Engineering Drawing and Design

Second Edition

CECIL JENSEN

Technical Director
R. S. McLaughlin Collegiate and
Vocational Institute
Oshawa, Ontario, Canada

JAY HELSEL

Vice President for Administrative Affairs
California State College
California, Pennsylvania

GREGG DIVISION
McGRAW-HILL BOOK COMPANY
New York • St. Louis • Dallas • San Francisco • Auckland • Bogotá
Düsseldorf • Johannesburg • London • Madrid • Mexico • Montreal • New Delhi
Panama • Paris • São Paulo • Singapore • Sydney • Tokyo • Toronto

Library of Congress Catalóging in Publication Data

Jensen, Cecil Howard, (date)
Engineering drawing and design.

Includes index.
1. Mechanical drawing. 2. Engineering design.
I. Helsel, Jay D., joint author. II. Title.
T353.J47 1979 604'.2 78-14799
ISBN 0-07-032516-2

ENGINEERING DRAWING AND DESIGN, Second Edition

1 2 3 4 5 6 7 8 9 0 V H V H 7 8 6 5 4 3 2 1 0 9

The editors for this book were Gerald O. Stoner and George McCloskey, the designer was Eileen Thaxton, the art supervisor was George T. Resch, the production supervisor was Kathleen Morrissey, the cover designer was Barbara Soll, and the cover photographers were Bellott/Wolfson. It was set in Times Roman by John C. Meyer & Son.
Printed and bound by Von Hoffman Press, Inc.

Contents

Preface

Engineering Drawing and Design is prepared for a two-semester course in engineering drawing. The contents are consistent with the trends and practices currently used in the preparation of engineering drawings.

Technical drafting, like all technical areas, is constantly changing. In *Engineering Drawing and Design*, the authors have made every effort to translate the most current technical information available into the most usable form from the standpoint of both teacher and student. The latest developments and current practices in all areas of graphic communication, functional drafting, materials representation, shop processes, numerical control, true positioning geometric tolerancing, and metrication have been incorporated into this text in a manner that synthesizes, simplifies, and converts complex drafting standards and procedures into understandable instructional units. Extensive author research and visits to drafting rooms throughout the country have resulted in a combination of current drafting practices and practical pedagogical techniques that produces the most efficient learning system yet designed for the instruction of engineering drawing.

Every chapter in *Engineering Drawing and Design* is divided into a number of single-concept units each with its own objective, instruction, examples, review and assignments. This organization provides the student with a logical sequence of experiences which can be adjusted to individual needs and also provides for maximum efficiency in learning essential concepts. Development of each unit is from the simple to the complex and from the familiar to the unfamiliar. Checkpoints are included to provide maximum reinforcement at each level.

Metric conventions as utilized on a practical level by American industry have been incorporated into this text.* Dual dimensioning, presenting both metric and customary units, is used throughout the text and in all problems. Each problem is stated in both conventional and metric form. Thus, the text may be used in a completely conventional (customary) course, in a complete metric-oriented course, or in a course which utilizes both metric and conventional systems. The teacher may also customize the course by selecting appropriate problems or materials to emphasize or deemphasize any degree of metrication.

A set of 11 x 17 inch worksheets is available for the completion of the problems. The worksheets include the problem in metric form on one side and in customary form on the reverse side. They are preprinted with light lines to provide the student with a beginning to each problem. Using these worksheets eliminates some of the initial work such as preparing borders, legends, data lines, and so forth. The worksheets also provide the student with the positioning of the drawing on each sheet, thus enabling the student to concentrate on the solution to the problem rather than on the mechanics of beginning the drawing. This focuses attention specifically on the concept under consideration and eliminates time wasted in nonessential aspects of the lesson. In earlier units, a certain amount of the work is completed for the student; in later units, however, fewer lines are provided.

A teacher's manual which provides complete solutions to all graphic problems, both metric and conventional, found in the text is also available from the publisher.

The authors wish to thank Fred Newman who has spent countless hours preparing the worksheets and solutions which accompany this text. Mr. Newman is head of the drafting department in a secondary school. He has had many years of experience in education and in industry, and he has brought these combined experiences to the preparation of the student assignments.

Cecil Jensen
Jay Helsel

*Since many major American corporations such as Caterpillar Tractor Company, Clark Equipment Company, Deere and Company, Ford Motor Company, General Motors, Honeywell, Inc., International Business Machines, and International Harvester Company now use the *re* spelling for metre and litre, this spelling has been adopted. If and when the official spelling is changed to meter, future printings of this text will reflect the latest standards.

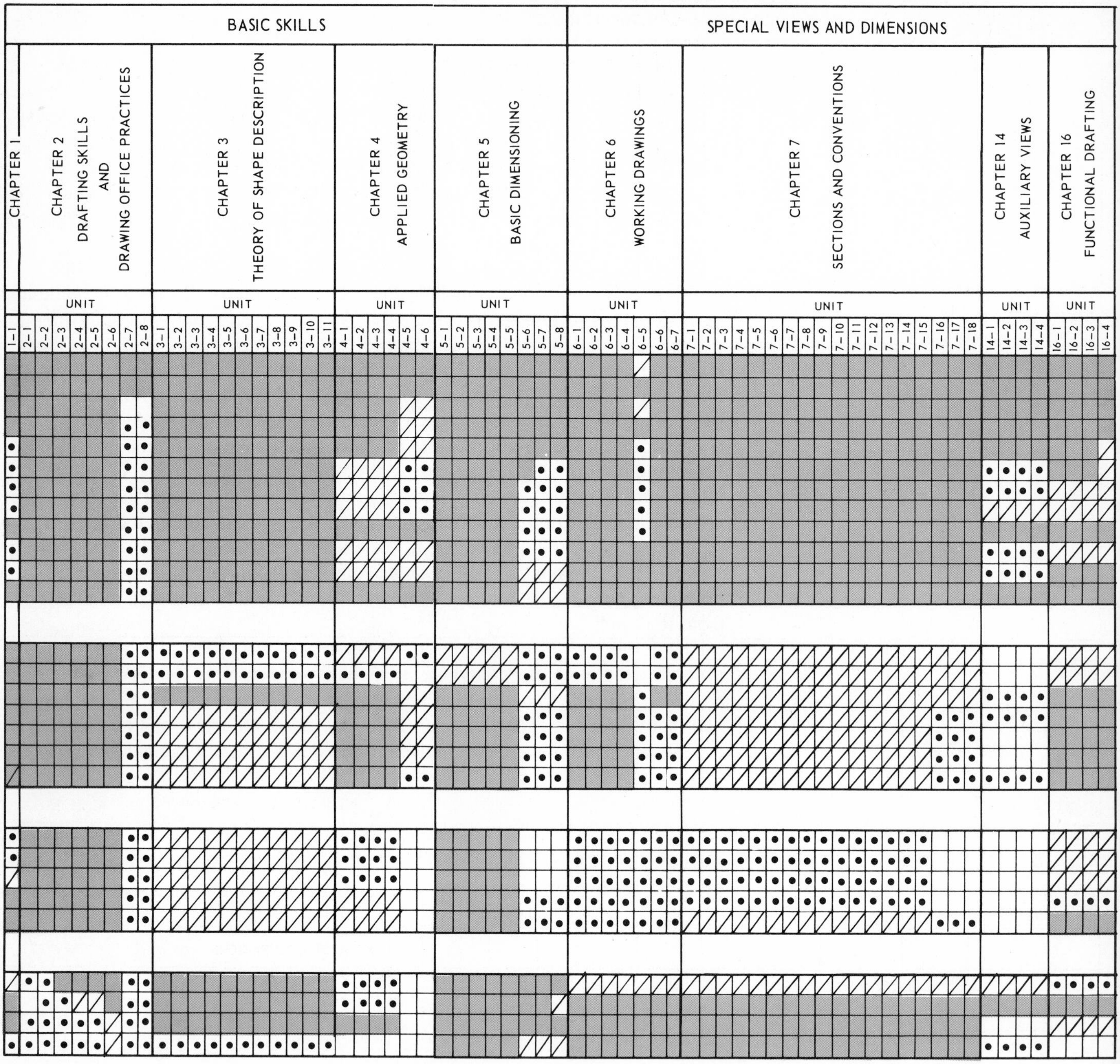
BASIC SKILLS
SPECIAL VIEWS AND DIMENSIONS
CHAPTER 1
CHAPTER 2 DRAFTING SKILLS AND DRAWING OFFICE PRACTICES
CHAPTER 3 THEORY OF SHAPE DESCRIPTION
CHAPTER 4 APPLIED GEOMETRY
CHAPTER 5 BASIC DIMENSIONING
CHAPTER 6 WORKING DRAWINGS
CHAPTER 7 SECTIONS AND CONVENTIONS
CHAPTER 14 AUXILIARY VIEWS
CHAPTER 16 FUNCTIONAL DRAFTING
UNIT
1–1
2–1 2–2 2–3 2–4 2–5 2–6 2–7 2–8
3–1 3–2 3–3 3–4 3–5 3–6 3–7 3–8 3–9 3–10 3–11
4–1 4–2 4–3 4–4 4–5 4–6
5–1 5–2 5–3 5–4 5–5 5–6 5–7 5–8
6–1 6–2 6–3 6–4 6–5 6–6 6–7
7–1 7–2 7–3 7–4 7–5 7–6 7–7 7–8 7–9 7–10 7–11 7–12 7–13 7–14 7–15 7–16 7–17 7–18
14–1 14–2 14–3 14–4
16–1 16–2 16–3 16–4

NOTE: SUMMARY BASED ON INDUSTRIAL SURVEY TAKEN IN 1976

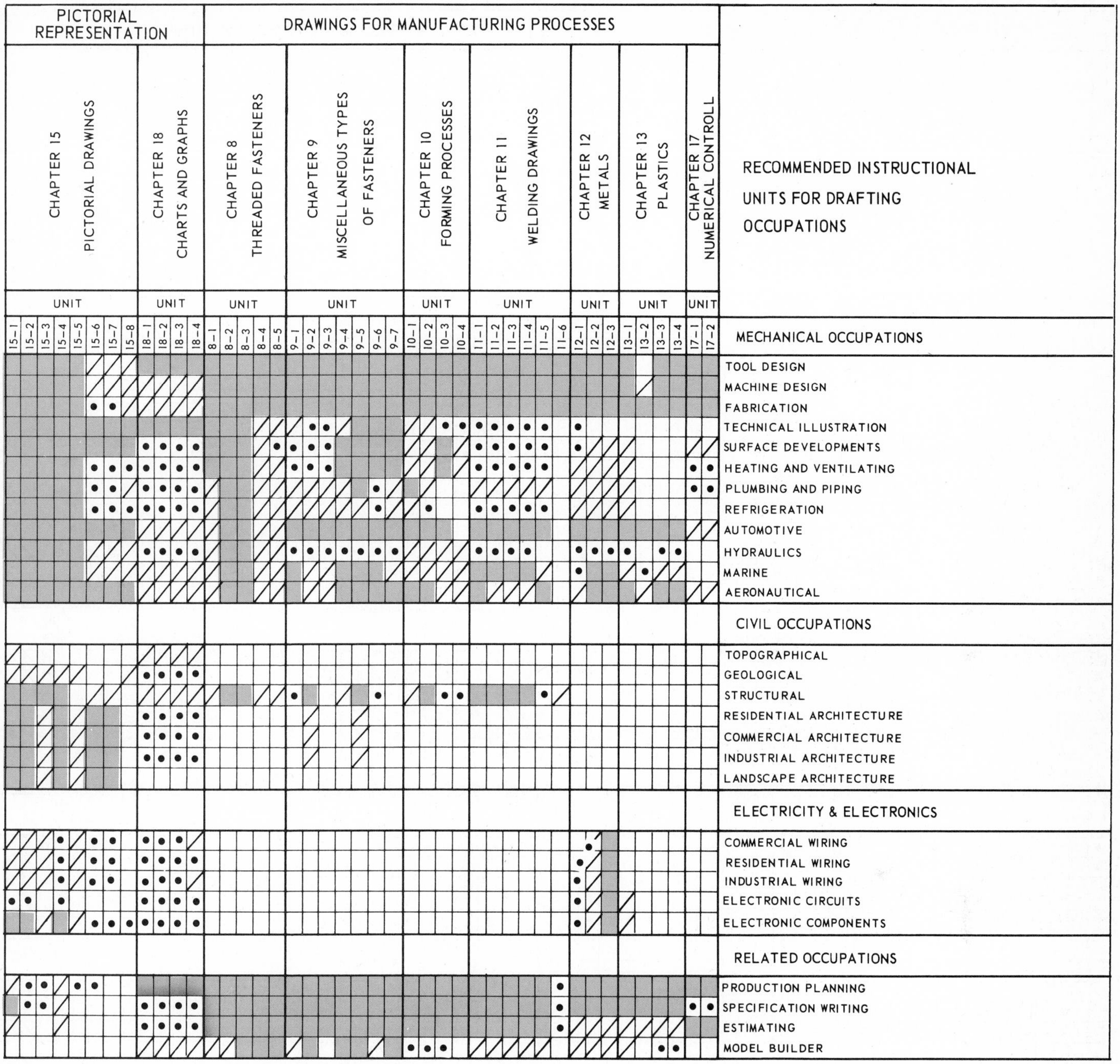

PICTORIAL REPRESENTATION
DRAWINGS FOR MANUFACTURING PROCESSES
CHAPTER 15 PICTORIAL DRAWINGS
CHAPTER 18 CHARTS AND GRAPHS
CHAPTER 8 THREADED FASTENERS
CHAPTER 9 MISCELLANEOUS TYPES OF FASTENERS
CHAPTER 10 FORMING PROCESSES
CHAPTER 11 WELDING DRAWINGS
CHAPTER 12 METALS
CHAPTER 13 PLASTICS
CHAPTER 17 NUMERICAL CONTROLL
UNIT
15-1 15-2 15-3 15-4 15-5 15-6 15-7 15-8
18-1 18-2 18-3 18-4
8-1 8-2 8-3 8-4 8-5
9-1 9-2 9-3 9-4 9-5 9-6 9-7
10-1 10-2 10-3 10-4
11-1 11-2 11-3 11-4 11-5 11-6
12-1 12-2 12-3
13-1 13-2 13-3 13-4
17-1 17-2
RECOMMENDED INSTRUCTIONAL UNITS FOR DRAFTING OCCUPATIONS
MECHANICAL OCCUPATIONS
TOOL DESIGN
MACHINE DESIGN
FABRICATION
TECHNICAL ILLUSTRATION
SURFACE DEVELOPMENTS
HEATING AND VENTILATING
PLUMBING AND PIPING
REFRIGERATION
AUTOMOTIVE
HYDRAULICS
MARINE
AERONAUTICAL
CIVIL OCCUPATIONS
TOPOGRAPHICAL
GEOLOGICAL
STRUCTURAL
RESIDENTIAL ARCHITECTURE
COMMERCIAL ARCHITECTURE
INDUSTRIAL ARCHITECTURE
LANDSCAPE ARCHITECTURE
ELECTRICITY & ELECTRONICS
COMMERCIAL WIRING
RESIDENTIAL WIRING
INDUSTRIAL WIRING
ELECTRONIC CIRCUITS
ELECTRONIC COMPONENTS
RELATED OCCUPATIONS
PRODUCTION PLANNING
SPECIFICATION WRITING
ESTIMATING
MODEL BUILDER

CODE

- ESSENTIAL
- OCCASIONALLY USED
- NICE TO KNOW
- NONESSENTIAL

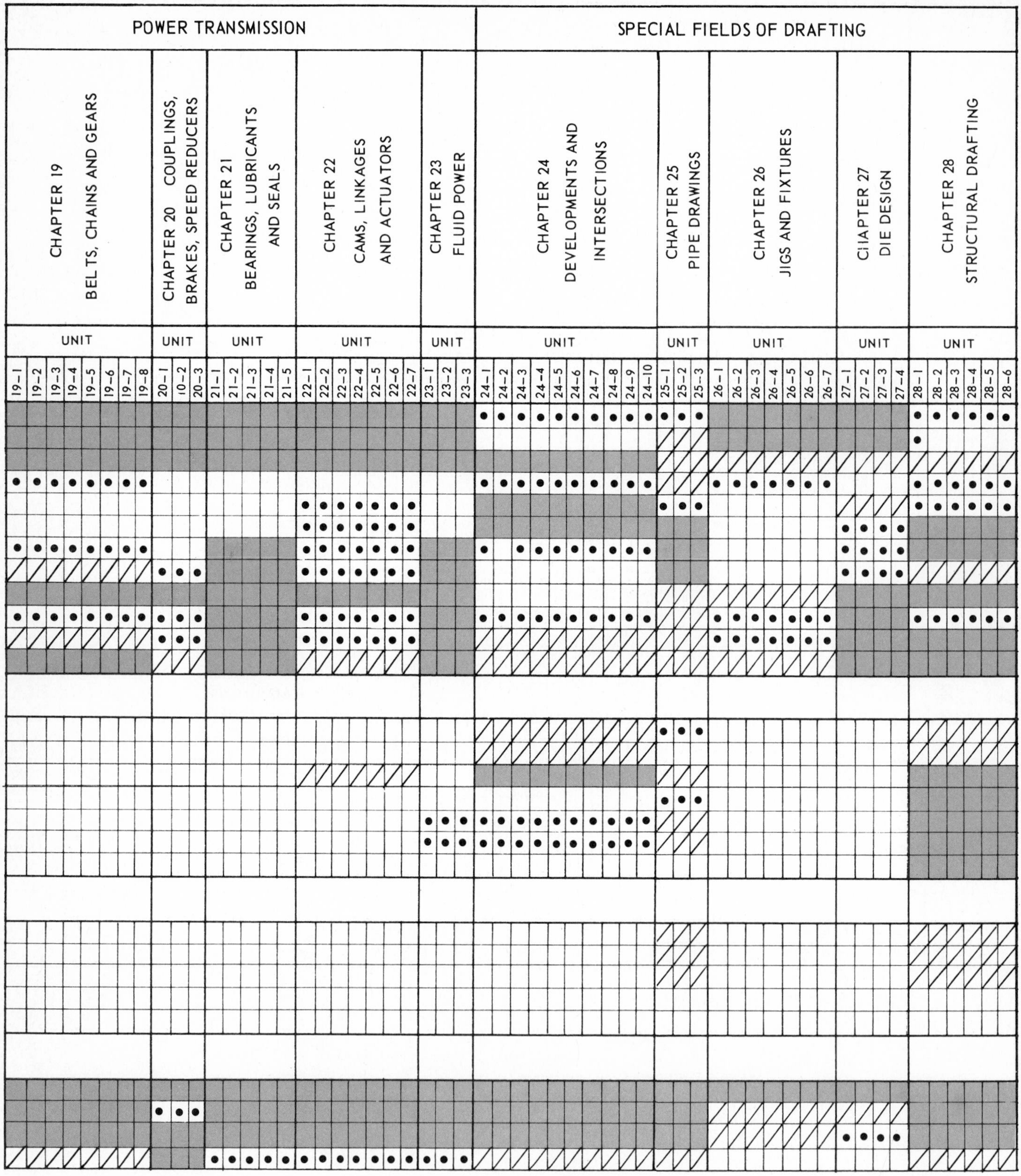
POWER TRANSMISSION
SPECIAL FIELDS OF DRAFTING
CHAPTER 19 BELTS, CHAINS AND GEARS
CHAPTER 20 COUPLINGS, BRAKES, SPEED REDUCERS
CHAPTER 21 BEARINGS, LUBRICANTS AND SEALS
CHAPTER 22 CAMS, LINKAGES AND ACTUATORS
CHAPTER 23 FLUID POWER
CHAPTER 24 DEVELOPMENTS AND INTERSECTIONS
CHAPTER 25 PIPE DRAWINGS
CHAPTER 26 JIGS AND FIXTURES
CHAPTER 27 DIE DESIGN
CHAPTER 28 STRUCTURAL DRAFTING
UNIT
19–1 19–2 19–3 19–4 19–5 19–6 19–7 19–8
20–1 10–2 20–3
21–1 21–2 21–3 21–4 21–5
22–1 22–2 22–3 22–4 22–5 22–6 22–7
23–1 23–2 23–3
24–1 24–2 24–3 24–4 24–5 24–6 24–7 24–8 24–9 24–10
25–1 25–2 25–3
26–1 26–2 26–3 26–4 26–5 26–6 26–7
27–1 27–2 27–3 27–4
28–1 28–2 28–3 28–4 28–5 28–6

NOTE: SUMMARY BASED ON INDUSTRIAL SURVEY TAKEN IN 1976

DESIGN CONCEPTS

RECOMMENDED INSTRUCTION UNITS FOR DRAFTING OCCUPATIONS

UNIT	CHAPTER 29 APPLIED MECHANICS					CHAPTER 30 STRENGTH OF MATERIALS								CHAPTER 31 ENGINEERING TOLERANCING											CHAPTER 32 DESCRIPTIVE GEOMETRY						CHAPTER 33 THE DESIGN PROCESS	
	29–1	29–2	29–3	29–4	29–5	30–1	30–2	30–3	30–4	30–5	30–6	30–7	30–8	31–1	31–2	31–3	31–4	31–5	31–6	31–7	31–8	31–9	31–10	31–11	32–1	32–2	32–3	32–4	32–5	32–6	33–1	33–2
MECHANICAL OCCUPATIONS																																
TOOL DESIGN	■	■	■	■	■	■	■	■	■	■	■	■	■	■	■	■	■	■	■	■	■	■	■	■	/	/	/	/	/	/	■	■
MACHINE DESIGN	■	■	■	■	■	■	■	■	■	■	■	■	■	■	■	■	■	■	■	■	■	■	■	■	/	/	/	/	/	/	■	■
FABRICATION	■	■	■	■	■	■	■	■	■	■	■	■	■	■	■	■	■	■	■	■	■	■	■	■	/	/	/	/	/	/	■	■
TECHNICAL ILLUSTRATION														■	■	■	■	■	■	■	■	■	■	■	■	■	■	■	■	■	•	•
SURFACE DEVELOPMENTS	•	•	•	•	•	/	/	/	/	/	/	/	/	■	■	■	■	■	■	■	■	■	■	■	■	■	■	■	■	■	•	•
HEATING AND VENTILATING	/	/	/	/	/	/	/	/	/	/	/	/	/	•	•	•	•	•	•	•	•	•	•	•	/	/	/	/	/	/	•	•
PLUMBING AND PIPING	/	/	/	/	/	/	/	/	/	/	/	/	/	•	•	•	•	•	•	•	•	•	•	•	/	/	/	/	/	/	•	•
REFRIGERATION	/	/	/	/	/	/	/	/	/	/	/	/	/		•	•	•	•	•	•	•	•	•	•	/	/	/	/	/	/	•	•
AUTOMOTIVE	■	■	■	■	■	■	■	■	■	■	■	■	■	■	■	■	■	■	■	■	■	■	■	■	■	■	■	■	■	■	■	■
HYDRAULICS	/	/	/	/	/	/	/	/	/	/	/	/	/	•	•	•	•	•	•	•	•	•	•	•	/	/	/	/	/	/	■	■
MARINE	■	■	■	■	■	■	■	■	■	■	■	■	■	/	/	/	/	/	/	/	/	/	/	/	/	/	/	/	/	/	■	■
AERONAUTICAL	■	■	■	■	■	■	■	■	■	■	■	■	■	■	■	■	■	■	■	■	■	■	■	■	■	■	■	■	■	■	■	■
CIVIL OCCUPATIONS																																
TOPOGRAPHICAL						/	/	/	/	/	/	/	/	•	•	•	•	•	•	•	•	•	•	•	/	/	/	/	/	/	•	•
GEOLOGICAL						/	/	/	/	/	/	/	/	•	•	•	•	•	•	•	•	•	•	•	/	/	/	/	/	/	•	•
STRUCTURAL	■	■	■	■	■	■	■	■	■	■	■	■	■	•	•	•	•	•	•	•	•	•	•	•	/	/	/	/	/	/	■	■
RESIDENTIAL ARCHITECTURE	/	/	/	/	/	/	/	/	/	/	/	/	/												/	/	/	/	/	/	/	/
COMMERCIAL ARCHITECTURE						/	/	/	/	/	/	/	/												/	/	/	/	/	/	/	/
INDUSTRIAL ARCHITECTURE						/	/	/	/	/	/	/	/												/	/	/	/	/	/	/	/
LANDSCAPE ARCHITECTURE						/	/	/	/	/	/	/	/												/	/	/	/	/	/	/	/
ELECTRICITY & ELECTRONICS																																
COMMERCIAL WIRING						•	•	•	•	•	•	•	•																		•	•
RESIDENTIAL WIRING						•	•	•	•	•	•	•	•																			
INDUSTRIAL WIRING						/	/	/	/	/	/	/	/																		•	•
ELECTRONIC CIRCUITS						•	•	•	•	•	•	•	•																			
ELECTRONIC COMPONENTS	/	/	/	/	/	/	/	/	/	/	/	/	/																			
RELATED OCCUPATIONS																																
PRODUCTION PLANNING	■	■	■	■	■	■	■	■	■	■	■	■	■	■	■	■	■	■	■	■	■	■	■	■	•	•	•	•	•	•	•	•
SPECIFICATION WRITING	•	•	•	•	•	•	•	•	•	•	•	•	•	■	■	■	■	■	■	■	■	■	■	■							•	•
ESTIMATING	•	•	•	•	•	•	•	•	•	•	•	•	•	•	•	•	•	•	•	•	•	•	•	•	■	■	■	■	■	■	•	•
MODEL BUILDER	/	/	/	/	/	/	/	/	/	/	/	/	/																		•	•

CODE

■ ESSENTIAL

/ OCCASIONALLY USED

• NICE TO KNOW

(blank) NONESSENTIAL

ABOUT THE AUTHORS

CECIL JENSEN, a technical director in the education system of the Province of Ontario, has had over twenty-five years teaching experience in mechanical drafting. He is the author of many successful technical books, including *Engineering Drawing and Design, Interpreting Engineering Drawings, Drafting Fundamentals,* and *Home Planning and Design.* Before entering the teaching profession, Mr. Jensen gained several years of design experience in industry. He has also been responsible for the supervision of adult education in Durham County in Ontario, Canada, and the supervision of the teaching of technical courses for General Motors apprentices in Oshawa, Canada. He is a member of the Canadian (CSA) Committee on Engineering Drawings and has represented Canada at several world conferences on the standardization of engineering drawings.

JAY HELSEL is a professor of industrial arts and vice president for administrative affairs at California State College in Pennsylvania. He completed his undergraduate work in industrial arts at California State College and was awarded a master's degree from Pennsylvania State University. He has done advanced graduate work at West Virginia and at the University of Pittsburgh, where he completed a doctoral degree in educational communications and technology. In addition, Dr. Helsel holds a certificate in airbrush techniques and technical illustration from the Pittsburgh Art Institute.

He has worked in industry and has taught drafting, metalworking, woodworking, and a variety of laboratory and professional courses at both the secondary and college levels. During the past fifteen years, he has worked as a freelance artist and illustrator. His work appears in a great variety of technical publications.

Dr. Helsel is coauthor of *Engineering Drawing and Design, Programmed Blueprint Reading, Mechanical Drawing, Reading Engineering Drawings Through Conceptual Sketching, Drawing and Blueprint Reading Transparencies, Architectural Drafting Transparencies, Woodworking Transparencies,* and *Automotive Transparencies.* He is also the author of a series of *Mechanical Drawing Film Loops.*

Part 1
Basic Drawing Design

Chapter 1
The Language of Industry

UNIT 1-1
THE LANGUAGE OF INDUSTRY

Since earliest times people have used drawings to communicate and record ideas so that they would not be forgotten. The earliest forms of writing, such as the Egyptian hieroglyphics, were picture forms.

The word *graphic* means dealing with the expression of ideas by lines or marks impressed on a surface. A drawing is a graphic representation of a real thing. Drafting, therefore, is a graphic language, because it uses pictures to communicate thoughts and ideas. Because these pictures are understood by people of different nations, drafting is referred to as a "universal language."

Drawing has developed along two distinct lines, with each form having a different purpose. On the one hand, artistic drawing is concerned mainly with the expression of real or imagined ideas of a cultural nature. Technical drawing, on the other hand, is concerned with the expression of technical ideas or ideas of a practical nature, and it is the method used in all branches of technical industry.

Even highly developed word languages are inadequate for describing the size, shape, and relationship of physical objects. For every manufactured object there are drawings that describe its physical shape completely and accurately, communicating the drafter's ideas to the worker. For this reason, drafting is referred to as the "language of industry."

Drafters translate the ideas, rough sketches, specifications, and calculations of engineers, architects, and designers into working plans which are used in making a product. See Figs. 1-1-3 through 1-1-8. Drafters may calculate the strength, reliability, and cost of materials. In their drawings and specifications, they describe exactly what materials workers are to use on a particular job. To prepare their drawings, drafters use instruments such as compasses, dividers, protractors, templates, and set squares, as well as machines that combine the functions of several devices. They also may use engineering handbooks, tables, and calculators to assist in solving technical problems.

Drafters are often classified according to their type of work or their level of responsibility. Senior drafters take the preliminary information provided by engineers and architects to prepare design "layouts" (drawings made to scale of the object to be built). Detailers make drawings of each part shown on the layout, giving dimensions, material, and any other information necessary to make the

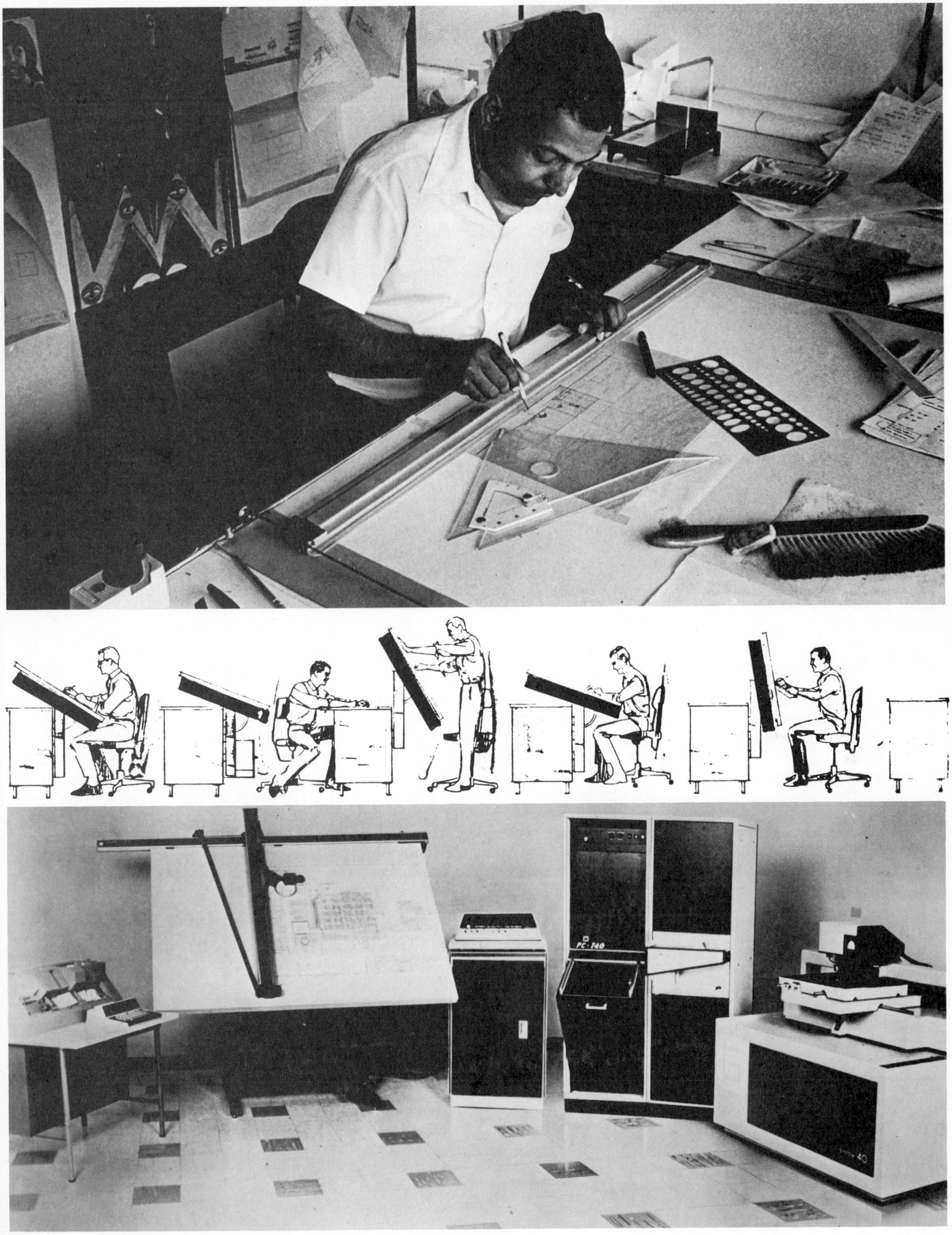

Fig. 1-1-1 Drafting—today and tomorrow.

detailed drawing clear and complete. Checkers carefully examine drawings for errors in computing or recording dimensions and specifications. Under the supervision of drafters, tracers make minor corrections and prepare drawings for reproductions by tracing them on transparent cloth, paper, or plastic film.

Drafters also may specialize in a particular field of work, such as mechanical, electrical, electronic, aeronautical, structural, or architectural drafting.

Changing Times[1]

Fifty years have brought great changes to the drafting room. Its physical appearance, furnishings, even its drafters and engineers have moved quickly from their battered domain of old into the Space Age.

These changes were brought about largely by the recognition of many factors that affect the performances of working people. Because designing and drafting are specialized technical fields today that require a high level of precision, personnel efficiency in these areas has been closely linked to the working atmosphere.

A constant reappraisal of this atmosphere should be a prime responsibility of every chief engineer and chief drafter. With an eye to improving working conditions, thereby increasing efficiency and bettering performance, they should reevaluate periodically the tables, boards, seating arrangements, drafting machines and tools, lighting, reference materials, and file units assigned to their department.

Drafting room technology has progressed at the same rapid pace as the economy of our country. Many changes have taken place in the modern drafting room, shown in Fig. 1-1-2, as compared to a typical drafting room scene before the turn of the century. Not only are there far more tools, but they are of much higher quality. From electric erasers to automated drafting machines and from combination reference tables with adjustable drawing boards to drawing media that contain all the desired qualities for reproduction. Noteworthy progress has been made and continues to be made as our expanding technology takes giant steps forward in this modern age.

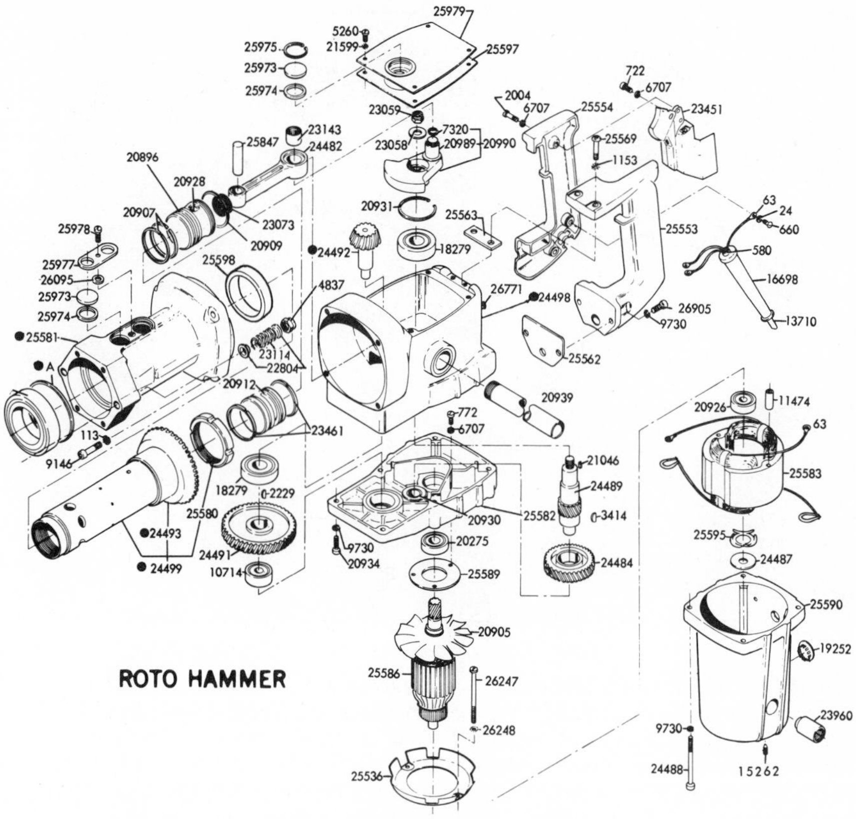
ROTO HAMMER

Fig. 1-1-3 Pictorial drawings. (Skill Corp. [Canada] Ltd.)

Drawing Standards

Throughout the long history of drafting, many drawing conventions, terms, abbreviations, and practices have come into common use. It is essential that different drafters use the same practices if drafting is to serve as a reliable means of communicating technical theories and ideas.

In the interest of efficient communica-

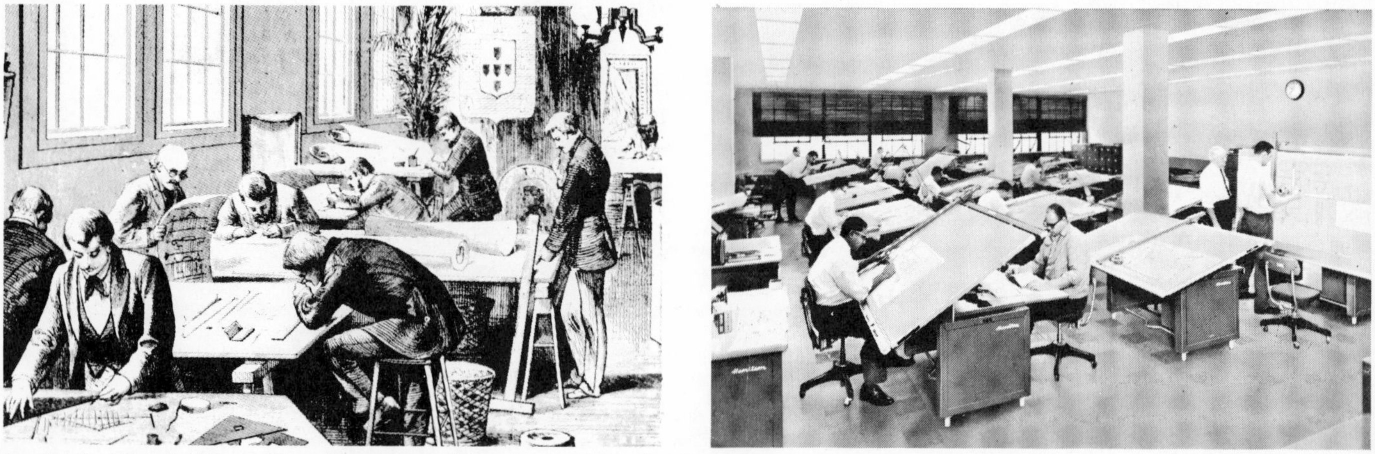

Fig. 1-1-2 The drafting office. (Bettman Archive, Inc. and Charles Bruning Company [Canada] Limited.)

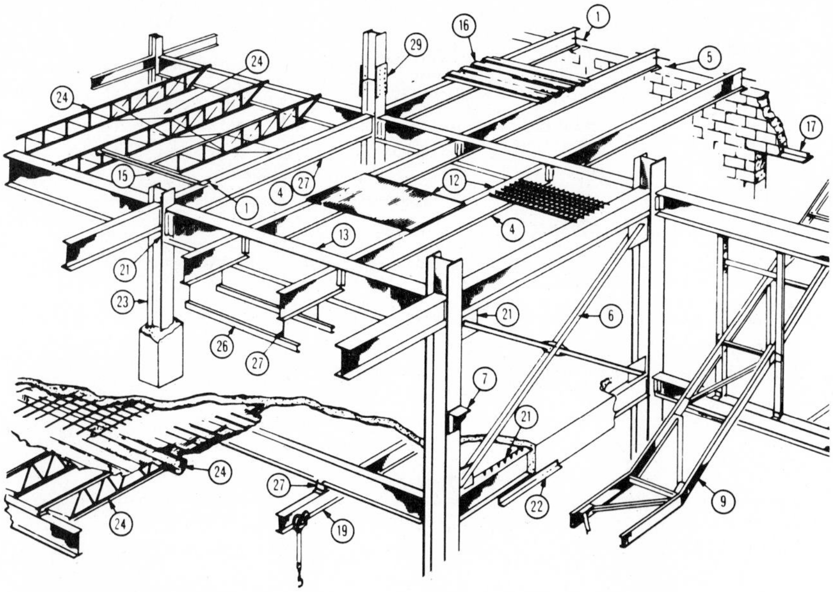

Fig. 1-1-4 **Structural drawings.** (Canadian Institute of Steel Construction.)

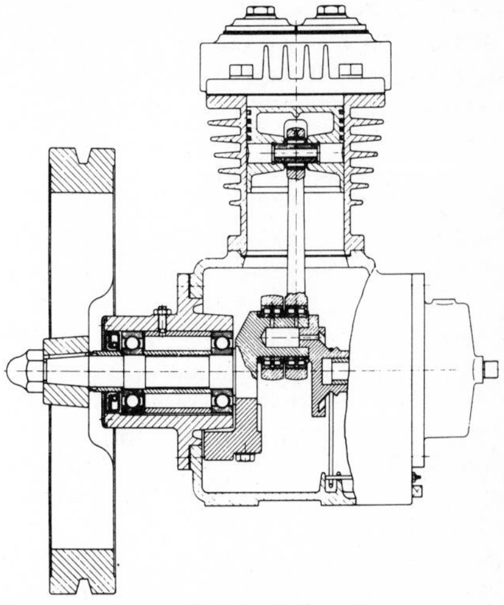

Fig. 1-1-6 Mechanical drawings.

tion, the American National Standards Institute (ANSI) and the Canadian Standards Association (CSA) have adopted a set of drafting standards which are recommended for drawing practice in all fields of engineering and are used and explained throughout this text. These standards apply primarily to end product drawings, which usually consist of detail or part drawings and assembly or subassembly drawings, but are not intended to fully cover other supplementary drawings such as checklists, parts lists, schematic diagrams, electrical wiring diagrams, flowcharts, installation drawings, process drawings, architectural drafting, and pictorial drawing.

The information and illustrations shown have been revised to reflect current industrial practices in the preparation and handling of engineering documents. The increased use of reduced-size copies of engineering drawings made from microfilm and the reading of microfilm require the proper preparation of the original engineering document. All future drawings should be prepared suitable for eventual photographic reduction or reproduction. The observance of the drafting practices described in this text will contribute substantially to the improved quality of photographically reproduced engineering drawings.

Places of Employment[2]

There are over 400,000 people working in drafting positions in the United States and Canada. Approximately 4 percent are women. About 9 out of 10 drafters are employed in private industry. Manufacturing industries that employ large numbers are those making machinery, electrical equipment, transportation equipment, and fabricated metal products. Nonmanufacturing industries employing large numbers are engineering and architectural consulting firms, construction companies, and public utilities.

Over 25,000 drafters work for the governments; the majority work for the armed services. Drafters employed by state and local governments work chiefly for highway and public works departments. Several thousand drafters are employed by colleges and universities and by nonprofit organizations.

Training, Qualifications, Advancement

Young persons interested in becoming drafters can acquire the necessary training from a number of sources, including technical institutes, junior and community colleges, extension divisions of universities, vocational and technical high schools, and correspondence schools. Others may qualify for drafting positions

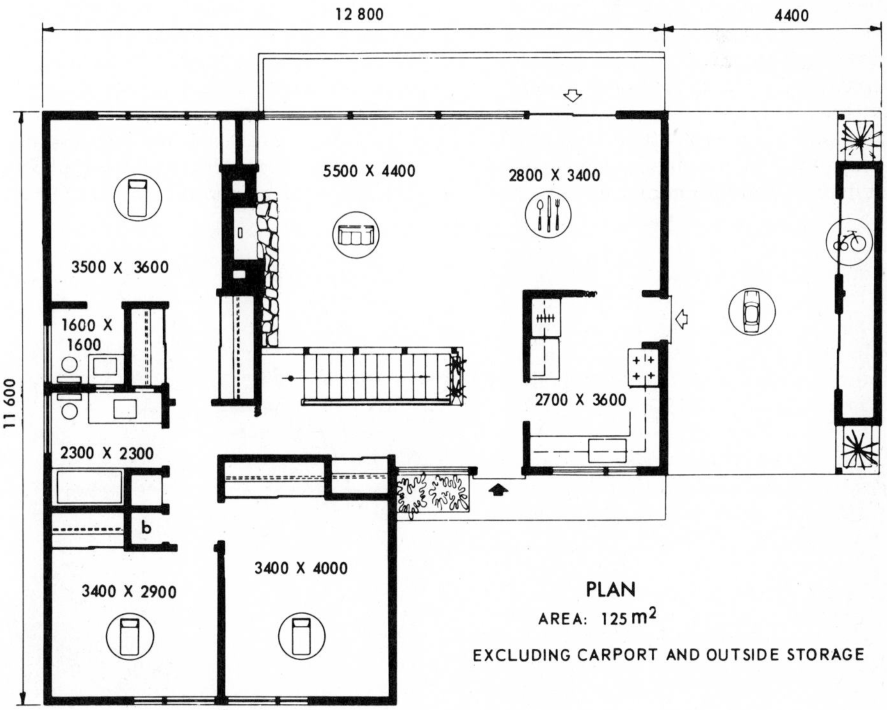
12 800
4400
11 600
5500 X 4400
2800 X 3400
3500 X 3600
1600 X 1600
2700 X 3600
2300 X 2300
b
3400 X 4000
3400 X 2900
PLAN
AREA: 125 m2
EXCLUDING CARPORT AND OUTSIDE STORAGE

Fig. 1-1-5 Architectural drawings.

through on-the-job training programs combined with part-time schooling or through three- or four-year apprenticeship programs.

The prospective drafter's training, whether obtained in high school or post-high school drafting programs, should include courses in mathematics and physical sciences, as well as in mechanical drawing and drafting. Studying shop practices and learning some shop skills also are helpful, since many higher-level drafting jobs require knowledge of manufacturing or construction methods. Many technical schools offer courses in structural design, strength of materials, and physical metallurgy.

Young people having only high school drafting training usually start out as tracers, or detailers. Those having some formal post-high school technical training can often qualify as junior drafters. As drafters gain skill and experience, they may advance to higher-level positions as checkers, detailers, senior drafters, or supervisors of other drafters. Drafters who take courses in engineering and mathematics are sometimes able to transfer to engineering positions.

Qualifications for success as a drafter may include the ability both to visualize objects in three dimensions and to do freehand drawing. Although such artistic ability is not generally required, it may be helpful in some specialized fields.

Drafting work also requires good eyesight (corrected or uncorrected), eye-hand coordination, and manual dexterity.

Employment Outlook

Employment opportunities for drafters are expected to be favorable in the future. Prospects will be best for those having post-high school drafting training. Well-qualified high school graduates who have had only high school drafting, however, also will be in demand for some types of jobs.

Employment of drafters is expected to rise rapidly as a result of the increasingly complex design problems of modern products and processes. In addition, as engineering and scientific occupations continue to grow, more drafters will be needed as supporting personnel. On the other hand, photoreproduction of drawings and expanding use of electronic drafting equipment and computers are eliminating some routine tasks done by drafters. This development will probably reduce the need for some less skilled drafters.

REFERENCES

1. Charles Bruning Co. (Canada) Ltd.
2. *Occupational Outlook Handbook*.

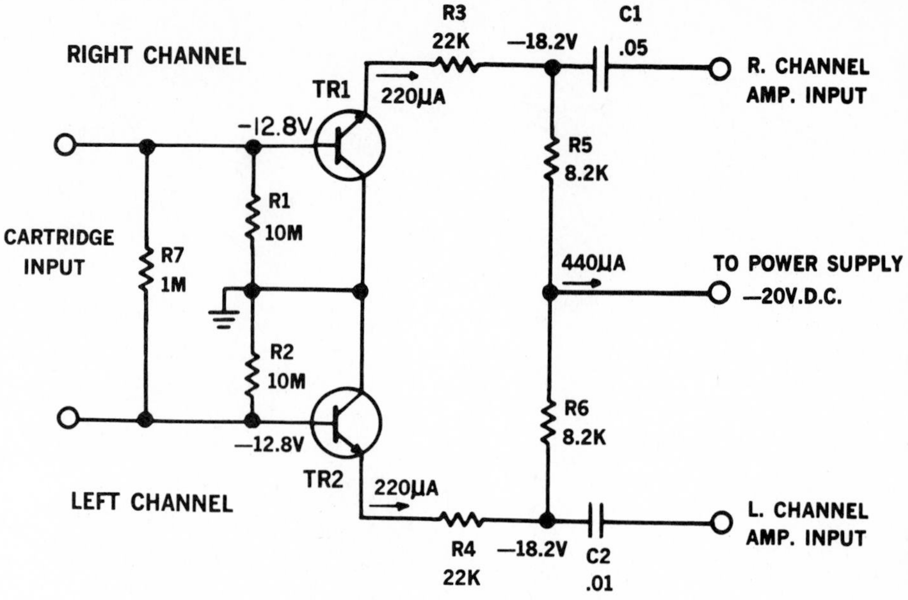

RIGHT CHANNEL
TR1
-12.8V
R3
22K
220μA
-18.2V
C1
.05
R. CHANNEL AMP. INPUT
R5
8.2K
CARTRIDGE INPUT
R7
1M
R1
10M
440μA
TO POWER SUPPLY
-20V.D.C.
R2
10M
R6
8.2K
-12.8V
TR2
LEFT CHANNEL
220μA
R4
22K
-18.2V
C2
.01
L. CHANNEL AMP. INPUT

ELEMENTARY DIAGRAM OF A PREAMPLIFIER

Fig. 1-1-7 Electrical drawings.

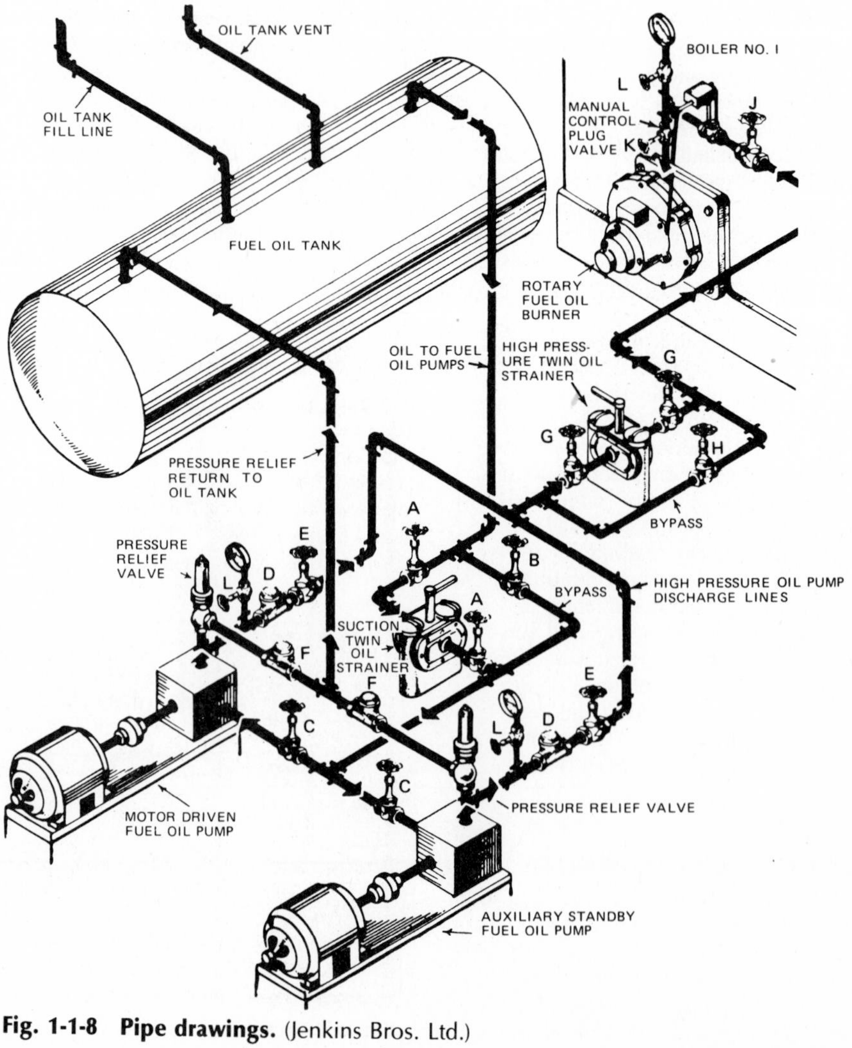

OIL TANK VENT
OIL TANK FILL LINE
BOILER NO. I
MANUAL CONTROL PLUG VALVE
FUEL OIL TANK
ROTARY FUEL OIL BURNER
OIL TO FUEL OIL PUMPS
HIGH PRESSURE TWIN OIL STRAINER
PRESSURE RELIEF RETURN TO OIL TANK
BYPASS
PRESSURE RELIEF VALVE
SUCTION TWIN OIL STRAINER
HIGH PRESSURE OIL PUMP DISCHARGE LINES
MOTOR DRIVEN FUEL OIL PUMP
PRESSURE RELIEF VALVE
AUXILIARY STANDBY FUEL OIL PUMP

Fig. 1-1-8 Pipe drawings. (Jenkins Bros. Ltd.)

Chapter 2
Drafting Skills and Drawing Office Practices

UNIT 2-1

THE DRAFTING OFFICE

The drafting office is the starting point for all engineering work. Its product, the *engineering drawing*, is the main method of communication between all persons concerned with the design and manufacture of parts. Therefore the drafting office must provide accommodations and equipment for the drafters, from designer and checker to detailer or tracer; for the personnel who make copies of the drawings and file the originals; and for the secretarial staff who assist in the preparation of the drawings (Fig. 2-1-1).

Fig. 2-1-1 Drafting office. (Charles Bruning Co. [Canada] Limited.)

Most engineering departments still rely on manual drafting needs. In the majority of cases, this is all that is necessary. Equipment for manual drafting is varied and is steadily being improved. Where a high volume of finished or repetitive work is not necessary, this equipment does the job adequately and inexpensively, and most designers are accustomed to working with it.

A growing number of companies have turned to automated drafting. The reason is not simply to speed the drafting process. Automatic drafting can serve as a full partner in the design process, enabling the designer to do jobs that are simply not possible or feasible with manual equipment.

UNIT 2-2

MANUAL DRAFTING EQUIPMENT AND SUPPLIES

Over the years, the designer's chair and drafting table have evolved into a drafting station which provides a comfortable, integrated work area. Yet much of the equipment and supplies employed years ago is still in use today, although it has been vastly improved.

Drafting Furniture

Special tables and desks are manufactured for use in single-station or multi-station design offices. Typical are desks with attached drafting boards (Fig. 2-2-1). The boards may be used by the occupant of the desk to which it is attached, in which case it may swing out of the way when not in use, or may be reversed for use by the person in the adjoining station.

In addition to such special "work stations," a variety of individual desks, chairs, tracing tables, filing cabinets, and special storage devices for drawings are available.

Fig. 2-2-1 Drafting work station. (Charles Bruning Co. [Canada] Limited and Keuffel & Esser Co.)

The simplest manually adjustable tables typically consist of a hinged surface riding on a vertical rod secured by a setscrew. The setscrew is loosened, the top is set at the desired angle, and the setscrew is retightened.

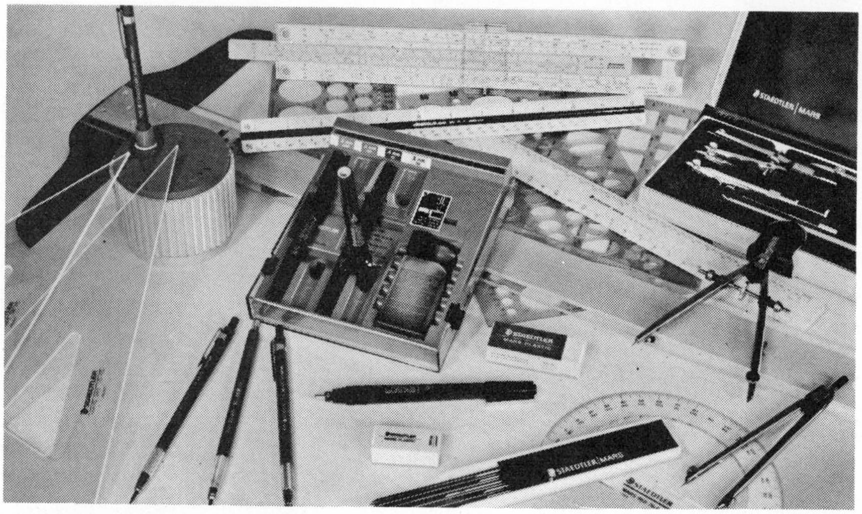

Fig. 2-2-2 Drafting equipment. (Staedtler.)

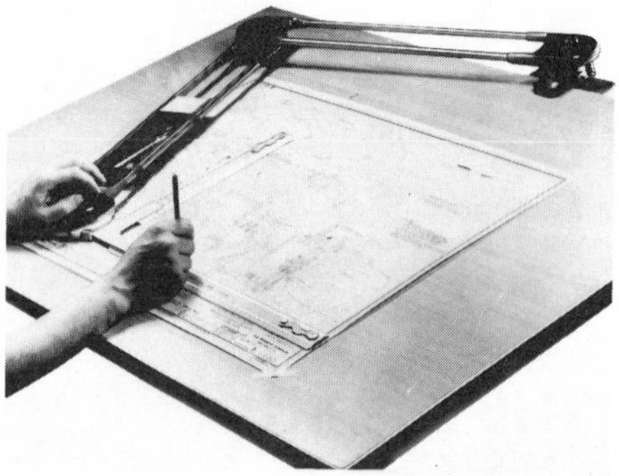

Fig. 2-2-3 Arm type drafting machine. (Keuffel and Esser Co.)

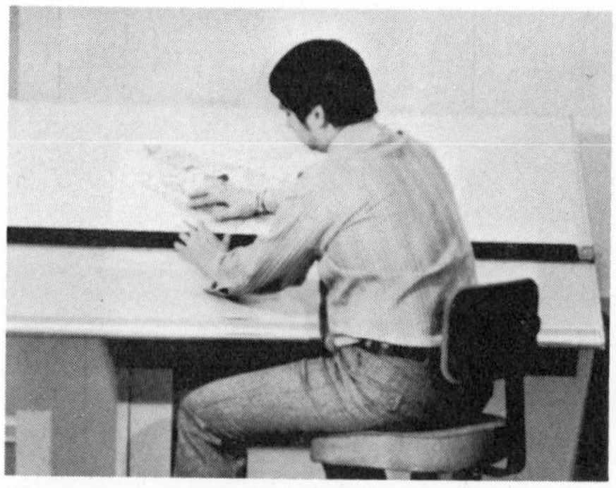

Fig. 2-2-4 Drafting table with straightedge. (Addressograph Multigraph Corp.)

Drafting Equipment

DRAFTING MACHINES

In the well-equipped engineering department, where the designer is expected to do accurate drafting, the T square has been replaced largely by the drafting machine. This device, which combines the functions of T square, set squares, scale, and protractor, is estimated to save up to 50 percent of the user's time. All positioning is done with one hand, while the other is free to draw.

Drafting machines may be attached to any drafting board or table. Two types are currently available. In the track type, a vertical beam carrying the drafting instruments rides along a horizontal beam fastened to the top of the table. In the arm, or elbow type (Fig. 2-2-3), two arms pivot from the top of the machine and relative to each other.

Track-type drafting machines are especially suitable for long-line work and large drawings.

STRAIGHTEDGES

The straightedge is used in drawing horizontal lines and for supporting set squares when vertical and sloping lines are being drawn. See Fig. 2-2-4. It is fastened on each end to cords, which pass over pulleys. This arrangement permits movement of the straightedge up and down the board while maintaining the straightedge in a horizontal position.

T SQUARES

The T square (Fig. 2-2-5) performs the same function as the parallel straightedge. While you are using the T square, keep the head of the instrument firmly against the side of the board to ensure that the lines you draw will be parallel. The head will be on the left edge of the board if you are right-handed and on the right if you are left-handed.

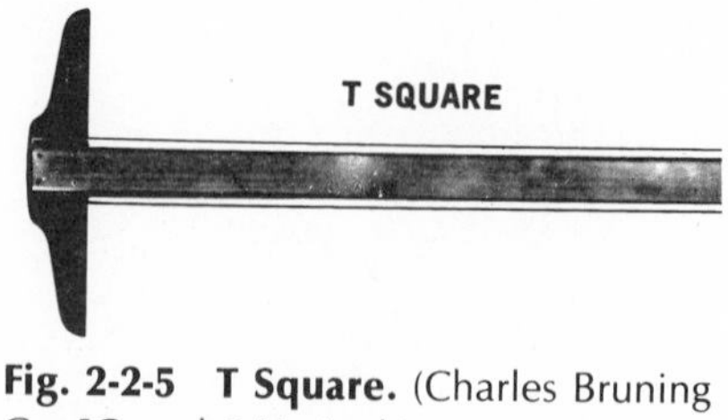
T SQUARE

Fig. 2-2-5 T Square. (Charles Bruning Co. [Canada] Limited.)

SET SQUARES

Set squares are used together with the parallel straightedge or T square when you are drawing vertical and sloping lines (Fig. 2-2-6). The set squares most commonly used are the 60–30° and the 45° set squares. Singly or in combination,

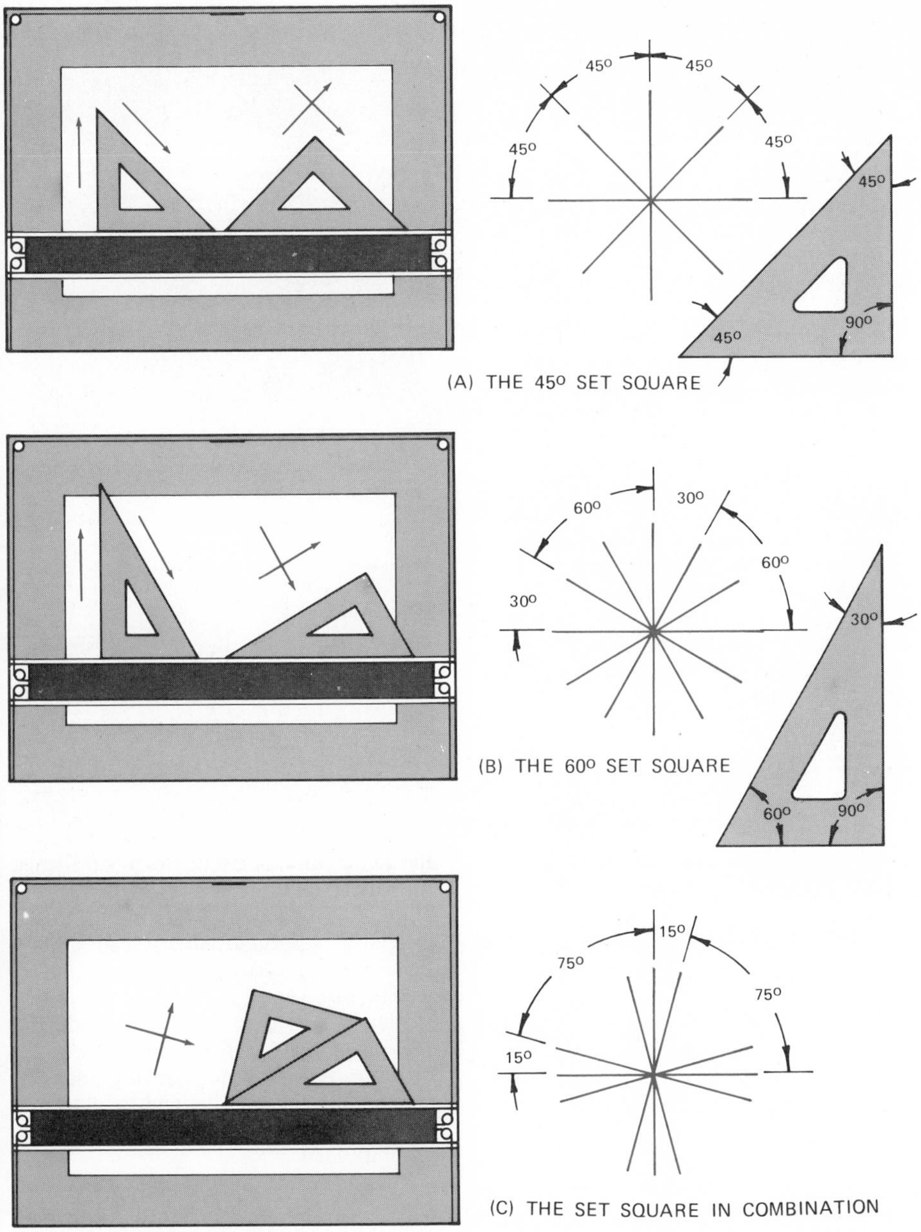

45°
45°
45°
45°
45°
45°
90°
(A) THE 45° SET SQUARE
60°
30°
60°
30°
30°
60°
90°
(B) THE 60° SET SQUARE
15°
75°
75°
15°
(C) THE SET SQUARE IN COMBINATION

Fig. 2-2-6 The set squares.

these set squares can be used to form angles in all the multiples of 15°. For other angles, the protractor is used. All angles can be drawn with the adjustable set square (Fig. 2-2-7); this instrument replaces the two common set squares and the protractor.

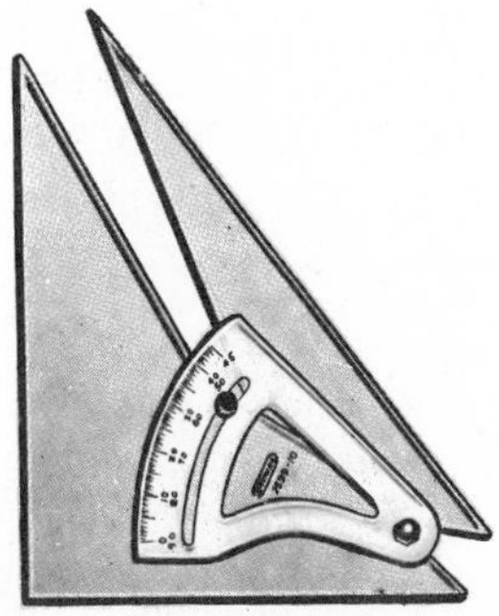

Fig. 2-2-7 Adjustable set square. (Charles Bruning Co. [Canada] Limited.)

SCALES

Shown in Fig. 2-2-8 are the common shapes of scales used by drafters to make measurements on their drawings. Scales are used only for measuring and are not to be used for drawing lines. It is important that drafters draw accurately to scale. The scale to which the drawing is made must be indicated in the title block or strip.

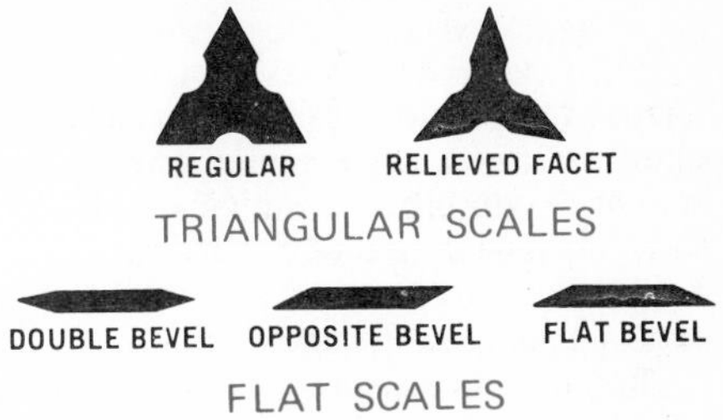

REGULAR
RELIEVED FACET
TRIANGULAR SCALES
DOUBLE BEVEL
OPPOSITE BEVEL
FLAT BEVEL
FLAT SCALES

Fig. 2-2-8 End view shapes of scales.

When objects are drawn at their actual size, the drawing is called *full scale* or *scale 1:1*. Many objects, however, such as buildings, ships, or airplanes, are too large to be drawn full scale, so they must be drawn to a reduced scale. An example would be the drawing of a house to a scale of 1:50 (metric) or, in the inch-pound scale, 1:48 [¼ inch (in.) = 1 foot (ft)]. See Fig. 2-2-9.

Frequently, objects such as small watch parts are drawn larger than their actual size so that their shape can be seen clearly. Such a drawing has been drawn to an enlarged scale. The minute hand of a wristwatch, for example, could be drawn to scale 4:1.

Many mechanical parts are drawn to half scale, 1:2, and fifth scale, 1:5. Notice that the scale of the drawing is expressed as an equation. The left side of the equation represents a unit of the size drawn; on the right side, a unit of the actual drawing equals five units of measurement of the actual object.

Scales are made with a variety of combined scales marked on their surfaces. This combination of scales spares the drafter the necessity of calculating the sizes to be drawn when working to a scale other than full size.

Metric Scales. The linear unit of measurement for mechanical drawings is the millimetre. Scale multipliers and divisors of 2 and 5 are recommended, which gives the scales shown in the accompanying table.

ENLARGED	SIZE AS	REDUCED
1000 : 1	1 : 1	1 : 2
500 : 1		1 : 5
200 : 1		1 : 10
100 : 1		1 : 20
50 : 1		1 : 50
20 : 1		1 : 100
10 : 1		1 : 200
5 : 1		1 : 500
2 : 1		1 : 1000

Fig. 2-2-9 Metric scales.

The numbers shown indicate the difference in size between the drawing and the actual part. For example, the ratio 10:1 shown on the drawing means that the drawing is 10 times the actual size of the part, whereas a ratio of 1:5 on the drawing means the object is 5 times as large as it is shown on the drawing.

The units of measurement for architectural drawings are the metre and millimetre. The same scale multipliers and divisors as used for mechanical drawings are used for architectural drawings.

Inch Scales. There are three types of scales which show various values that are equal to 1 in. They are the decimal inch scale, the fractional inch scale, and the scale which has divisions of 10, 20, 30, 40, 50, 60, and 80 parts to the inch. This last scale is known as the *civil engineer's scale*.

On fractional inch scales, multipliers or divisors of 2, 4, 8, and 16 are used, offering such scales as full size, half size, quarter size, etc.

Foot and Inch Scales. These scales are used mostly in architectural work. They differ from the inch scales in that each major division represents 1 ft, not 1 in., and the end units are subdivided into inches or parts of an inch. The more common scales are the ⅛ in. = 1 ft, ¼ in. = 1 ft, 1 in. = 1 ft, and 3 in. = 1 ft.

COMPASSES

The compass is used for drawing circles and arcs. Several basic types and sizes are available (Fig. 2-2-10).

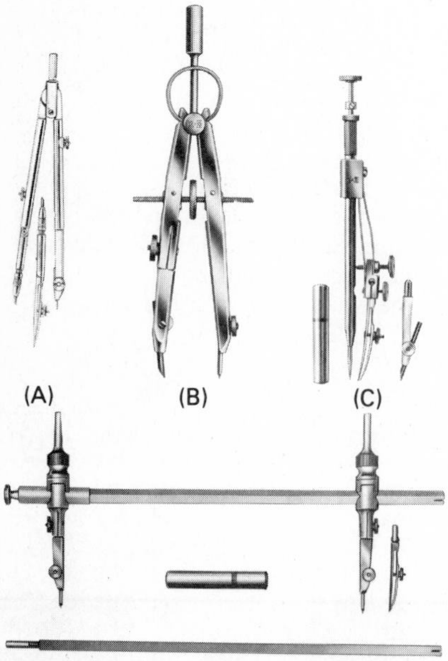
(A)
(B)
(C)
(D)

Fig. 2-2-10 Compasses. (a) Friction head, (b) Bow, (c) Drop bow, (d) Beam. (Keuffel & Esser Co.)

- *Friction head compass,* standard in most drafting sets.
- *Bow compass,* which operates on jackscrew or ratchet principle by turning a large knurled nut.
- *Drop bow compass,* mostly used for drawing small circles. The center rod contains the needle point and remains stationary while the pencil or pen leg revolves around it.
- *Beam compass,* a bar with an adjustable needle and pencil-and-pen attachment for drawing large arcs or circles.
- *Circuit scribing instrument,* a modified drop bow compass, used to cut terminal pads and prepare printed-circuit layouts on scribe coat film.

DIVIDERS

Used for the laying out of transferring measurements, dividers have a steel pin insert in each leg and come in a variety of sizes and designs, similar to the compass. See Fig. 2-2-11. The compass can be used as a divider by replacing the lead point with a steel pin.

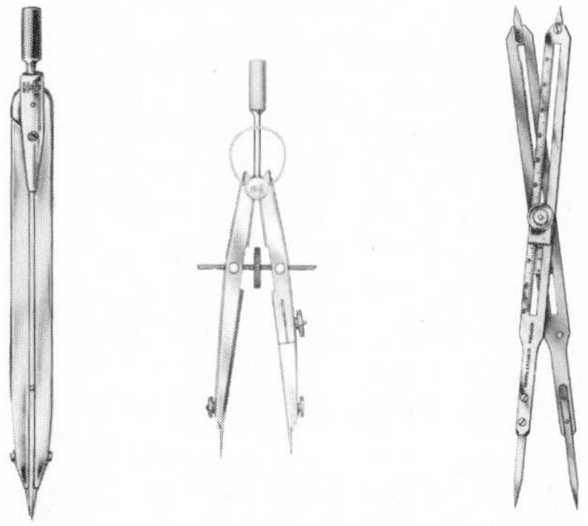

Fig. 2-2-11 Dividers. (Keuffel & Esser Co.)

DRAWING INSTRUMENT SETS

Many drafters have a complete drawing set, which usually includes several compasses and dividers with extension attachments for making inked drawings (Fig. 2-2-12).

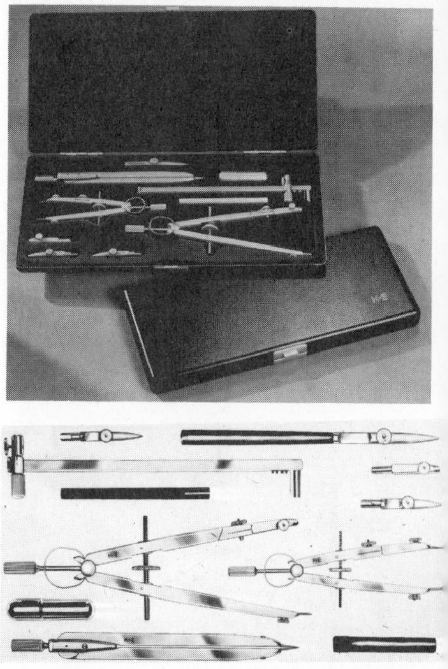

Fig. 2-2-12 Drawing instrument sets. (Keuffel & Esser Co.)

DRAFTING LEADS AND PENCILS

Leads. Because of the drawing media used and the type of reproduction required, pencil manufacturers have marketed three types of lead for the preparation of engineering drawings.

Graphite Lead. This is the conventional type of lead which has been used for years. It is made from graphite, clay, and resin. It is available in a variety of grades or hardness—9H, 8H, 7H, and 6H (hard); 5H and 4H (medium hard); 3H and 2H (medium); H and F (medium soft); and HB, B, 2B to 6B (very soft), the latter not being recommended for use on paper, vellum, or initial layout on cloth. The selection of the proper grade of lead is important. A hard lead might penetrate the drawing while a soft lead will smear. The next two types of drafting leads were developed as a result of the introduction of film as a drawing medium. A limited number of grades are available in these leads, and they do not correspond to the grades used for graphite lead.

Plastic Lead. This type of lead is designed for use on film only. It has good microfilm reproduction characteristics.

Plastic-Graphite Lead. As the name implies, this lead is made of plastic and graphite. There are two basic types: fired and extruded. They are similar in material content to the plastic-graphite fired lead, but they are processed differently. They are designed for use on film only, erase well, do not readily smear, and produce a good opaque line which is suitable for microfilm reproduction. The main drawback with this type of lead is that it does not hold a point well.

Drafting Pencils. The leads are held either in the conventional wood-bonded cases known as wooden pencils or in metal or plastic cases known as mechanical pencils. See Figs. 2-2-13 and 2-2-14. With the latter, the lead is ejected to the desired length of projection from the clamping chuck and then pointed in the

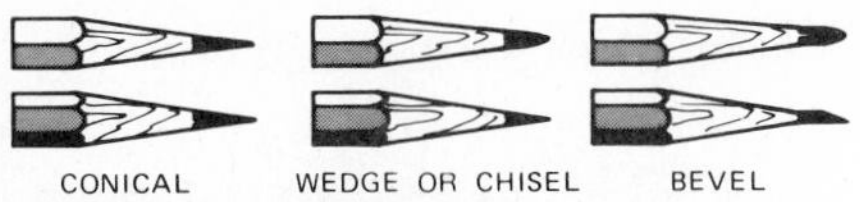
CONICAL
WEDGE OR CHISEL
BEVEL

Fig. 2-2-13 Pencil point shapes.

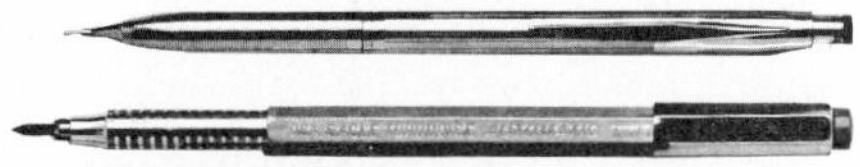

Fig. 2-2-14 Mechanical pencil-ejector type. (Berol Corp.)

same manner as the wood-bonded pencil. Recently, disposable mechanical pencils became available. These operate just as any mechanical pencil, but they are discarded after the lead has been used.

Lead Pointers. A fast, convenient means of putting a clean drafter's point on refillable pencils and on the exposed leads of wood-cased pencils is with a lead pointer. See Fig. 2-2-15. On some models short, medium, or long tapered lead points can be made by merely adjusting the length of lead. A sandpaper block or file is used to keep sharp points on wooden pencils.

ERASERS AND CLEANERS

Erasers. A variety of erasers have been designed to do special jobs—remove surface dirt, minimize surface damage on cloth or film, and remove ink or pencil lines.

Erasing Machines. In erasing, hardly any pressure is required because the high-speed rotation of the shaft actually does the clean-up job rapidly and flawlessly. These machines make erasures with pinpoint accuracy, and no drawing damage can occur because the motor is designed to stall when too much erasing pressure is applied (Fig. 2-2-16).

Cleaners. An easy way to clean tracings is to sprinkle them lightly with gum eraser particles

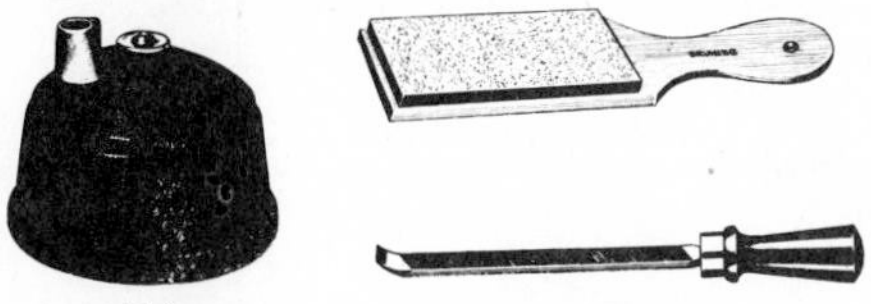

(A) (B)

Fig. 2-2-15 Lead pointers. (a) Mechanical pointer (Addressograph Multigraph Corp.). **(b) Sanding block and file** (Charles Bruning Company [Canada] Limited.)

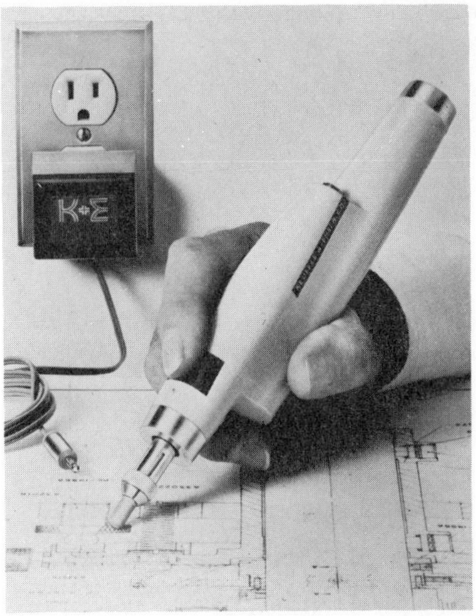
K+E

Fig. 2-2-16 Erasing machine. (Keuffel & Esser Co.)

squares, scales, etc., stay spotless and clean the surface automatically as they are moved back and forth. They contain no grit or abrasive, and will actually improve the ink-taking quality of the drafting surface.

Inking Powders. Surfaces of tracing cloths and prints made on reproduction papers and cloths require preparation before they are drawn on with ink. This is done by sprinkling an inking powder on the surface and rubbing it with a felt pad. The powder is then completely removed from the drawing surface.

Erasing Shields. These thin pieces of metal have a variety of openings to permit the erasure of fine detail lines or note work without disturbing nearby work that is to be left on the drawing. See Fig. 2-2-17. Through the use of this device, erasures can be performed quickly and accurately.

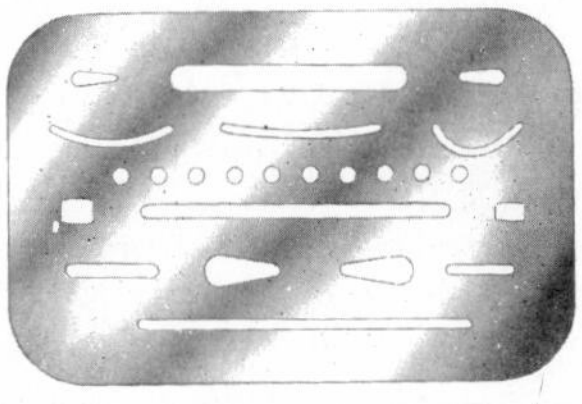

Fig. 2-2-17 Erasing shield. (Charles Bruning Co. [Canada] Limited.)

BRUSHES

A light brush (Fig. 2-2-18) is used to keep the drawing area clean. By using a brush to remove eraser particles and any accumulated dirt, the drafter avoids smudging the drawing.

Fig. 2-2-18 Drafter's brush. (Charles Bruning Co. [Canada] Limited.)

LETTERING AIDS

Lettering sets or guides (Fig. 2-2-19) are also used when it is desirable to have more uniform and accurate letters and numerals than can be obtained by the freehand method. Lettering sets contain a number of guide templates that give a variety of letter shapes and sizes, as well as different slope angles.

Instant Lettering is a new method of dry transfer lettering which offers a wide variety of lettering at good quality and speed. It adheres firmly to paper, wood, glass, and metal and is available in different colors. In case of errors, letters can be removed with cellophane tape or a pencil eraser.

Lettering typewriters have been used in drafting offices for some time for the lettering of bills of material and typing on appliques. But now small, movable typewriters can letter anywhere on the drawing without the need for removing the drawing from the board.

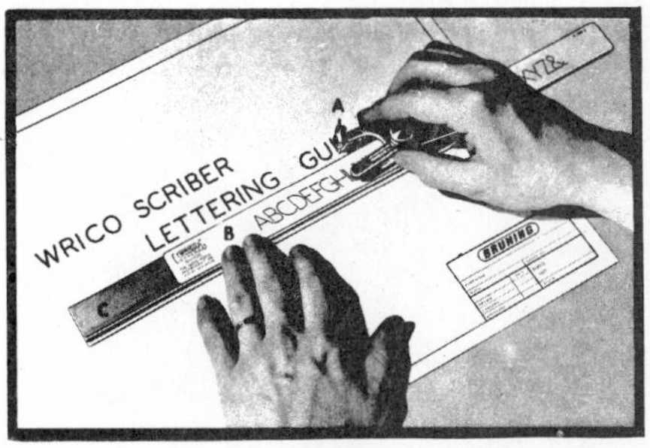
WRICO SCRIBER
LETTERING GU
BRUNING

(A)

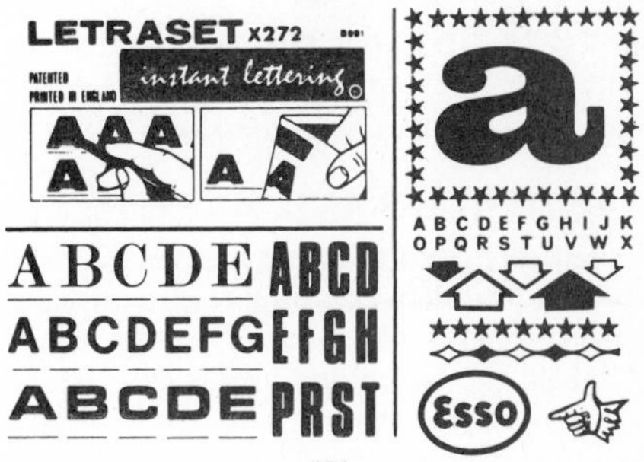
LETRASET X272
instant lettering
AAA
A
A
a
ABCDEFGHIJK
OPQRSTUVWX
ABCDE ABCD
ABCDEFG EFGH
ABCDE PRST
Esso

(B)

Fig. 2-2-19 Lettering aids. (a) Mechanical lettering (Addressograph Multigraph Corp.). **(b) Appliques** (Letraset Canada Limited.)

TEMPLATES

To save time, many drafters now use templates for drawing small circles and arcs. Templates are also available for drawing standard square, hexagonal, triangular, and elliptical shapes and standard electrical and architectural symbols. See Fig. 2-2-20.

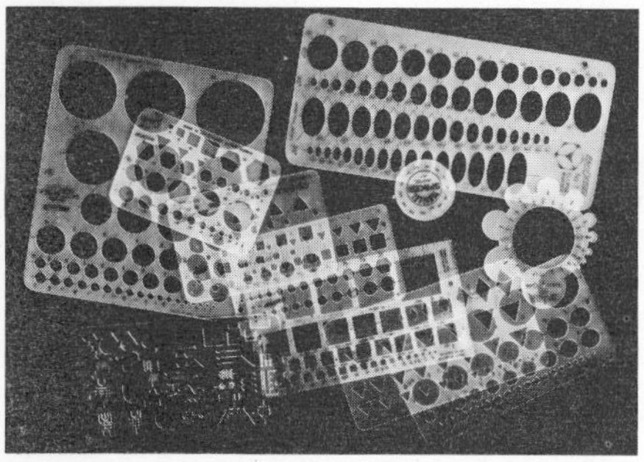

Fig. 2-2-20 Templates. (Rapid Design.)

IRREGULAR CURVES

For drawing curved lines in which, unlike circular arcs, the radius of curvature is not constant, an instrument known as an irregular or French curve (Fig. 2-2-21) is used. The patterns for these curves are based on various combinations of ellipses, spirals, and other mathematical curves. The curves are available in a variety of shapes and sizes. Generally, the drafter plots a series of points of intersection along the desired path and then uses the French curve to join these points so that a smooth-flowing curve results.

Fig. 2-2-21 Irregular curves. (Teledyne Post.)

CURVED RULES AND SPLINES

Curved rules and splines (Fig. 2-2-22) solve the problem of ruling a smooth curve through a given set of points. They lie flat on the board and are as easy to use as a set square; yet they can be bent to fit any contour to a 75 mm (3 in.) minimum radius and will hold the position without support. A clear, plastic ruling edge stands away from the board just far enough to prevent ink lines from smearing.

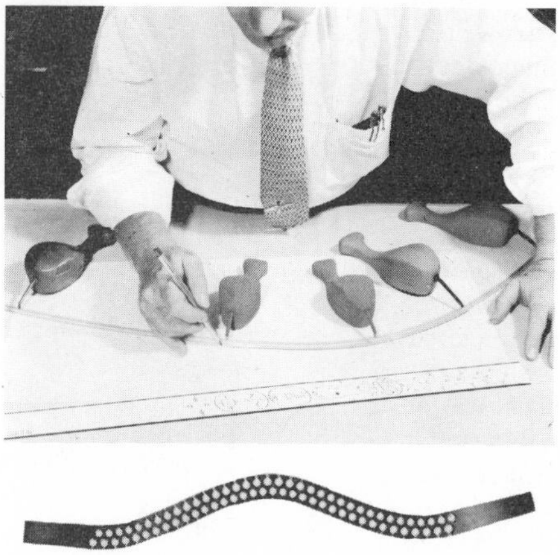

Fig. 2-2-22 Curved rules and splines. (Keuffel & Esser Co.)

INKING EQUIPMENT

Although most production drawings are drawn with pencil, in the last few years ink drawings have been on the increase. The use of this type of drawing for technical illustrations and the demand for good, clear drawings for microfilming have brought about the introduction of new and improved inking methods and techniques. Typical ink equipment is shown in Fig. 2-2-23.

Inking Pens. Two types of pens are used to produce ink lines. The ruling pen with an adjustable blade for drawing different-width lines has been available for many years. It is filled by a dropper cap, squeeze bottle, or cartridge tube. Inking compasses are available with permanent or detachable blades, which interchange with the lead attachment part of the compass. Specially designed pens for drawing curved, multiple broken or hidden lines are also available.

The second type, called a *technical fountain pen*, is a needle-in-tube type pen, is relatively new in comparison with the ruling pen, and has gained wide acceptance with drafters because the line widths are fixed. Also it is suitable for drawing both lines and letters. Different-size needle points are available which produce different-width lines. Several types of technical fountain pens and other needle-in-tube type pens now provide compass attachments so that these may be clamped to, or inserted on, a standard compass leg, thus providing the advantages of the compass while eliminating the inconvenience of the blade pen.

The use of a technical drawing pen with a circular template has, in many instances, replaced the drawing of circles and arcs by means of a compass. For best results, the pen should be perpendicular to the paper.

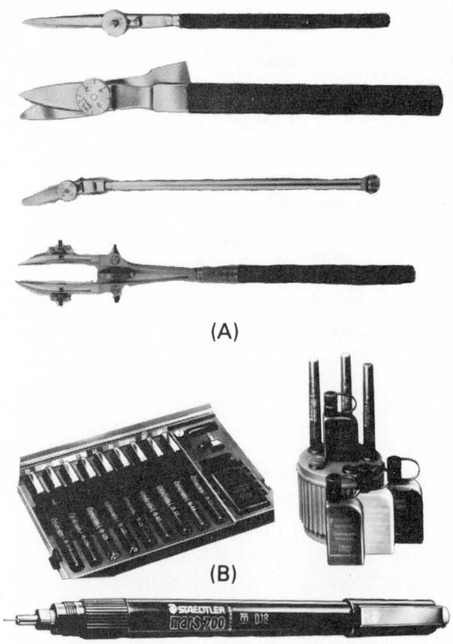
(A)
(B)

Fig. 2-2-23 Inking instruments. (a) Blade type (Charles Bruning Co. [Canada] Limited). **(b) Needle-in-tube-type pen** (J. S. Staedtler, Inc.)

CALCULATORS

Calculators, such as those shown in Fig. 2-2-24, are used by drafters to make fast mathematical calculations, using division, multiplication, and extractions of square roots, and to solve problems involving areas, volumes, masses, strengths of materials, pressures, etc. Because of cost, time, accuracy, and ease of operation, the calculator has replaced the slide rule in the engineering office.

Fig. 2-2-24 Calculators.

UNIT 2-3

AUTOMATED DRAFTING[1]

Most automatic drafting machines are digitally controlled. See Fig. 2-3-1. They were developed from, and resembled in their basic mode of operation, numerically controlled machine tools. The last few years have seen few earth-shaking developments in automatic drafting equipment. The price/performance ratio has improved considerably, and high-quality, relatively inexpensive equipment has become available. Manufacturers have introduced modular systems which can be easily upgraded.

Most automatic drafting is still done with line polo plotters. Dot-matrix plotters, such as the electrostatic machines, although very fast, don't usually have the accuracy needed to meet drafting requirements.

All machines can be made to automatically generate common shapes, such as squares, circles, ellipses, and other second-degree curves. Many can generate complex shapes, can do high-speed contouring and curve fitting, and can carry out interpolations, extrapolations, and fairing. Some systems can rotate the shape about any axis, scale down or blow up, and produce mirror-image drawings.

Any symbol or shape—circuit mechanism, rivet, screw, etc.—can be stored in memory and called out at any time. It is then drawn at the specified point. When a drawing must be updated, the ADM can make a new drawing, leaving blank those sections which must be revised; changes can be entered manually through the keyboard.

Now systems are available that can take a designer's rough sketch and transform it into finished drawings, photo-ready artwork, numerical control (NC) tapes, and various other output forms. These systems are simple to use and require virtually no programming skill.

One of the heaviest and most successful uses of automated drafting has been in electric circuit design. This is mainly because circuits are highly repetitive. A relatively small number of standard symbols are used over and over; these symbols, stored digitally in the machine's memory, can be called out quickly.

In addition to drawing electrical symbols where the designer indicates, the automatic plotter can route wiring properly and place components in their optimum relationship. This can be extremely useful, especially in designing multilayer printed-circuit boards (Fig. 2-3-2). In fact, as circuits get smaller and smaller, such tasks become nearly impossible except through computerized methods. After the circuit has been designed, the plotter can (assuming that it is sufficiently accurate) draw the artwork masters necessary to produce the printed circuits.

Under certain conditions, automated drafting is justified for certain types of mechanical drawing where a number of standard components such as standard shapes—screws, bolts, letters, dimensions, and other symbols—are involved. Such applications are adaptable to automated drafting, as are cases where the machine's logic can be used to position components, calculate dimensions, or do some other time-consuming task.

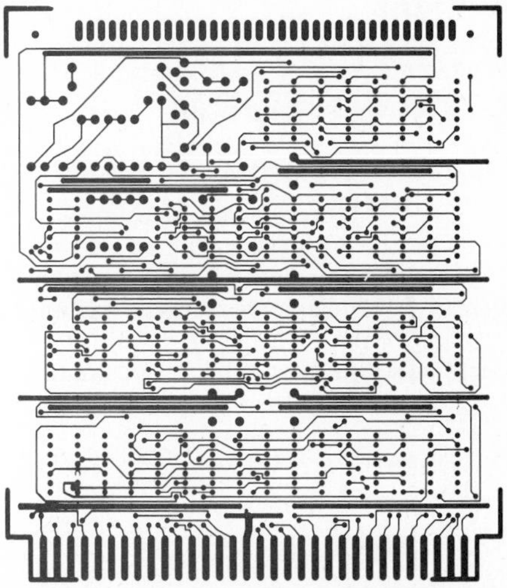

Fig. 2-3-2 Printed circuit produced by automated drafting machine in 25 minutes. (Gerber Scientific Instrument Co.)

Interactive drafting, as shown in Fig. 2-3-3, allows the user to see the final version of the sketch at any point. Typically, the user enters the rough sketch and calls for all or part of it to be displayed. He or she can then make any changes by light pen, digitizer, or teletypewriter. The user can also manipulate the sketch or zoom in on any particular part of it for a closer look. Once satisfied, the user can instruct the drafting machine to make a finished drawing.

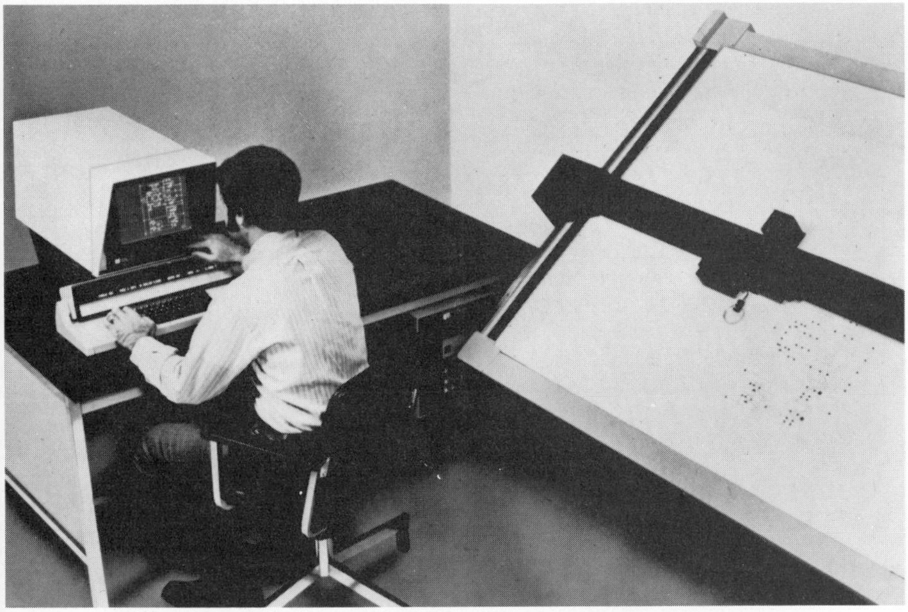

Fig. 2-3-1 Automated drafting equipment. (Gerber Scientific Instrument Co.)

REFERENCE

1. *Machine Design,* July 1971.

Fig. 2-3-3 Interactive drafting. (Systems Engineering Laboratories.)

UNIT 2-4

DRAWING AND LAYOUT FORM

Standard Drawing Sizes

Metric. Metric drawing sizes are based on the A0 size, having an area of 1 square metre (m^2) and a length-to-width ratio of $1:\sqrt{2}$. Each smaller size has an area half of the preceding size, and the length-to-width ratio remains constant. See Figs. 2-4-1 and 2-4-2.

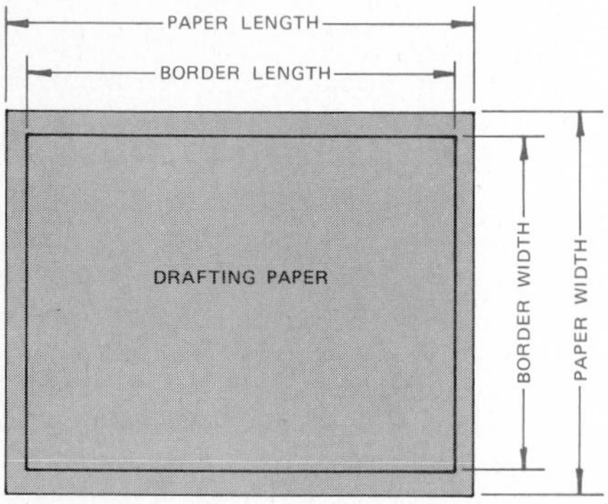

PAPER LENGTH
BORDER LENGTH
DRAFTING PAPER
BORDER WIDTH
PAPER WIDTH

METRIC DRAWING SIZES (MILLIMETRES)		
DRAWING SIZE	BORDER SIZE	OVERALL PAPER SIZE
A4	195 x 282	210 x 297
A3	277 x 400	297 x 420
A2	400 x 574	420 x 594
A1	574 x 821	594 x 841
A0	811 x 1159	841 x 1189

INCH DRAWING SIZES		
DRAWING SIZE	BORDER SIZE	OVERALL PAPER SIZE
A	8.00 x 10.50	8.50 x 11.00
B	10.50 x 16.50	11.00 x 17.00
C	16.25 x 21.25	17.00 x 22.00
D	21.00 x 33.00	22.00 x 34.00
E	33.00 x 43.00	34.00 x 44.00

Fig. 2-4-1 Standard drawing paper sizes.

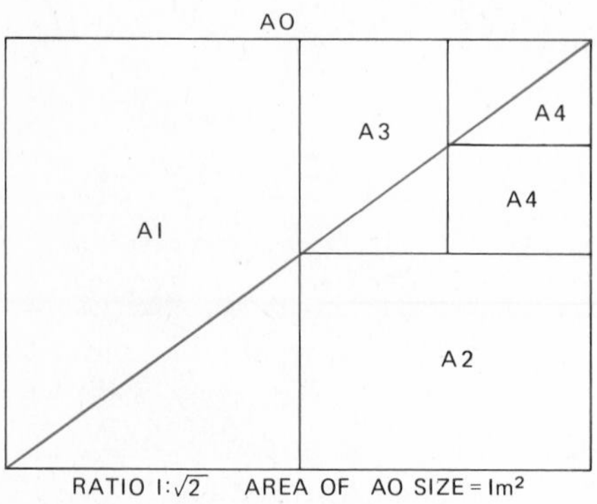

AO
A3
A4
A4
A1
A2
RATIO 1:√2
AREA OF AO SIZE = 1m²

Fig. 2-4-2 Metric drawing paper.

Inches. Drawing sizes in the inch system are based on dimensions of commercial letterheads, 8.5 × 11 in., and standard rolls of paper, cloth, or film 36 and 42 in. wide. They can be cut from these standard rolls with a minimum of waste.

Drawing Format

A general format for drawings is shown in Fig. 2-4-3, which illustrates a drawing trimmed to size. It is recommended that preprinted drawing forms be made to the trimmed size and have rounded corners, as shown, to minimize dog-ears and tears.

ZONING SYSTEM

Drawings larger than A3 size may be zoned for easy reference, by dividing the space between the trimmed size and the inside border into zones measuring 105 × 143.5 mm. These zones are numbered horizontally and lettered vertically, with uppercase letters, from the lower RH (right-hand) corner, as in Fig. 2-4-3, so that any area of the drawing can be identified by a letter and a number, such as B3.

MARGINAL MARKING[1]

In addition to zone identification, the margin may also carry fold marks to enable folding and a graphical scale to facilitate reproduction to a specific size. In the process of microfilming, it is necessary to center the drawing within rather close limits in order to meet standards. To facilitate this operation, it has become common practice to put a centering arrow or mark on at least three sides of the drawing. Most practices include the arrows on each of the four sides. If three sides are used, the arrows should be on the two sides and on the bottom. This helps the camera operator to align the drawing properly since the copyboard usually contains cross hairs through the center of the board at right angles. With any three arrows aligned on the cross hairs, centering is automatic. The arrows should be on the center of the border which outlines the information area of the drawing, not at the edge of the sheet on which the drawing is made.

Title Blocks and Tables

TITLE BLOCK

The title block is located in the lower right-hand corner. The arrangement and size of the title block are optional, but the following information should be included:

1. Drawing number
2. Name of firm or organization
3. Title or description
4. Scale

Provision may also be made within the title block for the date of issue, signatures, approvals, sheet number, drawing size, job, order, or contract number; references to this or other documents; and standard notes such as tolerances or finishes. An example of a typical title block is shown in Fig. 2-4-4. In classrooms, a title strip is often used on A3- and A4-size drawings, such as shown in Fig. 2-4-5.

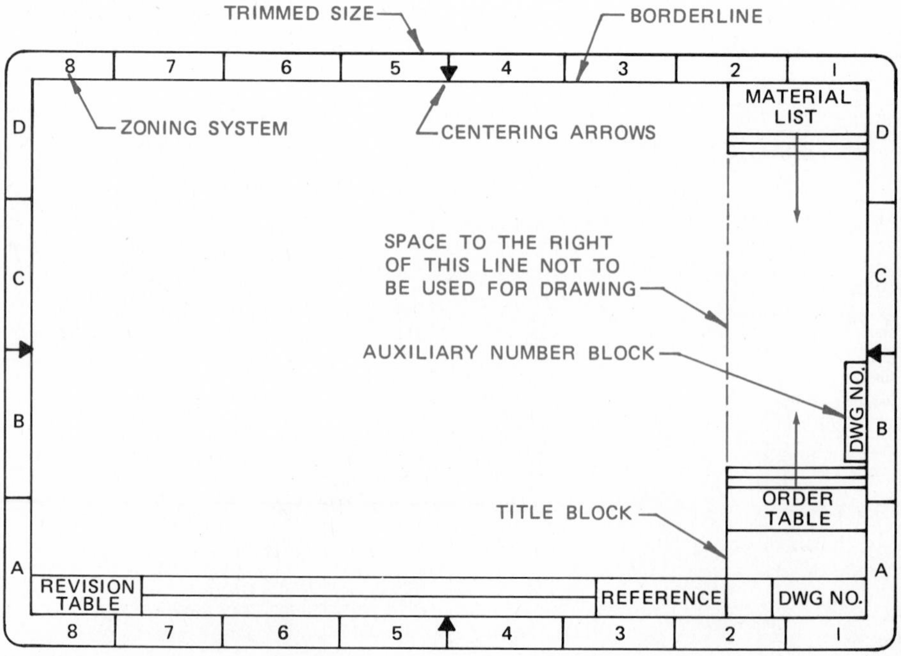

TRIMMED SIZE
BORDERLINE
8 7 6 5 4 3 2 1
D C B A
ZONING SYSTEM
CENTERING ARROWS
MATERIAL LIST
SPACE TO THE RIGHT OF THIS LINE NOT TO BE USED FOR DRAWING
AUXILIARY NUMBER BLOCK
DWG NO.
ORDER TABLE
TITLE BLOCK
REVISION TABLE
REFERENCE
DWG NO.

Fig. 2-4-3 Drawing paper format.

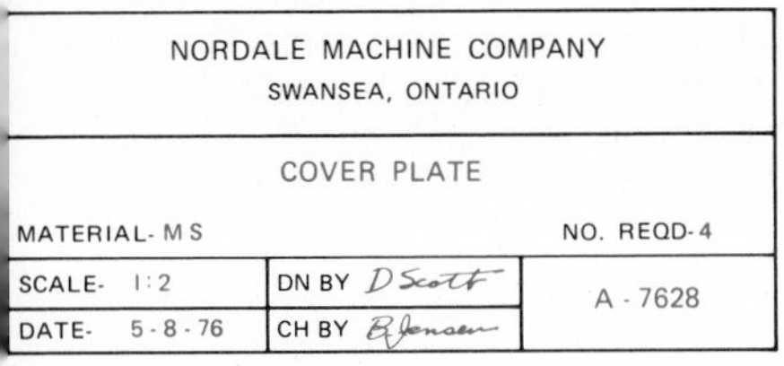
NORDALE MACHINE COMPANY
SWANSEA, ONTARIO
COVER PLATE
MATERIAL- M S
NO. REQD- 4
SCALE- 1:2
DN BY D Scott
A - 7628
DATE- 5-8-76
CH BY

Fig. 2-4-4 **Title block.**

AUXILIARY NUMBER BLOCKS

An auxiliary number block, approximately 50 × 10 mm, is placed above the title block so that when prints are folded, the number will appear close to the top RH corner of the print, as in Fig. 2-4-7. This is done to facilitate identification when the folded prints are filed on edge.

Auxiliary number blocks are usually placed within the inside border, but they may be placed in the margin outside the border line if space permits.

REFERENCE

1. National Microfilm.

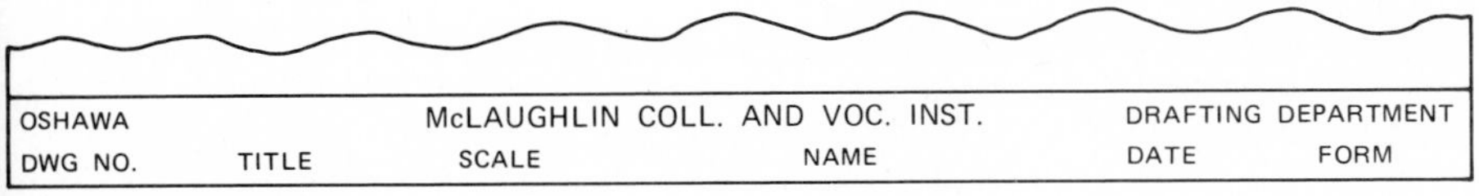
OSHAWA
McLAUGHLIN COLL. AND VOC. INST.
DRAFTING DEPARTMENT
DWG NO.
TITLE
SCALE
NAME
DATE
FORM

Fig. 2-4-5 **Title strip.**

MATERIAL LIST AND ORDER TABLE

The whole space above the title block, with the exception of the auxiliary number block, should be reserved for tabulating materials, charge of order, and revision; drawing in this space should be avoided. On preprinted forms, the right-hand inner border may be graduated to facilitate ruling, as shown in Fig. 2-4-6.

CHANGE OR REVISION TABLE

All drawings should carry a change or revision table, either down the right-hand side or across the bottom of the drawing. In addition to the description of drawing changes, provision may be made for recording a revision symbol zone location, issue number, date, and approval of the change. Typical revision tables are shown in Fig. 2-4-7.

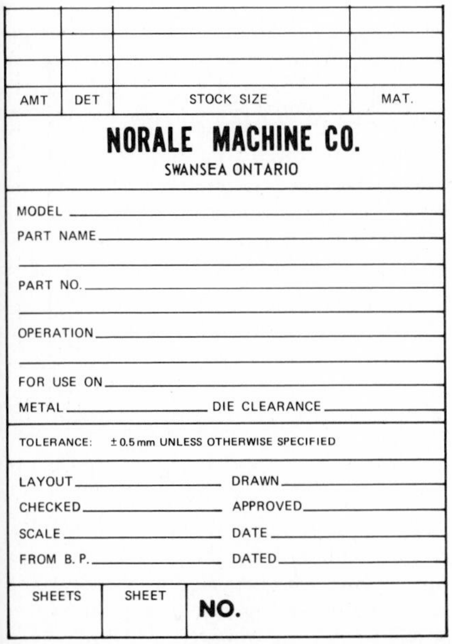
AMT
DET
STOCK SIZE
MAT.
NORALE MACHINE CO.
SWANSEA ONTARIO
MODEL
PART NAME
PART NO.
OPERATION
FOR USE ON
METAL
DIE CLEARANCE
TOLERANCE: ±0.5mm UNLESS OTHERWISE SPECIFIED
LAYOUT
DRAWN
CHECKED
APPROVED
SCALE
DATE
FROM B. P.
DATED
SHEETS
SHEET
NO.

Fig. 2-4-6 **Combined title block, order table, and material list.**

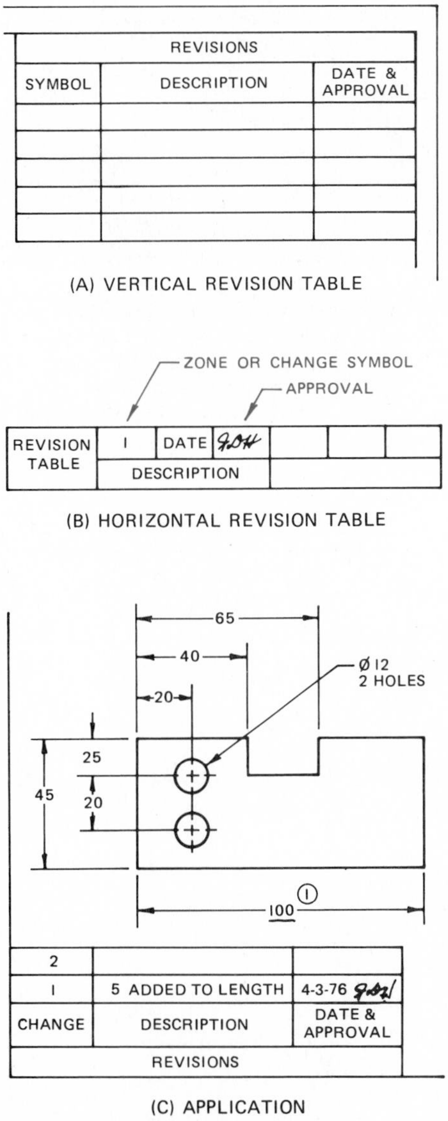
REVISIONS
SYMBOL
DESCRIPTION
DATE & APPROVAL
(A) VERTICAL REVISION TABLE
ZONE OR CHANGE SYMBOL
APPROVAL
REVISION TABLE
1
DATE
DESCRIPTION
(B) HORIZONTAL REVISION TABLE
65
40
20
Ø 12
2 HOLES
25
45
20
1
100
2
1
5 ADDED TO LENGTH
4-3-76
CHANGE
DESCRIPTION
DATE & APPROVAL
REVISIONS
(C) APPLICATION

Fig. 2-4-7 **Revision tables.**

UNIT 2-5

DRAWING MEDIA[1]

The three families of drafting and tracing materials—paper, cloth, and film—differ sufficiently between and within themselves to provide a wide choice of qualities and characteristics for selection of the perfect material for drawing requirements.

Paper

Drafting papers come with a wide range of qualities—strength, erasability, permanence, translucency, etc. The distinguishing feature between drawing and tracing paper is translucency.

Opaque papers are used primarily as drawing papers. Because of a change in drafting practices, the need for drawing papers today is limited to the two extremes in the scale of quality—the very highest grade, permanent drawing papers with the best possible erasing quality for maps or master drawings, which are later photographed, and the inexpensive school type of papers for the educational market. Since master drawings often must be revised and corrected, the major consideration in high-grade drawing papers is erasability.

The more transparent papers are used as tracing papers. Formerly, they were used almost exclusively for the ink tracing of pencil drawings made on opaque papers. Transparency was the prime requirement.

Today, the usual drafting practice is to develop the master drawing directly on transparent paper from which reproductions can be made, thereby saving the time, expense, and checking involved in the tracing process. This practice means that, in addition to good actinic transparency, the modern high-grade tracing paper must be able to withstand considerable handling. The paper should retain these qualities for a long time to avoid the eventual necessity of redrawing, either manually or photographically.

Cloth

Cloth combines the advantages of transparency, surface quality, strength, and permanence. It can withstand repeated erasures without major loss of surface quality, and aging does not materially affect its drawing and reproduction qualities. Though paper may exceed cloth in one or more of these characteristics, it does not equal its combined advantages.

Film[2]

The most recent drafting medium is film. The advantages of polyester film as a drawing material are many. Raw polyester has natural dimensional stability, great tearing strength, high transparency, age and heat resistance, and nonsolubility and waterproofness. The outstanding virtue of film over any other drafting medium is that film is almost indestructible. Its amazing permanence safeguards the important investment in engineering drawings and records. It is permanently transparent; is waterproof; is unaffected by aging; is superb for pencil drafting, ink work, and typewriting; and has unequaled erasability.

Polyester materials, however, present some problems. The material must be sufficiently dense to avoid reflection from the copyboard in microfilming, but translucent enough for backlighting or contact printing.

Preprinted Grid Sheets[3]

This type of drafting medium makes the job of engineering and freehand drawings easier and quicker. Cross-sectional guidelines can be preprinted directly on the paper or used as a liner beneath the drawing paper to provide an accurate guide for all drawing work. These cross-sectional guides are available in several grid sizes. The squared and pictorial styles (isometric, perspective, and oblique) are the more common preprinted grid paper used by drafters. These grids, when applied directly on the paper or cloth, as shown in Fig. 2-5-1, or on film with a special nonreproducible ink, will not appear on the prints when the drawing is reproduced by diazo and phototype methods.

For freehand work, whether rough sketches in the field or in the drafting room, whether these are preliminary or to be used as the finished drawing, grid line papers are an invaluable aid. The cross-sectional patterns serve as ready-made guides for base lines, dimensioning, and angles. See Fig. 2-5-2.

REFERENCES

1. Keuffel and Esser Co.
2. *Machine Design* and National Microfilm.
3. Eastman Kodak Company.

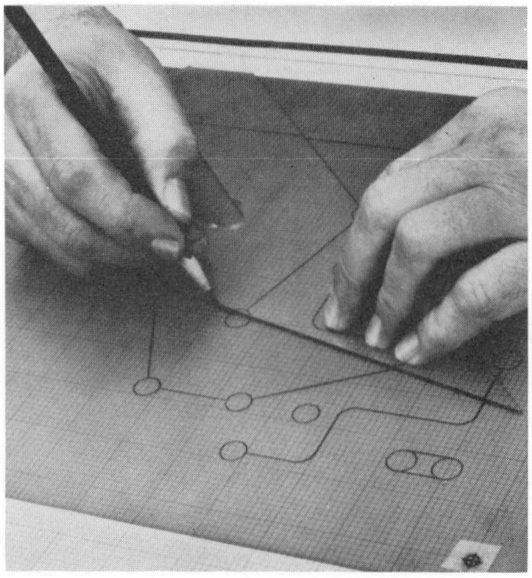

Fig. 2-5-2 Preprinted grid sheet being used as an underlay. (Keuffel & Esser Co.)

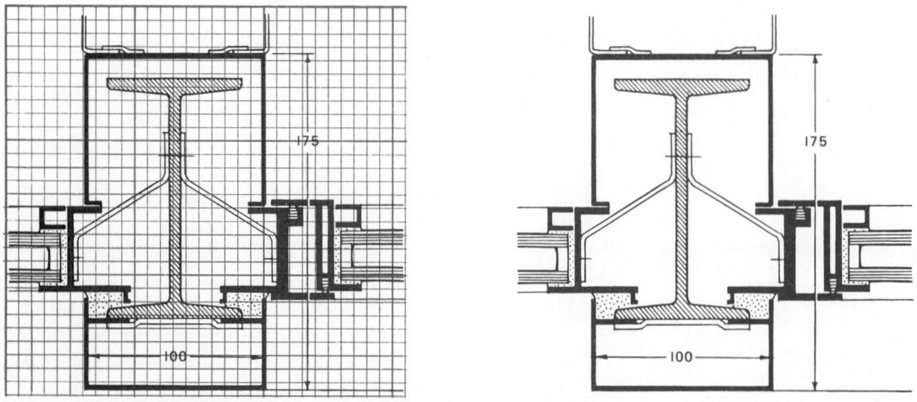

175
100
175
100

Fig. 2-5-1 Drawing directly on preprinted grid sheet.

UNIT 2-6

BASIC DRAFTING SKILLS

Lettering

SINGLE-STROKE GOTHIC LETTERING

The most important requirements for lettering are legibility, reproducibility, and ease of execution. These are particularly important because of the increased use of microfilming which requires optimum clarity and adequate size of all details and lettering. It is recommended that all drawings be made to conform to these requirements and that particular attention be paid to avoid the following common faults:

1. Unnecessarily fine detail
2. Poor spacing of details
3. Carelessly drawn figures and letters
4. Inconsistent delineation
5. Incompleted erasures that leave ghosted images
6. Use of differing densities, such as pencil, ink, and typescript on the same drawing

These requirements are met in the recommended single-stroke Gothic characters shown in Fig. 2-6-1, or adaptions thereof, which improve reproduction legibility. One such adaptation by the National Microfilm Association is the vertical Gothic-style Microfront alphabet (Fig. 2-6-2) intended for general usage.

Either inclined or vertical lettering is permissible, but only one style of lettering should be used throughout a drawing. The preferred slope for the inclined characters is 2 in 5, or approximately 68° with horizontal.

Uppercase letters should be used for all lettering on drawings unless lowercase letters are required to conform with other established standards, equipment nomenclature, or marking.

Lettering for titles, subtitles, drawing numbers, and other uses may be made freehand, by typewriter, or with the aid of mechanical lettering devices such as templates and lettering machines. Regardless of the method used, all characters are to conform, in general, with the recommended Gothic style and must be legible in full- or reduced-size copy by any accepted method of reproduction.

The recommended minimum freehand and mechanical letter heights for various applications are given in Fig. 2-6-3. So that lettering will be uniform and of proper height, light guidelines properly spaced, are drawn first and then the lettering is drawn between these lines.

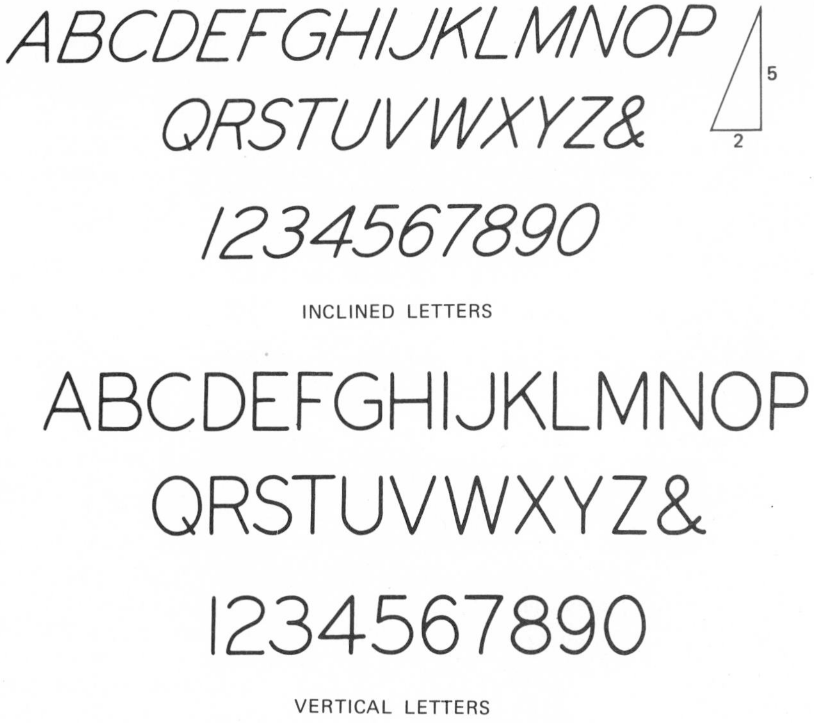
ABCDEFGHIJKLMNOP
QRSTUVWXYZ&
1234567890
5
2
INCLINED LETTERS
ABCDEFGHIJKLMNOP
QRSTUVWXYZ&
1234567890
VERTICAL LETTERS

Fig. 2-6-1 Approved lettering for engineering drawings.

ABCDEFGHIJKLMNO
PQRSTUVWXYZ
1234567890

Fig. 2-6-2 Microfont letters. (National Microfilm Assoc.)

USE	MINIMUM LETTER HEIGHT				DRAWING	
	FREEHAND		MECHANICAL			
	mm	INCHES	mm	INCHES	mm	INCHES
DRAWING NUMBERS IN TITLE BLOCK	8	.30	9	.35	ALL	
DRAWING TITLE AND SECTION AND TABULATION LETTERS	6	.25	6	.24		
ZONE LETTERS AND NUMBERS IN BORDER	5	.20	4	.17		
DIMENSION, TOLERANCES, LIMITS, NOTES, SUBTITLES FOR SPECIAL VIEWS, TABLES, REVISIONS, AND ZONE LETTERS FOR THE BODY OF THE DRAWING	3	.12	3	.12	UP TO AND INCLUDING 420 x 594	17 x 22
	4	.15	4	.15	LARGER THAN 420 x 594	17 x 22

Fig. 2-6-3 Recommended lettering heights.

Letters in words should be spaced so that the background areas between the letters are approximately equal, and words are to be clearly separated by a space equal to the height of the lettering. See Fig. 2-6-4. The vertical space between lines of lettering should be no more than the height of the lettering and no less than half the height of the lettering.

Notes should be placed horizontally on drawings and separated vertically by spaces at least equal to double the height of the character size used, to maintain the identity of each note.

Decimal points must be uniform, dense, and large enough to be clearly visible on approved types of reduced copy. Decimal points should be placed in line with the bottom of the associated digits and be given adequate space.

Lettering should not be underlined except when special emphasis is required. The underlining should not be less than 1.5 mm (0.06 in.) below the lettering.

When drawings are being made for microfilming, the size of the lettering is an important consideration. A drawing may be reduced to half size when microfilmed at 30X reduction and blown back at 15X magnification. (Most microfilm engineering readers and blowback equipment have a magnification of 15X. If a drawing is microfilmed at 30X reduction, the enlarged blown-back image is 50 percent, at 24X, it is 62 percent of its original size.) A consensus of engineers and drafters was taken to find out what they considered to be the smallest size lettering on drawings. Objections were strongly stressed on sizes under 2 mm in height.

Standards generally do not allow characters smaller than 3 mm (.12 in.) for drawings to be reduced 30X, and the

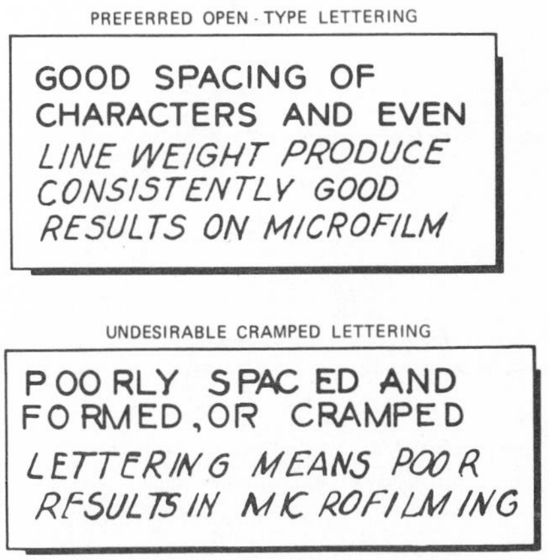
PREFERRED OPEN - TYPE LETTERING
GOOD SPACING OF CHARACTERS AND EVEN
LINE WEIGHT PRODUCE CONSISTENTLY GOOD RESULTS ON MICROFILM
UNDESIRABLE CRAMPED LETTERING
POORLY SPACED AND FORMED, OR CRAMPED
LETTERING MEANS POOR RESULTS IN MICROFILMING

Fig. 2-6-4 Spacing of lettering. (National Microfilm Assoc.)

trend is toward larger characters. Figure 2-6-5 shows the proportionate size of letters after reduction and enlargement.

The lettering heights, spacing, and proportions in Fig. 2-6-5 normally provide acceptable reproduction or camera reduction and blowback. However, manually, mechanically, optimechanically, or electromechanically applied lettering (typewriter, etc.) with heights, spacing, and proportions less than those recommended are acceptable when the reproducibility requirements of the accepted industry or military reproduction specifications are met.

Line Work

The various lines used in drawing form the alphabet of the drafting language: like letters of the alphabet, they are different in appearance. See Figs. 2-6-6 and 2-6-11. The distinctive features of all lines that form a permanent part of the drawing are the differences in their thickness and construction. Lines must be clearly visible and stand out in sharp contrast to one another. This line contrast is necessary if the drawing is to be clear and easily understood.

The drafter first draws very light construction lines, setting out the main shape of the object in various views. Since these first lines are very light, they can be erased easily should changes or corrections be necessary. When the drafter is satisfied that the layout is accurate, the construction lines are then changed to their proper type, according to the "alphabet" of lines. Guidelines, used to ensure uniform lettering, are also drawn very lightly.

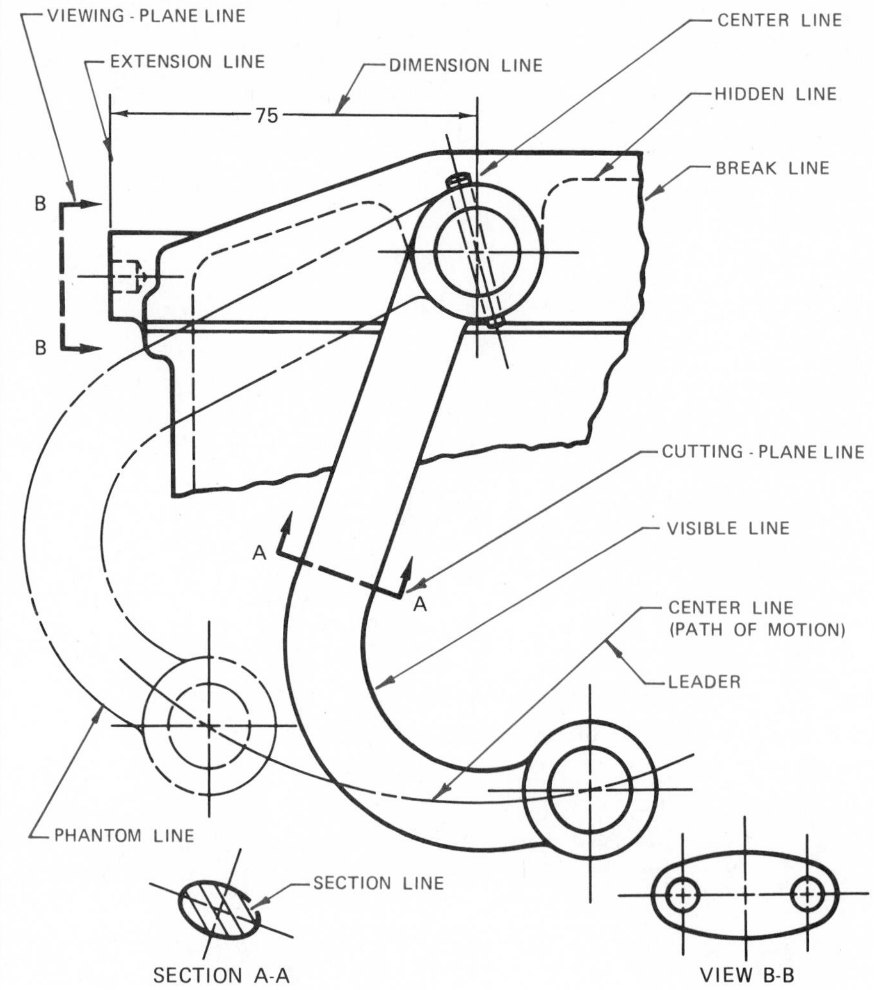

VIEWING - PLANE LINE
EXTENSION LINE
DIMENSION LINE
CENTER LINE
HIDDEN LINE
BREAK LINE
75
B
B
CUTTING - PLANE LINE
VISIBLE LINE
A
A
CENTER LINE (PATH OF MOTION)
LEADER
PHANTOM LINE
SECTION LINE
SECTION A-A
VIEW B-B

Fig. 2-6-6 **Application of lines.** (ANSI Y14.2, 1973.)

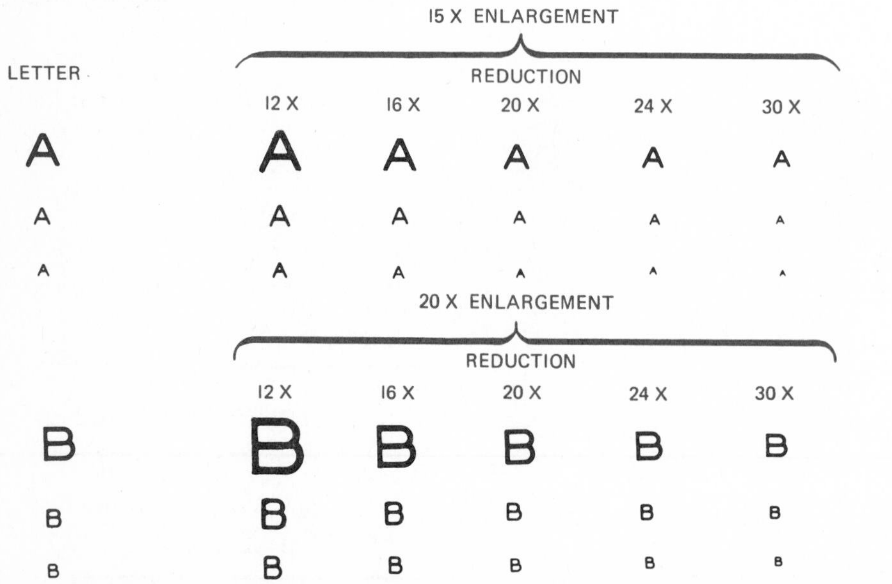

15 X ENLARGEMENT
LETTER
REDUCTION
12 X
16 X
20 X
24 X
30 X
A
20 X ENLARGEMENT
REDUCTION
12 X
16 X
20 X
24 X
30 X
B

Fig. 2-6-5 Proportionate size of letters after reduction and enlargement. (National Microfilm Assoc.)

LINE WIDTHS

Two widths of lines, thick and thin, as shown in Fig. 2-6-8, are recommended for use on drawings. Thick lines are 0.5 to 0.8 mm wide, thin lines between 0.3 and 0.5 mm wide. The actual width of each line is governed by the size and style of the drawing and the smallest size to which it is to be reduced. All lines of the same type should be uniform throughout the drawing. Spacing between parallel lines should be such that there is no fill-in when the copy is reproduced by available photographic methods. Spacing of no less than 3 mm (.12 in.) normally meets reproduction requirements.

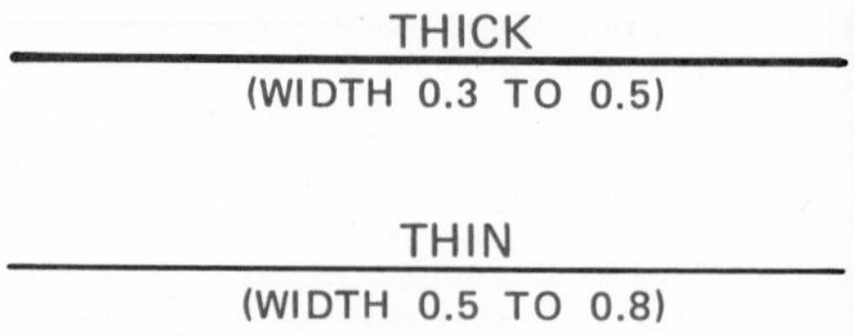

THICK
(WIDTH 0.3 TO 0.5)
THIN
(WIDTH 0.5 TO 0.8)

Fig. 2-6-7 Line widths.

All lines should be clean-cut, opaque, uniform, and properly spaced for legible reproduction by all commonly used methods, including microfilming in accordance with industry and government requirements. There should be a distinct contrast between the two widths of lines.

TYPES OF LINES

The types of lines used on engineering drawings are illustrated in Figs. 2-6-6 and 2-6-11. All lines must be clear and dense to obtain good reproduction. When additions or revisions are made to existing drawings, the line thicknesses and density should match the original work.

Visible Lines. The visible lines should be used for representing visible edges or contours of objects. Visible lines should be drawn so that the views they outline clearly stand out on the drawing with a definite contrast between these lines and secondary lines.

Hidden Lines. Hidden lines consist of short, evenly spaced dashes and are used to show the hidden features of an object. See Fig. 2-6-8. The lengths of the dashes may vary slightly in relation to the size of the drawing. Hidden lines should always begin and end with a dash, in contrast with the visible or hidden line from which they start, except when such a dash would form a continuation of a visible line. Dashes should join at corners, and arcs should start with dashes at tangent points. Hidden lines should be omitted when they are not required for the clarity of the drawing.

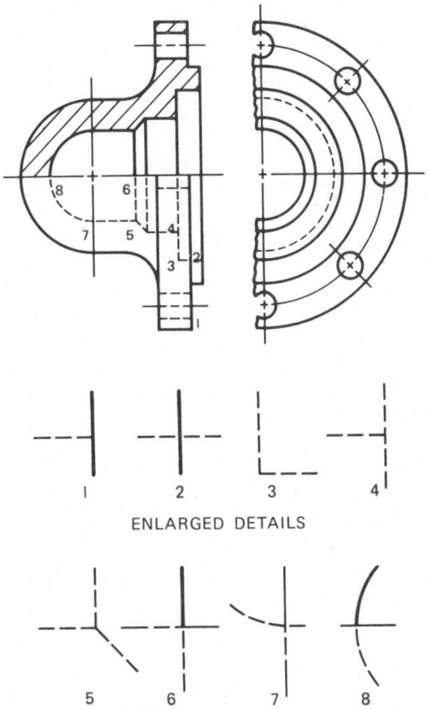

8
6
7
5
4
3
2
1
1
2
3
4
ENLARGED DETAILS
5
6
7
8

Fig. 2-6-8 Hidden-line technique.

Although features located behind transparent materials may be visible, they should be treated as concealed features and shown with hidden lines.

Center Lines. Center lines consist of alternating long and short dashes (Fig. 2-6-9). They are used to represent the axis of symmetrical parts and features, bolt circles, and paths of motion. The long dashes of the center lines may vary in length, depending upon the size of the drawing. Center lines should start and end with long dashes and should not intersect at the spaces between dashes. Center lines should extend uniformly and distinctly a short distance beyond the object or feature of the drawing unless a longer extension is required for dimensioning or for some other purpose. They should not terminate at other lines of the drawing, nor should they extend through the space between views. Very short center lines may be unbroken if no confusion results with other lines.

Extension and Dimension Lines. These lines are used when dimensioning a drawing and are explained in detail in Chap. 5.

Leaders. Leaders are used to indicate the part of a drawing to which a note refers. See Chap. 5 for further details.

Break Lines. Break lines are shown in Fig. 2-6-11. They are used to shorten the view of long uniform sections or when only a partial view is required. Such lines are used on both detail and assembly drawings. The thin line with freehand zigzags is recommended for long breaks, the thick freehand line for short breaks, and the jagged line for wood parts.

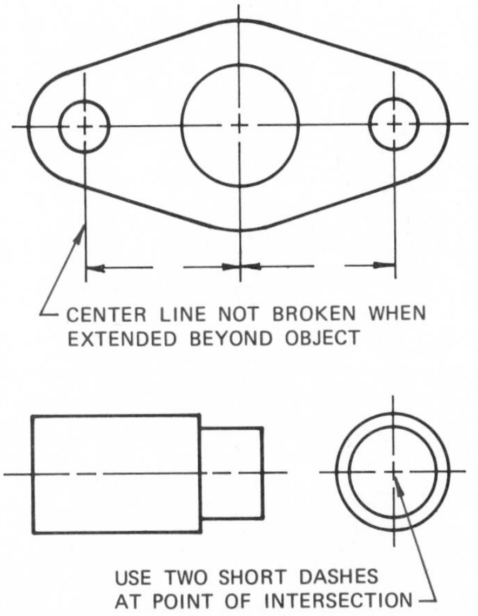

CENTER LINE NOT BROKEN WHEN EXTENDED BEYOND OBJECT
USE TWO SHORT DASHES AT POINT OF INTERSECTION

Fig. 2-6-9 Center-line technique.

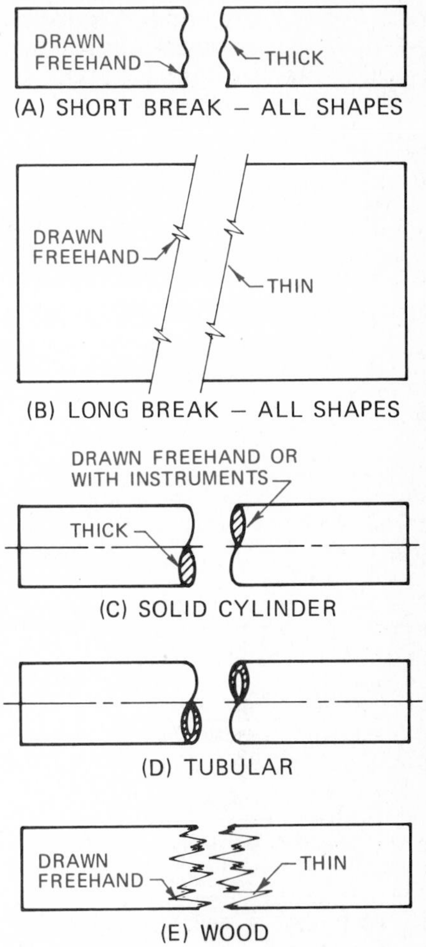

DRAWN FREEHAND
THICK
(A) SHORT BREAK – ALL SHAPES
DRAWN FREEHAND
THIN
(B) LONG BREAK – ALL SHAPES
DRAWN FREEHAND OR WITH INSTRUMENTS
THICK
(C) SOLID CYLINDER
(D) TUBULAR
DRAWN FREEHAND
THIN
(E) WOOD

Fig. 2-6-10 Conventional break lines.

The special breaks shown for cylindrical and tubular parts are useful when an end view is not shown; otherwise, the thick break line is adequate.

Cutting-Plane Lines. Cutting-plane lines are used to indicate the location of cutting planes for sectional views and are explained in detail in Chap. 6.

Section Lining. Section lining is used to indicate the surface in the section view imagined to be cut along the cutting plane line. Refer to Chap. 6 for further details.

Viewing Plane Lines. Viewing plane lines are used to indicate the viewing position for removed partial views.

Phantom Lines. Phantom lines consist of long dashes separated by pairs of short dashes. See Fig. 2-6-12. The long dashes may vary in length, depending on the size of the drawing. Phantom lines are used to indicate alternate positions of moving parts, adjacent positions of related parts, and repeated detail. These lines are also used for features such as bosses and lugs (later removed), for delineating machining stock and blanking

TYPE OF LINE	APPLICATION	DESCRIPTION
VISIBLE LINE THICK	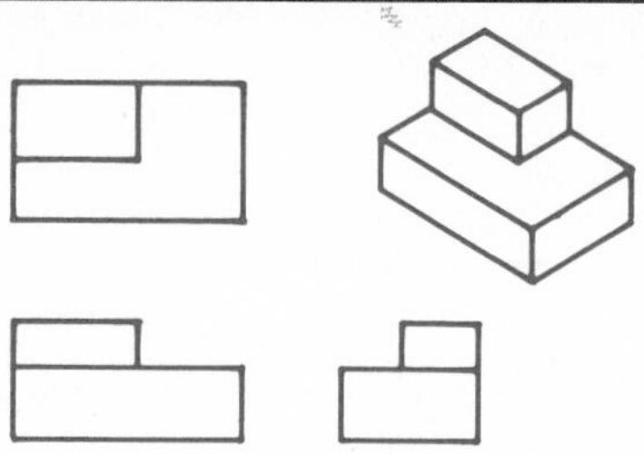	THE OBJECT LINE IS USED TO INDICATE ALL VISIBLE EDGES OF AN OBJECT. THEY SHOULD STAND OUT CLEARLY IN CONTRAST TO OTHER LINES SO THAT THE SHAPE OF AN OBJECT IS APPARENT TO THE EYE.
HIDDEN LINE THIN DASHES 2 TO 5 mm .10 TO .20 IN. SPACES 1 TO 2 mm .03 TO .06 IN.	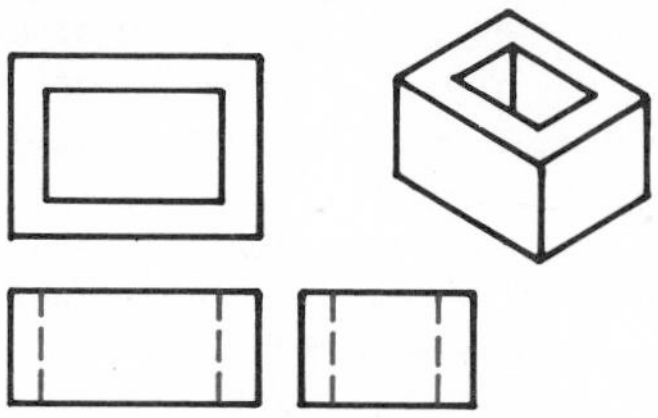	THE HIDDEN OBJECT LINE IS USED TO SHOW SURFACES, EDGES, OR CORNERS OF AN OBJECT THAT ARE HIDDEN FROM VIEW.
CENTER LINE THIN ALTERNATE LONG AND SHORT DASHES THIN SOLID USED IN SMALL AREAS	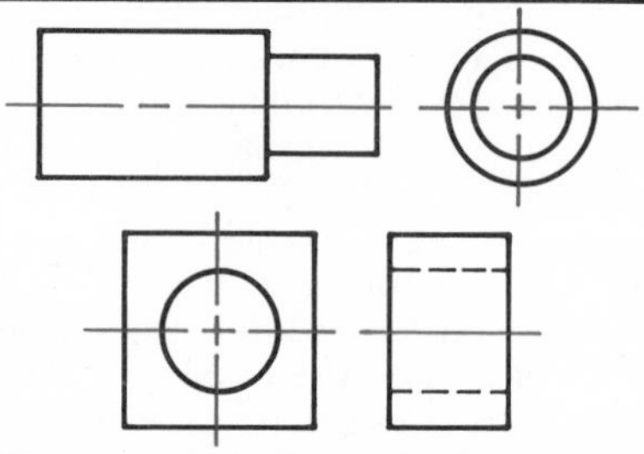	CENTER LINES ARE USED TO SHOW THE CENTER OF HOLES AND SYMMETRICAL OBJECTS.
EXTENSION AND DIMENSION LINE THIN DIMENSION LINE EXTENSION LINE	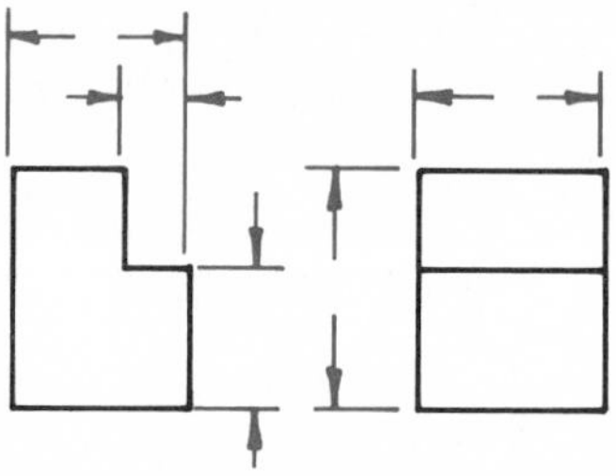	EXTENSION AND DIMENSION LINES ARE USED WHEN DIMENSIONING AN OBJECT.
LEADERS THIN	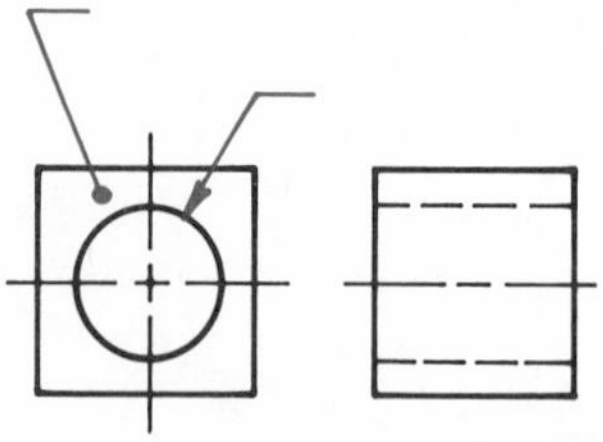	LEADERS ARE USED TO INDICATE THE PART OF THE DRAWING TO WHICH A NOTE REFERS. ARROWHEADS TOUCH THE OBJECT LINES WHILE THE DOT RESTS ON A SURFACE.
BREAK LINES THIN LONG BREAK THICK SHORT BREAK	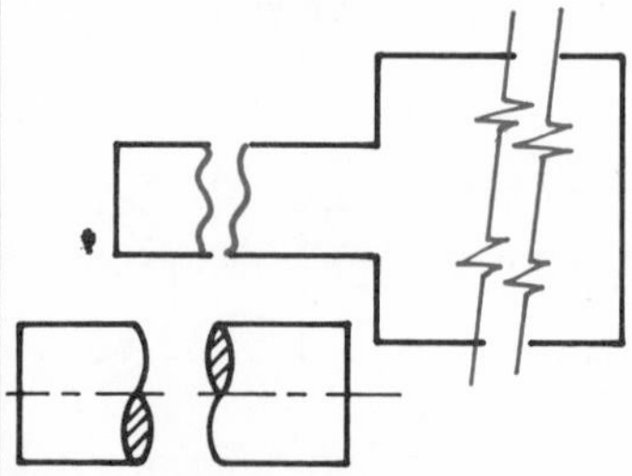	BREAK LINES ARE USED WHEN IT IS DESIRABLE TO SHORTEN THE VIEW OF A LONG PART.

Fig. 2-6-11 Types of lines.

TYPE OF LINE	APPLICATION	DESCRIPTION
CUTTING-PLANE LINE 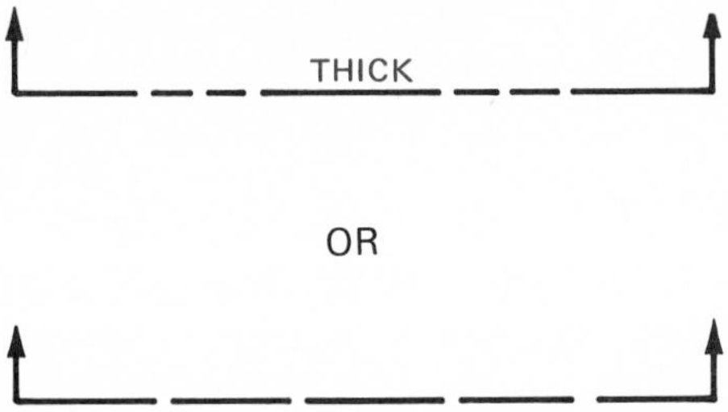 THICK OR	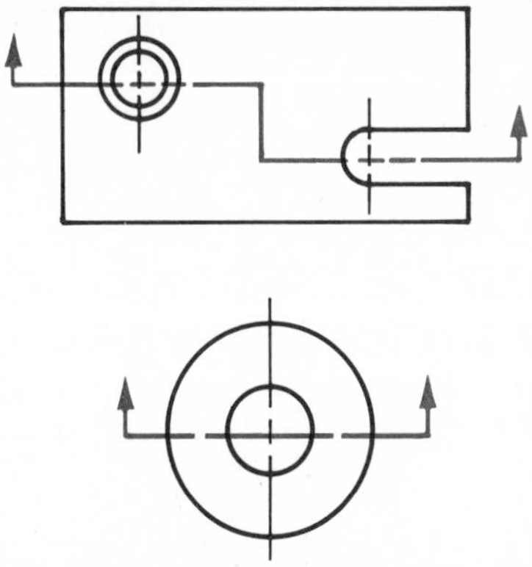	THE CUTTING-PLANE LINE IS USED TO DESIGNATE WHERE AN IMAGINARY CUTTING TOOK PLACE.
SECTION LINES 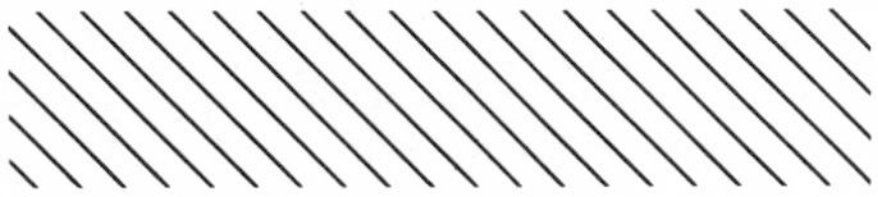 THIN LINES	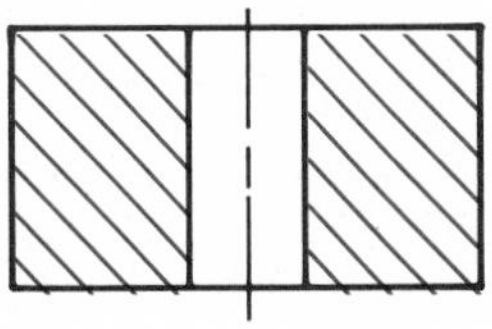	SECTION LINING IS USED TO INDICATE THE SURFACE IN THE SECTION VIEW IMAGINED TO HAVE BEEN CUT ALONG THE CUTTING-PLANE LINE.
VIEWING-PLANE LINE 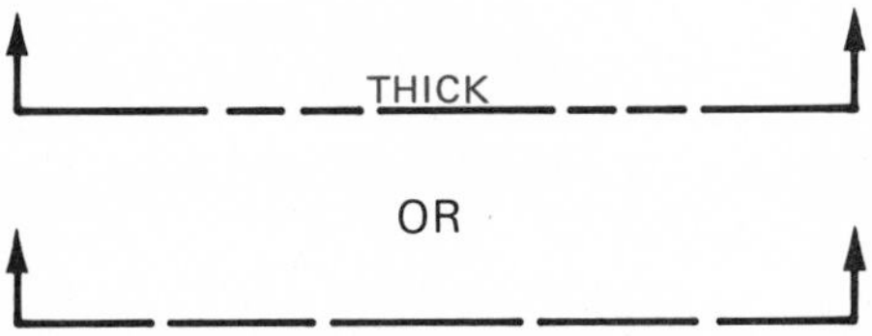 THICK OR	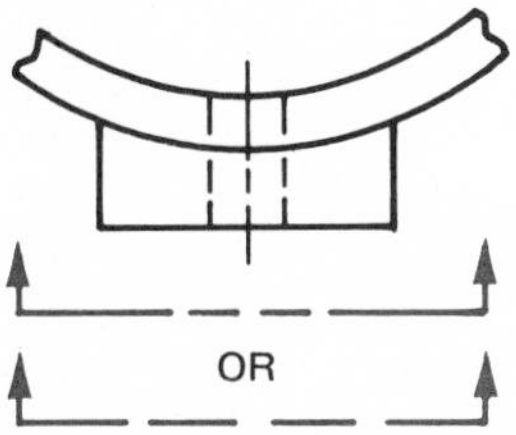 OR	THE VIEWING-PLANE LINE IS USED TO INDICATE DIRECTION OF SIGHT WHEN A PARTIAL VIEW IS USED.
PHANTOM LINE  THIN	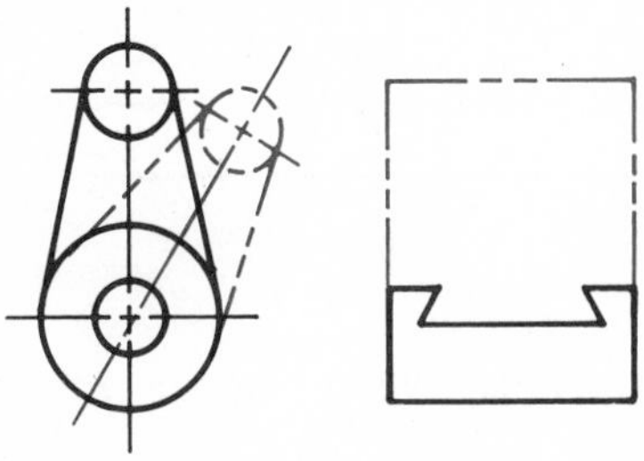	PHANTOM LINES ARE USED TO INDICATE THE ALTERNATE POSITION OF A PART OR TO SHOW THE POSITION OF AN ADJACENT PART.
STITCH LINE  THIN	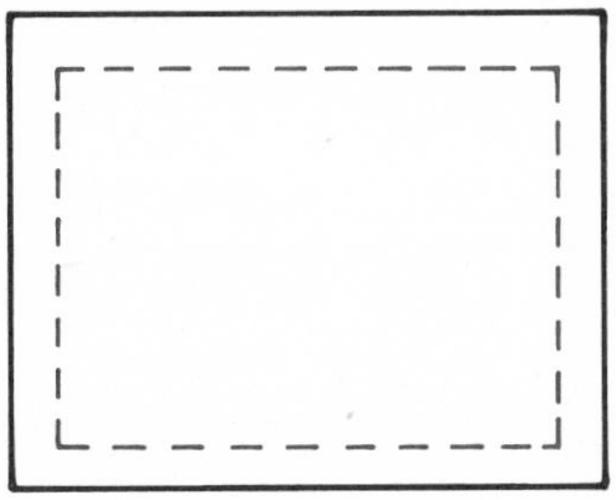	STITCH LINES ARE USED FOR INDICATING A SEWING OR STITCHING PROCESS.

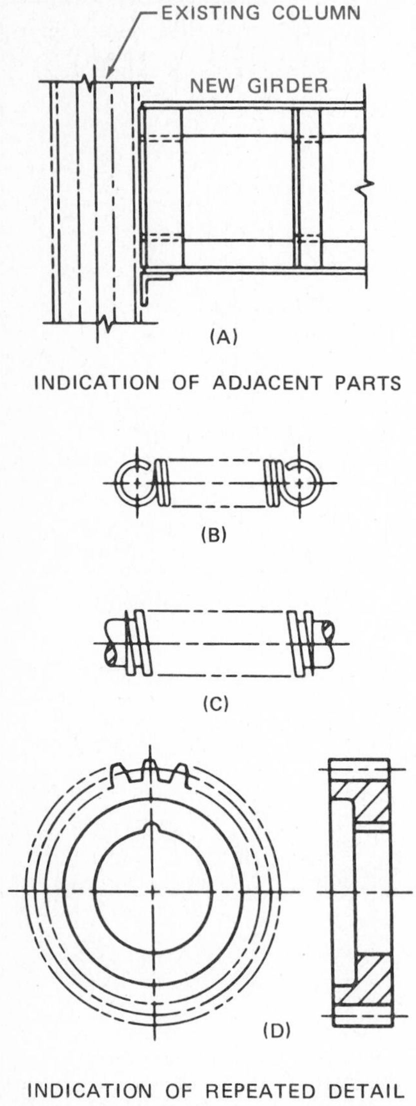
EXISTING COLUMN
NEW GIRDER
(A)
INDICATION OF ADJACENT PARTS
(B)
(C)
(D)
INDICATION OF REPEATED DETAIL

Fig. 2-6-12 Phantom-line application.

developments, for piece parts in jigs and fixtures, and for mold lines on drawings or formed metal parts. Phantom lines should start and end with long dashes.

Stitch Lines. Stitch lines consist of short dashes and spaces of equal lengths. Stitch lines are used for indicating a sewing or stitching process.

Drawing Lines

STRAIGHT LINES

A general rule to follow when drawing straight lines is this: lean the pencil in the direction of the line which you are about to draw. A right-handed person would lean the pencil to the right and draw horizontal lines from left to right. See Figs. 2-6-13 and 2-6-14. The left-handed person would reverse this procedure and draw horizontal lines from right to left. When drawing vertical lines, lean the pencil away from yourself, toward the top of the drafting board, and draw lines from bottom to top. Lines sloping from the bottom to the top right are drawn from bottom to top; lines sloping from the bottom to the left top are drawn from top to bottom. This procedure for sloping lines would be reversed for a left-handed person.

When using a conical-shaped lead, rotate the pencil slowly between your thumb and your forefinger when drawing lines. This keeps the lines uniform in width and the pencil sharp. Do not rotate a pencil having a bevel or wedge-shaped lead.

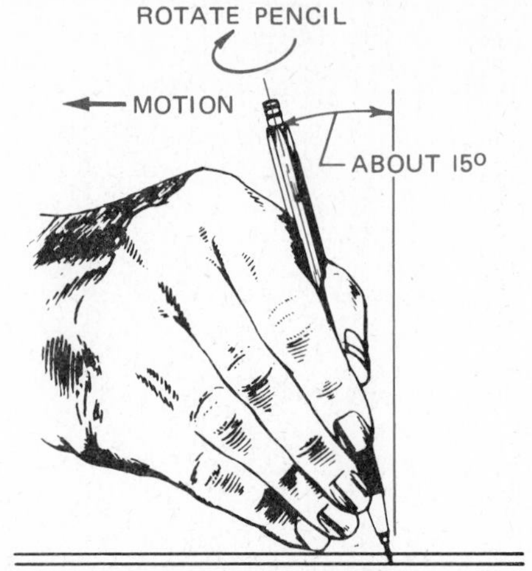
ROTATE PENCIL
MOTION
ABOUT 15°

Fig. 2-6-13 Drawing pencil lines.

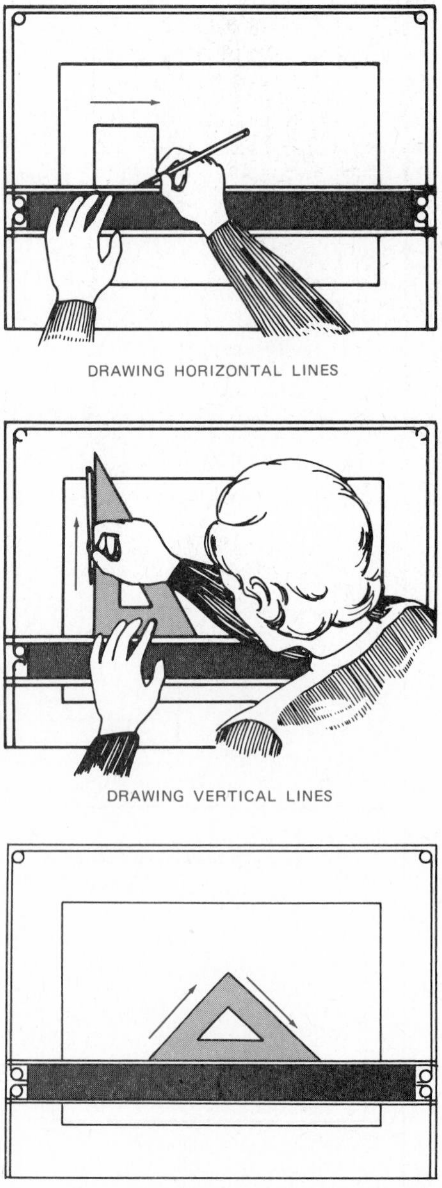
DRAWING HORIZONTAL LINES
DRAWING VERTICAL LINES
DRAWING SLOPING LINES

Fig. 2-6-14 Drawing straight lines.

CIRCLES AND ARCS

In drawing circles and arcs with a compass, it is recommended that the compass lead be softer and blacker than the pencil lead being used on the same drawing. For example, if you are drawing with a 2H or 3H pencil, use an H compass lead. This will produce a drawing having similar line work since it is necessary to compensate for the weaker impression left on the drawing medium by the compass lead as compared with the stronger direct pressure of the pencil point. For drawing circles and arcs, see Figs. 2-6-15 and 2-6-16.

It is essential that the compass lead be reasonably sharp at all times in order to ensure proper line width. The compass lead should be sharpened to a level point, with the top rounded off as shown in Fig. 2-6-17. The pencil lead is slightly shorter than the needle point, and the needle point must be readjusted after each sharpening.

For most drawings today, drafters find it much easier and faster to use plastic templates. There are sets which contain all common sizes and shapes of holes that most drafters are ever called upon to draw.

CURVED LINES

Curved lines may be drawn with the aid of irregular curves, flexible curves, and

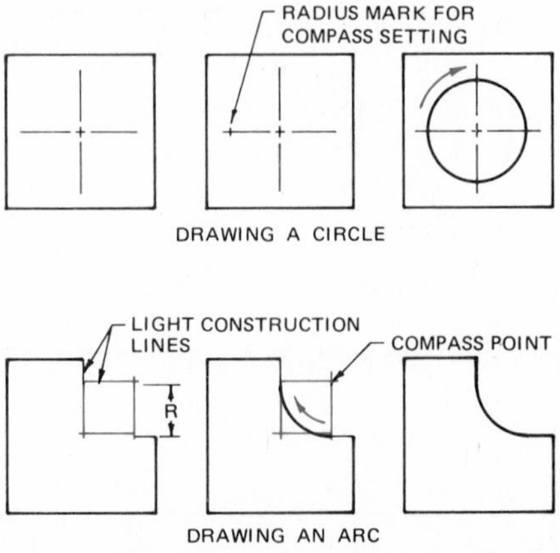
RADIUS MARK FOR COMPASS SETTING
DRAWING A CIRCLE
LIGHT CONSTRUCTION LINES
COMPASS POINT
R
DRAWING AN ARC

Fig. 2-6-15 Drawing circles and arcs.

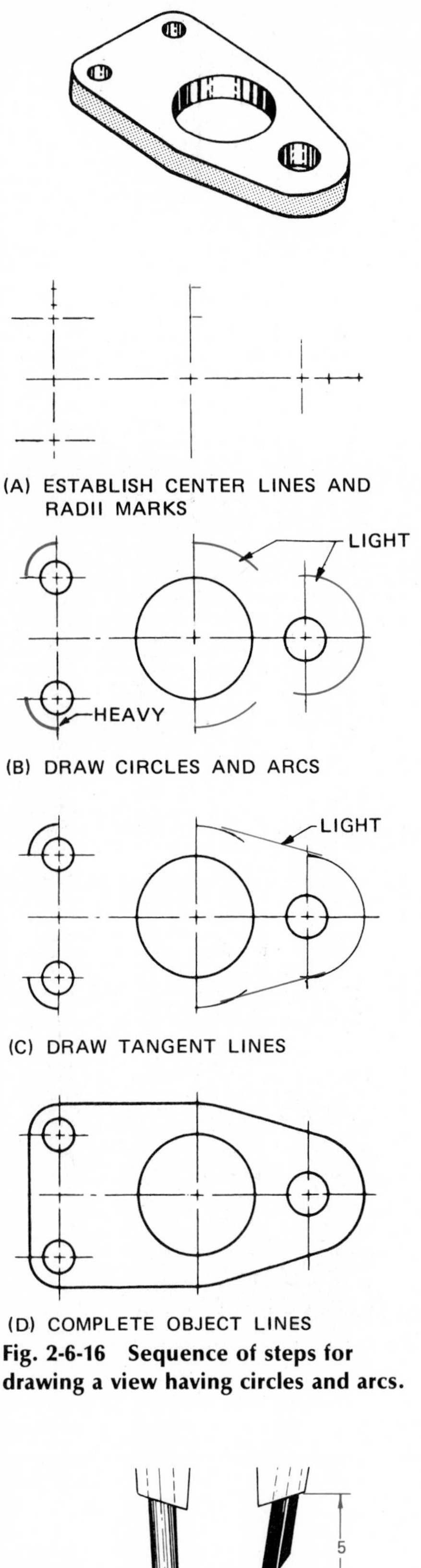
(A) ESTABLISH CENTER LINES AND RADII MARKS
LIGHT
HEAVY
(B) DRAW CIRCLES AND ARCS
LIGHT
(C) DRAW TANGENT LINES
(D) COMPLETE OBJECT LINES

Fig. 2-6-16 Sequence of steps for drawing a view having circles and arcs.

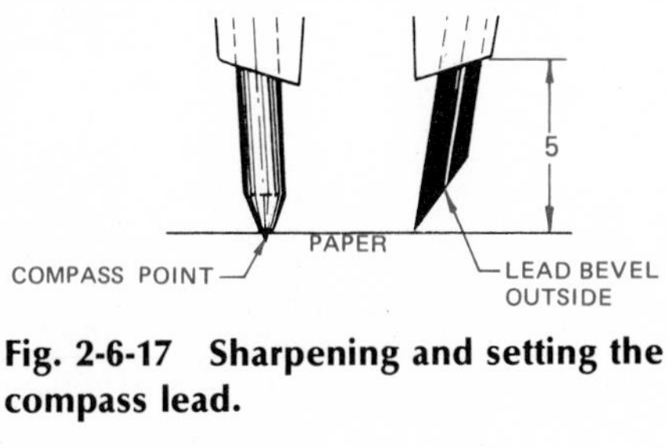
5
PAPER
COMPASS POINT
LEAD BEVEL OUTSIDE

Fig. 2-6-17 Sharpening and setting the compass lead.

elliptical templates (Fig. 2-6-18). After you have established the points through which the curved line passes, draw a light freehand line through these points. Next fit the irregular curve or other instrument by trial against a part of the curved line and draw a portion of the line. Move the curve to match the next portion, and so forth. Each new position should fit enough of the part just drawn (overlap) to ensure continuing a smooth line. It is very important to notice whether the radius of the curved line is increasing or decreasing and to place the irregular curve in the same way. If the curved line is symmetrical about the axis, the position of the axis may be marked on the irregular curve with a pencil for one side and then reversed to match and draw the other side.

ERASING TECHNIQUES[1]

Revision or change practice is inherent in the method of making engineering drawings. It is much more economical to introduce changes or additions on an original drawing than to redraw the entire drawing. Consequently, erasing has become a science all its own. Proper erasing is extremely important since some drawings are revised a great number of times. Consequently, good techniques and materials must be used which permit repeated erasures on the same area. Some recommendations follow.

1. Avoid damaging the surface of the drawing medium by selecting the proper eraser.

2. Lines not thoroughly erased produce ghostlike images on prints, resulting in reduced legibility.

3. A hard, smooth surface, such as a set square, placed under the lines being removed makes erasing easier.

4. Using an erasing shield protects the adjacent lines and lettering and also eliminates wrinkling. See Fig. 2-6-19.

5. Use an eraser on the back of the drawing medium as well as on the drawing side.

6. Be sure to completely remove erasure crumbs from the drawing surface.

7. When extensive changes are required, it may be more economical to cut and paste or make an intermediate drawing.

8. When you are erasing, use no more pressure than necessary.

9. The drawing quality of the drawing medium which may have been damaged by erasing may be improved by sprinkling an inking powder on the surface and rubbing it with a cloth.

In addition to these suggestions, it is necessary to match the density of the surrounding background when erasures are made. Often, the erased area is much cleaner than the rest of the drawing. If the change is made on this clean area, the contrast between line and background is different and that area presents a problem in reproduction.

It is usual practice to "dirty" the erased area so that it looks about the same shade as the surrounding area.

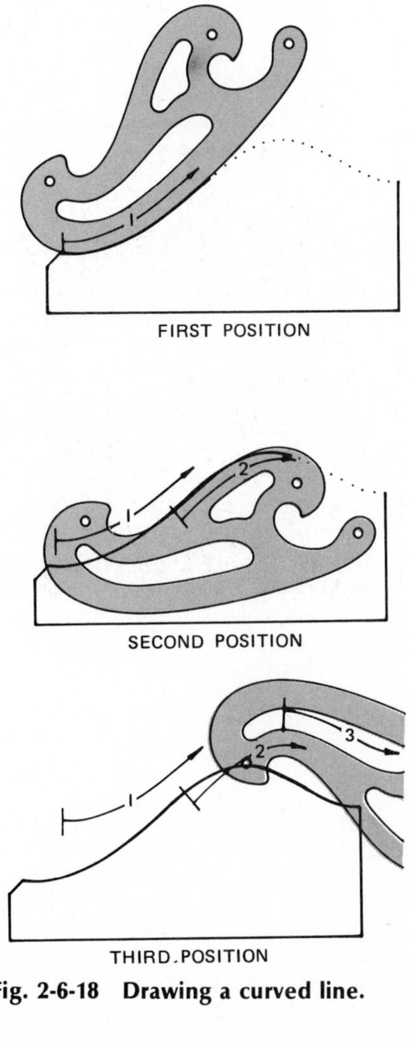
FIRST POSITION
2
SECOND POSITION
2
3
THIRD.POSITION

Fig. 2-6-18 Drawing a curved line.

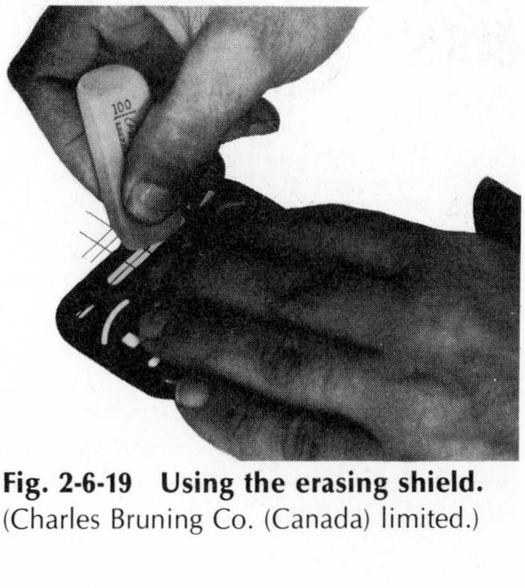

Fig. 2-6-19 Using the erasing shield. (Charles Bruning Co. (Canada) limited.)

REMOVING LINES ON FILM[2]

Erasers. Lines on photoreproduction film fall into two classifications: photographic lines and pencil-and-ink lines. All these lines can be removed easily so that the erased area can be used for further drafting. Here are some tips for removing lines.

There are three basic types of erasers: rubber, plastic, and liquid. Rubber and plastic erasers may tend to cause a shine on the drafting surface. This is not necessarily detrimental. Good drafting lines can be drawn easily over areas from which lines have been erased many times. A good general rule to follow is to use a soft, nonabrasive eraser and only enough pressure to remove the line. See Fig. 2-6-20. Liquid erasers do not put a shine on the drafting surface and can be used to make many erasures in the same place. When plastic erasers are used, the shiny appearance actually may be transparentizing because of the plasticizers used in their manufacture rather than wear on the drafting surface. The transparentizing effect is not detrimental since it does not reduce the ability of the surface to take a pencil line.

When the drafting surface is affected by excessive erasures, it can be repaired by rolling a regular typewriter eraser across the smooth area or by rubbing a small amount of drafting powder into the area with a finger.

Pencil Lines. Pencil lines can be removed from all film with a soft, nonabrasive rubber or plastic eraser or with liquid eraser. To keep the drafting surface from becoming too shiny, avoid excessive pressure.

Fastening Paper to the Board

The drafter in industry usually fastens the drawing paper to the board with staples or tape, and he or she does not remove it from the board until the drawing has been completed. In drafting classrooms, drawings are fastened to and removed from the boards so frequently that it is desirable to use fastening methods that permit easy removal of drawings without marking the board surface. Masking tape and spring clips are two methods that meet this requirement.

When you are fastening paper to the board, line up the bottom or top edge of the paper with the top horizontal edge of the T square, parallel straight edge, or horizontal scale of the drafting machine. See Fig. 2-6-21. When you are refastening a partially completed drawing, use lines rather than the edge of the paper for alignment.

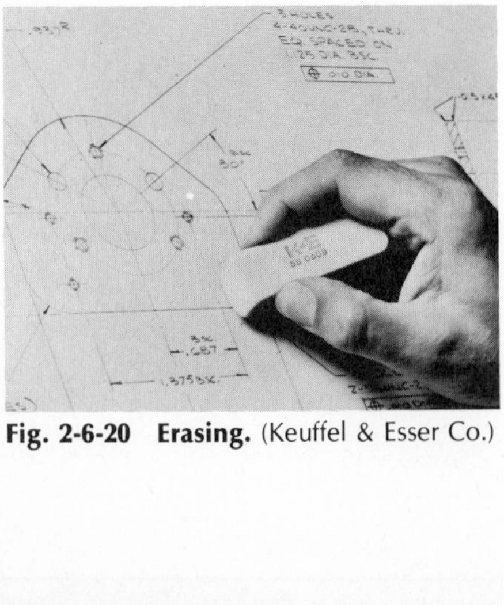

Fig. 2-6-20 Erasing. (Keuffel & Esser Co.)

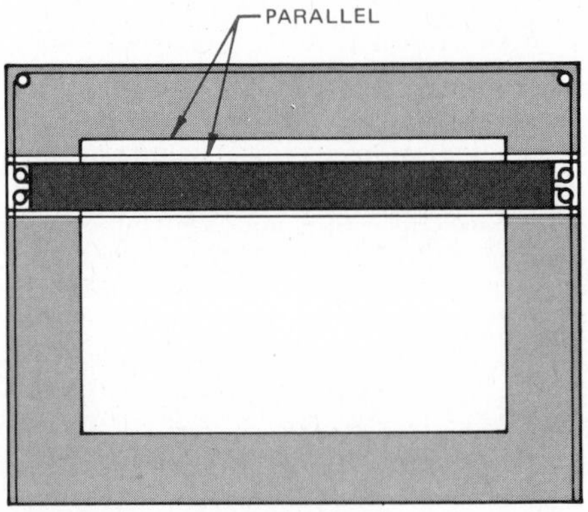
PARALLEL

Fig. 2-6-21 Positioning the paper on the board.

Sketching

Freehand sketching is a necessary part of drafting because the drafter in industry frequently sketches ideas and designs prior to making instrumental drawings.

The drafter may also use sketches to explain thoughts and ideas to other people in discussions of mechanical parts and mechanisms. Sketching, therefore, is an important method of communication. Practice in sketching helps the student to develop a good sense of proportion and accuracy of observation. It can be used to advantage when you are learning the fundamentals of drafting practice and procedures.

A fairly soft (HB, F, or H) pencil should be used for preliminary practice. Many types of graph or ordinate paper are available and can be used to advantage when close accuracy to scale or proportion is desirable. The directions in which horizontal, vertical, and oblique lines are sketched are illustrated in Fig. 2-6-22.

Since the shapes of objects are made up of flat and curved surfaces, the lines forming views of objects will be both straight and curved. Do not attempt to draw long lines with one continuous stroke. First plot points along the desired line path; then connect these points with a series of light strokes.

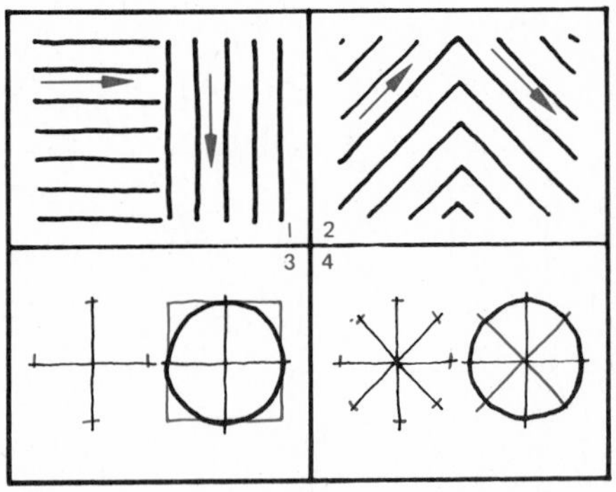
1
2
3
4

Fig. 2-6-22 Sketching lines and circles.

When you are sketching a view (or views), first lightly sketch the overall size as a rectangular or square shape, estimating its proportions carefully. Then add lines for the details of the shape, and thicken all lines forming a part of the view. See Fig. 2-6-23.

Figure 2-6-24 shows two methods of sketching circles. Figure 2-6-25 illustrates, both pictorially and orthographically, the use of graph paper for the sketching of a machine part. Coordinate and isometric sketching paper are shown in Fig. 2-6-26.

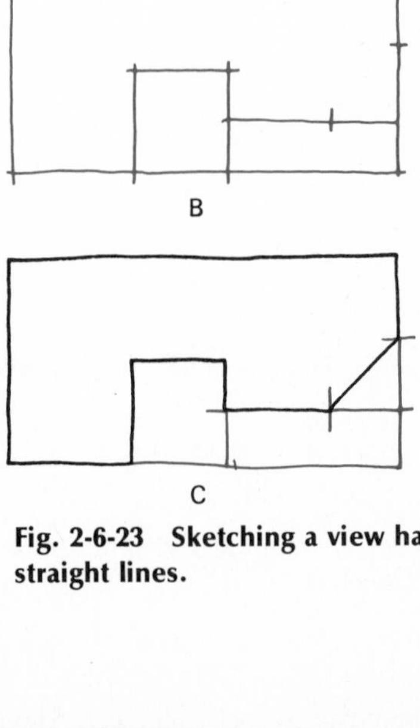
A
B
C

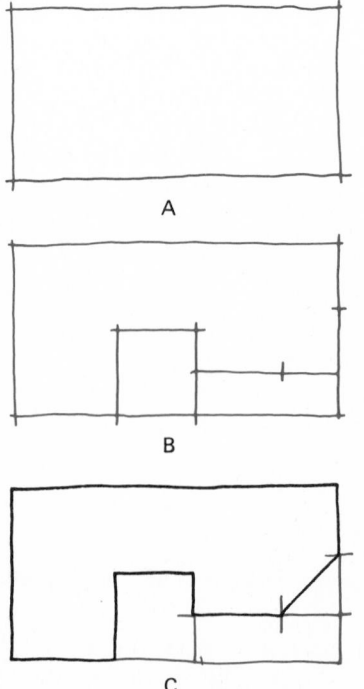

Fig. 2-6-23 Sketching a view having straight lines.

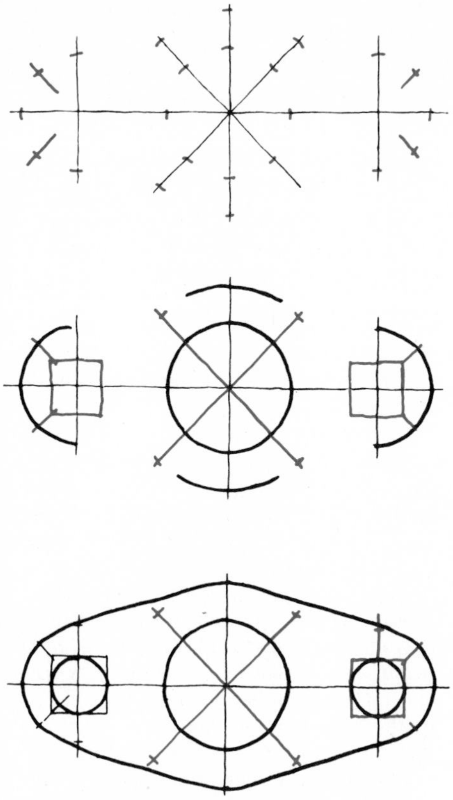

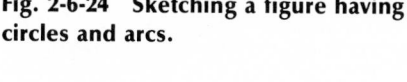

Fig. 2-6-24 Sketching a figure having circles and arcs.

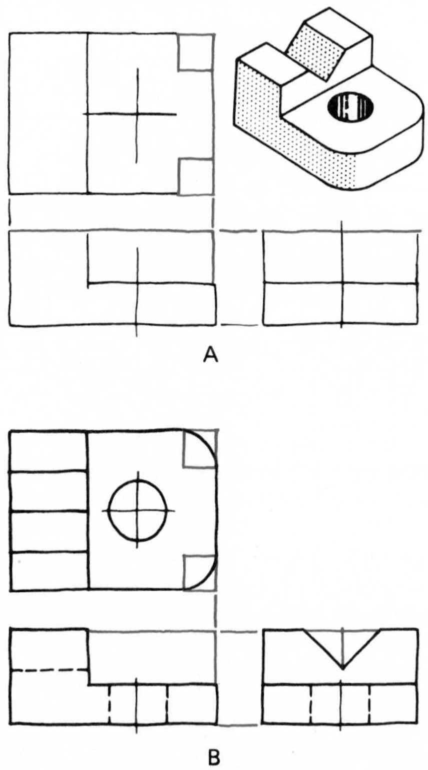
A
B

Fig. 2-6-25 Usual procedure for sketching three views.

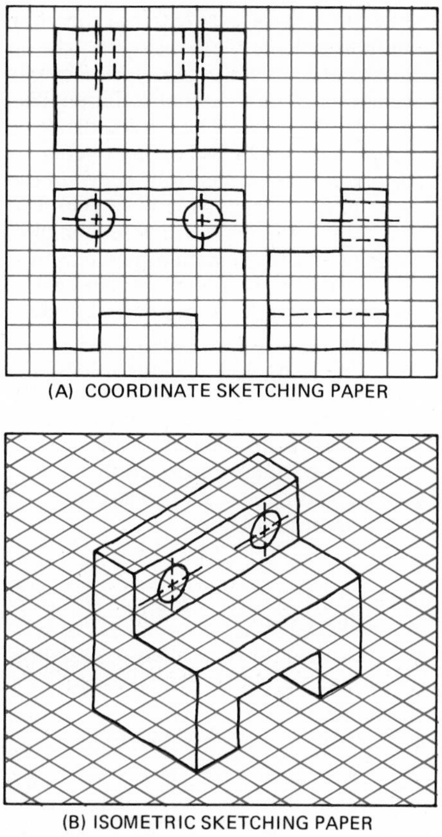
(A) COORDINATE SKETCHING PAPER
(B) ISOMETRIC SKETCHING PAPER

Fig. 2-6-26 Sketching paper.

Inking[3]

In inking, technique is an important factor. The pen should be moved as it touches the film or paper at the beginning of a line and as it leaves the film or paper at the end of the line. This will minimize belling, or spreading of ink, at the line ends.

Blade-type ruling pens should be inclined at an angle up to 15° from the vertical in the direction of motion. Needle-in-tube type pens should be held nearly vertical and moved with a light touch across the drawing medium. See Fig. 2-6-27.

There are several inks designed specially for use on drawing media. They can be used with ruling pens, capillary-type pens, lettering pens, or pen nibs. When you are inking on film, ruling pens should be filled about half-way, as is customary for drawing on vellum or cloth.

Normal handling of drawing media is bound to soil them. Ink lines applied over soil areas do not adhere well and may be chipped off or flake in time. It is always good practice to keep the drawing paper, film, or cloth clean. Soiled areas can be cleaned effectively with a cleaner.

The use of a needle-in-tube pen with a circular template has, in many instances, replaced the drawing of circles by means of the compass with its blade-type pen,

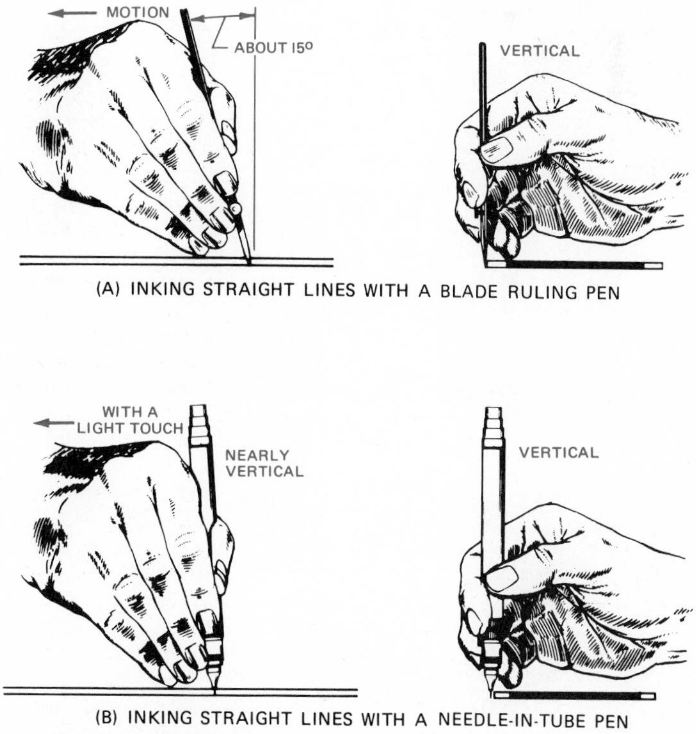
MOTION
ABOUT 15°
VERTICAL
(A) INKING STRAIGHT LINES WITH A BLADE RULING PEN
WITH A LIGHT TOUCH
NEARLY VERTICAL
VERTICAL
(B) INKING STRAIGHT LINES WITH A NEEDLE-IN-TUBE PEN

Fig. 2-6-27 Drawing ink lines.

which must be frequently refilled with ink (Fig. 2-6-28). When you are inking circles by template, the only thing to look out for is to keep the template above the drawing media. This is accomplished by inserting another template or other thin material between the template and the drawing surface in order to avoid blotting through capillary undertow of ink.

Scribing

Scribing is a drafting technique which in some areas has replaced pen-and-pencil drafting. It has already done so for some types of close-tolerance work. With regular drafting, using pencil or pen, you draw on top of the work. With scribing, you incise, or cut, lines into a special surface with scribing tools, making lines that are sharp and clean, never vary in width, and can't be smudged. Scribe lines also produce the sharpest prints you can get, direct from the scribed original. See Fig. 2-6-29.

Scribing is used for a variety of applications—as a drafting medium for the tools and templates from which fabricated parts are produced; in aircraft lofting and automobile layouts; in mapping; and for close-tolerance electronic circuitry and microcircuit layout.

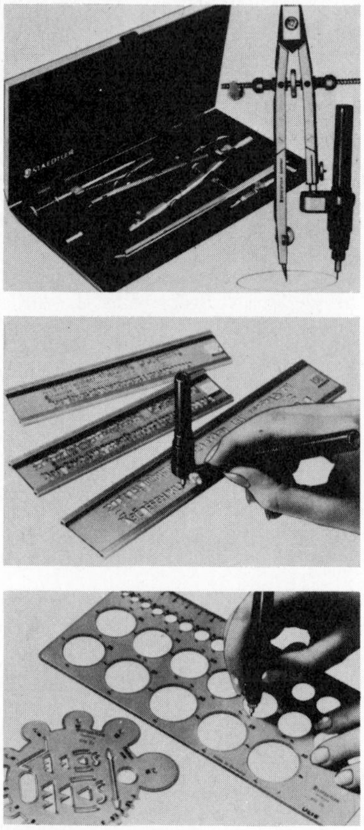

Fig. 2-6-28 Inking. (J. S. Staedtler, Inc.)

In the past, glass, steel, and aluminum were considered the only stable base materials for scribing because of their extremely high dimensional stability. These materials were adequate but difficult to handle and file. In subsequent years, glass cloth was developed, and it became quite popular as a stable base material. At present, the most useful and stable drafting and reproduction medium is polyester film. It is available in a wide range of surfaces for the precision drafting and reproduction requirements of such industries as mapping, electronics, and engineering. There are specially prepared scribe surfaces for the critical line work of mapping and undimensioned drafting—coatings which can be cut with a blade and then peeled back to the transparent film base.

REFERENCES

1. National Microfilming Association and Keuffel & Esser Co.
2. Keuffel & Esser Co.
3. Eastman Kodak Company.

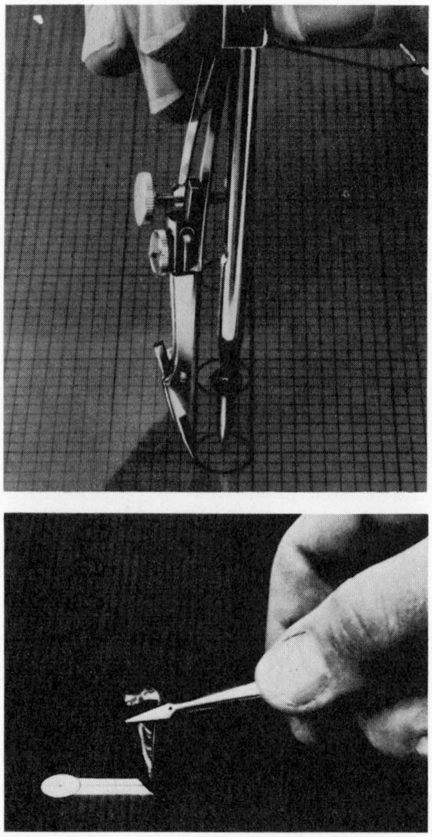

Fig. 2-6-29 Scribing. (Keuffel & Esser Co.)

UNIT 2-7

DRAWING REPRODUCTION[1]

A revolution in reproduction technologies and methods began in the 1940s and 1950s. It brought with it new equipment and supplies which have made quick copying commonplace. The new technologies make it possible to apply improved systems approaches and new information handling techniques to all types of files ranging from small documents to large engineering drawings. See Fig. 2-7-1.

The pressures on business and government for greater efficiency, space savings, cost reductions, lower investment costs, and equally important factors provided a fertile field for the new reproduction technologies. There is no reason to believe that such pressures will diminish; in fact, as the years go by, it is certain that more and more improvements will occur, newer and better reproduction and information handling equipment and methods will be discovered, and the advantages which they offer will find ever-widening application.

Reproduction Equipment[2]

Studies of reproduction facilities, existing or proposed, should first consider the nature of the demand for this service, then the processes which best satisfy the demand, and finally the particular machines which employ the processes. Factors to consider at these stages of study include:

- *Input originals*—sizes, paper mass, color, artwork
- *Quality of output copies*—depending on expected use and degree of legibility required
- *Size of copies*—same size, enlarged, reduced
- *Color*—copy paper and ink
- *Registration*—in multiple-run work
- *Volume*—numbers of orders and copies per order
- *Speed*—machine productivity, convenient start-stop, and load-unload
- *Cost*—direct labor, direct material, overhead
- *Future requirements*

Two general kinds of reproduction are recognized: copying and duplicating.

Copying machines are suitable for both line work and pictorials, often on large sheets. They operate at slow speeds, and so the cost per copy is relatively high.

In contrast, duplicating processes are characterized by high speed, high volume, and low cost per copy. They can be used with a wide variety of papers in

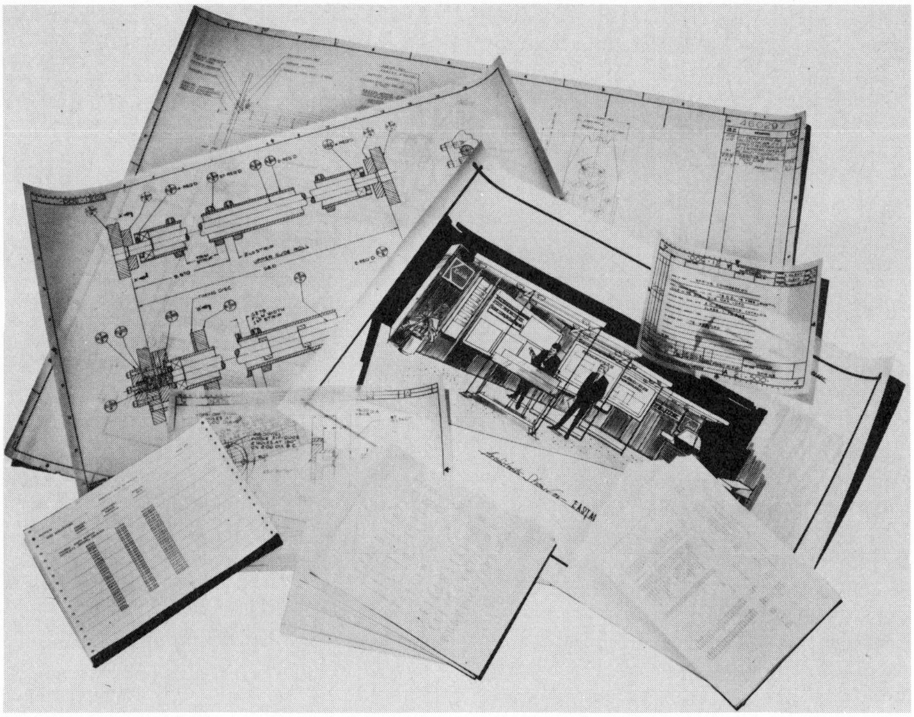

Fig. 2-7-1 Drawing reproduction. (Eastman Kodak Co.)

many sizes, masses, and finishes. Although duplicators—spirit, stencil, and offset—are used in conjunction with copying machines in the engineering and drafting offices, they will not be covered in this text.

COPIERS

The principal kinds of copiers are diazo, electrostatic, thermographic, and photographic devices.

Blueprinting has been the classical process for copying engineering drawings. It provided white lines on a blue background. Prints were made on a continuous roll-fed machine and were not intended for reproduction. The process was wet, and the scale of prints was poor. In recent years, blueprinting has been replaced, mainly by diazo.

DIAZO (WHITEPRINT)

In this process (Figs. 2-7-2 and 2-7-3), paper or film coated with a photosensitive diazonium salt is exposed to light passing through an original of translucent paper or film. The exposed coated sheet is then developed by an alkaline agent such as ammonia vapor. Where the light passes through the clear areas of the master, it decomposes the diazonium salt, leaving a clear area on the copy. Where markings on the original block the light, the ammonia and the unexposed coating produce an opaque die image of the original markings. A positive original makes a positive copy, and a negative original makes a negative copy.

Fig. 2-7-2 A whiteprint machine in use. (Addressograph Multigraph Corp.)

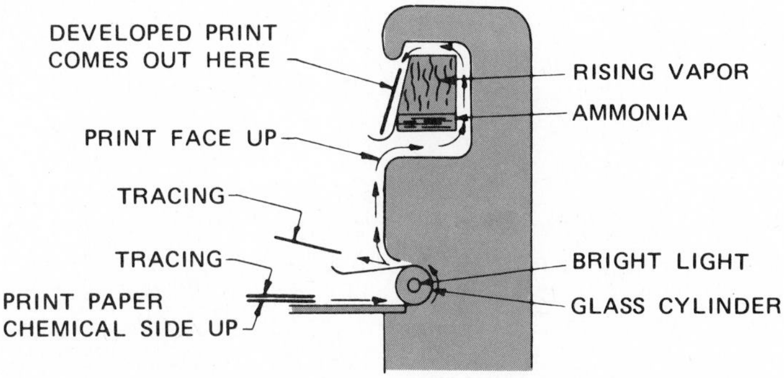
DEVELOPED PRINT
COMES OUT HERE
RISING VAPOR
AMMONIA
PRINT FACE UP
TRACING
TRACING
BRIGHT LIGHT
PRINT PAPER
CHEMICAL SIDE UP
GLASS CYLINDER

ROLLERS MOVE THE TRACING AND PRINT AROUND THE LIGHT, AND MOVE THE PRINT PAST THE RISING AMMONIA VAPOR.

Fig. 2-7-3 The diazo printing process.

Therefore, the polarity is said to be *nonreversing*. Three diazo processes currently used differ mainly in the way the ammonia is introduced to the diazo coat. These are ammonia vapor, moist developing, and pressure developing.

Perhaps the most significant characteristic of the diazo process is that it allows reproduction of fine detail. Diazo is a high-contrast process and thus ideal for document reproduction.

ELECTROSTATIC

Electrostatic reproduction, one form of which is xerography, is a dry copying process which uses electrostatic force to deposit dry powder on copy paper.

Electrostatic transfer enables printing on plain paper, offset paper masters, or transparent materials. Some machines copy only at the same size as the originals, while others can reduce or enlarge. See Fig. 2-7-4.

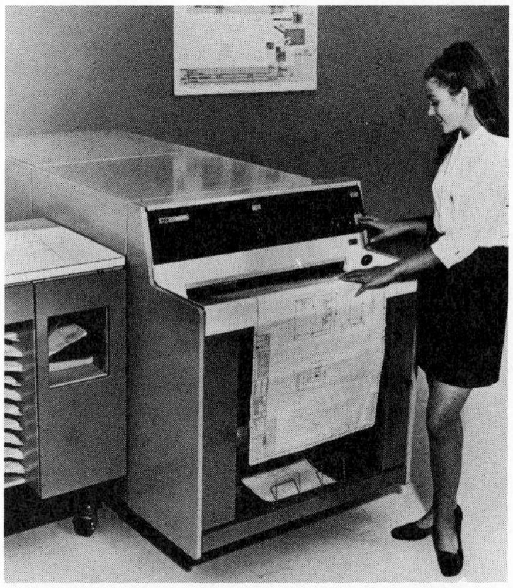

Fig. 2-7-4 Electrostatic printer. This machine will reproduce, reduce, fold, and sort prints. (Xerox Corp.)

PHOTOGRAPHIC

Photography is the process of creating latent images on light-sensitive silver halide material by exposure to light. The images are made visible and permanent by developing and fixing techniques. A camera provides for enlargement or reduction of the image size.

Contact printing and projection printing are the two principal methods of making photographic prints—contact for prints the same size as originals and projection for reduced or enlarged prints.

Microfilm

Microfilming of engineering drawings is now an established practice in many drafting offices. See Fig. 2-7-5. This has come about because of the primary savings in lower transportation, labor, and storage costs of microfilm.

Microfilmed prints are A3 size, regardless of their original size, and are much easier to handle and store. From the drafting room, drawings are taken to a camera, photographed, and stored in rolls or on cards.

FORMS OF FILM

One way to classify microfilm is according to the physical forms, called *microforms*, in which it is used.

Roll Film. This is the form of the film after it has been removed from the camera and developed. Microfilm comes in four different widths—16, 35, 70, and 105 mm—and is stored on magazines.

Aperture Cards. Perhaps the simplest of the flat microforms is finished roll film cut into separate frames, each mounted on a card having a rectangular hole as shown in Fig. 2-7-6. Aperture cards are available in many sizes, but the most widely used is 83 × 188 mm.

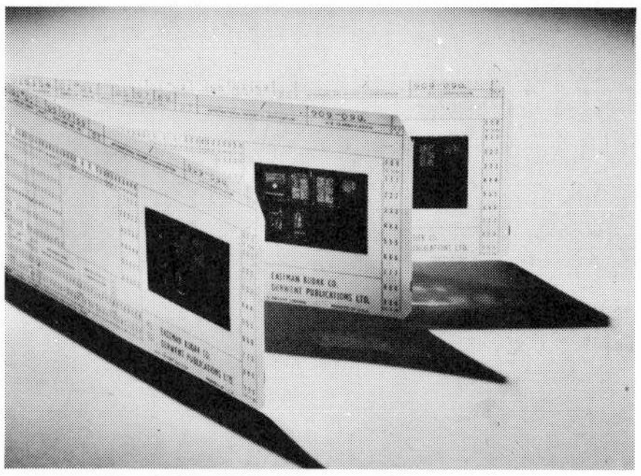

Fig. 2-7-6 Aperture Cards. (Eastman Kodak Co.)

Jackets. Jackets are made of thin, clear plastic and have channels into which short strips of microfilm are inserted. They come in a variety of film channel combinations for 16- and/or 35-mm microfilm. Like aperture cards, jackets can be viewed easily.

Microfiche. A microfiche is a sheet of clear film containing a number of microimages arranged in rows. A common size, 100 × 150 mm, frequently is arranged to contain 98 images. Microfiches are especially well suited for quantity distribution of standard information, parts, and service lists.

READERS AND VIEWERS

Microfilm readers magnify film images large enough to be read and project the images onto a translucent or opaque screen. Some readers accommodate only one microform (rolls, jackets, microfiches, or aperture cards), while others can be used with two or more. Scanning-type readers, having a variable-type magnification, are used when frames containing a large drawing (36 × 48 in.) are viewed. Only parts of the drawing can be viewed at one time.

Fig. 2-7-5 Microfilming. (3M Co. and American Motors Co.)

READER-PRINTERS

Two kinds of equipment are used to make enlarged prints from microfilm: reader-printers and enlarger-printers. The reader-printer, as illustrated in Fig. 2-7-7, is a reader which incorporates a means of making hard copy from the projected image. The enlarger-printer is designed only for copying and does not include the means for reading.

REFERENCES

1. National Microfilming Association.
2. *Machine Design*, July 1971.

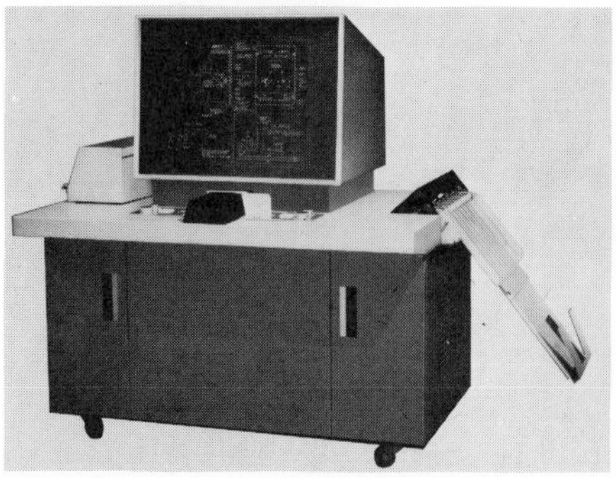

Fig. 2-7-7 Reader-printer. (Eastman Kodak Co.)

UNIT 2-8

FILING SYSTEMS[1]

One of the most common and difficult problems facing an engineering department is how to set up and maintain an efficient engineering filing area. Normal office file methods are not considered satisfactory for engineering drawings. To properly serve its function, an engineering filing area must meet two important criteria: accessibility of information and protection of valuable documentation.

For this kind of a system to be effective, drawings must be readily accessible. The degree of accessibility, of course, is dependent on whether drawings are considered active, semiactive, or inactive.

Filing Original Drawings

Unless a company has developed a full microfilming system, the original drawings which the drafter produced must be kept and filed for future use or reference. Unlike the prints, the originals must *not* be folded. They are filed in either a flat or rolled position. See Fig. 2-8-1.

In determining what type of equipment to use for engineering files, it should be remembered that different types of drawings require different kinds of files. Also,

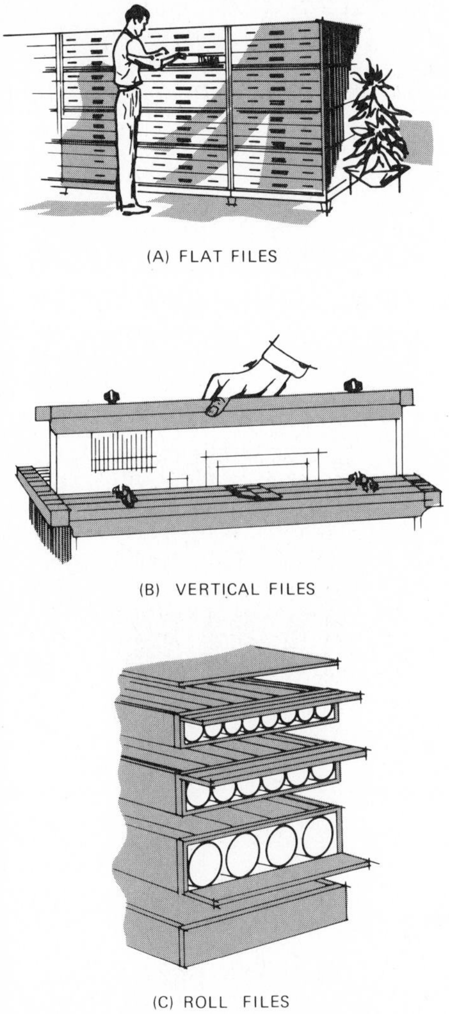
(A) FLAT FILES
(B) VERTICAL FILES
(C) ROLL FILES

Fig. 2-8-1 Filing systems for original drawing masters and prints. (Keuffel & Esser)

in planning a filing system, keep in mind that filing requirements are always increasing; unlike normal office files that can be purged each year, the more drawings are produced, the more need to be stored. Therefore, any filing system must have the flexibility of being easily expanded, and generally in a minimum of space.

MICROFILM FILING SYSTEMS

It seems logical that reducing drawings to tiny images on film would make them more difficult to locate. However, this is not the case, for while they are reduced in size, they are made more uniform. This results in improved file arrangements.

Roll Film. Roll film can be coded in several ways for visual or automatic retrieval. The more common methods used are flash cards, code lines, sequential numbering image control, and binary code patterns.

Aperture Cards. In many respects, aperture cards have the same filing and retrieval capabilities as jackets and microfiches. However, there is one important difference. It is possible to use aperture cards in machine records handling systems. They can be printed, punched, and sorted by machine.

Jackets and Microfiches. Both these microforms are basically the same with respect to retrieval. See Fig. 2-8-2. Each has a title header for identifying numbers and titles. Each jacket or microfiche contains a group of images arranged in a logical sequence so that the particular images can be readily found.

Folding of Prints

To facilitate handling, mailing, and filing, prints should be folded to letter size, 210 × 297 mm (8.5 × 11 in.), in such a way that the title block and auxiliary number always appear on the front face and the last fold is always at the top. In filing, this prevents other drawings from being pushed into the folds of filed prints.

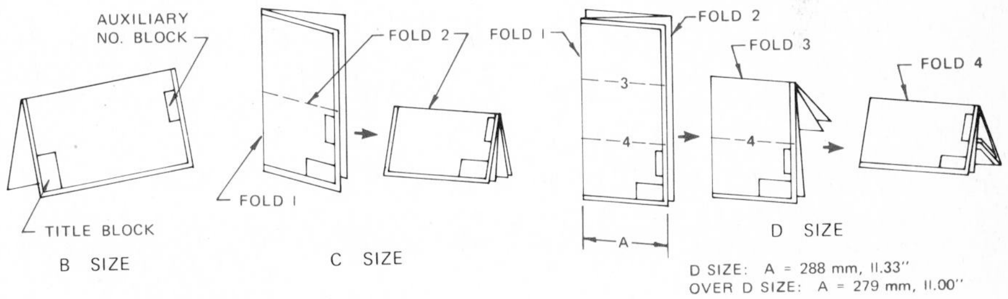
AUXILIARY NO. BLOCK
TITLE BLOCK
B SIZE
FOLD 2
FOLD 1
FOLD 1
C SIZE
FOLD 2
FOLD 3
FOLD 4
3
4
4
A
D SIZE
D SIZE: A = 288 mm, 11.33"
OVER D SIZE: A = 279 mm, 11.00"

Fig. 2-8-3 Folding prints.

Recommended methods of folding standard-size prints are illustrated in Fig. 2-8-3.

On preprinted forms, it is recommended that fold marks be included in the margin of the drawings on size B and larger and be identified by number, for example, "fold 1," "fold 2." In zoned prints, the fold lines will coincide with zone boundaries; but they should, nevertheless, be identified.

To avoid loss of clarity by frequent folding, important details should not be placed close to fold areas.

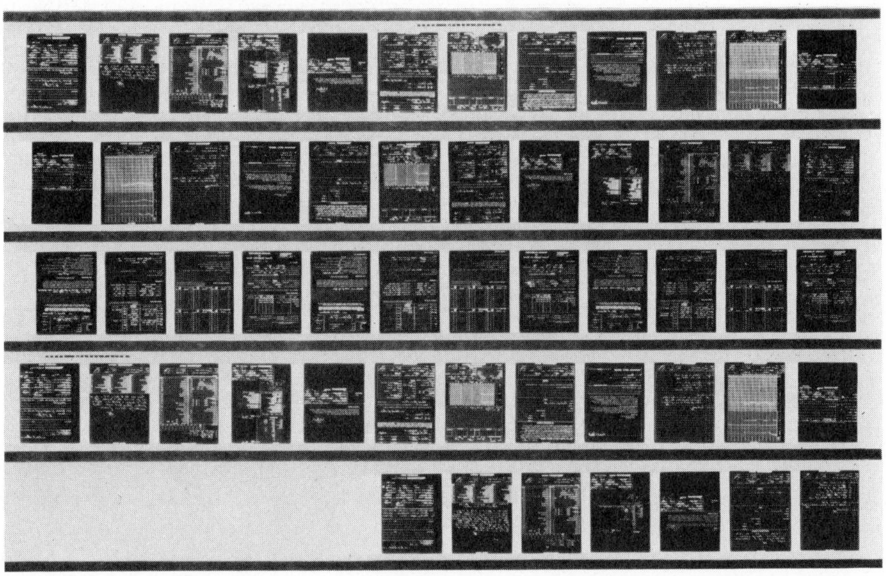

Fig. 2-8-2 Microfilm jacket. (Eastman Kodak Co.)

REFERENCE

1. Eastman Kodak Company and "Setting Up and Maintaining an Effective Drafting Filing System," *Reprographics*, March 1975.

Chapter 3
Theory of Shape Description

UNIT 3-1
THEORY OF SHAPE DESCRIPTION

Chapter 2 illustrated many simple parts that required only one view to completely describe them. However, in industry, the majority of parts that have to be drawn are more complicated than the ones previously described. More than one view of the object is required to show all the construction features.

Pictorial (three-dimensional) drawings of objects are sometimes used, but the majority of drawings used in mechanical drafting for completely describing an object are multiview drawings as shown in Fig. 3-1-1.

Pictorial projections, such as axonometric, oblique, and perspective projection, are useful for illustrative purposes and are frequently employed in installation and maintenance drawings and design sketches.

As a result of new drawing techniques and equipment, pictorial drawings are becoming a popular form of communication, especially with people not trained to read engineering drawings. Practically all drawings of do-it-yourself projects for the general public or of assembly-line instructions for nontechnical personnel are done in pictorial form.

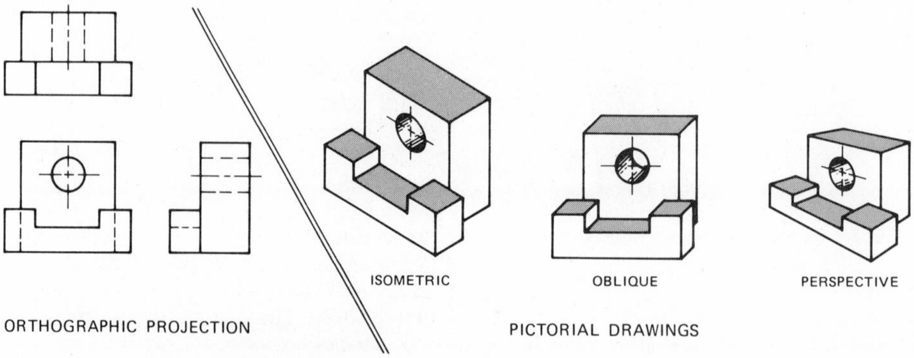

ISOMETRIC
OBLIQUE
PERSPECTIVE
ORTHOGRAPHIC PROJECTION
PICTORIAL DRAWINGS

Fig. 3-1-1 Types of projection used in drafting.

Orthographic Projection

The drafter must represent the part which appears as three-dimensional (width, height, depth) to the eye on the flat plane of the drawing paper. Different views of the object—front, side, and top views—are systematically arranged on the drawing paper to convey the necessary information to the reader (Fig. 3-1-2). Features are projected from one view to another. This type of drawing is called an *orthographic projection*. The word *orthographic* is derived from two Greek words: *orthos*, meaning straight, correct, at right angles to; and *graphikus*, meaning to write or describe by drawing lines.

The principles of orthographic projection can be applied in four different "angles" or systems: first-, second-, third-, and fourth-angle projection (Fig. 3-1-3).

However, only two systems—first- and third-angle projection—are used. Third-angle projection is used in the United States, Canada, and many other countries throughout the world. First-angle projection, which will be described in detail in Unit 3-6, is used mainly in many European and Asiatic countries. As world trade has brought about the exchange of engineering drawings as well as the end products, drafters are now called upon to communicate in, as well as understand, both types of orthographic projection.

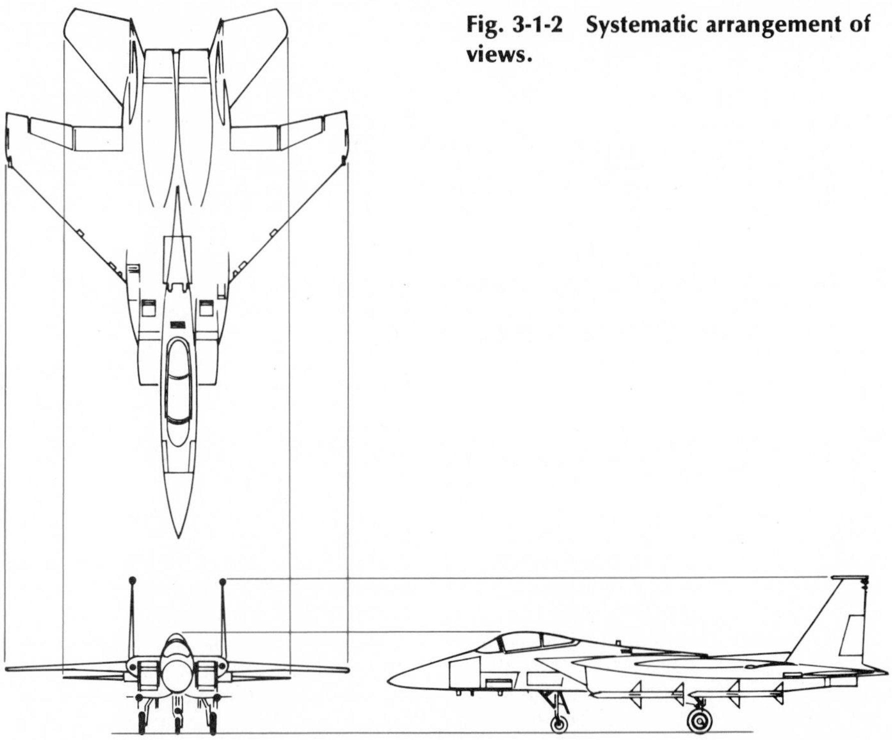

Fig. 3-1-2 Systematic arrangement of views.

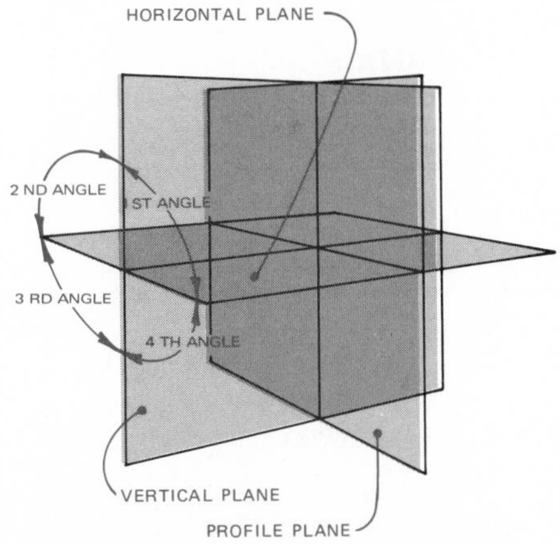
HORIZONTAL PLANE
2 ND ANGLE
1 ST ANGLE
3 RD ANGLE
4 TH ANGLE
VERTICAL PLANE
PROFILE PLANE

Fig. 3-1-3 The three planes used in orthographic projection.

ISO PROJECTION SYMBOL

With two types of projection being used on engineering drawings, a method of identifying the type of projection is necessary. The International Standards Organization, known as ISO, has recommended that the symbol shown in Fig. 3-1-4 be shown on all drawings and located preferably in the lower right-hand corner of the drawing, adjacent to the title block (Fig. 3-1-5). To aid the reader in learning the language of industry, objects throughout this text have been drawn in first- as well as third-angle projection. The ISO symbol will indicate the type of projection used.

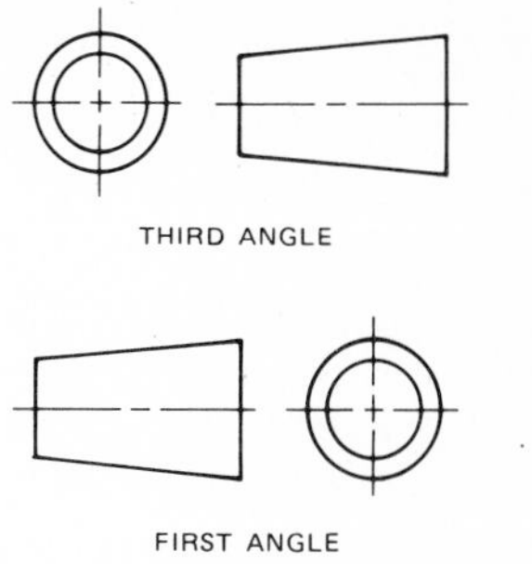
THIRD ANGLE
FIRST ANGLE

Fig. 3-1-4 ISO projection symbol.

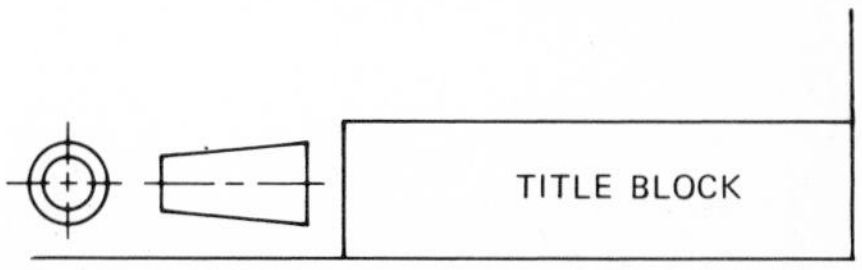
TITLE BLOCK

Fig. 3-1-5 Locating ISO symbol on drawing paper.

THIRD-ANGLE PROJECTION

In third-angle projection, the object is positioned in the third-angle quadrant, as shown in Fig. 3-1-6. The person viewing the object does so from six different positions, namely, from the top, front, right side, left side, rear, and bottom. The views or pictures seen from these positions are then recorded or drawn on the plane located between the viewer and the

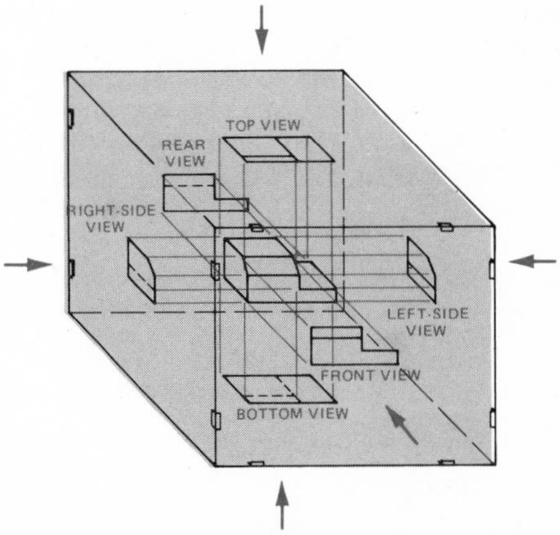
TOP VIEW
REAR VIEW
RIGHT-SIDE VIEW
LEFT-SIDE VIEW
FRONT VIEW
BOTTOM VIEW

Fig. 3-1-6 Relationship of object with viewing planes in third-angle projection.

object. These six viewing planes are then rotated or positioned so that they lie in a single plane, as shown in Fig. 3-1-7. Rarely are all six views used. Only the views which are necessary to fully describe the object are drawn. Simple objects, such as a gasket, can be described sufficiently by one view alone. However, in mechanical drafting two or three view drawings of objects are more common, the rear, bottom, and one of the two side views being rarely used. Figure 3-1-8 shows a simple object drawn in orthographic and pictorial form.

To fully appreciate the shape and detail of views drawn in third-angle orthographic projection, the next five drawing assignments have been chosen according to the types of surfaces generally found on objects. These surfaces can be divided into flat surfaces parallel to the viewing planes; flat surfaces which appear inclined in one plane and parallel to the other two principal reference planes (called *incline* surfaces); flat surfaces which are inclined in all three reference planes (called *oblique* surfaces); and surfaces which have diameters or radii. These drawings are so designed that only the top, front, and right side views are required.

All Surfaces Parallel to the Viewing Planes and All Edges and Lines Visible. When a surface is parallel to the viewing planes, that surface will show as a surface on one view and a line on the other views. The lengths of these lines are the same as the lines shown on the

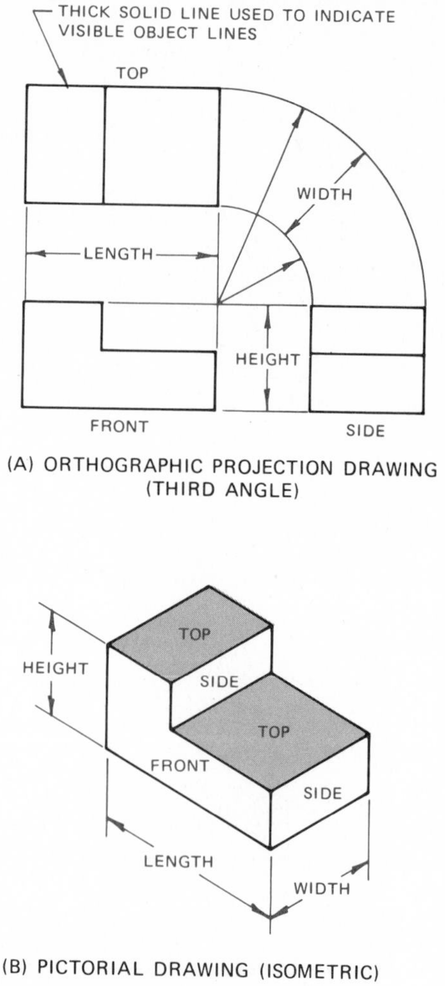
THICK SOLID LINE USED TO INDICATE VISIBLE OBJECT LINES
TOP
WIDTH
LENGTH
HEIGHT
FRONT
SIDE
(A) ORTHOGRAPHIC PROJECTION DRAWING (THIRD ANGLE)
HEIGHT
TOP
SIDE
TOP
FRONT
SIDE
LENGTH
WIDTH
(B) PICTORIAL DRAWING (ISOMETRIC)

Fig. 3-1-8 A simple object shown in orthographic projection and pictorial form.

surface view. The drawing has been made showing each side to represent the exact shape and size of the object and the relationship of the three views to one another. Figure 3-1-9 shows additional examples.

Assignment

On two A4-size sheets of preprinted grid paper (10-mm or 0.25-in. grids) sketch three views of each of the objects shown in Figs. 3-1-A and 3-1-B. Draw three objects on each sheet. Each square shown on the objects represents one square on the grid paper. Allow one grid space between views and a minimum of two grid spaces between objects. Identify the type of projection used by placing the appropriate ISO projection symbol on the bottom of the drawing.

REVIEW FOR ASSIGNMENT

Unit 2-6 Sketching
Unit 2-6 Visible Objects Lines
Unit 2-2 Scales

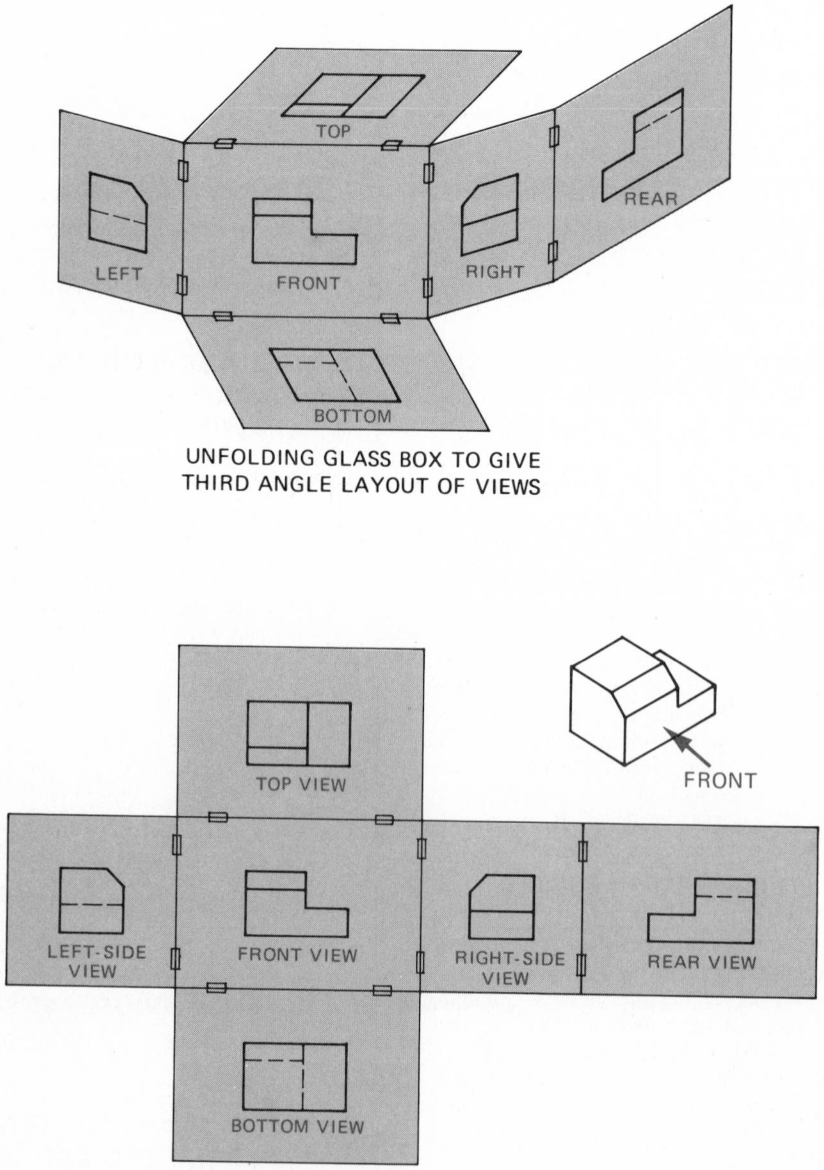
TOP
LEFT
FRONT
RIGHT
REAR
BOTTOM
UNFOLDING GLASS BOX TO GIVE THIRD ANGLE LAYOUT OF VIEWS
FRONT
TOP VIEW
LEFT-SIDE VIEW
FRONT VIEW
RIGHT-SIDE VIEW
REAR VIEW
BOTTOM VIEW

Fig. 3-1-7 Third-angle orthographic projection.

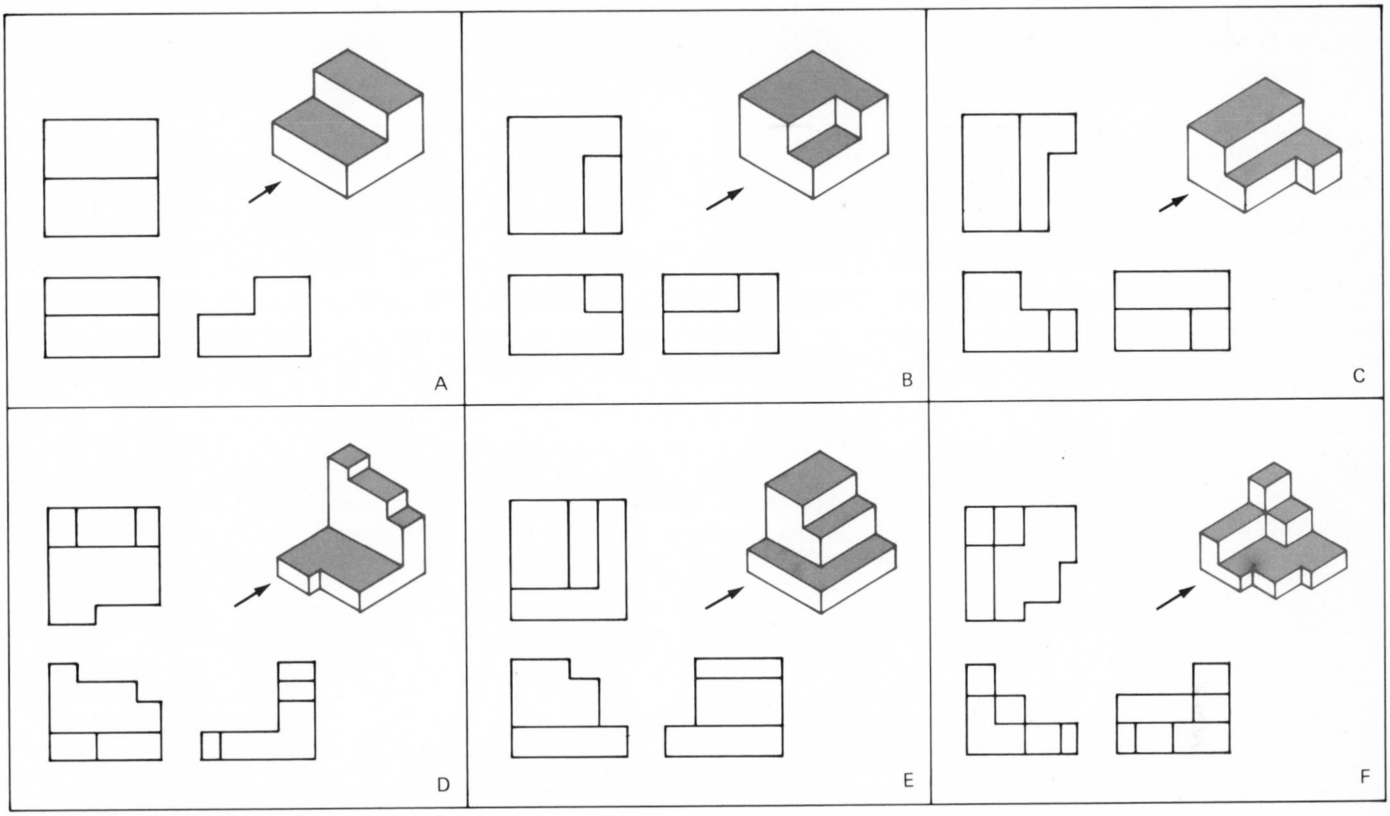
A
B
C
D
E
F

NOTE: ARROWS INDICATE DIRECTION OF SIGHT WHEN LOOKING AT THE FRONT VIEW.

Fig. 3-1-9 Illustrations of objects drawn in third-angle orthographic projection.

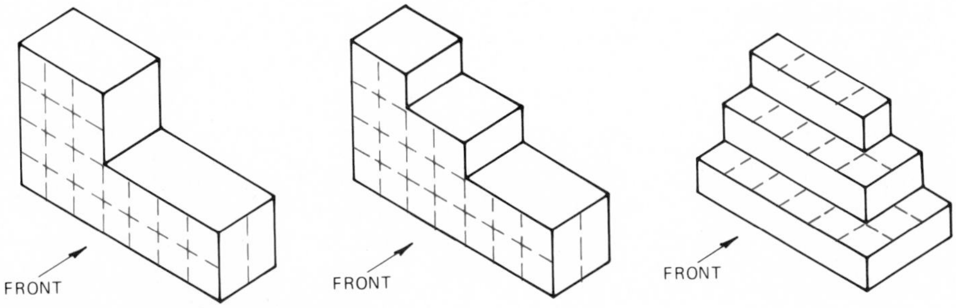
FRONT
FRONT
FRONT

Fig. 3-1-A Drawing assignment.

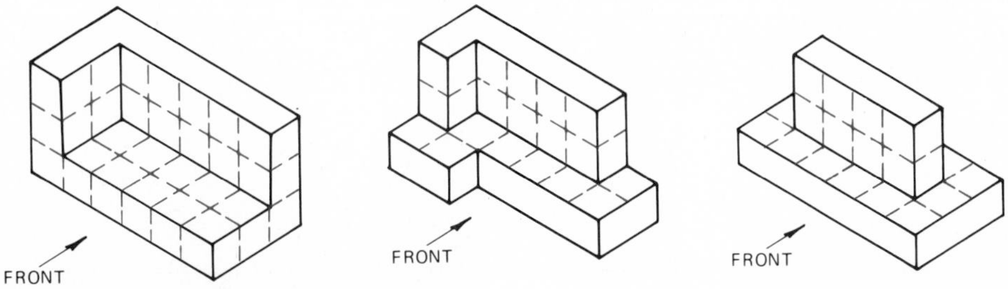
FRONT
FRONT
FRONT

Fig. 3-1-B Drawing assignment.

UNIT 3-2

ALL SURFACES PARALLEL TO THE VIEWING PLANES WITH SOME EDGES AND SURFACES HIDDEN

Most objects drawn in engineering offices are more complicated than the ones shown in Fig. 3-2-1. Many features

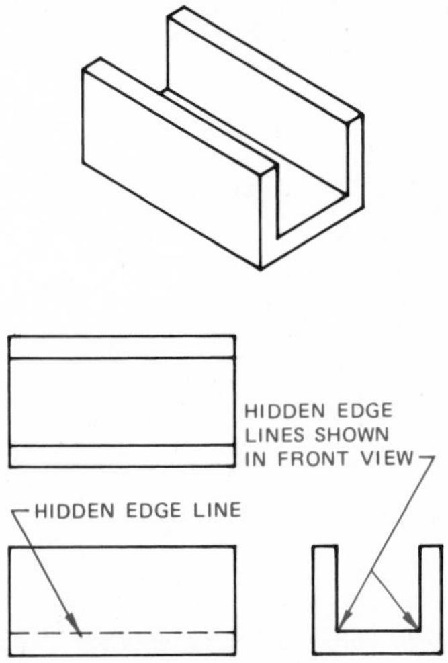
HIDDEN EDGE
LINES SHOWN
IN FRONT VIEW
HIDDEN EDGE LINE

Fig. 3-2-1 Hidden lines.

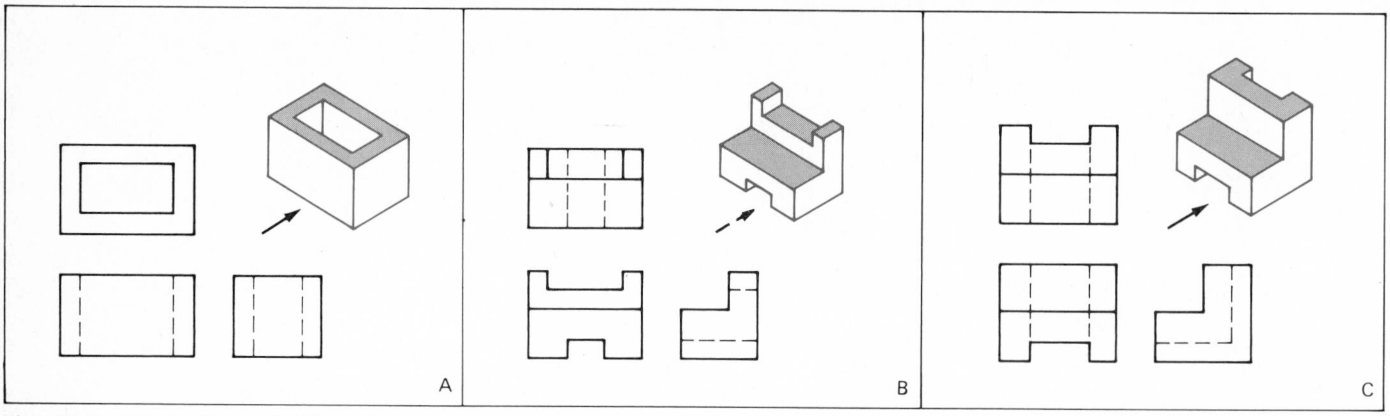
A
B
C

Fig. 3-2-2 **Illustrations of objects having hidden features.**

(lines, holes, etc.) cannot be seen when viewed from outside the piece. These hidden edges are called *hidden lines* and are normally required on the drawing to show the true shape of the object. Figure 3-2-2 shows additional examples of objects requiring hidden lines.

Assignment

On two A4-size sheets of preprinted grid paper (10-mm or 0.25-in. grids) sketch three views of each of the objects shown in Figs. 3-2-A and 3-2-B. Draw three objects on each sheet. Each square shown on the objects represents one square on the grid paper. Allow one grid space between views and a minimum of two spaces between objects. Identify the type of projection by placing the ISO projection symbol on the bottom of the drawing.

REVIEW FOR ASSIGNMENT

Unit 2-6 Hidden Lines
Unit 3-1 ISO Projection Symbol

UNIT 3-3

INCLINED SURFACES

If the surfaces of an object lie in either a horizontal or a vertical position, the surfaces appear in their true shapes in one of the three views, and these surfaces appear as a line in the other two views.

When a surface is inclined or sloped in only one direction, then that surface is not seen in its true shape in the top, front, or side view. It is, however, seen in two views as a distorted surface. On the third view it appears as a line.

The true length of surfaces A and B in Fig. 3-3-1 is seen in the front view only. In the top and side views, only the width of surfaces A and B appears in its true size. The length of these surfaces is foreshortened. Figure 3-3-2 shows additional examples.

Where an inclined surface has important features that must be shown clearly and without distortion, an *auxiliary* or helper view must be used. This type of view will be discussed in detail in Chapter 14. Figure 3-3-2 shows additional examples of objects having inclined surfaces.

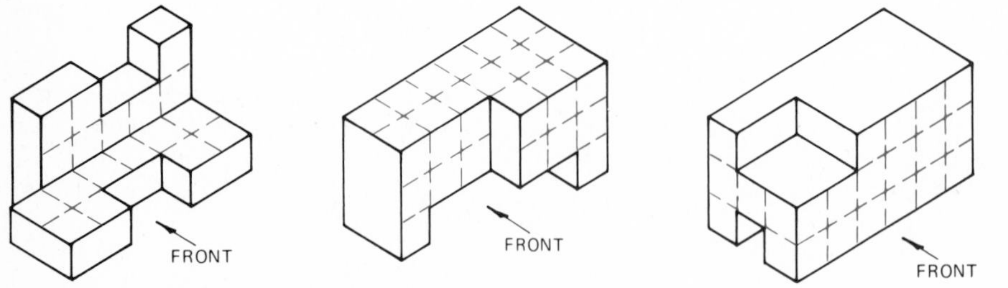
FRONT
FRONT
FRONT

Fig. 3-2-A **Drawing assignment.**

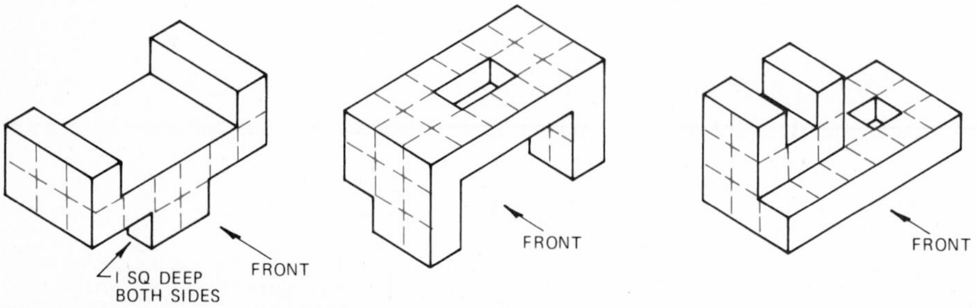
1 SQ DEEP
BOTH SIDES
FRONT
FRONT
FRONT

Fig. 3-2-B **Drawing assignment.**

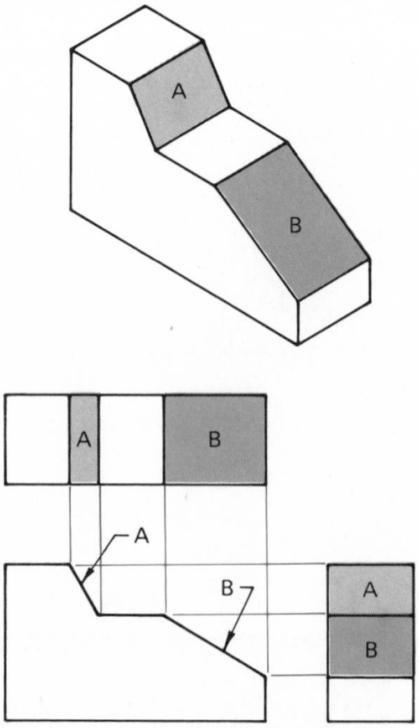
A
B
A
B
A
B
A
B

NOTE: THE TRUE SHAPE OF SURFACES A AND B DO NOT APPEAR ON THE TOP OR SIDE VIEWS.

Fig. 3-3-1 **Sloping surfaces.**

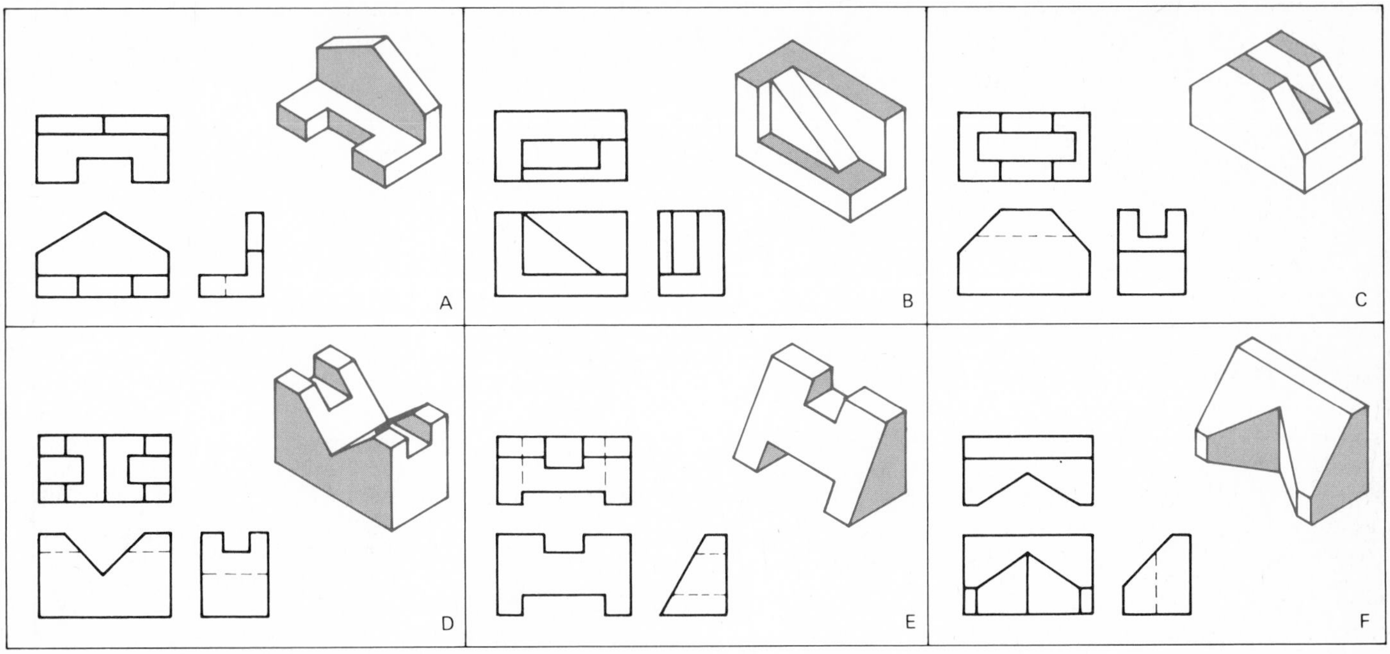
A
B
C
D
E
F

Fig. 3-3-2 Illustrations of objects having sloping surfaces.

Assignment

On two A4-size sheets of preprinted grid paper (10-mm or 0.25-in. grids) sketch three views of each of the objects shown in Figs. 3-3-A and 3-3-B. Draw three objects on each sheet. Each square shown on the objects represents one square on the grid paper. Allow one grid space between views and a minimum of two grid spaces between objects. The sloped (inclined) surfaces on each of the three objects are identified by a letter. Identify the sloped surfaces on each of the three views with a corresponding letter. Also identify the type of projection used by placing the appropriate ISO symbol on the bottom of the drawing.

REVIEW FOR ASSIGNMENT

Unit 2-6 Sketching
Unit 3-1 ISO Projection Symbol

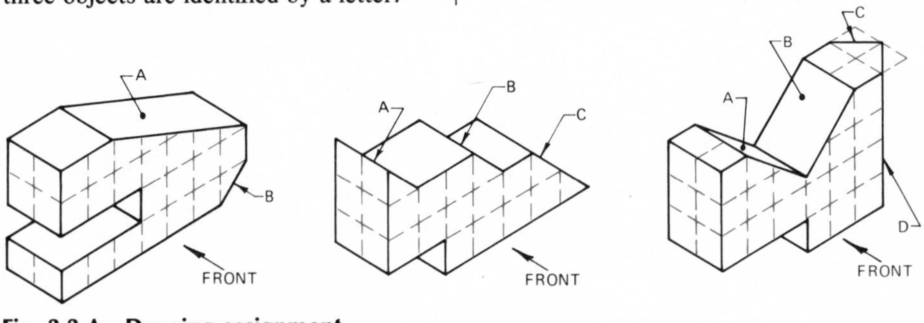
A
B
FRONT
A
B
C
FRONT
A
B
C
D
FRONT

Fig. 3-3-A Drawing assignment.

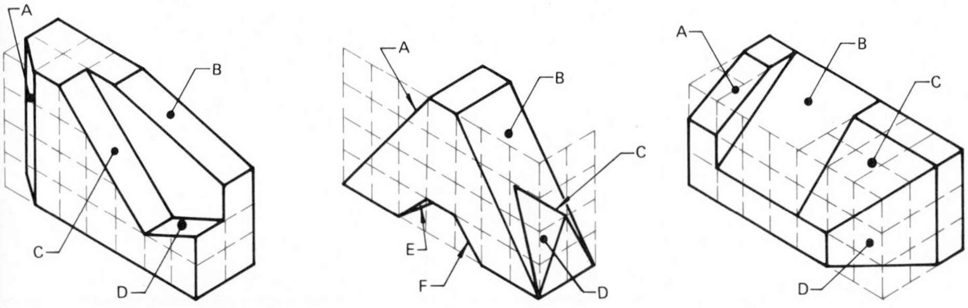
A
B
C
D
A
B
C
D
E
F
A
B
C
D

Fig. 3-3-B Drawing assignment.

UNIT 3-4

OBLIQUE SURFACES

When a surface is sloped so that it is not perpendicular to any of the three viewing planes, it will appear as a surface in all three views but never in its true shape. This is referred to as an *oblique surface* (Fig. 3-4-1). Since the oblique surface is not perpendicular to the viewing planes, it cannot be parallel to them and consequently appears foreshortened. If a true view is required for this surface,

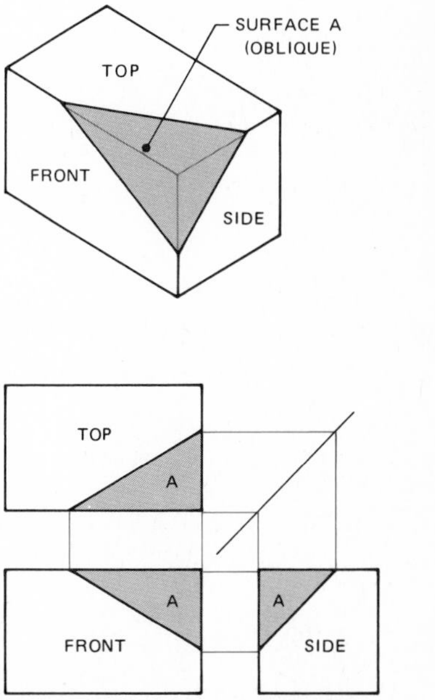
SURFACE A
(OBLIQUE)
TOP
FRONT
SIDE
TOP
A
A
A
FRONT
SIDE

Fig. 3-4-1 Oblique surface A not true shape in any of the three views.

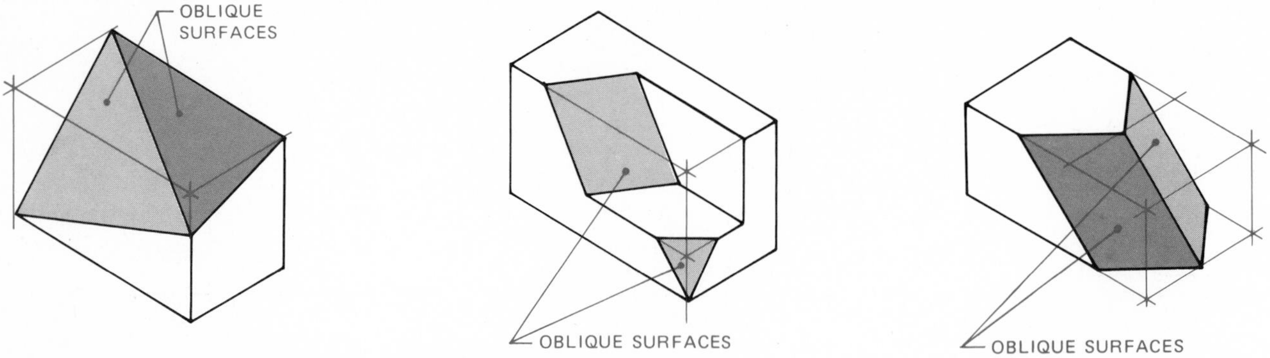
OBLIQUE SURFACES
OBLIQUE SURFACES
OBLIQUE SURFACES

Fig. 3-4-2 Illustrations of objects having oblique surfaces.

two auxiliary views—a primary and a secondary view—need to be drawn. This is discussed in detail under Secondary Auxiliary Views in Unit 14-4. Figure 3-4-2 shows additional examples of objects having oblique surfaces.

Assignment

On two A4-size sheets of preprinted grid paper (10-mm or 0.25-in. grids) sketch three views of each of the objects shown in Figs. 3-4-A and 3-4-B. Draw three objects on each sheet. Each square on the objects represents one square on the grid paper. Allow one grid space between views and a minimum of two grid spaces between objects. The oblique surfaces on the objects are identified by a letter. Identify the oblique surfaces on each of the three views with a corresponding letter. Also identify the type of projection used by placing the appropriate ISO symbol on the bottom of the drawing.

REVIEW FOR ASSIGNMENT

Unit 2-6 Sketching
Unit 3-1 ISO Projection Symbol

UNIT 3-5

CIRCULAR FEATURES

Typical parts with circular features are illustrated in Fig. 3-5-2. Note that the circular feature appears circular in one view only and that no line is used to indicate where a curved surface joins a flat surface. Hidden circles, like hidden flat surfaces, are represented on drawings by a hidden line.

CENTER LINES

A center line is drawn as a thin, broken line of long and short dashes, spaced alternately. They may be used to indicate center points, axes of cylindrical parts, and axes of symmetry, as shown in Fig. 3-5-1. Solid center lines are often used when the circular features are small. Center lines should project for a short distance beyond the outline of the part or feature to which they refer. They may be extended for use as extension lines for dimensioning purposes, but in this case the extended portion is not broken.

On views showing the circular features, the point of intersection of the two center lines is shown by the two intersecting short dashes.

Assignment

On two A4-size sheets of preprinted grid paper (10-mm or 0.25-in. grids) sketch three views of each of the objects shown in Figs. 3-5-A and 3-5-B. Draw three objects on each sheet. Each square shown on the objects represents one square on the grid paper. Allow one grid space between views and a minimum of two grids spaces between objects. Identify the type of projection used by placing the appropriate ISO projection symbol on the bottom of the drawing.

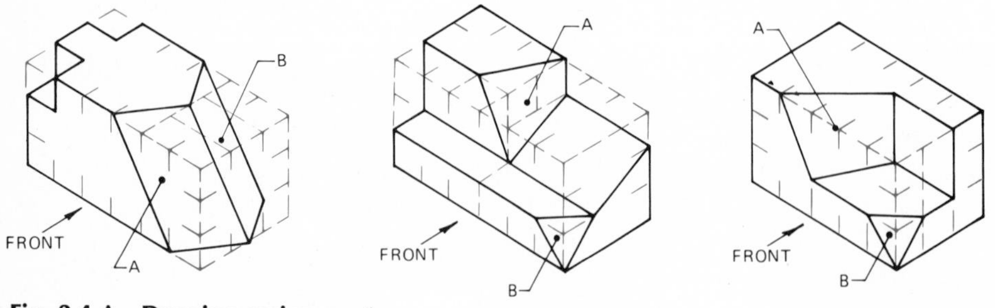
B
A
FRONT
A
B
FRONT
A
B
FRONT

Fig. 3-4-A Drawing assignment.

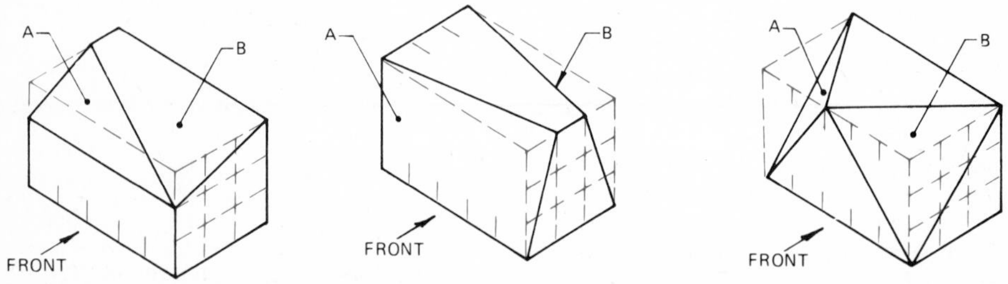
A
B
FRONT
A
B
FRONT
A
B
FRONT

Fig. 3-4-B Drawing assignment.

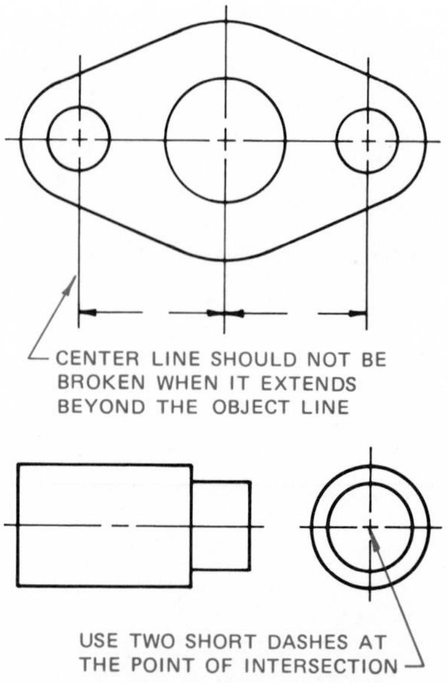
CENTER LINE SHOULD NOT BE BROKEN WHEN IT EXTENDS BEYOND THE OBJECT LINE
USE TWO SHORT DASHES AT THE POINT OF INTERSECTION

Fig. 3-5-1 Center line application.

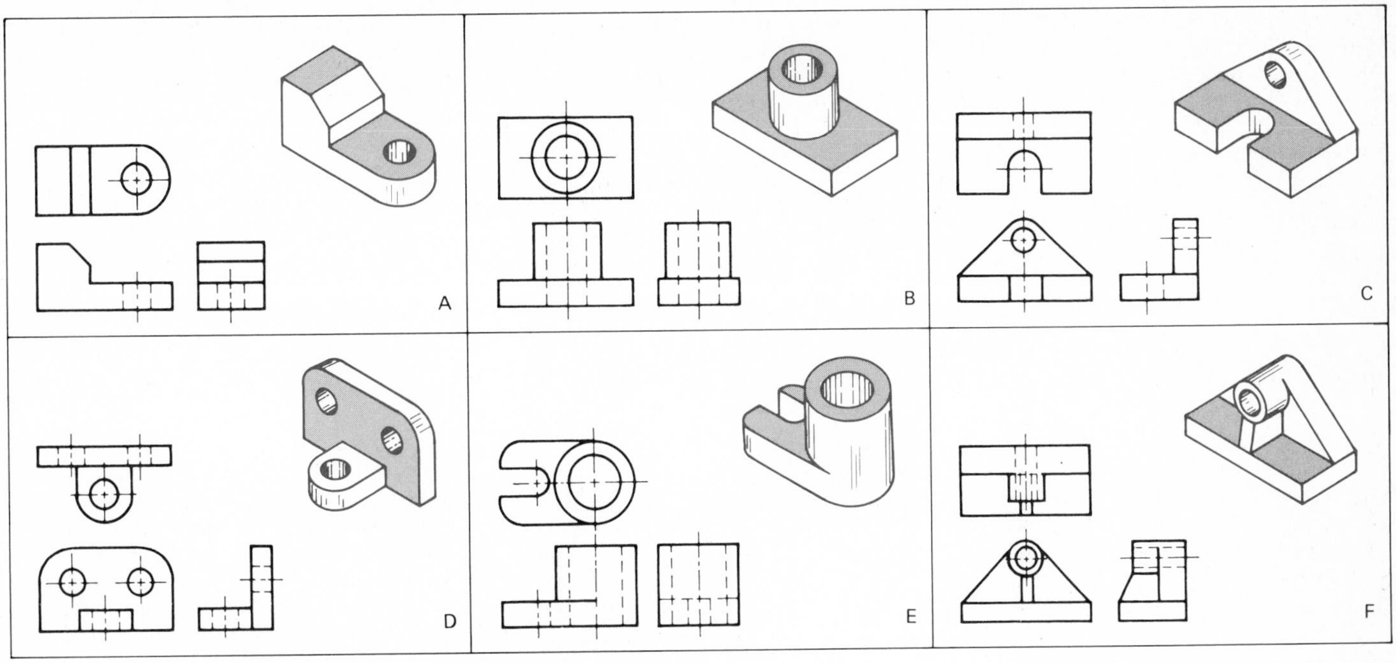
A
B
C
D
E
F

Fig. 3-5-2 Illustrations of objects having circular features.

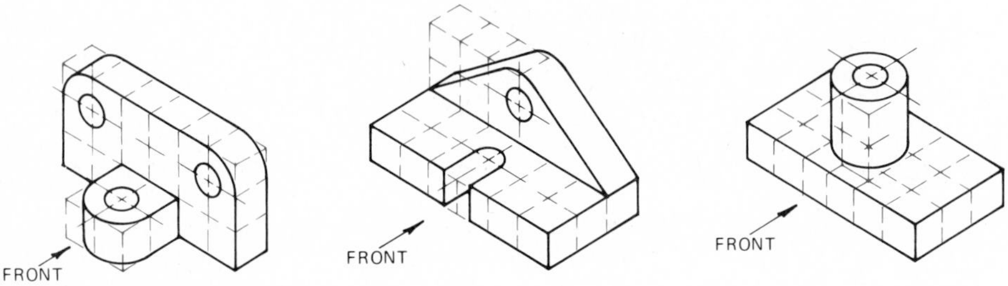
FRONT
FRONT
FRONT

Fig. 3-5-A Drawing assignment.

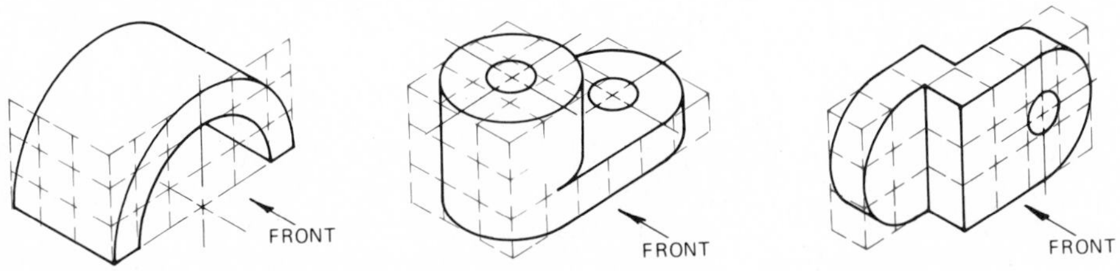
FRONT
FRONT
FRONT

Fig. 3-5-B Drawing assignment.

REVIEW FOR ASSIGNMENT

Unit 2-6 Center Lines
Unit 2-6 Sketching Circular Features
Unit 3-1 ISO Projection Symbol

UNIT 3-6

FIRST-ANGLE ORTHOGRAPHIC PROJECTION

As mentioned previously, first-angle orthographic projection is used in many countries throughout the world. Today with global marketing and the interchange of drawings with different countries, drafters are called upon to prepare and interpret drawings in both first- and third-angle projection. In first-angle projection, all the views are projected onto the planes located behind the objects rather than onto the planes lying between the objects and the viewer, as in third-angle projection. This is shown in Fig. 3-6-2. The unfolding and positioning of the views in one plane are shown in Fig. 3-6-3. Note that the views are on opposite sides of the front view with the exception of the rear view. A comparison

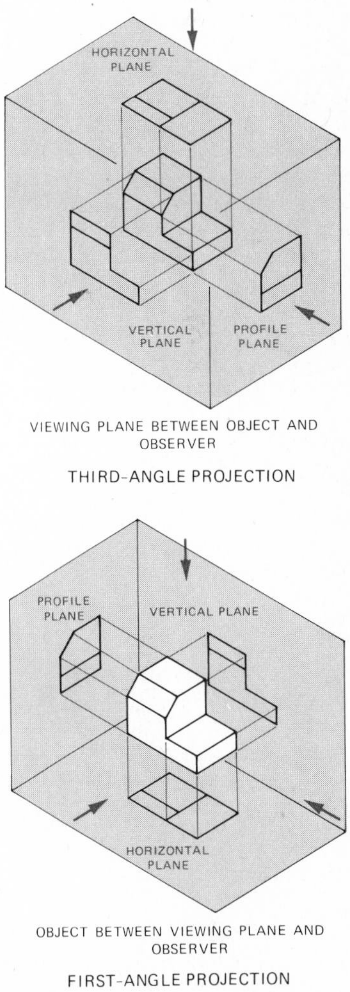
HORIZONTAL PLANE
VERTICAL PLANE
PROFILE PLANE
VIEWING PLANE BETWEEN OBJECT AND OBSERVER
THIRD-ANGLE PROJECTION
PROFILE PLANE
VERTICAL PLANE
HORIZONTAL PLANE
OBJECT BETWEEN VIEWING PLANE AND OBSERVER
FIRST-ANGLE PROJECTION

Fig. 3-6-1 A comparison between third-and-first-angle projection.

between the views of first- and third-angle projections is shown in Figs. 3-6-1 and 3-6-4. Remember that the views are identical in shape and detail, and only their location in reference to the front view has changed.

Assignment

On two A4-size sheets of preprinted grid paper (10-mm or 0.25-in. grids) sketch three views in first-angle orthographic projection of each of the objects shown in Figs. 3-6-A and 3-6-B. Draw three objects on each sheet. Each square on the objects represents one square on the grid paper. Allow one grid space between views and a minimum of two grid spaces between objects. Identify the type of projection used by placing the ISO projection symbol on the bottom of the drawing.

REVIEW FOR ASSIGNMENT

Unit 2-6 Sketching
Unit 3-1 ISO Projection Symbol

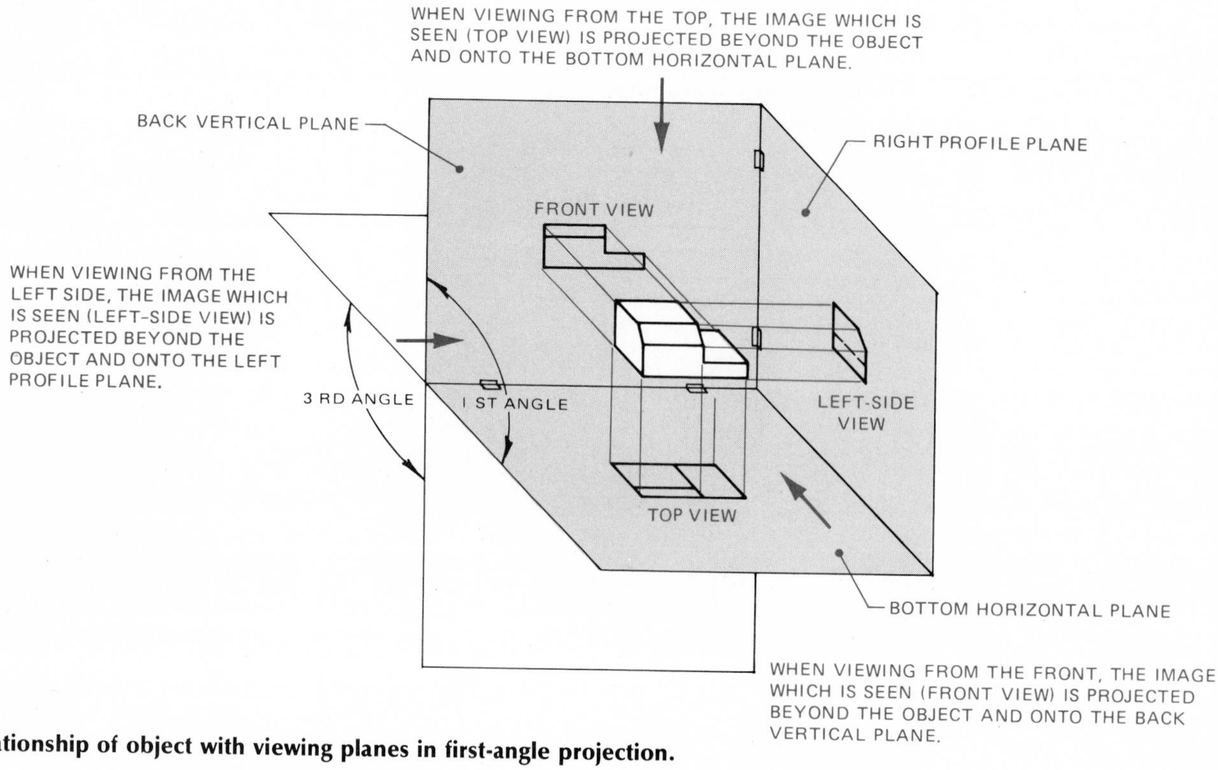

WHEN VIEWING FROM THE TOP, THE IMAGE WHICH IS SEEN (TOP VIEW) IS PROJECTED BEYOND THE OBJECT AND ONTO THE BOTTOM HORIZONTAL PLANE.
BACK VERTICAL PLANE
RIGHT PROFILE PLANE
FRONT VIEW
WHEN VIEWING FROM THE LEFT SIDE, THE IMAGE WHICH IS SEEN (LEFT-SIDE VIEW) IS PROJECTED BEYOND THE OBJECT AND ONTO THE LEFT PROFILE PLANE.
3 RD ANGLE
I ST ANGLE
LEFT-SIDE VIEW
TOP VIEW
BOTTOM HORIZONTAL PLANE
WHEN VIEWING FROM THE FRONT, THE IMAGE WHICH IS SEEN (FRONT VIEW) IS PROJECTED BEYOND THE OBJECT AND ONTO THE BACK VERTICAL PLANE.

Fig. 3-6-2 Relationship of object with viewing planes in first-angle projection.

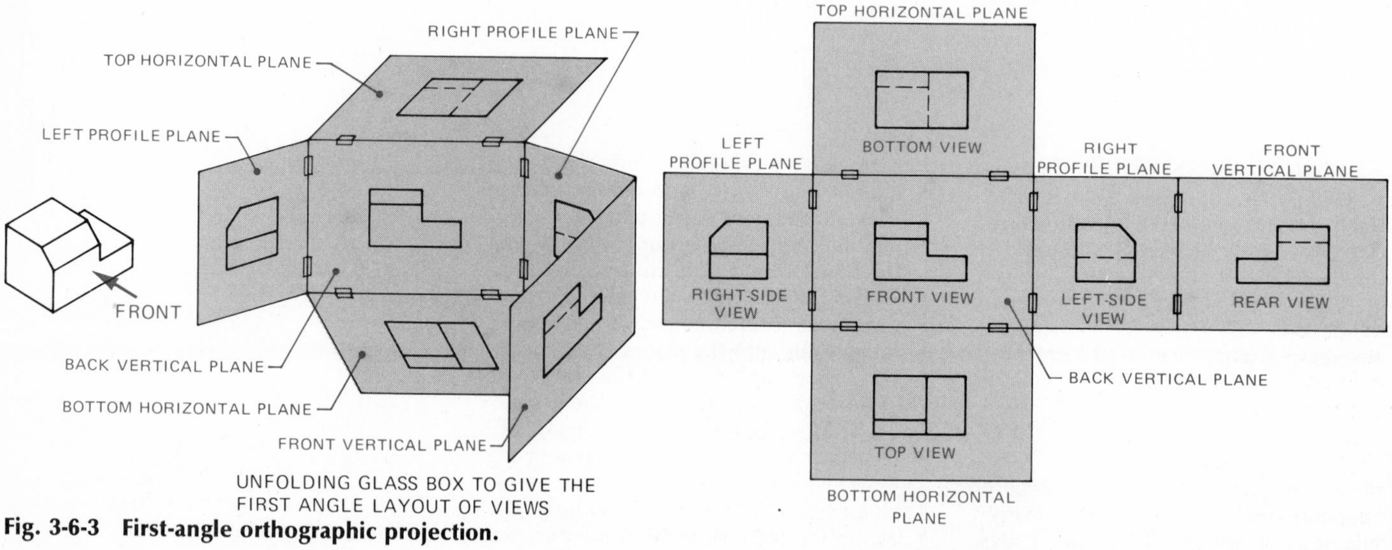

RIGHT PROFILE PLANE
TOP HORIZONTAL PLANE
LEFT PROFILE PLANE
FRONT
BACK VERTICAL PLANE
BOTTOM HORIZONTAL PLANE
FRONT VERTICAL PLANE
UNFOLDING GLASS BOX TO GIVE THE FIRST ANGLE LAYOUT OF VIEWS
TOP HORIZONTAL PLANE
BOTTOM VIEW
LEFT PROFILE PLANE
RIGHT PROFILE PLANE
FRONT VERTICAL PLANE
RIGHT-SIDE VIEW
FRONT VIEW
LEFT-SIDE VIEW
REAR VIEW
BACK VERTICAL PLANE
TOP VIEW
BOTTOM HORIZONTAL PLANE

Fig. 3-6-3 First-angle orthographic projection.

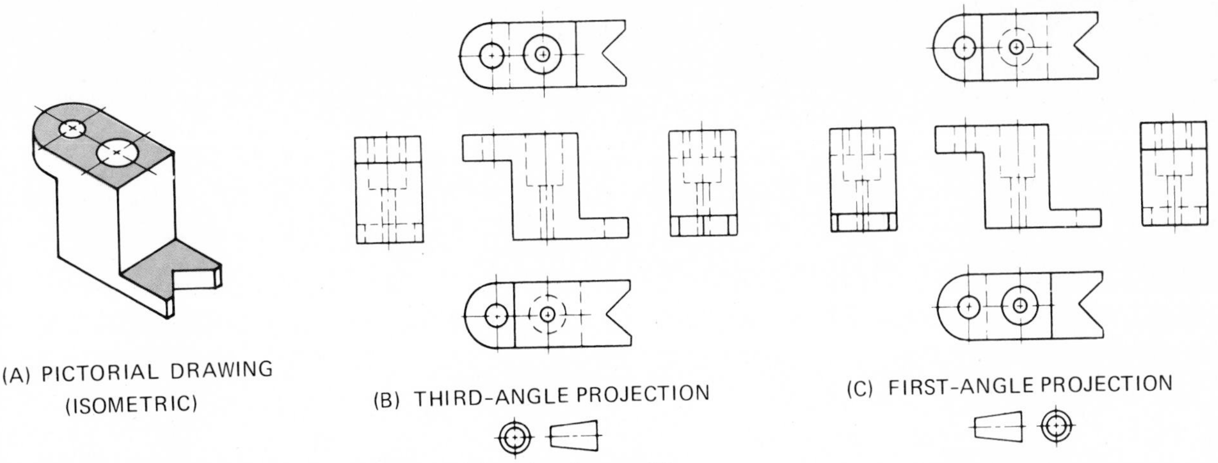
(A) PICTORIAL DRAWING
(ISOMETRIC)
(B) THIRD-ANGLE PROJECTION
(C) FIRST-ANGLE PROJECTION

Fig. 3-6-4 A simple object shown in pictorial and orthographic.

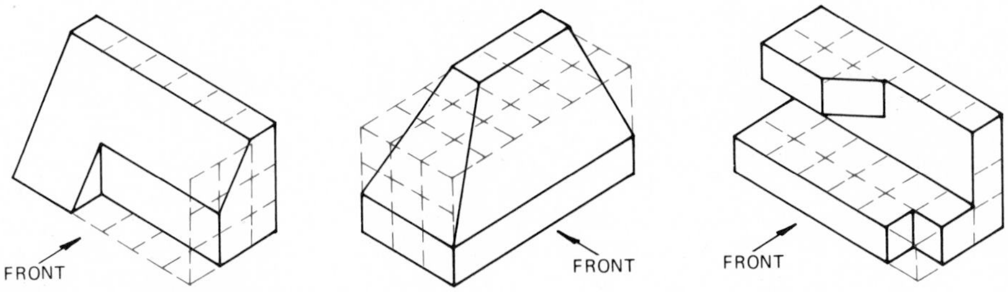
FRONT
FRONT
FRONT

Fig. 3-6-A Drawing assignment.

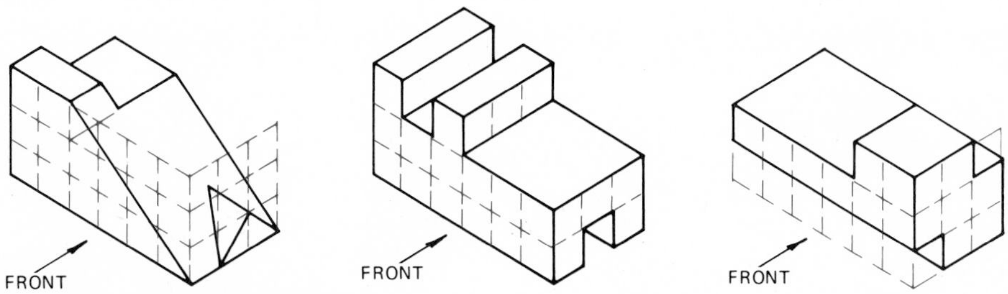
FRONT
FRONT
FRONT

Fig. 3-6-B Drawing assignment.

UNIT 3-7

ONE- AND TWO-VIEW DRAWINGS

View Selection

Views should be chosen that will best describe the object to be shown. Only the minimum number of views that will completely portray the size and shape of the part should be used. They should also be chosen to avoid hidden feature lines whenever possible, as shown in Fig. 3-7-1.

Except for complex objects of irregular shape, it is seldom necessary to draw more than three views. For representing simple parts, one- or two-view drawings will often be adequate.

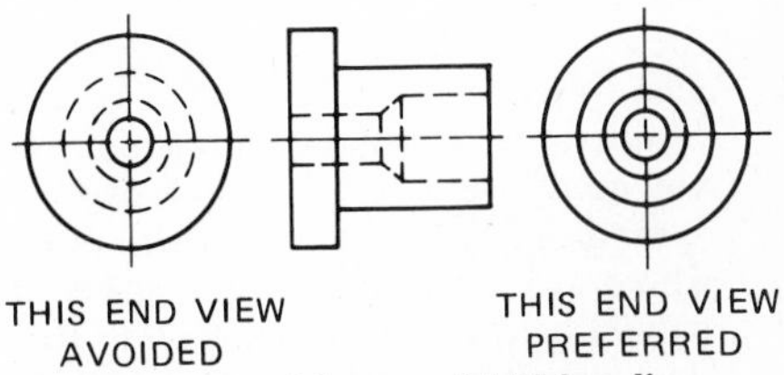
THIS END VIEW
AVOIDED
THIS END VIEW
PREFERRED

Fig. 3-7-1 Avoidance of hidden-line features.

One-View Drawings

In one-view drawings, the third dimension, such as thickness, may be expressed by a note or by descriptive words or abbreviations, such as DIA, ϕ, or HEXAGON ACROSS FLATS. Square sections may be indicated by light crossed diagonal lines. This applies whether the face is parallel or inclined to the drawing plane. These are illustrated in Fig. 3-7-2.

When cylindrical-shaped surfaces include special features such as a keyway, a side view (often called an *end view*) is required.

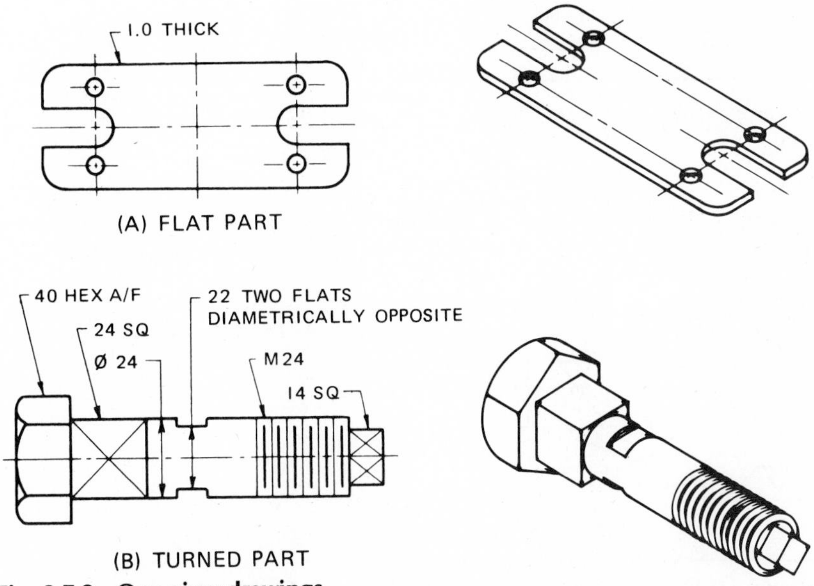
1.0 THICK
(A) FLAT PART
40 HEX A/F
24 SQ
Ø 24
22 TWO FLATS
DIAMETRICALLY OPPOSITE
M24
14 SQ
(B) TURNED PART

Fig. 3-7-2 One-view drawings.

Two-View Drawings

Frequently the drafter will decide that only two views are necessary to explain fully the shape of an object (Fig. 3-7-3). For this reason, some drawings consist of two adjacent views, such as the top and front views only, or front and right side views only. Two views are usually sufficient to explain fully the shape of cylindrical objects; if three views were used, two of them would be identical, depending on the detail structure of the part.

Assignment

On an A3-size sheet, select any four of the objects shown in Fig. 3-7-A or 3-7-B and draw only the necessary views in orthographic third-angle projection which will completely describe each part. Use symbols or abbreviations where possible. The drawings need not be to scale but should be drawn in proportion to the illustrations shown.

REVIEW FOR ASSIGNMENT

Unit 2-6 Line Work
Unit 2-6 Drawing Circles and Arcs

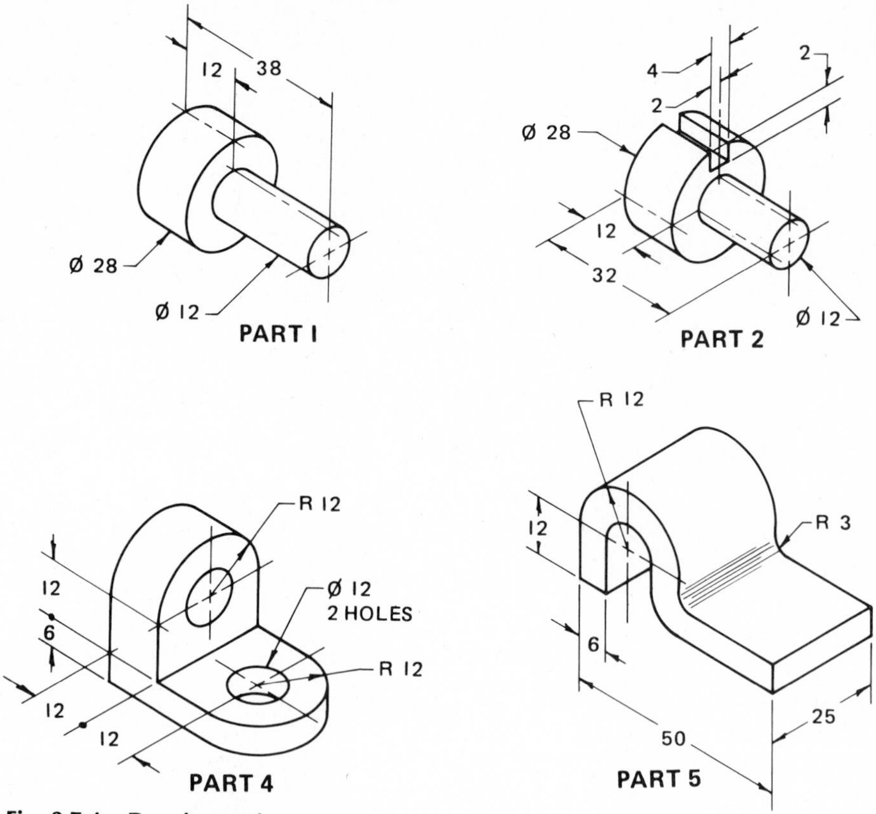

12
38
Ø 28
Ø 12
PART I
4
2
2
Ø 28
12
32
Ø 12
PART 2
R 12
Ø 12
2 HOLES
R 12
12
6
12
12
PART 4
R 12
12
R 3
6
50
25
PART 5

Fig. 3-7-A Drawing assignment.

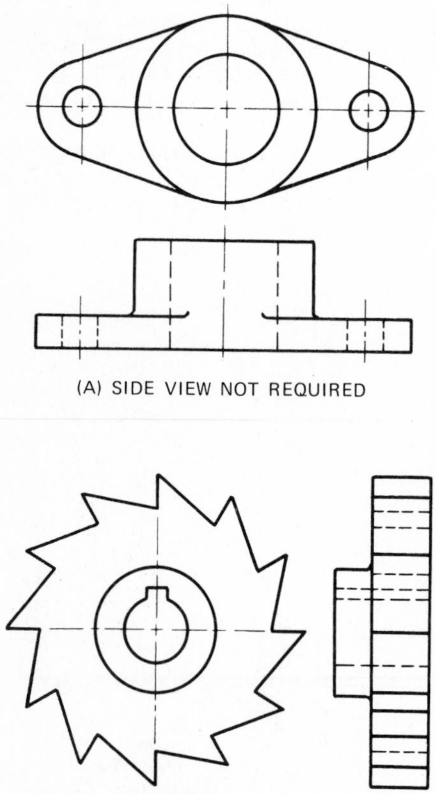

(A) SIDE VIEW NOT REQUIRED

(B) TOP VIEW NOT REQUIRED

Fig. 3-7-3 Two-view drawings.

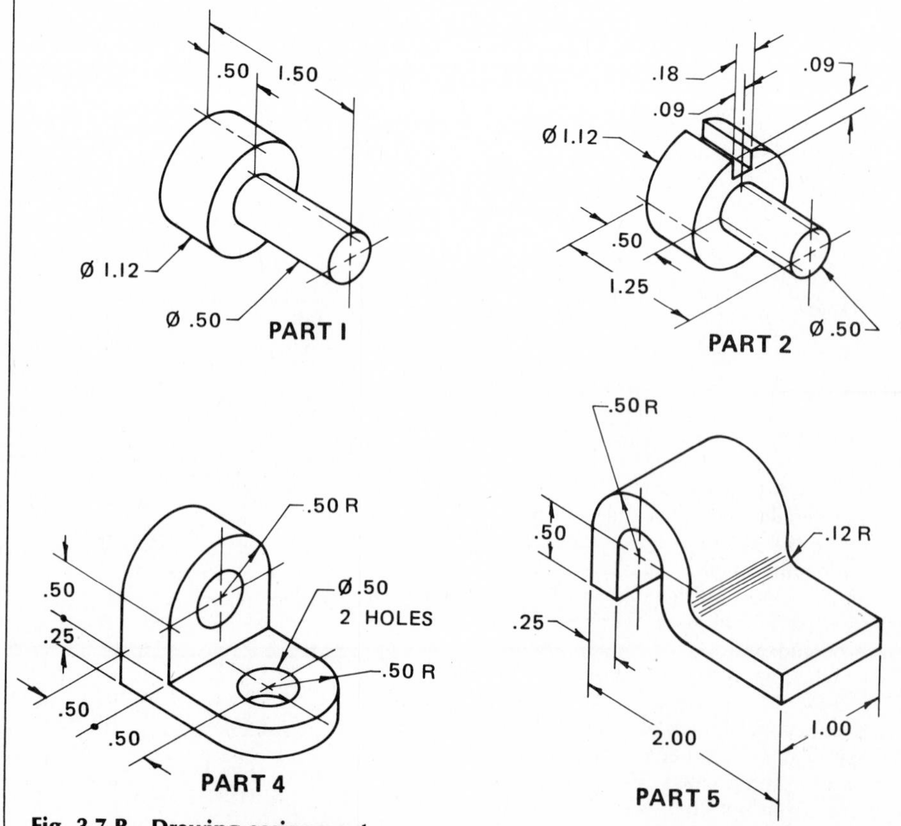

.50
1.50
Ø 1.12
Ø .50
PART I
.18
.09
.09
Ø 1.12
.50
1.25
Ø .50
PART 2
.50 R
Ø .50
2 HOLES
.50 R
.50
.25
.50
.50
PART 4
.50 R
.50
.12 R
.25
2.00
1.00
PART 5

Fig. 3-7-B Drawing assignment.

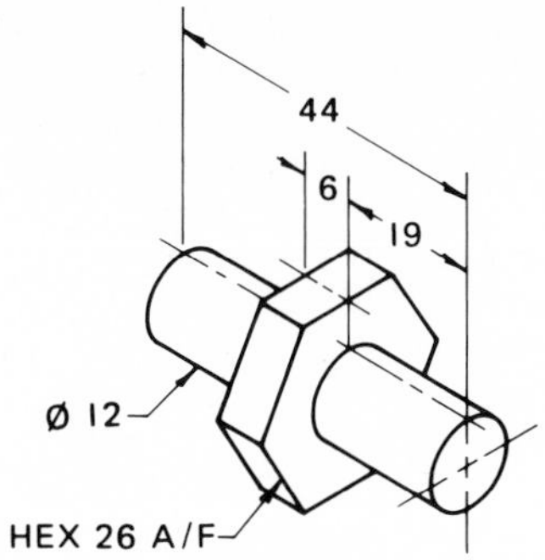
44
6
19
Ø 12
HEX 26 A/F

PART 3

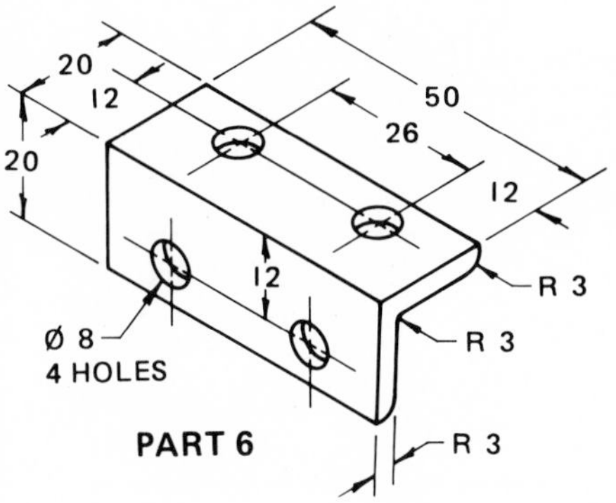
20
12
50
26
20
12
12
R 3
Ø 8
4 HOLES
R 3
R 3

PART 6

UNIT 3-8

PARTIAL VIEWS

Symmetrical objects may often be adequately portrayed by half views (Fig. 3-8-1). Partial views, which show only a limited portion of the object with remote details omitted, should be used, when necessary, to clarify the meaning of the drawing (Fig. 3-8-1). Such views are used to avoid the necessity of drawing many hidden features.

On drawings of objects where two side views can be used to better advantage than one, each need not be complete if together they depict the shape. Show only the hidden lines of features immediately behind the view (Fig. 3-8-1).

Assignment

On an A3-size sheet, select any one of the objects shown in Figs. 3-8-A to 3-8-D and draw only the necessary views (full and partial) which will completely describe each part. Add dimension and machining symbols.

REVIEW FOR ASSIGNMENT

Unit 2-6 Line Work
Unit 2-6 Drawing Circles and Arcs
Unit 5-7 Machining Symbols

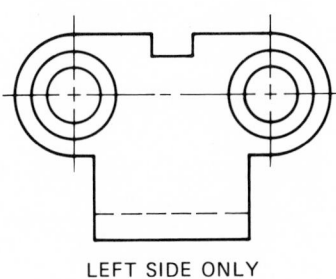

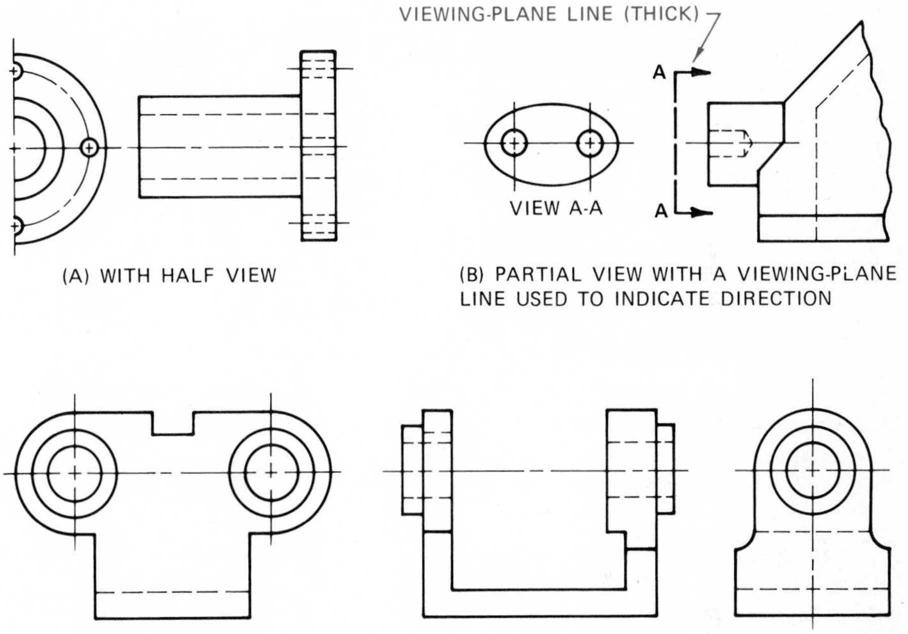
VIEWING-PLANE LINE (THICK)
A
A
VIEW A-A
(A) WITH HALF VIEW
(B) PARTIAL VIEW WITH A VIEWING-PLANE LINE USED TO INDICATE DIRECTION
LEFT SIDE ONLY
RIGHT SIDE ONLY
(C) PARTIAL SIDE VIEWS

Fig. 3-8-1 Partial views.

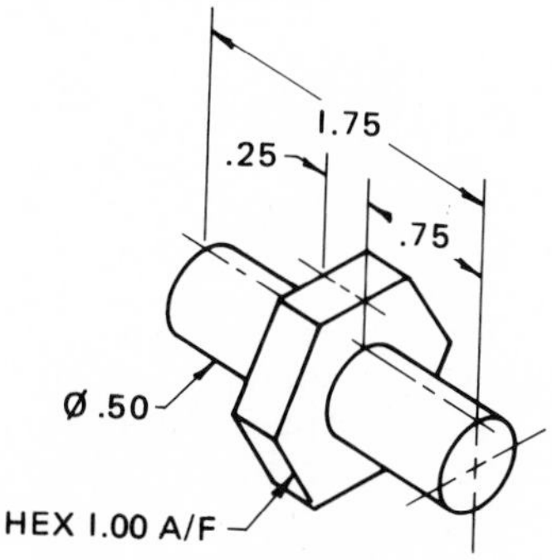
1.75
.25
.75
Ø .50
HEX 1.00 A/F

PART 3

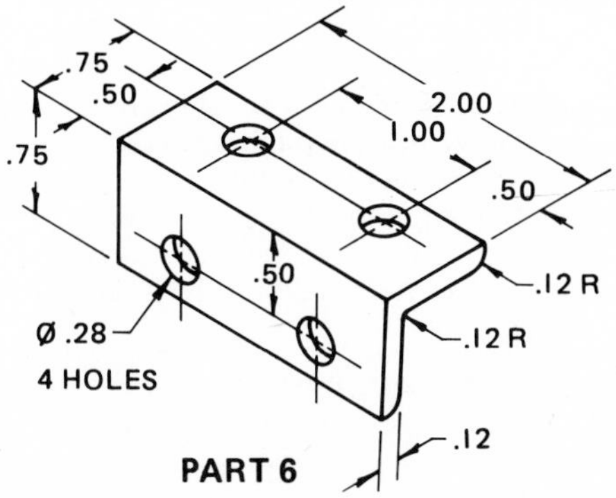
.75
.50
2.00
1.00
.75
.50
.50
.12 R
Ø .28
4 HOLES
.12 R
.12

PART 6

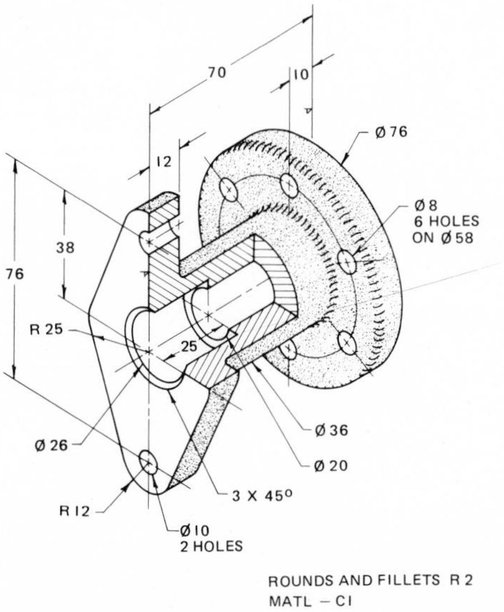
70
10
12
Ø 76
Ø 8
6 HOLES
ON Ø 58
38
76
R 25
25
Ø 36
Ø 26
Ø 20
3 X 45°
R 12
Ø 10
2 HOLES
ROUNDS AND FILLETS R 2
MATL – CI

Fig. 3-8-A Flanged coupling.

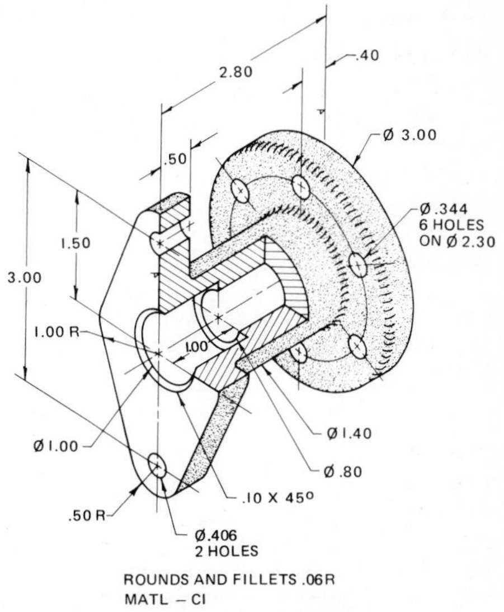
2.80
.40
.50
Ø 3.00
Ø .344
6 HOLES
ON Ø 2.30
1.50
3.00
1.00 R
1.00
Ø 1.00
Ø 1.40
Ø .80
.10 X 45°
.50 R
Ø.406
2 HOLES
ROUNDS AND FILLETS .06R
MATL – CI

Fig. 3-8-B Flanged coupling.

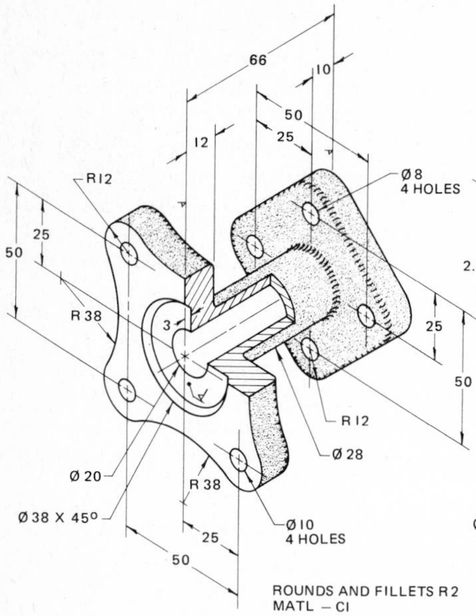

66
10
50
25
12
R12
Ø 8
4 HOLES
25
50
R 38
3
25
50
R 12
Ø 28
Ø 20
R 38
Ø 38 X 45°
Ø 10
4 HOLES
25
50
ROUNDS AND FILLETS R 2
MATL – CI

Fig. 3-8-C Flanged adapter.

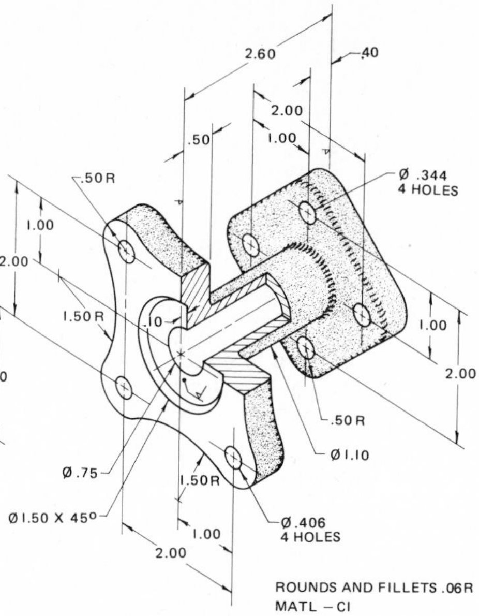

2.60
.40
2.00
1.00
.50
.50 R
Ø .344
4 HOLES
1.00
2.00
1.50 R
.10
1.00
2.00
.50 R
Ø 1.10
Ø .75
1.50 R
Ø 1.50 X 45°
Ø .406
4 HOLES
1.00
2.00
ROUNDS AND FILLETS .06R
MATL – CI

Fig. 3-8-D Flanged adapter.

UNIT 3-9

REAR VIEWS AND ENLARGED VIEWS

Placement of Views

When views are placed in the relative positions shown in Fig. 3-6-3, it is rarely necessary to identify them. When they are placed in other than the regular projected position, the removed view must be clearly identified.

Whenever appropriate, the orientation of the main view on a detail drawing should be the same as on the assembly drawing. To avoid the crowding of dimensions and notes, ample space must be provided between views.

Enlarged Views

Enlarged views are used when it is desirable to show a feature in greater detail or to eliminate the crowding of details or dimensions (Fig. 3-9-1). The enlarged view should be oriented in the same manner as the main view. However, if an enlarged view is rotated, state the direction and the amount of rotation of the detail. The scale of enlargement must be shown, and both views should be identified by one of the three methods shown.

Rear Views

Rear views are normally projected to the right or left. When this projection is not practical, because of the length of the part, particularly for panels and mounting plates, the rear view must not be projected up or down. Doing so would result in the part's being shown upside down. Instead, the view should be drawn as if it were projected sideways but located in some other position, and it should be clearly labeled REAR VIEW REMOVED (Fig. 3-9-2).

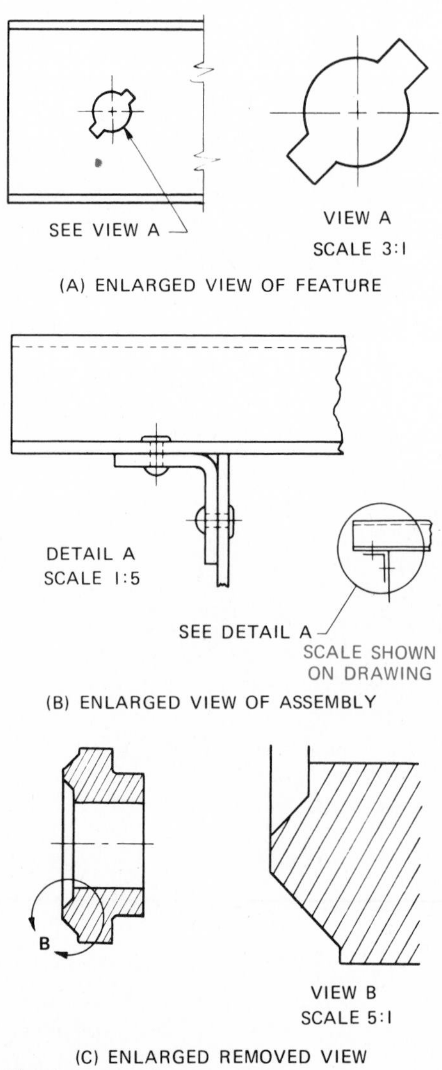

SEE VIEW A
VIEW A
SCALE 3:1
(A) ENLARGED VIEW OF FEATURE
DETAIL A
SCALE 1:5
SEE DETAIL A
SCALE SHOWN
ON DRAWING
(B) ENLARGED VIEW OF ASSEMBLY
B
VIEW B
SCALE 5:1
(C) ENLARGED REMOVED VIEW

Fig. 3-9-1 Enlarged views.

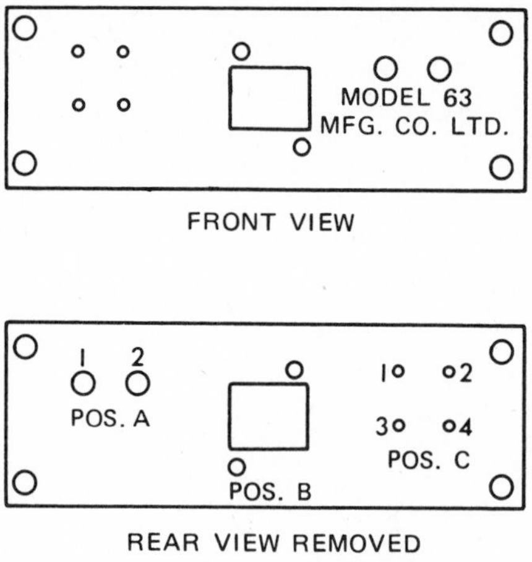

MODEL 63
MFG. CO. LTD.
FRONT VIEW
1 2
POS. A
POS. B
1 2
3 4
POS. C
REAR VIEW REMOVED

Fig. 3-9-2 Removed rear view.

Assignment

On an A3-size sheet, select one of the panels shown in Figs. 3-9-A to 3-9-D and make a detail drawing of the part. Panels such as these, where labeling is used to identify the terminals, are used extensively in the electrical and electronics industry.

In addition to the detail drawing, you are required to select a plastic material for the part. Design requirements dictate that the part be strong, have good electrical properties, be opaque, and, because of the large quantity required, be capable of being molded. Prepare a report to accompany the drawing which lists a minimum of three choices of materials that could be specified. The report should include comments on the above properties and your reasons for your choice.

REVIEW FOR ASSIGNMENT

Unit 13-1 Kinds of Plastics
Unit 13-2 Forming Methods
Unit 13-3 Design Consideration—Parts

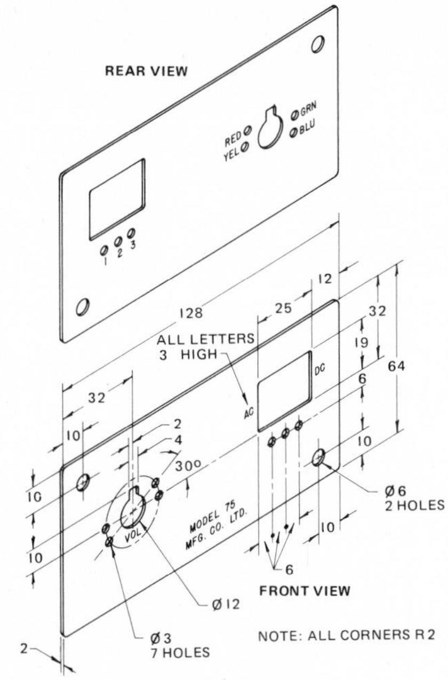
REAR VIEW
RED
YEL
GRN
BLU
128
25
12
32
19
6
64
10
ALL LETTERS
3 HIGH
AC
DC
32
10
2
4
30°
Ø6
2 HOLES
MODEL 75
MFG. CO. LTD.
VOL
10
10
6
FRONT VIEW
Ø12
Ø3
7 HOLES
NOTE: ALL CORNERS R2
2

Fig. 3-9-A Radio cover plate.

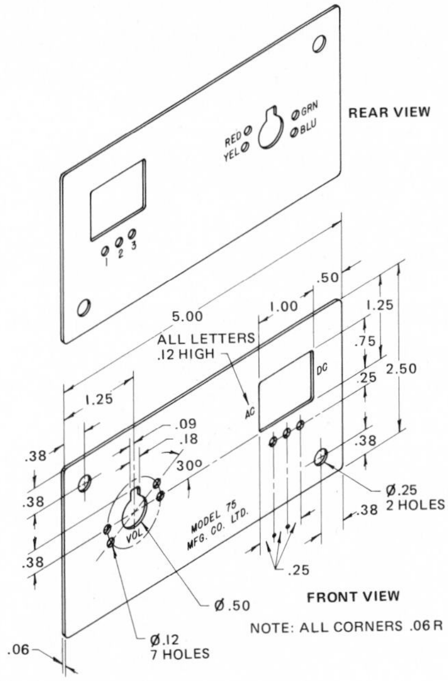
REAR VIEW
RED
YEL
GRN
BLU
5.00
1.00
.50
1.25
.75
.25
2.50
ALL LETTERS
.12 HIGH
AC
DC
1.25
.38
.09
.18
30°
.38
.38
.38
Ø.25
2 HOLES
MODEL 75
MFG. CO. LTD.
VOL
.25
Ø.50
FRONT VIEW
NOTE: ALL CORNERS .06R
Ø.12
7 HOLES
.06

Fig. 3-9-B Radio cover plate.

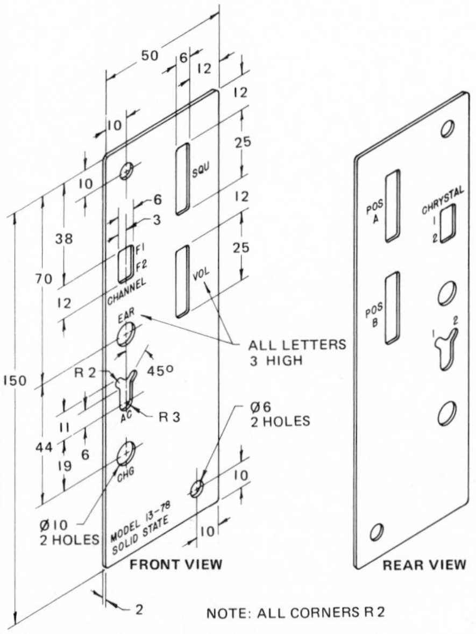
50
6
12
12
25
12
25
10
10
6
3
38
70
12
150
R2
45°
R3
11
44
6
19
Ø10
2 HOLES
SQU
VOL
CHANNEL
F1
F2
EAR
AC
CHG
MODEL 13-78
SOLID STATE
ALL LETTERS
3 HIGH
Ø6
2 HOLES
10
10
FRONT VIEW
REAR VIEW
POS A
POS B
CHRYSTAL
2
NOTE: ALL CORNERS R2

Fig. 3-9-C Transceiver cover plate.

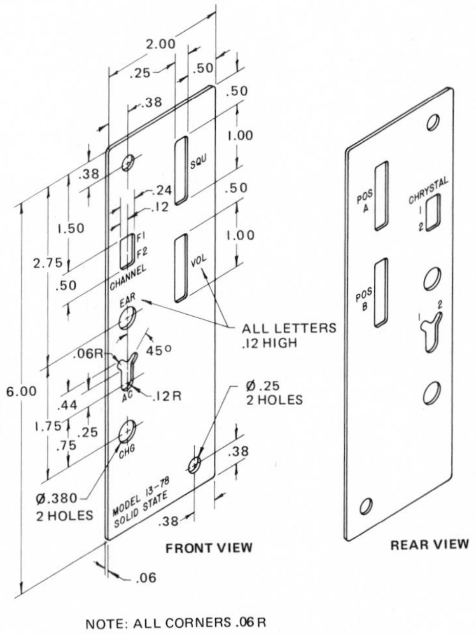
2.00
.25
.50
.38
.50
1.00
.50
1.00
.38
.24
.12
1.50
2.75
.50
6.00
.06R
45°
.12R
.44
1.75
.25
.75
Ø.380
2 HOLES
SQU
VOL
CHANNEL
F1
F2
EAR
AC
CHG
MODEL 13-78
SOLID STATE
ALL LETTERS
.12 HIGH
Ø.25
2 HOLES
.38
.38
FRONT VIEW
REAR VIEW
POS A
POS B
CHRYSTAL
.06
NOTE: ALL CORNERS .06R

Fig. 3-9-D Transceiver cover plate.

UNIT 3-10

KEY PLANS AND OPPOSITE HAND VIEWS

Key Plan

A method particularly applicable to structural work is to include on each sheet of a series of drawings a small key plan indicating in bold lines the relationship of the detail on that sheet to the whole work, as in Fig. 3-10-1.

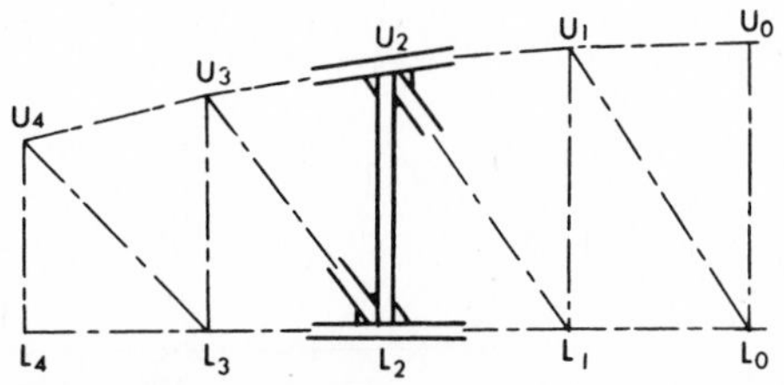
U4
U3
U2
U1
U0
L4
L3
L2
L1
L0

Fig. 3-10-1 Key plan.

Opposite-Hand Views

Where parts are symmetrically opposite, such as for right- and left-hand usage, one part is drawn in detail and the other is indicated by a note such as PART B SAME EXCEPT OPPOSITE HAND. It is preferable to show both part numbers on the same drawing (Fig. 3-10-2).

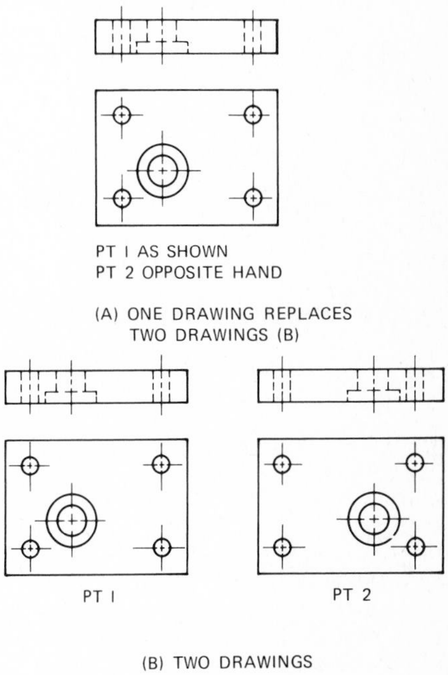
PT 1 AS SHOWN
PT 2 OPPOSITE HAND
(A) ONE DRAWING REPLACES
TWO DRAWINGS (B)
PT 1
PT 2
(B) TWO DRAWINGS

Fig. 3-10-2 Opposite hand views.

Assignment

With most truss drawings, the scale used on the overall assembly is such that intricate detail cannot be clearly shown. As a result, enlarged detail views are added. With this type of assembly, many parts are opposite hand to their counterparts.

The primary problem in the assembly of large-span multimember space structures has been to find some simple, inexpensive, and repetitive way to connect many members into and through a typical joint. The approach shown in this assignment is to shop-fabricate as much as possible, to ship these subassemblies to the site, and finally to complete the assembly by bolting the subassemblies together. The number, size, and strength of bolt is calculated by the loading requirements at each connection.

Benefits of shop prefabrication of large units include minimization of field erection time and less possible error in the field resulting from the greater tolerance control in the shop.

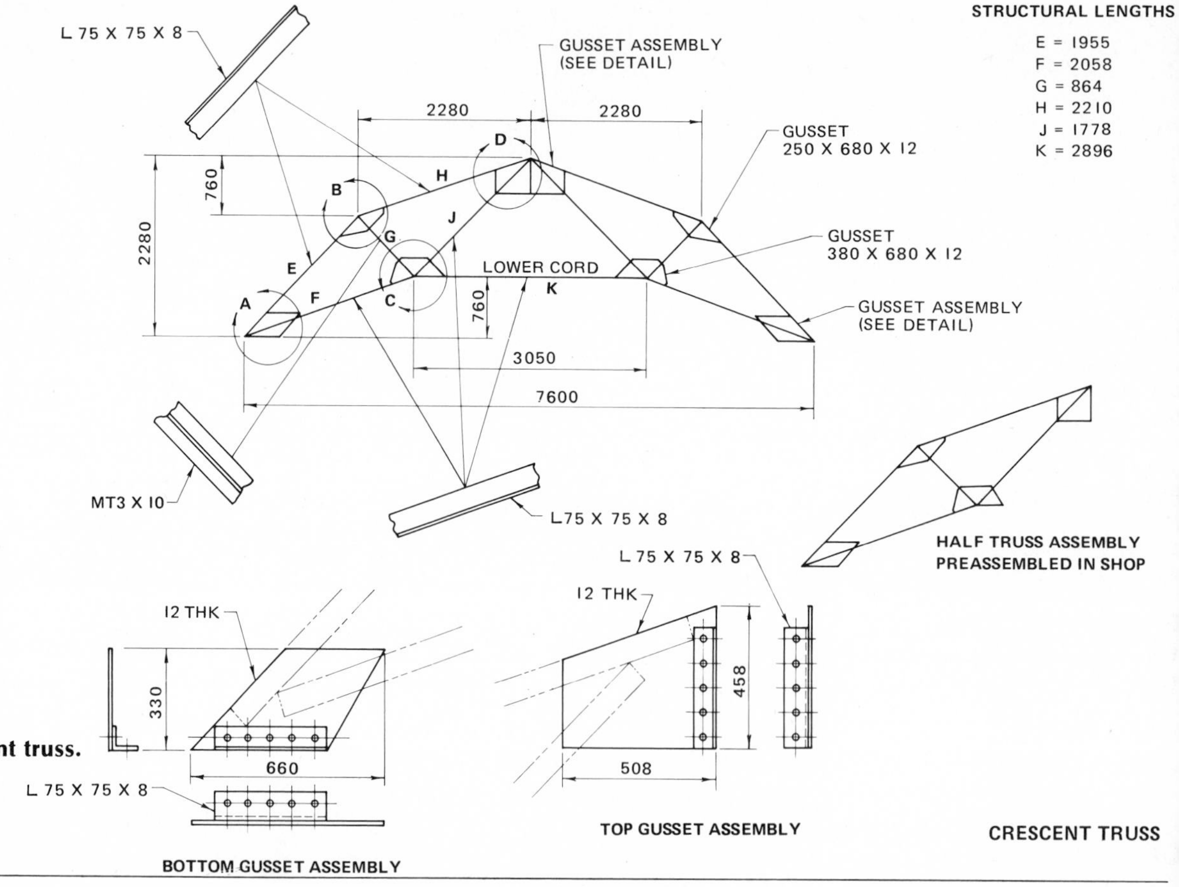
L 75 X 75 X 8
GUSSET ASSEMBLY
(SEE DETAIL)
2280
2280
GUSSET
250 X 680 X 12
STRUCTURAL LENGTHS
E = 1955
F = 2058
G = 864
H = 2210
J = 1778
K = 2896
760
2280
D
B
H
J
G
E
A
F
C
LOWER CORD
K
760
GUSSET
380 X 680 X 12
GUSSET ASSEMBLY
(SEE DETAIL)
3050
7600
MT3 X 10
L75 X 75 X 8
HALF TRUSS ASSEMBLY
PREASSEMBLED IN SHOP
L 75 X 75 X 8
12 THK
12 THK
330
458
660
508
L 75 X 75 X 8
TOP GUSSET ASSEMBLY
CRESCENT TRUSS
BOTTOM GUSSET ASSEMBLY

Fig. 3-10-A Crescent truss.

Fig. 3-10-B Crescent truss.

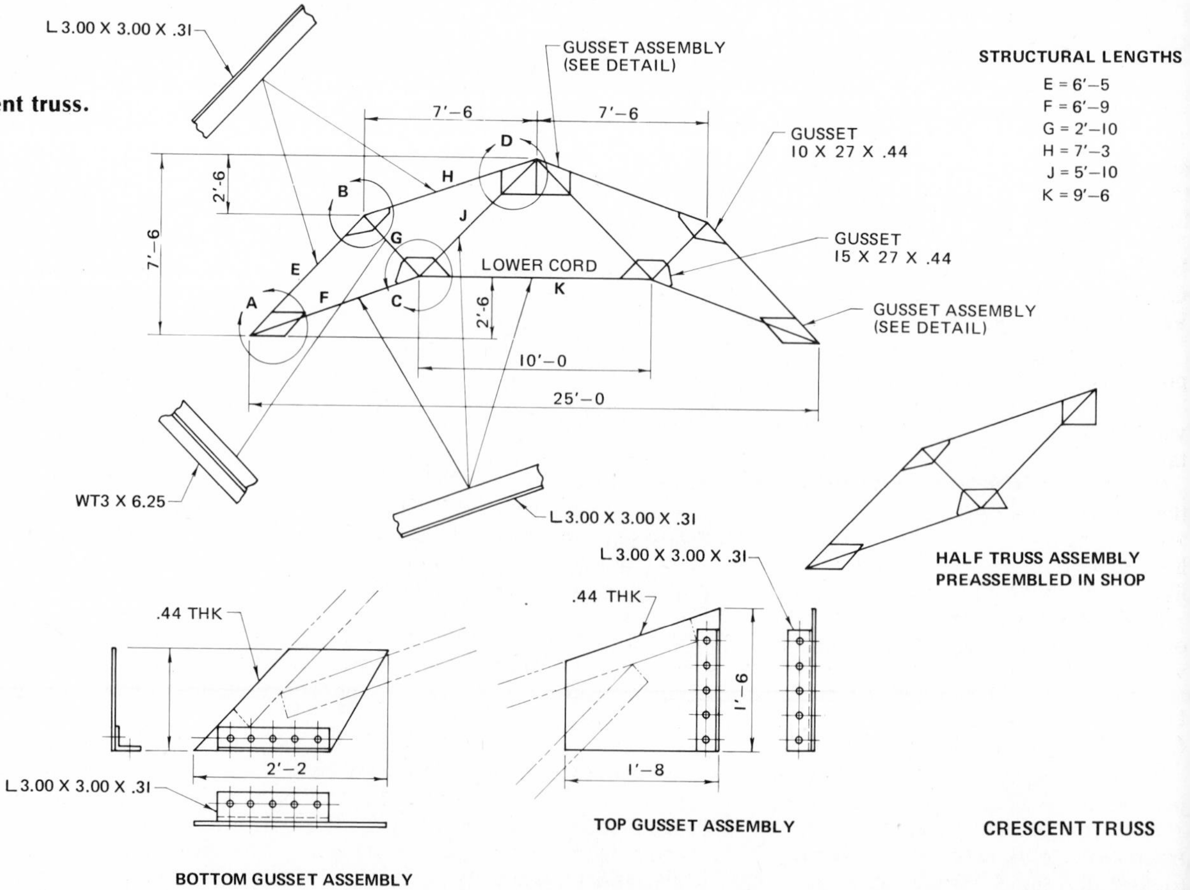
L 3.00 X 3.00 X .31
GUSSET ASSEMBLY
(SEE DETAIL)
STRUCTURAL LENGTHS
E = 6′–5
F = 6′–9
G = 2′–10
H = 7′–3
J = 5′–10
K = 9′–6
7′–6
7′–6
GUSSET
10 X 27 X .44
2′-6
7′–6
D
B
H
J
G
E
A
F
C
LOWER CORD
K
2′-6
GUSSET
15 X 27 X .44
GUSSET ASSEMBLY
(SEE DETAIL)
10′–0
25′–0
WT3 X 6.25
L 3.00 X 3.00 X .31
L 3.00 X 3.00 X .31
HALF TRUSS ASSEMBLY
PREASSEMBLED IN SHOP
.44 THK
.44 THK
1′–6
2′–2
1′–8
L 3.00 X 3.00 X .31
TOP GUSSET ASSEMBLY
CRESCENT TRUSS
BOTTOM GUSSET ASSEMBLY

Select one of the assignments from Figs. 3-10-A and 3-10-B. On an A3-size sheet, draw the enlarged views of the gusset assemblies to a scale of 1:10 or 1 in. = 1 ft. On a second sheet, draw up the complete bill of material necessary to make one complete truss. The bolts used to hold the truss to the walls are not to be included.

Shop Bolting Data. All structural members will be bolted to the gusset plate with four M10 (0.375-in.) high-strength bolts. Spacing is 40 mm (1.60 in.) from end and 80 mm (3.10 in.) center to center. Bracketed sizes are for inch-size fasteners.

Field Bolting Data. All connections are to be made with five M10 (0.375-in.) high-strength bolts. Spacing is 40 mm (1.60 in.) from end and 80 mm (3.10 in.) center to center. Bracketed sizes are for inch-size fasteners.

REVIEW FOR ASSIGNMENT

Unit 8-3 Bolt Specifications
Unit 6-4 Bill of Materials

UNIT 3-11

MITER LINES AND SPACING OF VIEWS

Use of a Miter Line

The use of a miter line provides a fast and accurate method of constructing the third view once two views are established (Fig. 3-11-1).

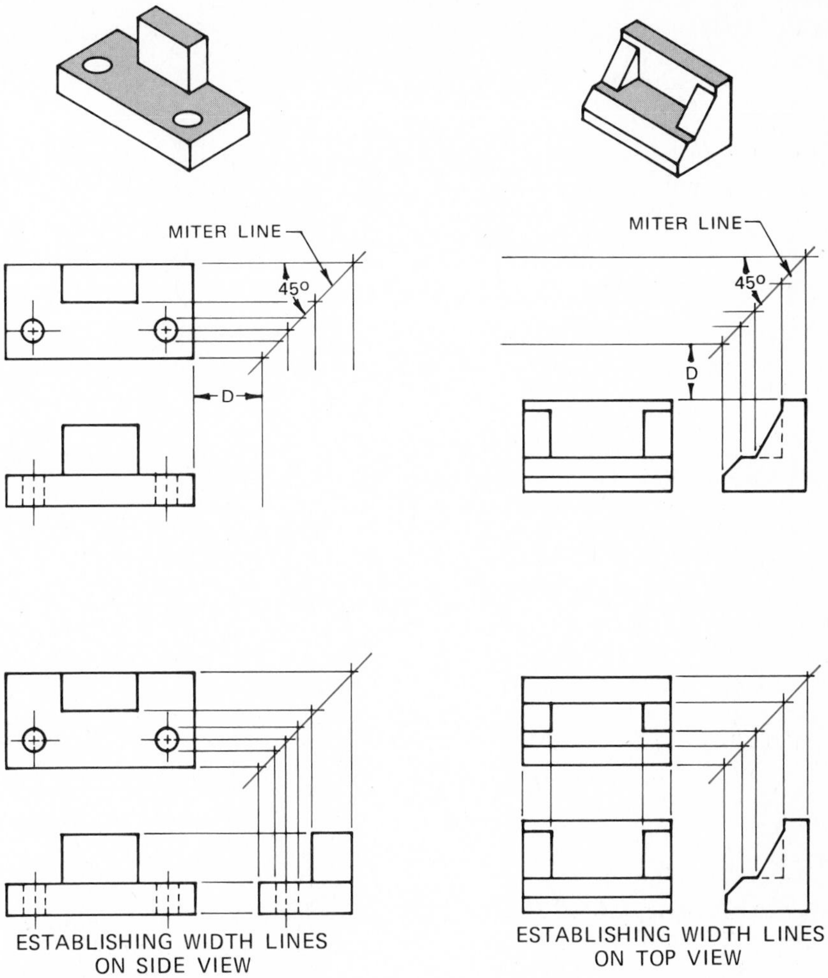
MITER LINE
45°
D
MITER LINE
45°
D
ESTABLISHING WIDTH LINES
ON SIDE VIEW
ESTABLISHING WIDTH LINES
ON TOP VIEW

Fig. 3-11-1 Use of miter line.

USING A MITER LINE TO CONSTRUCT THE RIGHT SIDE VIEW

1. Given the top and front views, project lines to the right of the top view.

2. Establish how far from the front view the side view is to be drawn (distance *D*).

3. Construct the miter line at 45° to the horizon.

4. Where the horizontal projection lines of the top view intersect the miter line, drop vertical projection lines.

5. Project horizontal lines to the right of the front view and complete the side view.

USING A MITER LINE TO CONSTRUCT THE TOP VIEW

1. Given the front and side views, project vertical lines up from the side view.

2. Establish how far away from the front view the top view is to be drawn (distance *D*).

3. Construct the miter line at 45° to the horizon.

4. Where the vertical projection lines of the side view intersect the miter lines, project horizontal lines to the left.

5. Project vertical lines up from the front view and complete the top view.

SPACING THE VIEWS

It is important for clarity and good appearance that the views be well balanced on the drawing paper, whether the drawing shows one view, two views, three views, or more. The drafter must anticipate the approximate space required. This is determined from the size of the object to be drawn, the number of views, the scale used, and the space between views. Ample space should be provided between views to permit placement of dimensions on the drawing without crowding. Space should also be allotted for notes without crowding. However, space between views should not be excessive. Once the size of paper, scale, and number of views are established, the balancing of the three views is relatively simple (Fig. 3-11-2). A popular method of doing so is shown. In this example, a distance of 40 mm is left between views. Remember that the spacing suitable between parallel dimension lines on most drawings is approximately 8 mm. Between the outline of the object and the nearest dimension line the space is usually about 10 mm. For the beginning drafter, between 30 and 40 mm is recommended as the distance between views.

Assignment

On an A3- or B-size sheet, select one of the parts shown in Figs. 3-11-A and 3-11-B and make a three-view working drawing. Balance the views on the paper, and with the use of a miter line project all lines and points from one view to the other. Dimensions are to be converted to millimetres. Only the dovetail and T slot dimensions are critical and must be taken

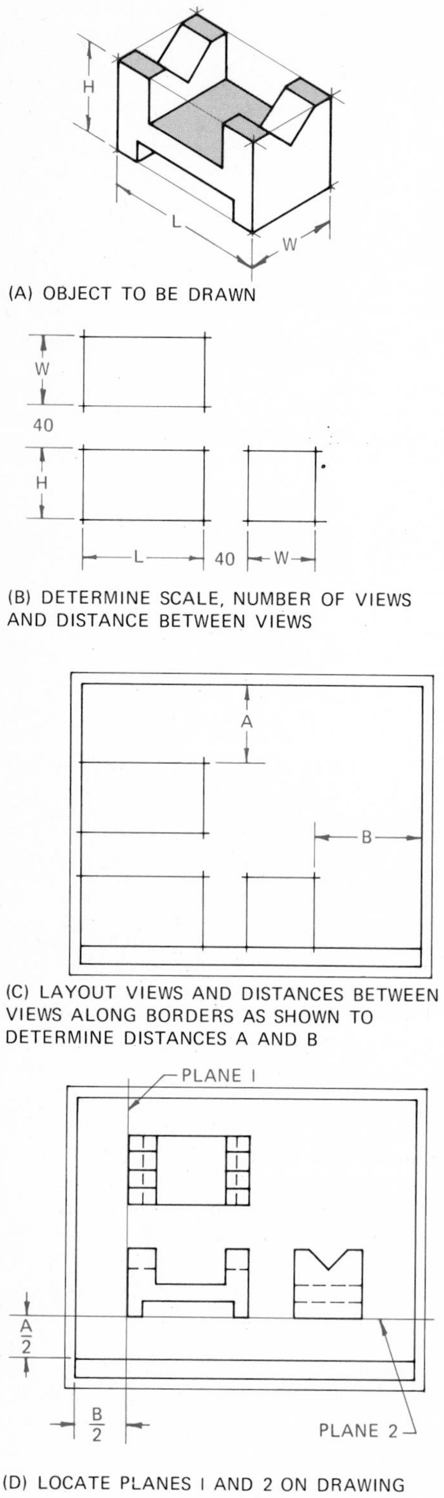
(A) OBJECT TO BE DRAWN
(B) DETERMINE SCALE, NUMBER OF VIEWS AND DISTANCE BETWEEN VIEWS
(C) LAYOUT VIEWS AND DISTANCES BETWEEN VIEWS ALONG BORDERS AS SHOWN TO DETERMINE DISTANCES A AND B
PLANE 1
PLANE 2
(D) LOCATE PLANES 1 AND 2 ON DRAWING

Fig. 3-11-2 Balancing the drawing on the paper.

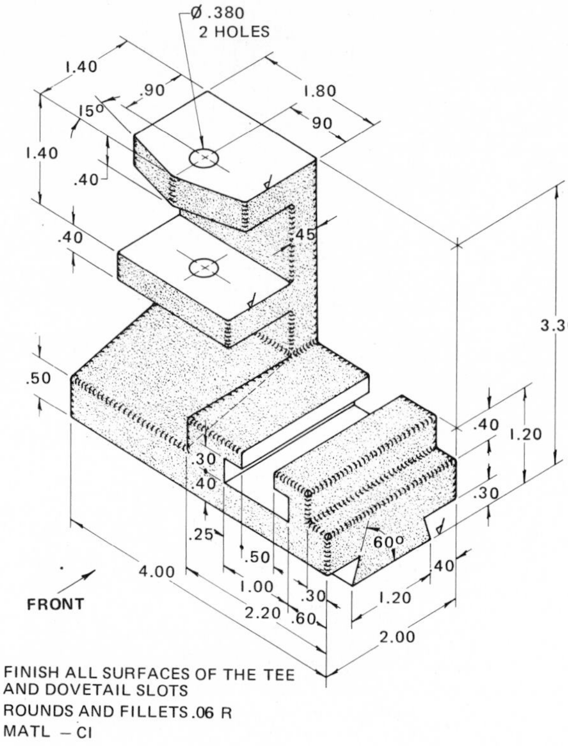
Ø .380
2 HOLES
FRONT
FINISH ALL SURFACES OF THE TEE AND DOVETAIL SLOTS
ROUNDS AND FILLETS .06 R
MATL – CI

Fig. 3-11-A Locating stand.

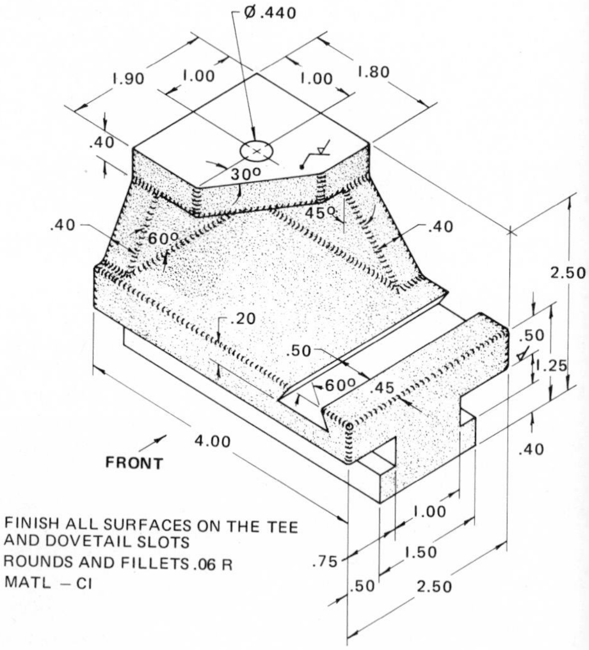
Ø .440
FRONT
FINISH ALL SURFACES ON THE TEE AND DOVETAIL SLOTS
ROUNDS AND FILLETS .06 R
MATL – CI

Fig. 3-11-B Cross slide.

to an accuracy of one point beyond the decimal point. All other dimensions are to be rounded off to whole numbers.

REVIEW FOR ASSIGNMENT

Unit 5-1 Basic Dimensioning
Unit 5-2 Dimensioning Circular Features
Unit 5-3 Dimensioning Common Features
Unit 5-7 Machining Allowance Symbols
Unit 10-1 Sand Castings

Chapter 4
Applied Geometry

UNIT 4-1
STRAIGHT LINES

Most of the lines forming the views on mechanical drawings can be drawn using the instruments and equipment described in Chap. 2. However, geometric constructions have important uses, both in making drawings and in solving problems by graphs and diagrams. Sometimes it is necessary to use geometric constructions, particularly if the drafter does not have the advantages afforded by a drafting machine, an adjustable set square, or templates for drawing hexagonal and elliptical shapes.

TO DRAW A LINE OR LINES PARALLEL TO AND AT A GIVEN DISTANCE FROM AN OBLIQUE LINE

1. Given line *AB* (Fig. 4-1-1), erect a perpendicular *CD* to *AB*.

2. Space the given distance from the line *AB* by scale measurement or by an arc along line *CD*.

3. Position a set square, using a second set square or a T square as base, so that one side of the set square is parallel with the given line.

4. Slide this set square along the base to the point at the desired distance from the given line, and draw the required line.

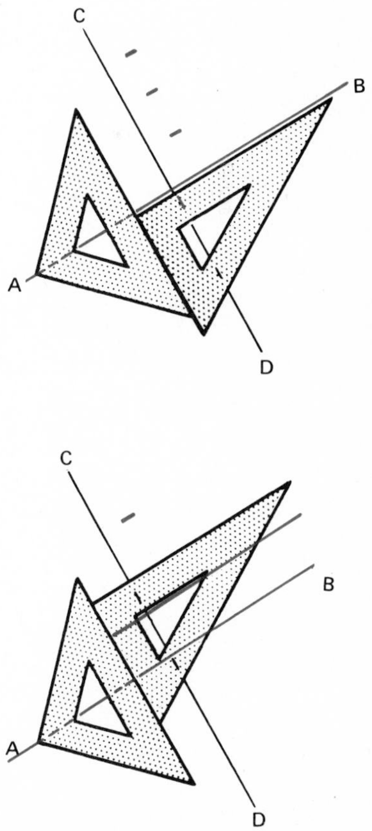
C
B
A
D
C
B
A
D

Fig. 4-1-1 Drawing parallel lines with the side of set squares.

TO DRAW A STRAIGHT LINE TANGENT TO TWO CIRCLES

Place a T square or straightedge so that the top edge just touches the edges of the circles, and draw the tangent line (Fig. 4-1-2). Perpendiculars to this line from the centers of the circles give the tangent points T_1 and T_2.

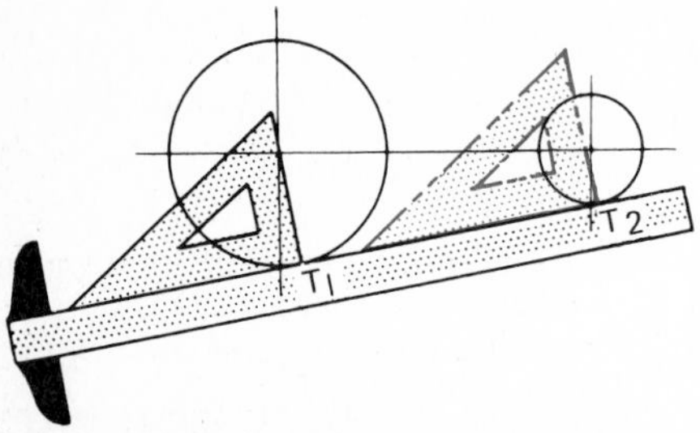

T1
T2

Fig. 4-1-2 Drawing a straight line tangent to two circles.

TO BISECT A STRAIGHT LINE

1. Given line *AB* (Fig. 4-1-3), set the compass to a radius greater than ½ *AB*.

2. Using centers at *A* and *B*, draw intersecting arcs above and below line *AB*. A line *CD* drawn through the intersections will divide *AB* into two equal parts and will be perpendicular to line *AB*.

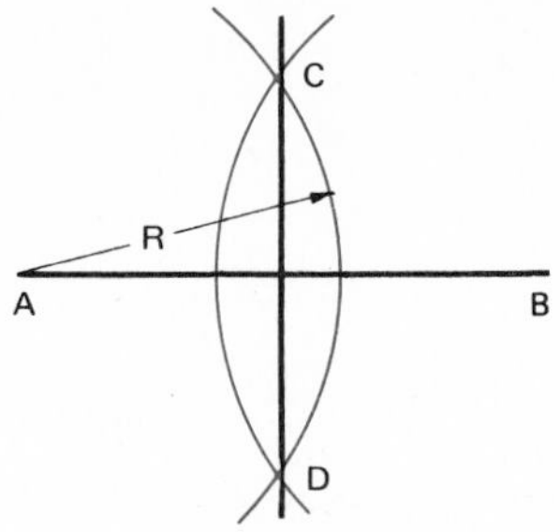

C
R
A
B
D

Fig. 4-1-3 Bisecting a line.

TO BISECT AN ARC

1. Given arc *AB* (Fig. 4-1-4), set the compass to a radius greater than ½ *AB*.

2. Using points *A* and *B* as centers, draw intersecting arcs above and below arc *AB*. A line drawn through the intersections *C* and *D* will bisect the arc *AB* into two equal parts.

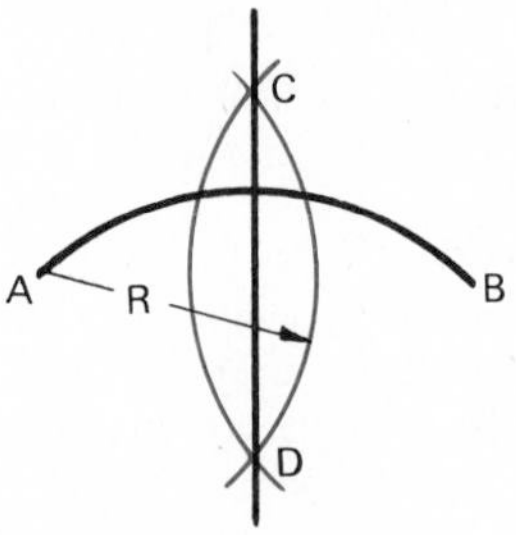

C
A
R
B
D

Fig. 4-1-4 Bisecting an arc.

TO BISECT AN ANGLE

1. Given angle *ABC* with center *B* and a suitable radius (Fig. 4-1-5), draw an arc to cut *BC* at *D* and *BA* at *E*.

2. With centers *D* and *E* and equal radii, draw arcs to intersect at *F*.

3. Join *BF* and produce to *G*. Line *BG* is the required bisector.

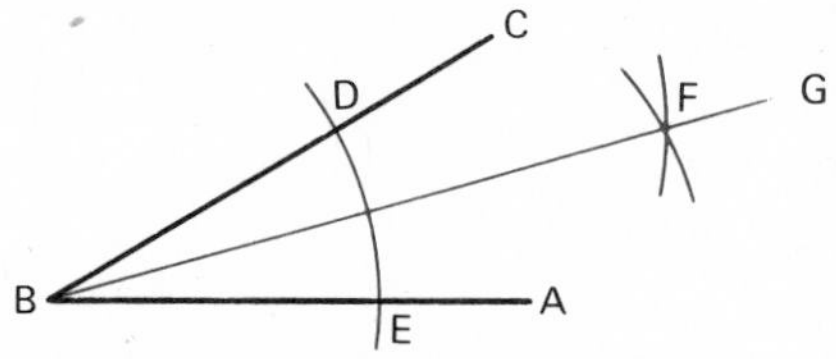

C
D
F
G
B
E
A

Fig. 4-1-5 Bisecting an angle.

TO DIVIDE A LINE INTO A GIVEN NUMBER OF EQUAL PARTS

1. Given line *AB* and the number of equal divisions desired (12, for example), draw a perpendicular from *A*.

2. Place the scale so that the desired number of equal divisions is conveniently included between *B* and the perpendicular. Then mark these divisions, using short vertical marks from the scale divisions as shown in Fig. 4-1-6.

3. Draw perpendiculars to line *AB* through the points marked, dividing the line *AB* as required.

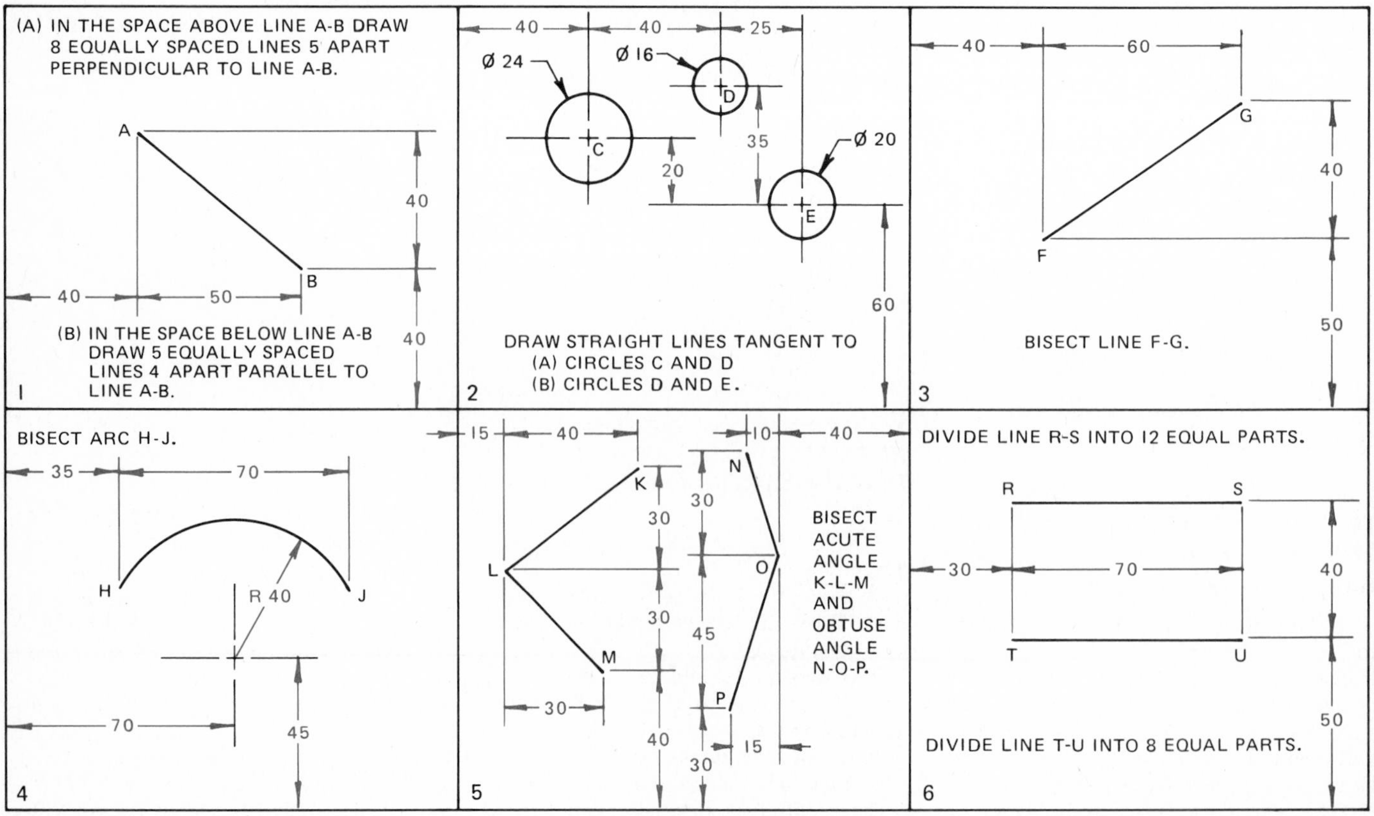

(A) IN THE SPACE ABOVE LINE A-B DRAW 8 EQUALLY SPACED LINES 5 APART PERPENDICULAR TO LINE A-B.
(B) IN THE SPACE BELOW LINE A-B DRAW 5 EQUALLY SPACED LINES 4 APART PARALLEL TO LINE A-B.
A
B
40
50
40
40
1
Ø 24
Ø 16
Ø 20
C
D
E
40
40
25
20
35
60
DRAW STRAIGHT LINES TANGENT TO
(A) CIRCLES C AND D
(B) CIRCLES D AND E.
2
40
60
G
F
40
50
BISECT LINE F-G.
3
BISECT ARC H-J.
35
70
H
J
R 40
70
45
4
15
40
10
40
K
L
M
N
O
P
30
30
30
45
40
30
15
30
BISECT ACUTE ANGLE K-L-M AND OBTUSE ANGLE N-O-P.
5
DIVIDE LINE R-S INTO 12 EQUAL PARTS.
R
S
T
U
30
70
40
50
DIVIDE LINE T-U INTO 8 EQUAL PARTS.
6

Fig. 4-1-A Straight-line construction.

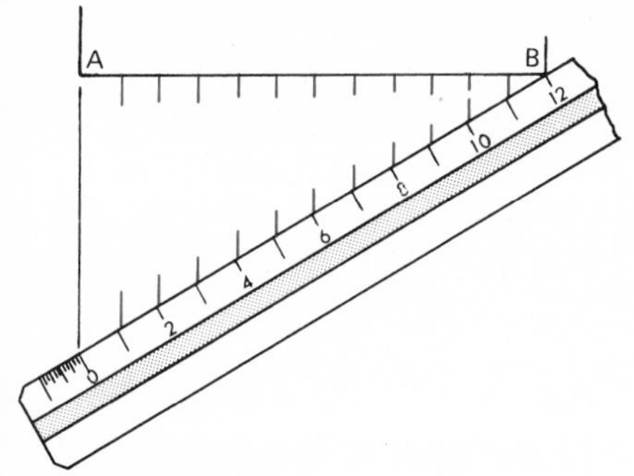
A
B

Fig. 4-1-6 Dividing a straight line into equal parts.

Assignment

Divide an A3-size sheet as shown in Fig. 4-1-A. In the designated areas draw the geometric constructions.

UNIT 4-2

ARCS AND CIRCLES

TO DRAW AN ARC TANGENT TO TWO LINES AT RIGHT ANGLES TO EACH OTHER

Given radius *R* of the arc (Fig. 4-2-1).

1. Draw an arc having radius *R* with center at *B*, cutting the lines *AB* and *BC* at *D* and *E* respectively.

2. With *D* and *E* as centers and with the same radius *R*, draw arcs intersecting at 0.

3. With center 0 draw the required arc. The tangent points are *D* and *E*.

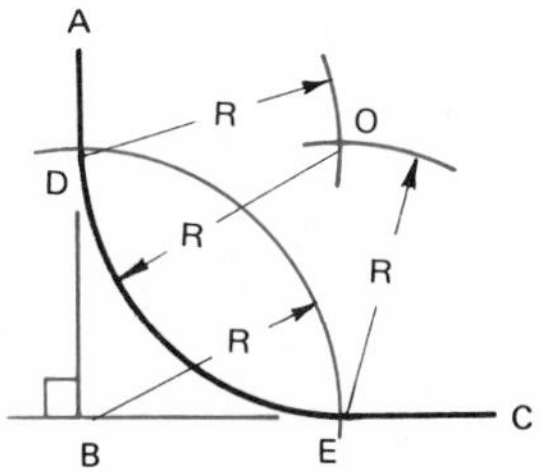
A
R
O
D
R
R
R
B
E
C

Fig. 4-2-1 Arc tangent to two lines at right angles to each other.

TO DRAW AN ARC TANGENT TO THE SIDES OF AN ACUTE ANGLE

Given radius *R* of the arc (Fig. 4-2-2).

1. Draw lines inside the angle, parallel to the given lines, at distance *R* away from the given lines. The center of the arc will be at *C*.

2. Set the compass to radius *R*, and with center *C* draw the arc tangent to the given sides. The tangent points *A* and *B* are found by drawing perpendiculars through point *C* to the given lines.

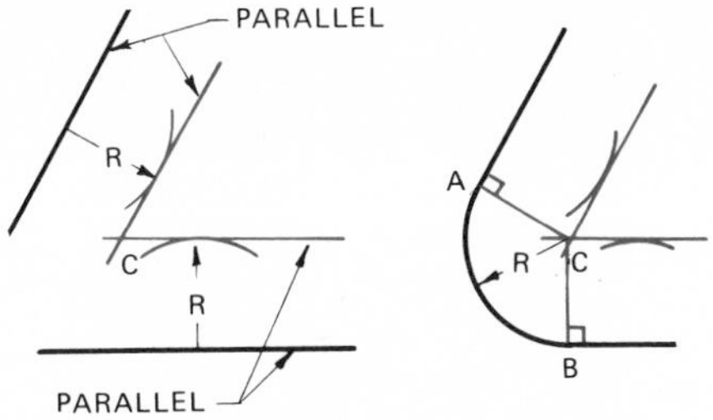
PARALLEL
R
C
R
PARALLEL
A
R
C
B

Fig. 4-2-2 Drawing an arc tangent to the sides of an acute angle.

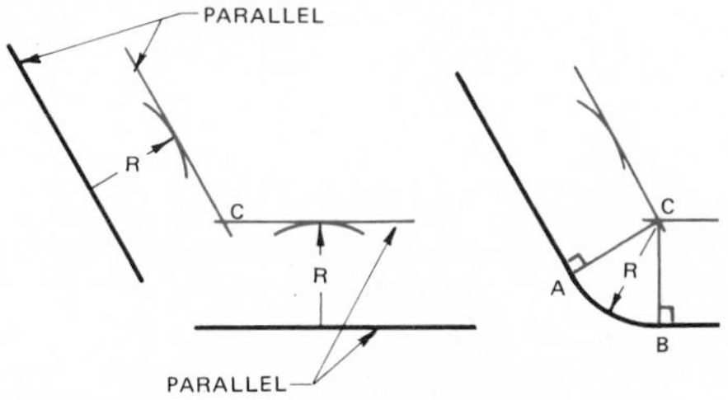
PARALLEL
R
C
R
PARALLEL
C
R
A
B

Fig. 4-2-3 Drawing an arc tangent to the sides of an obtuse angle.

TO DRAW AN ARC TANGENT TO TWO SIDES OF AN OBTUSE ANGLE

Follow the same procedure as for an acute angle (Fig. 4-2-3).

TO DRAW A CIRCLE ON A REGULAR POLYGON

1. Given the size of the polygon (Fig. 4-2-4), bisect any two sides; for example, *BC* and *DE*. The center of the polygon is where bisectors *F*0 and *G*0 intersect at point 0.

2. The inner circle radius is 0*H*, and the outer circle radius is 0*A*.

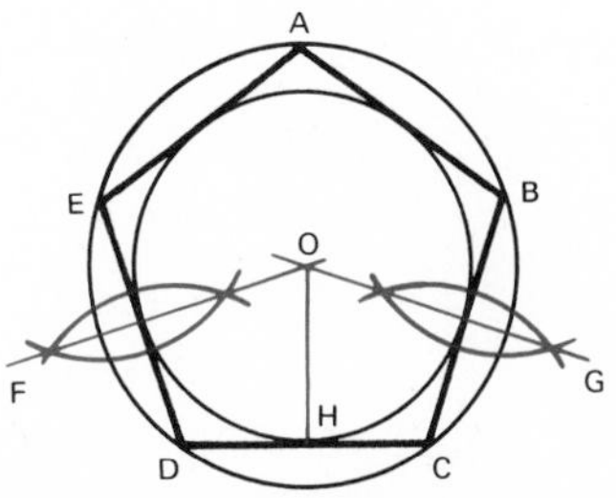
A
E
B
O
F
G
H
D
C

Fig. 4-2-4 Drawing a circle on a regular polygon.

TO DRAW A REVERSE, OR OGEE, CURVE CONNECTING TWO PARALLEL LINES

1. Given two parallel lines *AB* and *CD* and distances *X* and *Y* (Fig. 4-2-5), join points *B* and *C* with a line.

2. Erect a perpendicular to *AB* and *CD* from points *B* and *C* respectively.

3. Select point *E* on line *BC* where the curves are to meet.

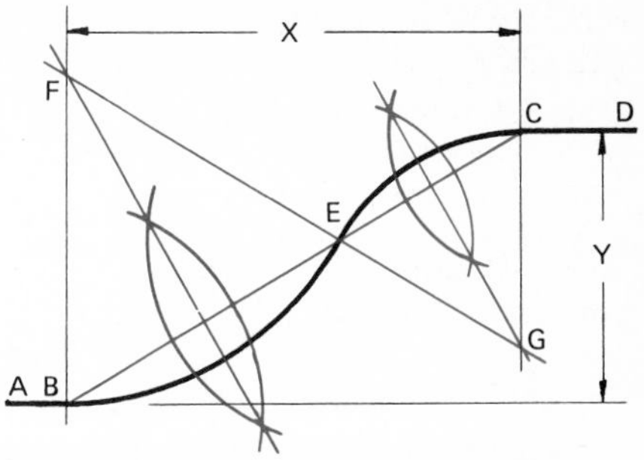
X
F
C
D
E
Y
G
A B

Fig. 4-2-5 Drawing a reverse (ogee) curve connecting two parallel planes.

4. Bisect *BE* and *EC*.

5. Points *F* and *G* where the perpendiculars and bisectors meet are the centers for the arcs forming the ogee curve.

TO DRAW AN ARC TANGENT TO A GIVEN CIRCLE AND STRAIGHT LINE

1. Given *R*, the radius of the arc (Fig. 4-2-6), draw a line parallel to the given straight line between the circle and the line at distance *R* away from the given line.

2. With the center of the circle as center and radius R_1 (radius of the circle plus *R*), draw an arc to cut the parallel straight line at *C*.

3. With center *C* and radius *R*, draw the required arc tangent to the circle and the straight line.

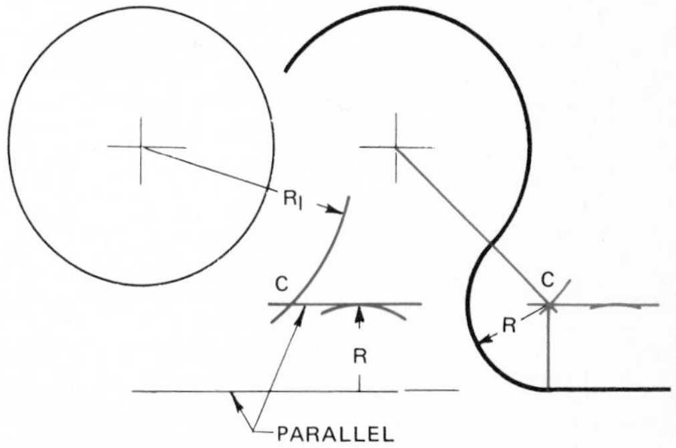
R_1
C
R
C
R
PARALLEL

Fig. 4-2-6 Drawing an arc tangent to a circle and a straight line.

TO DRAW AN ARC TANGENT TO TWO CIRCLES (FIG. 4-2-7)

Figure 4-2-7*a*

1. Given the radius of arc *R*, with the center of circle *A* as center and radius R_2 (radius of circle *A* plus *R*), draw an arc in the area between the circles.

2. With the center of circle *B* as center and radius R_2 (radius of circle *B* plus *R*), draw an arc to cut the other arc at *C*.

3. With center *C* and radius *R*, draw the required arc tangent to the given circles.

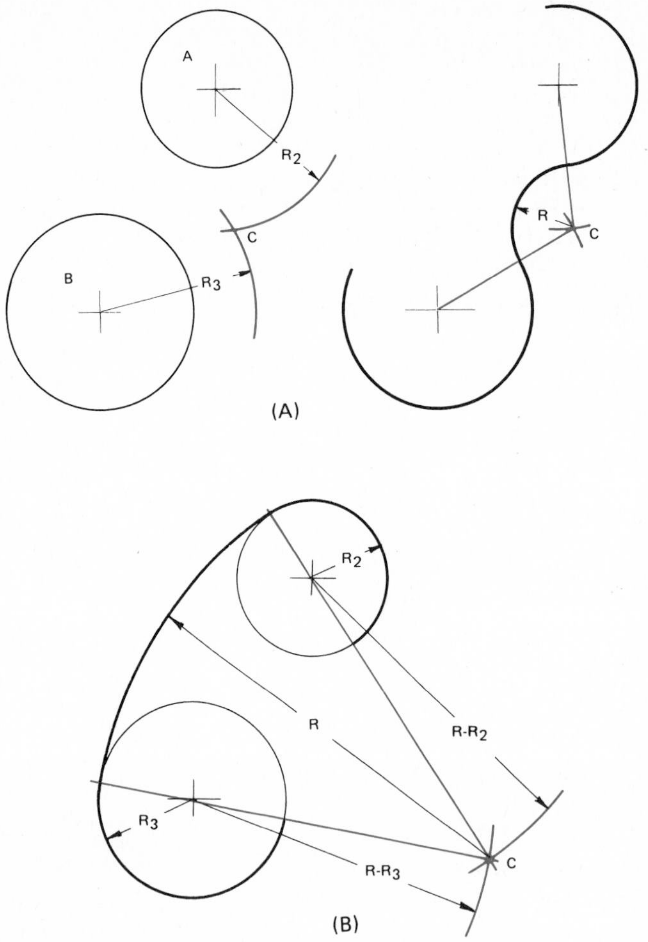
A
R2
C
B
R3
R
C
(A)
R2
R
R-R2
R3
R-R3
C
(B)

Fig. 4-2-7 Drawing an arc tangent to two circles.

Figure 4-2-7*b*

1. Given radius of arc R, with the center of circle A as center and radius $R - R_2$, draw an arc in the area between the circles.

2. With the center of circle B as center and radius $R - R_3$, draw an arc to cut the other arc at C.

3. With center C and radius R, draw the required arc tangent to the given circles.

TO DRAW AN ARC OR CIRCLE THROUGH THREE POINTS NOT IN A STRAIGHT LINE

1. Given points A, B, and C (Fig. 4-2-8), join points A, B, and C as shown.

2. Bisect lines AB and BC and extend bisecting lines to intersect at 0. Point 0 is the center of the required circle or arc.

3. With center 0 and radius 0A draw an arc.

Assignment

Divide a B-size sheet as shown in Fig. 4-2-A. In the designated areas draw the geometric constructions.

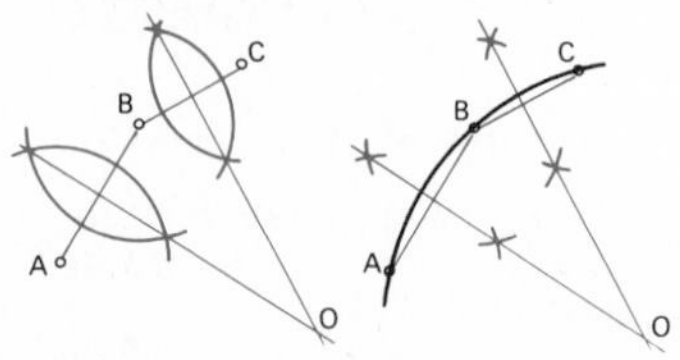
C
B
A
O
C
B
A
O

Fig. 4-2-8 Drawing an arc or circle through three points not in a straight line.

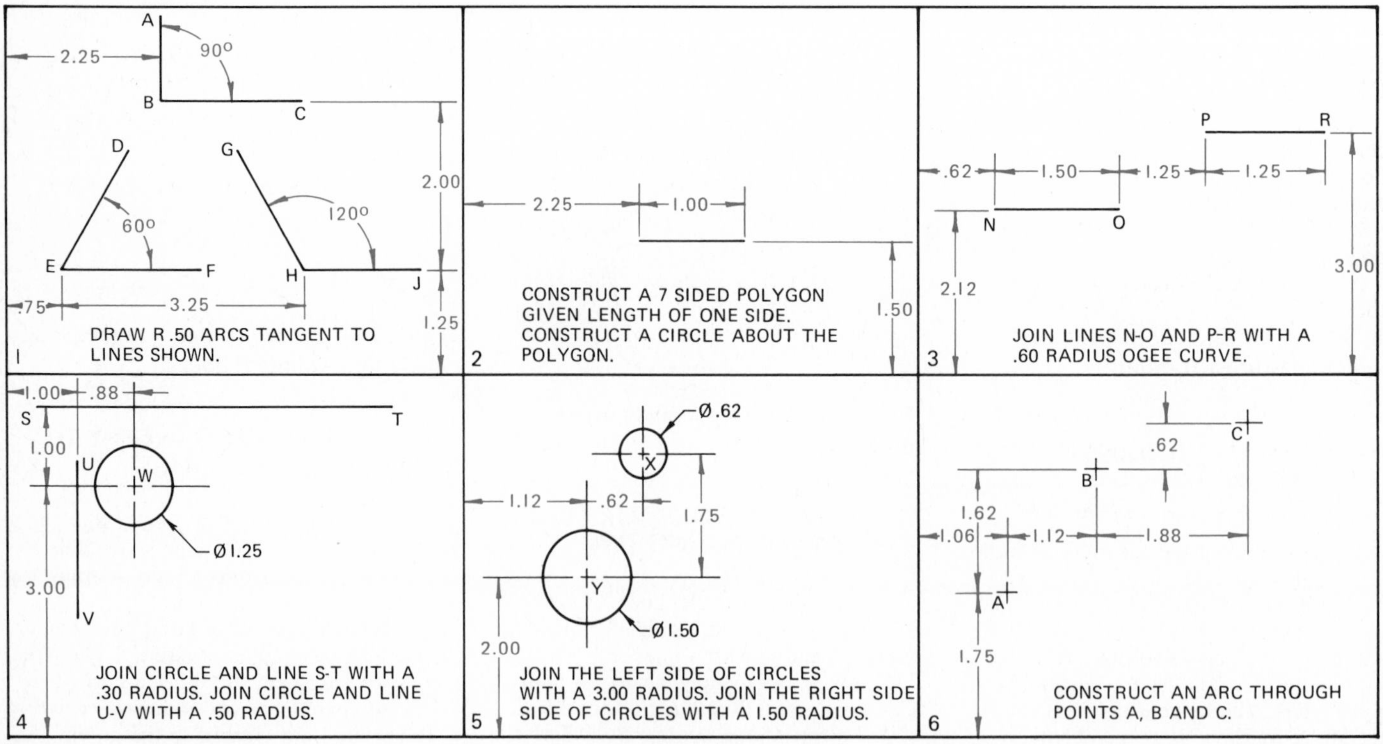
A
2.25
90°
B
C
D
G
2.00
60°
120°
E
F
H
J
.75
3.25
1.25
DRAW R .50 ARCS TANGENT TO LINES SHOWN.
1
2.25
1.00
CONSTRUCT A 7 SIDED POLYGON GIVEN LENGTH OF ONE SIDE. CONSTRUCT A CIRCLE ABOUT THE POLYGON.
1.50
2
P
R
.62
1.50
1.25
1.25
N
O
3.00
2.12
JOIN LINES N-O AND P-R WITH A .60 RADIUS OGEE CURVE.
3
1.00
.88
S
T
1.00
U
W
Ø1.25
3.00
V
JOIN CIRCLE AND LINE S-T WITH A .30 RADIUS. JOIN CIRCLE AND LINE U-V WITH A .50 RADIUS.
4
Ø.62
X
1.12
.62
1.75
Y
Ø1.50
2.00
JOIN THE LEFT SIDE OF CIRCLES WITH A 3.00 RADIUS. JOIN THE RIGHT SIDE SIDE OF CIRCLES WITH A 1.50 RADIUS.
5
C
.62
B
1.62
1.06
1.12
1.88
A
1.75
CONSTRUCT AN ARC THROUGH POINTS A, B AND C.
6

Fig. 4-2-A Assignment.

UNIT 4-3

POLYGONS

A regular polygon is a plane figure bounded by straight lines of equal length and contains angles of equal size.

TO DRAW A HEXAGON (FIG. 4-3-1), GIVEN THE DISTANCE ACROSS THE FLATS

1. Establish horizontal and vertical center lines for the hexagon.

2. Using the intersection of these lines as center, with radius one-half the distance across the flats, draw a light construction circle.

3. Using the 60° set square, draw six straight lines, equally spaced, passing through the center of the circle.

4. Draw tangents to these lines at their intersection with the circle.

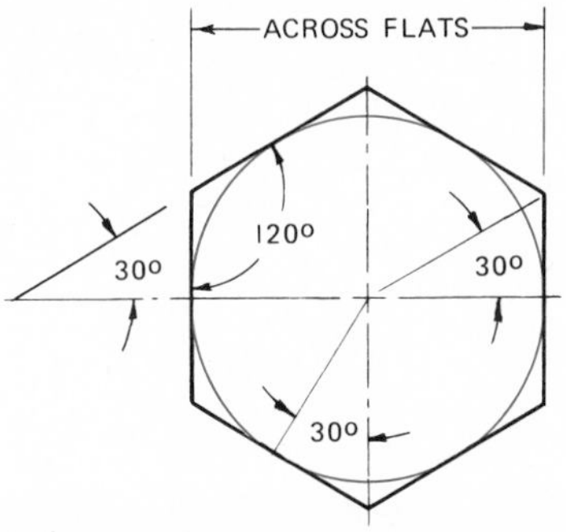
ACROSS FLATS
120°
30°
30°
30°

Fig. 4-3-1 Constructing a hexagon, given distance across flats.

TO DRAW A HEXAGON, GIVEN THE DISTANCE ACROSS THE CORNERS (FIG. 4-3-2)

1. Establish horizontal and vertical center lines, and draw a light construction circle with radius one-half the distance across the corners.

2. With a 60° set square, establish points on the circumference 60° apart.

3. Draw straight lines connecting these points.

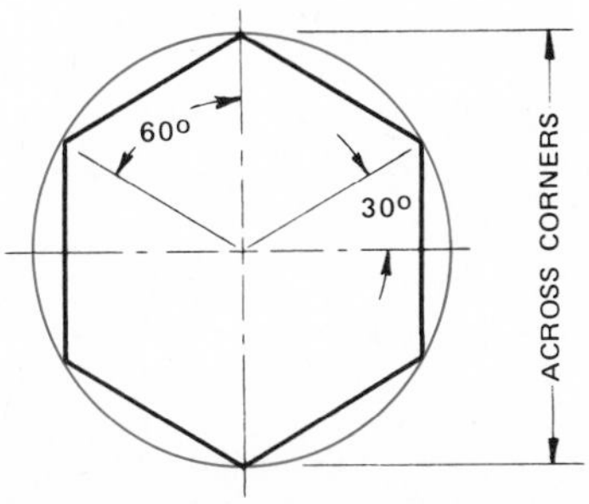
60°
30°
ACROSS CORNERS

Fig. 4-3-2 Constructing a hexagon, given distance across corners.

TO DRAW AN OCTAGON, GIVEN THE DISTANCE ACROSS THE FLATS (FIG. 4-3-3)

1. Establish horizontal and vertical center lines and draw a light construction circle with radius one-half the distance across the flats.

2. Draw horizontal and vertical lines tangent to the circle.

3. Using the 45° set square, draw lines tangent to the circle at a 45° angle from the horizontal.

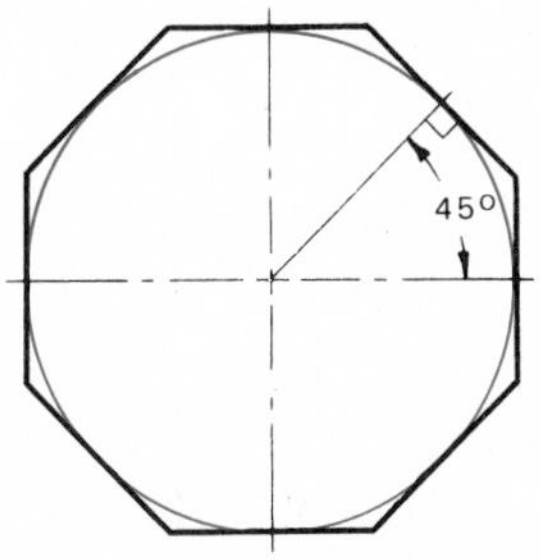
45°

Fig. 4-3-3 Constructing an octagon, given distance across flats.

TO DRAW AN OCTAGON, GIVEN THE DISTANCE ACROSS THE CORNERS (FIG. 4-3-4)

1. Establish horizontal and vertical center lines and draw a light construction circle with radius one-half the distance across the corners.

2. With the 45° set square, establish points on the circumference between the horizontal and vertical center lines.

3. Draw straight lines connecting these points to the points where the center lines cross the circumference.

TO DRAW A REGULAR POLYGON, GIVEN THE LENGTH OF THE SIDES

Let the polygon have seven sides.

1. Given the length of side *AB* (Fig. 4-3-5), with radius *AB* and *A* as center, draw a semicircle and divide it into seven equal parts using a protractor.

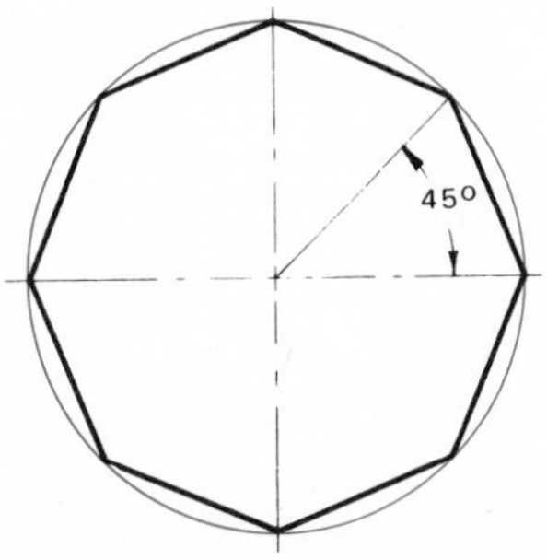
45°

Fig. 4-3-4 Constructing an octagon, given distance across corners.

2. Through the second division from the left, draw radial line *A*2.

3. Through points 3, 4, 5, and 6 extend radial lines as shown.

4. With *AB* as radius and *B* as center, cut line *A*6 at *C*. With the same radius and *C* as center, cut line *A*5 at *D*. Repeat at *E* and *F*.

5. Connect these points with straight lines.

These steps can be followed in drawing a regular polygon with any number of sides.

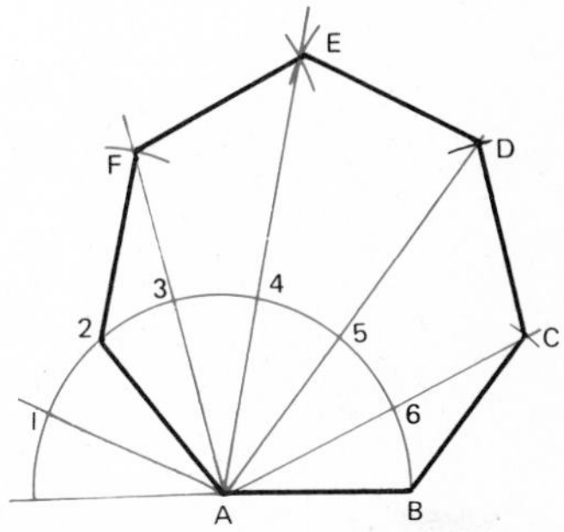
E
F
D
C
1
2
3
4
5
6
A
B

Fig. 4-3-5 Constructing a regular polygon, given length of one side.

TO INSCRIBE A REGULAR PENTAGON IN A GIVEN CIRCLE

1. Given circle with center 0 (Fig. 4-3-6), draw the circle with diameter *AB*.

2. Bisect line 0*B* at *D*.

3. With center *D* and radius *DC*, draw arc *CE* to cut the diameter at *E*.

4. With *C* as center and radius *CE*, draw arc *CF* to cut the circumference at *F*. Distance *CF* is one side of the pentagon.

5. With radius *CF* as a chord, mark off the remaining points on the circle. Connect the points with straight lines.

Assignment

Divide an A3-size sheet as shown in Fig. 4-3-A. In the designated areas, draw the geometric constructions.

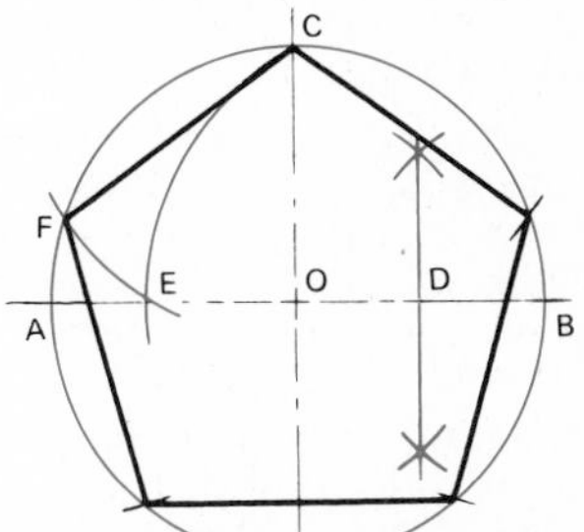
C
F
E
O
D
A
B

Fig. 4-3-6 Inscribing a regular pentagon in a given circle.

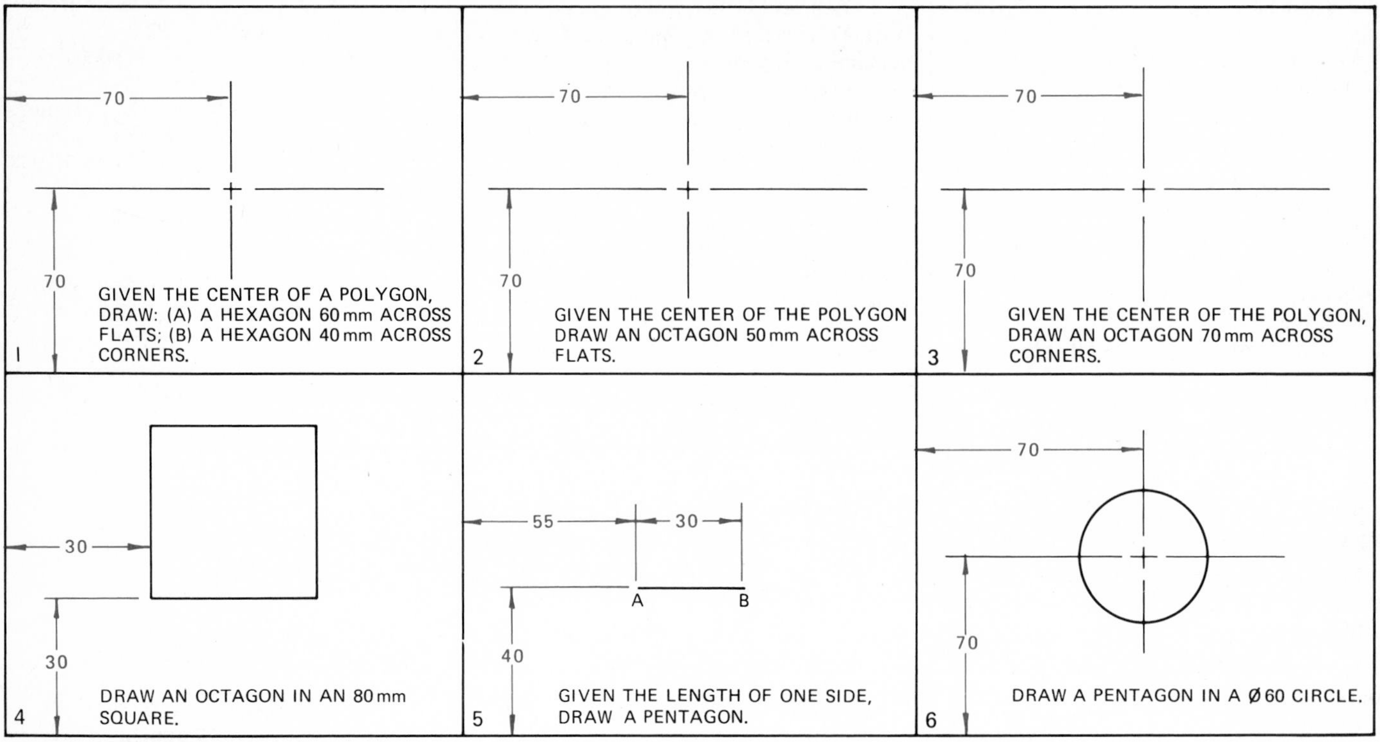

70
70
GIVEN THE CENTER OF A POLYGON, DRAW: (A) A HEXAGON 60 mm ACROSS FLATS; (B) A HEXAGON 40 mm ACROSS CORNERS.
1
70
70
GIVEN THE CENTER OF THE POLYGON DRAW AN OCTAGON 50 mm ACROSS FLATS.
2
70
70
GIVEN THE CENTER OF THE POLYGON, DRAW AN OCTAGON 70 mm ACROSS CORNERS.
3
30
30
DRAW AN OCTAGON IN AN 80 mm SQUARE.
4
55
30
A
B
40
GIVEN THE LENGTH OF ONE SIDE, DRAW A PENTAGON.
5
70
70
DRAW A PENTAGON IN A Ø60 CIRCLE.
6

Fig. 4-3-A Drawing assignment.

UNIT 4-4

THE ELLIPSE

The *ellipse* is the plane curve generated by a point moving so that the sum of the distances from any point on the curve to two fixed points, called foci, is a constant.

Often a drafter is called upon to draw oblique and inclined holes and surfaces which take the form of an ellipse. Several methods, true and approximate, are used for its construction. The terms *major diameter* and *minor diameter* will be used in place of *major axis* and *minor axis* so the reader won't become confused with the mathematical *X* and *Y* axes.

TO DRAW AN ELLIPSE—TWO-CIRCLE METHOD

1. Given the major and minor diameters (Fig. 4-4-1), construct two concentric circles with diameters equal to *AB* and *CD*.

2. Divide the circles into a convenient number of equal parts. Figure 4-4-1 shows 12.

3. Where the radial lines intersect the outer circle, as at 1, draw lines parallel to line *CD* inside the outer circle.

4. Where the same radial line intersects the inner circle, as at 2, draw a line parallel to axis *AB* away from the inner circle. The intersection of these lines, as at 3, gives points on the ellipse.

5. Draw a smooth curve through these points.

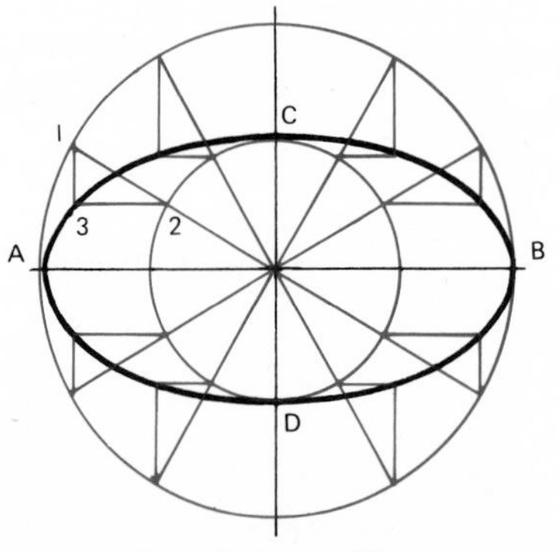

C
1
3
2
A
B
D

Fig. 4-4-1 Drawing an ellipse—two circle method.

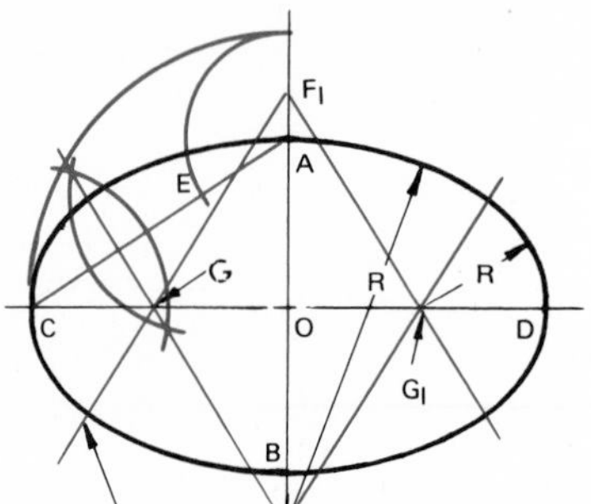

F_1
A
E
G
R
R
C
O
D
G_1
B
F
TANGENT POINT

Fig. 4-4-2 Drawing an ellipse—four center method.

TO DRAW AN ELLIPSE—FOUR-CENTER METHOD

1. Given the major diameter *CD* and the minor diameter *AB* (Fig. 4-4-2), join points *A* and *C* with a line.

2. Lay off *AE* equal to *CO-AO*.

3. Draw the right bisector of *CE* locating point *G* on line *CO* and point *F* on line *AB*. (Line *AB* may have to be extended.)

4. Make *OF* equal to OF_1 and *OG* equal to OG_1.

5. Points *F*, F_1, *G*, and G_1 are the centers for the two large and two small arcs forming the ellipse.

TO DRAW AN ELLIPSE—PARALLELOGRAM METHOD

1. Given the major diameter *CD* and minor diameter *AB* (Fig. 4-4-3), construct a parallelogram.

2. Divide *CO* into a number of equal parts. Divide *CE* into the same number of equal parts. Number the points from *C*.

3. Draw a line from *B* to point 1 on line *CE*. Draw a line from *A* through point 1 on *CO*, intersecting the previous line. The point of intersection will be one point on the ellipse.

4. Proceed in the same manner to find other points on the ellipse.

5. Draw a smooth curve through these points.

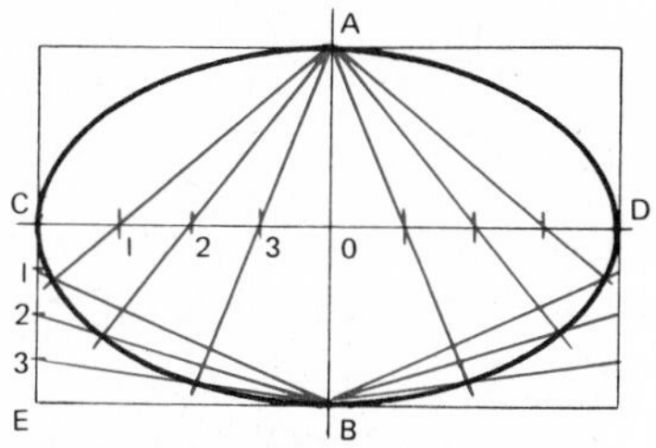
A
C
D
1
2
3
0
1
2
3
E
B

Fig. 4-4-3 Drawing an ellipse-parallelogram method.

TO DRAW AN ELLIPSE—THE TRAMMEL METHODS

First Method. Given the major diameter *CD* and the minor diameter *AB* (Fig. 4-4-4).

1. On a strip of paper or cardboard mark the distances *ao* and *co* as shown.

2. Placing point *A* on line *CD* and point *c* on line *AB*, mark point *o*.

3. Reposition the strip to obtain various positions of point *o* which are located on the ellipse.

4. Connect the points using an irregular curve.

Second Method. Given the major diameter *CD* and the minor diameter *AB*.

1. On a strip of paper or cardboard mark the distances *ao* and *co* as shown.

2. Placing point *a* on line *CD* and point *c* on line *AB*, mark point *o*.

3. Reposition the strip to obtain various positions of point *o*, which are located on the ellipse.

4. Draw a smooth curve through these points.

Assignment

Divide an A3- or B-size sheet as shown in Fig. 4-4-A. In the designated areas draw the geometric constructions.

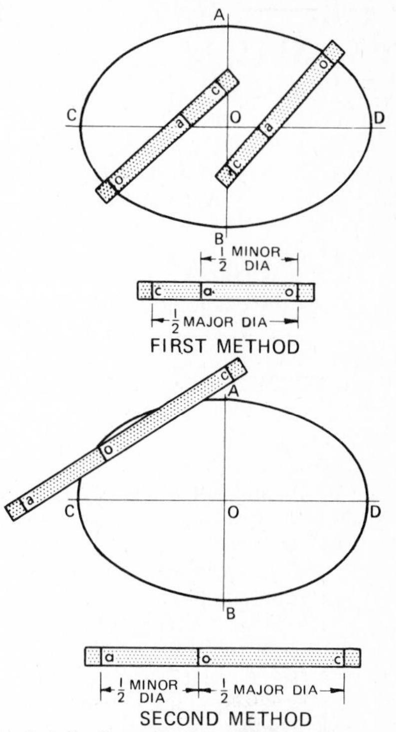
A
C
O
D
B
$\frac{1}{2}$ MINOR DIA
$\frac{1}{2}$ MAJOR DIA
FIRST METHOD
A
C
O
D
B
$\frac{1}{2}$ MINOR DIA
$\frac{1}{2}$ MAJOR DIA
SECOND METHOD

Fig. 4-4-4 Drawing an ellipse—the trammel methods.

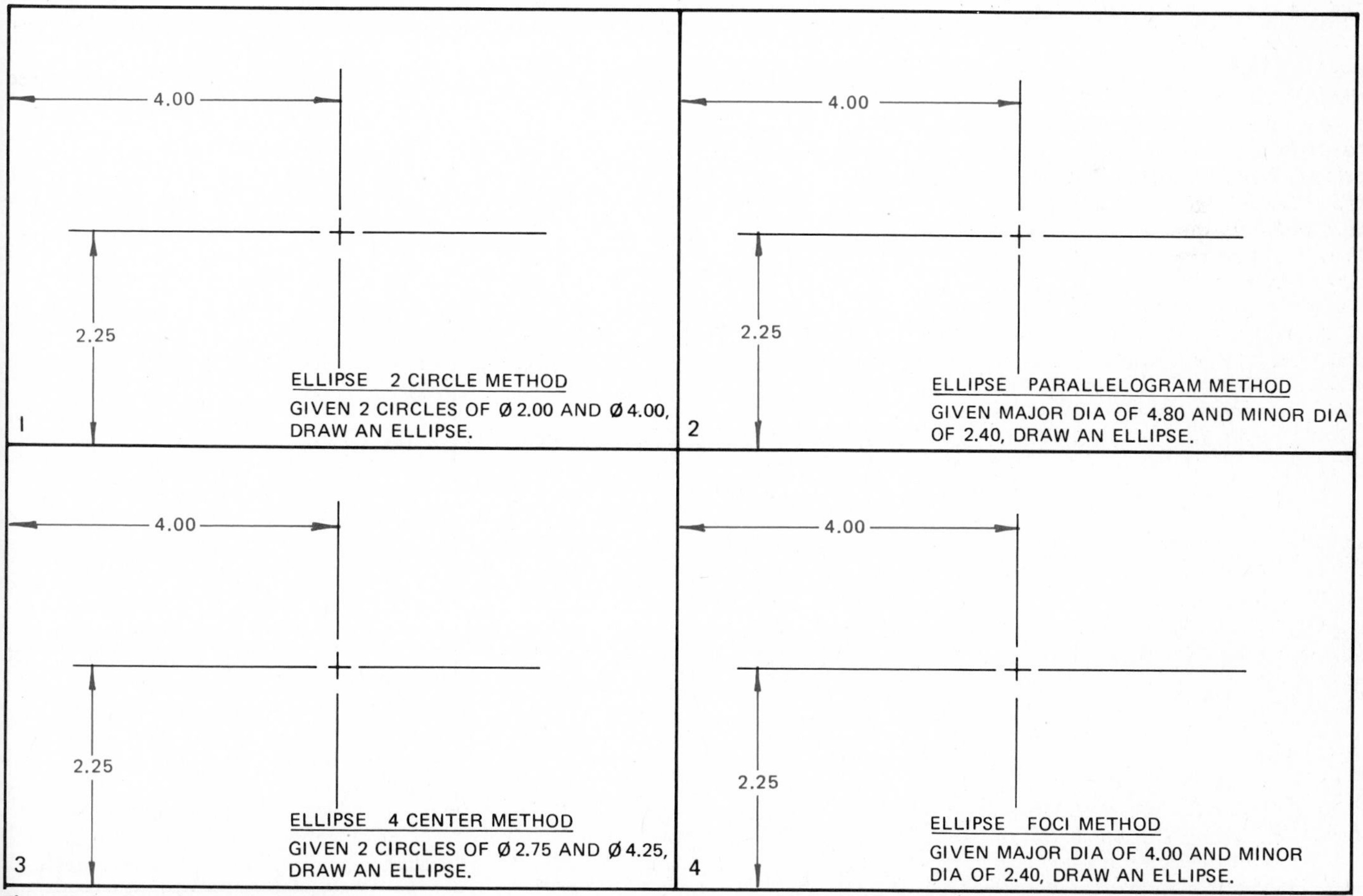
4.00
2.25
ELLIPSE 2 CIRCLE METHOD
GIVEN 2 CIRCLES OF Ø 2.00 AND Ø 4.00, DRAW AN ELLIPSE.
1
4.00
2.25
ELLIPSE PARALLELOGRAM METHOD
GIVEN MAJOR DIA OF 4.80 AND MINOR DIA OF 2.40, DRAW AN ELLIPSE.
2
4.00
2.25
ELLIPSE 4 CENTER METHOD
GIVEN 2 CIRCLES OF Ø 2.75 AND Ø 4.25, DRAW AN ELLIPSE.
3
4.00
2.25
ELLIPSE FOCI METHOD
GIVEN MAJOR DIA OF 4.00 AND MINOR DIA OF 2.40, DRAW AN ELLIPSE.
4

Fig. 4-4-A Drawing assignment.

UNIT 4-5

THE HELIX

The *helix* is the curve generated by a point that revolves uniformly around and up or down the surface of a cylinder. The *lead* is the vertical distance that the point rises or drops in one complete revolution.

TO DRAW A HELIX

1. Given the diameter of the cylinder and the lead (Fig. 4-5-1), draw the top and front views.
2. Divide the circumference (top view) into a convenient number of parts (use 12) and label them.
3. Project lines down to the front view.
4. Divide the lead into the same number of equal parts and label them as shown in Fig. 4-5-1.
5. The points of intersection of lines with corresponding numbers lie on the helix. *Note:* Since points 8 to 12 lie on the back portion of the cylinder, the helix curve starting at point 7 and passing through points 8, 9, 10, 11, 12 to point 1 will appear as a hidden line.
6. If the development of the cylinder is drawn, the helix will appear as a straight line on the development.

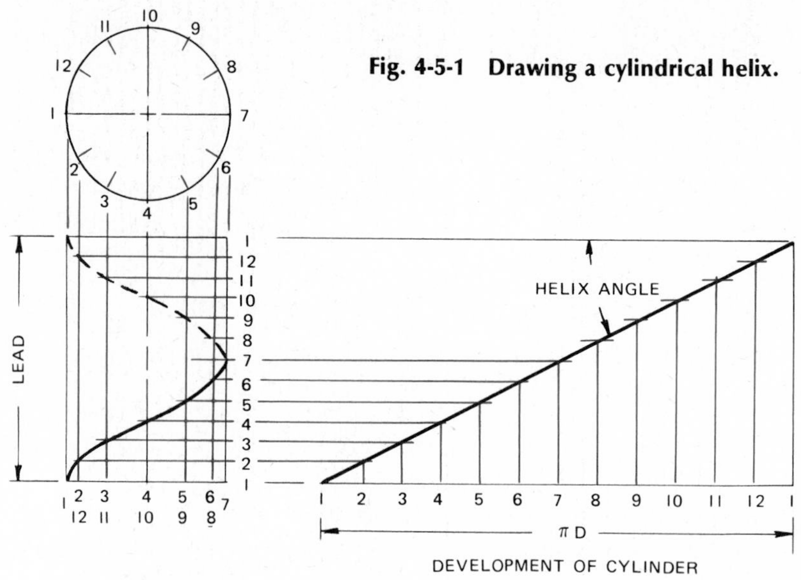
LEAD
HELIX ANGLE
πD
DEVELOPMENT OF CYLINDER

Fig. 4-5-1 Drawing a cylindrical helix.

TO DRAW A CONICAL HELIX

1. Given the diameter and height of the cone and the lead (Fig. 4-5-2), draw the top and front views.
2. Divide the circumference of the base (top view) into a convenient number of parts (use eight) and label them.
3. Project these points down to the base of the front view and label them.
4. Draw light lines from these points to the apex of the cone.
5. Divide the lead into the same number of equal parts and label them as shown in Fig. 4-5-2.
6. The points of intersection of lines with corresponding numbers lie on the conical helix. *Note:* Since points 6 to 8 lie on the back portion of the cone, the conical helix curve starting at point 5 and passing through points 6, 7, 8 to point 1 will appear as a hidden line.

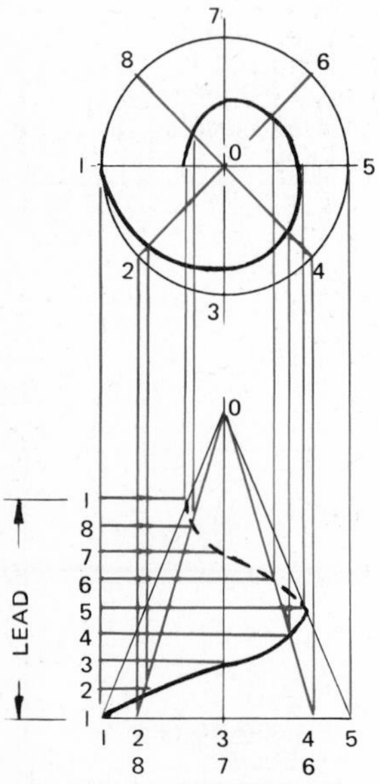
LEAD

Fig. 4-5-2 Drawing a conical helix.

Assignment

Divide an A3- or B-size sheet as shown in Fig. 4-5-A. In the designated areas, draw the geometric constructions.

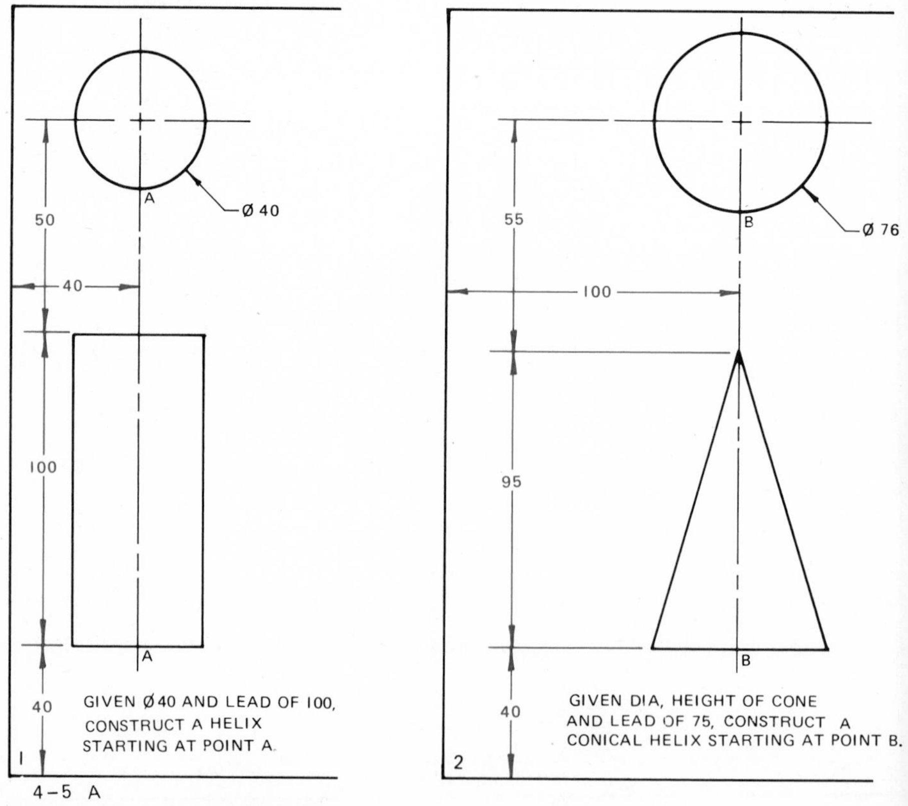
Ø 40
50
40
100
40
A
GIVEN Ø 40 AND LEAD OF 100, CONSTRUCT A HELIX STARTING AT POINT A.
1
4–5 A
Ø 76
55
100
95
40
B
GIVEN DIA, HEIGHT OF CONE AND LEAD OF 75, CONSTRUCT A CONICAL HELIX STARTING AT POINT B.
2

Fig. 4-5-A Drawing assignment.

UNIT 4-6
THE PARABOLA

The *parabola* is a plane curve generated by a point that moves along a path equidistant from a fixed line (directrix) and a fixed point (focus).

TO CONSTRUCT A PARABOLA—TRUE METHOD

1. Given the focus *F* and directrix *AB* (Fig. 4-6-1*a*), draw a line (axis) perpendicular to *AB* through focus *F*.
2. Draw a line parallel to line *AB*, intersecting the axis at any point 2.
3. With *F* as the center and a radius equal to 0-2, strike an arc above and below the axis intersecting the vertical line which passes through point 2. The intersections of these lines, points *C* and *D*, are points on the parabola.
4. Proceed in the same manner to find other points on the parabola.
5. Connect these points using an irregular curve.
6. To draw a tangent at any point *E*, draw *EG* parallel to the axis to bisect angle *GEF*.

TO CONSTRUCT A PARABOLA—PARALLELOGRAM METHOD

1. Given the sizes of the enclosing rectangle, distances *AB* and *AC* (Fig. 4-6-1*b*), construct a parallelogram.
2. Divide *AC* into a number of equal parts. Divide *A*0 into the same number of equal parts. Number the points as shown.
3. Draw a line from 0 to point 1 on line *AC*. Draw a line parallel to the axis through point 1 on line *A*0, intersecting the previous line 0-1. The point of intersection will be one point on the parabola.
4. Proceed in the same manner to find other points on the parabola.
5. Connect the points using an irregular curve.

TO CONSTRUCT A PARABOLA—OFFSET METHOD

1. Given the sizes of the enclosing rectangle, distances *AB* and *AC* (Fig. 4-6-1*c*), construct a parallelogram.
2. Divide 0*A* into four equal parts.
3. The offsets vary in length as the square of their distances from 0. Since 0*A* is divided into four equal parts, distance *AC* will be divided into 4^2, or 16, equal divisions. Thus since 01 is one-fourth the length of 0*A*, the length of line 1-1_1 will be $(\frac{1}{4})^2$, or $\frac{1}{16}$, the length of *AC*.
4. Since distance 02 is one-half the length of 0*A*, the length of line 2-2_1 will be $(\frac{1}{2})^2$, or $\frac{1}{4}$, the length of *AC*.

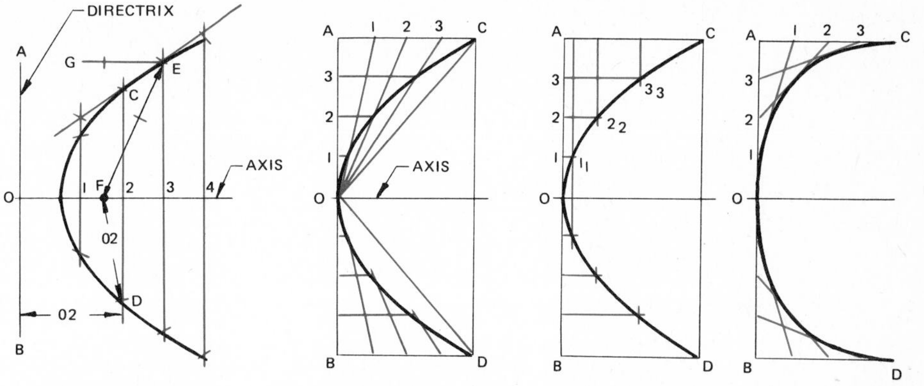
DIRECTRIX
AXIS
A
B
C
D
E
F
G
O
02
1 2 3 4
3 3
2 2
1 1
(A) TRUE METHOD (B) PARALLELOGRAM METHOD (C) OFFSET METHOD (D) ENVELOPE METHOD

Fig. 4-6-1 Common methods used to construct a parabola.

5. Since distance 03 is three-fourths the length of 0*A*, the length of line 3-3_1 will be $(\frac{3}{4})^2$, or $\frac{9}{16}$, the length of *AC*.
6. Complete the parabola by joining the points with an irregular curve.

TO CONSTRUCT A PARABOLA—ENVELOPE METHOD

1. Given the sizes of the enclosing rectangle, distances *AB* and *AC* (Fig. 4-6-1*d*), construct a parallelogram.
2. Divide *AC* into a number of equal parts. Divide *A*0 into the same number of equal parts.
3. Connect the corresponding numbers.
4. Draw the parabolic curve tangent to these lines.

Assignment

Divide an A3- or B-size sheet, as shown in Fig. 4-6-A. In the designated areas, draw the geometric constructions.

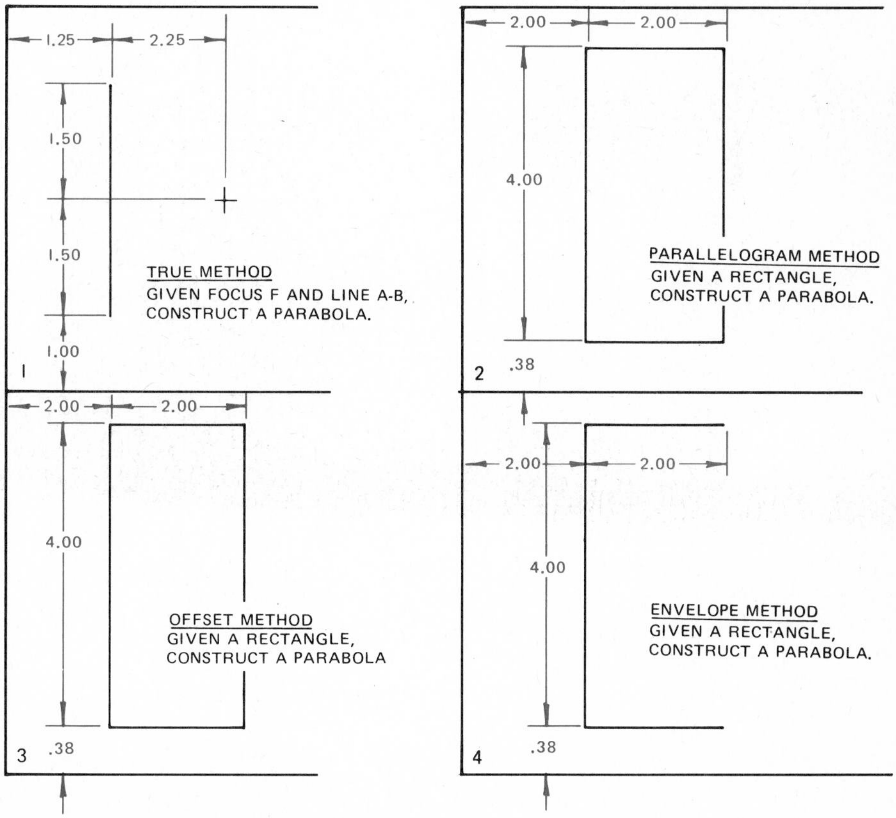
1.25
2.25
1.50
1.50
1.00
1
TRUE METHOD
GIVEN FOCUS F AND LINE A-B, CONSTRUCT A PARABOLA.
2.00
2.00
4.00
.38
2
PARALLELOGRAM METHOD
GIVEN A RECTANGLE, CONSTRUCT A PARABOLA.
2.00
2.00
4.00
.38
3
OFFSET METHOD
GIVEN A RECTANGLE, CONSTRUCT A PARABOLA
2.00
2.00
4.00
.38
4
ENVELOPE METHOD
GIVEN A RECTANGLE, CONSTRUCT A PARABOLA.

Fig. 4-6-A Drawing assignment.

Chapter 5
Basic Dimensioning

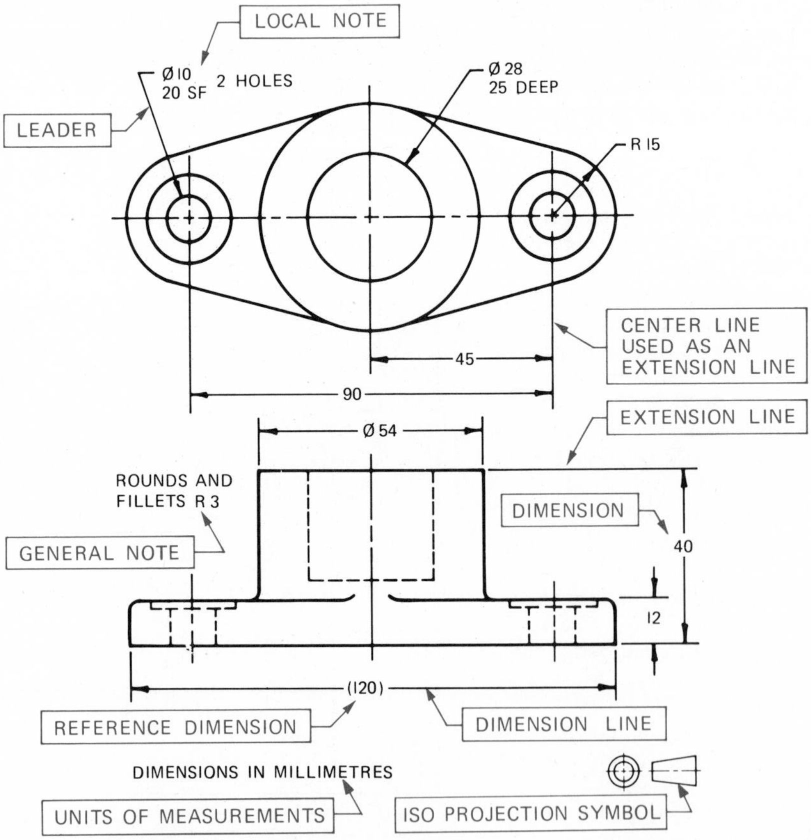

LOCAL NOTE
Ø10
20 SF
2 HOLES
Ø28
25 DEEP
LEADER
R15
CENTER LINE
USED AS AN
EXTENSION LINE
45
90
EXTENSION LINE
Ø54
ROUNDS AND
FILLETS R3
DIMENSION
40
GENERAL NOTE
12
(120)
REFERENCE DIMENSION
DIMENSION LINE
DIMENSIONS IN MILLIMETRES
UNITS OF MEASUREMENTS
ISO PROJECTION SYMBOL

Fig. 5-1-1 Basic dimensioning elements.

UNIT 5-1

BASIC DIMENSIONING[1,2]

A working drawing is one from which a tradesperson can produce a part. The drawing must be a complete set of instructions, so that it will not be necessary to give further information to the person(s) making the object. A working drawing, then, consists of the views necessary to explain the shape, the dimensions needed by the tradesperson, and required specifications, such as material and quantity needed. The latter information may be found in the notes on the drawing, or it may be located in the title block.

Dimensioning

Dimensions are indicated on drawings by extension lines, dimension lines, leaders, arrowheads, figures, notes, and symbols, to define geometric characteristics such as lengths, diameters, angles, and locations (Fig. 5-1-1). The lines used in dimensioning are thin in contrast to the outline of the object. The dimension must be clear and concise and permit only one interpretation. In general, each surface, line, or point is located by only one set of dimensions. Deviations from the approved rules for dimensioning should be made only in exceptional cases, when they will improve the clarity of the dimensioning. An exception to these rules is for arrowless and tabular dimensioning, which is discussed in Chap. 16.

In general, each surface, line, or point, where possible, is located by only one set of dimensions. These dimensions are not duplicated in other views, except for the purpose of identification, the improvement of clarity, or both.

Drawing for industry requires some form of tolerancing on dimensions so that components can be properly assembled and manufacturing and production requirements can be met. This chapter deals only with basic dimensioning and tolerancing techniques. Modern engineering tolerancing, such as true positioning and tolerance of form, is covered in detail in Chap. 31.

DIMENSION AND EXTENSION LINES

Dimension lines are used to indicate the extent and direction of dimensions, and they are terminated by neatly made, uniform arrowheads, as shown in Fig. 5-1-2. Arrowheads are usually drawn freehand, and the recommended length and the width should be in a ratio of 3:1 (Fig. 5-1-3*b*). The length of the arrowhead should be equal to the height of the dimension numerals. Where space is limited, a small circular dot may be used in lieu of an arrowhead (Fig. 5-1-3*d*). Normally, a break is made near the center of the dimension line for the insertion of the dimension which indicates the distance between the extension lines. When several dimension lines are directly above or next to one another, it is good practice to stagger the dimensions in order to improve the clarity of the drawing. In special cases, such as dual dimensioning that combines inch and metric units of measure on the same engineering drawing, or as a simplified drafting practice, the dimension line may be unbroken. The spacing suitable for most drawings between parallel dimension lines is 8 mm, and the spacing between the outline

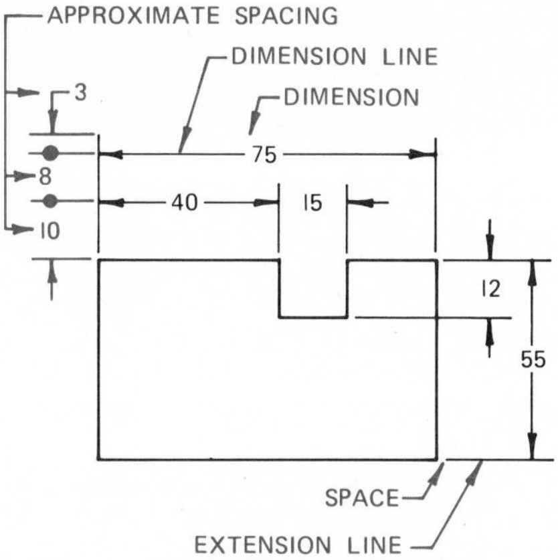

APPROXIMATE SPACING
DIMENSION LINE
DIMENSION
3
8
10
75
40
15
12
55
SPACE
EXTENSION LINE

Fig. 5-1-2 Dimension and extension lines.

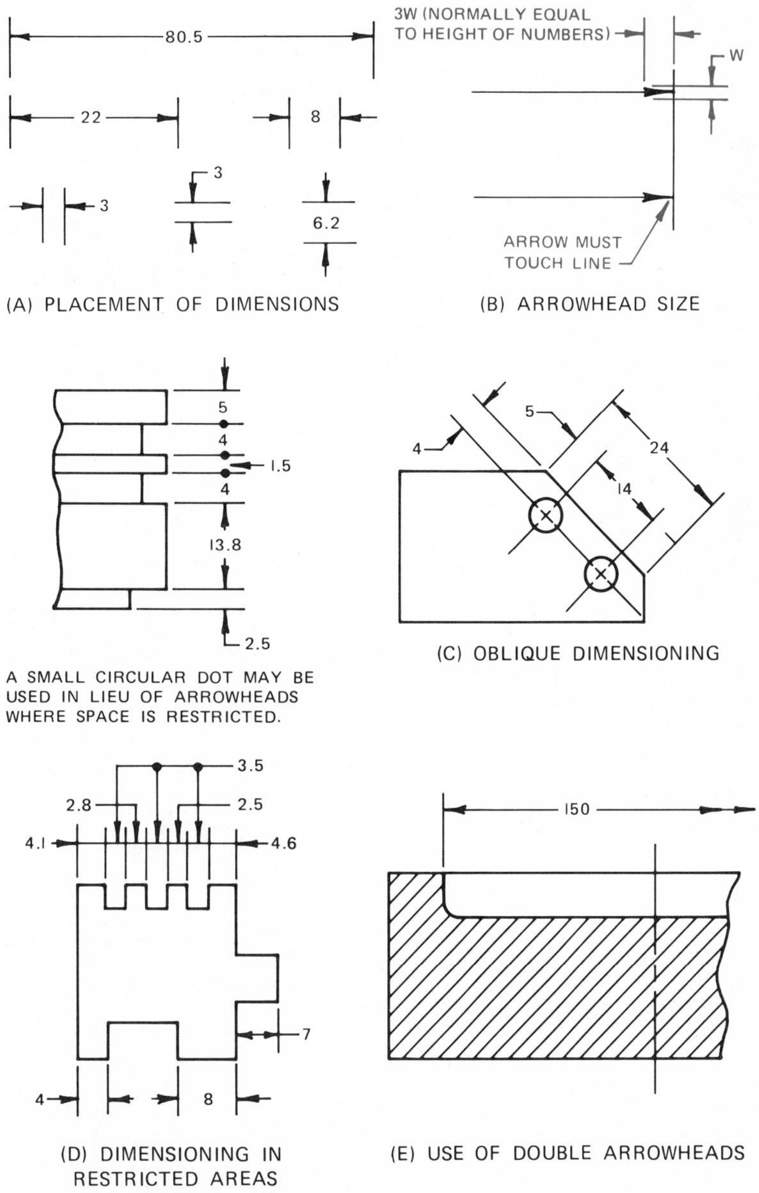

80.5
22
8
3
3
6.2
(A) PLACEMENT OF DIMENSIONS
3W (NORMALLY EQUAL TO HEIGHT OF NUMBERS)
W
ARROW MUST TOUCH LINE
(B) ARROWHEAD SIZE
5
4
1.5
4
13.8
2.5
A SMALL CIRCULAR DOT MAY BE USED IN LIEU OF ARROWHEADS WHERE SPACE IS RESTRICTED.
5
4
24
14
(C) OBLIQUE DIMENSIONING
3.5
2.8
2.5
4.1
4.6
7
4
8
(D) DIMENSIONING IN RESTRICTED AREAS
150
(E) USE OF DOUBLE ARROWHEADS

Fig. 5-1-3 Dimension lines.

of the object and the nearest dimension line should be approximately 10 mm. When the space between the extension lines is too small to permit the placing of the dimension line complete with arrowheads and dimension, then the alternate methods of placing the dimension line, dimension, or both outside the extension lines are used (Fig. 5-1-3). Center lines should never be used for dimension lines. Every effort should be made to avoid crossing dimension lines. This is accomplished by placing the shortest dimension closest to the outline.

Dimension lines should be placed outside the view where possible and should extend to extension lines rather than object lines. However, when readability is improved by avoiding either extra-long extension lines (Fig. 5-1-4) or the crowding of dimensions, placing of dimensions on views is permissible. Avoid dimensioning to hidden lines. In order to do so, it may be necessary to use a sectional view or a broken-out section. When the termination for a dimension is not included, as when used on partial views, a double arrow is used. For symmetrical features the dimension line should extend beyond the center line before the double arrowheads are added.

Extension (witness) lines are used to indicate the point or line on the drawing to which the dimension applies (Fig. 5-1-5). A small gap of about 1 mm (.04 in.) is left between the extension line and the outline to which it refers, and the extension line extends about 3 mm (.12 in.) beyond the outermost dimension line. However, when extension lines refer to points, as in Fig. 5-1-5*e*, they should extend through the points. Extension lines are usually drawn perpendicular to dimension lines. However, to improve clarity or when there is overcrowding, extension lines may be drawn at an oblique angle but must clearly illustrate where they apply.

Center lines may be used as extension lines in dimensioning. The center line extending past the circle is not broken.

Where extension lines cross other extension lines, dimension lines, or object lines, they are not broken. However, if extension lines cross arrowheads or dimension lines close to arrowheads, a break in the extension line is recommended.

LEADERS

Leaders are used to direct notes, dimensions, symbols, item numbers, or part numbers to features on the drawing (Fig. 5-1-6). A leader should generally be a single straight inclined line (not vertical or horizontal) except for a short horizontal portion extending to the center of the height of the first or last letter or digit of the note. According to ANSI Y14.5, 1973, a horizontal bar must be used. The leader is terminated by an arrowhead or a dot of at least 1.5 mm (.06 in.) in diameter. Arrowheads should always terminate on a line; dots should be used within the outline of the object and rest on a surface. Leaders should not be bent in any way unless it is unavoidable. Leaders should not cross one another, and two or more leaders adjacent to one

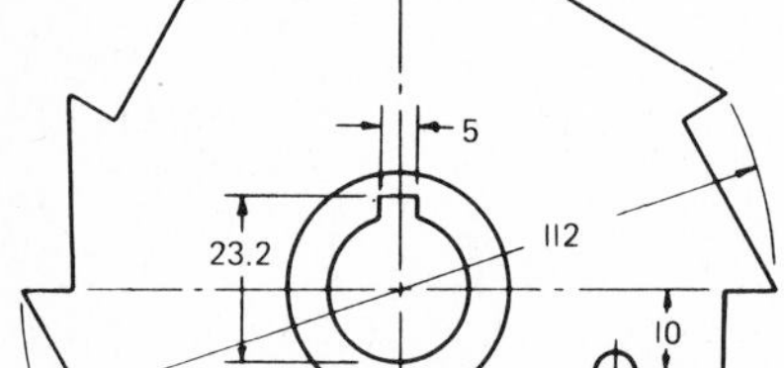
7.6
10
10
7.6
18
16
10
15
10

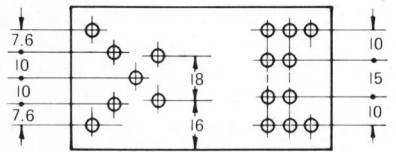

(A) IMPROVING READABILITY OF DRAWING

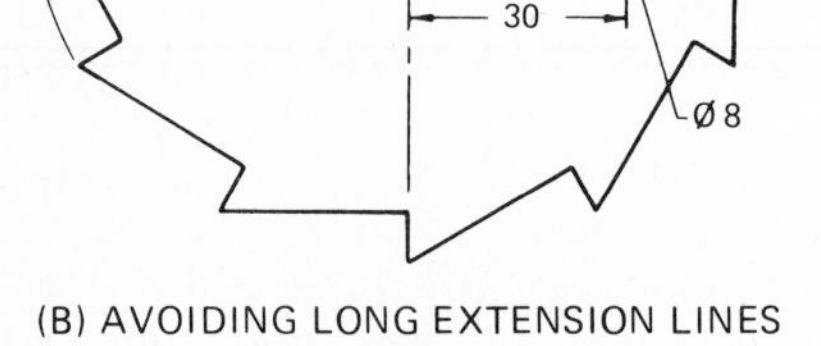
5
23.2
112
10
30
Ø 8

(B) AVOIDING LONG EXTENSION LINES

Fig. 5-1-4 Placing dimensions on view.

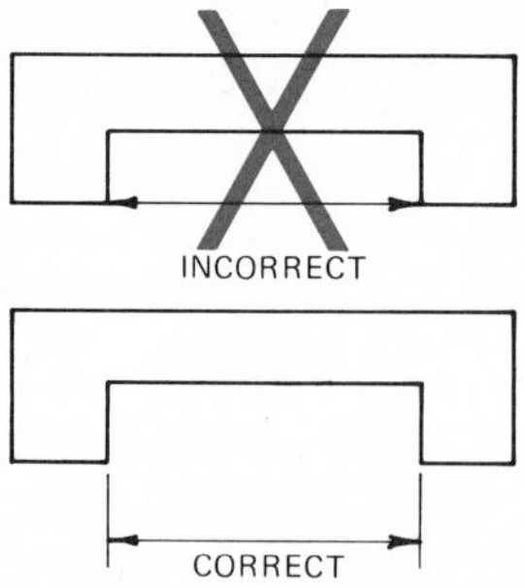
INCORRECT
CORRECT

(A) USE OF EXTENSION LINES

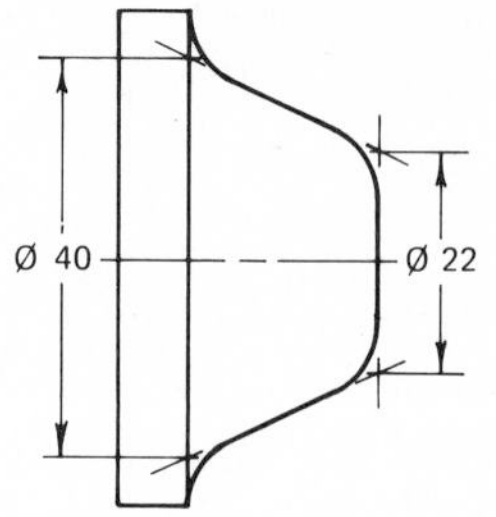
Ø 40
Ø 22

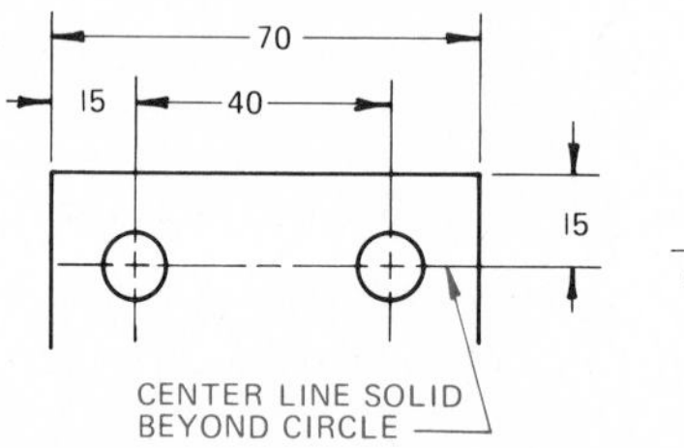
70
15
40
15
CENTER LINE SOLID
BEYOND CIRCLE

(B) CENTER LINE USED AS EXTENSION LINE

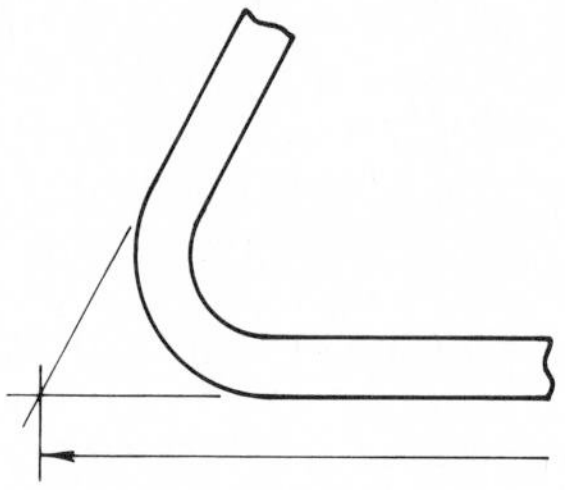

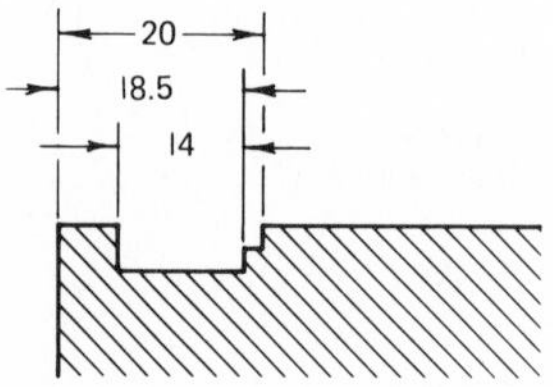
20
18.5
14

(C) BREAK IN EXTENSION LINES

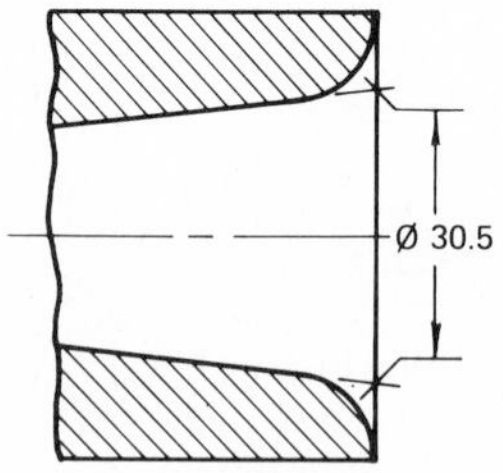
Ø 30.5

(E) EXTENSION LINE FROM POINTS

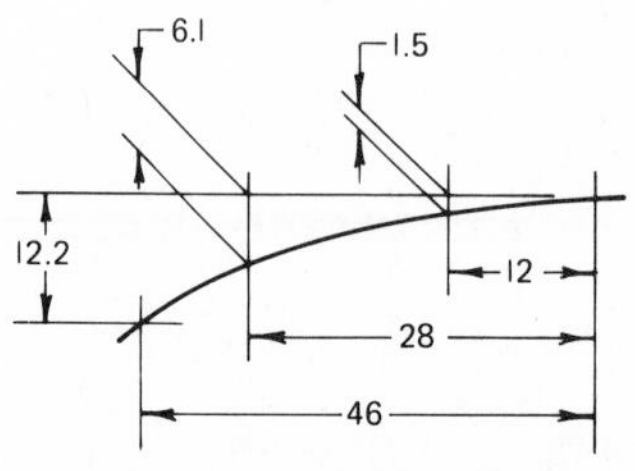
6.1
1.5
12.2
12
28
46

(D) OBLIQUE EXTENSION LINES

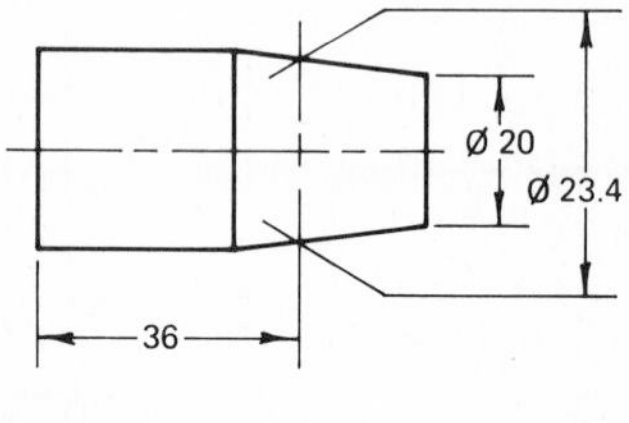
Ø 20
Ø 23.4
36

(F) EXTENSION LINE FROM POINTS

Fig. 5-1-5 Extension lines.

another should be drawn parallel if practicable. It is better to repeat dimensions or references than to use long leaders.

Where a leader is directed to a circle or circular arc, its direction should point to the center of the arc or circle. Regardless of the reading direction used, aligned or undirectional, all notes and dimensions used with leaders are placed in a horizontal position.

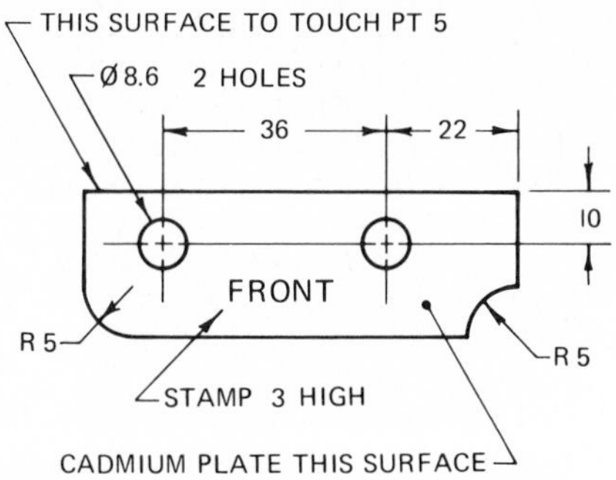
THIS SURFACE TO TOUCH PT 5
Ø8.6 2 HOLES
36
22
10
FRONT
R 5
R 5
STAMP 3 HIGH
CADMIUM PLATE THIS SURFACE

Fig. 5-1-6 Leaders.

NOTES

Notes are used to simplify or complement dimensioning by indicating information on a drawing in a condensed and systematic manner. They may be general or local notes, and should be in the present or future tense.

General Notes. These refer to the part or drawing as a whole. They should be shown in a central position below the view to which they apply or placed in a general note column. Typical examples of this type of note are

- FINISH ALL OVER
- ROUNDS AND FILLETS R 2
- REMOVE ALL SHARP EDGES

Local Notes. These apply to local requirements only and are connected by a leader to the point to which the note applies. Typical examples are

- ϕ6, 4 HOLES
- 2 × 45°
- ϕ3
  ϕ11.5 × 86° CSK
- M12 × 1.25

Units of Measurements

Although the metric system of dimensioning has become the official standard of measurement, many drawings in current use are still dimensioned in inches or feet and inches. For this reason, drafters should be familiar with all the dimensioning systems which they may encounter.

SI METRIC UNITS OF MEASUREMENT

The standard metric units on engineering drawings are the millimetre (mm) for linear measure and micrometre (μm) for surface roughness (Fig. 5-1-7). For architectural drawings, metre and millimetre units are used. Unless otherwise specified, the dimensions in this book are in millimetres.

Whole numbers from 1 to 9 will be shown without a zero to the left of the number or a zero to the right of the decimal point.

2 *not* 02 or 2.0

A millimetre value of less than 1 is shown with a zero to the left of the decimal point.

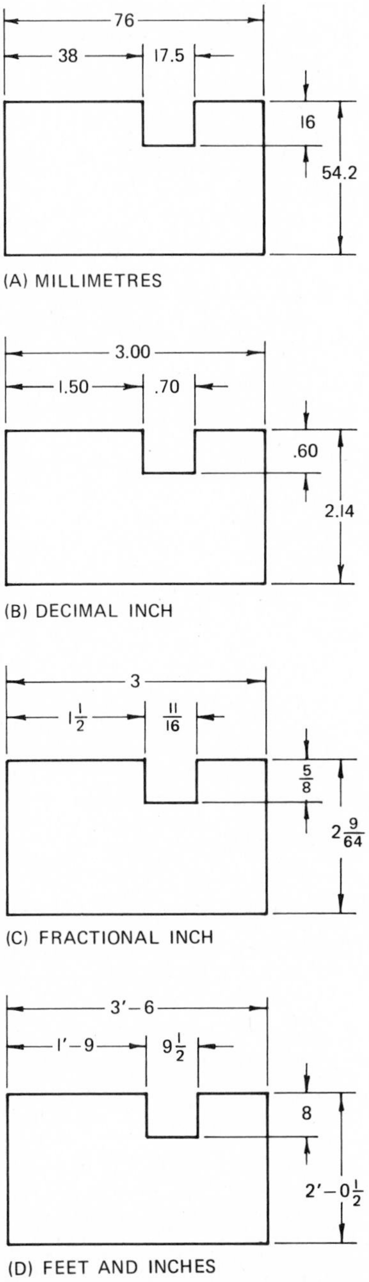
76
38
17.5
16
54.2
(A) MILLIMETRES
3.00
1.50
.70
.60
2.14
(B) DECIMAL INCH
3
1½
11/16
5/8
2 9/64
(C) FRACTIONAL INCH
3′−6
1′−9
9½
8
2′−0½
(D) FEET AND INCHES

Fig. 5-1-7 Dimensioning units.

0.2 *not* .2 or .20

0.26 *not* .26

Decimal points should be uniform and large enough to be clearly visible on reduced-size prints. They should be placed in line with the bottom of the associated numbers and be given adequate space.

Commas should not be used to separate groups of three numbers in either inch or metric values. A space should be used in place of the comma.

32 541 *not* 32,541

2.562 827 6 *not* 2.562827

Identification. A metric drawing should include a general note, such as UNLESS OTHERWISE SPECIFIED DIMENSIONS ARE IN MILLIMETRES and be identified by the word METRIC prominently displayed near the title block.

INCH UNITS OF MEASUREMENT

Decimal Inch System. Parts are designed in basic decimal increments, preferably 0.02 in., and are expressed with a minimum of two figures to the right of the decimal point. Using the 0.02 module, the second decimal place (hundredths) is an even number or zero. By using the design modules having an even number for the last digit, dimensions can be halved for center distances without increasing the number of decimal places. Decimal dimensions which are not multiples of 0.02, such as 0.01, 0.03, and 0.15, should be used only when it is essential to meet design requirements such as to provide clearance, strength, smooth curves, etc. When greater accuracy is required, sizes are expressed as three- or four-place decimal numbers, for example, 1.875.

Whole dimensions will show a minimum of two zeros to the right of the decimal point.

24.00 *not* 24

An inch value of less than 1 is shown without a zero to the left of the decimal point.

.44 *not* 0.44

In cases where parts have to be aligned with existing parts or commercial products, which are dimensioned in fractions, it may be necessary to use decimal equivalents of fractional dimensions.

Fractional Inch System. In this system, parts are designed in basic units of common fractions down to 1/64 in. Decimal dimensions are used when finer divisions than 1/64 in. must be made.

Common fractions are used for specifying the size of holes that may be produced by drills ordinarily stocked in fraction sizes and for the sizes of standard screw threads.

When common fractions are used on drawings, the fraction bar must not be omitted and should be horizontal except when applied with a typewriting machine which does not have a horizontal fraction bar.

When a dimension intermediate between 1/64 increments is necessary, it is expressed in decimals, such as .30, .257, or .2575.

The inch marks (″) should not be shown with the dimensions. A note such as DIMENSIONS ARE IN INCHES should be clearly shown on the drawing. The exception is when the dimension "1 in." is shown on the drawing. The 1 should then be followed by the inch marks—1″, *not* 1.

Foot and Inch System. Feet and inches are often used for installation drawings, drawings of large objects, and floor plans associated with architectural work. In this case, all dimensions 12 in. or greater are specified in feet and inches. Parts of an inch are usually expressed as common fractions, rather than as decimals.

The inch marks (″) are not shown. The drawing should carry a note such as DIMENSIONS ARE IN INCHES UNLESS NOT OTHERWISE SPECIFIED.

A dash and space should be left between the foot and inch values. For example, 1′ - 3, not 1′3.

Whole numbers of inches are expressed without decimals or fractions. For example, 24 in. is expressed as 2′ - 0, and 27 in. is expressed as 2′ - 3.

UNITS COMMON TO EITHER SYSTEM

Some measurements can be stated so that the call-out will satisfy the units of both systems. For example, tapers such as 0.006 mm per millimetre and 0.006 in. per inch can both be expressed simply as the ratio 0.006:1 or in a note TAPER 0.006:1. Angular dimensions are also specified the same in both inch and metric systems.

STANDARD ITEMS

Fasteners and Threads. Metric fasteners and threads should be used if possible. Refer to the Appendix and Chap. 9 for additional information.

Hole Sizes. The chart of metric drill sizes shown in Table 8 in the Appendix may be used as a reference in determining standard hole sizes.

Dual Dimensioning

With a great exchange of drawings taking place between North America and the rest of the world, at one time it became advantageous to show drawings in both inches and millimetres. As a result, many companies adopted a dual system of dimensioning. Today, however, this type of dimensioning should be avoided if possible. However, when it is necessary or desirable to give dimensions in both inches and millimetres on the same drawing, the following guidelines, as illustrated in Fig. 5-1-8, should be observed.

Show the millimetre dimension above the inch dimension separated by a horizontal line, or to the left of the inch dimension separated by a slash (oblique) line, or by enclosing the millimetre or inch dimension in brackets.

When a note with a leader is used, the units of measurement are separated by an oblique line or enclosing the latter dimension in brackets. The converted values are not shown below the measurements.

The method(s) used on the drawing should be clearly stated on the drawing.

A note or illustration should be located near the title block or strip to identify the inch and millimetre dimensions.

MILLIMETRE
———
INCH

MILLIMETRE/INCH

MILLIMETRE [INCH]

Examples of dual-dimensioned drawings are shown in Figs. 5-1-9 and 5-1-10.

Dual dimensioning, as described above, has some disadvantages. More space is required to place both sets of dimensions, errors may occur when dual dimensions are manually converted and placed on the drawing, and the time spent by the drafter to calculate the dual dimensions may be costly. Some drawing offices have added the dual dimensions to the drawing in chart form. The drawing is dimensioned to the units (either inches or millimetres) by the drafter. Then it is given to an assistant who feeds all the dimensions into a computer which accurately tabulates the dual dimensions. The results are printed on an adhesive translucent material which is attached to the drawing.

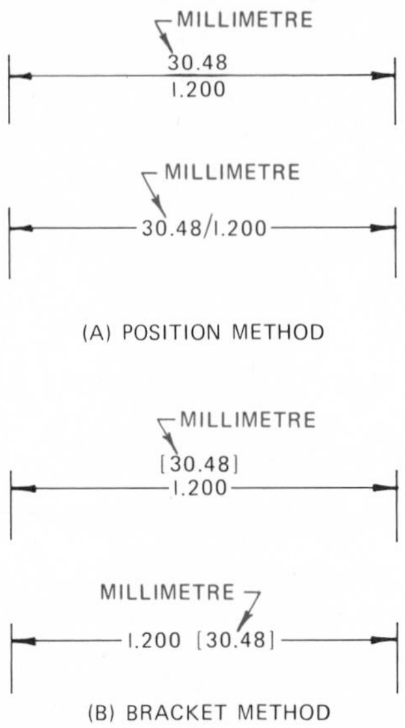
MILLIMETRE
30.48
1.200
MILLIMETRE
30.48/1.200
(A) POSITION METHOD
MILLIMETRE
[30.48]
1.200
MILLIMETRE
1.200 [30.48]
(B) BRACKET METHOD

Fig. 5-1-8 Dual dimensioning.

Angular Units

Angles are measured in degrees. The decimal degree is now preferred to the use of degrees, minutes, and seconds. For example, the use of 50.5° is preferred to the use of 60°30′. Where only minutes or seconds are specified, the number of minutes or seconds is preceded by 0°, or 0°0′, as applicable. Some examples follow.

Decimal Degree	Degrees, Minutes, and Seconds
10° ± 0.5°	10° ± 0°30′
0.75°	0°45′
0.004°	0°0′15″
90° ± 1.0°	90° ± 1°
25.6° ± 0.2°	25°36′ ± 0°12′
25.51°	25°30′40″

The dimension line of an angle is an arc drawn with the apex of the angle as the center point for the arc, wherever practicable. The position of the dimension varies according to the size of the angle and appears in a horizontal position. Recommended arrangements are shown in Fig. 5-1-11.

Reading Direction

Dimensions on drawings are either unidirectional or aligned (Fig. 5-1-12). Unidirectional is preferred. With this method, dimensions are placed to be read from the bottom of the drawing. Aligned dimensions are placed perpendicular to their dimension lines. Numerals are read from the bottom or right side of the drawing.

In both methods, angular dimensions and dimensions and notes shown with leaders should be aligned with the bottom of the drawing.

Basic Rules for Dimensioning

Refer to Fig. 5-1-13.

- Place dimensions between the views when possible.

- Place the dimension line for the shortest length, width, or height nearest the outline of the object. Parallel dimension lines are placed in order of their size, making the longest dimension line the outermost.
- Place dimensions near the view that best shows the characteristic contour or shape of the object. In following this rule, dimensions will not always be between views.
- On large views, dimensions can be placed on the view to improve clarity.
- Use only one system of dimensions, either the unidirectional or the aligned, on any one drawing.
- Dimensions should not be duplicated in other views.
- Dimensions should be selected so that it will not be necessary to add or subtract dimensions in order to define or locate a feature.

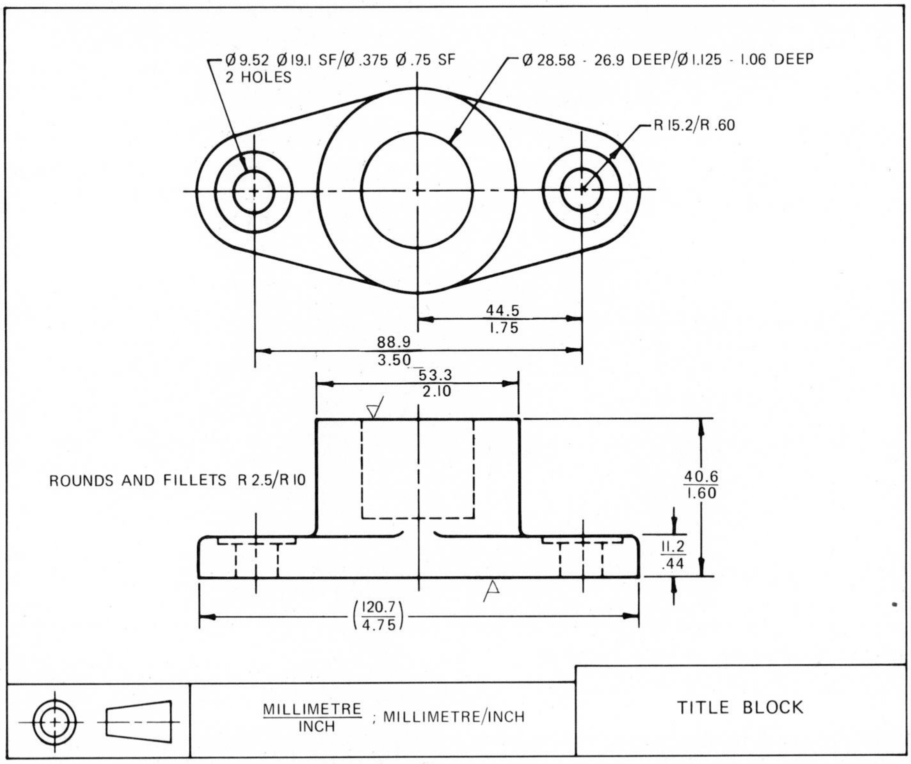
Ø 9.52 Ø 19.1 SF/Ø .375 Ø .75 SF
2 HOLES
Ø 28.58 - 26.9 DEEP/Ø 1.125 - 1.06 DEEP
R 15.2/R .60
44.5
1.75
88.9
3.50
53.3
2.10
ROUNDS AND FILLETS R 2.5/R 10
40.6
1.60
11.2
.44
120.7
4.75
MILLIMETRE
INCH
; MILLIMETRE/INCH
TITLE BLOCK

Fig. 5-1-9 Dual dimensioned drawing.

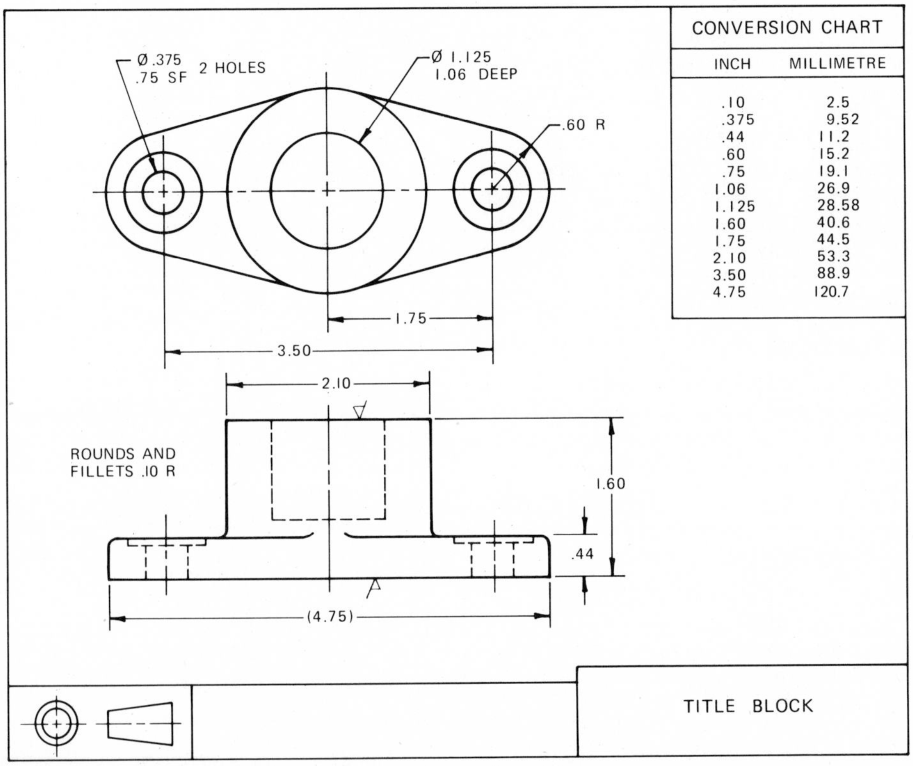
Ø .375
.75 SF
2 HOLES
Ø 1.125
1.06 DEEP
.60 R
1.75
3.50
2.10
ROUNDS AND
FILLETS .10 R
1.60
.44
(4.75)
TITLE BLOCK

CONVERSION CHART	
INCH	MILLIMETRE
.10	2.5
.375	9.52
.44	11.2
.60	15.2
.75	19.1
1.06	26.9
1.125	28.58
1.60	40.6
1.75	44.5
2.10	53.3
3.50	88.9
4.75	120.7

Fig. 5-1-10 Drawing with millimetre conversion chart.

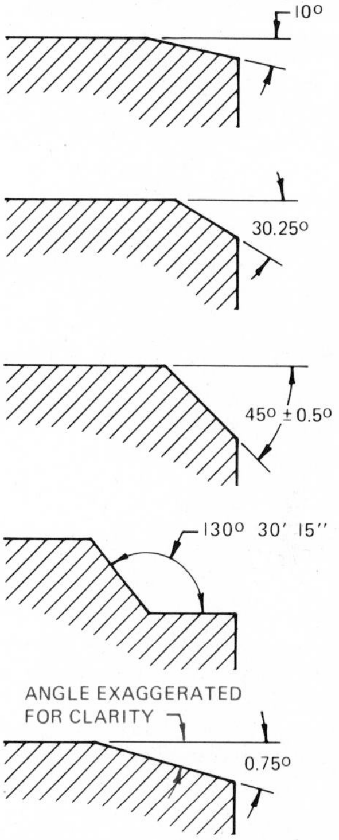
10°
30.25°
45° ± 0.5°
130° 30′ 15″
ANGLE EXAGGERATED
FOR CLARITY
0.75°

Fig. 5-1-11 Angular units.

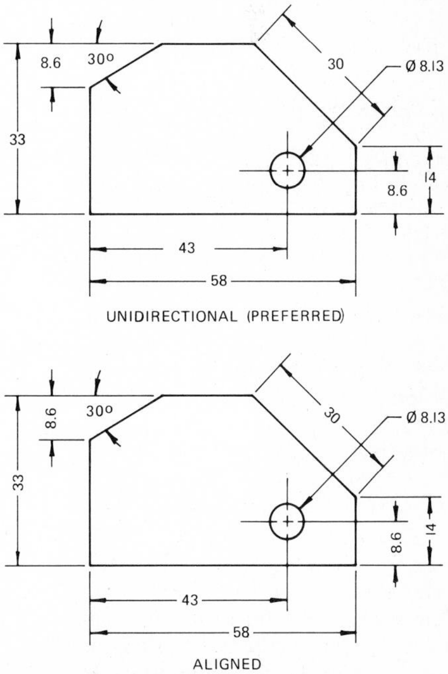
8.6
30°
30
Ø 8.13
33
14
8.6
43
58
UNIDIRECTIONAL (PREFERRED)
8.6
30°
30
Ø 8.13
33
8.6
14
43
58
ALIGNED

Fig. 5-1-12 Reading direction of dimensions.

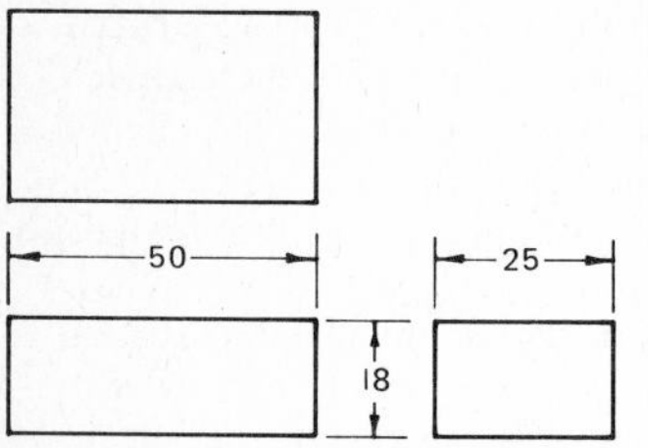

(A) PLACE DIMENSIONS BETWEEN VIEWS

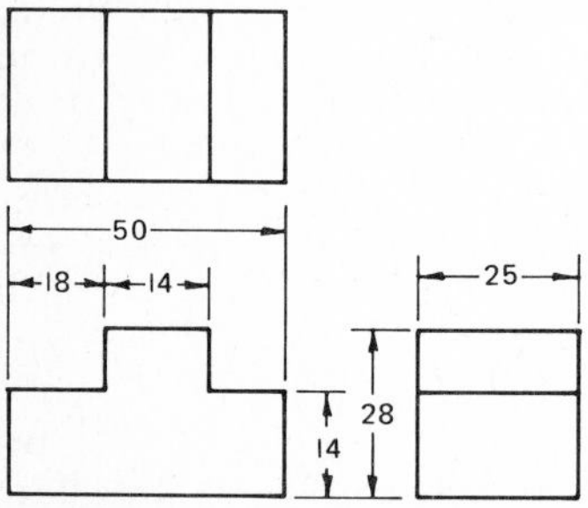

(B) PLACE SMALLEST DIMENSION NEAREST THE VIEW BEING DIMENSIONED

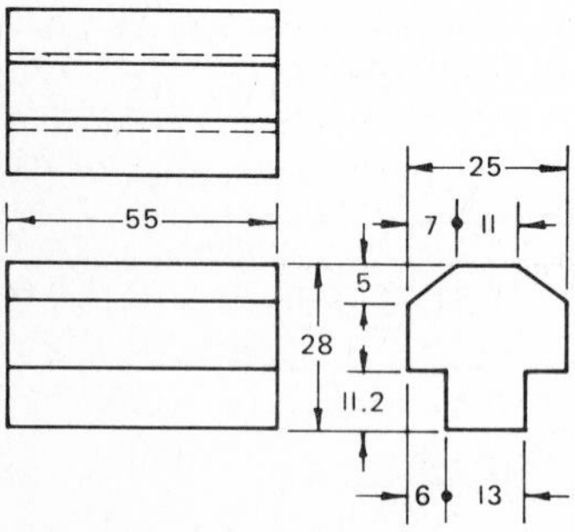

(C) DIMENSION THE VIEW THAT BEST SHOWS THE SHAPE

Fig. 5-1-13 Basic dimensioning rules.

REFERENCE DIMENSIONS

A reference dimension is shown for information only, and it is not required for manufacturing or inspection purposes. It is enclosed in parentheses, as shown in Fig. 5-1-14. Formerly the abbreviation REF was used to indicate a reference dimension.

NOT-TO-SCALE DIMENSIONS

When a dimension on a drawing is altered, making it not to scale, it should be underlined with a freehand line (Fig. 5-1-15), except when the condition is clearly indicated by break lines.

Operational Names

The use of operational names of dimensions, such as turn, bore, grind, ream, tap, and thread, should be avoided. While the drafter should be aware of the methods by which a part can be produced, the method of manufacture is better left to the producer. If the completed part is adequately dimensioned and has surface texture symbols showing finish quality desired, it remains a shop problem to meet the drawing specifications.

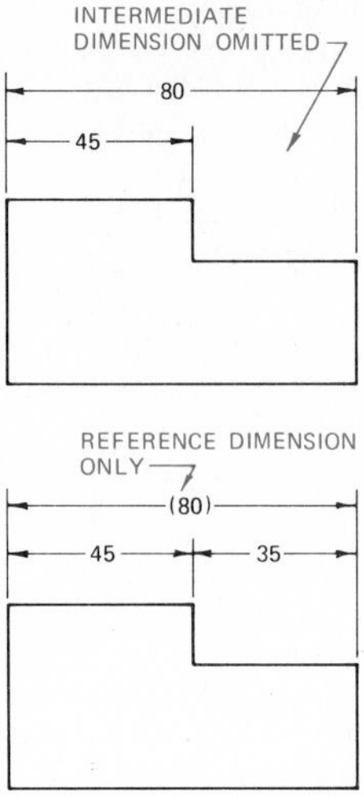
INTERMEDIATE DIMENSION OMITTED
80
45
REFERENCE DIMENSION ONLY
(80)
45
35

Fig. 5-1-14 Reference dimensions.

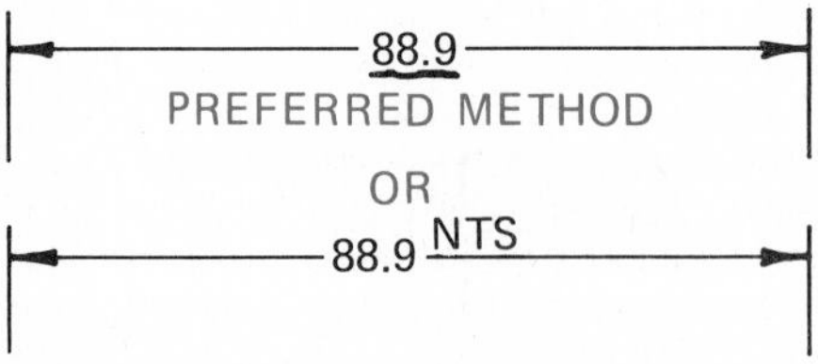
88.9
PREFERRED METHOD
OR
88.9 NTS

Fig. 5-1-15 Not to scale dimensions.

Abbreviations

Abbreviations and symbols are used on drawings to conserve space and time, but only where their meanings are quite clear. Therefore, only commonly accepted abbreviations such as those shown in the Appendix should be used on drawings.

REFERENCES

1. ANSI Y14.5, *Dimensioning and Tolerancing*.
2. CSA B78.2, *Dimensioning and Tolerancing of Mechanical Engineering Drawings*.

Assignment

Select one of the template drawings (Figs. 5-1-A to 5-1-D) and on an A3- or B-size sheet make a one-view drawing, complete with dimensions, of the part. Scale is 1:1.

REVIEW FOR ASSIGNMENT

Unit 2-6 Drafting Skills

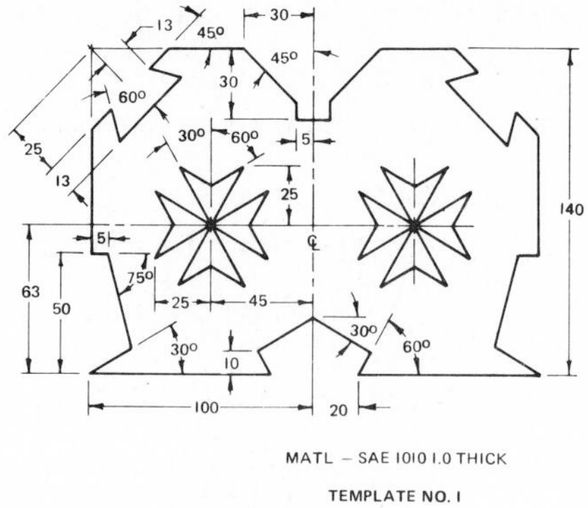
MATL – SAE 1010 1.0 THICK
TEMPLATE NO. 1

Fig. 5-1-A Template no. 1.

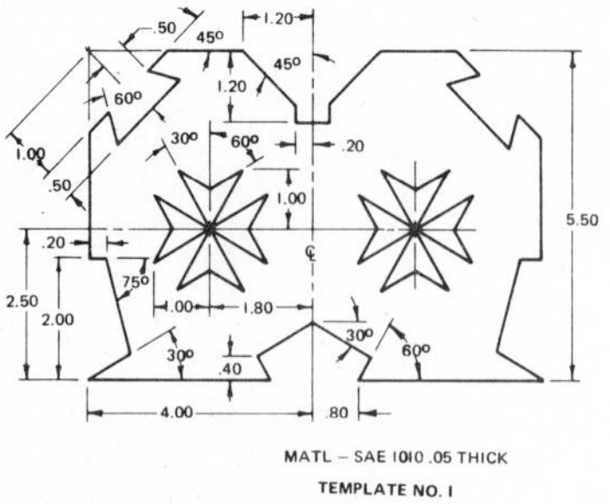
MATL – SAE 1010 .05 THICK
TEMPLATE NO. 1

Fig. 5-1-B Template no. 1.

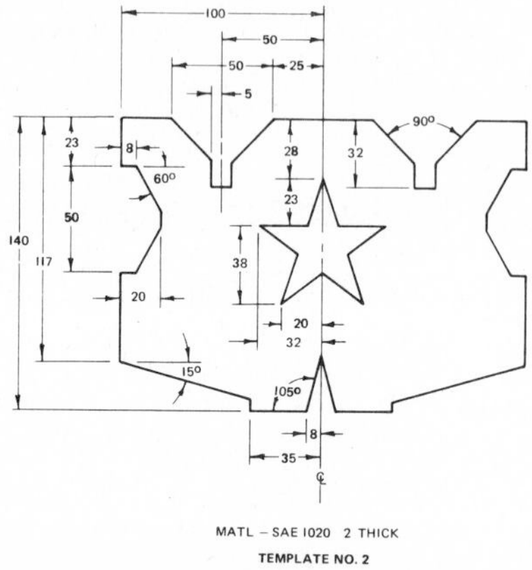
MATL – SAE 1020 2 THICK
TEMPLATE NO. 2

Fig. 5-1-C Template no. 2.

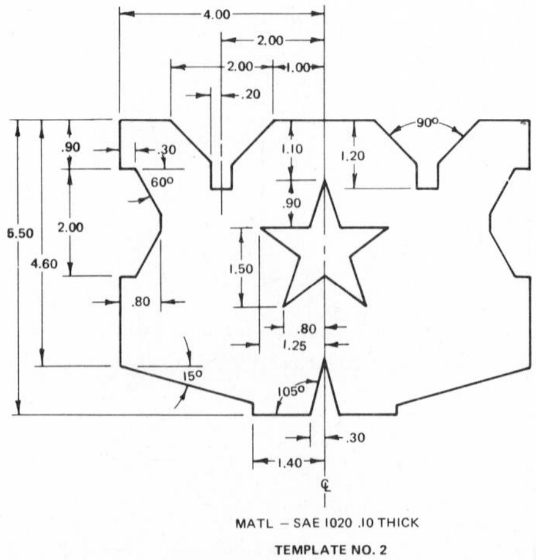
MATL – SAE 1020 .10 THICK
TEMPLATE NO. 2

Fig. 5-1-D Template no. 2.

UNIT 5-2

DIMENSIONING CIRCULAR FEATURES[1,2]

Diameters

Where the diameter of a single feature or the diameters of a number of concentric cylindrical features are to be specified, it is recommended that they be shown on the longitudinal view (Fig. 5-2-1). Where it is obvious from the views that the dimension refers to the diameter and a dimension line is used, the symbol ϕ or the abbreviation DIA may be omitted. It is recommended that the diameter symbol or abbreviation be used in conjunction with a leader.

Where the diameter is dimensioned on a single-view drawing which does not illustrate the circularity of the feature, the dimension indicating the diameter is preceded by the symbol ϕ or followed by the abbreviation DIA. The symbol is the preferred method since it is faster to draw, takes less space, and is universal in use.

Where space is restricted or when only a partial view is used, diameters may be dimensioned as illustrated in Fig. 5-2-2.

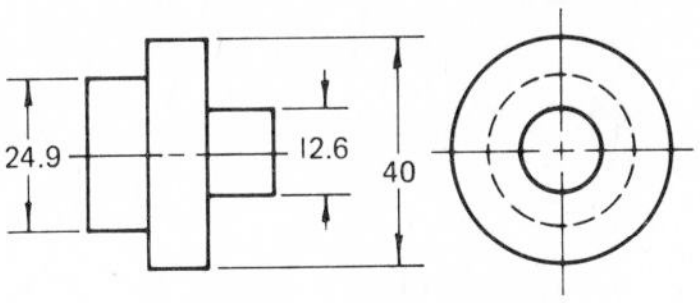
24.9
12.6
40

(A) TWO-VIEW DRAWING

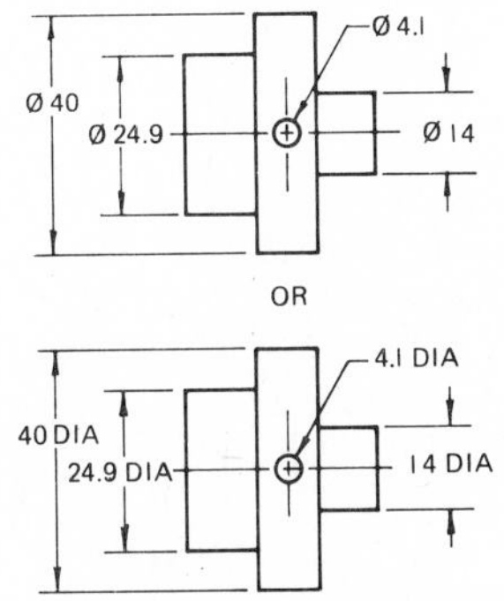
Ø 4.1
Ø 40
Ø 24.9
Ø 14
OR
4.1 DIA
40 DIA
24.9 DIA
14 DIA

(B) ONE-VIEW DRAWING

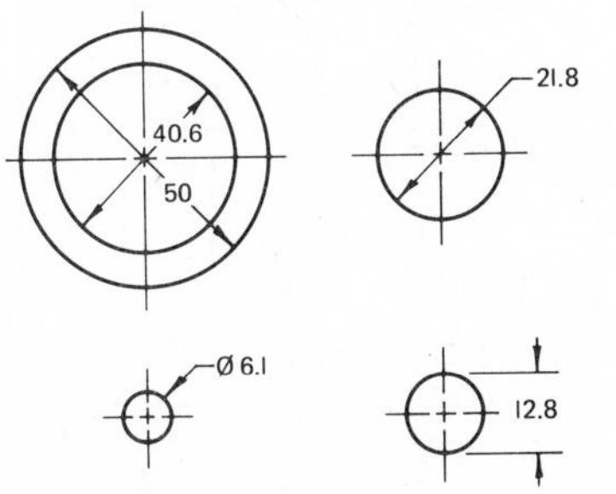
40.6
50
21.8
Ø 6.1
12.8

(C) DIMENSIONING DIAMETERS ON END VIEW

Fig. 5-2-1 Diameters.

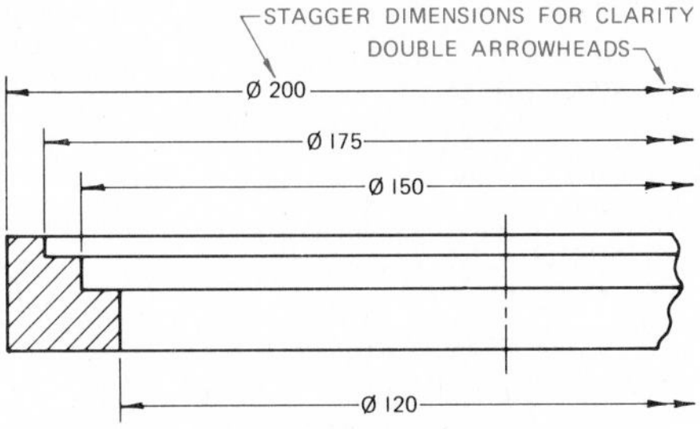
STAGGER DIMENSIONS FOR CLARITY
DOUBLE ARROWHEADS
Ø 200
Ø 175
Ø 150
Ø 120

Fig. 5-2-2 Dimensioning diameters where space is rest restricted.

Radii

The general method of dimensioning a circular arc is by giving its radius. A radius dimension line passes through, or is in line with, the radius center and terminates with an arrowhead touching the arc (Fig. 5-2-3). An arrowhead is never used at the radius center. The size of the dimension is preceded by the abbreviation R for metric dimensioning and followed by the abbreviation R for feet and inch dimensions. Where space is limited, as for a small radius, the radial dimension line may extend through the radius center. Where it is inconvenient to place the arrowhead between the radius center and the arc, it may be placed outside the arc, or a leader may be used (Fig. 5-2-3*a*).

Where a dimension is given to the center of the radius, a small cross should be drawn at the center (Fig. 5-2-3*b*). Extension lines and dimension lines are used to locate the center. Where the location of the center is unimportant, a radial arc may be located by tangent lines (Fig. 5-2-3*e*).

Where the center of a radius is outside the drawing or interferes with another view, the radius dimension line may be foreshortened (Fig. 5-2-3*d*). The portion of the dimension line next to the arrowhead should be radial relative to the curved line. Where the radius dimension line is foreshortened and the center is located by coordinate dimensions, the dimensions locating the center should be shown as foreshortened or a dimension not to scale.

Simple fillet and corner radii may also be dimensioned by use of a general note, such as ALL ROUNDS AND FILLETS UNLESS OTHERWISE SPECIFIED R 5 or ALL RADII R 5.

ROUNDED ENDS

Overall dimensions should be used for parts or features having rounded ends. For fully rounded ends, the radius is indicated but not dimensioned (Fig. 5-2-4*a*). For parts with partially rounded ends, the radius is dimensioned (Fig. 5-2-4*b*). Where a hole and radius have the same center and the hole location is more critical than the location of a radius, then either the radius or the overall length should be shown as a reference dimension (Fig. 5-2-4*c*).

SPHERICAL FEATURES

Spherical surfaces may be dimensioned as diameters or radii, but the dimension should be used with the abbreviations R

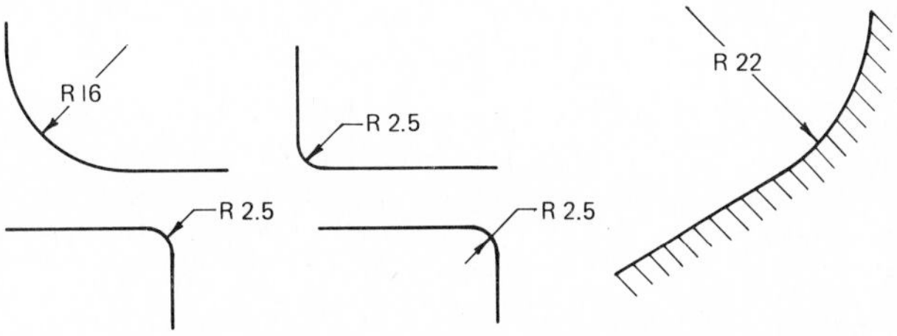
R 16
R 2.5
R 2.5
R 2.5
R 22

(A) RADII WHICH NEED NOT HAVE THEIR CENTERS LOCATED

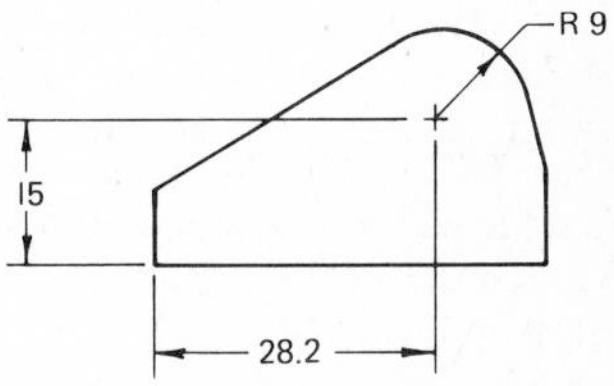
R 9
15
28.2

(B) LOCATING RADIUS CENTER

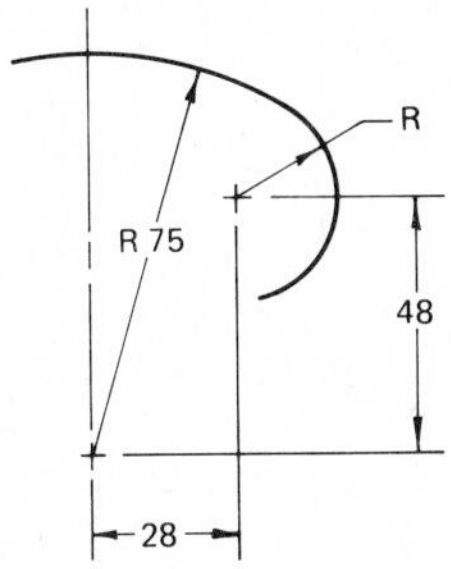
R
R 75
48
28

(C) RADII WITH COMMON TANGENT POINTS

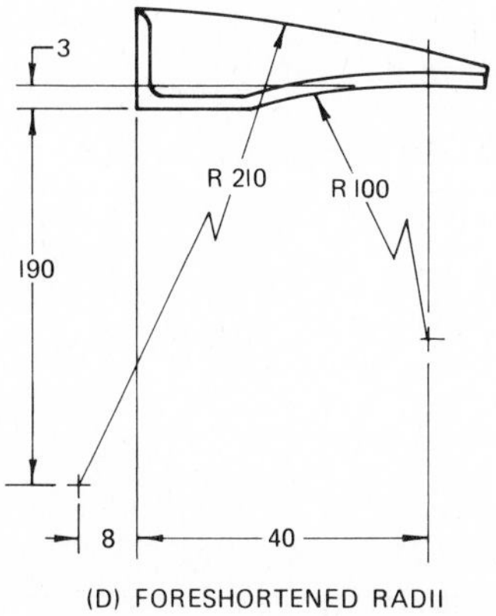
3
R 210
R 100
190
8
40

(D) FORESHORTENED RADII

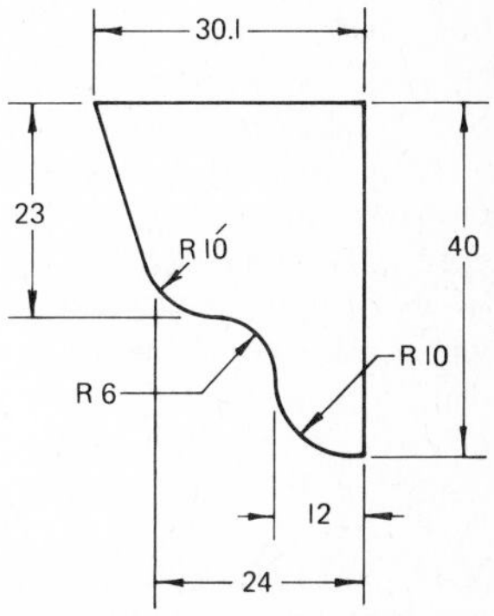
30.1
23
R 10
40
R 10
R 6
12
24

(E) RADII LOCATED BY TANGENTS

Fig. 5-2-3 Radii.

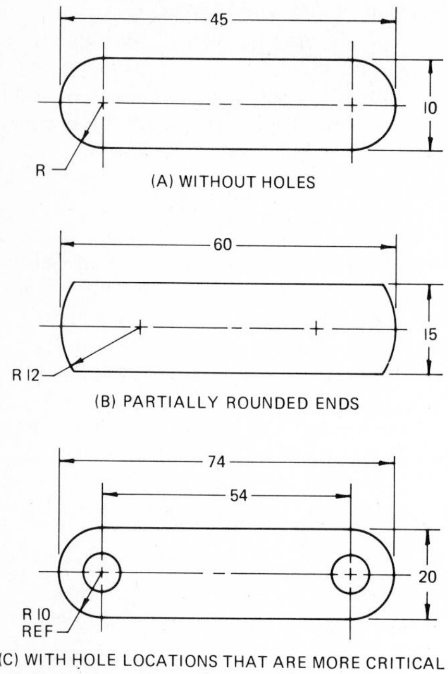

45
10
R
(A) WITHOUT HOLES
60
15
R 12
(B) PARTIALLY ROUNDED ENDS
74
54
20
R 10
REF
(C) WITH HOLE LOCATIONS THAT ARE MORE CRITICAL

Fig. 5-2-4 External surfaces with rounded ends.

SPHER, ϕ SPHER, or DIA SPHER. The symbol ϕ or the abbreviation DIA may be omitted when a diameter is clearly indicated by the view and with the use of extension lines (Fig. 5-2-5).

CYLINDRICAL HOLES

Plain, round holes are dimensioned in various ways, depending upon design and manufacturing requirements (Fig. 5-2-6). However, the leader is the method most commonly used. When a leader is used to specify diameter sizes, as with small holes, the dimension is identified as a diameter by preceding the numerical value with the diameter symbol ϕ or by adding the abbreviation DIA after the numerical value.

The size, quantity, and depth may be shown on a single line, or on several lines if preferable. For through holes, the abbreviation THRU should follow the dimension if the drawing does not make this clear. The depth dimension of a blind hole is the depth of the full diameter and is normally included as part of the dimensioning note.

When more than one hole of a size is required, the number of holes should be specified. However, care must be taken to avoid placing two values together without adequate spacing. For example, it may be better to show the note on two or more lines than to use a line note which might be misread (Fig. 5-2-6*d*).

SLOTTED HOLES

Elongated holes and slots are used to compensate for inaccuracies in manufacturing and to provide for adjustment (Fig. 5-2-7). The method selected to locate the slot would depend on how the slot was made. The method shown in Fig. 5-2-7*b* is used when the slot is punched out and the location of the punch is given. Figure 5-2-7*a* is the method of dimensioning used when the slot is machined out.

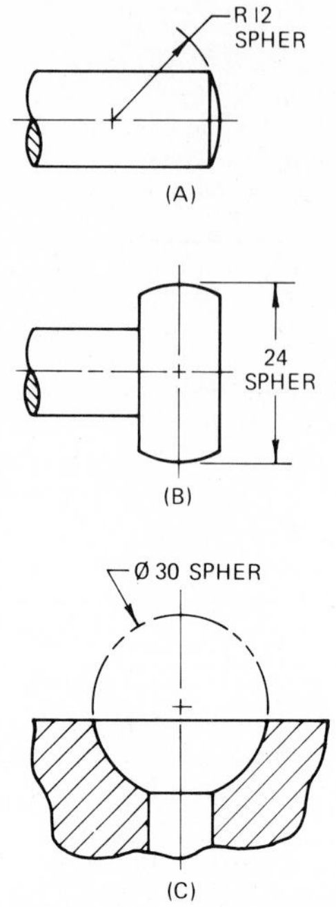

R 12
SPHER
(A)
24
SPHER
(B)
Ø 30 SPHER
(C)

Fig. 5-2-5 Spherical surfaces.

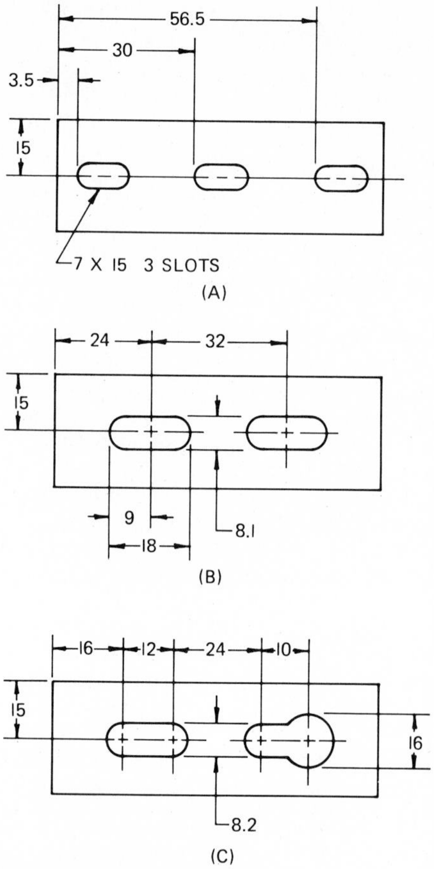

56.5
30
3.5
15
7 X 15 3 SLOTS
(A)
24
32
15
9
18
8.1
(B)
16
12
24
10
15
16
8.2
(C)

Fig. 5-2-7 Slotted holes.

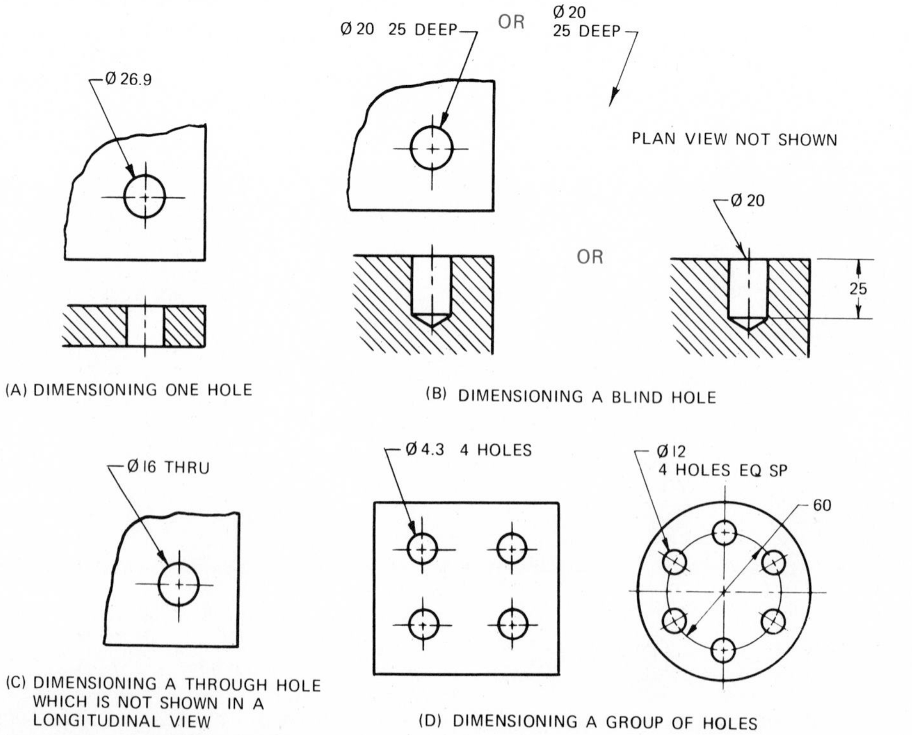

Ø 26.9
(A) DIMENSIONING ONE HOLE
Ø 20 25 DEEP
OR
Ø 20
25 DEEP
PLAN VIEW NOT SHOWN
Ø 20
OR
25
(B) DIMENSIONING A BLIND HOLE
Ø 16 THRU
(C) DIMENSIONING A THROUGH HOLE WHICH IS NOT SHOWN IN A LONGITUDINAL VIEW
Ø 4.3 4 HOLES
Ø 12
4 HOLES EQ SP
60
(D) DIMENSIONING A GROUP OF HOLES

Fig. 5-2-6 Cylindrical holes.

COUNTERSINK, COUNTERBORES, AND SPOTFACES

The abbreviations CSK, CBORE, and SF for countersink, counterbore, and spotface, respectively, indicate the form of the surface only and do not restrict the methods used to produce that form. The dimensions for them are usually given as a note, preceded by the size of the through hole (Fig. 5-2-8).

A countersink is an angular-sided recess to accommodate the head of flathead screws, rivets, and similar items. The diameter at the surface and the included angle are given. When the depth of the tapered section of the countersink is critical, this depth is specified in the note or by dimension. For counterdrilled holes, the diameter, depth, and included angle of the counterdrill are given.

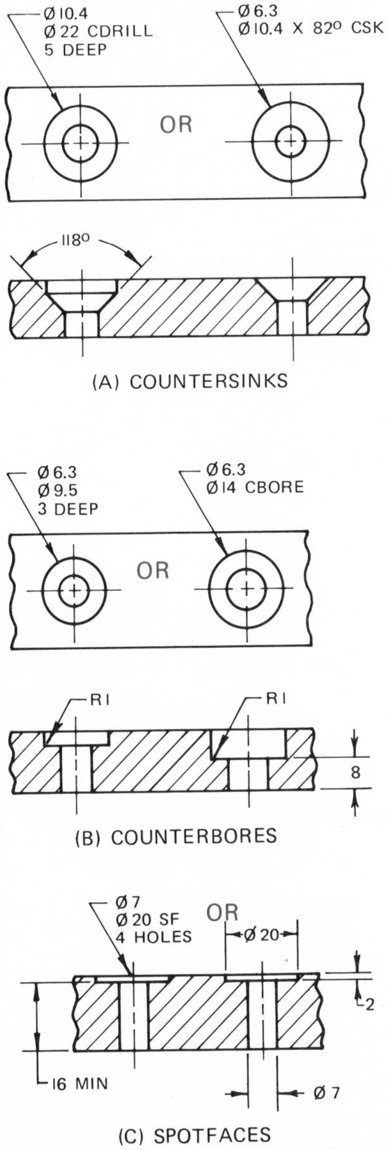

Ø 10.4
Ø 22 CDRILL
5 DEEP
Ø 6.3
Ø 10.4 X 82° CSK
OR
118°
(A) COUNTERSINKS
Ø 6.3
Ø 9.5
3 DEEP
Ø 6.3
Ø 14 CBORE
OR
R 1
R 1
8
(B) COUNTERBORES
Ø 7
Ø 20 SF
4 HOLES
OR
Ø 20
2
16 MIN
Ø 7
(C) SPOTFACES

Fig. 5-2-8 Countersinks, counterbores, and spotfaces.

A counterbore is a flat-bottomed, cylindrical recess which permits the head of a fastening device, such as a bolt, to lie recessed into the part. The diameter, depth, and corner radius are specified in the note. In some cases, the thickness of the remaining stock may be dimensioned rather than the depth of the counterbore.

A spotface is an area where the surface is machined just enough to provide smooth, level seating for a bolthead, nut, or washer.

The diameter of the faced area and either the depth or the remaining thickness are given. A spotface may be specified by a note only, and not delineated on the drawing. If no depth or remaining thickness is specified, it is implied that the spot-facing is the minimum depth necessary to clean up the surface to the specified diameter.

REFERENCES AND SOURCE MATERIALS

1. ANSI Y14.5, *Dimensioning and Tolerancing.*
2. CSA B78.2, *Dimensioning and Tolerancing of Mechanical Engineering Drawings.*

Assignment

Select one of the problems shown in Figs. 5-2-A through 5-2-C. On an A3 or B-size sheet make a one-view drawing, complete with dimensions, of the part. Scale is 1:1.

REVIEW FOR ASSIGNMENT

Unit 2-6 Drafting Skills

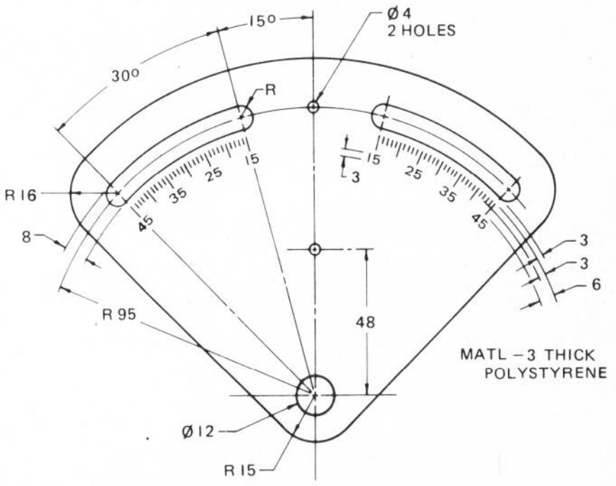

15°
Ø 4
2 HOLES
30°
R
R 16
8
R 95
48
3
3
6
MATL – 3 THICK
POLYSTYRENE
Ø 12
R 15

Fig. 5-2-A Dial indicator.

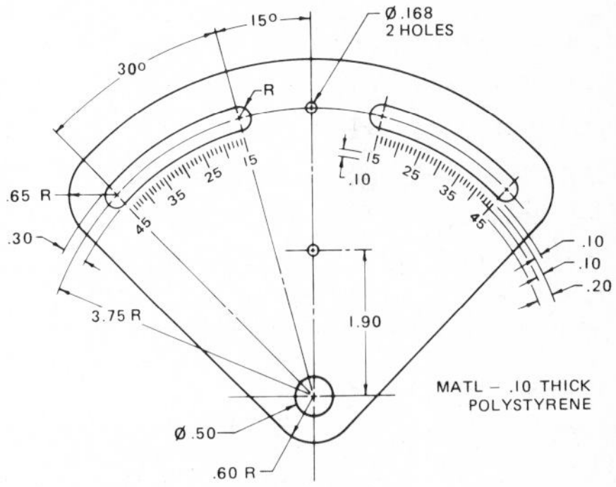

15°
Ø .168
2 HOLES
30°
R
.65 R
.30
3.75 R
1.90
.10
.10
.20
MATL – .10 THICK
POLYSTYRENE
Ø .50
.60 R

Fig. 5-2-B Dial indicator.

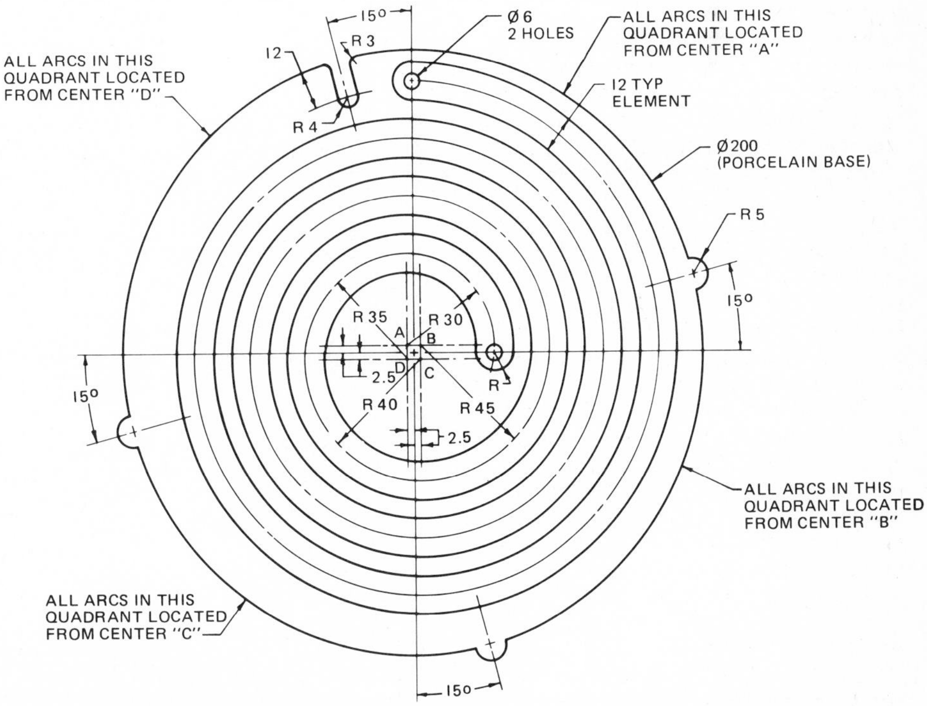

15°
R 3
12
R 4
ALL ARCS IN THIS
QUADRANT LOCATED
FROM CENTER "D"
Ø 6
2 HOLES
ALL ARCS IN THIS
QUADRANT LOCATED
FROM CENTER "A"
12 TYP
ELEMENT
Ø 200
(PORCELAIN BASE)
R 5
15°
R 35
R 30
A
B
D
C
2.5
R 40
R 45
R
2.5
15°
ALL ARCS IN THIS
QUADRANT LOCATED
FROM CENTER "B"
ALL ARCS IN THIS
QUADRANT LOCATED
FROM CENTER "C"
15°

Fig. 5-2-C Element plate.

UNIT 5-3

DIMENSIONING COMMON FEATURES

TAPERS

Circular Tapers. Taper shanks are used on many small tools, such as drills, reamers, counterbores, and spotfaces, to hold them accurately in the machine spindle (Fig. 5-3-1). *Taper* means the difference in diameter or width in a given length. There are many standard tapers; the Morse taper and the Brown and Sharpe taper are the most common.

The following dimensions may be used, in suitable combinations, to define the size and form of tapered features:

- The diameter (or width) at one end of the tapered feature
- The length of the tapered feature
- The rate of taper
- The included angle
- Taper ratio

In dimensioning a taper by means of the taper ratio, the taper symbol —▷ should precede the ratio figures.

Flat Tapers. Flat tapers are used as locking devices such as taper keys and adjusting shims. The methods recommended for dimensioning flat tapers are shown in Fig. 5-3-2.

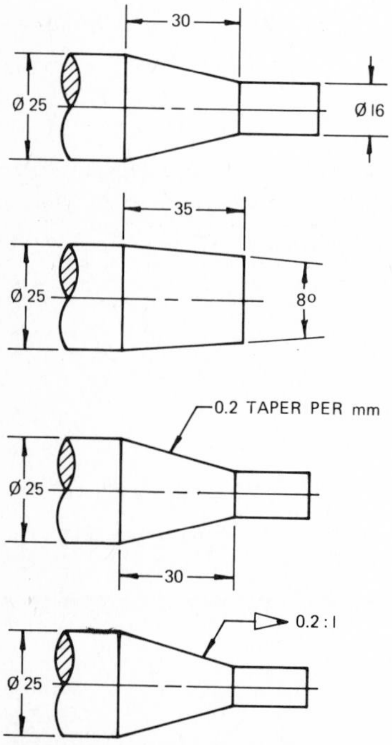
30
Ø 25
Ø 16
35
Ø 25
8°
0.2 TAPER PER mm
Ø 25
30
0.2 : 1
Ø 25

Fig. 5-3-1 Dimensioning circular tapers.

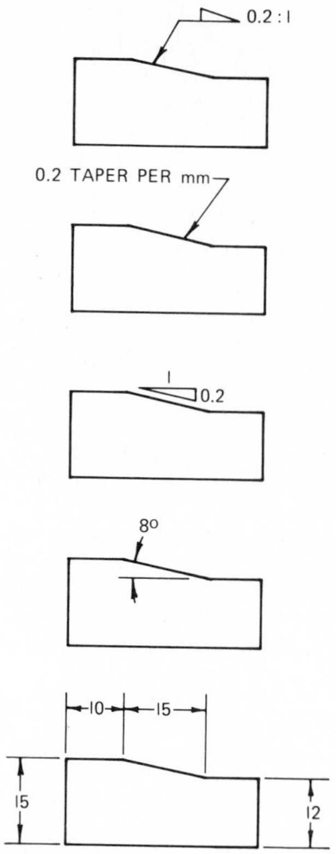
0.2 : 1
0.2 TAPER PER mm
1
0.2
8°
10
15
15
12

Fig. 5-3-2 Dimensioning flat tapers.

CHAMFERS

The process of chamfering, that is, cutting away the inside or outside piece, is done to facilitate assembly. Chamfers are normally dimensioned by giving their angle and length (Fig. 5-3-3). When the chamfer is 45°, it may be specified as a note.

When a very small chamfer is permissible, primarily to break a sharp corner, it may be dimensioned but not drawn, as in Fig. 5-3-3*c*. If not otherwise specified, an angle of 45° is understood.

Internal chamfers may be dimensioned in the same manner, but it is often desirable to give the diameter over the chamfer. The angle may also be given as the included angle if this is a design requirement. This type of dimensioning is generally necessary for larger diameters, especially those over 50 mm (2 in.), whereas chamfers on small holes are usually expressed as countersinks. Chamfers are never measured along the angular surface.

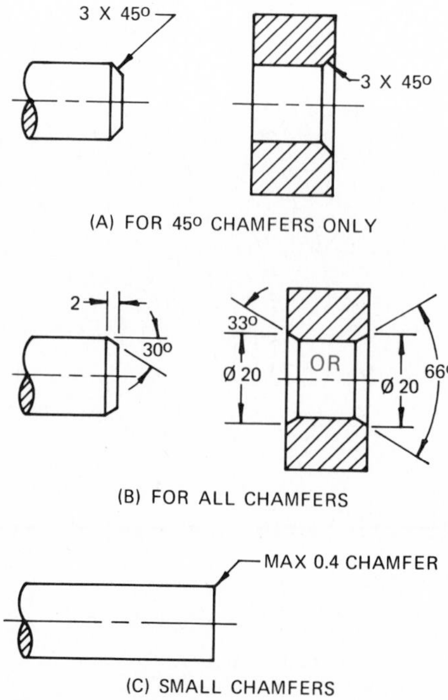
3 X 45°
3 X 45°
(A) FOR 45° CHAMFERS ONLY
2
30°
33°
Ø 20
OR
Ø 20
66°
(B) FOR ALL CHAMFERS
MAX 0.4 CHAMFER
(C) SMALL CHAMFERS

Fig. 5-3-3 Dimensioning chamfers.

KNURLS

Knurling is specified in terms of type, pitch, and diameter before and after knurling (Fig. 5-3-4). The letter P precedes the pitch number. Where control is not required, the diameter after knurling is omitted. Where only portions of a feature require knurling, axial dimensions must be provided. Where required to provide a press fit between parts, knurling is specified by a note on the drawing which includes the type of knurl required, the pitch, the toleranced diameter of the feature prior to knurling, and the minimum acceptable diameter after knurling. Commonly used types are straight, diagonal, spiral, convex, raised diamond, depressed diamond, and radial. The pitch is usually expressed in terms of teeth per millimetre and may be the straight pitch, circular pitch, or diametral pitch. For cylindrical surfaces, the latter is preferred.

The knurling symbol is optional and is used only to improve clarity on working drawings.

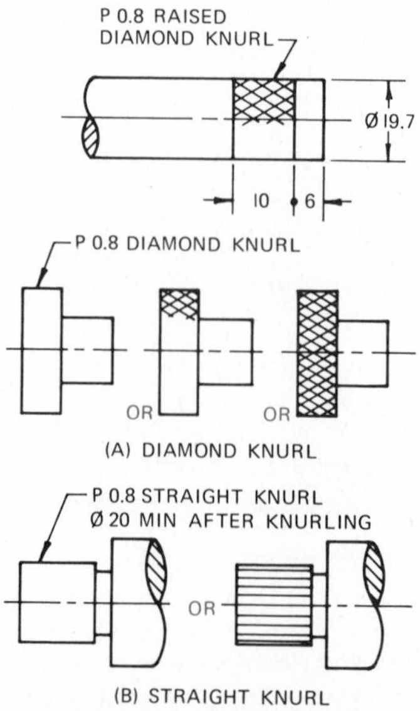
P 0.8 RAISED
DIAMOND KNURL
Ø 19.7
10
6
P 0.8 DIAMOND KNURL
OR
OR
(A) DIAMOND KNURL
P 0.8 STRAIGHT KNURL
Ø 20 MIN AFTER KNURLING
OR
(B) STRAIGHT KNURL

Fig. 5-3-4 Dimensioning knurls.

UNDERCUTS

The operation of undercutting or necking, that is, cutting a recess in a diameter, is done to permit two parts to come together, as illustrated in Fig. 5-3-5*a*. It is indicated on the drawing by a note listing the width first and then the diameter. If the radius is shown at the bottom of the undercut, it will be assumed that the radius is equal to half the width unless otherwise specified, and the diameter will apply to the center of the undercut. When the size of the undercut is unimportant, the dimension may be left off the drawing.

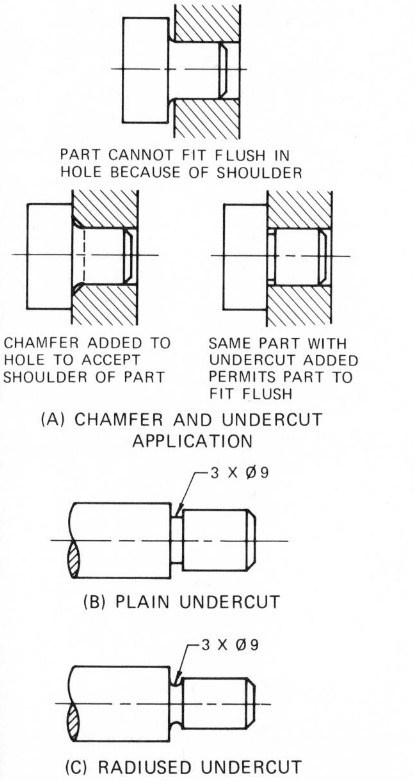
PART CANNOT FIT FLUSH IN HOLE BECAUSE OF SHOULDER
CHAMFER ADDED TO HOLE TO ACCEPT SHOULDER OF PART
SAME PART WITH UNDERCUT ADDED PERMITS PART TO FIT FLUSH
(A) CHAMFER AND UNDERCUT APPLICATION
3 X Ø9
(B) PLAIN UNDERCUT
3 X Ø9
(C) RADIUSED UNDERCUT

Fig. 5-3-5 Dimensioning undercuts.

FORMED PARTS

In dimensioning formed parts, the inside radius is usually specified, rather than the outside radius, but all forming dimensions should be shown on the same side if possible. Dimensions apply to the side on which the dimensions are shown unless otherwise specified (Fig. 5-3-6).

SYMMETRICAL OUTLINES

Symmetrical outlines may be dimensioned on one side of the axis of symmetry only (Fig. 5-3-7). Where only part of the outline is shown, because of functional drafting procedures, the size of the part, or space limitations, symmetrical shapes may be shown by only one-half of the outline, and the symmetry is indicated by a note. In such cases, the outline of the part should extend slightly beyond the center line and terminate with a break line. Note the dimensioning method of extending the dimension lines to act as extension lines for the perpendicular dimensions.

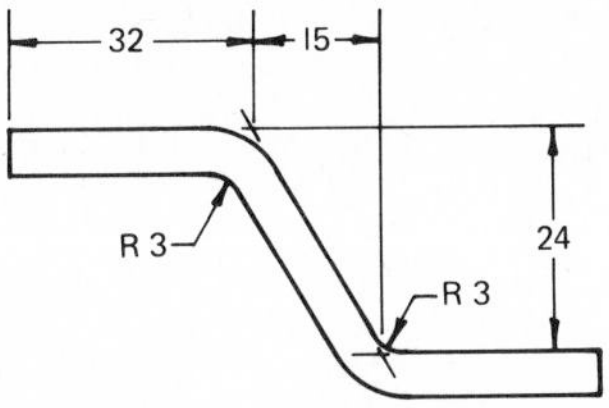
32
15
R 3
24
R 3

Fig. 5-3-6 Dimensioning theoretical points of intersection.

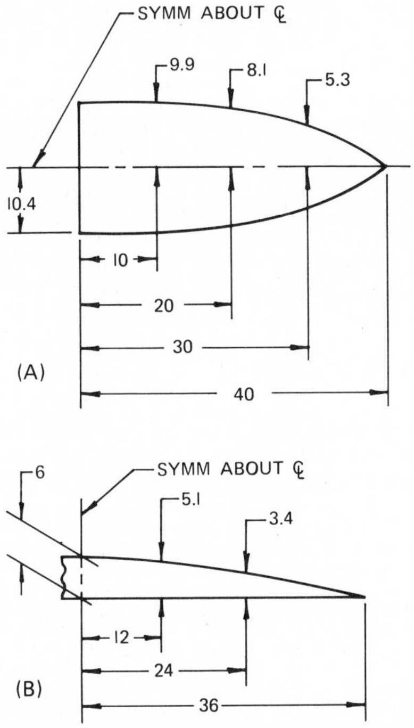
SYMM ABOUT ℄
9.9
8.1
5.3
10.4
10
20
30
40
(A)
6
SYMM ABOUT ℄
5.1
3.4
12
24
36
(B)

Fig. 5-3-7 Dimensioning symmetrical features.

REPETITIVE DETAIL

Where a series of holes or other features are spaced equally, the methods shown in Fig. 5-3-8 may be used.

WIRE, SHEET METAL, AND DRILL ROD

Wire, sheet metal, and drill rod, which are manufactured to gage or code sizes, should be shown by their decimal dimensions; but gage numbers, drill letters, etc., may be shown in parentheses following those dimensions.

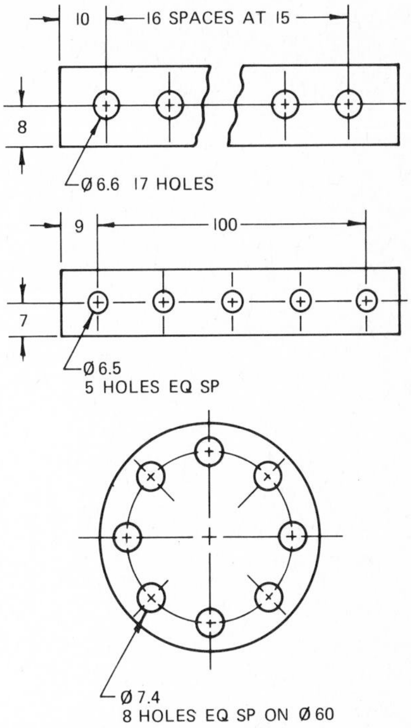
10
16 SPACES AT 15
8
Ø 6.6 17 HOLES
9
100
7
Ø 6.5
5 HOLES EQ SP
Ø 7.4
8 HOLES EQ SP ON Ø 60

Fig. 5-3-8 Dimensioning repetitive details.

Examples:
Sheet —3.57 (No. 10 USS GA)
—2.05 (No. 12 B & S GA)

SOURCE MATERIALS

1. ANSI Y14.5, *Dimensioning and Tolerancing*.
2. CSA B78.2, *Dimensioning and Tolerancing of Mechanical Engineering Drawings*.

Assignment

Select one of the problems shown in Figs. 5-3-A to 5-3-D. On an A3- or B-size sheet, make a working drawing, complete with dimensions, of the part. Scale 1:1.

REVIEW FOR ASSIGNMENT

Unit 2-6 Drafting Skills

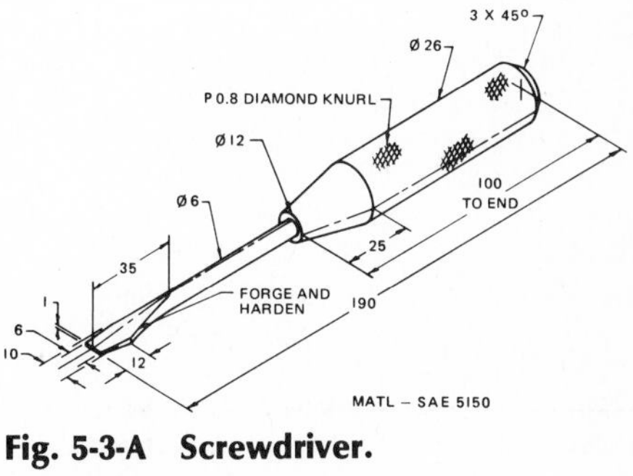
3 X 45°
Ø 26
P 0.8 DIAMOND KNURL
Ø 12
Ø 6
100
TO END
25
35
FORGE AND HARDEN
190
6
10
12
MATL – SAE 5150

Fig. 5-3-A Screwdriver.

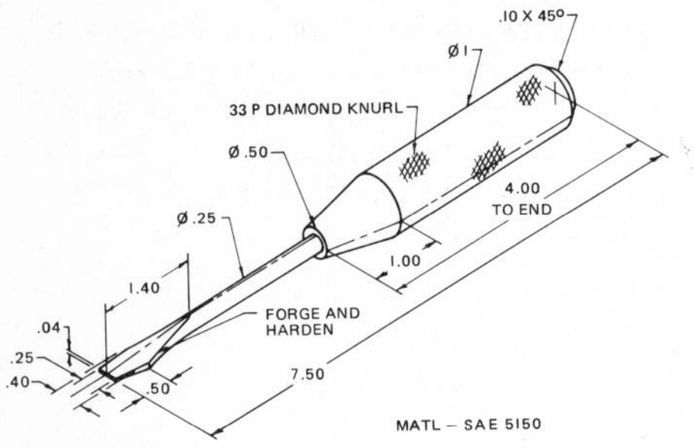

.10 X 45°
Ø1
33 P DIAMOND KNURL
Ø.50
Ø.25
4.00
TO END
1.00
1.40
.04
.25
.40
FORGE AND
HARDEN
.50
7.50
MATL – SAE 5150

Fig. 5-3-B Screwdriver.

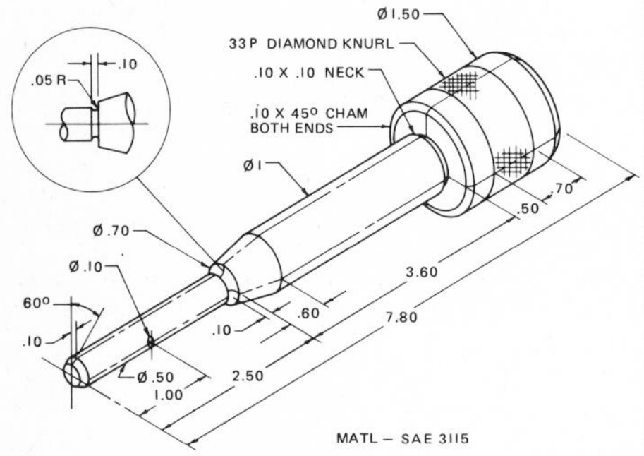

Ø1.50
33 P DIAMOND KNURL
.10 X .10 NECK
.10 X 45° CHAM
BOTH ENDS
Ø1
.05 R
.10
Ø.70
Ø.10
60°
.10
.70
.50
3.60
.60
7.80
.10
Ø.50
1.00
2.50
MATL – SAE 3115

Fig. 5-3-C Indicator rod.

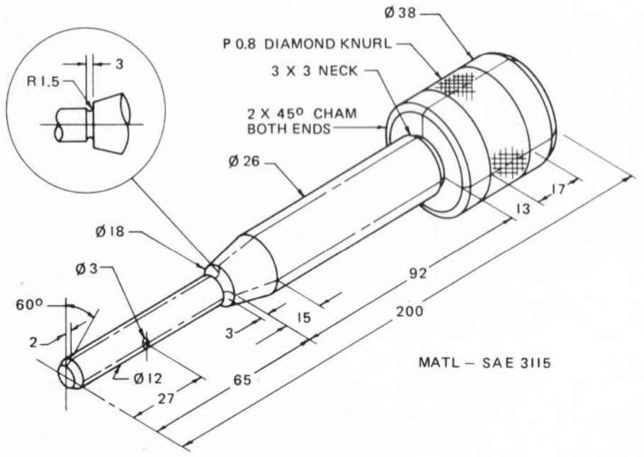

Ø38
P 0.8 DIAMOND KNURL
3 X 3 NECK
2 X 45° CHAM
BOTH ENDS
Ø26
R1.5
3
Ø18
Ø3
60°
2
17
13
92
200
15
3
Ø12
27
65
MATL – SAE 3115

Fig. 5-3-D Indicator rod.

UNIT 5-4

DIMENSIONING METHODS[1,2]

The choice of the most suitable dimensions and dimensioning methods will depend, to some extent, on how the part will be produced and whether the drawings are intended for unit or mass production.

Unit production refers to cases where each part is to be made separately, using general-purpose tools and machines.

Mass production refers to parts produced in quantity, where special tools and gages are usually provided.

Either linear or angular dimensions may locate features with respect to one another (point-to-point) or from a datum. Point-to-point dimensions may be adequate for describing simple parts. Dimensions from a datum may be necessary if a part with more than one critical dimension must mate with another part.

The following systems of dimensioning are used more commonly for engineering drawings.

THE RECTANGULAR COORDINATE DIMENSIONING SYSTEM

This is a method for indicating distance, location, and size by means of linear dimensions measured parallel or perpendicular to reference axes or datum planes that are perpendicular to one another.

Coordinate dimensioning with dimension lines must clearly identify the datum features from which the dimensions originate (Fig. 5-4-1).

TRUE POSITION DIMENSIONING

True position dimensioning (Fig. 5-4-2) has many advantages over the coordinate dimensioning system. Because of its newness to engineering drawings and its scope, it is covered as a complete topic in Chap. 31.

POLAR COORDINATE DIMENSIONING

Polar coordinate dimensioning is commonly used in circular planes or circular configurations of features. It is a method of indicating the position of a point, line, or surface by means of a linear dimension and an angle, other than 90°, that is implied by the vertical and horizontal center lines (Fig. 5-4-3).

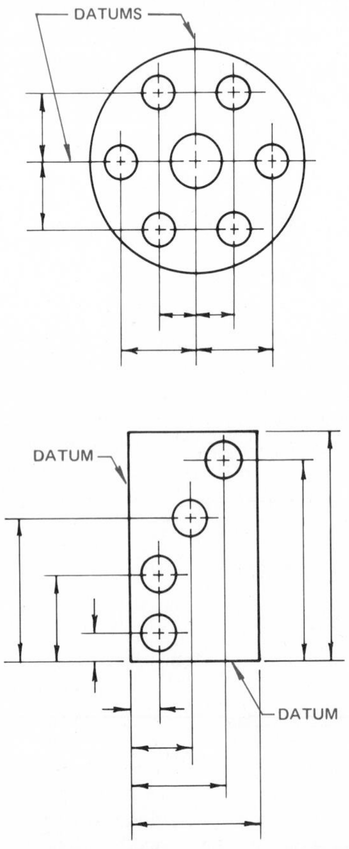

DATUMS
DATUM
DATUM

Fig. 5-4-1 Rectangular coordinate dimensioning.

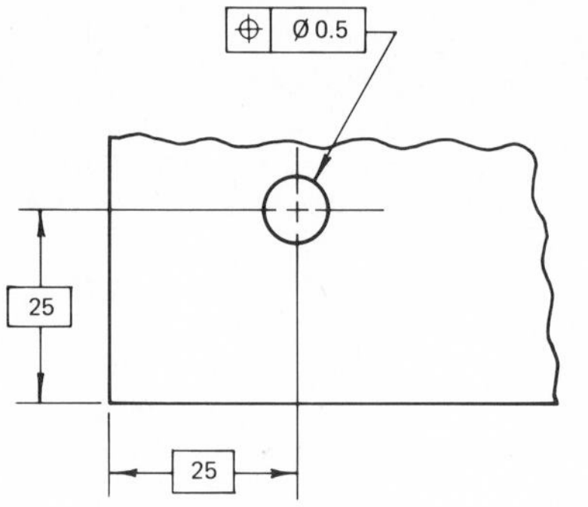

Ø0.5
25
25

Fig. 5-4-2 True position dimensioning.

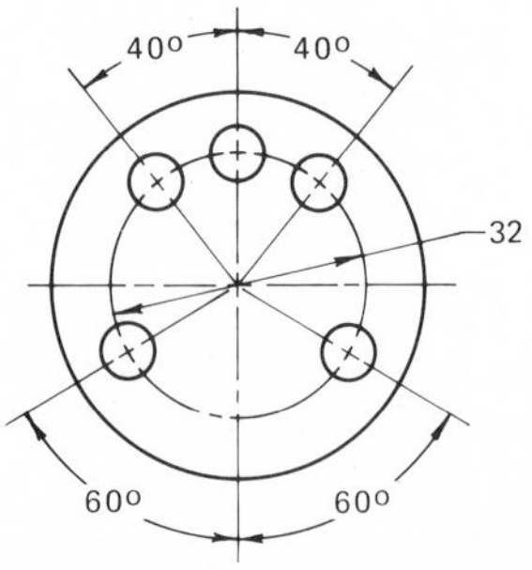

40°
40°
32
60°
60°

Fig. 5-4-3 Polar coordinate dimensioning.

CHORDAL DIMENSIONING SYSTEM

The chordal dimensioning system may also be used for the spacing of points on the circumference of a circle relative to a datum, where manufacturing methods indicate that this will be more convenient (Fig. 5-4-4).

ARROWLESS DIMENSIONING

Arrowless dimensioning is coordinate dimensioning without dimension lines (Fig. 5-4-5). The dimensions originate

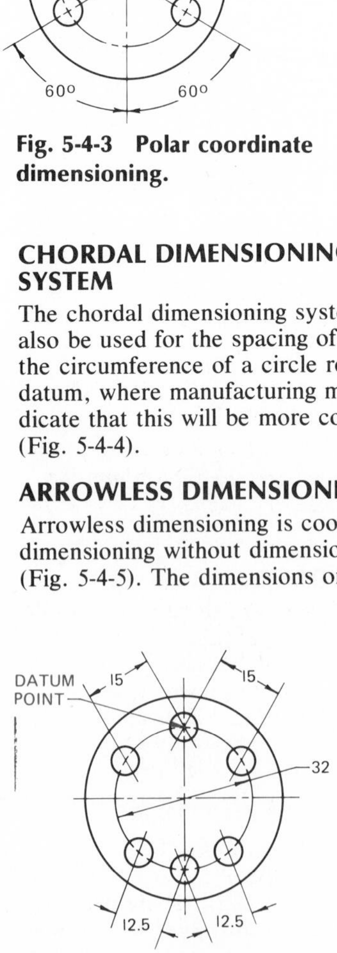

DATUM
POINT
15
15
32
12.5
12.5

Fig. 5-4-4 Chordal dimensioning.

HOLE SYMBOL	HOLE SIZE
A	6.4
B	4.8
C	4
D	3.2

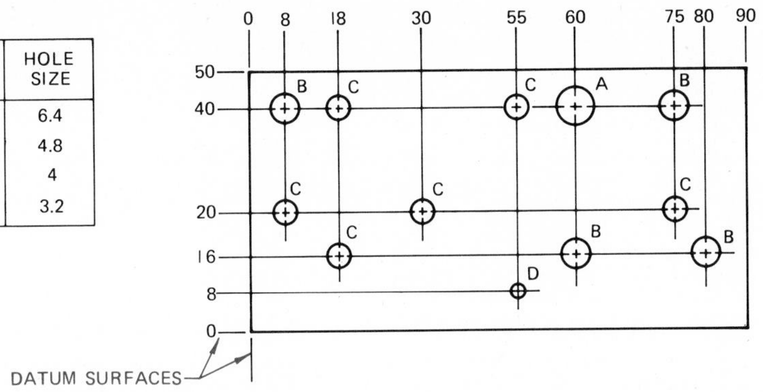
DATUM SURFACES

Fig. 5-4-5 Arrowless dimensioning.

from datum planes which are indicated by zero coordinates. Dimensions from them are shown on extension lines. There should never be more than one zero line in each direction. For further details, refer to Chap. 16.

TABULAR DIMENSIONING

Tabular dimensioning is a type of rectangular coordinate dimensioning in which dimensions from mutually perpendicular datum planes are listed in a table on the drawing rather than on the pictorial delineation (Fig. 5-4-6). Tables may be prepared in any suitable manner which will adequately locate the features. For further details, refer to Chaps. 16 and 17.

CHAIN DIMENSIONING

When a series of dimensions is applied on a point-to-point basis, it is called *chain dimensioning* (Fig. 5-4-7). A possible disadvantage of this system is that it may result in an undesirable accumulation of tolerances between individual features (see Unit 5-5).

DATUM OR COMMON-POINT DIMENSIONING

When several dimensions emanate from a common reference point or line, the method is called *common-point* or *datum dimensioning*.

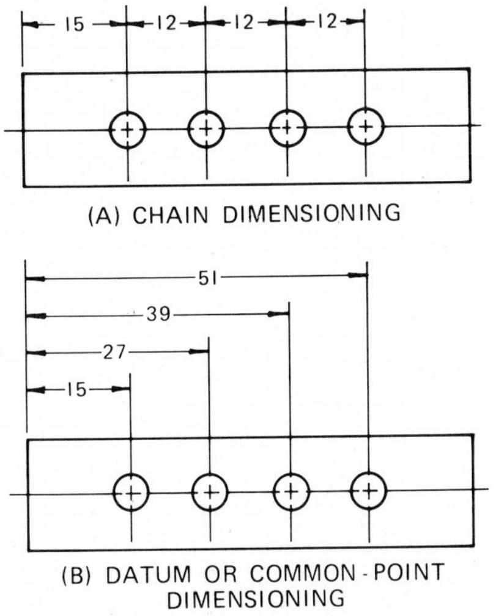
(A) CHAIN DIMENSIONING
(B) DATUM OR COMMON-POINT DIMENSIONING

Fig. 5-4-7 A comparison between chain and datum dimensioning.

REFERENCES

1. ANSI, Y14.5, *Dimensioning and Tolerancing*.
2. CSA B78.2, *Dimensioning and Tolerancing of Mechanical Engineering Drawings*.

Assignment

Select one of the problems shown in Figs. 5-4-A to 5-4-D, and on an A3- or B-size sheet make a working drawing of the part. The arrowless dimensioning shown is to be replaced with rectangular coordinate dimensioning and has the following dimensioning changes. For Figs. 5-4-A and 5-4-B,

- Holes A, E, and D are located from the zero coordinates.
- Hole B is located from center of hole E.
- Hole C is located from center of hole D.

For Figs. 5-4-C and 5-4-D,

- Holes E and D are located from left and bottom edges.
- Holes A and C are located from center of hole D.
- Hole B is located from center of hole E.

For the sake of clarity, some dimensions may best be shown on the part. Scale for drawings is 1:1.

REVIEW FOR ASSIGNMENT

Unit 2-6 Drafting Skills

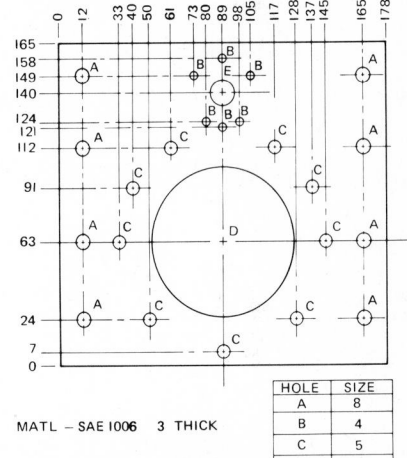

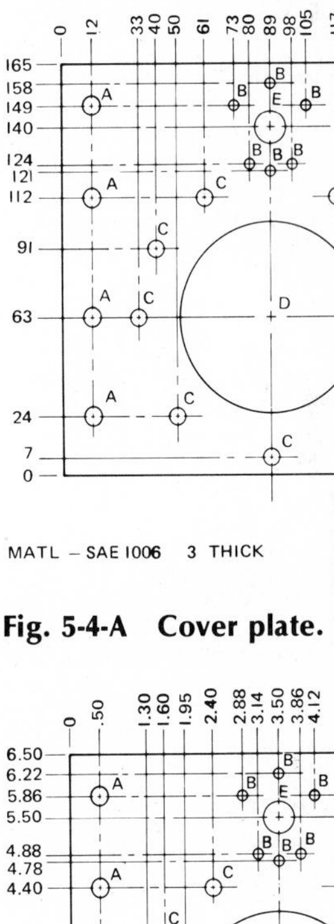
MATL – SAE 1006 3 THICK

HOLE	SIZE
A	8
B	4
C	5
D	76
E	12

Fig. 5-4-A Cover plate.

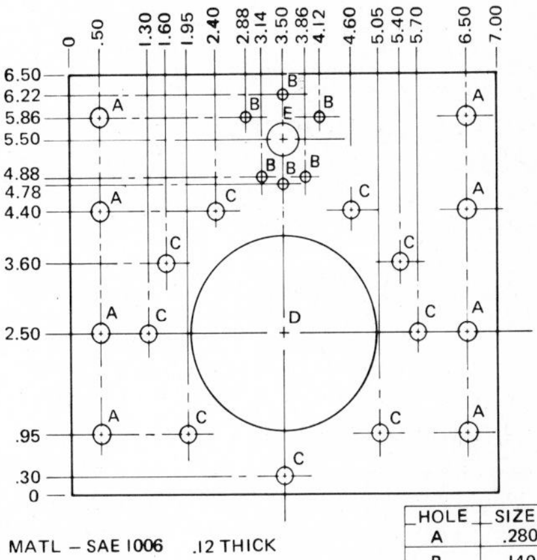
MATL – SAE 1006 .12 THICK

HOLE	SIZE
A	.280
B	.140
C	.200
D	3.00
E	.50

Fig. 5-4-B Cover plate.

HOLE SYMBOL	HOLE DIA	LOCATION X →	Y ↑
A_1	5.6	60	40
B_1	4.8	10	40
B_2	4.8	75	40
B_3	4.8	60	16
B_4	4.8	80	16
C_1	4	18	40
C_2	4	55	40
C_3	4	10	20
C_4	4	30	20
C_5	4	75	20
C_6	4	18	16
D_1	3.2	55	8

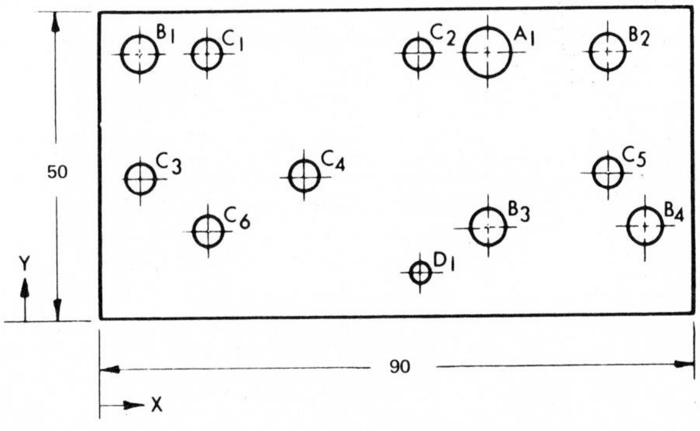

Fig. 5-4-6 Tabular dimensioning.

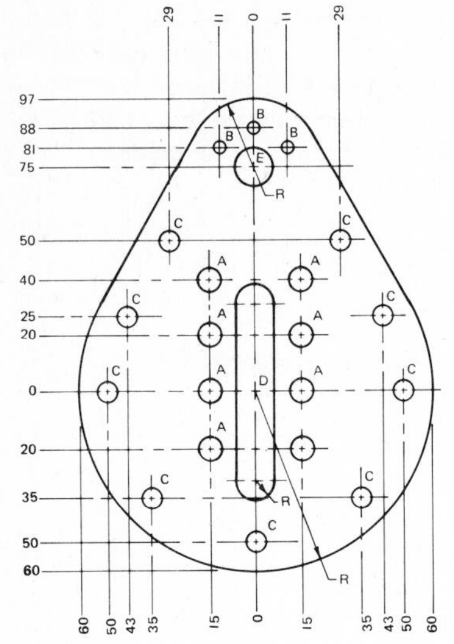

MATL – SAE 1008 3 THICK

HOLE	SIZE
A	8
B	4
C	6
D	10 X 70
E	12

Fig. 5-4-C Transmission cover.

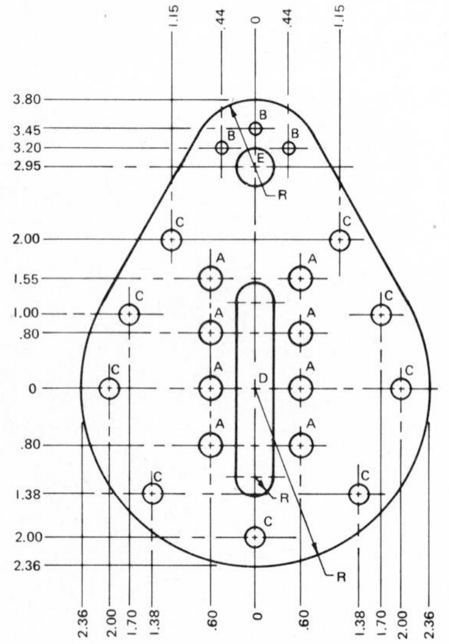

MATL – SAE 1008 .12 THICK

HOLE	SIZE
A	.30
B	.16
C	.24
D	.40 X 2.75
E	.50

Fig. 5-4-D Transmission cover.

UNIT 5-5
LIMITS AND TOLERANCES[1,2]

In the 6000 years of the history of engineering drawing as a means for the communication of engineering information, it seems inconceivable that such an elementary practice as the tolerancing of dimensions, which we take so much for granted today, was introduced for the first time about 60 years ago.

Apparently, engineers and workers came to the realization in a very gradual manner that exact dimensions and shapes could not be attained in the shaping of physical objects, such as the manufacture of materials and products.

The skilled tradespersons of old prided themselves on being able to work to exact dimensions. What they really meant was that they dimensioned objects with a degree of accuracy greater than that with which they could measure. The use of modern measuring instruments would have shown the deviations from the sizes which they called exact.

As soon as it was realized that variations in the sizes of parts had always been present, that such variations could be restricted but not avoided, and also that slight variations in the size which a part was originally intended to have could be tolerated without its correct functioning being impaired, it was evident that interchangeable parts need not be identical parts, but that it would be sufficient if the significant sizes which controlled their fits lay between definite limits. Accordingly, the problem of interchangeable manufacture evolved from the making of parts to a would-be exact size, to the holding of parts between two limiting sizes lying so closely together that any intermediate size would be acceptable.

Tolerances are the permissible variations in the specified form, size, or location of individual features of a part from that shown on the drawing. The finished form and size into which material is to be fabricated are defined on a drawing by various geometric shapes and dimensions.

As mentioned previously, the worker cannot be expected to produce the exact size of parts as indicated by the dimensions on a drawing; so a certain amount of variation on each dimension must be tolerated. For example, a dimension given as 40 ± 0.1 mm means that the manufactured part can be anywhere between 39.9 mm and 40.1 mm and that the tolerance permitted on this dimension is 0.2 mm. The largest and smallest permissible sizes (40.1 and 39.9 mm, respectively) are known as the *limits*.

Greater accuracy costs more money; and since economy in manufacturing would not permit all dimensions to be held to the same accuracy, a system for dimensioning must be used (Fig. 5-5-1). Generally, most parts require only a few dimensions to be accurate.

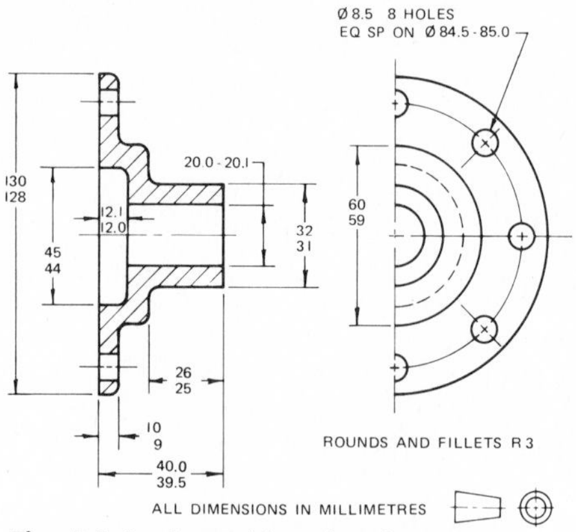

Ø8.5 8 HOLES
EQ SP ON Ø84.5-85.0
ROUNDS AND FILLETS R3
ALL DIMENSIONS IN MILLIMETRES

Fig. 5-5-1 A working drawing.

In order that assembled parts may function properly and to allow for interchangeable manufacturing, it is necessary to permit only a certain amount of tolerance on each of the mating parts and a certain amount of allowance between them.

In order to calculate limit dimensions, the following definitions should be clearly understood (refer to Fig. 5-5-2).

Basic Size. The *basic size* of a dimension is the theoretical size from which the limits for that dimension are derived, by the application of the allowance and tolerance.

Limits of Size. These limits are the maximum and minimum sizes permissible for a specific dimension.

Tolerance. The *tolerance* on a dimension is the total permissible variation in its size. The tolerance is the difference between the limits of size.

Allowance. An *allowance* is an intentional difference in correlated dimensions of mating parts. It is the minimum clear-

BASIC SIZE	40
BASIC SIZE WITH TOLERANCE ADDED	40 ±0.1
HALF OF TOTAL TOLERANCE	
LIMITS– LARGEST AND SMALLEST SIZES PERMITTED	40.1 39.9
TOLERANCE – DIFFERENCE BETWEEN MIN AND MAX LIMITS	0.2

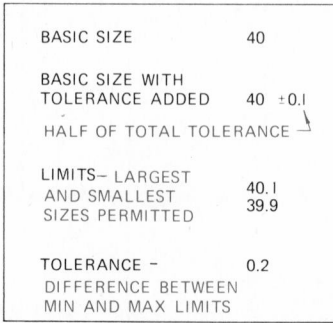

Fig. 5-5-2 Limit and tolerance terminology.

ance (positive allowance) or maximum interference (negative allowance) between such parts.

Maximum Material Size. The *maximum material size* is that limit of size of a feature which results in the part containing the maximum amount of material. Thus it is the maximum limit of size for a shaft or an external feature, or the minimum limit of size for a hole or internal feature.

Tolerancing[3,4]

All dimensions required in the manufacture of a product have a tolerance, except those identified as reference, basic, and gage. Tolerances may be expressed in one of the following ways:

- As specified limits of tolerances shown directly on the drawing for a specified dimension

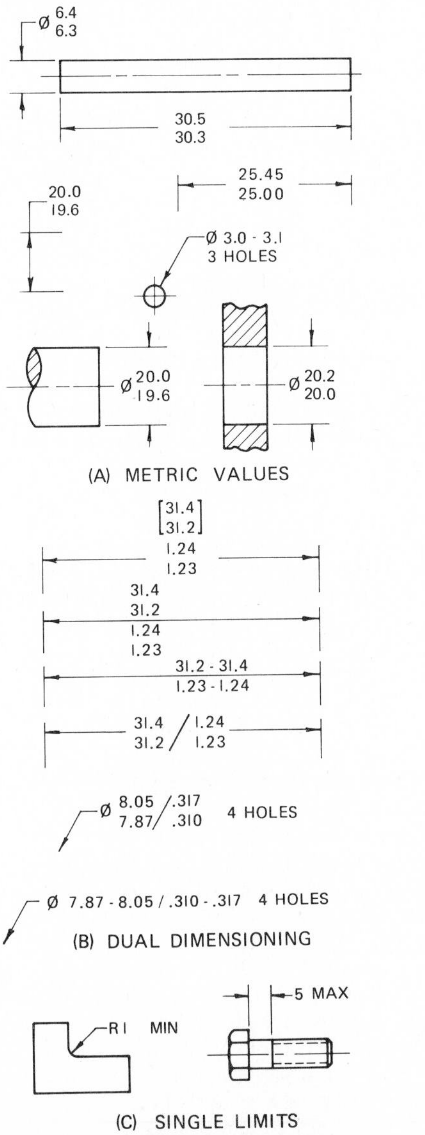
Ø 6.4
6.3
30.5
30.3
20.0
19.6
25.45
25.00
Ø 3.0 - 3.1
3 HOLES
Ø 20.0
19.6
Ø 20.2
20.0
(A) METRIC VALUES
[31.4]
[31.2]
1.24
1.23
31.4
31.2
1.24
1.23
31.2 - 31.4
1.23 - 1.24
31.4 / 1.24
31.2 / 1.23
Ø 8.05 / .317
7.87 / .310
4 HOLES
Ø 7.87 - 8.05 / .310 - .317 4 HOLES
(B) DUAL DIMENSIONING
R1 MIN
5 MAX
(C) SINGLE LIMITS

Fig. 5-5-3 Limit dimensioning.

- In a general tolerance note, referring to all dimensions on the drawing for which tolerances are not otherwise specified
- In the form of a special note referring to specific dimensions

Tolerances on dimensions that locate features may be applied directly to the locating dimensions or by the positional tolerancing method described in Chap. 31. Tolerancing applicable to the control of form and runout, referred to as *geometric tolerancing*, is also covered in detail in Chap. 31.

TOLERANCING METHODS

A tolerance applied directly to a dimension may be expressed in two ways.

Limit Dimensioning. For this method, the high limit (maximum value) is placed above the low limit (minimum value). When it is expressed in a single line, the low limit should precede the high limit and they should be separated by a dash, as shown in Figs. 5-5-3 and 5-5-4.

Where limit dimensions are used and where either the maximum or minimum dimension has digits to the right of the decimal point, the other value should have the zeros added so that both the limits of size are expressed to the same number of decimal places.

Plus and Minus Tolerancing. (Refer to Fig. 5-5-5.) For this method the dimension of the specified size is given first

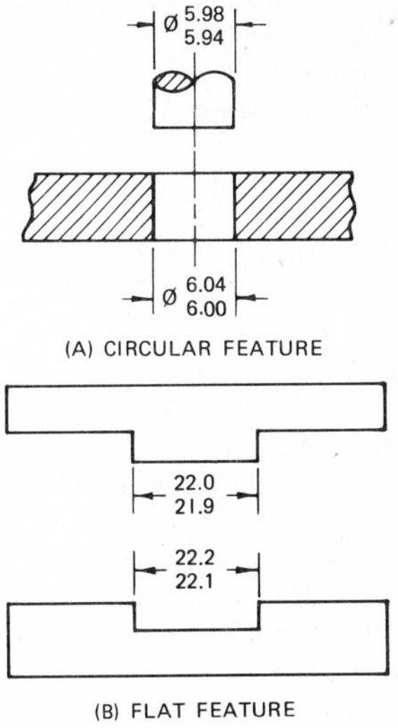
Ø 5.98
5.94
Ø 6.04
6.00
(A) CIRCULAR FEATURE
22.0
21.9
22.2
22.1
(B) FLAT FEATURE

Fig. 5-5-4 Limit dimensioning application.

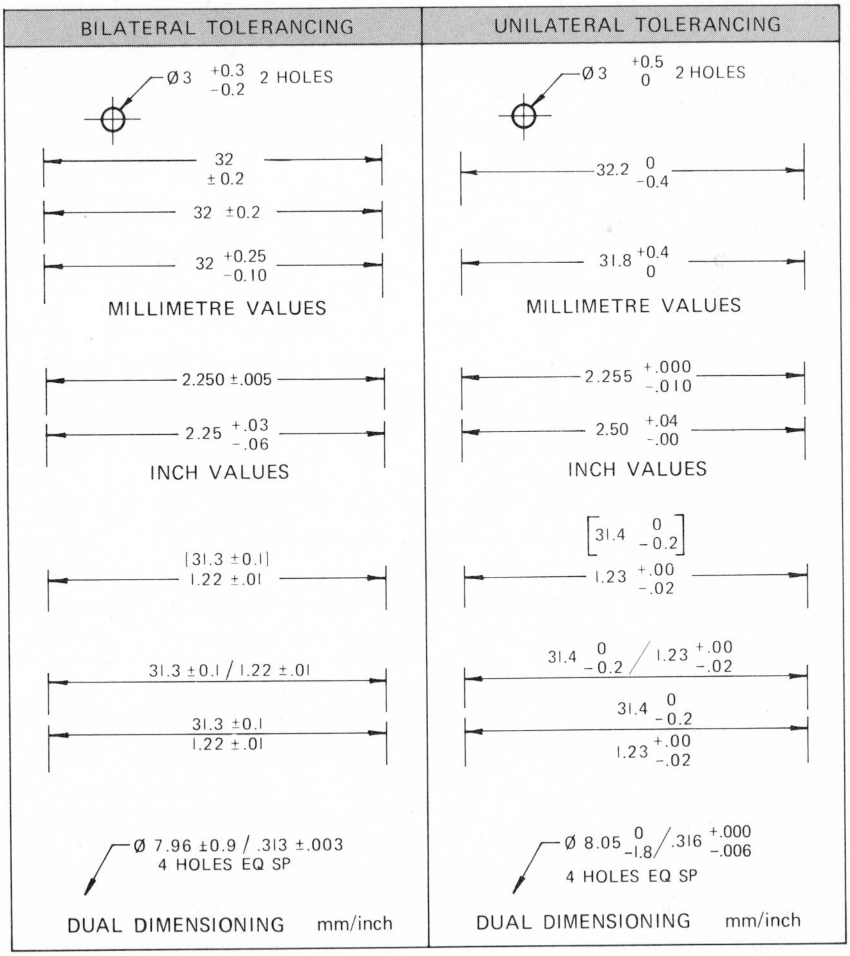
BILATERAL TOLERANCING
UNILATERAL TOLERANCING
Ø 3 +0.3 -0.2 2 HOLES
Ø 3 +0.5 0 2 HOLES
32 ± 0.2
32 ±0.2
32 +0.25 -0.10
32.2 0 -0.4
31.8 +0.4 0
MILLIMETRE VALUES
MILLIMETRE VALUES
2.250 ±.005
2.25 +.03 -.06
2.255 +.000 -.010
2.50 +.04 -.00
INCH VALUES
INCH VALUES
[31.3 ±0.1]
1.22 ±.01
[31.4 0 -0.2]
1.23 +.00 -.02
31.3 ±0.1 / 1.22 ±.01
31.4 0 -0.2 / 1.23 +.00 -.02
31.3 ±0.1
1.22 ±.01
31.4 0 -0.2
1.23 +.00 -.02
Ø 7.96 ±0.9 / .313 ±.003
4 HOLES EQ SP
Ø 8.05 0 -1.8 / .316 +.000 -.006
4 HOLES EQ SP
DUAL DIMENSIONING mm/inch
DUAL DIMENSIONING mm/inch

Fig. 5-5-5 Plus and minus tolerancing.

and is followed by a plus or minus expression of tolerancing. The plus value should be placed above the minus value. This type of tolerancing can be broken down into bilateral and unilateral tolerancing. In a bilateral tolerance, the plus and minus tolerances should generally be equal, but special design considerations may sometimes dictate unequal values (Fig. 5-5-6). The specified size is the design size, and the tolerance represents the desired control of quality and appearance. The dimension need not be shown to the same number of decimal places as its tolerance. For example:

1.5± 0.04 *not* 1.50± 0.04
10± 0.1 not 10.0± 0.1

The unilateral tolerance is generally used to establish the position of a feature, as shown in Fig. 5-5-7, whenever the ideal position of a feature is midway in the allowable tolerance range.

Where dual tolerancing is required for plus or minus tolerancing, the preferred methods are to have the millimetre dimension placed above or to the left of the inch dimensions. When a leader is used, the preferred method is to place

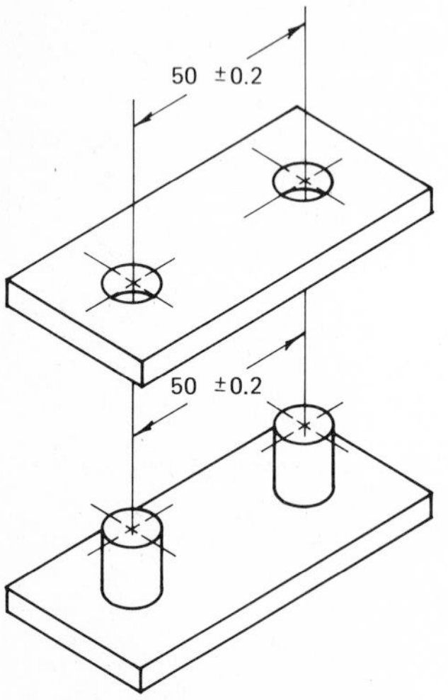
50 ±0.2
50 ±0.2

EQUAL BILATERAL TOLERANCES

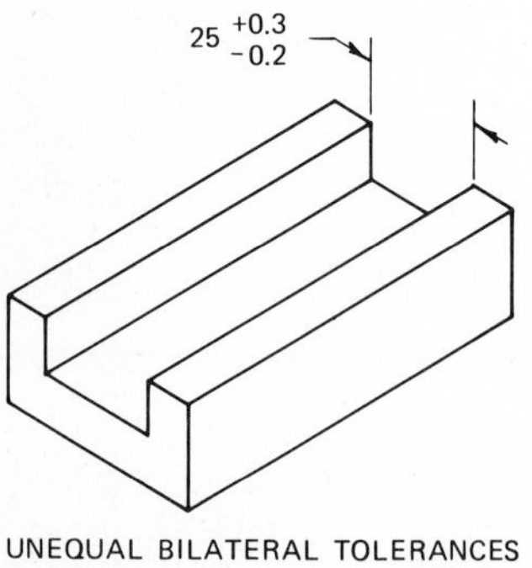
25 +0.3 −0.2

UNEQUAL BILATERAL TOLERANCES

Fig. 5-5-6 Application of bilateral tolerances.

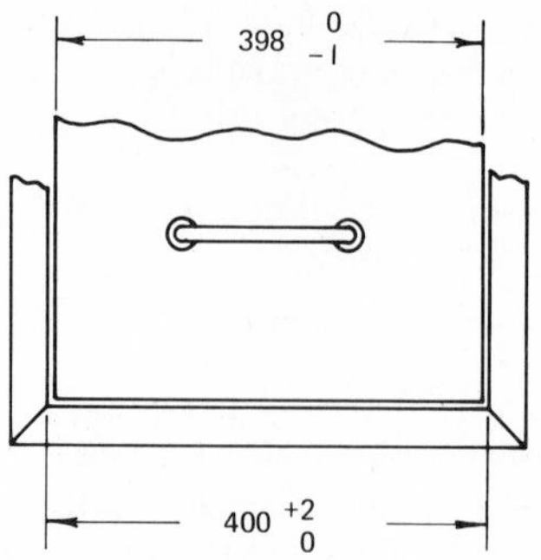
398 0 −1
400 +2 0

Fig. 5-5-7 Application of unilateral tolerances.

the metric value to the left of the inch dimension. A slash line is used to separate the values. Conversion charts for tolerances are shown in Fig. 5-5-8.

TOTAL TOLERANCE IN INCHES		MILLIMETRE CONVERSION ROUNDED TO
AT LEAST	LESS THAN	
.00004	.0004	4 DECIMAL PLACES
.0004	.004	3 DECIMAL PLACES
.004	.04	2 DECIMAL PLACES
.04	.4	1 DECIMAL PLACE
.4 AND OVER		WHOLE mm

TOTAL TOLERANCE IN MILLIMETRES		INCH CONVERSION ROUNDED TO
AT LEAST	LESS THAN	
0.002	0.02	5 DECIMAL PLACES
0.02	0.2	4 DECIMAL PLACES
0.2	2	3 DECIMAL PLACES
2 AND OVER		2 DECIMAL PLACES

Fig. 5-5-8 Conversion charts for tolerances.

Where a hole location is more critical than the location of a radius from the same center, the hole and radius are dimensioned and toleranced separately, as shown in Fig. 5-5-9.

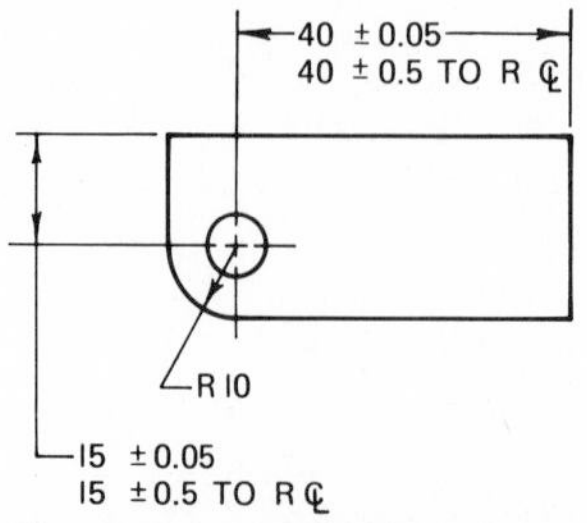
40 ±0.05
40 ±0.5 TO R ℄
R 10
15 ±0.05
15 ±0.5 TO R ℄

Fig. 5-5-9 Multiple tolerance.

General Tolerance Notes. The use of general tolerance notes greatly simplifies the drawing and saves much layout in its preparation. The following examples illustrate the wide field of application of this system. The values given in the examples are typical.

Example 1. EXCEPT WHERE STATED OTHERWISE, TOLERANCES ON FINISHED DECIMAL DIMENSIONS ±0.1.

Example 2. EXCEPT WHERE STATED OTHERWISE, TOLERANCES ON FINISHED DIMENSIONS TO BE AS FOLLOWS:

Dimension (mm)	Tolerance
Up to 100	± 0.1
From 101 to 300	± 0.2
From 301 to 600	± 0.5
Over 600	± 1

A comparison between the tolerancing methods described is shown in Fig. 5-5-10.

TOLERANCE GRADES

Standard tolerances in millimetres and inches are given in the Appendix. The millimetre and inch values are intended to be similar, but in all cases they have been rounded off to provide values more practical for normal use.

Grades 1 to 4 are very precise grades intended primarily for gage making and similar precision work, although grade 4 can also be used for very precise production work.

Grades 5 to 16 represent a progressive series suitable for cutting operations, such as turning, boring, grinding, milling, and sawing. Grade 5 is the most precise grade, obtainable by fine grinding and lapping, while 16 is the coarsest grade for rough sawing and machining.

Grades 110 to 117 are intended for manufacturing operations such as cold heading, pressing, rolling, and other forming operations. Grades 110 to 116 are identical to grades 10 to 16 in the smallest size range, but are based on a formula which provides for a more rapid increase in relation to size than that used for grades 10 to 16. Grade 117 is a coarser grade, suitable for very rough work.

Grades 210 to 217 are similar to grades 110 to 117, but provide a further increase of tolerance in relation to size, to be more consistent with operations where temperature change and shrinkage occur, since such changes are almost in direct proportion to size. These tolerances are therefore suitable for processes such as hot forging, casting, and molding.

As a guide to the selection of tolerances, Fig. 5-5-11 has been prepared to show grades which may be expected to be held by various manufacturing processes for work in metals. For work in other materials, such as plastics, it may be necessary to use coarser tolerance grades for the same process.

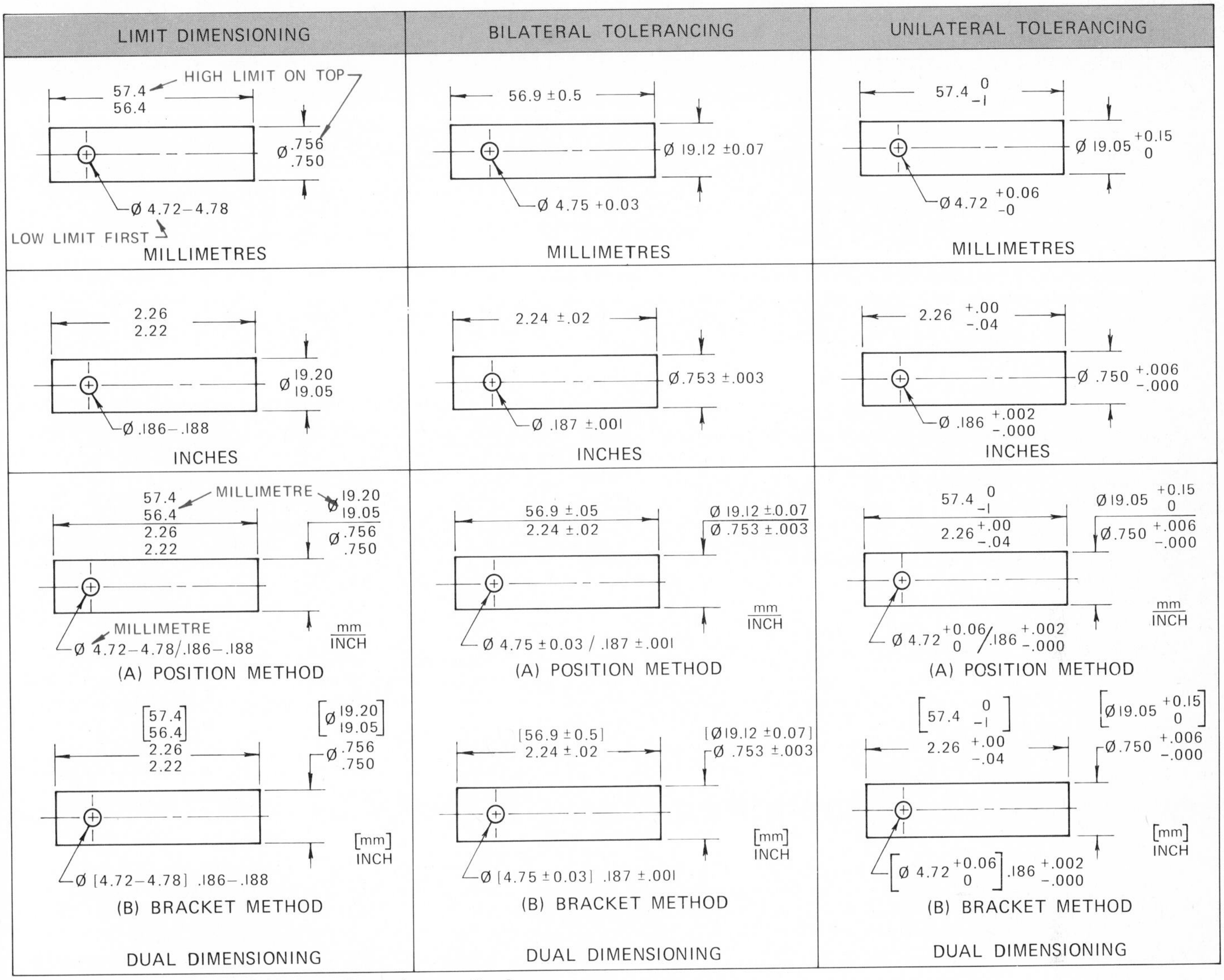
LIMIT DIMENSIONING
BILATERAL TOLERANCING
UNILATERAL TOLERANCING
HIGH LIMIT ON TOP
57.4
56.4
Ø .756
.750
Ø 4.72–4.78
LOW LIMIT FIRST
MILLIMETRES
56.9 ±0.5
Ø 19.12 ±0.07
Ø 4.75 +0.03
MILLIMETRES
57.4 0 -1
Ø 19.05 +0.15 0
Ø 4.72 +0.06 -0
MILLIMETRES
2.26
2.22
Ø 19.20
19.05
Ø .186–.188
INCHES
2.24 ±.02
Ø .753 ±.003
Ø .187 ±.001
INCHES
2.26 +.00 -.04
Ø .750 +.006 -.000
Ø .186 +.002 -.000
INCHES
57.4
56.4
2.26
2.22
MILLIMETRE
Ø 19.20
19.05
Ø .756
.750
MILLIMETRE
Ø 4.72–4.78/.186–.188
mm
INCH
(A) POSITION METHOD
56.9 ±.05
2.24 ±.02
Ø 19.12 ±0.07
Ø .753 ±.003
Ø 4.75 ±0.03 / .187 ±.001
mm
INCH
(A) POSITION METHOD
57.4 0 -1
2.26 +.00 -.04
Ø 19.05 +0.15 0
Ø .750 +.006 -.000
Ø 4.72 +0.06 0 / .186 +.002 -.000
mm
INCH
(A) POSITION METHOD
[57.4
56.4]
2.26
2.22
[Ø 19.20
19.05]
Ø .756
.750
[mm]
INCH
Ø [4.72–4.78] .186–.188
(B) BRACKET METHOD
[56.9 ±0.5]
2.24 ±.02
[Ø19.12 ±0.07]
Ø .753 ±.003
[mm]
INCH
Ø [4.75 ±0.03] .187 ±.001
(B) BRACKET METHOD
[57.4 0 -1]
2.26 +.00 -.04
[Ø19.05 +0.15 0]
Ø .750 +.006 -.000
[mm]
INCH
[Ø 4.72 +0.06 0] .186 +.002 -.000
(B) BRACKET METHOD
DUAL DIMENSIONING
DUAL DIMENSIONING
DUAL DIMENSIONING

Fig. 5-5-10 A comparison of the tolerancing methods.

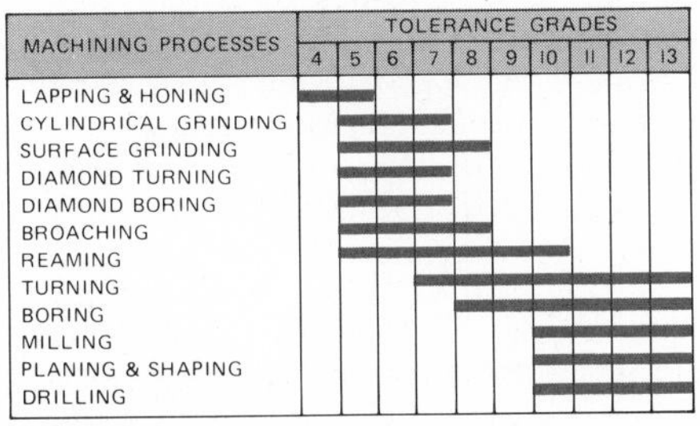
MACHINING PROCESSES
TOLERANCE GRADES
4 5 6 7 8 9 10 11 12 13
LAPPING & HONING
CYLINDRICAL GRINDING
SURFACE GRINDING
DIAMOND TURNING
DIAMOND BORING
BROACHING
REAMING
TURNING
BORING
MILLING
PLANING & SHAPING
DRILLING

Fig. 5-5-11 Tolerancing grades for machining processes.

TOLERANCE ACCUMULATION

It is necessary also to consider the effect of each tolerance with respect to other tolerances, and not to permit a chain of tolerances to build up a cumulative tolerance between surfaces or points that have an important relation to one another. Where the position of a surface in any one direction is controlled by more than one tolerance, then the tolerances are cumulative. Figure 5-5-12 compares the tolerance accumulation resulting from three different methods of dimensioning.

Chain Dimensioning. The maximum variation between any two features is equal to the sum of the tolerances on the intermediate distances. This results in the greatest tolerance accumulation, as illustrated by the ± 2 variation between holes X and Y, as shown in Fig. 5-5-12.

Datum Dimensioning. The maximum variation between any two features is equal to the sum of the tolerances on the two dimensions from the datum to the feature. This reduces the tolerance accumulation, as illustrated by the ± 1 variation between holes X and Y.

Direct Dimensioning. The maximum variation between any two features is controlled by the tolerance on the dimension between the features. This results in the least tolerance accumulation, as illustrated by the ± 0.5 variation between holes X and Y.

REFERENCES

1. ANSI Y14.5, *Dimensioning and Tolerancing*.

2. CSA B78.2, *Dimensioning and Tolerancing for Mechanical Engineering Drawings*.

3. CSA B97.1, *Standard Tolerances for Linear Dimensions, Inch and Metric*.

4. F. L. Spalding, "The Development of Standards for Dimensioning and Tolerancing," *Graphic Science*, vol. 8, no. 2, 1966.

Assignment

Calculate the sizes and tolerances for one of the drawings shown in Fig. 5-5-A.

REVIEW FOR ASSIGNMENT

Unit 2-6 Drafting Skills

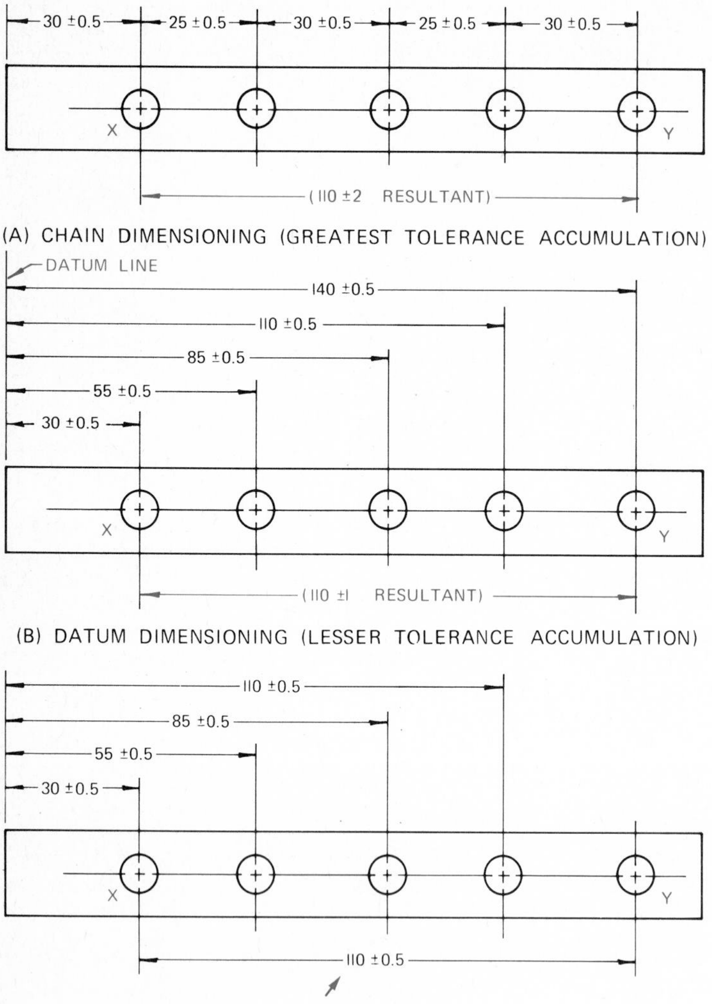
30 ±0.5
25 ±0.5
30 ±0.5
25 ±0.5
30 ±0.5
X
Y
(110 ±2 RESULTANT)
(A) CHAIN DIMENSIONING (GREATEST TOLERANCE ACCUMULATION)
DATUM LINE
140 ±0.5
110 ±0.5
85 ±0.5
55 ±0.5
30 ±0.5
(110 ±1 RESULTANT)
(B) DATUM DIMENSIONING (LESSER TOLERANCE ACCUMULATION)
110 ±0.5
85 ±0.5
55 ±0.5
30 ±0.5
110 ±0.5
(C) DIRECT DIMENSIONING (LEAST TOLERANCE ACCUMULATION)

Fig. 5-5-12 Dimensioning method comparison.

UNIT 5-6

FITS AND ALLOWANCES

Fits[1,2]

Fit is the general term used to signify the range of tightness or looseness which may result from the application of a specific combination of allowance and tolerances in the design of mating parts. Fits are of three general types: clearance, interference, and transition.

The following definitions cover terms of particular significance to fits (refer to Fig. 5-6-1).

Clearance fits are those having limits of size so prescribed that a clearance always results when mating parts are assembled.

Interference fits are those having limits of size so prescribed that an interference always results when mating parts are assembled.

Transition fits are those having limits of size so prescribed that either a clearance or an interference may result when mating parts are assembled.

Nominal size is the designation used for the purpose of general identification.

Basic size of a dimension is the theoretical size from which the limits for that dimension are derived, by the application of the allowance and tolerance.

Allowance is a prescribed difference in correlated dimensions of mating parts. It is the minimum clearance or maximum interference between such parts.

Basic hole system of fits is a system in which the minimum limit of size of each hole is basic. The fit desired is obtained

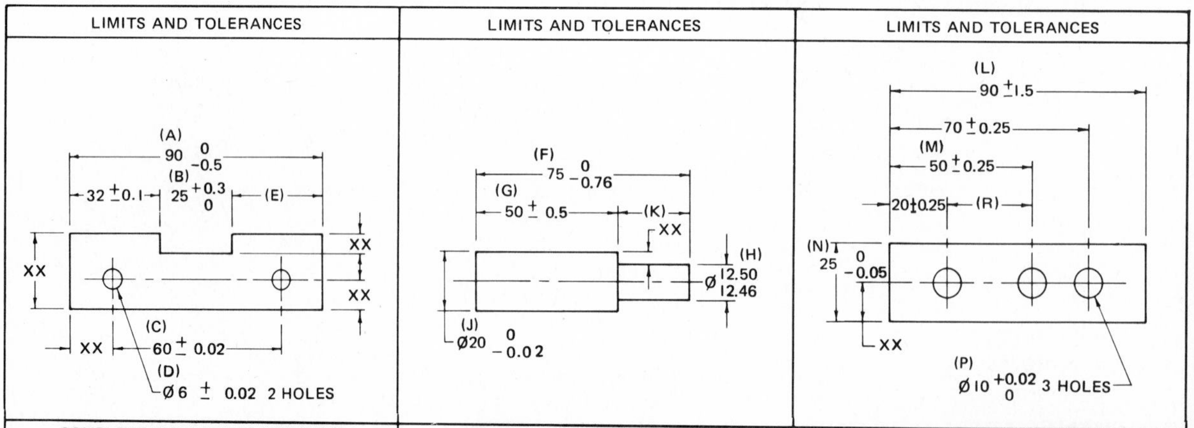
LIMITS AND TOLERANCES
(A)
90 0 -0.5
(B)
32 ±0.1
25 +0.3 0
(E)
XX
(C)
60 ± 0.02
(D)
Ø6 ± 0.02 2 HOLES
LIMITS AND TOLERANCES
(F)
75 0 -0.76
(G)
50 ± 0.5
(K)
XX
(H)
Ø 12.50 12.46
(J)
Ø20 0 -0.02
LIMITS AND TOLERANCES
(L)
90 ±1.5
70 ±0.25
(M)
50 ±0.25
20±0.25
(R)
(N)
25 0 -0.05
XX
(P)
Ø 10 +0.02 0 3 HOLES

COMPLETE THIS CHART FROM THE INFORMATION GIVEN ABOVE

		A	B	C	D	E
BASIC SIZE						X
TOLERANCE						X
LIMITS OF SIZE	MAX					
	MIN					

COMPLETE THIS CHART FROM THE INFORMATION GIVEN ABOVE

		F	G	H	J	K
BASIC SIZE						X
TOLERANCE						X
LIMITS OF SIZE	MAX					
	MIN					

COMPLETE THIS CHART FROM THE INFORMATION GIVEN ABOVE

		L	M	N	P	R
BASIC SIZE						X
TOLERANCE						X
LIMITS OF SIZE	MAX					
	MIN					

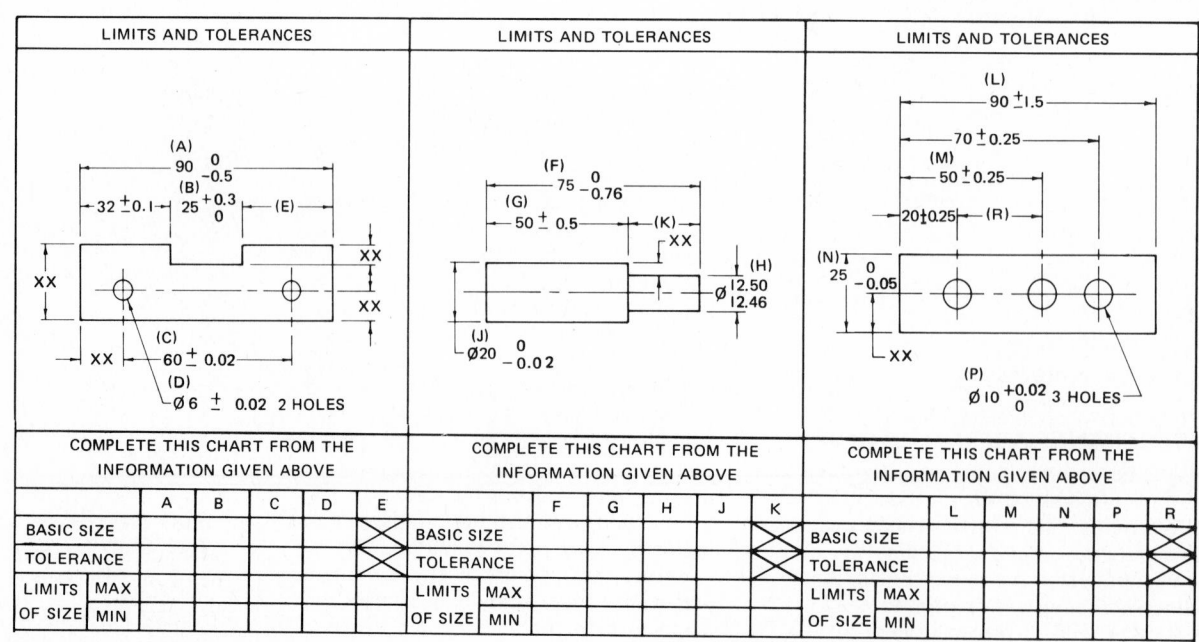

Fig. 5-5-A Limits and tolerances.

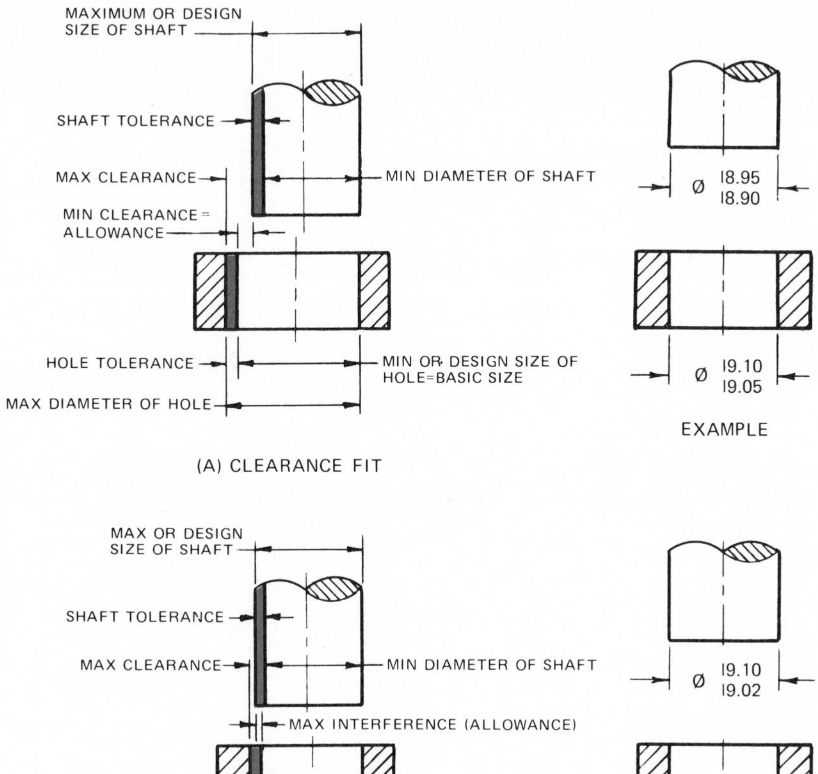

MAXIMUM OR DESIGN SIZE OF SHAFT
SHAFT TOLERANCE
MAX CLEARANCE
MIN DIAMETER OF SHAFT
MIN CLEARANCE = ALLOWANCE
HOLE TOLERANCE
MIN OR DESIGN SIZE OF HOLE=BASIC SIZE
MAX DIAMETER OF HOLE
Ø 18.95 18.90
Ø 19.10 19.05
EXAMPLE
(A) CLEARANCE FIT
MAX OR DESIGN SIZE OF SHAFT
SHAFT TOLERANCE
MAX CLEARANCE
MIN DIAMETER OF SHAFT
MAX INTERFERENCE (ALLOWANCE)
HOLE TOLERANCE
MIN OR DESIGN SIZE OF HOLE=BASIC SIZE
MAX DIAMETER OF HOLE
Ø 19.10 19.02
Ø 19.13 19.05
EXAMPLE
(B) TRANSITION FIT
MAX OR DESIGN SIZE OF SHAFT
SHAFT TOLERANCE
MIN DIAMETER OF SHAFT
MAX INTERFERENCE= ALLOWANCE
MIN INTERFERENCE
HOLE TOLERANCE
MIN OR DESIGN SIZE OF HOLE=BASIC SIZE
MAX DIAMETER OF HOLE
Ø 19.05 19.02
Ø 19.00 18.97
EXAMPLE
(C) INTERFERENCE FIT

Fig. 5-6-1 Types of fits.

by varying the allowance of the shaft and the tolerances of the mating parts.

In a *basic shaft system* of fits the maximum limit of size of each shaft is basic. The fit desired is obtained by varying the allowance of the hole and the tolerances of the mating parts.

Dimensions for individual members of the mating parts were commonly expressed in fractions along with the notations RUNNING FIT, PRESS FIT, FREE FIT, etc., to indicate the assembly requirements. See Fig. 5-6-2. This practice was acceptable because skilled tradespersons in the shop who were familiar with all parts had had the responsibility for assembling the parts properly. They would select their own tolerances and allowances with a reasonable degree of accuracy.

Increased demand for manufactured products led to the development of new production techniques. Interchangeability of parts became the basis for mass-production, low-cost manufacturing, and it brought about the refinement of machinery, machine tools, and measuring devices. Today it is possible and generally practical to design for 100 percent interchangeability.

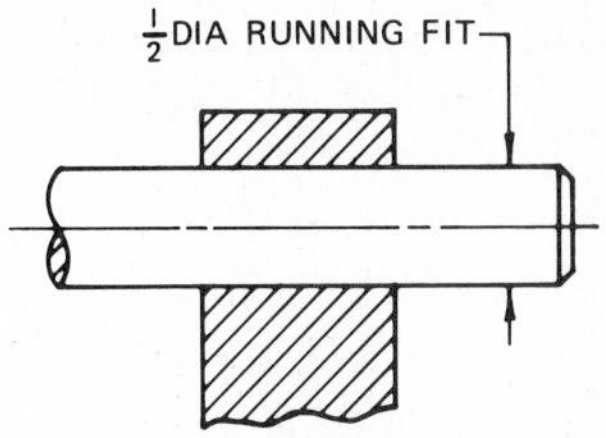

1/2 DIA RUNNING FIT

Fig. 5-6-2 Original method of specifying fits.

No part can be manufactured to exact dimensions. Tool wear, machine variations, and the human factor all contribute to some degree of deviation from perfection. It is therefore necessary to determine the deviation and permissible clearance, or interference, to produce the desired fit between parts.

Modern industry has adopted three basic approaches to manufacturing.

1. *The completely interchangeable assembly.* Any and all mating parts of a design are toleranced to permit them to assemble and function properly without the need for machining or fitting at assembly.
2. *The fitted assembly.* Mating features of a design are fabricated either simultaneously or with respect to one another. Individual members of mating features are not interchangeable.
3. *The selected assembly.* All parts are mass-produced, but members of mating features are individually selected to provide the required relationship with one another.

Standard Fits[3]

Both ANSI and CSA have adopted the basic hole system of fits. In this system of fits the minimum limit of size of each hole is basic. The fit desired is obtained by varying the allowance of the shaft and the tolerances of the mating parts. Tables shown in the Appendix have been developed to give a series of standard types and classes of fits on a unilateral hole basis, such that the fit produced by mating parts in any one class will produce an approximately similar performance throughout the range of sizes. These tables prescribe the fit for any given size or type of fit; they also prescribe the standard limits for the mating parts which will produce the fit.

In developing these tables, it has been recognized that any fit will usually be required to perform one of three functions, as indicated by the three general types of fit: running fits, locational fits, and force fits.

BASIC HOLE SYSTEM

In the basic hole system (Fig. 5-6-3*a*), which is recommended for general use, the basic size will be the design size for the hole. The design size for the shaft will be the basic size minus the smallest limit of clearance, or plus the maximum limit of interference. For example, for a 25-mm RC7 fit, the limits will be as follows:

- Basic size = 25
- Design size of hole = 25
- Table 38 gives values for a 25 mm RC7 fit of

  +0.05   0.06   −0.03
- Hole limits = $\phi 25^{+0.05}_{5} = \phi^{25.05}_{25.00}$
- Minimum clearance = 0.06 (allowance)
- Design size of shaft = 25 − 0.06 = 24.94
- Shaft limits = $\phi 24.94^{\;0}_{-0.03} = \phi^{24.94}_{24.91}$
- Maximum clearance = 25.11 − 24.97 = 0.14

BASIC SHAFT SYSTEM

Fits are sometimes required on a basic shaft system (Fig. 5-6-3*b*), especially in cases where two or more fits are required on the same shaft. This is designated for design purposes by a letter S following the fit symbol, for example, RC7S.

Tolerances for holes and shaft are identical with those for a basic hole system. However, the basic size becomes

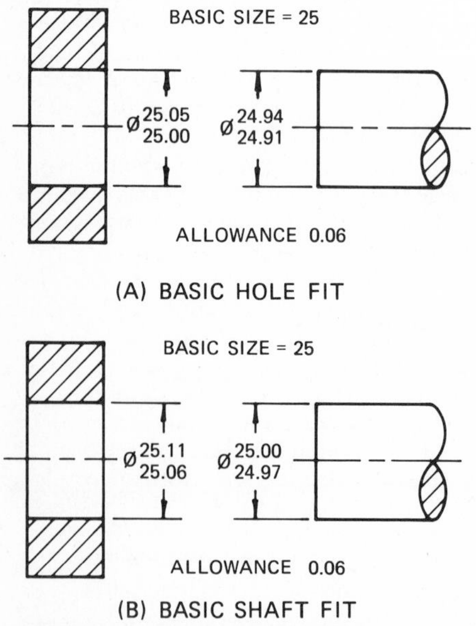
BASIC SIZE = 25
Ø25.05 25.00
Ø24.94 24.91
ALLOWANCE 0.06
(A) BASIC HOLE FIT
BASIC SIZE = 25
Ø25.11 25.06
Ø25.00 24.97
ALLOWANCE 0.06
(B) BASIC SHAFT FIT

Fig. 5-6-3 Comparison between basic hole and basic shaft fits.

the design size for the shaft, and the design size for the hole is found by adding the minimum clearance or subtracting the maximum interference from the basic size.

For example, for a 25-mm RC7S fit, the limits will be as follows:

- Basic size = 25
- Design size of shaft = 25
- Table 38 gives values for a 25 mm RC7 fit of

  +0.05   0.06   −0.03
- Tolerance on shaft size = 0.03
- Shaft limits = $25^{\;0}_{-0.03} = \phi^{25.00}_{24.97}$
- Minimum clearance = 0.06 (allowance)
- Design size of hole = largest shaft size + minimum clearance = 25 + 0.06 = 25.06
- Tolerance on hole = $^{+0.05}_{0}$
- Hole limits = $\phi 25.06^{+0.05}_{0} = \phi^{25.11}_{25.06}$
- Maximum clearance = 25.11 − 24.97 = 0.14

In using tables shown in the Appendix, the standard units for hole and shaft are applied algebraically to the basic size to obtain the limits of size for the parts.

DESIGNATION OF STANDARD FITS

Standard fits are designated for design purposes in specifications and on design sketches by means of the symbols shown in Fig. 5-6-4. These symbols, however, are not intended to be shown directly on shop drawings; instead, design size and tolerances are specified on the drawing.

The letter symbols used are as follows:

- RC   Running and sliding fit
- LC   Locational clearance fit
- LT   Locational transition fit
- LN   Locational interference fit
- FN   Force or shrink fit

These letter symbols are used in conjunction with numbers representing the class of fit; FN4 represents a class 4, force fit.

Description of Fits

RUNNING OR SLIDING FITS

These fits, for which tolerances and clearances are given (see Fig. 5-6-5), represent a special type of clearance fit. These are intended to provide a similar running performance, with suitable lubrication allowance, throughout the range of sizes.

These fits may be described as follows (refer to Fig. 5-6-5).

RC1 precision sliding fit is intended for the accurate location of parts which must assemble without perceptible play, for high-precision work such as gages.

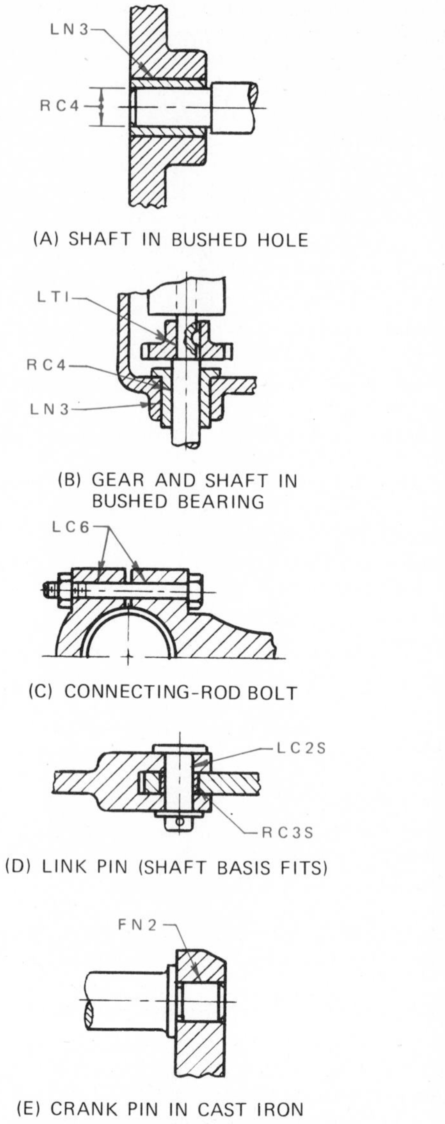
LN3
RC4
(A) SHAFT IN BUSHED HOLE
LT1
RC4
LN3
(B) GEAR AND SHAFT IN BUSHED BEARING
LC6
(C) CONNECTING-ROD BOLT
LC2S
RC3S
(D) LINK PIN (SHAFT BASIS FITS)
FN2
(E) CRANK PIN IN CAST IRON

Fig. 5-6-4 Typical design sketches showing classes of fits.

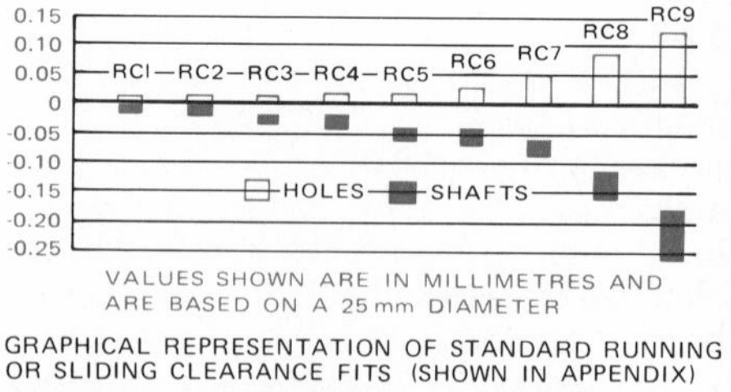
0.15
0.10
0.05
0
-0.05
-0.10
-0.15
-0.20
-0.25
RC1
RC2
RC3
RC4
RC5
RC6
RC7
RC8
RC9
HOLES
SHAFTS
VALUES SHOWN ARE IN MILLIMETRES AND ARE BASED ON A 25 mm DIAMETER
GRAPHICAL REPRESENTATION OF STANDARD RUNNING OR SLIDING CLEARANCE FITS (SHOWN IN APPENDIX)

Fig. 5-6-5 Graphical representation of running and sliding clearance fits.

RC2 sliding fit is intended for accurate location, but with greater maximum clearance than class RC1. Parts made to this fit move and turn easily but are not intended to run freely, and in the larger sizes they may seize with small temperature changes.

RC3 precision running fit is about the closest fit which can be expected to run freely. It is intended for precision work for oil-lubricated bearings at slow speeds and light journal pressures, but is not suitable where appreciable temperature differences are likely to be encountered.

RC4 close running fit is intended chiefly as a running fit for grease- or oil-lubricated bearings on accurate machinery with moderate surface speeds and journal pressures, where accurate location and minimum play are desired.

RC5 and RC6 medium running fits are intended for higher running speeds and/or for heavy journal pressures, where temperature variations are likely to be encountered.

RC7 free running fit is intended for use where accuracy is not essential, and/or where large temperature variations are likely to be encountered.

RC8 and RC9 loose running fits are intended for use where materials made to commercial tolerances are involved such as cold-rolled shafting, tubing, etc.

LOCATIONAL FITS

Locational fits are intended to determine only the location of the mating parts; they may provide rigid or accurate location, as with interference fits, or some freedom of location, as with clearance fits. Accordingly, they are divided into three groups: clearance fits, transition fits, and interference fits.

Locational clearance fits are intended for parts which are normally stationary but which can be freely assembled or disassembled. They run from snug fits for parts requiring accuracy of location, through the medium clearance fits for parts such as ball, race, and housing, to the looser fastener fits where freedom of assembly is of prime importance. These are classified as follows (refer to Fig. 5-6-6).

LC1 to LC4. These fits have a minimum zero clearance, but in practice the probability is that the fit will always have a clearance. These fits are suitable for location of nonrunning parts and spigots, although classes LC1 and LC2 may also be used for sliding fits.

LC5 and LC6. These fits have a small minimum clearance, intended for close location fits for nonrunning parts. LC5 can also be used in place of RC2 as a free slide fit, and LC6 may be used as a medium running fit having greater tolerances than RC5 and RC6.

LC7 to LC11. These fits have progressively larger clearances and tolerances and are useful for various loose clearances for assembly of bolts and similar parts.

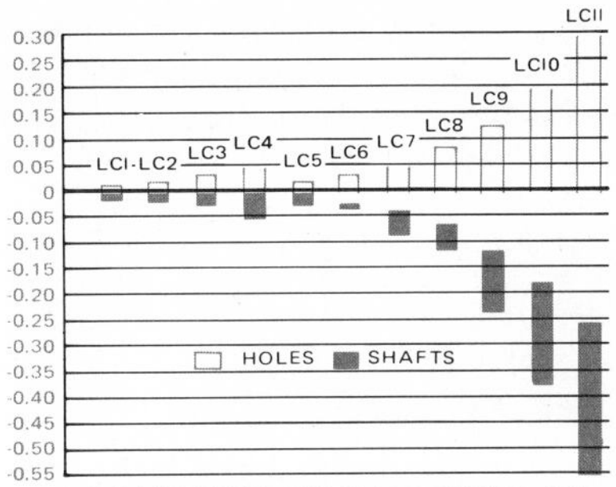
0.30
0.25
0.20
0.15
0.10
0.05
0
-0.05
-0.10
-0.15
-0.20
-0.25
-0.30
-0.35
-0.40
-0.45
-0.50
-0.55
LCI-LC2
LC3
LC4
LC5
LC6
LC7
LC8
LC9
LC10
LCII
HOLES
SHAFTS

VALUES SHOWN ARE IN MILLIMETRES AND ARE BASED ON A 25 mm DIAMETER

GRAPHICAL REPRESENTATION OF STANDARD LOCATIONAL CLEARANCE FITS (SHOWN IN APPENDIX)

Fig. 5-6-6 Graphical representation of locational clearance fits.

Locational transition fits are a compromise between clearance and interference fits, for application where accuracy of location is important but a small amount of either clearance or interference is permissible. These are classified as follows (refer to Fig. 5-6-7).

LT1 and LT2. These fits average a slight clearance, giving a light push fit, and are intended for use where the maximum clearance must be less than for the LC1 to LC3 fits and where slight interference can be tolerated for assembly by pressure or light hammer blows.

LT3 and LT4. These fits average virtually no clearance and are for use where some interference can be tolerated, for example, to eliminate vibration. These are sometimes referred to as an "easy keying fit" and are used for shaft keys and ball-race fits. Assembly is generally by pressure or hammer blows.

LT5 and LT6. These fits average a slight interference, although appreciable assembly force will be required when extreme limits are encountered, and selective assembly may be desirable. These fits are useful for heavy keying, for ball-race fits subject to heavy duty and vibration, and as light press fits for steel parts.

Locational interference fits are used where accuracy of location is of prime importance and for parts requiring rigidity and alignment with no special requirements for bore pressure. Such fits are not intended for parts designed to transmit frictional loads from one part to another by virtue of the tightness of fit, for these conditions are covered by force fits. These are classified as follows (refer to Fig. 5-6-8).

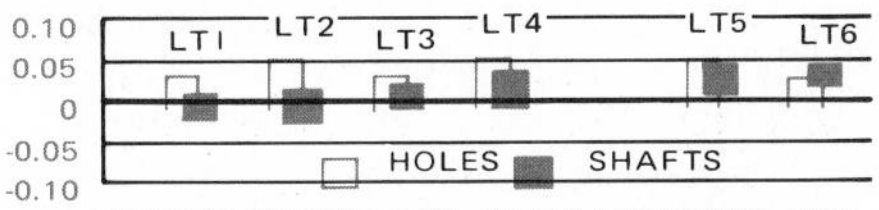
0.10
0.05
0
-0.05
-0.10
LT1
LT2
LT3
LT4
LT5
LT6
HOLES
SHAFTS

VALUES SHOWN ARE IN MILLIMETRES AND ARE BASED ON A 25 mm DIAMETER

GRAPHICAL REPRESENTATION OF STANDARD LOCATIONAL TRANSITION FITS (SHOWN IN APPENDIX)

Fig. 5-6-7 Graphical representation of locational transition fits.

LN1 and LN2. These are light press fits, with very small minimum interference, suitable for parts such as dowel pins, which are assembled with an arbor press in steel, cast iron, or brass. Parts can normally be dismantled and reassembled, since the interference is not likely to overstrain the parts, but the interference is too small for satisfactory fits in elastic materials or light alloys.

LN3. This is suitable as a heavy press fit in steel and brass or as a light press fit in more elastic materials and light alloys.

LN4 to LN6. While LN4 can be used for permanent assembly of steel parts, these fits are primarily intended as press fits for more elastic or soft materials, such as light alloys and the more rigid plastics.

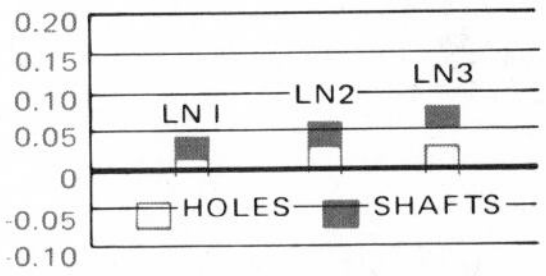
0.20
0.15
0.10
0.05
0
-0.05
-0.10
LN1
LN2
LN3
HOLES
SHAFTS

VALUES SHOWN ARE IN MILLIMETRES AND ARE BASED ON A 25 mm DIAMETER

GRAPHICAL REPRESENTATION OF STANDARD LOCATIONAL INTERFERENCE FITS (SHOWN IN APPENDIX)

Fig. 5-6-8 Graphical representation of locational interference fits.

FORCE OR SHRINK FITS

Force or shrink fits constitute a special type of interference fit, normally characterized by maintenance of constant bore pressures throughout the range of sizes. The interference therefore varies almost directly with diameter, and the difference between its minimum and maximum values is small, to maintain the resulting pressures within reasonable limits. These are classified as follows (refer to Fig. 5-6-9).

FN1 Light Drive Fit. This requires light assembly pressure and produces more or less permanent assemblies. It is suitable for thin sections or long fits or in cast-iron external members.

FN2 Medium Drive Fit. This is suitable for ordinary steel parts or as a shrink fit on light sections. It is about the tightest fit that can be used with high-grade cast-iron external members.

FN3 Heavy Drive Fit. This is suitable for heavier steel parts or as a shrink fit in medium sections.

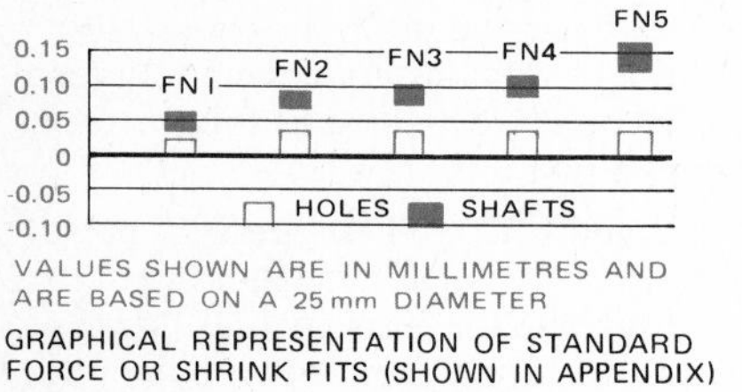
FN5
0.15
0.10
0.05
0
-0.05
-0.10
FN 1
FN2
FN3
FN4
HOLES
SHAFTS
VALUES SHOWN ARE IN MILLIMETRES AND ARE BASED ON A 25 mm DIAMETER
GRAPHICAL REPRESENTATION OF STANDARD FORCE OR SHRINK FITS (SHOWN IN APPENDIX)

Fig. 5-6-9 Graphical representation of force or shrink fits.

FN4 and FN5 Force Fits. These are suitable for parts which can be highly stressed and/or for shrink fits where the heavy pressing forces required are impractical.

REFERENCES

1. ANSI B4.1, *Preferred Limits and Fits for Cylindrical Parts.*
2. CSA B97.1, *Limits and Fits for Engineering and Manufacturing.*
3. CSA B97.3, *Standard Fits for Mating Parts, Inch Sizes.*

Assignment

Using the tables of fits located in the Appendix, calculate the missing dimensions located in the charts shown on Figs. 5-6-A and 5-6-B.

REVIEW FOR ASSIGNMENT

Unit 2-6 Drafting Skills

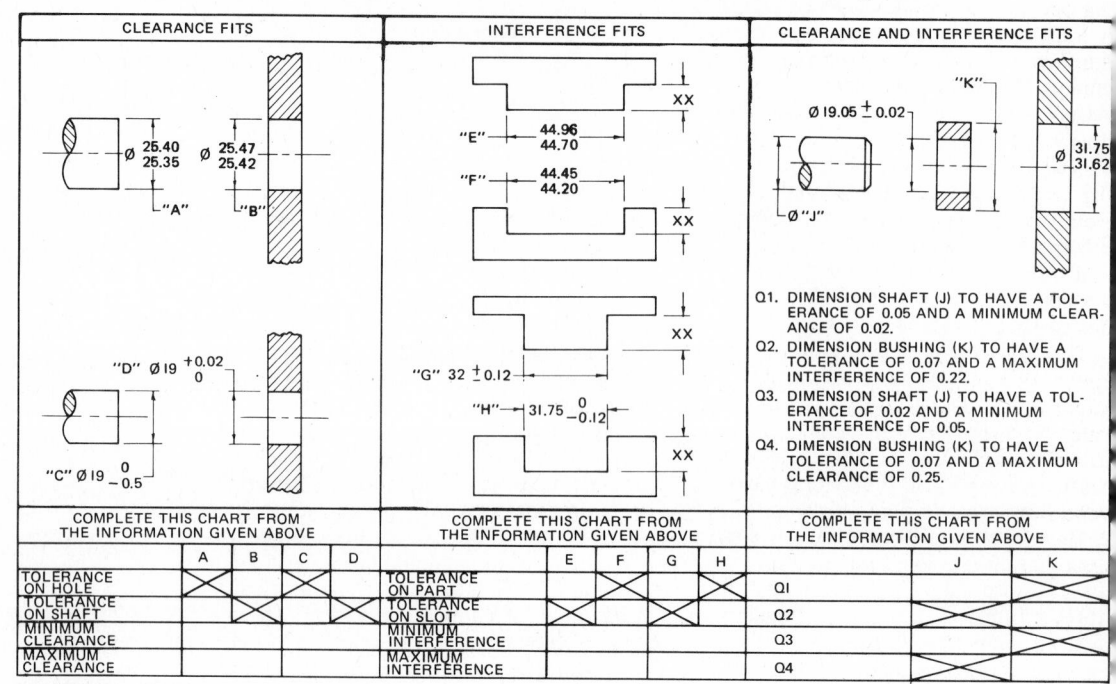

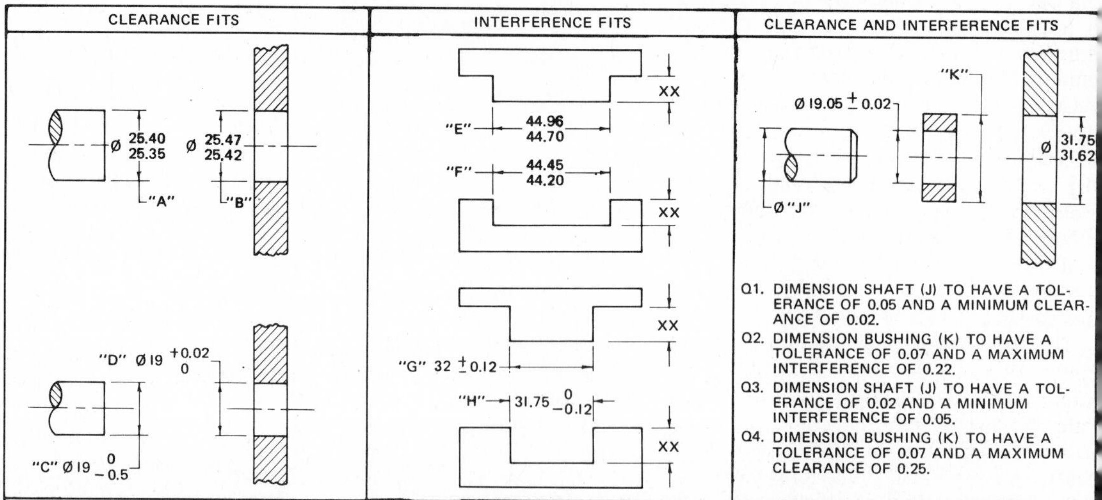
CLEARANCE FITS
Ø 25.40 25.35
Ø 25.47 25.42
"A"
"B"
"D" Ø 19 +0.02 0
"C" Ø 19 0 −0.5
INTERFERENCE FITS
XX
"E" 44.96 44.70
"F" 44.45 44.20
"G" 32 ± 0.12
"H" 31.75 0 −0.12
CLEARANCE AND INTERFERENCE FITS
"K"
Ø 19.05 ± 0.02
Ø 31.75 31.62
Ø "J"
Q1. DIMENSION SHAFT (J) TO HAVE A TOLERANCE OF 0.05 AND A MINIMUM CLEARANCE OF 0.02.
Q2. DIMENSION BUSHING (K) TO HAVE A TOLERANCE OF 0.07 AND A MAXIMUM INTERFERENCE OF 0.22.
Q3. DIMENSION SHAFT (J) TO HAVE A TOLERANCE OF 0.02 AND A MINIMUM INTERFERENCE OF 0.05.
Q4. DIMENSION BUSHING (K) TO HAVE A TOLERANCE OF 0.07 AND A MAXIMUM CLEARANCE OF 0.25.

COMPLETE THIS CHART FROM THE INFORMATION GIVEN ABOVE

	A	B	C	D
TOLERANCE ON HOLE				
TOLERANCE ON SHAFT				
MINIMUM CLEARANCE				
MAXIMUM CLEARANCE				

COMPLETE THIS CHART FROM THE INFORMATION GIVEN ABOVE

	E	F	G	H
TOLERANCE ON PART				
TOLERANCE ON SLOT				
MINIMUM INTERFERENCE				
MAXIMUM INTERFERENCE				

COMPLETE THIS CHART FROM THE INFORMATION GIVEN ABOVE

	J	K
Q1		
Q2		
Q3		
Q4		

Fig. 5-6-A Fits.

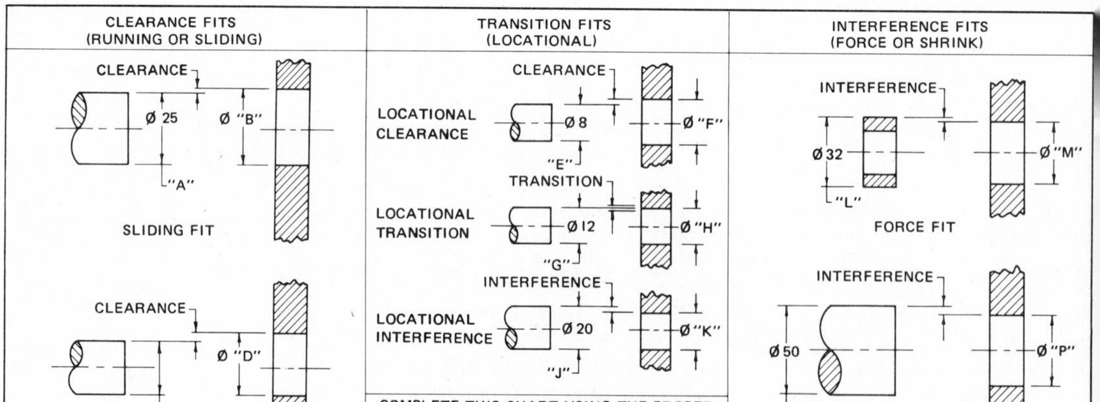
CLEARANCE FITS (RUNNING OR SLIDING)
CLEARANCE
Ø 25
Ø "B"
"A"
SLIDING FIT
CLEARANCE
Ø "D"
"C" Ø 19
RUNNING FIT
TRANSITION FITS (LOCATIONAL)
CLEARANCE
LOCATIONAL CLEARANCE
Ø 8
Ø "F"
"E"
TRANSITION
LOCATIONAL TRANSITION
Ø 12
Ø "H"
"G"
INTERFERENCE
LOCATIONAL INTERFERENCE
Ø 20
Ø "K"
"J"
INTERFERENCE FITS (FORCE OR SHRINK)
INTERFERENCE
Ø 32
Ø "M"
"L"
FORCE FIT
INTERFERENCE
Ø 50
Ø "P"
"N"
SHRINK FIT

COMPLETE THIS CHART USING THE PROPER LIMIT AND FIT TABLES

	Ø A	Ø B	Ø C	Ø D
RC I				
RC6				

COMPLETE THIS CHART USING THE PROPER LIMIT AND FIT TABLES

	LC 6	LT 3	LN 2
Ø E			
Ø F			
Ø G			
Ø H			
Ø J			
Ø K			

COMPLETE THIS CHART USING THE PROPER LIMIT AND FIT TABLES

	Ø L	Ø M	Ø N	Ø P
FN I				
FN 2				

Fig. 5-6-B Fits.

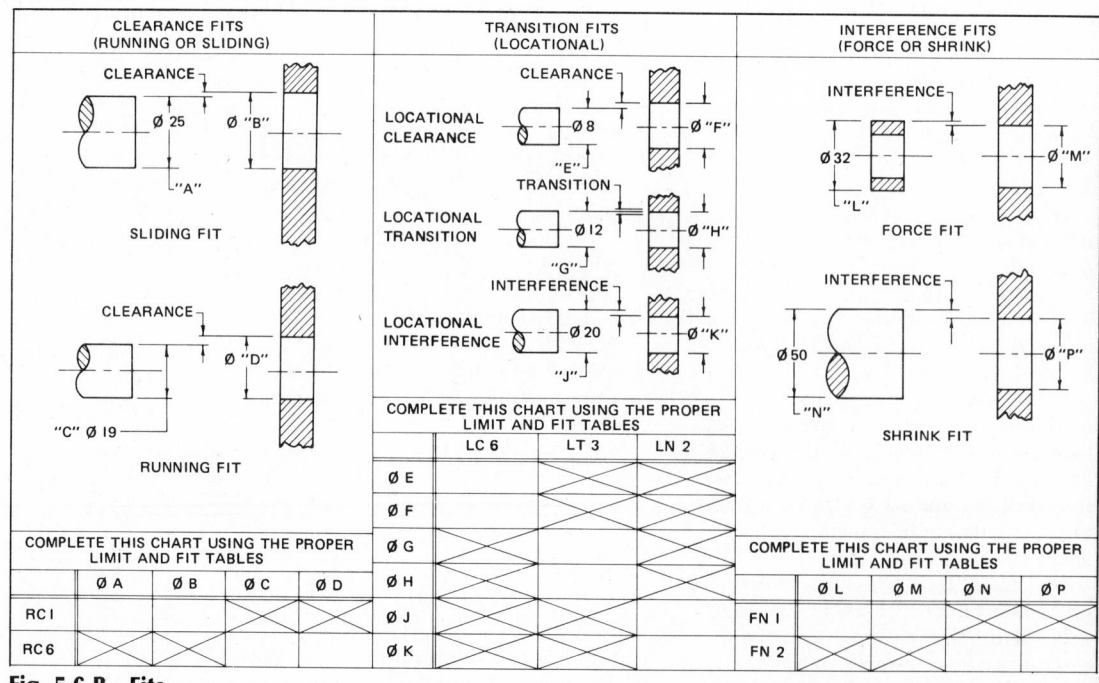

UNIT 5-7

MACHINING SYMBOLS (FINISH MARKS)

In preparing working drawings or parts to be cast, molded, or forged, the drafter must indicate the surfaces on the drawing which will require machining or finishing. The symbol √ identifies those surfaces which are produced by machining operations. It indicates that material is to be provided for removal by machining. Where all the surfaces are to be machined, a general note such as FINISH ALL OVER may be used, and the symbols on the drawing may be omitted. Where space is restricted, the machining symbol may be placed on an extension line.

The machining symbol is not to be construed as giving any indication of the surface finish quality. This is defined and discussed in Unit 5-8.

The former machining symbol, as shown in Fig. 5-7-1, may be found on many drawings in use today. When called upon to make changes or revisions to a drawing already in existence, a drafter must adhere to the drawing conventions shown on that drawing.

Machining symbols, like dimensions, are not normally duplicated. They should be used on the same view as the dimensions that give the size or location of the surfaces concerned. The symbol is placed on the line representing the surface or, where desirable, on the extension line locating the surface. Figures 5-7-2 and 5-7-3 show examples of the use of machining symbols.

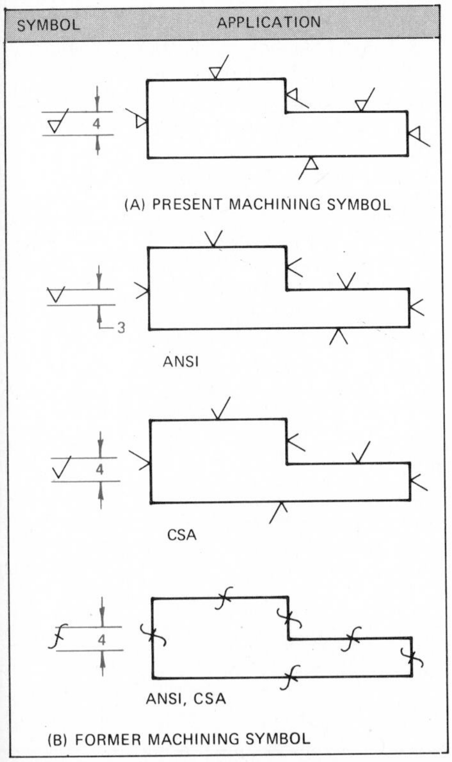
SYMBOL
APPLICATION
4
(A) PRESENT MACHINING SYMBOL
3
ANSI
4
CSA
4
ANSI, CSA
(B) FORMER MACHINING SYMBOL

Fig. 5-7-1 Machining symbols.

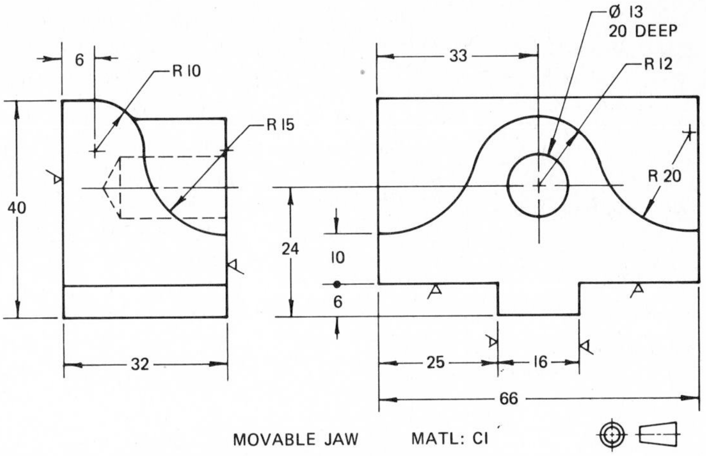
6
R 10
R 15
40
32
24
10
6
33
Ø 13
20 DEEP
R 12
R 20
25
16
66
MOVABLE JAW
MATL: CI

Fig. 5-7-2 Application of machining symbols.

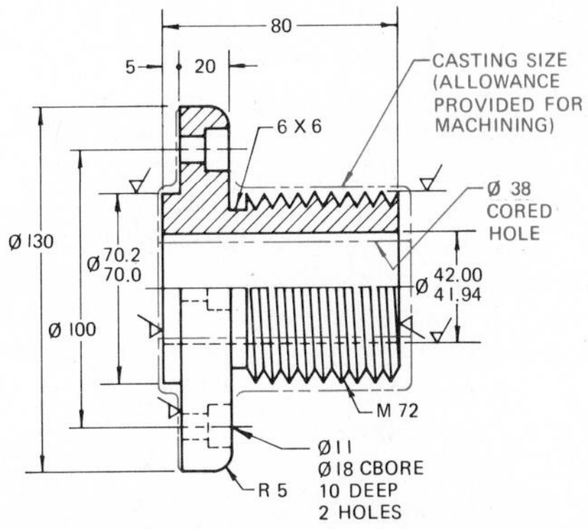
80
5
20
6 X 6
CASTING SIZE
(ALLOWANCE
PROVIDED FOR
MACHINING)
Ø 38
CORED
HOLE
Ø 130
70.2
Ø
70.0
42.00
Ø
41.94
Ø 100
M 72
Ø 11
Ø 18 CBORE
10 DEEP
2 HOLES
R 5

Fig. 5-7-3 Extra metal allowance for machined surfaces.

MATERIAL REMOVAL ALLOWANCE

When it is desirable to indicate the amount of material to be removed, the amount of material in millimetres is shown to the left of the symbol. Illustrations showing material removal allowance are shown in Figs. 5-7-4 and 5-7-5.

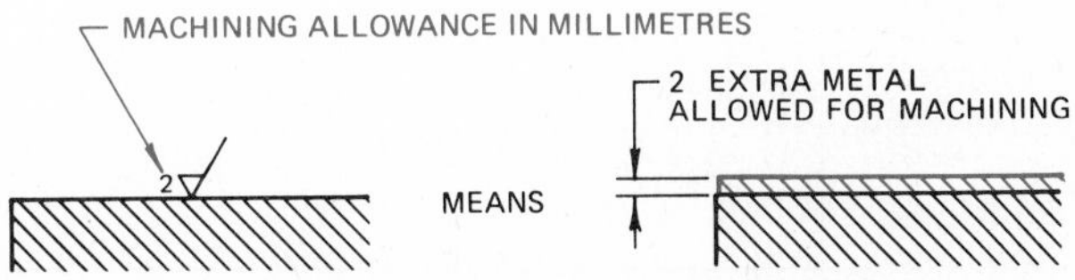
MACHINING ALLOWANCE IN MILLIMETRES
2
MEANS
2 EXTRA METAL
ALLOWED FOR MACHINING

Fig. 5-7-4 Indication of machining allowance.

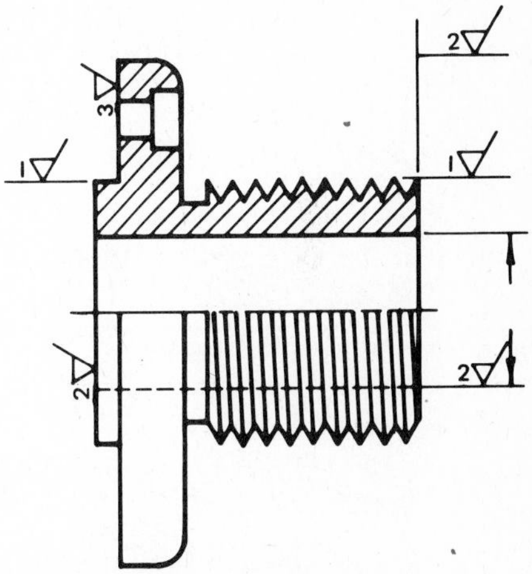
2
3
1
1
2
2

Fig. 5-7-5 Indicating machining allowance on drawings.

MATERIAL REMOVAL PROHIBITED

When it is necessary to indicate that a surface must be produced without material removal, the machining prohibited symbol shown in Fig. 5-7-6 must be used.

REFERENCES AND SOURCE MATERIALS

ANSI Y14.36 Surface Texture Symbols

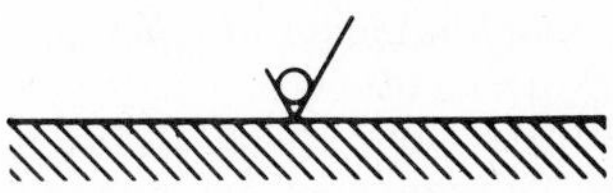

Fig. 5-7-6 Symbol for removal of material not permitted.

Assignments

1. On an A3- or B-size sheet make a working drawing of the adjustable base plate shown in Figs. 5-7-A or 5-7-B. The amount of material to be removed on the surfaces requiring machining is 2 mm or .08 in. The center hole is to be dimensioned for an LN4 fit with plain bearings. Scale 1:1.

REVIEW FOR ASSIGNMENTS

Unit 5-1 Basic Dimensioning
Unit 5-2 Dimensioning Circular Features
Unit 5-6 Fits

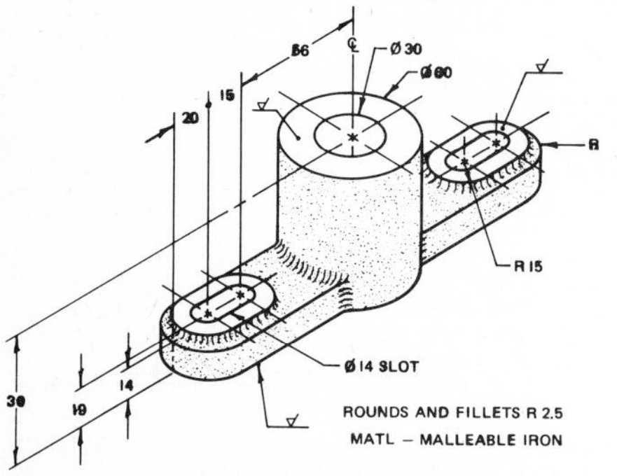
Ø 30
Ø 60
R
R 15
Ø 14 SLOT
ROUNDS AND FILLETS R 2.5
MATL — MALLEABLE IRON

Fig. 5-7-A Adjustable base plate.

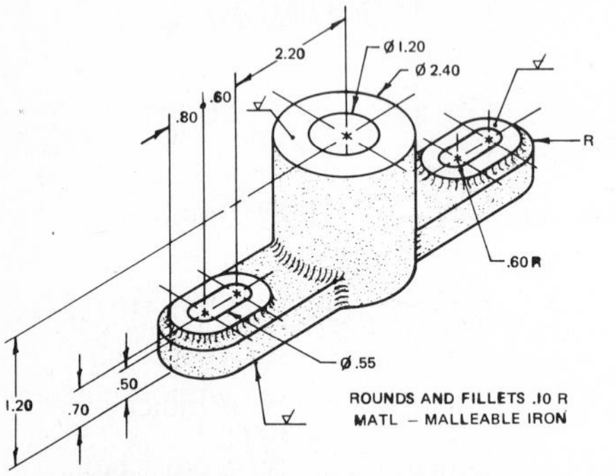
2.20
Ø 1.20
Ø 2.40
.60
.80
R
.60 R
Ø .55
1.20
.70
.50
ROUNDS AND FILLETS .10 R
MATL — MALLEABLE IRON

Fig. 5-7-B Adjustable base plate.

UNIT 5-8
SURFACE TEXTURE

Modern development of high-speed machines has resulted in higher loadings and increased speeds of moving parts. To withstand these more severe operating conditions with minimum friction and wear, a particular surface finish is often essential, making it necessary for the designer to accurately describe the required finish to the persons who are actually making the parts.

For accurate machines it is no longer sufficient to indicate the surface finish by various grind marks, such as "g," "f," or "fg." It becomes necessary to define surface finish and take it out of the "opinion" or "guesswork" class.

All surface finish control starts in the drafting room. The designer has the responsibility of specifying the right surface to give maximum performance and service life at the lowest cost. In selecting the required surface finish for any particular part, the designer bases her or his decision on past experience with similar parts, on field service data, or on engineering tests. Such factors as size and function of the parts, type of loading, speed and direction of movement, operating conditions, physical characteristics of both materials on contact, whether it is subjected to stress reversals, type and amount of lubricant, contaminants, temperature, etc., influence the choice.

There are two principal reasons for surface finish control:

1. To reduce friction
2. To control wear

Whenever a film of lubricant must be maintained between two moving parts, the surface irregularities must be small enough so they will not penetrate the oil film under the most severe operating conditions. Bearings, journals, cylinder bores, piston pins, bushings, pad bearings, helical and worm gears, seal surfaces, machine ways, and so forth, are examples where this condition must be fulfilled.

Surface finish is also important to the wear of certain pieces which are subject to dry friction, such as machine tool bits, threading dies, stamping dies, rolls, clutch plates, brake drums, etc.

Smooth finishes are essential on certain high-precision pieces. In mechanisms such as injectors and high-pressure cylinders, smoothness and lack of waviness are essential to accuracy and pressure-retaining ability.

Surfaces, in general, are very complex in character. Only the height, width, and direction of surface irregularities will be covered in this section since these are of practical importance in specific applications.

SURFACE TEXTURE CHARACTERISTICS

Refer to Fig. 5-8-1.

Micrometre. A micrometre is one millionth of a metre (0.000 001 m). For written specifications or reference to surface roughness requirements, micrometres may be abbreviated as μm.

Microinch. A microinch is one millionth of an inch (0.000 001 in.). For written specifications or reference to surface roughness requirements, microinches may be abbreviated as μin.

Roughness. Roughness consists of the finer irregularities in the surface texture usually including those which result from the inherent action of the production process. These are considered to include traverse feed marks and other irregularities within the limits of the roughness-width cutoff.

Roughness-Height Value. Roughness-height value is rated as the arithmetic average (AA) deviation expressed in micrometres or microinches measured normal to the center line. ISO and many European countries use the term *CLA* (center line average) in lieu of AA. Both have the same meaning.

Roughness Spacing. Roughness spacing is the distance parallel to the nominal surface between successive peaks or ridges which constitute the predominant pattern of the roughness. Roughness spacing is rated in millimetres or inches.

Roughness-Width Cutoff. The greatest spacing of repetitive surface irregularities is included in the measurement of average roughness height. Roughness-width cutoff is rated in millimetres or inches and must always be greater than the roughness width in order to obtain the total roughness height rating.

Waviness. Waviness is the usually widely spaced component of surface texture and is generally of wider spacing than the roughness-width cutoff. Waviness may result from such factors as machine or work deflections, vibration, chatter, heat treatment, or warping strains. Roughness may be considered as

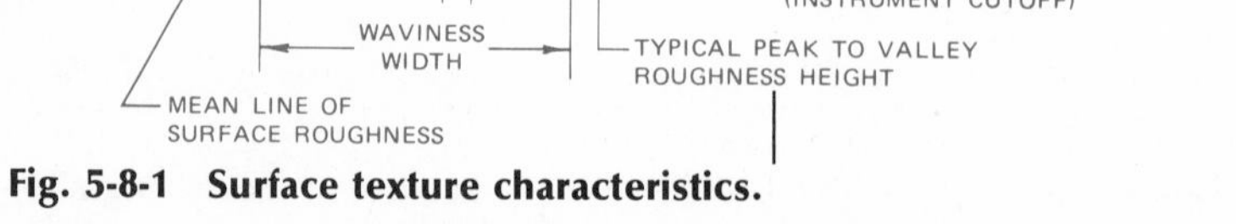
TYPICAL FLAW (SCRATCH)
WAVINESS HEIGHT (TRUE)
LAY (DIRECTION OF DOMINANT PATTERN)
TYPICAL ROUGHNESS WIDTH
SAMPLING LENGTH ROUGHNESS-WIDTH CUTOFF (INSTRUMENT CUTOFF)
WAVINESS WIDTH
TYPICAL PEAK TO VALLEY ROUGHNESS HEIGHT
MEAN LINE OF SURFACE ROUGHNESS

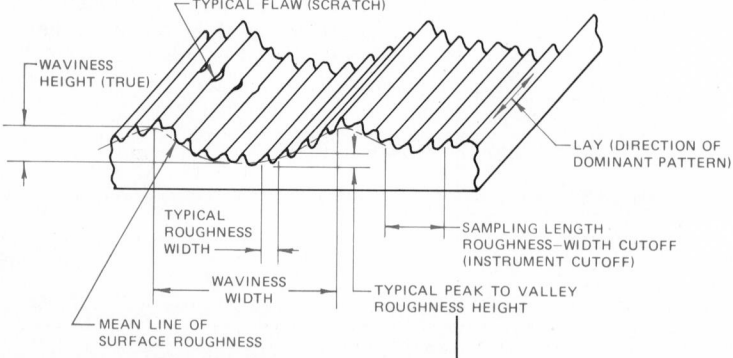

Fig. 5-8-1 Surface texture characteristics.

superimposed on a "wavy" surface.

Lay. The direction of the predominant surface pattern, ordinarily determined by the production method used, is the *lay*. Lay symbols are specified as shown in Fig. 5-8-6.

Flaws. Flaws are irregularities which occur at one place or at relatively infrequent or widely varying intervals in a surface. Flaws include such defects as cracks, blow holes, checks, ridges, scratches, etc. Unless otherwise specified, the effect of flaws is not included in the roughness height measurements.

SURFACE TEXTURE SYMBOL

Surface characteristics of roughness, waviness, and lay may be controlled by applying the desired values to the surface texture symbol, shown in Figs. 5-8-2 and 5-8-3, in a general note, or both. Where only the roughness value is indicated, the horizontal extension line on the symbol may be omitted. The horizontal bar is used whenever any surface characteristics are placed above the bar or to the right of the symbol. The point of the symbol should be located on the line indicating the surface, on an extension line from the surface, or on a leader pointing to the surface or extension line (Fig. 5-8-4). The symbol should be in an upright position in order to be readable from the bottom or right side. This means that the long leg and extension line are always on the right. The symbol applies to the entire surface, unless otherwise specified.

Like dimension, the symbol for the same surface should not be duplicated on other views.

APPLICATION

Plain (Unplated or Uncoated) Surfaces. Surface texture values specified on plain surfaces apply to the completed surface unless otherwise noted.

Plated or Coated Surfaces. Drawings or specifications for plated or coated parts must indicate whether the surface texture value applies before, after, or both before and after plating or coating.

Surface Texture Ratings. The roughness value rating is indicated at the left of the long leg of the symbol (Fig. 5-8-4). The specification of only one rating indicates the maximum value, and any lesser value is acceptable. The specification of two ratings indicates the minimum and maximum values, and anything lying within that range is acceptable (Fig. 5-8-5). The maximum value is placed over the minimum.

Waviness-height rating is indicated in millimetres or inches and is located above the horizontal extension of the symbol (Fig. 5-8-3). Any lesser value is acceptable.

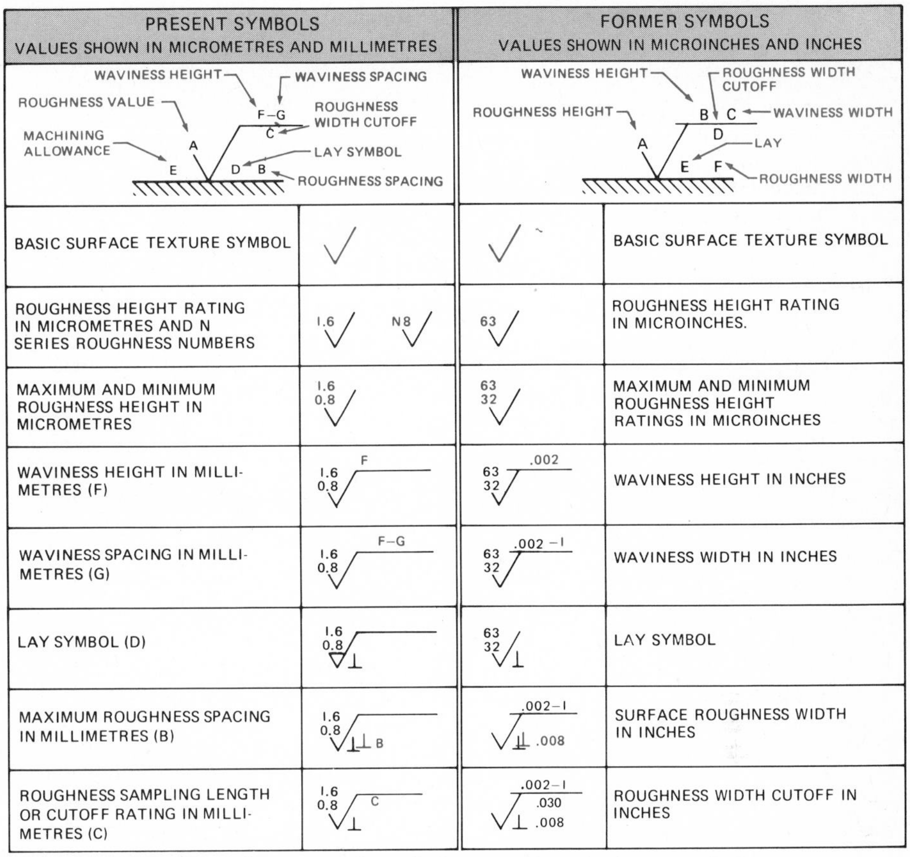
PRESENT SYMBOLS
VALUES SHOWN IN MICROMETRES AND MILLIMETRES
WAVINESS HEIGHT
WAVINESS SPACING
ROUGHNESS VALUE
ROUGHNESS WIDTH CUTOFF
MACHINING ALLOWANCE
LAY SYMBOL
ROUGHNESS SPACING
FORMER SYMBOLS
VALUES SHOWN IN MICROINCHES AND INCHES
WAVINESS HEIGHT
ROUGHNESS WIDTH CUTOFF
ROUGHNESS HEIGHT
WAVINESS WIDTH
LAY
ROUGHNESS WIDTH
BASIC SURFACE TEXTURE SYMBOL
ROUGHNESS HEIGHT RATING IN MICROMETRES AND N SERIES ROUGHNESS NUMBERS
ROUGHNESS HEIGHT RATING IN MICROINCHES.
MAXIMUM AND MINIMUM ROUGHNESS HEIGHT IN MICROMETRES
MAXIMUM AND MINIMUM ROUGHNESS HEIGHT RATINGS IN MICROINCHES
WAVINESS HEIGHT IN MILLIMETRES (F)
WAVINESS HEIGHT IN INCHES
WAVINESS SPACING IN MILLIMETRES (G)
WAVINESS WIDTH IN INCHES
LAY SYMBOL (D)
LAY SYMBOL
MAXIMUM ROUGHNESS SPACING IN MILLIMETRES (B)
SURFACE ROUGHNESS WIDTH IN INCHES
ROUGHNESS SAMPLING LENGTH OR CUTOFF RATING IN MILLIMETRES (C)
ROUGHNESS WIDTH CUTOFF IN INCHES

Fig. 5-8-3 **Location of notes and symbols on surface texture symbols.**

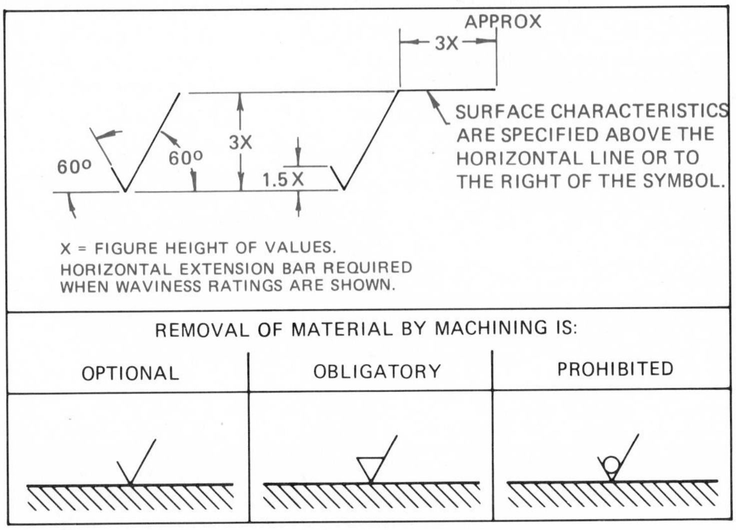
APPROX
3X
SURFACE CHARACTERISTICS ARE SPECIFIED ABOVE THE HORIZONTAL LINE OR TO THE RIGHT OF THE SYMBOL.
60°
60°
3X
1.5 X
X = FIGURE HEIGHT OF VALUES.
HORIZONTAL EXTENSION BAR REQUIRED WHEN WAVINESS RATINGS ARE SHOWN.
REMOVAL OF MATERIAL BY MACHINING IS:
OPTIONAL
OBLIGATORY
PROHIBITED

Fig. 5-8-2 **Basic surface texture symbol.**

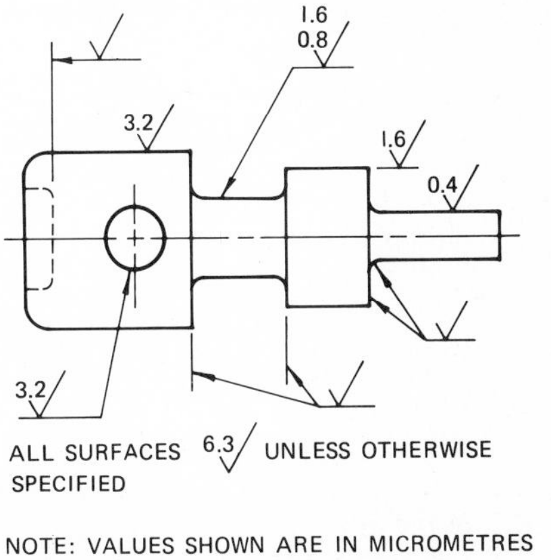
1.6
0.8
3.2
1.6
0.4
3.2
ALL SURFACES 6.3 UNLESS OTHERWISE SPECIFIED
NOTE: VALUES SHOWN ARE IN MICROMETRES

Fig. 5-8-4 **Application of surface texture symbols and notes.**

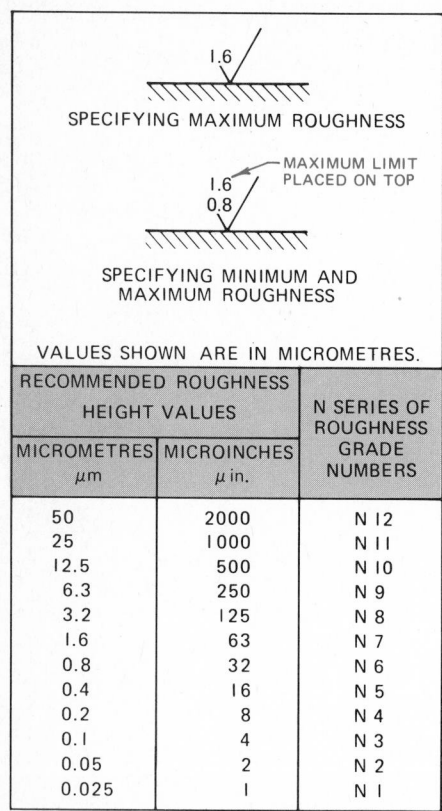

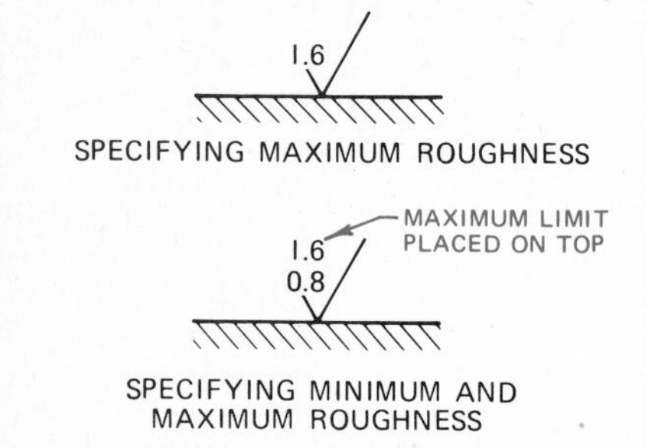
1.6
SPECIFYING MAXIMUM ROUGHNESS
1.6
0.8
MAXIMUM LIMIT PLACED ON TOP
SPECIFYING MINIMUM AND MAXIMUM ROUGHNESS

VALUES SHOWN ARE IN MICROMETRES.

RECOMMENDED ROUGHNESS HEIGHT VALUES		N SERIES OF ROUGHNESS GRADE NUMBERS
MICROMETRES μm	MICROINCHES μ in.	
50	2000	N 12
25	1000	N 11
12.5	500	N 10
6.3	250	N 9
3.2	125	N 8
1.6	63	N 7
0.8	32	N 6
0.4	16	N 5
0.2	8	N 4
0.1	4	N 3
0.05	2	N 2
0.025	1	N 1

Fig. 5-8-5 Roughness height ratings.

Waviness spacing is indicated in millimetres or inches and is located above the horizontal extension and to the right, separated from the waviness-height rating by a dash (Fig. 5-8-3). Any lesser value is acceptable. If the waviness value is a minimum, the abbreviation MIN should be placed after the value.

Lay symbols which indicate the directional pattern of the surface texture are shown in Fig. 5-8-6. The symbol is located to the right of the long leg of the symbol. On surfaces having parallel or perpendicular lay designated, the lead resulting from machine feeds may be objectionable. In these cases, the symbol should be supplemented by the words NO LEAD.

Roughness sampling length or cutoff rating is in millimetres or inches and is located below the horizontal extension (Fig. 5-8-3). Unless otherwise specified, roughness sampling length is 0.8 mm (.03 in.). See Fig. 5-8-7.

NOTES

Notes relating to surface roughness can be local or general.

General Note. Normally, a general note is used where a given roughness requirement applies to the whole part or the major portion. Any exceptions to the general note are indicated locally. See Fig. 5-8-8. Figures 5-8-9 and 5-8-10 sup-

SYMBOL	DESIGNATION	EXAMPLE
=	LAY PARALLEL TO THE LINE REPRESENTING THE SURFACE TO WHICH THE SYMBOL IS APPLIED.	DIRECTION OF TOOL MARKS
⊥	LAY PERPENDICULAR TO THE LINE REPRESENTING THE SURFACE TO WHICH THE SYMBOL IS APPLIED.	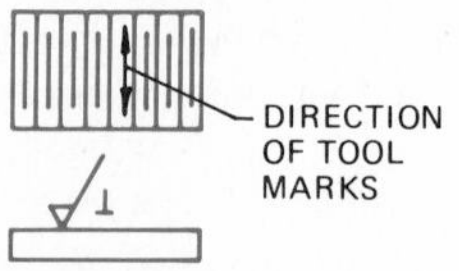DIRECTION OF TOOL MARKS
X	LAY ANGULAR IN BOTH DIRECTIONS TO LINE REPRESENTING THE SURFACE TO WHICH SYMBOL IS APPLIED.	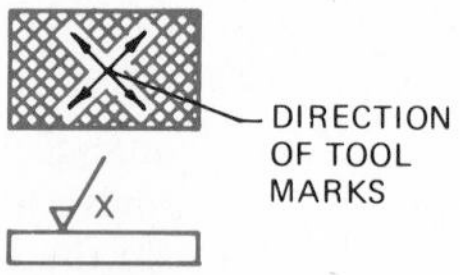DIRECTION OF TOOL MARKS
M	LAY MULTIDIRECTIONAL.	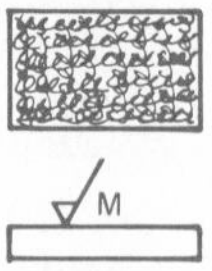
C	LAY APPROXIMATELY CIRCULAR RELATIVE TO THE CENTER OF THE SURFACE TO WHICH THE SYMBOL IS APPLIED.	
R	LAY APPROXIMATELY RADIAL RELATIVE TO THE CENTER OF THE SURFACE TO WHICH THE SYMBOL IS APPLIED.	
P	LAY NONDIRECTIONAL, PITTED OR PROTUBERANT.	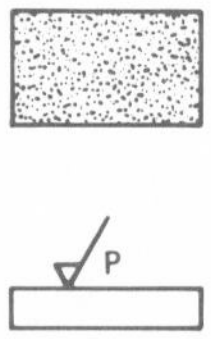

Fig. 5-8-6 Lay symbols.

PRESENT SYMBOL	FORMER SYMBOL
LAY SYMBOLS	

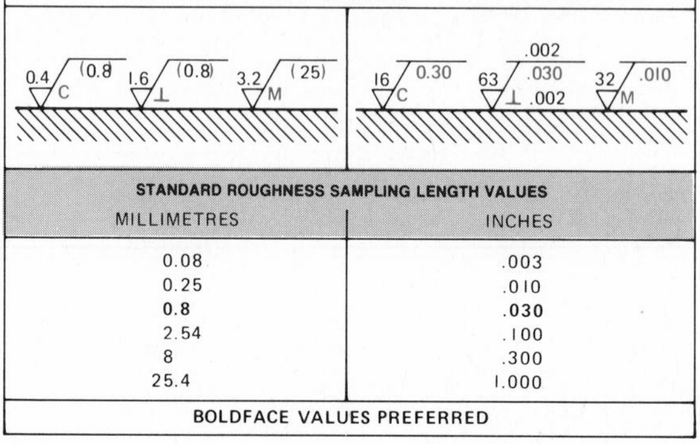
0.4 (0.8 C
1.6 (0.8) ⊥
3.2 (25) M
16 0.30 C
.002
63 .030 ⊥ .002
32 .010 M

STANDARD ROUGHNESS SAMPLING LENGTH VALUES	
MILLIMETRES	INCHES
0.08	.003
0.25	.010
0.8	**.030**
2.54	.100
8	.300
25.4	1.000
BOLDFACE VALUES PREFERRED	

Fig. 5-8-7 Lay and roughness sampling length specifications.

ply further data on roughness value applications.

REFERENCES AND SOURCE MATERIALS

1. ANSI Y14.36 Surface Texture Symbols
2. CSA B95 *Surface Texture Symbols*
3. GAR
4. General Motors

Assignments

1. On an A3- or B-size sheet make a working drawing of the column bracket

LOCAL NOTES

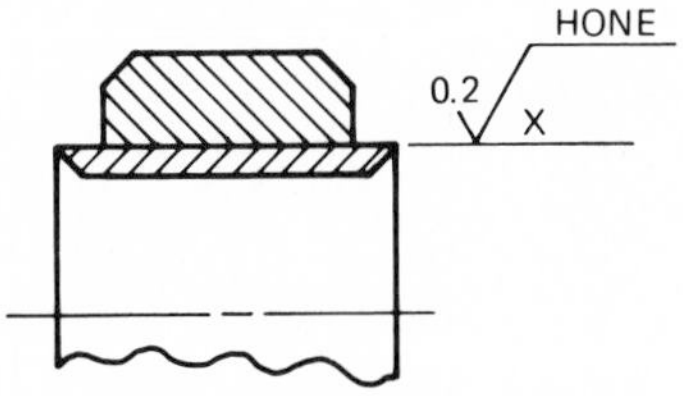
HONE
0.2
X

*ROUGHNESS VALUE SHOWN IN MICROMETRES.

GENERAL NOTES

(A) ALL SURFACES xx√ (. . . . . .)*

(B) ALL SURFACES xx√ (. . . . . .)* UNLESS OTHERWISE SPECIFIED.

(C) ALL (. . . . . .)** SURFACE xx√ (. . . . . .)* UNLESS OTHERWISE SPECIFIED.

(D) SURFACES MARKED √ TO BE xx√.

* SOCIETY, MILITARY, COMPANY OR OTHER APPLICABLE SPECIFICATIONS MAY BE ADDED THUSLY "IN ACCORDANCE WITH SPECIFICATIONS XXXX".

** INSERT THE APPLICABLE WORD "MACHINED", "CAST", "FORGED", ETC.

Fig. 5-8-8 Surface texture notes.

MICROMETRES AA RATING	MICROINCHES AA RATING	APPLICATION
25.2√	1000√	ROUGH, LOW GRADE SURFACE RESULTING FROM SAND CASTING, TORCH OR SAW CUTTING, CHIPPING OR ROUGH FORGING. MACHINE OPERATIONS ARE NOT REQUIRED AS APPEARANCE IS NOT OBJECTIONABLE. THIS SURFACE, RARELY SPECIFIED, IS SUITABLE FOR UNMACHINED CLEARANCE AREAS ON ROUGH CONSTRUCTION ITEMS.
12.5√	500√	ROUGH, LOW GRADE SURFACE RESULTING FROM HEAVY CUTS AND COARSE FEEDS IN MILLING TURNING, SHAPING, BORING, AND ROUGH FILING, DISC GRINDING AND SNAGGING. IT IS SUITABLE FOR CLEARANCE AREAS ON MACHINERY, JIGS, AND FIXTURES. SAND CASTING OR ROUGH FORGING PRODUCES THIS SURFACE.
6.3√	250√	COARSE PRODUCTION SURFACES, FOR UNIMPORTANT CLEARANCE AND CLEANUP OPERATIONS, RESULTING FROM COARSE SURFACE GRIND, ROUGH FILE, DISC GRIND, RAPID FEEDS IN TURNING, MILLING, SHAPING, DRILLING, BORING, GRINDING, ETC., WHERE TOOL MARKS ARE NOT OBJECTIONABLE. THE NATURAL SURFACES OF FORGINGS, PERMANENT MOLD CASTINGS, EXTRUSIONS, AND ROLLED SURFACES ALSO PRODUCE THIS ROUGHNESS. IT CAN BE PRODUCED ECONOMICALLY AND IS USED ON PARTS WHERE STRESS REQUIREMENTS, APPEARANCE, AND CONDITIONS OF OPERATIONS AND DESIGN PERMIT.
3.2√	125√	THE ROUGHEST SURFACE RECOMMENDED FOR PARTS SUBJECT TO LOADS, VIBRATION, AND HIGH STRESS. IT IS ALSO PERMITTED FOR BEARING SURFACES WHEN MOTION IS SLOW AND LOADS LIGHT OR INFREQUENT. IT IS A MEDIUM COMMERCIAL MACHINE FINISH PRODUCED BY RELATIVELY HIGH SPEEDS AND FINE FEEDS TAKING LIGHT CUTS WITH SHARP TOOLS. IT MAY BE ECONOMICALLY PRODUCED ON LATHES, MILLING MACHINES, SHAPERS, GRINDERS, ETC., OR ON PERMANENT MOLD CASTINGS, DIE CASTINGS, EXTRUSION, AND ROLLED SURFACES.
1.6√	63√	A GOOD MACHINE FINISH PRODUCED UNDER CONTROLLED CONDITIONS USING RELATIVELY HIGH SPEEDS AND FINE FEEDS TO TAKE LIGHT CUTS WITH SHARP CUTTERS. IT MAY BE SPECIFIED FOR CLOSE FITS AND USED FOR ALL STRESSED PARTS, EXCEPT FAST ROTATING SHAFTS, AXLES, AND PARTS SUBJECT TO SEVERE VIBRATION OR EXTREME TENSION. IT IS SATISFACTORY FOR BEARING SURFACES WHEN MOTION IS SLOW AND LOADS LIGHT OR INFREQUENT. IT MAY ALSO BE OBTAINED ON EXTRUSIONS, ROLLED SURFACES, DIE CASTINGS AND PERMANENT MOLD CASTINGS WHEN RIGIDLY CONTROLLED.
0.8√	32√	A HIGH-GRADE MACHINE FINISH REQUIRING CLOSE CONTROL WHEN PRODUCED BY LATHES, SHAPERS, MILLING MACHINES, ETC., BUT RELATIVELY EASY TO PRODUCE BY CENTERLESS, CYLINDRICAL OR SURFACE GRINDERS. ALSO, EXTRUDING, ROLLING, OR DIE CASTING MAY PRODUCE A COMPARABLE SURFACE WHEN RIGIDLY CONTROLLED. THIS SURFACE MAY BE SPECIFIED IN PARTS WHERE STRESS CONCENTRATION IS PRESENT. IT IS USED FOR BEARINGS WHEN MOTION IS NOT CONTINUOUS AND LOADS ARE LIGHT. WHEN FINER FINISHES ARE SPECIFIED, PRODUCTION COSTS RISE RAPIDLY; THEREFORE, SUCH FINISHES MUST BE ANALYZED CAREFULLY.
0.4√	16√	A HIGH QUALITY SURFACE PRODUCED BY FINE CYLINDRICAL GRINDING, EMERY BUFFING, COARSE HONING OR LAPPING. IT IS SPECIFIED WHERE SMOOTHNESS IS OF PRIMARY IMPORTANCE, SUCH AS RAPIDLY ROTATING SHAFT BEARINGS, HEAVILY LOADED BEARINGS AND EXTREME TENSION MEMBERS.
0.2√	8√	A FINE SURFACE PRODUCED BY HONING, LAPPING, OR BUFFING. IT IS SPECIFIED WHERE PACKINGS AND RINGS MUST SLIDE ACROSS THE DIRECTION OF THE SURFACE GRAIN, MAINTAINING OR WITHSTANDING PRESSURES, OR FOR INTERIOR HONED SURFACES OF HYDRAULIC CYLINDERS. IT MAY ALSO BE REQUIRED IN PRECISION GAGES AND INSTRUMENT WORK, OR SENSITIVE VALUE SURFACES, OR ON RAPIDLY ROTATING SHAFTS AND ON BEARINGS WHERE LUBRICATION IS NOT DEPENDABLE.
0.1√	4√	A COSTLY REFINED SURFACE PRODUCED BY HONING, LAPPING, AND BUFFING. IT IS SPECIFIED ONLY WHEN THE REQUIREMENTS OF DESIGN MAKE IT MANDATORY. IT IS REQUIRED IN INSTRUMENT WORK, GAGE WORK, AND WHERE PACKINGS AND RINGS MUST SLIDE ACROSS THE DIRECTION OF SURFACE GRAIN SUCH AS ON CHROME-PLATED PISTON RODS, ETC., WHERE LUBRICATION IS NOT DEPENDABLE.
0.05√ 0.025√	2√ 1√	COSTLY REFINED SURFACES PRODUCED ONLY BY THE FINEST OF MODERN HONING, LAPPING, BUFFING, AND SUPERFINISHING EQUIPMENT. THESE SURFACES MAY HAVE A SATIN OR HIGHLY POLISHED APPEARANCE DEPENDING ON THE FINISHING OPERATION AND MATERIAL. THESE SURFACES ARE SPECIFIED ONLY WHEN DESIGN REQUIREMENTS MAKE IT MANDATORY. THEY ARE SPECIFIED ON FINE OR SENSITIVE INSTRUMENT PARTS OR OTHER LABORATORY ITEMS, AND CERTAIN GAGE SURFACES, SUCH AS ON PRECISION GAGE BLOCKS.

Fig. 5-8-9 Typical surface roughness height applications.

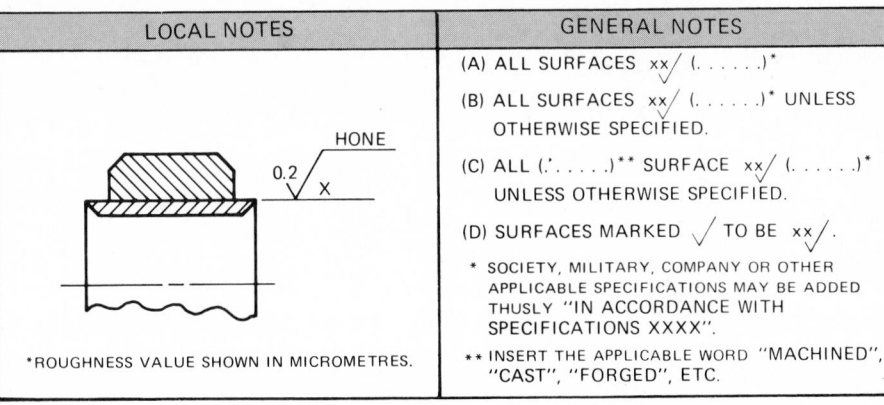

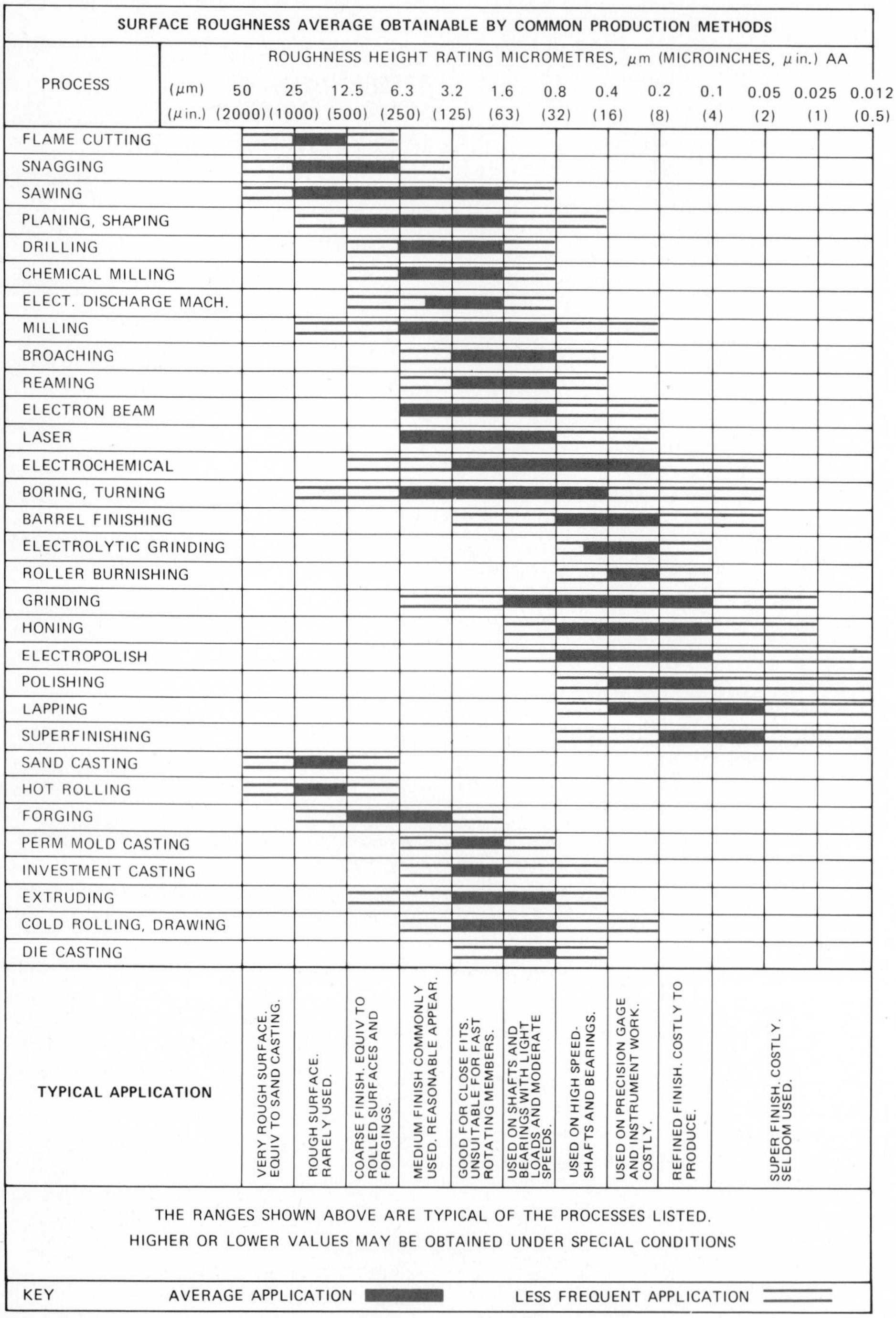
SURFACE ROUGHNESS AVERAGE OBTAINABLE BY COMMON PRODUCTION METHODS
PROCESS
ROUGHNESS HEIGHT RATING MICROMETRES, μm (MICROINCHES, μin.) AA
(μm) 50 25 12.5 6.3 3.2 1.6 0.8 0.4 0.2 0.1 0.05 0.025 0.012
(μin.) (2000) (1000) (500) (250) (125) (63) (32) (16) (8) (4) (2) (1) (0.5)
FLAME CUTTING
SNAGGING
SAWING
PLANING, SHAPING
DRILLING
CHEMICAL MILLING
ELECT. DISCHARGE MACH.
MILLING
BROACHING
REAMING
ELECTRON BEAM
LASER
ELECTROCHEMICAL
BORING, TURNING
BARREL FINISHING
ELECTROLYTIC GRINDING
ROLLER BURNISHING
GRINDING
HONING
ELECTROPOLISH
POLISHING
LAPPING
SUPERFINISHING
SAND CASTING
HOT ROLLING
FORGING
PERM MOLD CASTING
INVESTMENT CASTING
EXTRUDING
COLD ROLLING, DRAWING
DIE CASTING
TYPICAL APPLICATION
VERY ROUGH SURFACE. EQUIV TO SAND CASTING.
ROUGH SURFACE. RARELY USED.
COARSE FINISH. EQUIV TO ROLLED SURFACES AND FORGINGS.
MEDIUM FINISH COMMONLY USED. REASONABLE APPEAR.
GOOD FOR CLOSE FITS. UNSUITABLE FOR FAST ROTATING MEMBERS.
USED ON SHAFTS AND BEARINGS WITH LIGHT LOADS AND MODERATE SPEEDS.
USED ON HIGH SPEED-SHAFTS AND BEARINGS.
USED ON PRECISION GAGE AND INSTRUMENT WORK. COSTLY.
REFINED FINISH. COSTLY TO PRODUCE.
SUPER FINISH. COSTLY. SELDOM USED.
THE RANGES SHOWN ABOVE ARE TYPICAL OF THE PROCESSES LISTED.
HIGHER OR LOWER VALUES MAY BE OBTAINED UNDER SPECIAL CONDITIONS
KEY AVERAGE APPLICATION LESS FREQUENT APPLICATION

Fig. 5-8-10 Surface roughness range for common production.

shown in Figs. 5-8-A or 5-8-B. The following surface texture information is to be added to the drawing of Fig. 5-8-A.

The bottom of the base is to have a maximum roughness value of 3.2 μm and a machining allowance of 3 mm.

The tops of the bosses are to have a maximum roughness value of 6.3 μm and a machining allowance of 2 mm.

The end surfaces of the hubs supporting the shafts are to have maximum and minimum roughness values of 1.6 and 0.8 μm and a machining allowance of 2 mm.

The large hole is to be dimensioned for an RC4 fit. The small hole is to be dimensioned for an LN3 fit with plain bearings.

For Fig. 5-8-B use inch-size equivalents. Scale 1:1.

REVIEW FOR ASSIGNMENTS

Unit 5-1 Basic Dimensioning
Unit 5-2 Dimensioning Circular Features
Unit 5-6 Fits
Unit 5-7 Machining Symbols

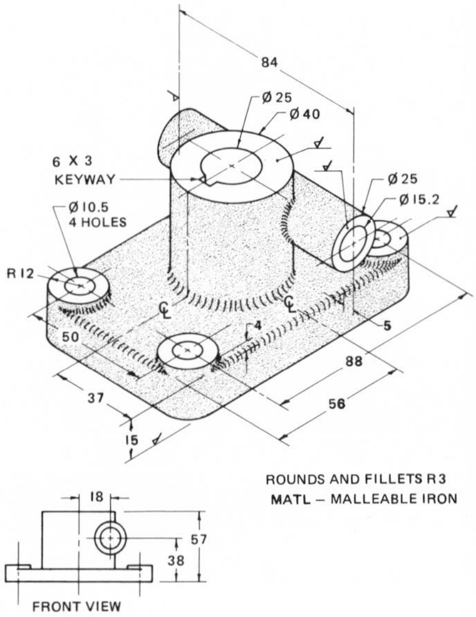
84
Ø 25
Ø 40
6 X 3
KEYWAY
Ø 10.5
4 HOLES
Ø 25
Ø 15.2
R 12
50
4
5
88
37
56
15
ROUNDS AND FILLETS R 3
MATL – MALLEABLE IRON
18
57
38
FRONT VIEW

Fig. 5-8-A Column bracket.

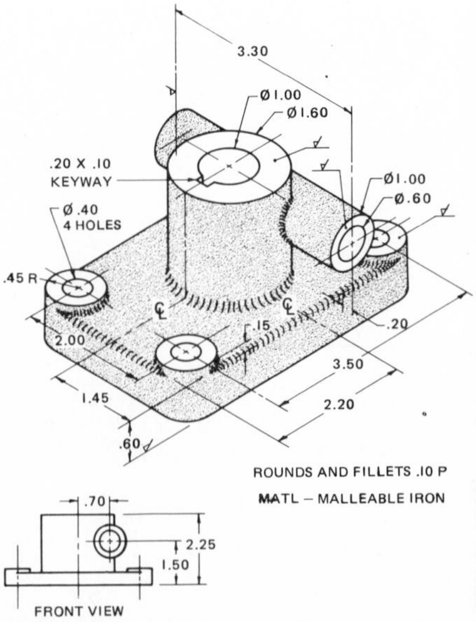
3.30
Ø 1.00
Ø 1.60
.20 X .10
KEYWAY
Ø .40
4 HOLES
Ø 1.00
Ø .60
.45 R
2.00
.15
.20
3.50
1.45
2.20
.60
ROUNDS AND FILLETS .10 R
MATL – MALLEABLE IRON
.70
2.25
1.50
FRONT VIEW

Fig. 5-8-B Column bracket.

Chapter 6
Working Drawings

UNIT 6-1
WORKING DRAWINGS

A *working drawing* is a drawing that supplies information and instructions for the manufacture or construction of machines or structures. Generally, working drawings may be classified into two groups: detail drawings, which provide the necessary information for the manufacture of the parts, and assembly drawings, which supply the necessary information for their assembly.

Since working drawings may be sent to other companies to make or assemble the parts, the drawings should conform with the drawing standards of that company. For this reason, most companies follow the drawing standards of their country. The drawing standards recommended by the American National Standards Institute (ANSI) and the Canadian Standards Association (CSA), which is very similar to ANSI, have been adopted by the majority of industries in North America.

Detail Drawings

A detail drawing (Fig. 6-1-1) must supply the complete information for the construction of a part. This information may be classified under three headings: shape description, size description, and specifications.

Shape Description. This term refers to the selection and number of views to show or describe the shape of the part. The part may be drawn in either pictorial or orthographic projection, the latter being used more frequently. Sectional views, auxiliary views, and enlarged detail views may be added to the drawing in order to provide a clearer image of the part.

Size Description. Dimensions which show the size and location of the shape features are then added to the drawing. The manufacturing process will influence the selection of some dimensions, such as datum features. Tolerances are then selected for each dimension.

Specifications. This term refers to general notes, material, heat treatment, finish, general tolerances, and number required. This information is located on or near the title block or strip.

Additional Drawing Information. In addition to the information pertaining to the part, a detail drawing includes additional information such as drawing number, scale, method or projection, date, name of part or parts, and the drafter's name.

The selection of paper size is determined by the number of views selected, the number of general notes required,

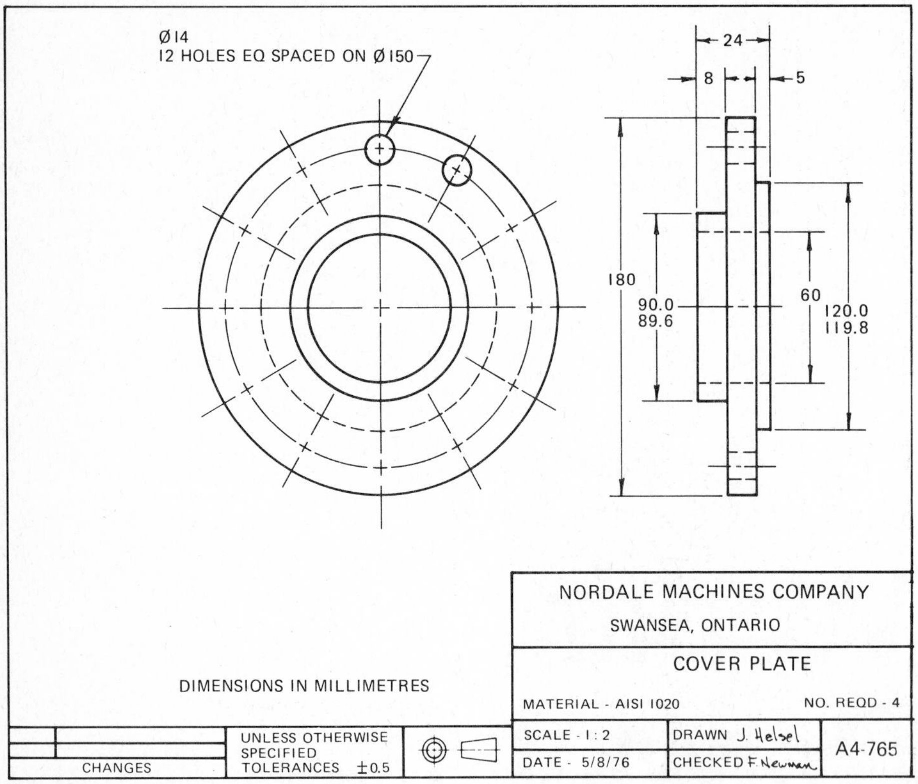
Ø14
12 HOLES EQ SPACED ON Ø150
24
8
5
180
90.0
89.6
60
120.0
119.8
NORDALE MACHINES COMPANY
SWANSEA, ONTARIO
COVER PLATE
DIMENSIONS IN MILLIMETRES
MATERIAL - AISI 1020
NO. REQD - 4
UNLESS OTHERWISE SPECIFIED TOLERANCES ±0.5
CHANGES
SCALE - 1 : 2
DATE - 5/8/76
DRAWN J. Helsel
CHECKED F. Newman
A4-765

Fig. 6-1-1 A simple detail drawing.

and the drawing scale used. If the drawing is to be microfilmed, then the lettering size would be another factor to consider. The drawing number usually carries a prefix or suffix number or letter to indicate the sheet size, such as 4-571 or A-571; the number 4 indicates that the drawing is made on a 210 × 297 mm sheet, and the letter A indicates that the drawing is made on an 8.50 × 11.00 in. sheet.

Drawing Checklist

As an added precaution against errors occurring on a drawing, many companies have provided checklists for drafters to follow before a drawing is issued to the shop. A typical checklist may be as follows:

1. *Dimensions.* Is the part fully dimensioned, and are the dimensions clearly positioned? Is the drawing dimensioned to avoid unnecessary shop calculations?
2. *Scale.* Is the drawing to scale? Is the scale shown?
3. *Tolerances.* Are the clearances and tolerances indicated by the linear and angular dimensions and by local, general, or title block notes suitable for proper functioning? Are they realistic? Can they be liberalized?
4. *Standards.* Have standard parts, design, materials, processes, or other items been used where possible?
5. *Surface Texture.* Have surface roughness values been shown where required? Are the values shown compatible with overall design requirements?
6. *Material.* Have proper material and heat treatment been specified?

QUALIFICATIONS OF A DETAILER

The detailer should have a thorough understanding of materials, shop processes, and operations in order to properly dimension the part and call for the correct finish and material. In addition, the detailer must have a thorough knowledge of how the part functions in order to provide the correct data and tolerances for each dimension.

The detailer may be called upon to work from a complete set of instructions and drawings, or he or she may be required to make working drawings of parts which involve the design of the part. Design considerations are limited in this chapter but are covered in detail in Chap. 33.

Manufacturing Methods

The type of manufacturing process will influence the selection of material and detailed feature of a part (see Fig. 6-1-2).

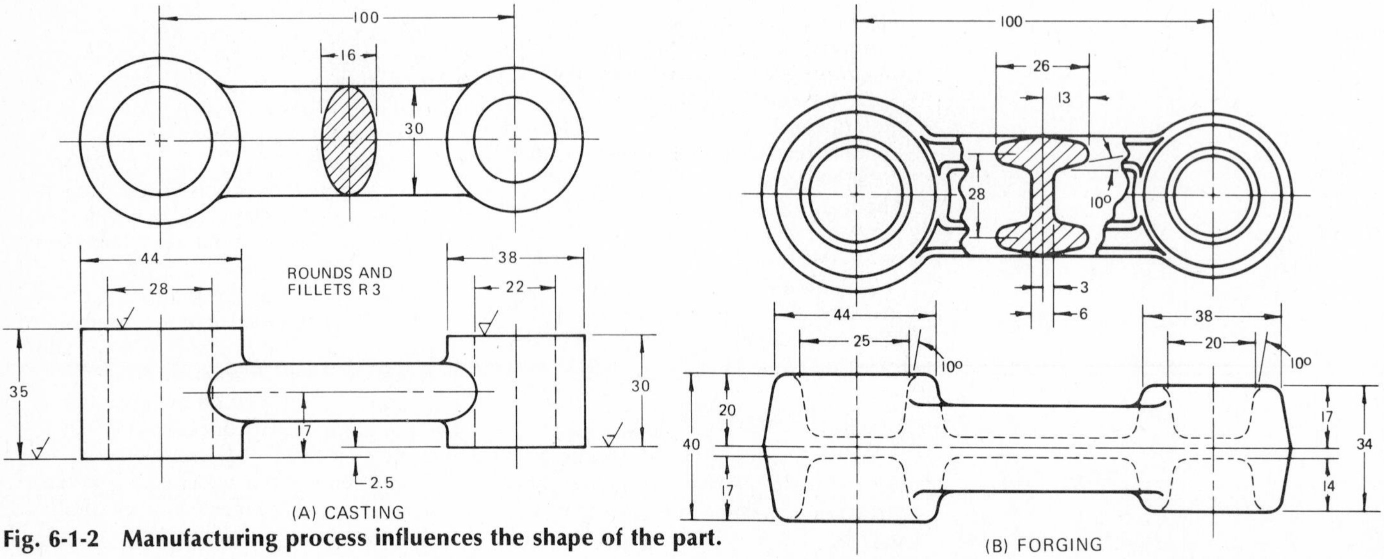
100
16
30
44
28
ROUNDS AND FILLETS R 3
38
22
35
30
17
2.5
(A) CASTING
100
26
13
28
10°
3
6
44
25
10°
38
20
10°
20
40
17
17
34
14
(B) FORGING

Fig. 6-1-2 Manufacturing process influences the shape of the part.

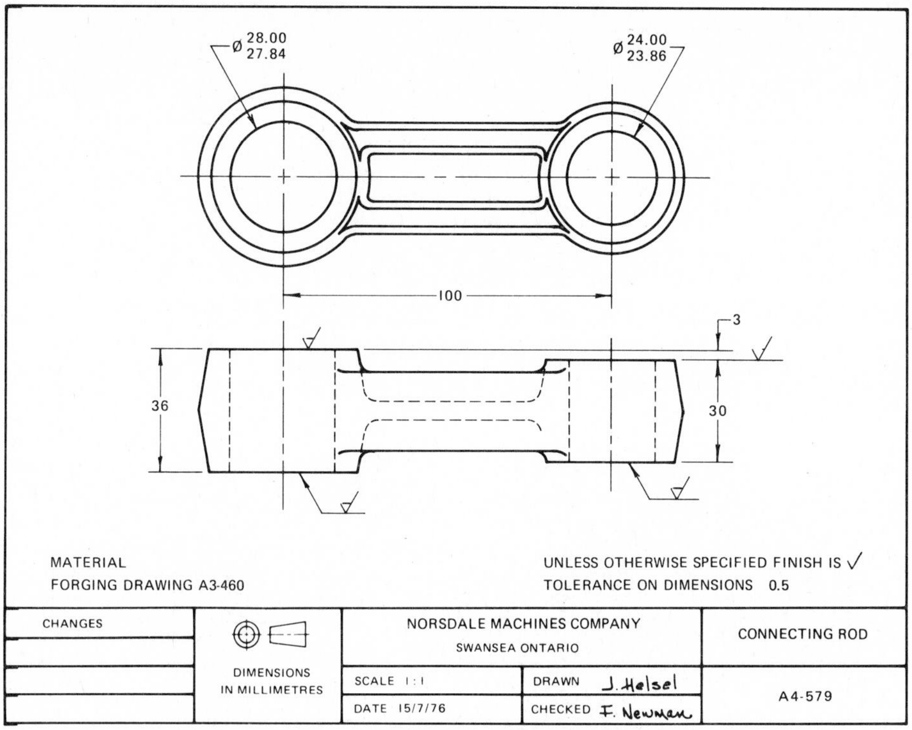
Ø 28.00
27.84
Ø 24.00
23.86
100
3
36
30
MATERIAL
FORGING DRAWING A3-460
UNLESS OTHERWISE SPECIFIED FINISH IS √
TOLERANCE ON DIMENSIONS 0.5
CHANGES
DIMENSIONS IN MILLIMETRES
NORSDALE MACHINES COMPANY
SWANSEA ONTARIO
CONNECTING ROD
SCALE 1:1
DATE 15/7/76
DRAWN J. Helsel
CHECKED F. Newman
A4-579

Fig. 6-1-3 Machining drawing for forged part shown in Fig. 6-1-2*b*.

For example, if it has been decided that the part is to be cast, then rounds and fillets will be added to the part. Additional material will also be required where surfaces are to be finished.

The more common manufacturing processes are machining from standard stock; prefabrication which includes welding, riveting, soldering, brazing, and gluing; forming from sheet stock; and casting and forging. The latter two processes can be justified only when large quantities are required and for specially designed parts. All these processes are described in detail in other chapters.

Several drawings may be made for the same part, each one giving only the information necessary for a particular step in the manufacture of the part. A part which is to be produced by forging, for example, may have one drawing showing the forged part and one detail of the finished forged part (see Figs. 6-1-2*b* and 6-1-3).

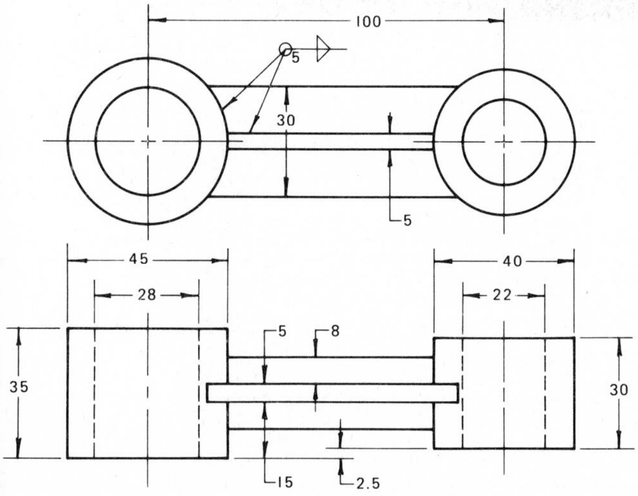
100
5
30
5
45
28
40
22
35
5
8
30
15
2.5

(C) WELDMENT

Assignment

On an A3- or B-size sheet, make a working drawing of two of the parts shown in Figs. 6-1-A and 6-1-B. Select appropriate views and dimensions and add to the drawing the information needed so that the parts can be completely manufactured. Scale is to suit.

REVIEW FOR ASSIGNMENT

Chapter 5 Basic Dimensioning
Unit 3-2 Shape Description

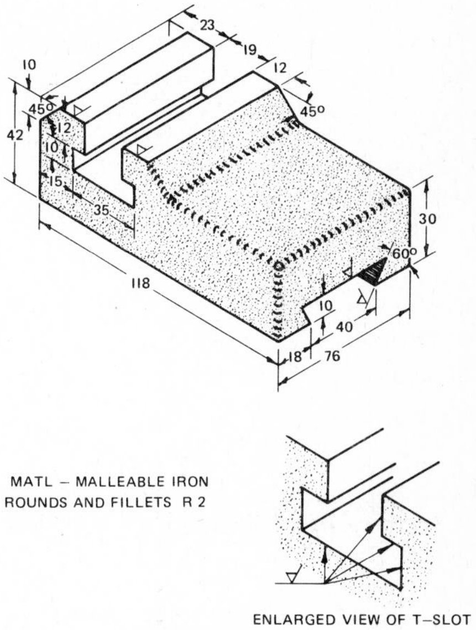
23
19
12
10
45°
45°
42
12
10
15
35
30
118
60°
10
40
18
76
MATL – MALLEABLE IRON
ROUNDS AND FILLETS R 2
ENLARGED VIEW OF T–SLOT

Fig. 6-1-A Cross slide.

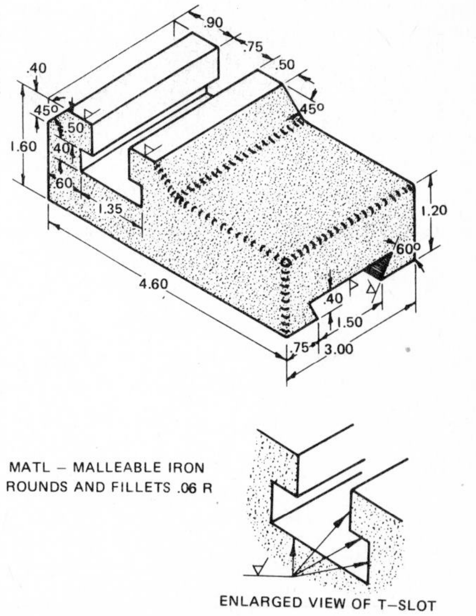
.90
.75
.50
.40
45°
45°
1.60
.50
.40
.60
1.35
1.20
4.60
60°
.40
1.50
.75
3.00
MATL – MALLEABLE IRON
ROUNDS AND FILLETS .06 R
ENLARGED VIEW OF T–SLOT

Fig. 6-1-B Cross slide.

UNIT 6-2

MULTIPLE DETAIL DRAWINGS

Detail drawings may be shown on separate sheets, or they may be grouped on one or more large sheets.

Often the detailing of parts is grouped according to the department in which they are made. For example, wood, fiber, and metal parts are used in the assembly of a transformer. Three separate detail sheets—one for wood parts, one for fiber parts, and the third for the metal parts—may be drawn. These parts would be made in the different shops and sent to another area for assembly. In order to facilitate assembly, each part is given an identification part number which is shown on the assembly drawing. A typical detail drawing showing multiple parts is illustrated in Fig. 6-2-1.

If the details are few, the assembly drawing may appear on the same sheet or sheets.

Assignments

1. On an A3- or B-size sheet, make detail drawings of all the parts shown of one of the assemblies in Figs. 6-2-A to 6-2-D. Since time is money, select only the views necessary to describe each part. Below each part show the following information: part number, name of part, material, number required. Scale is 1:1.

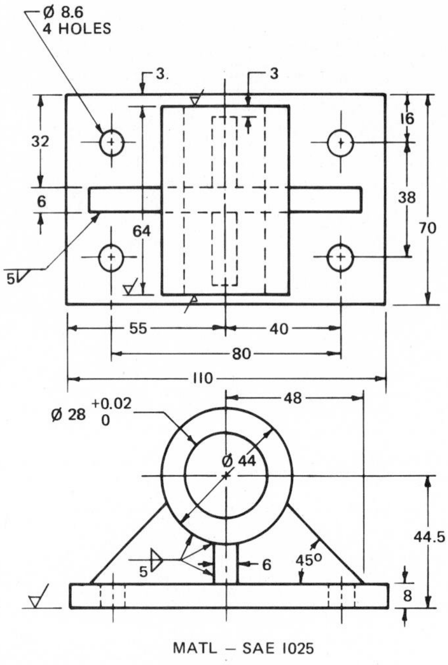
Ø 8.6
4 HOLES
MATL – SAE 1025

Fig. 6-2-A Shaft support.

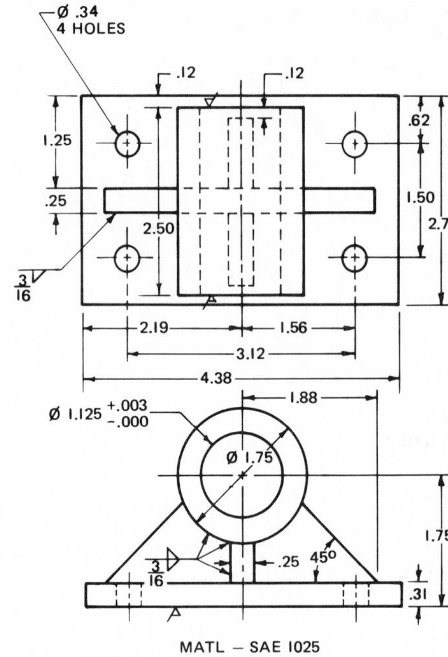
Ø .34
4 HOLES
MATL – SAE 1025

Fig. 6-2-B Shaft support.

REVIEW FOR ASSIGNMENT

Unit 3-2 Shape Description
Chapter 5 Basic Dimensioning

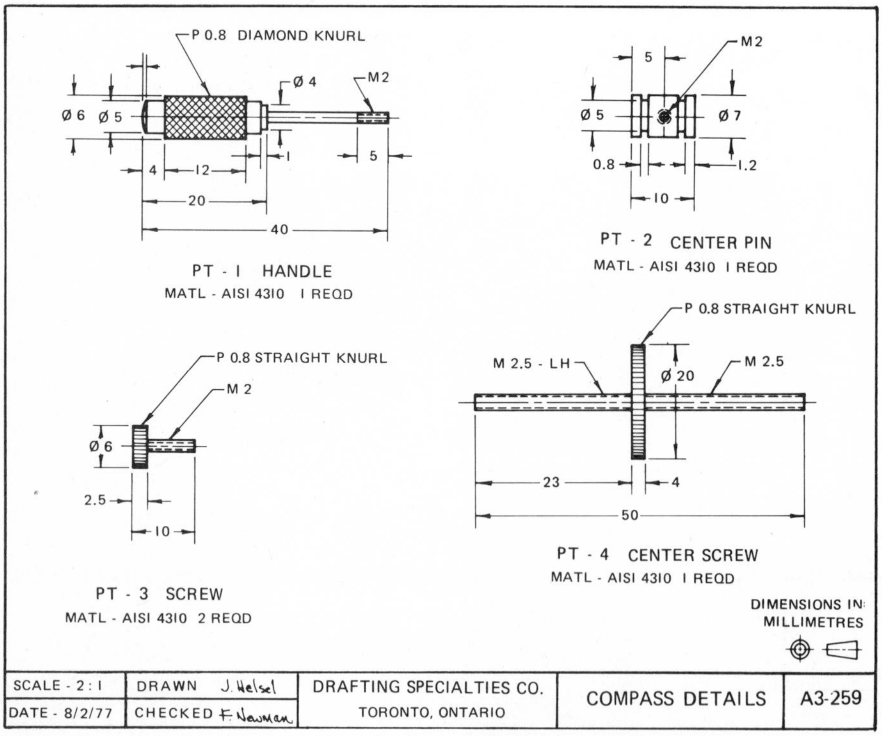
P 0.8 DIAMOND KNURL
PT - 1 HANDLE
MATL - AISI 4310 1 REQD
PT - 2 CENTER PIN
MATL - AISI 4310 1 REQD
P 0.8 STRAIGHT KNURL
PT - 3 SCREW
MATL - AISI 4310 2 REQD
P 0.8 STRAIGHT KNURL
PT - 4 CENTER SCREW
MATL - AISI 4310 1 REQD
DIMENSIONS IN MILLIMETRES
SCALE - 2 : 1
DATE - 8/2/77
DRAWN J. Helsel
CHECKED F. Newman
DRAFTING SPECIALTIES CO.
TORONTO, ONTARIO
COMPASS DETAILS
A3-259

Fig. 6-2-1 Detail drawing containing many details on one drawing.

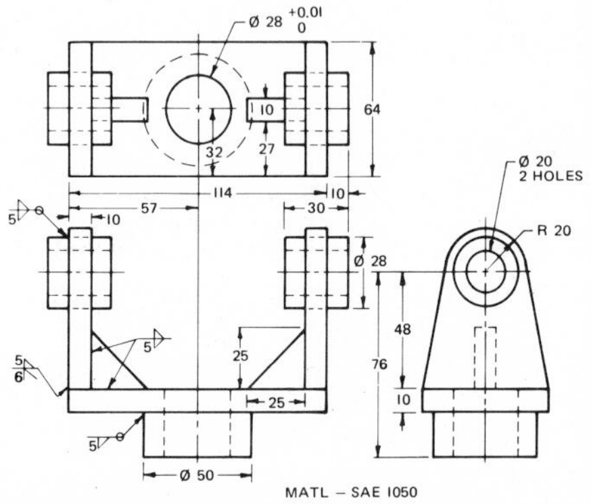
Ø 20
2 HOLES
R 20
MATL – SAE 1050

Fig. 6-2-C Shaft pivot support.

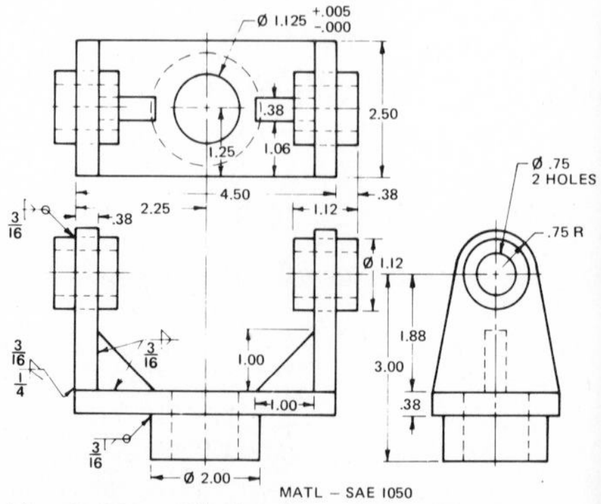
Ø .75
2 HOLES
.75 R
MATL – SAE 1050

Fig. 6-2-D Shaft pivot support.

UNIT 6-3

DRAWING REVISIONS[1,2]

Revisions are made to an existing drawing when manufacturing methods are improved, to reduce cost, to correct errors, and to improve design. A clear record of these revisions must be registered on the drawing.

All drawings must carry a change or revision table, either down the right-hand side or across the bottom of the drawing. In addition to a description of drawing changes, provision may be made for recording a revision symbol, a zone location, an issue number, a date, and the approval of the change. Should the drawing revision cause a dimension or dimensions to be other than the scale indicated on the drawing, then the dimensions that are not to scale should be indicated by one of the methods shown in Fig. 5-1-15. Typical revision tables are shown in Fig. 6-3-1.

At times, when there are a large number of revisions to be made, it may be more economical to make a new drawing. When this is done, the words REDRAWN and REVISED should appear in the revision column of the new drawing. A new date is also shown for updating old prints.

REFERENCES

1. ANSI Y14.5, *Dimensions and Tolerancing*.
2. CSA B78.2, *Dimensioning and Tolerancing of Mechanical Engineering Drawings*.

Assignment

Select two of the drawings shown in Figs. 6-3-A through 6-3-D and make appropriate revisions to these drawings, recording the changes in the drawing revision column.

REVIEW FOR ASSIGNMENT

Unit 5-1 Not-to-Scale Drawings

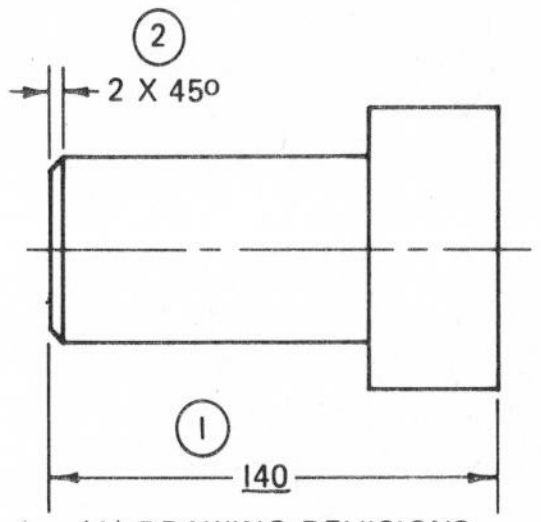

2
2 X 45°
1
140

(A) DRAWING REVISIONS

REVISIONS		
SYMBOL	DESCRIPTION	DATE & APPROVAL
1	LENGTH WAS 150	J. Helsel 12-4-76
2	CHAMFER ADDED	F. Newman 2-2-77

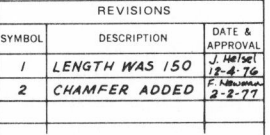

(B) VERTICAL REVISION BLOCK

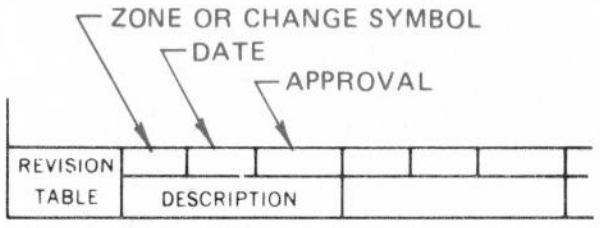

ZONE OR CHANGE SYMBOL
DATE
APPROVAL
REVISION TABLE
DESCRIPTION

(C) HORIZONTAL REVISION BLOCK

Fig. 6-3-1 Drawing revisions.

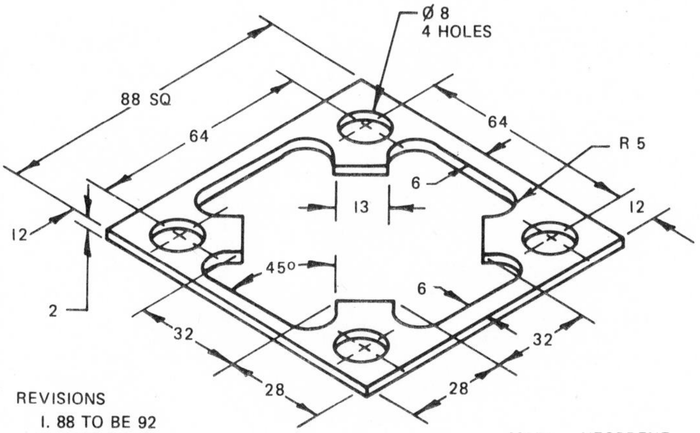

Ø 8
4 HOLES
88 SQ
64
64
R 5
6
13
12
12
45°
6
2
32
32
28
28

REVISIONS
1. 88 TO BE 92
2. 12 TO BE 14
3. Ø 8 TO BE Ø 10

MATL – NEOPRENE

Fig. 6-3-A Gasket.

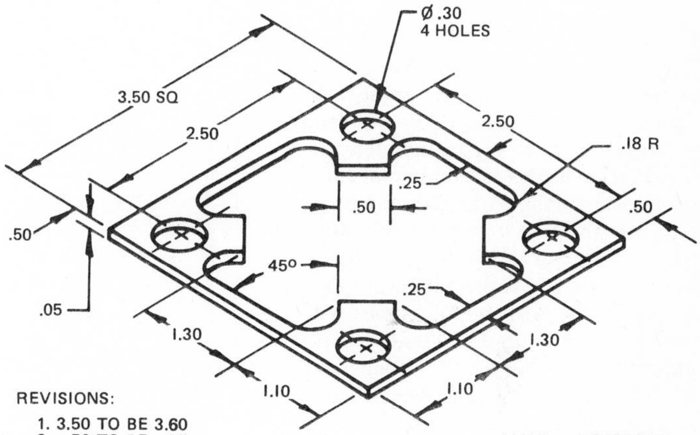

Ø .30
4 HOLES
3.50 SQ
2.50
2.50
.18 R
.25
.50
.50
.50
45°
.25
.05
1.30
1.30
1.10
1.10

REVISIONS:
1. 3.50 TO BE 3.60
2. .50 TO BE .55
3. Ø .30 TO BE Ø .40

MATL – NEOPRENE

Fig. 6-3-B Gasket.

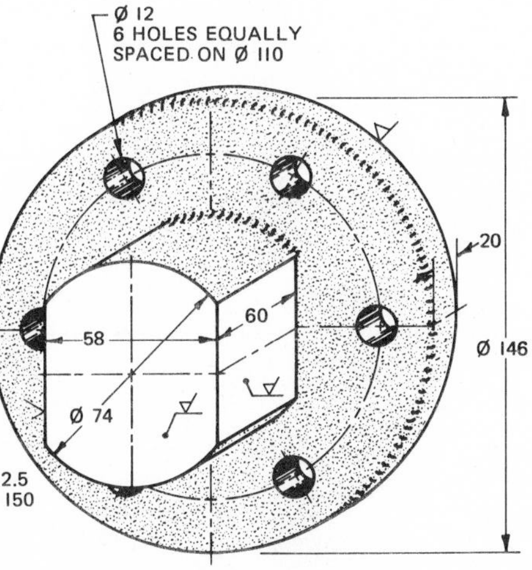

Ø 12
6 HOLES EQUALLY
SPACED ON Ø 110
20
60
58
Ø 146
Ø 74

REVISIONS:
1. Ø 12 TO BE Ø 12.5
2. Ø 146 TO BE Ø 150
3. 58 TO BE 60
4. 60 TO BE 50

ROUNDS AND FILLETS R 2
MATL – MALLEABLE IRON

Fig. 6-3-C Axle cap.

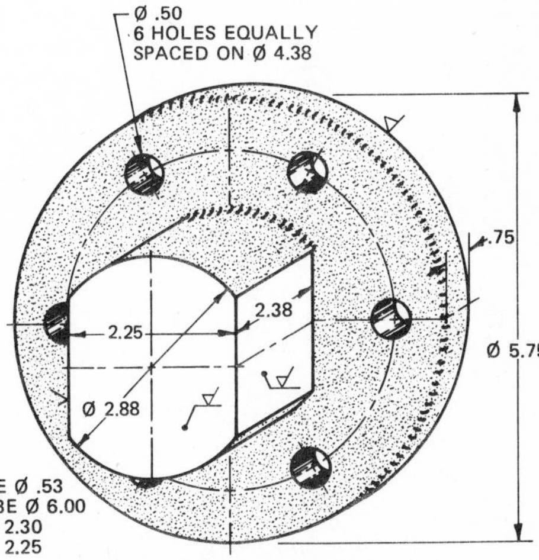

Ø .50
6 HOLES EQUALLY
SPACED ON Ø 4.38
.75
2.38
2.25
Ø 5.75
Ø 2.88

REVISIONS:
1. Ø .50 TO BE Ø .53
2. Ø 5.75 TO BE Ø 6.00
3. 2.25 TO BE 2.30
4. 2.38 TO BE 2.25

ROUNDS AND FILLETS .06 R
MATL – MALLEABLE IRON

Fig. 6-3-D Axle cap.

UNIT 6-4
ASSEMBLY DRAWINGS

All machines and mechanisms are composed of numerous parts. A drawing showing the product in its completed state is called an *assembly drawing*.

Assembly drawings vary greatly in the amount and type of information they give, depending on the nature of the machine or mechanism they depict. The primary functions of the assembly drawing are to show the product in its completed shape, to indicate the relationship of its various components, and to designate these components by a part or detail number. Other information that might be given includes overall dimensions, capacity dimensions, relationship dimensions between parts (necessary information for assembly), operating instructions, and data on design characteristics.

DESIGN ASSEMBLY DRAWINGS

When a machine is designed, an assembly drawing or a design layout is first drawn to clearly visualize the performance, shape, and clearances of the various parts. From this assembly drawing, the detail drawings are made and each part is given a part number.

To assist in the assembling of the machine, part numbers of the various details are placed on the assembly drawing. Small circles, 8 to 12 mm in diameter, that contain the part number are then attached to the corresponding part with a leader, as illustrated in Fig. 6-4-1. It is important that the detail drawings not use identical numbering schemes when several bills of material are used.

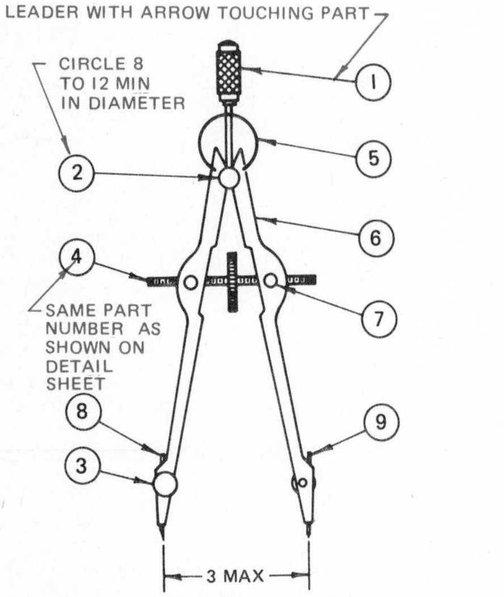
LEADER WITH ARROW TOUCHING PART
CIRCLE 8 TO 12 MIN IN DIAMETER
SAME PART NUMBER AS SHOWN ON DETAIL SHEET
3 MAX
1
2
3
4
5
6
7
8
9

Fig. 6-4-1 Identification numbers on assembly drawings.

INSTALLATION ASSEMBLY DRAWINGS

This type of assembly drawing is used when many unskilled persons are employed to mass-assemble parts. Since these people are not normally trained to read technical drawings, simplified pictorial assembly drawings similar to the one shown in Fig. 6-4-2 are used.

ASSEMBLY DRAWINGS FOR CATALOGS

Special assembly drawings are prepared for company catalogs. These assembly drawings show only pertinent details and dimensions that would interest the potential buyer. Often one drawing, having letter dimensions accompanied by a chart, is used to cover a range of sizes, such as the pillow block shown in Fig. 6-4-3*b*.

BILLS OF MATERIAL

A *bill of material* is an itemized list of all the components shown on an assembly drawing or a detail drawing (see Fig. 6-4-4). Often, a bill of material is placed on a separate sheet for ease of handling and duplicating. Since the bill of material is used by the purchasing department to order the necessary material for the design, the bill of material should show the raw material size, rather than the finished size of the part.

For castings, a pattern number should

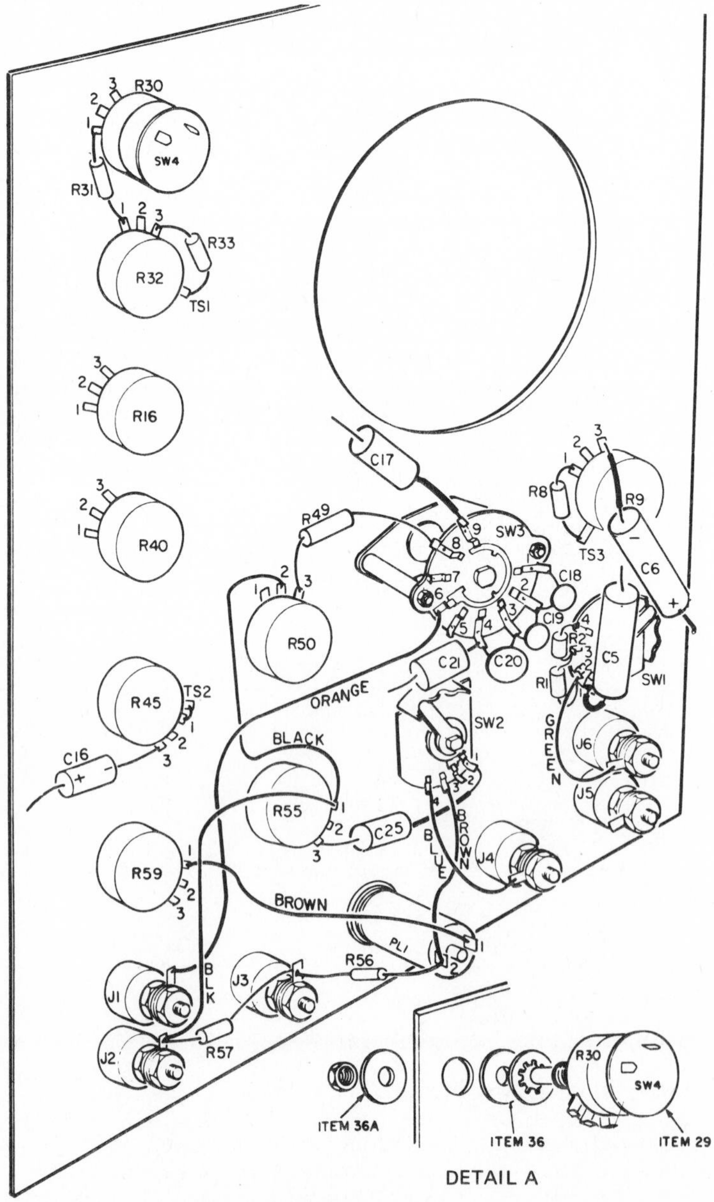
R30
SW4
R31
R33
R32
TS1
R16
R40
C17
R49
R8
R9
SW3
TS3
C6
C18
C19
C5
R50
C21
C20
R1
SW1
R45
TS2
ORANGE
SW2
GREEN
J6
C16
BLACK
J5
R55
C25
BLUE
BROWN
J4
R59
BROWN
PL1
R56
J1
J3
BLK
J2
R57
ITEM 36A
ITEM 36
ITEM 29
DETAIL A

Fig. 6-4-2 Installation assembly drawings.

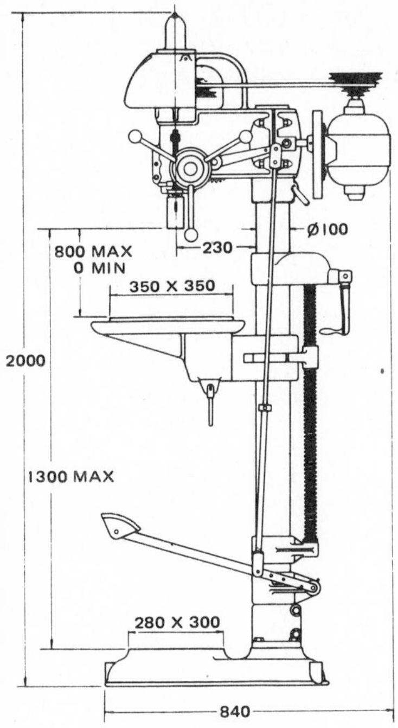
800 MAX
0 MIN
230
Ø100
350 X 350
2000
1300 MAX
280 X 300
840

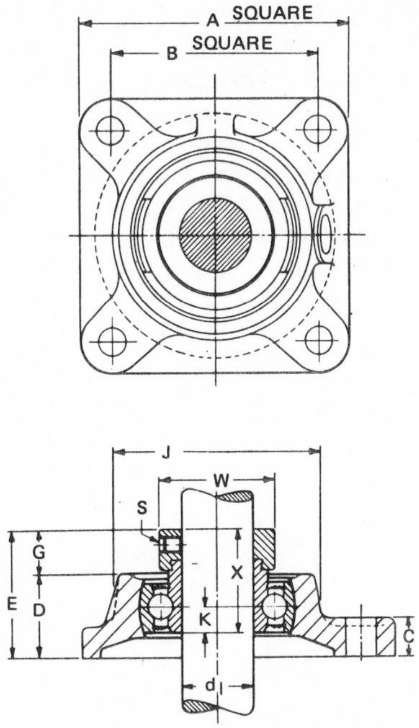
A SQUARE
B SQUARE
J
W
S
G
E
D
X
K
C
d1

A CHART IS USED WITH THIS TYPE OF DRAWING TO COVER A RANGE IN SIZES

(A) DRILL PRESS (B) PILLOW BLOCK

Fig. 6-4-3 Assembly drawings used in catalogs.

QTY	ITEM	MATL	DESCRIPTION	PT NO.
1	BASE	CI	PATTERN #A3154	1
1	CAP	CI	PATTERN #B7156	2
1	SUPPORT	MS	10 X 50 X 110	3
1	BRACE	MS	5.0 X 25 X 50	4
1	COVER	ST	3.57 (#10 GA USS) X 150 X 180	5
1	SHAFT	CRS	Ø 25 X 160	6
2	BEARINGS	SKF	RADIAL BALL #6200Z	7
2	RETAINING CLIP	TRUARC	N5000-725	8
1	KEY	ST	WOODRUFF #608	9
1	SET SCREW	CUP POINT	HEX SOCKET M6 X 10	10
4	BOLT–HEX HD–REG	SEMI-FIN	M10 X 40	11
4	NUT–REG HEX	ST	M10	12
4	LOCK WASHER–SPRING	ST	10 MED	13
				14

(A) TYPICAL BILL OF MATERIAL PARTS 7 TO 13 ARE PURCHASED ITEMS

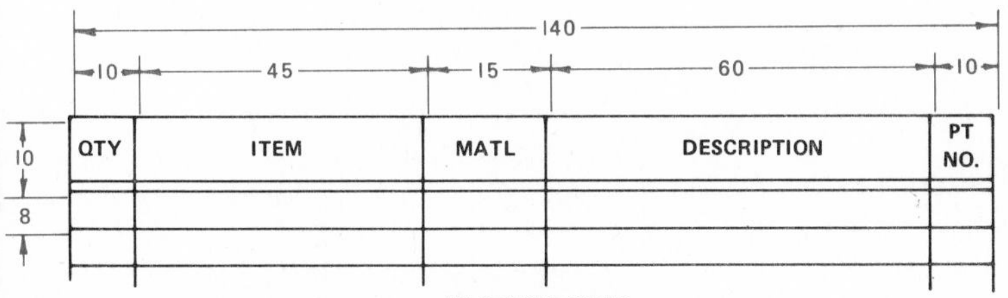
140
10
45
15
60
10
QTY
ITEM
MATL
DESCRIPTION
PT NO.
10
8

(B) SAMPLE SIZES

Fig. 6-4-4 Bill of material.

appear in the size column in lieu of the physical size of the part.

Standard components, which are purchased rather than fabricated, such as bolts, nuts, and bearings, should have a part number and appear on the bill of material. Information in the descriptive column should be sufficient for the purchasing agent to order these parts.

Parts lists for bills of material placed on the bottom of the drawing should read from bottom to top, while bills of material placed on the top of the drawings should read from top to bottom. This practice allows additions to be made at a later date.

Assignments

On an A3- or B-size sheet, make a one-view assembly drawing of one of the assemblies shown in Figs. 6-4-A and 6-4-B. Show a round bar (25 mm or 1 in.) in phantom being held in position.

Include on the drawing a bill of material and identification part numbers. Scale is 1:1.

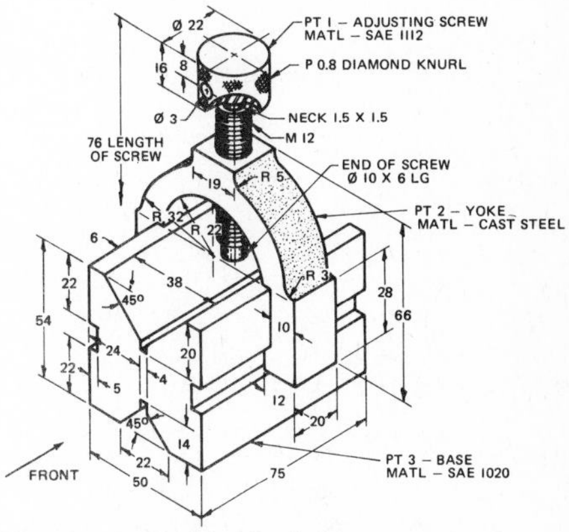
PT 1 – ADJUSTING SCREW
MATL – SAE 1112
P 0.8 DIAMOND KNURL
NECK 1.5 X 1.5
M 12
END OF SCREW
Ø 10 X 6 LG
PT 2 – YOKE
MATL – CAST STEEL
76 LENGTH OF SCREW
FRONT
PT 3 – BASE
MATL – SAE 1020

Fig. 6-4-A V-block clamp.

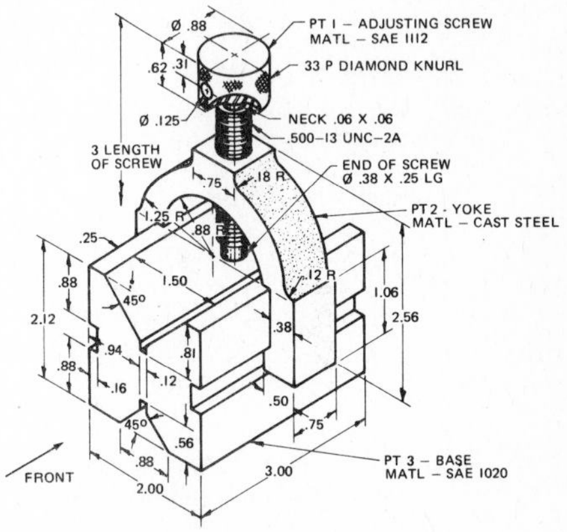
PT 1 – ADJUSTING SCREW
MATL – SAE 1112
33 P DIAMOND KNURL
NECK .06 X .06
.500–13 UNC–2A
END OF SCREW
Ø .38 X .25 LG
PT2 - YOKE
MATL – CAST STEEL
3 LENGTH OF SCREW
FRONT
PT 3 – BASE
MATL – SAE 1020

Fig. 6-4-B V-block clamp.

UNIT 6-5
EXPLODED ASSEMBLY DRAWINGS

In many instances parts must be identified or assembled by persons unskilled in the reading of engineering drawings. Examples are found in the appliance repair industry, which relies on assembly drawings for repair work and for reordering parts. Exploded assembly drawings, like that shown in Fig. 6-5-1 are used extensively in these cases, for they are easier to read. This type of assembly drawing is also used frequently by companies that manufacture do-it-yourself assembly kits, such as model-making kits.

For this type of drawing, the parts are aligned in position. Frequently, shading techniques are used to make the drawings appear more realistic.

Assignment

On an A3- or B-size sheet, make an exploded assembly in orthographic projection of one of the assemblies shown in Figs. 6-5-A and 6-5-B. Use center lines to align parts and holes. To make the parts appear more realistic, shading techniques are recommended. Scale is 1:1.

REVIEW FOR ASSIGNMENT

Unit 15-1 Pictorial Drawings
Unit 15-8 Technical Illustration

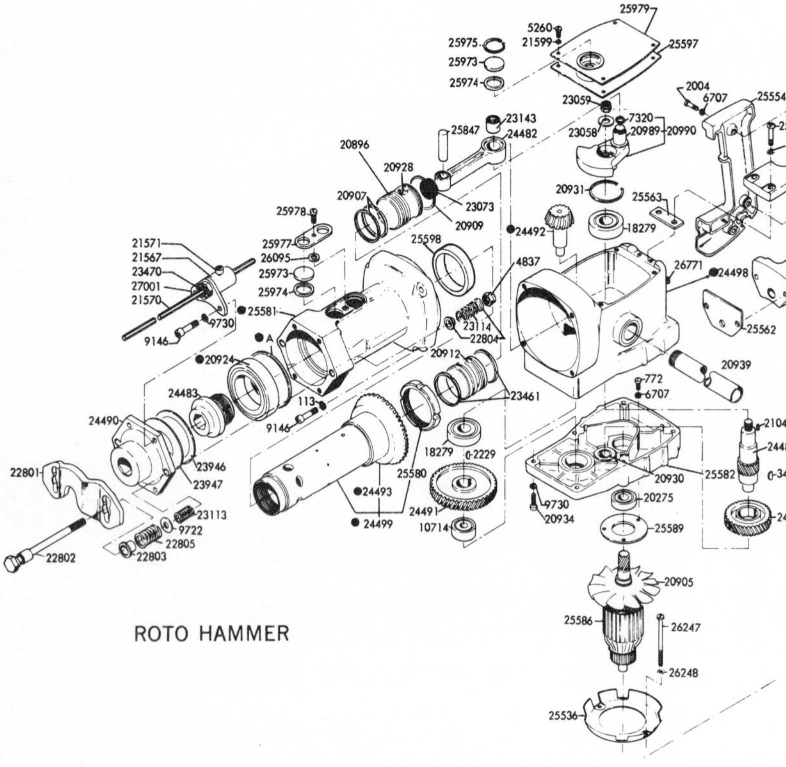
ROTO HAMMER

(A) PICTORIAL EXPLODED ASSEMBLY

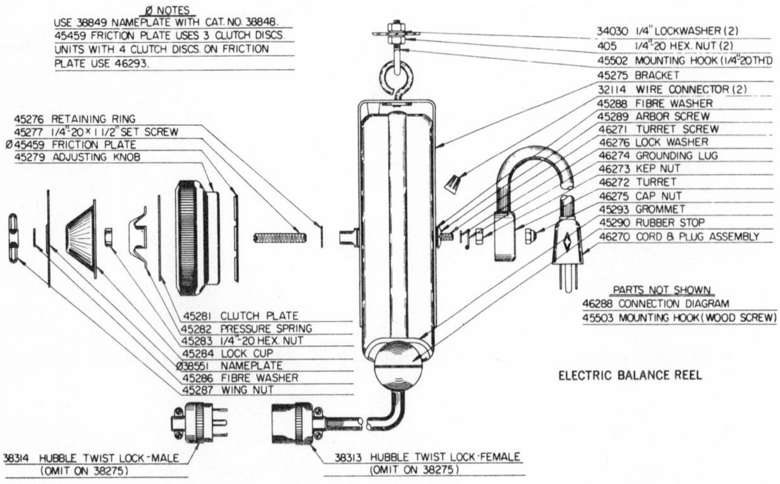
Ø NOTES
USE 38849 NAMEPLATE WITH CAT. NO. 38848.
45459 FRICTION PLATE USES 3 CLUTCH DISCS.
UNITS WITH 4 CLUTCH DISCS. ON FRICTION PLATE USE 46293.
45276 RETAINING RING
45277 1/4"-20 × 1 1/2" SET SCREW
Ø45459 FRICTION PLATE
45279 ADJUSTING KNOB
45281 CLUTCH PLATE
45282 PRESSURE SPRING
45283 1/4"-20 HEX. NUT
45284 LOCK CUP
Ø38551 NAMEPLATE
45286 FIBRE WASHER
45287 WING NUT
38314 HUBBLE TWIST LOCK-MALE (OMIT ON 38275)
38313 HUBBLE TWIST LOCK-FEMALE (OMIT ON 38275)
34030 1/4" LOCKWASHER (2)
405 1/4"-20 HEX. NUT (2)
45502 MOUNTING HOOK (1/4"-20 TH'D
45275 BRACKET
32114 WIRE CONNECTOR (2)
45288 FIBRE WASHER
45289 ARBOR SCREW
46271 TURRET SCREW
46276 LOCK WASHER
46274 GROUNDING LUG
46273 KEP NUT
46272 TURRET
46275 CAP NUT
45293 GROMMET
45290 RUBBER STOP
46270 CORD & PLUG ASSEMBLY
PARTS NOT SHOWN
46288 CONNECTION DIAGRAM
45503 MOUNTING HOOK (WOOD SCREW)
ELECTRIC BALANCE REEL

(B) ORTHOGRAPHIC EXPLODED ASSEMBLY

Fig. 6-5-1 Exploded assembly drawings.

UNIT 6-6
DETAILED ASSEMBLY DRAWINGS

Often these are made for fairly simple objects, such as pieces of furniture, when the parts are few in number and are not intricate in shape. All the dimensions and information necessary for the construction of each part and for the assembly of the parts are given directly on the assembly drawing. Separate views of specific parts, in enlargements showing the fitting together of parts, may also be drawn in addition to the regular assembly drawing. Note that in Fig. 6-6-1 the enlarged views are drawn in picture form, not as regular orthographic views. This method is peculiar to the cabinetmaking trade and is not normally used in mechanical drawing.

Assignment

On an A3- or B-size sheet, make a detailed assembly drawing of one of the assemblies shown in Figs. 6-6-A and 6-6-B. Include on the drawing the method of assembly (i.e., nailing, wood screws, dowling, etc.) and a bill of material. Include in the bill of material the assembly materials. Scale is 1:10 for Fig. 6-6-A. Scale is 1:8 for their inch equivalents (Fig. 6-6-B).

REVIEW FOR ASSIGNMENT

Unit 6-4 Assembly Drawings
Unit 6-1 Detail Drawings

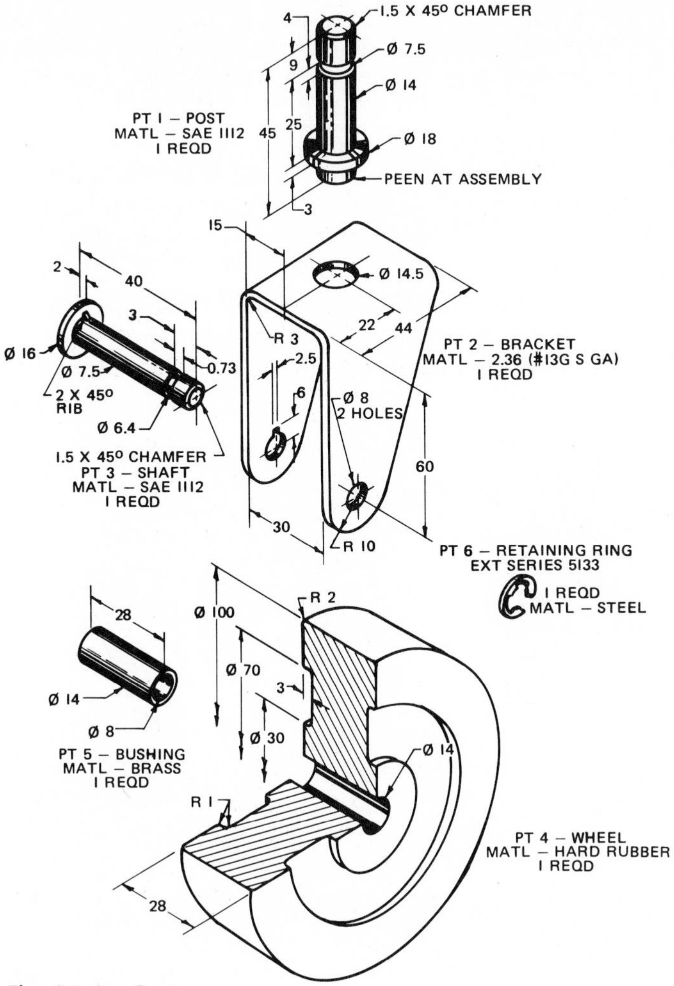
1.5 X 45° CHAMFER
Ø 7.5
Ø 14
Ø 18
PEEN AT ASSEMBLY
PT 1 – POST
MATL – SAE 1112
1 REQD
4
9
25
45
3
15
Ø 14.5
22
44
PT 2 – BRACKET
MATL – 2.36 (#13G S GA)
1 REQD
R 3
2.5
6
Ø 8
2 HOLES
60
30
R 10
2
40
3
0.73
Ø 16
Ø 7.5
2 X 45°
RIB
Ø 6.4
1.5 X 45° CHAMFER
PT 3 – SHAFT
MATL – SAE 1112
1 REQD
PT 6 – RETAINING RING
EXT SERIES 5133
1 REQD
MATL – STEEL
28
Ø 100
R 2
Ø 70
3
Ø 30
Ø 14
Ø 8
PT 5 – BUSHING
MATL – BRASS
1 REQD
Ø 14
R 1
PT 4 – WHEEL
MATL – HARD RUBBER
1 REQD
28

Fig. 6-5-A Caster.

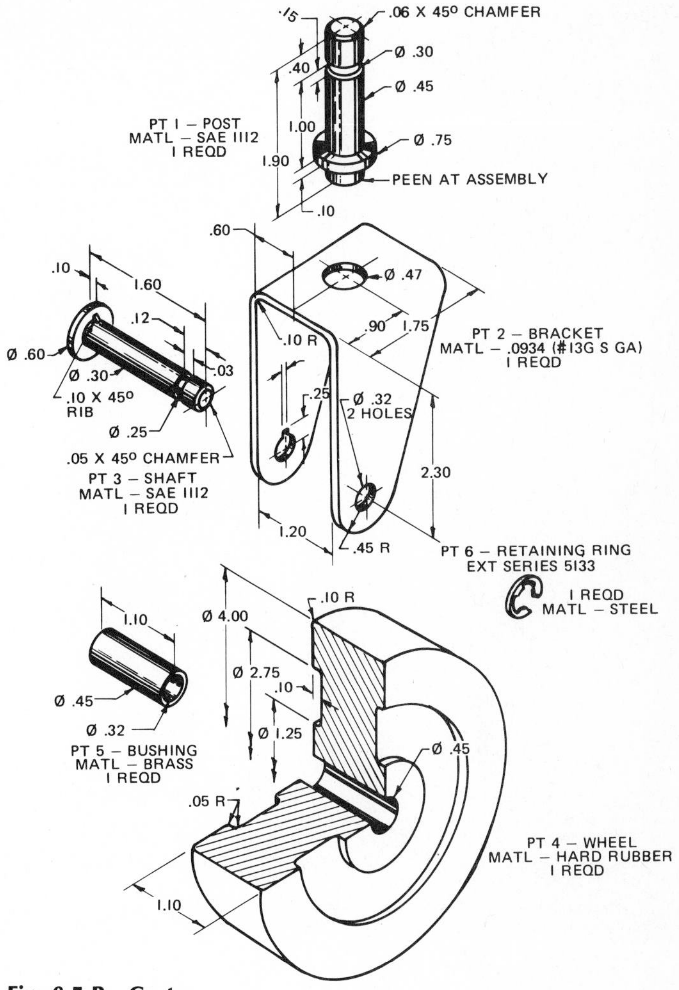
.06 X 45° CHAMFER
Ø .30
Ø .45
Ø .75
PEEN AT ASSEMBLY
.15
.40
1.00
1.90
.10
PT 1 – POST
MATL – SAE 1112
1 REQD
.60
Ø .47
.90
1.75
PT 2 – BRACKET
MATL – .0934 (#13G S GA)
1 REQD
.10 R
.25
Ø .32
2 HOLES
2.30
1.20
.45 R
.10
1.60
.12
.03
Ø .60
Ø .30
.10 X 45°
RIB
Ø .25
.05 X 45° CHAMFER
PT 3 – SHAFT
MATL – SAE 1112
1 REQD
PT 6 – RETAINING RING
EXT SERIES 5133
1 REQD
MATL – STEEL
1.10
Ø 4.00
.10 R
Ø 2.75
.10
Ø 1.25
Ø .45
Ø .45
Ø .32
PT 5 – BUSHING
MATL – BRASS
1 REQD
.05 R
PT 4 – WHEEL
MATL – HARD RUBBER
1 REQD
1.10

Fig. 6-5-B Caster.

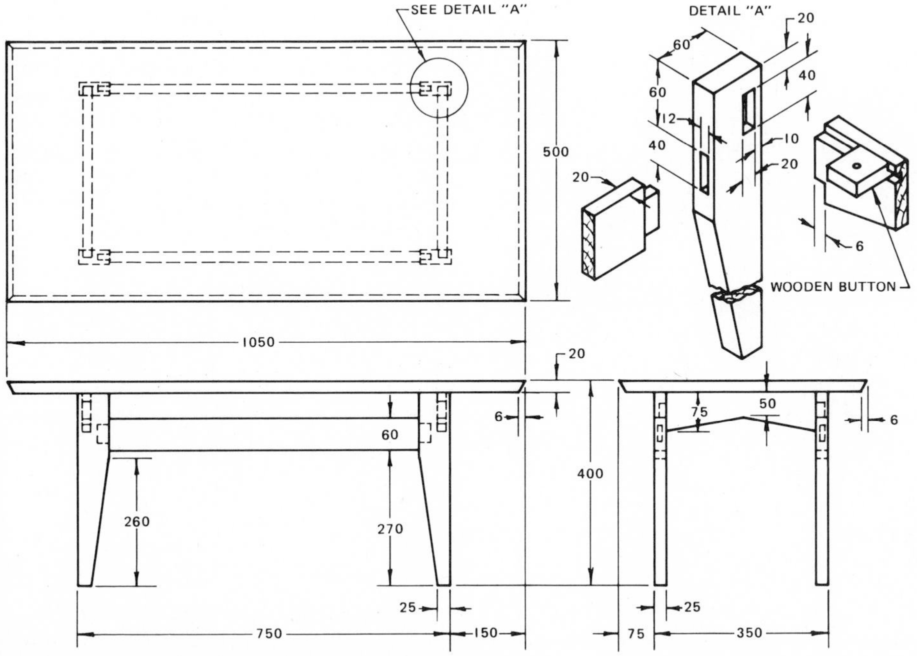
SEE DETAIL "A"
DETAIL "A"
500
1050
20
60
60
12
40
20
40
10
20
6
WOODEN BUTTON
20
6
60
260
270
400
75
50
6
25
25
750
150
75
350

Fig. 6-6-1 Detailed assembly drawing.

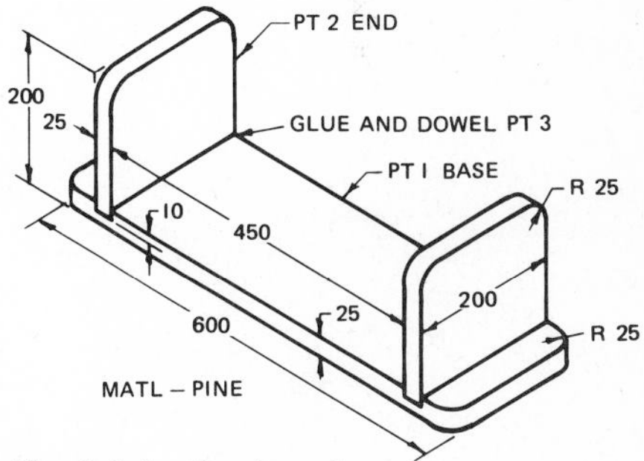
PT 2 END
200
25
GLUE AND DOWEL PT 3
PT 1 BASE
R 25
10
450
600
25
200
R 25
MATL – PINE

Fig. 6-6-A Book rack.

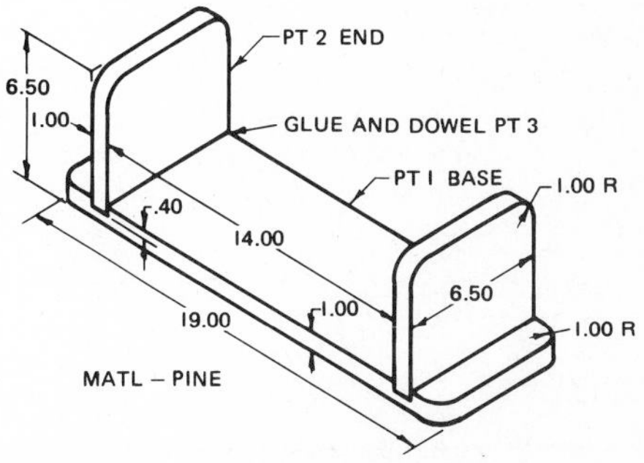
PT 2 END
6.50
1.00
GLUE AND DOWEL PT 3
PT 1 BASE
1.00 R
.40
14.00
19.00
1.00
6.50
1.00 R
MATL – PINE

Fig. 6-6-B Book rack.

UNIT 6-7

SUBASSEMBLY DRAWINGS

Many completely assembled items, such as a car and television set, are assembled with many preassembled components as well as individual parts. These preassembled units are referred to as *subassemblies*. The assembly drawings of a transmission for an automobile and the transformer for a television set are typical examples of subassembly drawings.

Subassemblies are designed to simplify final assembly as well as permit the item to be either assembled in a more suitable area or purchased from an outside source. This type of drawing shows only those dimensions which would be required for the completed assembly. Examples are size of the mounting holes and their location, shaft locations, and overall sizes. This type of drawing is found frequently in catalogs.

Assignment

On an A3- or B-size sheet make a one-view subassembly drawing of one of the assemblies shown in Figs. 6-7-A and 6-7-B. A broken-out or partial section view is recommended to show the interior features. Include on the drawing pertinent dimensions, identification numbers on assembly drawing, a bill of material, and a phantom outline of the adjoining part or features. Scale is 1:1.

REVIEW FOR ASSIGNMENT

Unit 6-4 Assembly Drawings
Unit 2-6 Phantom Lines

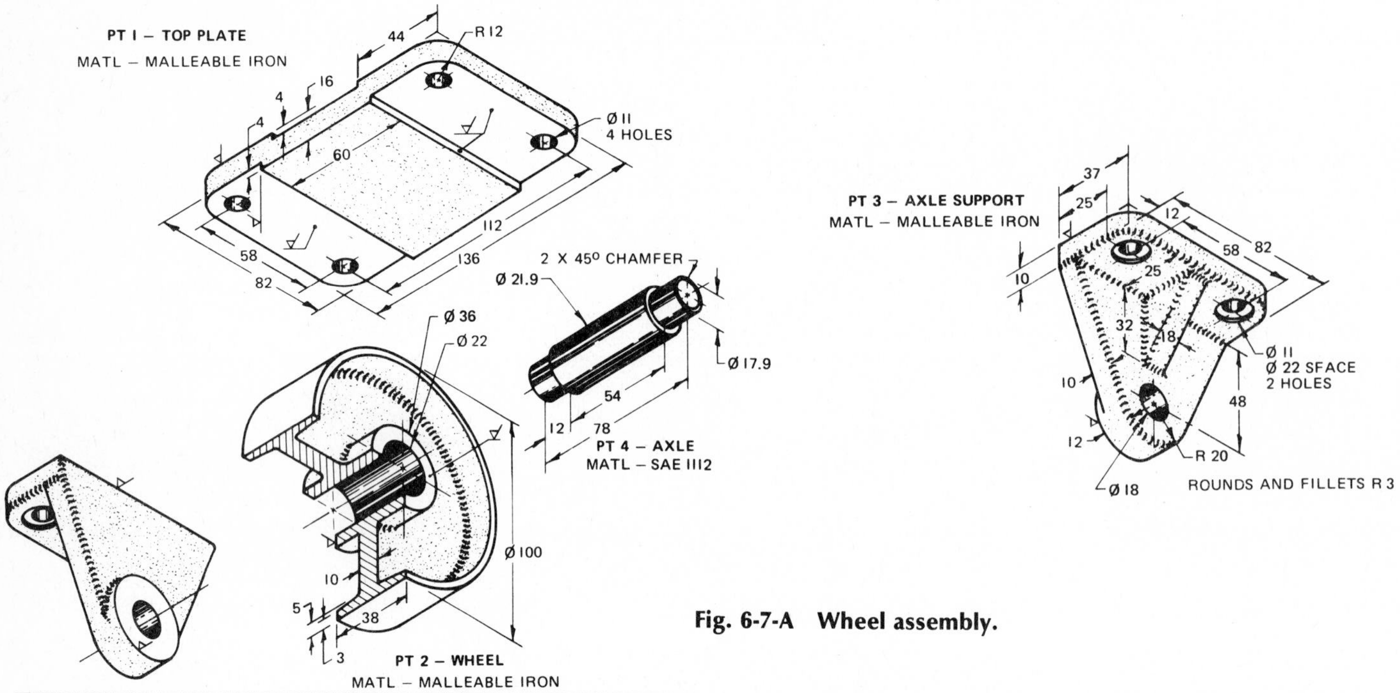

PT 1 – TOP PLATE
MATL – MALLEABLE IRON
44
R 12
16
4
4
60
Ø 11
4 HOLES
112
58
82
136
2 X 45° CHAMFER
Ø 21.9
Ø 17.9
54
12
78
PT 4 – AXLE
MATL – SAE 1112
Ø 36
Ø 22
Ø 100
10
5
38
3
PT 2 – WHEEL
MATL – MALLEABLE IRON
PT 3 – AXLE SUPPORT
MATL – MALLEABLE IRON
37
25
12
58
82
10
25
32
18
Ø 11
Ø 22 SFACE
2 HOLES
10
48
12
R 20
Ø 18
ROUNDS AND FILLETS R 3

Fig. 6-7-A Wheel assembly.

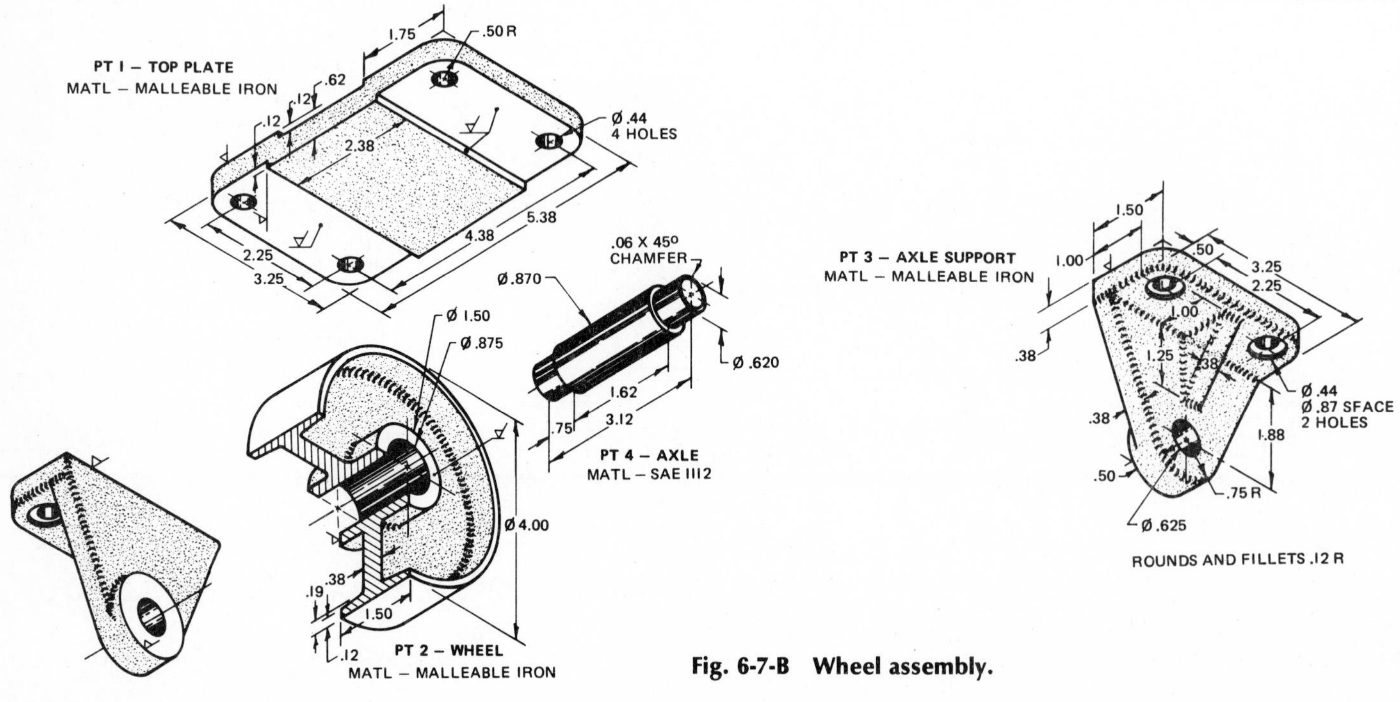

PT 1 – TOP PLATE
MATL – MALLEABLE IRON
1.75
.50 R
.62
.12
.12
2.38
Ø .44
4 HOLES
5.38
4.38
2.25
3.25
.06 X 45°
CHAMFER
Ø .870
Ø .620
1.62
.75
3.12
PT 4 – AXLE
MATL – SAE 1112
Ø 1.50
Ø .875
Ø 4.00
.38
.19
1.50
.12
PT 2 – WHEEL
MATL – MALLEABLE IRON
PT 3 – AXLE SUPPORT
MATL – MALLEABLE IRON
1.50
1.00
.50
3.25
2.25
1.00
.38
1.25
.38
Ø .44
Ø .87 SFACE
2 HOLES
.38
1.88
.50
.75 R
Ø .625
ROUNDS AND FILLETS .12 R

Fig. 6-7-B Wheel assembly.

Chapter 7
Sections and Conventions

UNIT 7-1
SECTIONAL VIEWS

Sectional views, commonly called *sections*, are used to show interior detail that is too complicated to be shown clearly by regular views containing many hidden lines. For some assembly drawings, they indicate a difference in materials. A sectional view is obtained by supposing the nearest part of the object to be cut or broken away on an imaginary cutting plane. The exposed or cut surfaces are identified by section lining or cross-hatching. Hidden lines and details behind the cutting-plane line are usually omitted unless they are required for clarity or dimensioning. It should be understood that only in the sectional view is any part of the object shown as having been removed.

A sectional view frequently replaces one of the regular views. For example, a regular front view is replaced by a front view in section, as shown in Fig. 7-1-1.

Whenever practical, except for revolved sections, sectional views should be projected perpendicular to the cutting plane and be placed in the normal position for third-angle projection.

When the preferred placement is not practical, the sectional view may be removed to some other convenient position on the drawing, but it must be clearly identified, usually by two capital letters, and labeled.

CUTTING-PLANE LINES

Cutting-plane lines (Fig. 7-1-2) are used to indicate the location of cutting planes for sectional views and the viewing position for removed partial views. Two

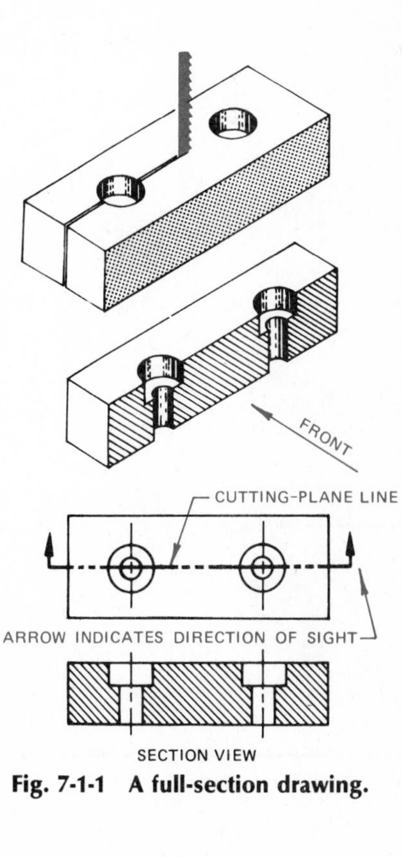

FRONT
CUTTING-PLANE LINE
ARROW INDICATES DIRECTION OF SIGHT
SECTION VIEW

Fig. 7-1-1 A full-section drawing.

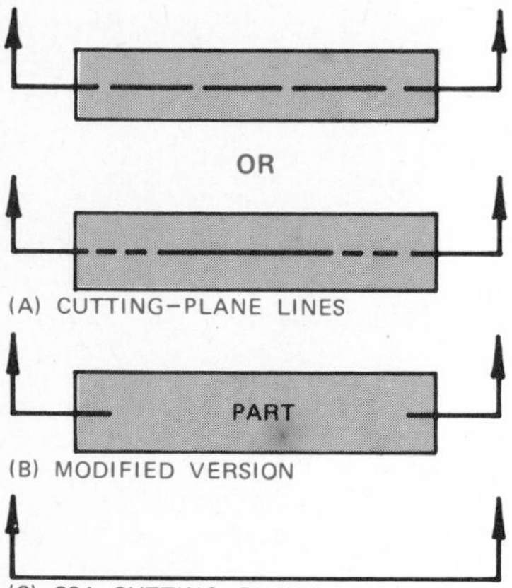
OR
(A) CUTTING–PLANE LINES
PART
(B) MODIFIED VERSION
(C) CSA CUTTING–PLANE LINE FOR SIMPLE PARTS

Fig. 7-1-2 Cutting-plane lines.

forms of cutting-plane lines are approved for general use.

The first form consists of evenly spaced dashes with arrowheads (Fig. 7-1-2*a*). The second form consists of alternating long dashes and pairs of short dashes (Fig. 7-1-2*a*). The long dashes may vary in length, depending on the size of the drawing.

Both forms of cutting-plane lines should be drawn to stand out clearly on the drawing. The ends of the lines are bent at 90° and terminated by bold arrowheads to indicate the direction of sight for viewing the section.

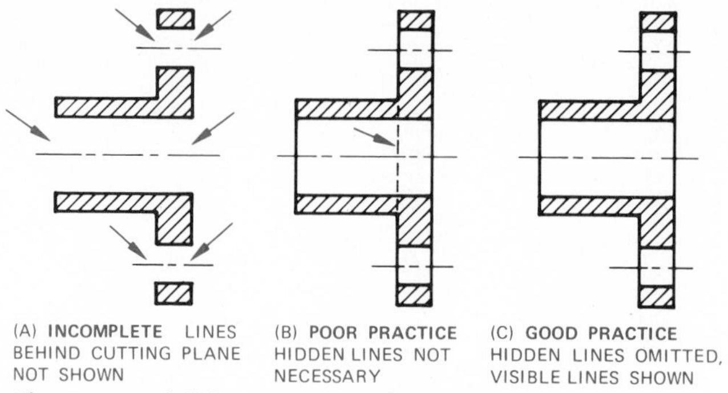
(A) INCOMPLETE LINES BEHIND CUTTING PLANE NOT SHOWN
(B) POOR PRACTICE HIDDEN LINES NOT NECESSARY
(C) GOOD PRACTICE HIDDEN LINES OMITTED, VISIBLE LINES SHOWN

Fig. 7-1-4 Visible and hidden lines on section views.

The cutting-plane line can be omitted when it corresponds to the center line of the part and it is obvious where the cutting plane lies. On drawings with a high density of line work, cutting-plane lines may be modified by omitting the dashes between the line ends for the purpose of obtaining clarity, as shown in Fig. 7-1-2*b* and *c*.

FULL SECTIONS

When the cutting plane extends entirely through the object in a straight line and the front half of the object is theoretically removed, a full section is obtained (Figs. 7-1-3 and 7-1-4). This type of section is used for both detail and assembly drawings. When the section is on an axis of symmetry, it is not necessary to indicate its location (Fig. 7-1-5). However, it may be identified and indicated in the normal manner to increase clarity, if so desired.

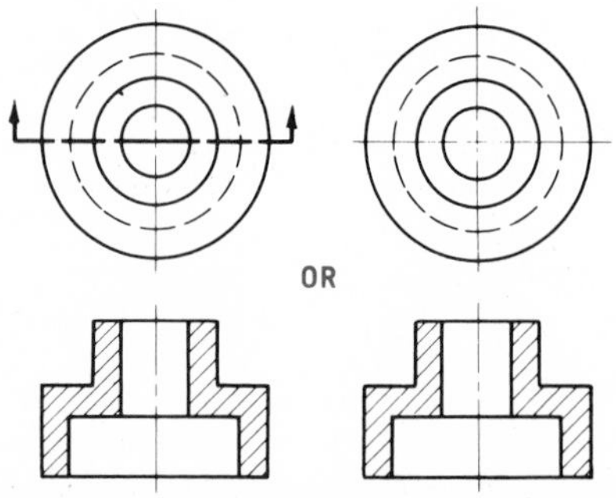
OR

Fig. 7-1-5 Cutting-plane line may be omitted when it corresponds with the center line.

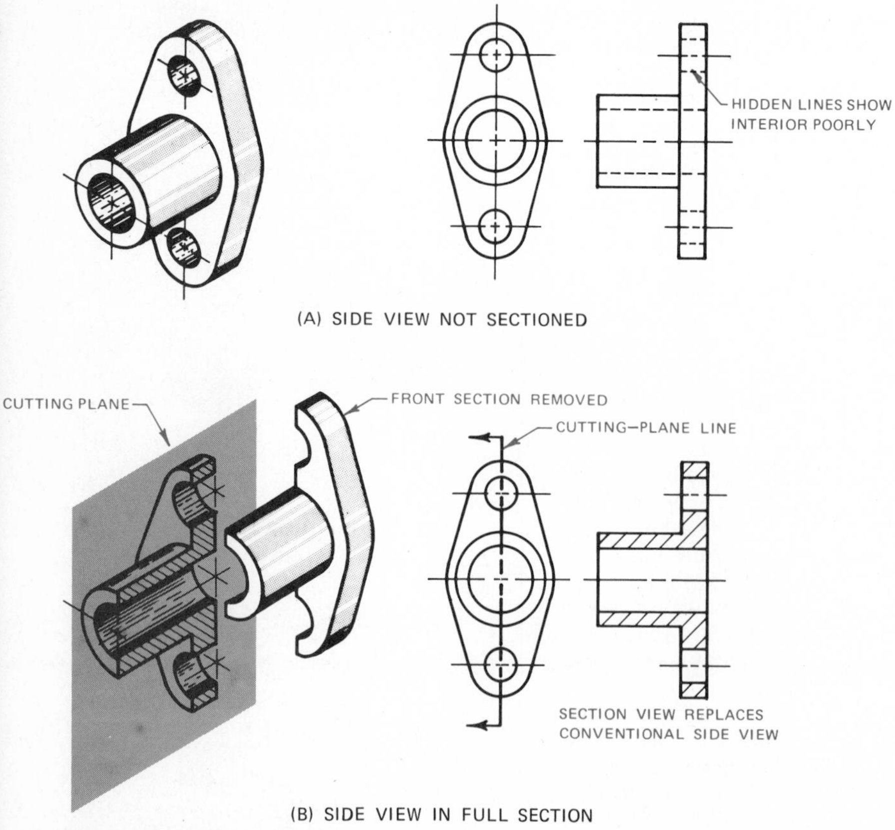
HIDDEN LINES SHOW INTERIOR POORLY
(A) SIDE VIEW NOT SECTIONED
CUTTING PLANE
FRONT SECTION REMOVED
CUTTING–PLANE LINE
SECTION VIEW REPLACES CONVENTIONAL SIDE VIEW
(B) SIDE VIEW IN FULL SECTION

Fig. 7-1-3 Full-section view.

SECTION LINING

Section lining, sometimes referred to as cross-hatching, can serve a double purpose. It indicates the surface that has been theoretically cut and makes it stand out clearly, thus helping the observer to understand the shape of the object. Section lining may also indicate the material from which the object is to be made, when the section lining symbols shown in Fig. 7-1-6 are used.

Section Lining for Detail Drawings. Since the exact material specifications for a part are usually given elsewhere on the drawing, the general-purpose section lining symbol is recommended for most detail drawings. An exception may be made for wood when it is desirable to show the direction of the grain.

The lines for section lining are thin and are usually drawn at an angle of 45° to the major outline of the object. The same angle is used for the whole "cut" surface of the object. If the part shape would cause section lines to be parallel, or nearly so, to one of the sides of the part, then some angle other than 45° should be

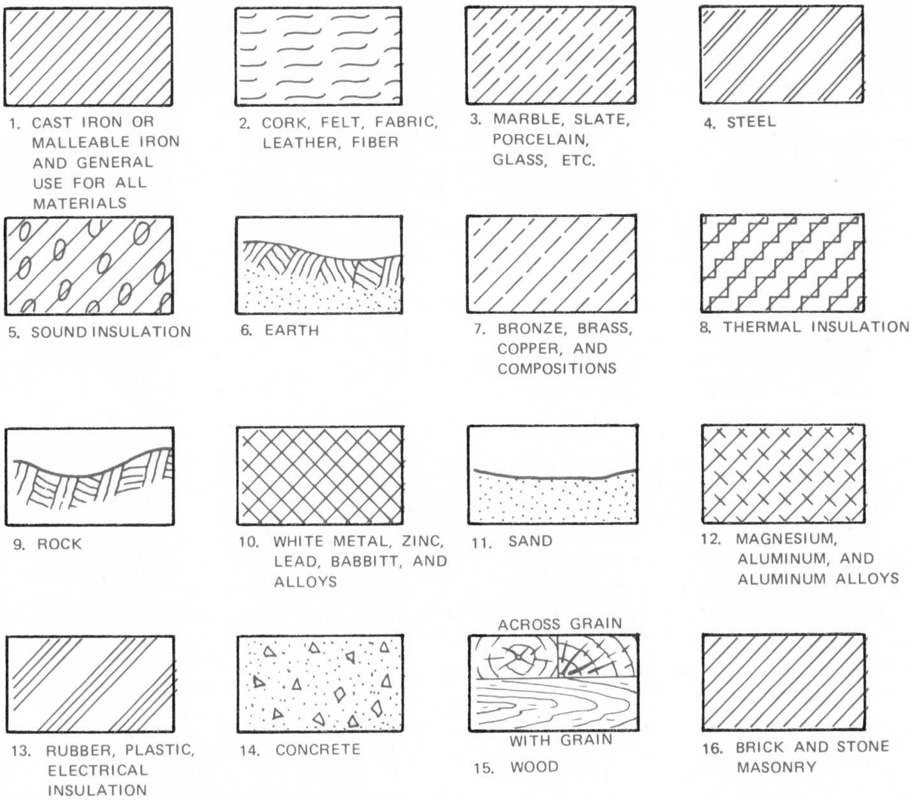
1. CAST IRON OR MALLEABLE IRON AND GENERAL USE FOR ALL MATERIALS
2. CORK, FELT, FABRIC, LEATHER, FIBER
3. MARBLE, SLATE, PORCELAIN, GLASS, ETC.
4. STEEL
5. SOUND INSULATION
6. EARTH
7. BRONZE, BRASS, COPPER, AND COMPOSITIONS
8. THERMAL INSULATION
9. ROCK
10. WHITE METAL, ZINC, LEAD, BABBITT, AND ALLOYS
11. SAND
12. MAGNESIUM, ALUMINUM, AND ALUMINUM ALLOYS
13. RUBBER, PLASTIC, ELECTRICAL INSULATION
14. CONCRETE
ACROSS GRAIN
WITH GRAIN
15. WOOD
16. BRICK AND STONE MASONRY

Fig. 7-1-6 Symbolic section lining.

chosen (Fig. 7-1-7). The spacing of the hatching lines should be reasonably uniform to give a good appearance to the drawing. The pitch, or distance between lines, normally varies between 1 and 3 mm depending on the size of the area to be sectioned.

Large areas shown in section need not be entirely section-lined (Fig. 7-1-8). Section lining around the outline will usually be sufficient, providing clarity is not sacrificed.

Dimensions or other lettering should not be placed in sectioned areas. When this is unavoidable, the section lining should be omitted for the numerals or lettering (Fig. 7-1-9).

Sections which are too thin for effective section lining, such as sheet-metal items, packing, and gaskets, may be shown without section lining; or the area may be filled in completely (Fig. 7-1-10).

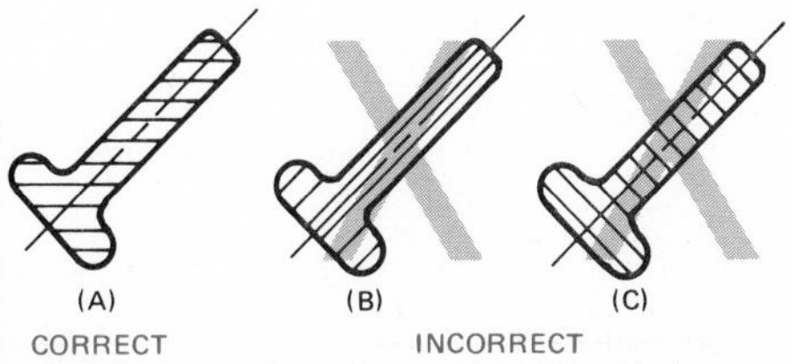
(A)
(B)
(C)
CORRECT
INCORRECT

Fig. 7-1-7 Direction of section lining.

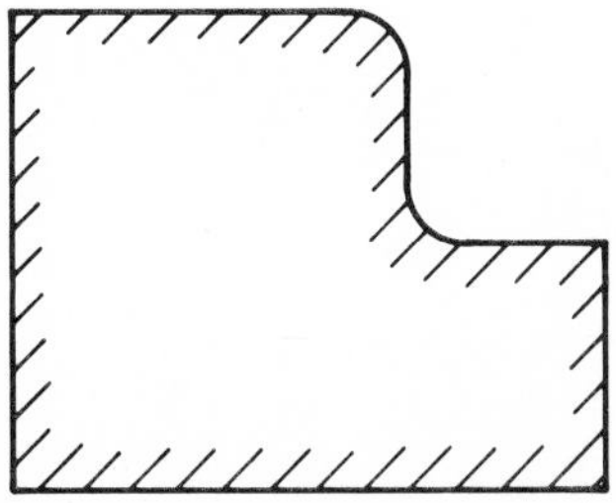

Fig. 7-1-8 Outline section lining.

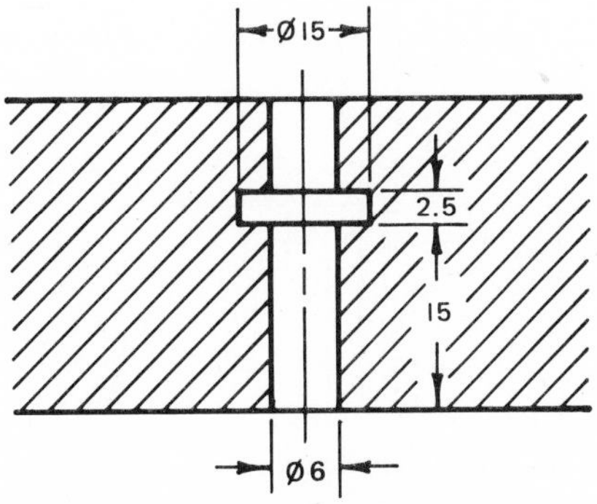
Ø15
2.5
15
Ø6

Fig. 7-1-9 Section lining omitted to accommodate dimensions.

Assignment

Select one of the problems shown in Figs. 7-1-A and 7-1-B and on an A3- or B-size sheet make a two-view working drawing of the part, showing one of the views in full section. Scale is 1:1.

REVIEW FOR ASSIGNMENT

Unit 10-1 Sand Castings
Unit 5-7 Finish Marks
Unit 5-2 Spotfacing
Unit 2-6 Drawing Circles and Arcs

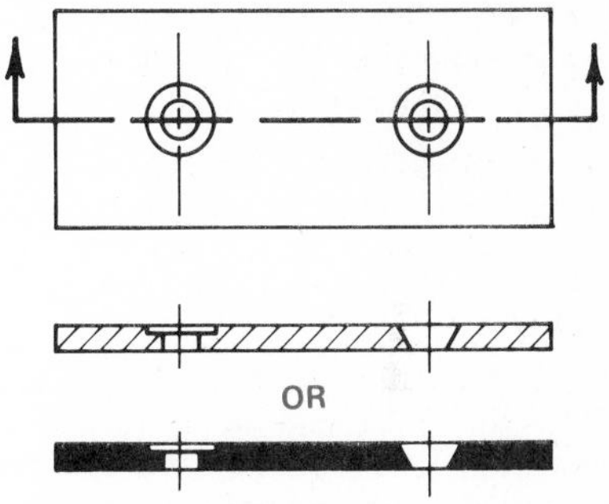
OR

Fig. 7-1-10 Thin parts in section.

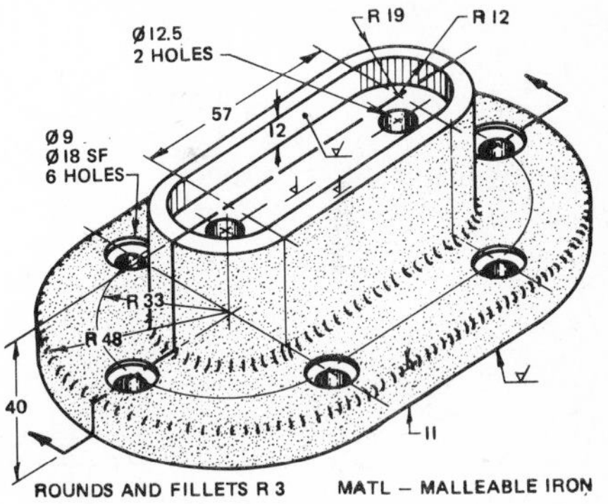
Ø12.5
2 HOLES
R 19
R 12
57
12
Ø9
Ø18 SF
6 HOLES
R 33
R 48
40
11
ROUNDS AND FILLETS R 3
MATL – MALLEABLE IRON

Fig. 7-1-A Shaft base.

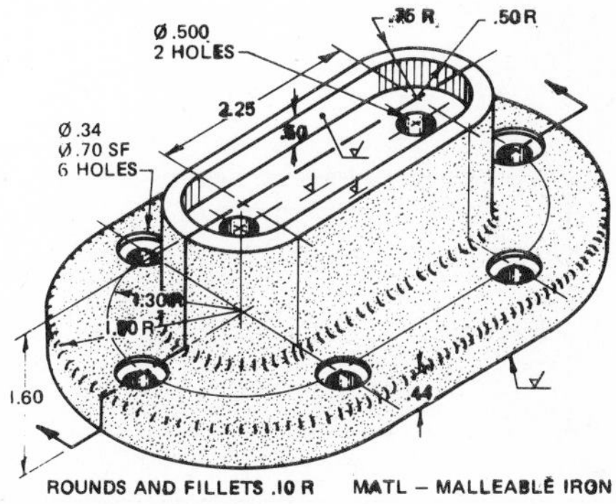
Ø .500
2 HOLES
.50R
2.25
Ø .34
Ø .70 SF
6 HOLES
1.60
ROUNDS AND FILLETS .10 R
MATL – MALLEABLE IRON

Fig. 7-1-B Shaft base.

UNIT 7-2

TWO OR MORE SECTIONAL VIEWS ON ONE DRAWING

If two or more sections appear on the same drawing, the cutting-plane lines are identified by two identical large, single-stroke, Gothic letters, one at each end of the line, placed behind the arrowhead so that the arrow points away from the letter. The identification letters should not include I, O, and Q (Fig. 7-2-1).

Sectional view subtitles are given when identification letters are used and appear directly below the view, incorporating the letters at each end of the cutting-plane line thus: SECTION A-A, or abbreviated, SECT. B-B. When the scale is different from the main view, it is stated below the subtitle thus:

SECTION A-A
SCALE 1:5

Assignment

Select one of the problems shown in Figs. 7-2-A and 7-2-B and on an A3- or B-size sheet make a working drawing of the part showing the appropriate views in sections. Refer to the Appendix for taper sizes. Scale is 1:1.

REVIEW FOR ASSIGNMENT

Unit 9-2 Taper Pins
Unit 5-2 Countersink, Counterbores, and Spotfaces
Unit 4-3 Drawing a Hexagon
Unit 5-3 Dimensioning Tapers

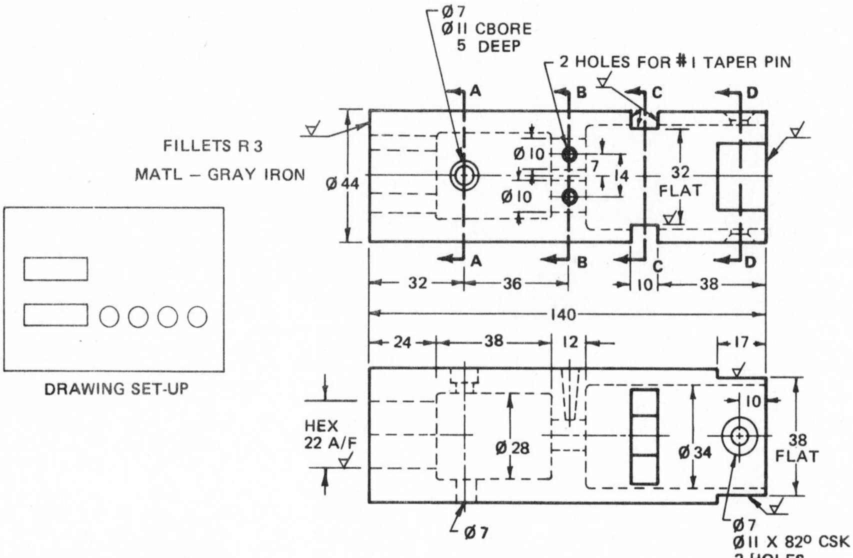

Ø 7
Ø 11 CBORE
5 DEEP
2 HOLES FOR # 1 TAPER PIN
FILLETS R 3
MATL – GRAY IRON
Ø 44
Ø 10
Ø 10
32
FLAT
DRAWING SET-UP
140
HEX
22 A/F
Ø 28
Ø 34
38
FLAT
Ø 7
Ø 7
Ø 11 X 82° CSK
2 HOLES

Fig. 7-2-A Housing.

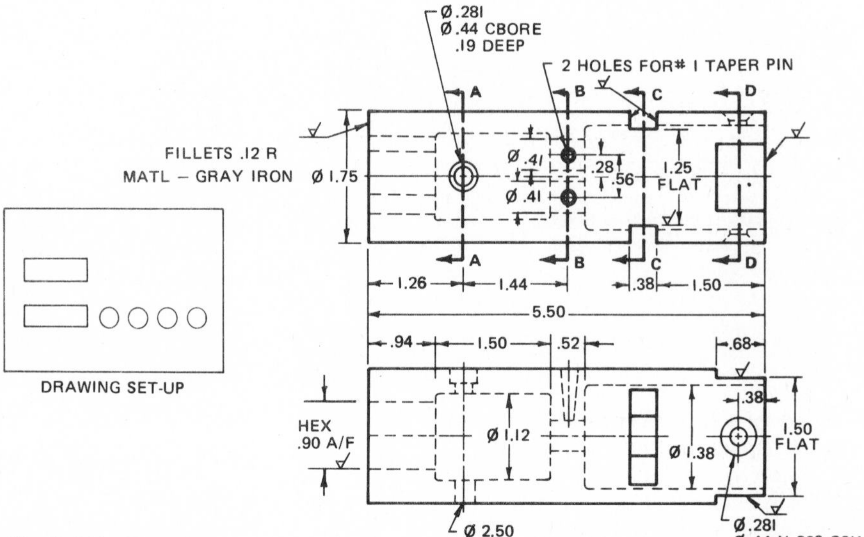

Ø .281
Ø .44 CBORE
.19 DEEP
2 HOLES FOR # 1 TAPER PIN
FILLETS .12 R
MATL – GRAY IRON
Ø 1.75
Ø .41
Ø .41
1.25
FLAT
DRAWING SET-UP
5.50
HEX
.90 A/F
Ø 1.12
Ø 1.38
1.50
FLAT
Ø 2.50
Ø .281
Ø .44 X 82° CSK
2 HOLES

Fig. 7-2-B Housing.

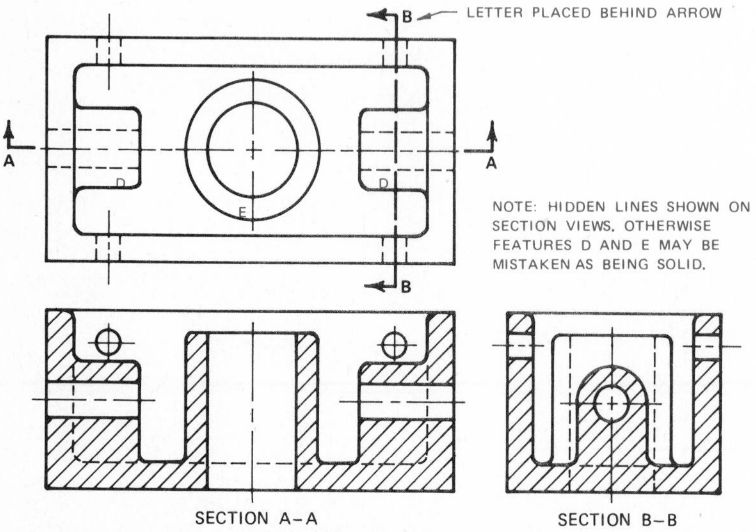

LETTER PLACED BEHIND ARROW
NOTE: HIDDEN LINES SHOWN ON SECTION VIEWS. OTHERWISE FEATURES D AND E MAY BE MISTAKEN AS BEING SOLID.
SECTION A–A
SECTION B–B

Fig. 7-2-1 Detail drawing having two section views.

UNIT 7-3

HALF SECTIONS

A *half section* is a view of an assembly or object, usually symmetrical, showing one half of the view in section. See Figs. 7-3-1 and 7-3-2. Two cutting-plane lines, perpendicular to each other, extend halfway through the view, and one quarter of the view is considered removed and the interior is exposed to view.

Similar to the practice followed for full-section drawings, the cutting-plane line need not be drawn for half sections when it is obvious where the cutting plane took place. Instead, center lines may be used. When a cutting plane is used, the common practice is to show only one end of the cutting-plane line, terminating with an arrow to indicate the direction of sight for viewing the section.

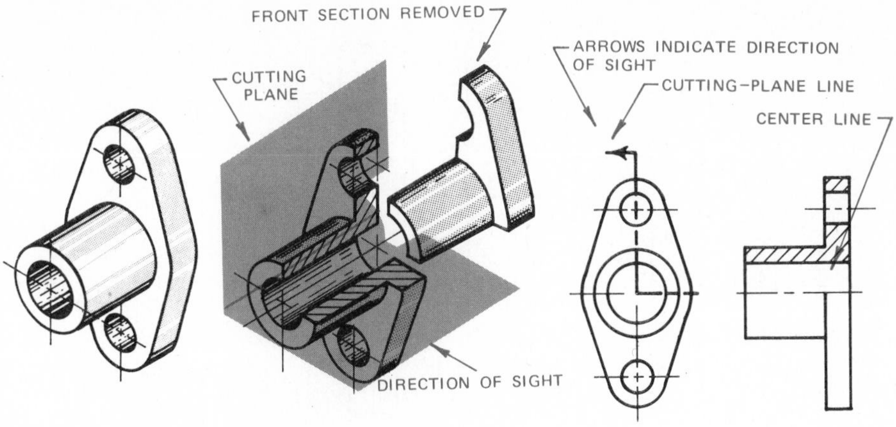
FRONT SECTION REMOVED
CUTTING PLANE
ARROWS INDICATE DIRECTION OF SIGHT
CUTTING-PLANE LINE
CENTER LINE
DIRECTION OF SIGHT

Fig. 7-3-1 A half-section drawing.

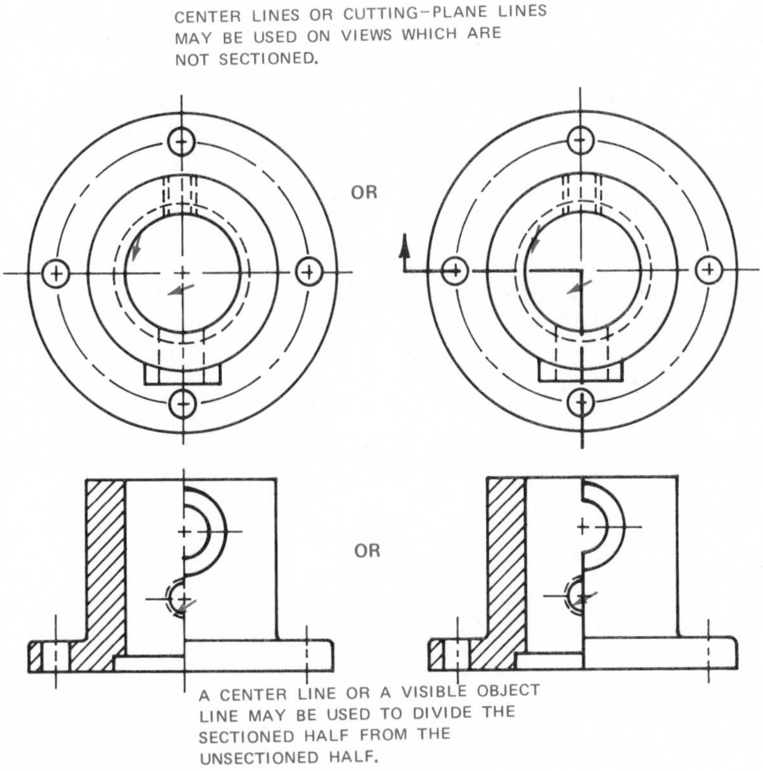
CENTER LINES OR CUTTING-PLANE LINES MAY BE USED ON VIEWS WHICH ARE NOT SECTIONED.
OR
OR
A CENTER LINE OR A VISIBLE OBJECT LINE MAY BE USED TO DIVIDE THE SECTIONED HALF FROM THE UNSECTIONED HALF.

Fig. 7-3-2 Half-section views.

On the sectional view a center line or a visible object line may be used to divide the sectioned half from the unsectioned half of the drawing. This type of sectional drawing is best suited for assembly drawings where both internal and external construction is shown on one view and where only overall and center-to-center dimensions are required. The main disadvantage of using this type of sectional drawing for detail drawings is the difficulty in dimensioning internal features without adding hidden lines. However, hidden lines may be added for dimensioning, as shown in Fig. 7-3-3.

Assignment

Select one of the problems shown in Figs. 7-3-A and 7-3-B. On an A3- or B-size sheet, make a two-view working drawing of the part, showing the side view in half section.

REVIEW FOR ASSIGNMENT

Unit 19-1 Vee-Belt Drive
Unit 9-1 Dimensioning Keyways
Unit 9-1 Keys
Unit 5-2 Double-Arrow Dimensioning

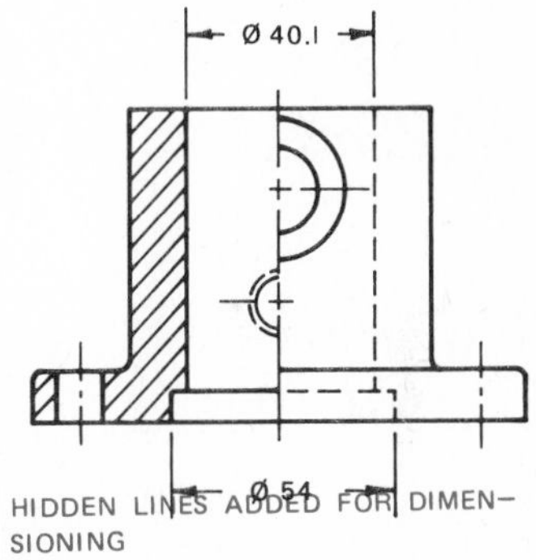
Ø 40.1
Ø 54
HIDDEN LINES ADDED FOR DIMENSIONING

Fig. 7-3-3 Dimensioning half-section view.

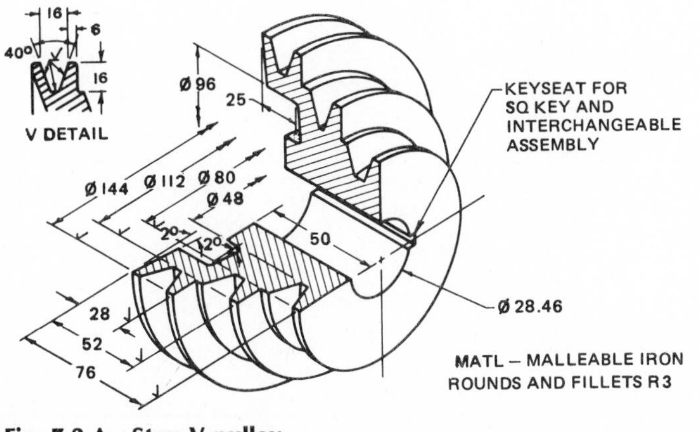
16
6
40°
16
V DETAIL
Ø 96
25
KEYSEAT FOR SQ KEY AND INTERCHANGEABLE ASSEMBLY
Ø 144
Ø 112
Ø 80
Ø 48
20
120
50
Ø 28.46
28
52
76
MATL – MALLEABLE IRON
ROUNDS AND FILLETS R3

Fig. 7-3-A Step-V pulley.

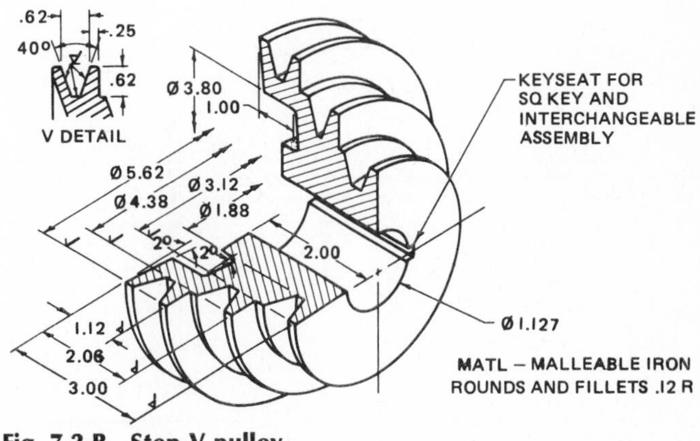
.62
.25
40°
.62
V DETAIL
Ø 3.80
1.00
KEYSEAT FOR SQ KEY AND INTERCHANGEABLE ASSEMBLY
Ø 5.62
Ø 4.38
Ø 3.12
Ø 1.88
20
120
2.00
Ø 1.127
1.12
2.06
3.00
MATL – MALLEABLE IRON
ROUNDS AND FILLETS .12 R

Fig. 7-3-B Step-V pulley.

UNIT 7-4
THREADS IN SECTION

True representation of a screw thread is seldom provided on working drawings because it would require very laborious and accurate drawing involving repetitious development of the helix curve of the thread. A symbolic representation of threads is now standard practice.

Three types of conventions are in general use for screw thread representation (Fig. 7-4-1). These are known as pictorial, schematic, and simplified representation. Simplified representation should be used whenever it will clearly portray the requirements. Schematic and pictorial representations require more drafting time, but are sometimes necessary to avoid confusion with other parallel lines or to more clearly portray particular aspects of the threads.

THREADED ASSEMBLIES

Any of the thread conventions shown here may be used for assemblies of threaded parts, and two or more methods may be used on the same drawing, as shown in Fig. 7-4-2. In sectional views, the externally threaded part is always shown covering the internally threaded part, as illustrated in Fig. 7-4-3.

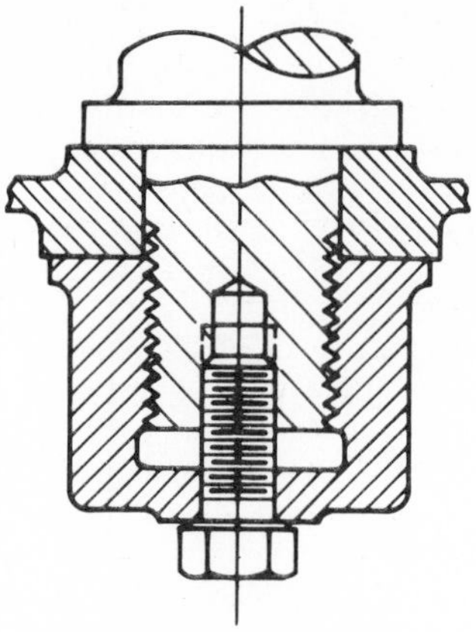

Fig. 7-4-2 Threaded assembly.

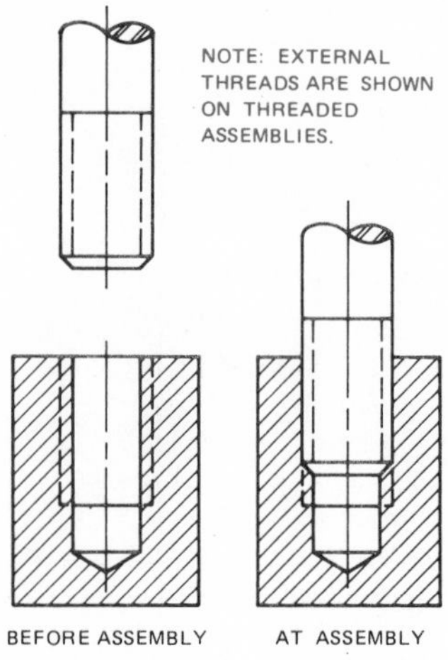

NOTE: EXTERNAL THREADS ARE SHOWN ON THREADED ASSEMBLIES.
BEFORE ASSEMBLY
AT ASSEMBLY

Fig. 7-4-3 Drawing threads in assembly drawings.

Assignment

Select one of the problems shown in Figs. 7-4-A and 7-4-B. On an A3- or B-size sheet make a working drawing of the part. Determine the number of views and the best type of section which will clearly describe the part. Scale is 1:1.

REVIEW FOR ASSIGNMENT

Unit 8-2	Thread Conventions
Unit 8-2	Pipe Threads
Unit 8-1	Threading Considerations
Unit 7-18	Intersection of Unfinished Surfaces
Unit 5-7	Machining Symbols

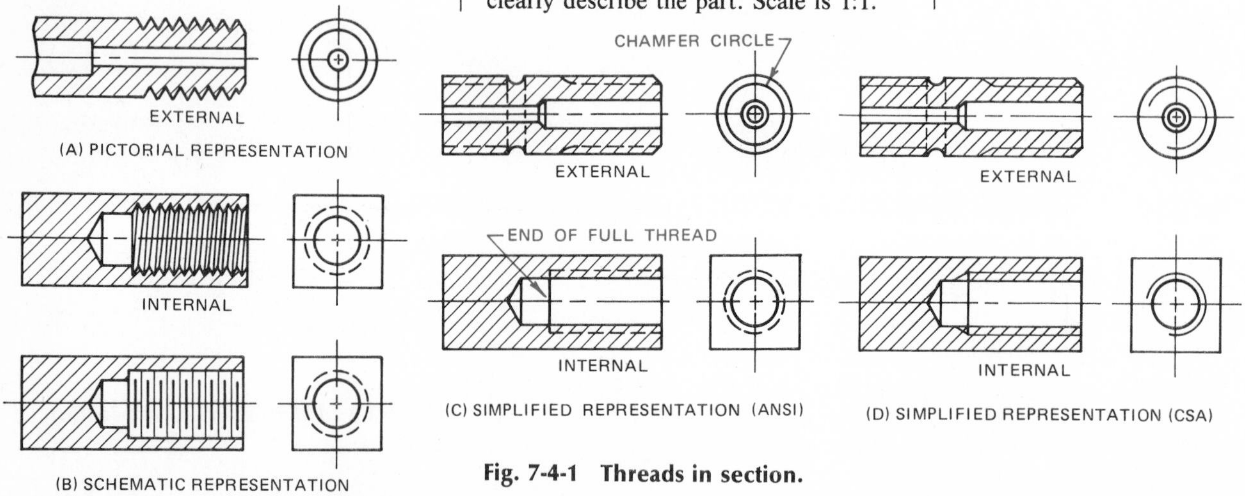

EXTERNAL
(A) PICTORIAL REPRESENTATION
INTERNAL
(B) SCHEMATIC REPRESENTATION
CHAMFER CIRCLE
EXTERNAL
END OF FULL THREAD
INTERNAL
(C) SIMPLIFIED REPRESENTATION (ANSI)
EXTERNAL
INTERNAL
(D) SIMPLIFIED REPRESENTATION (CSA)

Fig. 7-4-1 Threads in section.

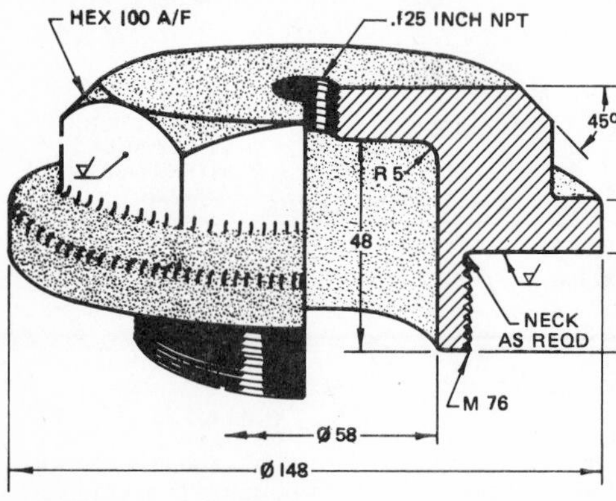

HEX 100 A/F
.125 INCH NPT
45°
R 5
64
12
48
22
NECK AS REQD
M 76
Ø 58
Ø 148
ROUNDS AND FILLETS R 3
MATL – MALLEABLE IRON

Fig. 7-4-A Pipe plug.

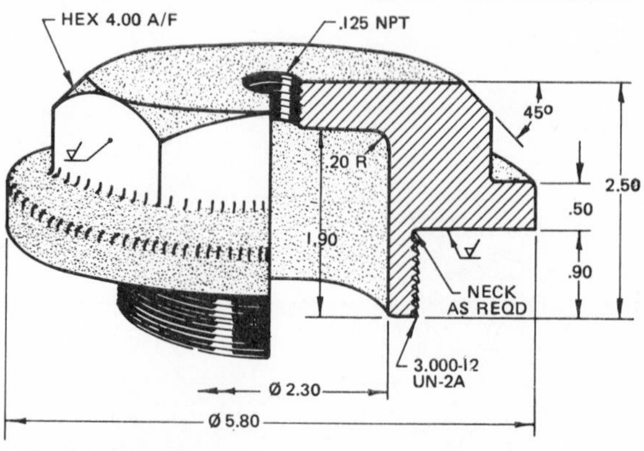

HEX 4.00 A/F
.125 NPT
45°
20 R
2.50
.50
1.90
.90
NECK AS REQD
3.000-12 UN-2A
Ø 2.30
Ø 5.80
ROUNDS AND FILLETS .10R
MATL – MALLEABLE IRON

Fig. 7-4-B Pipe plug.

UNIT 7-5

ASSEMBLIES IN SECTION

Section Lining on Assembly Drawings

General-purpose section lining is recommended for most assembly drawings, especially if the detail is small. Symbolic section lining is generally not recommended for drawings that will be microfilmed.

General-purpose section lining should be drawn at an angle of 45° with the main outlines of the view. On adjacent parts, the section lines should be drawn in the opposite direction, as shown in Figs. 7-5-1 and 7-5-2.

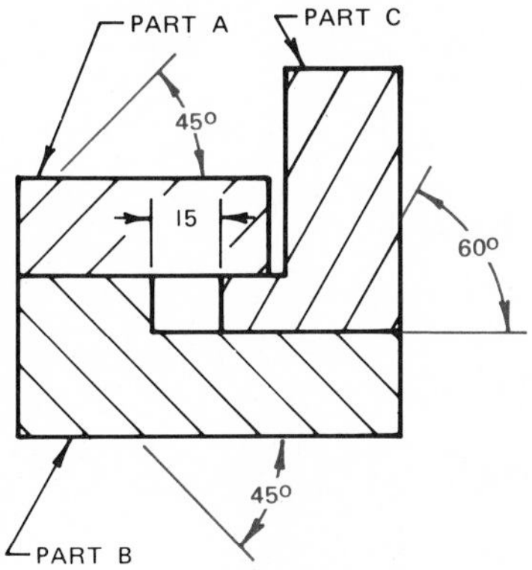
PART A
PART C
45°
15
60°
45°
PART B

Fig. 7-5-1 Direction of section lining.

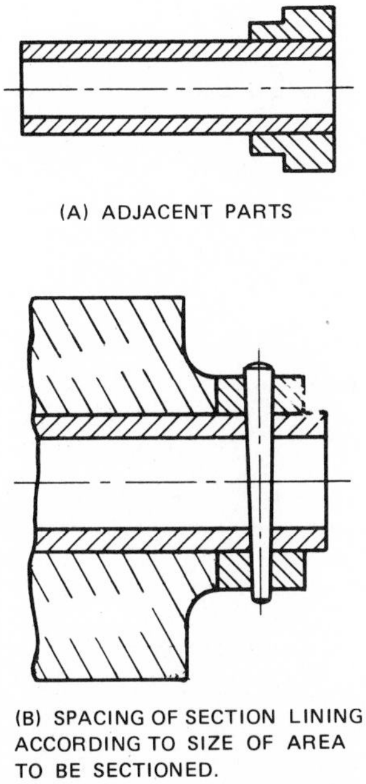
(A) ADJACENT PARTS
(B) SPACING OF SECTION LINING ACCORDING TO SIZE OF AREA TO BE SECTIONED.

Fig. 7-5-2 Arrangement of section lining.

For additional adjacent parts, any suitable angle may be used to make each part stand out separately and clearly. Section lines should not be purposely drawn to meet at common boundaries.

When two or more thin adjacent parts are filled in, a space is left between them, as shown in Fig. 7-5-3.

Symbolic section lining is used on special-purpose assembly drawings such as illustrations for parts catalogs, display assemblies, promotional materials, etc., when it is desirable to distinguish between different materials (Fig. 7-1-6).

All assemblies and subassemblies pertaining to one particular set of drawings should use the same symbolic conventions.

Shafts, Bolts, Pins, Keyways, etc., in Section. Shafts, bolts, nuts, rods, rivets, keys, pins, and similar solid parts, the axes of which lie in the cutting plane, should not be sectioned except that a broken-out section of the shaft may be used to indicate clearly the key, keyseat, or pin (Fig. 7-5-4).

Assignment

On an A3- or B-size sheet, make a one-view section assembly drawing of one of the problems shown in Figs. 7-5-A and 7-5-B. Include on your drawing a bill of materials and identify the parts on the assembly. Assuming that this drawing will be used in a catalog, place on the drawing the dimensions and information required by the potential buyer. Scale is 1:1.

REVIEW FOR ASSIGNMENT

Unit 8-3 Bolt Sizes
Unit 6-4 Assembly Drawings
Unit 6-4 Bill of Material
Unit 5-2 Spotfacing
Unit 8-3 Bolted Assemblies
Unit 21-6 Plain Bearings
Unit 2-6 Phantom Lines
Unit 7-14 Conventional Breaks
Unit 5-6 Classification of Fits

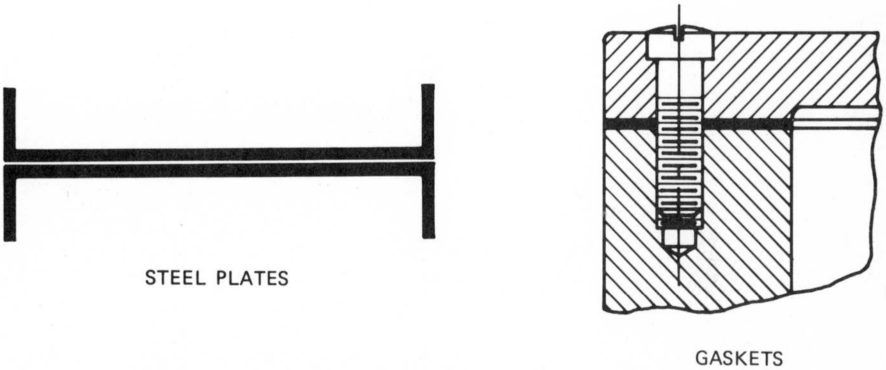
STEEL PLATES
GASKETS

Fig. 7-5-3 Assembly of thin parts in section.

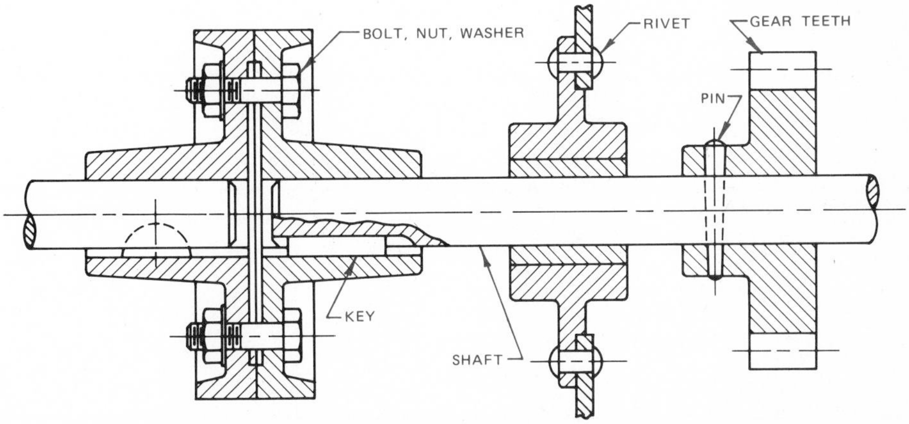
BOLT, NUT, WASHER
RIVET
GEAR TEETH
PIN
KEY
SHAFT

Fig. 7-5-4 Section lining used to distinguish between stationary and moving parts.

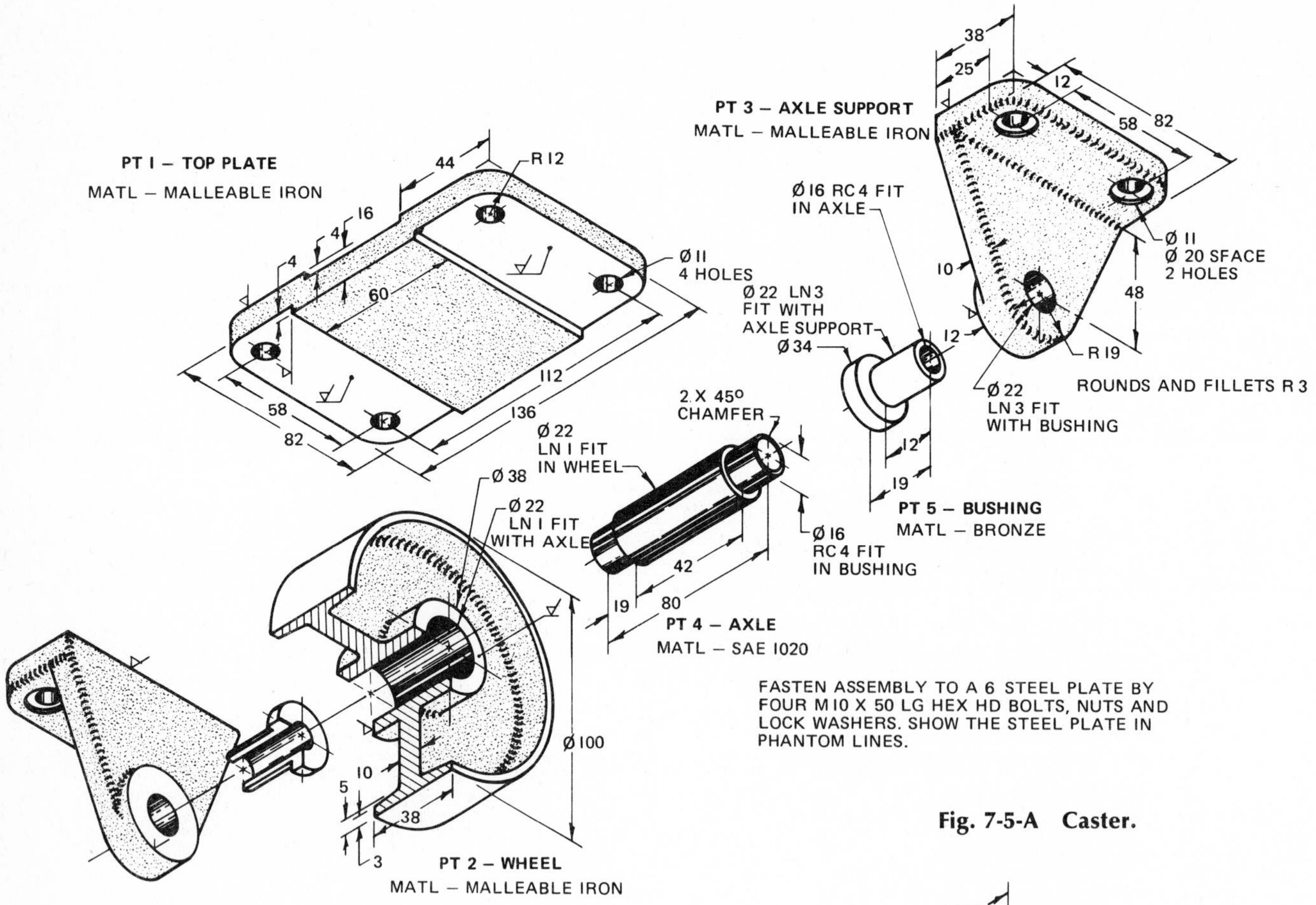
PT 1 – TOP PLATE
MATL – MALLEABLE IRON
44
R 12
16
4
4
60
Ø 11
4 HOLES
112
136
58
82
PT 3 – AXLE SUPPORT
MATL – MALLEABLE IRON
38
25
12
58
82
Ø 16 RC 4 FIT
IN AXLE
Ø 11
Ø 20 SFACE
2 HOLES
10
48
Ø 22 LN 3
FIT WITH
AXLE SUPPORT
Ø 34
12
R 19
Ø 22
LN 3 FIT
WITH BUSHING
ROUNDS AND FILLETS R 3
2 X 45°
CHAMFER
Ø 22
LN 1 FIT
IN WHEEL
12
19
PT 5 – BUSHING
MATL – BRONZE
Ø 38
Ø 22
LN 1 FIT
WITH AXLE
Ø 16
RC 4 FIT
IN BUSHING
42
19
80
PT 4 – AXLE
MATL – SAE 1020
Ø 100
10
5
38
3
PT 2 – WHEEL
MATL – MALLEABLE IRON

FASTEN ASSEMBLY TO A 6 STEEL PLATE BY FOUR M10 X 50 LG HEX HD BOLTS, NUTS AND LOCK WASHERS. SHOW THE STEEL PLATE IN PHANTOM LINES.

Fig. 7-5-A Caster.

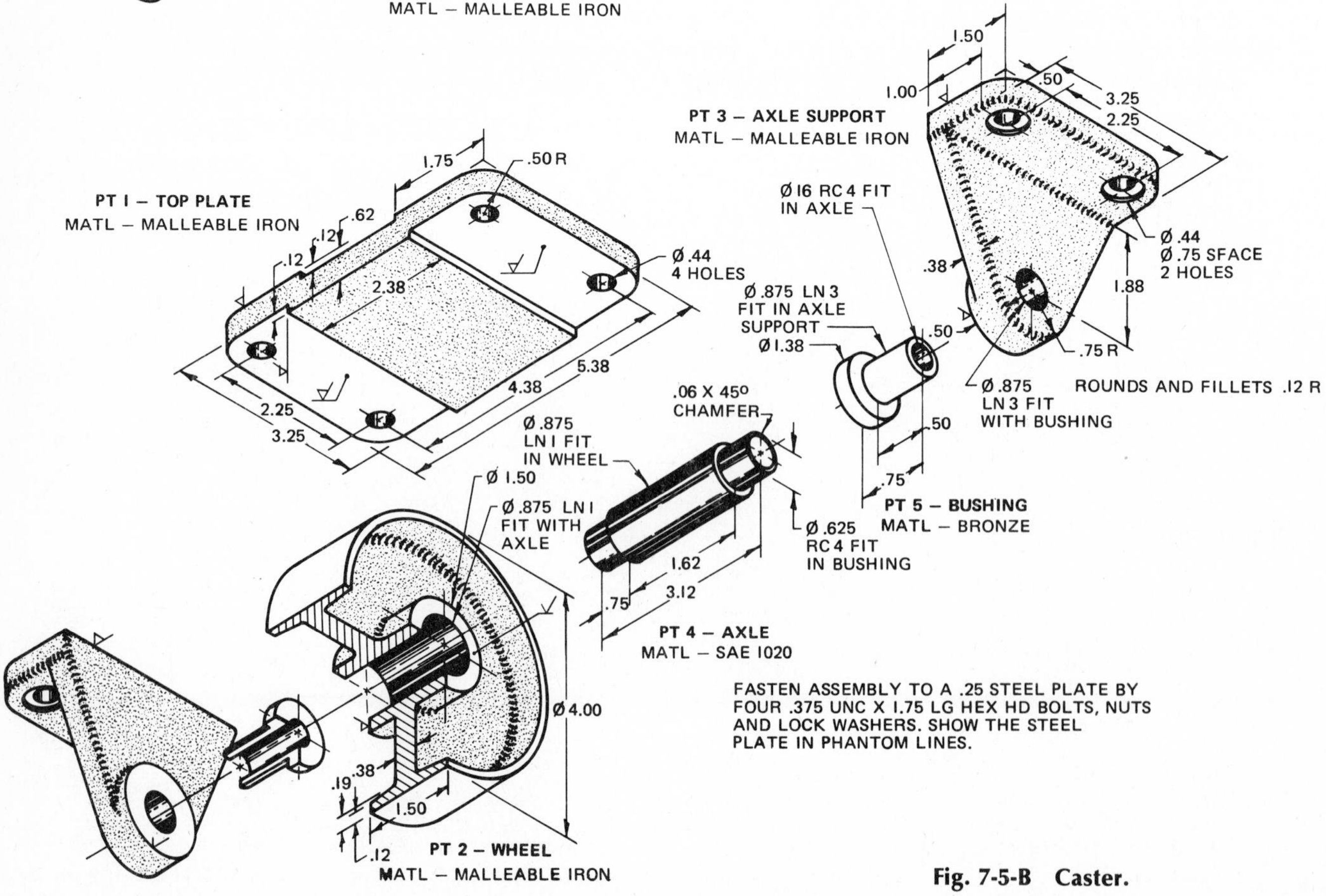
PT 1 – TOP PLATE
MATL – MALLEABLE IRON
1.75
.50 R
.62
.12
.12
2.38
Ø .44
4 HOLES
5.38
4.38
2.25
3.25
PT 3 – AXLE SUPPORT
MATL – MALLEABLE IRON
1.50
1.00
.50
3.25
2.25
Ø 16 RC 4 FIT
IN AXLE
Ø .44
Ø .75 SFACE
2 HOLES
.38
1.88
Ø .875 LN 3
FIT IN AXLE
SUPPORT
Ø 1.38
.50
.75 R
Ø .875
LN 3 FIT
WITH BUSHING
ROUNDS AND FILLETS .12 R
.06 X 45°
CHAMFER
Ø .875
LN 1 FIT
IN WHEEL
.50
.75
PT 5 – BUSHING
MATL – BRONZE
Ø 1.50
Ø .875 LN 1
FIT WITH
AXLE
Ø .625
RC 4 FIT
IN BUSHING
1.62
.75
3.12
PT 4 – AXLE
MATL – SAE 1020
Ø 4.00
.19
.38
1.50
.12
PT 2 – WHEEL
MATL – MALLEABLE IRON

FASTEN ASSEMBLY TO A .25 STEEL PLATE BY FOUR .375 UNC X 1.75 LG HEX HD BOLTS, NUTS AND LOCK WASHERS. SHOW THE STEEL PLATE IN PHANTOM LINES.

Fig. 7-5-B Caster.

UNIT 7-6
OFFSET SECTIONS

In order to include features that are not in a straight line, the cutting plane may be offset or bent, so as to include several planes or curved surfaces (Figs. 7-6-1 and 7-6-2).

An offset section is similar to a full section in that the cutting-plane line extends through the object from one side to the other. The change in direction of the cutting-plane line is not shown in the sectional view.

Assignment

Select one of the problems shown in Figs. 7-6-A and 7-6-B. On an A3- or B-size sheet, make a working drawing of the part. Scale is 1:1.

REVIEW FOR ASSIGNMENT

Unit 5-4 Arrowless and Tabular Dimensioning
Unit 16-2 Simplified Drafting
Unit 17-1 Drawings for Numerical Control

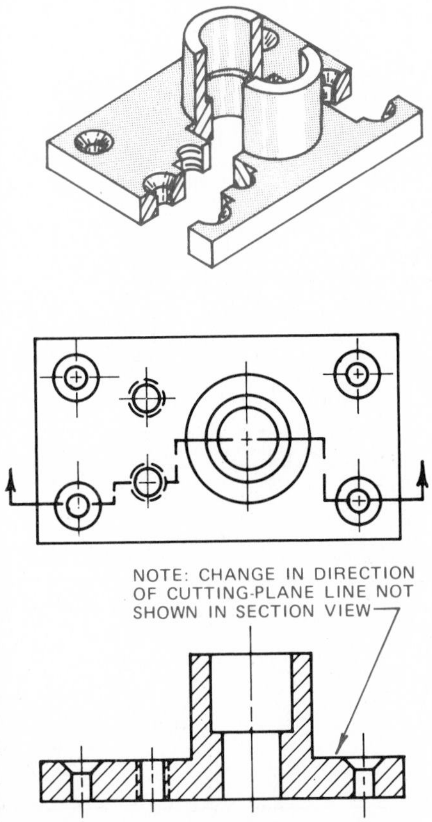
NOTE: CHANGE IN DIRECTION OF CUTTING-PLANE LINE NOT SHOWN IN SECTION VIEW

Fig. 7-6-1 An offset section.

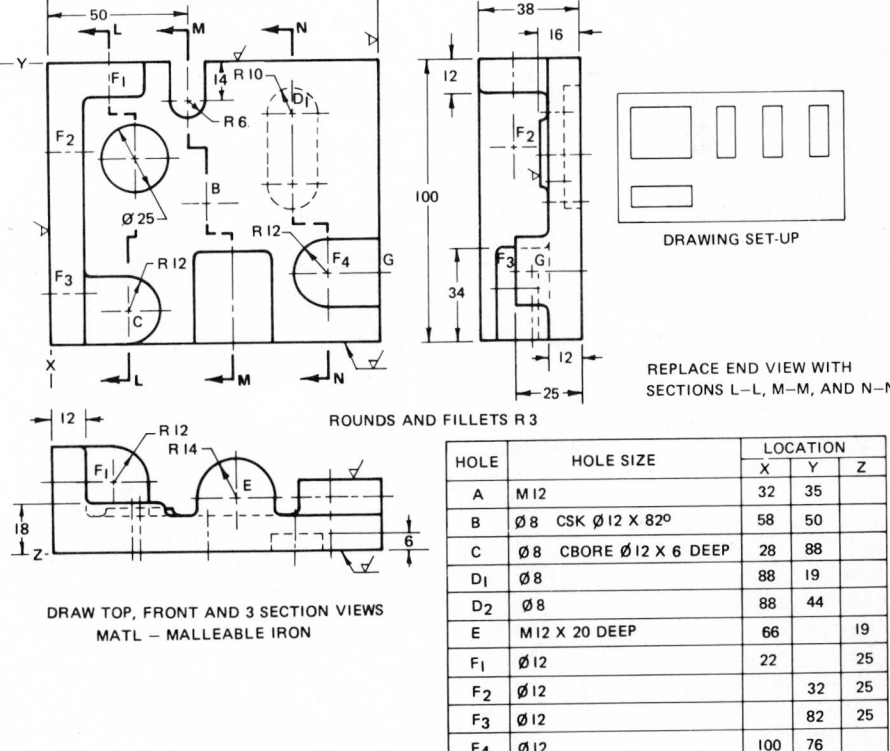

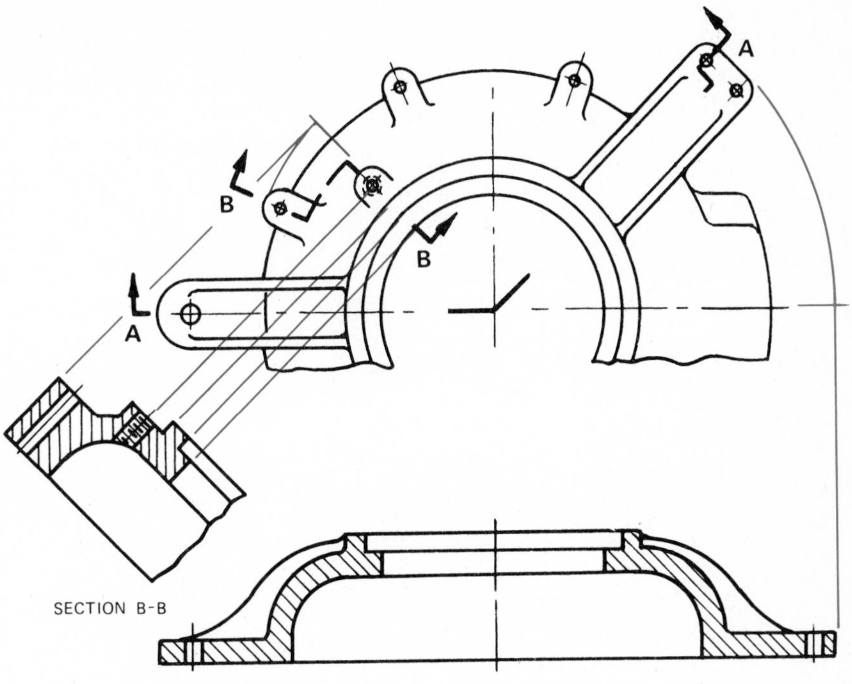
A
B
B
A
SECTION B-B
SECTION A-A

Fig. 7-6-2 Positioning offset sections.

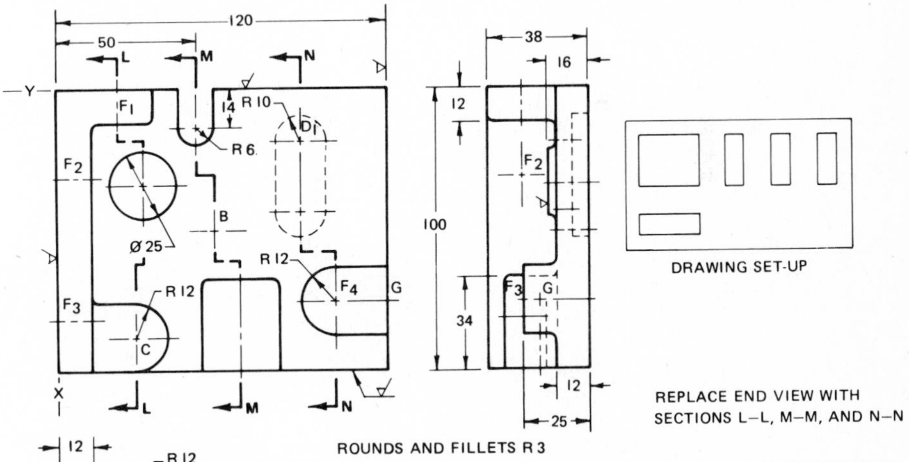
120
50
38
16
12
100
34
25
R 10
R 6
R 12
Ø 25
DRAWING SET-UP
REPLACE END VIEW WITH SECTIONS L–L, M–M, AND N–N
ROUNDS AND FILLETS R 3

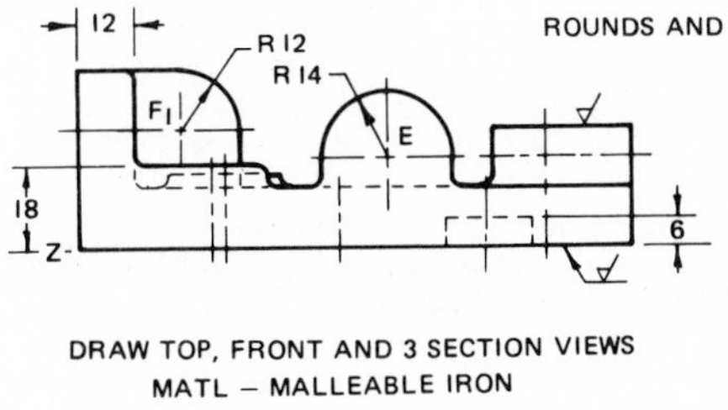
R 12
R 14
18
6
DRAW TOP, FRONT AND 3 SECTION VIEWS
MATL – MALLEABLE IRON

HOLE	HOLE SIZE	LOCATION X	Y	Z
A	M 12	32	35	
B	Ø 8 CSK Ø 12 X 82°	58	50	
C	Ø 8 CBORE Ø 12 X 6 DEEP	28	88	
D_1	Ø 8	88	19	
D_2	Ø 8	88	44	
E	M 12 X 20 DEEP	66		19
F_1	Ø 12	22		25
F_2	Ø 12		32	25
F_3	Ø 12		82	25
F_4	Ø 12	100	76	
G	Ø 3 THROUGH		76	19

Fig. 7-6-A Base plate.

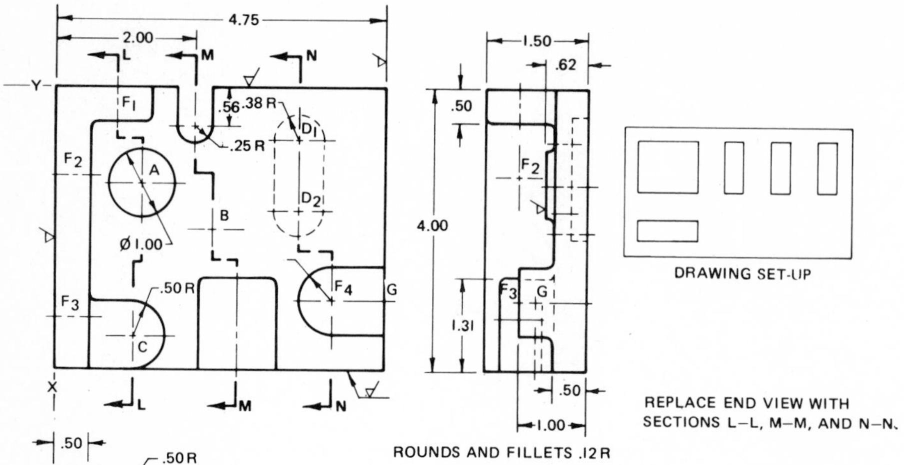

4.75
2.00
L M N
Y
F1
.56
.38 R
D1
.25 R
F2
A
D2
B
Ø 1.00
.50 R
F4
G
F3
C
X
L M N
1.50
.62
.50
F2
4.00
F3
G
1.31
.50
1.00
DRAWING SET-UP
REPLACE END VIEW WITH SECTIONS L–L, M–M, AND N–N.
ROUNDS AND FILLETS .12 R

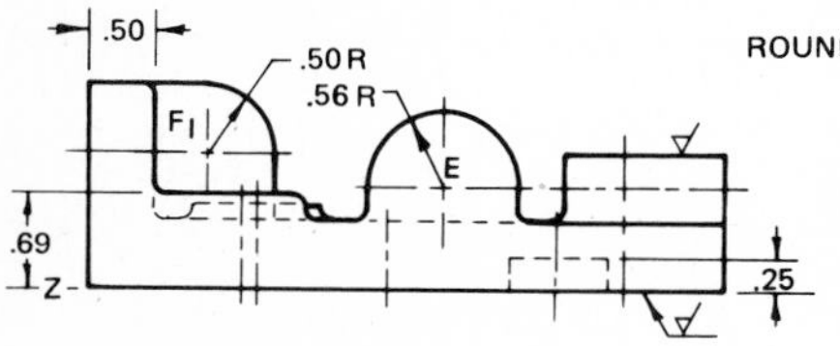

.50
.50 R
.56 R
F1
E
.69
Z
.25

HOLE	HOLE SIZE	LOCATION		
		X	Y	Z
A	.500-13 UNC-2B	1.25	1.38	
B	Ø .281 CSK Ø .50 X 82°	2.25	1.94	
C	Ø .312 CBORE Ø .50 X .25 DP	1.12	3.50	
D_1	Ø .312	3.50	.75	
D_2	Ø .312	3.50	1.75	
E	.500–13 UNC-2B X .75 DEEP	2.62		.75
F_1	Ø .500	.88		1.00
F_2	Ø .500		1.25	1.00
F_3	Ø .500		3.25	1.00
F_4	Ø .500	4.00	3.00	
G	Ø .125 THROUGH		3.00	.75

DRAW TOP, FRONT AND 3 SECTION VIEWS

MATL – MALLEABLE IRON

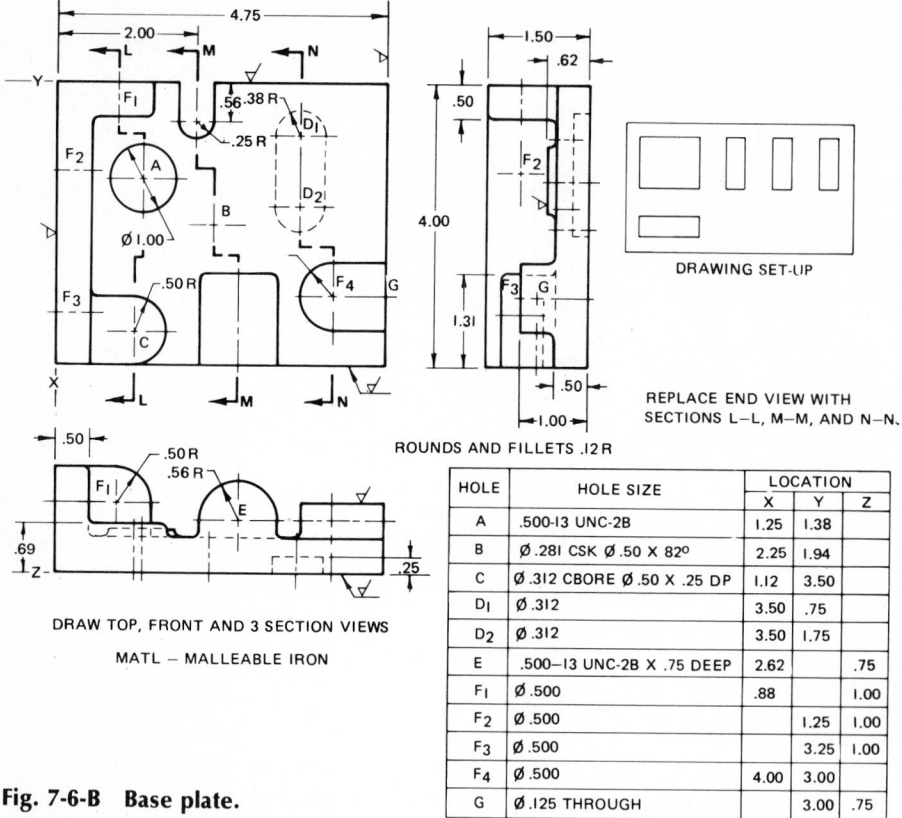

Fig. 7-6-B Base plate.

UNIT 7-7
RIBS, HOLES, AND LUGS IN SECTION

Ribs in Sections

A true-projection sectional view of a part, such as shown in Fig. 7-7-1, would be misleading when the cutting plane passes longitudinally through the center of the rib. To avoid this impression of solidity, a section not showing the ribs section-lined or cross-hatched is preferred. When there is an odd number of ribs, such as those shown in Fig. 7-7-1*b*, the top rib is aligned with the bottom rib to show its true relationship with the hub and flange. If the rib is not aligned or revolved, it appears distorted on the sectional view and is therefore misleading.

At times it may be necessary to use an alternative method of identifying ribs in a sectional view. Figure 7-7-2 shows a base and a pulley in section. If rib A of the base were not sectioned as previously mentioned, it would appear exactly like rib B in the sectional view and would be misleading. Similarly, rib C shown on the pulley may be overlooked. To distinguish between the ribs on the base and the ribs and spaces on the pulley, alternate section lining on the ribs is used. The line between the rib and solid portions is shown as a broken line.

Holes in Sections

Holes, like ribs, are aligned as shown in Fig. 7-7-1 to show their true relationship to the rest of the part.

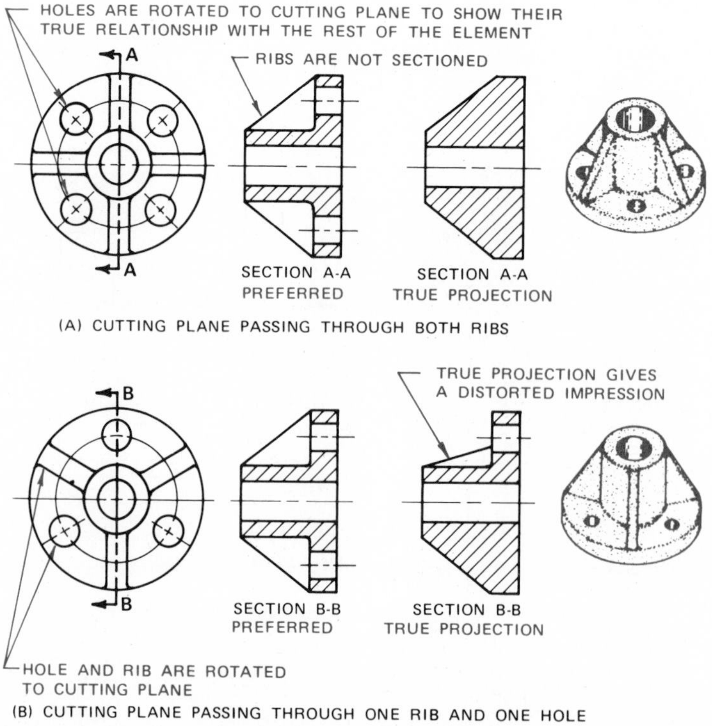

HOLES ARE ROTATED TO CUTTING PLANE TO SHOW THEIR TRUE RELATIONSHIP WITH THE REST OF THE ELEMENT
A
RIBS ARE NOT SECTIONED
A
SECTION A-A PREFERRED
SECTION A-A TRUE PROJECTION
(A) CUTTING PLANE PASSING THROUGH BOTH RIBS
TRUE PROJECTION GIVES A DISTORTED IMPRESSION
B
B
SECTION B-B PREFERRED
SECTION B-B TRUE PROJECTION
HOLE AND RIB ARE ROTATED TO CUTTING PLANE
(B) CUTTING PLANE PASSING THROUGH ONE RIB AND ONE HOLE

Fig. 7-7-1 Preferred and true projection through ribs and holes.

Lugs in Section

Lugs, like ribs and spokes, are also aligned to show their true relationship to the rest of the part, because true projection may be misleading. Figure 7-7-3 shows several examples of lugs in section. Note how the cutting-plane line is bent or offset so that the features may be clearly shown in the sectional view.

Some lugs are shown in section, and some are not. When the cutting plane passes through the lug crosswise, the lug is sectioned; otherwise, the lugs are treated in the same manner as ribs.

Assignment

Select one of the problems shown in Figs. 7-7-A and 7-7-B. On an A3- or B-size sheet make a three-view working drawing of the part showing the front and side views in section. Both parts are to be used here and abroad, so a dual dimensioning system must be used. Scale is 1:1.

REVIEW FOR ASSIGNMENT

Unit 10-1 Bosses
Unit 5-2 Clearance Holes for Bolted Assemblies
Unit 5-1 Dual Dimensioning
Unit 7-17 Foreshortened Projection
Unit 7-18 Intersection of Unfinished Surfaces

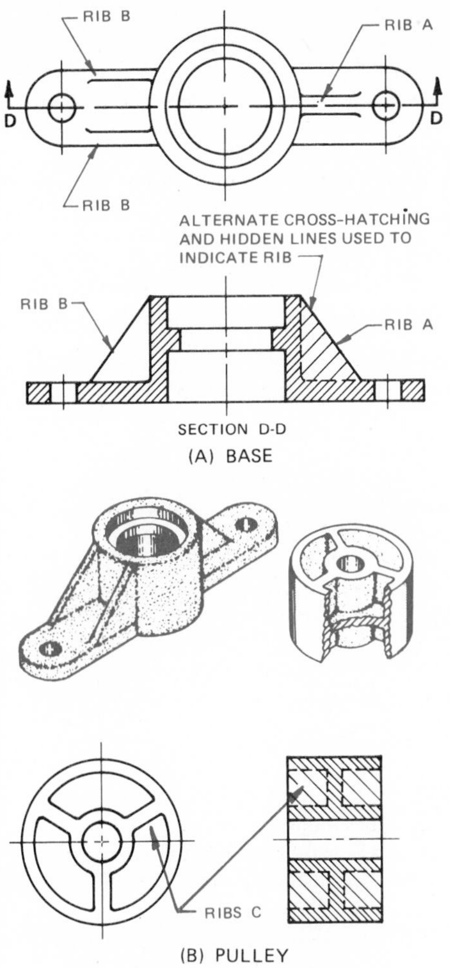
RIB B
RIB A
D
D
RIB B
ALTERNATE CROSS-HATCHING AND HIDDEN LINES USED TO INDICATE RIB
RIB B
RIB A
SECTION D-D
(A) BASE
RIBS C
(B) PULLEY

Fig. 7-7-2 Alternate method of showing ribs in section.

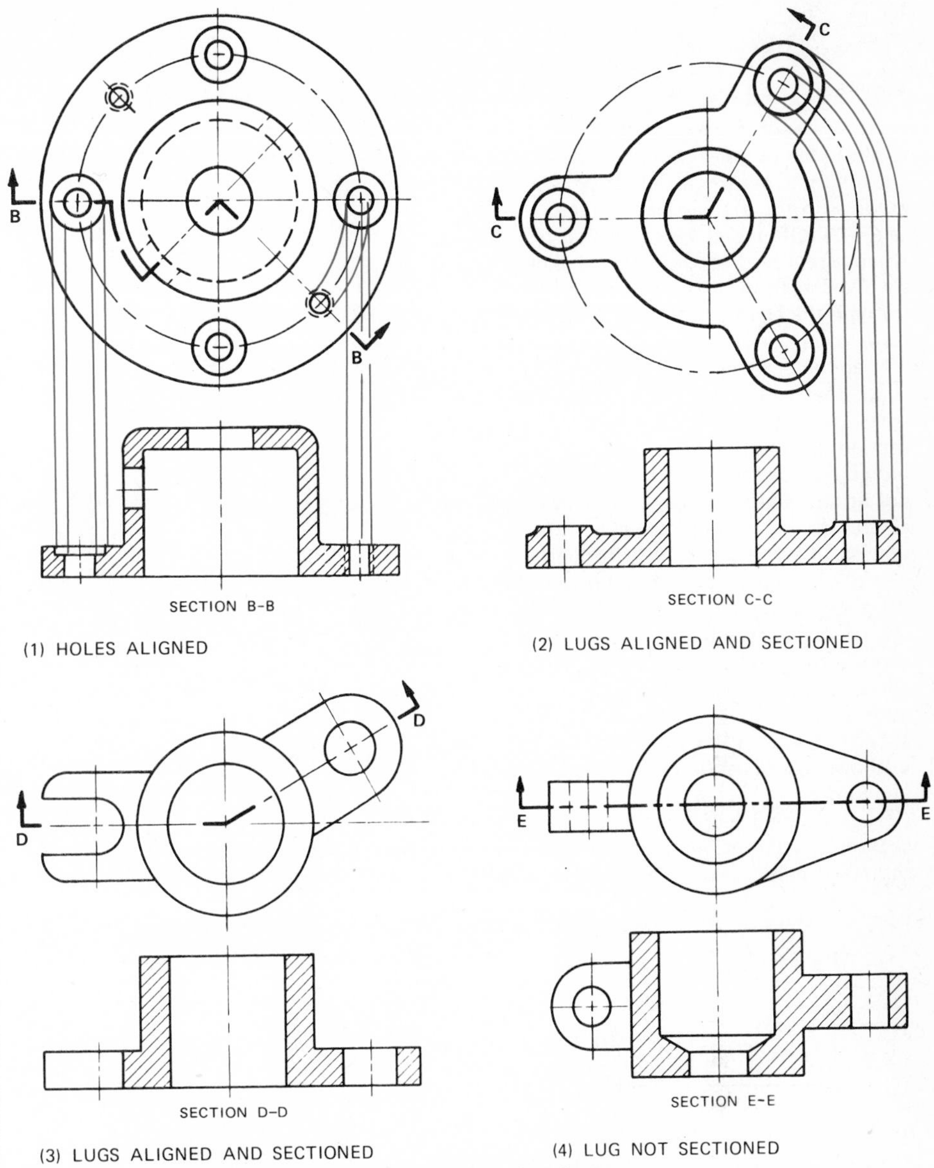
B
B
C
C
SECTION B-B
SECTION C-C
(1) HOLES ALIGNED
(2) LUGS ALIGNED AND SECTIONED
D
D
E
E
SECTION D–D
SECTION E-E
(3) LUGS ALIGNED AND SECTIONED
(4) LUG NOT SECTIONED

Fig. 7-7-3 Aligning holes and lugs in section drawings.

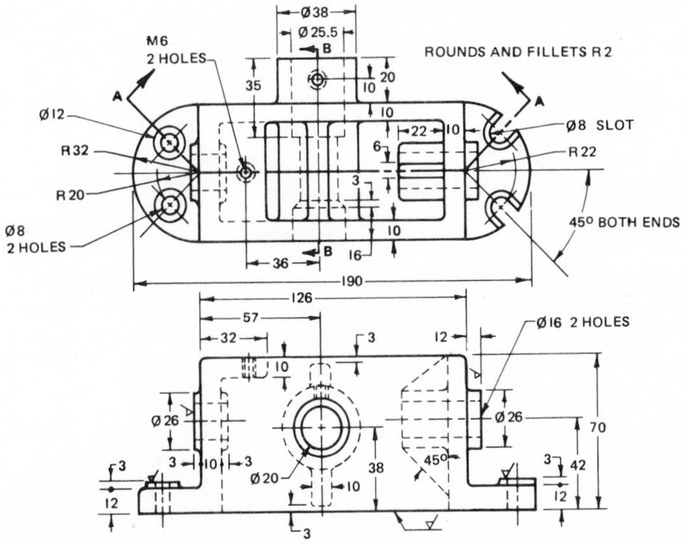
M6
2 HOLES
ROUNDS AND FILLETS R2
Ø12
R32
R20
Ø8
2 HOLES
Ø8 SLOT
R22
45° BOTH ENDS
Ø38
Ø25.5
190
126
57
32
Ø16 2 HOLES
Ø26
Ø20
MATL – MALLEABLE IRON

Fig. 7-7-A Shaft support.

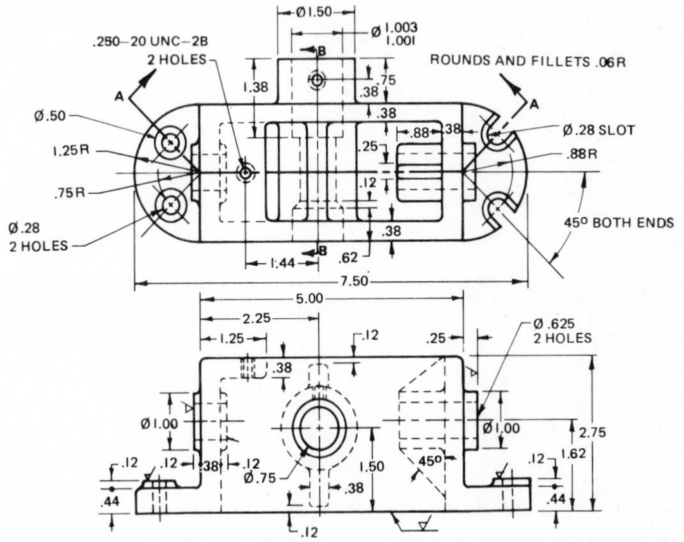
.250–20 UNC–2B
2 HOLES
ROUNDS AND FILLETS .06R
Ø.50
1.25R
.75R
Ø.28
2 HOLES
Ø.28 SLOT
.88R
45° BOTH ENDS
Ø1.50
Ø 1.003
1.001
7.50
5.00
2.25
1.25
Ø.625
2 HOLES
Ø1.00
Ø.75
MATL – MALLEABLE IRON

Fig. 7-7-B Shaft support.

UNIT 7-8

REVOLVED AND REMOVED SECTIONS

Revolved and removed sections are used to show the cross-sectional shape of ribs, spokes, or arms when the shape is not obvious in the regular views (Figs. 7-8-1 to 7-8-3). Often end views are not needed when a revolved section is used. For a revolved section, draw a center line through the shape on the plane to be described, imagine the part to be rotated 90°, and superimpose on the view the shape that would be seen when rotated (Figs. 7-8-1 and 7-8-3). If the revolved section does not interfere with the view on which it is revolved, then the view is not broken unless it would provide for clearer dimensioning. When the revolved section interferes or passes through lines on the view on which it is revolved, then the general practice is to break the view (Fig. 7-8-3). Often the break is used to shorten the length of the object. In no circumstances should the lines on the view pass through the section. When superimposed on the view, the outline of the revolved section is a thin, continuous line.

The removed section differs in that the section, instead of being drawn right on the view, is removed to an open area on the drawing (Fig. 7-8-2). Frequently the removed section is drawn to an enlarged scale for clarification and easier dimensioning. Removed sections of symmetrical parts should be placed, whenever possible, on the extension of the center line (Fig. 7-8-2*b*).

On complicated drawings where the placement of the removed view may be some distance from the cutting plane, auxiliary information, such as reference zone location (Fig. 7-8-4), may be helpful.

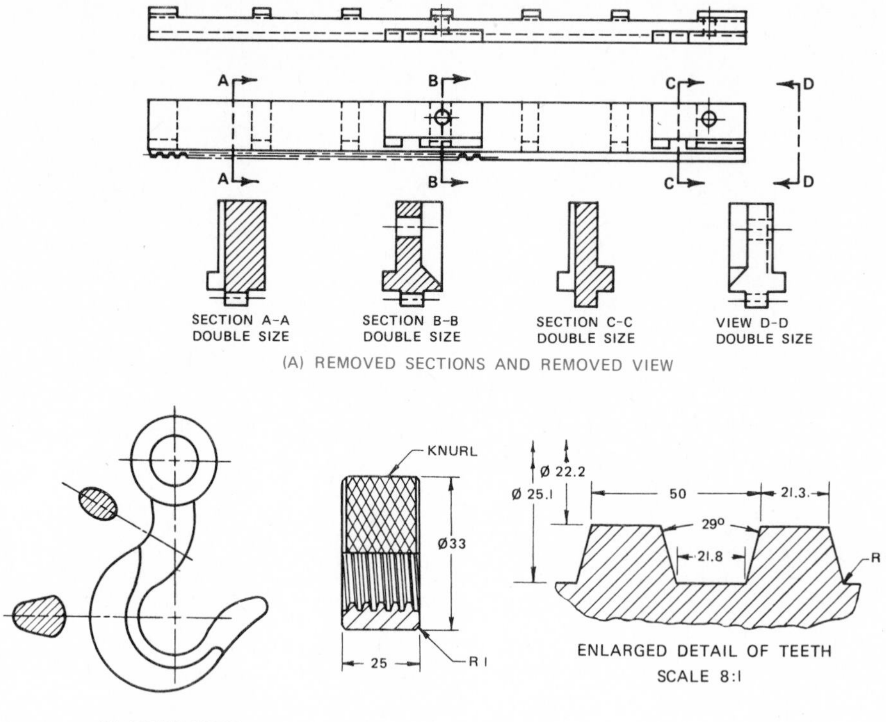

A
B
C
D
SECTION A–A
DOUBLE SIZE
SECTION B–B
DOUBLE SIZE
SECTION C–C
DOUBLE SIZE
VIEW D–D
DOUBLE SIZE
(A) REMOVED SECTIONS AND REMOVED VIEW
KNURL
Ø33
25
R 1
Ø 22.2
Ø 25.1
50
21.3
29°
21.8
R 0.5
ENLARGED DETAIL OF TEETH
SCALE 8:1
(B) CRANE HOOK
(C) NUT

Fig. 7-8-2 Removed sections.

PLACEMENT OF SECTIONAL VIEWS

Whenever practical, except for revolved sections, section views should be projected perpendicular to the cutting plane and be placed in the normal position for third-angle projection (Fig. 7-8-5).

When the preferred placement is not practical, the sectional view may be removed to some other convenient position

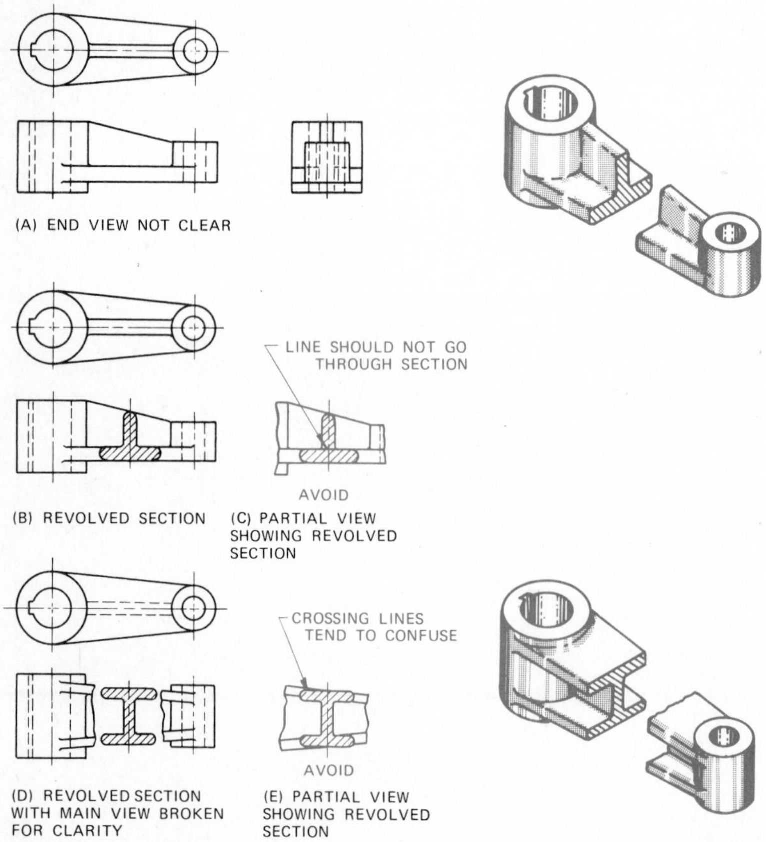

(A) END VIEW NOT CLEAR
LINE SHOULD NOT GO THROUGH SECTION
AVOID
(B) REVOLVED SECTION
(C) PARTIAL VIEW SHOWING REVOLVED SECTION
CROSSING LINES TEND TO CONFUSE
AVOID
(D) REVOLVED SECTION WITH MAIN VIEW BROKEN FOR CLARITY
(E) PARTIAL VIEW SHOWING REVOLVED SECTION

Fig. 7-8-1 Revolved sections.

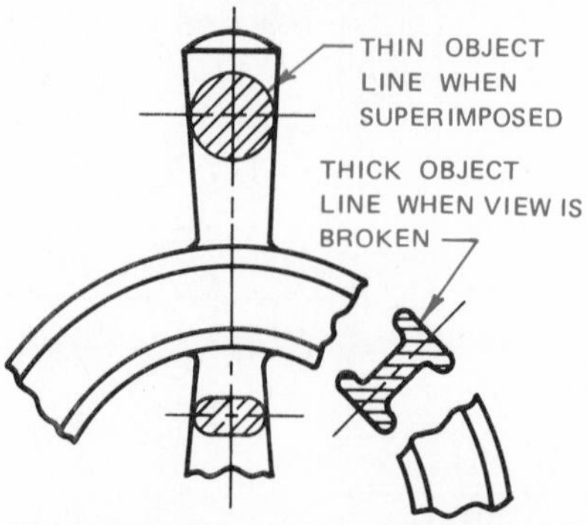

THIN OBJECT LINE WHEN SUPERIMPOSED
THICK OBJECT LINE WHEN VIEW IS BROKEN

Fig. 7-8-3 Revolved (superimposed) sections.

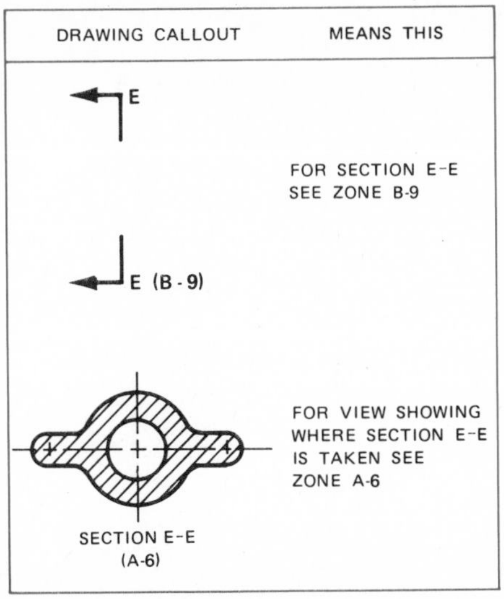
DRAWING CALLOUT
MEANS THIS
E
FOR SECTION E-E
SEE ZONE B-9
E (B-9)
FOR VIEW SHOWING
WHERE SECTION E-E
IS TAKEN SEE
ZONE A-6
SECTION E-E
(A-6)

Fig. 7-8-4 Reference zone location.

on the drawing, but it must be clearly identified, usually by two capital letters, excluding I, O, Q, and Z, and be labeled.

Assignment

Select one of the problems shown in Figs. 7-8-A and 7-8-B and on an A3- or B-size sheet make a working drawing of the part. For clarity it is recommended that an enlarged removed view be used to show the detail of the small hole. Scale is 1:1.

REVIEW FOR ASSIGNMENT

Unit 7-4 Conventional Breaks
Unit 7-18 Intersection of Unfinished Surfaces
Unit 5-7 Machining Symbols

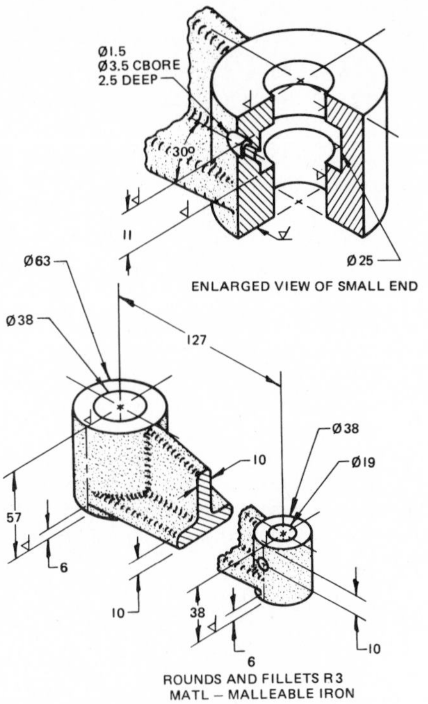
Ø1.5
Ø3.5 CBORE
2.5 DEEP
30°
11
Ø63
Ø25
ENLARGED VIEW OF SMALL END
Ø38
127
Ø38
10
Ø19
57
6
10
38
6
10
ROUNDS AND FILLETS R3
MATL – MALLEABLE IRON

Fig. 7-8-A Shaft support.

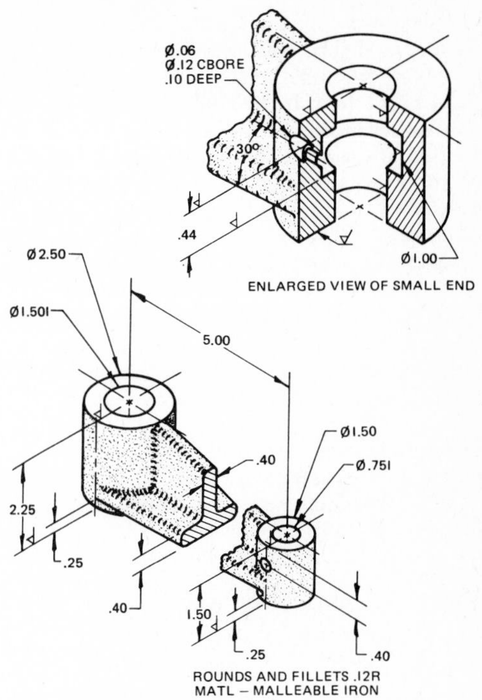
Ø.06
Ø.12 CBORE
.10 DEEP
30°
.44
Ø2.50
Ø1.00
ENLARGED VIEW OF SMALL END
Ø1.501
5.00
Ø1.50
.40
Ø.751
2.25
.25
.40
1.50
.25
.40
ROUNDS AND FILLETS .12R
MATL – MALLEABLE IRON

Fig. 7-8-B Shaft support.

UNIT 7-9

SPOKES AND ARMS IN SECTION

A comparison of the true projection of the wheel with spokes and the wheel with a web is made in Figs. 7-9-1*a* and *b*. This comparison shows that a preferred section for the wheel and spokes is desirable so that it will not appear to be a wheel with a solid web. In preferred sectioning, any part that is not solid or continuous around the hub is drawn without the section lining, even though the cutting plane passes through the spoke. When there is an odd number of spokes,

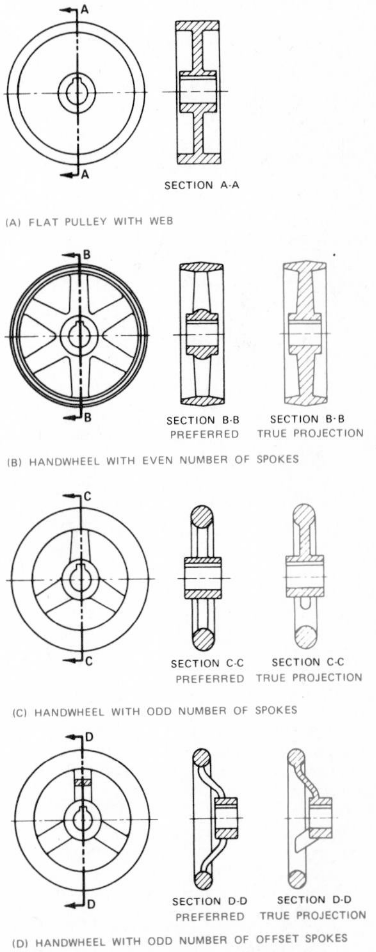
A
A
SECTION A-A
(A) FLAT PULLEY WITH WEB
B
B
SECTION B-B
PREFERRED
SECTION B-B
TRUE PROJECTION
(B) HANDWHEEL WITH EVEN NUMBER OF SPOKES
C
C
SECTION C-C
PREFERRED
SECTION C-C
TRUE PROJECTION
(C) HANDWHEEL WITH ODD NUMBER OF SPOKES
D
D
SECTION D-D
PREFERRED
SECTION D-D
TRUE PROJECTION
(D) HANDWHEEL WITH ODD NUMBER OF OFFSET SPOKES

Fig. 7-9-1 Preferred and true projection through spokes.

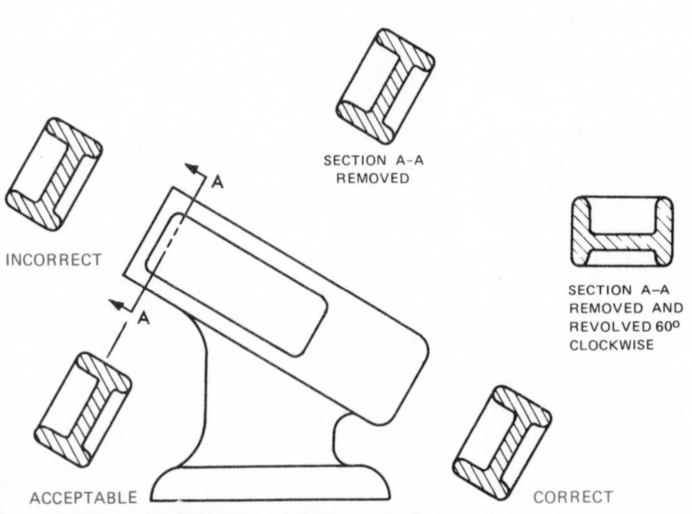
SECTION A-A
REMOVED
A
INCORRECT
A
SECTION A-A
REMOVED AND
REVOLVED 60°
CLOCKWISE
ACCEPTABLE
CORRECT

Fig. 7-8-5 Placement of sectional views.

as shown in Fig. 7-9-1*c*, the bottom spoke is aligned with the top spoke to show its true relationship to the wheel and to the hub. If the spoke were not revolved or aligned, it would appear distorted in the sectional view.

Assignment

Select one of the problems shown in Figs. 7-9-A and 7-9-B and make on an A3- or B-size sheet a two-view working drawing. Draw the side view in full section, and show a revolved section of the spoke in the front view. Scale is 1:1.

REVIEW FOR ASSIGNMENT

Unit 9-2 Fasteners, Pins, etc.
Unit 7-18 Intersection of Unfinished Surfaces

Fig. 7-9-A Offset handwheel.

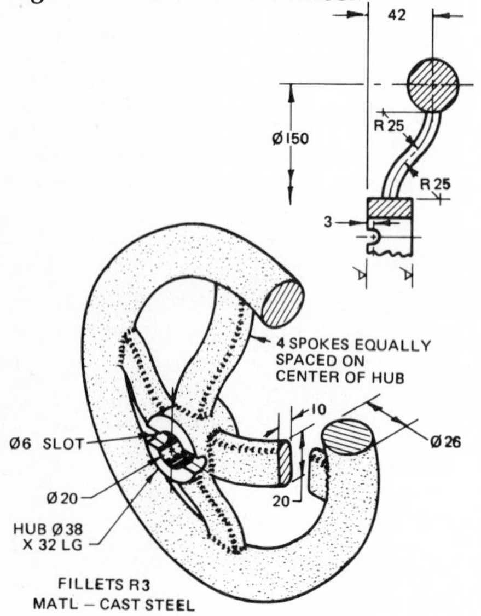

42
Ø150
R25
R25
3
4 SPOKES EQUALLY SPACED ON CENTER OF HUB
10
Ø6 SLOT
Ø26
Ø20
20
HUB Ø38 X 32 LG
FILLETS R3
MATL – CAST STEEL

Fig. 7-9-B Offset handwheel.

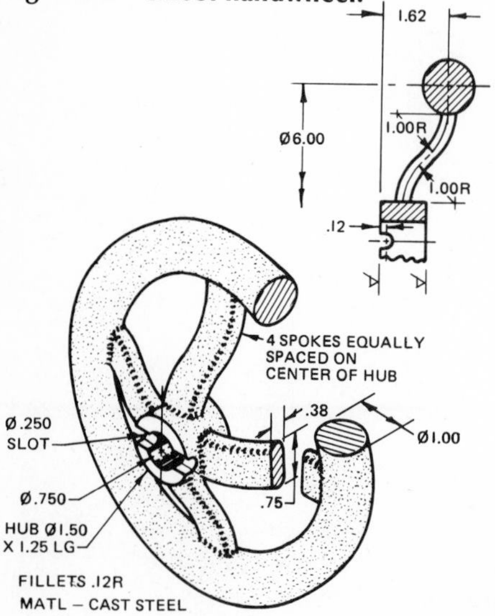

1.62
Ø6.00
1.00R
1.00R
.12
4 SPOKES EQUALLY SPACED ON CENTER OF HUB
.38
Ø.250 SLOT
Ø1.00
Ø.750
.75
HUB Ø1.50 X 1.25 LG
FILLETS .12R
MATL – CAST STEEL

UNIT 7-10

PARTIAL OR BROKEN-OUT SECTIONS

Where a sectional view of only a portion of the object is needed, partial sections may be used (Fig. 7-10-1). An irregular break line is used to show the extent of the section. With this type of section, a cutting-plane line is not required.

Assignment

Select one of the problems shown in Figs. 7-10-A and 7-10-B and make a two-view working drawing on an A3- or B-size sheet. Use partial sections where clarity of drawing can be achieved. Scale is 1:1.

REVIEW FOR ASSIGNMENT

Unit 9-3 Retaining Rings

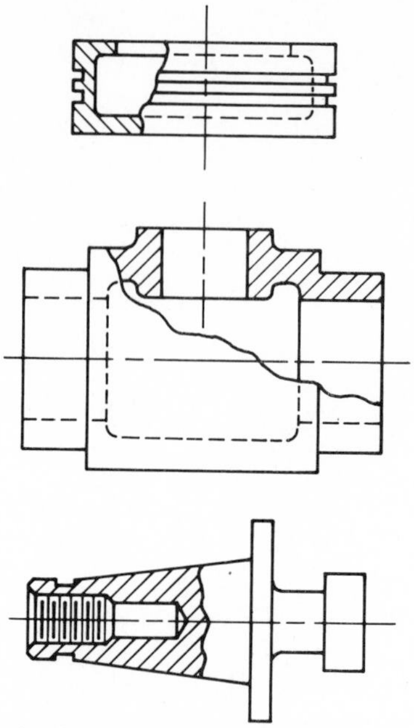

Fig. 7-10-1 Broken-out or partial sections.

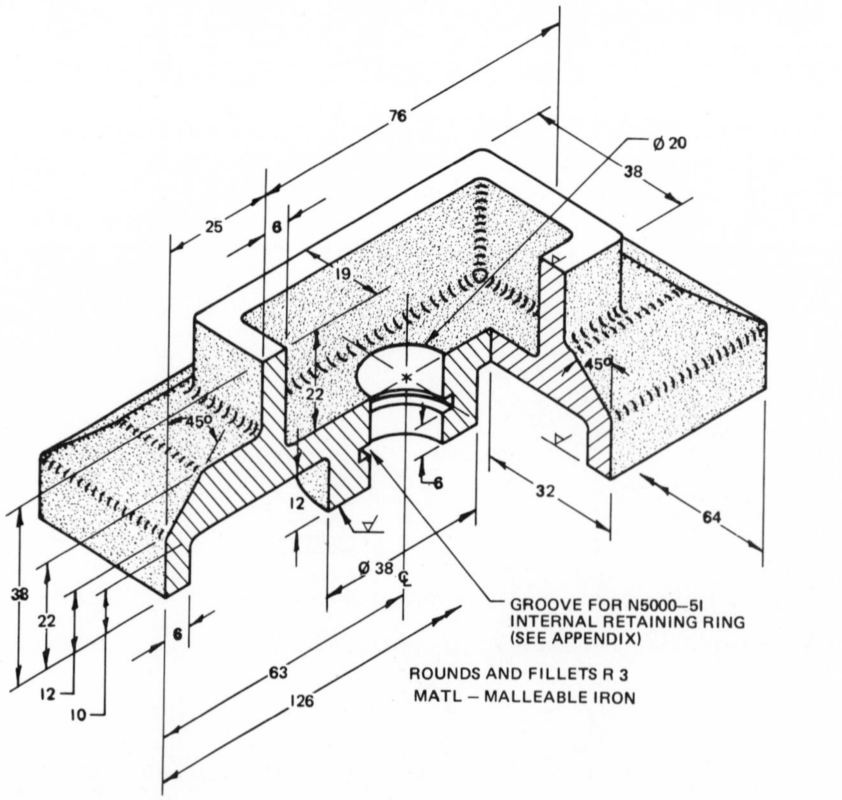

76
Ø20
38
25
8
19
22
45°
45°
6
12
32
64
Ø38
38
22
12
10
6
63
126
GROOVE FOR N5000–51 INTERNAL RETAINING RING (SEE APPENDIX)
ROUNDS AND FILLETS R 3
MATL – MALLEABLE IRON

Fig. 7-10-A Hold-down bracket.

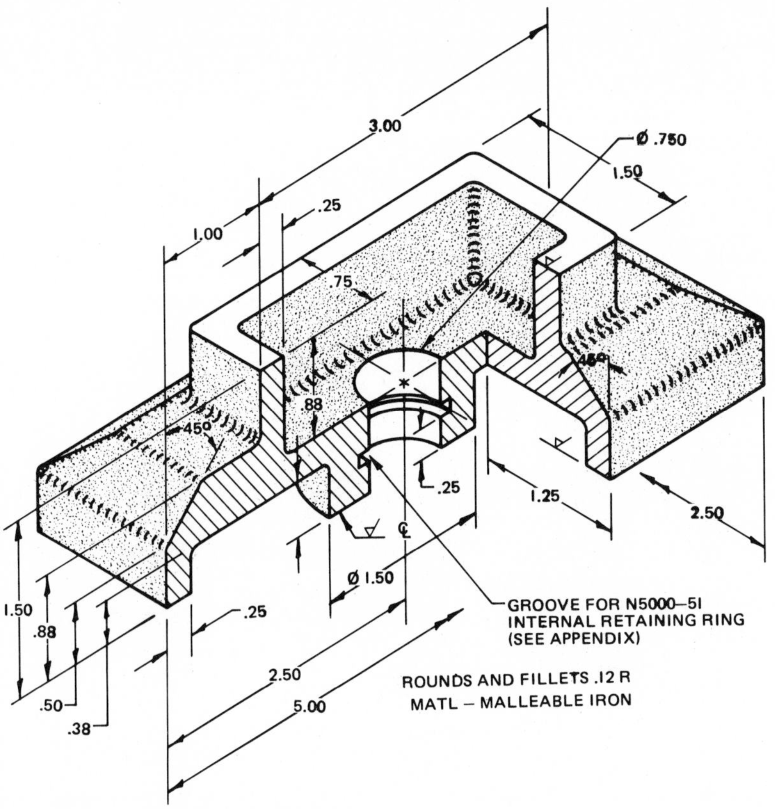
3.00
Ø .750
1.50
.25
1.00
.75
45°
.88
45°
.25
1.25
2.50
Ø 1.50
1.50
.88
.25
.50
.38
2.50
5.00
GROOVE FOR N5000–51
INTERNAL RETAINING RING
(SEE APPENDIX)
ROUNDS AND FILLETS .12 R
MATL – MALLEABLE IRON

Fig. 7-10-B Hold-down bracket.

UNIT 7-11

PHANTOM OR HIDDEN SECTIONS

A phantom section is used to show the typical interior shapes of an object in one view when the part is not truly symmetrical in shape, as well as to show mating parts in an assembly drawing (Fig. 7-11-1). It is a sectional view superimposed on the regular view without the removal of the front portion of the object. The section lining used for phantom sections consists of light, evenly spaced, broken lines.

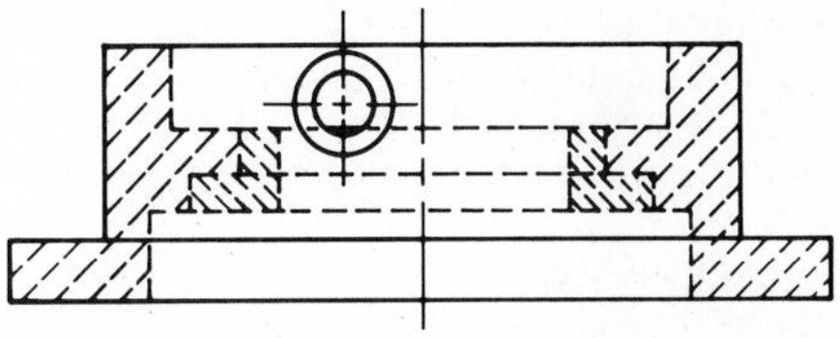

Fig. 7-11-1 Phantom or hidden section.

Assignment

On an A3- or B-size sheet, make a two-view assembly drawing of one of the assemblies shown in Figs. 7-11-A and 7-11-B. The front view is to be drawn as a phantom section drawing. Scale is 1:1. Show only the bushing hole sizes and their locations on the drawing.

REVIEW FOR ASSIGNMENT

Unit 5-3 Necking
Unit 5-7 Machining Symbols
Unit 5-6 Standard Fits
Unit 6-4 Part Number Identification

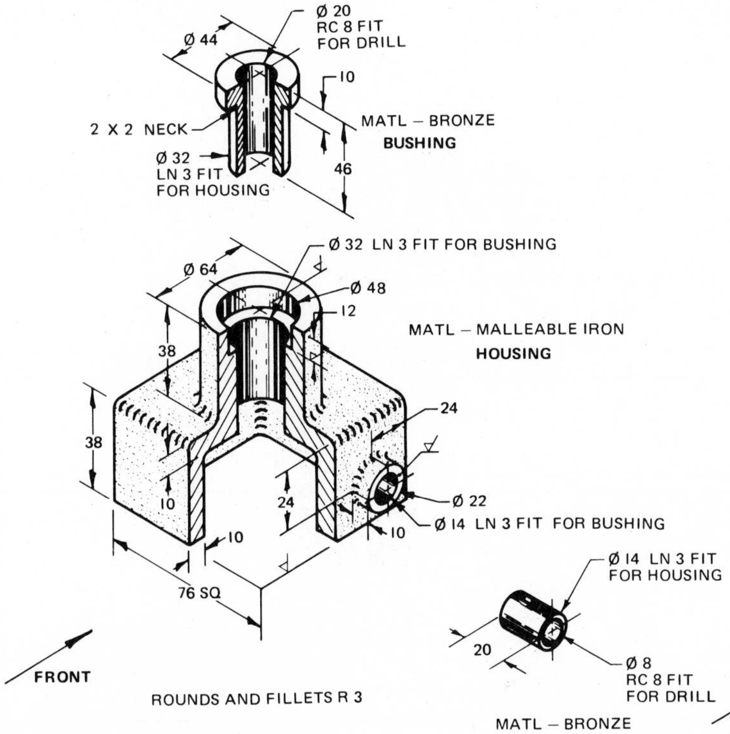
Ø 44
Ø 20
RC 8 FIT
FOR DRILL
10
2 X 2 NECK
MATL – BRONZE
BUSHING
Ø 32
LN 3 FIT
FOR HOUSING
46
Ø 32 LN 3 FIT FOR BUSHING
Ø 64
Ø 48
12
38
MATL – MALLEABLE IRON
HOUSING
24
38
10
24
Ø 22
Ø 14 LN 3 FIT FOR BUSHING
10
10
76 SQ
Ø 14 LN 3 FIT
FOR HOUSING
20
Ø 8
RC 8 FIT
FOR DRILL
FRONT
ROUNDS AND FILLETS R 3
MATL – BRONZE
BUSHING

Fig. 7-11-A Drill-jig assembly.

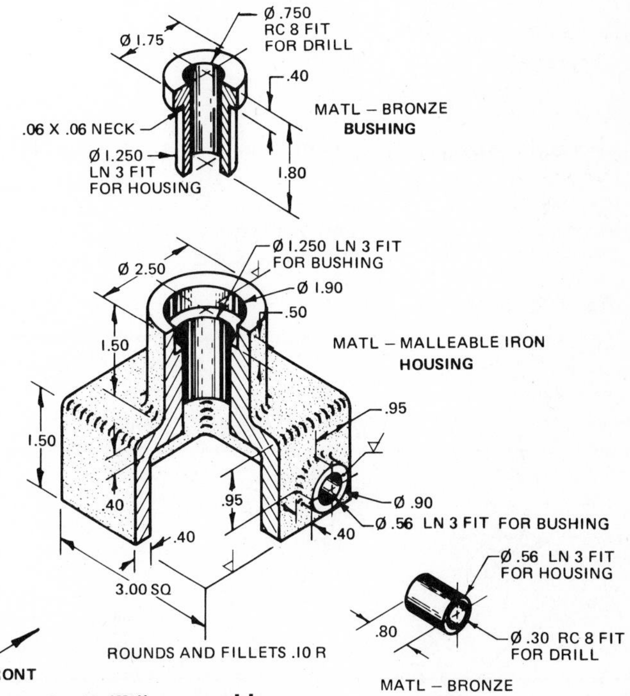
Ø 1.75
Ø .750
RC 8 FIT
FOR DRILL
.40
MATL – BRONZE
BUSHING
.06 X .06 NECK
Ø 1.250
LN 3 FIT
FOR HOUSING
1.80
Ø 2.50
Ø 1.250 LN 3 FIT
FOR BUSHING
Ø 1.90
.50
1.50
MATL – MALLEABLE IRON
HOUSING
.95
1.50
.40
.95
Ø .90
Ø .56 LN 3 FIT FOR BUSHING
.40
.40
3.00 SQ
Ø .56 LN 3 FIT
FOR HOUSING
.80
Ø .30 RC 8 FIT
FOR DRILL
FRONT
ROUNDS AND FILLETS .10 R
MATL – BRONZE
BUSHING

Fig. 7-11-B Drill-jig assembly.

UNIT 7-12

SECTIONAL DRAWING REVIEW

In Units 7-1 through 7-11 the different types of sectional views have been explained and drawing problems have been assigned with each type of section drawing.

In the drafting office, it is the drafter who must decide which views are required to fully explain the part to be made. In addition, the drafter must select the proper scale(s) which will show the features clearly.

This unit has been designed to review the sectional-view options open to the drafter.

Assignment

On an A3- or B-size sheet, make a three-view working drawing of one of the parts shown in Figs. 7-12-A and 7-12-B. From the information on section drawings found in Units 7-1 to 7-11, select appropriate sectional views which will improve the clarity of the drawing. Scale is 1:1.

REVIEW FOR ASSIGNMENT

Units 7-1 to 7-11	Section Drawings
Unit 6-1	Working Drawings

UNIT 7-13

CONVENTIONAL REPRESENTATION OF COMMON FEATURES

To simplify the representation of common features, a number of conventional drawing practices are used. Many conventions are deviations from true projection for the purpose of clarity; others are used to save drafting time. These conventions must be executed carefully, for clarity is even more important than speed.

Many drawing conventions such as those used on thread, gear, and spring drawings appear in various chapters throughout the text. Only the conventions not described in these chapters appear here.

REPETITIVE DETAILS

Repetitive features, such as gear and spline teeth, are shown by drawing a partial view, showing two or three of these features, with a phantom line or lines to indicate the extent of the remaining features (Fig. 7-13-1*a* and *b*). Alternatively, gears and splines may be shown with a solid line representing the basic outline of the part and a lighter line representing the root of the teeth. This is essentially the same convention that is used for screw threads. The pitch line may be added by using the standard center line.

KNURLS

Knurling is an operation which puts patterned indentations in the surface of a metal part to provide a good finger grip (Fig. 7-13-1*c* and *d*). Commonly used types of knurls are straight, diagonal, spiral, convex, raised diamond, depressed diamond, and radial. The pitch refers to the distance between corresponding indentations, and it may be a straight pitch, a circular pitch, or a diametral pitch. For cylindrical surfaces, the latter is preferred. The pitch of the teeth for coarse knurls (measured parallel to the axis of the work) is about 2 mm (.08 in.); for medium knurls, about 1.2 mm (.05 in.); and for fine knurls, 0.8 mm (.03 in.). The medium-pitch knurl is the most commonly used.

As a time-saver, the knurl symbol is shown on only a part of the surface being knurled.

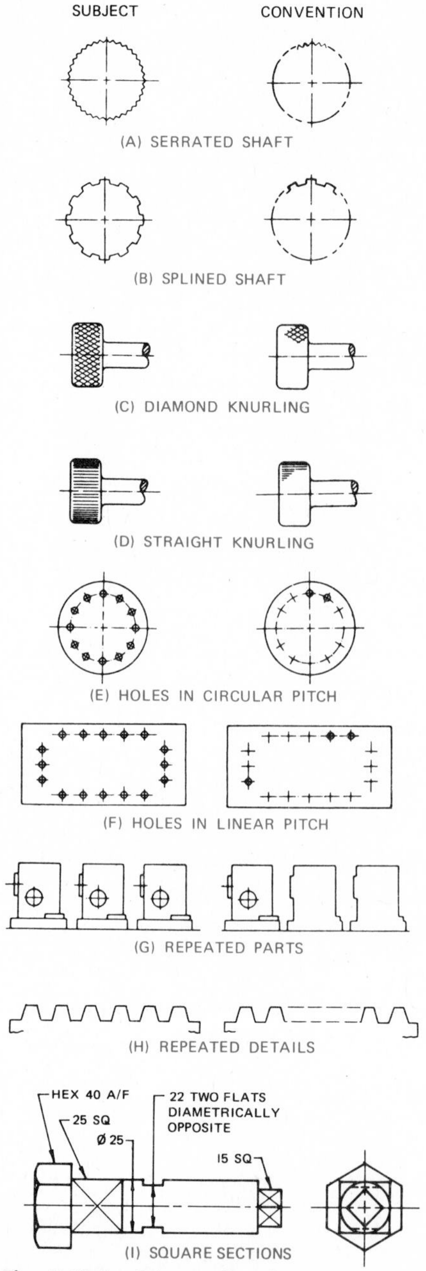
SUBJECT
CONVENTION
(A) SERRATED SHAFT
(B) SPLINED SHAFT
(C) DIAMOND KNURLING
(D) STRAIGHT KNURLING
(E) HOLES IN CIRCULAR PITCH
(F) HOLES IN LINEAR PITCH
(G) REPEATED PARTS
(H) REPEATED DETAILS
HEX 40 A/F
25 SQ
Ø 25
22 TWO FLATS DIAMETRICALLY OPPOSITE
15 SQ
(I) SQUARE SECTIONS

Fig. 7-13-1 Conventional representation of common features.

Fig. 7-12-A Shaft-support base.

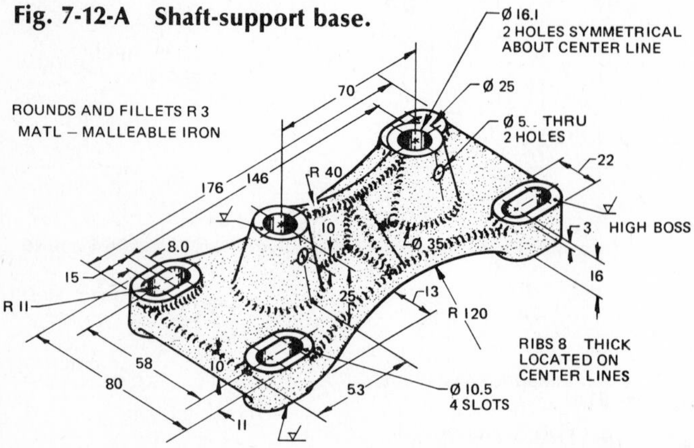
Ø 16.1
2 HOLES SYMMETRICAL ABOUT CENTER LINE
Ø 25
Ø 5. THRU
2 HOLES
ROUNDS AND FILLETS R 3
MATL – MALLEABLE IRON
70
176
146
R 40
22
10
3 HIGH BOSS
8.0
Ø 35
15
16
R 11
25
13
R 120
58
10
RIBS 8 THICK LOCATED ON CENTER LINES
80
53
Ø 10.5
4 SLOTS
11

Fig. 7-12-B Shaft-support base.

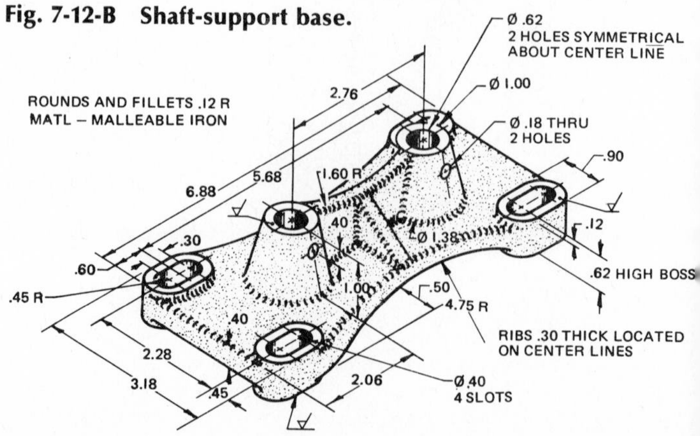
Ø .62
2 HOLES SYMMETRICAL ABOUT CENTER LINE
Ø 1.00
2.76
ROUNDS AND FILLETS .12 R
MATL – MALLEABLE IRON
Ø .18 THRU
2 HOLES
.90
5.68
1.60 R
6.88
40
Ø 1.38
.12
.30
.60
.62 HIGH BOSS
1.00
.50
.45 R
4.75 R
40
RIBS .30 THICK LOCATED ON CENTER LINES
2.28
3.18
.45
2.06
Ø .40
4 SLOTS

HOLES

A series of similar holes is indicated by drawing one or two holes and showing only the center for the others (Fig. 7-13-1*e* and *f*).

REPETITIVE PARTS

Repetitive parts, or intricate features, are shown by drawing one in detail and the others in simple outline only. A covering note is added to the drawing (Fig. 7-13-1*g* and *h*).

SQUARE SECTIONS

Square sections on shafts and similar parts may be indicated by thin, crossed, diagonal lines, as shown in Fig. 7-13-1*i*.

Assignment

On an A3- or B-size sheet, make a working drawing of one of the parts shown in Figs. 7-13-A and 7-13-B. Wherever possible, simplify the drawing by using conventional representation of features.

REVIEW FOR ASSIGNMENT

Unit 2-2 Enlarged Scales
Unit 3-7 One- and Two-View Drawings
Unit 5-3 Dimensioning Common Features

UNIT 7-14

CONVENTIONAL BREAKS

Long, simple parts such as shafts, bars, tubes, and arms need not be drawn to their entire length. Conventional breaks located at a convenient position may be used and the true length indicated by a dimension (Fig. 7-14-1). Often a part can be drawn to a larger scale to produce a clearer drawing if a conventional break is used. The breaks used on circular objects, known as "S breaks," may be drawn freehand, with an irregular curve, template, or compass. The procedure for drawing an S break using a compass is shown in Fig. 7-14-2. Freehand drawing of S breaks is not recommended on shafts 20 mm (.75 in.) in diameter or larger. The suggested radius used for the break varies between the radius of the shaft to 0.7 times the size of shaft radius.

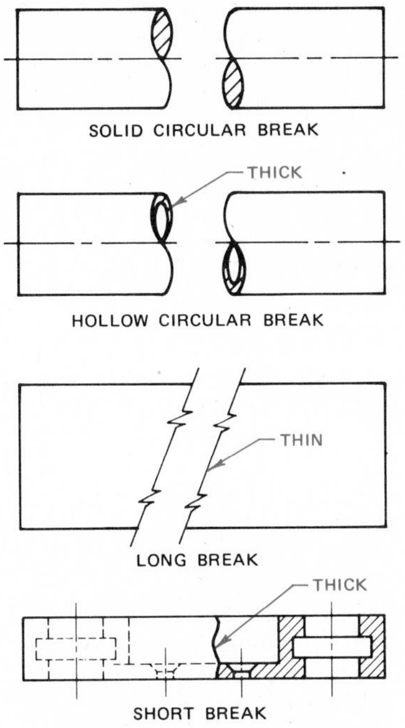
SOLID CIRCULAR BREAK
THICK
HOLLOW CIRCULAR BREAK
THIN
LONG BREAK
THICK
SHORT BREAK

Fig. 7-14-1 Conventional breaks.

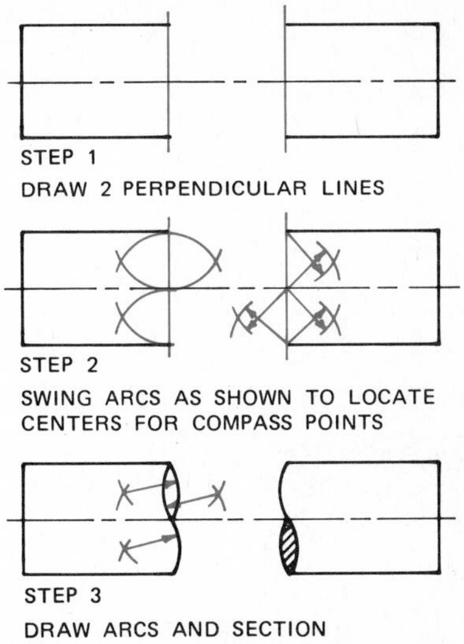
STEP 1
DRAW 2 PERPENDICULAR LINES
STEP 2
SWING ARCS AS SHOWN TO LOCATE CENTERS FOR COMPASS POINTS
STEP 3
DRAW ARCS AND SECTION

Fig. 7-14-2 Steps for drawing the S break for circular parts using a compass.

Assignment

On an A3- or B-size sheet, make a working drawing of one of the parts shown in Figs. 7-14-A and 7-14-B. Use conventional breaks to shorten the length of the part. An enlarged view is also recommended where the detail cannot be clearly shown at full scale. Scale is 1:1.

REVIEW FOR ASSIGNMENT

Unit 4-3 Drawing a Hexagon
Unit 3-9 Enlarged Views

Fig. 7-13-A Clock stem.

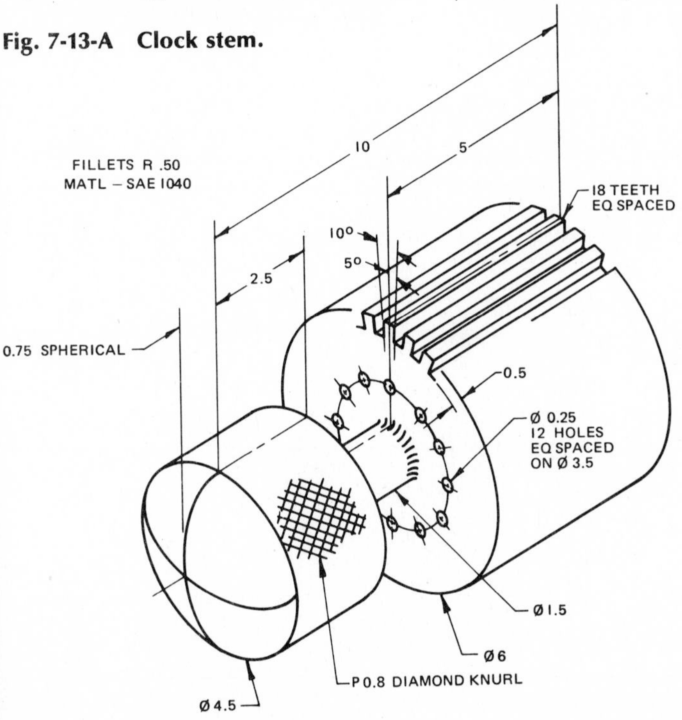
FILLETS R .50
MATL – SAE 1040
10
5
18 TEETH
EQ SPACED
10°
5°
2.5
0.75 SPHERICAL
0.5
Ø 0.25
12 HOLES
EQ SPACED
ON Ø 3.5
Ø 1.5
Ø 6
P 0.8 DIAMOND KNURL
Ø 4.5

Fig. 7-13-B Clock stem.

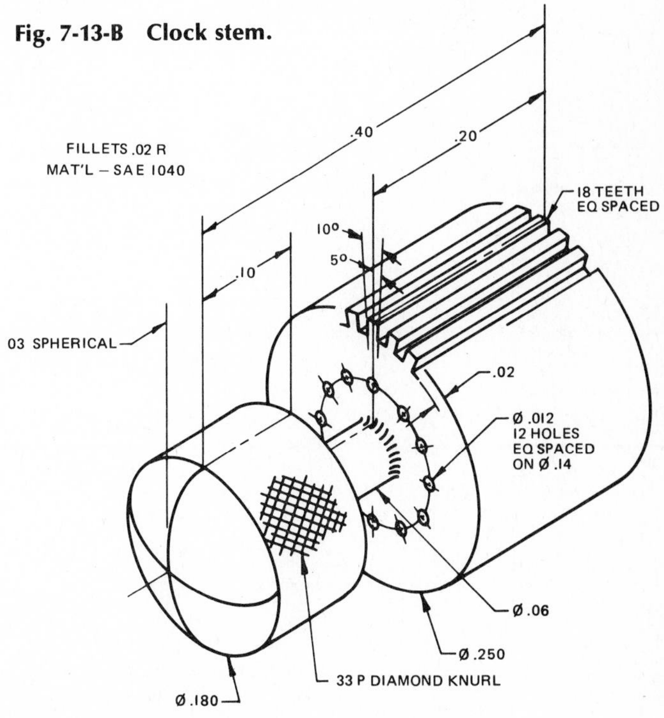
FILLETS .02 R
MAT'L – SAE 1040
.40
.20
18 TEETH
EQ SPACED
10°
5°
.10
03 SPHERICAL
.02
Ø .012
12 HOLES
EQ SPACED
ON Ø .14
Ø .06
Ø .250
33 P DIAMOND KNURL
Ø .180

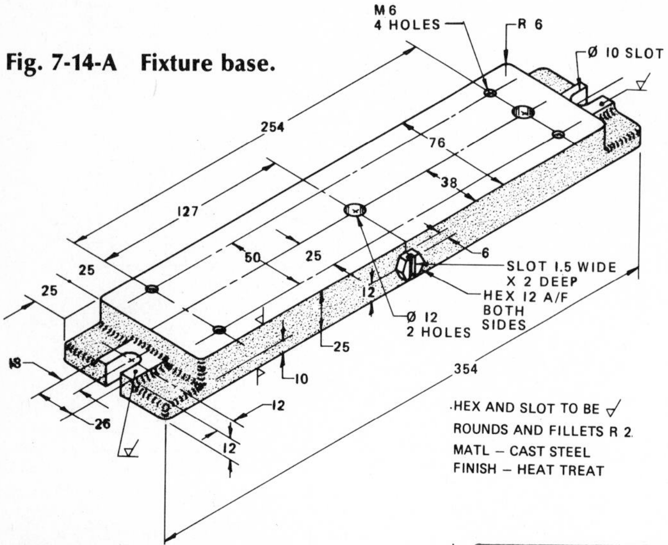
M6
4 HOLES
R 6
Ø 10 SLOT
254
76
38
127
50
25
25
25
6
SLOT 1.5 WIDE
X 2 DEEP
HEX 12 A/F
BOTH
SIDES
Ø 12
2 HOLES
12
25
18
10
26
12
12
354
HEX AND SLOT TO BE
ROUNDS AND FILLETS R 2
MATL – CAST STEEL
FINISH – HEAT TREAT

Fig. 7-14-A Fixture base.

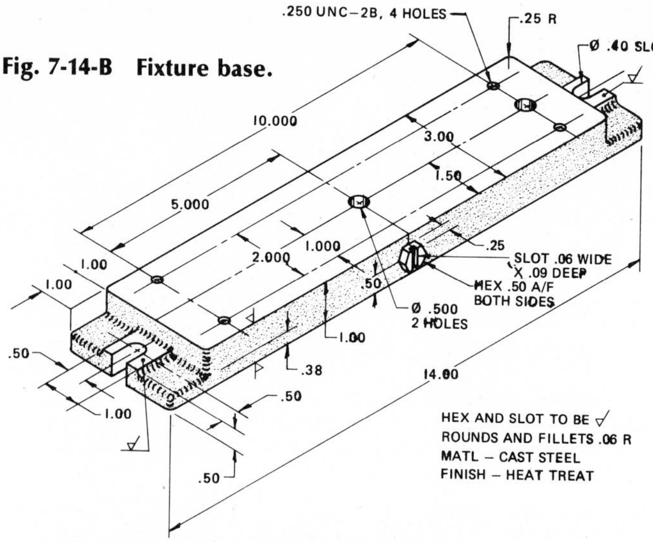
.250 UNC–2B, 4 HOLES
.25 R
Ø .40 SLOT
10.000
3.00
1.50
5.000
1.000
2.000
1.00
1.00
.25
SLOT .06 WIDE
X .09 DEEP
HEX .50 A/F
BOTH SIDES
.50
Ø .500
2 HOLES
1.00
.50
.38
.50
1.00
14.00
.50
HEX AND SLOT TO BE
ROUNDS AND FILLETS .06 R
MATL – CAST STEEL
FINISH – HEAT TREAT

Fig. 7-14-B Fixture base.

UNIT 7-15

MATERIALS OF CONSTRUCTION

Symbols used to indicate materials in sectional views are shown in Fig. 7-1-6. Those shown for concrete, wood, and transparent materials are also suitable for outside views. Other symbols which may be used to indicate areas of different materials are shown in Fig. 7-15-1. It is not necessary to cover the entire area affected with such symbolic lining, as long as the extent of the area is shown on the drawing.

TRANSPARENT MATERIALS

These should generally be treated in the same manner as opaque materials; i.e., details behind them are shown with hidden lines if such detail is necessary.

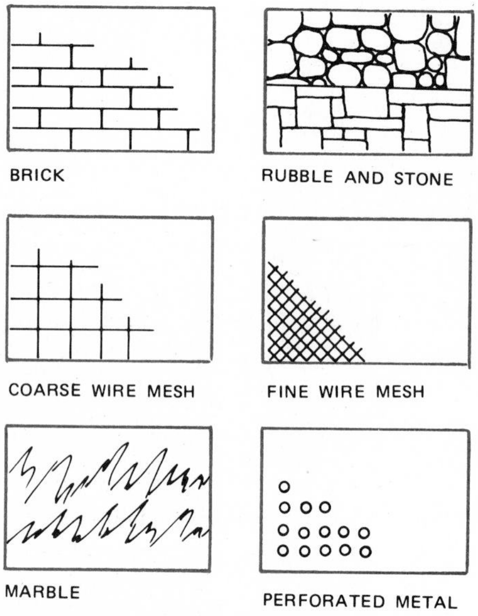
BRICK
RUBBLE AND STONE
COARSE WIRE MESH
FINE WIRE MESH
MARBLE
PERFORATED METAL

Fig. 7-15-1 Symbols to indicate materials of construction.

Assignment

On an A3- or B-size sheet, make a detailed assembly drawing of one of the assemblies shown in Figs. 7-15-A and 7-15-B. Enlarged details are recommended for the steel mesh and joints. Use conventional breaks to shorten the length. Scale is 1:5 (metric) or 1:4 (inches).

REVIEW FOR ASSIGNMENT

Unit 2-2	Enlarged Views
Unit 6-6	Detail Assembly Drawings
Unit 7-14	Conventional Breaks
Appendix	Sheet-Metal Gage Sizes

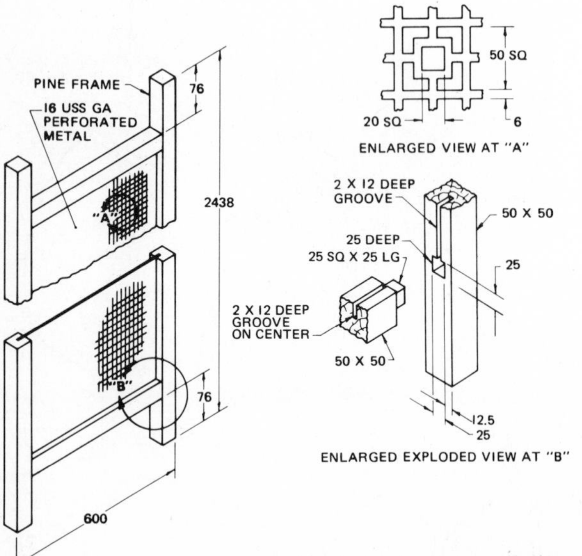
PINE FRAME
16 USS GA
PERFORATED
METAL
76
2438
"A"
"B"
76
600
50 SQ
20 SQ
6
ENLARGED VIEW AT "A"
2 X 12 DEEP
GROOVE
50 X 50
25 DEEP
25 SQ X 25 LG
25
2 X 12 DEEP
GROOVE
ON CENTER
50 X 50
12.5
25
ENLARGED EXPLODED VIEW AT "B"

Fig. 7-15-A Room divider.

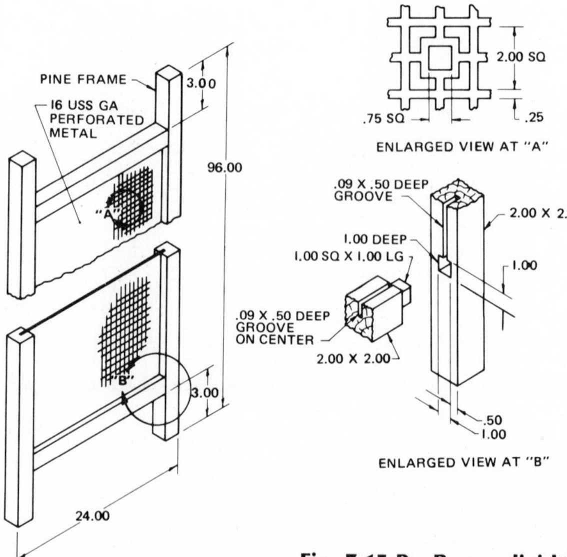
PINE FRAME
16 USS GA
PERFORATED
METAL
3.00
96.00
"A"
"B"
3.00
24.00
2.00 SQ
.75 SQ
.25
ENLARGED VIEW AT "A"
.09 X .50 DEEP
GROOVE
2.00 X 2.00
1.00 DEEP
1.00 SQ X 1.00 LG
1.00
.09 X .50 DEEP
GROOVE
ON CENTER
2.00 X 2.00
.50
1.00
ENLARGED VIEW AT "B"

Fig. 7-15-B Room divider.

UNIT 7-16
CYLINDRICAL INTERSECTIONS

The intersections of rectangular and circular contours, unless they are very large, are shown conventionally as in Figs. 7-16-1 and 7-16-2. The same convention may be used to show the intersection of two cylindrical contours, or the curve of intersection may be shown as a circular arc.

Assignment

On an A3- or B-size sheet, make a working drawing of one of the parts shown in Figs. 7-16-A and 7-16-B. A bushing is to be pressed (LN4 fit) into the large hole and the stepped smaller hole is to have a running fit (RC4) with its respective shaft. These sizes are to be given as limit dimensions. All other finished surfaces are to have 3.2μm or equivalent finish. Use your judgment in selecting the number of views required and deciding whether some form of sectional view would be desirable to improve the readability of the drawing. Scale is 1:1.

REVIEW FOR ASSIGNMENT

Unit 5-6 Fits
Unit 5-8 Surface Texture Symbols
Unit 7-10 Partial Views
Unit 7-8 Revolved Sections

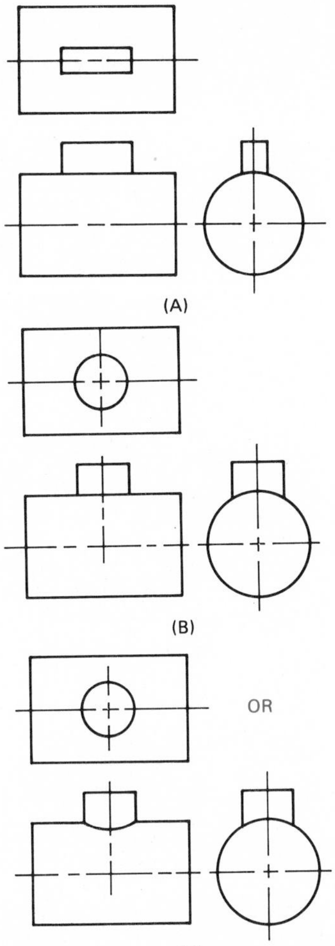
(A)
(B)
OR
(C)

Fig. 7-16-1 Conventional representation of external intersections.

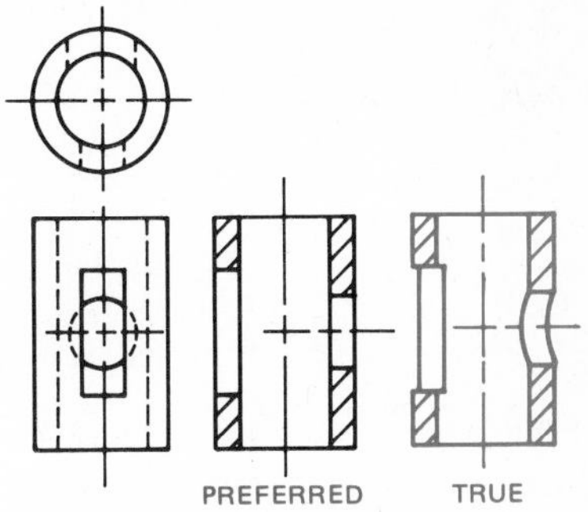
PREFERRED
TRUE

Fig. 7-16-2 Conventional representation of cylindrical intersections.

UNIT 7-17
FORESHORTENED PROJECTION

When the true projection of a feature would result in confusing foreshortening, it should be rotated until it is parallel to the line of the section or projection (Fig. 7-17-1).

HOLES REVOLVED TO SHOW TRUE DISTANCE FROM CENTER

Drilled flanges in elevation or section should show the holes at their true distance from the center, rather than the true projection.

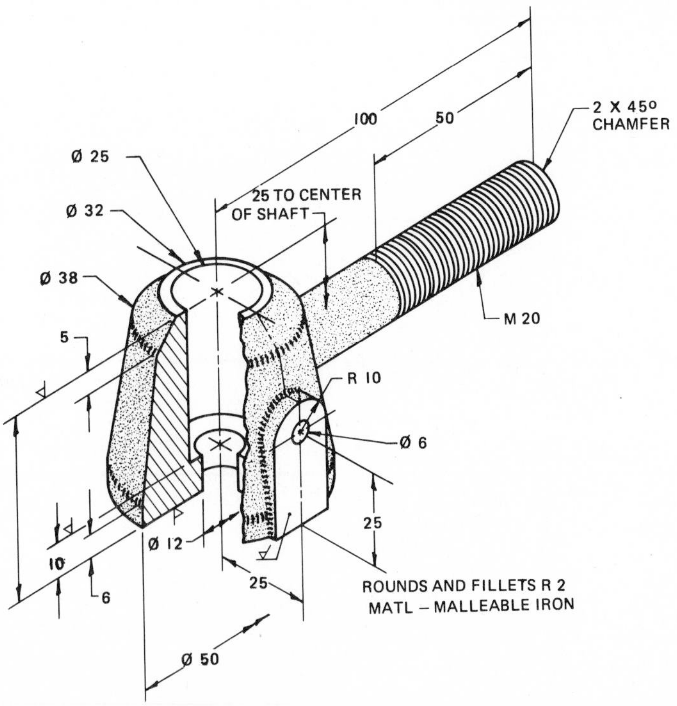
100
50
2 X 45°
CHAMFER
Ø 25
Ø 32
Ø 38
25 TO CENTER
OF SHAFT
M 20
5
R 10
Ø 6
25
10
6
Ø 12
25
ROUNDS AND FILLETS R 2
MATL – MALLEABLE IRON
Ø 50

Fig. 7-16-A Steering knuckle.

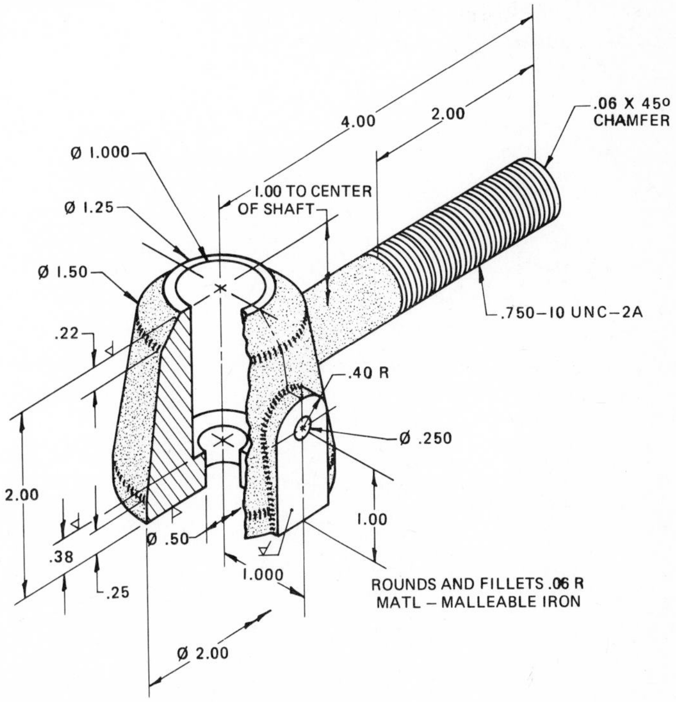
4.00
2.00
.06 X 45°
CHAMFER
Ø 1.000
Ø 1.25
1.00 TO CENTER
OF SHAFT
Ø 1.50
.750–10 UNC–2A
.22
.40 R
Ø .250
2.00
1.00
Ø .50
.38
.25
1.000
ROUNDS AND FILLETS .06 R
MATL – MALLEABLE IRON
Ø 2.00

Fig. 7-16-B Steering knuckle.

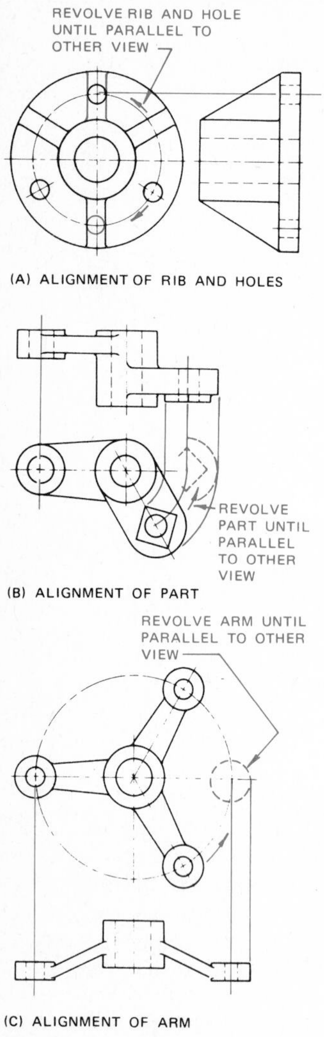
REVOLVE RIB AND HOLE UNTIL PARALLEL TO OTHER VIEW
(A) ALIGNMENT OF RIB AND HOLES
REVOLVE PART UNTIL PARALLEL TO OTHER VIEW
(B) ALIGNMENT OF PART
REVOLVE ARM UNTIL PARALLEL TO OTHER VIEW
(C) ALIGNMENT OF ARM

Fig. 7-17-1 Alignment of parts and holes to show true relationship.

Assignment

On an A3- or B-size sheet, make a detail drawing of one of the parts shown in Figs. 7-17-A and 7-17-B. All surface finishes are to be 1.6μm. Keyed holes will have RC6 fits with shafts, and keyways will be dimensioned for interchangeable assembly. Where required, rotate the features to show their true distances from the centers and edges. To show the true shape of the ribs or arms, a revolved section is recommended. Scale is 1:1.

REVIEW FOR ASSIGNMENT

Unit 5-6 Fits
Unit 5-8 Surface Finish
Unit 9-1 Keys and Keyways
Unit 7-8 Revolved Sections

UNIT 7-18

INTERSECTIONS OF UNFINISHED SURFACES

The intersections of unfinished surfaces that are rounded or filleted may be indicated conventionally by a line coinciding with the theoretical line of intersection. The need for this convention is demonstrated by the examples shown in Fig. 7-18-1, where the upper top views are shown in true projection. Note that in each example the true projection would be misleading. In the case of the large radius, such as shown in Fig. 7-18-1*d*, no line is drawn. Members such as ribs and arms that blend into other features terminate in curves called *runouts*. Small runouts are usually drawn freehand. Large runouts are drawn with an irregular curve, template, or compass. (See Fig. 7-18-2.)

Assignment

On an A3- or B-size sheet, make a three-view detail drawing of one of the problems shown in Figs. 7-18-A through 7-18-D. Scale is 1:1. Surface finish requirements are essential for both parts. For Fig. 7-18-A the T slot surfaces should have a maximum roughness of 0.8 μm and a maximum waviness of 0.05 mm for a 25-mm length. The back surface should have a maximum roughness of 3.2 μm with no restrictions or waviness. For Fig. 7-18-C the back surface and notch should have the same control as the T slot in Fig. 7-18-A. The faces on the boss should have a maximum roughness of 3.2 μm with no restrictions on waviness. Use inch-size equivalents for Figs. 7-18-B and 7-18-D.

REVIEW FOR ASSIGNMENT

Unit 5-2 Dimensioning Circular Features
Unit 5-8 Surface Texture Symbols
Unit 3-11 Miter Lines

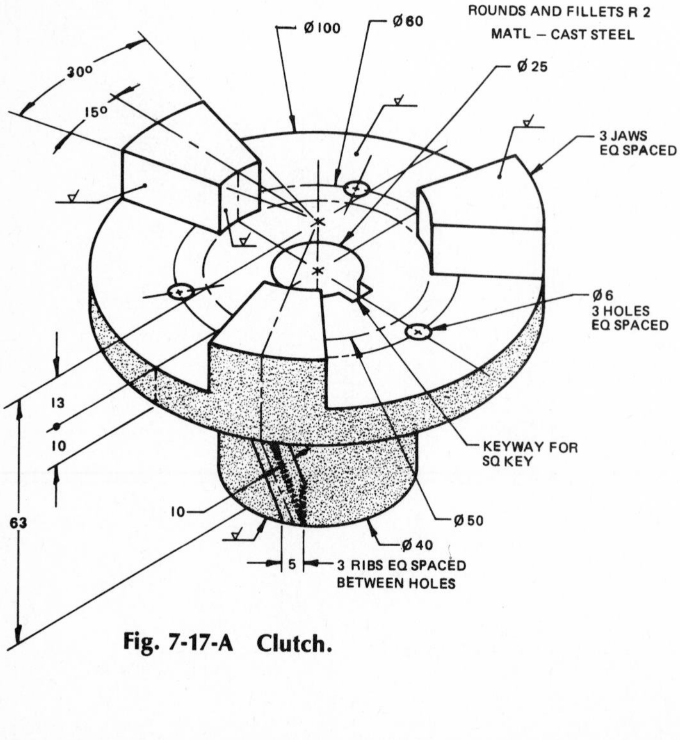
ROUNDS AND FILLETS R 2
MATL – CAST STEEL
Ø 100
Ø 60
Ø 25
30°
15°
3 JAWS EQ SPACED
Ø 6
3 HOLES EQ SPACED
13
10
63
10
KEYWAY FOR SQ KEY
Ø 50
Ø 40
5
3 RIBS EQ SPACED BETWEEN HOLES

Fig. 7-17-A Clutch.

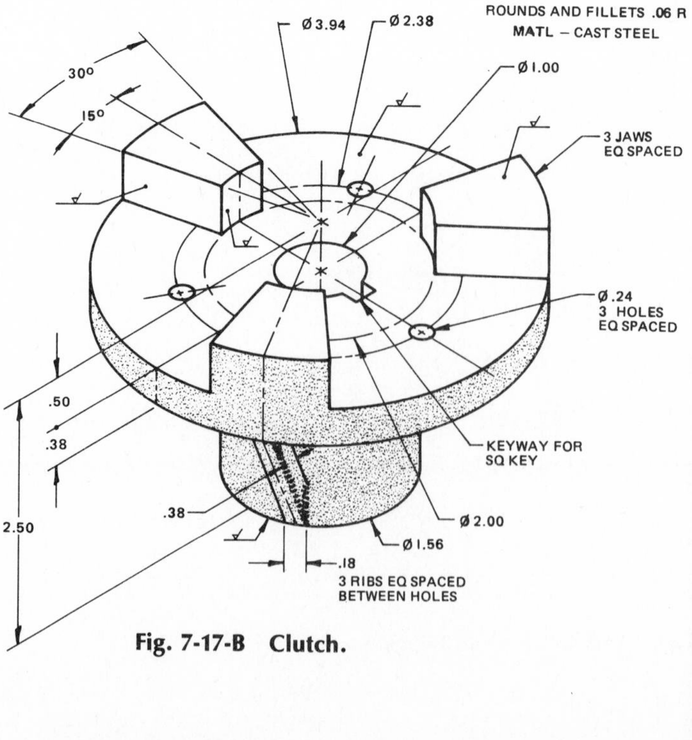
ROUNDS AND FILLETS .06 R
MATL – CAST STEEL
Ø 3.94
Ø 2.38
Ø 1.00
30°
15°
3 JAWS EQ SPACED
Ø .24
3 HOLES EQ SPACED
.50
.38
2.50
.38
KEYWAY FOR SQ KEY
Ø 2.00
Ø 1.56
.18
3 RIBS EQ SPACED BETWEEN HOLES

Fig. 7-17-B Clutch.

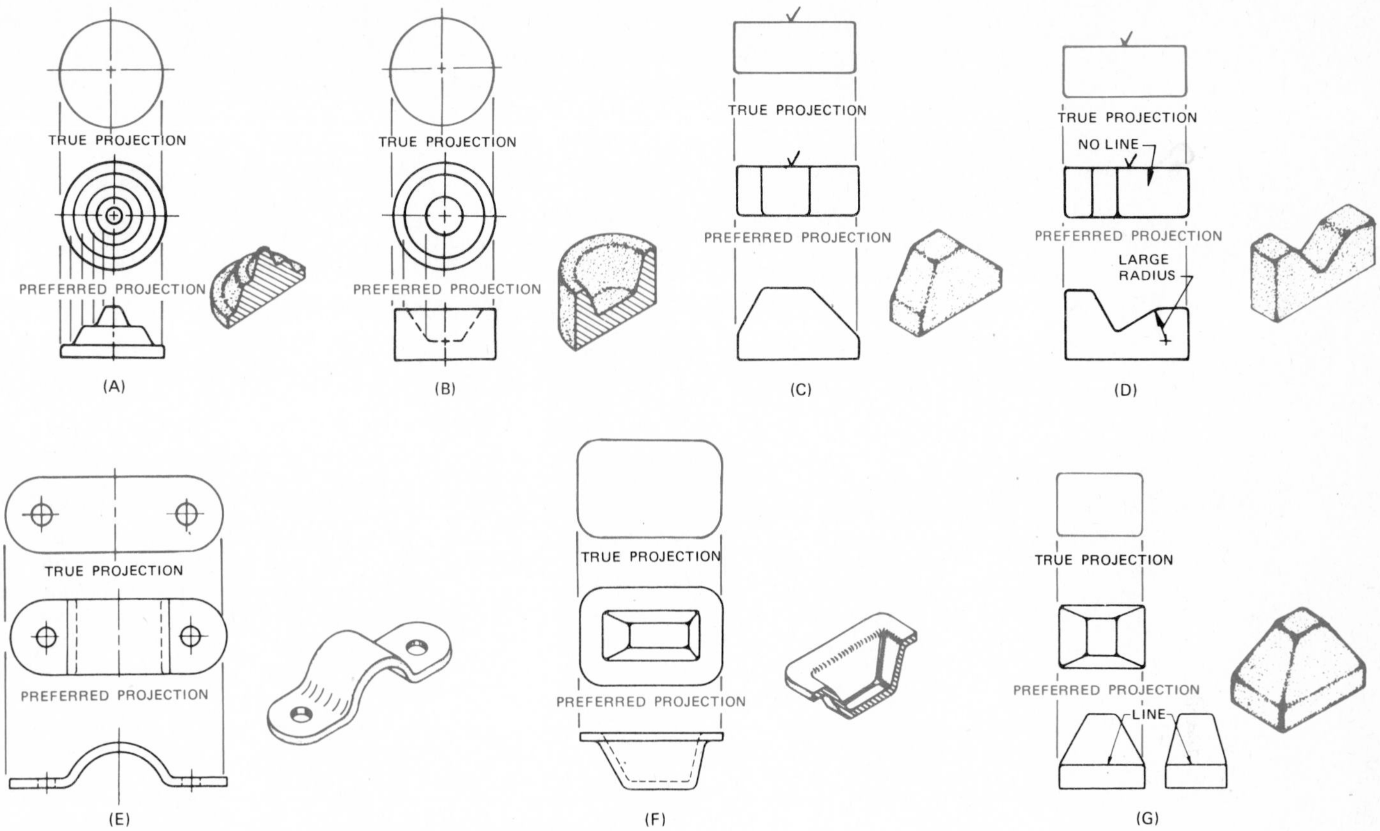

TRUE PROJECTION
PREFERRED PROJECTION
(A)
TRUE PROJECTION
PREFERRED PROJECTION
(B)
TRUE PROJECTION
PREFERRED PROJECTION
(C)
TRUE PROJECTION
NO LINE
PREFERRED PROJECTION
LARGE RADIUS
(D)
TRUE PROJECTION
PREFERRED PROJECTION
(E)
TRUE PROJECTION
PREFERRED PROJECTION
(F)
TRUE PROJECTION
PREFERRED PROJECTION
LINE
(G)

Fig. 7-18-1 Conventional representation of rounds and fillets.

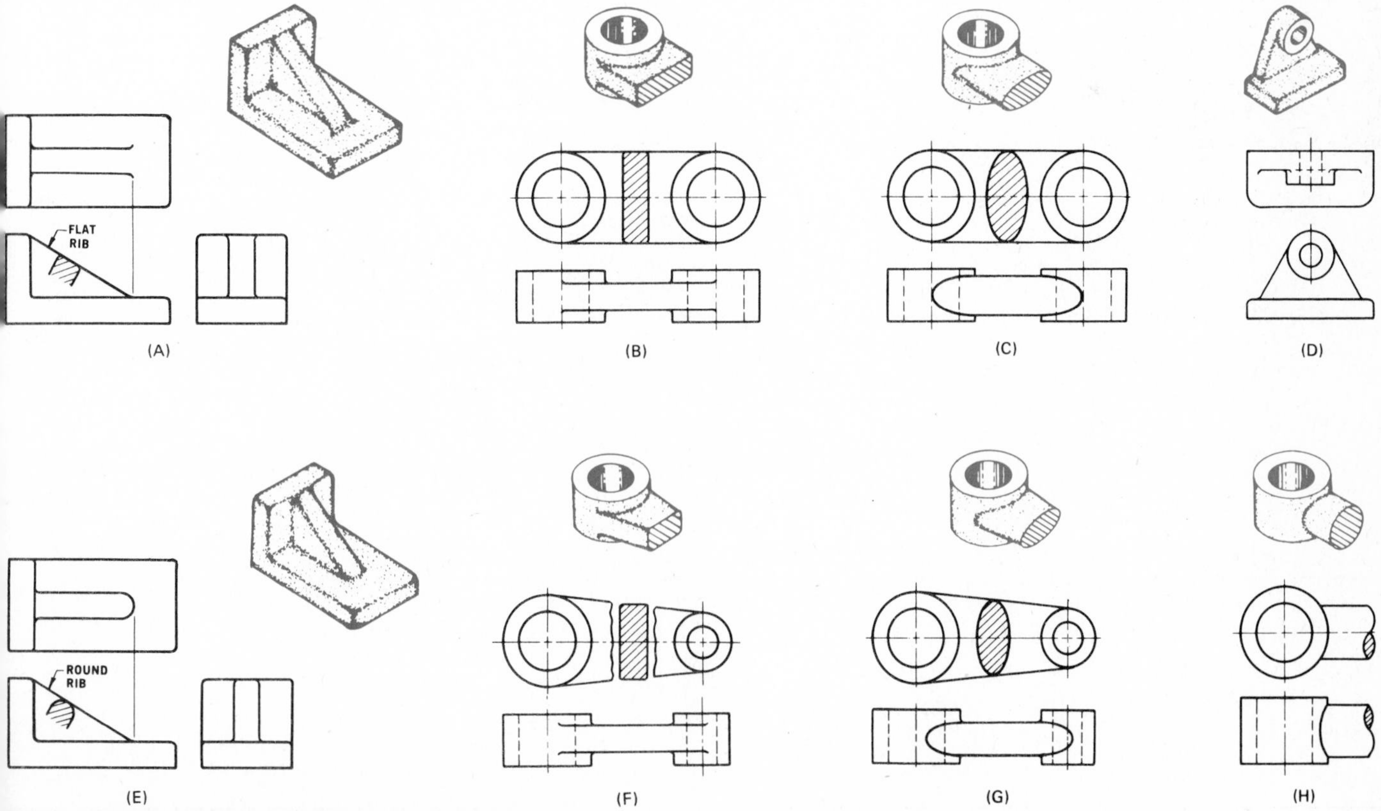

FLAT RIB
(A)
(B)
(C)
(D)
ROUND RIB
(E)
(F)
(G)
(H)

Fig. 7-18-2 Conventional representation of runouts.

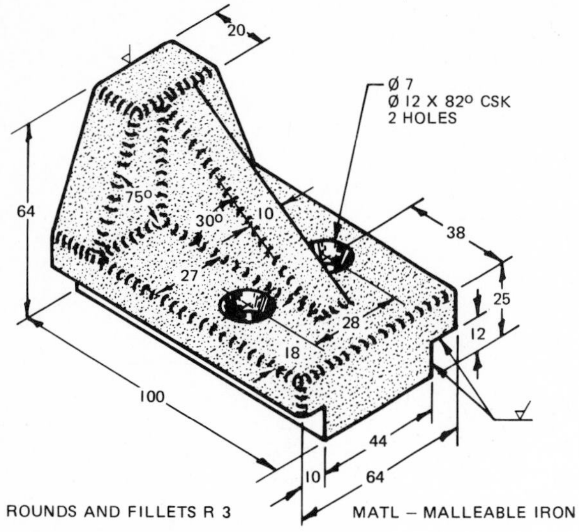
Ø 7
Ø 12 X 82° CSK
2 HOLES
ROUNDS AND FILLETS R 3
MATL – MALLEABLE IRON

Fig. 7-18-A Cut-off stop.

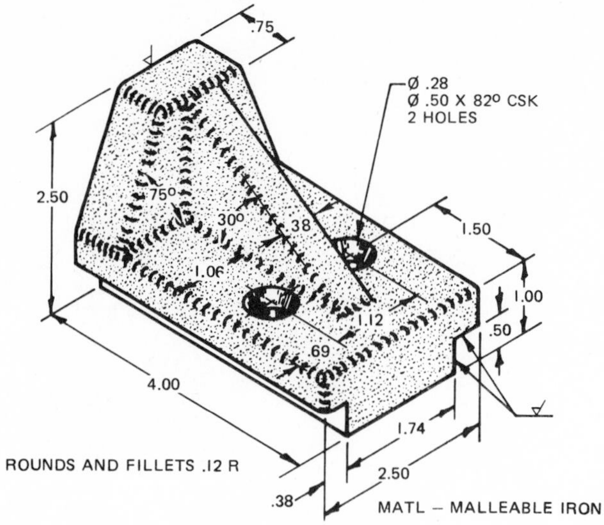
Ø .28
Ø .50 X 82° CSK
2 HOLES
ROUNDS AND FILLETS .12 R
MATL – MALLEABLE IRON

Fig. 7-18-B Cut-off stop.

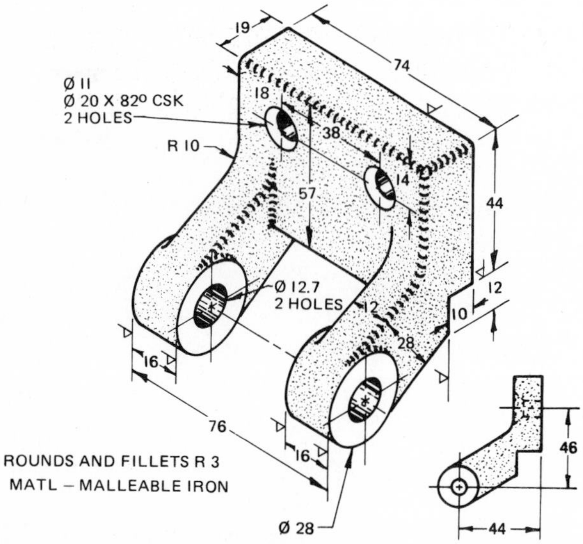
Ø 11
Ø 20 X 82° CSK
2 HOLES
Ø 12.7
2 HOLES
ROUNDS AND FILLETS R 3
MATL – MALLEABLE IRON
Ø 28

Fig. 7-18-C Sparker bracket.

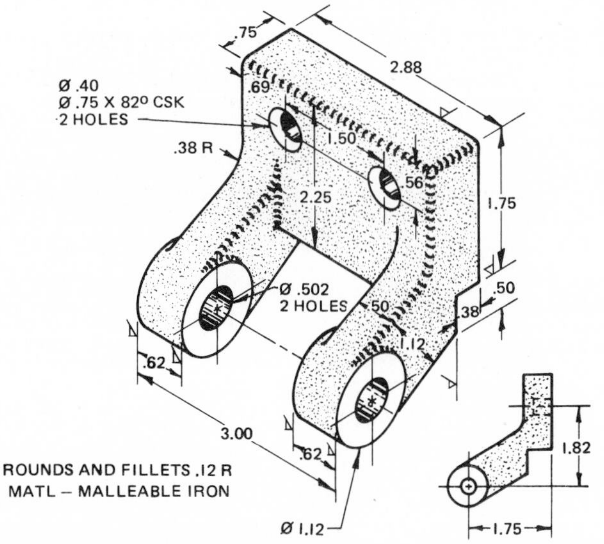
Ø .40
Ø .75 X 82° CSK
2 HOLES
Ø .502
2 HOLES
ROUNDS AND FILLETS .12 R
MATL – MALLEABLE IRON
Ø 1.12

Fig. 7-18-D Sparker bracket.

Part 2
Fasteners, Materials, and Forming Processes

Chapter 8 Threaded Fasteners

UNIT 8-1

THREAD FORMS[1,2]

Fastening devices are important in the construction of manufactured products, in the machines and devices used in manufacturing processes, and in the construction of all types of buildings. Fastening devices are used in the smallest watch to the largest ocean liner. See Fig. 8-1-1.

There are two basic kinds of fasteners: *permanent* and *removable*. Rivets and welds are permanent fasteners. Bolts, screws, studs, nuts, pins, rings, and keys are removable fasteners. As industry progressed, fastening devices became standardized, and they developed definite characteristics and names. A thorough knowledge of the design and graphic representation of the more common fasteners is an essential part of drafting.

The cost of fastening, once considered only incidental, is fast becoming recognized as a critical factor in total product cost. "It's the in-place cost that counts, not the fastener cost" is an old saying of fastener design. The art of holding down fastener cost is not learned simply by scanning a parts catalog. More subtly, it entails weighing such factors as standardization, automatic assembly, tailored fasteners, and joint preparation.

Standardization. A favorite cost-reducing method, standardization not only cuts the cost of parts but also reduces paperwork and simplifies inventory and quality control. By standardizing on type and size, it may be possible to reach the level of usage required to make power tools or automatic assembly feasible.

Screw Threads

A *screw thread* is a ridge of uniform section either in the form of a helix on the external or internal surface of a cylinder or in the form of a conical spiral on the external or internal surface of a cone or frustum of a cone. See Fig. 8-1-2.

The *pitch* of a thread **P** is the distance from a point on the thread form to the corresponding point on the next form, measured parallel to the axis (Fig. 8-1-3). The *lead* **L** is the distance the threaded part would move parallel to the axis during one complete rotation in relation to a fixed mating part (the distance a screw would enter a threaded hole in one turn).

Thread Forms

Figure 8-1-6 shows some of the more common thread forms in use today. The ISO metric thread shown in Fig. 8-1-6 has now replaced all the former V-shaped metric and inch threads. As for the other thread forms shown, the proportions will be the same for both metric- and inch-size threads.

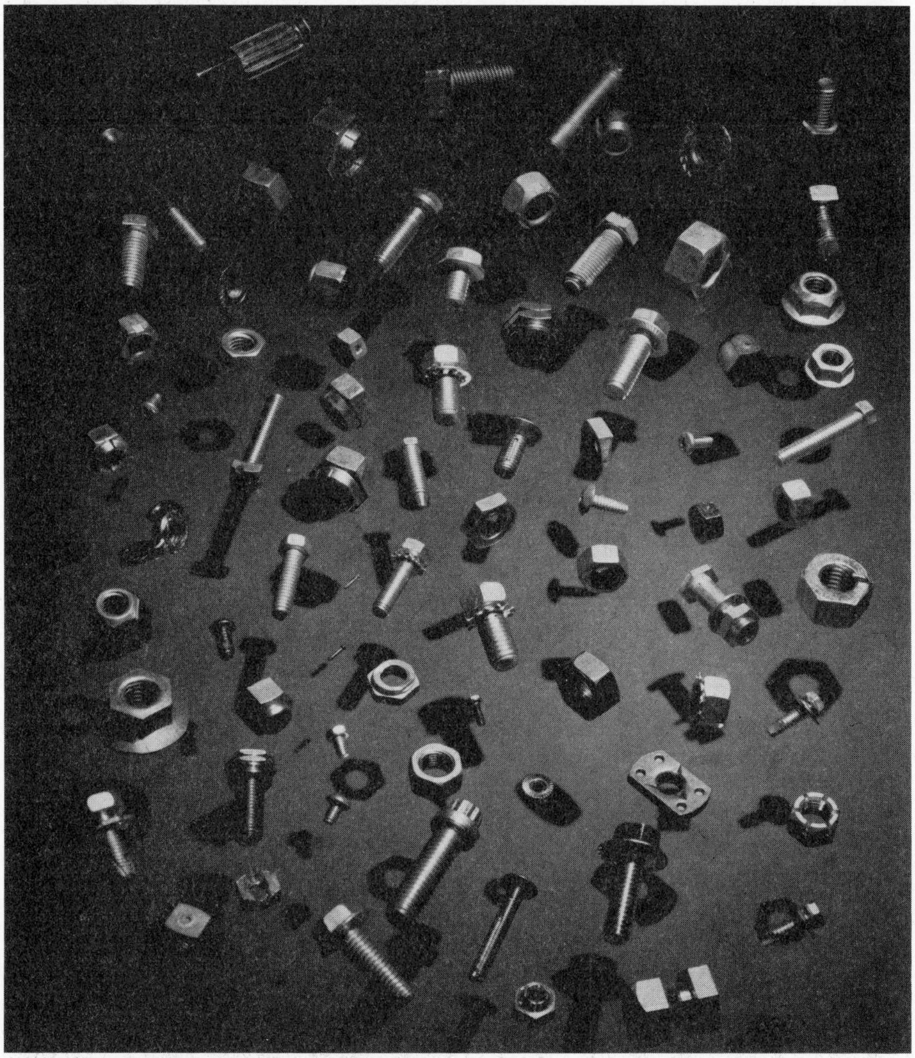

Fig. 8-1-1 Fasteners. (Industrial Fasteners Institute)

The knuckle thread is usually rolled or cast. A familiar example of this form is seen on electric light bulbs and sockets. See Fig. 8-1-4. The square and acme forms are designed to transmit motion or power, as on the lead screw of a lathe. The buttress thread takes pressure in only one direction—against the surface perpendicular to the axis.

Thread Representation

True representation of a screw thread is seldom provided on working drawings because it would require very laborious and accurate drawing involving repetitious development of the helix curve of the thread. See Figs. 8-1-3 and 8-1-5. A symbolic representation of threads is now standard practice.

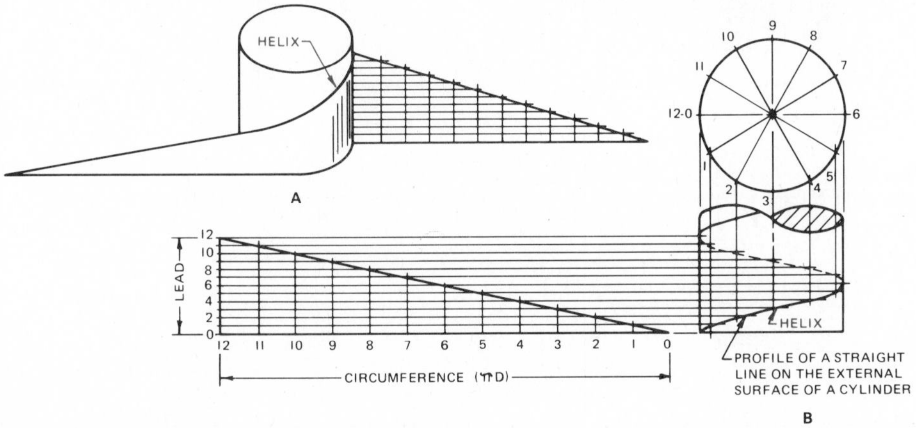
HELIX
A
12
10
8
6
4
2
0
LEAD
12 11 10 9 8 7 6 5 4 3 2 1 0
CIRCUMFERENCE (πD)
10 9 8
11 7
12-0 6
1 5
2 3 4
HELIX
PROFILE OF A STRAIGHT LINE ON THE EXTERNAL SURFACE OF A CYLINDER
B

Fig. 8-1-2 The helix.

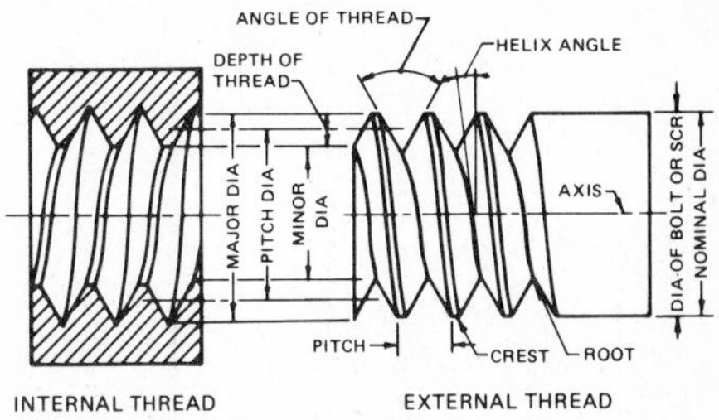
ANGLE OF THREAD
DEPTH OF THREAD
HELIX ANGLE
MAJOR DIA
PITCH DIA
MINOR DIA
AXIS
DIA OF BOLT OR SCR
NOMINAL DIA
PITCH
CREST
ROOT
INTERNAL THREAD
EXTERNAL THREAD

Fig. 8-1-3 Screw-thread terms.

Fig. 8-1-4 Application of a knuckle thread.

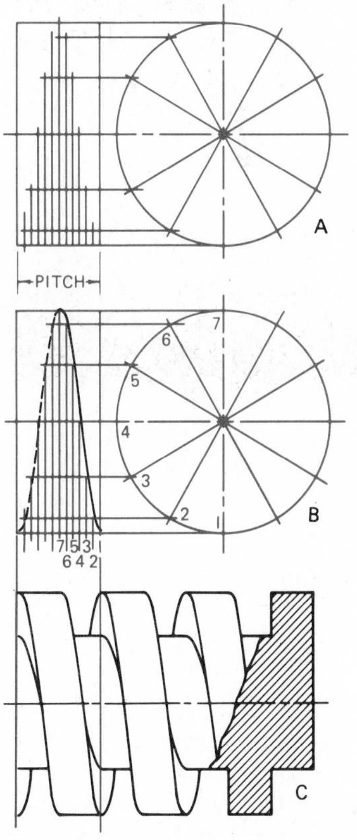
A
PITCH
7
6
5
4
3
2
1
B
7 5 3 1
6 4 2
C

Fig. 8-1-5 The helix of a square thread.

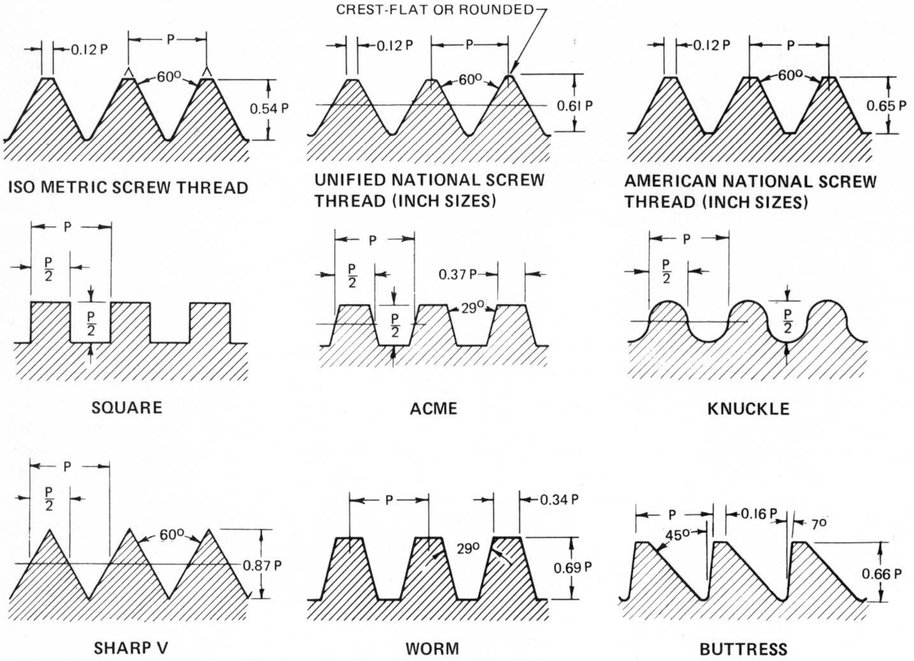
CREST-FLAT OR ROUNDED
0.12 P
P
60°
0.54 P
ISO METRIC SCREW THREAD
0.61 P
UNIFIED NATIONAL SCREW THREAD (INCH SIZES)
0.65 P
AMERICAN NATIONAL SCREW THREAD (INCH SIZES)
P/2
SQUARE
0.37 P
29°
ACME
KNUCKLE
0.87 P
SHARP V
0.34 P
0.69 P
WORM
45°
0.16 P
7°
0.66 P
BUTTRESS

Fig. 8-1-6 Common thread forms and proportions.

Pictorial Representation of Screw Threads. The detailed representation uses the sharp-V profile. Straight lines are used to represent the helices of the crest and root lines.

The order of drawing the V-form thread is shown in Fig. 8-1-8. The pitch is seldom drawn to scale; generally it is approximated and drawn to look good. Lay off the pitch *P* and the half-pitch *P*/2, as shown. Adjust the set square to the slope, and draw the crest lines (if a drafting machine is used, set the ruling arm to the slope of the crest line). At *b* draw the V profile for one thread, top and bottom, locating the root diameter. Draw light construction lines for the root diameter. At *c* set the ruling face of the 30° set square and draw one side of the remaining V's (thread profile). Reverse the set square and draw the other sides of the V's, completing the thread profile. At *d*, draw the root lines, which complete the pictorial representation of the threads.

PICTORIAL REPRESENTATION

A close approximation to the actual appearance of the screw thread is shown in Figs. 8-1-7 and 8-1-8. It is simplified in that crests and roots for full threads are shown sharp, with a single straight line, instead of a double curved line which would be required for flat crests and roots.

Pictorial representation should be used only for enlarged detail and other special applications.

Pictorial Representation of Square Threads. The depth of the square thread is one-half the pitch. In Fig. 8-1-7*a*, lay off spaces equal to *P*/2 along the diameter and add light lines which indicate the depth of thread. At *b* draw the crest lines. At *c* draw the root lines, as shown. At *d* the internal square thread is drawn in section. Note the reverse direction of the lines.

Pictorial Representation of Acme Threads. The depth of the acme thread is one-half the pitch (Fig. 8-1-7). The stages in drawing acme threads are shown at *e*. The pitch diameter is midway between the outside diameter and the root diameter and locates the pitch line. On the pitch line, lay off half-pitch spaces and draw the root lines to complete the view. The construction shown at *f* is enlarged.

A sectional view of an internal acme thread is shown at *g*. Other representations used for internal threads are hidden lines and sections. These are shown at *h*.

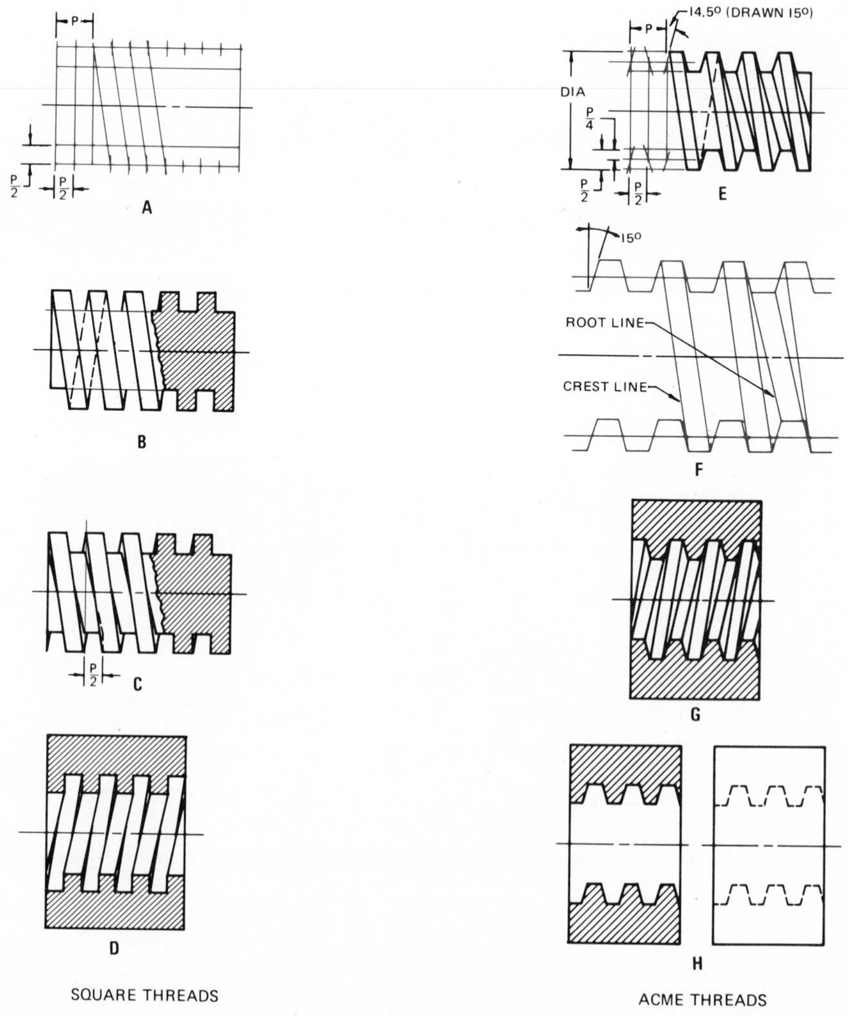
P
P/2
A
B
C
D
SQUARE THREADS
14.5° (DRAWN 15°)
DIA
P/4
E
15°
ROOT LINE
CREST LINE
F
G
H
ACME THREADS

Fig. 8-1-7 Steps in drawing pictorial representation of square and acme threads.

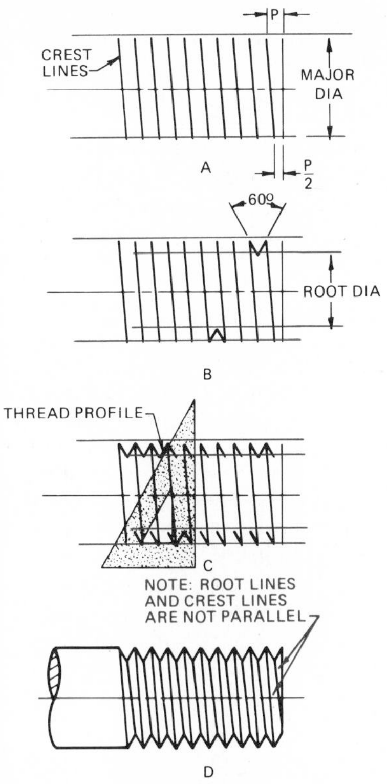
P
CREST LINES
MAJOR DIA
A
P/2
60°
ROOT DIA
B
THREAD PROFILE
C
NOTE: ROOT LINES AND CREST LINES ARE NOT PARALLEL
D

Fig. 8-1-8 Steps in drawing pictorial representation of screw thread.

Right- and Left-Hand Threads

Unless designated otherwise, threads are assumed to be right-hand. A bolt being threaded into a tapped hole would be turned in a right-hand (clockwise) direction. See Fig. 8-1-9. For some special applications, such as turnbuckles, left-hand threads are required. When such a thread is necessary, the letters LH are added after the thread designation.

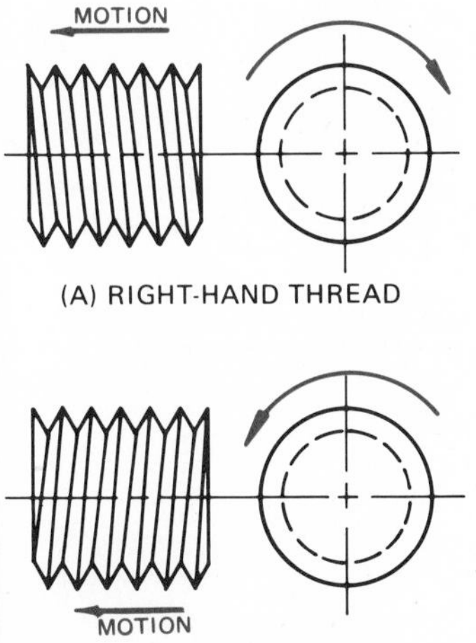
MOTION
(A) RIGHT-HAND THREAD
MOTION
(B) LEFT-HAND THREAD

Fig. 8-1-9 Right- and left-hand threads.

Single and Multiple Threads

Most screws have single threads; one would assume that unless the thread is designated otherwise, it is a single thread. The single thread has a single ridge in the form of a helix. See Fig. 8-1-10. The lead of a thread is the distance traveled parallel to the axis in one rotation of a part in relation to a fixed mating part (the distance a nut would travel along the axis of a bolt with one rotation of the nut). In single threads, the lead is equal to the pitch. A double thread has two ridges, started 180° apart, in the form of helices; and the lead is twice the pitch. A triple thread has three ridges, started 120° apart, in the form of helices; and the lead is 3 times the pitch. Multiple threads are used where fast movement is desired with a minimum number of rotations, such as on threaded mechanisms for opening and closing windows.

Threaded Assemblies

It is often desirable to show threaded assembly drawings in pictorial form, such as in preparing presentation or catalog drawings. Hidden lines are normally omitted on these drawings, as they do nothing to add to the clarity of the drawing. See Fig. 8-1-11.

Fastener Materials[3]

Because fasteners are available in virtually any material, design choices are practically unlimited. The key to material selection for fasteners is in knowing what job the fastener has to do and then specifying a material to meet those requirements. Consider these questions:

1. Will the fastener be subjected to a corrosive condition?
2. Will the fastener be used in high temperatures?
3. Is mass important?
4. Should the material be nonmagnetic?
5. Will the fastener be subjected to high vibrations or cyclic fatigue stresses?
6. Does it need good heat or electrical conductivity?

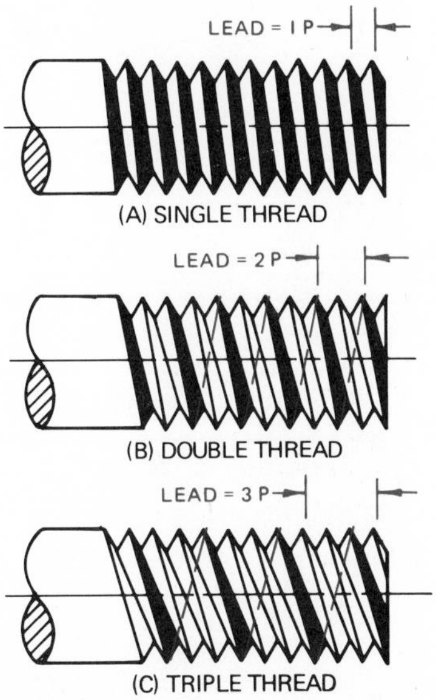
LEAD = 1 P
(A) SINGLE THREAD
LEAD = 2 P
(B) DOUBLE THREAD
LEAD = 3 P
(C) TRIPLE THREAD

Fig. 8-1-10 Single and multiple threads.

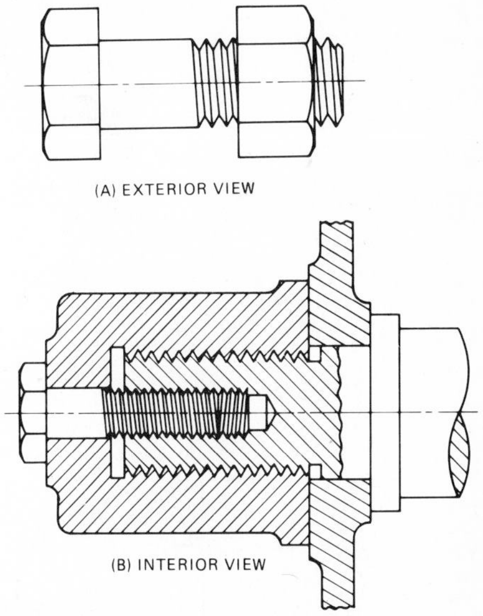
(A) EXTERIOR VIEW
(B) INTERIOR VIEW

Fig. 8-1-11 Pictorial threaded assembly.

REFERENCES AND SOURCE MATERIAL

1. ANSI Y14.6.
2. CSA.
3. R. H. Akers et al., "Materials," *Machine Design*, vol. 37, no. 6, 1965.

Assignments

1. Divide an A3- or B-size sheet into four sections, and draw the four threaded parts shown in Fig. 8-1-A or 8-1-B. Use pictorial representation for the threads. Use conventional breaks to shorten the lengths of the guide rod and jack screw.

REVIEW FOR ASSIGNMENTS

Unit 12-2	Steel Specifications
Unit 7-14	Conventional Breaks
Unit 7-13	Conventional Representation of Common Features
Unit 4-2	Arcs Tangent to Circles

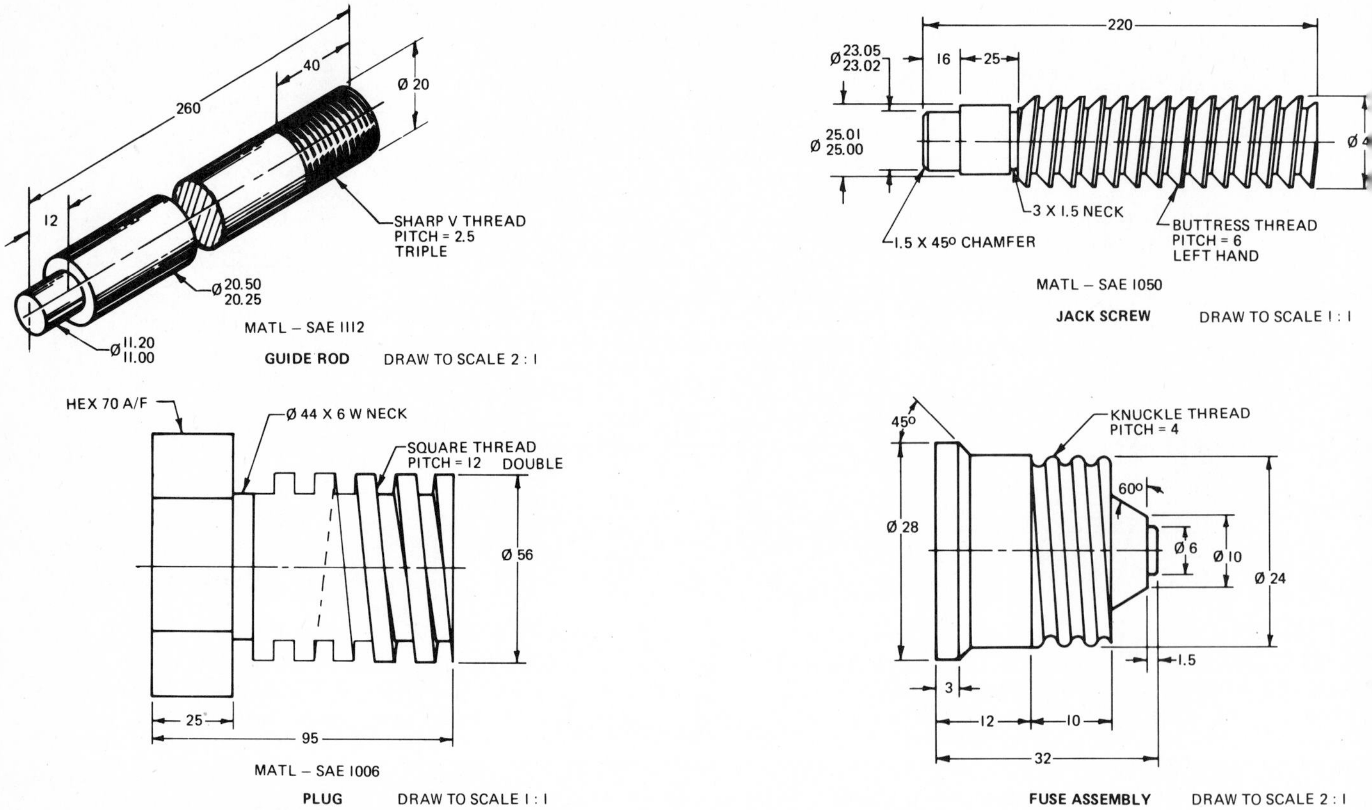

260
40
Ø 20
12
SHARP V THREAD
PITCH = 2.5
TRIPLE
Ø 20.50 20.25
Ø 11.20 11.00
MATL – SAE 1112
GUIDE ROD
DRAW TO SCALE 2 : 1
220
Ø 23.05 23.02
16
25
Ø 25.01 25.00
3 X 1.5 NECK
1.5 X 45° CHAMFER
BUTTRESS THREAD
PITCH = 6
LEFT HAND
MATL – SAE 1050
JACK SCREW
DRAW TO SCALE 1 : 1
HEX 70 A/F
Ø 44 X 6 W NECK
SQUARE THREAD
PITCH = 12 DOUBLE
Ø 56
25
95
MATL – SAE 1006
PLUG
DRAW TO SCALE 1 : 1
45°
KNUCKLE THREAD
PITCH = 4
60°
Ø 28
Ø 6
Ø 10
Ø 24
1.5
3
12
10
32
FUSE ASSEMBLY
DRAW TO SCALE 2 : 1

Fig. 8-1-A Pictorial representation of threaded parts.

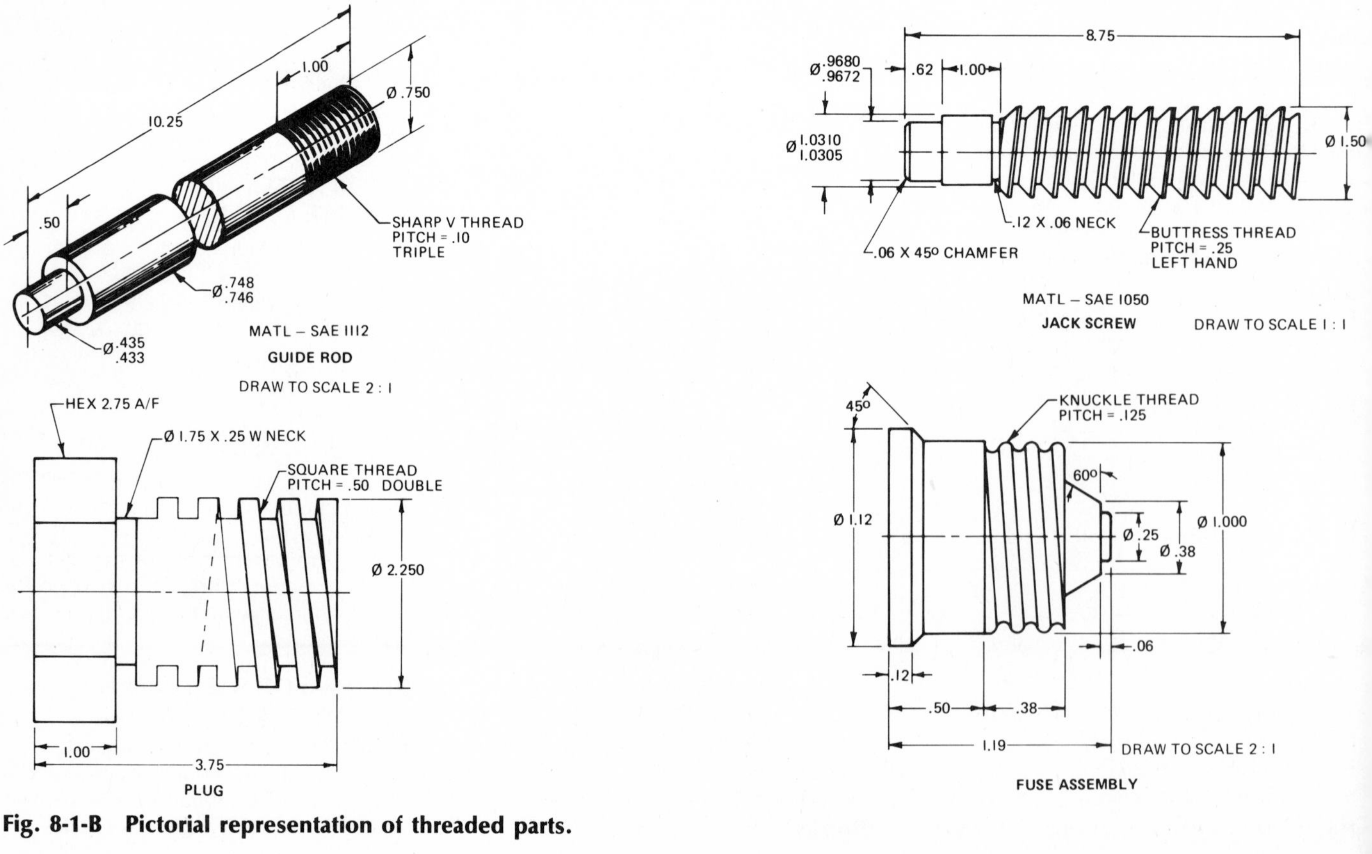

10.25
1.00
Ø .750
.50
SHARP V THREAD
PITCH = .10
TRIPLE
Ø .748 .746
Ø .435 .433
MATL – SAE 1112
GUIDE ROD
DRAW TO SCALE 2 : 1
8.75
Ø .9680 .9672
.62
1.00
Ø 1.0310 1.0305
Ø 1.50
.12 X .06 NECK
.06 X 45° CHAMFER
BUTTRESS THREAD
PITCH = .25
LEFT HAND
MATL – SAE 1050
JACK SCREW
DRAW TO SCALE 1 : 1
HEX 2.75 A/F
Ø 1.75 X .25 W NECK
SQUARE THREAD
PITCH = .50 DOUBLE
Ø 2.250
1.00
3.75
PLUG
45°
KNUCKLE THREAD
PITCH = .125
60°
Ø 1.12
Ø .25
Ø .38
Ø 1.000
.06
.12
.50
.38
1.19
DRAW TO SCALE 2 : 1
FUSE ASSEMBLY

Fig. 8-1-B Pictorial representation of threaded parts.

UNIT 8-2

SIMPLIFIED AND SCHEMATIC REPRESENTATION OF THREADS[1,2]

Three types of conventions are in general use for screw-thread representation. These are known as pictorial (as explained in Unit 8-1), schematic, and simplified representations. Simplified representation should be used whenever it will clearly portray the requirements. Schematic and pictorial representations require more drafting time, but are sometimes necessary to avoid confusion with other parallel lines or to more clearly portray particular aspects of the threads.

Simplified Representation

Simplified representation, as shown in Figs. 8-2-1*c* and 8-2-2*c*, should be used whenever it conveys the required information without confusion, because it requires the least amount of drafting effort. In this system the thread crests, except in hidden views, are represented by a thick outline and the thread roots by a thin broken line (ANSI) or a thin continuous line (ISO). The end of the full-form thread is indicated by a thick line across the part, and imperfect or runout threads are shown beyond this line by running the root line at an angle to meet the crest line. If the length of runout threads is unimportant, this portion of the convention may be omitted.

Schematic Representation

Schematic representation, as shown in Figs. 8-2-1*b* and 8-2-2*b*, should be used instead of the simplified representation when it is necessary to give more emphasis to the threaded aspects of a part or when the simplified representation might become confused with other parallel lines, grooves, slots, or adjacent features. When it is desired to distinguish between full threads and incomplete or runout threads, the latter are indicated by a crest and root line extending halfway across the part, as shown.

Threaded Assemblies

Any of the thread conventions shown here may be used for assemblies of threaded parts, and two or more methods may be used on the same drawing, as shown in Fig. 8-2-3. In sectional views,

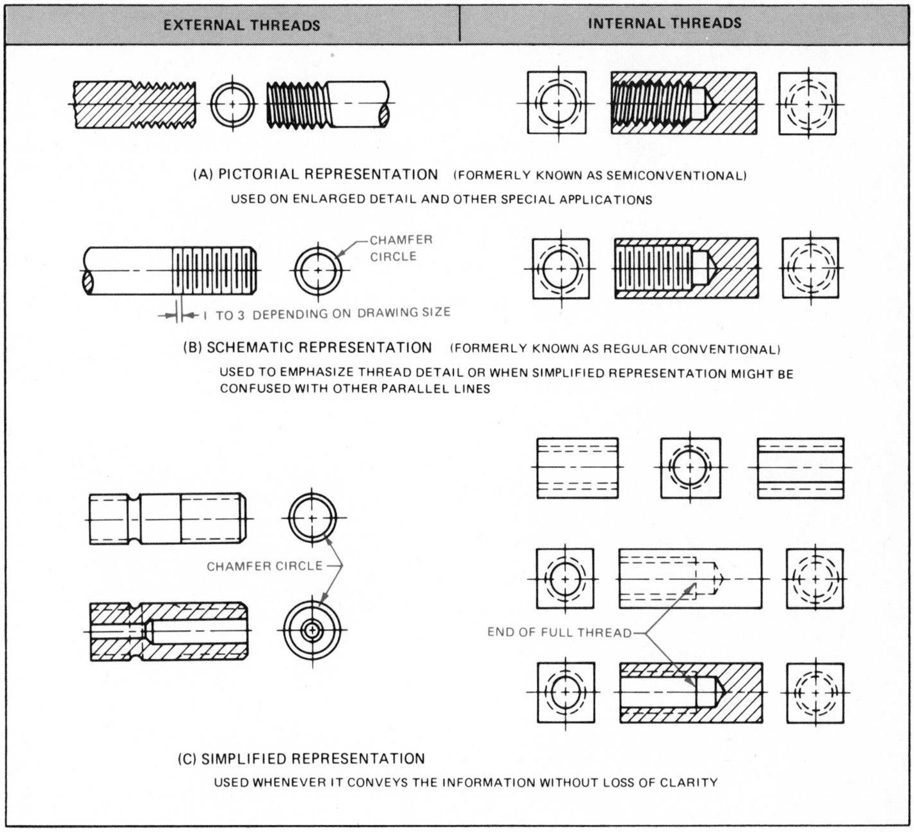
EXTERNAL THREADS
INTERNAL THREADS
(A) PICTORIAL REPRESENTATION (FORMERLY KNOWN AS SEMICONVENTIONAL)
USED ON ENLARGED DETAIL AND OTHER SPECIAL APPLICATIONS
CHAMFER CIRCLE
1 TO 3 DEPENDING ON DRAWING SIZE
(B) SCHEMATIC REPRESENTATION (FORMERLY KNOWN AS REGULAR CONVENTIONAL)
USED TO EMPHASIZE THREAD DETAIL OR WHEN SIMPLIFIED REPRESENTATION MIGHT BE CONFUSED WITH OTHER PARALLEL LINES
CHAMFER CIRCLE
END OF FULL THREAD
(C) SIMPLIFIED REPRESENTATION
USED WHENEVER IT CONVEYS THE INFORMATION WITHOUT LOSS OF CLARITY

Fig. 8-2-1 American National Standard Thread conventions. (ANSI)

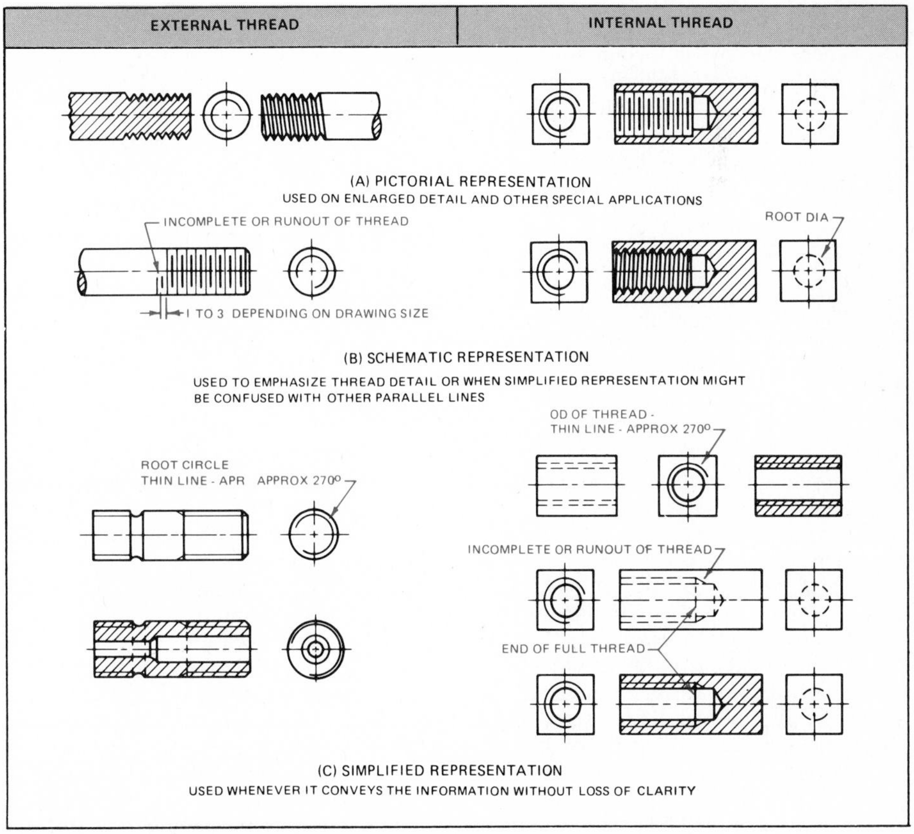
EXTERNAL THREAD
INTERNAL THREAD
(A) PICTORIAL REPRESENTATION
USED ON ENLARGED DETAIL AND OTHER SPECIAL APPLICATIONS
INCOMPLETE OR RUNOUT OF THREAD
ROOT DIA
1 TO 3 DEPENDING ON DRAWING SIZE
(B) SCHEMATIC REPRESENTATION
USED TO EMPHASIZE THREAD DETAIL OR WHEN SIMPLIFIED REPRESENTATION MIGHT BE CONFUSED WITH OTHER PARALLEL LINES
OD OF THREAD - THIN LINE - APPROX 270°
ROOT CIRCLE THIN LINE - APR APPROX 270°
INCOMPLETE OR RUNOUT OF THREAD
END OF FULL THREAD
(C) SIMPLIFIED REPRESENTATION
USED WHENEVER IT CONVEYS THE INFORMATION WITHOUT LOSS OF CLARITY

Fig. 8-2-2 ISO Standard Thread conventions. (Adopted by CSA)

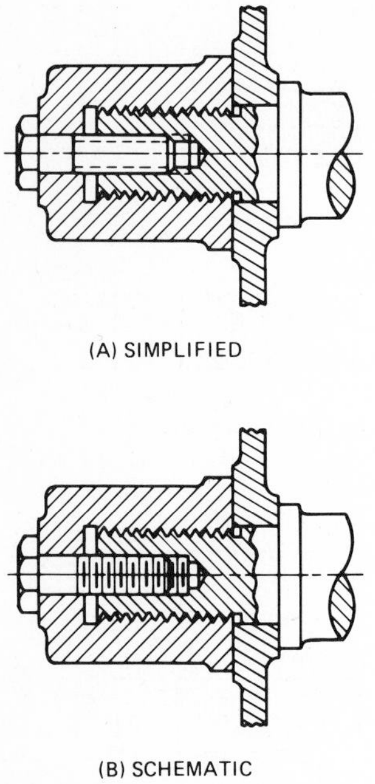
(A) SIMPLIFIED
(B) SCHEMATIC

Fig. 8-2-3 Simplified and schematic representation of threads in assembly drawings.

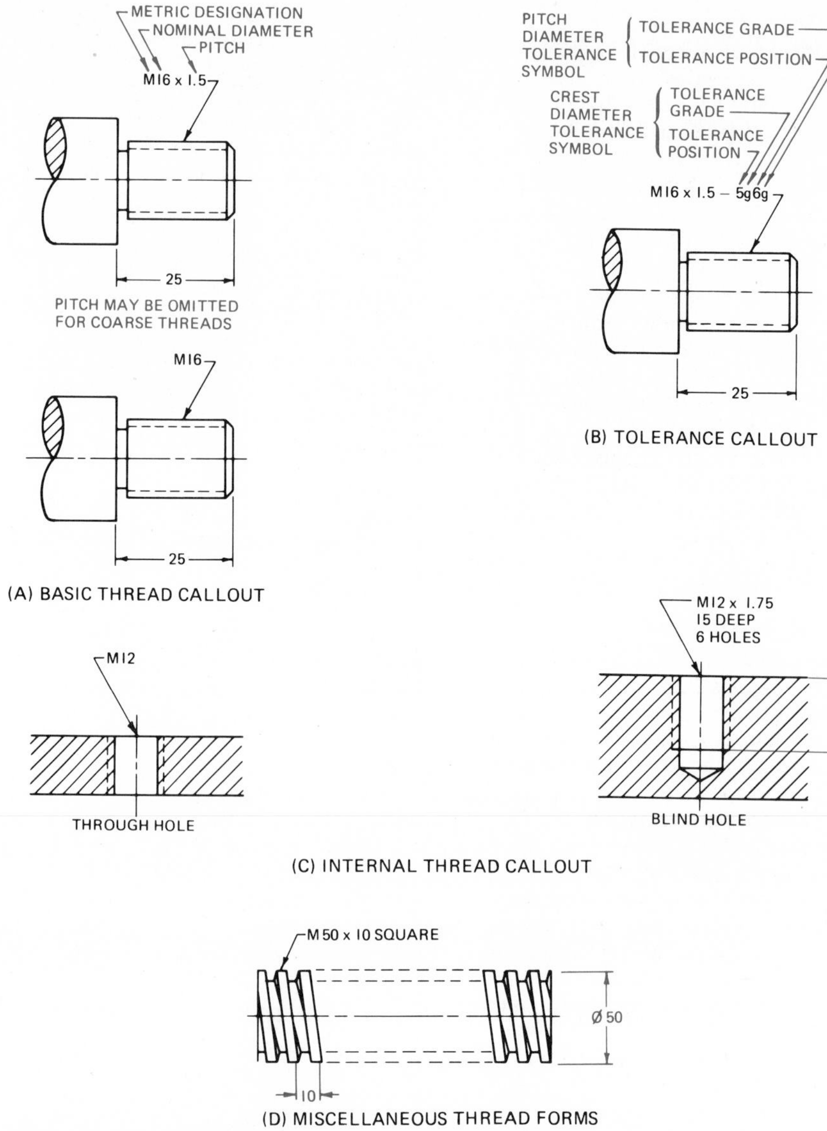
METRIC DESIGNATION
NOMINAL DIAMETER
PITCH
M16 x 1.5
25
PITCH MAY BE OMITTED FOR COARSE THREADS
M16
25
(A) BASIC THREAD CALLOUT
PITCH DIAMETER TOLERANCE SYMBOL
TOLERANCE GRADE
TOLERANCE POSITION
CREST DIAMETER TOLERANCE SYMBOL
TOLERANCE GRADE
TOLERANCE POSITION
M16 x 1.5 – 5g6g
25
(B) TOLERANCE CALLOUT
M12
THROUGH HOLE
M12 x 1.75
15 DEEP
6 HOLES
15
BLIND HOLE
(C) INTERNAL THREAD CALLOUT
M 50 x 10 SQUARE
Ø 50
10
(D) MISCELLANEOUS THREAD FORMS

Fig. 8-2-4 Thread specifications for metric threads.

the externally threaded part is always shown covering the internally threaded part.

Thread Standards

With the progress and growth of industry, a need has grown for uniform, interchangeable threaded fasteners. The factors that influence standards, aside from the threaded forms previously mentioned, are the pitch of the thread and the major diameter.

Metric Threads

Metric threads are grouped into diameter-pitch combinations distinguished from one another by the pitch applied to specific diameters. (See Fig. 8-2-4.) The *pitch* for metric threads is the distance between corresponding points on adjacent teeth. In addition to a coarse- and fine-pitch series, a series of constant pitches is available.

Coarse-Thread Series. This series is intended for use in general engineering work and commercial applications.

Fine-Thread Series. The fine-thread series is for general use where a finer thread than the coarse-thread series is desirable. In comparison with a coarse-thread screw, the fine-thread screw is stronger in both tensile and torsional strength and is less likely to loosen under vibration.

Thread Grades and Classes. The *fit* of a screw thread is the amount of clearance between the internal and external threads when they are assembled.

For each of the two main thread elements—pitch diameter and crest diameter—a number of tolerance grades have been established. The number of the tolerance grades reflects the size of the tolerance. For example, grade 4 tolerances are smaller than grade 6 tolerances, and grade 8 tolerances are larger than grade 6 tolerances.

In each case, grade 6 tolerances should be used for medium-quality length-of-engagement applications. The tolerance grades below grade 6 are intended for applications involving fine quality and/or short lengths of engagement. Tolerance grades above grade 6 are intended for coarse quality and/or long lengths of engagement.

In addition to the tolerance grade, a positional tolerance is required. This tolerance defines the maximum-material limits of the pitch and crest diameters of the external and internal threads and indicates their relationship to the basic profile.

In conformance with current coating (or plating) thickness requirements and the demand for ease of assembly, a series of tolerance positions reflecting the application of varying amounts of allowance has been established as follows:

For external threads:

- Tolerance position e (large allowance)
- Tolerance position g (small allowance)
- Tolerance position h (no allowance)

For internal threads:

- Tolerance position G (small allowance)
- Tolerance position H (no allowance)

THREAD DESIGNATION

ISO metric screw threads are defined by the nominal size (basic major diameter) and pitch, both expressed in millimetres. An "M" specifying an ISO metric screw thread precedes the nominal size, and an "x" separates the nominal size from the pitch. For the coarse-thread series only, the pitch is not shown unless the dimension for the length of the thread is required. In specifying the length of thread, an "x" is used to separate the length of thread from the rest of the designations. For external threads, the length of thread may be given as a dimension on the drawing.

For example, a 10-mm-diameter, 1.25 pitch, fine-thread series is expressed as M10 × 1.25. A 10-mm-diameter, 1.5 pitch, coarse-thread series is expressed as M10; the pitch is not shown unless the length of thread is required. If the latter thread were 25 mm long and this information were required on the drawing, then the thread callout would be M10 × 1.5 × 25.

A complete designation for an ISO metric screw thread comprises, in addition to the basic designation, an identification for the tolerance class. The tolerance class designation is separated from the basic designation by a dash and includes the symbol for the pitch diameter tolerance followed immediately by the symbol for crest diameter tolerance. Each of these symbols consists of a numeral indicating the grade tolerance followed by a letter indicating the tolerance position (a capital letter for internal threads and a lowercase letter for external threads). Where the pitch and crest diameter symbols are identical, the symbol need be given only once and not repeated.

For external threads, the length of thread is given as a dimension on the drawing. The length given is to be the minimum length of full thread. For threaded holes that go all the way through the part, the term THRU is sometimes added to the note. If no depth is given, the hole is assumed to go all the way through. For threaded holes that do not go all the way through, the depth is given in the note, for example, M12 × UNF × 20 deep. The depth given is the minimum depth of full thread.

Inch Threads

Up to 1976, practically all threaded assemblies on this continent were designed using inch-sized threads. In this system the *pitch* is equal to

$$\frac{1}{\text{Number of threads per inch}}$$

The number of threads per inch is set for different diameters in what is called a thread *series*. For the Unified National system there is the coarse-thread series and the fine-thread series. See Fig. 8-2-5.

In addition, there is an extra-fine-thread series, UNEF, for use where a small pitch is desirable, such as on thin-walled tubing. For special work and for diameters larger than those specified in the coarse and fine series, the Unified Thread system has three series that provide for the same number of threads per inch regardless of the diameter. These are the 8-thread series, the 12-thread series, and the 16-thread series.

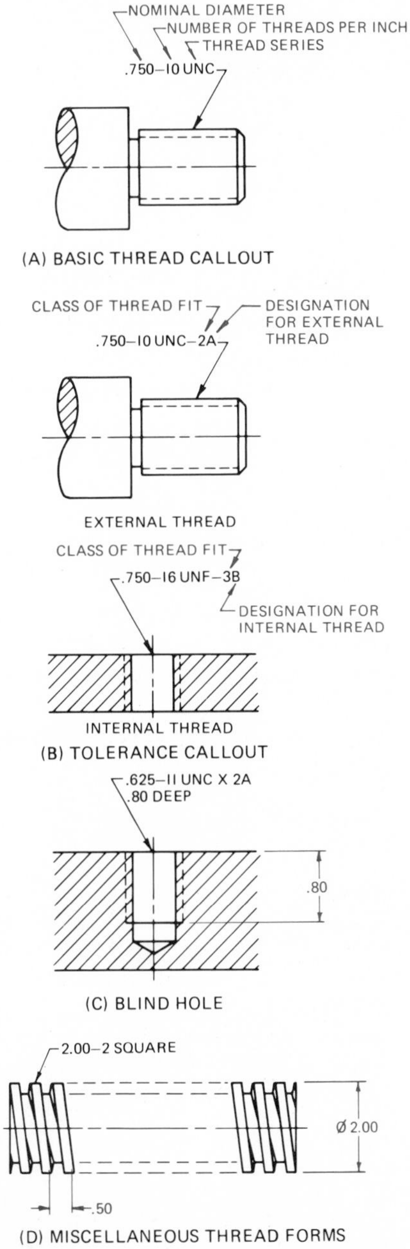
NOMINAL DIAMETER
NUMBER OF THREADS PER INCH
THREAD SERIES
.750–10 UNC
(A) BASIC THREAD CALLOUT
CLASS OF THREAD FIT
DESIGNATION FOR EXTERNAL THREAD
.750–10 UNC–2A
EXTERNAL THREAD
CLASS OF THREAD FIT
.750–16 UNF–3B
DESIGNATION FOR INTERNAL THREAD
INTERNAL THREAD
(B) TOLERANCE CALLOUT
.625–11 UNC X 2A
.80 DEEP
.80
(C) BLIND HOLE
2.00–2 SQUARE
Ø 2.00
.50
(D) MISCELLANEOUS THREAD FORMS

Fig. 8-2-5 Thread specifications for inch-size threads.

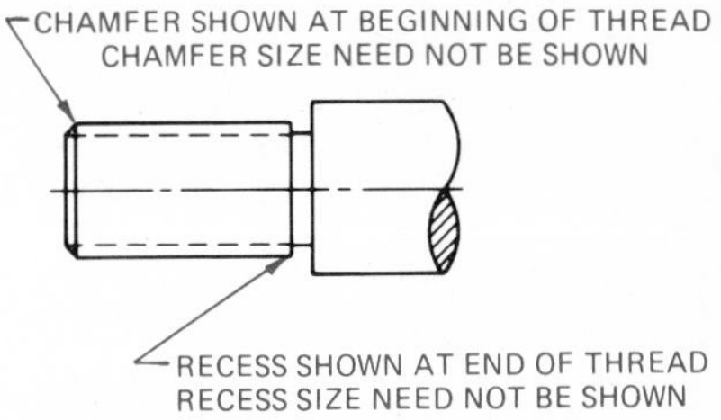
CHAMFER SHOWN AT BEGINNING OF THREAD
CHAMFER SIZE NEED NOT BE SHOWN
RECESS SHOWN AT END OF THREAD
RECESS SIZE NEED NOT BE SHOWN

Fig. 8-2-6 Omission of thread information on detail drawings.

THREAD CLASS

Three classes of external thread (classes 1A, 2A, and 3A) and three classes of internal thread (classes 1B, 2B, and 3B) are provided. These classes differ in the amount of allowances and tolerances provided in each class. A comparison of thread sizes is shown in Fig. 8-2-7.

The general characteristics and uses of the various classes are as follows.

Classes 1A and 1B. These classes produce the loosest fit, that is, the greatest amount of play in assembly. They are useful for work where ease of assembly and disassembly is essential, such as for some ordnance work and for stove bolts and other rough bolts and nuts.

Classes 2A and 2B. These classes are designed for the ordinary good grade of commercial products, such as machine screws and fasteners, and for most interchangeable parts.

Classes 3A and 3B. These classes are intended for exceptionally high-grade commercial products, where a particularly close or snug fit is essential and the high cost of precision tools and machines is warranted.

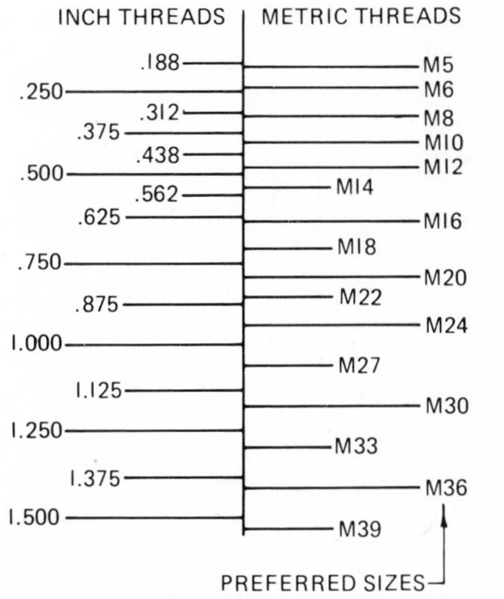
INCH THREADS
METRIC THREADS
.188
.250
.312
.375
.438
.500
.562
.625
.750
.875
1.000
1.125
1.250
1.375
1.500
M5
M6
M8
M10
M12
M14
M16
M18
M20
M22
M24
M27
M30
M33
M36
M39
PREFERRED SIZES

Fig. 8-2-7 Comparison of thread sizes.

THREAD DESIGNATION

Thread designation for inch threads, whether external or internal, is expressed in this order: diameter (nominal or major diameter), number of threads per inch, thread form and series, and class of fit.

Pipe Threads

The pipe universally used is the inch-size pipe and, as such, will not be changed. When pipe is ordered, the outside diameter and wall thickness (in millimetres) are given. In calling for the size of thread, the note used is similar to that for screw threads. See Fig. 8-2-8.

Example 1. 4 × 8NPT
where 4 = nominal diameter of pipe, in inches
8 = number of threads per inch
N = American Standard
P = pipe
T = taper thread

REFERENCES AND SOURCE MATERIAL

1. American National Standards Institute.
2. Canadian Standards Association.

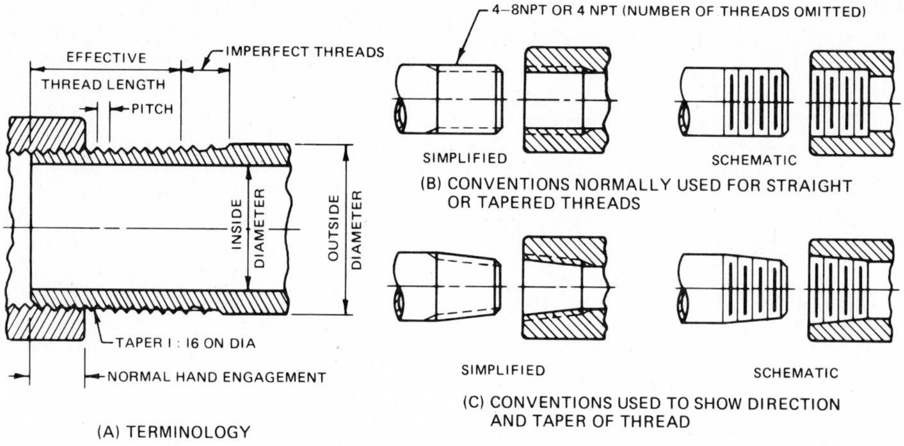

4–8NPT OR 4 NPT (NUMBER OF THREADS OMITTED)
EFFECTIVE
THREAD LENGTH
IMPERFECT THREADS
PITCH
INSIDE DIAMETER
OUTSIDE DIAMETER
TAPER 1 : 16 ON DIA
NORMAL HAND ENGAGEMENT
(A) TERMINOLOGY
SIMPLIFIED
SCHEMATIC
(B) CONVENTIONS NORMALLY USED FOR STRAIGHT OR TAPERED THREADS
SIMPLIFIED
SCHEMATIC
(C) CONVENTIONS USED TO SHOW DIRECTION AND TAPER OF THREAD

Fig. 8-2-8 Pipe thread terminology and conventions.

Assignments

1. On an A3- or B-size sheet, make a two-view assembly drawing of the parallel clamps shown in Fig. 8-2-A or 8-2-B. Use simplified thread connections and include on your drawing a bill of material calling for all the parts. The only dimension required on the drawing is the maximum opening of the jaws. Identify the parts on the assembly. Scale is 1:1.

REVIEW FOR ASSIGNMENTS

Unit 6-4	Bills of Materials
Unit 2-6	Phantom Line Application
Unit 7-13	Knurled Parts
Unit 6-4	Assembly Drawings
Unit 9-4	Springs
Appendix	Number Drill Sizes

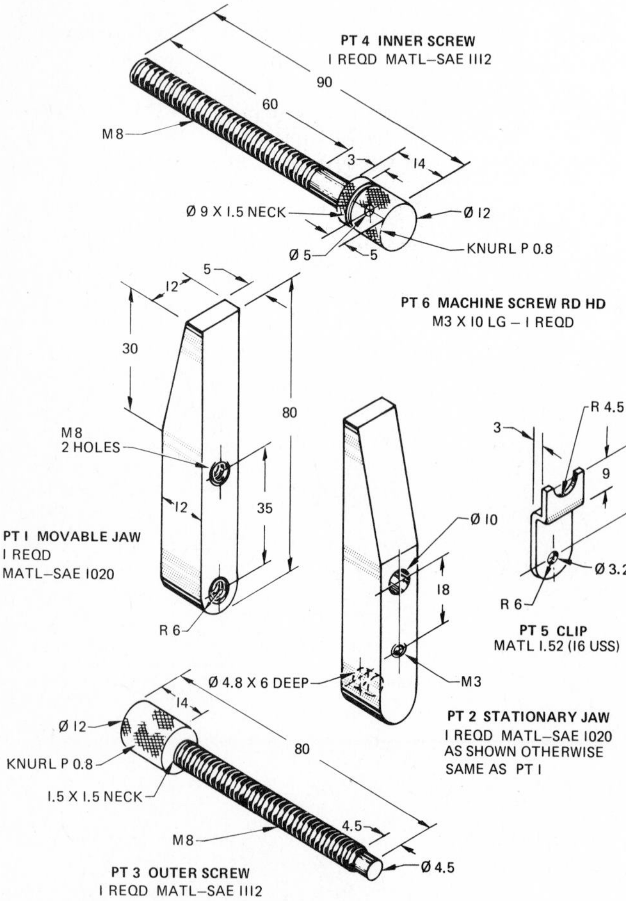

PT 4 INNER SCREW
1 REQD MATL–SAE 1112
90
60
M8
3
14
Ø 9 X 1.5 NECK
Ø 12
Ø 5
5
KNURL P 0.8
PT 6 MACHINE SCREW RD HD
M3 X 10 LG – 1 REQD
12
5
30
80
M8
2 HOLES
35
12
PT 1 MOVABLE JAW
1 REQD
MATL–SAE 1020
R 6
R 4.5
3
9
18
Ø 10
Ø 3.2
18
R 6
PT 5 CLIP
MATL 1.52 (16 USS)
Ø 4.8 X 6 DEEP
M3
PT 2 STATIONARY JAW
1 REQD MATL–SAE 1020
AS SHOWN OTHERWISE
SAME AS PT 1
Ø 12
14
KNURL P 0.8
80
1.5 X 1.5 NECK
M8
4.5
Ø 4.5
PT 3 OUTER SCREW
1 REQD MATL–SAE 1112

Fig. 8-2-A Parallel clamps.

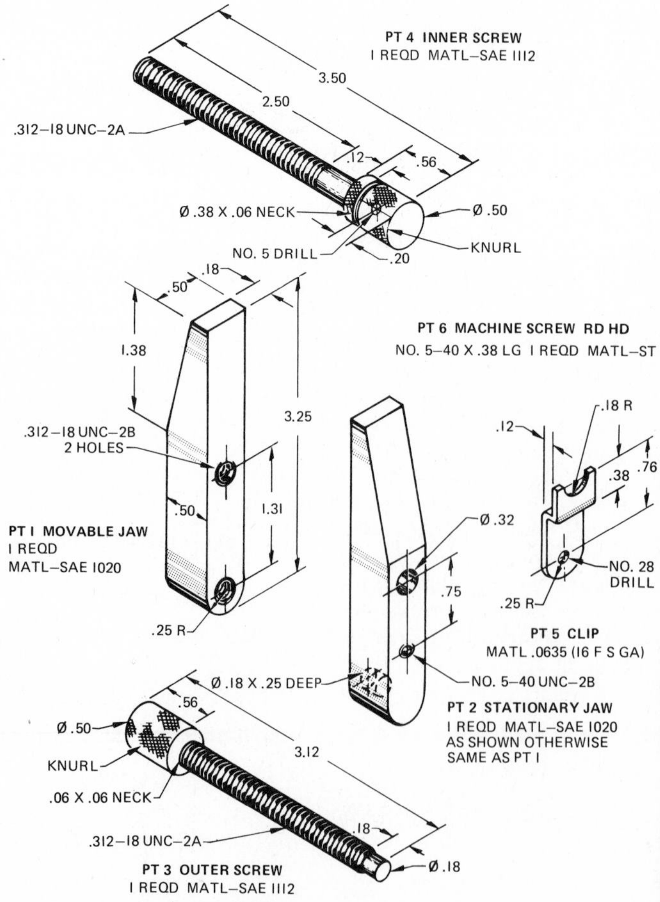

PT 4 INNER SCREW
1 REQD MATL–SAE 1112
3.50
2.50
.312–18 UNC–2A
.12
.56
Ø .38 X .06 NECK
Ø .50
NO. 5 DRILL
KNURL
.20
.18
.50
1.38
3.25
PT 6 MACHINE SCREW RD HD
NO. 5–40 X .38 LG 1 REQD MATL–ST
.312–18 UNC–2B
2 HOLES
.50
1.31
PT 1 MOVABLE JAW
1 REQD
MATL–SAE 1020
.25 R
.12
.18 R
.38
.76
Ø .32
NO. 28
DRILL
.75
.25 R
PT 5 CLIP
MATL .0635 (16 F S GA)
Ø .18 X .25 DEEP
NO. 5–40 UNC–2B
PT 2 STATIONARY JAW
1 REQD MATL–SAE 1020
AS SHOWN OTHERWISE
SAME AS PT 1
.56
Ø .50
3.12
KNURL
.06 X .06 NECK
.18
.312–18 UNC–2A
Ø .18
PT 3 OUTER SCREW
1 REQD MATL–SAE 1112

Fig. 8-2-B Parallel clamps.

UNIT 8-3

COMMON THREADED FASTENERS

Fastener Selection[1]

Fastener manufacturers agree that product selection must begin at the design stage. For it is here, when a product is still a figment of someone's imagination, that the best interests of the designer, production manager, and purchasing agent can be served. Designers, naturally, want optimum performance; production people are interested in the ease and economics of assembly; purchasing agents are keen to minimize initial costs and stocking costs.

The answer, pure and simple, is to determine the objectives of the particular fastening job and then consult with a fastener supplier. These technical experts can often shed light on the situation and then recommend the right item, at the best, in-place cost.

Machine screws are among the most common fasteners in industry. See Figs. 8-3-1 and 8-3-2. They are the easiest to install and remove. They are also among the least understood. To obtain maximum machine-screw efficiency, thorough knowledge of the properties of both the screw and the materials to be fastened together is required.

For a given application, a designer should know the load which the screw must withstand, whether the load is one of tension or shear, and whether the assembly will be subject to impact shock or that perennial nemesis, vibration. Once these factors have been determined, then the size, strength, head shape, and thread type can be selected.

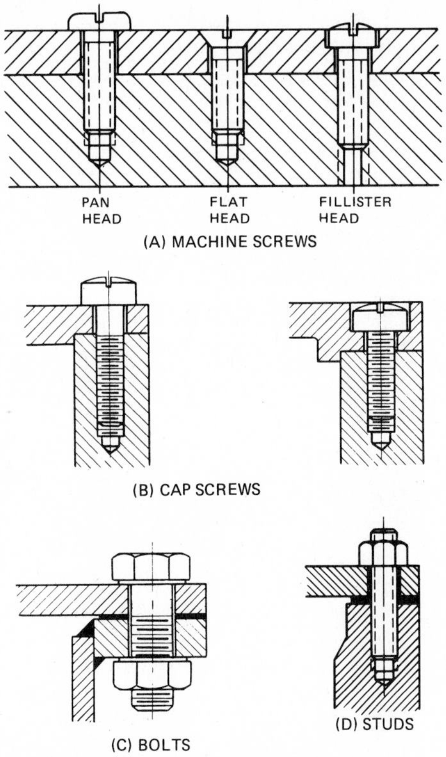
PAN HEAD
FLAT HEAD
FILLISTER HEAD
(A) MACHINE SCREWS
(B) CAP SCREWS
(C) BOLTS
(D) STUDS

Fig. 8-3-2 Fastener application.

Fastener Definitions

Machine Screws. Machine screws have either fine or coarse threads and are available in a variety of heads. They may be used in tapped holes or with nuts.

Cap Screws. A *cap screw* is a threaded fastener which joins two or more parts by passing through a clearance hole in one part and screwing into a tapped hole in the other. A cap screw is tightened or released by torquing the head.

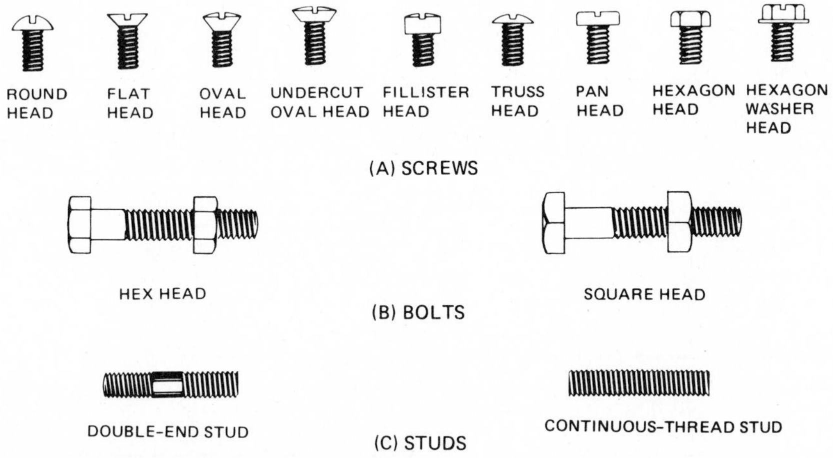
ROUND HEAD
FLAT HEAD
OVAL HEAD
UNDERCUT OVAL HEAD
FILLISTER HEAD
TRUSS HEAD
PAN HEAD
HEXAGON HEAD
HEXAGON WASHER HEAD
(A) SCREWS
HEX HEAD
SQUARE HEAD
(B) BOLTS
DOUBLE-END STUD
CONTINUOUS-THREAD STUD
(C) STUDS

Fig. 8-3-1 Common threaded fasteners.

Cap screws range in size from 6 mm in diameter and are available in five basic types of head.

Bolts. A *bolt* is a threaded fastener which passes through clearance holes in assembled parts and threads into a nut. Bolts and nuts are available in a variety of shapes and sizes. The square and hexagon head are the two most popular designs.

Studs. *Studs* are shafts threaded at both ends, and they are used in assemblies. One end of the stud is threaded into one of the parts being assembled; and the other assembly parts, such as washers and covers, are guided over the studs through clearance holes and are held together by means of a nut which is threaded over the exposed end of the stud.

EXPLANATORY DATA

A bolt is designed for assembly with a nut. A screw has features in its design which make it capable of being used in a tapped or other preformed hole in the work. Because of basic design, it is possible to use certain types of screws in combination with a nut. Any externally threaded fastener which has a majority of the design characteristics that assist its proper use in a tapped or other preformed hole is a *screw*, regardless of how it is used in its service application.

The Change to Metric Fasteners

In the United States, the Industrial Fasteners Institute (IFI) has undertaken a major compilation of standards in its 296-page volume *Metric Fastener Standards*. It covers more than 30 metric fastener standards which have been developed by the IFI over the past few years.

HEAD STYLES

Figure 8-3-3 shows some of the common head styles generic to bolts and screws alike. Of these basic styles, five (the pan, round, binding, truss, and hex cap) are designed to do the same job.

The pan, however, is the most versatile. That is why the Industrial Fasteners Institute recommended that it replace round and binding head styles. They also suggest that pan heads can substitute for truss and fillister heads in most assemblies. Economically, it makes good sense. A switch to a uniform head style means that fewer types of screws can be manufactured.

Drive Configurations. Figure 8-3-4 indicates five popular driving designs.

Shoulders and Necks. The *shoulder* of a fastener is the enlarged portion of the

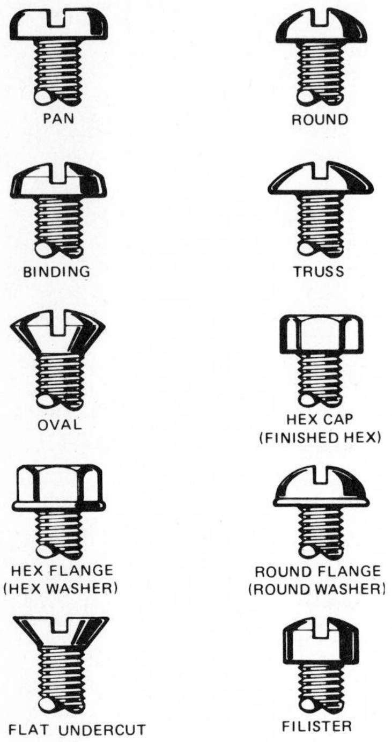
PAN
ROUND
BINDING
TRUSS
OVAL
HEX CAP
(FINISHED HEX)
HEX FLANGE
(HEX WASHER)
ROUND FLANGE
(ROUND WASHER)
FLAT UNDERCUT
FILISTER

Fig. 8-3-3 Common head styles.

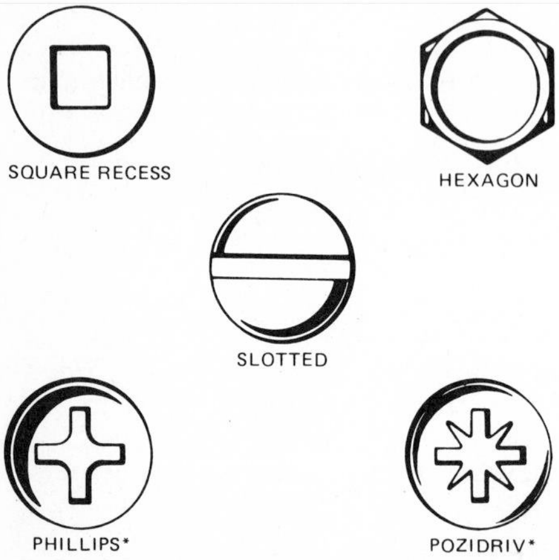
SQUARE RECESS
HEXAGON
SLOTTED
PHILLIPS*
POZIDRIV*

Fig. 8-3-4 Drive configurations.

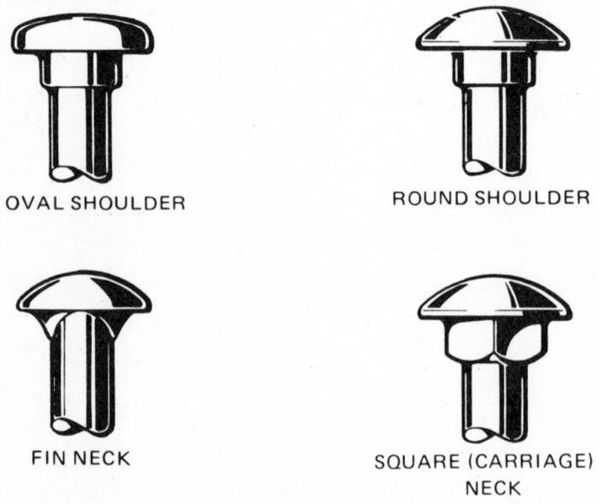
OVAL SHOULDER
ROUND SHOULDER
FIN NECK
SQUARE (CARRIAGE)
NECK

Fig. 8-3-5 Shoulders and necks.

body of a threaded fastener or the shank of an unthreaded fastener. See Fig. 8-3-5.

Point Designs. The *point* of a fastener is the configuration of the end of the shank of a headed fastener, or of each end of a headless fastener. Points are usually either roll-formed or cold-headed. They may also be formed by pinching or milling. See Fig. 8-3-6.

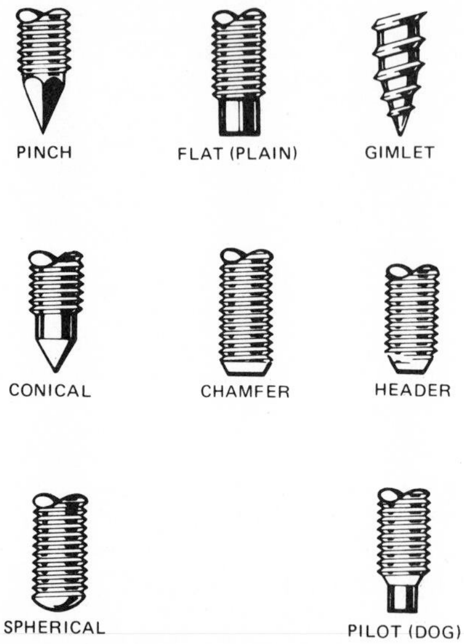
PINCH
FLAT (PLAIN)
GIMLET
CONICAL
CHAMFER
HEADER
SPHERICAL
PILOT (DOG)

Fig. 8-3-6 Point styles.

Property Classes of Metric Fasteners[2]

BOLTS, SCREWS, AND STUDS

For mechanical and material requirements, metric fasteners are classified under a number of property classes. Bolts, screws, and studs have seven property classes of steel suitable for general engineering applications. The property classes are designated by numbers where increasing numbers generally represent increasing tensile strengths. The designation symbol consists of two parts: the first numeral of a two-digit symbol or the first two numerals of a three-digit symbol approximates one-hundredth of the minimum tensile strength in millipascals; and the last numeral approximates one-tenth of the ratio expressed as a percentage of minimum yield strength and minimum tensile strength.

Example 1. A property class 4.8 fastener (see Fig. 8-3-7) has a minimum tensile strength of 420 MPa and a minimum yield strength of 340 MPa. One percent of 420 is 4.2. The first digit is 4. The minimum yield strength of 340 MPa is equal to approximately 80 percent of the minimum tensile strength of 420 MPa. One-tenth of 80 percent is 8. The last digit of the property class is 8.

PROPERTY CLASS	NOMINAL DIAMETER	MINIMUM TENSILE STRENGTH MPa	MINIMUM YEILD STRENGTH MPa
4.6	M5 THRU M36	400	240
4.8	M1.6 THRU M16	420	340
5.8	M5 THRU M24	520	420
8.8	M16 THRU M36	830	660
9.8	M1.6 THRU M16	900	720
10.9	M5 THRU M36	1040	940
12.9	M1.6 THRU M36	1220	1100

Fig. 8-3-7 Mechanical requirements for bolts, screws, and studs.

Example 2. A property class 10.9 fastener (see Fig. 8-3-7) has a minimum tensile strength of 1040 MPa and a minimum yield strength of 940 MPa. One percent of 1040 is 10.4. The first two numerals of the three-digit symbol are 10. The minimum yield strength of 940 MPa is equal to approximately 90 percent of the minimum tensile strength of 1040 MPa. One-tenth of 90 percent is 9. The last digit of the property class is 9.

Other styles of bolts and cap screws are shown in Fig. 8-3-8.

Machine screws are normally available only in classes 4.8 and 9.8; other bolts, screws, and studs are available in all classes within the specified product size limitations given in Fig. 8-3-7.

For guidance purposes only, to assist designers in selecting a property class, the following information may be used.

- Class 4.6 is approximately equivalent to SAE grade 1 and ASTM A307, grade A.
- Class 5.8 is approximately equivalent to SAE grade 2.
- Class 8.8 is approximately equivalent to SAE grade 5 and ASTM A449.
- Class 9.8 has properties approximately 9 percent stronger than SAE grade 5 and ASTM A449.
- Class 10.9 is approximately equivalent to SAE grade 8 and ASTM A354 grade BD.

Figure 8-3-9, shown for comparison as well as for identification purposes, lists the mechanical requirements of inch-size threaded fasteners.

Fastener Markings. Slotted and crossed recessed screws of all sizes and other screws and bolts of sizes M4 and smaller need not be marked. All other bolts and screws of sizes M5 and larger are marked to identify the property class and the manufacturer. The property class symbols are shown in Fig. 8-3-10. The symbol is located on the top of the bolt head or screw. Alternatively, for hex-head products, the markings may be indented on the side of the head.

All studs of size M5 and larger are identified by the property class symbol.

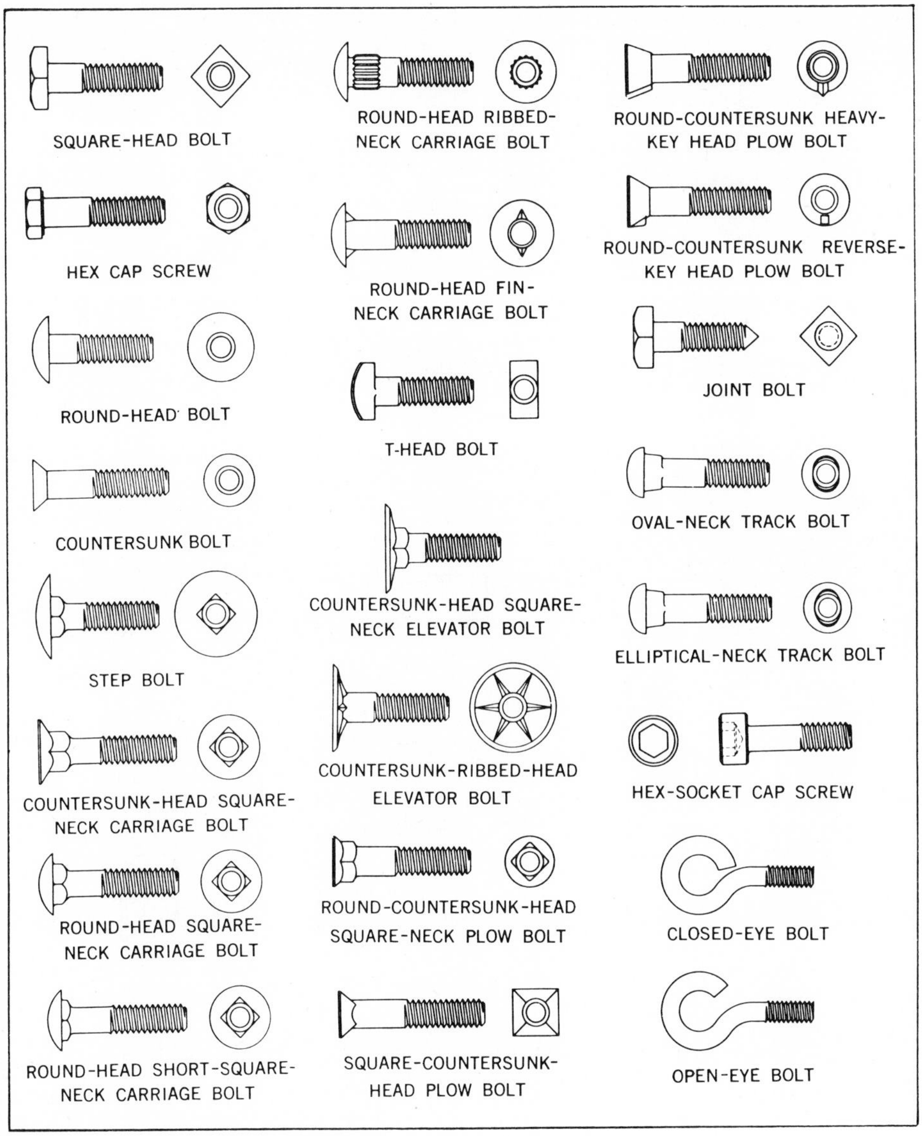
SQUARE-HEAD BOLT
HEX CAP SCREW
ROUND-HEAD BOLT
COUNTERSUNK BOLT
STEP BOLT
COUNTERSUNK-HEAD SQUARE-NECK CARRIAGE BOLT
ROUND-HEAD SQUARE-NECK CARRIAGE BOLT
ROUND-HEAD SHORT-SQUARE-NECK CARRIAGE BOLT
ROUND-HEAD RIBBED-NECK CARRIAGE BOLT
ROUND-HEAD FIN-NECK CARRIAGE BOLT
T-HEAD BOLT
COUNTERSUNK-HEAD SQUARE-NECK ELEVATOR BOLT
COUNTERSUNK-RIBBED-HEAD ELEVATOR BOLT
ROUND-COUNTERSUNK-HEAD SQUARE-NECK PLOW BOLT
SQUARE-COUNTERSUNK-HEAD PLOW BOLT
ROUND-COUNTERSUNK HEAVY-KEY HEAD PLOW BOLT
ROUND-COUNTERSUNK REVERSE-KEY HEAD PLOW BOLT
JOINT BOLT
OVAL-NECK TRACK BOLT
ELLIPTICAL-NECK TRACK BOLT
HEX-SOCKET CAP SCREW
CLOSED-EYE BOLT
OPEN-EYE BOLT

Fig. 8-3-8 Bolts and cap screws.

Head Designation	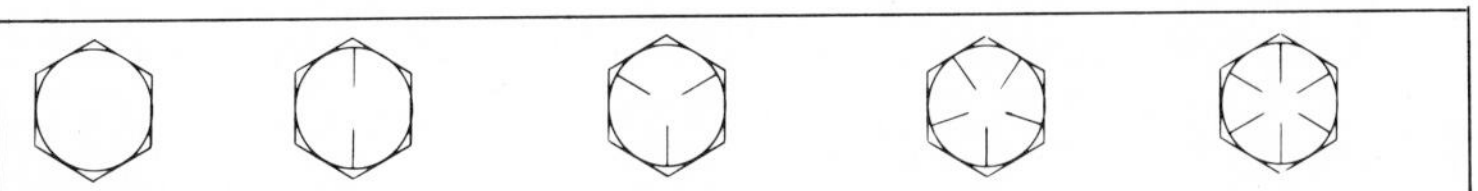				
Grade	GRADES 0, 1, 2	GRADE 3	GRADE 5	GRADE 7	GRADE 8
Minimum Tensile Strength Kips	0—No requirements 1—55 2—69 64 55	110 100	120 115 105	133	150

Fig. 8-3-9 Mechanical requirements for inch-size threaded fasteners.

The marking is located on the extreme end of the stud. For studs with an interference fit thread the markings are located at the nut end. Studs smaller than M12 use different identification symbols.

NUTS

The terms ***regular*** and ***thick*** for describing nut thicknesses have been replaced by the terms ***style 1*** and ***style 2***. The design of style 1 and 2 steel nuts shown in Fig. 8-3-11 is based on providing sufficient nut strength to reduce the possibility of thread stripping. There are three property classes of steel nuts. They are designated by the numbers 5, 9, and 10.

Hex Nuts. Class 5 nuts are non-heat-treated, have style 1 dimensions, and are designed for use with bolts and screws, property class 5.8 and lower strengths. Class 9 nuts are non-heat-treated, have style 2 dimensions, and are designed for use with bolts and screws of property class 9.8, 8.8, and lower strengths. Class 10 nuts are heat-treated, have style 1 dimensions, and are designed for use with bolts and screws of property class 10.9 and lower strengths.

Hex Slotted Nuts. Style 1 hex slotted nuts conform with either property class 5 or property class 10. Style 2 hex slotted nuts conform with property class 9.

PROPERTY CLASS	IDENTIFICATION SYMBOL	
	BOLTS, SCREWS AND STUDS	STUDS SMALLER THAN M12
4.6	4.6	–
4.8	4.8	–
5.8	5.8	–
8.8 (1)	8.8	○
9.8 (1)	9.8	+
10.9 (1)	10.9	□
12.9	12.9	△

NOTE 1: PRODUCTS MADE OF LOW CARBON MARTENSITE STEEL SHALL BE ADDITIONALLY IDENTIFIED BY UNDERLINING THE NUMERALS.

Fig. 8-3-10 Property class identification symbols for bolts, screws, and studs.

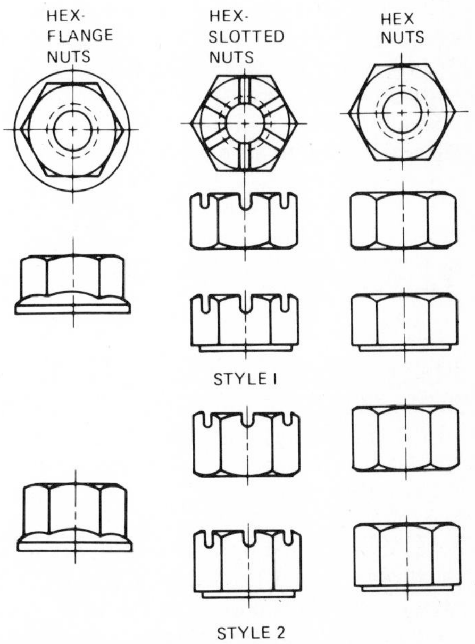
HEX FLANGE NUTS
HEX SLOTTED NUTS
HEX NUTS
STYLE 1
STYLE 2

Fig. 8-3-11 Hex-nut styles.

Hex Flanged Nuts. These nuts are intended for general use in applications requiring a large bearing contact area, such as where the nut face is to bear directly on the work of overslotted or oversized clearance holes or against material of relatively low compressive strength, or where the conditions might normally necessitate the use of a flat washer under a hex nut. The two styles of flanged hex nuts differ dimensionally in thickness only. The standard property classes for hex flanged nuts are identical to the hex nuts.

The three property classes of hex nuts are available in the sizes specified in Fig. 8-3-12. The two property classes of the hex flange nuts are designated by the numbers 9 and 10, and are available in sizes M6.3 to M20 inclusive.

PROPERTY CLASS	NOMINAL NUT SIZE	SUGGESTED PROPERTY CLASS OF MATING BOLT, SCREW OR STUD
5	M5 THRU M36	4.6, 4.8, 5.8
9	M5 THRU M16 M20 THRU M36	5.8, 9.8 5.8, 8.8
10	M6.3 THRU M36	10.9

Fig. 8-3-12 Nut selection for bolts, screws, and studs.

Markings. All nuts are marked to identify the property class in one of the following ways:

1. Property class number is located on any external surface.

2. Insert retaining stake marks for insert-type hex nuts: two equally spaced stake marks for class 5, three equally spaced stake marks for class 9, and four equally spaced stake marks for class 10.

3. Dots are located on the top surface of the nut. A dot is located either on the chamfered surface at one corner or normal to one of the wrenching flats for class 5. A dot is located either on the chamfered surface at each of the two adjacent corners or normal to each of two adjacent wrench flats for class 9. A dot is located either on the chamfered surface at each of the three adjacent corners or normal to each of three adjacent wrenching flats.

Drawing a Bolt and Nut

Bolts and nuts are not normally drawn on detail drawings unless they are of a special size or have been modified slightly. On some assembly drawings it may be necessary to draw the nut and bolt. Approximate nut and bolt sizes are shown in Fig. 8-3-13. Actual sizes are found in the Appendix. Nut and bolt templates are also available and are recommended as a cost-saving device. Conventional drawing practice is to draw the nuts and bolt heads in the across-corners position in all views.

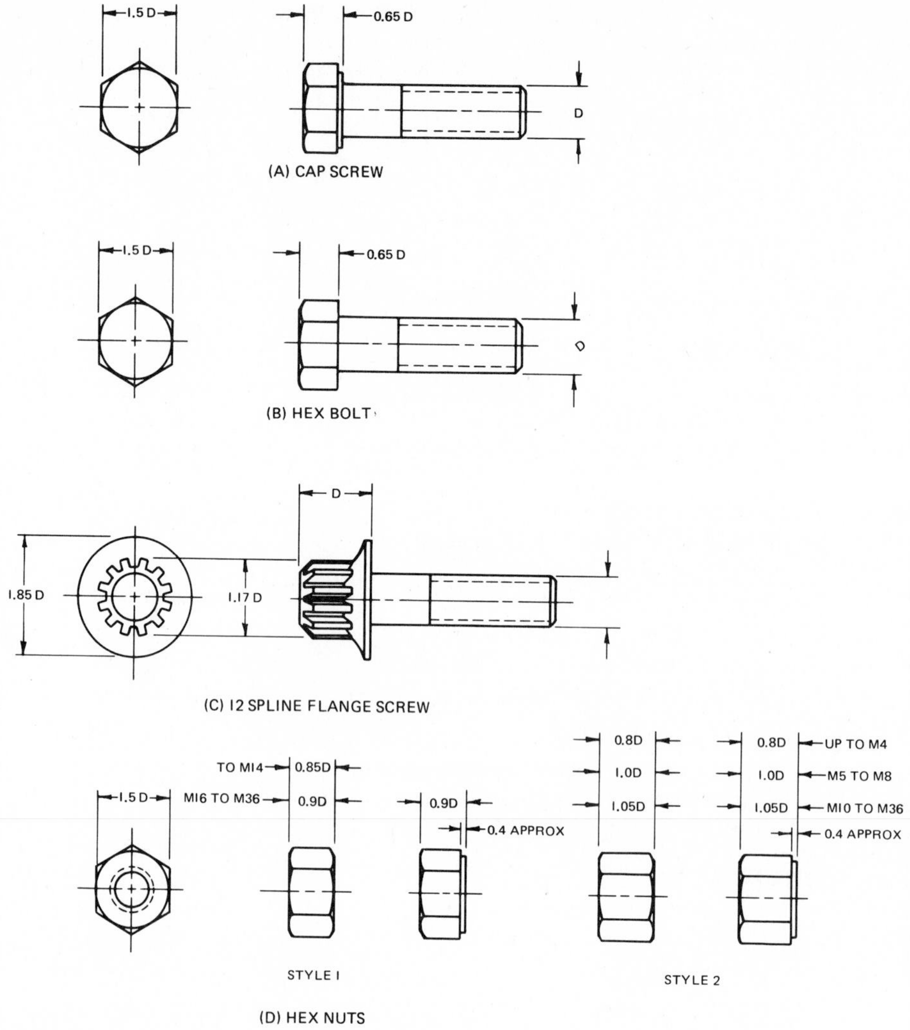
1.5 D
0.65 D
D
(A) CAP SCREW
1.5 D
0.65 D
D
(B) HEX BOLT
D
1.85 D
1.17 D
(C) 12 SPLINE FLANGE SCREW
TO M14
0.85D
1.5 D
M16 TO M36
0.9D
0.9D
0.4 APPROX
0.8D
1.0D
1.05D
0.8D
UP TO M4
1.0D
M5 TO M8
1.05D
M10 TO M36
0.4 APPROX
STYLE 1
STYLE 2
(D) HEX NUTS

Fig. 8-3-13 Approximate head proportions for hex-head cap screws, bolts, and nuts.

Studs[3]

Studs, as shown in Fig. 8-3-14, are still used in large quantities to best fulfill the needs of certain design functions and for overall economy.

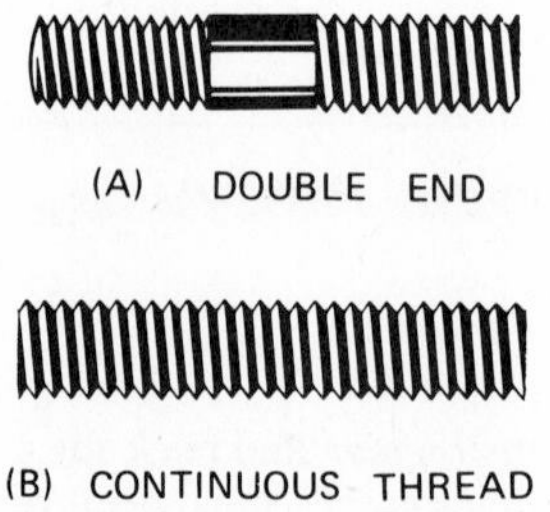
(A) DOUBLE END
(B) CONTINUOUS THREAD

Fig. 8-3-14 Studs.

STUD STANDARDS

There are four basic types of studs.

Type 1: Unfinished. These studs have an unfinished body with no specific body tolerances.

Type 2: Finished, Full or Undersize Body. These studs have a maximum body diameter equal to the basic major diameter of the thread and a minimum body diameter equal to the rolled thread blank size.

Type 3: Finished, Full Body. These studs have a maximum body diameter equal to basic major diameter of the thread and a minimum body diameter equal to the specified minimum major diameter of the thread.

Type 4: Finished, Close Body. These studs have body diameter tolerances as specified by the purchaser.

DESIGNATION

Double-End Studs. To avoid possible misunderstanding in specifying double-end studs, it is recommended that they be designated in the following sequence: Type and name; nominal size × thread pitch; stud length; material, including grade identification; and finish (plating or coating) if required.

Example: TYPE 2 DOUBLE-END STUD, M10 × 1.5 × 100, STEEL CLASS 9.8, CADMIUM PLATED

Continuous-Thread Studs. To avoid possible misunderstanding in specifying continuous thread studs, it is recommended that they be designated in the following sequence: Product name, nominal size and thread pitch, stud length, material and finish (plating or coating) if required.

Example: TYPE 3 CONTINUOUS THREAD STUD, M24 × 3 × 200, STEEL CLASS 8.8, ZINC PHOSPHATE AND OIL

Washers[4]

Washers are one of the most common forms of hardware and perform many varied functions in mechanically fastened assemblies. They may only be required to span an oversize clearance hole, to give better bearing for nuts or screw faces, or to distribute loads over a greater area. Often, they serve as locking devices for threaded fasteners. They are also used to maintain a spring-resistance pressure, to guard surfaces against marring, and to provide a seal.

CLASSIFICATION OF WASHERS

Washers are commonly the elements which are added to screw systems to keep them tight, but not all washers are locking types. Many washers serve other functions, such as calibrated and uncalibrated load distribution, surface protection, insulation, sealing, electrical connection, and spring-tension take-up devices.

Flat Washers. Plain, or flat, washers are used primarily to provide a bearing surface for a nut or a screw head, to cover large clearance holes, and to distribute fastener loads over a larger area—particularly on soft materials such as aluminum or wood. See Fig. 8-3-15.

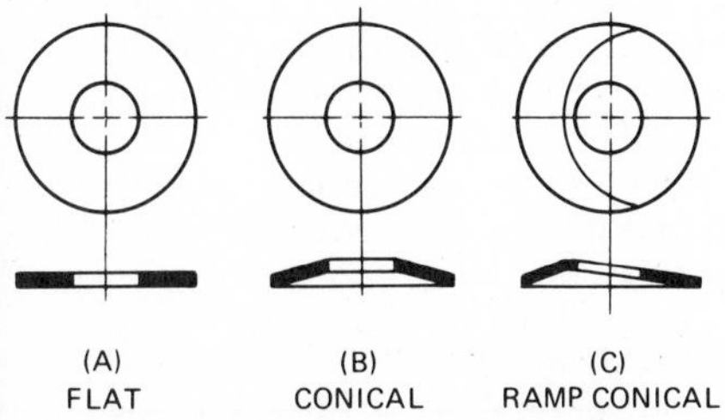
(A)
FLAT
(B)
CONICAL
(C)
RAMP CONICAL

Fig. 8-3-15 Flat and conical washers.

Conical Washers. These washers are used with screws to effectively add spring take-up to the screw elongation. They are usually made of hardened spring steel.

Belleville is a term commonly, but incorrectly, applied to coned (conical) washers. Cone washers do not have any auxiliary locking features other than friction, and in the flattened position, they are the equivalent of any flat washer as far as locking action is concerned.

Helical Spring Washers. These washers are made of slightly trapezoidal wire formed into a helix of one coil so that the free height is approximately twice the thickness of the washer section. See Fig. 8-3-16.

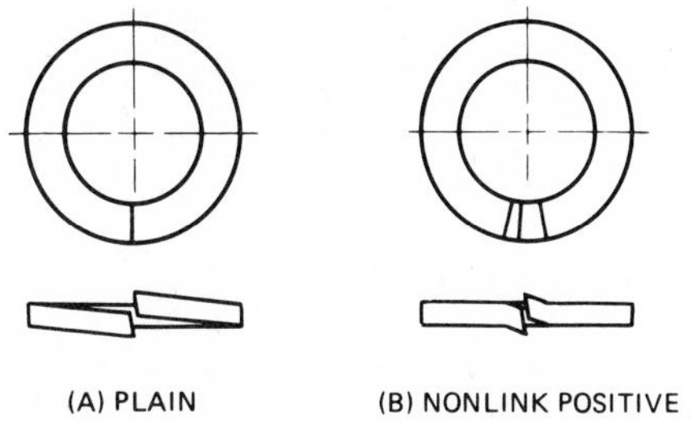
(A) PLAIN
(B) NONLINK POSITIVE

Fig. 8-3-16 Helical spring washers.

Tooth Lock Washers. Made of hardened carbon steel, a tooth lock washer has teeth that are twisted or bent out of the plane of the washer face so that sharp cutting edges are presented to both the workpiece and the bearing face of the screw head or nut. See Fig. 8-3-17.

Spring Washers. There are no standard designs for spring washers. See Fig. 8-3-18. They are made in a great variety of sizes and shapes and are usually selected from a manufacturer's catalog for some specific purpose.

Special-Purpose Washers. Molded or stamped nonmetallic washers are available in many materials and may be used as seals, electrical insulators, or for protection of the surface of assembled parts.

Many plain, cone, or tooth washers are available with special mastic sealing compounds firmly attached to the washer. These washers are used for sealing and vibration isolation in high-production industries.

TERMS RELATED TO THREADED FASTENERS

The *tap drill size* for a threaded (tapped) hole is a diameter equal to the minor diameter of the thread. The *clearance drill size*, which permits the free passage of a bolt, is a diameter slightly greater than

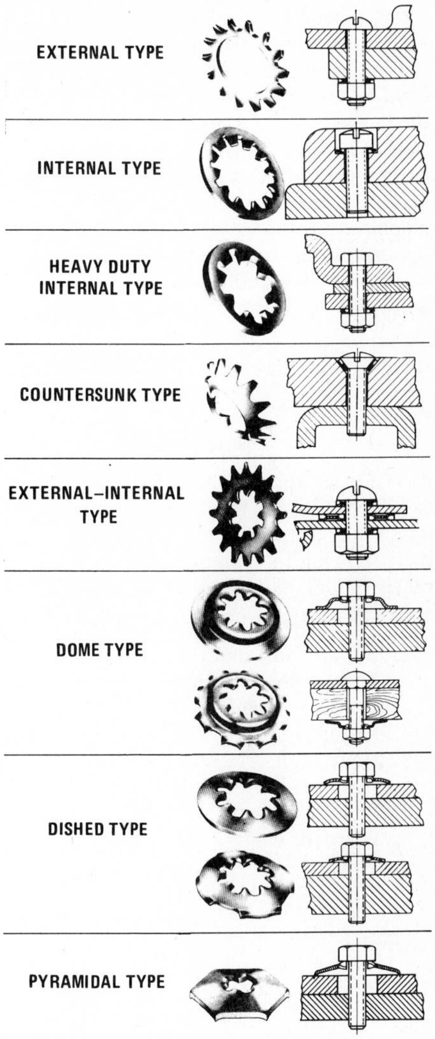
EXTERNAL TYPE
INTERNAL TYPE
HEAVY DUTY INTERNAL TYPE
COUNTERSUNK TYPE
EXTERNAL–INTERNAL TYPE
DOME TYPE
DISHED TYPE
PYRAMIDAL TYPE

Fig. 8-3-17 Tooth lock washers.

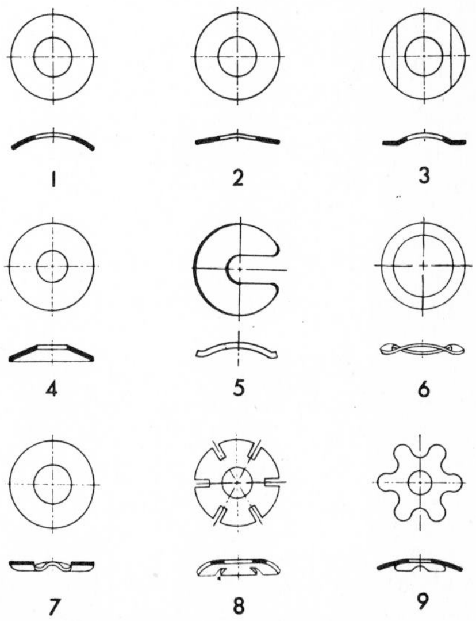
1
2
3
4
5
6
7
8
9

Fig. 8-3-18 Typical spring washers design.

the major diameter of the bolt. See Fig. 8-3-19. A *counterbored hole* is a circular, flat-bottomed recess that permits the head of a bolt or cap screw to rest below the surface of the part. A *countersunk hole* is an angular-sided recess that accommodates the shape of a flat-head cap screw or machine screw or an oval-head machine screw. *Spot-facing* is a machine operation that provides a smooth, flat surface where a bolt head or a nut will rest.

Specifying Fasteners

In order for the purchasing department to properly order the fastening device which has been selected in the design, the following information is required. *(Note: The information listed will not apply to all types of fasteners):*

1. Type of fastener
2. Thread specifications
3. Fastener length
4. Material
5. Head style
6. Type of driving recess
7. Point type (setscrews only)
8. Property class
9. Finish

Examples:

M8 × 1.25 × 100, 9.8 HEX BOLT, ZINC PLATED

M10 × 1.5 × 50, 9.8 12-SPLINE FLANGE SCREW, CADMIUM PLATED

TYPE 2 DOUBLE-END STUD, M10 × 1.5 × 100, STEEL CLASS 9.8, CADMIUM PLATED

NUT, HEX, STYLE 1, M12 STEEL

MACH SCREW, PHILLIPS ROUND HD, M4 × 0.7 × 25 LG, BRASS

WASHER, FLAT 8.4 ID × 17 OD × 2 THK, STEEL LOCKWASHER.

REFERENCES AND SOURCE MATERIAL

1. "Fastening Technology '77," Design Engineering and Staff of Stelco's "Fastener Facts."
2. CSA-B33.1, *Square and Hexagon Bolts and Nuts, Studs and Wrench Openings*.
3. W. G. Waltermire, "Studs," *Machine Design*, vol. 37, no. 6, 1965.
4. T. P. Hurst and D. F. Wagner, "Washers," *Machine Design*, vol. 37, no. 6, 1965.

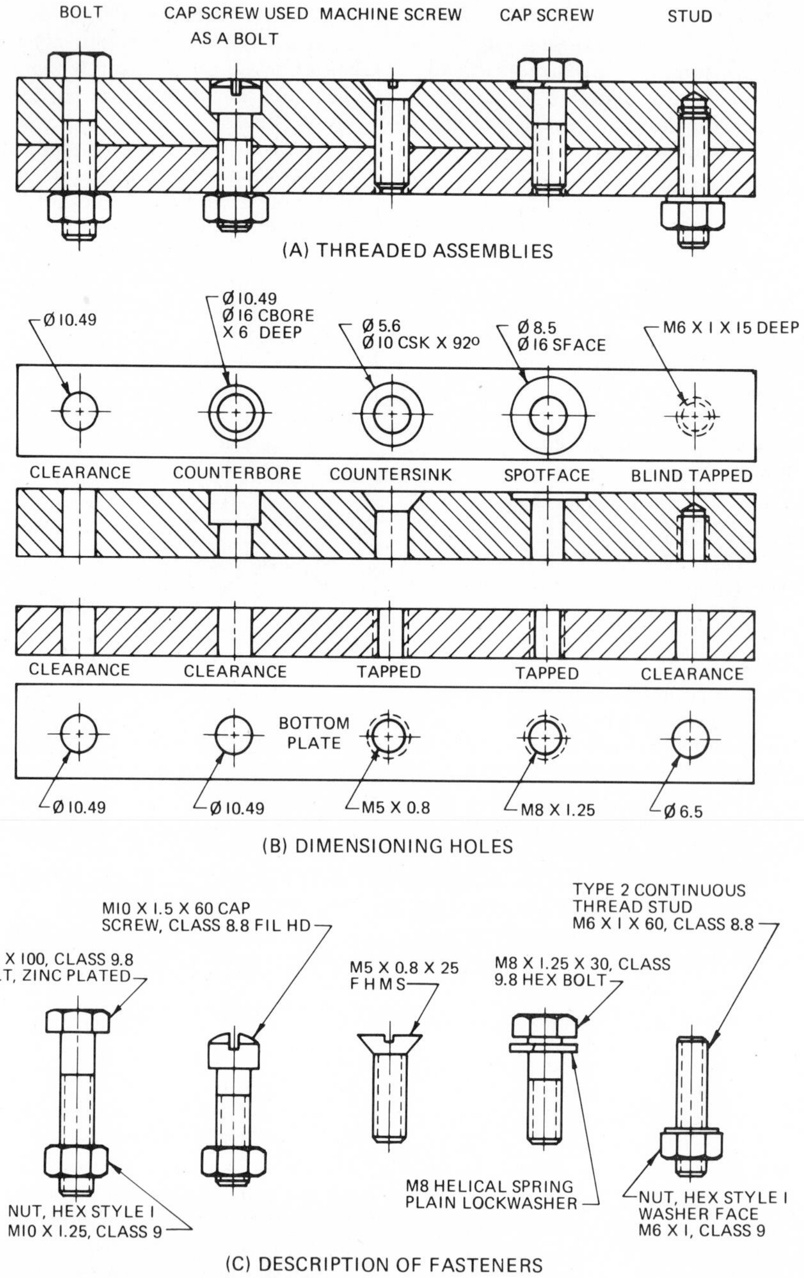

BOLT
CAP SCREW USED AS A BOLT
MACHINE SCREW
CAP SCREW
STUD
(A) THREADED ASSEMBLIES
Ø 10.49
Ø 10.49
Ø 16 CBORE X 6 DEEP
Ø 5.6
Ø 10 CSK X 92°
Ø 8.5
Ø 16 SFACE
M6 X 1 X 15 DEEP
CLEARANCE
COUNTERBORE
COUNTERSINK
SPOTFACE
BLIND TAPPED
CLEARANCE
CLEARANCE
TAPPED
TAPPED
CLEARANCE
BOTTOM PLATE
Ø 10.49
Ø 10.49
M5 X 0.8
M8 X 1.25
Ø 6.5
(B) DIMENSIONING HOLES
M10 X 1.5 X 60 CAP SCREW, CLASS 8.8 FIL HD
TYPE 2 CONTINUOUS THREAD STUD M6 X 1 X 60, CLASS 8.8
M10 X 1.5 X 100, CLASS 9.8 HEX BOLT, ZINC PLATED
M5 X 0.8 X 25 F H M S
M8 X 1.25 X 30, CLASS 9.8 HEX BOLT
NUT, HEX STYLE 1 M10 X 1.25, CLASS 9
M8 HELICAL SPRING PLAIN LOCKWASHER
NUT, HEX STYLE 1 WASHER FACE M6 X 1, CLASS 9
(C) DESCRIPTION OF FASTENERS

Fig. 8-3-19 Specifying threaded fasteners and holes.

Assignments

1. On an A3- or B-size sheet, draw the five standard fastener assemblies in full section shown in Fig. 8-3-A or 8-3-B. Use simplified thread symbols. Scale is 1:1. Dimension both the clearance and the threaded holes. If desired, a top view of the fasteners may be drawn.

REVIEW FOR ASSIGNMENTS

Unit 7-1	Full Sections
Unit 4-3	Drawing a Hexagon
Appendix	Nut and Bolt Sizes

UNIT 8-4

SPECIAL FASTENERS

Setscrews[1]

Setscrews are used as semipermanent fasteners to hold a collar, sheave, or gear on a shaft against rotational or translational forces. In contrast to most fastening devices, the setscrew is essentially a compression device. Forces developed by the screw point on tightening produce a strong clamping action that resists relative motion between assembled parts. The basic problem in setscrew selection is to find the best combination of

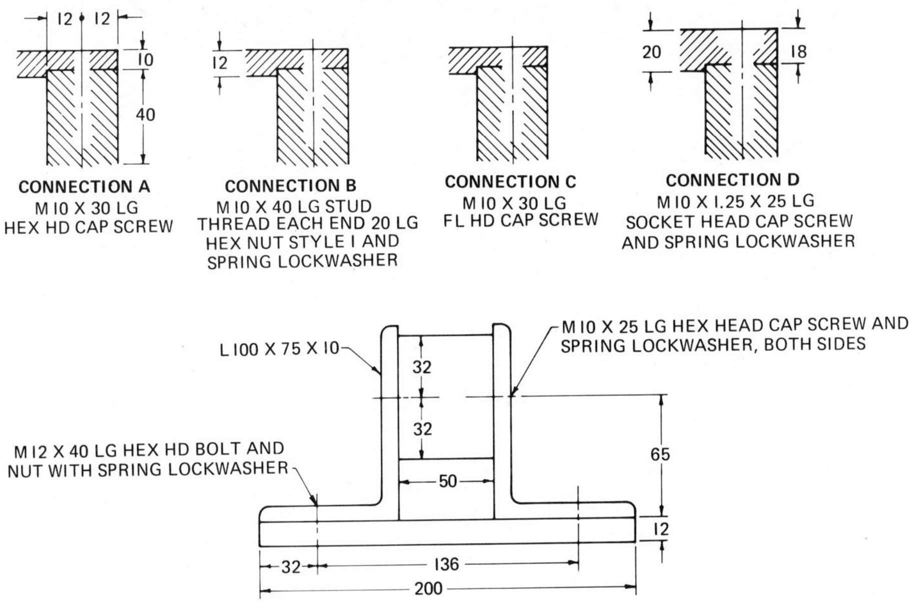
12
12
10
40
12
20
18
CONNECTION A
M 10 X 30 LG
HEX HD CAP SCREW
CONNECTION B
M 10 X 40 LG STUD
THREAD EACH END 20 LG
HEX NUT STYLE I AND
SPRING LOCKWASHER
CONNECTION C
M 10 X 30 LG
FL HD CAP SCREW
CONNECTION D
M 10 X 1.25 X 25 LG
SOCKET HEAD CAP SCREW
AND SPRING LOCKWASHER
L 100 X 75 X 10
M 10 X 25 LG HEX HEAD CAP SCREW AND
SPRING LOCKWASHER, BOTH SIDES
32
32
M 12 X 40 LG HEX HD BOLT AND
NUT WITH SPRING LOCKWASHER
65
50
12
32
136
200

Fig. 8-3-A Standard fasteners.

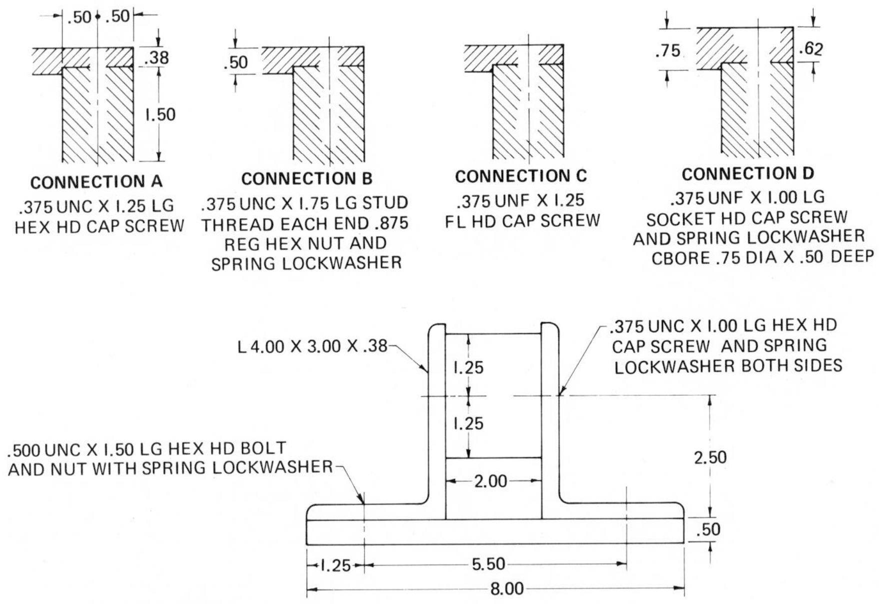
.50
.50
.38
1.50
.50
.75
.62
CONNECTION A
.375 UNC X 1.25 LG
HEX HD CAP SCREW
CONNECTION B
.375 UNC X 1.75 LG STUD
THREAD EACH END .875
REG HEX NUT AND
SPRING LOCKWASHER
CONNECTION C
.375 UNF X 1.25
FL HD CAP SCREW
CONNECTION D
.375 UNF X 1.00 LG
SOCKET HD CAP SCREW
AND SPRING LOCKWASHER
CBORE .75 DIA X .50 DEEP
L 4.00 X 3.00 X .38
.375 UNC X 1.00 LG HEX HD
CAP SCREW AND SPRING
LOCKWASHER BOTH SIDES
1.25
1.25
.500 UNC X 1.50 LG HEX HD BOLT
AND NUT WITH SPRING LOCKWASHER
2.50
2.00
.50
1.25
5.50
8.00

Fig. 8-3-B Standard fasteners.

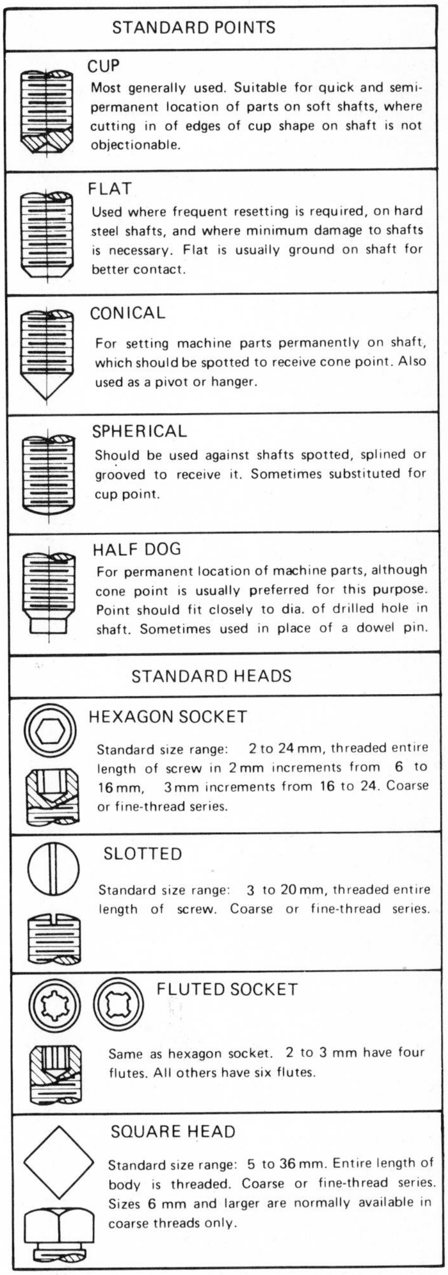
STANDARD POINTS
CUP
Most generally used. Suitable for quick and semi-permanent location of parts on soft shafts, where cutting in of edges of cup shape on shaft is not objectionable.
FLAT
Used where frequent resetting is required, on hard steel shafts, and where minimum damage to shafts is necessary. Flat is usually ground on shaft for better contact.
CONICAL
For setting machine parts permanently on shaft, which should be spotted to receive cone point. Also used as a pivot or hanger.
SPHERICAL
Should be used against shafts spotted, splined or grooved to receive it. Sometimes substituted for cup point.
HALF DOG
For permanent location of machine parts, although cone point is usually preferred for this purpose. Point should fit closely to dia. of drilled hole in shaft. Sometimes used in place of a dowel pin.
STANDARD HEADS
HEXAGON SOCKET
Standard size range: 2 to 24 mm, threaded entire length of screw in 2 mm increments from 6 to 16 mm, 3 mm increments from 16 to 24. Coarse or fine-thread series.
SLOTTED
Standard size range: 3 to 20 mm, threaded entire length of screw. Coarse or fine-thread series.
FLUTED SOCKET
Same as hexagon socket. 2 to 3 mm have four flutes. All others have six flutes.
SQUARE HEAD
Standard size range: 5 to 36 mm. Entire length of body is threaded. Coarse or fine-thread series. Sizes 6 mm and larger are normally available in coarse threads only.

Fig. 8-4-1 Setscrews.

setscrew form, size, and point style that provides the required holding power.

Setscrews can be categorized in two ways: by their forms and by the point style desired. See Fig. 8-4-1. Each of the standardized setscrew forms is available in any one of six point styles.

Conventional approach to setscrew selection is usually based on a rule-of-thumb procedure: The setscrew diameter should be roughly equal to one-half the shaft diameter. This rule of thumb often gives satisfactory results, but its range of usefulness is limited.

Setscrews and Keyways. When a setscrew is used in combination with a key, the screw diameter should be equal to the width of the key. In this combination the setscrew is locating the parts in an axial direction only. Torsional load on the parts is carried by the key.

The key should be tight-fitting, so that no motion is transmitted to the screw. Under high reversing or alternating loads, a poorly fitted key will cause the screw to back out and lose its clamping force.

Keeping Fasteners Tight[2]

Fasteners are inexpensive, but the cost of installing them can be substantial. Probably the simplest way to cut assembly costs is to make sure that, once installed, fasteners stay tight.

The American National Standards Institute has identified three basic locking methods: free-spinning, prevailing-torque, and chemical locking. Each has its own advantages and disadvantages. See Fig. 8-4-2.

Free-spinning devices include toothed and spring lockwashers and screws and bolts with washerlike heads. With these arrangements, the fasteners spin free in the clamping direction, which makes them easy to assemble, and the break-loose torque is greater than the seating

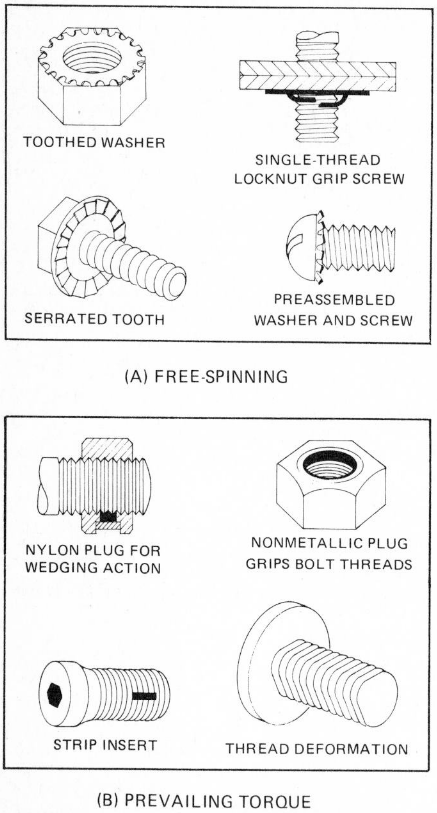
TOOTHED WASHER
SINGLE-THREAD LOCKNUT GRIP SCREW
SERRATED TOOTH
PREASSEMBLED WASHER AND SCREW
(A) FREE-SPINNING
NYLON PLUG FOR WEDGING ACTION
NONMETALLIC PLUG GRIPS BOLT THREADS
STRIP INSERT
THREAD DEFORMATION
(B) PREVAILING TORQUE

Fig. 8-4-2 Typical setscrew installation locking fasteners.

torque. However, once break-loose torque is exceeded, free-spinning washers have no prevailing torque to prevent further loosening.

Prevailing-torque methods make use of increased friction between nut and bolt. Metallic types usually have deformed threads or contoured thread profiles that jam the threads on assembly. Nonmetallic types make use of nylon or polyester insert elements that produce interference fits on assembly.

Chemical locking is achieved by coating the fastener with an adhesive. These adhesive chemicals can be formulated for different strengths, can be used on any size fastener, and can reduce the number of fasteners that must be carried in inventory.

Locknuts[3]

A *locknut* is a nut with special internal means for gripping a threaded fastener or connected material to prevent rotation in use. Generally it has the dimensions, mechanical requirements, and other specifications of a standard nut, but with a locking feature added.

Locknuts are divided into two general classifications: prevailing-torque and

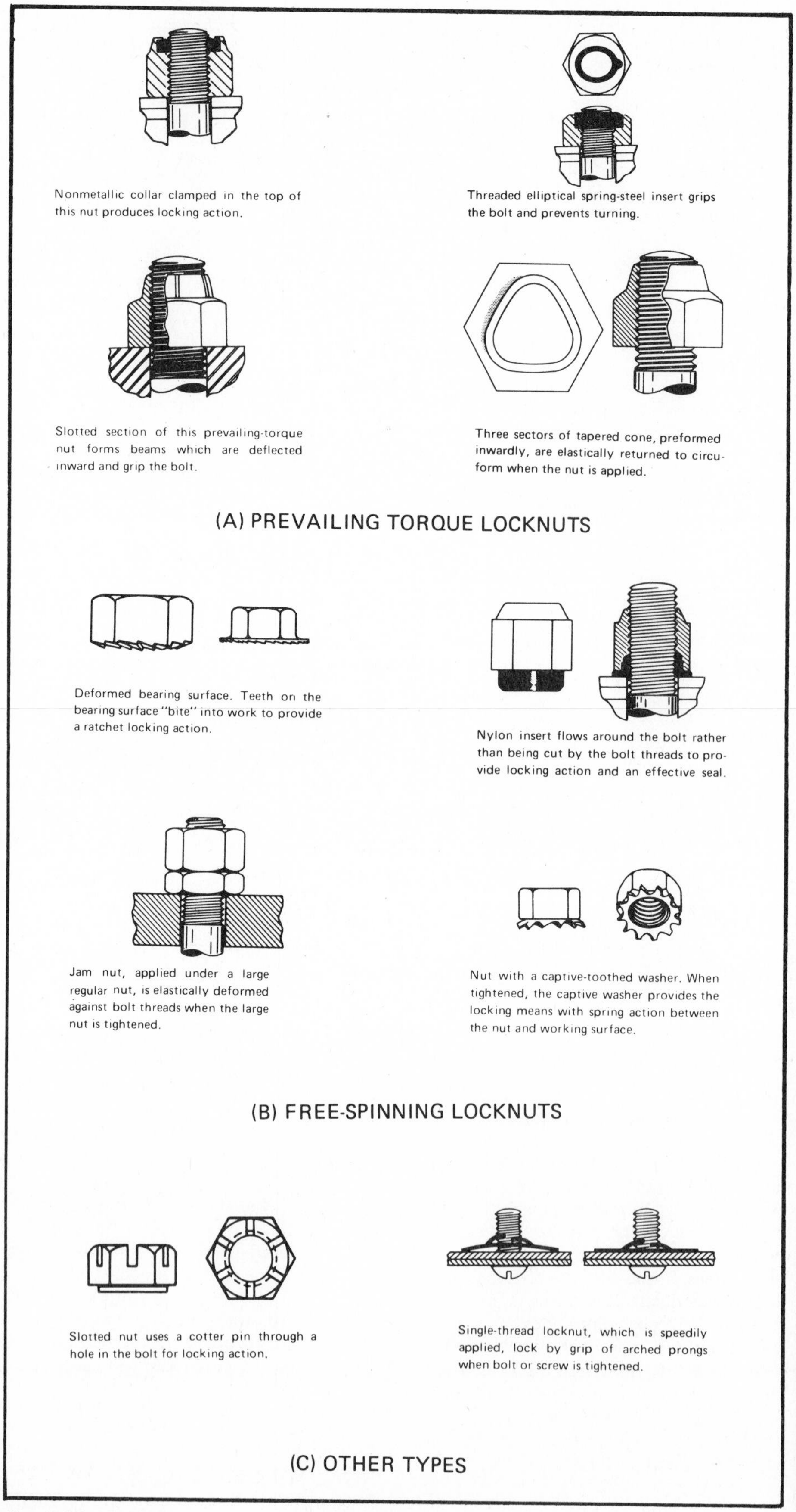
Nonmetallic collar clamped in the top of this nut produces locking action.
Threaded elliptical spring-steel insert grips the bolt and prevents turning.
Slotted section of this prevailing-torque nut forms beams which are deflected inward and grip the bolt.
Three sectors of tapered cone, preformed inwardly, are elastically returned to circuform when the nut is applied.
(A) PREVAILING TORQUE LOCKNUTS
Deformed bearing surface. Teeth on the bearing surface "bite" into work to provide a ratchet locking action.
Nylon insert flows around the bolt rather than being cut by the bolt threads to provide locking action and an effective seal.
Jam nut, applied under a large regular nut, is elastically deformed against bolt threads when the large nut is tightened.
Nut with a captive-toothed washer. When tightened, the captive washer provides the locking means with spring action between the nut and working surface.
(B) FREE-SPINNING LOCKNUTS
Slotted nut uses a cotter pin through a hole in the bolt for locking action.
Single-thread locknut, which is speedily applied, lock by grip of arched prongs when bolt or screw is tightened.
(C) OTHER TYPES

Fig. 8-4-3 Locknuts.

free-spinning types. However, there are some types which do not fall exactly into these categories. These are shown in Fig. 8-4-3.

PREVAILING-TORQUE LOCKNUTS

Prevailing-torque locknuts spin freely for a few turns, and then must be wrenched to final position. The maximum holding and locking power is reached as soon as the threads and the locking feature are engaged. Locking action is maintained until the nut is removed. Prevailing-torque locknuts are classified by basic design principles:

1. Thread deflection causes friction to develop when the threads are mated; thus the nut resists loosening.
2. Out-of-round top portion of the tapped nut grips the bolt threads and resists rotation.
3. Slotted section of locknut is pressed inward to provide a spring frictional grip on the bolt.
4. Inserts, either nonmetallic or of soft metal, are plastically deformed by the bolt threads to produce a frictional interference fit.
5. Spring wire or pin engages the bolt threads to produce a wedging or ratchet-locking action.

FREE-SPINNING LOCKNUTS

Free-spinning locknuts are free to spin on the bolt until seated. Additional tightening locks the nuts.

Free-spinning locknuts are often specified when long travel of nut on bolt is unavoidable. Since most free-spinning locknuts depend on clamping force for their locking action, they are usually not recommended for joints that might relax through plastic deformation or for fastening materials that might crack or crumble when subjected to preload.

OTHER LOCKNUT TYPES

Jam nuts are thin nuts used under full-sized nuts to develop locking action. The large nut has sufficient strength to elastically deform the lead threads of the bolt and jam nut. Thus, a considerable resistance against loosening is built up. The use of jam nuts is decreasing; a one-piece, prevailing-torque locknut usually is used instead at a savings in assembled cost.

The jam nut is considered ideal for assemblies where long travel of nut on bolt under load is necessary to bring mating parts into position.

Slotted and castle nuts have slots which receive a cotter pin that passes through a drilled hole in the bolt and thus serves as the locking member. These nuts are essentially free-spinning nuts with the locking feature added after the preload condition is developed. Castle nuts differ from slotted nuts in that they have a circular crown of a reduced diameter.

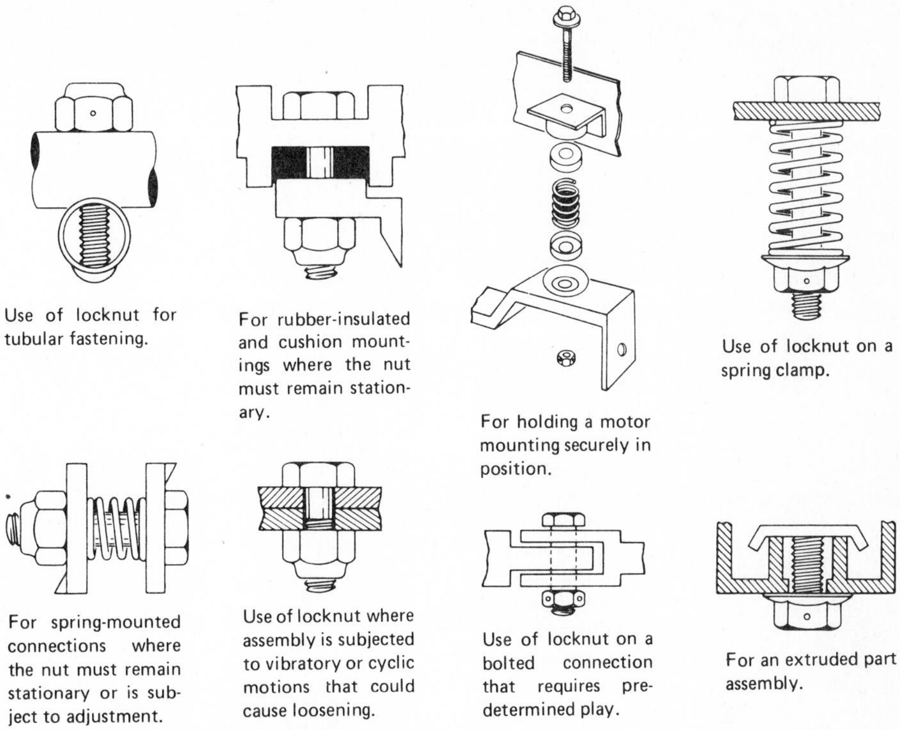
Use of locknut for tubular fastening.
For rubber-insulated and cushion mountings where the nut must remain stationary.
For holding a motor mounting securely in position.
Use of locknut on a spring clamp.
For spring-mounted connections where the nut must remain stationary or is subject to adjustment.
Use of locknut where assembly is subjected to vibratory or cyclic motions that could cause loosening.
Use of locknut on a bolted connection that requires predetermined play.
For an extruded part assembly.

Fig. 8-4-4 Typical locknut applications.

From an assembled-cost viewpoint, these nut forms are expensive because of the extra operations involved in their assembly. Usually a one-piece, prevailing-torque locknut is a better solution.

Single-thread engaging locknuts are spring steel fasteners which may be speedily applied. Locking action is provided by the grip of the thread-engaging prongs and the reaction of the arched base. Their use is limited to nonstructural assemblies and usually to screw sizes below 6 mm in diameter. Compared with multiple-thread locknuts, these types are less expensive and lighter in mass. The spring steel provides a resilient assembly which will absorb a certain amount of motion. Tightening torques for these nuts, however, must be lower than for multiple-thread locknuts.

Captive or Self-Retaining Nuts[4]

Self-retained or captive nuts provide a permanent, strong, multiple-threaded fastener in many types of thin materials. See Fig. 8-4-5. They are especially good where there are blind locations, and they can generally be attached without damaging finishes. Methods of attaching these types of nuts vary and tools required for assembly are generally uncomplicated and inexpensive. In this section, the self-retained nuts are grouped according to four means of attachment:

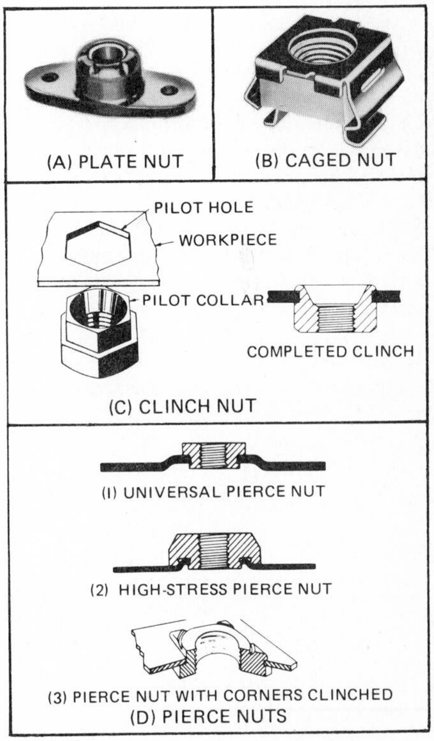
(A) PLATE NUT
(B) CAGED NUT
PILOT HOLE
WORKPIECE
PILOT COLLAR
COMPLETED CLINCH
(C) CLINCH NUT
(1) UNIVERSAL PIERCE NUT
(2) HIGH-STRESS PIERCE NUT
(3) PIERCE NUT WITH CORNERS CLINCHED
(D) PIERCE NUTS

Fig. 8-4-5 Captive or self-retaining nuts.

1. Plate or anchor nuts: These nuts have mounting lugs which can be riveted, welded, or screwed to the part.

2. Caged nuts: A spring-steel cage retains a standard nut. The cage snaps into a hole or clips over an edge to hold the nut in position.

3. Clinch nuts: They are specially designed nuts with pilot collars which are clinched or staked into the parent part through a precut hole.

4. Self-piercing nuts: A form of clinch nut that cuts its own hole.

PLATE NUTS

Plate nuts, commonly referred to as *anchor* nuts, have one or more lugs projecting from the base of the threaded body. The nuts are attached by riveting or welding the lugs to the work surface.

Plate nuts are divided into two general categories: nonfloating (fixed) and floating. Fixed plate nuts are generally one-piece units. The physical requirements of the application will dictate the most suitable lug configuration.

Floating plate nuts are multipiece units which permit permanent retainer mounting while allowing the nut to move radially within the retainer.

CAGED NUTS

A multiple-threaded nut enclosed within a spring-steel retainer is a fastening device that has the high-strength characteristics associated with multiple-threaded fasteners and the versatility and self-retaining features of spring-steel fasteners.

Multiple-thread nut retainers are particularly useful in blind fastening locations. Their self-retaining feature eliminates the need for welding, clinching, or staking nuts in place. They can be snapped into place at any convenient spot along the production line.

CLINCH NUTS

A clinch nut is a solid nut having a pilot or other feature designed to be inserted into a preformed hole and permanently clinched to the parent material. Clinch nuts are used to provide a strong multiple-threaded fastener in materials too thin to be extruded or tapped and are well adapted for inaccessible (blind) assembly locations.

SELF-PIERCING NUTS

A pierce nut is a one-piece metal nut with multiple internal threads, an external undercut section, and a work-hardened steel body for punching its own mounting hole. The nut is installed in a workpiece by an insertion head mounted in a punch press or brake.

The piercing impact forces metal from the workpiece into the nut undercuts and locks the nut securely in place. The result is an assembly in which the fastener is precisely positioned as an integral part of the component. Pierce nuts can either pierce and be clinched simultaneously or be clinched into prepierced holes.

Single-Thread Engaging Nuts[5]

Single-thread engaging nuts are formed by stamping a thread-engaging impression in a flat piece of metal. The stamped impression can take a number of shapes; for example, shear-formed helical prongs engage and lock on the screw-thread root diameter (Fig. 8-4-6*a*), or a protruding truncated cone (Fig. 8-4-6*b*) stamped into the metal provides a ramp that the screw climbs as it turns. Another type of impression has the thread-engaging elements spirally formed to match the pitch of the screw threads (Fig. 8-4-6*c*).

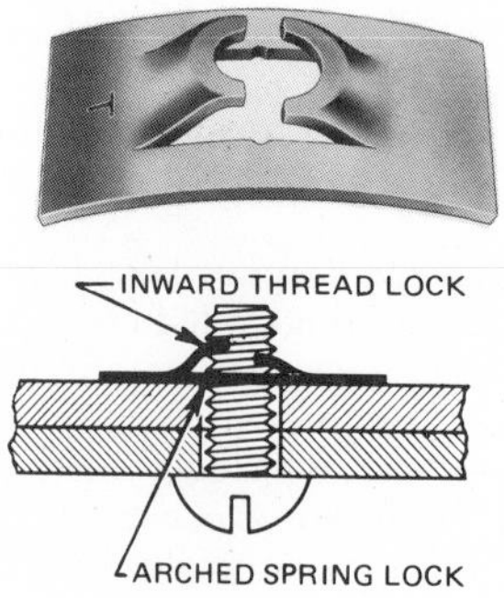
INWARD THREAD LOCK
ARCHED SPRING LOCK
(A) FLAT TYPE

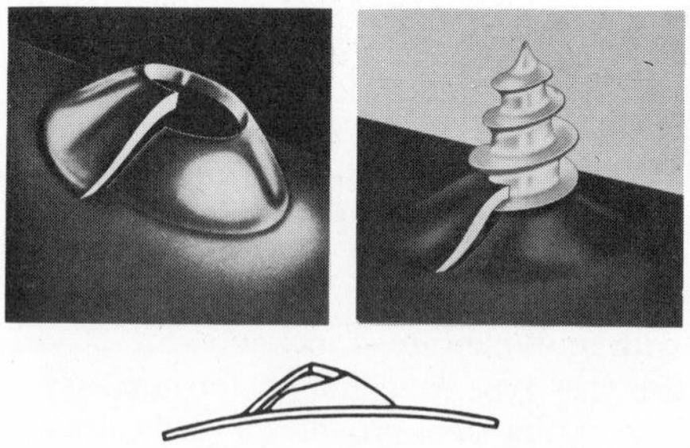
(B) FLAT-TYPE CONICAL THREAD

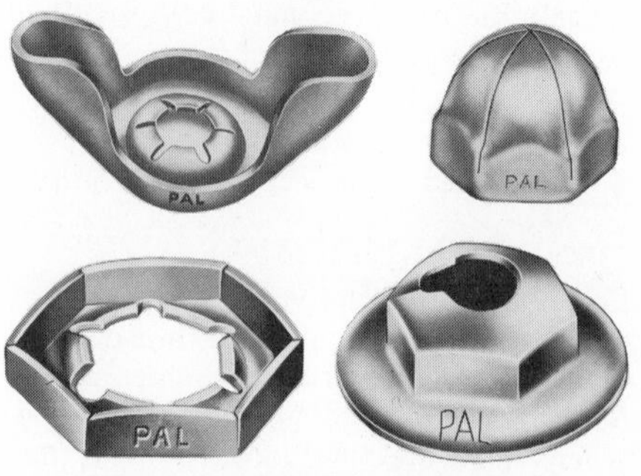
PAL
PAL
PAL
PAL
(C) SPIRAL-FORMED THREAD

Fig. 8-4-6 Single-thread engaging nuts.

Inserts[6]

Inserts are a special form of nut designed to serve the function of a tapped hole in blind or through-hole locations. See Fig. 8-4-7. They are sometimes referred to as *solid bushings*. Another basic type of screw-thread insert consists of precision-formed wire, spirally coiled to provide threads of proper form for installation in a tapped hole. It is known as a *wire insert*.

The internal thread of the insert is usually of standard size and thread form as governed by the screw-thread industry. The external configuration of the insert is designed to suit its particular purpose and may incorporate threads, flanges, grooves, knurls, or other shapes for retention purposes.

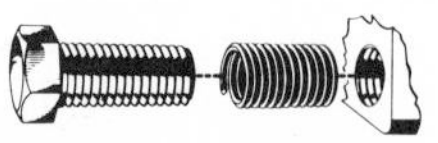
(A) EXTERNAL-INTERNAL THREADED INSERT

(B) SELF-TAPPING INSERT

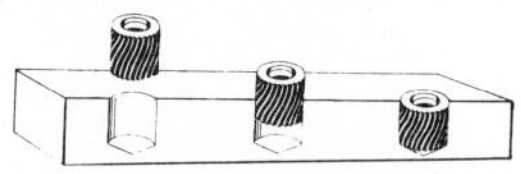
(C) PRESSED-IN INSERT

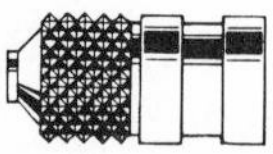
(D) MOLDED-IN INSERT

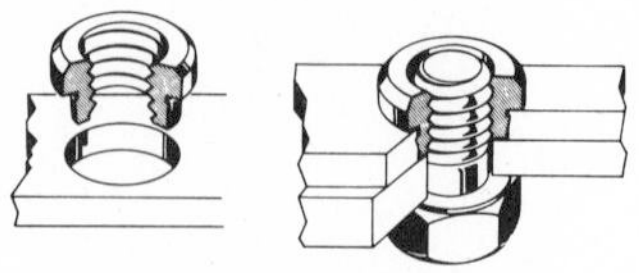
(E) THIN MATERIAL INSERT

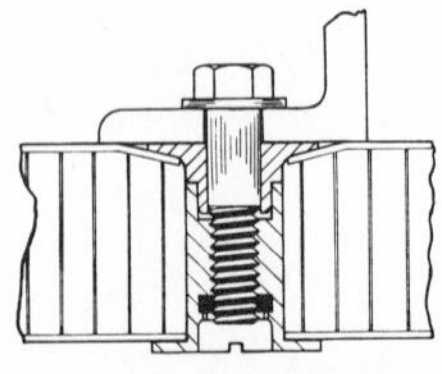
(F) SANDWICH PANEL INSERT

Fig. 8-4-7 Inserts.

Sealing Fasteners[7]

Fasteners hold two or more parts together, but they can perform other functions as well. One important auxiliary function is that of sealing gases and liquids against leakage.

Two types of sealed-joint constructions are possible with fasteners. See Fig. 8-4-8. In one approach, the fasteners enter the sealed medium and are separately sealed. A number of fastener designs with built-in sealing elements have been developed for this purpose.

The second approach uses a separate sealing element which is held in place by the clamping forces produced by conventional fasteners, such as rivets or bolts.

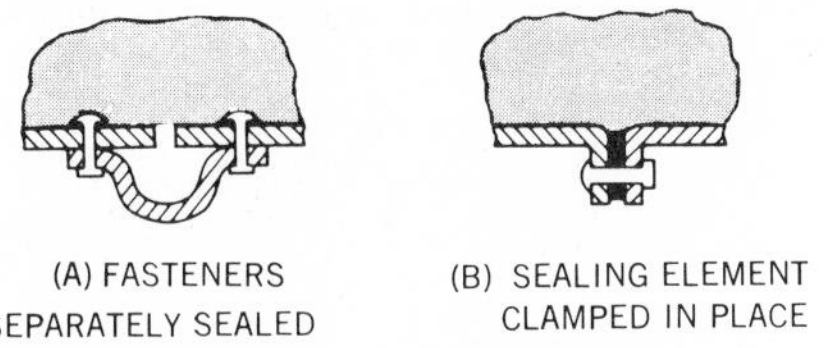
(A) FASTENERS SEPARATELY SEALED
(B) SEALING ELEMENT CLAMPED IN PLACE

Fig. 8-4-8 Types of sealed-joint construction.

SEALING FASTENER TYPES

There are many methods of obtaining a seal using sealing fasteners, as shown in Fig. 8-4-9. The more common sealing fasteners are screws, rivets, nuts, and washers.

Screws. Most sealing screws provide a sealing action by the use of a resilient material added to the screw. It may be rubber, plastic, mastic, or metal. In addition, some screws make use of an interference fit in the threads.

Rivets. Sealing rivets are used extensively in aircraft and ships. They seal integral tanks, pressurized cabins, and pontoons and are used in other critical applications.

Nuts. Starting with the simple device of adding a soft metal washer under a cap nut to effect a seal, the range of sealing nuts available today has steadily increased. Mastic, rubber, plastic, and soft metals are all used as seals in a wide variety of types.

Washers. Sealing washers are used under either the screw head or the nut. Cut rubber rings, O-rings, or molded-in sections are most commonly used. More recently, flowed-in sealants and nylon rings or sleeves have been made available.

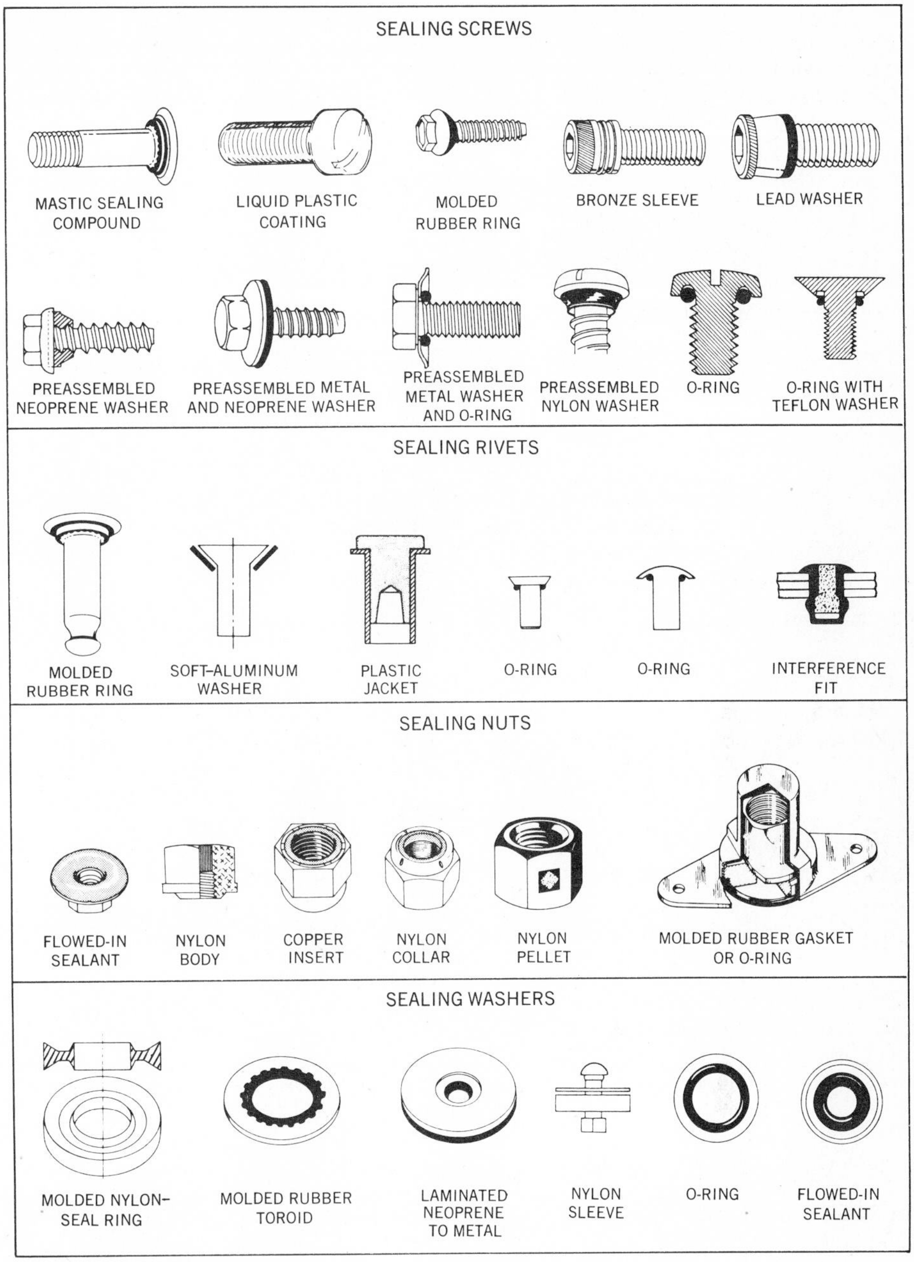
SEALING SCREWS
MASTIC SEALING COMPOUND
LIQUID PLASTIC COATING
MOLDED RUBBER RING
BRONZE SLEEVE
LEAD WASHER
PREASSEMBLED NEOPRENE WASHER
PREASSEMBLED METAL AND NEOPRENE WASHER
PREASSEMBLED METAL WASHER AND O-RING
PREASSEMBLED NYLON WASHER
O-RING
O-RING WITH TEFLON WASHER
SEALING RIVETS
MOLDED RUBBER RING
SOFT-ALUMINUM WASHER
PLASTIC JACKET
O-RING
O-RING
INTERFERENCE FIT
SEALING NUTS
FLOWED-IN SEALANT
NYLON BODY
COPPER INSERT
NYLON COLLAR
NYLON PELLET
MOLDED RUBBER GASKET OR O-RING
SEALING WASHERS
MOLDED NYLON-SEAL RING
MOLDED RUBBER TOROID
LAMINATED NEOPRENE TO METAL
NYLON SLEEVE
O-RING
FLOWED-IN SEALANT

Fig. 8-4-9 Sealing fasteners.

REFERENCES AND SOURCE MATERIAL

1. F. R. Kull, "Set Screws," *Machine Design*, vol. 37, no. 6, 1965.
2. R. Batson et al., "Keeping Fasteners Tight," *Machine Design*, vol. 47, no. 23, 1975.
3. R. B. Belford et al., "Locknuts," *Machine Design*, vol. 37, no. 6, 1965.
4. M. Mihaly; W. C. Seitz; S. Petrus; P. D. Massey; and J. H. Stewart, "Capture or Self-Retaining Nuts," *Machine Design*, vol. 37, no. 6, 1965.
5. W. L. Seitz and S. Petrus, "Single-Thread Engaging Nuts," *Machine Design*, vol. 37, no. 6, 1965.
6. A. H. Mussgnug, "Inserts," *Machine Design*, vol. 37, no. 6, 1965.
7. R. B. Belford et al., "Sealing Fasteners," *Machine Design*, vol. 37, no. 6, 1965.

Assignments

1. On an A3- or B-size sheet, make a two-view assembly drawing of the crane hook shown in Fig. 8-4-A or 8-4-B. The hook is to be held to the U-frame with a slotted locknut. A spring pin is inserted through the locknut slots to prevent the nut from turning. A clevis pin with washer and cotter pin holds the pulley to the frame. Include on the drawing a bill of material. Scale is 1:1.

REVIEW FOR ASSIGNMENTS

Unit 7-5 Shafts in Section
Unit 9-2 Pin Fasteners

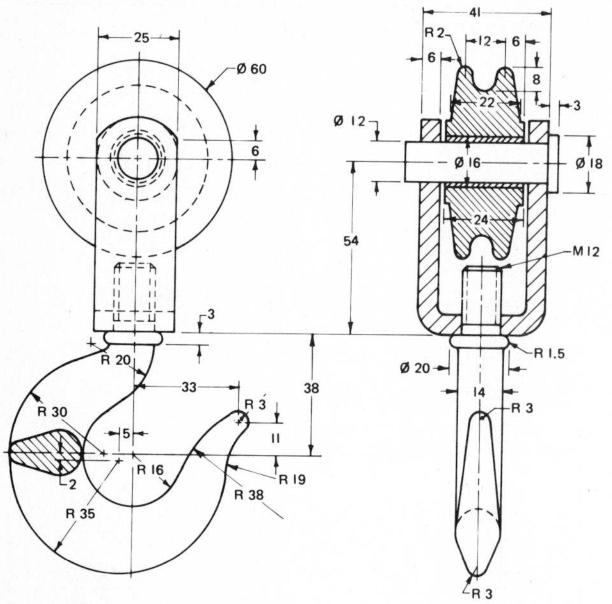

25
Ø 60
6
3
R 20
33
R 30
R 3
38
5
11
R 16
R 19
2
R 38
R 35
41
R 2
12
6
8
22
3
Ø 12
Ø 16
Ø 18
54
24
M12
R 1.5
Ø 20
14
R 3
R 3

Fig. 8-4-A Crane hook.

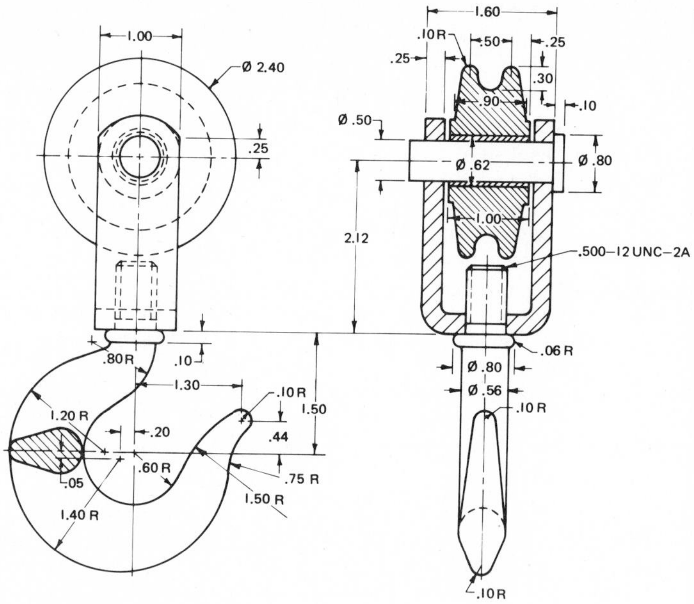

1.00
Ø 2.40
.25
.80R
.10
1.30
.10 R
1.20 R
1.50
.20
.44
.05
.60 R
.75 R
1.50 R
1.40 R
1.60
.10R
.50
.25
.25
.30
.90
.10
Ø .50
Ø .62
Ø .80
1.00
2.12
.500–12 UNC–2A
.06 R
Ø .80
Ø .56
.10 R
.10R

Fig. 8-4-B Crane hook.

UNIT 8-5

FASTENERS FOR LIGHT-GAGE METAL, PLASTIC, WOOD, AND MASONRY

Tapping Screws[1]

Tapping screws cut or form a mating thread when driven into drilled or cored holes. These one-piece fasteners permit rapid installation, since nuts are not used and access is required from only one side of the joint. The mating thread produced by the tapping screw fits the screw threads closely, and no clearance is necessary. This close fit usually keeps the screw tight, even under vibrating conditions. See Fig. 8-5-1.

Tapping screws are practically all case-hardened and, therefore, can be driven tight and have a relatively high ultimate torsional strength. The screws are used in steel, aluminum (cast, extruded, rolled, or die-formed) die castings, cast iron, forgings, plastics, reinforced plastics, asbestos, and resin-impregnated plywood. See Fig. 8-5-2.

Types C, D, F, G, and T tapping screws are available in both coarse- and fine-thread series. Coarse threads should be used with weak materials. Fine threads are recommended if two or more full threads of engagement must be above the top of the cutting slot but the thickness of materials is insufficient to allow two full threads of the coarse-thread series.

Self-drilling tapping screws, types BSD and CSD, have special points for drilling and then tapping their own holes (Fig. 8-5-3). These eliminate drilling or punching, but they must be driven by a power screwdriver. Once the self-drilling screw pierces the metal, it forms or cuts threads the same way standard tapping screws do.

SPECIAL TAPPING SCREWS

Typical specials are self-captive tapping screws and double-thread combinations for limited drive. Self-captive screws combine a coarse-pitched starting thread (similar to type B) with a finer pitch (machine-screw thread) farther along the screw shank. Initially, the coarse threads cut the hole, but then the fine threads retap the hole and change the thread pitch. Thus, attempts to back the screw out will cause the coarse thread to interfere with the fine threads, and consequently the screw is held captive.

Sealing tapping screws, with preassembled washers or O-rings (Fig. 8-5-4*b*) can be used to control leaks, squeaks, crazing of enamel, and electrolysis in all types of metal structures and assemblies.

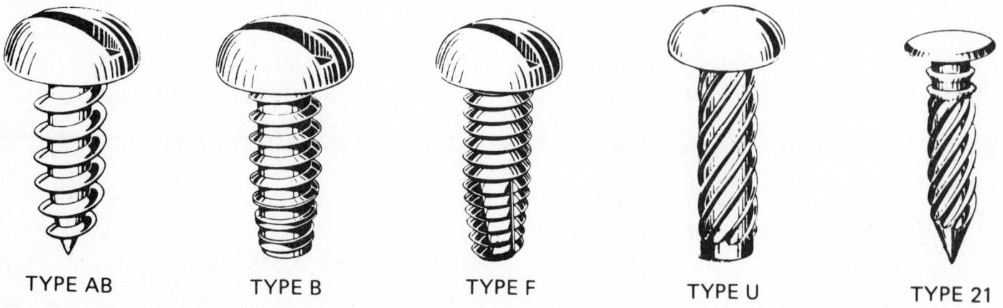

TYPE AB
TYPE B
TYPE F
TYPE U
TYPE 21

Fig. 8-5-1 Self-tapping screws.

Masonry Fasteners[2]

Masonry anchoring devices or methods for securing things to concrete, brick, stone, or plaster constitute a critically important element in all modern construction. Concrete anchors are the mainstay in securing practically everything that must be fastened to the concrete skeleton of a building—pipes, sprinkler systems, curtain walls, suspended ceilings—just to name several of the scores of vital applications. See Fig. 8-5-5.

Anchoring devices requiring predrilled holes involve the use of ordinary hand tools such as hammers and drills or portable power tools such as electric drills and impact hammers. The great majority of anchors used today consist of variations of this type.

Tapping Screw Assembly. The steel post is fastened to the panel by two rows of tapping screws. The steel strap is held

HEAVY GAGE SHEET METAL AND STRUCTURAL STEEL.
USE TYPES B, BP, C, U, D, F, G, T.

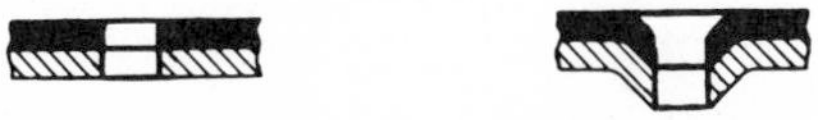

Holes may be drilled or clean-punched.

Two parts may have pierced holes to nest burrs. This results in a stronger joint.

Use a pierced hole in workpiece if clearance hole is needed in part to be fastened.

Extruded hole may also be used in workpiece if clearance hole is needed in fastened part.

LIGHT GAGE SHEET METAL.
USE TYPES AB, B, BP, C.

Holes may be drilled or clean-punched the same size in both sheet metal parts. For thicker sheet metal and structural steel, a clearance hole should be provided in the part to be fastened. Hole size depends on thickness of the workpiece.

Notes:
1. Use hex-head on Type B and BP screws.
2. Type C screw for sheet metal only; maximum thickness, 3.5 mm.
3. With Type U screws, material should be thick enough to permit sufficient thread engagement—at least one screw diameter.

PLASTICS.
USE TYPES B, BP, U, D, F, G, T, BF, BT.

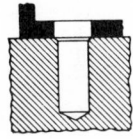

Screw holes may be molded or drilled. If material is brittle or friable, molded holes should be formed with a rounded chamfer, and drilled holes should be machine chamfered. Provide a clearance in the part to be fastened. Depth of penetration should be held within the "minimum and maximum" limits recommended. The hole should be deeper than the screw penetration to allow for chip clearance.

CASTINGS AND FORGINGS.
USE TYPES B, BP, U, D, F, G, T, BF, BT.

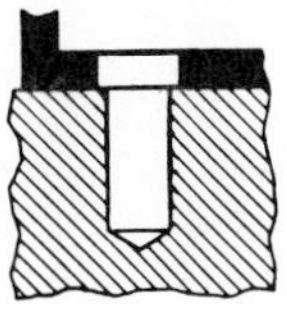

Holes may be cored if it is practical to maintain close tolerances. Otherwise blind drill holes to recommended hole size. Provide a clearance hole for screw in the part to be fastened. The hole in the casting,, if it is a blind hole, should be deeper than the screw penetration to allow for chip clearance.

Notes:
1. Hole in fastened part may be the same size as workpiece hole for Type U screws.
2. Types B, BP, BF, BT are only suitable for use in nonferrous castings.

Fig. 8-5-2 Tapping-screw application chart.

to the post by a single tapping screw which has the equivalent strength (body area) of at least three of the other tapping screws.

Wood Fasteners. The steel post is held to the wood furring with no. 8 RHWS. For safety reasons, the screw holding the steel strap must be flush with the face of the strap. A no. 10 wood screw is required.

Masonry Fasteners. The hooks must be anchored to the masonry wall at a

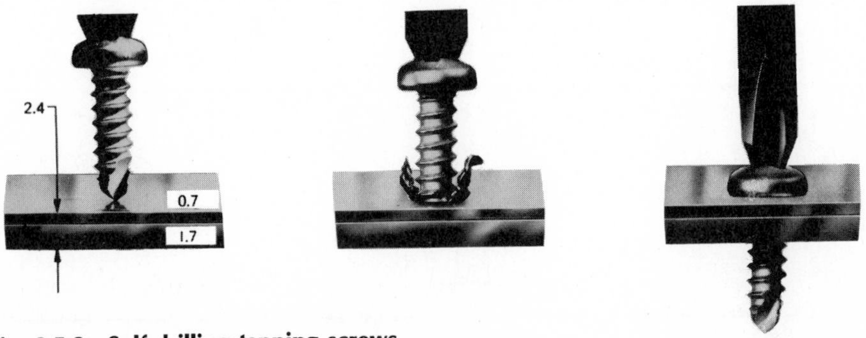
2.4
0.7
1.7

Fig. 8-5-3 Self-drilling tapping screws.

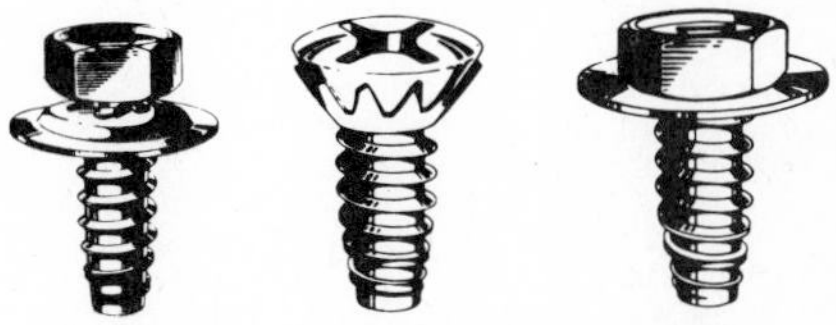

(A) TAPPING SCREWS WITH PREASSEMBLED WASHERS. THESE ARE AVAILABLE IN A GREAT VARIETY OF THREAD FORMS, HEAD STYLES, AND WASHER CONFIGURATIONS.

(B) TAPPING SCREWS WITH PREASSEMBLED SEALING WASHERS OR COMPOUNDS.

Fig. 8-5-4 Special tapping screws.

spacing which will have hooks anchored into hollow masonry wall. The diameter of the masonry fasteners should be 6 mm or its equivalent.

REFERENCES AND SOURCE MATERIAL

1. R. B. Belford et al., "Tapping Screws," *Machine Design*, vol. 37, no. 6, 1965.
2. Industrial Fasteners Institute.

Assignments

1. Divide an A3- or B-size sheet into three parts, and draw appropriate section views of the assemblies shown in Fig. 8-5-A or 8-5-B. In each assembly indicate the hole and fastener sizes.

REVIEW FOR ASSIGNMENTS

Unit 7-1 Section Drawings

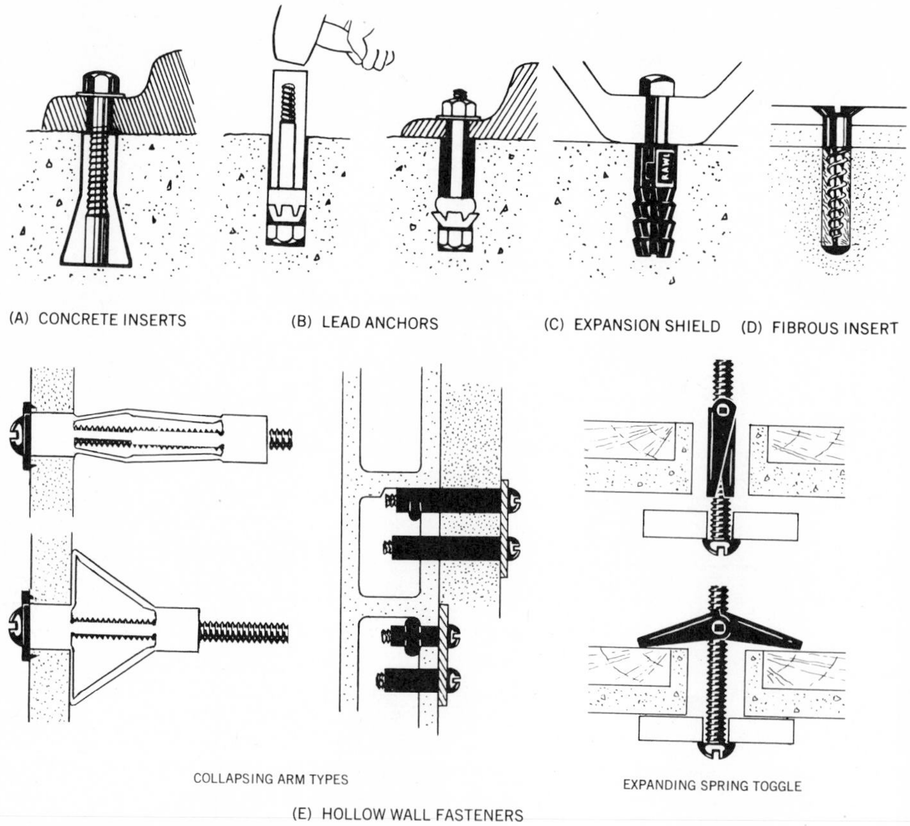

(A) CONCRETE INSERTS
(B) LEAD ANCHORS
(C) EXPANSION SHIELD
(D) FIBROUS INSERT
COLLAPSING ARM TYPES
EXPANDING SPRING TOGGLE
(E) HOLLOW WALL FASTENERS

Fig. 8-5-5 Typical masonry fasteners.

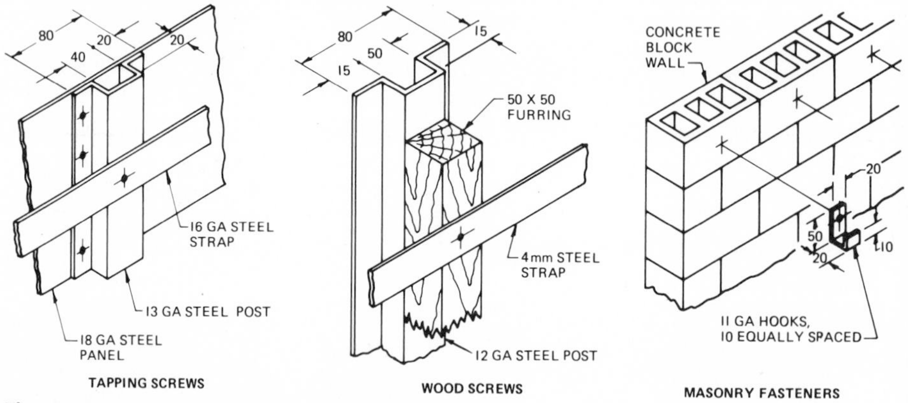

80
40
20
20
16 GA STEEL STRAP
13 GA STEEL POST
18 GA STEEL PANEL
TAPPING SCREWS
80
50
15
15
50 X 50 FURRING
4 mm STEEL STRAP
12 GA STEEL POST
WOOD SCREWS
CONCRETE BLOCK WALL
20
50
20
10
11 GA HOOKS, 10 EQUALLY SPACED
MASONRY FASTENERS

Fig. 8-5-A Special fastener problems.

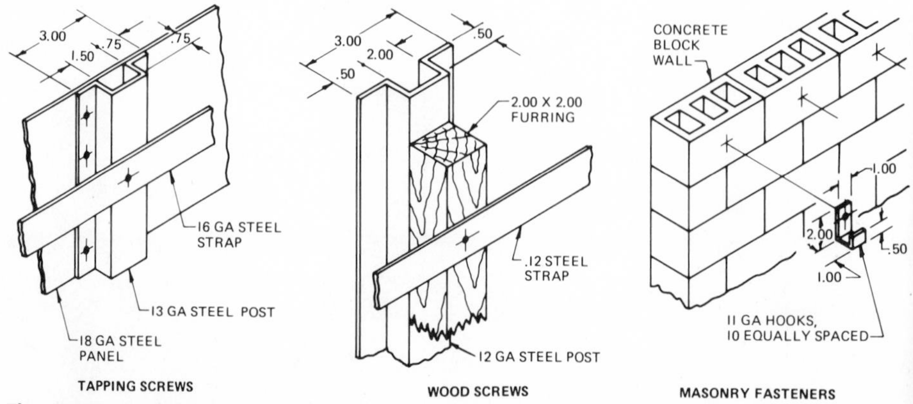

3.00
1.50
.75
.75
16 GA STEEL STRAP
13 GA STEEL POST
18 GA STEEL PANEL
TAPPING SCREWS
3.00
2.00
.50
.50
2.00 X 2.00 FURRING
.12 STEEL STRAP
12 GA STEEL POST
WOOD SCREWS
CONCRETE BLOCK WALL
1.00
2.00
.50
1.00
11 GA HOOKS, 10 EQUALLY SPACED
MASONRY FASTENERS

Fig. 8-5-B Special fastener problems.

Chapter 9
Miscellaneous Types of Fasteners

UNIT 9-1
KEYS, SPLINES, AND SERRATIONS

Keys

A *key* is a piece of steel lying partly in a groove in the shaft, called a *keyseat*, and extending into another groove, called a *keyway*, in the hub. It is used to secure gears, pulleys, cranks, handles, and similar machine parts to shafts, so that the motion of the part is transmitted to the shaft, or the motion of the shaft to the part, without slippage. The key may also

Fig. 9-1-1 Miscellaneous types of fasteners. (*Design Engineering*, Oct. 1968.)

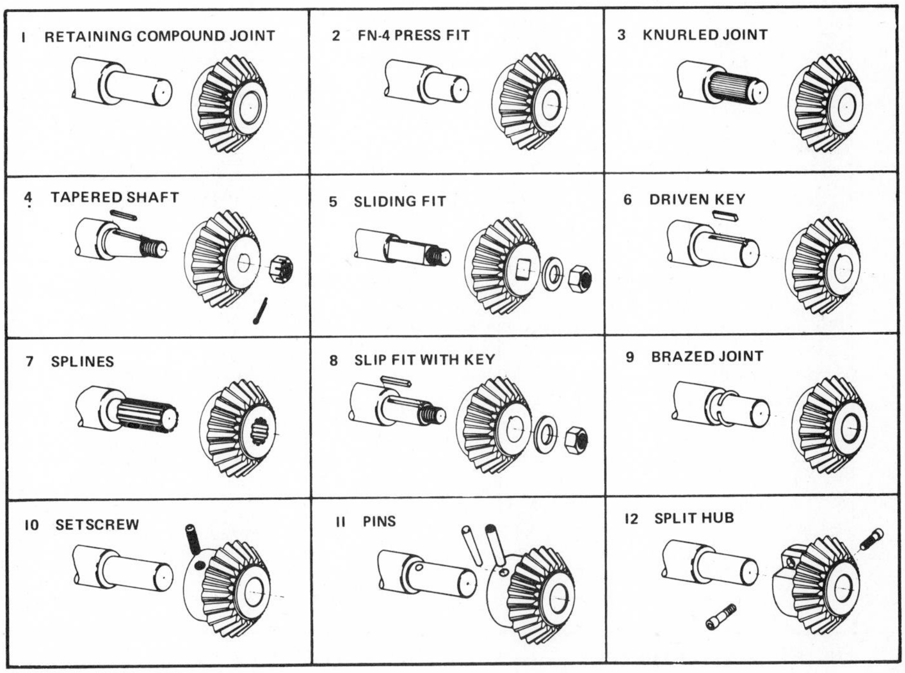

1 RETAINING COMPOUND JOINT
2 FN-4 PRESS FIT
3 KNURLED JOINT
4 TAPERED SHAFT
5 SLIDING FIT
6 DRIVEN KEY
7 SPLINES
8 SLIP FIT WITH KEY
9 BRAZED JOINT
10 SETSCREW
11 PINS
12 SPLIT HUB

act in a safety capacity; its size is generally calculated so that when overloading takes place, the key will shear or break before the part or shaft breaks.

There are many kinds of keys. The most common types are shown in Fig. 9-1-2. Square and flat keys are widely used in industry. The width of the square and flat key should be approximately one-quarter the shaft diameter, but for proper key selection refer to the Appendix. These keys are also available with a 1:100 taper on their top surfaces and are known as *square taper* or *flat tapered* keys. The keyway, not the keyseat, is tapered to accommodate the taper on the key.

The gibhead key is the same as the square or flat tapered key but has a head added for easy removal.

The Pratt and Whitney key is rectangular with rounded ends. Two-thirds of this key sits in the shaft, one-third sits in the hub.

The Woodruff key is semicircular and fits into a semicircular keyseat in the shaft and a rectangular keyway in the hub. The width of the key should be approximately one-quarter the diameter of the shaft, and its diameter should approximate the diameter of the shaft. Half the width of the key extends above the shaft and into the hub. Refer to the Appendix for exact sizes. Woodruff keys are identified by a number which indicates the nominal dimensions of the key. The numbering system which originated many years ago is identified with the fractional-inch system of measurement. The last two digits of the number give the normal diameter in eighths of an inch, and the digits preceding the last two give the nominal width in thirty-seconds of an inch. For example, a no. 1210 Woodruff key indicates a key $^{12}/_{32} \times {}^{10}/_{8}$ in., or a $^{3}/_{8} \times 1^{1}/_{4}$ in. key.

In calling up keys on a bill of material, only the information shown in the column "Specifications" in Fig. 9-1-2 need be given.

DIMENSIONING OF KEYWAYS AND KEYSEATS

All dimensions of keyways and keyseats for square and flat keys, with the exception of the length of the flat portion of the keyseat which is given by a direct dimension on the drawing, are shown on the drawing by a note specifying first the width and then the depth. This type of dimensioning is the standard method used for unit production where the machinist is expected to fit the key into the keyway and keyseat.

The width of all keyways and keyseats is the nominal width. The depth of all keyways and keyseats, as given on the drawing, may vary with the type of key, but it is based on the nominal key height.

The depth of all keyseats in shafts as shown on the drawing is the nominal depth $H/2$. See Fig. 9-1-3.

The depth of plain parallel keyways in hubs, which is shown on the drawing, is the nominal depth $H/2$, plus an allowance.

For interchangeable-assembly and mass-production purposes, keyway and keyseat dimensions are given in limit dimensions to ensure proper fits and are located from the opposite side of the hole or shaft. See Fig. 9-1-4.

Tapered Keyways. The depth of tapered keyways in hubs, which is shown on the drawing, is the nominal depth $H/2$ minus an allowance. This is always the depth at the large end of the tapered keyway and is indicated on the drawing by the abbreviation LE.

The radii of fillets, when required, must be dimensioned on the drawing, for example, 12 × 6 × RI.

Since standard milling cutters for Woodruff keys have the same appropriate number, it is possible to call for a Woodruff keyway or keyseat by the number only.

Where it is desirable to detail Woodruff keyseats on a drawing, all dimensions are given in the form of a note in the following order: width, depth, and radius of cutter.

Woodruff keyways may alternately be dimensioned in the same manner as for square and flat keys, specifying first the width and then the depth. See Fig. 9-1-5.

TYPE OF KEY	ASSEMBLY SHOWING KEY, SHAFT AND HUB	SPECIFICATION	DIMENSIONING OF	
			KEY SEAT	KEYWAY*
SQUARE 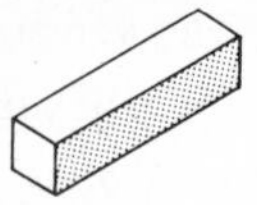	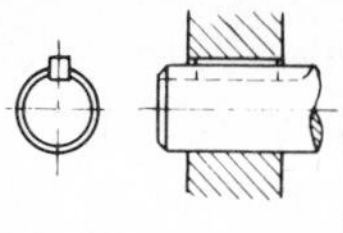	6 SQUARE KEY, 30 LG OR 6 SQUARE TAPERED KEY, 30 LG	6 x 3 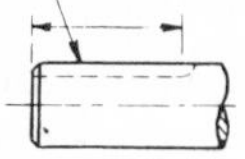	6 x 3 OR LE 6 x 3 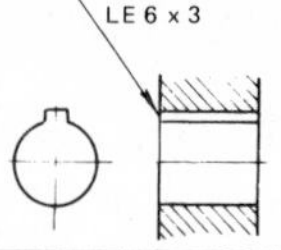
FLAT 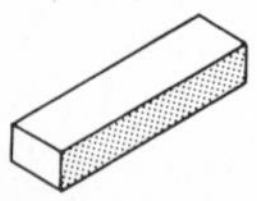	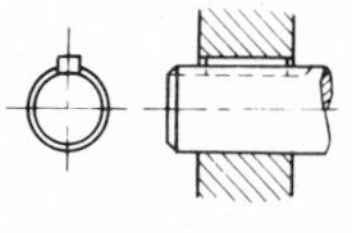	5 x 3 FLAT KEY, 25 LG OR 5 x 3 FLAT TAPERED KEY, 25 LG	5 x 1.5 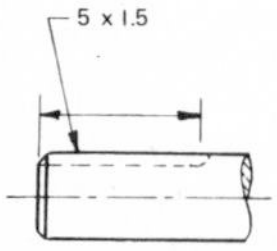	5 x 1.5 OR LE 5 x 1.5 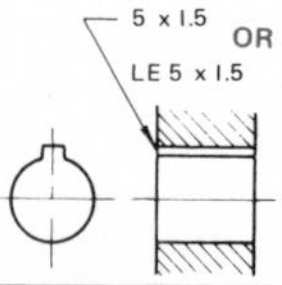
GIB-HEAD 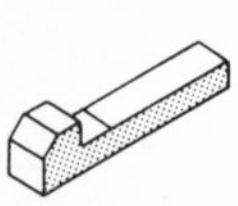	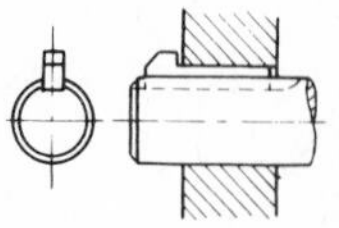	10 SQUARE GIB-HEAD KEY, 50 LG	10 x 5 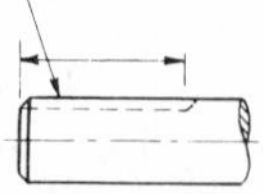	LE 10 x 4.5 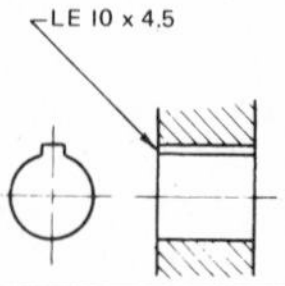
PRATT AND WHITNEY 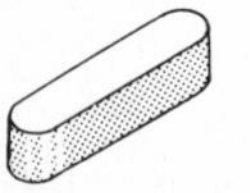	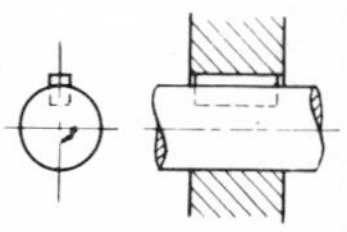	NO. 15 PRATT AND WHITNEY KEY	6 x 6 FOR NO. 15 PRATT & WHITNEY KEY 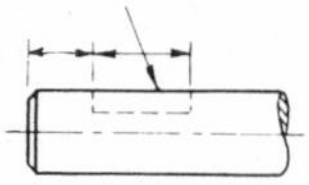	6 x 3 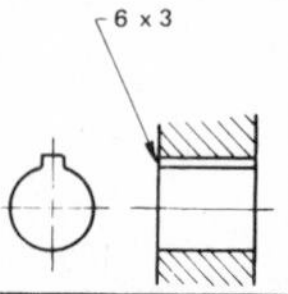
WOODRUFF 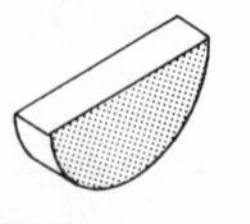	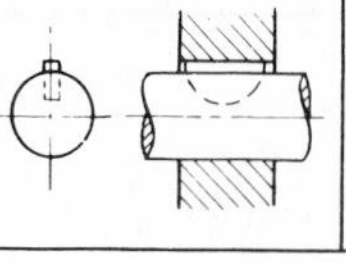	NO. 1210 WOODRUFF KEY	#1210 WOODRUFF KEYSEAT 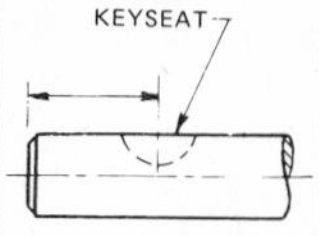	#1210 WOODRUFF KEYWAY 

*NOTE: LE REFERS TO LARGE END FOR TAPERED KEYS

Fig. 9-1-2 Common keys.

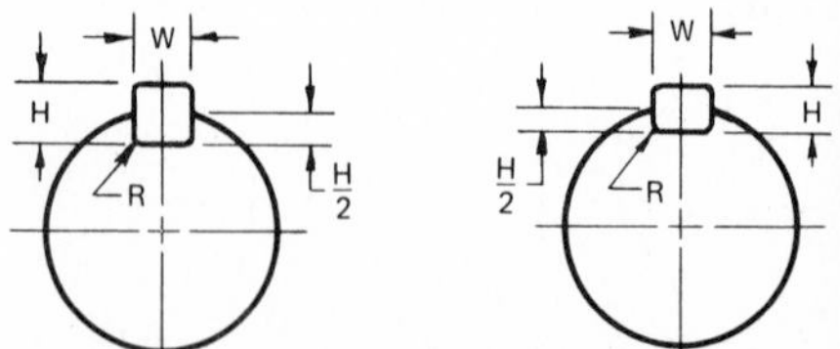
W
H
R
H/2
W
H
H/2
R

Fig. 9-1-3 Methods of establishing nominal depth of keyways and keyseats.

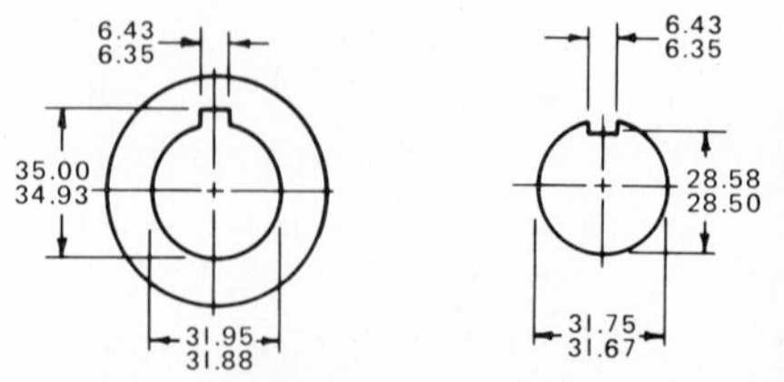
6.43
6.35
35.00
34.93
31.95
31.88
6.43
6.35
28.58
28.50
31.75
31.67

Fig. 9-1-4 Dimensioning keyways and keyseats for interchangeable assembly.

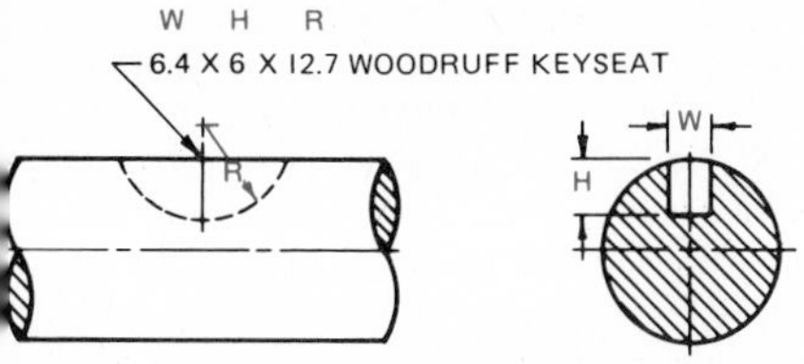
W H R
6.4 X 6 X 12.7 WOODRUFF KEYSEAT
W
H
R

Fig. 9-1-5 Alternate method of detailing a woodruff keyseat.

Splines and Serrations

A *spline shaft* is a shaft having multiple grooves, or keyseats, cut around its circumference for a portion of its length, in order that a sliding engagement may be made with corresponding internal grooves of a mating part.

Spline shafts are capable of carrying heavier loads than keys, permit lateral movement of a part while maintaining positive rotation, and allow the attached part to be indexed or changed to another angular position.

Splines have either straight-sided teeth or curved-sided teeth. The latter type is known as an *involute spline*.

Involute Splines. These splines are similar in shape to involute gear teeth but have pressure angles of 30, 37.5, or 45°. There are two types of fits, the *side fit* and the *major-diameter fit*. See Fig. 9-1-6.

In the side fit, the sides of the teeth make contact and the major and minor diameters have clearances. In the major-diameter fit, the mating parts make contact at the major diameter, and the sides of the teeth act as drivers.

The involute splines are becoming more popular for two reasons: they have greater torque-transmitting capacity, and they can be produced in a manner similar to that of cutting gear teeth.

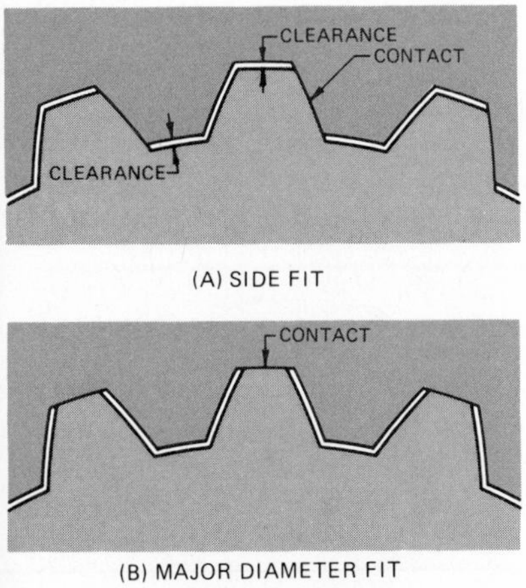
CLEARANCE
CONTACT
CLEARANCE
(A) SIDE FIT
CONTACT
(B) MAJOR DIAMETER FIT

Fig. 9-1-6 Involute splines.

Parallel-Side Splines. The most popular are the SAE parallel-side splines, as shown in Fig. 9-1-7. They have been used in many applications in the automotive and machine industry.

Serrations are shallow, involute splines with 45° pressure angles. They are primarily used for holding parts, such as plastic knobs on steel shafts.

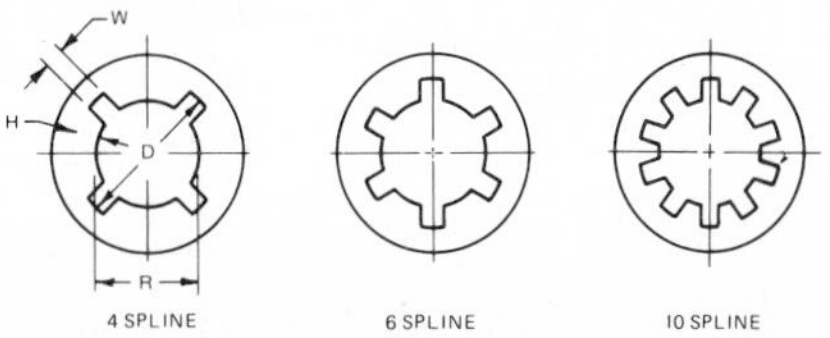
W
H
D
R
4 SPLINE
6 SPLINE
10 SPLINE

NUMBER OF SPLINES	W FOR ALL FITS	PERMANENT FIT		TO SLIDE WITHOUT LOAD		TO SLIDE UNDER LOAD	
		H	R	H	R	H	R
4	0.241 D	0.075 D	0.85 D	0.125 D	0.75 D		
6	0.250 D	0.050 D	0.90 D	0.075 D	0.85 D	0.100 D	0.80 D
10	0.156 D	0.045 D	0.91 D	0.070 D	0.86 D	0.095 D	0.81 D
16	0.098 D	0.045 D	0.91 D	0.070 D	0.86 D	0.095 D	0.81 D

Fig. 9-1-7 Sizes of SAE parallel-side splines.

DRAWING DATA

It is essential that a uniform system of drawing and specifying splines and serrations be used on drawings. The conventional method of showing splines on a drawing is shown in Fig. 9-1-8. The drawing callout is shown in Fig. 9-1-9. Distance *L* does not include the cutter runout. The drawing callout shows the symbol indicating the type of spline followed by the type of fit, the pitch diameter, number of teeth and pitch for involute splines, and number of teeth and outside diameter for straight-sided teeth.

Assignment

On an A3- or B-size sheet, lay out the four fastener assemblies shown in Fig. 9-1-A or 9-1-B. The following fasteners are used:

- Assembly A: square key
- Assembly B: flat key
- Assembly C: Woodruff key
- Assembly D: serrations

Refer to the Appendix and manufacturers' catalogs for sizes and use your judgment for dimensions not shown. Show the dimensions for the keyseats, keyways, and serrations. Scale is 1:1.

REVIEW FOR ASSIGNMENT

Unit 7-5 Assemblies in Section

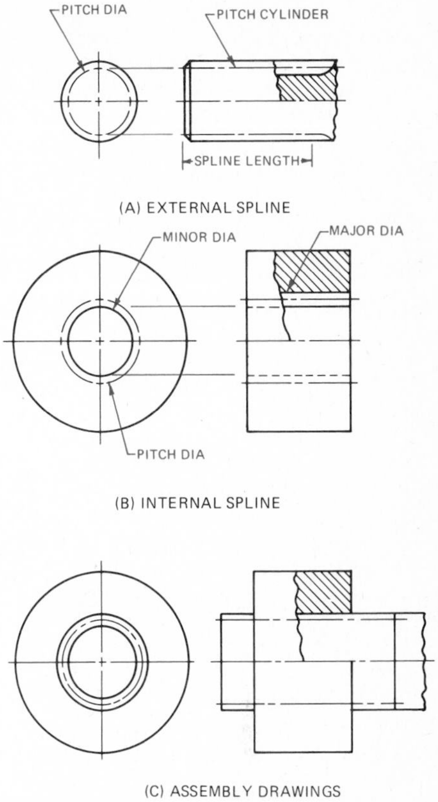
PITCH DIA
PITCH CYLINDER
SPLINE LENGTH
(A) EXTERNAL SPLINE
MINOR DIA
MAJOR DIA
PITCH DIA
(B) INTERNAL SPLINE
(C) ASSEMBLY DRAWINGS

Fig. 9-1-8 Representing splines on drawings.

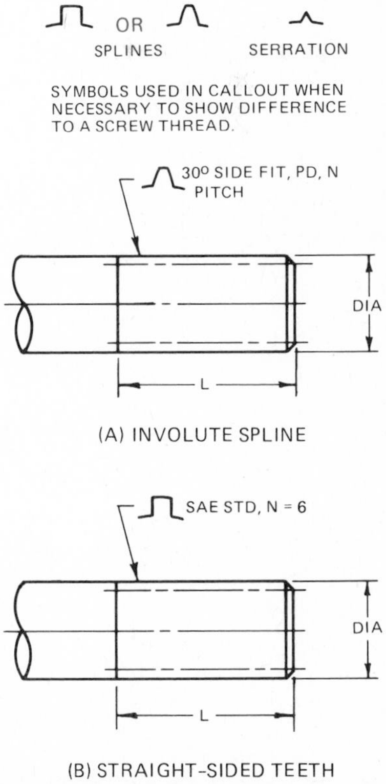
OR
SPLINES
SERRATION
SYMBOLS USED IN CALLOUT WHEN NECESSARY TO SHOW DIFFERENCE TO A SCREW THREAD.
30° SIDE FIT, PD, N PITCH
DIA
L
(A) INVOLUTE SPLINE
SAE STD, N = 6
DIA
L
(B) STRAIGHT-SIDED TEETH

Fig. 9-1-9 Drawing callout for splines.

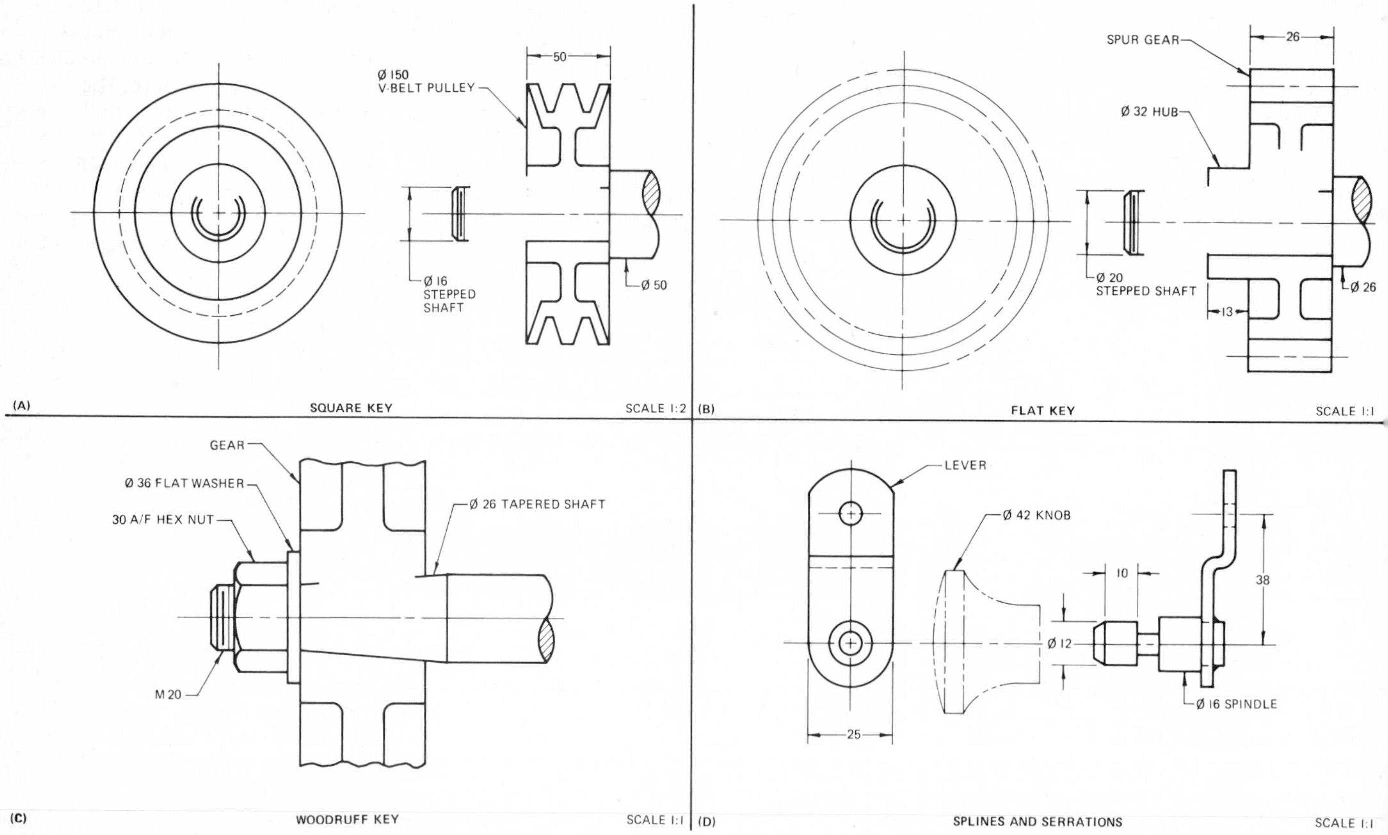
Ø 150
V-BELT PULLEY
50
Ø 16
STEPPED
SHAFT
Ø 50
(A)
SQUARE KEY
SCALE 1:2
SPUR GEAR
26
Ø 32 HUB
Ø 20
STEPPED SHAFT
13
Ø 26
(B)
FLAT KEY
SCALE 1:1
GEAR
Ø 36 FLAT WASHER
30 A/F HEX NUT
Ø 26 TAPERED SHAFT
M 20
(C)
WOODRUFF KEY
SCALE 1:1
LEVER
Ø 42 KNOB
10
38
Ø 12
Ø 16 SPINDLE
25
(D)
SPLINES AND SERRATIONS
SCALE 1:1

Fig. 9-1-A Key and serration fasteners.

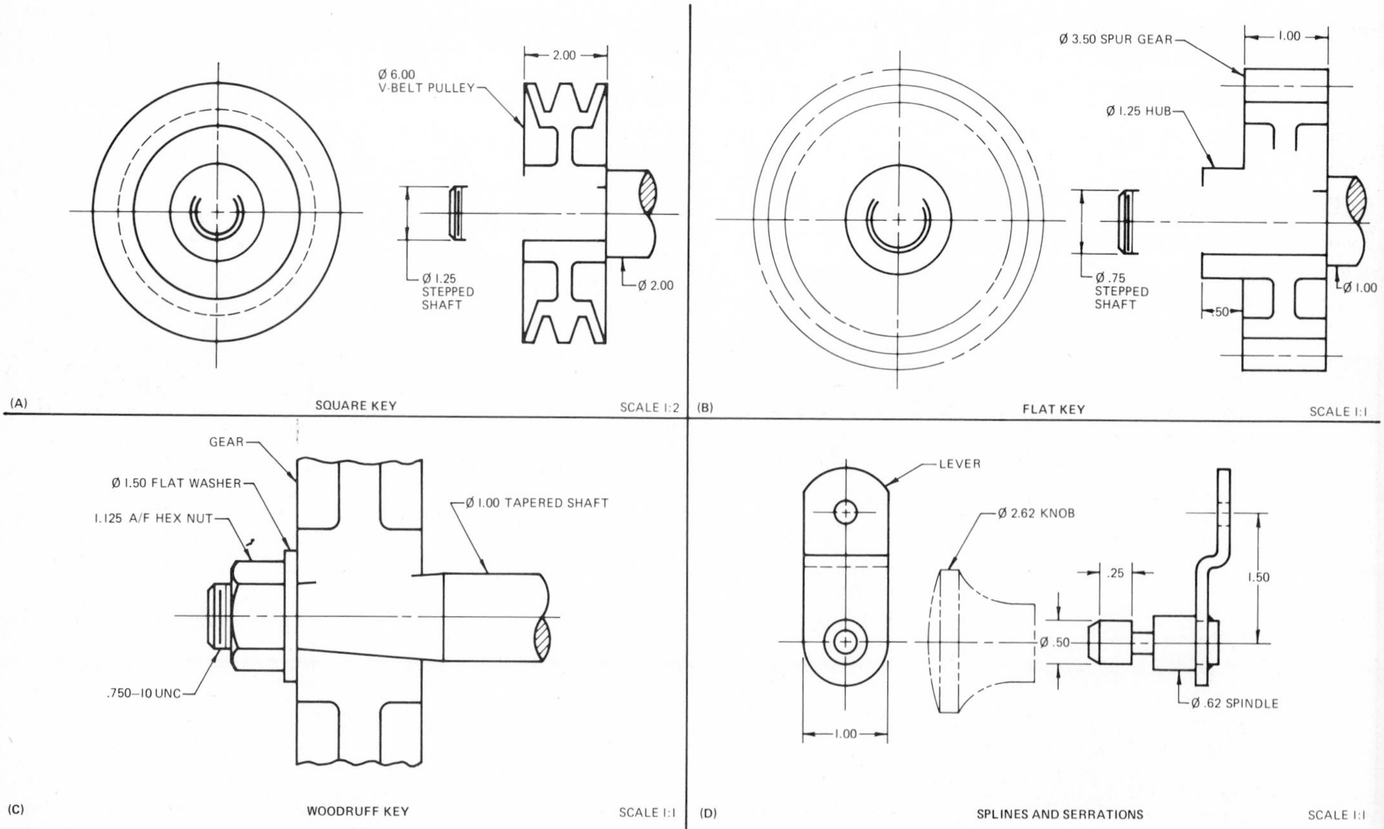
Ø 6.00
V-BELT PULLEY
2.00
Ø 1.25
STEPPED
SHAFT
Ø 2.00
(A)
SQUARE KEY
SCALE 1:2
Ø 3.50 SPUR GEAR
1.00
Ø 1.25 HUB
Ø .75
STEPPED
SHAFT
.50
Ø 1.00
(B)
FLAT KEY
SCALE 1:1
GEAR
Ø 1.50 FLAT WASHER
1.125 A/F HEX NUT
Ø 1.00 TAPERED SHAFT
.750–10 UNC
(C)
WOODRUFF KEY
SCALE 1:1
LEVER
Ø 2.62 KNOB
.25
1.50
Ø .50
Ø .62 SPINDLE
1.00
(D)
SPLINES AND SERRATIONS
SCALE 1:1

Fig. 9-1-B Key and serration fasteners.

UNIT 9-2

PIN FASTENERS[1]

Pin fasteners are an inexpensive and effective approach to assembly where loading is primarily in shear. They can be separated into two groups: *semipermanent* and *quick-release*.

Semipermanent Pins

Semipermanent pin fasteners require application of pressure or the aid of tools for installation or removal. The two basic types are machine pins and radial-locking pins.

The following general design rules apply to all types of semipermanent pins:

- Avoid conditions where the direction of vibration parallels the axis of the pin.
- Keep the shear plane of the pin a minimum distance of 1 diameter from the end of the pin.
- In applications where engaged length is at a minimum and appearance is not critical, allow pins to protrude the length of the chamfer at each end for maximum locking effect.

MACHINE PINS

Four types are generally considered to be most important: *hardened and ground dowel pins and commercial straight pins*, *taper pins*, *clevis pins*, and *standard cotter pins*. Descriptive data and recommended assembly practices for these four traditional types of machine pins are presented in Fig. 9-2-1. For proper size selection of cotter pins, refer to Fig. 9-2-2.

RADIAL LOCKING PINS

Two basic pin forms are employed: *solid with grooved surfaces* and *hollow spring pins*, which may be either slotted or spiral-wrapped. In assembly, radial forces produced by elastic action at the pin surface develop a secure, frictional locking grip against the hole wall. These pins are reusable and can be removed and reassembled many times without appreciable loss of fastening effectiveness. Spring action at the pin surface also prevents loosening under shock and vibration loads and accommodates hole-size variations.

Figure 9-2-3 shows six of the grooved-pin constructions that have been standardized.

NOMINAL THREAD SIZE (mm)	NOMINAL COTTER PIN SIZE (mm)	COTTER PIN HOLE (mm)	END CLEARANCE*
6	1.5	1.9	3
8	2	2.4	3
10	2.5	2.8	4
12	3	3.4	5
14	3	3.4	5
16	4	4.5	6
20	4	4.5	7
24	5	5.6	8
27	5	5.6	8
30	6	6.3	10
36	6	6.3	11
42	6	6.3	12
48	8	8.5	14

*DISTANCE FROM EXTREME POINT OF BOLT OR SCREW TO CENTER OF COTTER PIN HOLE.

Fig. 9-2-2 Recommended cotter pin sizes.

SOLID WITH GROOVED SURFACES

TYPE A

Full-length grooves. Used for general purpose fastening.

TYPE B

Grooves extend half length of the pin. Used as a hinge or linkage "bolt" but also can be employed for other functions in through-drilled holes where a locking fit over only part of the pin length is required.

TYPE E

Half-length groove section centered along the pin surface. Used as a cotter pin or in similar functions where an artificial shoulder or a locking fit over the centre portion of the pin is required.

TYPE A3

Full-length grooves with pilot section at one end to facilitate assembly. Expanded dimension of this pin is held to a maximum over the full grooved length to provide uniform locking action. It is recommended for applications subject to severe vibration or shock loads where maximum locking effect is required.

TYPE D

Reverse tapered grooves extend half the pin length. It is the counterpart of the Type B pin for assembly in blind holes.

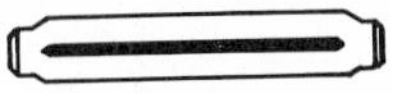

TYPE U

Full-length grooves with pilot section at both ends for hopper feeding. Same as Type C.

HOLLOW SPRING PINS

SPIRAL WRAPPED

SLOTTED TUBULAR

Fig. 9-2-3 Radial locking pins.

HARDENED AND GROUND DOWEL PIN	TAPER PIN	CLEVIS PIN	COTTER PIN
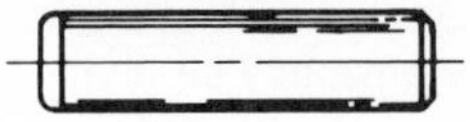		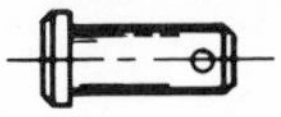	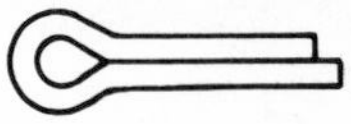
Standardized in nominal diameters ranging from 3 to 22 mm. 1. Holding laminated sections together with surfaces either drawn up tight or separated in some fixed relationship. 2. Fastening machine parts where accuracy of alignment is a primary consideration. 3. Locking components on shafts, in the form of transverse pin key.	Standard pins have a taper of 1:48 measured on the diameter. Basic dimension is the diameter of the large end. Used for light duty service in the attachment of wheels, levers and similar components to shafts. Torque capacity is determined on the basis of double shear, using the average diameter along the tapered section in the shaft for area calculations.	Standard nominal diameters for clevis pins range from 5 to 25 mm. Basic function of the clevis pin is to connect mating yoke, or fork, and eye members in knuckle-joint assemblies. Held in place by a small cotter pin or other fastening means, it provides a mobile joint construction, which can be readily disconnected for adjustment or maintenance.	Sizes have been standardized in nominal diameters ranging from 1 to 20 mm. Locking device for other fasteners. Used with a castle or slotted nut on bolt, screws, or studs, it provides a convenient, low-cost locknut assembly. Hold standard clevis pins in place. Can be used with or without a plain washer as an artificial shoulder to lock parts in position on shafts.

Fig. 9-2-1 Machine pins.

GROOVED STRAIGHT PINS

Locking action of the groove pin is provided by parallel, longitudinal grooves uniformly spaced around the pin surface. Rolled or pressed into solid pin stock, the grooves expand the effective pin diameter. When the pin is driven into a drilled hole corresponding in size to nominal pin diameter, elastic deformation of the raised groove edges produces a secure force-fit with the hole wall. For typical grooved pin applications and size selection, refer to Figs. 9-2-4 and 9-2-5.

HOLLOW SPRING PINS

Resilience of hollow cylinder walls under radial compression forces is the principle of spiral-wrapped and slotted tubular pins. Both pin forms are made to controlled diameters greater than the holes into which they are pressed. Compressed when driven into the hole, the pins exert spring pressure against the hole wall along their entire engaged length to develop locking action.

SHAFT DIA (mm)	TRANSVERSE KEY PIN DIA (mm)	TRANSVERSE KEY TAPER PIN NO.	LONGITUDINAL KEY PIN DIA (mm)
5	1.5	7/0	–
5.5	2	6/0	–
6	2.5	5/0	1.5
8	3	4/0	2
10	3	3/0	2.5
12	4	0	3
14	5	2	4
16	5	2	4
18	6	3	5
20	6	4	5
22	6	4	6
24	8	5	6
26	8	6	6
28	10	7	–
30	10	7	–
32	10	7	8
34	11	7	10
36	11	7	–
38	12	8	11

Fig. 9-2-5 Recommended cotter pin sizes.

Locking force of a spiral-wrapped pin is a function of length of engagement, pin diameter, and wall thickness.

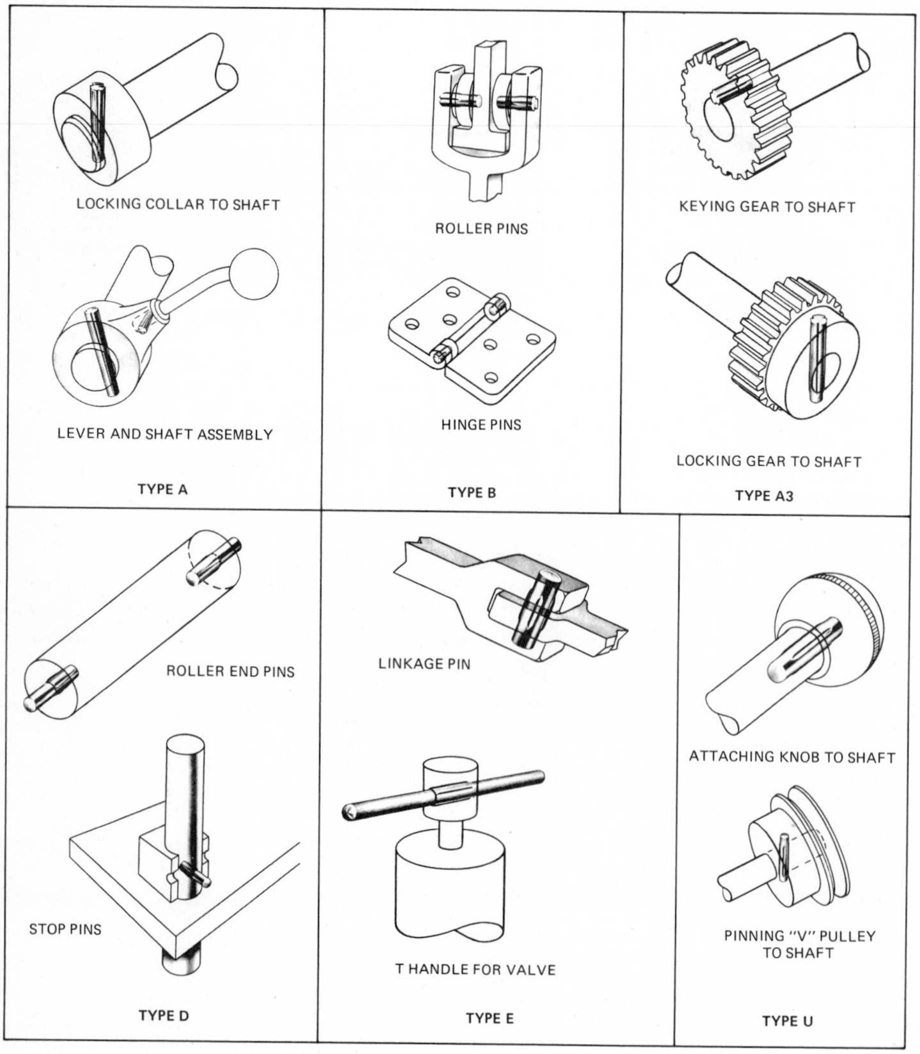
LOCKING COLLAR TO SHAFT
LEVER AND SHAFT ASSEMBLY
TYPE A
ROLLER PINS
HINGE PINS
TYPE B
KEYING GEAR TO SHAFT
LOCKING GEAR TO SHAFT
TYPE A3
ROLLER END PINS
STOP PINS
TYPE D
LINKAGE PIN
T HANDLE FOR VALVE
TYPE E
ATTACHING KNOB TO SHAFT
PINNING "V" PULLEY TO SHAFT
TYPE U

Fig. 9-2-4 Grooved pin application.

Standard slotted tubular pins are designed so that several sizes can be used inside one another. In such combinations, shear strengths of the individual pins are additive. For spring pin application refer to Fig. 9-2-6.

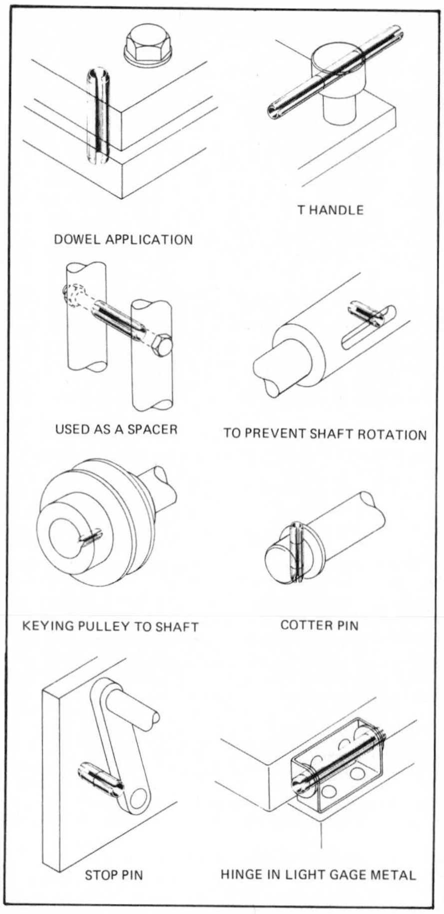
DOWEL APPLICATION
T HANDLE
USED AS A SPACER
TO PREVENT SHAFT ROTATION
KEYING PULLEY TO SHAFT
COTTER PIN
STOP PIN
HINGE IN LIGHT GAGE METAL

Fig. 9-2-6 Spring pin application. (Drive Lok).

Quick-Release Pins

Commercially available, quick-release pins vary widely in head styles, types of locking and release mechanisms, and range of pin lengths. See Fig. 9-2-7.

Quick-release pins may be divided into two basic types: *push-pull* and *positive-locking* pins. The positive-locking pins can be further divided into three categories: heavy-duty cotter pins, single-acting pins, and double-acting pins.

PUSH-PULL PINS

These pins are made with either a solid or a hollow shank, containing a detent assembly in the form of a locking lug, button, or ball backed up by some type of resilient core, plug, or spring. The detent member projects from the surface of

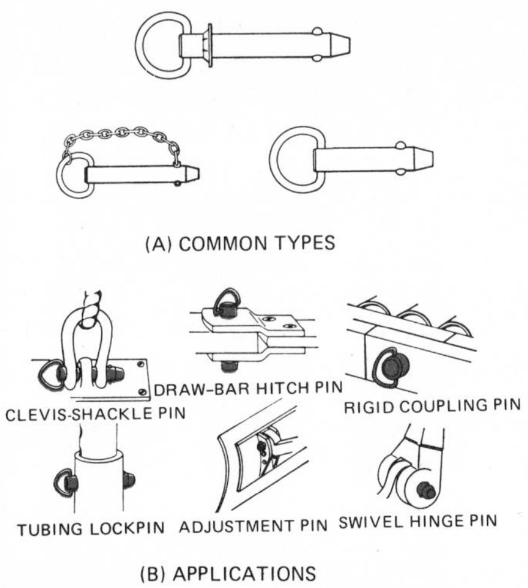
(A) COMMON TYPES
CLEVIS-SHACKLE PIN
DRAW-BAR HITCH PIN
RIGID COUPLING PIN
TUBING LOCKPIN
ADJUSTMENT PIN
SWIVEL HINGE PIN
(B) APPLICATIONS

Fig. 9-2-7 Quick-release pins.

the pin body until sufficient force is applied in assembly or removal to cause it to retract against the spring action of the resilient core and release the pin for movement.

POSITIVE-LOCKING PINS

For some quick-release fasteners, the locking action is independent of insertion and removal forces. As in the case of push-pull pins, these pins are primarily suited for shear-load applications. However, some degree of tension loading usually can be tolerated without affecting the pin function.

Positive-locking pins can be divided into three categories: heavy-duty cotter pins, single-acting pins, and double-acting pins.

Heavy-duty cotter pins employ a forged, high-carbon-steel body to replace the conventional split-cotter construction. Locking action is provided by a tempered-steel snap ring mounted on the head of the pin.

Single-acting pins have locking action controlled by a plunger-actuated locking mechanism. In normal (locked) position, the locking element projects beyond the surface of the pin shank to provide a positive lock. Either ball- or pin-type locking elements may be employed.

Double-acting pins are a modification of single-acting types and have a bidirectional, spring-located plunger. Movement of the plunger in either direction along its barrel releases the locking balls.

REFERENCE AND SOURCE MATERIAL

1. F. W. Braendel, "Pin Fasteners," *Machine Design*, vol. 45, no. 28, 1973.

Assignment

On an A3- or B-size sheet, complete the pin assemblies shown in Fig. 9-2-A or 9-2-B, given the following information:

- *Assembly A*. A type E groove pin holds the roller to the bracket. A washer and cotter pin are used to fasten the bracket to the pushrod.
- *Assembly B*. A type A3 groove pin holds the V-belt pulley to the shaft.
- *Assembly C*. Slotted tubular spring pins are used to fasten the cap and handle to the shaft.
- *Assembly D*. A clevis pin whose area is equal to the four rivets is used to fasten the trailer hitch to the tractor draw bar.

Refer to manufacturers' catalogs for pin sizes and provide the complete information to order each fastener. Scale is as shown.

REVIEW FOR ASSIGNMENT

Unit 8-3 Washers
Unit 7-5 Assemblies in Section

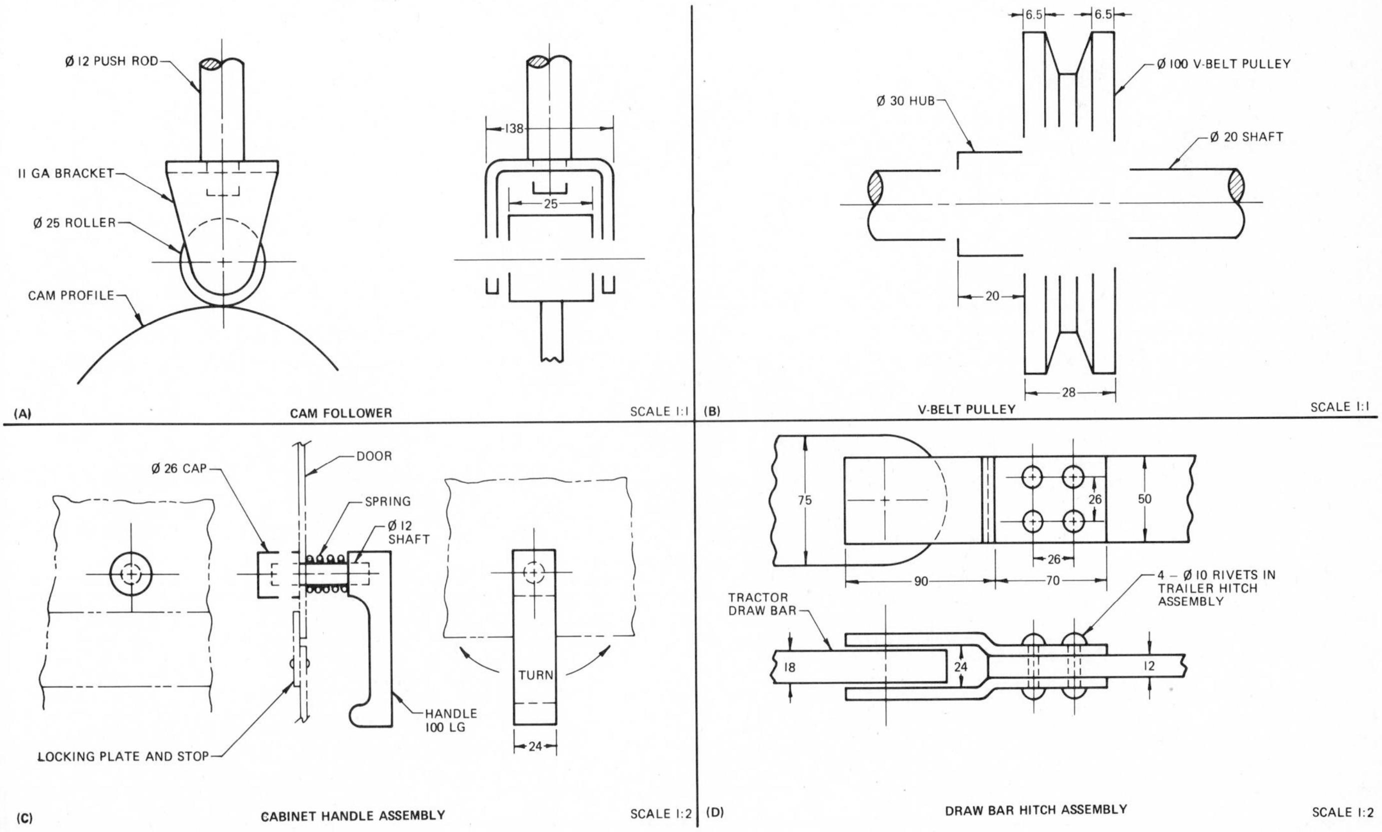
Ø 12 PUSH ROD
11 GA BRACKET
Ø 25 ROLLER
CAM PROFILE
138
25
(A)
CAM FOLLOWER
SCALE 1:1
6.5
6.5
Ø 100 V-BELT PULLEY
Ø 30 HUB
Ø 20 SHAFT
20
28
(B)
V-BELT PULLEY
SCALE 1:1
Ø 26 CAP
DOOR
SPRING
Ø 12 SHAFT
TURN
HANDLE 100 LG
24
LOCKING PLATE AND STOP
(C)
CABINET HANDLE ASSEMBLY
SCALE 1:2
75
26
50
26
90
70
4 – Ø 10 RIVETS IN TRAILER HITCH ASSEMBLY
TRACTOR DRAW BAR
18
24
12
(D)
DRAW BAR HITCH ASSEMBLY
SCALE 1:2

Fig. 9-2-A Pin fasteners.

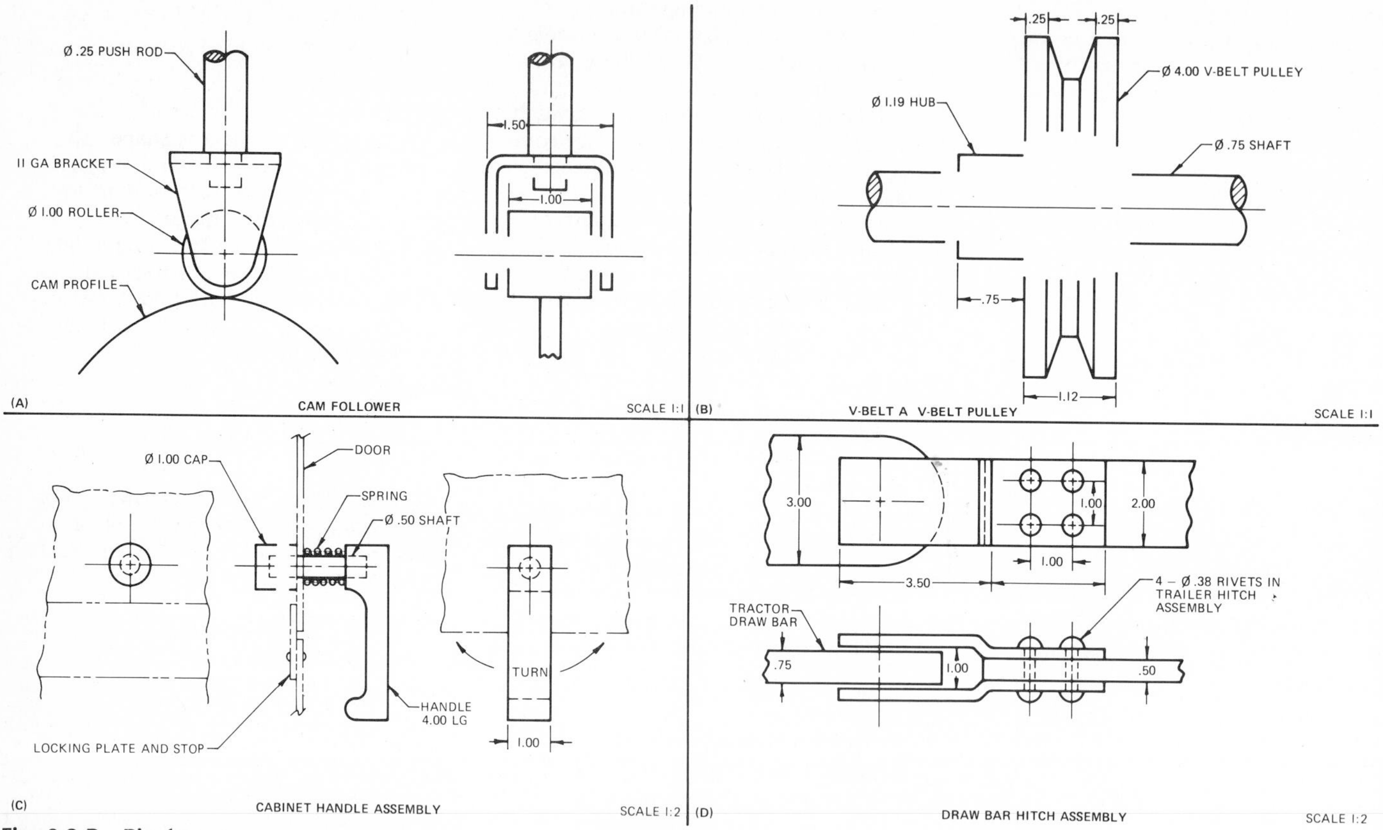

Ø .25 PUSH ROD
II GA BRACKET
Ø 1.00 ROLLER
CAM PROFILE
1.50
1.00
(A)
CAM FOLLOWER
SCALE 1:1
.25
.25
Ø 4.00 V-BELT PULLEY
Ø 1.19 HUB
Ø .75 SHAFT
.75
1.12
(B)
V-BELT A V-BELT PULLEY
SCALE 1:1
Ø 1.00 CAP
DOOR
SPRING
Ø .50 SHAFT
TURN
HANDLE
4.00 LG
1.00
LOCKING PLATE AND STOP
(C)
CABINET HANDLE ASSEMBLY
SCALE 1:2
3.00
1.00
2.00
1.00
3.50
4 – Ø .38 RIVETS IN
TRAILER HITCH
ASSEMBLY
TRACTOR
DRAW BAR
.75
1.00
.50
(D)
DRAW BAR HITCH ASSEMBLY
SCALE 1:2

Fig. 9-2-B Pin fasteners.

UNIT 9-3
RETAINING RINGS[1]

Retaining rings, or *snap rings*, are used to provide a removable shoulder to accurately locate, retain, or lock components on shafts and in bores of housings. See Fig. 9-3-1. They are easily installed and removed, and since they are usually made of spring steel, retaining rings have a high shear strength and impact capacity. In addition to fastening and positioning, a number of rings are designed for taking up end play caused by accumulated tolerances or wear in the parts being retained. In general, these devices can be placed into three categories which describe the type and method of fabrication: stamped retaining rings, wire-formed rings, and spiral-wound retaining rings.

Stamped Retaining Rings

Stamped retaining rings, in contrast to wire-formed rings with their uniform cross-sectional area, have a tapered radial width which decreases symmetrically from the center section to the free ends. The tapered construction permits the rings to remain circular when they are

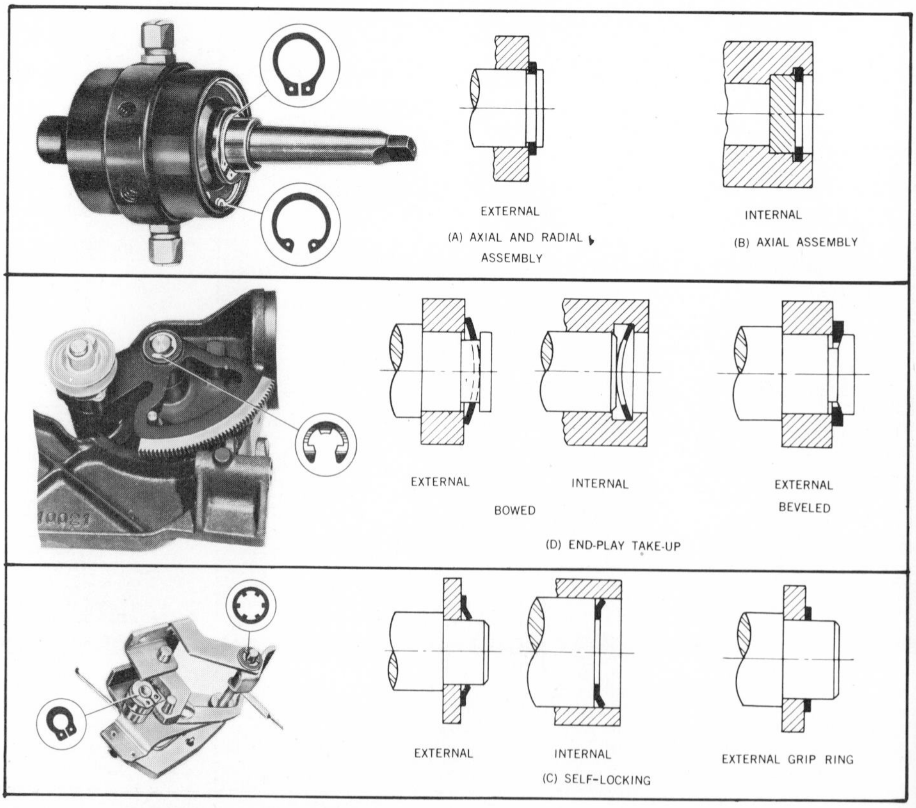

EXTERNAL
(A) AXIAL AND RADIAL ASSEMBLY
INTERNAL
(B) AXIAL ASSEMBLY
EXTERNAL
INTERNAL
BOWED
EXTERNAL
BEVELED
(D) END-PLAY TAKE-UP
EXTERNAL
INTERNAL
(C) SELF-LOCKING
EXTERNAL GRIP RING

Fig. 9-3-1 Retaining ring application.

xpanded for assembly over a shaft or ontracted for insertion into a bore or ousing. This constant circularity en- ures maximum contact surface with the ottom of the groove.

Stamped retaining rings can be clas- ified into three groups: axially assem- led rings, radially assembled rings, and elf-locking rings which do not require rooves. *Axially assembled rings* slip ver the ends of shafts or down into ores, while *radially assembled rings* ave side openings which permit the ings to be snapped directly into grooves n a shaft.

Most axially assembled rings have holes in the lugs at the free ends. Special retaining ring pliers fit into these holes and expand or contract the rings for installation or removal. Radially assembled rings are installed and removed with a screwdriver or other simple hand tool. Several varieties of self-locking rings, which do not require grooves, are available for assemblies in which the fastener need not absorb any sizable thrust, but instead serves mainly as a positioning and locking device. Commonly used types of stamped retaining rings are illustrated and compared in Fig. 9-3-2.

Wire-Formed Retaining Rings

The *wire-formed retaining ring* is a split ring formed and cut from spring wire of uniform sectional size and shape. The wire is cold-drawn or rolled into shape from a continuous coil or bar. Then the gap ends are cut into various configurations for ease of application and removal.

Rings are available in many cross-sectional shapes, but the most commonly used are the rectangular and circular, or round, cross sections.

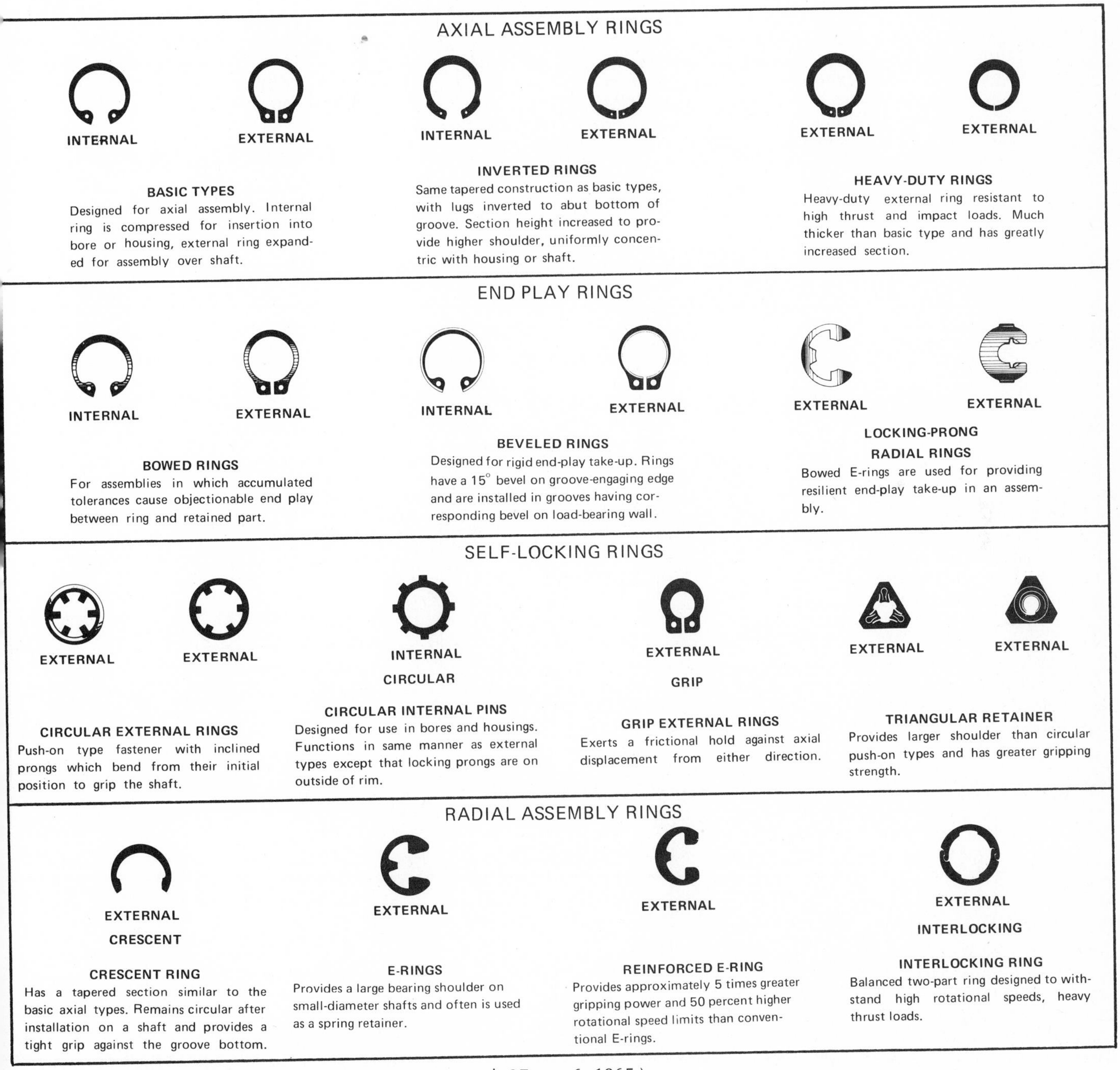

AXIAL ASSEMBLY RINGS
INTERNAL
EXTERNAL
BASIC TYPES
Designed for axial assembly. Internal ring is compressed for insertion into bore or housing, external ring expanded for assembly over shaft.
INTERNAL
EXTERNAL
INVERTED RINGS
Same tapered construction as basic types, with lugs inverted to abut bottom of groove. Section height increased to provide higher shoulder, uniformly concentric with housing or shaft.
EXTERNAL
EXTERNAL
HEAVY-DUTY RINGS
Heavy-duty external ring resistant to high thrust and impact loads. Much thicker than basic type and has greatly increased section.
END PLAY RINGS
INTERNAL
EXTERNAL
BOWED RINGS
For assemblies in which accumulated tolerances cause objectionable end play between ring and retained part.
INTERNAL
EXTERNAL
BEVELED RINGS
Designed for rigid end-play take-up. Rings have a 15° bevel on groove-engaging edge and are installed in grooves having corresponding bevel on load-bearing wall.
EXTERNAL
EXTERNAL
LOCKING-PRONG RADIAL RINGS
Bowed E-rings are used for providing resilient end-play take-up in an assembly.
SELF-LOCKING RINGS
EXTERNAL
EXTERNAL
CIRCULAR EXTERNAL RINGS
Push-on type fastener with inclined prongs which bend from their initial position to grip the shaft.
INTERNAL
CIRCULAR
CIRCULAR INTERNAL PINS
Designed for use in bores and housings. Functions in same manner as external types except that locking prongs are on outside of rim.
EXTERNAL
GRIP
GRIP EXTERNAL RINGS
Exerts a frictional hold against axial displacement from either direction.
EXTERNAL
EXTERNAL
TRIANGULAR RETAINER
Provides larger shoulder than circular push-on types and has greater gripping strength.
RADIAL ASSEMBLY RINGS
EXTERNAL
CRESCENT
CRESCENT RING
Has a tapered section similar to the basic axial types. Remains circular after installation on a shaft and provides a tight grip against the groove bottom.
EXTERNAL
E-RINGS
Provides a large bearing shoulder on small-diameter shafts and often is used as a spring retainer.
EXTERNAL
REINFORCED E-RING
Provides approximately 5 times greater gripping power and 50 percent higher rotational speed limits than conventional E-rings.
EXTERNAL
INTERLOCKING
INTERLOCKING RING
Balanced two-part ring designed to withstand high rotational speeds, heavy thrust loads.

Fig. 9-3-2 Stamped retaining rings. (*Machine Design*, vol. 37, no. 6, 1965.)

Spiral-Wound Retaining Rings

Spiral-wound retaining rings consist of two or more turns of rectangular material, wound on edge to provide a continuous crimped or uncrimped coil. This design has several advantages:

1. A simple, gapless ring which can be adapted easily to specific applications.
2. Since the coil material is thin in section, a flexible retaining ring is obtained which allows ease of assembly and disassembly.
3. Varying-thickness rings can be fabricated, either by varying the thickness of the material or by increasing the number of turns, to overcome problems of machining narrow and deep grooves or to provide optimum properties.
4. The laminated section permits a large deflection under thrust without overstressing the ring.
5. The use of thin section material permits the forming of tabs, prongs, and other devices for assembly and disassembly while maintaining optimum ring thickness.

REFERENCE AND SOURCE MATERIAL

1. H. Wurzel, R. J. Munsey, and H. E. McCormick, "Retaining Rings," *Machine Design*, vol. 37, no. 6, 1965.

Assignment

Divide an A3- or B-size sheet into four sections, and draw the assemblies shown in Figs. 9-3-A and 9-3-B. Complete the assemblies by adding suitable retaining rings as per the information supplied below. Refer to manufacturers' catalogs and show on the drawing the catalog number for the retaining ring. Add ring and groove sizes. Scale is 1:1. Use your judgment for dimensions not shown.

- *Assembly A*. An external radial retaining ring mounted on the shaft is to act as a shoulder for the shaft support. An external axial retaining ring is required to hold the gear on the shaft.
- *Assembly B*. The plunger and punch are held into the punch holder by internal retaining rings.
- *Assembly C*. External self-locking retaining rings hold the roller shaft in position on the bracket.
- *Assembly D*. An external self-locking ring holds the plastic housing to the viewer case. An internal self-locking ring holds the lens in position.

REVIEW FOR ASSIGNMENT

Unit 7-5 Assemblies in Section

UNIT 9-4

SPRINGS[1,2]

Springs may be classified into three general groups according to their application.

Controlled Action Springs. Controlled action springs have a well-defined function, or a constant range of action for each cycle of operation. Examples are valve, die, and switch springs.

Variable-Action Springs. Variable-action springs have a changing range of action because of the variable conditions imposed upon them. Examples are suspension, clutch, and cushion springs.

Static Springs. Static springs exert a comparatively constant pressure or tension between parts. Examples are packing or bearing pressure, antirattle, and seal springs.

Types of Springs

The type or name of a spring is determined by characteristics such as function, shape of material, application, or design. Figure 9-4-1 illustrates common springs in use. Figure 9-4-2 designates spring nomenclature.

COMPRESSION SPRINGS

A *compression spring* is an open-coil, helical spring that offers resistance to a compressive form. It has a wide variety

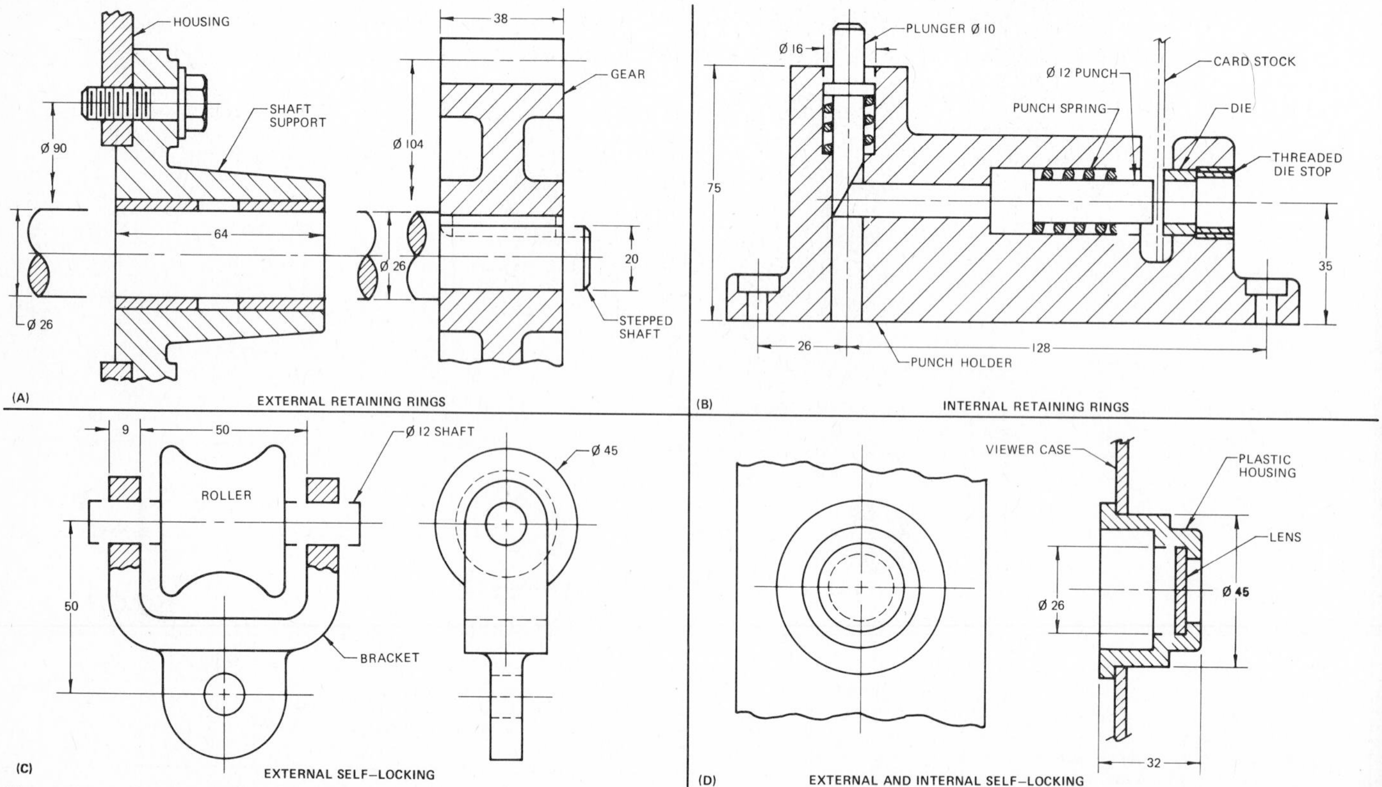
HOUSING
SHAFT SUPPORT
Ø 90
64
Ø 26
38
GEAR
Ø 104
Ø 26
20
STEPPED SHAFT
(A)
EXTERNAL RETAINING RINGS
PLUNGER Ø 10
Ø 16
Ø 12 PUNCH
CARD STOCK
PUNCH SPRING
DIE
THREADED DIE STOP
75
35
26
128
PUNCH HOLDER
(B)
INTERNAL RETAINING RINGS
9
50
Ø 12 SHAFT
Ø 45
ROLLER
50
BRACKET
(C)
EXTERNAL SELF–LOCKING
VIEWER CASE
PLASTIC HOUSING
LENS
Ø 45
Ø 26
32
(D)
EXTERNAL AND INTERNAL SELF–LOCKING

Fig. 9-3-A Retaining ring fasteners.

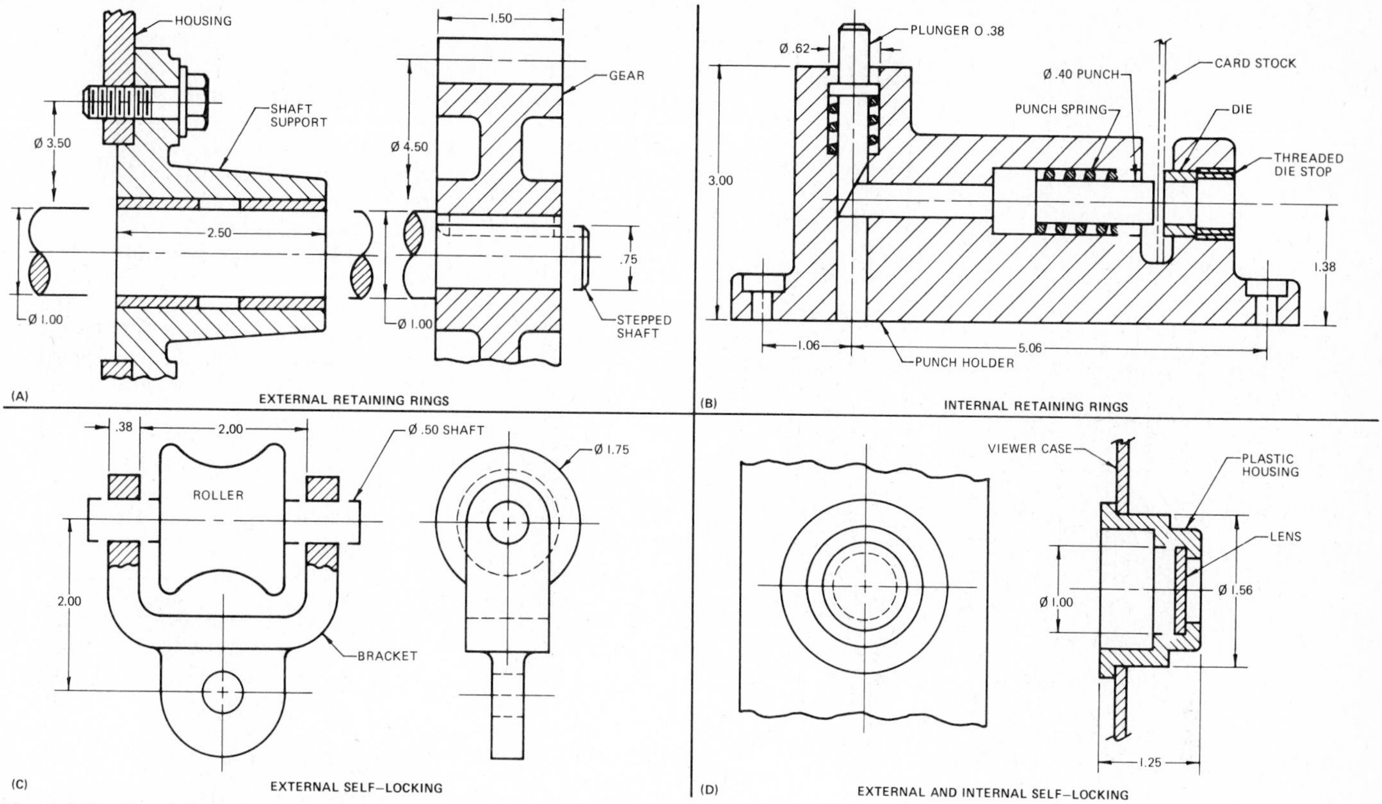
HOUSING
SHAFT SUPPORT
Ø 3.50
2.50
Ø 1.00
1.50
GEAR
Ø 4.50
.75
Ø 1.00
STEPPED SHAFT
(A)
EXTERNAL RETAINING RINGS
PLUNGER Ø .38
Ø .62
Ø .40 PUNCH
CARD STOCK
PUNCH SPRING
DIE
THREADED DIE STOP
3.00
1.38
1.06
5.06
PUNCH HOLDER
(B)
INTERNAL RETAINING RINGS
.38
2.00
Ø .50 SHAFT
Ø 1.75
ROLLER
2.00
BRACKET
(C)
EXTERNAL SELF-LOCKING
VIEWER CASE
PLASTIC HOUSING
LENS
Ø 1.00
Ø 1.56
1.25
(D)
EXTERNAL AND INTERNAL SELF-LOCKING

Fig. 9-3-B Retaining ring fasteners.

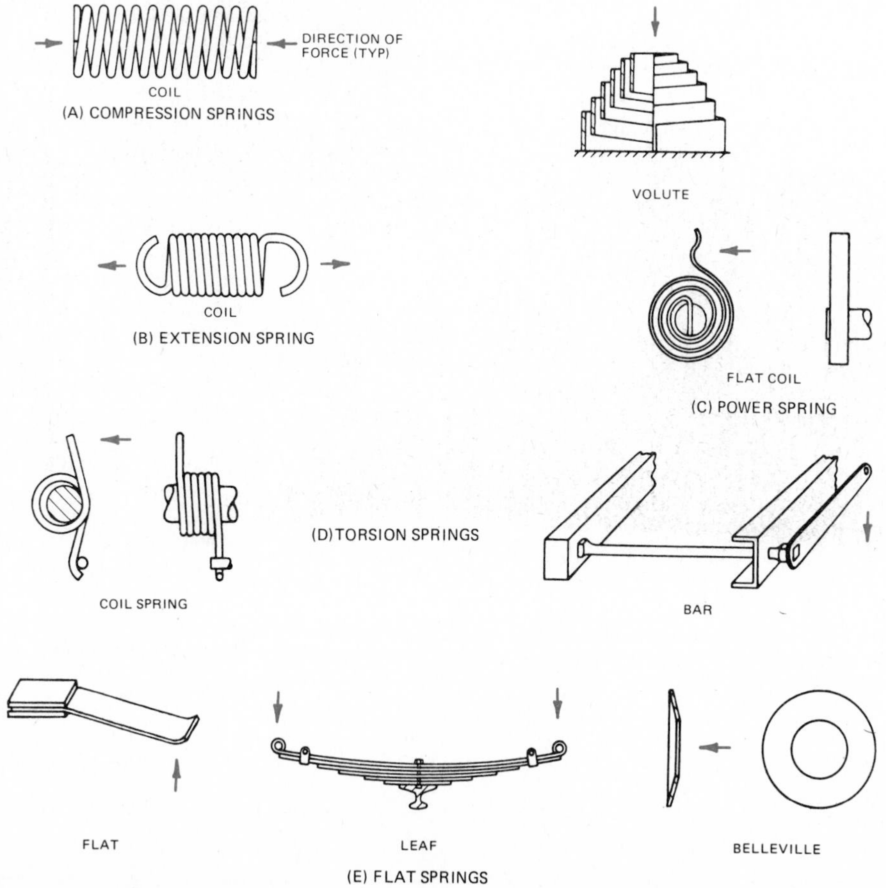
DIRECTION OF FORCE (TYP)
COIL
(A) COMPRESSION SPRINGS
VOLUTE
COIL
(B) EXTENSION SPRING
FLAT COIL
(C) POWER SPRING
(D) TORSION SPRINGS
COIL SPRING
BAR
FLAT
LEAF
BELLEVILLE
(E) FLAT SPRINGS

Fig. 9-4-1 Types of springs.

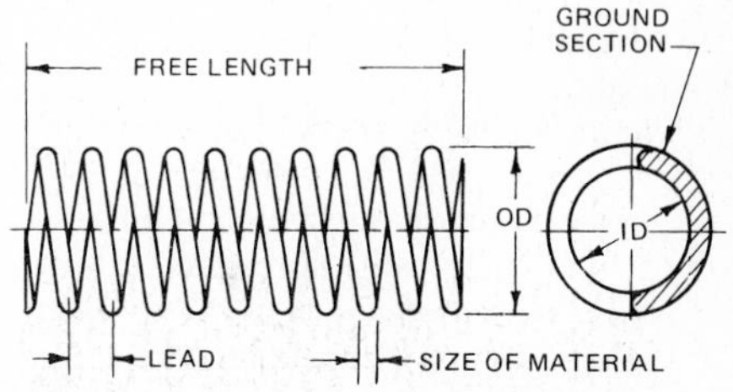
FREE LENGTH
GROUND SECTION
OD
ID
LEAD
SIZE OF MATERIAL

Fig. 9-4-2 Spring nomenclature.

of uses and is made in various forms and from different shapes of wire, depending upon its application.

The most common form of this type is the same diameter through its entire length, and it is known as a *straight spring*. Tapered and cone-shaped springs are used quite extensively, and sometimes a combination of straight and tapered sections works to good advantage.

Compression Spring Ends. Figure 9-4-3*a* shows the ends commonly used on compression springs.

Plain open ends are produced by straight cutoff with no reduction of helix angle. Though they are the most economical to make, their limited use is due to inherent buckling tendencies. The number of active coils approximates the total number of coils. See Fig. 9-4-4. The spring should be guided on a rod or in a hole to operate satisfactorily.

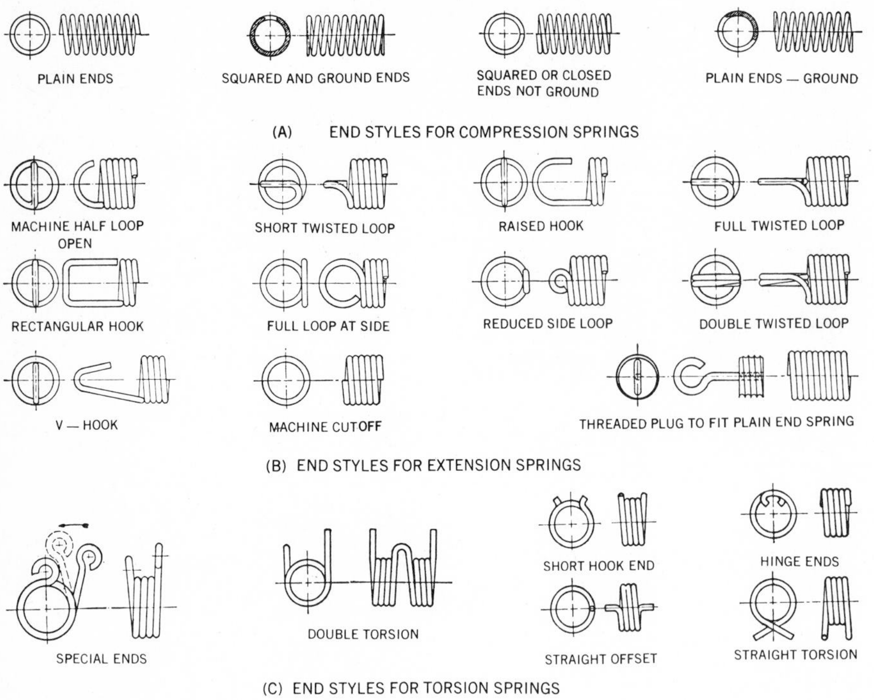
PLAIN ENDS
SQUARED AND GROUND ENDS
SQUARED OR CLOSED ENDS NOT GROUND
PLAIN ENDS — GROUND
(A) END STYLES FOR COMPRESSION SPRINGS
MACHINE HALF LOOP OPEN
SHORT TWISTED LOOP
RAISED HOOK
FULL TWISTED LOOP
RECTANGULAR HOOK
FULL LOOP AT SIDE
REDUCED SIDE LOOP
DOUBLE TWISTED LOOP
V — HOOK
MACHINE CUTOFF
THREADED PLUG TO FIT PLAIN END SPRING
(B) END STYLES FOR EXTENSION SPRINGS
SHORT HOOK END
HINGE ENDS
SPECIAL ENDS
DOUBLE TORSION
STRAIGHT OFFSET
STRAIGHT TORSION
(C) END STYLES FOR TORSION SPRINGS

Fig. 9-4-3 End styles for helical springs.

Ground open ends are produced by parallel grinding of open-end coil springs. Advantages of this type of end are improved stability and a larger number of total coils.

Plain closed ends are produced with a straight cutoff and with reduction of helix angle to obtain closed-end coils, resulting in a more stable spring.

Ground closed ends are produced by parallel grinding of closed-end coil springs, resulting in maximum stability.

Direction of Helix. In certain applications, such as one spring working inside another, it is necessary to coil the springs in opposite directions. If a spring operates over a screw thread, the direction of the helix should be opposite to that of the thread.

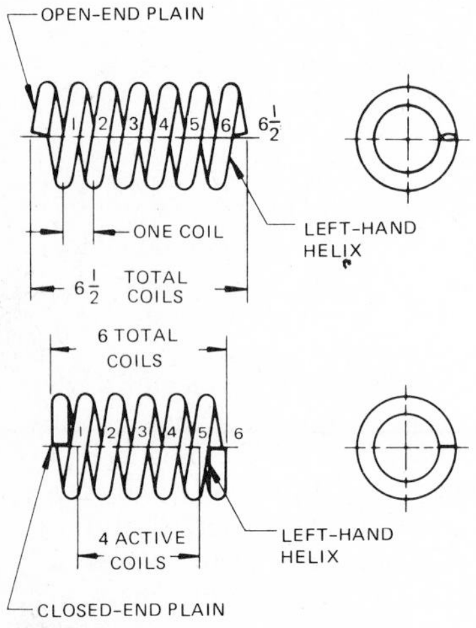
OPEN-END PLAIN
1 2 3 4 5 6 $6\frac{1}{2}$
ONE COIL
LEFT-HAND HELIX
$6\frac{1}{2}$ TOTAL COILS
6 TOTAL COILS
1 2 3 4 5 6
4 ACTIVE COILS
LEFT-HAND HELIX
CLOSED-END PLAIN

Fig. 9-4-4 Coil definitions.

VOLUTE SPRINGS

A *volute spring* is a conically shaped compression spring produced by winding a flat material upon itself, with the wide dimension parallel to the axis of the helix, in such a manner that the inner coils telescope within the outer coils when compressed. It is used for compactness, high forces, and high spring rates, where relatively small deflection and increasing rate toward the end of the stroke are found.

EXTENSION SPRINGS

An *extension spring* is a close-coiled helical spring that offers resistance to a pulling force. It is made from round and square wire. Coils are usually close wound and in contact with one another. It is different from a compression spring from a loading standpoint, inasmuch as the coils may be wound so tightly together that an effort is required to pull them apart.

Extension Spring Ends. The end of an extension spring is usually the most highly stressed part. Thus, proper consideration should be given to its selection. The types of ends shown in Fig. 9-4-3*b* are most commonly used on extension springs. Different types of ends can be used on the same spring.

TORSION SPRINGS

Springs exerting pressure along a path which is a circular arc, or, in other words, providing a torque, are called torsion springs, springs, motor springs, power springs, etc. The term *torsion spring* is usually applied to a helical spring of round, square, or rectangular wire, loaded by torque.

One of the most common uses of springs of this type is the ordinary spring hinge. Many other cases could be cited where they are used to rotate parts, or to cushion shocks on rotating parts, such as oven doors on gas and electric ranges, brush-holder springs on motors and generators, etc.

The variation in ends used is almost limitless, but a few of the more common types are illustrated in Fig. 9-4-3*c*.

Torsion Bar Springs. A torsion bar spring is a relatively straight bar anchored at one end, on which a torque may be exerted at the other end, thus tending to twist it about its axis. A torsion bar is sometimes favored because of its high efficiency in material utilization.

POWER SPRINGS

Clock or Motor Type. A *flat coil spring*, also known as a clock or motor spring, consists of a strip of tempered steel wound up on an arbor and usually confined in a case or drum. Such a spring is capable of storing up the energy applied in winding it and delivering such energy in the form of torque either through its central shaft or arbor or through the drum by which it is confined.

Hairspring. The success or failure of many of the ideas which can contribute so much to the future often depends on a precisely formed wire coil, many times finer than a human hair. Whether it be one of these finer sizes or a more sturdy version of a hairspring, great accuracy and good and careful design are vital to the proper functioning of the instrument for which it is required. A hairspring can control the release of a more powerful source of energy, as in a watch or clock, and commonly performs this duty 18 000 times every hour, year in and year out. A hairspring may measure a varying force, such as in an automobile speedometer or an electrical meter.

FLAT SPRINGS

Flat springs are made of flat material formed in such a manner as to apply force in the desired direction when deflected in the opposite direction.

Most flat springs may be classified as beams, either clamped at one end with force applied at the other end (cantilever springs) or supported at each end with a force applied in the center (elliptical springs).

Leaf Springs. A *leaf spring* is composed of a series of flat springs nested together and arranged so as to provide approximately uniform distribution of stress throughout its length. Springs may be used in multiple arrangements, as shown in Fig. 9-4-5.

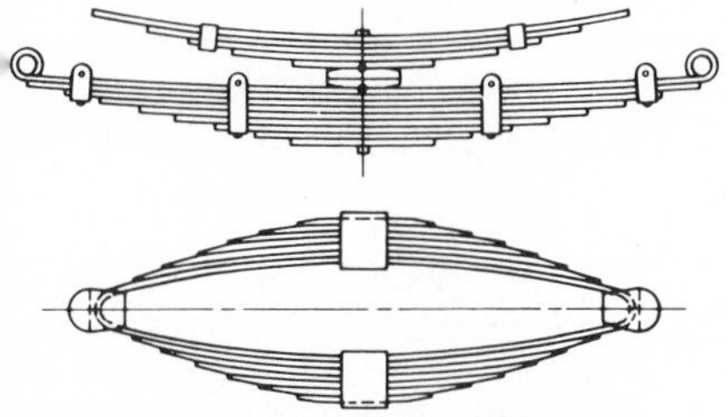

Fig. 9-4-5 Leaf springs—multiple arrangements.

Belleville Springs. Belleville springs are washer-shaped, made in the form of a short, truncated cone.

Belleville washers may be assembled in series to accommodate greater deflections, in parallel to resist greater forces, or in combination of series and parallel, as shown in Fig. 9-4-6.

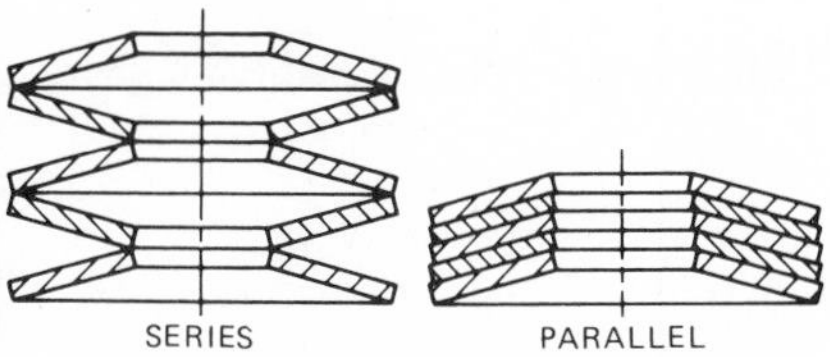
SERIES
PARALLEL

Fig. 9-4-6 Belleville spring installation.

Spring Drawings

On working drawings, a schematic drawing of a helical spring is recommended in order to save drafting time. See Fig. 9-4-7. Like screw-thread representation, straight lines are used in place of the helical curves. On assembly drawings, springs are normally shown in section, and either cross-hatching lines or solid black shading is recommended, depending on the size of the wire's diameter. See Fig. 9-4-8.

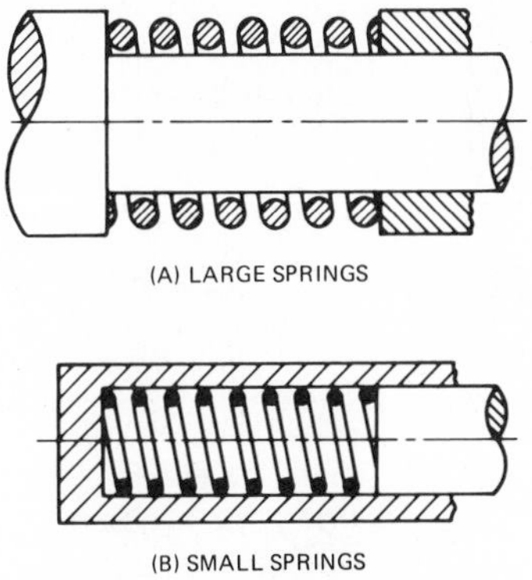
(A) LARGE SPRINGS
(B) SMALL SPRINGS

Fig. 9-4-8 Showing helical springs on assembly drawings.

Dimensioning Springs. In using single-line representation, the dimensions should state the applicable side of material as required to ensure correct interpretation on such features as inside diameter, outside diameter, and end loops. See Fig. 9-4-9.

Spring Clips[3]

Spring clips are a relatively new class of industrial fasteners. They perform multiple functions, eliminate the handling of several small parts, and thus reduce assembly costs. See Fig. 9-4-10.

The *spring clip* is generally self-retaining, requiring only a flange, panel edge, or mounting hole to clip to. It is a one-piece, self-sufficient fastener that requires no secondary fastening devices such as rivets, studs, or screws to make an attachment.

Basically, spring clips are light-duty fasteners and serve the same function as small bolts and nuts, self-tapping screws, clamps, spot welding, and formed retaining plates.

Generally, a spring clip may be said to have two parts: the legs, which fasten the clip to the panel or flange, and the arms, which hold the component being attached. Spring clips can be classified by the application for which they are designed, the function of the arms, the way they attach to the panel, or the configuration of the legs.

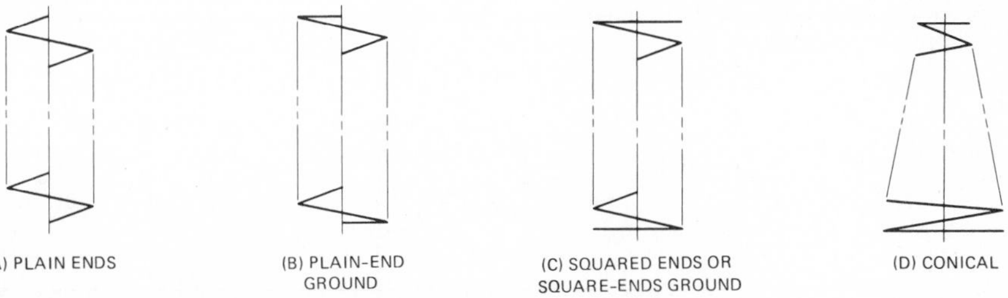
(A) PLAIN ENDS
(B) PLAIN-END GROUND
(C) SQUARED ENDS OR SQUARE-ENDS GROUND
(D) CONICAL

Fig. 9-4-7 Schematic drawing of compression springs.

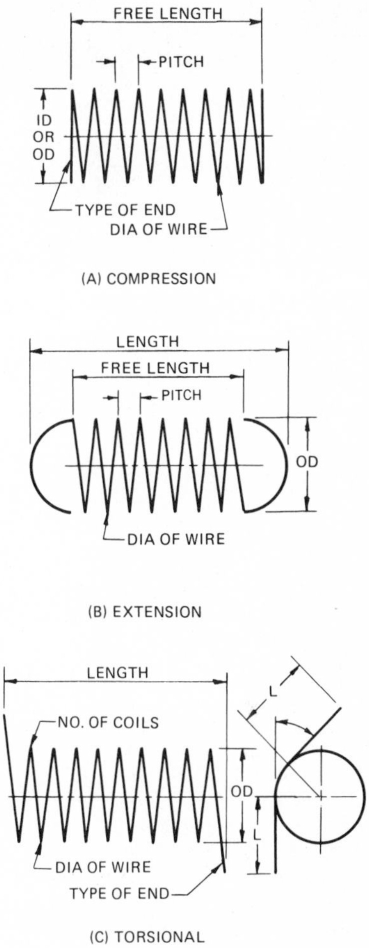
FREE LENGTH
PITCH
ID OR OD
TYPE OF END
DIA OF WIRE
(A) COMPRESSION
LENGTH
FREE LENGTH
PITCH
OD
DIA OF WIRE
(B) EXTENSION
LENGTH
NO. OF COILS
L
OD
L
DIA OF WIRE
TYPE OF END
(C) TORSIONAL

Fig. 9-4-9 Dimensioning springs.

DART-TYPE SPRING CLIPS

Dart-shaped panel retaining elements have hips to engage within panel or component holes. The top or arms of the fastener can be formed in any shape to perform unlimited fastening functions.

Dart-type spring clips are commonly used for securing two panel surfaces together as in refrigerator door liners, electronic printed-circuit attachment, cardboard facing, and cabinet shelving. Most dart-type clips are easily removable and minimize assembly damage with delicate components. The dart-type fastener requires a mounting hole and space behind the mounting panel to accommodate the dart section.

STUD RECEIVER CLIPS

There are three basic types of stud receivers: push-ons, tubular types, and self-threading fasteners. All are designed to make attachments to unthreaded studs, rivets, pins, or rods of metal or plastic.

Push-on stud receivers are made in flat and round styles with two or more

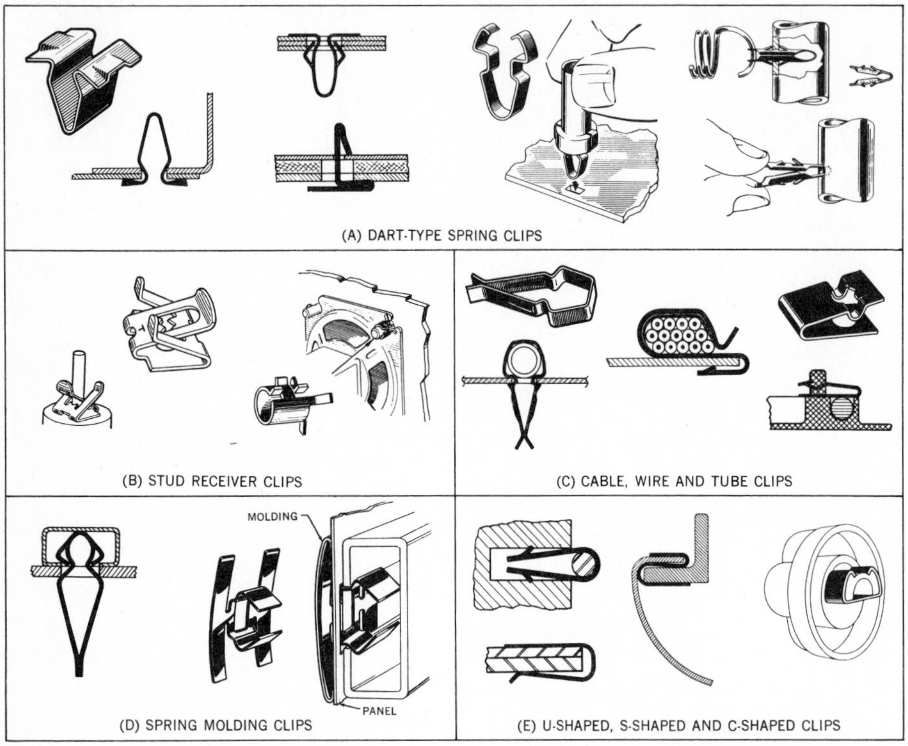
(A) DART-TYPE SPRING CLIPS
(B) STUD RECEIVER CLIPS
(C) CABLE, WIRE AND TUBE CLIPS
MOLDING
PANEL
(D) SPRING MOLDING CLIPS
(E) U-SHAPED, S-SHAPED AND C-SHAPED CLIPS

Fig. 9-4-10 Spring clips.

prongs that allow the fastener to be forced down on a stud and locked in place. Any back pressure against the fastener causes the prongs to bite deeper into the stud and prevent loosening.

CABLE, WIRE, AND TUBE CLIPS

These fasteners incorporate self-retaining elements for engaging panel holes or mounting on panel edges and flanges.

Many of the wire clips can be used on existing parts and assemblies without altering the part design. This is particularly true of those which clip on panel edges and flanges.

Spring-clip cable, wire, and tubing fasteners are front-mounting devices, requiring no access to the back of the panel.

SPRING MOLDING CLIPS

Molding retaining clips are formed with legs that hold the clips to a panel and arms that positively engage the flanges of various sizes and shapes of trim molding and pull the molding tightly to the attaching panel.

Dart-type retaining elements are probably used more for retaining molding clips to panels than any other means.

U-SHAPED, S-SHAPED, AND C-SHAPED SPRING CLIPS

These spring clips get their names from their shapes. The fastening function is accomplished by using inward compressive spring force to secure assembly components or provide self-retention after installation.

U-Shaped Clips. These are used to assemble cover panels, breaker trim, flange and channel assemblies, wire and cables, glass panels, armored cables, hinged components, and rubber and fiber materials.

S-Shaped Clips. Panels and flanges are assembled in line with one another, or at an angle with these clips. They are particularly useful in assembling plastics and other soft materials.

C-Shaped and Compression-Ring Clips. The compressive action of these clips holds plastic knobs firmly on steel shafts and permits their ready removal.

REFERENCES AND SOURCE MATERIAL

1. General Motors Corporation.
2. The Wallace Barnes Company Limited.
3. W. L. Seitl and S. Petrus, "Spring Clips," *Machine Design*, vol. 37, no. 6, 1965.

Assignment

On an A3- or B-size sheet, lay out the four assembly drawings as shown in Fig. 9-4-A or 9-4-B. Complete the drawings from the information supplied below, and make detail drawings of the springs in the spaces provided. Use your judgment for sizes not given.

- *Assembly A*. An extension spring controls the lever. The spring is fastened to the neck in the pin and through the hole in the lever. Scale is 1:1.
- *Assembly B*. A compression spring mounted on the shaft of the handle provides sufficient pressure to hold the lever in position, thus maintaining the door against the panel. To open the door, the handle is pushed in and turned. This action compresses the spring and forces the lever away from the notch in the panel edge, thus permitting the lever to turn. Scale is 1:1.
- *Assembly C*. The license plate holder is held to the frame of the car by a hinge. A torsion spring is required to keep the plate holder in position. The torsion spring is slipped over the hinge pin during assembly, and one end of the spring passes through the hole in the bumper. The other end of the spring is locked into the spring-retaining notch in the license plate holder. Scale is 1:2.
- *Assembly D*. Flat springs are positioned in openings C and D in the tape deck player. These springs hold the cassette against the bottom, and the locating pin positioned in the left side of the tape deck. Scale is 1:2.

REVIEW FOR ASSIGNMENT

Unit 7-5 Assemblies in Section

UNIT 9-5

RIVETS

Standard Rivets[1,2]

Riveting is a popular method of fastening and joining, primarily because of its simplicity, dependability, and low cost. Myriad manufactured products and structures, both small and large, are held together by these fasteners. Rivets are classified as permanent fastenings, as distinguished from removable fasteners, such as bolts and screws.

Basically, a *rivet* is a ductile metal pin which is inserted through holes in two or more parts, and the ends are formed over to securely hold the parts.

Another important reason for riveting is versatility, with respect to both the properties of rivets as fasteners and the method of clinching.

- Part materials: Rivets can be used to join dissimilar materials, metallic or nonmetallic, in various thicknesses.
- Rivet materials and finishes: Rivets can be made of any material that can

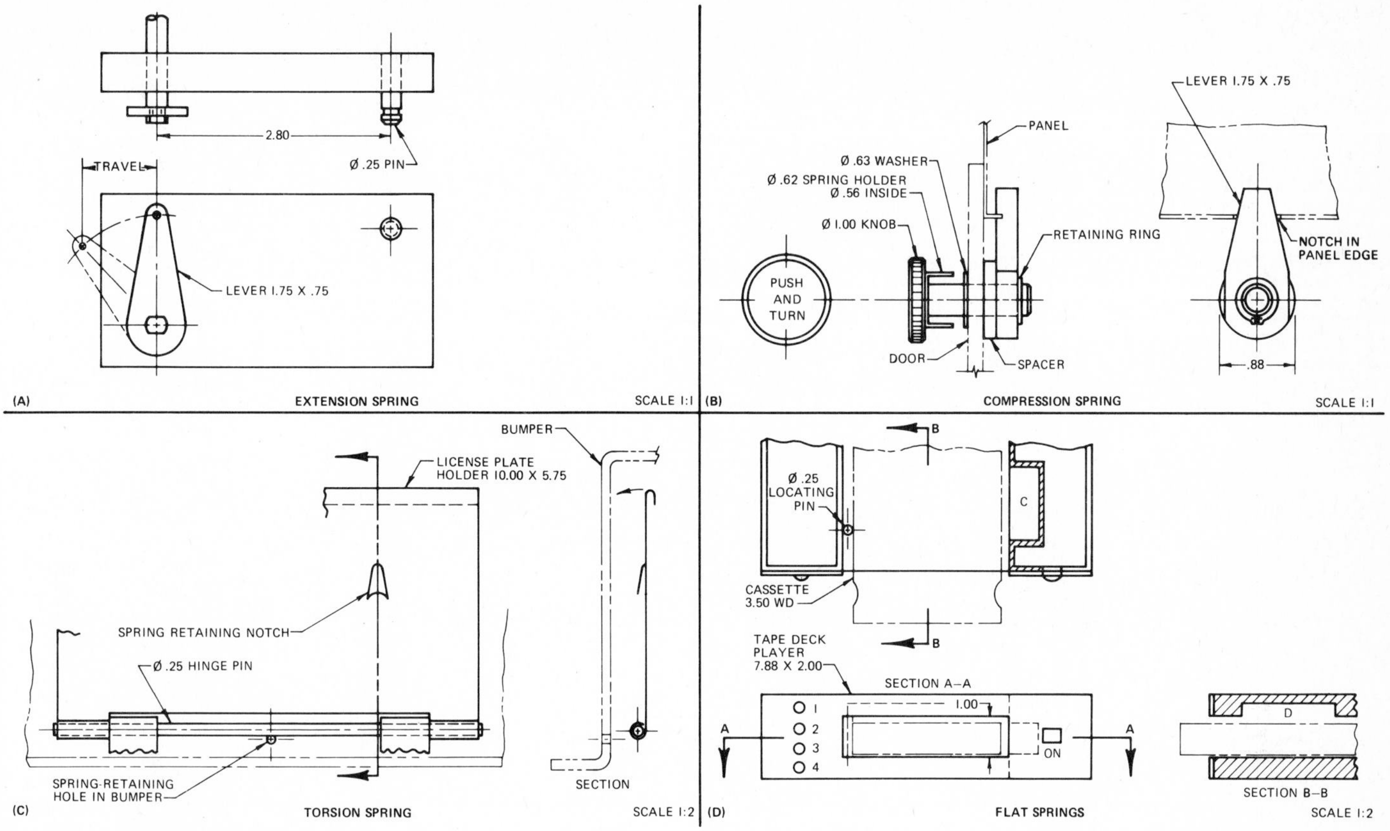
2.80
Ø .25 PIN
TRAVEL
LEVER 1.75 X .75
(A)
EXTENSION SPRING
SCALE 1:1
LEVER 1.75 X .75
PANEL
Ø .63 WASHER
Ø .62 SPRING HOLDER
Ø .56 INSIDE
Ø 1.00 KNOB
RETAINING RING
NOTCH IN
PANEL EDGE
PUSH
AND
TURN
DOOR
SPACER
.88
(B)
COMPRESSION SPRING
SCALE 1:1
BUMPER
LICENSE PLATE
HOLDER 10.00 X 5.75
SPRING RETAINING NOTCH
Ø .25 HINGE PIN
SPRING-RETAINING
HOLE IN BUMPER
SECTION
(C)
TORSION SPRING
SCALE 1:2
B
Ø .25
LOCATING
PIN
C
CASSETTE
3.50 WD
B
TAPE DECK
PLAYER
7.88 X 2.00
SECTION A–A
1.00
A
1
2
3
4
ON
A
D
SECTION B–B
(D)
FLAT SPRINGS
SCALE 1:2

Fig. 9-4-A Spring fasteners.

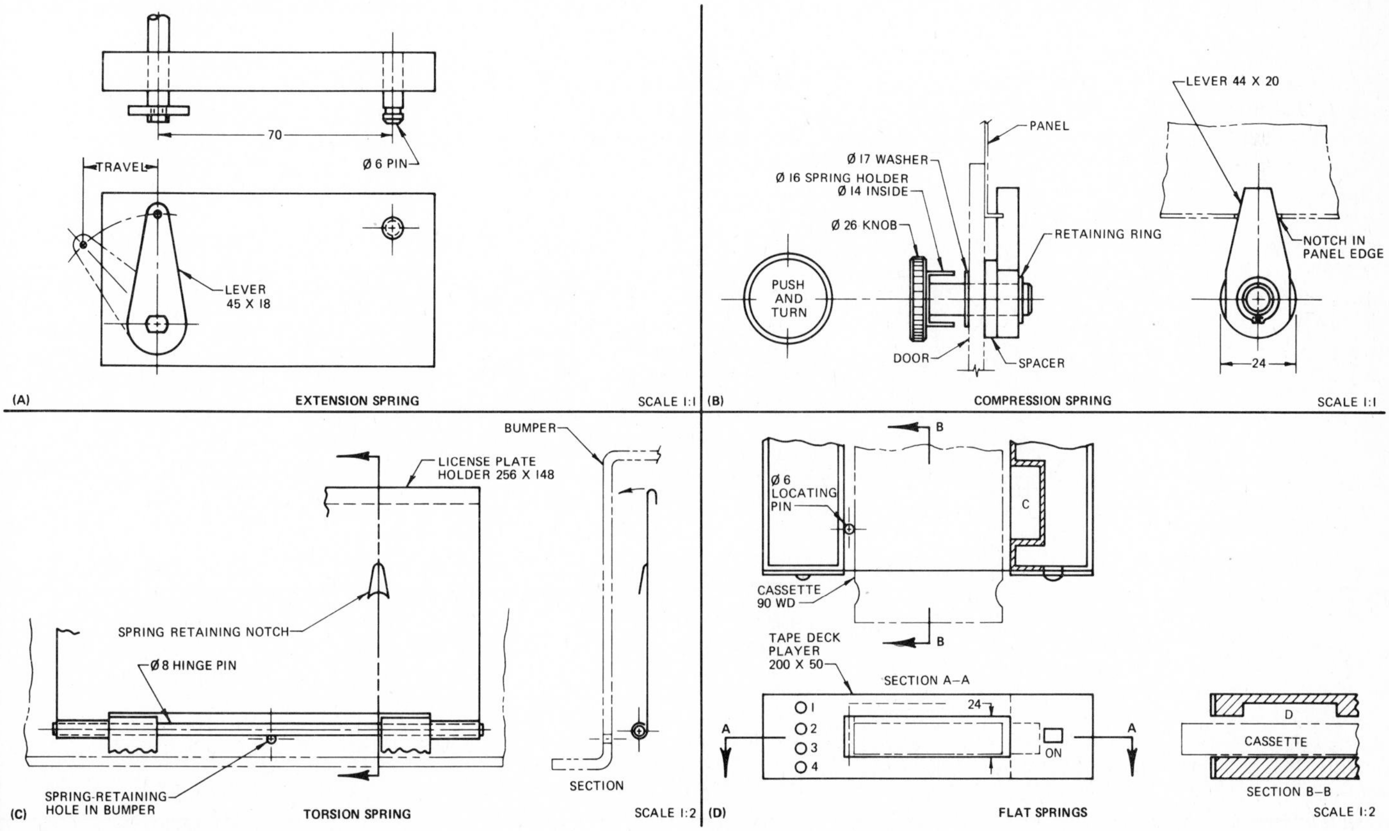
70
Ø 6 PIN
TRAVEL
LEVER
45 X 18
(A)
EXTENSION SPRING
SCALE 1:1
LEVER 44 X 20
PANEL
Ø 17 WASHER
Ø 16 SPRING HOLDER
Ø 14 INSIDE
Ø 26 KNOB
RETAINING RING
NOTCH IN
PANEL EDGE
PUSH
AND
TURN
DOOR
SPACER
24
(B)
COMPRESSION SPRING
SCALE 1:1
BUMPER
LICENSE PLATE
HOLDER 256 X 148
SPRING RETAINING NOTCH
Ø 8 HINGE PIN
SPRING-RETAINING
HOLE IN BUMPER
SECTION
(C)
TORSION SPRING
SCALE 1:2
B
Ø 6
LOCATING
PIN
C
CASSETTE
90 WD
B
TAPE DECK
PLAYER
200 X 50
SECTION A–A
24
A
1
2
3
4
ON
A
D
CASSETTE
SECTION B–B
(D)
FLAT SPRINGS
SCALE 1:2

Fig. 9-4-B Spring fasteners.

be cold-worked. They can have a variety of finishes such as plating, parkerizing, or paint.

- Shape flexibility: As long as there are flat, parallel surfaces for the rivet clinch and head, plus adequate space for the rivet driver during clinching, riveting can be adapted to fasten almost any part shape.
- Multiple functions: Rivets can serve as fasteners, pivot shafts, spacers, electric contacts, stops, or inserts.
- Fastening finished parts: Rivets can be used to fasten parts that have already received a final painting or other finishing.

Riveted joints are neither watertight nor airtight, although such a joint may be attained at some added cost by using a sealing compound. The riveted parts cannot be disassembled for maintenance or replacement without knocking the rivet out and clinching a new one in place for reassembly. Common riveted joints are shown in Fig. 9-5-1.

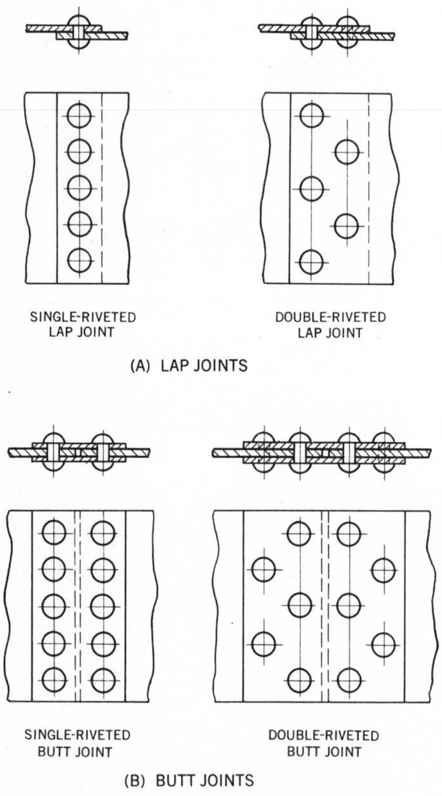

SINGLE-RIVETED LAP JOINT
DOUBLE-RIVETED LAP JOINT
(A) LAP JOINTS
SINGLE-RIVETED BUTT JOINT
DOUBLE-RIVETED BUTT JOINT
(B) BUTT JOINTS

Fig. 9-5-1 Common riveted joints.

LARGE RIVETS

Large rivets are used in structural work of buildings and bridges. Today, however, high-strength bolts have almost completely replaced rivets in field connections because of cost, strength, and noise factor. Rivet joints are of two types: butt and lapped. The more common types are shown in Fig. 9-5-2. In order to show the difference between *shop rivets* (rivets that are put in the structure at the shop) and *field rivets* (rivets that are used on the site), two types of symbols are used. In drawing shop rivets, the diameter of the rivet head is shown on the drawings. For field rivets, the shaft diameter is used. Figure 9-5-3 shows the conventional rivet symbols adopted by the American and Canadian Institutes of Steel Construction.

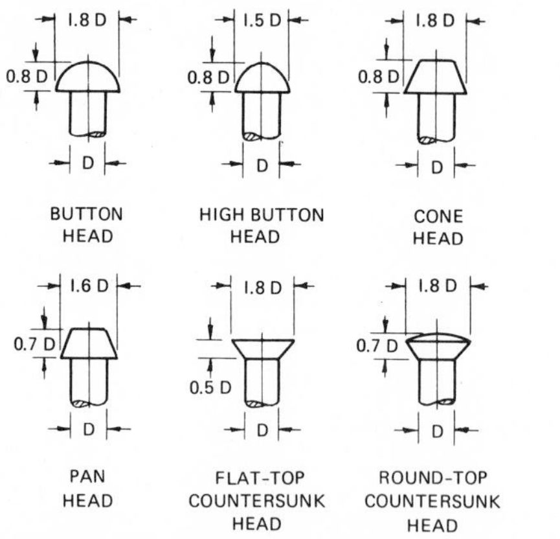

1.8 D
0.8 D
D
BUTTON HEAD
1.5 D
0.8 D
D
HIGH BUTTON HEAD
1.8 D
0.8 D
D
CONE HEAD
1.6 D
0.7 D
D
PAN HEAD
1.8 D
0.5 D
D
FLAT-TOP COUNTERSUNK HEAD
1.8 D
0.7 D
D
ROUND-TOP COUNTERSUNK HEAD

Fig. 9-5-2 Approximate sizes and types of large rivets 12 mm and up.

SMALL RIVETS[3]

Design of small rivet assemblies is influenced by two major considerations:

1. The joint itself, its strength, appearance, and configuration
2. The final riveting operation, in terms of equipment capabilities and production sequence

Since the basic advantage of riveting is that it permits automatic feeding of the fastener to assume high production rates and low assembly costs, the product design should be related to the riveting equipment and the assembly sequence if optimum results are to be obtained.

TYPES OF SMALL RIVETS (Fig. 9-5-4)

Semitubular. This is the most widely used type of small rivet. The depth of the hole in the rivet, measured along the wall, does not exceed 112 percent of the mean shank diameter. The hole may be extruded (straight or tapered) or drilled (straight), depending on the manufacturer and/or rivet size.

Full Tubular. This rivet has a drilled shank with a hole depth more than 112 percent of the mean shank diameter. It can be used to punch its own hole in fabric, some plastic sheets, and other soft materials, eliminating a preliminary punching or drilling operation.

Bifurcated (Split). The rivet body is sawed or punched to produce a pronged shank that punches its own hole through fiber, wood, or plastic. With a few exceptions, punched shanks are more suitable than sawed shanks for piercing operations on nonmetallic materials.

Compression. This rivet consists of two elements: the solid or blank rivet and the deep-drilled tubular member. Pressed together, these form an interference fit. Because the heads of both members can be produced to close tolerances, these rivets are commonly used when appearance from both sides of the work must be uniform and heads must be flush to prevent accumulation of dirt or

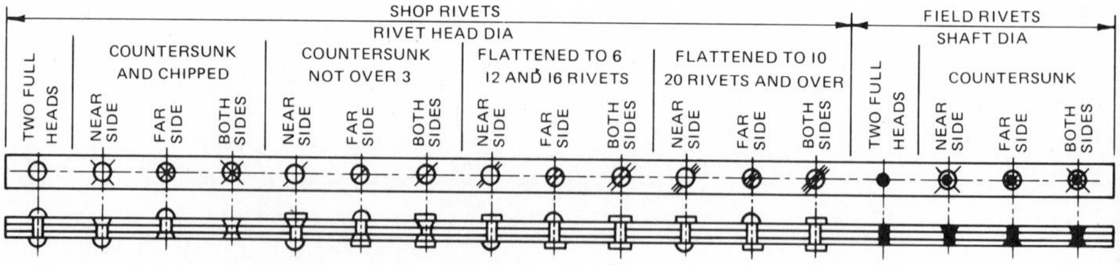

SHOP RIVETS
RIVET HEAD DIA
FIELD RIVETS
SHAFT DIA
TWO FULL HEADS
COUNTERSUNK AND CHIPPED
COUNTERSUNK NOT OVER 3
FLATTENED TO 6 12 AND 16 RIVETS
FLATTENED TO 10 20 RIVETS AND OVER
COUNTERSUNK
NEAR SIDE
FAR SIDE
BOTH SIDES

Fig. 9-5-3 Conventional rivet symbols.

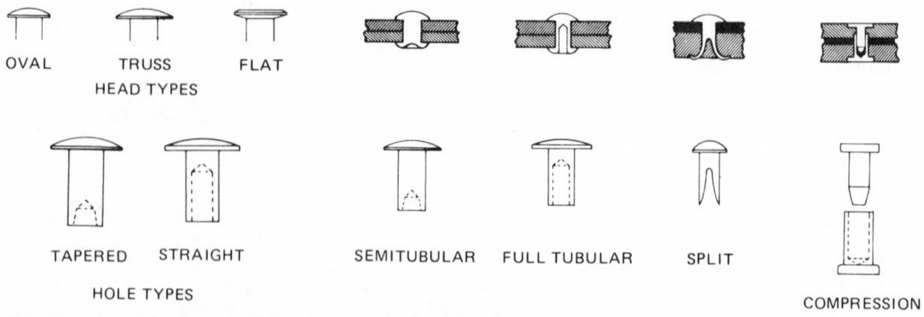

OVAL
TRUSS
FLAT
HEAD TYPES
TAPERED
STRAIGHT
HOLE TYPES
SEMITUBULAR
FULL TUBULAR
SPLIT
COMPRESSION

Fig. 9-5-4 Basic types of small rivets.

waste. They can be used in wood, brittle plastics, or other materials with little danger of splitting during setting.

DESIGN RECOMMENDATIONS (Fig. 9-5-5)

Select the Right Rivets. Basic types are covered in Fig. 9-5-4. Rivet standards for all types but compression rivets have been published by the Tubular and Split Rivet Council.

Use Proper Type of Clinch. There are two basic types of clinches for full tubular and semitubular rivets: roll set and star set.

The *roll set clinch* forms a rolled edge around the end of the rivet shank that clinches onto the work. A *star set clinch* cuts the wall of the shank and sets the rivet in the form of a star.

Rivet Diameters. The optimum rivet diameter is determined not by performance requirements but by economics—the costs of the rivet and the labor to install it. On the other hand, the required rivet shank length is fixed by the amount of rivet material needed for clinching and the total material thickness.

The most important factor in minimizing the initial cost of a rivet is to specify one of standard diameter, length increment, and tolerances.

Rivet Positioning. The location of the rivet in the assembled product influences both joint strength and clinching requirements. The important dimensions are edge distance and pitch distance.

Edge distance is the interval between the edge of the part and the center line of the rivet. If the edge distance in the direction of shear load is too small, then bearing strength of the material is significantly reduced and the joint might fail. Edge distance should be at least twice the rivet shank diameter.

The recommended edge distance for plastic materials, either solid or laminated, is between 2 and 3 diameters, depending on the thickness and inherent strength of the material.

Pitch distance—the interval between center lines of adjacent rivets—should not be too small. Unnecessarily high stress concentrations in the riveted material and buckling at adjacent empty holes can result if the pitch distance is less than 3 times the diameter of the largest rivet in the assembly (metal parts) or 5 times the diameter (plastic parts).

Blind Rivets[4,5]

Blind riveting is a technique for setting a rivet without access to the reverse side of the joint. However, blind rivets are used in many applications in which both sides of the joint are actually accessible.

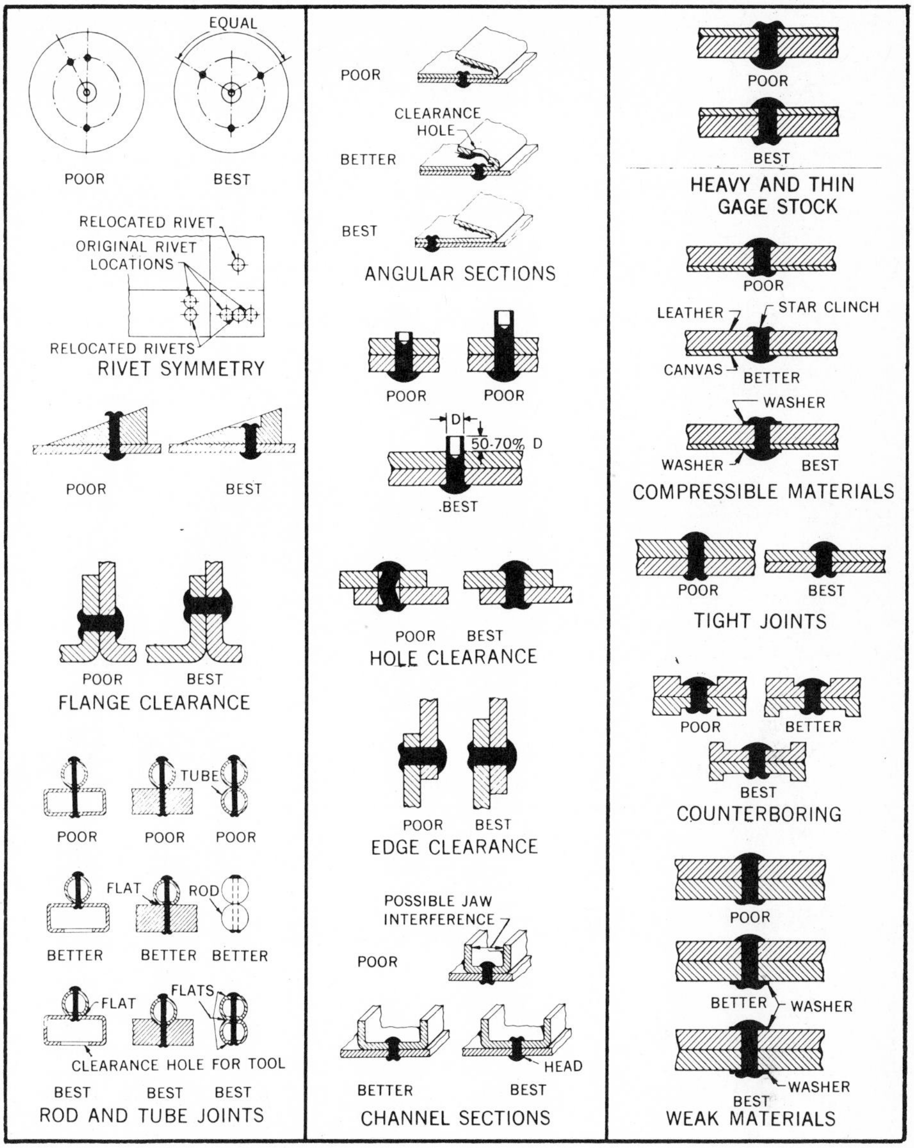

EQUAL
POOR
BEST
RELOCATED RIVET
ORIGINAL RIVET LOCATIONS
RELOCATED RIVETS
RIVET SYMMETRY
POOR
BEST
POOR
BEST
FLANGE CLEARANCE
TUBE
POOR
POOR
POOR
FLAT
ROD
BETTER
BETTER
BETTER
FLAT
FLATS
CLEARANCE HOLE FOR TOOL
BEST
BEST
BEST
ROD AND TUBE JOINTS
POOR
CLEARANCE HOLE
BETTER
BEST
ANGULAR SECTIONS
POOR
POOR
D
50-70% D
BEST
POOR
BEST
HOLE CLEARANCE
POOR
BEST
EDGE CLEARANCE
POSSIBLE JAW INTERFERENCE
POOR
HEAD
BETTER
BEST
CHANNEL SECTIONS
POOR
BEST
HEAVY AND THIN GAGE STOCK
POOR
LEATHER
STAR CLINCH
CANVAS
BETTER
WASHER
WASHER
BEST
COMPRESSIBLE MATERIALS
POOR
BEST
TIGHT JOINTS
POOR
BETTER
BEST
COUNTERBORING
POOR
BETTER
WASHER
WASHER
BEST
WEAK MATERIALS

Fig. 9-5-5 Design types for small rivets.

Blind rivets should be considered when the fasteners will not have to be removed for maintenance purposes, will have to function in a high-vibration environment, will serve only as tack or temporary fasteners, or will be used as repair fasteners by untrained operators in the field.

Blind rivets are classified according to the methods with which they are set: pull-mandrel, drive-pin, and chemically expanded. See Fig. 9-5-6.

DESIGN CONSIDERATIONS (Fig. 9-5-7)

Type of Rivet. Selection depends on a number of factors, such as speed of assembly, clamping capacity, available sizes, adaptability to the assembly, ease of removal, cost, and structural integrity of the joint.

Joint Design. Factors that must be known include allowable tolerances of rivet length versus assembly thickness, hole clearance, joint configuration, and type of loading.

Speed of Installation. The fastest, most efficient installation is done with power tools—air, hydraulic, or electric. Manual tools, such as special pliers, can be used efficiently with practically no training.

In-Place Costs. Blind rivets often have lower in-place costs than solid rivets or tapping screws because of low tooling investment, high installation speed, and single-operator requirements.

Loading. A blind-rivet joint is usually in compression or shear, both of which the rivets can support somewhat better than tensile loading.

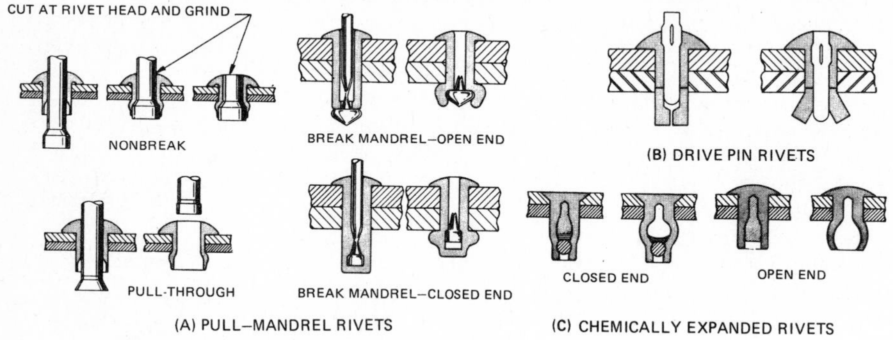
CUT AT RIVET HEAD AND GRIND
NONBREAK
BREAK MANDREL—OPEN END
(B) DRIVE PIN RIVETS
PULL-THROUGH
BREAK MANDREL—CLOSED END
CLOSED END
OPEN END
(A) PULL—MANDREL RIVETS
(C) CHEMICALLY EXPANDED RIVETS

Fig. 9-5-6 Basic types of blind rivets and methods of setting. (*Machine Design*, vol. 45, no. 28, 1973.)

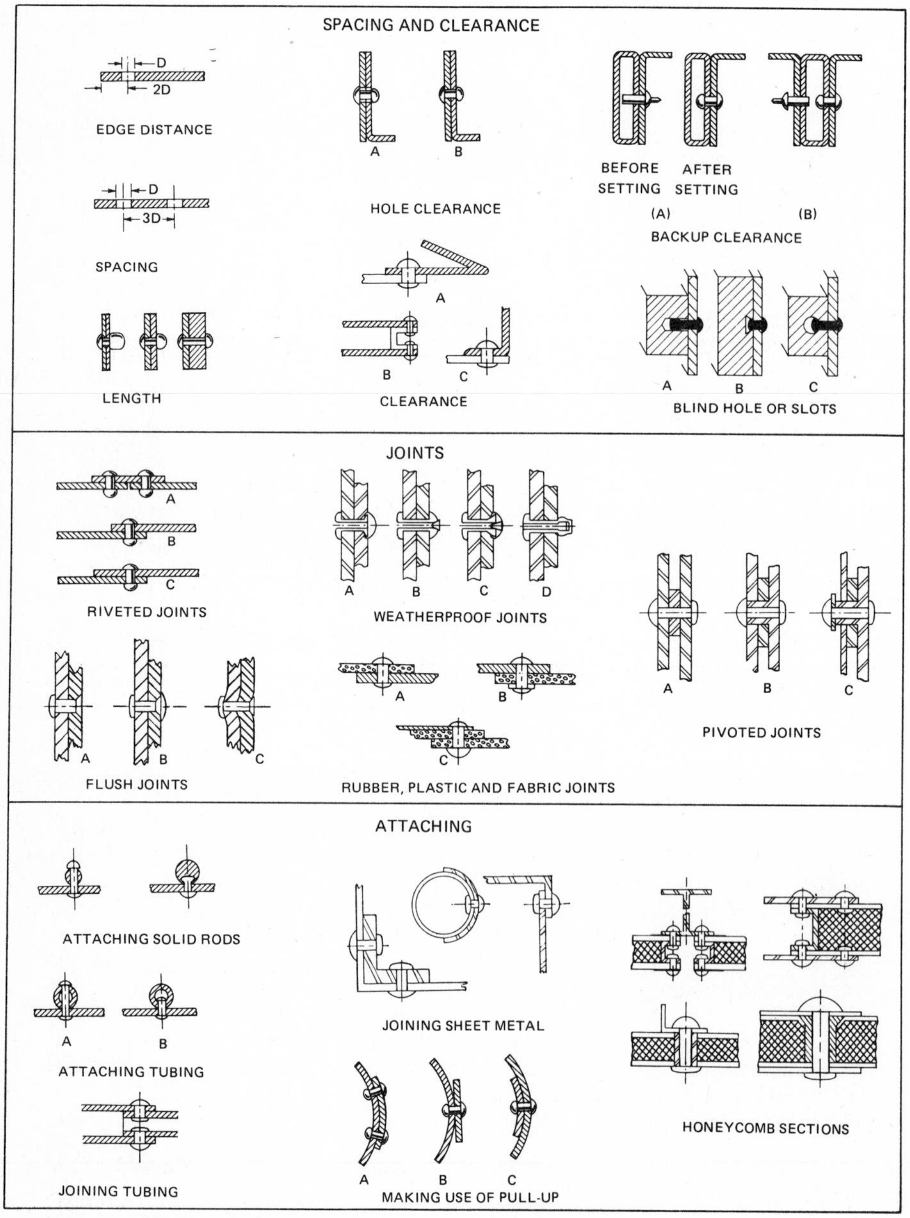
SPACING AND CLEARANCE
D
2D
EDGE DISTANCE
D
3D
SPACING
LENGTH
A
B
HOLE CLEARANCE
A
B
C
CLEARANCE
BEFORE SETTING
AFTER SETTING
(A)
(B)
BACKUP CLEARANCE
A
B
C
BLIND HOLE OR SLOTS
JOINTS
A
B
C
RIVETED JOINTS
A
B
C
D
WEATHERPROOF JOINTS
A
B
C
FLUSH JOINTS
A
B
C
RUBBER, PLASTIC AND FABRIC JOINTS
A
B
C
PIVOTED JOINTS
ATTACHING
ATTACHING SOLID RODS
A
B
ATTACHING TUBING
JOINING TUBING
JOINING SHEET METAL
A
B
C
MAKING USE OF PULL-UP
HONEYCOMB SECTIONS

Fig. 9-5-7 Blind rivet design data.

Material Thickness. Some rivets can be set in materials as thin as 0.5 mm if the head is properly formed and shank expansion is carefully controlled during setting. If possible, however, it is best to form the blind head against the thicker sheet of a combination. Also, if one component is of compressible material, rivets with extra-large head diameter should be used.

Edge Distance. The average recommended edge distance is twice the diameter of the rivet. In lightly loaded structures where the rivet performs only a holding function, this can be decreased to 1.5 diameters, and in heavily loaded structures it may be necessary to increase it to as much as 3 diameters to develop required joint strength.

Spacing. Rivet pitch should be 3 times the diameter of the rivet. Depending upon the nature of load, it may be desirable to decrease or increase this distance. However, it is generally considered that 3 diameters overcome the tendency of the material to fail and concentrate the load on the rivet.

Length. The amount of length needed for clinching action varies greatly and depends on material being fastened, necessary strength, and the method of riveting. Most rivet manufacturers provide data on grip ranges of their rivets to simplify selection for the user.

Hole Clearance. For vibration-proof construction or high shear strength, hole tolerances can be held very tight (B).

Clearance. An angular section is riveted to a straight member, A. If the angle projects above the rivet head, it may be difficult to put the tool in to upset the rivet. One method of solving this problem is to pull the rivet from the underside of the work. In riveting the U-channel, B, to the outer members, the rivet must be fed from the outer side. On this joint, allow enough back-up clearance to prevent interference. At C, not only tool clearance but also rivet head clearance is critical. A large-diameter head may interfere with the wall.

Back-Up Clearance. Full entry of the rivet is essential for tightly clinched joints. Sufficient back-up clearance must be provided to accommodate the full length of the unclinched rivet, A. In the case of two opposed rivets, the minimum allowable space accommodates one clinched and one unclinched rivet, B.

Blind Holes or Slots. A useful application of a blind rivet is in fastening members in a blind hole. At A, the formed head bears against the side of the hole only. As could be expected, this joint is not as strong as the other two (B and C), but the clinching action of the rivet head provides sufficient strength for light tension loads and moderate shear loads.

Riveted Joints. Riveted cleat or batten holds a butt joint, A. The simple lap joint, B, must have sufficient material beyond the hole for strength. Excessive material beyond rivet hole, C, may curl up or vibrate or cause interference problems, depending on the installation.

Flush Joints. Generally, flush joints are made by countersinking one of the sections and using a rivet with a countersunk head, A. Some rivets are available in either a round or a flat-top countersunk head, but not both. The rivet at B is capped.

Another popular method of providing flush assembly and gaining additional bearing strength is to dimple the sheet by forming a conical projection on the back of the sheets with a die, C.

Weatherproof Joints. A hollow-core rivet can be sealed by capping it, A; by plugging it, B; or by using both a cap and a plug, C. To obtain a true seal, however, a gasket or mastic should be used between the sections and perhaps under the rivet head. An ideal solution is to use a closed-end rivet, D.

Rubber, Plastic, and Fabric Joints. Some plastics, such as reinforced, molded Fiberglas, or polystyrene, which are reasonably rigid, present no problem for most small rivets. However, when the material is very flexible or is a fabric, set the rivet as shown at A or B, with the upset head against the solid member. If this practice is not possible, use a back-up strip as shown at C.

Pivoted Joints. There are a number of ways of producing a pivoted assembly, three of which are shown.

Attaching Solid Rod. In attaching a rod to other members, the usual practice is to pass the rivet completely through the rod.

Attaching Tubing. Attaching tubing is an application for which the blind rivet is ideally suited. At A, the rivet goes completely through the tubing. A very long rivet is required, but little advantage is obtained over the design shown at B, where a shorter length is used.

Joining Tubing. This tubing joint is a common form of blind riveting, used for both structural and low-cost power transmission assemblies.

Making Use of Pull-Up. By judicious positioning of rivets and parts that are to be assembled with rivets, the setting force can sometimes be used to pull together unlike parts.

Honeycomb Sections. Inserts should be employed to strengthen the section and provide a strong joint. Otherwise, setting the rivet may deform the section and cause a structural weakness that may later result in failure.

REFERENCES AND SOURCE MATERIAL

1. T. E. Weissman, "Guide to Fastening Devices," *Canadian Machinery and Metalworking*, vol. 73, no. 12, 1962.

2. Technical Committee, "Tubular and Split Rivet Council," *Machine Design*, vol. 45, no. 28, 1973.

3. A. Hartwell et al., "Small Rivets," *Machine Design*, vol. 37, no. 6, 1965.

4. T. R. Freeman, "Blind Rivets," *Machine Design*, vol. 37, no. 6, 1965.

5. Technical Committee, "Blind Rivets and Bolts," Industrial Fasteners Institute, *Machine Design*, vol. 45, no. 28, 1973.

Assignment

On an A3- or B-size sheet, draw the four assembly drawings shown in Fig. 9-5-A or 9-5-B. Complete the drawings from the information supplied below. Refer to manufacturers' catalogs for rivet type and sizes, and on each assembly show the callout for the rivets. Scale is as shown. Use your judgment for sizes not given.

- *Assembly A*. The grill is held to the panel by four truss-head full tubular rivets.
- *Assembly B*. The support is held to the plywood panel by drive rivets uniformly spaced on the gage lines. Two rivets hold the bracket to the support.
- *Assembly C*. Padlock brackets are riveted to the locker door and door frame with two blind rivets in each bracket.
- *Assembly D*. The roof truss is assembled in the shop with five evenly spaced ϕ12-mm (or 0.50-in.) rivets in each angle.

REVIEW FOR ASSIGNMENT

Unit 7-5 Assemblies in Section

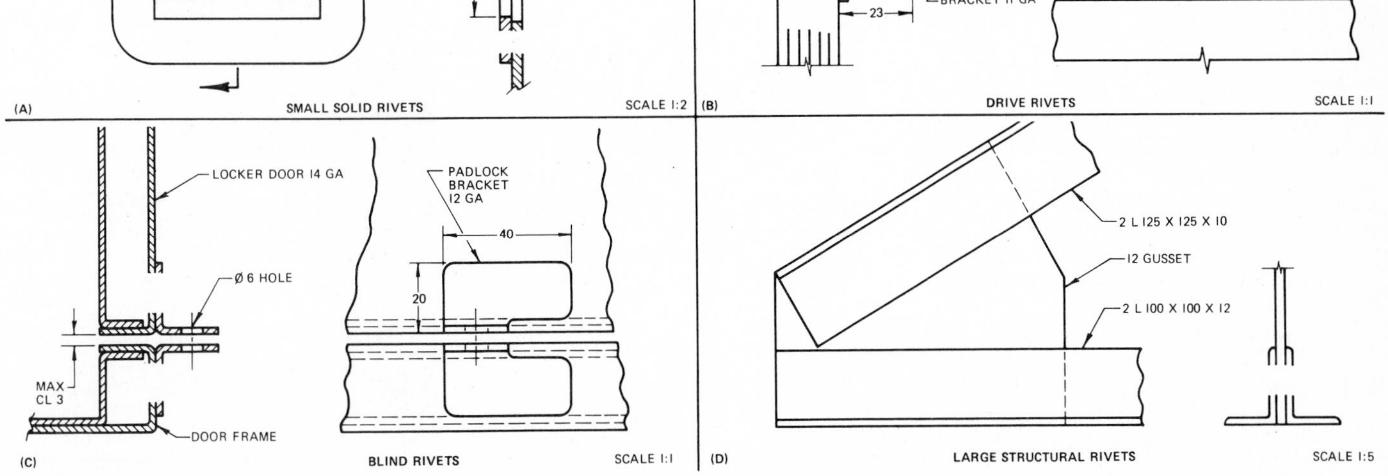
PLASTIC GRILL
76 X 76 X 4
PLASTIC PANEL
PANEL
4 THICK
50
(A)
SMALL SOLID RIVETS
SCALE 1:2
PLYWOOD
RIVET GAGE LINES
SUPPORT 10 GA
BRACKET 11 GA
23
26
6
38
(B)
DRIVE RIVETS
SCALE 1:1
LOCKER DOOR 14 GA
Ø 6 HOLE
MAX
CL 3
DOOR FRAME
PADLOCK
BRACKET
12 GA
40
20
(C)
BLIND RIVETS
SCALE 1:1
2 L 125 X 125 X 10
12 GUSSET
2 L 100 X 100 X 12
(D)
LARGE STRUCTURAL RIVETS
SCALE 1:5

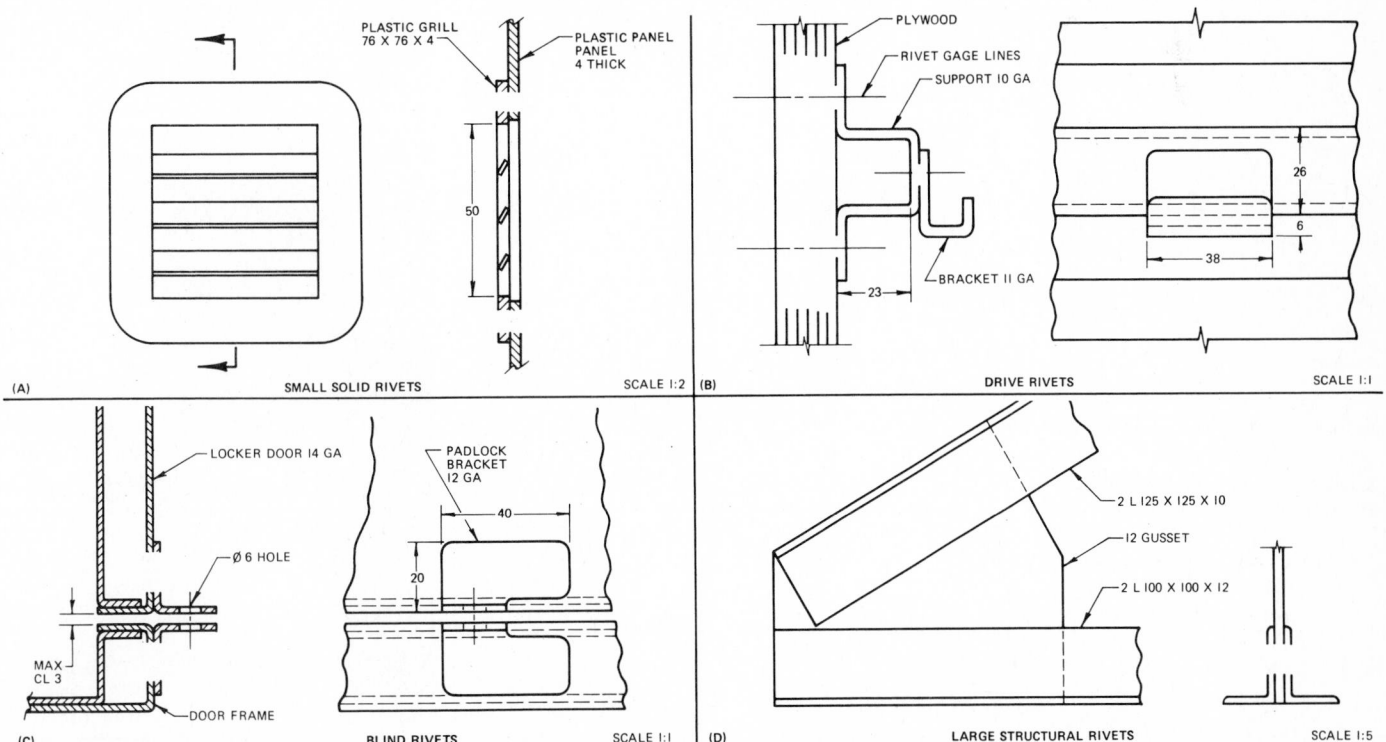

Fig. 9-5-A Rivet fasteners.

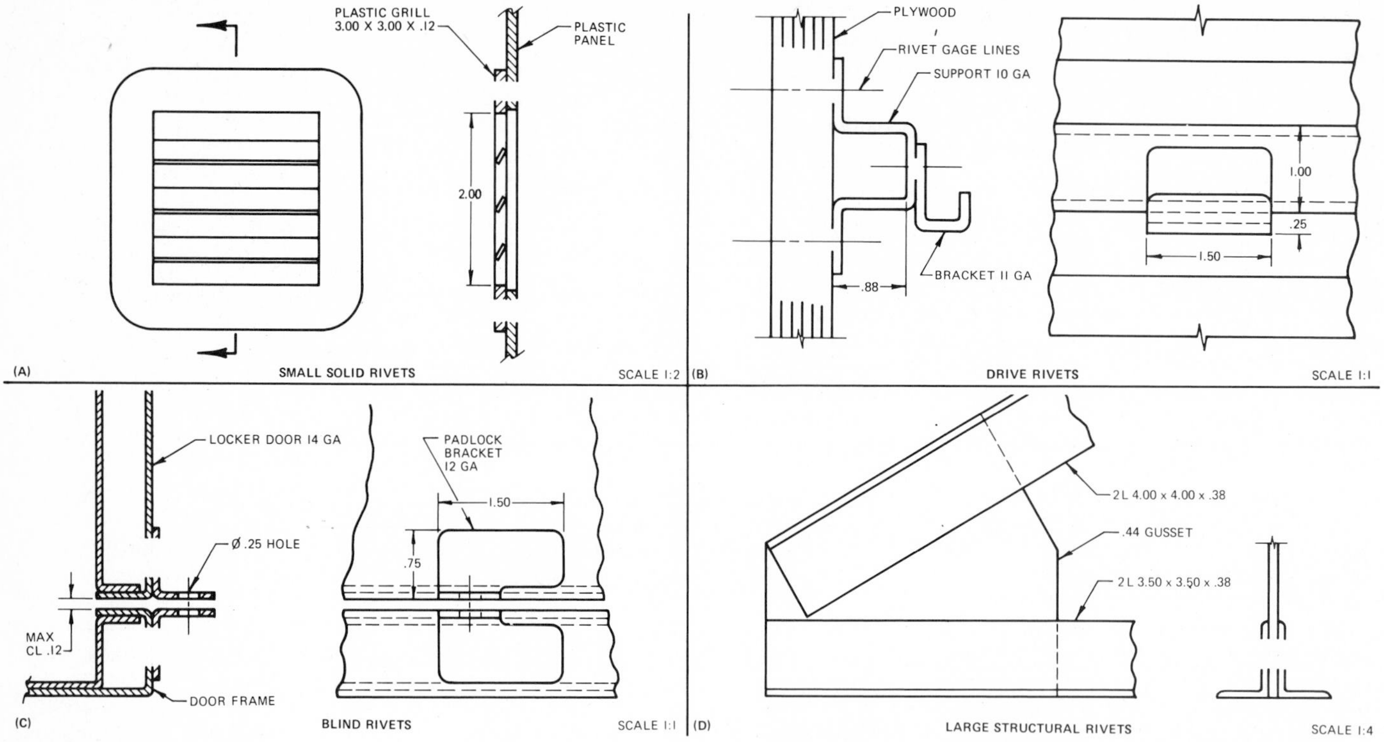
PLASTIC GRILL
3.00 X 3.00 X .12
PLASTIC PANEL
2.00
(A)
SMALL SOLID RIVETS
SCALE 1:2
PLYWOOD
RIVET GAGE LINES
SUPPORT 10 GA
BRACKET 11 GA
.88
1.00
.25
1.50
(B)
DRIVE RIVETS
SCALE 1:1
LOCKER DOOR 14 GA
Ø .25 HOLE
MAX CL .12
DOOR FRAME
PADLOCK BRACKET 12 GA
1.50
.75
(C)
BLIND RIVETS
SCALE 1:1
2L 4.00 x 4.00 x .38
.44 GUSSET
2L 3.50 x 3.50 x .38
(D)
LARGE STRUCTURAL RIVETS
SCALE 1:4

Fig. 9-5-B Rivet fasteners.

UNIT 9-6

WELDED FASTENERS

The most common forms of welded fasteners are screws and nuts. Welded pins or unthreaded studs generally serve as locating or bearing surfaces, rather than as fasteners. In this section, welded fasteners are grouped into ***resistance-welded*** threaded fasteners and ***arc-welded*** studs.

Resistance-Welded Fasteners[1]

Simply defined, a *resistance-welded fastener* is an externally or internally threaded metal part designed to be fused permanently in place by standard production welding equipment. Two methods of resistance welding are used to attach these fasteners (Fig. 9-6-1): projection welding or spot welding.

With either method, the fusion of the fastener to the metal-part surface is the result of the natural resistance of metal to a controlled current under pressure.

The most common forms of weld fasteners are screws and nuts. Pins are also available but generally serve as a locating or bearing surface when applied as a welded component rather than as a fastener (Fig. 9-6-2.)

DESIGN CONSIDERATIONS

Before fasteners can be used, four basic requirements must be met. See Figs. 9-6-3 and 9-6-4.

1. The materials to be joined, both part and fastener, must be suitable for resistance welding. The material most commonly used for weld fasteners is low-carbon steel, such as SAE 1010.

2. The parts to be welded must be portable enough to be carried to the welder. Use of portable welding equipment with these types of fasteners is generally not recommended.

3. Production volume should be great enough to justify tooling costs. As a general rule of thumb, a quantity of at least 1000 parts is generally considered necessary to make the application of weld fasteners feasible from a cost standpoint.

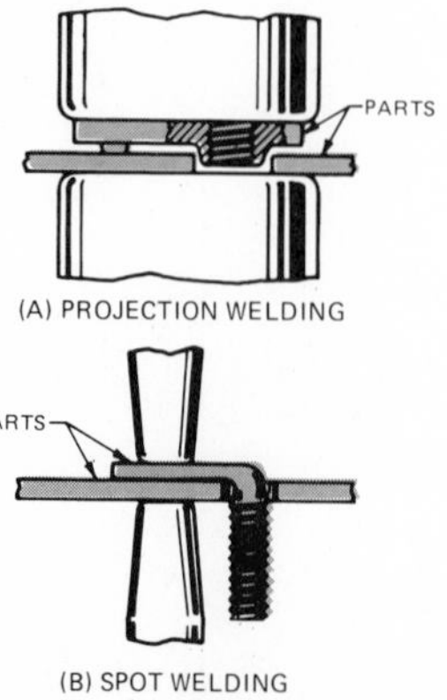
PARTS
(A) PROJECTION WELDING
PARTS
(B) SPOT WELDING

Fig. 9-6-1 Basic methods for attaching resistance-welded fasteners.

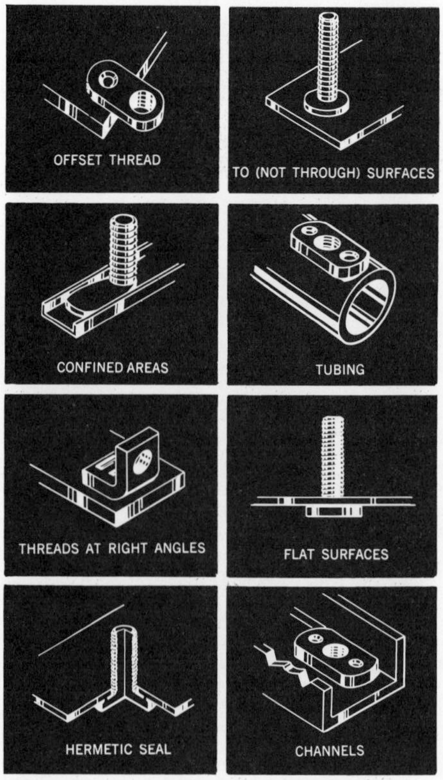
OFFSET THREAD
TO (NOT THROUGH) SURFACES
CONFINED AREAS
TUBING
THREADS AT RIGHT ANGLES
FLAT SURFACES
HERMETIC SEAL
CHANNELS

Fig. 9-6-2 Application of resistance-welded fasteners. (Ohio Nut and Bolt Co.)

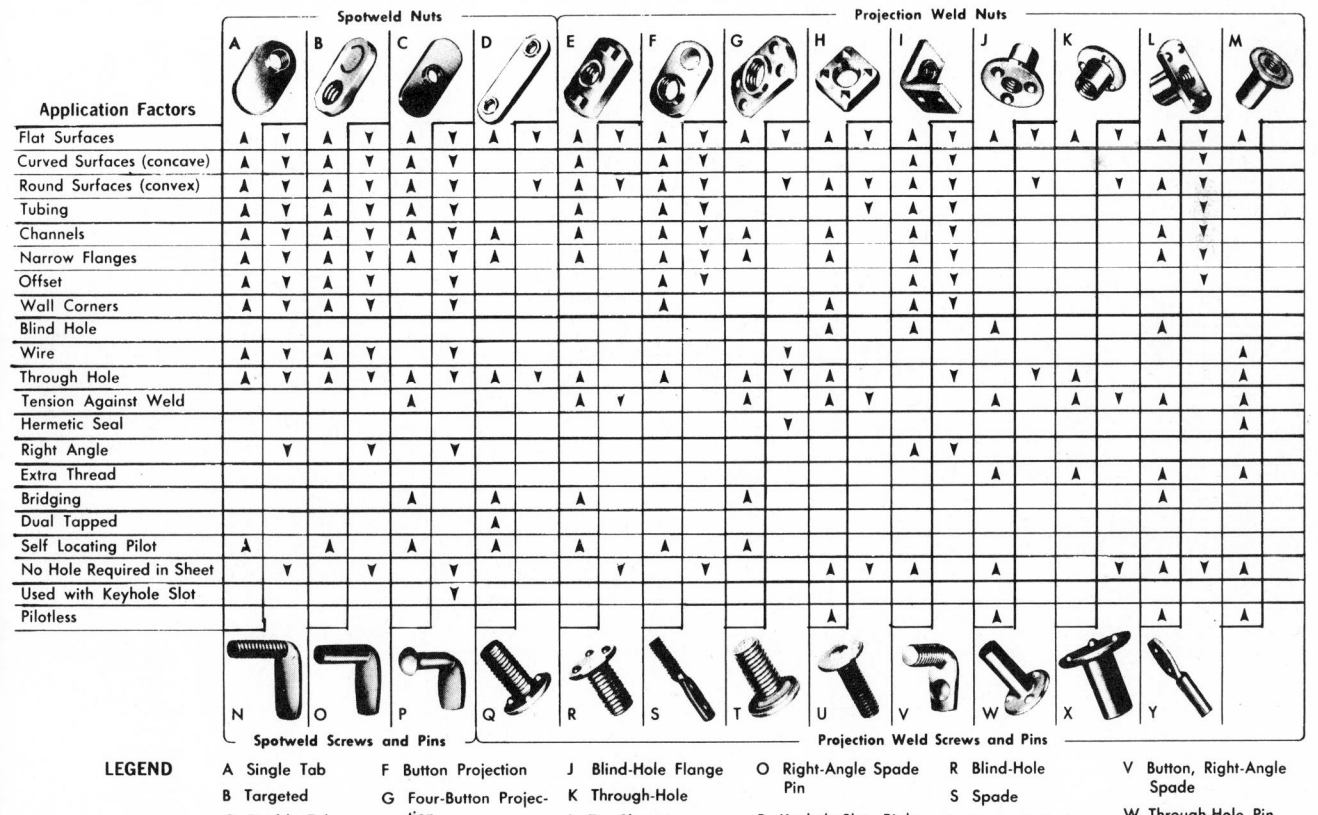

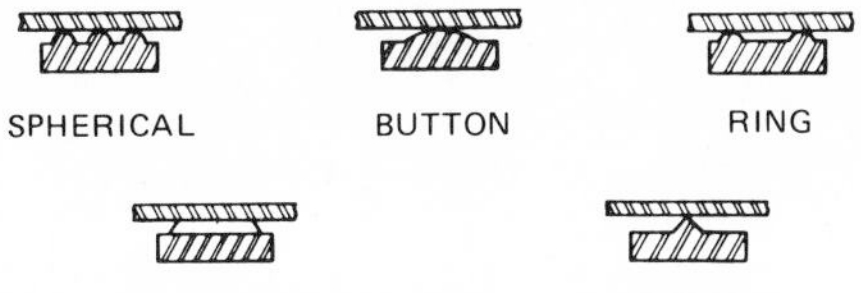
SPHERICAL
BUTTON
RING
RIB
PYRAMIDAL

Fig. 9-6-3 Basic types of weld projection.

USE PROJECTION-WELD FASTENERS WHEN	USE SPOT-WELD FASTENERS WHEN
• Suitable projection welding equipment is available.	• Suitable rocker-arm welding equipment is available.
• Appearance is an important consideration. Projection welding does not mark the surface on the opposite side of the weld.	• Appearance of the part surface opposite the weld is not critical. Spot welding leaves a slight indentation from the electrode tips.
• Simultaneous welding of multiple fasteners is required.	• Other spot welds are being performed on parts of the assembly.
• Spacing between fasteners must be kept close.	• Length of production run without maintenance is not too important. Spot-welding electrode tips will mushroom to some extent in production welding. Shorter runs before refacing or redressing must be expected.
• Fasteners must be welded to part sections of varying thicknesses.	• Dissimilar materials, such as aluminum, copper, or magnesium, are being welded.
• Fasteners must be welded to parts of unusual shape or a watertight weld joint is required.	• Shape, size, or space requirements do not permit use of projection welded fasteners.
• Welding fixtures can be used for easier locating or automatic feeding.	
• Length of production run without maintenance is critical.	

Fig. 9-6-4 Guide to weld fastener selection.

Figure 9-6-5 shows typical resistance-welded fasteners.

Arc-Welded Studs[2]

There are two basic stud welding processes: *electric-arc* and *capacitor-discharge*.

Electric-Arc Stud Welding. The more widely used stud welding process is a semiautomatic electric-arc process in which the heat for end welding the studs is from a motor generator or transformer-rectifier supplying dc current which passes through an arc from the stud (electrode) to the plate (work).

Electric-arc stud welding is used to best advantage when the base plate is heavy enough to support the full strength of the welded fastener; however, lighter-gage materials are often arc-welded. As a rule of thumb, to avoid burn-through, the plate thickness should be at least one-fifth the weld base diameter. To develop full fastener strength, base-plate thickness should be a minimum of one-third the weld base diameter.

Capacitor-Discharge Stud Welding. The second basic stud welding process derives its heat from an arc produced by a rapid discharge of stored electrical energy, with pressure applied during or immediately following the electrical discharge. Like arc stud welding, the heat for end welding of studs is developed by passage of current through an arc from the stud (electrode) to the plate (work).

Spotweld Nuts (A–D) — Projection Weld Nuts (E–M)

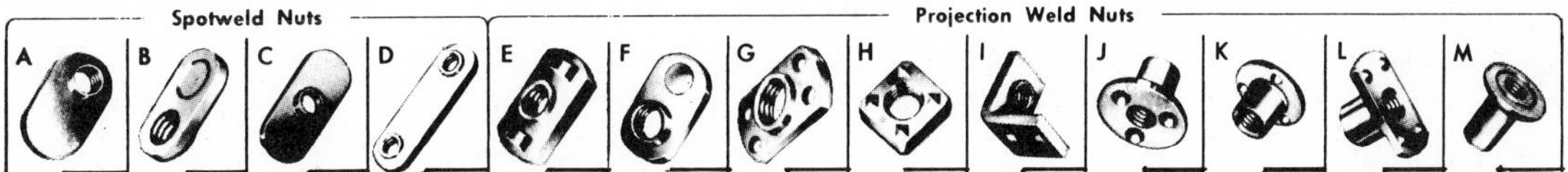

Application Factors	A	N	B	O	C	P	D	Q	E	R	F	S	G	T	H	U	I	V	J	W	K	X	L	Y	M
Flat Surfaces	▲	▼	▲	▼	▲	▼	▲	▼	▲	▼	▲	▼	▲	▼	▲	▼	▲	▼	▲	▼	▲	▼	▲	▼	▲
Curved Surfaces (concave)	▲	▼	▲	▼	▲	▼			▲		▲	▼					▲	▼						▼	
Round Surfaces (convex)	▲	▼	▲	▼	▲	▼		▼	▲	▼	▲	▼		▼	▲	▼	▲	▼		▼		▼	▲	▼	
Tubing	▲	▼	▲	▼	▲	▼			▲		▲	▼				▼	▲	▼						▼	
Channels	▲	▼	▲	▼	▲	▼	▲		▲		▲	▼	▲		▲		▲	▼					▲	▼	
Narrow Flanges	▲	▼	▲	▼	▲	▼	▲		▲		▲	▼	▲		▲		▲	▼					▲	▼	
Offset	▲	▼	▲	▼		▼					▲	▼					▲	▼						▼	
Wall Corners	▲	▼	▲	▼		▼					▲				▲		▲	▼							
Blind Hole															▲		▲		▲				▲		
Wire	▲	▼	▲	▼		▼								▼											▲
Through Hole	▲	▼	▲	▼	▲	▼	▲	▼	▲		▲		▲	▼	▲			▼		▼	▲				▲
Tension Against Weld					▲				▲	▼			▲		▲	▼			▲		▲	▼	▲		▲
Hermetic Seal														▼											▲
Right Angle		▼		▼		▼											▲	▼							
Extra Thread																			▲		▲		▲		▲
Bridging					▲		▲		▲				▲										▲		
Dual Tapped							▲																		
Self Locating Pilot	▲		▲		▲		▲		▲		▲		▲												
No Hole Required in Sheet		▼		▼		▼				▼		▼			▲	▼	▲		▲			▼	▲	▼	▲
Used with Keyhole Slot						▼																			
Pilotless															▲				▲				▲		▲

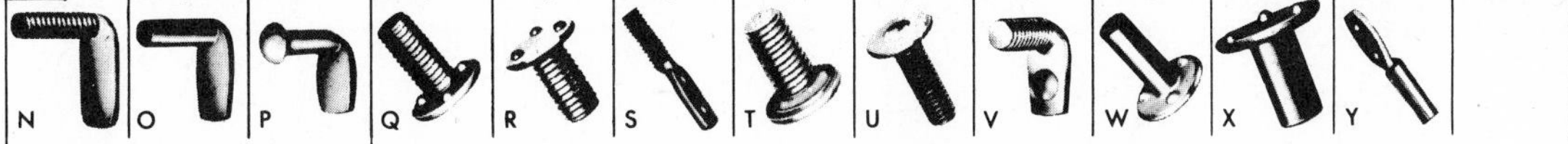

Spotweld Screws and Pins (N–P) — Projection Weld Screws and Pins (Q–Y)

LEGEND

A Single Tab	F Button Projection	J Blind-Hole Flange	O Right-Angle Spade Pin	R Blind-Hole	V Button, Right-Angle Spade
B Targeted	G Four-Button Projection	K Through-Hole	P Keyhole-Slot, Right-Angle Spade Pin	S Spade	W Through-Hole Pin
C Double Tab	H Pilotless	L Tee-Shape	Q Through-Hole	T Hermetic Seal	X Blind-Hole Pin
D Dual Tapped	I Right-Angle Bracket	M Hermetic Seal		U Button-Projection, Blind-Hole	Y Spade Pin
E Dual Projection		N Right-Angle Spade			

Fig. 9-6-5 Resistance-welded fasteners.

DESIGN CONSIDERATIONS

The following considerations are important in obtaining maximum economy in stud welding design.

Select the Proper Welding Method to Be Used. In most instances, the thickness of the plate for stud attachment will determine the stud welding process. Electric-arc stud welding is generally used for fasteners 8 mm and larger. The two capacitor-discharge methods are used for smaller diameters. Figure 9-6-6 is a cross reference for the two welding processes for such criteria as strength, plate thickness, plate material, fastener material, and stud diameter and shape.

Select Standard Stud Types. Standard studs cost less than special designs and are available for immediate delivery. A minor compromise on diameter, or length, which permits use of a standard stud, will often result in considerable savings without loss in function or strength of the assembly.

Standardize on Diameter and Length. Changes in diameter or length of a stud for a different application usually require adjustments in the welding tool and resetting of the timer and current controls.

REFERENCES AND SOURCE MATERIAL

1. G. F. Grey, "Resistance Welded Fasteners," *Machine Design*, vol. 37, no. 6, 1965.
2. R. C. Singleton, "Arc Welded Fasteners," *Machine Design*, vol. 45, no. 28, 1973.

MATERIAL DATA	ELECTRIC-ARC STUD WELDING	CAPACITOR DISCHARGE STUD WELDING
Fastener Shape		
Round	A	A
Square	A	A
Rectangular	A	A
Irregular	A	A
Fastener Diameter or Area		
1.5 to 3 mm dia.	D	A
3 to 6 mm dia.	C	A
6 to 12 mm dia.	A	D
12 to 25 mm dia.	A	D
Up to 30 mm²	C	A
Over 30 mm²	A	D
Fastener Material		
Carbon steel	A	A
Stainless steel	A	A
Alloy steel	B	C
Aluminum	B	A
Brass	C	A
Plate Material		
Carbon steel	A	A
Stainless steel	A	A
Alloy steel	B	A
Aluminum	B	A
Brass	C	A
Plate Thickness		
Under 0.4 mm	D	A
0.4 to 1.5 mm	C	A
1.5 to 3 mm	B	A
Over 3 mm	A	B

A—Applicable without special procedures, equipment, etc.
B—Applicable with special techniques on specific applications which justify preliminary trials or testing to develop welding procedure and weld technique.
C—Limited application.
D—Not recommended. Welding methods not developed at this time.

Fig. 9-6-6 Arc-welded stud process selection chart.

Assignment

On an A3- or B-size sheet, draw the four assemblies shown in Fig. 9-6-A or 9-6-B. Refer to manufacturers' catalogs and the Appendix for standard fastener components. Complete the drawings from the information supplied below. Use your judgment for sizes not given.

- *Assembly A*. A spot-weld nut is to be attached to the panel. A hole in the clamp permits a machine screw to fasten the pipe clamp to the nut.
- *Assembly B*. A right-angle bracket is to be fastened (projection welding) to the bottom plate. The vertical plate is secured to the bracket by a machine screw and lockwasher.
- *Assembly C*. Two resistance-welded threaded fasteners, one on each side of the pipe, are required. The bracket drops over the fasteners, and lockwashers and nuts secure the bracket to the pipe.
- *Assembly D*. A leakproof attaching method (stud welding) is required to hold the adapter to the panel.

REVIEW FOR ASSIGNMENT

Unit 7-10 Partial Sections

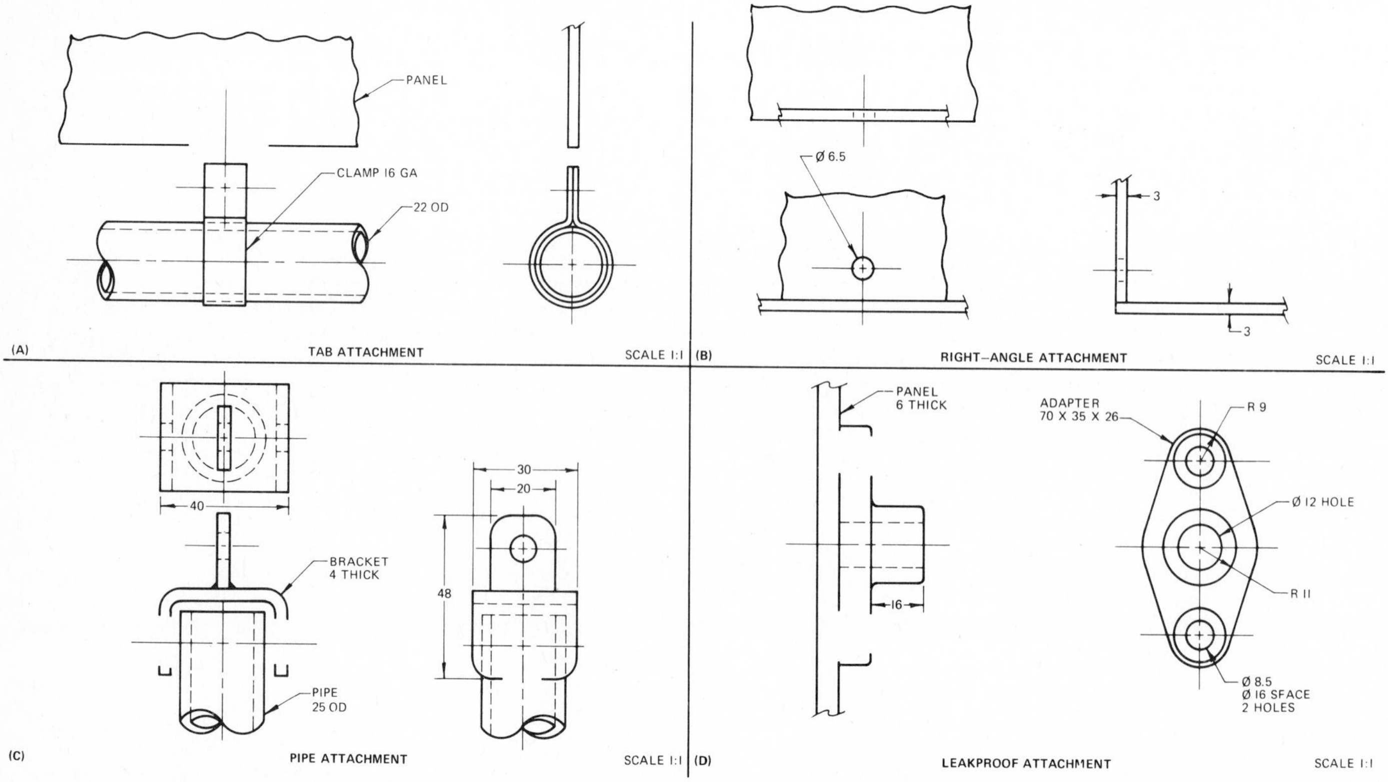

PANEL
CLAMP 16 GA
22 OD
(A)
TAB ATTACHMENT
SCALE 1:1
(B)
Ø 6.5
3
3
RIGHT–ANGLE ATTACHMENT
SCALE 1:1
40
BRACKET
4 THICK
PIPE
25 OD
30
20
48
(C)
PIPE ATTACHMENT
SCALE 1:1
PANEL
6 THICK
16
ADAPTER
70 X 35 X 26
R 9
Ø 12 HOLE
R 11
Ø 8.5
Ø 16 SFACE
2 HOLES
(D)
LEAKPROOF ATTACHMENT
SCALE 1:1

Fig. 9-6-A Welded fasteners.

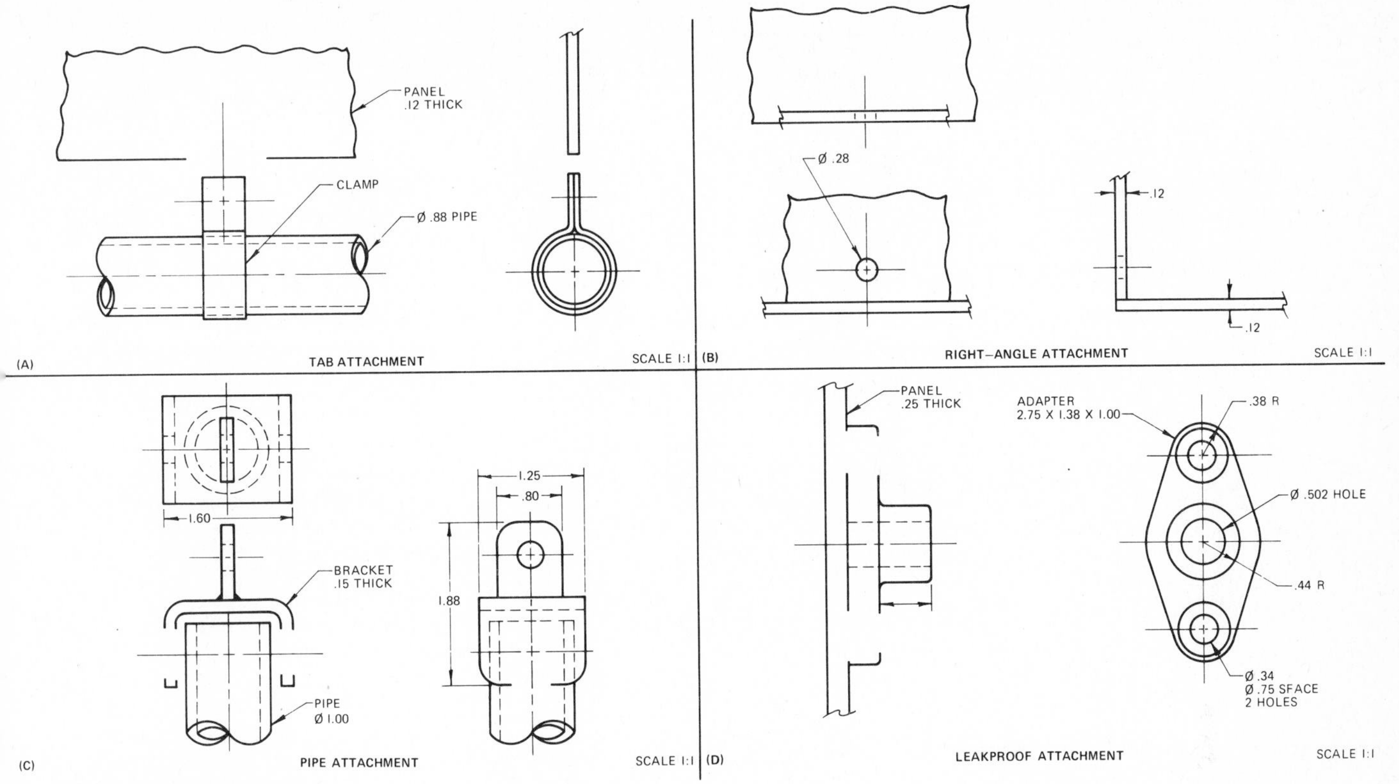
PANEL
.12 THICK
CLAMP
Ø .88 PIPE
(A)
TAB ATTACHMENT
SCALE 1:1
(B)
Ø .28
.12
.12
RIGHT–ANGLE ATTACHMENT
SCALE 1:1
1.60
1.25
.80
1.88
BRACKET
.15 THICK
PIPE
Ø 1.00
(C)
PIPE ATTACHMENT
SCALE 1:1
(D)
PANEL
.25 THICK
ADAPTER
2.75 X 1.38 X 1.00
.38 R
Ø .502 HOLE
.44 R
Ø .34
Ø .75 SFACE
2 HOLES
LEAKPROOF ATTACHMENT
SCALE 1:1

Fig. 9-6-B Welded fasteners.

UNIT 9-7

ADHESIVE FASTENINGS[1]

Today, industrial designers and manufacturers are relying on adhesives more than ever before. They allow greater versatility in design, styling, and materials. They can cut costs. However, as with any engineering tool, there are limitations as well as advantages.

Adhesion versus Stress

Adhesion is the force that holds together materials. There are three basic types:

1. *Specific Adhesion* is the molecular attraction between contacting surfaces.
2. *Mechanical Adhesion* occurs when an adhesive is applied to rough or porous surfaces.
3. *Effective Adhesion* combines specific and mechanical adhesion for optimum strength.

Stress, on the other hand, is the force pulling materials apart. See Fig. 9-7-1. The basic types of stress in adhesive technology are:

1. *Tensile*. Pull is exerted equally over the entire joint. Pull direction is straight and away from the adhesive bond. All adhesive contributes to bond strength.
2. *Shear*. Pull direction is across the adhesive bond. The bonded materials are being forced to slide over one another.
3. *Cleavage*. Pull is concentrated at one edge of the joint and exerts a prying force on the bond. The other edge of the joint is theoretically under zero stress.
4. *Peel*. One surface must be flexible. Stress is concentrated along a thin line at the edge of the bond. This line is the exact point where an adhesive would separate if the flexible surface were peeled away from its mating surface.

Resistance to stress is one general reason for the rapid increase in the use of adhesive for product assembly. The following points elaborate on stress resistance and the other advantages of adhesive technology.

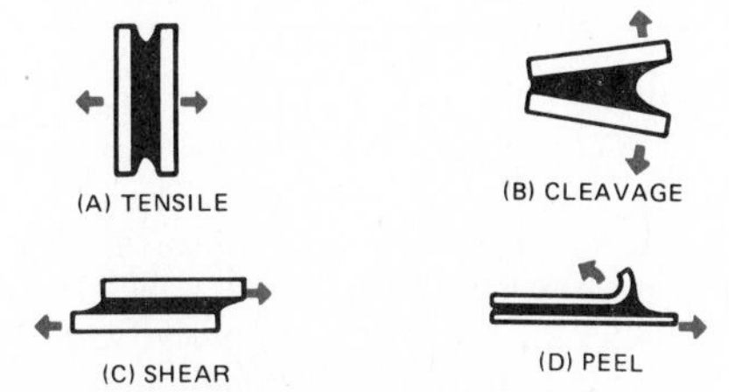
(A) TENSILE
(B) CLEAVAGE
(C) SHEAR
(D) PEEL

Fig. 9-7-1 Basic types of stress on bonded joints. (3M Co.)

ADVANTAGES

1. Adhesives allow uniform distribution of stress over the entire bond area. See Fig. 9-7-2. This eliminates stress concentration caused by rivets, bolts, spot welds, and similar fastening techniques. Lighter, thinner materials can be used without sacrificing strength.
2. Adhesives can effectively bond dissimilar materials. Laminates of dissimilar material can often produce combinations superior in strength and performance to either adherent alone.
3. Continuous contact between mating surfaces effectively bonds and seals against many environmental conditions.
4. Adhesives maintain the integrity of the bonded material. They eliminate holes needed for mechanical fasteners and surface marks resulting from spot welding, brazing, etc.

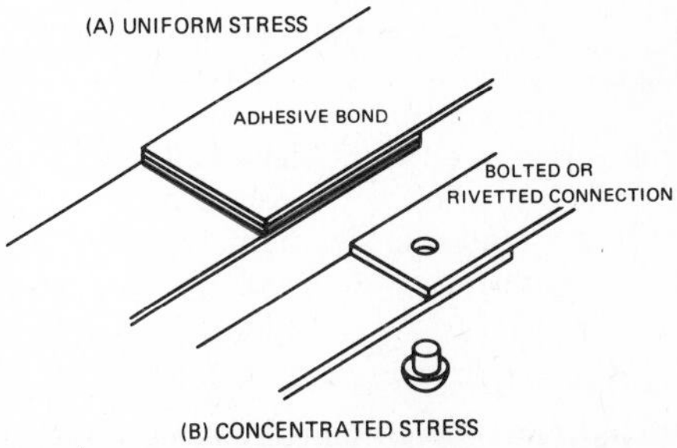
(A) UNIFORM STRESS
ADHESIVE BOND
BOLTED OR
RIVETTED CONNECTION
(B) CONCENTRATED STRESS

Fig. 9-7-2 Stresses caused by fasteners.

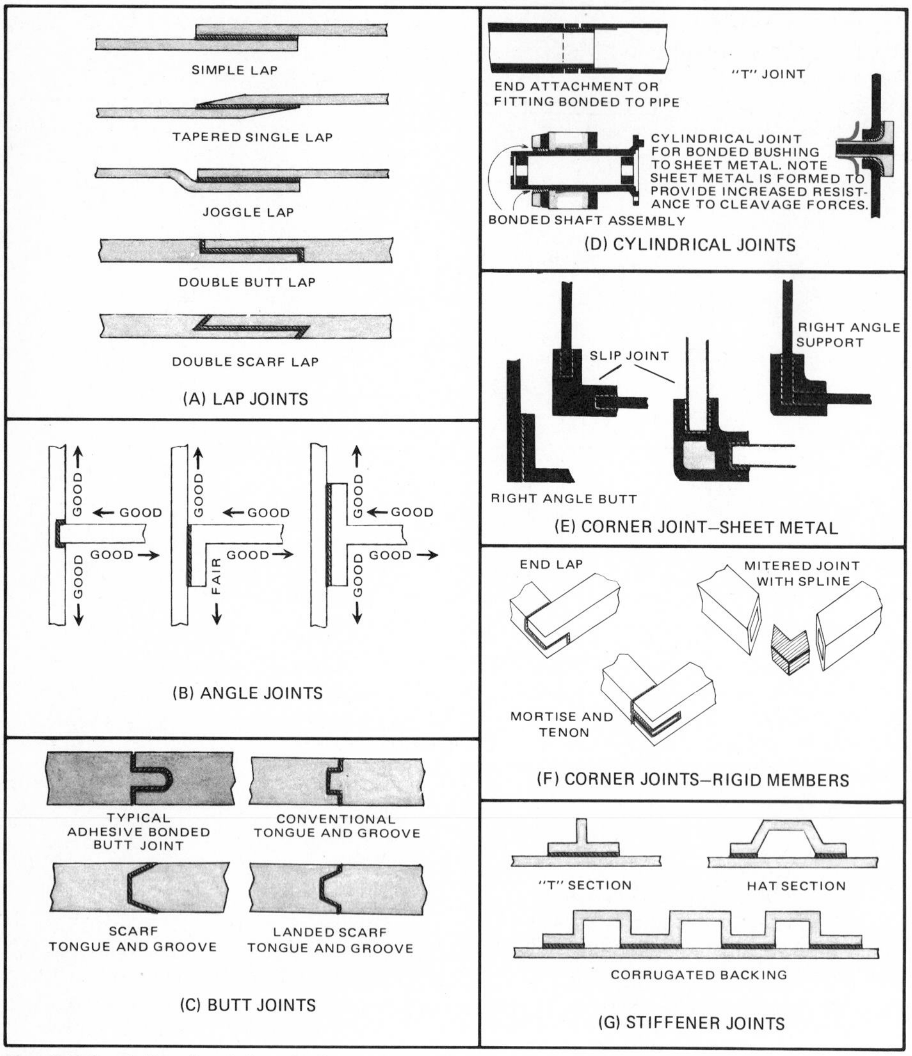

SIMPLE LAP
TAPERED SINGLE LAP
JOGGLE LAP
DOUBLE BUTT LAP
DOUBLE SCARF LAP
(A) LAP JOINTS
GOOD
GOOD
GOOD
GOOD
GOOD
GOOD
GOOD
FAIR
GOOD
GOOD
GOOD
GOOD
(B) ANGLE JOINTS
TYPICAL ADHESIVE BONDED BUTT JOINT
CONVENTIONAL TONGUE AND GROOVE
SCARF TONGUE AND GROOVE
LANDED SCARF TONGUE AND GROOVE
(C) BUTT JOINTS
END ATTACHMENT OR FITTING BONDED TO PIPE
"T" JOINT
CYLINDRICAL JOINT FOR BONDED BUSHING TO SHEET METAL. NOTE SHEET METAL IS FORMED TO PROVIDE INCREASED RESIST-ANCE TO CLEAVAGE FORCES.
BONDED SHAFT ASSEMBLY
(D) CYLINDRICAL JOINTS
SLIP JOINT
RIGHT ANGLE SUPPORT
RIGHT ANGLE BUTT
(E) CORNER JOINT–SHEET METAL
END LAP
MITERED JOINT WITH SPLINE
MORTISE AND TENON
(F) CORNER JOINTS–RIGID MEMBERS
"T" SECTION
HAT SECTION
CORRUGATED BACKING
(G) STIFFENER JOINTS

Fig. 9-7-3 Adhesive joint design. (3M Co.)

LIMITATIONS

1. Adhesive bonding can be slow or require critical processing. This is particularly true in mass production. Some adhesives require heat and pressure, or special jigs and fixtures, to establish the bond.

2. Adhesives are sensitive to surface conditions. Special surface preparation may be required for optimum bonding results.

3. Some adhesive solvents present hazards. Special ventilation may be required to protect employees from toxic vapors.

4. Environmental conditions can reduce bond strength of some adhesives. Some do not hold well when exposed to low temperatures, high humidity, severe heat, chemicals, water, etc.

Joint Design

Joints should be specifically designed for use with structural adhesives. This is largely a matter of common sense and experience. Two basic factors should be the design guidelines. First, structural joints should be designed so that all the bonded area shares the load equally. Second, the joint configuration should be designed so that basic stress is primarily in shear or tensile, with cleavage and peel minimized or eliminated.

The following structural joints and their advantages and disadvantages illustrate some typical design alternatives. They are not, of course, the limit of possible adhesive bonded joints. See Fig. 9-7-3.

LAP JOINTS

Lap joints are most practical and applicable in bonding thin materials. The *simple lap joint* is offset. This can result in cleavage and peel stress under load when thin materials are used. A *tapered single lap joint* is more efficient than a simple lap joint. The tapered edge allows bending of the joint edge under stress. The *joggle lap joint* gives more uniform stress distribution than either the simple or tapered lap joint. The joint can be formed by simple metal-forming operations. The curing pressure is easily applied.

The *double-butt lap joint* gives more uniform stress distribution in the load-bearing area than the above joints. This type of joint, however, requires machining which is not always feasible with thinner-gage metals. *Double-scarf lap joints* have better resistance to bending forces than double-butt joints. This type of joint, however, also requires machining.

ANGLE JOINTS

Angle joints give rise to either peel or cleavage stress depending on the gage metal. Typical approaches to the reduction of cleavage are illustrated.

BUTT JOINTS

A straight butt joint has poor resistance to cleavage. The following recessed butt joints are recommended: landed scarf tongue and groove, conventional tongue and groove, and scarf tongue and groove. Landed scarf tongue and groove joints act as stops which can control adhesive line thickness. Tongue and groove are self-aligning during assembly and act as a reservoir for void-filling-type adhesives.

CYLINDRICAL JOINTS

The *T joint* and *overlap slip joint* are typical for bonding cylindrical parts such as tubing, bushings, and shafts. With adhesive bonding, all available contact area contributes to carrying the load.

CORNER JOINTS—SHEET METAL

Corner joints can be assembled with adhesives by using simple supplementary attachments. This permits joining and sealing in a single operation. Typical designs are right-angle butt joints, slip joints, and right-angle support joints.

CORNER JOINTS—RIGID MEMBERS

Corner joints, as in storm doors or decorative frames, can be adhesive-bonded. End lap joints are the simplest design type, although they require machining. Adhesives requiring pressure during curing may be utilized in such designs. Mortise and tenon joints are excellent from a design standpoint, but they also require machining. Mitered joint with spline is best if members are hollow extrusions.

STIFFENER JOINTS

Deflection and flutter of thin metal sheets can be minimized with adhesive-bonded stiffeners. When such assemblies are flexed, peel stresses are exerted on the adhesives. If the flanges on the stiffening section can bend with the sheet, minimum peel stress on the bond will result. Increasing sheet gage or decreasing the gage of the stiffener flange will give equivalent results.

REFERENCE AND SOURCE MATERIAL

1. 3M Company.

Assignment

On an A3- or B-size sheet, draw the four adhesive-bonded assemblies shown in Fig. 9-7-A or 9-7-B. Complete the drawings from the information supplied below and the adhesive chart in the Appendix. List the adhesive product number and state the method of application you would recommend. Use your judgment for sizes not shown, and dimension the joint. Scale is to suit.

- *Assembly A*. Three pieces of wood are to be assembled into the shape shown. Joint design has not been shown.
- *Assembly B*. The riveted joint shown is to be replaced by a joggle lap joint. Must meet specification requirements of MMM-A-121.
- *Assembly C*. The riveted joint shown is to be replaced by a tongue and groove joint. It must be fast-drying.
- *Assembly D*. The sheet-metal corner joint shown is to be replaced by a slip joint. It must be water-resistant.

REVIEW FOR ASSIGNMENT

Unit 7-5 Assemblies in Section

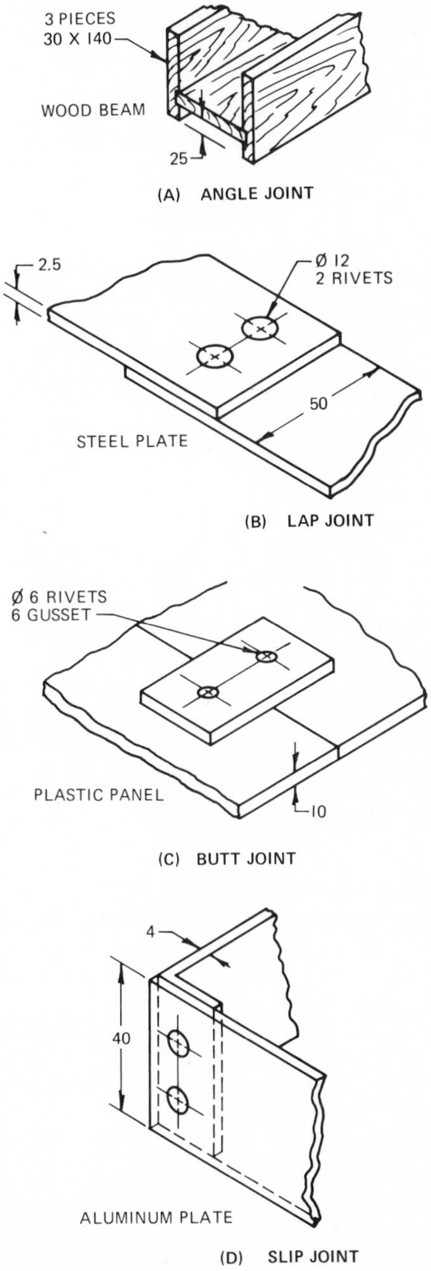
3 PIECES
30 X 140
WOOD BEAM
25
(A) ANGLE JOINT
2.5
Ø 12
2 RIVETS
50
STEEL PLATE
(B) LAP JOINT
Ø 6 RIVETS
6 GUSSET
PLASTIC PANEL
10
(C) BUTT JOINT
4
40
ALUMINUM PLATE
(D) SLIP JOINT

Fig. 9-7-A Adhesive fastenings.

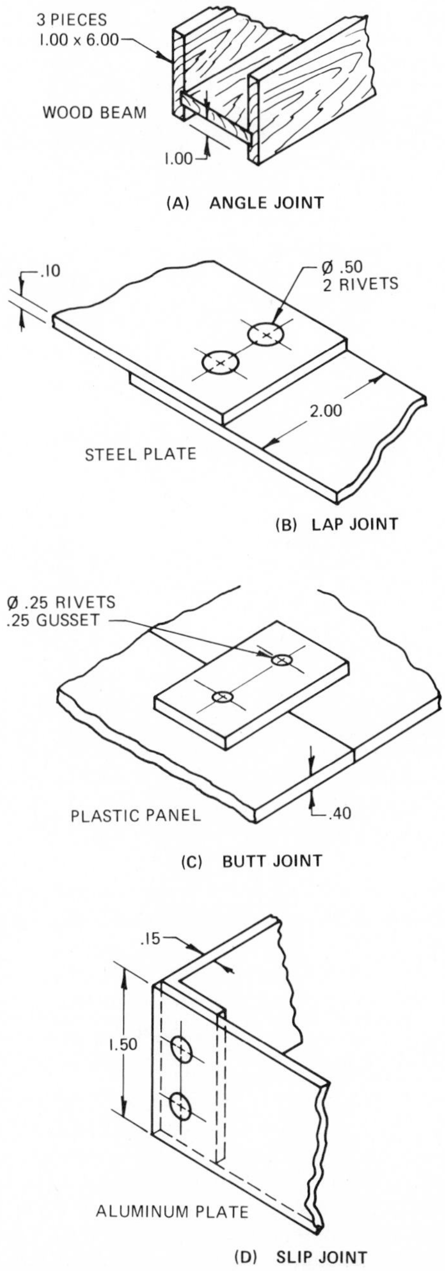
3 PIECES
1.00 x 6.00
WOOD BEAM
1.00
(A) ANGLE JOINT
.10
Ø .50
2 RIVETS
2.00
STEEL PLATE
(B) LAP JOINT
Ø .25 RIVETS
.25 GUSSET
PLASTIC PANEL
.40
(C) BUTT JOINT
.15
1.50
ALUMINUM PLATE
(D) SLIP JOINT

Fig. 9-7-B Adhesive fastenings.

Chapter 10
Forming Processes

UNIT 10-1
CASTINGS

Forming Processes[1]

When a component of a machine takes shape on the drawing board of the designer, the method of its manufacture may still be entirely open. The number of possible manufacturing processes is increasing day by day, and the optimum process is found only by carefully weighing technological advantages and drawbacks in relation to the economy of production.

The choice of the manufacturing process depends on the size and shape of the component.

Manufacturing processes are therefore important to the engineer and drafter in order to properly design a part. They must be familiar with the advantages, disadvantages, costs, and machines necessary for manufacturing. Since the cost of the part is influenced by the method of producing the part, such as welding or casting, the designer must be able to choose wisely the method which will reduce the overall cost of production. In some cases it may be necessary to recommend the purchase of a new or different machine in order to produce the part at a competitive price.

This means the designer should design the part for the process as well as for the function. Most of all, unnecessarily close tolerances on nonfunctional dimensions should be avoided.

This chapter covers the following manufacturing processes: casting, forging, extruding, cold heading, drawings, spinning, rolling, electroforming, high-energy-rate forming, and powder metallurgy. Forming by means of welding and stamping (punches and dies) is covered in Units 11 and 27, respectively.

Casting Processes[2,3]

Casting is the process whereby parts are produced by pouring molten metal into a mold. A typical cast part is shown in Fig. 10-1-1. Casting processes for metals can be classified by either the type of mold or pattern or the pressure or force

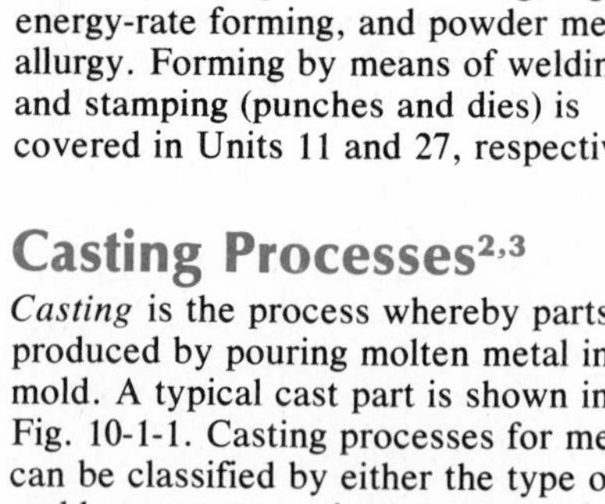

Fig. 10-1-1 Typical cast part. (General Motors Corp.)

used to fill the mold. Conventional sand, shell, and plaster molds utilize a permanent pattern, but the mold is used only once. Permanent molds and die-casting dies are machined in metal or graphite sections and are employed for a large number of castings. Investment casting and the relatively new full-mold process involve both an expendable mold and an expendable pattern.

Casting metals are usually alloys or compounds of two or more metals. They are generally classed as ferrous or nonferrous metals. *Ferrous* metals are those which contain iron, the most common being gray iron, steel, and malleable iron. *Nonferrous* alloys are those which contain no iron, such as aluminum, magnesium, brass, and bronze.

SAND MOLD CASTING

The most widely used casting process for metals uses a permanent pattern of metal or wood that shapes the mold cavity when loose molding material is compacted around the pattern. This material consists of a relatively fine sand, which serves as the refractory aggregate, plus a binder.

A typical sand mold, with the various provisions for pouring the molten metal and compensating for contraction of the solidifying metal, and a sand core for forming a cavity in the casting are shown in Fig. 10-1-2. Sand molds consist of two or more sections: bottom (*drag*), top (*cope*), and intermediate sections (*cheeks*) when required. The sand is contained in flasks equipped with pins and plates to ensure the alignment of the cope and drag.

Molten metal is poured into the sprue, and connecting runners provide flow channels for the metal to enter the mold cavity through gates. Riser cavities are located over the heavier sections of the casting. A vent is usually added to permit the escape of gases, which are formed during the pouring of metal.

When a hollow casting is required, a form called a *core* is usually used. Cores occupy that part of the mold which is intended to be hollow in the casting. Cores, like molds, are formed of sand and placed in the supporting impressions or core prints in the molds. The core prints ensure positive location of the core in the mold and, as such, should be placed so that they support the mass of the core uniformly to prevent shifting or sagging. Metal core supports called *chaplets*, which are used in the mold cavity and which fuse into the casting, are sometimes used by the foundry in addition to core prints. See Fig. 10-1-3. Chaplets and their locations are not usually specified on drawings.

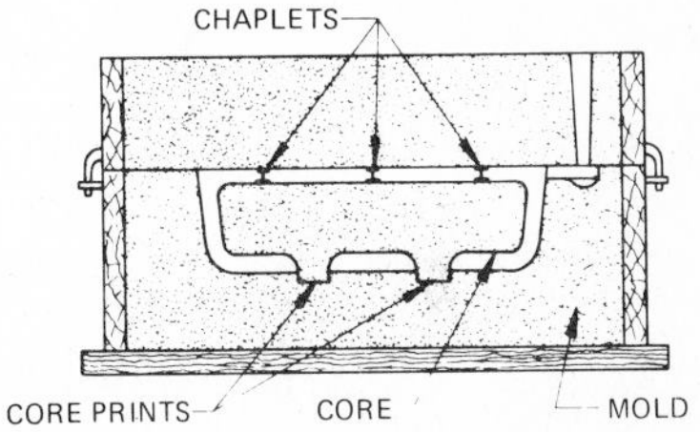
CHAPLETS
CORE PRINTS
CORE
MOLD

Fig. 10-1-3 Core prints and chaplets. (General Motors Corp.)

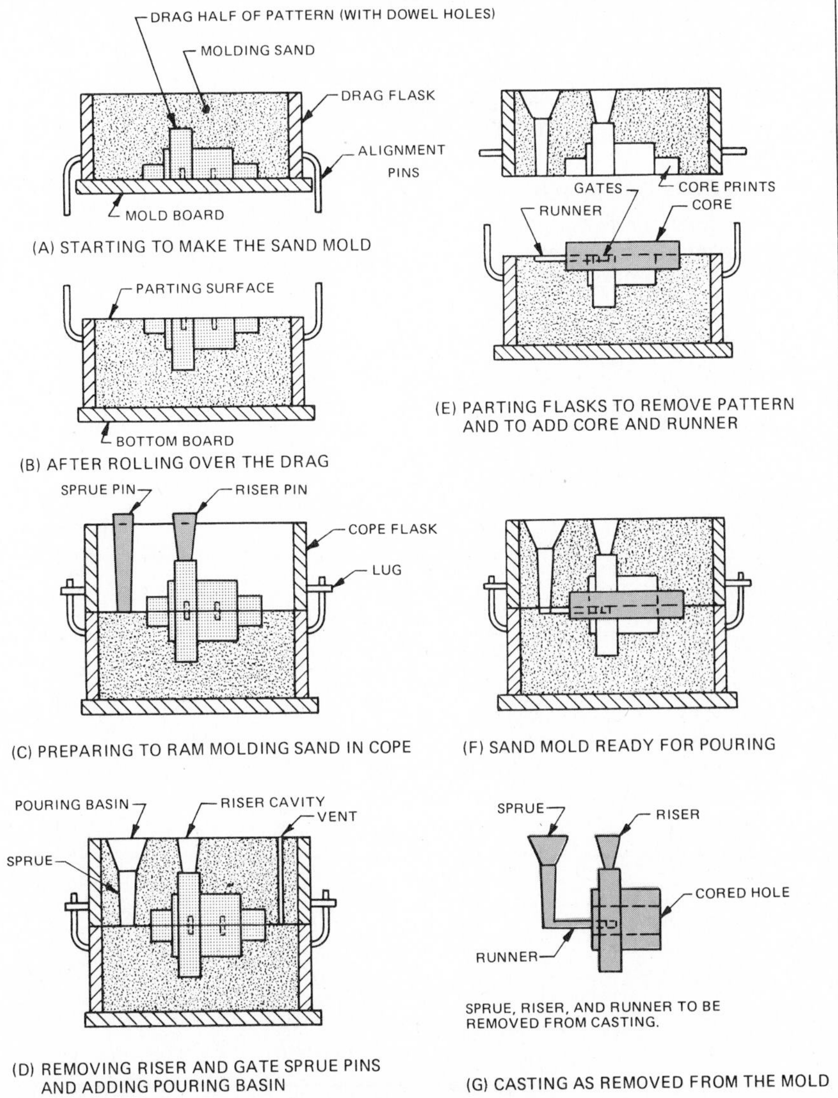
DRAG HALF OF PATTERN (WITH DOWEL HOLES)
MOLDING SAND
DRAG FLASK
ALIGNMENT PINS
MOLD BOARD
(A) STARTING TO MAKE THE SAND MOLD
PARTING SURFACE
BOTTOM BOARD
(B) AFTER ROLLING OVER THE DRAG
SPRUE PIN
RISER PIN
COPE FLASK
LUG
(C) PREPARING TO RAM MOLDING SAND IN COPE
POURING BASIN
RISER CAVITY
VENT
SPRUE
(D) REMOVING RISER AND GATE SPRUE PINS AND ADDING POURING BASIN
GATES
CORE PRINTS
RUNNER
CORE
(E) PARTING FLASKS TO REMOVE PATTERN AND TO ADD CORE AND RUNNER
(F) SAND MOLD READY FOR POURING
SPRUE
RISER
CORED HOLE
RUNNER
SPRUE, RISER, AND RUNNER TO BE REMOVED FROM CASTING.
(G) CASTING AS REMOVED FROM THE MOLD

Fig. 10-1-2 Sequence in preparing a sand casting.

In producing sand molds, a metal or wooden pattern must first be made. The pattern, normally made in two parts, is slightly larger in every dimension than the part to be cast, to allow for shrinkage when the casting cools. This is known as *shrinkage allowance*, and the pattern maker allows for it by using a shrink rule for each of the cast metals.

Drafts or slight tapers are also placed on the pattern to allow for easy withdrawal from the sand mold. Since shrinkage and draft are taken care of by the pattern maker, they are of no concern to the designer.

In the construction of patterns for castings in which various points on the surface of the casting must be machined, sufficient excess metal should be provided for all machined surfaces. The allowance, commonly called *finish*, depends on the metal used, the shape and size of the part, the tendency to warp, the machining method, and setup.

The molten metal is poured into the pouring basin and runs down the sprue to a runner and into the mold cavity.

When the metal has hardened, the sand is broken and the casting removed. Next the excess metal, gates, and risers are removed and remelted.

SHELL MOLD CASTING[4]

The refractory sand used in shell molding is bonded by a thermostatic resin that forms a relatively thin shell mold. A heated, reusable metal pattern plate (Fig. 10-1-4) is used to form each half of the mold by either dumping a sand-resin mixture on top of the heated pattern or by blowing a resin-coated sand under air pressure against the pattern.

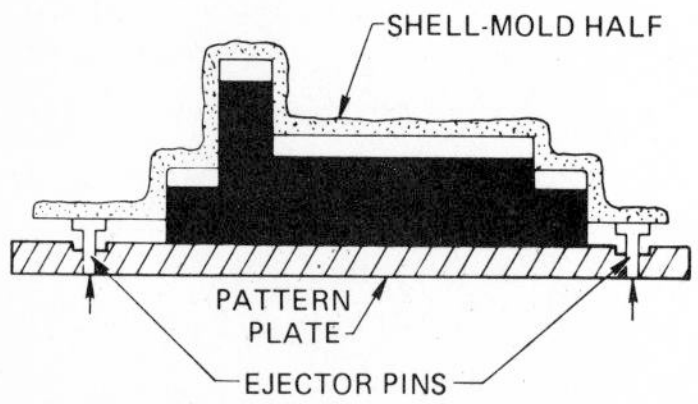
SHELL-MOLD HALF
PATTERN PLATE
EJECTOR PINS

Fig. 10-1-4 Shell mold being stripped from pattern.

PLASTER MOLD CASTING

Plaster of Paris and fillers are mixed with water and setting-control agents to form a *slurry*. This slurry is poured around a reusable metal or rubber pattern and sets to form a gypsum mold (Fig. 10-1-5). The molds are then dried, assembled, and filled with molten (nonferrous) metals. Plaster molding is ideal for producing thin, sound walls. As in sand mold casting, a new mold is required for each casting. Castings made by this process have smoother finish, finer detail, and greater dimensional accuracy than sand castings.

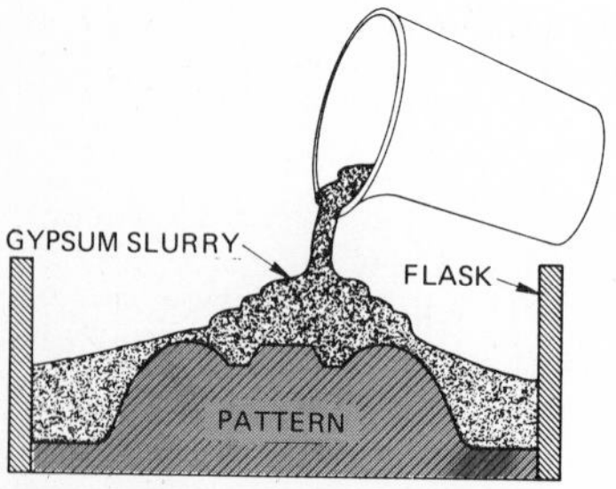
GYPSUM SLURRY
FLASK
PATTERN

Fig. 10-1-5 Pouring slurry over a plaster-mold pattern.

PERMANENT MOLD CASTING

This process makes use of a metal mold, similar to a die, which is utilized to produce many castings from each mold. See Fig. 10-1-6. It is used to produce some ferrous alloy castings. But due to rapid deterioration of the mold caused by the high pouring temperatures of these alloys, and the attendant high mold cost, the process is confined largely to production of nonferrous alloy castings.

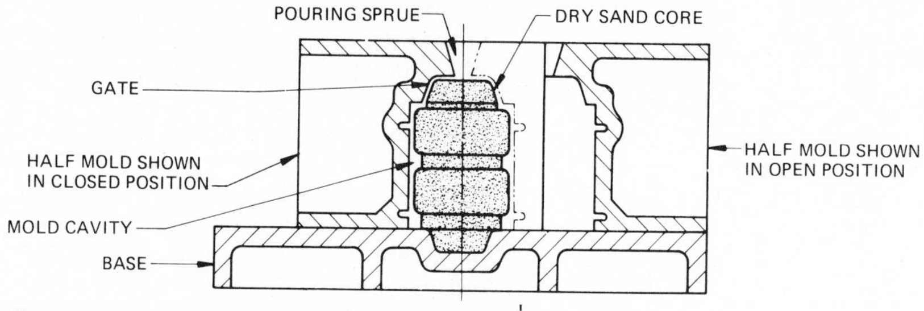
POURING SPRUE
DRY SAND CORE
GATE
HALF MOLD SHOWN IN CLOSED POSITION
HALF MOLD SHOWN IN OPEN POSITION
MOLD CAVITY
BASE

Fig. 10-1-6 Permanent mold. (General Motors Corp.)

INVESTMENT MOLD CASTING[5]

Investment castings have been better known in the past by the term *lost wax castings*. The term *investment* refers to the refractory material used to encase the wax patterns.

This process uses both an expendable pattern and an expendable mold. Patterns of wax, plaster, or frozen mercury are cast in metal dies. The molds are formed either by pouring a slurry of a refractory material around the pattern positioned in a flask or by building a thick layer of shell refractory on the pattern by repeated dipping into slurries and drying. The arrangement of the wax patterns in the flask method is shown in Fig. 10-1-7.

FULL MOLD CASTING[6]

The characteristic feature of the full mold process is the use of lost patterns made of foamed plastic. These are not extracted from the mold, but are vaporized by the molten metal.

The full mold process is suitable for individual castings and for small series of up to five castings. The advantages it offers are obvious: it is very economical and reduces the delivery time required for prototypes, articles urgently needed for repair jobs, or individual large machine parts.

CENTRIFUGAL CASTING

In the centrifugal casting process, commonly applied to cylindrical casting of either ferrous or nonferrous alloys, a permanent mold is rotated rapidly about the axis of the casting while a measured amount of molten metal is poured into the mold cavity. See Fig. 10-1-8. The centrifugal force is used to hold the metal against the outer walls of the mold with the volume of metal poured determining the wall thickness of the casting.

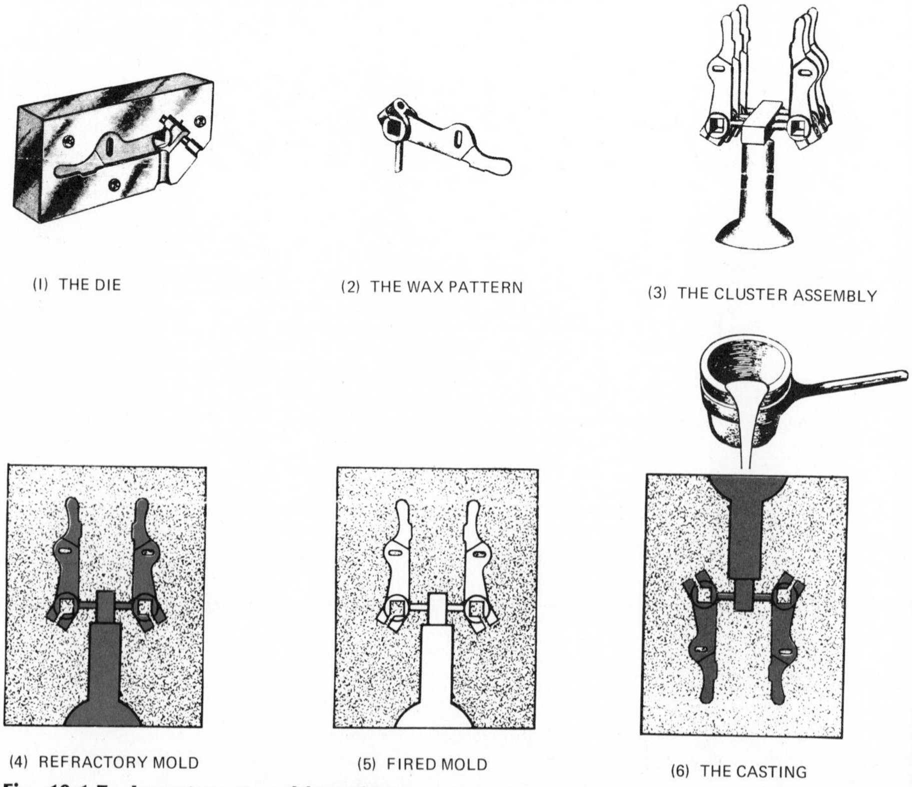
(1) THE DIE
(2) THE WAX PATTERN
(3) THE CLUSTER ASSEMBLY
(4) REFRACTORY MOLD
(5) FIRED MOLD
(6) THE CASTING

Fig. 10-1-7 Investment mold casting.

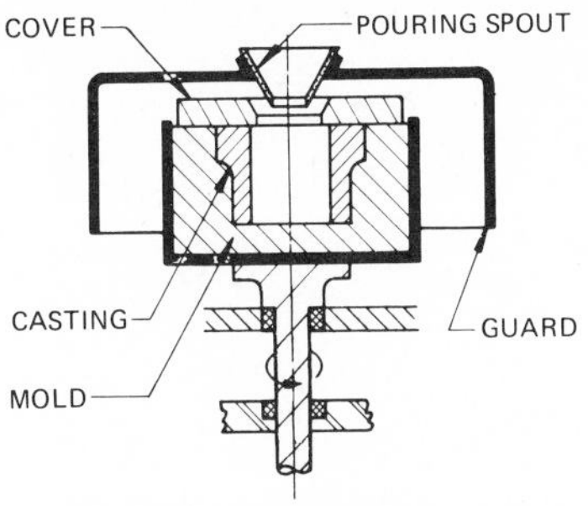
COVER
POURING SPOUT
CASTING
GUARD
MOLD

Fig. 10-1-8 Centrifugal mold equipment. (General Motors Corp.)

Rotation speed is rapid enough to form the central hole without a core. Castings made by this method are smooth, sound, and clean on the outside because impurities, being lighter than the metal, work toward the inner surface of the molten metal while rotating. The impurities can be removed by machining.

CONTINUOUS CASTING

Continuous casting produces semifinished shapes as uniform section rounds, ovals, squares, rectangles, and plates. These shapes are cast from nearly all ferrous and nonferrous metals by continuously pouring the molten metal into a water-jacketed mold. The metal solidifies in the mold, and the solid billet exits continuously into a water spray. These sections are processed further by rolling, drawing, or extruding into smaller, more intricate shapes. Iron bars cast by this process are finished by machining.

DIE CASTING

One of the cheapest, fastest, and most efficient processes used in the production of metal parts is die casting. Die castings are made by forcing molten metal into a die or mold. Large quantities, accurately cast, can be produced with a die-casting die, thus eliminating or cutting down machining costs. Many parts are completely finished when taken from a die. Since die castings can be accurate to within 0.02 mm of size, internal and external threads, gear teeth, and lugs can readily be cast.

Die casting has its limitations. Only nonferrous alloys can be die-cast economically because of the lack of a suitable die material to withstand the higher temperatures required for steel and iron.

Die-casting machines are of two types: the *submerged-plunger* type for low-melting alloys such as zinc, tin, lead, etc., and the *cold chamber* type for high-melting nonferrous alloys such as aluminum and magnesium. See Fig. 10-1-9.

Selection of Process

Selection of the most feasible casting process for a given part requires an evaluation of the type of metal, the number of castings required, their shape and size, the dimensional accuracy required, and the casting finish required. When the casting can be produced by a number of methods, selection of the process is based on the most economical production of the total requirement. Since final cost of the part, rather than price of the rough casting, is the significant factor, the number of finishing operations necessary on the casting is also considered. Those processes that provide the closest dimensions, the best surface finish, and the most intricate detail generally require the smallest number of finishing operations.

A direct comparison of the capabilities, production characteristics, and limitations of several processes is indicated in Fig. 10-1-10.

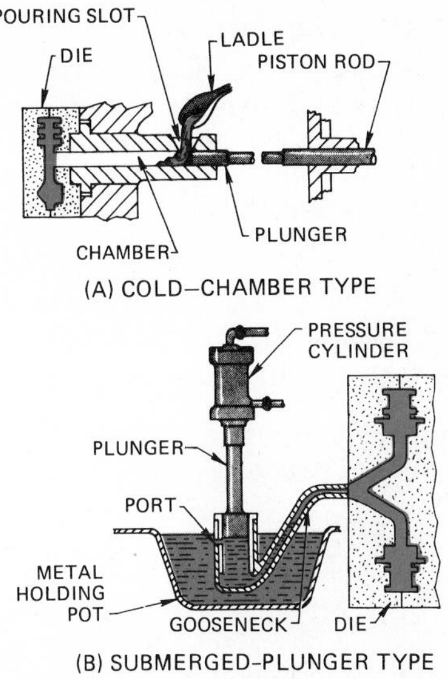
POURING SLOT
DIE
LADLE
PISTON ROD
CHAMBER
PLUNGER
(A) COLD–CHAMBER TYPE
PRESSURE CYLINDER
PLUNGER
PORT
METAL HOLDING POT
GOOSENECK
DIE
(B) SUBMERGED-PLUNGER TYPE

Fig. 10-1-9 Die-casting machines.

Process	Metals Cast	Usual Mass Range	Minimum Production Quantities	Relative Setup Cost	Casting Detail Feasible	Minimum Thickness (mm)	Dimensional Tolerances (mm)	Surface Finish, RMS (μm)
Sand (Green, Dry, and Core)	All ferrous and nonferrous	Less than 0.5 kg to several tonnes.	3, without mechanization	Very low to high depending on mechanization	Fair	3 to 6	± 0.8	10
CO_2 Sand						2.5 to 6	± 0.5	6
Shell	All ferrous and nonferrous	0.2 to 15 kg	50	Moderate to high depending on mechanization	Fair to good	0.8 to 2.5	± 0.4	5
Plaster	Al, Mg, Cu, and Zn alloys	Less than 0.5 kg to 1350 kg	1	Moderate	Excellent	0.8 to 2	± 0.2	2.5
Investment	All ferrous and nonferrous	Less than 30g to 25 kg	25	Moderate	Excellent	0.5 to 1.5	± 0.1	2
Permanent Mold Metal Mold	Nonferrous and cast iron	0.5 to 20 kg	100	Moderate to high	Poor	4.5 to 6	± 0.5	5
Graphite Mold	Steel	2 to 150 kg	100			6	± 0.8	5
Die	Sn, Pb, Zn, Al, Mg, and Cu alloys	Less than 0.5 kg to 10 kg	1000	High	Excellent	1.2 to 2	± 0.5	1.5

* Values listed are primarily for aluminum alloys, but data applies generally to other metals also.

† Depends on surface area. Double if dimension is across parting line.

Fig. 10-1-10 General characteristics of casting processes.

Design Considerations[7]

The advantages of using castings for engineering components are well appreciated by designers. Of major importance is the fact that they can produce shapes of any degree of complexity and of virtually any size.

SOLIDIFICATION OF METAL IN A MOLD

While this is not the first step in the sequence of events, it is of such fundamental importance that it forms the most logical point of beginning in understanding the making of a casting.

Consider a few simple shapes transformed into mold cavities and filled with molten metal.

In a sphere, heat dissipates from the surface through the mold while solidification commences from the outside and proceeds progressively inward, in a series of layers. As liquid metal solidifies, it contracts in volume, and unless feed metal is supplied, a shrinkage cavity will be formed in the center. See Fig. 10-1-11.

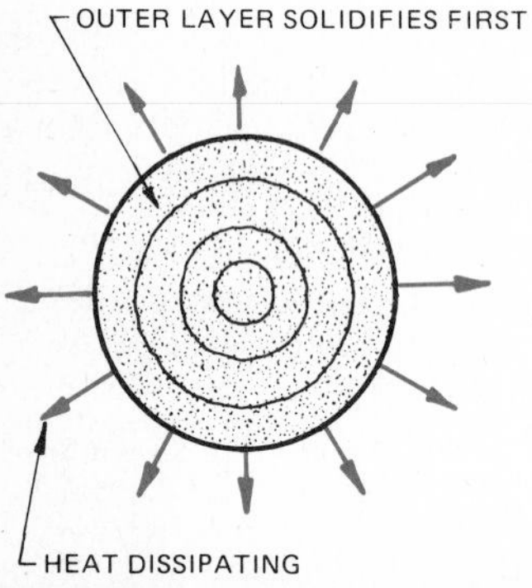
OUTER LAYER SOLIDIFIES FIRST
HEAT DISSIPATING

Fig. 10-1-11 Circular mold cavity filled with molten metal.

The designer must realize that a shrinkage problem exists and that the foundry worker will have to attach risers to the casting or resort to other means to overcome it.

When the simple sphere has solidified further, it continues to contract in volume, so that the final casting is smaller than the mold cavity.

Consider a shape with a square cross section such as shown in Fig. 10-1-12*a*. Here again, cooling proceeds at right angles to the surface and is necessarily faster at the corners of the casting. Thus, solidification proceeds more rapidly at the corners.

The resulting hot spot prolongs solidification, promoting solidification shrinkage and lack of density in this area. The only logical solution, from the designer's viewpoint, is the provision of very generous fillets or radii at these internal corners. Additionally, the relation of the two sections forming the corner to each other is of importance. If they are materially different, as in Fig. 10-1-12*d*, contraction in the lighter member will occur at a different time from that in the heavier member. Differential contraction is the major cause of casting stress, warping, and cracking.

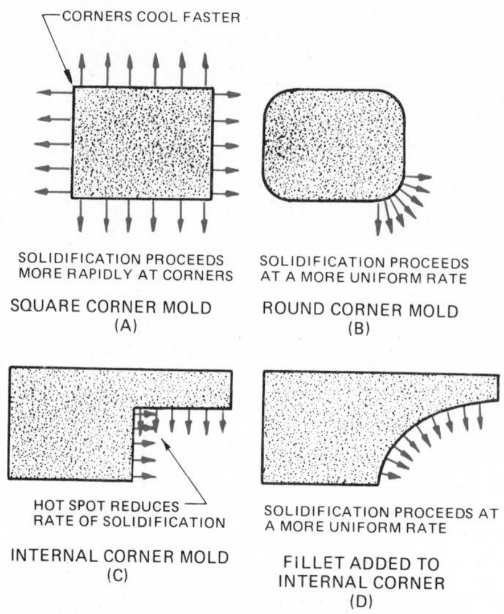
CORNERS COOL FASTER
SOLIDIFICATION PROCEEDS MORE RAPIDLY AT CORNERS
SQUARE CORNER MOLD (A)
SOLIDIFICATION PROCEEDS AT A MORE UNIFORM RATE
ROUND CORNER MOLD (B)
HOT SPOT REDUCES RATE OF SOLIDIFICATION
INTERNAL CORNER MOLD (C)
SOLIDIFICATION PROCEEDS AT A MORE UNIFORM RATE
FILLET ADDED TO INTERNAL CORNER (D)

Fig. 10-1-12 Cooling effect on mold cavities filled with molten metal. (Meehanite Metal Corp.)

GENERAL DESIGN RULES

Design for Casting Soundness. See Fig. 10-1-13. Most metals and alloys shrink when they solidify. Therefore, the design must be such that all members of the parts increase in dimension progressively to one or more suitable locations where feeder heads can be placed to offset liquid shrinkage.

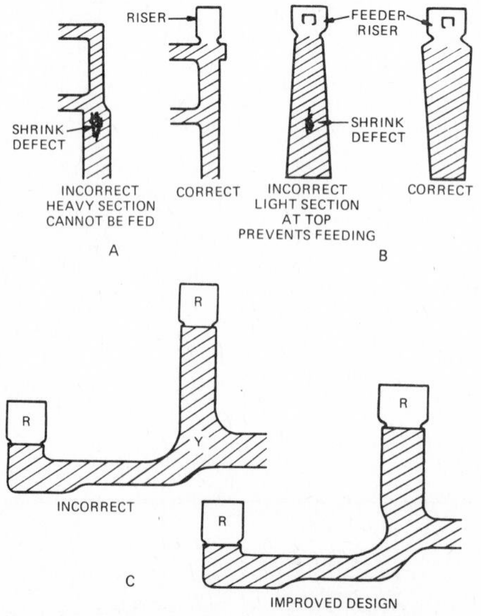
RISER
SHRINK DEFECT
INCORRECT HEAVY SECTION CANNOT BE FED
CORRECT
A
FEEDER RISER
SHRINK DEFECT
INCORRECT LIGHT SECTION AT TOP PREVENTS FEEDING
CORRECT
B
R
Y
INCORRECT
C
IMPROVED DESIGN

Fig. 10-1-13 Design members so that all parts increase progressively to feeder risers. (Meehanite Metal Corp.)

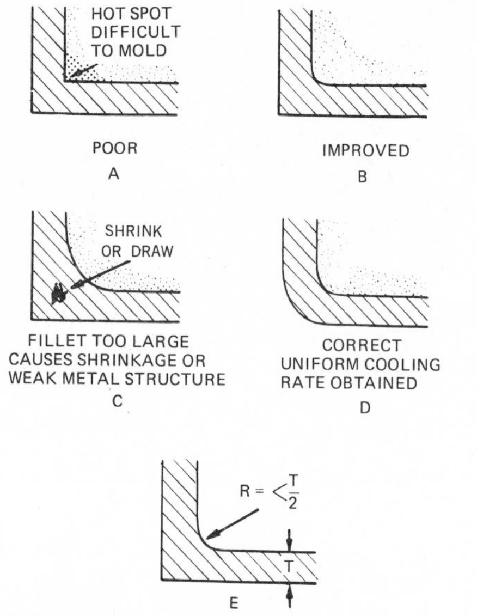
HOT SPOT DIFFICULT TO MOLD
POOR
A
IMPROVED
B
SHRINK OR DRAW
FILLET TOO LARGE CAUSES SHRINKAGE OR WEAK METAL STRUCTURE
C
CORRECT UNIFORM COOLING RATE OBTAINED
D
R = <T/2
T
E

Fig. 10-1-14 Fillet all-sharp angles. (Meehanite Metal Corp.)

Fillet All-Sharp Angles. See Fig. 10-1-14. Fillets have three functional purposes: to reduce stress concentration in the casting in service; to eliminate cracks, tears, and draws at reentry angles; and to make corners more moldable to eliminate hot spots.

Bring the Minimum Number of Adjoining Sections Together. See Fig. 10-1-15. A well-designed casting brings the minimum number of sections together and avoids acute angles.

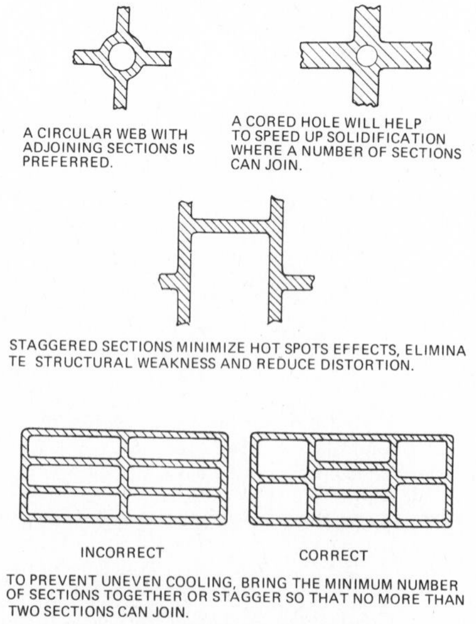
A CIRCULAR WEB WITH ADJOINING SECTIONS IS PREFERRED.
A CORED HOLE WILL HELP TO SPEED UP SOLIDIFICATION WHERE A NUMBER OF SECTIONS CAN JOIN.
STAGGERED SECTIONS MINIMIZE HOT SPOTS EFFECTS, ELIMINATE STRUCTURAL WEAKNESS AND REDUCE DISTORTION.
INCORRECT
CORRECT
TO PREVENT UNEVEN COOLING, BRING THE MINIMUM NUMBER OF SECTIONS TOGETHER OR STAGGER SO THAT NO MORE THAN TWO SECTIONS CAN JOIN.

Fig. 10-1-15 Bring the minimum number of adjoining sections together. (Meehanite Metal Corp.)

Design All Sections as Nearly Uniform in Thickness as Possible. Shrink defects and casting strains existed in the casting illustrated in Fig. 10-1-16. Redesigning eliminated excessive metal and resulted in a casting that was free from defects, was lighter in mass, and prevented the development of casting strains in the light radial veins.

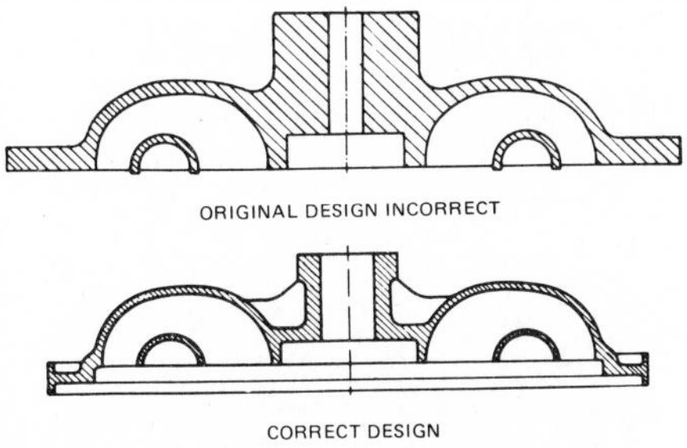
ORIGINAL DESIGN INCORRECT
CORRECT DESIGN

Fig. 10-1-16 Design all sections as nearly uniform in thickness as possible. (Meehanite Metal Corp.)

Avoid Abrupt Section Changes—Eliminate Sharp Corners at Adjoining Sections. See Fig. 10-1-17. The difference in the relative thickness of adjoining sections should be a minimum and not exceed a 2:1 ratio.

When a change of thickness is less than 2:1, it may take the form of a fillet; where the difference is greater, the form recommended is that of a wedge.

Wedge-shaped changes in wall thickness are to be designated with a taper not exceeding 1 in 4.

Design Ribs for Maximum Effectiveness. See Fig. 10-1-18. Ribs have two functions: to increase stiffness and to reduce the mass. If too shallow in depth or too widely spaced, they are ineffectual.

Bosses and Pads Should Not Be Used unless Absolutely Necessary. Bosses and pads increase metal thickness, create hot spots, and cause open grain or draws. Blend these into the casting by tapering or flattening the fillets. Bosses should not be included in casting design when the surface to support bolts, etc., may be obtained by milling or countersinking.

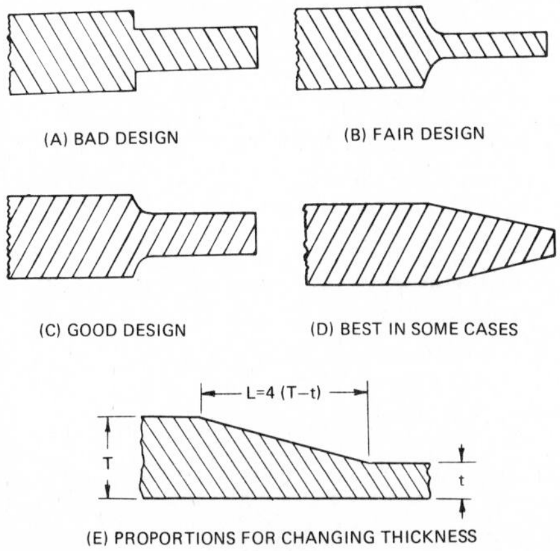
(A) BAD DESIGN
(B) FAIR DESIGN
(C) GOOD DESIGN
(D) BEST IN SOME CASES
L=4 (T−t)
T
t
(E) PROPORTIONS FOR CHANGING THICKNESS

Fig. 10-1-17 Avoid abrupt changes. (Meehanite Metal Corp.)

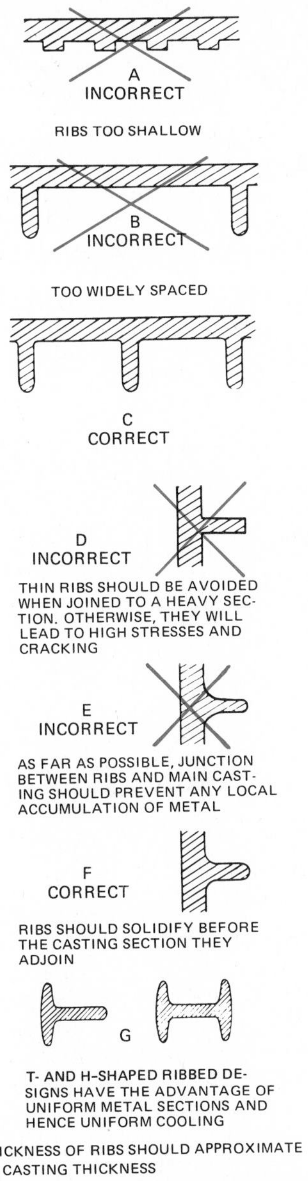
A
INCORRECT
RIBS TOO SHALLOW
B
INCORRECT
TOO WIDELY SPACED
C
CORRECT
D
INCORRECT
THIN RIBS SHOULD BE AVOIDED WHEN JOINED TO A HEAVY SECTION. OTHERWISE, THEY WILL LEAD TO HIGH STRESSES AND CRACKING
E
INCORRECT
AS FAR AS POSSIBLE, JUNCTION BETWEEN RIBS AND MAIN CASTING SHOULD PREVENT ANY LOCAL ACCUMULATION OF METAL
F
CORRECT
RIBS SHOULD SOLIDIFY BEFORE THE CASTING SECTION THEY ADJOIN
G
T- AND H-SHAPED RIBBED DESIGNS HAVE THE ADVANTAGE OF UNIFORM METAL SECTIONS AND HENCE UNIFORM COOLING
THICKNESS OF RIBS SHOULD APPROXIMATE 0.8 CASTING THICKNESS

Fig. 10-1-18 Design ribs for maximum effectiveness. (Meehanite Metal Corp.)

Spoked Wheels. See Fig. 10-1-19. A curved spoke is preferred to a straight one. It will tend to straighten slightly, thereby offsetting the dangers of cracking.

Use an odd number of spokes. A wheel having an odd number of spokes will not have the same direct tensile stress along the arms as one having an even number and will have more resiliency to casting stresses.

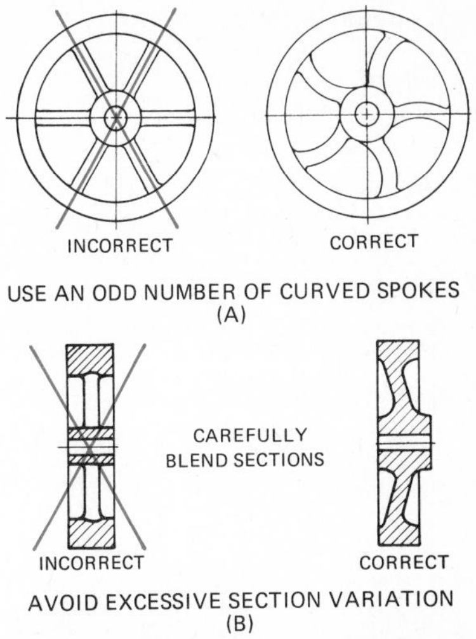
INCORRECT
CORRECT
USE AN ODD NUMBER OF CURVED SPOKES
(A)
CAREFULLY BLEND SECTIONS
INCORRECT
CORRECT
AVOID EXCESSIVE SECTION VARIATION
(B)

Fig. 10-1-19 Spoked-wheel design. (Meehanite Metal Corp.)

Wall Thicknesses. Walls should be of minimum thickness, consistent with good foundry practice, and should provide adequate strength and stiffness. The least expensive design is one which is light enough, but still presents the fewest hazards and difficulties in manufacturing. Wall thicknesses in castings should be uniform wherever the design permits. Wall thicknesses for different materials are as follows:

1. Walls of gray-iron castings and aluminum sand castings should not be less than 4 mm thick.
2. Walls of malleable iron and steel castings should not be less than 4.8 mm.
3. Walls of bronze, brass, or magnesium castings should not be less than 2.4 mm.

Parting Lines. A *parting line* is a line along which the pattern is divided for molding, or along which the sections of a mold separate. Selection of a parting line depends on a number of factors:

- Shape of the casting
- Elimination of machining on draft surfaces
- Method of supporting cores
- Location of gates and feeders

Holes in Castings. Small holes usually are drilled and not cored.

DRAWING PRACTICES[8]

It is important that a detail drawing give complete information on all cast parts, such as

- Machining allowances
- Surface texture
- Draft angles
- Limits on cast surfaces that must be controlled
- Locating points
- Parting lines

On small, simple parts all casting information is included on the finished drawing (see Fig. 10-1-20). On more complicated parts, it may be necessary to show additional casting views and sections to completely illustrate the construction of the casting. These additional views should show the rough casting outline in phantom lines and the finished contour in solid lines.

Material. In the selection of material for any particular application, the designer is influenced primarily by the physical characteristics such as strength, hardness, density, resistance to wear, mass, antifrictional properties, conductivity, corrosion resistance, shrinkage, and melting point. Often the demand is for a combination of a number of those qualities in the same casting.

Machining Allowance. In the construction of patterns for castings in which various points on the surface of the casting must be machined, sufficient excess metal should be provided for all machined surfaces. Unless otherwise specified, Fig. 10-1-21 may be used as a guide to machine finish allowance. When separate drawings are provided for cast and machined parts, suitable machining allowances must be selected and included in the casting dimensions.

CASTING ALLOY	DIMENSIONS WITHIN THIS RANGE	EXTERNAL SURFACE FINISH
CAST IRON ALUMINUM BRONZE, ETC. SAND CASTINGS	UP TO 200	2
	201 TO 400	2.5
	401 TO 600	3.6
	601 TO 800	4.6
	OVER 800	6
PEARLITIC MALLEABLE AND STEEL SAND CASTINGS	UP TO 200	2
	201 TO 400	3
	401 TO 600	4.6
	OVER 600	6
PERMANENT AND SEMIPERMANENT MOLD CASTINGS	UP TO 300	2
	301 TO 600	2.5
	OVER 600	4.6
PLASTER MOLD CASTING	UP TO 200	1
	201 TO 300	1.5
	OVER 300	2.5

Fig. 10-1-21 Guide to machining allowance for castings.

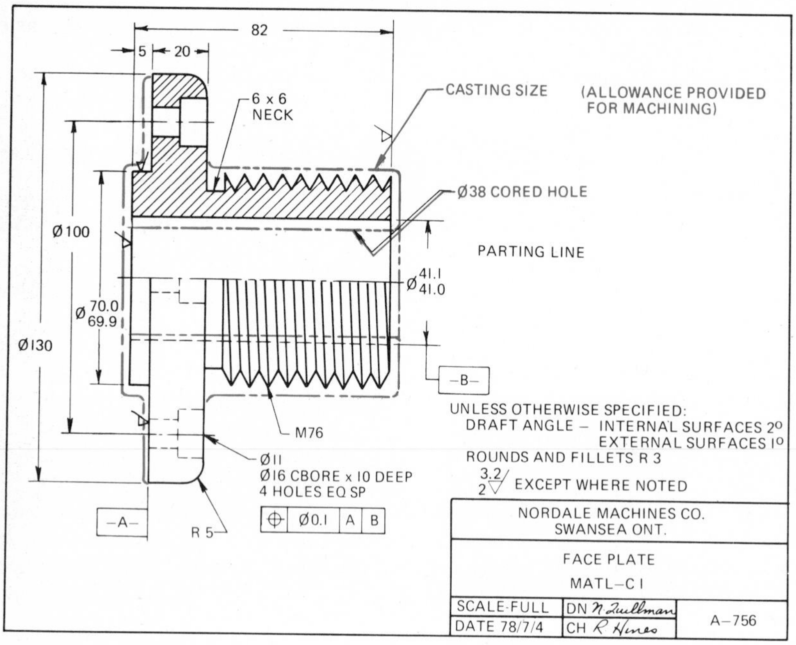
82
5
20
6 x 6
NECK
CASTING SIZE
(ALLOWANCE PROVIDED
FOR MACHINING)
Ø38 CORED HOLE
PARTING LINE
Ø100
Ø130
41.1
41.0
-B-
M76
Ø11
Ø16 CBORE x 10 DEEP
4 HOLES EQ SP
Ø0.1 A B
-A-
R 5
UNLESS OTHERWISE SPECIFIED:
DRAFT ANGLE – INTERNAL SURFACES 2°
EXTERNAL SURFACES 1°
ROUNDS AND FILLETS R 3
EXCEPT WHERE NOTED
NORDALE MACHINES CO.
SWANSEA ONT.
FACE PLATE
MATL–C I
SCALE-FULL
DATE 78/7/4
A–756

(A) WORKING DRAWING OF A CAST PART

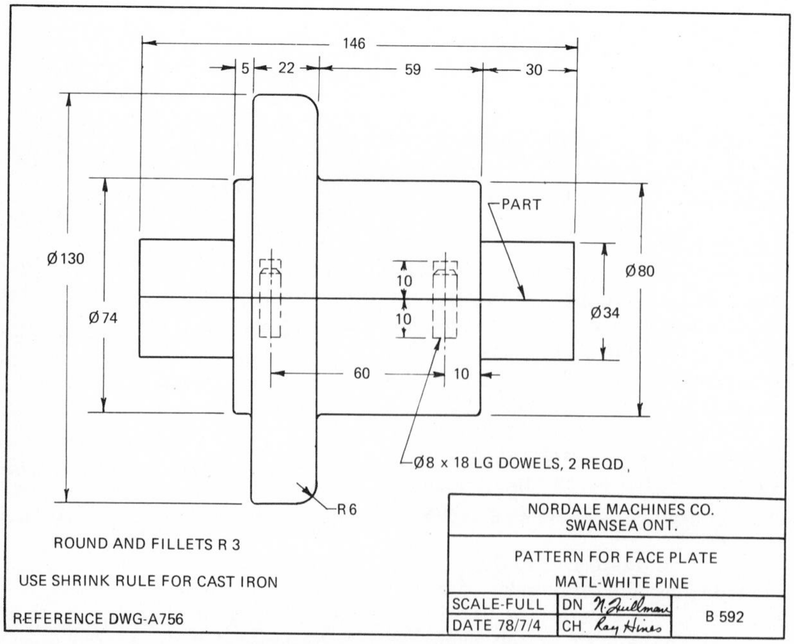
146
5
22
59
30
PART
Ø130
Ø80
Ø74
Ø34
10
10
60
10
Ø8 x 18 LG DOWELS, 2 REQD
R 6
ROUND AND FILLETS R 3
USE SHRINK RULE FOR CAST IRON
REFERENCE DWG-A756
NORDALE MACHINES CO.
SWANSEA ONT.
PATTERN FOR FACE PLATE
MATL-WHITE PINE
SCALE-FULL
DATE 78/7/4
B 592

(B) PATTERN DRAWING FOR THE CAST PART SHOWN IN (A)

Fig. 10-1-20 Cast part drawings.

Fillets and Radii. Generous fillets and radii should be provided on cast corners and must be specified on the drawing. This is best done by selecting a common radius for the majority of corners and fillets, to be specified in a note, and specifying separate radii in all cases where a different radius is required.

Casting Tolerances. A great many factors contribute to the dimensional variations of castings. However, the standard drawing tolerances specified in Fig. 10-1-22 can be satisfactorily attained in the production of average castings.

TYPE OF CASTING	DIMENSIONS WITHIN THIS RANGE	STANDARD DRAWING TOLERANCE (±)
IRON AND ALUMINUM SAND CASTINGS	UP TO 200	0.8
	201 TO 400	1.3
	401 TO 600	1.8
	601 TO 800	2.3
	OVER 800	3
PEARLITIC, MALLEABLE IRON AND STEEL SAND CASTINGS	UP TO 200	0.8
	201 TO 400	1.5
	401 TO 600	2.3
	OVER 600	3
PERMANENT MOLD CASTING (SEMIPERMANENT MOLD CASTING)	UP TO 130	0.8
	131 TO 300	0.8
	301 TO 600	1.3
	OVER 600	2.3
PLASTER MOLD CASTING	UP TO 100	0.5
	101 TO 200	0.5
	201 TO 300	0.8
	OVER 300	1.3
CENTRIFUGAL PRECISION CASTING	UP TO 13	0.5
	14 TO 130	0.5
	OVER 130	0.5

Fig. 10-1-22 Guide to casting tolerances.

Draft. All casting methods require a draft or taper on all surfaces perpendicular to the parting line, to facilitate removal of the pattern and ejection of the casting. The permissible draft must be specified on the drawing, in either degrees of taper for each surface or millimetres of taper per millimetre of length.

Suitable draft angles for general use, for both sand and die castings, are 1° for external surfaces and 2° for internal surfaces, as shown in Fig. 10-1-23.

The drawing must always clearly indicate whether the draft should be added to, or subtracted from, the casting dimensions.

Casting Datums[9]

It is recognized that in many cases a drawing is made of the fully machined end product; and casting dimensions, draft, and machining allowances are left entirely to the pattern maker or foundryman. However, for mass-production purposes it is generally advisable to make a separate casting drawing, with carefully selected datums, to ensure that parts will fit into machining jigs and fixtures and will meet final requirements after machining. Under these circumstances, dimensioning requires the selection of two sets of datum surfaces, lines, or points—one for the casting and one for the machining—to provide common reference points for measuring, machining, and assembly. To select suitable datums, it will be necessary to know how the casting is to be made, where the parting line or lines are to be, and how the part is going to fit into machining jigs and fixtures.

The first step in dimensioning is to select a primary datum surface for the casting, which is sometimes referred to as the *base* surface, and to label it DATUM A. See Fig. 10-1-24. This primary datum should be a surface which meets the following criteria as closely as possible:

1. It must be a surface, or datum targets on a surface (see Fig. 10-1-25),

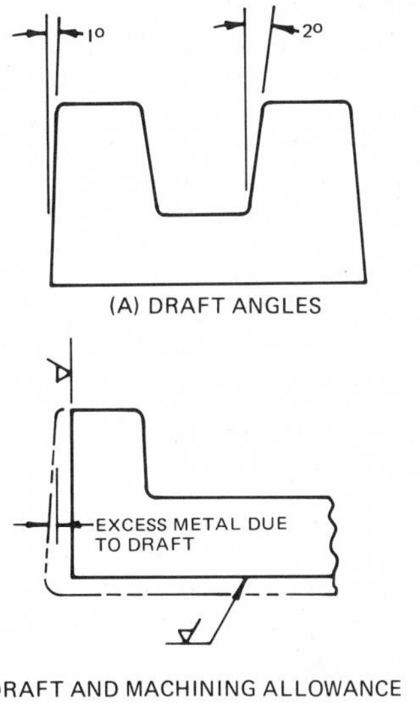

1°
2°
(A) DRAFT ANGLES
EXCESS METAL DUE TO DRAFT
(B) DRAFT AND MACHINING ALLOWANCE

Fig. 10-1-23 Draft for removal of pattern from mold.

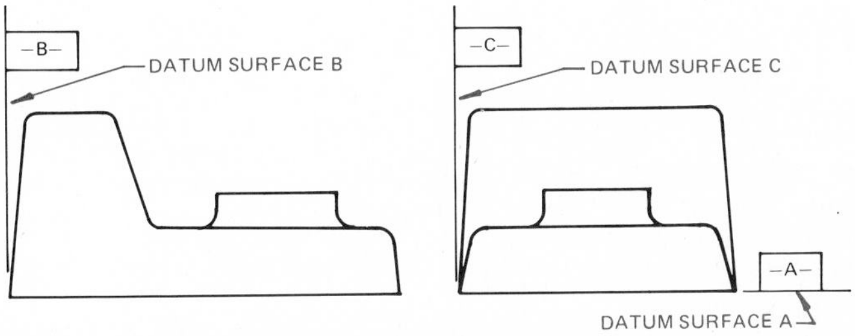

–B–
DATUM SURFACE B
–C–
DATUM SURFACE C
–A–
DATUM SURFACE A

Fig. 10-1-24 Casting datums.

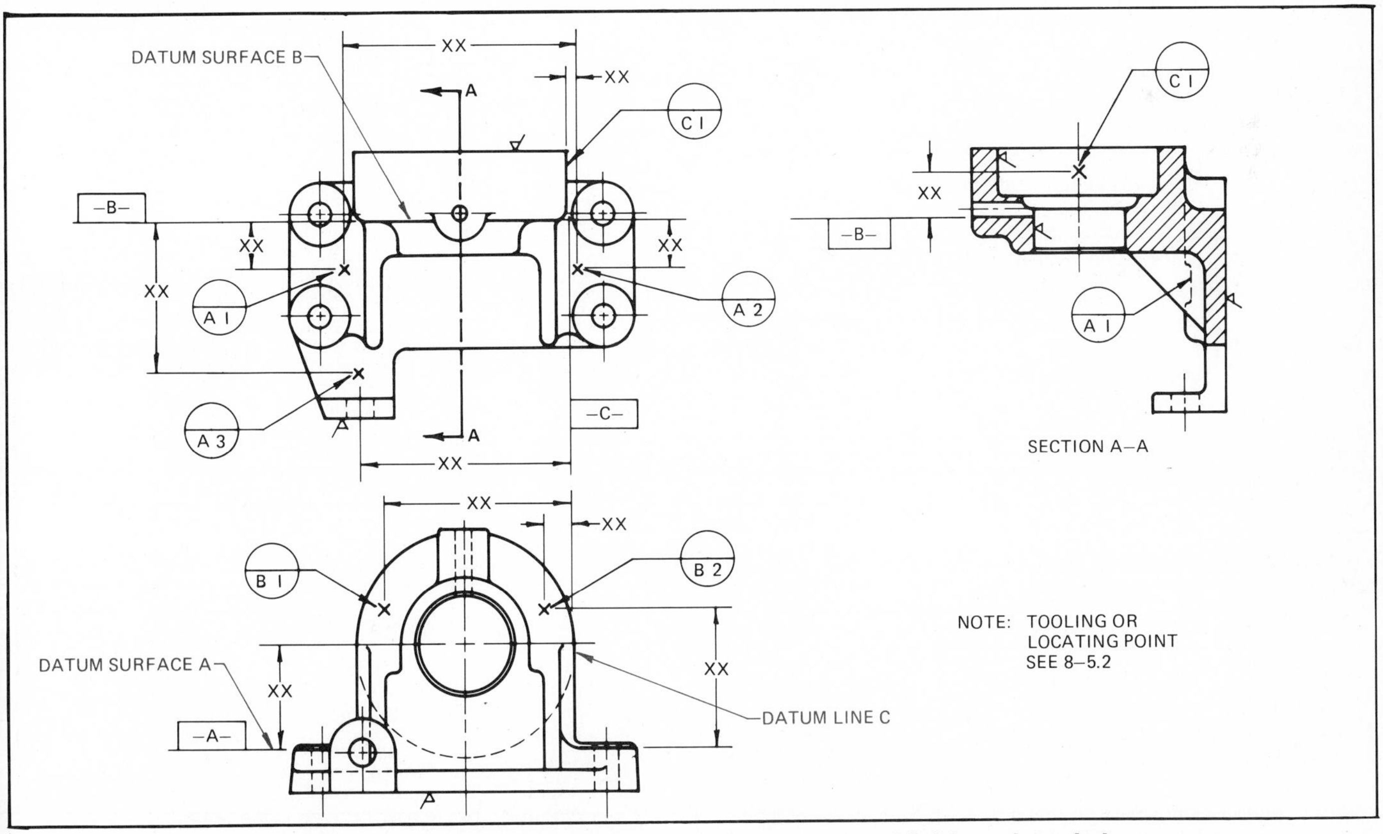

DATUM SURFACE B
XX
A
C 1
–B–
A 1
A 2
A 3
–C–
SECTION A–A
B 1
B 2
DATUM SURFACE A
–A–
DATUM LINE C
NOTE: TOOLING OR LOCATING POINT SEE 8–5.2

Fig. 10-1-25 Machined casting drawing illustrating datum lines, set-up points, and finish mark symbols.

which can be used as the basis for measuring the casting and which can later be used for mounting and locating the part in a jig or fixture, for the purpose of machining the primary datum for the finished part.

2. It should be a surface which will not be removed by machining, so that control of material to be removed is not lost, and can be checked at final inspection.

3. It should be parallel with the top of the mold, or parting line, that is, a surface which has no draft or taper.

4. It should be integral with the main body of the casting, so that measurements from it to the main surfaces of the casting will be least affected by cored surfaces, parting lines, or gated surfaces.

5. It should be a surface, or target areas on a surface, on which the part can be clamped without causing any distortion, so that the casting will not be under a distortional stress for the first machining operation.

6. It should be a surface which will provide locating points as far apart as possible, so that the effect of any flatness error will be minimized.

The second step is to select two other planes to serve as secondary and tertiary datums. These planes should be at right angles to one another and to the primary datum. They probably will not coincide with actual surfaces, because of taper or draft, except at one point, usually a point adjacent to the primary datum. These are labeled DATUM B and DATUM C, respectively, as shown in Fig. 10-1-26.

In the case of a circular part, the end-view center lines may be selected as secondary and tertiary datums, as shown in Fig. 10-1-27. In this case, unless otherwise specified, the center lines represent the center of the outside or overall diameter of the part.

Machining Datums

The first step in dimensioning the machined or finished part is to select a primary datum surface for machining and to label it DATUM D. This surface is the first surface on the casting to be machined and is thereafter used as the datum for all other machining operations. It should be selected to meet the following criteria:

1. It is generally preferable, though not essential, that it be a surface which is parallel to the primary casting datum.

2. It may be a large, flat, machined surface or several small areas of surfaces in the same or parallel planes.

3. If the primary casting datum surface is smooth and does not require machining, as in die castings, or if suitable target areas have been selected, the same surface may be used as the machining datum.

4. If the primary casting datum surface of sand castings appears to be the only suitable surface, it is recommended that three or four pads be provided, which can be machined to form the machining datum, as shown in Fig. 10-1-28.

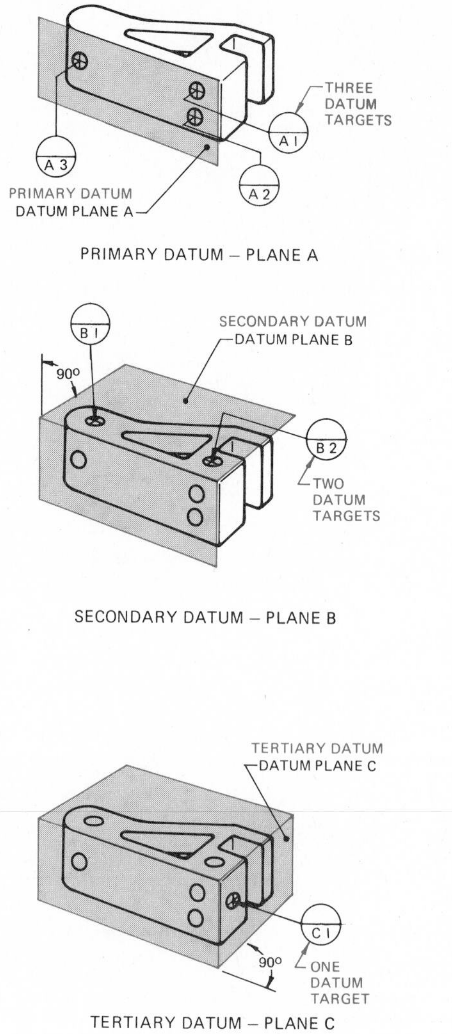
THREE DATUM TARGETS
A 1
A 2
A 3
PRIMARY DATUM
DATUM PLANE A
PRIMARY DATUM – PLANE A
B 1
90°
SECONDARY DATUM
DATUM PLANE B
B 2
TWO DATUM TARGETS
SECONDARY DATUM – PLANE B
TERTIARY DATUM
DATUM PLANE C
C 1
90°
ONE DATUM TARGET
TERTIARY DATUM – PLANE C

Fig. 10-1-26 Datum planes and datum targets.

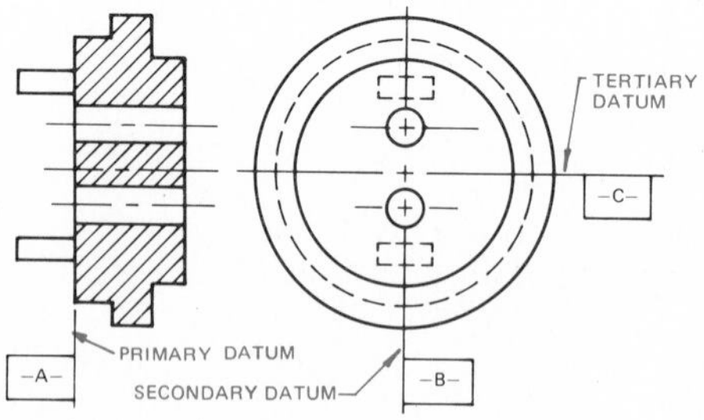
TERTIARY DATUM
-C-
-A-
PRIMARY DATUM
SECONDARY DATUM
-B-

Fig. 10-1-27 Datums for circular casting.

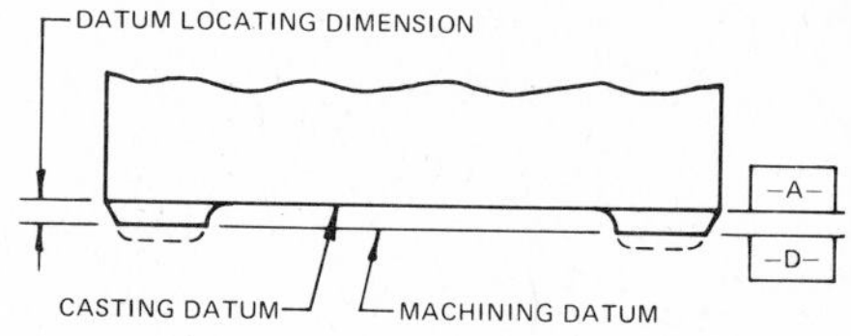
DATUM LOCATING DIMENSION
-A-
-D-
CASTING DATUM
MACHINING DATUM

Fig. 10-1-28 Primary machining datum.

5. When pads or small target areas are selected, they should be placed as far apart as possible and located where the part can be readily clamped in jigs or fixtures without distorting it or interfering with other machining operations.

The second step is to select two other surfaces to serve as secondary and tertiary datums. If these datums are required only for locating and dimensioning purposes, and not for clamping in a jig or fixture, some suitable datums other than flat, machined surfaces may be chosen. These could be the same datums as used for casting, if the locating point in each case is clearly defined and is not removed in machining. For circular parts, a hole drilled in the center, or a turned diameter other than the outside diameter, may provide suitable center lines for use as secondary datums, as shown in Fig. 10-1-29.

The third step is to specify the *datum-locating dimension*, that is, the dimension between each casting datum and the corresponding machining datum. See Fig. 10-1-28. There shall never be more than one such dimension from each casting datum surface.

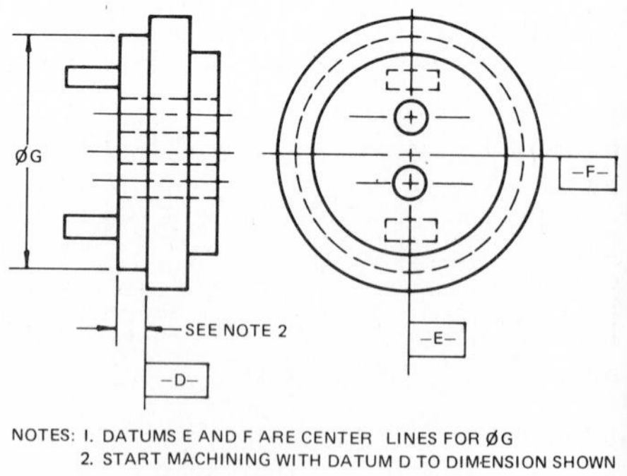
ØG
-F-
SEE NOTE 2
-E-
-D-
NOTES: 1. DATUMS E AND F ARE CENTER LINES FOR ØG
2. START MACHINING WITH DATUM D TO DIMENSION SHOWN

Fig. 10-1-29 Machining datums for circular parts.

DIMENSIONS

When suitable datums have been selected, with datum-locating dimensions for the machined casting drawing, dimensioning may proceed, with dimensions being specified directly from the datums to all main surfaces. However, where it is necessary to maintain a particular relationship between two or more surfaces or features, regular point-to-point dimensioning is usually the preferred method. This will generally include all such items as thickness of ribs, height of bosses, projections, depth of grooves, most diameters and radii, and center distances between holes and similar features. Whenever possible, specify dimensions to surfaces or surface intersections, rather than to radii centers or nonexistent center lines.

Dimensions given on the casting drawing should not be repeated, except as reference dimensions, on the machined part drawing.

When combined casting and machined part drawings are prepared, it is necessary to show both sets of datums and both sets of dimensions. However, since such drawings are used only for simple parts, there will be very few machined surfaces, and quite frequently the same datums will serve for both purposes.

REFERENCES AND SOURCE MATERIAL

1. American Iron and Steel Institute, "Principles of Forging Design."
2. General Motors Corporation.
3. J. F. Wallace, "Casting," *Machine Design*, vol. 37, no. 21, 1965.
4. General Motors Corporation.
5. Designed Precision Castings Ltd.
6. Full Mold Process (Canada) Limited.
7. Meehanite Metal Corporation.
8. General Motors Corporation.
9. General Motors Corporation.

Assignments

1. On an A3- or B-size sheet, complete the detail drawing of the base for the adjustable shaft support assembly shown in Fig. 10-1-A or 10-1-B. Cored holes are to be used for the shaft holes. Scale is 1:1.

2. On an A3- or B-size sheet, complete the detail drawing of the fork for the hinged pipe vise assembly shown in Fig. 10-1-C. Use your judgment for dimensions not given. Scale is 1:1.

REVIEW FOR ASSIGNMENTS

Unit 7-8 Removed and Revolved Sections
Unit 7-4 Threads in Section
Unit 6-1 Detail Drawings
Unit 5-7 Machine Symbols

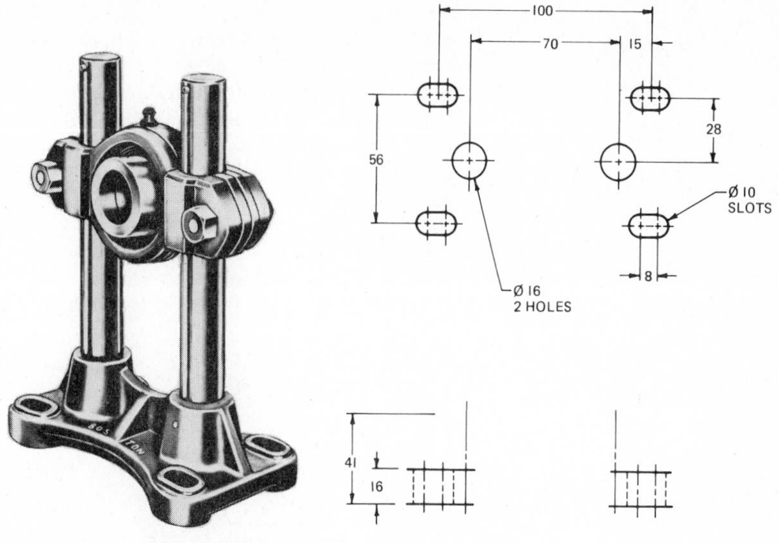

100
70
15
56
28
Ø 10 SLOTS
8
Ø 16 2 HOLES
41
16

Fig. 10-1-A Bearing stand.

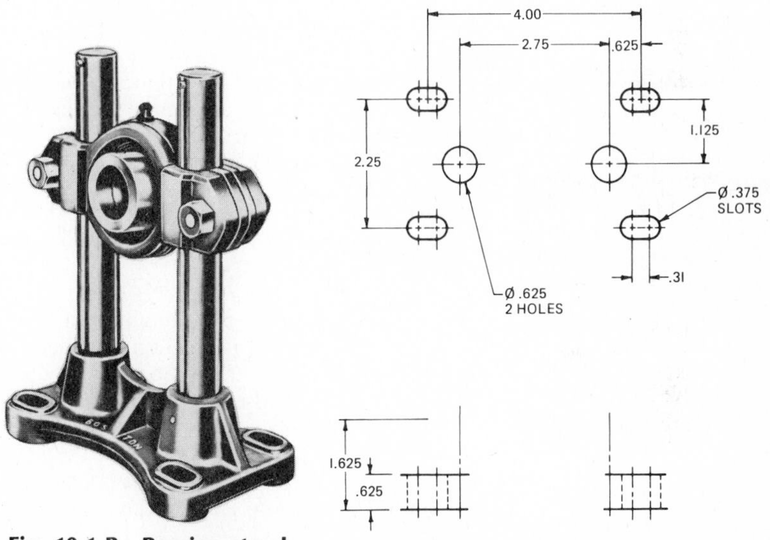

4.00
2.75
.625
2.25
1.125
Ø .375 SLOTS
.31
Ø .625 2 HOLES
1.625
.625

Fig. 10-1-B Bearing stand.

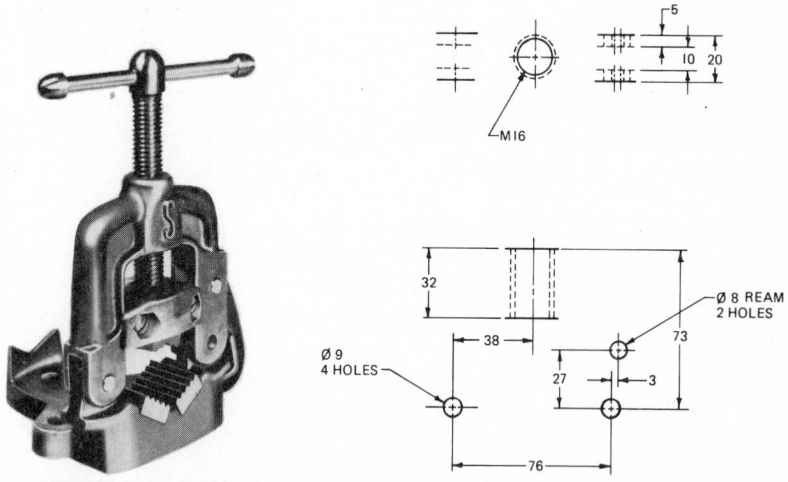

5
10
20
M16
32
Ø 8 REAM 2 HOLES
38
73
Ø 9 4 HOLES
27
3
76

Fig. 10-1-C Pipe vise base.

UNIT 10-2

FORGING, COLD HEADING, AND POWDER METALLURGY

Forging[1,2]

Forging consists of plastically deforming, either by a squeezing pressure or sharp blows, a cast or sintered ingot, a wrought bar or billet, or a powder-metal shape, to produce a desired shape with good mechanical properties. Practically all ductile metals can be forged. See Fig. 10-2-1.

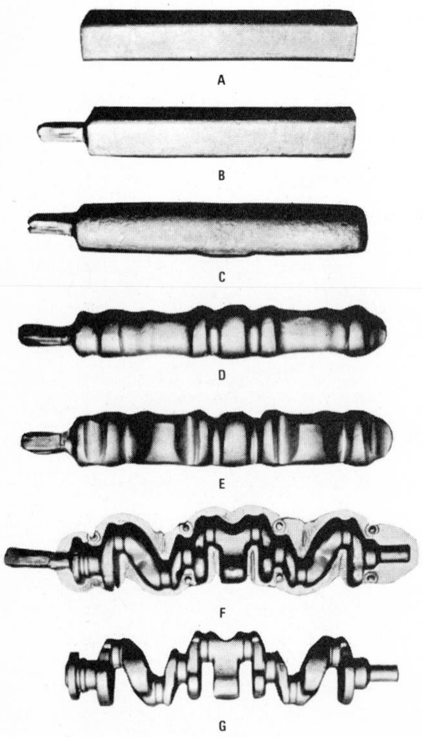
A
B
C
D
E
F
G

(A) BILLET; (B) TONGHOLD IS FIRST FORGED; (C) (D) (E) PRE-FORMING IMPRESSIONS; (F) BLOCKING AND FINISHING; (G) AFTER TRIMMING CRANKS ARE TWISTED INTO POSITION.

Fig. 10-2-1 The forging of a crankshaft. (Wyman-Gordon Co.)

TYPES OF FORGINGS

Closed-Die Forgings. Closed-die forgings are made by hammering or pressing metal until it conforms closely to the shape of the enclosing dies. Grain flow in the closed-die-forged parts can be oriented in the direction requiring greatest strength.

Three-dimensional control of the material to be forged requires a closed die, a simple and common form of which is the impression die.

In the simplest example of impression die forging (Fig. 10-2-2) the workpiece is cylindrical and is placed in the bottom-half die. On closing the top die, the cylinder undergoes elastic compression until its enlarged sides touch the side walls of the die impression. At this point, a small amount of excess material begins to form the flash between the two die faces. In the further course of die approach, this flash is gradually thinned.

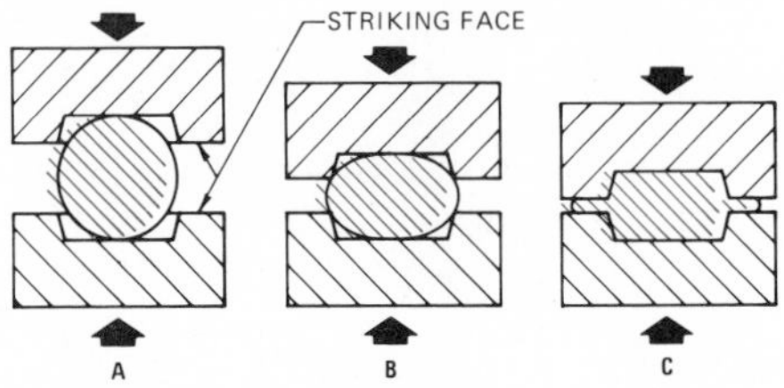
STRIKING FACE
A
B
C

Fig. 10-2-2 Compression in impression dies.

The forging impression die gives control over all three directions, except when the die is similar to that shown in Fig. 10-2-2, and the deforming forging machine tool has an unlimited stroke (e.g., a hammer or hydraulic press). In the latter case, the die must be shaped to allow complete closing of the striking faces at the end of the stroke (Fig. 10-2-3).

In practice, *closed-die forging* has become the term applied to all forging operations involving three-dimensional control. However, it is seen that the die is only closed by virtue of flash formation.

Forging dies can be divided into three main classes: single-impression, double-impression, and interlocking. See Fig. 10-2-4.

Single-impression dies have the impression of the desired forging entirely in one half of the die.

Double-impression dies have part of the impression of the desired forging sunk in each die in such a manner that no part of the die projects past the parting line into the other die. This type is the most common class of forging.

Interlock dies are the type in which a projection from either die extends past the parting line into the mating die. Interlock dies are generally more expensive because of the cost of making the projection of the die.

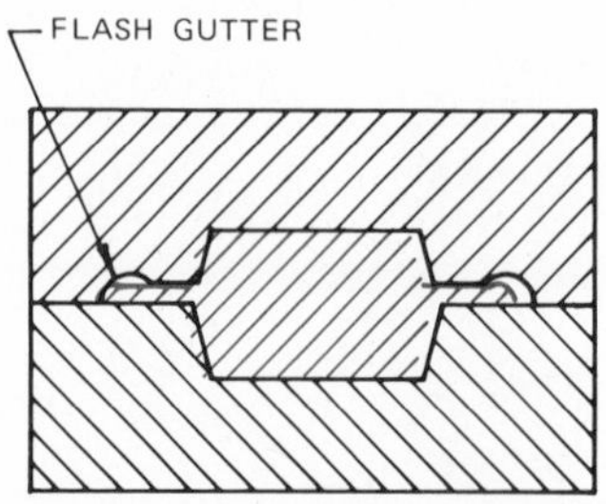
FLASH GUTTER

Fig. 10-2-3 Forging die with flesh gutter.

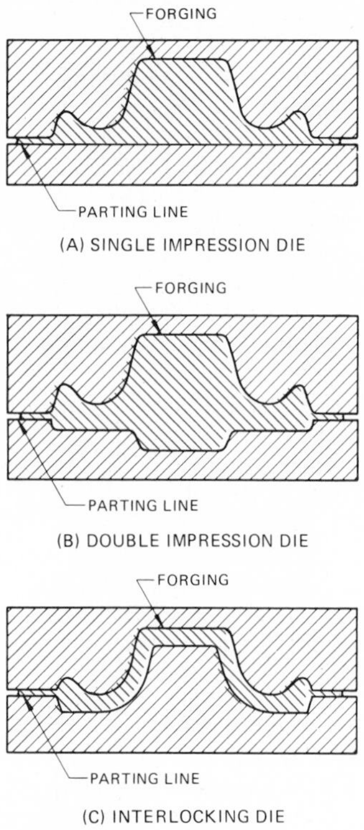
FORGING
PARTING LINE
(A) SINGLE IMPRESSION DIE
FORGING
PARTING LINE
(B) DOUBLE IMPRESSION DIE
FORGING
PARTING LINE
(C) INTERLOCKING DIE

Fig. 10-2-4 Forging dies. (General Motors Corp.)

TRIMMING

Because the quantity of forging metal is generally in excess of the space in the die cavity, space is provided between the die surfaces for the escape of the excess metal. This space is called the *flash space*, and the excess metal which flows into it is called *flash*. The flash thickness is proportionate to the mass of the forging.

The flash is removed from forgings by trimming dies which are formed to the outline of the part. These dies shear or punch the forging out of the flash, leaving a scar or mark, the width and raggedness of which depends on the quality and state of wear of the dies. See Fig. 10-2-5.

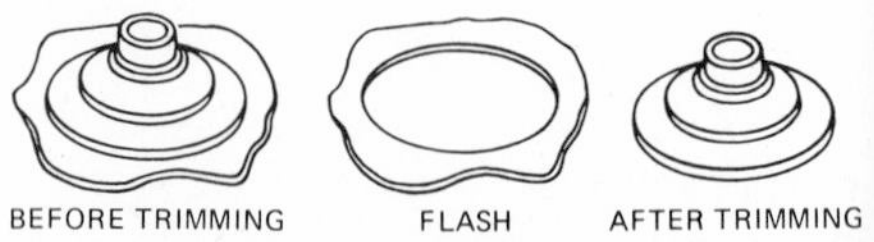
BEFORE TRIMMING
FLASH
AFTER TRIMMING

Fig. 10-2-5 Flash trimming.

DESIGN CONSIDERATIONS

Corner and Fillet Radii. It is important in forging design to use correct radii where two surfaces meet. Corner and fillet radii on forgings should be sufficient to facilitate the flow of metal for sound forgings and to permit economical manufacture.

Stress concentrations resulting from abrupt changes in section thickness or direction are minimized by corner and fillet radii of correct size. Any radius larger than recommended will increase die life. Any radius smaller than recommended will decrease die life. See Fig. 10-2-6 for recommendations.

Sharp fillets cause the formation of cold shuts. In a forging, a cold shut is a lap where two surfaces of metal have folded against each other, forming an undesirable flow of metal. A cold shut causes a weak spot that may be opened into a crack by heat treatment. Cold shuts are most likely to form at fillets in deep depressions or in deep sections, especially where the metal is confined. See Fig. 10-2-7. In these cases larger fillets are required, as shown in Fig. 10-2-6.

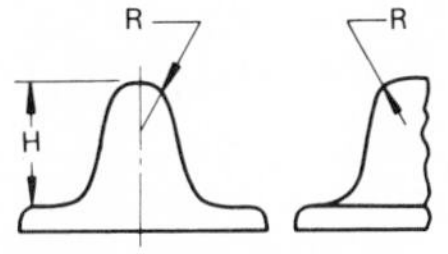
R
R
H

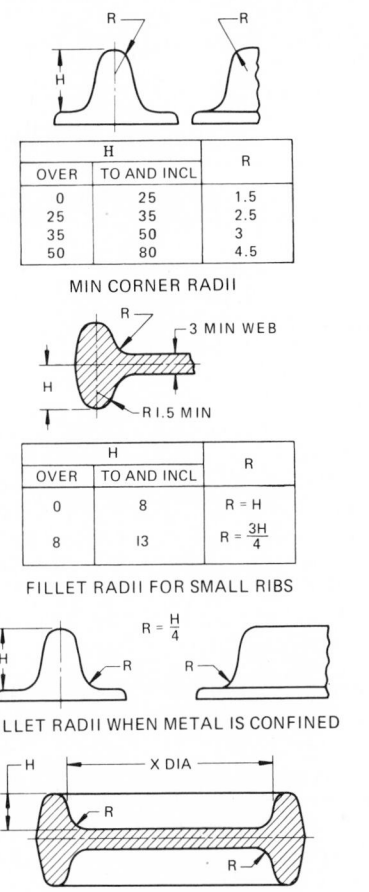

H		R
OVER	TO AND INCL	
0	25	1.5
25	35	2.5
35	50	3
50	80	4.5

MIN CORNER RADII

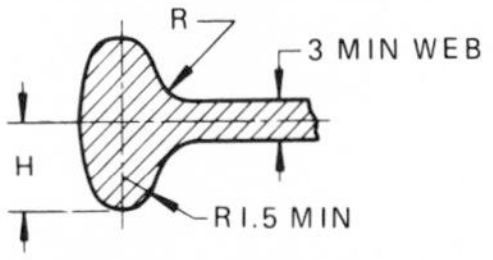
R
3 MIN WEB
H
R 1.5 MIN

H		R
OVER	TO AND INCL	
0	8	R = H
8	13	$R = \frac{3H}{4}$

FILLET RADII FOR SMALL RIBS

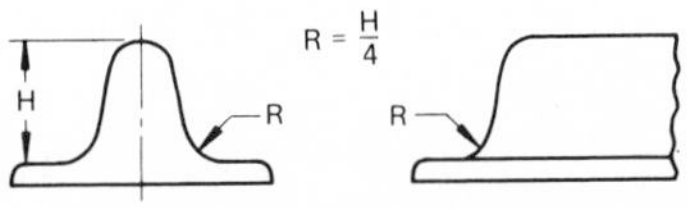
$R = \frac{H}{4}$
H
R
R

FILLET RADII WHEN METAL IS CONFINED

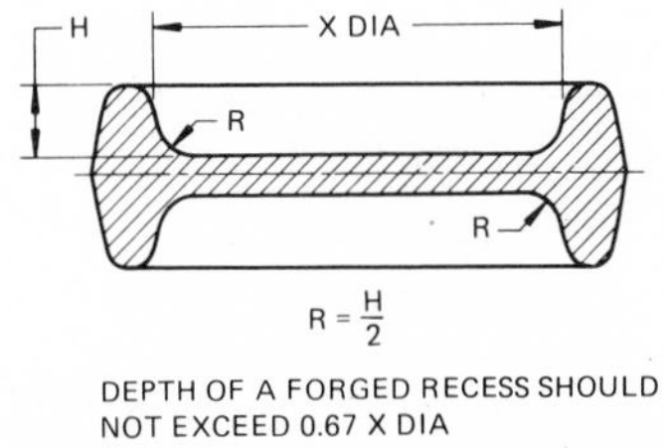
H
X DIA
R
R
$R = \frac{H}{2}$
DEPTH OF A FORGED RECESS SHOULD NOT EXCEED 0.67 X DIA

FILLET RADII WHEN METAL IS NOT CONFINED

Fig. 10-2-6 Corner and fillet radii. (General Motors Corp.)

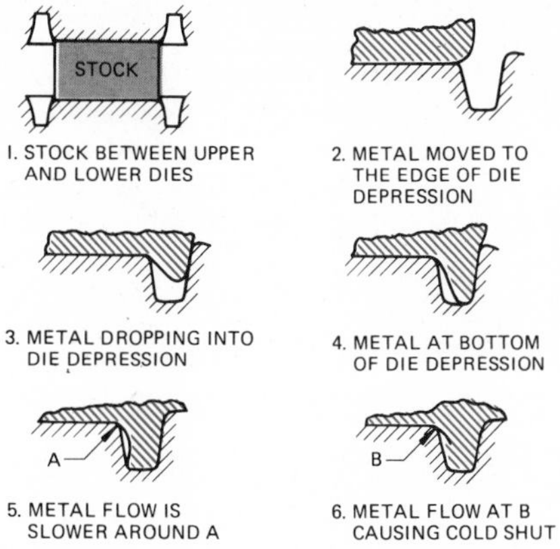
STOCK
1. STOCK BETWEEN UPPER AND LOWER DIES
2. METAL MOVED TO THE EDGE OF DIE DEPRESSION
3. METAL DROPPING INTO DIE DEPRESSION
4. METAL AT BOTTOM OF DIE DEPRESSION
A
B
5. METAL FLOW IS SLOWER AROUND A
6. METAL FLOW AT B CAUSING COLD SHUT

Fig. 10-2-7 Cold shut. (General Motors Corp.)

Draft Angle. Draft is one of the first factors to be considered in designing a forged part. See Fig. 10-2-8. *Draft* is defined as the angle of taper given to the side walls of the die in order to facilitate removal of the forging. Where little or no draft is allowed, stripper or ejection mechanisms must be used. The usual amount of draft for exterior contours is 7° and for interior contours 10°.

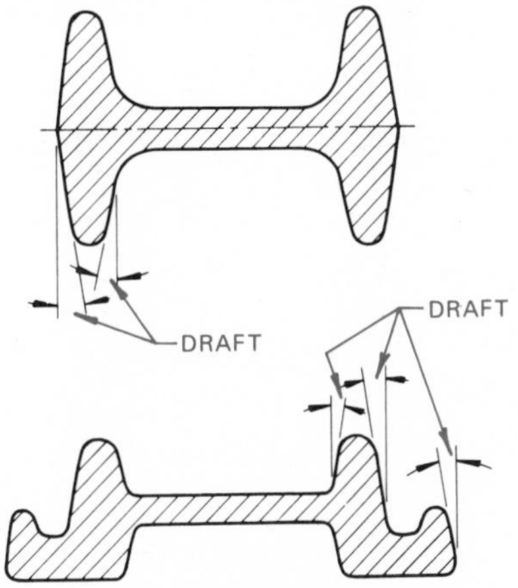
DRAFT
DRAFT

Fig. 10-2-8 Draft application.

Die draft equivalent is the amount of offset that results from draft. Figure 10-2-9 shows the draft equivalents for varying angles and depth of draft.

Parting Line. The surfaces of dies that meet in forgings are the striking surfaces. The line of meeting is the parting line. The parting line of the forging must be established in order to determine the amount of draft and its location.

The location and the type of parting as applied to simple forgings are shown in Fig 10-2-10.

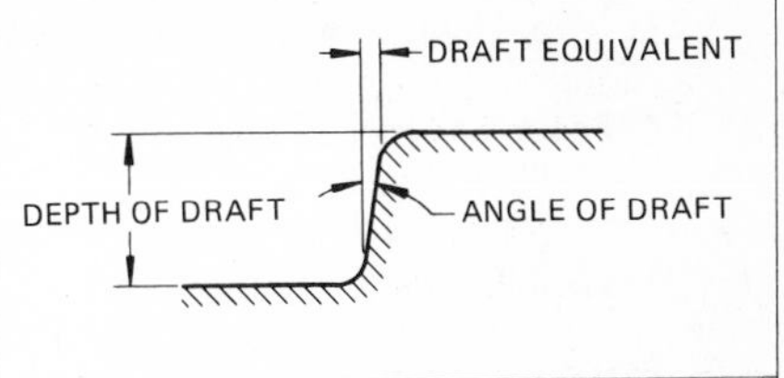
DRAFT EQUIVALENT
DEPTH OF DRAFT
ANGLE OF DRAFT

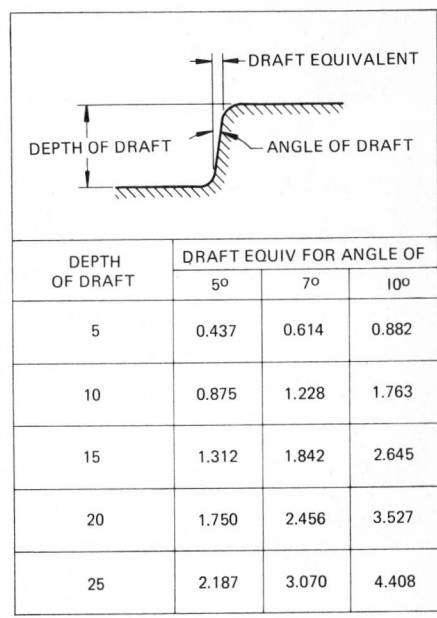

DEPTH OF DRAFT	DRAFT EQUIV FOR ANGLE OF		
	5°	7°	10°
5	0.437	0.614	0.882
10	0.875	1.228	1.763
15	1.312	1.842	2.645
20	1.750	2.456	3.527
25	2.187	3.070	4.408

Fig. 10-2-9 Die draft equivalent. (General Motors Corp.)

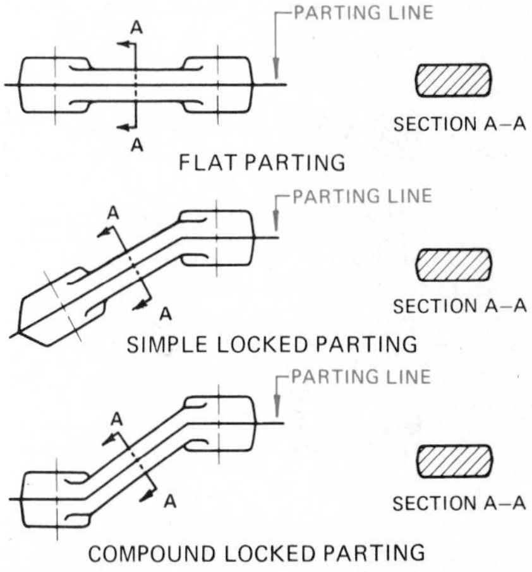
PARTING LINE
A
A
SECTION A–A
FLAT PARTING
PARTING LINE
A
A
SECTION A–A
SIMPLE LOCKED PARTING
PARTING LINE
A
A
SECTION A–A
COMPOUND LOCKED PARTING

Fig. 10-2-10 Parting line application. (General Motors Corp.)

DRAWING PRACTICES

In preparing forging drawings, it is important to consider drawing practices which may be peculiar to forgings, such as the following:

- Dimensioning
- Draft angles and parting lines
- Corner and fillet radii
- Forging tolerances
- Allowance for machining
- Material specifications
- Heat treatment
- Location of trademark, part number, and vendor specification

Dimensioning. It is generally desirable to apply dimensions to the forged part of the depths of the die impressions. Draft is additive to these dimensions and should be expressed in degrees or linear dimensions.

When the depth of the die impression is located, only one dimension should originate from the parting line. This surface should then be used to establish other dimensions, as shown in Fig. 10-2-11.

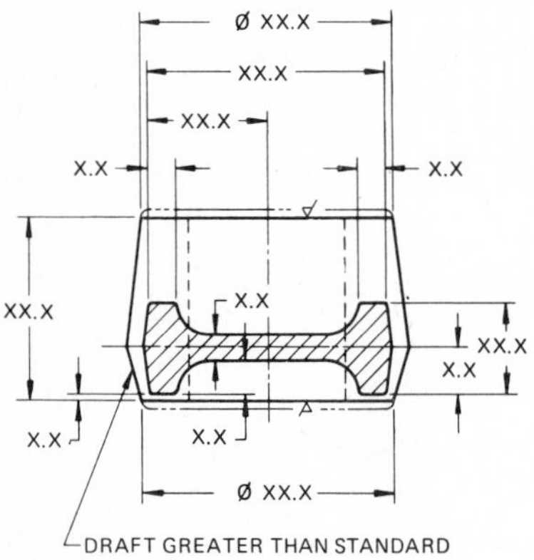
Ø XX.X
XX.X
XX.X
X.X
X.X
XX.X
X.X
XX.X
X.X
X.X
X.X
Ø XX.X
DRAFT GREATER THAN STANDARD

Fig. 10-2-11 Dimensioning. (General Motors Corp.)

Allowance for Machining. When a forging is to be machined, allowance must be made for metal to be removed.

Composite Drawings. Generally, a forged part should be shown on one drawing with the forging outline shown in phantom lines, as in Fig. 10-2-12. Forging outlines for machining allowance should not be dimensioned unless the amount of finish cannot be controlled by the machining symbol.

Separate drawings for rough forgings should be made only when the part is complicated and the outline of the rough forging cannot be clearly visualized, or where the outline of the rough forging must be maintained for tooling purposes.

Where both the forging and machining drawings are shown on the same sheet, as in Fig. 10-2-13, place the headings FORGING DRAWING and MACHINING DRAWING directly under the corresponding views.

Cold Heading[3]

Cold heading was developed primarily to make nuts, bolts, screws, and other nail-like shapes. Today it is used to form a variety of more complex shapes which retain only a rough family resemblance to nails.

The process is simply one in which wire is cut from a coil, fed into a die, then worked by a series of die motions so that it assumes diameters of varying

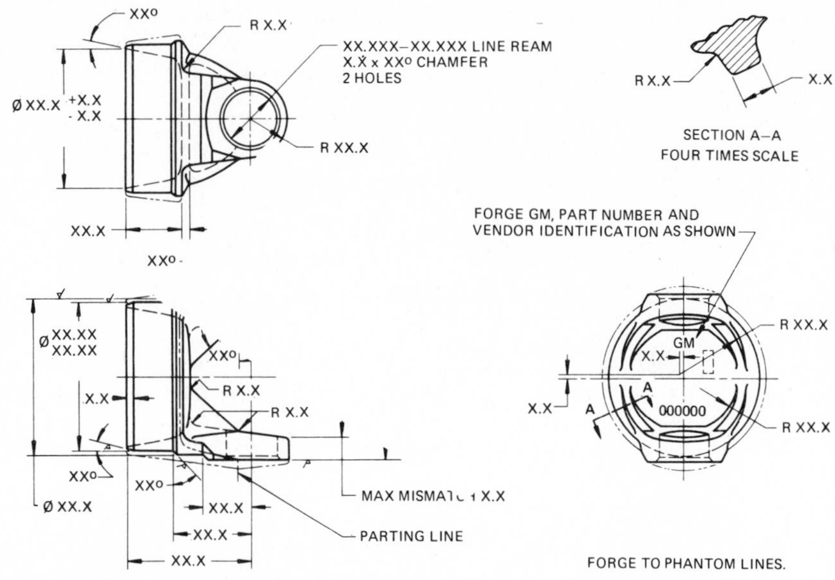
XX°
R X.X
XX.XXX–XX.XXX LINE REAM
X.X x XX° CHAMFER
2 HOLES
Ø XX.X +X.X -X.X
R XX.X
XX.X
XX°
R X.X
X.X
SECTION A–A
FOUR TIMES SCALE
FORGE GM, PART NUMBER AND
VENDOR IDENTIFICATION AS SHOWN
Ø XX.XX XX.XX
XX°
R X.X
X.X
R X.X
XX°
XX°
Ø XX.X
XX.X
XX.X
XX.X
MAX MISMATCH X.X
PARTING LINE
R XX.X
GM
X.X
A
000000
X.X
A
R XX.X
FORGE TO PHANTOM LINES.

Fig. 10-2-12 Composite forging drawing. (General Motors Corp.)

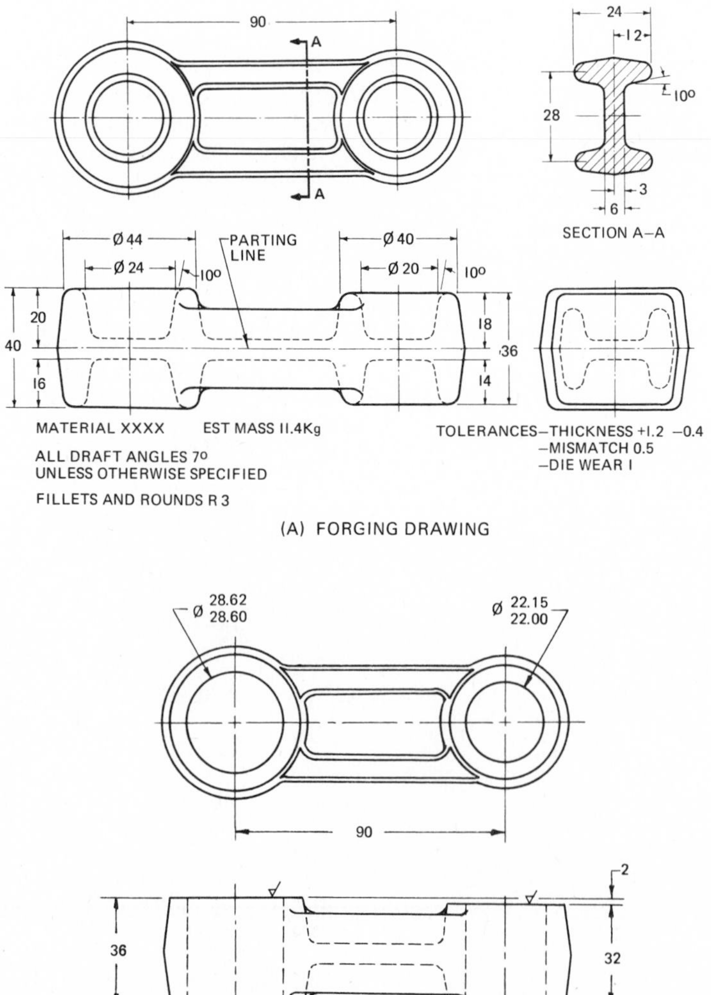
90
A
A
24
12
28
10°
3
6
SECTION A–A
Ø 44
Ø 24
10°
PARTING LINE
Ø 40
Ø 20
10°
20
40
16
18
36
14
MATERIAL XXXX
EST MASS 11.4Kg
ALL DRAFT ANGLES 7°
UNLESS OTHERWISE SPECIFIED
FILLETS AND ROUNDS R 3
TOLERANCES–THICKNESS +1.2 –0.4
–MISMATCH 0.5
–DIE WEAR 1
(A) FORGING DRAWING
Ø 28.62 28.60
Ø 22.15 22.00
90
2
36
32
(B) MACHINING DRAWING

Fig. 10-2-13 Separate forging and machining drawings.

thicknesses while retaining its basic cylindrical shape. Each forming strike is called a *blow*, and most commercial parts are made on two-blow machines.

The rules governing recommended design practice are simple (Fig. 10-2-14). Shoulders can be included in a headed part, but corners should have generous fillets and radii.

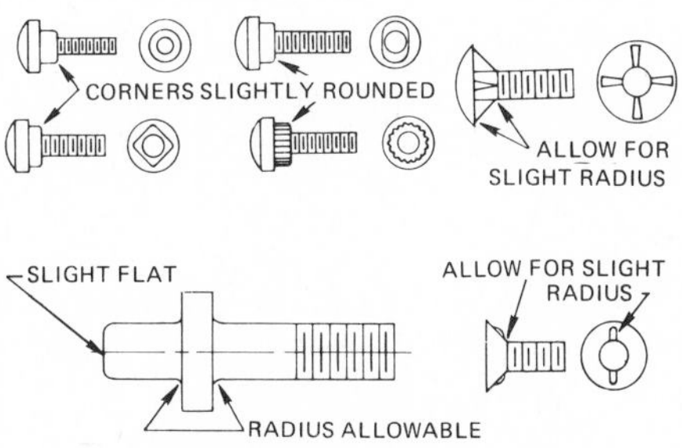
CORNERS SLIGHTLY ROUNDED
ALLOW FOR SLIGHT RADIUS
SLIGHT FLAT
ALLOW FOR SLIGHT RADIUS
RADIUS ALLOWABLE

Fig. 10-2-14 Recommended design practice for radii.

Powder Metallurgy[4]

Powder metallurgy is the process of making parts by compressing and sintering various metallic and nonmetallic powders into shape. See Fig. 10-2-15.

Dies and presses known as *briquetting machines* are used to compress the powders into shape. These briquets or compacts are then sintered or heated in an atmosphere-controlled furnace, bonding the powdered materials.

DESIGN CONSIDERATIONS

The following should be considered when powder metal parts are designed in order to realize the maximum benefits from the powder metallurgy process. This process is most applicable to the production of cylindrical, rectangular, or irregular shapes that do not have large variations in cross-sectional dimensions. Surface indentations or projections can be formed on either end or both ends of a part. Splines, gear teeth, axial holes, counterbores, straight knurls, serrations, slots, and keyways present few problems.

Ejection from the Die. The shape of the part must permit ejection from the die. The design requirements for some parts can be achieved only by subsequent machining, as in some corner relief designs, reverse tapers, holes at right angles to the direction of pressing, diamond knurls, and undercuts.

Corner Reliefs. Corner reliefs can be molded or machined. A molded corner relief will save machining. See Fig. 10-2-16.

Reverse Tapers. Reverse tapers cannot be molded. They can be machined. See Fig. 10-2-17.

Holes at Right Angles to the Direction of Pressing. Right-angle holes must be machined. See Fig. 10-2-18.

Knurls. Straight knurls can be molded; diamond knurls cannot. See Fig. 10-2-19.

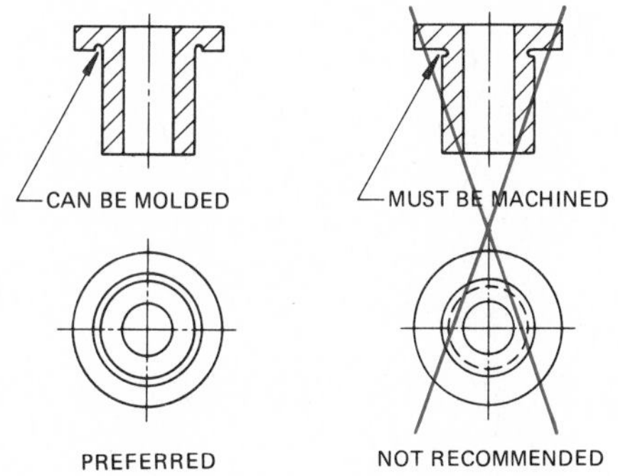
CAN BE MOLDED
MUST BE MACHINED
PREFERRED
NOT RECOMMENDED

Fig. 10-2-16 Corner relief. (General Motors Corp.)

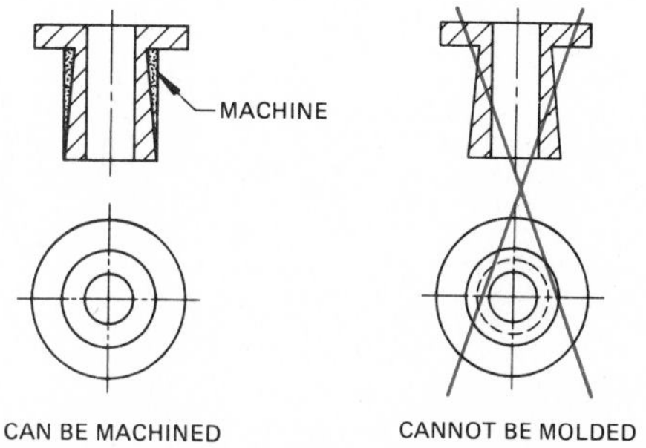
MACHINE
CAN BE MACHINED
CANNOT BE MOLDED

Fig. 10-2-17 Reverse taper. (General Motors Corp.)

Undercuts. Undercuts must be machined. See Fig. 10-2-20.

Wall Thickness. In general, sidewalls bordering a depression or hole should be a minimum of 0.8 mm thick. See Fig. 10-2-21.

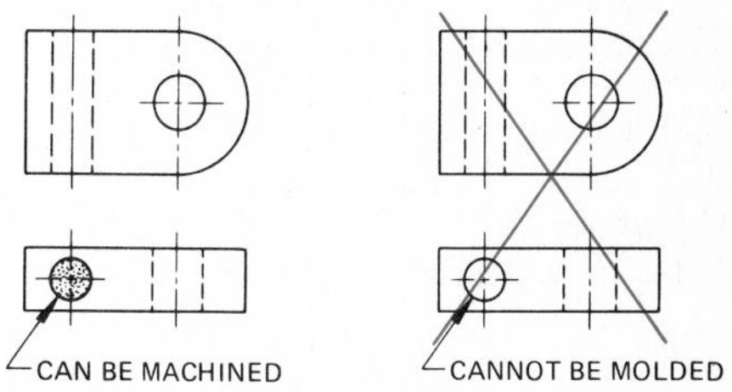
CAN BE MACHINED
CANNOT BE MOLDED

Fig. 10-2-18 Right-angle hole. (General Motors Corp.)

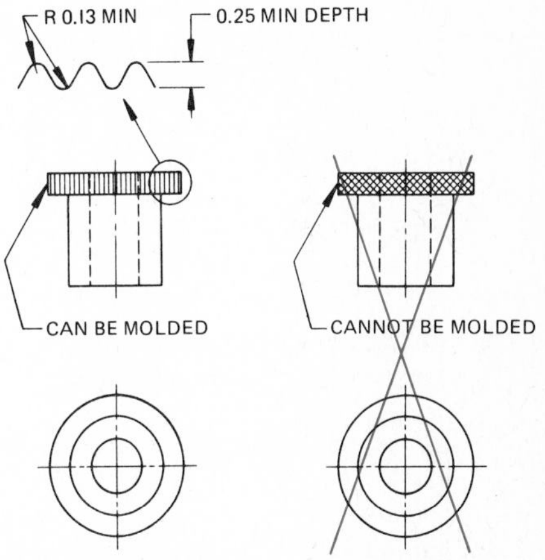
R 0.13 MIN
0.25 MIN DEPTH
CAN BE MOLDED
CANNOT BE MOLDED

Fig. 10-2-19 Knurls. (General Motors Corp.)

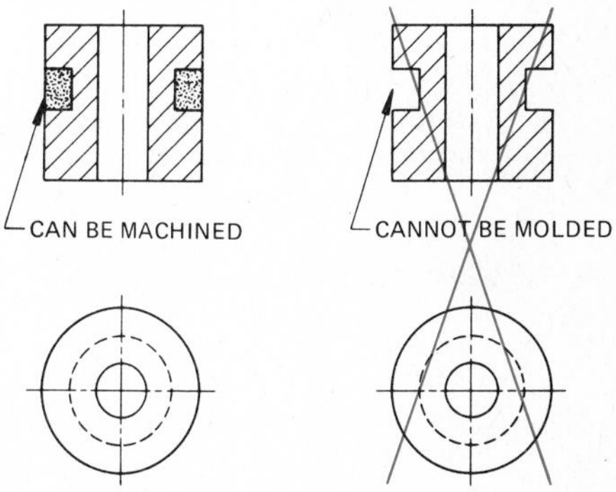
CAN BE MACHINED
CANNOT BE MOLDED

Fig. 10-2-20 Undercuts. (General Motors Corp.)

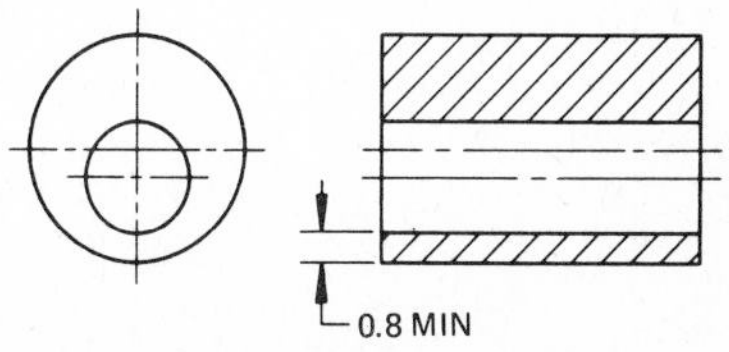
0.8 MIN

Fig. 10-2-21 Wall thickness. (General Motors Corp.)

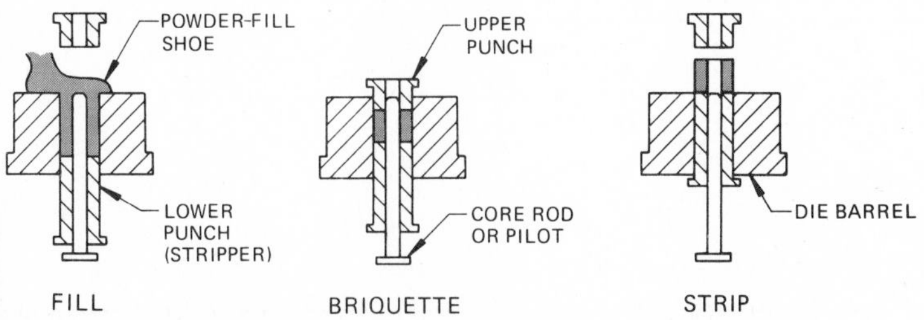
POWDER-FILL SHOE
UPPER PUNCH
LOWER PUNCH (STRIPPER)
CORE ROD OR PILOT
DIE BARREL
FILL
BRIQUETTE
STRIP
EJECT

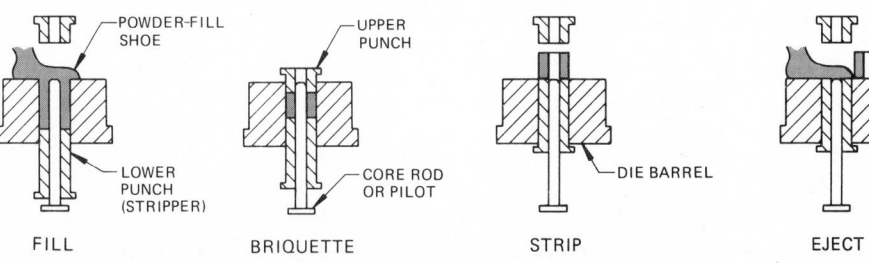

Fig. 10-2-15 Compacting sequence. (General Motors Corp.)

Corners. A fillet radius must be provided under the flange on a flanged part. It allows uniform powder flow in the die and produces a high-strength part. See Fig. 10-2-22.

Flanges. A 1.5 mm minimum flange overhang is desired to provide longer tool life. See Fig. 10-2-23.

Blind Holes. If a flange is opposite the blind end of the hole, the part must be modified to allow powder to fill in the die. See Fig. 10-2-24.

Changes in Cross Section. Large changes in cross section should be avoided because they cause density variation. Warping and cracking are likely to occur during sintering. See Fig. 10-2-25.

Chamfers. Care in the design of chamfers minimizes sharp edges on tools and improves tool life. Where a chamfer is required, a flat land should be provided to strengthen tools. Chamfer angles of less than 45° are attainable but should be avoided. See Fig. 10-2-26.

Holes. A variety of odd-shaped holes can be produced economically by the powder metallurgy process. The use of round holes, instead of odd-shaped holes, will simplify tooling, strengthen the part, and reduce costs. Depending on the material used, hole diameters should not be less than 20 to 25 percent of their length with a 2 mm diameter being considered the practical minimum. See Fig. 10-2-27.

Multilevel Parts. Design parts with a minimum of changes in section thickness. Where too many levels exist, density varies considerably and quality control becomes a problem. Design for pressing as many levels as practical, enabling uniform density to be maintained, and then machine the remaining levels. Lower strength may result where all steps are molded. See Fig. 10-2-28.

Axial Variations. Slots having a depth greater than one-fourth the axial length of the part require multiple punch action and result in higher production costs. See Fig. 10-2-29.

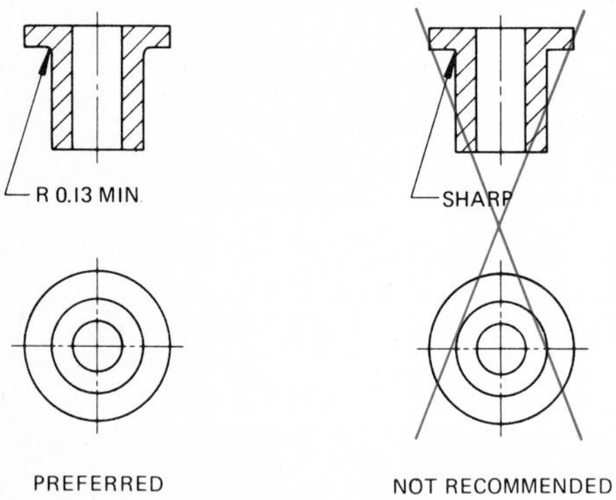
R 0.13 MIN
SHARP
PREFERRED
NOT RECOMMENDED

Fig. 10-2-22 Corners. (General Motors Corp.)

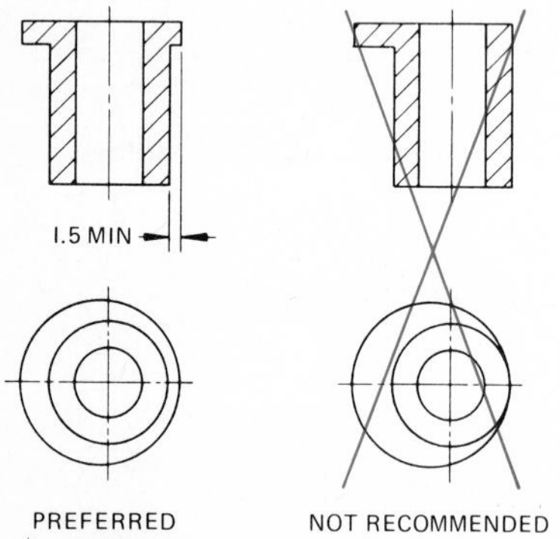
1.5 MIN
PREFERRED
NOT RECOMMENDED

Fig. 10-2-23 Flanges. (General Motors Corp.)

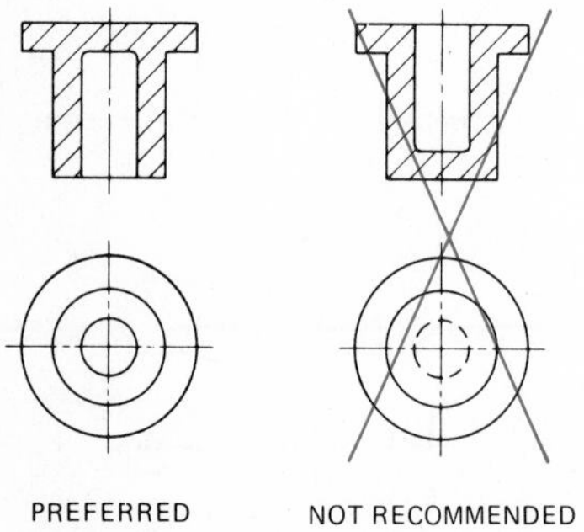
PREFERRED
NOT RECOMMENDED

Fig. 10-2-24 Blind holes. (General Motors Corp.)

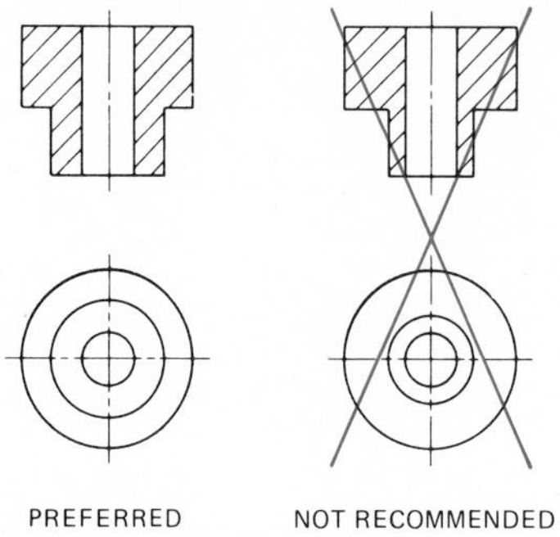
PREFERRED
NOT RECOMMENDED

Fig. 10-2-25 Change in cross section. (General Motors Corp.)

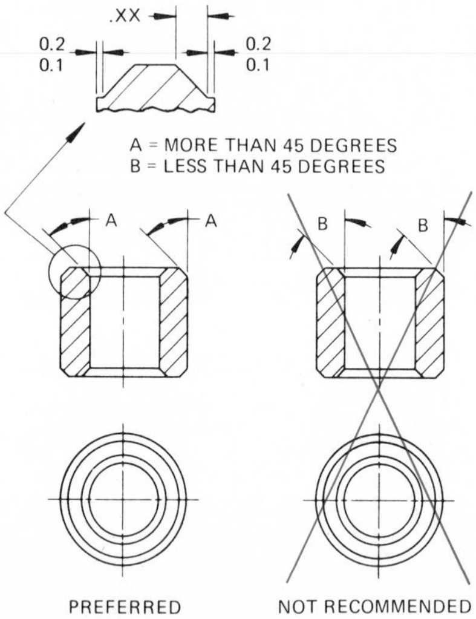
.XX
0.2
0.1
0.2
0.1
A = MORE THAN 45 DEGREES
B = LESS THAN 45 DEGREES
A
A
B
B
PREFERRED
NOT RECOMMENDED

Fig. 10-2-26 Chamfers with flat hand. (General Motors Corp.)

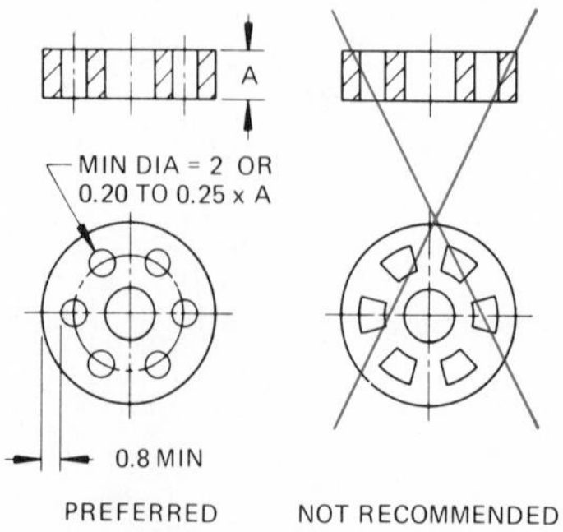
A
MIN DIA = 2 OR
0.20 TO 0.25 x A
0.8 MIN
PREFERRED
NOT RECOMMENDED

Fig. 10-2-27 Holes. (General Motors Corp.)

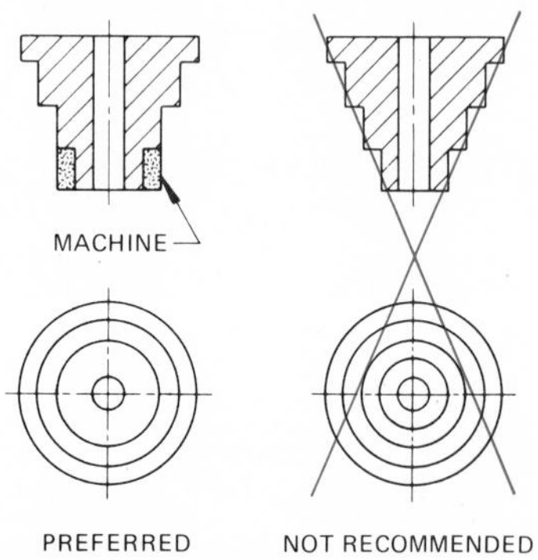
MACHINE
PREFERRED
NOT RECOMMENDED

Fig. 10-2-28 Multilevel parts. (General Motors Corp.)

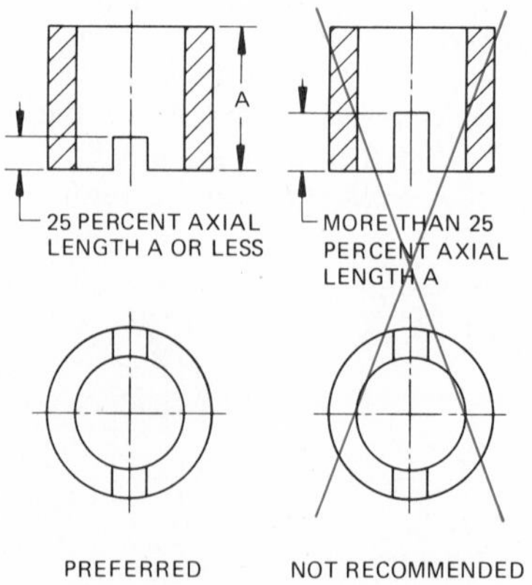
A
25 PERCENT AXIAL
LENGTH A OR LESS
MORE THAN 25
PERCENT AXIAL
LENGTH A
PREFERRED
NOT RECOMMENDED

Fig. 10-2-29 Axial variations. (General Motors Corp.)

REFERENCES AND SOURCE MATERIAL

1. Frank Burbank, "Forging," *Machine Design*, vol. 37, no. 21, 1965.
2. General Motors Corporation.
3. J. Havlis, "Cold Heading," *Machine Design*, vol. 37, no. 21, 1965.
4. General Motors Corporation.

Assignments

1. On an A3- or B-size sheet, prepare a forging drawing of the open-end wrench shown in Fig. 10-2-A or 10-2-B. Scale is 1:1

2. On an A3- or B-size sheet, prepare both the casting and the machining drawings for the connector shown in Fig. 10-2-C or 10-2-D. Draw a one-view full section, complete with the necessary dimensions for each drawing. Scale is 1:1

REVIEW FOR ASSIGNMENTS

Unit 6-1 Detail Drawings
Unit 7-1 Full Sections

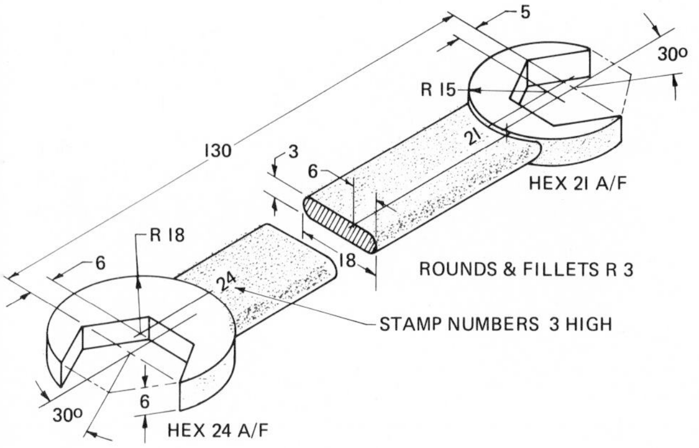
5
30°
R 15
130
3
6
21
HEX 21 A/F
R 18
6
18
24
ROUNDS & FILLETS R 3
STAMP NUMBERS 3 HIGH
30°
6
HEX 24 A/F

Fig. 10-2-A Open-end wrench.

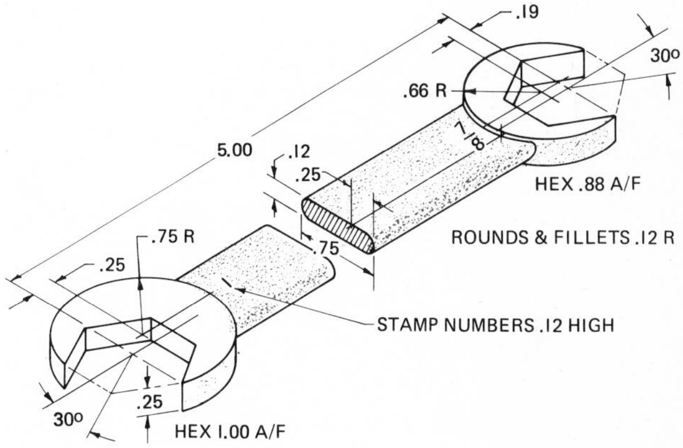
.19
30°
.66 R
5.00
.12
.25
7/8
HEX .88 A/F
.75 R
.25
.75
ROUNDS & FILLETS .12 R
STAMP NUMBERS .12 HIGH
30°
.25
HEX 1.00 A/F

Fig. 10-2-B Open-end wrench.

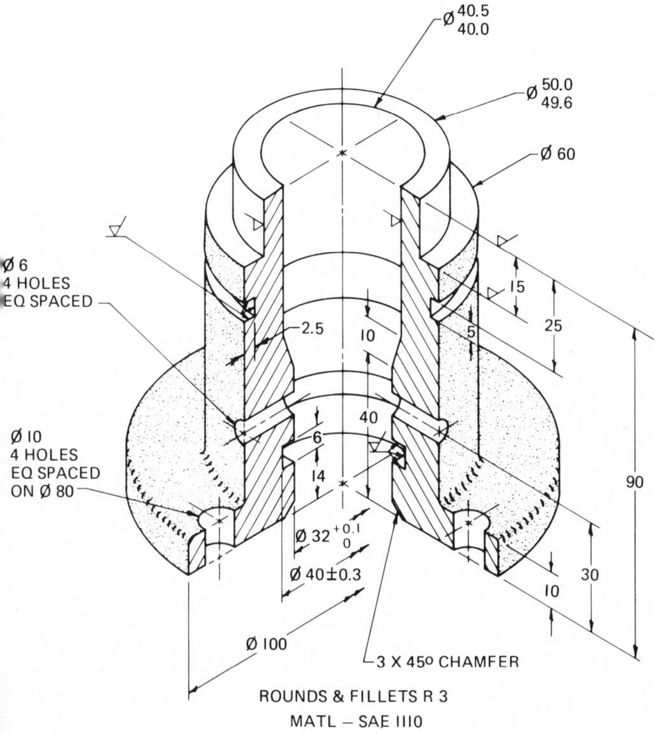
Ø 40.5 40.0
Ø 50.0 49.6
Ø 60
Ø 6
4 HOLES
EQ SPACED
15
25
2.5
10
5
Ø 10
4 HOLES
EQ SPACED
ON Ø 80
40
6
14
90
Ø 32 +0.1 0
Ø 40±0.3
30
10
Ø 100
3 X 45° CHAMFER
ROUNDS & FILLETS R 3
MATL – SAE 1110

Fig. 10-2-C Connector.

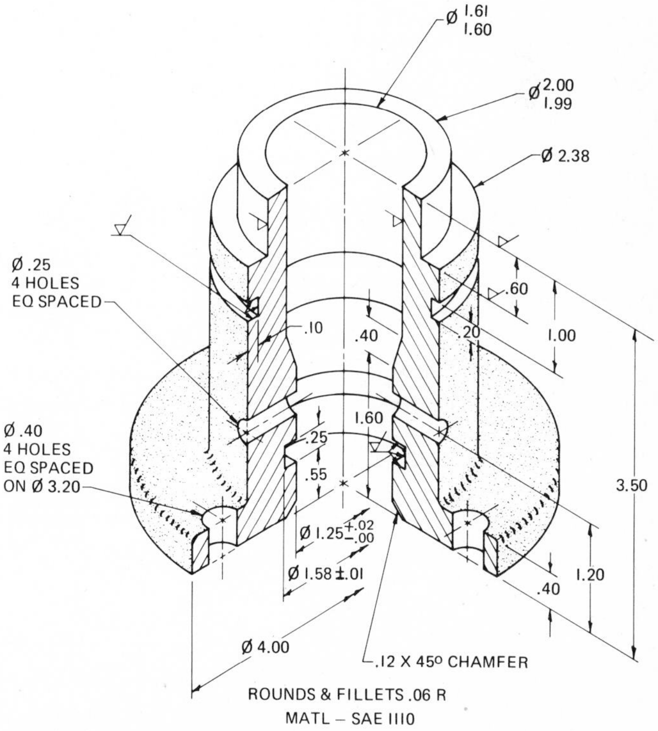
Ø 1.61 1.60
Ø 2.00 1.99
Ø 2.38
Ø .25
4 HOLES
EQ SPACED
.60
.10
.40
.20
1.00
Ø .40
4 HOLES
EQ SPACED
ON Ø 3.20
1.60
.25
.55
3.50
Ø 1.25 +.02 -.00
Ø 1.58 ±.01
.40
1.20
Ø 4.00
.12 X 45° CHAMFER
ROUNDS & FILLETS .06 R
MATL – SAE 1110

Fig. 10-2-D Connector.

UNIT 10-3

EXTRUDING, DEEP DRAWING, SPINNING, AND ROLL FORMING

Extruding[1]

In the *extrusion process*, hot metal is forced through a die to form a shape having a cross section essentially the same as the opening in the die. Extrusion presses are usually horizontal and hydraulically actuated. An extrusion press consists of a heated container to hold a billet, a ram that applies pressure to the billet, a means of supporting the die at the exit end, and a shear.

A typical tooling setup for extruding a solid shape and a setup for extruding tubing are shown in Fig. 10-3-1.

Extrusions can be hollow or solid. Contours can be irregular or nonsymmetrical. See Fig. 10-3-2. Shapes that cannot be rolled, or shapes for which the required quantities do not justify rolls, may be produced by extrusion.

COLD EXTRUDING[2]

Cold extruding is a process in which a metal slug is shaped in a die by a punch. Most parts formed by the process are characterized by fairly long tubelike or constant-section regions.

The process is also referred to by a number of other names, including *impact extruding* and *cold forging*. Parts formed in soft, non-heat-treatable aluminum or copper are often called *impacts*, while steels and higher-strength, nonferrous parts are referred to as *cold forgings*. See Fig. 10-3-3.

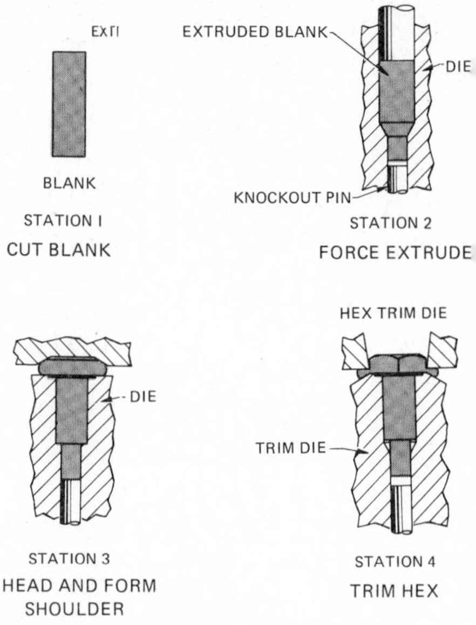
EXTRUDED BLANK
DIE
BLANK
KNOCKOUT PIN
STATION I
CUT BLANK
STATION 2
FORCE EXTRUDE
HEX TRIM DIE
DIE
TRIM DIE
STATION 3
HEAD AND FORM SHOULDER
STATION 4
TRIM HEX

Fig. 10-3-3 Progressive header operations to form a hex-head shoulder nut.

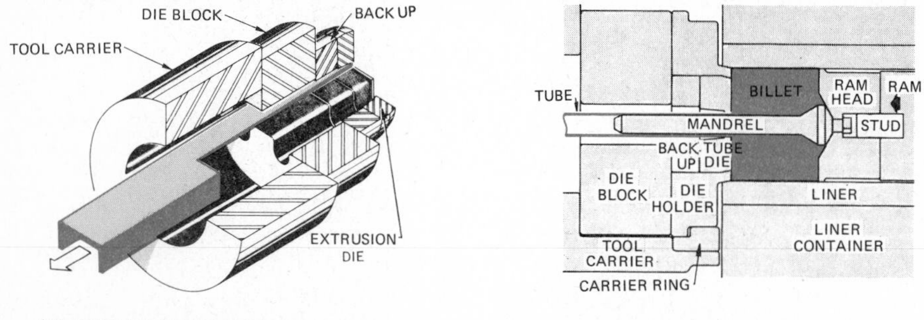
DIE BLOCK
BACK UP
TOOL CARRIER
EXTRUSION DIE
TUBE
BILLET
RAM HEAD
RAM
MANDREL
STUD
BACK-UP
TUBE DIE
DIE BLOCK
DIE HOLDER
LINER
LINER CONTAINER
TOOL CARRIER
CARRIER RING

(A) EXTRUSION OF A CHANNEL SHAPE

(B) EXTRUSION OF TUBING

Fig. 10-3-1 Extrusions.

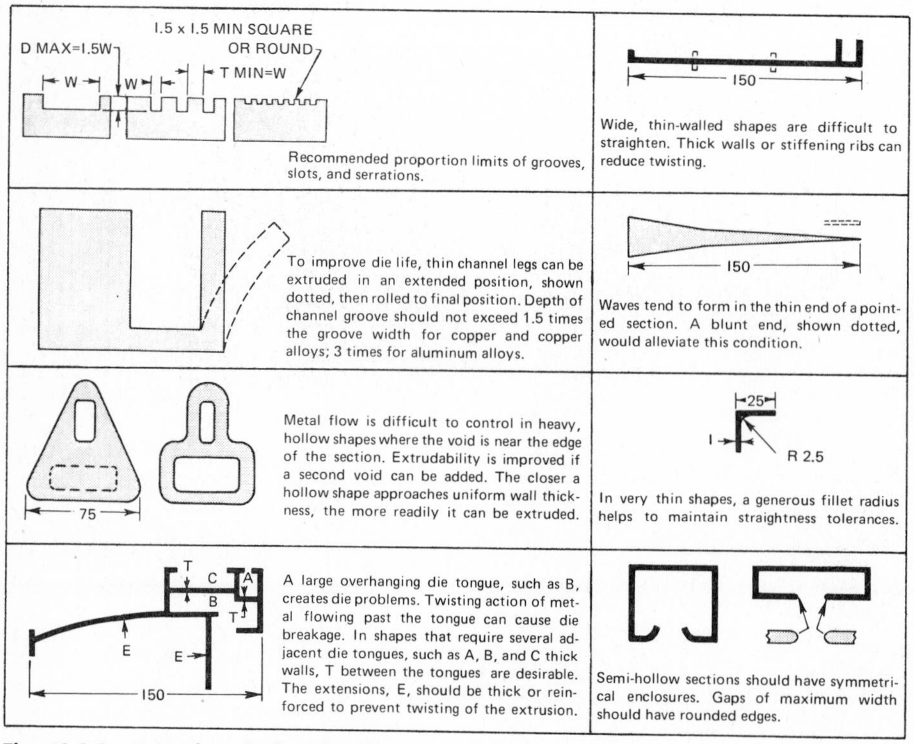
D MAX=1.5W
1.5 x 1.5 MIN SQUARE OR ROUND
W
W
T MIN=W
Recommended proportion limits of grooves, slots, and serrations.
150
Wide, thin-walled shapes are difficult to straighten. Thick walls or stiffening ribs can reduce twisting.
To improve die life, thin channel legs can be extruded in an extended position, shown dotted, then rolled to final position. Depth of channel groove should not exceed 1.5 times the groove width for copper and copper alloys; 3 times for aluminum alloys.
150
Waves tend to form in the thin end of a point-ed section. A blunt end, shown dotted, would alleviate this condition.
75
Metal flow is difficult to control in heavy, hollow shapes where the void is near the edge of the section. Extrudability is improved if a second void can be added. The closer a hollow shape approaches uniform wall thick-ness, the more readily it can be extruded.
25
1
R 2.5
In very thin shapes, a generous fillet radius helps to maintain straightness tolerances.
T
C
A
B
T
E
E
150
A large overhanging die tongue, such as B, creates die problems. Twisting action of met-al flowing past the tongue can cause die breakage. In shapes that require several ad-jacent die tongues, such as A, B, and C thick walls, T between the tongues are desirable. The extensions, E, should be thick or rein-forced to prevent twisting of the extrusion.
Semi-hollow sections should have symmetri-cal enclosures. Gaps of maximum width should have rounded edges.

Fig. 10-3-2 Extrusion design details.

Deep Drawing[3]

Deep drawing and *stamping* are terms often used interchangeably, yet the processes are basically different. In *deep drawing*, metal is held between a pair of draw rings and then made to flow over a punch of the desired contour (Fig. 10-3-4).

Deep drawing is usually specified when parts must have no mechanical joints or seams; where dimensional stability is required in long runs; where uniformity and closely controlled size are required; where freedom from defects is necessary; and where appearance is important.

Spinning[4]

Spinning is one of the simpler methods of metal forming for sheet and plate. Essentially, it involves shaping a circular disk of material over a mandrel. The mandrel is mounted on the headstock of a heavy-duty lathe, the disk of material is clamped to the mandrel by tailstock pressure, and forming is done by manipulating an antifriction-type roller to shape the metal tightly against the mandrel. The finished piece is necessarily concentric about an axis. See Fig. 10-2-5.

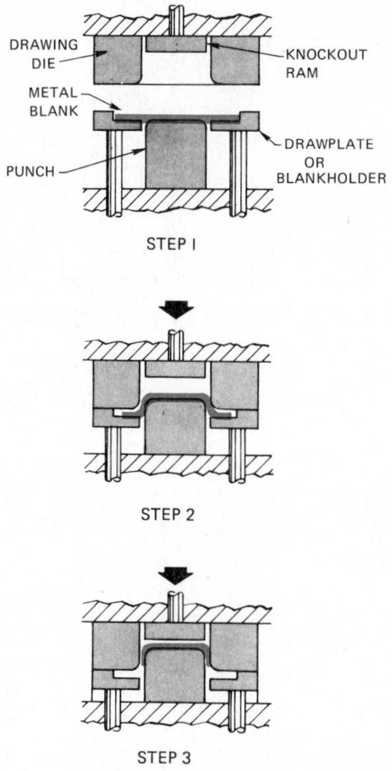
DRAWING DIE
KNOCKOUT RAM
METAL BLANK
DRAWPLATE OR BLANKHOLDER
PUNCH
STEP I
STEP 2
STEP 3

PUNCH IS SHAPED TO PRODUCE DESIRED CONFIGURATION ON INTERIOR OF CUP. DRAWPLATE, OR BLANKHOLDER, AND DRAWING DIE HOLD THE METAL BLANK. DRAWING BEGINS IN STEP 2 AS BLANKHOLDER AND DRAWING DIE MOVE DOWNWARD AND STRETCH THE BLANK OVER THE PUNCH. COMPLETED PART IN STEP 3 IS EJECTED FROM THE DRAWING DIE BY THE KNOCKOUT RAM AFTER THE TOOLING HAS RETURNED TO THE STARTING POSITION.

Fig. 10-3-4 Drawing sequences for cylindrical cup.

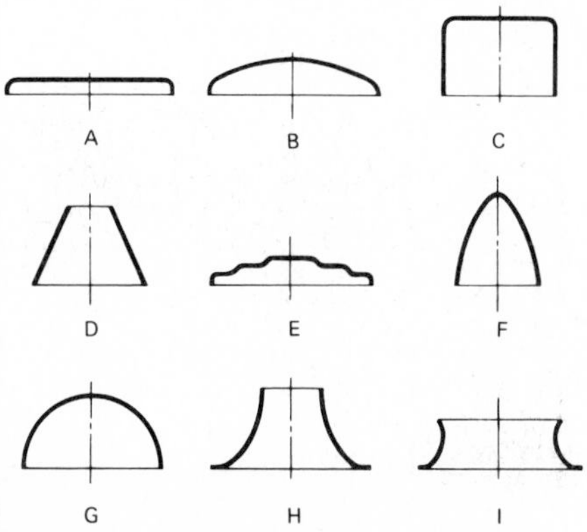
A
B
C
D
E
F
G
H
I

Fig. 10-3-5 Representative shapes suitable for spinning. (General Motors Corp.)

Roll Forming[5]

Cold roll forming is a continuous, high-production process that shapes a metal sheet or strip by means of progressive forming rolls. The process produces shapes of uniform cross section with close dimensional tolerances. Production is rapid, and secondary operations such as notching, slotting, punching, embossing, or straightening can be incorporated into the line with little or no time loss.

Sheet or strip stock is fed into a roll-forming machine, where it is transformed gradually as it passes through contoured rolls. Only bending takes place; the metal gage remains essentially constant.

The number of roll stands needed to form a given part depends on configuration of the final shape and material used. Simple shapes may require only three or four roll stands; as many as 30 stands may be needed for complex sections. The spring clip shown in Fig. 10-3-6 is produced in a 14-roll-stand machine.

Piercing, slotting, trimming, and embossing are performed on equipment located ahead of the first forming pass or between roll passes. Printing or other identification can be done at either end.

Other operations include spot welding, seam welding, straightening, curving, and coiling. Parts such as automotive headlamp rims and other glass-retaining rings are produced from roll-formed coils.

REFERENCES AND SOURCE MATERIAL

1. D. Cullen and J. J. Barrett, "Extruding," *Machine Design*, vol. 37, no. 21, 1965.

2. L. Schiller and H. Isbit, "Cold Extruding," *Machine Design*, vol. 37, no. 21, 1965.

3. H. Bartle, "Deep Drawing," *Machine Design*, vol. 37, no. 21, 1965.

4. W. A. Wenman, "Spinning," *Machine Design*, vol. 37, no. 21, 1965.

5. E. J. Vanderploeg, "Roll Forming," *Machine Design*, vol. 37, no. 21, 1965.

Assignments

1. On an A3- or B-size sheet, make proposed extrusion designs for the two moldings shown in Fig. 10-3-A or 10-3-B. Variations in sizes other than those shown are permitted, and rounds and fillets are to be added wherever desirable. List below the proposed extrusion design any operations which may be required on the extrusion in order to complete the part. Scale is 2:1.

REVIEW FOR ASSIGNMENTS

Unit 6-6 Detail Assembly Drawings
Unit 6-4 Bill of Materials
Unit 9-5 Rivets

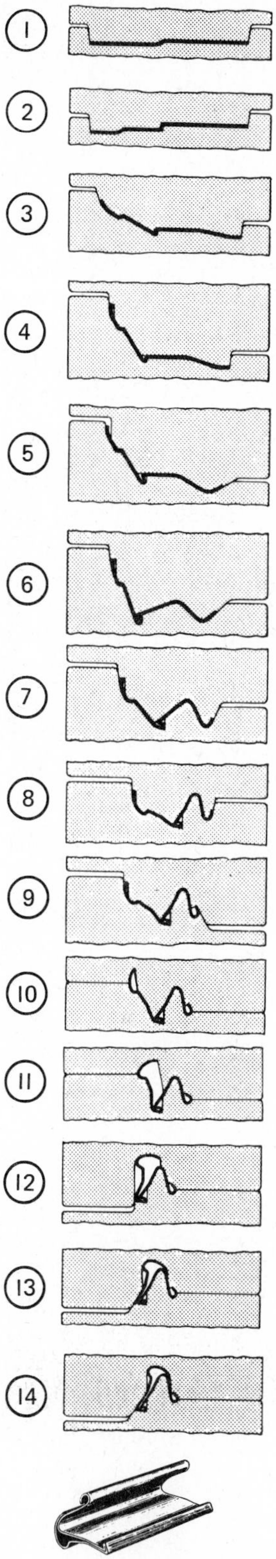
1
2
3
4
5
6
7
8
9
10
11
12
13
14

Fig. 10-3-6 Roll forming.

Fig. 10-3-A Extrusions.

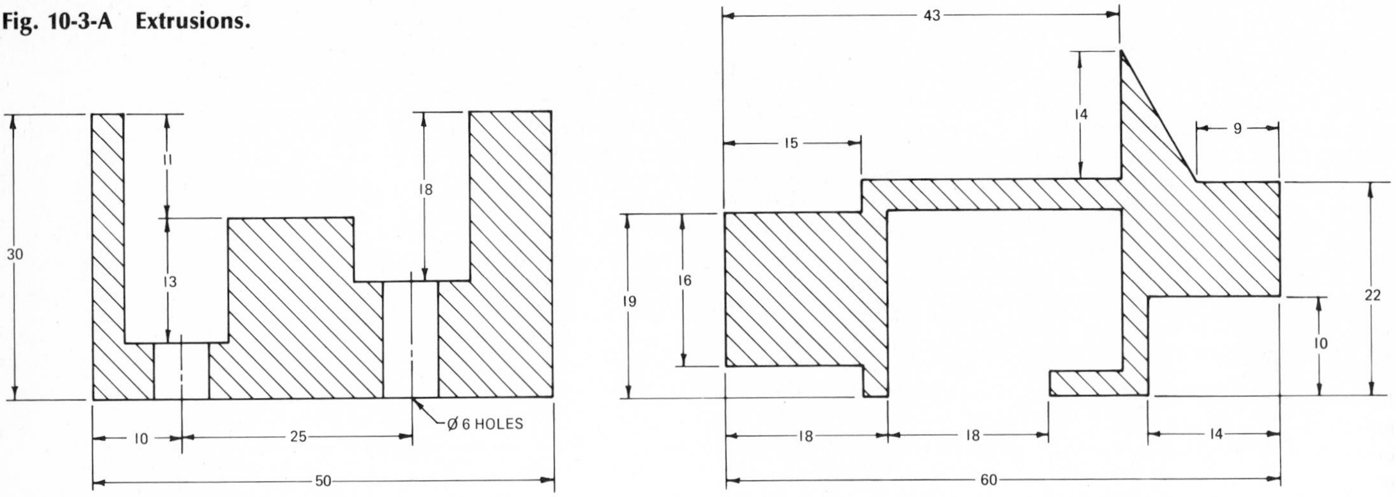
30
11
13
18
10
25
50
Ø 6 HOLES
43
15
14
9
19
16
22
10
18
18
14
60

INITIAL SKETCH OF PLASTIC MOLDED PART.

INITIAL SKETCH OF ALUMINUM MOLDED PART.

Fig. 10-3-B Extrusions.

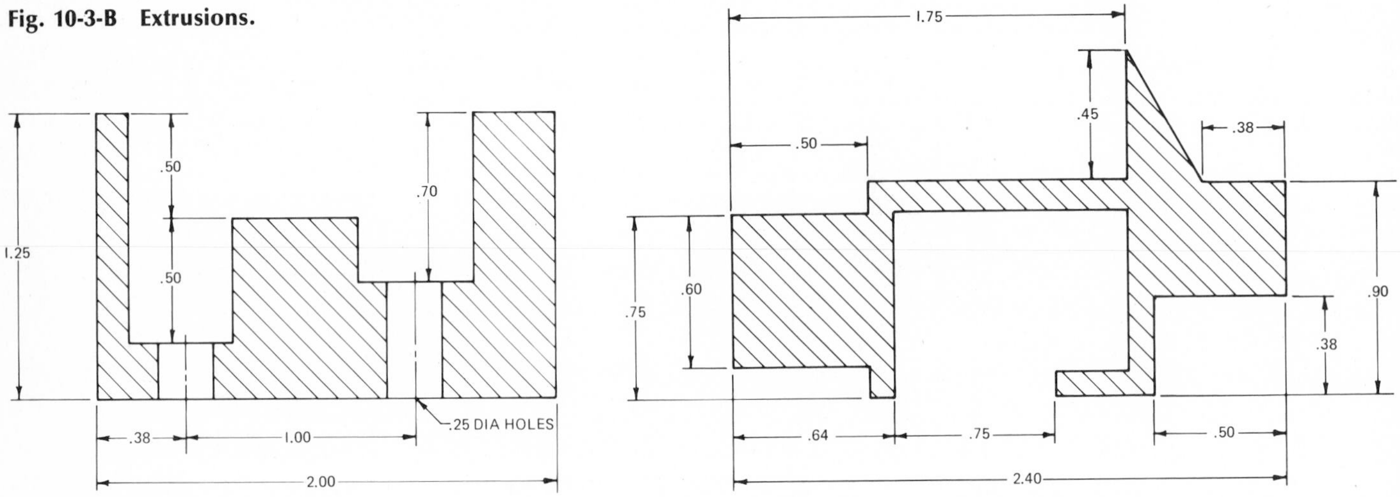
1.25
.50
.50
.70
.38
1.00
2.00
.25 DIA HOLES
1.75
.50
.45
.38
.75
.60
.90
.38
.64
.75
.50
2.40

INITIAL SKETCH OF PLASTIC MOLDED PART.

INITIAL SKETCH OF ALUMINUM MOLDED PART.

UNIT 10-4

ELECTRIC AND HIGH-ENERGY-RATE FORMING

Electroforming[1]

Electroforming is similar to electroplating except that the thickness of deposit is 10 to 200 times as great. It is used to manufacture a wide range of precision parts such as antennas, dies, venturi and wind tunnel nozzles, heating ducts, stereophonic records, musical instrument components, and liners.

The advantages of electroforming are as follows:

1. It can produce dense, high-purity copper deposits.
2. It needs no costly dies or presses, nor does it require heavy plates or castings.
3. It can exercise control over quality (hardness and metallurgy) throughout processing, and any layer found defective can be readily removed.
4. It needs little machining to reach final specifications.

At present, commercial electroforming is limited chiefly to the use of nickel, silver, copper, and their combinations.

Basically, electroforming consists of three steps:

1. A mandrel of proper shape and size is prepared; its surface is a mirror image of the surface desired on the finished part. Almost any material will serve for the mandrel provided it can be accurately shaped and readily separated from the deposit built on it by electroforming.
2. The mandrel is made cathode in an electrolytic bath, and metal is plated onto it to a specified deposit thickness.
3. After the plating step has been completed, the mandrel is separated from the deposited shape.

Nonconductive mandrels are also used where the mandrel will be part of the finished product. Plating over plastics is one example of the use of nonconductive mandrels. Complicated shapes can readily be electroformed.

DESIGN CONSIDERATIONS

Combined Assemblies. Joining parts by electroforming is often more practical and less costly than other techniques such as brazing or soldering.

Wall Variations. Normal application is to have the outside surface follow the same contour as the inside surface. Where desirable, any part of the external surface may be altered in thickness by masking off that surface, preventing any further buildup by electroforming.

Reproduction Accuracy. Reproduction within 0.001 mm can be obtained.

Corners and Recesses. Sharp corners and recesses should be avoided.

Tolerances and Surface Quality. Internal tolerances of ±0.01 on parts electroformed on permanent mandrels are readily obtainable. On external dimensions, tolerances such as ±0.02 are easily held in production.

Electric-Discharge Machining[2]

Electric-discharge machining is the removal of metal from a surface by spark erosion. It permits more economical machining of hardened tool steels and carbide, and makes it possible to machine certain configurations which cannot be done in any other way.

In principle, electric-discharge machining is removal of metal by an electric discharge or spark in the presence of a dielectric, the sparking action being performed with a predetermined frequency over a gap maintained between the electrode and the workpiece. This gap is the *overcut* of resulting clearance and can be controlled. See Fig. 10-4-1.

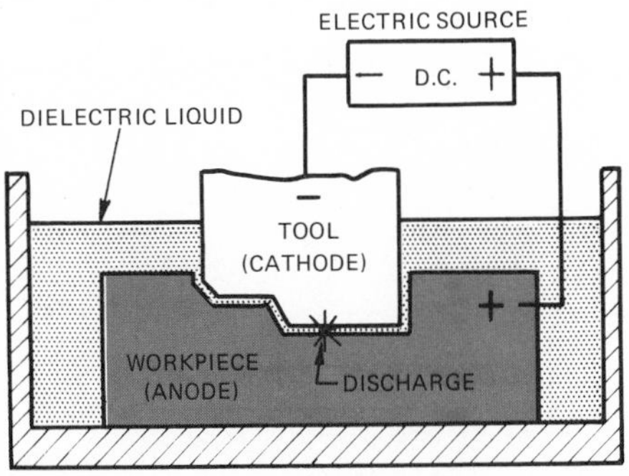
ELECTRIC SOURCE
D.C.
DIELECTRIC LIQUID
TOOL
(CATHODE)
WORKPIECE
(ANODE)
DISCHARGE

Fig. 10-4-1 Electric discharging process.

High-Energy-Rate Forming[3]

High-energy-rate forming is the plastic shaping of materials through the sudden exertion of forces.

Four major processes use the HVF principle:

1. Explosive forming
2. Electrohydraulic forming
3. Magnetic forming
4. Pneumatic-mechanical forming

EXPLOSIVE FORMING[4]

This metalworking process makes use of the enormous energies instantaneously developed when explosives are detonated. For metal forming, the energy of detonation of explosives is usually harnessed by one of three possible methods: the totally closed die, the open die, or the free die.

The open die is used with high explosives and a pressure-transmitting medium. See Fig. 10-4-2*a*. The male die is replaced by the pressure-transmitting medium, usually water, which applies a uniform pressure over the whole area of the female die. The blank is held to the die by a retaining ring, or merely by adhesive tape. The die is evacuated and the whole assembly, together with a charge located at a suitable distance from the blank, is lowered under a metre head of water. The water not only provides a pressure-transmitting medium, but also confines the charge.

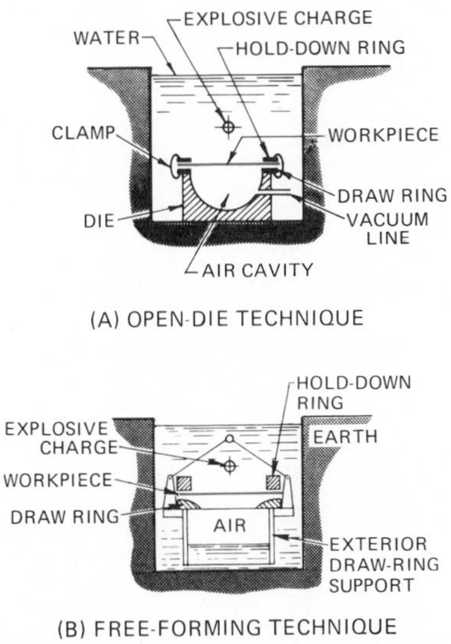
EXPLOSIVE CHARGE
WATER
HOLD-DOWN RING
CLAMP
WORKPIECE
DRAW RING
DIE
VACUUM LINE
AIR CAVITY
(A) OPEN-DIE TECHNIQUE
HOLD-DOWN RING
EXPLOSIVE CHARGE
EARTH
WORKPIECE
DRAW RING
AIR
EXTERIOR DRAW-RING SUPPORT
(B) FREE-FORMING TECHNIQUE

Fig. 10-4-2 Deformation velocities for high-velocity-forming techniques.

ELECTROHYDRAULIC FORMING[5]

With *electrohydraulic forming*, a shock wave and flow field form a workpiece in a manner similar to explosive forming, except that the forming forces are caused by the discharge of electrical energy stored in a capacitor. The energy is discharged either through a spark between electrodes or through a fine metal wire joining the electrodes is used as a shock medium.

MAGNETIC FORMING

Magnetic forming shapes workpieces by utilizing forces of repulsion caused by eddy currents induced in the workpieces.

PNEUMATIC-MECHANICAL FORGING

Except for forming speed, the *pneumatic-mechanical method* is similar to forging. The process uses a ram that is driven by a compressed gas to velocities of about 20 to 35 metres per second (m/s). The ram induces plastic flow in a metal workpiece so that it conforms to a die. The gas is compressed by hydraulic power or by a high-pressure gas compressor. Usually, the ram strikes a stationary workpiece, but a variation of the process moves the ram and workpiece toward each other.

REFERENCES AND SOURCE MATERIAL

1. P. Sherwood, "Electroformed Precision Parts Demand Right Techniques," *Canadian Machinery and Metalworking*, vol. 75, no. 12, 1964.
2. "E. D. M. Proves Boon to Montreal Firm," *Canadian Machinery and Metalworking*, vol. 75, no. 12, 1964.
3. L. Zernow, "High-Energy-Rate Forming," *Machine Design*, vol. 37, no. 21, 1965.
4. G. G. Goyer, "Explosive Forming Studies in Quebec Laboratory," *Canadian Machinery and Manufacturing News*, vol. 71, March 1960.
5. L. Zernow, "High-Energy-Rate Forming," *Machine Design*, vol. 37, no. 21, 1965.

Assignments

1. On an A3- or B-size sheet, design a storage tank from the information shown in Fig. 10-4-A or 10-4-B and the following data:

(A) The two ends are semispherical and are formed by explosive forming. The tank is made in four sections and welded.

(B) The tank is supported at two positions, and the supports are bolted to a concrete supporting pad.

(C) A fill and drain system is provided as shown. Flanged connections for 4-in. nominal pipe is welded to the tank to accommodate a drain valve at the bottom connection and a pipe nipple and cap at the top (fill) position.

(D) A manhole must be provided at the top of the tank for maintenance purposes.

REVIEW FOR ASSIGNMENTS

Unit 11-1	Design for Welding
Unit 11-2	Welding Symbols
Unit 11-3	Fillet Welds
Unit 11-4	Groove Welds
Unit 29-1	Structural Steel
Unit 8-3	Bolts
Appendix	Pipe Sizes

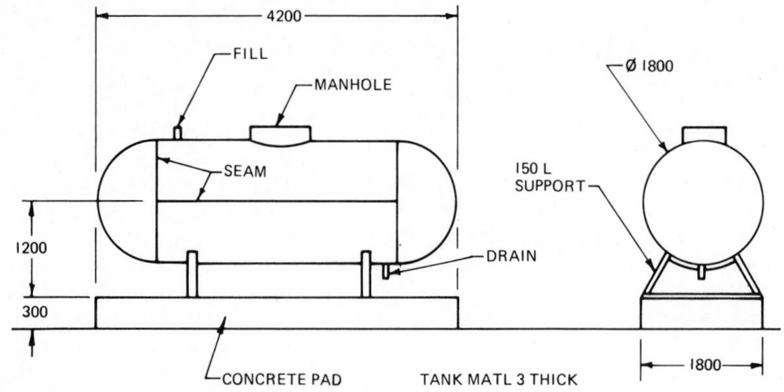
4200
FILL
MANHOLE
Ø 1800
SEAM
150 L
SUPPORT
1200
DRAIN
300
CONCRETE PAD
TANK MATL 3 THICK
1800

Fig. 10-4-A Storage tank base.

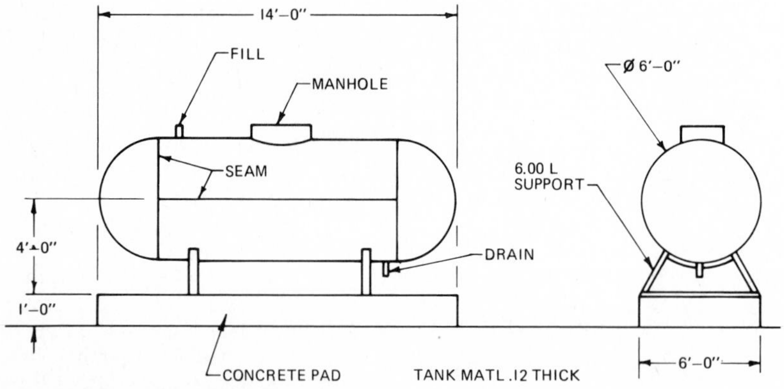
14'–0"
FILL
MANHOLE
Ø 6'–0"
SEAM
6.00 L
SUPPORT
4'–0"
DRAIN
1'–0"
CONCRETE PAD
TANK MATL .12 THICK
6'–0"

Fig. 10-4-B Storage tank base.

Chapter 11
Welding Drawings

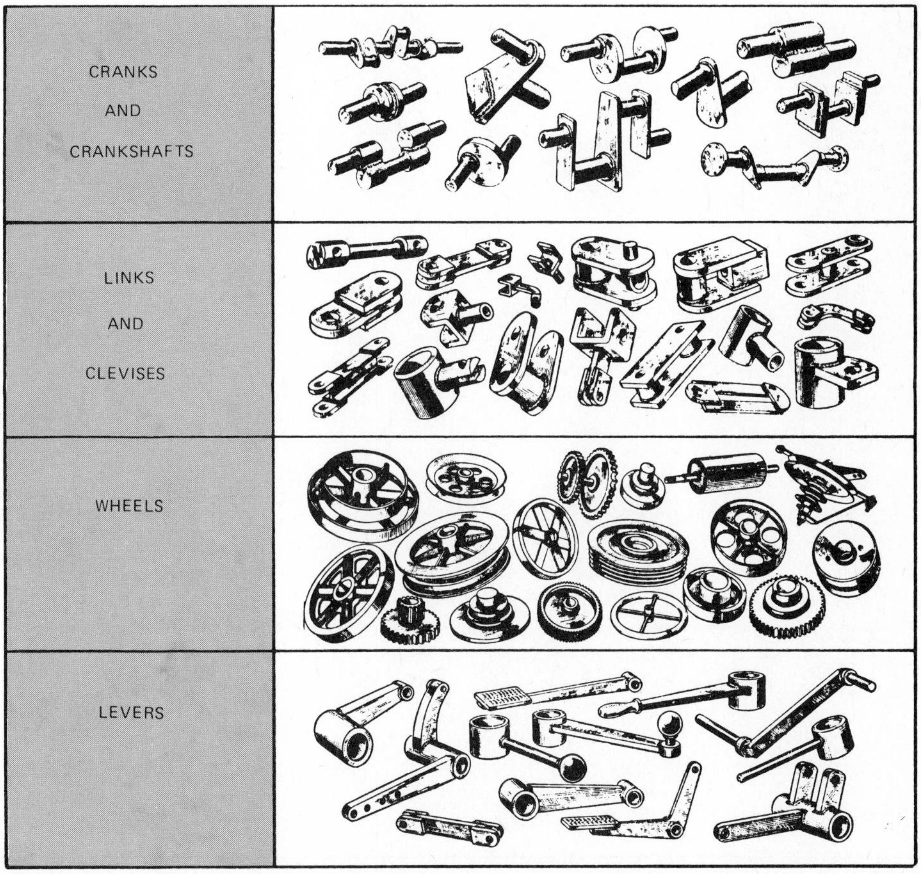
CRANKS
AND
CRANKSHAFTS
LINKS
AND
CLEVISES
WHEELS
LEVERS

Fig. 11-1-1 A variety of weldments. (James F. Lincoln Arc Welding Foundation.)

UNIT 11-1

DESIGNING FOR WELDING[1,2]

The primary importance of welding is to unite various pieces of metal so that they will operate as a unit structure to support the loads to be carried. In order to design such a structure, which will be both economical and efficient, the drafter must have a knowledge of the basic principles of welding practice and an understanding of the advantages and limitations of the process.

In order to produce an economical and pleasing design, the designer should endeavor to utilize the method of construction which is clearly the most advantageous for the application under consideration. This may mean a combination of welding and bolting, or even the incorporation of pressings, forgings, or even castings where they may be advantageous. The possibility of using structural steel shapes and tubes should also be kept in mind. See Figs. 11-1-1 and 11-1-2.

Of more than 40 welding processes used in industry today, only a handful are industrially important. Arc welding, gas welding, and resistance welding are the three most important types of welding.

The workpieces are melted along a common edge or surface so that their molten metal—and usually a filler metal also—is allowed to form a common pool or puddle. The pieces are fused when the puddle solidifies. See Figs. 11-1-3 and 11-1-4.

Gas welding, the most common form of which is oxyacetylene welding, gets its heat from the burning of flammable gases. Welding skills for this process are slow compared to other modern welding methods, so gas welding is normally confined to repair and maintenance work rather than being a major mass-production technique. See Fig. 11-1-5.

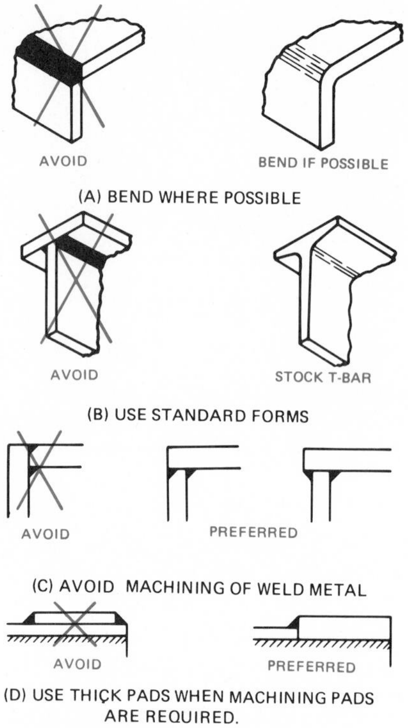
AVOID
BEND IF POSSIBLE
(A) BEND WHERE POSSIBLE
AVOID
STOCK T-BAR
(B) USE STANDARD FORMS
AVOID
PREFERRED
(C) AVOID MACHINING OF WELD METAL
AVOID
PREFERRED
(D) USE THICK PADS WHEN MACHINING PADS ARE REQUIRED.

Fig. 11-1-3 Preferred welding design. (Canadian Welding Bureau.)

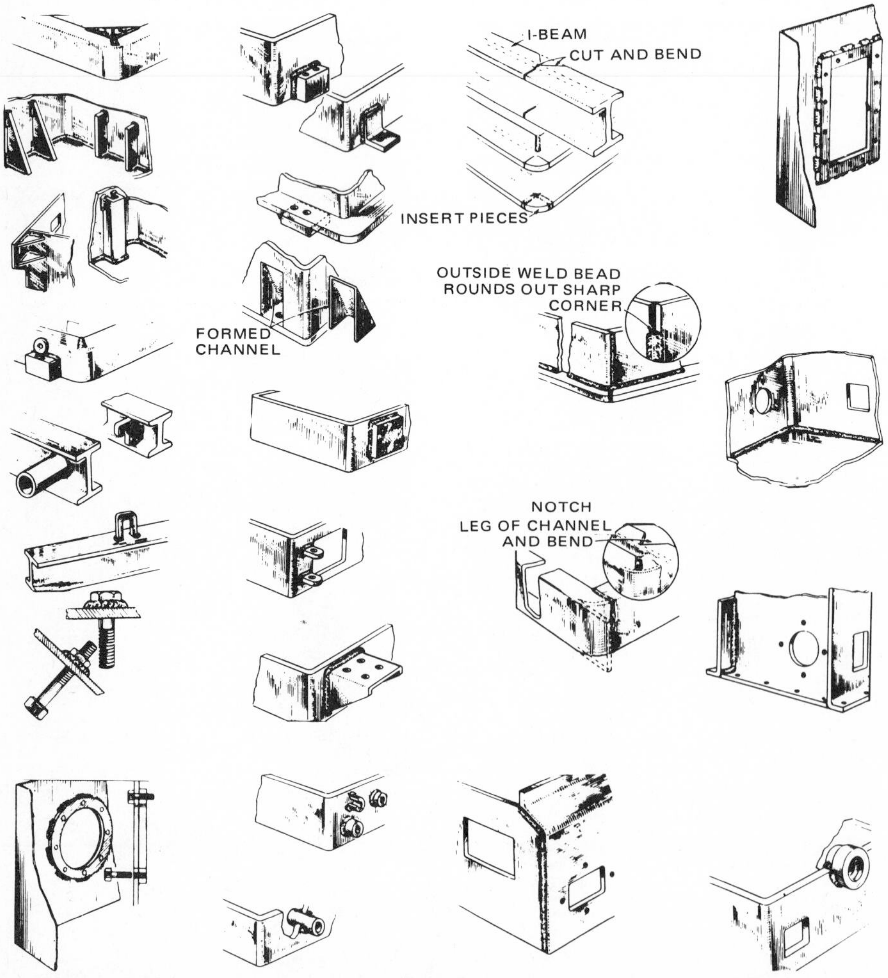
I-BEAM
CUT AND BEND
INSERT PIECES
OUTSIDE WELD BEAD ROUNDS OUT SHARP CORNER
FORMED CHANNEL
NOTCH LEG OF CHANNEL AND BEND

Fig. 11-1-2 Design ideas for fabricated parts. (James F. Lincoln Arc Welding Foundation.)

Design of Welded Structures

In designing machine frames and similar structures for fabrication by welding, the main considerations, apart from ensuring that the part will fulfill its intended function, are usually confined to the necessity for evolving an article which will be pleasing in appearance and which can be economically produced. In this respect the drafter has more scope for the application of his or her inventiveness than in other branches of welding design. The drafter should try to avoid being unduly influenced by the design principles which have been developed for other methods of construction. For example, in designing machinery parts for fabrication, especially when they are intended to replace or supersede castings and forgings, it is generally essential for the drafter to avoid any tendency to design on the basis of making the weldment look like a casting or forging. Otherwise, an uneconomical and excessively heavy product is sure to result. Instead, full advantage should be taken of the manipulative possibilities and physical characteristics of mild steel, the efficiency and neatness of the welded joint, the possibilities of mass saving, reducing machining and material costs, and the greater flexibility of design.

TYPE OF JOINT	APPLICABLE WELDS
T-JOINT 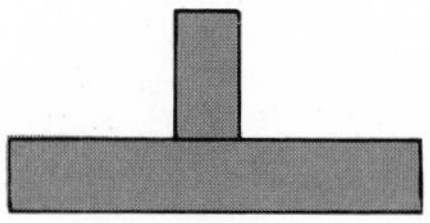	FILLET PLUG SLOT SQUARE-GROOVE BEVEL-GROOVE J-GROOVE FLARE-BEVEL-GROOVE SPOT PROJECTION SEAM
BUTT JOINT 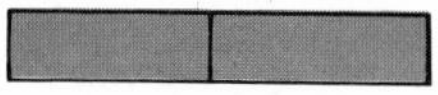	SQUARE-GROOVE V-GROOVE BEVEL-GROOVE U-GROOVE J-GROOVE FLARE-V-GROOVE FLARE-BEVEL-GROOVE EDGE-FLANGE
CORNER JOINT 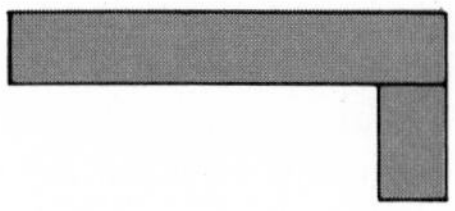	FILLET SQUARE-GROOVE V-GROOVE BEVEL-GROOVE U-GROOVE J-GROOVE FLARE-V-GROOVE FLARE-BEVEL-GROOVE EDGE-FLANGE CORNER-FLANGE SPOT PROJECTION SEAM
LAP JOINT 	FILLET PLUG SLOT BEVEL-GROOVE J-GROOVE FLARE-BEVEL-GROOVE SPOT PROJECTION SEAM
EDGE JOINT 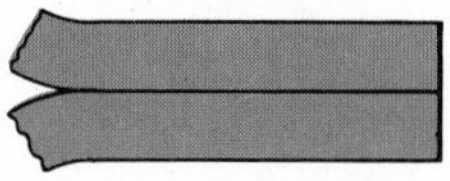	PLUG SLOT SQUARE-GROOVE BEVEL-GROOVE V-GROOVE U-GROOVE J-GROOVE EDGE-FLANGE CORNER-FLANGE SPOT PROJECTION SEAM EDGE

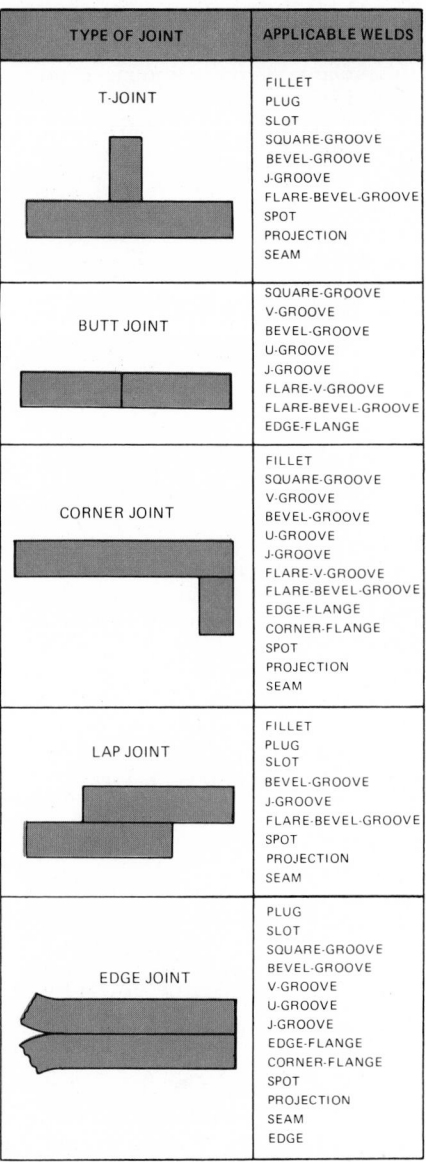

Fig. 11-1-4 Basic welding joints.

METAL OR ALLOY	GAS	ARC
ALUMINUM		
—COMMERCIALLY PURE	X	X
—Al-Mn ALLOY	X	X
BRASS, COMMERCIAL	X	
BRONZE, COMMERCIAL	X	
COPPER (DEOXIDIZED)	X	
IRON		
—GRAY AND ALLOY	X	
—MALLEABLE		
LEAD	X	
MAGNESIUM ALLOYS	X	
NICKEL AND NICKEL ALLOYS	X	X
STEELS, CARBON		
—LOW AND MEDIUM CARBON	X	X
—HIGH CARBON		X
—TOOL STEEL		
STEEL, CAST	X	X
STEELS, STAINLESS		
—CHROMIUM		X
CHROMIUM-NICKEL	X	X

Fig. 11-1-5 Weldability of various metals and alloys.

MASS SAVING

When castings are to be superseded by weldments, the higher labor costs of the weldment must be offset by simplifying the design and reducing the mass. With a casting some extra thickness usually has been provided to allow for defective metal and maybe for shifting cores. With steel there is practically no risk of defective material so that this surplus can be eliminated. Moreover, since cast iron has less than half the tensile strength of steel, the mass or size of a steel part can be reduced proportionately. For example, for the same overall dimensions, because of the higher stresses that can be allowed, a steel section need be no more than half the thickness of a cast-iron one. This is shown in the drawing of a pump base in Fig. 11-1-6.

Savings in mass are not only favorable in reducing the manufacturing cost of the weldment, but they also reduce transportation and handling costs.

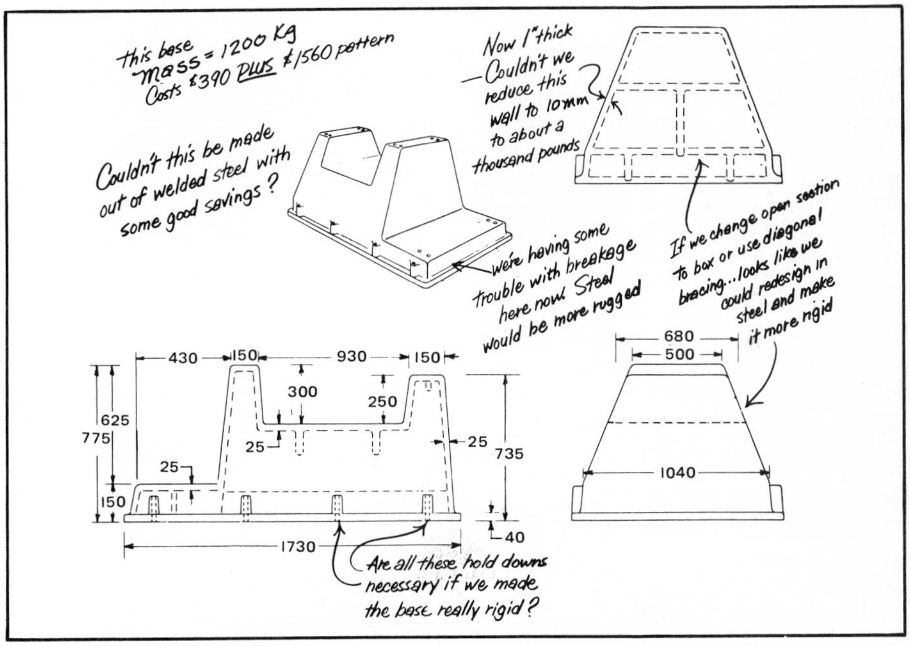
This base mass = 1200 Kg
Costs $390 PLUS $1560 pattern
Couldn't this be made out of welded steel with some good savings?
Now 1" thick — Couldn't we reduce this wall to 10 mm to about a thousand pounds
We're having some trouble with breakage here now. Steel would be more rugged
If we change open section to box or use diagonal bracing...looks like we could redesign in steel and make it more rigid
Are all these hold downs necessary if we made the base really rigid?
430
150
930
150
300
250
625
775
25
25
735
150
1730
40
680
500
1040

(A) ORIGINAL PUMP BASE SUBJECT OF COST STUDY

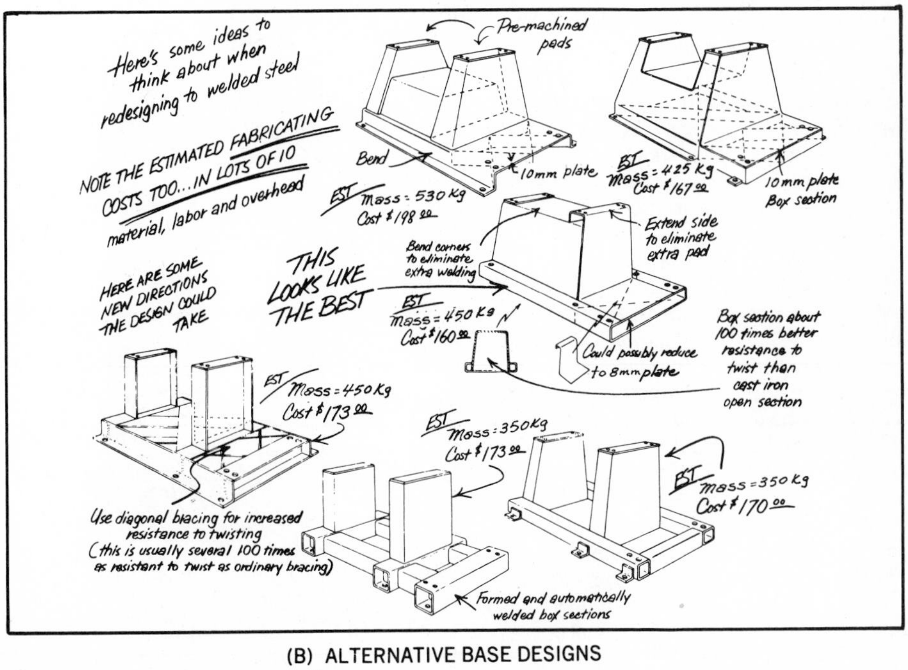
Here's some ideas to think about when redesigning to welded steel
NOTE THE ESTIMATED FABRICATING COSTS TOO...IN LOTS OF 10 material, labor and overhead
Pre-machined pads
Bend
EST mass = 530 Kg Cost $198.00
10mm plate
EST mass = 425 Kg Cost $167.00
10mm plate Box section
HERE ARE SOME NEW DIRECTIONS THE DESIGN COULD TAKE
THIS LOOKS LIKE THE BEST
Bend corners to eliminate extra welding
EST mass = 450 Kg Cost $160.00
Extend side to eliminate extra pad
Could possibly reduce to 8mm plate
Box section about 100 times better resistance to twist than cast iron open section
EST mass = 450 Kg Cost $173.00
EST mass = 350 Kg Cost $173.00
EST mass = 350 Kg Cost $170.00
Use diagonal bracing for increased resistance to twisting (this is usually several 100 times as resistant to twist as ordinary bracing)
Formed and automatically welded box sections

(B) ALTERNATIVE BASE DESIGNS

Fig. 11-1-6 Design of pump base.

QUANTITIES AND TIME FACTORS

The next important point concerns those factors influencing the cost of the weldment. When compared, for example, with castings, often fabrication by welding is particularly suited to the small-quantity type of job. For many cases this alone is an outstanding advantage over casting, since the pattern cost, which may be prohibitive for small quantities of castings, is entirely eliminated. Moreover, a weldment can often be produced in a fraction of the time required for the production of a casting.

CONCLUSION

To a large extent, the ultimate cost of the job is usually the yardstick by which the advantages of any type of construction are measured. The drafter should, therefore, review those factors which are contributory to the cost of a weldment. Although the cost of steel is low compared with that of cast iron or cast steel, and generally it is possible to use less metal in a weldment than in an equivalent casting, it is essential to remember that there are more operations involved in the production of a weldment than there are in the case of a casting. The plate or section must be prepared for welding, the various components must then be assembled and fitted; finally, there is the actual welding which may be followed by stress relieving.

CHOICE OF RAW MATERIALS

In this type of design, the drafter has a wide choice of raw materials, plates, structural shapes, forgings, tubes, castings, etc. Careful consideration of the function of the various components of the structure is desirable in order to enable the most suitable raw material to be selected to ensure efficiency, economy, and pleasing appearance. Steel plates will no doubt provide the basic element in the majority of cases, and by flamecutting there is no limit to the variety of shapes that can be produced.

Steel plate surfaces are usually flat and smooth enough to be used as seating or bolting surfaces without further machining. However, where bearings in a plate are required or should a machined surface be considered desirable for the seating of bolt heads, collars, washers, etc., it is often not essential to weld on bosses such as would ordinarily be employed on a casting. By making the plate a little thicker than normal, the machined areas can be spotfaced into the plate surface. The spotfacing costs about the same as a boss machining operation, but the work in preparation and welding on of bosses is eliminated. See Fig. 11-1-7 for types of boss attachments.

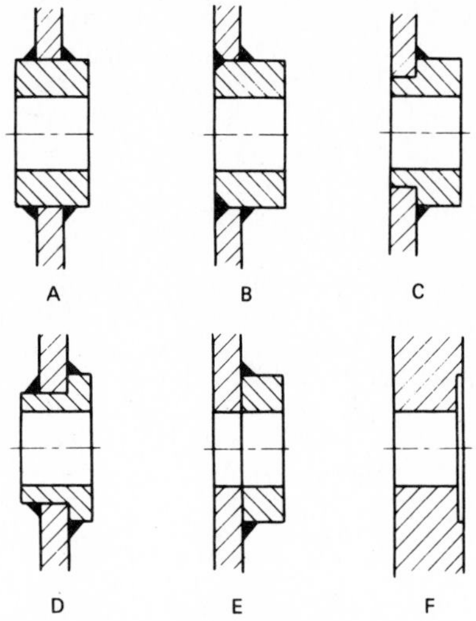
A
B
C
D
E
F

FIGURE "F" SUGGESTS HOW BOSSES MAY BE AVOIDED BY THICKENING THE PLATE AND SPOT FACING.

Fig. 11-1-7 Various types of boss attachments. (Canadian Welding Bureau.)

STRESS RELIEVING

Any weldment which is subjected to dynamic loading or which must maintain a high standard of accuracy in service (for example, machine tool beds, some types of engine frames, etc.) should be stress-relieved in order to remove any residual stresses. Where welded subassemblies are employed, it may be desirable to stress-relieve each subassembly before incorporation in the complete structure which will then be stress-relieved after the completion of the welding.

REFERENCES AND SOURCE MATERIAL

1. Canadian Welding Bureau.
2. *Machine Design*, fastening and joining reference issue, Nov. 22, 1973.

Assignment

On an A3- or B-size sheet, redesign the cast parts shown in Fig. 11-1-A or 11-1-B for fabrication by welding, using standard steel sizes and shapes. Make a detail assembly drawing. Welding symbols or sizes are not required. Include on the drawing a bill of material and identify each part on the assembly. Scale is 1:1.

REVIEW FOR ASSIGNMENT

Unit 6-4 Bill of Material
Unit 6-6 Detailed Assembly Drawing

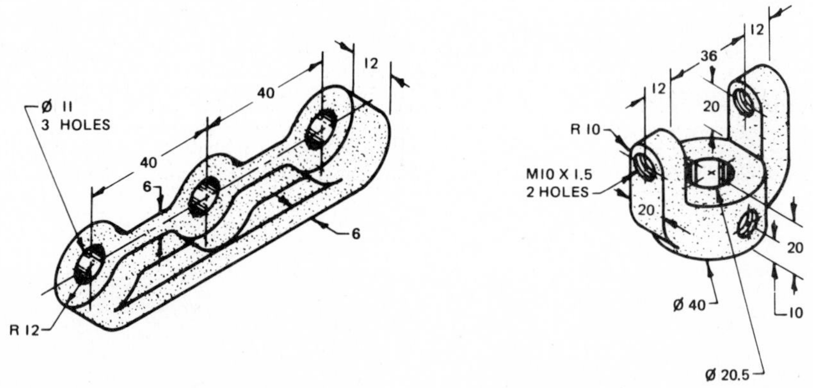
12
40
Ø 11
3 HOLES
40
6
6
R 12
12
36
12
20
R 10
M10 X 1.5
2 HOLES
20
20
Ø 40
10
Ø 20.5

Fig. 11-1-A Cast parts.

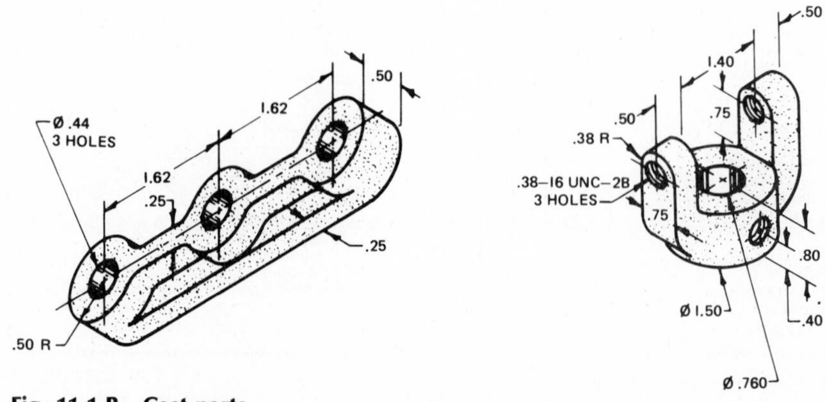
.50
1.62
Ø .44
3 HOLES
1.62
.25
.25
.50 R
.50
1.40
.50
.75
.38 R
.38–16 UNC–2B
3 HOLES
.75
.80
Ø 1.50
.40
Ø .760

Fig. 11-1-B Cast parts.

UNIT 11-2

WELDING SYMBOLS[1]

The introduction of welding symbols enables the designer to indicate clearly the type and size of weld required to meet design requirements, and it is becoming increasingly important for the designer to indicate the required type of weld correctly. Points which must be made clear are the type of weld, the joint preparation, the weld size, the root gap (if any), and the degree of penetration required. These points can be clearly indicated on the drawing by the welding symbol. See Fig. 11-2-1.

Welding symbols are a shorthand language. They save time and money and ensure understanding and accuracy. It is desirable that they should be a universal language; and for this reason the symbols of the American Welding Society and the Canadian Welding Bureau, already well established, have been adopted (Figs. 11-2-2 and 11-2-3).

The use of the words *far side* and *near side* in the past has led to confusion because when joints are shown in section, all welds are equally distant from the reader, and the words *near* and *far* are meaningless. In the present system the joint is the basis of reference. Any joint, the welding of which is indicated by a

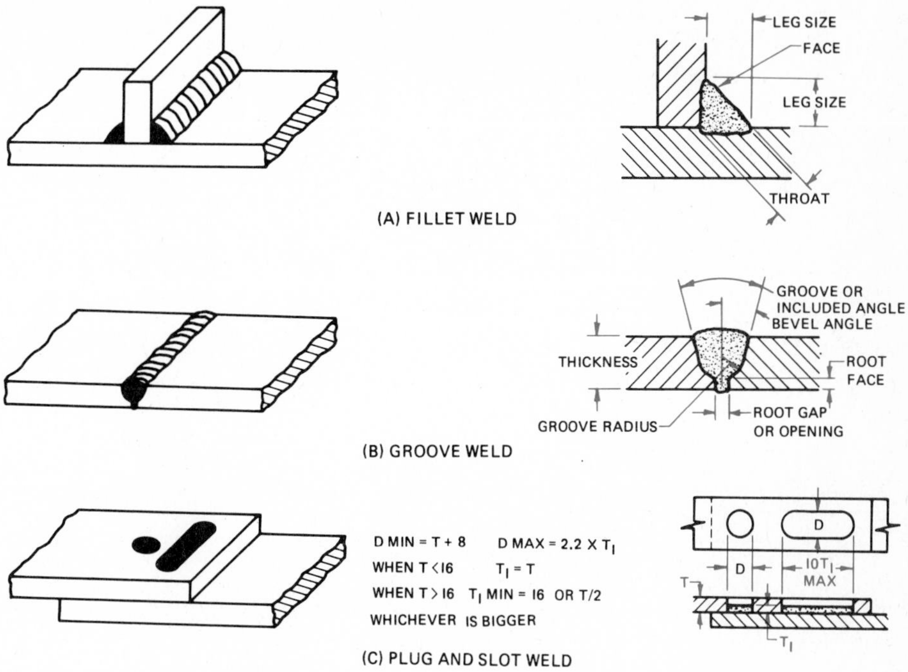

LEG SIZE
FACE
LEG SIZE
THROAT
(A) FILLET WELD
GROOVE OR
INCLUDED ANGLE
BEVEL ANGLE
THICKNESS
ROOT
FACE
ROOT GAP
OR OPENING
GROOVE RADIUS
(B) GROOVE WELD
D MIN = T + 8 D MAX = 2.2 X T_I
WHEN T < 16 T_I = T
WHEN T > 16 T_I MIN = 16 OR T/2
WHICHEVER IS BIGGER
D
D
10T_I MAX
T
T_I
(C) PLUG AND SLOT WELD

Fig. 11-2-2 Basic weld terminology.

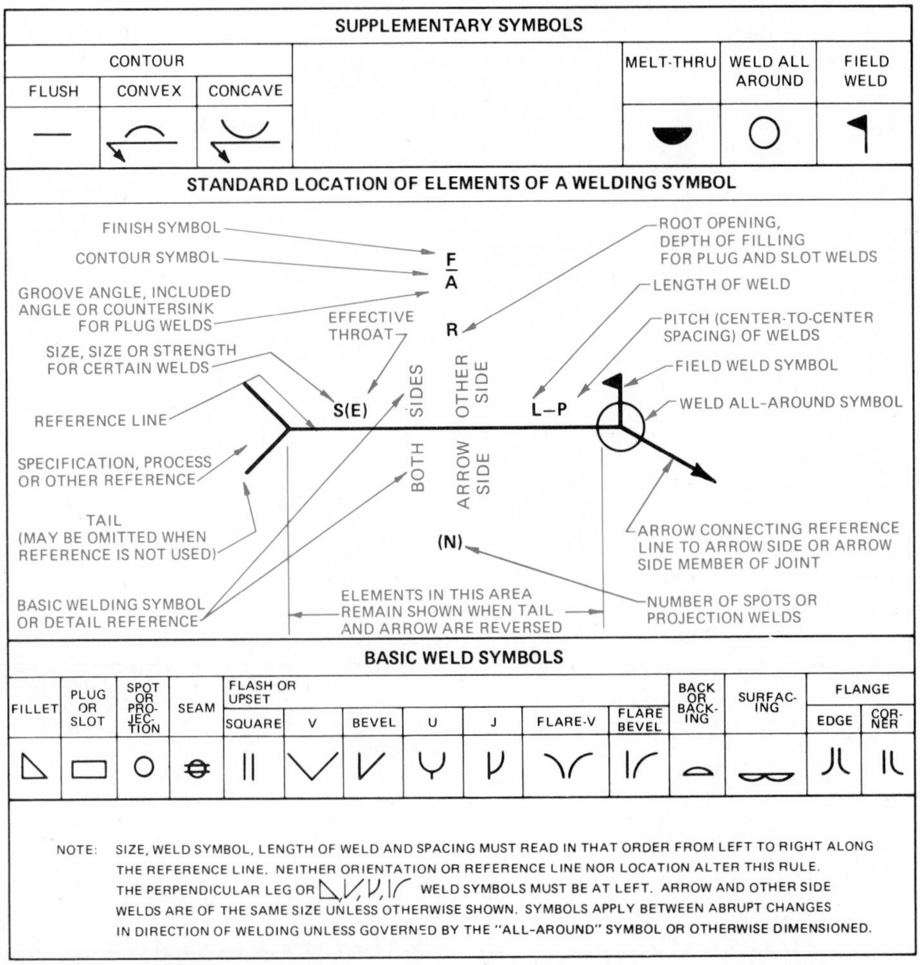

SUPPLEMENTARY SYMBOLS
CONTOUR
FLUSH
CONVEX
CONCAVE
MELT-THRU
WELD ALL AROUND
FIELD WELD
STANDARD LOCATION OF ELEMENTS OF A WELDING SYMBOL
FINISH SYMBOL
CONTOUR SYMBOL
GROOVE ANGLE, INCLUDED ANGLE OR COUNTERSINK FOR PLUG WELDS
SIZE, SIZE OR STRENGTH FOR CERTAIN WELDS
REFERENCE LINE
SPECIFICATION, PROCESS OR OTHER REFERENCE
TAIL (MAY BE OMITTED WHEN REFERENCE IS NOT USED)
BASIC WELDING SYMBOL OR DETAIL REFERENCE
EFFECTIVE THROAT
F
A
R
S(E)
BOTH SIDES
OTHER SIDE
ARROW SIDE
L–P
(N)
ROOT OPENING, DEPTH OF FILLING FOR PLUG AND SLOT WELDS
LENGTH OF WELD
PITCH (CENTER-TO-CENTER SPACING) OF WELDS
FIELD WELD SYMBOL
WELD ALL-AROUND SYMBOL
ARROW CONNECTING REFERENCE LINE TO ARROW SIDE OR ARROW SIDE MEMBER OF JOINT
NUMBER OF SPOTS OR PROJECTION WELDS
ELEMENTS IN THIS AREA REMAIN SHOWN WHEN TAIL AND ARROW ARE REVERSED
BASIC WELD SYMBOLS
FILLET
PLUG OR SLOT
SPOT OR PROJECTION
SEAM
FLASH OR UPSET
SQUARE
V
BEVEL
U
J
FLARE-V
FLARE BEVEL
BACK OR BACKING
SURFACING
FLANGE
EDGE
CORNER
NOTE: SIZE, WELD SYMBOL, LENGTH OF WELD AND SPACING MUST READ IN THAT ORDER FROM LEFT TO RIGHT ALONG THE REFERENCE LINE. NEITHER ORIENTATION OR REFERENCE LINE NOR LOCATION ALTER THIS RULE. THE PERPENDICULAR LEG OR WELD SYMBOLS MUST BE AT LEFT. ARROW AND OTHER SIDE WELDS ARE OF THE SAME SIZE UNLESS OTHERWISE SHOWN. SYMBOLS APPLY BETWEEN ABRUPT CHANGES IN DIRECTION OF WELDING UNLESS GOVERNED BY THE "ALL-AROUND" SYMBOL OR OTHERWISE DIMENSIONED.

Fig. 11-2-1 Welding symbols.

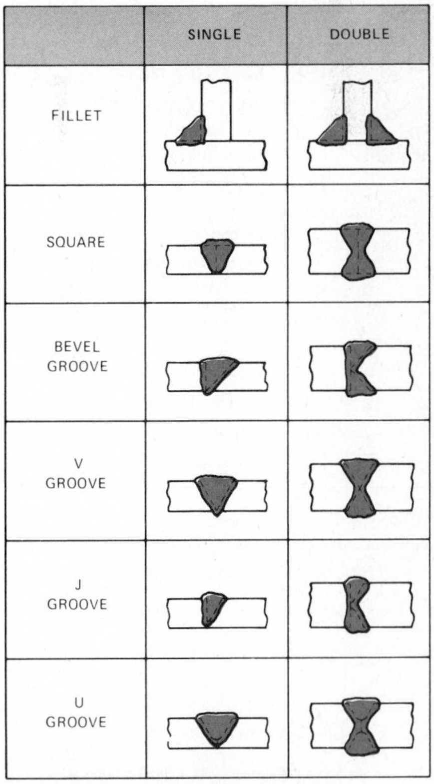

SINGLE
DOUBLE
FILLET
SQUARE
BEVEL GROOVE
V GROOVE
J GROOVE
U GROOVE

Fig. 11-2-3 Basic types of welds.

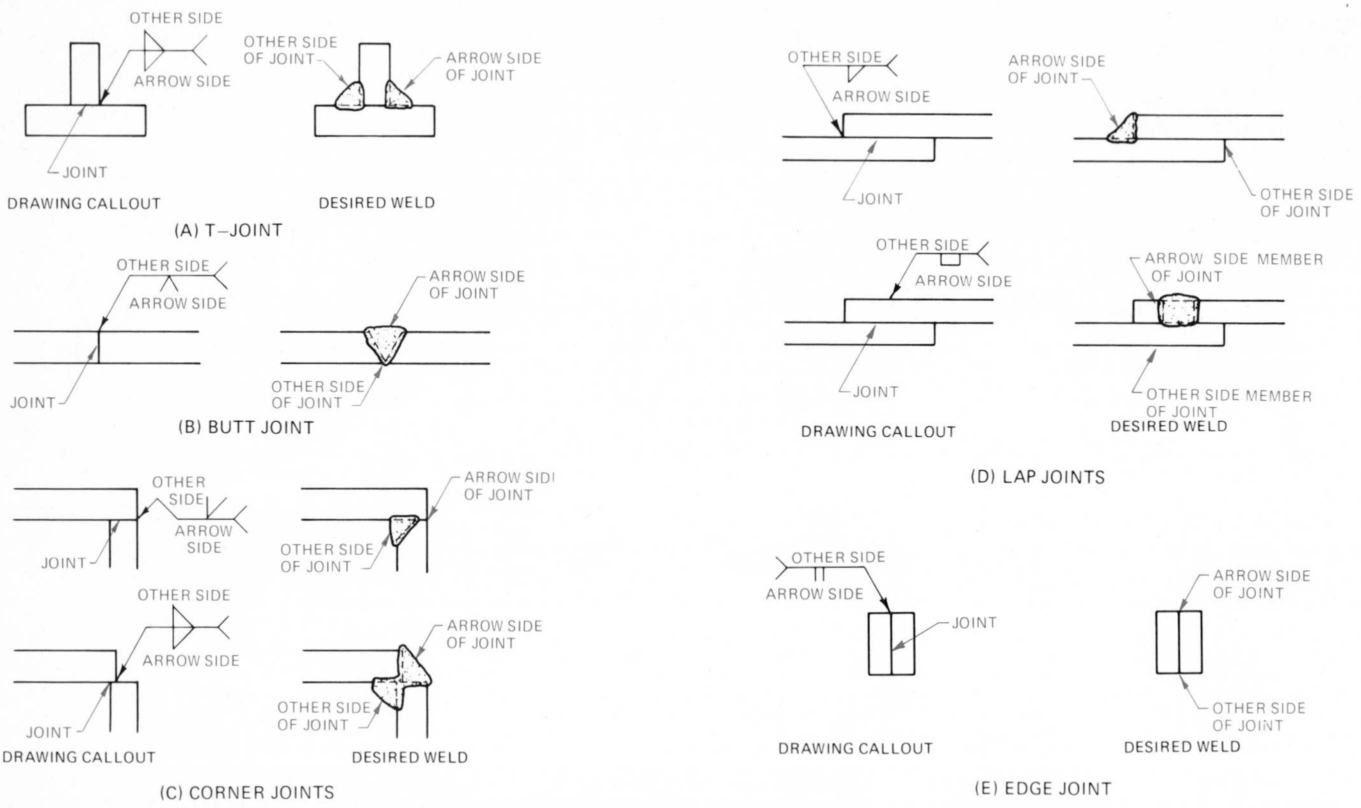
OTHER SIDE
ARROW SIDE
OTHER SIDE OF JOINT
ARROW SIDE OF JOINT
JOINT
DRAWING CALLOUT
DESIRED WELD
(A) T–JOINT
OTHER SIDE
ARROW SIDE
ARROW SIDE OF JOINT
JOINT
OTHER SIDE OF JOINT
(B) BUTT JOINT
OTHER SIDE
ARROW SIDE
ARROW SIDE OF JOINT
JOINT
OTHER SIDE OF JOINT
OTHER SIDE
ARROW SIDE
ARROW SIDE OF JOINT
JOINT
OTHER SIDE OF JOINT
DRAWING CALLOUT
DESIRED WELD
(C) CORNER JOINTS
OTHER SIDE
ARROW SIDE
ARROW SIDE OF JOINT
JOINT
OTHER SIDE OF JOINT
OTHER SIDE
ARROW SIDE
ARROW SIDE MEMBER OF JOINT
JOINT
OTHER SIDE MEMBER OF JOINT
DRAWING CALLOUT
DESIRED WELD
(D) LAP JOINTS
OTHER SIDE
ARROW SIDE
JOINT
ARROW SIDE OF JOINT
OTHER SIDE OF JOINT
DRAWING CALLOUT
DESIRED WELD
(E) EDGE JOINT

Fig. 11-2-4 Identification of arrow side and other side of joints.

symbol, will always have an arrow side and another side. Accordingly, the words *arrow side*, *other side*, and *both sides* are used here to locate the weld with respect to the joint. See Fig. 11-2-4.

The tail of the symbol is used for designating the welding specifications, procedures, or other supplementary information to be used in the making of the weld (Fig. 11-2-5). The notation to be placed on the tail of the symbol is to indicate the process, the type of filler metal to be used, and whether peening or root chipping is required. If notations are not used, the tail of the symbol may be omitted.

The use of letters can designate different welding and cutting processes. See Figs. 11-2-6 and 11-2-7.

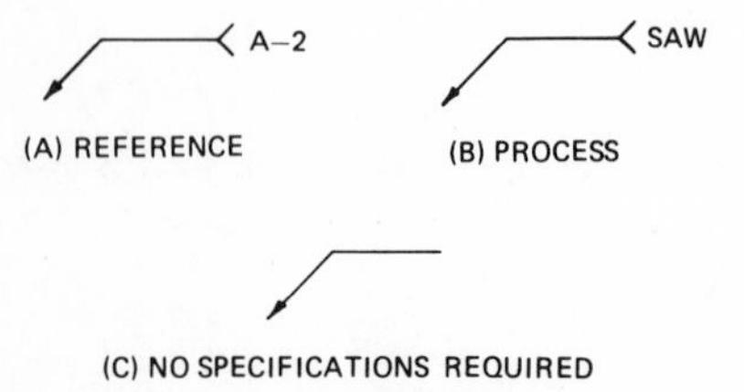
A–2
SAW
(A) REFERENCE
(B) PROCESS
(C) NO SPECIFICATIONS REQUIRED

Fig. 11-2-5 Location of reference and processes on welding symbols.

DESIGNATION	WELDING PROCESS
CAW	Carbon-Arc Welding
CW	Cold Welding
DB	Dip Brazing
DFW	Diffusion Welding
EBW	Electron Beam Welding
EW	Electrostag Welding
EXW	Explosion Welding
FB	Furnace Brazing
FCAW	Flux Cord Arc Welding
FOW	Forge Welding
FRW	Friction Welding
FW	Flash Welding
GMAW	Gas Metal-Arc Welding
GTAW	Gas Tungsten-Arc Welding
IB	Induction Brazing
IRB	Infrared Brazing
IW	Induction Welding
LBW	Laser Beam Welding
OAW	Oxyacetylene Welding
OHW	Oxyhydrogen Welding
PAW	Plasma-Arc Welding
PEW	Percussion Welding
PGW	Pressure Gas Welding
RB	Resistance Brazing
RPW	Projection Welding
RSEW	Resistance Seam Welding
RSW	Resistance Spot Welding
SAW	Submerged Arc Welding
SMAW	Shielded Metal Arc Welding
SW	Stud Welding
TB	Torch Brazing
TW	Thermit Welding
USW	Ultrasonic Welding
UW	Upset Welding

Fig. 11-2-6 Designation of welding processes by letters.

LOCATION SIGNIFICANCE OF ARROW

1. In the case of fillet and groove welding symbols, the arrow connects the welding symbol reference line to one side of the joint, and this side is considered the *arrow side* of the joint. The side opposite the arrow side of the joint is considered the *other side* of the joint.

2. When a joint is depicted by a single line on the drawing and the arrow of a welding symbol is directed to this line,

DESIGNATION	CUTTING PROCESS
AAC	Air-Carbon Arc Carbon
AC	Arc Cutting
AOC	Oxygen Arc Cutting
CAC	Carbon Arc Cutting
FOC	Chemical Flux Cutting
MAC	Metal-Arc Cutting
OC	Oxygen Cutting
PAC	Plasma Arc Cutting
POC	Metal Powder Cutting

Fig. 11-2-7 Designation of cutting processes by letters.

the arrow side of the joint is considered the *near side* of the joint.

3. In the case of plug, slot, arc-spot, arc-seam, resistance-spot, resistance-seam, and projection welding symbols, the arrow connects the welding symbol reference line to the outer surface of one of the members of the joint at the center line of the desired weld. The member to which the arrow points is the *arrow-side* member. The remaining member of the joint is considered the *other member*.

4. When a joint is depicted as an area parallel to the plane of projection in the drawing and the arrow of a welding symbol is directed to that area, the arrow-side member of the joint is considered the *near member* of the joint.

LOCATION OF WELD SYMBOL WITH RESPECT TO JOINT

1. Welds on the arrow side of the joint are shown by placing the weld symbol on the bottom side of the reference line.

2. Welds on the other side of the joint are shown by placing the weld symbol on the top side of the reference line.

3. Welds on both sides of the joint are shown by placing the weld symbol on both sides of the reference line.

USE OF WELD ALL-AROUND SYMBOL

A weld extending completely around a joint is indicated by means of a weld-all-around symbol placed at the intersection of the reference line and the arrow. See Fig. 11-2-8.

USE OF FIELD WELD SYMBOL

Field welds (welds not made in a shop or at the place of initial construction) are indicated by means of the field weld symbol placed at the intersection of the reference line and the arrow. The flag always points toward the tail of the arrow. See Fig. 11-2-9.

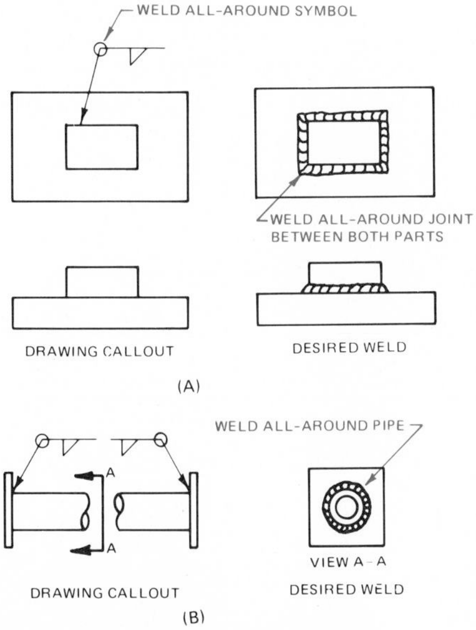
WELD ALL-AROUND SYMBOL
WELD ALL-AROUND JOINT BETWEEN BOTH PARTS
DRAWING CALLOUT
DESIRED WELD
(A)
WELD ALL-AROUND PIPE
A
A
VIEW A–A
DRAWING CALLOUT
DESIRED WELD
(B)

Fig. 11-2-8 Application of weld all-around symbol.

COMBINED WELDING SYMBOLS

For joints having more than one weld, a symbol shall be shown for each weld. See Fig. 11-2-10.

FINISHING OF WELDS

Finishing of welds, other than cleaning, is indicated by suitable contour and finish symbols (Fig. 11-2-11). The following finishing symbols indicate the method, not the degree, of finish:

- C = Chipping
- G = Grinding
- M = Machining
- R = Rolling
- H = Hammering

For methods of indicating degree of finish see ANSI B46.1, *Surface Texture*.

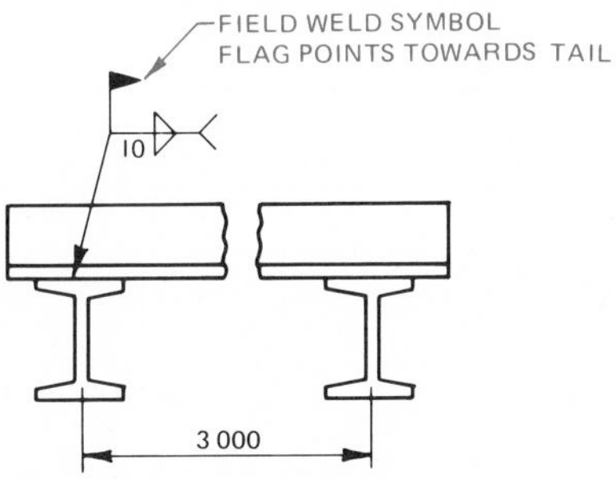
FIELD WELD SYMBOL
FLAG POINTS TOWARDS TAIL
10
3 000

Fig. 11-2-9 Application of field weld symbol.

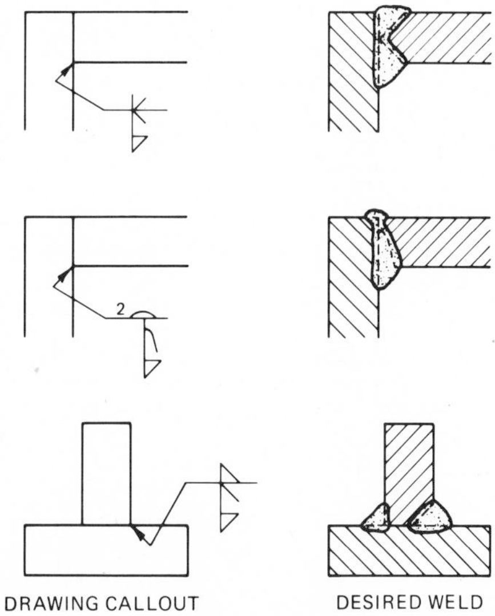
2
DRAWING CALLOUT
DESIRED WELD

Fig. 11-2-10 Combined welding symbols.

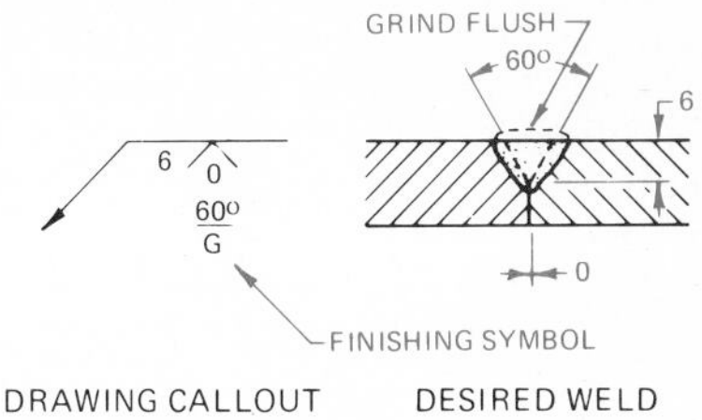
GRIND FLUSH
60°
6
6
0
60°
G
0
FINISHING SYMBOL
DRAWING CALLOUT
DESIRED WELD

Fig. 11-2-11 Welding finishing symbols.

The Design of Welded Joints[2]

Since loads are transferred from one member to another through the welds on a fabricated assembly, the type of joint and weld is specified by the designer. Figure 11-2-4 shows basic joint and weld types. Specifying the joint does not by itself describe the type of weld to be used. Several types of welds may be used for making a joint.

The fillet weld, requiring no groove penetration, is one of the most commonly used welds. Corner welds are also widely used in machine design. The corner-to-corner joint, shown in Fig. 11-2-12*a*, is difficult to assemble because neither plate can be supported by the other. The joint also requires a larger amount of weld than the other joints illustrated. The corner joint shown in Fig. 11-2-12*b* is easy to assemble and requires half the amount of weld metal as the joint in Fig. 11-2-12*a*. However, by using half the weld size, but placing two welds, one outside, as in Fig. 11-2-12*c*, it is possible to obtain the same total throat as with the first weld. Only half the weld metal is required.

With thick plates, a partial-penetration groove joint, as in Fig. 11-2-12*d*, is used. This requires beveling. For a deeper joint, a J preparation, as in Fig. 11-2-12*e*, may be used in preference to a bevel. The fillet weld in Fig. 11-2-12*f* is out of sight and makes a neat and economical corner.

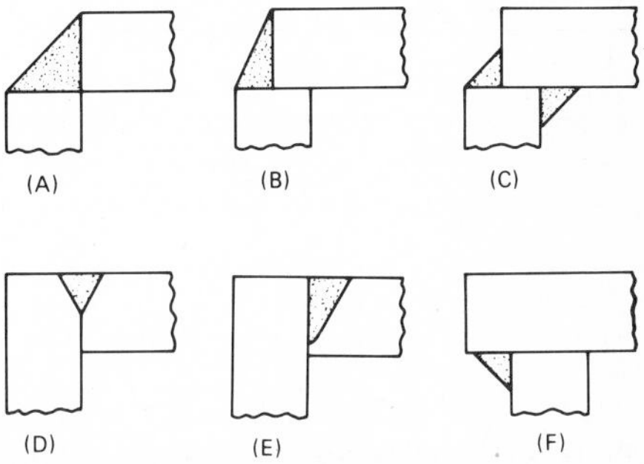
(A)
(B)
(C)
(D)
(E)
(F)

Fig. 11-2-12 Corner joints.

The size of the weld should always be designed with reference to the size of the thinner member. The joint cannot be made any stronger by using the thicker member for the weld size, and much more weld metal may be required, as illustrated in Fig. 11-2-13.

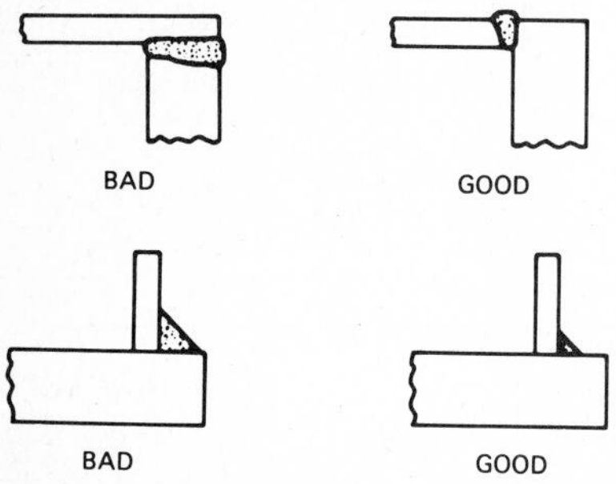
BAD
GOOD
BAD
GOOD

Fig. 11-2-13 Size of weld determined by thinner member.

The designer is frequently faced with the question of whether to use fillet or groove welds. Here cost becomes a major consideration. The fillet welds in Fig. 11-2-14*a* are easy to apply and require no special plate preparation. They can be made using large-diameter electrodes with high welding currents; as a consequence, the deposit rate is high.

In comparison, the double-bevel groove weld in Fig. 11-2-14*b* has about one-half the weld area of the fillet welds. However, it requires extra preparation and the use of smaller-diameter electrodes with lower welding currents to place the initial pass without burning through. As plate thickness increases, this initial low-deposition region becomes a less important factor, and the higher cost factor decreases in significance.

Refer to Fig. 11-2-14*c*. It will be noted that the single-bevel groove weld requires about the same amount of weld metal as the fillet welds deposited in Fig. 11-2-14*a*. Thus, there is no apparent economic advantage. There are some disadvantages, though. The single-bevel joint requires bevel preparation and initially a lower deposit rate at the root of the joint. From a design standpoint, however, it offers a direct transfer of force through the joint, which means that it is probably better under fatigue loading. Although the illustrated full-strength fillet welds, having leg sizes equal to 75 percent of the plate thickness, would be sufficient, some codes have lower allowable limits for fillet welds and may require a leg size equal to the plate thickness. In this case, the cost of the fillet-welded joint may exceed the cost of a single-bevel groove in thicker plates. Also, if the joint is so positioned that the weld can be made in a flat position, a single-bevel groove weld would be less expensive than if the fillet welds were specified. As can be seen in Fig. 11-2-15, one of the fillets would have to be made in the overhead position—a costly operation.

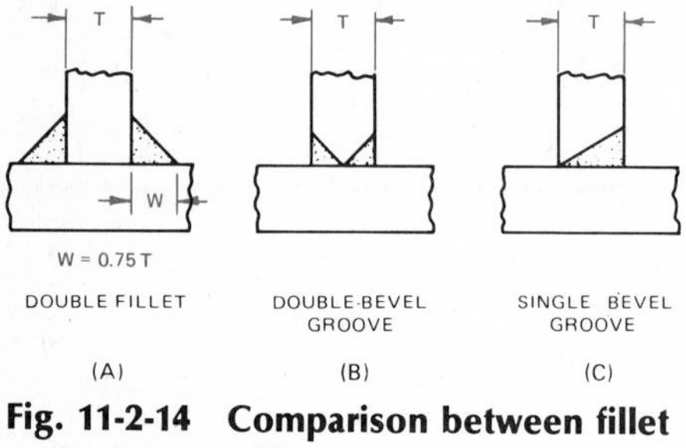
T
T
T
W
W = 0.75 T
DOUBLE FILLET
DOUBLE-BEVEL GROOVE
SINGLE BEVEL GROOVE
(A)
(B)
(C)

Fig. 11-2-14 Comparison between fillet and groove welds.

REFERENCES AND SOURCE MATERIAL

1. American Welding Society and Canadian Welding Bureau.
2. The Lincoln Electric Company.

Assignments

1. On a A3- or B-size sheet, complete the enlarged view of the welded joint for the drawing callouts shown in Figs. 11-2-A and 11-2-B. Use notes to explain any additional welding requirements.
2. On an A3- or B-size sheet, add the information shown above Figs. 11-2-C and 11-2-D to the seven welding symbols shown in these assignments.

REVIEW FOR ASSIGNMENTS

Unit 3-9 Enlarged Views
Unit 7-5 Direction of Section Lining

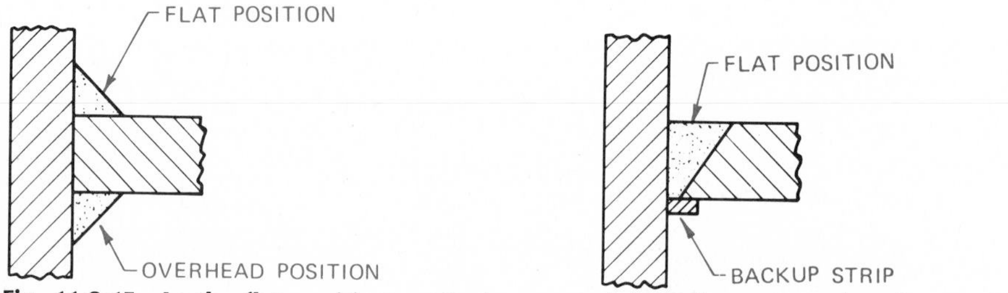
FLAT POSITION
OVERHEAD POSITION
FLAT POSITION
BACKUP STRIP

Fig. 11-2-15 In the flat position, a single-groove joint is less expensive than two fillet welds.

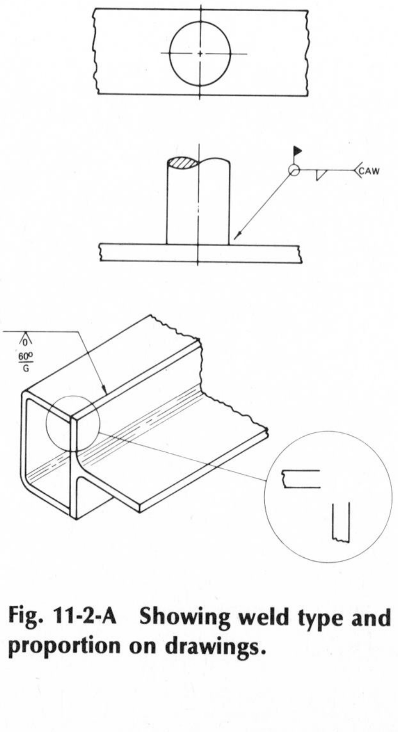
CAW
60°
G

Fig. 11-2-A Showing weld type and proportion on drawings.

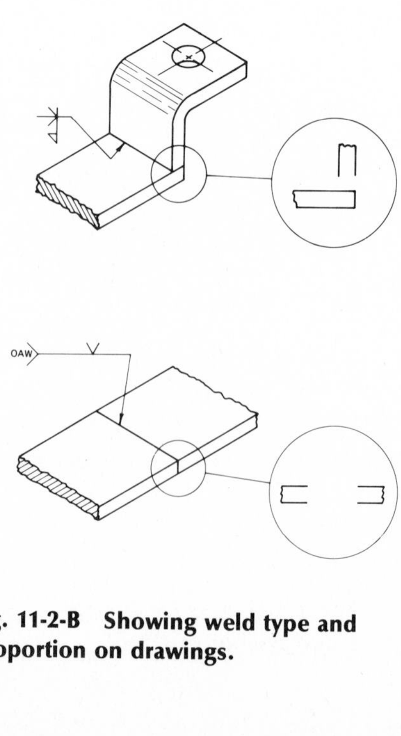
OAW

Fig. 11-2-B Showing weld type and proportion on drawings.

Weld	Welding Process Required	Type of Weld	Additional Requirements
1	Carbon-Arc Welding	Bevel	
2	Oxyacetylene Welding	Double Fillet	Both Sides Field Weld
3	Oxyacetylene Welding	Double Fillet	Both Sides
4	No Specifications Required	J Groove	
5	Carbon-Arc Welding	Fillet	All Around
6	Carbon-Arc Welding	Fillet	All Around Field Weld
7	Gas Metal-Arc Welding	Double Fillet	

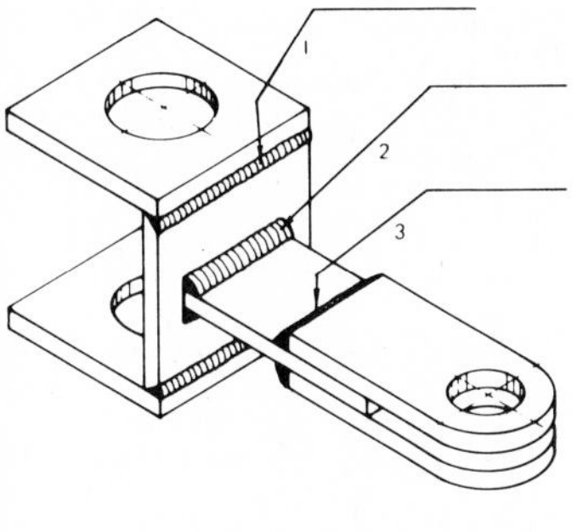

BEVEL AND FILLET WELDS

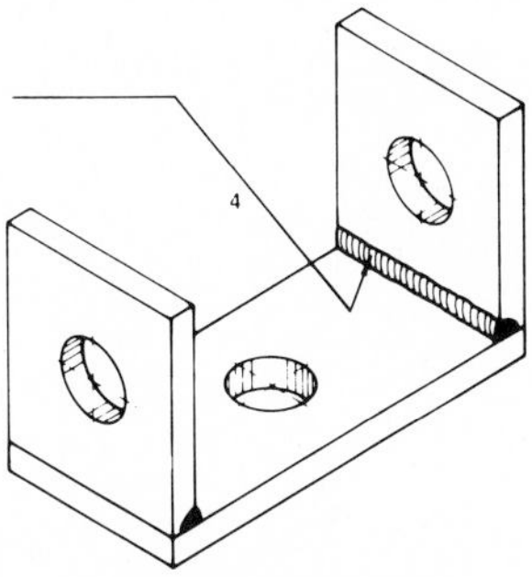

J-GROOVE WELD

Fig. 11-2-C Indicating welding symbols on drawings.

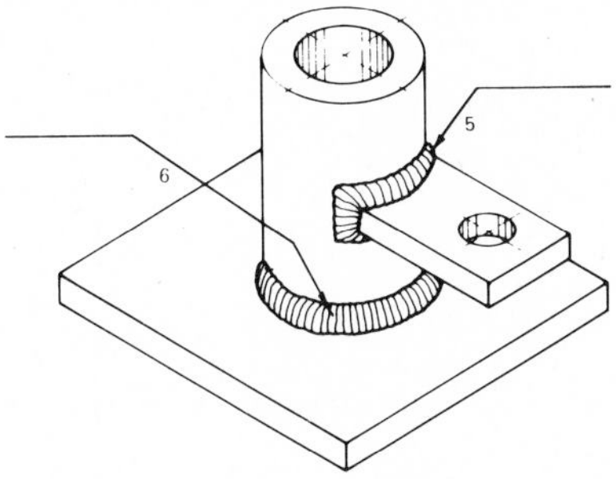

FILLET WELD ALL AROUND

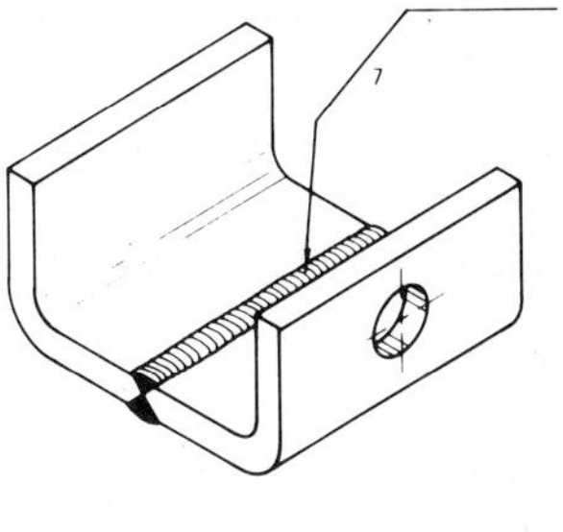

DOUBLE V-GROOVE WELD

Fig. 11-2-D Indicating welding symbols on drawings.

UNIT 11-3

FILLET WELDS

Fillet Weld Symbols[1]

1. Dimensions of fillet welds are shown on the same side of the reference line as the weld symbol.

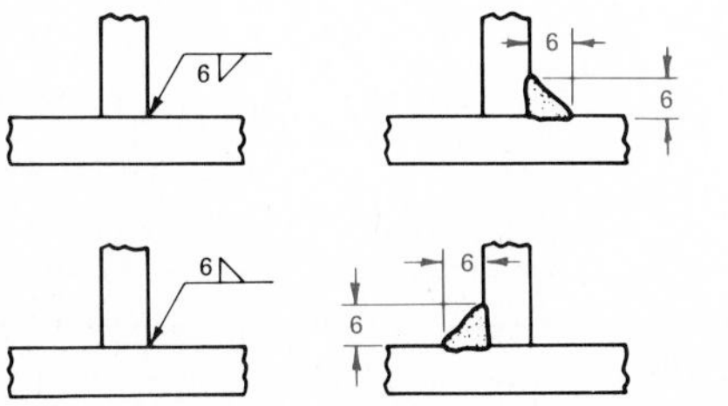

2. When both sides of a joint have same-size fillet welds, one or both may be dimensioned.

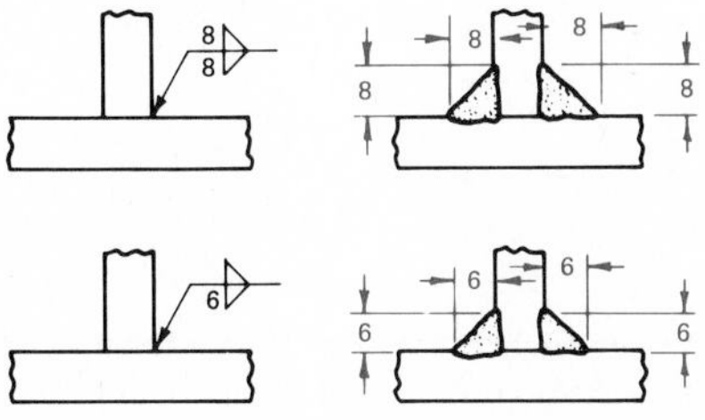

3. When both sides of a joint have different-size fillet welds, both are dimensioned.

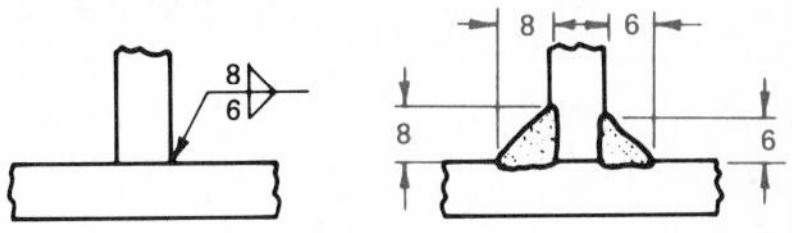

4. When there appears on a drawing a general note governing the dimension of fillet welds, such as ALL FILLET WELDS 6 mm UNLESS OTHERWISE NOTED and all the welds have dimensions governed by the note, the dimension need not be shown.

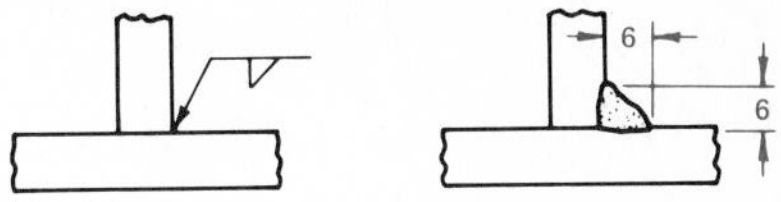

5. When the dimensions of one or both welds differ from the dimensions given in the general note, both welds are dimensioned.

6. The size of a fillet weld is shown to the left of the weld symbol.

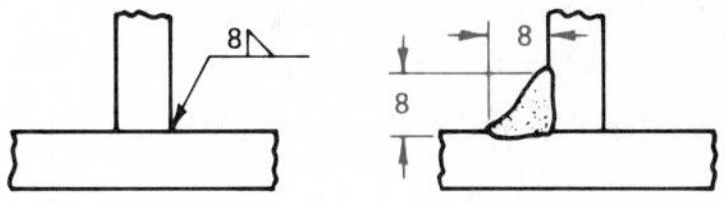

LOCATION SIGNIFICANCE	SYMBOL
ARROW SIDE	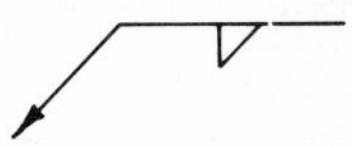
OTHER SIDE	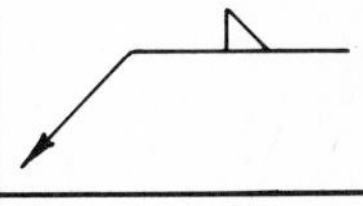
BOTH SIDES	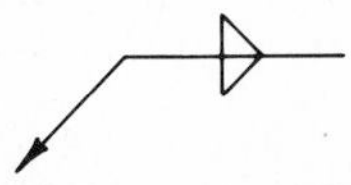

Fig. 11-3-1 Fillet weld symbol and its location significance.

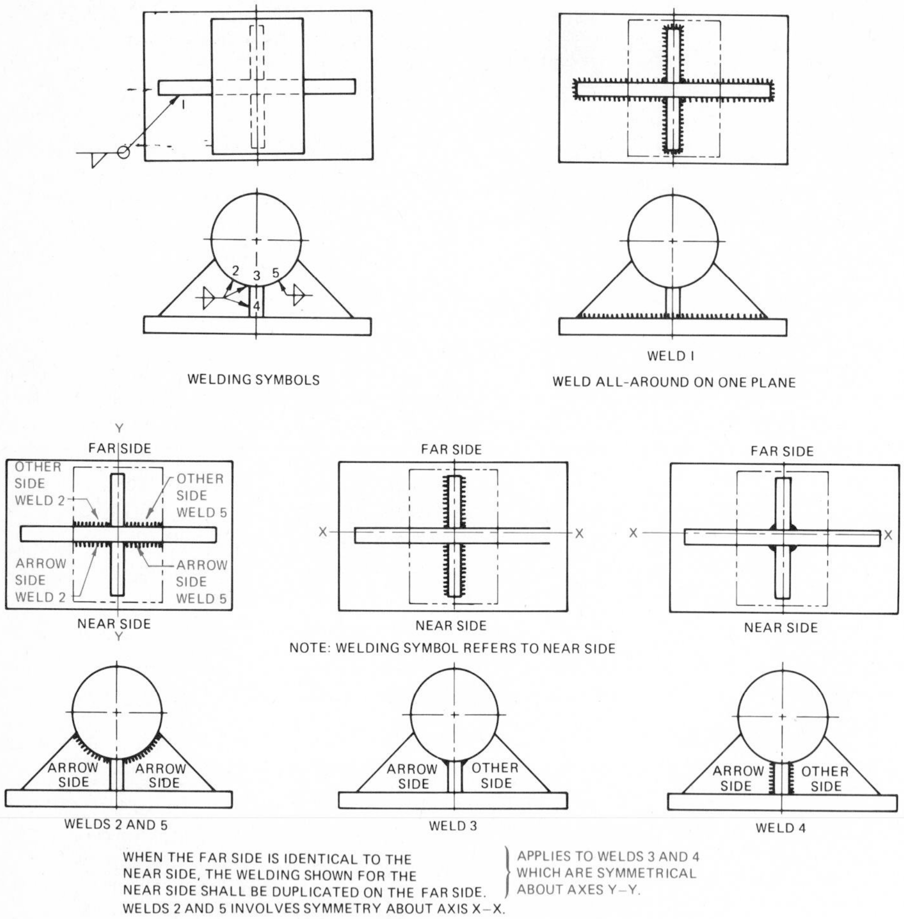

2
3
5
4
WELDING SYMBOLS
WELD I
WELD ALL-AROUND ON ONE PLANE
Y
FAR SIDE
OTHER SIDE WELD 2
OTHER SIDE WELD 5
ARROW SIDE WELD 2
ARROW SIDE WELD 5
NEAR SIDE
Y
FAR SIDE
X
X
NEAR SIDE
NOTE: WELDING SYMBOL REFERS TO NEAR SIDE
FAR SIDE
X
X
NEAR SIDE
ARROW SIDE
ARROW SIDE
WELDS 2 AND 5
ARROW SIDE
OTHER SIDE
WELD 3
ARROW SIDE
OTHER SIDE
WELD 4
WHEN THE FAR SIDE IS IDENTICAL TO THE NEAR SIDE, THE WELDING SHOWN FOR THE NEAR SIDE SHALL BE DUPLICATED ON THE FAR SIDE.
WELDS 2 AND 5 INVOLVES SYMMETRY ABOUT AXIS X–X.
APPLIES TO WELDS 3 AND 4 WHICH ARE SYMMETRICAL ABOUT AXES Y–Y.

Fig. 11-3-2 Application of fillet welds for shaft support.

7. When a fillet weld is required to go completely around a part, a weld-all-around symbol is used.

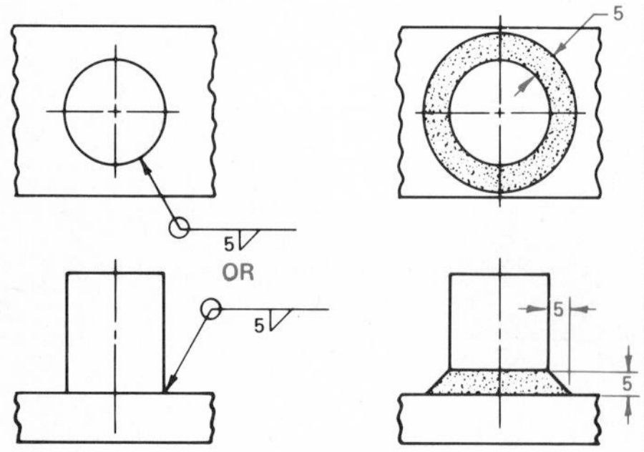

5
5
OR
5
5
5

8. The size of a fillet weld with unequal legs is shown in parentheses to the left of the weld symbol. Weld orientation is not shown by the symbol. It is shown on the drawing when necessary.

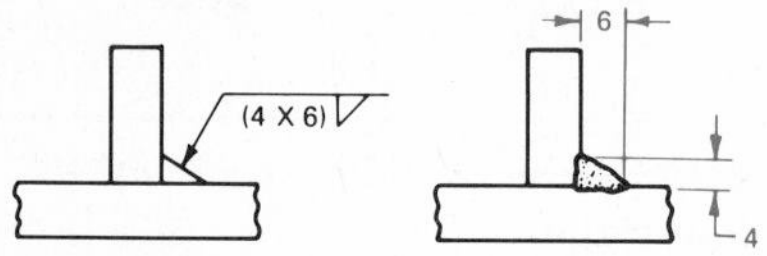

(4 X 6)
6
4

9. The length of a fillet weld, when indicated on the welding symbol, is shown to the right of the weld symbol.

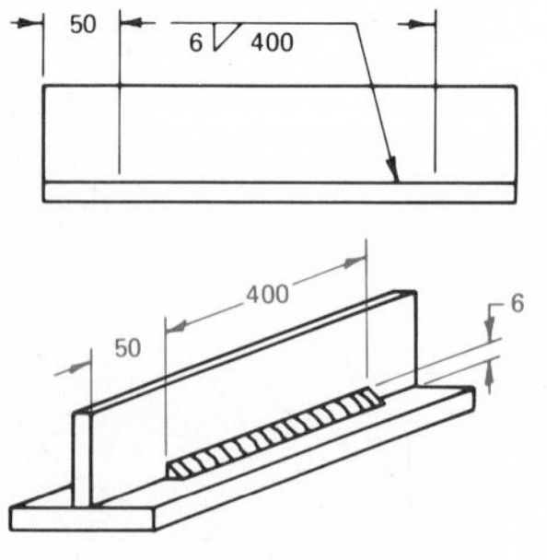

50
6
400
400
50
6

10. Specific lengths of fillet welds may be indicated by symbols in conjunction with dimension lines.

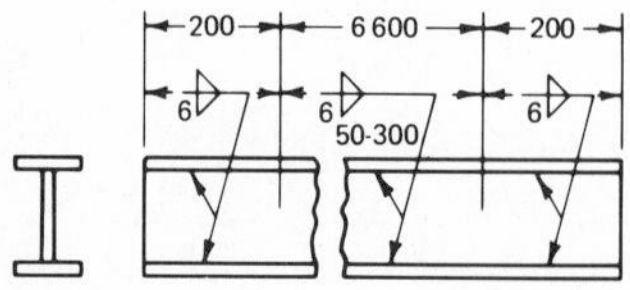

200
6 600
200
6
6
50-300
6

11. The pitch (center-to-center spacing) of intermittent fillet welding is shown as the distance between centers of increments on one side of the joint. It is shown to the right of the length dimension.

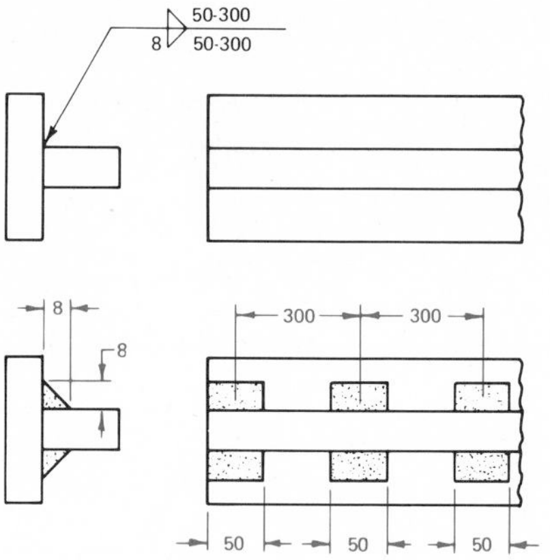

50-300
8
50-300
8
8
300
300
50
50
50

12. Staggered intermittent fillet welds are shown with the weld symbols staggered.

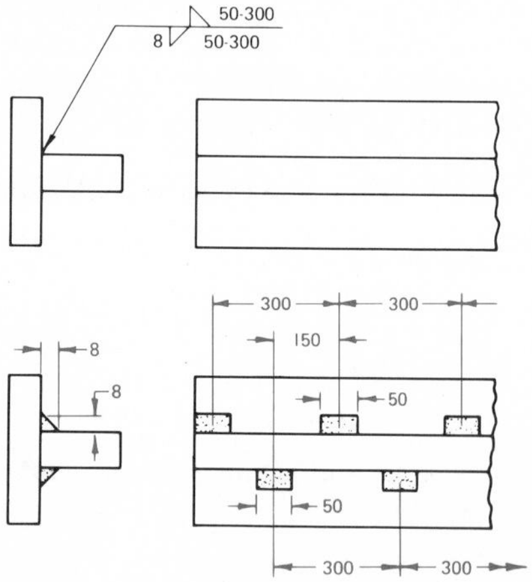

50-300
8
50-300
300
300
150
8
8
50
50
300
300

13. Fillet welds that are to be welded approximately flat-faced without recourse to any method of finishing are shown by adding the flush-contour symbol to the weld symbol.

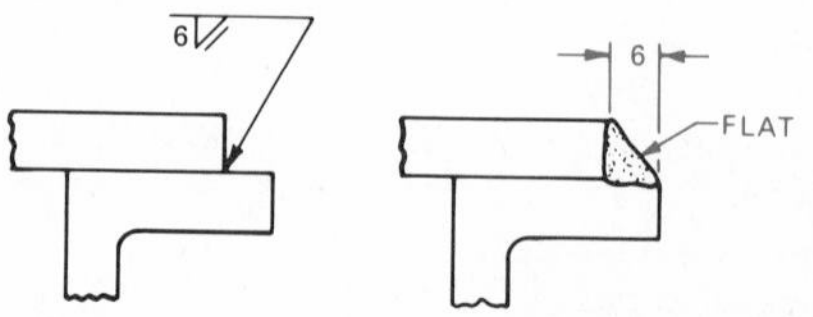

6
6
FLAT

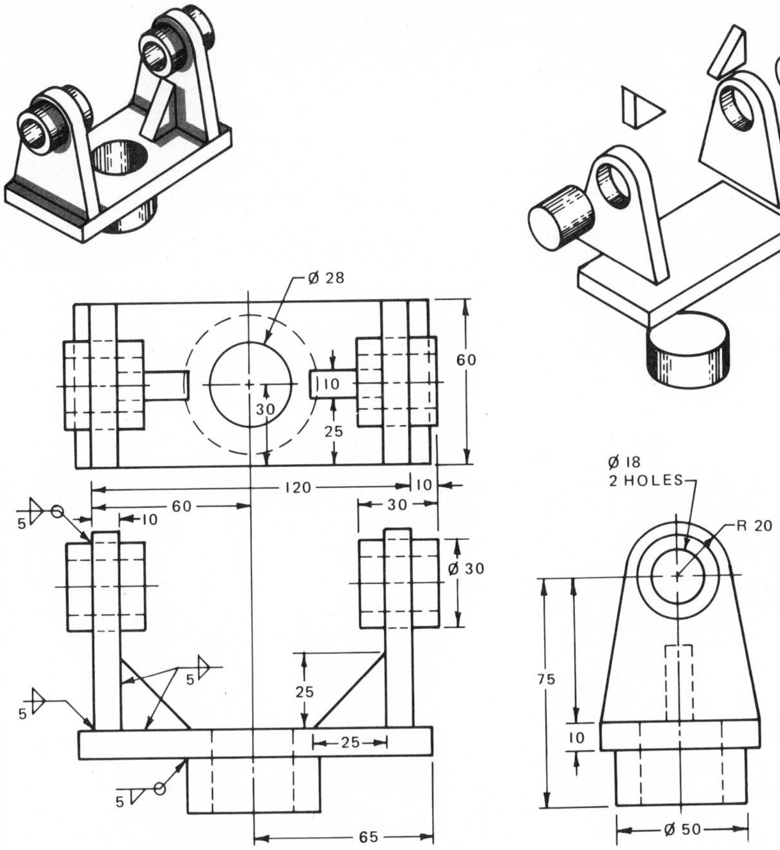
Ø 28
60
10
30
25
120
10
60
10
30
5
Ø 30
5
25
5
25
5
65
Ø 18
2 HOLES
R 20
75
10
Ø 50

Fig. 11-3-3 Welded-steel shaft support.

A very practical method is to design the weld for the thinner plate, making it sufficient to carry one-third to one-half the carrying capacity of the plate. This means that if the plate were stressed one-third to one-half its usual value, the weld would be of sufficient size. Most rigidity designs are stressed much below these values. However, any reduction in weld size below one-third the full-strength value would give a weld too small in appearance for general acceptance.

Example 1. What size fillet weld is required to match the strength of the fabricated design shown in Fig. 11-3-5*a*?

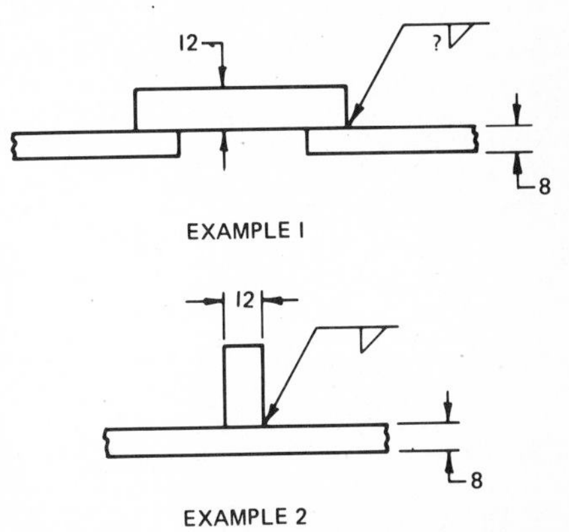
12
?
8
EXAMPLE I
12
8
EXAMPLE 2

Fig. 11-3-5 Calculating fillet-weld size.

14. Fillet welds that are to be made flat-faced by mechanical means are shown by adding both the flush-contour symbol and the user's standard finish symbol.

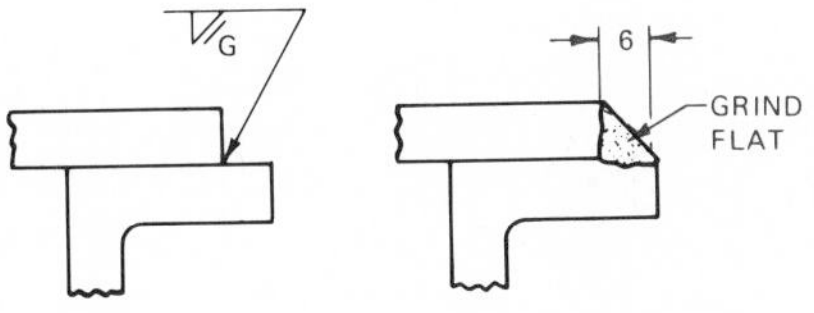
G
6
GRIND
FLAT

Size of Fillet Welds[2]

Figure 11-3-4 gives the sizing of fillet welds for rigidity designs at various strengths and plate thicknesses, where the strength of the weld metal matches the plate.

In machine design work, where the primary design requirement is rigidity, members are often made with extra-heavy sections, so that movement under load would be within very close tolerances. The question arises as how to determine the weld size for these types of rigidity designs.

DIMENSIONS IN MILLIMETRES				DIMENSIONS IN INCHES			
	Strength Design	Rigidity Design			Strength Design	Rigidity Design	
Plate Thickness (t)	Full-strength Weld = 0.75 T	50% of Full-strength Weld = 0.38 T	33% of Full-strength Weld = 0.25 T	Plate Thickness (t)	Full-strength Weld = .75 T	50% of Full-strength Weld = .38 T	33% of Full-strength Weld = .25 T
6	3	3	3	Less than .25	.12	.12	.12
6	5	5	5	.25	.19	.19	.19
8	6	5	5	.31	.25	.19	.19
10	8	5	5	.38	.31	.19	.19
11	10	5	5	.44	.38	.19	.19
12	10	5	5	.50	.38	.19	.19
14	11	6	6	.56	.44	.25	.25
16	12	6	6	.62	.50	.25	.25
20	14	8	6	.75	.56	.31	.25
22	16	10	8	.88	.62	.38	.31
25	16	10	8	1.00	.62	.38	.31
28	22	11	8	1.12	.88	.44	.31
32	25	12	8	1.25	1.00	.50	.31
35	25	12	10	1.38	1.00	.50	.38
38	28	14	10	1.50	1.12	.56	.38

Fig. 11-3-4 Rule-of-thumb fillet-weld sizes where the strength of the weld metal matches the plate.

Solution With reference to Fig. 11-3-4, a full-strength weld is required. Thinner plate = 8. Fillet weld required = 6 mm.

Example 2. What size fillet weld is required to hold the rib to the plate shown in Fig. 11-3-5*b*? Weld design is for rigidity only, and only 33 percent of full-strength weld is required.

Solution Thinner plate = 8. With reference to Fig. 11-3-4, the weld size under rigidity design, 33 percent opposite 8 mm, is 5.

REFERENCES AND SOURCE MATERIAL

1. Canadian Welding Bureau and the American Welding Society.
2. The Lincoln Electric Company.

Assignment

On an A3- or B-size sheet, select one of the problems shown in Figs. 11-3-A to 11-3-D and make a two-view working drawing complete with dimensions and welding symbols. Include on the drawing a bill of material, and identify each part on the assembly. Use full-strength welds. Scale is 1:1.

REVIEW FOR ASSIGNMENT

Unit 6-4 Bill of Material
Unit 6-6 Detailed Assembly Drawing
Unit 11-2 Welding Symbols

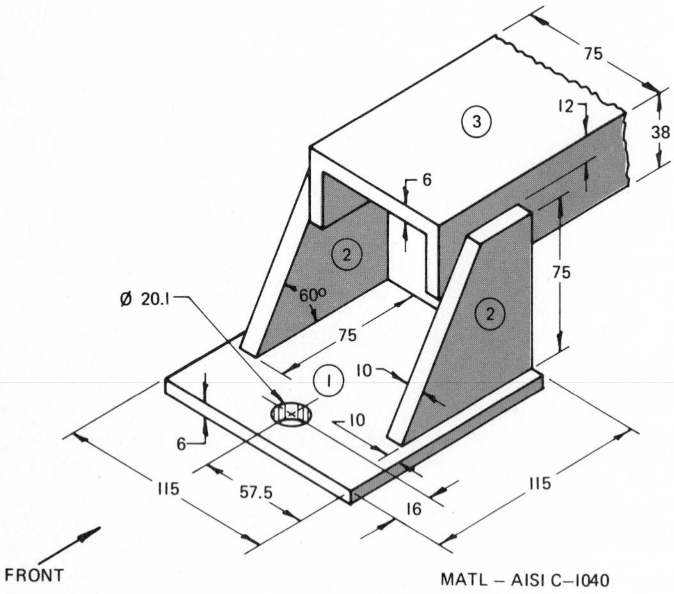

75
12
38
6
75
Ø 20.1
60°
75
10
10
6
115
57.5
16
115
FRONT
MATL – AISI C–1040

Fig. 11-3-A Step bracket.

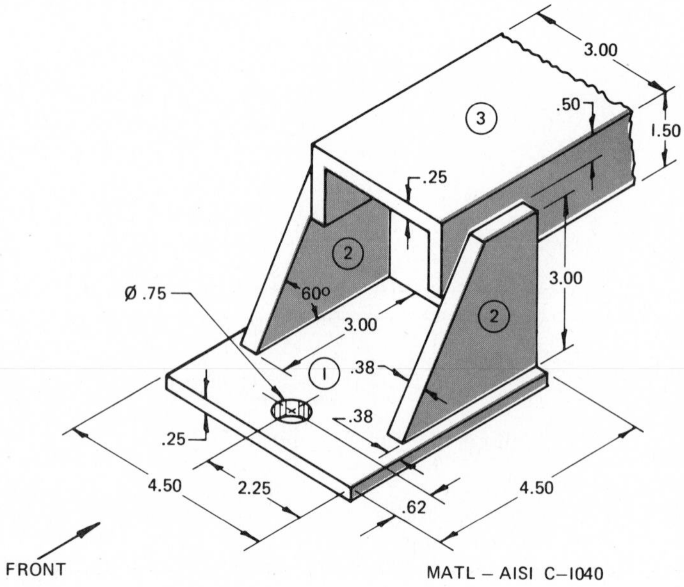

3.00
.50
1.50
.25
3.00
Ø .75
60°
3.00
.38
.38
.25
4.50
2.25
.62
4.50
FRONT
MATL – AISI C–1040

Fig. 11-3-B Step bracket.

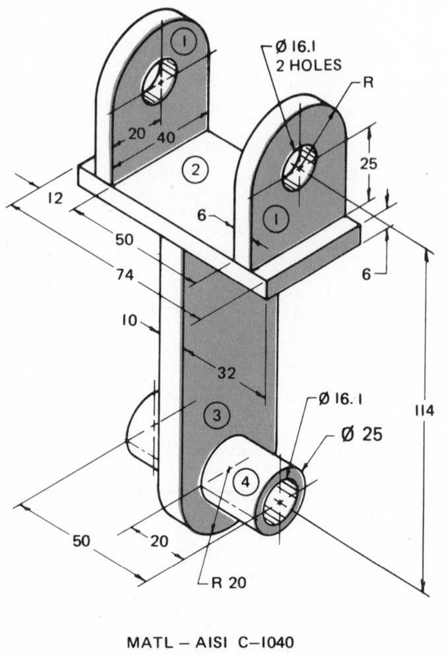

Ø 16.1
2 HOLES
R
20
40
25
12
6
50
74
6
10
32
114
Ø 16.1
Ø 25
50
20
R 20
MATL – AISI C–1040

Fig. 11-3-C Swing bracket.

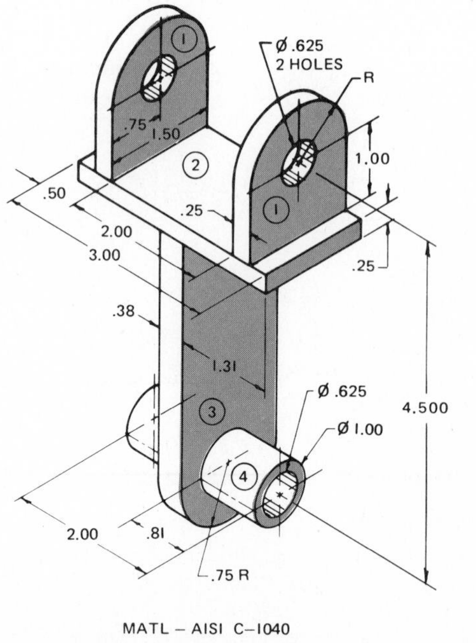

Ø .625
2 HOLES
R
.75
1.50
1.00
.50
.25
2.00
3.00
.25
.38
1.31
Ø .625
Ø 1.00
4.500
2.00
.81
.75 R
MATL – AISI C–1040

Fig. 11-3-D Swing bracket.

UNIT 11-4

GROOVE WELDS[1]

Use of Break-in Arrow of Bevel and J-Groove Welding Symbols

When a bevel or J-groove weld symbol is used, the arrow points with a definite break toward the member which is to be chamfered. In cases where the member to be chamfered is obvious, the break in the arrow may be omitted. See Figs. 11-4-1 through 11-4-3.

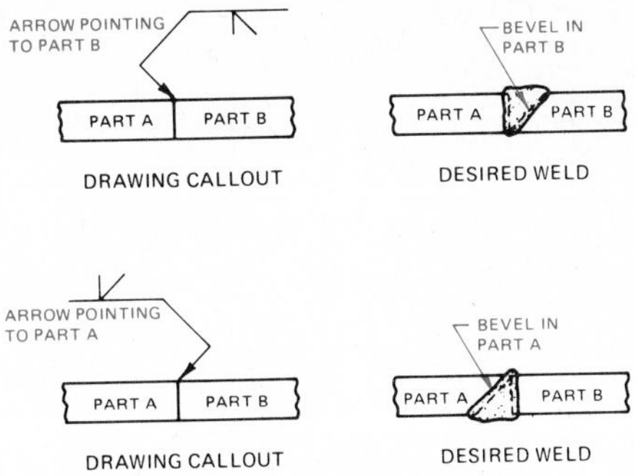
ARROW POINTING TO PART B
PART A
PART B
DRAWING CALLOUT
BEVEL IN PART B
DESIRED WELD
ARROW POINTING TO PART A
BEVEL IN PART A

Fig. 11-4-2 Use of break in arrow.

Groove Weld Symbols

1. Dimensions of groove welds are shown on the same side of the reference line as the weld symbol.

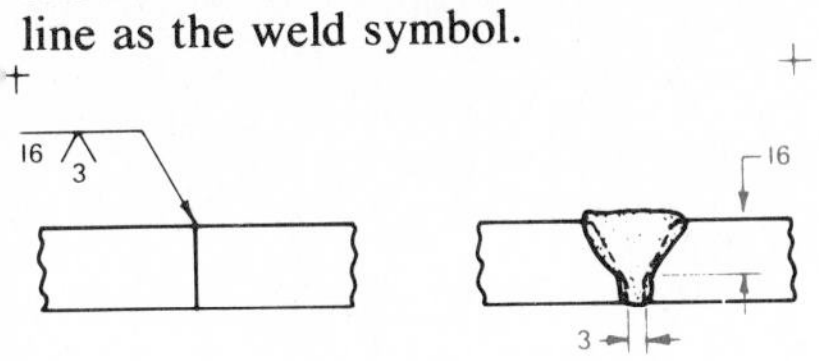

2. When both sides of a double-groove weld have the same dimensions, one or both may be dimensioned.

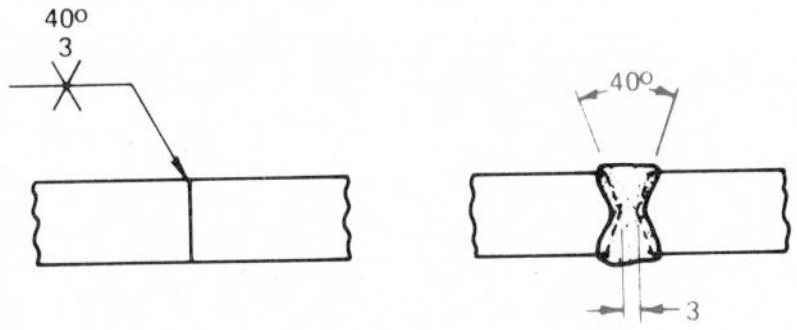

3. When both sides of a double-groove weld differ in dimensions, both are dimensioned.

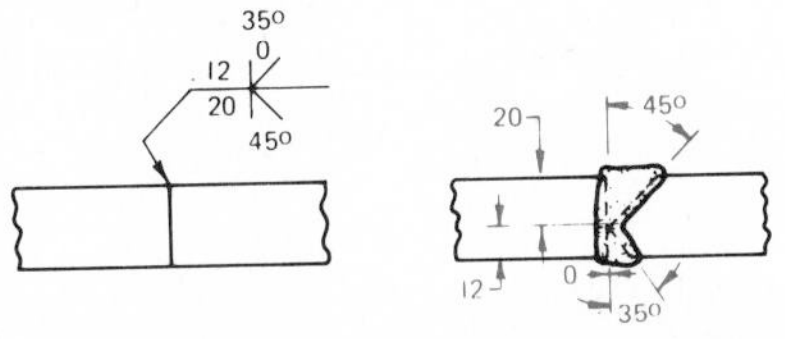

4. When there appears on the drawing a general note governing the dimensions of groove welds, such as ALL V-GROOVE WELDS ARE TO HAVE A 60° ANGLE UNLESS OTHERWISE NOTED, groove welds need not be dimensioned.

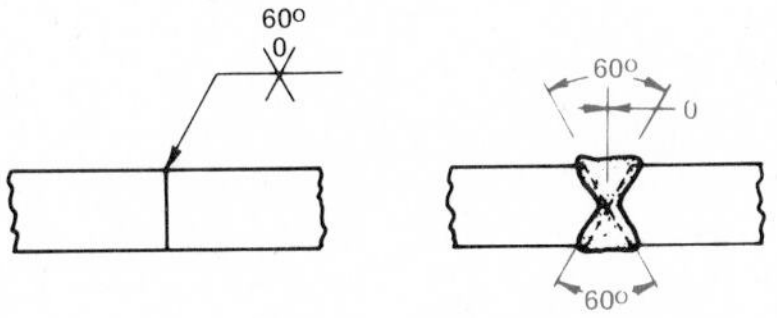

5. For bevel and groove welds, the arrow points with a definite break toward the member being beveled.

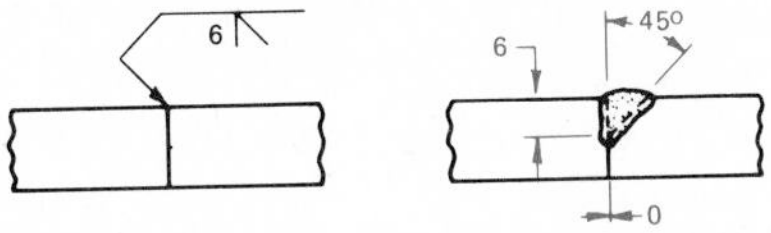

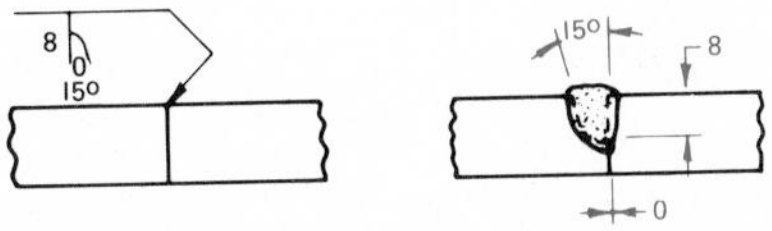

6. When the dimensions of one or both welds differ from the dimensions given in the general note, both welds are dimensioned.

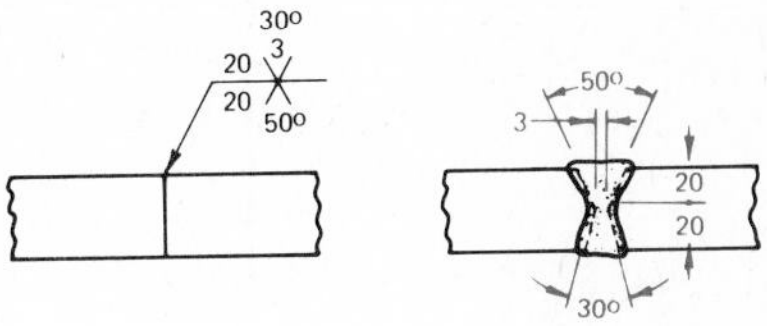

7. The size of groove welds is shown to the left of the weld symbol.

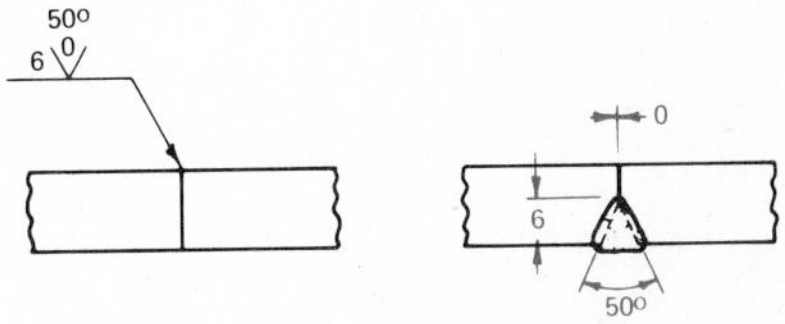

8. When the single-groove and symmetrical double-groove welds extend completely through the member or members being joined, the size of the weld need not be shown on the welding symbol.

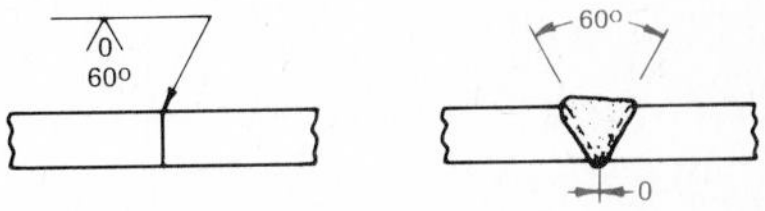

9. When the groove welds extend only partly through the member being joined, the size of the weld is shown on the welding symbol.

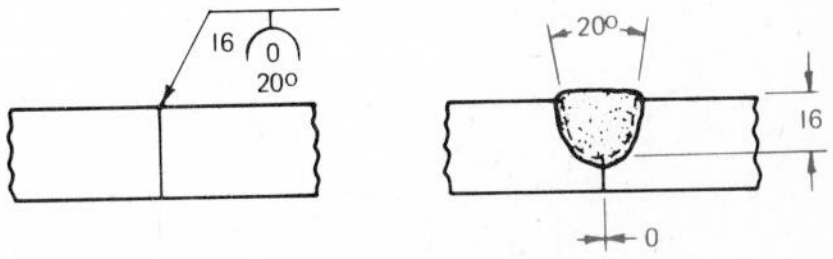

10. The size of groove welds with specified root penetration, except square-groove welds, is indicated by showing both the depth of chamfering and the root penetration to the left of the weld symbol. The size of the square-groove welds is indicated by showing only the root penetration. The depth of chamfering and the root penetration, enclosed in brackets, are read in that order from left to right along the reference line.

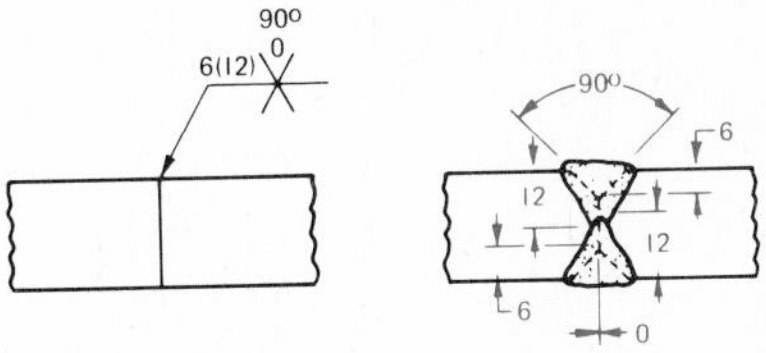

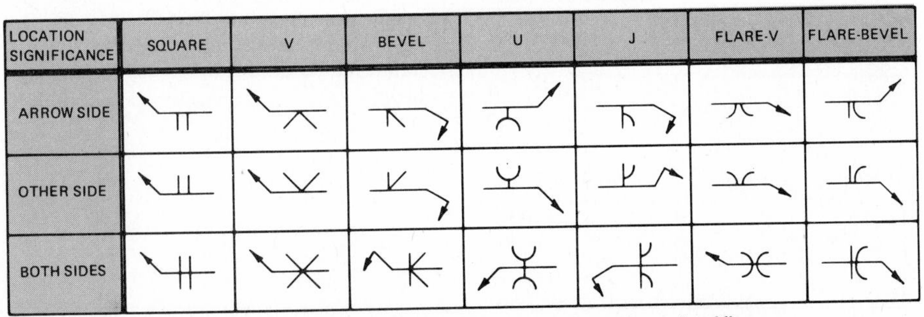
LOCATION SIGNIFICANCE
SQUARE
V
BEVEL
U
J
FLARE-V
FLARE-BEVEL
ARROW SIDE
OTHER SIDE
BOTH SIDES

Fig. 11-4-1 Basic groove welding symbols and their location significance.

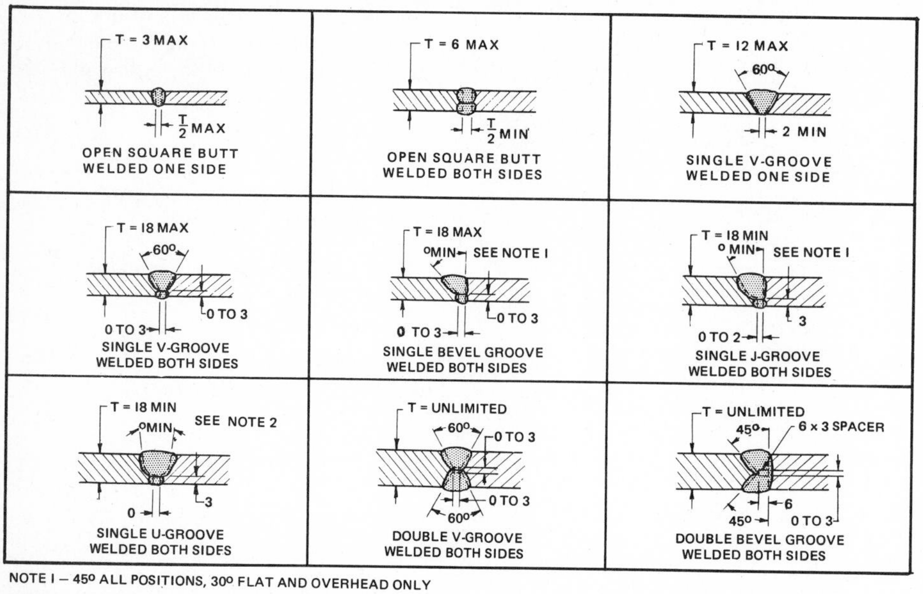
T = 3 MAX
OPEN SQUARE BUTT WELDED ONE SIDE
T = 6 MAX
OPEN SQUARE BUTT WELDED BOTH SIDES
T = 12 MAX
60°
2 MIN
SINGLE V-GROOVE WELDED ONE SIDE
T = 18 MAX
60°
0 TO 3
SINGLE V-GROOVE WELDED BOTH SIDES
T = 18 MAX
SEE NOTE 1
0 TO 3
SINGLE BEVEL GROOVE WELDED BOTH SIDES
T = 18 MIN
SEE NOTE 1
0 TO 2
3
SINGLE J-GROOVE WELDED BOTH SIDES
T = 18 MIN
SEE NOTE 2
3
SINGLE U-GROOVE WELDED BOTH SIDES
T = UNLIMITED
60°
0 TO 3
DOUBLE V-GROOVE WELDED BOTH SIDES
T = UNLIMITED
45°
6 x 3 SPACER
6
0 TO 3
DOUBLE BEVEL GROOVE WELDED BOTH SIDES

NOTE 1 – 45° ALL POSITIONS, 30° FLAT AND OVERHEAD ONLY
NOTE 2 – 45° ALL POSITIONS, 20° FLAT AND OVERHEAD ONLY

Fig. 11-4-3 Spacing and material thickness for common butt joints.

11. The size of flare-groove welds is considered as extending only to the tangent points. The extension beyond the point of tangency is treated as an edge or lap joint. See Fig. 11-4-4.

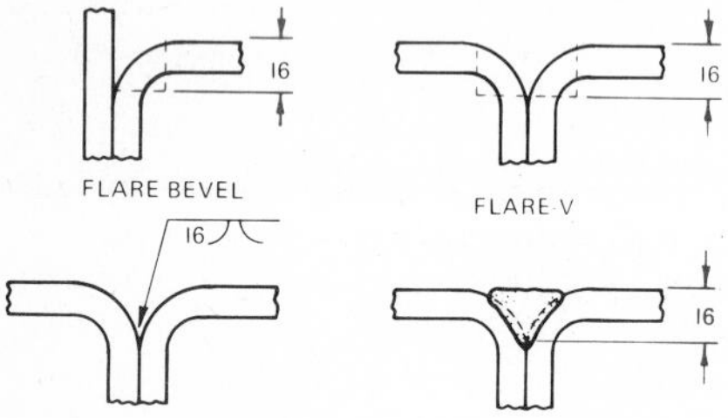
16
FLARE BEVEL
16
FLARE V
16
16

12. Root opening of groove welds is the user's standard unless otherwise indicated. Root opening of groove welds, when not the user's standard, is shown inside the weld symbol.

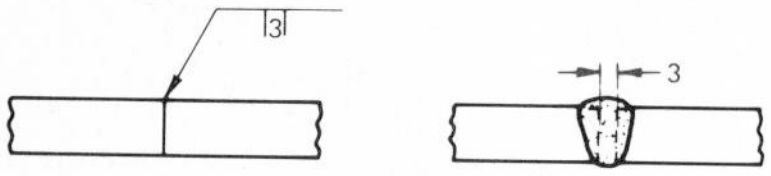
3
3

13. Groove angle of groove welds is the user's standard, unless otherwise indicated. Groove angle of groove welds, when not the user's standard, is shown.

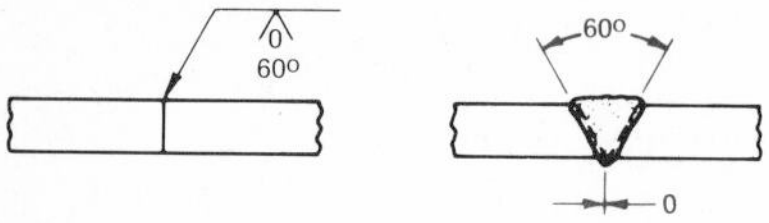
0
60°
60°
0

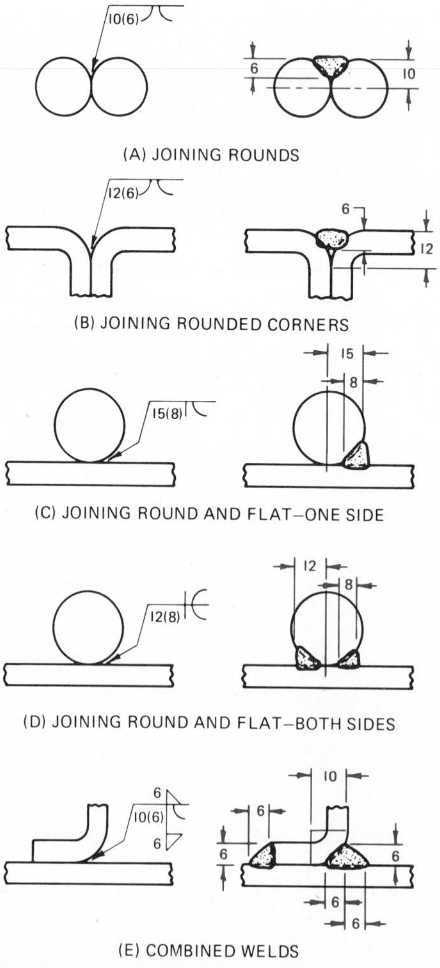
10(6)
6
10
(A) JOINING ROUNDS
12(6)
6
12
(B) JOINING ROUNDED CORNERS
15(8)
15
8
(C) JOINING ROUND AND FLAT—ONE SIDE
12(8)
12
8
(D) JOINING ROUND AND FLAT—BOTH SIDES
6
10(6)
6
10
6
6
6
6
6
(E) COMBINED WELDS

Fig. 11-4-4 Application of flare-V and flare-bevel welds.

14. Groove radii and root faces of U- and J-groove welds are the user's standard, unless otherwise indicated. When groove radii and root faces of U- and J-groove welds are not the user's standard, the weld is shown by a cross section, detail, or other data, with a reference thereto on the welding symbol, observing the usual locational significance.

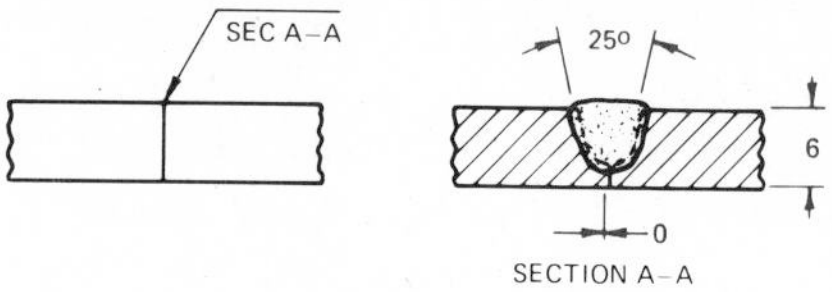
SEC A–A
25°
6
0
SECTION A–A

15. Bead-type back and backing welds of single-groove welds are shown by means of the back or backing weld symbol.

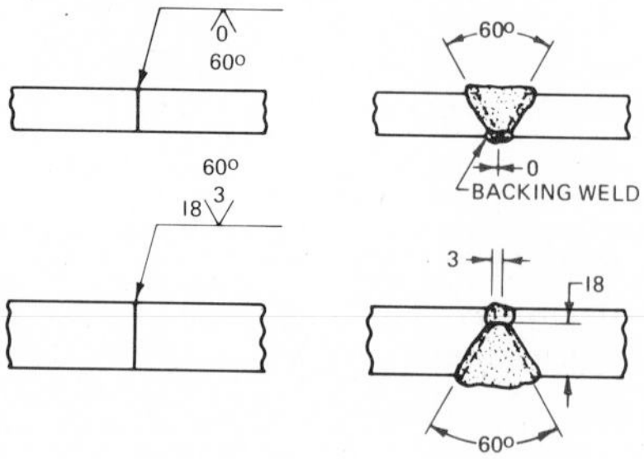
0
60°
60°
0
BACKING WELD
60°
3
18
3
18
60°

16. Groove welds that are welded approximately flush without recourse to any method of finishing are shown by adding the flush-contour symbol to the weld symbol, observing the usual locational significance.

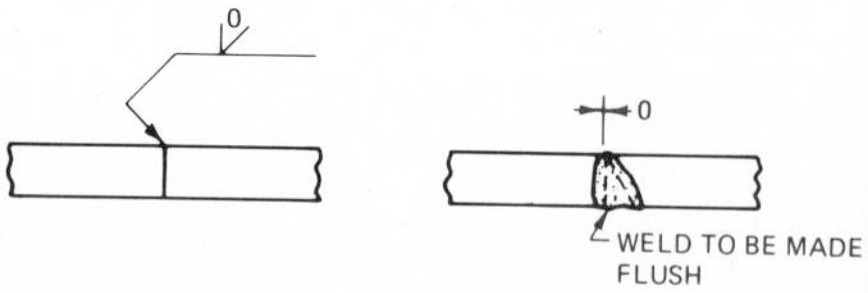
0
0
WELD TO BE MADE FLUSH

17. Groove welds that are to be made flush by mechanical means are shown by adding both the flush-contour symbol and the user's standard finish symbol to the weld symbol, observing the usual locational significance. C = chipping, G = grinding, M = machining, R = rolling, H = hammering.

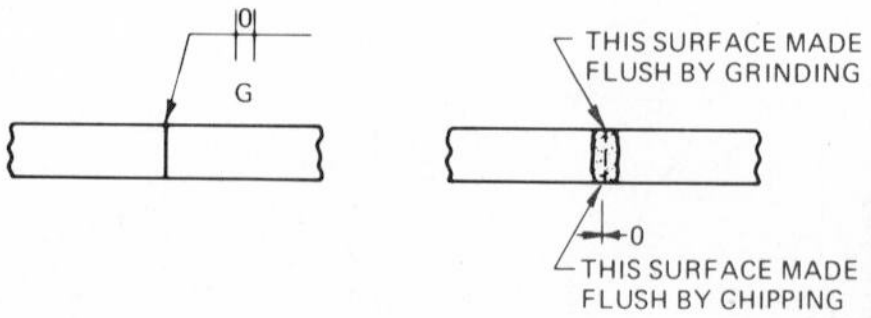
C
0
G
THIS SURFACE MADE FLUSH BY GRINDING
0
THIS SURFACE MADE FLUSH BY CHIPPING

Groove Joint Design[2]

Figure 11-4-5 indicates that the root opening (R) is the separation between the members to be formed. A root opening is used for electrode accessibility to the base or root of the joint. The smaller the angle of bevel, the larger the root opening must be to get good fusion at the root. If the root opening is too small, root fusion is more difficult to obtain, and smaller electrodes must be used, thus slowing down the welding process.

If the root opening is too large, weld quality does not suffer but more weld metal is required. This increases welding cost and will tend to increase distortion.

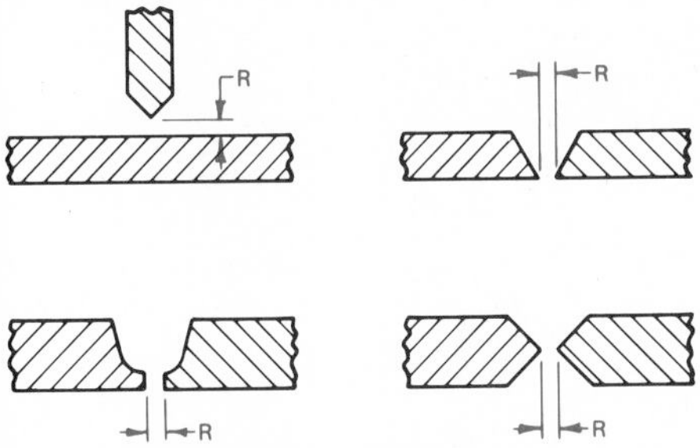
R
R
R
R

Fig. 11-4-5 Root openings.

Figure 11-4-6 indicates how the root opening must be increased as the included angle of the bevel is decreased. Backup strips are used on larger root openings. All three preparations are acceptable; all are conducive to good welding procedure and good weld quality. Selection, therefore, is usually based on cost.

Root opening and joint preparation will directly affect weld cost, and selection should be made with this in mind. Joint preparation involves the work required on plate edges prior to welding and includes beveling and providing a root face.

Using a double-groove joint in preference to a single-groove, Fig. 11-4-7 halves the amount of welding. This reduces distortion and makes possible alternating the weld passes on each side of the joint, again reducing distortion.

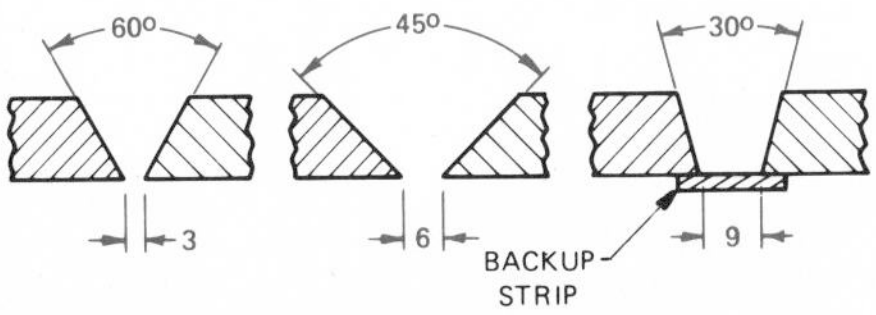
60°
45°
30°
3
6
9
BACKUP STRIP

Fig. 11-4-6 Root opening increases as the angle decreases.

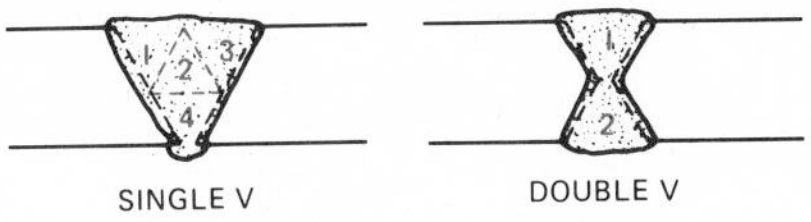
SINGLE V
DOUBLE V

Fig. 11-4-7 Single-V weld uses twice as much weld material as double-V weld.

The main purpose of a root face is to provide an additional thickness of metal, as opposed to a featheredge, in order to minimize any burn-through tendency.

Plate edges are beveled to permit accessibility to all parts of the joint and ensure good fusion. Accessibility can be gained by compromising between maximum bevel and minimum root opening (Fig. 11-4-8). The degree of bevel may be dictated by the importance of maintaining proper electrode angle in confined quarters (Fig. 11-4-9). For the joint illustrated, the minimum recommended bevel is 45°.

J and U preparations are excellent to work with, but economically they may have little to offer because preparation requires machining as opposed to simple torch cutting. Also J and U grooves require a root face and thus back-gouging.

REFERENCES AND SOURCE MATERIAL

1. American Welding Society and the Canadian Welding Bureau.
2. Lincoln Electric Company.

Assignment

On an A3- or B-size sheet, select one of the problems shown in Figs. 11-4-A and 11-4-B and make a working drawing complete with dimensions and welding symbols. Include on the drawing a bill of material and identify each part on the assembly. Use full-strength welds. Scale is as shown.

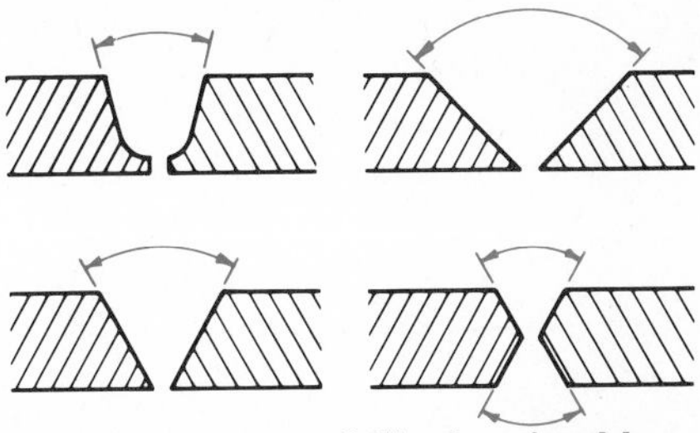

Fig. 11-4-8 Accessibility is gained by compromising between bevel and root opening.

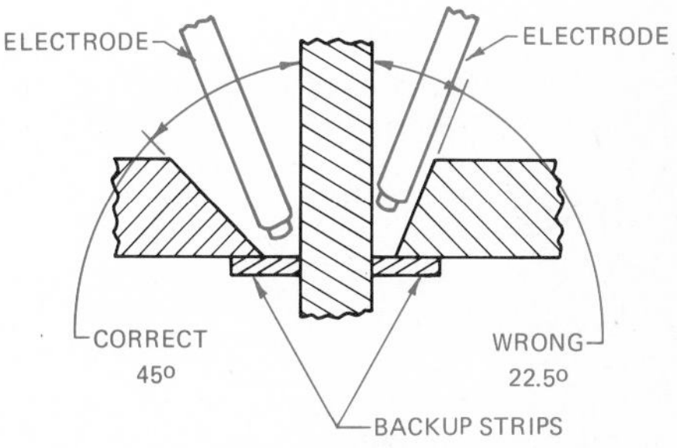
ELECTRODE
ELECTRODE
CORRECT
45°
WRONG
22.5°
BACKUP STRIPS

Fig. 11-4-9 Degree of bevel may be dictated by the need to obtain proper electrode angle.

REVIEW FOR ASSIGNMENT

Unit 6-4	Bill of Material
Unit 6-6	Detailed Assembly Drawings
Unit 11-2	Welding Symbols
Unit 28-1	Structural Steel Shapes and Sizes

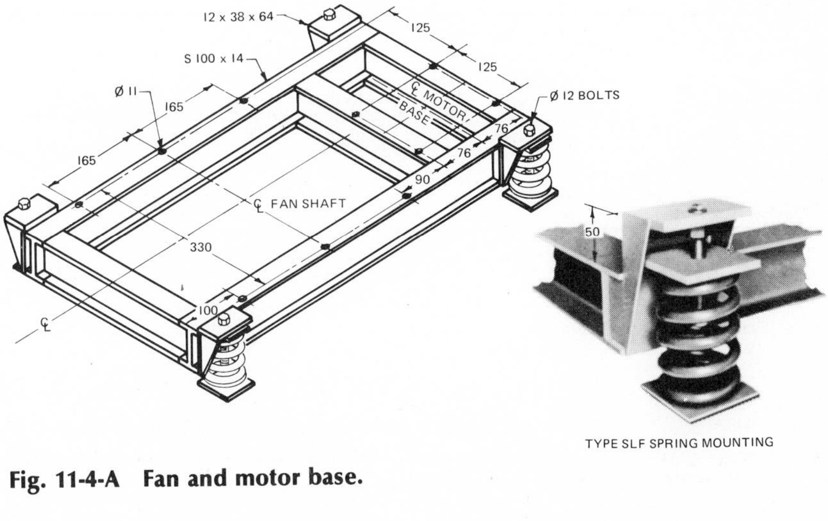
12 x 38 x 64
S 100 x 14
Ø 11
165
165
125
125
Ø 12 BOLTS
℄ MOTOR BASE
76
76
90
℄ FAN SHAFT
330
100
℄
50
TYPE SLF SPRING MOUNTING

Fig. 11-4-A Fan and motor base.

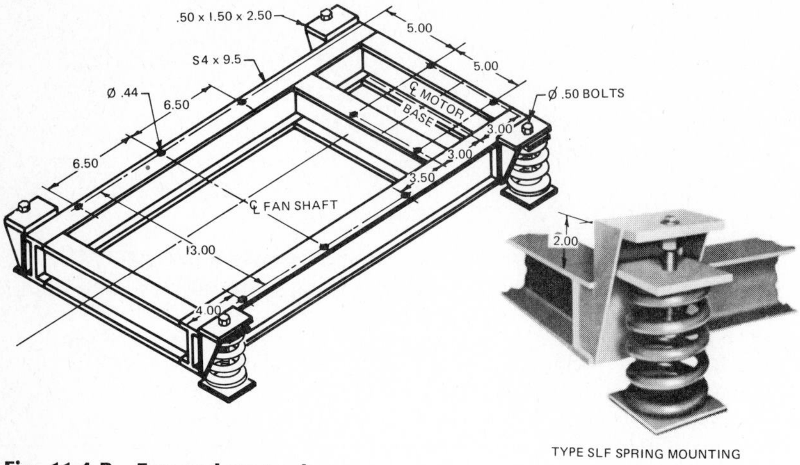
.50 x 1.50 x 2.50
S4 x 9.5
Ø .44
6.50
6.50
5.00
5.00
MOTOR BASE
Ø .50 BOLTS
3.00
3.00
3.50
FAN SHAFT
13.00
4.00
2.00
TYPE SLF SPRING MOUNTING

Fig. 11-4-B Fan and motor base.

UNIT 11-5

OTHER BASIC WELDS[1]

In order for engineers to keep abreast of national and international thinking and to reduce the complexity inherent in providing symbols for a variety of ways of making the same type of weld, new symbols for spot and seam welds have been established (Fig. 11-5-2). As a result, the old symbols shown in Fig. 11-5-1 should no longer be used. They have been included in this text in order to acquaint the drafter with welding symbols found on existing drawings.

ARC-SEAM OR ARC-SPOT	RESISTANCE SPOT	PROJECTION	RESISTANCE SEAM	FLASH OR UPSET

USE PREFERRED SYMBOL WITH PROCESS REFERENCE IN THE TAIL

Fig. 11-5-1 Former welding symbols.

Plug Welds

1. Holes in the arrow-side member of a joint for plug welding are indicated by placing the weld symbol on the side of the reference line toward the reader.

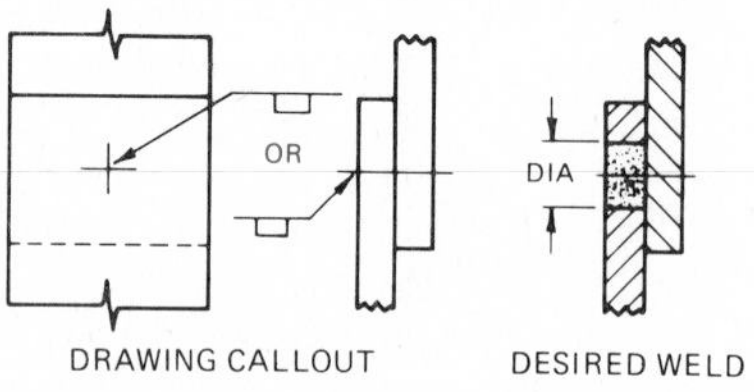
OR
DIA
DRAWING CALLOUT
DESIRED WELD

2. Holes in the other-side member of a joint for plug welding are indicated by placing the weld symbol on the side of the reference line away from the reader.

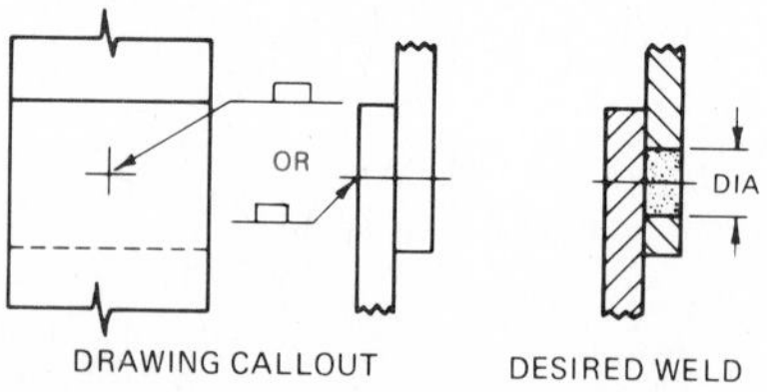
OR
DIA
DRAWING CALLOUT
DESIRED WELD

3. The size of a plug weld is shown on the same side and to the left of the weld symbol.

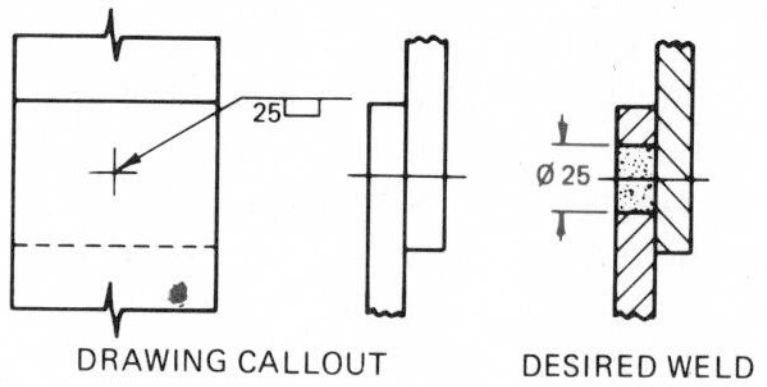
25
Ø 25
DRAWING CALLOUT
DESIRED WELD

4. The included angle of countersink of plug welds is the user's standard, unless otherwise indicated. Included angle, when not the user's standard, is shown.

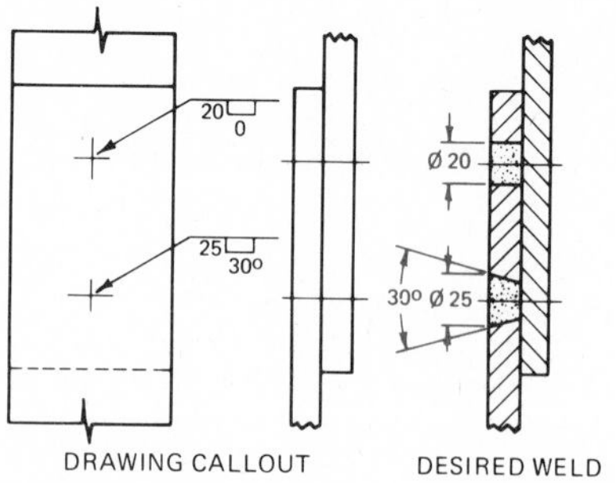
20
0
25
30°
Ø 20
30° Ø 25
DRAWING CALLOUT
DESIRED WELD

5. The depth of filling of plug welds is complete unless otherwise indicated. When the depth of filling is less than complete, the depth of filling in millimetres is shown inside the weld symbol.

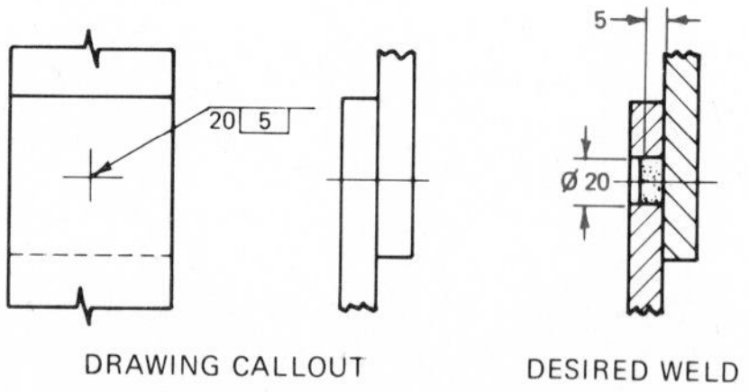
20 5
5
Ø 20
DRAWING CALLOUT
DESIRED WELD

6. Pitch (center-to-center spacing) of plug welds is shown to the right of the weld symbol.

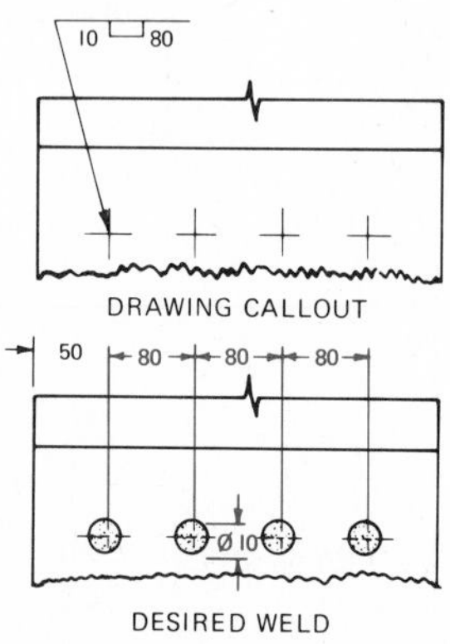
10 80
DRAWING CALLOUT
50
80
80
80
Ø 10
DESIRED WELD

LOCATION SIGNIFICANCE	PLUG OR SLOT	SPOT OR PROJECTION	SEAM	BACK OR BACKING	SURFACING	FLANGE EDGE	FLANGE CORNER	MELT THRU
ARROW SIDE				GROOVE WELD SYMBOL				
OTHER SIDE				GROOVE WELD SYMBOL	NOT USED			
BOTH SIDES	NOT USED	NOT USED	NOT USED	NOT USED	NOT USED	NOT USED	NOT USED	NOT USED
NO ARROW SIDE OR OTHER SIDE SIGNIFICANCE	NOT USED			NOT USED	NOT USED	NOT USED	NOT USED	NOT USED

Fig. 11-5-2 Other basic welding symbols and their location significance.

7. Plug welds that are to be welded pproximately flush without recourse to ny method of finishing are shown by dding the flush-contour symbol to the veld symbol.

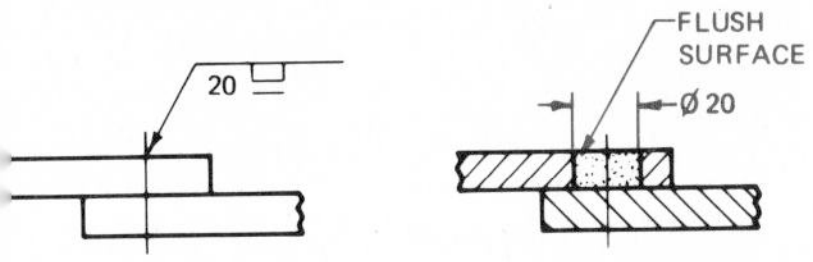
20
FLUSH SURFACE
Ø 20

8. Plug welds that are to be made lush by mechanical means are shown by dding both the flush-contour symbol nd the user's standard finish symbol to he weld symbol.

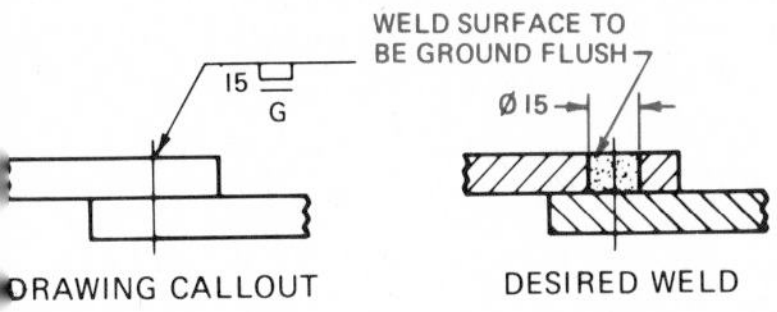
15
G
WELD SURFACE TO BE GROUND FLUSH
Ø 15
RAWING CALLOUT
DESIRED WELD

Slot Welds

1. Slots in the arrow-side member of a joint for slot welding are indicated by placing the weld symbol on the side of the reference line toward the reader. Slot orientation must be shown on the drawing.

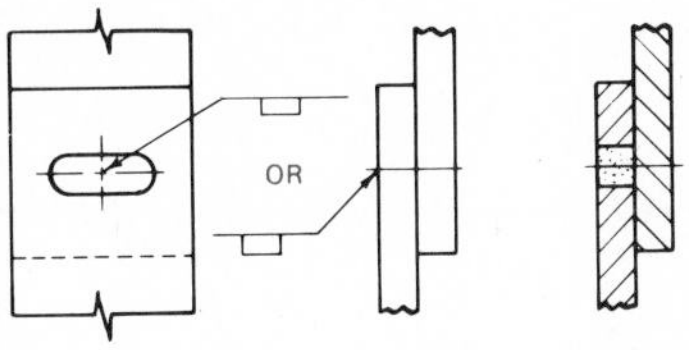
OR

2. Slots in the other-side member of a joint for slot welding are indicated by placing the weld symbol on the side of the reference line away from the reader.

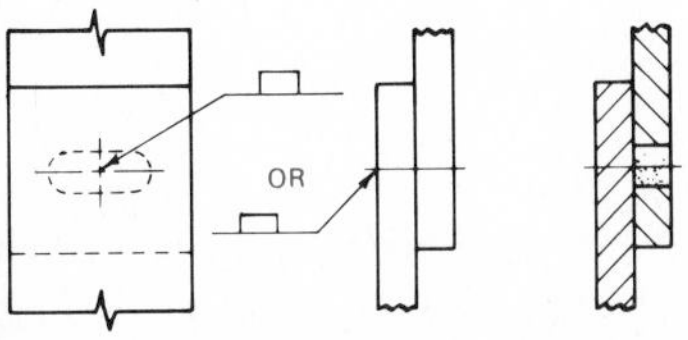
OR

3. Depth of filling of slot welds is complete unless otherwise indicated. When the depth of filling is less than complete, the depth of filling, in millimetres, is shown inside the welding symbol.

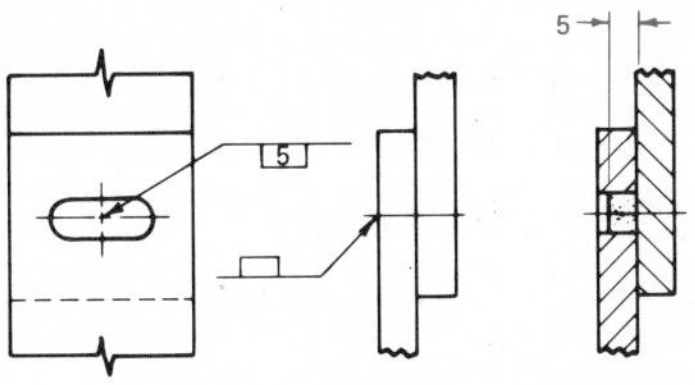
5
5

4. Length, width, spacing, included angle of countersink, orientation, and location of slot welds should be shown on the drawing or by a detail with reference to it on the welding symbol, observing the usual locational significance.

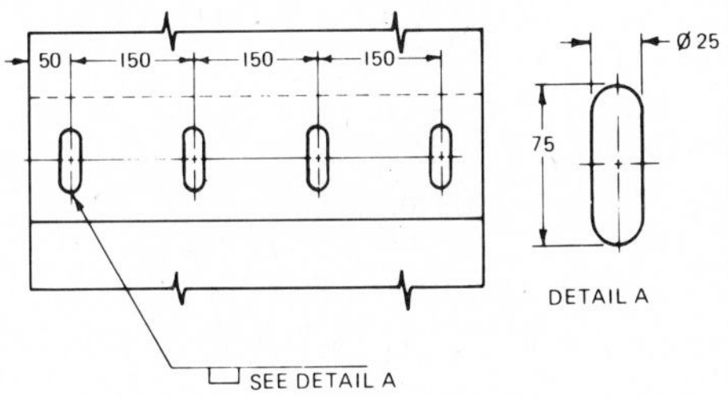
50
150
150
150
Ø 25
75
DETAIL A
SEE DETAIL A

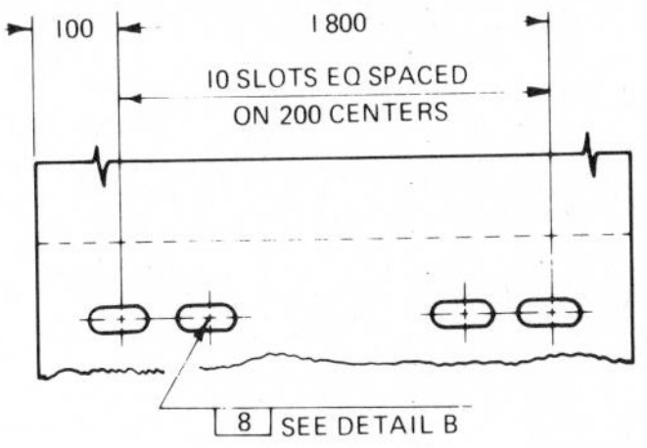
100
1 800
10 SLOTS EQ SPACED
ON 200 CENTERS
8
SEE DETAIL B

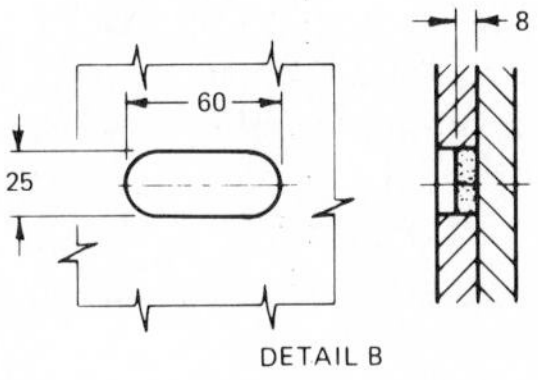
8
60
25
DETAIL B

5. Slot welds that are to be welded approximately flush without recourse to any method of finishing are shown by adding the flush-contour symbol to the welding symbol.

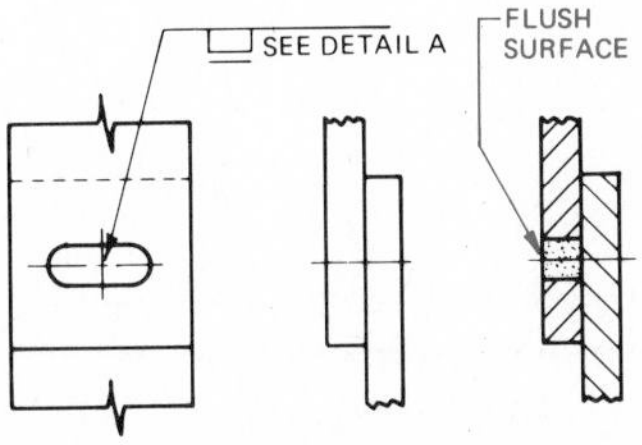
SEE DETAIL A
FLUSH SURFACE

6. Slot welds that are to be made flush by mechanical means are shown by adding both the flush-contour symbol and the user's standard finish symbol to the weld symbol.

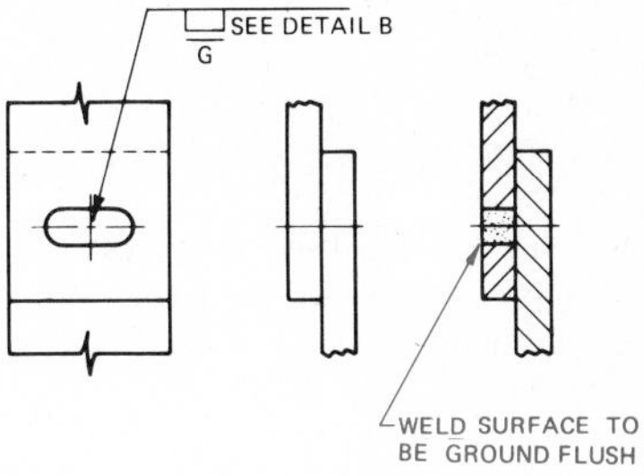
SEE DETAIL B
G
WELD SURFACE TO BE GROUND FLUSH

Spot or Projection Welds

The symbol for all spot or projection welds is a circle, regardless of the welding process used. There is no attempt to provide symbols for different ways of making a spot weld, such as resistance, arc, and electron beam welding.

The symbol for a spot weld is a circle placed:

1. Below the reference line, indicating arrow side
2. Above the reference line, indicating other side
3. On the reference line, indicating that there is no arrow or other side

SPOT- OR PROJECTION-WELD APPLICATION

1. Dimensions of spot welds are shown on the same side of the reference line as the weld symbol, or on either side when the symbol is located astride the reference line and has no arrow-side or other-side significance. They are dimensioned by either the size or the strength. The size is designated as the diameter of the weld and is shown to the left of the weld symbol. The strength of the spot weld is designated in newtons (or pounds) per spot and is shown to the left of the weld symbol.

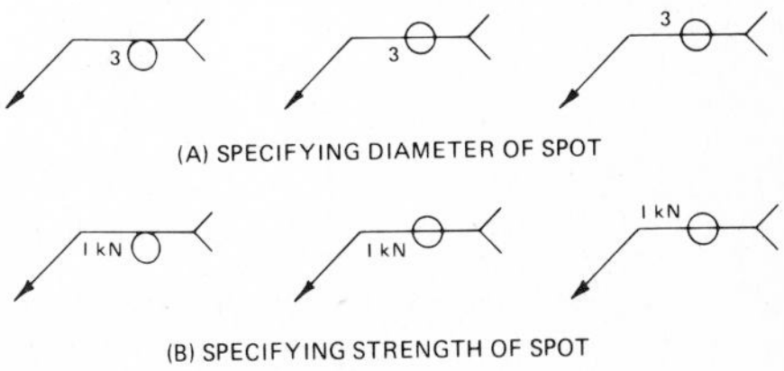
3
3
3
(A) SPECIFYING DIAMETER OF SPOT
1 kN
1 kN
1 kN
(B) SPECIFYING STRENGTH OF SPOT

2. The pitch (center-to-center spacing) is shown to the right of the weld symbol.

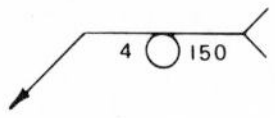

4
150

3. When spot welding extends less than the distance between abrupt changes in the direction of the welding or less than the full length of the joint, the extent is dimensioned.

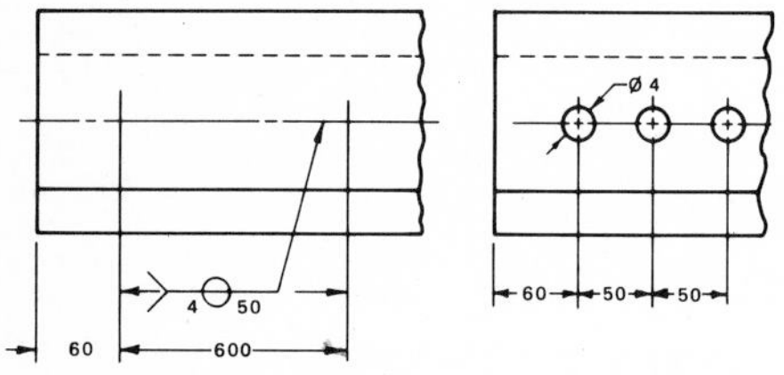

Ø 4
4
50
60
600
60
50
50

4. When projection welding is used, the projection-welding process shall be referenced in the tail of the welding symbol. The spot-weld symbol is placed either above or below (not on) the reference line to designate in which member the embossment is placed.

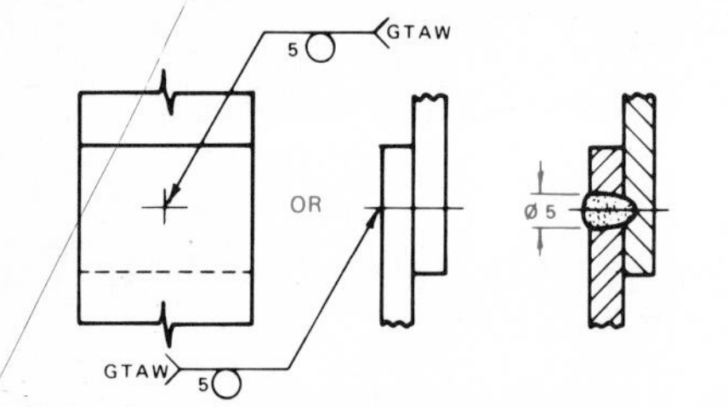

5
GTAW
OR
Ø 5
GTAW
5

ARROW SIDE SPOT WELD SYMBOL (GAS TUNGSTEN-ARC SPOT)

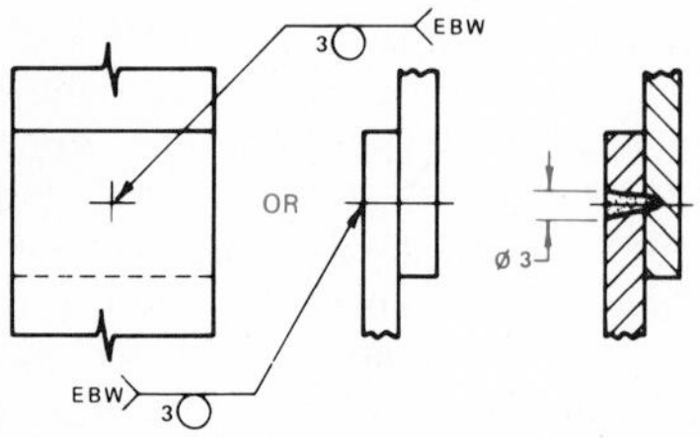

3
EBW
OR
Ø 3
EBW
3

ARROW SIDE SPOT WELD SYMBOL (ELECTRON BEAM SPOT)

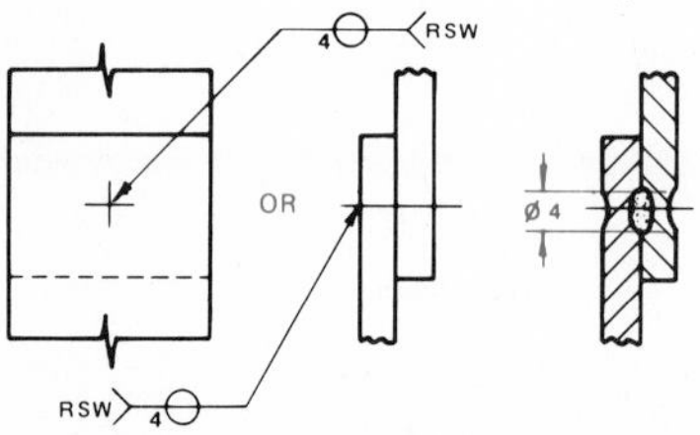

4
RSW
OR
Ø 4
RSW
4

NO ARROW OR OTHER SIDE REFERENCE OR SIGNIFICANCE (RESISTANCE SPOT)

5. When the exposed surface of one member of a spot-welded joint is to be flush, that surface is indicated by adding the flush-contour symbol to the weld symbol, observing the usual locational significance.

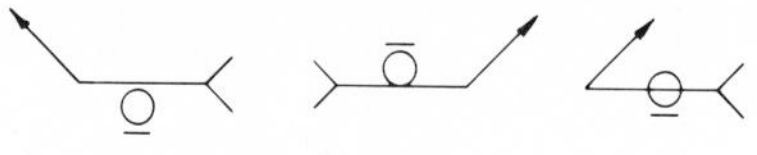

6. When a definite number of spot welds is desired in a certain joint, the number is shown in parentheses either above or below the weld symbol.

1.5 kN
(6)
(6)
1.5 kN

Seam Welds

The symbol for all seam welds is a circle transversed by two horizontal parallel lines. This symbol is used for all seam welds regardless of the way they are made. The seam-weld symbol is placed (1) below the reference line to indicate arrow side, (2) above the reference line to indicate other side, and (3) on the reference line to indicate that there is no arrow or other side.

SEAM-WELD APPLICATION

1. Dimensions of seam welds are shown on the same side of the reference line as the weld symbol. They are dimensioned by either size or strength. The size of seam welds is designated as the width of the weld and is shown to the left of the weld symbol. The strength of seam welds is designated in newtons per millimetre (N/mm) and is shown to the left of the weld symbols.

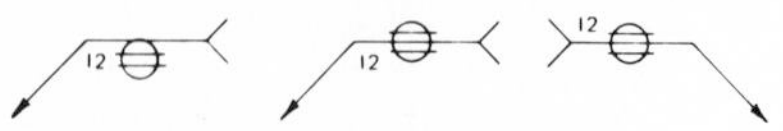

12
12
12

(A) SPECIFYING WIDTH OF THE WELD

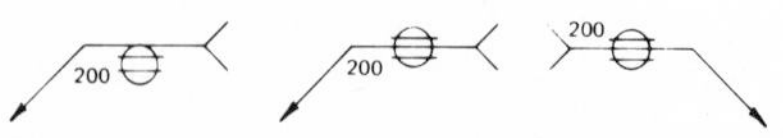

200
200
200

(B) SPECIFYING STRENGTH OF WELD IN NEWTONS PER MILLIMETRE

2. The length of a seam weld, when indicated on the welding symbol, is shown to the right of the weld symbol. When seam welding extends for the full distance between abrupt changes in the direction of the welding, no length dimension needs to be shown on the welding symbol. When a seam weld extends less than the full length of the joint, the extent of the weld should be shown.

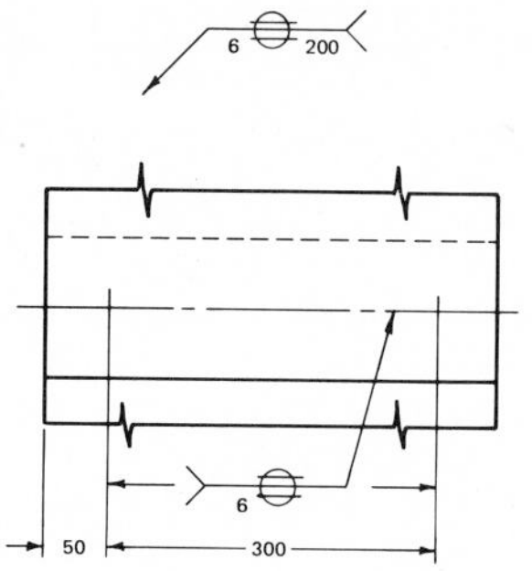

6
200
6
50
300

3. The pitch of intermittent seam welds is shown as the distance between centers of the weld increments. The pitch is shown to the right of the length dimension. Unless otherwise indicated, intermittent seam welds are interpreted as having the length and pitch measured parallel to the axis of the weld.

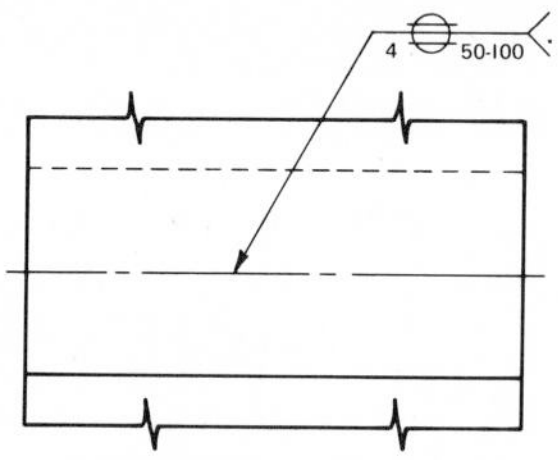

4
50-100

DRAWING CALLOUT

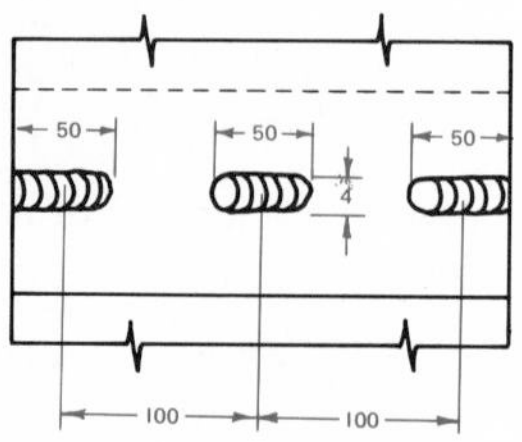

50
50
50
4
100
100

INTERPRETATION

4. When the exposed surface of one member of a seam-welded joint is to be flush, that surface is indicated by adding the flush-contour symbol to the weld symbol, observing the usual locational significance.

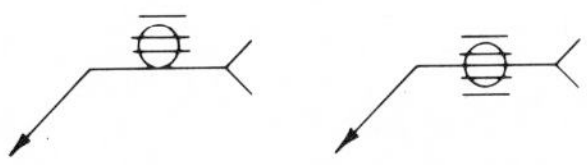

5. When the pitch or length of seam is not parallel to the axis of the weld, then it must be shown on the drawing.

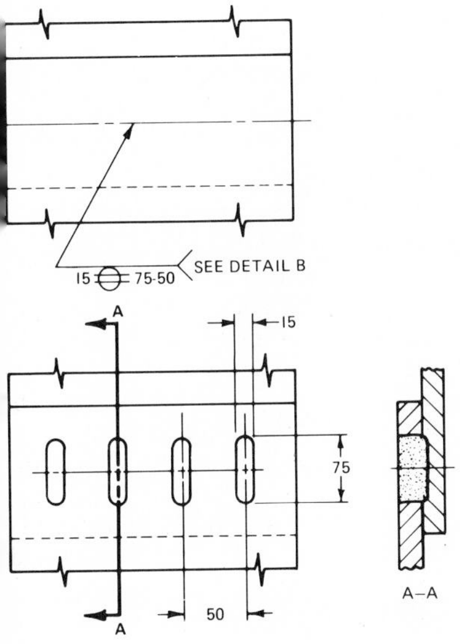
15
75-50
SEE DETAIL B
A
15
75
50
A
A–A

DETAIL B

6. The seam-weld process is referenced in the tail of the welding symbol.

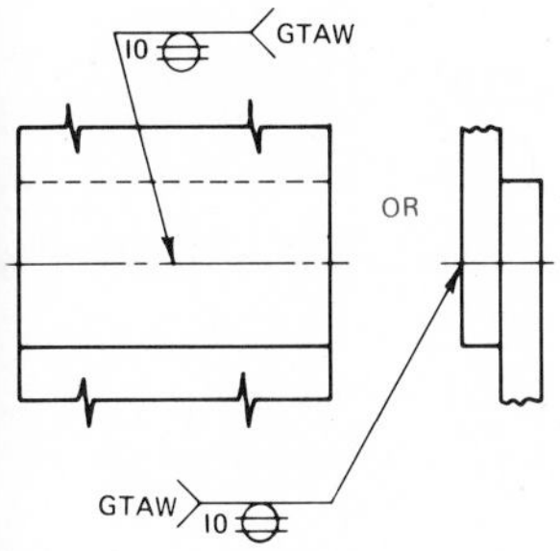
10
GTAW
OR
GTAW
10

DRAWING CALLOUT

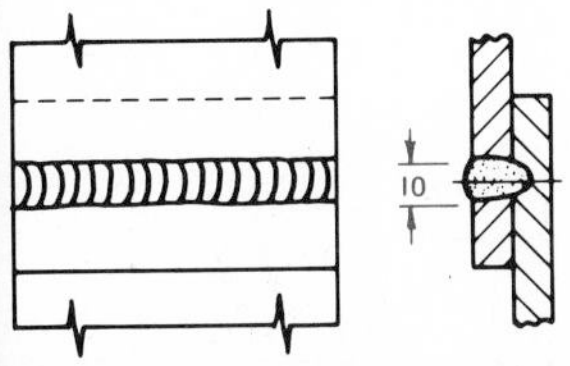
10

INTERPRETATION

Back or Backing Welds

The back or backing weld symbol is used to indicate bead-type back or backing welds of single-groove welds.

1. Back or backing welds of single-groove welds are shown by placing a back or backing weld symbol on the side of the reference line opposite the groove-weld symbol. Dimensions of back or backing welds are not shown on the welding symbol. If it is desired to specify these dimensions, they are shown on the drawing.

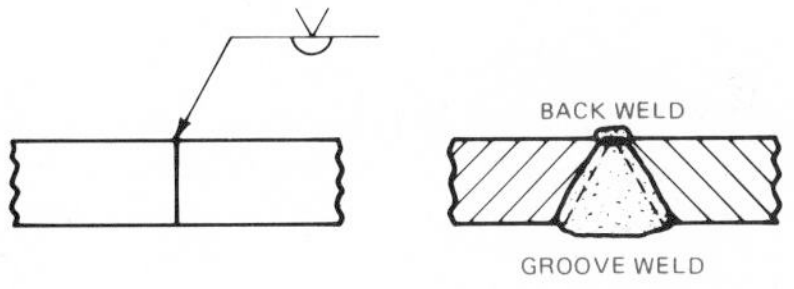
BACK WELD
GROOVE WELD

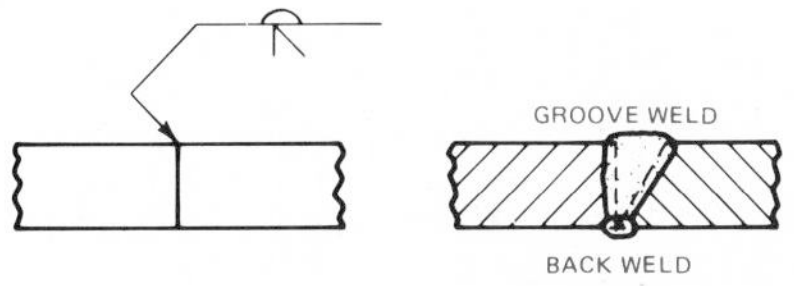
GROOVE WELD
BACK WELD

2. Back or backing welds that are to be welded approximately flush, without recourse to any method of finishing, are shown by adding the flush-contour symbol to the back or backing weld symbol.

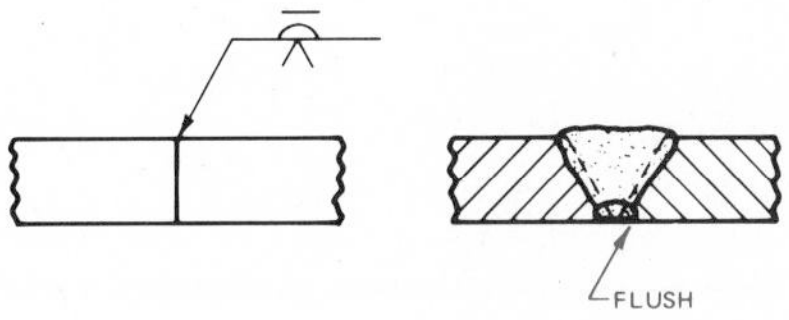
FLUSH

3. Back or backing welds that are made flush by mechanical means are shown by adding both the flush-contour symbol and the user's standard finish symbol to the back or backing weld symbol.

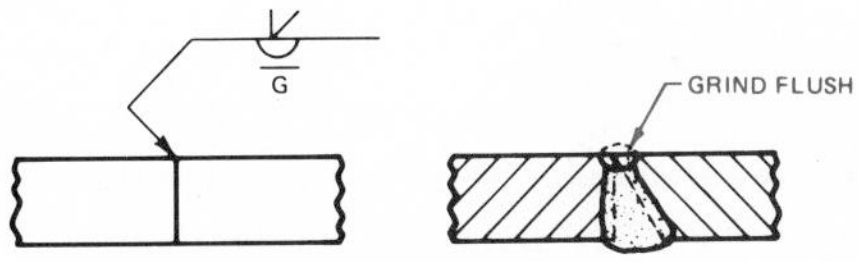
G
GRIND FLUSH

Surfacing Welds

1. The surfacing-weld symbol is used to indicate surfaces built up by welding. Surfaces built up by welding, whether by single- or multiple-pass surfacing welds, are shown by the surfacing-weld symbol. The surfacing-weld symbol does not indicate the welding of a joint, and hence has no arrow or other-side significance. This symbol is drawn on the side of the reference line toward the reader, and the arrow must point clearly to the surface on which the weld is to be deposited.

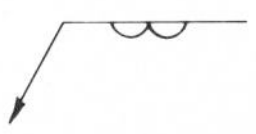

2. Dimensions used in conjunction with the surfacing-weld symbol are shown on the same side of the reference line as the weld symbol. The size or height of the surface built up by welding is indicated by showing the minimum height of the weld deposit to the left of the weld symbol. When no specific height of weld deposit is desired, no size dimension need be shown on the welding symbol. When the entire area of a plane or curved surface is to be built up by welding, no dimension other than size (height of deposit) need be shown on the welding symbol.

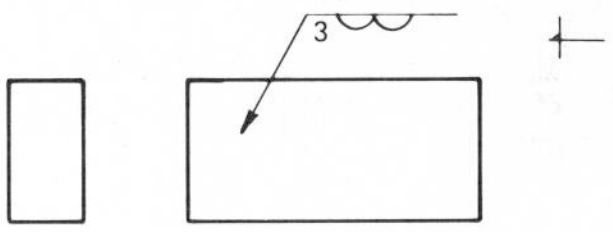
3

DRAWING CALLOUT

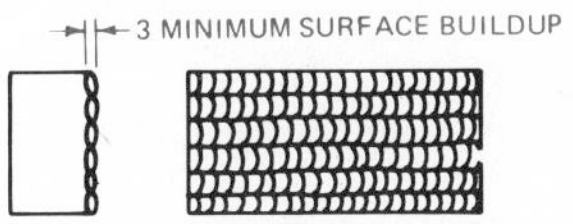
3 MINIMUM SURFACE BUILDUP

INTERPRETATION

SURFACE BUILDUP OF A FLAT SURFACE

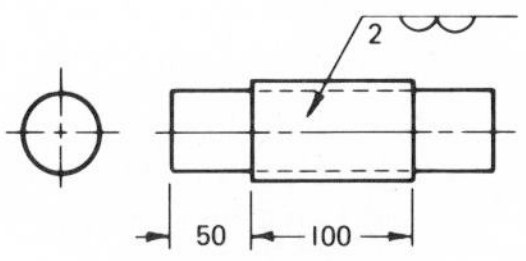
2
50
100

DRAWING CALLOUT

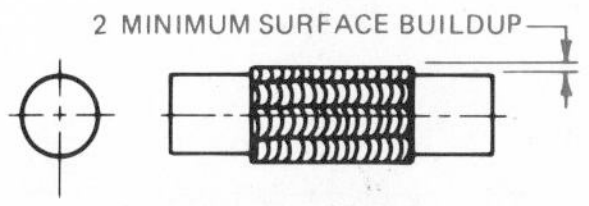
2 MINIMUM SURFACE BUILDUP

INTERPRETATION

Flanged Welds

The following welding symbols are intended to be used for light-gage metal joints involving the flaring or flanging of the edges to be joined.

1. Edge-flange welds are shown by the edge-flange-weld symbol. This symbol has no both-sides significance.

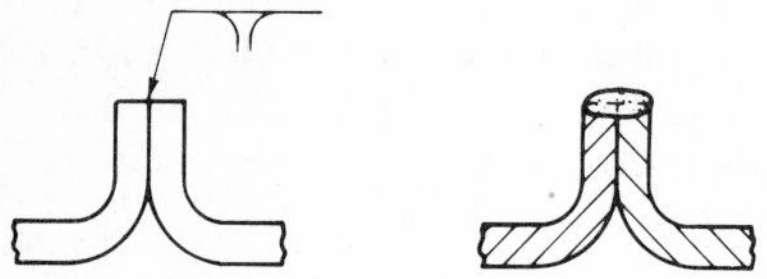

2. Corner-flange welds are shown by the corner-flange-weld symbol. This symbol has no both-sides significance.

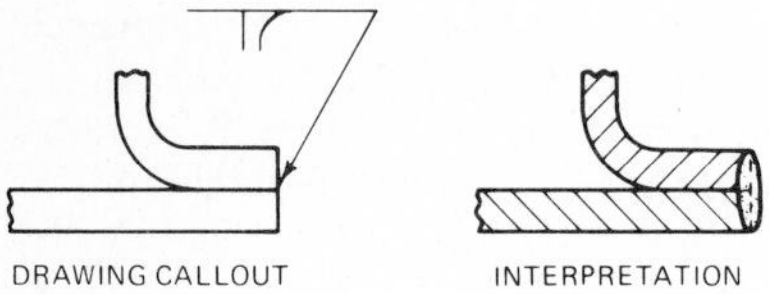
DRAWING CALLOUT
INTERPRETATION

3. Dimensions of flange welds shall be shown on the same side of the reference line as the weld symbol. The radius and the height above the point of tangency are indicated by showing both the radius and the height separated by a plus mark, and placed to the left of the weld symbol. The radius and the height read in that order from left to right along the reference line.

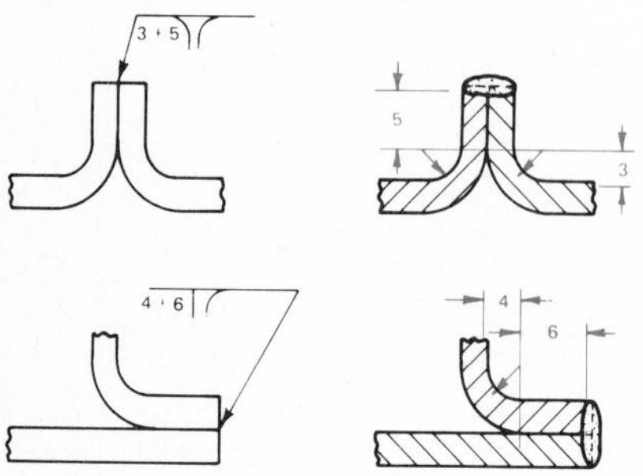
3 + 5
4 + 6

4. The size of flange welds is shown by a dimension placed outward of the flanged dimensions.

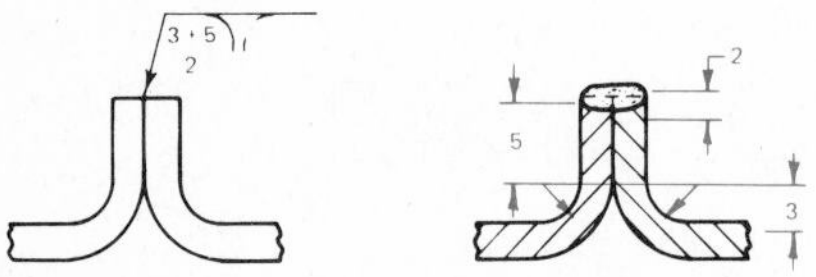

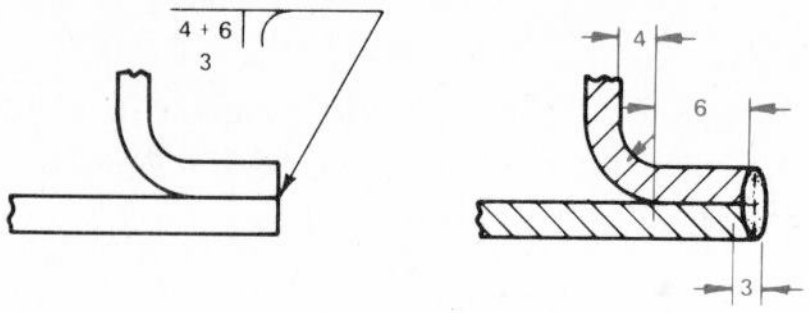

5. Root opening of flange welds is not shown on the welding symbol. If it is desired to specify this dimension, it is shown on the drawing.

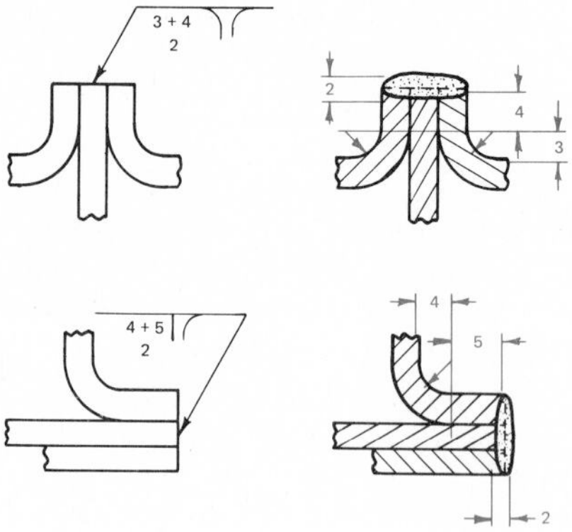

6. For flange welds, when one or more pieces are inserted between the two outer pieces, the same welding symbol as for the two outer pieces is used regardless of the number of pieces inserted.

Melt-Through Welds

1. The melt-through symbol is used where at least 100 percent joint penetration of the weld through the material is required in welds made from one side only. Melt-through welds are shown by placing the melt-through weld symbol on the side of the reference line opposite the groove- or flange-weld symbol. No dimensions of melt-through, except height of reinforcement, are shown on the welding symbol. If it is desired to specify height of reinforcement, it is shown to the left of the melt-through symbol.

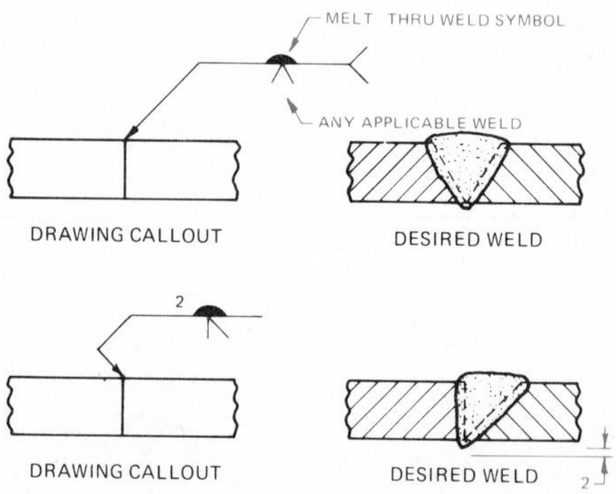
MELT THRU WELD SYMBOL
ANY APPLICABLE WELD
DRAWING CALLOUT
DESIRED WELD
DRAWING CALLOUT
DESIRED WELD

2. Melt-through welds that are to be made flush by mechanical means are shown by adding both the flush-contour symbol and the user's standard finish symbol to the melt-through weld symbol

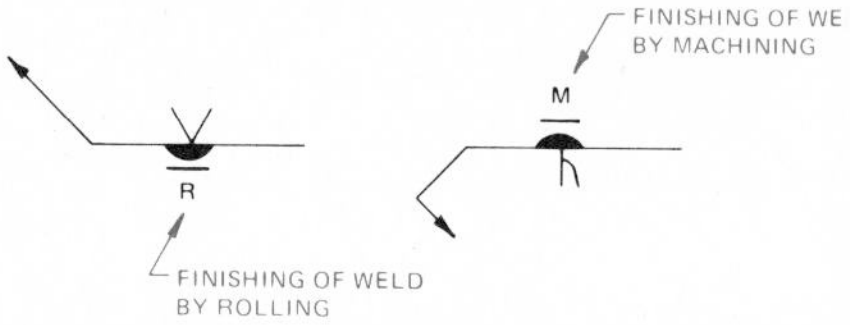
FINISHING OF WE
BY MACHINING
FINISHING OF WELD
BY ROLLING

3. Melt-through welds that are to be mechanically finished to a convex contour are shown by adding the convex-contour symbol to the melt-through weld symbol.

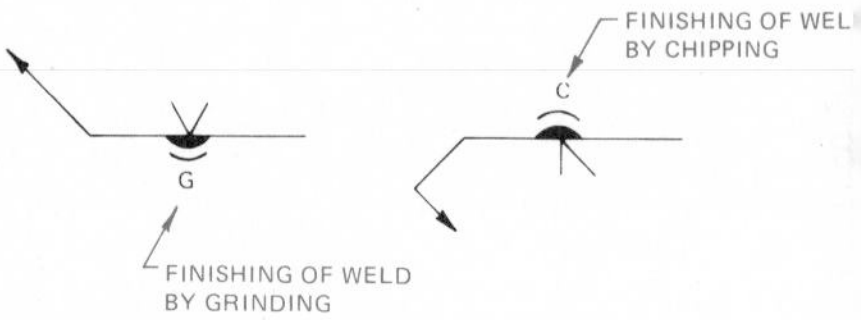
FINISHING OF WEL
BY CHIPPING
FINISHING OF WELD
BY GRINDING

REFERENCE AND SOURCE MATERIAL

1. Canadian Welding Bureau and the American Welding Society.

Assignment

On an A3- or B-size sheet, draw the parts and welds shown in Fig. 11-5-A or 11-5-B and add the weld-size dimensions.

REVIEW FOR ASSIGNMENT

Unit 7-5 Assemblies in Section

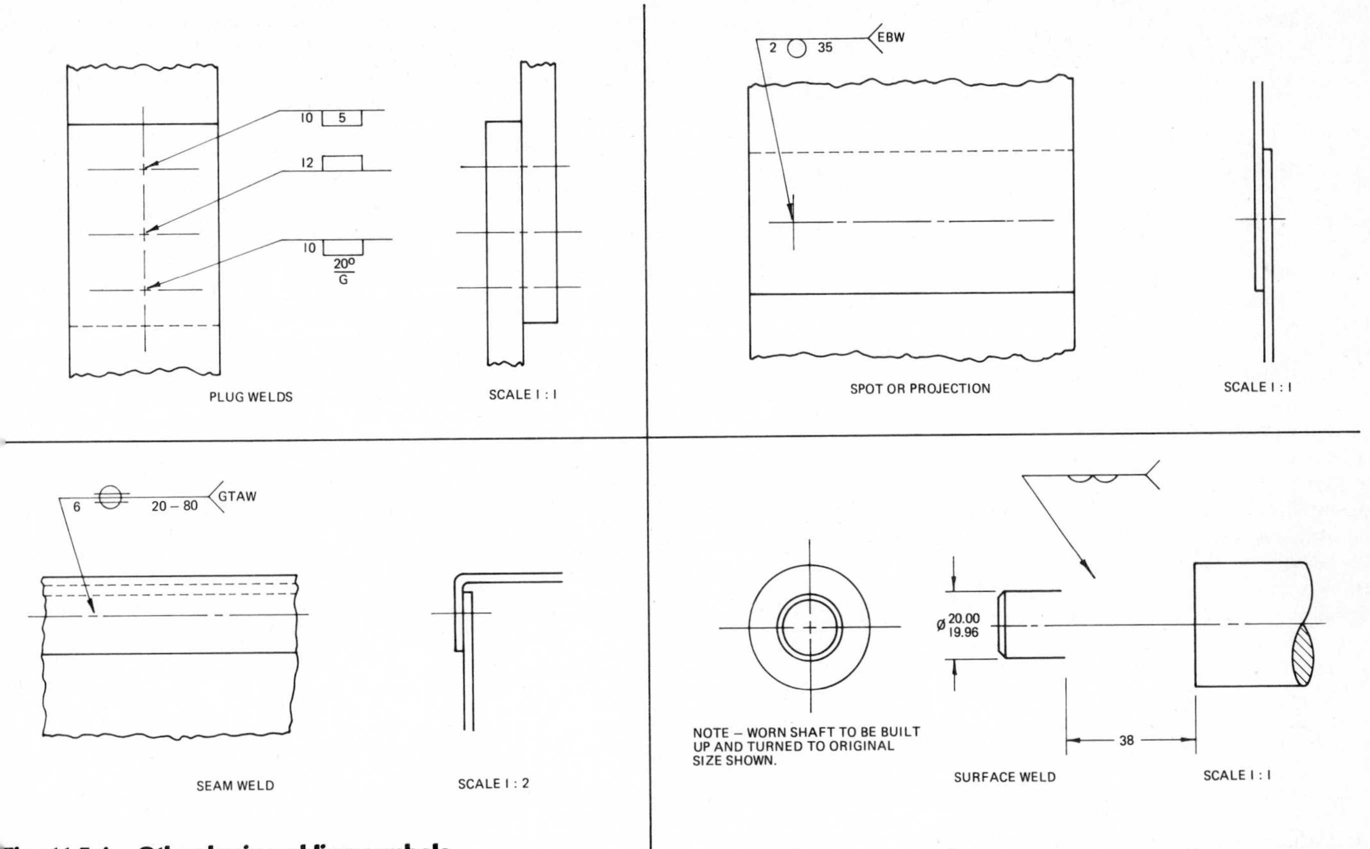
10 5
12
10 20° G
PLUG WELDS
SCALE 1 : 1
2 35 EBW
SPOT OR PROJECTION
SCALE 1 : 1
6 20 – 80 GTAW
SEAM WELD
SCALE 1 : 2
Ø 20.00 19.96
38
NOTE – WORN SHAFT TO BE BUILT UP AND TURNED TO ORIGINAL SIZE SHOWN.
SURFACE WELD
SCALE 1 : 1

Fig. 11-5-A Other basic welding symbols.

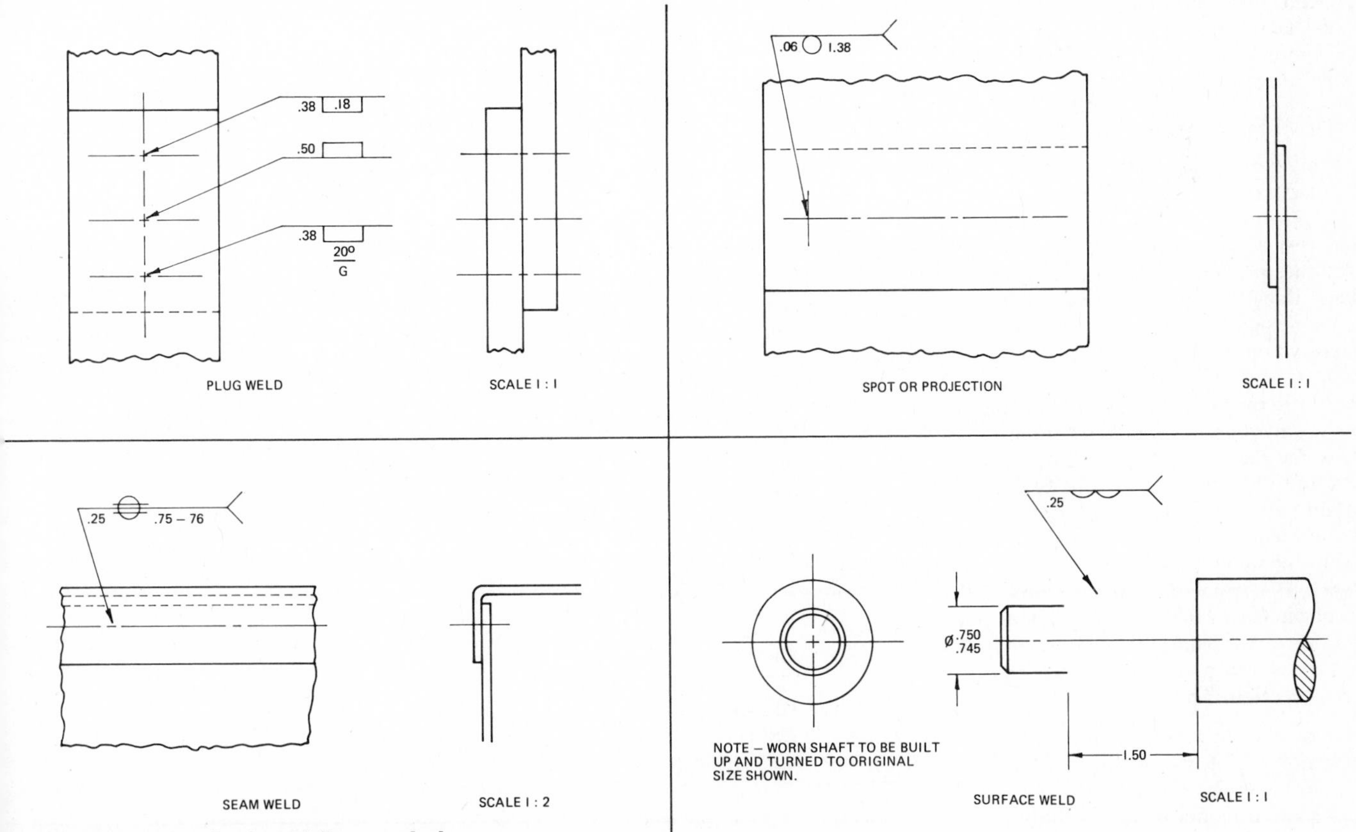
.38 .18
.50
.38 20° G
PLUG WELD
SCALE 1 : 1
.06 1.38
SPOT OR PROJECTION
SCALE 1 : 1
.25 .75 – 76
SEAM WELD
SCALE 1 : 2
.25
Ø .750 .745
1.50
NOTE – WORN SHAFT TO BE BUILT UP AND TURNED TO ORIGINAL SIZE SHOWN.
SURFACE WELD
SCALE 1 : 1

Fig. 11-5-B Other basic welding symbols.

UNIT 11-6

REDESIGNING BY MEANS OF EQUIVALENT SECTIONS[1]

Although it is preferable in most cases to design a machine on the basis of calculated loading, at times the designer wants to convert from a casting or a forging to fabricated steel in the simplest way possible.

The *equivalent sections* concept is aimed at this direct conversion from one material to another. Possibly a single member or assembly will be redesigned for steel and must be functional within an overall machine design still based on cast iron. Or, the decision may be to lean heavily on the plant's casting experience and pretty much duplicate the original machine in steel rather than go into stress analysis. Or, the need may be to convert from one steel design to another in order to take advantage of new manufacturing techniques.

The basic three-step approach to converting a casting into a steel weldment by means of equivalent sections is this:

1. Determine the type of loading under requirements of strength or rigidity of each member.
2. Determine the critical property of this cast member.
3. Determine the required property for the steel member.

Use of Equivalent Tables

The equivalent tables were developed from a simplification of traditional engineering formulas. The tables permit the designer to be concerned only with an appropriate property of the section. The designer does not have to work directly with the design loads, which would be the case when using the traditional formulas.

Step 1: Determine the Type of Loading under the Requirements of Strength or Rigidity for Each Member. All parts of a structure must have basic jobs to do: maintain sufficient strength or rigidity, withstand loads applied in tension, compression, bending, or torsion.

Step 2: Determine the Critical Properties of the Cast Member. The ability of the part to withstand the above loadings is measured by certain properties of its cross section. These are:

A = area of the cross section
I = moment of inertia, for resistance to bending
S = section modulus, for flexural strength
J = polar moment of inertia, for resistance to twisting
J/C = polar section modulus, for strength under torsion

Step 3: Determine the Required Properties for the Steel Member. If these properties of a cast part or member are known, equivalent tables facilitate determining the corresponding properties of the steel member or part that will have equal rigidity or equal strength (Fig. 11-6-1). It is necessary only to multiply the known properties of the casting by the factor obtained from the appropriate equivalent table. For instance, to determine how much area must be provided in a steel tension member to equal the rigidity of a gray-iron casting, ASTM 20, refer to Fig. 11-6-1, which shows that the steel member need have only 40 percent as much area.

Example. To see how the system is applied to an actual problem, consider the gray-iron shaft support in Fig. 11-6-2. The redesign objective is the conversion of this support to lower-cost, welded steel members of equal rigidity.

1. Determine the type of loading under requirements of strength and rigidity of each member. Since the pillow block is used in conjunction with a bronze bushing to support a shaft, we may assume the main consideration would be rigidity. Since the pillow block could be mounted in any position, both tension and compression should be considered.

2. It appears that the base and the ribs could be reduced in thickness. However, certain dimensions must not be altered as the welded steel design may be used as a replacement for a broken cast-iron pillow block. The 44.5 mm height is an example of a dimension that should not be altered.

3. Determine the required thickness for the steel members.

(*a*) Thickness of steel base: a steel base need only be 50 percent as thick as a gray-iron ASTM 30 base. See Fig. 11-6-1. Therefore steel base thickness = 50 percent of 16 = 8.

(*b*) Thickness of ribs: the same reduction is permissible for the ribs. Therefore rib thickness = 50 percent of 12 = 6.

REFERENCE AND SOURCE MATERIAL

1. The James F. Lincoln Arc Welding Foundation.

Assignment

On an A3- or B-size sheet, redesign by means of equivalent sections one of the castings shown in Figs. 11-6-A to 11-6-D to welded steel construction. Completely

		RIGIDITY		STRENGTH			
Step 1. Determining the type of loading		**Tension**	**Short Column Compression**	**Tension**	**Short Column Compression**	**Bending**	**Torsion**
Step 2. Determine this property of the cast member		Area (A)	Area (A)	Area (A)	Area (A)	Section Modulus S	Polar Section Modulus $\frac{J}{C}$
Step 3. Equivalent factors		Multiply the above property of the cast member by the following factor to get the equivalent value for steel.					
Gray Iron	ASTM 20	40%	40%	21%	94%	21%	28%
	ASTM 30	50	50	31	123	31	42
	ASTM 40	63	63	42	136	42	56
	ASTM 50	67	67	52	156	52	70
	ASTM 60	70	70	63	167	63	83
Malleable	A47-33 35018	83	83	68		68	76
	A47-33 32510	83	83	54		54	70
Meehanite	GRADE GE	40	40	31	125	31	42
	GRADE GD	48	48	36	136	36	49
	GRADE GC	57	57	44	164	44	58
	GRADE GB	60	60	49	174	49	64
	GRADE GA	67	67	57	199	57	73
Cast Steel	0.12 to 0.2%	100	100	75	75	75	75

Fig. 11-6-1 Equivalent rigidity and strength factors for steel selection equal to cast iron.

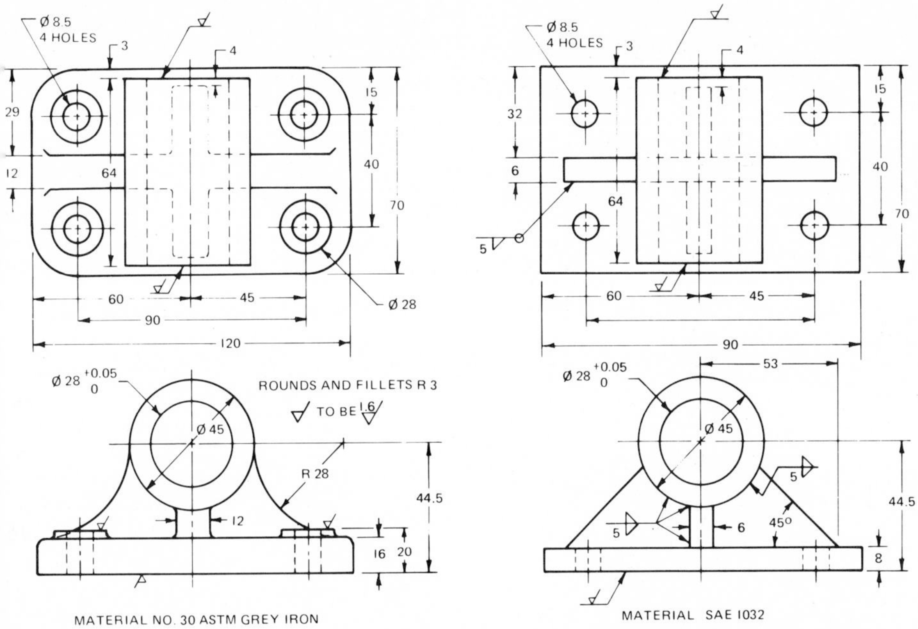
ROUNDS AND FILLETS R 3
MATERIAL NO. 30 ASTM GREY IRON
MATERIAL SAE 1032
Ø 8.5
4 HOLES

Fig. 11-6-2 Comparison of a cast shaft support with a welded-steel shaft support.

dimension and include a bill of material. Select suitable welds and refer to Fig. 11-3-4 for proper weld sizes (use 50 percent of full-strength weld). Scale is 1:2. Material—ASTM 40 gray iron.

REVIEW FOR ASSIGNMENT

Unit 6-6 Detailed Assembly Drawings
Unit 11-1 Design for Welding
Unit 11-2 Welding Symbols
Unit 11-3 Fillet Welds
Unit 11-4 Groove Welds

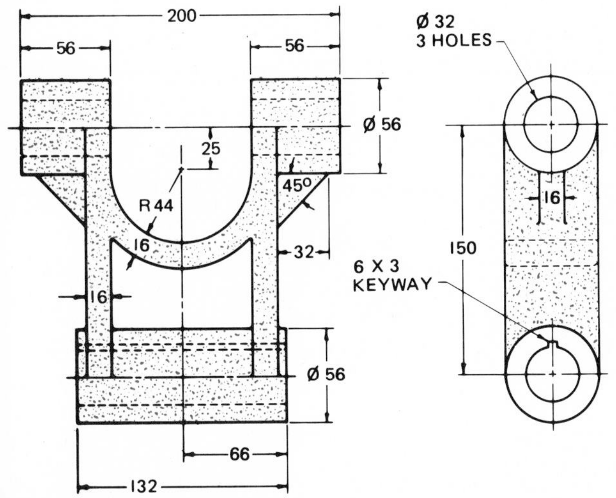
Ø 32
3 HOLES
6 X 3
KEYWAY

Fig. 11-6-A Swing bracket.

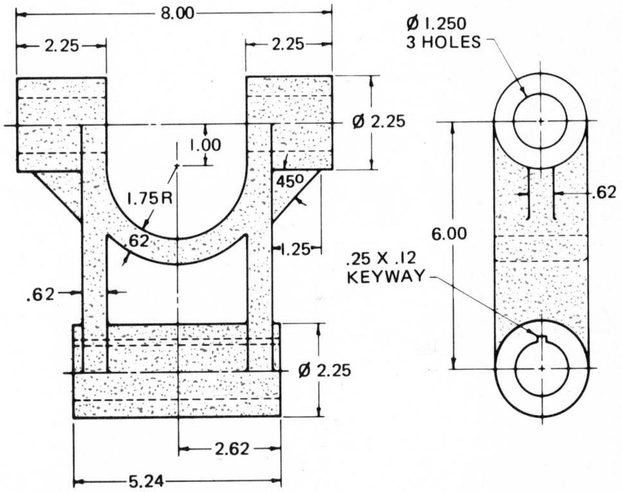
Ø 1.250
3 HOLES
.25 X .12
KEYWAY

Fig. 11-6-B Swing bracket.

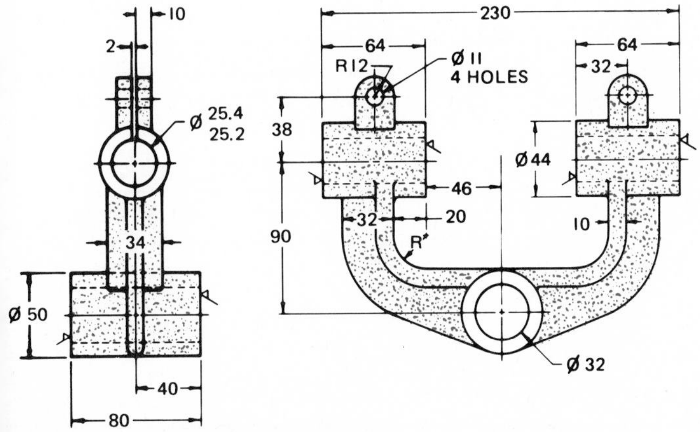
Ø 11
4 HOLES

Fig. 11-6-C Connecting bracket.

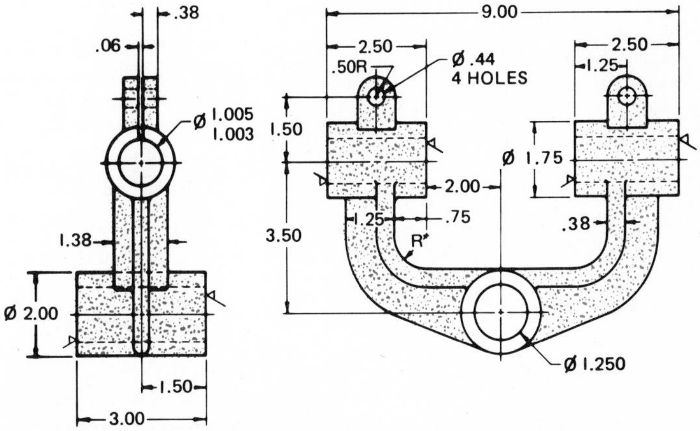
Ø .44
4 HOLES

Fig. 11-6-D Connecting bracket.

Chapter 12
Metals

UNIT 12-1
CAST IRONS

This chapter is a complete up-to-date reference manual on ferrous and nonferrous metals. It provides the engineer or designer with basic reference material for solving the majority of practical, everyday problems encountered in machine design.

Ferrous Metals

Iron and the large family of iron alloys called steel are the most frequently specified metals. Iron is abundant (iron ore constitutes about 5 percent of the earth's crust), easy to convert from ore to a useful form, and iron and steel are sufficiently strong and stable for most engineering applications.

All commercial forms of iron and steel contain carbon, which is an integral part of the metallurgy of iron and steel.

Cast Iron

Because of its low cost, cast iron is often considered a simple metal to produce and to specify. Actually, the metallurgy of cast iron is more complex than that of steel and other familiar design materials. Whereas most other metals are usually specified by a standard chemical analysis, the same analysis of cast iron can produce several entirely different types of iron, depending upon rate of cooling, thickness of the casting, and how long the casting remains in the mold. By controlling these variables, the foundry can produce a variety of irons for heat or wear-resistant uses, or for high-strength components. See Fig. 12-1-1.

The single most important element in cast iron is carbon. It represents approximately 2 to 6 percent of the molten metal when it is poured. Less than 2 percent of the carbon can remain in solution after the iron is set. The excess carbon separates out during the solidification process and forms free graphite, or in white iron, the excess carbon forms iron carbide. The excess carbon is the key to high fluidity, low shrinkage, high damping capacity, and good machinability of the cast irons.

TYPES OF CAST IRON[1,2]

Ductile or Nodular Iron. Ductile or nodular iron is not as available as gray iron, and it is more difficult to control in production.

However, ductile iron can be used where higher ductility or strength is required than is available in gray iron.

Ductile iron is specified by its tensile properties. A three-number symbol is used to designate the various grades.

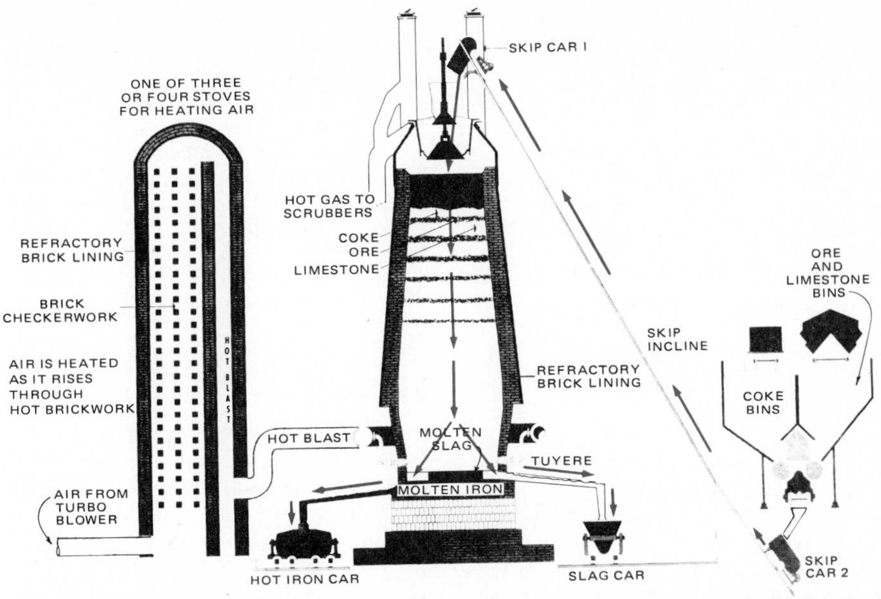
SKIP CAR I
ONE OF THREE
OR FOUR STOVES
FOR HEATING AIR
HOT GAS TO
SCRUBBERS
COKE
ORE
LIMESTONE
REFRACTORY
BRICK LINING
BRICK
CHECKERWORK
AIR IS HEATED
AS IT RISES
THROUGH
HOT BRICKWORK
HOT BLAST
ORE
AND
LIMESTONE
BINS
SKIP
INCLINE
REFRACTORY
BRICK LINING
COKE
BINS
HOT BLAST
MOLTEN
SLAG
TUYERE
MOLTEN IRON
AIR FROM
TURBO
BLOWER
SKIP
CAR 2
HOT IRON CAR
SLAG CAR

Fig. 12-1-1 Schematic diagram of a blast furnace, hot blast stone, and skiploader. (American Iron and Steel Institute.)

These numbers refer to tensile strength, yield strength, and percent elongation. See Fig. 12-1-2. The more common specifications range from the high-ductility grade, 55-40-18, to the 120-90-02 high-strength grade.

The designation of a typical alloy, 60-45-18 for example, specifies a minimum tensile strength of 410 MPa (60 000 psi), a minimum yield strength of 275 MPa (40 000 psi), and 18 percent elongation in 50 mm. Ductile iron, sometimes called nodular iron, is used in applications such as crankshafts because of its good machinability, fatigue strength, and high modulus of elasticity; heavy-duty gears because of its high-yield strength and wear resistance; and automobile door hinges because of its ductility.

Gray Iron. Gray iron is a supersaturated solution of carbon in an iron matrix. The excess carbon precipitates out in the form of graphite flakes. Gray iron is specified by a single-digit designation. Class 20, for example, specifies that the material must have a minimum tensile strength of 140 MPa. In addition, however, because the strength of gray iron is highly sensitive to cross section (the smaller the cross section, the faster the cooling rate, and thus, the higher the strength), gray iron is also specified by the cross section and minimum strength of a special test bar. Typical applications of gray iron include automotive block, flywheels, brake disks and drums, the machine bases, and gears. Generally, gray iron serves well in many machinery applications because of its fatigue resistance.

White Iron. White iron is produced by a process called "chilling" which prevents graphitic carbon from precipitating out. Either gray or ductile iron can be chilled to produce a surface of white iron. In castings that are white iron throughout, however, the composition of iron is selected according to part size to insure that the volume of metal involved can chill rapidly enough to produce white iron.

Because of their extreme hardness, white irons are used primarily for applications requiring wear and abrasion resistance such as mill liners and shot-blasting nozzles. Other uses include railroad brake shoes, rolling-mill rolls, clay-mixing and brick-making equipment, and crushers and pulverizers. Generally, plain (unalloyed) white iron costs less than other cast irons.

The principal disadvantage of white iron is that it is very brittle.

High-Alloy Irons. High-alloy irons are ductile, gray, or white irons that contain over 3 percent alloy content. These irons have properties that are significantly different from the unalloyed irons and are usually produced by specialized foundries.

White high-alloy irons containing nickel and chromium yield a very high hardness with extreme wear and abrasion resistance.

Gray-iron high-nickel and high-tin alloys have excellent vibration damping and are used for precision machine tools, gages, and optical devices.

Malleable Iron. Malleable iron is white iron that has been converted to a "malleable" condition by a two-stage heat-treating process.

		DUCTILE				WHITE	GRAY						MALLEABLE						
		0-55-06	60-40-18	100-70-03	120-90-02		20	25	30	40	50	60	32510	35018	40010	45006	50005	70003	90001
Yield Strength	MPa	410-520	310-410	520-620	620-860								220	240	275	310	345	485	620
	10^3 PSI	60-75	45-60	75-90	90-125								32	35	40	45	50	70	90
Tensile Strength	MPa	620-760	410-550	690-825	860-1035	140-345	140-170	170-205	205-240	275-330	345-390	415-455	345	365	415	450	480	585	725
	10^3 PSI	90-110	60-80	100-120	125-150	20-50	20-25	25-30	30-35	40-48	50-57	60-66	50	53	60	65	70	85	105
Elongation in 50 mm (%)		3-10	10-25	6-10	2-7	—	1%	1%	1%	0.8%	0.5%	0.5%	10	18	10	6	5	3	1
Modulus of Elasticity	10^3 MPa	150-170	150-170	150-170	150-170	8 —	83	90	103	117	131	138	172	172	180	180	180	180-193	180-193
	10^6 PSI	22-25	22-25	22-25	22-25	—	12	13	15	17	19	20	25	25	26	26	26	26-28	26-28

Fig. 12-1-2 Mechanical properties of cast iron.

It is a commercial cast material which is similar to steel in many respects. It is strong and ductile, has good impact and fatigue properties, and has excellent machining characteristics.

Malleable iron differs from other cast irons in the shape of the contained graphite. It is used because of its impact and fatigue resistance, wear resistance, and good machinability. Malleable iron castings generally cost slightly less than ductile iron castings.

The two basic types of malleable iron are ferritic and pearlitic. Ferritic grades are more machinable and ductile, whereas the pearlitic grades are stronger and harder.

REFERENCES AND SOURCE MATERIAL

1. C. F. Walton, "Gray, Ductile, and High-Alloy Iron," *Machine Design,* vol. 37, no. 21, 1965.

2. H. J. Heine, "Malleable Iron," *Machine Design,* vol. 37, no. 21, 1965.

Assignments

1. On an A3- or B-size sheet make a two-view working drawing of the door closer arm shown in Fig. 12-1-A or 12-1-B. Use a revolved section to show the center section of the arm. Select a suitable cast iron for the part. Scale 1:1.

REVIEW FOR ASSIGNMENTS

Unit 4-2	Arcs Tangent to Two Lines
Unit 5-1	Basic Dimensioning
Unit 5-7	Machining Symbols
Unit 6-1	Detail Drawings
Unit 7-8	Revolved Sections
Unit 7-18	Intersection of Unfinished Surfaces

UNIT 12-2

CARBON STEEL

Carbon steel is essentially an iron-carbon alloy with small amounts of other elements (either intentionally added or unavoidably present) such as silicon, magnesium, copper, and sulphur. Steels can be either cast to shape or wrought into various mill forms from which finished parts can be machined, forged, formed, stamped, or otherwise generated.

First wrought steel is either poured into ingots or is sand-cast. After solidification the metal is reheated and hot rolled—often in several steps—into the finished wrought form. Hot-rolled steel is characterized by a scaled surface and a decarburized skin. Hot-rolled bars may be subsequently cold finished by acid pickling or shot blasting to remove scale, and then by drawing through a die and restraightening to improve surface properties and strength or for closer size control. Also, hot-rolled steel may be cold finished by metal-removal processes such as turning or grinding.

Carbon and Low-Alloy Cast Steels[1]

Carbon and low-alloy cast steels lend themselves to the formation of streamlined, intricate parts with high strength and rigidity.

A number of advantages favor steel casting as a method of construction:

1. The metallographic structure of steel castings is uniform in all directions. It is free from the directional variations in properties of wrought-steel products.

2. Cast steels are available in a wide range of mechanical properties depending on the compositions and heat treatments.

3. Steel castings can be annealed, normalized, tempered, hardened, or carburized.

4. Steel castings are as easy to machine as wrought steels.

5. Most compositions of carbon and low-alloy cast steels are easily welded because carbon content is under 0.45 percent.

The making of steel is illustrated in Fig. 12-2-1.

High-Alloy Cast Steels[2]

The term *high alloy* is applied arbitrarily to steel castings containing a minimum of 8 percent nickel and/or chromium. Such castings are used mostly to resist corrosion or provide strength at temperatures above 650°C.

Carbon Steels[3]

Carbon steels are the workhorse of product design. They account for over 90 percent of total steel production. More carbon steels are used in product manufacturing than all other metals combined.

A thorough understanding of the selection and specification criteria for all types of steel requires knowledge of what is implied by carbon-steel mill forms, qualities, grades, tempers, finishes, edges, heat treatments; and how and where these terms relate to dimensions, tolerances, physical and mechanical properties, and manufacturing requirements.

The designer's specification job really begins the instant that molten steel hits the mold. The conditions under which steel solidifies have a significant effect

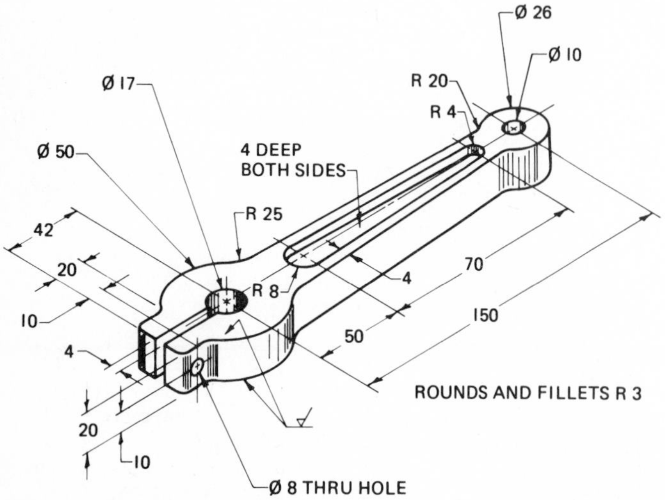
Ø 26
Ø 10
Ø 17
R 20
R 4
Ø 50
4 DEEP
BOTH SIDES
R 25
42
20
10
R 8
4
70
150
50
4
ROUNDS AND FILLETS R 3
20
10
Ø 8 THRU HOLE

Fig. 12-1-A Door closer arm.

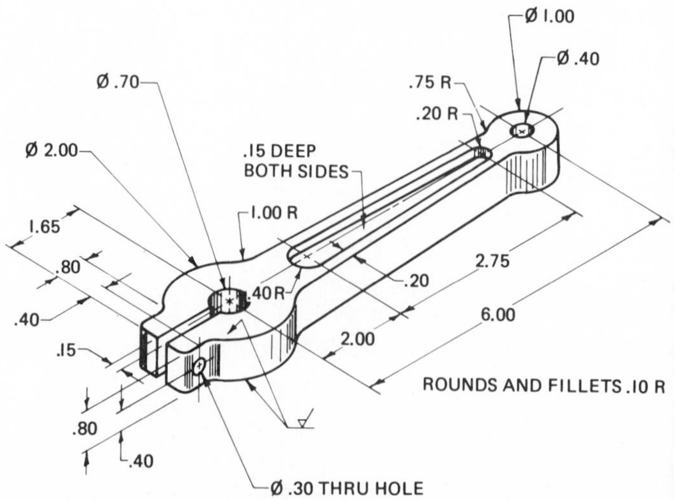
Ø 1.00
Ø .40
Ø .70
.75 R
.20 R
Ø 2.00
.15 DEEP
BOTH SIDES
1.00 R
1.65
.80
.40R
.20
2.75
.40
6.00
2.00
.15
ROUNDS AND FILLETS .10 R
.80
.40
Ø .30 THRU HOLE

Fig. 12-1-B Door closer arm.

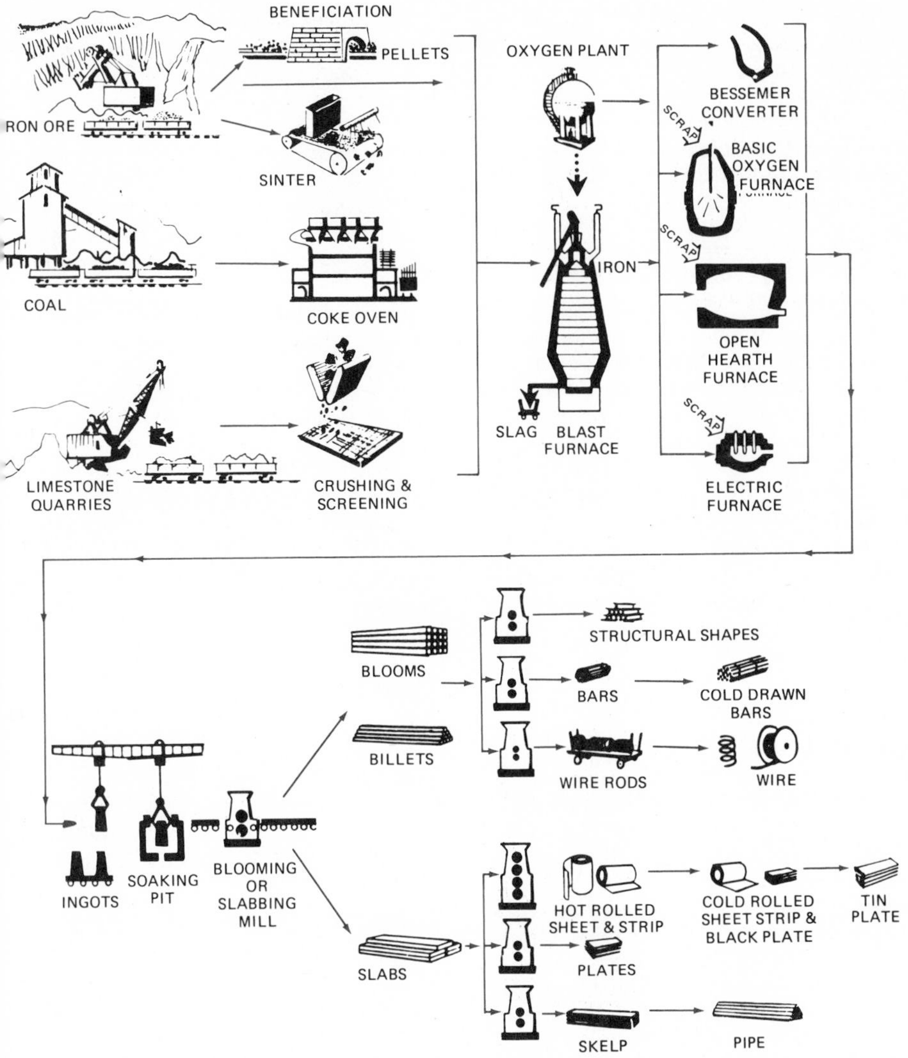
BENEFICIATION
PELLETS
IRON ORE
SINTER
COAL
COKE OVEN
LIMESTONE QUARRIES
CRUSHING & SCREENING
OXYGEN PLANT
IRON
SLAG
BLAST FURNACE
SCRAP
BESSEMER CONVERTER
BASIC OXYGEN FURNACE
OPEN HEARTH FURNACE
ELECTRIC FURNACE
INGOTS
SOAKING PIT
BLOOMING OR SLABBING MILL
BLOOMS
BILLETS
SLABS
STRUCTURAL SHAPES
BARS
COLD DRAWN BARS
WIRE RODS
WIRE
HOT ROLLED SHEET & STRIP
COLD ROLLED SHEET STRIP & BLACK PLATE
TIN PLATE
PLATES
SKELP
PIPE

Fig. 12-2-1 Flow chart of steelmaking.

on production and on performance of subsequent mill products.

Steel Specification

Several ways are used to identify a specific steel: by chemistry, by mechanical properties, by its ability to meet a standard specification or industry-accepted practice, or by its ability to be fabricated into an identified part.

SPECIFYING BY CHEMISTRY

The producer can be instructed to produce a desired composition in one of three ways:

1. By a maximum limit.
2. By a minimum limit.
3. By an acceptable range.

The following are some commonly specified elements.

Carbon. The principal hardening element in steel. As carbon content is increased to about 0.85 percent, hardness and tensile strength increase, but ductility and weldability decrease.

Manganese. A lesser contributor to hardness and strength. Properties depend on carbon content. Increasing manganese increases the rate of carbon penetration during carburizing. It is beneficial to surface finish in all carbon ranges.

Phosphorus. Large amounts increase strength and hardness but reduce ductility and impact toughness, particularly in the higher-carbon grades. Phosphorus in low carbon, free-machining steels improves machinability.

Sulphur. Increased sulphur content reduces transverse ductility, notch impact toughness, and weldability. Sulphur is added to improve machinability of steels.

Silicon. A principal deoxidizer in the steel industry. Silicon increases strength and hardness but to a lesser extent than manganese.

Copper. Improves atmospheric corrosions resistances when present in excess of 0.15 percent. (Minimum of 0.2 percent is usually specified.) Copper affects forge welding adversely but not arc or acetylene welding.

Lead. Improves machinability of steel.

CLASSIFICATION BODIES

The specifications covering the composition of these metals have been issued by various classification bodies. They serve as a selection guide and provide a means for the buyer to conveniently specify certain known and recognized requirements. The main classification bodies are:

CSA—Canadian Standards Association.

SAE—Society of Automotive Engineers.

AISI—American Iron and Steel Institute. This is an association of steel producers which issues steel specifications for the steel-making industry and cooperates with the SAE in using the same numbers for the same steel.

ASME—American Society of Mechanical Engineers. This group is interested in the steel used in boilers and other mechanical equipment. The ASME issues steel plate specifications cooperating closely with the ASTM using the same designations for the same plate.

ASTM—American Society for Testing Materials. This group is interested in materials of all kinds and writes specifications. The ASTM steel specifications for steel plate and structural shapes used by all steel makers in North America.

The best known and most used steel-bar specifications are those issued by the SAE and the AISA. In recent years the AISI has become the more widely used specification. Since both use the same basic system it is difficult to differentiate between them.

The ASTM has several specifications covering structural steel while the CSA has two specifications, for Canadian use. ASTM appears to have precedence since both the AISI and AISC (American Institute of Steel Construction) refer to ASTM specifications.

These various bodies tend to dominate their respective fields. The advantages of standardization of the different product fields are obvious; and, through standardization, the clashing of the various groups would be eliminated. Thus the ASME and the ASTM cooperatively developed specifications for steel plate that are generally quoted and used while the automotive engineers have a major interest in steel bars and have consequently developed specifications for this product that have come into general use.

SAE and AISI—Systems of Steel Identification. The specifications for steel bar are based on a code that indicates the composition of each type of steel covered. They include both plain carbon and alloy steels. The code is a 4-number system. See Figs. 12-2-2 and 12-2-3. Each figure in the number has the following specific function: the first or left-side figure indicates the major class of steel, the second figure represents a subdivision of the major class; for example, the series having *one* as the left-hand figure covers the carbon steels. The second figure breaks this class up into normal low sulphur steels, the high sulphur free machining grades, and another grade having higher than normal manganese.

Class 1	
Carbon steels	1xxx
Basic open hearth and acid Bessemer carbon steels nonsulphurized and non-phosphorized	10xx
Basic open hearth and acid Bessemer carbon steels, sulphurized but not phosphorized	11xx
Basic open hearth carbon steels phosphorized	12xx

Originally this second figure indicated the percentage of the major alloying element present, and this is true of many of the alloy steels. However, this had to be varied in order to care for all the steels that are available.

The third and fourth figures indicate carbon content in hundredths of 1 percent, thus the figure xx15 indicates 0.15 of 1 percent carbon.

Example. SAE 2335 is a nickel steel containing 3.5 percent nickel and 0.35 of 1 percent carbon.

Most steel product specifications indicate mechanical properties as well as composition. This is not true of SAE numbers and constitutes the major difference between specifications such as those issued by the ASTM, ASME, and the SAE or AISI numbers. Despite this the SAE numbers are convenient, widely employed, and very useful as long as they are thoroughly understood and their limitations appreciated. Typical properties of rolled carbon steels are shown in Fig. 12-2-4.

CARBON STEEL SHEET

Flat-rolled carbon steel sheets are made from heated slabs that are progressively reduced in size as they move through a series of rolls.

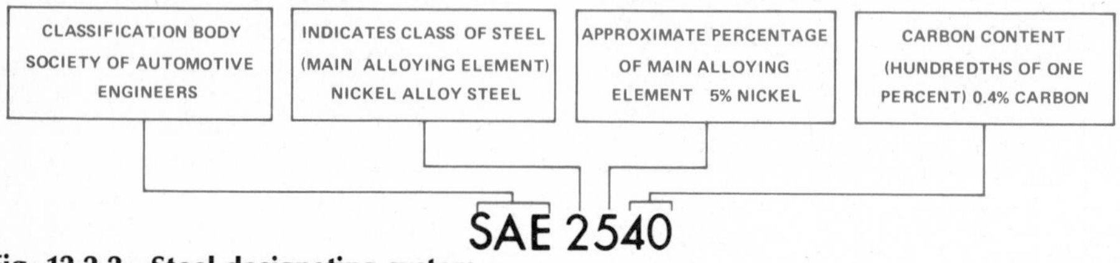
CLASSIFICATION BODY
SOCIETY OF AUTOMOTIVE ENGINEERS
INDICATES CLASS OF STEEL
(MAIN ALLOYING ELEMENT)
NICKEL ALLOY STEEL
APPROXIMATE PERCENTAGE
OF MAIN ALLOYING
ELEMENT 5% NICKEL
CARBON CONTENT
(HUNDREDTHS OF ONE
PERCENT) 0.4% CARBON
SAE 2540

Fig. 12-2-2 Steel designating system.

TYPE OF CARBON STEEL	NUMBER SYMBOL	PRINCIPAL PROPERTIES	COMMON USES
—Plain Carbon	10XX		
—Low-Carbon Steel (0.6% to 0.20% Carbon)	1006 to 1020	Toughness and Less Strength	Chains, Rivets, Shafts, Pressed Steel Products
—Medium-Carbon Steel (0.20% to 0.50% Carbon)	1020 to 1050	Toughness and Strength	Gears, Axles, Machine Parts, Forgings, Bolts and Nuts
—High-Carbon Steel (Over 0.50% Carbon)	1050 and over	Less Toughness and Greater Hardness	Saws, Drills, Knives, Razors, Finishing Tools, Music Wire
—Sulphurized (Free Cutting)	11XX	Improves Machinability	Threads, Splines, Machined Parts
—Phosphorized	12XX	Increases Strength and Hardness but Reduces Ductility	
—Manganese Steels	13XX	Improves Surface Finish	

Fig. 12-2-3 Carbon steel designations, properties, and uses of steel.

Hot-Rolled Sheets. Hot-rolled sheets are produced to three principal qualities.

Commercial Quality. Surface quality is of secondary importance, so its applications are those where oxide and normal surface imperfections are not objectionable.

Drawing Quality. Drawing quality (DQ) sheet is used for applications where surface appearance is a secondary consideration.

Physical Quality. Produced for applications requiring specific mechanical properties other than the bend characteristics of commercial-quality sheet.

Cold-Rolled Sheets. Cold-rolled sheets are made from hot-rolled coils which are pickled, then cold reduced to the desired thickness. The commercial quality of cold-rolled sheets is normally produced with a matte finish suitable for painting or enameling but not suitable for electroplating.

CARBON STEEL PLATES

Carbon steel plates are produced (in rectangular plates or in coils) by hot rolling direct from the ingot or slab. Plate thickness ranges from 4 mm and thicker for plates up to 1200 mm wide, and from 6 mm and thicker for plates wider than 1200 mm. Thickness is specified in millimetres. It can also be specified by mass/m^2.

CARBON STEEL BARS

Hot-Rolled Bars. Hot-rolled carbon-steel bars are produced from blooms or billets in a variety of cross sections and sizes. See Figs. 12-2-5 and 12-2-6. They are produced in straight lengths or, for some sizes and sections, in coils. Hot-rolled carbon-steel bars have two standard quality designations.

Merchant Quality. Merchant-quality steel bars are produced for structural and other applications requiring mild hot and cold bending operations, punching, and welding. They are used in noncritical parts of buildings, ships, mobile equipment, and many types of machinery.

Special Quality. Special-quality hot-rolled bars are produced for applications where the end use requires forging, heat treatment, cold drawing, or machining.

Cold-Finished Bars. Cold-finished carbon-steel bars are produced from hot-rolled steel by a cold finishing process which improves surface finish, dimensional accuracy, and alignment. Cold drawing and cold rolling also increase the yield and tensile strength. For machinability ratings of cold drawn carbon steel see Fig. 12-2-7.

STEEL WIRE

Steel wire is made from hot-rolled rods produced in continuous-length coils.

		AISI STEEL											
		1015/1020/1022			1035/1040			1045/1050			1095		
		Hot Rolled	Cold Drawn	Annealed	Hot Rolled	Cold Drawn	Quenched and Tempered	Hot Rolled	Cold Drawn	Quenched and Tempered	Hot Rolled	Cold Drawn and Annealed	Quenched and Tempered
Yield Strength	MPa	270	350	295	290	440	435-660	335	580	470-800	455	525	580-1050
	10^3 PSI	40	51	42	42	71	63-96	49	84	68-117	66	76	80-152
Tensile Strength	MPa	450	420	415	525	585	660-895	620	690	725-945	895	680	895-1490
	10^3 PSI	65	61	60	76	85	96-130	90	100	105-137	130	99	130-216
Elongation in 50 mm %	%	25	15	38	18	12	17-24	15	10	25-15	9	13	10-84

Fig. 12-2-4 Typical properties of rolled carbon steel.

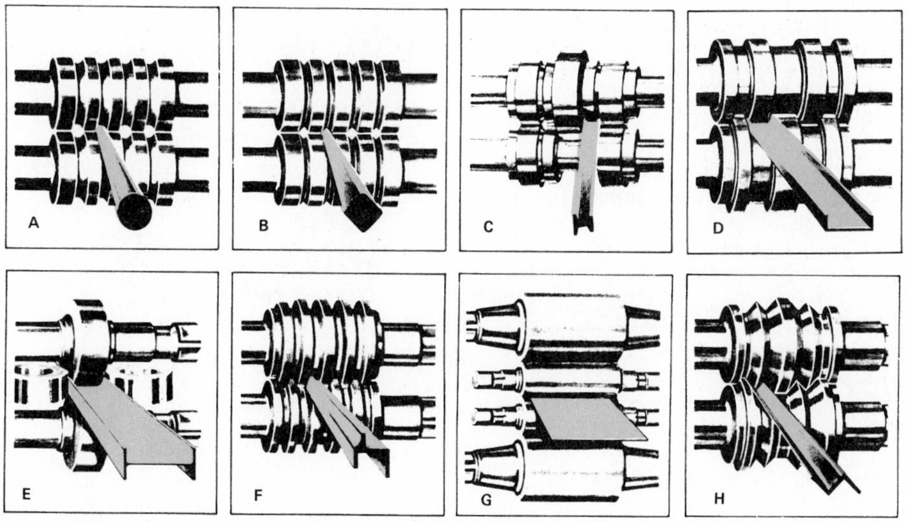
A
B
C
D
E
F
G
H

Fig. 12-2-5 Standard stock. (American Iron and Steel Institute.)

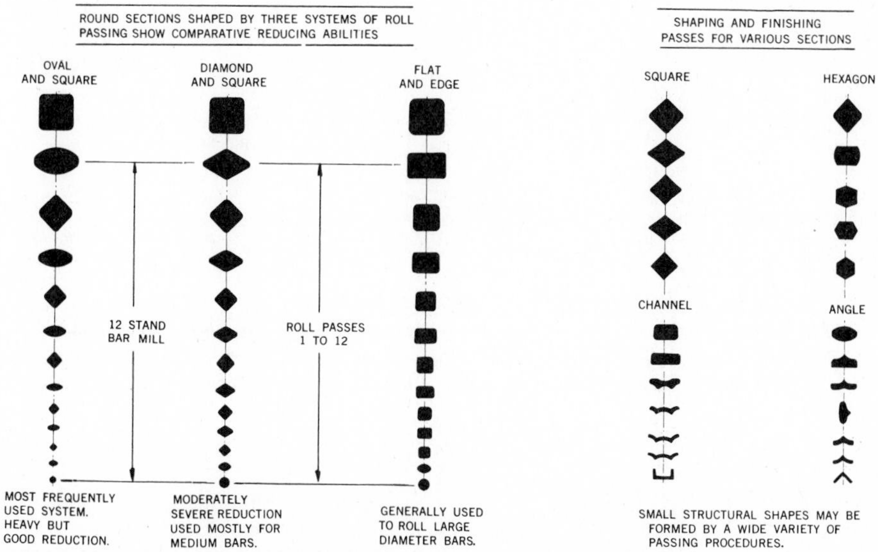
ROUND SECTIONS SHAPED BY THREE SYSTEMS OF ROLL PASSING SHOW COMPARATIVE REDUCING ABILITIES
OVAL AND SQUARE
DIAMOND AND SQUARE
FLAT AND EDGE
12 STAND BAR MILL
ROLL PASSES 1 TO 12
MOST FREQUENTLY USED SYSTEM. HEAVY BUT GOOD REDUCTION.
MODERATELY SEVERE REDUCTION USED MOSTLY FOR MEDIUM BARS.
GENERALLY USED TO ROLL LARGE DIAMETER BARS.
SHAPING AND FINISHING PASSES FOR VARIOUS SECTIONS
SQUARE
HEXAGON
CHANNEL
ANGLE
SMALL STRUCTURAL SHAPES MAY BE FORMED BY A WIDE VARIETY OF PASSING PROCEDURES.

Fig. 12-2-6 Bar-mill roll passes. (American Iron and Steel Institute.)

Most wire is drawn, but some special shapes are rolled. Before the drawing operation, the rods are descaled, cleaned, and coated with a lubricant.

PIPE AND TUBING

Pipe and tubing range from the familiar plumber's black pipe to high-precision mechanical tubing for bearing races. Pipe and tubing may contain fluids, support structures, or be a primary shape from which products are fabricated. The range of end uses for steel pipe and tubing is almost equaled by the types available.

Welded Tubular Products. Welded tubular products are made from hot-rolled or cold-rolled flat steel coils.

Pipe. Pipe is produced of carbon or alloy steel to nominal dimensions. Nominal pipe sizes are expressed in inch sizes, but the outside diameter and the wall thickness are expressed in millimetres. The outside diameter is often much larger than the nominal size. For example, a .75 in. standard-weight pipe has an outside diameter of 26.7 mm (1.050 in.). The outside diameter of nom-

AISI No.	RATING*
12L14	195
1213	137
1215	137
1212	100
1211	91
1117	89
1114	85
1137	72
1141	69
1018	66
1045	55

* Based on 1212 = 100%

Fig. 12-2-7 Machinability rating of cold-drawn carbon steel.

inal size pipe always remains the same and the mass or wall thickness changes. The wall thickness can be designated as *standard weight, extra strong, or double-extra strong*.

Nominal pipe size designation stops at 12 in. Pipe 14 in. and over is listed on basis of outside diameter and wall thickness.

Tubing. Tubing is usually specified by a combination of either outside diameter, inside diameter, or wall thickness. Sizes range from approximately 6 to 125 mm, in increments of 3 mm. Wall thickness is usually specified in millimetres or by gage number.

STRUCTURAL STEEL SHAPES

A large tonnage of structural-steel shapes goes into manufactured products rather than buildings. The frame of a truck, railroad car, or earth-moving equipment is a structural design problem, just as is a high-rise building.

Size Designations. Several ways are used to describe structural sections in a specification, depending primarily on its shape:

1. Beams and channels are measured by the depth of the section in millimetres and by mass per metre in kilograms.

2. Angles are described by length of legs and thickness in millimetres or, more commonly, by lengths of legs and mass per metre. The longest leg is always stated first.

3. Tees are specified by width of flange, overall depth of stem, and mass per metre, in that order.

4. Zees are specified by width of flange and thickness in millimetres, or by depth, width across flange, and mass per metre.

5. Wide-flange sections: Described by depth, width across flange, and mass per metre.

High-Strength Low-Alloy Steels

The properties of high-strength low-alloy (HSLA) steels generally exceed those of conventional carbon structural steels. These low-alloy steels are usually chosen for their high ratios of yield to tensile strength, resistance to puncturing, abrasion resistance, corrosion resistance, and toughness.

ASTM SPECIFICATIONS

ASTM has six specifications covering high-strength low-alloy steels. These are:

ASTM A-94. Used primarily for riveted and bolted structures and for special structural purposes.

ASTM A-242. Used primarily for structural members where light mass and durability are important.

ASTM A-374. Used where high strength is required and where resistance to atmospheric corrosion must be at least equal to that of plain copper-bearing steel.

ASTM A-375. This specification differs slightly from ASTM A-374 in that material can be specified in the annealed or normalized condition.

ASTM A-440. This covers high-strength intermediate-manganese steels for nonwelded applications.

ASTM A-441. This covers the intermediate manganese HSLA steels which are readily weldable when proper welding procedures are used.

Low- and Medium-Alloy Steels

There are two basic types of alloy steel: *through hardenable* and *surface hardenable*. Each type contains a broad family of steels whose chemical, physical, and mechanical properties make them suitable for specific product applications. See Fig. 12-2-9.

Stainless Steels

Stainless steels have many industrial uses because of their desirable corrosion resistance and strength properties. High resistance to corrosion in a broad spectrum of environments, high strength to withstand processing stresses, excellent oxidation resistance and strength at elevated temperature, and good fabricability are characteristics of most of the stainless steels.

FREE-MACHINING STEELS

A whole family of free-machining steels has been developed for fast and economical machining. See Fig. 12-2-10. These steels are available in bar stock in various compositions, some standard and some proprietary. When utilized properly, they lower the cost of machining by reducing metal removal time.

REFERENCES AND SOURCE MATERIAL

1. C. W. Briggs, "Carbon and Low-Alloy Steels," *Machine Design*, vol. 37, no. 21, 1965.

TYPE OF STEEL	ALLOY SERIES	APPROXIMATE ALLOY CONTENT	PRINCIPAL PROPERTIES	COMMON USES
Manganese Steel	13XX	Mn 1.6-1.9	Improve Surface Finish	
Molybdenum Steel	40XX	Mo 0.15-0.3		
	41XX	Cr 0.4-1.1; Mo 0.08-0.35		Axles,
	43XX	Ni 1.65-2; Cr 0.4-0.9; Mo 0.2-0.3	High	Forgings, Gears,
	44XX	Mo 0.45-0.6	Strength	Cams,
	46XX	Ni 0.7-2; Mo 0.15-0.3		Mechanical
	47XX	Ni 0.9-1.2; Cr 0.35-0.55; Mo 0.15-0.4		Parts
	48XX	Ni 3.25-3.75; Mo 0.2-0.3		
Chromium Steels	50XX	Cr 0.3-0.5	Hardness,	Gears,
	51XX	Cr 0.7-1.15	Great Strength	Shafts,
	E51100	C 1.0; Cr 0.9-1.15	and Toughness	Bearings, Spring
	E52100	C 1.0; Cr 0.9-1.15		Connecting Rod
Chromium Vanadium Steel	61XX	Cr 0.5-1.1; V 0.1-0.15	Hardness and Strength	Punches and Di Piston Rods, Gears, Axles
Nickel-Chromium-Molybdenum Steels	86XX	Ni 0.4-0.7; Cr 0.4-0.6; Mo 0.15-0.25	Rust Resistance, Hardness and Strength	Food containers Surgical Equipment
	87XX	Ni 0.4-0.7; Cr 0.4-0.6; Mo 0.2-0.3		
	88XX	Ni 0.4-0.7; Cr 0.4-0.6; Mo 0.3-0.4		
Silicon-Manganese Steels	92XX	Si 1.8-2.2	Springiness and Elasticity	Springs

Fig. 12-2-8 AISI designation system for alloy steel.

		12L13/12L14 12L15		1211/1212/ 1213		1117/1118/1119			1137			1141/1144		
		Hot Rolled	Cold Drawn	Hot Rolled	Cold Drawn	Hot Rolled	Cold Drawn	Quenched and Tempered	Hot Rolled	Cold Drawn	Quenched and Tempered	Hot Rolled	Cold Drawn	Quenched and Tempered
Yield Strength	MPa	235	416-550	225	400	235-315	350-470	345-525	330	565	335	350	620	1120
	10^3 PSI	34	60-80	33	58	34-46	51-68	50-76	48	82	136	51	90	163
Tensile Strength	MPa	390	480-620	380	517	425-525	475-535	615-780	605	675	1080	650	690	1310
	10^3 PSI	57	70-90	55	75	62-76	69-78	89-113	88	98	157	94	100	190
Elongation in 50 mm	%	22	10-18	25	10	23-33	15-20	17-22	15	10	5	15	10	9
Machinability	(B1212 = 100%)	195-296		91-137		89-100			71			69		

Fig. 12-2-9 Typical mechanical properties of free-machining carbon steels.

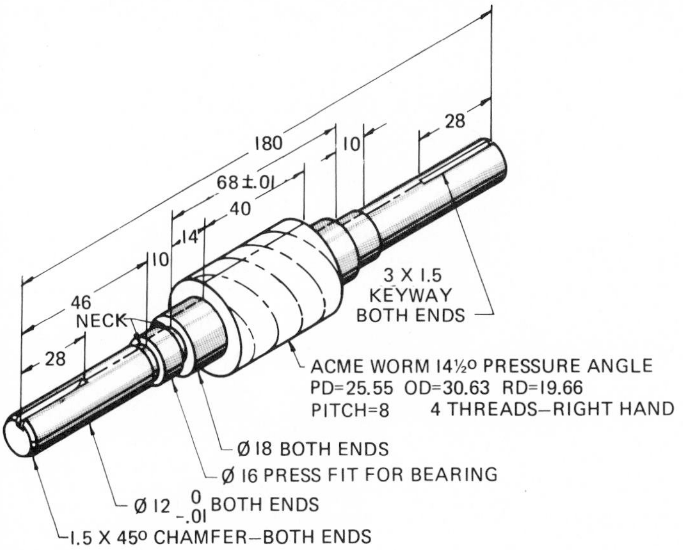
180
68 ±.01
40
10
28
14
10
46
NECK
28
3 X 1.5
KEYWAY
BOTH ENDS
ACME WORM 14½° PRESSURE ANGLE
PD=25.55 OD=30.63 RD=19.66
PITCH=8 4 THREADS–RIGHT HAND
Ø 18 BOTH ENDS
Ø 16 PRESS FIT FOR BEARING
Ø 12 0 -.01 BOTH ENDS
1.5 X 45° CHAMFER–BOTH ENDS

Fig. 12-2-A Worm for gear jack.

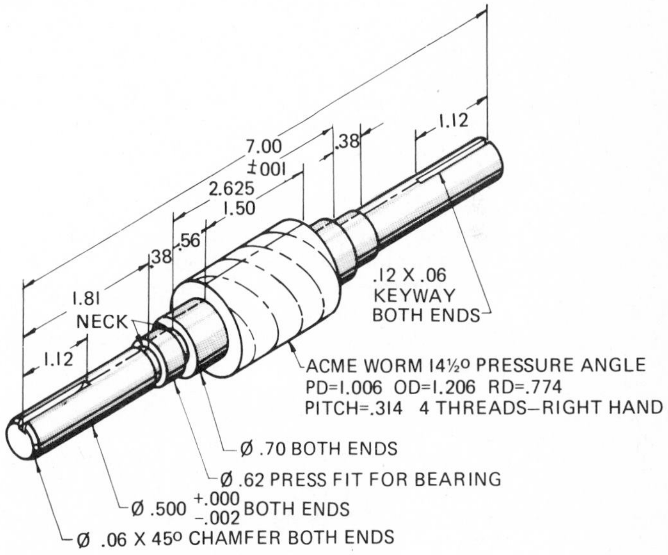
7.00
±.001
.38
1.12
2.625
1.50
.56
.38
1.81
NECK
1.12
.12 X .06
KEYWAY
BOTH ENDS
ACME WORM 14½° PRESSURE ANGLE
PD=1.006 OD=1.206 RD=.774
PITCH=.314 4 THREADS–RIGHT HAND
Ø .70 BOTH ENDS
Ø .62 PRESS FIT FOR BEARING
Ø .500 +.000 -.002 BOTH ENDS
Ø .06 X 45° CHAMFER BOTH ENDS

Fig. 12-2-B Worm for gear jack.

2. E. A. Schoeper, "High-Alloy Steels," *Machine Design*, vol. 37, no. 21, 1965.

3. C. N. Parker, "Carbon Steels," *Machine Design*, vol. 37, no. 21, 1965.

Assignments

1. On an A3- or B-size sheet make a working drawing of the worm for the gear jack shown in Fig. 12-2-A or 12-2-B. Show the worm threads in pictorial form. Scale 1:1. Select a suitable steel for the part.

REVIEW FOR ASSIGNMENTS

Unit 5-3 Dimensioning Common Features
Unit 6-1 Detail Drawings
Unit 8-1 Pictorial Representation and Thread Forms
Unit 9-1 Keys

UNIT 12-3
NONFERROUS METALS

Although ferrous alloys are specified for more engineering applications than all nonferrous metals combined, the large family of nonferrous metals offers a wider variety of characteristics and mechanical properties. For example, the lightest metal is lithium, 0.53 gm/cm³; the heaviest is osmium with a mass of 22.5 gm/cm³—nearly twice the mass of lead. Mercury melts at around −39°C, while tungsten, the highest-melting metal, liquefies at 3410°C.

Availability, abundance, and the cost to convert the metal into useful forms all play an important part in selecting a nonferrous metal. Although nearly 80 percent of all elements are called "metals," only about two dozen of these are used as structural engineering materials. Of the balance, however, many are used not structurally but as coatings, in electronic devices, as nuclear materials, and as minor constituents in other systems.

One of the most important aspects in selecting a material for a mechanical or structural application is how easily the material can be shaped into the finished part—and how its properties can be either intentionally or inadvertently altered in the process. See Fig. 12-3-1. Frequently, metals are simply cast into the finished part. In other cases, metals are cast into an intermediate form (such as an ingot) then worked or "wrought" by rolling, forging, extruding, or other deformation processes. While this applies to ferrous as well as nonferrous metals and alloys, the reaction of nonferrous metals to these forming processes is often more severe—properties may differ remarkably between the cast and wrought forms of the same metal or alloy.

Manufacturing with Metals (Fig. 12-3-1)

Machining. Most metals can be machined. Machinability is best for metals that allow easy chip removal with minimum tool wear.

PM Compacting. Parts can be made from most metals and alloys by PM compacting, although only a few are economically justified. Iron and iron-copper alloys are most commonly used.

Casting. Theoretically, any metal that can be melted and poured can be cast. However, economic limitations usually narrow down the number of ways metals are cast commercially.

Extruding and Forging. Metals to be forged or extruded must be ductile and not work harden at working temperature. Some metals show these characteristics at room temperature and can be cold worked; others must be heated.

Stamping and Forming. Most metals, except brittle alloys, can be press worked.

Cold Heading. Metals must be ductile and should not work harden rapidly. Annealing should restore ductility and softness in cold-heading alloys.

Deep Drawing. Deep drawing involves severe deformation and the metal is usually stretched over the die.

Aluminum[1]

The density of aluminum is about one-third that of steel, brass, nickel, or copper. Yet, some alloys of aluminum are stronger than structural steel. Under most service conditions, aluminum has high resistance to corrosion and forms no colored salts which might stain or discolor adjacent components.

Aluminum and its alloys are available in practically all commercial forms in which metals are commonly used: wire, foil, powder, sheet, plate, extrusions, bar, forgings, and castings. See Fig. 12-3-2.

MAJOR ALLOYING ELEMENT	DESIGNATION
Aluminum (99% or more)	1XXX
Copper	2XXX
Manganese	3XXX
Silicon	4XXX
Magnesium	5XXX
Magnesium and Silicon	6XXX
Zinc	7XXX
Other elements	8XXX
Unused Series	9XXX

Fig. 12-3-2 Wrought aluminum alloy designations.

Copper[2]

Copper alloys, approximately 250 of them, are fabricated in rod, sheet tube, and wire form. Each of these alloys has some property or combination of properties which makes it unique. They can be grouped into several general headings, such as coppers, brasses, leaded brasses, phosphor bronzes, aluminum bronzes, silicon bronzes, beryllium coppers, cupro nickels, and nickel silvers. See Fig. 12-3-3.

Copper alloys are used where one or more of the following properties are needed: thermal or electrical conductivity, corrosion resistance, strength, ease of forming, ease of joining, and color. Volume-for-volume, copper has the best conductivity of any commercially priced metal. Copper and its alloys have excellent corrosion resistance. Most of the systems are readily fabricated, cold or hot, and can be joined by conventional methods.

The major alloy usages are:

1. Copper in pure form as a conductor in the electrical industry

2. Copper or alloy tubing for water, drainage, air conditioning, and refrigeration lines

3. Brasses, phosphor bronzes, and nickel silvers as springs or in construction of equipment if corrosive conditions are too severe for iron or steel

An advantage of copper and its alloys, offered by no other metals, is the wide range of colors available. Brasses can be obtained in shades from yellow to reddish brown; bronzes from copper to gold, copper nickels and nickel silvers, from pink to silver.

	Al	Cu	Fe	Pb	Mg	MI	Ag, Au, Pt	Cu Mo Ta,W	Steel	Sn	Ti	Zn
Casting												
Centrifugal	✓	✓	✓			✓			✓			
Continuous	✓	✓	✓	✓					✓			
Ceramic mold	✓	✓	✓		✓	✓			✓			✓
Investment	✓	✓			✓	✓	✓		✓			
Permanent mold	✓	✓	✓	✓	✓	✓			✓	✓		✓
Sand	✓	✓	✓	✓	✓	✓			✓	✓		✓
Shell mold	✓	✓	✓			✓			✓			
Die casting	✓	✓		✓	✓					✓		✓
Cold heading	✓	✓		✓		✓	✓		✓			
Deep drawing	✓	✓			✓	✓			✓		✓	✓
Extruding	✓	✓		✓	✓	✓		✓	✓	✓	✓	
Forging	✓	✓			✓	✓		✓	✓		✓	
Machining	✓	✓	✓		✓	✓	✓	✓	✓		✓	✓
PM compacting	✓	✓	✓			✓	✓	✓	✓		✓	
Stamping and forming	✓	✓			✓	✓	✓	✓	✓		✓	✓

Fig. 12-3-1 Common methods of forming metals.

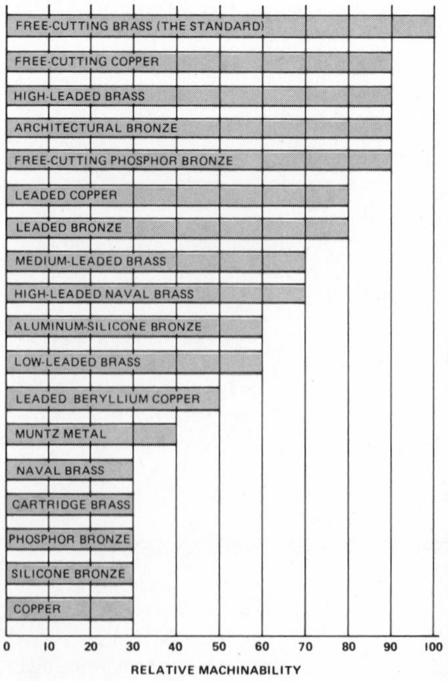
FREE-CUTTING BRASS (THE STANDARD)
FREE-CUTTING COPPER
HIGH-LEADED BRASS
ARCHITECTURAL BRONZE
FREE-CUTTING PHOSPHOR BRONZE
LEADED COPPER
LEADED BRONZE
MEDIUM-LEADED BRASS
HIGH-LEADED NAVAL BRASS
ALUMINUM-SILICONE BRONZE
LOW-LEADED BRASS
LEADED BERYLLIUM COPPER
MUNTZ METAL
NAVAL BRASS
CARTRIDGE BRASS
PHOSPHOR BRONZE
SILICONE BRONZE
COPPER
0 10 20 30 40 50 60 70 80 90 100
RELATIVE MACHINABILITY

Fig. 12-3-3 Free-machining copper alloys.

Nickel[3]

Commercially pure wrought nickel is a grayish-white metal capable of taking a high polish. Because of its combination of attractive mechanical properties, corrosion resistance, and formability, nickel or its alloys are used in a variety of structural applications usually requiring specific corrosion resistance.

Nickel is important as an electroplated coating offering good resistance to corrosion and oxidation, favorable physical and mechanical properties, and ease of application.

Magnesium[4]

Magnesium, with density of only 1.74 g/cm^3, is the world's lightest structural metal. The combination of low density and good mechanical strength makes possible alloys with a high strength-to-weight ratio.

Alloys of magnesium are the easiest of all structural metals to machine, and they can be readily fabricated by most metal-working processes. They are easily welded and have high weld joint efficiency.

Zinc[5]

Zinc is a relatively inexpensive metal which has moderate strength and toughness and outstanding corrosion resistance in many types of service.

The principal characteristics which influence the selection of zinc alloys for die castings include the dimensional accuracy obtainable, castability of thin sections, smooth surface, dimensional stability, and adaptability to a wide variety of finishes.

Titanium[6]

Titanium is a light metal (4.43 g/cm^3); being 60 percent heavier than aluminum (2.77 g/cm^3) but 45 percent lighter than alloy steel (7.92 g/cm^3). It is the fourth most abundant metallic element in the earth's crust and the ninth most common element.

Titanium-base alloys are much stronger than aluminum alloys and superior in many respects to most alloy steels. They are superior to all the usual engineering metals and alloys in strength-mass ratio at temperatures ranging from −258°C to +500°C.

Beryllium[7]

Beryllium has a strength-mass ratio comparable to high-strength steel, yet it is lighter than aluminum. Its melting point is 1285°C and it has excellent thermal conductivity. It is nonmagnetic and a good conductor of electricity.

Refractory Metals[8]

Refractory metals are those metals with melting points above 2000°C. Among these, the best known and most extensively used are tungsten, tantalum, molybdenum, and columbium.

Refractory metals are characterized by high-temperature strength corrosion resistance and high melting points.

TANTALUM AND COLUMBIUM

Tantalum and columbium are usually discussed together, since most of their working operations are identical. Unlike molybdenum and tungsten, tantalum and columbium can be worked at room temperatures. The major differences between tantalum and columbium are in density, nuclear cross section, and corrosion resistance. The density of tantalum is almost twice that of columbium.

The only acids which will attack tantalum are hydrofluoric, fuming sulphuric, and phosphoric; the last two acids only at high concentration and temperatures.

MOLYBDENUM

Molybdenum is widely used in missiles, electronic tubes, industrial furnaces, and nuclear projects. Its melting point is lower than that of tantalum and tungsten. Molybdenum has a high strength-to-mass ratio, a low vapor pressure, is a good conductor of heat and electricity, has a high modulus of elasticity, and a low coefficient of expansion. In general, molybdenum has properties similar to those of tungsten. However, it is more ductile and easier to fabricate.

Molybdenum wire is widely used for grids and support members and for anodes, cathodes, and filament supports in electronic tubes. It is also used extensively for heating elements and heat shields in furnaces.

TUNGSTEN

Tungsten is the only refractory metal that has the combination of excellent corrosion resistance, good electrical and thermal conductivity, a low coefficient of expansion, and high strength at elevated temperatures.

Precious Metals[9]

Gold costs over 8000 times an equal mass of iron; rhodium costs nearly 32 000 times more than copper. With prices such as these, why are precious metals ever specified?

In some cases, precious metals are used for their unique surface characteristics. They reflect light better than other metals. Gold, for example, is specified as a surface for heat reflectors, insulators, and collectors because of its outstanding ability to reflect ultraviolet radiation.

Precious metals are excellent conductors of heat and electricity. Although pure-gold contacts weld together under high currents, precious-metal alloys have been developed for contacts that neither oxidize nor weld—even under severe atmospheric conditions at high loads.

The family of metals called precious metals can be divided into three subgroups: silver and silver alloys; gold and gold alloys; and the so-called "platinum" metals, which are platinum, palladium, rhodium, ruthenium, iridium, and osmium. The six platinum metals are so grouped because they occur naturally in the same ore.

Most precious metals are available as sheet, tape, foil, wire, tubing, gauze, disks, electrodes, cathodes, crucibles, catalysts, and salts or solutions for plating and coating.

Often the properties of the platinum-group metals and their alloys are the only materials capable of meeting the required service conditions. As a result they may become, in many cases, the most economical choice of metals.

GOLD

Gold is an extremely soft, ductile metal which undergoes very little work hardening under deformation. Pure gold is too soft and dense for large free-standing components and is used chiefly for linings or electrodeposits.

SILVER

Silver is the least costly of the precious metals. It is very soft when fully annealed (heated to just above 200°C) but work hardens appreciably during fabrication.

PLATINUM

Platinum is white, malleable, ductile, and takes a high, permanent polish. When heated to redness, it softens and can be easily worked. It is nearly nonoxidizable, and is soluble only in liquids generating free chlorine, such as aqua regia.

REFERENCES AND SOURCE MATERIAL

1. H. J. Rowe, W. E. King, and E. V. Blackmun, "Aluminum," *Machine Design*, vol. 37, no. 21, 1965.
2. G. C. Strubell, "Copper," *Machine Design*, vol. 37, no. 21, 1965.
3. A. M. Hall, "Nickel," *Machine Design*, vol. 37, no. 21, 1965.

4. J. D. Hanawalt and W. H. Gross, "Magnesium," *Machine Design*, vol. 37, no. 21, 1965.

5. E. W. Horvick, "Zinc," *Machine Design*, vol. 37, no. 21, 1965.

6. E. F. Erbin, "Titanium," *Machine Design*, vol. 37, no. 21, 1965.

7. J. A. Hawk, Jr., "Beryllium," *Machine Design*, vol. 37, no. 21, 1965.

8. J. Chelius, "Refractory Metals," *Machine Design*, vol. 37, no. 21, 1965.

9. H. K. Lake, "Precious Metals," *Machine Design*, vol. 37, no. 21, 1965.

Assignments

1. On an A3- or B-size sheet make a two-view working drawing of the outdoor motor clamp shown in Figs. 12-3-A or 12-3-B. Use lines or surfaces marked *A*, *B*, and *C* as the zero lines and use arrowless dimensioning. Scale 1:1. Select a suitable material noting that the part must be water resistant, have a painted finish, have moderate strength, and have a light mass.

2. On an A3- or B-size sheet recommend the material for each of the parts shown in Figs. 12-3-C and 12-3-D. State your reasons for each of the materials selected.

REVIEW FOR ASSIGNMENTS

Unit 4-2 Arcs Tangent to Two Lines
Unit 12-1 Cast Irons
Unit 12-2 Steels
Unit 16-2 Arrowless Dimensioning

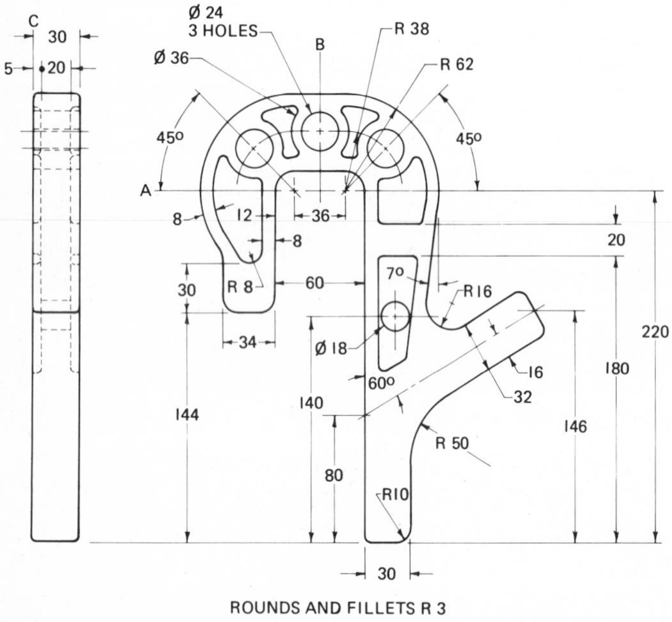

C
30
5
20
Ø 24
3 HOLES
B
R 38
Ø 36
R 62
45°
45°
A
8
12
36
8
20
30
R 8
60
7°
R16
34
Ø 18
16
60°
32
144
140
180
220
146
R 50
80
R10
30
ROUNDS AND FILLETS R 3

Fig. 12-3-A **Outboard motor clamp.**

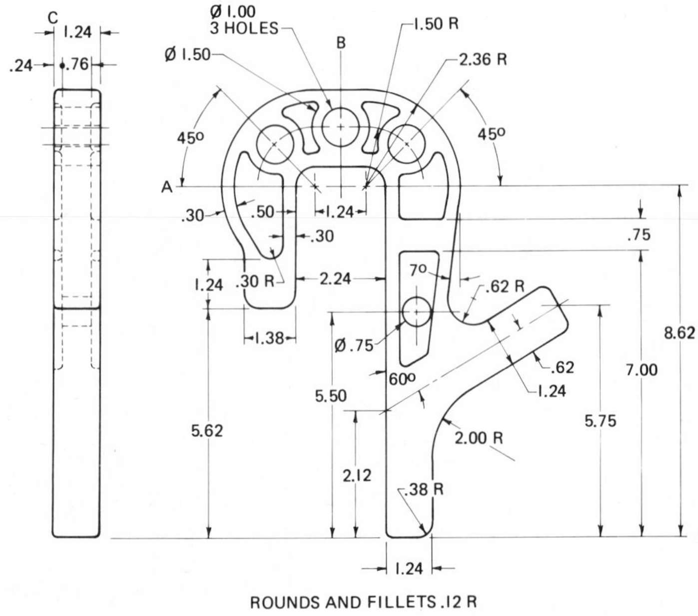

C
1.24
.24
.76
Ø 1.00
3 HOLES
B
1.50 R
Ø 1.50
2.36 R
45°
45°
A
.30
.50
1.24
.30
.75
1.24
.30 R
2.24
7°
.62 R
1.38
Ø .75
.62
8.62
60°
1.24
7.00
5.50
5.62
5.75
2.00 R
2.12
.38 R
1.24
ROUNDS AND FILLETS .12 R

Fig. 12-3-B Outboard motor clamp.

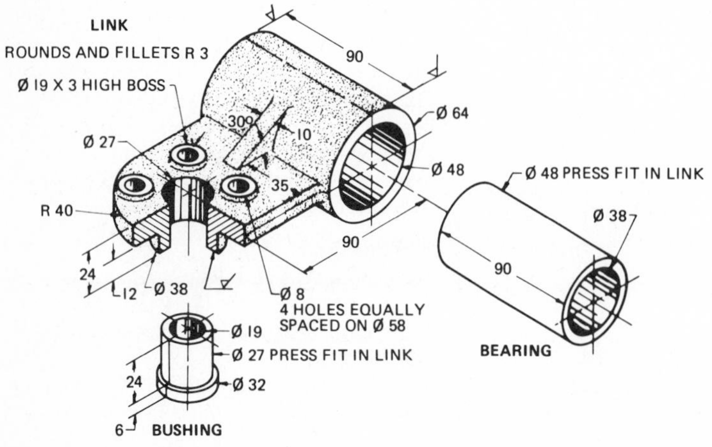

LINK
ROUNDS AND FILLETS R 3
Ø 19 X 3 HIGH BOSS
90
Ø 64
30°
10
Ø 27
Ø 48
Ø 48 PRESS FIT IN LINK
35
R 40
Ø 38
90
90
24
12
Ø 38
Ø 8
4 HOLES EQUALLY SPACED ON Ø 58
Ø 19
Ø 27 PRESS FIT IN LINK
BEARING
24
Ø 32
6
BUSHING

Fig. 12-3-C Connecting link.

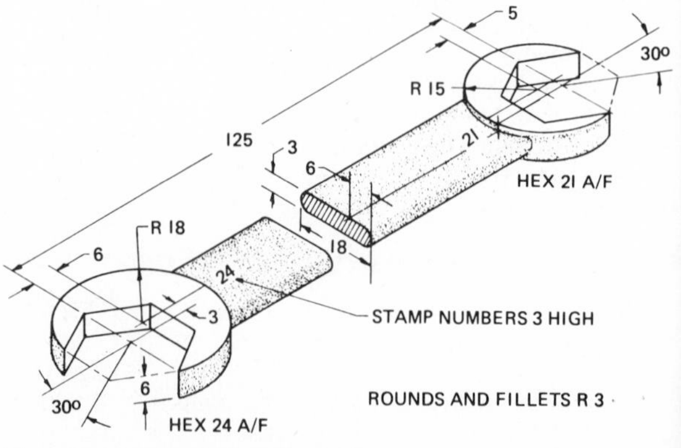

5
30°
R 15
125
3
6
21
HEX 21 A/F
R 18
6
18
24
3
STAMP NUMBERS 3 HIGH
6
30°
ROUNDS AND FILLETS R 3
HEX 24 A/F

Fig. 12-3-D Open-end wrench.

Chapter 13
Plastics

UNIT 13-1
KINDS OF PLASTICS[1,2,3]

This chapter is intended to acquaint drafters with the general characteristics of commercially available plastics so that they may make proper use of plastics in products.

Plastics may be defined as nonmetallic materials capable of being formed or molded with the aid of heat, pressure, chemical reaction, or a combination of these. See Fig. 13-1-1. Many plastics comprise a base material whose properties have been modified by the incorporation of plasticizers and fillers. Each base material is the foundation for a

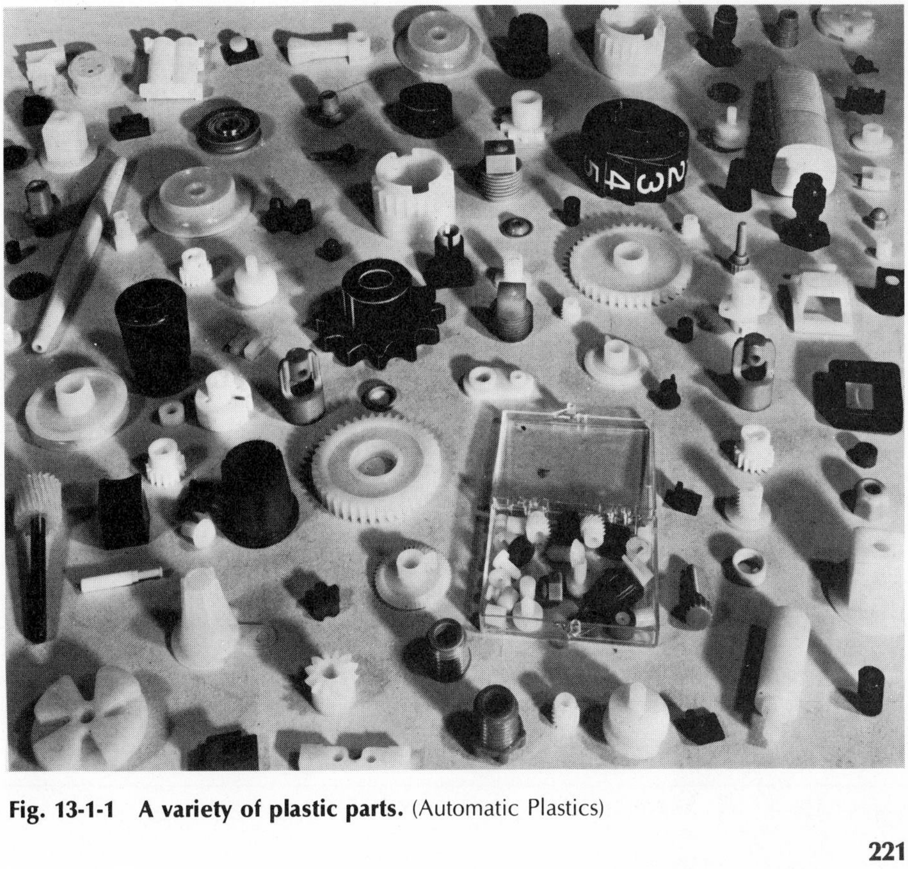

Fig. 13-1-1 A variety of plastic parts. (Automatic Plastics)

group of compositions related in general behavior but differing from one another in individual properties. Base materials are modified to improve characteristics, such as temperature and moisture resistance and mechanical and chemical properties.

Plastics are strong, tough, durable materials which solve many problems in machine and equipment design. Metals, it is true, are hard and rigid. This means that they can be machined, to very close tolerances, into cams, bearings, bushings, and gears, which will work smoothly under heavy loads for long periods. Although some come close, no plastic has the hardness and creep resistance of, say, steel. However, metals have many weaknesses which engineering plastics do not. Metals corrode or rust, they must be lubricated, their working surfaces wear readily, they cannot be used as electrical or thermal insulators, they are opaque and noisy, and where they must flex, metals fatigue rapidly.

Plastics can solve these weaknesses, though not necessarily all with one material. The engineering plastics are resistant to most chemicals; fluorocarbon is one of the most chemically inert substances known. None of the engineering plastics corrode or rust; acetal resin and fluorocarbon are unaffected even when continuously immersed in water. Engineering plastics can be run at low speeds and loads without lubrication and are among the world's slipperiest solids, as a comparison, equivalent to that of ice. Engineering plastics are resilient; therefore, they run more quietly and smoothly than equivalent metal ones, and they are able to stand periodic overloads without harmful effects.

Plastics are a family of materials—not a single material—each member of which has its special advantages.

Being manufactured, plastics raw materials are capable of being variously combined to give almost any property desired in an end product. But these are controlled variations unlike those of nature's products. Some thermoplastics can be sterilized.

The widespread and growing use of plastics in almost every phase of modern living can be credited in large part to their unique combinations of advantages. These advantages are light mass, range of color, good physical properties, adaptability to mass-production methods, and, often, lower cost.

Aside from the range of uses attributable to the special qualities of different plastics, these materials achieve still greater variety through the many forms in which they can be produced.

They may be made into definite shapes like dinnerware and electric switchboxes. They may be made into flexible film and sheeting such as shower curtains and upholstery. Plastics may be made into sheets, rods, and tubes that are later shaped or machined into internally lighted signs, airplane blisters. They may be made into filaments for use in household screening, industrial strainers, and sieves. Plastics may be used as a coating on textiles and paper. They may be used to bind such materials as fibers of glass and sheets of paper or wood to form boat hulls, airplane wing tips, and tabletops. Plastics may also be used as adhesives and in lacquers and paints, uses which are but mentioned in this chapter and which deal specifically with solid forms of plastics.

Plastics are usually classified as either *thermoplastic* or *thermosetting*. Some other materials may be formed in a similar manner but are not normally referred to as plastics. These include rubber, glass, some forms of ceramics, and compressed powders.

Thermoplastics

These materials soften, or liquefy, and flow when heat is applied. Removal of the heat causes these materials to set or solidify. They may be reheated and reformed or reused. In this group fall the acrylics, the cellulosics, nylon (polyamide), polyethylene, polystyrene or styrene, polyfluorocarbons, the vinyls, polyvinylidene, ABS, acetal resin, polypropylene, and polycarbonates. See Fig. 13-1-2.

Thermosetting

These materials undergo an irreversible chemical change when heat is applied or when a catalyst or reactant is added. They become hard, insoluble, and infusible, and they do not soften upon reapplication of heat. Thermosetting plastics include phenolics, amino plastics (melamine and urea), cold-molded polyesters, epoxies, silicones, alkyds, allylics, and casein. See Fig. 13-1-3.

Selection of Materials

The properties of various materials influence the shape of the part. The essential properties data needed are static strength values in tension, compression,

THERMOPLASTICS

Name of Plastic	Properties	Forms and Methods of Forming	Uses
ABS (Acrylonitrile Butadiene-Styrene)	Strong, tough, good electrical properties.	Available in powder or granules for injection molding, extrusion, and calendering and as sheet for vacuum forming.	Pipe, wheels, football helmets, battery cases, radio cases, children's skates, tote boxes.
ACETAL RESIN	Rigid without being brittle, tough, resistant to extreme temperatures, good electrical properties.	Produced in powder form for molding and extrusion. Available in rod, bar, tube, strip, slab.	Automobile instrument clusters, gears, bearings, bushings, door handles, plumbing fixtures, threaded fasteners, cams.
ACRYLICS	Exceptional clarity and good light transmission. Strong, rigid, and resistant to sharp blows. Excellent insulator, colorless or full range of transparent, translucent, or opaque colors.	Available in sheet, rod, tube, and molding powders. Plastic products can be produced by fabricating of sheets, rods, and tubes, hot forming of sheets, injection and compression molding of powder, extrusion, casting.	Airplane canopies and windows, television and camera viewing lenses, combs, costume jewelry, salad bowls, trays, lamp bases, scale models, automobile tail lights, outdoor signs.

Fig. 13-1-2 Thermoplastics. (The Society of Plastics Industry, Inc.)

THERMOPLASTICS (Continued)

Name of Plastic	Properties	Forms and Methods of Forming	Uses
CELLULOSICS (A) Cellulose Acetate		Available in pellets, sheets, film, rods, tubes, strips, coated cord. Can be made into products by injection, compression molding, extrusion, blow molding, and vacuum forming, or sheets and coating.	Spectacle frames, toys, lamp shades, combs, shoe heels.
(B) Cellulose Acetate Butyrate	Among the toughest of plastics. Retains a lustrous finish under normal wear. Transparent, translucent, or opaque in wide variety of colors and in clear transparent. Good insulators.	Available in pellets, sheets, rods, tubes, strips, and as a coating. Can be made into products by injection, compression molding, extrusion, blowing and drawing of sheet, laminating, coating.	Steering wheels, radio cases, pipe and tubing, tool handles, playing cards.
(C) Cellulose Propionate		Available in pellets for injection extrusion or compression molding.	Appliance housing, telephone hand sets, pens and pencils.
(D) Ethyl Cellulose		Available in granules, flake, sheet, rod, tube, film, or foil. Can be made into finished products by injection, compression molding, extrusion, drawing.	Edge moldings, flashlights, electrical parts.
(E) Cellulose Nitrate		Available in rods, tubes, sheets for machining and as a coating.	Shoe heel covers, fabric coating.
FLUOROCARBONS	Low coefficient of friction, resistant to extreme heat and cold. Strong, hard, and good insulators.	Available as powder and granules in resin form. Sheet, rod, tube, film, tape, and dispersions. Molded, extruded, and machined.	Valve seats, gaskets, coatings, linings, tubings.
NYLON (Polyamides)	Resistant to extreme temperatures. Strong and long-wearing range of soft colors.	Available as a molding powder, in sheets, rods, tubes, and filaments. Injection, compression, blow molding, and extrusion.	Tumblers, faucet washers, gears. As a filament, it is used as brush bristles, fishing line.
POLYCARBONATE	High impact strength, resistant to weather, transparent.	Primarily a molding material, may take form of film, extrusion, coatings, fibers, or elastomers.	Parts for aircraft, automobiles, business machines, gages, safety-glass lenses.
POLYETHYLENE	Excellent insulating properties, moisture proof. Clear, transparent, translucent.	Available in pellet, powder, sheet, film, filament, rod, tube, and foamed. Injection, compression, blow molding, extrusion, coating, and casting.	Ice cube trays, tumblers, dishes, bottles, bags, balloons, toys, moisture barriers.
POLYSTYRENE	Clear, transparent, translucent, or opaque. All colors. Water and weather resistant. Resistance to heat or cold.	Available in molding powders or granules, sheets, rods, foamed blocks, liquid solution, coatings, and adhesives. Injection, compression molding, extrusion, laminating, machining.	Kitchen items, food containers, wall tile, toys, instrument panels.
POLYPROPYLENES	Good heat resistance. High resistance to cracking. Light range of color.	Processed by injection molding, blow molding, and extrusion.	Thermal dishware, washing machine agitators, pipe and pipe fittings, wire and cable insulation, battery boxes, packaging film and sheets.
URETHANES	Tough and shock resistant for solid materials. Flexible for foamed material, can be foamed in place.	Solid type—starting two reactants, final article can be extruded, molded, calendered, or cast. Foamed type—can be made by either a prepolymer or one-shot process, in either slab stock or molded form.	Mattresses, cushioning, padding, toys, rug underlays, crashpads, sponges, mats, adhesion, thermal insulation, industrial tires.
VINYLS	Strong and abrasion-resisting. Resistant to heat and cold. Wide color range.	Available in molding powder, sheet, rod, tube, granules, powder. It can be formed by extrusion, casting, calendering, compression, and injection molding.	Raincoats, garment bags, inflatable toys, hose, records, floor and wall tile, shower curtains, draperies, pipe, paneling.

Fig. 13-1-2 (cont'd.)

THERMOSETTING PLASTICS			
Name of Plastic	**Properties**	**Forms and Methods of Forming**	**Uses**
Alkyds	Excellent dielectric strength, heat resistance, and resistance to moisture.	Available in molding powder and liquid resin. Finished molded products are produced by compression molding.	Light switches, electric motor insulator and mounting cases, television tuning devices and tube supports. Enamels and lacquers for automobiles, refrigerators, and stoves are typical uses for the liquid form.
Allylics	Excellent dielectric strength and insulation resistance. No moisture absorption; stain resistant. Full range of opaque and transparent colors.	Available in the form of monomers, prepolymers, and powders. Finished articles may be made by transfer or compression molding, lamination, coating, or impregnation.	Electrical connectors, appliance handles, knobs, etc. Laminated overlays or coatings for plywood, hardboard, and other laminated materials needing protection from moisture.
Amino (melamine and urea)	Full range of translucent and opaque colors. Very hard, strong, but not unbreakable. Good electrical qualities.	Available as molding powder or granules, as a foamed material in solution, and as resins. Finished products can be made by compression, transfer, plunger molding, and laminating with wood, paper, etc.	Melamine—tableware, buttons, distributor cases, tabletops, plywood adhesive, and as a paper and textile treatment. Urea—scale housing, radio cabinets, electrical devices, appliance housings, stove knobs in resin form as baking enamel coatings, plywood adhesive and as a paper and textile treatment.
Casein	Excellent surface polish. Wide range of near transparent and opaque colors. Strong, rigid, affected by humidity and temperature changes.	Available in rigid sheets, rods, and tubes, as a powder and liquid. Finished products are made by machining of the sheets, rods, and tubes.	Buttons, buckles, beads, game counters, knitting needles, toys, and adhesives.
Cold-Molded 3 Types: Bitumin Phenolic Cement-Asbestos	Resistance to high heat, solvents, water, and oil.	Available in compounds. Finished articles produced by molding and curing.	Switch bases and plugs, insulators, small gears, handles and knobs, tiles, jigs and dies, toy building blocks.
Epoxy	Good electrical properties; water and weather resistance.	Available as molding compounds, resins, foamed blocks, liquid solutions, adhesives, coatings, sealants.	Protective coating for appliances, cans, drums, gymnasium floors, and other hard-to-protect surfaces. They firmly bond metals, glass, ceramics, hard rubber and plastics, printed circuits, laminated tools and jigs, and liquid storage tanks.
Phenolics	Strong and hard. Heat and cold resistant; excellent insulators.	Cast and molded.	Radio and TV cabinets, washing machine agitators, juke box housings, jewelry, pulleys, electrical insulation.
Polyesters (Fiberglass)	Strong and tough, bright and pastel colors, high dielectric qualities.	Produced as liquids, dry powders, premix molding compounds, and as cast sheets, rods, and tubes. They are formed by molding, casting, impregnating, and premixing.	Used to impregnate cloth or mats of glass fibers, paper, cotton, and other fibers in the making of reinforced plastic for use in boats, automobile bodies, luggage.
Silicones	Heat resistant, good dielectric properties.	Available as molding compounds, resins, coatings, greases, fluids, and silicon rubber. Finished by compression and transfer molding, extrusion, coating, calendering, casting, foaming, and impregnating.	Coil forms, switch parts, insulation for motors, and generator coils.

Fig. 13-1-3 Thermosetting plastics. (The Society of Plastics Industry, Inc.)

PRODUCTION REPORT					
Part Name	Material 1st Choice	Material 2nd Choice	Reason for Selection	Machining Required	Color
Telephone Case	ABS	Cellulosics	Good impact strength Good range of colors Light mass. Excellent surface finish Good electrical properties Variety of forming methods	None	Green Blue White Tan Red

Fig. 13-1-4 Selection of material.

shear, and flexure and mechanical constants, such as moduli of elasticity and shear. Most thermoplastics have flow properties which cause degradation of fastening torque retention. Composition and fabrication methods influence the mechanical properties of plastics and, therefore, the working stresses. The conditions under which the plastics are used may also influence their mechanical behavior and, consequently, the design stresses.

Machining

Practically all thermoplastics and thermosets can be satisfactorily machined on standard equipment with adequate tooling. The nature of the plastic will determine whether heat should be applied, as in some laminates, or avoided, as in buffing some thermoplastics. Standard machining operations can be used, such as turning, drilling, tapping, milling, blanking, and punching.

Material Selection

One of the first decisions a designer makes is the choice of materials. The choice is influenced by many factors, such as the end use of the product and the properties of the selected material. No attempt is made at this point to discuss the engineering approach to selection of materials. This is covered in Chap. 33.

However, a basic examination and selection of a plastic material at this time will help acquaint the drafter with the wide range of plastics available. For instance, the preliminary production report for the material selection of the telephone case shown in Fig. 13-1-4 is an example of the type of research required in selecting a material.

REFERENCES AND SOURCE MATERIAL

1. The Society of the Plastics Industry, Inc.
2. Crystaplex Plastics.
3. General Motors Corporation.

Assignments

1. On an A3- or B-size sheet, prepare a production report for the selection of materials for the parts shown in Fig. 13-1-A or 13-1-B. Convert as many parts as possible to plastic. The crane hook assembly is to be used in a water dip tank operation. Assume that the production run is such that all forming processes can be considered. Include with your report a bill of material.

REVIEW FOR ASSIGNMENT

Unit 6-4 Bill of Material

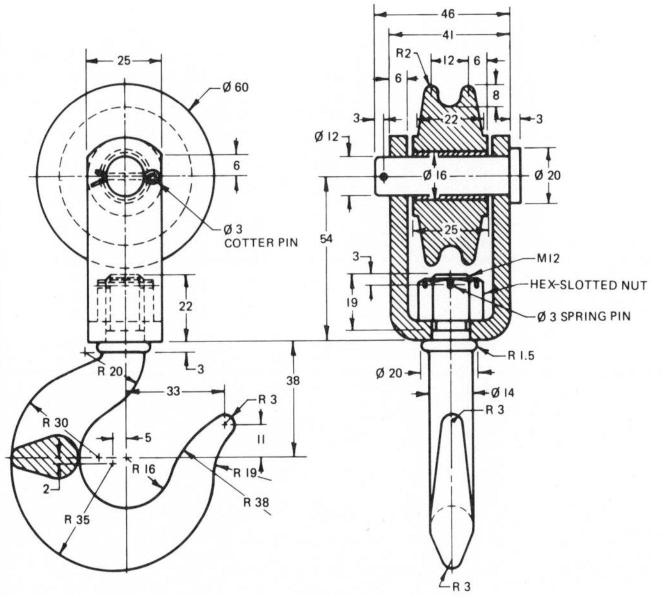

Ø 60
Ø 3 COTTER PIN
Ø 12
Ø 16
Ø 20
M12
HEX-SLOTTED NUT
Ø 3 SPRING PIN
R 1.5
Ø 14
Ø 20
R 20
R 30
R 16
R 19
R 38
R 35
R 3

Fig. 13-1-A Crane hook assembly.

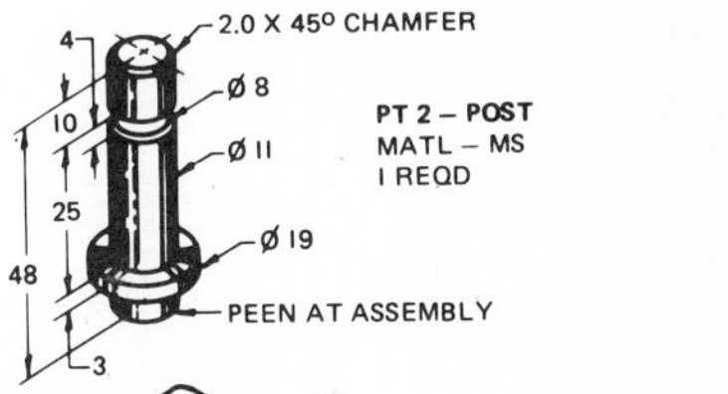

2.0 X 45° CHAMFER
Ø 8
Ø 11
Ø 19
PEEN AT ASSEMBLY
PT 2 – POST
MATL – MS
I REQD

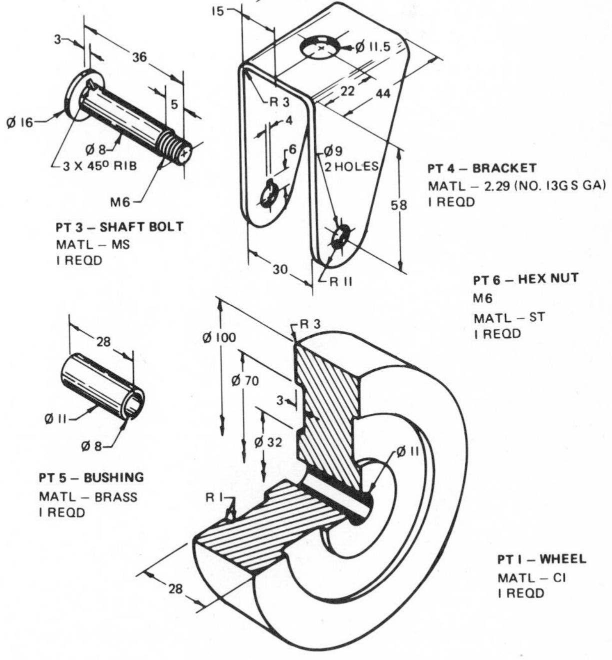

Ø 16
Ø 8
3 X 45° RIB
M6
PT 3 – SHAFT BOLT
MATL – MS
I REQD
Ø 11.5
Ø 9
2 HOLES
PT 4 – BRACKET
MATL – 2.29 (NO. 13G S GA)
I REQD
PT 6 – HEX NUT
M6
MATL – ST
I REQD
Ø 11
Ø 8
PT 5 – BUSHING
MATL – BRASS
I REQD
Ø 100
Ø 70
Ø 32
Ø 11
PT I – WHEEL
MATL – CI
I REQD

Fig. 13-1-B Caster assembly.

UNIT 13-2

FORMING METHODS[1,2]

The most common methods of forming plastic parts are by the application of heat and pressure, casting, and machining. Heat and pressure may be applied by different methods, the most common of which are compression molding, injection molding, transfer molding, extrusion, blow and vacuum forming, and embossing. Other methods include laminating or layup and cold forming and embossing. See Fig. 13-2-1. The method of forming is governed by the material, part, part design, and cost.

Injection Molding

Injection is the principal method of forming thermoplastic materials. Modifications of the injection process are sometimes used for thermosetting plastics.

In *injection molding*, plastic material is put into a hopper which feeds into a heating chamber. A plunger pushes the plastic through this long heating chamber where the material is softened to a fluid state. At the end of this chamber there is a nozzle which abuts firmly against an opening into a cool, closed mold. The fluid plastic is forced at high pressure through this nozzle into the cold mold. As soon as the plastic cools to a solid state, the mold opens and the finished plastic piece is ejected from the press. See Fig. 13-2-2.

Extruding

Extruding is a process generally used with the thermoplastic materials, although it is applicable to the thermosetting plastics. It lends itself to forming of shapes having uniform sections such as sheets, rods, tubes, and filaments. Extrusion is accomplished by forcing material, which has been softened in a heating chamber, through dies of the desired shape, using the pressure created from a screw or hydraulic ram. Extrusions may be conducted continuously and rapidly with a screw-type machine, as shown in Fig. 13-2-3.

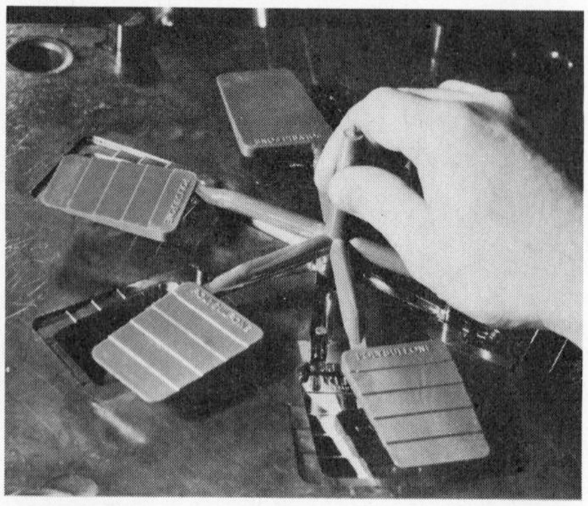

Fig. 13-2-1 Plastic parts being removed from an injection mold. (Union Carbide Corp.)

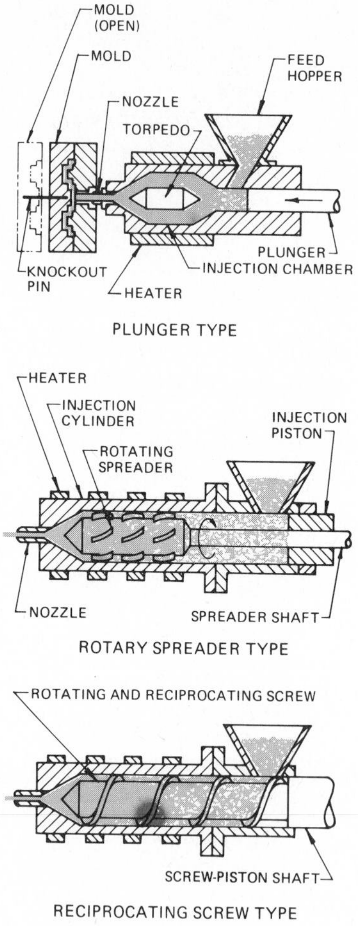
MOLD (OPEN)
MOLD
FEED HOPPER
NOZZLE
TORPEDO
PLUNGER
INJECTION CHAMBER
KNOCKOUT PIN
HEATER
PLUNGER TYPE
HEATER
INJECTION CYLINDER
INJECTION PISTON
ROTATING SPREADER
NOZZLE
SPREADER SHAFT
ROTARY SPREADER TYPE
ROTATING AND RECIPROCATING SCREW
SCREW-PISTON SHAFT
RECIPROCATING SCREW TYPE

Fig. 13-2-2 Injection molding machines. (General Motors Corp.)

Blow Molding

Blow molding is a method of forming used with thermoplastic materials. Basically, blow molding consists of stretching and then hardening a plastic against a mold. See Fig. 13-2-4.

Compression Molding

The molding material is placed in the lower portion of a heated mold cavity having no top and with sides high enough to retain the material. See Fig. 13-2-5*a*. The upper portion of the mold is forced down over the material, usually with high pressure. The combination of heat and pressure causes the molding material to liquefy and flow, filling the mold. In

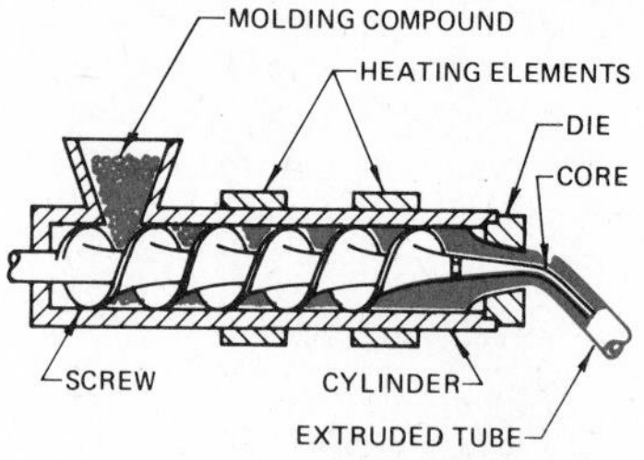
MOLDING COMPOUND
HEATING ELEMENTS
DIE
CORE
SCREW
CYLINDER
EXTRUDED TUBE

Fig. 13-2-3 Extruding machine. (General Motors Corp.)

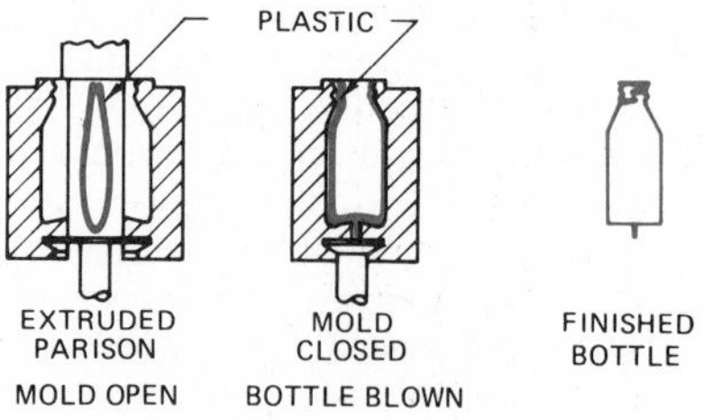
PLASTIC
EXTRUDED PARISON
MOLD OPEN
MOLD CLOSED
BOTTLE BLOWN
FINISHED BOTTLE

Fig. 13-2-4 Blow molding.

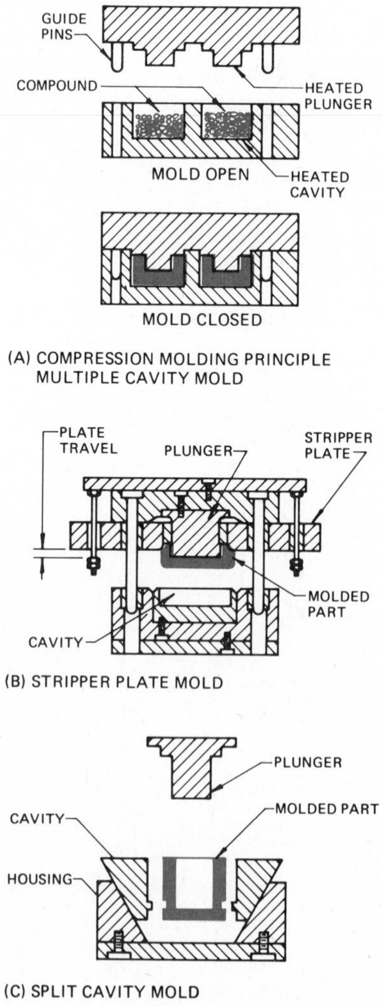
GUIDE PINS
COMPOUND
HEATED PLUNGER
MOLD OPEN
HEATED CAVITY
MOLD CLOSED
(A) COMPRESSION MOLDING PRINCIPLE MULTIPLE CAVITY MOLD
PLATE TRAVEL
PLUNGER
STRIPPER PLATE
MOLDED PART
CAVITY
(B) STRIPPER PLATE MOLD
PLUNGER
CAVITY
MOLDED PART
HOUSING
(C) SPLIT CAVITY MOLD

Fig. 13-2-5 Compression molding. (General Motors Corp.)

the case of thermoplastics, the closed mold must be cooled to permit the molded part to solidify before removal. Thermosetting materials may be removed from the hot mold.

Transfer Molding

Transfer molding is most generally used for thermosetting plastics. This method is like compression molding in that the plastic is cured into an infusible state in a mold under heat and pressure. It differs from compression molding in that the plastic is heated to a point of plasticity before it reaches the mold and is forced into a closed mold by means of a hydraulically operated plunger. See Fig. 13-2-6.

Pulp Molding

Thermosetting plastics are used in *pulp molding*. In this process a porous form, approximating the shape of a finished article, is lowered into a tank containing a mixture of pulp, plastic resins, and water. The water is drawn off through the porous form by a vacuum. This causes the pulp and resin mixture to be drawn to the form and adhere to it. When a sufficient thickness of pulp has been drawn onto the form, it is removed and then molded into final shape.

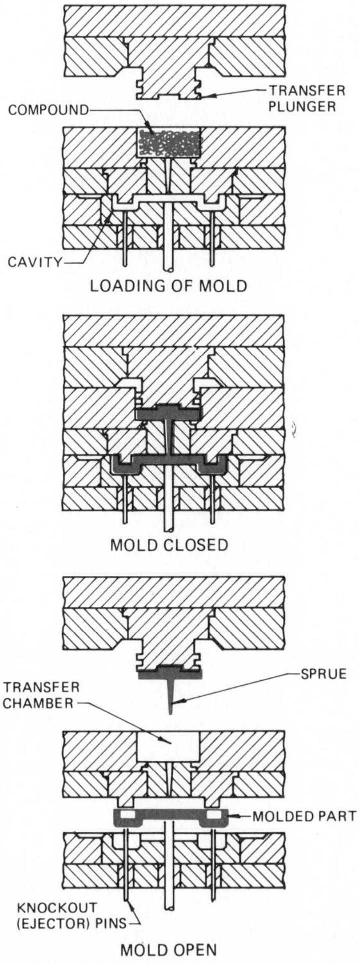
COMPOUND
TRANSFER PLUNGER
CAVITY
LOADING OF MOLD
MOLD CLOSED
TRANSFER CHAMBER
SPRUE
MOLDED PART
KNOCKOUT (EJECTOR) PINS
MOLD OPEN

Fig. 13-2-6 Transfer molding. (General Motors Corp.)

Solvent Molding

Thermoplastic materials are used in this process of forming. *Solvent molding* is based on the fact that when a mold is immersed in a solution and withdrawn, or when it is filled with a liquid plastic and then emptied, a layer of plastic film adheres to the sides of the mold.

Some articles thus formed, like a bathing cap or vial, are removed from the molds. Other solvent moldings remain permanently on the form as, for example, a plastic coating on a metal tube.

Casting

Casting may be employed for both thermoplastic and thermosetting materials in making special shapes, rigid sheets, film and sheeting, rods, and tubes.

The essential difference between casting and molding is that no pressure is used in casting as it is in molding.

In casting, the plastic material is heated to a fluid mass, poured into either open or closed molds, cured at varying temperatures depending on the plastic used, and removed from the molds.

Calendering

Calendering can be used to process thermoplastics into film and sheeting and to apply a plastic coating to textiles or other supporting materials. See Fig. 13-2-7. *Film* refers to thicknesses up to and including 0.2 mm while sheeting includes thicknesses over 0.2 mm.

Coating

Thermosetting and thermoplastic materials may both be used as a coating. The materials to be coated may be metal, wood, paper, fabric, leather, glass, concrete, ceramics, or other plastics.

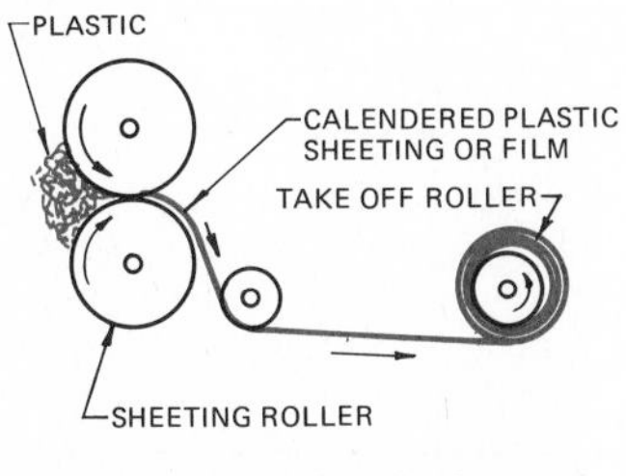
PLASTIC
CALENDERED PLASTIC SHEETING OR FILM
TAKE OFF ROLLER
SHEETING ROLLER

Fig. 13-2-7 Calendering.

Methods of coating are varied and include knife or spread coating, spraying, roller coating, dipping, and brushing. See Fig. 13-2-8. Calendering of a film to a supporting material, described under calendering, is also a form of coating. A new method for applying plastics to metals is called the *fluidized bed process*.

High-Pressure Laminating

Thermosetting plastics are most generally used in *high-pressure laminating* which is distinguished by the use of high heat and pressure. See Fig. 13-2-9. These plastics are used to hold together the reinforcing materials that comprise the body of the finished product. The reinforcing materials may be cloth, paper, wood, fibers of glass.

The end product of high-pressure laminating may be plain flat sheets, decorative sheets as in countertops, or rods, tubes, or formed shapes.

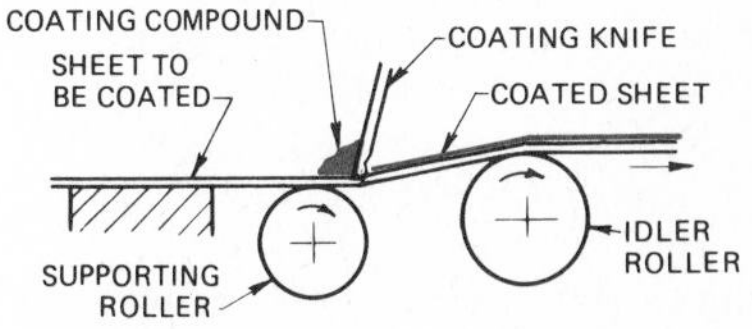
COATING COMPOUND
SHEET TO BE COATED
COATING KNIFE
COATED SHEET
SUPPORTING ROLLER
IDLER ROLLER

Fig. 13-2-8 Coating.

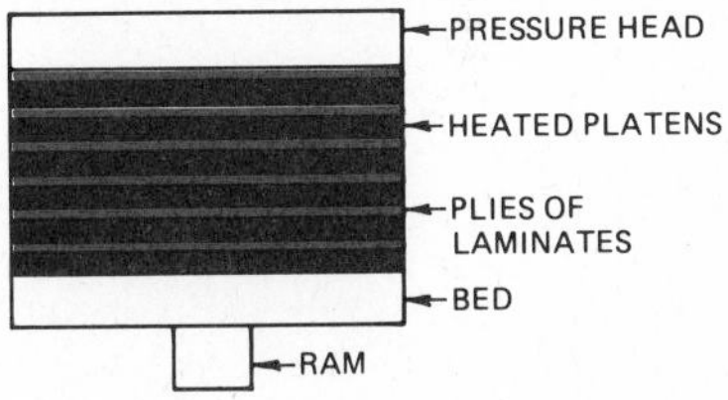
PRESSURE HEAD
HEATED PLATENS
PLIES OF LAMINATES
BED
RAM

Fig. 13-2-9 High-pressure laminating.

Reinforcing

Reinforced plastics mostly employ thermoset plastics, though some thermoplastics are used.

Reinforced plastics differ from high-pressure laminates in that very low or no pressure is used in the processing. The two methods are alike in that the plastic is used to bind together the cloth, paper, or glass fiber reinforcing material used for the body of the product. The reinforcing materials may be in sheet or mat form, and their selection depends on the qualities desired in the end product. See Fig. 13-2-10.

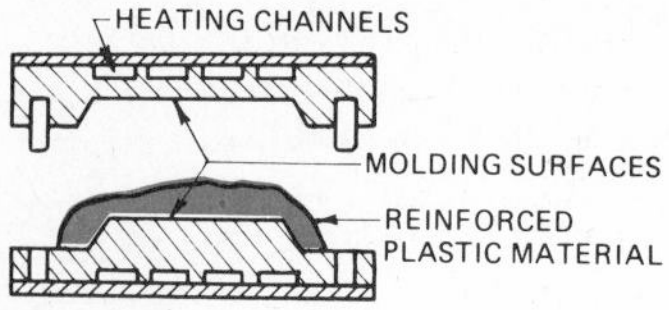
HEATING CHANNELS
MOLDING SURFACES
REINFORCED PLASTIC MATERIAL

Fig. 13-2-10 Reinforcing.

Fabricating

Fabricating covers operations on sheet, rod, tube, sheeting, film, and special shapes to make them into finished products. The materials may be thermosetting or thermoplastic.

Fabricating divides into three broad categories: machining; cutting, sewing, and sealing of film and sheeting; and forming.

Machining. Machining is used on rigid sheets, rods, tubes, and special shapes. The various operations include grinding, turning on a lathe, sawing, reaming, milling, routing, drilling, tapping.

Cutting, Sewing, and Sealing of Film and Sheeting. In this category of fabricating fall all the operations involved in fashioning plastic film and sheeting into finished articles like inflatable toys, garment bags, aprons, raincoats, luggage.

For all these articles, the film or sheeting must first be cut to pattern.

Forming. In working with flexible thermoplastic sheets, the first step is usually the cutting or blanking out of sections roughly approximating the dimensions of the finished article. This blank may be beaded for added strength, creased, and folded into final form, a box for example; or deep drawing may be employed, using male and female molds to shape the plastic blank.

Finishing

The finishing of plastics includes the different methods of adding either decorative or functional surface effects to a plastic product.

REFERENCES AND SOURCE MATERIAL

1. The Society of the Plastics Industry, Inc.
2. General Motors Corporation.

Assignment

Prepare a report to your supervisor recommending the method of manufacture for any one of the parts shown in Figs. 13-2-A and 13-2-B. The report must be concise and brief. The manner of presentation and the analysis arriving at your conclusion could well be a deciding factor for future promotion within a company.

REVIEW FOR ASSIGNMENT

Unit 13-1 Kinds of Plastic

UNIT 13-3

DESIGN CONSIDERATIONS—FOR SINGLE PARTS[1,2]

The design of molded parts involves several factors not normally encountered with machine-fabricated and assembled parts. It is important that the designer take these factors into consideration.

Shrinkage. *Shrinkage* is defined as the difference between dimensions of the mold and the corresponding dimensions of the molded part. Normally the mold designer is more concerned with shrinkage than the molded part designer. Shrinkage does, however, affect dimensions, warpage, residual stress, and moldability. The amount of shrinkage varies with the material, the part design, and the curing cycle.

Section Thickness. Solidification is a function of heat transfer from or to the mold for both thermoplastics and thermosets. Each material has a fixed rate of heat transfer. Therefore, where section thickness varies, areas within a molded part will solidify at different rates. The varying rates will cause irregular shrinkage, sink marks, additional strain, and warpage. For these reasons, uniform section thickness is important and may be maintained by adding holes or depressions, as shown in Figs. 13-3-1 and 13-3-2.

Molded Holes. A through hole is more advantageous than a blind hole since it is more accurate and economical. Blind holes should not be more than twice as deep as the diameter, as shown in Fig. 13-3-3. Avoid placing holes at an-

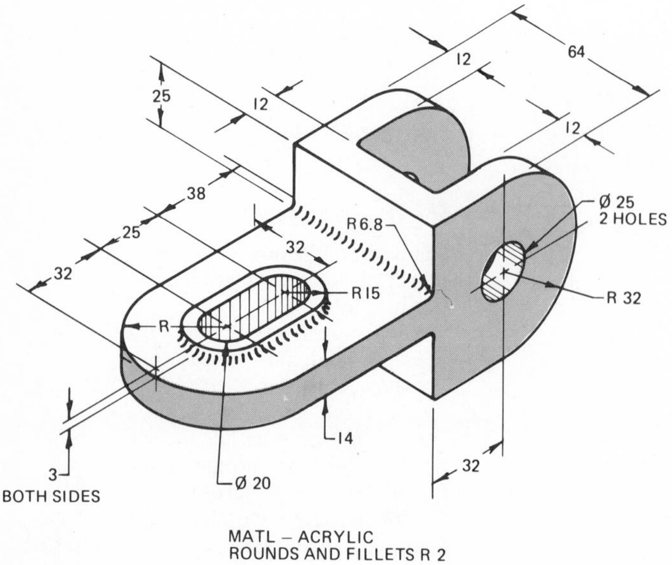
64
12
12
12
25
38
25
32
R6.8
32
R15
R
Ø 25
2 HOLES
R 32
14
3
BOTH SIDES
Ø 20
32
MATL – ACRYLIC
ROUNDS AND FILLETS R 2

Fig. 13-2-A Swing bracket.

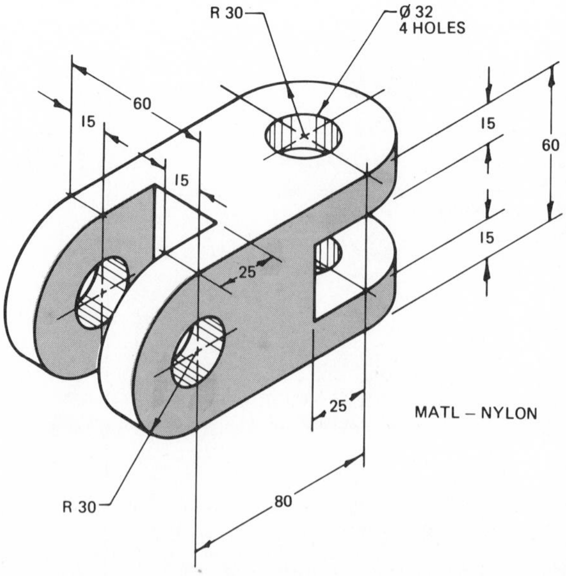
R 30
Ø 32
4 HOLES
60
15
15
15
60
15
25
25
MATL – NYLON
R 30
80

Fig. 13-2-B Coupling.

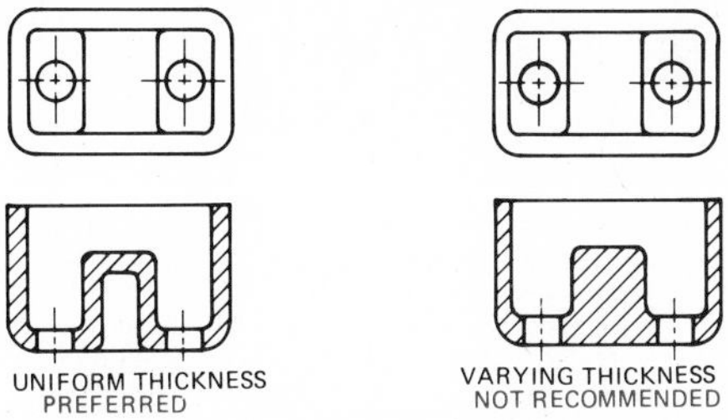
UNIFORM THICKNESS
PREFERRED
VARYING THICKNESS
NOT RECOMMENDED

Fig. 13-3-1 Section thickness.

MATERIAL	ABSOLUTE MINIMUM (mm)	RECOMMENDED MINIMUM (mm)
THERMOPLASTICS		
ABS	0.6	1.6
ACRYLIC	0.6	2.4
CELLULOSIC	0.6	1.9
FLUOROCARBON	0.2	0.3
POLYAMIDE	0.4	1.5
POLYCARBONATE	0.6	2.4
POLYETHYLENE	0.9	1.6
PLOYSTYRENE	0.8	1.6
POLYVINYL CHLORIDE	2.4	2.4
THERMOSETS		
EPOXY	1.6	3.2
POLYESTER		
GLASS FILLED	1.0	4.7
MINERAL FILLED	1.0	3.2
PHENOLICS		
GENERAL PURPOSE	1.3	3.2
FABRIC FILLED	1.6	4.7
MINERAL FILLED	3.2	4.7
UREA AND MALAMINES		
GENERAL PURPOSE	0.9	2.5
FABRIC FILLED	1.3	3.2
MINERAL FILLED	1.0	4.7

Fig. 13-3-2 Section thickness for various plastics. (General Motors Corp.)

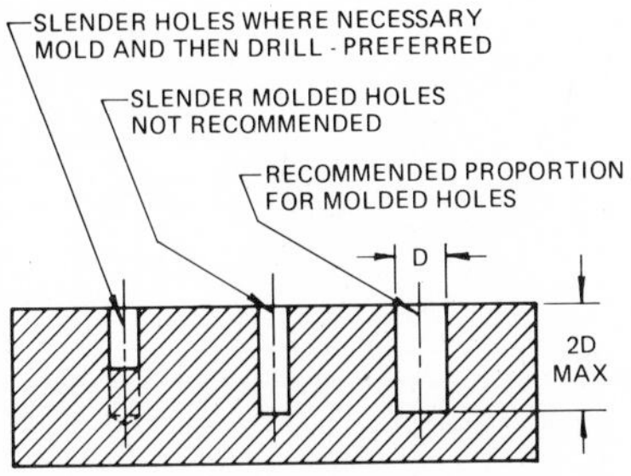
SLENDER HOLES WHERE NECESSARY
MOLD AND THEN DRILL - PREFERRED
SLENDER MOLDED HOLES
NOT RECOMMENDED
RECOMMENDED PROPORTION
FOR MOLDED HOLES
D
2D
MAX

Fig. 13-3-3 Blind holes.

gles other than perpendicular to the flash line. If such holes are necessary, consider using a drilled hole to maintain simple molding. See Fig. 13-3-4.

Threads. External and internal threads can be easily molded by means of loose-piece inserts and rotating core pins. External threads may be formed by placing the cavity so that the threads are formed in the mold platen. This method requires only a sliding side core. Internal threads may be formed either with a rotating core pin or by unthreading the part from the core pin.

Gates. Gate location should be anticipated during the design stage. Avoid gating into areas subjected to high stress levels, fatigue, or impact. To optimize molding, locate gates in the heaviest section of the part. See Fig. 13-3-5.

Internal and External Draft. Draft is necessary on all rigid molded articles to facilitate removal of the part from the mold. Draft may vary from 0.25 to 4° per side, depending upon the length of the vertical wall, surface area, finish, kind of material, and the mold or method of ejection used.

Parting or Flash Line. *Flash* is that portion of the molding material which flows or exudes from the mold parting line during molding. Any mold which is made of two or more parts may produce flash at the line of junction of the mold parts. The thickness of flash usually varies between 0.05 and 0.40 mm, depending upon the accuracy of the mold, type of material, and the process used. See Fig. 13-3-6.

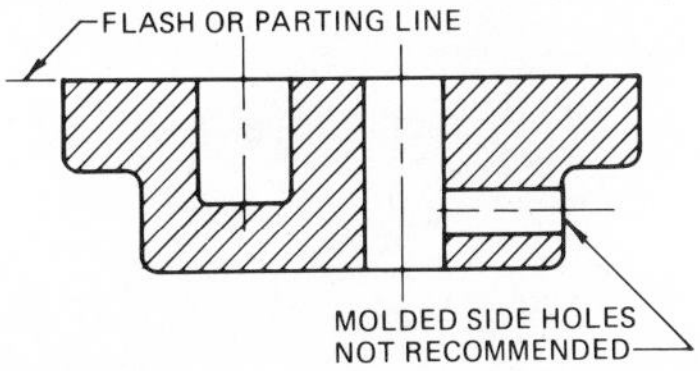
FLASH OR PARTING LINE
MOLDED SIDE HOLES
NOT RECOMMENDED

Fig. 13-3-4 Avoiding molded holes which are perpendicular to parting line.

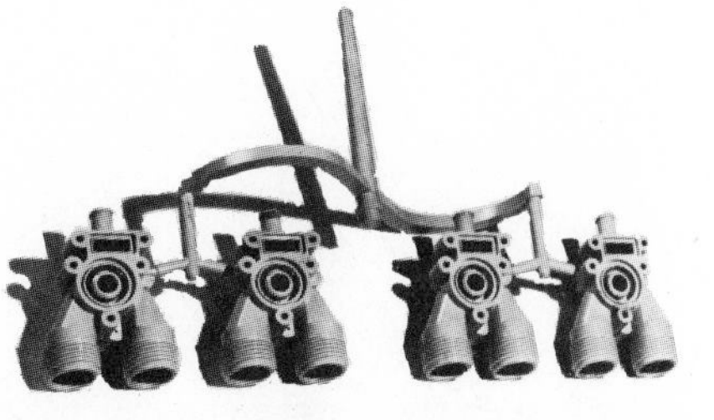

Fig. 13-3-5 Gating. (Union Carbide Corp.)

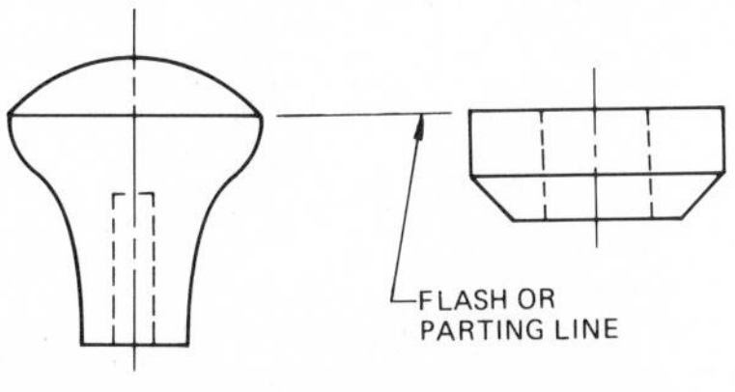
FLASH OR
PARTING LINE

Fig. 13-3-6 Parting or flash line. (General Motors Corp.)

Fillets and Radii. The principal functions of fillets and radii are to ease the flow of plastic within the mold, to facilitate ejection of the part, and to distribute stress in the part in service. During molding, the material is liquefied, but it is a heavy, viscous liquid which does not easily flow around sharp corners. The liquid tends to bend around corners; therefore, rounded corners permit the liquid plastic to flow smoothly and easily through the mold. For recommended radii, see Fig. 13-3-7.

Undercuts. Parts with undercuts should be avoided. Normally, parts with external undercuts cannot be withdrawn from a one-piece mold. Internal undercuts are considered impractical and should be avoided. If an internal undercut is essential, it may be achieved by machining or by use of a flexible material for part or mold core. See Fig. 13-3-8.

Ribs and Bosses. Ribs increase rigidity of a molded part without increasing wall thickness and sometimes facilitate flow during molding. Bosses reinforce small, stressed areas, providing sufficient strength for assembly with inserts or screws. Recommended proportions for ribs and bosses are shown in Fig. 13-3-9*a*.

REFERENCES AND SOURCE MATERIAL

1. General Motors Corporation.
2. General Electric Company.

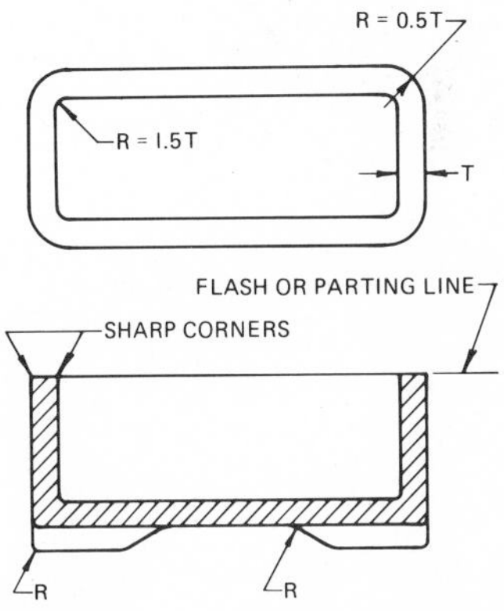
R = 0.5T
R = 1.5T
T
FLASH OR PARTING LINE
SHARP CORNERS
R
R

Fig. 13-3-7 Fillets and radii. (General Motors Corp.)

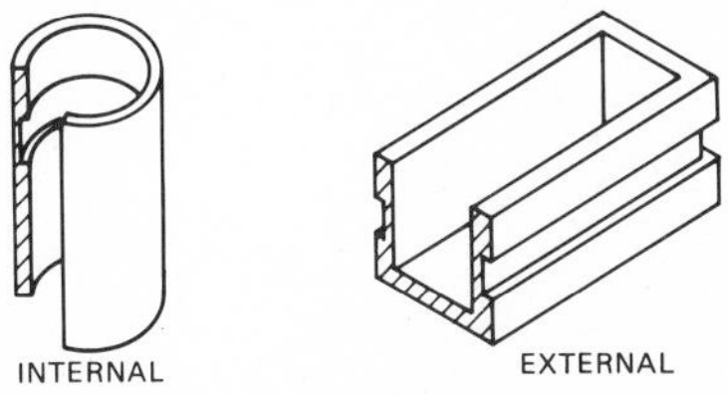
INTERNAL
EXTERNAL

Fig. 13-3-8 Undercuts. (General Motors Corp.)

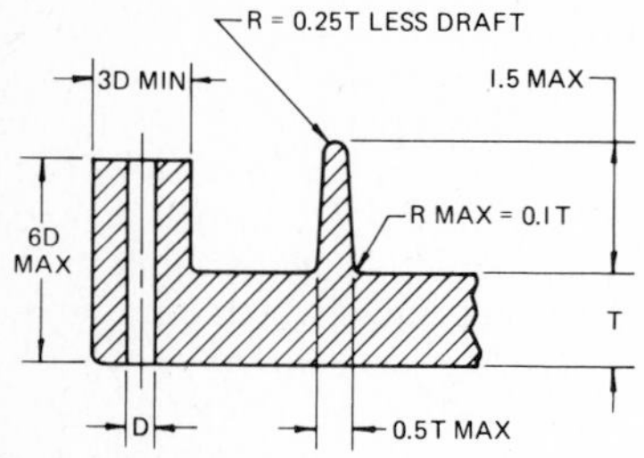
R = 0.25T LESS DRAFT
3D MIN
1.5 MAX
6D MAX
R MAX = 0.1T
T
D
0.5T MAX

(A) RIB AND BOSS PROPORTION

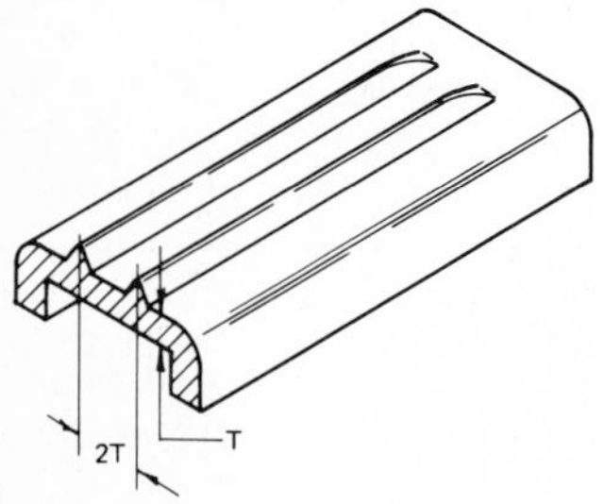
2T
T

(B) RIB SPACING

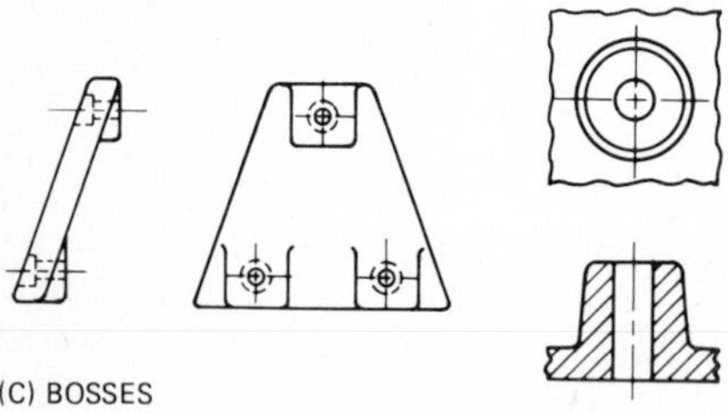

(C) BOSSES

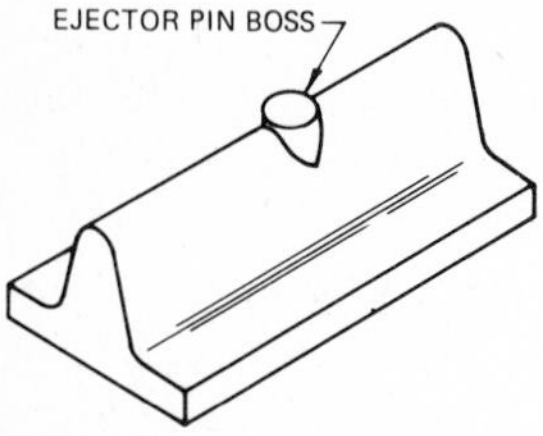
EJECTOR PIN BOSS

(D) EJECTOR PIN BOSSES

Fig. 13-3-9 Ribs and bosses. (General Motors Corp.)

Assignment

On an A3- or B-size sheet, prepare a casting drawing for one of the parts shown in Figs. 13-3-A to 13-3-C. Refer to the casting recommendations shown in this unit and indicate the parting line on the drawing. Use your judgment for dimensions not given. Scale is 1:1.

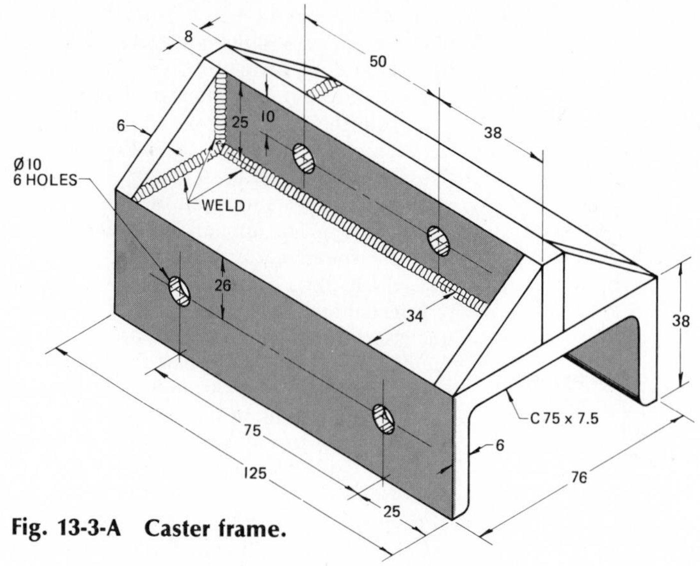
8
50
6
25
10
38
Ø 10
6 HOLES
WELD
26
34
38
C 75 x 7.5
75
6
125
76
25

Fig. 13-3-A Caster frame.

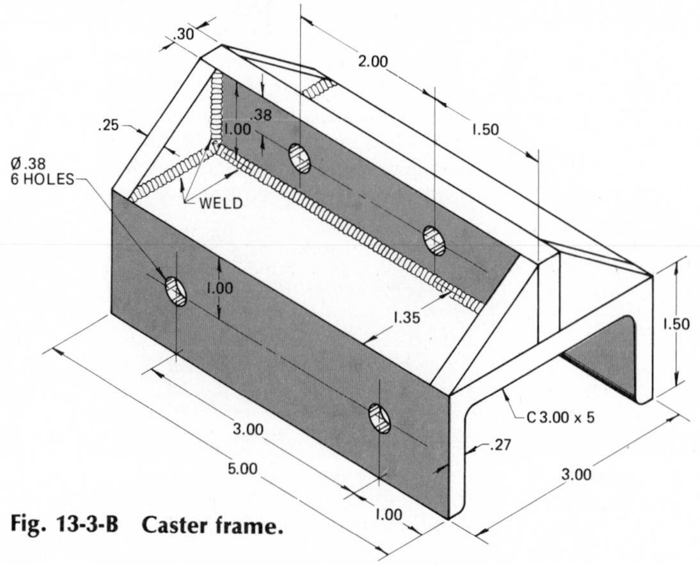
.30
2.00
.25
.38
1.00
1.50
Ø .38
6 HOLES
WELD
1.00
1.35
1.50
C 3.00 x 5
3.00
.27
5.00
3.00
1.00

Fig. 13-3-B Caster frame.

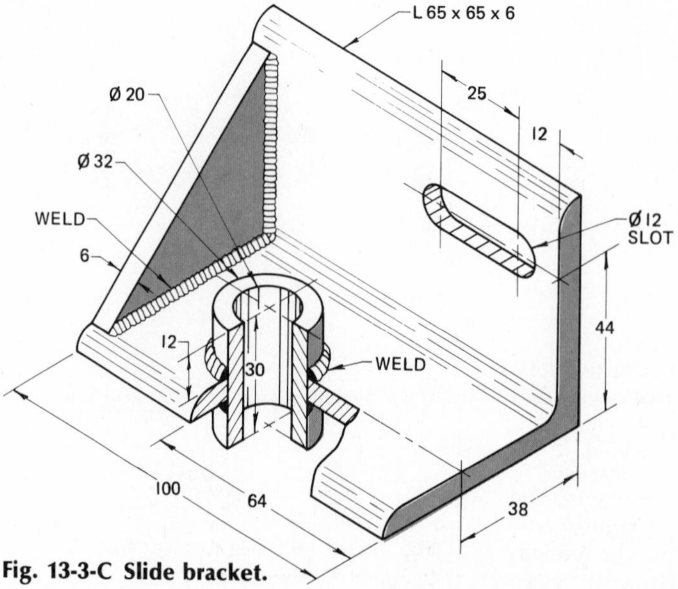
L 65 x 65 x 6
Ø 20
25
12
Ø 32
WELD
Ø 12
SLOT
6
44
12
30
WELD
100
64
38

Fig. 13-3-C Slide bracket.

UNIT 13-4

DESIGN CONSIDERATIONS—FOR ASSEMBLIES[1,2]

The design of molded parts which are to be assembled with typical fastening methods involves factors different from those normally encountered with metal.

Holes and Threads

Mechanical fasteners, in general, depend upon a hole of some type. Holes should be designed and located to provide maximum strength and minimum molding problems. Any straight hole, molded or machined, should have between it and an adjacent hole, or side wall, an amount of material equal to or greater than the diameter or width of the hole. Any threaded hole, molded or tapped, should have between it and an adjacent hole, or side wall, an amount of material at least 3 times the outside diameter of the thread. Spacing may be reduced, however, by proper use of bosses.

Drilled holes are often more accurate and easier to produce than molded holes, even though they require a second operation. In order to locate holes consistently, drill jigs may be used or spot points may be molded directly into the part.

Tapped holes provide an economical means of joining a molded part to its assembly. The designer should avoid threads with a pitch of less than 0.8 mm. Holes which are to be tapped should be countersunk to prevent chipping when the tap is inserted.

External and internal threads can be molded integral with the part. Molded threads are generally more expensive to form than other threads because either a method of unscrewing the part from the mold must be provided or a split mold must be used. Molded threads should be used only for diameters of 8 mm or greater and should not have a pitch less than 0.8 mm. Thread length of 1.25 turns will provide adequate holding power with minimum material usage.

Inserts

After the molding material has been determined, the insert should be designed. The molded part should be designed around the insert. Inserts, usually metallic, may be divided into two general groups:

1. Decorative, protective, or structural inserts please the eye, protect from bodily injury, add strength, or control shrinkage.

2. Attaching or conductive inserts assist assembly or provide an electrical or heat-conducting path.

Inserts of round rod stock, coarse diamond knurled, and grooved provide the strongest anchorage under torque and tension. A large single groove with knurling on each end, as in Fig. 13-4-1, is superior to two or more grooves with smaller knurled surface areas. See also examples of inserts in Fig. 13-4-2.

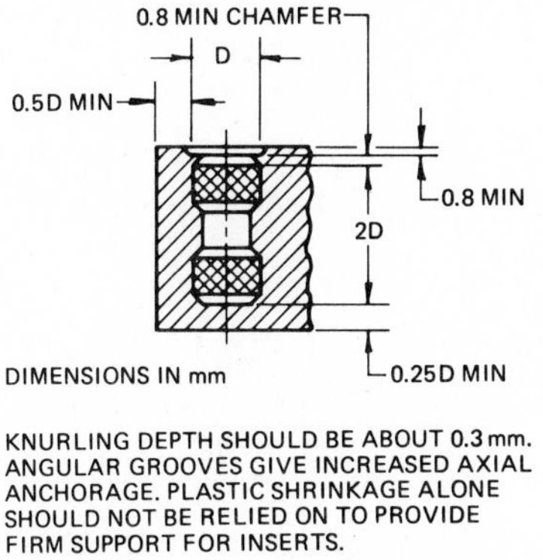
0.8 MIN CHAMFER
D
0.5D MIN
0.8 MIN
2D
DIMENSIONS IN mm
0.25D MIN
KNURLING DEPTH SHOULD BE ABOUT 0.3 mm.
ANGULAR GROOVES GIVE INCREASED AXIAL ANCHORAGE. PLASTIC SHRINKAGE ALONE SHOULD NOT BE RELIED ON TO PROVIDE FIRM SUPPORT FOR INSERTS.

Fig. 13-4-1 Supporting inserts. (General Motors Corp.)

Press and Shrink Fits

Inserts may be secured by a press fit, or the plastic molding material may be assembled to a larger part by a shrink fit, as shown in Fig. 13-4-3. Both methods rely on shrinkage of the material, which is greatest immediately after removal from the mold.

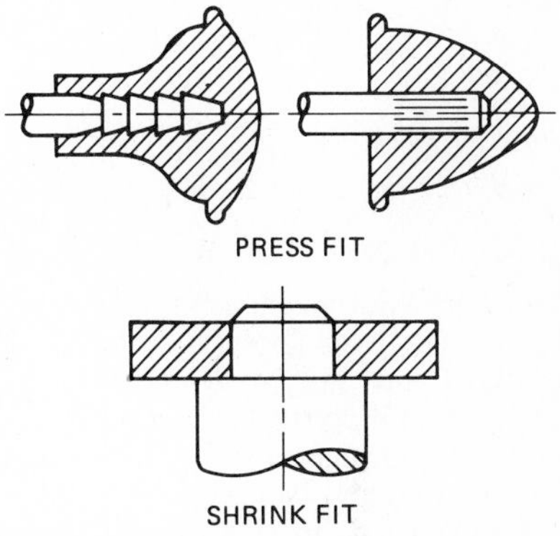
PRESS FIT
SHRINK FIT

Fig. 13-4-3 Press and shrink fits. (General Motors Corp.)

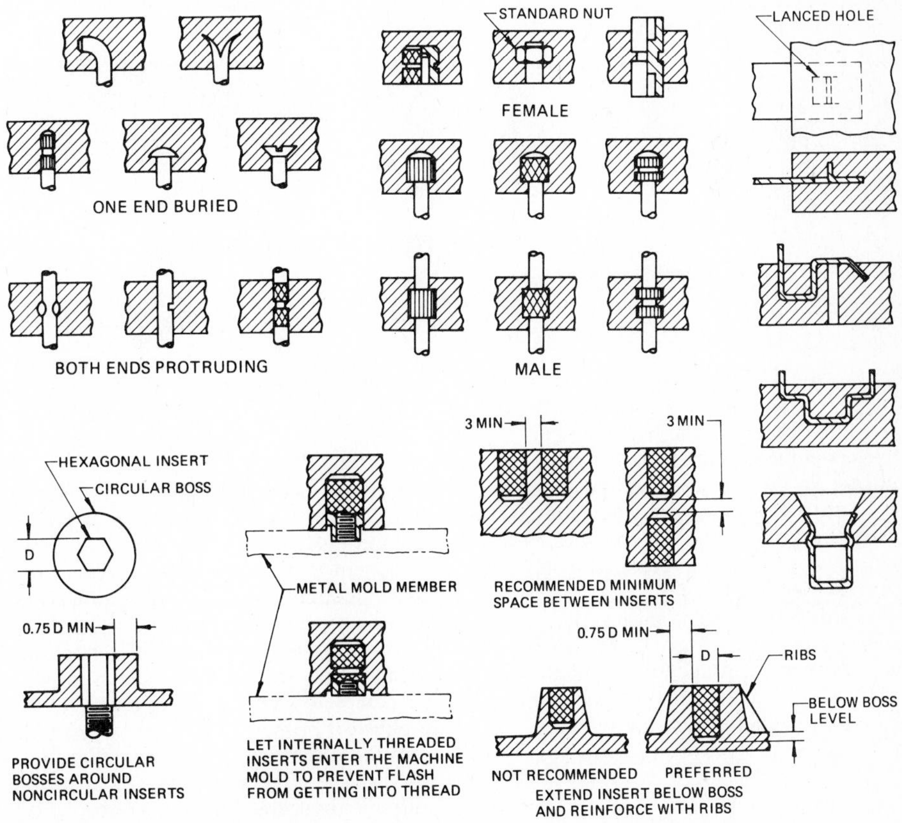
ONE END BURIED
BOTH ENDS PROTRUDING
STANDARD NUT
FEMALE
MALE
LANCED HOLE
HEXAGONAL INSERT
CIRCULAR BOSS
D
0.75 D MIN
PROVIDE CIRCULAR BOSSES AROUND NONCIRCULAR INSERTS
METAL MOLD MEMBER
LET INTERNALLY THREADED INSERTS ENTER THE MACHINE MOLD TO PREVENT FLASH FROM GETTING INTO THREAD
3 MIN
3 MIN
RECOMMENDED MINIMUM SPACE BETWEEN INSERTS
0.75 D MIN
D
RIBS
BELOW BOSS LEVEL
NOT RECOMMENDED
PREFERRED
EXTEND INSERT BELOW BOSS AND REINFORCE WITH RIBS

Fig. 13-4-2 Insert application. (General Motors Corp.)

Heat Forming and Heat Sealing

Most thermoplastics can be reformed by the application of heat and pressure, as in Fig. 13-4-4. This reforming often eliminates the need for other assembly methods, such as adhesive bonding and mechanical fasteners. This method cannot be used with thermosetting materials.

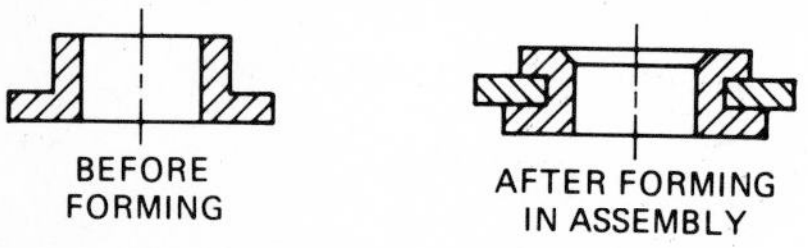
BEFORE
FORMING
AFTER FORMING
IN ASSEMBLY

Fig. 13-4-4 Heat forming.

Mechanical Fastening

Various designs of mechanical fasteners are commercially available. Spring-type metal hinges and clips, speed clips or nuts, and expanding rivets are a few of these designs. Design of the parts for assembly requires that molded parts have sufficient sectional strength to withstand the stresses that will be encountered with fasteners. A strengthening of the area which will receive the brunt of these applied stresses is usually required. See Fig. 13-4-5.

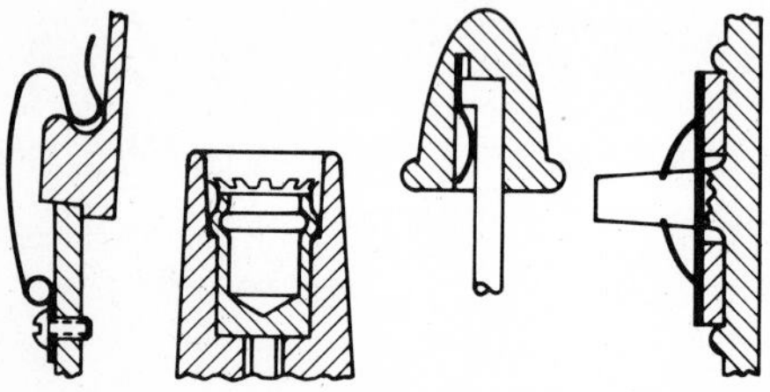

Fig. 13-4-5 Mechanical fasteners.

Rivets

Conventional riveting equipment and procedures can be used with plastics. Care must be exercised to minimize stresses induced during the fastening operation. To do this, the rivet head should be 2.5 to 3 times the shank diameter. Also, rivets should be backed with either plates or washers to avoid high localized stresses. See Fig. 13-4-6.

Drilled holes rather than punched holes are preferred for fasteners. If possible, fastener clearance in the hole should be at least 0.3 mm to maintain a plane stress condition at the fastener.

Boss Caps

A boss cap is a cup-shaped metal ring which is pressed onto the boss by hand, with an air cylinder, or with a light-duty press. It is designed to reinforce the boss against the expansion force exerted by tapping screws. See Fig. 13-4-7.

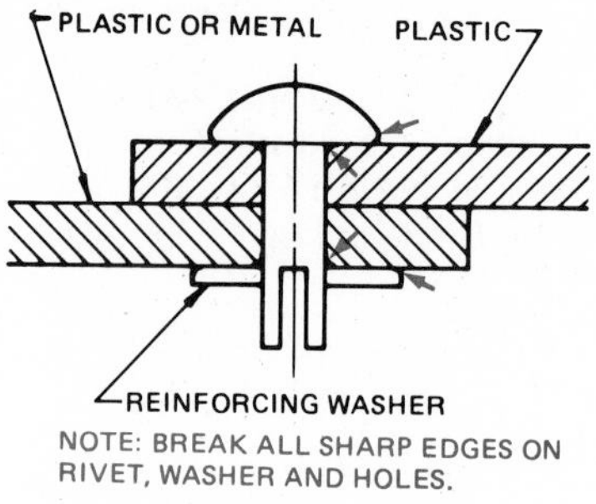
PLASTIC OR METAL
PLASTIC
REINFORCING WASHER
NOTE: BREAK ALL SHARP EDGES ON
RIVET, WASHER AND HOLES.

Fig. 13-4-6 Recommended riveting procedure.

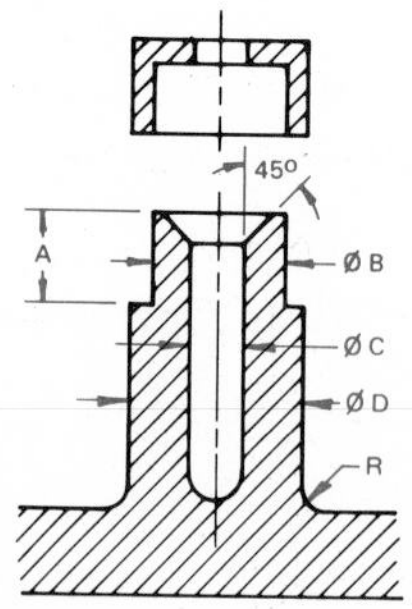
45°
A
ØB
ØC
ØD
R

THREAD SIZE	DIMENSIONS				
	A	B	C	D	R
M 3.5	3.5	5.3	2.5	7	0.5
M 4	4	6.3	3.3	8.5	0.5
M 5	4.5	7.3	4.2	10	0.5

Fig. 13-4-7 Boss cap design.

Adhesive Bonding

When two or more parts are to be joined into an assembly, adhesives permit a strong, durable fastening between similar materials and often are the only fastening method available for joining dissimilar materials. Structural adhesives are made from the same basic resins as many plastics and thus react to their operating environment in a similar manner. In order to provide maximum strength, adhesives must be applied as a liquid to thoroughly wet the surface of the part. The bonding surface must be chemically clean to permit complete wetting. Basic plastics vary in physical properties, so adhesives made from these materials also vary. (See Fig. 13-4-8.) The effects of applied loads to an adhesive joint are similar to those applied to a molded or formed plastic part. A proper adhesive joint will cause the tensile failure to occur in the part or the adhesive, but not at the bonded surface. The tensile strength of the parts, as well as the tensile strength of the adhesive in bulk, is important in predicted joint strength. In order to minimize the weak points of adhesives, the geometry of the joint should be properly designed. Size and shape of the parts to be joined are dictated by the required function and may not allow the most effective joint design.

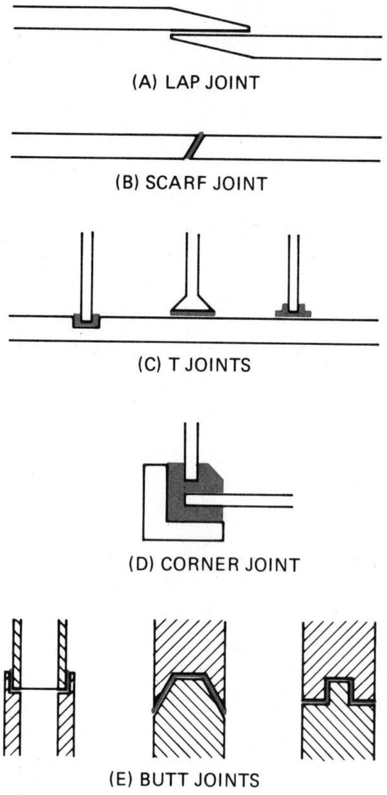
(A) LAP JOINT
(B) SCARF JOINT
(C) T JOINTS
(D) CORNER JOINT
(E) BUTT JOINTS

Fig. 13-4-8 Adhesive bonding.

Ultrasonic Bonding

Ultrasonic bonding often is used instead of solvent cementing to bond plastic parts. By using this technique, irregularly shaped parts can be bonded in 2 seconds or less. The bonded parts may be handled and used at reasonable temperatures within minutes after joining.

Only one of the mating parts comes in contact with the horn. The part transmits the ultrasonic vibration to small, hidden bonding areas, resulting in fast, perfect welds. Both mating halves remain cool except at the seam, where the energy is quickly dissipated.

This technique is not recommended where high impact strength is required in the bond area. See Fig. 13-4-9.

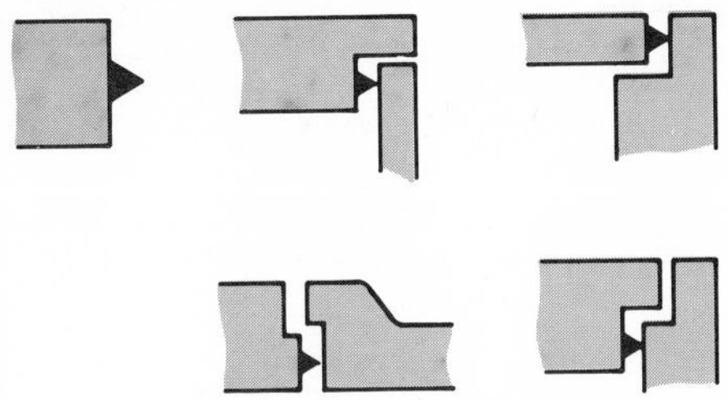

Fig. 13-4-9 Design joints for ultrasonic bonding.

Ultrasonic Staking

Ultrasonic staking frequently involves the assembly of metal parts. In this technique, a stud molded into the plastic part protrudes through a hole in the metal part. The surface of the stud is vibrated with a horn having high amplitude and a relatively small contact area. The vibration causes the stud to melt and reform in the configuration of the horn tip. See Fig. 13-4-10.

Friction or Spin Welding

This welding technique is limited to parts with circular joints. It is especially useful for large parts where ultrasonic welding or chemical bonding is impractical.

In friction or spin welding, the faces to be joined are pressed together while one part is spun and the other is held fixed. Frictional heat produces a molten zone that becomes a weld when spinning stops. See Fig. 13-4-11.

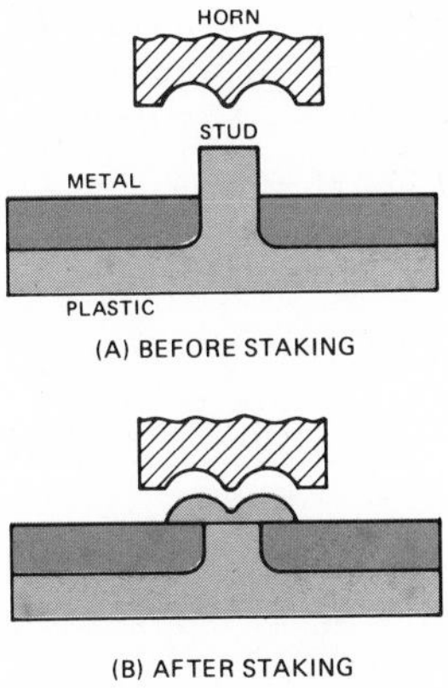
HORN
STUD
METAL
PLASTIC
(A) BEFORE STAKING
(B) AFTER STAKING

Fig. 13-4-10 Typical ultrasonic staking operation.

DRAWINGS

In addition to the usual considerations, the following points should be taken into account when a detail drawing of a plastic part is made:

1. Can the part be removed from the mold?
2. Is location of flash line consistent with design requirements?
3. Is section thickness consistent? Are there thick sections? thin sections? Could greater uniformity of section thickness be maintained?
4. Has the material been correctly specified?
5. Is each feature in accordance with the thinking of competent materials engineers and molders?

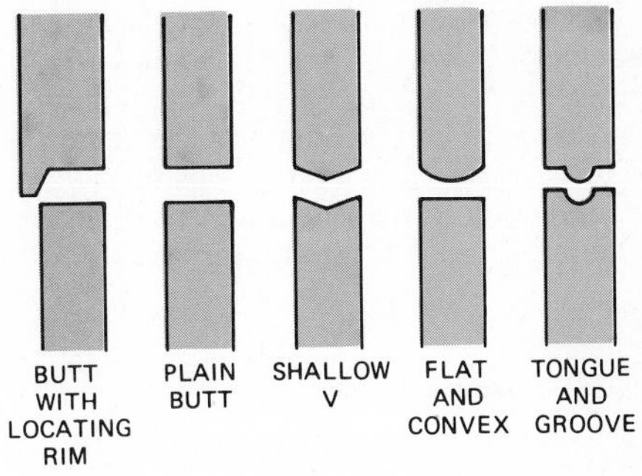
BUTT WITH LOCATING RIM
PLAIN BUTT
SHALLOW V
FLAT AND CONVEX
TONGUE AND GROOVE

Fig. 13-4-11 Joint shapes for spin welding.

6. Have close tolerance requirements been reviewed with responsible engineers?
7. Have marking requirements been specified to inform field service people of the material from which the part is fabricated?

REFERENCES AND SOURCE MATERIAL

1. General Motors Corporation.
2. General Electric Company.

Assignments

1. On an A3- or B-size sheet, add a threaded insert to one of the parts shown in Fig. 13-4-A or 13-4-B. Use your judgment for dimensions not shown and the type and number of views required. Scale is 1:1.

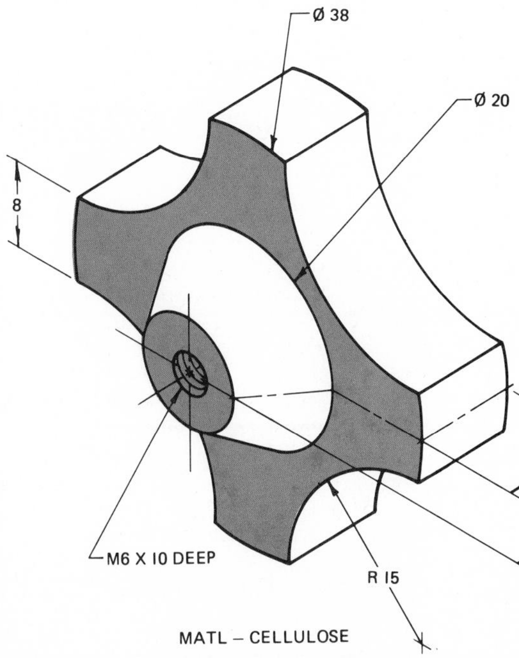
Ø 38
Ø 20
8
10
16
M6 X 10 DEEP
R 15
MATL – CELLULOSE

Fig. 13-4-A Lamp adjusting knob.

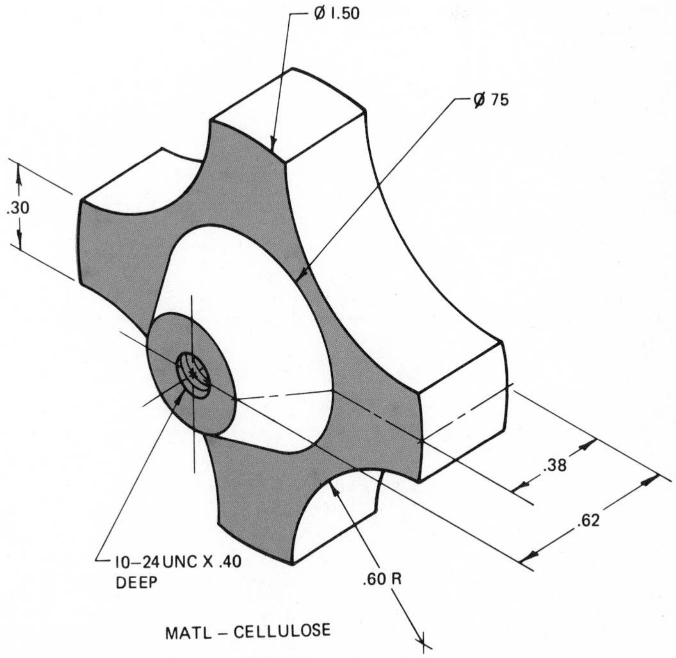
Ø 1.50
Ø 75
.30
.38
.62
10–24 UNC X .40 DEEP
.60 R
MATL – CELLULOSE

Fig. 13-4-B Lamp adjusting knob.

2. On an A3- or B-size sheet, make an assembly drawing of the parts shown in either Fig. 13-4-C or 13-4-D. The retaining ring is to be positioned in the center of the part and molded into position. Modification to the retaining ring may be required to prevent the ring from turning in the wheel. Scale is 5:1. Show a top view and a full-section view. Dimension the finished assembly.

REVIEW FOR ASSIGNMENTS

Unit 6-1 Working Drawings
Unit 7-1 Full Sections

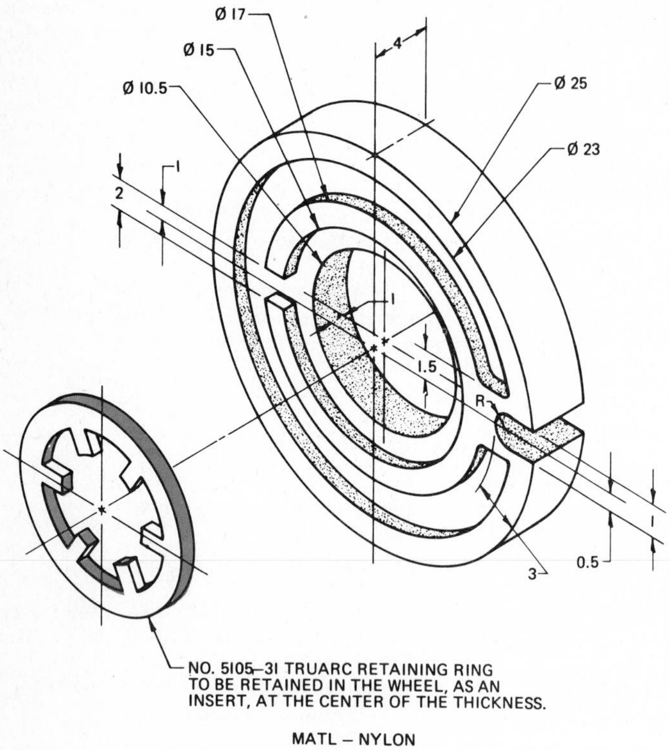

Ø 17
Ø 15
Ø 10.5
4
Ø 25
Ø 23
2
1
1
1.5
R
1
0.5
3
NO. 5105–31 TRUARC RETAINING RING
TO BE RETAINED IN THE WHEEL, AS AN
INSERT, AT THE CENTER OF THE THICKNESS.
MATL – NYLON

Fig. 13-4-C Cassette-tape drive wheel.

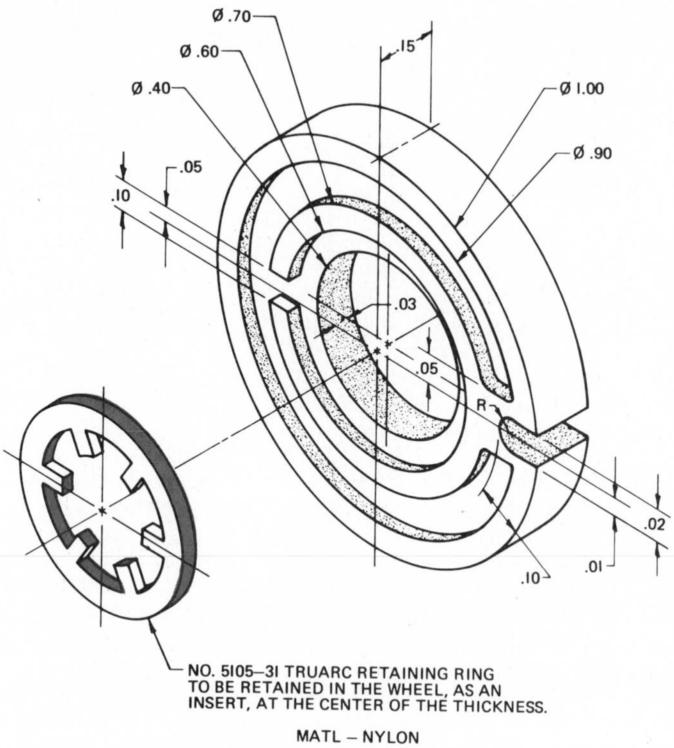

Ø .70
Ø .60
Ø .40
.15
Ø 1.00
Ø .90
.05
.10
.03
.05
R
.02
.01
.10
NO. 5105–31 TRUARC RETAINING RING
TO BE RETAINED IN THE WHEEL, AS AN
INSERT, AT THE CENTER OF THE THICKNESS.
MATL – NYLON

Fig. 13-4-D Cassette-tape drive wheel.

Part 3
Intermediate Drawing Design

Chapter 14
Auxiliary Views

UNIT 14-1
PRIMARY AUXILIARY VIEWS

Many machine parts have surfaces that are not perpendicular, or at right angles, to the plane of projection. These are referred to as *sloping* or *inclined* surfaces. In the regular orthographic views, such surfaces appear to be foreshortened, and their true shape is not shown. When an inclined surface has important characteristics that should be shown clearly and without distortion, an auxiliary view is used so that the drawing completely and clearly explains the shape of the object. In many cases, the auxiliary view will replace one of the regular views on the drawing, as illustrated in Fig. 14-1-1.

One of the regular orthographic views will have an edge line representing the inclined surface. The auxiliary view is projected from this edge line, at right angles, and is drawn parallel to the edge line.

Only the true shape features on the views need be drawn, as shown in Fig. 14-1-2. Since the auxiliary view shows only the true shape and detail of the inclined surface or features, a partial auxiliary view is all that is necessary. Likewise, the distorted features on the regular views may be omitted. Hidden lines are usually omitted unless required for clarity. This procedure is recommended for functional and production drafting where drafting costs are an important consideration. However, the drafter may be called upon to draw the complete views of the part. This type of drawing is often used for catalog and standard parts drawings.

Dimensioning Auxiliary Views

One of the basic rules of dimensioning is to dimension the feature where it can be seen in its true shape and size. Thus the auxiliary view will show only the dimensions pertaining to those features for which the auxiliary view was drawn. The recommended dimensioning method for engineering drawings is the unidirectional system. Not only is this method of dimensioning easier to prepare and read, but it is readily adaptable to mechanical lettering. See Fig. 14-1-3.

Assignments

1. On an A3- or B-size sheet, make a working drawing of the cross slot bracket shown in Fig. 14-1-A or 14-1-B. Replace the side view with an auxiliary view. Only partial views need be drawn, and hidden lines may be added to improve clarity. Scale is 1:1.

2. On an A3- or B-size sheet, make a working drawing of the angle bracket shown in Fig. 14-1-C or 14-1-D. Replace the top view with an auxiliary view. Draw complete views with hidden lines. Scale is 1:1.

REVIEW FOR ASSIGNMENTS

Unit 5-7 Machining Symbols
Unit 7-18 Rounds and Fillets

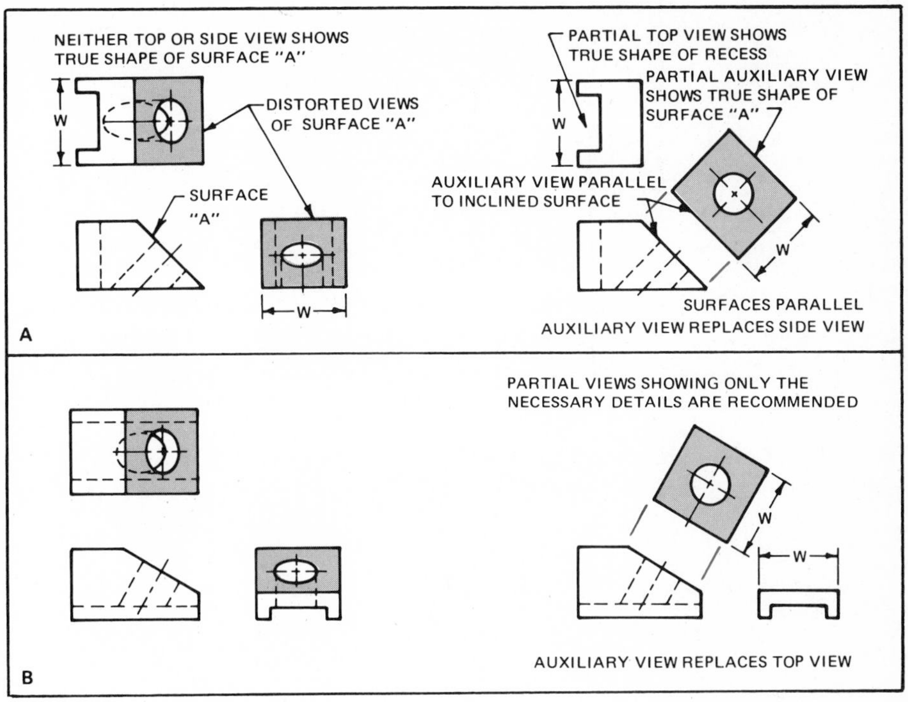
NEITHER TOP OR SIDE VIEW SHOWS TRUE SHAPE OF SURFACE "A"
DISTORTED VIEWS OF SURFACE "A"
SURFACE "A"
W
PARTIAL TOP VIEW SHOWS TRUE SHAPE OF RECESS
PARTIAL AUXILIARY VIEW SHOWS TRUE SHAPE OF SURFACE "A"
AUXILIARY VIEW PARALLEL TO INCLINED SURFACE
SURFACES PARALLEL
AUXILIARY VIEW REPLACES SIDE VIEW
A
PARTIAL VIEWS SHOWING ONLY THE NECESSARY DETAILS ARE RECOMMENDED
AUXILIARY VIEW REPLACES TOP VIEW
B

Fig. 14-1-1 Auxiliary views replacing regular views.

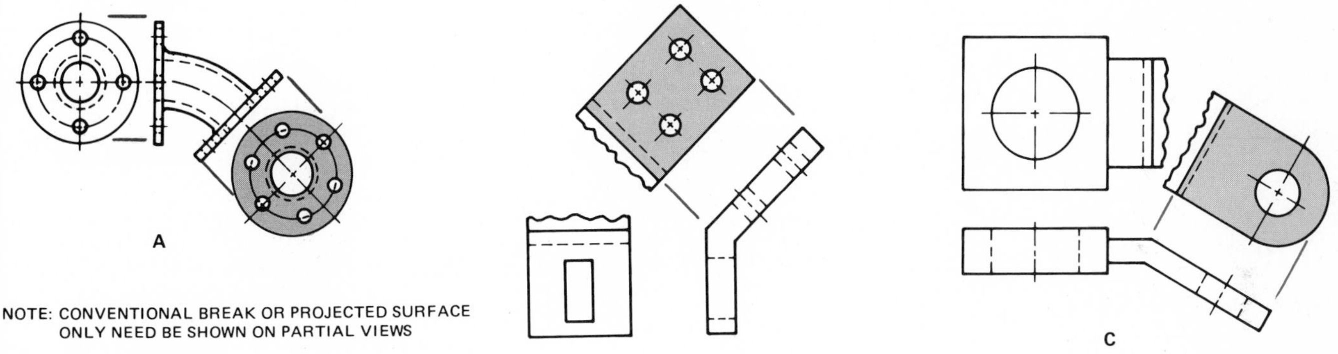
A
B
C

NOTE: CONVENTIONAL BREAK OR PROJECTED SURFACE ONLY NEED BE SHOWN ON PARTIAL VIEWS

Fig. 14-1-2 Examples of auxiliary view drawings.

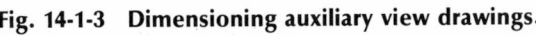

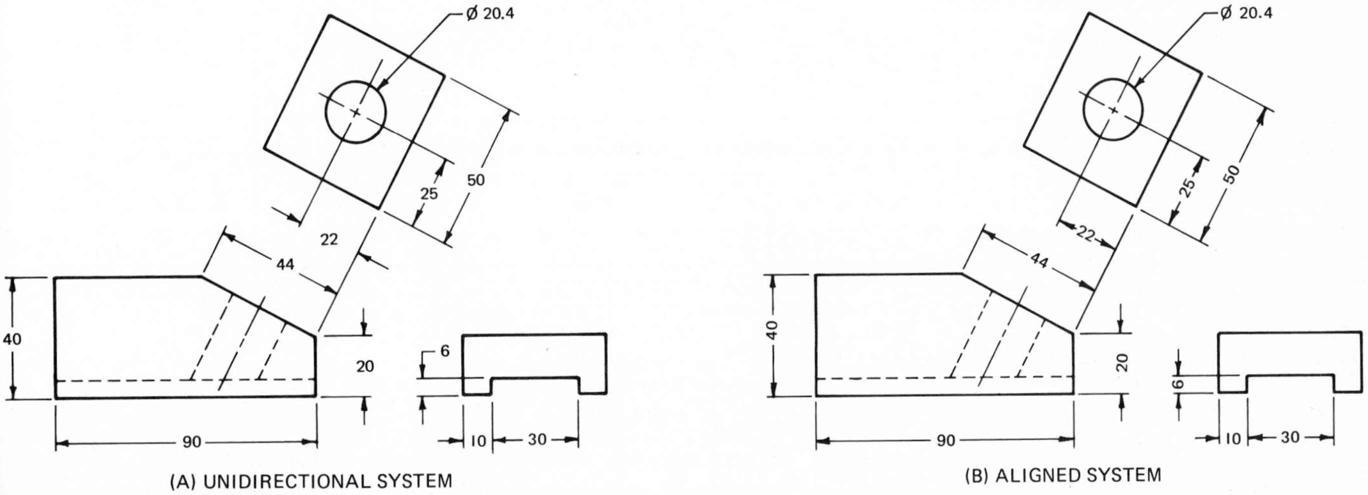
Ø 20.4
50
25
22
44
40
20
6
90
10
30
(A) UNIDIRECTIONAL SYSTEM
(B) ALIGNED SYSTEM

Fig. 14-1-3 Dimensioning auxiliary view drawings.

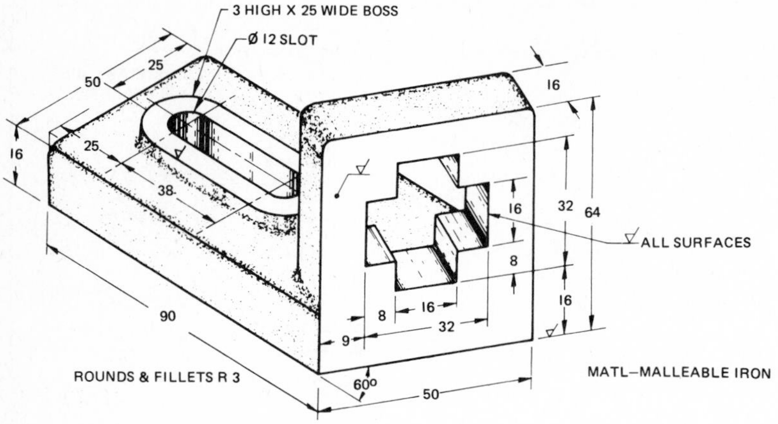
3 HIGH X 25 WIDE BOSS
Ø 12 SLOT
25
50
16
25
38
16
ALL SURFACES
16
32
64
8
16
8
16
32
9
90
ROUNDS & FILLETS R 3
60°
50
MATL—MALLEABLE IRON

Fig. 14-1-A **Cross-slide bracket.**

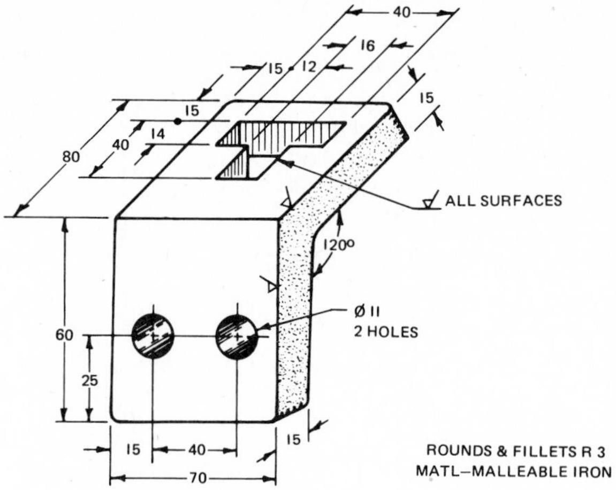
40
16
15
12
15
15
14
40
80
ALL SURFACES
120°
Ø 11
2 HOLES
60
25
15
40
15
70
ROUNDS & FILLETS R 3
MATL—MALLEABLE IRON

Fig. 14-1-C **Angle bracket.**

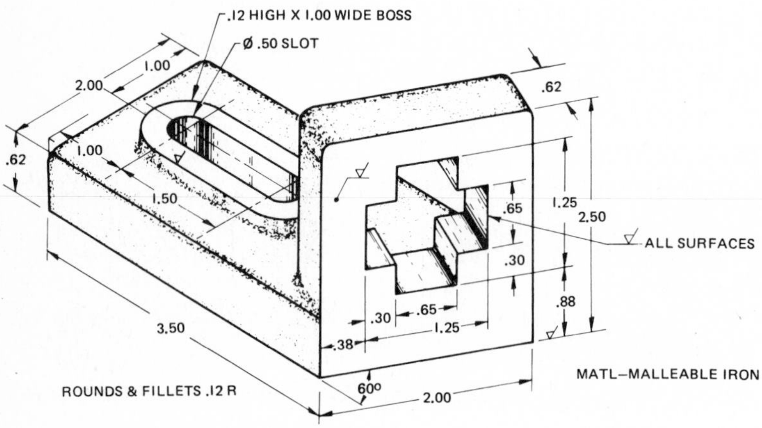
.12 HIGH X 1.00 WIDE BOSS
Ø .50 SLOT
1.00
2.00
.62
1.00
1.50
.62
.65
1.25
2.50
ALL SURFACES
.30
.88
.30
.65
1.25
.38
3.50
ROUNDS & FILLETS .12 R
60°
2.00
MATL—MALLEABLE IRON

Fig. 14-1-B **Cross-slide bracket.**

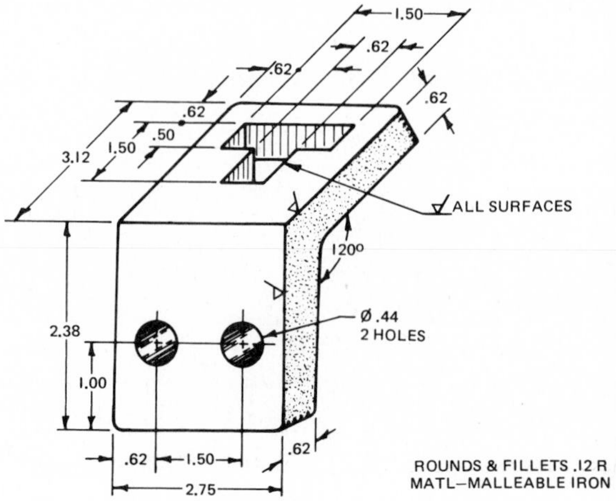
1.50
.62
.62
.62
.62
.50
1.50
3.12
ALL SURFACES
120°
Ø .44
2 HOLES
2.38
1.00
.62
1.50
.62
2.75
ROUNDS & FILLETS .12 R
MATL—MALLEABLE IRON

Fig. 14-1-D **Angle bracket.**

UNIT 14-2
CIRCULAR FEATURES IN AUXILIARY PROJECTION

As mentioned in Unit 14-1, at times it is necessary to show the complete views of an object. If circular features are involved in auxiliary projection, then the surfaces appear elliptical, not circular, in one of the views.

The method most commonly used to draw the true-shape projection of the curved surface is the plotting of a series of points on the line, the number of points being governed by the accuracy of the curved line required.

Figure 14-2-1 illustrates an auxiliary view of a truncated cylinder. The shape seen in the auxiliary view is an *ellipse*. This shape is drawn by plotting *lines of intersection*. The perimeter of the circle in the top view is divided to give a number of equally spaced points—in this case, 12 points, *A* to *M*, spaced 30° apart (360°/12 = 30°). These points are projected down to the edge line on the front view, then at right angles to the edge line to the area where the auxiliary view will be drawn. A center line for the auxiliary view is drawn parallel to the edge line, and width settings taken from the top view are transferred to the auxiliary view. Note width setting *R* for point *L*. Because the illustration shows a true cylinder and the point divisions in the top view are all equal, the width setting *R* taken at *L* is also the correct width setting for *C*, *E*, and *J*. Width setting *S* for *B* is also the correct width setting for *F*, *H*, and *M*. When all the width settings have been transferred to the auxiliary view, the resulting points of intersection are connected with the use of a French curve to give the desired elliptical shape.

It is often necessary to construct the auxiliary view first in order to complete the regular views. This is shown in Fig. 14-2-2.

Assignments

1. On an A3- or B-size sheet, make a working drawing of the link shown in Fig. 14-2-A or 14-2-B. Draw the top, front, and auxiliary views. Show the complete top view. Add hidden lines when required for clarity.

REVIEW FOR ASSIGNMENTS

Unit 5-7	Machining Symbols
Unit 14-1	Dimensioning Auxiliary Views
Unit 7-18	Intersection of Unfinished Surfaces

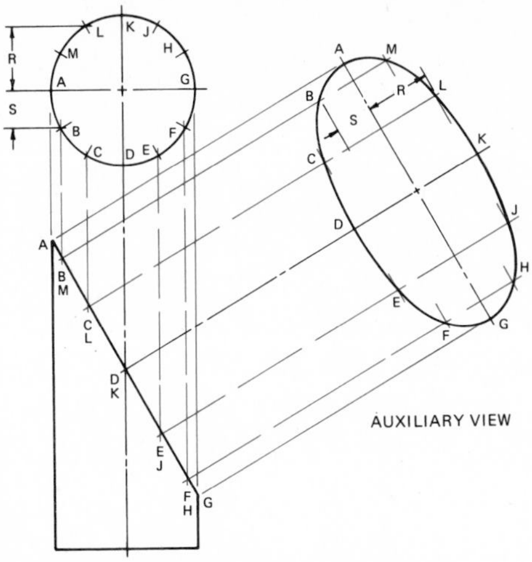
AUXILIARY VIEW

Fig. 14-2-1 Truncated cylinder and auxiliary view.

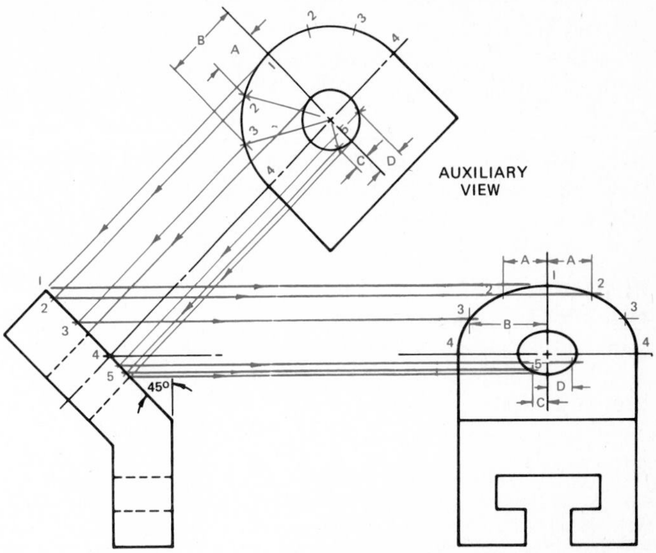
AUXILIARY
VIEW
45°

Fig. 14-2-2 Constructing the true shape of a curved surface by the plotting method.

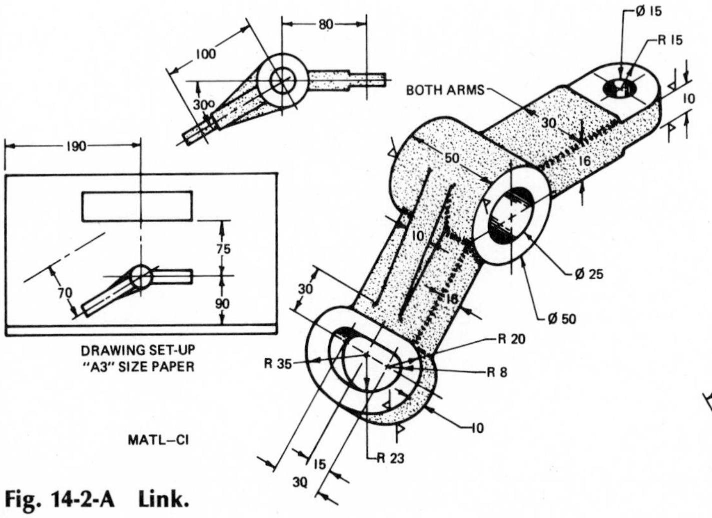
100
80
30°
190
75
70
90
DRAWING SET-UP
"A3" SIZE PAPER
MATL—CI
BOTH ARMS
Ø 15
R 15
10
30
50
16
Ø 25
Ø 50
R 20
R 8
R 35
10
R 23
15
30

Fig. 14-2-A Link.

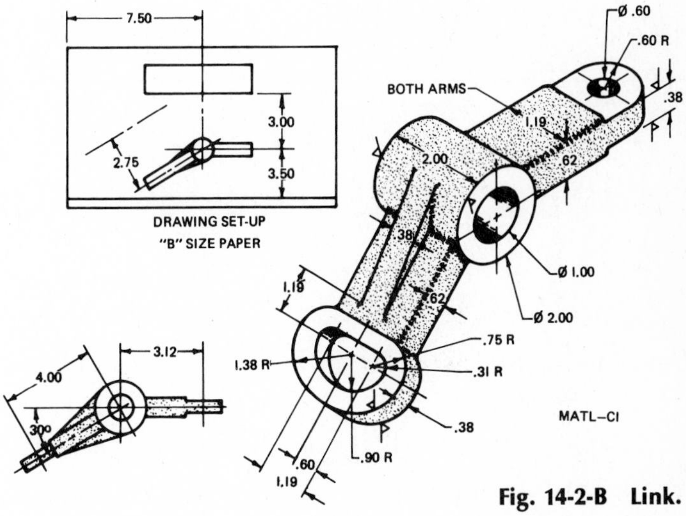
7.50
3.00
2.75
3.50
DRAWING SET-UP
"B" SIZE PAPER
4.00
3.12
30°
BOTH ARMS
Ø .60
.60 R
.38
1.19
2.00
.62
Ø 1.00
Ø 2.00
.75 R
.31 R
1.38 R
.38
.90 R
.60
1.19
MATL—CI

Fig. 14-2-B Link.

UNIT 14-3

MULTI-AUXILIARY-VIEW DRAWINGS

Some objects have more than one surface not perpendicular to the plane of projection. In preparing working drawings of these objects, an auxiliary view may be required for each surface. Naturally, this would depend upon the amount and type of detail lying on these surfaces. This type of drawing is often referred to as a *multi-auxiliary-view* drawing. See Fig. 14-3-1.

One can readily see the advantage of using the unidirectional system of dimensioning for dimensioning an object such as shown in Fig. 14-3-2.

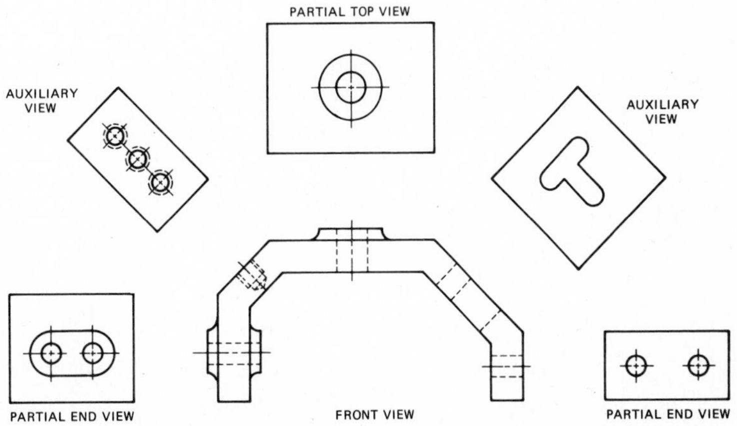
PARTIAL TOP VIEW
AUXILIARY
VIEW
AUXILIARY
VIEW
PARTIAL END VIEW
FRONT VIEW
PARTIAL END VIEW

Fig. 14-3-1 Auxiliary views added to regular views to show true shape of features.

Assignments

1. On an A3- or B-size sheet, make a working drawing of the connecting bar shown in Fig. 14-3-A or 14-3-B. The selection and placement of views are shown beside the drawing. Only partial views need be drawn except where noted and hidden lines may be added to improve clarity. Scale is 1:1.

REVIEW FOR ASSIGNMENTS

Unit 5-7 Machining Symbols
Unit 7-18 Intersection of Unfinished Surfaces
Unit 14-1 Dimensioning Auxiliary Views

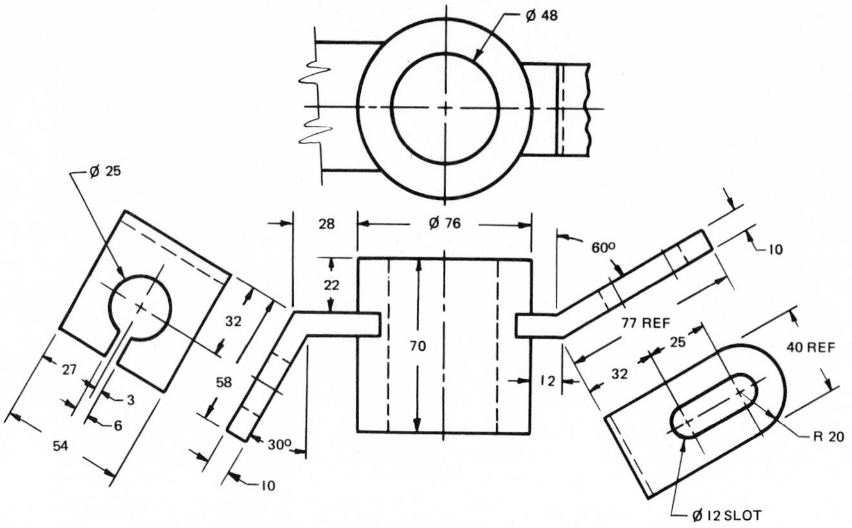
Ø 48
Ø 25
28
Ø 76
60°
10
22
32
70
77 REF
25
40 REF
27
58
12
32
3
6
54
30°
10
R 20
Ø 12 SLOT

Fig. 14-3-2 Dimensioning a multi-auxiliary-view drawing.

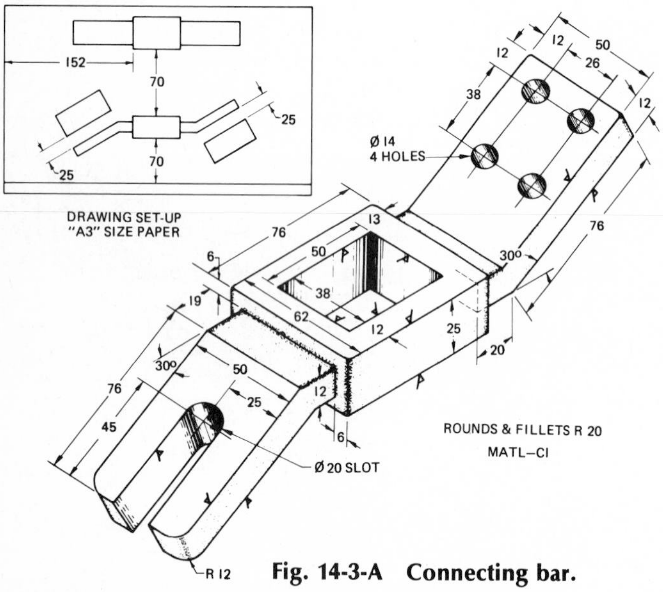
152
70
70
25
25
DRAWING SET-UP
"A3" SIZE PAPER
Ø 14
4 HOLES
12
12
50
26
38
12
76
76
13
6
50
38
62
12
19
30°
25
20
30°
50
25
12
76
45
6
Ø 20 SLOT
ROUNDS & FILLETS R 20
MATL–CI
R 12

Fig. 14-3-A Connecting bar.

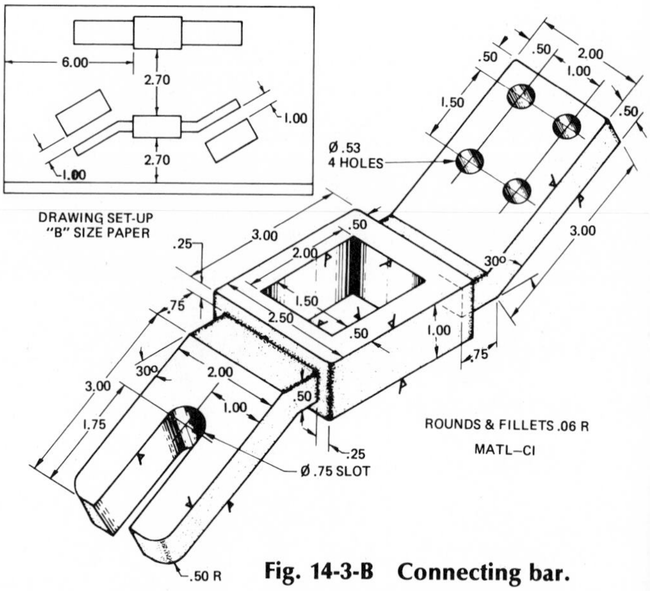
6.00
2.70
2.70
1.00
1.00
DRAWING SET-UP
"B" SIZE PAPER
Ø .53
4 HOLES
.50
.50
2.00
1.00
1.50
.50
3.00
3.00
.25
.50
2.00
1.50
2.50
.50
.75
30°
1.00
.75
30°
2.00
1.00
3.00
1.75
.50
.25
Ø .75 SLOT
ROUNDS & FILLETS .06 R
MATL–CI
.50 R

Fig. 14-3-B Connecting bar.

UNIT 14-4

SECONDARY AUXILIARY VIEWS

Some objects, because of their shape, require a secondary auxiliary view to show the true shape of the surface or feature. See Fig. 14-4-1. The surface or feature is usually oblique to the principal planes of projection. In order to draw a secondary auxiliary view, the primary auxiliary view is first constructed so that it is perpendicular to the inclined surface and to one of the principal planes. The secondary auxiliary view is then projected from the primary auxiliary view, perpendicular to it. Figure 14-4-2 shows the procedure for drawing a secondary auxiliary view.

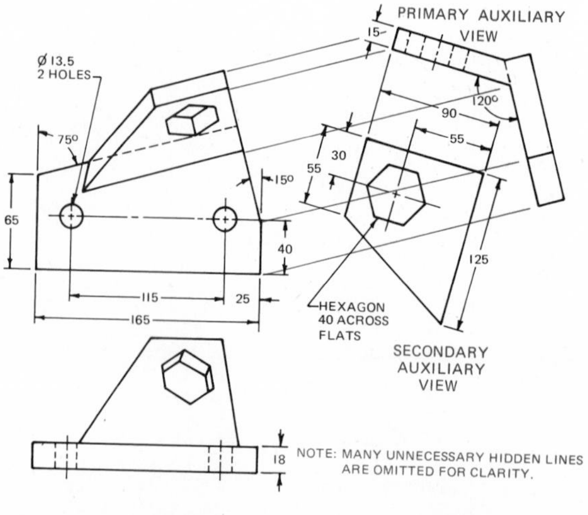
PRIMARY AUXILIARY
VIEW
15
Ø 13.5
2 HOLES
120°
90
55
75°
30
55
15°
65
40
125
115
25
165
HEXAGON
40 ACROSS
FLATS
SECONDARY
AUXILIARY
VIEW
18
NOTE: MANY UNNECESSARY HIDDEN LINES
ARE OMITTED FOR CLARITY.

Fig. 14-4-1 Secondary auxiliary-view drawing.

Assignments

1. On an A3- or B-size sheet, make a working drawing of the dovetail bracket shown in Fig. 14-4-A or 14-4-B. The selection and placement of views are shown beside the drawing. Only partial auxiliary views need be drawn, and hidden lines may be added to improve clarity. Scale is 1:1.

REVIEW FOR ASSIGNMENTS

Unit 14-1 Dimensioning Auxiliary Views

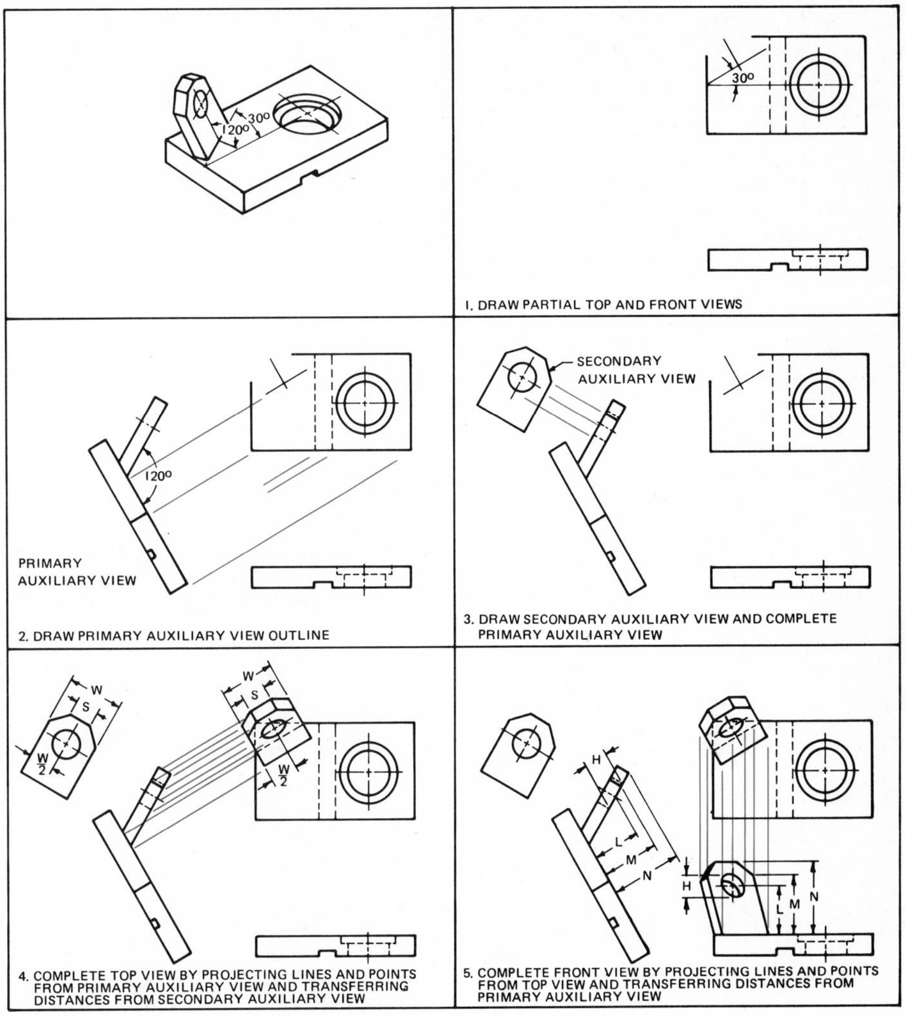

30°
120°
1. DRAW PARTIAL TOP AND FRONT VIEWS
PRIMARY AUXILIARY VIEW
2. DRAW PRIMARY AUXILIARY VIEW OUTLINE
SECONDARY AUXILIARY VIEW
3. DRAW SECONDARY AUXILIARY VIEW AND COMPLETE PRIMARY AUXILIARY VIEW
W
S
W/2
4. COMPLETE TOP VIEW BY PROJECTING LINES AND POINTS FROM PRIMARY AUXILIARY VIEW AND TRANSFERRING DISTANCES FROM SECONDARY AUXILIARY VIEW
H
L
M
N
5. COMPLETE FRONT VIEW BY PROJECTING LINES AND POINTS FROM TOP VIEW AND TRANSFERRING DISTANCES FROM PRIMARY AUXILIARY VIEW

Fig. 14-4-2 Steps in drawing a secondary auxiliary view.

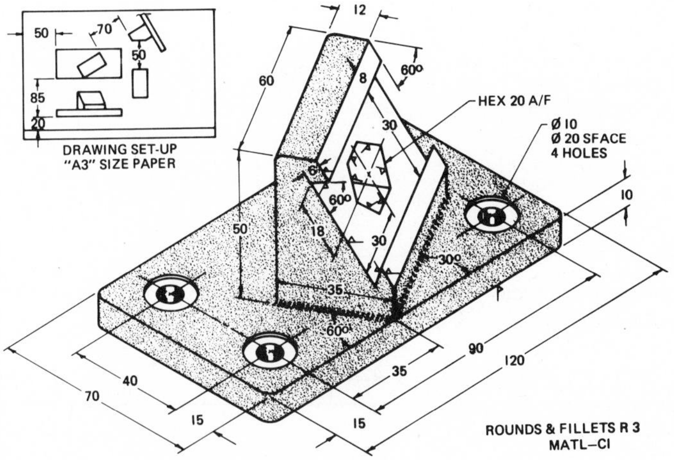

50
70
50
85
20
DRAWING SET-UP "A3" SIZE PAPER
12
60
8
60°
HEX 20 A/F
30
Ø 10
Ø 20 SFACE
4 HOLES
60°
50
18
30
10
30°
35
60°
35
90
120
70
40
15
15
ROUNDS & FILLETS R 3
MATL–CI

Fig. 14-4-A Dovetail bracket.

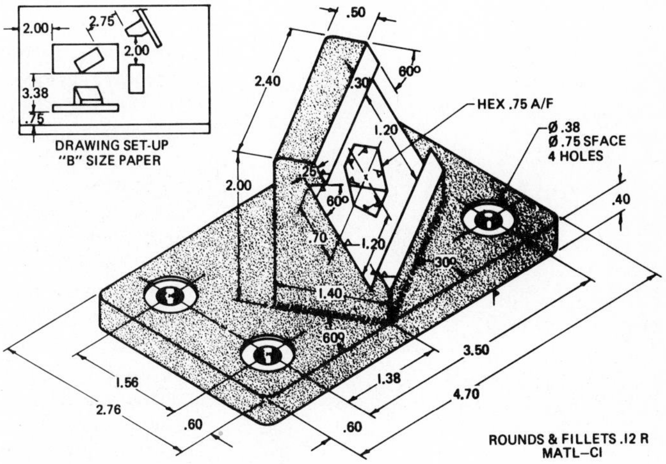

2.00
2.75
2.00
3.38
.75
DRAWING SET-UP "B" SIZE PAPER
.50
2.40
.30
60°
HEX .75 A/F
1.20
Ø .38
Ø .75 SFACE
4 HOLES
2.00
.25
60°
.40
.70
1.20
30°
1.40
60°
3.50
1.56
1.38
4.70
2.76
.60
.60
ROUNDS & FILLETS .12 R
MATL–CI

Fig. 14-4-B Dovetail bracket.

Chapter 15
Pictorial Drawings

UNIT 15-1
PICTORIAL DRAWINGS

Pictorial drawing is the oldest written method of communication known, but the character of pictorial drawing has continually changed with the advance of civilization. In this text only those kinds of pictorial drawings commonly used by the engineer and drafter are considered. Pictorial drawings are useful in design, construction or production, erection or assembly, service or repairs, and sales. They are used to explain complicated engineering drawings to people who do not have the ability or training to read the conventional multiview drawings; to help the designer work out problems in space, including clearances and interferences; to train new employees in the shop; to speed up and clarify the assembly of a machine or the ordering of new parts; to transmit ideas from one person to another, from shop to shop, or from salesperson to purchaser; and as an aid in developing the power of visualization. The type of pictorial drawing used depends on the purpose for which it is drawn.

There are three general types into which pictorial drawings may be divided: axonometric, oblique, and perspective. These three differ from one another in the fundamental scheme of projection, as shown in Fig. 15-1-1.

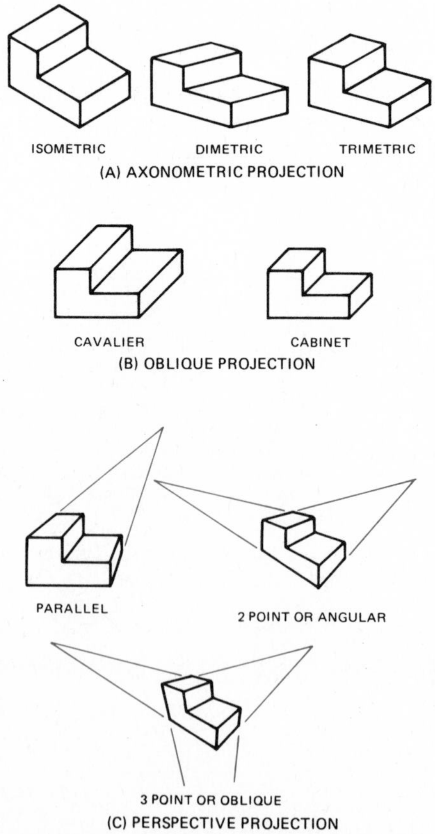
ISOMETRIC
DIMETRIC
TRIMETRIC
(A) AXONOMETRIC PROJECTION
CAVALIER
CABINET
(B) OBLIQUE PROJECTION
PARALLEL
2 POINT OR ANGULAR
3 POINT OR OBLIQUE
(C) PERSPECTIVE PROJECTION

Fig. 15-1-1 Types of pictorial drawings.

Axonometric Projection[1]

A projected view in which the lines of sight are perpendicular to the plane of projection, but in which the three faces of a rectangular object are all inclined to the plane of projection, is called an *axonometric projection*. See Fig. 15-1-2. The projections of the three principal axes may make any angle with one another except 90°. Axonometric drawings, as shown in Figs. 15-1-3 and 15-1-4, are classified into three forms: *isometric drawings*, where the three principal faces and axes of the object are equally inclined to the plane of projection; *dimetric drawings*, where two of the three principal faces and axes of the object are equally inclined to the plane of projection; and *trimetric drawings*, where all three faces and axes of the object make different angles with the plane of projection. The most popular form of axonometric projection is the isometric.

Isometric Drawings[2]

This method is based on a procedure of revolving the object at an angle of 45° to the horizontal, so that the front corner is toward the viewer, then tipping the object up or down at an angle of 35° 16′. See Fig. 15-1-5. When this is done to a cube, the three faces visible to the viewer appear equal in shape and size, and the side faces are at an angle of 30° to the horizontal. If the isometric view were actually projected from a view of the object in the tipped position, the lines in the isometric view would be foreshortened and would, therefore, not be seen in their true length. To simplify the drawing of an isometric view, the actual measurements of the object are used. Although the object appears slightly larger without the allowance for foreshortening, the proportions are not affected. All isometric drawings are

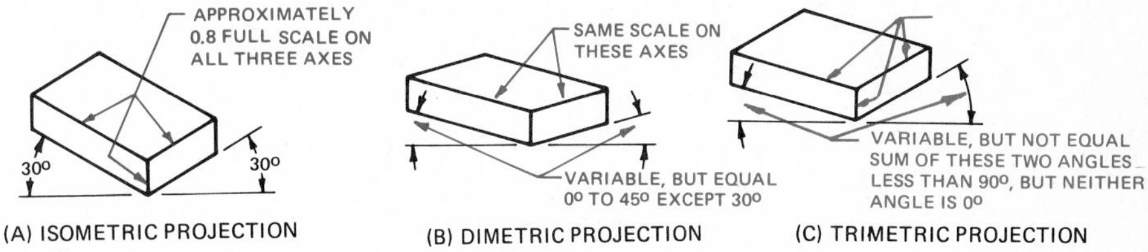
APPROXIMATELY 0.8 FULL SCALE ON ALL THREE AXES
30°
30°
(A) ISOMETRIC PROJECTION
SAME SCALE ON THESE AXES
VARIABLE, BUT EQUAL 0° TO 45° EXCEPT 30°
(B) DIMETRIC PROJECTION
VARIABLE, BUT NOT EQUAL SUM OF THESE TWO ANGLES LESS THAN 90°, BUT NEITHER ANGLE IS 0°
(C) TRIMETRIC PROJECTION

Fig. 15-1-3 Types of axonometric projection. (Graphic Standard Instruments Co.)

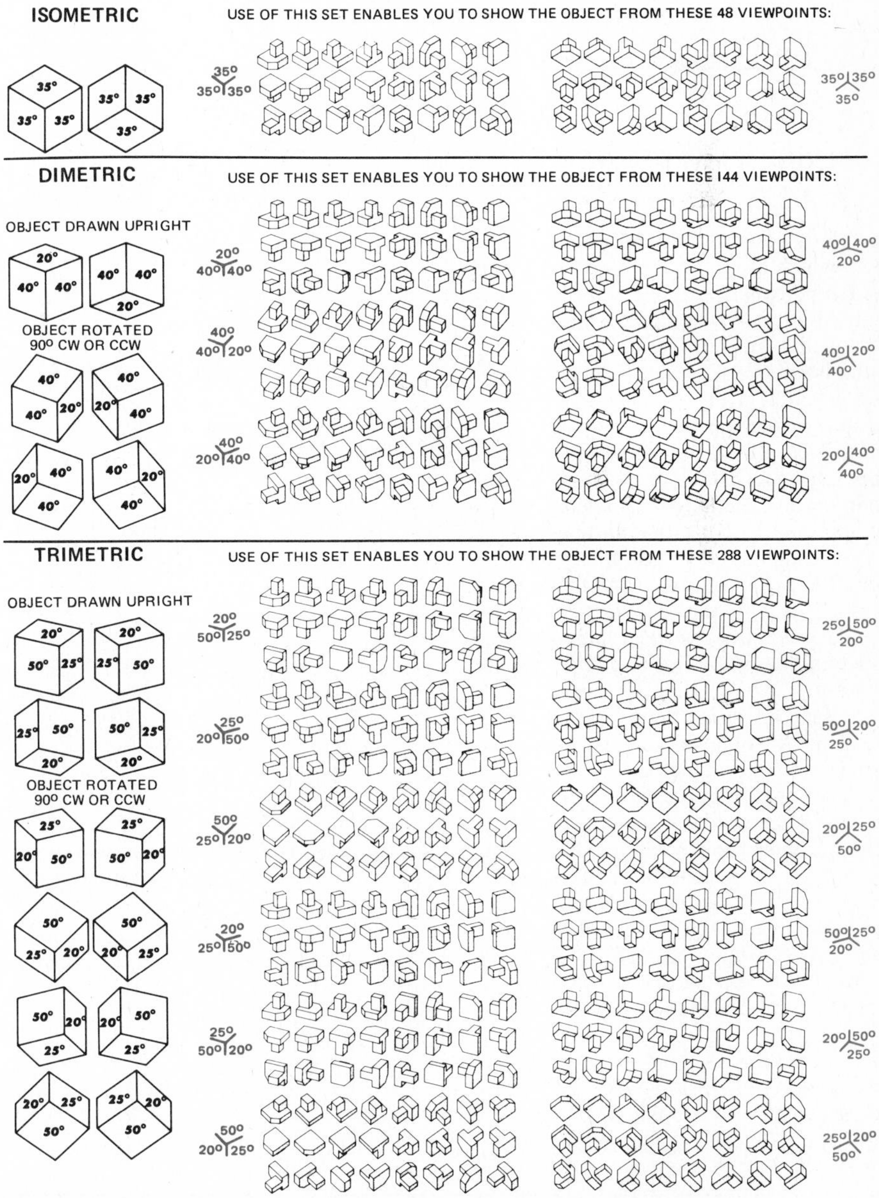
ISOMETRIC
USE OF THIS SET ENABLES YOU TO SHOW THE OBJECT FROM THESE 48 VIEWPOINTS:
DIMETRIC
OBJECT DRAWN UPRIGHT
OBJECT ROTATED 90° CW OR CCW
USE OF THIS SET ENABLES YOU TO SHOW THE OBJECT FROM THESE 144 VIEWPOINTS:
TRIMETRIC
OBJECT DRAWN UPRIGHT
OBJECT ROTATED 90° CW OR CCW
USE OF THIS SET ENABLES YOU TO SHOW THE OBJECT FROM THESE 288 VIEWPOINTS:

Fig. 15-1-4 Axonometric projection. (Graphic Standard Instruments Co.)

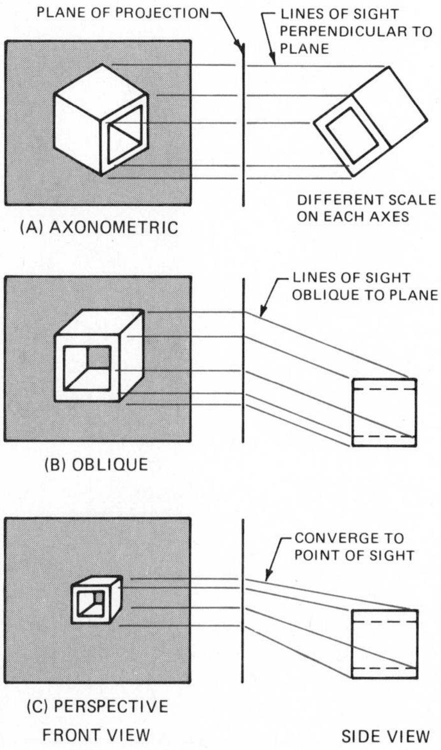
PLANE OF PROJECTION
LINES OF SIGHT PERPENDICULAR TO PLANE
DIFFERENT SCALE ON EACH AXES
(A) AXONOMETRIC
LINES OF SIGHT OBLIQUE TO PLANE
(B) OBLIQUE
CONVERGE TO POINT OF SIGHT
(C) PERSPECTIVE
FRONT VIEW
SIDE VIEW

Fig. 15-1-2 Kinds of projection.

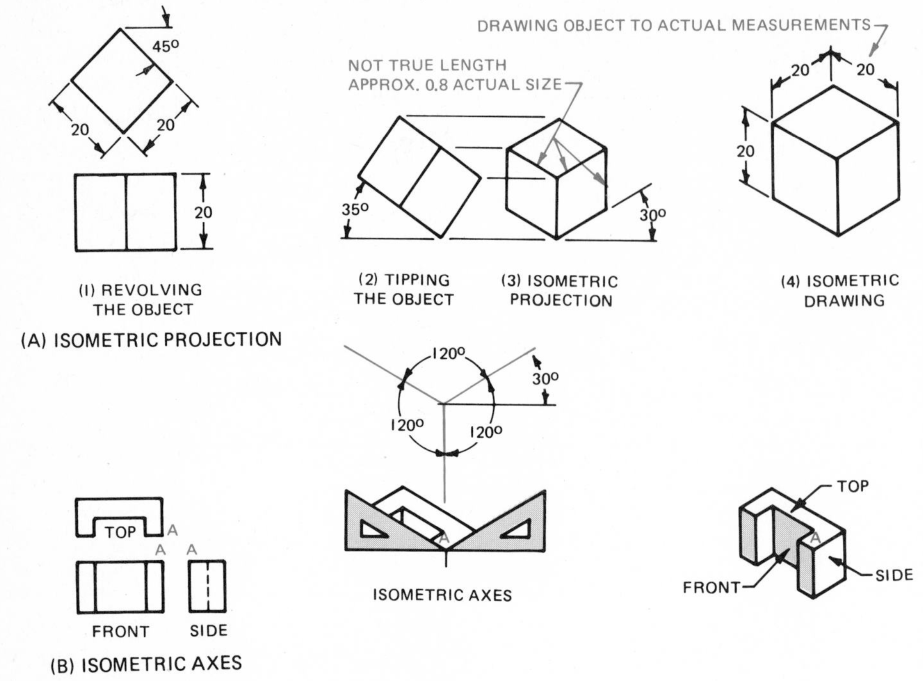
45°
20
20
20
DRAWING OBJECT TO ACTUAL MEASUREMENTS
NOT TRUE LENGTH
APPROX. 0.8 ACTUAL SIZE
35°
30°
20
20
20
(I) REVOLVING THE OBJECT
(2) TIPPING THE OBJECT
(3) ISOMETRIC PROJECTION
(4) ISOMETRIC DRAWING
(A) ISOMETRIC PROJECTION
120°
30°
120°
120°
TOP
A
A A
FRONT
SIDE
(B) ISOMETRIC AXES
A
ISOMETRIC AXES
TOP
A
FRONT
SIDE

Fig. 15-1-5 Isometric axes and projections.

started by constructing the isometric axes, which are a vertical line for heights and isometric lines to left and right, at an angle of 30° from the horizontal, for lengths and widths. The three faces seen in the isometric view are the same faces that would be seen in the normal orthographic views: top, front, and side. Figure 15-1-5*b* illustrates the selection of the front corner (*A*), the construction of the isometric axes, and the completed isometric view. Note that all lines are drawn to their true length, measured along the isometric axes, and that hidden lines are usually omitted. Vertical edges are represented by vertical lines, and horizontal edges by lines at 30° to the horizontal. Two techniques can be used for making an isometric drawing of irregularly shaped objects, as illustrated in Fig. 15-1-6. In one method the object is divided mentally into a number of sections, and the sections are drawn one at a time in their proper relationship to one another. In the second method, a box is drawn with the maximum height, width, and depth of the object; then the parts of the box that are not part of the object are removed, leaving the pieces that form the total object.

NONISOMETRIC LINES

Many objects have sloping surfaces that are represented by sloping lines in the orthographic views. In isometric drawing, sloping surfaces appear as *nonisometric* lines. To draw them, locate their endpoints, found on the ends of isometric lines, and join them with a straight line. Figures 15-1-7 and 15-1-8 illustrate examples in the construction of nonisometric lines.

DIMENSIONING ISOMETRIC DRAWINGS

At times, an isometric drawing of a simple object may serve as a working drawing. In such cases, the necessary dimensions and specifications are placed on the drawing.

Dimension lines, extension lines, and the line being dimensioned should be in the same plane. Arrowheads, which should be long and narrow, should be in the plane of the dimension and extension lines. See Fig. 15-1-9.

There are two acceptable methods of dimensioning an isometric drawing: pictorial plane dimensioning and unidirectional dimensioning. In *pictorial plane dimensioning*, the lettering should lie in one of the pictorial planes. This also applies to notes, which should be located outside the view whenever possible. An example of pictorial plane dimensioning is shown in Fig. 15-1-10*a*.

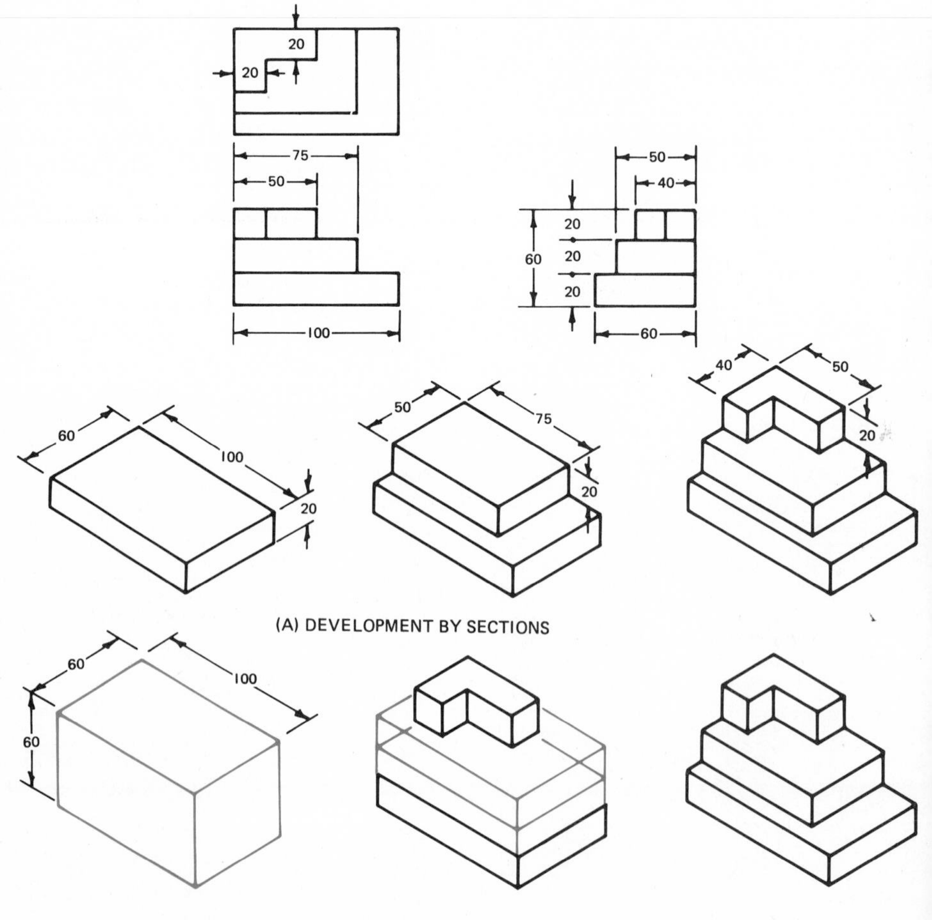
20
20
75
50
50
40
20
60
20
20
100
60
60
100
20
50
75
20
40
50
20
(A) DEVELOPMENT BY SECTIONS
60
100
60
(B) BOX CONSTRUCTION

Fig. 15-1-6 Developing an isometric drawing.

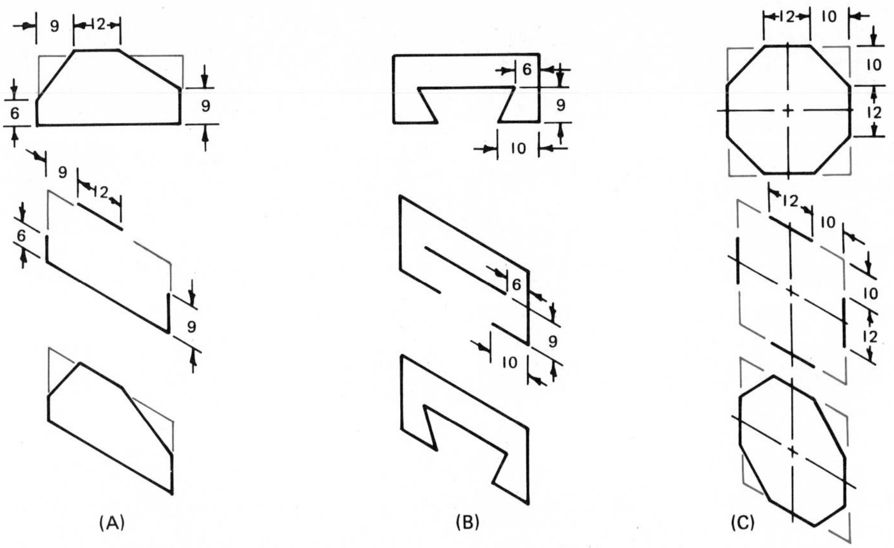

Fig. 15-1-7 **Examples in the construction of nonisometric lines.**

In *unidirectional dimensioning*, the letters and numbers should be vertical and read from the bottom of the sheet. An example of this type of dimensioning is shown in Fig. 15-1-10*b*. This is the preferred method.

Since the isometric is a one-view drawing, it is not usually possible to avoid placing dimensions on the view or across dimension lines. However, this practice should be avoided whenever possible.

ISOMETRIC GRID PAPER

Isometric grid sheets are another time-saving device. Designers and engineers frequently use isometric grid paper on which they sketch their ideas and designs. See Fig. 15-1-11. Many companies, such as those which prepare pipe drawings, have large drawing sheets made with nonreproducible isometric grid lines.

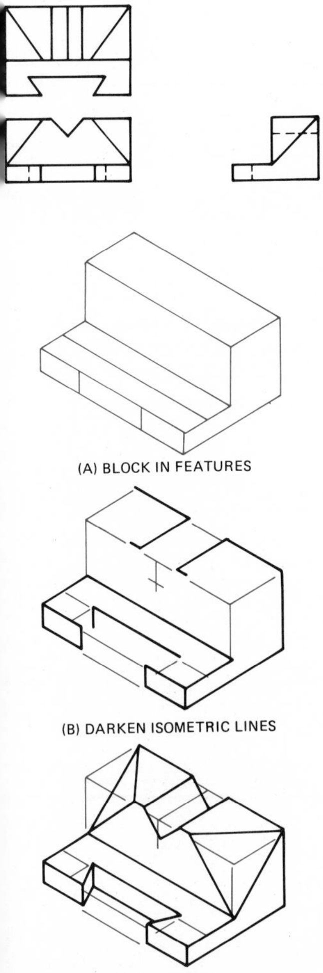

(A) BLOCK IN FEATURES

(B) DARKEN ISOMETRIC LINES

(C) COMPLETE NONISOMETRIC LINES

Fig. 15-1-8 **Sequence in drawing an object having nonisometric lines.**

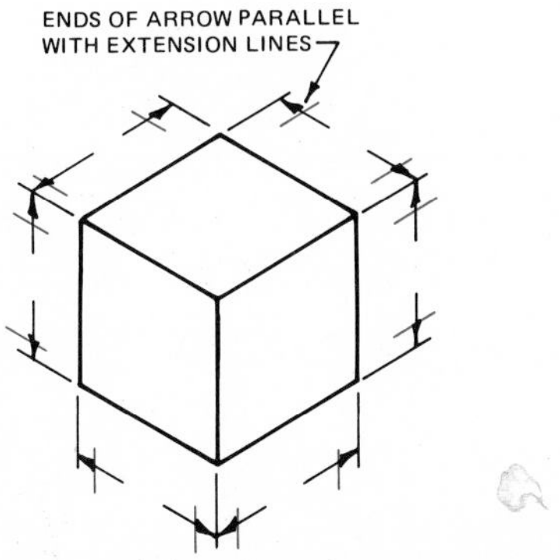
ENDS OF ARROW PARALLEL WITH EXTENSION LINES

Fig. 15-1-9 **Orienting the dimension line, arrowhead, and extension line.**

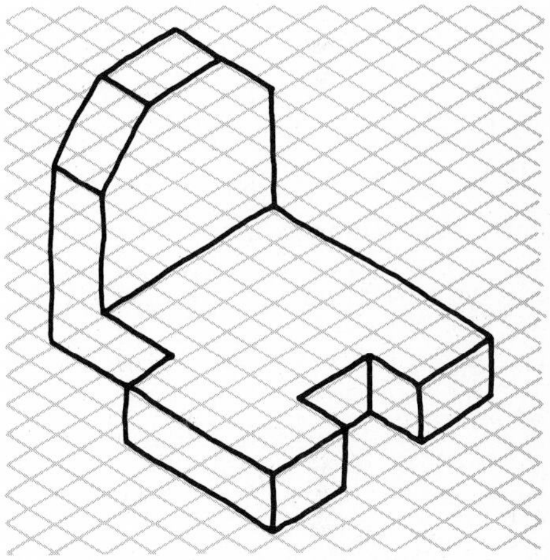

Fig. 15-1-11 **Isometric grid paper.**

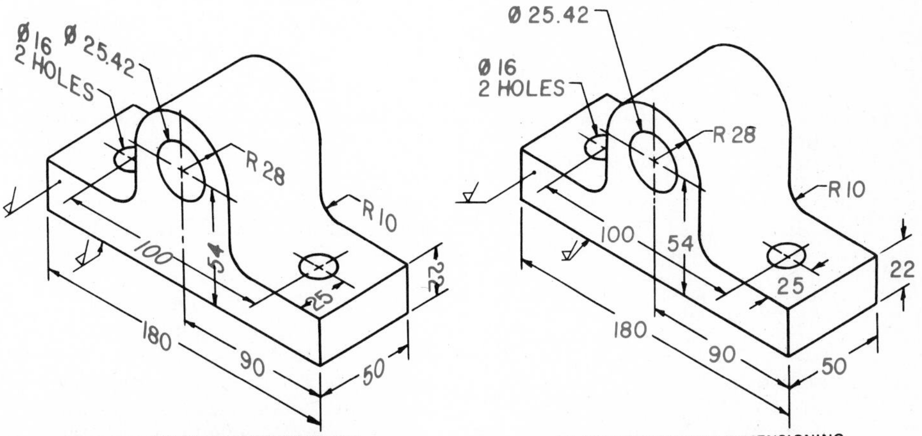
Ø16 2 HOLES
Ø25.42
R28
R10
100
54
25
22
180
90
50

(A) PICTORIAL PLANE DIMENSIONING

(B) UNIDIRECTIONAL DIMENSIONING

Fig. 15-1-10 **Isometric dimensioning.**

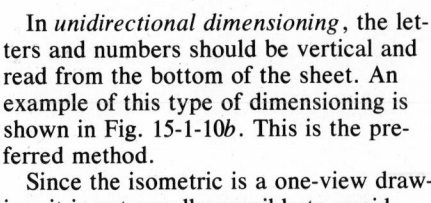

REFERENCES AND SOURCE MATERIAL

1. Extracted from American Drafting Standards Manual, Pictorial Drawing (ANSI Y14.4-1957), with the permission of the publisher, The American Society of Mechanical Engineers, 345 East 47th Street, New York, N.Y. 10017.

2. General Motors Corporation.

Assignments

1. On isometric graph paper sketch the four parts shown in Fig. 15-1-A. Do not show hidden lines. Each square shown on the drawing represents one isometric square on the graph paper.

2. On isometric graph paper sketch the four parts shown in Fig. 15-1-B. Do not show hidden lines. Each square shown on the drawing represents one isometric square on the graph paper.

3. On an A3- or B-size sheet, make an isometric drawing, complete with dimensions, of either the base plate or step block shown in Fig. 15-1-C or 15-1-D. Scale is 1:1.

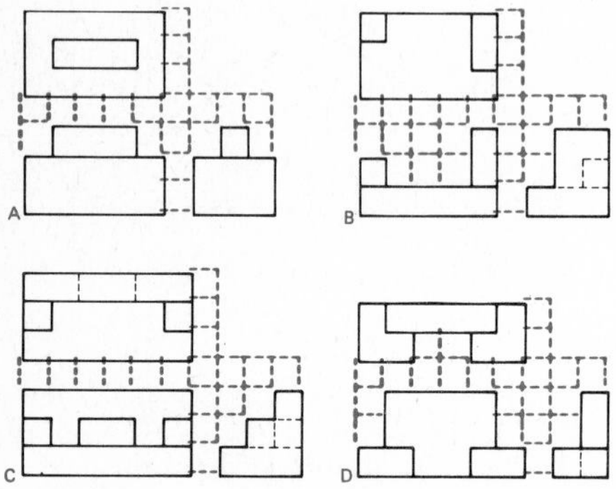
A
B
C
D

Fig. 15-1-A Sketching problems.

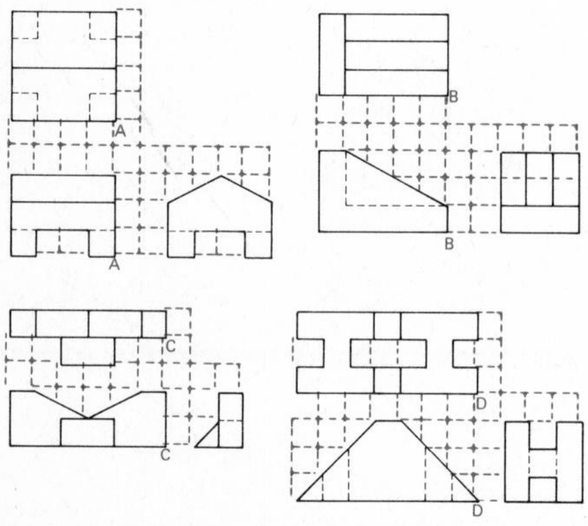
A
B
C
D

Fig. 15-1-B Sketching problems—sloped surfaces.

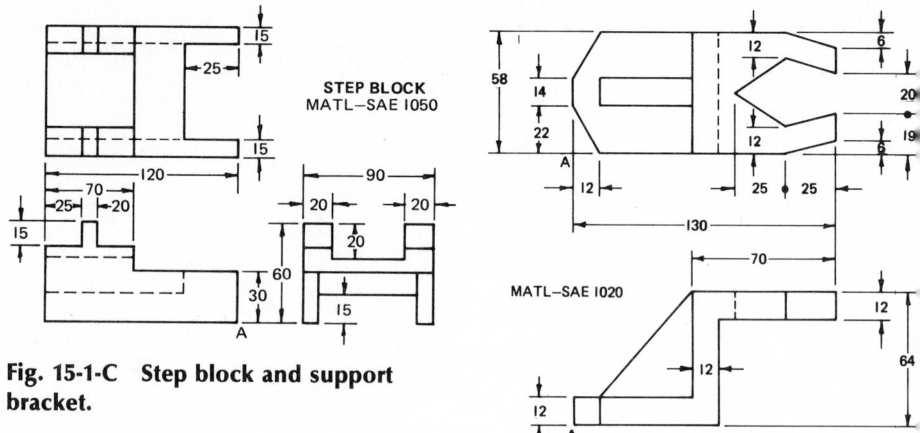
STEP BLOCK
MATL—SAE 1050
MATL—SAE 1020
SUPPORT BRACKET

Fig. 15-1-C Step block and support bracket.

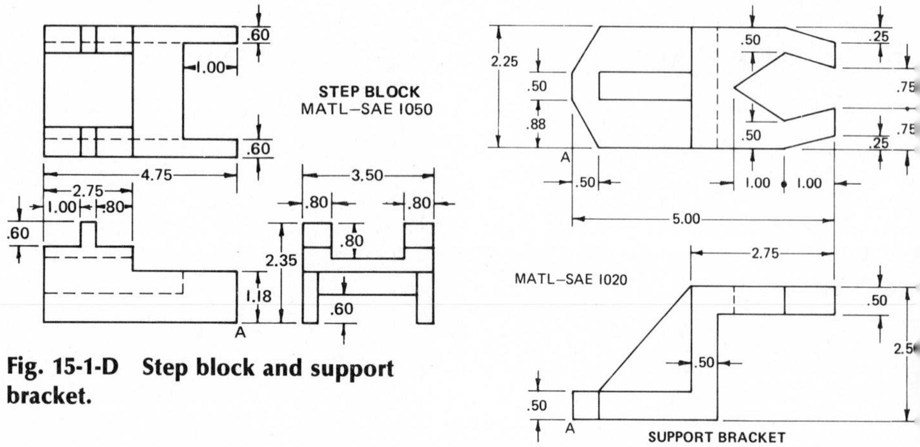
STEP BLOCK
MATL—SAE 1050
MATL—SAE 1020
SUPPORT BRACKET

Fig. 15-1-D Step block and support bracket.

UNIT 15-2
CURVED SURFACES IN ISOMETRIC

Circles and Arcs in Isometric

A circle on any of the three faces of an object drawn in isometric has the shape of an ellipse. See Fig. 15-2-1. Figure 15-2-2 illustrates the steps in drawing circular features on isometric drawings.

1. Draw the center lines and a square, with sides equal to the circle diameter, in isometric.

2. Using the obtuse-angled (120°) corners as centers, draw arcs tangent to the sides forming the obtuse-angled corners, stopping at the points where the center lines cross the sides of the square.

3. Draw construction lines from these same points to the opposite obtuse-angled corners. The points at which these construction lines intersect are the centers for arcs drawn tangent to the sides forming the acute-angled corners, meeting the first arcs.

In drawing concentric circles, each circle must have its own set of centers for the arcs, as shown in Fig. 15-2-3.

The same technique is used for drawing part-circles (arcs). See Fig. 15-2-4.

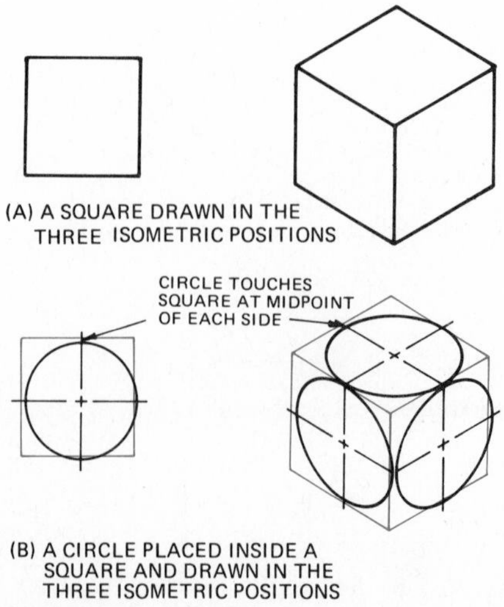
(A) A SQUARE DRAWN IN THE THREE ISOMETRIC POSITIONS
CIRCLE TOUCHES SQUARE AT MIDPOINT OF EACH SIDE
(B) A CIRCLE PLACED INSIDE A SQUARE AND DRAWN IN THE THREE ISOMETRIC POSITIONS

Fig. 15-2-1 Circles in isometric.

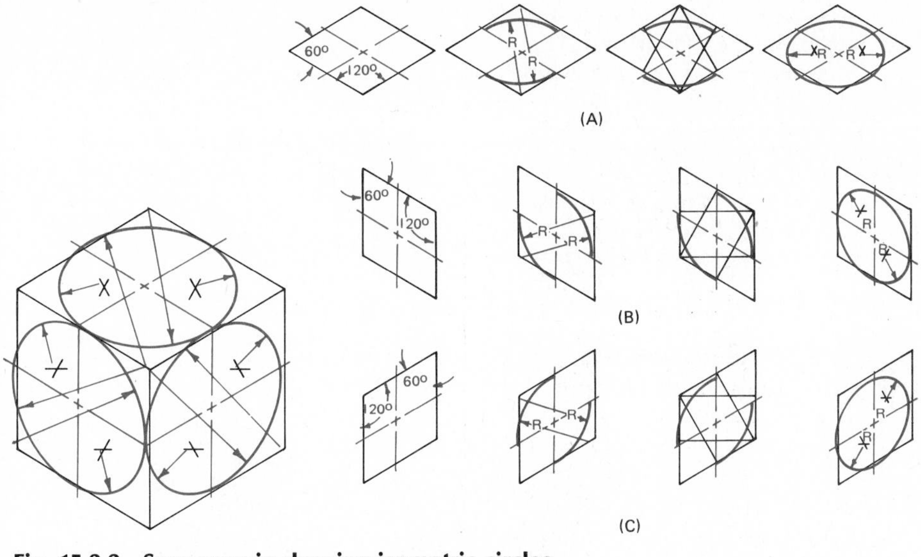
60°
120°
R
(A)
(B)
(C)
Fig. 15-2-2 Sequence in drawing isometric circles.

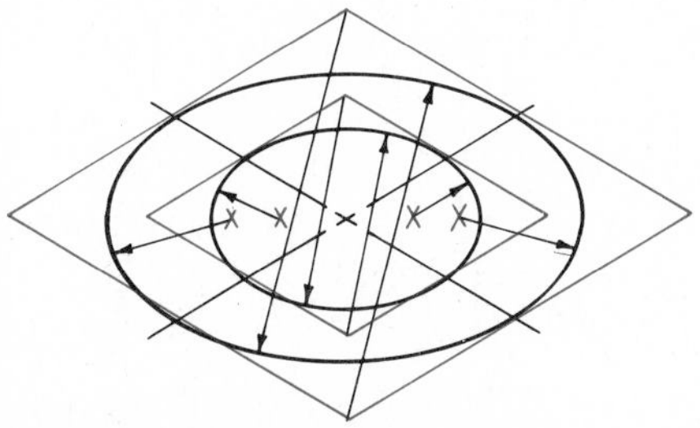
Fig. 15-2-3 Drawing concentric isometric circles.

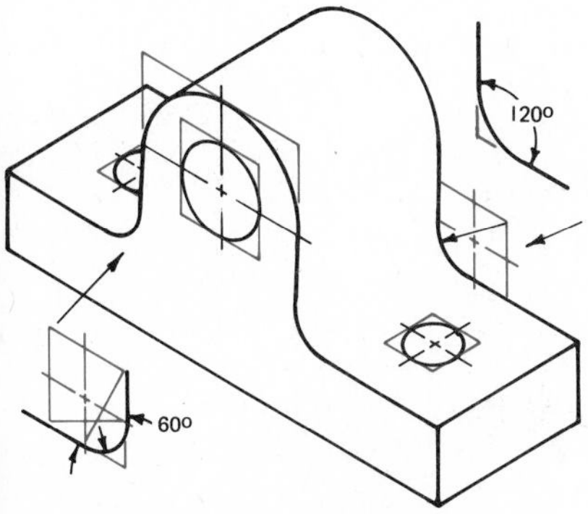
120°
60°
Fig. 15-2-4 Drawing isometric arcs.

Construct an isometric square with sides equal to twice the radius, and draw that portion of the ellipse necessary to join the two faces. When these faces are parallel, draw half of an ellipse (one long radius and one short radius); when they are at an obtuse angle (120°), draw one long radius; and when they are at an acute angle (60°), draw one short radius.

Isometric Templates

For convenience and time saving, isometric ellipse templates should be used whenever possible. A wide variety of elliptical templates are available. The template shown in Fig. 15-2-5 combines ellipse, scales, and angles. Markings on the ellipses coincide with the center lines of the holes which speeds up drawing circles and arcs. Figure 15-2-6 shows the same part as shown in Fig. 15-2-4 but with the arcs and circles being constructed with a template.

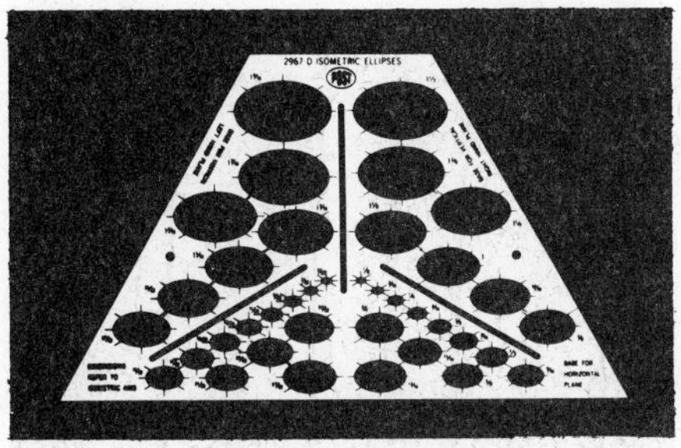
Fig. 15-2-5 Isometric ellipse template.

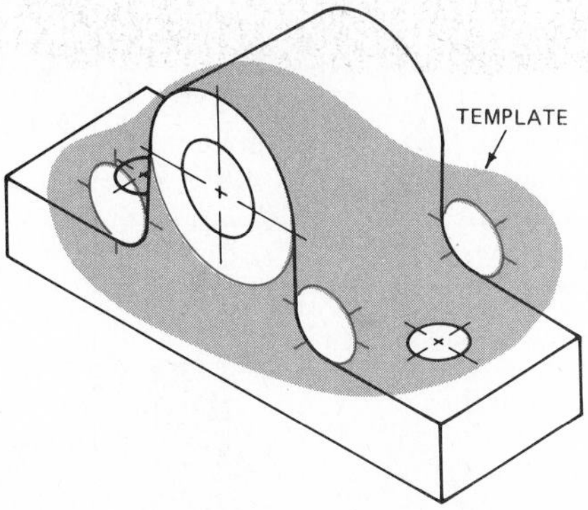
TEMPLATE
Fig. 15-2-6 Circles and arcs drawn with isometric ellipse templates.

Sketching Circles and Arcs

In sketching circles and arcs on isometric grid paper, locate the center lines first, then lightly sketch in construction boxes (isometric squares) where the circles and arcs should be. See Fig. 15-2-7. Sketch the ellipse (isometric circle) just touching the center of each of the four sides of the square.

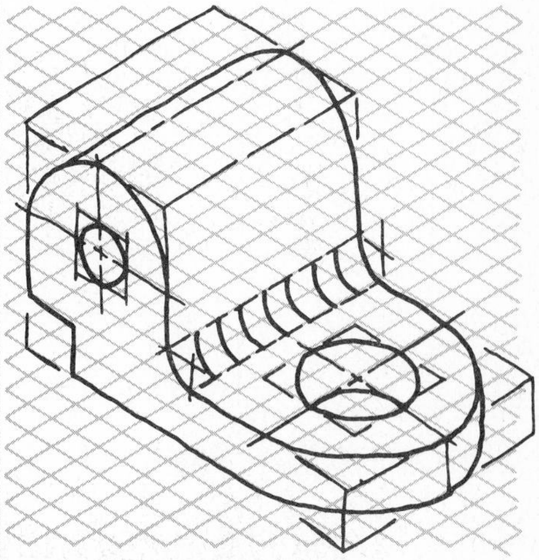
Fig. 15-2-7 Isometric sketching paper.

Drawing Irregular Curves in Isometric

To draw curves other than circles or arcs, the plotting method shown in Fig. 15-2-8 is used:

1. Draw an orthographic view, and divide the area enclosing the curved line into equal squares.
2. Produce an equivalent area on the isometric drawing, showing the offset squares.
3. Take positions relative to the squares from the orthographic view, and plot them on the corresponding squares on the isometric view.
4. Draw a smooth curve through the established points with the aid of an irregular curve.

Assignments

1. On isometric grid paper sketch the four parts shown in Fig. 15-2-A. Each square shown on the figure represents

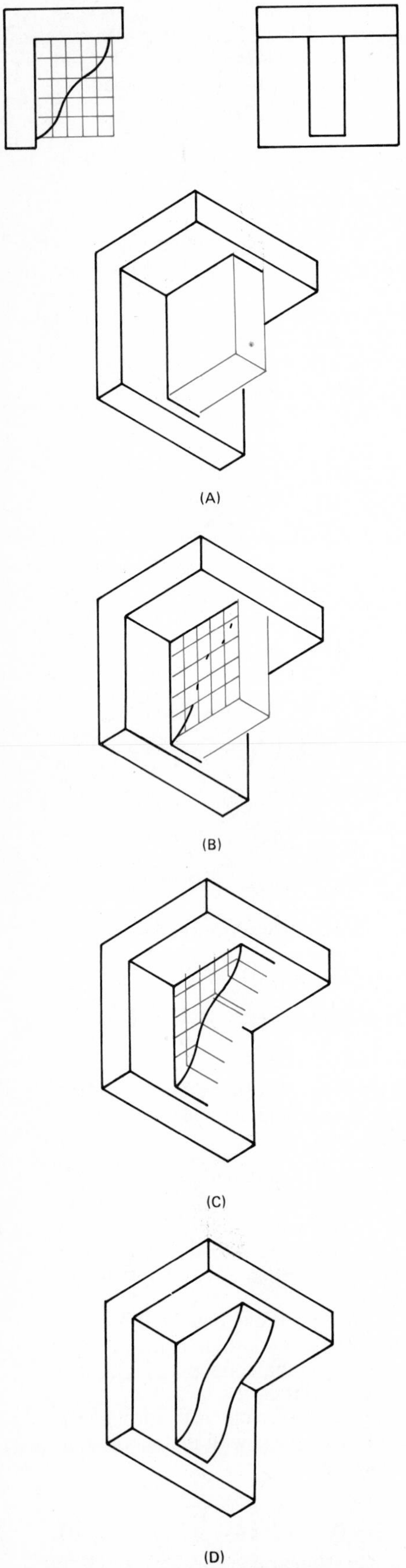
(A)
(B)
(C)
(D)

Fig. 15-2-8 Curves drawn in isometric by means of offset measurements.

one square on the isometric grid. Hidden lines may be omitted for clarity.

2. On isometric grid paper sketch the four parts shown in Fig. 15-2-B. Each square shown on the figure represents one square on the isometric grid. Hidden lines may be omitted for clarity.

3. On an A3- or B-size sheet, make an isometric drawing complete with dimensions of either the link or rocker arm shown in Fig. 15-2-C or 15-2-D. Scale is 1:1.

REVIEW FOR ASSIGNMENTS

Unit 15-1 Dimensioning Isometric Drawings

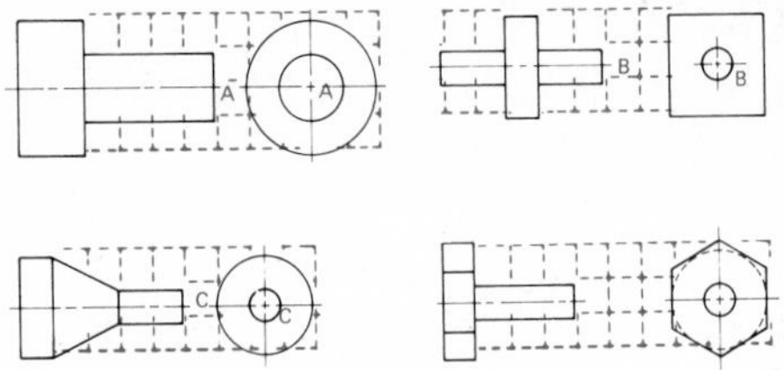

Fig. 15-2-A Sketching problems.

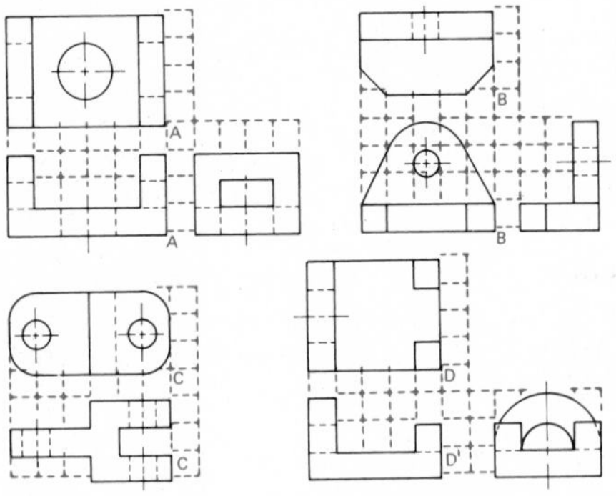

Fig. 15-2-B Sketching problems.

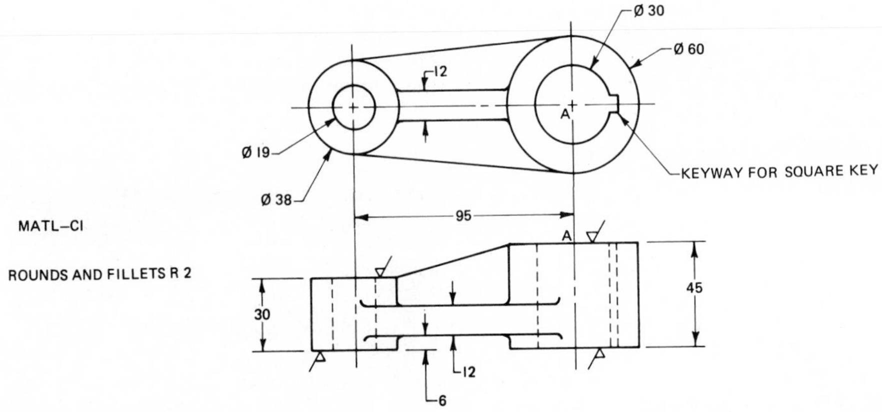
Ø 30
Ø 60
12
A
KEYWAY FOR SQUARE KEY
Ø 19
Ø 38
95
MATL—CI
ROUNDS AND FILLETS R 2
A
45
30
12
6

Fig. 15-2-C Link.

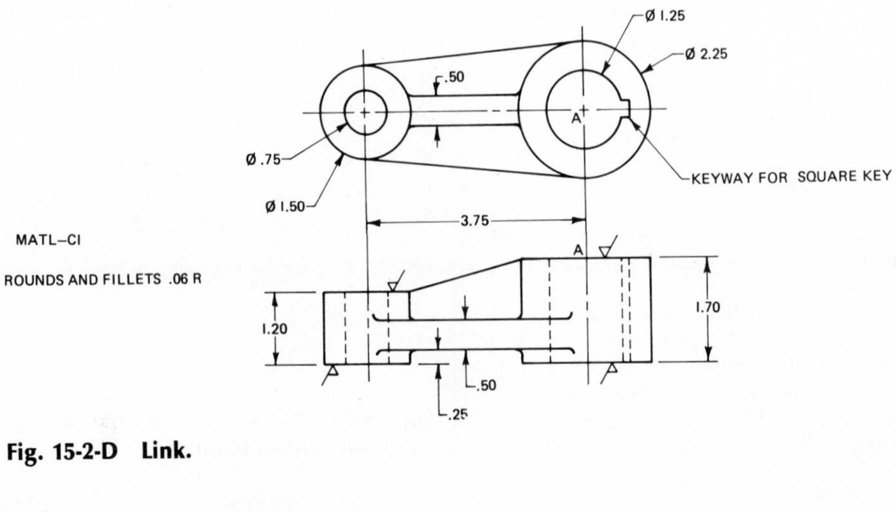
Ø 1.25
Ø 2.25
.50
A
KEYWAY FOR SQUARE KEY
Ø .75
Ø 1.50
3.75
MATL—CI
ROUNDS AND FILLETS .06 R
A
1.70
1.20
.50
.25

Fig. 15-2-D Link.

UNIT 15-3

COMMON FEATURES IN ISOMETRIC

Isometric Sectioning

Isometric drawings are generally made as outside views, but sometimes a sectional view is needed. The section is taken on an isometric plane, that is, on a plane parallel to one of the faces of the cube. Figure 15-3-1 shows isometric full sections taken on a different plane for each of three objects. Note the construction lines indicating the part that has been cut away. Isometric half sections are illustrated in Fig. 15-3-2. Notice the outlines of the cut surfaces in *a* and *b*. The cut method is to draw the complete outside view and the isometric cutting plane.

When a section drawing is drawn in isometric, the section lines are drawn at an angle of 60° with the horizontal or in a horizontal position, depending on where the cutting-plane line is located. In half sections, the section lines are sloped in opposite directions, as shown in Fig. 15-3-2.

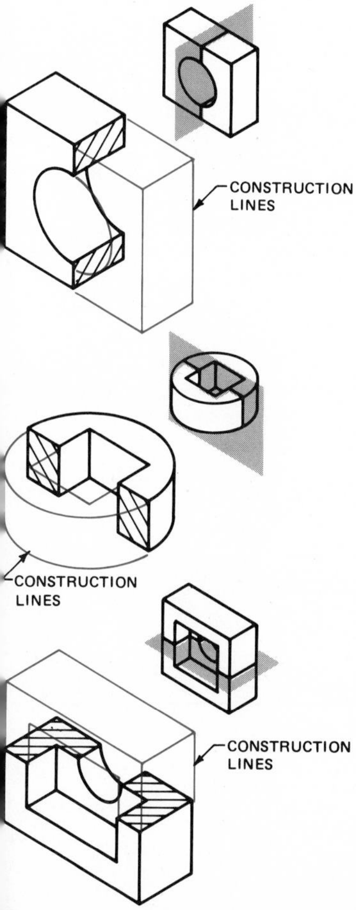
CONSTRUCTION LINES
CONSTRUCTION LINES
CONSTRUCTION LINES

Fig. 15-3-1 Examples of isometric full sections.

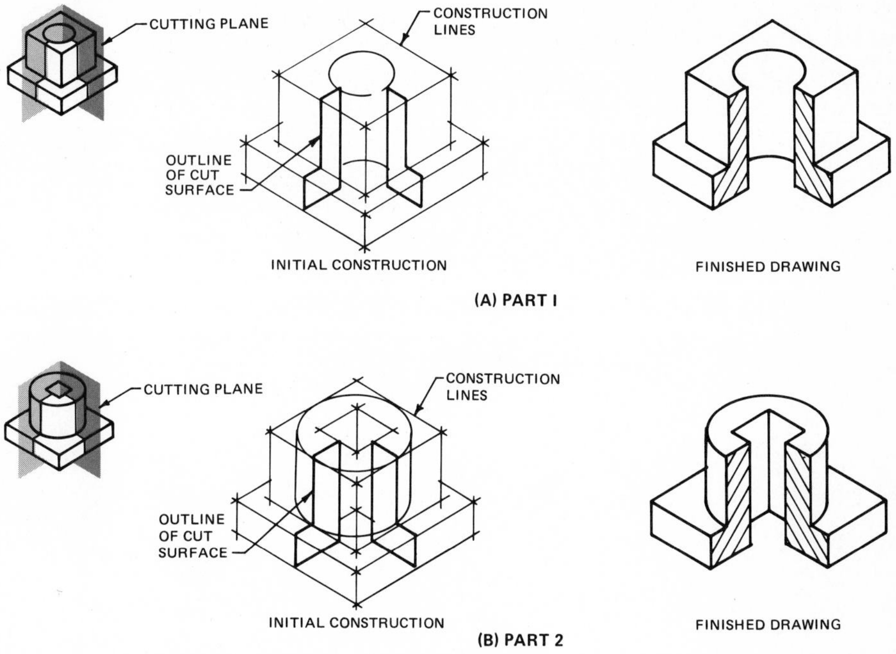
CUTTING PLANE
CONSTRUCTION LINES
OUTLINE OF CUT SURFACE
INITIAL CONSTRUCTION
FINISHED DRAWING
(A) PART I
CUTTING PLANE
CONSTRUCTION LINES
OUTLINE OF CUT SURFACE
INITIAL CONSTRUCTION
(B) PART 2
FINISHED DRAWING

Fig. 15-3-2 Examples of isometric half sections.

Fillets and Rounds

For most isometric drawings of parts having small fillets and rounds, the adopted practice is to draw the corners as sharp features. However, when it is desirable to represent the part, normally a casting, as having a more realistic appearance, either of the methods shown in Fig. 15-3-3 may be used.

Threads

The conventional method for showing threads in isometric is shown in Fig. 15-3-4. The threads are represented by a series of ellipses uniformly spaced along the center line of the thread. The spacing of the ellipses need not be the spacing of the actual pitch.

Break Lines

For long parts, break lines should be used to shorten the length of the drawing. Freehand breaks are preferred, as shown in Fig. 15-3-5.

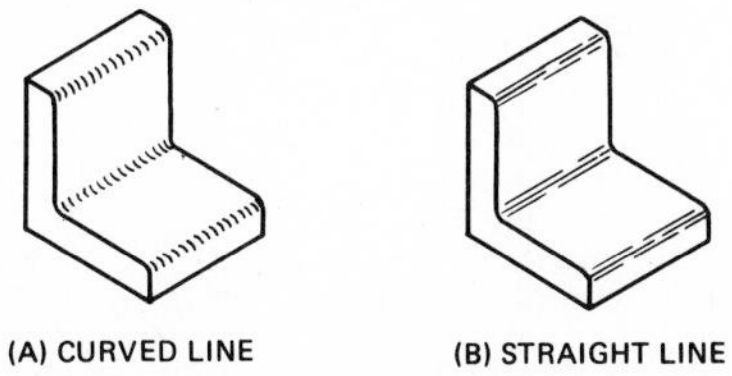
(A) CURVED LINE
(B) STRAIGHT LINE

Fig. 15-3-3 Representation of fillets and rounds.

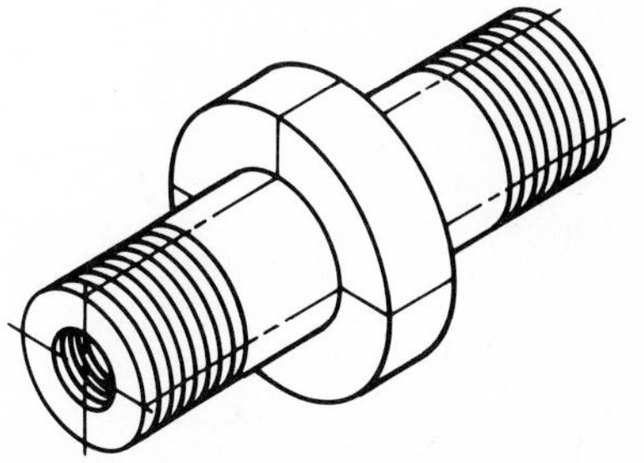

Fig. 15-3-4 Representation of threads in isometric.

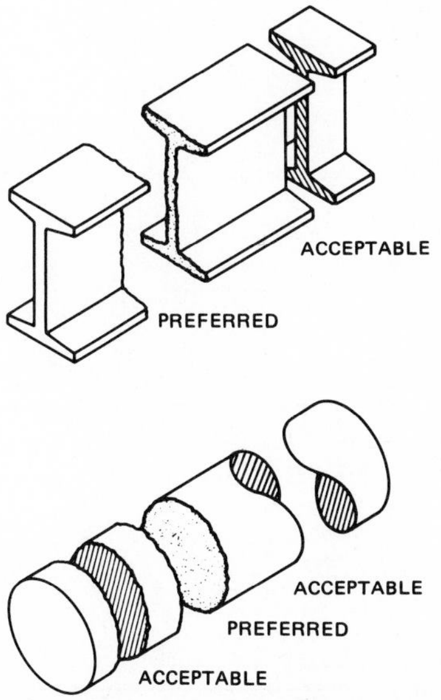
ACCEPTABLE
PREFERRED
ACCEPTABLE
PREFERRED
ACCEPTABLE

Fig. 15-3-5 Conventional breaks.

Isometric Assembly Drawings

Regular or exploded assembly drawings are frequently used in catalogs and sales literature, as illustrated by Figs. 15-3-6 and 15-3-7.

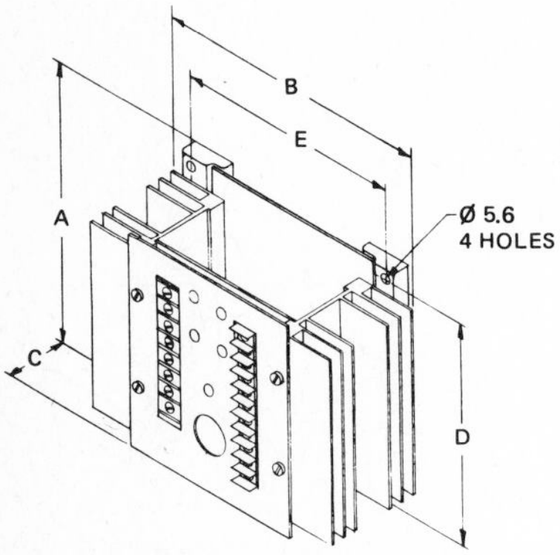

A
B
C
D
E
Ø 5.6
4 HOLES

Fig. 15-3-6 Isometric assembly drawings.

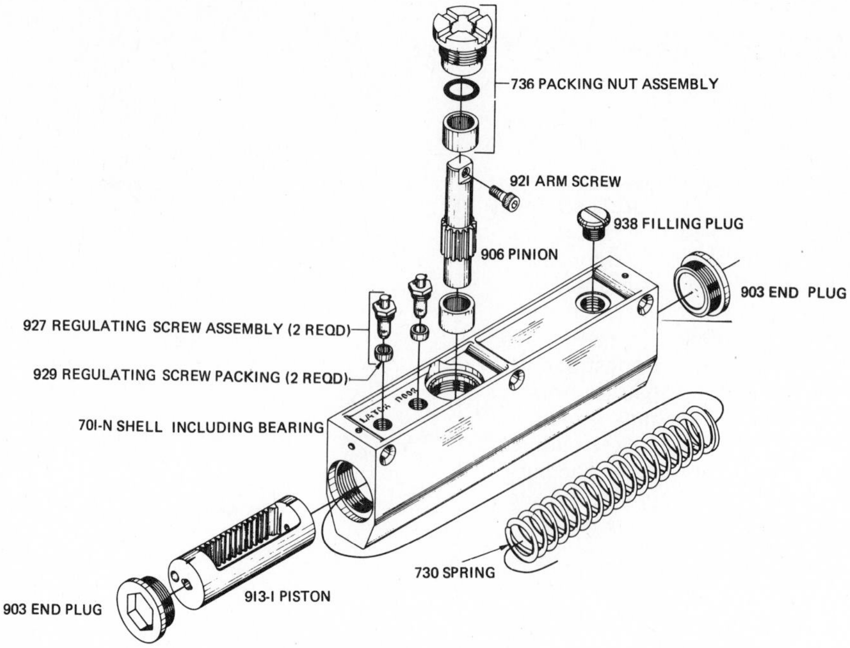

736 PACKING NUT ASSEMBLY
921 ARM SCREW
938 FILLING PLUG
906 PINION
903 END PLUG
927 REGULATING SCREW ASSEMBLY (2 REQD)
929 REGULATING SCREW PACKING (2 REQD)
701-N SHELL INCLUDING BEARING
730 SPRING
913-1 PISTON
903 END PLUG

Fig. 15-3-7 Exploded isometric assembly drawing. (Yale and Towne Inc.)

Assignments

1. On an A3- or B-size sheet, draw a full-section view of the pencil holder shown in Fig. 15-3-A or 15-3-B. The section view is to be taken at cutting plane *BB*. Do not dimension. Scale is 1:1.

2. On an A3- or B-size sheet, draw an isometric assembly drawing of the two-post die set, model 302, shown in Fig. 15-3-C or 15-3-D. Allow 50 mm [2 in.] between the top and base. Scale is 1:2. Do not dimension. Include on the drawing a bill of material. Using part numbers, identify the parts on the assembly.

3. On an A3- or B-size sheet, draw an isometric exploded assembly drawing of the book rack shown in Fig. 15-3-E or 15-3-F. Scale is 1:2. Do not dimension. Include on the drawing a bill of material. Using part numbers, identify the parts on the assembly.

REVIEW FOR ASSIGNMENTS

Unit 6-4 Bill of Materials
Unit 6-4 Assembly Drawings

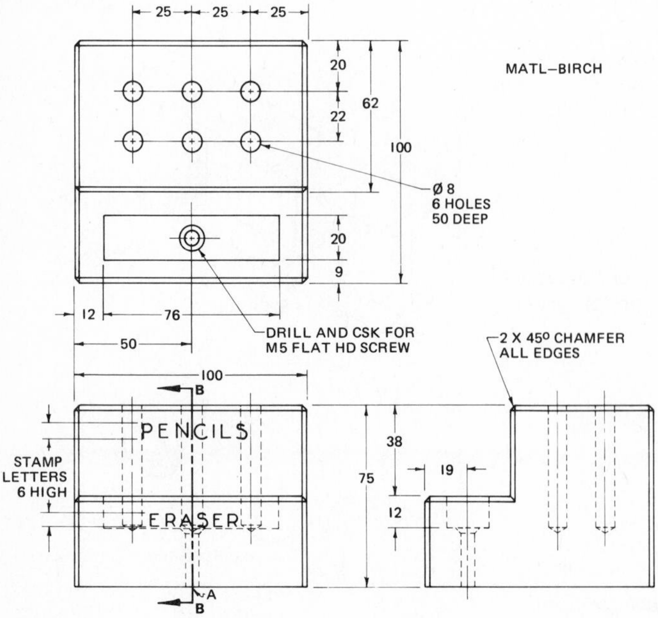

25
25
25
MATL—BIRCH
20
22
62
100
Ø 8
6 HOLES
50 DEEP
20
9
12
76
50
DRILL AND CSK FOR
M5 FLAT HD SCREW
2 X 45° CHAMFER
ALL EDGES
100
B
PENCILS
STAMP
LETTERS
6 HIGH
ERASER
38
75
19
12
A
B

Fig. 15-3-A Pencil holder.

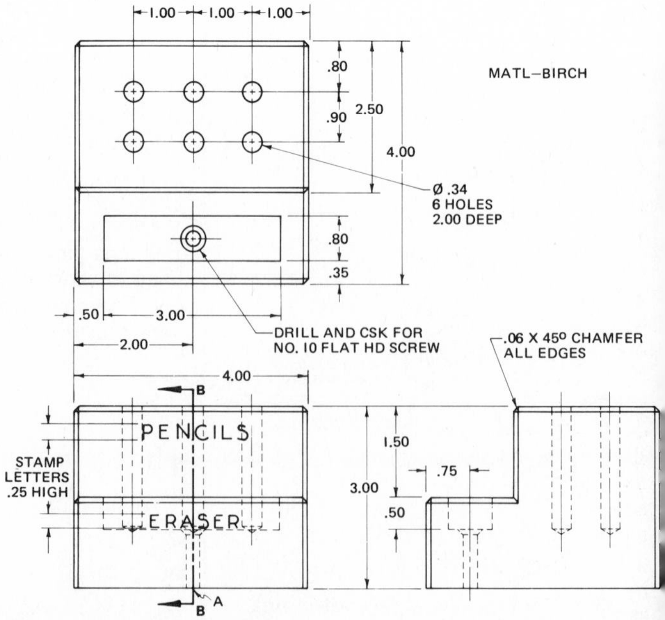

1.00
1.00
1.00
MATL—BIRCH
.80
.90
2.50
4.00
Ø .34
6 HOLES
2.00 DEEP
.80
.35
.50
3.00
2.00
DRILL AND CSK FOR
NO. 10 FLAT HD SCREW
.06 X 45° CHAMFER
ALL EDGES
4.00
B
PENCILS
STAMP
LETTERS
.25 HIGH
ERASER
1.50
.75
3.00
.50
B
A

Fig. 15-3-B Pencil holder.

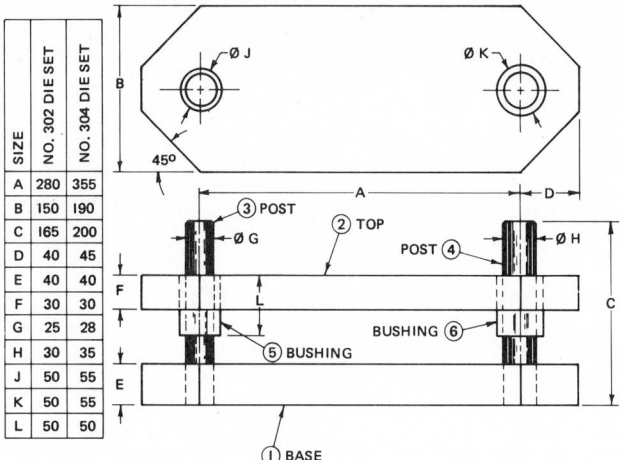

SIZE	NO. 302 DIE SET	NO. 304 DIE SET
A	280	355
B	150	190
C	165	200
D	40	45
E	40	40
F	30	30
G	25	28
H	30	35
J	50	55
K	50	55
L	50	50

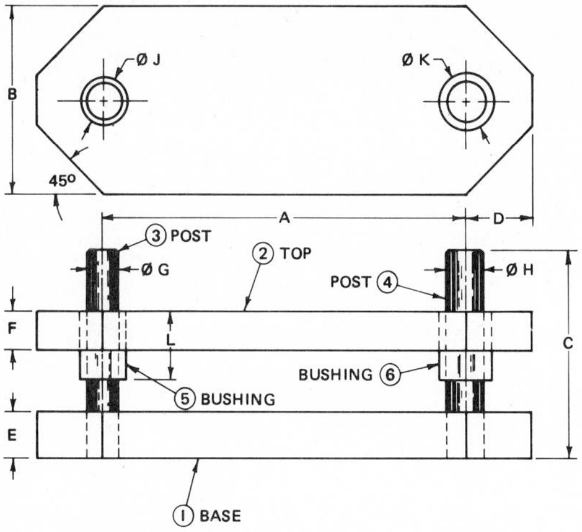

Ø J
Ø K
B
45°
A
D
3 POST
2 TOP
POST 4
Ø G
Ø H
F
L
C
BUSHING 6
5 BUSHING
E
1 BASE

Fig. 15-3-C **Two-post die set.**

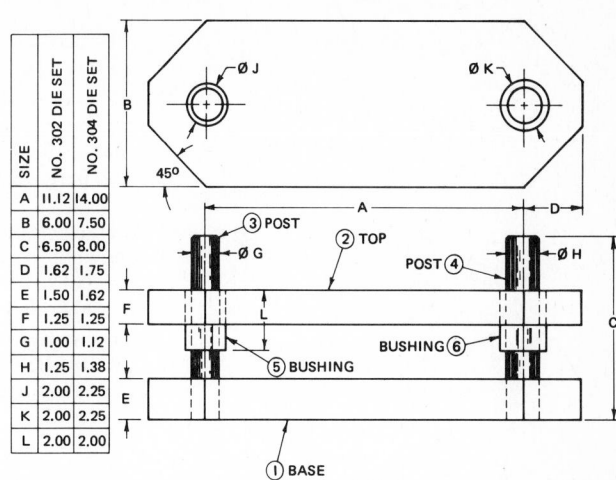

SIZE	NO. 302 DIE SET	NO. 304 DIE SET
A	11.12	14.00
B	6.00	7.50
C	6.50	8.00
D	1.62	1.75
E	1.50	1.62
F	1.25	1.25
G	1.00	1.12
H	1.25	1.38
J	2.00	2.25
K	2.00	2.25
L	2.00	2.00

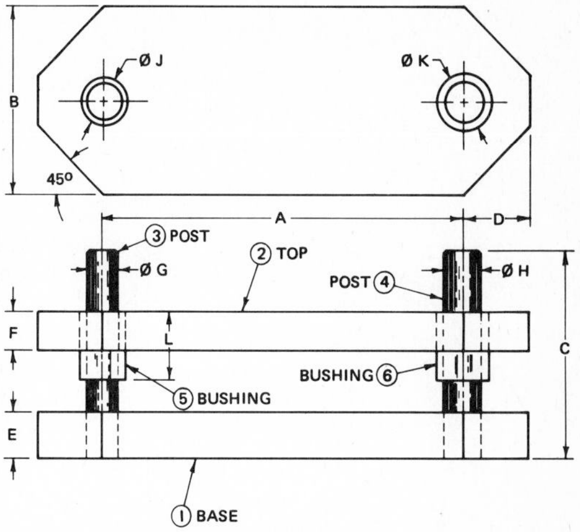

Ø J
Ø K
B
45°
A
D
3 POST
2 TOP
POST 4
Ø G
Ø H
F
L
C
BUSHING 6
5 BUSHING
E
1 BASE

Fig. 15-3-D **Two-post die set.**

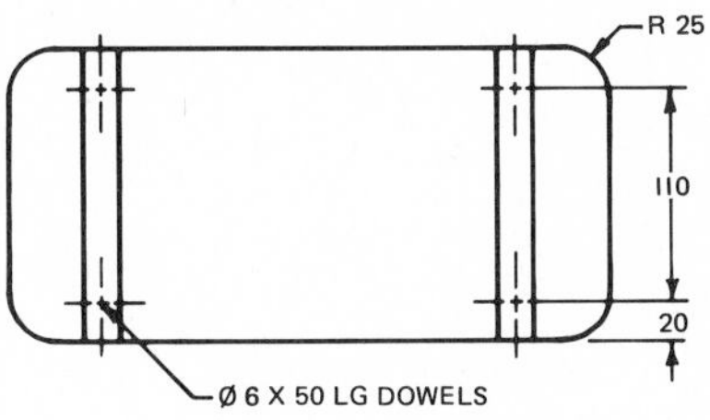

R 25
110
20
Ø 6 X 50 LG DOWELS

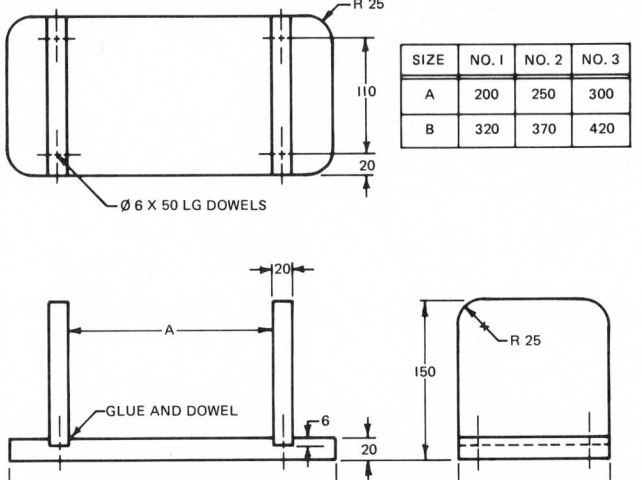

SIZE	NO. 1	NO. 2	NO. 3
A	200	250	300
B	320	370	420

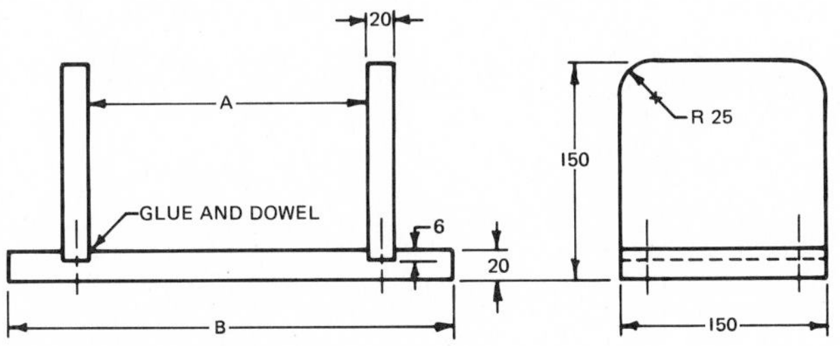

20
A
GLUE AND DOWEL
6
20
B
150
R 25
150

Fig. 15-3-E **Book rack.**

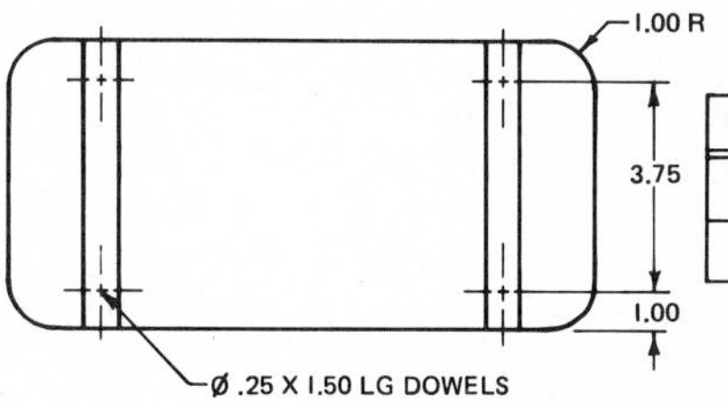

1.00 R
3.75
1.00
Ø .25 X 1.50 LG DOWELS

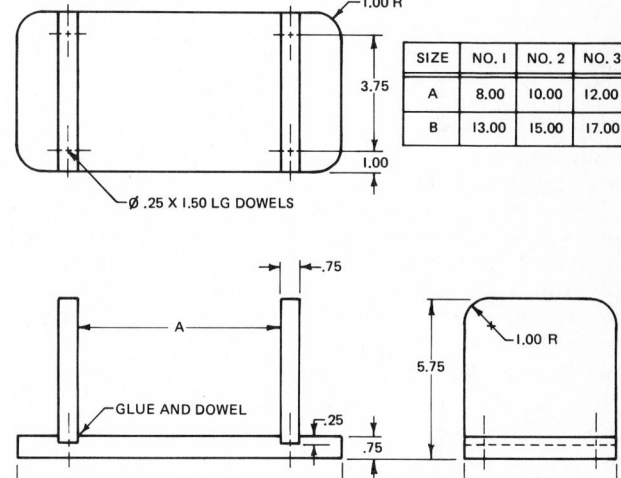

SIZE	NO. 1	NO. 2	NO. 3
A	8.00	10.00	12.00
B	13.00	15.00	17.00

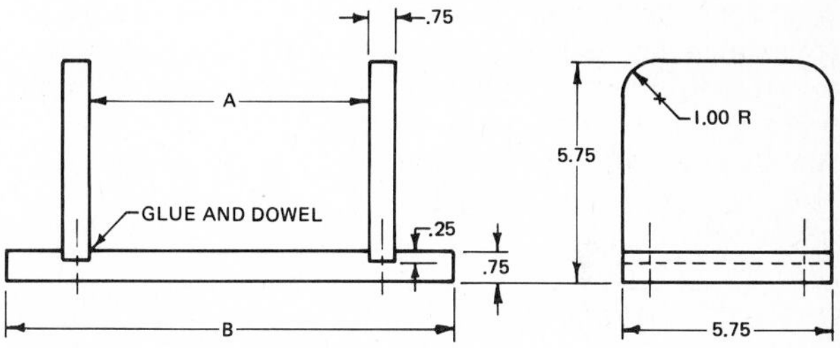

.75
A
GLUE AND DOWEL
.25
.75
B
5.75
1.00 R
5.75

Fig. 15-3-F **Book rack.**

UNIT 15-4

OBLIQUE PROJECTION

This method of pictorial drawing is based on the procedure of placing the object with one face parallel to the frontal plane and placing the other two faces on oblique (or receding) planes, to left or right, top or bottom, at a convenient angle. The three axes of projection are *vertical*, *horizontal*, and *receding*. Figure 15-4-1 illustrates a cube drawn in typical positions with the receding axis at 60, 45, and 30°. This form of projection has the advantage of showing one face of the object without distortion. The face with the greatest irregularity of outline or contour, or the face with the longest dimension, faces the front. See Fig. 15-4-2.

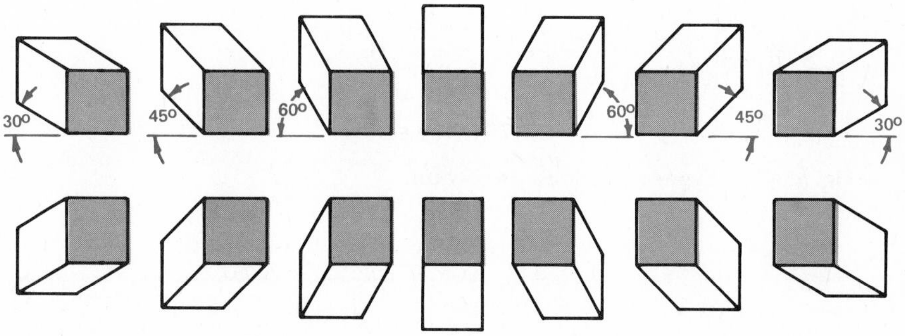

30°
45°
60°
60°
45°
30°

Fig. 15-4-1 **Typical positions of receding axes for oblique projections.**

Two types of oblique projection are used extensively. In *cavalier oblique*, all lines are drawn to their true length, measured on the axes of the projection. In *cabinet oblique*, the lines on the receding axis are shortened by one-half their true length to compensate for distortion and to approximate more closely what the

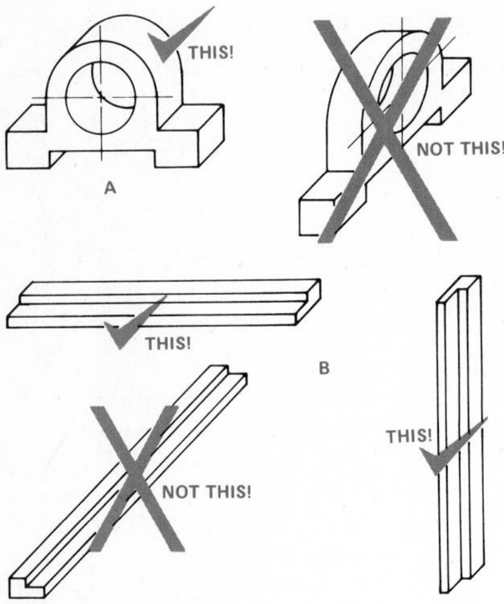
THIS!
NOT THIS!
A
THIS!
B
NOT THIS!
THIS!

Fig. 15-4-2 Two general rules for oblique drawings.

human eye would see. For this reason, and because of the simplicity of projection, cabinet oblique is a commonly used form of pictorial representation, especially when circles and arcs are to be drawn. Figure 15-4-3 shows a comparison of cavalier and cabinet oblique. Note that hidden lines are omitted unless required for clarity. Many of the drawing techniques for isometric projection apply to oblique projection. Figure 15-4-4 illustrates the construction of an irregularly shaped object by the box method.

Inclined Surfaces

Angles which are parallel to the picture plane are drawn as their true size. Other angles can be laid off by locating the ends of the inclined line.

A part with notched corners is shown in Fig. 15-4-5*a*. An oblique drawing with the angles parallel to the picture plane is shown at *b*. In Fig. 15-4-5*c* the angles are parallel to the profile plane. In each case the angle is laid off by measurement parallel to the oblique axes, as shown by the construction lines. Since the part, in each case, is drawn in cabinet oblique, the receding lines are shortened by one-half their true length.

Oblique Sketching

Specially designed oblique sketching paper with 45° lines is available and, like isometric sketching paper, is used extensively by engineers and drafters. See Fig. 15-4-6.

Dimensioning an Oblique Drawing

Dimension lines are drawn parallel to the axes of projection. Extension lines are projected from the horizontal and vertical object lines whenever possible.

The dimensioning of an oblique drawing is similar to that of an isometric drawing. The recommended method is unidirectional dimensioning, which is

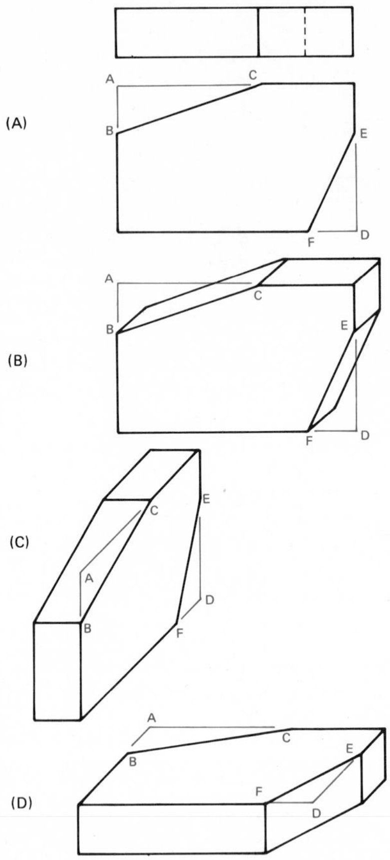
(A)
(B)
(C)
(D)
A
B
C
D
E
F

Fig. 15-4-5 Drawing inclined surfaces.

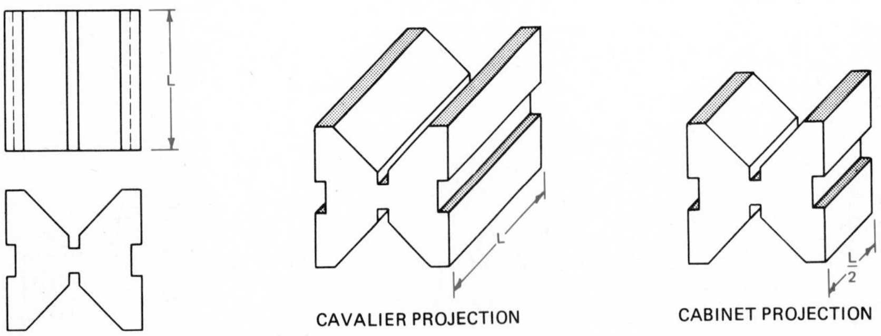
L
CAVALIER PROJECTION
L/2
CABINET PROJECTION

Fig. 15-4-3 Types of oblique projection.

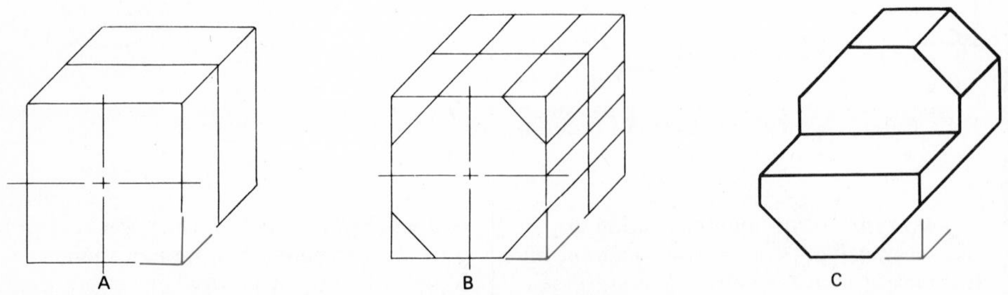
A
B
C

Fig. 15-4-4 Oblique construction by the box method.

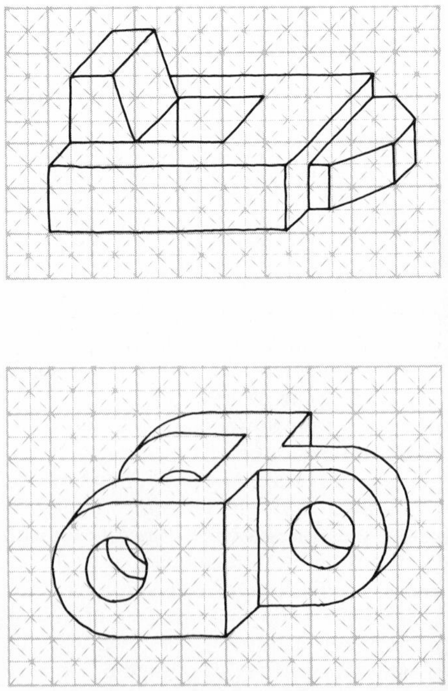

Fig. 15-4-6 Oblique sketching paper.

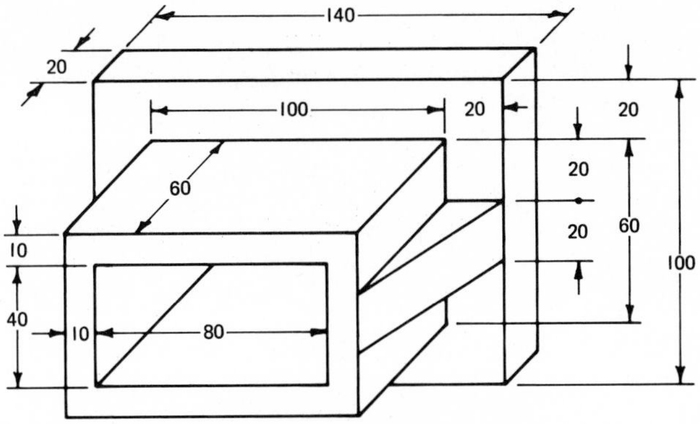
140
20
100
20
20
20
60
20
60
100
10
40
10
80

Fig. 15-4-7 Dimensioning an oblique drawing.

shown in Fig. 15-4-7. As in isometric dimensioning, usually it is necessary to place some dimensions directly on the view.

Assignments

1. On coordinate grid paper make oblique sketches of the four parts shown in Fig. 15-4-A. Each square shown on the figure represents one square on the grid paper. Hidden lines may be omitted to improve clarity.

2. On coordinate grid paper make oblique sketches of the four parts shown in Fig. 15-4-B. Each square shown on the figure represents one square on the grid paper. Hidden lines may be omitted to improve clarity.

3. On an A3- or B-size sheet, make an oblique drawing, complete with dimensions, of the dovetail guide shown in Fig. 15-4-C or 15-4-D. Scale is 1:1.

REVIEW FOR ASSIGNMENTS

Unit 15-1 Dimensioning Isometric Drawings

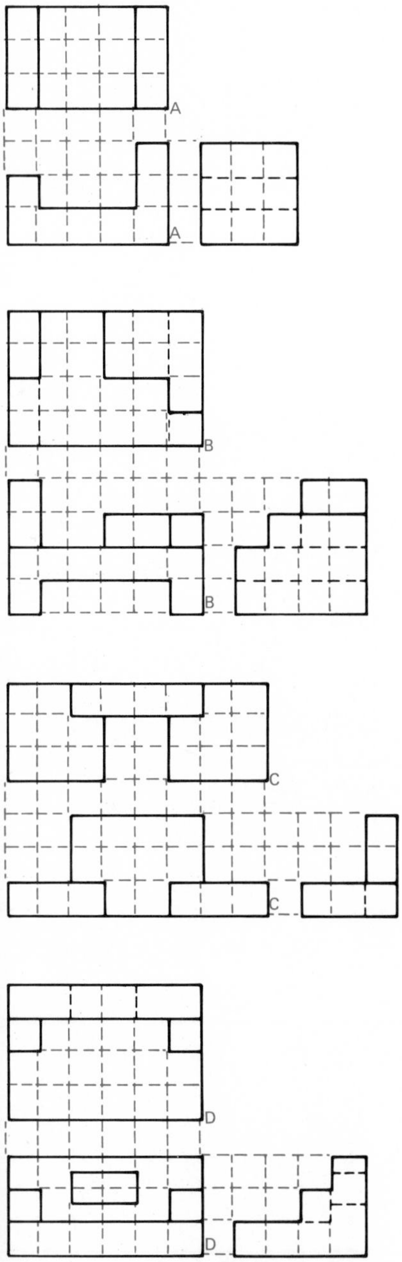
A
A
B
B
C
C
D
D

Fig. 15-4-A Sketching problems.

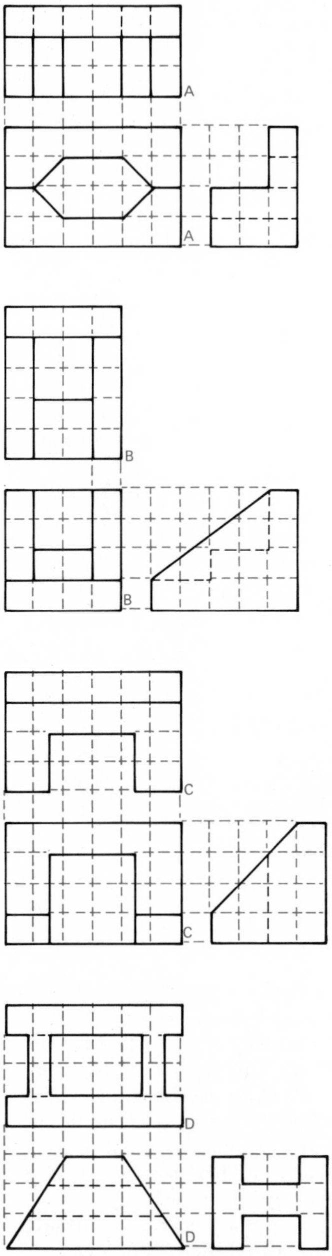
A
A
B
B
C
C
D
D

Fig. 15-4-B Sketching problems.

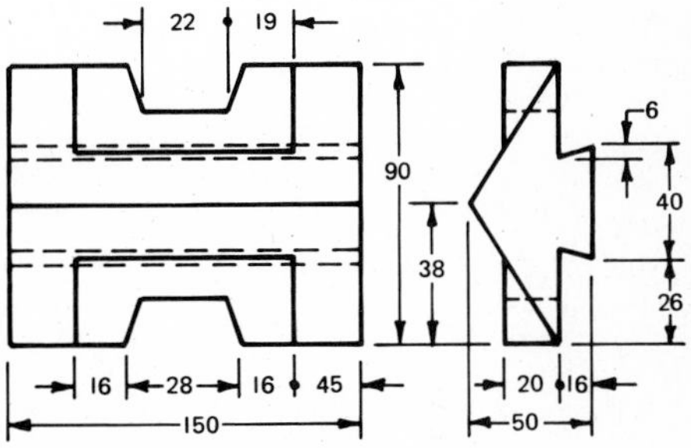
22
19
90
38
6
40
26
16
28
16
45
150
20
16
50

Fig. 15-4-C Dovetail guide.

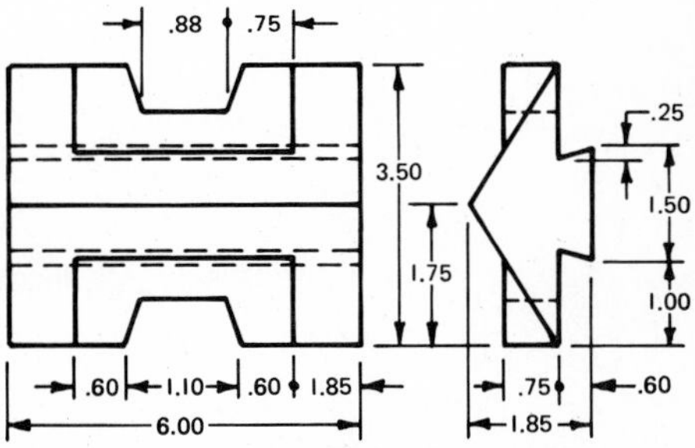
.88
.75
3.50
1.75
.25
1.50
1.00
.60
1.10
.60
1.85
6.00
.75
.60
1.85

Fig. 15-4-D Dovetail guide.

UNIT 15-5

COMMON FEATURES IN OBLIQUE

Circles and Arcs

Whenever possible, the face of the object having circles or arcs should be selected as the *front* face, so that such circles or arcs can be easily drawn in their true shape. See Fig. 15-5-1. When circles or arcs must be drawn on one of the oblique faces, the offset measurement method illustrated in Fig. 15-5-2 may be used:

1. Draw an oblique square about the center lines, with sides equal to the diameter.
2. Draw a true circle adjacent to the oblique square, and establish equally spaced points about its circumference.
3. Project these point positions to the edge of the oblique square, and draw lines on the oblique axis from these positions. Similarly spaced lines are drawn on the other axis, forming offset squares and giving intersection points for the oval shape.

Another method used when circles or arcs must be drawn on one of the oblique surfaces is the *four-center method*. In Fig. 15-5-3*a* a circle is shown as it would be drawn on a front plane, a side plane, and a top plane.

In Fig. 15-5-3*b*, the oblique drawing has some arcs in a horizontal plane. In Fig. 15-5-3*c*, the oblique drawing shown has some arcs in a profile plane.

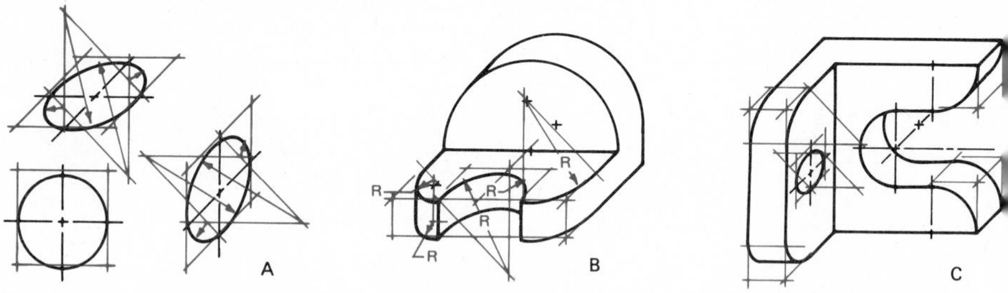
A
B
C
R

Fig. 15-5-3 Circles parallel to the picture plane are true circles; on other planes, ellipses.

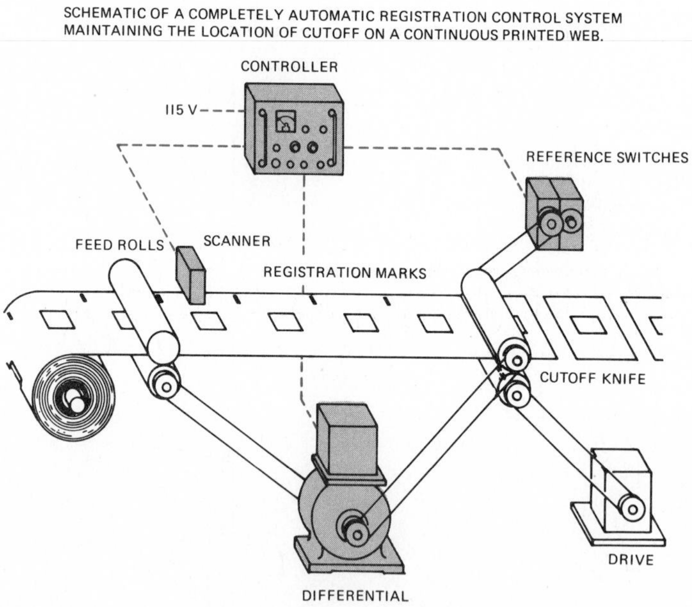
SCHEMATIC OF A COMPLETELY AUTOMATIC REGISTRATION CONTROL SYSTEM
MAINTAINING THE LOCATION OF CUTOFF ON A CONTINUOUS PRINTED WEB.
CONTROLLER
115 V
REFERENCE SWITCHES
FEED ROLLS
SCANNER
REGISTRATION MARKS
CUTOFF KNIFE
DRIVE
DIFFERENTIAL

Fig. 15-5-1 Application of oblique drawing.

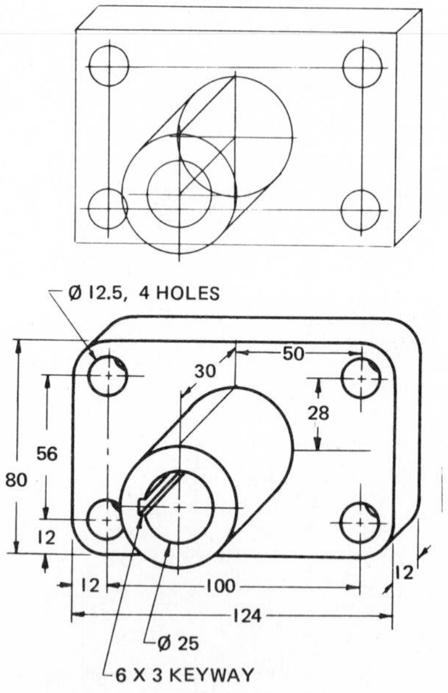
Ø 12.5, 4 HOLES
30
50
28
56
80
12
12
100
124
12
Ø 25
6 X 3 KEYWAY

Fig. 15-5-4 Construction and dimensioning of an oblique object.

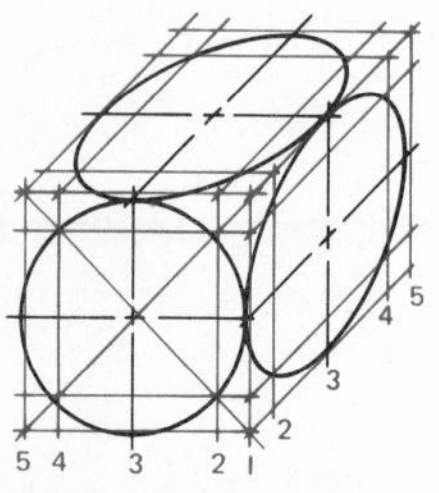
5
4
3
2
5 4 3 2 1

Fig. 15-5-2 Drawing oblique circles by means of offset measurements.

Circles not parallel to the picture plane when drawn by the approximate method are not pleasing but are satisfactory for some purposes. Ellipse templates, when available, should be used because they reduce drawing time and give much better results. If a template is used, the oblique circle should first be blocked in as an oblique square in order to locate the proper position of the circle. Blocking in the circle first also helps the drafter select the proper size and shape of the ellipse. The construction and dimensioning of an oblique part are shown in Fig. 15-5-4.

Oblique Sectioning

Oblique drawings are generally made as outside views, but sometimes a sectional view is necessary. The section is taken on a plane parallel to one of the faces of an oblique cube. Figure 15-5-5 shows an oblique full section and an oblique half

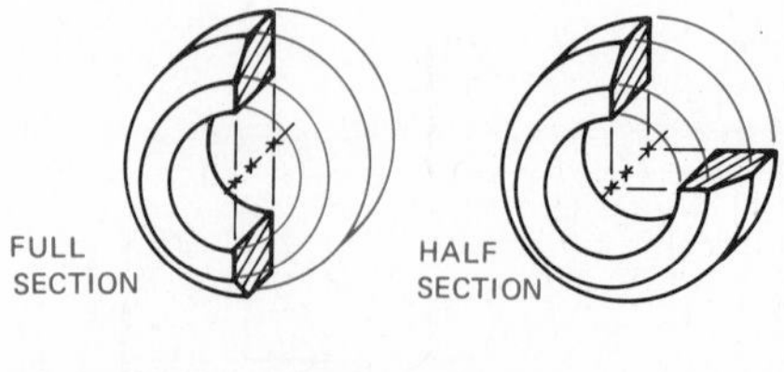
FULL
SECTION
HALF
SECTION

Fig. 15-5-5 Oblique full section and an oblique half section.

section. Note the construction lines which indicate the part that has been cut away.

Treatment of Conventional Features

Fillets and Rounds. Small fillets and rounds normally are drawn as sharp corners. When it is desirable to show the corners rounded, then either of the methods shown in Fig. 15-5-6 is recommended.

Threads. The conventional method of showing threads in oblique is shown in Fig. 15-5-7. The threads are represented by a series of circles uniformly spaced along the center of the thread. The spacing of the circles need not be the spacing of the pitch.

Breaks. Figure 15-5-8 shows the conventional method for representing breaks.

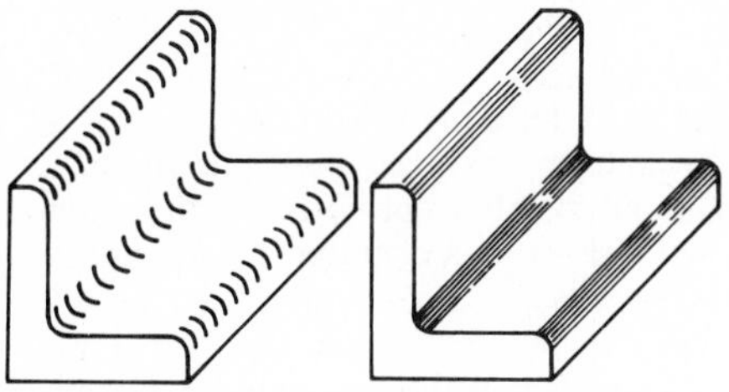

(A) CURVED LINE (B) STRAIGHT LINE

Fig. 15-5-6 Representing rounds and fillets.

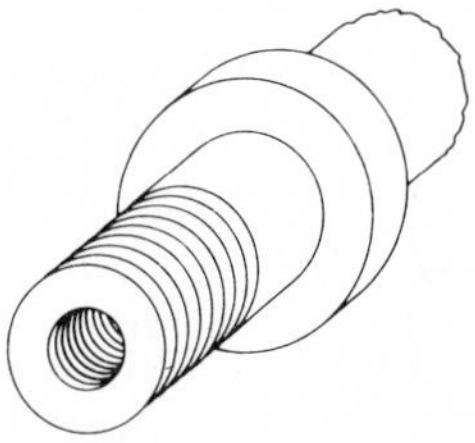

Fig. 15-5-7 Representation of threads in oblique.

Assignments

1. On an A3- or B-size sheet, make an oblique drawing, complete with dimensions, of the bearing support shown in Fig. 15-5-A or 15-5-B. Scale is 1:1.

2. On an A3- or B-size sheet, make an oblique drawing, complete with dimensions, of the end bracket shown in Fig. 15-5-C or 15-5-D. Use the straight-line method of showing the rounds and fillets. Scale is 1:1.

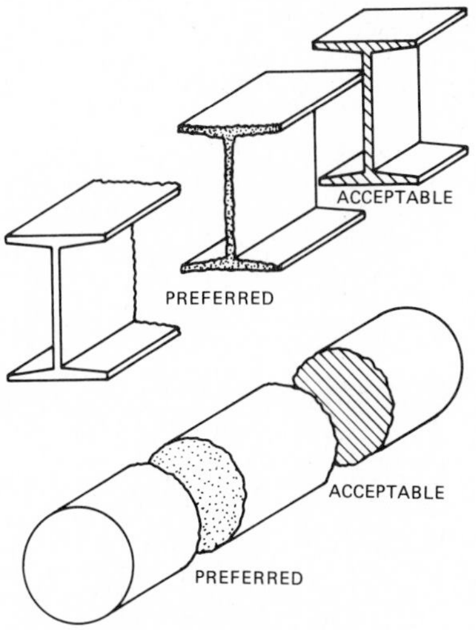
ACCEPTABLE
PREFERRED
ACCEPTABLE
PREFERRED

Fig. 15-5-8 Conventional breaks.

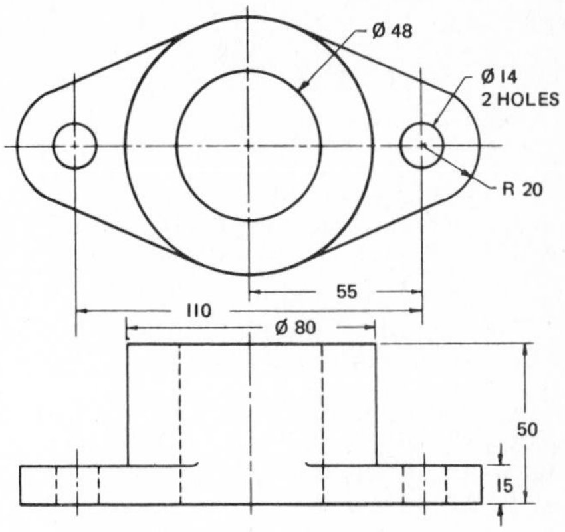
Ø 48
Ø 14
2 HOLES
R 20
55
110
Ø 80
50
15

Fig. 15-5-A Bearing support.

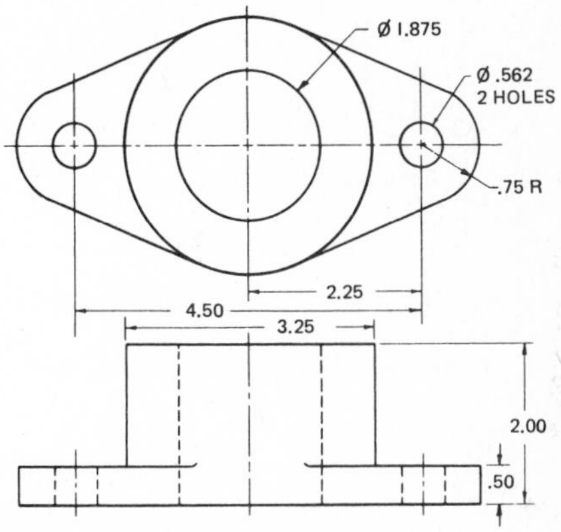
Ø 1.875
Ø .562
2 HOLES
.75 R
2.25
4.50
3.25
2.00
.50

Fig. 15-5-B Bearing support.

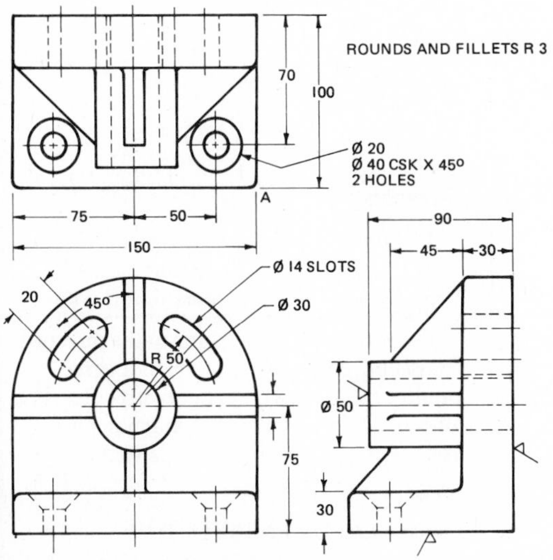
ROUNDS AND FILLETS R 3
70
100
Ø 20
Ø 40 CSK X 45°
2 HOLES
A
75
50
150
90
45
30
Ø 14 SLOTS
Ø 30
20
45°
R 50
Ø 50
75
30

Fig. 15-5-C End bracket.

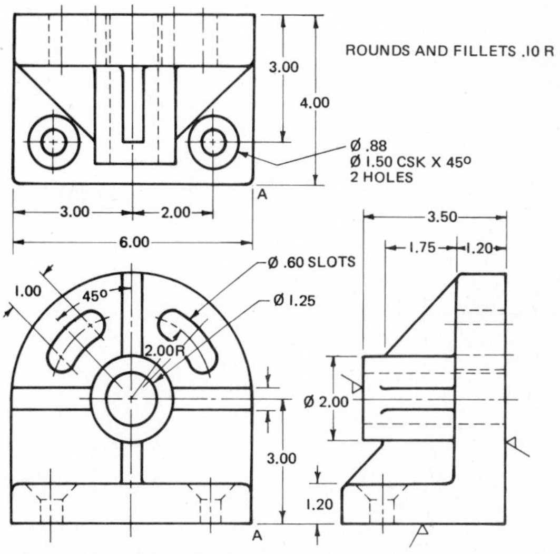
ROUNDS AND FILLETS .10 R
3.00
4.00
Ø .88
Ø 1.50 CSK X 45°
2 HOLES
A
3.00
2.00
6.00
3.50
1.75
1.20
Ø .60 SLOTS
Ø 1.25
1.00
45°
2.00R
Ø 2.00
3.00
1.20
A

Fig. 15-5-D End bracket.

UNIT 15-6

PERSPECTIVE PROJECTION[1,2]

Perspective is a method of drawing that depicts a three-dimensional object on a flat plane as it appears to the eye. See Fig. 15-6-1. A pictorial drawing made by the intersection of the picture plane with lines of sight converging from points on the object to the point of sight, which is located at a finite distance from the picture plane, is called a *perspective*. See Fig. 15-6-2.

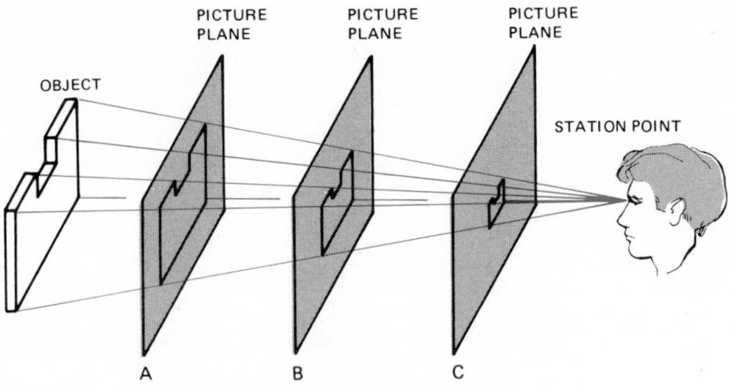
PICTURE PLANE
PICTURE PLANE
PICTURE PLANE
OBJECT
STATION POINT
A
B
C

Fig. 15-6-3 Location of the picture plane.

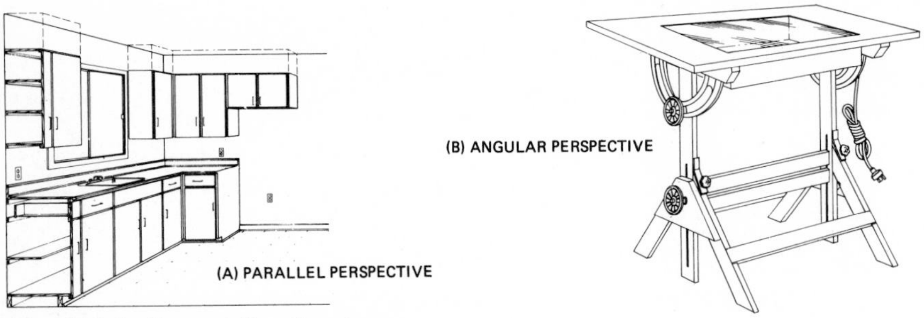
(B) ANGULAR PERSPECTIVE
(A) PARALLEL PERSPECTIVE

Fig. 15-6-1 Perspective drawings.

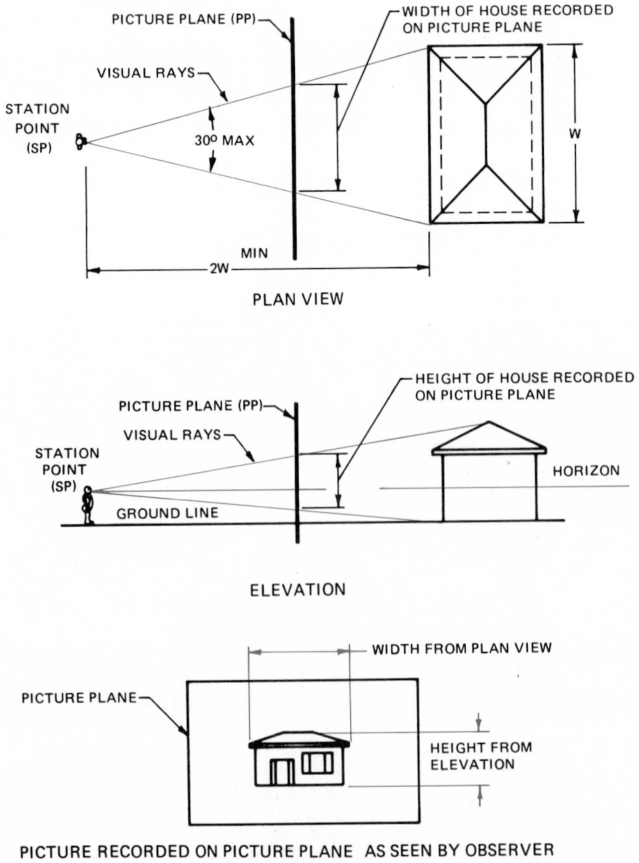
PICTURE PLANE (PP)
WIDTH OF HOUSE RECORDED ON PICTURE PLANE
VISUAL RAYS
STATION POINT (SP)
30° MAX
W
MIN 2W
PLAN VIEW
HEIGHT OF HOUSE RECORDED ON PICTURE PLANE
PICTURE PLANE (PP)
VISUAL RAYS
STATION POINT (SP)
HORIZON
GROUND LINE
ELEVATION
WIDTH FROM PLAN VIEW
PICTURE PLANE
HEIGHT FROM ELEVATION
PICTURE RECORDED ON PICTURE PLANE AS SEEN BY OBSERVER

Fig. 15-6-2 Perspective drawings.

Perspective drawings are more realistic than axonometric or oblique drawings because the object is shown as the eye would see it. Since they are far more difficult to draw than the other types of pictorial drawings, their use in drafting is limited mainly to production illustrations and illustrations of proposed structures by architects.

The main elements of a perspective drawing are the *picture plane* (plane of projection), the *station point* (the position of the observer's eye when he or she is viewing the object), the *horizon* (an imaginary horizontal line taken at eye level), the *vanishing point or points* (a point or points on the horizon where all the receding lines converge), and the *ground line* (the base line of the picture plane and object).

To avoid undue distortion in perspective, the point of sight should be located so that the cone of rays from the observer's eye has an angle at the apex not greater than 30°. This would place the station point a distance away from the outside portion of the object of approximately twice the width of the object being viewed (see Figs. 15-6-2 and 15-6-3).

Types of Perspective Drawings

There are three types of perspective drawings:

1. Parallel: One vanishing point
2. Angular: Two vanishing points
3. Oblique: Three vanishing points

In industry they are normally referred to as one-point, two-point, and three-point perspectives, respectively. See Fig. 15-6-4. Only parallel and angular perspectives are covered in this text.

PARALLEL, OR ONE-POINT, PERSPECTIVE

Parallel-perspective drawings are similar to oblique drawings, except the receding

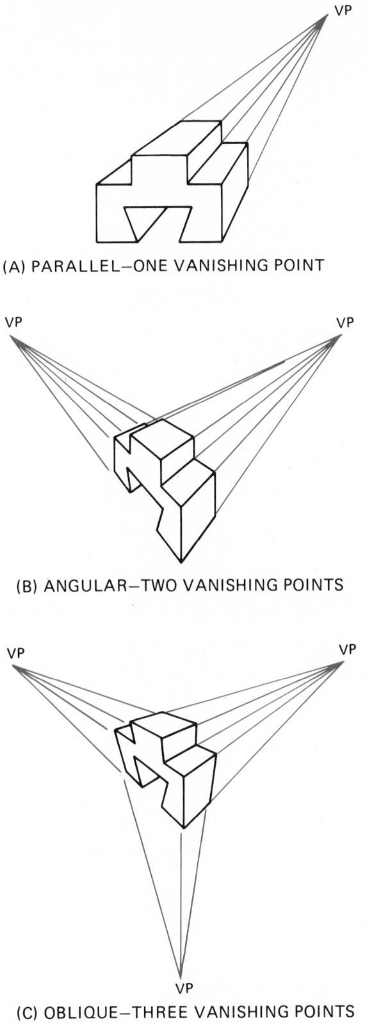
VP
(A) PARALLEL—ONE VANISHING POINT
VP
VP
(B) ANGULAR—TWO VANISHING POINTS
VP
VP
VP
(C) OBLIQUE—THREE VANISHING POINTS

Fig. 15-6-4 Perspective drawings.

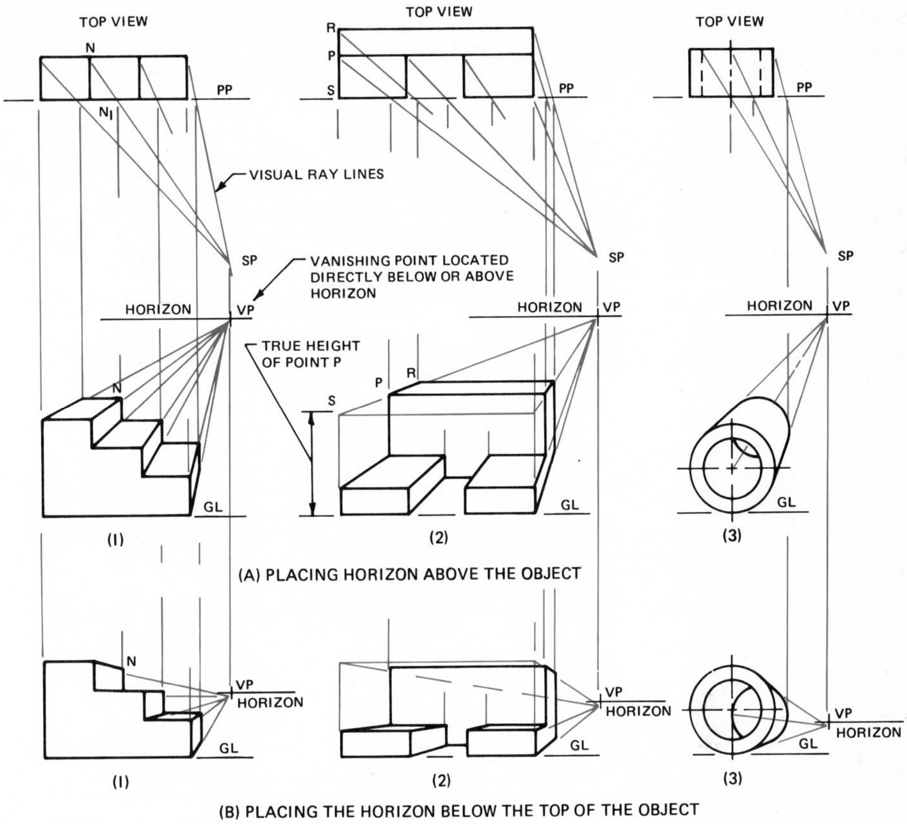
TOP VIEW
N
PP
N_1
VISUAL RAY LINES
SP
HORIZON
VP
N
GL
(1)
TOP VIEW
R
P
S
PP
VANISHING POINT LOCATED DIRECTLY BELOW OR ABOVE HORIZON
SP
HORIZON
VP
TRUE HEIGHT OF POINT P
P
R
S
GL
(2)
TOP VIEW
PP
SP
HORIZON
VP
GL
(3)
(A) PLACING HORIZON ABOVE THE OBJECT
N
VP
HORIZON
GL
(1)
VP
HORIZON
GL
(2)
VP
HORIZON
GL
(3)
(B) PLACING THE HORIZON BELOW THE TOP OF THE OBJECT

Fig. 15-6-5 Parallel or one-point perspective.

lines all converge at one point on the horizon. In drawing a parallel-perspective drawing, it is desirable to place one face of the object on the picture plane line so that it will be drawn in its true size and shape, as shown in Fig. 15-6-5. The *PP* line shown in the top view represents the picture plane line, and point *SP* is the position of the observer. The lines of the object, which are not on the picture plane, are found by projecting lines down from the top view from the point of intersection of the visual ray and the picture plane, as shown by point *N* in Fig. 15-6-5*a*.

Where the true height of a line or a point does not lie on the picture plane, such as point *P* in Fig. 15-6-5*b*, then the true height may be found by extending line *PR* to point *S* on the picture plane. Since point *S* lies on the picture plane and is the same height as point *P*, it may readily be found on the perspective drawing. Point *P* will lie on the receding line joining point *S* to the line *VP*.

In drawing a one-point perspective, a side or front view and a top view are normally drawn first—the top view to locate the part with respect to the picture plane and the side or front view to obtain the height of the various features. Figure 15-6-6 shows a simple, one-point perspective drawing on which the construction lines are left.

One of the most common uses of an angular-perspective drawing is for representing the interior of a building. With this type of drawing, the vanishing point is located inside the room and is normally at eye level. See Fig. 15-6-7.

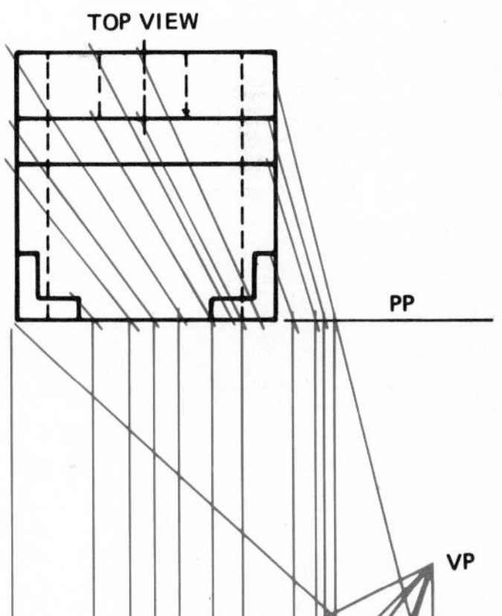
TOP VIEW
PP
VP
SP

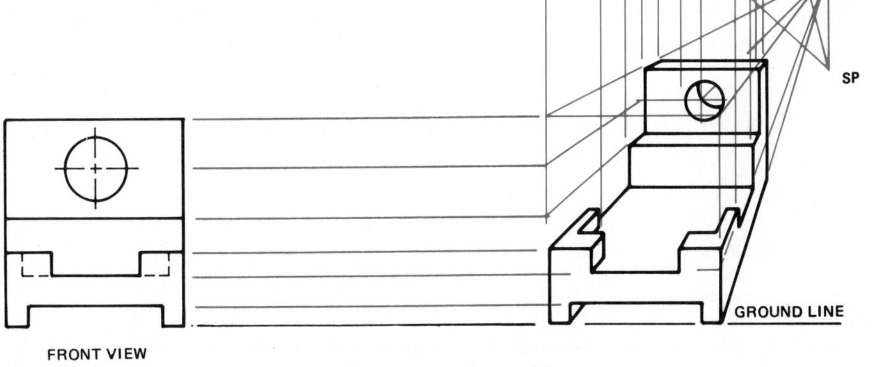
GROUND LINE
FRONT VIEW

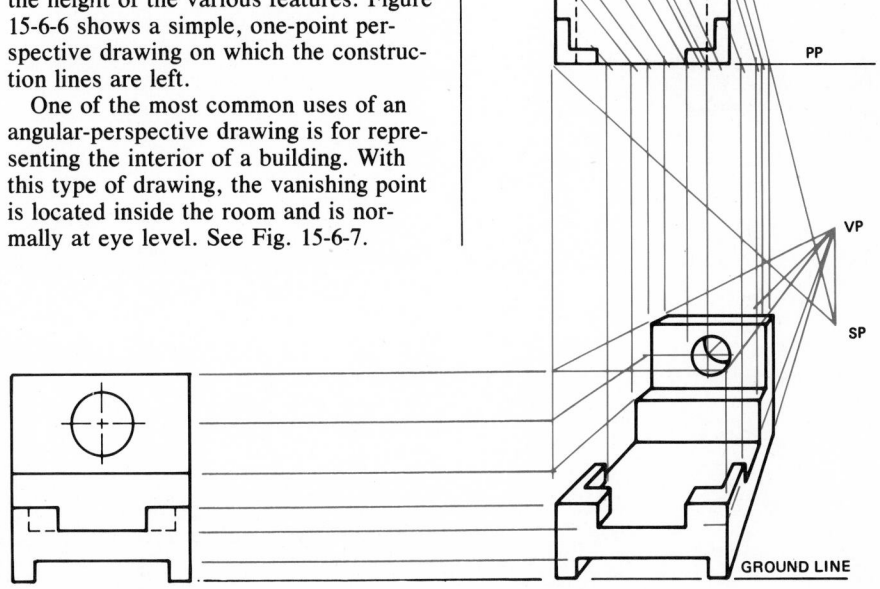

Fig. 15-6-6 Construction of a one-point perspective.

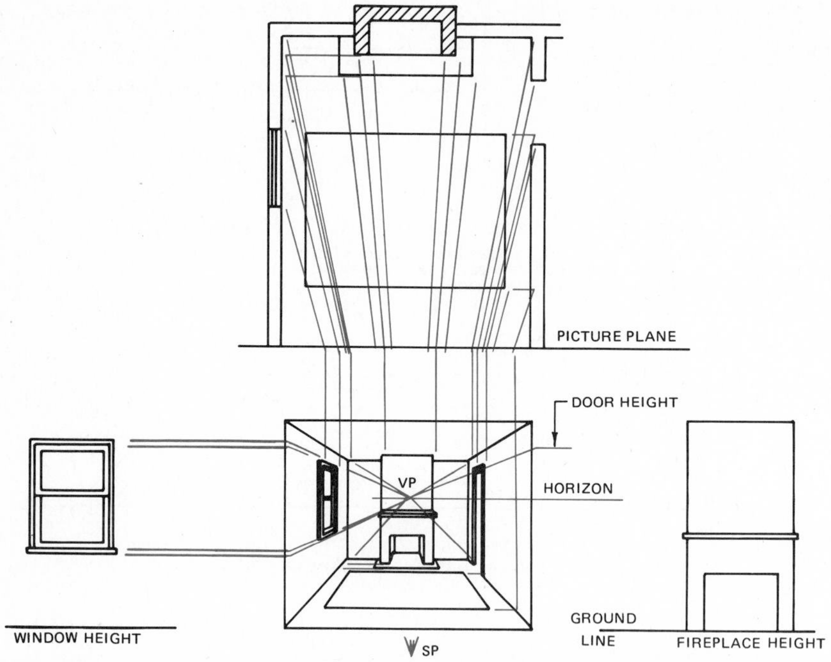
PICTURE PLANE
DOOR HEIGHT
VP
HORIZON
GROUND LINE
WINDOW HEIGHT
SP
FIREPLACE HEIGHT

Fig. 15-6-7 **Parallel-perspective drawing of an interior of a house.**

PARALLEL-PERSPECTIVE GRID

A variety of perspective grid sheets are available, which enables the drafter to produce perspective drawings in less time than by the conventional manner. Using a grid eliminates the tedious effort of establishing and projecting from the vanishing points for each individual feature. It also eliminates the problem of having the vanishing points located, in many instances, beyond the drawing area.

The cube grid, which is most widely used, has two basic variations: an exterior grid and an interior grid. See Fig. 15-6-8. The grid sizes are dependent on the desired scale of the parts to be drawn. The height and width planes are subdivided into identical increments, each increment representing any convenient size, such as 10, 100, or 1000 mm. The plane or surface representing the depth is subdivided into increments which are proportionately foreshortened as it recedes from the picture plane and thus creates the perspective illusion. See Figs. 15-6-9 and 15-6-10.

REFERENCES AND SOURCE MATERIAL

1. Extracted from American Drafting Standards Manual, Pictorial Drawing (ASA Y14.4—1957), with the permission

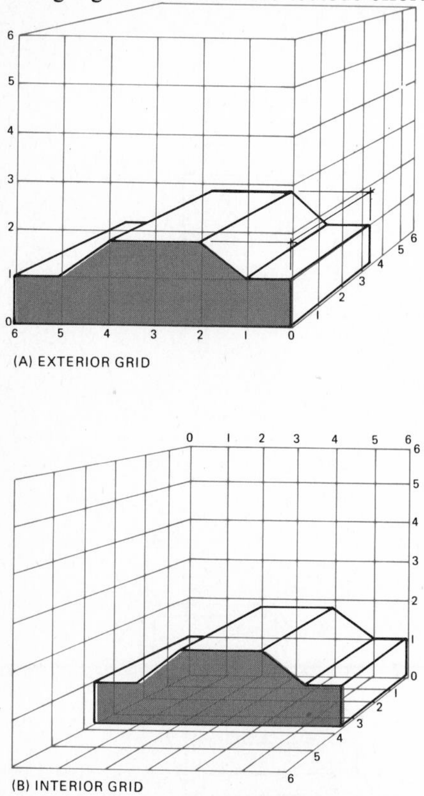
(A) EXTERIOR GRID
(B) INTERIOR GRID

Fig. 15-6-8 Parallel-perspective grid types.

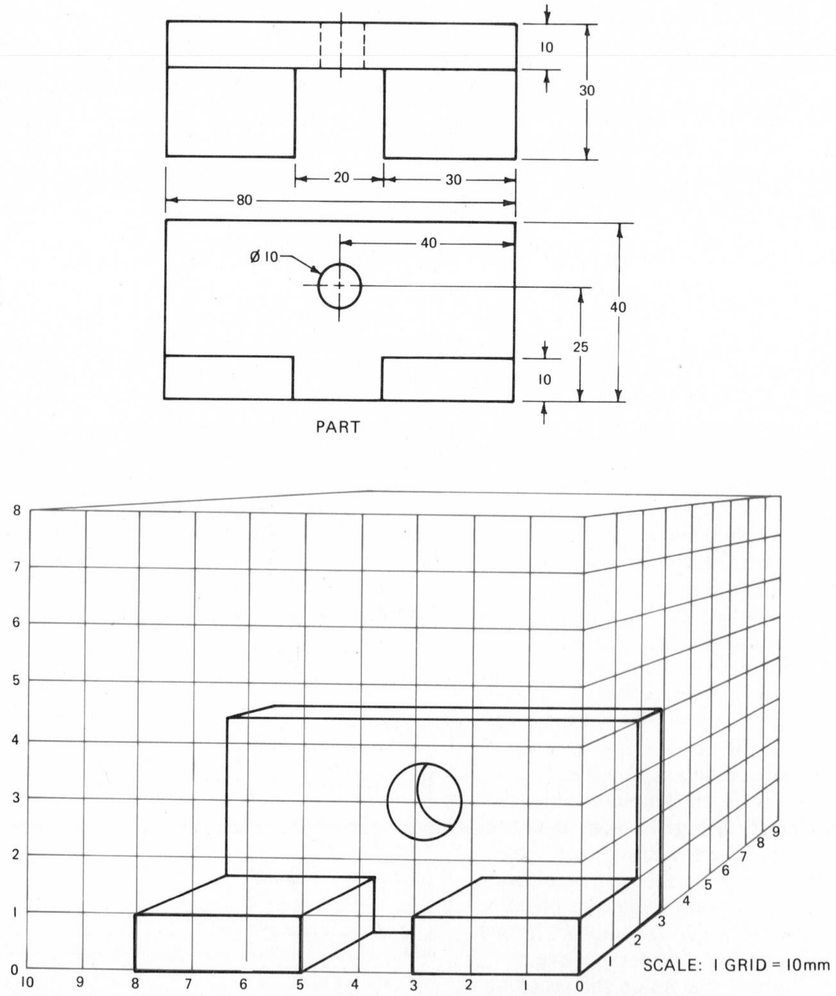
10
30
20
30
80
Ø 10
40
40
25
10
PART
SCALE: 1 GRID = 10mm

Fig. 15-6-9 Part drawn on parallel-perspective paper—exterior grid.

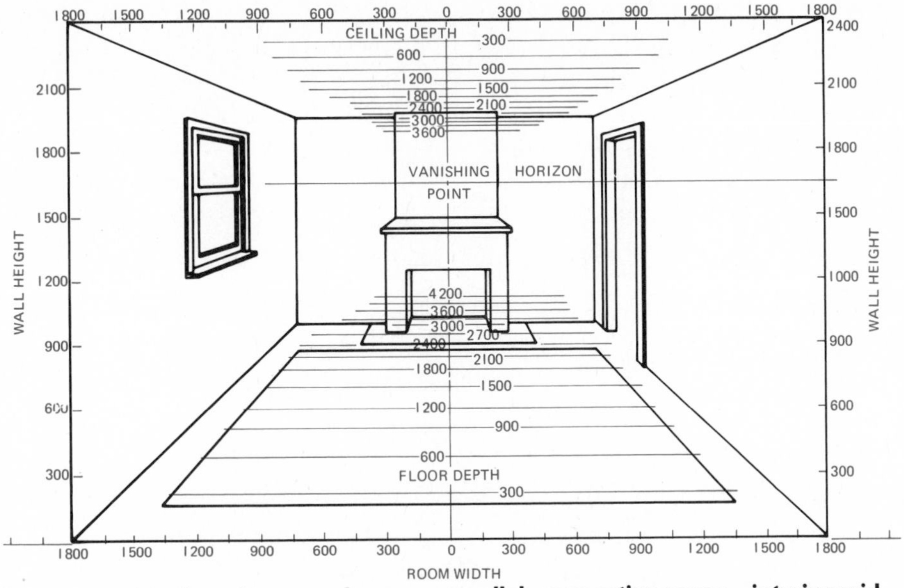
CEILING DEPTH
VANISHING POINT
HORIZON
WALL HEIGHT
WALL HEIGHT
FLOOR DEPTH
ROOM WIDTH

Fig. 15-6-10 Interior of a room drawn on parallel-perspective paper—interior grid.

of the publisher, The American Society of Mechanical Engineers, 29 W. 39th Street, New York, N.Y.

2. General Motors Corporation.

Assignments

1. On an A3- or B-size sheet, make a parallel-perspective drawing of the vise base shown in Fig. 15-6-A or 15-6-B. Scale is 1:1.

2. With the aid of a parallel-perspective grid, make a perspective drawing of the triple bookcase shown in Fig. 15-6-C. Scale is as shown. Each unit is 900W x 300D x 1800H (metric) or 36″ W x 12″ D x 72″ H (conventional). Use your judgment for sizes not shown.

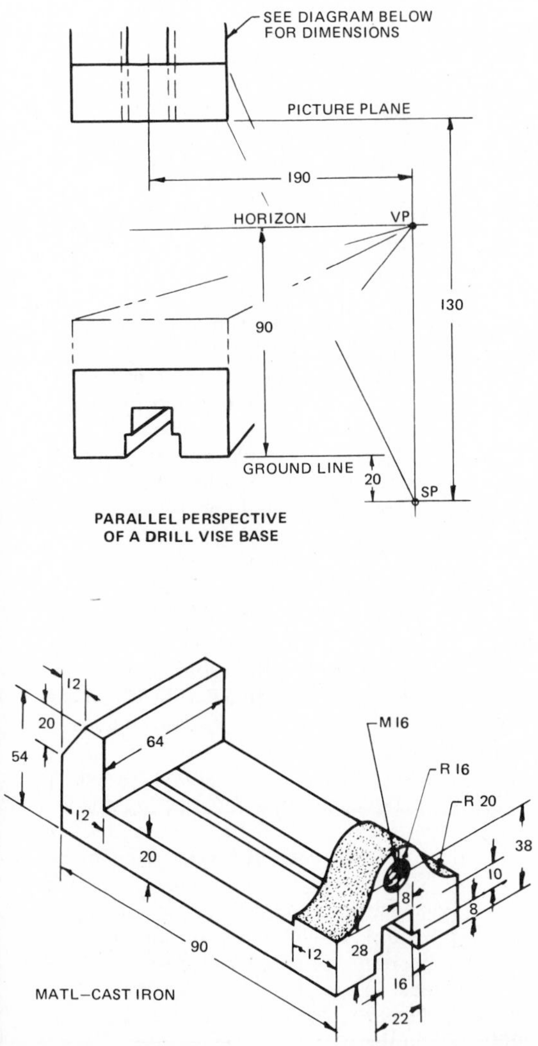
SEE DIAGRAM BELOW FOR DIMENSIONS
PICTURE PLANE
190
HORIZON
VP
130
90
GROUND LINE
20
SP
PARALLEL PERSPECTIVE OF A DRILL VISE BASE
M 16
R 16
R 20
MATL–CAST IRON

Fig. 15-6-A Vise base.

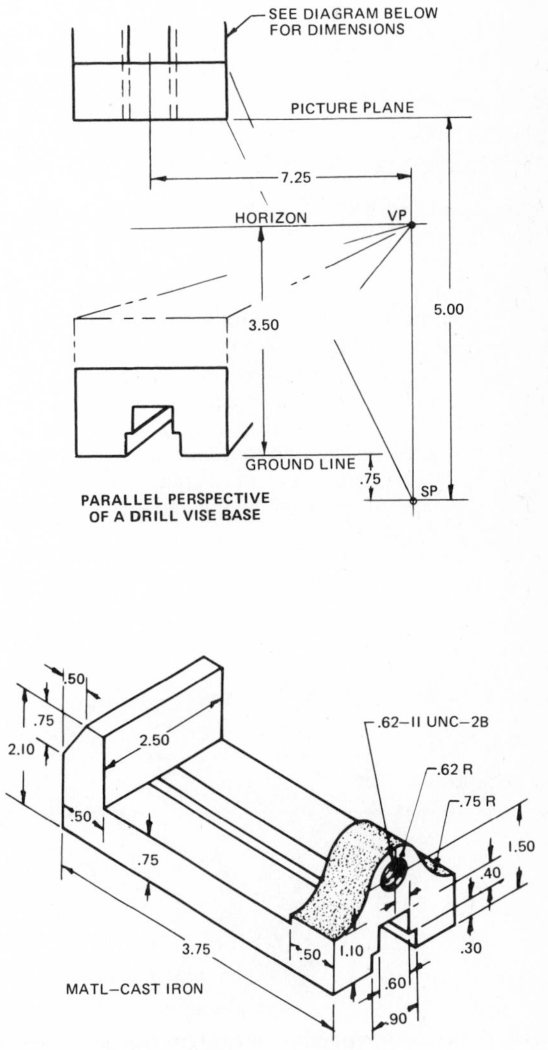
SEE DIAGRAM BELOW FOR DIMENSIONS
PICTURE PLANE
7.25
HORIZON
VP
5.00
3.50
GROUND LINE
.75
SP
PARALLEL PERSPECTIVE OF A DRILL VISE BASE
.62–11 UNC–2B
.62 R
.75 R
MATL–CAST IRON

Fig. 15-6-B Vise base.

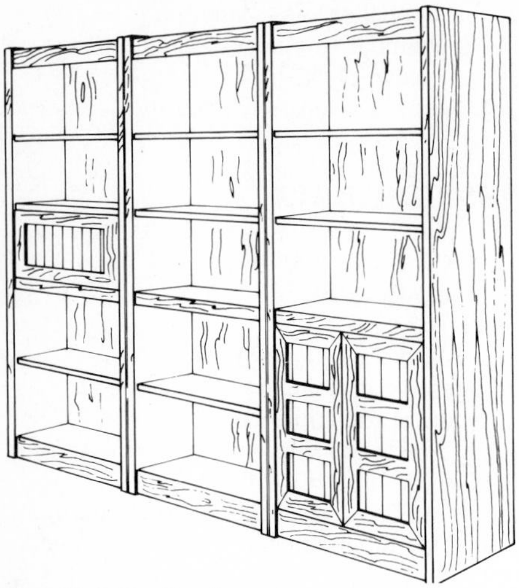

Fig. 15-6-C Bookcase

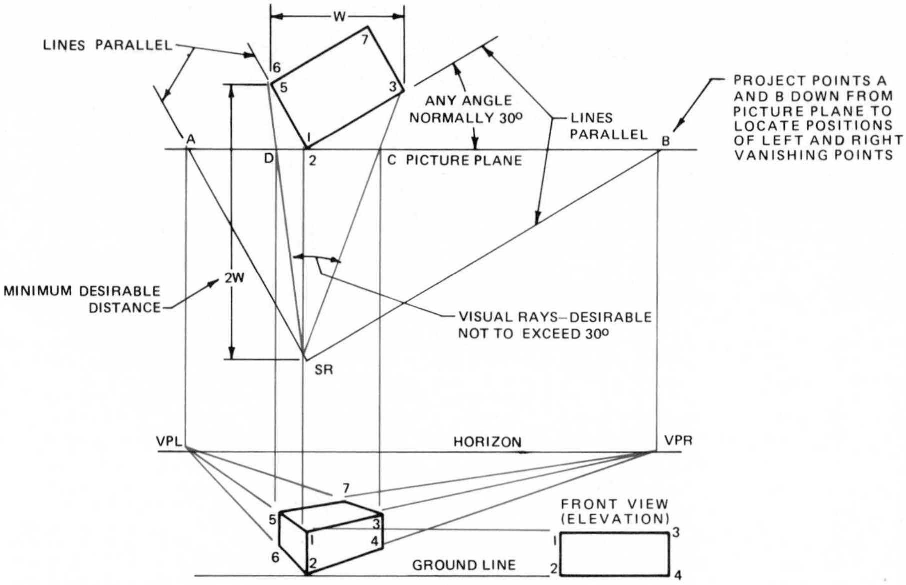
W
LINES PARALLEL
ANY ANGLE
NORMALLY 30°
LINES
PARALLEL
PROJECT POINTS A
AND B DOWN FROM
PICTURE PLANE TO
LOCATE POSITIONS
OF LEFT AND RIGHT
VANISHING POINTS
A
B
D
C PICTURE PLANE
2W
MINIMUM DESIRABLE
DISTANCE
VISUAL RAYS–DESIRABLE
NOT TO EXCEED 30°
SR
VPL
HORIZON
VPR
FRONT VIEW
(ELEVATION)
GROUND LINE

Fig. 15-7-2 Angular-perspective drawing of a prism.

UNIT 15-7

ANGULAR, OR TWO-POINT, PERSPECTIVE[1]

Two-point perspective is used quite extensively for architectural and product illustration, as shown in Fig. 15-7-1. *Angular-perspective* drawings are similar to axonometric drawings except that the receding lines converge at two vanishing points located on the horizon. Normally the height or vertical lines are parallel to the picture plane, and the length and width lines recede.

The construction for a simple prism is shown in Fig. 15-7-2. Since line 1-2 rests on the picture plane, it will appear as its true height on the perspective drawing and will be located directly below line 1-2 on the top view. The next step is to join points 1 and 2 with light, receding lines to both vanishing points. These receding lines represent the width and length lines of the prism; the width lines recede to *VPL* and the length lines recede to *VPR*. Since line 3-4 on the top view does not rest on the picture plane, it will not appear in its true height nor at its true distance from line 1-2 in the perspective. To find its position on the perspective drawing, join line 3-4, which appears as a point in the top view, to *SP* with a visual ray line. Where this visual ray line intersects the picture plane at *C*, project a vertical line down to the perspective view until it intersects the receding lines 1-*VPR* and 2-*VPR* at points 3 and 4 respectively. Line 5-6 may be found in the same manner. Next join point 3 to *VPL* and point 5 to *VPR* with light receding lines. The intersection of these lines is point 7.

Lines Not Touching on the Picture Plane. Figure 15-7-3 illustrates the construction of a perspective drawing where none of the object lines touches the picture plane.

All these lines can be constructed by using the following procedure, which locates the position and size of lines 1-2 and 3-4. Extend line 1-3 in the top view to intersect the picture plane at point *C*. Project a line down from *C* to intersect horizontal lines 1-*D* and 2-*E* at *D* and *E* respectively. Had line 1-2 been located at *C* in the top view, it would have appeared as its true height and at *D-E* on the perspective. Join points *D* to *VPR* and *E* to *VPR* with light receding lines. Somewhere along these lines are points 1, 2, 3, and 4. Next join lines 1-2 and 3-4 in the top view to *SP* with visual ray lines. Where these visual ray lines intersect the picture plane at *F* and *G* respectively, project vertical lines down to the perspective view intersecting line *D-VPR* at 1 and 3 and *E-VPR* at 2 and 4.

Construction of Circles and Curves in Perspective. Circles and curves may be constructed in perspective, as illustrated in Fig. 15-7-4. Using orthographic projections, oriented with respect to the subject in the plan and side views, plot and label the desired points (using numbers) on the curved surfaces. From the plan view project these points to the picture

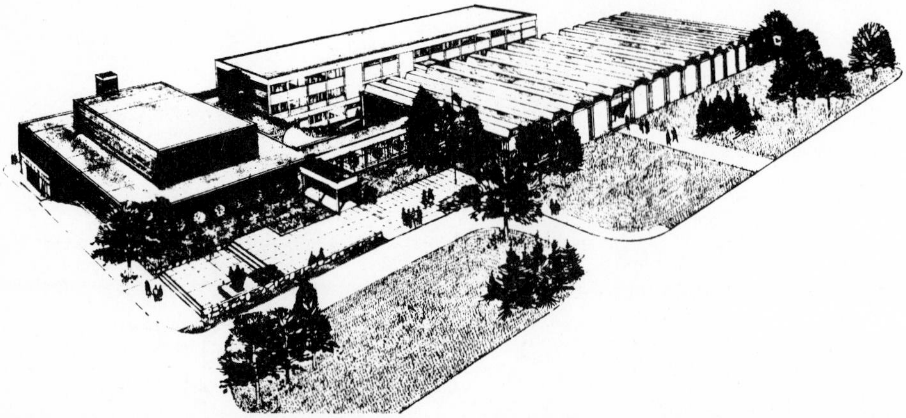

Fig. 15-7-1 Perspective drawing of the R. S. McLaughlin Collegiate and Vocational Institute, Oshawa, Canada.

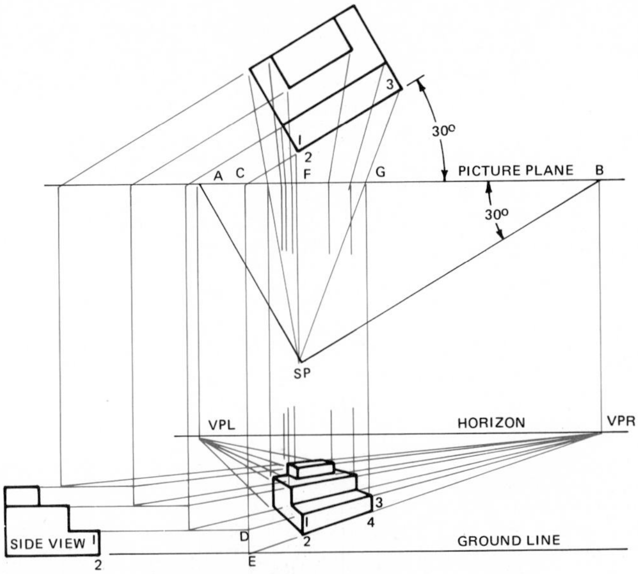
30°
PICTURE PLANE
A C F G B
30°
SP
VPL
HORIZON
VPR
SIDE VIEW
GROUND LINE
D
E

Fig. 15-7-3 Angular-perspective view of an object which does not touch the picture plane.

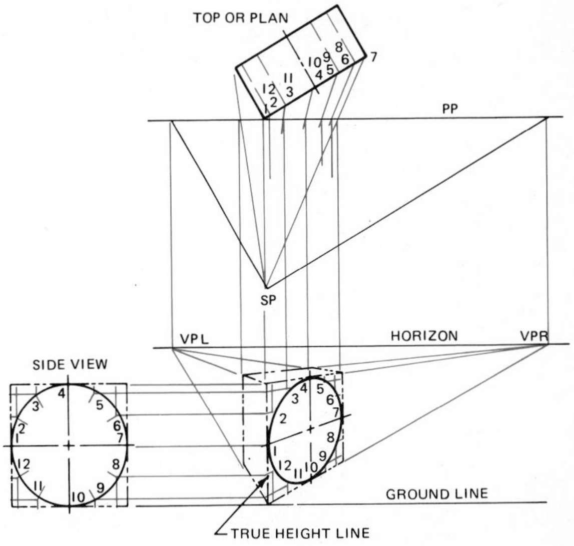
TOP OR PLAN
PP
SP
VPL
HORIZON
VPR
SIDE VIEW
GROUND LINE
TRUE HEIGHT LINE

Fig. 15-7-4 Constructing a circle in angular perspective.

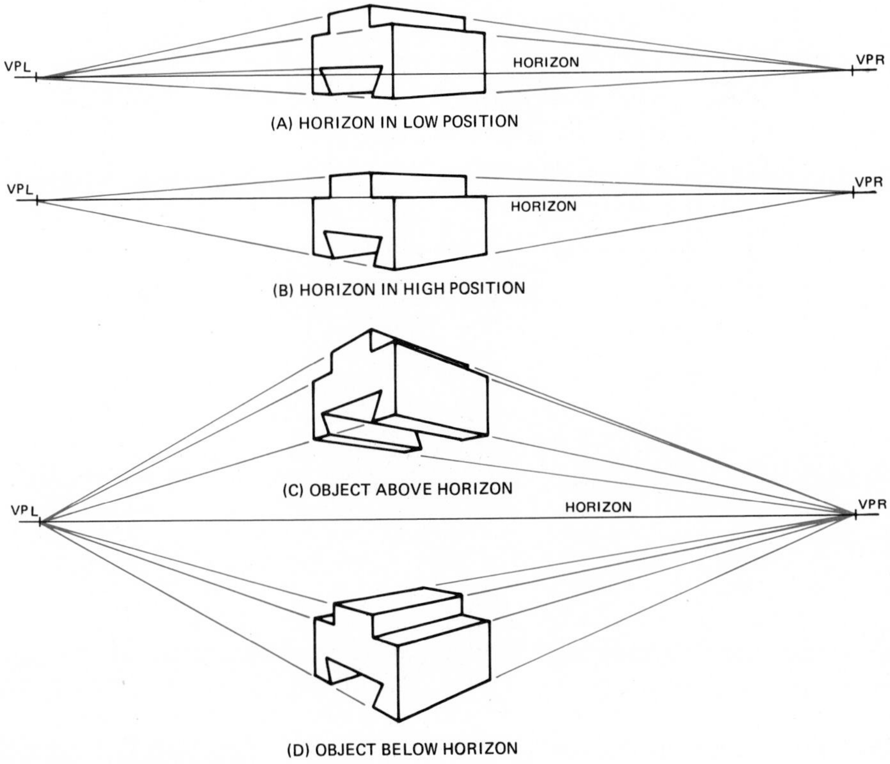
VPL
HORIZON
VPR
(A) HORIZON IN LOW POSITION
VPL
HORIZON
VPR
(B) HORIZON IN HIGH POSITION
(C) OBJECT ABOVE HORIZON
VPL
HORIZON
VPR
(D) OBJECT BELOW HORIZON

Fig. 15-7-5 Horizon lines.

plane, then vertically down to the perspective view. Project horizontally from the side view to the true-height line in the perspective view, the height of the plotting numbers. The position of the plotting numbers may now be located on the perspective view. Locate the points of intersection of the lines projected down from the picture plane with the visual ray lines receding to the right vanishing point from the appropriate numbers on the true-height line.

Horizon Line. Figure 15-7-5 illustrates different effects produced by repositioning the object with respect to the horizon.

Angular-Perspective Grids

Exterior Grid. When the three adjacent exterior planes of the cube are developed, the resultant image is referred to as an *exterior grid*. In using this grid, the points are projected from the top plane downward and from the picture planes away from the observer. See Figs. 15-7-6 and 15-7-7.

Interior Grid. When the three adjacent interior planes of the cube are exposed and developed, the resultant image is referred to as an *interior grid*. In using this grid, the points are projected from the base plane upward and from the picture planes toward the observer. The choice of usage of either variation is a matter of individual preference. Each produces the same results.

Two further variations of both the exterior and interior grids are known as the *bird's-eye* and *worm's-eye grids*.

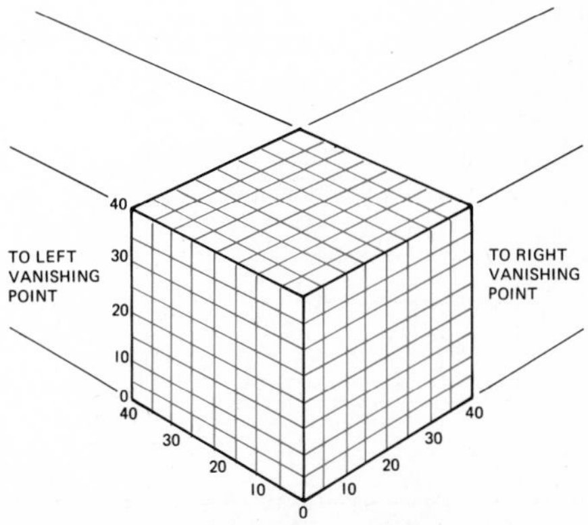
TO LEFT VANISHING POINT
TO RIGHT VANISHING POINT
40
30
20
10
0

Fig. 15-7-6 Angular-perspective grid.

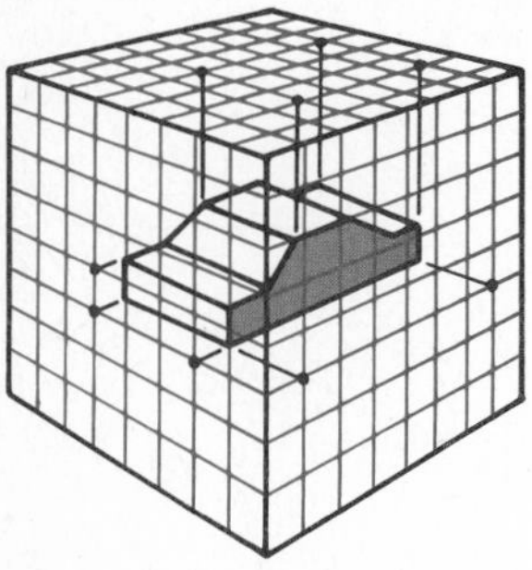

(A) EXTERIOR GRID

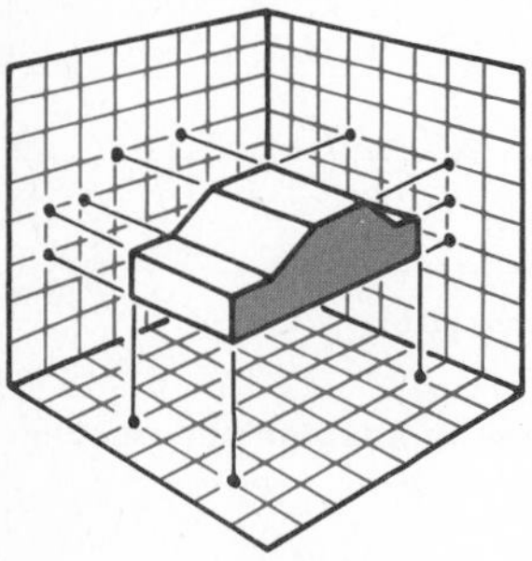

(B) INTERIOR GRID

Fig. 15-7-7 Angular-perspective grid types.

These effects are achieved by rotating the vertical plane of the grid about the horizon line.

Objects drawn in the bird's-eye grid appear as if they were being viewed from above the horizon line, as seen in Fig. 15-7-8. Objects drawn in the worm's-eye grid appear as if they were being viewed from below the horizon line.

GRID INCREMENTS

The three surfaces or planes of the grid are subdivided into multiple, vertical, and horizontal increments. Each increment is proportionately foreshortened as it recedes from the picture plane and

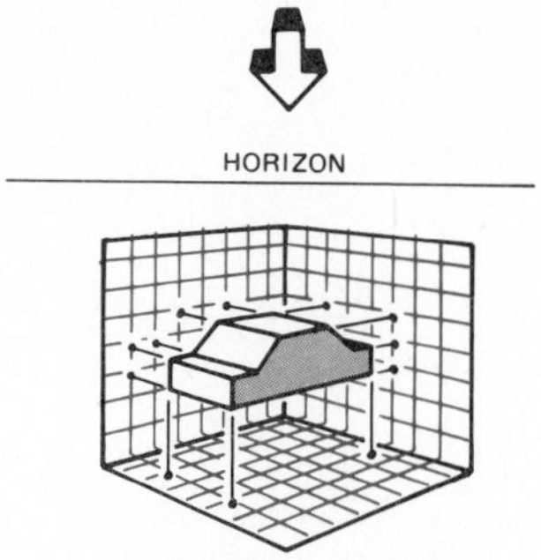
HORIZON

(A) BIRDS-EYE VIEW

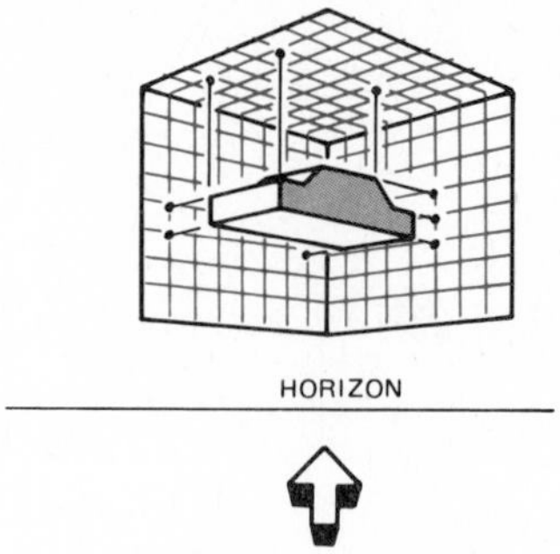
HORIZON

(B) WORMS-EYE VIEW

Fig. 15-7-8 Grid variations.

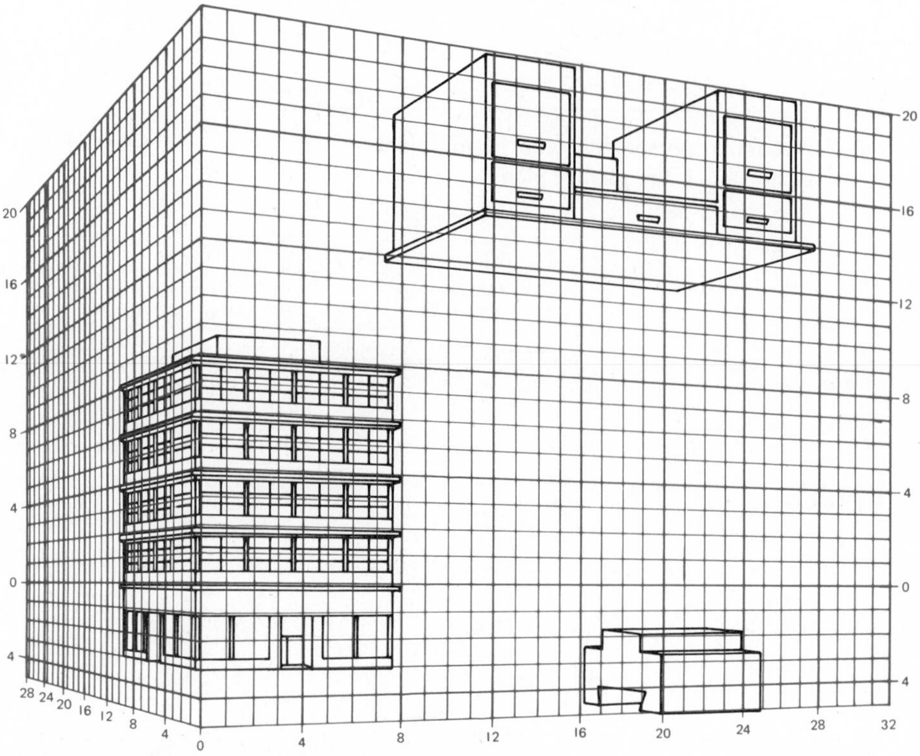
20
16
12
8
4
0
4
28 24 20 16 12 8 4 0
4 8 12 16 20 24 28 32

Fig. 15-7-9 Angular-perspective grid application.

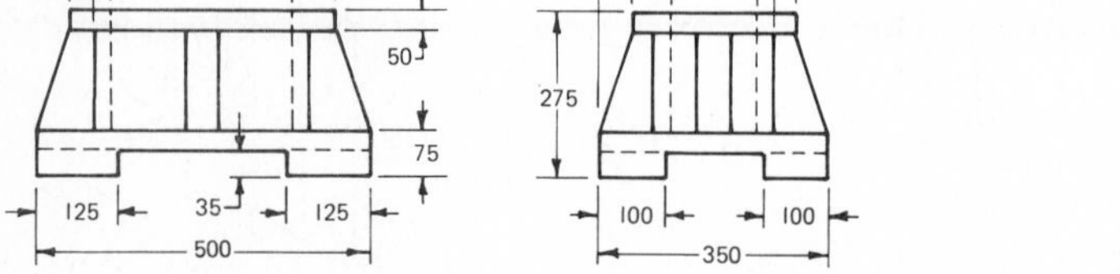
PP
30°
PLACE HORIZON
400 ABOVE
GROUND LINE
675
SP
100
50
150
225
50
400
50
25
300
50
50
275
75
125
35
125
500
250
50
150
25
100
100
350

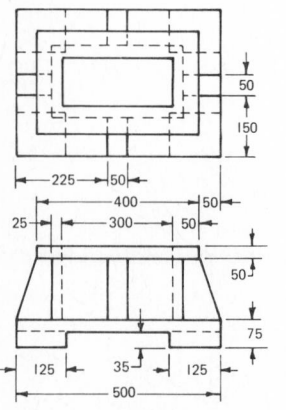

Fig. 15-7-A Planter box.

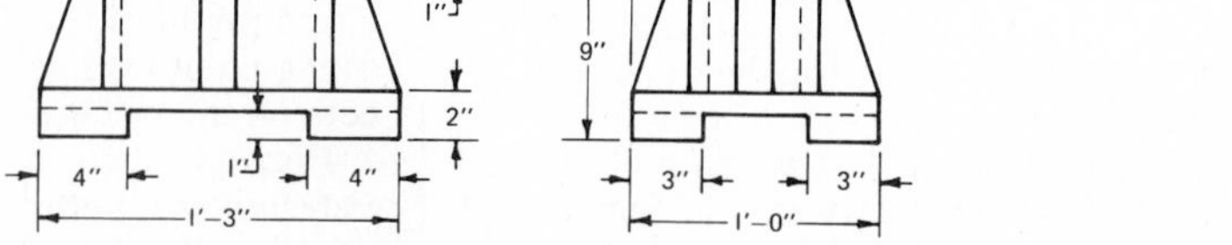
PP
30°
PLACE HORIZON
12" ABOVE
GROUND LINE
21"
SP
3"
1 1/2"
5"
6 3/4"
1 1/2"
1'–0"
1 1/2"
3/4"
9"
1 1/2"
1"
2"
4"
1"
4"
1'–3"
9"
6"
3/4"
9"
3"
3"
1'–0"

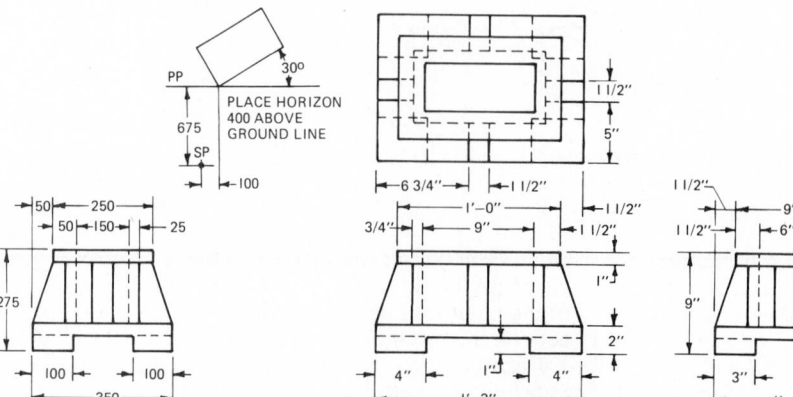

Fig. 15-7-B Planter box.

thus creates the perspective illusion. The grid increments can be any size desired. See Fig. 15-7-9.

REFERENCE AND SOURCE MATERIAL

1. General Motors Corporation.

Assignments

1. On an A3- or B-size sheet, make an angular-perspective drawing of the *planter box* shown in Fig. 15-7-A or 15-7-B. Scale is as shown.
2. On angular-perspective grid paper make a bird's-eye perspective view of the cross slide shown in Fig. 15-7-C or 15-7-D. Scale is to suit.

UNIT 15-8

TECHNICAL ILLUSTRATION[1]

Technical illustrations have an important place in all phases of engineering drawing. They form an essential part of technical manuals and catalogs, as well as illustrations appearing in technical magazines. However, this unit will not cover technical illustration techniques beyond the scope of the general drafting office.

Technical illustration drawings vary from simple sketches to rather extensive shaded drawings. They may be based upon any of the pictorial methods: isometric, oblique, or perspective, as shown in Fig. 15-8-1. The project being illustrated may be an assembly, a single part, or just a portion of a part. It may take the form of an exterior view, a sectional view, or a phantom view. The purpose in all cases is to provide a clear and easily understood drawing. See Fig. 15-8-2.

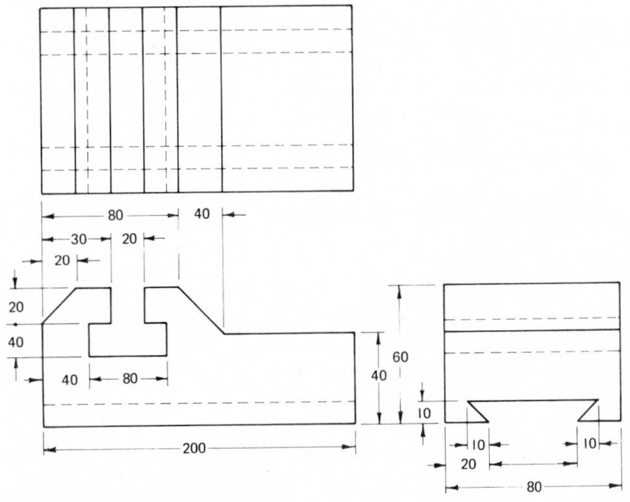
80
40
30
20
20
20
40
40
80
60
40
10
200
10
10
20
80

Fig. 15-7-C Cross slide.

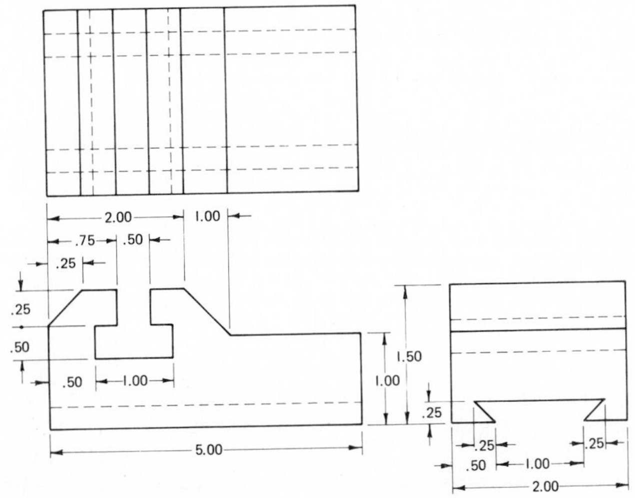
2.00
1.00
.75
.50
.25
.25
.50
.50
1.00
1.50
1.00
.25
5.00
.25
.25
.50
1.00
2.00

Fig. 15-7-D Cross slide.

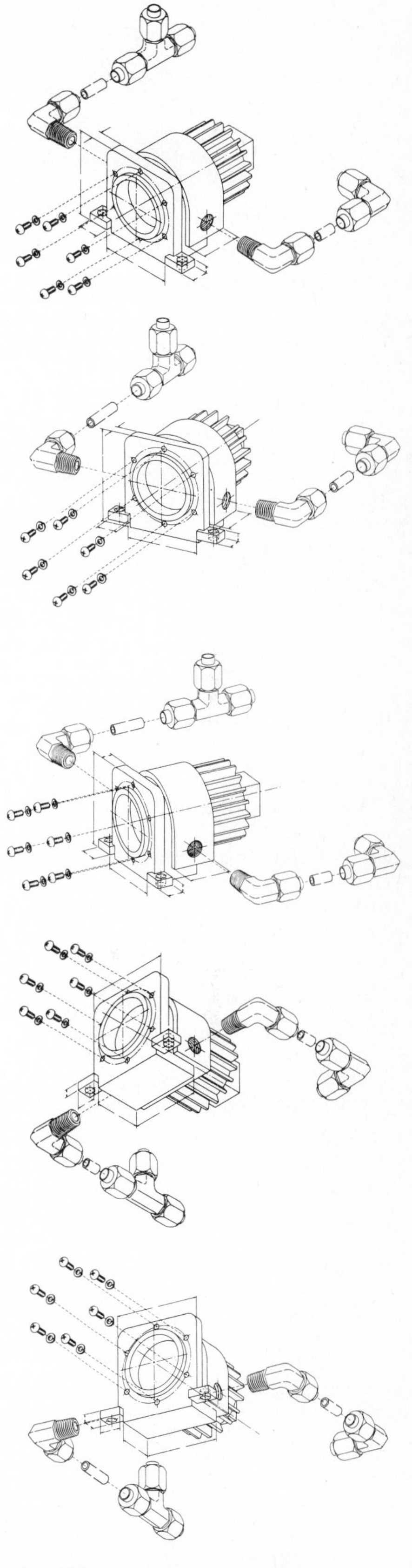

Fig. 15-8-1 A variety of pictorial methods used in technical illustrating. (Graphic Standard Instruments Co.)

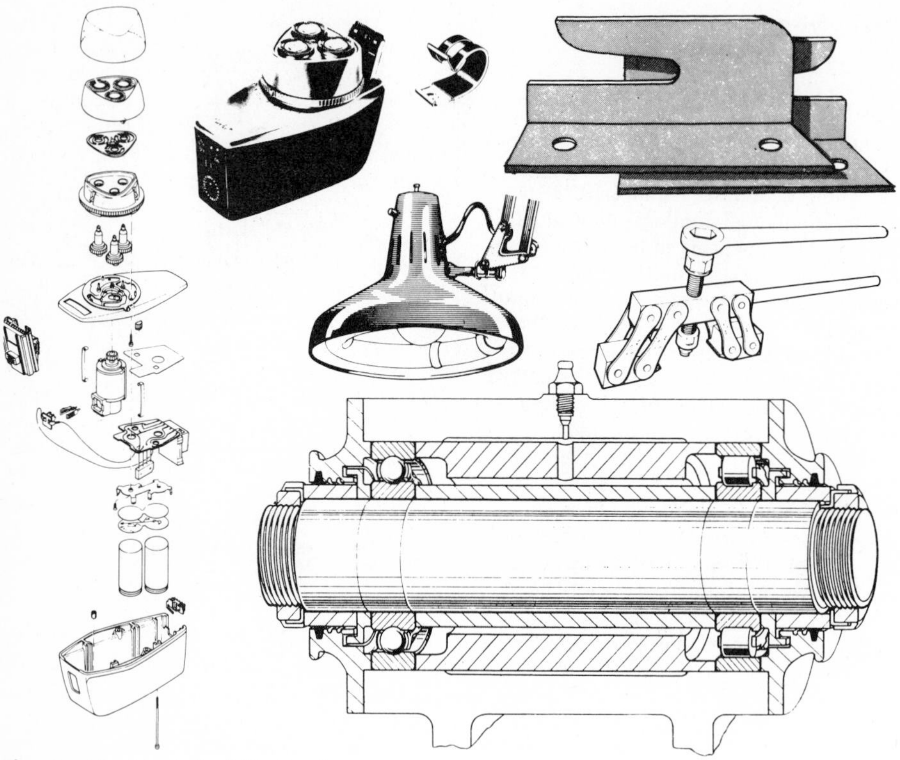

Fig. 15-8-2 **Technical illustrations.**

In many cases the drafting department may do the simple illustrations. However, for most purposes the special requirements of such drawings call for work by a professional technical illustrator.

Pictorial Line Drawings

Since nearly all technical illustrations are basically pictorial line drawings, a complete understanding of the various types and their applications is necessary.

While any type of pictorial drawing can be used as the basis for a technical illustration, some types are more suitable than others. This is especially true if the illustration is to be rendered. Figure 15-8-3 shows a V block drawn in various types of pictorial form. Notice the differences in appearance of each. Isometric is the least natural in appearance; perspective is the most natural. This might suggest, then, that all technical illustrations should be drawn in perspective. This is not necessarily true. While perspective is the more natural in appearance, it takes more time to draw if perspective grid sheets are not used. Thus, it could be more costly.

The shape of the object and how the drawing is to be used also influence the type of pictorial drawing chosen. If the object is circular, it will be easier to draw in oblique rather than isometric, if elliptical templates are not available. If the illustration is to be used in a publication such as a journal, operator's manual, technical publication, etc., dimetric, trimetric, or perspective may be the best choice.

Line Application

Line thickness in a pictorial drawing has an illusionary meaning. A pictorial drawing created with lines only, that is, without shading, must rely on converging lines and subtle line thicknesses to appear three-dimensional. See Fig. 15-8-4*a*. Basically, there are three types of lines required to create a good linear pictorial drawing. They are thick, medium, and thin.

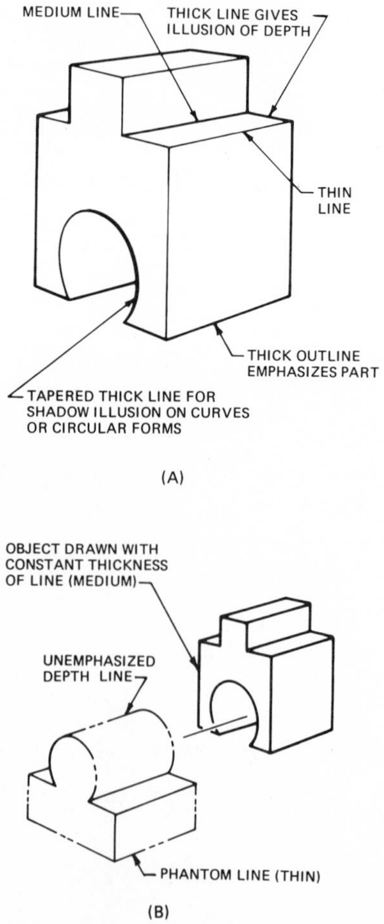
MEDIUM LINE
THICK LINE GIVES ILLUSION OF DEPTH
THIN LINE
THICK OUTLINE EMPHASIZES PART
TAPERED THICK LINE FOR SHADOW ILLUSION ON CURVES OR CIRCULAR FORMS
(A)
OBJECT DRAWN WITH CONSTANT THICKNESS OF LINE (MEDIUM)
UNEMPHASIZED DEPTH LINE
PHANTOM LINE (THIN)
(B)

Fig. 15-8-4 **Line usage.**

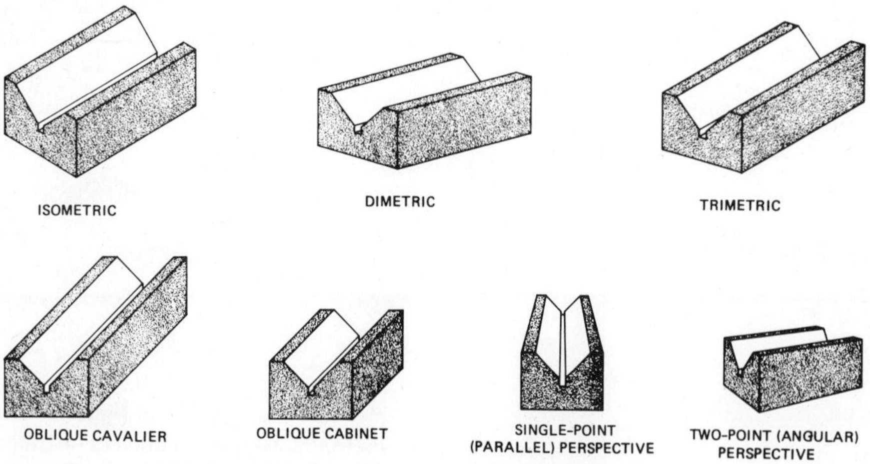
ISOMETRIC
DIMETRIC
TRIMETRIC
OBLIQUE CAVALIER
OBLIQUE CABINET
SINGLE-POINT (PARALLEL) PERSPECTIVE
TWO-POINT (ANGULAR) PERSPECTIVE

Fig. 15-8-3 **V-block in various types of pictorial drawings.**

Thick Lines. Thick or heavy lines are used to emphasize a part in a pictorial drawing. Also, an illusion of depth is created when certain lines are thicker than others. Similarly, a tapered line is used to illustrate depth on a curved surface.

Medium Lines. Medium-thickness lines are used as secondary emphasizing lines in the main object of a pictorial. They are also used as depth lines in an unemphasized or background object, as shown in Fig. 15-8-4*b*.

Thin Lines. Thin or fine lines are used as front-edge lines on the surface of the object nearest to the light source. This gives a feeling of light on this edge which further enhances the illusion of perspective. A second use of thin lines is in subduing or deemphasizing an object because it is of secondary importance in the pictorial. Phantom and broken lines are also used to illustrate background or secondary parts.

Figure 15-8-5 illustrates technical illustrations featuring line application. Notice that only the necessary detail is shown and that just enough shading is added to emphasize and give form to the parts.

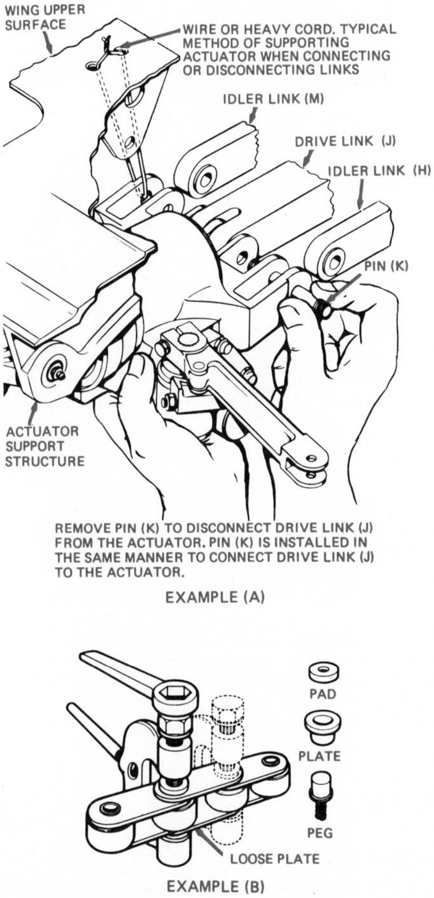
WING UPPER SURFACE
WIRE OR HEAVY CORD. TYPICAL METHOD OF SUPPORTING ACTUATOR WHEN CONNECTING OR DISCONNECTING LINKS
IDLER LINK (M)
DRIVE LINK (J)
IDLER LINK (H)
PIN (K)
ACTUATOR SUPPORT STRUCTURE
REMOVE PIN (K) TO DISCONNECT DRIVE LINK (J) FROM THE ACTUATOR. PIN (K) IS INSTALLED IN THE SAME MANNER TO CONNECT DRIVE LINK (J) TO THE ACTUATOR.
EXAMPLE (A)
PAD
PLATE
PEG
LOOSE PLATE
EXAMPLE (B)

Fig. 15-8-5 Line application.

Identification Illustrations

Pictorial drawings are very useful in identifying parts. They help save time when the parts are manufactured or assembled in place and are useful for illustrating operating instruction manuals and parts catalogs.

Identification illustrations usually take the form of exploded views. If parts are few, they can be identified by names and leaders. The identification illustration in Fig. 15-8-6 is an example showing numbers for the parts and a tabulated parts list. This method is especially recommended where identification of parts is desirable and the viewer is not trained to read technical drawings.

Rendering

For certain purposes or where shapes are difficult to read, surface shading or rendering of some kind may be desirable. For most industrial illustrations, accurate descriptions of shapes and positions are more important than fine artistic effects. Desired results can often be obtained without any shading. In general, surface shading should be limited to the least amount necessary to define the shapes illustrated. Different ways of rendering technical illustrations include line shading, screen tints, and special appliqués and pencil shading (smudge).

Some shaded surfaces are indicated in Fig. 15-8-7. An unshaded view is shown at *a* for comparison. Ruled-surface shading is shown at *b*, freehand shading at *c*, and appliqué shading at *d*.

Line Shading. Line shading is a simple, fast, and effective method of defining form. The technique may use straight or curved lines, as shown in Figs. 15-8-8 and 15-8-9.

With the light rays coming in the usual conventional direction, as in Fig. 15-8-10*a*, the top and front surfaces would be lighted and the right-hand surface would be shaded, as in Fig. 15-8-10*b*. The front surface can have light shading with heavy shading on the right-hand side, as in *c*.

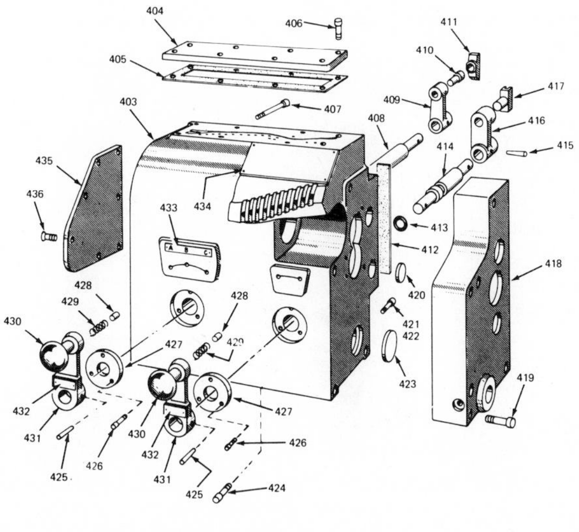

PART NO.	PART NAME	QTY.
403	QUICK CHANGE BOX	1
404	COVER, TOP	1
405	GASKET, COVER	1
406	SCREW, SOCKET HEAD CAP	8
407	SCREW	2
408	SHAFT, SHIFTER	1
409	LINK, SHIFTER	1
410	PIN	1
411	SHOE, SHIFTER	1
412	GASKET (MAKE IN PATTERN SHOP-BOX TO BED)	2
413	"O" RING	2
414	SHAFT, SHIFTER	1
415	PIN, TAPER	2
416	LINK, SHIFTER	1
417	SHOE, SHIFTER	1
418	COVER, SLIP GEAR	1
419	SCREW	4
420	PLUG	2
421	SCREW	3
422	SCREW	1
423	PLUG (NOT USED WITH SCREW REVERSE)	1
424	SCREW	3
425	PIN	2
426	SCREW	6
427	COLLAR	2
428	PLUNGER	2
429	SPRING	2
430	KNOB	2
431	LEVER	2
432	PLATE, FEED-THD.	1
433	PLATE, COMPOUND	1
434	PLATE, ENGLISH INDEX	1
435	COVER	1
436	SCREW	7

Fig. 15-8-6 Exploded view.

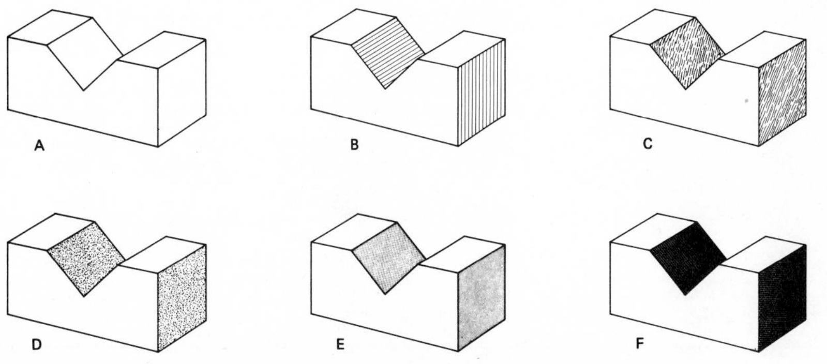
A
B
C
D
E
F

Fig. 15-8-7 Examples of various kinds of rendering.

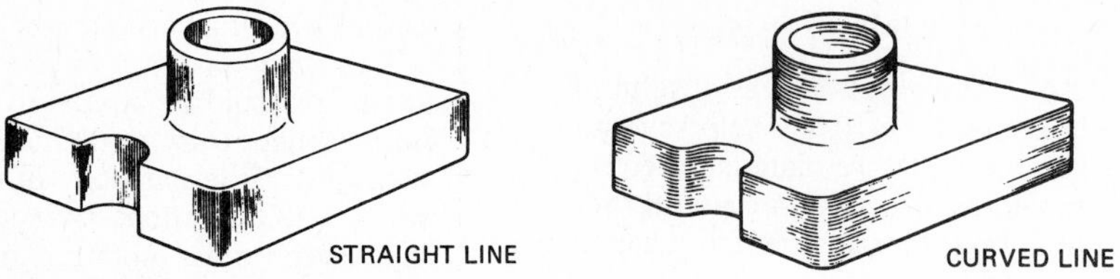
STRAIGHT LINE
CURVED LINE

Fig. 15-8-8 Line shading.

TORQUE CONVERTER
10g
100g

Fig. 15-8-9 Application of line shading.

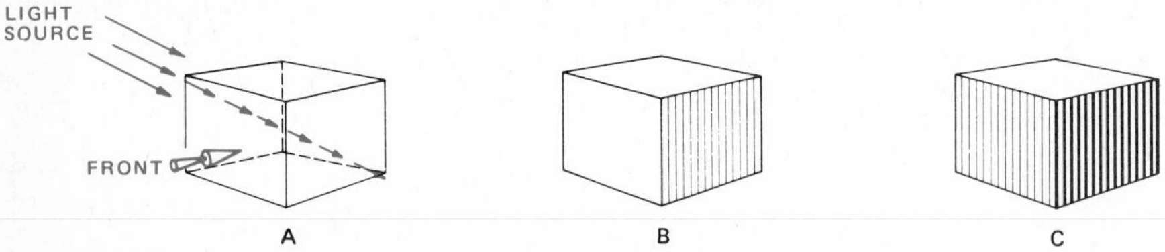
LIGHT SOURCE
FRONT
A
B
C

Fig. 15-8-10 Line rendering the faces of a cube.

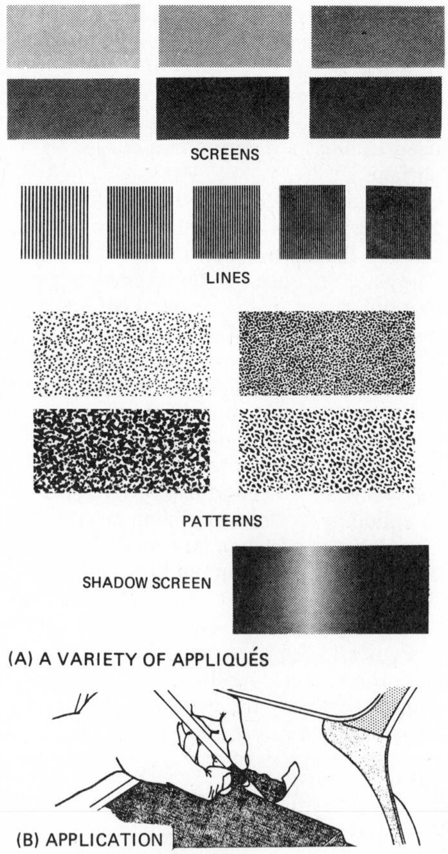
SCREENS
LINES
PATTERNS
SHADOW SCREEN
(A) A VARIETY OF APPLIQUÉS
(B) APPLICATION

Fig. 15-8-11 Appliqués.

Appliqué Shading. Commercial products of varied patterns are used to render areas which require distinction. See Fig. 15-8-11. These products are self-adhering and are easily cut out to match the areas to be covered. Shadow effects are achieved by the addition of another layer of the same material or of a contrast pattern. Both methods are illustrated in Fig. 15-8-12.

Stippling consists of dots, short crooked lines, or similar treatment to produce a shaded effect. It is a good method when it is well done.

Screens (measured in the number of dots per area) are available in a great variety of patterns and are very effective. A combination of line and appliqué rendering is shown in Fig. 15-8-13.

Tonal Pencil Rendering. Tonal pencil rendering, while in limited use, produces a finished drawing of more professional quality. This method is usually used with diazo reproduction of limited distribution. See Fig. 15-8-14.

Special Features

Screw Threads. Both internal and external threads are conveniently shown by a series of ellipses uniformly spaced

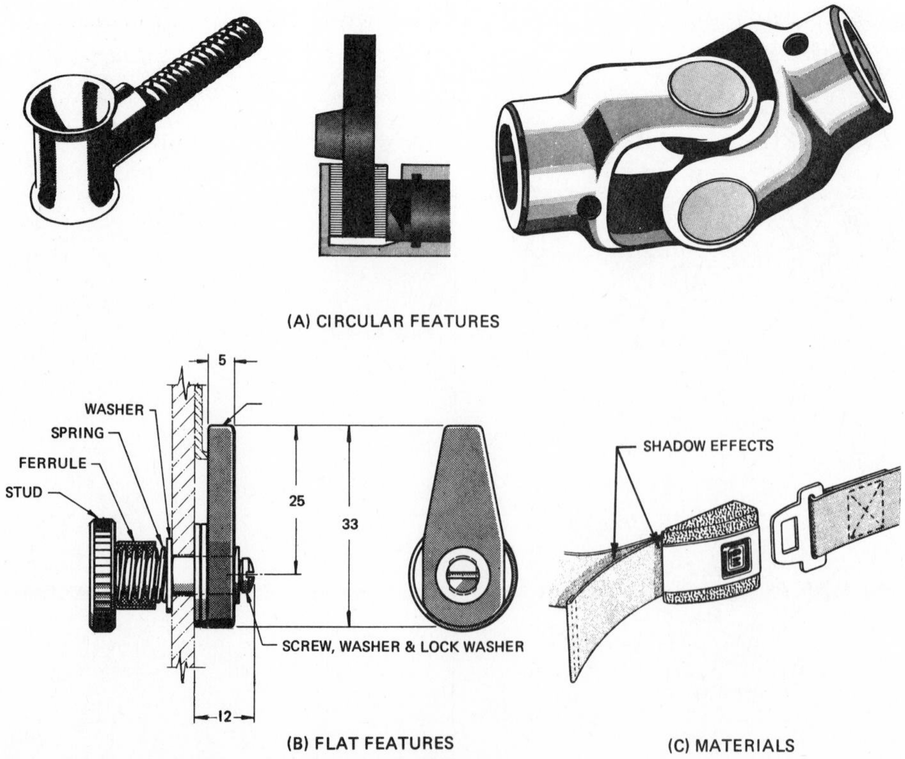
(A) CIRCULAR FEATURES
5
WASHER
SPRING
FERRULE
STUD
25
33
SCREW, WASHER & LOCK WASHER
12
(B) FLAT FEATURES
SHADOW EFFECTS
(C) MATERIALS

Fig. 15-8-12 Examples of appliqué rendering.

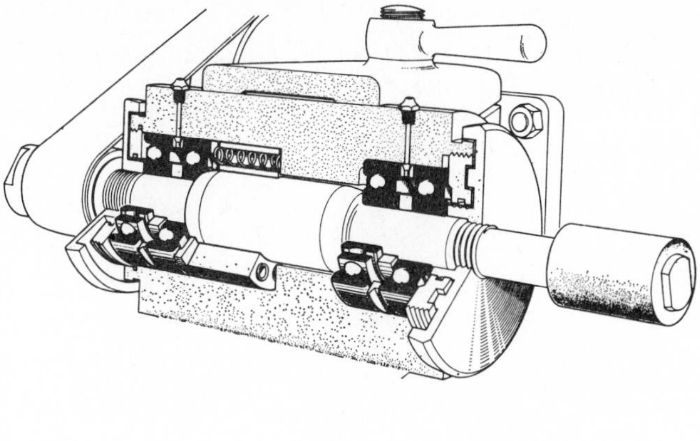

(A)

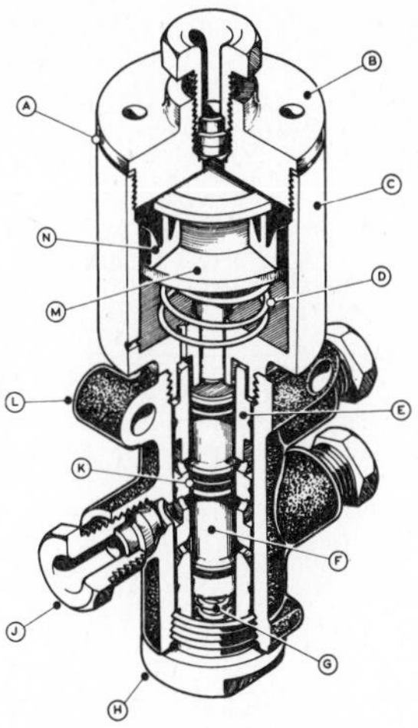

(B)

Fig. 15-8-13 Combination of line and appliqué rendering. (A—Renold Canada Ltd., B—Holman Bros. [Canada] Ltd.)

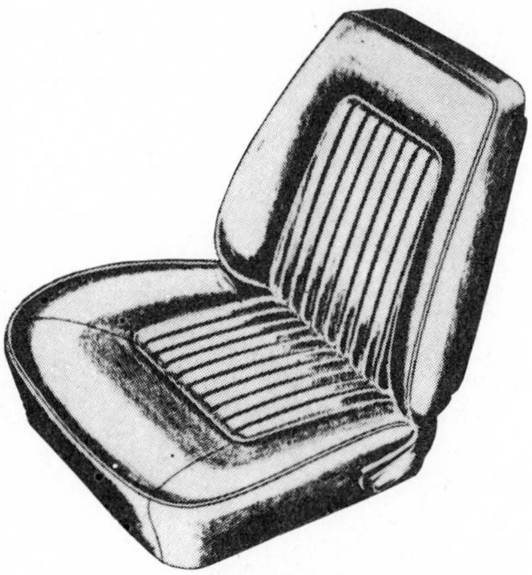

Fig. 15-8-14 Tonal pencil rendering. (General Motors Corp.)

along the axis of the thread. The spacing of these lines may be greater than the actual thread pitch to allow room for the effective line shading. See Fig. 15-8-15.

REFERENCE AND SOURCE MATERIAL

1. General Motors Corporation.

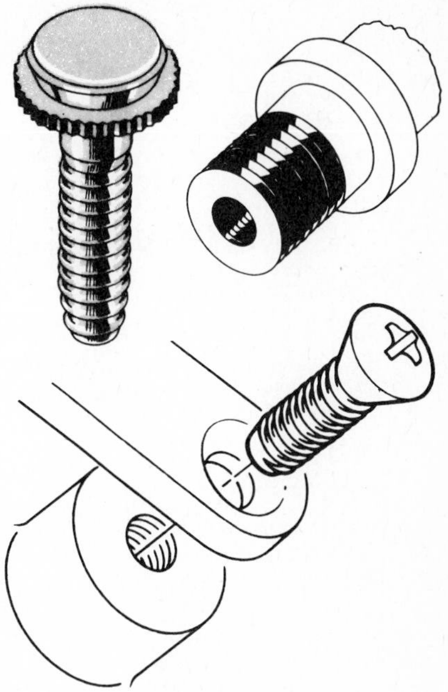

Fig. 15-8-15 Screw threads.

Assignments

1. On an A3- or B-size sheet, prepare two identical pictorial drawings of the cast swing bracket shown in Fig. 15-8-A or 15-8-B. On the first drawing use line shading; on the second, use appliqués. Do not dimension. Select any pictorial method you wish. Scale is 1:1.

2. On an A3- or B-size sheet, make an exploded isometric assembly drawing of the universal joint shown in Fig. 15-8-C or 15-8-D. Scale is 1:1. Do not dimension. Use a rendering technique of your choice to improve the appearance of the drawing.

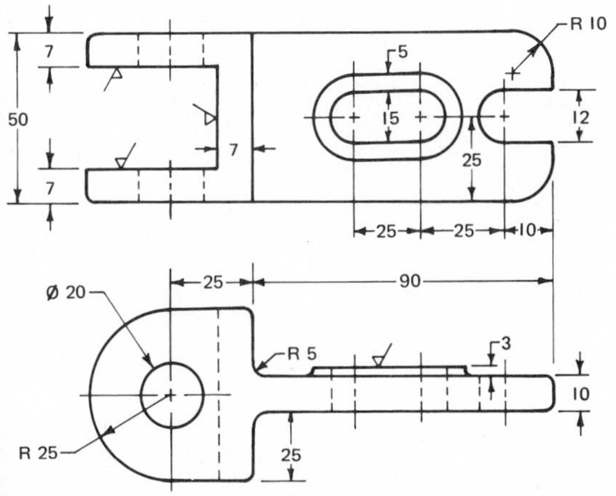
R 10
5
7
50
15
12
7
25
7
25
25
10
Ø 20
25
90
R 5
3
10
R 25
25

Fig. 15-8-A Swing bracket.

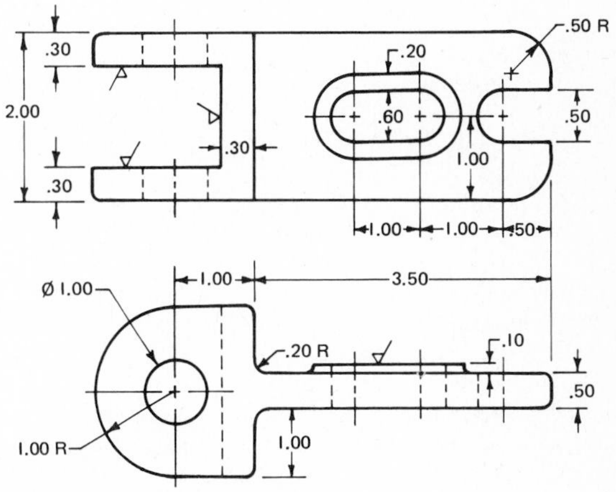
.50 R
.30
.20
2.00
.60
.50
.30
1.00
.30
1.00
1.00
.50
Ø 1.00
1.00
3.50
.20 R
.10
.50
1.00 R
1.00

Fig. 15-8-B Swing bracket.

Fig. 15-8-C Universal joint details.

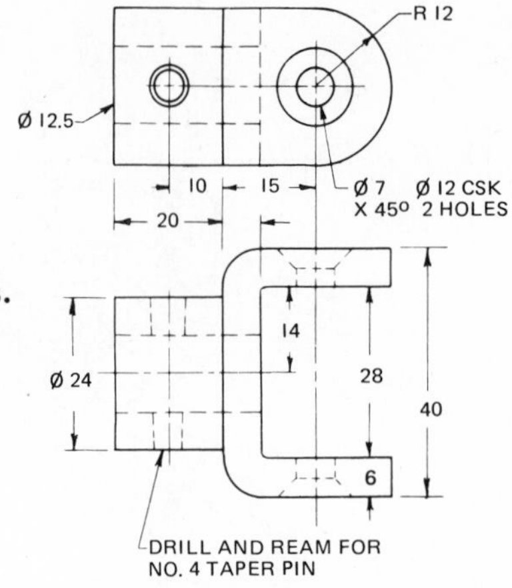
R 12
Ø 12.5
10
15
20
Ø 7 Ø 12 CSK
X 45° 2 HOLES
14
Ø 24
28
40
6
DRILL AND REAM FOR
NO. 4 TAPER PIN

PT I – FORK – 2 REQD

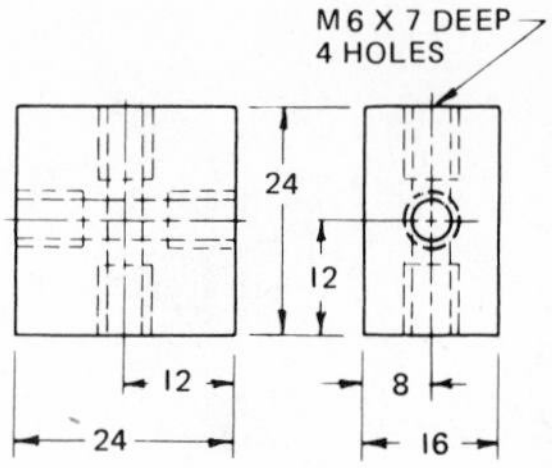
M 6 X 7 DEEP
4 HOLES
24
12
12
24
8
16

PT 2 – RING – I REQD

PT 3 – NO. 4 TAPER PIN – 2 REQD

PT 4 – M6 – FHMS I5 LG – 4 REQD

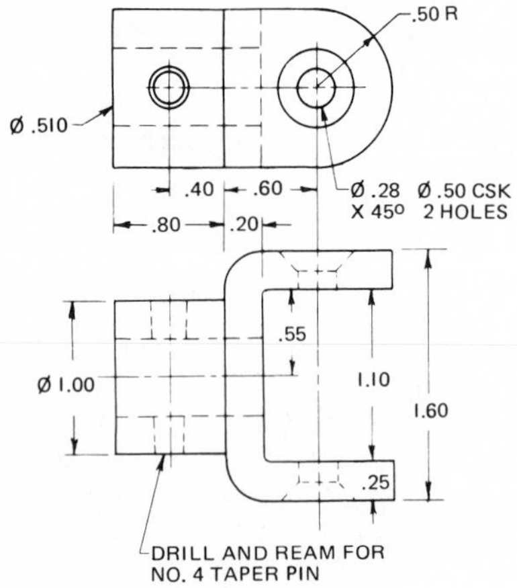
.50 R
Ø .510
.40
.60
.80
.20
Ø .28 Ø .50 CSK
X 45° 2 HOLES
.55
Ø 1.00
1.10
1.60
.25
DRILL AND REAM FOR
NO. 4 TAPER PIN

PT I – FORK – 2 REQD

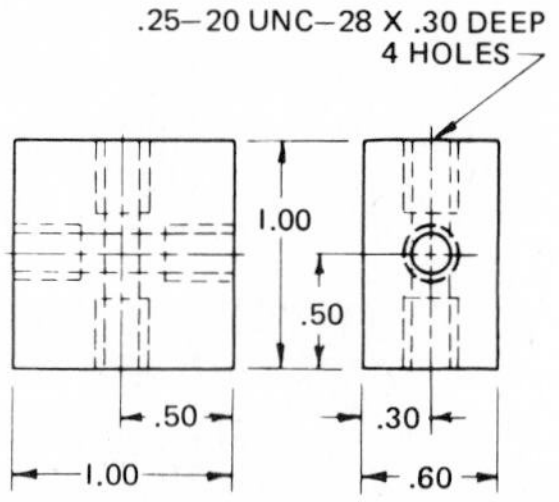
.25–20 UNC–28 X .30 DEEP
4 HOLES
1.00
.50
.50
1.00
.30
.60

PT 2 – RING – I REQD

PT 3 – NO. 4 TAPER PIN – 2 REQD

PT 4 – .25–20 FHMS .60 LG – 4 REQD

Fig. 15-8-D Universal joint details.

Chapter 16 Functional Drafting

UNIT 16-1

FUNCTIONAL DRAFTING[1,2]

Since the basic function of the drafting department is to provide sufficient information to produce or assemble parts, functional drafting must embrace every possible means to communicate this information in the least expensive manner. Functional drafting also applies to any method which would lower the cost of producing the part. New technological developments have provided many new ways of producing drawings at lower costs and/or in less time. This means that the drafting office must be prepared to discard some of the old, traditional methods in lieu of these newer means of communication.

There are many ways in which to reduce the drafting time in preparing a drawing. These drawing shortcuts, when collectively used, are of prime importance in an effective drafting system.

These newer techniques cannot be blindly applied, however, but must be carefully evaluated to make certain that the benefits outweigh the potential disadvantages. This evaluation should answer the following questions:

- What is its purpose?
- Is it a personal preference disguised as a project requirement?
- Does it meet contractual requirements?
- Will the shortcut increase costs in other areas such as manufacturing, purchasing, or inspection?
- Is it an effective communication link?
- How much training or education is required to make effective use of it?
- Are facilities available to implement it?
- Does the shortcut bypass a real bottleneck?

As each of these categories is examined, the advantages of the shortcuts will become apparent.

Procedural Shortcuts

There are a number of procedural shortcuts which, if properly applied and carefully managed, can shorten the drawing preparation cycle and result in savings.

Streamlined Approval Requirements. It is obvious that the more signatures required on a drawing, the greater the delays in releasing data. The decision as to who will approve drawings and drawing changes must be carefully considered to make certain that all necessary functions have been taken into account (checkers, responsible engineers, important technical specialists, etc.) without imposing undue restrictions. Project ground rules and contractual requirements also play an important part in this decision.

Eliminating Drawing Check from the Preparation Cycle. One of the most common suggested shortcuts, proposed usually when a project is behind schedule or exceeding its budget or when experienced personnel are involved, is to eliminate checking from the drawing preparation cycle.

Using Standard and Existing Drawings. Every year numerous drawings of parts are prepared which are repetitions of existing drawings. If the drafter were to incorporate in the new design parts that were already drawn, many drawing hours would be saved. Good drawing application records and an efficient multiple-use drawing system can eliminate a great deal of duplication. Standard tabulated drawings may be used to eliminate hundreds of drawings. See Figs. 16-1-1 and 16-1-2. For additional information on using existing drawings refer to Unit 16-3 on scissors and paste-up drafting.

Standard Drafting Practices. Standard drafting practices are obviously the backbone of efficient drafting room operations. The best way to establish and implement these practices is through a good drafting room manual, whose requirements must be strictly observed by all affected personnel. See Fig. 16-1-3.

The drafting room manual should contain data on the use and preparation of specific types of drawings, drawing and part number requirements, standard and special drafting practices, rules for dimensioning and tolerancing, specifications for associated lists, and company procedures for the preparation, handling, release, and control of drawings.

Team Drafting. Many engineering departments have turned out drawings by the method of one drafter to one drawing. Team drafting involves a number of people producing one drawing. While this may seem uneconomical, it is an expeditious means with visible cost savings over the traditional method.

Some firms are using team drafting because it is a better utilization of skill levels. It is a training program through which drafting skills are taught and semiskilled people are given an opportunity to gain experience.

Data Retrieval. The use of microfilm reader-printers in the drafting room provides quick and ready access to standard drawings and parts. However, for this method to be effective, a full-time librarian is needed. The use of microfilm cards is becoming popular, because they can hold up to 70 pages of information.

Standard Parts and Design Standard Information. Encouraging the use of standard parts and standard approaches to design will not only result in drafting time saved, but will also cut costs in areas such as purchasing, material control, manufacturing, etc. The odd-size cutout that requires special tooling, the design that calls for nonstandard hardware, and the equipment that uses a wide variety of fasteners when only one or two would suffice are typical cases where properly applied standards would reduce both time and cost.

Training Programs. To provide drafters with standard procedures and technical information is not enough; they must be trained in their use. New drafters are frequently overwhelmed by a strange environment, while old employees fail to keep up with new requirements or properly use the services available. Training programs for the indoctrination of new personnel and the updating of long-serviced employees are rewarded by more efficient and versatile operation.

Copying Machines. One of the most important time-saving devices, which should be available in every drafting area, is a copying machine for reference copies, check prints of work in preparation, and similar uses. See Fig. 16-1-4. When a drafter needs a copy, work is delayed until the copy is made available. Therefore a good copying machine will soon pay for itself in drawing hours saved.

GM
GENERAL MOTORS
ENGINEERING STANDARDS
ENGINEERING STANDARDS

Fig. 16-1-3 General Motors drawing standards manual.

QTY	PART	MATL	DESCRIPTION	PT NO.
2	CABLE SUPPORT	MAPLE	A–5374 PT 1	1
2	CABLE SUPPORT	MAPLE	A–5374 PT 2	2
3	CABLE SUPPORT	MAPLE	A–5374 PT 4	3

(A) DRAWING CALLOUT

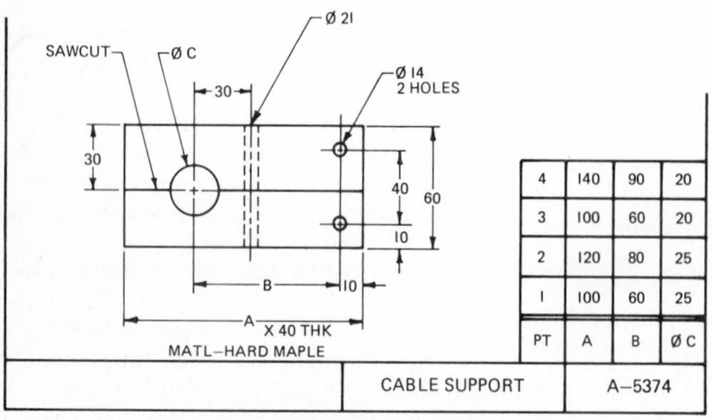
SAWCUT
Ø C
Ø 21
Ø 14
2 HOLES
30
30
40
60
10
B
10
A
X 40 THK
MATL–HARD MAPLE
4 140 90 20
3 100 60 20
2 120 80 25
1 100 60 25
PT A B Ø C
CABLE SUPPORT
A–5374

(B) STANDARD PART

Fig. 16-1-1 Standard tabulated drawings.

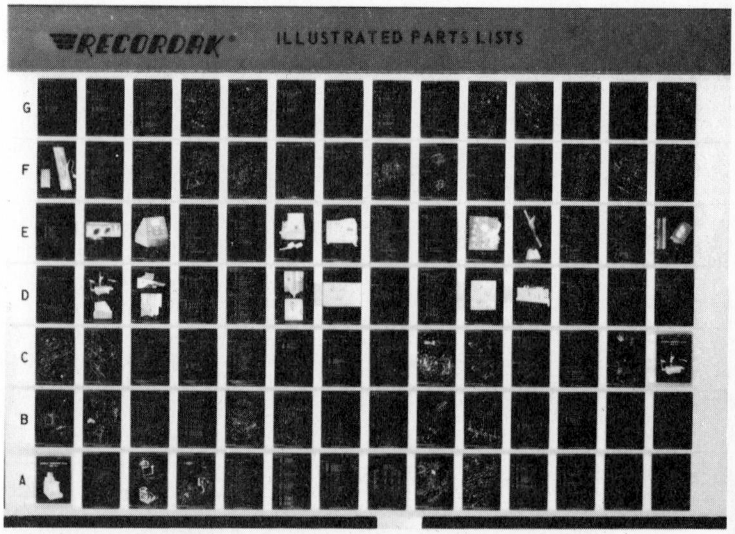
RECORDAK
ILLUSTRATED PARTS LISTS
G
F
E
D
C
B
A

Fig. 16-1-2 Standard parts drawings stored on microfilm. (Eastman Kodak Co.)

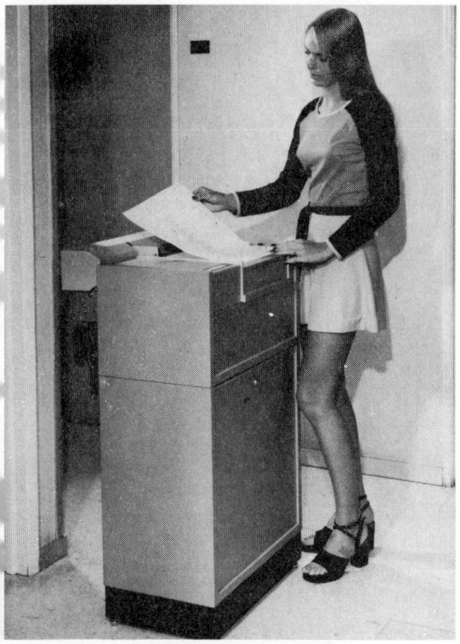

Fig. 16-1-4 Copying machine.

Drafting Equipment and Materials

The quality of the material and supplies used in the preparation of drawings is as important as the quality of the instruments used in fabrication. Leads which break easily, vellum which ghosts or smudges, and inks that crack and chip are some of the many material factors which contribute to increased drawing preparation time and decreased drawing life. Drafting materials must be carefully evaluated before orders are placed. Purchasing must be aware of the importance of procuring the specific materials requested and cautioned against the irresistible urge to "save" by purchasing drafting supplies from the lowest bidder at a sacrifice to quality.

Numerous time-saving devices are available: templates for every application, "pens" for easier linework, and more application of tape to artwork, transfer-type lettering, etc. Since drafting applications vary so widely, only the drafting supervisor can determine which devices will increase the drafting production.

Drafting aids are designed to facilitate the making of drawings by removing or reducing some of the more tedious aspects of drafting.

Templates. Templates, such as shown in Fig. 16-1-5, play an important part in functional drafting, for they save a great deal of time in drawing common shapes of details such as rounds, squares, hexagons, and ellipses. In addition to common shapes, templates have been made for standard parts such as nuts, bolt heads, electrical symbols, outline of tools and equipment, and many other outlines which are often repeated.

Mechanical Lettering. The results of a recent survey disclosed that mechanical lettering is generally replacing hand lettering. When mechanical lettering is required, it should be performed, whenever possible, by a subordinate. Mechanical lettering is particularly effective when it is used for general notes, particularly when preprinted standard notes on adhesive material are used.

Reducing the Number of Drawings Required

The cost of a project is, to some extent, directly related to the number of drawings which must be prepared. Therefore, careful planning to reduce the number of drawings required can result in significant savings. Some ways to reduce the number of drawings are explained below.

Detail Assembly Drawings. Detail assembly drawings, in which parts are detailed in place on the assembly (see Fig. 16-1-6), and multidetail assembly drawings, in which there are separate detail views for the assembly and each of its parts, will reduce the number of drawings required. However, they must be used with extreme care. They can easily become too complicated and confusing to be an effective means of communication.

Selecting the Most Suitable Type of Projection to Describe the Part. The selection of the type of projection (orthographic, isometric, or oblique) can greatly increase the ease with which some drawings can be read and, in many cases, reduce drafting time. For example, a single-line piping drawing drawn in isometric projection simplifies an otherwise difficult drawing problem in orthographic projection. See Fig. 16-1-7.

REFERENCES AND SOURCE MATERIAL

1. George R. Beck, "Reorganization in the Drafting Room," *Graphic Science*, October 1966.
2. Xerox Corp.

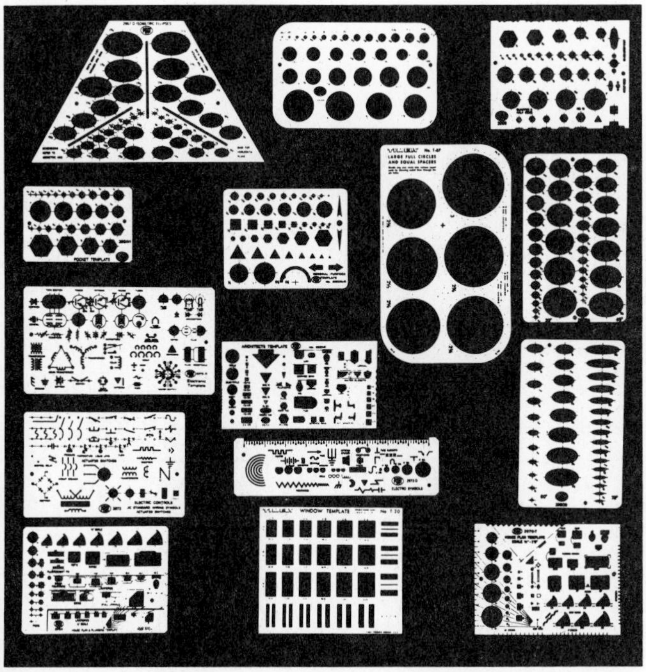

Fig. 16-1-5 Templates are made for all possible uses and save a lot of time.

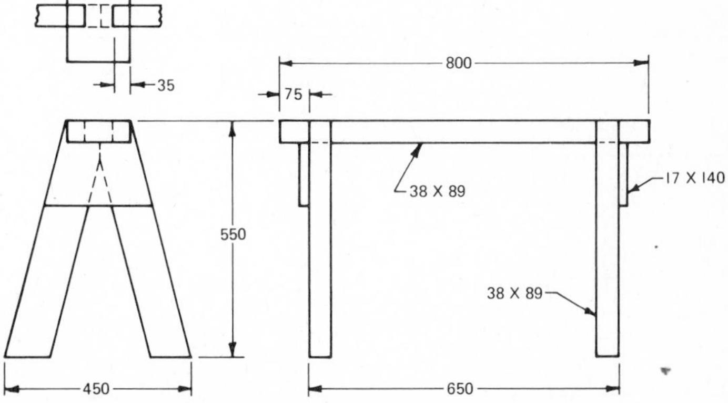
35
800
75
550
38 X 89
17 X 140
38 X 89
450
650

Fig. 16-1-6 Detail assembly drawing of a sawhorse.

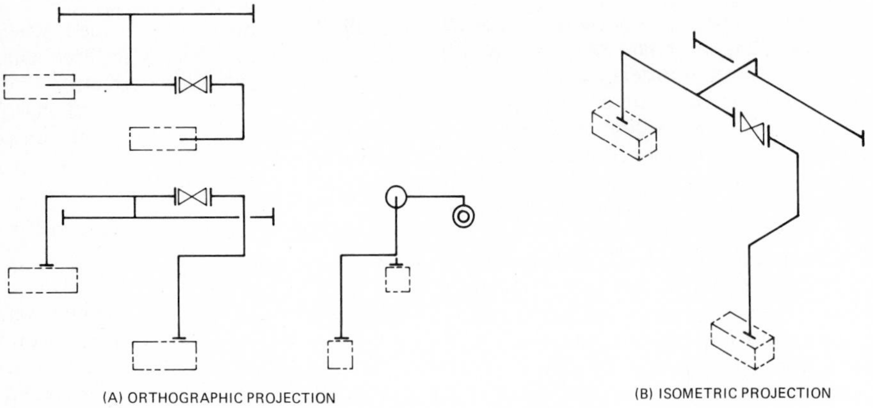
(A) ORTHOGRAPHIC PROJECTION
(B) ISOMETRIC PROJECTION

Fig. 16-1-7 Selecting the most suitable type of projection.

Assignments

1. After the number of drawings made over the last 6 months was reviewed, it was discovered that a great number of cable straps, shown in Fig. 16-1-A or 16-1-B, were being made which were similar in design. On an A3- or B-size sheet, prepare a standard tabulated drawing similar to Fig. 16-1-1, reducing the number of standard parts to 4. Scale is 1:1.

2. On an A3- or B-size sheet, the rod guide shown in Fig. 16-1-C or 16-1-D is to be drawn twice and the drawing time for each recorded. First, on plain paper make an isometric drawing of the part, using a compass to draw the circles and

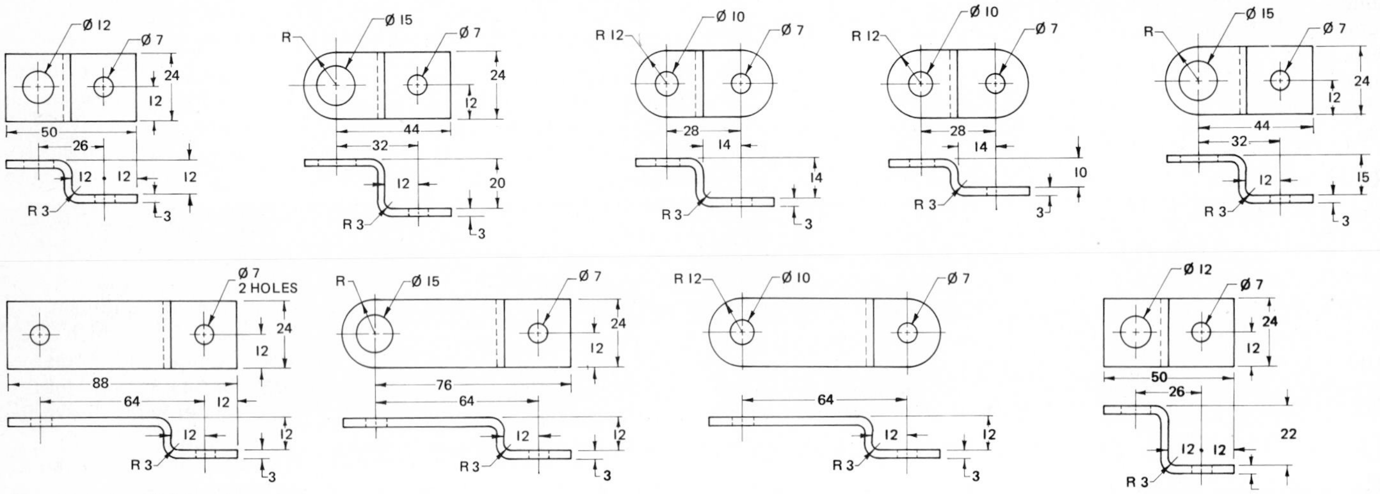
Ø 12
Ø 7
Ø 15
Ø 10
R 12
R 3
2 HOLES

Fig. 16-1-A Cable straps.

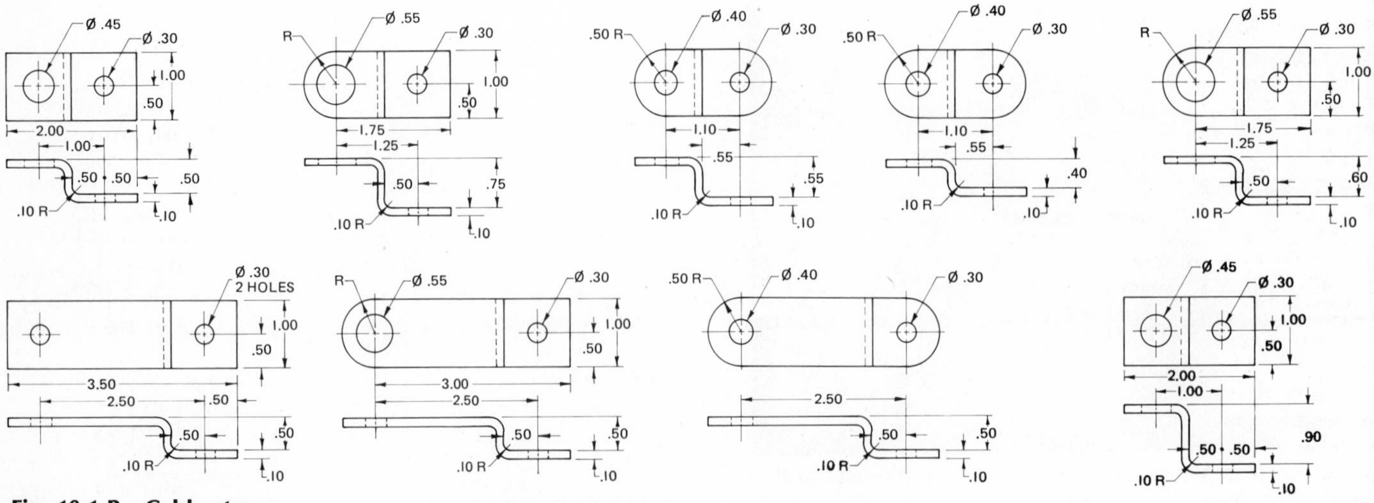
Ø .45
Ø .30
Ø .55
Ø .40
.50 R
.10 R
2 HOLES

Fig. 16-1-B Cable straps.

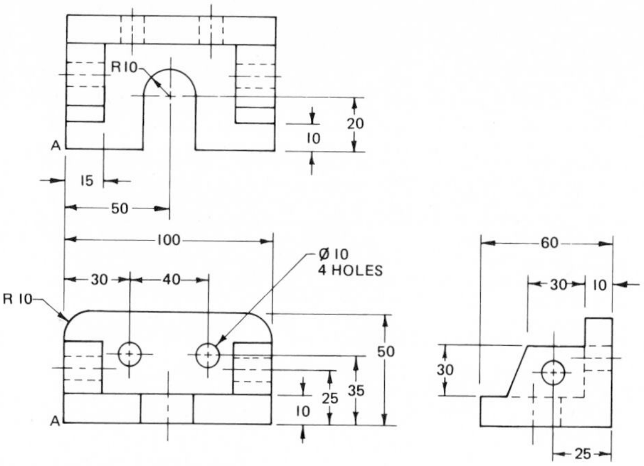
R10
A
20
10
15
50
100
Ø 10
4 HOLES
30
40
R 10
50
25
35
10
60
30
10
30
25

Fig. 16-1-C Rod guide.

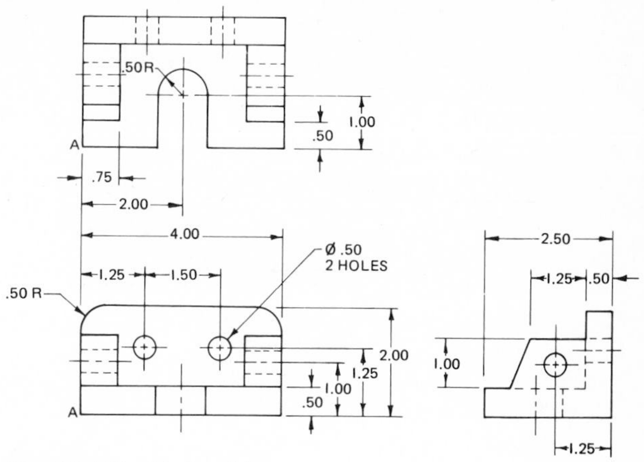
.50R
A
1.00
.50
.75
2.00
4.00
Ø .50
2 HOLES
1.25
1.50
.50 R
2.00
1.25
1.00
.50
2.50
1.25
.50
1.00
1.25

Fig. 16-1-D Rod guide.

arcs. Next repeat the drawing, only this time use isometric grid paper and a template for drawing the circle and arcs. From the drawing times recorded, state in percentage the time saved by the use of grid paper and templates. Scale is 1:1. Do not dimension.

REVIEW FOR ASSIGNMENTS

Unit 15-2 Isometric Circles

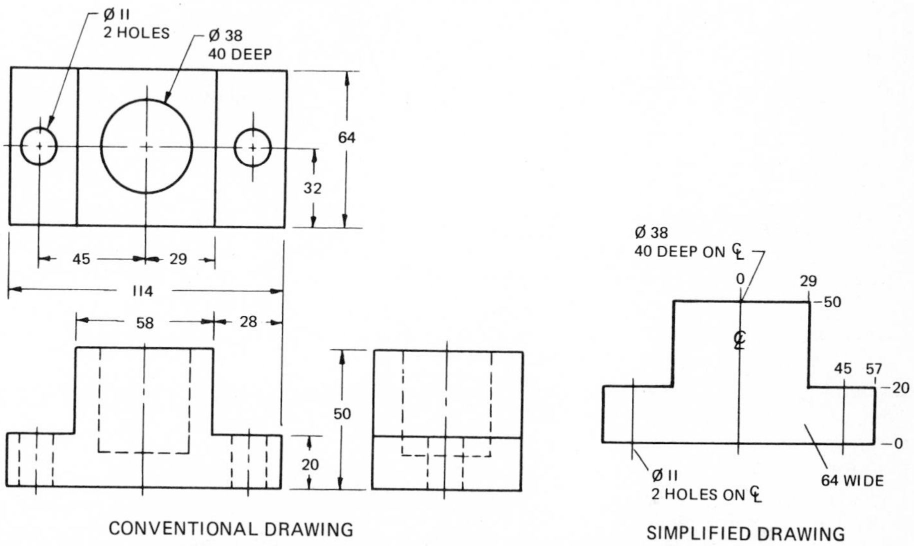
Ø 11
2 HOLES
Ø 38
40 DEEP
64
32
45
29
114
58
28
50
20
Ø 38
40 DEEP ON ℄
0
29
50
45
57
20
0
Ø 11
2 HOLES ON ℄
64 WIDE
CONVENTIONAL DRAWING
SIMPLIFIED DRAWING

Fig. 16-2-1 A comparison between conventional and simplified drafting.

UNIT 16-2
SIMPLIFIED DRAFTING

The challenge of modern industry is to produce more and better goods at competitive prices. Drafting, like all other branches of industry, must share in the responsibility for making this increased productivity possible. The old concept of drafting—that of producing an elaborate and beautiful drawing, complete with all the lines, projected views, and sections—must give way to a simplified method. The new, simplified method of drafting must embrace many modern economical drafting practices but surrender nothing in either clarity of presentation or accuracy of dimensioning. Drafting stripped of its frills is the new standard. See Fig. 16-2-1.

Although simplified drafting was pursued largely to reduce drafting costs and to improve the drafting product, it had a beneficial side effect when the drawing had to be microfilmed. Even more important, the worker on the shop floor has an "easier to understand" print. Much of the fine, difficult-to-produce detail, such as unnecessary frills and curlicues, has been eliminated and replaced by simplified or symbolic presentation.

Drawings cluttered by repetitive detail are now shown simplified and are more suitable for making quality microfilm reproduction and reduced-size drawings.

Three most effective practices used in simplified drafting are

- Simplification of dimensioning
- Simplification of detail drawing
- Extensive use of freehand sketching

Simplification of Dimensioning[1,2]

Simplification of dimensioning not only reduces drafting time but also avoids cluttering a drawing with unnecessary lines, thereby making it easier to read.

No other single factor has more influence on the use of drawings than dimensioning. It is not sufficient to have dimensions numerically correct; it is equally important to dimension a drawing properly so that computation of sizes is unnecessary.

- Use arrowless dimensioning.
- Use tabular dimensioning.
- Use abbreviations and symbols.

Arrowless Dimensioning. To avoid having a large number of dimensions extending away from the part, arrowless or ordinate dimensioning may be used. See Fig. 16-2-2. In this system, the "zero" lines represent the vertical and horizontal datum lines, and each of the dimensions without an arrowhead indicates the distance from the zero line. There is never more than one zero line in each direction.

Tabular Dimensioning. When there is a very large number of holes or repetitive features, such as in a chassis or a printed circuit board, and where the multitude of center lines would make a drawing difficult to read, tabular dimensioning

HOLE SYMBOL	HOLE DIA
A	6
B	5
C	4
D	3

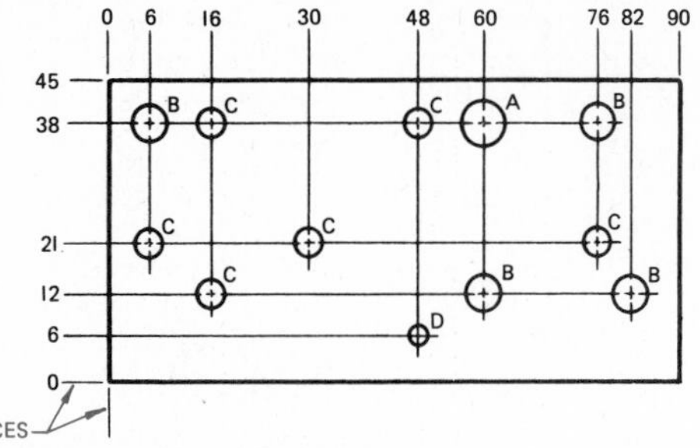
0 6 16 30 48 60 76 82 90
45
38
21
12
6
0
A B C D
DATUM SURFACES

Fig. 16-2-2 **Arrowless dimensioning.**

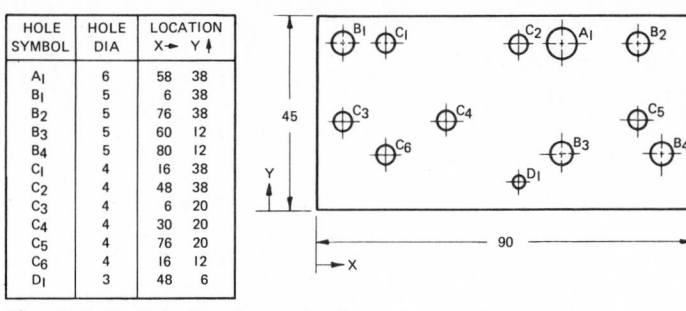

HOLE SYMBOL	HOLE DIA	LOCATION X	LOCATION Y
A_1	6	58	38
B_1	5	6	38
B_2	5	76	38
B_3	5	60	12
B_4	5	80	12
C_1	4	16	38
C_2	4	48	38
C_3	4	6	20
C_4	4	30	20
C_5	4	76	20
C_6	4	16	12
D_1	3	48	6

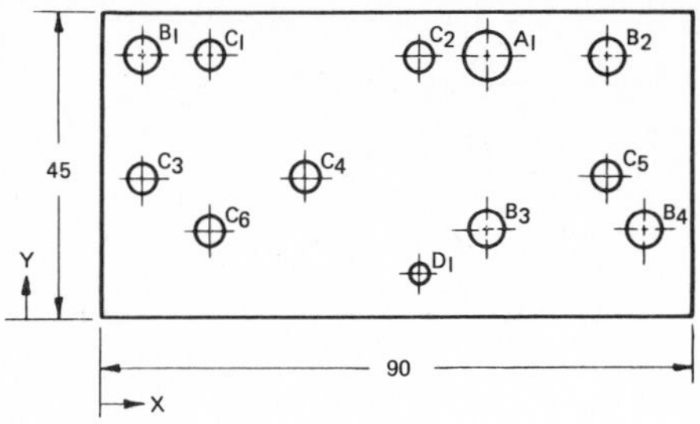
45
Y
90
X

Fig. 16-2-3 **Tabular dimensioning.**

is recommended. See Fig. 16-2-3. In this system each hole or feature is assigned a letter, or a letter and numeral subscript. The feature dimensions and the feature location along the *X* and *Y* axes are given in a table.

Abbreviations and Symbols. Abbreviations and symbols are shortened forms of words or expressions used to conserve drafting time and drawing space. Refer to the Appendix for commonly used abbreviations and symbols.

Simplification of Detail Drawing

1. Complicated parts are best described by means of a drawing. However, explanatory notes can complement the drawing, thereby eliminating views that are time-consuming to draw. See Figs. 16-2-4 and 16-2-5.

2. Use simplified drawing practices as described throughout this text, especially on threads and common features.

3. Avoid unnecessary views. In many cases one or two views are sufficient to explain the part fully.

4. When a large number of holes of similar size are to be made in a part, there is a chance that the person producing the part may misinterpret a conventional drawing. To simplify the drawing and reduce the chance of error, hole symbols such as shown in Figs. 16-2-6 and 16-2-7 are recommended.

5. The use of the abbreviation ℄ indicates that all dimensions are symmetrical about the line indicated.

6. A simplified assembly drawing should be for assembly purposes only. Some means of simplification are:

- Standard parts such as nuts, bolts, and washers need not be drawn.
- Reference part circles and arrowheads on leaders can be omitted.
- Small fillets and rounds on cast parts need not be shown.
- Phantom outline of complicated details can often be used.

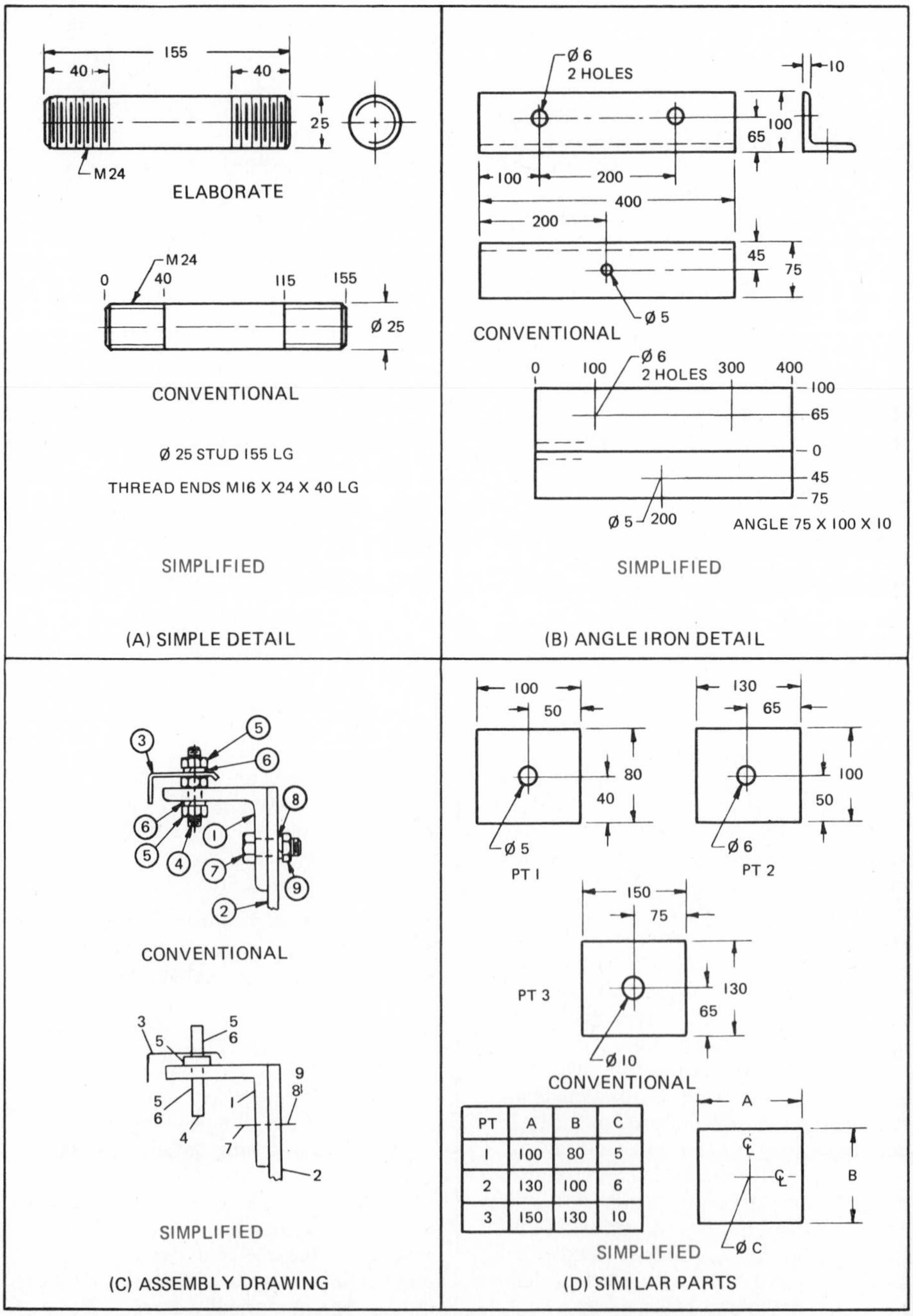
155
40
40
25
M24
ELABORATE
M24
0
40
115
155
Ø 25
CONVENTIONAL
Ø 25 STUD 155 LG
THREAD ENDS M16 X 24 X 40 LG
SIMPLIFIED
(A) SIMPLE DETAIL
Ø 6
2 HOLES
10
100
65
100
200
400
200
45
75
Ø 5
CONVENTIONAL
0
100
Ø 6
2 HOLES
300
400
100
65
0
45
75
Ø 5
200
ANGLE 75 X 100 X 10
SIMPLIFIED
(B) ANGLE IRON DETAIL
CONVENTIONAL
SIMPLIFIED
(C) ASSEMBLY DRAWING
100
50
80
40
Ø 5
PT 1
130
65
100
50
Ø 6
PT 2
150
75
PT 3
130
65
Ø 10
CONVENTIONAL
A
B
Ø C
SIMPLIFIED
(D) SIMILAR PARTS

PT	A	B	C
1	100	80	5
2	130	100	6
3	150	130	10

Fig. 16-2-4 **Comparison between conventional and simplified drawings.**

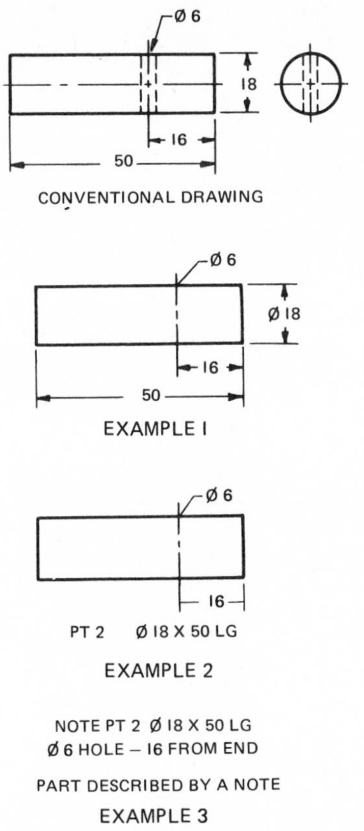
Ø 6
18
16
50
CONVENTIONAL DRAWING
Ø 6
Ø 18
16
50
EXAMPLE I
Ø 6
16
PT 2 Ø 18 X 50 LG
EXAMPLE 2
NOTE PT 2 Ø 18 X 50 LG
Ø 6 HOLE – 16 FROM END
PART DESCRIBED BY A NOTE
EXAMPLE 3

Fig. 16-2-5 Simplified drafting practices for detailed parts.

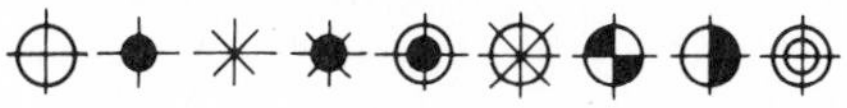

Fig. 16-2-6 Recommended hole symbols in order of preference.

7. Use templates where possible.
8. Within limits a small drawing is made more quickly than a large drawing.
9. Eliminate hidden lines which do not add clarification.
10. Show only partial views of symmetrical objects. See Fig. 16-2-8.
11. Avoid the use of elaborate pictorial or repetitive detail.
12. Eliminate repetitive data by use of general notes or phantom lines.
13. Eliminate views where the shape or dimension can be given by description, for example, hex, sq., ϕ, on ℄, thk, etc.

Freehand Sketching

Most shops care little whether the drawing is freehand, whether one view is shown, or whether the drawing is to scale, as long as the proportions are approximate. They are interested in having the necessary information clearly shown. Freehand sketches and drawings made with instruments can be shown on one sheet. However, it must be clearly understood that the use of freehand sketching does not give the drafter a license to turn out sloppy work.

Savings as high as 30 percent in the preparation of working drawings have been attributed to the use of freehand sketches as opposed to instrument-produced drawings. However, freehand sketching has its limitations. It is highly effective on simple detailed parts, for small radii, such as rounds and fillets, and for small holes. In many cases the term *freehand* is not entirely correct. For instance, templates may be used to draw circles, resistors, or other common features, or a straightedge may be used to produce long lines since it is faster and more accurate than freehand sketching. But short lines are drawn faster freehand.

Drawing paper with nonreproducible grid lines is ideal for freehand sketching. See Fig. 16-2-9. For this reason many companies have their drawing paper made with nonreproducible grid lines over the entire drawing area. Other advantages of having the grid lines on the paper are that (1) they may serve as guidelines in lettering notes and dimensions and (2) they may be used for measuring distances, thereby reducing the number of times the scale is used for measuring.

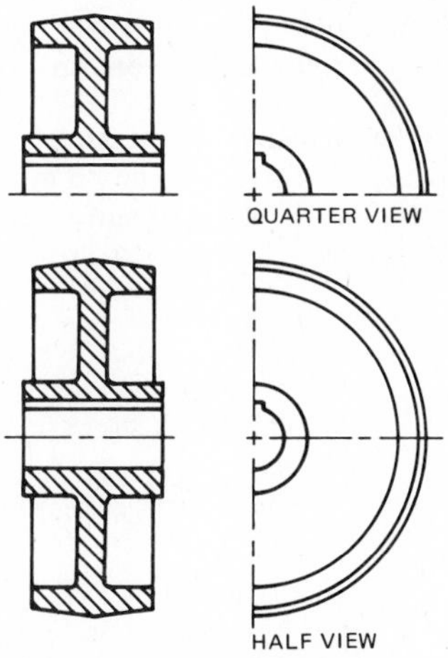
QUARTER VIEW
HALF VIEW

Fig. 16-2-8 Partial views.

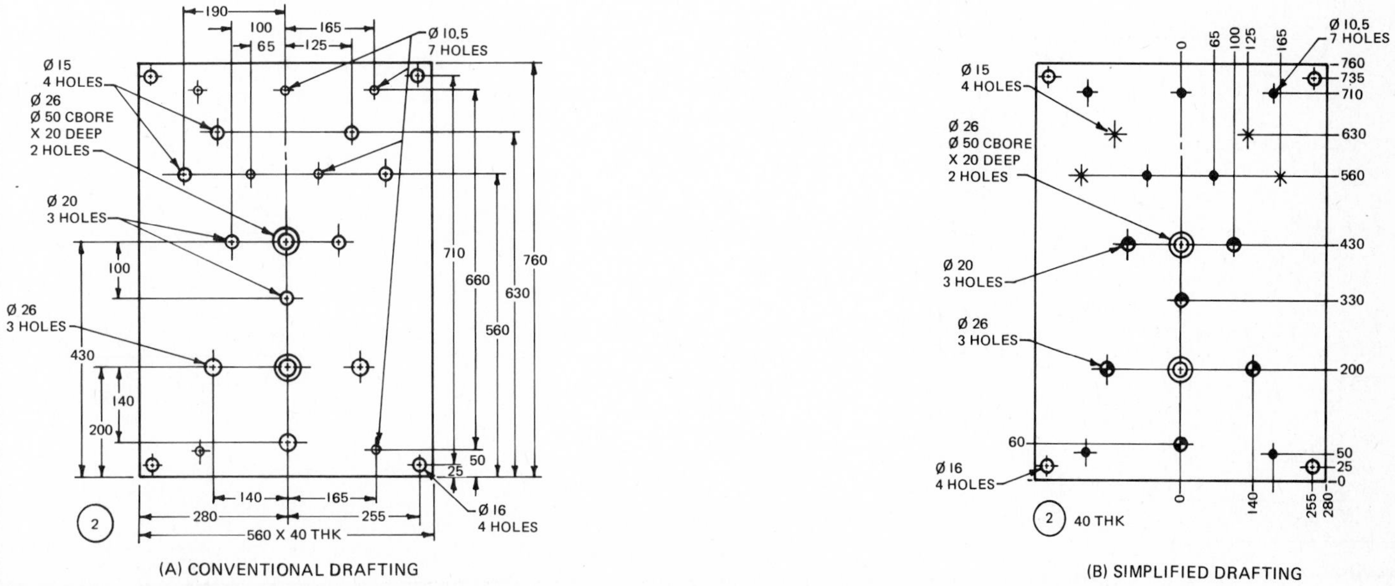
Ø 15
4 HOLES
Ø 26
Ø 50 CBORE
X 20 DEEP
2 HOLES
Ø 20
3 HOLES
Ø 26
3 HOLES
Ø 10,5
7 HOLES
Ø 16
4 HOLES
560 X 40 THK
(A) CONVENTIONAL DRAFTING
40 THK
(B) SIMPLIFIED DRAFTING

Fig. 16-2-7 Application hole symbols and arrowless dimensioning.

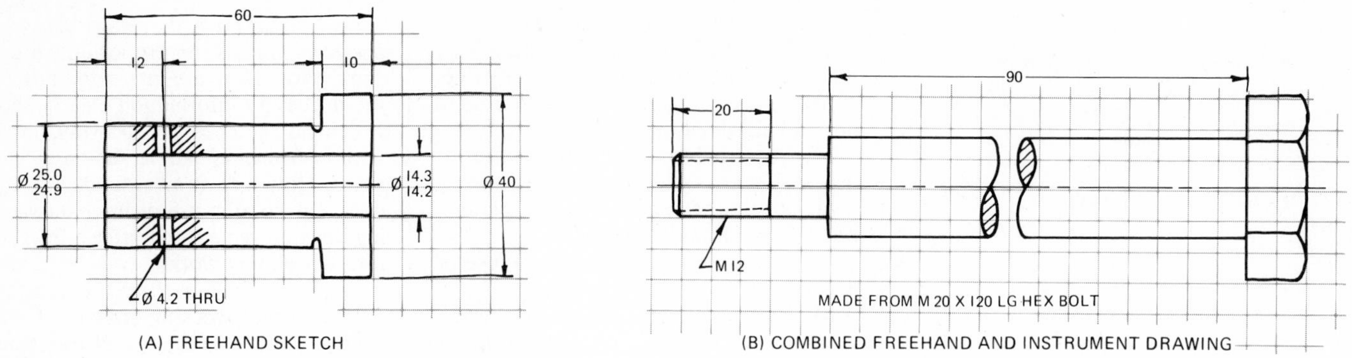

60
12
10
Ø 25.0
24.9
Ø 14.3
14.2
Ø 40
Ø 4.2 THRU
(A) FREEHAND SKETCH
90
20
M 12
MADE FROM M 20 X 120 LG HEX BOLT
(B) COMBINED FREEHAND AND INSTRUMENT DRAWING

Fig. 16-2-9 Sketching of parts on coordinate paper.

REFERENCES AND SOURCE MATERIAL

1. ANSI Y14.5.
2. CSA B78.2.

Assignments

1. On an A3- or B-size sheet, redraw the two parts shown in Fig. 16-2-A or 16-2-B using arrowless dimensioning and simplified drawing practices. Scale is 1:2. For the cover plate use the bottom and left-hand edge for the datum surfaces. For the back plate use the bottom and the center of the part for the datum surfaces.

2. On an A3- or B-size sheet, make simplified drawings of the parts shown in Fig. 16-2-C or 16-2-D. Scale is to suit.

REVIEW FOR ASSIGNMENTS

Unit 2-6	Sketching
Unit 6-1	Detail Drawings
Appendix	Keys and Keyways

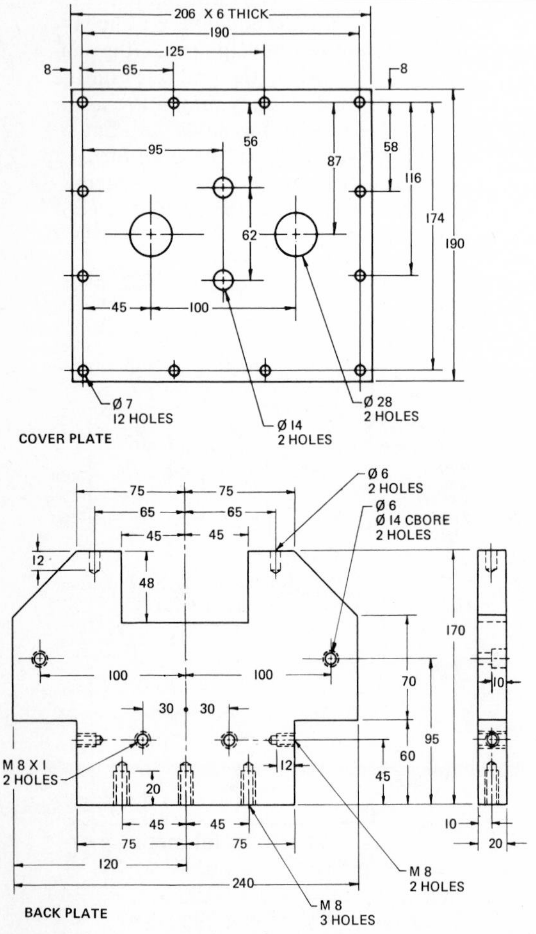

206 X 6 THICK
190
125
65
8
8
95
56
87
58
116
174
190
62
45
100
Ø 7
12 HOLES
Ø 14
2 HOLES
Ø 28
2 HOLES
COVER PLATE
75
75
65
65
45
45
12
48
Ø 6
2 HOLES
Ø 6
Ø 14 CBORE
2 HOLES
170
100
100
70
30
30
95
60
45
12
10
M 8 X 1
2 HOLES
20
45
45
75
75
120
240
20
M 8
2 HOLES
M 8
3 HOLES
BACK PLATE

Fig. 16-2-A Arrowless dimensioning assignment.

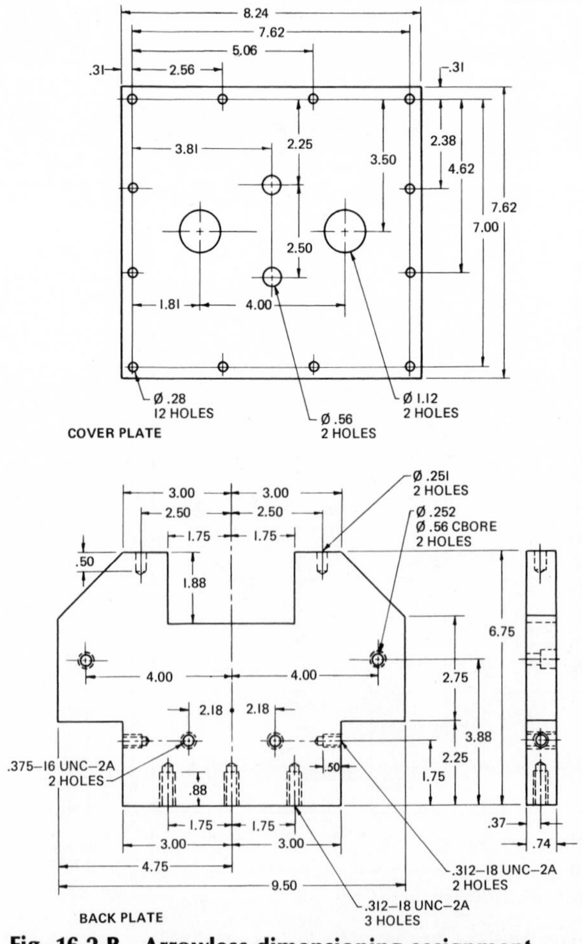

8.24
7.62
5.06
.31
2.56
.31
3.81
2.25
3.50
2.38
4.62
7.62
7.00
2.50
1.81
4.00
Ø .28
12 HOLES
Ø .56
2 HOLES
Ø 1.12
2 HOLES
COVER PLATE
3.00
3.00
2.50
2.50
1.75
1.75
Ø .251
2 HOLES
Ø .252
Ø .56 CBORE
2 HOLES
.50
1.88
6.75
4.00
4.00
2.75
2.18
2.18
3.88
.375–16 UNC–2A
2 HOLES
.50
2.25
1.75
.88
1.75
1.75
3.00
3.00
4.75
9.50
.37
.74
.312–18 UNC–2A
2 HOLES
.312–18 UNC–2A
3 HOLES
BACK PLATE

Fig. 16-2-B Arrowless dimensioning assignment.

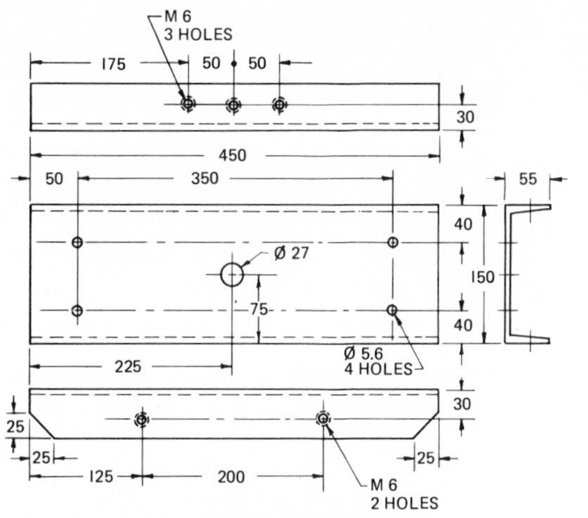
M 6
3 HOLES
175
50
50
30
450
50
350
55
40
Ø 27
150
75
40
225
Ø 5.6
4 HOLES
30
25
25
25
125
200
M 6
2 HOLES

CHANNEL CLAMP

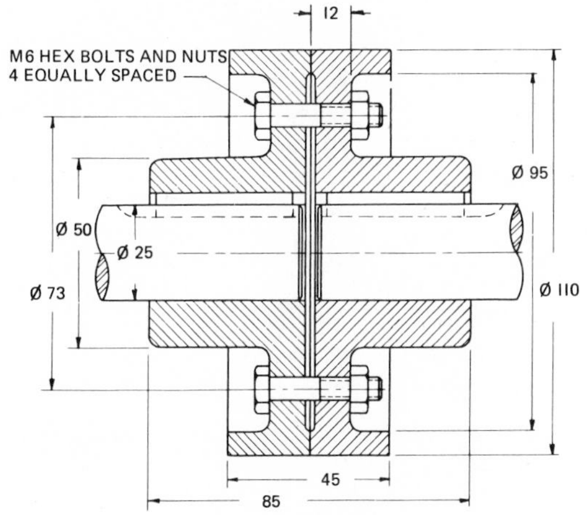
12
M6 HEX BOLTS AND NUTS
4 EQUALLY SPACED
Ø 95
Ø 50
Ø 25
Ø 73
Ø 110
45
85

FLANGED COUPLING

Fig. 16-2-C Simplification of detail assignment.

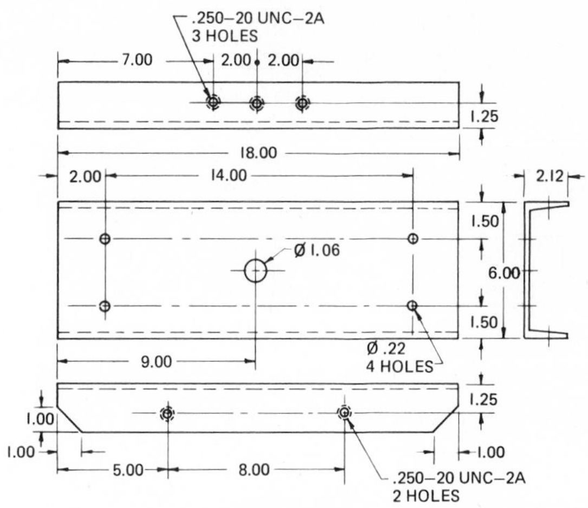
.250–20 UNC–2A
3 HOLES
7.00
2.00
2.00
1.25
18.00
2.00
14.00
2.12
1.50
Ø 1.06
6.00
1.50
9.00
Ø .22
4 HOLES
1.25
1.00
1.00
5.00
8.00
1.00
.250–20 UNC–2A
2 HOLES

CHANNEL CLAMP

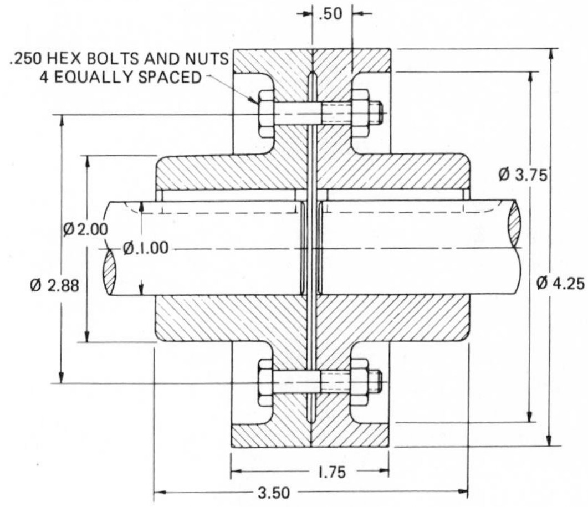
.50
.250 HEX BOLTS AND NUTS
4 EQUALLY SPACED
Ø 3.75
Ø 2.00
Ø 1.00
Ø 2.88
Ø 4.25
1.75
3.50

FLANGED COUPLING

Fig. 16-2-D Simplification of detail assignment.

UNIT 16-3

REPRODUCTION SHORTCUTS[1]

In the past few years, a number of reproduction techniques have been developed which, if properly used, can greatly reduce drawing preparation time. An understanding of available techniques and their limitations, supported by the close cooperation of a reproduction group familiar with drafting operation, can help the drafting supervisor to make significant cost savings.

Reproducibles from Existing Drawings

When a new drawing is to be made from an existing drawing with few changes, reproducibles will save a great deal of preparation time. This procedure involves making a translucent or transparent print from the original drawing, removing unwanted material from this print, and adding the new information to the drawing. The main drawback to this method is that the existing drawing may not conform to the latest drawing standard practice.

Scissors and Paste-up Drafting[2,3]

No matter how original a design may be, a great number of part features are repetitive. With the aid of modern reproduction methods, drawings can be created by using unchanged portions of existing drawings. Transferring them from one drawing to the next is accomplished by scissors and paste-up drafting. It provides a way of using all or parts of existing drawings, notes, charts, drawing forms, to revise existing drawings and to create new drawings. Through the utilization of existing drawings much valuable drafting time is freed for creative design drafting rather than hand copying.

Finished prints can be made on paper, acetate, or vellum. They can be the same size or reduced to five different sizes, depending on the reproduction equipment being used.

Another important advantage of cut-and-paste drafting is that materials copied from existing drawings do not have to be rechecked minutely, as must be done with new drawings. Rechecking time is reduced. See Fig. 16-3-1.

Appliqués

One of the most successful methods of reducing drawing time is the use of appliqués. When parts, shapes, symbols, or notes are used repeatedly, then appliqués should be considered. These

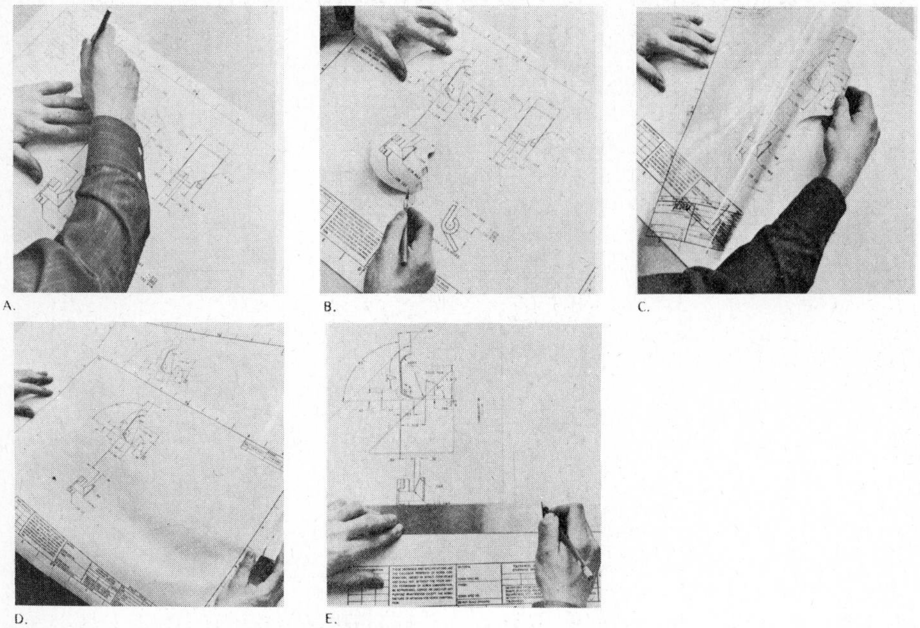
A.
B.
C.
D.
E.

Fig. 16-3-1 Cut and paste drafting. (Xerox Corp.)

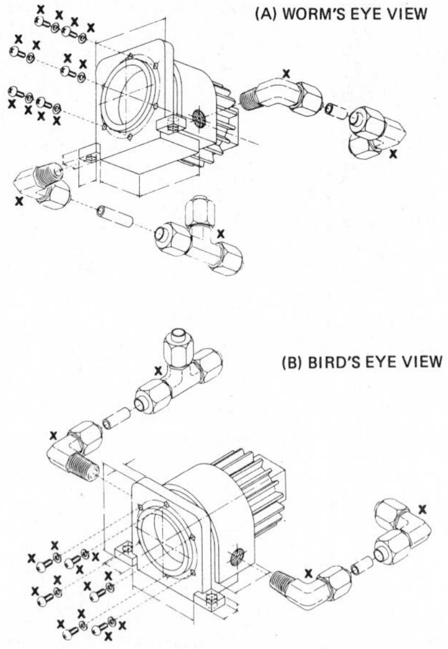
(A) WORM'S EYE VIEW
(B) BIRD'S EYE VIEW

Fig. 16-3-3 Application of appliqués shown in Fig. 16-3-2. (Graphic Standard Instruments Co.)

pressure-sensitive overlays may be printed or plain on transparent or translucent sheets with an adhesive backing.

Appliqués are available in a great variety of standard symbols or patterns (Fig. 16-3-2) and in blank (unprinted) sheets. A matte surface on the blank sheet will accept typewriter copy as well as pencil or ink lines. This material is often used for making corrections on drawings and for adding materials lists or detailed notes which can be typed faster than they can be lettered. Figure 16-3-3 shows a drawing which used many of the appliqués shown in Fig. 16-3-2.

Appliqués are available in two basic types: ***cutout*** and ***transfer***. Cutout appliqués are applied by positioning the desired image in the correct position on the drawing, burnishing (rubbing) the image area, and cutting around it to remove the portion not wanted. The transfer-type pressure-sensitive appliqué works on a somewhat different principle. The carrier is removed from the translucent image sheet, and the area to be transferred is placed in position on the drawing. The image to be transferred is then rubbed over the top surface of the transfer sheet with a burnishing stick or other blunt instrument. When the transfer sheet is lifted, the image remains on the drawing sheet. After the image has been transferred, the carrier sheet is placed over the image and reburnished.

The combined use of cut-and-paste drafting and appliqués for new drawings is found extensively in industry, especially in the electronics and piping fields. See Fig. 16-3-4.

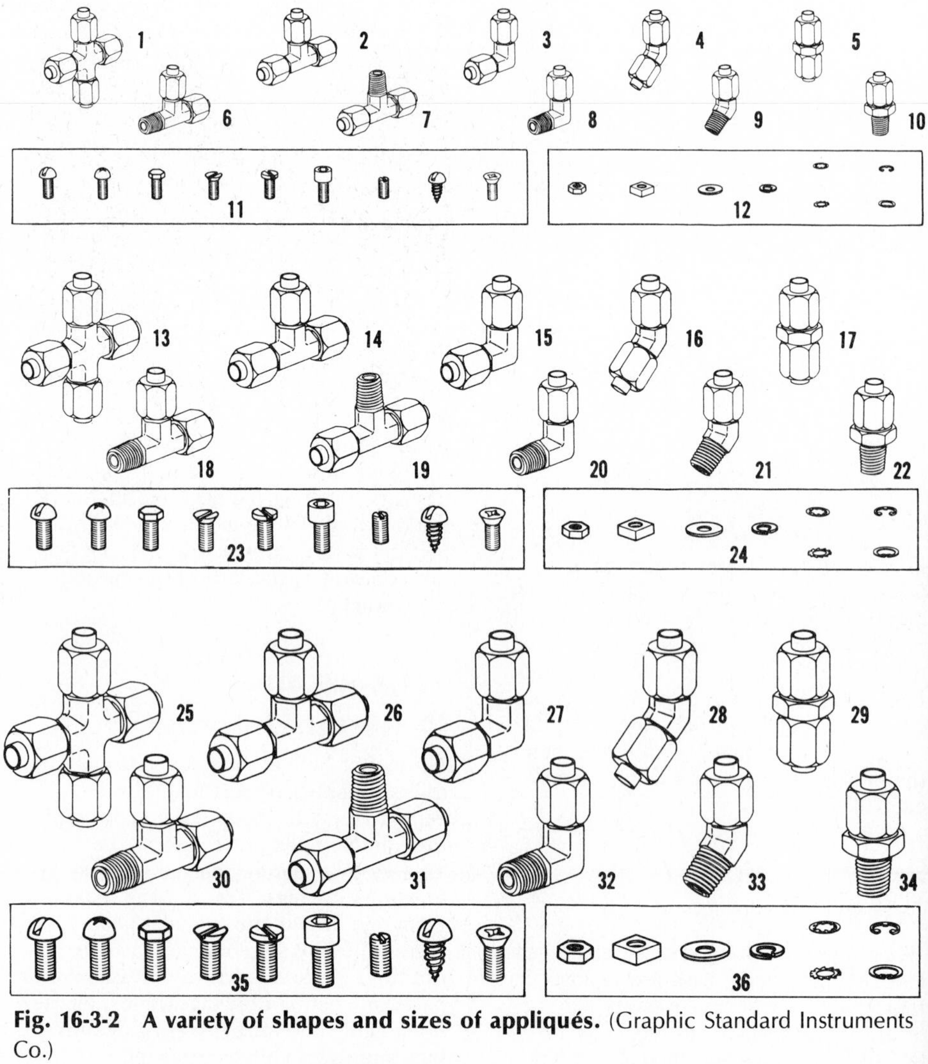
1
2
3
4
5
6
7
8
9
10
11
12
13
14
15
16
17
18
19
20
21
22
23
24
25
26
27
28
29
30
31
32
33
34
35
36

Fig. 16-3-2 A variety of shapes and sizes of appliqués. (Graphic Standard Instruments Co.)

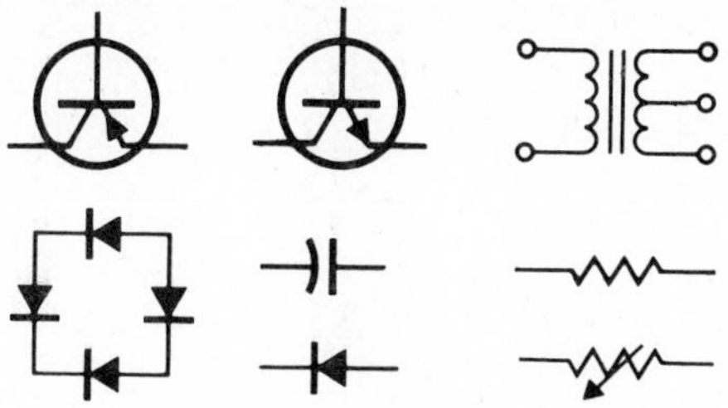

(A) TYPICAL APPLIQUES

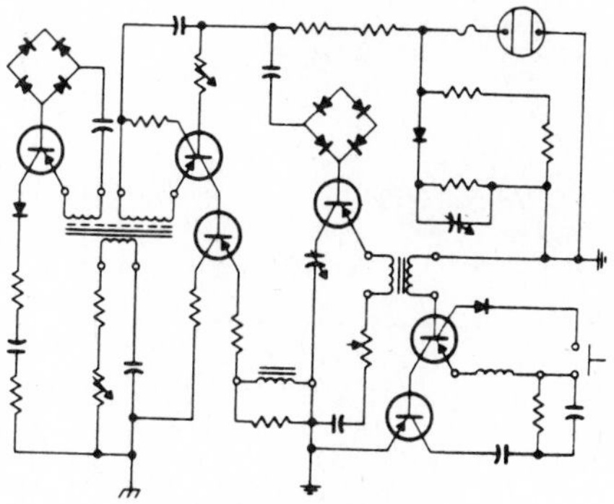

(B) APPLICATION

Fig. 16-3-4 Electronics appliqués. (Bishop Industries Corp.)

REFERENCES AND SOURCE MATERIAL

1. George R. Beck, "Reorganization in the Drafting Room," *Graphic Science*, October 1966.
2. Eastman Kodak Company.
3. Xerox.

Assignments

1. An exploded isometric assembly drawing of the wheel puller shown in Fig. 16-3-A or 16-3-B is urgently required. Time does not permit one drafter to do the entire drawing; thus three drafters will be required to draw the parts. Scale is 1:2. On plain paper draw all the parts in isometric. When all the parts have been drawn, cut them from the paper and assemble them in the exploded position on an A3- or B-size sheet. Use glue or rubber cement. Make a suitable print of the exploded assembly from a copying machine.

2. On an A3- or B-size sheet, make the electronic layout drawing shown in Fig. 16-3-C. Appliqués and templates are to be used. If appliqués of the electronic components are not available, make your own by photostating Fig. 16-3-4 and cutting them out and gluing them to your drawing. There is no scale.

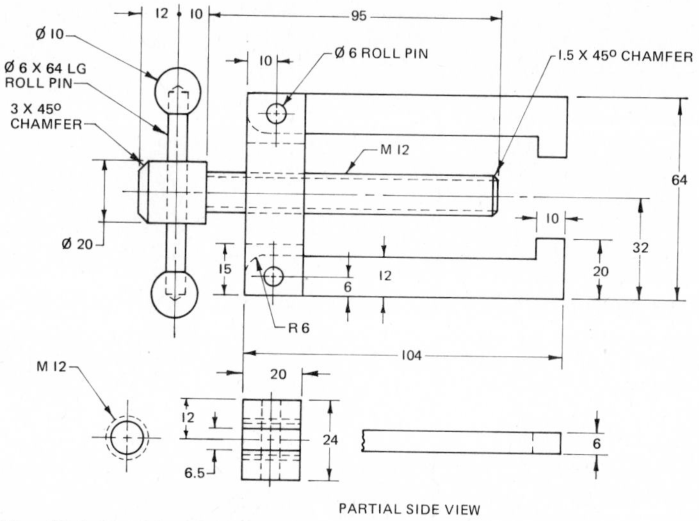

12
10
95
Ø 10
Ø 6 X 64 LG ROLL PIN
3 X 45° CHAMFER
10
Ø 6 ROLL PIN
1.5 X 45° CHAMFER
M 12
64
10
32
Ø 20
15
12
20
6
R 6
104
M 12
20
12
24
6
6.5
PARTIAL SIDE VIEW

Fig. 16-3-A Wheel puller. (Cut and paste)

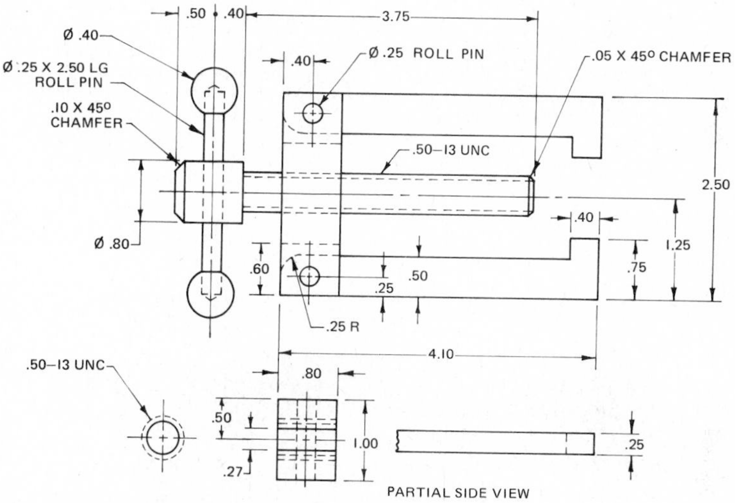

.50
.40
3.75
Ø .40
Ø .25 X 2.50 LG ROLL PIN
.10 X 45° CHAMFER
.40
Ø .25 ROLL PIN
.05 X 45° CHAMFER
.50–13 UNC
2.50
.40
1.25
Ø .80
.60
.75
.50
.25
.25 R
4.10
.50–13 UNC
.80
.50
1.00
.25
.27
PARTIAL SIDE VIEW

Fig. 16-3-B Wheel puller. (Cut and paste)

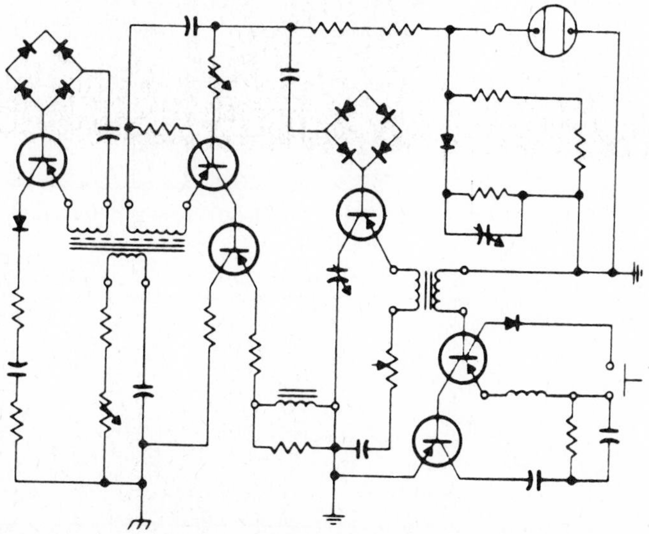

Fig. 16-3-C Electronics diagram. (Appliqué)

UNIT 16-4

PHOTODRAWINGS[1]

Photodrawings, that is, engineering drawings into which one or more photographs are incorporated, have increased in popularity because they can present a subject sometimes even more clearly than conventional drawings. Photodrawings supplement rather than replace conventional engineering drawings by eliminating much of the tedious and time-consuming effort involved when the subject is difficult to draw. They are particularly useful for assembly drawings, piping diagrams, large machine installations, switchboards, etc., provided, of course, that the subject of the drawings exists so that it may be photographed.

Photodrawings are also a comprehensive means of clearly transmitting technical information; they free the drafter from having to draw things which already exist. See Fig. 16-4-1.

Photodrawings have several other advantages. They are easy to make and usually take much less time to prepare than an equivalent amount of conventional drafting.

Background. Any photodrawing must begin with a photograph of an object, a part or assembly, a building, a model, or whatever else may be the subject of the drawing.

Photography. The best photographic angle usually is one which shows the subject in a flat view with as little perspective as possible. (If the situation calls for a perspective, then select the angle that best tells the story.) Make certain that all the parts important to the photodrawing are in view of the camera.

REFERENCE AND SOURCE MATERIAL

1. Eastman Kodak Company.

Assignment

On an A3- or B-size sheet, make a photostat of the sprocket (exploded halftone) shown in Fig. 16-4-A. The photostat, which is to service as a photodrawing, is to replace the two-view drawing, and a new chart listing metric sizes is to replace the existing chart. Leaders, dimensions, and dimension lines, which are to be inked, are to be added to the photodrawing.

REVIEW FOR ASSIGNMENT

Unit 2-6 Ink Drawings

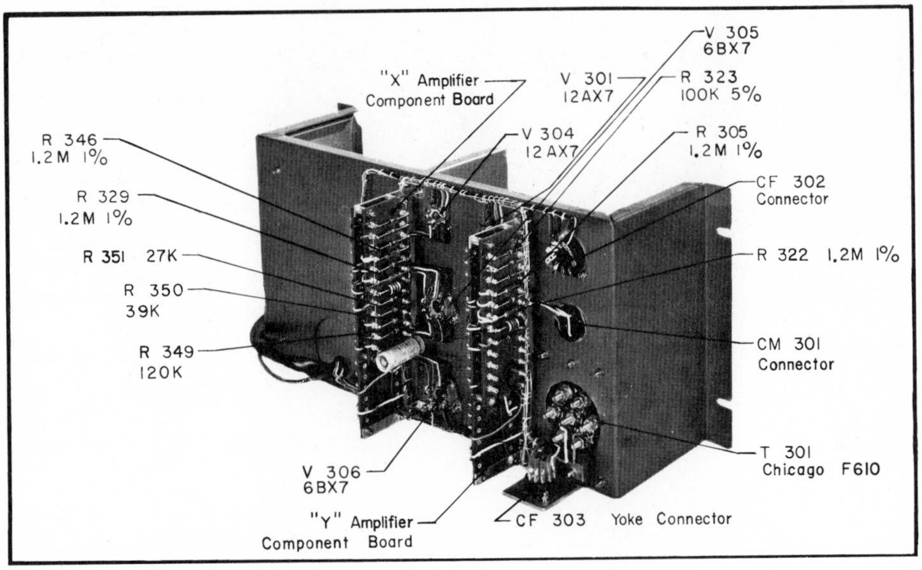

V 305
6BX7
"X" Amplifier
Component Board
V 301
12AX7
R 323
100K 5%
R 346
1.2M 1%
V 304
12 AX7
R 305
1.2M 1%
R 329
1.2M 1%
CF 302
Connector
R 351 27K
R 322 1.2M 1%
R 350
39K
R 349
120K
CM 301
Connector
T 301
Chicago F610
V 306
6BX7
"Y" Amplifier
Component Board
CF 303
Yoke Connector

Fig. 16-4-1 Photodrawing. (Eastman Kodak Co.)

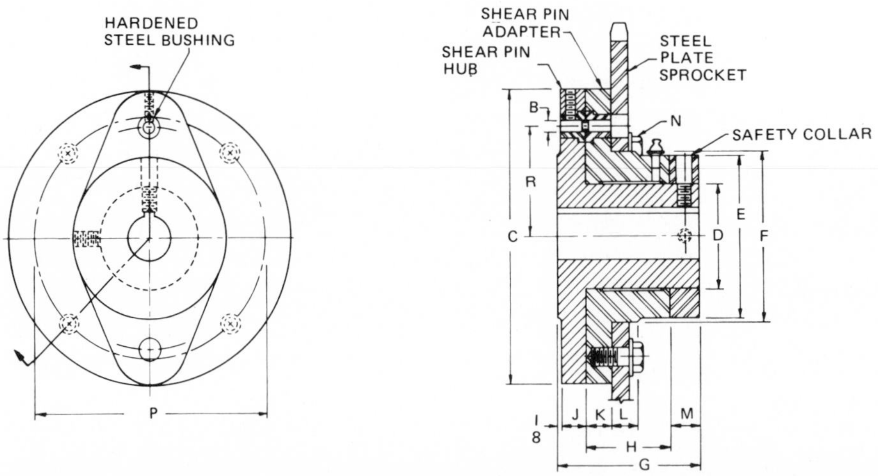

HARDENED
STEEL BUSHING
SHEAR PIN
ADAPTER
SHEAR PIN
HUB
STEEL
PLATE
SPROCKET
B
N
SAFETY COLLAR
R
E
C
D
F
P
J
K
L
M
H
G

Hub Bore Range	Shear Pin Assembly Number	Shear Pin Radius R	Shear Pin Pin Dia. B	Diameters Flange C	Diameters Shear Pin Hub D	Diameters Adapt Hub & Collar E	Diameters Sprocket Seat F	Length Thru Shear Pin Hub G	Length Thru Adapt H	Length Thru Collar M	Hub Flange Thickness J	Adapt Flange Thickness K	Sprocket Seat Width L	Bolts Number & Size N	Bolts Bolt Circle P
1.00 & under	SP-17	1.80	.25	5.25	1.75	2.50	2.62	2.44	1.38	.38	.56	.56	.44	4-.38	4.00
1.06-1.25	SP-18	2.18	.25	6.00	2.25	3.25	3.38	2.94	1.75	.50	.56	.56	.56	4-.38	4.75
1.30-1.50	SP-19	2.56	.30	6.75	2.75	4.00	4.12	3.56	2.12	.62	.68	.68	.68	4-.50	5.50
1.56-1.75	SP-20	3.00	.38	7.75	3.25	4.75	4.88	4.18	2.50	.75	.80	.80	.68	4-.50	6.25
1.80-2.00	SP-21	3.30	.45	8.75	3.75	5.25	5.38	4.80	2.88	.88	.94	.94	.94	4-.62	7.00
2.06-2.25	SP-22	3.80	.50	9.75	4.25	6.25	6.38	5.18	3.00	1.00	1.06	1.06	1.18	4-.62	8.00
2.30-2.50	SP-23	4.00	.50	10.00	4.50	6.50	6.62	5.68	3.50	1.00	1.06	1.06	1.38	4-.62	8.25
2.56-2.75	SP-24	4.40	.55	11.50	5.00	7.00	7.12	6.30	3.88	1.12	1.18	1.18	1.38	4-.62	9.25
2.80-3.00	SP-25	4.90	.62	12.50	5.50	8.00	8.12	6.94	4.25	1.25	1.30	1.30	1.38	6-.62	10.25

Fig. 16-4-A Bolt-on shear pin sprocket.

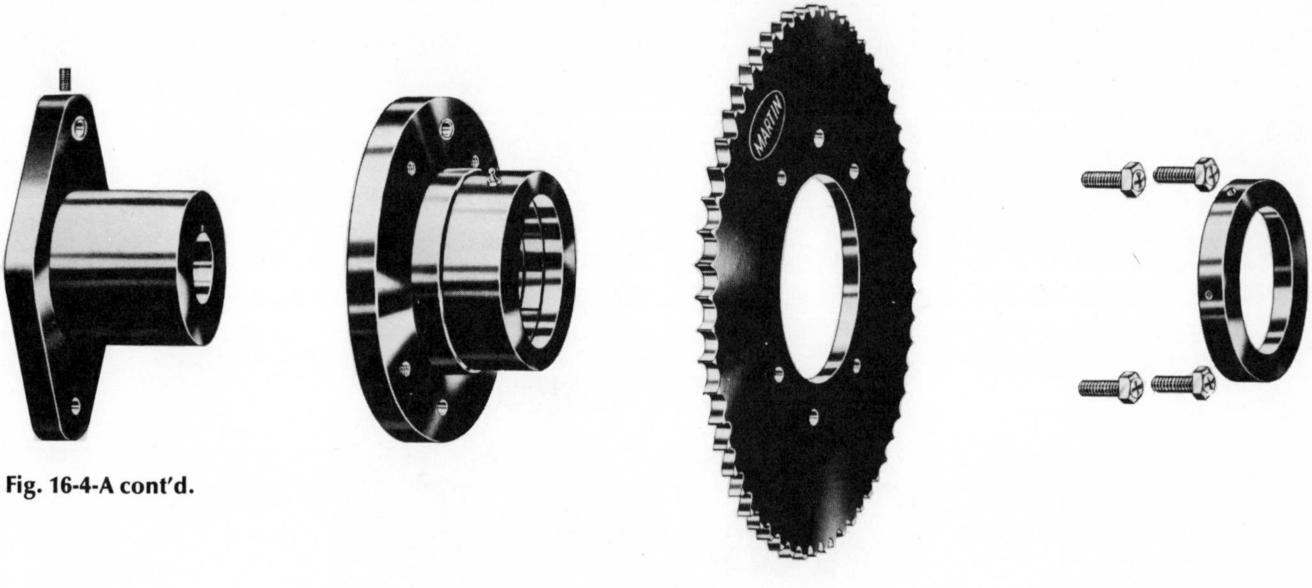
MARTIN

Fig. 16-4-A cont'd.

Chapter 17 Drawing for Numerical Control

UNIT 17-1

Numerical control is a means of directing some of or all the functions of a machine automatically from instructions. The instructions are generally stored on tape and are fed to the controller through a tape reader. The controller interprets the coded instructions and directs the machine through the required operations.

It has been established that because of the consistent high accuracy of numerically controlled machines, and because human errors have been almost entirely eliminated, scrap has been considerably reduced.

Because both setup and tape preparation times are short, numerically controlled machines produce a part faster than manually controlled machines.

When changes become necessary on a part, they can easily be implemented by changing the original tapes. The process takes very little time and expense in comparison to the alterations of a jig or fixture.

Another area where numerically controlled machines are better is in the quality or accuracy of the work. In many cases a numerically controlled machine can produce parts more accurately at no additional cost, resulting in reduced assembly time and better interchangeability of parts. This latter fact is especially important when spare parts are required.

Acceptance of numerical control processes is being accelerated by the development of computer-aided design and computer-aided manufacturing techniques.

Dimensioning for Numerical Control

Common guidelines have been established that enable dimensioning and tolerancing practices to be used effectively in delineating parts for both numerical control and conventional fabrication.

Coordinate System

The numerical control concept is based on the system of rectangular or cartesian coordinates in which any position can be described in terms of distance from an origin point along either two or three mutually perpendicular axes. Two dimensional coordinates (*X*, *Y*) define points in a plane. See Fig. 17-1-1.

The *X* axis is horizontal and is considered the first and basic reference axis.

Distances to the right of the zero *X* axis are considered positive *X* values and to the left of the zero *X* axis as negative *X* values.

The *Y* axis is vertical and perpendicular to the *X* axis in the plane of a drawing showing *XY* relationships. Distances above the zero *Y* axis are considered positive *Y* values and below the zero *Y* axis as negative *Y* values. The position where the *X* and *Y* axes cross is called the *origin*, or *zero point*.

For example, four points lie in a plane, as shown in Fig. 17-1-1. The plane is divided into four quadrants. Point *A* lies in quadrant 1 and is located at position (6, 5), with the *X* coordinate first, followed by the *Y* coordinate. Point *B* lies in quadrant 2 and is located at position (−4, 3). Point *C* lies in quadrant 3 and is located at position (−5, −4). Point *D* lies in quadrant 4 and is located at position (3, −2).

Designing for numerical control would be greatly simplified if all work were done in the first quadrant because all the values would be positive and the plus and minus signs would not be required. However, any of the four quadrants may be used, and, as such, programming in any of the quadrants should be understood.

Some numerically controlled machines are designed for locating points in only the *X* and *Y* directions. These are called *two-axis machines*.

The function of these machines is to move the machine table or tool to a specified position in order to perform work, as shown in Fig. 17-1-2. With the fixed spindle and movable table as shown in Fig. 17-1-2*b*, hole *A* is drilled; then the table moves to the left, positioning point *B* below the drill. This is the most frequently used method. With the fixed table and movable spindle as shown in Fig. 17-1-2*c*, hole *A* is drilled; then the spindle moves to the right, positioning the drill above point *B*. This changes the direction of the motion, but the movement of the cutter as related to the work remains the same.

ZERO POINT

As previously mentioned, the zero point is the point where the *X* and *Y* axes intersect. It is the point where all coordinate dimensions are measured. Many machines have a fixed zero point built in.

Two examples of fixed zero points on machine tables are shown in Fig. 17-1-3. In Fig. 17-1-3*a* all points are located in the first quadrant, resulting in positive *X*

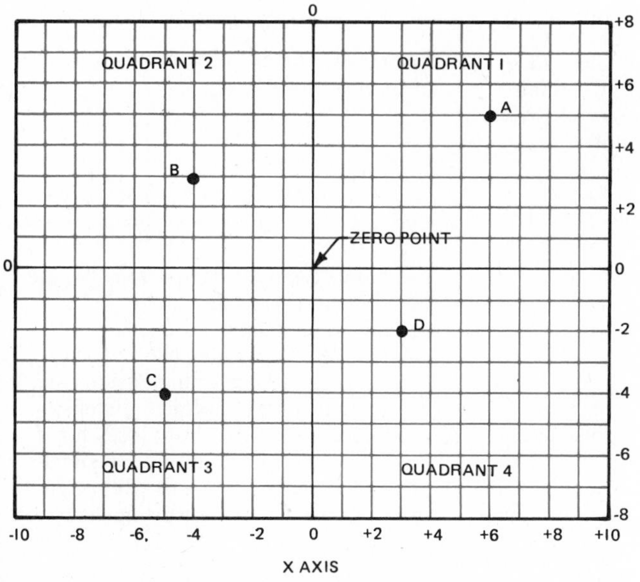

0
QUADRANT 2
QUADRANT I
A
B
ZERO POINT
D
C
QUADRANT 3
QUADRANT 4
+8
+6
+4
+2
0
-2
-4
-6
-8
-10
-8
-6,
-4
-2
0
+2
+4
+6
+8
+10
Y AXIS
X AXIS

Fig. 17-1-1 Two dimensional coordinates. (X and Y)

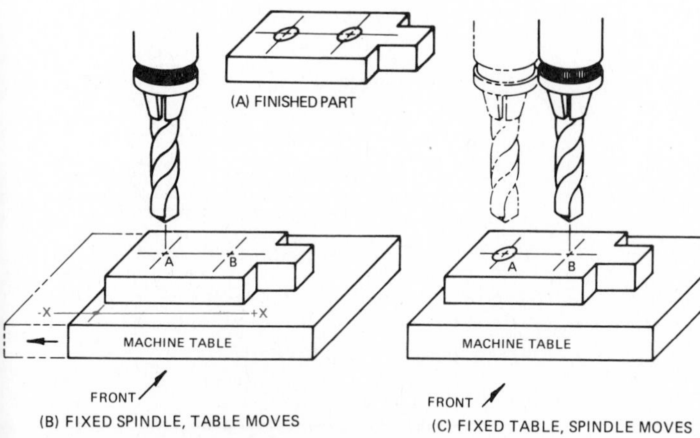

(A) FINISHED PART
A
B
-X
+X
MACHINE TABLE
FRONT
(B) FIXED SPINDLE, TABLE MOVES
MACHINE TABLE
FRONT
(C) FIXED TABLE, SPINDLE MOVES

Fig. 17-1-2 Positioning the work.

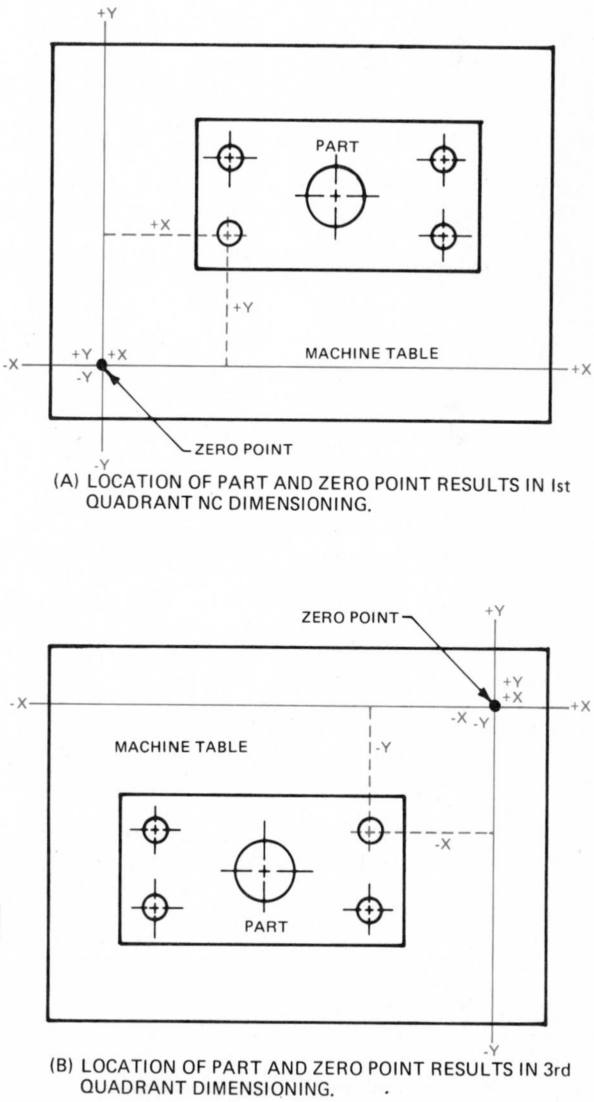

+Y
PART
+X
+Y
-X
+Y
+X
-Y
MACHINE TABLE
+X
ZERO POINT
-Y
(A) LOCATION OF PART AND ZERO POINT RESULTS IN Ist QUADRANT NC DIMENSIONING.
ZERO POINT
+Y
-X
+Y
+X
+X
-X
-Y
MACHINE TABLE
-Y
-X
PART
-Y
(B) LOCATION OF PART AND ZERO POINT RESULTS IN 3rd QUADRANT DIMENSIONING.

Fig. 17-1-3 Zero point location.

and *Y* values. In Fig. 17-1-3*b* all points are located in the third quadrant, resulting in negative *X* and *Y* values.

SETUP POINT

The setup point is located on the part or the fixture holding the part. It may be the intersection of two finished surfaces, the center of a previously machined hole in the part, or a feature of the fixture. It must be accurately located in relation to the zero point, as shown in Fig. 17-1-4.

Point-to-Point Programming

Point-to-point programming is the most common type of positioning system. With this system each new positioning is given from the last position. To compute the next position wanted, it is necessary to establish the sequence in which the work is to be done.

An example of this type of dimensioning is shown in Fig. 17-1-4*a*. The distance between the left edge of the part and hole 1 is given as 20 mm. From hole 1 to hole 4 the dimension shown is 120 mm (*X* axis), and from hole 1 to hole 2 the dimension shown is 40 mm (*Y* axis). These dimensions give the distance from the last hole drilled to the next drilled hole. Assume the holes are to be drilled in the sequence shown in the figure. Hole 1 is located (70, 70) from the zero point. After hole 1 has been drilled, the drill spindle is positioned above the center of hole 2. Hole 2 has the same *X*-coordinate dimension as hole 1, making the *X* increment zero. Since the vertical distance between holes 1 and 2 is 40 mm, the *Y* increment becomes +40.

After hole 2 is drilled, the drill spindle is positioned above the center of hole 3. Since the horizontal distance between holes 2 and 3 is 120 mm, the *X* increment is +120. Hole 3 has the same *Y*-coordinate dimension as hole 2, making the *Y* increment zero.

From hole 3 the drill spindle is positioned above the center of hole 4. Hole 4 has the same *X*-coordinate dimension as hole 3, making the *X* increment zero. Since the vertical distance between hole 3 and hole 4 is 40 mm, the *Y* increment is −40.

Figure 17-1-5 lists the distance between holes and indicates the direction of motion by plus and minus signs. It can be seen that each pair of coordinates is the distance between the two locations.

Coordinate Programming

Many machines use coordinate programming instead of the point-to-point method of dimensioning. With this type of dimensioning all dimensions are taken from the zero point; as such, base-line dimensioning, as shown in Fig. 17-1-4*b*, is used. For example, after hole 1 is drilled, then the drill spindle has to be positioned above the center of hole 2. The coordinates for hole 2 are (70, 110). Figure 17-1-6 shows the coordinate dimensions of the holes shown in Fig. 17-1-4.

Assignments

1. On an A3- or B-size sheet, prepare a chart listing the *X* and *Y* (coordinate) locations and the quadrant for the points *A* to *V* shown on Fig. 17-1-A. The grid is 10× 10 to the centimetre or inch.

HOLE	X	Y
1	+70	+70
2	0	40
3	+120	0
4	0	−40

Fig. 17-1-5 Point-to-point dimensioning of holes shown in Fig. 17-1-4.

HOLE	X	Y
1	+70	+70
2	+70	+110
3	+190	+110
4	+190	+70

Fig. 17-1-6 Coordinate dimensioning of holes shown in Fig. 17-1-4.

2. On an A3- or B-size sheet, prepare two drawings of the cover plate shown in Fig. 17-1-B or 17-1-C. One drawing is to use point-to-point dimensioning for the 10 holes; the other drawing is to use coordinate dimensioning. Only the dimensions locating the holes need be shown. The radial and angular dimensions are to be replaced with coordinate dimensions and taken to two decimal places for millimetre dimensions and three decimal places for inch dimensions. Below each drawing prepare a chart listing each hole

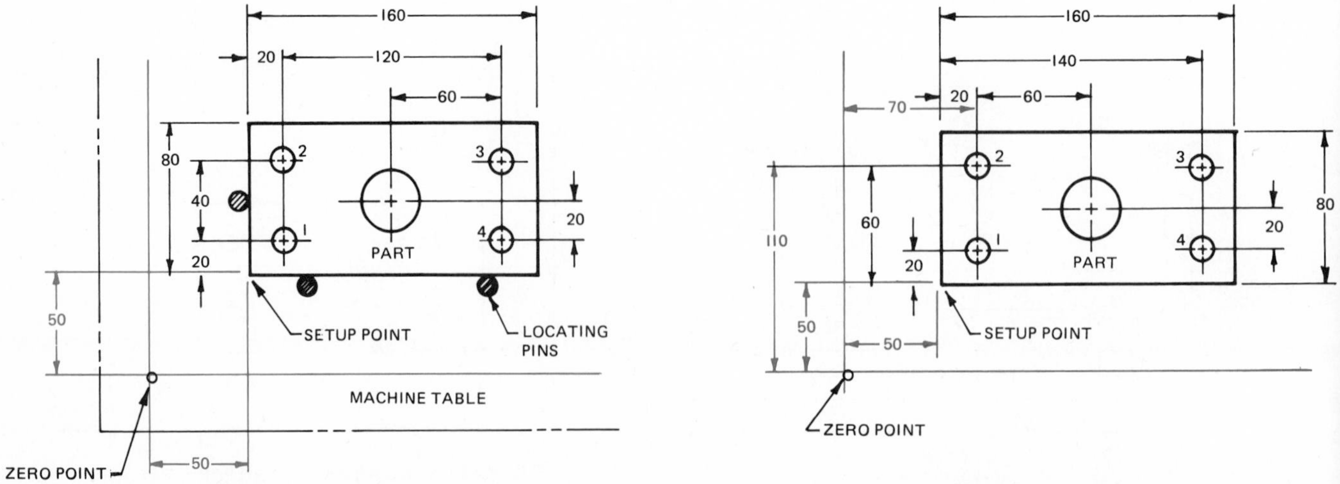
160
20
120
60
80
40
20
2
3
1
4
20
PART
50
SETUP POINT
LOCATING PINS
MACHINE TABLE
ZERO POINT
50
160
140
70
20
60
2
3
1
4
110
60
20
50
50
PART
20
80
SETUP POINT
ZERO POINT

(A) POINT-TO-POINT DIMENSIONING (B) BASE LINE DIMENSIONING

Fig. 17-1-4 Dimensioning for N/C.

and their *X* and *Y* coordinates. The letters shown on the holes show the sequence in which they are to be drilled. The center of the part is the zero point. Scale is 1:1.

REVIEW FOR ASSIGNMENTS

Appendix Trigonometric Functions

POINT	X AXIS	Y AXIS	QUADRANT
A			
B			
C			

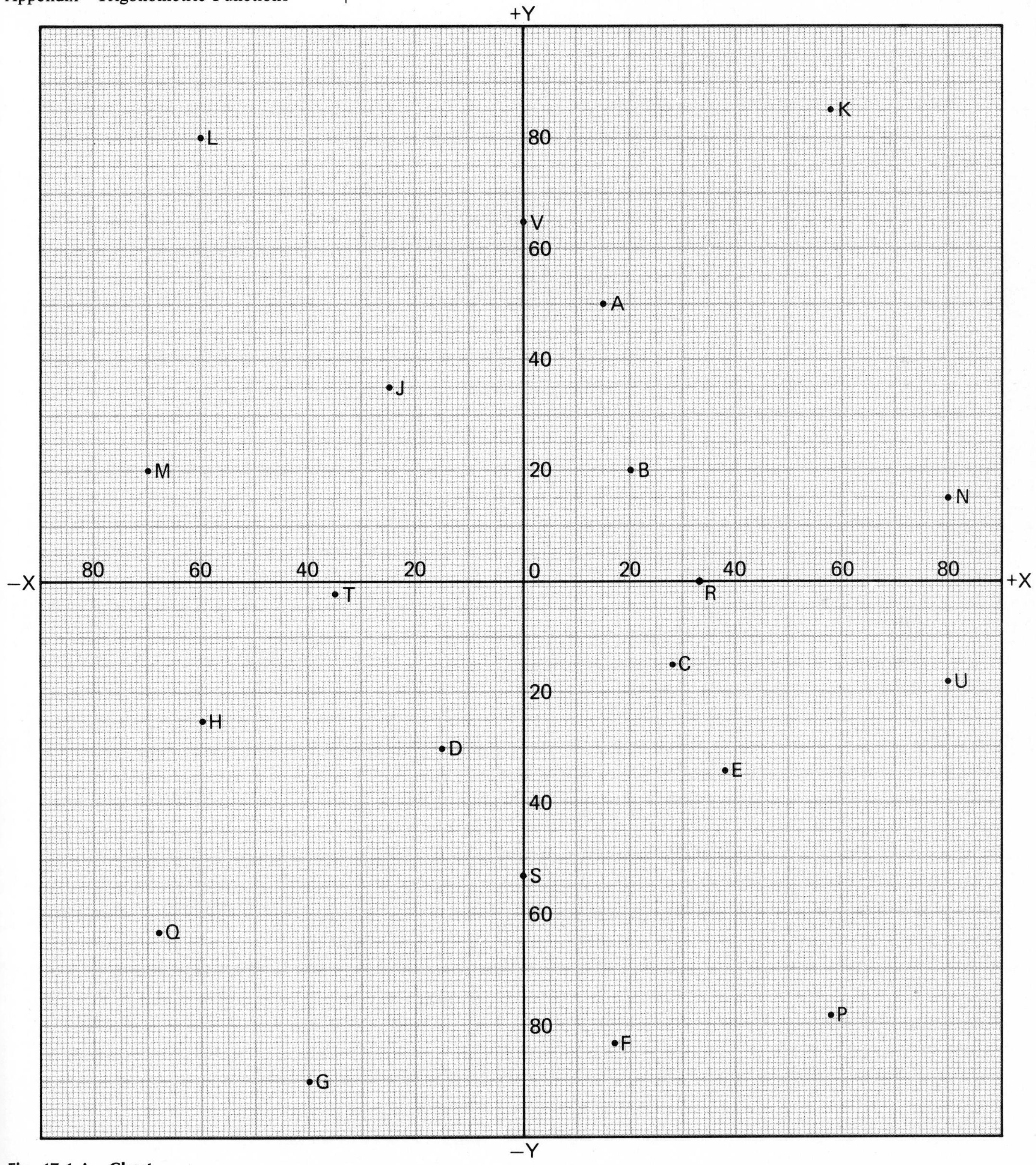
+Y
−Y
−X
+X
80
60
40
20
0
20
40
60
80
K
L
V
A
J
M
B
N
T
R
C
U
H
D
E
S
Q
P
F
G

Fig. 17-1-A Chart.

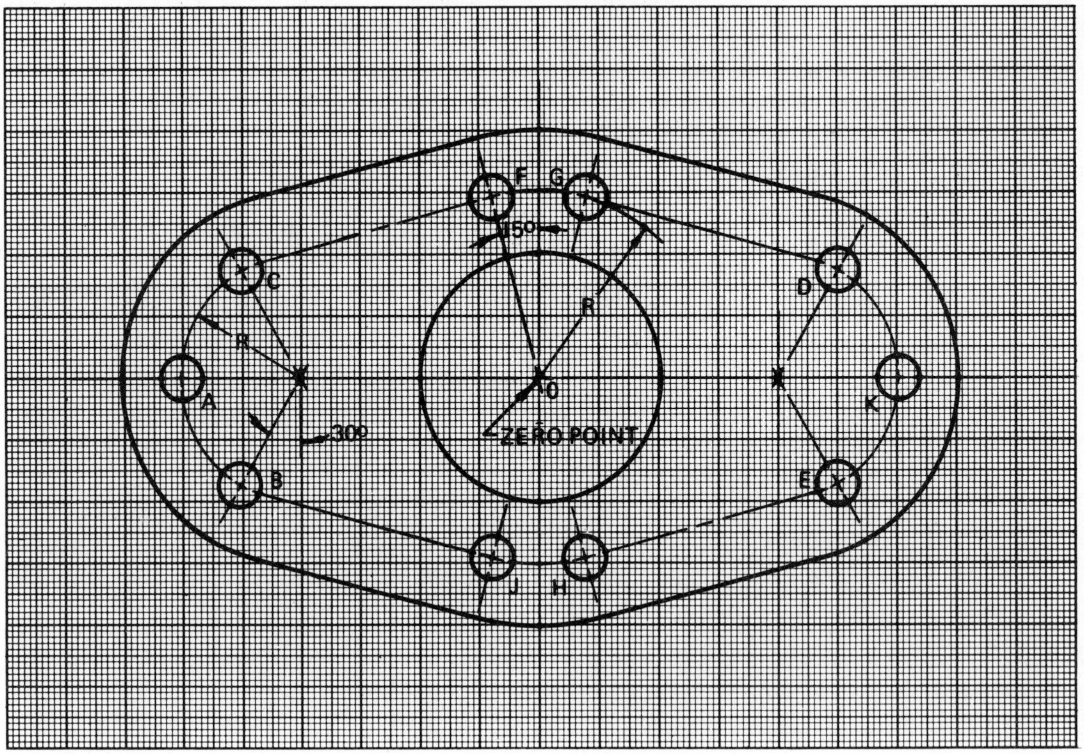

F
G
15°
C
D
R
R
A
0
K
30°
ZERO POINT
B
E
J
H

NOTE: GRID 10 X 10 TO THE CM

Fig. 17-1-B Cover plate

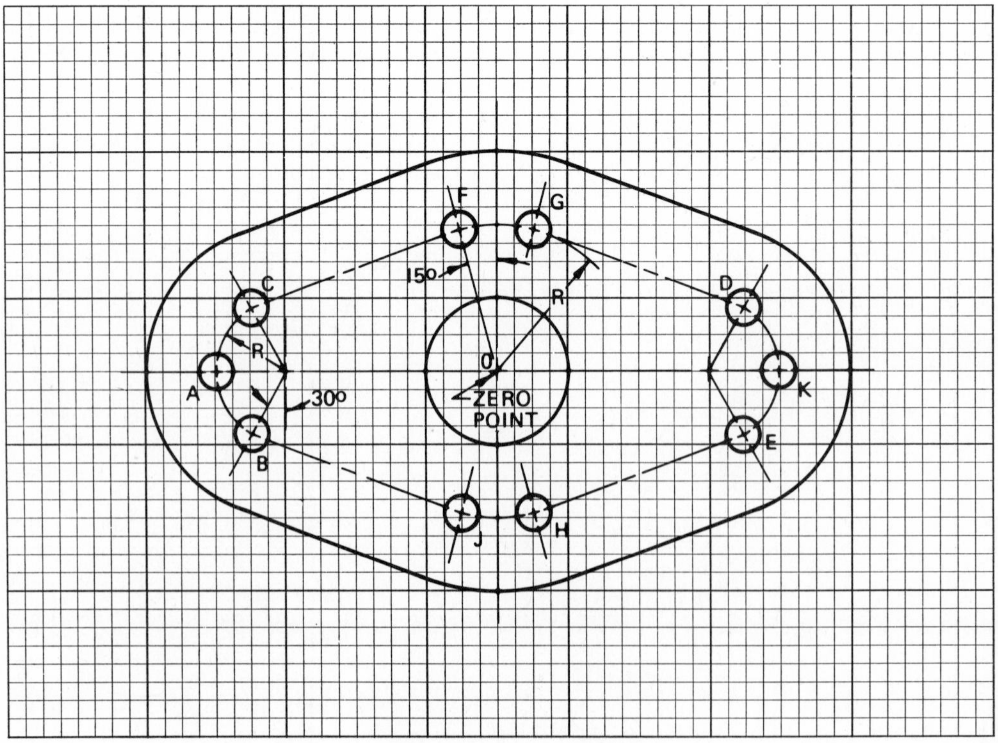

F
G
15°
C
D
R
R
0
A
30°
ZERO POINT
K
B
E
J
H

NOTE: GRID 8 X 8 TO THE INCH

Fig. 17-1-C Cover plate.

UNIT 17-2
THREE-AXIS CONTROL SYSTEMS

Many numerically controlled machines operate in three directions, the table and carriage moving in the *X* and *Y* directions, as explained in Unit 17-1, and the tool spindle, such as a turret drill, traveling in an up-and-down direction. A vertical line taken through the center of the machine spindle is referred to as the *Z* axis and is perpendicular to the plane formed by the *X* and *Y* axes. See Fig. 17-2-1.

Thus, a point in space can be described by its *X*, *Y*, and *Z* coordinates. For example, P_1 in Fig. 17-2-2 can be described by its (*X*, *Y*, *Z*) coordinates as (4, 3, 5) and P_2 as (11, 2, 8).

A popular system used on many numerically controlled machines, such as the turret drill, is to establish the *Z* zero reference plane above the workpiece. Each tool is then adjusted and calibrated to the *Z* zero reference plane.

For example, Fig. 17-2-3 shows a part requiring three drilled holes. As the center hole is drilled through, the part is raised by gage blocks so that the drill does not touch the machine table. The height of the gage blocks is determined by the distance the drill passes through the workpiece plus clearance, or $2 + 0.3D + 4$. See Fig. 17-2-4. If a 20 mm drill were used, the gage block height would be $2 + 6 + 4 = 12$ mm.

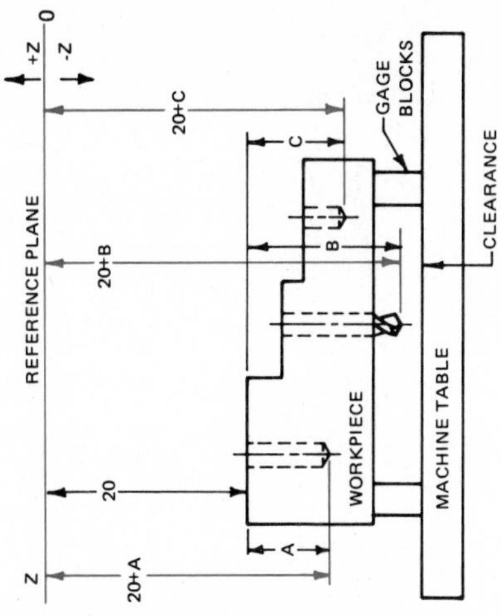
+Z
-Z
0
20+C
GAGE BLOCKS
C
CLEARANCE
REFERENCE PLANE
20+B
B
WORKPIECE
MACHINE TABLE
20
A
Z
20+A

Fig. 17-2-3 Calculating Z distance.

If the distance from the top of the workpiece to the *Z* zero reference plane is set at 20 mm, the *Z* coordinates for the three holes shown are $-(20 + A)$, $-(20 + B)$, and $-(20 + C)$.

Dimensioning and Tolerancing

Recommended guidelines for dimensioning and tolerancing practices for use in defining parts for numerical control fabrication are as follows:

1. When the basic coordinate system is established, the setup point should be placed at an appropriate location on the part itself.

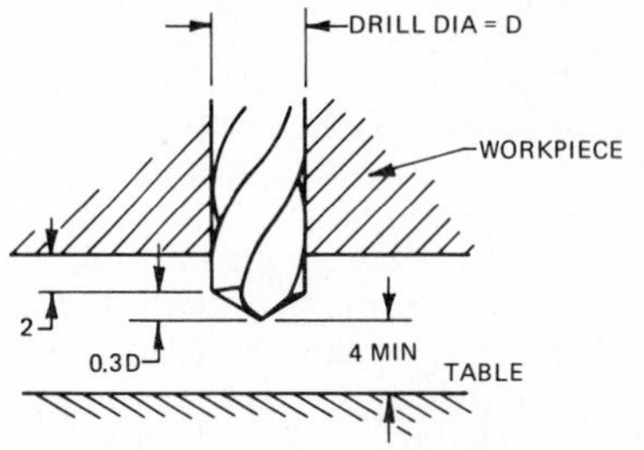
DRILL DIA = D
WORKPIECE
2
0.3D
4 MIN
TABLE

Fig. 17-2-4 Determining gage block height.

2. Any number of subcoordinate systems may be used to define features of a part as long as these systems can be related to the basic coordinate system of the given part.

3. Define part surfaces in relation to three mutually perpendicular reference planes. Establish these planes along part surfaces which parallel the machine axes if these axes can be predetermined.

4. Dimension the part precisely so that the physical shape can be readily determined. Dimension to points on the part surfaces.

5. Regular geometric contours such as ellipses, parabolas, hyperbolas, etc., may be defined on the drawing by mathematical formulas. The numerically controlled machinery can easily be programmed to

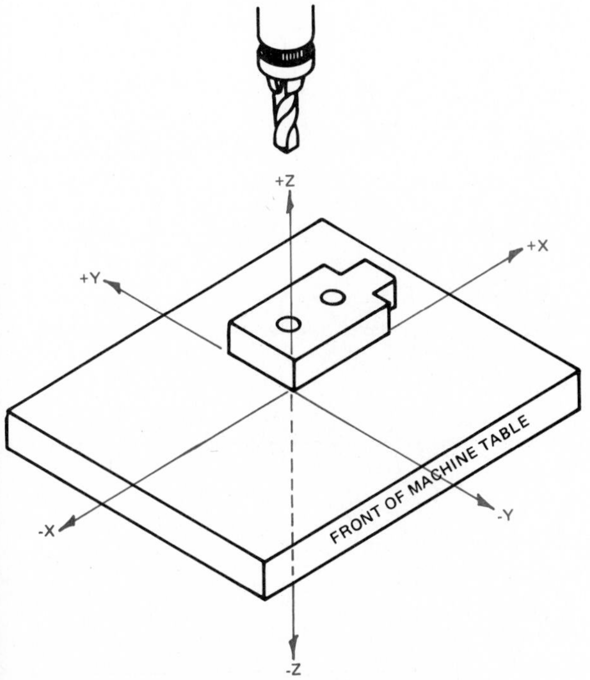
+Z
+X
+Y
-X
-Y
-Z
FRONT OF MACHINE TABLE

Fig. 17-2-1 X, Y, and Z axes.

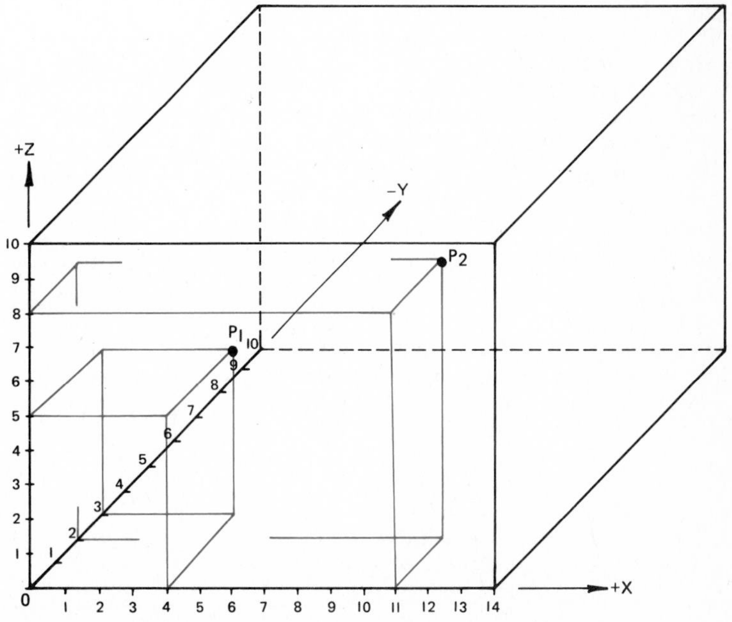
+Z
-Y
+X
P1
P2
0
1
2
3
4
5
6
7
8
9
10
11
12
13
14

Fig. 17-2-2 Points in space.

approximate these curves by linear interpolation, that is, as a series of short, straight lines whose endpoints are close enough together to ensure meeting the required tolerances for contour. In the case of arbitrary curves, the drawing should specify appropriate points on the curve by coordinate dimensions or a table of coordinates. Consideration should be given to the number of points needed to define the curve; however, one should keep in mind the fact that the tighter the tolerance or the smaller the radius of curvature, the closer together the points should be. Such terms as *blend smoothly* and *faired curve* are not used. Curves may also be defined by other coordinates, such as polar, spherical, or cylindrical, as applicable.

6. Changes in contour should be unambiguously defined with prime consideration for design intent.

7. Holes in a circular pattern should be located with coordinate dimensions preferably.

8. Express angular dimensions, where possible, relative to the *X* axis in degrees and decimal parts of a degree.

9. Use plus and minus tolerances, not limit dimensions. Preferably, the tolerance should be equally divided bilaterally.

10. Positional tolerancing, form tolerancing, and datum referencing should be used where applicable. Datum features specified on the drawing in proper sequence will clearly indicate their usage for setup.

11. Where profile tolerances are specified, the geometric boundary should be equally disposed bilaterally along the true profile. Avoid profile tolerances applied unilaterally along the true profile. Include no less than four defined points along the profile.

12. Tolerances are specified only on the basis of actual design requirements. The accuracy capability of numerically controlled equipment is not a basis for specifying more restrictive tolerances than functionally required.

Assignments

1. On an A3- or B-size sheet make a two-view drawing of the end plate shown in Fig. 17-2-A or 17-2-B. Point-to-point dimensioning is to be used, and only the dimensions locating the holes need be shown. Below the drawing prepare a chart listing each hole and its *X*, *Y*, and *Z* coordinates. The letters on the holes show the sequence in which they are to be drilled. Calculating the *Z* coordinate is to be done in the same manner as for the part shown in Fig. 17-2-3. Scale is 1:1. *Note:* Programming will be for the tap-drill holes and the six through holes shown. Zero point for the *X* and *Y* coordinates is the center of the end plate.

2. On an A3- or B-size sheet, locate points *P*1 to *P*10 on the coordinate chart 1 shown in Fig. 17-2-C. On chart 2 points *P*1 to *P*10 are located, but only one of their coordinates is known. Accurately lay out the missing coordinates on the chart and record their values in the table. Scale—as shown.

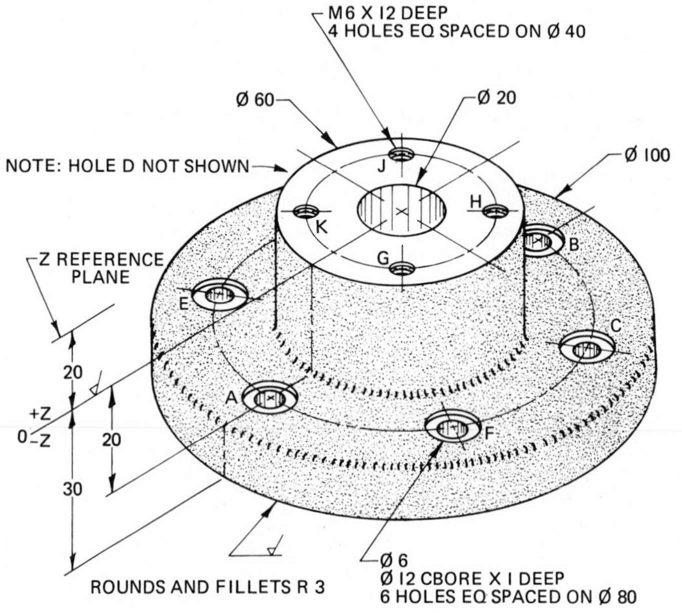
M6 X 12 DEEP
4 HOLES EQ SPACED ON Ø 40
Ø 60
Ø 20
Ø 100
NOTE: HOLE D NOT SHOWN
J
H
K
G
B
E
C
A
F
Z REFERENCE PLANE
20
+Z
0
-Z
20
30
ROUNDS AND FILLETS R 3
Ø 6
Ø 12 CBORE X 1 DEEP
6 HOLES EQ SPACED ON Ø 80

Fig. 17-2-A End plate.

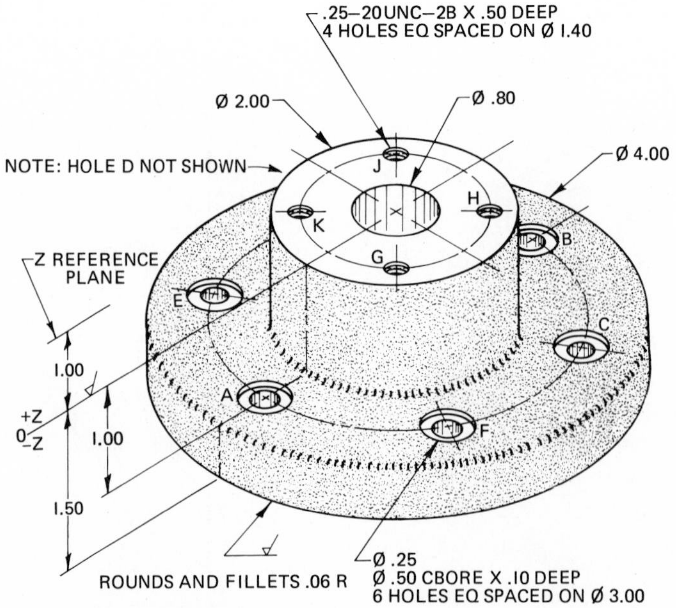
.25–20UNC–2B X .50 DEEP
4 HOLES EQ SPACED ON Ø 1.40
Ø 2.00
Ø .80
Ø 4.00
NOTE: HOLE D NOT SHOWN
J
H
K
G
B
E
C
A
F
Z REFERENCE PLANE
1.00
+Z
0
-Z
1.00
1.50
ROUNDS AND FILLETS .06 R
Ø .25
Ø .50 CBORE X .10 DEEP
6 HOLES EQ SPACED ON Ø 3.00

Fig. 17-2-B End plate.

CHART I

LOCATE POINTS PI TO PI0 ON CHART

NOTE: ALL COORDINATES ARE +

POINT	X AXIS	Y AXIS	Z AXIS
PI	10	20	60
P2	20	70	70
P3	0	20	30
P4	100	60	75
P5	40	30	20
P6	40	60	10
P7	70	20	0
P8	50	50	50
P9	85	65	30
PI0	60	65	15

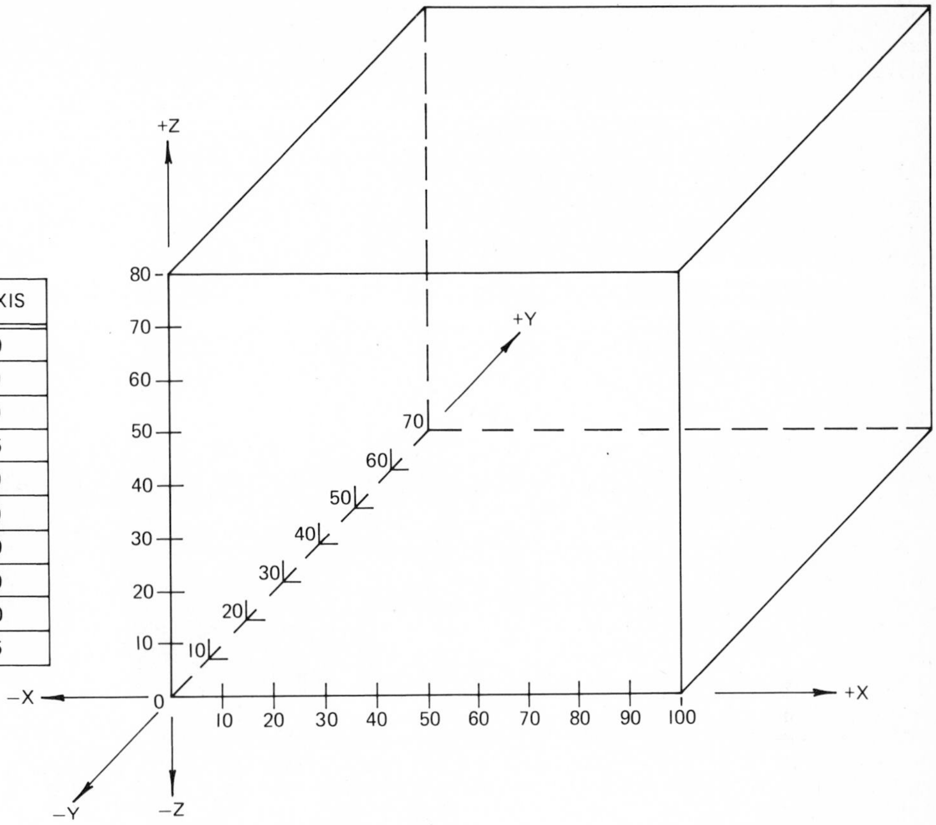
+Z
80
70
60
50
40
30
20
10
0
+Y
10 20 30 40 50 60 70
-X
+X
10 20 30 40 50 60 70 80 90 100
-Y
-Z

CHART 2

LOCATE COORDINATES AND RECORD IN TABLE

NOTE: ALL COORDINATES ARE +

POINT	X AXIS	Y AXIS	Z AXIS
PI	20		
P2			55
P3		60	
P4	30		
P5		30	
P6			0
P7			40
P8	65		
P9		30	
PI0			55

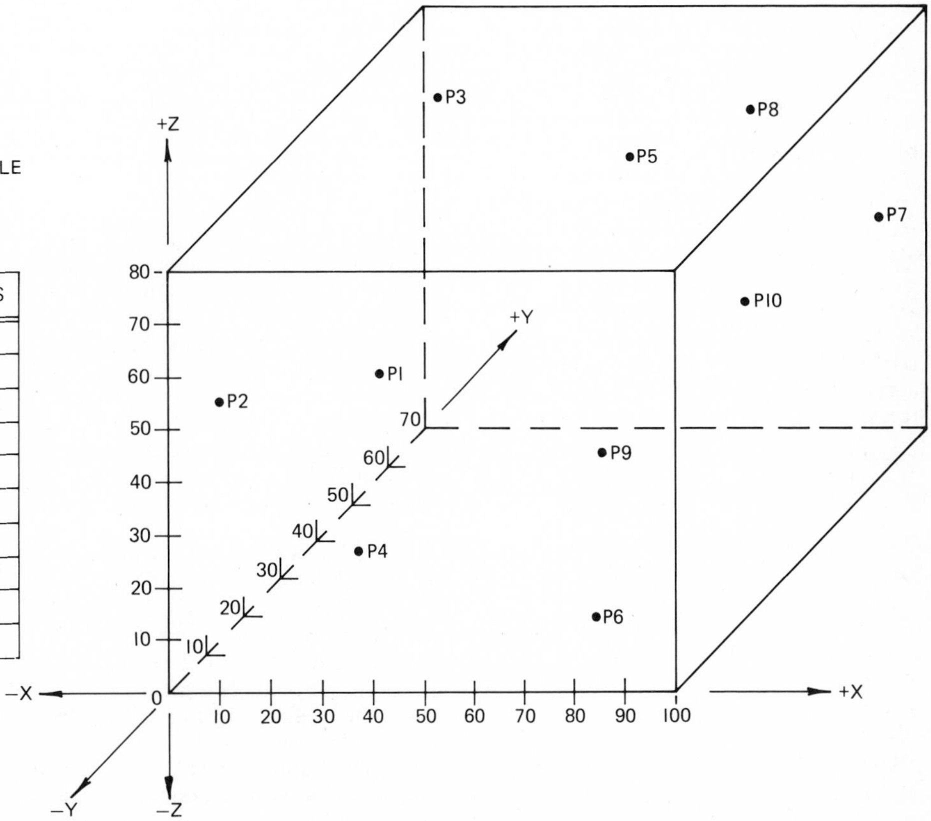
+Z
P3
P8
P5
P7
PI0
PI
P2
P9
P4
P6
80
70
60
50
40
30
20
10
0
+Y
10 20 30 40 50 60 70
-X
+X
10 20 30 40 50 60 70 80 90 100
-Y
-Z

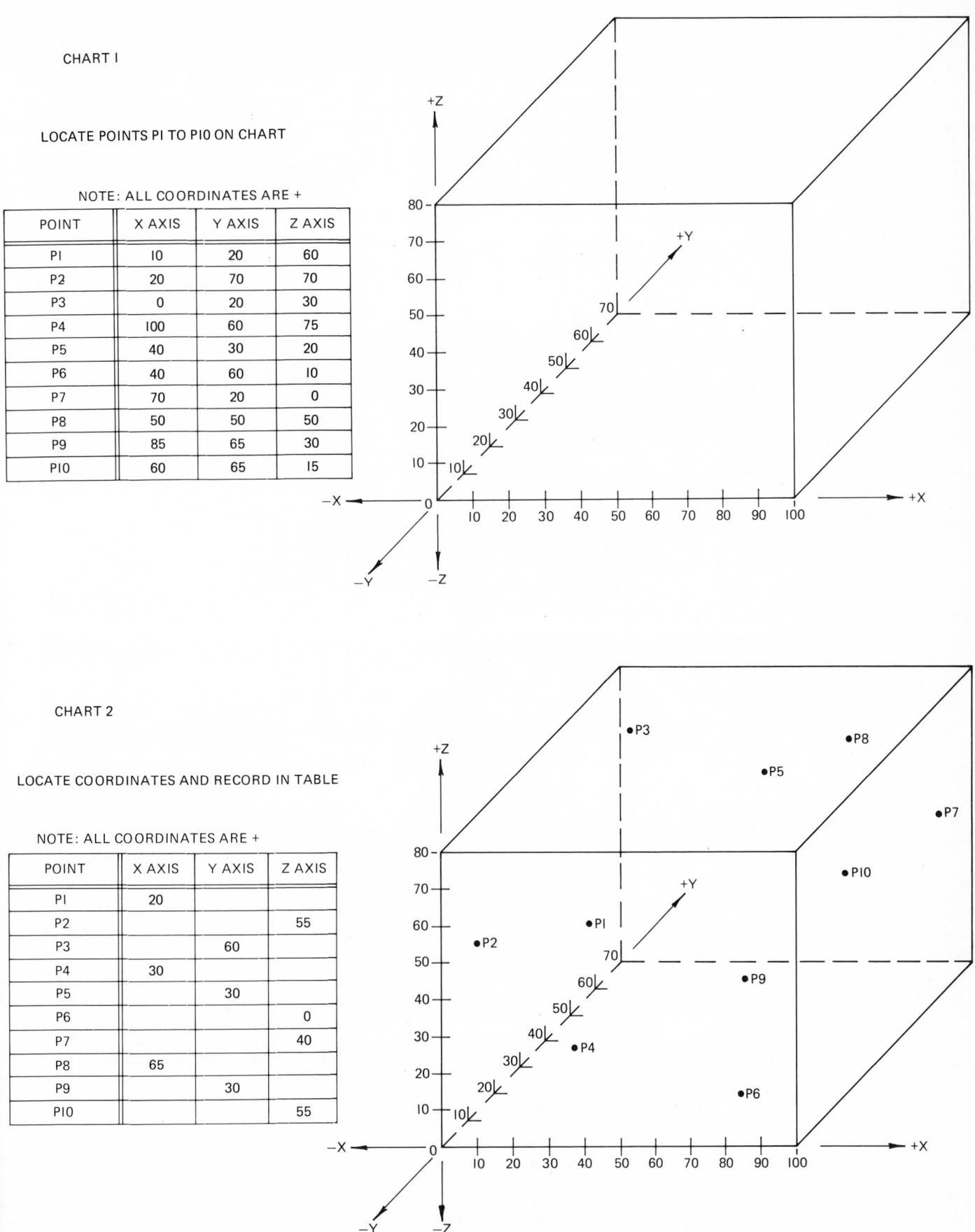

Fig. 17-2-C Chart.

Chapter 18
Charts and Graphs

UNIT 18-1
CHARTS AND GRAPHS

Charts and graphs are an important part of technical drawing. They are important to scientists, engineers, mathematicians, and nearly everyone else in everyday life.

Scientists use charts and graphs to record and study the results of research. Engineers use them to record information about materials and conditions. Mathematicians use them to record facts and trends of numerical information. Doctors use charts to record body temperature, heart action, and other body functions. Most people use and read charts and graphs to see about the weather, the stock market, where the tax dollar goes, and for many other purposes. Because Fig. 18-1-1 is in chart form, it can be seen at a glance that driver reaction distance and automobile braking distance increase as speed is increased. Figure 18-1-2 has the same information but requires a great deal more study time to obtain it. In addition, the graphic (pictorial) aspect of Fig. 18-1-1 makes the relationship of speed and distance more easily understood.

Graphic charts and graphs are used pictorially to show information. They show trends, such as whether the cost of living and wages are rising or falling over a period or whether one is rising and the other is falling.

Charts can show ratios, such as speed versus distance, or they may show percentages of a whole, as in a bar chart or a pie chart. Charts can also be used to explain information which is not numerical. A flowchart shows sequential information, that is, which operation is first, second, third, and so on. Graphic charts may also be used to solve various kinds of mathematical problems and assist in making predictions.

Chart Elements

An almost endless variety of graphic charts may be made for visual communication. This chapter has suggested the general character of a few and how they are used. The drawings throughout this chapter will indicate some of the elements of chart making and suggest the ways in which these elements affect the chart.

Every chart should have a suitable title, well lettered and placed within the area of the chart. Every chart should also have a key to tell the reader what the elements represent.

TAPES

Tapes are used as a drafting aid in the preparing of charts and graphs. Adhesive

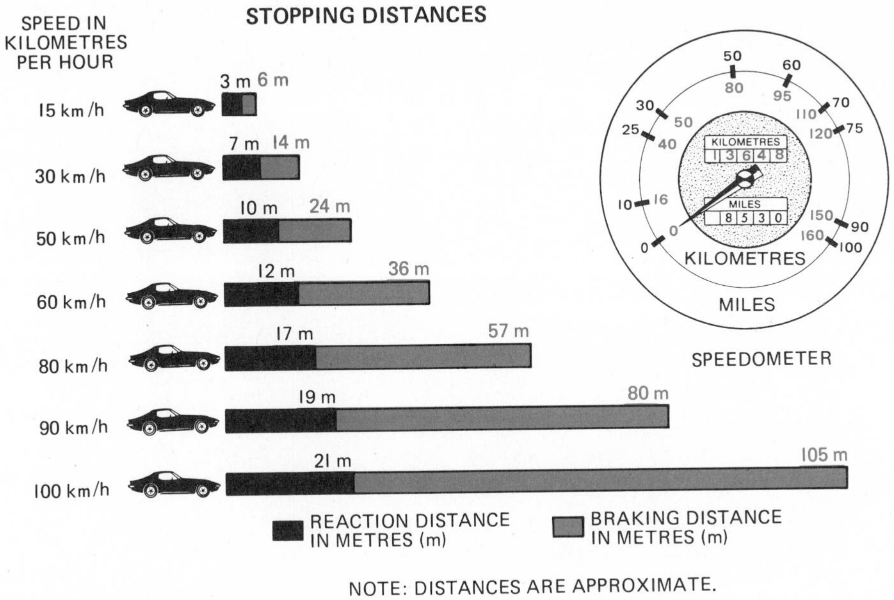
STOPPING DISTANCES
SPEED IN KILOMETRES PER HOUR
15 km/h
30 km/h
50 km/h
60 km/h
80 km/h
90 km/h
100 km/h
3 m 6 m
7 m 14 m
10 m 24 m
12 m 36 m
17 m 57 m
19 m 80 m
21 m 105 m
REACTION DISTANCE IN METRES (m)
BRAKING DISTANCE IN METRES (m)
NOTE: DISTANCES ARE APPROXIMATE.
KILOMETRES
MILES
SPEEDOMETER

Fig. 18-1-1 **One-column bar chart.**

STOPPING DISTANCES AT DIFFERENT SPEEDS FOR AUTOMOBILES

Kilometres Per Hour	Distance in Metres		
	Reaction Distance	**Braking Distance**	**Total Distance**
15	3	3	6
30	7	7	14
50	10	14	24
60	12	24	36
80	17	40	57
90	19	61	80
100	21	84	105

Fig. 18-1-2 Information shown in Fig. 18-1-1 in table form.

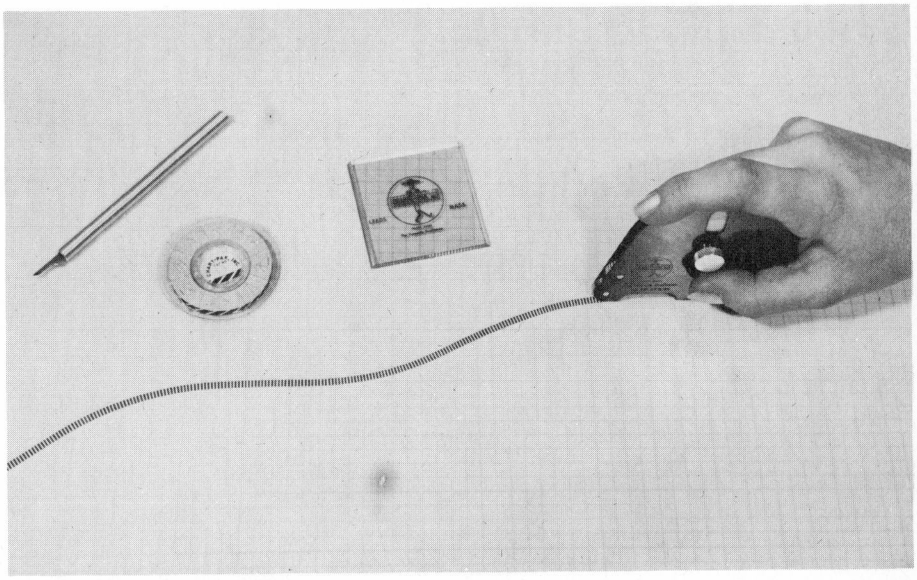

Fig. 18-1-3 Applying adhesive tape to a graphic chart. (Chartpak, Inc.)

tape comes in many colors, designs, and widths. It is applied from a roll dispenser (Fig. 18-1-3) and pressed onto the chart in the desired position. Tapes of selected widths provide a quick and simple way of making bar charts.

THE USE OF COLOR

Black-and-white charts are often used by scientists and mathematicians for recording data. Newspapers use black-and-white charts, and they usually serve the particular purpose rather well. However, the use of color has become quite common in the preparation of charts and diagrams for magazines, books, brochures, and other publications where it is economically practical. Color is also used extensively in making charts for display purposes.

The use of color adds a great deal to appearance, readability, and emphasis. See Fig. 18-1-4. Color may be added in a variety of ways. Colored pencils, felt-tipped pens, water colors, or other similar materials are easy to use and can probably be found in the drafting room, art room, or at home. Commercially prepared pressure-sensitive materials are available at art and engineering supply stores.

Bar Charts

Bar charts are probably the most familiar and most easily read and understood kind of graphic charts. A bar chart consists of a single rectangle representing 100 percent or several rectangular bars of varying sizes, the lengths or heights

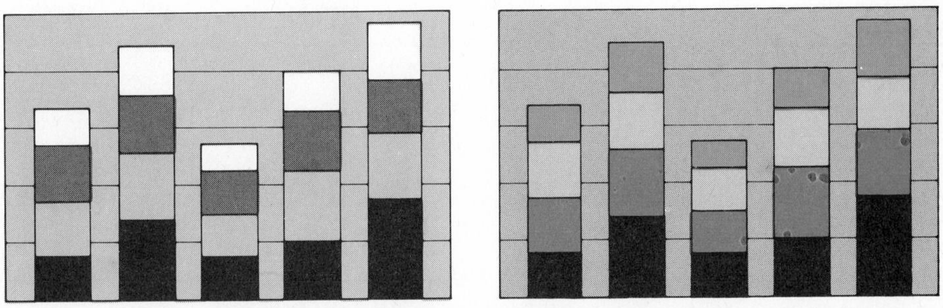

Fig. 18-1-4 A multiple bar chart in black and white and the same one in color.

being in proportion to the values being depicted. The one-column bar chart shown in Fig. 18-1-5 represents the total number of games played during a season. It is divided to show the proportional part, or percentage, of the total number of games won, lost, and tied. Multiple-bar charts are also useful when comparisons are to be made.

PERCENTAGE BAR CHART

A *percentage bar chart* is made by following steps A through F shown in Fig. 18-1-5.

A. Prepare and list the information to be presented as a percentage.
B. Lay off the long side, using a convenient scale equal to 100 units.
C. Lay off a suitable width and complete the rectangle.
D. Lay off the percentage of the parts and draw lines parallel to the base.
E. Crosshatch, shade, or color the various parts, as shown.
F. Letter all necessary information in or near the parts so that it can be read easily.

MULTIPLE-COLUMN BAR CHART

A multiple-column bar chart is made by following steps A through D.

A. Prepare and list the information to be presented (Fig. 18-1-6*a*).
B. Select a suitable scale and lay off the vertical and horizontal axes. Lay off the scale divisions (Fig. 18-1-6*b*).
C. Block in the bars, using the information gathered in Fig. 18-1-6*a*. Allow enough space between bars for all necessary lettering. Make the bars any convenient width for best appearance (Fig. 18-1-6*c*).
D. Complete the bar chart by adding shading or color to the bars, lettering, and any other lines and information needed to make the chart easily understood (Fig. 18-1-6*d*).

COMPOUND BAR CHART

A *compound bar chart* is shown in Fig. 18-1-1, where the total length of the bars is made up of two parts. The black part is the distance traveled at a given speed before the driver starts to apply the brakes. The red part is the distance traveled in coming to a stop. The black

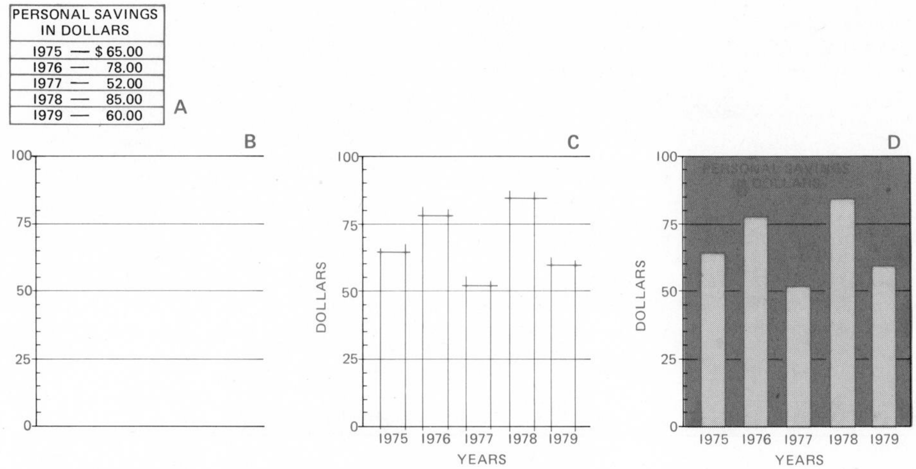
PERSONAL SAVINGS IN DOLLARS
1975 — $ 65.00
1976 — 78.00
1977 — 52.00
1978 — 85.00
1979 — 60.00
A
B
C
D
100
75
50
25
0
DOLLARS
1975 1976 1977 1978 1979
YEARS
PERSONAL SAVINGS IN DOLLARS

Fig. 18-1-6 Steps in drawing a multiple-column bar chart.

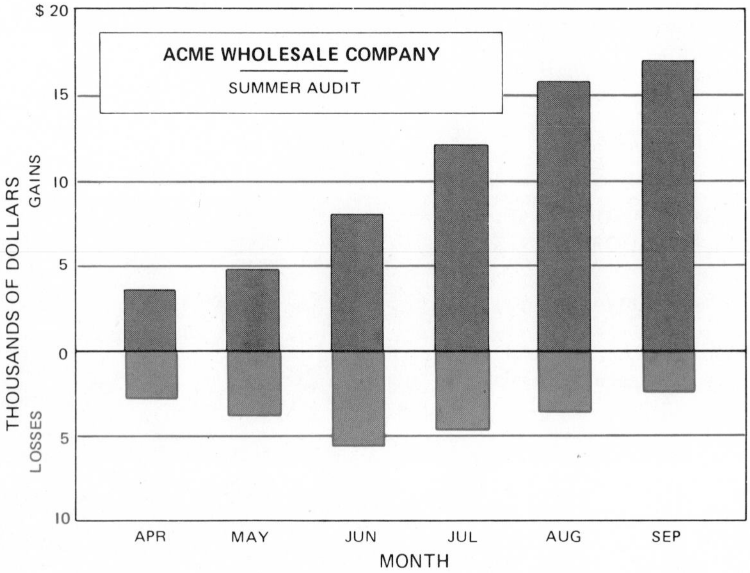
ACME WHOLESALE COMPANY
SUMMER AUDIT
$ 20
15
10
5
0
5
10
THOUSANDS OF DOLLARS
GAINS
LOSSES
APR
MAY
JUN
JUL
AUG
SEP
MONTH

Fig. 18-1-7 A multiple-column bar chart in which the bars have plus and minus values.

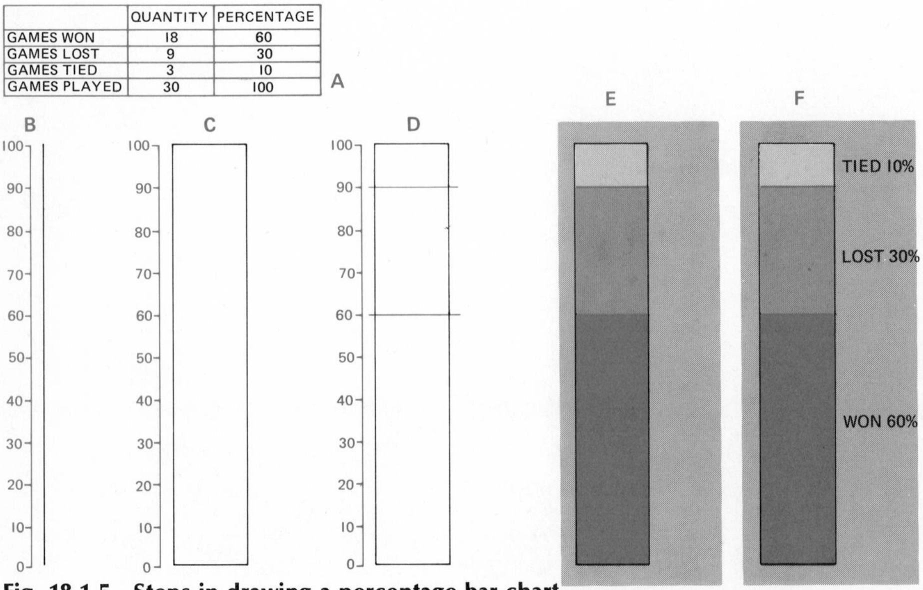
QUANTITY
PERCENTAGE
GAMES WON 18 60
GAMES LOST 9 30
GAMES TIED 3 10
GAMES PLAYED 30 100
A
B
C
D
E
F
100
90
80
70
60
50
40
30
20
10
0
TIED 10%
LOST 30%
WON 60%

Fig. 18-1-5 Steps in drawing a percentage bar chart.

and the red parts are added graphically to give the total distance. With this type of compound bar chart, one end of the bars lines up on a datum or starting line.

Another type of compound bar chart is shown in Fig. 18-1-7. With this type of chart, the datum or starting line is located in the middle of the bars, and plus values are shown above the datum line while minus values are shown below the datum line and in a second color.

PROGRESSIVE BAR CHART

Another type of bar chart with horizontal bars is shown in Fig. 18-1-8. It gives speed ranges for Caterpillar tractors. Note that the bars do not start at a common line because they show different speed ranges. This is called a *progressive bar chart*.

The shape of the bars is not restricted to rectangles alone. Three-dimensional and irregular shapes may be used, as shown in Fig. 18-1-9, to convey the information in the best possible manner.

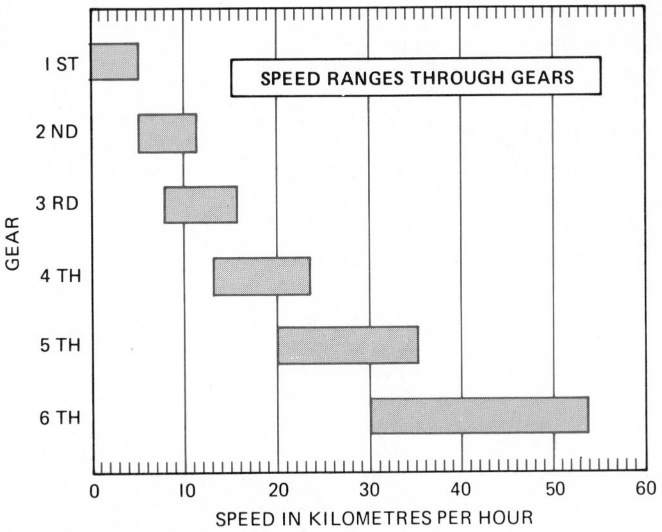
SPEED RANGES THROUGH GEARS
GEAR
1 ST
2 ND
3 RD
4 TH
5 TH
6 TH
0
10
20
30
40
50
60
SPEED IN KILOMETRES PER HOUR

Fig. 18-1-8 Progressive bar chart.

Assignments

1. On an A3- or B-size sheet, design a multiple-column bar chart showing the spending of the following companies for the year 1978. Use the following data: Dodge Company, $127 million; Heinen A. Ltd., $113 million; Bruning Ltd., $108 million; American Gas, $94 million; United Brass, $85 million; Textile Oils, $83 million; Solar Energy Electric, $79 million; Reactors Ltd., $56 million.

2. On an A3- or B-size sheet, design a multiple-column bar chart showing the breakdown of where the home-building dollar goes today. Place the construction bar on the right-hand side; divide it into proper proportions, making it a percentage bar chart which shows how the construction costs are made up. Use the following data:

Total Home Costs	Construction Breakdown
Land—22 percent	—Wood cabinetry —18 percent
Profit—13 percent	—Finishes —13 percent
Sales and marketing —6 percent	—Roofing — 4 percent
Construction —59 percent	—Electric — 3 percent
	—Concrete and masonry — 3 percent
	—Miscellaneous —18 percent
	59 percent

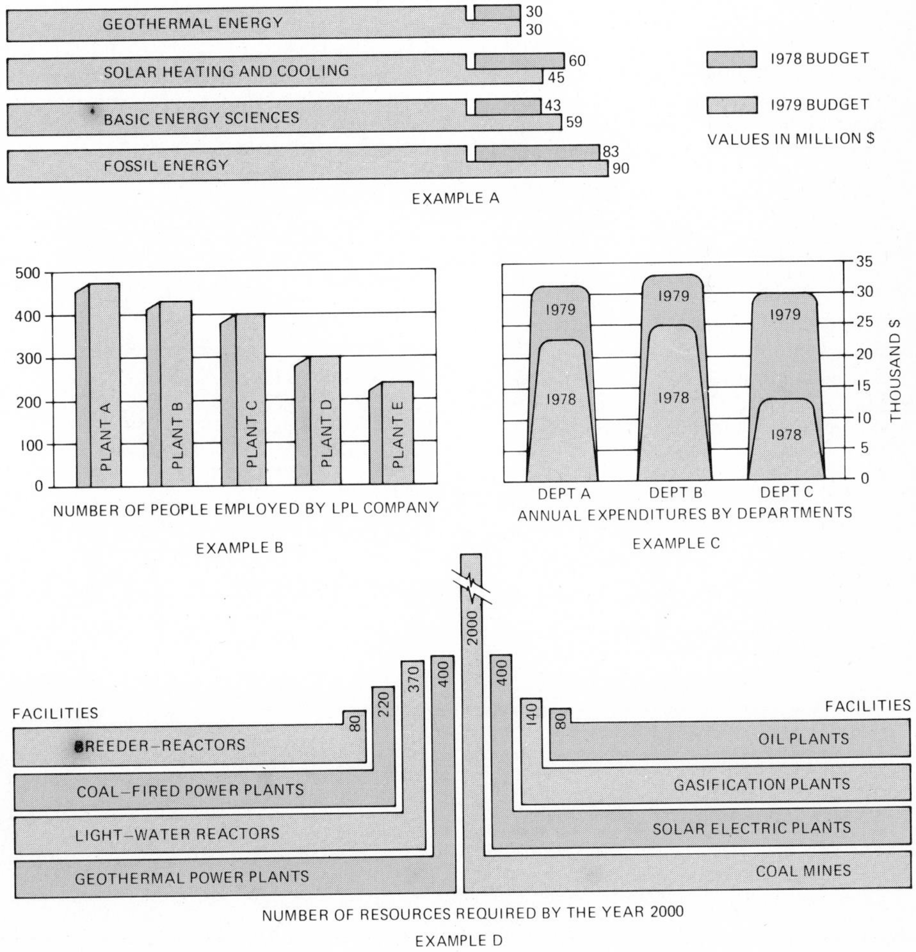
GEOTHERMAL ENERGY
30
30
SOLAR HEATING AND COOLING
60
45
BASIC ENERGY SCIENCES
43
59
FOSSIL ENERGY
83
90
1978 BUDGET
1979 BUDGET
VALUES IN MILLION $
EXAMPLE A
500
400
300
200
100
0
PLANT A
PLANT B
PLANT C
PLANT D
PLANT E
NUMBER OF PEOPLE EMPLOYED BY LPL COMPANY
EXAMPLE B
1979
1978
DEPT A
DEPT B
DEPT C
35
30
25
20
15
10
5
0
THOUSAND $
ANNUAL EXPENDITURES BY DEPARTMENTS
EXAMPLE C
FACILITIES
BREEDER–REACTORS
COAL–FIRED POWER PLANTS
LIGHT–WATER REACTORS
GEOTHERMAL POWER PLANTS
80
220
370
400
2000
400
140
80
FACILITIES
OIL PLANTS
GASIFICATION PLANTS
SOLAR ELECTRIC PLANTS
COAL MINES
NUMBER OF RESOURCES REQUIRED BY THE YEAR 2000
EXAMPLE D

Fig. 18-1-9 Bar chart styles.

3. On an A3- or B-size sheet, design a multiple-column bar chart showing the pattern of family expenditure of families with an annual income of $15,000. Use the following data which are shown as percentages: food, 16.1; shelter, 13.8; household operations, 3.9; furnishings, 3.7; appliances, 1.1; clothing, 8.6; personal care, 2.1; medical and health, 2.8; transportation, 13.1; miscellaneous, 6.6; recreation and holidays, 3.8; personal taxes, 16.8; insurance, 5.1; gifts and contributions, 2.5. Compute the dollar value of each column and mark it on the chart. Be sure the figures total $15,000.

4. On an A3- or B-size sheet, design a suitable bar chart showing the number of calories and grams of carbohydrates in a 100 g (grams) serving of each of the following common foods.

Food	Calories	Carbohydrates
Ice cream	150	14
Peas	75	14
Pizza	260	29
Milk	85	6
Strawberries	30	6

5. On an A3- or B-size sheet, design a progressive bar chart to show the operational sequence of a standard washing machine, using the following information:

- Complete cycling operation—30 minutes
- Fill—4 minutes
- Wash—11 minutes
- Idle—1 minute
- Spin (water removal)—2 minutes
- Fill—3 minutes
- Rinse—3 minutes
- Idle—1 minute
- Spin (water removal)—5 minutes

UNIT 18-2

PIE CHARTS

This is a form of 100 percent chart in which a circle represents 100 percent and sectors represent parts of the whole (Fig. 18-2-1).

To Draw a Pie Chart. A pie chart is made by following steps A through D, as shown in Fig. 18-2-2.

A. Prepare and list the information to be presented.

B. Draw a circle of the desired size. Lay off and draw the radial lines representing the amount or percentage for each part on the circumference of the circle. If a protractor is used, 3.6° = 1 percent. Thirty percent is 30 × 3.6°, or 108°, and so on. If a circle is to be divided into a 24-hour day, each hour represents 15° on the circle. This can be drawn by using a straightedge and the two set squares.

C. Crosshatch, shade, or color the various parts, as shown.

D. Complete the pie chart by adding all necessary information.

Three-Dimensional Pie Chart

This type of chart can be used effectively to show comparisons such as illustrated in Fig. 18-2-3. With this type of chart, two or more pie charts are drawn. Instead of a circle an ellipse is drawn and divided up in percentages. Another factor, such as quantity, is added to the chart in the form of the height of the pie chart.

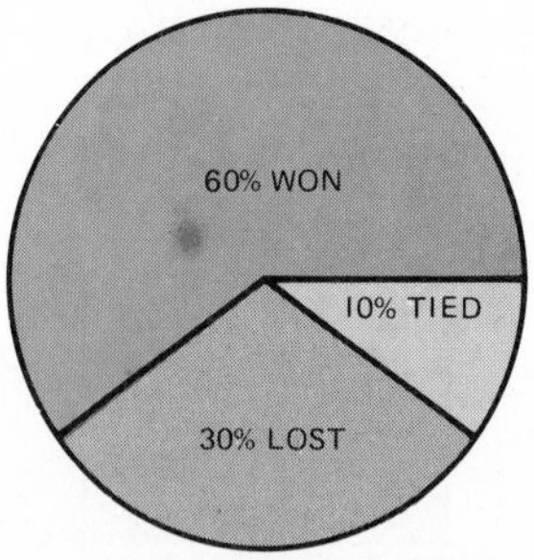
60% WON
10% TIED
30% LOST

FOOTBALL SEASON RECORD

RELATIONSHIP OF GAMES WON, LOST, AND TIED

Fig. 18-2-1 Pie chart.

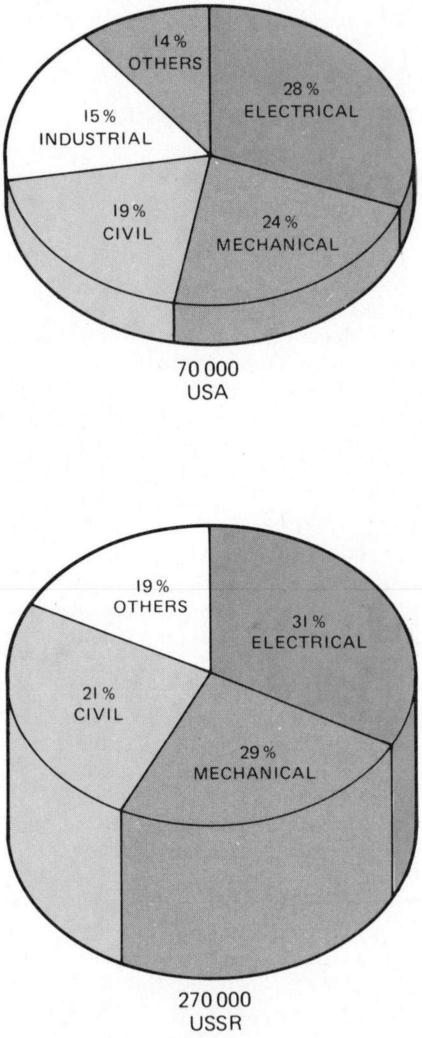
14 %
OTHERS
28 %
ELECTRICAL
15 %
INDUSTRIAL
19 %
CIVIL
24 %
MECHANICAL
70 000
USA
19 %
OTHERS
31 %
ELECTRICAL
21 %
CIVIL
29 %
MECHANICAL
270 000
USSR

COMPARISON OF GRADUATE ENGINEERS

Fig. 18-2-3 Three-dimensional pie chart.

DISTRIBUTION OF CLASS TREASURY		
ITEM	COST	%
DANCE	$ 72.00	40
PARTY	36.00	20
PICNIC	32.40	18
CLASS PLAY	21.60	12
PHOTOGRAPHS	18.00	10
TOTAL	$180.00	100 %

A

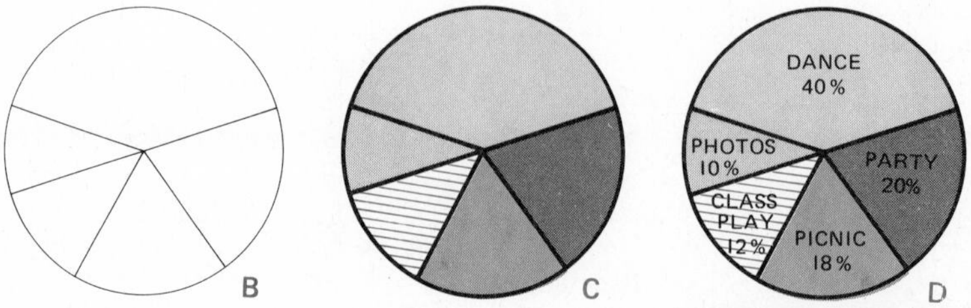
B
C
DANCE
40 %
PHOTOS
10 %
PARTY
20%
CLASS
PLAY
12 %
PICNIC
18 %
D

DISTRIBUTION OF CLASS TREASURY

Fig. 18-2-2 Steps in drawing a pie chart.

Assignments

1. On an A3- or B-size sheet, draw a pie chart to show the population distribution in the following four regions of the United States:

• Northeast	24.2 percent
• North Central	27.8 percent
• South	31.2 percent
• West	16.8 percent

Use color or crosshatching for contrast.

2. On an A3- or B-size sheet, draw a pie chart showing a breakdown of the expenditure of each dollar by the federal government. Use the following information:

• National Defense	\$0.31
• Income Security	0.27
• Interest	0.08
• Health	0.07
• Transportation and Housing	0.06
• Veterans	0.05
• Education	0.04
• Agriculture	0.03
• Others (Miscellaneous)	0.09

3. On an A3- or B-size sheet, draw three-dimensional pie charts showing the comparisons between urban and suburban accident statistics involving children. Use color or crosshatching for contrast.

Location	Urban	Suburban
At Home	25%	32%
Between Home and School	8%	12%
On School Grounds	15%	14%
In School Buildings	21%	18%
In Other Places	31%	24%
Number of Accidents (in Thousands)	84	38

4. On an A3- or B-size sheet, prepare three charts showing a breakdown on the source of each dollar received by the federal government, using the following information:

• Individual income tax	\$0.38
• Employment tax	0.26
• Corporate income tax	0.14
• Borrowing	0.10
• Excise tax	0.07
• Other (Miscellaneous)	0.05

The three charts to be drawn are a percentage bar chart, a multiple-column bar chart, and a pie chart.

UNIT 18-3

LINE CHARTS

Line charts are used most often to show *trends* or changes. For example, changes in the weather, ups and downs in sales, or trends in population growth may be plotted and shown graphically on a line chart. A line chart may contain one or several curves. A conversion chart with one curve (Fig. 18-3-1) is often convenient for changing from one value to another. This one is used to convert Fahrenheit to Celsius. Figure 18-3-2 contains four curves comparing sales in various parts of the country.

Graphic charts may also be used to solve various kinds of mathematical problems. The following problems can easily be solved by studying the chart in Fig. 18-3-3:

1. What is the normal water level in the lake?
2. What is the maximum rise?
3. When does the highest level occur?

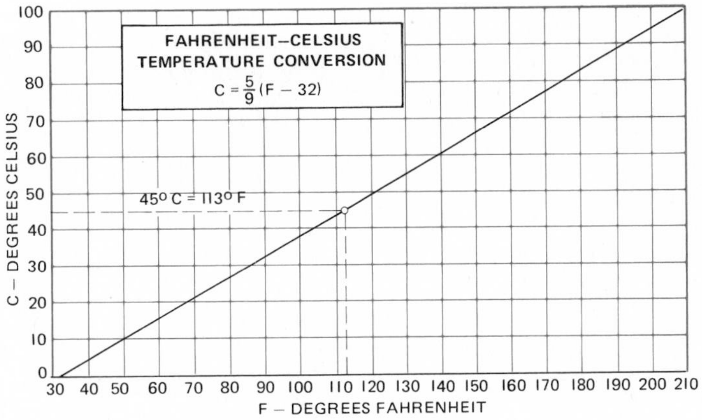
FAHRENHEIT–CELSIUS
TEMPERATURE CONVERSION
$C = \frac{5}{9}(F - 32)$
45° C = 113° F
C – DEGREES CELSIUS
0 10 20 30 40 50 60 70 80 90 100
30 40 50 60 70 80 90 100 110 120 130 140 150 160 170 180 190 200 210
F – DEGREES FAHRENHEIT

Fig. 18-3-1 A conversion chart.

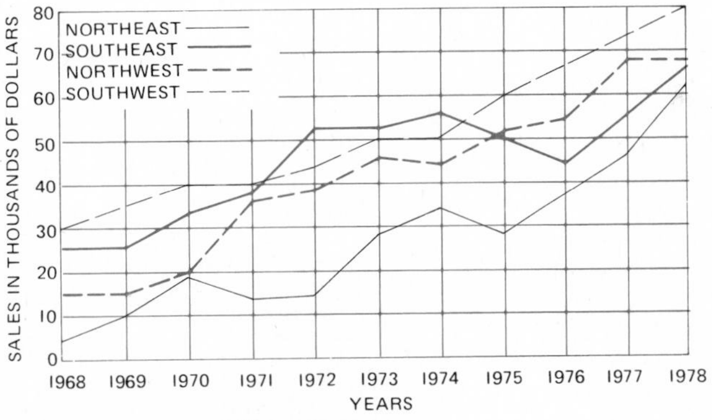
NORTHEAST
SOUTHEAST
NORTHWEST
SOUTHWEST
SALES IN THOUSANDS OF DOLLARS
0 10 20 30 40 50 60 70 80
1968 1969 1970 1971 1972 1973 1974 1975 1976 1977 1978
YEARS

Fig. 18-3-2 A multiline chart.

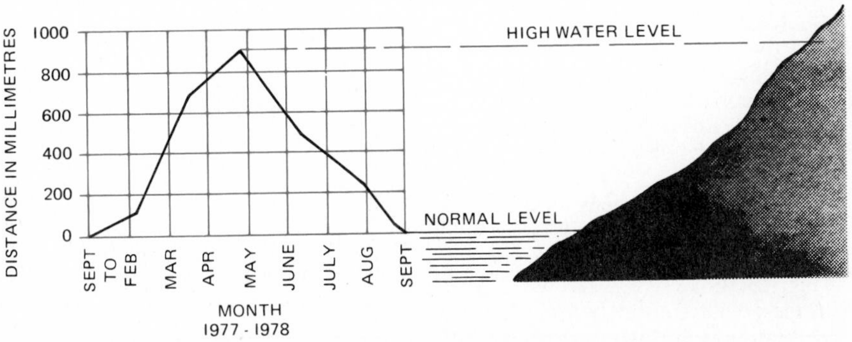
HIGH WATER LEVEL
NORMAL LEVEL
DISTANCE IN MILLIMETRES
0 200 400 600 800 1000
SEPT TO FEB
MAR
APR
MAY
JUNE
JULY
AUG
SEPT
MONTH
1977 - 1978

Fig. 18-3-3 Line chart showing rise and fall of water level on a lake.

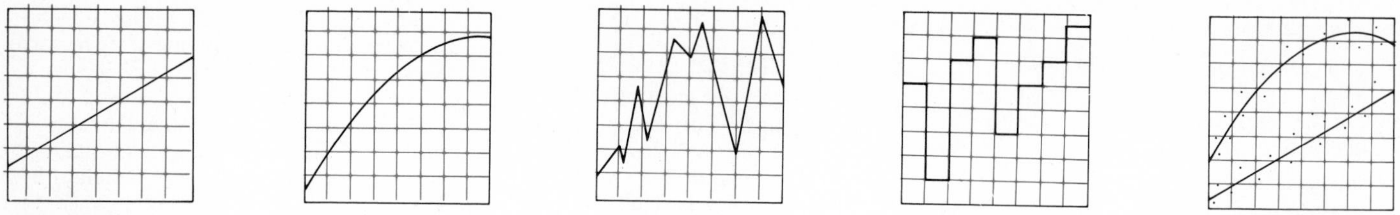
STRAIGHT LINE
CURVED LINE
BROKEN LINE
STEPPED LINE
CURVED OR STRAIGHT LINE ADJUSTED TO PLOTTED POINTS

Fig. 18-3-4 Lines on graphs may have different forms.

The line on a chart is not necessarily straight. As shown in Fig. 18-3-4, a line on a chart may be a straight line, a curved line, a broken line, a stepped line, or a straight line or curved line adjusted to plotted points.

The selection of the proper scales, the vertical and horizontal grid, is important. The vertical and horizontal scales must be such that a true pictorial impression is given by the angle of slope of the line. In Fig. 18-3-5, notice the different impressions given by the three charts. In Fig. 18-3-5*a* it appears as though there is a very abrupt change; at *b*, a normal change; at *c*, a very slow or gradual change. The scale chosen should be the one that gives the most accurate pictorial impression.

Printed grid or graph paper is available in many forms. It may be purchased with lines ruled (drawn or printed) with a variety of uniform spaces between the lines and in many other forms (Fig. 18-3-6). Graph paper may also have certain lines printed heavier to make it easier to plot points and to read the finished chart. Figure 18-3-6*b* shows every tenth line printed heavier. In this case, the heavy lines are 10 mm apart. The lines on grid paper may form squares or rectangles, as shown in Fig. 18-3-5.

To Draw a Line Chart. A line chart is made by following steps 1 through 9.

1. Prepare and list the information to be presented (Fig. 18-3-7).
2. Select plain or ready-ruled graph paper.
3. Select a suitable size and proportion to give the desired result.
4. Select the proper scale.
5. If graph paper is not used, lay off and draw thin horizontal lines, called *X* axis, or abscissa, lines, as shown in Fig. 18-3-8*a*. The intersection of *X* and *Y* is zero.
6. Lay off the scale divisions on the *X* axis and the *Y* axis (Fig. 18-3-8*b*).
7. Mark the scale values on the *X* and *Y* axes (Fig. 18-3-8*b*).
8. Plot the points accurately from the listed information (Fig. 18-3-7). It is usually better to use small, open circles, triangles, or squares rather than crosses or solid dots for plotting purposes (Fig. 18-3-8*c*).
9. Connect the points to complete the line chart (Fig. 18-3-8*d*).

If more than one curve is drawn on a chart, use different types of lines or different colors for each curve (Fig. 18-3-9). Use a full, continuous line of the brightest color for the most important curve. In general, the scales and other identifying notes, or captions, are placed below the *X* axis and to the left of the *Y*

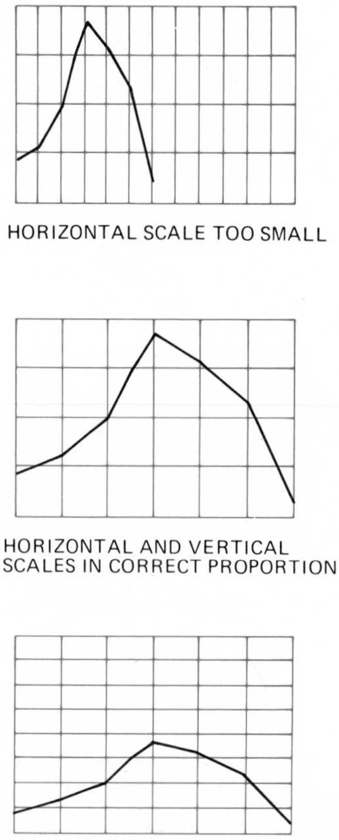
(A) HORIZONTAL SCALE TOO SMALL
(B) HORIZONTAL AND VERTICAL SCALES IN CORRECT PROPORTION
(C) VERTICAL SCALE TOO SMALL

Fig. 18-3-5 Selecting the correct scales to obtain proper proportions.

SCORING INFORMATION
EASTERN HIGH SCHOOL

Game Number	Points Scored
1	38
2	20
3	50
4	40
5	40
6	30
7	10
8	40
9	55
10	45

Fig. 18-3-7 Information to be presented in a line chart.

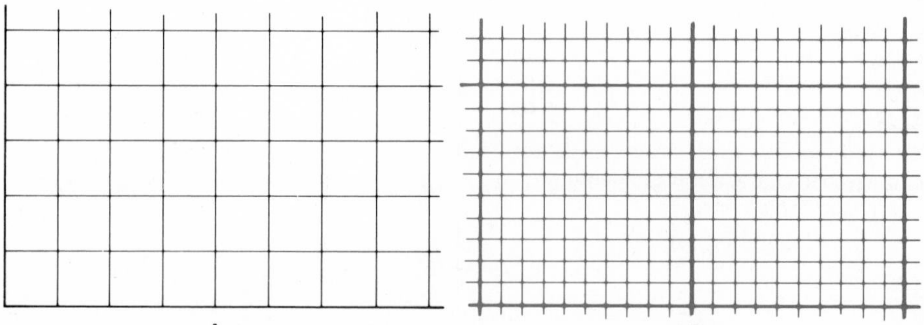
A
B

Fig. 18-3-6 Graph paper is available in many forms and colors.

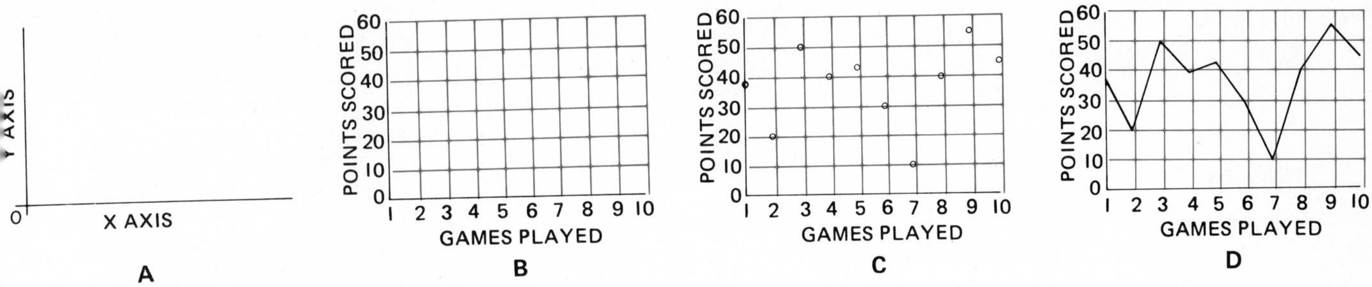

Y AXIS
X AXIS
0
POINTS SCORED
60
50
40
30
20
10
0
1 2 3 4 5 6 7 8 9 10
GAMES PLAYED
A
B
C
D

Fig. 18-3-8 Steps in drawing a line chart.

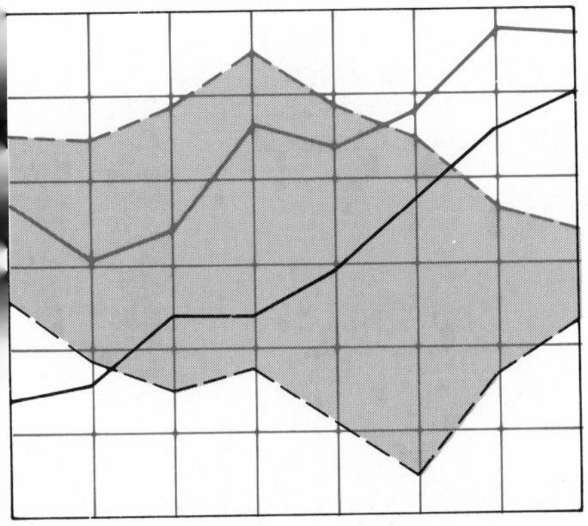

Fig. 18-3-9 The use of different types of lines on multiline charts.

axis. For large charts it is sometimes desirable to show the Y-axis scales at both the right and the left and the X-axis scales at the top and bottom for convenience in reading.

Assignments

1. On an A3- or B-size sheet, prepare a line chart to show the information in Fig. 18-3-10. Shade in the area between the quartiles.

2. On an A3- or B-size sheet, draw a line chart to show the consumption of electricity for one year from the following data:

Month	Kilowatt Hours Used
January	900
February	885
March	800
April	783
May	722
June	600
July	494
August	478
September	525
October	650
November	735
December	820

3. The average cost of meat has varied over a period of 10 years as shown in the chart at right. On an A3- or B-size sheet prepare a line chart showing the relationship of the three types of meat. Extend the chart to 1980, and assuming that the trend will remain, predict the prices of the meat in 1980. Convert the prices to cost per kilogram. Note: 1 kilogram (kg) = approximately 2.2 lb. Cost shown is per kilogram.

Year	Cost per Pound		
	Beef	Pork	Poultry
1969	$0.78	$0.70	$0.35
1970	0.80	0.80	0.40
1971	0.78	0.85	0.33
1972	0.95	0.90	0.38
1973	1.35	1.10	0.45
1974	1.40	1.28	0.48
1975	1.55	1.50	0.61
1976	1.72	1.67	0.71
1977	1.87	1.74	0.79
1978*	4.51	4.18	1.87

*Cost shown is per kilogram.

SALARIES FOR ENGINEERING TECHNOLOGISTS — 1977

Years of Experience	Average Monthly Salary in Dollars For				
				Quarterlies	
	Top 10%	Bottom 10%	Median	25% Above Median	25% Below Median
1	1490	890	1220	1360	1040
2	1470	885	1180	1340	1050
3	1500	1070	1290	1380	1160
4	1510	1110	1225	1350	1115
5	1650	1100	1300	1425	1100
6	1840	1180	1450	1630	1300
7	1925	1200	1530	1690	1330
8	1830	1180	1490	1690	1310
9	2190	1170	1500	1820	1290
10	2000	1300	1630	1870	1450
11	2090	1240	1550	1830	1350
12	2200	1320	1680	1925	1500
13	2200	1410	1730	1950	1525
14	2170	1340	1740	2030	1520
15	2330	1330	1750	1990	1535
16	2040	1300	1680	1880	1500
17	2020	1340	1600	1870	1450
18	2200	1355	1670	1980	1500
19	2420	1345	1830	2000	1580
20	2210	1315	1690	1945	1490
21	2420	1500	1845	2100	1590
22	2330	1190	1630	1990	1380
23	2430	1380	1740	1980	1530
24	2100	1250	1800	2030	1600
25	1990	1420	1730	1830	1530

Fig. 18-3-10 Salaries for engineering technologies—1977.

UNIT 18-4
OTHER CHARTS

Flowcharts

A flowchart may show the path or series of operations in manufacturing a product or in producing a material, such as the flowchart of steelmaking (Fig. 18-4-1).

Organization Charts

Organization charts are of many kinds but have the features of a flowchart. Figure 18-4-2 is an example. It shows the path, or flow, of drawings from the top engineer to the shop. It also shows the organization of the drafting department.

Engineering Charts

Experimental data may be plotted from tests and used to obtain an unknown value. In Fig. 18-4-3 the results of tests have been plotted, and they form a straight-line curve when drawn in an adjusted position. (Ω is the Greek letter omega.) Values can be taken from two points and inserted in the formula to obtain and check the value of the unknown resistance. Notice that tests were made on two occasions, and the results were plotted on the chart. A straight-line curve was drawn along the center of the path made by the dots.

Nomograms are charts that show the solutions to problems containing three or more variables (kinds of information).

Figure 18-4-4 is an example of this type of chart. A straightedge from values on the outside scales will cross the inside scale, where the solution to the equation may be read. Nomography is a special division of chart construction which requires more than simple mathematics.

Pictorial Charts, or Pictographs

Pictorial charts, or pictographs, are, in effect, bar charts which use pictures or symbols in place of bars. Figure 18-4-5 illustrates a pictorial chart. Each figure represents 100 people; it could represent 1000 or any assigned number. The chart is then drawn in the same manner as a

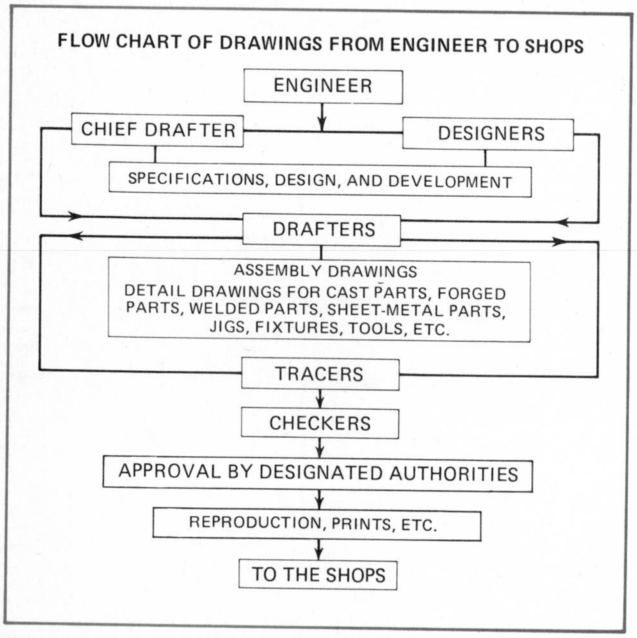

FLOW CHART OF DRAWINGS FROM ENGINEER TO SHOPS
ENGINEER
CHIEF DRAFTER
DESIGNERS
SPECIFICATIONS, DESIGN, AND DEVELOPMENT
DRAFTERS
ASSEMBLY DRAWINGS
DETAIL DRAWINGS FOR CAST PARTS, FORGED PARTS, WELDED PARTS, SHEET-METAL PARTS, JIGS, FIXTURES, TOOLS, ETC.
TRACERS
CHECKERS
APPROVAL BY DESIGNATED AUTHORITIES
REPRODUCTION, PRINTS, ETC.
TO THE SHOPS

Fig. 18-4-1 An organization flow chart.

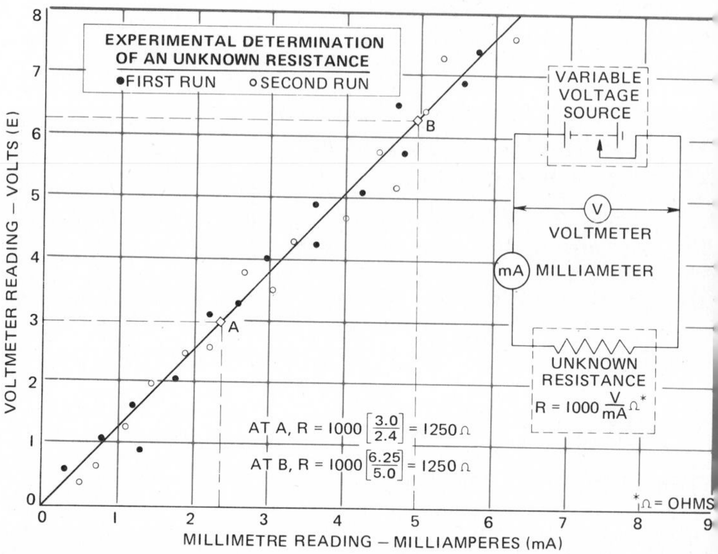

EXPERIMENTAL DETERMINATION OF AN UNKNOWN RESISTANCE
•FIRST RUN ○SECOND RUN
VARIABLE VOLTAGE SOURCE
VOLTMETER
MILLIAMETER
UNKNOWN RESISTANCE
R = 1000 V/mA Ω*
AT A, R = 1000 [3.0/2.4] = 1250 Ω
AT B, R = 1000 [6.25/5.0] = 1250 Ω
*Ω = OHMS
VOLTMETER READING – VOLTS (E)
MILLIMETRE READING – MILLIAMPERES (mA)

Fig. 18-4-3 An engineering test chart.

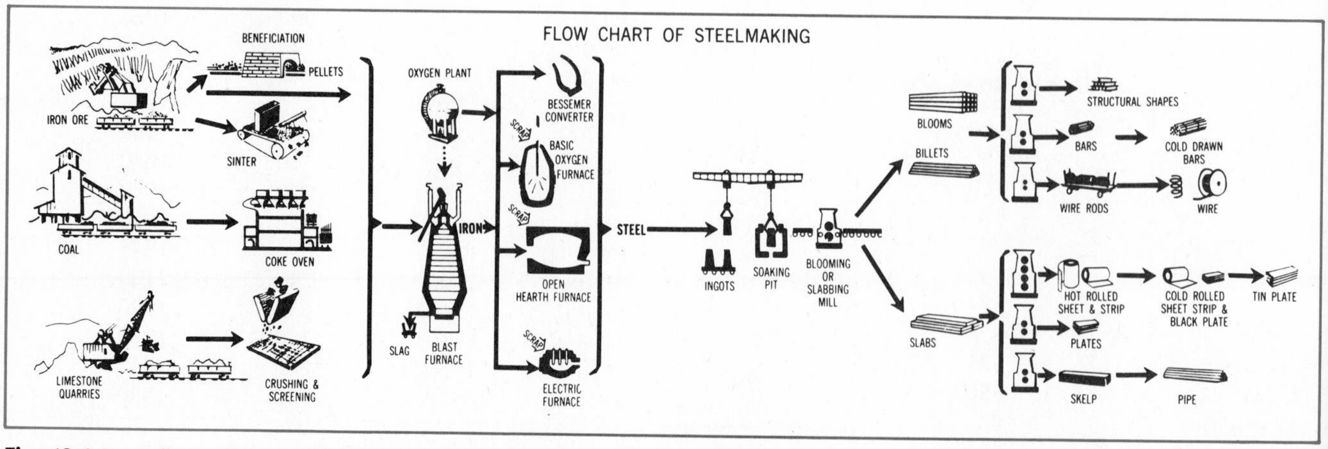

FLOW CHART OF STEELMAKING
BENEFICIATION
PELLETS
IRON ORE
SINTER
COAL
COKE OVEN
LIMESTONE QUARRIES
CRUSHING & SCREENING
OXYGEN PLANT
IRON
SLAG
BLAST FURNACE
BESSEMER CONVERTER
BASIC OXYGEN FURNACE
OPEN HEARTH FURNACE
ELECTRIC FURNACE
SCRAP
STEEL
INGOTS
SOAKING PIT
BLOOMING OR SLABBING MILL
BLOOMS
BILLETS
SLABS
STRUCTURAL SHAPES
BARS
COLD DRAWN BARS
WIRE RODS
WIRE
HOT ROLLED SHEET & STRIP
COLD ROLLED SHEET STRIP & BLACK PLATE
TIN PLATE
PLATES
SKELP
PIPE

Fig. 18-4-2 A flow chart of steelmaking. (American Iron and Steel Institute)

multiple-column bar chart in which a different individual is used for each column. Adhesive symbols are available in many forms (Fig. 18-4-6). These make pictorial charts easy to construct and understand.

Assignments

1. On an A3- or B-size sheet, prepare a flowchart showing the steps taken when a cast-iron part is manufactured and designed by a manufacturing firm for a customer.

2. On an A3- or B-size sheet, prepare an organization chart showing the administrative organization of your school. Consult with your instructor for details.

3. On an A3- or B-size sheet, make a pictorial chart showing the enrollment of technical programs in your school. Your instructor will supply the information.

4. The number of cars on the highways is steadily increasing. On an A3- or B-size sheet, prepare a pictorial chart to show the growth and anticipated increase from the following data:

Year	Cars per Kilometre of Road
1930	9
1940	11
1950	15
1960	20
1970	26
1980	38

5. A local manufacturer has requested that your firm design a set of fuel tanks. The diameters selected are 1000, 1500, 2000, 2500, and 3000 mm. The tank capacities are 0.5, 1, 2, 5, 10, and 20 km. On an A3- or B-size sheet, prepare a table from the nomogram shown in Fig. 18-4-4 and the data given above listing 25 different sizes of tanks that you would recommend. Height sizes are to be designed in multiples of 100 mm; the tank capacities may be slightly larger, but not less, than the capacities shown.

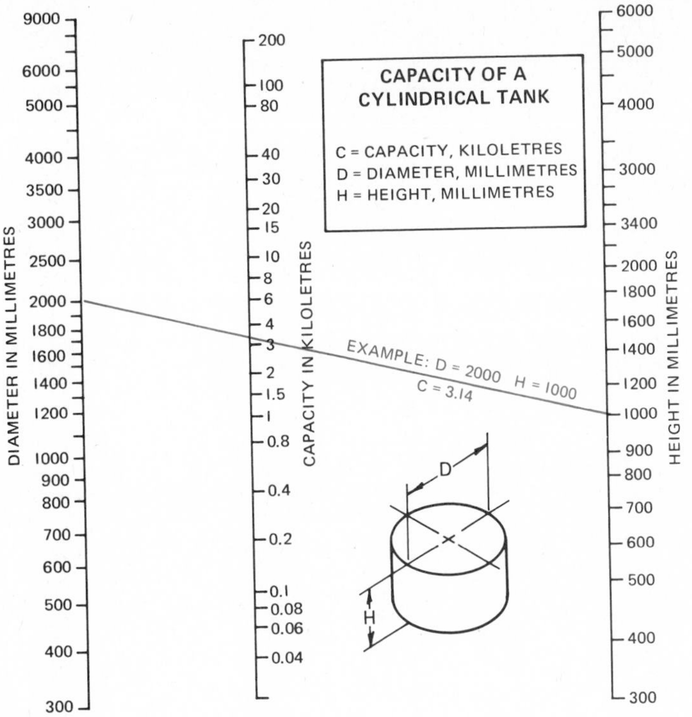
CAPACITY OF A CYLINDRICAL TANK
C = CAPACITY, KILOLETRES
D = DIAMETER, MILLIMETRES
H = HEIGHT, MILLIMETRES
DIAMETER IN MILLIMETRES
CAPACITY IN KILOLETRES
HEIGHT IN MILLIMETRES
EXAMPLE: D = 2000 H = 1000
C = 3.14
D
H

Fig. 18-4-4 **A nomogram.**

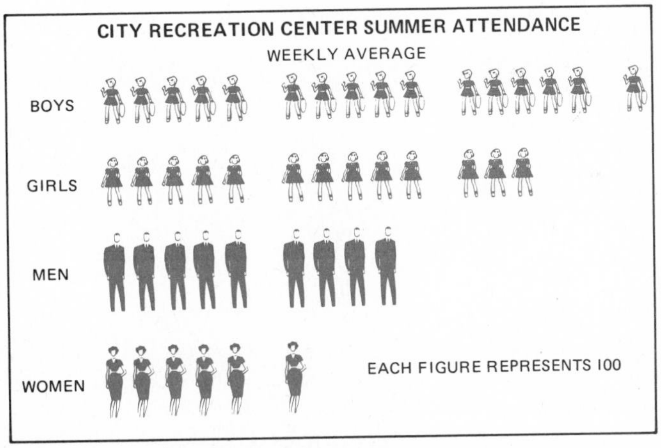
CITY RECREATION CENTER SUMMER ATTENDANCE
WEEKLY AVERAGE
BOYS
GIRLS
MEN
WOMEN
EACH FIGURE REPRESENTS 100

Fig. 18-4-5 **A pictorial graphic chart.**

Fig. 18-4-6 **Many styles of adhesive symbols are available for use on graphic charts.** (Chartpak, Inc.)

Part 4
Power Transmissions

Chapter 19
Belts, Chains, and Gears

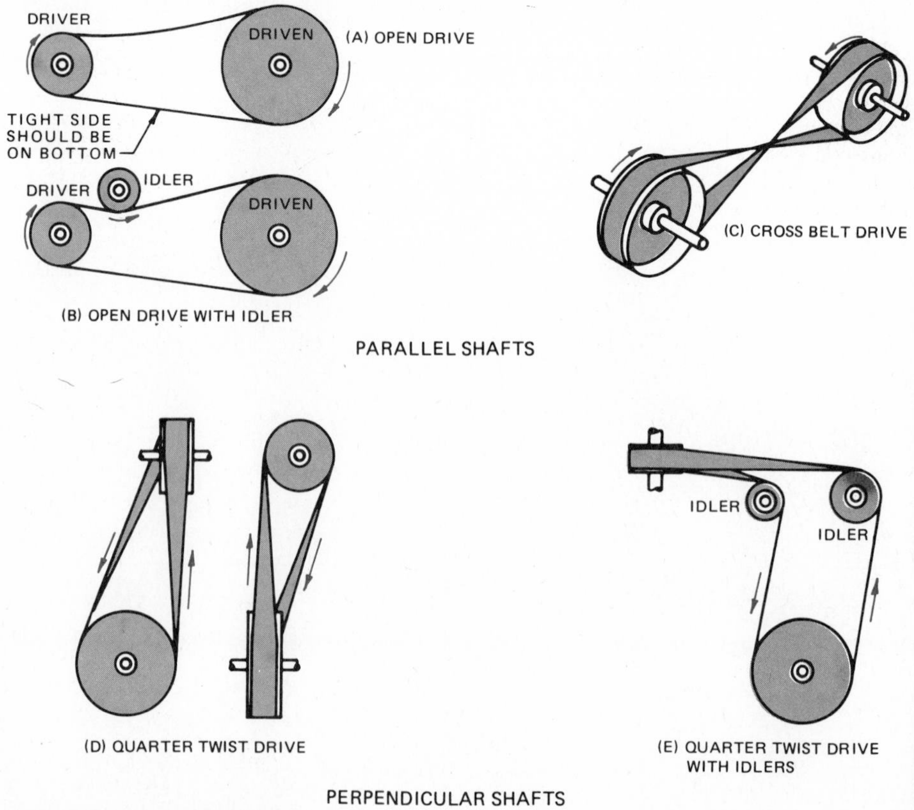
DRIVER
DRIVEN
(A) OPEN DRIVE
TIGHT SIDE
SHOULD BE
ON BOTTOM
DRIVER
IDLER
DRIVEN
(B) OPEN DRIVE WITH IDLER
(C) CROSS BELT DRIVE
PARALLEL SHAFTS
(D) QUARTER TWIST DRIVE
IDLER
IDLER
(E) QUARTER TWIST DRIVE
WITH IDLERS
PERPENDICULAR SHAFTS

Fig. 19-1-1 Flat-belt drives.

UNIT 19-1

BELT DRIVES

Flat Belts[1]

Flat-belt drives offer flexibility, shock absorption, efficient power transmission at high speeds, resistance to abrasive atmospheres, and comparatively low cost. The belts can operate on relatively small pulleys and can be spliced or connected for endless operation. However, because they require high tension, they also impose high bearing loads. They are sometimes noisier than other belt drives, will slip, and have comparatively low efficiency at moderate speeds. See Fig. 19-1-1.

Flat belts for power transmission can be broken down into three classes:

1. *Conventional*: plain flat belt without teeth, grooves, or serrations.
2. *Grooved or serrated*: basic flat belt modified to provide the advantages of another type of transmission product, such as V-belts.
3. *Positive drive*: basic flat belt modified to eliminate the need for frictional force for power transmission.

Conventional belts are available in two types: *reinforced*, which utilizes a tensile member to obtain strength, and *nonreinforced*, which depends upon the tensile strength of the basic material for its strength.

Longitudinally grooved or serrated belts use a flat belt as the tensile section and a series of adjacent V-shaped grooves as compression section and tracking means. These are generally known as *poly-V belts* (Fig. 19-1-2).

Positive-drive belts use a flat belt as the tensile section and a series of evenly spaced teeth on the bottom surface.

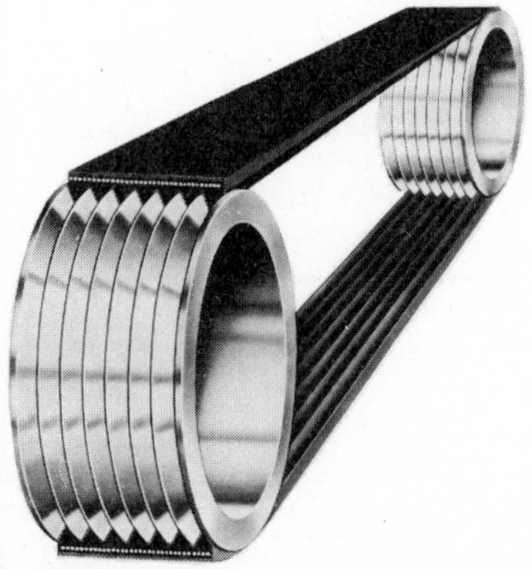

Fig. 19-1-2 Grooved or poly-V belt. (Raybestos-Manhattan [Canada])

These teeth engage a similarly grooved pulley to achieve positive mesh. These are also known as *timing belts*. See Fig. 19-1-3.

Fig. 19-1-3 Timing belt. (Morse Chain Co.)

CONVENTIONAL FLAT BELTS

Conventional flat belts are available either as endless belts or as belting which can be spliced or fastened to length. An endless belt is best, of course, since it has no weak points caused by splices or connectors, and will usually operate much more smoothly. Either a vulcanized splice or mechanical fastener may be used.

Conventional belts are normally available in five basic materials:

1. Leather
2. Rubberized fabric or cord
3. Nonreinforced rubber or plastic
4. Reinforced leather
5. Fabric

Leather. Most leather belts are made of plies of belting bonded together. They provide excellent coefficient of friction, flexibility, long life, and are easily repaired. On the other hand, the initial cost is high, they must be cleaned, and they require belt dressing. They also stretch and shrink, depending on atmospheric conditions.

This type of belt is used primarily for slow to moderate speeds, with a maximum of 30 metres per second (m/s); for medium to heavy loads, up to 375 kilowatts (kW), ratios of 16:1 are normally possible. It has good shock-absorbing qualities.

Rubberized Fabric or Cord. Many types and grains of rubberized belting are presently available. Almost all are moisture-, acid- and alkali-resistant. They consist of rubber-impregnated fabric or cord.

Rubberized Fabric. This is the least expensive type of flat belting. See Fig. 19-1-4. It is made up of plies of cotton or

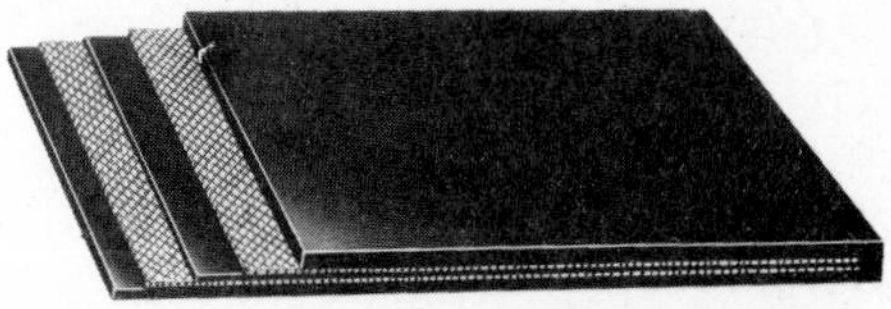

Fig. 19-1-4 Flat belt—rubberized-fabric type. (American Beltrite Rubber Co.)

synthetic duck, impregnated with rubber. Rubberized-fabric belts transmit less power per millimetre of width for the same thickness and have a shorter life than leather belts.

Rubberized Cord. These belts consist of a series of plies of rubber-impregnated cords. They offer high tensile strength for a modest size and mass. Such belts are generally furnished in widths from about 10 to 330 mm and in two to four plies. They are available only as endless belts and are not designed to be used with fasteners.

Nonreinforced Rubber or Plastic. For light-duty applications, flat belts are available in a number of unreinforced materials.

Rubber. Basically a simple strip of rubber, these belts are available in various compounds. They are designed specifically for low-kilowatt, low-speed drives. They are especially useful for fixed-center drives since they can be simply stretched into place over their pulleys.

Plastic. Unreinforced plastic belts transmit higher kilowatts than rubber belts. They are available in a number of plastic compounds.

Reinforced Leather. These belts consist of a plastic tensile member, generally reorientated nylon, and leather top and bottom layers. This combination offers the long life, high coefficient of friction, and flexibility of leather without excessive stretch.

Most of these belts are used where ordinary leather belts cannot handle the high kilowatts to be transmitted. Like leather, these belts can be spliced.

Fabric. These all-fabric belts may consist of a single piece of cotton or duck folded and sewn with rows of longitudinal stitches. Others are woven endless.

Fabric belts are either made plain or treated with a chemical or rubber solution to improve the coefficient of friction.

The major advantage of all-fabric belts is their ability to track uniformly and to operate at high speeds. Capacity depends on the number of plies of fabric, size of thread, and belt width. They are used typically in check-sorting machines.

GROOVED BELTS

This is basically a flat belt with a longitudinally ribbed underside. The *flat* belt section serves as the load-carrying component, and the ribs provide traction in the sheave grooves.

The belt, although it bears a resemblance to the conventional V-belt, operates on a different principle. Rather than depending on wedging action to transmit power, it depends solely on friction between sheave and belt. Power capacity depends on belt width; only a single belt, with a varying number of ribs, is used for each drive. Tension is somewhat greater than for conventional V-belts, but less than for conventional flat belts.

The ribbed belt has high efficiency when used on small sheaves and can be used on vertical and turned drives.

POSITIVE-DRIVE BELTS

Another variation of the flat belt is the positive-drive belt, commonly known as the *timing belt*. Basically a flat belt with a series of evenly spaced teeth on the inside circumference, it combines the advantages of the flat belt with the positive-grip features of chain and gears. The belt has high-strength steel or glass tensile members, with nylon-jacketed neoprene teeth. The belts are available in five stock pitches and in various widths. Special widths and pitches are available.

Positive-drive belts have many advantages. There is no slippage or speed variation, and a wide range of speed ratios is possible. Required belt tension is minimal, so that bearing loads are low. They are ideally suited for high-kilowatt fixed-center drives and can be used at belt speeds up to 80 m/s. They are especially recommended for drives where high mechanical efficiency, constant angular velocity, or positive synchronization is desired.

These belts are not recommended where pulleys are misaligned. Also, high-speed operation may cause some noise, but this is not generally a problem at normal operating speeds.

PULLEYS FOR FLAT BELTS

Different types of pulleys are used for flat, ribbed, and positive-drive belts.

Flat-Belt Pulleys. These are generally made of cast iron. However, they are also available in steel and in various rim and hub combinations. They may have solid, spoked, or split hubs as well as other modifications of the basic pulley.

Crowning. All power-transmission pulleys should be crowned or flanged. Crowned pulleys are more effective than flanged pulleys for flat belts, provided proper crowning is used. To prevent concentration of stresses in a narrow section of belt width, crown should be limited. The degree of crown varies with the application, but maximum recommended slope is 1:100 on the pulley face. See Fig. 19-1-5.

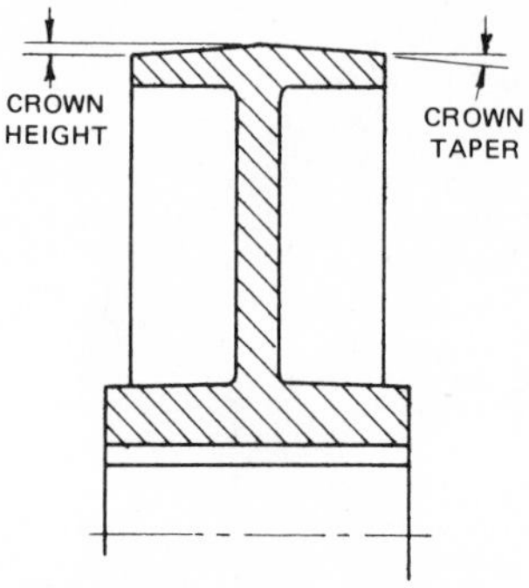
CROWN
HEIGHT
CROWN
TAPER

Fig. 19-1-5 Crown on pulley.

Other Types. Pulleys for ribbed and positive-drive belts are available in a variety of stock sizes and widths. Special sizes are also available. These pulleys must have the same groove dimension or pitch dimensions as the belt being used.

At least one pulley in a timing-belt drive must be flanged in order to keep the belt on the drive. For long-center drives, flanging both pulleys is recommended but not required. Idler pulleys should not be crowned. No flanging or crowning is required for ribbed belts.

V-Belts[2]

V-belts are presently available in a wide variety of standardized sizes and types, for transmitting almost any amount of kilowatts. In addition, special constructions are available for specific applications.

Normally, V-belt drives operate best at belt speeds between 8 and 30 m/s. For standard belts, ideal (peak capacity) speed is approximately 23 m/s. Narrow V-belts, however, will operate up to 50 m/s. A summary of belt characteristics is given in Fig. 19-1-6. For most drive applications, the maximum satisfactory speed ratio is approximately 7:1.

The efficiency of a V-belt ranges from 90 to 98 percent, with a generally accepted average of 95 percent. However, the more belts used on a single drive and the higher the speed ratio, the lower the efficiency.

Advantages. V-belt drives permit large speed ratios and provide long life (3 to 5 years). They are easily installed and removed, quiet, and low in maintenance; and they provide shock absorption between driver and driven shafts.

Limitations. Because they are subject to a certain amount of creep and slip, V-belts should not be used where synchronous speeds are required, such as in timing applications. Improper belt tensioning and mismatching of belt lengths can reduce service life. Belt life in temperatures above 80°C and below −50°C is significantly shortened. Centrifugal force prevents the use of V-belts at speeds above 50 m/s. At speeds below 5 m/s they are usually not economical.

STANDARD DIMENSIONS

To facilitate interchangeability and to ensure uniformity, V-belt manufacturers have developed industrial standards for the various types of V-belts. Most of these standards were first set by the Rubber Manufacturers Association (RMA) and the Mechanical Power Transmission Association (MPTA) and have been issued as American National Standards Institute (ANSI) standards. For the three major areas of application—

Type of Belt	Maximum Power (kilowatts)	Maximum Speed (m/s)	Belt Speed for Max. Power (m/s)	Max. Speed Ratio	Shock Absorption
Constant-Speed					
Light duty	5.6	25	18	8	Poor
Standard	260	30	23	7	Good
Super	375	30	25	7	Very Good
Cogged	375	30	25	8	Very Good
Steel cable	375	40	25	7	Poor
Narrow	200	50	38	7	Very Good
Variable-Speed					
Conventional	225	30	—	—	Good
Wide-range	55	30	—	—	Good

Fig. 19-1-6 V-belt characteristics.

*Stock items. Drives available to 1100 kilowatts.

industrial, automotive, and agricultural—the standards cover, among other factors, belt and sheave dimensions and the numerical factors and pertinent equations necessary to design the drive.

Cross Sections. Industrial and agricultural V-belts are always made to standard cross sections. This is not true in automotive applications, where a number of special sizes are produced. However, even in this case most cross sections are standard. Because of differences in construction and manufacturing methods, cross-section shapes, dimensions, and the included angle between the sidewall may differ slightly with different manufacturers. However, all standard cross sections will operate interchangeably in standard grooves. See Fig. 19-1-7.

Industrial. These are made in two types: heavy duty (conventional, narrow) and light duty. Conventional belts are available in A, B, C, D, and E sections. See Fig. 19-1-8. Narrow belts are made in 3V, 5V, and 8V sections. Light-duty (less than 0.7 kW) belts come in 2L, 3L, 4L, and 5L sections. Conventional sections are also available in the double-V form for use where a grooved sheave must ride on the back of the belt. These are available in AA, BB, CC, and DD sections.

Open-end belting is available in A, B, C, and D sections. Link-V belting, which is not covered by a standard, is made in A, B, C, D, and E sections, and in some sizes for small kilowatt applications.

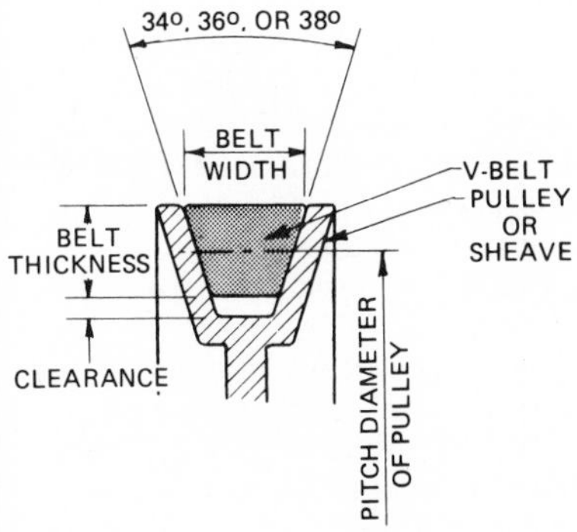
34°, 36°, OR 38°
BELT WIDTH
V-BELT PULLEY OR SHEAVE
BELT THICKNESS
CLEARANCE
PITCH DIAMETER OF PULLEY

Fig. 19-1-7 V-belt and pulley.

Wide-range V-belts, used for variable-speed drives, are available in Q, P, R, T, and W sections.

Agricultural. These belts are made in the same sections as conventional belts. They are designated as HA, HB, HC, HD, and HE; in double-V sections HAA, HBB, HCC, and HDD are available. Agricultural belts differ from industrial belts mainly in construction.

Automotive. Belts for automotive applications are made in six SAE-designated cross sections identified by the nominal top widths 10, 12, 17, 19, 22, and 25 mm.

Length. Although endless V-belts can be manufactured in any length within a fairly wide range, manufacturers have standardized on certain lengths which are produced for stock. These lengths will not necessarily represent the only sizes considered as stock by a manufacturer.

BELT-SIZE DESIGNATION

For the different types of V-belts, the same basic method is used to designate belt size. In some cases, this method is specified in the standard. Basically, the belt-size specification consists of the symbol for the cross section accompanied by the accepted length designation.

The following are examples for each type of belt.

For a conventional V-belt, a B230 belt has a B cross section and a 2300 mm standard length designation. For a narrow belt, 5V355 indicates a 5V cross section and a belt with a 3550 mm effective outside length. For a light-duty V-belt, a 2L203 belt has a 2L cross section and an effective outside length of 2030 mm. For a wide-range belt, 3030V93 indicates a 30-mm belt; 30 (the two numbers preceding the V) is the sheave groove angle in degrees, V denotes a wide-range belt, 93 is the belt length, 930 mm. There is no standard method for designating an SAE automotive belt or an agricultural belt.

Basically, a V-belt consists of five component sections (Fig. 19-1-9):

1. Tensile members or load-carrying section
2. Low-durometer cushion section surrounding tensile members

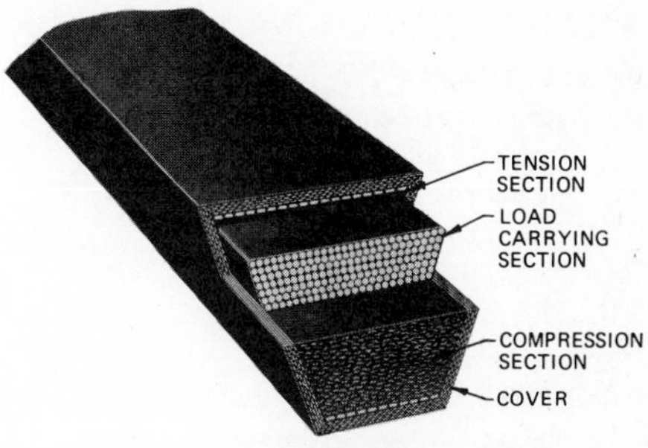
TENSION SECTION
LOAD CARRYING SECTION
COMPRESSION SECTION
COVER

Fig. 19-1-9 Basic V-belt construction. (American Beltrite Rubber Co.)

3. Flexible top section
4. Bottom compression section
5. Cover or jacket

SHEAVES AND HUBS

Most sheaves (the grooved wheels of pulleys) are made of cast iron, which is economical and stable and which provides long groove life. For light duty, sheaves may be of formed steel, cast iron, plastic, or die-cast. Formed-steel sheaves are used primarily in automotive and agricultural applications. For special applications they may be made of steel or aluminum alloy. Typical V-belt applications are shown in Figs. 19-1-10 and 19-1-11.

Cast-iron sheaves are generally limited to 33 m/s rim speeds. For speeds up to 50 m/s, aluminum, steel, and ductile iron are used.

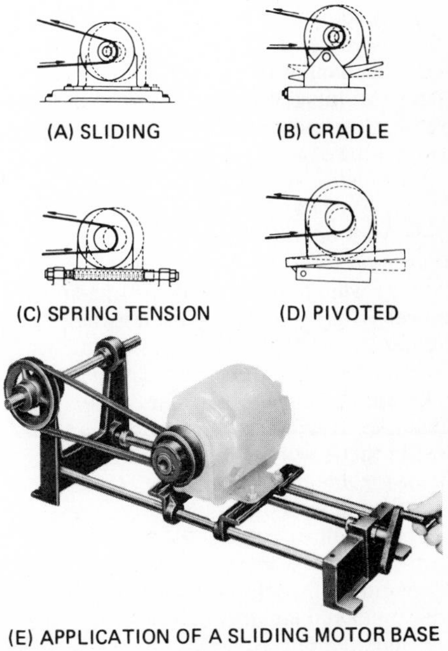
(A) SLIDING
(B) CRADLE
(C) SPRING TENSION
(D) PIVOTED
(E) APPLICATION OF A SLIDING MOTOR BASE

Fig. 19-1-10 Common types of motor bases. (T. B. Wood's Sons Co.)

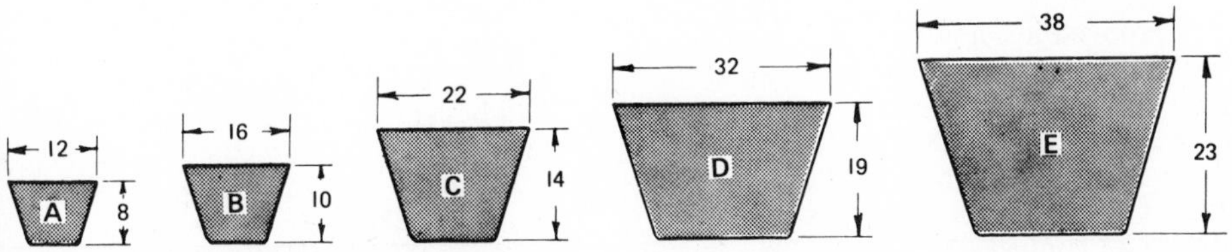
12
8
A
16
10
B
22
14
C
32
19
D
38
23
E

Fig. 19-1-8 Industrial V-belts.

(A) SINGLE PULLEY

(B) DOUBLE PULLEY

(C) SINGLE DRIVE

(D) MULTIPLE DRIVE

Fig. 19-1-11 Single- and multiple-belt drives.

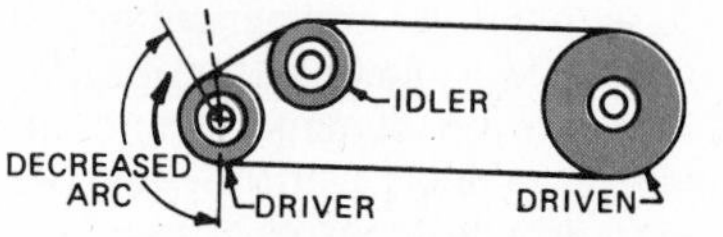
DECREASED ARC
IDLER
DRIVER
DRIVEN

(A) INSIDE IDLER PULLEY, AT LEAST AS LARGE AS THE SMALL SHEAVE, ON THE SLACK SIDE OF THE DRIVE

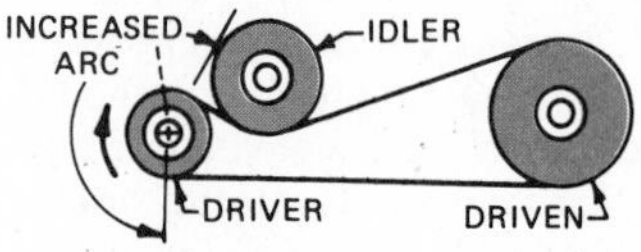
INCREASED ARC
IDLER
DRIVER
DRIVEN

(B) OUTSIDE IDLER PULLEY, AT LEAST 1.3 LARGER THAN THE SMALL SHEAVE

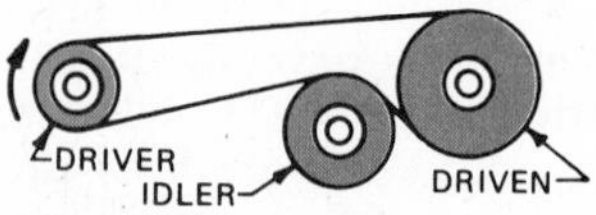
DRIVER
IDLER
DRIVEN

(C) OUTSIDE IDLER PULLEY ON THE TIGHT SIDE OF THE DRIVE

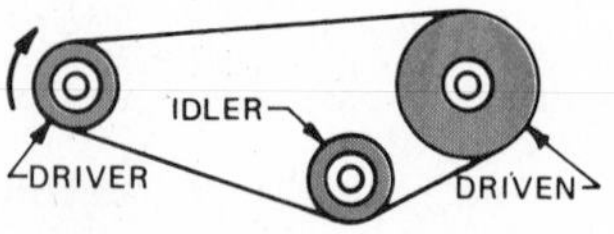
IDLER
DRIVER
DRIVEN

(D) INSIDE IDLER PULLEY ON THE TIGHT SIDE OF THE DRIVE

Fig. 19-1-12 Location of idler pulleys.

Sheaves are made with either regular or deep grooves. A deep-groove sheave is generally used when the V-belt enters the sheave at an angle, for example, in a quarter-turn drive, on vertical shaft drives, or wherever belt vibration may be a problem.

Formed-steel sheaves usually have an integral hub, but are sometimes available with removable bushings for various bore sizes. Multiple-V sheaves are generally stocked with detachable tapered bushings to fit most shaft sizes and to permit ease of installation and removal.

THE USE OF IDLER PULLEYS[3]

Idler pulleys are grooved sheaves or flat pulleys which do not serve to transmit power. Usually they are used as belt tighteners when it is not possible to move either shaft for belt installation and take-up, as between two line shafts, for example. It is better, and more economical in the long run, to provide movement for one of the shafts where it is at all possible to do so, rather than use idlers as belt tighteners. However, if idlers must be used, they are perfectly acceptable for multiple V-belt drives. Idler pulleys may also be used when the belt must be passed around some obstruction. Such a pulley may be run inside the drive or on the outside. An inside idler may be either a grooved sheave or a flat, uncrowned pulley, but an outside idler must be flat and without any crown.

An inside idler pulley invariably decreases the arc of contact of the belts on each loaded sheave of the drive. It should be at least as large as the small loaded sheave and located, preferably, on the slack slide of the drive. See Fig. 19-1-12*a*.

An outside idler pulley invariably increases the arc of contact of the belts on each loaded sheave of the drive, but the amount of take-up that can be obtained is definitely limited by the belts as they run on the opposite side. If you are designing a V-flat drive and cannot provide movement for one of the shafts for belt installations and take-up, an outside idler pulley should be used. It should be at least one-third larger than the small loaded sheave and located, preferably, on the slack side of the drive. See Fig. 19-1-12*b*.

A grooved inside idler pulley may be located at any point along the span of belts, preferably on the slack side of the drive. Since it will decrease the arc of contact of the belts on each loaded sheave, it can be located so as to make the arc of contact more nearly the same on each sheave.

A flat idler pulley, whether used inside or outside the drive, should be located as close as possible to the place where the belts leave the sheave. On the slack side of the drive, which is the preferred location, this means as close as possible to the driver sheave. See Fig. 19-1-12*a* and *b*. On the tight side of the drive, this means as close as possible to the driven sheave. See Fig. 19-1-12*c* and *d*.

A spring-loaded or weighted idler pulley is sometimes used to provide take-up and to maintain the necessary operating tension in the belts when the drive is running under normal load. This type of idler should be located on the slack side of the drive. It should not be used on a drive where the load may be reversed, that is, where the slack side may temporarily become the tight side.

A drive that uses an idler pulley should be laid out to scale, the extreme installation and take-up positions established, and the belts measured in each position to make sure that sufficient take-up for the length of the belts is available with the chosen arrangement.

How to Select a Light-Duty V-Belt Drive[4]

The proper selection of V-belt drives for light machinery has been simplified and condensed into three steps. Complete selection involves the proper choice of:

1. V-pulley size for driver shaft and belt cross section
2. V-pulley size for driven shaft
3. Belt length for required center distance

Proper duty classification helps to ensure maximum drive life. The following are typical duty classifications:

- **Light Duty**
  Household washers
  Household ironers
  Dish washers
  Fans and blowers
  Centrifugal pumps
- **Normal Duty**
  Oil burners
  Buffers
  Heating and ventilating fans
  Meat slicers
  Speed-up drives
  Drill presses
  Generators
  Power lawn mowers
- **Heavy Duty**
  Gasoline engine drives
  Metalworking machines
  Sanding machines
  Stokers
  Spray equipment
  Woodworking machines
  Lathes
  Industrial machines
  Refrigerators
  Compressors
  Piston or plunger pumps
  Grinders

The kilowatt ratings listed in Fig. 19-1-13 are suitable for normal-duty applications. For light duty, multiply the normal-duty kilowatt rating by 1.2. For heavy duty, multiply the normal kilowatt rating by 0.85.

KILOWATT RATINGS

Rev/min of Small Pulley	Outside Diameter of Small V-Pulley—Millimetres														
	38	**44**	**51**	**57**	**64**	**70**	**76**	**83**	**89**	**95**	**102**	**108**	**114**	**121**	**127**
200											0.13	0.16	0.18	0.21	0.22
400			0.04	0.06	0.09	0.11	0.13	0.16	0.19	0.23	0.26	0.31	0.34	0.39	0.42
600	0.03	0.05	0.06	0.09	0.13	0.16	0.20	0.24	0.27	0.33	0.38	0.43	0.49	0.54	0.60
800	0.04	0.06	0.08	0.11	0.16	0.21	0.25	0.31	0.34	0.41	0.48	0.55	0.60	0.69	0.75
1000	0.04	0.07	0.09	0.13	0.19	0.25	0.31	0.36	0.41	0.48	0.56	0.64	0.74	0.82	0.90
1160	0.05	0.08	0.11	0.16	0.22	0.28	0.34	0.40	0.46	0.51	0.63	0.73	0.80	0.92	1.01
1400	0.06	0.09	0.13	0.17	0.25	0.32	0.40	0.48	0.55	0.63	0.72	0.82	0.93	1.06	1.16
1600	0.06	0.10	0.14	0.19	0.27	0.36	0.43	0.51	0.60	0.67	0.76	0.90	1.01	1.14	1.25
1750	0.06	0.11	0.15	0.19	0.28	0.38	0.47	0.55	0.63	0.72	0.81	0.93	1.07	1.20	1.33
2000	0.07	0.12	0.16	0.21	0.31	0.41	0.51	0.60	0.69	0.78	0.87	1.01	1.15	1.29	1.42
2200	0.07	0.13	0.18	0.23	0.33	0.43	0.54	0.64	0.74	0.84	0.94	1.05	1.20	1.34	1.48
2400	0.07	0.13	0.19	0.24	0.34	0.46	0.57	0.68	0.78	0.89	0.98	1.08	1.23	1.39	1.51
2600	0.07	0.14	0.19	0.26	0.35	0.48	0.59	0.72	0.81	0.93	1.03	1.10	1.26	1.41	1.56
2800	0.08	0.14	0.21	0.27	0.36	0.49	0.62	0.74	0.85	0.95	1.06	1.10	1.28	1.42	1.57
3000	0.08	0.16	0.22	0.29	0.37	0.51	0.63	0.76	0.88	0.98	1.09	1.10	1.26	1.41	1.55
3200	0.08	0.16	0.22	0.29	0.38	0.52	0.66	0.78	0.90	1.01	1.12	1.12	1.25	1.39	1.51
3450	0.09	0.16	0.24	0.31	0.38	0.53	0.67	0.80	0.92	1.03	1.13	1.13	1.20	1.33	1.45
3600	0.09	0.16	0.25	0.31	0.39	0.54	0.68	0.81	0.93	1.04	1.15	1.15	1.15	1.28	1.38
3800	0.09	0.16	0.25	0.31	0.39	0.54	0.69	0.81	0.93	1.05	1.15	1.15	1.15	1.19	1.28
4000	0.09	0.16	0.25	0.33	0.40	0.54	0.69	0.82	0.94	1.05	1.15	1.15	1.13	1.13	1.16

FOR [light shaded] BACKGROUND USE A — FOR [white] BACKGROUND USE A — FOR [dark shaded] BACKGROUND USE A

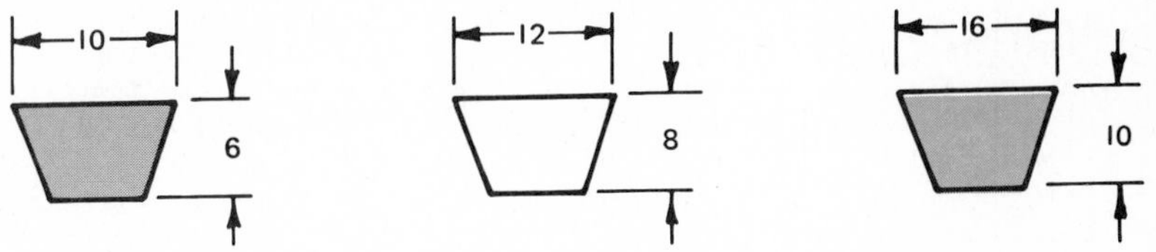
10
6
12
8
16
10

NOTE: THIS TABLE INCORPORATES A SERVICE FACTOR OF 1.3.
FOR HEAVY DUTY, MULTIPLY NORMAL DUTY KILOWATT RATING BY .85.
FOR LIGHT DUTY, MULTIPLY NORMAL DUTY KILOWATT RATING BY 1.20.

NOTE: SIZES SHOWN ARE INCH SIZES SOFT CONVERTED TO MILLIMETRES.

Fig. 19-1-13 Calculating pulley diameter of driver shaft and belt cross section.

FOLLOW THESE THREE EASY STEPS

Step 1: Selecting Driver V-Pulley Diameter and Belt Cross Section. First, classify the application and apply the proper service factor, as explained above. Refer to Fig. 19-1-13 for driver V-pulley diameter and belt cross section.

Step 2: Choosing Driven V-Pulley Diameter. Refer to Fig. 19-1-14 for the speed of the motor. Locate desired driven speed in driver V-pulley column; read driven V-pulley diameter in the first column.

Step 3: Finding Belt Length and Center Distance. Add the diameter of driver and driven V-pulley and refer to Fig. 19-1-15. Locate the sum of V-pulley diameters at the top of the chart, read down to the required centers, and read the belt length in the belt length column. **Note:** After calculating a center distance from a standard pitch length, make sure that the centers can be moved closer together to facilitate installation without injury to the belt. A minimum allowance of approximately 25 mm below standard center distance is required for the application of belts.

Although the amount of stretch in V-belts is relatively small, some adjustment between centers of pulleys is necessary to compensate for stretch and side wear on the belts and sheaves.

For larger V-belt drives, refer to manufacturers' catalogs and specifications.

The most common problem involves a tool or machine to be driven at a certain speed by an electric motor, gasoline engine, or some other power source usually operating at a higher speed.

To design a belt drive, the following information should be known:

1. The speed [revolutions per minute (r/min)] and kilowatts of the motor or driver unit
2. The speed (r/min) the driven shaft is to turn
3. The space available for the drive

Example 1. A 0.38-kW, 1750-r/min motor is to operate a drill press having a spindle of approximately 1200 r/min. The center distance between the motor shaft

DRIVEN SPEEDS FOR 1160 REV/MIN MOTORS

DriveN V-Pulley OD mm	DriveR V-Pulley OD—mm												
	38	**44**	**51**	**57**	**64**	**70**	**76**	**83**	**89**	**95**	**102**	**108**	**114**
38	1160	1392	1625	1855	2085	2325	2550	2785	3015	3250	3480	3715	
51	829	995	1160	1325	1490	1658	1825	1988	2150	2315	2485	2650	
64	645	774	903	1031	1160	1290	1418	1546	1675	1805	1933	2032	2190
76	528	634	739	845	950	1057	1160	1266	1370	1475	1580	1685	1793
89	447	536	625	715	804	894	982	1071	1160	1248	1340	1428	1518
102	387	465	542	620	696	775	851	929	1008	1082	1160	1238	1315
114	341	409	477	545	614	682	750	819	886	955	1022	1091	1160
127	305	366	427	488	549	610	671	732	794	854	915	976	1039
140	277	332	381	442	497	553	608	663	718	774	829	884	939
152	253	302	353	404	454	505	555	605	655	706	756	806	857
178	215	258	301	344	388	430	474	516	560	602	648	688	732
203	187	224	262	297	337	374	411	449	486	524	561	599	636
254	149	179	208	238	268	298	328	357	387	417	446	477	506
305	123	148	173	197	222	247	272	296	321	346	370	395	420

DRIVEN SPEEDS FOR 1750 REV/MIN MOTORS

DriveN V-Pulley OD mm	DriveR V-Pulley OD—mm												
	38	**44**	**51**	**57**	**64**	**70**	**76**	**83**	**89**	**95**	**102**	**108**	**114**
38	1750	2100	2450	2800	3150	3500	3850						
51	1250	1500	1750	2000	2250	2500	2750	3000	3250	3500	3750	4000	
64	974	1167	1360	1555	1750	1945	2140	2330	2530	2725	2915	3110	3305
76	797	955	1113	1272	1431	1590	1750	1910	2070	2225	2385	2545	2700
89	674	808	942	1077	1210	1346	1480	1615	1750	1885	2020	2155	2290
102	584	700	817	935	1050	1168	1283	1400	1518	1634	1750	1865	1985
114	516	618	720	824	926	1030	1131	1235	1339	1440	1543	1650	1750
127	462	554	646	737	830	922	1013	1105	1198	1290	1382	1473	1568
140	417	500	584	667	750	834	917	1000	1082	1167	1250	1333	1417
152	381	456	533	610	685	760	837	913	990	1065	1140	1217	1290
165	350	420	490	560	630	700	771	840	910	980	1050	1120	1190
178	324	389	454	518	584	648	713	778	843	907	973	1039	1102
203	282	339	394	451	507	564	620	676	734	789	845	902	959
229	250	300	350	400	450	500	550	600	650	700	750	800	850
254	224	270	315	360	405	450	495	540	585	630	675	720	765
279	203	244	285	326	366	407	448	488	530	570	610	652	692
305	186	224	261	298	336	373	410	446	485	522	560	596	634

DRIVEN SPEEDS FOR 3500 REV/MIN MOTORS

DriveN V-Pulley OD mm	DriveR V-Pulley OD—mm												
	38	**44**	**51**	**57**	**64**	**70**	**76**	**83**	**89**	**95**	**102**	**108**	**114**
38	3500	4200	4900	5600	6300	7000	7700						
51	2500	3000	3500	4000	4500	5000	5500	6000	6500	7000	7500	8000	
64	1948	2334	2720	3110	3500	3890	4280	4660	5060	5450	5830	6220	6610
76	1594	1910	2236	2544	2862	3180	3500	3820	4140	4450	4770	5090	5400
89	1348	1616	1884	2154	2420	2692	2960	3230	3500	3770	4040	4310	4580
102	1168	1400	1634	1870	2030	2336	2566	2800	3036	3268	3500	3730	3970
114	1032	1236	1440	1648	1852	2060	2262	2470	2678	2880	3086	3300	3500
127	924	1108	1292	1474	1660	1844	2026	2210	2396	2580	2764	2946	3136
140	834	1000	1168	1334	1500	1668	1834	2000	2164	2334	2500	2666	2834
152	762	912	1066	1220	1370	1520	1774	1826	1980	2130	2280	2434	2580
165	700	840	980	1120	1260	1400	1542	1680	1820	1960	2100	2240	2380
178	648	778	908	1036	1168	1296	1426	1556	1686	1814	1946	2078	2204
203	564	678	788	902	1014	1128	1240	1352	1468	1578	1690	1804	1918
229	500	600	700	800	900	1000	1100	1200	1300	1400	1500	1600	1750
254	448	540	630	720	810	900	990	1080	1170	1260	1350	1440	1530
279	406	488	570	652	732	814	896	976	1060	1140	1220	1304	1384
305	372	448	522	596	672	746	820	892	970	1044	1120	1192	1268

Fig. 19-1-14 Calculating r/min and diameter of driven pulley.

and spindle is approximately 500 mm. The type of drive required is V-belt.

Solution: Since drill press operations come under the classification of normal duty, no adjustment needs to be made to the kilowatt rating.

Step 1. Selecting Driver V-Pulley Diameter and Belt Cross Section (Fig. 19-1-13). Read down the extreme left column to the r/min figure nearest that of the speed of the motor, which is 1750 r/min. Read across this line to the figure closest to the design kilowatts of the drive. The closest kilowatt rating is found to be 0.38. Read up from the 0.38-kW figure. The figure at the top of the column is the outside diameter of the motor pulley in millimetres. The 0.38-kW figure is in the white area. The reference at the bottom of the chart indicates the size of the belt required.

Pulley size for motor = 70 mm

Belt section = 12 mm wide × 8 mm thick

Step 2. Choosing Driven V-Pulley Diameter (Fig. 19-1-14). Refer to the table for driven speeds for 1750 r/min motors. Read across the top of the table to the figure nearest the small pulley size. Column 70 corresponds exactly with the small pulley diameter. Read down this column to the figure nearest the desired speed (1200 r/min) of the driven shaft. The nearest figure is 1168. By reading to the left of this figure, the driven-pulley diameter is found to be 102 mm.

Step 3. Finding Belt Length and Center Distance (Fig. 19-1-15). Add the diameter of the pulleys and select the number in the top row that is nearest to this sum.

Motor pulley diameter = 70 mm

Spindle pulley diameter = 102 mm

Sum of diameter = 172 mm

The exact sum of the diameter is not shown on the top row; therefore use 180 mm. Read down this column to the figure indicated below the shaded area. The 500 mm distance is the ideal center distance. Other figures in this column indicate alternative center distances. Use 493 mm since the approximate center distance required is 500 mm. Follow along this line to the left to column Market Belt Length to obtain a belt length of 1270.

REFERENCES AND SOURCE MATERIAL

1. J. J. Zaiss, "Flat Belts," *Machine Design*, Vol. 37, No. 14, 1965.

2. E. L. Nuernberger, "V-Belts," *Machine Design*, Vol. 37, No. 14, 1965.

Installation Allowance	Take-Up	Belt Length	SUM OF BOTH V-BELT PULLEY DIAMETERS																			
			125	140	150	165	180	190	200	215	230	240	255	265	280	290	305	320	330	345	355	370
12	10	410	105																			
15	10	460	130	117																		
15	10	510	155	142	132																	
15	10	560	180	168	157	147																
15	10	610	205	193	183	173	160	147														
15	10	660	231	218	208	198	185	175	165													
15	10	710	257	269	234	224	213	201	193	180	168											
15	10	760	282	295	259	249	239	226	218	205	196	185										
15	10	810	308	295	284	274	264	254	244	231	221	213	203									
15	10	860	333	323	310	230	290	280	269	259	246	239	228	218								
15	10	910	358	348	335	325	315	305	295	284	272	264	254	244	229							
15	10	960	384	373	361	351	340	330	320	310	300	290	279	269	254	246	231					
20	10	1010	409	399	389	376	366	356	345	335	325	315	305	295	282	272	257	249				
20	10	1070	434	424	414	401	391	381	371	361	351	340	333	320	307	297	284	274	259			
20	10	1120	460	450	439	427	417	406	396	386	376	366	358	345	333	325	310	302	284	277		
20	10	1170	485	475	465	455	442	432	422	411	401	391	384	371	358	351	335	328	312	304	277	266
20	10	1220	511	500	490	480	467	457	450	437	427	417	409	396	384	376	363	353	338	330	304	295
20	10	1270	536	526	516	505	493	483	475	462	452	442	434	424	411	401	389	378	366	358	333	323
20	10	1320	513	551	541	531	518	508	500	488	478	467	460	450	437	427	414	404	391	381	358	351
20	10	1370	587	577	566	556	544	534	526	513	503	493	485	475	462	452	439	431	417	410	386	376
20	10	1420	612	602	592	582	569	559	551	538	528	518	511	500	488	478	465	457	442	434	411	404
20	10	1470	638	627	617	607	594	584	577	564	554	544	536	526	513	503	490	482	470	460	439	429
22	15	1520	663	653	643	632	622	610	602	589	579	569	561	551	538	528	518	508	495	485	465	457
22	15	1570	688	678	668	658	648	635	627	617	605	594	587	577	564	554	544	533	520	511	493	483
22	15	1630	714	704	693	683	673	660	653	643	630	620	612	602	589	582	569	559	546	536	518	508
22	15	1680	790	729	719	709	699	686	678	668	658	645	638	627	615	607	594	584	571	564	544	536
22	15	1730	765	754	744	734	724	714	704	693	683	671	663	653	640	632	620	610	597	589	569	561
22	15	1780	790	780	770	759	749	739	729	719	709	696	688	678	665	658	645	635	622	615	597	587
22	15	1830	815	805	798	785	765	765	754	744	734	721	714	704	691	683	671	660	647	640	622	612
22	15	1880	841	831	820	810	800	790	780	770	759	747	739	729	716	709	696	685	673	665	648	638
22	15	1930	866	886	846	836	826	815	805	795	785	772	765	754	742	734	721	711	701	691	673	665
22	15	1980	892	881	869	861	851	841	831	820	810	798	790	780	767	759	747	736	726	716	698	691
22	15	2030	917	907	897	886	876	866	856	846	836	823	815	805	795	785	772	762	752	742	726	716
22	15	2080	932	922	912	902	892	881	871	861	851	838	831	820	810	800	787	780	767	757	742	732
22	15	2130	968	958	947	937	927	917	907	897	886	874	866	856	846	836	823	815	803	792	777	767

Fig. 19-1-15 Determining V-belt length from pulley diameters and center distances.

3. The Gates Rubber Company, Denver, Colorado.
4. T. B. Wood's Sons Company.

Assignments

1. *V-Belt Motor Drive.* On an A3- or B-size sheet, lay out a 0.2 kW or 0.25 horsepower (hp) motor to drive shaft A between 815 and 835 r/min by means of a V-belt drive. Refer to Fig. 19-1-A or 19-1-B. Design for normal duty. Other details can be seen in Fig. 19-1-11. Draw top and front views. From manufacturers' catalogs select the belt and pulleys and call for them in a bill of material. Scale is 1:2.

2. *V-Belt Drive Problems.* Do any three.

(*a*) A 0.25 kW, 1750-r/min motor is to operate a furnace blower having a shaft speed of approximately 765 r/min. The center distance between the motor and blower shafts is approximately 340 mm. The type of drive required is V-belt.

(*b*) A 0.37 kW, 1160 r/min motor is used to operate a drill press. The spindle speed is 520 r/min, ±5 r/min. The center distance between the motor and blower shafts is approximately 550 mm. Select a suitable V-belt.

(*c*) A 1.1 kW, 1750 r/min motor is to operate a band saw whose flywheel turns at approximately 800 r/min. A pulley attached to the flywheel shaft connects, by means of a V-belt, to the pulley on the motor shaft. Center-to-center distance of shafts is 340 mm. Calculate the size of the V-belt required.

(*d*) A 0.37 kW, 1750 r/min motor drives a power hacksaw. The shaft on the hacksaw is to run at approximately 750 r/min, and the center-to-center distance of shafts is 400 mm. Calculate the size of V-belt required.

(*e*) A 0.6 kW, 1750 r/min motor is used to drive a punch machine whose flywheel turns at approximately 500 r/min. A pulley is attached to the flywheel shaft and connects to the motor pulley by means of a V-belt. Center-to-center distance is 430 mm. Calculate the size of V-belt required.

REVIEW FOR ASSIGNMENTS

Unit 6-4 Bill of Material

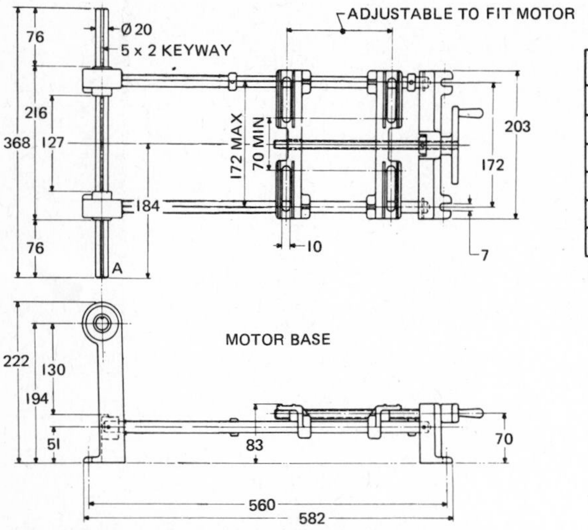
ADJUSTABLE TO FIT MOTOR
Ø 20
5 x 2 KEYWAY
76
216
368
127
184
76
A
172 MAX
70 MIN
203
172
10
7
MOTOR BASE
222
130
194
51
83
70
560
582

kW	C	L
0.1	274	118
0.12	274	118
0.20	286	130
0.25	286	144
0.37	320	166
0.56	344	188

115 VOLT 1750 REV/MIN
MOTOR DIMENSIONS

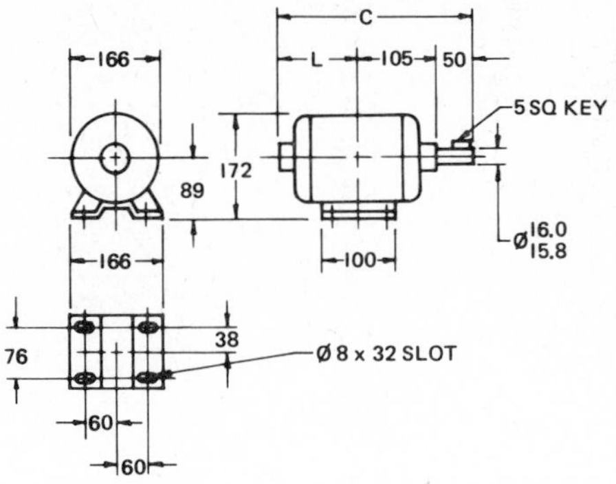
C
166
L
105
50
5 SQ KEY
89
172
Ø 16.0 15.8
166
100
76
38
Ø 8 x 32 SLOT
60
60

Fig. 19-1-A V-belt drive problems.

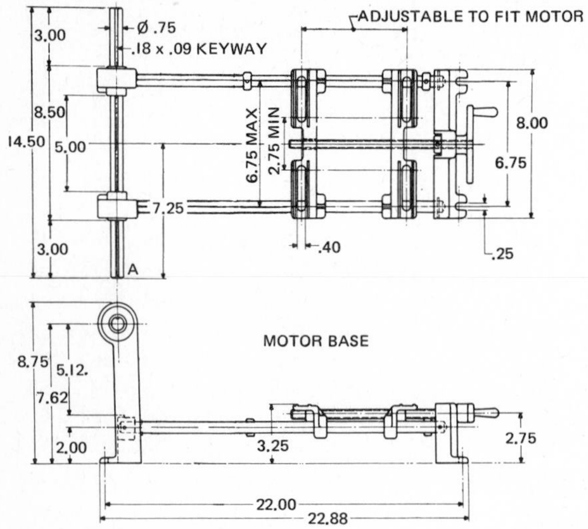
ADJUSTABLE TO FIT MOTOR
Ø .75
.18 x .09 KEYWAY
3.00
8.50
14.50
5.00
7.25
3.00
A
6.75 MAX
2.75 MIN
8.00
6.75
.40
.25
MOTOR BASE
8.75
5.12
7.62
2.00
3.25
2.75
22.00
22.88

HP	C	L
.12	10.75	4.62
.16	10.75	4.62
.25	11.25	5.12
.33	11.75	5.62
.50	12.62	6.50
.75	13.50	7.38

115 VOLT 1750 RPM
MOTOR DIMENSIONS

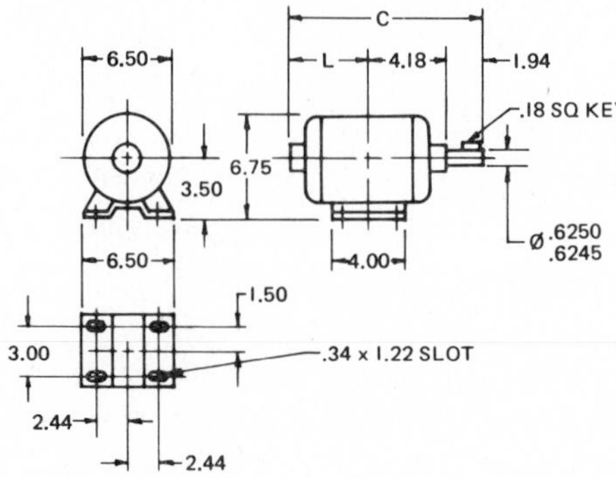
C
6.50
L
4.18
1.94
.18 SQ KEY
6.75
3.50
Ø .6250 .6245
6.50
4.00
1.50
3.00
.34 x 1.22 SLOT
2.44
2.44

Fig. 19-1-B V-belt drive problems.

UNIT 19-2
CHAIN DRIVES

Nearly all types of power-transmission chains have two basic components: side bars or link plates, and pin and bushing joints. The chain articulates at each joint to operate around a toothed sprocket. The *pitch* of the chain is the distance between centers of the articulating joints.

Power-transmission chains have several advantages: relatively unrestricted shaft center distances, compactness and ease of assembly, elasticity in tension with no slip or creep, and ability to operate in relatively high-temperature atmospheres. A typical application can be seen in Fig. 19-2-1.

Basic Types

There are six major types of power-transmission chains, with numerous modifications and special shapes for specific applications. A seventh type, the bead

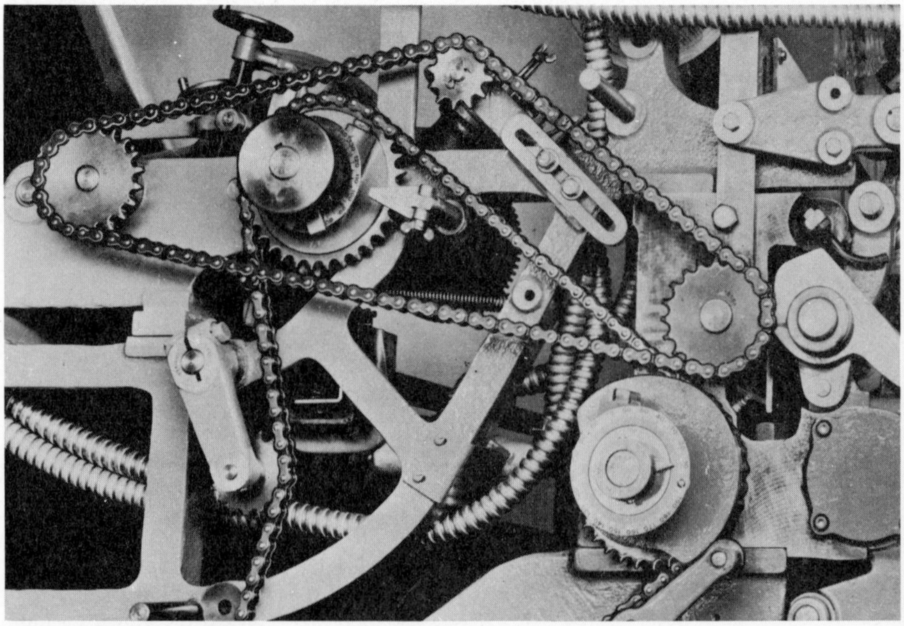

Fig. 19-2-1 Chain drives.

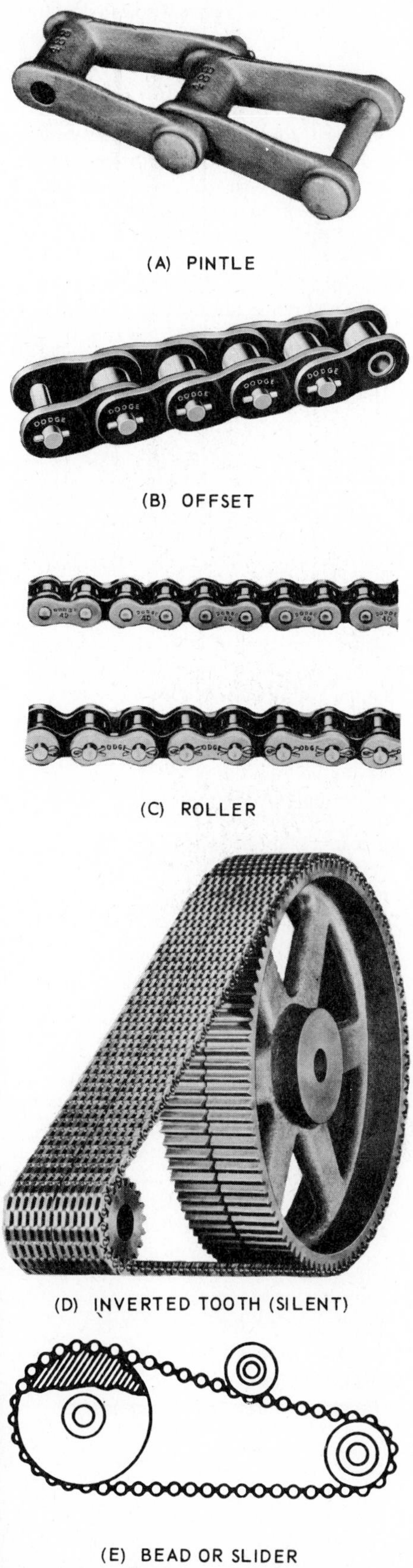
(A) PINTLE
(B) OFFSET
(C) ROLLER
(D) INVERTED TOOTH (SILENT)
(E) BEAD OR SLIDER

Fig. 19-2-2 Basic chain types. (A and D—Link Belt Ltd.; B and C—Dodge Mfg. Corp.; E—Machine De)

chain, is often used for light-duty applications. Figure 19-2-2 shows basic characteristics of the major types.

DETACHABLE

The malleable detachable chain is made in a range of sizes from 23 to 103 mm pitch and ultimate strength from 5 to 110 MPa (megapascals).

Of the same type is the steel detachable chain, covered by ANSI-B29.6. This chain is made in sizes from 23 mm to just under 76 mm in pitch, with ultimate strength from 35 MPa.

The ends of the detachable link are referred to as the *bar end* and the *hook end*.

Steel and malleable detachable chains are used mainly on farm machinery for speeds up to approximately 1.8 m/s and loads up to approximately 18 kW. These machines are operated seasonally, and the chains are not normally lubricated, since lubricants are apt to retain grit particles and thus abrade the joint wear surfaces they are intended to lubricate.

PINTLE

For slightly higher speeds (to about 2.2 m/s) and heavier loads, pintle chains are used. *Pintle chains* are made up of individual cast links having a full, round barrel end cast integral with offset sidebars. These links are intercoupled with steel pins. The ends of the pintle chain links are referred to as the *barrel end* and the *open end* of the links.

Many of these chains have been designed to operate over sprockets intended for detachable chain. Therefore, chains range from just over 25 mm up to 150 mm in pitch with ultimate strengths from 25 to 200 MPa. These chains are normally not lubricated. They are a bit more costly than detachable chain, and the mass is more per kilogram of useful load transmitted than precision chains.

OFFSET-SIDEBAR

Steel offset-sidebar chains are used extensively as drive chains on construction machinery. They operate at speeds to 5 m/s and transmit loads to approximately 185 kW.

Each link has two offset sidebars, one bushing, one roller, one pin, and, if the chain is detachable, a cotter pin. Some offset-sidebar chains are made without rollers. Lubrication of clean steel offset-sidebar chains will lengthen wear life appreciably.

These are precision chains, costing more than those previously described. They are designed to withstand infrequently applied overload without serious damage. (Through-hardened pins give the chains high ultimate strength.) The relatively large ratio of pitch to roller diameter makes operation over cast-tooth sprockets practical.

ROLLER

Transmission roller chain (Fig. 19-2-3) is available in pitches from 6 to 75 mm. In the single-width roller, the ultimate strength ranges from 6 to 900 MPa. It is also available in multiple widths. Small-pitch sprockets can operate at speeds as high as 10 000 r/min, and 750 to 900 kW drives are not unusual.

These chains are assembled from roller links and pin links. If the chain is detachable, cotter pins are used in the chain pin holes. For best joint wear life, these chains should be lubricated.

ANSI-B29.1 also covers a number of special types of roller chains. One is equipped with oil-impregnated, sintered, powdered metal bushings, for self-lubrication. This chain handles lighter loads at reduced speeds and is limited in application because it does not use rollers. Instead, it uses bushings of the same outside diameter as normal rollers. See Fig. 19-2-4.

Another approach to self-lubrication has been the use of special roller chains with plastic sleeves between the chain

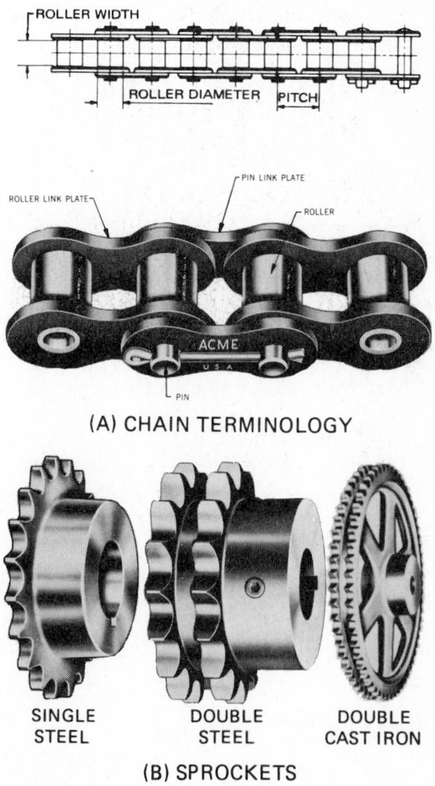
ROLLER WIDTH
ROLLER DIAMETER
PITCH
PIN LINK PLATE
ROLLER LINK PLATE
ROLLER
ACME
PIN
(A) CHAIN TERMINOLOGY
SINGLE STEEL
DOUBLE STEEL
DOUBLE CAST IRON
(B) SPROCKETS

Fig. 19-2-3 Chain terminology and sprockets.

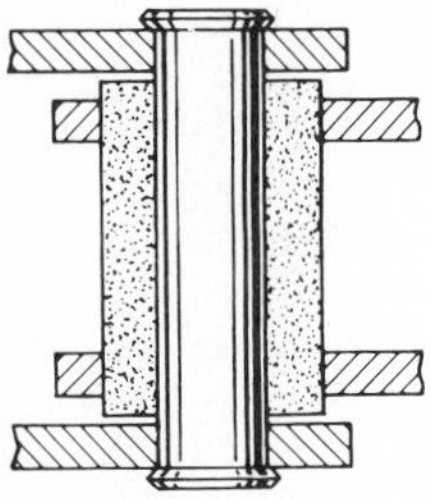

THIS TYPE USES OIL-IMPREGNATED SINTERED METAL BUSHING IN PLACE OF ROLLERS.

Fig. 19-2-4 Self-lubricating chain.

rivets and bushings. The plastic reduces joint friction. Also, a roller chain with Delrin stud bushings has been developed recently for light loads and slow speeds where the rigidity of metal is not required. This type, too, requires little or no lubrication; it has been used successfully in subzero temperatures. Still another recently developed chain is made entirely of self-lubricating plastic.

Another recent roller-chain development is one with extra joint clearances. These clearances permit side flexing of the chain so that it can be operated over sprockets that are not accurately aligned. The chain is as precise as standard roller chain, but loads and speeds should be reduced slightly.

DOUBLE-PITCH

These are basically the same as roller chains, except that the pitch is twice as long. Roller and double-pitch chains have the same diameter pins and rollers, the same width rollers, and the same thickness of link plates.

INVERTED-TOOTH SILENT

These are high-speed chains, used predominantly for prime-mover, power takeoff drives, such as on power cranes or shovels, machine tools, and pumps. Drives transmitting up to 900 kW are in use.

These chains are made up of a series of tooth links, alternately assembled with either pins or a combination of joint components in such a way that the joint articulates between adjoining pitches. Center-guide chain has guide links which engage a groove or grooves in the sprocket, while the side-guide chain has guides which engage the sides of the sprocket. Because of their high-speed operation (as high as 25 to 300 m/s), these chains are normally lubricated.

BEAD OR SLIDER

Bead chains are used as manually controlled or slow-speed drives in numerous products such as television tuners, radio tuners, computing devices, time recorders, air conditioners, toys, display drives, ventilator controls, and venetian blinds. They can be used on die-cast, sheet metal or nylon molded sprockets.

Four standard sizes, having a bead diameter of 2.4, 3.2, 4.8, and 6.4 mm, are available. These sizes have safe working loads of 70, 110, 180, and 330 N (newtons). Monel, carbon-steel, or stainless-steel chains provide extra strength. A new type, called *bead belt*, consists of solid plastic beads molded onto a cord. These plastic chains can be run at higher speeds than metal chains, but they do not offer as great strength. They are nonconductive. Bead diameters are 2.4 and 3.2 mm.

Sprockets

Basic sprocket types used with precision steel roller chains conform to ANSI standards.

Used for mounting on flanges, hubs, or other devices, the plate sprocket is a flat, hubless sprocket. It is made from bar stock or hot-rolled plate in either solid or split construction with plain, countersunk, or tapped holes.

Small- and medium-size hub sprockets are turned from bar stock or forgings or are made by welding a bar-stock hub to a hot-rolled plate. For small, low-load applications, only one hub extension may be needed. Large-diameter sprockets normally have two hub projections equidistant from the center plane of the sprocket. Thus, the line of action from chain pull reacts through the center of the hub, providing stability and ensuring an even distribution of stress on shaft and key.

Materials. Although normally machined from gray-iron castings, sprockets are also available in cast steel or welded hub construction.

Sprockets made of sintered powdered metal, and from nylon and other plastics, have become economical in large quantities. These sprockets offer many advantages. For example, plastic sprockets require minimum lubrication and are widely used where cleanliness is essential. Moreover, both powdered metal and plastics cost less in quantity than machined sprockets where milling and drilling are required.

Special Types. Several types of special-purpose sprockets are available. Split-type sprockets facilitate installation or replacement of sprockets mounted between bearings or at any intermediate location on the shaft; double-duty sprockets consist of a plate sprocket bolted to a cast-iron hub. They are used when frequent sprocket changes must be made quickly and economically. Jaw-clutch sprockets are used for slow- and moderate-speed drives when an inexpensive device is required for infrequent engagement and disengagement of a drive.

For built-in overload protection, the shear-pin type of sprocket is often used. It consists of a modified hub sprocket mounted on a cast-iron hub. Torque is transmitted from sprocket to hub through a shear pin. Other sprockets protect machinery from overloads with slip-clutch mechanisms which cushion the driving train by slipping until the torque drops to a preset level. The shear-pin sprocket requires a new pin in the event of an overload, whereas the slip-clutch requires occasional adjustment.

Design of Roller Chain Drives[1,2]

The design of a roller chain drive consists, primarily, of the selection of the chain and sprocket sizes. It also includes the determination of chain length, center distance, method of lubrication, and, in some cases, the arrangement of chain casings and idlers.

Unlike belt drives, which are based on lineal speeds in metres per minute, the limiting factor of chain drives is based on the rotative speed, or revolutions per minute, of the smaller sprocket, which in most installations is the driven member.

Design of chain drives is based not only on kilowatt and speeds but also on the following factors relative to broad service conditions:

1. Average kilowatts to be transmitted. See Fig. 19-2-5.
2. The revolutions per minute of the driving and driven members.
3. Shaft diameter.
4. Permissible diameters of sprockets.
5. Load characteristics, whether smooth and steady, pulsating, heavy-starting, or subject to peaks.
6. Lubrication, whether periodic, occasional, or copious. Where chains are exposed to dust, dirt, or injurious foreign matter, chain cases should be used.

CHAIN SPEED (m/s)	POWER (kilo-watts)	TYPE OF CHAIN
1.8	15	Detachable
2.2	30	Pintle
5	190	Offset-sidebar
12.5	1100	Roller
20	1850	Silent

Fig. 19-2-5 Tentative selection factors for chain drives.

7. Life expectancy: the amount of service required, or total life. It is much better to *overchain* rather than skimp on the size of the chain used.

In designing chain drives, it is of the utmost importance to consider and study the pitch or size of the chain used. The number of revolutions per minute and size of the smaller or faster-moving sprocket determine the pitch of chain that should be used.

Smaller-pitch chains in single or multiple width are adaptable for elevated speed drives, and also for any speed drives where smoother and quieter performance is essential.

Large-pitch chains are adaptable for slow- and medium-speed drives.

Multiple-width roller chains are becoming increasingly popular. They not only solve the problem of transmitting greater power at higher speeds, but because of their smoother action they also substantially reduce the noise factor. See Fig. 19-2-6.

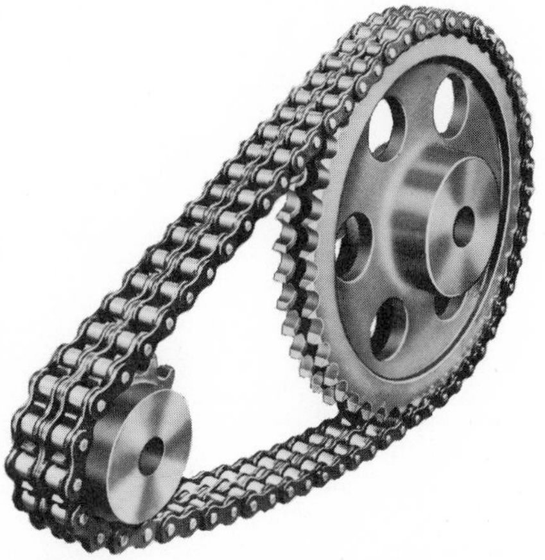

Fig. 19-2-6 Multiple-roller chain drive.

Because of the essential polygon action of roller chains, as the roller enters the sprocket tooth, an impact or hammer action occurs at the extreme end of functioning contact. The measure of impact is determined by the pitch and speed of chain used. Therefore, it is important to select the proper pitch chain to accommodate a given chain speed. In many instances it is more desirable to consider a multiple-width chain of smaller pitch but equal carrying capacity to the larger-pitch, single-width chain.

Size of Sprockets. It is general practice to use a minimum size sprocket of 17 teeth in order to obtain smooth operation at high speeds. Nineteen- or twenty-one-tooth sprockets should be considered from a standpoint of greater life expectancy, and smoother operation, because of the lessening of tooth impact. On slow-speed and special-purpose installations or where space limitations are involved, smaller than 17-tooth sprockets can be used.

The normal maximum number of teeth is 120. If more than 120 teeth are used, a small amount of permissible pitch elongation will result and cause the chain to ride high on the sprocket teeth long before the chain is actually worn out.

Ordinary practice indicates that the ratio of driver to driven sprockets should be no more than 6:1.

The recommended chain wrap on driver is 120°.

Center Distances. Center distances must be more than one-half the diameter of the smaller sprocket, plus one-half of the diameter of the larger sprocket; otherwise the sprocket teeth will touch. (When necessary, drives may be operated with a small amount of clearance between sprockets.) However, best results are obtained by using a center distance of 30 to 50 times the pitch of the chain used.

Eighty times the pitch is considered maximum.

For pulsating drives, the center distance should be shortened to 20 to 30 times the pitch.

Chain Tension. Chains should never run with both sides tight. Adjustable centers should be provided when possible to permit proper initial slack and to allow for periodic adjustment necessitated by natural chain wear. The chain sag should be equivalent to approximately 2 percent of the center distance. See Fig. 19-2-7.

Idler sprockets should be used as a means of taking up chain slack where it is not possible to provide adjustable centers. They are not needed for normal drives where other means of taking up the chain slack are available. Idler sprockets may be fixed or adjustable depending on their purpose. Adjustable idlers have the advantage in that chain tension can be controlled. Idler sprockets should also be applied against the slack side of the chain. When running inside the chain drive, idlers may be located somewhat from the middle of the drive toward the larger sprocket. When running outside the chain drive, idlers should be located toward the smaller sprocket. On drives with short centers, the idler on outside of chain should be as near as possible to the smaller sprocket.

Chain Length. The chain length is a function of the number of teeth in both sprockets and of the center distance. In addition, the chain must consist of an integral number of pitches, with an even

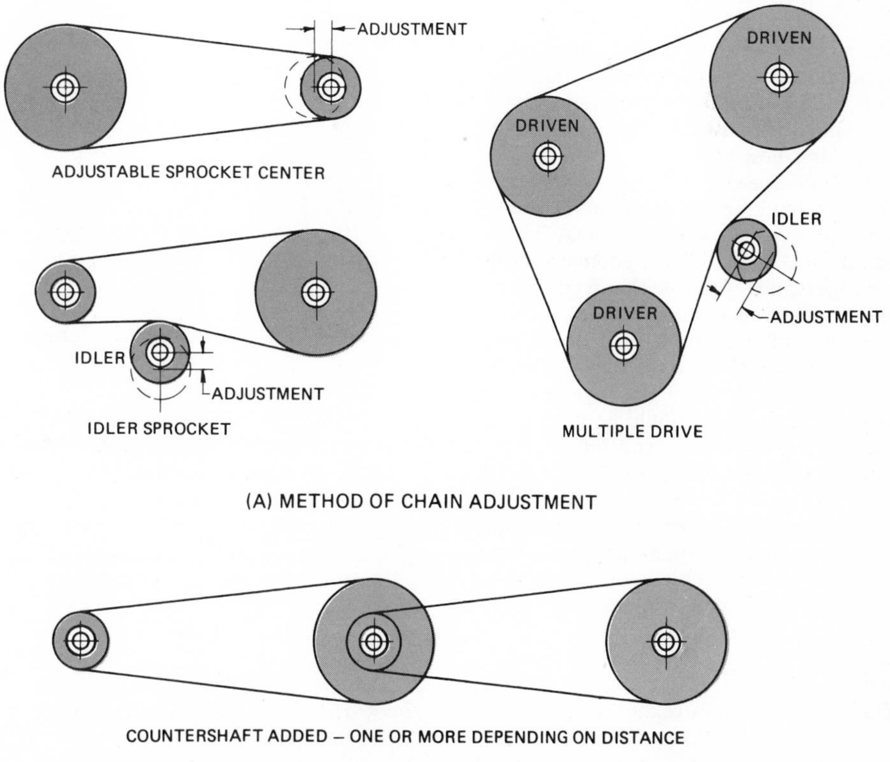
ADJUSTMENT
ADJUSTABLE SPROCKET CENTER
IDLER
ADJUSTMENT
IDLER SPROCKET
DRIVEN
DRIVEN
IDLER
DRIVER
ADJUSTMENT
MULTIPLE DRIVE
(A) METHOD OF CHAIN ADJUSTMENT
COUNTERSHAFT ADDED – ONE OR MORE DEPENDING ON DISTANCE
(B) CHAIN DRIVE WITH LONG CENTER DISTANCE

Fig. 19-2-7 Chain drives.

number preferable, in order to avoid the use of an offset link.

Chain Length Formula. For simplicity it is customary to compute the chain length in terms of chain pitches and then to multiply the result by the chain pitch to obtain the length in millimetres. The following formula is a quick and convenient method of finding the chain length in pitches (Fig. 19-2-8).

1. Divide center distance in millimetres by pitch of chain, obtaining C.
2. Add number of teeth in small sprocket to number of teeth in large sprocket, obtaining M.
3. Subtract number of teeth in small sprocket from number of teeth in large sprocket, obtaining value F to obtain the corresponding value of S.
4. Chain length in pitches equals

$$2C + \frac{M}{2} + \frac{S}{C}$$

A chain cannot contain the fractional part of a pitch. It is therefore necessary to increase the pitch to the next higher whole number, preferably an even number. The center distance must then be corrected.

5. Multiply number of pitches by chain pitch used in order to get chain length in millimetres.

LUBRICATION

Slow-Speed Drives. Correct lubrication is essential to obtain long life of the chain drive. At slow speeds good results are obtained by periodic lubrication. Mineral oil of medium consistency may be applied with a brush, when the chain is running slowly.

Medium-Speed Drives. Medium-speed drives should receive more lubrication than slow-speed drives. This may be accomplished by utilizing a good-size drip feed lubricator, fitted with a feed pipe and arranged in such a manner as to drop oil on the chain, so that the lubricant may penetrate to the chain bearing.

High-Speed Drives. For high-speed drives a chain case should be used with the chain just touching and running through oil in the bottom of the case. On very high-speed drives, an oil-slinging disk of larger diameter than the sprocket should be mounted on either shaft. The small amount of oil picked up is thrown against a baffle from which it drips on the inside of the chain.

DRIVE SELECTION

The kilowatt ratings relate to the speed of the smaller sprocket, and drive selections are made on this basis, whether the drive is speed-reducing or speed-increasing. In making drive selections,

F	S	F	S	F	S	F	S	F	S	F	S
1	.03	32	25.94	63	100.54	94	223.82	125	395.79	156	616.44
2	.10	33	27.58	64	103.75	95	228.61	126	402.14	157	624.37
3	.23	34	29.28	65	107.02	96	233.44	127	408.55	158	632.35
4	.41	35	31.03	66	110.34	97	238.33	128	415.01	159	640.38
5	.63	36	32.83	67	113.71	98	243.27	129	421.52	160	648.46
6	.91	37	34.68	68	117.13	99	248.26	130	428.08	161	656.59
7	1.24	38	36.58	69	120.60	100	253.30	131	434.69	162	664.77
8	1.62	39	38.53	70	124.12	101	258.39	132	441.36	163	673.00
9	2.05	40	40.53	71	127.69	102	263.54	133	448.07	164	681.28
10	2.53	41	42.58	72	131.31	103	268.73	134	454.83	165	689.62
11	3.06	42	44.68	73	134.99	104	273.97	135	461.64	166	698.00
12	3.65	43	46.84	74	138.71	105	279.27	136	468.51	167	706.44
13	4.28	44	49.04	75	142.48	106	284.67	137	475.42	168	714.92
14	4.96	45	51.29	76	146.31	107	290.01	138	482.39	169	723.46
15	5.70	46	53.60	77	150.18	108	295.45	139	489.41	170	732.05
16	6.48	47	55.95	78	154.11	109	300.95	140	496.47	171	740.60
17	7.32	48	58.36	79	158.09	110	306.50	141	503.59	172	749.37
18	8.21	49	60.82	80	162.11	111	312.09	142	510.76	173	758.11
19	9.14	50	63.33	81	166.19	112	317.74	143	517.98	174	766.90
20	10.13	51	65.88	82	170.32	113	323.44	144	525.25	175	775.74
21	11.17	52	68.49	83	174.50	114	329.19	145	532.57	176	784.63
22	12.26	53	71.15	84	178.73	115	334.99	146	539.94	177	793.57
23	13.40	54	73.86	85	183.01	116	340.84	147	547.36	178	802.57
24	14.59	55	76.62	86	187.34	117	346.75	148	554.83	179	811.61
25	15.83	56	79.44	87	191.73	118	352.70	149	562.36	180	820.70
26	17.12	57	82.30	88	196.16	119	358.70	150	569.93	181	829.85
27	18.47	58	85.21	89	200.64	120	364.76	151	577.56	182	839.04
28	19.86	59	88.17	90	205.18	121	370.86	152	585.23	183	848.29
29	21.30	60	91.19	91	209.76	122	377.02	153	592.96	184	857.58
30	22.80	61	94.25	92	214.40	123	383.22	154	600.73	185	866.93
31	24.34	62	97.37	93	219.08	124	389.48	155	608.56	—	—

STEP 1	Divide center distance which is given in millimetres by pitch of chain used obtaining C.
STEP 2	Add number of teeth in smaller sprocket to number of teeth in larger sprocket obtaining M.
STEP 3	Subtract number of teeth in smaller sprocket from number of teeth in larger sprocket which gives F in table above. Use Corresponding Constant S.
STEP 4	Chain length in pitches = $2C + \frac{M}{2} + \frac{S}{C}$.
STEP 5	Multiply number of pitches by chain pitch used in order to get chain length in millimetres.

Fig. 19-2-8 Determining chain length.

SERVICE FACTOR FOR SINGLE CHAINS
Kilowatt Ratings for Multiple-Strand Chains Equal Single Strand Ratings Multiplied by Multiple-Strand Factor

MULTIPLE-STRAND FACTOR

	Type of Input Power				
Type of Driven Load	**Internal Combustion Engine with Hydraulic Drive**	**Electric Motor or Turbine**	**Internal Combustion Engine with Mechanical Drive**	**Number of Strands**	**Multiple-Strand Factor**
Smooth	1.0	1.0	1.2	2	1.7
Moderate Shock	1.2	1.3	1.4	3	2.5
Heavy Shock	1.4	1.5	1.7	4	3.3

Fig. 19-2-9 Service- and multiple-strand factors for chain drives.

No. of Teeth on Small Spkt.	6.0 PITCH						NO. 25 ASA STANDARD ROLLER CHAIN										
	Revolutions Per Minute of Small Sprocket																
	100	500	900	1200	1800	2500	3000	3500	4000	4500	5000	5500	6000	6500	7000	7500	8000
17	.064	0.3	0.5	0.6	0.9	1.2	1.4	1.6	1.8	1.7	1.5	1.3	1.1	1.0	0.9	0.8	0.7
18	.069	0.3	0.5	0.6	0.9	1.2	1.5	1.7	1.9	1.9	1.6	1.4	1.2	1.1	1.0	0.9	0.8
19	.072	0.3	0.5	0.7	1.0	1.3	1.5	1.8	2.0	2.0	1.7	1.5	1.3	1.2	1.0	0.9	0.9
20	.077	0.3	0.6	0.7	1.0	1.4	1.6	1.9	2.1	2.2	1.9	1.6	1.4	1.3	1.1	1.0	0.9
21	0.1	0.3	0.6	0.8	1.1	1.5	1.7	2.0	2.2	2.3	2.0	1.7	1.5	1.3	1.2	1.1	1.0
22	0.1	0.4	0.6	0.8	1.1	1.5	1.8	2.1	2.3	2.5	2.1	1.9	1.6	1.4	1.3	1.2	1.1
23	0.1	0.4	0.6	0.8	1.2	1.6	1.9	2.2	2.5	2.7	2.3	2.0	1.7	1.5	1.4	1.2	1.1
24	0.1	0.4	0.7	0.9	1.3	1.7	2.0	2.3	2.6	2.9	2.4	2.1	1.9	1.6	1.5	1.3	1.2
25	0.1	0.4	0.7	0.9	1.3	1.8	2.1	2.4	2.7	3.0	2.6	2.2	2.0	1.7	1.6	1.4	1.3
28	0.1	0.5	0.8	1.0	1.5	2.0	2.3	2.7	3.0	3.4	3.1	2.7	2.3	2.1	1.9	1.7	1.5
30	0.1	0.5	0.9	1.1	1.6	2.1	2.5	2.9	3.3	3.6	3.4	3.0	2.6	2.3	2.1	1.9	1.7
32	0.1	0.5	0.9	1.2	1.7	2.3	2.7	3.1	3.5	3.9	3.8	3.3	2.9	2.5	2.3	2.0	1.9
35	0.1	0.6	1.0	1.3	1.9	2.5	3.0	3.4	3.9	4.3	4.3	3.7	3.3	2.9	2.6	2.3	2.1
40	0.2	0.7	1.2	1.5	2.2	2.9	3.5	4.0	4.5	5.0	5.3	4.6	4.0	3.5	3.2	2.9	2.6
45	0.2	0.8	1.3	1.7	2.5	3.3	3.9	4.5	5.1	5.7	6.2	5.4	4.8	4.2	3.8	3.4	3.1

No. of Teeth on Small Spkt.	10 PITCH						NO. 35 ASA STANDARD ROLLER CHAIN										
	Revolutions Per Minute of Small Sprocket																
	100	500	900	1200	1800	2500	3000	3500	4000	4500	5000	5500	6000	6500	7000	7500	8000
17	0.2	0.9	1.6	2.1	2.9	4.0	4.2	3.3	2.7	2.3	2.0	1.7	1.5	1.3	1.2	1.1	1.0
18	0.2	1.0	1.7	2.2	3.1	4.2	4.6	3.6	3.0	2.5	2.1	1.8	1.6	1.4	1.3	1.2	1.1
19	0.2	1.1	1.8	2.3	3.3	4.5	5.0	3.9	3.1	2.7	2.3	2.0	1.8	1.6	1.4	1.3	1.1
20	0.3	1.1	1.9	2.4	3.5	4.7	5.4	4.3	3.5	2.9	2.5	2.2	2.0	1.7	1.5	1.4	1.2
21	0.3	1.2	2.0	2.6	3.7	5.0	5.8	4.6	3.7	3.1	2.7	2.3	2.0	1.8	1.6	1.5	1.3
22	0.3	1.2	2.1	2.7	3.9	5.2	6.2	4.9	4.0	3.4	2.9	2.5	2.2	1.9	1.7	1.6	1.4
23	0.3	1.3	2.2	2.8	4.1	5.5	6.5	5.2	4.3	3.6	3.1	2.7	2.3	2.1	1.9	1.7	1.5
24	0.3	1.4	2.3	3.0	4.3	5.8	6.8	5.6	4.6	3.8	3.3	2.8	2.5	2.2	2.0	1.8	1.6
25	0.3	1.4	2.4	3.1	4.5	6.0	7.1	5.9	4.9	4.1	3.5	3.0	2.6	2.3	2.1	1.9	1.7
28	0.4	1.6	2.7	3.5	5.1	6.8	8.0	7.0	5.8	4.8	4.1	3.6	3.1	2.8	2.5	2.2	2.0
30	0.4	1.7	2.9	3.8	5.4	7.3	8.7	7.8	6.4	5.4	4.6	4.0	3.5	3.1	2.8	2.5	2.3
32	0.4	1.8	3.1	4.1	5.8	7.8	9.3	8.6	7.0	5.9	5.0	4.4	3.8	3.4	3.0	2.7	2.5
35	0.5	2.0	3.4	4.5	6.4	8.7	10.2	9.8	8.1	6.8	5.8	5.0	4.4	4.0	3.5	3.1	2.8
40	0.6	2.3	4.0	5.2	7.4	10.0	11.8	12.0	9.8	8.3	7.0	6.1	5.4	4.8	4.3	3.8	3.5
45	0.7	2.7	4.5	5.9	8.4	11.3	13.4	14.3	11.8	9.8	8.4	7.3	6.4	5.7	5.1	4.6	

No. of Teeth on Small Spkt.	13 PITCH						NO. 40 ASA STANDARD ROLLER CHAIN										
	Revolutions Per Minute of Small Sprocket																
	50	200	400	600	900	1200	1800	2400	3000	3500	4000	4500	5000	5500	6000	6500	7000
17	0.3	0.9	1.8	2.6	3.7	4.8	6.7	4.3	3.1	2.5	2.0	1.7	1.4	1.3	1.1	1.0	0.9
18	0.3	1.0	1.9	2.7	4.0	5.1	7.3	4.7	3.4	2.7	2.2	1.8	1.6	1.4	1.2	1.1	0.9
19	0.3	1.1	2.0	2.9	4.1	5.4	7.8	5.1	3.8	2.9	2.4	2.0	1.7	1.5	1.3	1.1	1.0
20	0.3	1.1	2.1	3.0	4.4	5.7	8.3	5.5	4.0	3.1	2.6	2.1	1.8	1.6	1.4	1.2	1.1
21	0.3	1.2	2.3	3.2	4.7	6.1	8.7	6.0	4.3	3.4	2.8	2.3	2.0	1.7	1.5	1.3	1.2
22	0.4	1.3	2.4	3.4	4.9	6.4	9.2	6.4	4.6	3.6	3.0	2.5	2.1	1.8	1.6	1.4	1.3
23	0.4	1.3	2.5	3.6	5.1	6.7	9.6	6.8	4.9	3.9	3.2	2.7	2.3	2.0	1.7	1.5	1.4
24	0.4	1.4	2.6	3.7	5.4	7.0	10.1	7.3	5.2	4.1	3.4	2.8	2.4	2.1	1.8	1.6	1.5
25	0.4	1.5	2.7	3.9	5.6	7.3	10.5	7.8	5.5	4.4	3.6	3.0	2.6	2.2	2.0	1.7	
28	0.5	1.6	3.0	4.4	6.4	8.3	11.9	9.2	6.6	5.2	4.3	3.6	3.1	2.6	2.3	2.1	
30	0.5	1.8	3.3	4.8	6.9	8.9	12.8	10.1	7.3	5.8	4.8	3.1	3.4	2.9	2.6		
32	0.5	1.9	3.5	5.1	7.4	9.5	13.7	11.2	8.1	6.4	5.2	4.4	3.7	3.2	2.8		
35	0.6	2.1	3.9	5.6	8.1	10.5	15.1	12.8	9.2	7.3	6.0	5.0	4.3	3.7			
40	0.7	2.4	4.5	6.5	9.0	12.2	17.5	15.7	11.2	8.9	7.3	6.1	5.2				
45	0.7	2.7	5.1	7.4	10.6	13.8	19.8	18.7	13.4	10.6	8.8	7.3					

Fig. 19-2-10 Kilowatt ratings for 6-, 10-, and 13-pitch single-strand roller chain.

16 PITCH — **NO. 50 ASA STANDARD ROLLER CHAIN**

No. of Teeth on Small Spkt.	Revolutions Per Minute of Small Sprocket															
	50	100	300	500	900	1200	1500	1800	2100	2400	2700	3000	3300	3500	4000	4500
17	0.54	1.00	2.7	4.2	7.2	9.3	10.7	8.0	6.3	5.2	4.3	4.0	3.2	3.0	2.4	2.0
18	0.57	1.07	2.9	4.5	7.7	10.0	11.6	8.7	6.9	5.7	4.7	4.0	3.6	3.2	2.7	2.2
19	0.60	1.13	3.0	4.8	8.1	10.5	12.6	9.4	7.5	6.1	5.1	4.4	3.9	3.4	2.9	2.4
20	0.64	1.2	3.2	5.1	8.7	11.2	13.6	10.2	8.1	6.6	5.6	4.8	4.1	3.8	3.1	2.6
21	0.67	1.3	3.4	5.3	9.1	11.8	14.4	11.0	8.7	7.1	6.0	5.1	4.4	4.1	3.3	3.0
22	0.71	1.3	3.6	5.6	9.5	12.4	15.1	11.7	9.3	7.6	6.4	5.5	4.8	4.3	3.6	3.0
23	0.75	1.4	3.7	5.9	10.0	13.0	15.9	12.6	10.0	8.1	7.0	5.9	5.1	4.7	3.8	3.2
24	0.78	1.5	3.9	6.1	10.5	13.6	16.6	13.4	10.6	8.7	7.2	6.2	5.4	5.0	4.0	3.4
25	0.81	1.5	4.1	6.4	11.0	14.2	17.4	14.2	11.3	9.3	7.8	6.7	5.8	5.2	4.3	3.6
28	0.90	1.7	4.6	7.3	12.4	16.1	19.6	17.0	13.4	11.0	9.1	7.9	6.9	6.2	5.1	0
30	0.99	1.8	5.0	7.8	13.4	17.3	21.2	18.7	14.8	12.2	10.2	8.8	7.6	7.0	5.7	
32	1.06	2.0	5.3	8.4	14.3	18.6	22.7	20.7	16.3	13.4	12.0	10.0	8.2	8.0	6.2	
35	1.17	2.2	5.9	9.2	15.8	20.4	25.0	23.7	18.7	15.3	13.0	11.0	10.0	9.0	7.1	
40	1.35	2.5	6.8	10.6	18.2	23.2	28.9	28.9	22.8	18.7	15.7	13.4	11.7	11.0	0	
45	1.54	2.9	7.7	12.2	20.7	26.9	32.7	34.4	27.2	22.3	19.0	16.0	14.0	0		

20 PITCH — **NO. 60 ASA STANDARD ROLLER CHAIN**

No. of Teeth on Small Spkt.	Revolutions Per Minute of Small Sprocket															
	50	100	200	500	700	900	1200	1400	1600	1800	2000	2200	2400	2600	2800	3000
17	0.9	1.7	3.2	7.3	10.0	12.5	16.2	13.6	11.0	9.3	7.9	6.9	6.0	5.3	4.8	4.3
18	1.0	1.8	3.4	7.8	10.5	13.3	17.2	14.8	12.0	10.1	8.6	7.5	6.5	5.8	5.2	4.7
19	1.0	1.9	3.6	8.3	11.2	14.0	18.2	16.0	13.1	11.0	9.3	8.1	7.1	6.3	5.7	5.1
20	1.1	2.1	3.8	8.8	11.9	14.9	19.2	17.3	14.1	11.9	10.1	8.7	7.7	6.8	6.1	5.5
21	1.2	2.2	4.0	9.2	12.5	15.7	20.3	18.6	15.1	12.8	10.8	9.4	8.3	7.3	6.6	5.9
22	1.2	2.3	4.2	9.7	13.1	16.5	21.3	20.0	16.3	13.8	11.6	10.1	8.9	7.8	7.0	6.3
23	1.3	2.4	4.5	10.1	13.7	17.3	22.4	21.3	17.4	14.7	12.5	10.7	9.5	8.4	7.5	6.8
24	1.3	2.5	4.7	10.6	14.4	18.1	23.4	22.7	18.5	15.6	13.2	11.5	10.1	9.0	8.0	7.2
25	1.4	2.6	4.9	11.1	15.1	19.0	24.6	24.2	19.7	16.7	14.1	12.2	10.8	9.5	8.5	7.7
28	1.6	3.0	5.5	12.6	17.0	21.4	27.7	28.7	23.3	19.7	16.8	14.5	12.7	11.3	10.1	9.1
30	1.7	3.2	6.0	13.6	18.4	23.1	30.0	31.8	25.9	21.8	18.6	16.0	14.1	12.5	11.2	10.1
32	1.8	3.4	6.4	14.5	19.7	24.7	32.0	35.0	28.5	24.0	20.4	17.0	15.6	13.8	12.3	11.1

25 PITCH — **NO. 80 ASA STANDARD ROLLER CHAIN**

No. of Teeth on Small Spkt.	Revolutions Per Minute of Small Sprocket															
	25	50	100	200	300	400	500	700	900	1000	1200	1400	1600	1800	2000	2200
17	1.2	2.1	4.0	7.5	10.8	14.0	17.1	23.1	29.0	28.0	21.3	16.9	13.9	11.6	9.9	8.6
18	1.2	2.3	4.3	8.0	11.5	14.8	18.2	24.6	30.8	30.6	23.3	18.5	15.1	12.7	10.8	9.4
19	1.3	2.4	4.5	8.4	12.2	15.7	19.2	26.1	32.7	33.2	25.3	20.1	16.4	13.7	11.7	10.1
20	1.4	2.6	4.8	9.0	12.8	16.6	20.4	27.6	34.6	35.9	27.3	21.6	17.8	14.8	12.7	11.0
21	1.4	2.7	5.0	9.4	13.6	17.6	21.5	29.1	36.5	38.6	29.4	23.3	19.1	16.0	13.7	11.9
22	1.5	2.8	5.3	9.9	14.2	18.5	22.6	30.6	38.3	41.4	31.5	25.0	20.4	17.2	14.6	12.7
23	1.6	3.0	5.6	10.4	14.9	19.4	23.6	32.1	40.2	44.2	33.6	26.7	21.9	18.4	15.7	13.6
24	1.7	3.1	5.8	10.9	15.7	20.3	24.8	33.6	42.1	46.3	35.9	28.4	23.3	19.5	16.6	14.5
25	1.7	3.3	6.1	11.3	16.3	21.2	26.0	35.1	44.0	48.4	38.1	30.3	24.8	20.7	17.8	15.4
28	2.0	3.7	6.9	12.8	18.5	23.9	29.3	39.7	49.7	54.7	45.2	35.9	29.4	24.6	21.0	18.2
30	2.1	4.0	7.4	13.8	19.9	25.8	31.6	42.7	53.6	58.9	50.1	39.8	32.5	27.3	23.3	18.3
32	2.3	4.3	8.0	14.8	21.3	27.7	33.8	45.8	57.4	63.2	55.2	43.8	35.9	30.1	25.7	

Fig. 19-2-11 Kilowatt ratings for 16-, 20-, and 25-pitch single-strand roller chain.

consideration is given to the loads imposed on the chain by the type of input power and the type of equipment to be driven. Service factors are used to compensate for these loads, and the required kilowatt rating of the chain is determined by the following equation:

Required kilowatt table rating (Fig. 19-2-9)

$$= \frac{\text{kW to be transmitted} \times \text{service factor}}{\text{multiple-strand factor}}$$

Figures 19-2-10 and 19-2-11 show the kilowatt ratings for just a few of the many roller chains available. For additional information refer to manufacturers' catalogs.

The kilowatt rating chart (Fig. 19-2-12) provides a quick means of determining the probable chain requirements without the need for a random search through the kilowatt rating charts which offer more accurate kilowatt values.

CHAIN DRIVE DESIGN

Example 1. Select an electric motor-driven roller chain drive to transmit 3.7 kW from a countershaft to the main shaft of a wire drawing machine. The countershaft is 38 mm in diameter and operates at 1200 r/min. The main shaft is also 38 mm and must operate between 378 and 382 r/min. Shaft centers, once established, are fixed and by initial calculations must be approximately 570 mm. The load on the main shaft is uneven and presents peaks which places it in the heavy shock load category. The driving head is fully enclosed, and all parts are pressure-lubricated from the central system; therefore, the drive will receive copious lubrication.

Solution

Step 1. Service Factor.
The corresponding service factor from

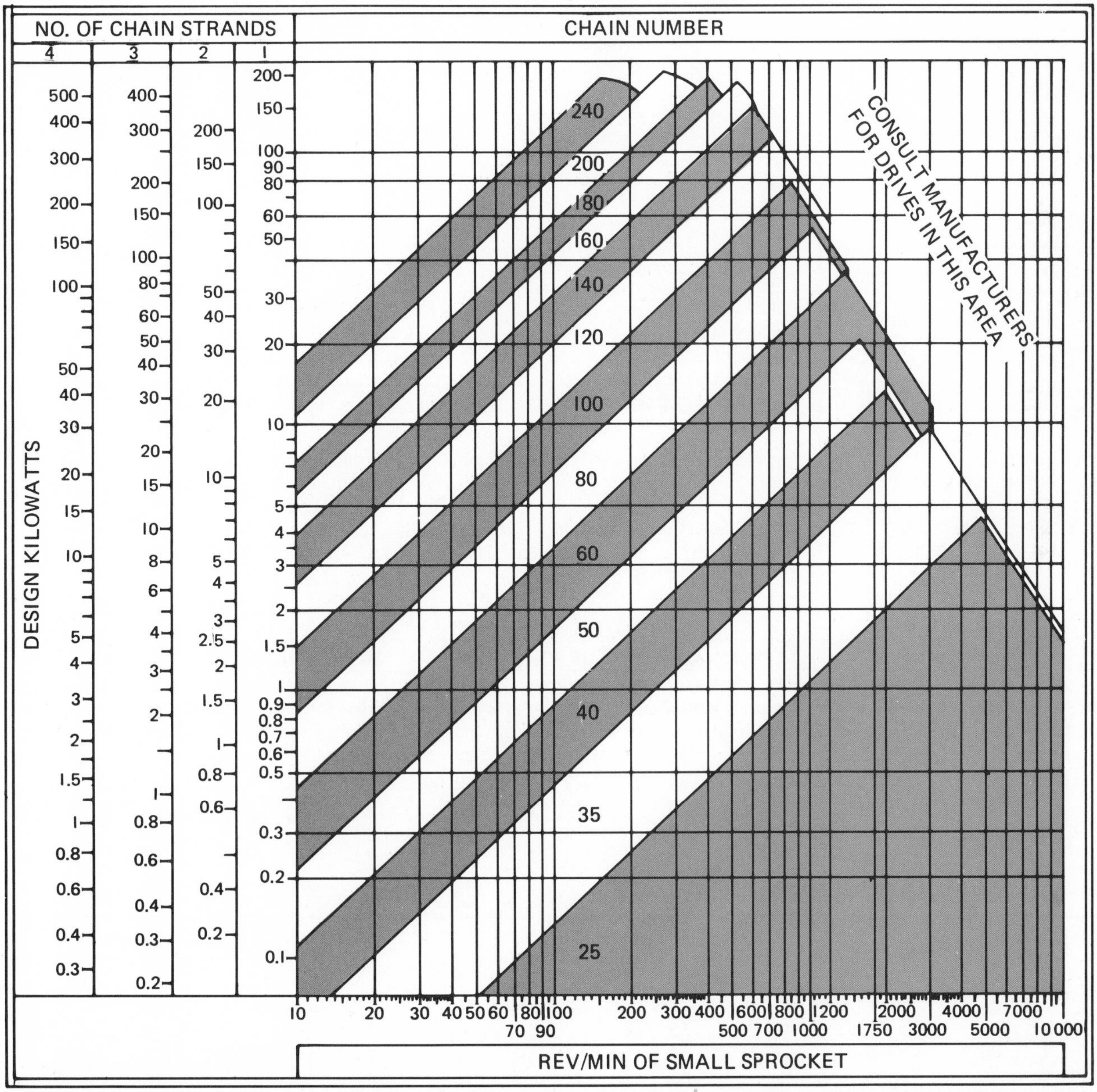
NO. OF CHAIN STRANDS
4
3
2
1
CHAIN NUMBER
DESIGN KILOWATTS
240
200
180
160
140
120
100
80
60
50
40
35
25
CONSULT MANUFACTURERS FOR DRIVES IN THIS AREA
REV/MIN OF SMALL SPROCKET

NOTE: THE MAXIMUM KILOWATT RATING SPECIFIED IN EACH OF THE STRAND COLUMNS IS NOT LIMITING FOR CHAIN DRIVES. CONSULT CHAIN MANUFACTURERS ON THOSE APPLICATIONS WHICH ARE ABOVE THE KILOWATT RANGE OF THE CHART.

Fig. 19-2-12 Kilowatt rating chart.

Fig. 19-2-9 for heavy shock load and electric motor is 1.5.

Step 2. Design Kilowatts.
Design kilowatts is $3.7 \times 1.5 = 5.6$ kW.

Step 3. Tentative Chain Selection.
On the kilowatt rating chart (Fig. 19-2-12) the suggested selection using a design of 5.6 kW and a 1200-r/min sprocket is a no. 40 (13-mm pitch) chain.

If a multiple-strand chain has been selected, determine the required kilowatt table rating per strand from the following equation:

$$\text{Required kilowatt table rating} = \frac{\text{design kilowatts}}{\text{multiple-strand factor}}$$

Refer to the RH columns shown in Fig. 19-2-9.

Step 4. Final Selection of Chain and Small Sprocket.
On the kilowatt rating table for a no. 40 chain, Fig. 19-2-10 at 1200 r/min, the computed design of 5.6 kW is realized with a 20-tooth sprocket. Follow down the column headed by the speed of the small sprocket and find the nearest value to the design kilowatts. Follow this line horizontally to the left to find the number of teeth for the small sprocket.

For intermediate speeds or sprocket sizes not tabulated, interpolate between the appropriate columns or lines. Check the maximum bore for the selected sprocket (Fig. 19-2-13). If the selected sprocket will not accommodate the shaft, use a larger sprocket or make a new sprocket and chain selection from the rating table for the next larger chain number. In this problem, the 20-tooth sprocket will accommodate the 38 mm shaft.

Step 5. Selection of the Large Sprocket.
Since the driver is to operate at 1200 r/min and the driven at a minimum of 378 r/min, the speed ratio = 1200/378 = 3.17:1 minimum. Therefore, the large sprocket should have 20×3.17 teeth = 63.4 teeth. Since the standard sprocket

NUMBER OF TEETH ON SPROCKET												
No. 25	No. 35		No. 40			No. 50			No. 60		No. 80	
9	9	48	8	34	59	9	34	59	9	34	9	34
10	10	54	9	35	60	10	35	60	10	35	10	35
11	11	60	10	36	70	11	36	70	11	36	11	36
12	12	70	11	37	72	12	37	72	12	37	12	37
13	13	72	12	38	80	13	38	80	13	38	13	38
14	14	80	13	39	84	14	39	84	14	39	14	39
15	15	84	14	40	96	15	40	96	15	40	15	40
16	16	96	15	41	112	16	41	112	16	41	16	41
17	17	112	16	42		17	42		17	42	17	42
18	18		17	43		18	43		18	43	18	43
19	19		18	44		19	44		19	44	19	44
20	20		19	45		20	45		20	45	20	45
21	21		20	46		21	46		21	46	21	46
22	22		21	47		22	47		22	47	22	47
24	23		22	48		23	48		23	48	23	48
25	24		23	49		24	49		24	49	24	54
26	25		24	50		25	50		25	50	25	60
28	26		25	51		26	51		26	51	26	
30	28		26	52		27	52		27	52	27	
32	30		27	53		28	53		28	53	28	
36	32		28	54		29	54		29	54	29	
40	35		30	55		30	55		30	60	30	
45	36		31	56		31	56		31	70	31	
48	40		32	57		32	57		32	72	32	
52	45		33	58		33	58		33	80	33	
60										84		

Fig. 19-2-14 Stock sprockets.

No. of Teeth	10 Pitch No. 25 Max. Bore	10 Pitch No. 25 Maximum Hub Dia.	13 Pitch No. 40 Max. Bore	13 Pitch No. 40 Maximum Hub Dia.	16 Pitch No. 50 Max. Bore	16 Pitch No. 50 Maximum Hub Dia.	20 Pitch No. 60 Max. Bore	20 Pitch No. 60 Maximum Hub Dia.	25 Pitch No. 80 Max. Bore	25 Pitch No. 80 Maximum Hub Dia.
11	15	22	20	30	25	37	32	45	41	50
12	16	25	22	34	29	42	33	51	45	69
13	20	28	25	38	33	47	38	57	51	76
14	22	32	30	42	33	53	45	64	58	85
15	22	35	32	46	39	58	45	70	61	93
16	25	38	32	50	43	63	50	76	69	101
17	28	40	35	54	45	68	56	82	71	109
18	31	44	39	58	48	73	58	88	79	118
19	32	47	43	62	52	78	62	94	84	126
20	33	50	45	67	57	83	68	100	89	134
21	34	53	45	71	58	88	71	106	95	142
22	36	56	49	75	62	94	75	111	98	150
23	40	59	53	79	67	99	79	119	106	158
24	43	62	57	83	71	104	83	124	116	167
25	45	65	58	87	72	109	86	131	119	175

Fig. 19-2-13 Maximum bore and hub diameters.

sizes near this number of teeth is either 60 or 70 teeth (Fig. 19-2-14), it may be more economical and time-saving to try to use a combination of standard sprockets. In rechecking the smaller sprocket, the 19-tooth sprocket would also be acceptable. This would require a large sprocket of 19 × 3.17 teeth = 60.2 teeth (use 60 teeth). Since the 19- and 60-tooth sprockets are acceptable and standard, it would be more economical to use this combination.

Step 6. Chain Length in Pitches.
Since 19- and 60-tooth sprockets are to be placed on 570 mm centers, calculations are as follows to determine chain length:

Chain length in pitches =

$$2C + \frac{M}{2} + \frac{S}{C}$$

where C = center distance ÷ pitch
= 570 ÷ 13 = 43.8 = 44

M = total number of teeth on both sprockets
= 19 + 60 = 79

S = value obtained from table (Fig. 19-2-8)
$F = 60 - 19 = 41$
$S = 42.58$

Substituting values for C, M, and S, we get

Chain length in pitches =

$$2 \times 44 + \frac{79}{2} + \frac{42.58}{44} = 128.47$$

Since the chain is to couple to an even number of pitches, we will use 128 pitches since the leeway on the 570 mm centers is not critical.

Step 7. Chain Length in Millimetres.

Length of chain = number of pitches × pitch
= 128 × 13 = 1664 mm

REFERENCES AND SOURCE MATERIAL

1. B. L. Pearce, "Chains," *Machine Design,* vol. 37, no. 14, 1965.
2. American Chain Association.

Assignments

The following problems provide practice in the design of roller chain drives. It should be remembered that more than one combination of chain and sprockets will give acceptable results.

Assignment 19-2-A

1. A tumbler barrel is to be driven at approximately 40 r/min by a speed reducer powered by a 3.7 kW electric motor. The reducer output speed is 100 r/min, and the output shaft is 44 mm diameter. The shaft diameter of the tumbling barrel is 50 mm. The shaft center distance is approximately 900 mm. Select a single chain (heavy shock).

2. This is the same as Problem 1, except that a double chain is to be used.

3. The head shaft of an apron conveyor, which handles rough castings from a shakeout, operates at 66 r/min and is driven by a gear motor whose output is 5.6 kW at 100 r/min. The head shaft has a 50 mm diameter, and the gear motor shaft has a 44 mm diameter. Shaft center distance should not exceed 1055 mm. Select a multiple chain (moderate shock).

Assignment 19-2-B

4. A gear-type lubrication pump located in the base of a large hydraulic press is to be driven at 860 r/min from a 32 mm diameter shaft operating at 1000 r/min. The pump is rated at 2.4 kW and has a 35 mm diameter shaft. Shaft center distance must not be less than 250 mm.

5. A 7.5 kW, 480 r/min electric motor is to drive a line shaft, which is subject to light service, at 160 r/min. The motor shaft will be in approximately the same horizontal plane as the line shaft. The diameters of the motor shaft and line shaft are 42 and 44 mm, respectively. A shaft distance of 1220 to 1520 mm will be acceptable. Select a triple chain.

6. A centrifugal fan is to be driven at 2800 r/min by a 7.5 kW electric motor. The motor speed is 1800 r/min, and the shaft has a 35 mm diameter. The compressor shaft has a 32 mm diameter. The center distance is to be approximately 500 mm. The overall drive must not exceed a 125 mm radius on the motor or a 75 mm radius at the fan.

UNIT 19-3
GEAR DRIVES

The function of a gear is to transmit motion, rotating or reciprocating, from one machine part to another and where necessary reduce or increase the revolutions per minute of a shaft. *Gears* are rolling cylinders or cones having teeth on their contact surfaces to ensure positive motion. See Figs. 19-3-1 and 19-3-2.

There are many kinds of gears, and they may be grouped according to the position of the shafts that they connect. *Spur gears* connect parallel shafts, *bevel gears* connect shafts whose axes intersect, and *worm gears* connect shafts whose axes do not intersect. A spur gear with a rack converts rotary motion to reciprocating or linear motion. The smaller of two gears is known as the *pinion*.

Gear design is very complicated, dealing with such problems as strength, wear, and material selection. Normally a drafter selects a gear from a catalog.

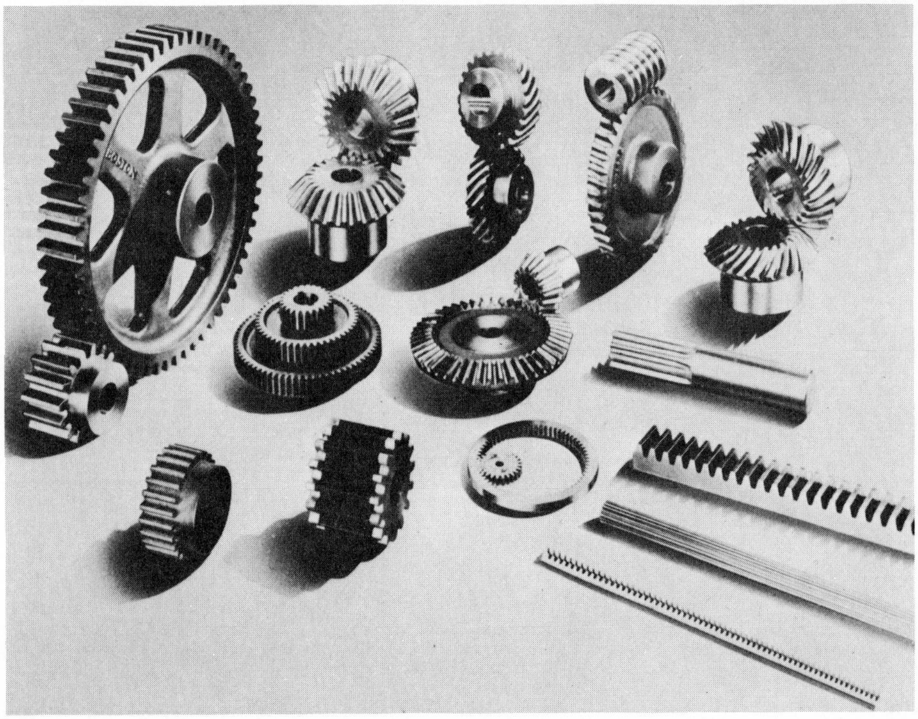

Fig. 19-3-1 Gears. (Boston Gear Works)

MODULE FOR METRIC SIZE GEARS	DIAMETRAL PITCH FOR INCH SIZE GEARS	14.5°	20°
6.35	4		
5.08	5		
4.23	6		
3.18	8		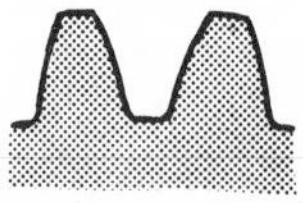
2.54	10		
2.17	12		
1.59	16		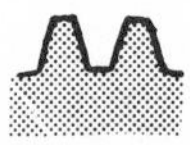
1.27	20		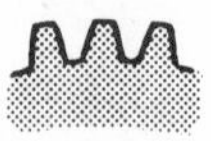
1.06	24		
0.79	32		

NOTE: MODULE SIZES SHOWN ARE CONVERTED INCH SIZES.

Fig. 19-3-2 Gear-teeth sizes.

Most gears are made of cast iron or steel, but brass, bronze, and plastic are used when factors such as wear or noise must be considered.

Spur Gears

Spur gear proportions and the shape of gear teeth are standardized, and the definitions, symbols, and formulas are given in Figs. 19-3-3 to 19-3-6.

Gears are used to transmit motion and power at constant angular velocity. The specific form of the gear which best produces this constant angular velocity is the *involute*. The action between the teeth of a pair of gears is called *mating*, or *conjugate*, action.

Classically, the involute is described as the curve traced by a point on a taut string unwinding from a circle. This circle is called the *base circle*.

Every involute gear has only one base circle from which all the involute surfaces of the gear teeth are generated. This base circle is not a physical part of the gear and cannot be measured directly. The contact between mating involutes takes place along a line which is always tangent to and crossing between the two base circles. This is the *line of action*. This line of action is a line on paper only. In reality, on spur, helical,

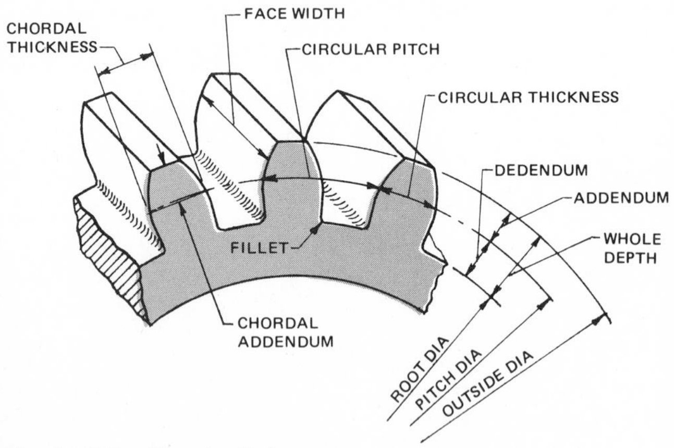
FACE WIDTH
CHORDAL THICKNESS
CIRCULAR PITCH
CIRCULAR THICKNESS
DEDENDUM
ADDENDUM
WHOLE DEPTH
FILLET
CHORDAL ADDENDUM
ROOT DIA
PITCH DIA
OUTSIDE DIA

Fig. 19-3-3 Gear-teeth terms.

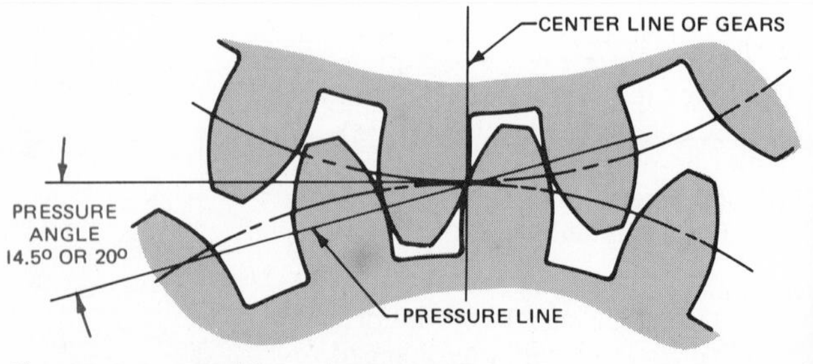
CENTER LINE OF GEARS
PRESSURE ANGLE 14.5° OR 20°
PRESSURE LINE

Fig. 19-3-4 Meshing of gear teeth.

worm, and all gears which are not paper-thin, it is a plane of action.

The 14.5° pressure angle has been used for many years and remains useful for duplicate or replacement gearing or in situations where the control of backlash is of primary importance.

The 20° pressure angle has become the standard for new gearing because of the smoother and quieter running characteristics, greater load-carrying ability, and the fewer number of teeth affected by undercutting.

Standard spur gears having a 14.5° pressure angle should have a minimum of

TERM AND SYMBOL	DEFINITION	FORMULA	
		METRIC GEARS	INCH GEARS
Pitch Diameter—D	The diameter of an imaginary circle on which the gear tooth is designed	D = Mo × N	D = N ÷ P
Number of Teeth—N	The number of teeth on a gear	N = D/Mo	N = D× P
Module—Mo	The length of pitch diameter per tooth	Mo = D/N	
Diametral Pitch—P	A ratio equal to the number of teeth on a gear for every inch of pitch diameter		P = N/D
Addendum—A	The radial distance from the pitch circle to the top of the tooth	14.5° or 20° A = Mo 20° stub A = 0.8 × Mo	14.5° or 20° A = I/P 20° stub A = 0.8/P
Dedendum—B	The radial distance from the pitch circle to the bottom of the tooth	14.5° or 20° B = 1.157 × Mo 20° stub B = Mo	14.5° or 20° B = 1.157/P 20° stub B = I/P
Whole Depth—WD	The overall height of the tooth	14.5° or 20° WD = 2.157 × Mo 20° stub WD = 1.8 × Mo	14.5° or 20° WD = 2.157/P 20° stub WD = 1.8/P
Clearance—C	The radial distance between the bottom of one tooth and the top of the mating tooth	14.5° or 20° C = 0.157 × Mo 20° stub C = 0.2 × Mo	14.5° or 20° C = .157/P 20° stub C × .2/P
Outside Diameter—OD	The overall diameter of the gear	14.5° or 20° OD = D + 2A = D + 2 Mo 20° stub OD = D + 2A = D + 1.6 Mo	14.5° or 20° OD = D + 2A = (N + 2) ÷ P 20° stub OD = D + 2A = (N + 1.6) ÷ P
Root Diameter—RD	The diameter at the bottom of the tooth	14.5° or 20° RD = D − 2B = D + 2.314 Mo 20° stub RD = D − 2B = D + 2 Mo	14.5° or 20° RD = D − 2B = (N − 2.314) ÷ P 20° stub RD = D − 2B = (N − 2) ÷ P
Base Circle—BC	The circle from which the involute curve of the tooth is formed	BC = D Cos PA	BC = D Cos PA
Pressure Angle—PA	The angle between the direction of pressure between contacting teeth and a line tangent to the pitch circle	14.5° or 20°	14.5° or 20°
Backlash	The clearance between the teeth of two meshing gears	—	—
Circular Pitch—CP	The distance measured from the point of one tooth to the corresponding point on the adjacent tooth on the circumference of the pitch diameter	CP = 3.1416 D/N = 3.1416 Mo	CP = 3.1416 D ÷ N = 3.1416 ÷ P
Circular Thickness—T	The thickness of a tooth or space measured on the circumference of the pitch diameter	T = 3.1416 D/2N = 1.57 D/N = 1.57 Mo	T = 3.1416 D ÷ 2N = 1.57 ÷ P
Chordal Thickness—Tc	The thickness of a tooth or space measured along a chord on the circumference of the pitch diameter	Tc = D Sin (90°/N)	Tc = D Sin (90° ÷ N)
Chordal Addendum—Ac	Chordal Addendum, also known as Corrected Addendum, is the perpendicular distance from chord to outside circumference of gear	Ac = A + T^2/4D	Ac = A + T^2 ÷ 4D

Fig. 19-3-5 Spur gear definitions and formulas.

16 teeth with at least 40 teeth in a mating pair. Gears with 20° pressure angle should have a minimum of 13 teeth with at least 26 teeth in a mating pair.

The formulas for the 14.5° and the 20° full-depth teeth are identical.

The 20° stub tooth differs from the 20° standard tooth depth. The stub tooth is shorter and stronger and therefore is preferred where maximum power transmission is required.

DRAWING GEAR TEETH

The teeth on a gear are not normally shown on the working drawings. Instead, they are represented by solid, broken, and hidden lines which will be discussed under working drawings. However, presentation or display drawings normally require the teeth to be shown.

Since the exact form of an involute tooth would require too much time to draw, approximate methods are used. The two most common methods are shown in Fig. 19-3-6. To draw the teeth using the approximate representation of involute spur-gear teeth, lay out the root, pitch, and outside circles. On the pitch circle mark off the circular thickness. Through the pitch point on the pitch circle draw the pressure line at an angle of 14.5° with the line tangent to the pitch circle for the 14.5° involute tooth (use 15° for convenience) or 20° for the 20° involute tooth. Draw the base circle tangent to this pressure line. With the compass set to a radius equal to one-eighth the pitch diameter and the compass point on the base circle, draw arcs passing through the circular thickness points established on the pitch diameter, starting at the base circle and ending at the top of the teeth. The part of the tooth profile below the base circle is drawn as a radial line ending in a small fillet at the root circle.

For a closer approximation of the involute tooth profile, the Grant's odontograph method is used. Lay out the outside, pitch, root, and base circles and circular thickness in the same manner as used in the approximation method. The top portion of the tooth profile from point *A* to point *B* is drawn with the radius *R*, and the portion of the tooth profile from point *B* to point *C* is drawn with the radius *r*.

The values of the radii *R* and *r* are found by multiplying the numbers in the table by the module for metric-size gears or by dividing the numbers found in the table by the diametral pitch for inch-size gears. The lower portion of the tooth from points *C* to *D* is drawn as a radial line ending in a small fillet at the root circle.

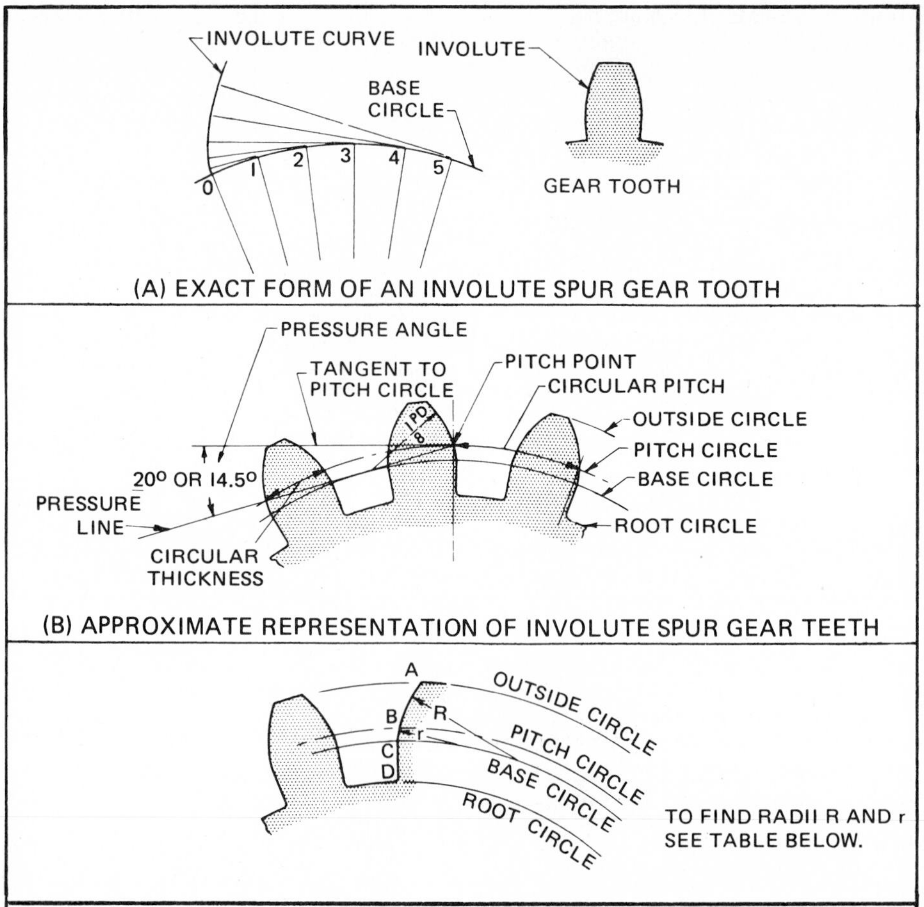
INVOLUTE CURVE
INVOLUTE
BASE CIRCLE
0
1
2
3
4
5
GEAR TOOTH
(A) EXACT FORM OF AN INVOLUTE SPUR GEAR TOOTH
PRESSURE ANGLE
TANGENT TO PITCH CIRCLE
PITCH POINT
CIRCULAR PITCH
OUTSIDE CIRCLE
PITCH CIRCLE
BASE CIRCLE
ROOT CIRCLE
20° OR 14.5°
PRESSURE LINE
CIRCULAR THICKNESS
(B) APPROXIMATE REPRESENTATION OF INVOLUTE SPUR GEAR TEETH
A
B
C
D
R
r
OUTSIDE CIRCLE
PITCH CIRCLE
BASE CIRCLE
ROOT CIRCLE
TO FIND RADII R AND r SEE TABLE BELOW.

METRIC GEARS Radii in Millimetres			INCH GEARS Radii in Inches	
RADIUS R Multiply No. by Mo	RADIUS r Multiply No. by Mo	No. of Teeth	RADIUS R Divide No. by P	RADIUS r Divide No. by P
2.51	0.96	12	2.51	0.96
2.62	1.09	13	2.62	1.09
2.72	1.22	14	2.72	1.22
2.82	1.34	15	2.82	1.34
2.92	1.46	16	2.92	1.46
3.02	1.58	17	3.02	1.58
3.12	1.69	18	3.12	1.69
3.22	1.79	19	3.22	1.79
3.32	1.89	20	3.32	1.89
3.41	1.98	21	3.41	1.98
3.49	2.06	22	3.49	2.06
3.57	2.15	23	3.57	2.15
3.64	2.24	24	3.64	2.24
3.71	2.33	25	3.71	2.33
3.78	2.42	26	3.78	2.42
3.85	2.50	27	3.85	2.50
3.92	2.59	28	3.92	2.59
3.99	2.67	29	3.99	2.67
4.06	2.76	30	4.06	2.76
4.13	2.85	31	4.13	2.85
4.20	2.93	32	4.20	2.93
4.27	3.01	33	4.27	3.01
4.33	3.09	34	4.33	3.09
4.39	3.16	35	4.39	3.16
4.20	3.23	36	4.20	3.23
4.45	4.20	37-40	4.45	4.20
4.63	4.63	41-45	4.63	4.63
5.06	5.06	46-51	5.06	5.06
5.74	5.74	52-60	5.74	5.74
6.52	6.52	61-70	6.52	6.52
7.72	7.72	71-90	7.72	7.72
9.78	9.78	91-120	9.78	9.78
13.38	13.38	121-180	13.38	13.38
21.62	21.62	181-360	21.62	21.62

Fig. 19-3-6 Methods of drawing involute spur gear teeth.

WORKING DRAWINGS OF SPUR GEARS

The working drawings of gears which are normally cut from blanks are not complicated. A sectional view is sufficient unless a front view is required to show web or arm details. Since the teeth are cut to shape by cutters, they need not be shown in the front view. See Figs. 19-3-7 and 19-3-8.

ANSI recommends the use of phantom lines for the outside and root circles. The CSA recommends that a solid line be used for the outside and root circles and a center line for the pitch circle. In the section view, the root and outside circles are shown as solid lines for both standards.

The dimensioning for the gear is divided into two groups, because the finishing of the gear blank and the cutting of the teeth are separate operations in the shop. The gear-blank dimensions are shown on the drawing while the gear tooth information is given in a table.

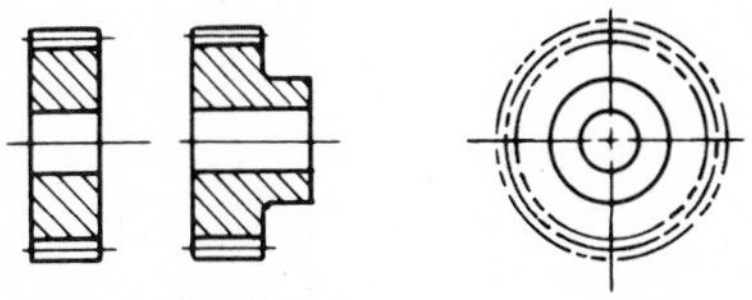

(A) PLAIN STYLE

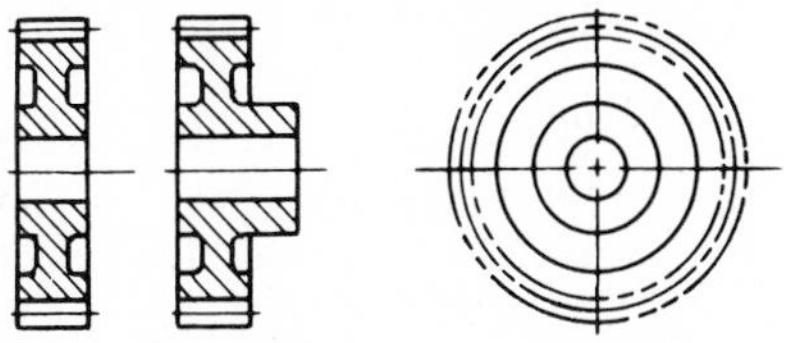

(B) WEBBED STYLE

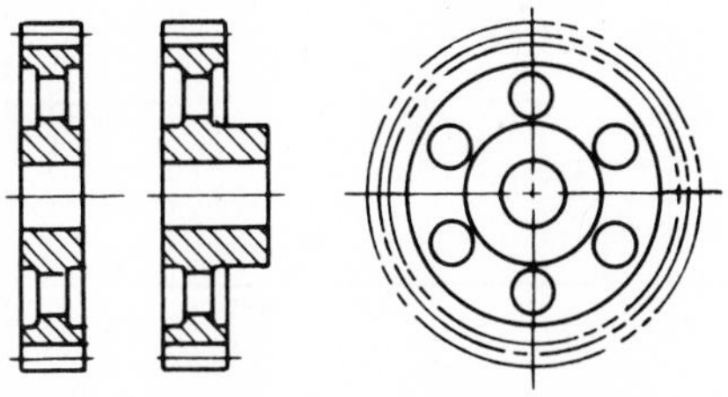

(C) WEBBED WITH CORED HOLES

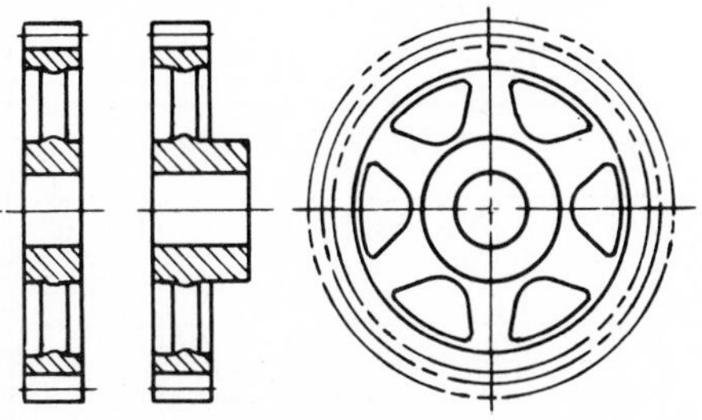

(D) SPOKED STYLE

Fig. 19-3-7 Stock spur gear styles.

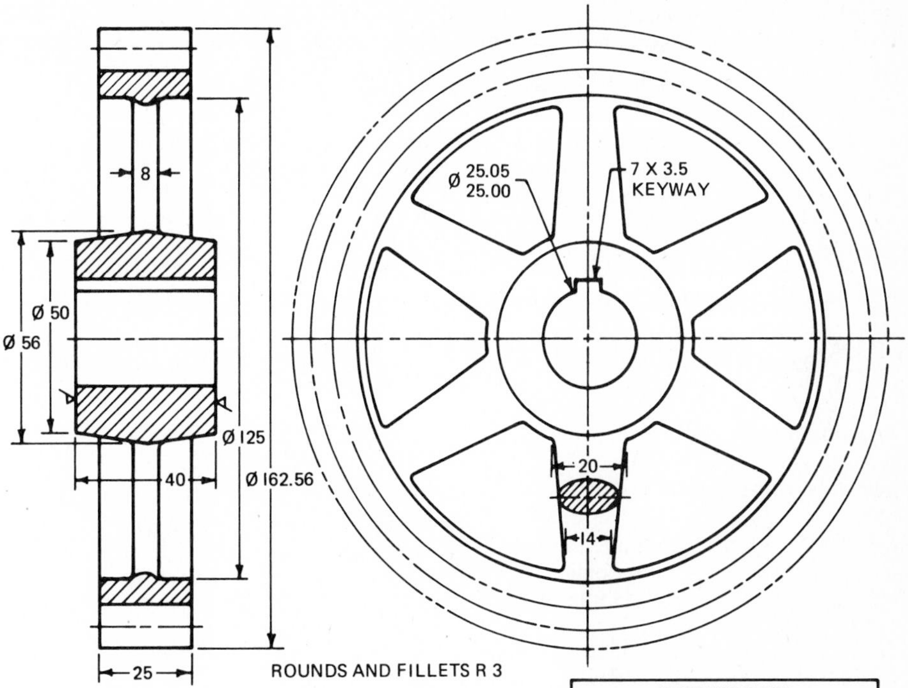
8
Ø 50
Ø 56
Ø 125
40
Ø 162.56
25
ROUNDS AND FILLETS R 3
Ø 25.05
25.00
7 X 3.5
KEYWAY
20
14

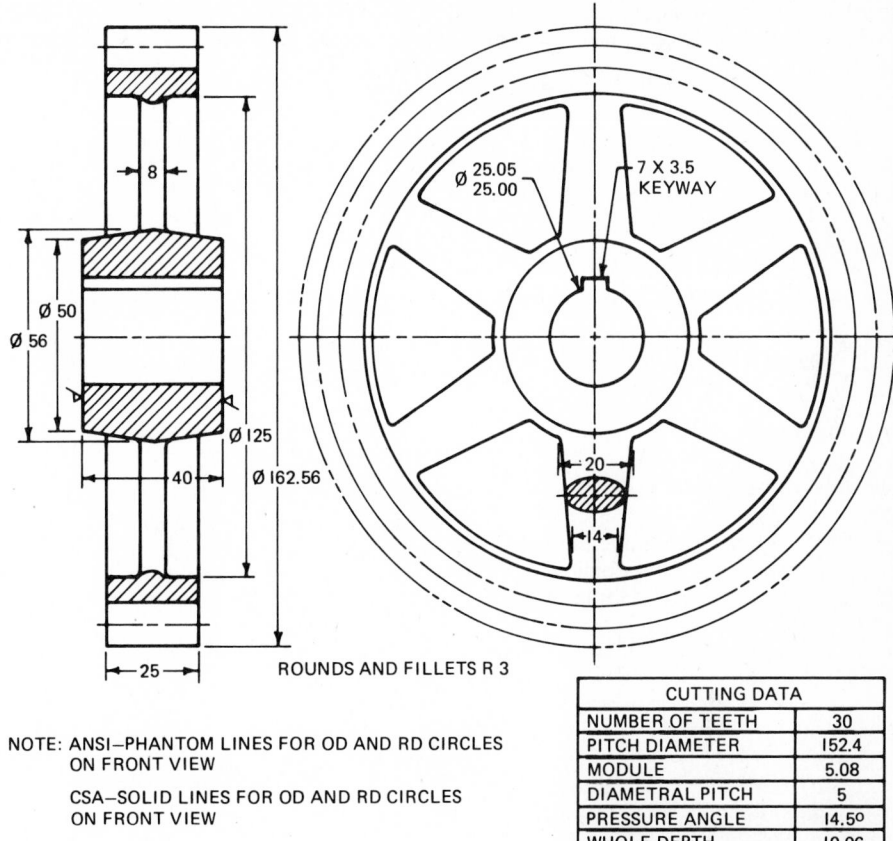

NOTE: ANSI–PHANTOM LINES FOR OD AND RD CIRCLES ON FRONT VIEW

CSA–SOLID LINES FOR OD AND RD CIRCLES ON FRONT VIEW

CUTTING DATA	
NUMBER OF TEETH	30
PITCH DIAMETER	152.4
MODULE	5.08
DIAMETRAL PITCH	5
PRESSURE ANGLE	14.5°
WHOLE DEPTH	10.96
CHORDAL ADDENDUM	5.18
CHORDAL THICKNESS	7.97

Fig. 19-3-8 Working drawing of a spur gear.

The only differences in terminology between metric-size and inch-size gear drawings are the terms *module* and *diametral pitch*. *Module* is the term used on metric gears. It is the length of pitch diameter per tooth measure in millimetres:

$$\text{Module} = \text{Mo} = \frac{D}{N}$$

For inch-size gears, the term *diametral pitch* is used instead of the term *module*. The diametral pitch is a ratio of the number of teeth to a unit length of pitch diameter:

$$\text{Diametral pitch} = \frac{N}{D}$$

From these definitions it can be seen that the module is equal to the reciprocal of the diametral pitch and thus is not its metric dimensional equivalent. If the diametral pitch is known, the module can be obtained as follows:

$$\text{Module} = 25.4 \div \text{diametral pitch}$$

Gears presently being used in North America are designed in the inch system and have a standard diametral pitch instead of a preferred standard module. Therefore it is recommended that the diametral pitch be referenced beneath the module on the drawing when gears designed with standard pitches are used. For gears designed with standard modules, the diametral pitch need not be referenced on the gear drawing.

SPUR-GEAR CALCULATIONS

Center Distance. The center distance between the two shaft centers is determined by adding the pitch diameter of the two gears and dividing the sum by 2.

Example 1. A 3.18 module, 24-tooth pinion mates with a 96-tooth gear. Find the center distance.

Pitch diameter of pinion
= number of teeth × module
= $24 \times 3.18 = \phi 76.3$
= $96 \times 3.18 = \phi 305.2$

Sum of the two pitch diameter = 76.3 + 305.2 = 381.5

Center distance = ½ sum of the two pitch diameter =

$$\frac{381.5}{2} = 190.75 \text{ mm}$$

Ratio. The ratio of gears is a relationship between any of the following:

1. Revolutions per minute of the gears
2. Number of teeth on the gears
3. Pitch diameter of the gears

The ratio is obtained by dividing the larger value of any of the three by the corresponding smaller value.

Example 1. A gear rotates at 90 r/min and the pinion at 360 r/min.

$$\text{Ratio} = \frac{360}{90} = 4 \text{ or ratio} = 4{:}1$$

Example 2. A gear has 72 teeth, the pinion 18 teeth.

$$\text{Ratio} = \frac{72}{18} = 4 \text{ or ratio} = 4{:}1$$

Example 3. A gear with a pitch diameter of 216 mm meshes with a pinion having a pitch diameter of 54 mm.

$$\text{Ratio} = \frac{D \text{ of gear}}{D \text{ of pinion}}$$

$$= \frac{216}{54} = 4$$

or $\text{Ratio} = 4{:}1$

Determining the Pitch Diameter and Outside Diameter. The pitch diameter of a gear can readily be found if the number of teeth and the module are known. The OD is equal to the pitch diameter plus two addendums. The addendum for a 14.5° spur-gear tooth is equal to the module.

Example 4. A 14.5° spur gear has a module of 6.35 and 34 teeth.

$$\text{Pitch diameter} = N \times Mo$$
$$= 34 \times 6.35 = 216 \text{ mm}$$
$$OD = D + 2A = 216 + 2(6.35)$$
$$= 228.7 \text{ mm}$$

Assignments

1. On an A3- or B-size sheet, make working drawings for the two gears described in either Fig. 19-3-A or Fig. 19-3-B. Gear no. 1 will require one view only and be drawn to the scale 1:1. Gear no. 2 will require two views and will be drawn to the scale 1:2. Select a proper key size and use your judgment for dimensions not given. Include with each gear drawing a cutting data block.

2. On an A3- or B-size sheet, prepare a working drawing of two gears in mesh from the information found in Fig. 19-3-C or 19-3-D. Show two views with three or four teeth shown in mesh. Add suitable keyways and use your judgment for dimensions not given. Include cutting data for each gear. Scale is 1:1.

3. On an A3- or B-size sheet, complete the missing information on the gear-train problems shown in Fig. 19-3-E or 19-3-F.

GEAR #1

Tooth Form—14.5°
D = 128
Module—6.35
Face Width—26
Web—10
Shaft—ϕ28
Hub—ϕ50 x 40 Lg
Matl—MI

GEAR #2

Tooth Form—20°
N = 44
Module—6.35
Face Width—46
Shaft—ϕ45
Hub—ϕ76 x 70.6 (Total Length)
6 Spokes—16 Thk, 40 Wide, Tapered to 30 Wide
Matl—MI

Fig. 19-3-A Single spur gears.

GEAR #1

Tooth Form—14.5°
D—ϕ5.00
P—4
Face Width—1.00
Web—40
Shaft—ϕ1.10
Hub—ϕ1.90 x 1.50 Lg
Matl—MI

GEAR #2

Tooth Form—20°
N—44
P—4
Face Width—1.75
Shaft—ϕ1.75
Hub—ϕ3.00 x 2.75 (Total Length)
6 Spokes—.60 Thk, 1.50 Wide, Tapered to 1.10 Wide
Matl—MI

Fig. 19-3-B Single spur gears.

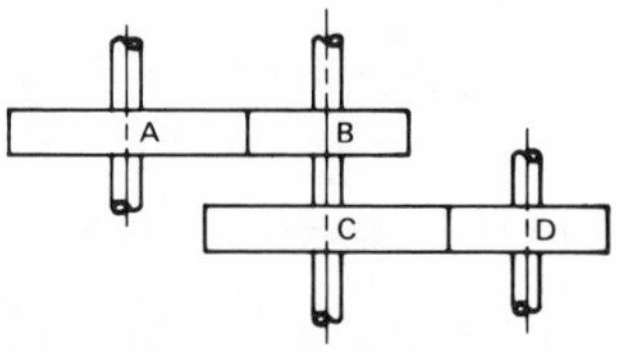
A
B
C
D

GEAR	D	N	Mo	DIRECTION	REV/MIN	CENTER DISTANCE
A			6.35	C'WISE	300	
B		12				
C						
D		12	3			

GEAR	D	N	Mo	DIRECTION	REV/MIN	CENTER DISTANCE
A		54			160	
B			6			
C			5			
D				C'WISE	2400	

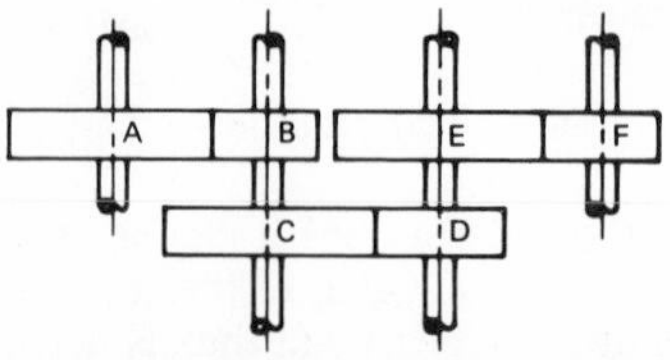
A
B
E
F
C
D

GEAR	D	N	Mo	DIRECTION	REV/MIN	CENTER DISTANCE
A			4	A'WISE	240	
B		18				
C						
D		16				
E			6			
F		40				

Fig. 19-3-E Gear-train calculations.

GEAR DATA

Tooth Form—14.5°
Center to Center Distance—127
Gear—N—24
Face Width—30
Shaft—ϕ32
Web—10
Hub—ϕ54 x 38 Lg
Matl—MI
Pinion—N—16
Shaft—ϕ30
Matl—Steel

Fig. 19-3-C Meshing spur gears.

GEAR DATA

Tooth Form—14.5°
Center to Center Distance—5.00
Gear—N—24
Face Width—1.10
Shaft—ϕ1.25
Web—.40
Hub—ϕ2.10 x 1.50 Lg
Matl—MI
Pinion—N—16
Shaft—ϕ1.10
Matl—Steel

Fig. 19-3-D Meshing spur gears.

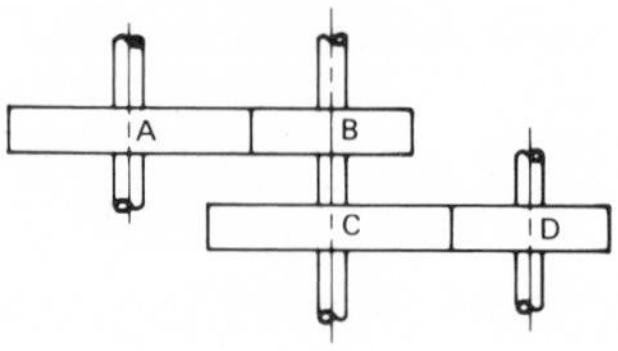
A
B
C
D

GEAR	D	N	P	DIRECTION	RPM	CENTER DISTANCE
A	7.0		4	C'WISE	300	
B		12				
C	6.0					
D		12	3			

GEAR	D	N	P	DIRECTION	RPM	CENTER DISTANCE
A		54			160	
B	3.0		6			
C			5			
D	2.4			C'WISE	2400	

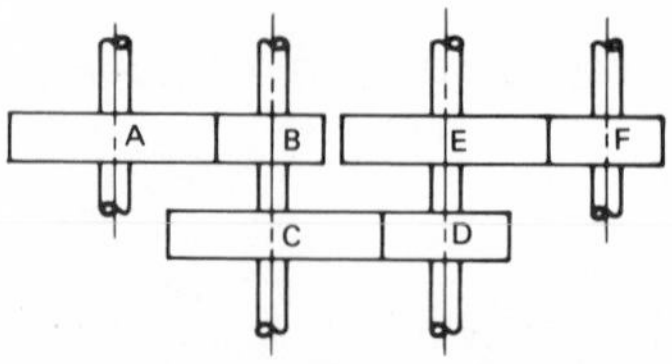
A
B
E
F
C
D

GEAR	D	N	P	DIRECTION	RPM	CENTER DISTANCE
A	7.5		4	A'WISE	240	
B		18				
C	10.0					
D	3.2	16				
E	8.0		6			
F		40				

Fig. 19-3-F Gear-train calculations.

UNIT 19-4

POWER-TRANSMITTING CAPACITY OF SPUR GEARS[1]

Gear drives are required to operate under such a wide variety of conditions that it is very difficult and expensive to determine the best gear set for a particular application.

The most economical procedure is to select standard stock gears with an adequate load rating for the application.

Approximate kilowatt ratings for spur gears of various sizes (numbers of teeth), at several operating speeds (revolutions per minute), are given in catalogs with the spur-gear listings.

The kilowatt ratings are for properly mounted, lubricated gears of catalog face widths and are based on the safe beam strength of gear teeth as calculated with the Lewis formula (Barth revision).

Ratings for gear sizes and/or speeds not listed may be estimated from the values shown in Fig. 19-4-1, based on the Lewis formula.

Pitch line velocities exceeding 5 m/s for 14.5° PA (pressure angle), or 6 m/s for 20° PA, are not recommended for metallic spur gears. Ratings are listed for speeds below these limits.

The ratings given (or calculated) should be satisfactory for gears used under normal operating conditions, that is, when they are properly mounted and lubricated, carrying a smooth load (without shock) for 8 to 10 hours a day.

The charts shown in Fig. 19-4-1 indicate the approximate kilowatt ratings of 16- and 20-tooth steel spur gears of several teeth sizes operating at various speeds. They may be used to determine the approximate module of a 16- or 20-tooth steel pinion that will carry the kilowatts required at the desired speed. The intersection of the lines representing values of revolutions per minute and kilowatts indicates the approximate gear module required.

Often the size of the pinion shaft is known and a pinion may be tentatively selected whose pitch diameter is at least twice the shaft diameter.

The number of teeth normally should not be less than 16 to 20 in a 14.5° pinion, nor less than 13 in a 20° pinion.

A trial module may be obtained by dividing the pinion *D* by the number of pinion teeth or by use of the module selection charts. Select gears by using the kilowatt tables normally found in gear catalogs.

Ratings shown for spur gears in catalogs normally are for class 1 service. For other classes of service, the service factors in Fig. 19-4-2 should be used.

Catalog ratings apply when gears are accurately mounted and properly lubricated with pitch line velocities not exceeding 5 m/s for 14.5° PA or 6 m/s for 20° PA.

SELECTING THE SPUR-GEAR DRIVE

1. Determine the class of service.

2. Multiply the kilowatts required for the application by the service factor.

3. Select spur-gear pinion with a catalog rating equal to or greater than the kilowatts determined in step 2.

4. Select driven spur gear with a catalog rating equal to or greater than the kilowatts determined in step 2.

Example 1. Select a pair of 20° spur gears which will drive a machine at 150 r/min. Size of driving motor = 20 kW, 600 r/min. Service factor = 1.

Solution: Since the service factor is 1, we do not need to increase or decrease the design kilowatts. Refer to the charts shown in Fig. 19-4-1*b*, which show design data for 20° spur gears hav-

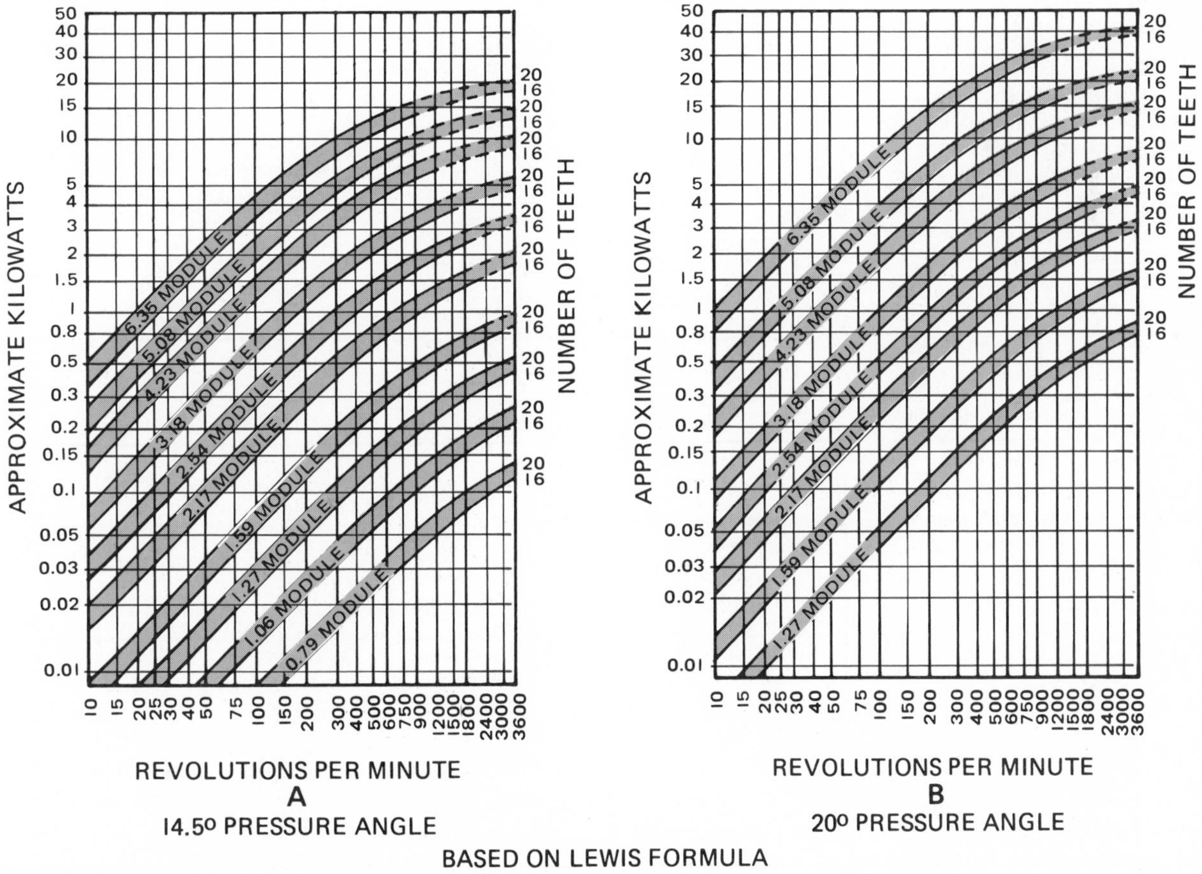
APPROXIMATE KILOWATTS
NUMBER OF TEETH
6.35 MODULE
5.08 MODULE
4.23 MODULE
3.18 MODULE
2.54 MODULE
2.17 MODULE
1.59 MODULE
1.27 MODULE
1.06 MODULE
0.79 MODULE
REVOLUTIONS PER MINUTE
A
14.5° PRESSURE ANGLE
B
20° PRESSURE ANGLE
BASED ON LEWIS FORMULA

Fig. 19-4-1 Module selection chart for 16- and 20-spur gears.

BGW SPUR SERVICE CLASS	OPERATING CONDITIONS	SERVICE FACTOR
Class I	Continuous 8 to 10 hrs. per day duty, with Smooth Load (No Shock).	1.0
Class II	Continuous 24 hr. duty, with Smooth Load, or 8 to 10 hrs. per day, with Moderate Shock.	1.2
Class III	Continuous 24 hr. duty with Moderate Shock Load.	1.3
Class IV	Intermittent duty, not over 30 min. per hr., with Smooth Load (No Shock).	0.7
Class V	Hand operation, Limited Duty, with Smooth Load (No Shock).	0.5

Heavy Shock loads and/or severe wear conditions require the use of higher Service Factors. Such conditions may require Factors of 1.5 to 2.0 or greater than required for Class I service.

Fig. 19-4-2 Service class and factor for spur gears.

ng 16 and 20 teeth. Selecting a pinion having 16 teeth and reading vertically on the column showing 600 r/min to kilowatt rating of 20, we find that the required module is 6.35. (Select the module equal to or greater than the kilowatts required.)

- Pinion: N = 16, Mo = 6.35, D = 16 × 6.35 = 101.6 mm
- Ratio 4:1
- Gear: N = 16 × 4 = 64, Mo = 6.35, D = 64 × 6.35 = 406.4 mm

Example 2. A 4 kW, 1200 r/min motor is used to drive a machine that runs 8 hours a day under moderate shock. If the machine is to run at 200 r/min and at the capacity of the motor, what spur gears would you select?

Solution The operating conditions of the machine are such that the machine fits into the service class 2 and requires a service factor of 1.2. Therefore

Kilowatts required for design purposes = 4 × 1.2 = 4.8 kW

The pinion will be mounted on the motor and runs at 1200 r/min. Selecting a 20° pinion having 16 teeth, refer to Fig. 19-4-1*b* to find the required module. Reading vertically on the 1200 r/min line and horizontally at 4.8 kW, we find that the required module is 3.18.

- Pinion: N = 16, Mo = 3.18, D = 16 × 3.18 = 50.88 mm
- Gear: N = 16 × 6 = 96, Mo = 3.18, D = 96 × 3.18 = 305.28 mm

Example 3. A 7.5 kW, 900 r/min motor is attached by means of 14.5° spur gears to a punch that operates 24 hours a day. The reduction in revolutions per minute is 4:1. Select a gear and pinion assuming that the punch is being operated at motor capacity.

Solution: The operating conditions of the machine are such that the machine fits into the service class 3 and requires a service factor of 1.3. Therefore,

Kilowatts required for design purposes = 7.5 × 1.3 = 9.75 kW

The pinion is mounted on the motor and runs at 900 r/min. Refer to Fig. 19-4-1*a*. Reading vertically on the 900 r/min line and horizontally at 9.75 kW, we may select either a pinion having a module of 5.08 and N of 20 or a module of 6.35 and N of 16. First, using a module of 5.08, we have

- Pinion: N = 20, Mo = 5.08, D = 101.6 mm. Gear travels at 225 r/min, or one-fourth of pinion revolutions per minute.
- Gear: N = 4 × 20 = 80, Mo = 5.08, D = 80 × 5.08 = 406.4 mm.

Second, using a module of 6.35, we have

- Pinion: N = 16, Mo = 6.35, D = 101.6 mm
- Gear: N = 64, Mo = 6.35, D = 406.4 mm

Since both sets of gears are of the same diameter, the overall size is not a factor. Checking on the cost per set, we find that there would be considerable savings by selecting the gear and pinion having the module of 5.08. Since the extra strength of the gear set having a module of 6.35 is not required in this instance, we would recommend the gear and pinion having a module of 5.08.

REFERENCE AND SOURCE MATERIAL

1. Boston Gear Works.

Assignments

1. On an A3- or B-size sheet, show your calculations for the design of a suitable pair of spur gears to operate the equipment shown in Assignment 19-4-A or 19-4-B.

2. On an A3- or B-size sheet, show your calculations for the design of a suitable pair of spur gears to operate the equipment shown in Assignment 19-4-C.

1. A 1200 r/min motor drives, by means of a spur gear and pinion, a machine rated at 7.5 kW and operating under moderate shock 8 hours a day. The reduction in rev/min is 3:1. Select a suitable pair of spur gears to transmit the power required.

2. An 1800 r/min motor drives a machine which is rated at 2 kW and runs at 450 rev/min under moderate shock 18 hours a day. Select a suitable pair of spur gears to transmit the power required.

Assignment 19-4-A Power transmitting problems.

1. A 1200 r/min motor drives, by means of a spur gear and pinion, a machine rated at 10 hp and operating under moderate shock 8 hours a day. The reduction in r/min is 3:1. Select a suitable pair of spur gears to transmit the power required.

2. An 1800 r/min motor drives a machine which is rated at 2.5 hp and runs at 450 r/min under moderate shock 18 hours a day. Select a suitable pair of spur gears to transmit the power required.

Assignment 19-4-B Power transmitting problems.

1. A 900 r/min motor drives an air compressor which operates between 15 to 20 minutes every hour. The compressor which is rated at 5 kW runs at 600 r/min, under smooth operating conditions. Select a suitable pair of spur gears to transmit the power required.

2. A machine which works under smooth operating conditions is used twice a day for about 10 minutes. It is manually operated and is rated at 6 kW and runs at 800 r/min. Two motors are in stock at the plant; one rated at 6 kW and 1200 r/min, the other at 4 kW and 750 r/min. Spur gears are to be used. Make a report to the plant engineer on the selection of motor and gears which you would recommend.

Assignment 19-4-C Power transmitting problems.

UNIT 19-5

RACK AND PINION

The *rack* is a straight bar having teeth which engage the teeth on a spur gear (Fig. 19-5-1). In theory, it is a spur gear having an infinite pitch diameter. Therefore, all circular dimensions become linear. The addendum, dedendum, and tooth thickness are the same as the mating spur gear. To draw the teeth on a rack, lay out the addendum and dedendum distances from the pitch line. Divide the pitch line into lineal pitch distances equal in size to the circular pitch on the gear. Divide each of these spaces in half to get the lineal thickness. Through these points draw the tooth faces at angles of 14.5 or 20° from the vertical lines. Darken in the top and bottom lines of the teeth and add the tooth fillets. The specifications for the teeth of the rack are given in the same manner as for spur gears. See Fig. 19-5-2.

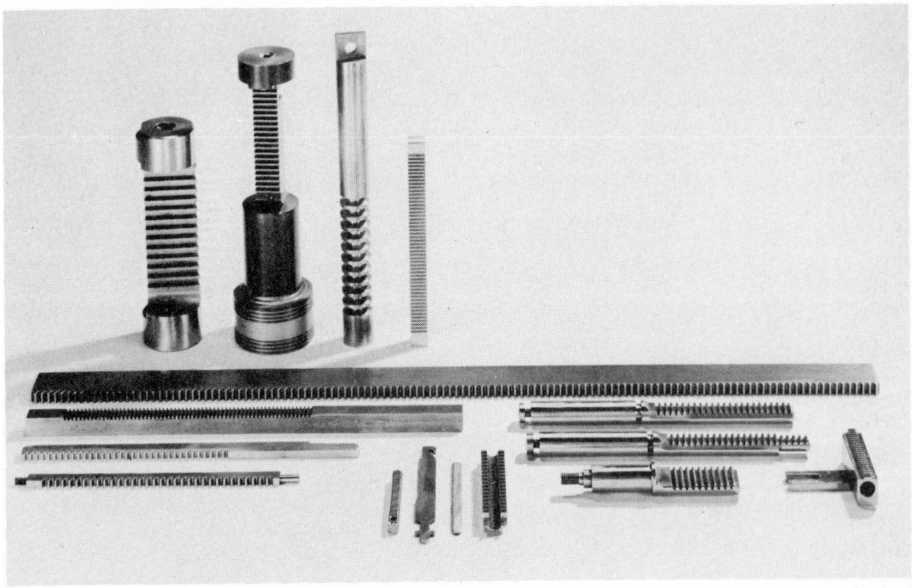

Fig. 19-5-1 Racks.

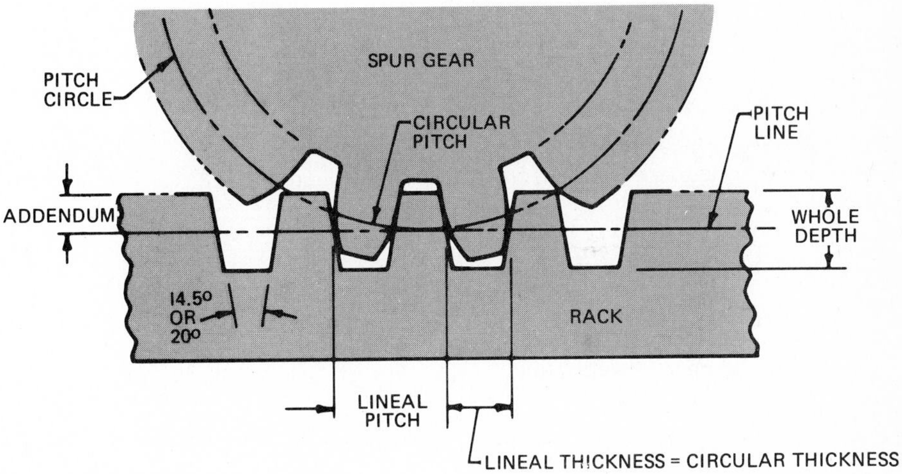
SPUR GEAR
PITCH CIRCLE
CIRCULAR PITCH
PITCH LINE
ADDENDUM
WHOLE DEPTH
14.5° OR 20°
RACK
LINEAL PITCH
LINEAL THICKNESS = CIRCULAR THICKNESS

Fig. 19-5-2 Rack and pinion.

Assignment

On an A3- or B-size sheet, make a working assembly drawing of the gear and rack shown in Fig. 19-5-A or 19-5-B. Use your judgment for dimensions not given. Show four or five teeth in mesh. Scale is 1:1.

REVIEW FOR ASSIGNMENT

Unit 19-3 Gear Formulas

GEAR AND RACK DATA

Gear— N—30
M_0—5.08
Tooth Form—14.5°
Web—11
Shaft—ϕ35
Hub—ϕ58 x 45 Lg
Face Width—32
Matl—MI
Rack—Matl—Steel

Fig. 19-5-A Gear and rack.

GEAR AND RACK DATA

Gear— N—30
P—5
Tooth Form—14.5°
Web—.44
Shaft—ϕ1.38
Hub—ϕ2.25 x 1.75 Lg
Face Width—1.25
Matl—MI
Rack—Matl—Steel

Fig. 19-5-B Gear and rack.

UNIT 19-6

BEVEL GEARS

Bevel gears are used to transmit power between two shafts whose axes intersect. The axes may intersect at any angle, but the most common is 90°. They are similar to rolling cones having the same apex. The teeth are the same shape as spur-gear teeth but taper toward the cone apex. Therefore, many spur-gear terms may apply to bevel gears. *Miter gears* are bevel gears having the same module, pressure angle, and number of teeth.

Figures 19-6-1 and 19-6-2 show bevel gear definitions and formulas.

TERM	FORMULA
Addendum, dedendum, whole depth, pitch diameter, module, diametral pitch, number of teeth, circular pitch, chordal thickness, circular thickness	Same as for spur gears
Pitch cone radius	$\dfrac{D}{2 \times \text{Sin of pitch angle}}$
Pitch cone angle	$\text{Tan pitch angle} = \dfrac{\text{D of gear}}{\text{D of pinion}} = \dfrac{\text{N of gear}}{\text{N of pinion}}$
Addendum angle	$\text{Tan addendum angle} = \dfrac{\text{Addendum}}{\text{Pitch cone radius}}$
Dedendum angle	$\text{Tan dedendum angle} = \dfrac{\text{Dedendum}}{\text{Pitch cone radius}}$
Face angle	Pitch cone angle plus addendum angle
Cutting angle	Pitch cone angle minus dedendum angle
Back angle	Same as pitch cone angle
Angular addendum	Cosine of pitch cone angle × addendum
Outside diameter	Pitch diameter plus two angular addendum
Crown height	Divide ½ the outside diameter by the tangent of the face angle
Face width	1½ to 2½ times the circular pitch
Chordal addendum	$\text{Addendum} + \dfrac{\text{circular thickness}^2 \times \text{Cos pitch cone angle}}{4D}$

Fig. 19-6-1 Bevel-gear formulas.

WORKING DRAWINGS OF BEVEL GEARS

The working drawings of bevel gears, like those of spur gears, give only the dimensions of the bevel gear blank. The cutting data for the teeth are given in a note or table. A single section view is normally used, unless a second view is required to show such details as spokes. Sometimes both the bevel gear and pinion are drawn together, showing their relationship. The dimensions and cutting data will depend on the method used in cutting the teeth, but the information shown in Fig. 19-6-3 is commonly used.

The actual gear teeth are often shown on assembly or display drawings. One of the most common conventions used for drawing the teeth is the Tredgold method, which is shown in Fig. 19-6-4.

An arc whose radius is taken on the back cone is used as the pitch circle, and a tooth is developed using standard spur-gear formulas. Tooth sizes taken on the OD and pitch diameter are transferred to the front view, and the profiles for the teeth are drawn. Radial lines from these points are taken, and the small end of the tooth is developed. The teeth on the side or section view are now drawn by projecting the teeth from the front view. Cast iron is normally used for large

Fig. 19-6-2 Bevel-gear nomenclature.

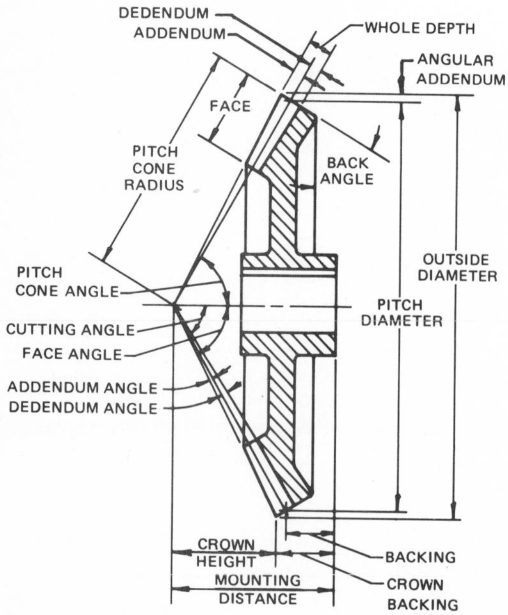

DEDENDUM
ADDENDUM
WHOLE DEPTH
ANGULAR ADDENDUM
FACE
PITCH CONE RADIUS
BACK ANGLE
OUTSIDE DIAMETER
PITCH CONE ANGLE
PITCH DIAMETER
CUTTING ANGLE
FACE ANGLE
ADDENDUM ANGLE
DEDENDUM ANGLE
CROWN HEIGHT
MOUNTING DISTANCE
BACKING
CROWN BACKING

Fig. 19-6-3 Working drawing of a bevel gear.

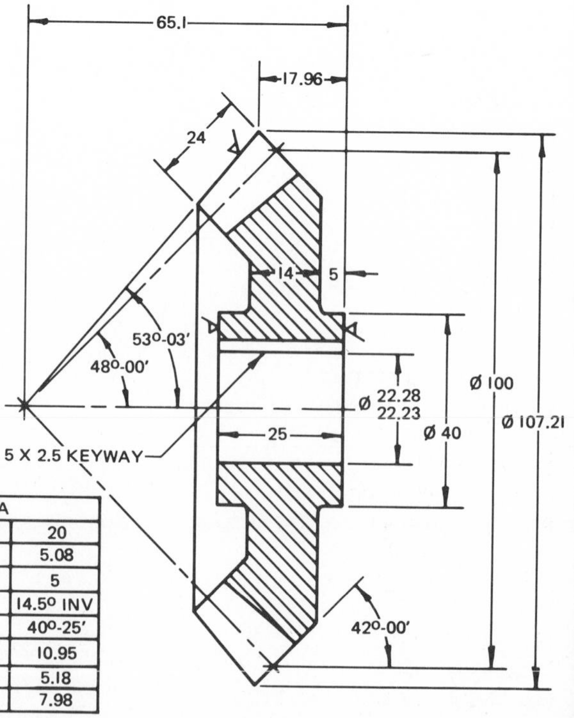

65.1
17.96
24
14
5
53°-03′
48°-00′
Ø 22.28
22.23
25
Ø 40
Ø 100
Ø 107.21
5 X 2.5 KEYWAY
42°-00′

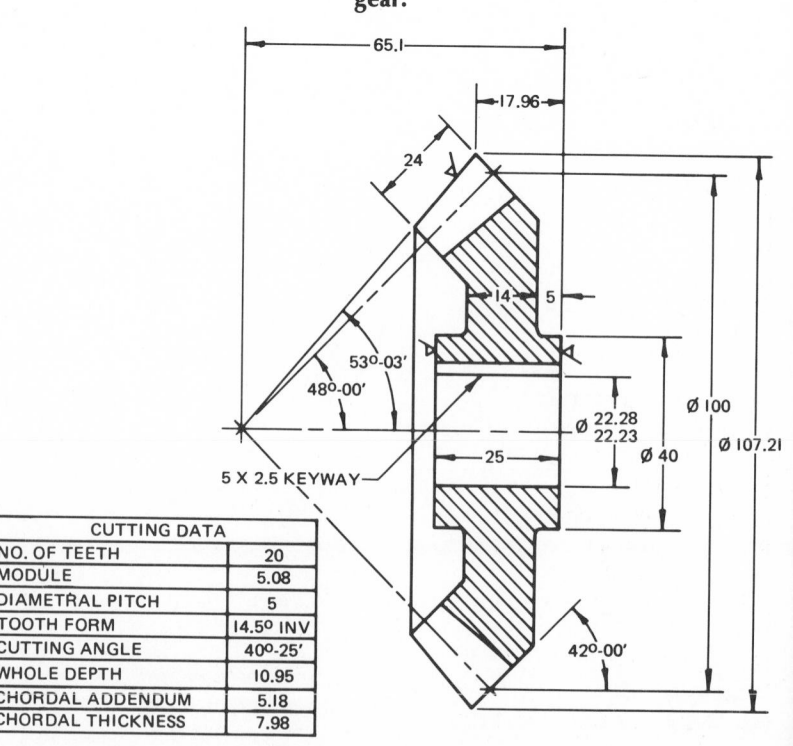

CUTTING DATA	
NO. OF TEETH	20
MODULE	5.08
DIAMETRAL PITCH	5
TOOTH FORM	14.5° INV
CUTTING ANGLE	40°-25′
WHOLE DEPTH	10.95
CHORDAL ADDENDUM	5.18
CHORDAL THICKNESS	7.98

gears and small gears that are not subjected to heavy duty. Often a gear and pinion are made of different materials for efficiency and durability. The pinion is made of a stronger material because the teeth on the pinion come into contact more times than the teeth on the gear. Common combinations are steel and cast iron, and steel and bronze.

For kilowatt ratings of bevel gears, refer to manufacturers' catalogs.

Assignments

1. On an A3- or B-size sheet, make a working drawing of a bevel gear from the data shown in Fig. 19-6-A or 19-6-B. Use your judgment for dimensions not given. Scale is 1:1.

2. On an A3- or B-size sheet, make an assembly working drawing of the gears described in Fig. 19-6-C or 19-6-D. Add to the drawing the cutting data for the gears. Use your judgment for dimensions not given. Scale is 1:1.

REVIEW FOR ASSIGNMENTS

Unit 6-4 Bills of Materials
Unit 7-1 Full-Section Views
Unit 9-3 Retaining Rings
Unit 9-1 Keys
Unit 21-1 Plain Bearings
Unit 21-4 Oil Seals

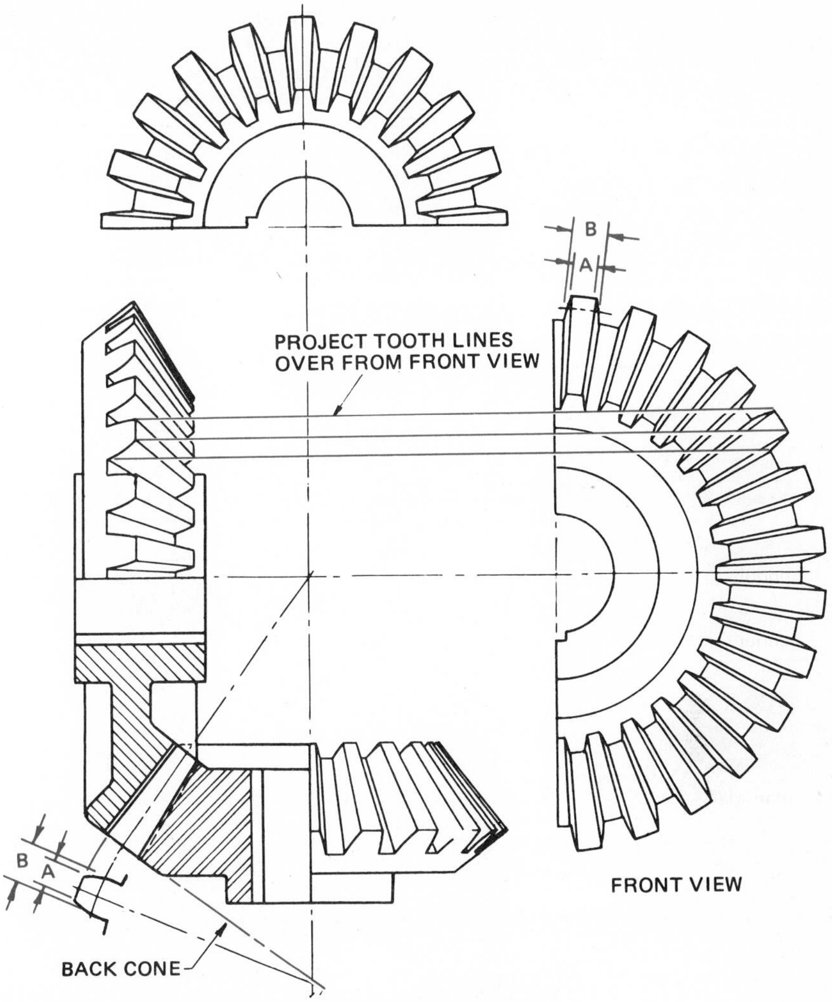
B
A
PROJECT TOOTH LINES OVER FROM FRONT VIEW
B
A
BACK CONE
FRONT VIEW

Fig. 19-6-4 Bevel gear assembly or display drawing.

GEAR DATA

D—114
Pitch Cone Angle—45°
Module—6.35
Tooth Form—14.5°
Face Width—32
Shaft—ϕ24
Hub—ϕ44 x 32 Lg
Web Thickness—16

Fig. 19-6-A Single bevel gear.

GEAR DATA

D—4.50
Pitch Cone Angle—45°
P—4
Tooth Form—14.5°
Face Width—1.25
Shaft—ϕ.94
Hub—ϕ1.75 x 1.25 Lg
Web Thickness—.62

Fig. 19-6-B Single bevel gear.

GEAR DATA

Module—6.35
Face Width—30
N—22
Shaft—ϕ25
Hub—ϕ48 x 38 Lg
Web—20
Matl—MI

PINION DATA

N—14
Shaft—ϕ20
Hub—32 Lg
Matl—Steel

Fig. 19-6-C Bevel gear assembly.

GEAR DATA

P—4
Face Width—1.10
N—22
Shaft—ϕ1.00
Hub—ϕ1.90 x 1.50 Lg
Web—.75
Matl—MI

PINION DATA

N—14
Shaft—ϕ.75
Hub—1.25 Lg
Matl—Steel

Fig. 19-6-D Bevel gear assembly.

UNIT 19-7

WORM AND WORM GEARS

Worm gears are used to transmit power between two shafts that are at right angles to each other and are nonintersecting. The teeth on the worm are similar to the teeth on the rack, and the teeth on the worm gear are curved to conform with the teeth on the worm. Thread terms such as *pitch* and *lead* are used on the worm.

As a single-thread worm gear in one revolution advances the worm gear only one tooth and space, a large reduction in velocity is obtained. Another feature of worm gearing is the high mechanical advantage acquired. The ratio of worm gear speed to the worm speed is the ratio between the number of teeth on the worm gear and the number of threads on the worm. A worm gear with 33 teeth and a worm with a multiple thread of three has a ratio of 11:1.

About 50:1 is the maximum ratio recommended. Since a single-thread worm has a low lead angle, it is inefficient and consequently not used to transmit power. The lead (or helix) angle should be between 25 and 45° for efficiency in transmitting power; as a result, multithread worms are used. The number of threads on a worm may vary from one to eight.

Steel worms and phosphor bronze

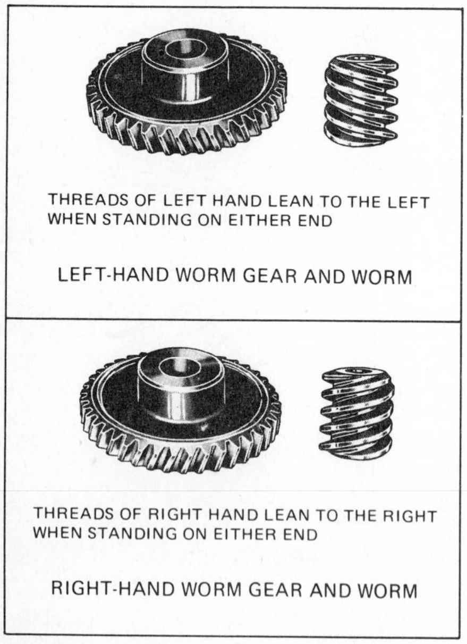
THREADS OF LEFT HAND LEAN TO THE LEFT WHEN STANDING ON EITHER END
LEFT-HAND WORM GEAR AND WORM
THREADS OF RIGHT HAND LEAN TO THE RIGHT WHEN STANDING ON EITHER END
RIGHT-HAND WORM GEAR AND WORM

Fig. 19-7-1 How to identify a left-hand and right-hand worm gear.

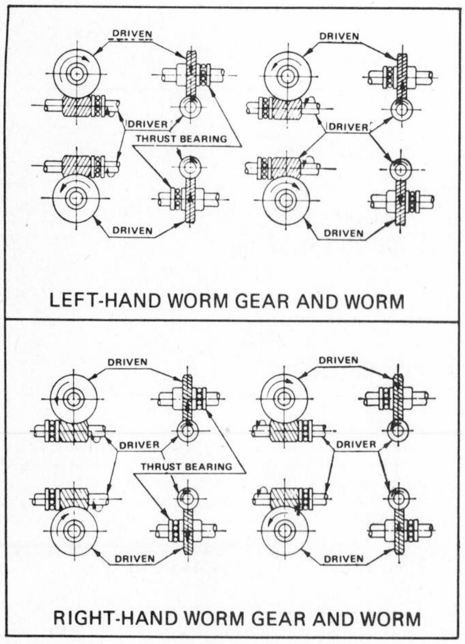
DRIVEN
DRIVER
THRUST BEARING
LEFT-HAND WORM GEAR AND WORM
RIGHT-HAND WORM GEAR AND WORM

Fig. 19-7-2 Location of bearings to absorb thrust load on worm and worm gear.

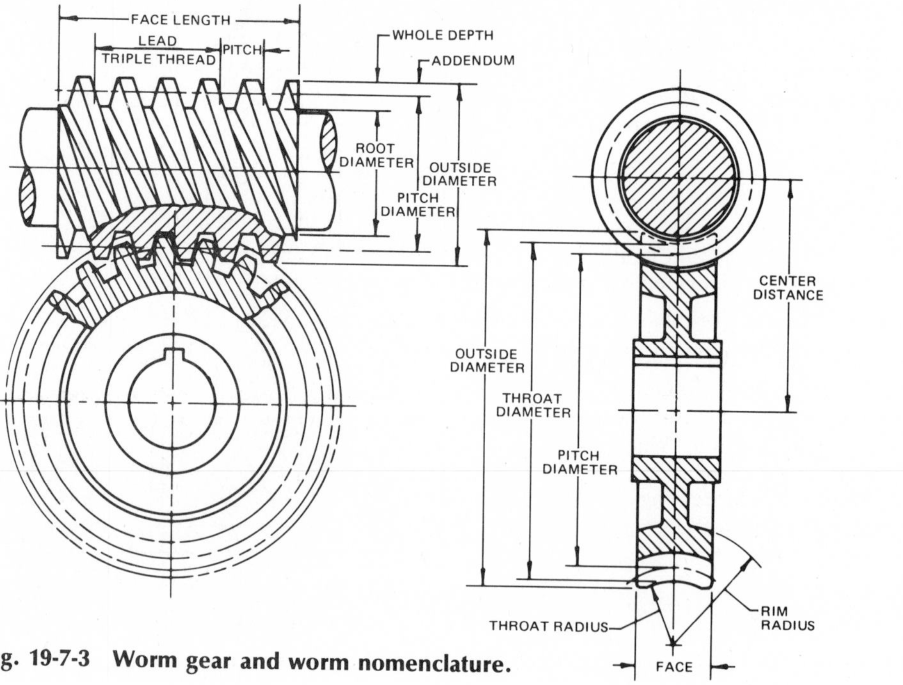
FACE LENGTH
LEAD
TRIPLE THREAD
PITCH
WHOLE DEPTH
ADDENDUM
ROOT DIAMETER
OUTSIDE DIAMETER
PITCH DIAMETER
CENTER DISTANCE
OUTSIDE DIAMETER
THROAT DIAMETER
PITCH DIAMETER
THROAT RADIUS
RIM RADIUS
FACE

Fig. 19-7-3 Worm gear and worm nomenclature.

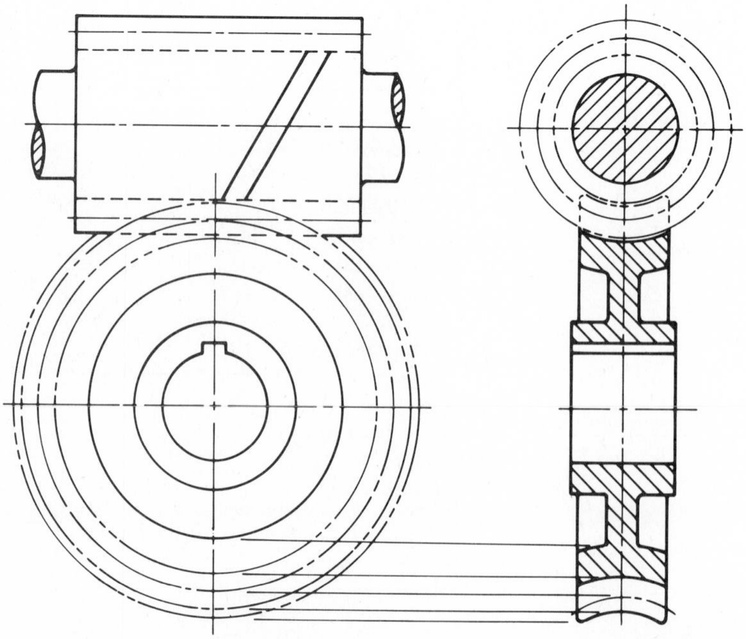

Fig. 19-7-4 Assembly drawing of a worm and worm gear assembly. (Dimensions not shown.)

gears are the most common materials used in worm gearing.

Figures 19-7-1 to 19-7-5 supply data on worm gear drawings and formulas.

WORKING DRAWINGS OF WORM AND WORM GEARS (FIG. 19-7-6)

These are similar to working drawings of other gears. A one-view section drawing is normally used for the worm gear. When a second view is required, the throat and root circles are shown as solid lines, and the outside circle is not shown on this view. As for the worm drawing, the root and the outside diameter are shown as solid lines, and a second view is not normally required.

When a worm and worm gear appear as an assembly drawing, both views are drawn and the conventional solid line for the OD of the worm and the throat diameter of the worm gear is shown as broken lines where the teeth mesh.

Assignments

1. On an A3- or B-size sheet, make a working drawing of a worm and the mating worm gear from the data given in Fig. 19-7-A or 19-7-B. Use your judg-

TERM	SYMBOL	FORMULA	DEFINITION
Pitch diameter of worm	Dw	Dw = 2C − Dg	
Pitch diameter of gear	Dg	Dg = 2C − Dw or $\frac{NP}{\pi}$	
Pitch	P	$P = \frac{L}{T}$ $P = \frac{(2C - Dw) \times \pi}{N}$	The distance from one tooth to the corresponding point on the next tooth measured parallel to the worm axis. It is equal to the circular pitch on the worm gear
Lead	L	L = Dg ÷ R L = P × T L = Tan La × πDw	The distance the thread advances axially in one revolution of the worm
Threads	T	$T = \frac{L}{P}$	The number of threads or starts on worm; e.g., 2 for double thread, 3 for triple thread
Gear teeth	N	$N = \frac{\pi Dg}{P}$	Number of teeth on worm gear
Ratio	R	$R = \frac{N}{T}$	Divide number of gear teeth by number of worm threads
Center distance	C	$C = \frac{Dw + Dg}{2}$	
Addendum	A	A = 0.318P	Single and double threads
		A = 0.286P	Triple and quadruple threads
Whole depth	WD	WD = 0.686P	Single and double threads
		WD = 0.623P	Triple and quadruple threads
Outside diameter, worm	ODw	ODw = Dw + 2A	
Outside diameter, gear	ODg	ODg = TD + 0.4775P	Single and double threads
		ODg = TD + 0.3183P	Triple and quadruple threads
Throat diameter	TD	TD = Dg + 2A	
Face width, gear	F	F = 2.38P + 6 (metric) F = 2.38P + .25 (inch)	Single and double threads
		F = 2.15P + 5 (metric) F = 2.15P + .20 (inch)	Triple and quadruple threads
Face length, worm	FL	FL = (0.02N + 115)P (metric) FL = (0.02N + 4.5)P (inch)	
Lead angle	La	$\text{Tan La} = \frac{L}{Dw \times 3.1416}$	Divide lead by circumference of pitch diameter of worm. Quotient is tangent of lead angle.
Throat radius	R_t	$R_t = \frac{Dw}{2} - A$	Subtract addendum from half of pitch diameter of worm
Rim radius	R_r	$R_r = \frac{Dw}{2} + P$	

Fig. 19-7-5 Worm and worm gear formulas.

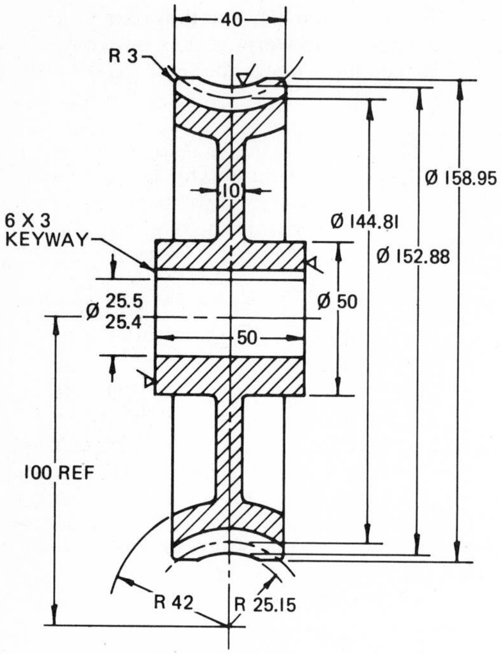
40
R 3
10
Ø 158.95
6 X 3
KEYWAY
Ø 144.81
Ø 152.88
Ø 25.5
25.4
Ø 50
50
100 REF
R 42
R 25.15

CUTTING DATA	
NO. OF TEETH	36
ADDENDUM	4.09
WHOLE DEPTH	8.71
NO. OF THREADS	2
PITCH (AXIAL)	12.7
PRESSURE ANGLE	14.5°
LEAD ANGLE	7°-53′
LEAD–RH	

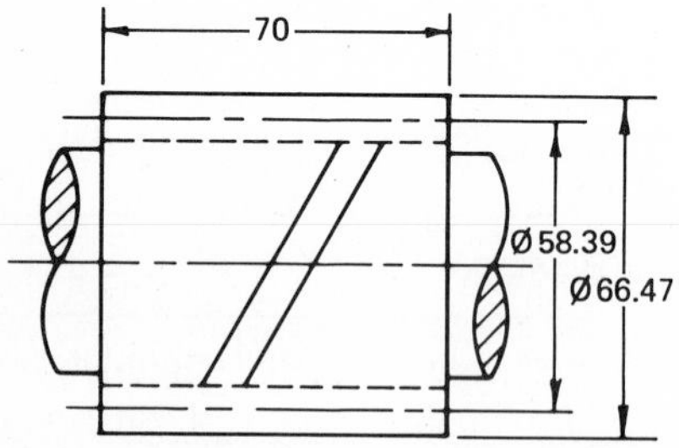
70
Ø58.39
Ø66.47

CUTTING DATA	
NO. OF THREADS	2
PITCH	12.7
PRESSURE ANGLE	14.5°
LEAD ANGLE	7°-53′
LEAD–RH	
WHOLE DEPTH	8.71
ADDENDUM	4.09

Fig. 19-7-6 Working drawing of a worm and worm gear.

ment for dimensions not given. Include on the drawing the cutting data.
Scale is 1:1.

2. On an A3- or B-size sheet, make a two-view detail assembly drawing of a worm and worm gear from the data given in Fig. 19-7-C or 19-7-D. Use your judgment for dimensions not given. Include the cutting data on the drawing.
Scale is 1:2.

REVIEW FOR ASSIGNMENTS

Unit 21-1 Bearings
Unit 21-4 Oil Seals
Unit 7-5 Assemblies in Section
Unit 6-4 Bills of Material
Unit 9-2 Pin Fasteners

A worm and worm gear have a pitch of 13.3. The gear, made of cast iron, has 30 teeth; shaft dia. = 22, hub dia. = 44, hub length = 48, face width = 25, web thickness = 10. The worm, made of hardened steel, is 88 long on a ϕ22 shaft, pitch dia. = 54, single thread, RH lead.

Fig. 19-7-A Working drawings of a worm and worm gear.

A worm and worm gear have a pitch of .5236 in. The gear, made of cast iron, has 30 teeth; shaft dia. = .88, hub dia. = 1.75, hub length = 1.90, face width = 1.00, web thickness = .40. The worm, made of hardened steel, is 3.50 long on a ϕ.88 shaft, pitch dia. = 2.12, single thread, RH lead.

Fig. 19-7-B Working drawings of a worm and worm gear.

A worm and worm gear have a pitch of 19. The gear, made of phosphor bronze, has 24 teeth, shaft dia. = 32, hub dia. = 58, hub length = 64, web thickness = 13. The worm, made of steel, has a pitch diameter of 64, left-hand double thread, shaft dia. = 26.

Fig. 19-7-C Assembly drawing of a worm and worm gear.

A worm and worm gear have a pitch of .75. The gear, made of phosphor bronze, has 24 teeth, shaft dia. = 1.25, hub dia. = 2.25, hub length = 2.50, web thickness = .50. The worm, made of steel, has a pitch diameter of 2.50, left-hand double thread, shaft dia. = 1.00.

Fig. 19-7-D Assembly drawing of a worm and worm gear.

UNIT 19-8

COMPARISON OF CHAIN, GEAR, AND BELT DRIVES[1]

Chains, gears, and belts are used for power drives between rotating shafts that cannot be directly coupled. In this unit the characteristics of these media are compared, and the conditions favorable to the use of each type of drive are discussed.

Chains

A *chain drive* consists of an endless chain whose links mesh with toothed wheels, called *sprockets*, which are keyed to the shafts of the driving and driven mechanisms.

Precision chains, of which roller chains and silent chains are examples, are made of parts that are finished to close tolerances. They engage the teeth of sprocket wheels, which are manufactured to equally precise standards, and provide smooth, efficient operation. Where the refinements of precision chains are not required, a wide variety of fabricated steel, forged, and cast link chains are available.

Roller Chains. The unique feature of a roller chain is its freedom of joint action during its engagement with the sprocket. This is accomplished by articulation of the pins of the bushings, while the rollers turn on the outside of the bushings, thus eliminating rubbing action between the rollers and the sprocket teeth.

Silent Chains. Comparable ease of joint action occurs in the engagement of the silent chain with the sprocket. When the chain is wrapped around the sprocket, the chain links simultaneously engage both sides of the sprocket teeth.

Gears

A *simple gear drive* consists of a toothed driving wheel meshing with a similar driven wheel. Tooth forms are designed to ensure uniform angular rotation of the driven wheel during tooth engagement. Gears are available with precision-cut teeth or with unfinished teeth.

Spur gears, which are used for drives between parallel shafts, have their teeth on the rim of the wheel. A pair of spur gears operates as though it consists of two cylindrical surfaces, called *pitch surfaces*, which are modified to form teeth. The teeth prevent slippage between the pitch surfaces and thus maintain a constant speed ratio between the driving and the driven wheel.

A modified form of spur gear, known as a *helical gear*, has the tooth faces cut in the form of spirals rather than cylindrical surfaces. When the two opposite-hand helices are cut on a single wheel, the gear is known as a *herringbone gear*.

Belts

A *belt drive* consists of an endless flexible belt that connects two wheels, or pulleys. Belt drives depend on friction between the belt and the pulley surfaces for the transmission of power.

The most common material used for flat belting is leather, either a single strip or two or more layers cemented together. In most cases, the belt is cut to the required length, and the two ends are joined by lacing them with leather thongs, by means of one of a variety of metal *belt lacings*, or by gluing.

Other belt materials include rubber, steel, canvas, and canvas impregnated with rubber.

Pulleys for flat belts are manufactured of cast iron, steel, wood, or various synthetic materials. The pulley surface is smooth and is usually crowned to help prevent the belt from running off the pulley.

In a *V-belt drive*, the belt has a trapezoidal cross section and runs in V-shaped grooves on the pulleys. These belts are made up of cords or cables, impregnated and covered with rubber or other organic compound. The covering is formed to produce the required cross section. V-belts are usually manufactured as endless belts, although open-end and link types are available.

In the case of V-belts, the friction for the transmission of the driving force is increased by the wedging of the belt into the grooves on the pulley.

V-belt drives are available in single or multiple strands for varying horsepower requirements.

Another type of belt has shallow teeth molded on the inside of the driving face. The pulleys have teeth for engagement with the belt teeth.

Chain Drives Compared with Gear Drives

ADVANTAGES OF CHAINS

Shaft center distances for chain drives are relatively unrestricted, whereas with gears, the center distance must be such that the pitch surfaces of the gears are tangent. This advantage often will result in a simpler, less costly, and more practical design.

Chains are easily installed. While all drive media require proper installation, the assembly tolerances for chain drives are not as restricted as those for gears; and the resultant savings in the time of installation may be an important item in meeting the production schedule required of the driven machine.

The ease of chain installation is a definite advantage where later changes in design, such as speed ratio, capacity, and centers, are anticipated.

The normal elasticity in tension of chain, plus the cushioning effect of the lubricant in the numerous chain joints, provides better shock absorption than is possible with the lubricated metal-to-metal contact of one or two gear teeth.

In gear trains, there is a combination of rolling and sliding action between the surfaces of the teeth in meshing, whereas the rollers of a chain engage the sprocket teeth with a rolling action. Wear is also reduced, since the chain load is distributed over a number of sprocket teeth simultaneously, in contrast to the concentration of load on the one or two teeth that are in mesh in the case of a gear.

ADVANTAGES OF GEARS

When space limitations require the shortest possible distance between shaft centers, a gear drive is usually preferable to a chain drive.

The maximum speed ratio for satisfactory operation of a gear drive is usually greater than that for chain drive.

Gears can be operated at higher rotative speeds than chain drives.

When combinations of extremely high speed and high kilowatts are encountered, gears are generally more practical. At medium and low speeds, either type of drive can be used, with some advantage in compactness on the side of gearing.

Chain Drives Compared with Belt Drives

ADVANTAGES OF CHAINS

Chain drives do not slip or creep as do belt drives. As a result, chains maintain a positive speed ratio between the driving and the driven shafts, and they are more efficient since no power is lost because of slippage.

A chain drive requires no tension on the slack side; therefore, it imposes lower loads on the shaft bearings than those occurring with a belt drive. Such lower loads reduce bearing maintenance as well as bearing friction losses.

Chain drives are more compact than belt drives. For a given capacity, a chain will be narrower than a belt, and sprockets will be smaller in diameter than pulleys; thus the chain drive will occupy less overall space.

Chains are easy to install. A chain can be installed by wrapping it around the sprockets and then slipping the pins of a connecting link into position. A belt, if it is already laced, must be pried over the pulleys with considerable difficulty. If the belt is not already laced, it must be laced on the pulleys with special equipment.

The required minimum arc of contact is smaller for chains than for belts. This advantage becomes more pronounced as the speed ratio increases and thus permits chain drives to operate on much shorter shaft center distances.

Where several shafts are to be driven from a single shaft, positive speed synchronism between the driven shafts is usually imperative. For such applications, chains are more suitable.

Chain drives, when used in a dusty atmosphere, eliminate the fire hazard which is possible with belt drives because of static electricity.

With a stalled, belt-driven pulley, excessive belt slippage will occur; and the resulting temperature rise may be sufficient to ignite atmospheric dust or other materials. Such a condition cannot develop with a chain drive.

Chains do not deteriorate with age; nor are they affected by sun, oil, and grease. Chains can operate at higher temperatures. Chain drives are more practical for low speeds.

Chain elongation resulting from normal wear is a slow process; the chain, therefore, requires infrequent adjustment. Belt stretch, however, necessitates frequent tightening by shaft adjustment, by idlers, or by shortening the belt.

ADVANTAGES OF BELTS

Since no metal-to-metal contact occurs between a belt and pulleys, belts require no lubrication, although leather belts need periodic applications of belt dressing to preserve their flexibility.

Generally speaking, a belt drive operates with less noise than a chain drive.

Flat belt drives can be used where extremely long center distances would make chain drives impractical.

In the extremely high-speed ranges, flat belts can be operated to better advantage than chains.

Conclusion

No one type of power drive is ideal for all types of service. This unit has discussed the relative merits of chain, gear, and belt drives, and should provide a guide to the selection of the best type for a given application.

REFERENCE AND SOURCE MATERIAL

1. American Chain Association.

Assignment

Your supervisor has asked you to submit a report recommending the type of power transmission best suited for the machinery layouts shown in either Fig. 19-8-A or Fig. 19-8-B. Make a detailed report that specifies the power transmission parts required to properly operate the equipment.

REVIEW FOR ASSIGNMENT

Unit 19-1 Belt Drives
Unit 19-2 Chain Drives
Unit 19-3 Gear Drives

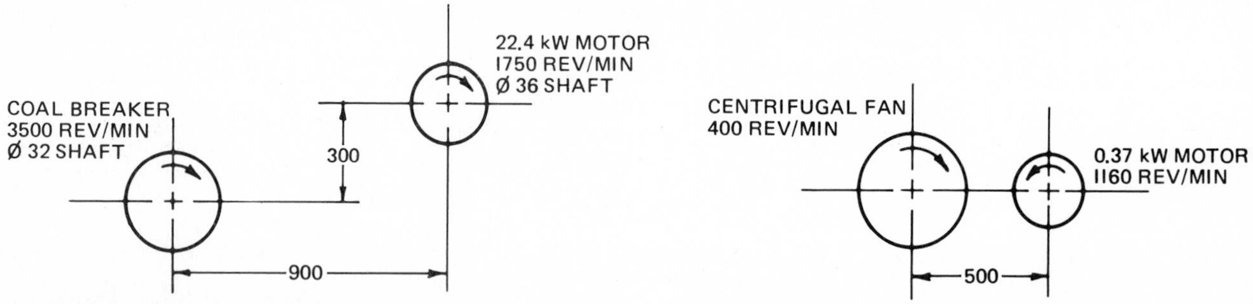
22.4 kW MOTOR
1750 REV/MIN
Ø 36 SHAFT
COAL BREAKER
3500 REV/MIN
Ø 32 SHAFT
300
900
CENTRIFUGAL FAN
400 REV/MIN
0.37 kW MOTOR
1160 REV/MIN
500

Fig. 19-8-A Power-transmission drive.

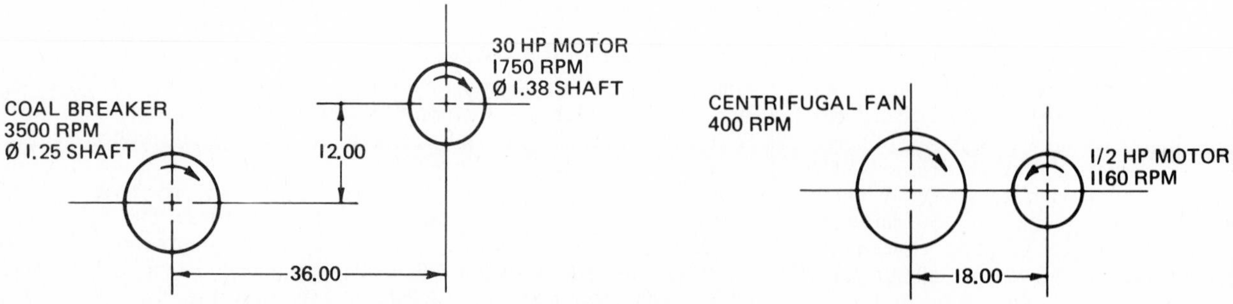
30 HP MOTOR
1750 RPM
Ø 1.38 SHAFT
COAL BREAKER
3500 RPM
Ø 1.25 SHAFT
12.00
36.00
CENTRIFUGAL FAN
400 RPM
1/2 HP MOTOR
1160 RPM
18.00

Fig. 19-8-B Power-transmission drive.

Chapter 20
Couplings, Clutches, Brakes, and Speed Reducers

UNIT 20-1

COUPLINGS AND FLEXIBLE SHAFTS

Couplings

Couplings, as the name implies, are used to couple or join shafts. There are two types of couplings: permanent couplings and clutches.

Permanent couplings are not normally disconnected except for assembly or disassembly purposes, while clutches permit shafts to be connected or disconnected at will.

PERMANENT COUPLINGS[1]

Permanent couplings can be divided into three main categories: solid, flexible, and universal.

Solid Couplings. Solid couplings should be used only when driving and driven shafts are mounted on a common rigid base, so that shafts can be perfectly aligned and will stay that way in service. If two shafts are not in exact alignment and are connected by a rigid coupling, excess bearing wear may occur on the bearings supporting the shaft. The steel sleeve coupling and the flanged coupling shown in Fig. 20-1-1 are solid couplings.

Flexible Couplings. These are intended to compensate for unintentional misalignments or transient misalignments such as those caused by thermal expansion or vibration. They also prevent shock from being transferred from one shaft to another and are recommended where power machines are connected on one shaft. See Figs. 20-1-2 and 20-1-3.

Because of the wide variety of low-cost flexible couplings now available, solid couplings are falling from favor. The time required for accurate alignment and the necessary shimming, or the

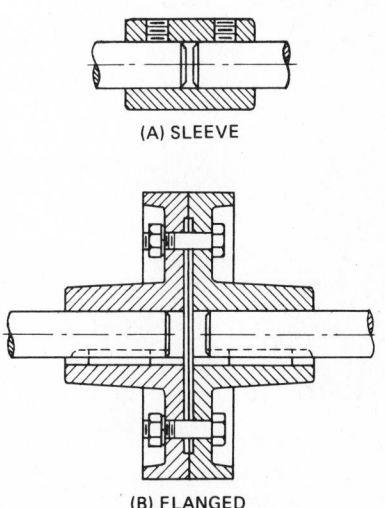

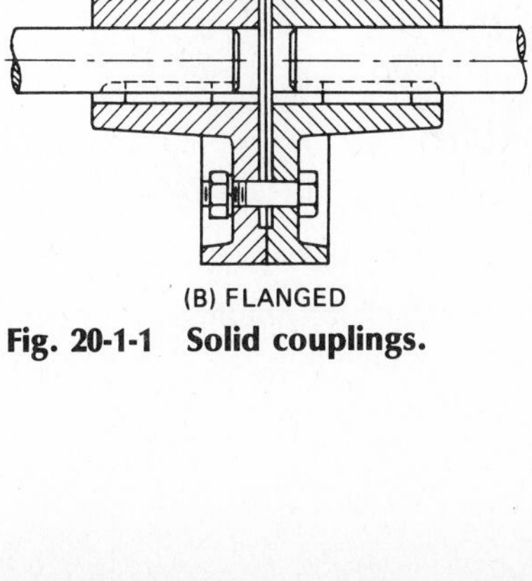
(A) SLEEVE
(B) FLANGED

Fig. 20-1-1 Solid couplings.

cost of close production tolerances, usually makes a flexible coupling more economical.

There are many types of flexible couplings, but all are similar in operation. There are two hubs, one on each shaft, connected by an intermediate part which may be flexible, floating, or both.

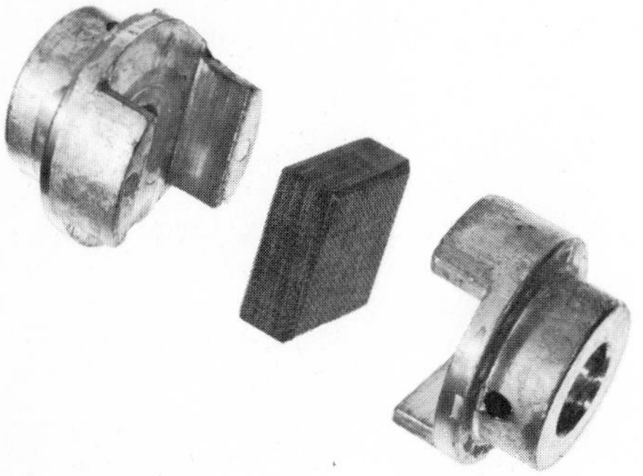

Fig. 20-1-2 Flexible coupling. (Oldham Principle)

Flexible couplings may also be divided into three main categories: those that use mechanical movement, those that depend on the flexing of materials, and those that combine mechanical movement with flexing.

The operational life therefore depends on the wear characteristics of the contacting materials, or the fatigue resistance of the flexing materials. Life will be shortened by the amount of angular displacement of the shafts, or the amount that parallel shafts are offset to one another.

Factors to consider before selection is made of a flexible coupling are as follows: power to be transmitted (revolutions per minute and torque); the amount of angular displacement and out-of-parallel of the two shafts; if varying end float is necessary; if backlash is permissible; torsional wind-up; vibration insulation; thermal insulation; electrical insulation; environment; accessibility for maintenance; whether periodic lubrication is permissible; whether replacement of the coupling without moving the shafts is necessary; and sometimes, balance and moment of inertia. For some installations, particularly for accurate instrumentation, constant velocity is also required.

The table shown in Fig. 20-1-4 lists the most common types of couplings and their main qualities. It should be used only as a guide, since special materials can affect qualities considerably; and, of course, there are exceptions to every rule. For most jobs, any one of several couplings may be suitable; then cost determines the final selection.

To aid in selecting the correct-size coupling, most manufacturers rate power transmitted in kilowatts per revolution per minute and give a maximum permissible revolutions per minute. The rating can be determined by the simple formula

$$\text{Kilowatts per 100 r/min} = \frac{\text{driving kilowatts} \times 100 \times \text{service factor}}{\text{coupling revolutions per minute}}$$

The service factor depends on the source of driving power and the type of duty. For smooth power sources such as an electric motor driving smooth loads like a centrifugal compressor, the factor is 1. It can be as high as 5 for reciprocating gasoline or diesel engines coupled to loads with cyclic torque variations such as a single-cylinder compressor without a flywheel. Not all manufacturers use exactly the same service factors; so when you are ordering couplings, the type of duty should be specified.

Universal Couplings. Commonly called universal joints, universal couplings are for applications where angular displacement of shafts is a design requirement. The selection of universal couplings is easier than for flexible couplings because there are fewer types. Most common is the Hooke's joint, which has a cross-type trunnion connected to driving and driven shafts by U-shaped endpieces. See Fig. 20-1-5. Its main disadvantage is that because the trunnion is always at right angles to the driven shaft, it gives a sinusoidal variation in angular velocity between shafts. Other disadvantages are that it cannot compensate for out-of-parallel and it does not compensate for changing distances between driving and driven points when the angle between shafts changes. These disadvantages disappear when two universals are used, one with a sliding spline, as in automotive systems using the Hotchkiss drive. Here, the transmission and differential pinion shafts are parallel, so rotational fluctuations are canceled out. When two joints are used

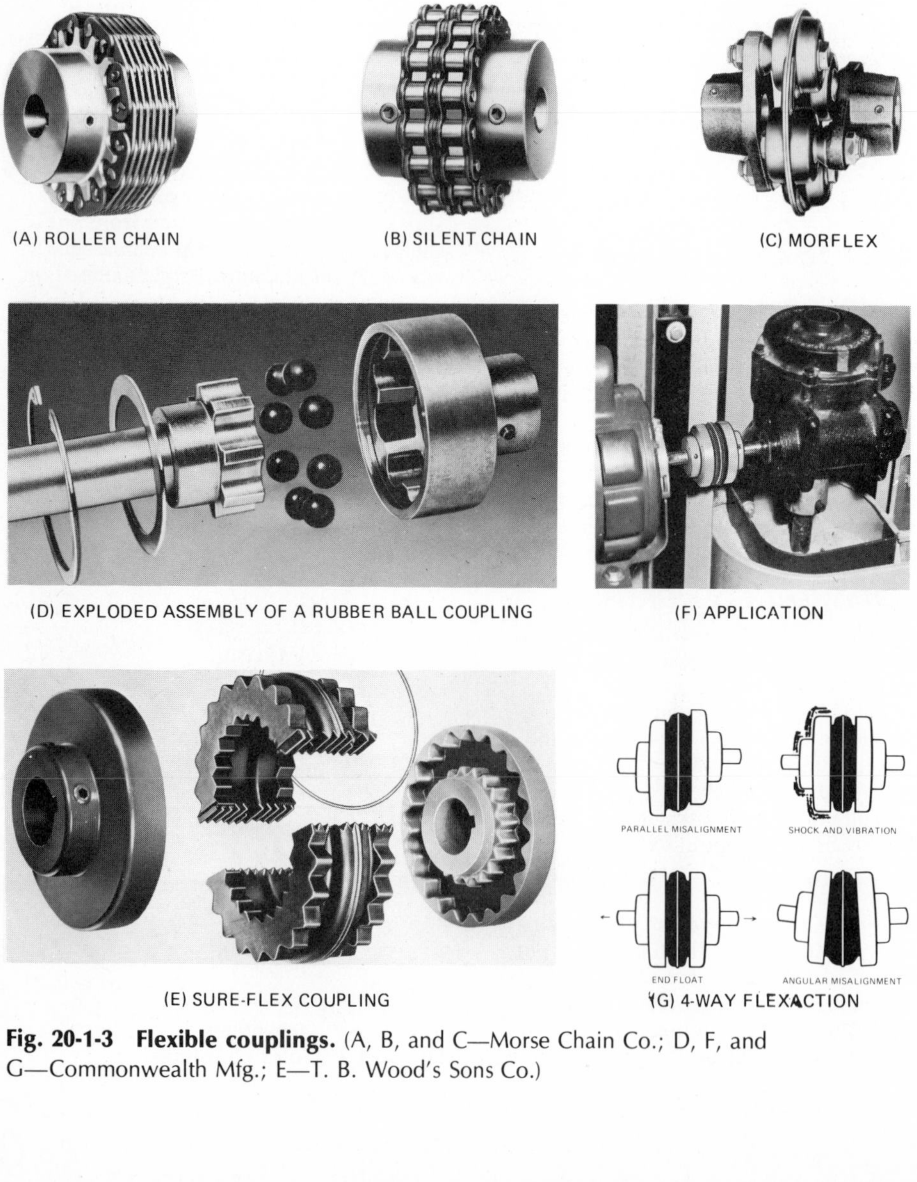
(A) ROLLER CHAIN
(B) SILENT CHAIN
(C) MORFLEX
(D) EXPLODED ASSEMBLY OF A RUBBER BALL COUPLING
(F) APPLICATION
PARALLEL MISALIGNMENT
SHOCK AND VIBRATION
END FLOAT
ANGULAR MISALIGNMENT
(E) SURE-FLEX COUPLING
(G) 4-WAY FLEXACTION

Fig. 20-1-3 Flexible couplings. (A, B, and C—Morse Chain Co.; D, F, and G—Commonwealth Mfg.; E—T. B. Wood's Sons Co.)

		Chain	Straight gear	Curved gear	Sliding insert	Metal ball	Metal disc	Plastic disc	Bellows	Helical spring	Rubber tire	Plastic insert	Rubber insert	Rubber ball	Radial spring	Grid spring
Angular misalignment	—high							*	*	*	*		*	*		
	—average			*	*	*	*					*			*	*
	—low	*	*													
Out-of-parallel misalignment	—high							*			*		*	*		
	—average				*	*			*			*			*	*
	—low	*	*	*			*			*						
End float	—high		*	*										*	*	*
	—average	*			*	*	*	*	*		*	*	*			
	—low									*						
RPM	—high	*	*	*			*			*						
	—average				*	*		*	*		*	*	*	*	*	*
	—low															
Torsional resilience	—high										*		*	*		
	—average							*		*		*			*	*
	—low	*	*	*	*	*	*		*							
Lubrication required		*	*	*	*	*										

Fig. 20-1-4 Basic features of common flexible couplings.

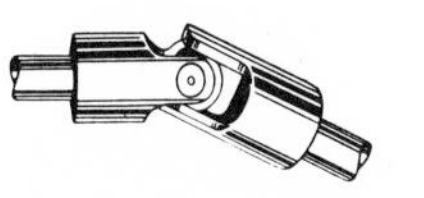

(A) SINGLE JOINT

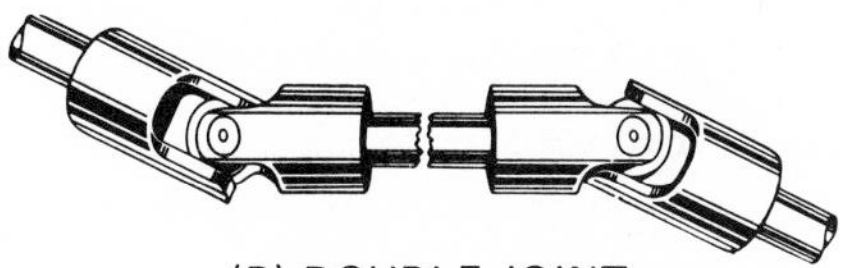

(B) DOUBLE JOINT

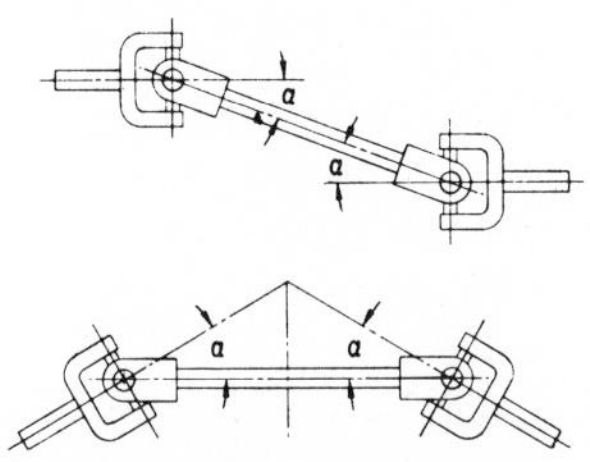

(C) CORRECT ARRANGEMENT
ANGLES MUST BE EQUAL

Fig. 20-1-5 Universal joints—Hooke's type. (Boston Gear Works)

in this manner, the U-shaped fittings on the drive-shaft ends must be parallel, or else the rotational fluctuations will be increased instead of canceling out.

If constant velocity is essential with only one universal, a special constant-velocity-type universal must be used. Most of these have some type of ball drive, where the driving points of contact bisect the driving angle. They are more complex than the Hooke's type and more expensive. The universal coupling shown in Fig. 20-1-6 is designed to transmit a constant velocity. The drive is through steel balls in races, designed so that the plane of contact between the balls and races always bisects the shaft angle. Flexible shafts also give constant velocity but are limited to transmitting relatively low power.

Flexible Shafts[2]

Flexible shafts are used to transmit power around corners and at various angles when driving and driven elements are not aligned. Speedometers, tachometers, and indicating and recording instruments are typical applications. In remote control applications, they operate parts which must be either rotated or both rotated and reciprocated.

Flexible shafts are used in applications where power shafts are subjected to continuous vibration and where starting and stopping cause surging load changes. When rigid elements transmit power, even slight misalignment causes vibration. Portable tools constitute a large field of application for flexible shafts.

The most recent advances in flexible-shaft application have been in the field of power-driven remote control where they provide intermittent service at high speeds in both directions. Typical uses are for power-operated automobile seats, windows, and convertible tops; for power-operated mechanisms for jet aircraft canopies and wing flaps; and for valve controls, recorders, etc.

Flexible shafts are constructed of helically wound wire and designed for transmission of rotary power and motion between two points located so that their relative positions preclude the use of solid shafts.

Principal parts are shown in Fig. 20-1-7:

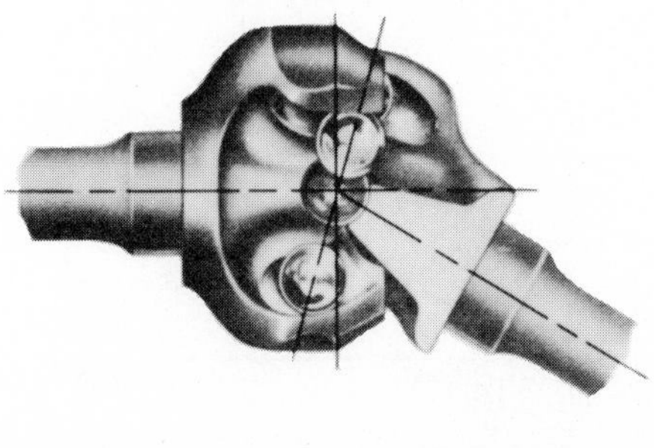

Fig. 20-1-6 Constant velocity universal coupling. (The Bendix Corp.)

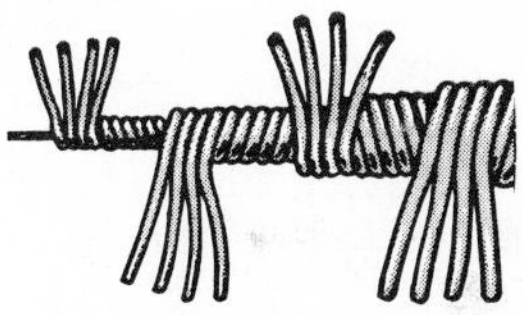

(A) CABLE

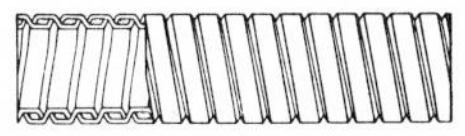

(B) CASING

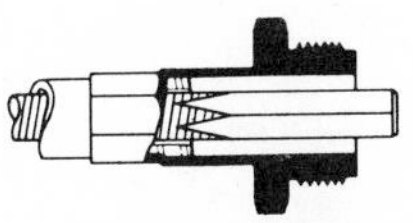

(1) LOOSE MALE NUT

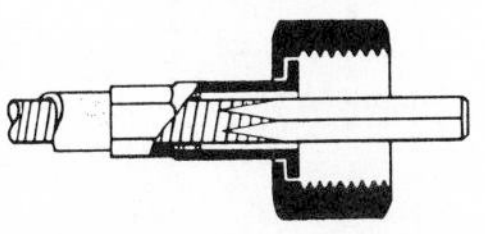

(2) LOOSE FEMALE NUT

(C) CASING END FITTINGS

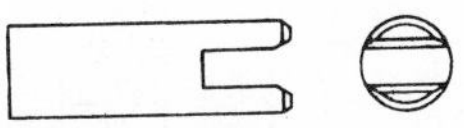

(D) END FITTING

Fig. 20-1-7 Principal parts of a flexible shaft.

1. *Flexible shaft,* the bare working element without end fittings, sometimes called the *core* or *cable.*

2. *Shaft end fittings* fastened to the ends of a flexible shaft to permit connection to driving and driven elements.

3. *Flexible casing,* a flexible, tubelike covering which acts as a runway or guide for the flexible shaft, protects it from dirt and injury and helps retain lubrication.

4. *Casing end fittings,* parts fastened to allow connection or coupling to the housing of the driving and driven elements.

REFERENCES AND SOURCE MATERIAL

1. J. Young, "Selection Factors for Shaft Couplings," *Design Engineering,* Vol. 11, No. 4, 1965.
2. F. Zambetti, "Flexible Shafts," *Machine Design,* Vol. 37, No. 14, 1965.

Assignments

1. On an A3- or B-size sheet, lay out the motor-to-pump drive assembly shown in Fig. 20-1-A or 20-1-B. Flexible couplings are required to connect the shafts, and the type of coupling specified is shown in the figure. Call for the proper couplings on the drawing. Scale is as specified.

2. On an A3- or B-size sheet, lay out the motor-to-gear-box drive unit shown in Fig. 20-1-C or 20-1-D. A flexible coupling is required to connect the shafts, and the type of coupling specified is shown in the figure. Call for the correct-size coupling on the drawing. Scale is as specified.

REVIEW FOR ASSIGNMENTS

Unit 6-4 Assembly Drawings
Unit 7-1 Full Sections
Unit 8-3 Drawing Nuts and Bolts
Unit 9-1 Keys

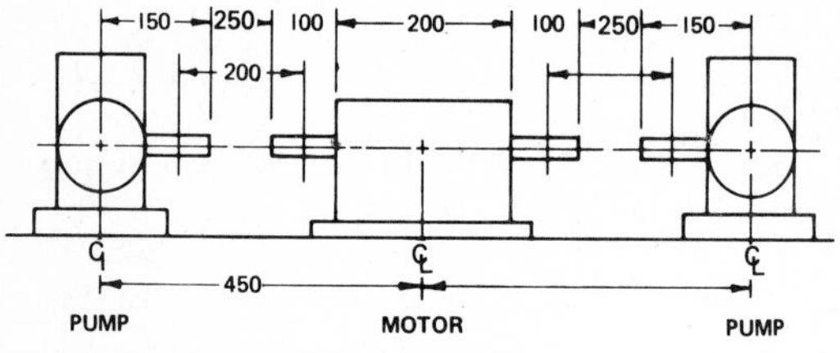

150
250
100
200
100
250
150
200
450
PUMP
MOTOR
PUMP

PUMP DRIVE LAYOUT

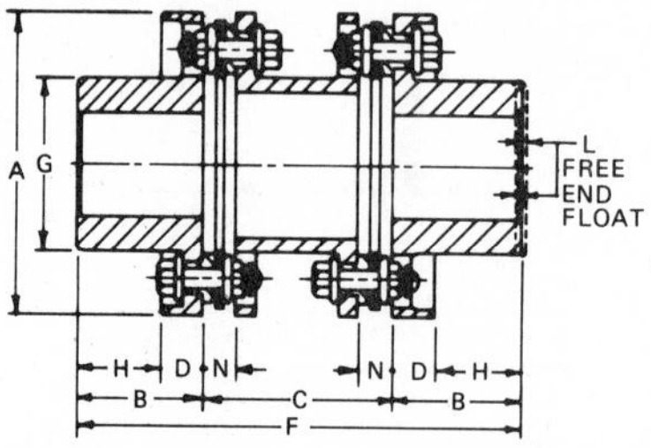

A
G
L
FREE
END
FLOAT
H
D
N
N
D
H
B
C
B
F

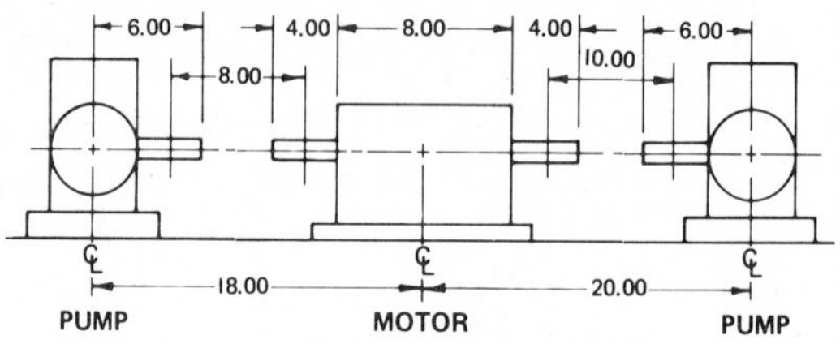

6.00
4.00
8.00
4.00
6.00
8.00
10.00
18.00
20.00
PUMP
MOTOR
PUMP

PUMP DRIVE LAYOUT

PART NO.	SHAFT SIZE	MAX rev/min	A	B	C	D	H	F	G	N	L
163 A	Ø 40	5000	120	50	100	20	30	200	60	10	3
163 B	Ø 40	5000	120	50	115	20	30	215	60	10	3
163 C	Ø 40	5000	120	50	130	20	30	230	60	10	3
163 D	Ø 40	5000	120	50	145	20	30	245	60	10	3
163 E	Ø 40	5000	120	50	160	20	30	260	60	10	3

COUPLING DATA

Fig. 20-1-A Motor to pump drive.

PART NO.	SHAFT SIZE	MAX RPM	A	B	C	D	H	F	G	N	L
163 A	1.50	5000	4.60	2.00	4.00	.70	1.30	8.00	2.50	.50	.10
163 B	1.50	5000	4.60	2.00	4.50	.70	1.30	8.50	2.50	.50	.10
163 C	1.50	5000	4.60	2.00	5.00	.70	1.30	9.00	2.50	.50	.10
163 D	1.50	5000	4.60	2.00	5.50	.70	1.30	9.50	2.50	.50	.10
163 E	1.50	5000	4.60	2.00	6.00	.70	1.30	10.00	2.50	.50	.10

COUPLING DATA

Fig. 20-1-B Motor to pump drive.

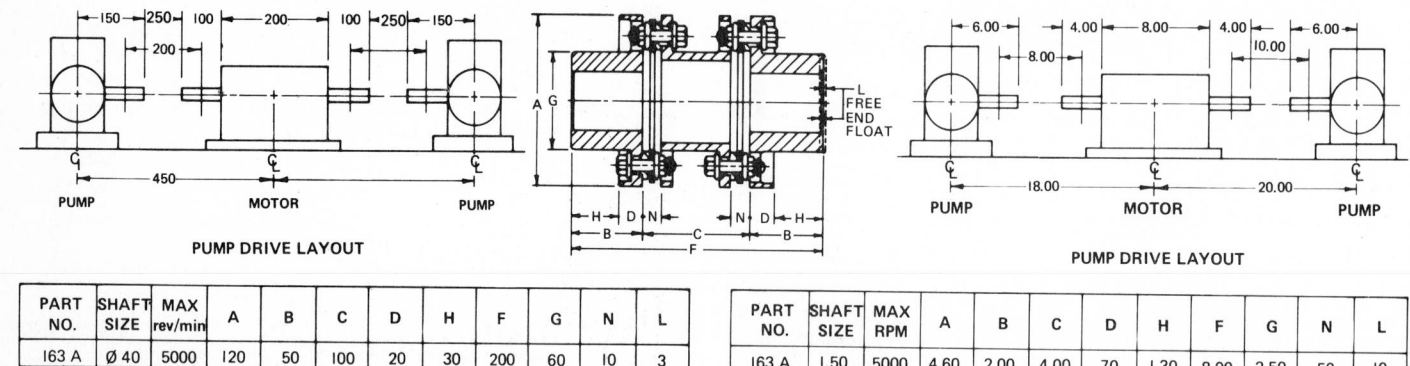

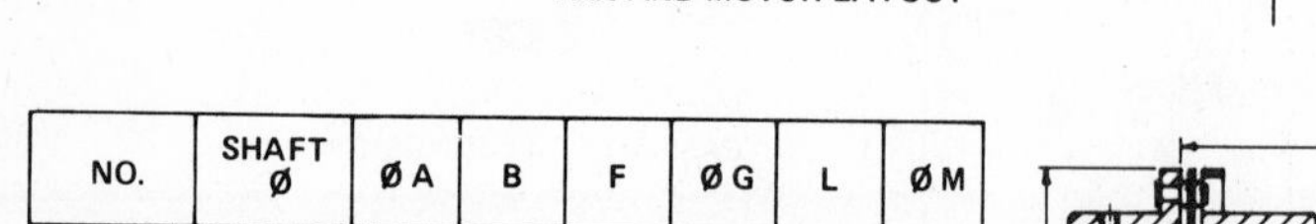

HUMID OR CORROSIVE
ATMOSPHERE
HOUSING
FAN
80
120
180
160
Ø 20
MOTOR
GEAR BOX
60

FAN AND MOTOR LAYOUT

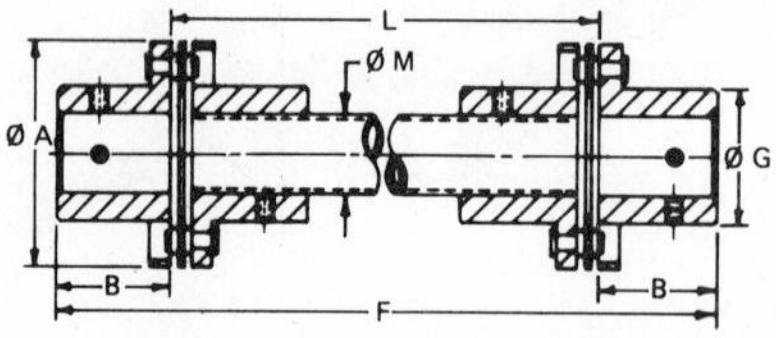

L
Ø M
Ø A
Ø G
B
F
B

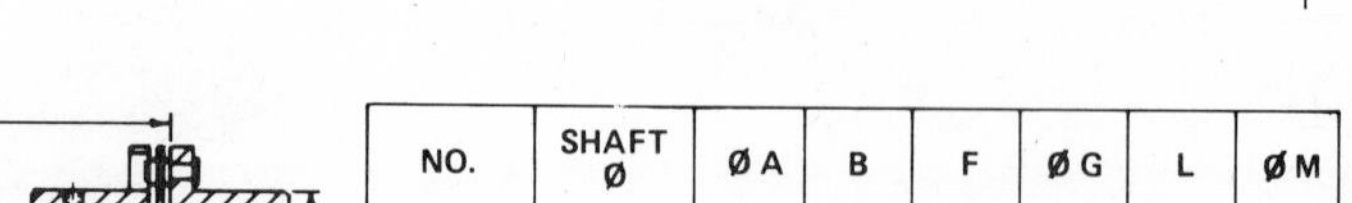

HUMID OR CORROSIVE
ATMOSPHERE
HOUSING
FAN
3.00
4.50
7.00
6.00
Ø .75
MOTOR
GEAR BOX
2.50

FAN AND MOTOR LAYOUT

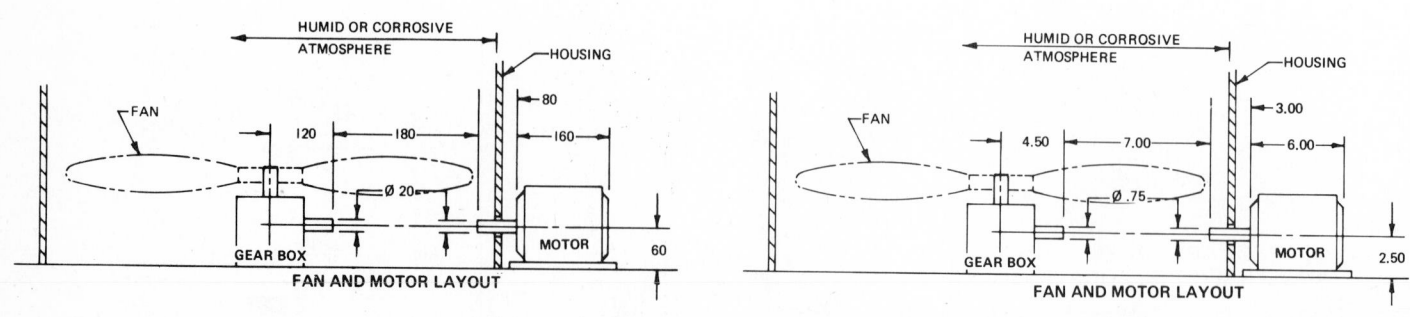

NO.	SHAFT Ø	Ø A	B	F	Ø G	L	Ø M
10 FN	10	38	18	VARIES WITH DIMENSION L	22	TO SUIT YOUR REQUIREMENTS	10
12 FN	12	42	25		24		12
16 FN	16	46	28		36		16
20 FN	20	50	30		40		20

FLEXIBLE COUPLING DATA

Fig. 20-1-C Motor to gearbox drive.

NO.	SHAFT Ø	Ø A	B	F	Ø G	L	Ø M
37 FN	.375	1.50	.70	VARIES WITH DIMENSION L	.88	TO SUIT YOUR REQUIREMENTS	.375
50 FN	.50	1.62	1.00		1.00		.50
62 FN	.62	1.75	1.10		1.38		.62
75 FN	.75	2.00	1.20		1.50		.75

FLEXIBLE COUPLING DATA

Fig. 20-1-D Motor to gearbox drive.

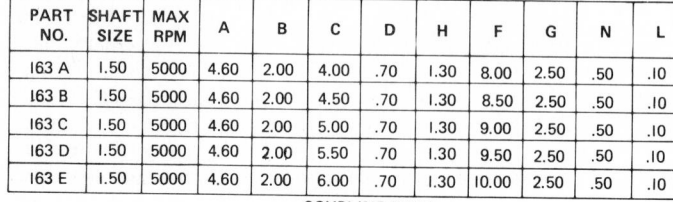

UNIT 20-2
CLUTCHES AND BRAKES

Clutches

The simplest use of a clutch is to start and stop a machine or rotating element without starting and stopping the prime mover. Clutching devices have been designed to do a variety of other jobs, such as to maintain constant speed, torque, and power or to limit torque. They are also used for automatic disconnection, quick starts and stops, gradual starts, and nonreversing and overrunning functions.

Three types of clutches currently in general use are mechanical, electrical, and hydraulic. These categories break down into numerous subtypes providing special characteristics and functional capabilities.

MECHANICAL CLUTCHES[1]

Mechanical clutches are basically of two types: positive and friction. Positive clutches operate by meshing the metal teeth or jaws of the driving member with corresponding elements on the driven member. Friction clutches press one or more driving members against corresponding driven members, such as disks, bands, or shoes. Positive clutches will break but cannot slip; friction clutches slip but do not break.

Positive Clutches. The positive clutch will not slip up to the point of destruction. Generally, for a comparable torque capacity this clutch is smaller, lighter, and less costly than a friction clutch. It is also simpler and does not require adjustment for wear.

Because it does not develop heat appreciably during use, it may be cycled frequently. On the other hand, it cannot be engaged at high speeds, although it can be run and sometimes disengaged at high speeds. Shock results from engagement at any speed above zero. Another disadvantage is that positive clutches cannot always be engaged when both drive and load are completely at rest: some relative motion usually is required to mesh the parts.

Although not used as commonly as friction clutches, positive clutches have important applications. For example, they are used in many special synchronized drives. They are used extensively in machine tools, business machines, and household appliances. In integral sizes they are used in power presses, construction machinery, and automotive transmissions.

Positive clutches are available in three basic types, in addition to numerous special designs.

The *square-jaw clutch*, also called the *claw clutch* (Fig. 20-2-1*a*), originally used teeth in the as-cast condition. Large clearances were necessary to engage the clutch.

Square-jaw clutches are compact, of simple construction, and low in cost. During operation they produce no heat. Although engagement speed is low, they can run at high speeds.

This clutch is not in wide use today because it cannot be safely engaged while moving. Also, it picks up the load at the instant of engagement: if load inertia is high, destructive shock may be transmitted to the drive.

Spiral-jaw clutches (Fig. 20-2-1*b*) have eliminated many of the undesirable features of the square-jaw type. They are usually used as-cast, where the load is to be disengaged while running (finished jaws should be specified) and where lubrication is provided.

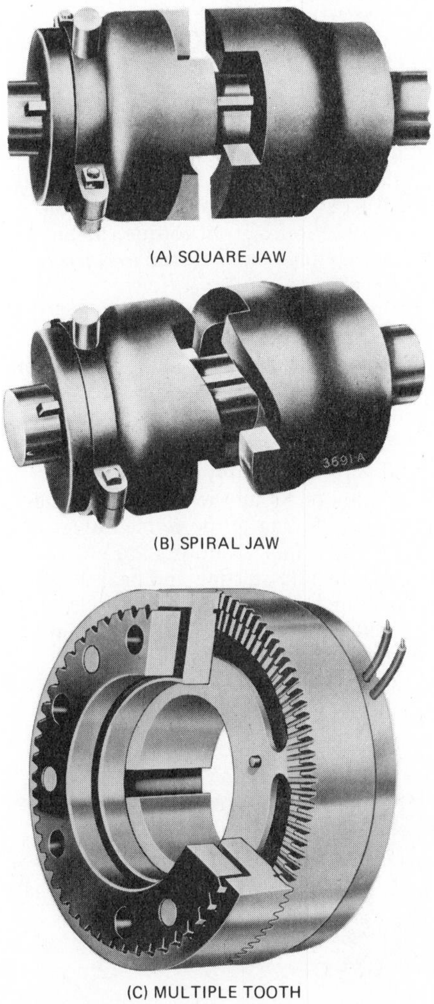
(A) SQUARE JAW
3691-A
(B) SPIRAL JAW
(C) MULTIPLE TOOTH

Fig. 20-2-1 Positive clutches. (A and B—Link Belt Ltd.; C—The Bendix Corp.)

The spiral-jaw clutch will engage at higher speeds than the square-jaw, but will shock-load the driver at engagement if load inertia is high. It drives in one direction only and has a tendency to freewheel.

Multiple-tooth clutches (Fig. 20-2-1*c*) have multiple teeth modified in various ways to provide strength or high-speed capability.

Friction Clutches. These clutches offer an extensive choice of characteristics. They can slip during engagement, which enables the drive to pick up and accelerate the load with minimum shock. They do not employ teeth or jaws, which permits their use at high engagement speeds. They need not be unloaded before engagement, and they will slip momentarily under shock loads, thus providing a cushioning action.

Friction clutches usually fall into two principal classifications: rim clutches, in which the contact pressure is applied, normally, to the shaft axis against a rim or drum; and axial clutches, in which the contact pressure is applied by shift movement along the shaft, as in cone or disk clutches. Axial-type clutches are virtually free from the effects of centrifugal force. They have several other advantages: large friction surfaces are possible for a given size and mass, the effective heat-dissipating surfaces are large, and pressure distribution is more uniform over the friction lining area.

One of the earliest of the axial-type clutches was the familiar cone clutch (Fig. 20-2-2*a*). These clutches are simple in construction and comparatively powerful for a given outside diameter and axial energizing force. The present tendency is to limit cone clutches to fairly low peripheral velocities; for instance, in machine-tool feeds, drill-press tapping attachments, synchronizing clutches in synchromesh transmissions, and other low-horsepower applications. Under such conditions the characteristic sudden pickup of load is not objectionable. Cone clutches are also used effectively for overload-release units.

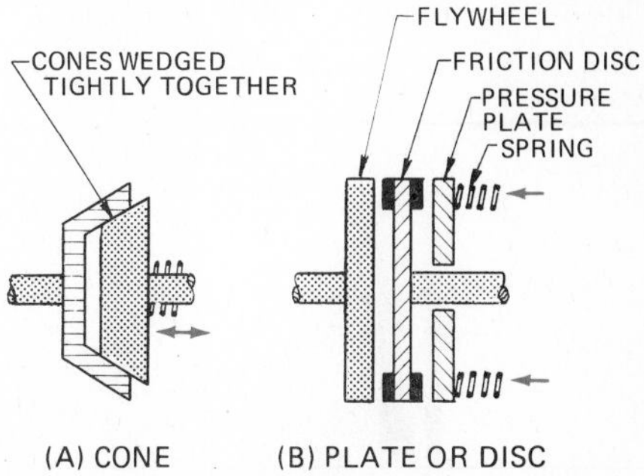
CONES WEDGED TIGHTLY TOGETHER
FLYWHEEL
FRICTION DISC
PRESSURE PLATE
SPRING
(A) CONE
(B) PLATE OR DISC

Fig. 20-2-2 Axial-type friction clutches.

In the disk type of clutch, one or more friction disks are clamped to metal plates (Fig. 20-2-2*b*). Because clamping movement is axial, centrifugal force cannot affect clamping pressure. Uniform pressure distribution can be achieved, and large friction areas are available for a given size and mass.

Overrun clutches drive in one direction only and overrun or freewheel in the other direction. The operation of the overrun clutch is very simple. As the shaft and inner race rotate in the direction of the arrow shown in Fig. 20-2-3, the rollers roll up to the high spots of the inner race, thus jamming the inner and outer races as one, causing the outer race to rotate.

However, if the inner race should travel at a lower speed than the outer race, stop, or reverse its direction, then the outer race would not turn with the inner race because the rollers would not engage the outer race.

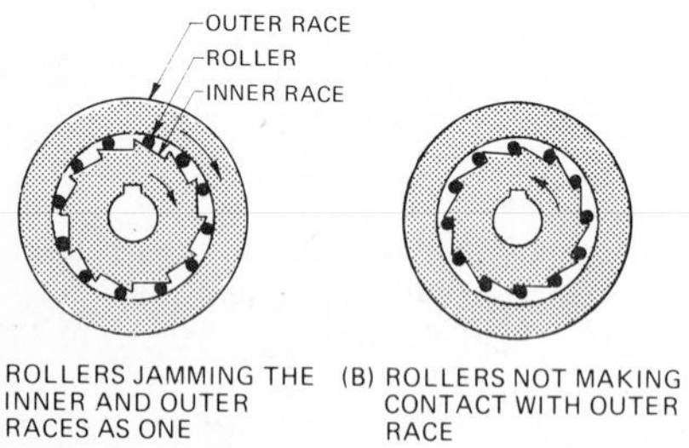
OUTER RACE
ROLLER
INNER RACE
(A) ROLLERS JAMMING THE INNER AND OUTER RACES AS ONE
(B) ROLLERS NOT MAKING CONTACT WITH OUTER RACE

Fig. 20-2-3 Overrun clutch.

ELECTRIC CLUTCHES[2]

Electric clutches perform the same functions as mechanical clutches but are controlled electromagnetically rather than mechanically. They range from the precision miniature clutch with a 12 mm diameter to units with 600 mm diameter.

All electric clutches have one thing in common—a magnetic field that determines torque. Remote control is thus both feasible and economical.

An electric clutch acts as either an on-off or a continuous-slip device. Whenever a machine operation involves starting or stopping a motor more than approximately 12 times per minute (or 4 to 5 times per minute for an enclosed motor), an electric clutch should be used to clutch the load in and out, thus allowing the motor to run continuously. The running torque of a motor, which is higher than the starting torque, can be used for starting high-inertia loads or for quick-starting by clutching in the motor after full speed is obtained. In other applications, the continuous-slip capability of the electric clutch provides overload protection and cushioned starting.

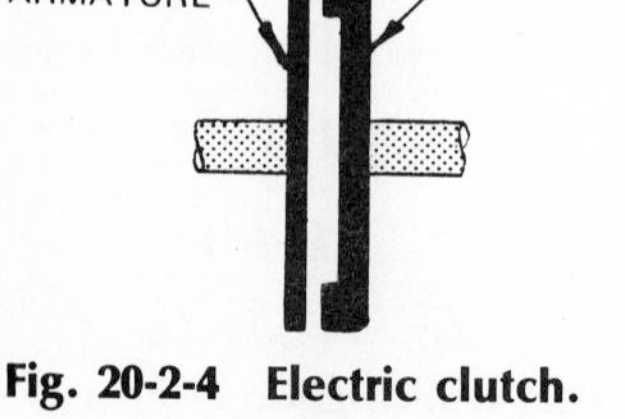
ARMATURE
MAGNET

Fig. 20-2-4 Electric clutch.

The simplest and most commonly used electric clutch is the single-disk friction clutch (Fig. 20-2-4). Its action is basically that of an electromagnet. When the electromagnet is energized, an armature is drawn into direct contact with the face of the electromagnet. In the single-disk clutch, the electromagnet is shaped like a doughnut, and the coil which creates the field is located inside the doughnut shell. Friction material is added to one side of the shell to form one of the clutch faces. The other face is that of the armature, a segmented iron disk.

HYDRAULIC CLUTCHES (FLUID COUPLINGS)[3]

Fluid couplings are similar to torque converters, except they do not have a stator between the impeller and the turbine (Fig. 20-2-5). They simply transmit input torque, whereas the torque converter multiplies input torque. Some fluid couplings can be used as variable-speed drives.

The fluid coupling performs much the same function as the centrifugal clutch. However, because it is never fully disengaged, it is torsionally softer. It has several advantages:

1. During starting and heavy running-load surges, it simply slips while it smoothly picks up the load. Thus, lugging and stalling of the prime mover are prevented. Special motors are not necessary to handle starting and peak loads.

2. Shock loads and torsional vibrations are dampened, so that power transmission is always smooth.

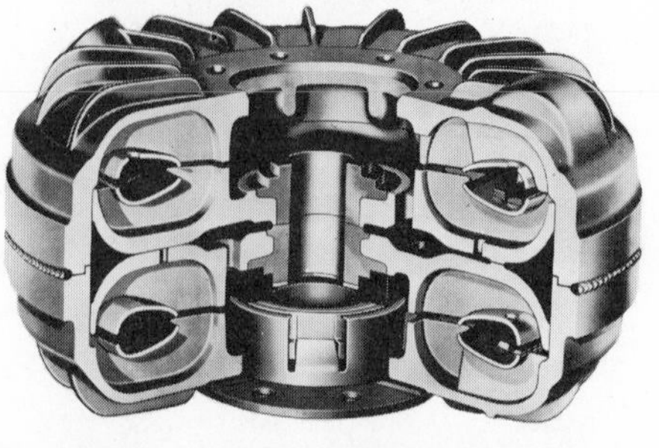

Fig. 20-2-5 Fluid coupling. (Twin Disc Clutch Co.)

3. In compound drives, fluid couplings equalize the load on the various prime movers.

4. Output torque is always the same as input torque; thus even when the driven load is stalled, the prime-mover torque is still transmitted to the drive shaft.

Operation. As the prime mover rotates the input shaft, a vortex is created between the impeller and the turbine, and energy is transmitted from one to the other. With the input shaft rotating at constant speed, the output shaft can rotate at any speed from 0 to within 2 percent of input speed. (In actual practice, speed is, of course, never exactly constant.)

In order for a fluid coupling to transmit torque, there must be slip between impeller and turbine. Torque capacity is 0 at no slip. As load is applied to the output shaft, slip increases. Assuming constant input speed, transmitted torque increases until maximum torque is attained at 100 percent slip; at this point the turbine is stalled.

Brakes

Basically, any brake is simply an extension of a clutch, in which one member is held stationary. Brakes may be used as on-off devices or as drags. Brakes may be classified into two major types: mechanical and electrical.

MECHANICAL BRAKES[4]

A mechanical brake is a frictional device that converts kinetic energy to heat and dissipates it into the atmosphere. It can be actuated mechanically, pneumatically, hydraulically, or electrically. The friction material used in brakes may be organic, metallic, ceramic, or a combination of any two or all three. Because the organic is the most versatile type of lining, it is used for most ordinary applications.

Two factors are of paramount importance in selecting and applying a brake: torque requirement, or force which the brake must resist; and amount of energy which the brake and drum (or caliper and disk) must absorb and dissipate repeatedly.

Most commercial brakes are band, drum, or disk types. In addition, a number of specialized configurations are also made. These are not in general use and will not be covered in this chapter.

Band Brakes. This is one of the oldest types of brakes. Basically, it consists of a flexible steel band lined with friction material. Normally the operating pull on the band is in the same direction as the resisted torque, so that there is a self-energization, or *self-servo,* effect. However, *negative-servo* band brakes are also used, although they are not common. They provide greater stability. Certain configurations may be designed so that they are equally effective in either direction of drum rotation (Fig. 20-2-6*a*). Also, band brakes may be used in combination with disks to form a band-disk brake (Fig. 20-2-6*b*), which is used primarily in farm machinery.

Tension is applied to the band through one of several mechanical linkages, depending on the application. Band brakes may also be spring-set and released hydraulically or otherwise.

Drum Brakes. Drum brakes are perhaps the best known type of mechanical brake, mainly because of their widespread use in automobiles. See Fig. 20-2-7. They are available in two basic configurations, external-contracting and internal-expanding, in sizes ranging from 125 × 25 mm to 750 × 300 mm and larger. Most drum brakes in use today are internal-expanding and are of the fixed-anchor type, in which the fulcrum points of the shoes are positively located with reference to the brake drum. This arrangement results in consistent performance and can provide the maximum amount of self-energization.

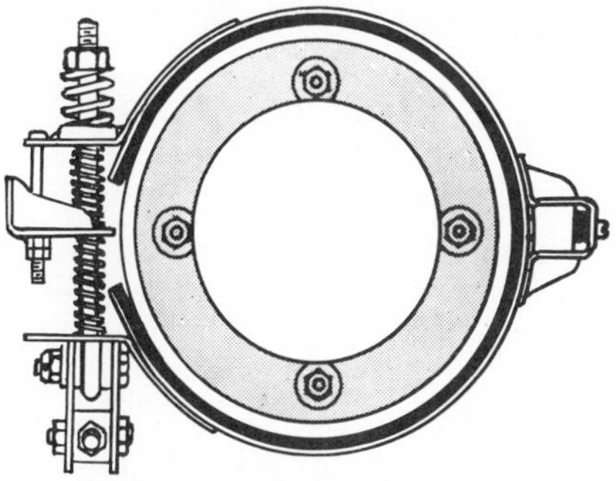

(A) AUXILIARY BRAKE FOR A TRUCK USUALLY MOUNTED ON THE TRANSMISSION

(B) COMBINATION BAND-DISC BRAKE USED MOSTLY IN FARM MACHINERY

Fig. 20-2-6 Band brakes. (The Bendix Corp.)

Fig. 20-2-7 Drum brakes. (The Bendix Corp.)

ELECTRIC BRAKES[5]

Basically, an electric brake is similar to an electric clutch, except that one element is rigidly held. Like clutches, electric brakes are available in four basic types: magnetic-particle, eddy-current, hysteresis, and friction.

The most widely used type of electric brake is the electromagnetic friction-disk type, in which a friction unit is electrically actuated or released. See Fig. 20-2-8. These are ordinarily used for on-off operation. For retarding, as in tensioning applications, pure-electrical brakes (magnetic-particle, hysteresis, or eddy-current) are normally used.

Electric clutches and brakes are similar enough that virtually all the basic information on electrical clutches can also be applied to electrical brakes.

(A) SPRING-ACTUATED ELECTRICALLY RELEASED

(B) ELECTRICALLY ACTUATED

Fig. 20-2-8 Electromagnet friction-disk brakes. (A—Dings Brakes Co.; B—Warner Electric Brake & Clutch Co.)

REFERENCES AND SOURCE MATERIAL

1. N. C. Harrison, "Mechanical Clutches," *Machine Design*, Vol. 37, No. 14, 1965.
2. J. F. Pech, "Electric Clutches," *Machine Design*, Vol. 37, No. 14, 1965.
3. F. J. Lavoie, "Fluid Couplings," *Machine Design*, Vol. 37, No. 14, 1965.
4. E. K. Dombeck, "Mechanical Brakes," *Machine Design*, Vol. 37, No. 14, 1965.
5. F. J. Lavoie, "Electric Brakes," *Machine Design*, Vol. 37, No. 14, 1965.

Assignments

1. On an A3- or B-size sheet, make a full-section assembly drawing of the friction clutch shown in Fig. 20-2-A or 20-2-B. Scale is 1:1. Use your judgment for dimensions not shown.

2. On an A3- or B-size sheet, make a half-section assembly drawing of a gear mounted on the outer race hub of the overrunning clutch, model 12, shown in Fig. 20-2-C or 20-2-D. Scale is 1:1. Use your judgment for dimensions not given.

REVIEW FOR ASSIGNMENTS

Unit 6-4 Assembly Drawings
Unit 7-1 Full Sections
Unit 7-2 Half Sections
Unit 8-3 Drawing Nuts and Bolts
Unit 9-1 Keys

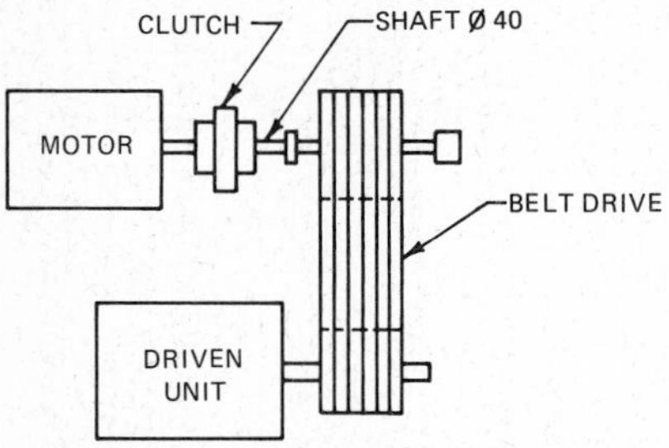
CLUTCH
SHAFT Ø 40
MOTOR
BELT DRIVE
DRIVEN UNIT

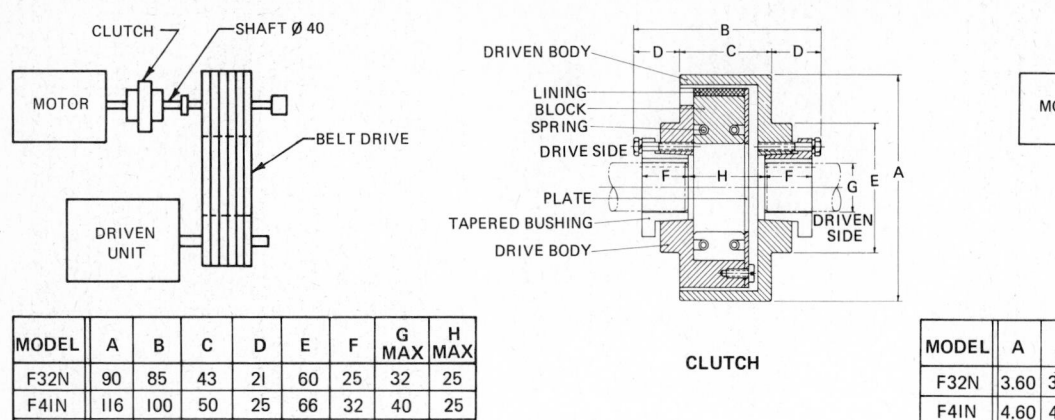

MODEL	A	B	C	D	E	F	G MAX	H MAX
F32N	90	85	43	21	60	25	32	25
F41N	116	100	50	25	66	32	40	25
F54N	140	125	55	35	90	32	45	50

Fig. 20-2-A Friction clutch.

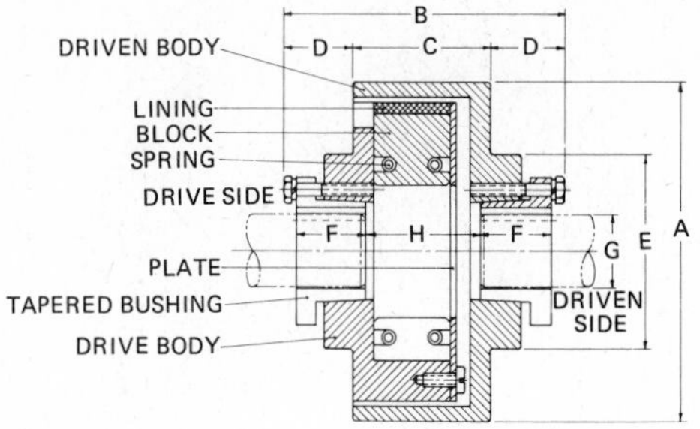
B
D
C
D
DRIVEN BODY
LINING
BLOCK
SPRING
DRIVE SIDE
F
H
F
G
E
A
PLATE
TAPERED BUSHING
DRIVEN SIDE
DRIVE BODY

CLUTCH

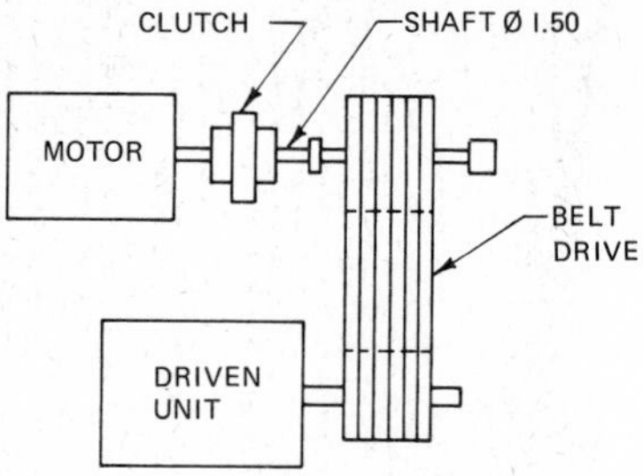
CLUTCH
SHAFT Ø 1.50
MOTOR
BELT DRIVE
DRIVEN UNIT

MODEL	A	B	C	D	E	F	G MAX	H MAX
F32N	3.60	3.40	1.70	.85	2.30	1.00	1.25	1.00
F41N	4.60	4.00	1.90	1.05	2.60	1.25	1.50	1.00
F54N	5.60	5.00	2.20	1.40	3.50	1.25	1.75	2.00

Fig. 20-2-B Friction clutch.

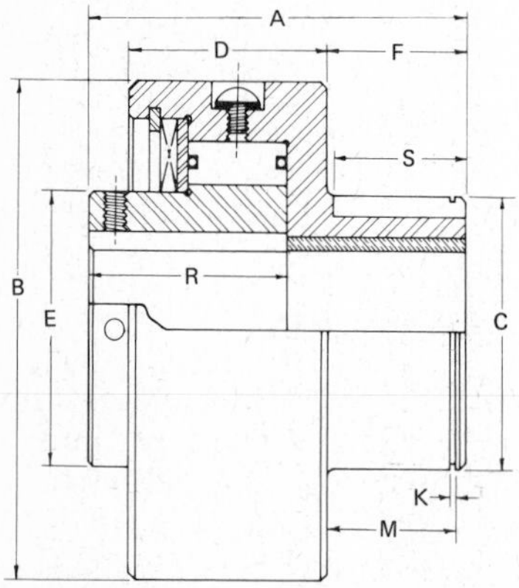
A
D
F
S
R
B
E
C
K
M

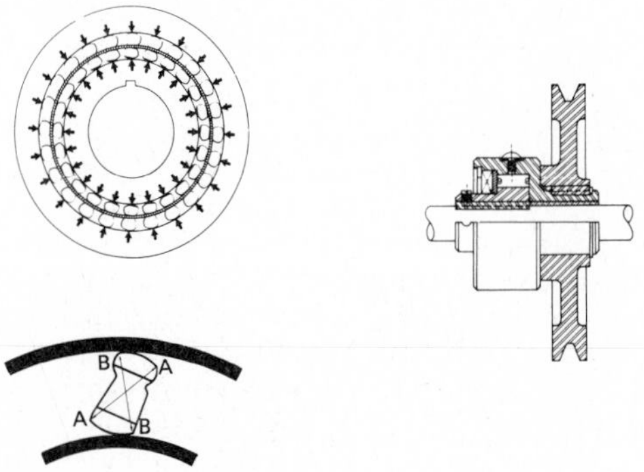
B
A
A
B

F-S SERIES MODEL NUMBER	5	6	8	10	12	14
STANDARD BORE SIZE	12 15	19	22 25	28 32	34 38	40 44
KEYWAY	5 x 2.5	5 x 2.5	6 x 3	7 x 3.5	10 x 5	11 x 5.5
STANDARD HUB KEYWAY	5 x 2.5	5 x 2.5	6 x 3	7 x 3.5	10 x 5	11 x 5.5
A	70	80	84	90	100	110
B	56	74	82	96	112	134
C	31.8 31.8	34.9 34.8	45.5 44.5	57.2 57.1	64.1 64.0	73.1 73.0
D	32	38	40	44	50	58
E	25	32	40	50	60	74
F	25	33	33	37	37	40
K	1.0 1.5	1.0 1.5	1.2 1.8	1.2 1.8	1.2 1.8	1.2 1.8
M	22.8 22.9	30.7 30.9	30.7 30.9	34.4 34.5	34.4 34.5	38.6 38.8
R	37	39	43	45	53	60
S	14	25	23	25	36	39
OIL HOLE	M6	M6	M6	M6	M6	M6

GEAR DATA: 20° SPUR GEAR, MODULE – 6.35 PITCH DIA – 152.4,
FACE WIDTH – 25, HUB PROJECTION ON ONE SIDE ONLY – 6,
HUB DIA – 90, SHAFT DIA – 35.

Fig. 20-2-C Overrun clutch.

A FULL COMPLEMENT OF SPRAGS BETWEEN CONCENTRIC INNER AND OUTER RACES TRANSMITS POWER FROM ONE RACE TO THE OTHER BY WEDGING ACTION OF THE SPRAGS WHEN EITHER RACE IS ROTATED IN THE DRIVING DIRECTION. ROTATION IN THE OPPOSITE DIRECTION FREES THE SPRAGS AND CLUTCH IS DISENGAGED OR "OVER-RUNS."

OVERRUN CLUTCH

F-S SERIES MODEL NUMBER	5	6	8	10	12	14
STANDARD BORE SIZE	.500 .625	.750	.875 1.000	1.125 1.250	1.375 1.500	1.625 1.750
KEYWAY	.10x.05 .20x.10	.20x.10	.24x.12	.30x.15	.30x.15 .40x.20	.44x.22
STANDARD HUB KEYWAY	.20x.10	.20x.10	.24x.12	.30x.15	.40x.20	.44x.22
A	2.75	3.20	3.30	3.60	3.90	4.40
B	2.20	2.90	3.25	3.75	4.45	5.25
C	1.250 1.249	1.375 1.374	1.750 1.749	2.250 2.249	2.500 2.499	2.875 2.874
D	1.25	1.45	1.60	1.75	2.00	2.25
E	1.00	1.25	1.60	2.00	2.40	2.90
F	1.00	1.30	1.30	1.45	1.45	1.60
K	.048 .068	.048 .068	.056 .076	.056 .076	.056 .076	.056 .076
M	.900 .905	1.215 1.220	1.215 1.220	1.340 1.345	1.340 1.345	1.525 1.530
R	1.45	1.55	1.70	1.80	2.10	2.35
S	.55	1.00	.90	1.00	1.40	1.55
OIL HOLE	.25-28	.25-28	.25-28	.25-28	.25-28	.25-28

GEAR DATA: 20° SPUR GEAR, PITCH – 4, PITCH DIA – 6.00,
FACE WIDTH – 1.00, HUB PROJECTION ONE SIDE ONLY – .25,
HUB DIA – 3.50, SHAFT DIA – 1.375.

Fig. 20-2-D Overrun clutch.

UNIT 20-3

ADJUSTABLE-SPEED DRIVES AND SPEED REDUCERS

Packaged Adjustable-Speed Drives

Packaged mechanical adjustable-speed drives are available for small machine tools to trucks. They may provide only a few selected speeds or be able to vary speed infinitely over a wide range. Efficiency is generally high; for many units, efficiency will be over 90 percent depending on the type of drive. An example of a variable-speed drive is shown in Fig. 20-3-1.

Adjustable-speed drives can be broken down into three major categories:

1. Stepped
2. Stepless, limited range
3. Stepless, infinite range

The five major types of mechanical *transmission* are covered in this unit. (The torque converter, although based on hydraulic principles, is here considered a mechanical drive.) In each case, the principle of operation and characteristics are discussed. The final step in deciding on a specific drive depends, of course, on which type meets the system requirements most closely and economically.

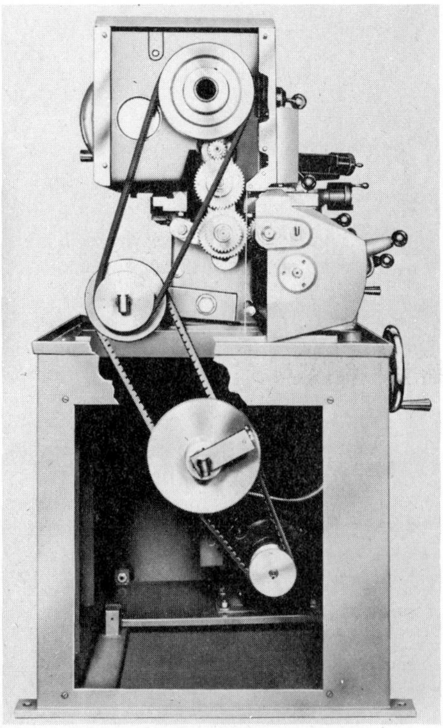

Fig. 20-3-1 Variable-speed drive for 350 mm lathe. (Rockwell Mfg. Co.)

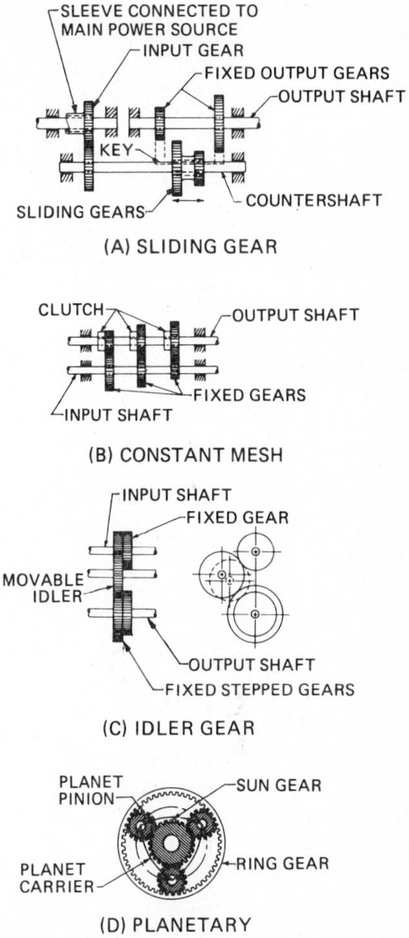
SLEEVE CONNECTED TO MAIN POWER SOURCE
INPUT GEAR
FIXED OUTPUT GEARS
OUTPUT SHAFT
KEY
SLIDING GEARS
COUNTERSHAFT
(A) SLIDING GEAR
CLUTCH
OUTPUT SHAFT
FIXED GEARS
INPUT SHAFT
(B) CONSTANT MESH
INPUT SHAFT
FIXED GEAR
MOVABLE IDLER
OUTPUT SHAFT
FIXED STEPPED GEARS
(C) IDLER GEAR
PLANET PINION
SUN GEAR
PLANET CARRIER
RING GEAR
(D) PLANETARY

Fig. 20-3-2 Gear-type adjustable-speed drives.

Gear Drives[1]

Multispeed gear transmissions provide exact shaft speed at high efficiency. They are used in machine tools, mobile equipment, and other applications where selective control of a number of fixed speeds is necessary.

Two broad categories of general transmissions are available. In one type, the speed is selected manually; in the other, speed changes occur automatically at predetermined points.

BASIC TYPES

The most common selective-speed transmissions are parallel-axis arrangements. These can be broadly classified into four types.

Sliding Gears. Speed adjustment is effected by sliding gears on one or more intermediate parallel shafts (Fig. 20-3-2*a*). Shifting is generally accomplished by disengaging the input shaft.

Sliding-gear transmissions are usually manually shifted by means of a lever or a handwheel. A variety of shaft arrangements and mountings are available.

Constant-Mesh. Several gears of different sizes mounted rigidly to one shaft mesh with mating gears free to rotate on the other shaft (Fig. 20-3-2*b*). Speed adjustment is obtained by locking different gears individually on the second shaft by means of splined clutches or sliding couplings.

Constant-mesh gears are used in numerous applications, among them heavy-duty industrial transmissions. This arrangement can use virtually any type of gearing, for example, spur, helical, herringbone, and bevel gears.

Manually shifted automotive transmissions combine two arrangements. The forward speeds are constant-mesh helical gears, whereas reverse speeds use sliding spur gears.

Idler. One shaft carries several different-size gears that are rigidly mounted (Fig. 20-3-2*c*). Speed adjustment is through an adjustable arm which carries an idler gear to connect with fixed or sliding gears on the other shaft. This arrangement is used to provide stepped speeds in small increments and is frequently found in machine tools.

A transmission of this type can be connected to another multispeed train to provide an extremely wide range of speeds with constant-speed input. Shifting is usually manual. The transmission is disengaged and allowed to come to a stop before the sliding idler gear is moved to the desired setting.

Planetary. This type of gearing (Fig. 20-3-2*d*) is the most versatile and compact gear arrangement for a given ratio range and torque capacity. At the same time, it is also the most expensive, because of the clutching and braking elements necessary to control the operation of the unit. In addition, practical ratios available with planetary sets are limited.

Most planetary-gear transmissions are of the *automatic* type, in which speed changes are carried out automatically at a selected speed or torque level.

Belt and Chain Drives[2]

Packaged belt and chain drives convert constant input speed into output that is steplessly variable within a certain range. They may contain integral motors and built-in gear reducers to obtain low output speeds. Electric motors are the most commonly used power-input devices.

Drive packages may be mounted horizontally, vertically, or on a 45° angle and are available in open or enclosed designs. Speed ranges of 10:1 to 2:1 can be obtained with most units—some are available for 16:1 ratios. A typical distribution of available output speeds is 4660 to 1.7 r/min, including drives with and without reducer gearing. Recent improvements in increaser gearing, offered as integrally mounted packages, provide speeds to 16 000 r/min.

Many accessories and modifications are available for packaged-belt and chain-drive units. For example, the integral-motor drive package may also include brakes, special clutches and couplings, special drive motors, and sophisticated controls.

All-Metal Belt. In this unit, power is transmitted by means of a self-forming laminated steel belt or chain which engages radial teeth on conical pulley flanges (Fig. 20-3-3). The steel laminations shuffle and reshuffle into driving contact with every meshing engagement of the pulley teeth. The principle of speed-changing operation is the same as for the other variable-pitch pulley designs.

V-Belt. A highly compact belt drive for wide speed variation has been achieved through use of a modified form of the compound floating-sheave principle. Speed adjustment is accomplished by changing the position of the movable flange on the bottom input sheave. This action is transmitted by belt pressure that automatically adjusts the upper sheaves and the bottom output sheave to correspond.

Friction and Traction Drives[3]

Friction or *traction drives* transmit rotary motion by friction generated at the point or line of contact. Speeds are changed by moving the contact point or line relative to the centers of rotation of the driving and driven members. The amount of friction between the parts determines the transmittable power. This friction, in turn, is determined by the force applied at the contact point. A successful design must handle these forces without excessive wear or early failure. Friction and traction drives are practically without vibration.

These drives operate best at constant load, with no sudden shocks to cause destructive slippage at the friction point. They have a lower tolerance for load variations that do adjustable-speed belt or chain drives. They also are less able to handle high-inertia loads.

Several types of packed friction or traction drives are commercially available.

Contact of metallic surfaces provides one practical method of obtaining stepless speed adjustment. See Fig. 20-3-4.

Present metallic-traction transmission units are efficient and reliable and offer a high degree of speed control. Because metal contact surfaces are hard and highly finished, parts can be small and the transmissions compact. Since metal parts (usually alloy steel) can run in oil, the speed-changing mechanism can be self-contained and splash-lubricated.

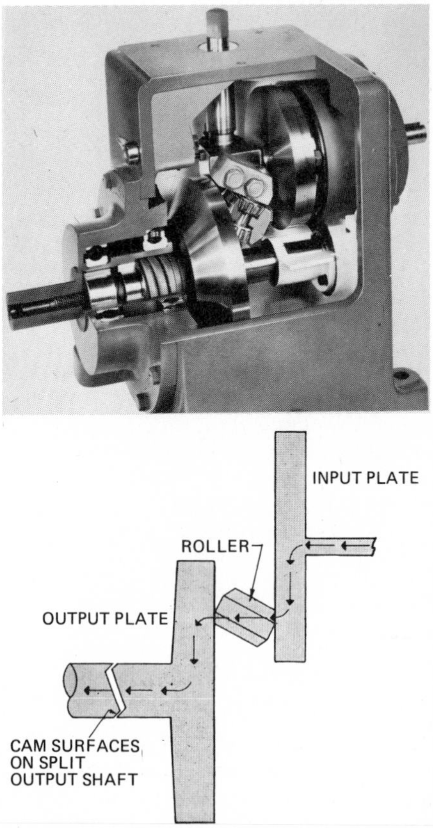
INPUT PLATE
ROLLER
OUTPUT PLATE
CAM SURFACES ON SPLIT OUTPUT SHAFT

Fig. 20-3-4 Friction-type adjustable-speed drive. (Cone Drive Gears)

Metallic systems are economically competitive with other mechanical units of equal power and range.

Impulse Drives[4]

Adjustable impulse drives provide infinitely adjustable output speeds, usually in low-speed ranges. They provide high-ratio speed reductions in a compact package. While most impulse drives may be adjusted to 0, typical output speeds range from 1.5 to 40 r/min. Speed variation is stepless, and output speed may be changed while the drive is operating. See Fig. 20-3-5.

The principle of operation is continuous indexing. The driving member engages the driven, moves it a predetermined distance, and then disengages. This is done through one-way clutches operating in phased sequence to minimize pulsation in the output.

Impulse drives are sometimes incorrectly called *ratcheting drives* because a ratchet has similar action; however, output in a ratchet drive is stepped. An adjustable impulse drive does not include a ratchet.

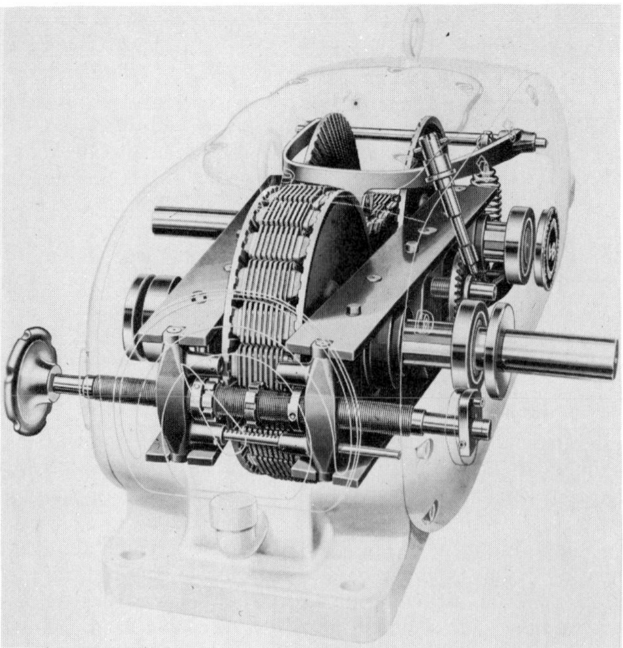

Fig. 20-3-3 Steel-belt variable drive. (Link Belt Ltd.)

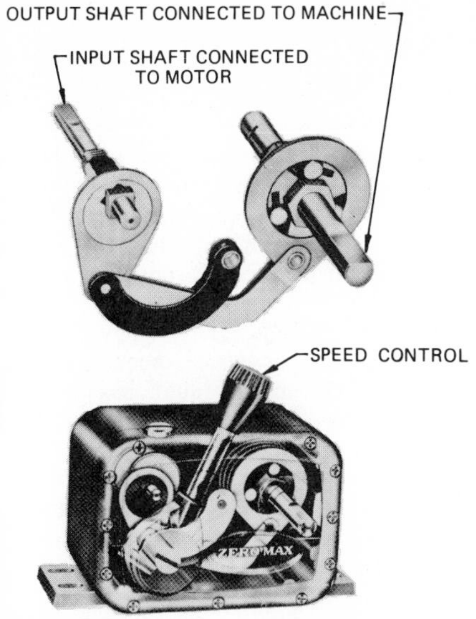
OUTPUT SHAFT CONNECTED TO MACHINE
INPUT SHAFT CONNECTED TO MOTOR
SPEED CONTROL

Fig. 20-3-5 Impulse-type variable-speed drive. (Zero-Max Ind. Inc.)

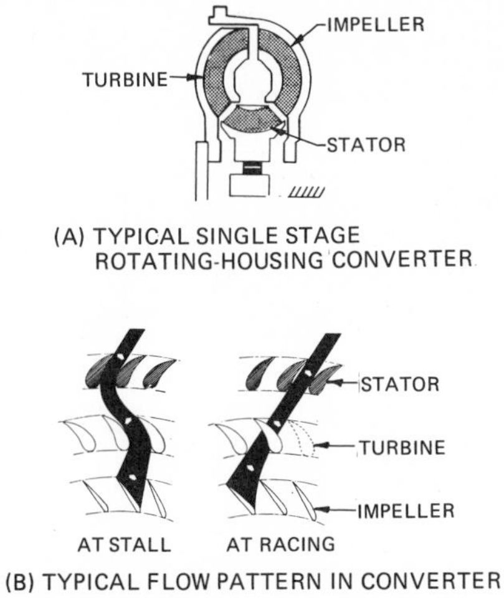
IMPELLER
TURBINE
STATOR
(A) TYPICAL SINGLE STAGE ROTATING-HOUSING CONVERTER
STATOR
TURBINE
IMPELLER
AT STALL
AT RACING
(B) TYPICAL FLOW PATTERN IN CONVERTER

Fig. 20-3-6 Torque converter adjustable-speed drive.

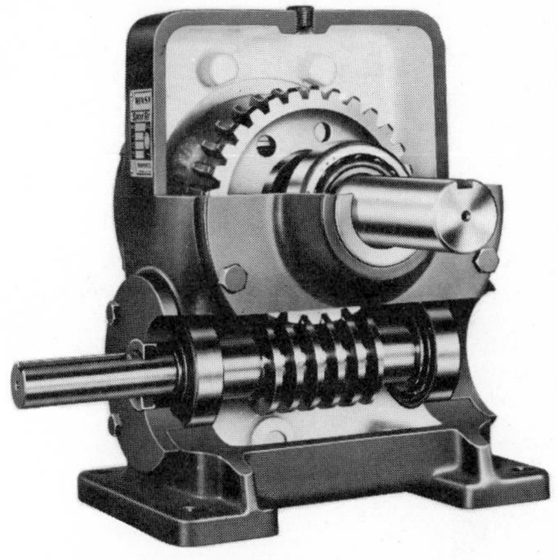

Fig. 20-3-7 Vase-mounted speed reducer. (Winsmith Div. of U.M.C.)

The one-way clutches in adjustable impulse drives provide stepless speed variation. Speed is changed by adjusting the distance the driven member is moved at each stroke. Output speed is changed by adjusting the angular relationship of bars in the linkages which actuate the one-way clutches. This changes the clutch stroke and, consequently, the output speed. With proper linkage design, output torque is constant.

Two commercial impulse drives are presently available: the Morse VID and Zero-Max.

Torque Converters[5]

The hydraulic torque converter provides infinitely variable torque within its limits, solely in response to load variation. It makes use of the kinetic energy of fluid in motion. The torque converter is similar to a fluid coupling except for the addition of a stator.

Figure 20-3-6*a* shows the schematic blade arrangement of a typical single-stage, rotating-housing converter. It consists of three parts: the impeller, the turbine, and the stator. The impeller, driven by the engine, imparts kinetic energy to the fluid which rotates the turbine and the output members. The fluid leaving the turbine passes through the stator blades, assuming the correct reentry angle to the impeller for recirculation. Direction of this fluid varies with load.

Under light load, the impeller causes the turbine to turn almost freely; fluid passes through the converter quickly and strikes the turbine blades and stator blades at a slight angle. Under load, the turbine slows down and the fluid strikes the turbine blades at a sharper angle, causing more torque to be transmitted. Figure 20-3-6*b* shows the typical flow pattern among the impeller, turbine, and stator at stall when maximum torque multiplication occurs, and at racing when no torque is transmitted.

Speed Reducers

Any device which reduces the speed of the driving unit is a speed reducer. However, this section considers only those packaged units—primarily gear-type—whose prime function is the reduction of speed. As defined here, a speed reducer has a fixed ratio, which cannot be readily changed. Although this section mentions specifically speed reducers, the considerations given here also apply to speed increasers; the same units are used for increasing as for reducing speeds.

Conventional geared speed reducers can be classified as either base-mounted or shaft-mounted.

Small reducers are frequently combined with a motor, to form a *gear-motor* package.

BASE-MOUNTED REDUCERS[6]

Base-mounted speed reducers are available in a number of gear types: helical, double-helical spur, spiral bevel, straight bevel, worm, double-enveloping worm, and herringbone. These may be used singly or in combination. Input-shaft arrangements include concentric, parallel-offset (vertical and horizontal), and right-angle. Single, double, triple, and, in some cases, quadruple reductions provide a wide range of reduction ratios.

In addition to the common geared reducers, many unconventional designs are available. Figure 20-3-7 shows a worm-type speed reducer.

SHAFT-MOUNTED REDUCERS[7]

A shaft-mounted speed reducer is an enclosed gear unit mounted on and supported by the input shaft of the driven machine. To prevent rotation of the housing, it is anchored by a torque-reaction member to a suitable rigid support.

These reducers are available as helical, herringbone, and spur-gear units, of single or multiple reduction stages (Fig. 20-3-8) with the output hub either concentric with or parallel to the input shaft of the speed reducer.

Antifriction bearings are used for both axial and radial loads, and all gears and bearings are splash-lubricated.

Shaft-mounted reducers are normally combined with a V-belt drive. Combinations of standard motor speeds, V-belt drives, and gear reduction ratios cover ranges of output speeds from 10 to 400 r/min power capacities to 135 kW, and torque capacities to 1000 megapascals (MPa).

All shaft-mounted speed reducers have fixed gear ratios. Output speed can be increased or decreased by a change of the V-belt drive ratio. If frequent output speed changes are required, a conventional variable-pitch sheave can be used. Interchangeable bushings are used to adapt the output hub of the reducer to standard shaft sizes.

REFERENCES AND SOURCE MATERIALS

1. R. P. Wadlington, "Gear Drives," *Machine Design,* Vol. 37, No. 14, 1965.

Fig. 20-3-8 Shaft-mounted reducer. (Dodge Mfg. Corp.)

2. G. Malcolm, "Belt and Chain Drives," *Machine Design,* Vol. 37, No. 14, 1965.

3. J. R. Burnett, "Friction and Traction Drives," *Machine Design,* Vol. 37, No. 14, 1965.

4. C. E. Hein, "Impulse Drives," *Machine Design,* Vol. 37, No. 14, 1965.

5. H. J. Wirry, "Torque Converters," *Machine Design,* Vol. 37, No. 14, 1965.

6. R. W. Whelan, "Base-Mounted Reducers," *Machine Design,* Vol. 37, No. 14, 1965.

7. J. Chung, "Shaft-Mounted Reducers," *Machine Design,* Vol. 37, No. 14, 1965.

Assignments

1. On an A3- or B-size sheet, make a full-section assembly drawing of the sliding gear speed reducer shown in Fig. 20-3-A or 20-3-B. Support the gears on journal bearings. The gears are held to the shafts by setscrews and keys. Gears *C* and *D* are combined into one part and slide on the countershaft. See Fig. 20-3-2*a*. Complete the chart on the figure showing the two speeds available (when gears *C* and *E* mesh and when gears *F* and *D* mesh) for the different motor inputs. Use your judgment for dimensions not shown. Scale is 1:1.

2. On an A3- or B-size sheet, make a drawing of the speed reduction assembly shown in Fig. 20-3-C or 20-3-D. The motor and worm gear reducer are mounted on a table. The coupling FC15 joins the two. A steel sprocket, mounted directly on the reducer shaft, is to move a chain at an approximate rate of 12 m per hour (m/h). Call out on the assembly drawing the catalog numbers for the coupling and sprocket selected. Scale is 1:2. Show the coupling and sprocket in full section.

REVIEW FOR ASSIGNMENTS

Unit 6-4	Assembly Drawings
Unit 7-1	Full Sections
Unit 19-3	Spur Gears
Unit 19-2	Chain Drives
Unit 21-1	Plain Bearings
Unit 9-1	Keys

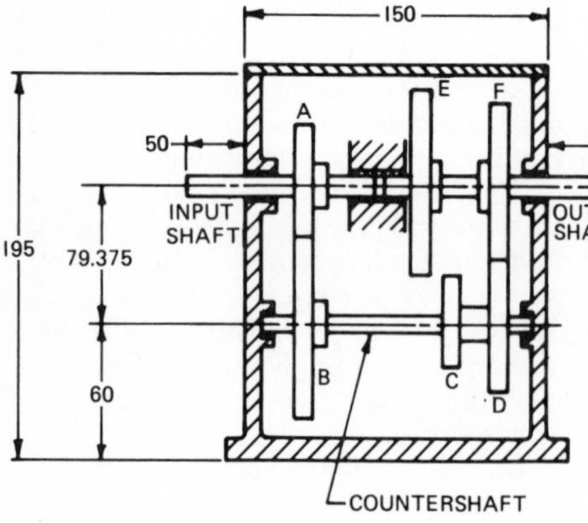
GEAR REDUCER
150
50
50
A
E
F
INPUT SHAFT
OUTPUT SHAFT
195
79.375
60
B
C
D
COUNTERSHAFT

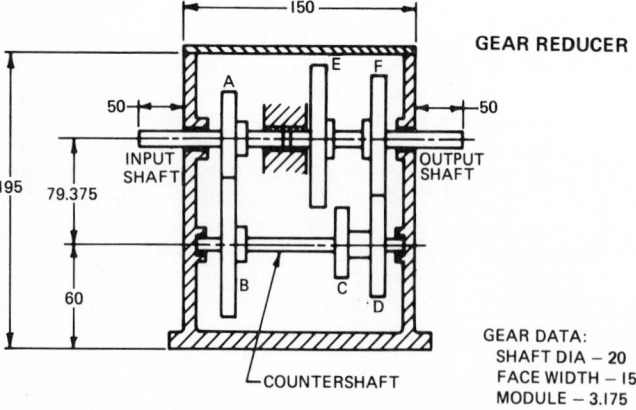

GEAR DATA:
SHAFT DIA – 20
FACE WIDTH – 15
MODULE – 3.175

GEAR	NUMBER OF TEETH
A	20
B	30
C	20
D	24
E	30
F	26

INPUT REV/MIN	OUTPUT REV/MIN	
1150		
1750		

Fig. 20-3-A Sliding gear speed reducer.

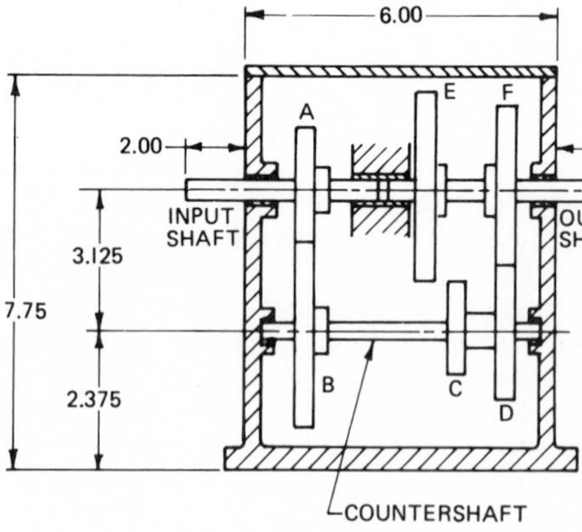
GEAR REDUCER
6.00
2.00
2.00
A
E
F
INPUT SHAFT
OUTPUT SHAFT
3.125
7.75
2.375
B
C
D
COUNTERSHAFT

GEAR DATA:
SHAFT DIA – .75
FACE WIDTH – .625
DIAMETRAL PITCH –

GEAR	NUMBER OF TEETH
A	20
B	30
C	20
D	24
E	30
F	26

INPUT RPM	OUTPUT RPM	
1150		
1750		

Fig. 20-3-B Sliding gear speed reducer.

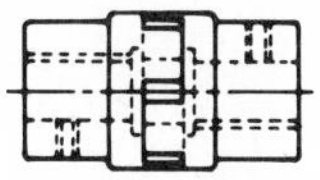

COUPLING

CATALOG NO.	HOLE DIA	HOLE LENGTH	O.D.	HUB DIA	HUB PROJ	LENGTH
FC 12	12.7	20	32	25	15	60
FC 15	12.7	25	38	32	20	70
FC 20	12.7	35	50	44	28	95

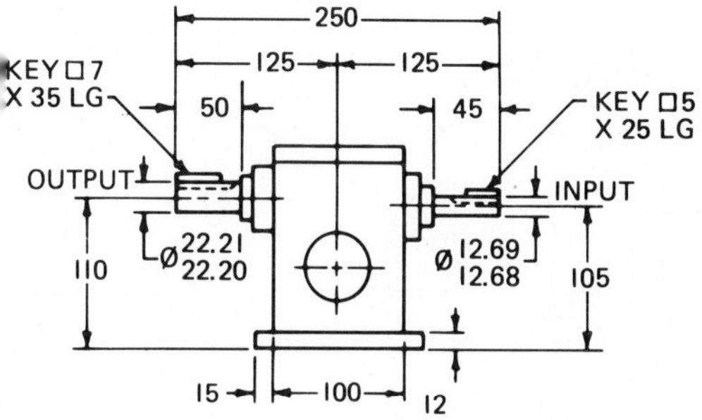
250
125
125
KEY □ 7
X 35 LG
50
45
KEY □ 5
X 25 LG
OUTPUT
INPUT
110
Ø 22.21
22.20
Ø 12.69
12.68
105
15
100
12

RATIO 1740 : 1
WORM GEAR REDUCER

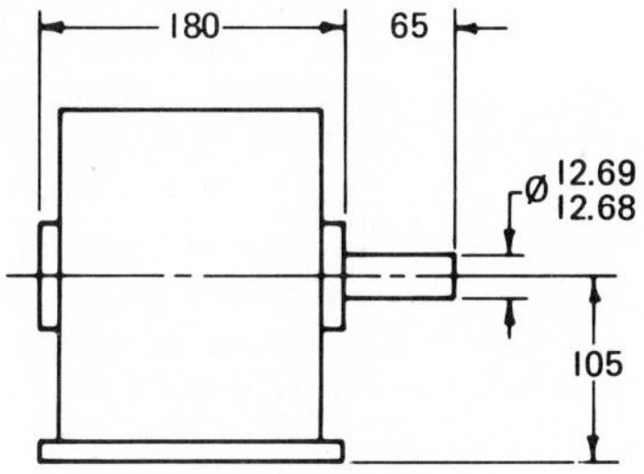
180
65
Ø 12.69
12.68
105

MOTOR 1750 REV/MIN

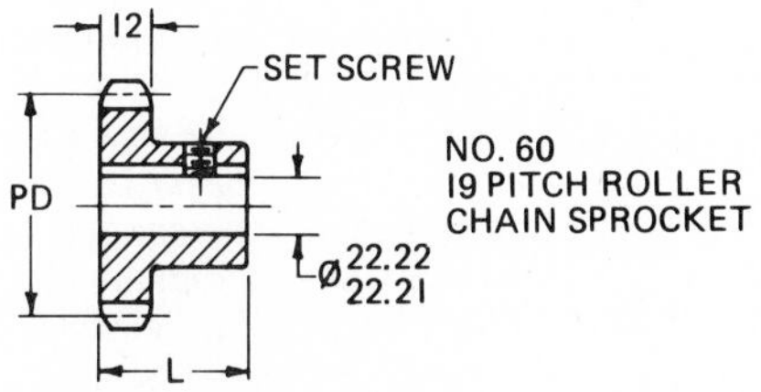
12
SET SCREW
PD
NO. 60
19 PITCH ROLLER
CHAIN SPROCKET
Ø 22.22
22.21
L

CATALOG NO.	P.D.	TEETH	HUB DIA	L
KS8	49.8	8	38	35
KS9	55.6	9	44	35
KS 10	61.7	10	50	35
KS 11	67.5	11	54	32
KS 12	73.7	12	54	32

SPEED REDUCTION ASSEMBLY

Fig. 20-3-C Speed reduction assembly.

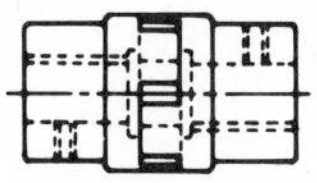

COUPLING

CATALOG NO.	HOLE DIA	HOLE LENGTH	O.D.	HUB DIA	HUB PROJ	LENGTH
FC 12	.50	.84	1.25	1.00	.60	2.30
FC 15	.50	1.00	1.50	1.25	.75	2.75
FC 20	.50	1.40	2.00	1.75	1.10	3.70

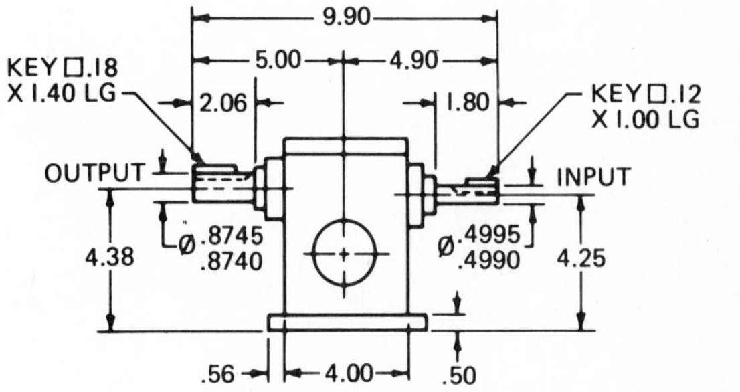
9.90
5.00
4.90
KEY □ .18
X 1.40 LG
2.06
1.80
KEY □ .12
X 1.00 LG
OUTPUT
INPUT
4.38
Ø .8745
.8740
Ø .4995
.4990
4.25
.56
4.00
.50

RATIO 1740 : 1
WORM GEAR REDUCER

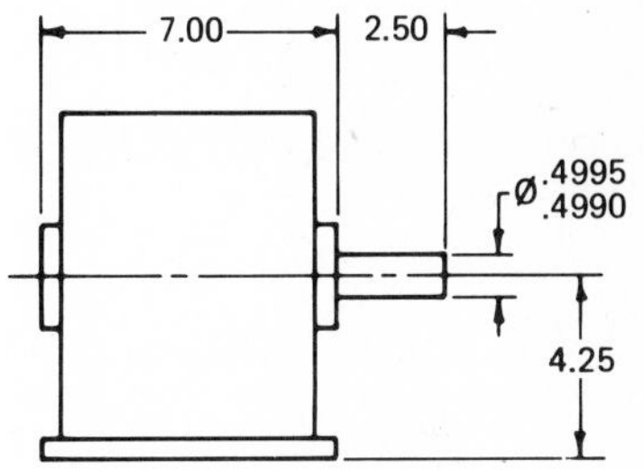
7.00
2.50
Ø .4995
.4990
4.25

MOTOR 1750 RPM

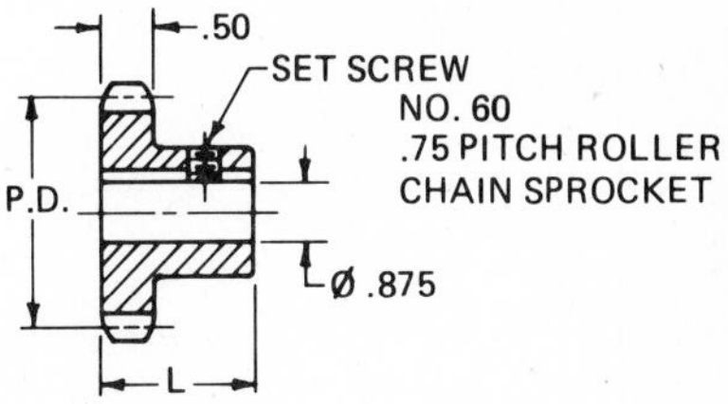
.50
SET SCREW
NO. 60
.75 PITCH ROLLER
CHAIN SPROCKET
P.D.
Ø .875
L

CATALOG NO.	P.D.	TEETH	HUB DIA	L
KS8	1.96	8	1.50	1.38
KS9	2.19	9	1.65	1.38
KS10	2.43	10	1.94	1.38
KS11	2.66	11	2.10	1.25
KS12	2.90	12	2.10	1.25

SPEED REDUCTION ASSEMBLY

Fig. 20-3-D Speed reduction assembly.

Chapter 21
Bearings, Lubricants, and Seals

UNIT 21-1

BEARINGS[1,2]

Bearings permit smooth, low-friction movement between two surfaces. The movement can be either rotary (a shaft rotating within a mount) or linear (one surface moving along another).

Bearings can employ either a sliding or a rolling action. In both cases, there is a strong attempt to provide enough lubrication to keep the bearing surfaces separated by a film of oil or other lubricant. It is this absence of physical contact that provides most bearings with long service lives.

Bearings based on rolling action are called *rolling-element bearings.* Those based on sliding action are called *plain bearings.*

Bearings are evaluated on the basis of how much load they can carry, at what speeds they can carry this load, and how long they will serve under the specified conditions. Friction, start-up torques or forces, ability to withstand impact or harsh environments, rigidity, size, cost, and complexity also are important design considerations.

The basic principles of design and application of antifriction bearings were conceived many centuries ago. They originated for one purpose only—to lessen friction. Through the ages people wanted to move heavy objects across the earth's surface. As far back as 1100 B.C., we know that such friction was reduced by the insertion of rollers between the object and the surface over which it was being moved. The Assyrians and Babylonians used rollers to move enormous stones for their monuments and palaces. Down through history are recorded many similar examples of people's war on friction.

Plain Bearings

A plain bearing is any bearing that works by sliding action, with or without lubricant. This group encompasses essentially all types other than rolling-element bearings.

Plain bearings are often referred to as *sleeve bearings* or *thrust bearings,* terms that designate whether the bearing is loaded axially or radially.

Plain bearings are generally less costly than rolling-element bearings. This is especially true in mass-production quantities. In moderate and small quantities, roller bearings are more competitive from a price standpoint, especially if a plain bearing requires special lubrication or demands special designs that cannot be supplied off the shelf.

Lubrication is critical to the operation of plain bearings, so their application and function are also often referred to according to the type of lubrication principle used. Thus, terms such as *hydrodynamic, fluid-film, hydrostatic, boundary-lubricated,* and *self-lubricated* are designations for particular types of plain bearings.

Although some materials have an inherent lubricity or can be lubricated by virtue of a film of slippery solid, most bearings operate with a fluid film—generally oil but sometimes a gas.

By far the largest number of bearings is oil-lubricated. The oil film can be maintained through pumping by a pressurization system, in which case the lubrication is termed *hydrostatic*. Or it can be maintained by a squeezing or wedging of lubricant produced by the rolling action of the bearing itself; this is termed *hydrodynamic* lubrication. The designs shown in Fig. 21-1-1 illustrate simple, effective arrangements for providing supplementary lubrication.

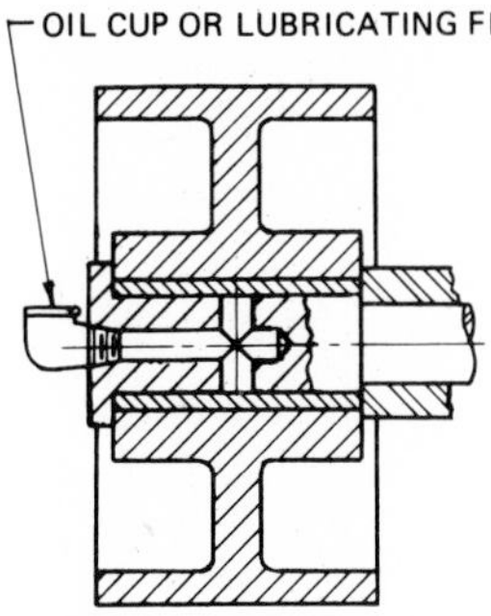
OIL CUP OR LUBRICATING FITTING

(A) OIL HOLE IN SHAFT

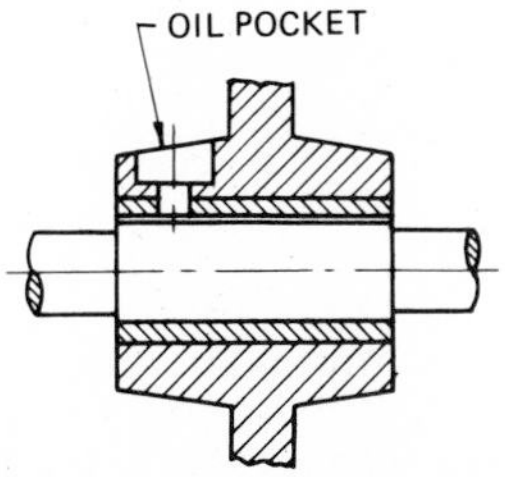
OIL POCKET

(B) OIL GROOVE IN BEARING

Fig. 21-1-1 Common methods of lubricating plain bearings.

BEARING TYPES

Journal or Sleeve Bearings. These are cylindrical or ring-shaped bearings designed to carry radial loads. See Fig. 21-1-2. The terms *sleeve* and *journal* are used more or less synonymously since *sleeve* refers to the general configuration while *journal* pertains to any portion of a shaft supported by a bearing. In another sense, however, the term *journal* may be reserved for two-piece bearings used to support the journals of an engine crankshaft.

The simplest and most widely used types of sleeve bearings are cast-bronze and porous-bronze (powdered-metal) cylindrical bearings. Cast-bronze bearings are oil- or grease-lubricated. Porous bearings are impregnated with oil and often have an oil reservoir in the housing.

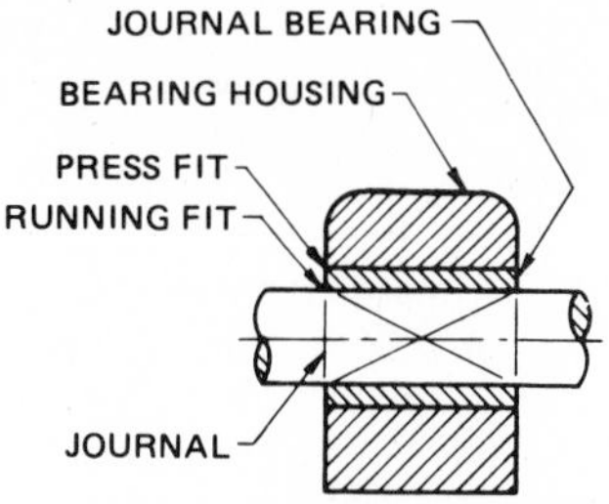
JOURNAL BEARING
BEARING HOUSING
PRESS FIT
RUNNING FIT
JOURNAL

Fig. 21-1-2 Journal or sleeve bearing.

Plastic bearings are being used increasingly in place of metal. Originally, plastic was used only in small, lightly loaded bearings where cost saving was the primary objective. More recently, plastics are being used because of functional advantages, including resistance to abrasion, and are being made in large sizes.

Thrust Bearings. This type of bearing differs from a sleeve bearing in that loads are supported axially rather than radially. See Fig. 21-1-3. Thin, disklike thrust bearings are called *thrust washers.*

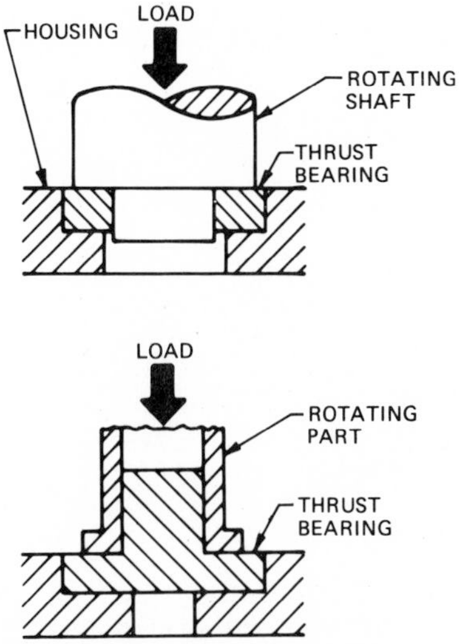
HOUSING
LOAD
ROTATING SHAFT
THRUST BEARING
LOAD
ROTATING PART
THRUST BEARING

Fig. 21-1-3 Thrust bearings.

BEARING MATERIALS

Babbitts. Tin and lead-base babbitts are among the most widely used bearing materials. They have an ability to embed dirt and have excellent compatibility properties under boundary lubrication conditions.

In bushings for small motors and in automotive engine bearings, babbitt is generally used as a thin coating over a steel strip. For larger bearings in heavy-duty equipment, thick babbitt is cast on a rigid backing of steel or cast iron. After machining, the babbitt layer is usually 3 to 6 mm thick.

Bronzes and Copper Alloys. Dozens of copper alloys are available as bearing materials. Most of these can be grouped into four classes: copper-lead, leaded-bronze, tin-bronze, and aluminum-bronze.

Aluminum. Aluminum bearing alloys have high wear resistance, load-carrying capacity, fatigue strength, thermal conductivity, and excellent corrosion resistance and low-cost. They are used extensively in connecting-rod and main bearings in internal-combustion engines; in hydraulic gear pumps, in oil-well pumping equipment, in roll-neck bearings in steel mills; and in reciprocating compressors and aircraft equipment. Aluminum alloys require sufficient lubrication, good surface finish, and hardened shafts.

Porous Metals. Sintered-metal self-lubricating bearings, often called *powder-metal bearings,* are simple and low-cost. They are widely used in home appliances, small motors, machine tools, business machines, and farm and construction equipment.

Most porous-metal bearings consist of either bronze or iron which has interconnecting pores. These voids take up 10 to 35 percent of the total volume. In operation, lubricating oil is stored in these voids and feeds through the interconnected pores to the bearing surface. Since these bearings can operate for long periods without additional supply of lubricant, they can be used in inaccessible or inconvenient places where lubrication would be difficult. Common methods used when supplementary lubrication for oil-impregnated bearings is needed are shown in Fig. 21-1-4.

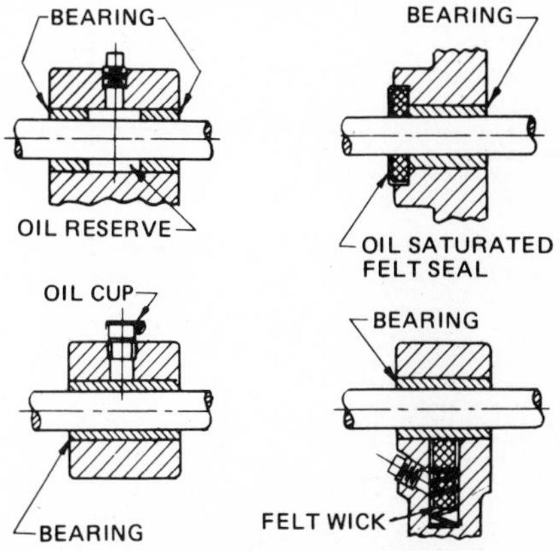
BEARING
OIL RESERVE
BEARING
OIL SATURATED FELT SEAL
OIL CUP
BEARING
BEARING
FELT WICK

Fig. 21-1-4 Supplementary lubrication for oil-impregnated bearings.

REFERENCES AND SOURCE MATERIAL

1. Canadian SKF Company Limited.
2. "Bearings," *Machine Design,* Vol. 50, No. 15, 1977.

Assignments

1. On an A3- or B-size sheet, complete the assembly drawings shown in Fig. 21-1-A or 21-1-B. The shaft, shown in the left view, is supported by two plain bearings press-fit into the shaft support and lubricated by means of an oil fitting. Select suitable bolts and fasten the shaft support to the mounting plate.

The shafts for the gear box are supported by two plain bearings press-fit into the gear box. Setscrew collars, as shown in the Appendix, are mounted on the shafts to prevent lateral movement. Select suitable 1.59 module (16-pitch), 20° spur gears from manufacturers' catalogs which will revolve the smaller shaft 4 times as fast as the larger shaft. Lock the gears to the shafts using setscrews and flats on the shaft. Scale is 1:1.

2. On an A3- or B-size sheet, complete the assembly drawings shown in Fig. 21-1-C or 21-1-D. For the assembly shown on the left, the largest portion of the shaft is positioned in the housing by a combination journal and thrust bearing. Complete the internal design of the housing to properly locate and support the bearing.

The shaft shown in the assembly on the right rests on a thrust bearing. Complete the housing detail, and show the bearing in position.

REVIEW FOR ASSIGNMENTS

Unit 7-5	Assembly Drawings in Section
Unit 19-3	Spur Gears
Appendix	Bolts, Setscrews

ASSIGNMENT	A	B	C
21-1-A METRIC	47.63	10	16
21-1-B INCH	1.875	.375	.625

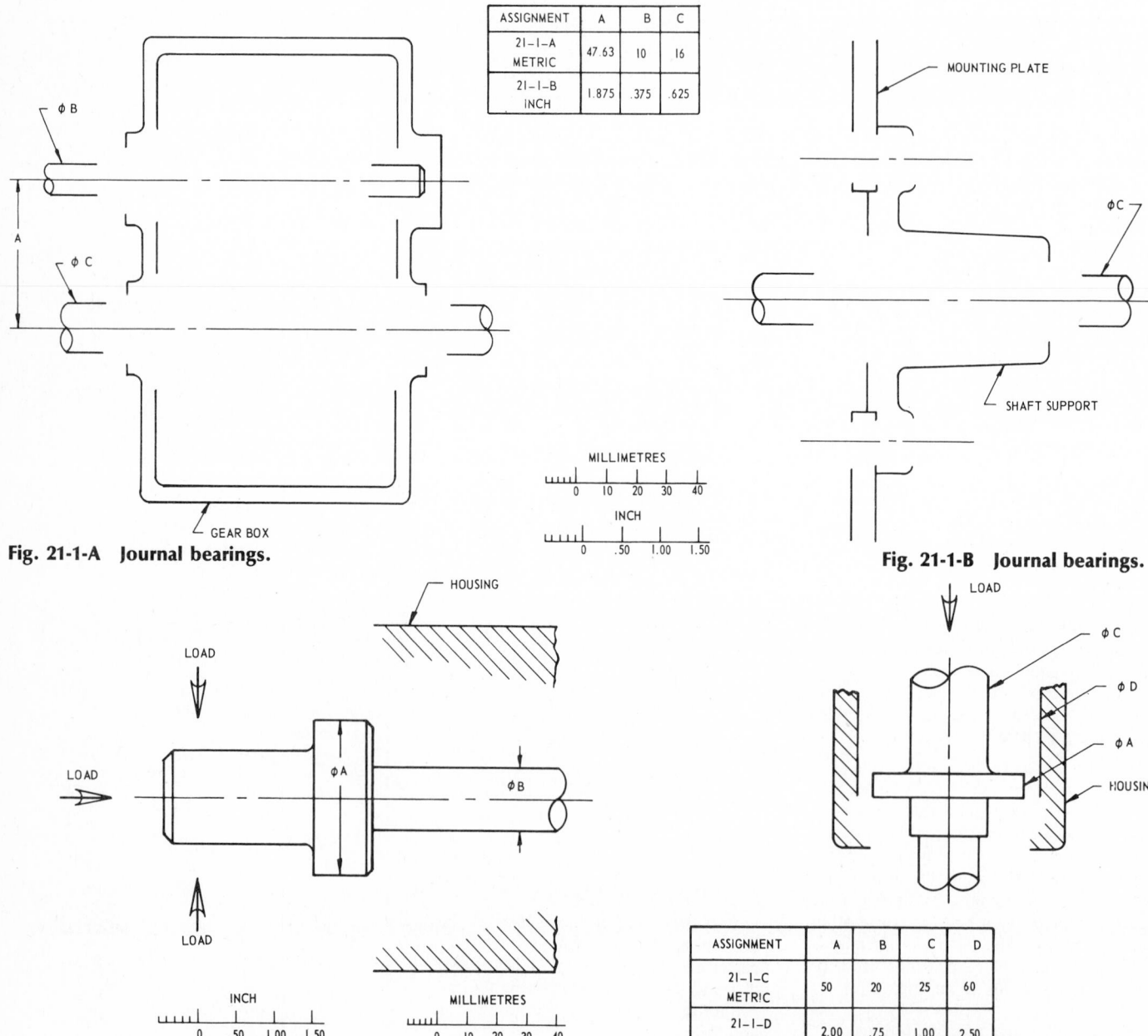

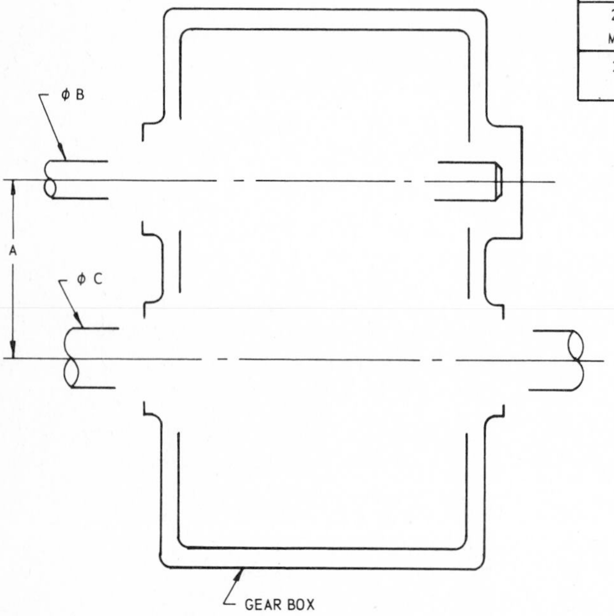
φB
A
φC
GEAR BOX

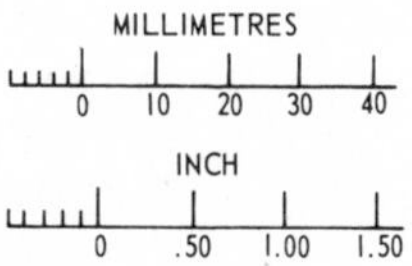
MILLIMETRES
0 10 20 30 40
INCH
0 .50 1.00 1.50

Fig. 21-1-A **Journal bearings.**

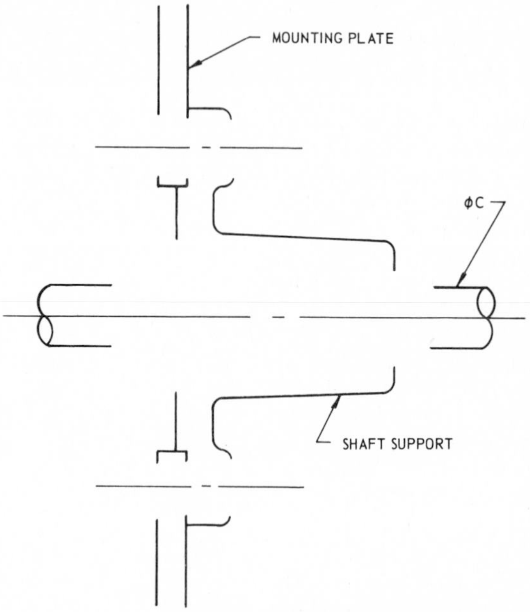
MOUNTING PLATE
φC
SHAFT SUPPORT

Fig. 21-1-B **Journal bearings.**

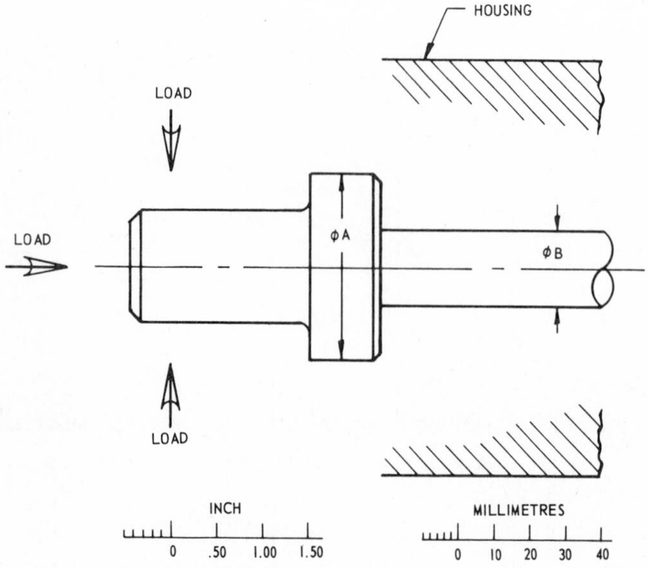
HOUSING
LOAD
LOAD
φA
φB
LOAD
INCH
0 .50 1.00 1.50
MILLIMETRES
0 10 20 30 40

Fig. 21-1-C Thrust and journal bearings.

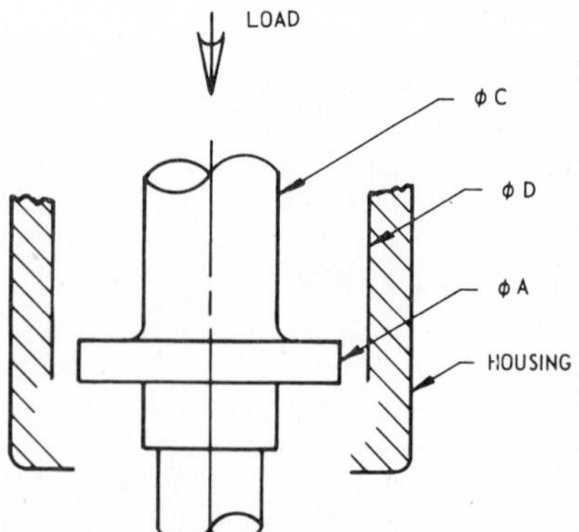
LOAD
φC
φD
φA
HOUSING

ASSIGNMENT	A	B	C	D
21-1-C METRIC	50	20	25	60
21-1-D INCH	2.00	.75	1.00	2.50

Fig. 21-1-D Thrust and journal bearings.

UNIT 21-2
ANTIFRICTION BEARINGS[1]

Ball, roller, and needle bearings are classified as antifriction bearings since friction has been reduced to a minimum. They may be divided into two main groups: radial bearings and thrust bearings. Except for special designs, ball and roller bearings consist of two rings, a set of rolling elements, and a cage. The cage separates the rolling elements and spaces them evenly around the periphery (circumference of the circle). The nomenclature of an antifriction bearing is given in Fig. 21-2-1.

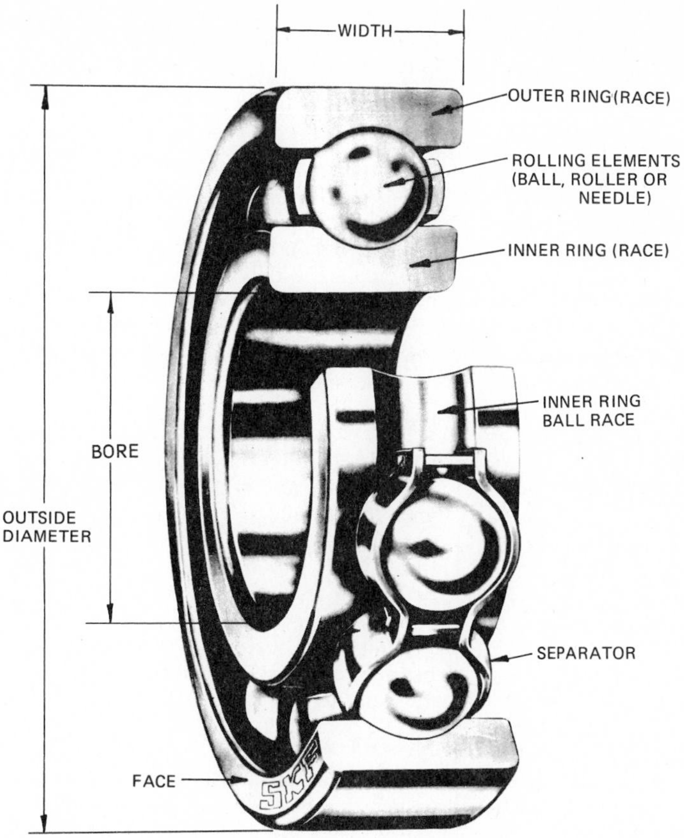
WIDTH
OUTER RING(RACE)
ROLLING ELEMENTS (BALL, ROLLER OR NEEDLE)
INNER RING (RACE)
INNER RING BALL RACE
BORE
OUTSIDE DIAMETER
SEPARATOR
FACE

Fig. 21-2-1 Antifriction bearing nomenclature.

Bearing Loads

Radial Load. Loads acting perpendicular to the axis of the bearing are called *radial loads*. See Fig. 21-2-2. Although radial bearings are designed primarily for straight radial service, they will withstand considerable thrust loads when deep ball tracks in the raceway are used.

Thrust Load. Loads applied parallel to the axis of the bearing are called *thrust loads*. These bearings are not designed to carry radial loads.

Combination Radial and Thrust Loads. When loads are exerted both parallel and perpendicular to the axis of the bearings, a combination radial and thrust bearing is used. The load ratings listed in the manufacturers' catalogs for this type of bearing are for either pure thrust loads or a combination of both radial and thrust loads.

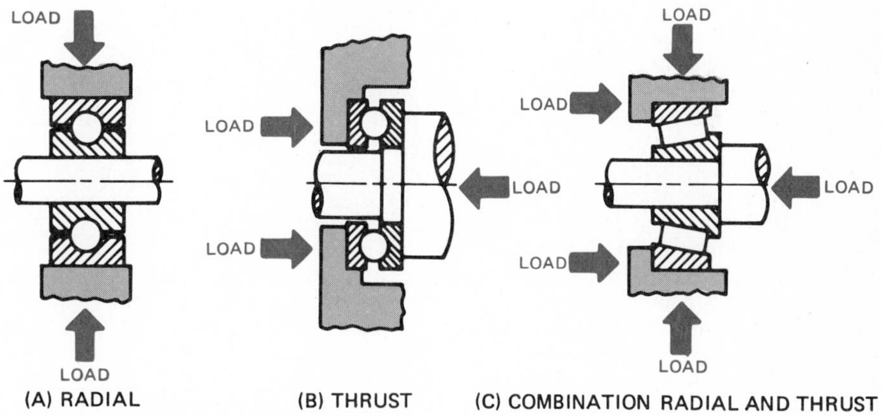
LOAD
LOAD
LOAD
LOAD
LOAD
LOAD
LOAD
LOAD
LOAD
LOAD
(A) RADIAL
(B) THRUST
(C) COMBINATION RADIAL AND THRUST

Fig. 21-2-2 Types of bearing loads.

Ball Bearings

Ball bearings fall roughly into three classes: *radial, thrust,* and *angular-contact*. Angular-contact bearings are used for combined radial and thrust loads and where precise shaft location is needed. Uses of the other two types are described by their names: radial bearings for radial loads and thrust bearings for thrust loads. See Fig. 21-2-3.

RADIAL BEARINGS

Deep-groove bearings are the most widely used ball bearings. In addition to radial loads, they can carry substantial thrust loads at high speeds, in either direction. They require careful alignment between shaft and housing.

Self-aligning bearings come in two types: internal and external. In internal bearings, the outer-ring ball groove is ground as a spherical surface. Such a surface makes the bearing insensitive

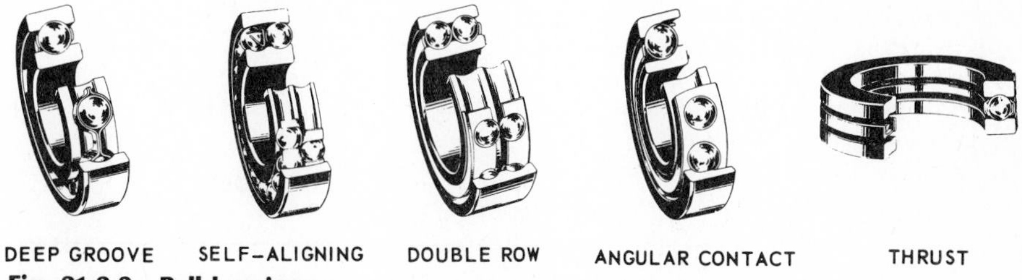
DEEP GROOVE
SELF-ALIGNING
DOUBLE ROW
ANGULAR CONTACT
THRUST

Fig. 21-2-3 Ball bearings.

to misalignment, but increases raceway contact stresses, which cuts load capacity.

Externally self-aligning bearings have a spherical surface on the outside of the outer ring, which matches a concave spherical housing. External bearings have better load capacity than internal, but demand greater radial space.

Double-row, deep-groove bearings embody the same principle of design as single-row bearings. However, grooves for the two rows of balls may be positioned so that the load lines through the balls converge either outwardly or inwardly. Inwardly converging bearings deflect more under moment loading but can stand more misalignment.

Double-row bearings can be used where high radial and thrust rigidity is needed and space is limited. They are about 60 to 80 percent wider than comparable single-row, deep-groove bearings, and they have about 50 percent more radial capacity.

Angular-contact thrust bearings can support a heavy thrust load in one direction, combined with a moderate radial load. High shoulders on the inner and outer rings provide steep contact angles for high thrust capacity and axial rigidity. Bearings with contact angles from 15 to 40° are available, but contact angles above 30° are not recommended for high-speed applications.

THRUST BEARINGS

In a sense, thrust bearings can be considered to be 90° angular-contact bearings. They support pure thrust loads at moderate speeds, but for practical purposes their radial load capacity is nil. Because they cannot support radial loads, ball thrust bearings must be used with radial bearings.

Flat-race bearings consist of a pair of flat washers separated by the ball complement and a shaft-piloted retainer, so load capacity is limited. Contact stresses are high, and torque resistance is low. The prime advantage of flat-race bearings is that they permit eccentricities, since there are no grooves to restrain one race from wandering slightly with respect to the other.

One-directional, grooved-race bearings have grooved races very similar to those in radial bearings. Thrust capacity is about twice that of the flat-race bearing, and limiting speed is about 3 times as high. However, limiting speed is still quite low because balls spin at high speeds.

Two-directional, grooved-race bearings consist of two stationary races, one rotating race, and two ball complements. Their characteristics are identical to those of the one-directional bearing, except that they can support thrust in both directions.

Roller Bearings

The principal types of roller bearings are *cylindrical, needle, tapered,* and *spherical.* In general, they have higher load capacities than ball bearings of the same size and are widely used in heavy-duty, moderate-speed applications. However, except for cylindrical bearings, they have lower speed capabilities than ball bearings. See Fig. 21-2-4.

CYLINDRICAL BEARINGS

Cylindrical roller bearings have high radial capacity and provide accurate guidance to the rollers. Their low friction permits operation at high speed, and thrust loads of some magnitude can be carried through the flange-roller end contacts.

Unlike ball bearings, cylindrical roller bearings are generally lubricated with oil; most of the oil serves as a coolant.

NEEDLE BEARINGS[2]

Needle bearings are roller bearings with rollers that have high length-to-diameter ratios. They are used in farm and construction equipment, automotive transmissions, small gasoline engines, gear pumps, small appliance and tool motors, alternators, helicopter rotors, and aircraft controls.

Compared with other roller bearings, needle bearings have much smaller rollers for a given bore size. They have the highest load capacity for a given radial space of all rolling-element bearings, but their use is limited to bore diameters less than 250 mm. All the various types of needle bearings are variations of two basic designs.

Loose-needle bearings are simply a full complement of needles in the annular space between two hardened machine components, which form the bearing raceways. They provide an effective and inexpensive bearing assembly with moderate speed capability, but they are sensitive to misalignment.

Caged assemblies are simply a roller complement with a retainer, placed between two hardened machine elements that act as raceways. Their speed capability is about 3 times higher than that of loose-needle bearings, but the smaller complement of needles reduces load capacity for the caged assemblies.

Thrust bearings are caged bearings with rollers assembled like the spokes of a wheel in a waferlike retainer. These bearings can operate at fairly high speeds, despite some continuous slippage between the roller and raceway surfaces, which occurs because the rollers are cylindrical rather than tapered.

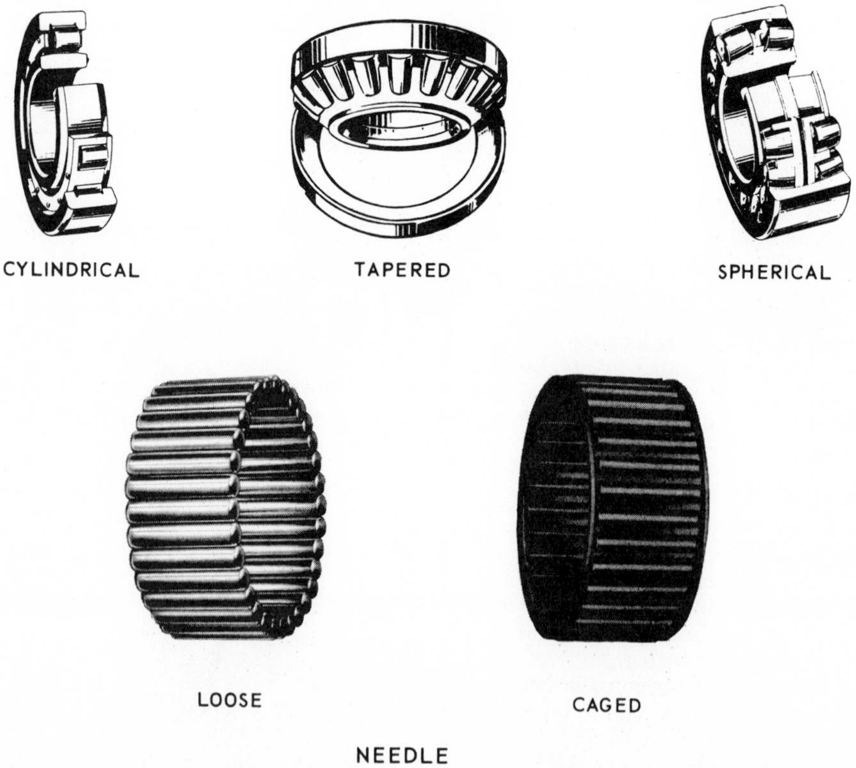
CYLINDRICAL
TAPERED
SPHERICAL
LOOSE
CAGED
NEEDLE

Fig. 21-2-4 Roller bearings. (The Torrington Co. & Orange Roller Bearing Co.)

TAPERED BEARINGS

Tapered roller bearings are widely used in roll-neck applications in rolling mills, transmissions, gear reducers, geared shafting, steering mechanisms, and machine-tool spindles. Where speeds are low, grease lubrication suffices, but high speeds demand oil lubrication—and very high speeds demand special lubricating arrangements.

SPHERICAL BEARINGS

Spherical roller bearings offer an unequaled combination of high load capacity, high tolerance to shock loads, and self-aligning ability, but they are speed-limited. Accordingly, they are used in vibrators, shakers, conveyors, speed reducers, transmissions, and other heavy machinery.

Single-row bearings are the most widely used tapered roller bearings. They have a high radial capacity and a thrust capacity about 60 percent of radial capacity.

Two-row bearings can replace two single-row bearings mounted back to back or face to face when the required capacity exceeds that of a single-row bearing. If the bearing contact lines converge away from the bearing axis, the bearing is more tolerant to misalignment. If they converge toward the axis, the bearing has higher resistance to moment loading but lower tolerance for misalignment.

Bearing Selection

Machine designers have a large variety of bearing types and sizes from which to choose. Each of these types has characteristics which make it best for a certain application. Although selection may sometimes present a complex problem requiring considerable experience, the following considerations are listed to serve as a general guide for conventional applications.

1. Generally, ball bearings are the less expensive choice in the smaller sizes and lighter loads, while roller bearings are less expensive for the larger sizes and heavier loads.

2. Roller bearings are more satisfactory under shock or impact loading than ball bearings.

3. If there is misalignment between housing and shaft, either a self-aligning ball or spherical roller bearing should be used.

4. Ball thrust bearings should be subjected to pure thrust loads only. At high speeds, a deep-groove or angular-contact ball bearing will usually be a better choice even for pure thrust loads.

5. Self-aligning ball bearings and cylindrical roller bearings have very low friction coefficients.

6. Deep-groove ball bearings are available with seals built into the bearings so that the bearing can be prelubricated and thus operate for long periods without attention.

Bearing Classifications

Because of standardization of boundary dimensions, it is possible to replace a bearing by another bearing produced by a different manufacturer without any modification to the existing assembly.

Ball and roller bearings are classified into various series: rigid ball journals, self-aligning ball journals, rigid roller journals, etc. Each series is subdivided into types—extra light, light, medium, and heavy—to meet varying load requirements. Each type is manufactured to a range of standard sizes which are usually represented by the diameter of the bore. Therefore, when a bearing is ordered, the series, type, and size are specified.

Figure 21-2-5 shows a range of bearings to a common bore at *a* and to a common outside diameter at *b*. A selection can therefore be made for a given shaft size or for a given housing diameter, and the series selected will depend on the load which is applied to the bearing.

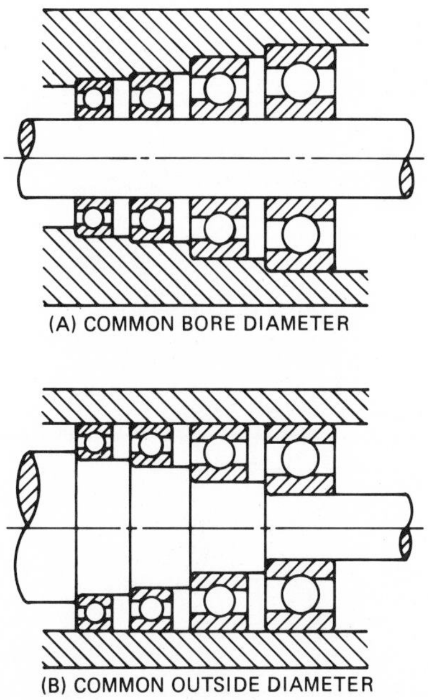
(A) COMMON BORE DIAMETER
(B) COMMON OUTSIDE DIAMETER

Fig. 21-2-5 Standard bearing sizes.

Shaft and Housing Fits[3]

If a ball or roller bearing is to function satisfactorily, both the fit between the inner ring and the shaft and the fit between the outer ring and the housing must be suitable for the application. The desired fits can be obtained by selecting the proper tolerances for the shaft diameter and the housing bore.

Bearings may be mounted directly on the shaft or on tapered adapter sleeves. When the bearing is mounted directly on the shaft, the inner ring should be located against a shaft shoulder of proper height. This shoulder must be machined square with the bearing seat, and a shaft fillet should be used. The radius of the fillet must clear the corner radius of the inner ring. See Fig. 21-2-6. This also applies when the outer ring is mounted in the housing.

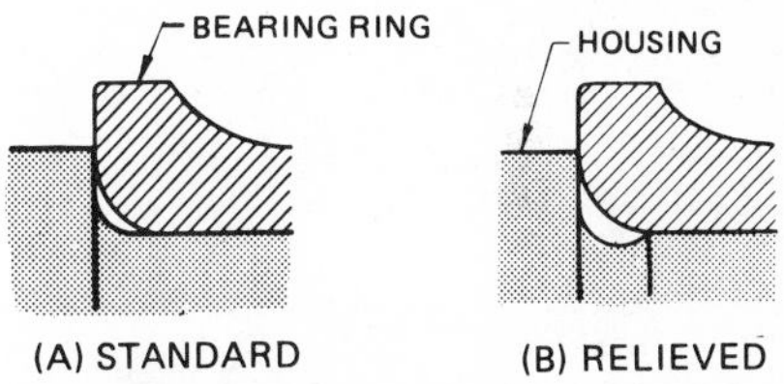
BEARING RING
HOUSING
(A) STANDARD
(B) RELIEVED

Fig. 21-2-6 Correct shaft and housing fillet radii.

To hold the bearing inner ring axially on the shaft, a locknut and lockwasher are commonly used. See Fig. 21-2-7. Not only is this method effective and convenient, but nuts and washers specially made for the purpose are also readily obtainable. A tab in the bore of the lockwasher engages a slot in the shaft, and one of the many tabs on the periphery of the washer is bent over into one of the slots in the nut OD.

Instead of a nut, a retaining ring fitted into a groove in the shaft can be used for simple bearing arrangements. See Fig. 21-2-8.

If another machine component, such as a gear or pulley, is fitted alongside the bearing, the inner ring is often secured by means of a spacing sleeve. A sleeve is often used for spacing the inner rings when the bearings are located reasonably close together.

Some bearings are merely mounted against a shoulder without other means of securing the inner ring axially. This is particularly the case where there are no axial forces tending to displace the bearings on the shaft. The housings for the two bearings are rigidly connected, and when thrust occurs, the bearing taking the load is pressed against its shoulder.

(A) LOCKWASHER LOCKNUT

(B) ADAPTOR SLEEVE

(C) WITHDRAWAL SLEEVE

Fig. 21-2-7 Locking devices.

On long, standard shafting it is impractical to apply bearings, with an interference fit, directly on the shaft. Therefore, they are applied with tapered adapter sleeves. The outer surface of the sleeve is tapered to match the tapered bore of the bearing inner ring. This will provide the required tight fit between the inner ring and the shaft. The adapter sleeve is slotted to permit easy contraction and is threaded at the small end to fit a locknut. When the sleeve is drawn up tight between the bearing and the shaft, a press fit is provided at both the shaft and the inner ring. When this type of mounting is employed, commercial grades of cold-finished shafting may be used without machining. For normal thrust loads the friction developed by the press fit is sufficient to prevent displacement of the sleeve along the shaft.

If the operating conditions are such that the outer rings can be mounted with a push fit in the housing and *closed bearings* (bearings capable of carrying thrust load in either direction) are used, axial location may be controlled, as shown in Fig. 21-2-9*a*. The outer ring of the held bearing has a clearance of only 0.05 to 0.1 mm with the housing shoulders, while the floating bearings (Fig. 21-2-9*b*) have a free displacement axially in the housing.

One of the most critical factors affecting bearing operation is the mounting fit of the bearing on the shaft and in the housing.

If there is any clearance or looseness between the shaft and the bore of the inner ring, then as the shaft rotates, it will roll along the bore of the inner ring. This rolling of the shaft in the bearing bore will cause the shaft to rapidly wear progressively looser, and soon it becomes too sloppy for further operation. The best way to prevent this rolling action and wear is to press-fit the inner ring on the shaft.

Similar reasoning applies to a bearing subject to a load which rotates in space with the inner ring. Here, if the outer ring has a clearance in the housing bore, it will roll around the housing bore and wear loose. In this case it would be necessary to have the outer ring press-fit in the housing.

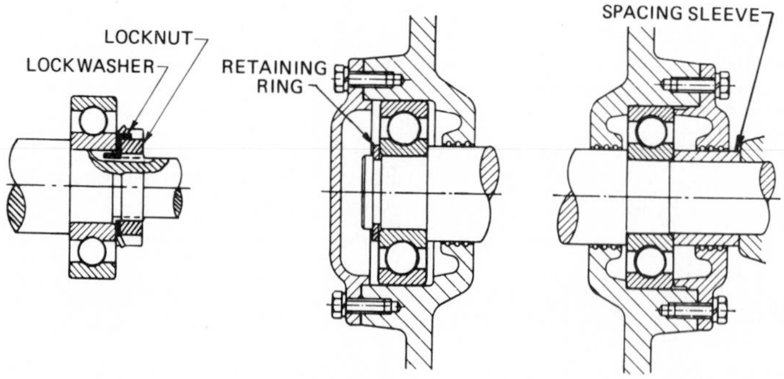
LOCKNUT
LOCKWASHER
RETAINING RING
SPACING SLEEVE

(A) LOCKNUT (B) FLOATING SNAP RING (C) FIXED SPACING SLEEVE

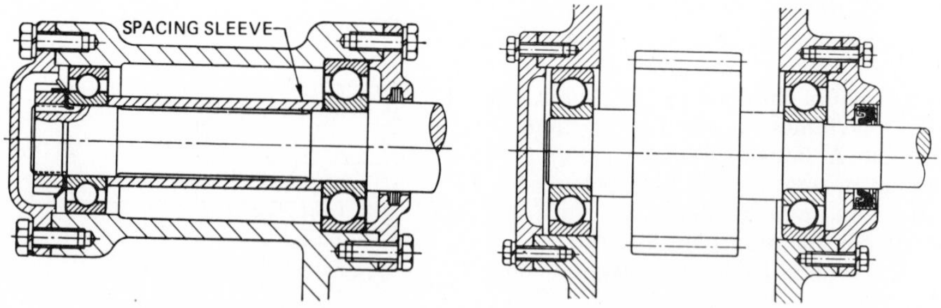
SPACING SLEEVE

(D) SPACING SLEEVE (E) SHOULDER MOUNTING

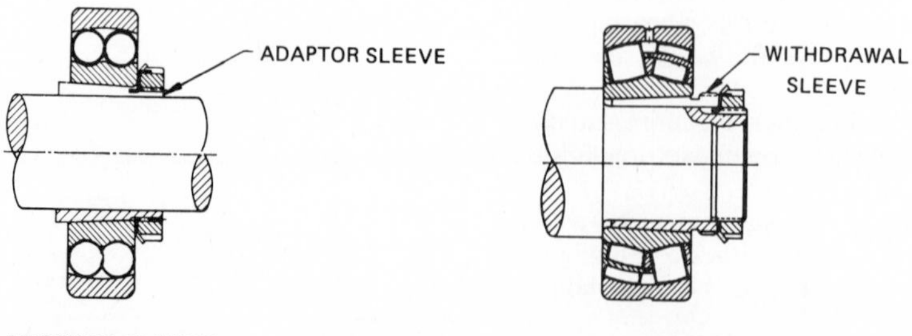
ADAPTOR SLEEVE
WITHDRAWAL SLEEVE

(F) ADAPTOR SLEEVE (G) WITHDRAWAL SLEEVE

Fig. 21-2-8 Axial mounting of inner rings.

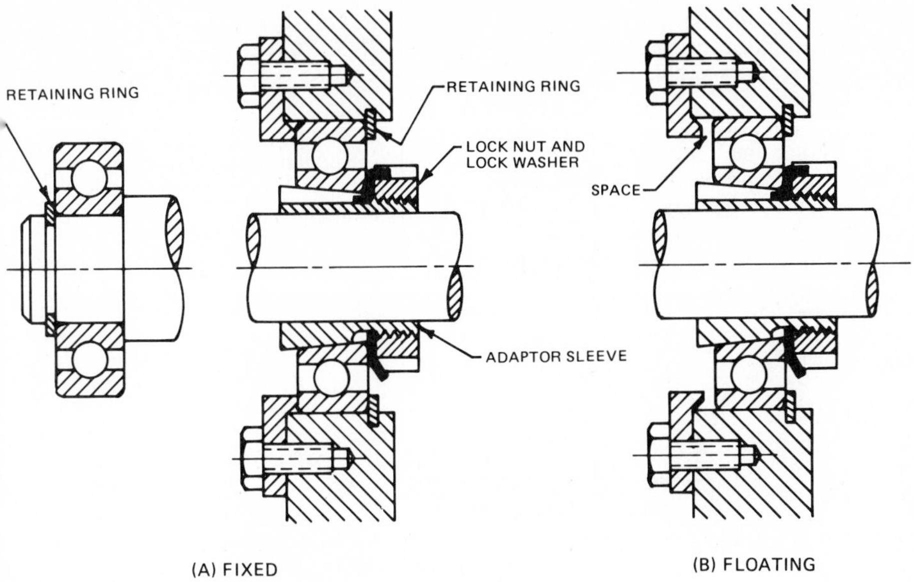
RETAINING RING
RETAINING RING
LOCK NUT AND
LOCK WASHER
SPACE
ADAPTOR SLEEVE
(A) FIXED
(B) FLOATING

Fig. 21-2-9 Outer ring mountings.

In all cases, it is necessary to press-fit the bearing ring which has relative rotation with respect to the direction of the radial load.

SEALS FOR GREASE LUBRICATION

In order that ball or roller bearings may operate properly, they must be protected against loss of lubricant and entrance of dirt and dust on the bearing surfaces. In its simplest and least space-requiring form, this is accomplished in some types of bearings by the use of a thin steel shield on one or both sides of the bearing, fastened in a groove in the outer ring and reaching almost to the inner ring as illustrated in Fig. 21-2-10. All other types of bearings require a seal between the bearing housing and the shaft, the types and designs of which are shown in Fig. 21-2-11. Other types of seals are explained in Units 21-4 and 21-5.

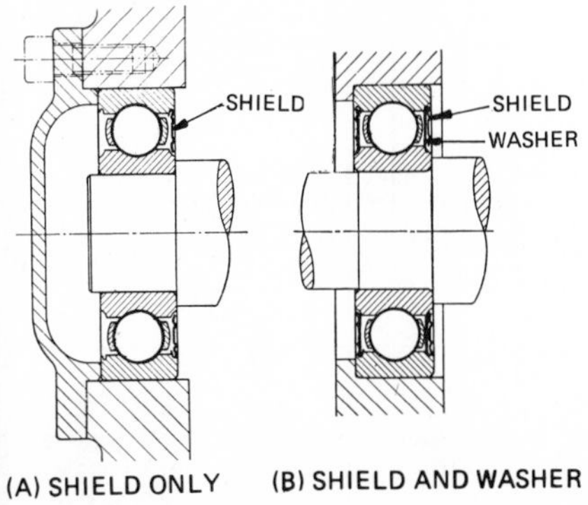
SHIELD
SHIELD
WASHER
(A) SHIELD ONLY
(B) SHIELD AND WASHER

Fig. 21-2-10 Bearing seals for grease lubrication.

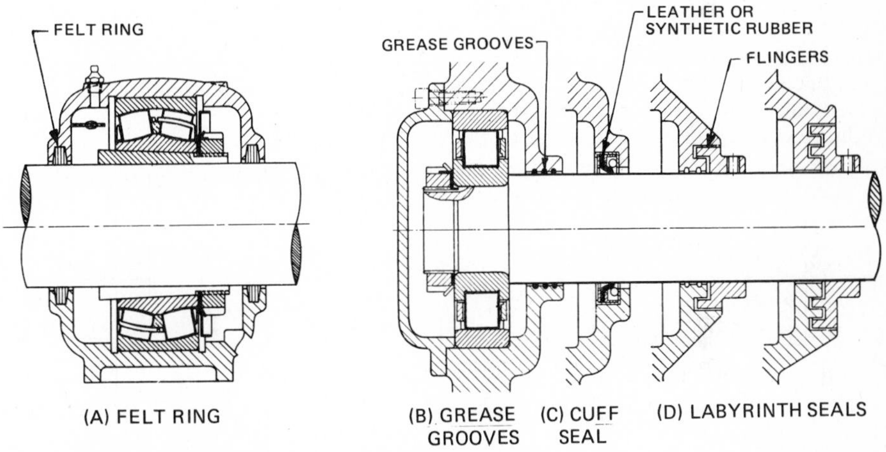
FELT RING
GREASE GROOVES
LEATHER OR
SYNTHETIC RUBBER
FLINGERS
(A) FELT RING
(B) GREASE
GROOVES
(C) CUFF
SEAL
(D) LABYRINTH SEALS

Fig. 21-2-11 Housing seals for grease lubrication.

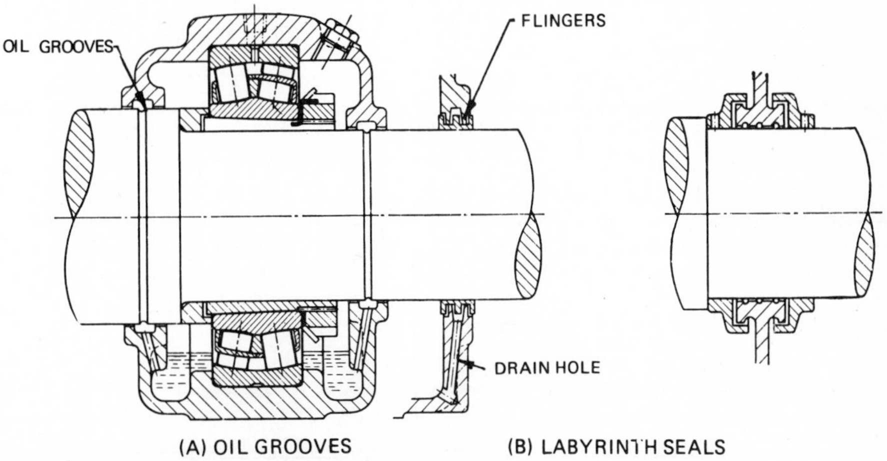
OIL GROOVES
FLINGERS
DRAIN HOLE
(A) OIL GROOVES
(B) LABYRINTH SEALS

Fig. 21-2-12 Housing seals for oil lubrication.

SEALS FOR OIL LUBRICATION

With oil lubrication, the sealing devices have the double function of protecting the bearing against contamination and retaining the lubricant in the housing. Protection is obtained by means of friction seals or flingers, as when grease lubrication is used. The essential feature for retaining the oil is a groove in the rotating shaft, or a rotating ring or collar from whose edges the oil is thrown by centrifugal force. The oil-groove seal shown in Fig. 21-2-12*a* retains the oil effectively but should be used only in dry and dust-free places where there is little danger of contamination. Figure 21-2-12*b* shows examples of labyrinth seals which retain the oil and protect against contamination.

Bearing Symbols

Simplified Representation. The simplified representation (general symbol) of rolling bearings (see Fig. 21-2-13)

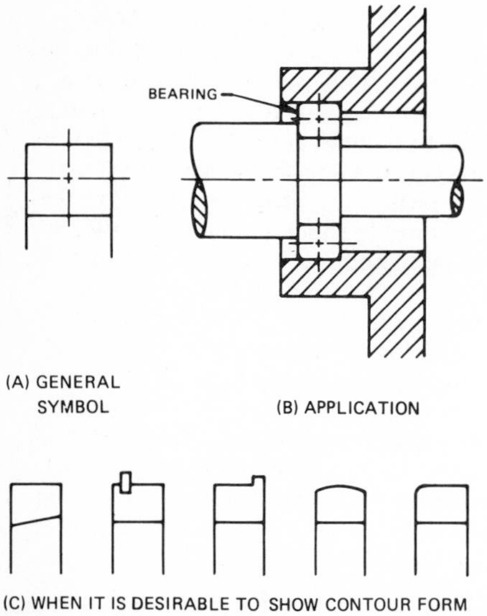
BEARING
(A) GENERAL SYMBOL
(B) APPLICATION
(C) WHEN IT IS DESIRABLE TO SHOW CONTOUR FORM

Fig. 21-2-13 Simplified representation of ball and roller bearings.

should be used in all types of technical drawings, wherever it is not necessary to show the exact form or size of the rolling bearings or details of their inner design.

Where it is desirable to show the functional principle of the set of rolling elements, symbols for the appropriate type of rolling element and raceway surface are added. See Fig. 21-2-14*b*.

Pictorial Representation. Pictorial representation of bearings, as shown in Fig. 21-2-14*a*, is used chiefly in catalogs and magazines. It is not recommended for production drawings because of the extra drafting time required.

Schematic Representation. Designers and engineers frequently use schematic layouts in their initial design layout. The schematic diagrams of bearing types and their application are shown in Figs. 21-2-14*c* and 21-2-15.

REFERENCES AND SOURCE MATERIAL

1. "Bearings," *Machine Design,* Vol. 50, No. 15, 1977.
2. P. R. Glazier, "Needle-Roller Bearings," *Machine Design,* Vol. 35, No. 14, 1963.
3. Canadian SKF Company, LTD.

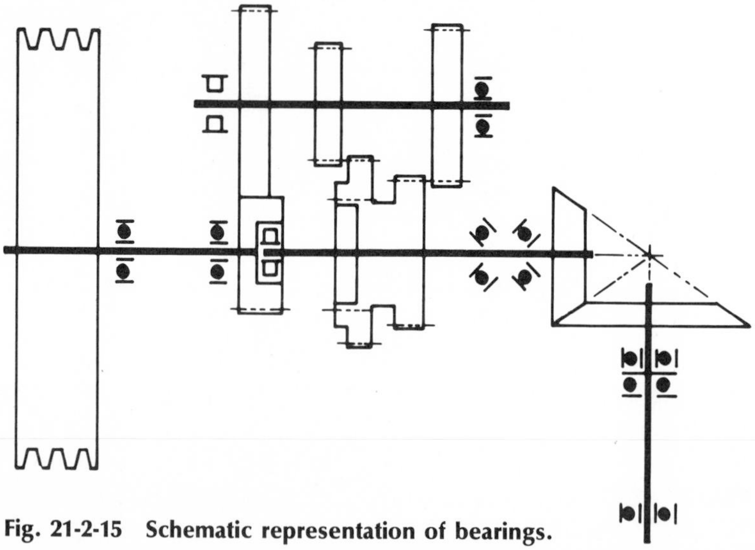

Fig. 21-2-15 Schematic representation of bearings.

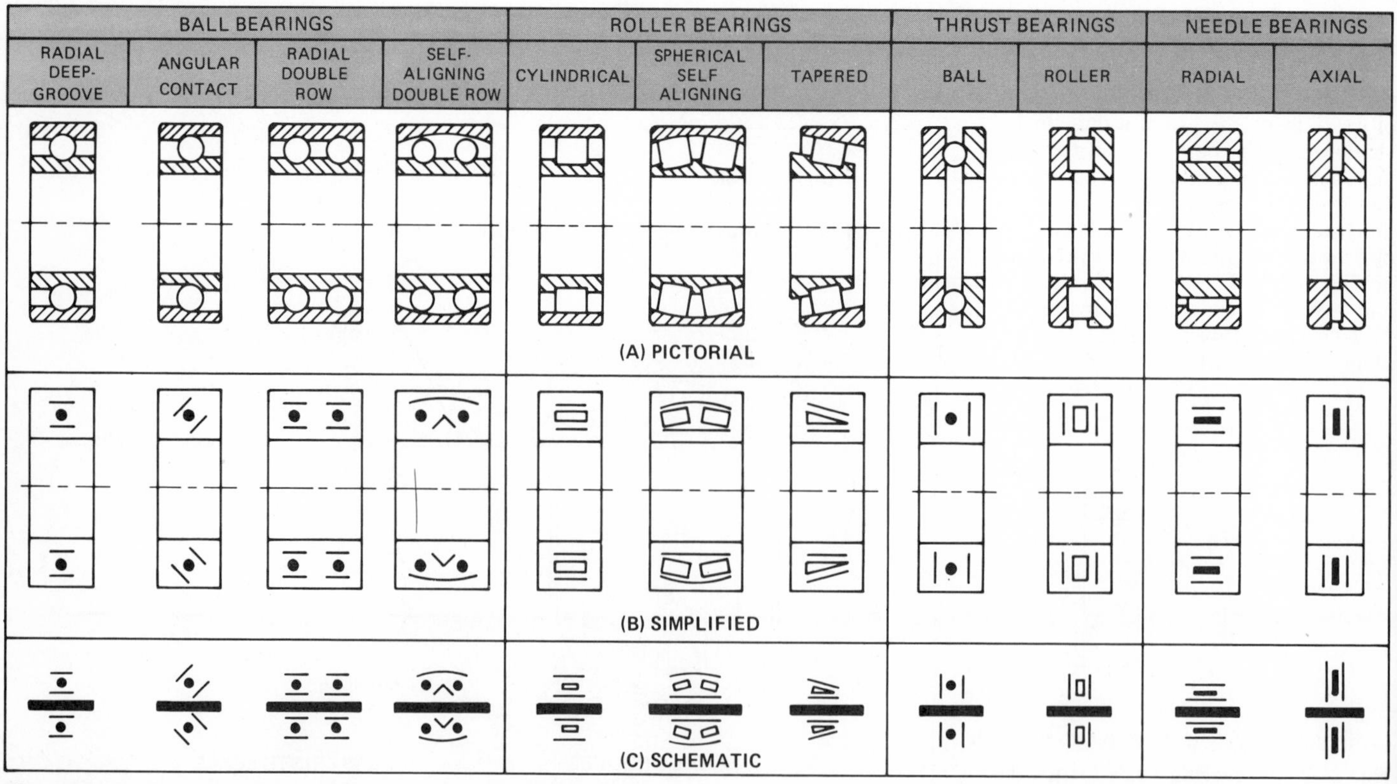
BALL BEARINGS
RADIAL DEEP-GROOVE
ANGULAR CONTACT
RADIAL DOUBLE ROW
SELF-ALIGNING DOUBLE ROW
ROLLER BEARINGS
CYLINDRICAL
SPHERICAL SELF ALIGNING
TAPERED
THRUST BEARINGS
BALL
ROLLER
NEEDLE BEARINGS
RADIAL
AXIAL
(A) PICTORIAL
(B) SIMPLIFIED
(C) SCHEMATIC

Fig. 21-2-14 Representation of bearings on drawings.

Assignments

1. On an A3- or B-size sheet, complete the gearbox assembly drawing shown in Fig. 21-2-A, B. Gears are mounted on shafts *A* and *B* and are positioned and held to the shafts by Woodruff keys and setscrews. The shafts are supported by radial ball bearings which are positioned on the shafts with retaining rings. The bearings are to be positioned and held to the housing by internal shoulders on the castings and by cover plates bolted to the housing. Each shaft will have one floating and one fixed outer ring mounting. The bearings will be purchased with seals on one side. From the information given, select suitable keys, bearings, retaining rings, and gears from the Appendix or manufacturers' catalogs. Note: shaft *A* must be able to be removed from the housing with the gear in position. Scale is 1:1.

2. On an A3- or B-size sheet, complete the gearbox assembly drawing

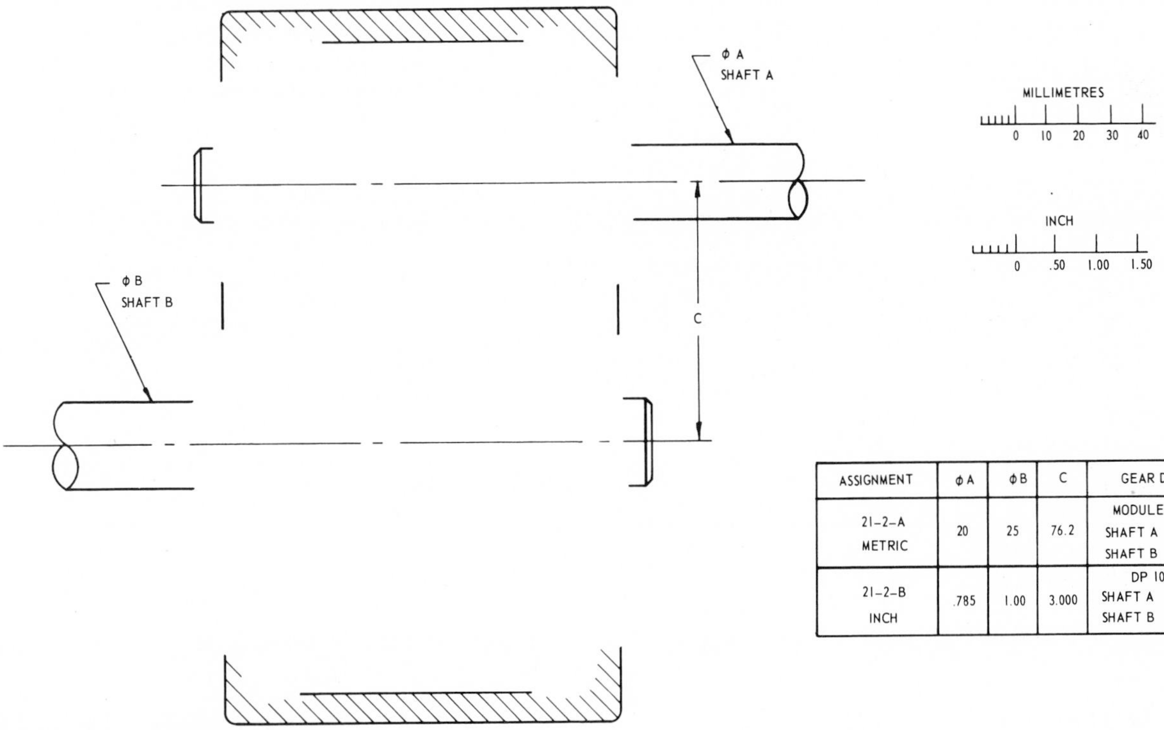
ϕ A
SHAFT A
MILLIMETRES
0 10 20 30 40
INCH
0 .50 1.00 1.50
ϕ B
SHAFT B
C

ASSIGNMENT	ϕ A	ϕ B	C	GEAR DATA
21-2-A METRIC	20	25	76.2	MODULE 2.54 SHAFT A N = 15 SHAFT B N = 45
21-2-B INCH	.785	1.00	3.000	DP 10 SHAFT A N = 15 SHAFT B N = 45

Fig. 21-2-A, B Ball bearings.

ASSIGNMENT	SHAFT	A	B	GEAR DATA								
21-2-C METRIC	PRIMARY	10	15	MODULE=2.54			20° SPUR GEAR			FACE WIDTH = 16		
	MAINSHAFT	10	20	GEAR	1	2	3	4	5	6	7	8
	LAYSHAFT	15	15	N	20	40	25	35	30	30	35	25
INCH	PRIMARY	.394	.591	PITCH = 10		20° SPUR GEAR				FACE WIDTH = .60		
	MAINSHAFT	.394	.787	GEAR	1	2	3	4	5	6	7	8
	LAYSHAFT	.591	.591	N	20	40	25	35	30	30	35	25

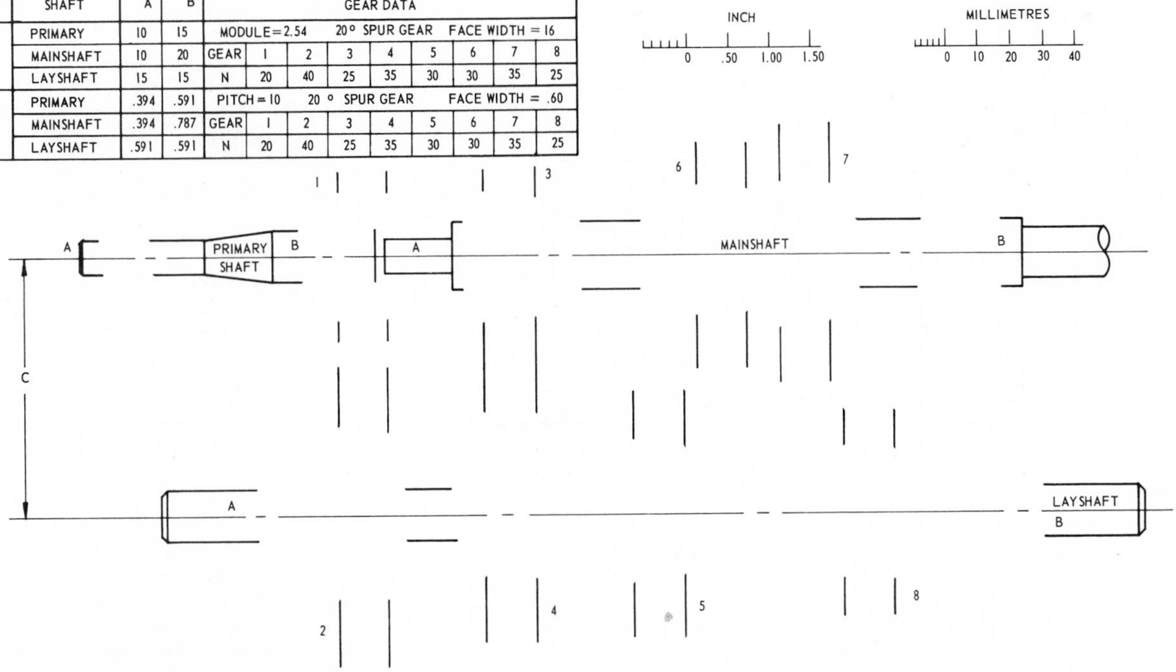
INCH
0 .50 1.00 1.50
MILLIMETRES
0 10 20 30 40
6
7
3
A
PRIMARY SHAFT
B
A
MAINSHAFT
B
C
A
LAYSHAFT
B
2
4
5
8

Fig. 21-2-C, D Gearbox.

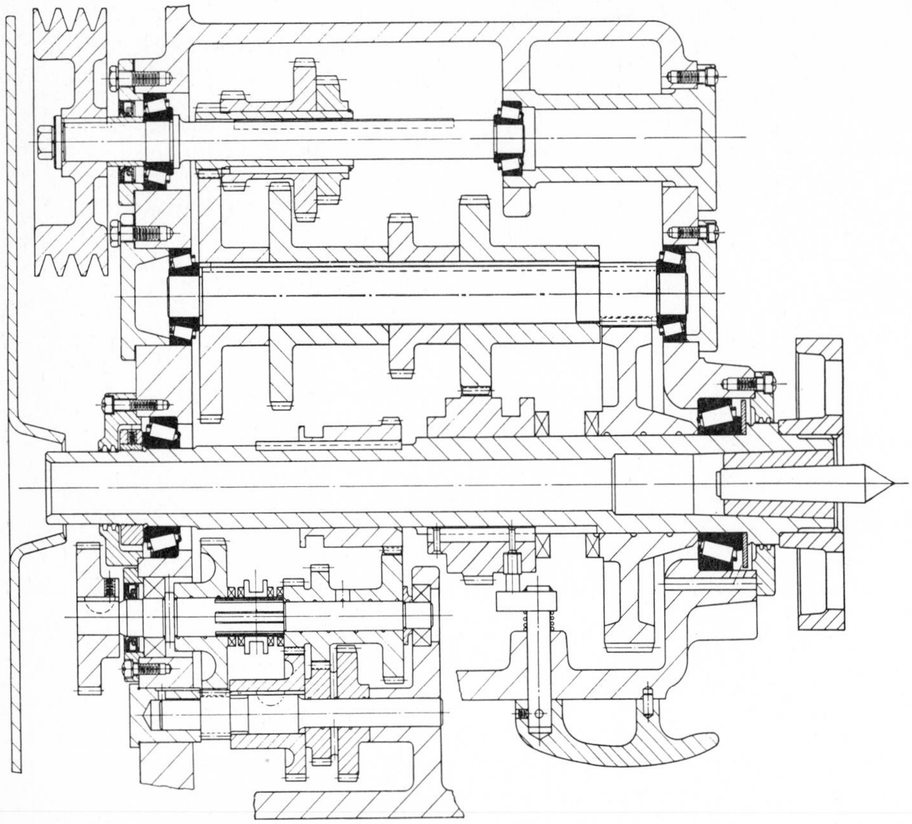

Fig. 21-2-E Lathe. (Timken Roller Bearing Co.)

shown in Fig. 21-2-C, D. Gear 1 and the shaft are cast as a single unit. Gears 2, 3, 6, and 7 are fastened to their respective shafts by keys and are held in location by retaining rings. Gears 4, 5, and 8 are formed as one part which slides along the lay-shaft meshing with gear 3, 6, or 7. Retaining rings located at each end of this sliding-gear assembly locate it in the three positions, and a key locks the assembly to the shaft. Radial ball bearings are positioned at points *A* and *B* on each shaft. Each shaft will have one floating and one fixed outer ring mounting. The gear end of the primary shaft must be designed to house bearing *A* of the mainshaft. Refer to manufacturers' catalogs or the Appendix for standard parts. Scale is 1:1.

3. On an A3- or B-size sheet, make a schematic presentation of bearings, gears, and belts, similar to Fig. 21-2-15, of one of the assemblies shown in Fig. 21-2-E or 21-2-F. Scale is to suit.

REVIEW FOR ASSIGNMENTS

Unit 19-3 Spur Gears
Unit 9-3 Retaining Rings
Unit 9-1 Keys

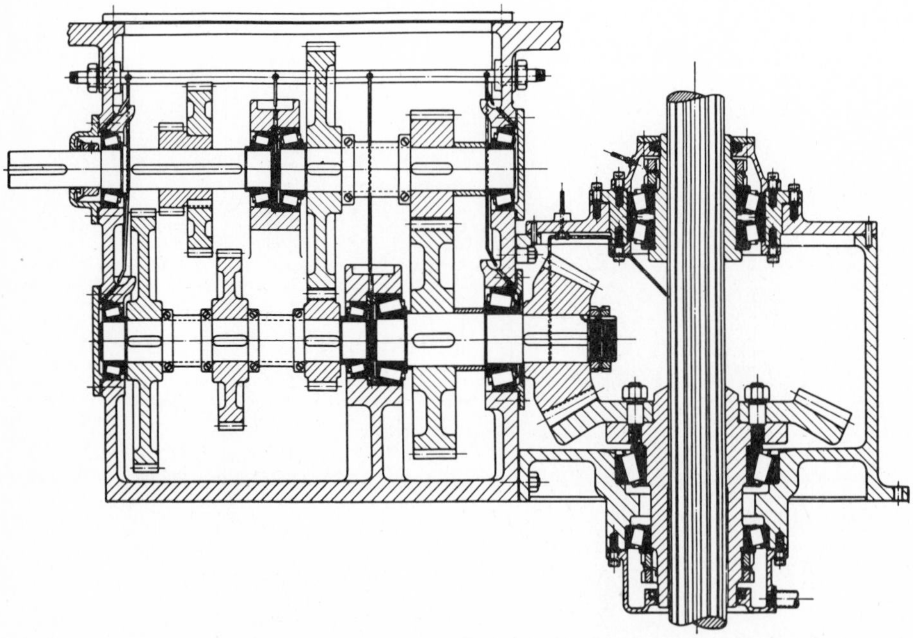

Fig. 21-2-F Honing gearbox. (Timken Roller Bearing Co.)

UNIT 21-3

PREMOUNTED BEARINGS[1]

Premounted bearing units consist of a bearing element and a housing, usually assembled to permit convenient adaptation to a machinery frame. All components are incorporated within a single unit to ensure proper protection, lubrication, and operation of the bearing. Both plain and rolling-element bearing units are available in a variety of housing designs and for a wide range of shaft sizes, as shown in Figs. 21-3-1 and 21-3-2.

Provision for lubrication is made within the units, and sealing elements retain the lubricant and exclude foreign materials. Some types are prelubricated and sealed at the factory.

Rigid and Self-Aligning Types. Rigid premounted units require accurate align-

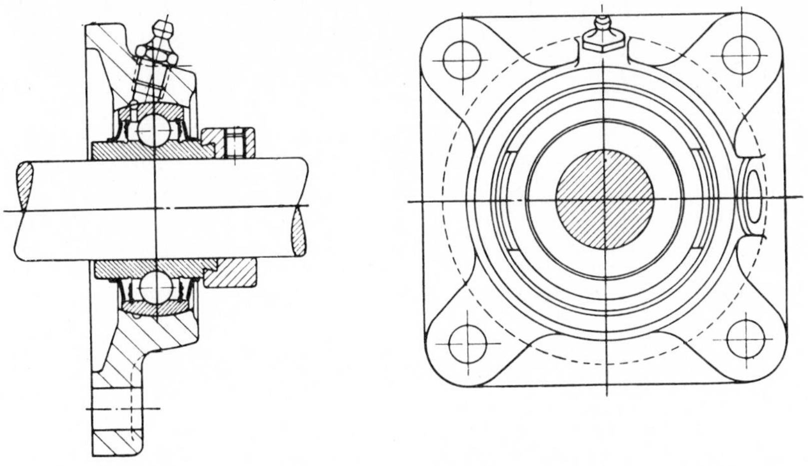

(A) FLANGED HOUSING SELF-ALIGNING-SEALED

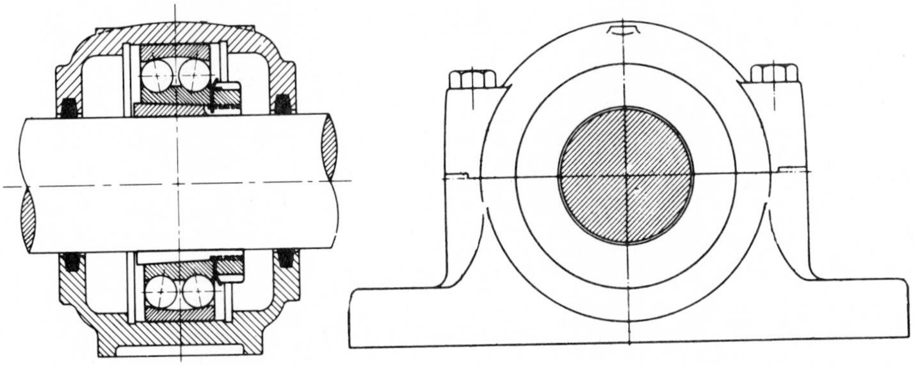

(B) PILLOW BLOCK SELF-ALIGNING-SEALED

Fig. 21-3-1 Premounted bearing units.

ment with the shaft. Although unit cost is usually less than that of self-aligning types, installed cost may be more, because of additional labor required during assembly.

Self-aligning units compensate for minor misalignment in mounting structures, shaft deflection, and changes which may occur after installation. Self-alignment in sleeve and in some rolling types is accomplished by the use of separate inner housings into which the bearing element is assembled. A ball-and-socket action between inner and outer housings accommodates the misalignments. Some rolling-bearing types have spherical outer races which permit self-alignment in the outer housing; others are self-aligning within the bearing element.

Expansion and Nonexpansion Types. Expansion bearings permit axial shaft movement. The principal application for expansion units is in equipment where shafts become heated and increase in length at a greater rate than the structure on which the bearings are mounted. Under such conditions, sleeve bearings permit the shaft to move through the bearing. Expansion-mounted rolling-bearing units provide for axial shaft movement in the housing design; the bearing or an inner housing moves within the outer housing.

Nonexpansion bearings restrict shaft movement relative to the mounting structure and keep shaft and attached components accurately positioned. These bearings also serve as thrust bearings within their capacity. Nonexpansion sleeve bearings usually require collars attached to the shaft at both ends of the housing. Nonexpansion rolling bearings are constructed with the bearing element or inner housing mounted in the outer housing to restrict lateral movement. The bearing element is secured to the shaft by conventional methods.

Pillow blocks provide a convenient means of mounting shafts parallel to the surface of a supporting structure. Bolt holes are provided, usually elongated, to permit alignment; and dowel holes are sometimes predrilled for use in maintaining final position on the supporting member. Pillow blocks are available with rigid or self-aligning bearings of expansion or nonexpansion types and with either sleeve or rolling bearings. Housings are either split or solid.

Fig. 21-3-2 Adjustable shaft support with journal bearings.

REFERENCE AND SOURCE MATERIAL

1. D. P. Lower, "Premounted Bearings," *Machine Design,* Vol. 35, No. 14, 1963.

Assignments

1. On an A3- or B-size sheet, make a two-view (front- and side-view) assembly drawing of the adjustable shaft support shown in Fig. 21-3-A or 21-3-B. Draw the front view in full section. Include on your drawing a bill of material. Scale is 1:1.

2. On an A3- or B-size sheet, make a one-view assembly drawing of the adjustable shaft support showing the bearing housing in its top position and a phantom outline of the bearing housing in its lowest position. Show only those dimensions that would be used for catalog purposes. Scale is 1:1.

REVIEW FOR ASSIGNMENTS

Unit 2-6 Phantom Outlines
Unit 6-4 Assembly Drawings
Unit 7-1 Full Sections

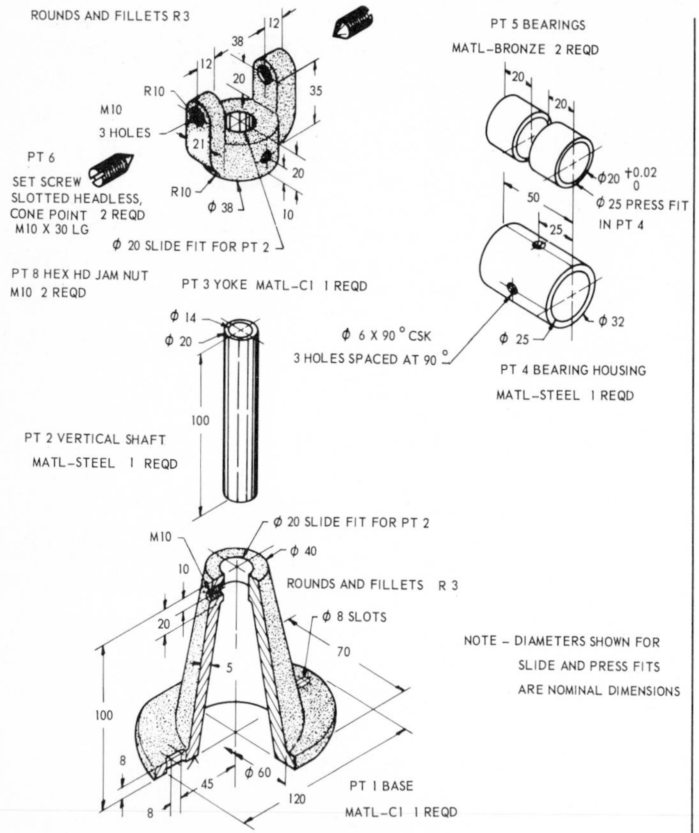
ROUNDS AND FILLETS R3
M10
3 HOLES
PT 6
SET SCREW
SLOTTED HEADLESS,
CONE POINT 2 REQD
M10 X 30 LG
Φ 20 SLIDE FIT FOR PT 2
PT 8 HEX HD JAM NUT
M10 2 REQD
PT 3 YOKE MATL–CI 1 REQD
PT 5 BEARINGS
MATL–BRONZE 2 REQD
Φ20 +0.02 0
Φ 25 PRESS FIT
IN PT 4
Φ 6 X 90° CSK
3 HOLES SPACED AT 90°
PT 4 BEARING HOUSING
MATL–STEEL 1 REQD
PT 2 VERTICAL SHAFT
MATL–STEEL 1 REQD
Φ 20 SLIDE FIT FOR PT 2
ROUNDS AND FILLETS R 3
Φ 8 SLOTS
NOTE – DIAMETERS SHOWN FOR
SLIDE AND PRESS FITS
ARE NOMINAL DIMENSIONS
PT 1 BASE
MATL–CI 1 REQD

Fig. 21-3-A Adjustable shaft support.

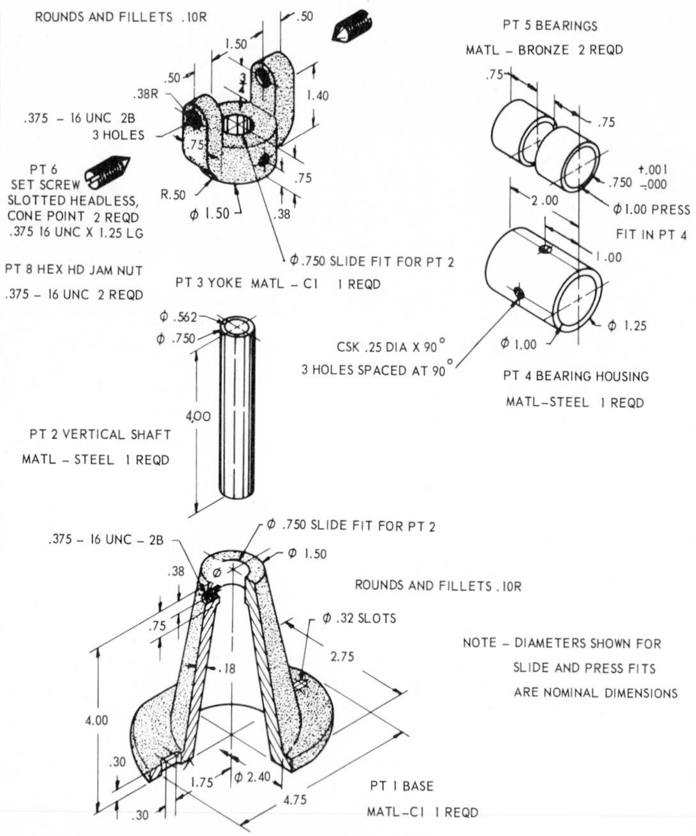
ROUNDS AND FILLETS .10R
.375 – 16 UNC 2B
3 HOLES
PT 6
SET SCREW
SLOTTED HEADLESS,
CONE POINT 2 REQD
.375 16 UNC X 1.25 LG
Φ .750 SLIDE FIT FOR PT 2
PT 8 HEX HD JAM NUT
.375 – 16 UNC 2 REQD
PT 3 YOKE MATL – CI 1 REQD
PT 5 BEARINGS
MATL – BRONZE 2 REQD
+.001 .750 –.000
Φ 1.00 PRESS
FIT IN PT 4
CSK .25 DIA X 90°
3 HOLES SPACED AT 90°
PT 4 BEARING HOUSING
MATL–STEEL 1 REQD
PT 2 VERTICAL SHAFT
MATL – STEEL 1 REQD
.375 – 16 UNC – 2B
Φ .750 SLIDE FIT FOR PT 2
ROUNDS AND FILLETS .10R
Φ .32 SLOTS
NOTE – DIAMETERS SHOWN FOR
SLIDE AND PRESS FITS
ARE NOMINAL DIMENSIONS
PT 1 BASE
MATL–CI 1 REQD

Fig. 21-3-B Adjustable shaft support.

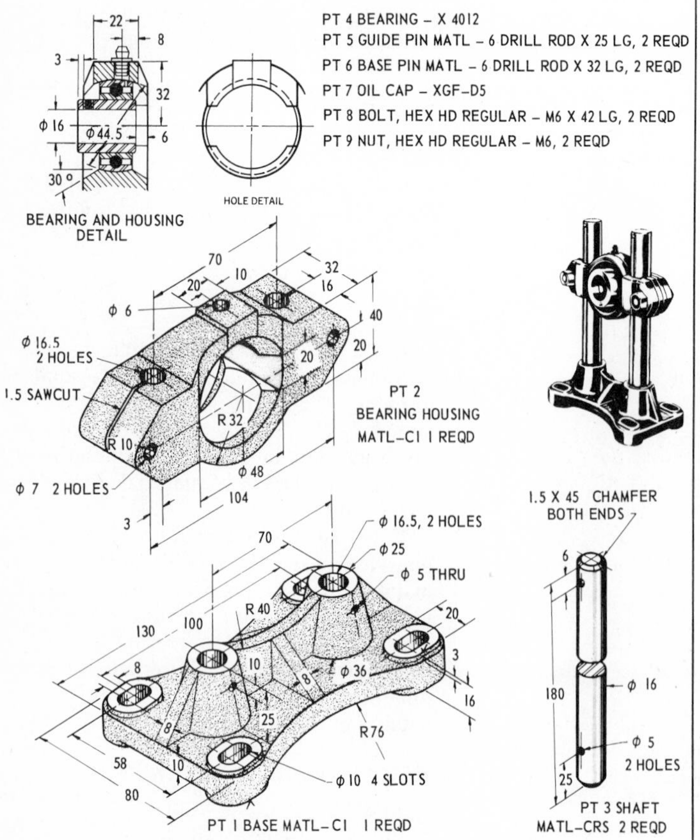
PT 4 BEARING – X 4012
PT 5 GUIDE PIN MATL – 6 DRILL ROD X 25 LG, 2 REQD
PT 6 BASE PIN MATL – 6 DRILL ROD X 32 LG, 2 REQD
PT 7 OIL CAP – XGF–D5
PT 8 BOLT, HEX HD REGULAR – M6 X 42 LG, 2 REQD
PT 9 NUT, HEX HD REGULAR – M6, 2 REQD
BEARING AND HOUSING
DETAIL
HOLE DETAIL
Φ 16.5
2 HOLES
1.5 SAWCUT
Φ 7 2 HOLES
PT 2
BEARING HOUSING
MATL–CI 1 REQD
Φ 16.5, 2 HOLES
Φ 5 THRU
Φ 10 4 SLOTS
PT 1 BASE MATL–CI 1 REQD
1.5 X 45 CHAMFER
BOTH ENDS
Φ 5
2 HOLES
PT 3 SHAFT
MATL–CRS 2 REQD

Fig. 21-3-C Adjustable shaft support.

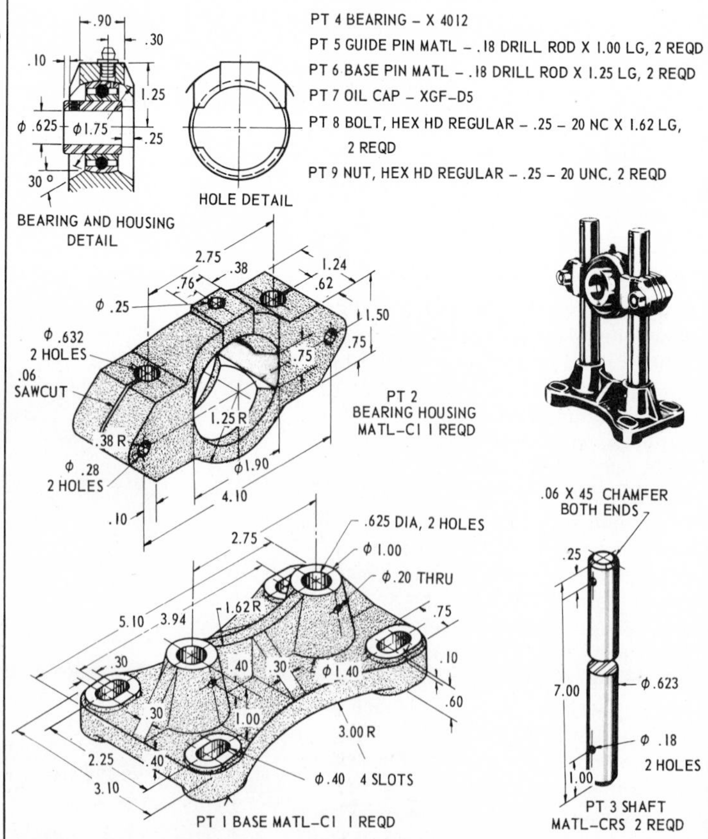
PT 4 BEARING – X 4012
PT 5 GUIDE PIN MATL – .18 DRILL ROD X 1.00 LG, 2 REQD
PT 6 BASE PIN MATL – .18 DRILL ROD X 1.25 LG, 2 REQD
PT 7 OIL CAP – XGF–D5
PT 8 BOLT, HEX HD REGULAR – .25 – 20 NC X 1.62 LG,
2 REQD
PT 9 NUT, HEX HD REGULAR – .25 – 20 UNC, 2 REQD
BEARING AND HOUSING
DETAIL
HOLE DETAIL
Φ .632
2 HOLES
.06
SAWCUT
Φ .28
2 HOLES
PT 2
BEARING HOUSING
MATL–CI 1 REQD
.625 DIA, 2 HOLES
Φ .20 THRU
Φ .40 4 SLOTS
PT 1 BASE MATL–CI 1 REQD
.06 X 45 CHAMFER
BOTH ENDS
Φ .18
2 HOLES
PT 3 SHAFT
MATL–CRS 2 REQD

Fig. 21-3-D Adjustable shaft support.

UNIT 21-4

LUBRICANTS AND RADIAL SEALS

Lubricants

There are two main reasons why lubricants are used in any bearing: (1) to reduce friction between rubbing surfaces and (2) as coolants to carry off heat which may be generated in bearings. Either or both of these functions may be required of a lubricant on a particular bearing.

As friction reducers, lubricants can be considered from two aspects. When a hydrodynamic bearing is started, for instance, metal-to-metal contact occurs. Here, the actual oiliness of the lubricant lowers the coefficient of friction between the two sliding surfaces. In slider bearings operating on full fluid-film lubrication, the lubricant separates the two sliding surfaces completely, and shearing of the lubricant is substituted for sliding friction.

Any system of rolling elements, like a ball bearing, should theoretically reduce friction radically. If balls and rollers were perfectly smooth and inelastic, friction would be very low. But materials deform, and rolling elements slip under load. Also, uncaged balls or rollers tend to rub or slide against one another. When a separator or cage is present, the rolling elements slide against it, and the cage itself rubs against any guiding flange surfaces. Because of this sliding, lubrication is needed to minimize wear and friction.

All lubricants can be grouped roughly into three general types: oils, greases, and solid-film lubricants.

OILS AND GREASES[1]

Whether to use oil or grease and what kind of oil or grease to use are questions that, for slider bearings, must usually be decided early in design phases, since bearing design depends on the lubricant and the type of lubrication selected.

Oils are slippery, hydrocarbon liquids. Grease is a semisolid, combining a fluid lubricant with a thickening agent, usually a soap. In the past, the soaps in greases were considered as storehouses for the oil. Pressure and temperature squeezed out the oil to lubricate bearing surfaces. This is probably only partly true.

Soap molecules are attracted to metal surfaces, their long hydrocarbon chains sticking out to separate the rubbing metal surfaces.

Both oil and grease are used to lubricate rolling- and sliding-contact bearings. In fact, either type of lubricant can be used in some applications, but each type has peculiar assets that equip it for certain types of applications.

Assets of Oil. These are some of the advantages of oils.

1. Oil is easier to drain and refill. This is important if lubricating intervals are close together. It is also easier to control the fill volume of the oil in the housing or reservoir.

2. An oil lubricant for a bearing might also be usable at many other points in the machine, even eliminating the need for a second grease-type lubricant.

3. Oil is more effective than grease in carrying heat away from bearing and housing surfaces. In addition, oils are available for a greater range of operating speeds and temperatures than greases.

4. Oil readily feeds into all areas of contact and can carry away dirt, water, and the products of wear.

Assets of Grease. Some of the advantages of greases are as follows.

1. Grease does not flow as readily as oil, so it can be more easily retained in a housing. Since grease is easily contained, leakproof designs are unnecessary.

2. Less maintenance is required. There is no oil level to maintain; regreasing is infrequent.

3. Grease has better sealing abilities than oil. This asset may help to keep dirt and moisture out of the housing.

Oil Selection. The term *oil* covers a broad class of fluid lubricants, each of which has various physical properties or characteristics. Many of these physical characteristics are important, but the most important is viscosity.

The characteristic of viscosity determines the oil's load-carrying capacity, the thickness of the lubricating film that will be maintained, the operating temperature of the bearing, and the amount of friction that will be generated. Therefore, this important characteristic is used almost universally to subclassify oil products.

SOLID-FILM LUBRICANTS[2]

Even the smoothest machined surfaces have microscopic roller-coaster profiles. When one such surface slides on another, the surface irregularities complicate lubrication. Under hydrodynamic conditions, a lubricating liquid may be interposed between the surfaces to prevent them from scraping one peak against another. However, when the critical surface load is exceeded or when sliding velocities are low, the peaks on each surface may break through the lubricant layer and collide, causing galling and seizing. Under these conditions, boundary lubrication must be provided.

Lubricant Selection. Basic requirements of a boundary lubricant are resistance to penetration by surface peaks under extreme loads and ability to shear more easily than base materials, by encouraging relative movement between the sliding surfaces to occur in the solid lubricant layer.

Secondary requirements, such as load capacity and temperature resistance, limit lubricant selection to a few types, principally graphite, molybdenum disulfide, tungsten disulfide, and, under low loads, TFE fluorocarbons and nylon.

When a pure lubricant or a mixture of lubricants is applied as dry powder, grease, or an oil suspension, it is called a *solid lubricant*. When the lubricant is applied in a uniformly thin layer, confining a high concentration of lubricant to a given area, it is called a *bonded dry film*. Inorganic or organic binders and solvents provide the vehicle for these films.

Selection of the solid lubricant and the best method of application will depend on wear life, load requirements, sliding velocity, cost, operating temperatures and environment, corrosiveness, and abrasiveness. High-vacuum and radioactive-environment applications also limit the choice of lubricants. Lubricant purity, optimum particle size, and methods of base metal pretreatment must also be considered.

LUBRICATING DEVICES[3]

Available lubricating devices range from simple fittings to completely automatic systems. Lubricating devices may be classified as internal or external. The bearings they serve can be lubricated individually or as a group.

Internal devices, or devices that use reservoirs contained within the housing of the bearing or group of bearings, are part of the original equipment design. External devices, or devices that rely on reservoirs in a separate housing, are often proprietary. Frequently, external devices are applied to bearings containing internal lubrication methods to provide additional lubricant.

Individual bearing devices include oil cups, hydraulic grease fittings, and drip oilers. Group methods generally supply lubricant under pressure through a distribution system to a number of bearings.

Hand Lubrication. *Hand lubrication* refers to the manual use of any portable or semiportable lubrication equipment for bearing-by-bearing application.

Small portable lubricators have reservoir capacities of approximately 0.5 kg. With the exception of the oil can or squirt oiler which has an open-end gooseneck extension, these devices have a short pipe or hose extension with a coupler that permits connection with bearing fittings in tight quarters.

Medium-size, portable lubricators have reservoir capacities of 10 to 15 kg. Many use original lubricant containers or drums as reservoirs.

Large lubricators have 50 kg, 200 kg, or larger reservoirs, which are stationary. Hoses that extend from spring-tensioned reels or hoses that connect to couplings located remotely from the reservoir are used.

For hand lubrication from portable devices, accessibly located fittings must be provided. For oil-lubricated bearings, the simplest provision is a drilled hole into which fluid lubricant is dripped. To avoid plugging or contamination, a tube or cup with a spring-loaded hinged lid is usually installed.

When the lubricant does not flow easily, fittings that connect with the coupling on the lubrication equipment to form a pressure-type union are used. A check valve in the fitting seals the entrance.

The hydraulic fitting is the most effective type. Jaws in the couplings snap over the ball on the fitting; lubricant pressure then seals the coupling.

Fittings must be accessible not only to the coupling on a portable lubricating device, but also for possible field installation of other types of lubrication equipment. Pipe-thread connections are the most universal.

If a fitting cannot be installed on a stationary part, it should be located so that its axis is common to the axis of rotation.

Individual Bearings. When life expectancy is satisfactory, certain bearings require no lubrication maintenance—prelubricated, permanently sealed bearings (rolling-element or plain), solid or dry-film lubricated plain bearings, and porous bushings. But the capabilities of even these three types of bearings can be improved by adding internal reservoirs or external lubricating devices.

Internal Reservoirs. The oldest method uses the direct contact of grease stored in the cavity of a rolling-element bearing or in the grooves of a plain bearing. This method is most satisfactory for widely separated, slow-moving bearings in locations which are subjected to external abuse and which are difficult or impossible to lubricate when the machine is in operation.

For oil lubrication, felt wicks or wool-waste packings are used to retain the oil and to transfer it to the moving surface by direct contact.

External Reservoirs. These devices for individual bearings include drop-feed, constant-level, thermal-expansion, bottle, wick-feed oilers, and pressure grease cups. Only the drop-feed or gravity oiler (Fig. 21-4-1) is in fairly common use. In this unit, a needle valve regulates the flow of oil from the reservoir, and a sight glass in the shank permits observation of flow rate.

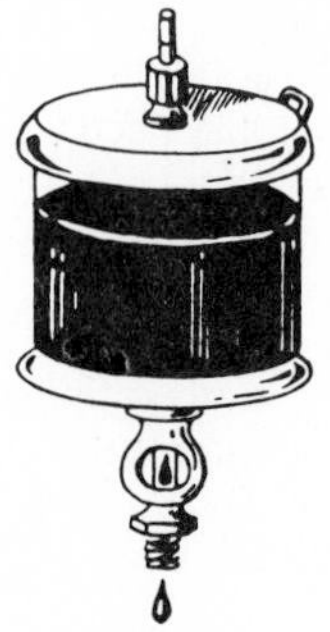

Fig. 21-4-1 External reservoir lubricating system.

Multiple Bearings. A suitable housing or enclosure is required for all internal-reservoir methods for multiple bearings. This enclosure maintains proper lubricant level and prevents loss of lubricant from within the internal complex. The enclosure must also prevent the entrance of contaminants. Three common types of internal lubricating systems are shown in Fig. 21-4-2.

Grease and Oil Seals

FACTORS IN THE SELECTION OF OIL SEALS[4]

Efficient oil seals are available today for every application. But if modern sealing techniques and advancements are to be utilized, oil seals must be selected, inspected, and installed correctly. There are a number of factors to be considered in carrying out these operations.

The environment in which the seal will operate is the most important factor in the selection of materials to be used for the seal. In particular, the user must consider:

1. Fluid to be sealed in
2. Fluid to be sealed out
3. Mean temperature of the environment
4. The shaft on which the sealing element runs
5. The sealing element designs for which tooling has been developed

The fluids to be sealed in are usually lubricants. The composition of fluid differs greatly, even within one classification.

Extreme-pressure gear lubricants have peculiar oil-seal problems. Most of them break down at temperatures of 100°C and deposit sludge between the seal lip and the shaft. This is the cause of many seal failures. In addition, many of them cause rapid hardening of nitrile rubber compounds and acrylic rubber compounds, and will rapidly disintegrate silicone rubber.

At high temperatures, the adverse effect of extreme-pressure gear oils on nitrile rubber seals is so severe that acrylic rubber compounds are usually used with these oils in spite of their generally less satisfactory abrasion resistance, greater low-temperature brittleness, and higher cost. Nitrile rubber seals can be used successfully when shaft speeds and environmental temperatures do not cause oil breakdown.

The medium to be sealed out is usually air containing varying amounts of dust, gravel, water, slate, etc. Dirt can radically shorten seal life and often dictates the selection of the oil-seal compound. Dirt and dust get under the sealing lip and cause heavy wear and scoring of both the shaft and seal lip. Water and salt cause rusting of the shaft surface, with consequent pitting of the shaft and rapid wear of the seal lip.

The mean temperature of the environment has a radical effect on the seal life and strongly influences the choice of seal compound. Most estimates of the temperature requirements for seals are based on maximum attainable temperatures

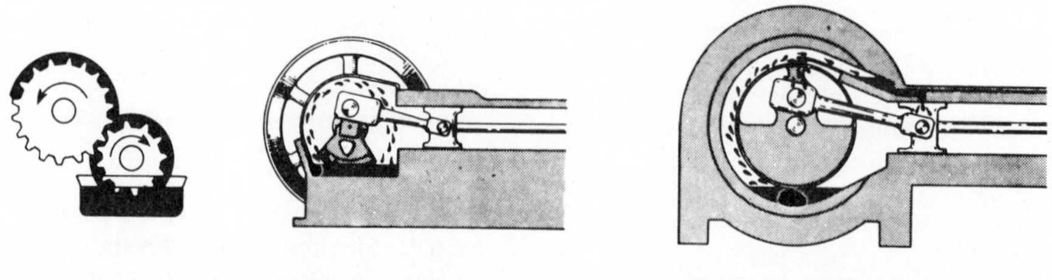

(A) BATH LUBRICATION (B) SPLASH LUBRICATION (C) SPLASH LUBRICATION WITH PRESSURE LUBRICATING SYSTEM

Fig. 21-4-2 Internal lubricating systems.

under the most extreme conditions. The designer's predilection for safety tends to make her or him estimate a mean temperature much higher than necessary. This results in the use of compounds with maximum resistance to high temperatures. It also means acceptance of the weaknesses of these compounds, such as poor abrasion resistance and low-temperature brittleness of the acrylics, the dimensional instability of the silicones, and the higher costs.

The shaft on which the seal must ride has some influence on the selection of the seal compound. If the shaft is hard and has a surface finish of better than 0.5 micrometres (μm) (one micrometre = one millionth metre), any of the compounds can be used. On rougher shafts, the acrylics and silicones wear too rapidly to be used. Even nitrile rubber compounds suffer excessive wear from rough shafts. Also, on rough shafts the lead of the grind may cause unpreventable leakage. The recent development of standardized wear sleeves, machined to a surface finish of 0.2 to 0.6 μm, eliminates many of the problems encountered with shaft finish. The wear sleeves are replaceable and can be installed each time the oil seal is changed, providing an undamaged surface of the seal without regrinding of the shaft.

Available tooling must be considered when seal designs and compounds are selected. Seal design is a conglomeration of little details that have been more or less perfected after years of cut-and-try artistry. Because tooling is expensive to replace, the compound must often meet the requirements of the tooling standards available.

Sealing Materials. The type of lubricant and the mean operating temperature usually govern the choice of the elastomer (any elastic substance resembling rubber) to be used for the seal compound. Since mean operating temperatures seldom exceed 105°C, nitrile rubber compounds are the most widely used sealing materials. They wear best, are easiest to mold, and are low in cost.

Silicone compounds are preferred for some applications. Not all silicone compounds are safe to use, however. Most will disintegrate rapidly in many automatic-transmission fluids and in some engine oils. Silicone compounds have been used occasionally in extreme-pressure gear oils, but only where operating temperatures are under 65°C. These compounds have poor abrasion resistance under dirty or dusty operating conditions. They perform well where shafts must run for short periods without lubrication, because silicone compounds absorb oil while running hot. On standing and cooling, this oil sweats and serves as a lubricant during dry periods. This same characteristic causes silicones to swell 5 to 20 percent during operation. Consequently, designs must compensate for this instability. Silicone seals cost slightly more than acrylic seals.

New fluorelastomer compounds, such as Viton, have a long life at very high temperatures in almost any lubricant. Their cost is high, however. They get quite stiff, but not brittle, at low temperatures. Because of their low coefficient of friction, they polish shafts rather than groove them. Viton tends to resist the deposit of carbon and varnish from the lubricant. The high cost of Viton is justified for only a few seal applications needing its resistance to high temperatures. Considerable progress is being made in the development of extenders, seal designs, and manufacturing techniques that will reduce the cost of Viton seals.

Radial Seals

Typical seal designs being used today feature both single- and dual-lip sealing elements bonded securely to metal cases that add strength and rigidity to the seals. The bonding of the sealing element to the case eliminates internal leakage resulting from clamping. The centering of the metal case within the mold at the time of molding ensures a precision product.

For several years, extensive tests and investigations have been conducted to evaluate the effect of various cross-sectional shapes for sealing elements. Several conditions must be satisfied in developing proper shapes, some of which conflict with others. The sealing element must be flexible enough to follow shaft runout but stiff enough to prevent collapse under operating conditions. Because of manufacturing variations and the use of elastomeric material, it is desirable for the spring to apply its force slightly behind the sealing lip. If the spring is ahead of the contact point and toward the oil, the sealing lip becomes unstable. A spring behind the contact point can cause lip sag and reduce sealing effectiveness.

The combination of angles between the trim surface and what is called the *approach angle* is critical. This is particularly true of the angle toward the oil. If an angle is too acute, an otherwise well-designed seal will perform poorly.

It is now possible to design a sealing element configuration by equations based on the desired shape, angles, and cross section. In this way, consistent variations from size to size can be made. This practice also makes possible the development of standards for all sizes of seals and can eliminate arduous calculations.

Installation. The bore must be round and smooth. It must have a proper lead-in chamfer, with a minimum of tool leads and marks, and no tool-return grooves. A bottom should be designed into the bore, and the bore should be concentric with the bearing retention surface.

The shaft should have a chamfer, and, generally speaking, the surface should be approximately 0.5 μm with above-C45 Rockwell hardness in abrasive applications and above-B80 Rockwell hardness when abrasive conditions are absent. Sleeves should be used over keyways and splines during installation, and the surface should have no nicks and chips caused by handling operations.

To prevent cocking and distortion, it is preferred that the seal be installed in the bore by hydraulic action or a hydraulic press. The tool used to press or drive the seal into the bore should concentrate its force at the periphery of the seal. The diameter of such a fixture should be only a few thousandths less than the outside diameter of the seal.

FELT RADIAL SEALS[5]

Felt is a built-up fabric made by interlocking fibers through a suitable combination of mechanical work, chemical action, moisture, and heat, without spinning, weaving, or knitting. It may consist of one or more classes of fibers: wool, reprocessed wool, or

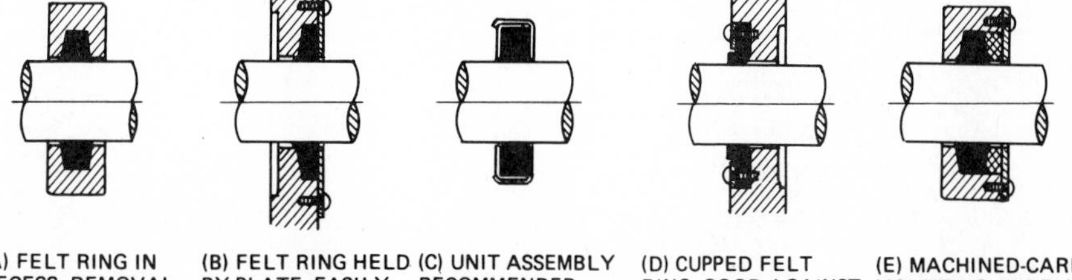

(A) FELT RING IN RECESS. REMOVAL OF SHAFT FOR SEAL REPLACEMENT
(B) FELT RING HELD BY PLATE. EASILY REPLACED
(C) UNIT ASSEMBLY RECOMMENDED WHERE SPACE IS CRITICAL
(D) CUPPED FELT RING. GOOD AGAINST GRIT AND DUST
(E) MACHINED-CARRIER MOUNTING. WIDELY USED WITH BALL AND ROLLER BEARING

Fig. 21-4-3 Felt seal designs.

reused wool, which are used alone or combined with animal, vegetable, and synthetic fibers.

Felt has long been used as an important material for sealing purposes. The main reasons are oil wicking, oil absorption, filtration, resiliency, low friction, polishing action, and cost. See Fig. 21-4-3.

RADIAL POSITIVE-CONTACT SEALS[6]

Radial positive-contact seals are dynamic rubbing seals. Operational effectiveness of a dynamic seal installation was once measured by an easy standard; if it did not leak too much too soon, it was a good seal. Today's operational concepts require sealing effectiveness with absolute minimal leakage over wide service parameters.

A radial positive-contact seal is a device which applies a sealing pressure to a mating cylindrical surface to retain fluids and, in some cases, to exclude foreign matter. Although this definition fits almost all dynamic contact seals, including packings and felt rubbing seals, attention is given in this chapter to the types of seals more commonly known as *oil seals* or *shaft seals* (Fig. 21-4-4).

The rotating-shaft application of the radial seal is most common. However, the radial seal is also used where shaft motion is oscillating or reciprocating. Where the application requires that the seal rotate with the shaft, external seals are available.

Among the factors which recommend a shaft seal over other possible sealing media are ease of installation and small space allocation necessary in design of equipment, relative low cost for high effectiveness, and ability to handle simultaneously a wide range of variables while providing a positive sealing effect throughout. The variables include numerous environments of oils and fluids over wide temperature gradients, coupled with an infinite variety of shaft speeds, moderate fluid pressures, eccentricities, and dynamic shaft runout.

Types. Because of the variety of applications, the radial seal is manufactured in numerous types and sizes. They are generally categorized as follows:

1. Cased seals, where the leather or synthetic sealing element is retained in a precision-manufactured metal case

2. Bonded seals, with the synthetic element permanently bonded to a flat washer or to a formed-metal case

Seals of both categories can be provided with spring-tension elements, either garter-spring or finger type, for sealing low-viscosity fluids or where either shaft speed or eccentricity demands higher seal contact pressures. Use of the bonded-type seal is often recommended where installation space is limited by external design requirements.

CLEARANCE SEALS[7]

Clearance seals limit leakage by closely controlling the angular clearance between the rotating or reciprocating shaft and the relatively stationary housing. There are two basic clearance-seal types: *labyrinth* and *bushing*, or *ring*. These seal types are employed when a small loss in efficiency because of leakage can be permitted. Clearance seals are used when pressure differentials are beyond the design limitations of contact seals (face and circumferential).

Labyrinths. Some advantages of labyrinth seals are reliability, simplicity, and flexibility in material selection. They are used mainly in heavy industrial, power, and aircraft applications where relatively high leakage rates may be tolerated and where design simplicity is an absolute necessity.

A labyrinth seal consists of one or more thin strips or knives which are attached to either the stationary housing or the rotating shaft. Design clearances between the knives and the housing, or the strips and the shaft, are dependent on bearing clearances, amplitude of shaft runout (vibration), and differential expansion. Final running clearances are frequently established by contact between the knife and the housing which rubs away or flares the knives. A simple labyrinth is shown in Fig. 21-4-5*a*.

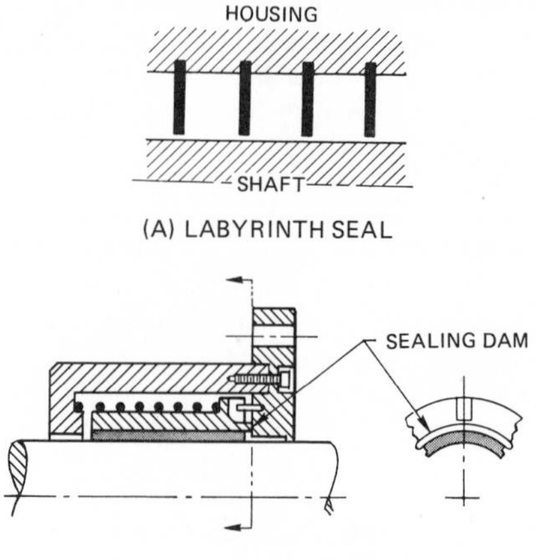
HOUSING
SHAFT
(A) LABYRINTH SEAL
SEALING DAM
(B) BUSHING AND RING SEAL

Fig. 21-4-5 Clearance seals.

Bushing and Ring Seals. The bushing-type seal is a close-fitting stationary sleeve within which the shaft rotates. Leakage from a high-pressure station at one end of the bushing to a region of low pressure at the other end is controlled by the restricted clearance between shaft and bushing. Ideally, the bushing and shaft are perfectly concentric, and no rubbing takes place. Leakage limitation depends upon the type of flow and the type of bushing used. There are four types of flow: compressible and incompressible, each of which may be either laminar or turbulent. Bushings are divided into two categories: fixed breakdown bushings and floating breakdown bushings, according to whether or not they are fixed with respect to the stationary housing.

SPLIT-RING SEALS[8]

Split rings are used for a large number of seal applications. See Fig. 21-4-6. Expanding split rings (piston rings) are used in compressors, pumps, and internal-combustion engines. Applications for straight-cut and seal-joint rings are common in industrial and aerospace hydraulic and pneumatic cylinders (linear actuators), where the ruggedness of piston rings is advantageous and where various degrees of leakage can be tolerated.

The use of contracting split rings or rod seals in linear actuators is increasing where high pressure, high temperature,

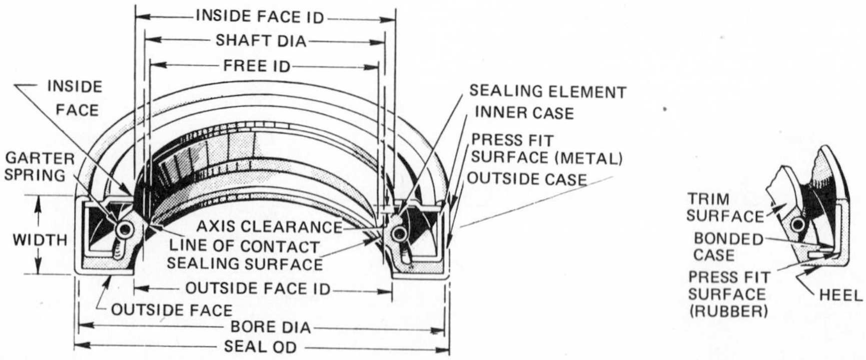
INSIDE FACE ID
SHAFT DIA
FREE ID
INSIDE FACE
GARTER SPRING
WIDTH
SEALING ELEMENT
INNER CASE
PRESS FIT SURFACE (METAL)
OUTSIDE CASE
AXIS CLEARANCE
LINE OF CONTACT
SEALING SURFACE
OUTSIDE FACE ID
OUTSIDE FACE
BORE DIA
SEAL OD
TRIM SURFACE
BONDED CASE
PRESS FIT SURFACE (RUBBER)
HEEL

Fig. 21-4-4 Metal-cased radial seal nomenclature.

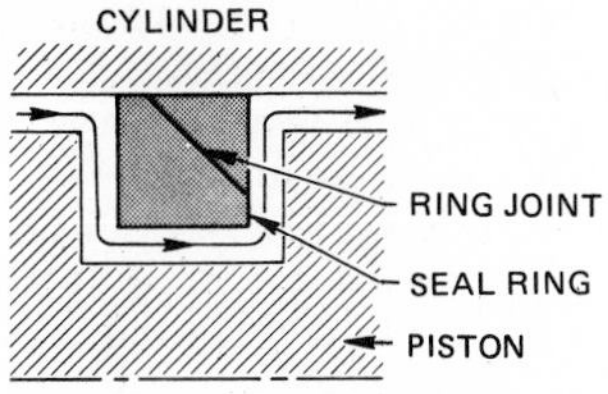
CYLINDER
RING JOINT
SEAL RING
PISTON

(A) ACTION OF MEDIUM ON SPLIT RING SEAL

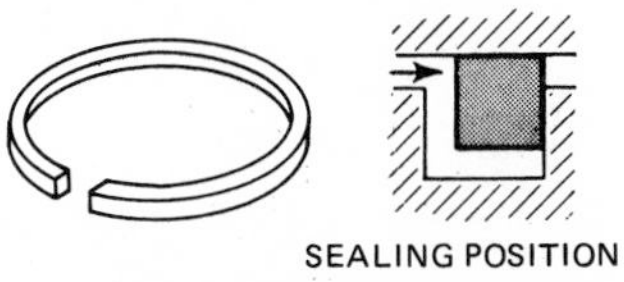
SEALING POSITION

(B) STRAIGHT-CUT SEAL RING

SEALING POSITION

(C) STEP SEAL RING

Fig. 21-4-6 Split-ring seals.

radiation, thermal fatigue, and the need for reliability prohibit the use of more conventional seals or make them less desirable.

AXIAL MECHANICAL SEALS

By convention, the term *axial mechanical seal,* or *end-face seal,* designates a sealing device which forms a running seal between flat, precision-finished surfaces. Used for rotating shafts, the sealing surfaces usually are located in a plane at a right angle to the shaft. Forces which hold the rubbing faces in contact are parallel to the shaft.

Axial mechanical seals replace conventional stuffing boxes where a fluid must be contained in spite of a substantial pressure head. These seals have many advantages:

1. Reduced friction and power losses
2. Elimination of wear on shaft or shaft sleeve
3. Zero or controlled leakage over a long service life
4. Relative insensitivity to shaft deflection or endplay
5. Freedom from periodic maintenance

Axial mechanical seals do have disadvantages. As a precision component, they demand careful handling and installation.

While differing in design detail, all mechanical seals make use of the following elements:

1. Rotating seal rings
2. Stationary seal rings
3. Spring-loading devices
4. Static seals

The rotating seal ring and the stationary seal ring are spring-loaded together by the spring-loading apparatus, and sealing takes place on the surfaces of these two rings, which rub together. The static seal component of an axial mechanical seal stops leakage of fluid past the juncture of the rotating seal ring and the shaft.

Since the rotating seal ring is stationary with respect to the turning shaft, sealing at their junction point is accomplished easily through the use of gaskets, O-rings, V-rings, cups, and so forth.

END-FACE SEALS[9]

The main advantage of an end-face seal is its low leakage rate. For example, the ratio of leakage between mechanical packings and end-face seals averages about 100:1. In addition, the end-face seal causes little wear of the sleeve or shaft on which it seals. Dynamic sealing is created on the seal faces in a vertical plane to the shaft. Very little relative movement occurs between the seal head and the sleeve or shaft on which it functions. Replacement of the sleeve or shaft underneath the seal is, therefore, seldom required. In most instances, the end-face seal either can be automatically assembled into a box or can be set into the unit by use of dimensional references. Manual adjustment does not have to be relied on, since uniform assembly can be easily accomplished.

The basic development of an end-face seal is shown in Fig. 21-4-7. A shaft with a simple O-ring as its sealing member is provided with a housing that incorporates one of the sealing faces. The housing encloses the O-rings and effects a preload on the shaft, thereby ensuring its sealing. A spring assembly is added to energize the end-face member axially, providing spring pressure against the end-face member to keep the faces together during periods of shutdown or lack of hydraulic pressure in the unit. To complete the basic seal, a stationary member is incorporated in the end cap of the unit.

A complete seal consists basically of two elements: the seal-head unit, which incorporates the housing, the end-face member, and the spring assembly; and the seal seat, which is the mating member that completes the precision-lapped face combination.

Shaft Sealing. Shaft sealing elements include the O-ring, V-ring, U-cup, wedge, and bellows (Fig. 21-4-8).

The first four of these elements constitute one category—the *pusher-type* seal. As the face wears, these sealing elements

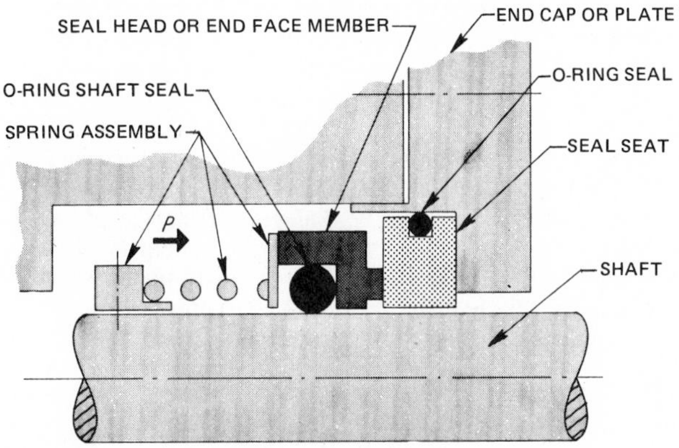
SEAL HEAD OR END FACE MEMBER
END CAP OR PLATE
O-RING SHAFT SEAL
O-RING SEAL
SPRING ASSEMBLY
SEAL SEAT
P
SHAFT

Fig. 21-4-7 Basic end-face seal design.

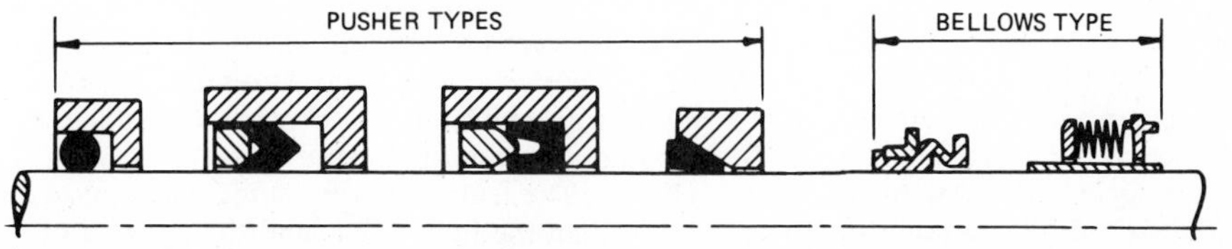
PUSHER TYPES
BELLOWS TYPE

(A) O-RING (B) V-RING (C) U-CUP (D) WEDGE (E) ELASTOMER (F) CONVOLUTION

Fig. 21-4-8 Shaft seal configurations.

are pushed forward along the shaft to maintain the seal.

Pusher-Type Elements. For the case of the O-ring, the hydraulic pressure and a mechanical preloading factor provide the sealing effect.

For the V-ring, U-cup, and wedge, the sealing function is created by mechanical and hydraulic means. Mechanical preloading to the shaft is provided by spring action incorporated in the seal design and by hydraulic pressure in the stuffing box.

The V-ring and U-cup designs seal at the shaft surface and at the mating surface of the housing. The sealing action is obtained from both spring force and hydraulic pressure acting against the spreader element, which reacts against the wings of the seal, spreading it in both directions.

Bellows-Type Elements. The bellows-shaped sealing member differs from the pusher type in that it forms a static seal between itself and the shaft. Hence, all axial movement is taken up by bellows flexure. Molded elastomers are used for one design, and plastic and corrugated metal or welded-disk forms are used for another. The static seal between the bellows and shaft is obtained by preloading the tail section of the elastomer bellows with a metal member.

MOLDED PACKINGS[10]

Molded packings are often called *automatic, hydraulic,* or *mechanical* packings. As a general group, these packings usually do not require any gland adjustment after installation. The fluid being sealed supplies the pressure needed to produce the force for sealing the packings against the wearing surface.

This general classification of packings can be subdivided into two categories: lip and squeeze types.

Lip-Type Packings. Lip-type packings of the flange, cup, U-cup, U-ring, and V-ring configurations are used almost exclusively for dynamic applications. Although rotary motions are encountered, the packings discussed here are used primarily for sealing during reciprocating motion. Hence, all the recommendations and designs mentioned apply to reciprocating service (see Fig. 21-4-9).

Flanged Packings. The flange, sometimes called the *hat*, is the least popular of all the lip-type packings. Primarily made of leather, it is also produced in fabricated and homogeneous materials. It is unbalanced, with sealing on the ID only, and is used for low-pressure, outside-packed installations where a V-ring or U-ring cannot be used because of lack of space. With a minimum cross-sectional width and a shallow gland depth, the flange can be used successfully for reciprocating motion. The flange is also used for rotary motion.

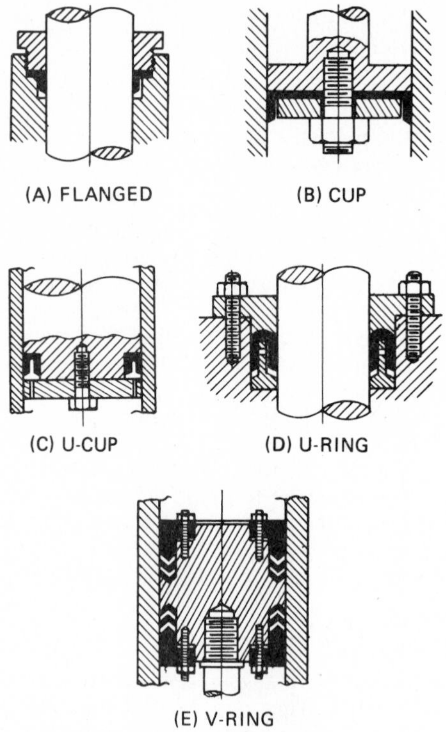
(A) FLANGED
(B) CUP
(C) U-CUP
(D) U-RING
(E) V-RING
Fig. 21-4-9 Lip-type packings.

Cup Packings. Leather cup packings, one of the oldest types of lip or mechanical packings, are used in large volume for both hydraulic and pneumatic service at low and high pressures. They are also available in homogeneous and fabricated snythetic rubber, but other materials such as TFE fluorocarbon resins, nylon, and various types of plastics are used for special applications.

Cups have only a single lip, as shown in Fig. 21-4-9*b*, and are classified as an unbalanced packing.

U-Cup Packings. This style of balanced packing is made primarily of synthetic rubber and is easily installed. It has low frictional characteristics and is primarily used for pressures under 10 000 kPa.

U-Ring Packings. This style packing in leather is sometimes called a *double cup* and in synthetic rubber a *U-cup*, although the correct and most common designation is *U-ring packing*.

All U-ring packings are balanced packings which seal on both the OD and ID. They are completely automatic in action with low frictional characteristics and are not stacked in nested sets.

The leather U-ring is more common than the fabricated, and the method of support, configuration of packing, and standard sizes vary between the two.

V-Ring Packings. These are one of the most popular of all the lip types and are used for low- or high-pressure applications. They can be installed on a piston (inside-packed) or in a gland (outside-packed). However, V-rings are primarily used for outside-packed applications with the U-ring being used primarily for inside-packed.

Also, small cross sections of this type can be used to reduce the stress on the gland rings. Properly installed, V-rings usually outperform other lip-type seals, particularly at pressures above 7000 kPa.

Squeeze-Type Packings.[11] Squeeze-type molded packings are made in a

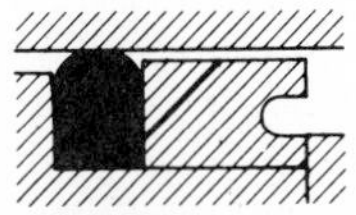
D-RING

D-SHAPED RING MAKES GOOD ROD SEAL FOR RECIPROCATING MOTION. PERFORMS EQUALLY WELL IN HYDRAULIC OR PNEUMATIC APPLICATIONS.

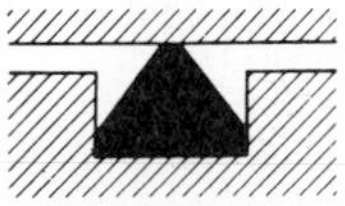
DELTA-RING

TRIANGLE-SHAPED RING. SOLVES TWISTING PROBLEM OF O-RING, BUT SINCE FRICTION IS GREATER, EXPECTED LIFE IS RELATIVELY SHORT. HAS LIMITED APPLICATIONS.

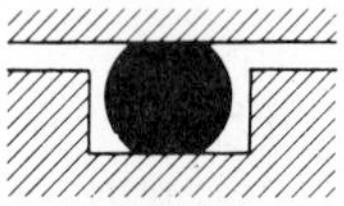
O-RING

O-RING IS MOST COMMON FORM OF SQUEEZE PACKING. SEALS IN BOTH DIRECTIONS. HAS LOW INITIAL COST. USED FOR RECIPROCATING, OSCILLATING AND ROTATING MOTION.

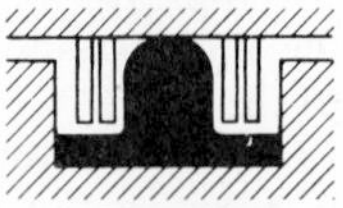
T-RING

T-SHAPED RING IS NOT SUSCEPTIBLE TO SPIRAL FAILURES. USED AS ROD OR PISTON SEAL FOR RECIPROCATING MOTION, CAN BE USED FOR OSCILLATING MOTION UNDER LOW PRESSURES.

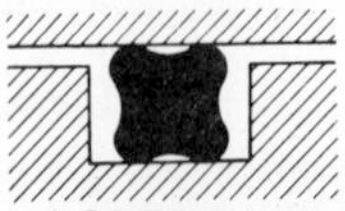
LOBED RING

SQUARE-SHAPED RING WITH FOUR ROUNDED LOBES. CAN BE USED IN CONVENTIONAL O-RING GROOVES FOR RECIPROCATING, ROTATING, AND OSCILLATING MOTION. SUPERIOR TO O-RING IN MOST ROTATING APPLICATIONS.

Fig. 21-4-10 Squeeze packings.

variety of sizes and shapes (see Fig. 21-4-10), but nearly all of them offer these advantages:

1. Low initial cost
2. Adaptability to limited space
3. Ease of installation
4. High efficiency
5. No need for adjustment
6. Tolerance to wide ranges of pressure, temperature, and fluids
7. Sealing in both directions
8. Relatively low friction

The squeeze-type packings are a relatively recent development and were not used extensively until recently because of the lack of suitable synthetic rubbers and inadequate precision molding techniques.

Squeeze-type packings are generally fitted to a rectangular groove, machined in a hydraulic or pneumatic mechanism. Nomenclature for the dimensions of a squeeze-packing seal groove is identified as that part which applies the *squeeze* to the cross section of the O-ring. The length of grooves for dynamic seals is from 25 to 50 percent longer than the O-ring cross section. Lengths of all other grooves are designed to allow for a 15 percent volume swell of the O-ring material.

Since the O-ring is the most commonly used form of squeeze-type packing and its application is similar to other types, this section will be devoted mainly to design considerations for O-rings.

O-Rings. O-ring seals work under the principle of controlled deformation. Some slight deformation is given the elastic O-ring in the form of diametral squeeze when it is installed (Fig. 21-4-11). But it is the pressure from the confined fluid that produces the deformation which causes the elastic O-ring to seal.

There are three types of applications for dynamic O-rings:

1. Reciprocating, where the sealing action is that of a piston ring or a seal around a piston rod.
2. Oscillating, where the seal rotates back and forth through a limited number of degrees or several complete turns. This may be combined with very short reciprocating strokes. The main difference between oscillation and rotation is the amount of motion involved.
3. Rotating, where a shaft turns inside the ID of the O-ring.

Squeeze-type packings are economical and easy to install and can be used whenever conditions permit.

One of the ideal applications of an O-ring is as a piston seal in a hydraulic-actuating cylinder. Another common application uses the O-ring as a valve seat or as a valve stem packing.

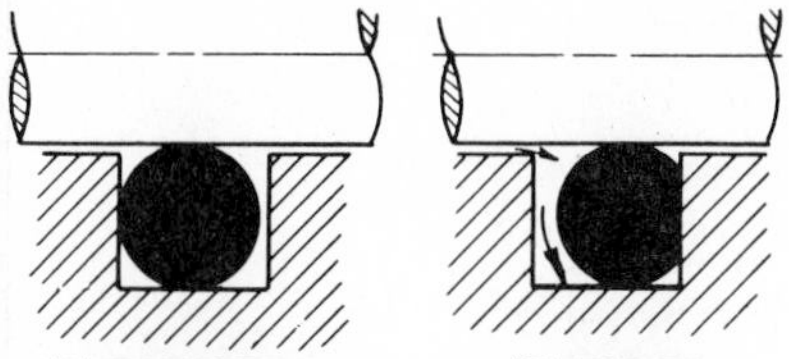

(A) DIAMETRAL SQUEEZE (B) UNDER PRESSURE

O-RINGS ARE FITTED INTO RECTANGULAR GROOVES IN HYDRAULIC MECHANISMS, AND SEALED BY BEING FORCED, BY PRESSURE, INTO CREVICES.

Fig. 21-4-11 O-rings.

Seal Symbols

The simplified representation of seals as shown in Fig. 21-4-12 is recommended for use on drawings, wherever it is not necessary to show the exact form and size of seals.

Where it is desirable to show the functional principle of the seal, symbols for the appropriate type of seal are added. See Fig. 21-4-13.

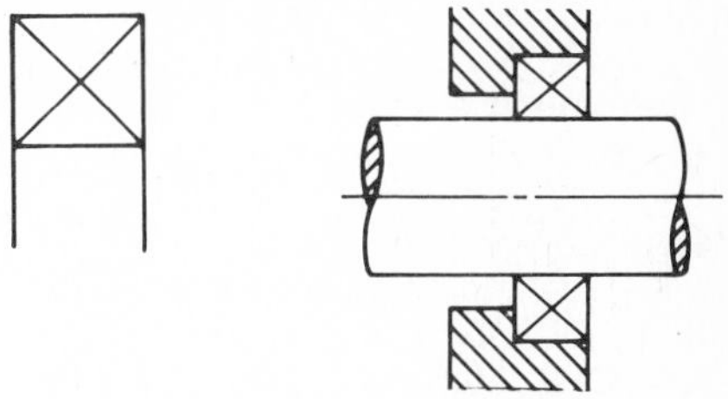

(A) GENERAL SYMBOL (B) APPLICATION

Fig. 21-4-12 Simplified representation of seals.

SHAFT—SEALS			GROOVE—SEALS		
1		ONE TONGUE, LEFT SIDE, SEALING INSIDE.	1		SEALING OUTSIDE, LEFT.
2		ONE TONGUE, RIGHT SIDE, SEALING INSIDE.	2		SEALING OUTSIDE, RIGHT.
3		ONE TONGUE, LEFT SIDE, SEALING OUTSIDE.	3		SEALING OUTSIDE, LEFT, WITH BACKRING.
4		ONE TONGUE, RIGHT SIDE, SEALING OUTSIDE.	4		SEALING OUTSIDE, RIGHT, WITH BACKRING.
5		ONE TONGUE, LEFT SIDE, WITH DUST TONGUE, SEALING INSIDE.	5		SEALING INSIDE, LEFT.
6		ONE TONGUE, RIGHT SIDE, WITH DUST TONGUE, SEALING INSIDE.	6		SEALING INSIDE, RIGHT.
7		ONE TONGUE, RIGHT SIDE, WITH DUST TONGUE, SEALING OUTSIDE.	7		SEALING INSIDE, LEFT, WITH BACKRING.
8		ONE TONGUE, LEFT SIDE, WITH DUST TONGUE, SEALING OUTSIDE.	8		SEALING INSIDE, RIGHT, WITH BACKRING.
9		TONGUES, LEFT AND RIGHT, SEALING OUTSIDE.	9		SEALING IN-AND OUTSIDE.
10		TONGUES, LEFT AND RIGHT, SEALING INSIDE.	10		SEALING LEFT AND RIGHT.

Fig. 21-4-13 Functional representation of seals.

REFERENCES AND SOURCE MATERIAL

1. B. M. Dunham, "Oils and Greases," *Machine Design,* Vol. 35, No. 14, 1963.
2. H. S. Gerstung, "Solid and Bonded-Film Lubricants," *Machine Design,* Vol. 35, No. 14, 1963.
3. J. R. Brehmer, "Lubricating Devices," *Machine Design,* Vol. 35, No. 14, 1963.
4. National Seal Division, Federal-Mogul Corporation.
5. E. A. Smith, "Felt Radial Seals," *Machine Design,* Vol. 36, No. 14, 1964.
6. C. R. McCray and H. G. Baeder, "Radial Positive-Contact Seals," *Machine Design,* Vol. 36, No. 14, 1964.
7. T. Kuchler, "Clearance Seals," *Machine Design,* Vol. 36, No. 14, 1964.
8. P. R. Shepler, "Split-Ring Seals," *Machine Design,* Vol. 36, No. 14, 1964.
9. H. Tankus, "Axial Mechanical Seals," *Machine Design,* Vol. 36, No. 14, 1964.
10. J. N. Smith, "Molded Packings," *Machine Design,* Vol. 36, No. 14, 1964.
11. M. H. Everett and H. G. Gillette, "Squeeze-Type," *Machine Design,* Vol. 36, No. 14, 1964.

Assignments

1. On an A3- or B-size sheet, complete the two assemblies shown in Fig. 21-4-A or 21-4-B given the following information.

Radial Oil-Seal Assembly. The inner ring of the tapered roller bearing is held laterally on the shaft by the shaft shoulder. A cover plate which is bolted to the housing by four socket-head cap screws has a stepped shoulder the same diameter as that of the outside diameter of the bearing. This shoulder serves two purposes: it locks the outer ring of the bearing in position, and it locates the cover plate radially on the shaft. The cover plate has a recess to accommodate a metal-cased radial seal. The shaft diameter for the oil seal should be slightly smaller than the diameter of the shaft for the bearing. The outside face of the cover plate and the oil seal should be flush. Scale is 1:1.

Oil Ring Seal Assembly. A magnetized ring press-fitted into the housing firmly holds the mating ring on the shaft element by magnetic force. The carbon ring in the face of the mating ring, in balanced contact with the lapped surface of the magnet, forms a permanent, self-adjusting face seal. O-rings in both the mating ring and the magnetic shaft element (between the element and housing) prevent leakage of confined fluids. Scale is 1:2.

2. On an A3- or B-size sheet, complete the hydraulic cylinder assembly shown in Fig. 21-4-C or 21-4-D. The sealing and fastener requirements are listed with the drawing. Add a bill of material calling out the standard sealing and fastener parts. Scale is 1:1.

REVIEW FOR ASSIGNMENTS

Unit 21-2	Antifriction Bearings
Unit 9-3	Retaining Rings
Appendix	Cap Screws

ASSIGNMENT	A	B	C	D
21-4-A METRIC	20	30	38	19
21-4-B INCH	.75	1.25	1.50	.75

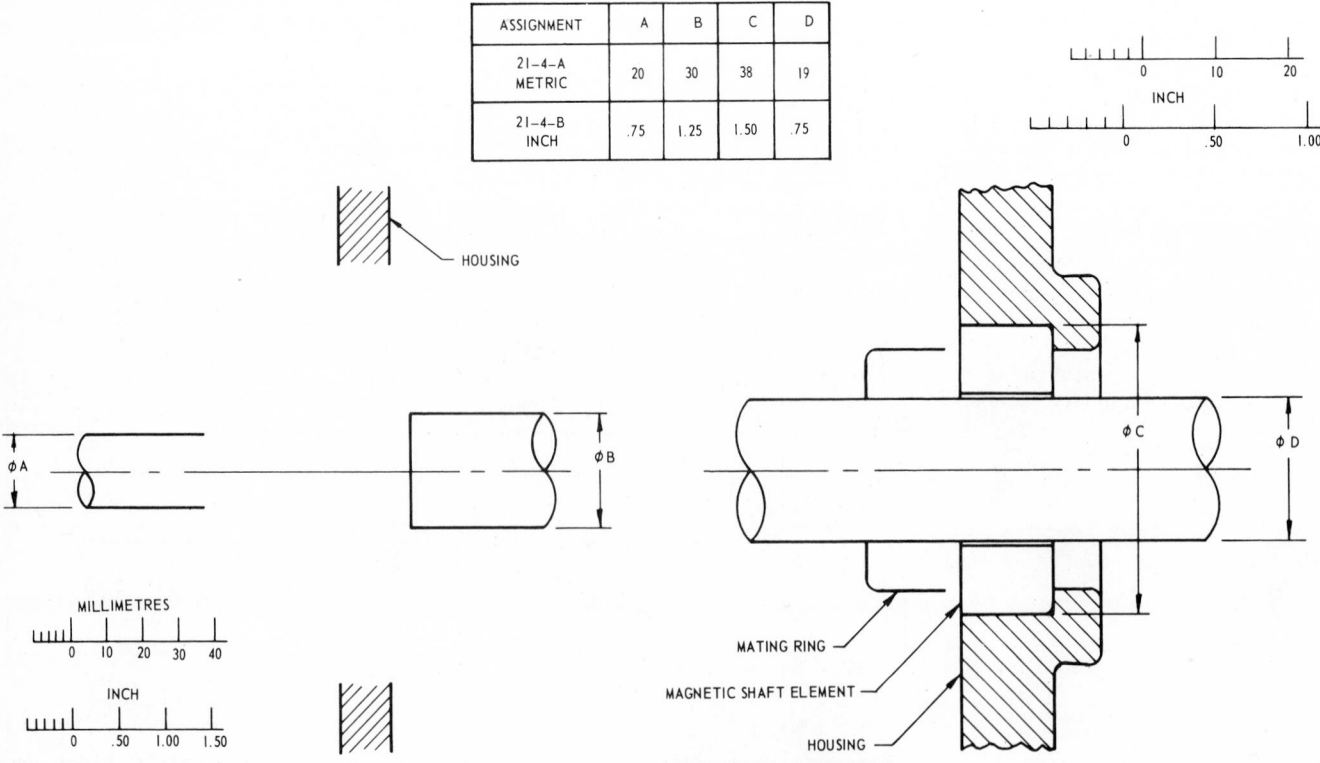

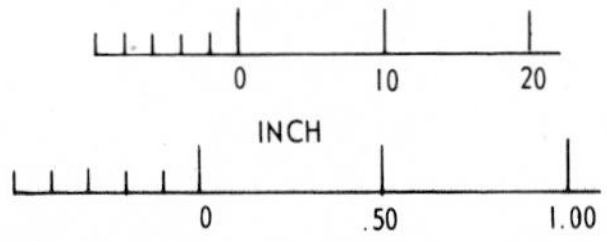

0
10
20
INCH
0
.50
1.00

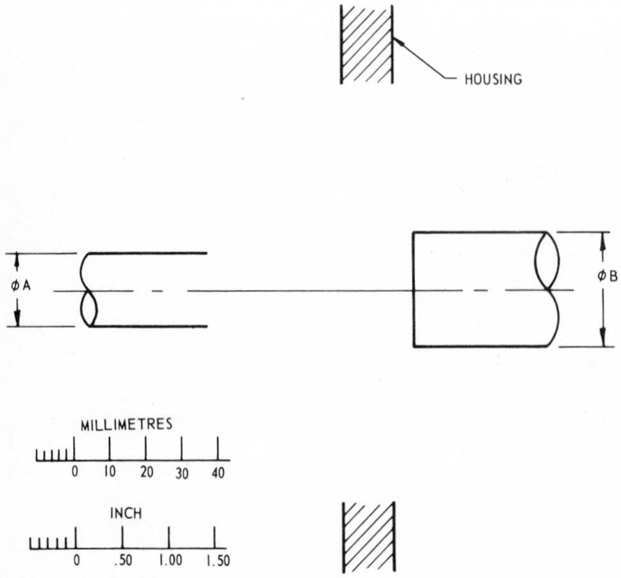

HOUSING
ϕA
ϕB
MILLIMETRES
0 10 20 30 40
INCH
0 .50 1.00 1.50

Fig. 21-4-A Oil seals.

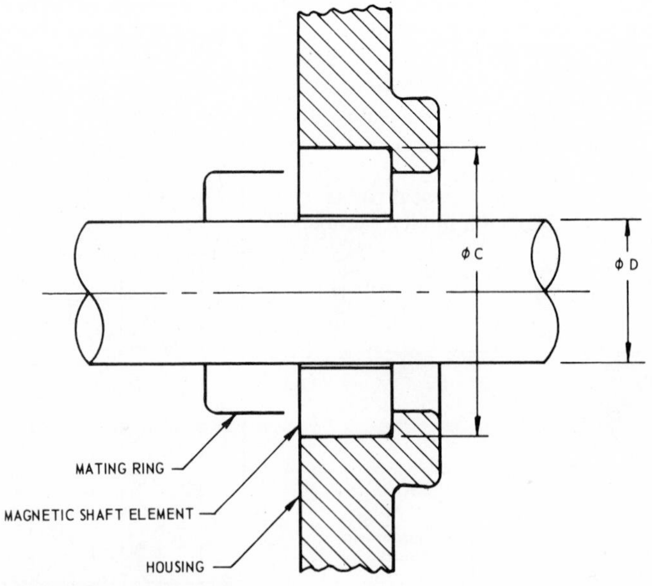

ϕC
ϕD
MATING RING
MAGNETIC SHAFT ELEMENT
HOUSING

Fig. 21-4-B Oil seals.

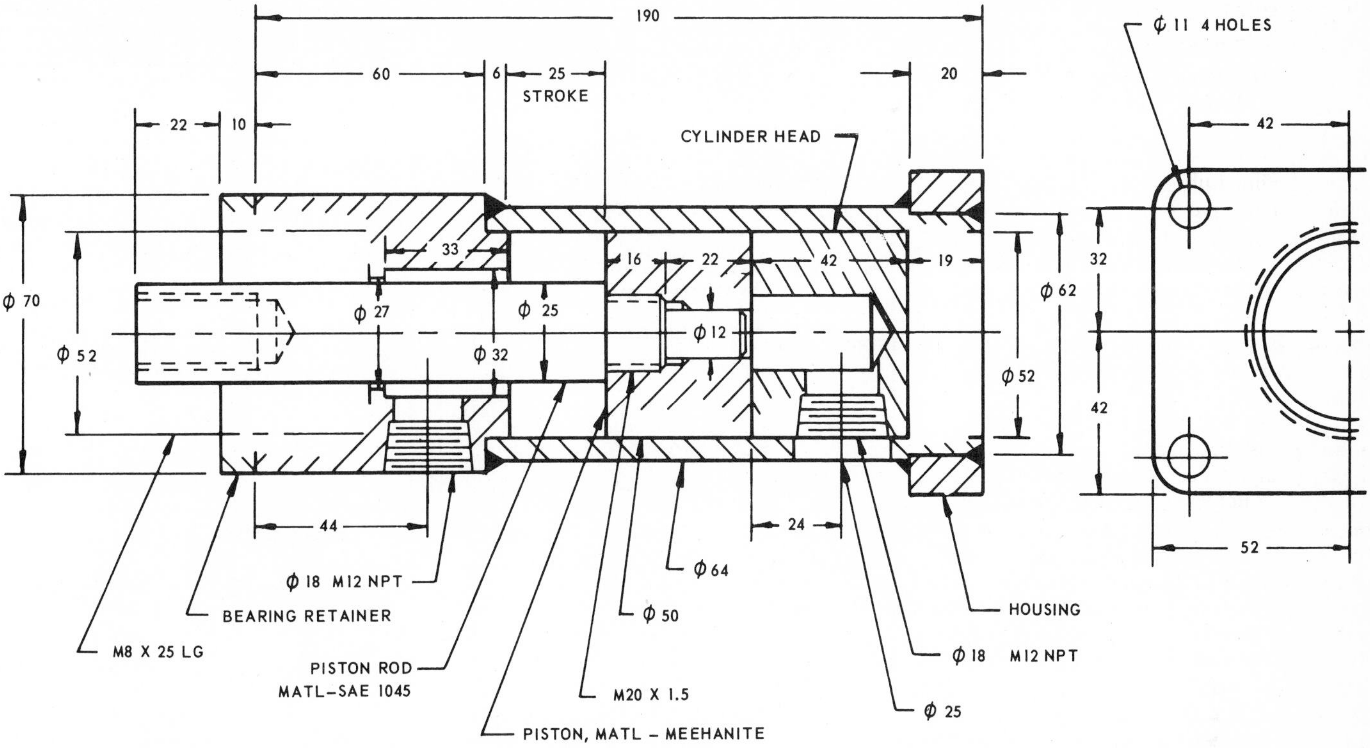

190
60
6
25
STROKE
22
10
20
ϕ 11 4 HOLES
42
CYLINDER HEAD
33
16
22
42
19
ϕ 70
ϕ 27
ϕ 25
ϕ 12
ϕ 62
ϕ 52
ϕ 32
ϕ 52
32
42
44
24
52
ϕ 18 M12 NPT
BEARING RETAINER
M8 X 25 LG
PISTON ROD
MATL–SAE 1045
ϕ 64
ϕ 50
M20 X 1.5
PISTON, MATL – MEEHANITE
HOUSING
ϕ 18 M12 NPT
ϕ 25

NOTE – ALL DIMENSIONS SHOWN ARE NOMINAL: CLEARANCES, TYPES OF FIT, O–RINGS, RETAINING RINGS, PACKING, SEALS, ETC. TO BE SELECTED BY STUDENT

Fig. 21-4-C Hydraulic cylinder.

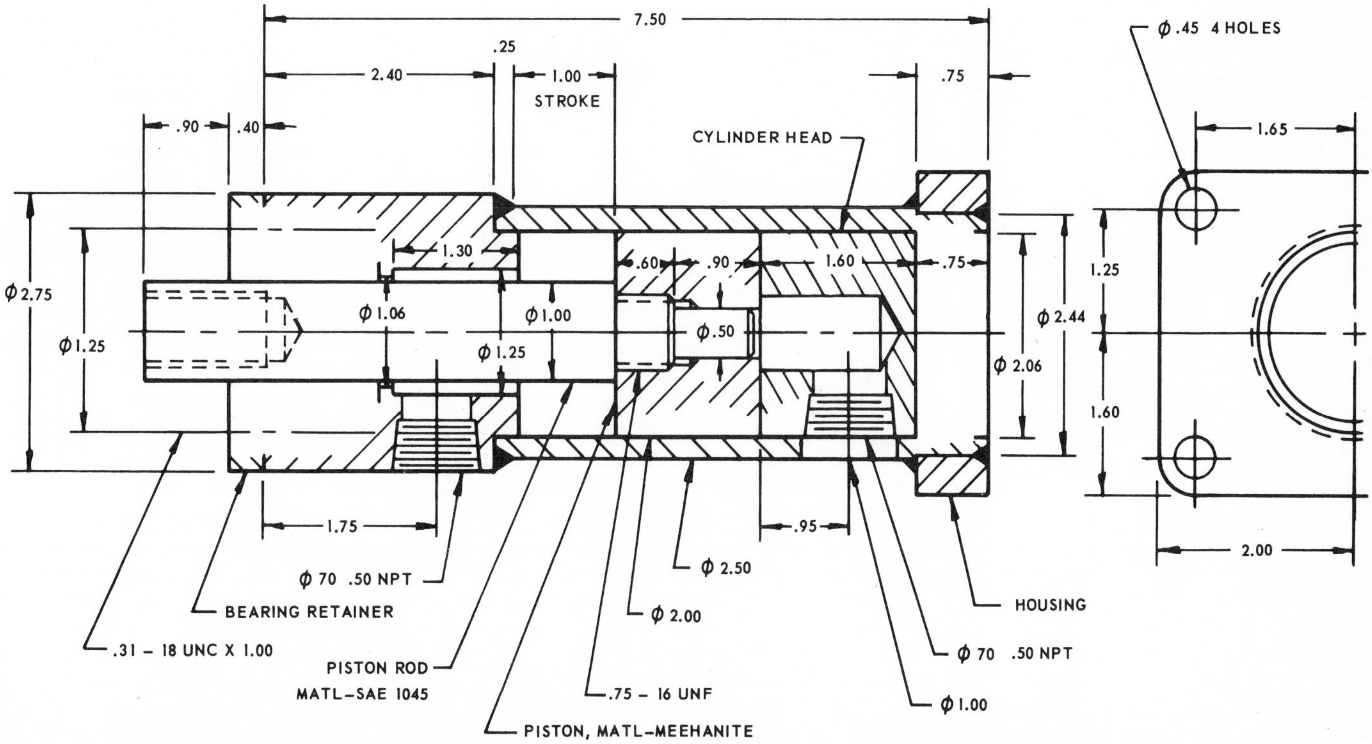

7.50
.25
2.40
1.00
STROKE
.90
.40
.75
ϕ .45 4 HOLES
1.65
CYLINDER HEAD
1.30
.60
.90
1.60
.75
ϕ 2.75
ϕ 1.06
ϕ 1.00
ϕ .50
ϕ 2.44
1.25
ϕ 1.25
ϕ 1.25
ϕ 2.06
1.60
1.75
.95
2.00
ϕ 70 .50 NPT
BEARING RETAINER
.31 – 18 UNC X 1.00
PISTON ROD
MATL–SAE 1045
ϕ 2.50
ϕ 2.00
.75 – 16 UNF
PISTON, MATL–MEEHANITE
HOUSING
ϕ 70 .50 NPT
ϕ 1.00

NOTE – ALL DIMENSIONS SHOWN ARE NOMINAL: CLEARANCES, TYPES OF FIT, O–RINGS, RETAINING RINGS, PACKING, SEALS, ETC. TO BE SELECTED BY STUDENT

Fig. 21-4-D Hydraulic cylinder.

UNIT 21-5

STATIC SEALS AND SEALANTS

O-Ring Seals[1]

All static O-ring seals are classified as gasket-type seals. Static O-ring seals are generally easier to design into a unit than dynamic O-ring seals. Wider tolerances and rougher surface finishes are allowed on metal mating members. The amount of squeeze applied to the O-ring cross section can also be increased. This type of nonmoving seal is used in flanges, flange fittings, flange unions and cylinder end caps, valve covers, plugs, etc.

When O-ring seals are used as gaskets for flanges and flange fittings, pressures as high as 170 000 kPa can be sealed, even with line vibration. Excessive bolt strain, often associated with common compression gaskets, is not encountered with static O-ring seals, because the nuts need only be tightened enough to obtain metal-to-metal surface contact. The flat flanges will not overstress or crack cast-iron fittings, which normally require special low-tension bolting. Nuts need not be tightened uniformly as with common gaskets, and pipe dopes are not required. The O-ring cannot shred or become nibbled or braided. Hence, contamination of filters and clogging of pumps are prevented. No periodic tightening of nuts is necessary to maintain a tight joint. Once the O-ring is installed, the operating pressure keeps the O-ring sealed against the metal clearance areas.

Groove Design. A rectangular groove is the most common for O-rings used as flange gaskets.

The rectangular groove can be machined half in the face plate and half in the flange, or the entire groove can be cut in one member. Various flange-type sealing designs and applications are shown in Fig. 21-5-1. In some flange-gasket designs, a triangular groove can be used to provide ease of machining and consequent reduced cost. Round-bottom grooves are also used.

Flat Nonmetallic Gaskets[2]

A gasket creates and maintains a tight seal between separable members of a mechanical assembly. Although a seal may be obtained without a gasket, the gasket promotes an efficient initial seal and prolongs the useful life of an assembly. In compensating for errors and irregularities in joined rigid members of an assembly, the gasket serves as a caulking compound. However, the forces which are brought to bear upon a gasket are quite complex and variable.

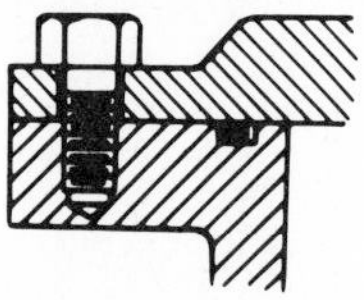

APPLICATIONS OF AN O-RING TO CYLINDER HEAD COVERS, BLIND FLANGES, ETC. THE HIGHER THE PRESSURE, THE TIGHTER THE SEAL. THIS DESIGN AUTOMATICALLY PRE-LOADS THE O-RING IN THE GROOVE. NUTS ARE TIGHTENED ONLY ENOUGH TO MAINTAIN METAL-TO-METAL SURFACE CONTACT. THIS TYPE OF INSTALLATION WILL SEAL HIGH PRESSURES WITHOUT THE EXCESSIVE BOLT STRESS NECESSARY WITH CONVENTIONAL GASKETED JOINTS.

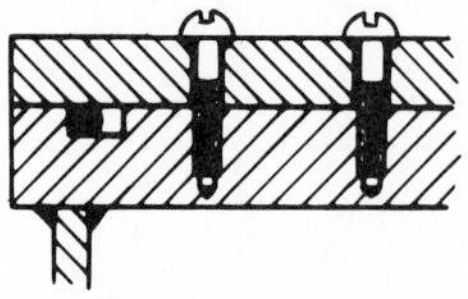

TWO O-RINGS SIZES USED FOR SEALING A RECTANGULAR PRESSURE CHAMBER. THE OUTSIDE O-RING IS STRETCHED IN A GROOVE AND THE CAPSCREW HEADS ARE SEALED BY SMALL O-RINGS IN COUNTERBORE. THIS DESIGN IS A SIMPLE AND EFFECTIVE METHOD OF SEALING X-RAY HEADS, GEAR PUMP END-PLATES AND OTHER APPLICATIONS WITHOUT REQUIRING LAPPED SURFACES.

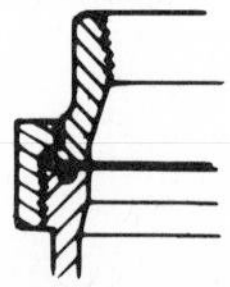

O-RING GASKET AS USED ON A FLANGE UNION. THE O-RING MAKES A TIGHT SEAL WHEN THE UNION IS SCREWED DOWN FINGER-TIGHT. THE ROUND-BOTTOM GROOVE HAS THE SAME DIAMETER AS THE ACTUAL CROSS SECTION OF THE O-RING. THE O-RING PROTRUDES .39 TO .79mm ABOVE THE FACE OF THE FLANGE. THE VOLUME OF THE GROOVE IS CALCULATED TO BE THE SAME AS THE MINIMUM VOLUME OF THE O-RING.

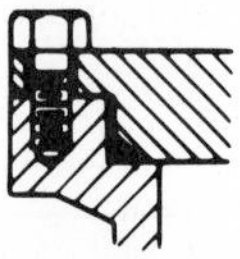

A MODIFICATION OF THE FLANGE-TYPE GROOVE. THIS TYPE OF TRIANGULAR GROOVE IS USED WHERE EASE OF MACHINING AND REDUCED COST ARE IMPORTANT. THIS TYPE OF DESIGN MAKES AN EFFECTIVE SEAL. HOWEVER, THE O-RING IS PERMANENTLY DEFORMED. PRESSURES ARE LIMITED ONLY BY THE CLEARANCE BETWEEN THE MATING METAL SURFACES AND THE STRENGTH OF THE METAL ITSELF.

Fig. 21-5-1 Flange-type static O-ring seal designs.

Basic flange joints (Fig. 21-5-2) are suitable for all kinds of flat gaskets, plain or jacketed. For moderate pressures up to 1400 kPa the simple flange joint is applicable. For higher fluid pressures, some variation of this joint is required. In the reduced-section joint, the gasket-wall section has been reduced to give a high gasket stress without change in bolt stress. The tongue-and-groove variation of the simple flange joint is useful for applying extremely high stress to gaskets and for holding high fluid pressures. Since it completely confines the gasket, it limits the flow that might otherwise occur in many materials, especially rubber and rubberlike materials.

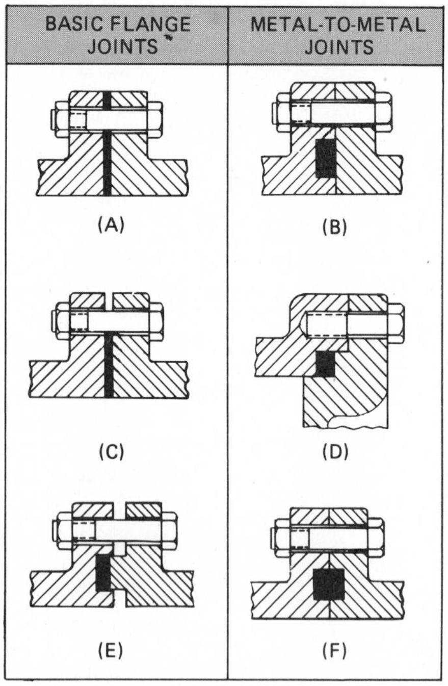
BASIC FLANGE JOINTS
METAL-TO-METAL JOINTS
(A)
(B)
(C)
(D)
(E)
(F)

Fig. 21-5-2 Flat gasket joints.

Metal-to-metal joints are particularly suitable for truly compressible materials, such as cork composition and cork-and-rubber. These joints bear a great similarity to joints designed for rubber O-rings. Rectangular-section rubber gaskets may be used if shape and volume are correct. Initial gasket volume must not exceed that of the cavity. But initial shape should be such that less than 20 to 25 percent deflection of the rubber part occurs as the joint is closed. These joints are particularly good where it is essential to maintain accurate internal clearances or alignments. They are also useful for limiting the amount of stress on the gasket. Gasket design considerations are shown in Fig 21-5-3.

Metallic Gaskets

Metallic gaskets are used for high pressures and for temperature extremes that cannot be handled by nonmetallic gaskets. They are available in a variety of shapes, and their temperature limits depend almost entirely on the metal chosen. The gaskets can withstand sealing pressures up to 70 000 kPa, but in that region the rings have heavy cross sections and require flange grooves for accurate location.

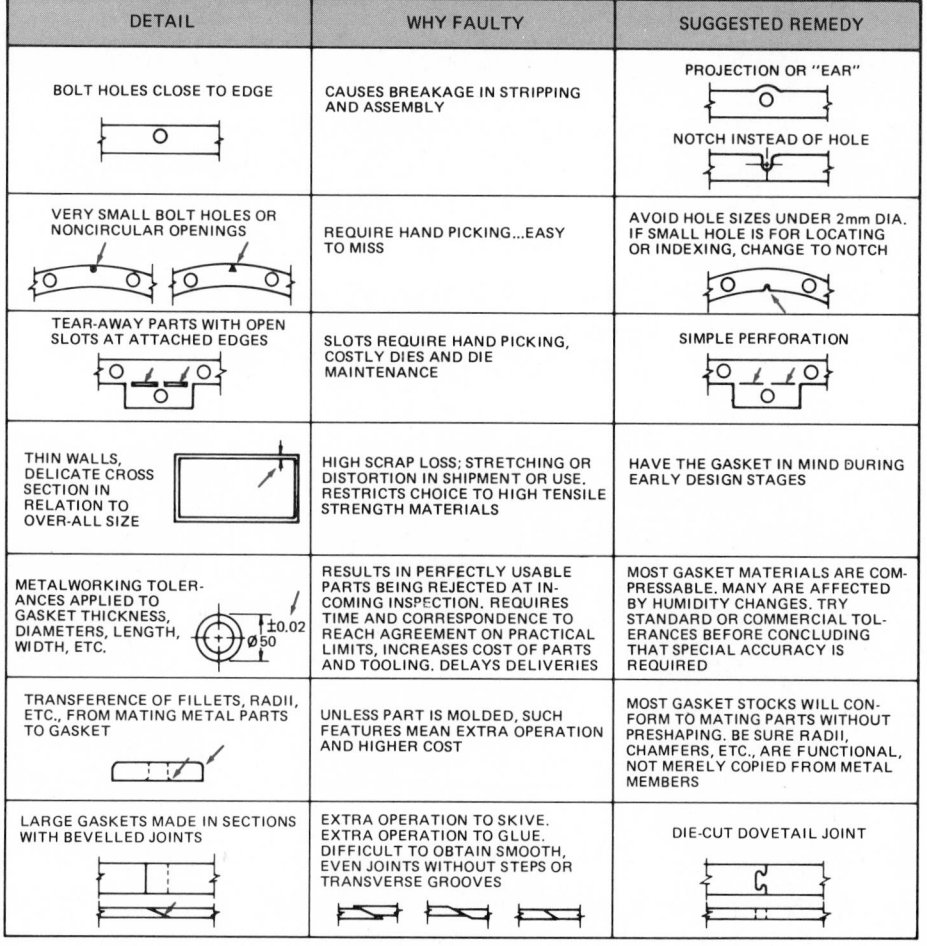

DETAIL	WHY FAULTY	SUGGESTED REMEDY
BOLT HOLES CLOSE TO EDGE	CAUSES BREAKAGE IN STRIPPING AND ASSEMBLY	PROJECTION OR "EAR" / NOTCH INSTEAD OF HOLE
VERY SMALL BOLT HOLES OR NONCIRCULAR OPENINGS	REQUIRE HAND PICKING...EASY TO MISS	AVOID HOLE SIZES UNDER 2mm DIA. IF SMALL HOLE IS FOR LOCATING OR INDEXING, CHANGE TO NOTCH
TEAR-AWAY PARTS WITH OPEN SLOTS AT ATTACHED EDGES	SLOTS REQUIRE HAND PICKING, COSTLY DIES AND DIE MAINTENANCE	SIMPLE PERFORATION
THIN WALLS, DELICATE CROSS SECTION IN RELATION TO OVER-ALL SIZE	HIGH SCRAP LOSS; STRETCHING OR DISTORTION IN SHIPMENT OR USE. RESTRICTS CHOICE TO HIGH TENSILE STRENGTH MATERIALS	HAVE THE GASKET IN MIND DURING EARLY DESIGN STAGES
METALWORKING TOLERANCES APPLIED TO GASKET THICKNESS, DIAMETERS, LENGTH, WIDTH, ETC. (±0.02, Ø50)	RESULTS IN PERFECTLY USABLE PARTS BEING REJECTED AT INCOMING INSPECTION. REQUIRES TIME AND CORRESPONDENCE TO REACH AGREEMENT ON PRACTICAL LIMITS, INCREASES COST OF PARTS AND TOOLING. DELAYS DELIVERIES	MOST GASKET MATERIALS ARE COMPRESSABLE. MANY ARE AFFECTED BY HUMIDITY CHANGES. TRY STANDARD OR COMMERCIAL TOLERANCES BEFORE CONCLUDING THAT SPECIAL ACCURACY IS REQUIRED
TRANSFERENCE OF FILLETS, RADII, ETC., FROM MATING METAL PARTS TO GASKET	UNLESS PART IS MOLDED, SUCH FEATURES MEAN EXTRA OPERATION AND HIGHER COST	MOST GASKET STOCKS WILL CONFORM TO MATING PARTS WITHOUT PRESHAPING. BE SURE RADII, CHAMFERS, ETC., ARE FUNCTIONAL, NOT MERELY COPIED FROM METAL MEMBERS
LARGE GASKETS MADE IN SECTIONS WITH BEVELLED JOINTS	EXTRA OPERATION TO SKIVE. EXTRA OPERATION TO GLUE. DIFFICULT TO OBTAIN SMOOTH, EVEN JOINTS WITHOUT STEPS OR TRANSVERSE GROOVES	DIE-CUT DOVETAIL JOINT

Fig. 21-5-3 Common faults in gasket design and suggested remedies.

Solid metal gaskets generally require thick flanges. Thinner flanges can be used with metal O-rings. The rings are made of thin-walled metal tube, bent and welded to form a continuous circle.

Sealants

Sealants are used to exclude dust, dirt, moisture, and chemicals or to contain a liquid or gas. They can also protect against mechanical or chemical attack, exclude noise, improve appearance, and act as an adhesive. See Fig. 21-5-4.

Sealants are generally used for less severe conditions of temperature and pressure than gaskets. Sealants are categorized as hardening and nonhardening. Whether a sealant is hardening or nonhardening depends upon its chemical composition and curing characteristics, rather than its initial form. Sealants generally come in nonsolid forms in a wide range of viscosities. Some epoxy sealers come in powdered form and must be melted to be applied. Certain asphalt-based sealers and waxes are solid and are applied by a hot-melt system. Thermosetting film adhesives used for sealing also come in tape form and require heat and pressure for curing.

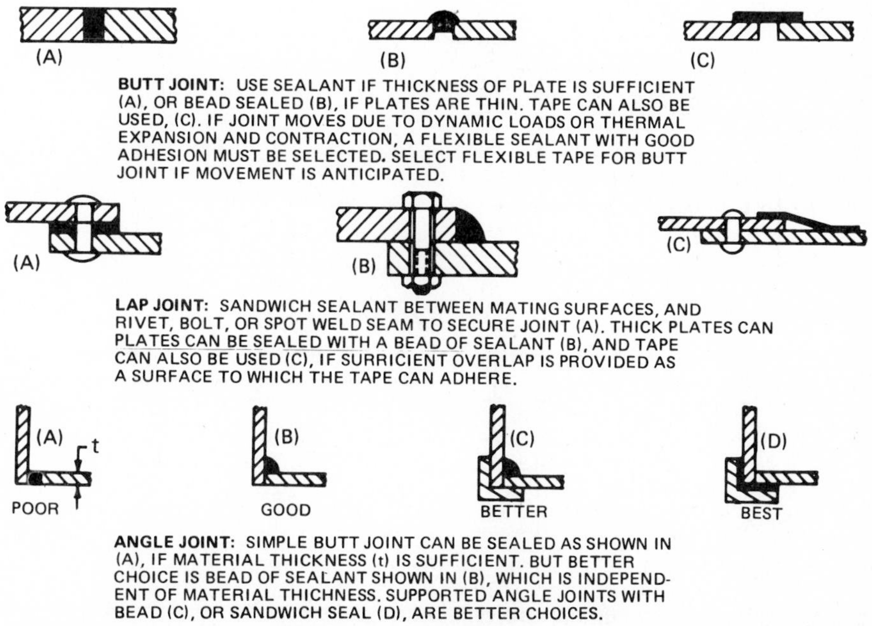
(A)
(B)
(C)
BUTT JOINT: USE SEALANT IF THICKNESS OF PLATE IS SUFFICIENT (A), OR BEAD SEALED (B), IF PLATES ARE THIN. TAPE CAN ALSO BE USED, (C). IF JOINT MOVES DUE TO DYNAMIC LOADS OR THERMAL EXPANSION AND CONTRACTION, A FLEXIBLE SEALANT WITH GOOD ADHESION MUST BE SELECTED. SELECT FLEXIBLE TAPE FOR BUTT JOINT IF MOVEMENT IS ANTICIPATED.
(A)
(B)
(C)
LAP JOINT: SANDWICH SEALANT BETWEEN MATING SURFACES, AND RIVET, BOLT, OR SPOT WELD SEAM TO SECURE JOINT (A). THICK PLATES CAN PLATES CAN BE SEALED WITH A BEAD OF SEALANT (B), AND TAPE CAN ALSO BE USED (C), IF SURRICIENT OVERLAP IS PROVIDED AS A SURFACE TO WHICH THE TAPE CAN ADHERE.
(A) t
POOR
(B)
GOOD
(C)
BETTER
(D)
BEST
ANGLE JOINT: SIMPLE BUTT JOINT CAN BE SEALED AS SHOWN IN (A), IF MATERIAL THICKNESS (t) IS SUFFICIENT. BUT BETTER CHOICE IS BEAD OF SEALANT SHOWN IN (B), WHICH IS INDEPENDENT OF MATERIAL THICHNESS. SUPPORTED ANGLE JOINTS WITH BEAD (C), OR SANDWICH SEAL (D), ARE BETTER CHOICES.

Fig. 21-5-4 Common methods for sealing joints.

Hardening sealants may be either rigid or flexible, depending on their composition. Nonhardening types are characterized by plasticizers that come to the surface continually, so that the sealant stays "wet" after application.

Exclusion Seals[3,4]

Exclusion seals are used to prevent the entry of foreign material into the moving parts of machinery. See Fig. 21-5-5. This

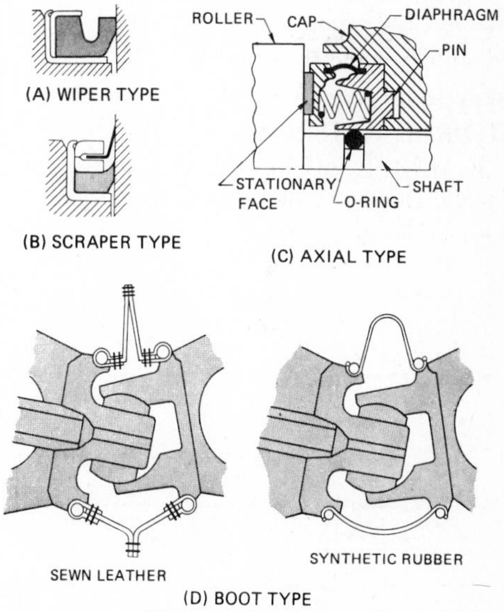
(A) WIPER TYPE
(B) SCRAPER TYPE
ROLLER
CAP
DIAPHRAGM
PIN
STATIONARY FACE
SHAFT
O-RING
(C) AXIAL TYPE
SEWN LEATHER
SYNTHETIC RUBBER
(D) BOOT TYPE

Fig. 21-5-5 Exclusion seals.

protection is necessary because foreign material entry contaminates the lubricant and accelerates wear and corrosion. Static joints are easily sealed by tight fits and gaskets. Sealing between parts having relative motion, such as between a housing and a moving shaft, is more difficult.

Sometimes seals designed only for inclusion are used to perform the inclusion and exclusion functions simultaneously. This is inadvisable, except under very light service conditions. Inclusion seals usually do a poor exclusion job and are damaged by even small amounts of abrasive material.

Exclusion seals can be classified into four general groups: wiper, scraper, axial, and boot seals.

REFERENCES AND SOURCE MATERIAL

1. M. H. Everett and H. G. Gillette, "Static O-Ring Seals," *Machine Design,* Vol. 36, No. 14, 1964.

2. E. M. Smoley and E. C. Frazier, "Nonmetallic Gaskets," *Machine Design,* Vol. 36, No. 14, 1964.

3. R. O. Isenbarger, "Exclusion Devices," *Machine Design,* Vol. 36, No. 14, 1964.

4. "Seals," *Machine Design,* Vol. 50, No. 15, 1977.

Assignments

1. On an A3- or B-size sheet, complete the two assemblies shown in Fig. 21-5-A or 21-5-B, given the following information.

Flanged Pipe Coupling. The flanges are fastened by hex bolts, nuts, and lockwashers. Alignment is accomplished by a tongue-and-groove joint similar to that shown in Fig. 21-5-2*e*, and a gasket positioned in the groove provides the seal.

Cylinder Head Cap. The cylinder head cap is fastened to the cylinder head by four hex-head cap screws equally spaced. Locking is accomplished by lockwashers. Spotfacing on the cast head cap is required because of the rough finish of the casting. An O-ring provides the seal.

2. On an A3- or B-size sheet, make a detail drawing of the gasket used with the domed cover shown in Fig. 21-5-C or 21-5-D. Material is Neoprene. Scale is 1:2.

REVIEW FOR ASSIGNMENTS

Unit 8-3 Spotfacing
Appendix Cap Screws

ASSIGNMENT	ϕ A	ϕ B	ϕ C	ϕ D	ϕ E	F	G	ϕ H	ϕ J	K	L	M	N	ϕ P	ϕ R
21–5–A METRIC	73	50	25	95	110	M6	45	120	90	M10	15	10	40	50	90
21–5–B INCH	2.90	2.00	1.00	3.75	4.50	.25	1.80	4.75	3.50	.375	.60	.40	1.50	2.00	3.50

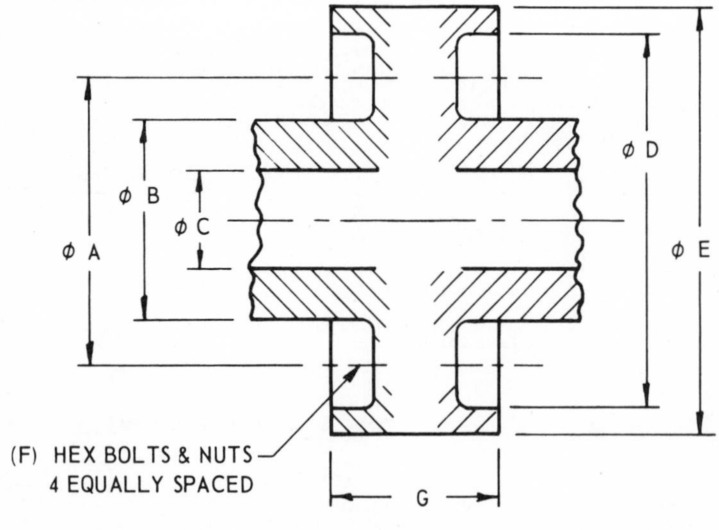
ϕ D
ϕ B
ϕ C
ϕ A
ϕ E
(F) HEX BOLTS & NUTS
4 EQUALLY SPACED
G

Fig. 21-5-A Flanged pipe coupling and cylinder-head cap.

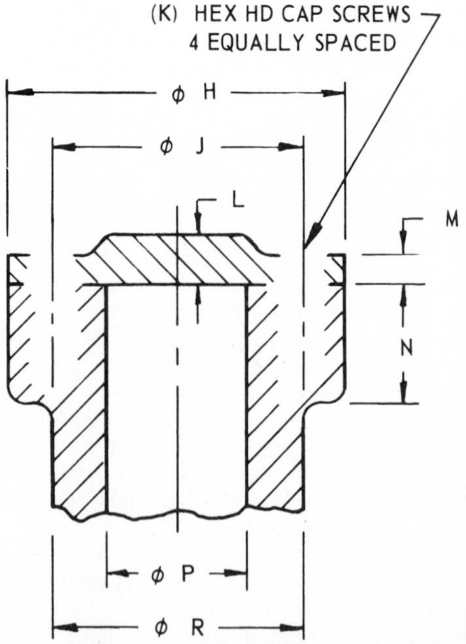
(K) HEX HD CAP SCREWS
4 EQUALLY SPACED
ϕ H
ϕ J
L
M
N
ϕ P
ϕ R

Fig. 21-5-B Flanged pipe coupling and cylinder-head cap.

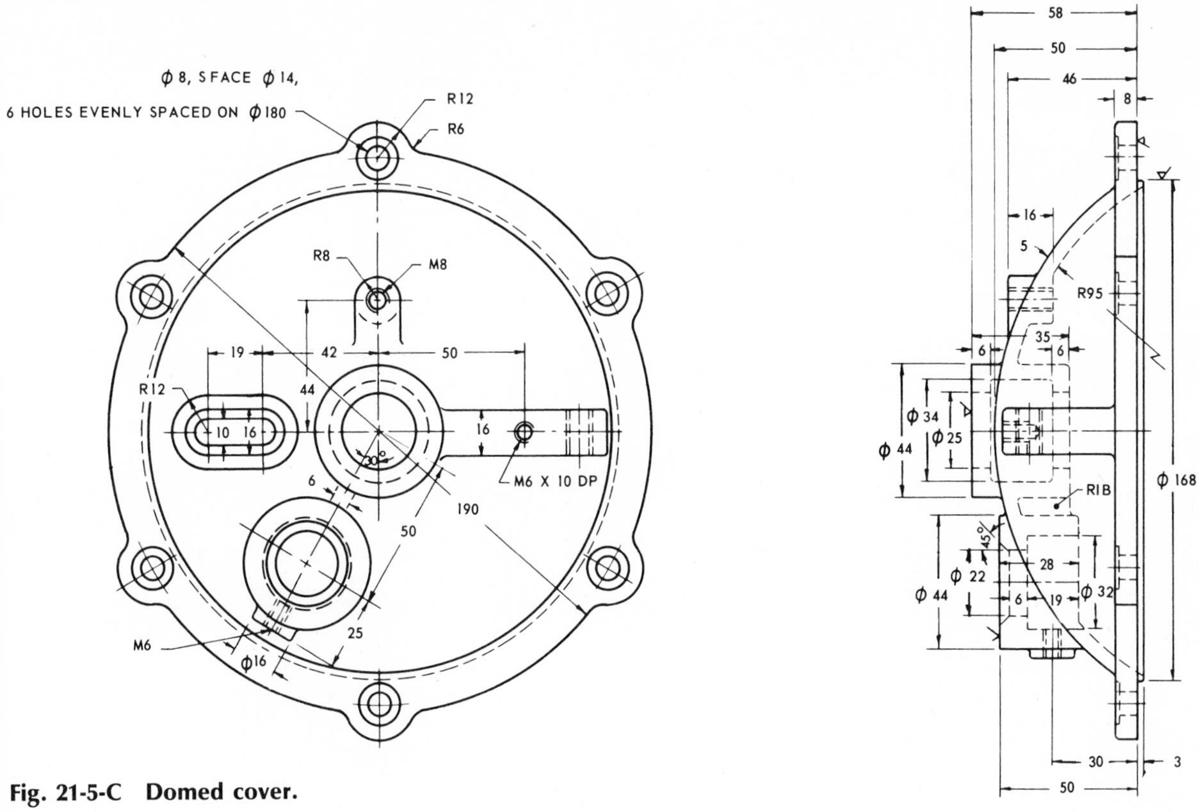

ϕ 8, SFACE ϕ 14,
6 HOLES EVENLY SPACED ON ϕ 180
R12
R6
R8
M8
19
42
50
44
R12
10
16
16
30°
6
M6 X 10 DP
190
50
M6
ϕ16
25
58
50
46
8
16
5
R95
35
6
6
ϕ 34
ϕ 25
ϕ 44
RIB
ϕ 168
45°
ϕ 22
28
ϕ 44
ϕ 32
6
19
30
3
50

Fig. 21-5-C Domed cover.

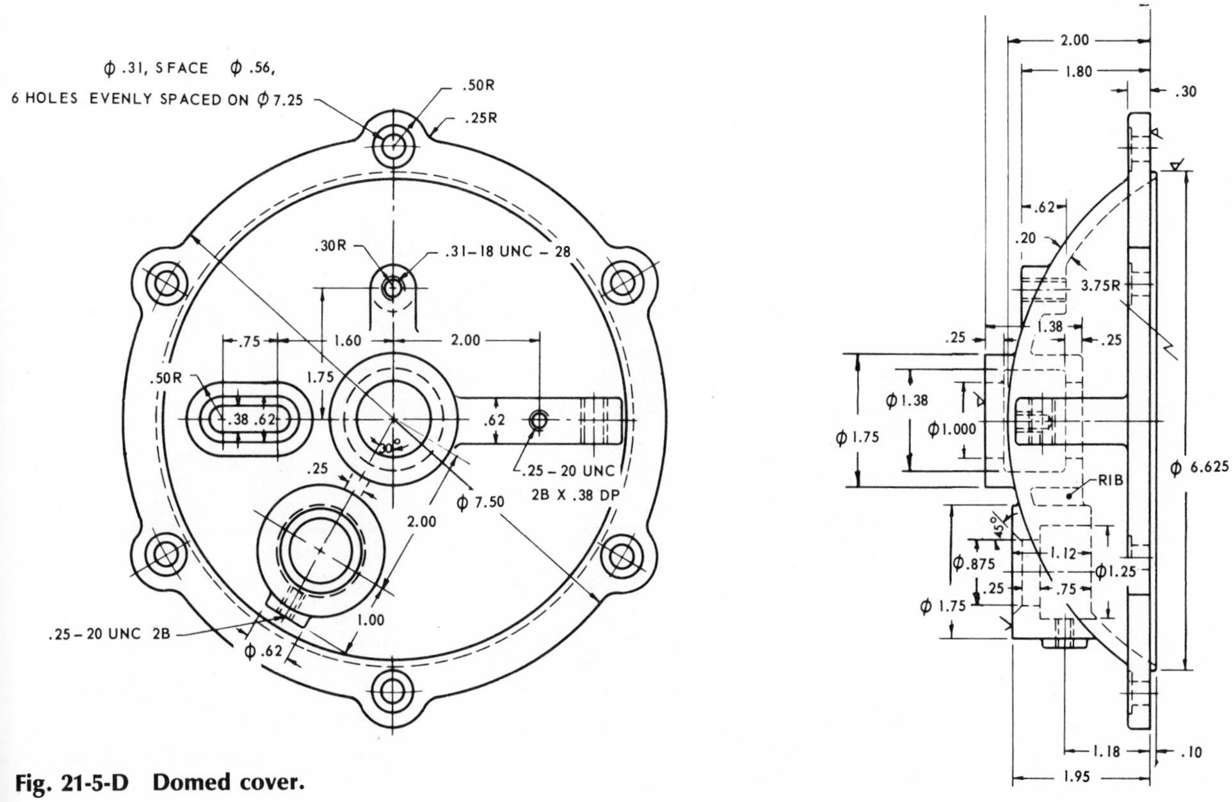

ϕ .31, SFACE ϕ .56,
6 HOLES EVENLY SPACED ON ϕ 7.25
.50R
.25R
.30R
.31–18 UNC – 28
.75
1.60
2.00
1.75
.50R
.38
.62
.62
30°
.25
.25 – 20 UNC
2B X .38 DP
ϕ 7.50
2.00
.25 – 20 UNC 2B
ϕ .62
1.00
2.00
1.80
.30
.62
.20
3.75R
.25
1.38
.25
ϕ 1.38
ϕ 1.000
ϕ 1.75
RIB
ϕ 6.625
45°
1.12
ϕ.875
ϕ1.25
.25
.75
ϕ 1.75
1.18
.10
1.95

Fig. 21-5-D Domed cover.

Chapter 22
Cams, Linkages, and Actuators

UNIT 22-1
CAMS[1]

A *cam* is a machine element designed to generate a desired motion to a follower by means of direct contact. Cams are generally mounted on rotating shafts, although they can be used so that they remain stationary and the follower moves about them. Cams may also produce oscillating motion, or they may convert motions from one form to another.

The shape of a cam is always determined by the motion of the follower. The cam is actually the end product of a desired follower movement. From the standpoint of engineering alone, cams have many decided advantages over the fundamental kinematic four-bar linkages (Fig. 22-1-1). Once they are understood, they are easier to design and the action produced by them can be most accurately forecast. For example, to cause the follower system to remain stationary

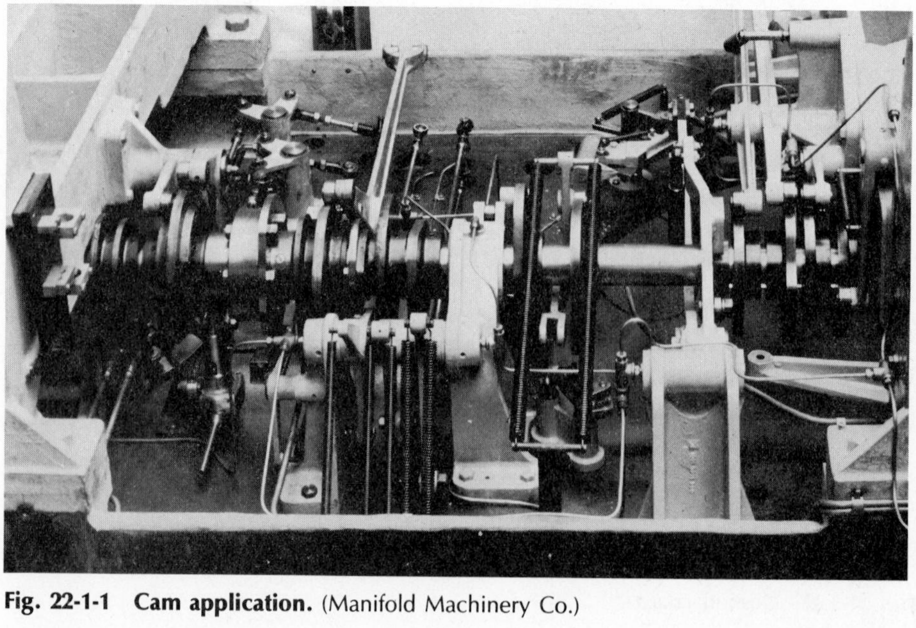

Fig. 22-1-1 Cam application. (Manifold Machinery Co.)

during a portion of a cycle is very difficult when linkages are used. With a cam this is accomplished by a contour surface which runs concentric with the center of rotation. To produce a given motion, velocity, or acceleration during a specific portion of a cycle is very difficult to do with linkages, whereas it is comparatively easy with standard cam motions, especially when the design is achieved with the aid of a computer (see Fig. 22-1-2).

By far the most popular types of cams are the OD or plate cam and the drum or cylinder cam. In the case of the OD cam, the body of the cam is usually shaped like a disk with the cam contour developed along its circumference. With these cams the line of action of a follower is generally perpendicular to the cam axis. With the drum cam, the cam track is generally machined around the circumference of the drum. In this type of cam the line of action is normally parallel to the cam axis. The level-winding mechanism on a fishing reel is an example of a drum cam. Other popular types of cams include the conjugate cam (more than one cam joined together); the face cam, where the cam track is cut into the face of the disk; and the index cam, which is similar to a drum cam except the motion of the follower passes in an arc over the cam itself.

As machine speeds increase, the need for properly designed quality cams becomes more evident. The essential specifications necessary to produce a cam of optimum quality are:

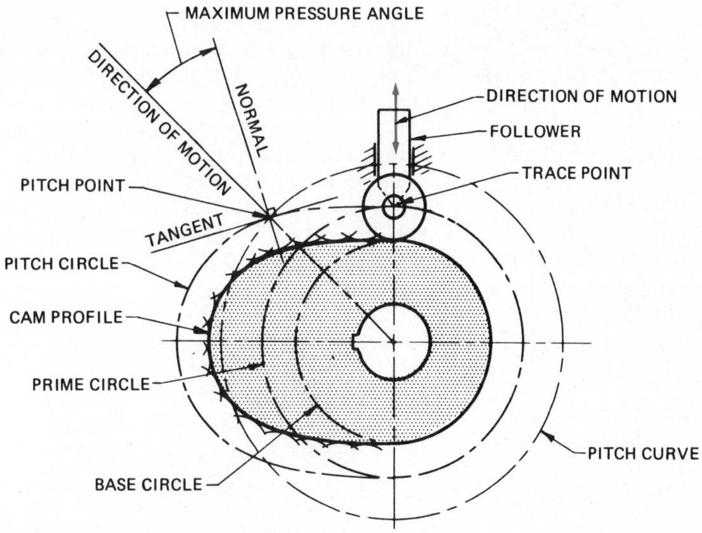
MAXIMUM PRESSURE ANGLE
DIRECTION OF MOTION
NORMAL
DIRECTION OF MOTION
FOLLOWER
TRACE POINT
PITCH POINT
TANGENT
PITCH CIRCLE
CAM PROFILE
PRIME CIRCLE
PITCH CURVE
BASE CIRCLE

Fig. 22-1-3 Cam nomenclature.

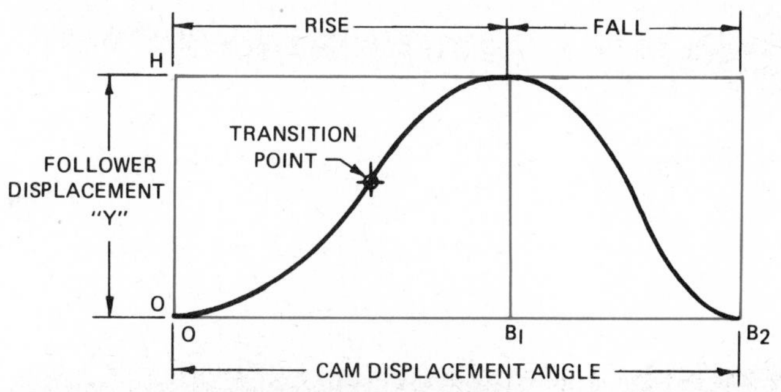
RISE
FALL
H
TRANSITION POINT
FOLLOWER DISPLACEMENT "Y"
O
O
B1
B2
CAM DISPLACEMENT ANGLE

Fig. 22-1-4 Cam displacement diagram.

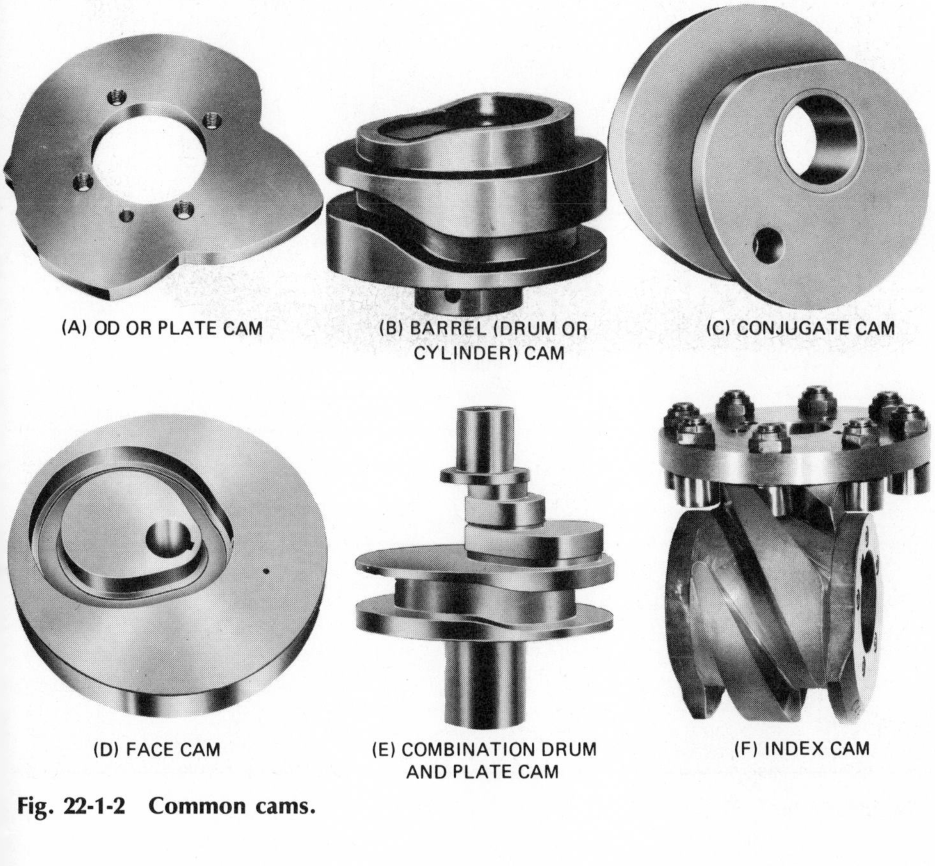
(A) OD OR PLATE CAM
(B) BARREL (DRUM OR CYLINDER) CAM
(C) CONJUGATE CAM
(D) FACE CAM
(E) COMBINATION DRUM AND PLATE CAM
(F) INDEX CAM

Fig. 22-1-2 Common cams.

1. Proper dynamic design which considers the velocity, acceleration, and jerk characteristics of the follower system. It includes vibration and shaft torque analysis.
2. Proper material selection which takes into account cost, wear, and surface stresses produced by the system.
3. Quality manufacturing which ensures that the required design is produced in the finished cam.

Cam Nomenclature

(Refer to Figs. 22-1-3 and 22-1-4.)

1. *Follower displacement* is generally defined as the position of the follower mechanism from a specific zero or rest position in relation to time or some fraction of the machine cycle (cam displacement) measured in degrees or millimetres.

2. *Cam displacement,* measured in degrees or millimetres, is the cam motion measured from a specific zero or rest position and relates to the follower mechanism as defined above.

3. *Cam profile* is the actual working surface contour of cam.

4. *Base circle* is the smallest circle drawn to the cam profile.

5. *Trace point* is the center line of the follower roller or its equivalent. When a flat follower is used, the cam profile is the envelope of successive positions of the flat follower.

6. *Pitch curve* is the locus of successive positions of the trace point as cam displacement takes place.

7. *Prime circle* is the smallest circle drawn in the pitch curve from the cam center. It is related to the base circle by the roller radius.

8. *Pressure angle* is the angle between the normal to the pitch curve and the instantaneous direction of motion of the follower.

9. *Pitch point* is the position on the pitch curve where the pressure angle is maximum.

10. *Pitch circle* is the circle which passes through the pitch point.

11. *Transition point* is the position of maximum velocity where acceleration changes from plus to minus (force on follower changes direction). In a closed cam this is sometimes referred to as the *crossover point*, where, because of the reversing acceleration, the follower roller leaves one cam profile and crosses over to the opposite (or conjugate) one.

Figure 22-1-5 illustrates typical cam and follower combinations used in machine design.

Cam Followers

The common types of cam followers are shown in Fig. 22-1-6. The roller follower is more suitable where high speeds, heat (friction), and wear are factors (Fig. 22-1-7).

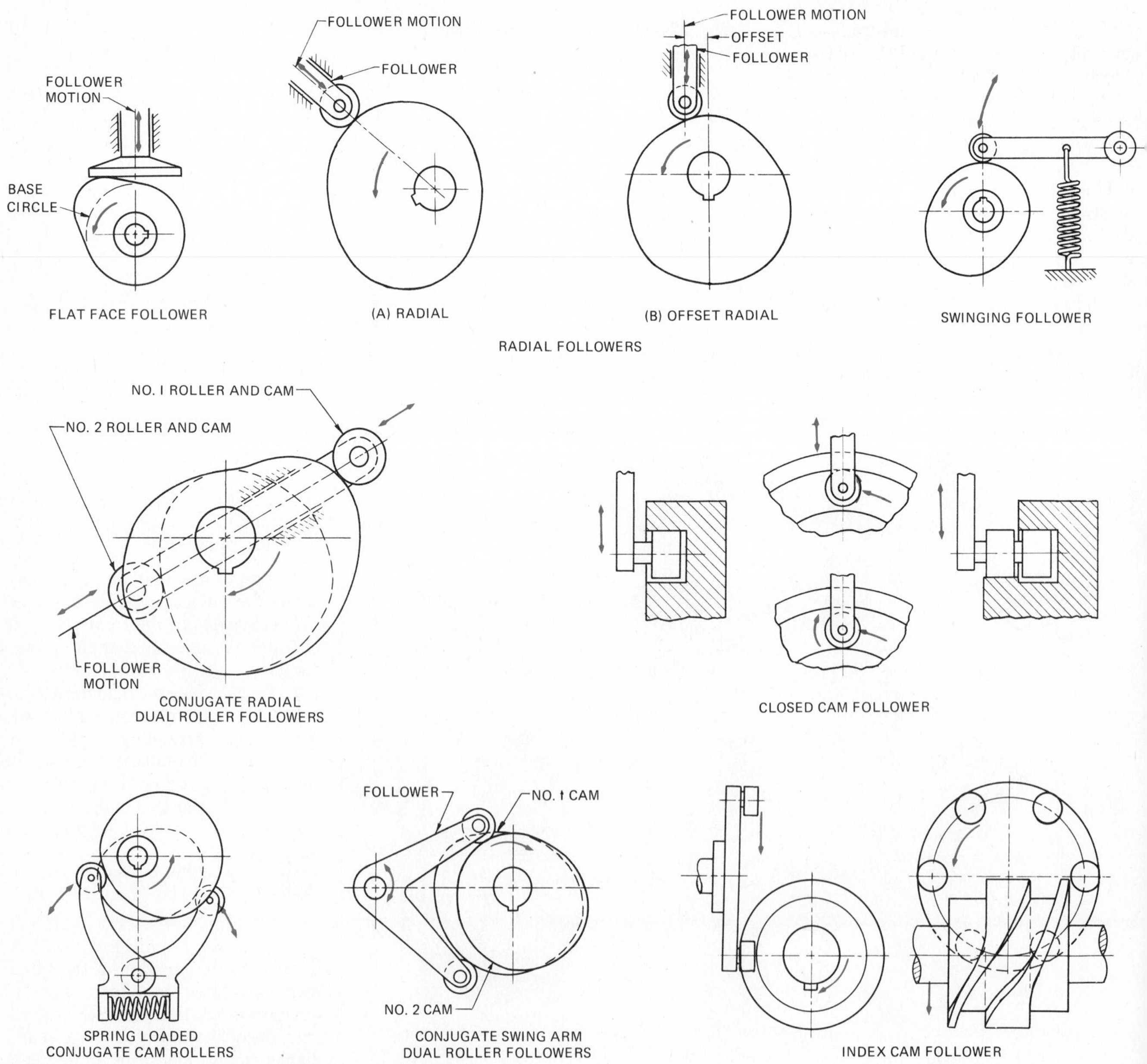

FOLLOWER MOTION
BASE CIRCLE
FLAT FACE FOLLOWER
FOLLOWER MOTION
FOLLOWER
(A) RADIAL
FOLLOWER MOTION
OFFSET
FOLLOWER
(B) OFFSET RADIAL
RADIAL FOLLOWERS
SWINGING FOLLOWER
NO. 1 ROLLER AND CAM
NO. 2 ROLLER AND CAM
FOLLOWER MOTION
CONJUGATE RADIAL DUAL ROLLER FOLLOWERS
CLOSED CAM FOLLOWER
SPRING LOADED CONJUGATE CAM ROLLERS
FOLLOWER
NO. 1 CAM
NO. 2 CAM
CONJUGATE SWING ARM DUAL ROLLER FOLLOWERS
INDEX CAM FOLLOWER

Fig. 22-1-5 Typical cam and follower combinations.

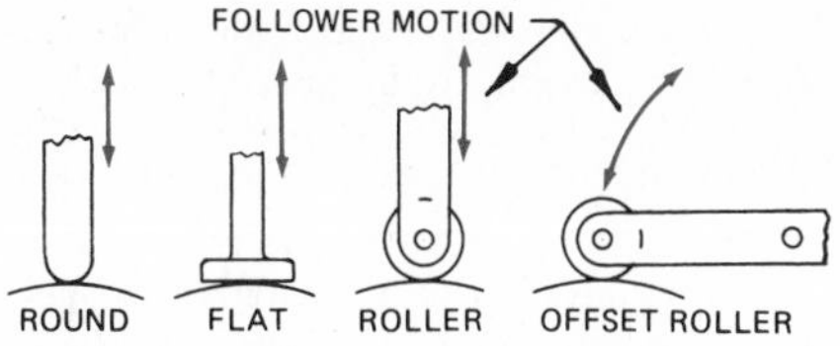
FOLLOWER MOTION
ROUND
FLAT
ROLLER
OFFSET ROLLER

Fig. 22-1-6 Types of cam followers.

Cam Motions[2]

In the early phases of the development of the cam mechanism, it is customary to work with only center lines to establish the desired motions. It is obvious that some data have been specified or determined from related parts of the design to establish the cam and linkage requirements and to provide base points from which to start the cam linkage design. These data will usually be the motion requirements and timing relationships of a particular part of the machine such as a feed slide, a folding mechanism, or a label applicator.

The choice of motion that the cam must produce will depend, first, on the cycle timing and, second, on the system or machine dynamics. For the purpose of showing cam layout techniques, cams producing the following motions will be discussed:

1. Uniform motion
2. Parabolic motion
3. Harmonic motion
4. Cycloidal motion
5. Modified trapezoid motion
6. Modified sine motion
7. Synthesized, modified sine-harmonic motion

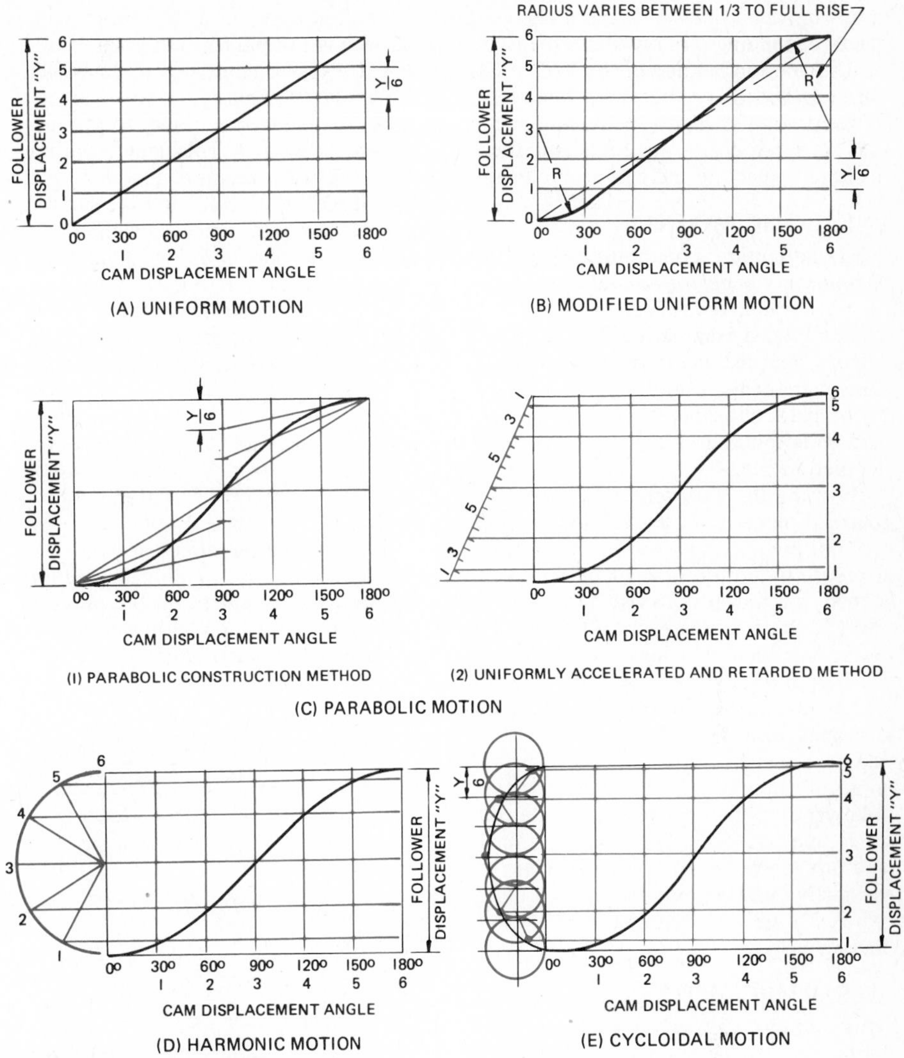
FOLLOWER DISPLACEMENT "Y"
CAM DISPLACEMENT ANGLE
(A) UNIFORM MOTION
RADIUS VARIES BETWEEN 1/3 TO FULL RISE
(B) MODIFIED UNIFORM MOTION
(1) PARABOLIC CONSTRUCTION METHOD
(2) UNIFORMLY ACCELERATED AND RETARDED METHOD
(C) PARABOLIC MOTION
(D) HARMONIC MOTION
(E) CYCLOIDAL MOTION

Fig. 22-1-8 Cam motions.

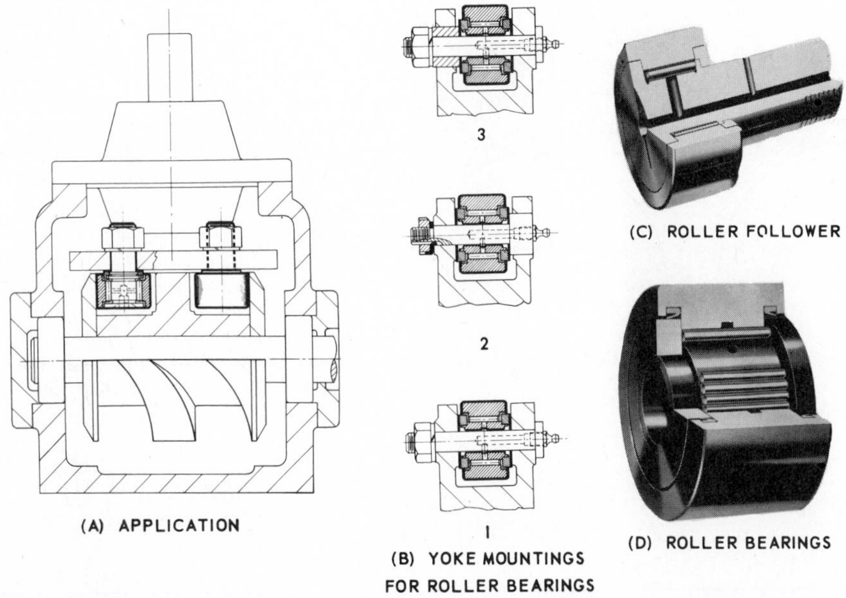
(A) APPLICATION
(B) YOKE MOUNTINGS FOR ROLLER BEARINGS
(C) ROLLER FOLLOWER
(D) ROLLER BEARINGS

Fig. 22-1-7 Cam-roller followers.

The first four are illustrated in Fig. 22-1-8.

UNIFORM MOTION (CONSTANT VELOCITY MOTION)

Uniform motion is used when the follower is required to rise and drop at a uniform rate of speed. If a follower is to rise 42 mm in one-half of a revolution, or 180° of the cam, then for every 30° of cam rotation the follower would rise one-sixth of 42 mm, or 7 mm.

This curve is also referred to as a *straight-line motion*, and it is most commonly used in connection with screw machines to control the feed of a cutting tool. If it were used with a dwell area in a cam, there would be a jerk at the start and stop of the motion.

Since this kind of motion starts and ends abruptly, it is often modified slightly to reduce the shock on the follower. A radius is used at the beginning and end of the motion, and a line tangent to these arcs is drawn. The size of radius varies between one-third to full rise height depending on how sharp the rise is. This motion is known as *modified uniform motion*. Since this type of motion is not desirable for high speeds, motions that start and end slowly, reaching their maximum speed in the center, are used.

PARABOLIC MOTION

Parabolic motion, commonly referred to as *uniformly accelerated and retarded motion*, or *constant acceleration*, is a curve found by taking the distance traveled and making it proportional to the square of the time. The construction of the parabolic curve as shown in Fig. 22-1-8*c* is found in the same manner as detailed in Fig. 4-6-1.

By using the uniformly accelerated and retarded method of construction for this motion, the divisions will increase and decrease by a ratio of 1:3:5:5:3:1. For instance, a follower is to rise 54 mm in 180°. Plotting points every 30° and using six proportional divisions of 1:3:5:5:3:1, we find in the first 30° the follower rises one-eighteenth of the total rise of 54 mm, or 3 mm; in the next 30° the follower rises three-eighteenths of the rise of 54 mm, or 9 mm, and in the third 30° the follower rises five-eighteenths of the rise of 54 mm, or 15 mm; the fourth, fifth, and last rises being 15, 9, and 3 mm, respectively. This motion would produce a jerk if used in connection with a cam having a dwell.

HARMONIC MOTION

This motion, often referred to as *crank motion*, is produced by a true eccentric operating against a flat follower whose surface is normal to the direction of linear displacement. Figure 22-1-9 illustrates this type of cam. However, it is more frequently necessary to produce a simple harmonic displacement with less than 360° of rotation of the cam, as illustrated in Fig. 22-1-10, and the ordinates for the cam pitch curve can then be determined as shown in Fig. 22-1-8*d*. It may be impossible to use a flat follower since the harmonic pitch curve usually has a reentrant or reversing curve and a flat follower would just *bridge* the hollow part. Since a roller follower is the most practical and reliable type, the development of the cam profile with this type follower is shown. This motion would also produce a jerk if used in connection with a cam having a dwell.

To illustrate the effect of cam displacement for a given cam size and follower displacement on the pressure angle, the return, or fall curve, has been shown with a much larger angle. Note that the maximum pressure angle has been considerably reduced.

CYCLOIDAL MOTION

Figure 22-1-8*e* illustrates the graphical method of laying out a cycloidal profile using a rolling circle, as shown on the left end of the illustration. This curve, when manufactured accurately, produces a very smooth jerk-free motion when used in a cam having a dwell. This curve is best suited for light loading at high speeds.

MODIFIED TRAPEZOID

The *modified trapezoid* is made by combining the cycloid and the constant-acceleration curve. Manufacturing accuracy requirements are less critical with the modified trapezoid than with the cycloidal curve. An advantage over the cycloid is lower acceleration, which means lower forces on output members (the follower system). High inertias can be handled more satisfactorily with the modified trapezoid than with the cycloid. This curve also is jerk-free when used in a cam having a dwell.

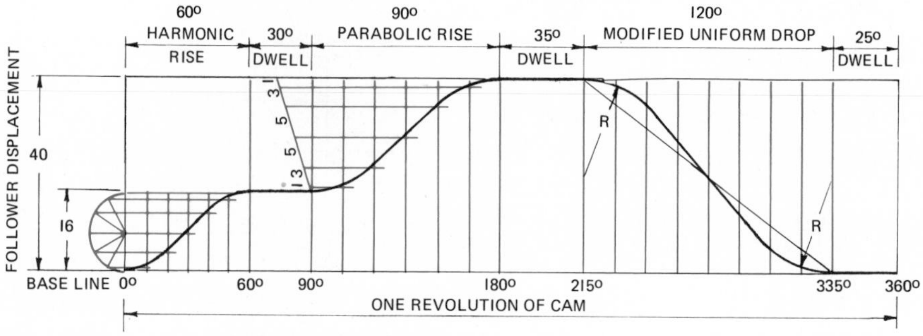
60°
HARMONIC
RISE
30°
DWELL
90°
PARABOLIC RISE
35°
DWELL
120°
MODIFIED UNIFORM DROP
25°
DWELL
FOLLOWER DISPLACEMENT
40
16
R
R
BASE LINE 0°
60°
90°
180°
215°
335°
360°
ONE REVOLUTION OF CAM

Fig. 22-1-10 Cam displacement diagram.

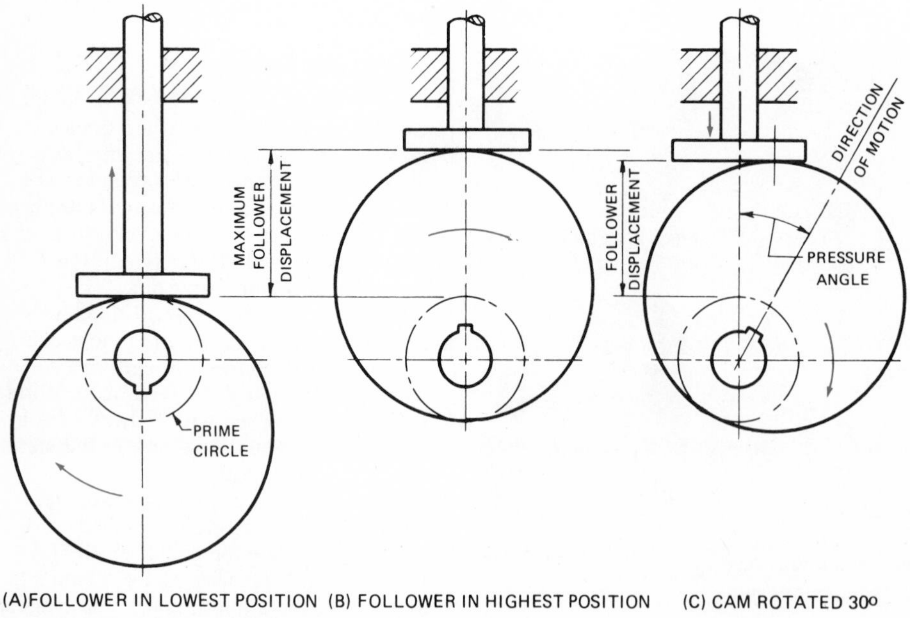
MAXIMUM
FOLLOWER
DISPLACEMENT
PRIME
CIRCLE
FOLLOWER
DISPLACEMENT
DIRECTION
OF MOTION
PRESSURE
ANGLE
(A)FOLLOWER IN LOWEST POSITION
(B) FOLLOWER IN HIGHEST POSITION
(C) CAM ROTATED 30°

Fig. 22-1-9 Eccentric plate cam.

MODIFIED SINE

The *modified sine* is a combination of cycloidal and harmonic curves. This curve will absorb more errors than the modified trapezoid or cycloidal curves. The torque change from positive to negative is 0.2 in the modified trapezoid and 0.4 in the modified sine. This means that we can stand a more flexible, or elastic, input drive with the modified sine than with the modified trapezoid. This curve (modified sine) is ideal for high inertia, as well as for reasonably high speed.

SYNTHESIZED, MODIFIED SINE-HARMONIC MOTION

Because of the complex makeup of the profiles of this curve, only the information shown in Fig. 22-1-11 is covered in this text. For further information see the references for this chapter.

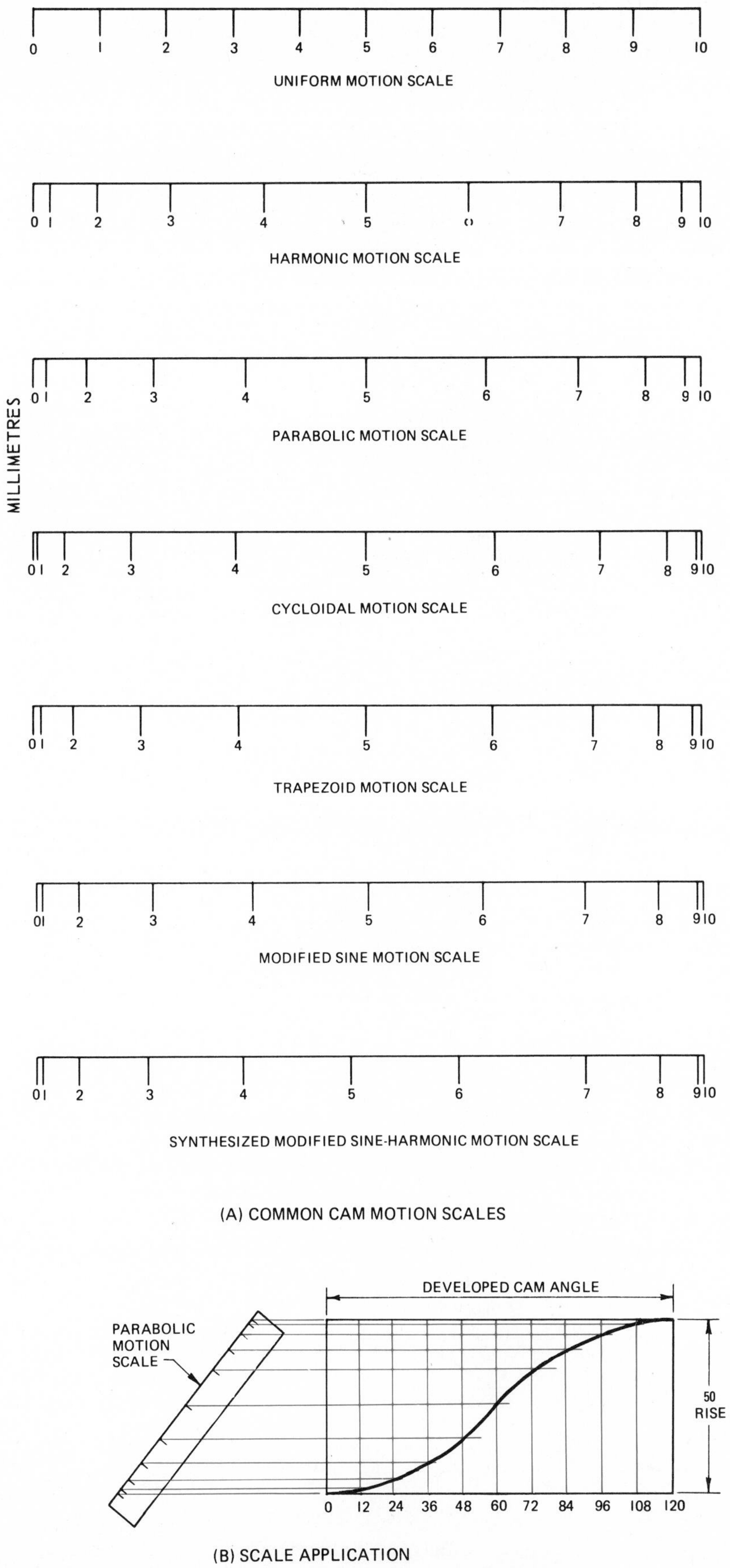
UNIFORM MOTION SCALE
HARMONIC MOTION SCALE
PARABOLIC MOTION SCALE
MILLIMETRES
CYCLOIDAL MOTION SCALE
TRAPEZOID MOTION SCALE
MODIFIED SINE MOTION SCALE
SYNTHESIZED MODIFIED SINE-HARMONIC MOTION SCALE
(A) COMMON CAM MOTION SCALES
DEVELOPED CAM ANGLE
PARABOLIC MOTION SCALE
50 RISE
0 12 24 36 48 60 72 84 96 108 120
(B) SCALE APPLICATION

Fig. 22-1-11 Simplified method of laying out cam motion.

Simplified Method of Laying Out a Cam Motion

The method shown in Fig. 22-1-11*b* is a quick and accurate method for laying out a cam motion. The divisions shown on the lines in Fig. 22-1-11*a* are accurately divided into the proper divisions for the various cam motions. For example, it is required to construct a 50 mm parabolic rise in 120° of cam rotation.

METHOD

1. Draw two parallel horizontal lines 50 mm apart representing the rise.
2. Select a suitable distance for the cam displacement and divide the 120° into 10 equal parts (12°, 24°, 36°, etc.).
3. Using the edge of a sheet of paper, mark on the divisions for the parabolic motion shown in Fig. 22-1-11*a*.
4. Using this marked paper as the scale, lay the scale between the base line and the top of the 50 mm rise, as shown in Fig. 22-1-11*b*, and transfer the points from the scale to the drawing.
5. Project these points horizontally to their respective cam divisions and draw the curve.

Cam Displacement Diagrams

In preparing cam drawings, a cam displacement diagram is drawn first to plot the motion of the follower. The curve on the drawing represents the path of the follower, not the face of the cam. The length of the diagram can be any convenient length, but often it is drawn equal to the circumference of the base circle of the cam, and the height is drawn equal to the follower displacement. The lines drawn on the motion diagram are shown as radial lines on the cam drawing, and sizes are transferred from the motion diagram to the cam drawing. Figure 22-1-10 shows a cam displacement diagram having three different types of motion plus three dwell periods. Most cam displacement diagrams have cam displacement angles of 360°.

REFERENCES AND SOURCE MATERIAL

1. Eonic Cams.
2. Commercial Cam and Machine Co.

Assignment

On an A3- or B-size sheet, draw displacement diagrams for each of the two cams shown in Fig. 22-1-A or Fig. 22-1-B.

REVIEW FOR ASSIGNMENT

Unit 4-1 Dividing a Line into Equal Parts

CAM 1

Draw the displacement diagram for a cam that has the following action:

Rise 50 mm in 120° with cycloidal motion
Dwell for 60°
Drop 10 mm in 90° with uniform motion
Drop 40 mm in 90° with uniformly accelerated and retarded motion.
Displacement diagram 50 mm high x 300 mm long.
Scale 1:1

CAM 2

Draw the displacement diagram for a cam that will raise a follower 30 mm with modified uniform motion during 90° of revolution; dwell for 60°; raise 20 mm with harmonic motion in 45°; drop 50 mm with parabolic motion in 120°; and dwell to the starting point.

Displacement diagram 50 mm high x 300 mm long.
Scale 1:1

Fig. 22-1-A Cam displacement problem.

CAM 1

Draw the displacement diagram for a cam that has the following action:

Rise 2.00 in 120° with cycloidal motion
Dwell for 60°
Drop .50 in 90° with uniform motion
Drop 1.50 in 90° with uniformly accelerated and retarded motion.
Displacement diagram 2.00 high x 12.00 long.
Scale 1:1

CAM 2

Draw a displacement diagram for a cam that will raise a follower 1.25 with modified uniform motion during 90° of revolution; dwell for 60°; raise .75 with harmonic motion for 45°; drop 2.00 with parabolic motion in 120°; and dwell to the starting point.

Displacement diagram 2.00 high x 12.00 long.
Scale 1:1

Fig. 22-1-B Cam displacement problem.

UNIT 22-2

PLATE CAMS

In preparing cam drawings, the radial ordinates should be laid out in the opposite direction to that in which the cam rotates.

In drawing plate cams, the prime circle is constructed first. This circle represents the face of a flat follower or the center line of a roller follower, whichever is used, in its lowest position. It also represents the base line on the motion diagram.

One of the simplest cams to produce is the eccentric-plate cam, as illustrated in Fig. 22-2-1. The shape of the cam is a perfect circle, and the offset distance for the cam shaft is equal to one-half the follower displacement. Radial lines are marked off on the cam drawing, with the center of the shaft as center. The distance between the prime circle and the center of the roller follower on the radial lines is transposed to the displacement diagram. The length of the displacement diagram can be any convenient figure, but often the circumference of the prime circle is chosen in order to keep it in the same scale as the follower displacement height. Since this type of cam does not provide a dwell period, it has limited applications.

Since most cams combine motions and dwells in their design, the drawing sequence is different than that for eccentric cams. The cam displacement diagram is drawn first. The ordinate lines constructed on the motion or displacement diagram are drawn on the cam drawing as radial lines, and the corresponding distances from the base line to the motion curve are transposed to the cam drawing, locating the path of the follower. With cams using a roller follower, the roller diameter is then drawn in several positions along the path of the follower in order to construct the profile of the cam face. When the follower has a flat surface, the paths of the follower and the cam face are one.

Figure 22-2-2 shows a plate cam which produces a simple harmonic displacement with less than 360° of rotation of the cam. The ordinates for the cam pitch curve are constructed as shown in Fig. 22-1-8*d*. It may be impossible to use a flat follower since the harmonic pitch curve usually has a reentrant or reversing curve and a flat follower would just bridge the hollow part. Since a roller follower is the most practical and reliable

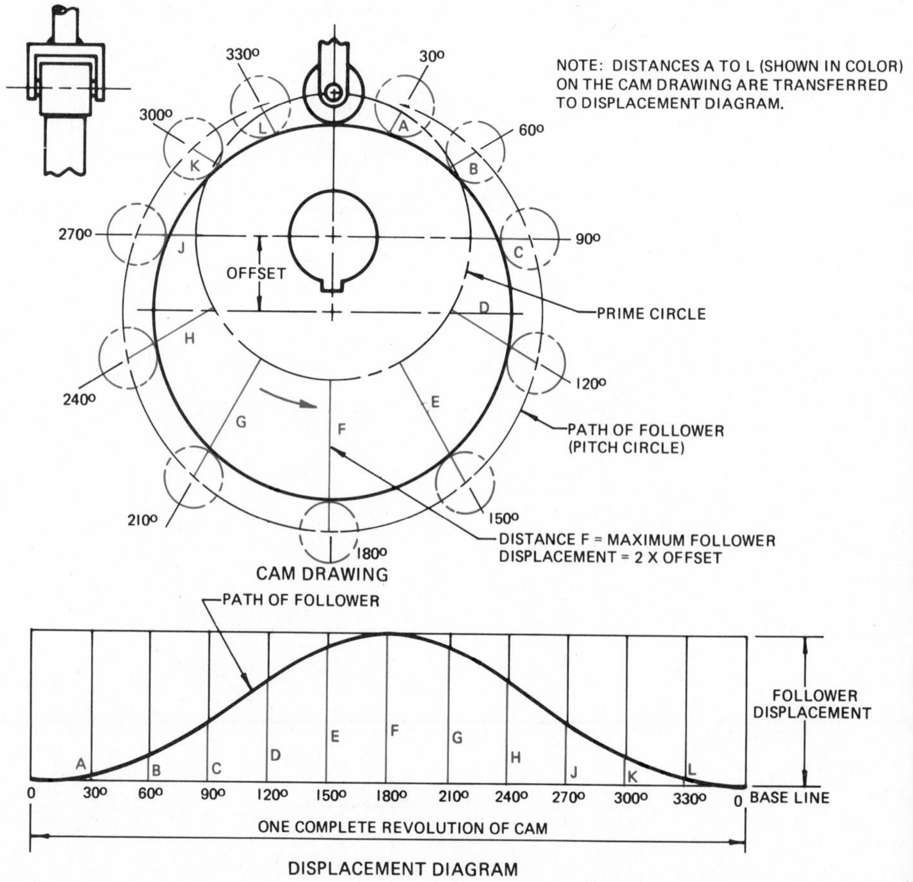

NOTE: DISTANCES A TO L (SHOWN IN COLOR) ON THE CAM DRAWING ARE TRANSFERRED TO DISPLACEMENT DIAGRAM.
330°
30°
300°
60°
270°
90°
OFFSET
PRIME CIRCLE
120°
240°
PATH OF FOLLOWER (PITCH CIRCLE)
210°
150°
180°
DISTANCE F = MAXIMUM FOLLOWER DISPLACEMENT = 2 X OFFSET
A
B
C
D
E
F
G
H
J
K
L
CAM DRAWING
PATH OF FOLLOWER
FOLLOWER DISPLACEMENT
0
30°
60°
90°
120°
150°
180°
210°
240°
270°
300°
330°
0
BASE LINE
ONE COMPLETE REVOLUTION OF CAM
DISPLACEMENT DIAGRAM

Fig. 22-2-1 Eccentric plate cam.

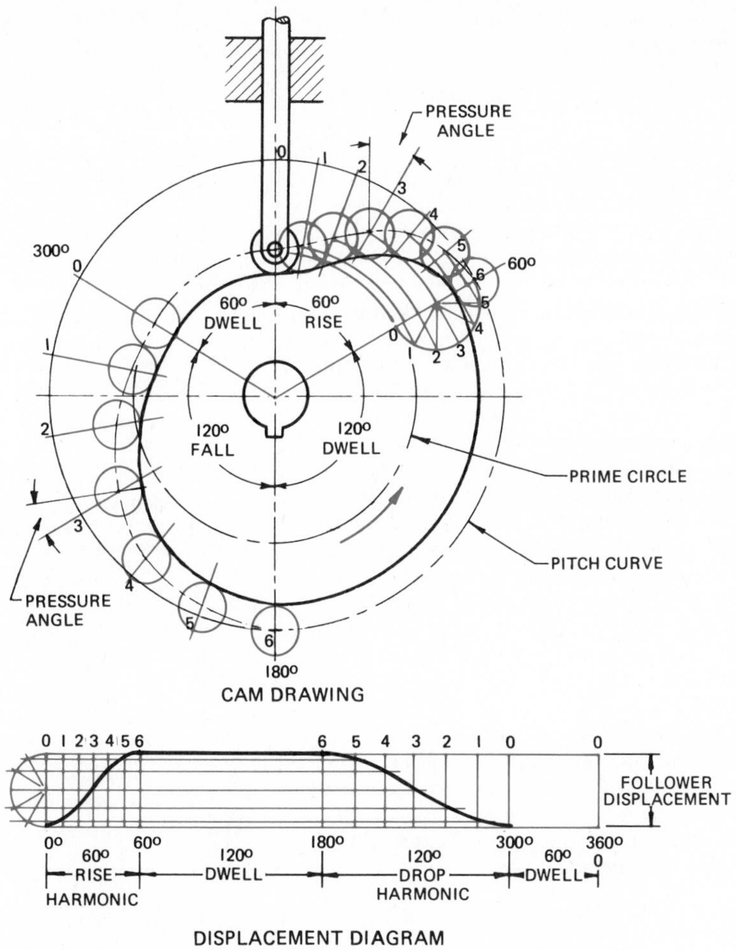
PRESSURE ANGLE
300°
60°
60° DWELL
60° RISE
120° FALL
120° DWELL
PRIME CIRCLE
PITCH CURVE
PRESSURE ANGLE
180°
CAM DRAWING
0 1 2 3 4 5 6
6 5 4 3 2 1 0
FOLLOWER DISPLACEMENT
0° 60° 120° 180° 300° 360°
60° RISE
120° DWELL
120° DROP
60° DWELL
HARMONIC
HARMONIC
DISPLACEMENT DIAGRAM

Fig. 22-2-2 Simple plate cam with harmonic motion.

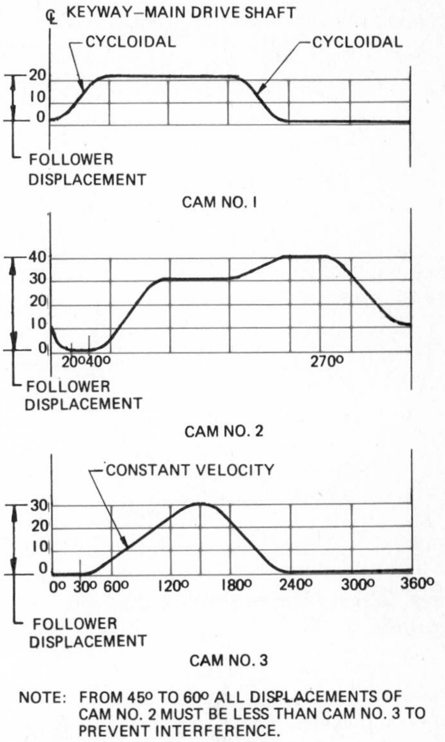
℄ KEYWAY–MAIN DRIVE SHAFT
CYCLOIDAL
CYCLOIDAL
20
10
0
FOLLOWER DISPLACEMENT
CAM NO. 1
40
30
20
10
0
20° 40°
270°
FOLLOWER DISPLACEMENT
CAM NO. 2
CONSTANT VELOCITY
30
20
10
0
0° 30° 60° 120° 180° 240° 300° 360°
FOLLOWER DISPLACEMENT
CAM NO. 3
NOTE: FROM 45° TO 60° ALL DISPLACEMENTS OF CAM NO. 2 MUST BE LESS THAN CAM NO. 3 TO PREVENT INTERFERENCE.

Fig. 22-2-4 Timing diagram.

type, the development of the cam profile with this type follower is shown.

Conjugate Cams

Conjugate cams are used when a desired motion cannot be obtained by a single cam (Fig. 22-2-3). Many indexing mechanisms use conjugate cams to obtain the necessary indexing. A displacement diagram is required for each cam.

Timing Diagrams

A convenient method of relating the movement of various machine members which are activated by cams is by the use of a timing diagram. Figure 22-2-4 shows the timing relationship for three cams. If displacements are plotted to scale, the diagram can be used for checking interferences. It can also be used for specifying the various types of transitions. If zero displacement is used to denote the prime circle radius, the timing diagram can be used by most manufacturers to produce finished cam data. The only additional data required would be a detailed drawing of the cam blank.

Dimensioning Cams[1]

The old method of developing cam contours on the drawing board by layout has definitely been outdated. In the past, a detailed cam was developed from an enlarged layout, using swung arcs and straight lines. One of the hardest tasks for a manufacturer is to make a quality cam that has been developed by this method.

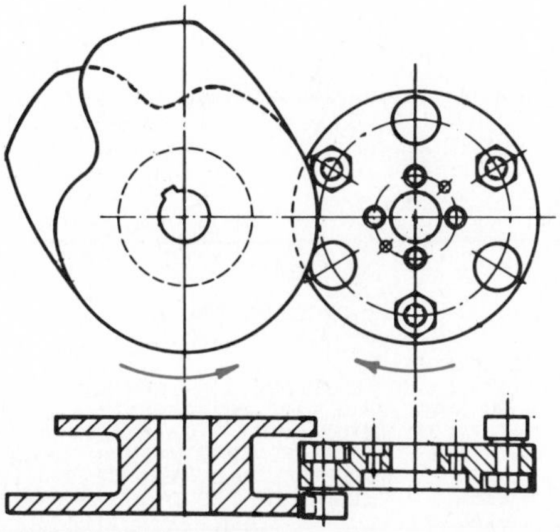

Fig. 22-2-3 Conjugate cam.

To produce a master cam or a single cam, a table of cam radii with corresponding cam angles must be supplied. The cam is then cut on a milling machine, or some other suitable machine tool, by point settings. The result is a surface with a series of ridges which must be filed down to a smooth profile. The cam radius, cutting radius, and frequency of machine setting determine the extent of filing and the final accuracy of the profile. For accurate master cams, settings must often be in 0.5° increments, calculated to seconds. The preparation of this table may require the solution of six or eight equations for each of these machine settings.

If a cam has been developed by layouts and has a profile as shown in Fig. 22-2-5, it may appear that the easiest way to describe this contour is to scale the angle R_1 from 0° and scale the displacement **D** from the center of the cam to the profile surface. Admittedly, this method would define the surface or cam profile.

However, the only method of manufacturing that could be used is to broach the cam with a very small point cutter. If a cutter radius is added to the displacement value and a cut is made, the adjacent contour is undercut. To properly cut the point *A* and maintain the

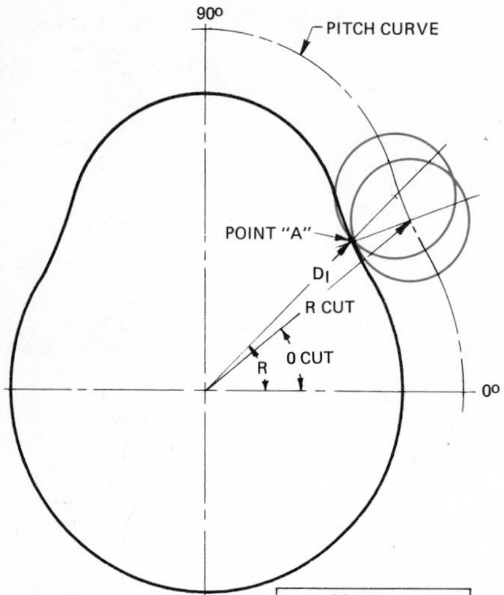
90°
PITCH CURVE
POINT "A"
DI
R CUT
R
0 CUT
0°

POINT "A"	
ANGLE	DISP
48°	82

Fig. 22-2-5 Dimensioning the cam profile.

contour, a new set of coordinates must be established. These are shown as R_{cut} and Θ_{cut}. To produce such data becomes a laborious and expensive task.

In describing a profile, *always* dimension to the pitch curve produced by the follower center. This holds true whether the cam is developed by layout or analytically. The actual follower location requires two physical dimensions: radial displacement and angular displacement. *Radial displacement* is expressed as a distance from the cam center or as a displacement from the prime circle. *Angular displacement* is measured in degrees from some zero reference, such as a keyway, dowel hole, or a timing hole, as shown in Fig. 22-2-6.

The easiest method of presenting these data is in tabular form, rather than dimensioning the detailed cam. Data should be given in at least 1° increments, although 0.5° increments are preferred. Increments of 0.5° allow the manufacturer to use discretion in selecting the intervals required to produce the finished cam.

Standard practice on tolerancing polar (angular) data is to hold the angles basic (zero tolerance) and to place all tolerance on the displacement value. The only angle that is toleranced is the one that relates the zero reference to some other point on the cam, such as a keyway or dowel hole. Figure 22-2-7 shows this method, plus one of the ways in which to tolerance the pitch curve contour.

Figure 22-2-8 is an alternate method of establishing the pitch-curve tolerances. Both these examples contain three fundamental specifications:

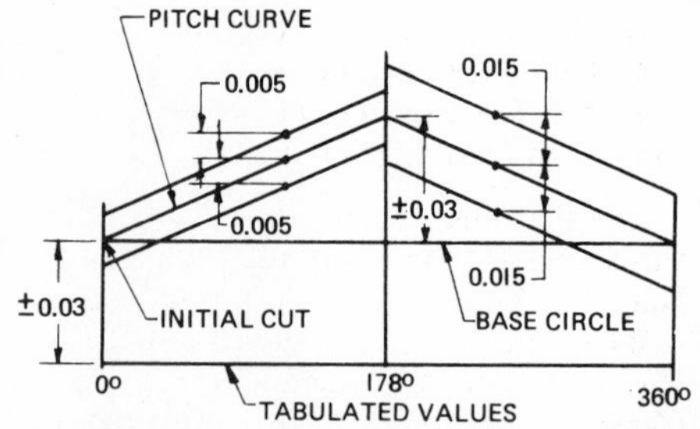
PITCH CURVE
0.005
0.015
0.005
±0.03
0.015
±0.03
INITIAL CUT
BASE CIRCLE
0°
178°
TABULATED VALUES
360°

Variations from the smooth curve representing the tabulated values shall not exceed the limits as follows:

A – Tabulated values are to the center of a 6.35 ± 0.005 diameter cutter.

B – The base circle, as established by the initial cut, shall establish a reference. This initial cut shall not vary by more than 0.03 from the tabulated values.

C – The values of all errors, in the interval from 0° through 178°, shall lie within bands allowing either cyclic or random variations not exceeding ± 0.005 from the (straight) center line of the band. The center line shall start at the initial cut point and shall end at 178° within 0.03 from the initial cut point.

D – In the interval from 178° through 360°, all errors shall lie within bands allowing either cyclic or random variations not exceeding 0.015 from the (straight) center line of the band. The center line of this interval shall start at the actual end point of the first interval center line at 178° and end at the initial cut position at 360°.

Fig. 22-2-8 Alternate method of establishing pitch-curve tolerances.

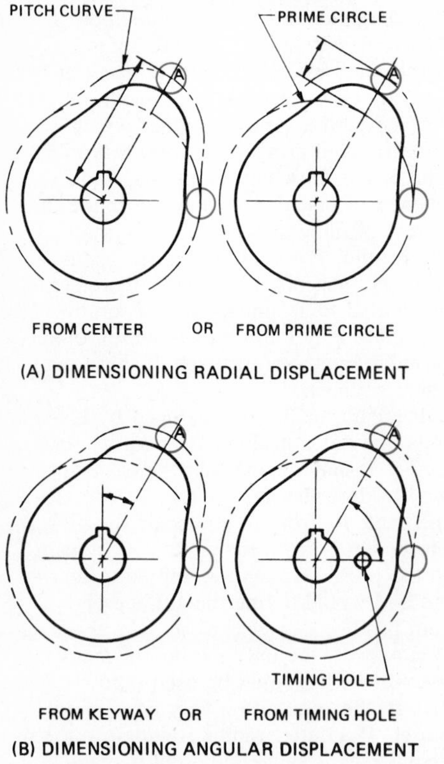
PITCH CURVE
PRIME CIRCLE
A
FROM CENTER
OR
FROM PRIME CIRCLE
(A) DIMENSIONING RADIAL DISPLACEMENT
TIMING HOLE
FROM KEYWAY
OR
FROM TIMING HOLE
(B) DIMENSIONING ANGULAR DISPLACEMENT

Fig. 22-2-6 Dimensioning point "A" on the cam profile.

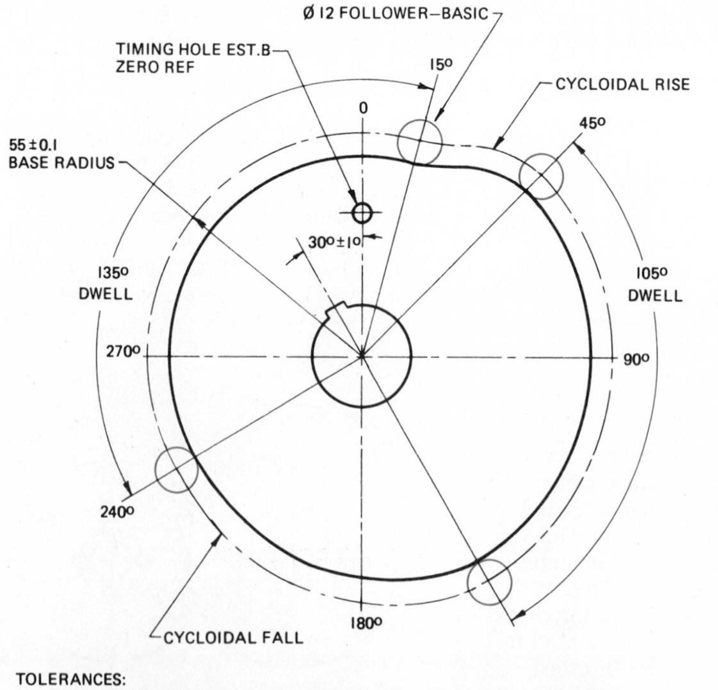
Ø 12 FOLLOWER–BASIC
TIMING HOLE EST.B
ZERO REF
15°
CYCLOIDAL RISE
0
45°
55 ±0.1
BASE RADIUS
30° ± 1°
135°
DWELL
105°
DWELL
270°
90°
240°
180°
CYCLOIDAL FALL

TOLERANCES:
1–TABULATED VALUES ARE TO THE CENTER OF A 12 DIAMETER FOLLOWER.
2– ALL ANGLES ARE BASIC AND IN RELATION TO TIMING HOLE.
3–TOTAL TRANSITION DWELL TO DWELL MAY VARY ±0.05 FROM TABULATED VALUES.
4–THE RELATIVE DIFFERENCE BETWEEN ANY TWO ADJACENT DEGREES MUST NOT EXCEED 0.01 FROM THE DIFFERENCE ESTABLISHED BY THE TABULATED DATA.
5–DWELLS TO BE CONCENTRIC TO ID OF CAM WITHIN 0.02 T.I.R. (TOTAL (INDICATOR) READING).

Fig. 22-2-7 Tolerancing polar data.

1. Tolerance on basic cam size
2. Tolerance on total transition
3. Tolerance on the pitch curve over some increment of cam angle

It is the third specification which ensures smooth continuity of the cam surface.

By far the simplest method of describing the contour is by denoting the type of transition. In this case, the type of dynamic curve which is chosen is called out on the detailed print, such as cycloidal, harmonic, modified-sine, etc. Most cam manufacturers are capable of producing their own incremental data. In most cases, the charge for this service is nominal. Figure 22-2-9 is an illustration of this type of cam detail drawing.

Figure 22-2-10 illustrates a detailed cam drawing used by cam manufacturers. The radial and displacement dimensions for the motions are shown in tabulated form.

Cam Size[2]

Cam size depends primarily on three factors: the pressure angle, the curvature of profile, and the camshaft size. Secondary factors which affect size and design are cam-follower stresses, available cam material, and available space.

It is always desirable to design the smallest possible cam consistent with available material, manufacturing accuracy, and overall economics. Judgment on the part of the designer is required because it is obviously impossible to cover every possible situation that may develop in the course of a machine design.

If a layout is made, such as shown in Fig. 22-2-11, it becomes obvious that the maximum pressure angle for a given cam and follower displacement becomes smaller as the cam-pitch circle becomes larger. It is advisable to limit this maximum pressure angle to 30 or 35°, even though with ball- or roller-bearing followers and slides higher angles have been successfully used. A large pressure angle puts high side loads on the follower linkage, increases operating torque, and increases wear.

Figure 22-2-11 also shows how cam curvature is related to cam size. For a given displacement **H**, cam rotation **B**, and roller radius r, the larger cam with pitch-curve radius Rp_2 has a much easier curve to manufacture. Note how the smaller cam with radius Rp_1 has much smaller radii of curvature near the high end of the displacement. Such a curve requires much more precise manufacturing methods. A given amount of error has a much greater effect on the follower displacement of the smaller cam than the

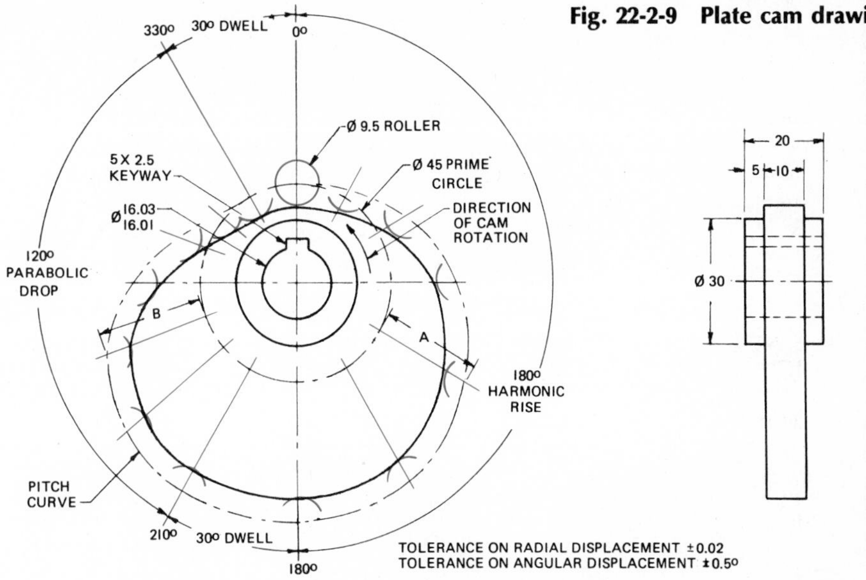
330° 30° DWELL 0°
Ø 9.5 ROLLER
5 X 2.5 KEYWAY
Ø 45 PRIME CIRCLE
Ø 16.03 16.01
DIRECTION OF CAM ROTATION
120° PARABOLIC DROP
B
A
180° HARMONIC RISE
PITCH CURVE
210° 30° DWELL
180°
20
5
10
Ø 30
TOLERANCE ON RADIAL DISPLACEMENT ±0.02
TOLERANCE ON ANGULAR DISPLACEMENT ±0.5°

ANGULAR DISPLACEMENT	RADIAL DISPLACEMENT FROM PRIME CIRCLE
0°	0
180°	31.75
210°	31.75
330°	0

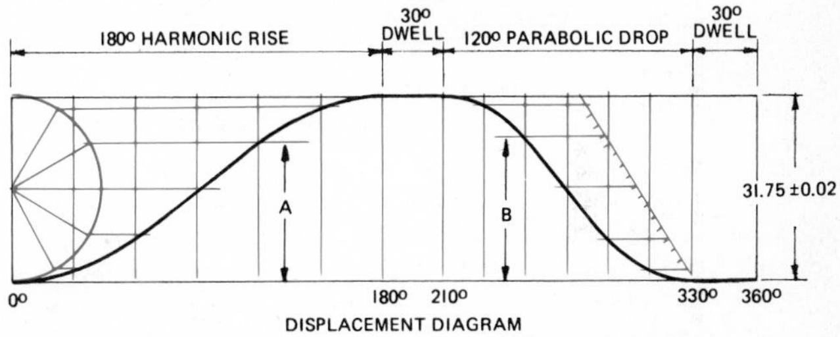
180° HARMONIC RISE
30° DWELL
120° PARABOLIC DROP
30° DWELL
A
B
31.75 ±0.02
0° 180° 210° 330° 360°
DISPLACEMENT DIAGRAM

NOTE: ANGULAR AND RADIAL DISPLACEMENT DIMENSIONS FOR MOTIONS SUPPLIED BY CAM MANUFACTURER.

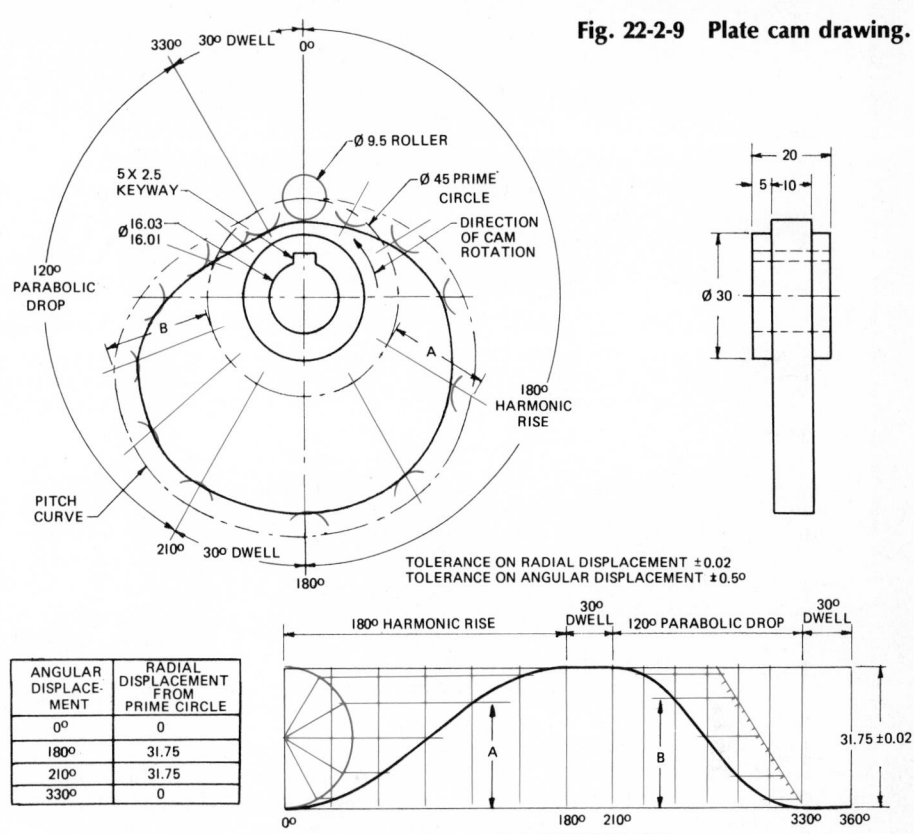

Fig. 22-2-9 Plate cam drawing.

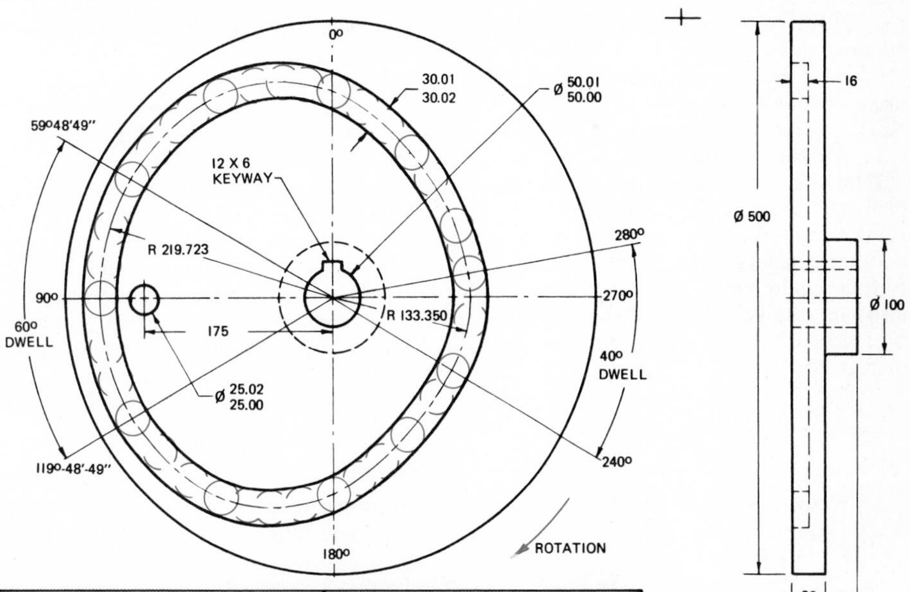
0°
30.01 30.02
Ø 50.01 50.00
59°48'49"
12 X 6 KEYWAY
R 219.723
280°
90°
270°
R 133.350
60° DWELL
175
40° DWELL
Ø 25.02 25.00
119°-48'-49"
240°
180°
ROTATION
16
Ø 500
Ø 100
30
60

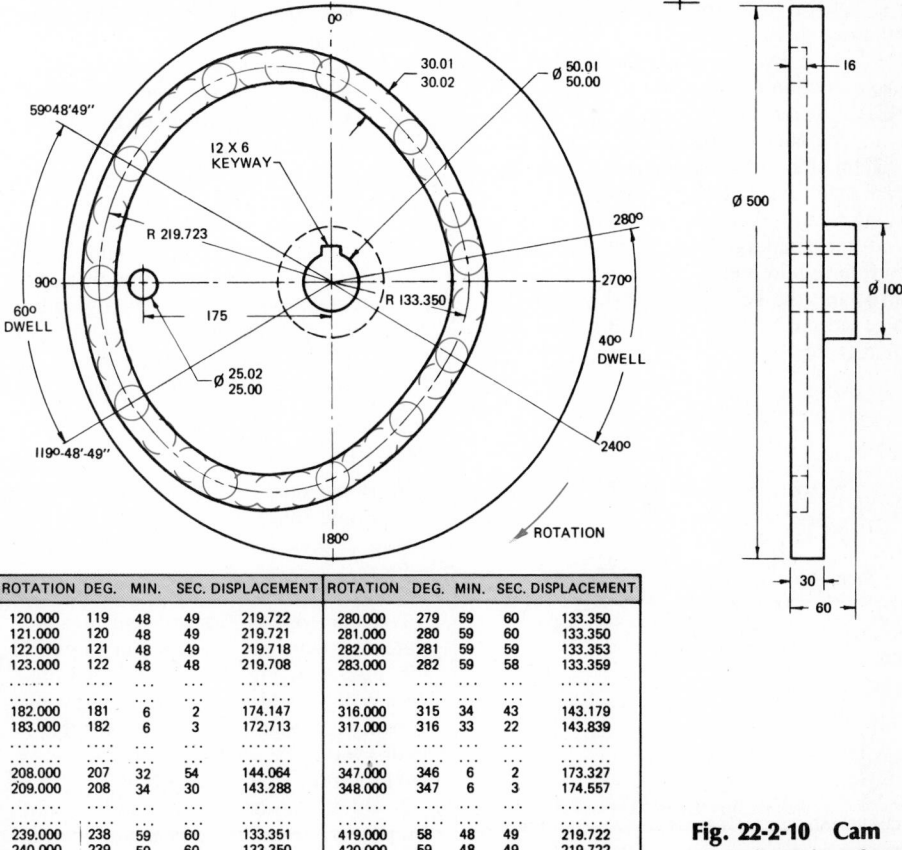

ROTATION	DEG.	MIN.	SEC.	DISPLACEMENT	ROTATION	DEG.	MIN.	SEC.	DISPLACEMENT
120.000	119	48	49	219.722	280.000	279	59	60	133.350
121.000	120	48	49	219.721	281.000	280	59	60	133.350
122.000	121	48	49	219.718	282.000	281	59	59	133.353
123.000	122	48	48	219.708	283.000	282	59	58	133.359
....	....	...	...	....	....	....	...	...	....
....	....	...	...	....	....	....	...	...	....
182.000	181	6	2	174.147	316.000	315	34	43	143.179
183.000	182	6	3	172.713	317.000	316	33	22	143.839
....	....	...	...	....	....	....	...	...	....
....	....	...	...	....	....	....	...	...	....
208.000	207	32	54	144.064	347.000	346	6	2	173.327
209.000	208	34	30	143.288	348.000	347	6	3	174.557
....	....	...	...	....	....	....	...	...	....
....	....	...	...	....	....	....	...	...	....
239.000	238	59	60	133.351	419.000	58	48	49	219.722
240.000	239	59	60	133.350	420.000	59	48	49	219.722

Fig. 22-2-10 Cam manufacturing data.

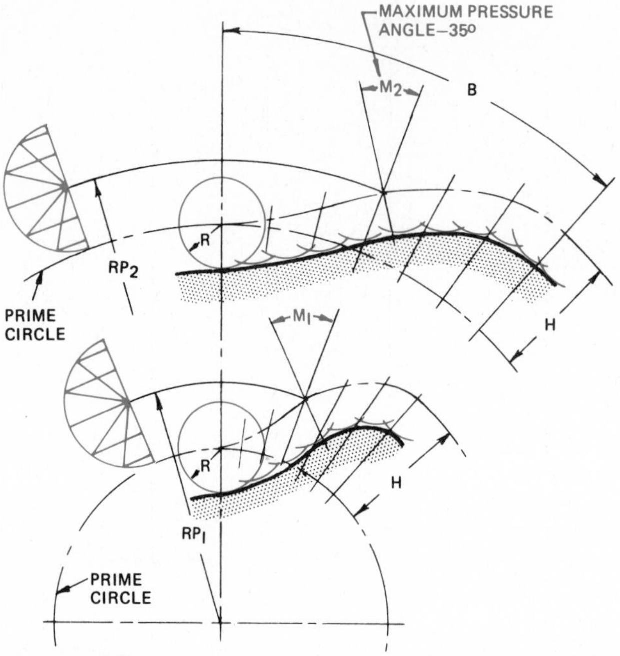
MAXIMUM PRESSURE
ANGLE—35°
M2
B
R
RP2
PRIME
CIRCLE
H
M1
R
H
RP1
PRIME
CIRCLE

Fig. 22-2-11 Increasing the cam size decreases the pressure angle.

larger one. Note how the points of tangency of the cam roll and the cam profile of the small cam are closely bunched at the upper end.

Figure 22-2-11 also shows the effect of roller diameter on the shape and accuracy of the cam profile. Always use the smallest possible roller consistent with the load it has to carry. Small rollers produce greater stress for a given load, so the designer must compromise with cam size and roller size.

Cam shaft size is a limiting factor on cam size, particularly if it is a shaft carrying a multiplicity of cams and gears and must be large because of the total torque requirement. The cam hub can also limit the base circle of the cam profile, although in face cams this need not be a factor.

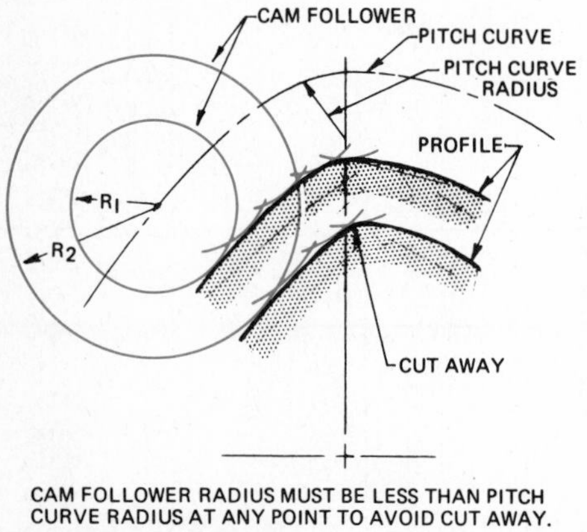
CAM FOLLOWER
PITCH CURVE
PITCH CURVE
RADIUS
PROFILE
R1
R2
CUT AWAY
CAM FOLLOWER RADIUS MUST BE LESS THAN PITCH
CURVE RADIUS AT ANY POINT TO AVOID CUT AWAY.

Fig. 22-2-12 Factors affecting cam size.

High-speed machinery is affected by unbalanced rotating components, and cams are a particularly difficult component to balance. Face cams can generally be balanced quite well. Care should always be taken to make the cams as small as possible, consistent with all the other requirements.

Another factor affecting cam size is the cutting away of a previously generated cam profile by virtue of too large a cam roll. This is illustrated in Fig. 22-2-12. Basically, the cam-follower-roll radius must be less than the pitch-curve radii at any point along the pitch curve.

Fortunately, modern machining methods make possible extremely accurate profiles, even on small cams, and it becomes the designer's responsibility to make the very best compromise with the specifications and materials at hand.

REFERENCES AND SOURCE MATERIAL

1. Eonic Inc.
2. Eonic Inc.

Assignments

1. On an A3- or B-size sheet, design a plate cam from the information shown in Fig. 22-2-A or Fig. 22-2-B. Add a suitable keyway. Add a chart to the drawing showing the angular and radial displacement every 15°, taking measurements from the prime circle. Scale is 1:1.

2. On an A3- or B-size sheet, lay out the parallel-drive indexing unit shown in Fig. 22-2-C or Fig. 22-2-D with the timing hole rotated to position *B*. Use your judgment for dimensions not shown. The angular and radial displacement values locate the center of the roller. Scale is 1:2.

REVIEW FOR ASSIGNMENTS

Unit 4-1 Dividing a Line into Equal Parts
Unit 9-1 Keys
Unit 22-1 Cam Motions

CAM DATA

Make a drawing of a plate cam that will raise a ϕ10 mm roller follower 40 mm at a uniformly accelerated and retarded motion in 150°, dwell 45°, drop with modified uniform motion in 120°, and dwell for the remainder. Prime circle = 70 mm, plate thickness = 10 mm, shaft = ϕ26, hub = ϕ44 x 32 long.

ANGULAR DISPLACEMENT	RADIAL DISPLACEMENT FROM PRIME CIRCLE
0°	
15°	
30°	
45°	
60°	
75°	
90°	
105°	
120°	
165°	
180°	
195°	
210°	
225°	
240°	
255°	
270°	
285°	
300°	
315°	
330°	
345°	
360°	

Fig. 22-2-A Plate cam.

CAM DATA

Make a drawing of a plate cam that will raise a ϕ.38 in. roller follower 1.50 in. at a uniformly accelerated and retarded motion in 150°, dwell 45°, drop with modified uniform motion in 120°, and dwell for the remainder. Prime circle = 2.75 in. plate, thickness = .38 in. shaft = ϕ1.00, hub = ϕ1.75 x 1.25 long.

Fig. 22-2-B Plate cam.

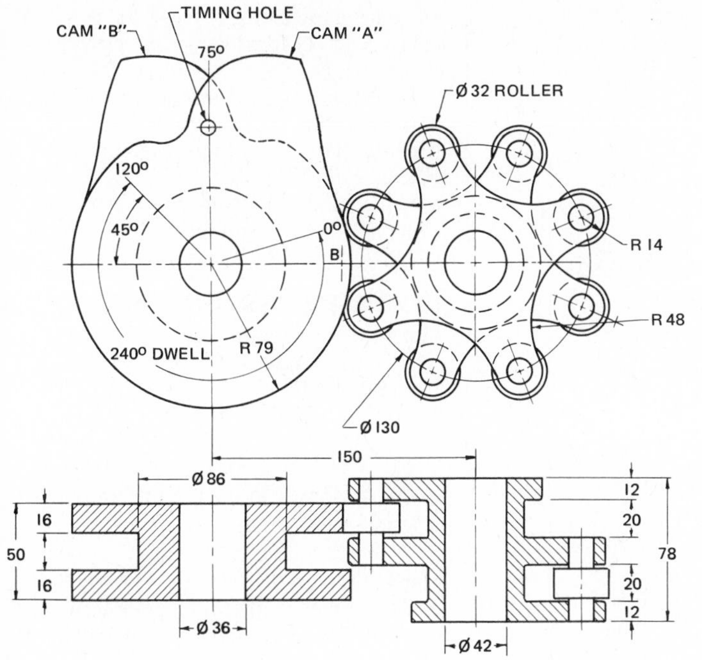
TIMING HOLE
CAM "B"
CAM "A"
75°
Ø 32 ROLLER
120°
45°
0°
B
R 14
R 48
240° DWELL
R 79
Ø 130
150
Ø 86
16
50
16
12
20
78
20
12
Ø 36
Ø 42

ANGULAR DISPLACEMENT AND RADIAL DISPLACEMENT FOR CAM "A". CAM "B" IS OPPOSITE HAND TO CAM "A". ANGULAR DIMENSIONS SHOWN ON DRAWING ARE FOR CAM "A".									
0°	94.9	25°	103.6	50°	139.9	75°	123.4	95°	88.1
5°	95.2	30°	109.7	55°	138.6	80°	116.5	100°	91.6
10°	95.5	35°	117.8	60°	136.3	85°	108.2	105°	93.4
15°	96.5	40°	125.7	65°	133.0	90°	96.0	110°	94.7
20°	99.0	45°	133.0	70°	128.5	93°	86.6	115° AND 120°	94.9

Fig. 22-2-C Parallel-drive indexing unit.

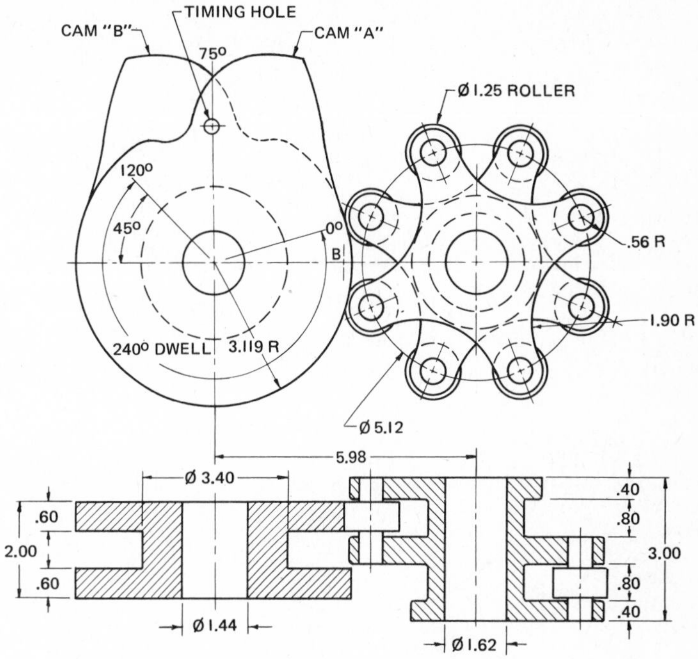
TIMING HOLE
CAM "B"
CAM "A"
75°
Ø 1.25 ROLLER
120°
45°
0°
B
.56 R
1.90 R
240° DWELL
3.119 R
Ø 5.12
5.98
Ø 3.40
.60
2.00
.60
.40
.80
3.00
.80
.40
Ø 1.44
Ø 1.62

ANGULAR DISPLACEMENT AND RADIAL DISPLACEMENT FOR CAM "A" CAM "B" IS OPPOSITE HAND TO CAM "A". ANGULAR DIMENSIONS SHOWN ON DRAWING ARE FOR CAM "A".									
0°	3.74	25°	4.08	50°	5.51	75°	4.86	95°	3.47
5°	3.75	30°	4.32	55°	5.46	80°	4.59	100°	3.61
10°	3.76	35°	4.64	60°	5.37	85°	4.26	105°	3.68
15°	3.80	40°	4.95	65°	5.24	90°	3.78	110°	3.73
20°	3.90	45°	5.24	70°	5.06	93°	3.41	115° AND 120°	3.74

Fig. 22-2-D Parallel-drive indexing unit.

UNIT 22-3

POSITIVE-MOTION CAMS

To ensure positive motion of the follower in both directions, positive-motion cams are employed. Two types are *face cams* and *cams with yoke-type followers*. Face cams are similar to plate cams, except the follower engages a groove on the face of the cam rather than on the outside edge of the cam. One disadvantage to this type of cam is that the outer edge of the cam groove tends to rotate the roller in the opposite direction to that of the inner edge, resulting in wear in both the cam and the roller. However, this is not serious at slow speeds. Yoke-type followers are used for operating light mechanisms. The follower surface is flat or tangent to the curvature of the cam. With this type of cam, only one-half of the cam displacement diagram need be drawn since the other half of the cam is identical to the first half. See Figs. 22-3-1 and 22-3-2.

Assignments

1. On an A3- or B-size sheet, make a two-view drawing of a face cam from the information given in Fig. 22-3-A or Fig. 22-3-B. Add a suitable keyway. Prepare a chart showing angular and radial dis-

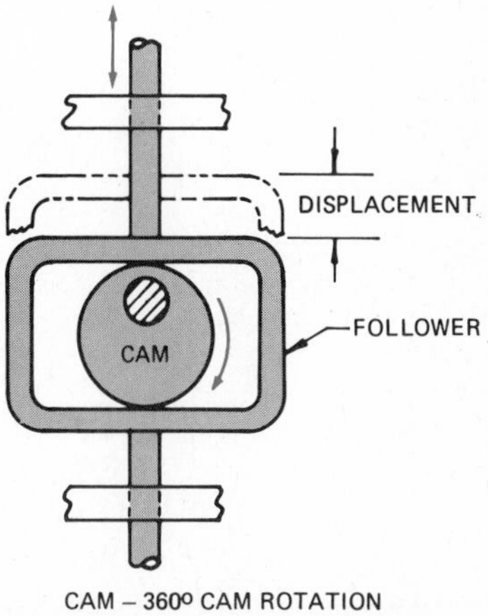
DISPLACEMENT
FOLLOWER
CAM
CAM – 360° CAM ROTATION

Fig. 22-3-1 Face cam drawing.

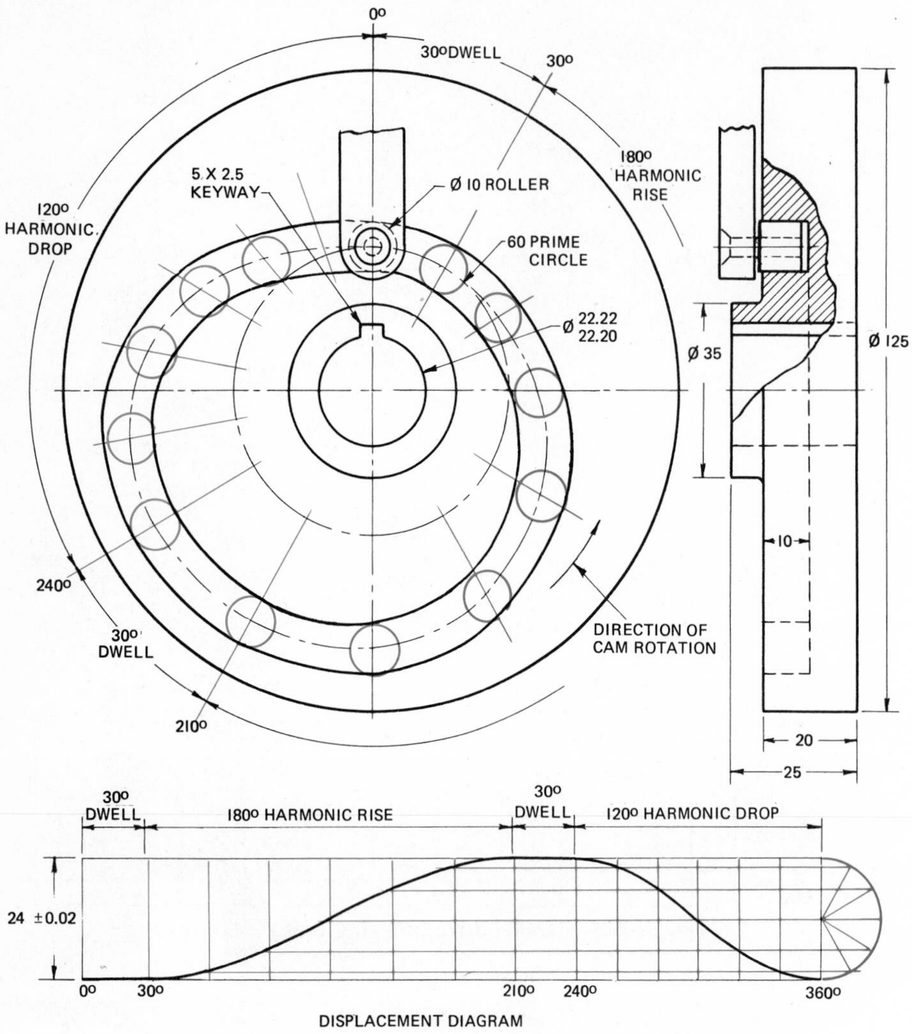
0°
30°DWELL
30°
180° HARMONIC RISE
5 X 2.5 KEYWAY
Ø 10 ROLLER
120° HARMONIC DROP
60 PRIME CIRCLE
Ø 22.22 22.20
Ø 35
Ø 125
10
240°
30° DWELL
DIRECTION OF CAM ROTATION
210°
20
25
30° DWELL
180° HARMONIC RISE
30° DWELL
120° HARMONIC DROP
24 ±0.02
0°
30°
210°
240°
360°
DISPLACEMENT DIAGRAM

TOLERANCE ON RADIAL DISPLACEMENT ±0.02

TOLERANCE ON ANGULAR DISPLACEMENT ±0.5°

NOTE: ANGULAR AND RADIAL DISPLACEMENT DIMENSIONS FOR MOTIONS SUPPLIED BY CAM MANUFACTURER

ANGULAR DISPLACEMENT FROM KEYWAY	RADIAL DISPLACEMENT FROM PRIME CIRCLE
0°	0
30°	0
210°	24
240°	24

Fig. 22-3-2 Cam with yoke follower.

placement for every 15°, taking measurements from the prime circle. Scale is 1:1.

2. On an A3- or B-size sheet, make a two-view drawing of a yoke cam from the information shown in Fig. 22-3-C or Fig. 22-3-D. From the cam drawing make a displacement diagram plotting points every 30°. Add a suitable keyway. Scale is 1:1.

REVIEW FOR ASSIGNMENTS

Unit 4-1	Dividing a Line into Equal Parts
Unit 9-1	Keys
Unit 22-1	Cam Motions

Make a drawing of a face cam that will raise a ϕ12 mm roller 24 mm with parabolic motion in 120°, dwell 45°, drop with cycloidal motion for the remainder. The face cam is to have an outside dia. of 160 mm, cam thickness 24 mm, and a groove depth of 12 mm.

Prime circle = ϕ80

Shaft = ϕ24, hub = ϕ42 x 28 long

Fig. 22-3-A Face cam.

Make a drawing of a face cam that will raise a ϕ.44 roller .88 with parabolic motion in 120°, dwell 45°, drop with cycloidal motion for the remainder. The face cam is to have an outside dia. of 6.25, cam thickness .88, and a groove depth of .50.

Prime circle = 3.25

Shaft = ϕ.88, hub = ϕ1.62 x 1.12 long

Fig. 22-3-B Face cam.

Make a two-view drawing of a yoke cam that will raise the yoke 35 mm. The cam is an eccentric cam having a dia. of 90 and a plate thickness of 20. Shaft = ϕ28, hub = ϕ44 x 30 long having the extension on one side only. The yoke is 10 thick and has a wall width of 20. A 6.0 x 30 steel guide bar is welded to the top and bottom of the yoke.

Fig. 22-3-C Yoke cam.

Make a two-view drawing of a yoke cam that will raise the yoke 1.40. The cam is an eccentric cam having a dia. of 3.50 and a plate thickness of .75. Shaft = ϕ1.06, hub = ϕ1.75 x 1.10 long having the extension on one side only. The yoke is .40 thick and has a wall width of .75. A .25 x 1.25 steel guide bar is welded to the top and bottom of the yoke.

Fig. 22-3-D Yoke cam.

UNIT 22-4

DRUM CAMS

The layout of a drum or cylinder cam starts, as with any cam, with the decision as to what profile and follower type will be used. Many cylinder cams are used with straight in-line followers so that the follower moves in a path parallel to the axis of the cam. The pitch surface is developed and shown as a rectangle (Fig. 22-4-1*a*), and the follower displacement plotted is for rectangular coordinates.

Theoretically, a tapered follower with its cone center on the cam axis should give the best results (Fig. 22-4-1*b*). Actually, straight rollers give excellent results as long as the roller length and diameter are not too large in relation to the cam cylinder diameter. Swinging followers are used on indexing-type cylindrical cams, as shown in Fig. 22-1-5.

For drum or cylindrical groove cams, the displacement diagram is replaced by the developed surface of the cam, as shown in Fig. 22-4-2. The groove shown in the front view of the cam is found by projection.

Points from the developed surface of the cam and their corresponding points on the top view are projected to the front view, as shown by the letter *A* at position 210°.

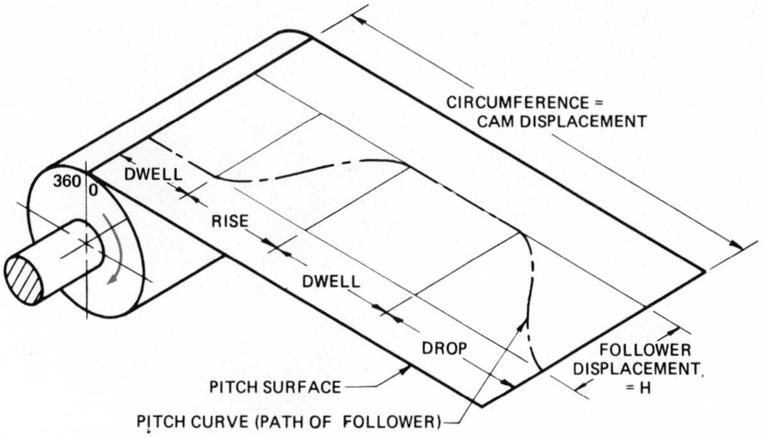

CIRCUMFERENCE =
CAM DISPLACEMENT
360
0
DWELL
RISE
DWELL
DROP
FOLLOWER
DISPLACEMENT
= H
PITCH SURFACE
PITCH CURVE (PATH OF FOLLOWER)

(A) CAM DISPLACEMENT DIAGRAM FOR DRUM CAM

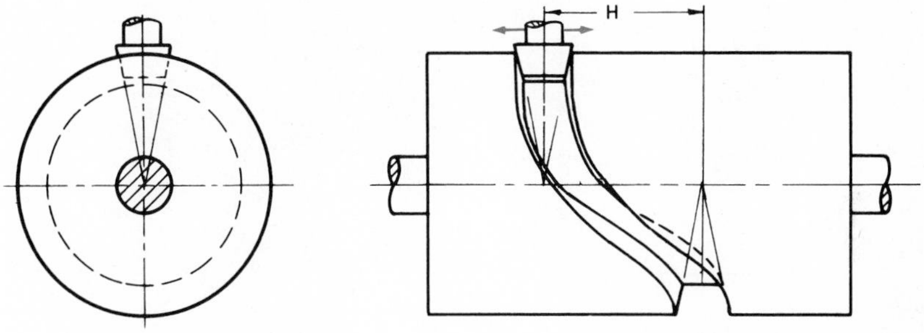

H

(B) TAPERED FOLLOWER

Fig. 22-4-1 Drum cam details.

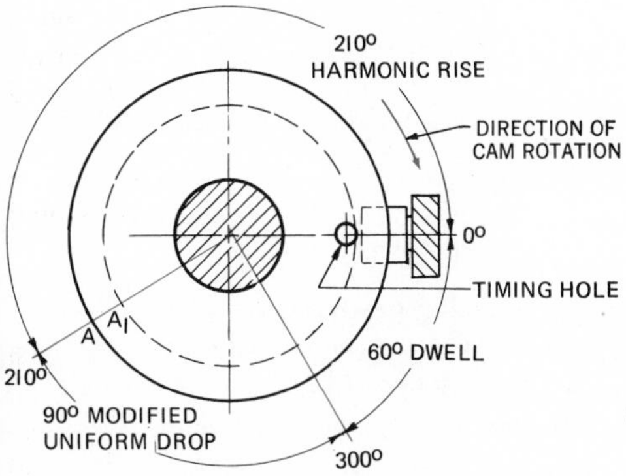

210°
HARMONIC RISE
DIRECTION OF
CAM ROTATION
0°
TIMING HOLE
60° DWELL
A
A1
210°
90° MODIFIED
UNIFORM DROP
300°

ANGULAR DISPLACEMENT FROM TIMING HOLE	DISPLACEMENT FROM BASE LINE
0°	0
210°	32
300°	0

TOLERANCE ON RADIAL DISPLACEMENT ±0.02
TOLERANCE ON ANGULAR DISPLACEMENT ±0.5°

NOTE: ANGULAR DISPLACEMENT AND DISPLACEMENT FROM BASE LINE SUPPLIED BY CAM MANUFACTURER.

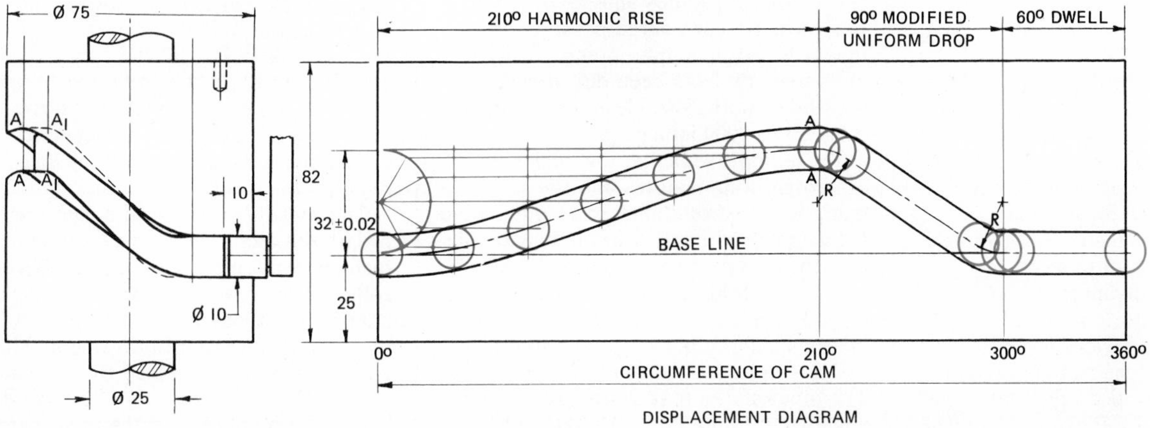

Ø 75
A
A1
A
A1
10
Ø 10
Ø 25
82
32 ±0.02
25
210° HARMONIC RISE
90° MODIFIED
UNIFORM DROP
60° DWELL
A
A
R
R
BASE LINE
0°
210°
300°
360°
CIRCUMFERENCE OF CAM
DISPLACEMENT DIAGRAM

Fig. 22-4-2 Drum cam drawing.

Assignment

On an A3- or B-size sheet, make a two-view drawing of a drum cam from the information given in Fig. 22-4-A or Fig. 22-4-B. Show the full development of the cam, which will serve as the motion diagram. Prepare a chart showing the angular and radial displacement from the base line for every 15°. Scale is 1:1.

CAM DATA

Make a drawing of a drum cam that will raise a ϕ14 roller follower 32 mm with harmonic motion in 150°, dwell 45°, drop with trapezoid motion in 120°, and dwell for the remainder. Cam = ϕ70 x 64 long, follower groove = 10 deep. Use your judgment for sizes not given.

ANGULAR DISPLACEMENT FROM TIMING HOLE	DISPLACEMENT FROM BASE LINE
0°	
15°	
30°	
60°	
315°	
360°	

Fig. 22-4-A Drum cam.

Make a drawing of a drum cam that will raise a ϕ.50 roller follower 1.25 with harmonic motion in 150°, dwell 45°, drop with trapezoid motion in 120°, and dwell for the remainder. Cam = ϕ2.75 x 2.50 long, follower groove = .40 deep. Use your judgment for sizes not given.

Fig. 22-4-B Drum cam.

(A) BARREL CAM

(B) BARREL CAM

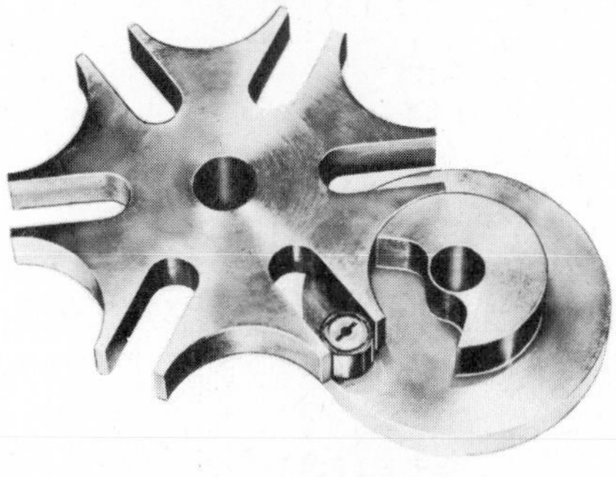

(C) 6-STATION DRIVE

(D) CONJUGATE CAMS

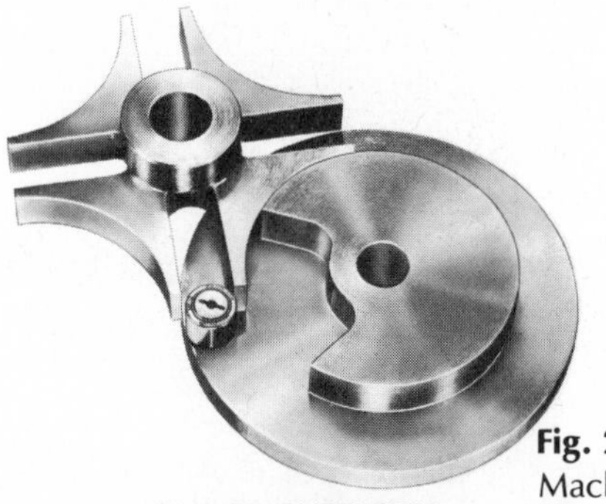

(E) 4-STATION DRIVE

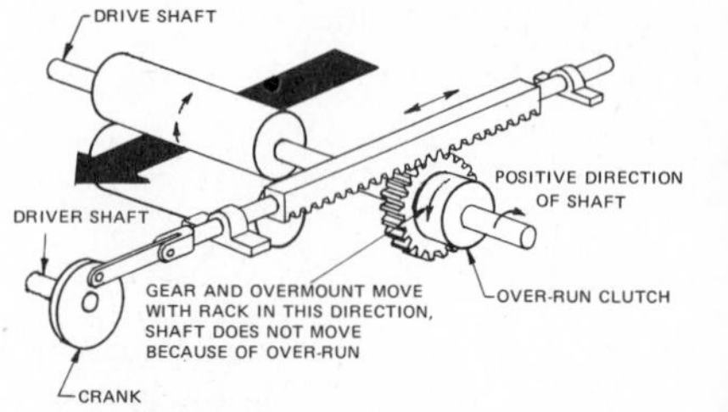
DRIVE SHAFT
DRIVER SHAFT
POSITIVE DIRECTION OF SHAFT
GEAR AND OVERMOUNT MOVE WITH RACK IN THIS DIRECTION, SHAFT DOES NOT MOVE BECAUSE OF OVER-RUN
OVER-RUN CLUTCH
CRANK

(F) OVERRUN CLUTCH, CRANK, AND GEAR AND RACK

Fig. 22-5-1 Indexing mechanisms. (A—Manifold Machinery Co. Ltd.; B—Commercial Cam & Machine Co.; C—Geneva Motions Corp.; D—Commercial Cam & Machine Co.; E—Geneva Motions Corp.)

UNIT 22-5

INDEXING[1,2]

Indexing is the conversion of a constant-speed rotary-input motion to an intermittent rotary-output motion. Press-feed tables, packaging machines, machine tools, switch gear, and feeding devices are but a few of the many machines found in industry that require indexing or intermittent motion.

In recent years, advances in the design and manufacture of positive intermittent-motion devices have improved significantly the smoothness and speed of indexing motions possible with Geneva-type drives. Geneva mechanisms currently being produced can operate at higher load capacities and speeds than was formerly considered possible or practical. For example, precision drive units recently have been developed to provide smooth, shockless operation at rates up to 2000 indices per minute.

These improvements, together with other operating features and the relatively low cost of commercial units, have led to renewed interest in the capabilities of Geneva-type drive systems.

The manifold indexing mechanism (Fig. 22-5-1*a*) consists of two basic elements: a *cam* attached to the input shaft and a *turret* attached to the output shaft. The input and output shafts are at right angles but do not lie in the same plane. The cam is of concave globoidal form with a track that engages roller followers which project radially from the edge of the turret disk. Part of the cam track is straight, so that when the roller followers engage it, no movement of the turret can take place. The angle of the cam occupied by this part is called the *dwell angle*.

The remaining part of the cam track progresses along the cam axis in helical fashion, thus rotating the turret. Before a roller leaves one end of the cam track, another roller enters the other end to maintain continuity of movement. The angle of the cam occupied by this part of the track is called the *cam index angle*. Thus, one revolution of the cam represents one indexing cycle, during which

DIMENSIONS		MODEL	A	B	C	D	DRIVER					WHEEL		
							E	F	G	H	K		M	N
4 STATIONS STATIONS TIME RATIO: 90° INDEX - 270° DWELL	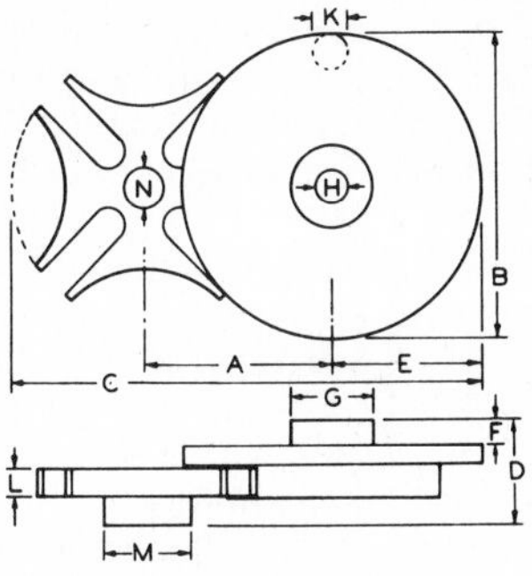	4S75	75	124	194	48	62	12	40	16	16	12	45	20
		4S90	90	150	224	48	75	12	40	16	16	12	45	20
		4S100	100	164	256	50	82	12	40	20	20	16	48	22
		4S115	115	190	287	50	95	12	40	20	20	16	48	22
		4S125	125	200	320	53	100	12	40	20	22	16	50	25
		4S140	140	224	350	53	112	12	40	20	22	16	50	25
		4S150	150	240	382	55	120	12	40	20	25	20	50	25
5 STATIONS STATIONS TIME RATIO: 108° INDEX - 252° DWELL	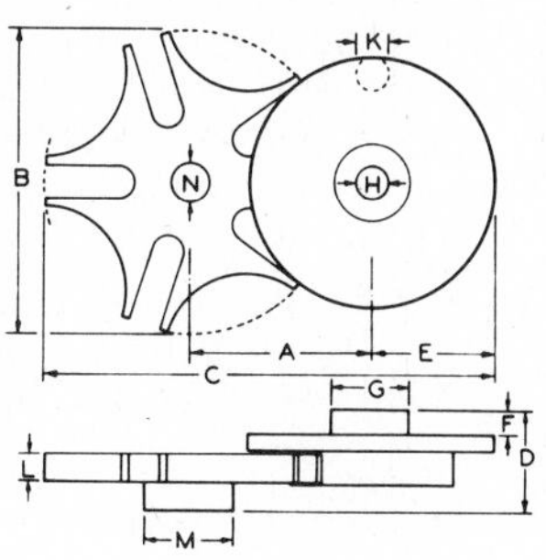	5S75	75	124	190	48	53	12	40	16	16	12	45	20
		5S90	90	150	225	48	60	12	40	16	16	12	45	20
		5S100	100	164	252	50	70	12	40	20	20	16	48	22
		5S115	115	190	287	50	77	12	40	20	20	16	48	22
		5S125	125	210	316	53	86	12	40	20	22	16	50	25
		5S140	140	234	351	53	94	12	40	20	22	16	50	25
		5S150	150	250	377	55	102	12	40	20	25	20	50	25
6 STATIONS STATIONS TIME RATIO: 120° INDEX - 240° DWELL	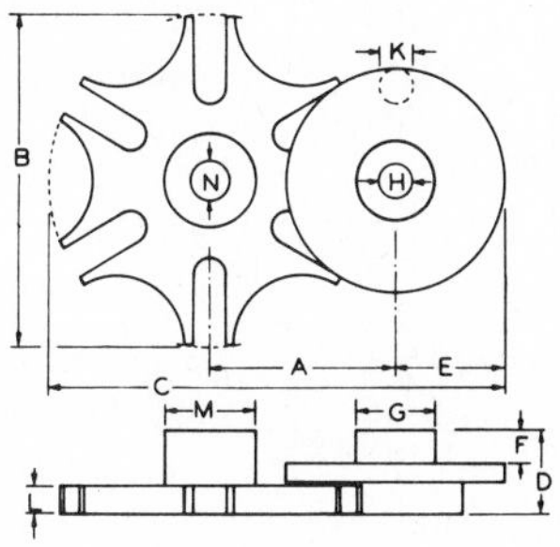	6S75	75	134	188	35	46	12	40	16	16	12	45	20
		6S90	90	160	222	35	52	12	40	16	16	12	45	20
		6S100	100	180	250	38	60	12	40	20	20	16	48	22
		6S115	115	204	274	38	67	12	40	20	20	16	48	22
		6S125	125	224	312	41	75	12	40	20	22	16	50	25
		6S140	140	250	346	41	81	12	40	20	22	16	50	25
		6S150	150	264	371	45	89	12	40	20	25	20	50	25
8 STATIONS STATIONS TIME RATIO: 135° INDEX - 225° DWELL	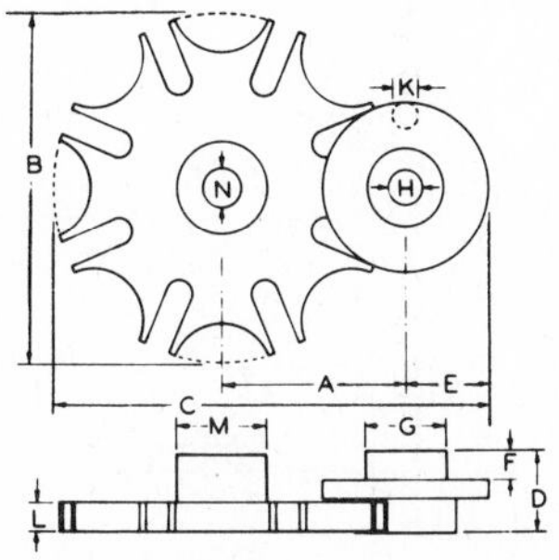	8S75	75	144	193	35	46	12	40	16	12	12	45	20
		8S90	90	170	215	35	40	12	40	16	12	12	45	20
		8S100	100	190	243	38	48	12	40	20	16	16	48	22
		8S115	115	214	274	38	52	12	40	20	16	16	48	22
		8S125	125	234	300	41	58	12	40	20	20	16	50	25
		8S140	140	270	339	41	64	12	40	20	20	16	50	25
		8S150	150	284	362	45	70	12	40	20	20	20	50	25

Fig. 22-5-2 Indexing drives.

the turret indexes from one station to the next and dwells for a specific period. The number of times that this takes place in one revolution of the turret is called the *number of stops*.

The tangent drive (Fig. 22-5-1*c* and *e*) consists of a constantly rotating driver and a driven wheel. The wheel may have four, five, six, or eight precision-machined radial slots. A matching cam follower, mounted on needle bearings on the driver, engages one of the slots on each revolution of the driver, thereby indexing the wheel. The concave section between the slots is precisely machined to mate with the locking hub of the driver to prevent movement of the wheel during dwell.

The tangent drive indexes over an angle equal to 360° divided by the number of slots or stations in its wheel. For example, each index of a four-station tangent drive is 90°; each index of a five-station drive is 72°.

The time ratio of a tangent drive is expressed by the arc (in degrees) of each revolution of the driver that the wheel is being indexed and the arc of each revolution that the wheel is at rest or dwell. The time ratio refers to each revolution of the driver and, therefore, remains constant, regardless of the driver speed. The actual speed of indexing is a function of the driver speed and is directly proportional to it.

The indexing application shown in Fig. 22-5-1*f* employs an overrun clutch and a rack and gear. The input or driver shaft is connected to a rack which converts rotary motion into reciprocating motion. The gear which is attached to an overrun clutch rotates in both directions. The overrun clutch drives the shaft in one direction but overruns or freewheels on the shaft in the other direction, producing an intermittent rotary motion.

REFERENCES AND SOURCE MATERIAL

1. Manifold Machinery Company, Limited.
2. Geneva Motions Corporation.

Assignments

1. On an A3- or B-size sheet, make a two-view drawing of the indexing drive 6S75 shown in Fig. 22-5-2. Use your judgment for dimensions not shown. Draw the cam displacement diagram, plotting points every 5° on the index cycle. Add suitable keyways. Scale is 1:1.

REVIEW FOR ASSIGNMENTS

Unit 9-1 Keys and Keyways
Unit 22-1 Cam Motions

UNIT 22-6

LINKAGES[1]

One of the ever-present problems in machine design is the mechanization of various interrelated motions. These motions are usually preassigned, often quite arbitrarily, and specify the relationships of moving parts or simply the end motion of a single part. Illustrations are present in all types of machinery. Typical examples are seen in textile machinery, packaging machinery, printing presses, valve mechanisms in steam locomotives, machine tools, automotive equipment, household articles, instruments, computing devices, and many other common mechanisms. Upon closer observation, it will be noticed that all these devices are simply combinations and arrangements of basic mechanical elements such as gear trains, cam actuators, cranks and links, sliders, bolts and pulleys, and other rotating and sliding parts. Combinations of the crank, link, and sliding elements are commonly termed *bar linkages*. These motion devices may be defined as either motion generators or function generators. *Motion generators* produce some particular motion relationships where the output may be restrained to one, two, or three degrees of freedom. Often, only the two endpoints and perhaps some intermediate conditions are specified, with the in-between motion having no particular importance. *Function generators,* on the other hand, have their entire motion specified and are restrained to one degree of freedom, either translation along a line or rotation about some axis. Mechanically there is no difference between the two types. The input-output relationship of a function generator may be specified by an explicit mathematical expression or by some tabular data, whose explicit function is not necessarily known. It is in this latter sense that the function generator will be considered, where the motion desired is not one for which a known linkage exists.

LOCUS OF A POINT

The *locus* of a point in a linkage or mechanism is the path traced out by that point as it moves according to certain controlled conditions. The study of loci is important in machine design to determine:

1. The position of various links and joints in the cycle of the mechanism
2. The relative speeds of different parts

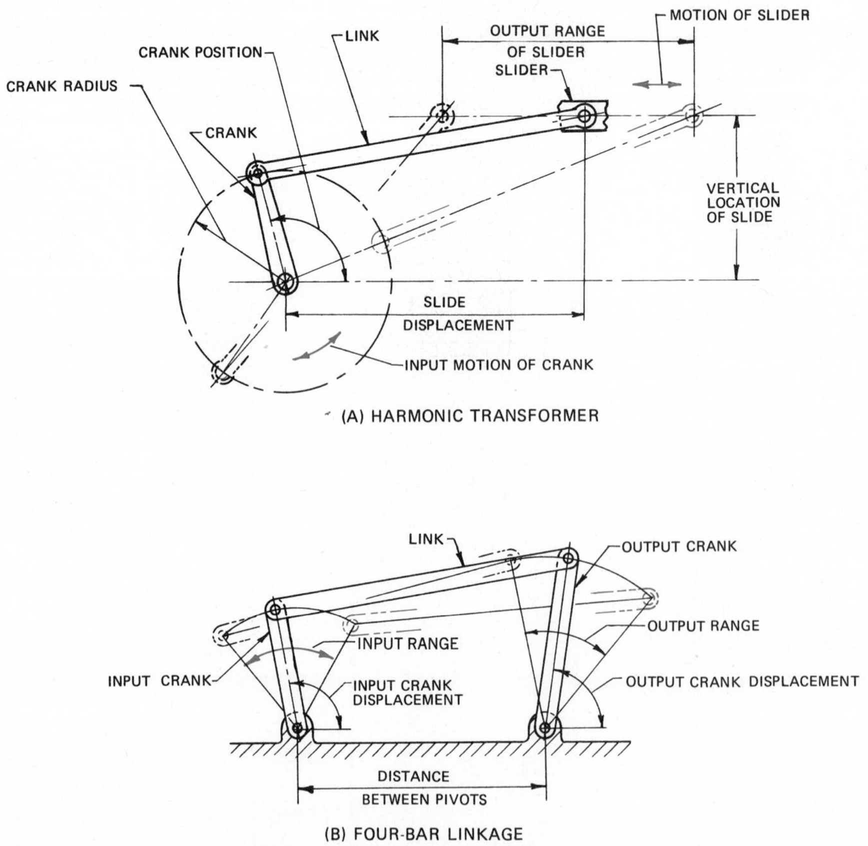
CRANK RADIUS
CRANK POSITION
LINK
OUTPUT RANGE OF SLIDER
SLIDER
MOTION OF SLIDER
CRANK
VERTICAL LOCATION OF SLIDE
SLIDE DISPLACEMENT
INPUT MOTION OF CRANK
(A) HARMONIC TRANSFORMER
LINK
OUTPUT CRANK
OUTPUT RANGE
INPUT RANGE
INPUT CRANK
INPUT CRANK DISPLACEMENT
OUTPUT CRANK DISPLACEMENT
DISTANCE BETWEEN PIVOTS
(B) FOUR-BAR LINKAGE

Fig. 22-6-1 Linkages used as function generators.

3. Forces exerted in the mechanism using applied machines in conjunction with diagrammatic layouts

The designer is often called upon to make these diagrammatic layouts to assist in the design of the container for linkages that will be of the most economical and space-saving nature, as well as to ensure that parts of adjacent linkages will not foul one another at any point of the movement of the machine. Further, a skillfully designed linkage can often accomplish a movement which would be difficult to produce using cams or gears. The designer may be required to plot a diagram or chart to show in graphical form the relative speed or displacement of different related parts of the linkage, including the input and output shafts. Applied mechanics, in conjunction with the linkage layout, can be used to construct vector diagrams to determine forces in links and pins. The latter components can be designed for strength using a knowledge of material strengths.

CAMS VERSUS LINKAGES[2]

The most well-known solution for a function generator is the cam: *flat cams* for functions of single variables and *barrel cams* for functions of two variables. Aside from the kinematics, cam design is very straightforward and consists simply of assigning coordinates corresponding to the specified functions. Perhaps the least known solution is the bar linkage. Both devices have their advantages and limitations.

Cams in general require less design effort and are able to mechanize discontinuous or jagged-shaped functions. Essentially, an infinite number of dimensions determine a cam profile.

As computing devices, linkage mechanisms enjoy a number of advantages over cams, with the one exception that the functions must be continuous. *Linkages* are essentially straight members jointed together. Only a small number of dimensions need to be held closely. The joints make use of standard bearings, and the links in effect form a solid chain and are not subject to undue acceleration limitations.

The harmonic transformer and the four-bar linkage shown in Fig. 22-6-1 are the two bar linkages most commonly used as function generators. The harmonic transformer consists of crank, connecting link, and slider. It may be driven from the crank end when a rotary input and linear output are desired or from the slide end when a linear input and rotary output are required. Two cranks and a connecting link form the four-bar linkage, whose input and output are both rotary. By assignment of correct values to the various parameters, these linkages will mechanize many single-variable functions. The selection of these values is termed a *linkage layout*. Typical linkage joints are shown in Fig. 22-6-2.

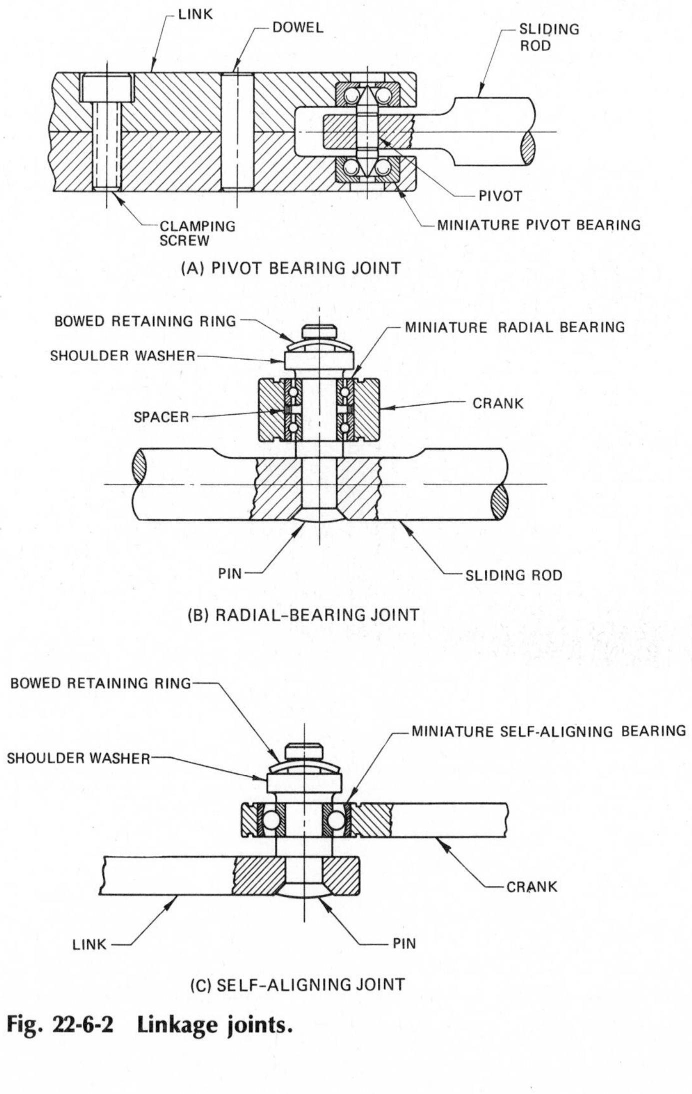
LINK
DOWEL
SLIDING ROD
PIVOT
CLAMPING SCREW
MINIATURE PIVOT BEARING
(A) PIVOT BEARING JOINT
BOWED RETAINING RING
SHOULDER WASHER
MINIATURE RADIAL BEARING
SPACER
CRANK
PIN
SLIDING ROD
(B) RADIAL-BEARING JOINT
BOWED RETAINING RING
SHOULDER WASHER
MINIATURE SELF-ALIGNING BEARING
CRANK
LINK
PIN
(C) SELF-ALIGNING JOINT

Fig. 22-6-2 Linkage joints.

Straight-Line Mechanism[3]

A *straight-line mechanism* is a linkage device used to guide a given point in an approximate straight line. Several such mechanisms use five or more links to produce exact straight-line motion of a given point. A four-link (or four-bar) mechanism, using finite links, can only approximate a straight line.

The four elements of the four-bar linkage (Fig. 22-6-3) are as follows:

1. A link which will be caused to move so that one point on it, called the *indicator*, travels along the desired path. This link is composed of two rigidly connected parts: the *connecting link*, joining the drive point and the control point, and the *indicator link*, connecting the indicator to the drive point.

2. A *drive crank*, to which a turning torque is applied to move the mechanism and which is connected to the link at the drive point.

3. A *control crank*, which serves only to guide the link control point in the proper path.

4. The *base* of the machine, to which the two cranks are pivotally attached.

For identification purposes, *indicator path* is the term used to describe the approximate straight-line path through which the indicator travels, and *straight line* refers to the desired theoretical straight line. The indicator path and the straight line will coincide at three or four places.

Systems Having Linkages and Cams[4]

A cam is of no value and can perform no useful function without a follower linkage. A *simple follower* is generally not

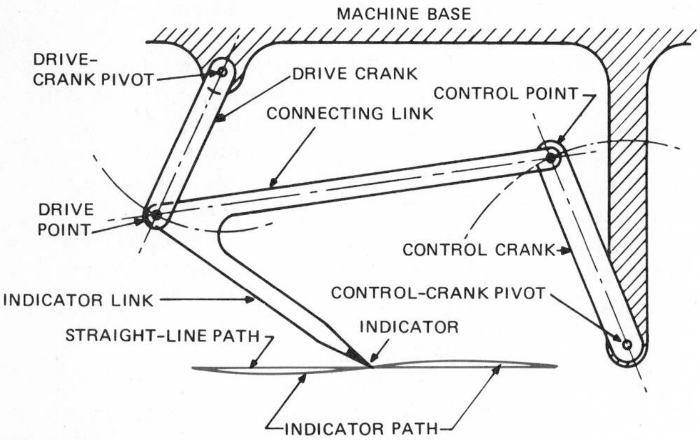
MACHINE BASE
DRIVE-CRANK PIVOT
DRIVE CRANK
CONTROL POINT
CONNECTING LINK
DRIVE POINT
CONTROL CRANK
INDICATOR LINK
CONTROL-CRANK PIVOT
STRAIGHT-LINE PATH
INDICATOR
INDICATOR PATH

Fig. 22-6-3 Terminology of a four-bar straight-line mechanism.

thought of as a linkage since it is usually a slide or plunger, such as an automotive valve assembly in a simple L-head engine. A *linkage* is generally considered to be a group of levers and links (Fig. 22-6-4).

Figure 22-6-5 shows a typical cam linkage comprised of a fairly heavy slide, a short link, and a bell crank. If the slide, which has the largest mass in the linkage, is to be moved with the most favorable accelerating forces, its displacement-to-time relationship must govern the shape of the cam profile. Since the bell crank and link swing about fixed and instantaneous centers during the stroke, the displacement increments of the slide and the cam profile do not bear the same relationship to the cam displacement. It is therefore necessary to plot the cam-pitch curve from the slide displacement at each increment of cam

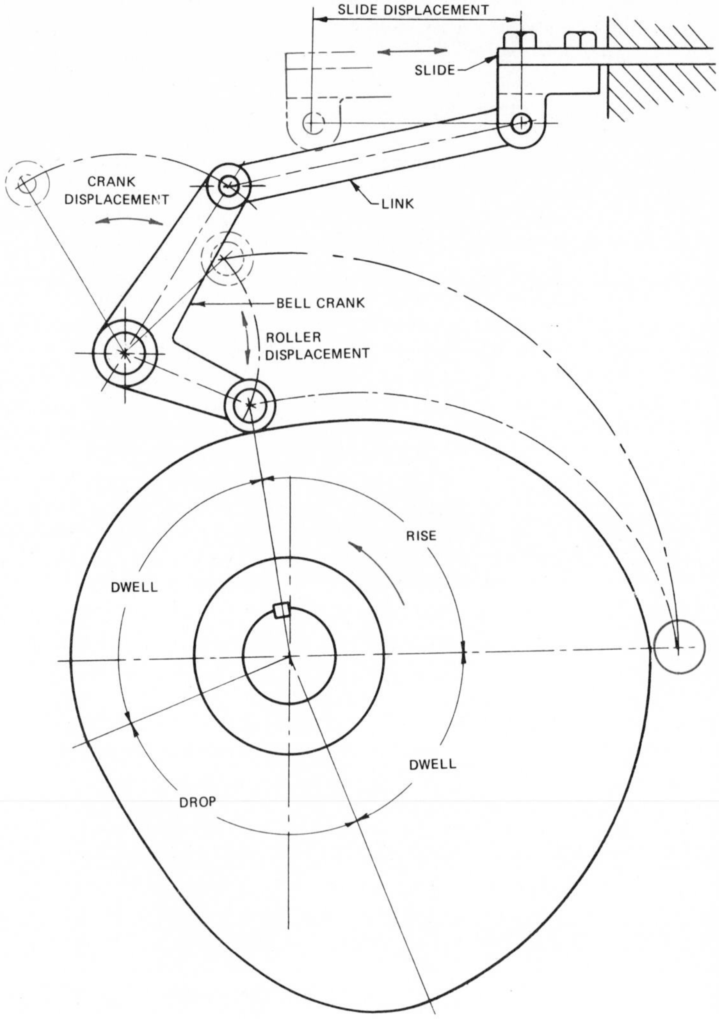
SLIDE DISPLACEMENT
SLIDE
CRANK DISPLACEMENT
LINK
BELL CRANK
ROLLER DISPLACEMENT
RISE
DWELL
DWELL
DROP

Fig. 22-6-5 Cam linkage.

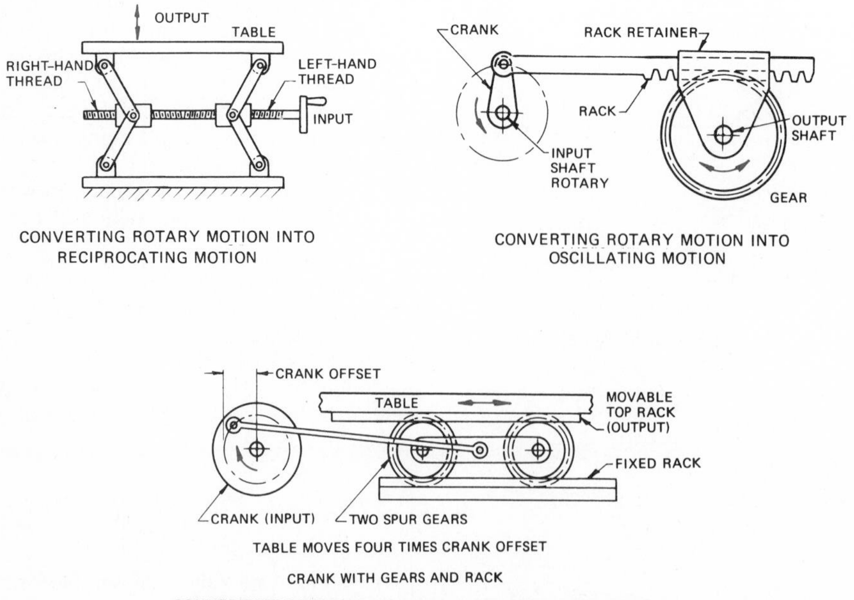
OUTPUT
TABLE
RIGHT-HAND THREAD
LEFT-HAND THREAD
INPUT
CONVERTING ROTARY MOTION INTO RECIPROCATING MOTION
CRANK
RACK RETAINER
RACK
OUTPUT SHAFT
INPUT SHAFT ROTARY
GEAR
CONVERTING ROTARY MOTION INTO OSCILLATING MOTION
CRANK OFFSET
TABLE
MOVABLE TOP RACK (OUTPUT)
FIXED RACK
CRANK (INPUT)
TWO SPUR GEARS
TABLE MOVES FOUR TIMES CRANK OFFSET
CRANK WITH GEARS AND RACK
CONVERTING ROTARY MOTION INTO RECIPROCATING MOTION

Fig. 22-6-4 Converting rotary motion into oscillating or reciprocating motion.

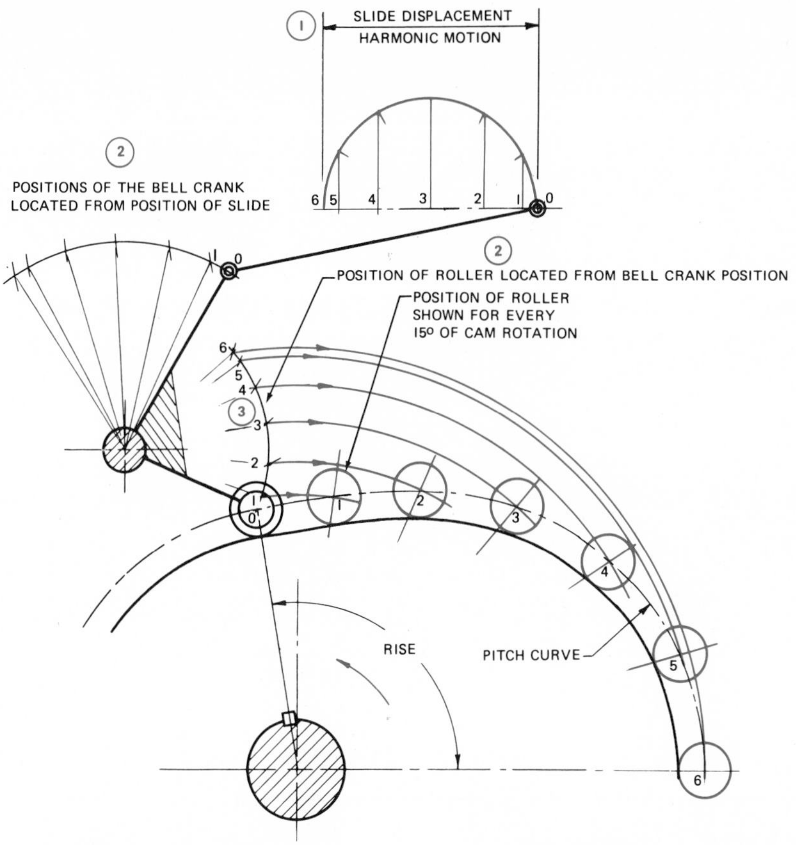
SLIDE DISPLACEMENT
HARMONIC MOTION
POSITIONS OF THE BELL CRANK
LOCATED FROM POSITION OF SLIDE
POSITION OF ROLLER LOCATED FROM BELL CRANK POSITION
POSITION OF ROLLER
SHOWN FOR EVERY
15º OF CAM ROTATION
RISE
PITCH CURVE

Fig. 22-6-6 Plotting the pitch curve from the slide displacement for the cam shown in Fig. 24-6-5.

displacement. Figure 22-6-6 shows how this is done.

For extra accuracy and for high-speed machinery, these displacements should be calculated.

REFERENCES AND SOURCE MATERIAL

1. P. T. Nickson, "A Simplified Approach to Linkage Design," *Machine Design*, Vol. 25, No. 12, 1953.
2. Percy Weir.
3. W. R. Kearney and M. G. Wright, "Straight-Line Mechanism," *Machine Design*, Vol. 26, No. 12, 1954.
4. Eonic Inc.

Assignments

1. *Simple Crank Mechanism*. On an A3- or B-size sheet, lay out the two linkages shown in Fig. 22-6-A or Fig. 22-6-B, and plot the paths at 15° intervals of the points indicated. Scale is 1:1.

REVIEW FOR ASSIGNMENTS

Unit 9-4 Springs
Unit 22-1 Cam Motions

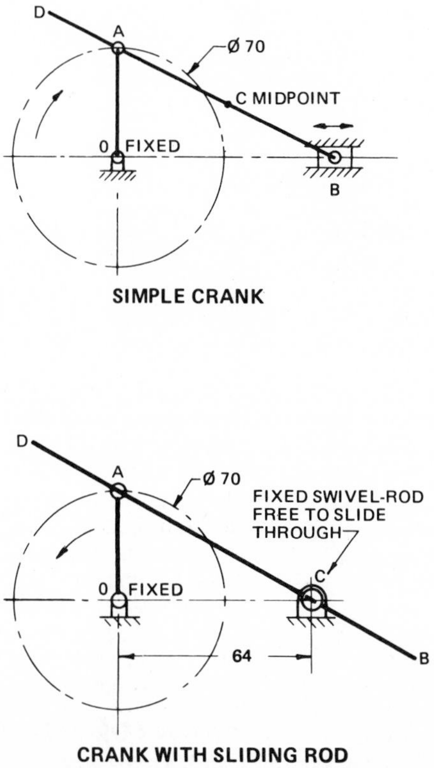
Ø 70
C MIDPOINT
FIXED
SIMPLE CRANK
Ø 70
FIXED SWIVEL-ROD
FREE TO SLIDE
THROUGH
FIXED
64
CRANK WITH SLIDING ROD

Fig. 22-6-A Simple crank mechanism.

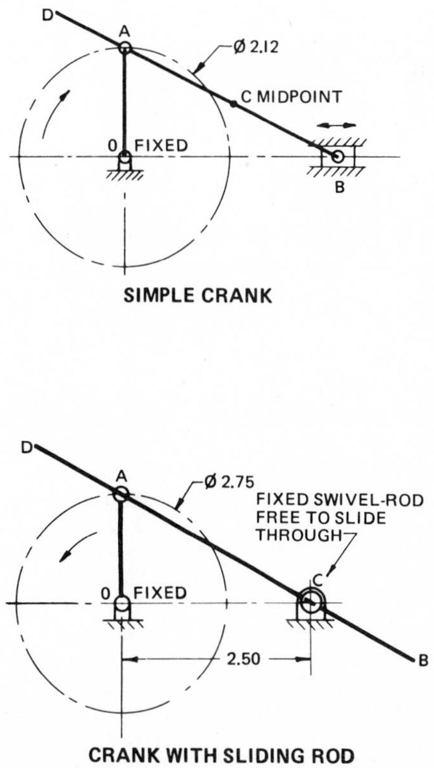
Ø 2.12
C MIDPOINT
FIXED
SIMPLE CRANK
Ø 2.75
FIXED SWIVEL-ROD
FREE TO SLIDE
THROUGH
FIXED
2.50
CRANK WITH SLIDING ROD

Fig. 22-6-B Simple crank mechanism.

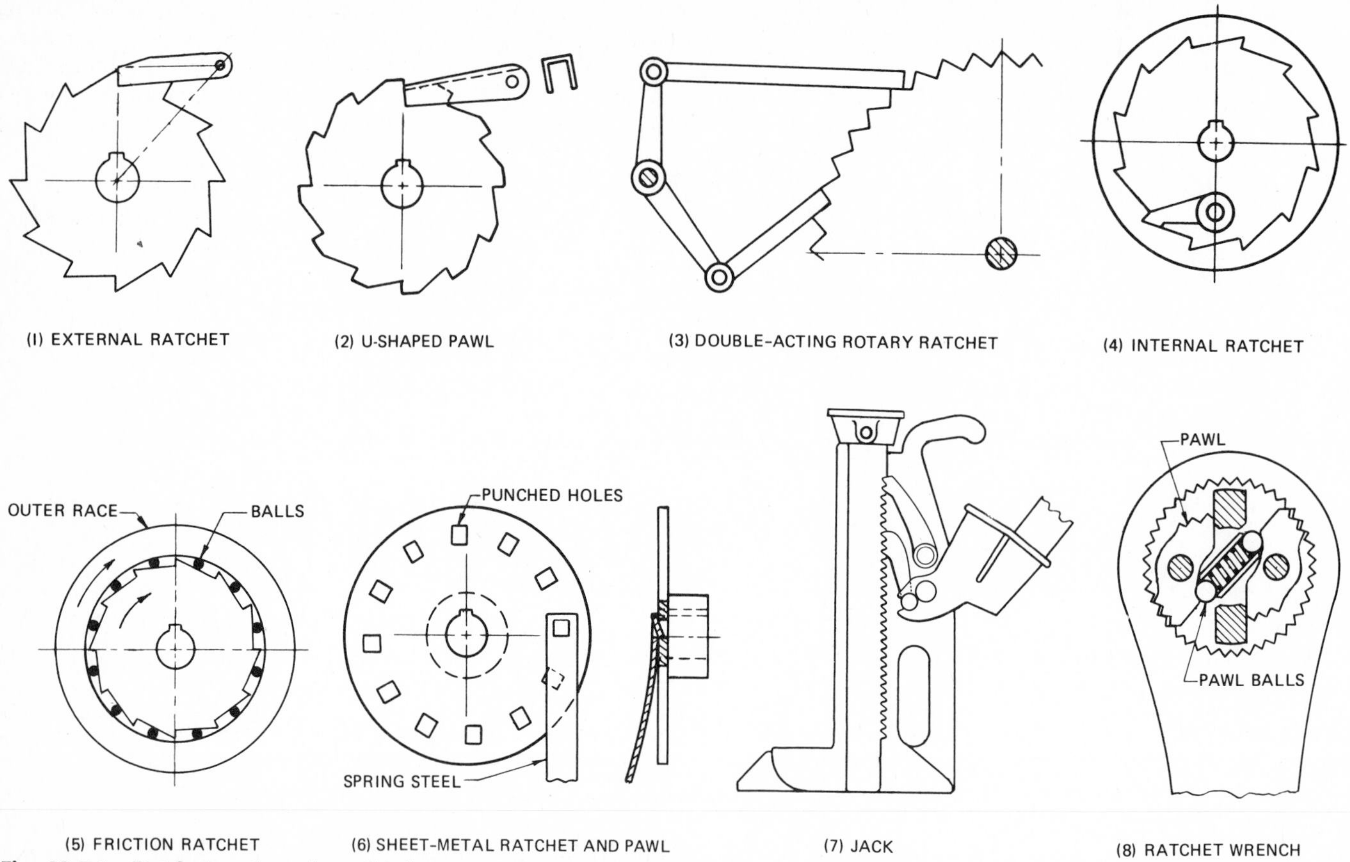
(1) EXTERNAL RATCHET
(2) U-SHAPED PAWL
(3) DOUBLE-ACTING ROTARY RATCHET
(4) INTERNAL RATCHET
OUTER RACE
BALLS
PUNCHED HOLES
SPRING STEEL
PAWL
PAWL BALLS
(5) FRICTION RATCHET
(6) SHEET-METAL RATCHET AND PAWL
(7) JACK
(8) RATCHET WRENCH

Fig. 22-7-1 Ratchet and pawl application.

UNIT 22-7
RATCHET WHEELS

Ratchet wheels are used to transform reciprocating or oscillatory motion into intermittent motion, to transmit motion in one direction only, or as an indexing device.

When a motion is to be transmitted at intervals rather than continuously and the loads are light, ratchets are ideal because of their low cost.

Common forms of ratchet and pawls are shown in Fig. 22-7-1. The teeth in the ratchet engage the teeth in the pawl, permitting rotation in one direction only. The pawl, which fits into the ratchet teeth as shown in Fig. 22-7-2, is pivoted at one end. A spring or counterweight is normally used to maintain contact between the wheel and pawl. In Fig. 22-7-1(7) and (8) lever or pawl balls are used to shaft the pawl to the alternative position so that the ratchet will work in reverse.

In friction ratchets [Fig. 22-7-1(5)], balls are used between the ratchet and the follower. As the ratchet rotates in the direction of the arrow shown, the balls roll up on the high spots on the teeth, wedging the ratchet and outer race together. If the direction of the ratchet is reversed, the balls roll to the low points on the teeth and disengage the outer roller. This principle is used on overrunning clutches.

Ratchet wheels and pawls are also used widely in hoisting equipment to control drum rotation.

In designing a ratchet wheel and pawl, lay out points *A*, *B*, and *C*, as shown in Fig. 22-7-2, on the same circle to ensure the smallest forces are acting on the system.

Another method to increase the number of stops made by the ratchet wheel without increasing the number of teeth is by the use of multiple pawls (Fig. 22-7-3). Adding another pawl of different length doubles the number of indexing positions.

Assignments

1. *Ratchet Wheel and Pawl* (Fig. 22-7-A or Fig. 22-7-B). On an A3- or B-size sheet, design a ratchet and a U-shaped pawl. Use your judgment for dimensions not given. Scale is 1:1. Show two views.

2. *Ratchet and Crank Mechanism*. On an A3- or B-size sheet, lay out a one-view drawing of the ratchet design shown in Fig. 22-7-C or Fig. 22-7-D. Two pawls are used, a drive pawl as shown and a holding pawl held in position by a spring. Using crank rotation positions every 22.5°, plot the path of the end of the drive pawl. Use your judgment for dimensions not shown. Scale is 1:1.

REVIEW FOR ASSIGNMENTS

Unit 9-4 Springs
Unit 9-1 Keys

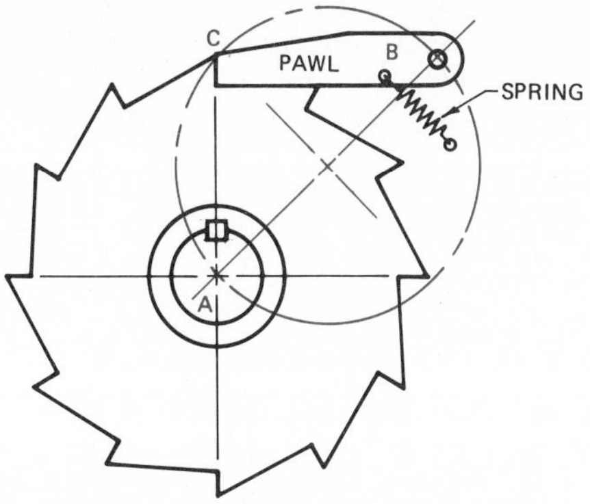
C
PAWL
B
SPRING
A

Fig. 22-7-2 Designing a ratchet wheel and pawl.

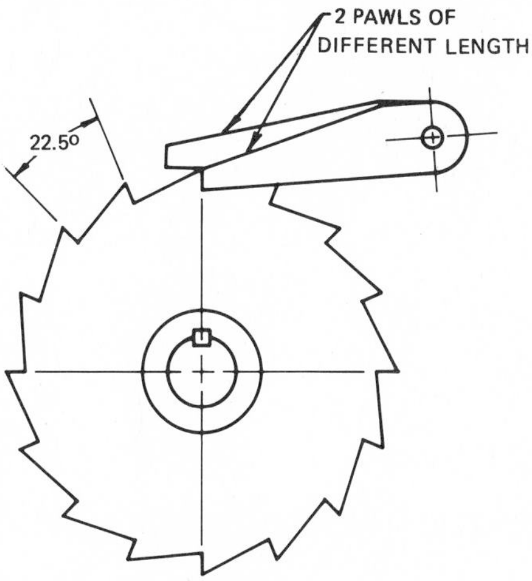
2 PAWLS OF
DIFFERENT LENGTH
22.5°

PAWLS LOCKS RATCHET WHEEL EVERY 11.25°

Fig. 22-7-3 Ratchet with two pawls.

Design a brake assembly using a U shaped pawl with a ratchet wheel. The ratchet wheel is to have 24 teeth O.D. of $\phi 146$; hub $\phi 48$; shaft $\phi 32.5$; keyway to suit; width of teeth 12; depth of teeth 10; the hub is extended 16 on one side. Scale 1:1. Show two views.

Fig. 22-7-A Ratchet wheel and pawls.

Design a brake assembly using a U shaped pawl with a ratchet wheel. The ratchet wheel is to have 24 teeth O.D. of $\phi 5.75$; hub $\phi 1.88$; shaft ϕ 1.250; keyway to suit; width of teeth .50; depth of teeth .38; the hub is extended .62 on one side. Scale 1:1. Show two views.

Fig. 22-7-B Ratchet wheel and pawls.

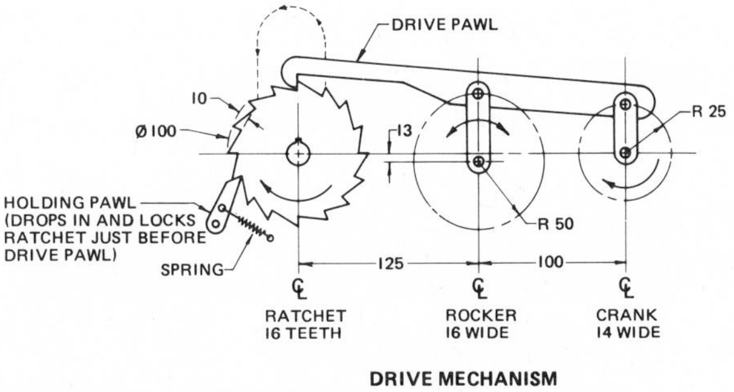
DRIVE PAWL
10
Ø 100
13
R 25
R 50
HOLDING PAWL
(DROPS IN AND LOCKS
RATCHET JUST BEFORE
DRIVE PAWL)
SPRING
125
100
℄
RATCHET
16 TEETH
℄
ROCKER
16 WIDE
℄
CRANK
14 WIDE
DRIVE MECHANISM

Fig. 22-7-C Ratchet and crank mechanism.

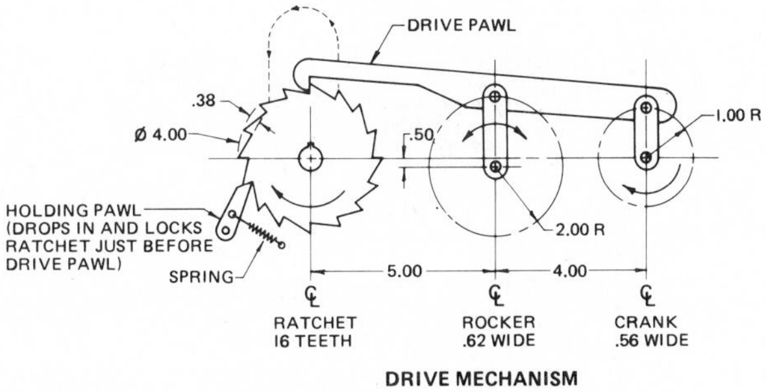
DRIVE PAWL
.38
Ø 4.00
.50
1.00 R
2.00 R
HOLDING PAWL
(DROPS IN AND LOCKS
RATCHET JUST BEFORE
DRIVE PAWL)
SPRING
5.00
4.00
℄
RATCHET
16 TEETH
℄
ROCKER
.62 WIDE
℄
CRANK
.56 WIDE
DRIVE MECHANISM

Fig. 22-7-D Ratchet and crank mechanism.

Chapter 23
Fluid Power

UNIT 23-1
HYDRAULICS[1,2]

The more complex industry becomes, the more vital becomes the role played by fluids in the industrial machine. One hundred years ago, water was the only important fluid which was conveyed from one point to another by pipe. Today, almost every conceivable fluid is handled in pipe during its production, processing, transportation, or utilization. The age of atomic energy and rocket power has added fluids such as liquid metals—for example, sodium, potassium, and bismuth—as well as liquid oxygen, nitrogen, etc., to the list of more common fluids such as oil, water, acids, and liquors that are being transported in pipe today. Nor is the transportation of fluids the only phase of hydraulics which warrants attention now. Hydraulic and pneumatic mechanisms are used extensively for the controls of modern aircraft, sea-going vessels, automotive equipment, machine tools, earth-moving and road-building machines, and even in scientific laboratory equipment where precise control of fluid flow is required.

So extensive are the applications of hydraulics and pneumatics that almost every designer has found it necessary to be familiar with at least the elementary laws of fluid flow.

Basic Principles

The science of hydraulics dates back several thousand years when water wheels, dams, and sluice gates were used to control the flow of water for irrigation and domestic use. Today, however, the term *hydraulics* commonly refers to *power hydraulics* in which fluid is used under controlled pressure to do work.

A fluid is infinitely flexible, yet as unyielding as steel. It can readily change its shape; it can be divided into parts to do work in different locations; it can move rapidly in one place and slowly in another; and it can transmit a force in any or all directions. No other medium except fluid combines the same degree of positiveness, accuracy, force, and pressure.

It is interesting and important to note that a hydraulic pump does not *pump* pressure. The pump merely produces flow. Pressure is generated only when a cylinder, motor, valve, or restriction tends to resist fluid flow. If the flow were to encounter only negligible resistance, the developed pressure would be slight.

Both force and pressure are primarily measures of effort. *Work*, however, is a measure of accomplishment. It describes the application of a force and flexibility of control, with the ability to transmit a maximum of power in a minimum of mass.

Although this unit is not aimed at an in-depth treatment of hydraulic components and circuits, it will nevertheless be helpful to start by considering some of the fundamental principles involved.

FORCE, PRESSURE, WORK, AND POWER

Force is defined as any cause which tends to produce or modify motion. To move an object, such as a machine tool head, force must be applied to it. The amount of force required depends on the object's inertia. Force is measured in newtons. The preferred units used in this text are *newtons* (N) for light forces, *kilonewtons* (kN) for intermediate forces, and *meganewtons* (MN) for high forces. (A newton is approximately one-quarter of a pound-force.)

Pressure is force per unit area, and it is measured in pascals. A *pascal* (Pa) is the pressure produced when a force of one newton is applied to an area of one square metre: Pa = N/sq m.

The pascal is a very small unit of measure, equivalent to approximately 0.001 pounds per square inch (psi). It is used for very low-pressure applications. In most instances the *kilopascal* (kPa) and *megapascal* (MPa) are used. In fluid power applications the kilopascal is the recommended unit of pressure. A 1 kPa of pressure is equal to approximately 20.9 lb/sq ft.

The earth's atmosphere provides an example of the relationship between force and pressure. The blanket of air enveloping the earth's surface is of such volume that its total mass could be measured in metric tons (megagrams). However, the force exerted by the mass of a column of air 1 m^2 in cross-sectional area is only 101.325 kPa at sea level. Thus atmospheric pressure at sea level is 101.325 kPa, but for low-accuracy work 100 kPa is used. The relationship between *force* (newtons), *pressure* (pascals), and *area* (square metres) is expressed mathematically as follows:

$$\text{Force} = \text{pressure} \times \text{area}$$
$$\text{Newtons} = \text{pascals} \times \text{square metres}$$
$$N = Pa \times m^2$$

As mentioned previously, work is a measure of accomplishment, describing the application of a force moving through a distance. It is expressed in joules. A *joule* (J) is the energy required or expended to move a force of one newton a distance of one metre. *Joule* is the special name given the derived unit newton-metre (N • m). The *kilojoule* (kJ) is the preferred unit. *Joule* (J) is used to express small quantities of energy, *megajoules* (MJ) large quantities. Thus, if a force of 500 N moves a ram 3 m, the work accomplished equals 1500 J, or 1.5 kJ. The formula is

$$\text{Work} = \text{force} \times \text{distance}$$
$$J = N \times m$$

The concept of work makes no allowance for the time factor. *Power* is work per unit of time, and the *watt* (W) is the standard unit for all forms of power: heat, electrical, mechanical, etc. A *watt* is energy per unit of time measured in joules per second (J/s). The preferred units are *watt* (W) for small values, *kilowatt* (kW) for intermediate values, and *megawatt* (MW) for large values. One watt is that amount of power necessary to move one newton one metre in one second.

BEHAVIORS AND MECHANICS OF FLUIDS

In the 17th century, Pascal formulated the fundamental law which forms the basis of modern hydraulics. Pascal's law states: *Pressure at any one point in a static liquid is the same in every direction and exerts equal force on equal areas.* Figure 23-1-1 illustrates this principle.

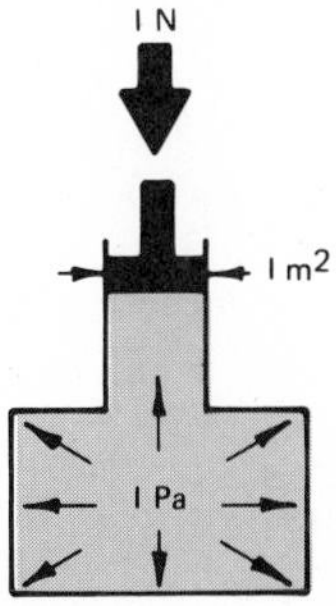
1 N
1 m2
1 Pa

Fig. 23-1-1 Pascal's law.

Since fluids are practically incompressible, mechanical forces may be transmitted, multiplied, or controlled by means of hydraulic fluids under pressure. This is demonstrated in Fig. 23-1-2. Assume that the masses of pistons *P* and *W* are equal. Assume also that the face of piston *P* has an area of 0.02 m^2, that the face of piston *W* has an area of 1 m^2, and that the two communicating containers are filled with liquid.

If we were to exert a downward force of 100 N on piston *P*, this force would be transmitted undiminished in every direction, acting with equal force on equal areas. Since the area of piston *P* is 0.02 m^2, the pressure at every point in the system is $(100 \times 1) / 0.02 = 5000$ Pa. When a pressure of 5000 Pa is applied to the lower surface of piston *W*, the resultant upward force is 5000×1, or 5000 N, because the area of piston *W* is 1 m^2 and force equals pressure times area.

In either case, we can multiply force only by sacrificing distance proportionally. If we move piston *P* downward a distance of 0.5 m, we have forced $0.5 \times 0.02 = 0.01$ m^3, or 0.01 kilolitre (kl), of liquid to the underside of piston *W*. One cubic metre equals one kilolitre. Since piston *W* has an area of 1 m, 0.01 kl of liquid can raise piston *W* only 0.01 m.

Characteristics of Flow. Pascal's law neglects the factor of friction because it deals with static fluids. When a fluid flows in a hydraulic circuit, friction results and heat is produced. Thus some of the energy being transferred is lost in the form of heat energy.

Although friction can never be eliminated entirely, it can be controlled to some extent. The four main causes of excess friction in hydraulic lines are excessive length of lines, excessive number of bends and fittings or improper (too sharp) bends, excessive fluid velocity

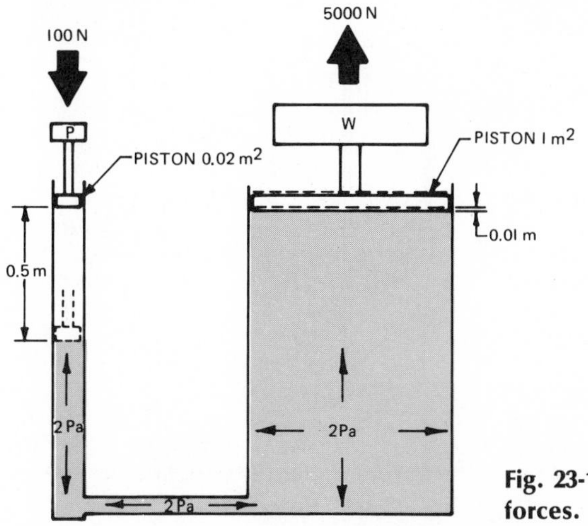
100 N
5000 N
P
W
PISTON 0.02 m2
PISTON 1 m2
0.5 m
0.01 m
2 Pa
2Pa
2 Pa

Fig. 23-1-2 Transmission of mechanical forces.

caused by undersized lines, and excessive viscosity of fluid.

Figure 23-1-3 illustrates the effect of friction upon pressure. Since pressure is a result of resistance to flow, the pressure at point *B* is zero. Assume that the mass of the fluid and the diameter of tube *A* are such that a static pressure of 100 kPa is created at point *C*. Then the flow from point *C* to point *B* has resulted in a pressure drop of 100 kPa.

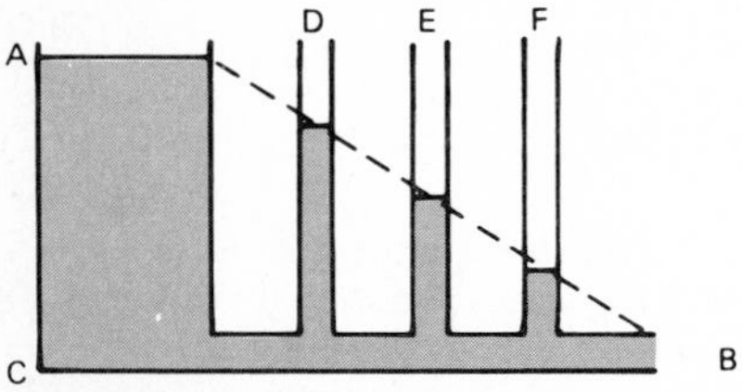
D
E
F
A
C
B

Fig. 23-1-3 Illustrating effect of friction upon pressure.

The potential energy at point *C* is completely dissipated in moving the fluid to point *B*. This potential energy has been converted into the heat energy produced by the friction of the fluid moving through the tubes.

The height of the fluid in tubes *D*, *E*, and *F* illustrates the action of friction in producing a pressure drop. In a moving fluid, pressure drop tends to increase and pressure tends to decrease as the distance from the source of pressure increases. Pressure drop through an orifice of a given size varies as the square of the flow passing through the orifice.

Helpful Information

Some of the points already mentioned are repeated below, along with several additional facts that may help in understanding machinery hydraulics.

1. Oil is the most commonly used hydraulic fluid because it serves as a lubricant for hydraulic components and is practically incompressible.

2. The mass of oil varies considerably with change in viscosity. However, 2.6 to 2.8 kPa covers the viscosity range of common hydraulic fluids.

3. Pressure at the bottom of a 1 m column of oil will be approximately 9 kPa. To find the approximate pressure at the bottom of any oil column, multiply the height in metres by 9.

4. There must be a pressure drop (pressure difference) across an orifice or restriction to cause flow through it. Conversely, if there is no flow, there will be no pressure drop.

5. A fluid is pushed into a pump. Atmospheric pressure supplies this push (in an unsupercharged pump) at 101.3 kPa at sea level.

6. The force exerted by a cylinder is dependent on the pressure applied and the piston area.

7. The speed of a cylinder is dependent on its piston area and the rate of fluid flow into it.

8. The flow velocity through a pipe varies inversely to the square of the inside diameter (ID). Doubling the ID increases the area 4 times.

Hydraulic Circuits

All *hydraulic circuits* are essentially the same regardless of the application (machine tools, airplanes, farm equipment, boats, etc.). There are four basic components required: a tank (reservoir) to hold the fluid, a pump to force the fluid through the system (the pump is driven by an electric motor or other power source), valves to control fluid pressure and flow and an actuator (a cylinder for linear or a motor for rotary motion) to convert the energy of fluid movement into mechanical force to do the work.

The complexity of hydraulic systems will vary, of course, depending on the application.

Hydraulic Equipment

RESERVOIRS

Although it primarily serves as a supply source for hydraulic system fluid, a well-designed reservoir serves other useful purposes. Its construction assists in the separation of air and contaminants from the fluid and helps to dissipate heat generated within the system.

A reservoir which conforms to Joint Industry Conference (JIC) specifications is illustrated in Fig. 23-1-4. The tank is constructed of hot-rolled steel plates with welded joints. Extensions of the tank ends support the tank and may be bolted to the floor. The tank bottom is concave and has a drain plug in the middle.

The large removable covers on each end permit easy access for cleaning. One of the covers has a fluid level indicator and a filler hole covered with a chained cap. Since the fluid level should be checked periodically, the indicator is located conveniently. A replaceable fine-mesh wire screen is installed in the filler hole to keep foreign matter out of the reservoir when new fluid is added.

STRAINERS AND FILTERS

To ensure long life and trouble-free performance of hydraulic components, it is important to keep the hydraulic fluid clean. When systems operate at high speeds and pressures, contaminants cause excessive wear and power loss and can cause malfunction of components. Adequate filtration pays for itself many times over in decreased maintenance and replacement costs.

Filters, strainers, and magnetic plugs can be used to remove foreign particles from the hydraulic fluid and are most effective as safeguards against contamination. Typical filters and strainers are shown in Fig. 23-1-5.

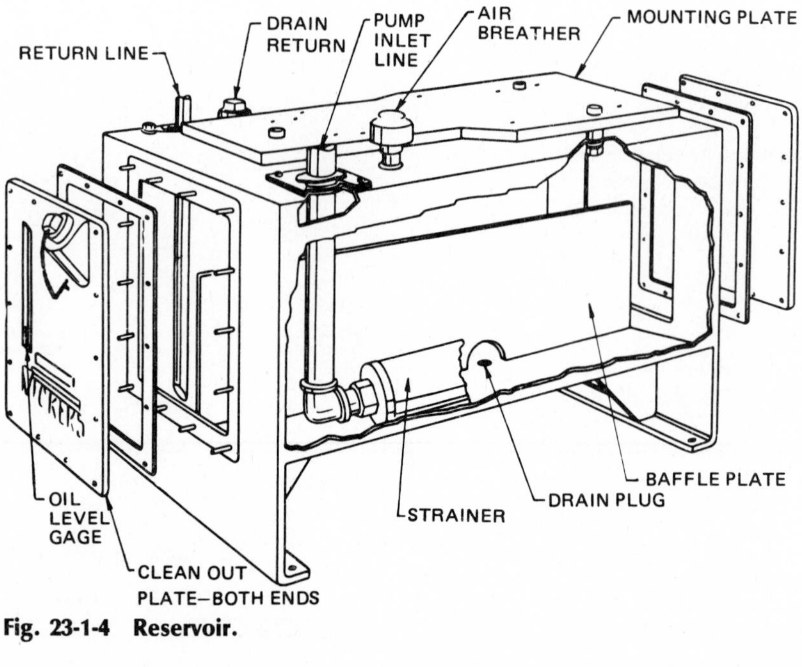
RETURN LINE
DRAIN RETURN
PUMP INLET LINE
AIR BREATHER
MOUNTING PLATE
BAFFLE PLATE
DRAIN PLUG
STRAINER
OIL LEVEL GAGE
CLEAN OUT PLATE—BOTH ENDS

Fig. 23-1-4 Reservoir.

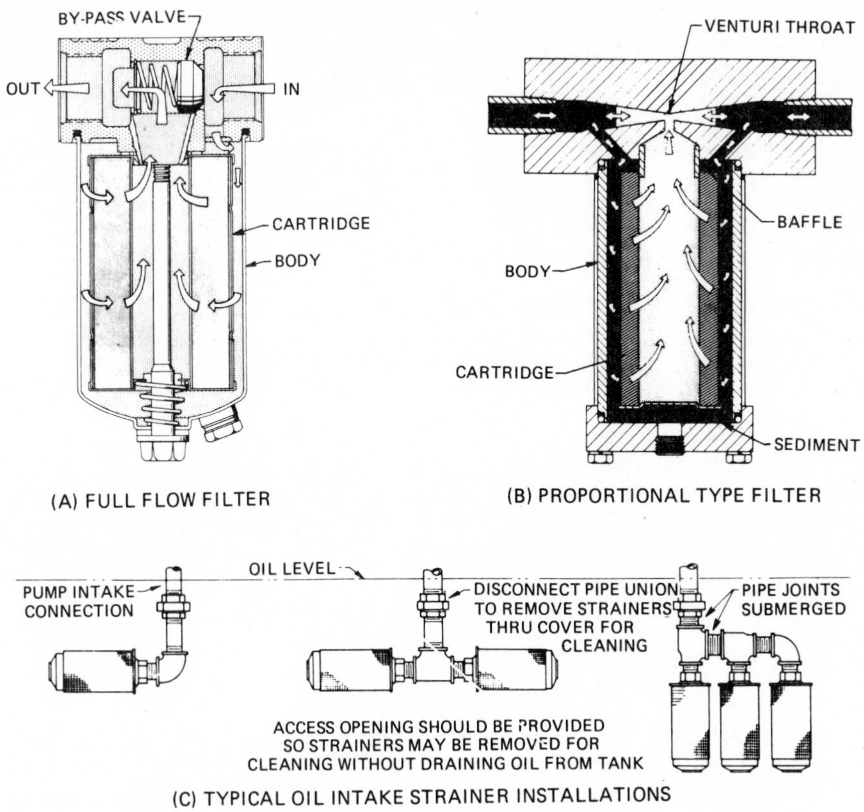
BY-PASS VALVE
OUT
IN
CARTRIDGE
BODY
(A) FULL FLOW FILTER
VENTURI THROAT
BAFFLE
BODY
CARTRIDGE
SEDIMENT
(B) PROPORTIONAL TYPE FILTER
OIL LEVEL
PUMP INTAKE CONNECTION
DISCONNECT PIPE UNION TO REMOVE STRAINERS THRU COVER FOR CLEANING
PIPE JOINTS SUBMERGED
ACCESS OPENING SHOULD BE PROVIDED SO STRAINERS MAY BE REMOVED FOR CLEANING WITHOUT DRAINING OIL FROM TANK
(C) TYPICAL OIL INTAKE STRAINER INSTALLATIONS

Fig. 23-1-5 Filters and strainers.

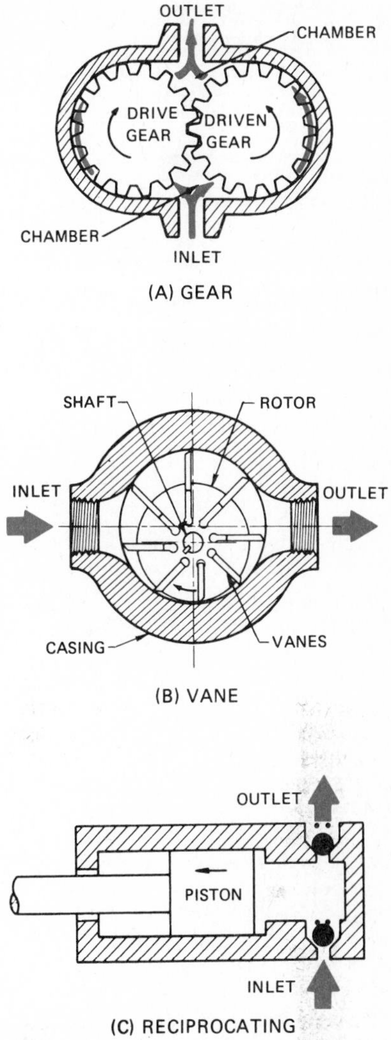
OUTLET
CHAMBER
DRIVE GEAR
DRIVEN GEAR
CHAMBER
INLET
(A) GEAR
SHAFT
ROTOR
INLET
OUTLET
CASING
VANES
(B) VANE
OUTLET
PISTON
INLET
(C) RECIPROCATING

Fig. 23-1-6 Positive-displacement pumps.

HYDRAULIC FLUIDS

Hydraulic fluid characteristics have an important effect on equipment performance and maintenance.

The use of a clean, high-quality fluid is the first step in achieving efficient hydraulic system operation. It is as important as any other component in the system and should be selected with the same care that the machine builder uses in selecting pumps, motors, valves, tubing, etc.

In addition to serving as a power transmitting medium, hydraulic fluid must keep wear to a minimum by providing good lubrication. It must maintain its body over a wide temperature range to prevent excessive slippage or leakage. It must have good lubricity to assure maximum life of pumps and motors.

Petroleum-base fluids are the most widely used. Well-refined oils, in themselves, are satisfactory for light service. Additives must be included to meet some of the preceding requirements, and properly compounded fluids should be used for normal service.

HYDRAULIC PUMPS

Hydraulic pumps are devices for converting mechanical energy into hydraulic energy. When a hydraulic pump is operated, it performs two functions. First, its mechanical action creates a partial vacuum at the pump inlet which enables atmospheric pressure in the reservoir to force liquid through the inlet line into the pump. Second, its mechanical action delivers this liquid to the pump outlet and forces it into the hydraulic system.

Pumps are broadly classified as either non-positive-displacement or positive-displacement units.

A *non-positive-displacement pump* produces a continuous flow. However, because of its design, there is no positive internal seal against leakage, and its output varies considerably as pressure varies.

Practically all pumps used in power hydraulic systems—whether on industrial machinery, mobile vehicles, or aircraft—are of the positive-displacement type.

A *positive-displacement pump* produces a pulsating flow, but since it provides a positive internal seal against leakage, its output is relatively unaffected by variations in system pressure. See Fig. 23-1-6.

ACTUATORS

Hydraulic actuators perform the opposite function of hydraulic pumps in that they convert hydraulic energy back to mechanical energy to perform useful work. In a typical circuit, the actuator is mechanically linked to the workload, and is actuated by fluid from the pump so that thrust or torque is transferred to the work.

Actuators can be classified broadly as either linear or rotary. A *linear actuator* such as a ram or cylinder is used for such operations as clamping, pressing, or traverse and feed motions. *Rotary actuator* applications include chucking, indexing, and turning.

Linear Actuators. The simplest linear actuator is the single-acting cylinder or ram (Fig. 23-1-7*a*) which applies force in only one direction. Fluid directed into the housing displaces the rod hydraulically; the retracting force can be gravity or some mechanical means, such as a spring. Two opposing rams are sometimes used in reciprocating grinder tables and similar applications where their simplicity permits the rams to be built right into the machine structure.

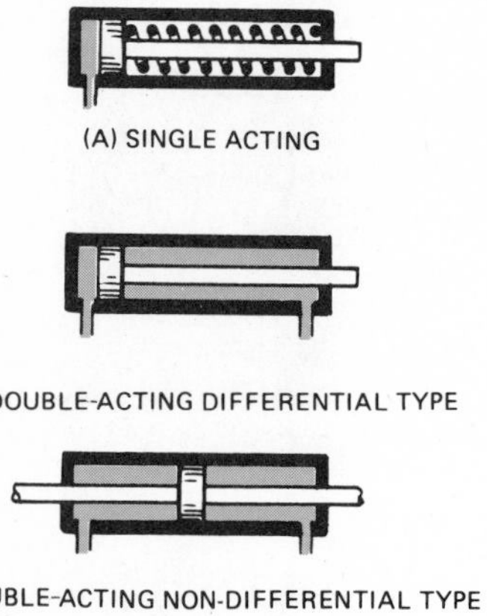
(A) SINGLE ACTING
(B) DOUBLE-ACTING DIFFERENTIAL TYPE
(C) DOUBLE-ACTING NON-DIFFERENTIAL TYPE

Fig. 23-1-7 Cylinders.

A double-acting cylinder (Fig. 23-1-7*b*) permits application of hydraulic pressure on either side of the piston to control linear motion in either of two opposite directions. The type shown is called a *differential cylinder* because the piston area at the left is larger, providing a slower, more powerful work stroke when fluid pressure is applied to the LH side. The return stroke will be faster due to the smaller piston area. Reciprocating motions of this type are required on machine tools such as shapers. If equal forces in both directions are required, the piston rod is designed to extend through both ends of the cylinder, as in Fig. 23-1-7*c*. This is a *non-differential-type cylinder*.

Rotary Actuators. Rotary actuators or motors, like rotary pumps, are of either the gear, vane, or piston design. The piston design is further divided into radial or axial types. Actually, many hydraulic pumps can be used as motors with little or no modification.

VALVES

Valves are used in hydraulic circuits to control pressure, as well as the direction and rate of fluid flow. They can be classified as pressure controls, directional controls, and flow controls.

Valves, like pipes, are generally rated according to size and pressure. The valve name may be based on its usual function (relief valve) or a feature of its construction (gate valve).

Pressure Control Valves. The most common type of pressure control valve is the *relief valve*. A relief valve may be used to provide overload protection for circuit components or to limit the force or torque exerted by a linear actuator or rotary motor. With a relief valve the flow is diverted back to the reservoir.

A relief valve of the simple type is shown in Fig. 23-1-8*a*. The valve is installed so that one port is connected to the pressure line, the other to the reservoir. The ball is held on its seat by thrust of the spring. Spring thrust can be changed by turning the adjusting screw.

When pressure at the valve inlet is insufficient to overcome spring force, the ball remains on its seat and the valve is closed as shown. The position of the ball prevents flow through the valve.

When pressure at the valve inlet exceeds the adjusted spring force, the ball is forced off its seat and the valve is opened. Liquid flows from the pressure line through the valve to the reservoir. This diversion of flow prevents further pressure increase in the pressure line. When pressure decreases below the valve setting, the spring reseats the ball and the valve is again closed.

Sequence valves, as shown in Fig. 23-1-8*b*, are used on machines requiring operations which must occur in a proper sequence. Sequence valves will not operate until the pressure of one unit has reached a certain level. Sequence valves differ from relief valves in that when a sequence valve is used the flow is diverted to another portion of the system to perform work.

Pressure-reducing valves are used to block or modulate flow at a preset pressure. Unlike the two valves discussed so far, reducing valves are normally open, the most common being a spool valve, which is shown in Fig. 23-1-8*c*.

Directional Control Valves. Although all *directional control valves* have a common function of controlling direction of fluid flow, they vary considerably in construction and operation.

Directional controls can be classified on the basis of principal characteristics: the type of internal valving element, the method of actuating the valving element, and the number of positions of the valving element, as well as the flow paths created in the various positions.

Check valves are the simplest of all directional control valves. They permit fluid to flow in one direction only. See Fig. 23-1-9.

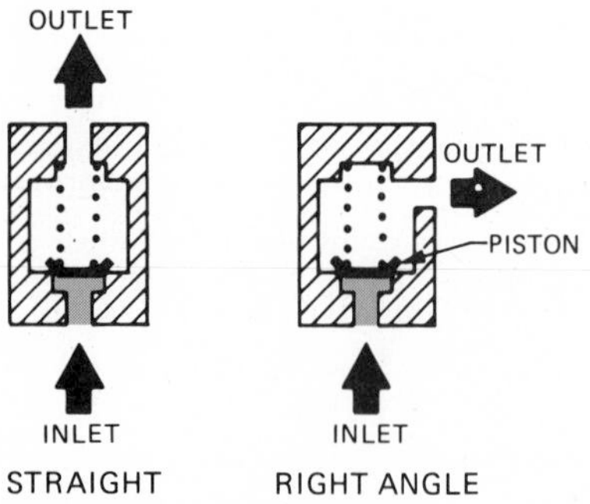
OUTLET
INLET
STRAIGHT
OUTLET
PISTON
INLET
RIGHT ANGLE

Fig. 23-1-9 Check valves.

Multiple-way valves are known as two-way, three-way, or four-way, depending on the number of primary parts the valve contains. See Figs. 23-1-10 through 23-1-12.

Many variations are possible when classifying valves according to the number of positions or flow paths. They

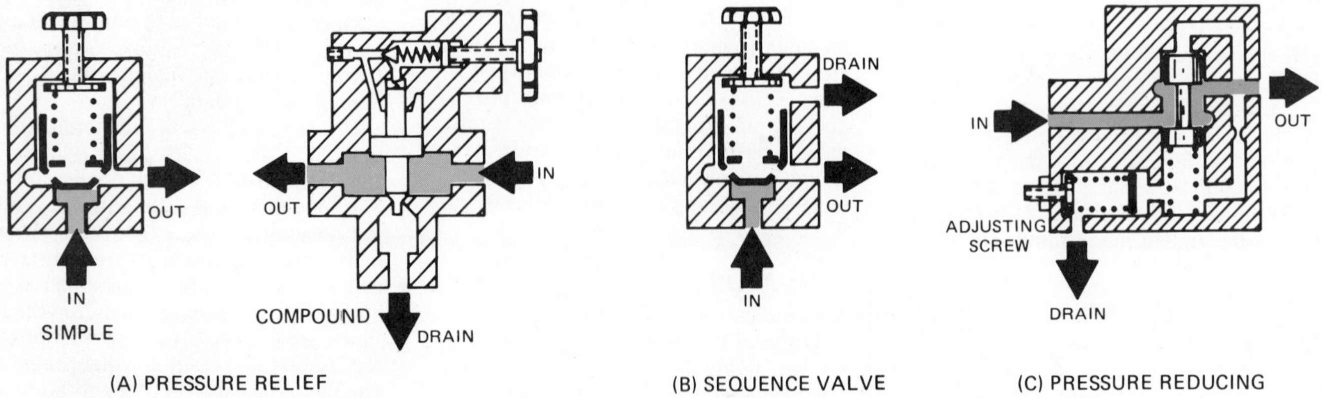
OUT
IN
SIMPLE
OUT
IN
COMPOUND
DRAIN
(A) PRESSURE RELIEF
DRAIN
OUT
IN
(B) SEQUENCE VALVE
IN
OUT
ADJUSTING SCREW
DRAIN
(C) PRESSURE REDUCING

Fig. 23-1-8 Pressure-control valves.

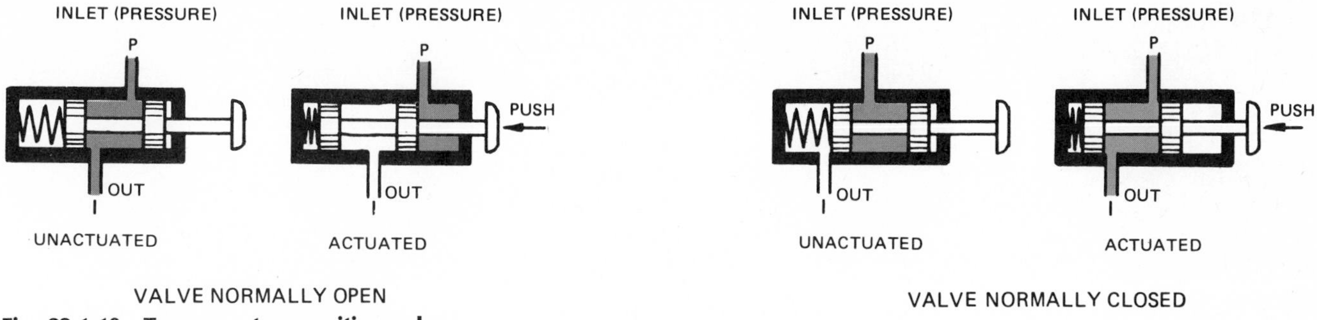
INLET (PRESSURE)
P
OUT
UNACTUATED
INLET (PRESSURE)
P
PUSH
OUT
ACTUATED
VALVE NORMALLY OPEN
INLET (PRESSURE)
P
OUT
UNACTUATED
INLET (PRESSURE)
P
PUSH
OUT
ACTUATED
VALVE NORMALLY CLOSED

Fig. 23-1-10 Two-way, two-position valves.

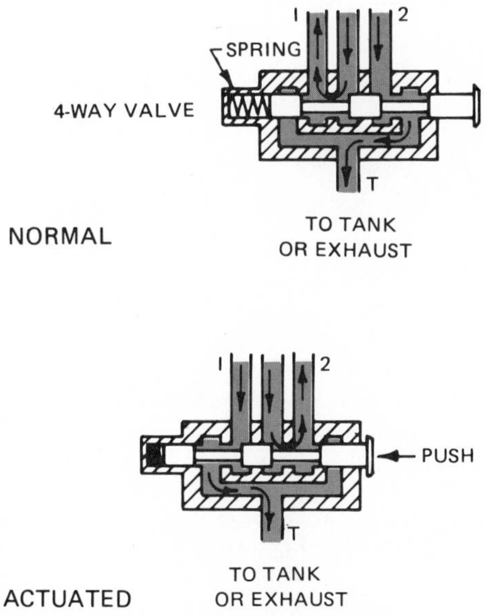
1
2
SPRING
4-WAY VALVE
T
NORMAL
TO TANK
OR EXHAUST
1
2
PUSH
T
TO TANK
OR EXHAUST
ACTUATED

Fig. 23-1-11 Four-way, two-position valve.

may be of the simple on-off variety or valves having a wide selection of flow paths through them.

The operation of a two-way sliding-spool valve is shown in Fig. 23-1-10.

Usually containing a sliding spool, two-way valves provide a choice of two flow paths exclusive of the valve's center position. In either shifted position, the pressure port is open to one cylinder port, but the opposite cylinder port is not open to the tank. (In a four-way valve, the opposite cylinder port would be open to the tank, making two flow paths in either shifted position.)

Flow Control Valves. The widespread use of hydraulic circuitry in machine tool applications stems in part from the ease with which traverse and feed rates can be controlled through the use of various types of flow controls.

The speed of an actuator, whether it be a hydraulic cylinder or a motor, is determined by its displacement and the amount of fluid available to it. Changing the displacement would change the operating speed, and would have an inverse effect on the force or torque output. For this and other reasons, such as the impracticability of changing the bore of a cylinder, speed control usually is accomplished by controlling the flow rate.

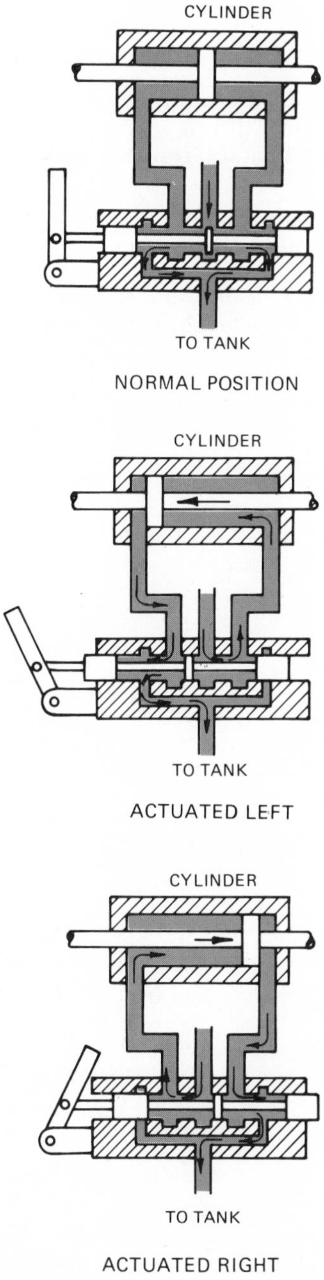
CYLINDER
TO TANK
NORMAL POSITION
CYLINDER
TO TANK
ACTUATED LEFT
CYLINDER
TO TANK
ACTUATED RIGHT

Fig. 23-1-12 Four-way, three-position valve.

With full pump delivery directed into a cylinder, it travels at its maximum speed. A larger pump would move it faster; a smaller one would move it slower. Logically, then, the use of a variable delivery pump would provide a relatively simple means of controlling actuator speed. Since operating pressure would vary with the workload, the volumetric efficiency of the pump would determine feed-rate accuracy. One advantage of this method is that the amount of fluid pumped would be only that required for feed purposes. If, however, flow is required at more than one location during the feed portion of the cycle, a separate source would be necessary.

The use of fixed delivery pumps and flow control valves makes a more flexible circuit since the flow that is not required for feed can be used elsewhere in the system for clamping, indexing, and cross-feeding. In such a circuit, some portion of the pump delivery will return to the tank through the relief valve at its setting during the feed cycle. Pressure at the actuator will change with workload changes. This will cause a varying pressure drop across the flow control valve and result in varying feed rates unless some means of pressure compensation is included.

A restrictor-type flow control valve (Fig. 23-1-13) has a compensator spool which is held in position by a light spring and pressure sensed beyond the throttle. In this type, the compensator spool is normally held in the open position and tends to shut off, not permitting fluid in excess of the throttle setting to enter the valve. The balance of the pump delivery is then available for other purposes.

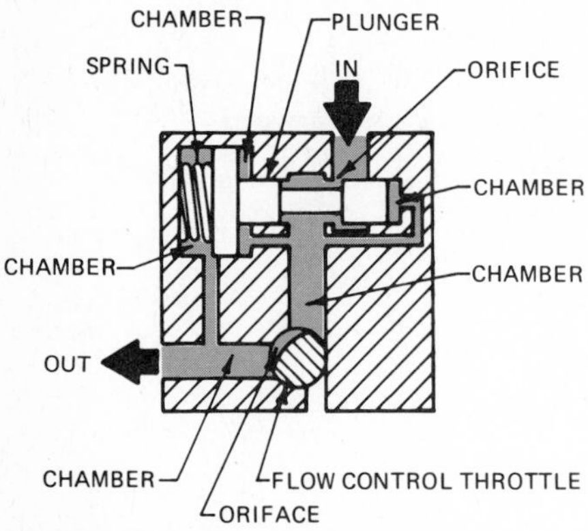
CHAMBER
PLUNGER
SPRING
IN
ORIFICE
CHAMBER
CHAMBER
CHAMBER
OUT
CHAMBER
FLOW CONTROL THROTTLE
ORIFACE

Fig. 23-1-13 Flow-control valve.

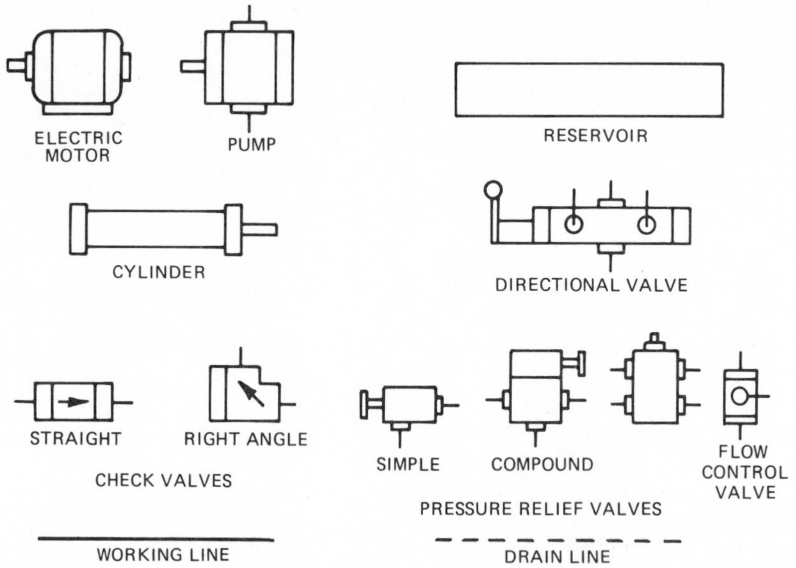
ELECTRIC MOTOR
PUMP
RESERVOIR
CYLINDER
DIRECTIONAL VALVE
STRAIGHT
RIGHT ANGLE
CHECK VALVES
SIMPLE
COMPOUND
PRESSURE RELIEF VALVES
FLOW CONTROL VALVE
WORKING LINE
DRAIN LINE

Fig. 23-1-15 Typical pictorial diagrams.

Fluid Circuit Diagrams

An accurate diagram of the fluid circuit is one of the most important pieces of literature accompanying a machine. The information shown in the circuit diagram is essential for an understanding of the operation of the machine and for installation or troubleshooting.

Not all circuit diagrams are complete. For instance, a diagram which is to be used only for piping would not need to show the sequence of operations or sizes and ratings of the components.

Fluid circuit diagrams are of four types, namely, pictorial, cutaway, graphical, and combination. The type used depends on the amount and kind of information required. Pictorial and cutaway diagrams are illustrated in Fig. 23-1-14.

PICTORIAL DIAGRAMS

A pictorial diagram is primarily used to show the piping arrangement of a circuit. The symbols are outline drawings showing the actual external shape of the components, with the piping shown to the various parts of the units. Because they show nothing of the internal construction or function of the components, pictorial diagrams have little value for instruction or troubleshooting. Pictorial symbols are difficult to standardize from a functional basis. Typical pictorial symbols are shown in Figs. 23-1-14*a* and 23-1-15.

CUTAWAY DIAGRAMS

Cutaway diagrams contain much information about the operation of a circuit and the construction and operation of its components. These diagrams are ideally suited for instruction and are widely used for that purpose. Because of the time and cost involved, they are seldom made for other purposes. Cutaway symbols are difficult to draw, and the part functions are not readily apparent. Figures 23-1-5 through 23-1-14*b* illustrate typical cutaway symbols for common hydraulic parts.

ARRANGEMENT OF SYMBOLS

Symbols are arranged in the diagram to facilitate the use of straight interconnecting lines. Where components have definite mechanical, functional, or otherwise important relationships to one another, their symbols should be so placed in the diagram. The conductors (interconnect-

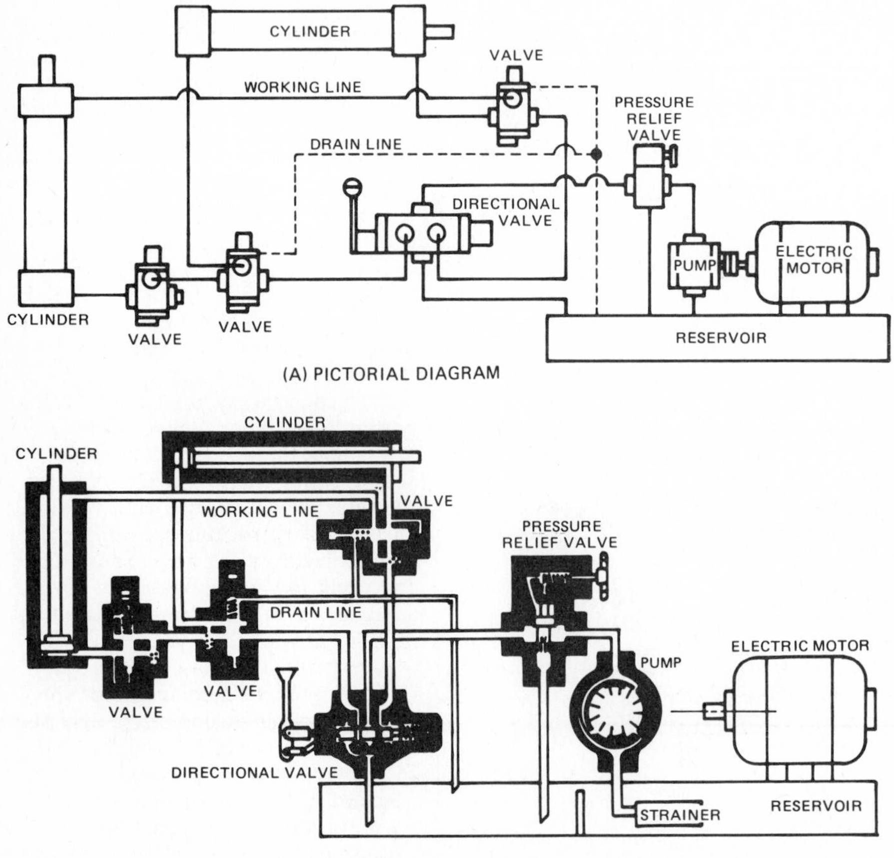
CYLINDER
VALVE
WORKING LINE
PRESSURE RELIEF VALVE
DRAIN LINE
DIRECTIONAL VALVE
PUMP
ELECTRIC MOTOR
CYLINDER
VALVE
VALVE
RESERVOIR
(A) PICTORIAL DIAGRAM
CYLINDER
CYLINDER
VALVE
WORKING LINE
PRESSURE RELIEF VALVE
DRAIN LINE
ELECTRIC MOTOR
PUMP
VALVE
VALVE
DIRECTIONAL VALVE
STRAINER
RESERVOIR
(B) CUTAWAY DIAGRAM

Fig. 23-1-14 Comparison between pictorial and cutaway diagrams.

ing lines) are drawn straight for clarity. They are not intended to indicate actual installation of pipe lines and are drawn as single or double lines depending on the type of diagram. Single lines are used on pictorial diagrams, double lines on cutaway diagrams.

REFERENCES AND SOURCE MATERIALS

1. Crane Canada, Limited.
2. Vickers, Incorporated.

Assignments

1. On an A3- or B-size sheet, make a pictorial diagram of the automobile brake hydraulic system shown in Fig. 23-1-A. Label the parts. Scale is to suit.

2. On an A3- or B-size sheet, make a cutaway diagram of the hydraulic circuit shown in Fig. 23-1-B. Label the parts. Scale is to suit.

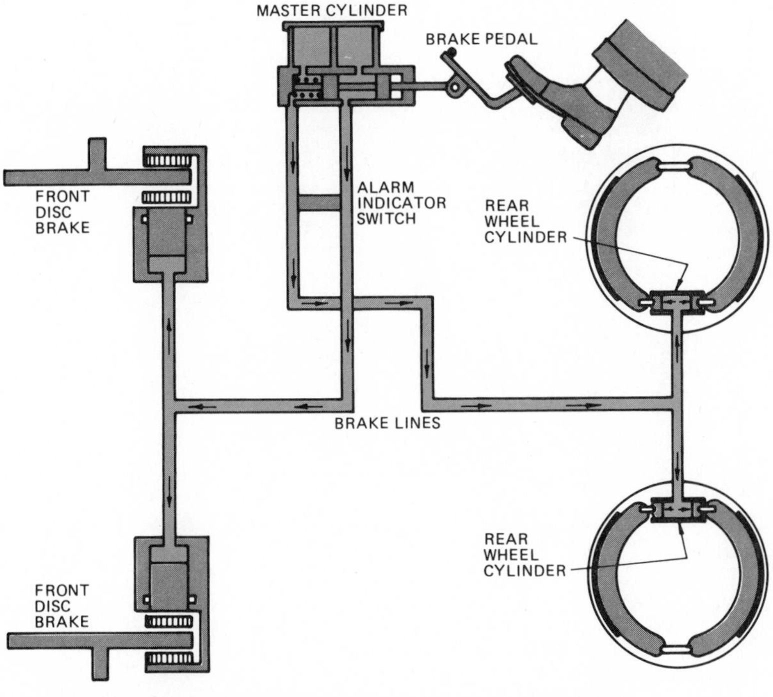
MASTER CYLINDER
BRAKE PEDAL
FRONT DISC BRAKE
ALARM INDICATOR SWITCH
REAR WHEEL CYLINDER
BRAKE LINES
REAR WHEEL CYLINDER
FRONT DISC BRAKE
AUTOMOBILE BRAKE HYDRAULIC SYSTEM

Fig. 23-1-A Brake hydraulic system.

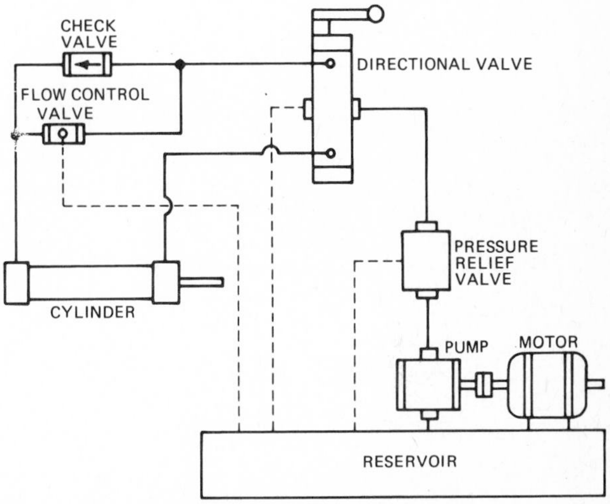
CHECK VALVE
FLOW CONTROL VALVE
DIRECTIONAL VALVE
CYLINDER
PRESSURE RELIEF VALVE
PUMP
MOTOR
RESERVOIR

Fig. 23-1-B Hydraulic circuit with one double-acting differential-type cylinder.

UNIT 23-2

GRAPHICAL DIAGRAMS[1]

Graphical diagrams are preferred by most application and service engineers for designing and troubleshooting hydraulic circuits. The graphical symbols are combinations of simple geometric figures which are easy to work with and clearly define the types and functions of the components. No attempt is made to show the shape or internal construction of a component. Note the differences in the diagrams shown in Fig. 23-2-1.

Some of the standard graphical symbols specified by the Joint Industry Conference (JIC) are shown in Fig. 23-2-2. The symbols can be combined in any form that is necessary to correctly depict a composite unit. Unless multiple diagrams are furnished showing the various phases of operation, the symbols will be shown in a diagram in their normal or neutral position. The reservoir symbol, like that of the electrical ground symbol, may be repeated several times on a graphical diagram.

Directional valves, because of various types of controls and flow paths, require a number of different symbols. A basic directional valve symbol consists of an envelope (square) for each position of the valve. The envelopes are drawn side by side in their correct order with each showing the flow path for its position. The method of actuating a valve is also shown symbolically. See Fig. 23-2-3.

Symbol Rules

1. Symbols show connections, flow paths, and function of the component represented. They do not indicate conditions occurring during transition from one flow-path arrangement to another. Further, they do not indicate construction or values such as pressure, flowrate, and other component settings.

2. Symbols do not indicate the location of ports, direction of shifting of spools, or position of control elements on actual components.

3. Symbols may be rotated or reversed without altering their meaning.

4. Line width does not alter meaning of symbols.

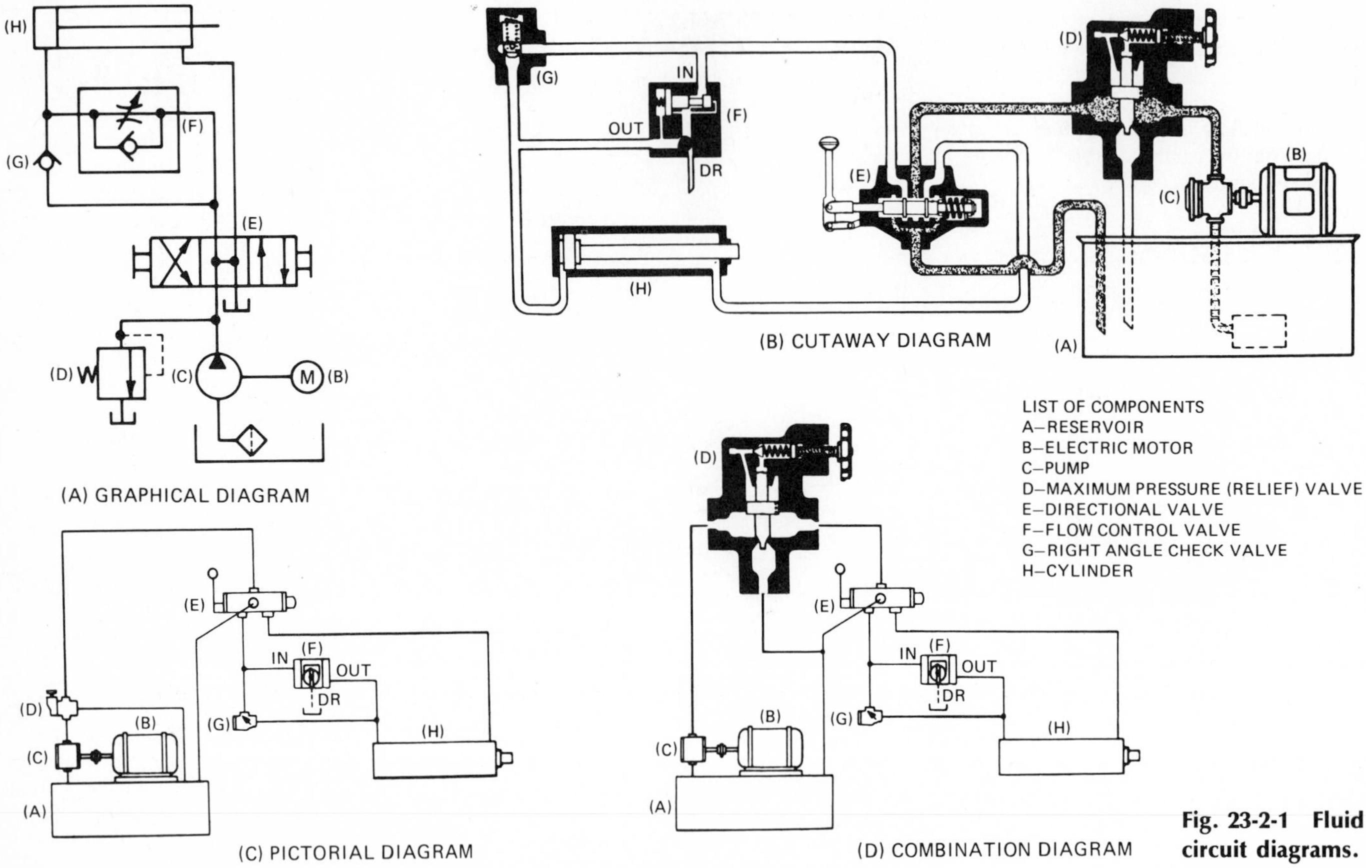

(H)
(F)
(G)
(E)
(D)
(C)
(M)
(B)
(A) GRAPHICAL DIAGRAM
(G)
IN
(F)
OUT
DR
(D)
(E)
(B)
(C)
(H)
(B) CUTAWAY DIAGRAM
(A)
LIST OF COMPONENTS
A–RESERVOIR
B–ELECTRIC MOTOR
C–PUMP
D–MAXIMUM PRESSURE (RELIEF) VALVE
E–DIRECTIONAL VALVE
F–FLOW CONTROL VALVE
G–RIGHT ANGLE CHECK VALVE
H–CYLINDER
(E)
IN
(F)
OUT
DR
(D)
(B)
(G)
(H)
(C)
(A)
(C) PICTORIAL DIAGRAM
(D)
(E)
IN
(F)
OUT
DR
(B)
(G)
(H)
(C)
(A)
(D) COMBINATION DIAGRAM

Fig. 23-2-1 Fluid circuit diagrams.

5. Symbols may be drawn any suitable size. Size may be varied for emphasis or clarity.

6. Letter combinations used as parts of graphical symbols are not necessarily abbreviations.

7. Where flow lines cross, a loop is used except within a symbol envelope. A loop may be used in this case, if clarity is improved by its use.

8. In multiple envelope symbols, the flow condition shown nearest a control symbol takes place when that control is caused or permitted to actuate.

9. Each symbol is drawn to show the normal or neutral condition of the component unless multiple circuit diagrams are furnished showing various phases of circuit operation.

10. Arrows should be used within symbol envelopes to show the direction of flow path in the component as used in the application represented. A double-end arrow is used to indicate reversing flow.

11. External ports are where flow lines connect to the basic symbol, except where component enclosure is used.

12. External ports are at the intersections of flow lines and the component enclosure symbol, when enclosure is used.

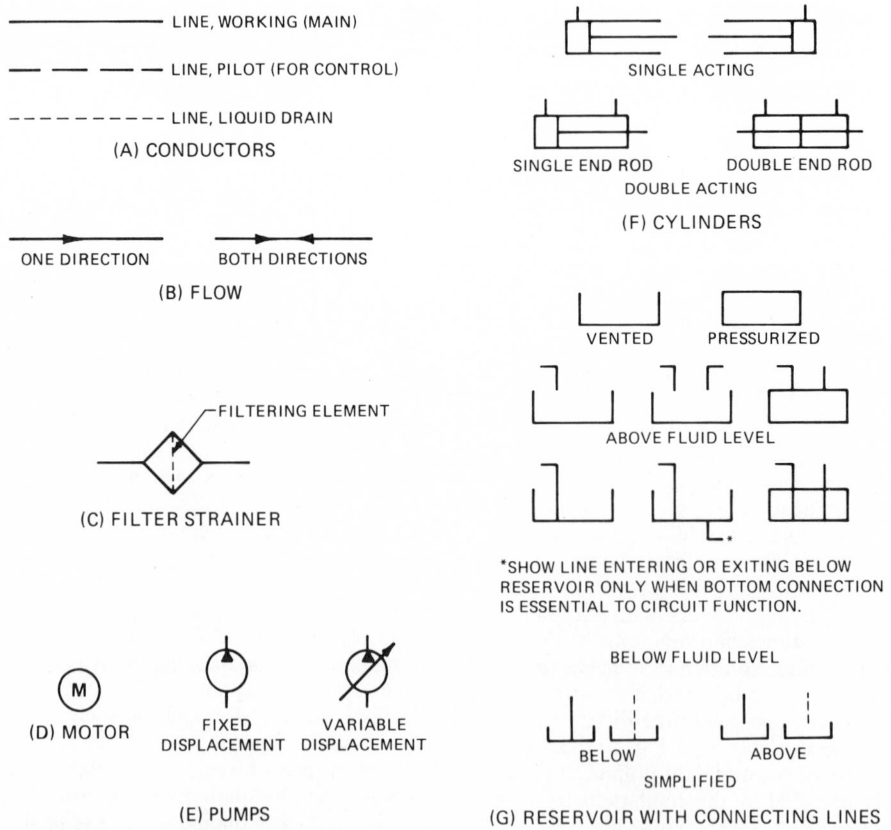

LINE, WORKING (MAIN)
LINE, PILOT (FOR CONTROL)
LINE, LIQUID DRAIN
(A) CONDUCTORS
ONE DIRECTION
BOTH DIRECTIONS
(B) FLOW
FILTERING ELEMENT
(C) FILTER STRAINER
M
(D) MOTOR
FIXED DISPLACEMENT
VARIABLE DISPLACEMENT
(E) PUMPS
SINGLE ACTING
SINGLE END ROD
DOUBLE END ROD
DOUBLE ACTING
(F) CYLINDERS
VENTED
PRESSURIZED
ABOVE FLUID LEVEL
*SHOW LINE ENTERING OR EXITING BELOW RESERVOIR ONLY WHEN BOTTOM CONNECTION IS ESSENTIAL TO CIRCUIT FUNCTION.
BELOW FLUID LEVEL
BELOW
ABOVE
SIMPLIFIED
(G) RESERVOIR WITH CONNECTING LINES

Fig. 23-2-2 Graphical symbols.

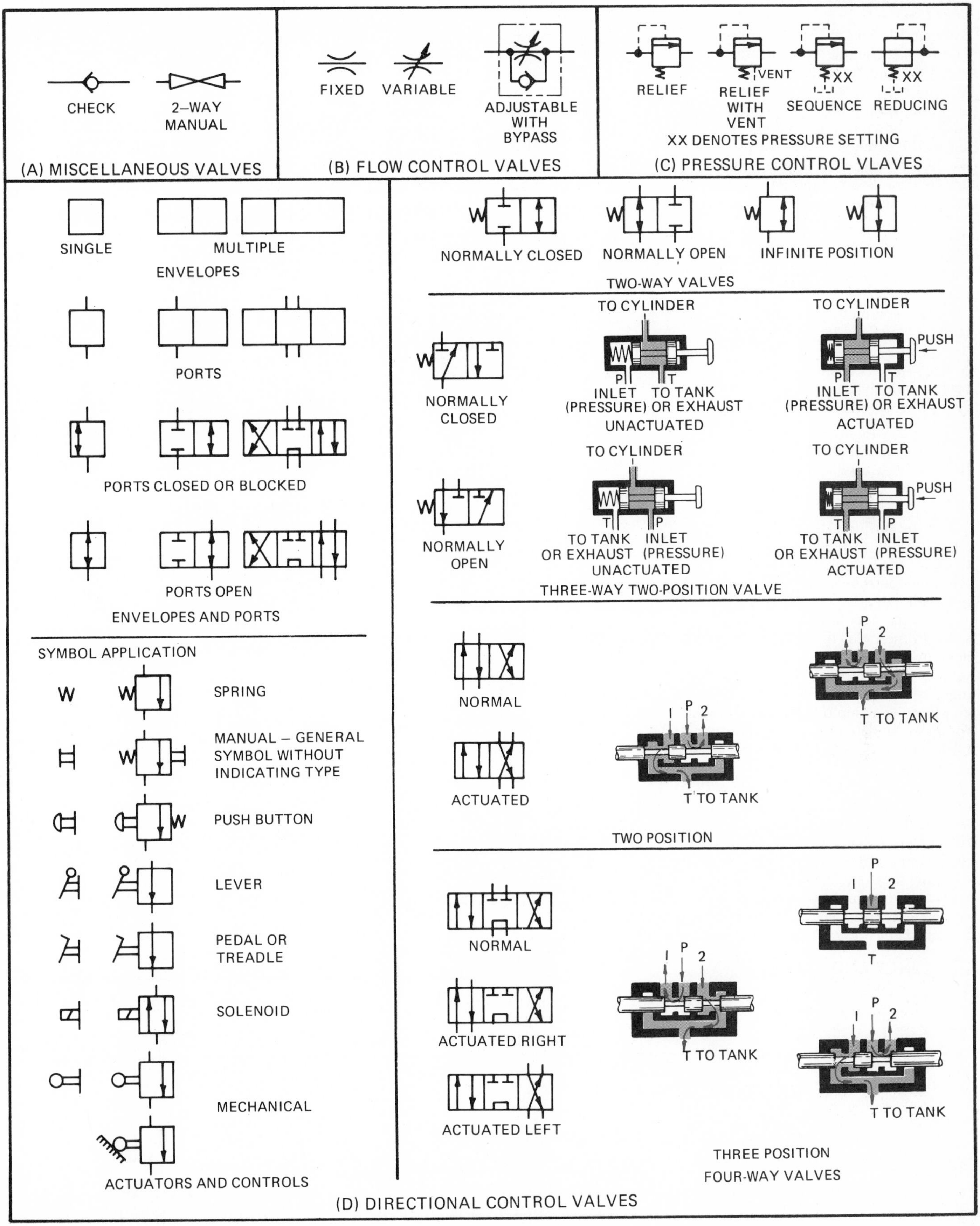
CHECK
2–WAY MANUAL
(A) MISCELLANEOUS VALVES
FIXED
VARIABLE
ADJUSTABLE WITH BYPASS
(B) FLOW CONTROL VALVES
RELIEF
VENT
RELIEF WITH VENT
XX
SEQUENCE
REDUCING
XX DENOTES PRESSURE SETTING
(C) PRESSURE CONTROL VLAVES
SINGLE
MULTIPLE
ENVELOPES
PORTS
PORTS CLOSED OR BLOCKED
PORTS OPEN
ENVELOPES AND PORTS
SYMBOL APPLICATION
W
SPRING
MANUAL – GENERAL SYMBOL WITHOUT INDICATING TYPE
PUSH BUTTON
LEVER
PEDAL OR TREADLE
SOLENOID
MECHANICAL
ACTUATORS AND CONTROLS
NORMALLY CLOSED
NORMALLY OPEN
INFINITE POSITION
TWO-WAY VALVES
TO CYLINDER
PUSH
P
T
INLET (PRESSURE)
TO TANK OR EXHAUST
UNACTUATED
ACTUATED
NORMALLY CLOSED
NORMALLY OPEN
THREE-WAY TWO-POSITION VALVE
NORMAL
ACTUATED
I
2
T TO TANK
TWO POSITION
NORMAL
ACTUATED RIGHT
ACTUATED LEFT
THREE POSITION FOUR-WAY VALVES
(D) DIRECTIONAL CONTROL VALVES

Fig. 23-2-3 Graphical valve symbols.

Hydraulic Circuits

Example 1. Figure 23-2-4 is a typical example of a simple hydraulic circuit. When the directional control valve (two-position, three-way, spring-operated, push-button control) is in the position shown in Fig. 23-2-4*a*, the cylinder piston extends. When the piston reaches the end of its stroke, the pump flow is then diverted through the pressure control (relief) valve.

When the push button is operated on the directional control valve, the port to the pump is blocked, as shown in Fig. 23-2-4*b*. The mass of the piston forces the hydraulic fluid from the cylinder through the directional control valve to the reservoir, and the piston retracts. Since pump flow through the directional control valve is blocked, the fluid is diverted through the pressure control (relief) valve back to the reservoir.

Example 2. Figure 23-2-5 shows a simple hydraulic circuit using a three-position, four-way valve controlling the movement of a double-acting cylinder. Figure 23-2-5*a* shows the control valve in its neutral position. Since pump flow through the directional control valve is blocked, the fluid is diverted through the pressure control (relief) valve back to the reservoir.

When the operator shifts the control valve to the right, fluid is directed to the head of the cylinder (Fig. 23-2-5*b*). The fluid at the other end of the cylinder is directed back to the reservoir. The pump continues to pump oil after the cylinder rod has completed its stroke. This excess oil is returned via the relief valve to the reservoir.

At the end of the forward stroke, the operator shifts the control valve to the left which causes the cylinder rod to retract (Fig. 23-2-5*c*). Fluid is directed to the rod end of the cylinder and the fluid from the head end returns to the reservoir. When the operator releases the control valve handle, the valve returns to its neutral position.

REFERENCE AND SOURCE MATERIAL

1. Vickers, Incorporated.

Assignments

1. On an A3- or B-size sheet, make a graphical diagram of the hydraulic circuit shown in Fig. 23-2-A.
2. On an A3- or B-size sheet, make a graphical diagram of the hydraulic circuit shown in Fig. 23-2-B.

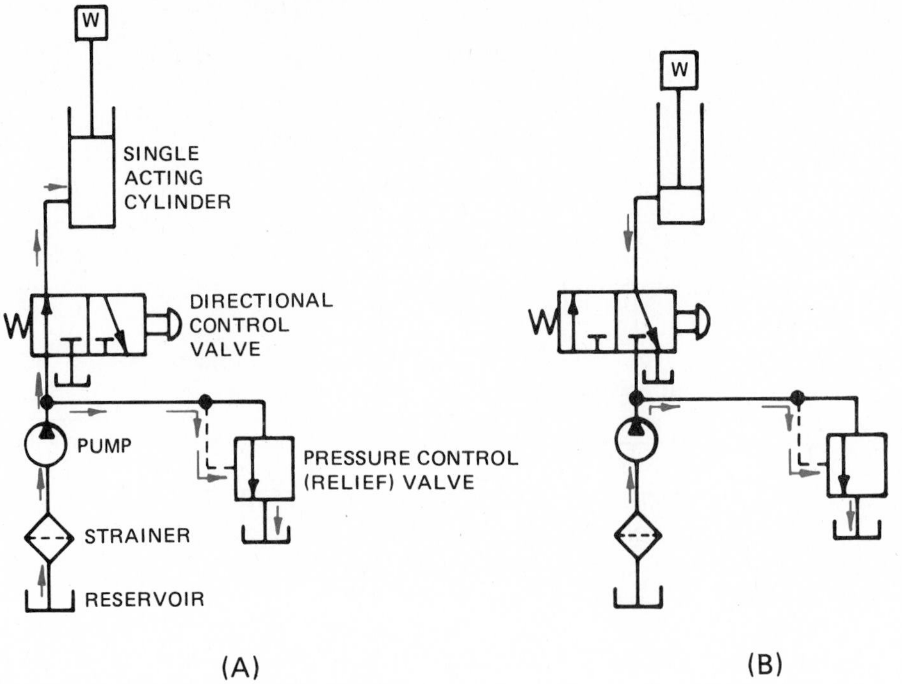
W
SINGLE ACTING CYLINDER
DIRECTIONAL CONTROL VALVE
PUMP
PRESSURE CONTROL (RELIEF) VALVE
STRAINER
RESERVOIR
(A)
(B)

Fig. 23-2-4 Simple hydraulic circuit.

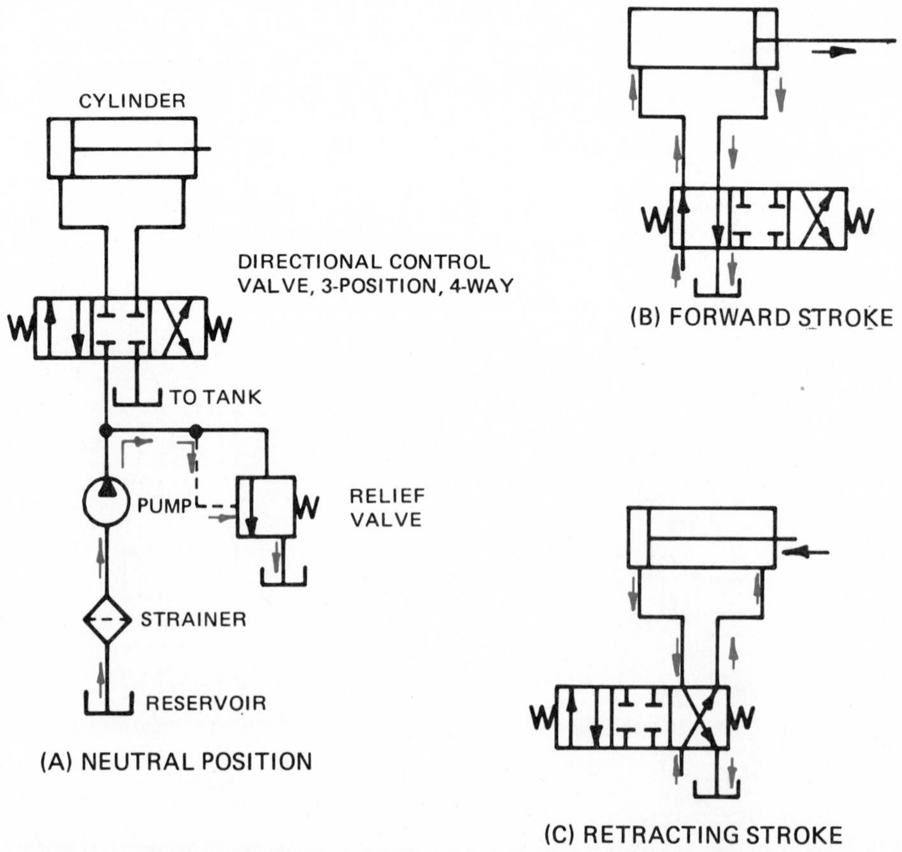
CYLINDER
DIRECTIONAL CONTROL VALVE, 3-POSITION, 4-WAY
TO TANK
PUMP
RELIEF VALVE
STRAINER
RESERVOIR
(A) NEUTRAL POSITION
(B) FORWARD STROKE
(C) RETRACTING STROKE

Fig. 23-2-5 Graphic diagram showing a three-position, four-way valve controlling a cylinder.

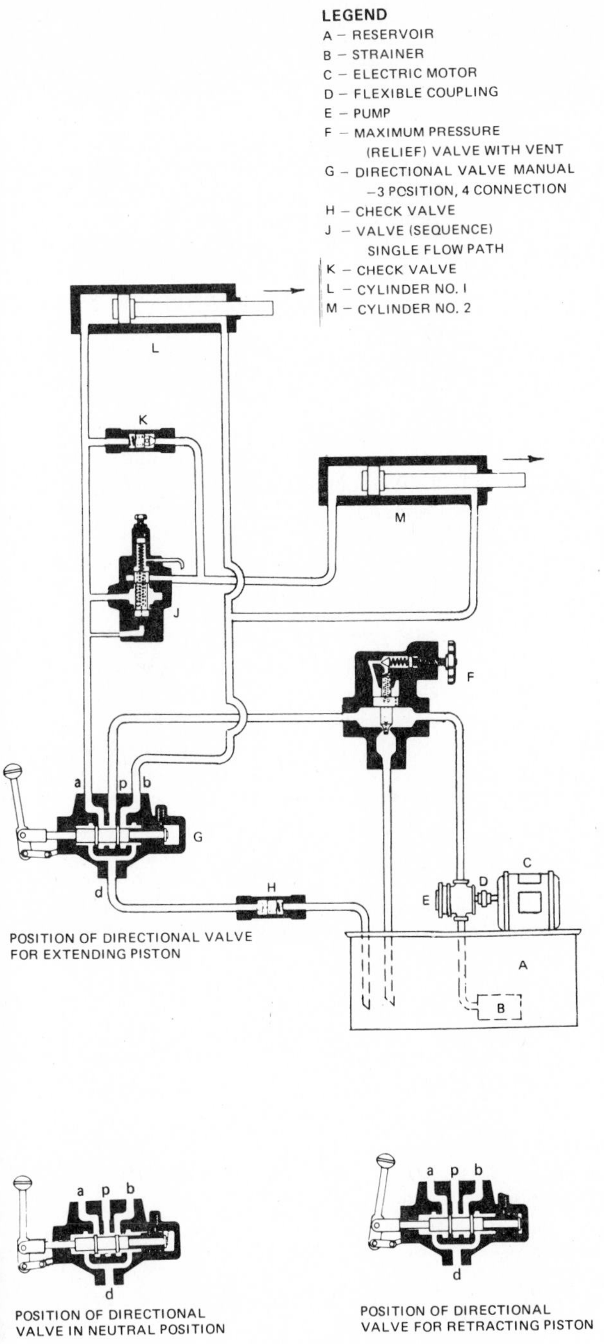
LEGEND
A – RESERVOIR
B – STRAINER
C – ELECTRIC MOTOR
D – FLEXIBLE COUPLING
E – PUMP
F – MAXIMUM PRESSURE (RELIEF) VALVE WITH VENT
G – DIRECTIONAL VALVE MANUAL –3 POSITION, 4 CONNECTION
H – CHECK VALVE
J – VALVE (SEQUENCE) SINGLE FLOW PATH
K – CHECK VALVE
L – CYLINDER NO. 1
M – CYLINDER NO. 2
POSITION OF DIRECTIONAL VALVE FOR EXTENDING PISTON
POSITION OF DIRECTIONAL VALVE IN NEUTRAL POSITION
POSITION OF DIRECTIONAL VALVE FOR RETRACTING PISTON

TO SEQUENCE THE ADVANCE OF TWO CYLINDERS, A SINGLE FLOW SEQUENCE VALVE IN THE SUPPLY LINE OF NO.2 CYLINDER (M) IS SET TO OPEN AT A PRESSURE IN EXCESS OF THAT REQUIRED TO ADVANCE NO.1 CYLINDER (L). AS THE DIRECTIONAL VALVE IS SHIFTED IN POSITION FOR EXTENDING PISTONS, FLOW IS DIRECTED FROM PORT (a). WHEN CYLINDER (L) COMPLETES ITS STROKE, SYSTEM PRESSURE BUILDS UP OPENING THE SEQUENCE VALVE (J) AND ALLOWING FLOW TO ADVANCE NO.2 CYLINDER (M). BY SHIFTING THE DIRECTIONAL CONTROL VALVE TO NEUTRAL, THE CYLINDERS CAN BE STOPPED AND HELD IN ANY POSITION WHILE PERMITTING THE PUMP OUTPUT TO RESERVOIR THROUGH PORT (d). WHEN THE DIRECTIONAL CONTROL VALVE IS SHIFTED TO THE RETRACTING POSITION FLOW IS DIRECTED FROM PORT (p) TO PORT (b). THE BY-PASS CHECK VALVE (K) ALLOWS NO.2 CYLINDER (M) TO RETRACT FREELY.

SEQUENCING THE ADVANCED STROKE OF TWO HYDRAULIC CYLINDERS

Fig. 23-2-A Sequencing the advanced stroke of two hydraulic cylinders.

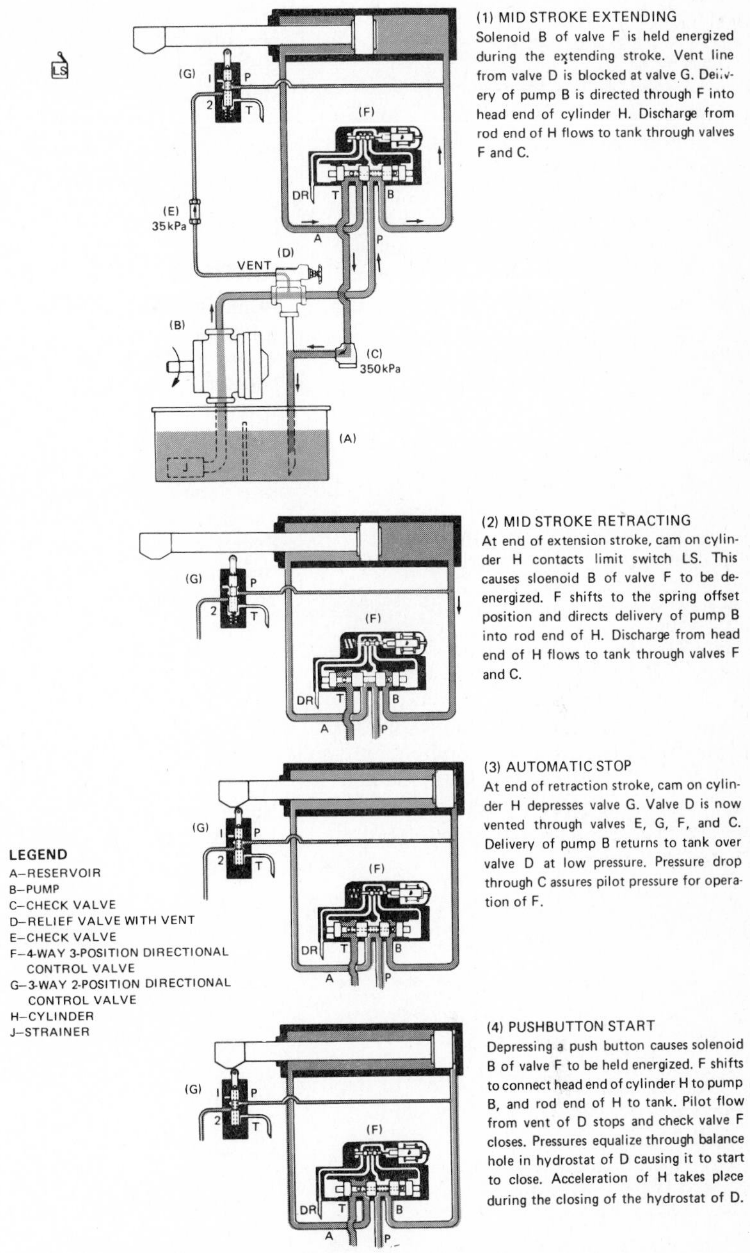
(1) MID STROKE EXTENDING
Solenoid B of valve F is held energized during the extending stroke. Vent line from valve D is blocked at valve G. Delivery of pump B is directed through F into head end of cylinder H. Discharge from rod end of H flows to tank through valves F and C.
(2) MID STROKE RETRACTING
At end of extension stroke, cam on cylinder H contacts limit switch LS. This causes sloenoid B of valve F to be de-energized. F shifts to the spring offset position and directs delivery of pump B into rod end of H. Discharge from head end of H flows to tank through valves F and C.
(3) AUTOMATIC STOP
At end of retraction stroke, cam on cylinder H depresses valve G. Valve D is now vented through valves E, G, F, and C. Delivery of pump B returns to tank over valve D at low pressure. Pressure drop through C assures pilot pressure for operation of F.
(4) PUSHBUTTON START
Depressing a push button causes solenoid B of valve F to be held energized. F shifts to connect head end of cylinder H to pump B, and rod end of H to tank. Pilot flow from vent of D stops and check valve F closes. Pressures equalize through balance hole in hydrostat of D causing it to start to close. Acceleration of H takes place during the closing of the hydrostat of D.
LEGEND
A–RESERVOIR
B–PUMP
C–CHECK VALVE
D–RELIEF VALVE WITH VENT
E–CHECK VALVE
F–4-WAY 3-POSITION DIRECTIONAL CONTROL VALVE
G–3-WAY 2-POSITION DIRECTIONAL CONTROL VALVE
H–CYLINDER
J–STRAINER
(E) 35kPa
(C) 350kPa
VENT

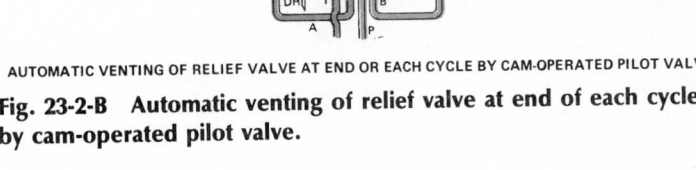

AUTOMATIC VENTING OF RELIEF VALVE AT END OR EACH CYCLE BY CAM-OPERATED PILOT VALVE

Fig. 23-2-B Automatic venting of relief valve at end of each cycle by cam-operated pilot valve.

UNIT 23-3

PNEUMATICS[1,2]

Compressed air has been used as a means of transmitting power for a long time, but in recent years pneumatics has taken on a new sophistication. With the availability of miniaturized valves and fluidic devices, air can now supply the nerves and brains, as well as the muscles, of complex automated equipment.

What is air? *Air* may be defined as a colorless, odorless, tasteless gas, chemically composed of nitrogen, oxygen, carbon dioxide, and water vapor.

Its molecules are widely separated and in constant motion, traveling in straight lines.

At sea level, air exists at a pressure of 101.3 kPa.

Air mixes readily with other fluids.

Basic Laws

There are several simple basic laws governing the behavior of air that are important to pneumatic system design.

Boyle's Law. This law defines the relationship between volume and pressure and states that if the temperature is constant, the volume of a given mass of gas varies inversely as its absolute pressure. It can be written

$$P_1V_1 = P_2V_2$$

where P_1 = the initial pressure (absolute)
V_1 = the initial volume
P_2 = the final pressure (absolute)
V_2 = the final volume

Charles' Law. This law refers to the behavior of a gas when changes in temperature take place and states that when there is no change in volume, the pressure of a gas varies directly with its absolute temperature.

Pascal's Law. This law states that the pressure of a gas in a container is transmitted undiminished in all directions and acts at right angles to the surfaces of the container (Fig. 23-1-1).

AIRFLOW

Air will only flow when a difference of pressure exists and flows *toward* the lower pressure. The rate of flow depends on the initial pressure, the difference in pressure, and the size, shape, and smoothness of the orifice. The shape and smoothness of the orifice are important. A straight or smoothly curved pipe assists airflow, and smooth inner walls reduce friction. Rough surfaces slow adjacent layers of air and reduce the effective area of the pipe.

CRITICAL BACK-PRESSURE RATIO

In an enclosed air system, one other factor must be considered. Figure 23-3-1 shows a supply tank at a constant pressure of 700 Pa connected through a stop valve to a receiver tank at atmospheric pressure (0 Pa). If the stop valve is opened, air will flow from *A* to *B* at a rate dependent on the orifice and pipe characteristics. Continuing flow will reduce the differential pressure between the supply and receiver tanks, but it is an observed phenomenon in gas behavior that the rate of flow does not change until the receiver tank pressure has reached 53 percent of the supply pressure, in this case, 371 Pa. Conversely, if the pressure in the receiver falls below 371 Pa, any further decrease in pressure will not increase airflow. This is called the *critical back-pressure ratio*, and can be important when selecting the size of a valve.

Pneumatic Equipment

AIR SUPPLY INSTALLATION

Before any pneumatic equipment can operate, it must have a supply of air at the correct pressure, in sufficient volume, and properly conditioned. See Fig. 23-3-2. The supply starts at the compressor which has a rated output pressure in pascals and a rated volume in cubic metres per second. This is often known as *free air delivered* (FAD). A standard cubic metre of air is the mass of air contained on 1 m^3 of space at 21°C and 101.3 kPa.

The compressor obeys Boyle's law by pushing air at atmospheric pressure into a smaller volume to increase the pressure. Unfortunately, Charles' law dictates that as pressure increases, so does temperature; and warm air contains more water vapor than cool air.

To decrease water content, a filter should be fitted immediately downstream of the compressor where moisture can be precipitated and drained off.

Although there is a definite trend toward oil-free compressors, industrial compressors are still predominantly of the oil-flooded type. Most have internal oil separators that are of quite high efficiency, but a certain amount of oil vapor does get through. This oil can contain wear particles from compressor parts, and also tends to be *burnt* due to the heat of compression. To prevent contamination of the air system, it is recommended that an efficient oil separator be fitted between the compressor discharge point and the air receiver.

AIR DISTRIBUTION SYSTEM

The air distribution system will be considered as beginning at the storage tank where a supply of air is available with most of the oil vapor and solids already removed. Between the tank outlet and the feeds to the individual systems, the biggest problem is how to get rid of water. A certain amount of precipitation will take place in the tank, so a drain is normally provided there. Methods of removing water in other parts of a pneumatic circuit are shown in Fig. 23-3-3.

CONDITIONING THE AIR

So far, the compression of the air, the removal of the moisture from the air, and the piping of the air to the supply points have been discussed. The next concern is to condition the air before it enters the tool, control, or power elements to do some useful work.

Filters. The average filter available (Fig. 23-3-4) is designed to do two jobs: remove moisture and remove dirt. When air enters the filter, its path is changed abruptly to a rotary direction, so centrifugal force hurls any particles of water

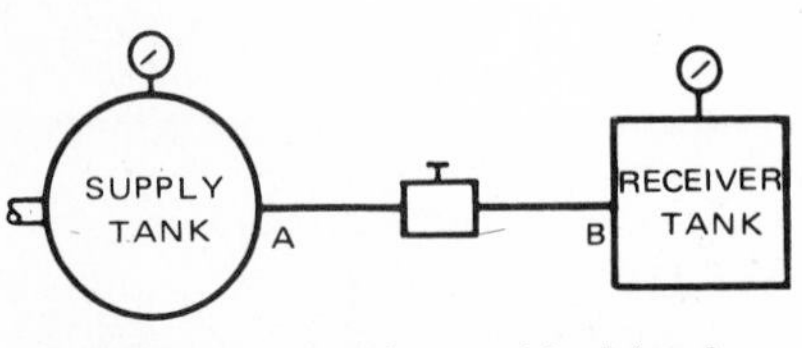
SUPPLY TANK
A
B
RECEIVER TANK

Fig. 23-3-1 Receiver critical back pressure is 53 percent of supply pressure.

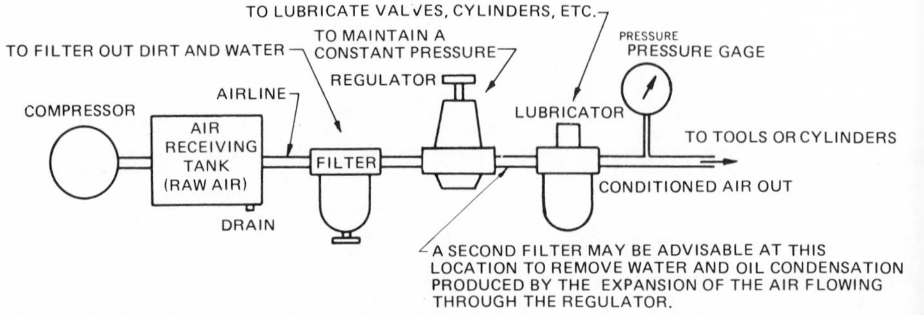
TO LUBRICATE VALVES, CYLINDERS, ETC.
TO MAINTAIN A CONSTANT PRESSURE
TO FILTER OUT DIRT AND WATER
PRESSURE
PRESSURE GAGE
REGULATOR
AIRLINE
COMPRESSOR
LUBRICATOR
AIR RECEIVING TANK (RAW AIR)
FILTER
TO TOOLS OR CYLINDERS
CONDITIONED AIR OUT
DRAIN
A SECOND FILTER MAY BE ADVISABLE AT THIS LOCATION TO REMOVE WATER AND OIL CONDENSATION PRODUCED BY THE EXPANSION OF THE AIR FLOWING THROUGH THE REGULATOR.

Fig. 23-3-2 Basic pneumatic components.

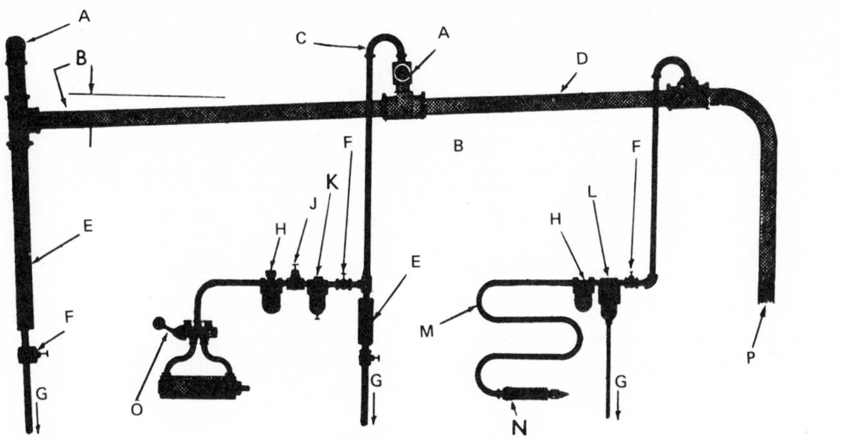
A
B
C
A
D
B
F
F
K
J
L
H
H
E
E
F
M
G
O
G
G
P
N
A—BRANCH MAIN
B—PITCH WITH FLOW
C—WIDE PATTERN RETURN BEND
D—MAIN
E—TRAP OR WATER LEG
F—STOP VALVE
G—DRAIN
H—LUBRICATOR
J—REGULATOR
K—FILTER
L—AUTOMATIC DRAIN AIR FILTER
M—FLEXIBLE HOSE
N—AIR DRILL
O—HAND-OPERATED VALVE AND AIR CYLINDER
P—FROM RECEIVER

Fig. 23-3-3 Air distribution diagram.

outward. Here they collect on the sides of the filter bowl and gravitate to the bottom where a baffle prevents air turbulence. This area is called the *quiet zone* and is drained by an automatic or manual valve.

Dirt exists to a greater or lesser extent in various forms in any plant system and is intercepted in most filters by a cartridge element. These elements are rated by the size, in microns, of particles they will intercept. A micron is 0.001 mm.

For work, or power air, a filter of 50 to 60 microns rating is normally sufficient, since this air comes in contact with contamination from cylinder or shaft packings. Control air is not normally subject to this, so should be passed through a secondary filter of about 5 microns. This finer filtration will add appreciably to the efficiency and life of the control section of the circuit.

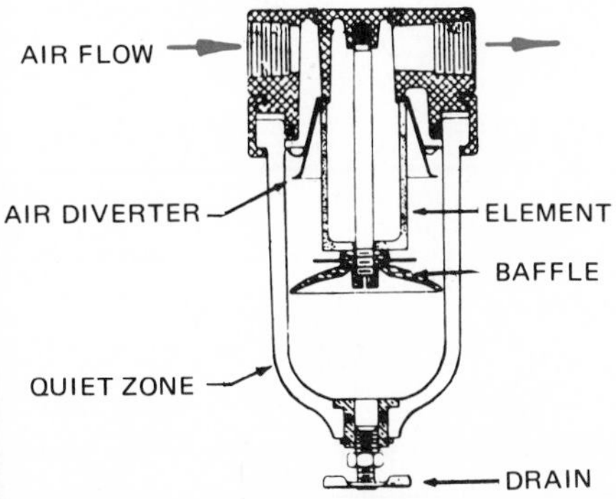
AIR FLOW
AIR DIVERTER
ELEMENT
BAFFLE
QUIET ZONE
DRAIN

Fig. 23-3-4 Typical air filter.

PRESSURE REGULATORS (Fig. 23-3-5)

The three main reasons for regulating air pressure are as follows:

1. To keep wasted air to a minimum
2. To achieve maximum consistency in circuit performance
3. To maintain an optimum balance between work output of components and their wear rate

A regulator must be chosen to give the correct working range, usually 0 to 1000 kPa.

In locating the regulator, always try to get it upstream of a lubricator, since air can be contaminated by the interaction of some oils on the regulator diaphragm.

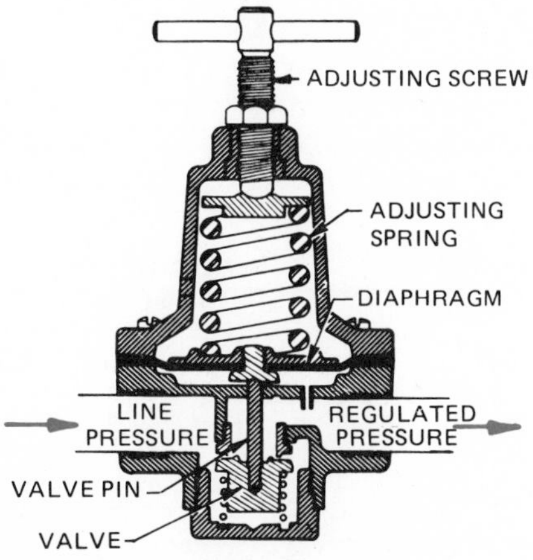
ADJUSTING SCREW
ADJUSTING SPRING
DIAPHRAGM
LINE PRESSURE
REGULATED PRESSURE
VALVE PIN
VALVE

Fig. 23-3-5 Simple spring-type regulator.

AIR LUBRICATORS

The standard type lubricator (Fig. 23-3-6) operates by creating a pressure differential between the lubricant container and a metering chamber. This causes the oil to enter the metering chamber and disperse into the pipeline as fog. Lubricators are designed to operate with a certain size pipe and airflow.

Common types of lubricators in use are the oil-fog type and the micro-fog type. The *oil-fog type* disperses relatively large drops of oil which have a tendency to fall out early, the normal carrying distance before fallout being about 5 m in a straight length of pipe.

The *micro-fog type* disperses much smaller oil particles which remain in suspension more easily. The suggested carrying distance before fallout for this type is 8 000 mm in a straight pipe.

In choosing the lubricant, it is preferable to use the grade recommended by the component manufacturers.

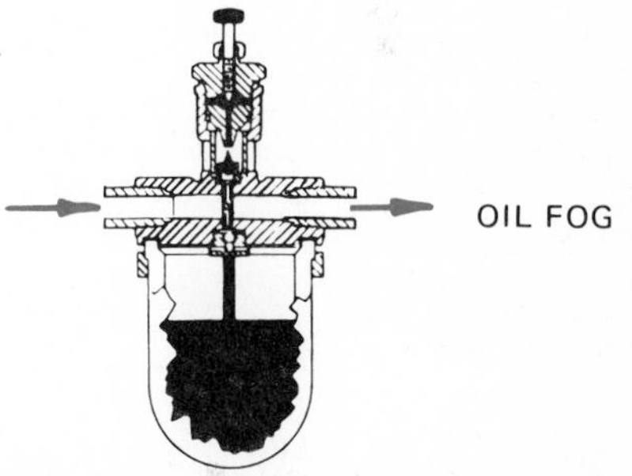
OIL FOG

Fig. 23-3-6 Air-lubricator oil-fog type.

Air Circuit Components

Air circuit elements can be considered under the three separate functions they perform: power, control, and signal.

POWER ELEMENTS

Cylinders. The most common power elements are cylinders, and many factors influence their choice.

If work is to be performed in one direction only, then a *single-acting cylinder* (Fig. 23-1-7*a*) may be used. This type is retracted by an internal spring, an external load or gravity, or bucking air. If work is to be done in both directions, i.e., if the return load exceeds cylinder friction, a *double-acting cylinder* (Fig. 23-1-7*b*) is needed.

The size of the cylinder will depend on the magnitude of load and the distance it must be moved. The bore can be found from the formula $A = F/P$, where F is the load and P is the air pressure available. The stroke should equal the distance to be moved.

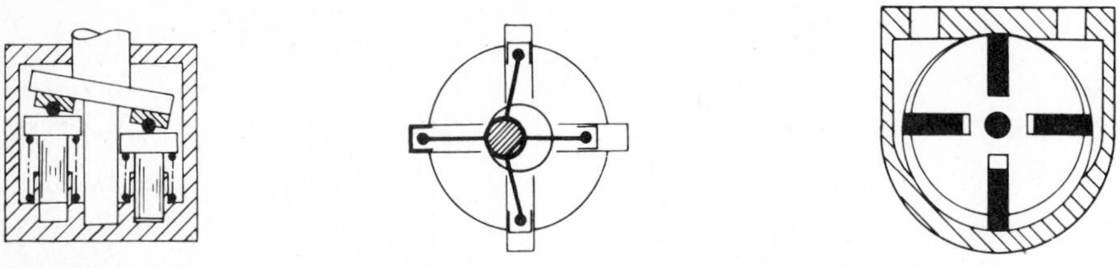
(A) AXIAL PISTON
(B) RADIAL PISTON
(C) VANE TYPE

Fig. 23-3-7 Air motors.

After it is decided which cylinder will be used, its air consumption must be calculated. This is the volume found by calculating the volume of the free end of the cylinder, adding the volume of the piston rod end if the piston is double-acting, and multiplying by the frequency in strokes per minute.

This calculation is important, but it is often neglected. It must be done not only to decide the size of lubricator required, but also to check if the plant system has the capacity to operate the circuit.

Air Motors. *Air motors* are used to convert the energy of compressed air into continuous torque. They are not the most efficient means of producing torque, since the average motor needs about 4 kW at the compressor to produce 1 kW at the motor, but they still have a lot in their favor.

With the growing popularity of pneumatic controls, air motors can make some machines or systems independent of electricity, particularly if the compressor is driven by an internal combustion engine. They cannot be harmed by stalling; they do not overheat, since the air expanding through the motor has a cooling effect; they have a very high power-to-mass ratio; and they are very reliable (seldom break down, only wear out slowly with lots of warning).

There are three main types (see Fig. 23-3-7). The two *piston types* are the work horses, giving high power at lower speeds. *Vane types* are the race horses, suitable for lighter loads at higher speeds, and they are more compact. The majority of low-kilowatt motors in use are of this type.

In selecting an air motor, it must be remembered that power is proportional to revolutions per minute and pressure. This is the pressure at the motor inlet, so air lines must be big enough to pass the required volume of air. If several motors are operating intermittently from the same air system, surge tanks or air reservoirs should be provided. The motor chosen should give the required power and revolutions per minute at about one-half the maximum pressure, so that there will be no power loss under adverse conditions.

CONTROL ELEMENTS

Power Valves. *Power valves* direct airflow to and from the work elements.

There are various designs of valve action, and it is helpful if the action requires low thrust to actuate the valve. More important, the required thrust should be constant and unaffected by variations in pressure or airflow through the valve or by changes in friction within the valve.

To eliminate unnecessary air wastage as the valve shifts, the action should block the air supply from the connecting flow paths in the valve as it moves through midposition.

In most applications, the valve action should be detented to keep it in its selected position in the event of failure in the controlling medium (electricity or air). The detent also safeguards against valve shift due to machine vibration and allows valve mounting in any position without the possibility of shifting due to gravity. It will also prevent valve creep due to the slow pressure buildup that sometimes occurs in air-pilot-actuated valves.

A good valve action provides a variety of flow paths by accepting and valving air at any port. This allows the use of the same valve for different circuit functions, such as normally open, normally closed, two-way, three-way, dual-pressure, and others.

There are three methods commonly used to actuate a power valve: mechanical, electrical, and air. In a two-position valve, any combination of these may be used.

The selection of power valves is governed by the airflow required, the flow paths needed, and the method of actuation.

Valves are termed by the number of flow paths they provide, either two- or three-position.

Two-position means that two flow conditions exist and are relative to the position of the valve.

Three-position is similar but has a third flow condition when the valve mechanism is centered.

One other factor must be known when describing a power valve. That is the number of *ways* it may be used and refers basically to the number of ports or connections in the valve body.

Figure 23-3-8 shows two types of two-way, two-position valves. Air connected to one port on the right valve will only flow out of the other port when the valve is actuated. This is known as a normally closed, two-way, two-position valve. On the left is a normally open, two-way, two-position valve, since air flows until the valve is actuated.

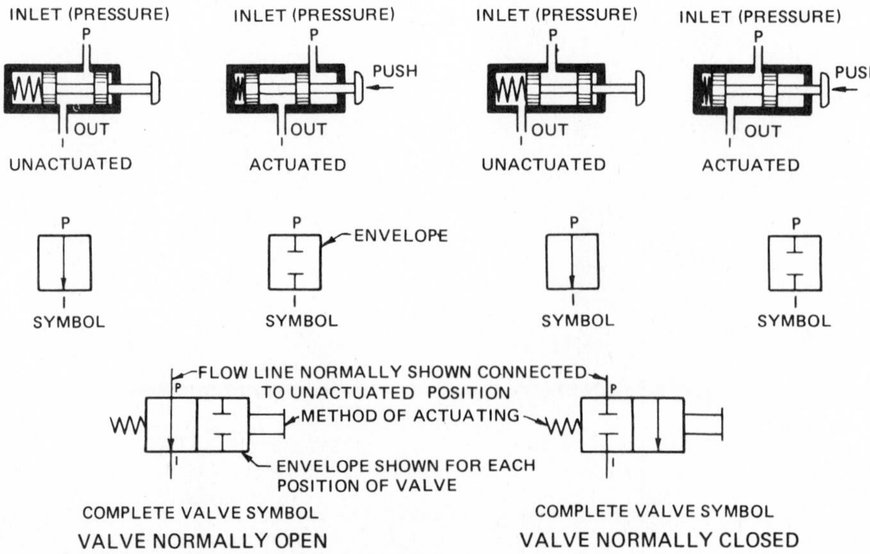
INLET (PRESSURE)
P
OUT
UNACTUATED
INLET (PRESSURE)
P
PUSH
OUT
ACTUATED
INLET (PRESSURE)
P
OUT
UNACTUATED
INLET (PRESSURE)
P
PUSH
OUT
ACTUATED
P
SYMBOL
P
ENVELOPE
SYMBOL
P
SYMBOL
P
SYMBOL
FLOW LINE NORMALLY SHOWN CONNECTED TO UNACTUATED POSITION
METHOD OF ACTUATING
ENVELOPE SHOWN FOR EACH POSITION OF VALVE
P
P
COMPLETE VALVE SYMBOL
VALVE NORMALLY OPEN
COMPLETE VALVE SYMBOL
VALVE NORMALLY CLOSED

Fig. 23-3-8 Two-way, two-position valves.

Pneumatic Circuit Diagrams

Pictorial and Cutaway Diagrams. The same type symbols as shown in hydraulic pictorial and cutaway diagrams also apply to pneumatic circuits. See Unit 23-1.

Graphic Symbols. With the exception of the symbols shown in Fig. 23-3-9, pneumatic graphic symbols are identical to those used in hydraulic circuits. See Figs. 23-2-2 and 23-2-3.

PNEUMATIC CIRCUITS

Example 1. See Fig. 23-3-10. When the operator shifts the control valve to the right, air is directed to the head end of the cylinder. The return air is directed back through the control valve to the atmosphere. At the end of the forward stroke, the operator shifts the control valve to the left. Air is directed to the rod end of the cylinder, and the air from the head end is directed to the atmosphere by way of the control valve. The pump continues to pump air after the cylinder rod has completed its stroke. This excess air is disposed of by the pressure regulator valve to the atmosphere. When the operator releases the control-valve handle, the spring-centered valve returns to neutral. The pressure on the cylinder is relieved while the pressure inlet remains closed, preventing the valve from draining the compressor. With the pressure removed from both ends of the cylinder, the cylinder rod can be moved readily. This action is called *floating* and is used in both air and hydraulic circuitry.

Example 2. A typical sequence circuit is shown in Fig. 23-3-11. The sequence of operation is (1) extend clamp cylinder, (2) extend work cylinder, (3) retract work cylinder, and (4) retract clamp cylinder. When the control valve is shifted, air is directed into the head of the clamp cylinder extending the piston and clamping the part. Air also flows to sequence valve 1, but no flow occurs through the valve because of the spring-loaded ball. After the clamp cylinder has completed its stroke, pressure builds up in the line and overcomes the spring tension in the valve, permitting air to pass through the valve to the head of the work cylinder. The piston of the work cylinder extends to perform the work. The pressure in the line is controlled by the pressure regulator valve.

When the control valve is released, air pressure flows into the rod end of the work cylinder retracting the piston. The air also flows to sequence valve 2 but is blocked by the spring-loaded valve. The air from the work cylinder head is forced through the sequence and control valves to the atmosphere.

After the work cylinder is fully retracted, the pressure again builds up and overcomes the spring tension in sequence valve 2, forcing air into the rod end of the clamp cylinder and thus retracting the piston and releasing the work. The air in the head of the work cylinder is forced out of the cylinder through the sequence and control valves to the atmosphere.

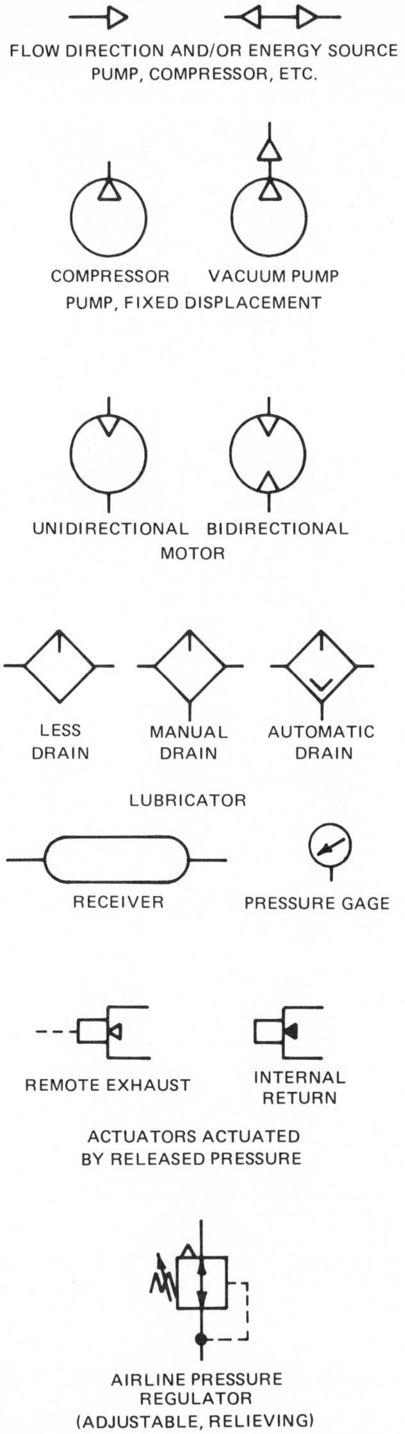
FLOW DIRECTION AND/OR ENERGY SOURCE
PUMP, COMPRESSOR, ETC.
COMPRESSOR
VACUUM PUMP
PUMP, FIXED DISPLACEMENT
UNIDIRECTIONAL
BIDIRECTIONAL
MOTOR
LESS DRAIN
MANUAL DRAIN
AUTOMATIC DRAIN
LUBRICATOR
RECEIVER
PRESSURE GAGE
REMOTE EXHAUST
INTERNAL RETURN
ACTUATORS ACTUATED BY RELEASED PRESSURE
AIRLINE PRESSURE REGULATOR (ADJUSTABLE, RELIEVING)

Fig. 23-3-9 Pneumatic graphic symbols.

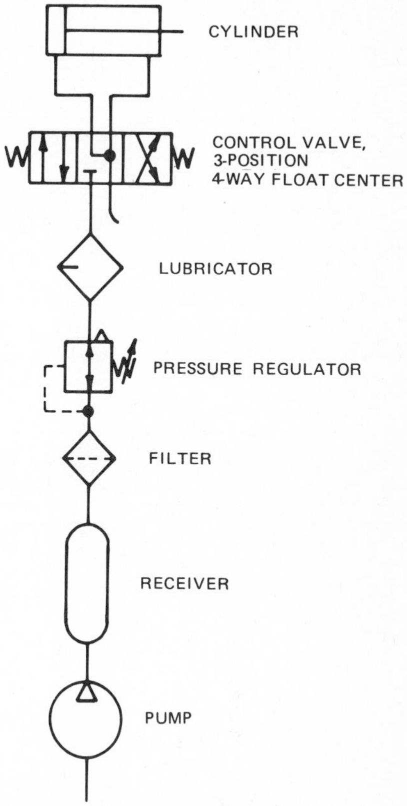
CYLINDER
CONTROL VALVE, 3-POSITION 4-WAY FLOAT CENTER
LUBRICATOR
PRESSURE REGULATOR
FILTER
RECEIVER
PUMP

Fig. 23-3-10 Simple pneumatic circuit.

REFERENCES AND SOURCE MATERIAL

1. J. Mooney, "Course on Basic Pneumatics," *Design Engineering*, Vol. 11, Nos. 11 and 12, 1965; Vol. 12, No. 1, 1966.

2. Holman Bros. (Canada), Limited, Maxam-Nopak Division.

Assignments

1. On an A3- or B-size sheet, make a graphical diagram of the pneumatic circuit shown in Fig. 23-3-A. The circuit operations are as follows.

All air entering the circuit passes through the lubro control unit, where it is filtered, lubricated, and delivered at a constant controllable pressure to the inlet port of valves 1, 3, 5, 6, 7, 8, 10, 12, 13, 14, and 15.

The continued operation of this circuit, which is automatic, depends entirely on the operation of valve 2. This valve is depressed only when a bar of sufficient length is in the machine and is positioned accordingly.

With this condition satisfied, the operator momentarily depresses valve 1. Valve 1 passes a feed through valve 2 to

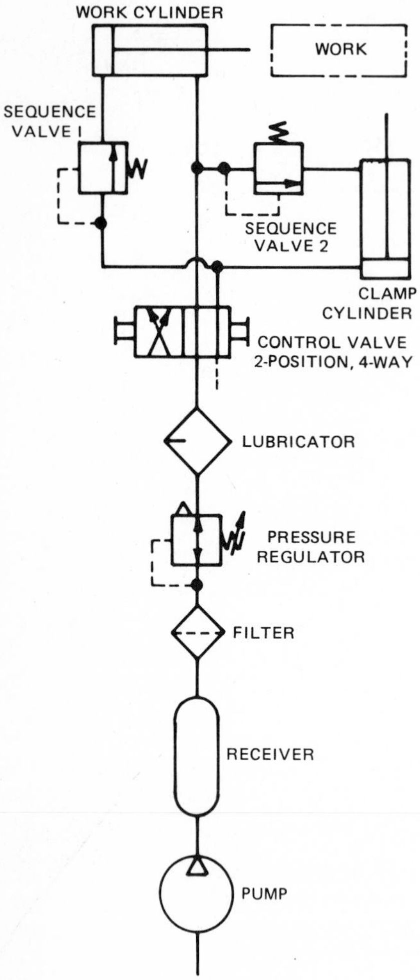

WORK CYLINDER
WORK
SEQUENCE VALVE 1
SEQUENCE VALVE 2
CLAMP CYLINDER
CONTROL VALVE 2-POSITION, 4-WAY
LUBRICATOR
PRESSURE REGULATOR
FILTER
RECEIVER
PUMP

Fig. 23-3-11 Sequence circuit.

the bottom end of valve 3, which operates, passing air unrestricted to the bottom end of cylinder A, thus extending the piston rod to lift the saw.

Toward the end of the upward stroke, valve 5 is momentarily tripped, passing air to the bottom end of valve 6, which operates, passing air to the top end of cylinder B, thus retracting the piston rod to open the vise. When nearing its retracted position, the piston rod momentarily trips valve 7 to pass air to the bottom end of valve 8, which operates, passing air to the top end of cylinder C, thus retracting the piston rod to open the clamp. When fully retracted, a pressure buildup operates sequence valve 9, which allows air to pass into the RH end of the cylinder D, thus retracting the piston rod. At an appropriate point (depending on the required length of the billets), valve 10 is momentarily tripped to pass air to the top end of valve 8, which operates, passing air to the bottom end of cylinder C, thus extending the piston rod to clamp the bar. When fully ex-

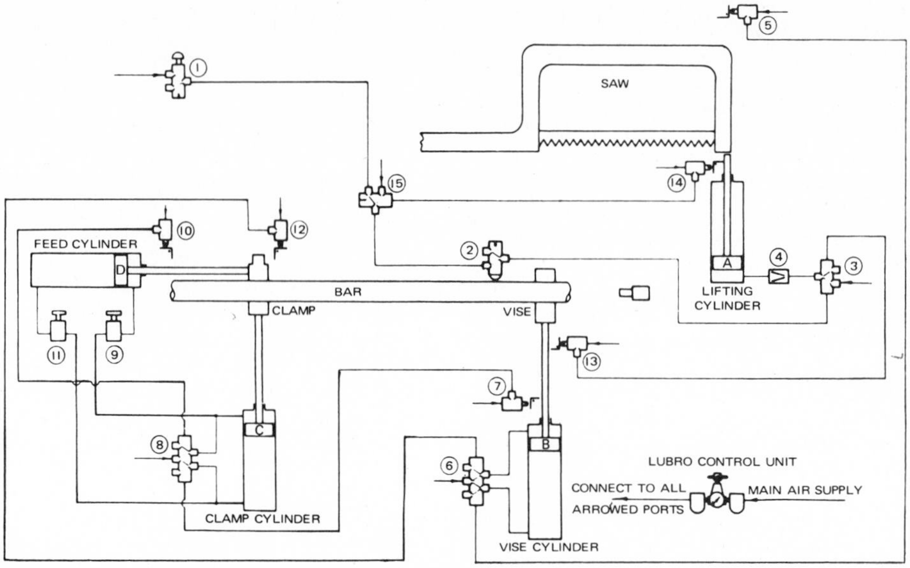

SAW
FEED CYLINDER
BAR
CLAMP
VISE
LIFTING CYLINDER
CLAMP CYLINDER
VISE CYLINDER
LUBRO CONTROL UNIT
CONNECT TO ALL ARROWED PORTS
MAIN AIR SUPPLY

Fig. 23-3-A Control circuit for automatic sawing machine.

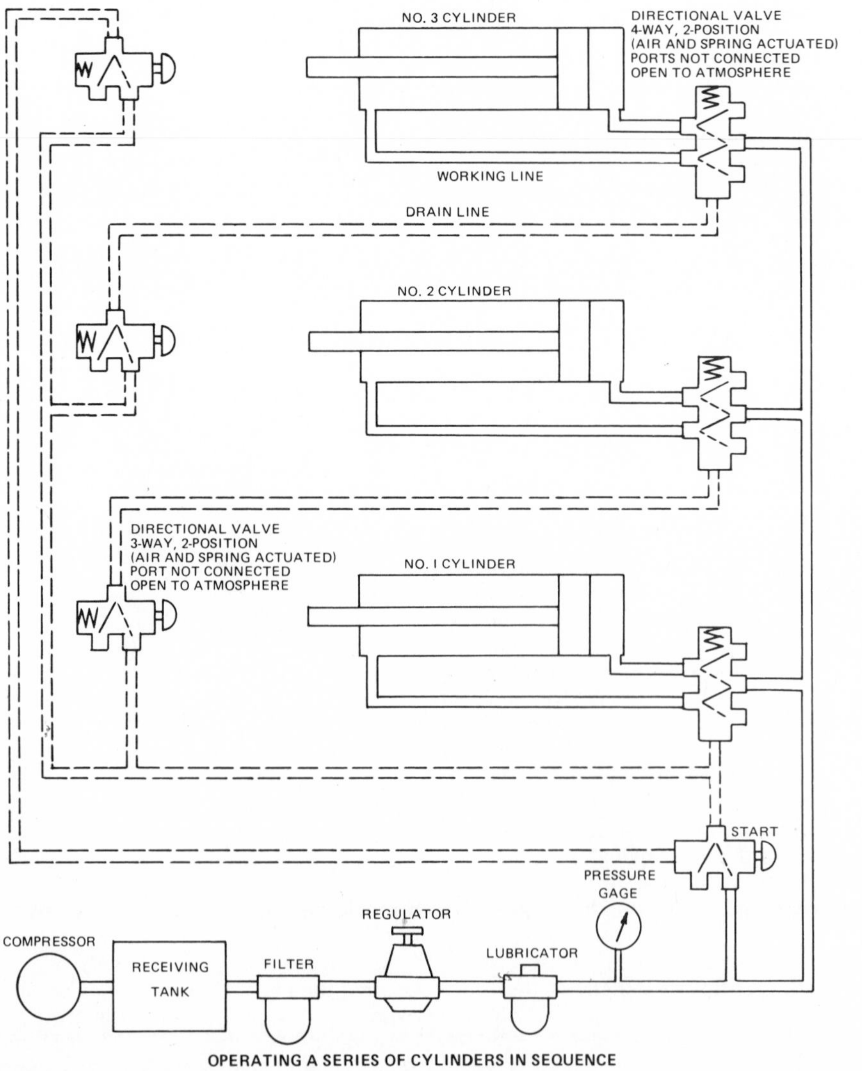

NO. 3 CYLINDER
DIRECTIONAL VALVE 4-WAY, 2-POSITION (AIR AND SPRING ACTUATED) PORTS NOT CONNECTED OPEN TO ATMOSPHERE
WORKING LINE
DRAIN LINE
NO. 2 CYLINDER
DIRECTIONAL VALVE 3-WAY, 2-POSITION (AIR AND SPRING ACTUATED) PORT NOT CONNECTED OPEN TO ATMOSPHERE
NO. 1 CYLINDER
START
PRESSURE GAGE
REGULATOR
COMPRESSOR
RECEIVING TANK
FILTER
LUBRICATOR
OPERATING A SERIES OF CYLINDERS IN SEQUENCE

Fig. 23-3-B Operating a series of cylinders in sequence.

tended, a pressure buildup operates sequence valve 11, which allows air to pass to the LH end of cylinder *D*, thus extending the piston rod to feed the required length of bar into the vise.

At a predetermined point in its travel, the piston rod of cylinder *D* momentarily trips valve 12 to pass air to the top end of valve 6, which operates, passing air to the bottom end of cylinder *B*, thus extending the piston rod to close the vise. As the vise closes, valve 13 is tripped to pass air to the top end of valve 3, which operates, exhausting the line to the bottom of cylinder *A*. The saw, therefore, descends under its own mass at a speed controlled by adjustment of restrictor valve 4. When the bar has been cut through, the saw trips valve 14 to pass air to the RH end of valve 15, which operates, passing air via valve 2 to the bottom end of valve 3, which operates to recommence the cycle.

2. On an A3- or B-size sheet, make a graphical diagram of the pneumatic circuit shown in Fig. 23-3-B.

3. On an A3- or B-size sheet, lay out the pneumatic system shown in Fig. 23-3-C using graphic symbols.

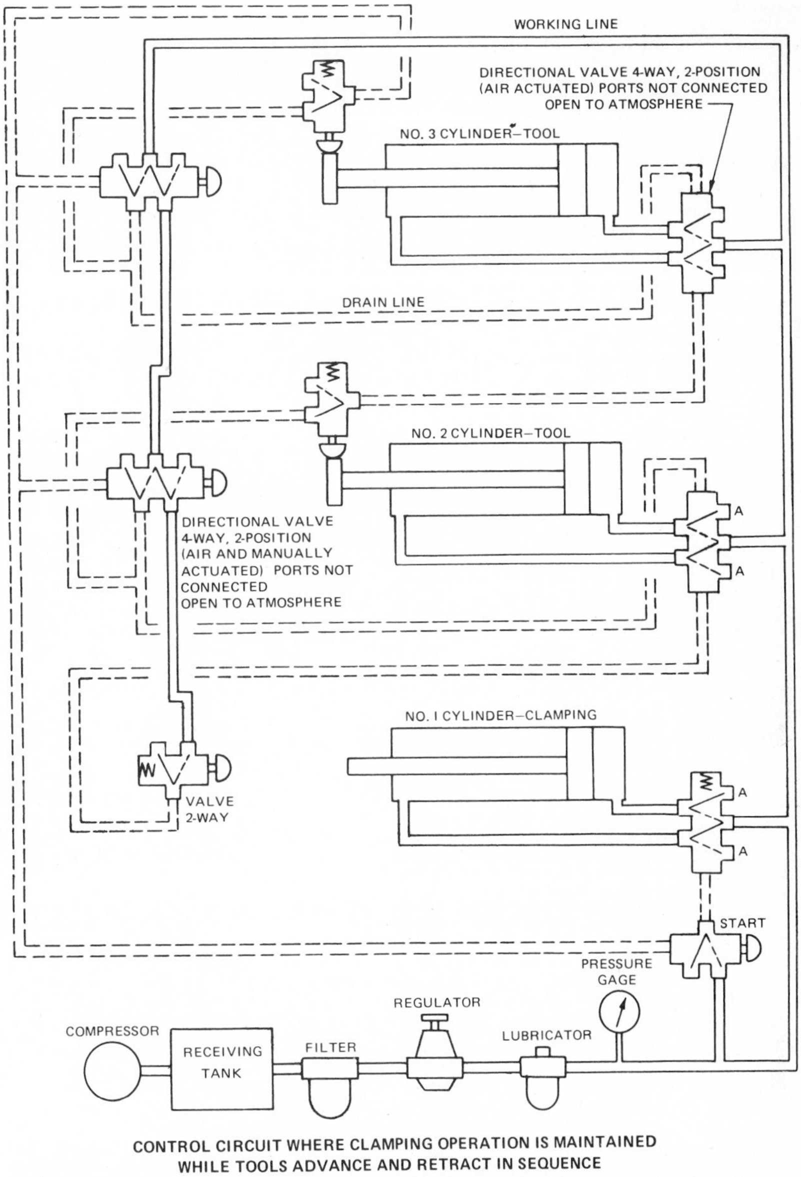
WORKING LINE
DIRECTIONAL VALVE 4-WAY, 2-POSITION
(AIR ACTUATED) PORTS NOT CONNECTED
OPEN TO ATMOSPHERE
NO. 3 CYLINDER–TOOL
DRAIN LINE
NO. 2 CYLINDER–TOOL
A
DIRECTIONAL VALVE
4-WAY, 2-POSITION
(AIR AND MANUALLY
ACTUATED) PORTS NOT
CONNECTED
OPEN TO ATMOSPHERE
A
NO. I CYLINDER–CLAMPING
A
VALVE
2-WAY
A
START
PRESSURE
GAGE
REGULATOR
COMPRESSOR
RECEIVING
TANK
FILTER
LUBRICATOR
CONTROL CIRCUIT WHERE CLAMPING OPERATION IS MAINTAINED
WHILE TOOLS ADVANCE AND RETRACT IN SEQUENCE

Fig. 23-3-C Control circuit where clamping operation is maintained while tools advance and retract in sequence.

Part 5
Special Fields of Drafting

Chapter 24 Development and Intersections

UNIT 24-1
SURFACE DEVELOPMENTS AND INTERSECTIONS

Surface Developments

Many objects, such as cardboard and metal boxes, "tin" cans, funnels, cake pans, furnace pipes, elbows, ducts, and eaves troughing, are made from flat sheet material that is cut so that, when folded, formed, or rolled, it will take the shape of an object. Since a definite shape and size are desired, a regular orthographic drawing of the object, such as shown in Fig. 24-1-1, is made first; then a development drawing is made to show the complete surface or surfaces laid out in a flat plane.

SHEET-METAL DEVELOPMENT

Surface development drawing is sometimes referred to as *pattern drawing*, because the layout, when made on heavy cardboard, thin metal, or wood, is used as a pattern for tracing out the developed shape on flat material. Such patterns are used extensively in sheet-metal shops.

When making a development drawing of an object which will be constructed of thin metal, such as a tin can or a dust pan, the drafter must be concerned not only with the developed surfaces but also with the joining of the edges of these surfaces and with exposed edges. An allowance must be made for the additional material necessary for such seams and edges. The drafter must also indicate where the material is bent. Several commonly used methods of showing bend lines are shown in Fig. 24-1-2. If the finished part is not shown with the development drawing, then bending instructions such as bend up 90°, bend down 180°, bend up 45°, are shown beside each bend line. Figure 24-1-3 shows a number of common methods for seaming and edging. Seams are used to join edges. The seams may be fastened together by lock seams, solder, rivets, adhesive, or welds. Exposed edges are folded or wired to give the edge added strength and to eliminate the sharp edge.

A surface is said to be *developable* if a thin sheet of flexible material, such as paper, can be wrapped smoothly about its surface. Objects that have plane, or flat surfaces or single-curved surfaces are developable; but if a surface is double-curved or warped, approximate methods must be used to develop the surface. The development of a spherical shape would thus be approximate, and the material would be stretched to compensate for small inaccuracies. For example, the coverings for a football or a basketball are made in segments, each segment cut

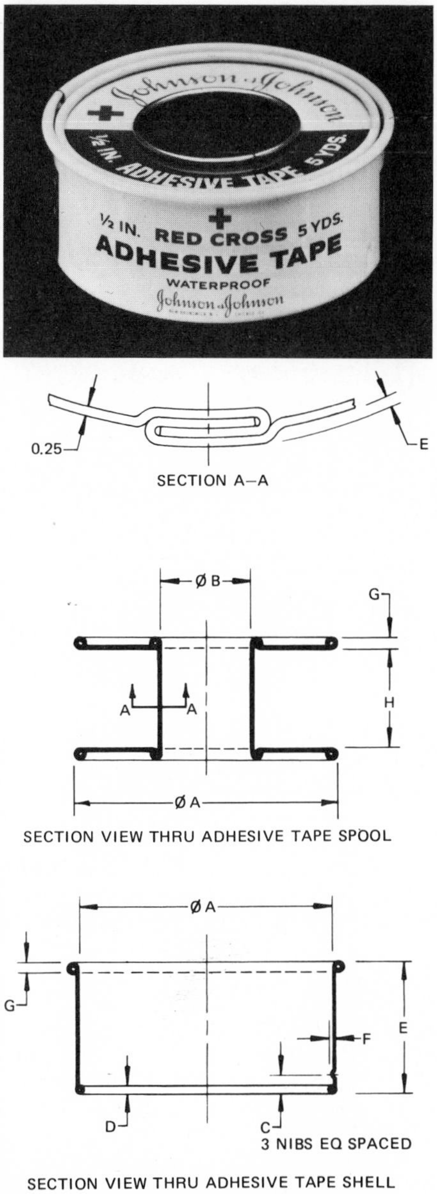
Johnson & Johnson
½ IN. ADHESIVE TAPE 5 YDS.
½ IN. RED CROSS 5 YDS.
ADHESIVE TAPE
WATERPROOF
Johnson & Johnson
0.25
E
SECTION A–A
Ø B
G
A
A
H
Ø A
SECTION VIEW THRU ADHESIVE TAPE SPOOL
Ø A
G
E
F
D
C
3 NIBS EQ SPACED
SECTION VIEW THRU ADHESIVE TAPE SHELL

Fig. 24-1-1 Sheet-metal application. (Johnson & Johnson)

to an approximate developed shape; the segments are then stretched and sewn together to give the desired shape.

SHEET-METAL SIZES

Metal thicknesses up to 6 mm are usually designated by a series of gage numbers, the more common gages being shown in Table 47 of the Appendix. Metal 6 mm and over is given in millimetre sizes. In calling for the material size of sheet-metal developments, customary practice is to give the gage number, type of gage, and its millimetre equivalent in brackets followed by the developed width and length. See Fig. 24-1-4.

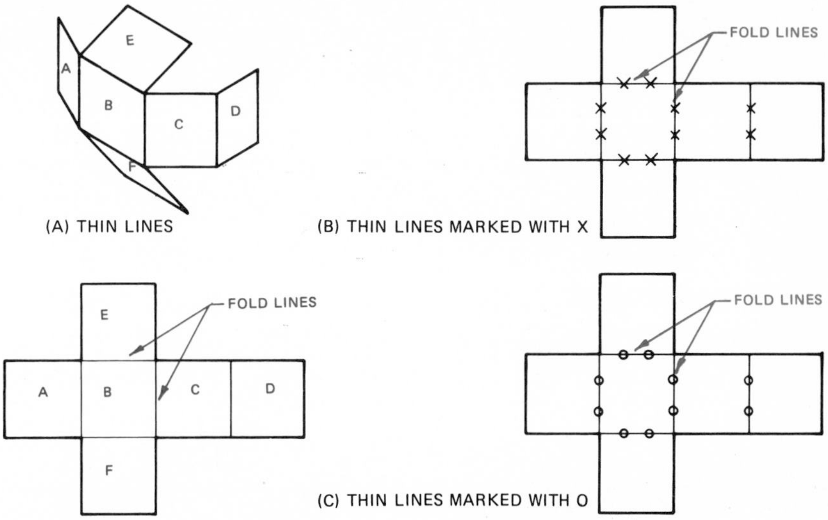
E
A
B
C
D
F
(A) THIN LINES
FOLD LINES
(B) THIN LINES MARKED WITH X
E
FOLD LINES
A
B
C
D
F
FOLD LINES
(C) THIN LINES MARKED WITH O

Fig. 24-1-2 Common methods used to identify fold or bend lines on development drawings.

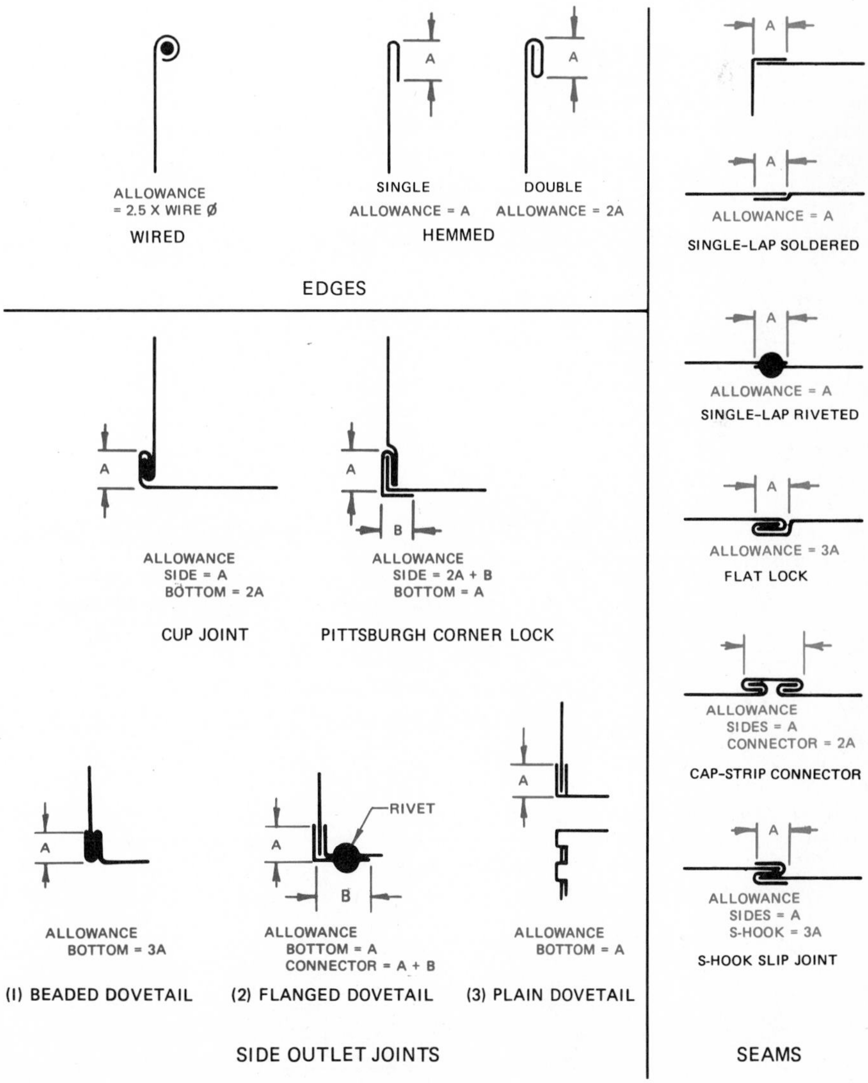
ALLOWANCE
= 2.5 X WIRE Ø
WIRED
A
SINGLE
ALLOWANCE = A
A
DOUBLE
ALLOWANCE = 2A
HEMMED
EDGES
A
ALLOWANCE
SIDE = A
BOTTOM = 2A
CUP JOINT
A
B
ALLOWANCE
SIDE = 2A + B
BOTTOM = A
PITTSBURGH CORNER LOCK
A
ALLOWANCE
BOTTOM = 3A
(1) BEADED DOVETAIL
RIVET
A
B
ALLOWANCE
BOTTOM = A
CONNECTOR = A + B
(2) FLANGED DOVETAIL
A
ALLOWANCE
BOTTOM = A
(3) PLAIN DOVETAIL
SIDE OUTLET JOINTS
A
A
ALLOWANCE = A
SINGLE-LAP SOLDERED
A
ALLOWANCE = A
SINGLE-LAP RIVETED
A
ALLOWANCE = 3A
FLAT LOCK
ALLOWANCE
SIDES = A
CONNECTOR = 2A
CAP-STRIP CONNECTOR
A
ALLOWANCE
SIDES = A
S-HOOK = 3A
S-HOOK SLIP JOINT
SEAMS

Fig. 24-1-3 Joints, seams, and edges.

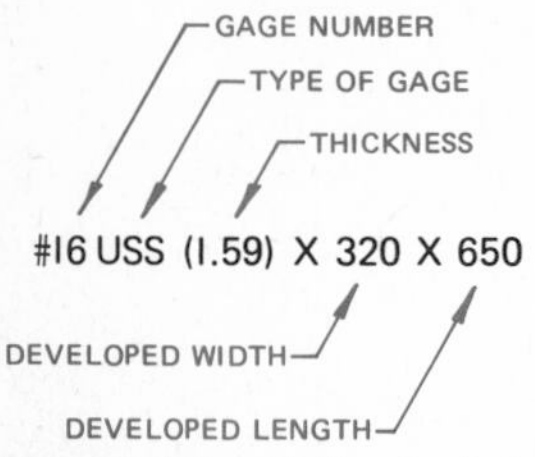
GAGE NUMBER
TYPE OF GAGE
THICKNESS
#16 USS (1.59) X 320 X 650
DEVELOPED WIDTH
DEVELOPED LENGTH

Fig. 24-1-4 Callout of sheet-metal sizes.

Straight-line Development

This is the term given to the development of an object that has surfaces on a flat plane of projection. The true size of each side of the object is known, and these sides can be laid out in successive order. Figure 24-1-5 shows the development of a simple rectangular box having a bottom and four sides. Note that in the development of the box, an allowance is made for lap seams at the corners and for a folded edge. The fold lines are shown as thin, unbroken lines. Note also that all lines for each surface are straight.

Figure 24-1-6 shows a development drawing with a complete set of folding instructions. Figure 24-1-7 shows a letter box development drawing where the back is higher than the front surface.

Assignment

On an A3- or B-size sheet, make a development drawing complete with bending instructions of one of the parts shown in Figs. 24-1-A to 24-1-D. Dimension the development drawing, showing the distance between bend lines, and show the overall sizes. Scale is 1:1.

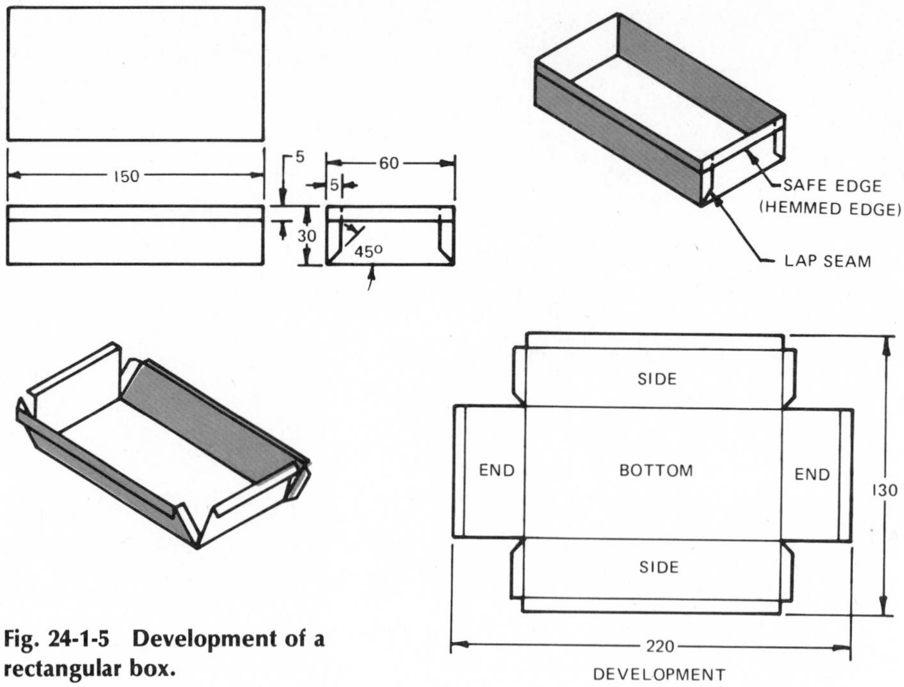
150
5
60
5
30
45°
SAFE EDGE
(HEMMED EDGE)
LAP SEAM
SIDE
END
BOTTOM
END
SIDE
130
220
DEVELOPMENT

Fig. 24-1-5 Development of a rectangular box.

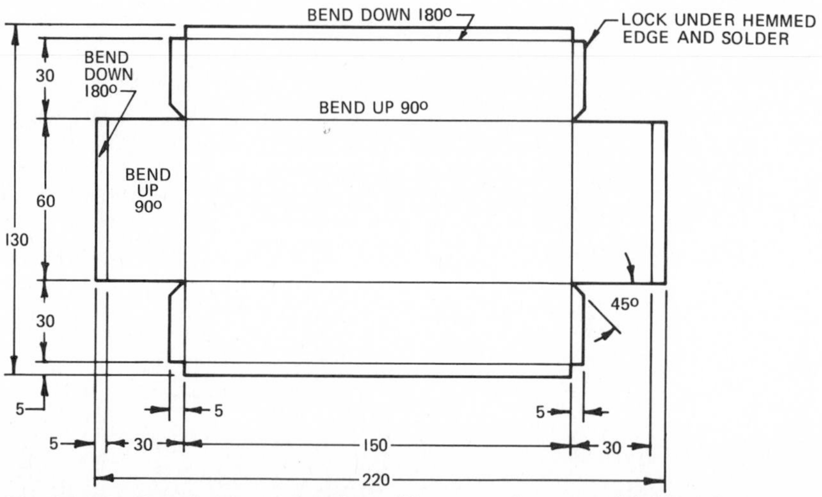
BEND DOWN 180°
LOCK UNDER HEMMED
EDGE AND SOLDER
BEND
DOWN
180°
BEND UP 90°
BEND
UP
90°
30
60
130
30
45°
5
5
5
5
30
150
30
220

Fig. 24-1-6 Development drawing with a complete set of folding instructions.

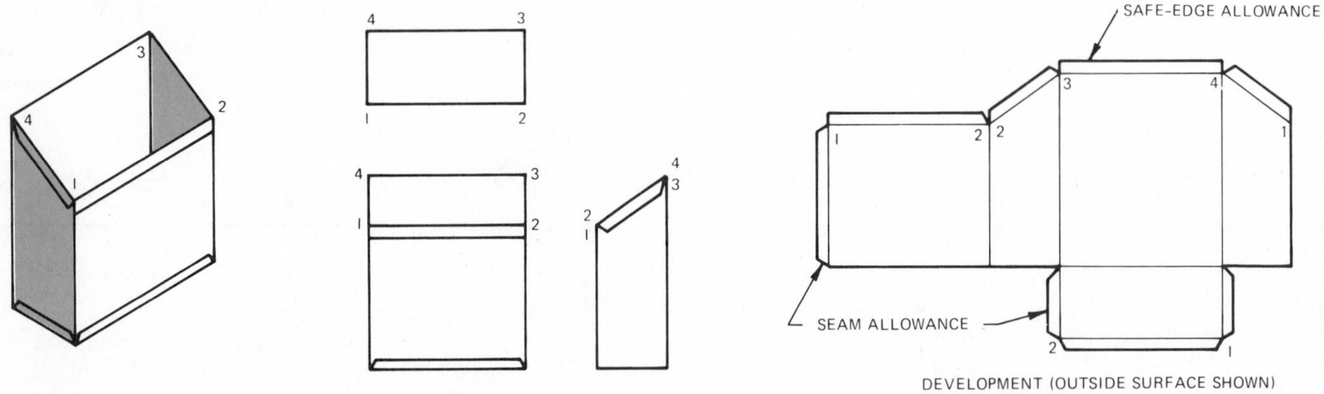
SAFE-EDGE ALLOWANCE
SEAM ALLOWANCE
DEVELOPMENT (OUTSIDE SURFACE SHOWN)

Fig. 24-1-7 Development drawing of a letter box.

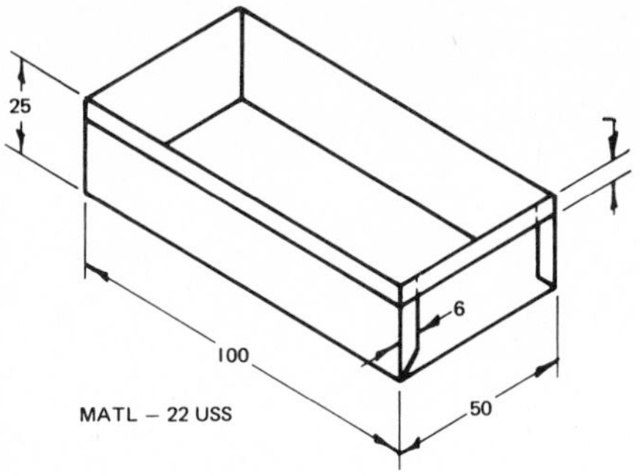
25
6
100
50
MATL – 22 USS

Fig. 24-1-A **Nail box.**

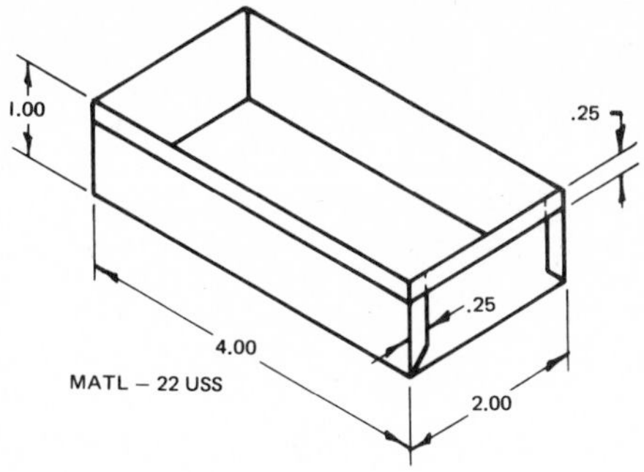
1.00
.25
.25
4.00
2.00
MATL – 22 USS

Fig. 24-1-B **Nail box.**

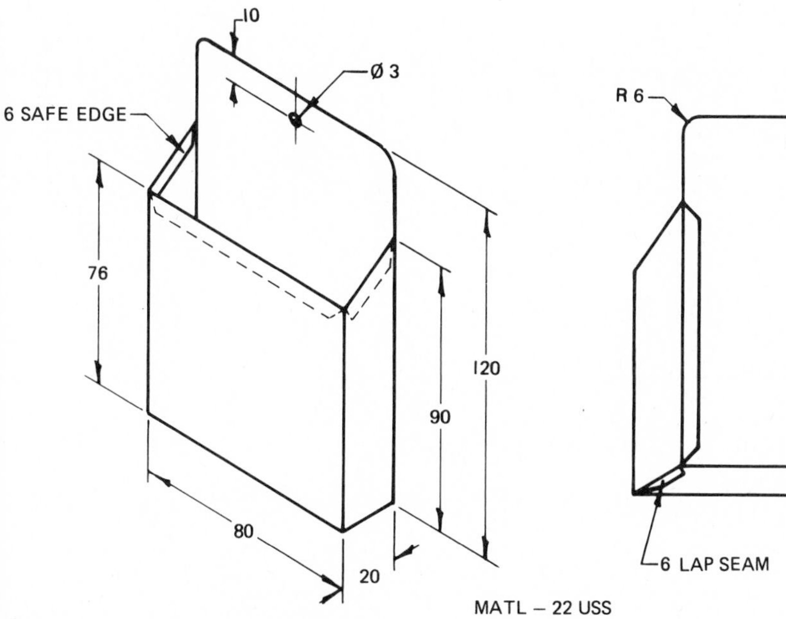
10
Ø 3
6 SAFE EDGE
76
120
90
80
20
R 6
6 LAP SEAM
MATL – 22 USS

Fig. 24-1-C **Memo pad holder.**

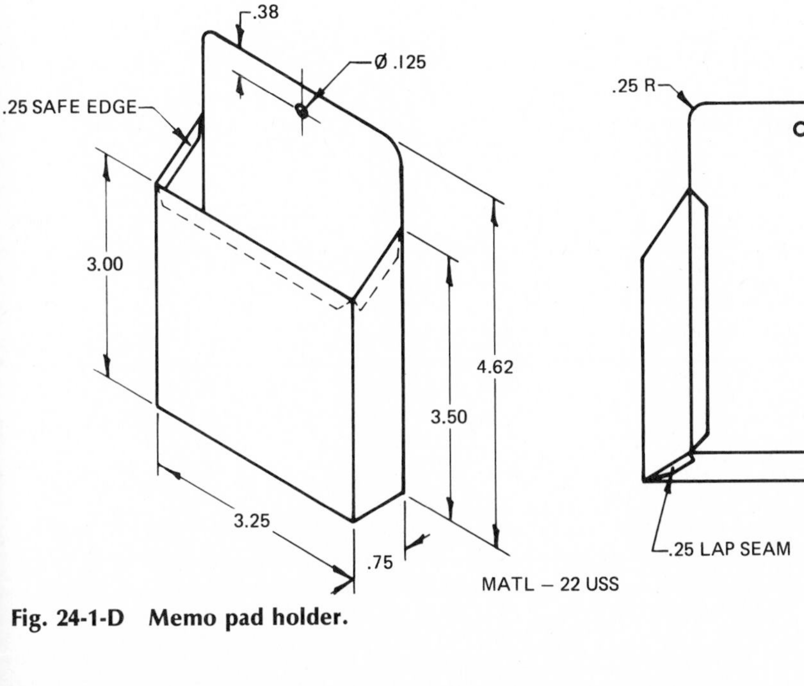
.38
Ø .125
.25 SAFE EDGE
3.00
4.62
3.50
3.25
.75
.25 R
.25 LAP SEAM
MATL – 22 USS

Fig. 24-1-D **Memo pad holder.**

UNIT 24-2
THE PACKAGING INDUSTRY

Packaging, which involves the principles of surface development, is one of the largest and most diversified industries in the world. Most products are packaged in metal, plastic, or cardboard containers. Small products, from candy-coated gum to large television sets, are packaged in cardboard containers. See Figs. 24-2-1 and 24-2-2. Such containers, often referred to as *cartons*, in many instances must be attractive as well as functional. They may be designed for sales appeal as

Fig. 24-2-1 Typical commercial containers. (Photo: Paul Newman)

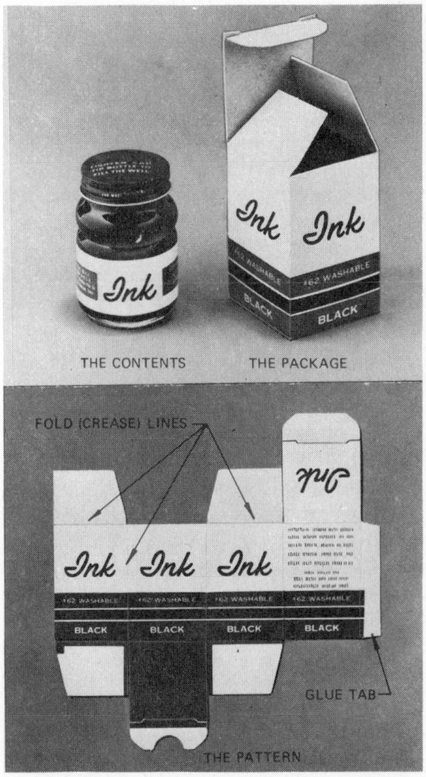
Ink
Ink
Ink
BLACK
THE CONTENTS
THE PACKAGE
FOLD (CREASE) LINES
Ink
Ink
Ink
Ink
BLACK
BLACK
BLACK
BLACK
GLUE TAB
THE PATTERN

Fig. 24-2-2 A familiar container made by cutting and folding a flat sheet.

well as for protection against contamination, shipping, and handling. They are also designed for temporary or permanent use.

Many cartons are printed, cut, creased, and sent to the customer in a flat position. See Fig. 24-2-3. They take less space to store and ship and are readily assembled. Locking devices such as tabs hold the box together. This type of container is used extensively by food chain operators such as Donut King, McDonalds, and Mr. Chicken. Other shapes, such as hexagons and octagons, as shown in Figs. 24-2-4 and 24-2-5, are becoming popular because of their novel form.

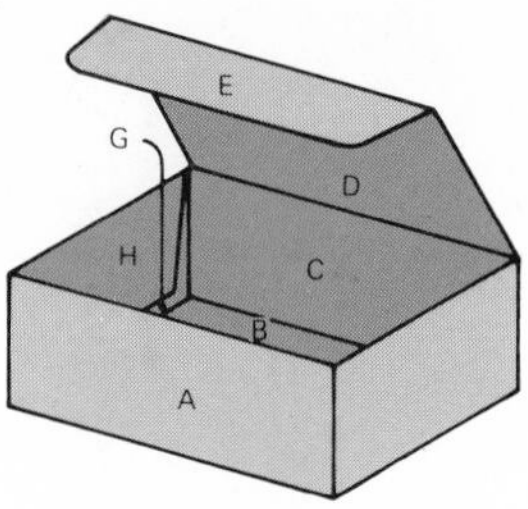
E
D
C
G
H
B
A

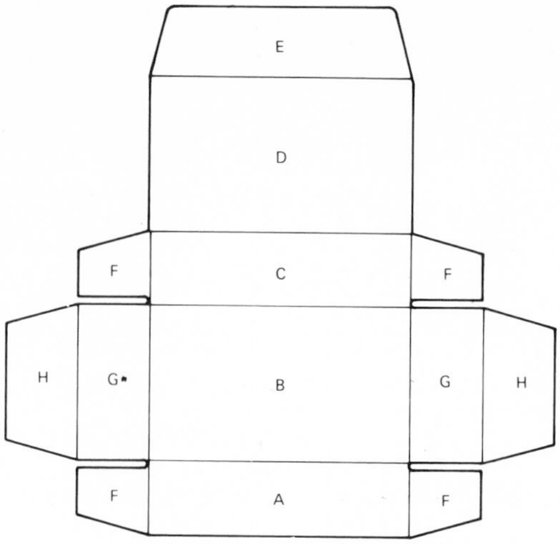
E
D
C
F
F
H
G
B
G
H
F
A
F

Fig. 24-2-3 Development of a one-piece carton with fold-down corners.

Assignments

1. On an A3- or B-size sheet, make a development drawing of one of the boxes shown in Figs. 24-2-A to 24-2-D. Scale is 1:1. After the development drawing has been checked by your instructor, add suitable seams and joint allowances. Then cut out the development, score on the bend lines, and form and glue the box together.

2. On an A3- or B-size sheet, make a development drawing of one of the pencil boxes or swim goggle boxes shown in Figs. 24-2-E to 24-2-H. On the exterior surface of the box, lay out a design which has eye appeal (color can be used) and which contains in the design the name of the item being sold, a company name, a slogan, and any other feature you believe would improve the salability of the article. Cut out the development, score on the bend lines, and glue together. *Note:* With reference to the swim goggle box, the box is completely sealed and must be broken to remove the goggles. Scale is 1:1.

3. Many containers are designed for a dual purpose. The main purpose is to accommodate the article being sold. The secondary purpose is to use the container as a novelty item after the article is removed. This next product is such a container. A company which produces cat and dog food wishes to have a dodecahedron- (12-sided) or icosahedron- (20-sided) shaped container having illustrations of different animals on the sides. The container can be used to hold small articles or as an art object (mobile) which can be hung from the ceiling.

On an A3- or B-size sheet, lay out one of the containers shown in Figs. 24-2-I to 24-2-L. One of the sides forms a lid. On the exterior surface of the container add a suitable design. Cut out the development, score on the bend lines, and glue together. Scale is 1:1.

REVIEW FOR ASSIGNMENTS

Unit 24-1 Surface Development
Unit 4-3 Constructing a Polygon

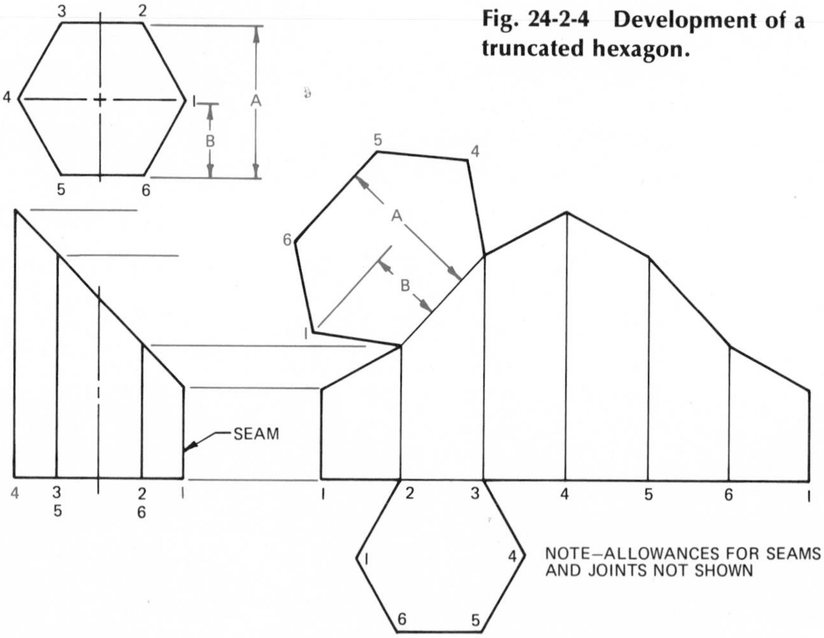
A
B
SEAM
NOTE—ALLOWANCES FOR SEAMS AND JOINTS NOT SHOWN

Fig. 24-2-4 Development of a truncated hexagon.

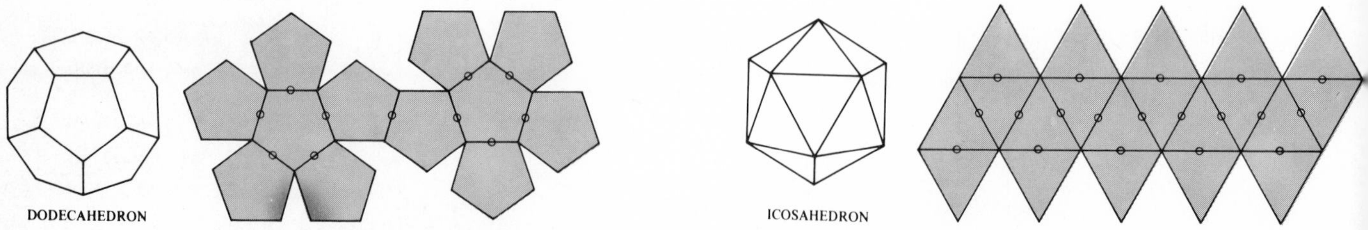
DODECAHEDRON
ICOSAHEDRON

Fig. 24-2-5 Twelve- and twenty-sided shapes.

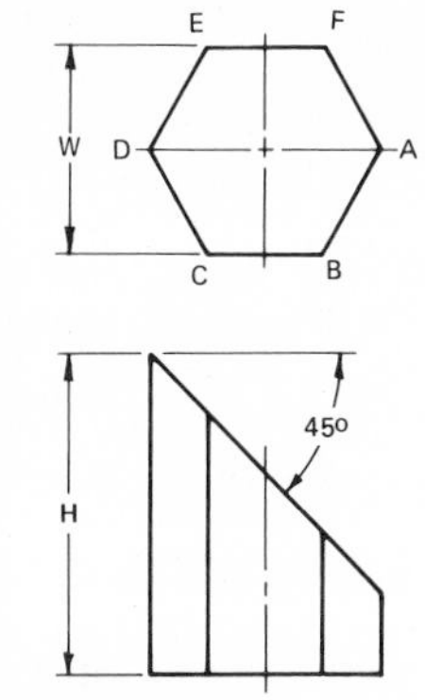
E
F
W
D
A
C
B
45°
H

NOTES:
SEAM IS AT A.
TOP AND BOTTOM TO HINGE AT C–D.

Fig. 24-2-A, B Hexagon box.

A. W=44 H=75 All seams 6 mm wide, placed on the inside and glued. Material 0.5 mm cardboard.

B. W=1.75 H=2.88 All seams .25 in. wide, placed on the inside and glued. Material 0.2 in. cardboard.

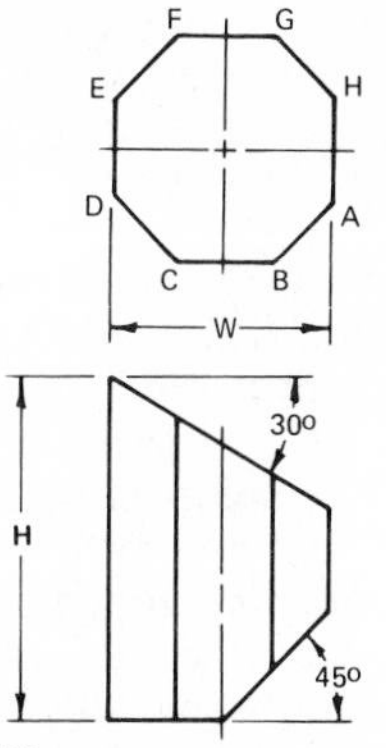
F
G
E
H
D
A
C
B
W
30°
H
45°

NOTES:
SEAM IS AT A.
TOP AND BOTTOM TO HINGE AT D–E.

Fig. 24-2-C, D Octagon box.

C. W=44 H=75 All seams 6 mm wide, placed on the inside and glued. Material 0.3 mm cardboard.

D. W=1.75 H=2.88 All seams .25 in. wide, placed on the inside and glued. Material .01 in. cardboard.

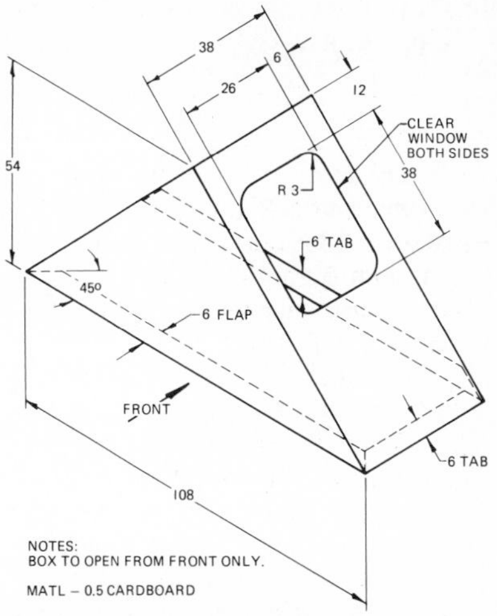
38
6
26
12
CLEAR WINDOW BOTH SIDES
38
54
R 3
6 TAB
45°
6 FLAP
FRONT
6 TAB
108
NOTES:
BOX TO OPEN FROM FRONT ONLY.
MATL – 0.5 CARDBOARD

Fig. 24-2-G Swim goggle box.

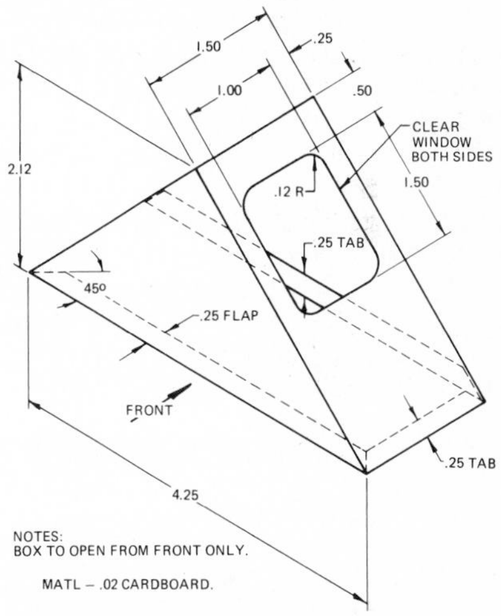
1.50
.25
1.00
.50
CLEAR WINDOW BOTH SIDES
1.50
2.12
.12 R
.25 TAB
45°
.25 FLAP
FRONT
.25 TAB
4.25
NOTES:
BOX TO OPEN FROM FRONT ONLY.
MATL – .02 CARDBOARD.

Fig. 24-2-H Swim goggle box.

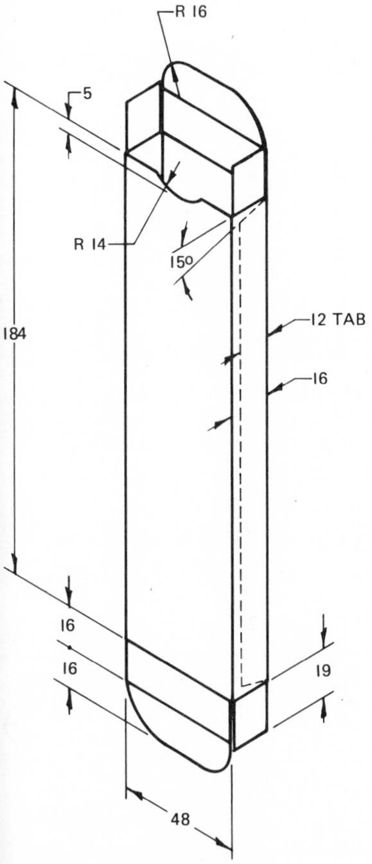
R 16
5
R 14
15°
12 TAB
16
184
16
16
19
48

MATL – 0.5 CARDBOARD

Fig. 24-2-E Pencil box.

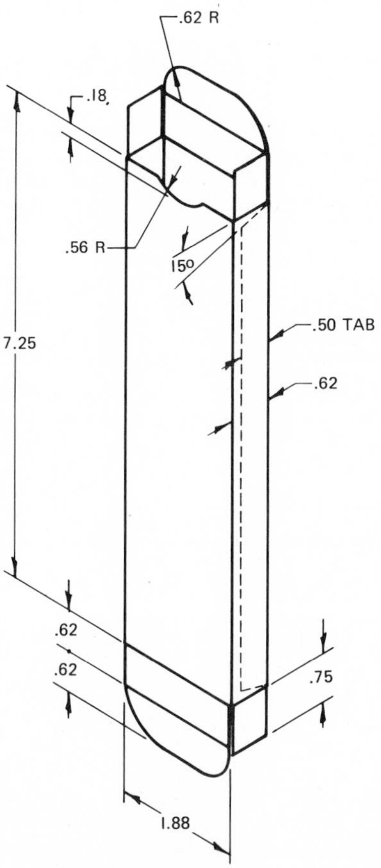
.62 R
.18
.56 R
15°
.50 TAB
.62
7.25
.62
.62
.75
1.88

MATL – .02 CARDBOARD

Fig. 24-2-F Pencil box.

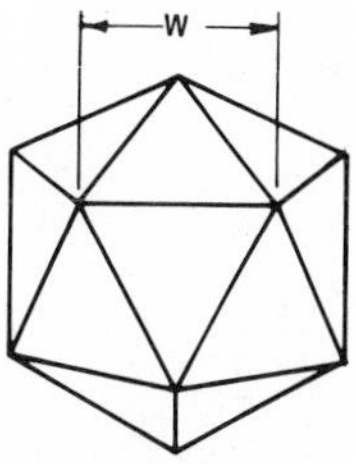
W

ICOSAHEDRON

Fig. 24-2-I, J Food container box.

I. W=50 All seams 5 mm wide, placed on the inside and glued. Material 0.5 mm cardboard.

J. W=2.00 All seams .20 in. wide, placed on the inside and glued. Material .02 in. cardboard.

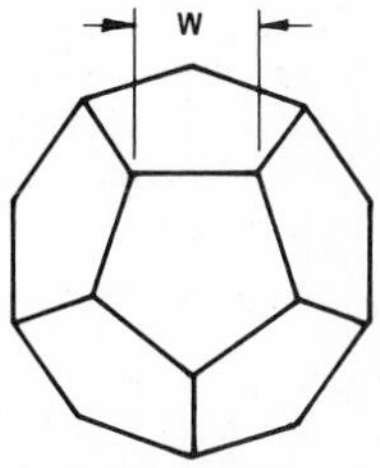
W

DODECAHEDRON

Fig. 24-2-K, L Food container box.

K. W=40 All seams 5 mm wide, placed on the inside and glued. Material 0.5 mm cardboard.

L. W=1.50 All seams .20 in. wide, placed on the inside and glued. Material .02 in. cardboard.

UNIT 24-3

RADIAL LINE DEVELOPMENT OF FLAT SURFACES

Development of a Right Pyramid with True Length of Edge Lines Shown. See Fig. 24-3-1. A right pyramid is a pyramid having all the lateral edges (from vertex to the base) of equal length. Since the true length of the lateral edges is shown in the front view (line 0-1 or 0-3) and the top view shows the true lengths of the edges of the base (lines 1-2, 2-3, etc.), the development may be constructed as follows: with 0 as center (corresponding to the apex) and with a radius equal to the true length of the lateral edges (line 0-1 in the front view), draw an arc as shown. Drop a perpendicular from 0 to intersect the arc at point 3. With a radius equal to the length of the edges of the base (line 1-2 on the top view), start at point 3 and step off the distances 3-2, 2-1, 3-4, and 4-1 on the large arc. Join these points with straight lines. These points are then connected to point 0 by straight lines to complete the development. Lines 0-2, 0-3, and 0-4 are the lines on which the development is folded to shape the pyramid. The base and seam allowances have been omitted for clarity.

In developing a truncated pyramid of this type, the procedure is the same as mentioned above, except that only a portion of lines 0-1, 0-2, 0-3, and 0-4 is required. The positions of points B and D in the top view are found by projecting lines horizontally from points B and D in the front view to intersect the true-length line 0-3 at B_1. Project a vertical line from point B_1 to intersect point B_2 in the top view. Rotate B_2 90° from point 0 to intersect line 0-2 at B and 0-4 at D. It will be noted that only lines A-1 and C-3 appear as their true length in the front view. The true length of lines B-2 and D-4 may be found by projecting a horizontal line from points B and D to point E on the true-length line 0-1.

To complete the development, step off distances 1-A on line 1-0, 2-B on line 2-0, 3-C on line 3-0, and 4-D on line 4-0. Join points A, B, C, D, A with straight lines. The top surface of the truncated pyramid may be added to the development as follows: with A as center and with a radius equal to distance AC in the front view, swing an arc. With B as center and with a radius equal to line BC on the development, swing an arc intersecting the first arc at C_1. Join point B to point C_1 with a straight line. With A as center and with a radius equal to line AB in the development, swing an arc. With B as center and with a radius equal to distance BD in the top view, swing an arc intersecting the first arc at D_1. Join AD_1 and D_1C_1 with straight lines.

Development of a Right Pyramid with True Length of Edge Lines Not Shown. See Fig. 24-3-2. In order to construct the development, the true length of the edge lines 0-1, 0-2, etc., must first be found. The true length of the edge lines would be equal to the hypotenuse of a right-angled triangle, having one leg equal in length to the projected edge line in the top view and the other leg equal to the height of the projected edge line in the front view. Since only one true-length line is required, it may be developed directly on the front view rather than by making a separate true-length diagram. With 0 in the top view as center and

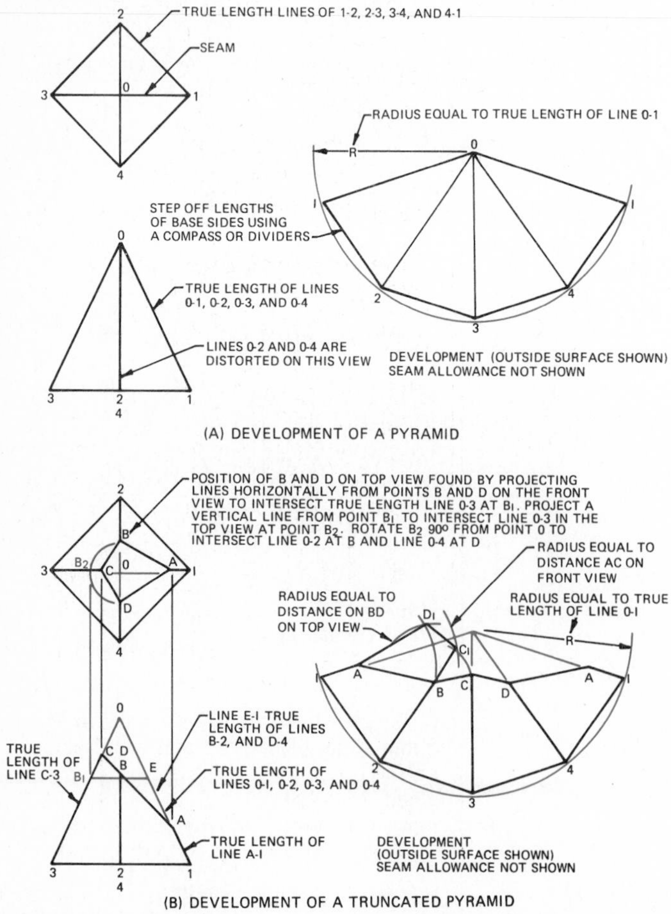
TRUE LENGTH LINES OF 1-2, 2-3, 3-4, AND 4-1
SEAM
RADIUS EQUAL TO TRUE LENGTH OF LINE 0-1
STEP OFF LENGTHS OF BASE SIDES USING A COMPASS OR DIVIDERS
TRUE LENGTH OF LINES 0-1, 0-2, 0-3, AND 0-4
LINES 0-2 AND 0-4 ARE DISTORTED ON THIS VIEW
DEVELOPMENT (OUTSIDE SURFACE SHOWN) SEAM ALLOWANCE NOT SHOWN
(A) DEVELOPMENT OF A PYRAMID
POSITION OF B AND D ON TOP VIEW FOUND BY PROJECTING LINES HORIZONTALLY FROM POINTS B AND D ON THE FRONT VIEW TO INTERSECT TRUE LENGTH LINE 0-3 AT B1. PROJECT A VERTICAL LINE FROM POINT B1 TO INTERSECT LINE 0-3 IN THE TOP VIEW AT POINT B2. ROTATE B2 90° FROM POINT 0 TO INTERSECT LINE 0-2 AT B AND LINE 0-4 AT D
RADIUS EQUAL TO DISTANCE AC ON FRONT VIEW
RADIUS EQUAL TO DISTANCE ON BD ON TOP VIEW
RADIUS EQUAL TO TRUE LENGTH OF LINE 0-1
LINE E-1 TRUE LENGTH OF LINES B-2, AND D-4
TRUE LENGTH OF LINE C-3
TRUE LENGTH OF LINES 0-1, 0-2, 0-3, AND 0-4
TRUE LENGTH OF LINE A-1
DEVELOPMENT (OUTSIDE SURFACE SHOWN) SEAM ALLOWANCE NOT SHOWN
(B) DEVELOPMENT OF A TRUNCATED PYRAMID

Fig. 24-3-1 Development of a right pyramid with true length of edge lines shown.

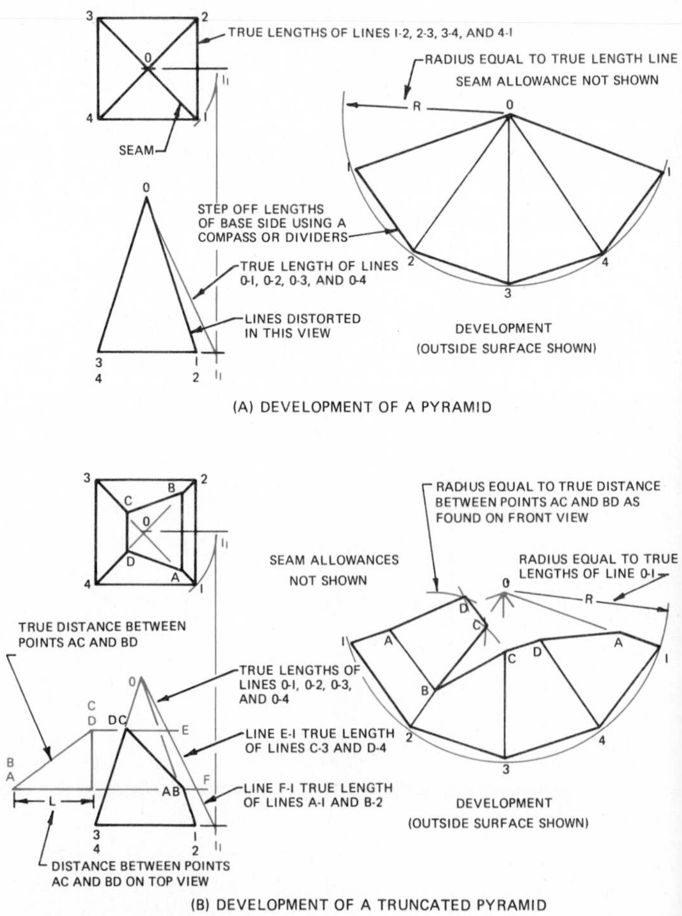
TRUE LENGTHS OF LINES 1-2, 2-3, 3-4, AND 4-1
RADIUS EQUAL TO TRUE LENGTH LINE
SEAM ALLOWANCE NOT SHOWN
SEAM
STEP OFF LENGTHS OF BASE SIDE USING A COMPASS OR DIVIDERS
TRUE LENGTH OF LINES 0-1, 0-2, 0-3, AND 0-4
LINES DISTORTED IN THIS VIEW
DEVELOPMENT (OUTSIDE SURFACE SHOWN)
(A) DEVELOPMENT OF A PYRAMID
RADIUS EQUAL TO TRUE DISTANCE BETWEEN POINTS AC AND BD AS FOUND ON FRONT VIEW
SEAM ALLOWANCES NOT SHOWN
RADIUS EQUAL TO TRUE LENGTHS OF LINE 0-1
TRUE DISTANCE BETWEEN POINTS AC AND BD
TRUE LENGTHS OF LINES 0-1, 0-2, 0-3, AND 0-4
LINE E-1 TRUE LENGTH OF LINES C-3 AND D-4
LINE F-1 TRUE LENGTH OF LINES A-1 AND B-2
DISTANCE BETWEEN POINTS AC AND BD ON TOP VIEW
DEVELOPMENT (OUTSIDE SURFACE SHOWN)
(B) DEVELOPMENT OF A TRUNCATED PYRAMID

Fig. 24-3-2 Development of a right pyramid with true length of edge lines not shown.

radius equal to distance 0-1 in the top view, swing an arc from point 1 until it intersects the center line at point 1_1. Project a vertical line down to the front view, intersecting the base line at point 1_1. Line 0-1_1 is the true length of the edge lines. The development may now be constructed in a similar manner to that outlined in the previous development.

In developing a truncated pyramid of this type, the procedure is the same except only the truncated edge lines are required. The true length of the truncated edge lines is required and may be found by projecting lines horizontally from points *A*, *B*, *C*, and *D* in the front view to intersect the true-length line 0-1_1 at points *F* and *E*, respectively. Line $F1_1$ is the true length of the truncated edge lines *A*1 and *B*1, and line $E1_1$ is the true length of the remaining truncated edge lines *C*3 and *D*4. The sides of the truncated pyramid may now be constructed in the development view. The top surface of the truncated pyramid may be added to the development as follows: with points *A* and *B* on the development as centers with a radius equal in length to line *BC* on the development, swing light arcs. With a radius equal in length to the true distance between points *A* and *C* or *B* and *D* (this is found on the true-length diagram constructed to the left of the front view) and with center *B*, swing an arc intersecting the first arc at *C*. Repeat, using point *A* as center and intersecting the other arc at point *C*. Join points *B*, *C*, *C*, and *A* with straight lines to complete the top surface. The base and seam lines have been omitted for clarity.

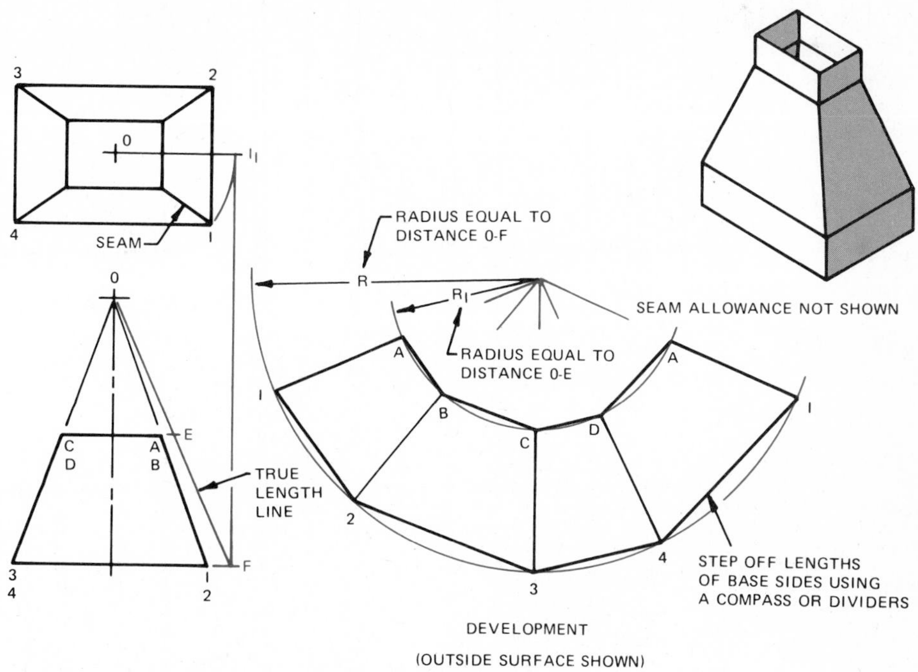

SEAM
TRUE LENGTH LINE
RADIUS EQUAL TO DISTANCE 0-F
RADIUS EQUAL TO DISTANCE 0-E
SEAM ALLOWANCE NOT SHOWN
STEP OFF LENGTHS OF BASE SIDES USING A COMPASS OR DIVIDERS
DEVELOPMENT
(OUTSIDE SURFACE SHOWN)

Fig. 24-3-4 Development of a transition piece.

Development of an Oblique Pyramid. See Fig. 24-3-3, an oblique pyramid having all its lateral edges of unequal length. The true length of each of these edges must first be found as shown in the true-length diagram. The development may now be constructed as follows: lay out base line 1-2 in the development view equal in length to the base line 1-2 found in the top view. With point 1 as center and radius equal in length to line 0-1 in the true-length diagram, swing an arc. With point 2 as center and radius equal in length to line 0-2 in the true-length diagram, swing an arc intersecting the first arc at 0. With point 0 as center and radius equal in length to line 0-3 in the true-length diagram, swing an arc. With point 2 as center and radius equal in length to base line 2-3 found in the top view, swing an arc intersecting the first arc at point 3. Locate point 4 and point 1 in a similar manner, and join these points, as shown, with straight lines. The base and seam lines have been omitted on the development drawing.

Development of a Transition Piece. See Fig. 24-3-4. The development of the transition piece is created in a similar manner to that of the development of the right pyramid (Fig. 24-3-2).

Development of an Offset Transition Piece. See Fig. 24-3-5. The offset transition piece is developed in a similar manner to that of the oblique pyramid (Fig. 24-3-3).

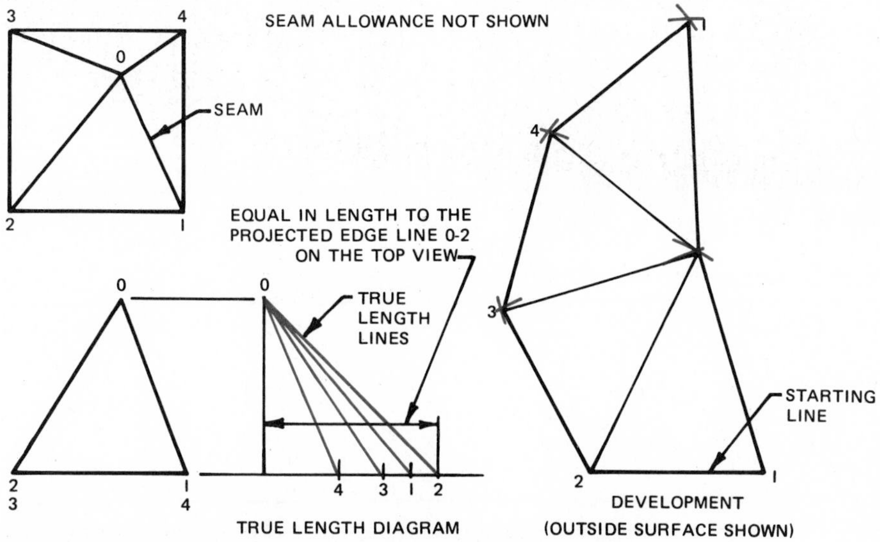

SEAM ALLOWANCE NOT SHOWN
SEAM
EQUAL IN LENGTH TO THE PROJECTED EDGE LINE 0-2 ON THE TOP VIEW
TRUE LENGTH LINES
STARTING LINE
TRUE LENGTH DIAGRAM
DEVELOPMENT
(OUTSIDE SURFACE SHOWN)

Fig. 24-3-3 Development of an oblique pyramid by triangulation.

Assignments

1. On an A3- or B-size sheet, make a development drawing of one of the concentric pyramids shown in Figs. 24-3-A to 24-3-D. Add suitable seams. Scale is 1:1.

2. On an A3- or B-size sheet, make a development drawing of the eccentric pyramid shown in Fig. 24-3-E. Add suitable seams. Scale 1:1.

REVIEW FOR ASSIGNMENTS

Unit 24-1 Surface Development

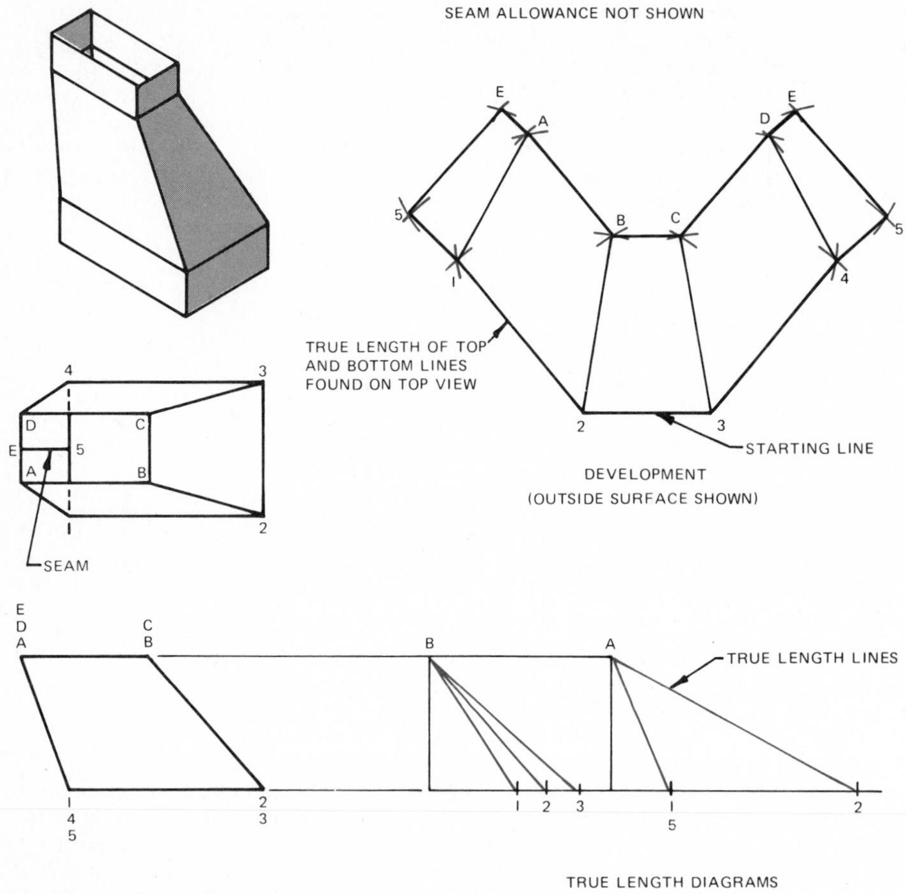
SEAM ALLOWANCE NOT SHOWN
TRUE LENGTH OF TOP AND BOTTOM LINES FOUND ON TOP VIEW
STARTING LINE
DEVELOPMENT
(OUTSIDE SURFACE SHOWN)
SEAM
TRUE LENGTH LINES
TRUE LENGTH DIAGRAMS

Fig. 24-3-5 Development of an offset transition piece.

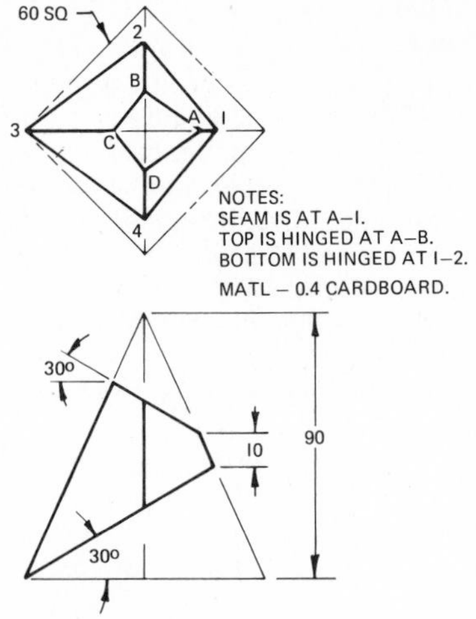
60 SQ
NOTES:
SEAM IS AT A–I.
TOP IS HINGED AT A–B.
BOTTOM IS HINGED AT I–2.
MATL – 0.4 CARDBOARD.
30°
10
90
30°

Fig. 24-3-C Truncated concentric pyramid.

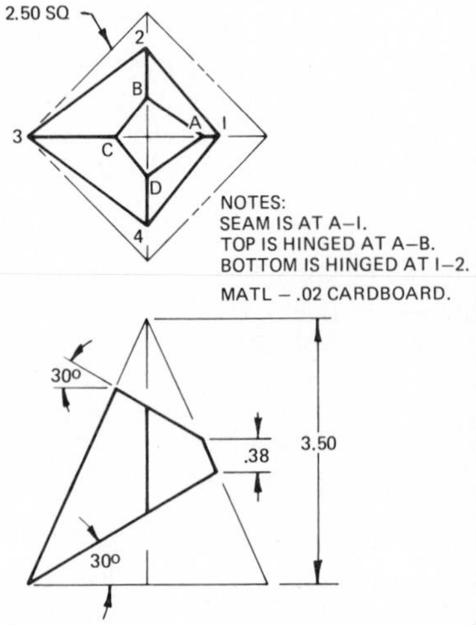
2.50 SQ
NOTES:
SEAM IS AT A–I.
TOP IS HINGED AT A–B.
BOTTOM IS HINGED AT I–2.
MATL – .02 CARDBOARD.
30°
.38
3.50
30°

Fig. 24-3-D Truncated concentric pyramid.

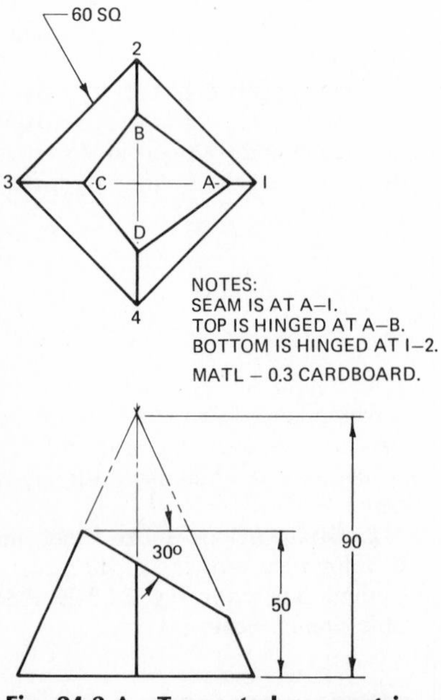
60 SQ
NOTES:
SEAM IS AT A–I.
TOP IS HINGED AT A–B.
BOTTOM IS HINGED AT I–2.
MATL – 0.3 CARDBOARD.
30°
90
50

Fig. 24-3-A Truncated concentric pyramid.

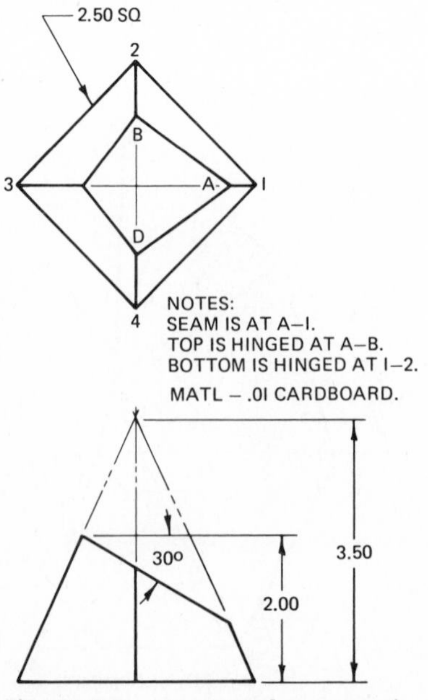
2.50 SQ
NOTES:
SEAM IS AT A–I.
TOP IS HINGED AT A–B.
BOTTOM IS HINGED AT I–2.
MATL – .01 CARDBOARD.
30°
3.50
2.00

Fig. 24-3-B Truncated concentric pyramid.

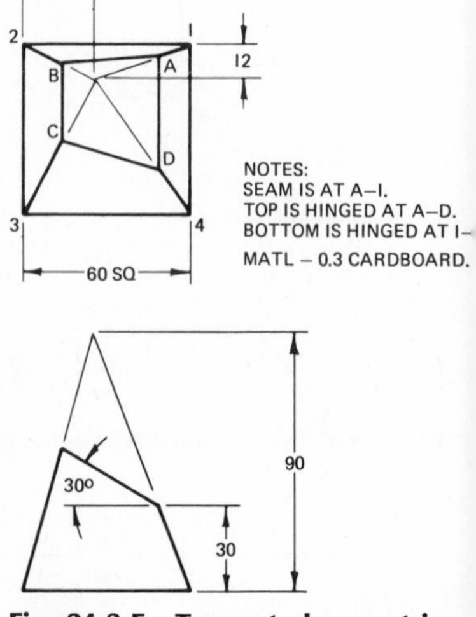
25
12
60 SQ
NOTES:
SEAM IS AT A–I.
TOP IS HINGED AT A–D.
BOTTOM IS HINGED AT I–
MATL – 0.3 CARDBOARD.
30°
90
30

Fig. 24-3-E Truncated eccentric pyramid.

UNIT 24-4

PARALLEL LINE DEVELOPMENT

The lateral, or curved, surface of a cylindrically shaped object, such as a "tin" can, is developable since it has a single-curved surface of one constant radius. The development technique used for such objects is called *parallel line development*. Figure 24-4-1*a* shows the development of the lateral surface of a simple hollow cylinder. The width of the development is equal to the height of the cylinder, and the length of the development is equal to the circumference of the cylinder (*d*) plus the seam allowance. Figure 24-4-1*b* shows the development of a cylinder with the top truncated at a 45° angle (one-half of a two-piece 90° elbow). Points of intersection are established to give the curved shape on the development. These points are derived from the intersection of a length location, representing a certain distance around the circumference from a starting point, and the height location at that same point on the circumference. The closer the points of intersection are to one another, the greater the accuracy of the development. An irregular curve is used to connect the points of intersection.

Figure 24-4-1*c* shows the development of the surface of a cylinder with both the top and the bottom truncated at an angle

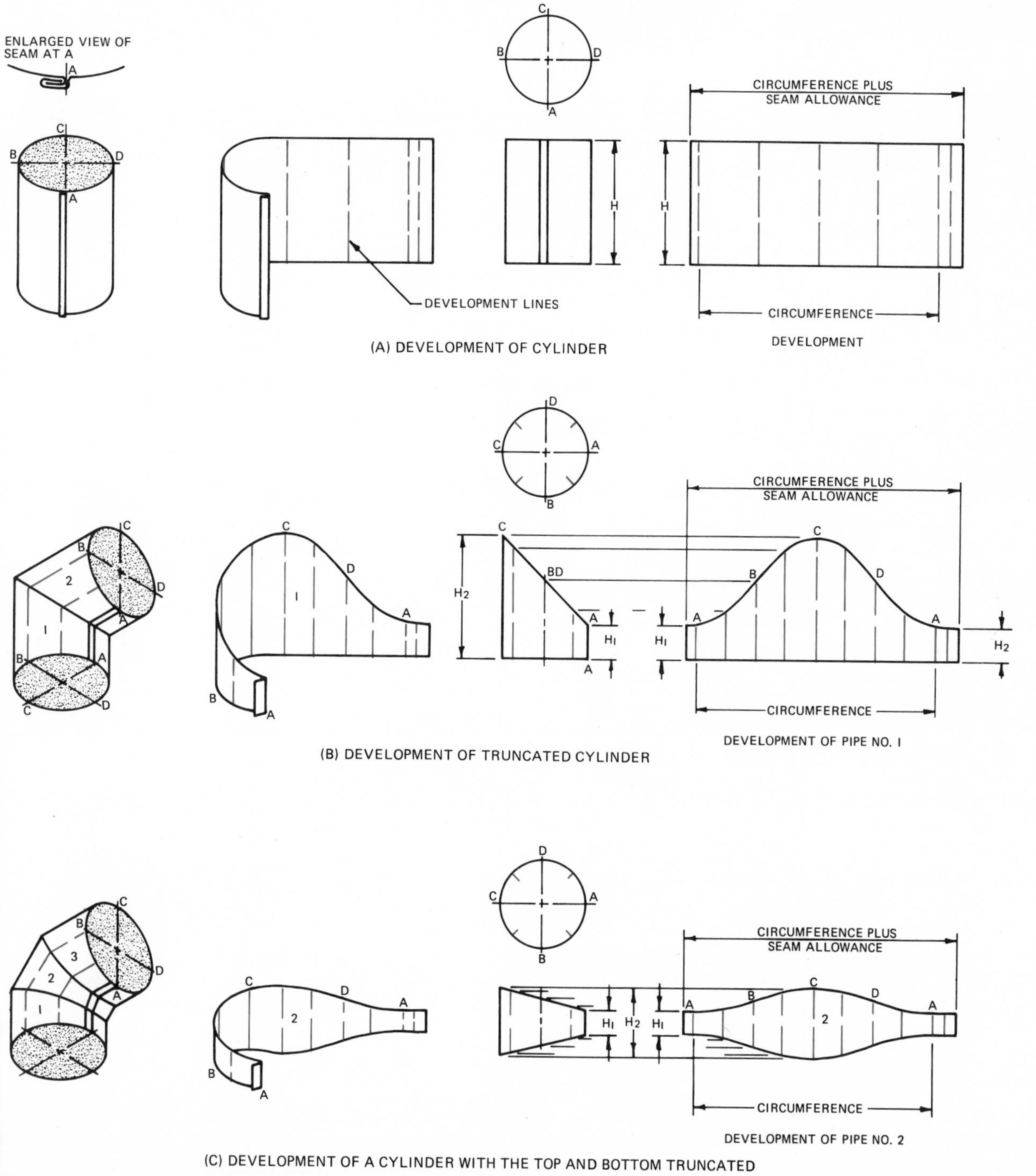
ENLARGED VIEW OF SEAM AT A
CIRCUMFERENCE PLUS SEAM ALLOWANCE
DEVELOPMENT LINES
CIRCUMFERENCE
DEVELOPMENT
(A) DEVELOPMENT OF CYLINDER
CIRCUMFERENCE PLUS SEAM ALLOWANCE
CIRCUMFERENCE
DEVELOPMENT OF PIPE NO. 1
(B) DEVELOPMENT OF TRUNCATED CYLINDER
CIRCUMFERENCE PLUS SEAM ALLOWANCE
CIRCUMFERENCE
DEVELOPMENT OF PIPE NO. 2
(C) DEVELOPMENT OF A CYLINDER WITH THE TOP AND BOTTOM TRUNCATED

Fig. 24-4-1 Development of cylinders.

of 22.5° (the center part of a three-piece elbow). It is normal practice in sheet-metal work to place the seam on the shortest side. In the development of elbows, however, this practice would result in considerable waste of material, as illustrated by Fig. 24-4-2*a*. To avoid this waste and to simplify cutting the pieces, the seams are alternately placed 180° apart, as illustrated by Fig. 24-4-2*b* for a two-piece elbow and by Fig. 24-4-2*c* for a three-piece elbow. Refer to Figs. 24-4-3 and 24-4-4 for complete developments of two- and four-piece elbows.

Assignments

1. On an A3- or B-size sheet, make a two-view and a development drawing of one of the elbows shown in Figs. 24-4-A to 24-4-D. Scale is 1:1.

2. On an A3- or B-size sheet, make a two-view and a development drawing of one of the sugar scoops shown in Figs. 24-4-E and 24-4-F. Scale is 1:1.

REVIEW FOR ASSIGNMENTS

Unit 24-1 Surface Development

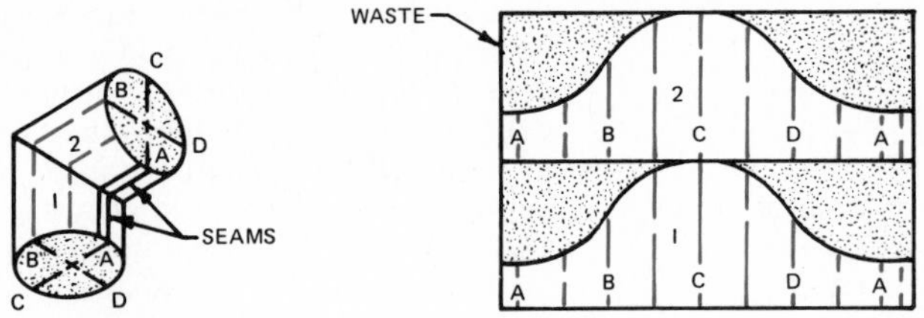
WASTE
SEAMS

(A) DEVELOPMENT OF A 2-PIECE ELBOW WITH BOTH SEAMS ON LINE A

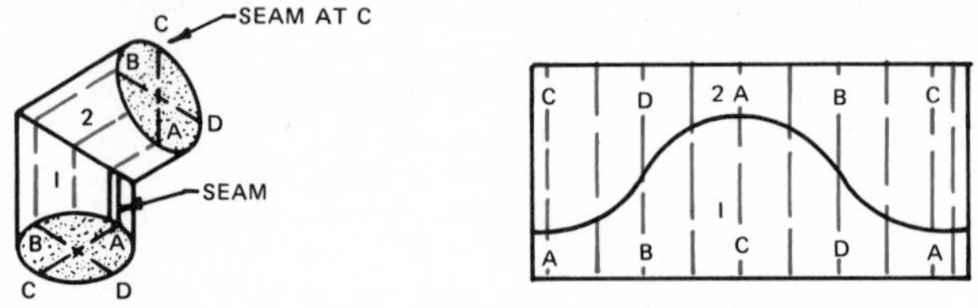
SEAM AT C
SEAM

(B) DEVELOPMENT OF A 2-PIECE ELBOW WITH SEAMS ON LINES A AND C

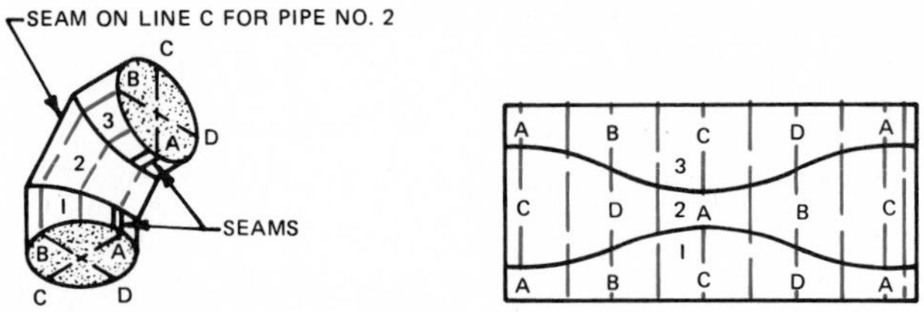
SEAM ON LINE C FOR PIPE NO. 2
SEAMS

(C) DEVELOPMENT OF A 3-PIECE ELBOW WITH SEAMS ALTERNATED ON LINES A AND C

Fig. 24-4-2 Location of seams on elbows.

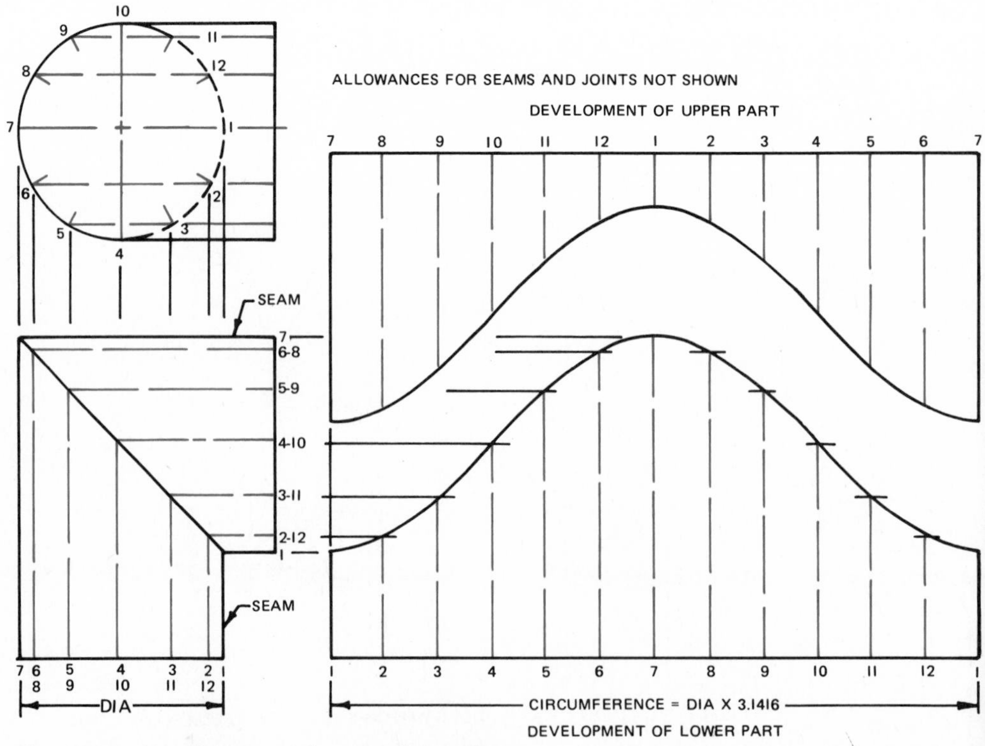
ALLOWANCES FOR SEAMS AND JOINTS NOT SHOWN
DEVELOPMENT OF UPPER PART
SEAM
SEAM
DIA
CIRCUMFERENCE = DIA X 3.1416
DEVELOPMENT OF LOWER PART

Fig. 24-4-3 Development of a two-piece elbow.

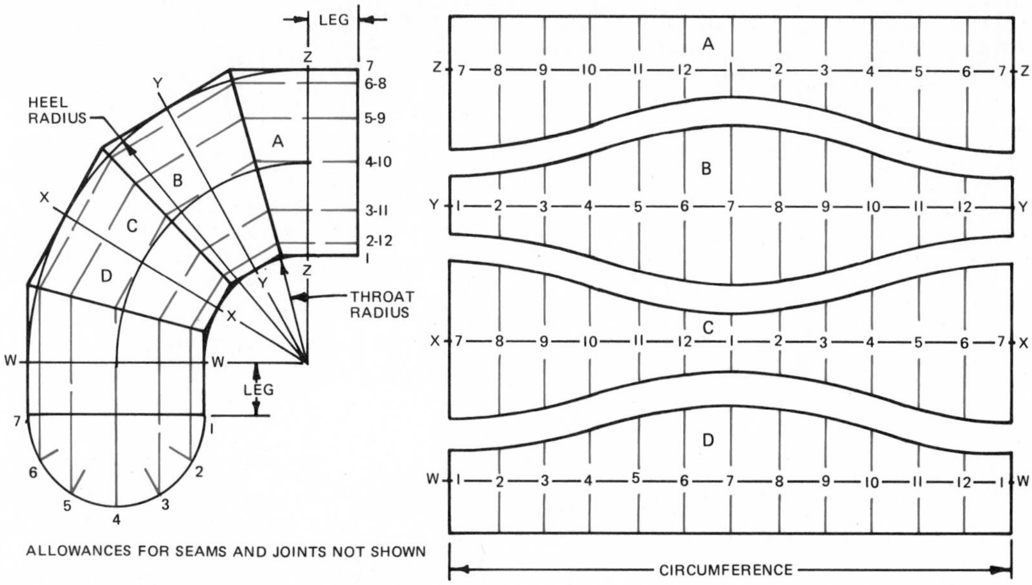
LEG
HEEL RADIUS
THROAT RADIUS
LEG
A
B
C
D
Z
Y
X
W
7
6-8
5-9
4-10
3-11
2-12
1
ALLOWANCES FOR SEAMS AND JOINTS NOT SHOWN
Z 7 8 9 10 11 12 1 2 3 4 5 6 7 Z
Y 1 2 3 4 5 6 7 8 9 10 11 12 1 Y
X 7 8 9 10 11 12 1 2 3 4 5 6 7 X
W 1 2 3 4 5 6 7 8 9 10 11 12 1 W
CIRCUMFERENCE

Fig. 24-4-4 Development of a four-piece elbow.

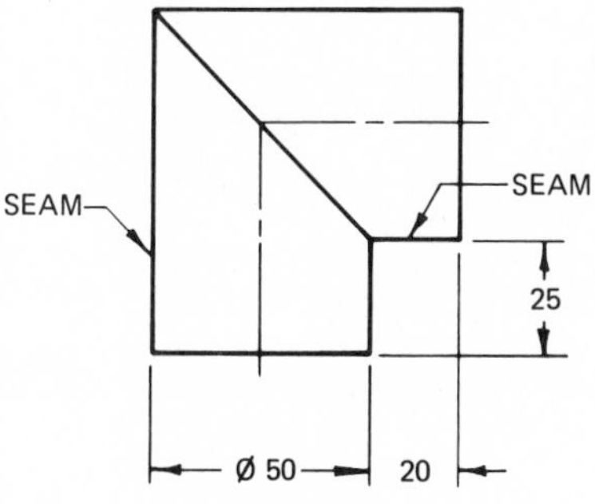
SEAM
SEAM
25
Ø 50
20

NOTES:
ALL SEAMS 6 WIDE.
MATL – 18 USS

Fig. 24-4-A Two-piece elbow.

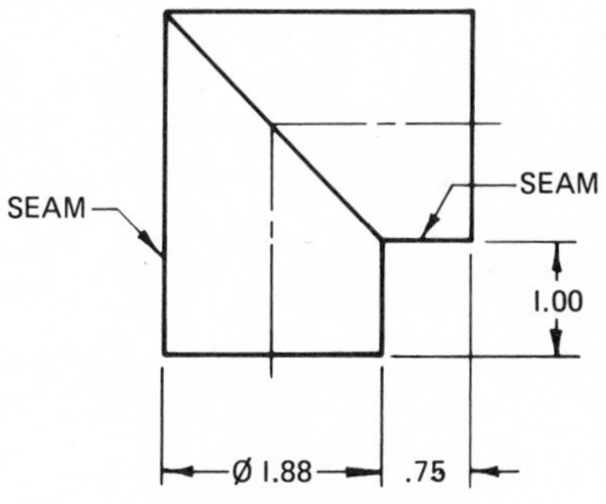
SEAM
SEAM
1.00
Ø 1.88
.75

NOTES:
ALL SEAMS .25 WIDE.
MATL – 18 USS

Fig. 24-4-B Two-piece elbow.

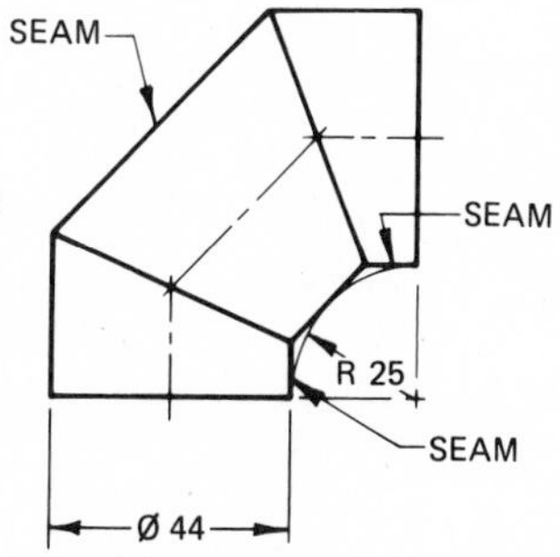
SEAM
SEAM
R 25
SEAM
Ø 44

NOTES:
ALL SEAMS 6 WIDE.
MATL – 18 USS

Fig. 24-4-C Three-piece elbow.

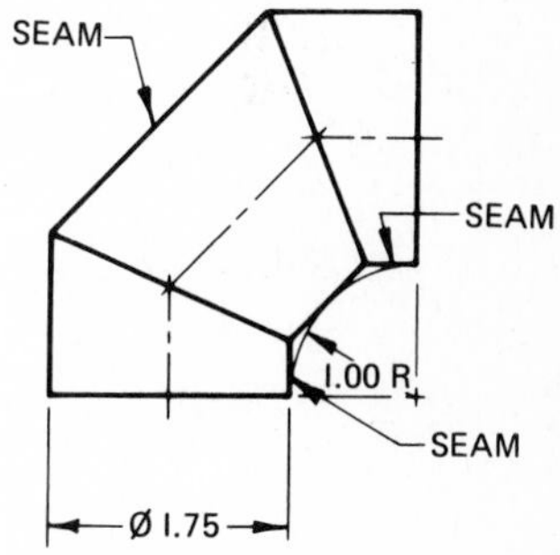
SEAM
SEAM
1.00 R
SEAM
Ø 1.75

NOTES:
ALL SEAMS .25 WIDE.
MATL – 18 USS

Fig. 24-4-D Three-piece elbow.

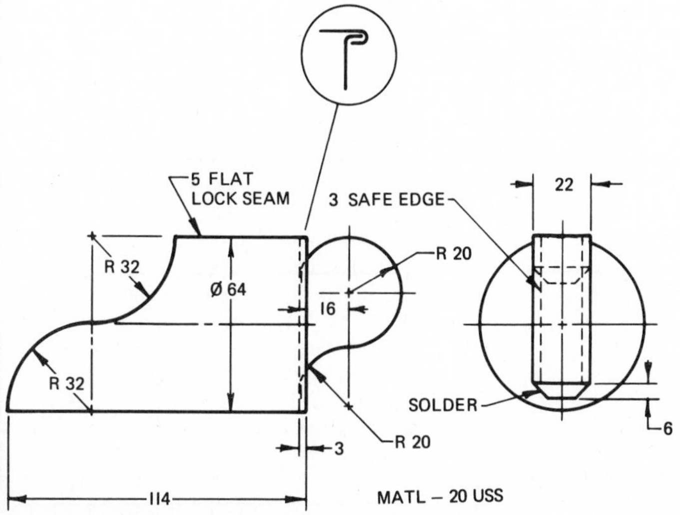
5 FLAT LOCK SEAM
3 SAFE EDGE
22
R 32
Ø 64
R 20
16
R 32
SOLDER
6
3
R 20
114
MATL – 20 USS

Fig. 24-4-E Sugar scoop.

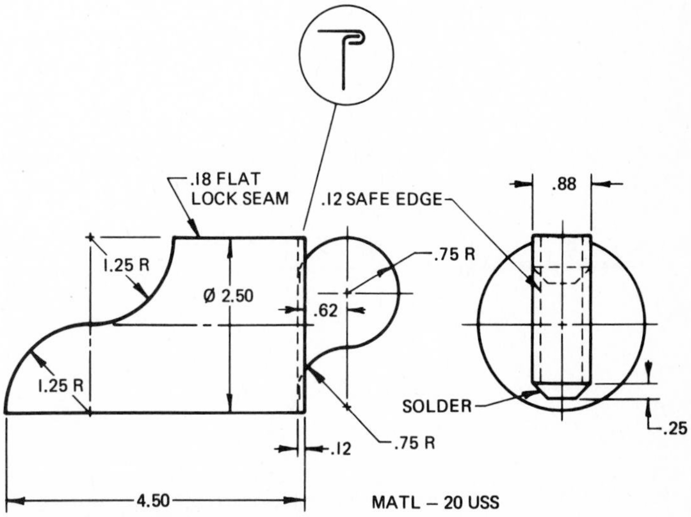
.18 FLAT LOCK SEAM
.12 SAFE EDGE
.88
1.25 R
Ø 2.50
.75 R
.62
1.25 R
SOLDER
.25
.12
.75 R
4.50
MATL – 20 USS

Fig. 24-4-F Sugar scoop.

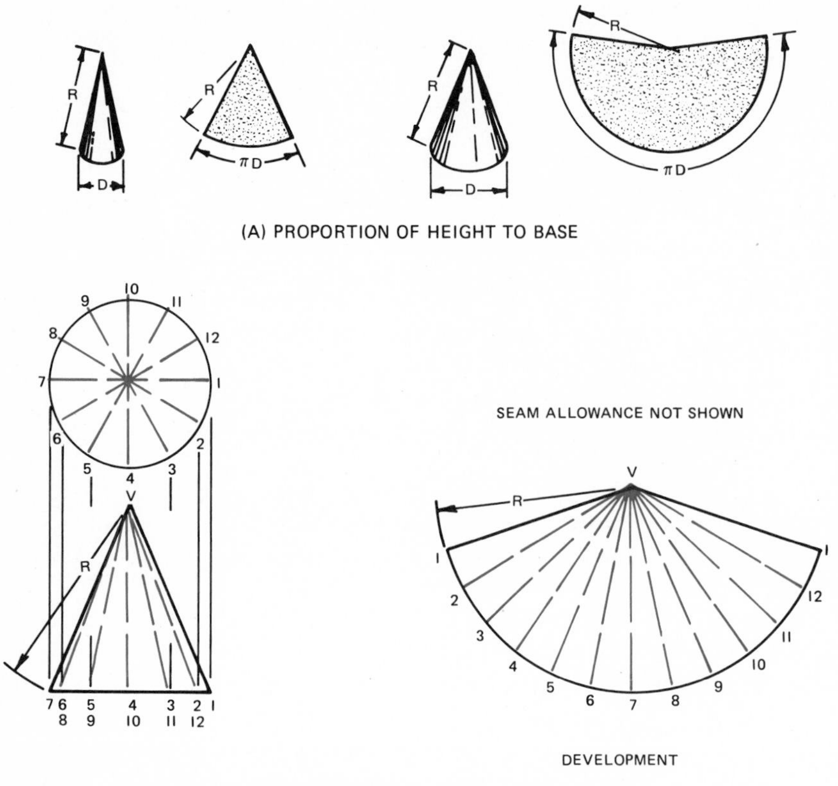

R
D
πD
(A) PROPORTION OF HEIGHT TO BASE
SEAM ALLOWANCE NOT SHOWN
V
DEVELOPMENT

(B) DEVELOPMENT PROCEDURE

Fig. 24-5-1 Development of a cone.

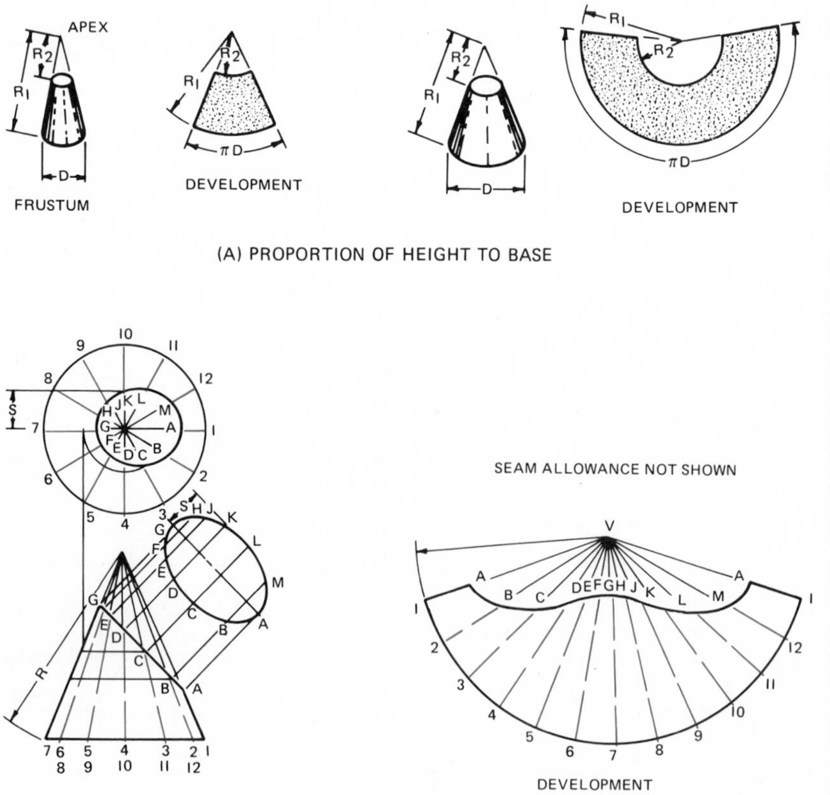

APEX
FRUSTUM
DEVELOPMENT
DEVELOPMENT
(A) PROPORTION OF HEIGHT TO BASE
SEAM ALLOWANCE NOT SHOWN
V
DEVELOPMENT

(B) DEVELOPMENT PROCEDURE

Fig. 24-5-2 Development of a truncated cone.

UNIT 24-5

RADIAL LINE DEVELOPMENT OF CONICAL SURFACES

Development of a Cone. The surface of a cone is developable, because a thin sheet of flexible material can be wrapped smoothly about it. The two dimensions necessary to make the development of the surface are the slant height of the cone and the circumference of its base. For a right circular cone (symmetrical about the vertical axis), the developed shape is a sector of a circle. The radius for this sector is the slant height of the cone, and the length around the perimeter of the sector is equal to the circumference of the base. The proportion of the height to the base diameter determines the size of the sector, as illustrated by Fig. 24-5-1*a*.

Figure 24-5-1*b* shows the steps in the development of a cone. The top view is divided into a convenient number of equal divisions, in this instance 12. The chordal distance between these points is used to step off the length of arc on the development. The radius *R* for the development is seen as the slant height in the front view. If a cone is truncated at an angle to the base, the inside shape on the development no longer has a constant radius; that is, it is an ellipse, which must be plotted by establishing points of intersection. The divisions made on the top view are projected down to the base of the cone in the front view. Element lines are drawn from these points to the apex of the cone. These element lines are seen in their true length only when the viewer is looking at right angles to them. Thus the points at which they cross the truncation line must be carried across, parallel to the base, to the outside element line, which is seen in its true length. The development is first made to represent the complete surface of the cone. Element lines are drawn from the step-off points about the circumference to the center point. True-length settings for each element line are taken from the front view and marked off on the corresponding element lines in the development. An irregular curve is used to connect these points of intersection, giving the proper inside shape.

Development of a Truncated Cone. The development of a frustum of a cone is the development of a full cone less the development of the part removed, as shown in Fig. 24-5-2. Note that, at all times, the radius setting, either R_1 or R_2, is a slant height, a distance taken on the surface of the cone.

When the top of a cone is truncated at an angle to the base, the top surface will not be seen as a true circle. This shape must also be plotted by establishing points of intersection. True radius settings for each element line are taken from the front view and marked off on the corresponding element line in the top view. These points are connected with an irregular curve to give the correct oval shape for the top surface. If the development of the sloping top surface is required, an auxiliary view of this surface shows its true shape.

Development of an Oblique Cone. See Fig. 24-5-3. The development of an oblique cone is generally accomplished by the triangulation method. The base of the cone is divided into a convenient number of equal parts and elements; 0-1, 0-2, etc., are drawn in the top view and projected down and drawn in the front view. The true lengths of the elements are not shown in either the top or front view but would be equal in length to the hypotenuse of a right-angle triangle, having one leg equal in length to the projected element in the top view and the other leg equal to the height of the projected element in the front view.

When it is necessary to find the true length of a number of edges, or elements, then a true-length diagram is drawn adjacent to the front view. This prevents the front view from being cluttered with lines.

Since the development of the oblique cone will be symmetrical, the starting line will be element 0-7. The development is constructed as follows: with 0 as center and radius equal to the true length of element 0-6, draw an arc. With 7 as center and radius equal to distance 6-7 in the top view, draw a second arc intersecting the first at point 6. Draw element 0-6 on the development. With 0 as center and the radius equal to the true length of element 0-5, draw an arc. With 6 as center and the radius equal to distance 5-6 in the top view, draw a second arc intersecting the first at point 5. Draw element 0-5 on the development. This is repeated until all the element lines are located on the development view. No seam allowance is shown on the development.

Assignment

On an A3- or B-size sheet, make development drawings of one of the assembled parts shown in Figs. 24-5-A to 24-5-D. Use your judgment for the other views required and add suitable seams. Scale is 1:1.

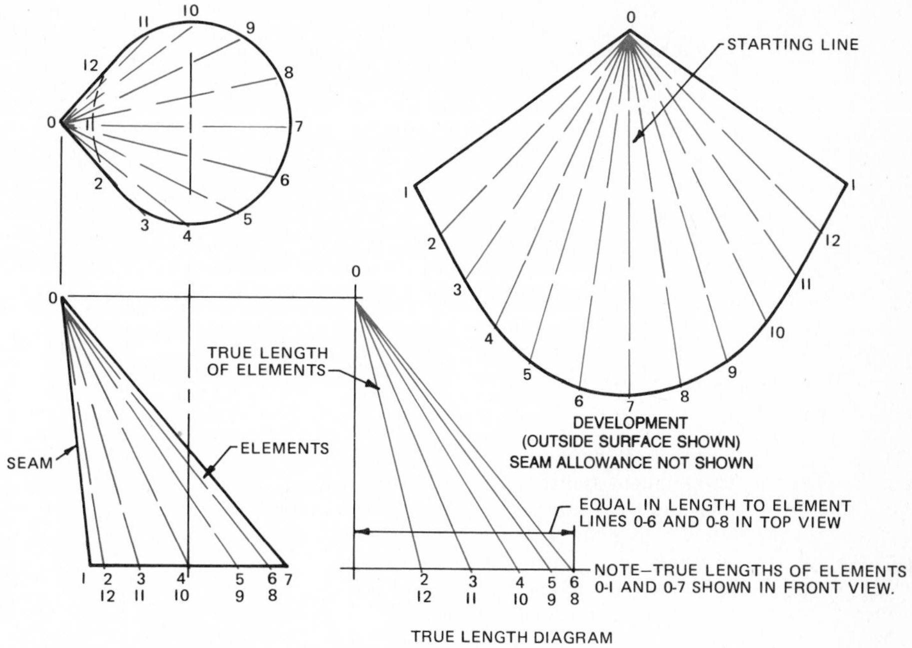
STARTING LINE
TRUE LENGTH OF ELEMENTS
ELEMENTS
SEAM
DEVELOPMENT
(OUTSIDE SURFACE SHOWN)
SEAM ALLOWANCE NOT SHOWN
EQUAL IN LENGTH TO ELEMENT LINES 0-6 AND 0-8 IN TOP VIEW
NOTE—TRUE LENGTHS OF ELEMENTS 0-I AND 0-7 SHOWN IN FRONT VIEW.
TRUE LENGTH DIAGRAM

Fig. 24-5-3 Development of an oblique cone.

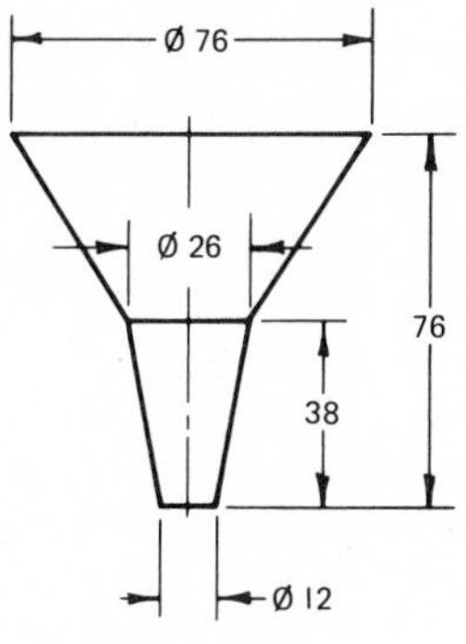
Ø 76
Ø 26
76
38
Ø 12
MATL – 24 USS

Fig. 24-5-A Funnel.

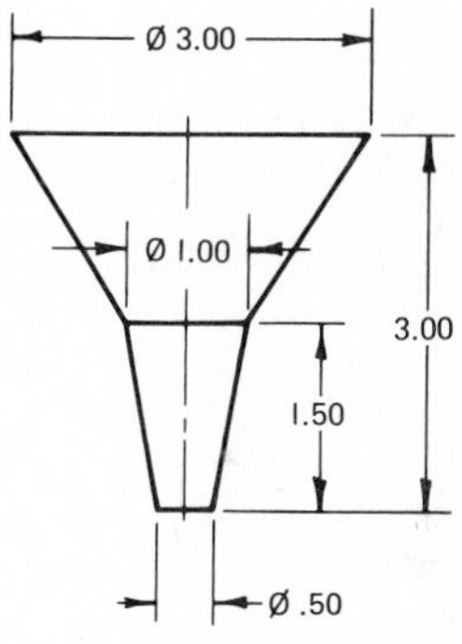
Ø 3.00
Ø 1.00
3.00
1.50
Ø .50
MATL – 24 USS

Fig. 24-5-B Funnel.

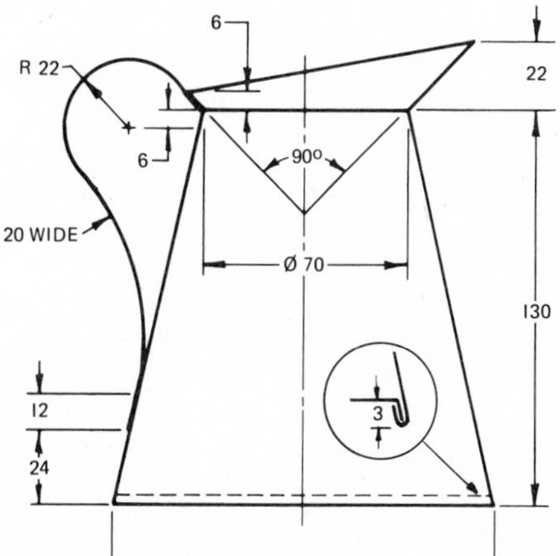
6
R 22
6
90°
22
20 WIDE
Ø 70
130
12
3
24
Ø 130
MATL – 22 USS

Fig. 24-5-C Measuring can.

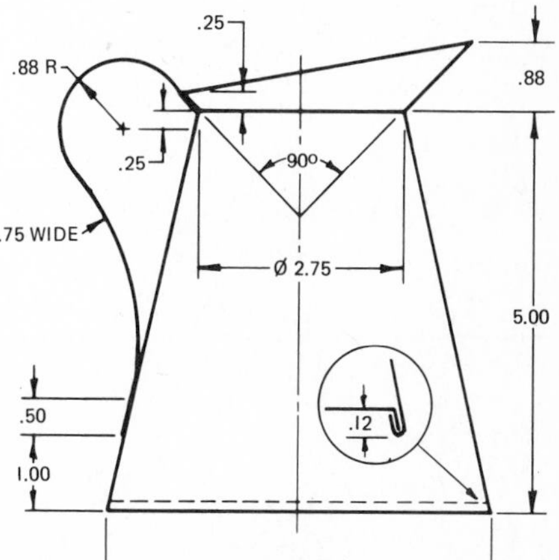
.25
.88 R
.25
90°
.88
.75 WIDE
Ø 2.75
5.00
.50
.12
1.00
Ø 5.00
MATL – 22 USS

Fig. 24-5-D Measuring can.

UNIT 24-6

DEVELOPMENT OF TRANSITION PIECES

Triangulation

Nondevelopable surfaces can be developed approximately by assuming the surface to be made up from a series of triangular surfaces laid side by side to form the development. This form of development is known as triangulation. Refer to Figs. 24-6-1 and 24-6-2.

Development of a Transition Piece—Square to Round. See Fig. 24-6-3. The transition piece shown is used to connect round and square pipes. It can be seen from both the development and the pictorial drawings that the transition piece is made of four isosceles triangles whose bases connect with the square duct and four parts of an oblique cone having the circle as the base and the corners of the square pipe as the vertices. To make the development, a true-length diagram is drawn first. When the true length of line 1*A* is known, the four equal isosceles triangles can be developed. After the triangle *G*-2-3 has been developed, then the partial developments of the oblique cone are added until points *D* and *K* have been located. Next the isosceles triangles *D*-1-2 and *K*-3-4 are added, then the partial cones, and, last, half of the isosceles triangle placed at each side of the development.

Development of an Offset Transition Piece—Rectangular to Round. See Fig. 24-6-4. The development of the transition piece shown is constructed in the same manner as the one previously developed, except that all the elements are of different lengths. To avoid confusion, four true-length diagrams are drawn and the true-length lines are clearly labeled.

Transition Piece Connecting Two Circular Pipes—Parallel Joints. See Fig. 24-6-5. The development of a transition piece connecting two circular pipes is

Fig. 24-6-1 Forming a square-to-round transition piece.

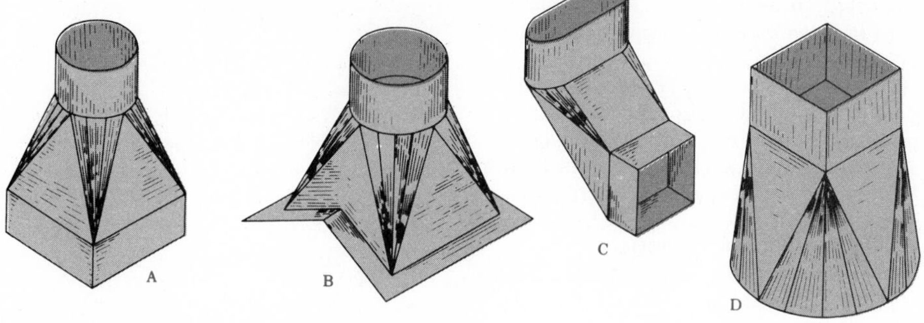
A
B
C
D

Fig. 24-6-2 Transition pieces.

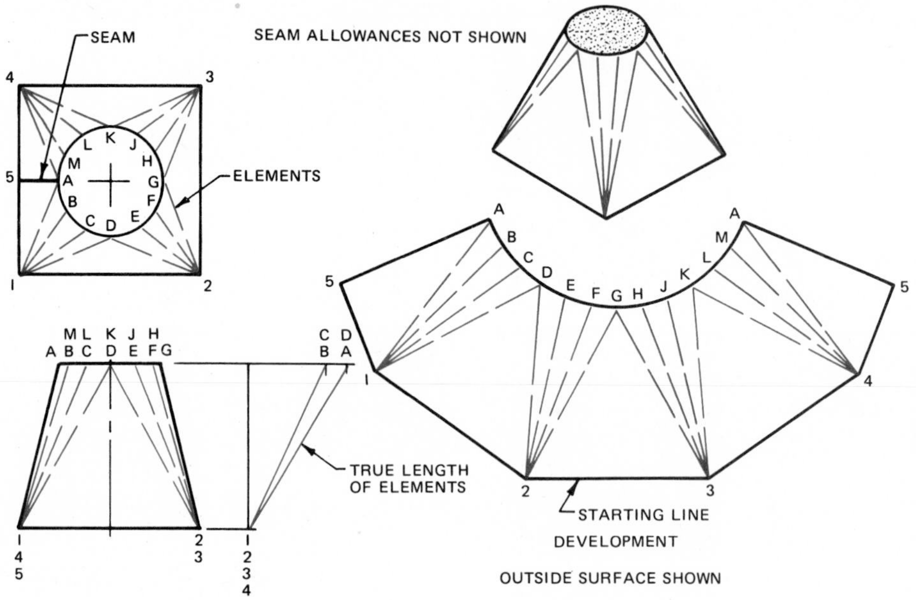
SEAM
SEAM ALLOWANCES NOT SHOWN
ELEMENTS
TRUE LENGTH OF ELEMENTS
STARTING LINE
DEVELOPMENT
OUTSIDE SURFACE SHOWN

Fig. 24-6-3 Development of a transition piece—square to round.

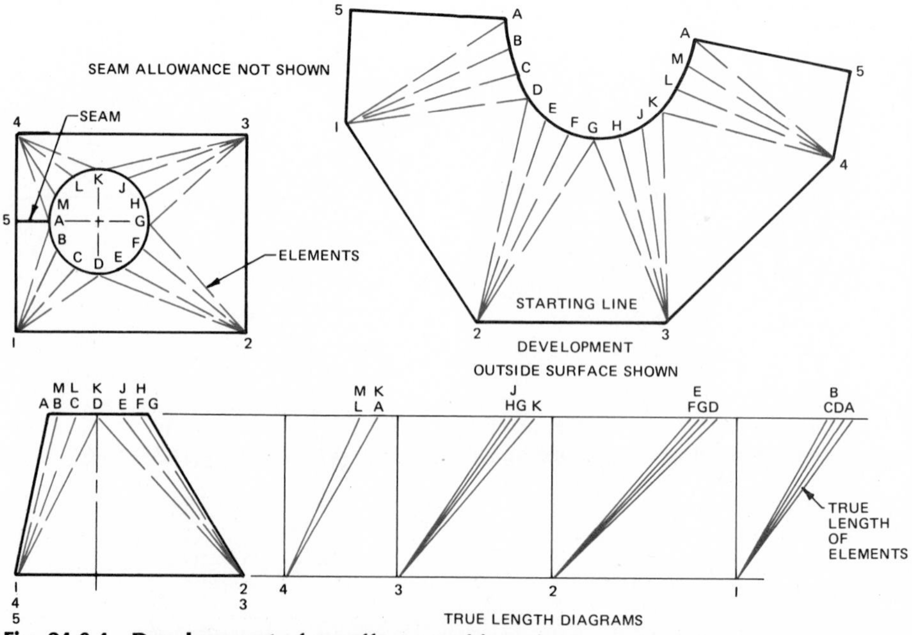
SEAM ALLOWANCE NOT SHOWN
SEAM
ELEMENTS
STARTING LINE
DEVELOPMENT
OUTSIDE SURFACE SHOWN
TRUE LENGTH OF ELEMENTS
TRUE LENGTH DIAGRAMS

Fig. 24-6-4 Development of an offset transition piece—rectangular to round.

similar to the development of an oblique cone (Fig. 24-5-3), except that the cone is truncated. The apex of the cone, 0, is located by drawing the two given pipe diameters in their proper positions and extending the radial lines $1\text{-}1_1$ and $7\text{-}7_1$ to intersect at point 0. First the development is made to represent the complete development of the cone, and then the top portion is removed. Radius settings for distances $0\text{-}2_1$ and $0\text{-}3_1$ on the development are taken from the true-length diagram.

Transition Piece Connecting Two Circular Pipes—Oblique Joints. See Fig. 24-6-6. When the joints between the pipe and transition piece are not perpendicular to the pipe axis, then the transition piece may be developed as shown. Since the top and bottom of the transition piece will be elliptical in shape, a partial auxiliary view is required to find the true length of the chords between the end points of the elements. The development is then constructed in a manner similar to that outlined for Fig. 24-6-5.

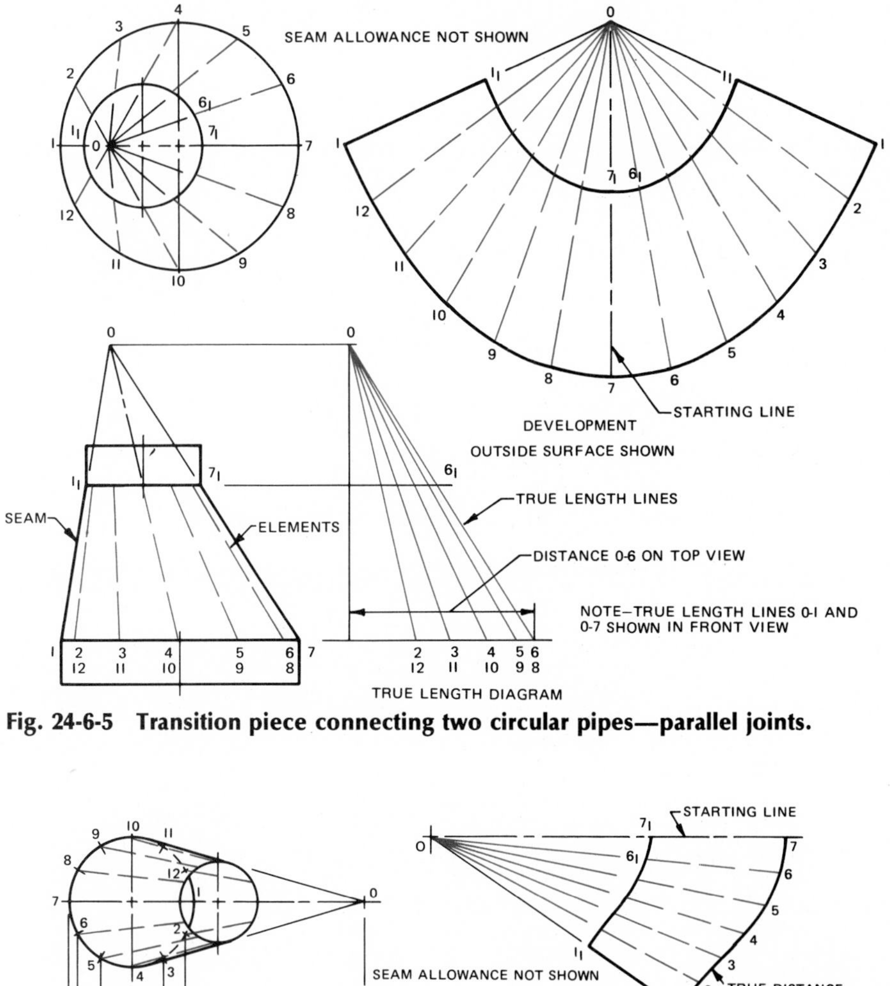
SEAM ALLOWANCE NOT SHOWN
STARTING LINE
DEVELOPMENT
OUTSIDE SURFACE SHOWN
TRUE LENGTH LINES
SEAM
ELEMENTS
DISTANCE 0-6 ON TOP VIEW
NOTE—TRUE LENGTH LINES 0-1 AND 0-7 SHOWN IN FRONT VIEW
TRUE LENGTH DIAGRAM

Fig. 24-6-5 Transition piece connecting two circular pipes—parallel joints.

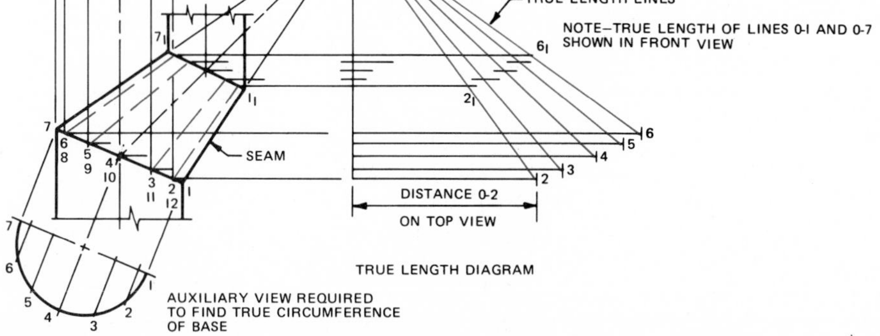
STARTING LINE
SEAM ALLOWANCE NOT SHOWN
HALF DEVELOPMENT
TRUE DISTANCE BETWEEN NUMBERS SHOWN IN AUXILIARY VIEW
TRUE LENGTH LINES
NOTE—TRUE LENGTH OF LINES 0-1 AND 0-7 SHOWN IN FRONT VIEW
SEAM
DISTANCE 0-2 ON TOP VIEW
TRUE LENGTH DIAGRAM
AUXILIARY VIEW REQUIRED TO FIND TRUE CIRCUMFERENCE OF BASE

Fig. 24-6-6 Transition piece connecting two circular pipes—oblique joints.

Assignment

On an A3-size sheet or a B-size sheet, make a two-view drawing plus a development drawing of the transition piece of one of the parts shown in Figs. 24-6-A to 24-6-D. Use your judgment for size and location of seams. Scale is 1:1.

REVIEW FOR ASSIGNMENT

Unit 24-5 Development of a Cone

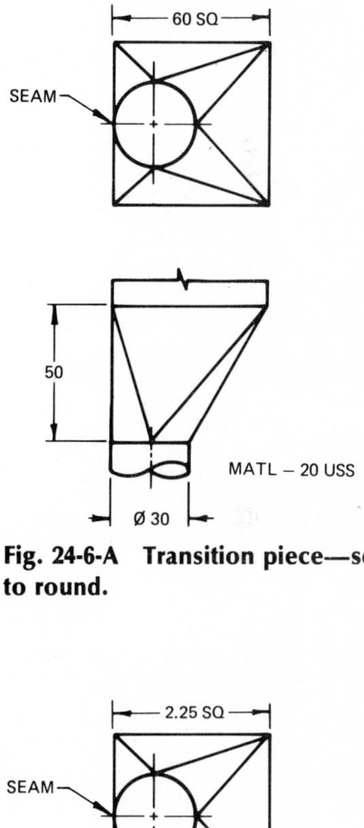
60 SQ
SEAM
50
MATL – 20 USS
Ø 30

Fig. 24-6-A Transition piece—square to round.

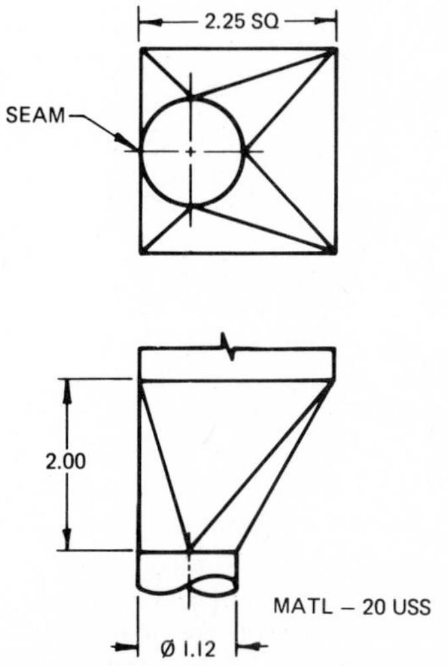
2.25 SQ
SEAM
2.00
MATL – 20 USS
Ø 1.12

Fig. 24-6-B Transition piece—square to round.

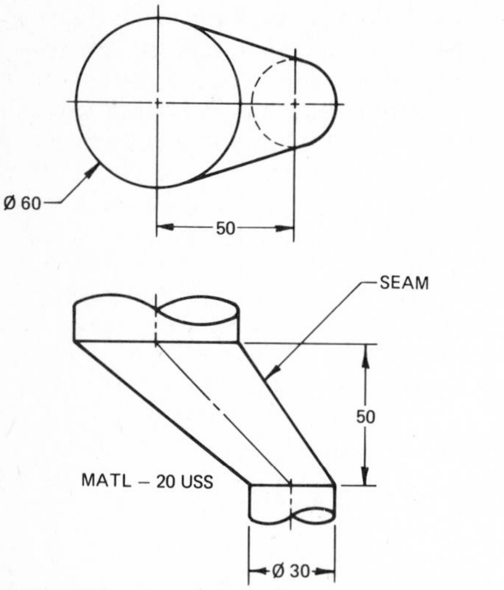
Ø 60
50
SEAM
50
MATL – 20 USS
Ø 30

Fig. 24-6-C Transition piece—round to round.

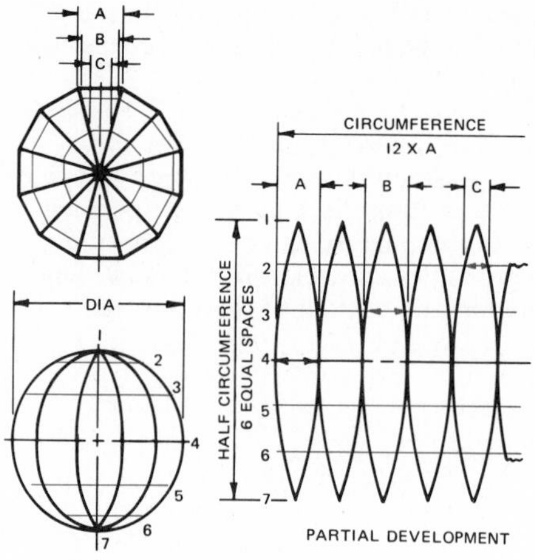
A
B
C
DIA
CIRCUMFERENCE
12 X A
HALF CIRCUMFERENCE
6 EQUAL SPACES
PARTIAL DEVELOPMENT

Fig. 24-7-1 Development of a sphere—gore method.

Assignment

On an A3- or B-size sheet, make a development drawing of one of the spheres shown in Figs. 24-7-A to 24-7-D. Use your judgment for the other views required. Either the gore or zone method may be used. Scale is 1:1.

Fig. 24-7-A Ball. Sphere—ϕ 80
Fig. 24-7-B Ball. Sphere—ϕ 3.00

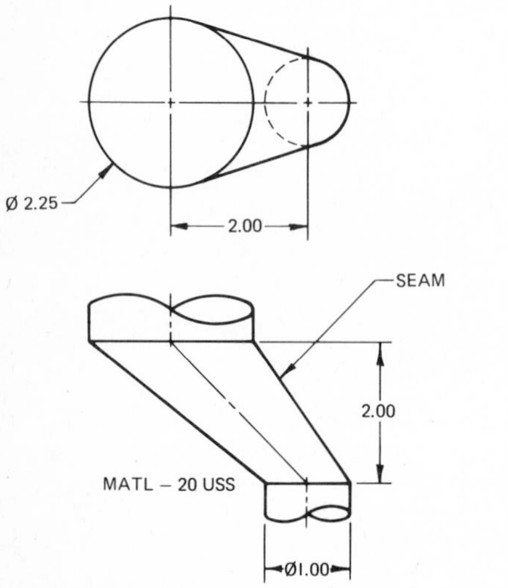
Ø 2.25
2.00
SEAM
2.00
MATL – 20 USS
Ø1.00

Fig. 24-6-D Transition piece—round to round.

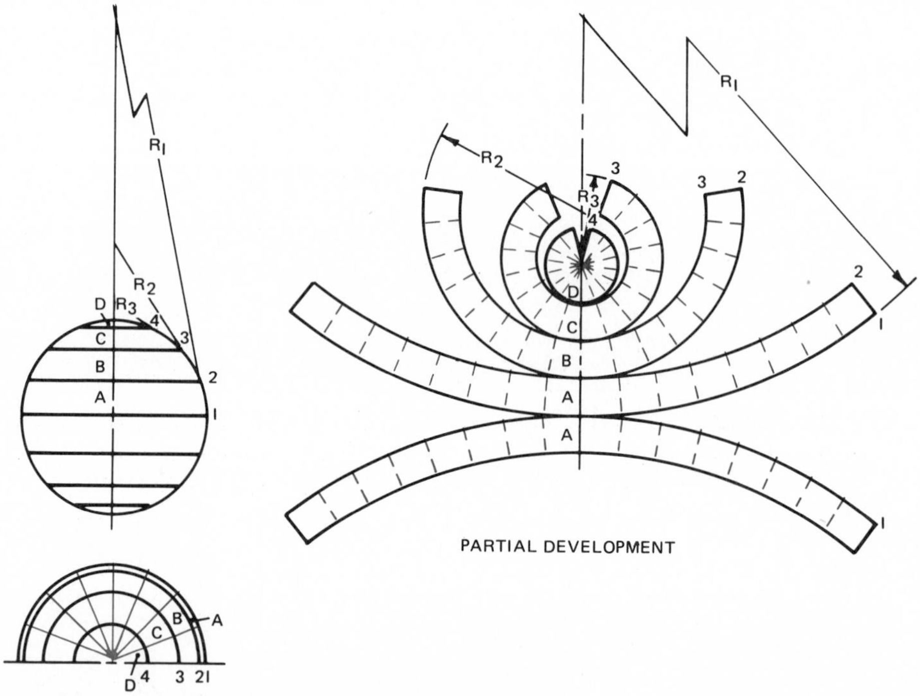
R1
R2
R3
R4
PARTIAL DEVELOPMENT

Fig. 24-7-2 Development of a sphere—zone method.

UNIT 24-7

DEVELOPMENT OF A SPHERE

Since the surface of a sphere is double-curved, it is not developable. However, the surface may be approximately developed by either the gore or the zone method. In the gore method (Fig. 24-7-1) the surface is divided into a number of equal sections, each section being considered as a section of a cylinder. Only one section need be developed, for it will serve as a pattern for the others. In the zone method (Fig. 24-7-2), the sphere is divided into horizontal zones and each zone is developed as a frustum of a cone.

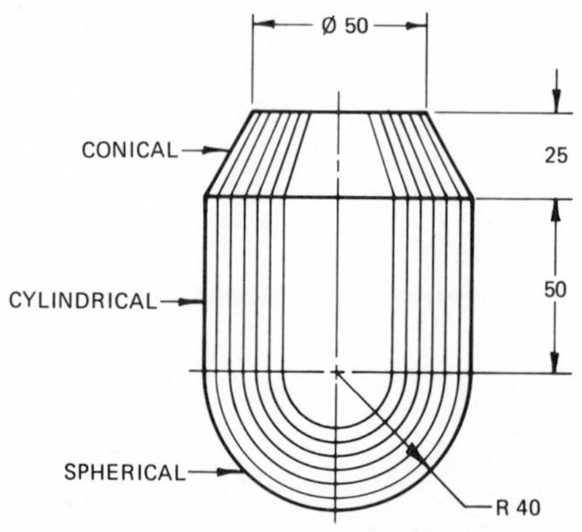
Ø 50
CONICAL
25
CYLINDRICAL
50
SPHERICAL
R 40

Fig. 24-7-C Thermos bottle liner.

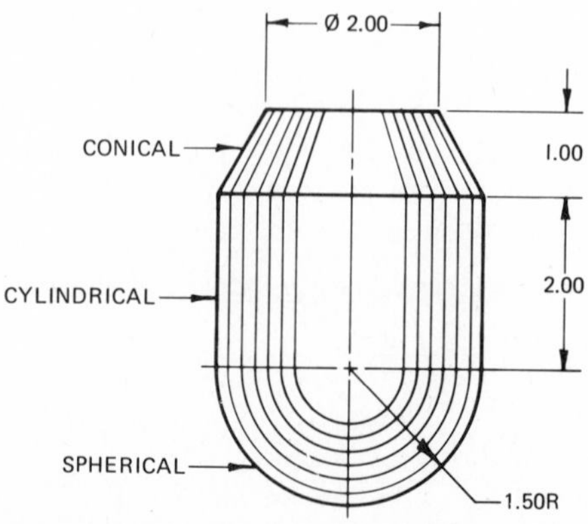
Ø 2.00
CONICAL
1.00
CYLINDRICAL
2.00
SPHERICAL
1.50R

Fig. 24-7-D Thermos bottle liner.

UNIT 24-8

INTERSECTION OF FLAT SURFACES —LINES PERPENDICULAR

Whenever two surfaces meet, there is a line common to both called the *line of intersection*. In making the orthographic drawing of objects that comprise two or more intersecting parts, the lines of intersection of these parts must be plotted on the orthographic views. Figures 24-8-1 and 24-8-2 illustrate this plotting technique for the intersection of flat-sided prisms. Figure 24-8-1 shows the development of the parts. A numbering technique is very valuable in plotting lines of intersection. In the illustrations shown, the lines of intersection appear in the front view. The end points for these lines are established by projecting the height position from the right side view to intersect the corresponding length position projected from the top view. When the prisms are flat-sided, the lines of intersection are straight, and the lines in the development are straight.

Intersecting Prisms—Triangle and Pyramid. See Fig. 24-8-3. In drawing the intersection of these two prisms, the points of intersection in the top view are found by projecting the points of intersection from the front view. If a development of the pyramid is required, true-length lines, which do not appear in either the top or the front view, must be found. Since only a few true-length lines are unknown, line 0-*D* on the front view serves as a true-length diagram. Lines $0\text{-}2_1$, $0\text{-}1_1$, and $0\text{-}5_1$ on line 0-*D* are the true lengths of lines 0-2, 0-1, and 0-5, respectively. Point 1 on surface 0-*EF* and 0-*BD* is located on the development as follows: draw a straight line through points 0 and 1 in the front view to intersect the base at point 7. Transfer distance *C*7 in the front view to the development view. Join points 0 and 7 with a straight line. With center 0 and radius equal to distance $0\text{-}1_1$ shown on the front view, swing an arc intersecting line 0-7 at point 1.

Points on the development of the triangular prism are found by projecting lines from the top view to the development view and transferring distances between points (numbers) and lines (letters)

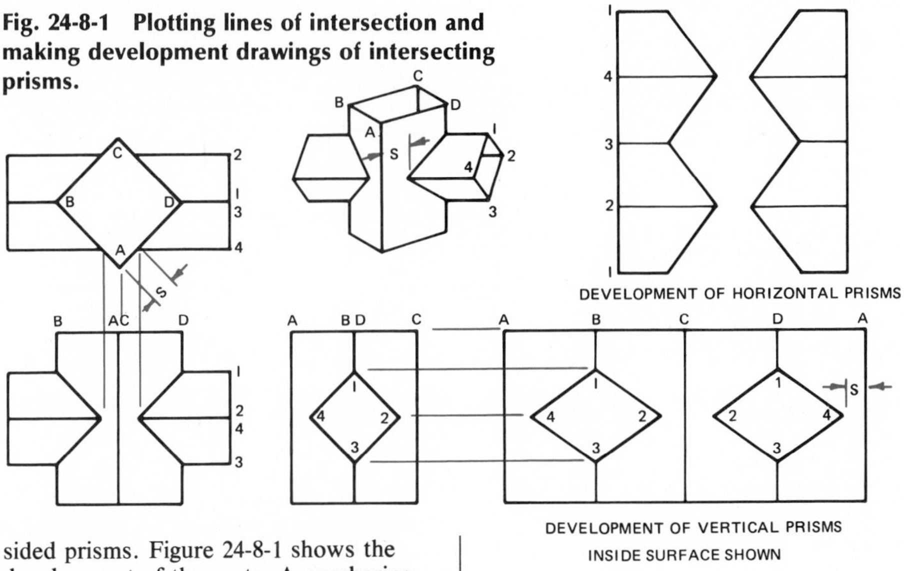
DEVELOPMENT OF HORIZONTAL PRISMS
DEVELOPMENT OF VERTICAL PRISMS
INSIDE SURFACE SHOWN

Fig. 24-8-1 Plotting lines of intersection and making development drawings of intersecting prisms.

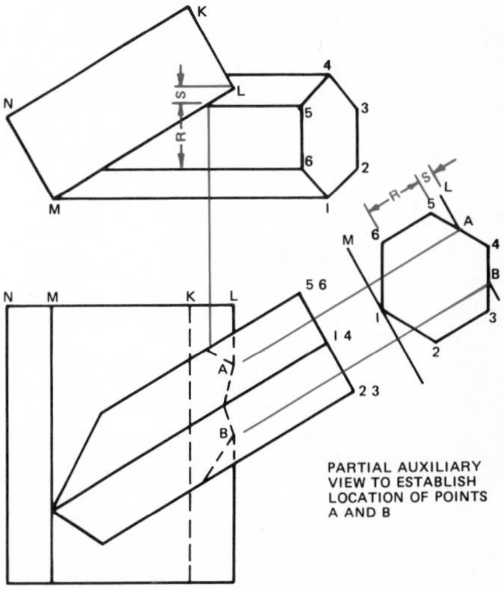
PARTIAL AUXILIARY VIEW TO ESTABLISH LOCATION OF POINTS A AND B

Fig. 24-8-4 Intersecting prisms not at right angles—hexagon and rectangle.

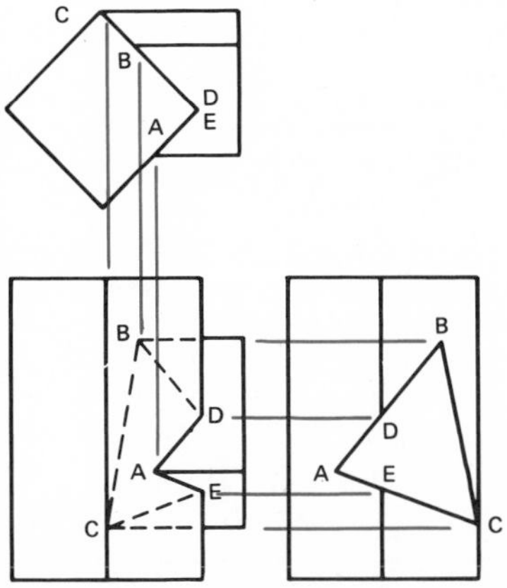

Fig. 24-8-2 Intersecting prisms at right angles.

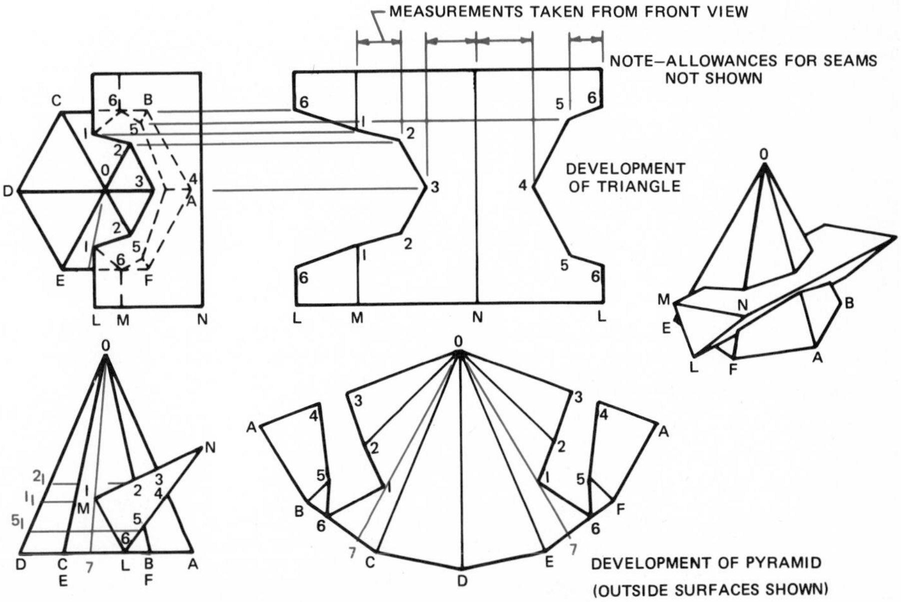
MEASUREMENTS TAKEN FROM FRONT VIEW
NOTE—ALLOWANCES FOR SEAMS NOT SHOWN
DEVELOPMENT OF TRIANGLE
DEVELOPMENT OF PYRAMID
(OUTSIDE SURFACES SHOWN)

Fig. 24-8-3 Intersecting prisms—triangle and pyramid.

from the front view to corresponding points on the development.

Intersecting Prisms Not at Right Angles—Hexagon and Rectangle. See Fig. 24-8-4. Often a partial auxiliary view is drawn to locate points of intersection such as points *A* and *B* on line *L*.

Intersecting Prisms Not at Right Angles—Hexagon and Triangle. See Fig. 24-8-5. An auxiliary view is required to locate points of intersection, such as points *A*, *B*, and *C*. To complete the side view, ends of lines *D*, *E*, and *F* and points of intersection *A*, *B*, and *C* are projected from the top view. Distances between the lines of the hexagon and triangle are transferred from either the top view or the auxiliary view.

Intersecting Prisms—Triangle and Pyramid. See Fig. 24-8-6. Another method commonly used to locate points of intersection of lines and surfaces is the use of vertical cutting planes located on the edges piercing the surface. Thus section *R-R* locates point *C*, section *S-S* locates point *A*, and section *T-T* locates point *B*. The sectional views shown are for illustrative purposes only and need not be drawn. To establish point *C*, extend line *LC* in the top view to intersect line 0-3 at C_2 and line 1-3 at C_1. Project vertical lines from C_1 and C_2 down the front view, locating points C_1 on base line 1-3 and C_2 on line 0-3. Join points C_1 and C_2 at point *C*. Extend a vertical line up from point *C* to the top view, intersecting line C_2L at point *C*. Repeat for points *A* and *B*.

Assignment

On an A3- or B-size sheet, select one of the assembled parts shown in Figs. 24-8-A to 24-8-D, complete the views, and make a development drawing of the vertical parts. In each case assume that the horizontal part passes through the vertical support. Add suitable seams. Scale is 1:1.

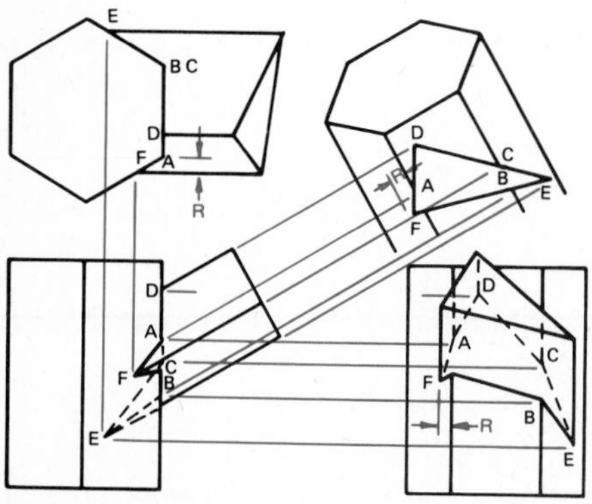
E
BC
D
F
A
R
C
B
E
F
D
A
C
B
E
F
E
R
B
E

Fig. 24-8-5 Intersecting prisms not at right angles—hexagon and triangle.

REVIEW FOR ASSIGNMENT

Unit 24-2 Development of Hexagons
Unit 24-3 Development of Pyramids

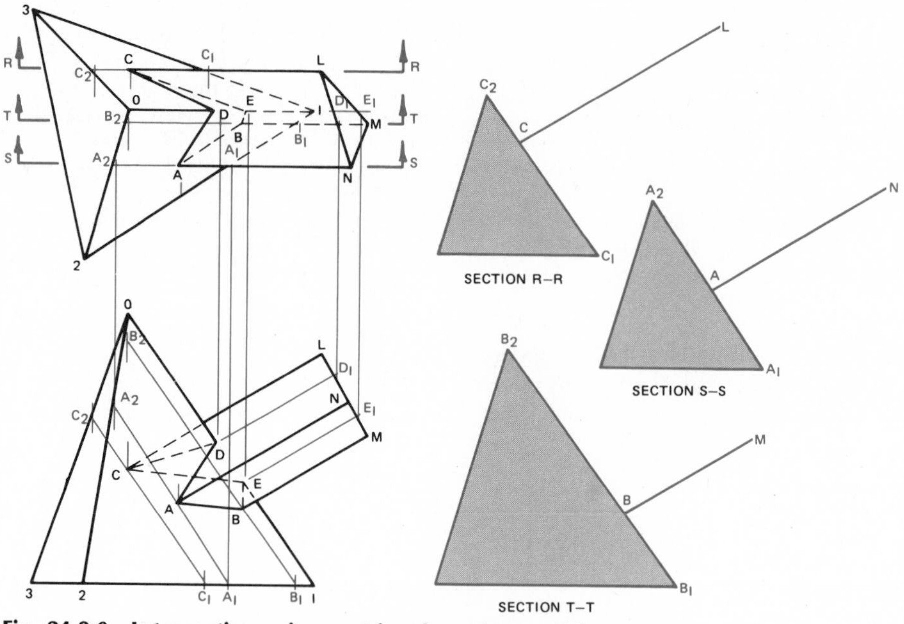
SECTION R–R
SECTION S–S
SECTION T–T

Fig. 24-8-6 Intersecting prisms—triangle and pyramid.

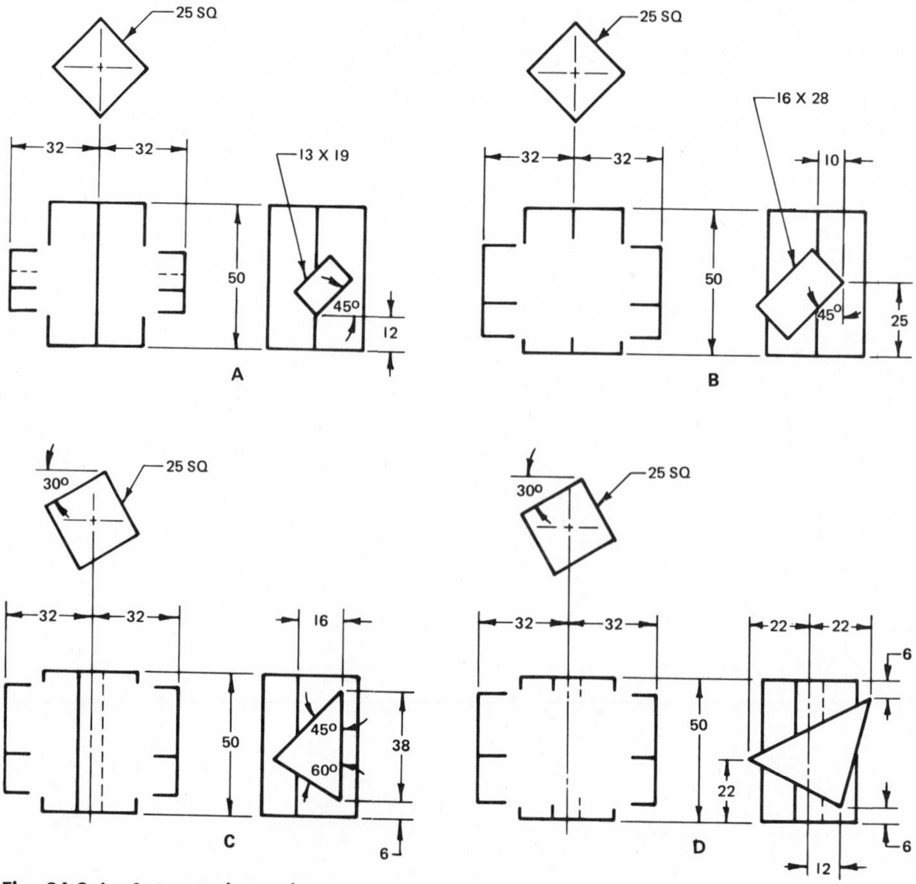
25 SQ
13 X 19
45°
50
12
A
16 X 28
10
25
B
30°
16
38
45°
60°
6
C
22
12
6
D

Fig. 24-8-A Intersecting prisms.

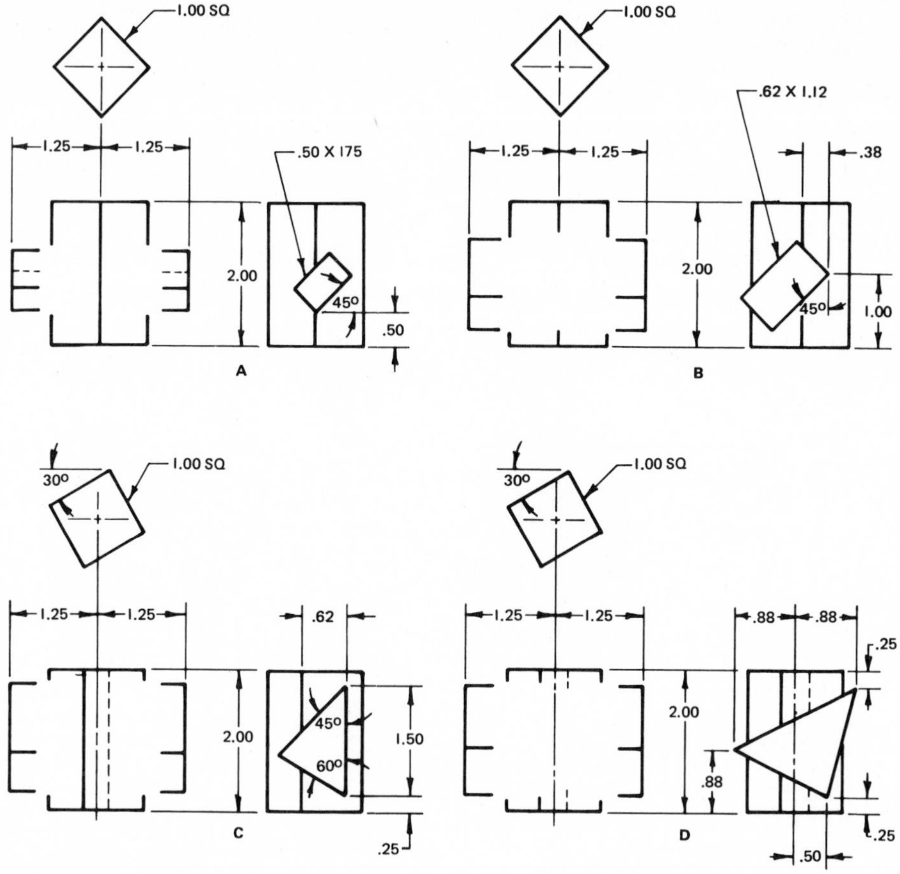
1.00 SQ
1.25
1.25
.50 X 175
2.00
45°
.50
A
1.00 SQ
.62 X 1.12
1.25
1.25
.38
2.00
45°
1.00
B
30°
1.00 SQ
1.25
1.25
.62
2.00
45°
60°
1.50
.25
C
30°
1.00 SQ
1.25
1.25
.88
.88
.25
2.00
.88
.25
.50
D

Fig. 24-8-B Intersecting prisms.

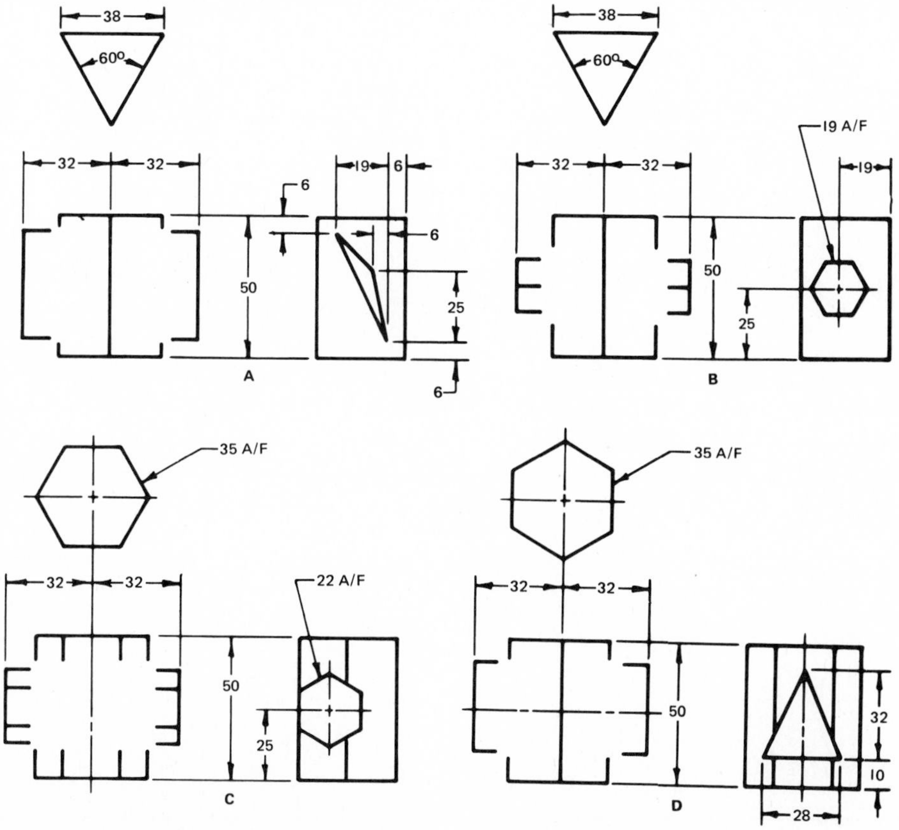
38
60°
32
32
19
6
6
6
50
25
6
A
38
60°
19 A/F
32
32
19
50
25
B
35 A/F
32
32
22 A/F
50
25
C
35 A/F
32
32
50
32
10
28
D

Fig. 24-8-C Intersecting prisms.

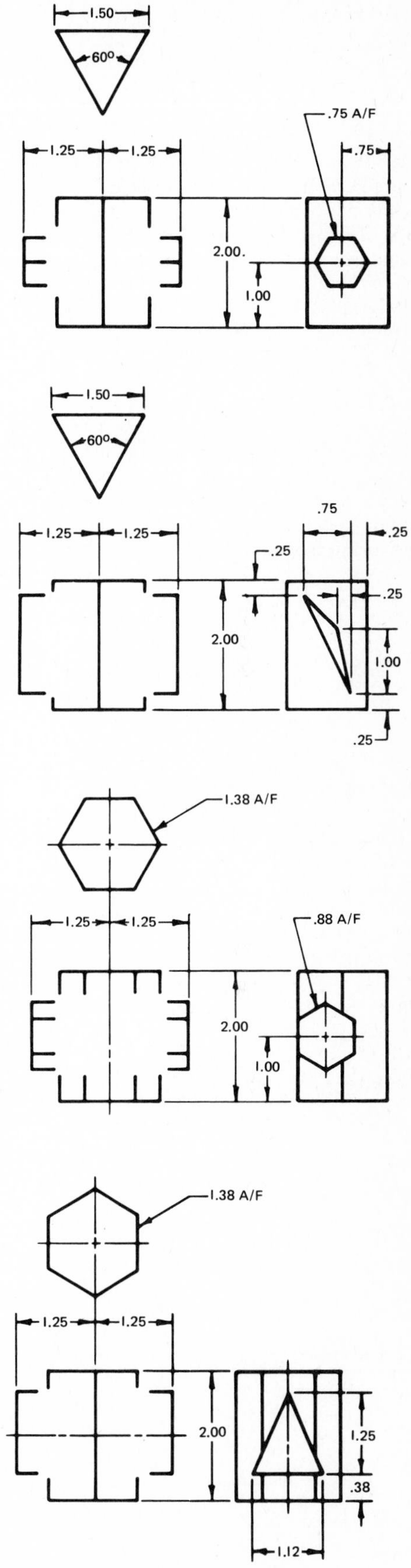
1.50
60°
.75 A/F
1.25
1.25
.75
2.00.
1.00
1.50
60°
.75
1.25
1.25
.25
.25
.25
2.00
1.00
.25
1.38 A/F
1.25
1.25
.88 A/F
2.00
1.00
1.38 A/F
1.25
1.25
2.00
1.25
.38
1.12

Fig. 24-8-D Intersecting prisms.

UNIT 24-9

INTERSECTION OF CYLINDRICAL SURFACES

90° Reducing Tee. See Fig. 24-9-1. This figure illustrates the plotting technique for the intersection of cylinders. Because there are no edges on the cylinders, element lines of reference are established about the cylinders in their orthographic views. In the top view, the element lines for the small cylinder are drawn to touch the surface of the large cylinder; for example, line 2 touches at *E*. This point location is then projected down to the front view to intersect the corresponding element line, establishing the height at that point. The points of intersection thus established are connected by an irregular curve to produce the line of intersection. The same points of reference used to establish the line of intersection are used to draw the development.

45° Reducing Tee. See Fig. 24-9-2. This figure illustrates the intersection of a small pipe at an angle of 45° to a large pipe. The same techniques of plotting reference points are used that were previously described for a 90° reducing tee.

Assignment

On an A3- or B-size sheet, select one of Figs. 24-9-A to 24-9-D and draw all four assemblies. Complete the views, and make a development drawing of the vertical parts. In each case assume the horizontal part passes through the horizontal support. Add the suitable seams. Scale is 1:1.

REVIEW FOR ASSIGNMENT

Unit 24-1 Development of Cylinders
Unit 24-8 Lines of Intersection

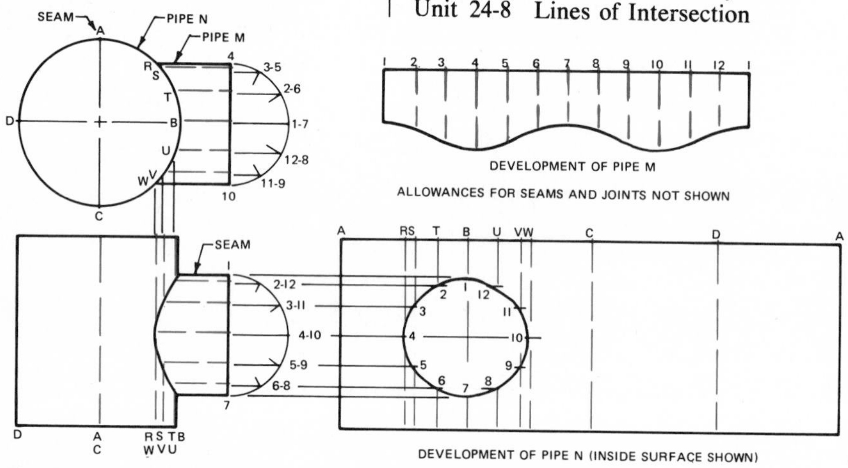
SEAM
PIPE N
PIPE M
DEVELOPMENT OF PIPE M
ALLOWANCES FOR SEAMS AND JOINTS NOT SHOWN
SEAM
DEVELOPMENT OF PIPE N (INSIDE SURFACE SHOWN)

Fig. 24-9-1 Plotting lines of intersection and making development drawings for a 90° reducing tee.

Fig. 24-9-2 Plotting lines of intersection making development drawings for a 45° reducing tee.

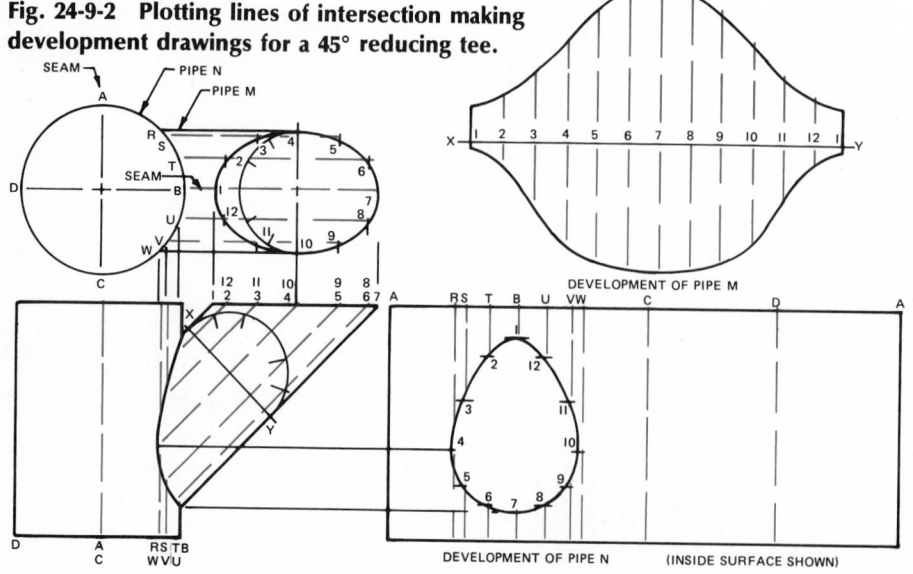

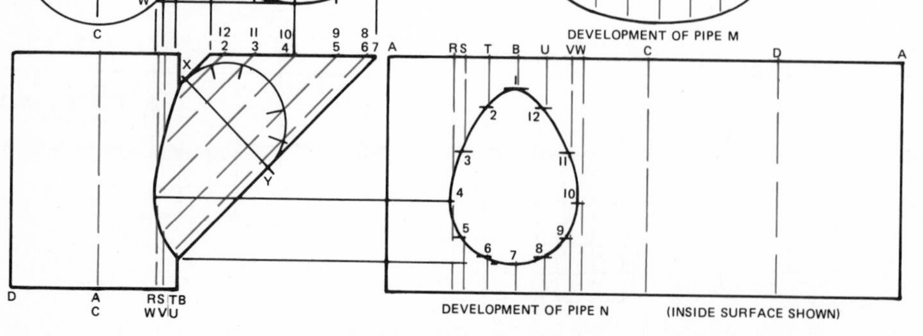
SEAM
PIPE N
PIPE M
SEAM
DEVELOPMENT OF PIPE M
DEVELOPMENT OF PIPE N
(INSIDE SURFACE SHOWN)
NOTE—ALLOWANCES FOR SEAMS AND JOINTS NOT SHOWN

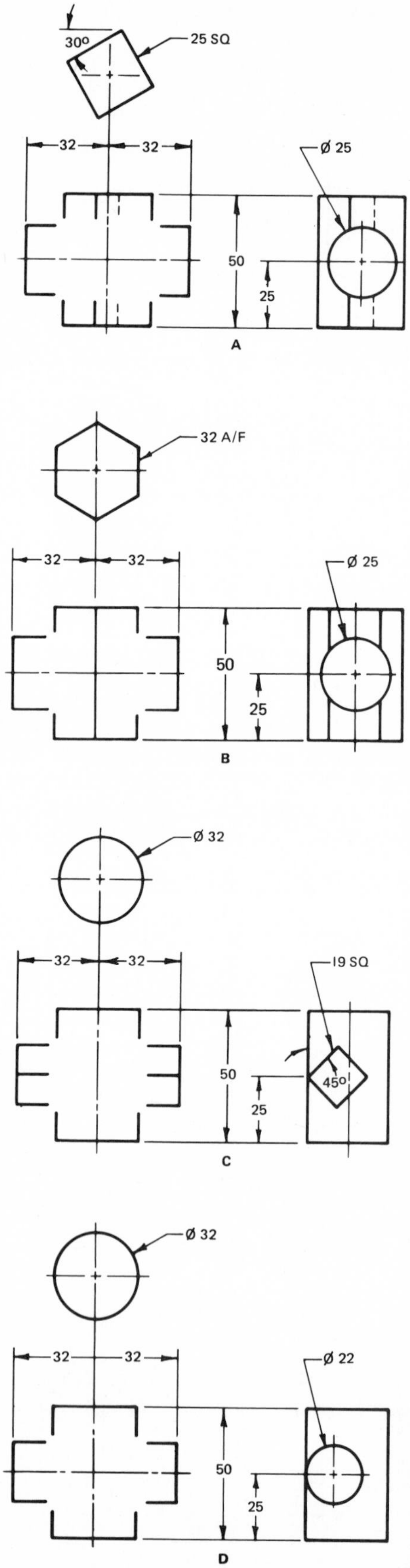
30°
25 SQ
Ø 25
A
32 A/F
Ø 25
B
Ø 32
19 SQ
45°
C
Ø 32
Ø 22
D

Fig. 24-9-A Intersecting prisms.

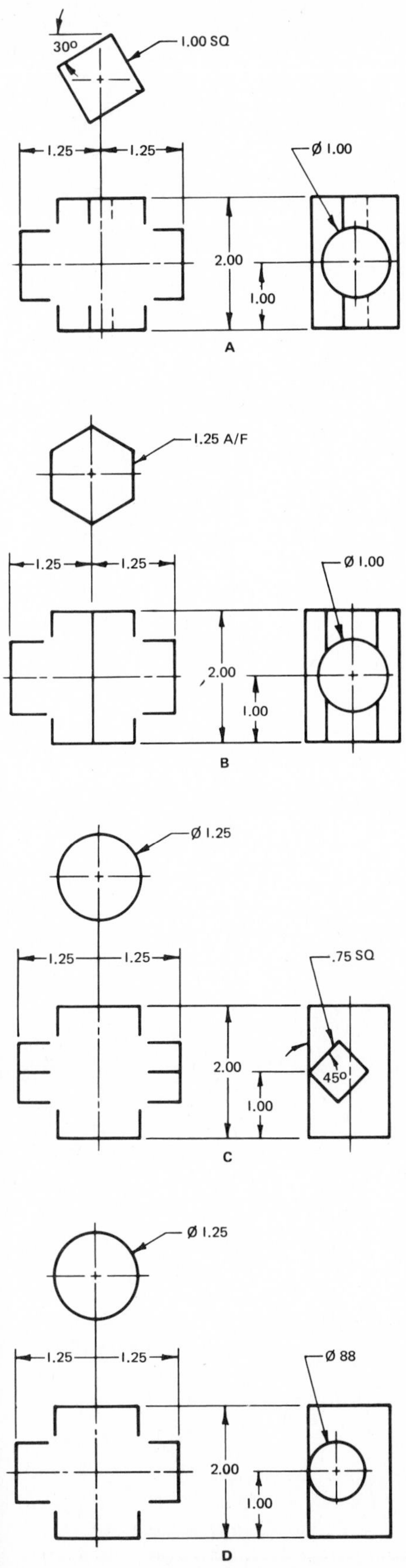
30°
1.00 SQ
1.25
1.25
Ø 1.00
2.00
1.00
A
1.25 A/F
1.25
1.25
Ø 1.00
2.00
1.00
B
Ø 1.25
1.25
1.25
.75 SQ
2.00
45°
1.00
C
Ø 1.25
1.25
1.25
Ø 88
2.00
1.00
D

Fig. 24-9-B Intersecting prisms.

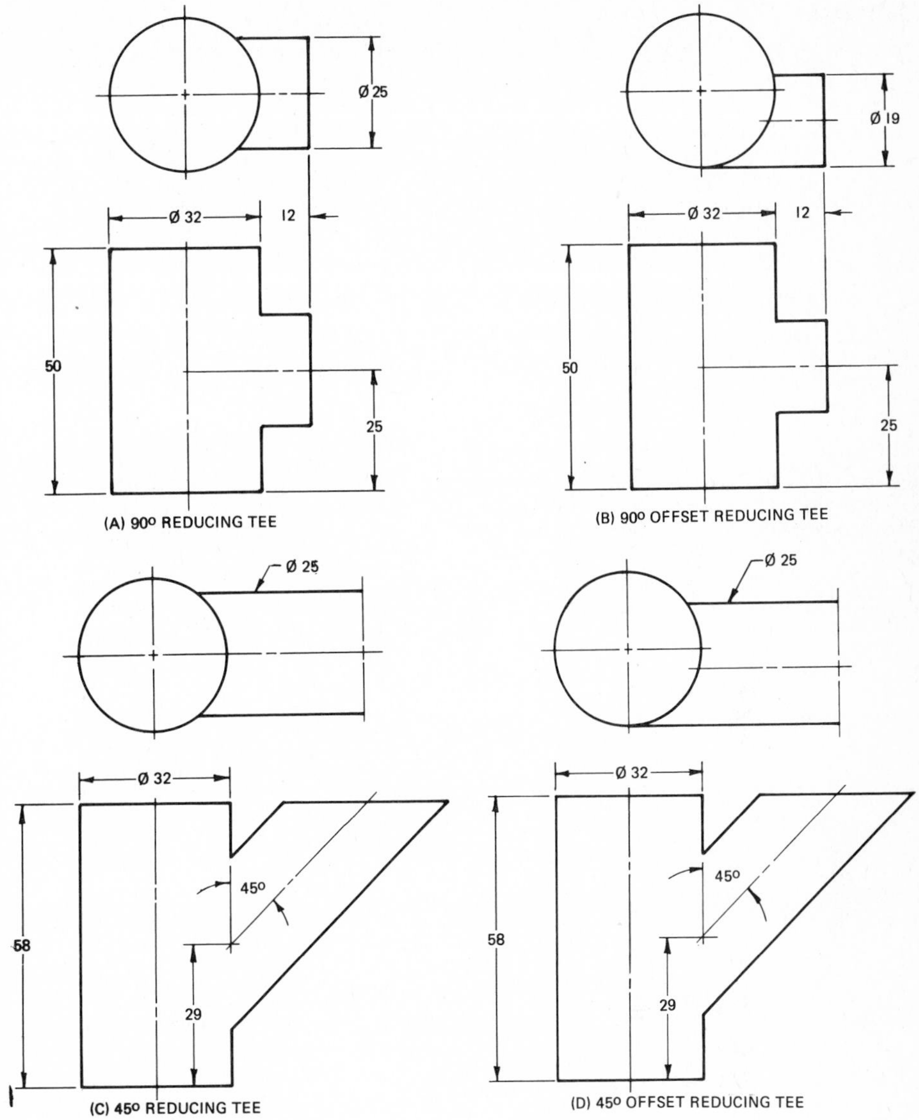
Ø 25
Ø 32
12
50
25
(A) 90° REDUCING TEE
Ø 19
Ø 32
12
50
25
(B) 90° OFFSET REDUCING TEE
Ø 25
Ø 32
45°
58
29
(C) 45° REDUCING TEE
Ø 25
Ø 32
45°
58
29
(D) 45° OFFSET REDUCING TEE

Fig. 24-9-C Tees and laterals.

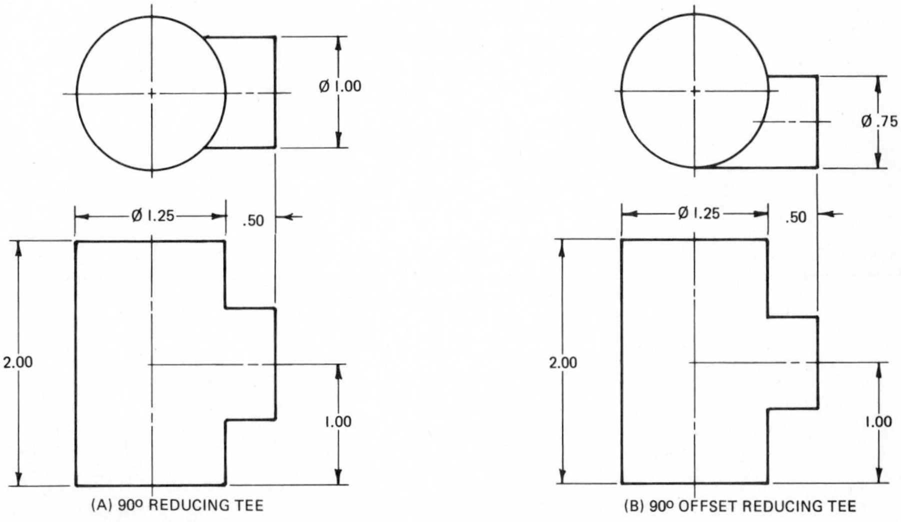
Ø 1.00
Ø 1.25
.50
2.00
1.00
(A) 90° REDUCING TEE
Ø .75
Ø 1.25
.50
2.00
1.00
(B) 90° OFFSET REDUCING TEE

Fig. 24-9-D Tees.

UNIT 24-10

INTERSECTING PRISMS

Intersecting Prisms—Hexagon and Cone. See Fig. 24-10-1. The lines of intersection between the hexagon and cone are developed as follows. Divide each side of the hexagon into four parts (lines every 15°). Through point *A* in the top view, swing an arc intersecting horizontal center line 0-4 at A_1. From point A_1 drop a vertical line down to the front view, intersecting line 0-*D* at point A_1. Draw a light horizontal line through A_1 in the front view. Drop vertical lines down from point *A* in the top view, intersecting this line at point *A*. Repeat locating point *B*. Point *C* may be located in the front view by extending a light horizontal line from point *C* on line 0-*D*. Join points *A*, *B*, and *C* with a line, which forms a hyperbolic curve.

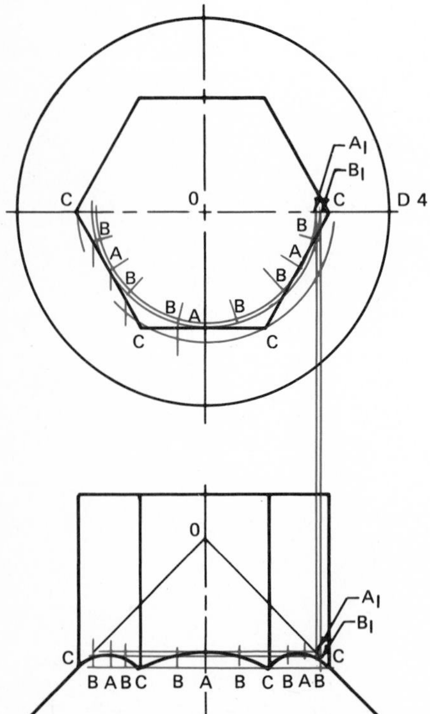
A1
B1
C
0
D 4
B A B C B A B C B A B
D

Fig. 24-10-1 Intersecting prisms—hexagon and cone.

Intersecting Prisms—Cone and Cylinder, Fig. 24-10-2. The intersections of the cone and cylinder elements in the top view are first found and are then projected down to the corresponding elements in the front view. A smooth curve is drawn through these points to produce the line of intersection.

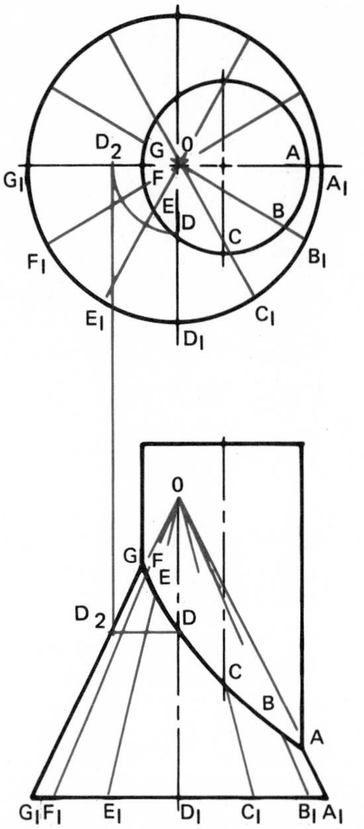
G1
F1
E1
D1
C1
B1
A1
D2
0
G F E D C B A
G1F1 E1 D1 C1 B1 A1

Fig. 24-10-2 Intersecting prisms—cone and cylinder.

Intersecting Prisms—Cone and Cylinder, Fig. 24-10-3. The line of intersection between the cone and the cylinder is found by assuming the front view to have a series of horizontal cutting planes passing through points 2, 3, 4, 5, and 6. The cutting-plane line passes through the

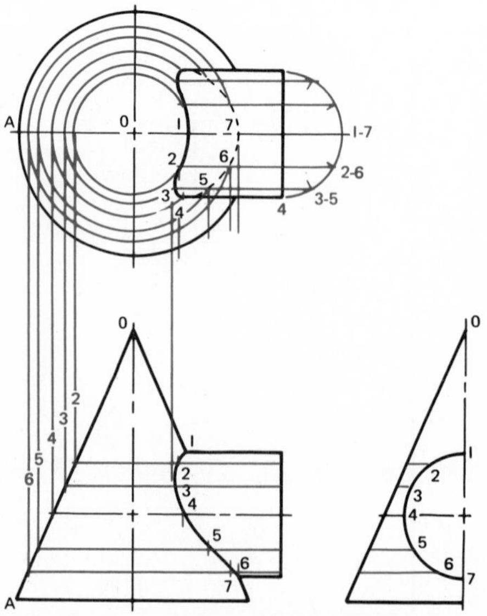
A
0
1-7
2-6
3-5
4

Fig. 24-10-3 Intersecting prisms—cone and cylinder.

intersection of the cone and cylinder. Each point on the line of intersection is developed in a manner similar to that for Fig. 24-10-2.

Intersecting Prisms—Pyramid and Cylinder. See Fig. 24-10-4. Cutting-plane lines taken horizonally through points 2, 3, 4, 5, and 6 in the front view are used to locate the lines of intersection. Point 5 on the line of intersection is located as follows. Draw a horizontal line through point 5 on the half circle in the front view to intersect line 0-*A* at point 5_1. Extend a vertical line from point 5_1 to the top view to intersect line 0-*A* at point 5_1. Extend a line from 5_1 in the top view at an angle of 45° to intersect line 0-*D*. From this intersection, extend a line at an angle of 45° in the direction of line 0-*C* to intersect a horizontal line passing through point 5 on the half circle in the top view. The intersection of these lines is point 5. To locate point 5 in the front view, drop a vertical line from point 5 in the top view to intersect the horizontal line passing through point 5 of the half

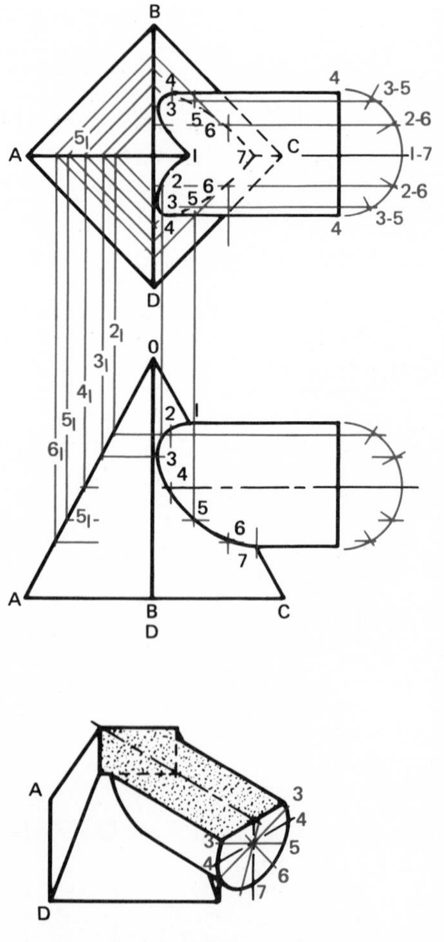
B
A
C
D
3-5
2-6
1-7
0
B
D
C

SECTION TAKEN ON PLANE 3

Fig. 24-10-4 Intersecting prisms—pyramid and cylinder.

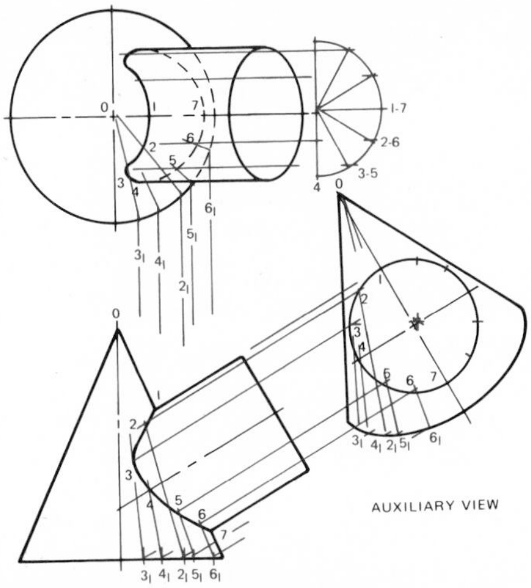
AUXILIARY VIEW

Fig. 24-10-5 Intersecting prisms—cone and oblique cylinder.

circle. Locate the other points in a similar manner, and connect them with a smooth, curved line.

Intersecting Prisms—Cone and Oblique Cylinder—Fig. 24-10-5. The cylinder shown in the auxiliary view is divided into 12 equal parts. Element lines are drawn from the apex of the cone, point 0, through points 2 to 6 to the base of the cone, establishing points 2_1 to 6_1 inclusive. These points are projected to corresponding points in the front view and the element lines are drawn.

The element lines in the top view are located by projecting vertical lines up from points 2_1 to 6_1 in the front view to intersect the base circle in the top view. The lines of intersection are then found by projecting lines from the circle to meet their corresponding element lines.

Assignment

On an A3- or B-size sheet, select one of Figs. 24-10-A to 24-10-D and draw all four assemblies. Complete the lines of intersections on the partially completed views, and make a development drawing of the vertical part(s). In each case, the other part is assumed to be passing through the vertical part; add suitable seams. Scale is 1:1.

REVIEW FOR ASSIGNMENT

Unit 24-3 Development of Pyramids
Unit 24-4 Development of Cylinders
Unit 24-5 Development of Cones

Fig. 24-10-A Intersecting prisms.

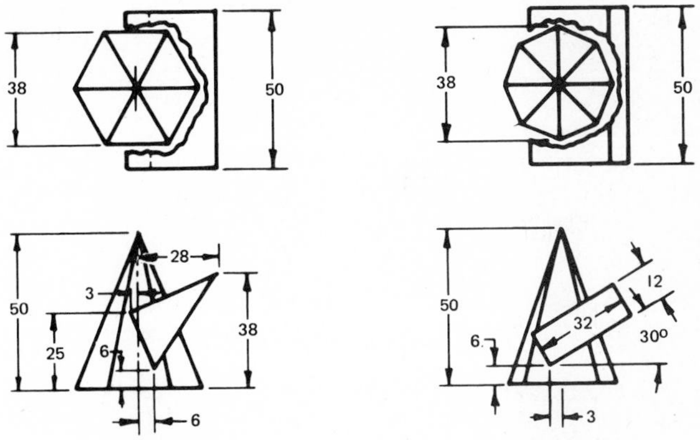

Fig. 24-10-B Intersecting prisms.

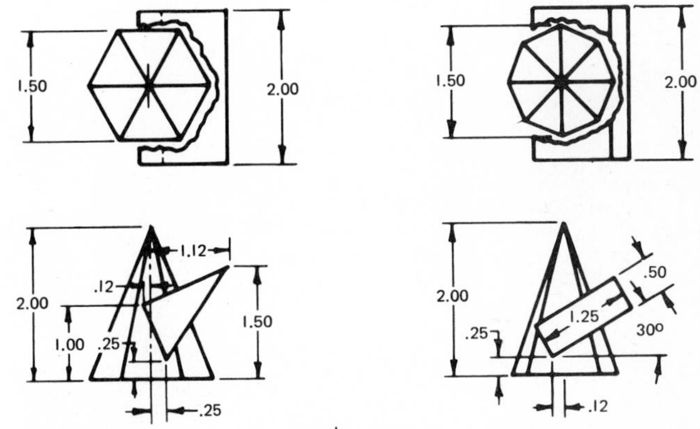

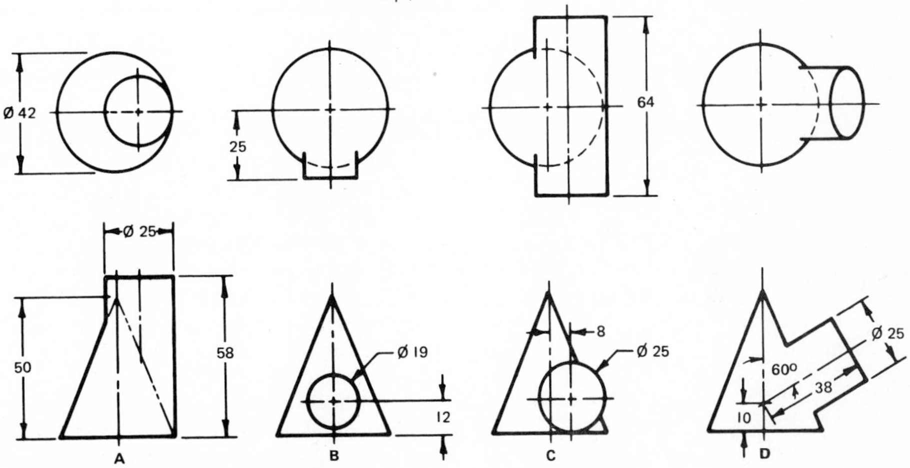
Ø 42
25
64
Ø 25
50
58
Ø 19
12
8
Ø 25
Ø 25
60°
38
10
A
B
C
D

Fig. 24-10-C Intersecting prisms.

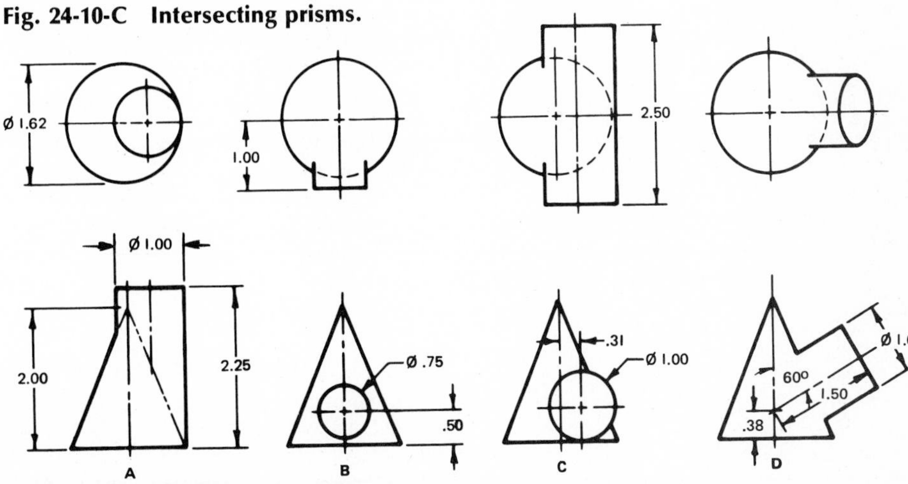
Ø 1.62
1.00
2.50
Ø 1.00
2.00
2.25
Ø .75
.50
.31
Ø 1.00
Ø 1.00
60°
1.50
.38
A
B
C
D

Fig. 24-10-D Intersecting prisms.

Chapter 25
Pipe Drawings

UNIT 25-1
PIPES

One hundred years ago, water was the only important fluid which was conveyed from one point to another in pipe. Today almost every conceivable fluid is handled in pipe during its production, processing, transportation, or utilization. The age of atomic energy and rocket power has added fluids such as liquid metals, sodium, and nitrogen to the list of more common fluids such as oil, water, gases, and acids that are being transported in pipe today. Nor is the transportation of fluids the only phase of hydraulics which warrants attention now. Hydraulic and pneumatic mechanisms are used extensively for the controls of machinery and numerous other equipment. Piping is also employed as a structural element such as in columns and handrails. It is for these reasons that drafters and engineers should familiarize themselves with pipe drawings.

Kinds of Pipe

STEEL AND WROUGHT-IRON PIPE

This type of pipe carries water, steam, oil, and gas and is commonly used where high temperatures and pressures are encountered. Standard steel and cast-iron pipe is specified by the nominal diameter, which is always less than the actual inner diameter (ID) of the pipe. This pipe was available up to recent times in only three wall thicknesses—standard, extra-strong, and double extra-strong. See Fig. 25-1-1. In order to use common fittings with these different wall thicknesses of pipe, the outer diameter (OD) of each of the different pipes remained the same, and the extra metal was added to the ID to increase the wall thickness of the extra strong and double extra strong pipe.

The nominal size of pipe is given in inch sizes, but the inside and outside diameters and wall thickness are given in millimetre sizes.

Because of the demand for a greater variety of pipe for increased pressure and temperature uses, the ANSI and CSA have now made available 10 different wall thicknesses of pipe, each desig-

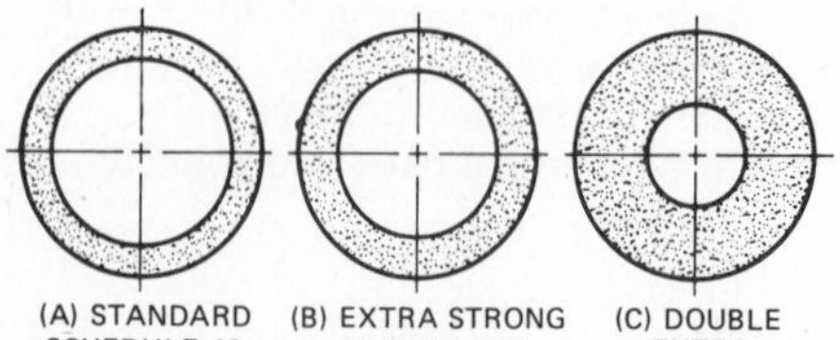

(A) STANDARD SCHEDULE 40
(B) EXTRA STRONG SCHEDULE 80
(C) DOUBLE EXTRA STRONG

Fig. 25-1-1 A comparison between steel pipe.

nated by a schedule number. Standard pipe is now referred to as *schedule 40 pipe* and extra strong pipe as *schedule 80*. Pipe over 12 in. is referred to as *OD pipe*, and the nominal size is the OD of the pipe.

CAST-IRON PIPE

Cast-iron pipe is often installed underground to carry water, gas, and sewage. It is also used for low-pressure steam connections. Cast-iron pipe joints are normally of the flanged type or the bell-and-spigot type.

SEAMLESS BRASS AND COPPER PIPE

These types of pipe are used extensively in plumbing because of their ability to withstand corrosion. They have the same nominal diameter as steel or iron pipe, but they have thinner wall sections.

COPPER TUBING

This pipe is used in plumbing and heating and where vibration and misalignment are factors, such as in automotive, hydraulic, and pneumatic design.

PLASTIC PIPE

This pipe or tubing, because of its corrosion and chemical resistance, is used extensively in the chemical industry. It is flexible and readily installed but is not recommended where heat or pressure is a factor.

Pipe Joints and Fittings

Parts that are joined to pipe are called *fittings*. They may be used to change size or direction and to join or provide branch connections. They fall into three general classes: screwed, welded, and flanged. See Fig. 25-1-2. Other methods are used for cast-iron pipe, copper, and plastic tubing.

Pipe fittings are specified by the nominal pipe size, the name of the fitting, and the material. Some fittings, such as tees, crosses, and elbows, are used to connect different sizes of pipe. These are called *reducing fittings*, and their nominal pipe sizes must be specified. The largest opening of the through run is given first, followed by the opposite end and the outlet. Figure 25-1-3 illustrates the method of designating sizes of reducing fittings.

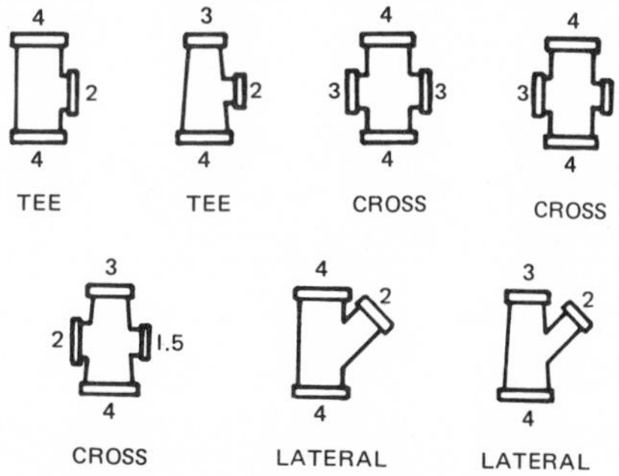
TEE
TEE
CROSS
CROSS
CROSS
LATERAL
LATERAL

Fig. 25-1-3 Order of specifying the openings of reducing fittings.

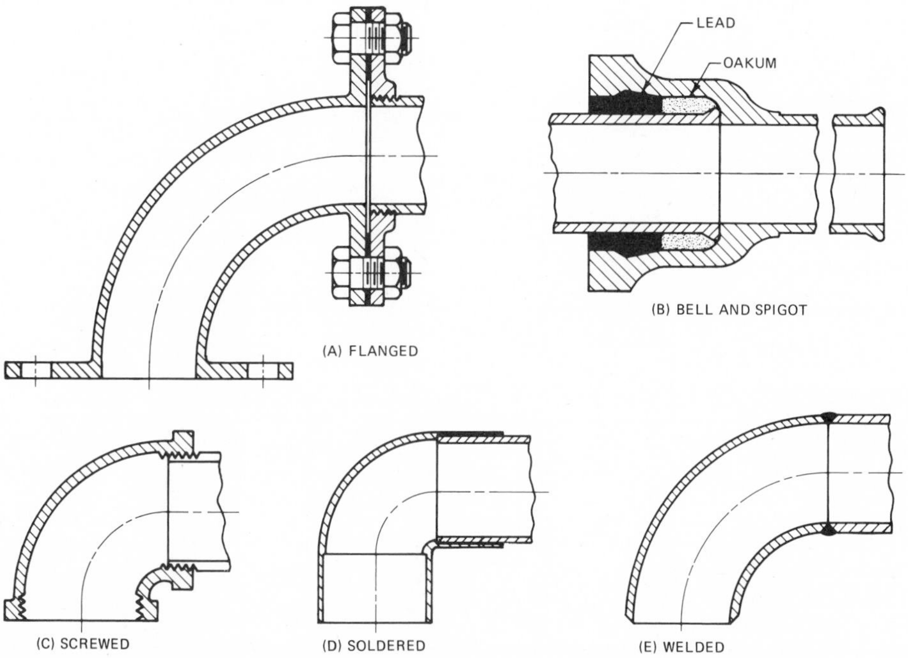
LEAD
OAKUM
(A) FLANGED
(B) BELL AND SPIGOT
(C) SCREWED
(D) SOLDERED
(E) WELDED

Fig. 25-1-2 Common types of pipe joints.

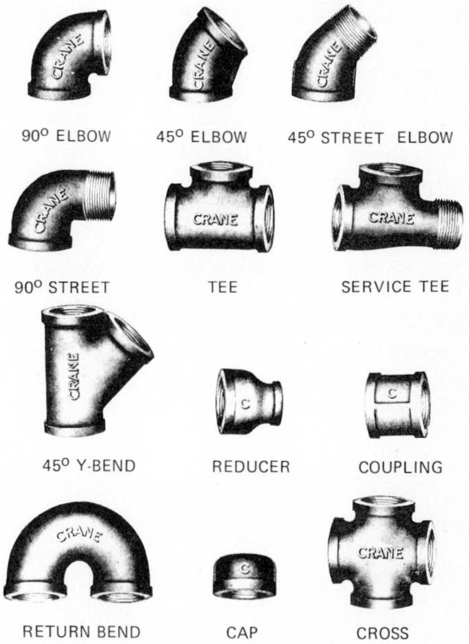
90° ELBOW
45° ELBOW
45° STREET ELBOW
90° STREET
TEE
SERVICE TEE
45° Y-BEND
REDUCER
COUPLING
RETURN BEND
CAP
CROSS

Fig. 25-1-4 Screwed fittings. (Crane Canada Ltd.)

SCREWED FITTINGS

Screwed fittings, as shown in Fig. 25-1-4, are generally used on small pipe design of 2.50 in. or less. Common practice is to use a pipe compound (a mixture of lead and oil) on the threaded connection to provide a lubricant and to seal any irregularities.

The American standard pipe thread is of two types—tapered or straight. The tapered thread, which is the more common, employs a 1:16 taper or diameter on both the external and internal threads. See Fig. 25-1-5. This fixes the distance to which the pipe enters the fitting and ensures a tight joint.

Straight threads are used for special applications which are listed in the ANSI handbook.

Both the taper and straight pipe threads have the same number of threads per inch of nominal pipe size, and a pipe

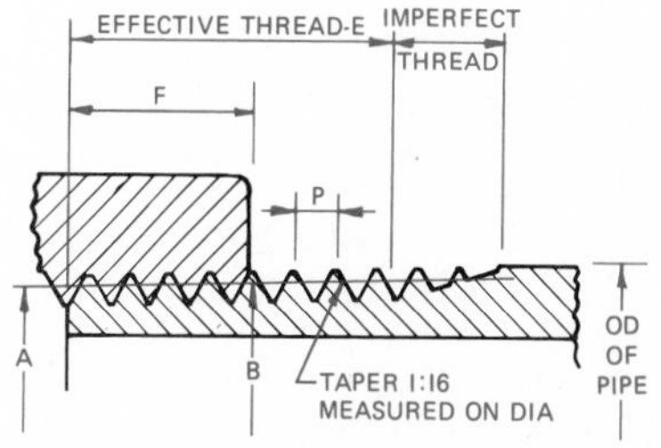
EFFECTIVE THREAD-E
IMPERFECT THREAD
F
P
A
B
TAPER 1:16 MEASURED ON DIA
OD OF PIPE

A = PITCH DIAMETER AT END OF PIPE D–(0.05D + 1.1)P
B = PITCH DIAMETER AT GAGING NOTCH (A + 0.0625F)
E = EFFECTIVE THREAD (0.8D + 6.8)P
F = NORMAL ENGAGEMENT BY HAND
P = PITCH
DEPTH OF THREAD = 0.8P

Fig. 25-1-5 American standard pipe thread.

with a tapered thread may thread into a fitting having a straight thread, resulting in a tight seal.

Tapered threads are designated on drawings as NPT (American Standard Pipe Taper Thread) and may be drawn with or without the taper, as shown in Fig. 25-1-6. When drawn in tapered form, the taper is exaggerated. Straight pipe threads are designated on drawings as NPS (American Standard Pipe Straight Thread), and standard thread symbols are used. All pipe threads are assumed to be tapered unless otherwise specified.

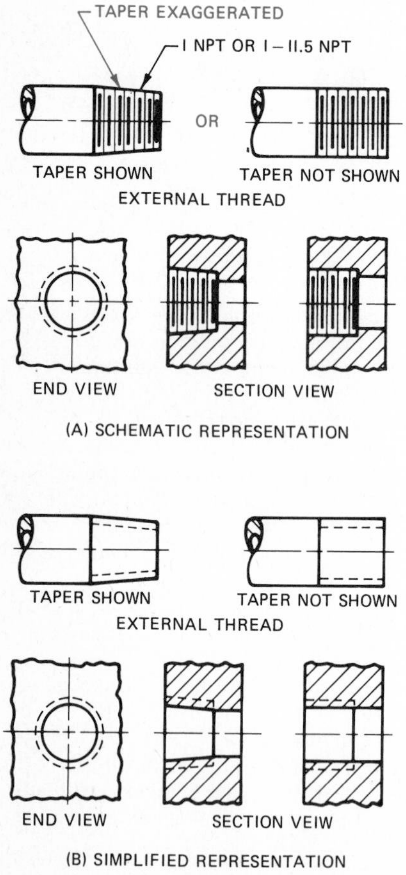
TAPER EXAGGERATED
I NPT OR I – II.5 NPT
OR
TAPER SHOWN
TAPER NOT SHOWN
EXTERNAL THREAD
END VIEW
SECTION VIEW
(A) SCHEMATIC REPRESENTATION
TAPER SHOWN
TAPER NOT SHOWN
EXTERNAL THREAD
END VIEW
SECTION VEIW
(B) SIMPLIFIED REPRESENTATION

Fig. 25-1-6 Pipe thread conventions.

WELDED FITTINGS

Welded fittings are used where connections are to be permanent and on high-pressure and -temperature lines. Other advantages over flanged or screwed fittings are that welded pipes are easier to insulate, they may be placed closer together, and they are lighter in mass. The ends of the pipe and pipe fittings are normally beveled, as shown in Fig. 25-1-7, to accommodate the weld. Joint rings may be used when welded pipe must be disassembled periodically.

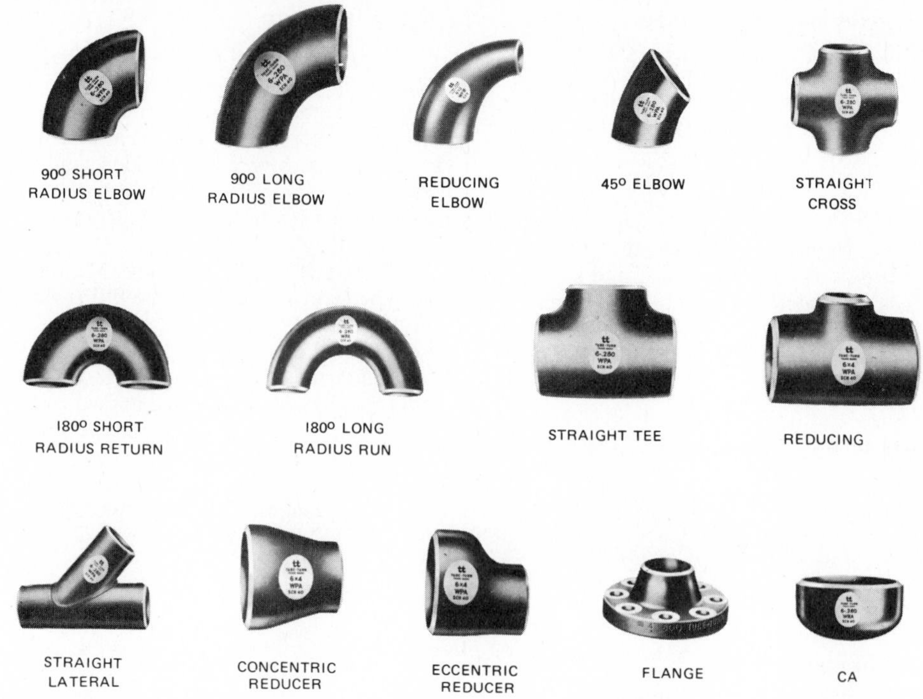
90° SHORT RADIUS ELBOW
90° LONG RADIUS ELBOW
REDUCING ELBOW
45° ELBOW
STRAIGHT CROSS
180° SHORT RADIUS RETURN
180° LONG RADIUS RUN
STRAIGHT TEE
REDUCING
STRAIGHT LATERAL
CONCENTRIC REDUCER
ECCENTRIC REDUCER
FLANGE
CA

Fig. 25-1-7 Welded fittings. (Tube Turns)

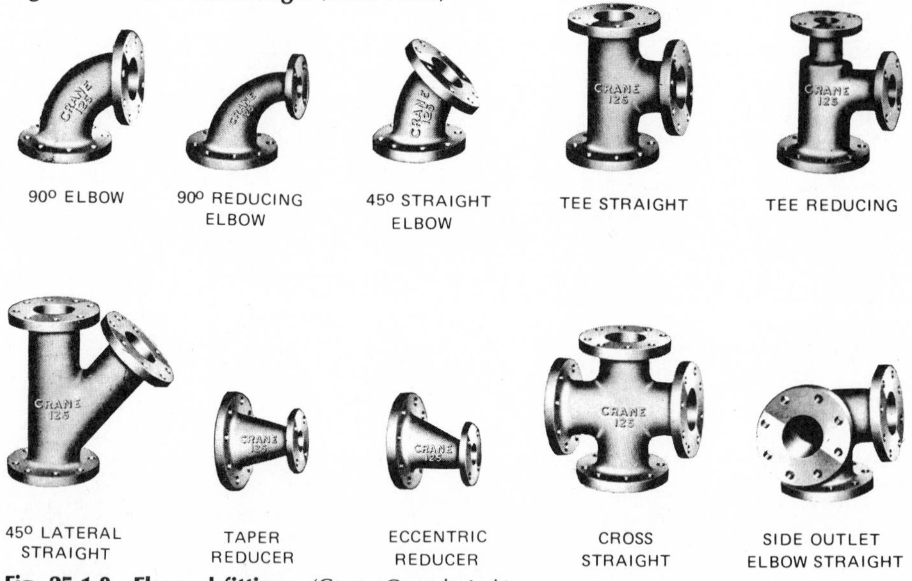
90° ELBOW
90° REDUCING ELBOW
45° STRAIGHT ELBOW
TEE STRAIGHT
TEE REDUCING
45° LATERAL STRAIGHT
TAPER REDUCER
ECCENTRIC REDUCER
CROSS STRAIGHT
SIDE OUTLET ELBOW STRAIGHT

Fig. 25-1-8 Flanged fittings. (Crane Canada Ltd.)

FLANGES

Flanges provide a quick means of disassembling pipe. Flanges are attached to the pipe ends by welding, screwing, or lapping. The flange faces are then drawn together by bolts, the size and spacing being determined by the size and working pressure of the joint. See Figs. 25-1-8 and 25-1-9.

Valves

Valves are used in piping systems to stop or to regulate the flow of fluids and gases. A few of the more common types are described here.

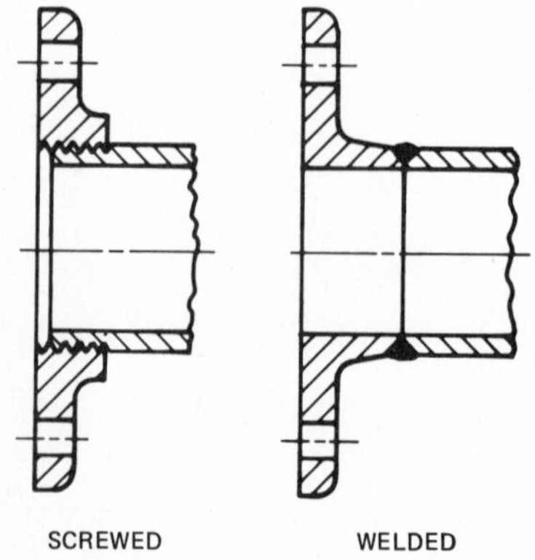
SCREWED
WELDED

Fig. 25-1-9 Flanges.

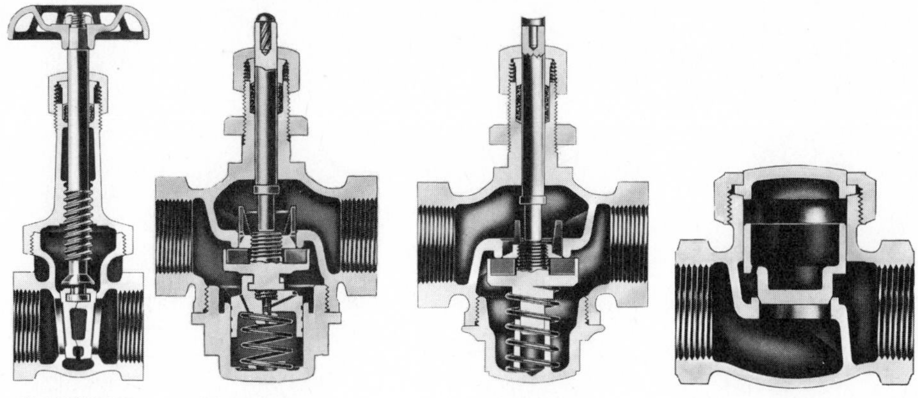

(A) GATE VALVE (B) GLOBE VALVE (C) GLOBE VALVE (D) CHECK VALVE

Fig. 25-1-10 Common valves. (Jenlins Bros. Ltd.)

GATE VALVES

Gate valves are used to control the flow of liquids. The wedge, or gate, lifts to allow full, unobstructed flow and lowers to stop it completely. See Fig. 25-1-10*a*. They are normally used where operation is infrequent and are not intended for throttling or close control.

GLOBE VALVES

These valves are used to control the flow of liquids or gases. The design of the globe valve requires two changes in the direction of flow, which slightly reduces the pressure in the system. The globe valve shown in Fig. 25-1-10*b* is installed so that the pressure is on the disk which assists the spring in the cap to make a tight closure. This type of valve is recommended for the control of air, steam, gas, or other compressibles where instantaneous on-and-off operation is essential. Figure 25-1-10*c* is recommended for the control of liquids such as hot or cold water, gasoline, oil, or solvents, where the sudden closure of a valve might cause objectionable and destructive water hammer. The cap is fitted with a spring-loaded piston dashpot arrangement which retards closure times and helps eliminate shock.

CHECK VALVES

As the name implies, check valves permit flow in one direction, but check all reverse flow. They are operated by pressure and velocity of line flow alone, and they have no external means of operation. See Fig. 25-1-10*d*.

Piping Drawings

The purpose of piping drawings is to show the size and location of pipes, fittings, and valves. Since these items may be purchased, a set of symbols has been developed to portray these features on a drawing.

There are two types of piping drawings in use—single-line and double-line drawings. See Fig. 25-1-11. Double-line drawings take more time to draw and therefore are not recommended for production drawings. They are, however, suitable for catalogs and other applications where the visual appearance is more important than the extra drafting time taken to make the drawing.

SINGLE-LINE DRAWINGS

Beyond dispute, single-line drawings, also known as *simplified representations*, of pipe lines are able to provide substantial savings without loss of clarity

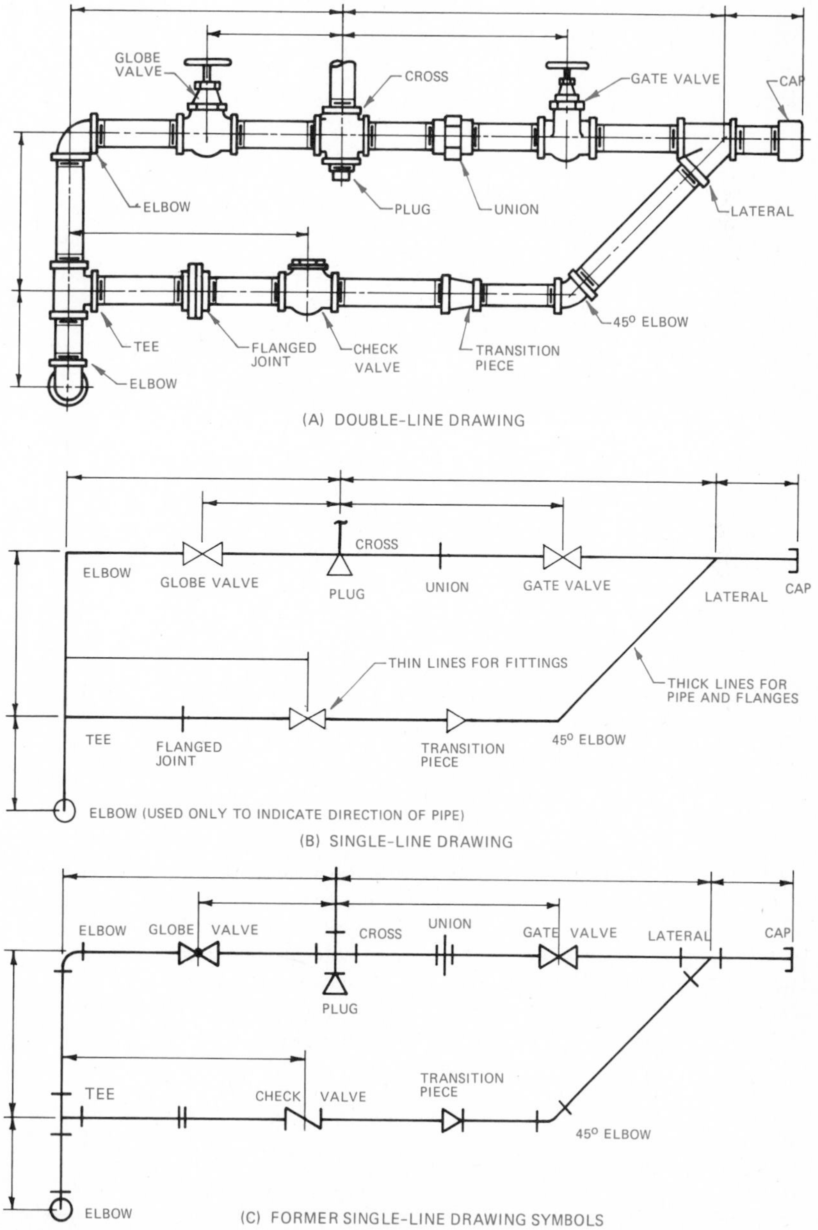
GLOBE VALVE
CROSS
GATE VALVE
CAP
ELBOW
PLUG
UNION
LATERAL
45° ELBOW
TEE
FLANGED JOINT
CHECK VALVE
TRANSITION PIECE
ELBOW
(A) DOUBLE-LINE DRAWING
CROSS
ELBOW
GLOBE VALVE
PLUG
UNION
GATE VALVE
LATERAL
CAP
THIN LINES FOR FITTINGS
THICK LINES FOR PIPE AND FLANGES
TEE
FLANGED JOINT
TRANSITION PIECE
45° ELBOW
ELBOW (USED ONLY TO INDICATE DIRECTION OF PIPE)
(B) SINGLE-LINE DRAWING
ELBOW
GLOBE VALVE
UNION
CROSS
GATE VALVE
LATERAL
CAP
PLUG
TEE
CHECK VALVE
TRANSITION PIECE
45° ELBOW
ELBOW
(C) FORMER SINGLE-LINE DRAWING SYMBOLS

Fig. 25-1-11 Piping drawing symbols.

or reduction of comprehensiveness of information. As such, the simplified method is used whenever possible.

Single-line piping drawings, as the name implies, use a single line to show the arrangement of the pipe and fittings. The center line of the pipe, regardless of pipe size, is drawn as a thick line to which the valve symbols are added. The size of the symbol is left to the discretion of the drafter. When the pipe lines carry different liquids, such as cold or hot water, a coded line symbol is often used.

Drawing Projection. Two methods of projection are used—orthographic and pictorial. Orthographic projection, as shown in Fig. 25-1-12, is recommended for the representation of single pipes either straight or bent in one plane only. However, this method is also used for more complicated pipings.

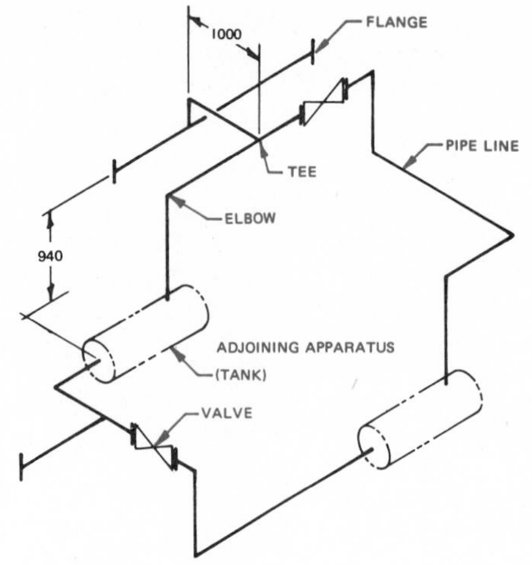
1000
FLANGE
PIPE LINE
TEE
ELBOW
940
ADJOINING APPARATUS
(TANK)
VALVE

Fig. 25-1-13 Single-line pictorial piping drawing.

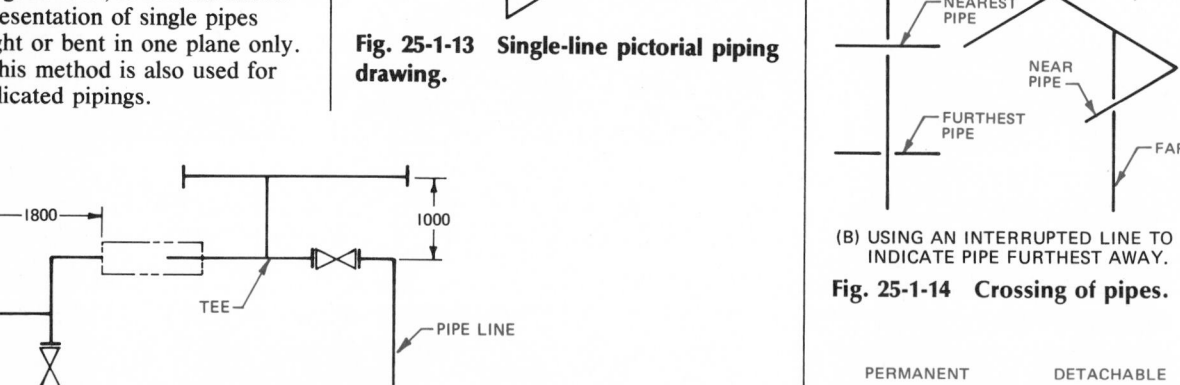

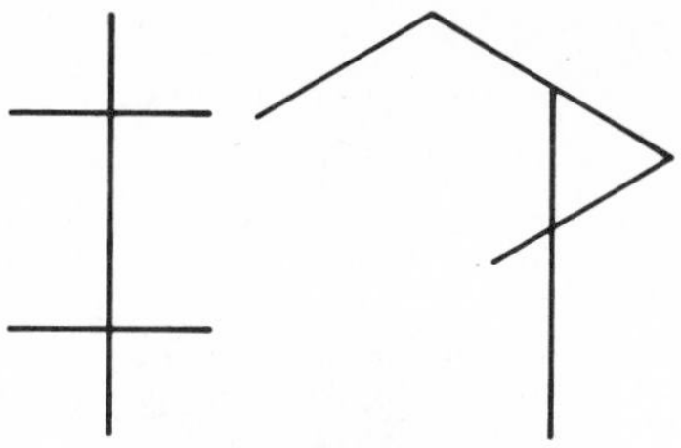

(A) CROSSING OF PIPE SHOWN WITHOUT INTERRUPTING THE PIPE PASSING BEHIND THE NEAREST PIPE.

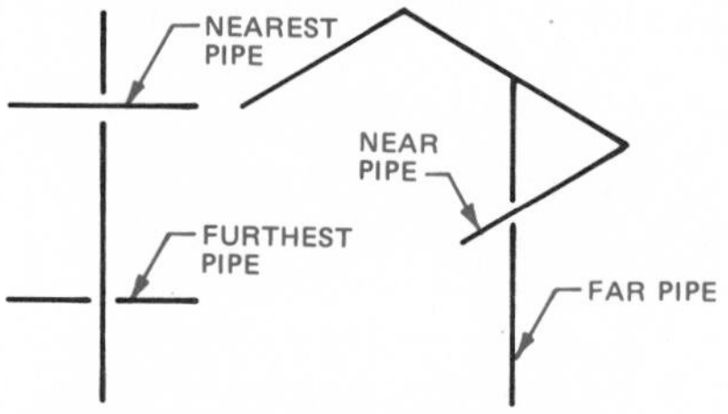
NEAREST PIPE
NEAR PIPE
FURTHEST PIPE
FAR PIPE

(B) USING AN INTERRUPTED LINE TO INDICATE PIPE FURTHEST AWAY.

Fig. 25-1-14 Crossing of pipes.

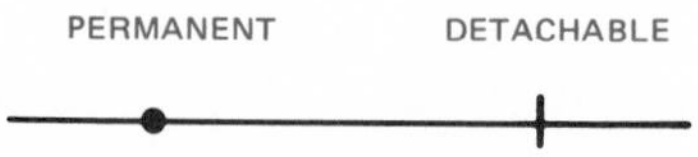
PERMANENT
DETACHABLE

Fig. 25-1-15 Pipe connection.

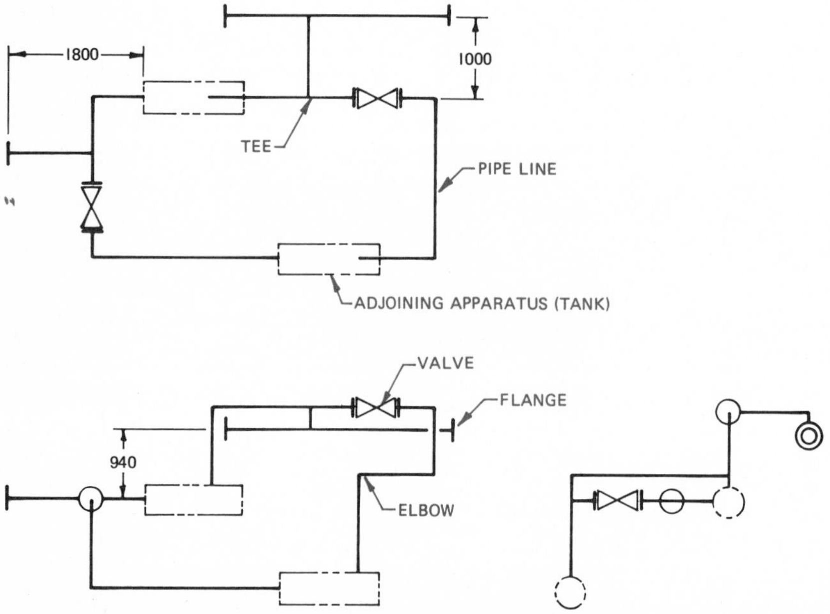
1800
1000
TEE
PIPE LINE
ADJOINING APPARATUS (TANK)
VALVE
FLANGE
940
ELBOW

Fig. 25-1-12 Single-line orthographic piping drawing.

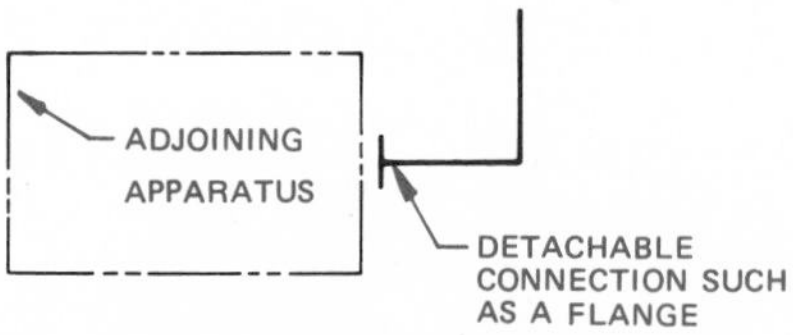
ADJOINING
APPARATUS
DETACHABLE
CONNECTION SUCH
AS A FLANGE

Fig. 25-1-16 Adjoining apparatus.

Pictorial projection, as shown in Fig. 25-1-13, is recommended for all pipes bent in more than one plane and for assembly and layout work, because the finished drawing is easier to understand.

Crossings. Crossing of pipes without connections are normally to be depicted without interrupting the line representing the hidden line (Fig. 25-1-14); but when it is desirable to indicate that one pipe must pass behind the other, the line representing the pipe farthest from the viewer will be shown with a break, or interruption, where the other pipe passes in front of it. For microfilming purposes, the break should not be less than 10 times the line thickness.

Connections. Permanent connections or junctions, whether made by welding or other processes such as gluing and soldering, are to be indicated on the drawing by a heavy dot (Fig. 25-1-15). A general note or specification may indicate the process used.

Detachable connections or junctions may be indicated by the use of a single thick line instead of a heavy dot, as shown in Fig. 25-1-16. The specifications, a general note, or the bill of materials will indicate the type of fitting such as flanges, union, or coupling and whether the fittings are flanged or threaded.

Fittings. If no specific symbols are standardized, fittings like tees, elbows, crosses, etc., are not specially drawn, but are represented, like pipe, by a continuous line. The circular symbol for a tee or elbow may be used when it is necessary to indicate whether the piping is coming toward or going away from the viewer, as shown in Fig. 25-1-18. Elbows on isometric drawings may be shown without the radius. However, the change of direction that the piping takes should be quite clear if this method is used.

Adjoining Apparatus. If needed, adjoining apparatus, such as tanks, machinery, etc., not belonging to the piping itself may be shown by an outline drawn with a thin phantom line. See Fig. 25-2-16.

DIMENSIONING

- Dimensions for pipe and pipe fittings are always given from center to center of pipe and to the outer face of the pipe end or flange (Fig. 25-1-17).
- Pipe lengths are not normally shown on the drawings, but left to the pipe fitter.
- Pipe and fitting sizes and general notes are placed on the drawing beside the part concerned or, where space is restricted, with a leader.
- A bill of materials is usually provided with the drawing.
- Pipes with bends are dimensioned from vertex to vertex.
- Radii and angles of bends may be indicated as shown in Fig. 25-1-17*b*. Whenever possible, the smaller of the supplementary angles is to be specified.

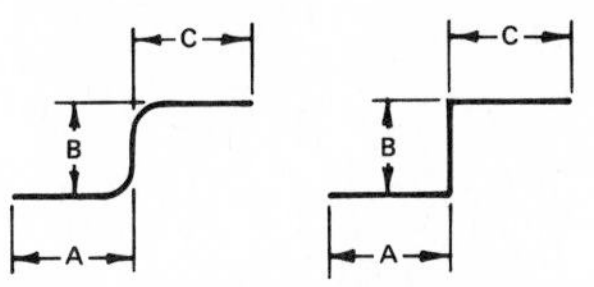

C
C
B
B
A
A

SHOWING THE RADIUS OF ELBOW OPTIONAL

(A) LINEAR DIMENSIONS

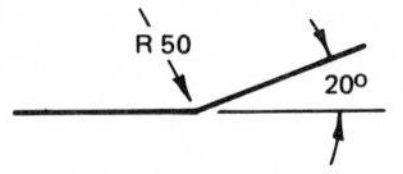

R 50
20°

(B) RADII AND ANGLES OF BENDS

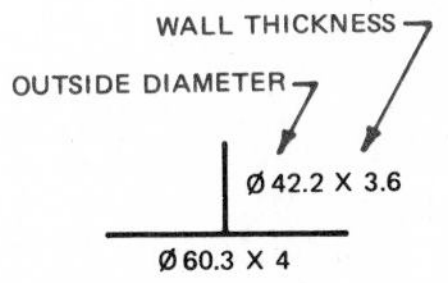

WALL THICKNESS
OUTSIDE DIAMETER
Ø 42.2 X 3.6
Ø 60.3 X 4

(C) PIPE SIZE INDICATED ON DRAWINGS

Fig. 25-1-17 Dimensioning piping drawings.

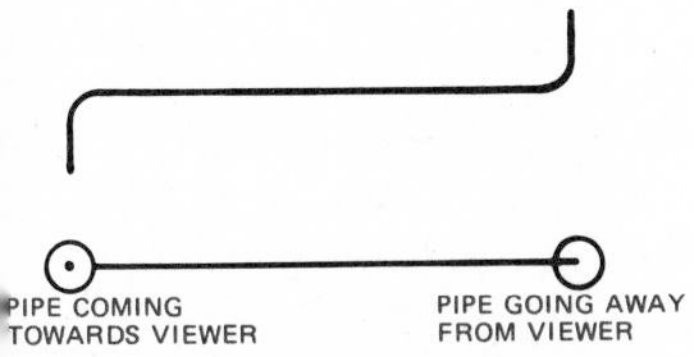

PIPE COMING TOWARDS VIEWER
PIPE GOING AWAY FROM VIEWER

(A) PIPE LINE WITHOUT FLANGES CONNECTED TO ENDS OF PIPE

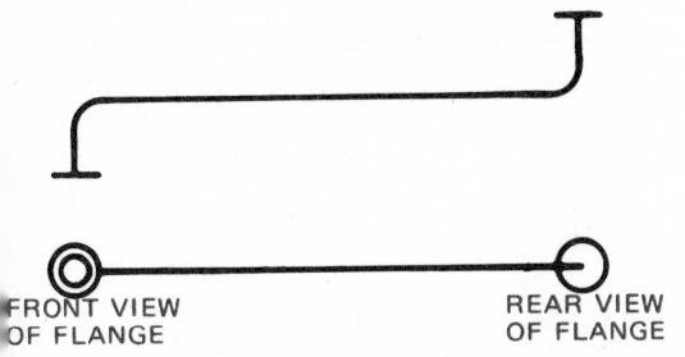

FRONT VIEW OF FLANGE
REAR VIEW OF FLANGE

(B) FLANGES CONNECTED TO ENDS OF PIPE LINE

Fig. 25-1-18 Indicating ends of pipe lines.

- The outer diameter and wall thickness of the pipe may be indicated on the line representing the pipe or elsewhere (parts list, general note, specification, etc.).

ORTHOGRAPHIC PIPING SYMBOLS

Pipe Symbols. If flanges are not attached to the ends of the pipelines when orthographic projections are drawn in, pipeline symbols indicating the direction of the pipe are required. If the pipeline is coming toward the front (or viewer), it will be shown by two concentric circles, the smaller one being solid. See Fig. 25-1-18*a*. If the pipeline is going toward the back (or away from the viewer), it will be shown by one circle. No extra lines are required on the other views.

Flange Symbols. See Fig. 25-1-18*b*. Flanges are to be represented, irrespective of their type and sizes, by two concentric circles in the front view, by one circle in the rear view, and by a short stroke in the side view, while lines of equal thickness, as chosen for the representation of pipes, are used.

Valve Symbols. Symbols representing valves are drawn with continuous thin lines (as opposed to thick lines for piping and flanges). The valve spindles should be shown only if it's necessary to define their positions. It will be assumed that unless otherwise indicated, the valve spindle is in the position shown in Fig. 25-1-19.

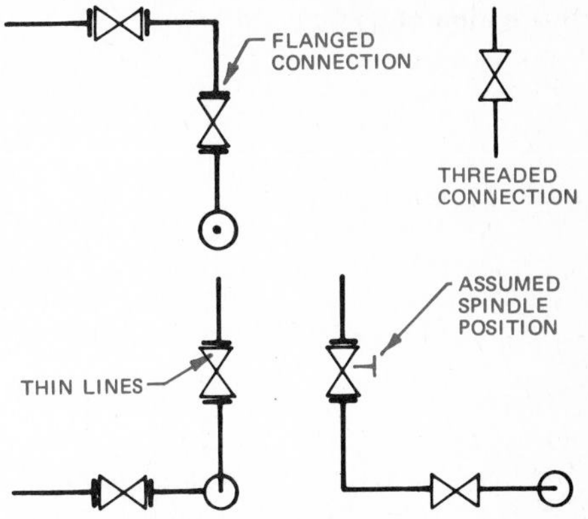

FLANGED CONNECTION
THREADED CONNECTION
ASSUMED SPINDLE POSITION
THIN LINES

NOTE: WHEN VALVE SPINDLES NOT SHOWN IT WILL BE ASSUMED THAT THEY WILL BE IN THE POSITIONS INDICATED ABOVE.

Fig. 25-1-19 Valve symbols.

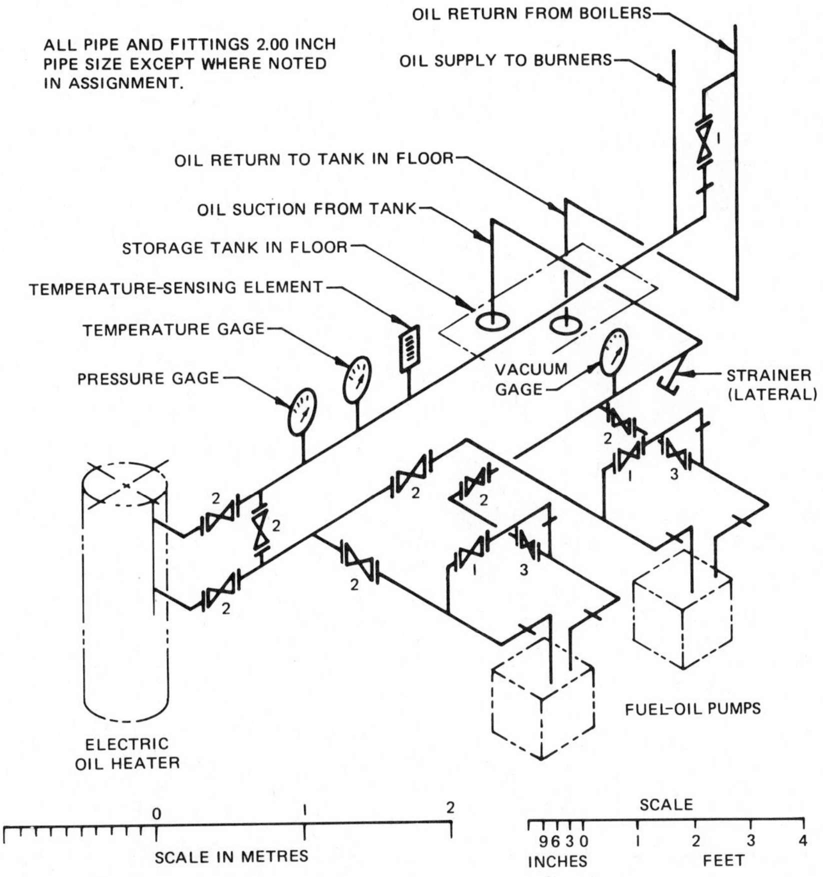

ALL PIPE AND FITTINGS 2.00 INCH PIPE SIZE EXCEPT WHERE NOTED IN ASSIGNMENT.
OIL RETURN FROM BOILERS
OIL SUPPLY TO BURNERS
OIL RETURN TO TANK IN FLOOR
OIL SUCTION FROM TANK
STORAGE TANK IN FLOOR
TEMPERATURE-SENSING ELEMENT
TEMPERATURE GAGE
PRESSURE GAGE
VACUUM GAGE
STRAINER (LATERAL)
FUEL-OIL PUMPS
ELECTRIC OIL HEATER
0
1
2
SCALE IN METRES
SCALE
9 6 3 0
INCHES
1 2 3 4
FEET

Fig. 25-1-A Fuel-oil-supply system.

SOURCE MATERIAL

- Crane Canada Limited
- Jenkins Bros. Limited

Assignments

1. On an A3- or B-size sheet, make a three-view drawing of the fuel-oil-supply system shown in Fig. 25-1-A. Include with the drawing a bill of materials calling for all the pipe fittings and valves. The following valves are used: (1) relief valves, (2) globe valves, (3) check valves. The numbers shown on the assignment correspond with the numbers listed above. Unions are used above the fuel oil pumps as detachable connections for ease of assembling and disassembling. The gages and temperature-sensing element have 1.0-in. pipe connections and, as such, necessitate the use of reducing tees. Scale is 1:20 (metric) or ½ in. = 1 ft (feet and inches).

2. Heated tanks must be provided for the storage of industrial heating oil in most plants using this fuel for boiler furnaces generating heat or power or for processing furnaces.

To ensure uninterrupted service when cleaning, or in the event of a breakdown of one of the systems, a duplicate installation of tanks is shown in this layout. Since circulation must be provided to keep the oil fluid, a return line as well as a suction line from the tanks is shown. A valve is provided directly to the suction line. Connections are provided for both high and low suction. High suction guards against difficulties from sediment, while the low suction is necessary when the fuel oil supply is extremely low.

The free blow shown on the steam-line connections to the heating coils in the tanks is important for testing for the presence of oil in the steam return line, since oil would indicate a leak.

Extra heavy globe valves of the regrind-renew type are recommended on the oil lines to ensure maximum safety in the transmission of hazardous fluid and to meet code requirements. Outside screw and yoke gate valves are suggested because they show at a glance if the valve is opened or closed.

On an A3- or B-size sheet, make a three-view drawing of the fuel-oil-storage connections with heating coil shown in Fig. 25-1-B. Include with the drawing a bill of materials calling for all the pipe fittings and valves. The oil lines are 1.50 in. pipe, and the steam line is 2 in. pipe. For the scale, see the drawing.

REVIEW FOR ASSIGNMENTS

Unit 6-4 Bill of Material

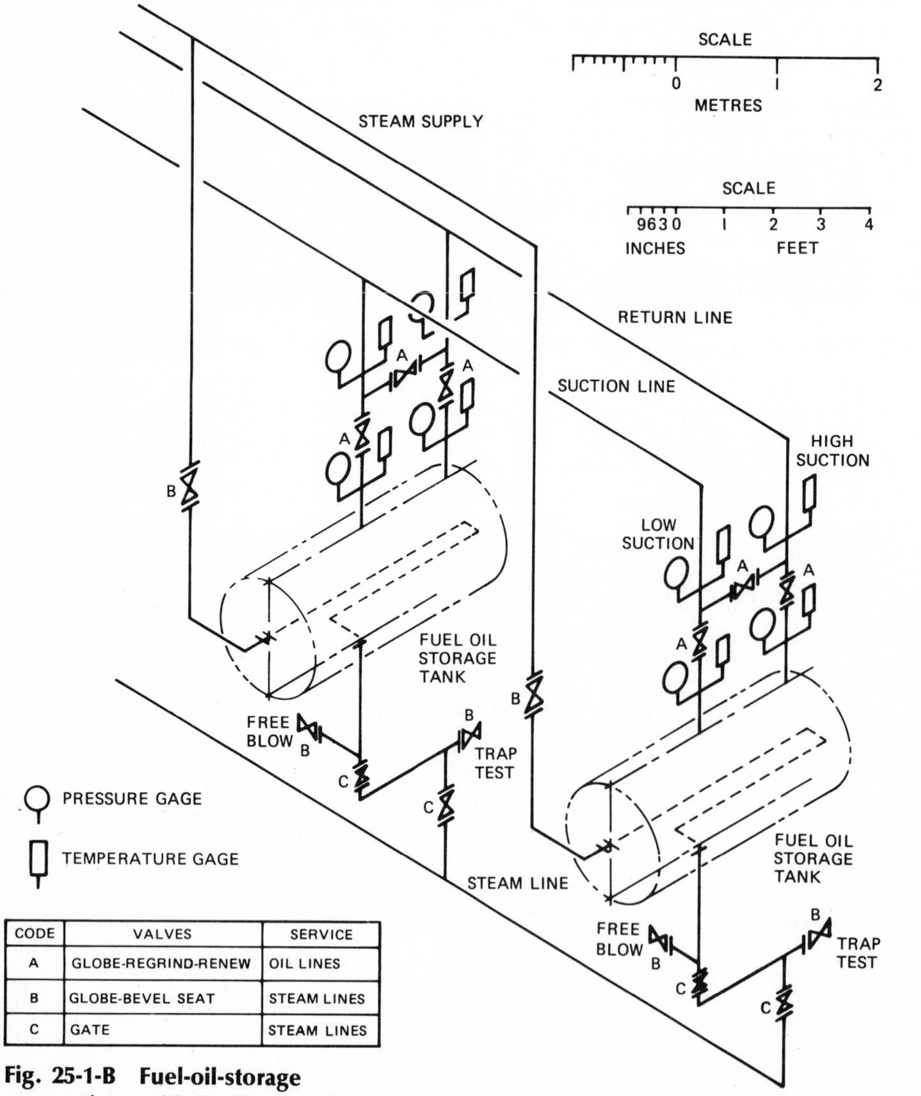
SCALE
0
1
2
METRES
SCALE
9 6 3 0
1
2
3
4
INCHES
FEET
STEAM SUPPLY
RETURN LINE
SUCTION LINE
HIGH SUCTION
LOW SUCTION
FUEL OIL STORAGE TANK
FREE BLOW
TRAP TEST
STEAM LINE
FUEL OIL STORAGE TANK
FREE BLOW
TRAP TEST
PRESSURE GAGE
TEMPERATURE GAGE

CODE	VALVES	SERVICE
A	GLOBE-REGRIND-RENEW	OIL LINES
B	GLOBE-BEVEL SEAT	STEAM LINES
C	GATE	STEAM LINES

Fig. 25-1-B Fuel-oil-storage connections with heating coils.

UNIT 25-2

ISOMETRIC PROJECTION OF PIPING DRAWINGS

The scale of the drawing applies to the dimension taken along the coordinate axes (isometric axes). In drawing to the

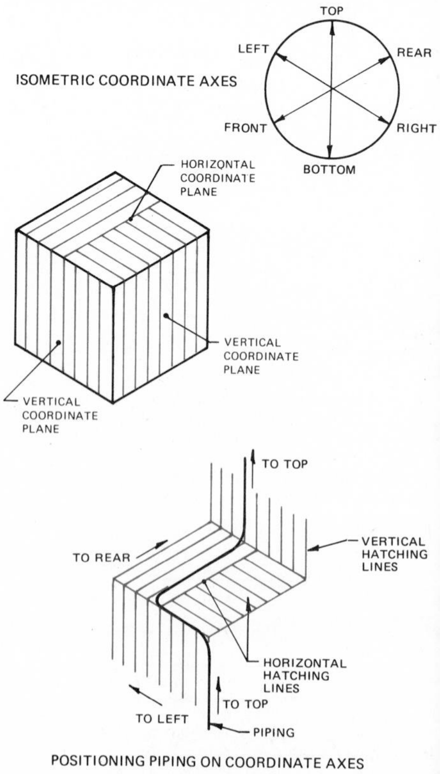
ISOMETRIC COORDINATE AXES
TOP
LEFT
REAR
FRONT
RIGHT
BOTTOM
HORIZONTAL COORDINATE PLANE
VERTICAL COORDINATE PLANE
VERTICAL COORDINATE PLANE
TO TOP
TO REAR
VERTICAL HATCHING LINES
HORIZONTAL HATCHING LINES
TO TOP
TO LEFT
PIPING
POSITIONING PIPING ON COORDINATE AXES

Fig. 25-2-1 Coordinate axes for piping drawings.

isometric projection method, the following rules should be followed.

- Parts of pipe that run parallel to the coordinate axes are drawn without any special indication of being parallel to the isometric axes. See Figs. 25-2-1 and 25-2-2.
- With reference to calculations or programming for computer drafting, it will probably be necessary to indicate the *x*, *y*, and *z* axes (coordinates) on the drawing.

FLANGES

Flanges are to be represented, irrespective of their type and sizes, by short strokes of equal-thickness lines as chosen for the representation of the pipes. See Fig. 25-2-3.

Flanges at the ends of vertical pipe parts should preferably be drawn to an angle of 30° to horizontal and flanges at the ends of horizontal pipe parts in a vertical direction.

VALVES

For isometric drawings it will be assumed that unless otherwise indicated, the valve spindle is in the position shown in Fig. 25-2-4. Valve spindles should be drawn only if it is necessary to define their positions. Deviations from these positions can be indicated by specifying the angle to which the valve is turned in a right-hand direction, looking in the direction of the positive *x*, *y*, or *z* axes (Fig. 25-2-5).

DIMENSIONING

There are two acceptable methods of dimensioning isometric pipe drawings: pictorial plane dimensioning and unidirectional dimensioning. In pictorial plane dimensioning, the lettering lies in one of the pictorial planes. In unidirectional dimensioning, which is preferred because of the ease in execution and reading, all the notes are read from the bottom of the sheet. See Fig. 25-2-6.

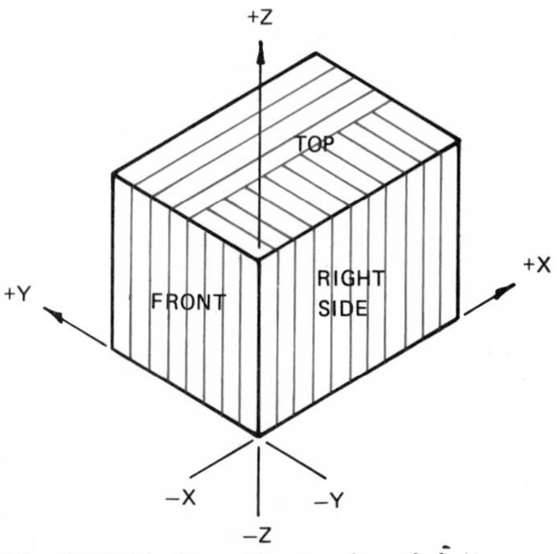

+Z
TOP
FRONT
RIGHT SIDE
+Y
+X
−X
−Y
−Z

Fig. 25-2-2 Coordinates for piping drawings.

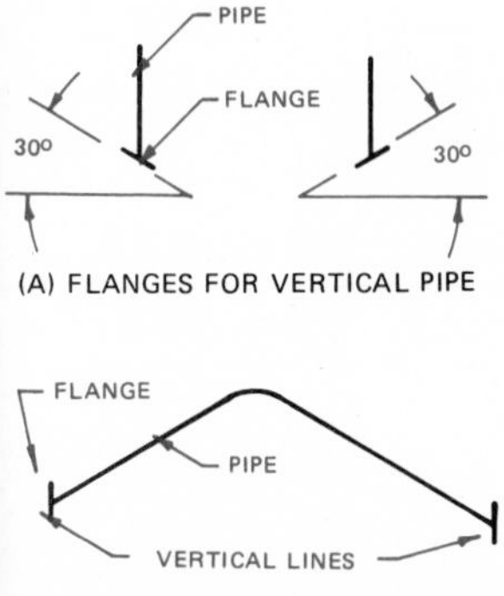

PIPE
FLANGE
30°
30°
FLANGE
PIPE
VERTICAL LINES

(A) FLANGES FOR VERTICAL PIPE

(B) FLANGES FOR HORIZONTAL PIPE

Fig. 25-2-3 Flange positioning for isometric drawings.

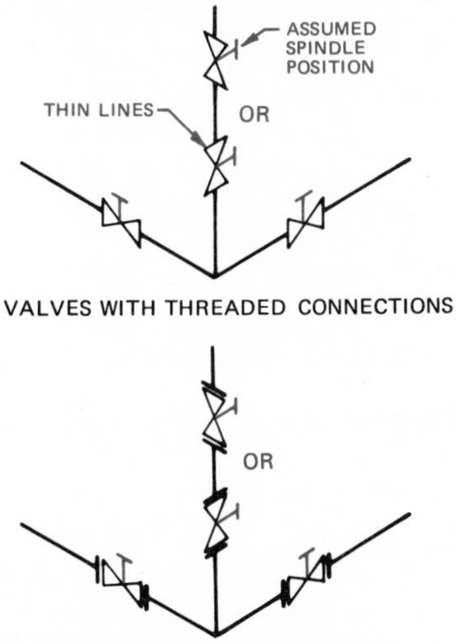

ASSUMED SPINDLE POSITION
THIN LINES
OR
OR

(A) VALVES WITH THREADED CONNECTIONS

(B) VALVES WITH FLANGE CONNECTIONS

Fig. 25-2-4 Valve symbols.

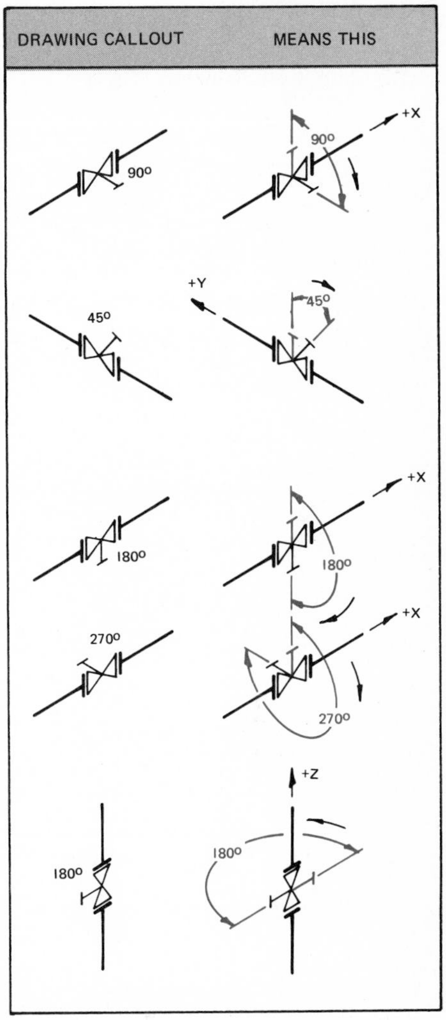

DRAWING CALLOUT
MEANS THIS
90°
90°
+X
+Y
45°
45°
180°
180°
+X
270°
270°
+X
+Z
180°
180°

Fig. 25-2-5 Deviations from normal position of valve spindle.

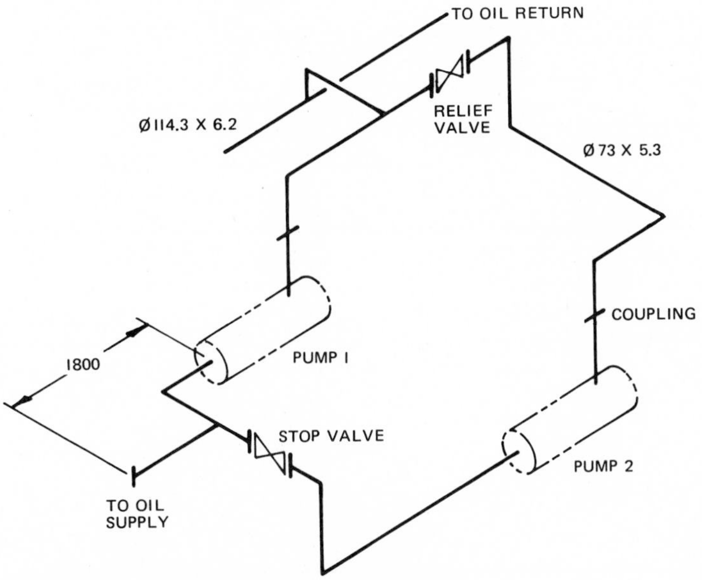

TO OIL RETURN
RELIEF VALVE
Ø 114.3 X 6.2
Ø 73 X 5.3
COUPLING
1800
PUMP 1
STOP VALVE
PUMP 2
TO OIL SUPPLY

Fig. 25-2-6 Unidirectional dimensioning.

SOURCE MATERIAL

- Crane Canada Limited
- Jenkins Bros. Limited

Assignments

1. For starting diesel engines, the most dependable and widely used method is an air system of the type illustrated in Fig. 25-2-A. With this starting system hooked up to diesel installations generating power and heat for such buildings as factories, hotels, large apartment houses, and stores, interruptions such as might occur through failure of electric supply or storage cells are avoided.

Safety valves are provided for the compressor and the air storage tanks. Check valves are installed on the air storage tank feed lines and the compressor discharge lines to prevent accidental discharge of the tanks.

Piping is arranged so that the compressor will either fill the storage tanks and/or pump directly to the engines. Any of the three storage tanks may be used for starting, and pressure gages indicate their readiness. The engines are fitted with quick-opening valves to admit air quickly at full pressure and shut it off the instant rotation is obtained. A bronze globe valve is installed to permit complete shutdown of the engine for repairs and for regulation of the air flow. Drains are provided at low points to remove condensate from the air storage tanks, lines, and engine feeds.

Globe valves are recommended throughout this hook-up except on the main shutoff lines where gate valves are used because of infrequent operation.

All valves connected to horizontal pipe lines 1800 mm (6 ft) or higher above the floor will have their spindles located on the underside for ease of operation. Other horizontally positioned valves will have their spindles in the upright position.

All valves connected to vertical pipes will have their valve spindles oriented to the front of the drawing.

Flanges are to be located on the top pipeline near the three starting air tanks and near the air compressor for assembly and disassembly purposes. Flanges are located on the starting diesel engines.

On an A3- or B-size sheet, make an isometric piping drawing for the diesel engine starting system shown in Fig. 25-2-A. Scale is 1:50 (metric) or ¼ in. = 1 ft. (conventional). Include on your drawing a bill of materials calling for the pipe fittings and valves. All the fittings are threaded, and 1.5 in. pipe is used throughout.

2. In the piping layout of a boiler room shown in Fig. 25-2-B, boilers 1, 3,

Fig. 25-2-A Diesel-engine starting air system.

CODE	VALVE	SERVICE
A	BRONZE GLOBE	AIR STORAGE TANK FEED LINES
B	BRONZE GLOBE	AIR STORAGE TANK DISCHARGE LINES
C	BRONZE GLOBE	DIESEL ENGINE SHUTOFF CONTROL
D	BRONZE GLOBE	AIR COMPRESSOR DISCHARGE
E	BRONZE GLOBE	PRESSURE GAGE SHUTOFF
F	BRONZE GLOBE	DRAIN VALVES
G	SPINDLE GATES	MAIN LINE SHUTOFF
H	BRONZE CHECK	AIR COMPRESSOR CHECK
J	BRONZE CHECK	AIR STORAGE TANK FEED LINES

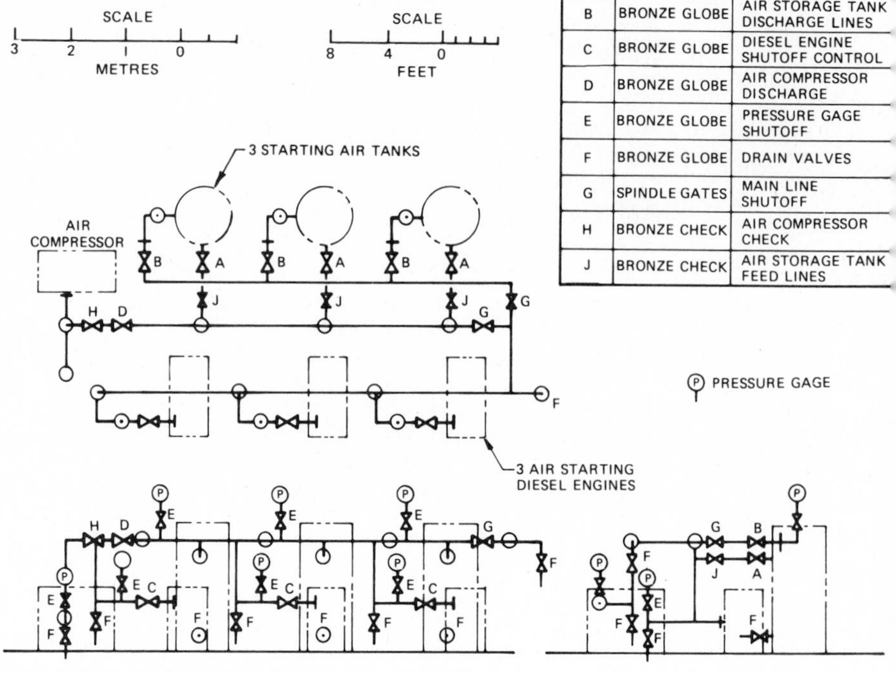
SCALE
3 2 1 0
METRES
SCALE
8 4 0
FEET
3 STARTING AIR TANKS
AIR COMPRESSOR
3 AIR STARTING DIESEL ENGINES
PRESSURE GAGE

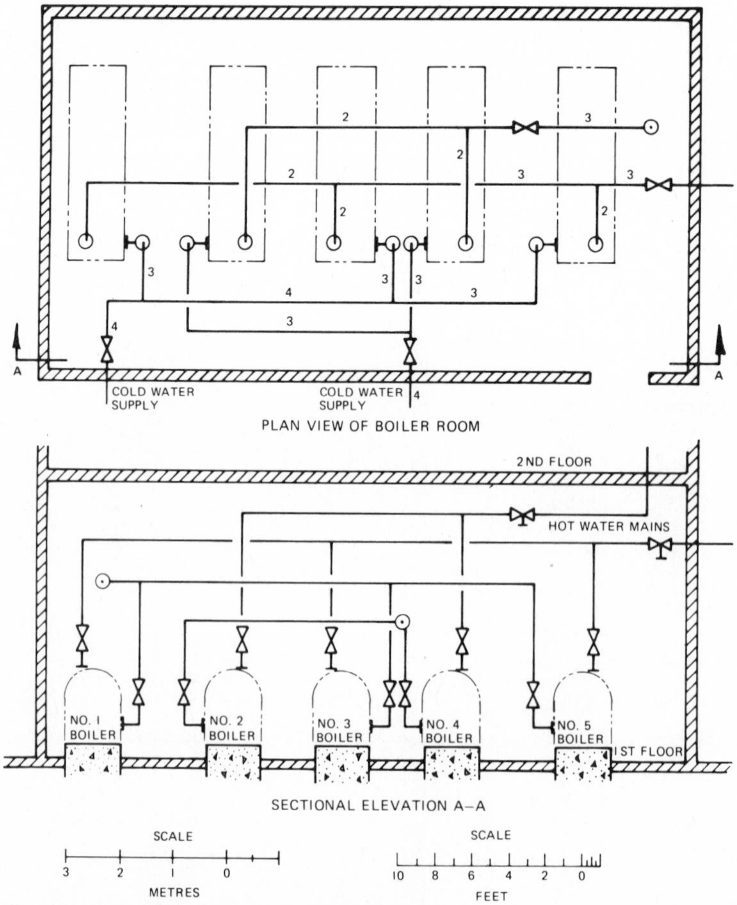
COLD WATER SUPPLY
COLD WATER SUPPLY
PLAN VIEW OF BOILER ROOM
2ND FLOOR
HOT WATER MAINS
NO. 1 BOILER
NO. 2 BOILER
NO. 3 BOILER
NO. 4 BOILER
NO. 5 BOILER
1ST FLOOR
SECTIONAL ELEVATION A–A
SCALE
3 2 1 0
METRES
SCALE
10 8 6 4 2 0
FEET

Fig. 25-2-B Piping layout of boiler room.

and 5 are connected to supply the hot water to rooms located on the first floor. Check valves are placed on the cold-water lines adjacent to each boiler to prevent the hot water from backing into the cold-water lines. Gate valves are used to shut off the main hot- and cold-water supply lines. Globe valves disconnecting the boilers are being made. Flanged connections are used at the boilers for ease of assembly and disassembly. The size of pipe is indicated on the plane view, and the scale is shown on the drawing.

On an A3- or B-size sheet, make an isometric drawing of the piping layout of a boiler room. Include on the drawing a bill of materials calling for the pipe fittings and valves. All fittings are of flanged type.

UNIT 25-3
SUPPLEMENTARY PIPING INFORMATION

Direction of Flow. The direction of flow may be indicated by an arrowhead on the line representing the piping, as shown in Fig. 25-3-1.

Level Indicators. Level indications in lieu of linear measurements may be used to show the height of pipelines and fittings. The preferred method of indicating levels is shown in Fig. 25-3-2.

Specifying Slope on Pipes. The direction of slope is indicated by an arrow located above the pipe pointing from the higher to the lower level. The amount of slope may be indicated by either a general note on the drawing or one of the methods shown in Fig. 25-3-3. However, with long piping runs, it may be useful to specify the slope by reference to a datum and level indication as shown in Fig. 25-3-3*c*.

Support and Hangers. Support and hangers are to be represented by their appropriate symbols, as shown in Fig. 25-3-4*a*. The representation of repetitive accessories may be simplified, as shown in Fig. 25-3-4*b*.

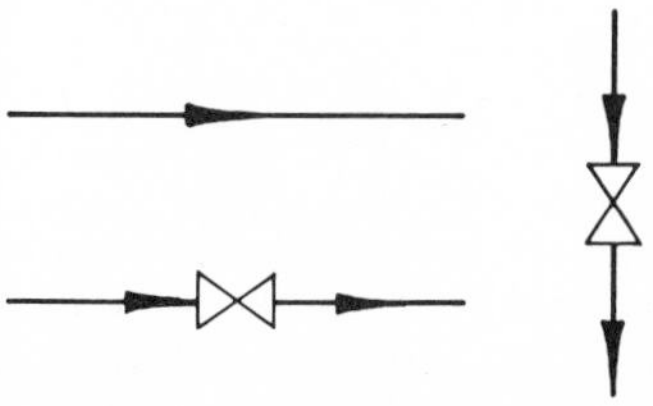

Fig. 25-3-1 Indicating direction of flow.

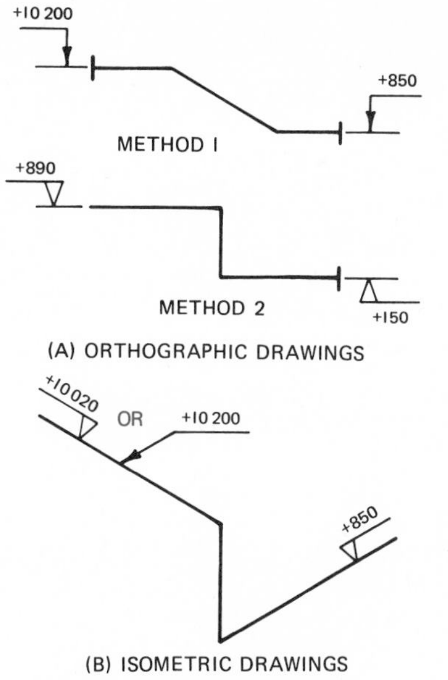
+10 200
+850
METHOD I
+890
METHOD 2
+150
(A) ORTHOGRAPHIC DRAWINGS
+10 020
OR
+10 200
+850
(B) ISOMETRIC DRAWINGS

Fig. 25-3-2 Level indicators.

Transition Pieces. Transition pieces for changing the cross section are indicated by the symbols shown in Fig. 25-3-5. The relevant nominal sizes are indicated above the symbols.

PIPE RUNS NOT PARALLEL WITH COORDINATE AXES

Deviations from the directions of the coordinate axes are to be indicated by means of hatched planes, as follows:

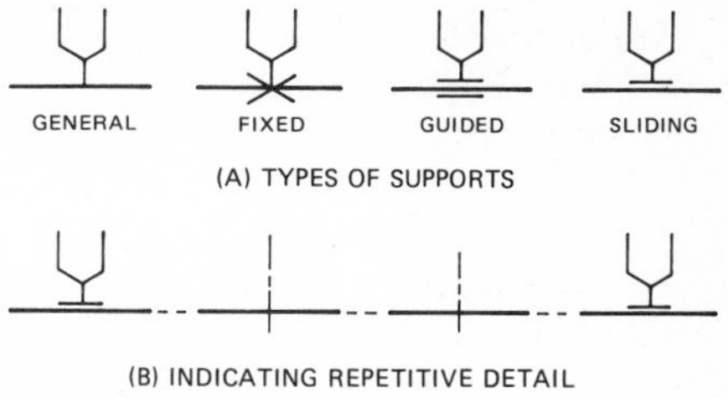
GENERAL
FIXED
GUIDED
SLIDING
(A) TYPES OF SUPPORTS
(B) INDICATING REPETITIVE DETAIL

Fig. 25-3-4 Supports and hangers.

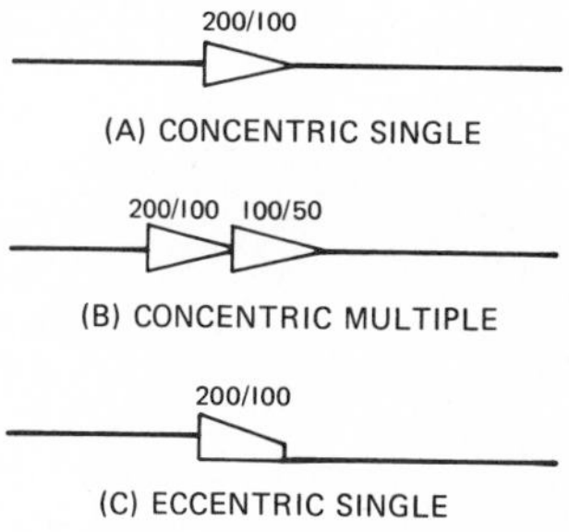
200/100
(A) CONCENTRIC SINGLE
200/100 100/50
(B) CONCENTRIC MULTIPLE
200/100
(C) ECCENTRIC SINGLE

Fig. 25-3-5 Transition pieces.

1. For a part of a pipe situated in a plane parallel to one of the vertical projection planes, vertical hatching lines are drawn to indicate the vertical projection plane (Fig. 25-3-6*a*).

2. For a part of a pipe situated in a plane parallel to the horizontal coordinate plane, horizontal hatching lines are

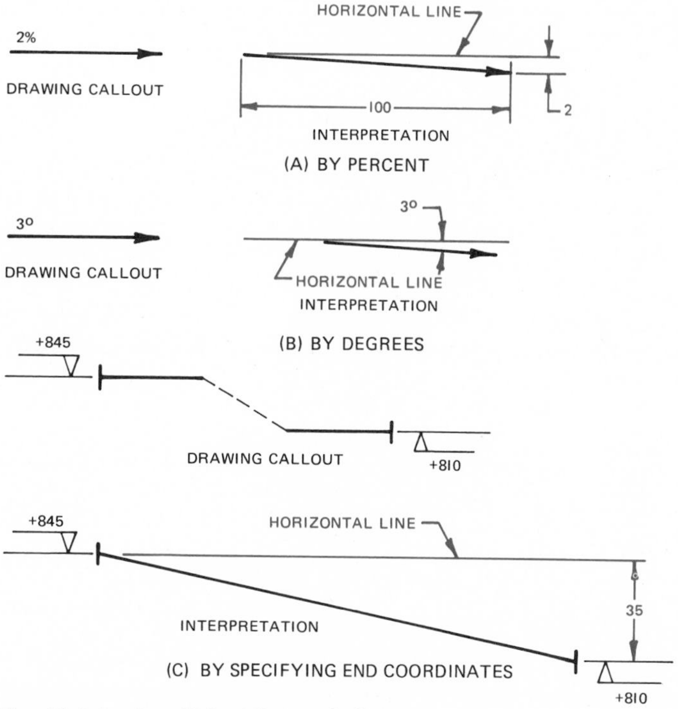
2%
DRAWING CALLOUT
HORIZONTAL LINE
100
2
INTERPRETATION
(A) BY PERCENT
3°
DRAWING CALLOUT
3°
HORIZONTAL LINE
INTERPRETATION
(B) BY DEGREES
+845
DRAWING CALLOUT
+810
+845
HORIZONTAL LINE
35
INTERPRETATION
(C) BY SPECIFYING END COORDINATES
+810

Fig. 25-3-3 Specifying slope of pipes.

drawn to indicate the horizontal projection plane (Fig. 25-3-6*b*).

3. For a part of a pipe not running parallel to any of the coordinate planes, both vertical and horizontal hatching lines are drawn to indicate the vertical and horizontal projection planes (Fig. 25-3-6*c*).

If desired, in addition to the coordinate planes, the prism of which the pipe part forms the diagonal may be shown in thin lines (Fig. 25-3-6*d*).

If such hatching is not convenient, for instance when using automated drafting equipment, it may be omitted but should be replaced with the thin-line rectangles or parallelopipeds whose diagonal coincides with the pipe. See Fig. 25-3-7. Application of projection planes is shown in Fig. 25-3-8.

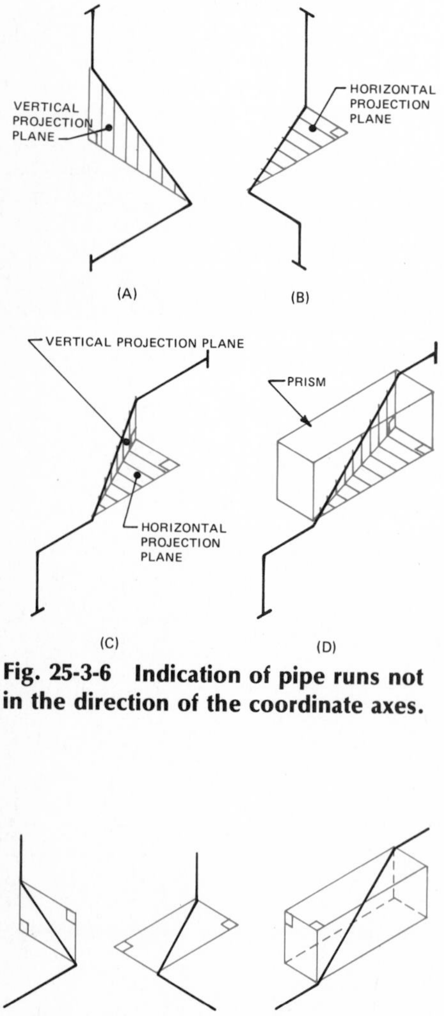
VERTICAL PROJECTION PLANE
HORIZONTAL PROJECTION PLANE
(A)
(B)
VERTICAL PROJECTION PLANE
PRISM
HORIZONTAL PROJECTION PLANE
(C)
(D)

Fig. 25-3-6 Indication of pipe runs not in the direction of the coordinate axes.

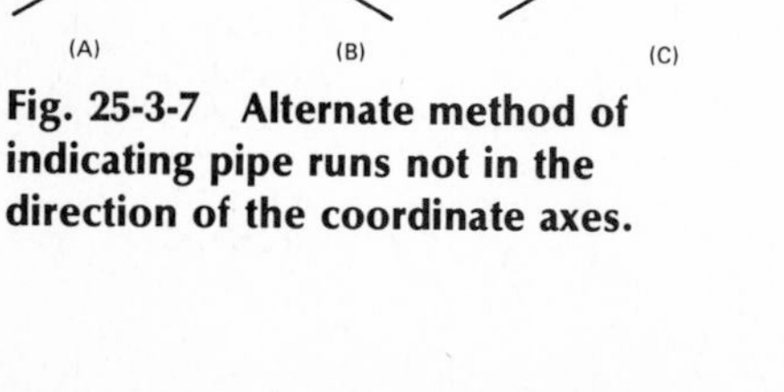
(A)
(B)
(C)

Fig. 25-3-7 Alternate method of indicating pipe runs not in the direction of the coordinate axes.

SOURCE MATERIAL

- Jenkins Bros. Limited

Assignments

1. The one-story "taxpayer" building (Fig. 25-3-A) has been developed and steadily improved as a result of the movement of shopping centers to suburban areas. This type of building, which is multiplying rapidly, is constructed either with or without basement. It houses retail stores, service establishments, amusement centers, restaurants, and offices.

Heat and plumbing services in such buildings are usually provided by the owner or operator; and for this reason, he or she might give careful consideration to low-cost and trouble-free installations. An oil- or gas-fired steam boiler with automatic control, and a separate gas-fired heater for hot-water supply, will generally meet these requirements.

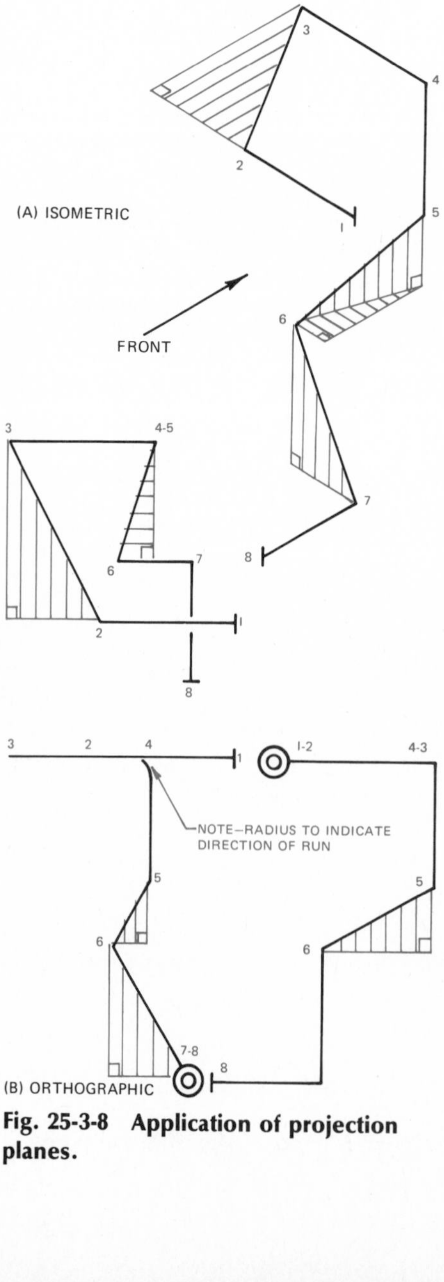
(A) ISOMETRIC
FRONT
3
4
2
1
5
6
7
8
4-5
(B) ORTHOGRAPHIC
1-2
4-3
NOTE–RADIUS TO INDICATE DIRECTION OF RUN
7-8

Fig. 25-3-8 Application of projection planes.

The two-pipe heating system, located in the basement in the installation illustrated, utilizes unit heaters with individual thermostatic controls. Valuable extra floor space is made available for tenants' use because the heaters are hung from the ceiling. Since the heaters in each store or each section of a store are automatically controlled, fuel savings are effected and even heating is ensured.

On an A3- or B-size sheet, make an isometric drawing of the piping layout shown. Include with the drawing a bill of materials, calling for all the pipe fittings and valves. Pipe hangers are required for every 2400 mm (8 ft) of piping. Direction of flow, level indicators for horizontal piping using the basement floor as zero, indication of pipe runs not in the direction of the coordinate axes (see Fig. 25-3-6), and a drainage slope of 1:20 are to be shown on the drawing. Scale is 1:100 (metric) and ⅛ in. = 1 ft. (conventional).

2. Light and medium fuel oils, numbers 1, 2, 3, and 5 (cold), which do not require preheating can be handled in a layout such as shown in Fig. 25-3-B. Since the expense of a preheater installation can be eliminated, this relatively simple system is economical and easy to operate. Similar systems are often installed in hotels, apartment houses, taxpayers' office buildings, large residences, and small industrial plants.

Fuel oil, stored in an unheated tank, flows through a large mesh, twin-type strainer to a motor-driven pump which provides the necessary oil pressure for satisfactory operation. The oil then passes through another fine-mesh strainer, which removes any small particles that might clog the burner. Oil flow to the burner is controlled by a burner control valve which opens or shuts according to the boiler pressure.

Although one fuel oil pump can adequately handle the maximum boiler demands, two are recommended to provide a second pump for standby service in case of breakdown. Each pump is provided with a pressure relief valve as a protection against excessive oil pressure, which might become high enough to cause leaks in the oil piping. Check valves in the relief lines prevent relieved oil from entering the idle standby pump.

Bronze valves are recommended throughout and must be of the appropriate pressure rating. The plug-type globe valve, recommended for the important individual burner shutoff, ensures positive tightness when closed and extremely close regulation of oil flow, both of which are essential to good oil burner operation.

The swing check valve indicated in this layout is exceptionally serviceable for the nonreturn control of steam, oil, water, and gas. It is generally used in connection with a gate valve, offering comparable full, free flow.

On an A3- or B-size sheet, make a single-line isometric drawing of the piping drawing shown. Design your own symbols for the indicators (gages, strainers, etc.) for which there are no standard symbols. The coding of these items should be shown clearly off the main drawing. Show the direction of flow and indicate on the drawing that all horizontal pipes require a slope of 1:20 for drainage purposes. Using the floor as zero elevation, scale the drawing and show by means of level indicator symbols the height of all horizontal pipelines. Include on your drawing a bill of materials, listing all the valves and fittings. Scale is 1:50 for Fig. 25-3-B (metric) and ¼ in. = 1 ft. (conventional).

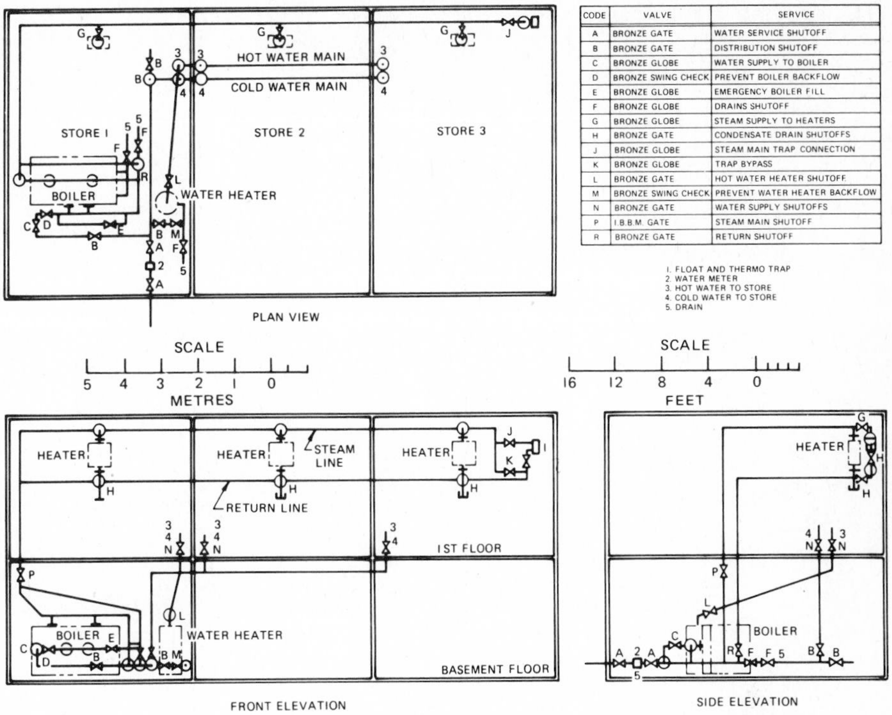

HOT WATER MAIN
COLD WATER MAIN
STORE 1
STORE 2
STORE 3
BOILER
WATER HEATER
PLAN VIEW
SCALE
5 4 3 2 1 0
METRES
SCALE
16 12 8 4 0
FEET
1. FLOAT AND THERMO TRAP
2. WATER METER
3. HOT WATER TO STORE
4. COLD WATER TO STORE
5. DRAIN
HEATER
STEAM LINE
RETURN LINE
1ST FLOOR
WATER HEATER
BOILER
BASEMENT FLOOR
FRONT ELEVATION
SIDE ELEVATION

CODE	VALVE	SERVICE
A	BRONZE GATE	WATER SERVICE SHUTOFF
B	BRONZE GATE	DISTRIBUTION SHUTOFF
C	BRONZE GLOBE	WATER SUPPLY TO BOILER
D	BRONZE SWING CHECK	PREVENT BOILER BACKFLOW
E	BRONZE GLOBE	EMERGENCY BOILER FILL
F	BRONZE GLOBE	DRAINS SHUTOFF
G	BRONZE GLOBE	STEAM SUPPLY TO HEATERS
H	BRONZE GATE	CONDENSATE DRAIN SHUTOFFS
J	BRONZE GLOBE	STEAM MAIN TRAP CONNECTION
K	BRONZE GLOBE	TRAP BYPASS
L	BRONZE GATE	HOT WATER HEATER SHUTOFF
M	BRONZE SWING CHECK	PREVENT WATER HEATER BACKFLOW
N	BRONZE GATE	WATER SUPPLY SHUTOFFS
P	I.B.B.M. GATE	STEAM MAIN SHUTOFF
R	BRONZE GATE	RETURN SHUTOFF

Fig. 25-3-A Piping connections for plumbing and heating in a small building.

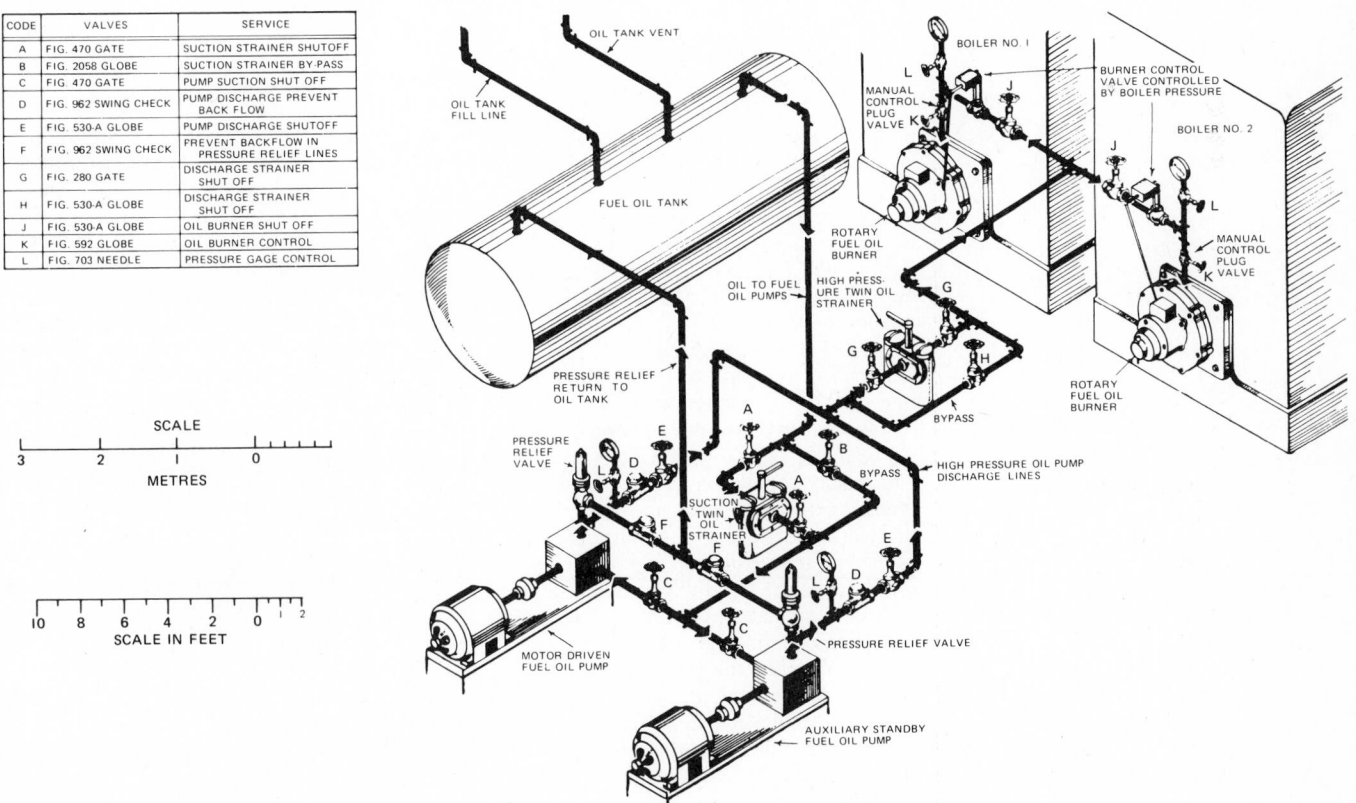

CODE	VALVES	SERVICE
A	FIG. 470 GATE	SUCTION STRAINER SHUTOFF
B	FIG. 2058 GLOBE	SUCTION STRAINER BY-PASS
C	FIG. 470 GATE	PUMP SUCTION SHUT OFF
D	FIG. 962 SWING CHECK	PUMP DISCHARGE PREVENT BACK FLOW
E	FIG. 530-A GLOBE	PUMP DISCHARGE SHUTOFF
F	FIG. 962 SWING CHECK	PREVENT BACKFLOW IN PRESSURE RELIEF LINES
G	FIG. 280 GATE	DISCHARGE STRAINER SHUT OFF
H	FIG. 530-A GLOBE	DISCHARGE STRAINER SHUT OFF
J	FIG. 530-A GLOBE	OIL BURNER SHUT OFF
K	FIG. 592 GLOBE	OIL BURNER CONTROL
L	FIG. 703 NEEDLE	PRESSURE GAGE CONTROL

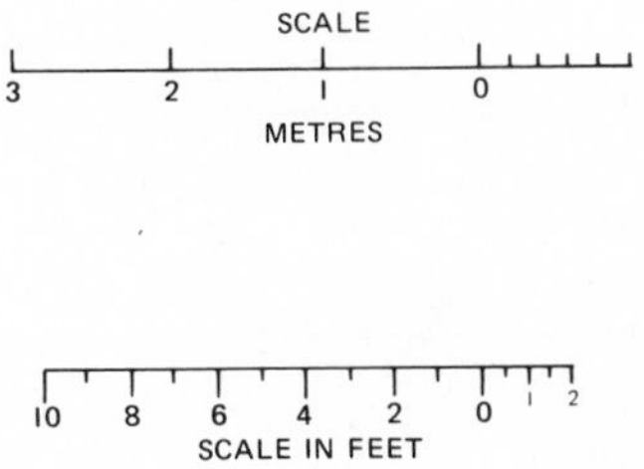

SCALE
3 2 1 0
METRES
10 8 6 4 2 0 1 2
SCALE IN FEET

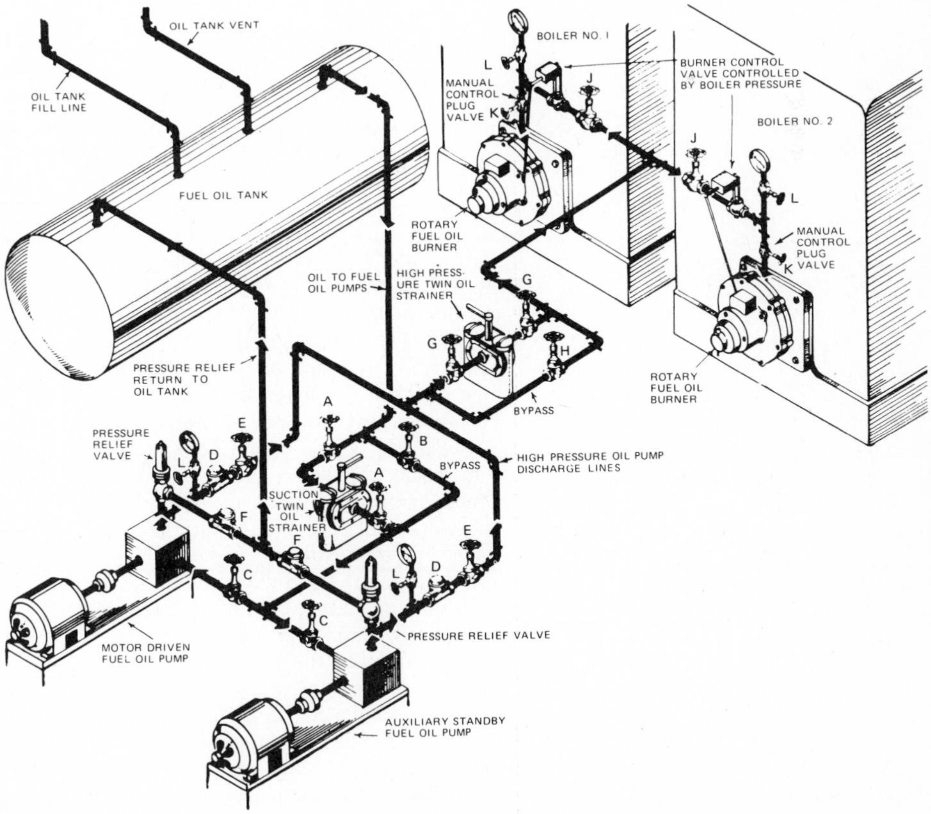

OIL TANK VENT
OIL TANK FILL LINE
FUEL OIL TANK
BOILER NO. 1
MANUAL CONTROL PLUG VALVE
BURNER CONTROL VALVE CONTROLLED BY BOILER PRESSURE
BOILER NO. 2
ROTARY FUEL OIL BURNER
OIL TO FUEL OIL PUMPS
HIGH PRESSURE TWIN OIL STRAINER
MANUAL CONTROL PLUG VALVE
PRESSURE RELIEF RETURN TO OIL TANK
BYPASS
ROTARY FUEL OIL BURNER
PRESSURE RELIEF VALVE
BYPASS
HIGH PRESSURE OIL PUMP DISCHARGE LINES
SUCTION TWIN OIL STRAINER
PRESSURE RELIEF VALVE
MOTOR DRIVEN FUEL OIL PUMP
AUXILIARY STANDBY FUEL OIL PUMP

Fig. 25-3-B Oil-burner piping for light oils.

Chapter 26
Jigs and Fixtures

UNIT 26-1

JIG AND FIXTURE DESIGN

With mass production and interchangeable assembly being used extensively in industry, it is imperative that components be machined and sized to identical standards. To do this, devices called *jigs* and *fixtures* are employed to hold and locate the work or to guide the tools while machining operations are being performed. Jigs and fixtures also cut down machining time, thus lowering production costs.

A jig is a device which holds the work and locates the path of the cutting tool. See Fig. 26-1-1. Generally, a jig may readily be moved about or repositioned. An example of this would be a drill jig which may reposition the work several times when many holes are required in the workpiece, the drill being located each time by a drill bushing located on the jig. Jigs are used extensively for drilling, reaming, tapping, and counterboring operations. A fixture, as the name implies, is fixed to the worktable of the machine and locates the work in an exact position relative to the cutting tool. The fixture does not guide the cutting tool. Fixtures are often employed when milling, grinding, welding, and honing operations are required.

Jigs

There are two main types of jigs: those used for machining purposes and those used for assembly purposes.

Fig. 26-1-1 Drill jig. (Ex-Cell-O Corp. of Canada Ltd.)

When a jig is used in conjunction with a machine tool, its function is to locate the component, hold it firmly, and guide the cutting tool during its operation. The jig need not be secured to the machine. The term thus used usually refers to drilling, reaming, tapping, and boring operations. See Fig. 26-1-2. The size of the jig in this case is limited by the proportions of the machine and the handling characteristic required of the jig. This is normally moved about frequently and is stored when not in use.

When a jig is used for assembly purposes, its function is to locate separate component parts and hold them rigidly in their correct relative positions to each other while they are being connected. These parts usually form large structural frameworks from which accurate locators are taken.

Frequently it is necessary to do some final drilling and reaming at the assembly stage, and this may also be catered for in the assembly jig. In this case, the size of the jig is determined by the proportions of the finished assembly, and in some cases, such as the aircraft industry, this may be very large indeed. This jig is normally fixed in one position, but if it is very small, it may be portable.

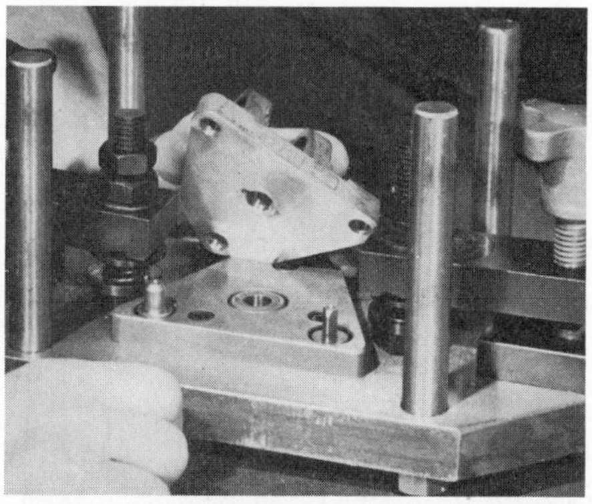

(A) LOADING THE WORKPIECE

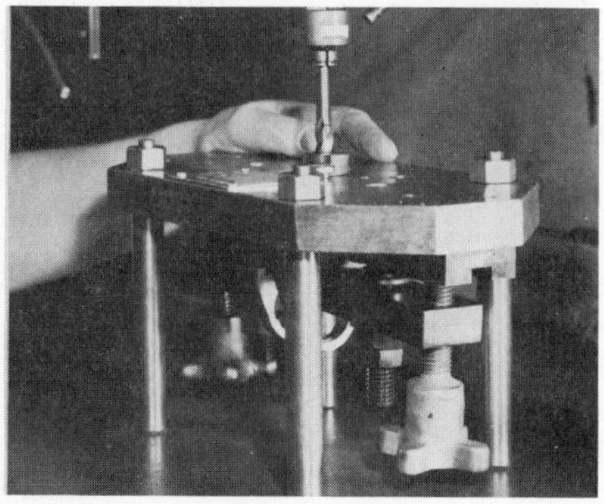

(B) TAPPING THE WORKPIECE

Fig. 26-1-2 Drilling and tapping jig. (Northwestern Tools)

THE DESIGN OF JIGS

The design of a jig is governed by five major factors:

1. The machining operation or operations involved
2. The number of parts to be produced
3. The degree of accuracy of the component
4. The state of the component
5. Any other relevant factors such as portability requirements and external locations

Machining Operation(s) Involved. As has been stated earlier, the term *jig* usually refers to a drilling, reaming, tapping, or boring device. More often than not, a jig may perform a combination of these functions—such as drill and ream, drill and tap, drill, ream and counterbore, etc. Drilling, reaming, and tapping jigs and their combinations are usually similar in construction since all these operations are performed on the one machine, namely, the drill press. Boring jigs are of a slightly different construction since they are used on the boring machine.

Number of Parts to Be Produced. The number of parts to be produced has an important effect on design. For example, in very large quantity production, the cost of an expensive clamping device may be recovered many times over, as a result of time saved through its use. Of course, in small quantity production, the cost of the device might not be recovered; thus a cheaper device should be used.

Degree of Accuracy of the Component. It is logical, of course, that if a component is required to be very accurate, the tool producing it will have to be even more accurate.

Stage of the Component. The designer must know the stage of manufacturing of the component so that he or she can use any available machined faces for location purposes.

Any Other Relevant Factors. Sometimes it is necessary to bolt a jig to the table of the machine. For example, when a large hole is being drilled or reamed, the designer must know what facilities for clamping may be available on the machine.

Before designing a jig, the designer must have or be able to find all the information such as that given above. The information is given to the designer in the form of a working drawing of the component, a process sheet showing the sequence of operations on the component, and the remainder is general information usually available in the department.

Having progressed this far, there are several principles that must also be considered before the designer can finally decide on the design of the jig:

1. The machine on which the operation is being performed
2. Loading and unloading of the component: (*a*) clearances necessary for locating the part; (*b*) methods of foolproofing against improper loading
3. Rapid methods of clamping the work
4. Chip clearance and chip removal
5. Allowance for observation of operation where possible
6. Safety in operation

There are many other considerations, of course, but those are the major ones.

If the following questions are asked and answered before the jig design is passed, much time and money (time *is* money) will be saved and trouble will be avoided.

JIGS IN GENERAL

1. Can a component be inserted and withdrawn without difficulty?

2. Should the component be located to ensure symmetry or balance, i.e., optical balance or material balance?

3. Have the best points of location been chosen with regard to the accuracy of location and the function of the component?

4. Are hardened location points provided where necessary?

5. Can locating points be adjusted where required to make allowance for wear of forging dies or patterns?

6. Are locations clear of flash or burrs?

7. Will the locating devices permit a commercial variation in the machining of the component without affecting the accuracy of location or causing it to bind in the jig?

8. Can the jig be easily cleared of metal shavings and grit, particularly the locating faces?

9. Are all the clamps strong enough?

10. Will any clamp-operating lever or nut be in a dangerous position, i.e., near the cutter?

11. Are all clamps and clamping screws in the most accessible and natural positions?

12. Can wrenches be eliminated by the use of ball or eccentric levers?

13. Will one wrench fit all clamp-operating bolts and nuts?

14. Is the component well supported against the actions of pressure of the cutter?

15. Is the jig foolproof? That is, can the component, tools, or bushes, etc., be wrongly inserted or used?

16. Has the operator an unobstructed view of the component, particularly at the points of location, clamping, and cut?

17. Is the jig as light as possible, consistent with the desired strength?

18. Can coolant, if used, reach the point of cut?

19. Have loose parts been eliminated wherever possible?

20. Have standard parts been used where circumstances allow?

21. Can the jig be designed to hold RH and LH or other similar or complementary components which may be required?

22. Where will burrs be formed, and is clearance for them arranged?

23. Are all corners and sharp edges that are likely to cut the operator shown well radiused on the drawing?

24. Are locating and other working faces and holes protected as far as possible from dirt and cuttings?

25. Will the jig as designed produce components within the required degree of accuracy?

DRILLING AND BORING JIGS

26. Do drills, tools, etc., enter the component at the face which directly adjoins the face of the component to which it fits?

27. Have all slip bushings necessary for reaming, spotfacing, tapping, counterboring, seating, etc., been arranged?

MILLING JIGS

28. Have clamps, etc., been designed to permit the use of the smallest possible diameter of cutter(s)?

29. Will the cutter mandrel clear all parts of the jig when it passes over?

30. Have means for setting the cutter(s) in the correct position been provided?

DRILL JIGS

Drill jigs are of two general types: open jigs and closed or box-type jigs. Open jigs are often referred to as *plate* or *template jigs*. Closed jigs will be discussed in Unit 26-5.

Open Jigs. The simplest tool used to locate holes for drilling is the plate jig or drill template. It consists of a plate with holes to guide the drills, and it has locating pins that locate the workpiece on the jig; or, the workpiece may be nested on the jig and then both are turned over for the drilling operation. Jigs of this type are generally without clamping devices. They are used where the cost of more elaborate tools would not be justified. The holes for guiding the drills are frequently bored into the jig plate, and the plate is sometimes hardened for longer wear; or the plate may have hardened steel drill bushings pressed into it.

A separate base is often used with the template or top plate, thus forming the sandwich type of jig. The base may have holes or grooves to provide clearance for the end of the drill as it breaks through the work.

In the drill jig shown in Fig. 26-1-3, the component is not clamped into or onto the jig. The jig rests upon the component. Since the center-to-center distance between the holes is probably more critical and held to closer tolerances than the distance between the holes and the edge of the part, a locking pin is used to ensure the center-to-center hole accuracy. After the first hole is drilled, the locking pin is inserted into the drill jig and workpiece.

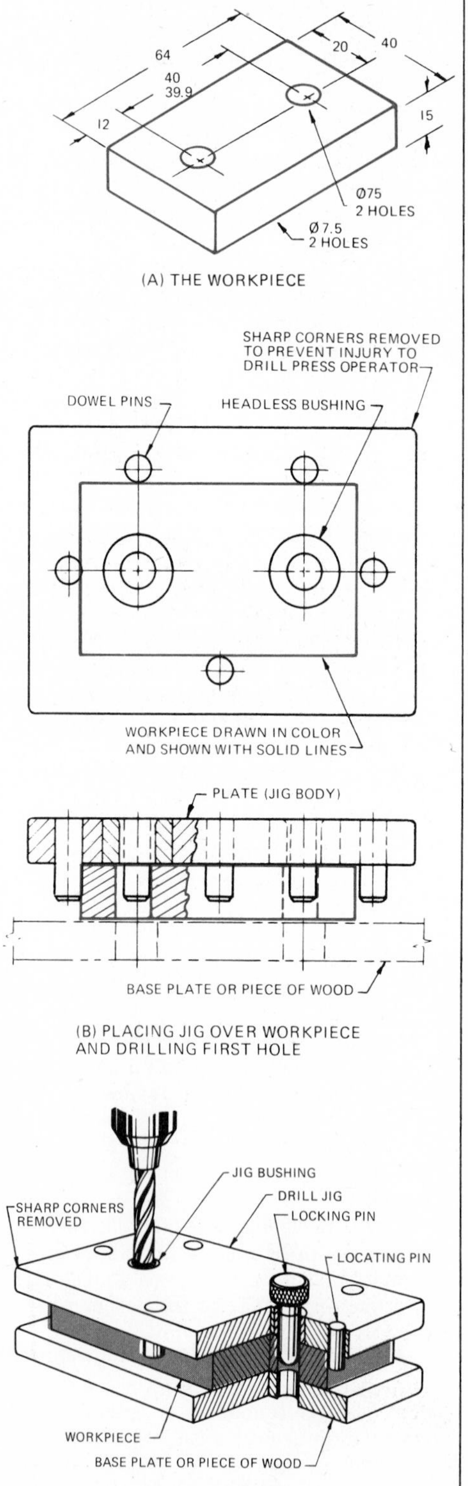
64
40
39.9
12
20
40
15
Ø75
2 HOLES
Ø7.5
2 HOLES
(A) THE WORKPIECE
SHARP CORNERS REMOVED TO PREVENT INJURY TO DRILL PRESS OPERATOR
DOWEL PINS
HEADLESS BUSHING
WORKPIECE DRAWN IN COLOR AND SHOWN WITH SOLID LINES
PLATE (JIG BODY)
BASE PLATE OR PIECE OF WOOD
(B) PLACING JIG OVER WORKPIECE AND DRILLING FIRST HOLE
JIG BUSHING
DRILL JIG
LOCKING PIN
SHARP CORNERS REMOVED
LOCATING PIN
WORKPIECE
BASE PLATE OR PIECE OF WOOD

Fig. 26-1-3 A simple plate jig.

DRILL BUSHINGS[1]

These are precision tools that guide cutting tools such as drills and reamers into precise locations in a workpiece. Assembled in a jig or fixture, drill bushings are capable of producing duplicate parts to extremely close tolerances on location and hole size.

A variety of bushings have been developed for a wide range of portable or machine drilling, reaming, and tapping operations. They include headless and head press-fit bushings, slip and fixed renewable bushings, headless and head liners, thin-wall bushings, and a number of embedment bushings for plastic or castable tooling, soft materials, and special applications.

Each type of bushing has its preferred use. Only proper selection can give the service that the manufacturer has built into the product.

To select the proper bushing, it is necessary to consider not only the function of the jig but also the quantity of production. Life of the average bushing is no more than 5000 to 10 000 pieces. This varies, depending on the operator, the cutting fluid, sharpness of cutting tools, and whether the operation is on multiple or automatic drilling machines or by hand.

The use of special alloy steels or carbides can increase bushing life up to 50 times that of an ordinary steel bushing. Other bushing materials include tool steels, graphitized steel, aluminum, bronze, brass, and other alloys that may be selected to fit the customer's needs.

Press-Fit Bushings. Press-fit bushings are available in two basic styles: headless (type P) and headed (type H). See Fig. 26-1-4. These bushings are permanently pressed into the jig plate or fixture. Press-fit bushings are recommended for use in limited production runs where replacement due to wear is not anticipated during the life of the tooling, and where a single operation, such as drilling only or reaming only, is performed. Headless press-fit bushings offer two advantages: they can be installed flush with the jig plate without counterboring the mounting hole, and they can be mounted closer together than headed bushings. However, where space permits, the use of head press-fit bushings is preferable in any application where heavy axial loads

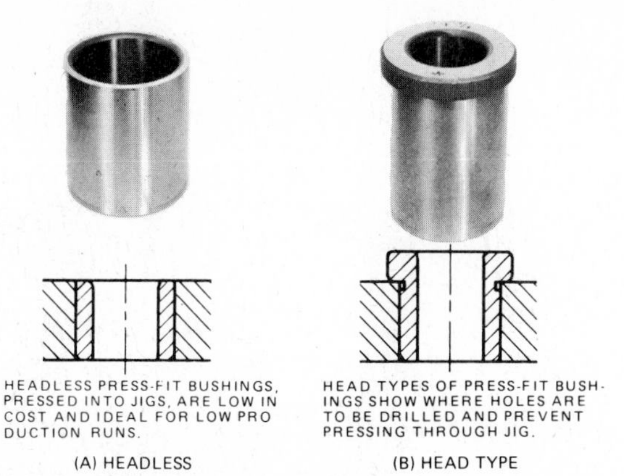
HEADLESS PRESS-FIT BUSHINGS, PRESSED INTO JIGS, ARE LOW IN COST AND IDEAL FOR LOW PRODUCTION RUNS.
HEAD TYPES OF PRESS-FIT BUSHINGS SHOW WHERE HOLES ARE TO BE DRILLED AND PREVENT PRESSING THROUGH JIG.
(A) HEADLESS
(B) HEAD TYPE

Fig. 26-1-4 Press-fit drill bushings. (American Drill Bushing Co.)

most applications because the abrasive action of metal particles will accelerate bushing wear. The recommended chip clearance for metals such as cast iron, which produces small chips, is equal to half the bushing ID. For other metals, such as cold-rolled steel, which produce long, stringy chips, the chip clearance should be at least equal to the bushing ID, but should not exceed 1.5 times the ID. Excessive chip clearance should be avoided. Because most cutting tools are slightly larger in diameter at the cutting end, excessive clearance reduces the guiding effect of the bushing and results in less accurate drilling.

may eventually force a headless bushing out of the jig plate. Typical sizes of standard press-fit drill jig bushings are shown in Fig. 26-1-5.

Renewable bushings are covered in Unit 26-3.

INSTALLATION

Chip Control. See Fig. 26-1-6. Sufficient clearance should be provided between the bushing and the workpiece to permit the removal of chips. The exception to this rule occurs in drilling operations requiring maximum precision where the bushings should be in direct contact with the workpiece. However, suitable chip clearance should be provided in

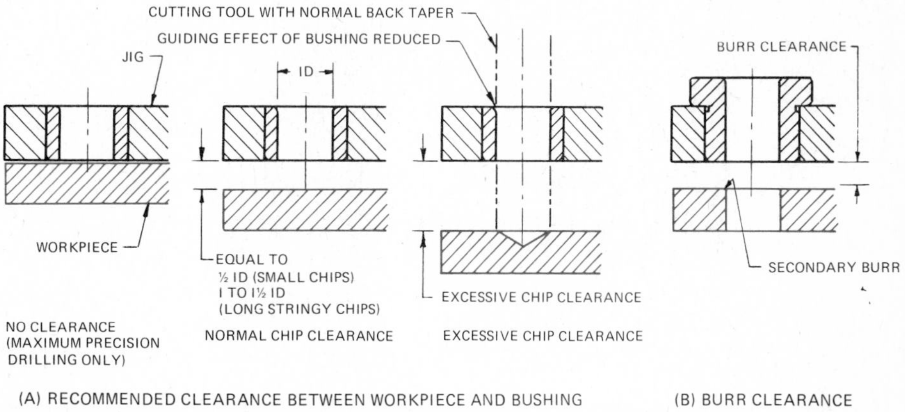
CUTTING TOOL WITH NORMAL BACK TAPER
GUIDING EFFECT OF BUSHING REDUCED
JIG
ID
WORKPIECE
EQUAL TO
½ ID (SMALL CHIPS)
1 TO 1½ ID
(LONG STRINGY CHIPS)
EXCESSIVE CHIP CLEARANCE
BURR CLEARANCE
SECONDARY BURR
NO CLEARANCE (MAXIMUM PRECISION DRILLING ONLY)
NORMAL CHIP CLEARANCE
EXCESSIVE CHIP CLEARANCE
(A) RECOMMENDED CLEARANCE BETWEEN WORKPIECE AND BUSHING
(B) BURR CLEARANCE

Fig. 26-1-6 Chip and burr clearance.

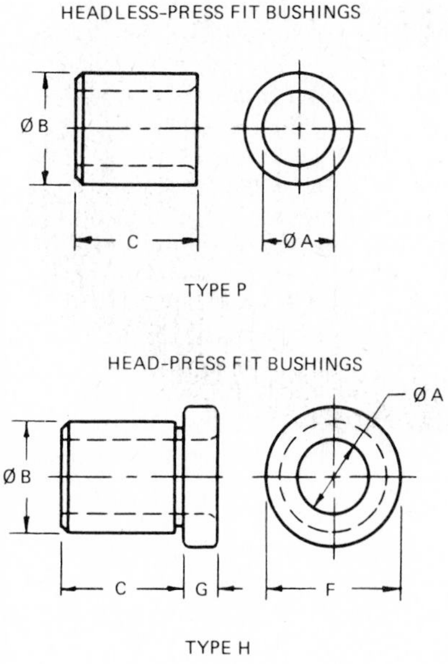
HEADLESS-PRESS FIT BUSHINGS
ØB
C
ØA
TYPE P
HEAD-PRESS FIT BUSHINGS
ØA
ØB
C
G
F
TYPE H

MILLIMETRE SIZES

A (ID Range) Diameter in Millimetres From	To	B	F	G	Bushing Length C 6	8	10	12	20	25	35	45
0.4	1.6	4	6	2.3	•	•	•	•	•	•	•	
0.4	2.5	5	8	2.3	•	•	•	•	•	•	•	
2.5	3.6	6.4	9.4	2.3	•	•	•	•	•	•	•	•
3	5	8	10.6	3	•	•	•	•	•	•	•	•
5	8	13	16	5.6	•	•	•	•	•	•	•	•
8	13.5	20	24	5.6	•	•	•	•	•	•	•	•
12.5	16.5	22	28	6		•	•	•	•	•	•	•
12.5	19.5	25	32	8				•	•	•	•	•
16	26	35	40	10				•	•	•	•	•
25	35	45	50	10					•	•	•	•
34	45	58	64	10						•	•	•

INCH SIZES

A (ID Range) Diameter in Inches From	To	B	F	G	Bushing Length C .25	.30	.40	.50	.75	1.00	1.40	1.75
.0135	.0625	.16	.25	.09	•	•	•	•	•	•	•	
.0135	.0995	.20	.30	.09	•	•	•	•	•	•	•	
.0980	.1406	.25	.36	.09	•	•	•	•	•	•	•	•
.125	.1935	.31	.42	.12	•	•	•	•	•	•	•	•
.188	.3160	.50	.61	.22	•	•	•	•	•	•	•	•
.312	.531	.75	.92	.22	•	•	•	•	•	•	•	•
.500	.656	.88	1.10	.25		•	•	•	•	•	•	•
.500	.766	1.00	1.24	.31				•	•	•	•	•
.625	1.031	1.38	1.60	.38				•	•	•	•	•
1.000	1.391	1.75	1.98	.38					•	•	•	•
1.375	1.766	2.25	2.48	.38						•	•	•

Fig. 26-1-5 Typical sizes of standard press-fit drill jig bushings. (American Drill Bushing Co.)

Burr Clearance. See Fig. 26-1-6*b*. Burr clearance should be provided between the bushing and the workpiece when wiry metals such as copper are drilled. Metals of this type tend to produce secondary burrs around the top of the drilled holes, which act to lift the jig from the workpiece and to cause difficulty in the removal of workpieces from side-loaded jigs. The recommended burr clearance is half the bushing ID.

REFERENCE

1. American Drill Bushing Co.

Assignment

On an A3- or B-size sheet, design a simple plate jig for drilling the holes in one of the parts shown in Figs. 26-1-A to 26-1-D. Scale is 1:1. State the sequence of operations and the time at which locking pins are employed.

REVIEW FOR ASSIGNMENT

Unit 7-5 Assemblies in Sections

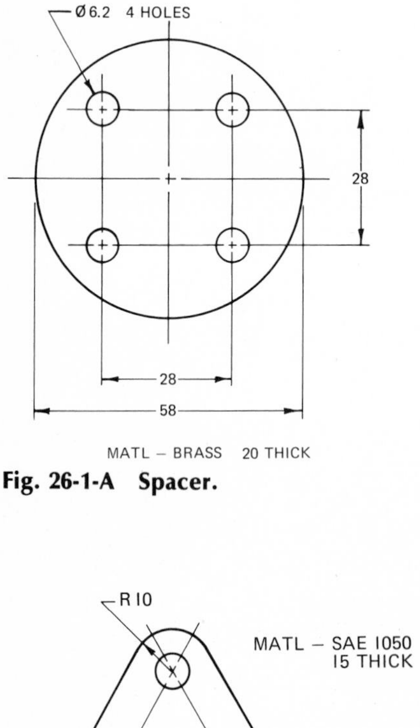
Ø6.2 4 HOLES
28
28
58
MATL – BRASS 20 THICK

Fig. 26-1-A Spacer.

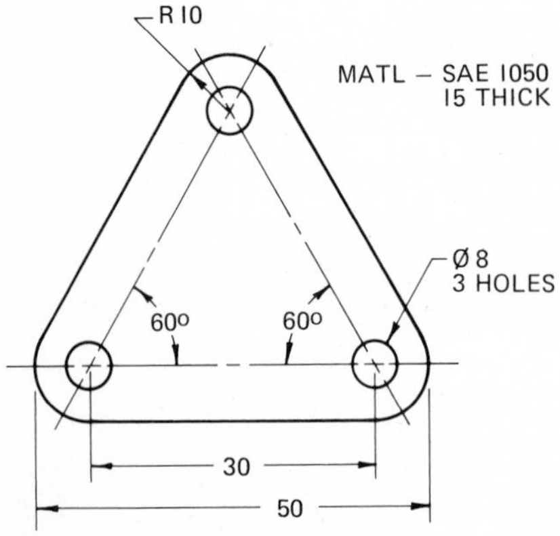
R10
MATL – SAE 1050
15 THICK
Ø8
3 HOLES
60°
60°
30
50

Fig. 26-1-C Cover plate.

UNIT 26-2
DRILL JIG COMPONENTS

JIG BODY

The frame which holds the various parts of a jig assembly is called the *jig body*. It may be in one piece or bolted or welded together. Rigid construction is necessary because of the accuracy required, yet the jig should be light enough to provide ease in handling. Sharp edges or burrs which may harm the operator should be removed. Supporting legs—a minimum of four being recommended—should be provided on the opposite side of each drilling surface. Standard shapes have been designed for jig bodies and are generally more economical than their fabrication in the shop. See Figs. 26-2-1 and 26-2-2.

CAP SCREWS AND DOWEL PINS

The purpose of cap screws in jig design is to hold together fabricated parts.

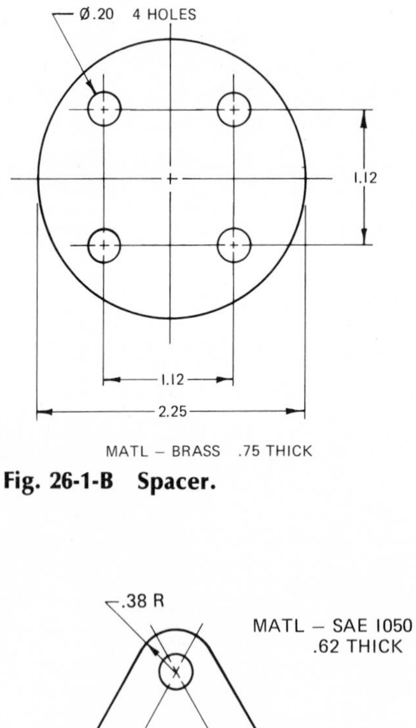
Ø.20 4 HOLES
1.12
1.12
2.25
MATL – BRASS .75 THICK

Fig. 26-1-B Spacer.

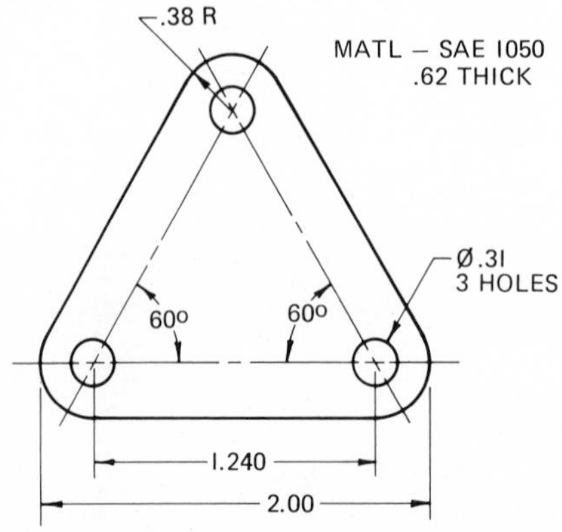
.38 R
MATL – SAE 1050
.62 THICK
Ø.31
3 HOLES
60°
60°
1.240
2.00

Fig. 26-1-D Cover plate.

Fig. 26-2-1 Machined cast iron and aluminum sections for construction of jigs and fixtures. (Standard Parts Co.)

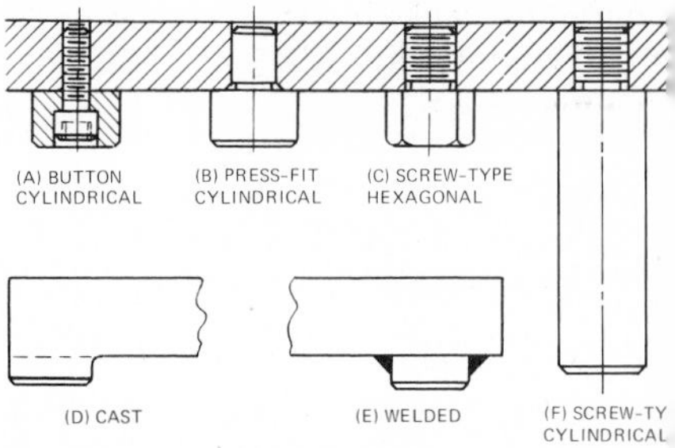
(A) BUTTON CYLINDRICAL
(B) PRESS-FIT CYLINDRICAL
(C) SCREW-TYPE HEXAGONAL
(D) CAST
(E) WELDED
(F) SCREW-TY CYLINDRICAL

Fig. 26-2-2 Typical jig feet.

Dowel pins provide the necessary alignment between the parts, a minimum of two being recommended. See Fig. 26-2-3. Wherever possible, cap screws should be recessed and have socket fillister heads. This type of screw cap can be tightened, with a greater amount of pressure providing better holding power. When thin stock is to be fastened together and counterboring is not possible, a hexagon-head cap screw is used. Flat-head machine screws are normally avoided where accuracy in alignment is required because the conical head forces the parts in a definite position. This type of screw may be used when alignment of parts is not critical and when the screw must be recessed in thin stock.

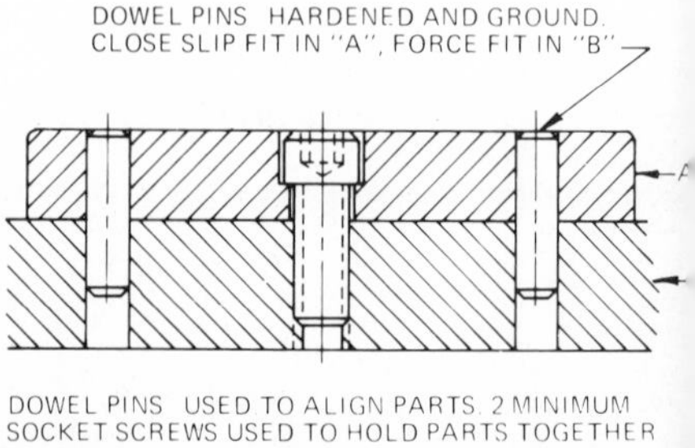
DOWEL PINS HARDENED AND GROUND.
CLOSE SLIP FIT IN "A", FORCE FIT IN "B"
DOWEL PINS USED TO ALIGN PARTS. 2 MINIMUM
SOCKET SCREWS USED TO HOLD PARTS TOGETHER

Fig. 26-2-3 Dowel pins and cap screws.

Dowel pins may be tapered or straight, the latter being used more frequently. A press fit into the two parts ensures the proper alignment required in jig design.

LOCATING DEVICES

The shape of the object determines the type of location best suited for the part. Pins, pads, and recesses are the more common methods used to locate the workpiece on the jig.

Internal Locating Devices. A machined recess in the jig plate (Fig. 26-2-4*a*), and a nesting ring (Fig. 26-2-4*b*), attached to the plate are two methods used to locate a part having a circular projection. The latter method is preferred because the part can be machined more readily and can be replaced when worn. Dowel pins—normally two—position the ring while fastening screws (the number being determined by the size of the ring) and secure it to the plate. For small cylindrical extensions, a headless bushing mounted flush with the locating surface may be used, provided that the shoulder of the workpiece rests on the locating surface, as shown in Fig. 26-2-4*c*. An example of a drilling jig that has an internal locating device is shown in Fig. 26-2-5.

External Locating Devices. Locating studs (Fig. 26-2-4*d*) provides an excellent means of locating workpieces with circular holes. When it is desirable to clamp the workpiece to the stud, the stud should be lengthened and fastened in place by a nut and washer (Fig. 26-2-4*e*). This secures the stud to the jig body and also provides for the interchanging of studs when necessary. Disk-type locators (Fig. 26-2-4*f*) are used when the locating diameter is over 50 mm. Dowels and fastening screws, the number determined by the size of the disk, locate and secure the disk to the plate.

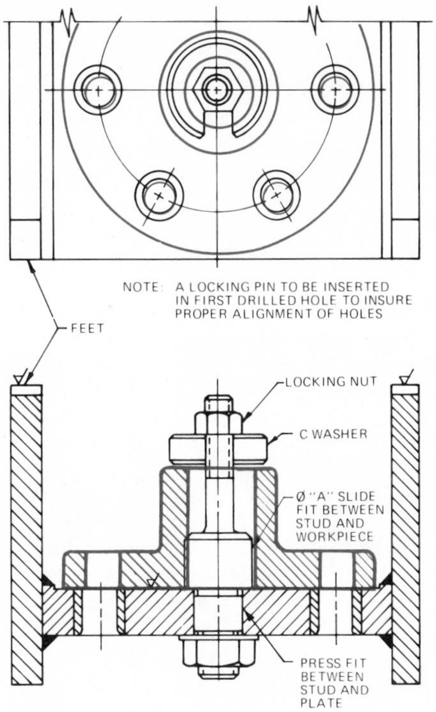
NOTE: A LOCKING PIN TO BE INSERTED IN FIRST DRILLED HOLE TO INSURE PROPER ALIGNMENT OF HOLES
FEET
LOCKING NUT
C WASHER
Ø "A" SLIDE FIT BETWEEN STUD AND WORKPIECE
PRESS FIT BETWEEN STUD AND PLATE

Fig. 26-2-5 Plate drill jig for drilling holes in flange.

Stops. When the workpiece cannot be located by recesses or projections as outlined above, locating stops are used. They are classified as either fixed or adjustable. See Fig. 26-2-6.

The commonest types of fixed stops are the stop pin, flatted shoulder plug, crowned shoulder plug, and stop pads. While stop pins (dowels) are the most economical, their main disadvantages are rapid wear and marring of the finished surface of the workpiece. Shoulder plugs, with one side of the head flattened, provide a greater bearing surface

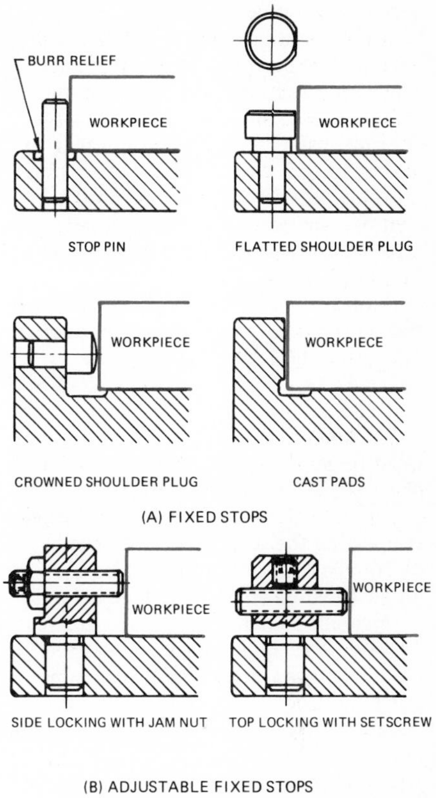
BURR RELIEF
WORKPIECE
STOP PIN
FLATTED SHOULDER PLUG
CROWNED SHOULDER PLUG
CAST PADS
(A) FIXED STOPS
SIDE LOCKING WITH JAM NUT
TOP LOCKING WITH SETSCREW
(B) ADJUSTABLE FIXED STOPS

Fig. 26-2-6 Fixed and adjustable stops.

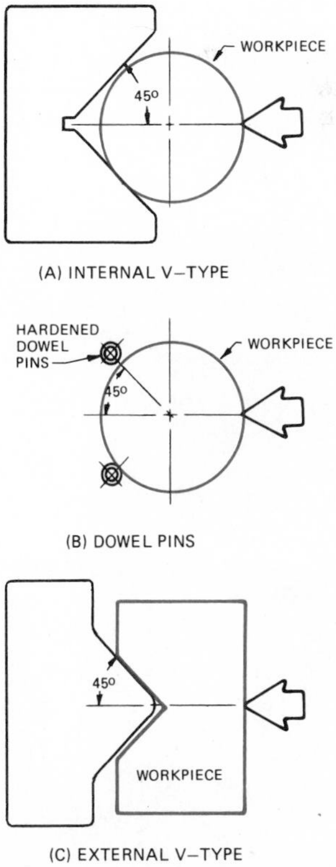
WORKPIECE
45°
(A) INTERNAL V–TYPE
HARDENED DOWEL PINS
(B) DOWEL PINS
(C) EXTERNAL V–TYPE

Fig. 26-2-7 Centralizers.

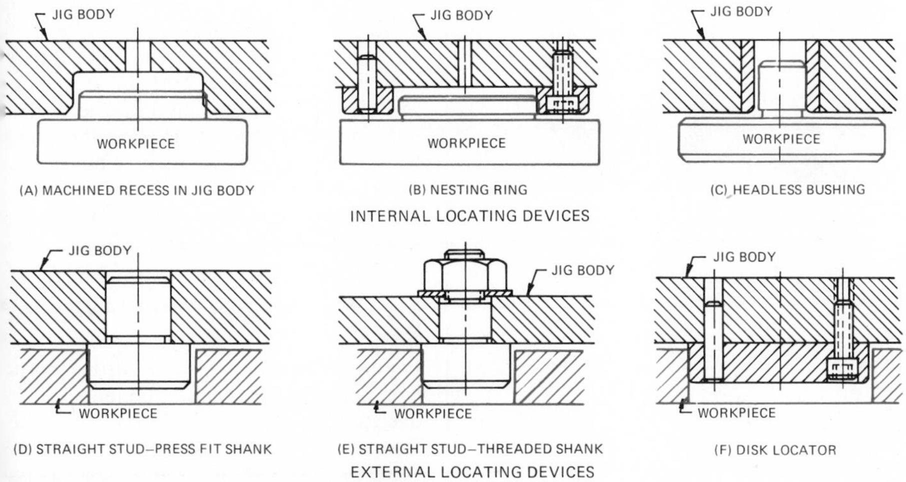
JIG BODY
WORKPIECE
(A) MACHINED RECESS IN JIG BODY
(B) NESTING RING
(C) HEADLESS BUSHING
INTERNAL LOCATING DEVICES
(D) STRAIGHT STUD–PRESS FIT SHANK
(E) STRAIGHT STUD–THREADED SHANK
(F) DISK LOCATOR
EXTERNAL LOCATING DEVICES

Fig. 26-2-4 Common locating devices.

and will not wear as readily. The crowned shoulder plug is similar to the flatted shoulder plug except that the pressure exerted on the plug is parallel to the plug axis. Stop pads provide large bearing surfaces which will not mar the surface of the workpiece.

Adjustable fixed stops are used with castings and forgings where variations in size occur on workpieces and where minor adjustment is necessary.

Centralizers. Circular workpieces, or flat workpieces with rounded or angled ends, may be located or centered by centralizers, as shown in Figs. 26-2-7 and 26-2-8.

Workpiece Supports. The workpiece must be supported so as to avoid distortion caused by either clamping or machining. See Fig. 26-2-9. The surfaces supporting the workpiece are called *workpiece supports* and are classified as either fixed or adjustable. They should be located, as nearly as possible, directly opposite the clamping force. It is recommended that four small work support areas be used in lieu of one large area, because the latter may produce a rocking condition.

The jig body with metal cut away and steel blocks called rest buttons are the more common types of fixed supports used. Rest buttons are preferred because, made from hardened tool steel, they offer an excellent wear-resistant surface. Another advantage is that they raise the workpiece sufficiently to reduce chip problems.

When variation in the workpiece is encountered, adjustable supports, commonly referred to as "jacks," are used

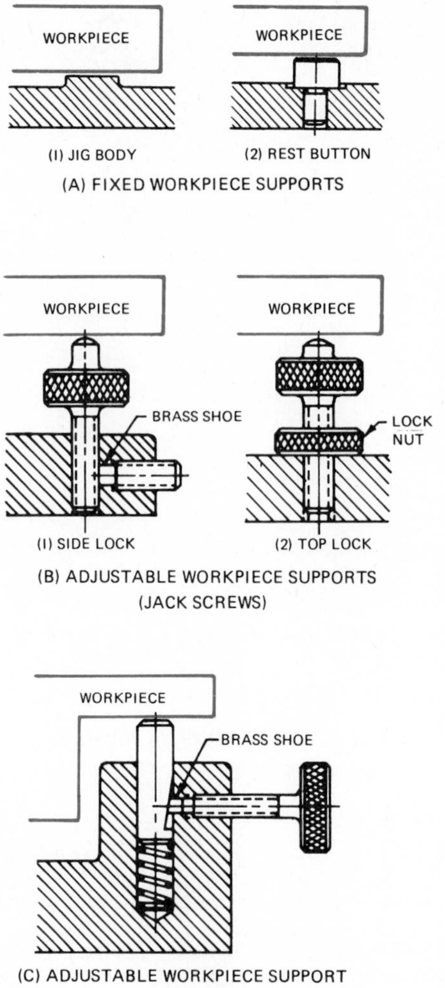
WORKPIECE
WORKPIECE
(1) JIG BODY
(2) REST BUTTON
(A) FIXED WORKPIECE SUPPORTS
WORKPIECE
WORKPIECE
BRASS SHOE
LOCK NUT
(1) SIDE LOCK
(2) TOP LOCK
(B) ADJUSTABLE WORKPIECE SUPPORTS (JACK SCREWS)
WORKPIECE
BRASS SHOE
(C) ADJUSTABLE WORKPIECE SUPPORT (JACK PIN)

Fig. 26-2-9 Workpiece supports.

to provide the necessary adjustment for proper alignment. The operator adjusts each support separately, as required.

CLAMPING DEVICES

The clamping components must be designed to securely hold the workpiece but not distort it, to be quickly and easily locked and unlocked, and to swing out of the way during loading and unloading. Some of the more common types of clamps are shown in Fig. 26-2-10.

Screw clamps are commonly used because they do not tend to loosen under vibration and they provide adequate clamping force. One of the simplest types of screw clamps is the cone-point setscrew. The incline on the screw tends to push the workpiece against the locating pads as well as against the stops. It is best suited for clamping unfinished surfaces such as castings because the point of the setscrew will mar the workpiece surface. The toggle-head type clamp provides a larger contact surface with the workpiece, thereby reducing the possibility of marring. It is also ideally suited for clamping workpieces having side drafts. Where only moderate clamping pressures are required, a knurled knob, lever nuts and thumbscrew may be used.

The two-direction clamp provides both side and top clamping. As pressure is exerted on the end of the screw thread, the clamp is pivoted about the pivot pin, producing a downward pressure at the top of the workpiece.

Since screw-thread type clamps are relatively slow, they are often used in combination with other devices to speed up the clamping and unclamping operations. The travel cam lock assembly and the hinged cam assembly clamps are two such types.

LOCKING PINS

A locking pin is used in jig design to lock or hold the workpiece securely to the jig plate while the second or subsequent holes are being drilled. After the first hole is drilled, the locking pin is inserted through the drill bushing into the drilled hole in the workpiece, locking the drill jig and workpiece together. When more than two holes are drilled, a second locking pin is used to maintain proper alignment. The use of a locking pin is illustrated in Fig. 26-1-3*c*.

MISCELLANEOUS STANDARD PARTS

Figure 26-2-11 shows some of the more common standard jig components. The designer should, wherever possible, use standard parts in the design in order to simplify the work and reduce the manufacturing cost.

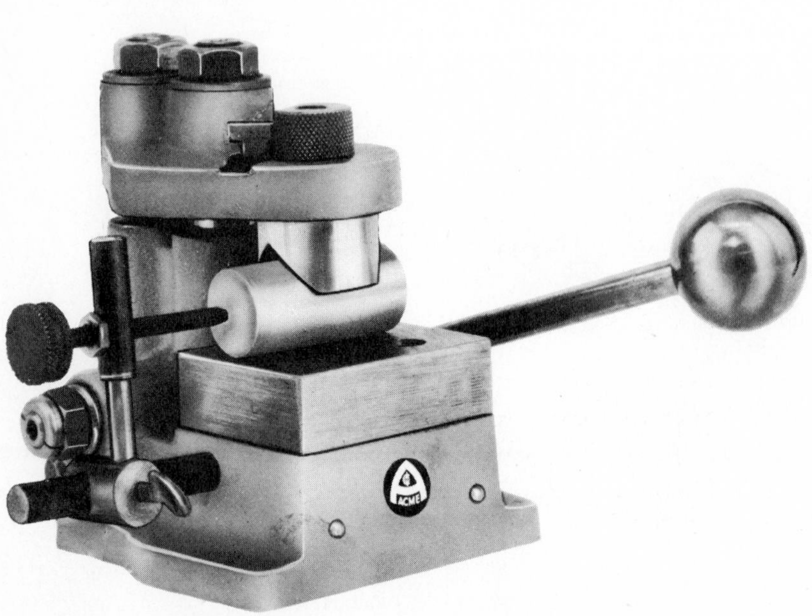

Fig. 26-2-8 V-bushing drill jig. (Acme Industrial Co.)

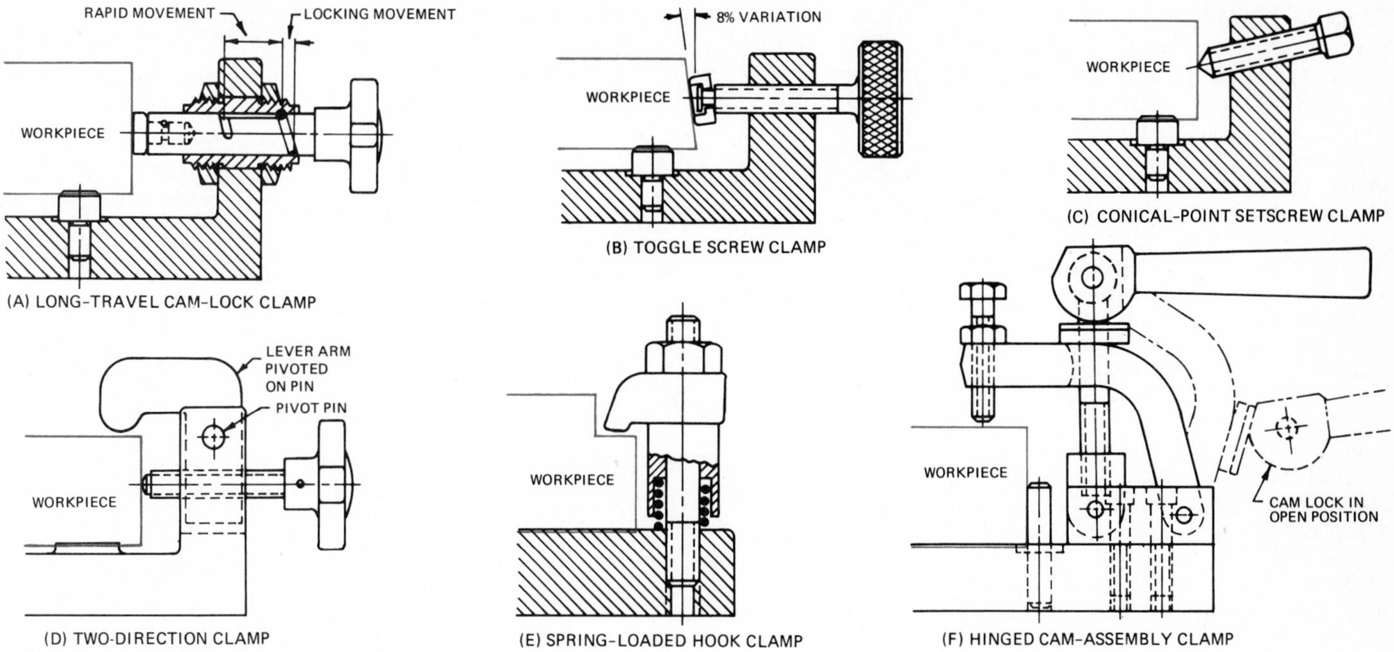
RAPID MOVEMENT
LOCKING MOVEMENT
WORKPIECE
(A) LONG-TRAVEL CAM-LOCK CLAMP
8% VARIATION
WORKPIECE
(B) TOGGLE SCREW CLAMP
WORKPIECE
(C) CONICAL-POINT SETSCREW CLAMP
LEVER ARM PIVOTED ON PIN
PIVOT PIN
WORKPIECE
(D) TWO-DIRECTION CLAMP
WORKPIECE
(E) SPRING-LOADED HOOK CLAMP
WORKPIECE
CAM LOCK IN OPEN POSITION
(F) HINGED CAM-ASSEMBLY CLAMP

Fig. 26-2-10 Common clamping devices.

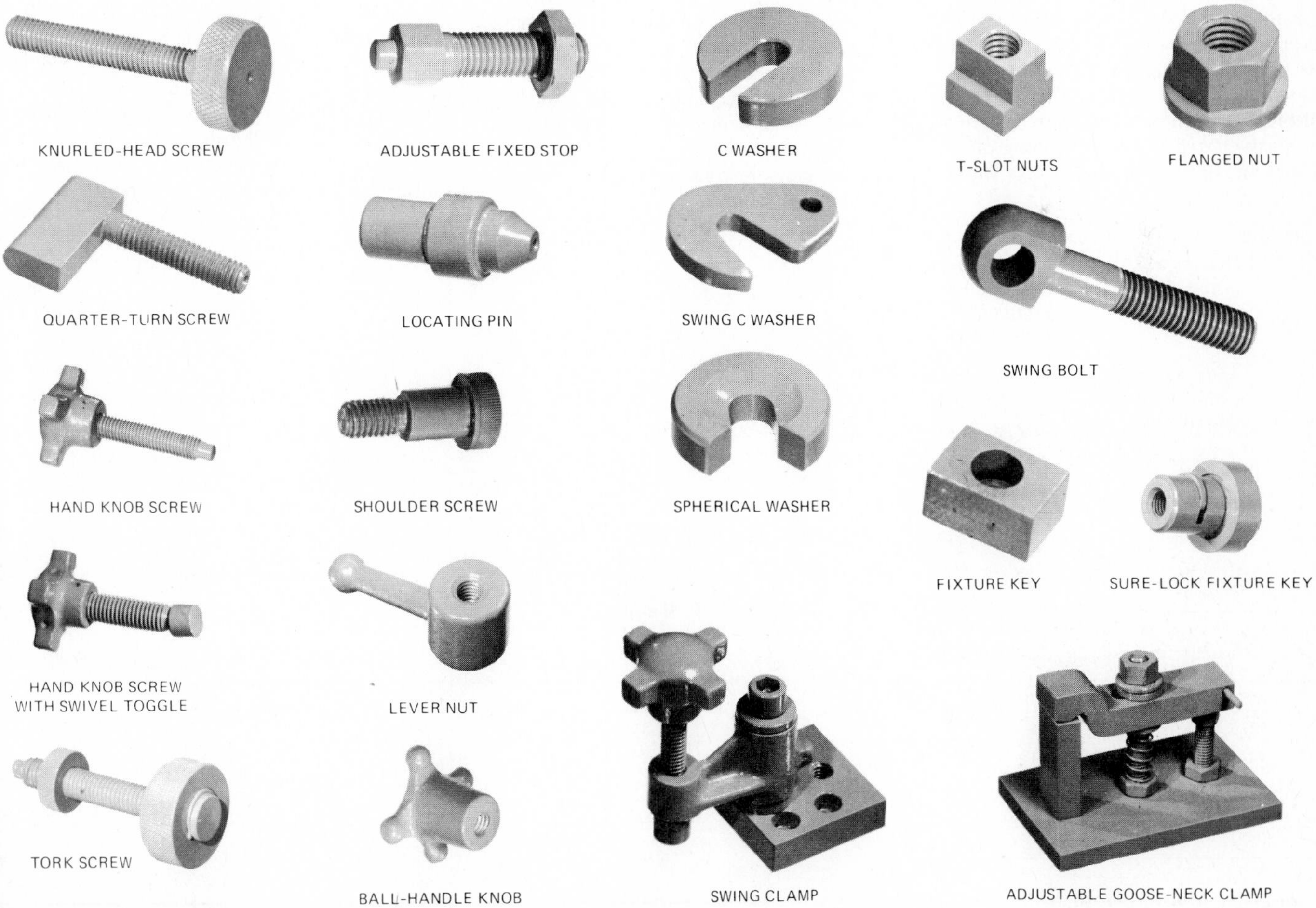
KNURLED-HEAD SCREW
ADJUSTABLE FIXED STOP
C WASHER
T-SLOT NUTS
FLANGED NUT
QUARTER-TURN SCREW
LOCATING PIN
SWING C WASHER
SWING BOLT
HAND KNOB SCREW
SHOULDER SCREW
SPHERICAL WASHER
FIXTURE KEY
SURE-LOCK FIXTURE KEY
HAND KNOB SCREW WITH SWIVEL TOGGLE
LEVER NUT
TORK SCREW
BALL-HANDLE KNOB
SWING CLAMP
ADJUSTABLE GOOSE-NECK CLAMP

Fig. 26-2-11 Standard jig and fixture parts. (Standard Parts Co.)

DESIGN EXAMPLES

Example 1. An alternative drill jig for the workpiece shown in Fig. 26-1-3 is shown in Fig. 26-2-12. This jig employs a lever arm and a knurled head screw which applies pressure on two sides of the workpiece, forcing it against the locating pins. This jig not only locates but also holds the workpiece in position. It is also recommended where more accurate tooling is required. Since this jig is turned over when the holes are to be drilled, feet must be added to provide stability and clearance between the clamping devices and the worktable.

In a more elaborate plate jig or drill template, clamps hold the workpiece underneath the plate, which contains the drill guiding holes or the drill bushings. The plate must then be provided with legs to raise both the clamps and the workpiece from the table and to support the jig. However, in all probability, the jig and component would be clamped as a unit onto the table of the drill press. This is done to prevent the jig from spiraling up the drill and cuttings from getting between the bottom face of the jig and the top face of the template. The above procedure is normally adopted when the sizes of the drills are large and/or the mass of the jig is small. The base plate or piece of wood is provided to prevent damage to the table when the drill breaks through. This jig is designed for short-run production. For long-run production, flatted shoulder plugs should be used in place of the dowels.

Example 2. For drilling a series of bolt holes in a flange, a drill jig, as shown in Fig. 26-2-5, may be used. The base surface of the flange and the diameter *A* of the workpiece, which were previously machined, were used as locating surfaces. The workpiece slips over the study and rests on the jig plate. The C-washer is then inserted over the workpiece, and the locking nut is screwed down to securely clamp the parts together. The size of the locknut was selected to clear diameter *A*. The body of the jig was designed to protect the threads on the stud from being damaged. The position shown is the loading and unloading position. For drilling, the jig must be inverted and the side walls which act as feet must be machined to level and true-up the jig. Notice that part of the sides has been machined away, leaving only the four small surfaces to act as jig feet. This type of jig is frequently used where only a limited number of workpieces are required. When a large number of parts are to be drilled, a quick-acting clamping device would be used.

Example 3. The screw latch clamp jig, similar to the one shown in Fig. 26-2-13, is frequently used because of its simple design and fast clamping action. All the parts shown, with the exception of the clamp plate, are standard items which were purchased. The standard cast-iron section used as the jig body was purchased at only a fraction of the cost of fabricating a similar type of body.

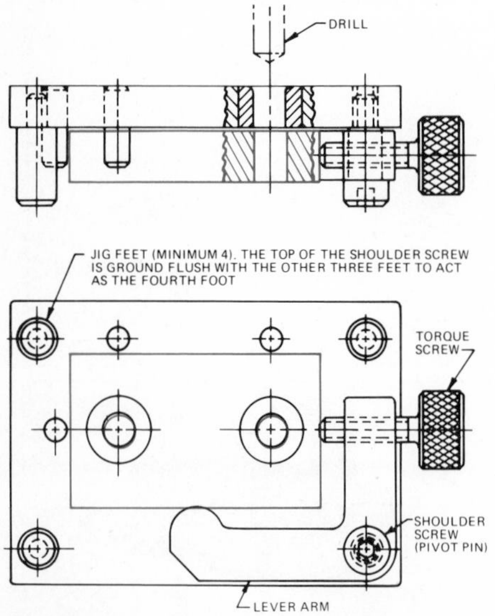
DRILL
JIG FEET (MINIMUM 4). THE TOP OF THE SHOULDER SCREW IS GROUND FLUSH WITH THE OTHER THREE FEET TO ACT AS THE FOURTH FOOT
TORQUE SCREW
SHOULDER SCREW (PIVOT PIN)
LEVER ARM

Fig. 26-2-12 Alternative plate jig for workpiece shown in Figure 26-1-3.

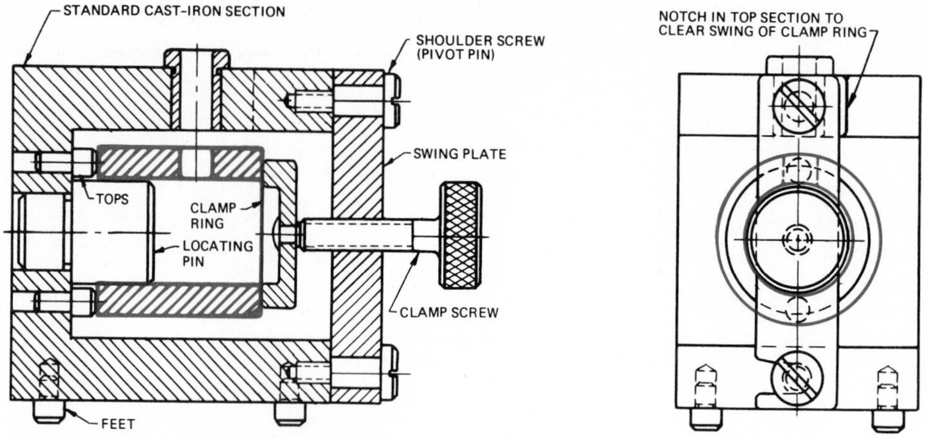
STANDARD CAST-IRON SECTION
SHOULDER SCREW (PIVOT PIN)
SWING PLATE
TOPS
CLAMP RING
LOCATING PIN
CLAMP SCREW
FEET
NOTCH IN TOP SECTION TO CLEAR SWING OF CLAMP RING

Fig. 26-2-13 Screw latch clamp jig.

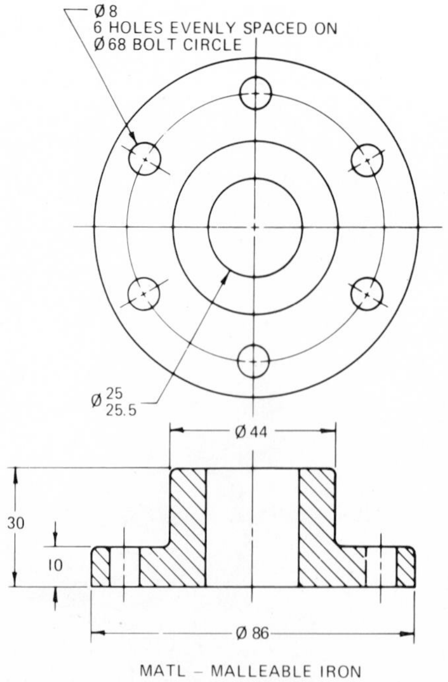
Ø8
6 HOLES EVENLY SPACED ON
Ø68 BOLT CIRCLE
Ø 25 25.5
Ø44
30
10
Ø86
MATL – MALLEABLE IRON

Fig. 26-2-A Flanged bracket.

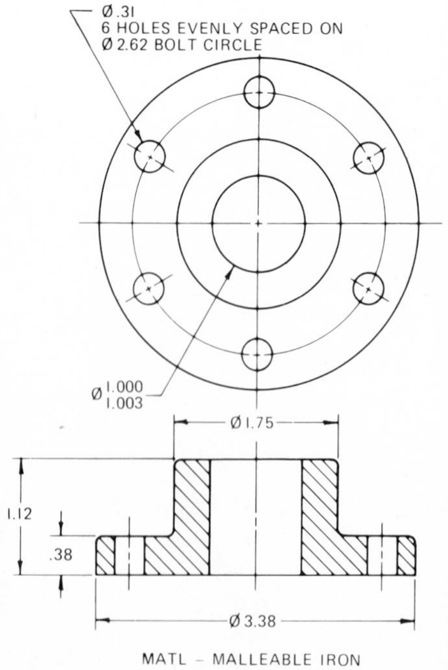
Ø.31
6 HOLES EVENLY SPACED ON
Ø2.62 BOLT CIRCLE
Ø 1.000 1.003
Ø1.75
1.12
.38
Ø3.38
MATL – MALLEABLE IRON

Fig. 26-2-B Flanged bracket.

Assignment

On an A3- or B-size sheet, design a jig for drilling the six holes in the part in either Fig. 26-2-A or Fig. 26-2-B. The large center hole and finished base should be the features used for locating the part in the jig. A locking pin is recommended for alignment after the first hole is drilled. Standard components should be used wherever possible.

An alternative to the above assignment is to design a drill jig for two of the three holes shown in Fig. 26-2-C or Fig. 26-2-D. The hole in the hub of the part and the finished surfaces should be the features used for locating the part in the jig. A locking pin is optional, depending on the design. Standard components should be used wherever possible. Scale is 1:1.

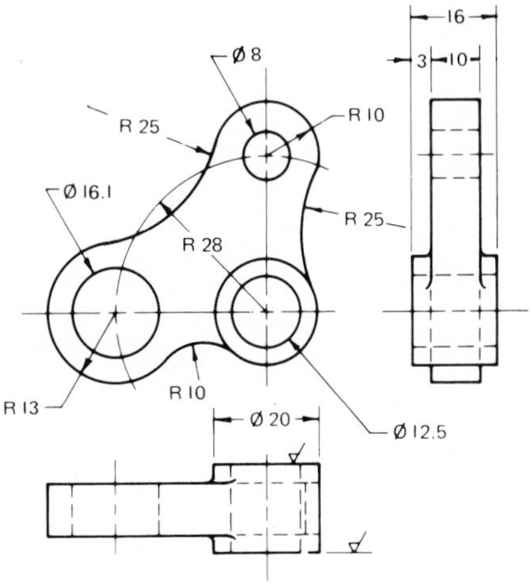

16
3
10
Ø 8
R 25
R 10
Ø 16.1
R 25
R 28
R 10
R 13
Ø 20
Ø 12.5

MATL SAE 1050

Fig. 26-2-C Connector.

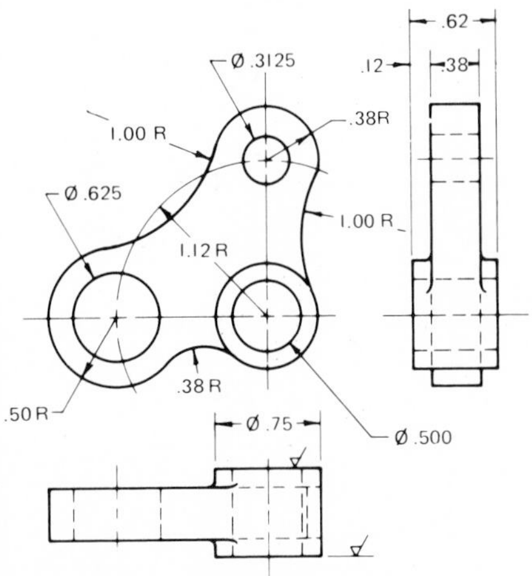

.62
.12
.38
Ø .3125
1.00 R
.38R
Ø .625
1.00 R
1.12 R
.38 R
.50 R
Ø .75
Ø .500

MATL SAE 1050

Fig. 26-2-D Connector.

REVIEW FOR ASSIGNMENT

Unit 7-5 Assemblies in Section
Unit 26-1 Jig Design

UNIT 26-3
RENEWABLE BUSHINGS[1]

Renewable bushings are designed to be easily replaced and are available in two styles—fixed renewable (type F) and slip renewable (type S). See Fig. 26-3-1. Both types are installed as slipfits in liners, are designed for long production runs, and are intended to remain fixed in the jig or fixture until worn out. In most applications, the liners (head or headless types) are mounted flush with the jig plate; however, projected mounting is sometimes used for head liners when the jig plate is too thin to accept a suitable counterbore or when the intended application does not warrant the machining costs involved.

Lock screws are suitable only for use with flush-mounted liners, usually in

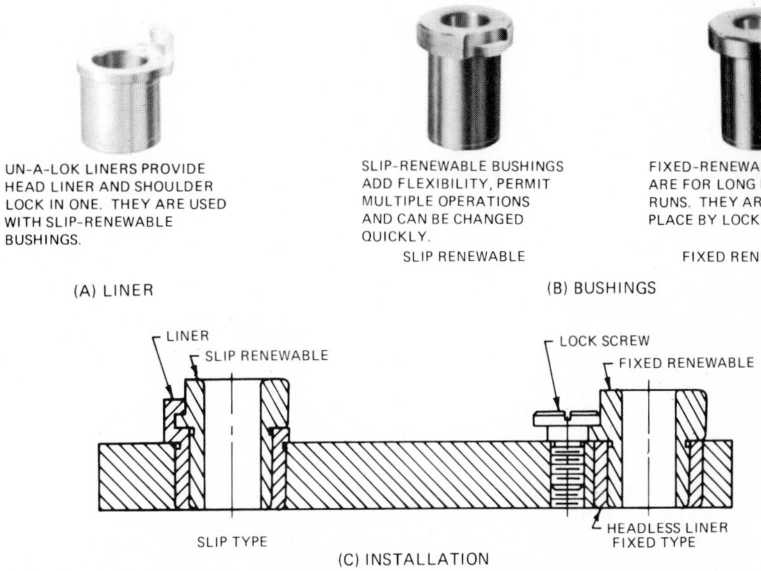

UN-A-LOK LINERS PROVIDE HEAD LINER AND SHOULDER LOCK IN ONE. THEY ARE USED WITH SLIP-RENEWABLE BUSHINGS.
SLIP-RENEWABLE BUSHINGS ADD FLEXIBILITY, PERMIT MULTIPLE OPERATIONS AND CAN BE CHANGED QUICKLY.
SLIP RENEWABLE
FIXED-RENEWABLE BUSHINGS ARE FOR LONG PRODUCTION RUNS. THEY ARE HELD IN PLACE BY LOCK SCREW.
FIXED RENEWABLE
(A) LINER
(B) BUSHINGS
LINER
SLIP RENEWABLE
LOCK SCREW
FIXED RENEWABLE
SLIP TYPE
HEADLESS LINER FIXED TYPE
(C) INSTALLATION

Fig. 26-3-1 Renewable bushings. (American Drill Bushing Co.)

FIXED RENEWABLE BUSHINGS		SLIP RENEWABLE BUSHINGS
FLUSH MOUNTED (HEAD OR HEADLESS LINERS)		
LOCK SCREW • RECOMMENDED FOR LIGHT DRILLING APPLICATIONS • SMALL HEAD DIAMETER PERMITS CLOSE BUSHING PLACEMENT	**ROUND CLAMP** • PROVIDES BETTER SECURITY THAN LOCK SCREW IN HEAVIER DUTY APPLICATIONS • DIAMETER SAME AS LOCK SCREW HEAD FOR CLOSE BUSHING PLACEMENT	**LOCK SCREW** • RECOMMENDED FOR LIGHT DRILLING OPERATIONS • SMALL HEAD DIAMETER PERMITS CLOSE BUSHING PLACEMENT
ROUND END CLAMP • FOR LOCKING STANDARD FIXED-RENEWABLE BUSHINGS • PROVIDES LARGE BEARING SURFACE AGAINST JIG PLATE	**FLAT CLAMP** • FOR TYPE FX BUSHINGS ONLY • PROVIDES MAXIMUM SECURITY AGAINST VIBRATION AND TORQUE	**ROUND END CLAMP** • FOR HEAVY DUTY APPLICATIONS • PROVIDES LARGE BEARING SURFACE AGAINST JIG PLATE
PROJECTED MOUNTED (HEAD LINERS ONLY)		
ROUND END CLAMP • FOR LOCKING STANDARD FIXED RENEWABLE BUSHINGS IN PROJECTED MOUNTED LINERS	**FLAT CLAMP** • FOR TYPE FX BUSHINGS ONLY • PROVIDES MAXIMUM SECURITY AGAINST VIBRATION AND TORQUE	**ROUND END CLAMP** • FOR ANY APPLICATION USING SLIP-RENEWABLE BUSHINGS INSTALLED IN PROJECTED MOUNTED HEAD LINERS

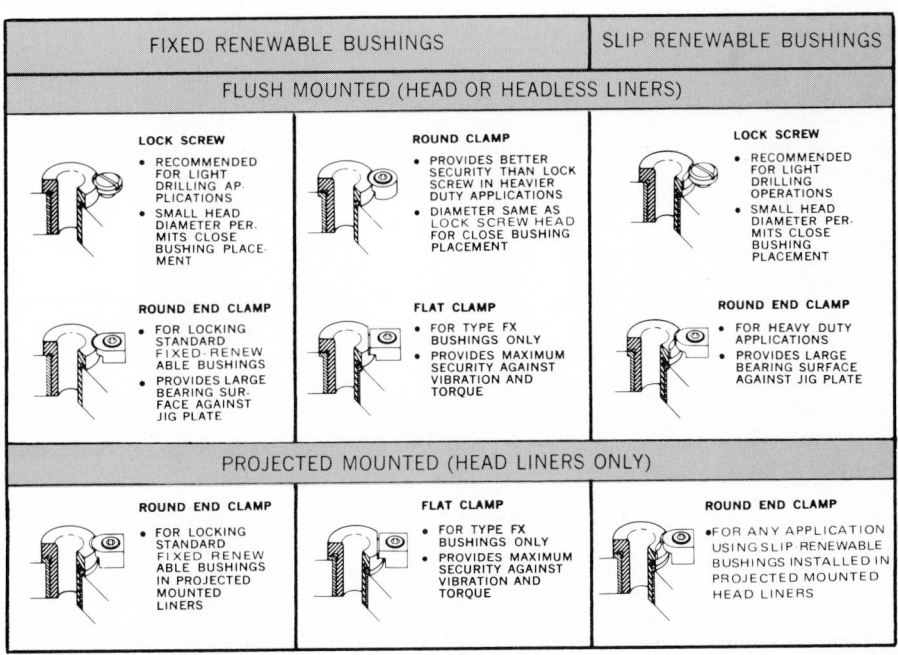

Fig. 26-3-2 Typical installations of renewable bushings. (American Drill Bushing Co.)

light-duty applications. For more heavy-duty applications, clamps provide a better means of locking the bushing against the effects of vibration and torque produced by drill rotation. Clamps provide a larger bearing surface against the jig plate and are secured by standard socket-head cap screws (SHCS).

Replacement of a fixed renewable bushing can be accomplished simply by removing its lock screw or clamp, without removing the jig from the production line. Slip renewable bushings are recommended for production runs of any length where more than one operation is performed in a hole, such as drilling, and then reaming or counterboring. This type is especially designed for fast change by merely turning and lifting the bushing. Slip renewable bushings may be interchanged in the same liner without affecting the centering accuracy. For example, suppose you wish to produce a hole in a part between 12 and 12.01 mm in diameter. Since a drill alone cannot come that close, the hole must be reamed. First, drill through a 11.5 mm ID slip renewable bushing. After drilling the hole, remove the bushing and replace it with a 12 mm slip renewable bushing for the reaming operation. Both bushings fit perfectly in the same 20 mm ID jig hole liner. This or other sequences may be endlessly repeated.

Since changing slip renewable bushings takes less than a minute, they are invaluable for high-speed production where machine down-time must be held to an absolute minimum. A slight turn of the milled head locks or unlocks the bushing without removing the lock screw. Typical installations and sizes of fixed and slip renewable bushings are shown in Figs. 26-3-2 and 26-3-3.

Liners

Liners are permanently pressed into the jig plate or fixture both to provide precision mounting holes or correct slipfit for renewable bushings and to prevent wear of soft jig plates caused by frequent replacement of renewable bushings. Liners are available in two basic styles: headless (type L) and headed (type HL). Headless and headed liners are similar to headless and headed bushings in their advantages and limitations. See Figs. 26-3-4 and 26-3-5.

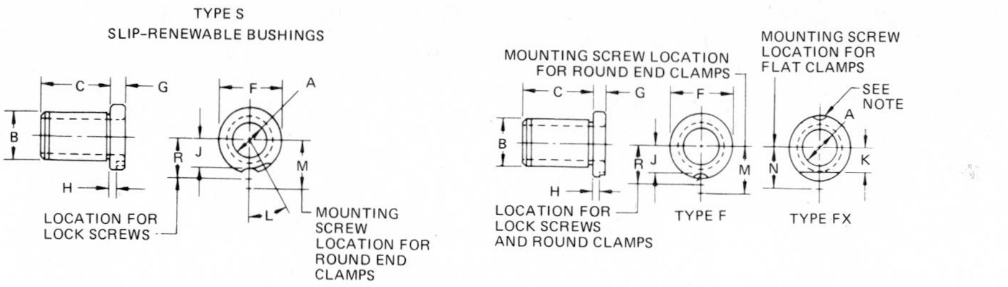
TYPE S
SLIP-RENEWABLE BUSHINGS
C
G
F
A
B
R
J
M
H
L
LOCATION FOR LOCK SCREWS
MOUNTING SCREW LOCATION FOR ROUND END CLAMPS
MOUNTING SCREW LOCATION FOR ROUND END CLAMPS
LOCATION FOR LOCK SCREWS AND ROUND CLAMPS
TYPE F
MOUNTING SCREW LOCATION FOR FLAT CLAMPS
SEE NOTE
N
K
TYPE FX

MILLIMETRE SIZES

A (ID Range) Millimetre Sizes From	To	B	F	G	H	J	K	L	M	N	R	Bushing Length C 6	8	10	12	16	20	25	35	45
0.34	1.6	3.2	8	5	2.4	2.4	—	55°	—	—	6.6	•	•	•	•	•	•			
0.34	4.92	8	14	10	3	4.3	4.3	65°	16	17	12.7		•	•	•	•	•	•	•	
3.57	8.8	12.5	20	11	3	7.6	6.6	65°	19	20	16		•	•	•		•	•	•	•
7.1	14.5	20	26	11	3	10.7	9.9	50°	22	20	19				•		•	•	•	•
11.5	20	25	36	11	5	15.2	12.7	35°	28	29	23.4				•		•	•	•	•
18	27	35	46	11	5	19.8	17.3	30°	33	34	28						•	•	•	•
24	36	45	58	16	5	25.4	22.4	30°	42	38.5	35.6						•	•	•	•

INCH SIZES

A (ID Range) Inch Sizes From	To	B	F	G	H	J	K	L	M	N	R	Bushing Length C .25	.31	.38	.50	.62	.75	1.00	1.38	1.75
.014	.062	.188	.31	.18	.09	.09	—	55°	—	—	.26	•	•	•	•	•	•			
.014	.194	.312	.55	.38	.12	.17	.17	65°	.62	.68	.50		•	•	•	•	•	•	•	
.141	.344	.500	.80	.44	.12	.30	.26	65°	.75	.78	.62		•	•	•		•	•	•	•
.281	.562	.750	1.05	.44	.12	.42	.39	50°	.88	.80	.75				•		•	•	•	•
.469	.781	1.000	1.42	.44	.18	.60	.50	35°	1.10	1.14	.92				•		•	•	•	•
.719	1.062	1.375	1.80	.44	.18	.78	.68	30°	1.30	1.33	1.10						•	•	•	•
.969	1.406	1.750	2.30	.62	.18	1.00	.88	30°	1.64	1.52	1.40						•	•	•	•

Fig. 26-3-3 Typical sizes for renewable drill jig bushings. (American Drill Bushing Co.)

MILLIMETRE SIZES

Part No.	A	B	C	E	F	Thread Metric	Thread Inch
LS-0	11	5	8	5	3	M4	8–32
LS-1	16	10	16	7	3.5	M8	.312–18
LS-2	22	10	16	10	5	M8	.312–18
LS-3	25	11	20	10	5	M10	.375–16
LS-4	27	11	20	11	5.5	M10	.375–16

FOR FLUSH MOUNTING OF LINERS

Part No.	A	B	D	E	F	G	H Socket Head Mounting Screw Metric	Inch
RE-1	11	20	16	8	5	3	M6	.250–20
RE-2	13	20	16	8	6	5	M6	.250–20
RE-3	16	25	20	10	8	5	M8	.312–18
RE-4	18	28	20	12	8	6	M10	.375–16

FOR PROJECTED MOUNTING OF SHOULDER LINERS

Part No.	A	B	D	E	F	G	H Mounting Screw Metric	Inch
RE-10	11	20	16	10	5	5	M6	.250–20
RE-12	12	20	16	14	6	8	M6	.250–20
RE-14	16	25	20	14	8	10	M10	.375–16
RE-15	18	28	20	16	8	10	M10	.375–16

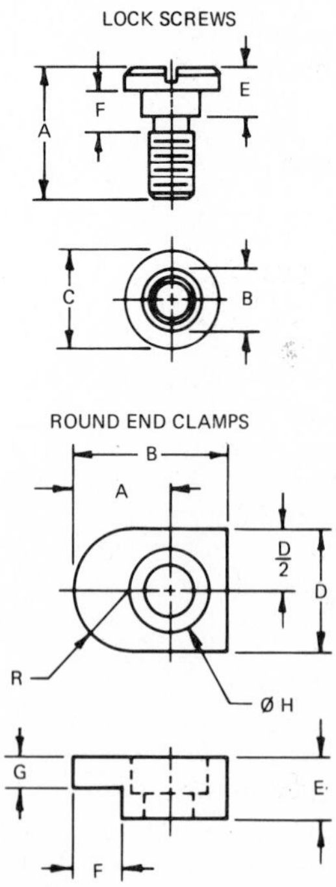
LOCK SCREWS
E
F
A
C
B
ROUND END CLAMPS
B
A
D/2
D
R
ØH
G
E
F

Fig. 26-3-4 Locking devices for renewable bushings. (American Drill Bushing Co.)

INSTALLATION

Multiple Operations. See Fig. 26-3-6*a*. In performing multiple operations such as drilling and reaming, slip renewable bushings of different lengths may be used to obtain the combined advantages of adequate chip removal and precise accuracy. The slip renewable bushing should be short enough to provide proper chip clearance during the drilling operation, while the reamer bushing may be long enough to contact or closely approach the workpiece, thus providing maximum guiding effect during the reaming operation.

Close-Hole Patterns. See Fig. 26-3-6*b*. For many applications requiring close center-to-center placement of bushings, thin-wall and miniature head series will prove helpful; however, for especially difficult close-hole patterns, it may be necessary to grind flats on the bushing

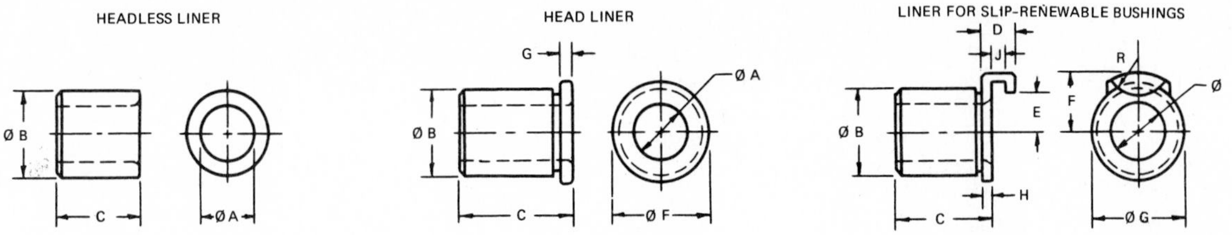
HEADLESS LINER
HEAD LINER
LINER FOR SLIP-RENEWABLE BUSHINGS
Ø B
C
Ø A
G
Ø F
D
J
R
E
F
Ø
H
Ø G

MILLIMETRE SIZES

A ID Size	B OD Size	C 8.0	C 10	C 12	C 20	C 25	C 35	D	E	F	G	H	J	R
5	8	•	•	•	•			5.3	2.8	6.4	9.2	1.5	2.3	4
8	12.5		•	•	•			7.9	4.8	9.6	14.2	2.3	3.3	8
12.5	19.2			•	•	•		7.9	7.9	12.7	20	2.3	3.3	8
20	25.5			•	•	•		9.3	11.2	16.8	27	3	3.3	8
25	35			•	•	•	•	11.1	15.5	22.4	36.6	3	5.1	8

INCH SIZES

A ID Size	B OD Size	C .31	C .38	C .50	C .75	C 1.00	C 1.38	D	E	F	G	H	J	R
.188	.312	•	•	•	•			.21	.11	.25	.36	.06	.09	.16
.312	.500		•	•	•			.31	.19	.38	.56	.09	.13	.31
.500	.750			•	•	•		.31	.31	.50	.81	.09	.13	.31
.750	1.000			•	•	•		.37	.44	.66	1.06	.12	.13	.31
1.000	1.375			•	•	•	•	.44	.61	.88	1.44	.12	.20	.31

MILLIMETRE SIZES

A ID Size	B OD Size	Liner Length (C) 6	8	10	12	16	20	25	35	45	F	G
5	8	•	•	•	•	•	•					
8	12.9	•	•	•	•	•	•	•	•		16	2.5
12.5	19.2		•	•	•		•	•	•	•	22	2.5
20	25				•		•	•	•	•	28	3
25	35				•		•	•	•	•	38	3

INCH SIZES

A ID Size	B OD Size	Liner Length (C) .25	.31	.38	.50	.62	.75	1.00	1.38	1.75	F	G
.188	.314	•	•	•	•	•	•					
.312	.502	•	•	•	•	•	•	•	•		.62	.09
.500	.752		•	•	•		•	•	•	•	.88	.09
.750	1.002				•		•	•	•	•	1.12	.12
1.000	1.337				•		•	•	•	•	1.50	.12

Fig. 26-3-5 Typical sizes for drill bushing liners. (American Drill Bushing Co.)

ODs and/or heads to achieve minimum spacing. When this technique is used, the bushing flats and mounting holes must be accurately machined.

Irregular Work Surfaces. See Fig. 26-3-6*c*. When bushings are adapted to suit applications involving irregular work surfaces, the ends of the bushings should be formed to the contour of the workpiece. In many applications of this nature, the drill point does not enter perpendicular to the work surface and has a tendency to skid or wander. For this reason, the distance between the bushing and the workpiece must be held to a minimum so that the full guiding effect of the bushing can be obtained. The side load exerted by drill in applications of this type is usually concentrated, causing bushing wear. Except in short production runs, the use of fixed renewable bushings simplifies the replacement of worn bushings and facilitates proper orientation of the bushing with respect to the contoured work surface. When press-fit bushings are used, bushing contours should be applied after bushings are installed in the jig plate to ensure proper contour placement with respect to the workpiece.

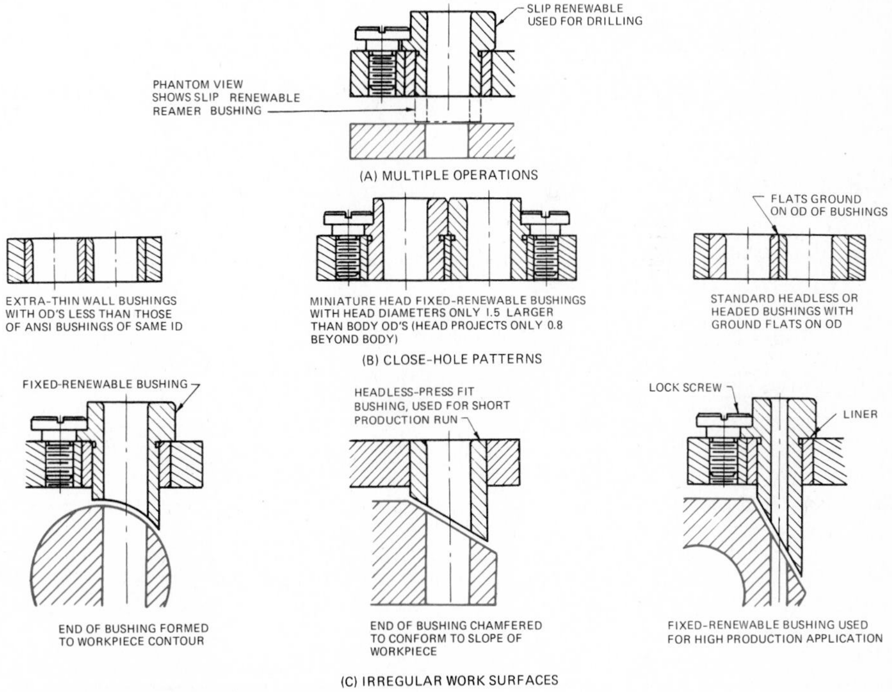
SLIP RENEWABLE USED FOR DRILLING
PHANTOM VIEW SHOWS SLIP RENEWABLE REAMER BUSHING
(A) MULTIPLE OPERATIONS
FLATS GROUND ON OD OF BUSHINGS
EXTRA-THIN WALL BUSHINGS WITH OD'S LESS THAN THOSE OF ANSI BUSHINGS OF SAME ID
MINIATURE HEAD FIXED-RENEWABLE BUSHINGS WITH HEAD DIAMETERS ONLY 1.5 LARGER THAN BODY OD'S (HEAD PROJECTS ONLY 0.8 BEYOND BODY)
STANDARD HEADLESS OR HEADED BUSHINGS WITH GROUND FLATS ON OD
(B) CLOSE-HOLE PATTERNS
FIXED-RENEWABLE BUSHING
HEADLESS-PRESS FIT BUSHING, USED FOR SHORT PRODUCTION RUN
LOCK SCREW
LINER
END OF BUSHING FORMED TO WORKPIECE CONTOUR
END OF BUSHING CHAMFERED TO CONFORM TO SLOPE OF WORKPIECE
FIXED-RENEWABLE BUSHING USED FOR HIGH PRODUCTION APPLICATION
(C) IRREGULAR WORK SURFACES

Fig. 26-3-6 Installation considerations.

DESIGN EXAMPLE

The jig shown in Fig. 26-3-7 is similar in design to the one shown in Fig. 26-2-5 except for the feet and bushings. Holes in the jig plate were provided for the feet, which are purchased items. The bushings are the fixed renewable type, designed for long production runs.

REFERENCE

1. American Drill Bushing Co.

Assignment

On an A3- or B-size sheet, design a jig using slip renewable bushings for one of the parts shown in Figs. 26-3-A to 26-3-D. For the jig for Fig. 26-3-D, only

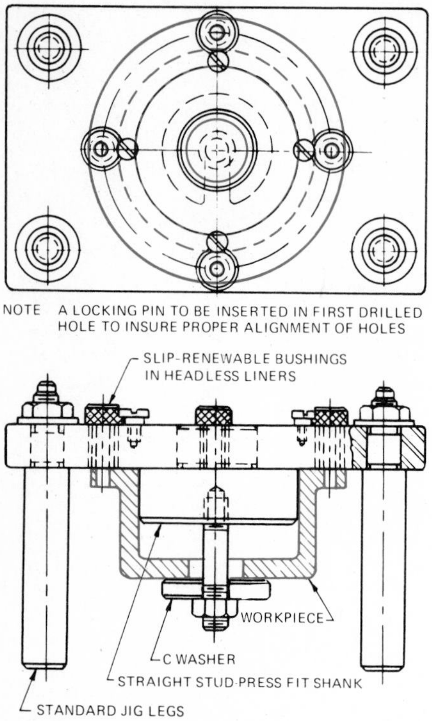
NOTE A LOCKING PIN TO BE INSERTED IN FIRST DRILLED HOLE TO INSURE PROPER ALIGNMENT OF HOLES
SLIP-RENEWABLE BUSHINGS IN HEADLESS LINERS
WORKPIECE
C WASHER
STRAIGHT STUD-PRESS FIT SHANK
STANDARD JIG LEGS

Fig. 26-3-7 Open jig. (Standard Parts Co.)

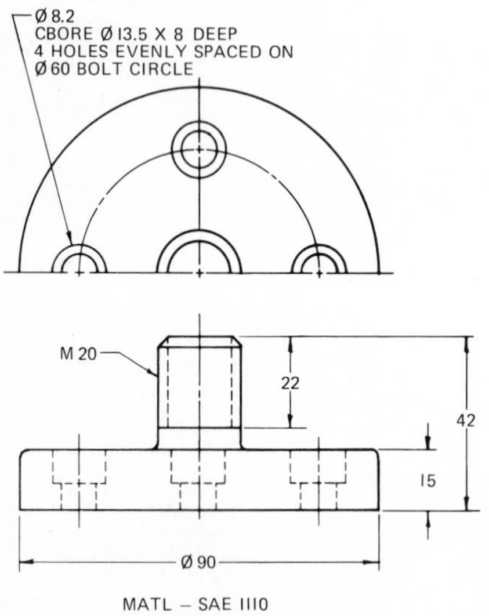
Ø 8.2
CBORE Ø 13.5 X 8 DEEP
4 HOLES EVENLY SPACED ON
Ø 60 BOLT CIRCLE
M 20
22
42
15
Ø 90
MATL – SAE 1110

Fig. 26-3-A End plate.

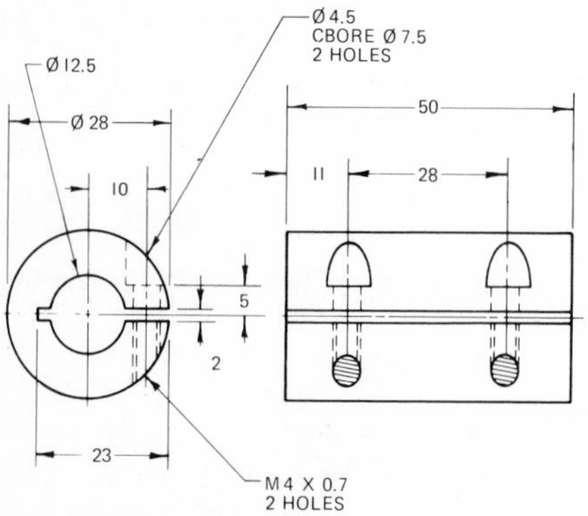
Ø 12.5
Ø 28
10
Ø 4.5
CBORE Ø 7.5
2 HOLES
50
11
28
5
2
23
M4 X 0.7
2 HOLES

Fig. 26-3-C Coupling.

two drilling operations are required—a tap drill hole and a clearance hole for the shank of the cap screw. The clearance hole for the head of the cap screw and the threading of the holes will each be done as a separate operation. Show the drill bushing for the largest hole in the assembly, and show a detail drawing of the slip bushings used for the smallest or smaller holes. Only nominal sizes need to be shown. Scale is 1:1.

REVIEW FOR ASSIGNMENT

Unit 7-5 Assemblies in Section
Unit 26-1 Jig Design
Unit 26-2 Drill Jig Components

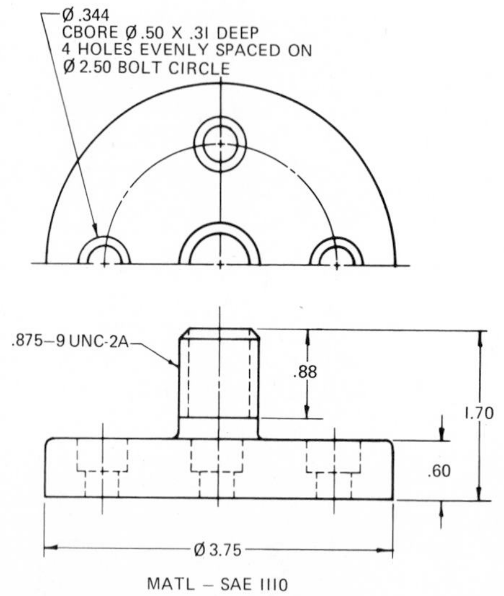
Ø .344
CBORE Ø .50 X .31 DEEP
4 HOLES EVENLY SPACED ON
Ø 2.50 BOLT CIRCLE
.875–9 UNC-2A
.88
1.70
.60
Ø 3.75
MATL – SAE 1110

Fig. 26-3-B End plate.

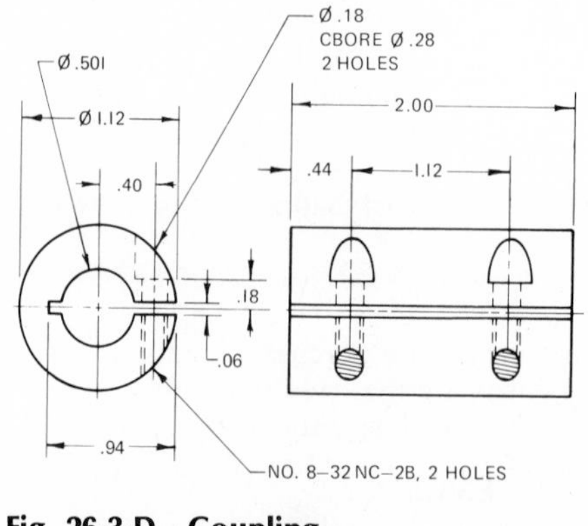
Ø .501
Ø 1.12
.40
Ø .18
CBORE Ø .28
2 HOLES
2.00
.44
1.12
.18
.06
.94
NO. 8–32 NC–2B, 2 HOLES

Fig. 26-3-D Coupling.

UNIT 26-4

DIMENSIONING JIG DRAWINGS

The finished detail drawing for the simple plate jig (Fig. 26-1-3) is shown in Fig. 26-4-1. A brief explanation of why the various dimensions were chosen is as follows.

1. *Distance between Holes.* The dimension between the ϕ 7.5 holes on the workpiece is 39.9–40.0. Therefore, the tolerance allowed on this dimension is 0.1 mm. It stands to reason that the distance between the drill bushings on the jig plate must be kept to a closer tolerance because of bushing wear. A tolerance of 0.04 was chosen for the center distance between the bushings, and the limits were placed midway between the workpiece limits. Thus the center-to-center distance was established at 39.93–39.97.

2. *Size of Bushing Holes.* In tool design it is general practice to show on the drawing only a note listing the nominal diameter of the hole and the part number of the mating part. It is the job of the machinist to select the proper diameters to ensure a press fit.

3. *Size of Dowel Pinholes.* Dowels are commercially available, at low cost, in a wide range of standard sizes. Standard commercial dowels are finished to 0.006 mm larger than the nominal diameter with a tolerance of $\pm$0.003 mm. The size of the dowel pins used is ϕ 6 or ϕ 6.002–6.009 × 20 long. One end of the dowel pin has a chamfer to facilitate pressing it into the jig plate and to permit easy loading onto the workpiece.

As mentioned above, a note calling for the nominal diameter and the part number of the mating part covers all the information required.

4. *Center Distance between Dowels.* Since the width and length of the workpiece are shown in nominal dimensions, the tolerance permitted on these dimensions is $\pm$0.05. Thus the size of the largest workpiece which would be permissible is 40.05 × 64.05. These are the workpiece sizes which are used in calculating the center-to-center distance between dowel pins. A clearance of 0.02 to 0.06 was decided on between the maximum workpiece size and the largest dowel pin. Thus the center-to-center distances between dowels were calculated to be 70.07–70.11 and 46.07–46.11.

5. *Center Distance, Bushing and Dowel.* The maximum limits were calculated by taking half the difference between the maximum limits of 70.11 and 39.97 for length and half the limit 46.11

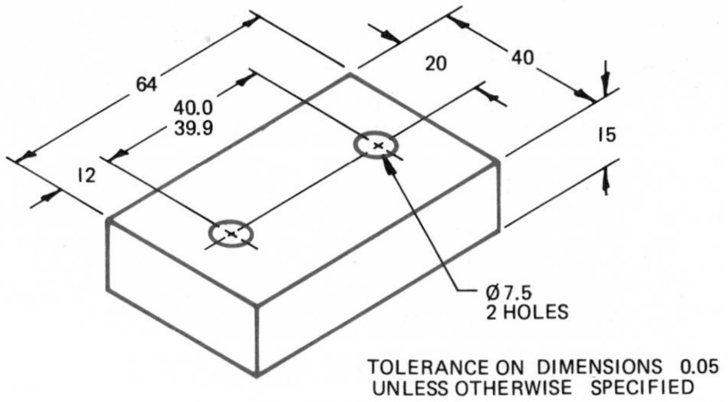
64
40.0
39.9
20
40
15
12
Ø7.5
2 HOLES
TOLERANCE ON DIMENSIONS 0.05
UNLESS OTHERWISE SPECIFIED

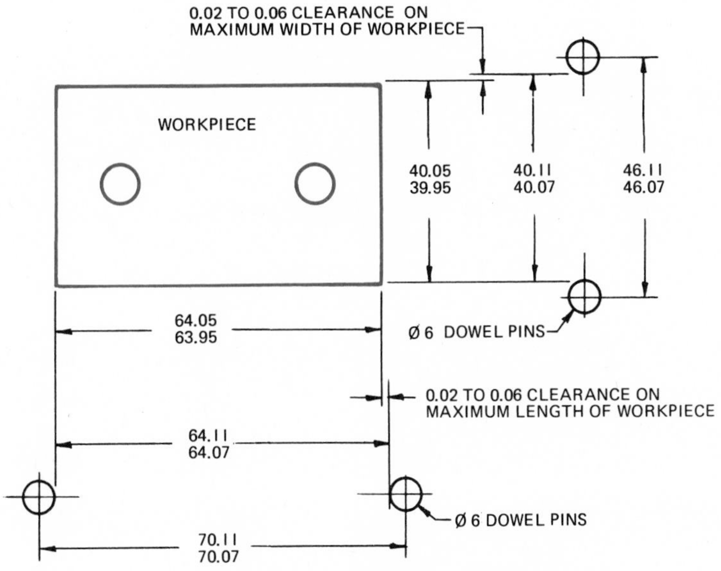
0.02 TO 0.06 CLEARANCE ON
MAXIMUM WIDTH OF WORKPIECE
WORKPIECE
40.05
39.95
40.11
40.07
46.11
46.07
64.05
63.95
Ø 6 DOWEL PINS
0.02 TO 0.06 CLEARANCE ON
MAXIMUM LENGTH OF WORKPIECE
64.11
64.07
Ø 6 DOWEL PINS
70.11
70.07

(A) CALCULATING DISTANCES BETWEEN DOWEL PINS

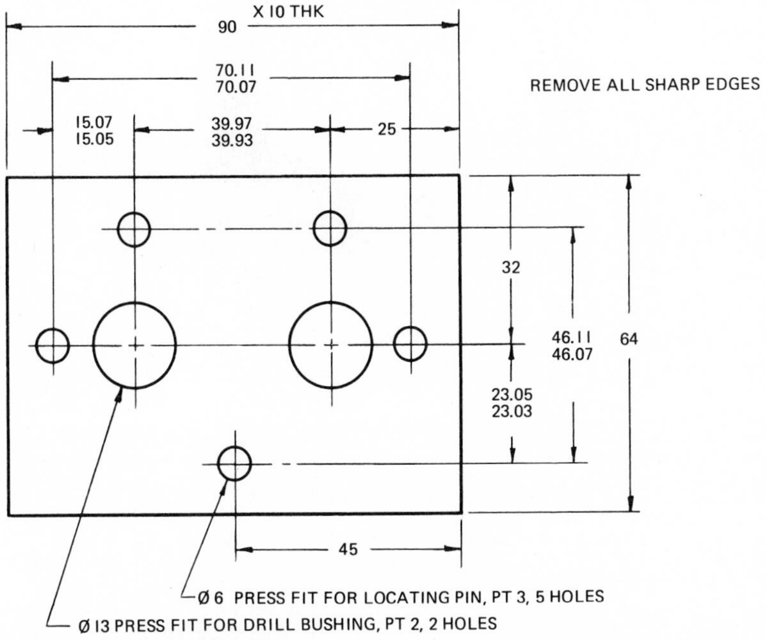
X 10 THK
90
70.11
70.07
REMOVE ALL SHARP EDGES
15.07
15.05
39.97
39.93
25
32
46.11
46.07
64
23.05
23.03
45
Ø 6 PRESS FIT FOR LOCATING PIN, PT 3, 5 HOLES
Ø 13 PRESS FIT FOR DRILL BUSHING, PT 2, 2 HOLES

(B) DIMENSIONING JIG PLATE SHOWN IN FIGURE 27-1-3

Fig. 26-4-1 Dimensioning jig drawings.

(rounded off to 23.05) for width. An allowance of −0.02 was given to these dimensions.

Assignment

On an A3- or B-size sheet, design a simple plate jig for drilling the holes in one of the parts shown in Figs. 26-4-A to 26-4-D. The size of the dowel pins used in the design is 6.006 ± 0.003. After the overall design has been approved by your instructor, dimension the jig plate as per procedures outlined in this unit. Scale is 1:1.

REVIEW FOR ASSIGNMENT

Unit 26-1 Jig Design
Unit 26-2 Drill Jig Components
Unit 26-3 Drill Bushing

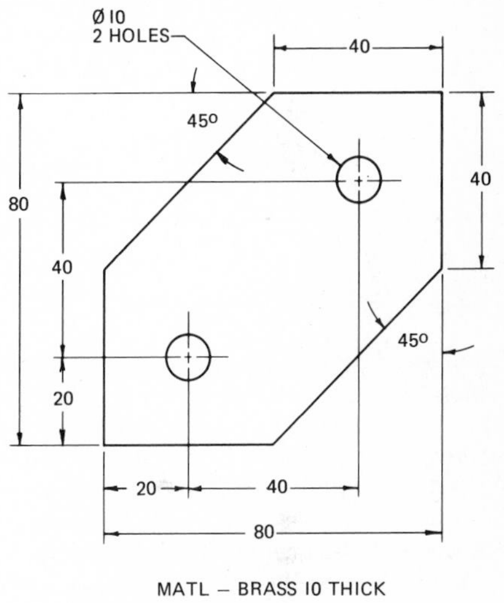
Ø 10
2 HOLES
40
45°
40
80
40
45°
20
20
40
80
MATL – BRASS 10 THICK

Fig. 26-4-A Locking plate.

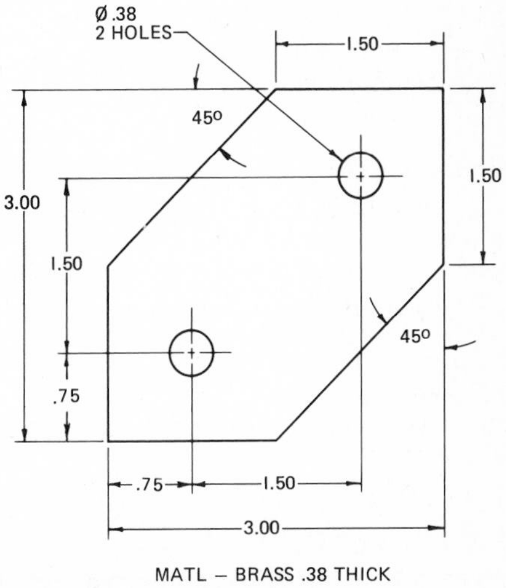
Ø .38
2 HOLES
1.50
45°
1.50
3.00
1.50
45°
.75
.75
1.50
3.00
MATL – BRASS .38 THICK

Fig. 26-4-B Locking plate.

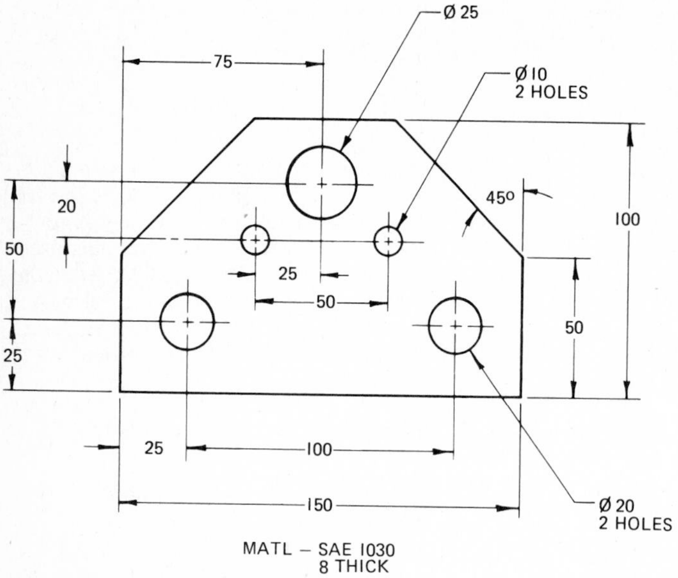
Ø 25
75
Ø 10
2 HOLES
45°
20
100
50
25
50
50
25
25
100
150
Ø 20
2 HOLES
MATL – SAE 1030
8 THICK

Fig. 26-4-C Trolley slide plate.

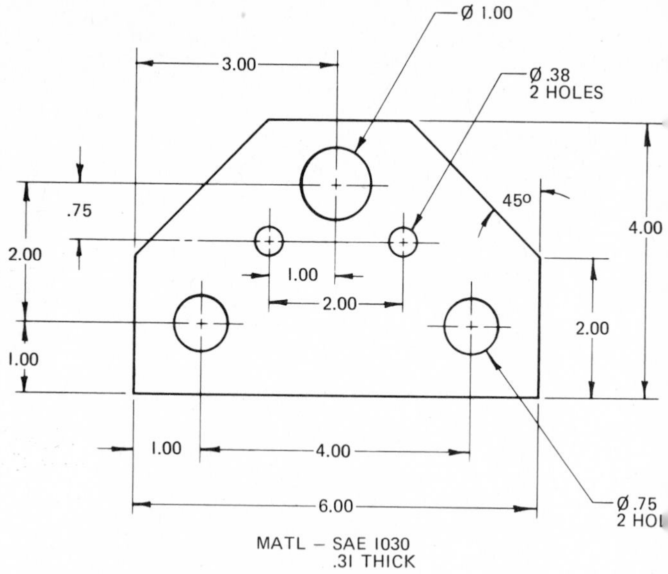
Ø 1.00
3.00
Ø .38
2 HOLES
45°
.75
4.00
2.00
1.00
2.00
2.00
1.00
1.00
4.00
6.00
Ø .75
2 HOLES
MATL – SAE 1030
.31 THICK

Fig. 26-4-D Trolley slide plate.

UNIT 26-5
CLOSED JIGS

The closed or box type of drill jig, within which the work is clamped, is usually used when holes have to be drilled in several directions. To firmly support the jig, sets of supporting legs or feet must be provided on the side of the box opposite each of the drilling faces. The jig is normally opened by swinging back a leaf or cover. The part to be drilled is placed within the box and accurately located and clamped with devices which are, as a rule, permanently attached to the jig body. The jig body frame, shown in Figs. 26-5-1 and 26-5-2, is typical of the type of jig body readily purchased. These are available in a wide variety of sizes, with or without the pop-up leaf.

DESIGN EXAMPLE

The box-type jig shown in Fig. 26-5-3 is designed to permit drilling from two sides. The complete design is made up of standard parts, and the job of the toolroom is assembling rather than manufacturing and assembling. The use of adjustable stops reduces machining time and permits adjustment and reuse at a later date.

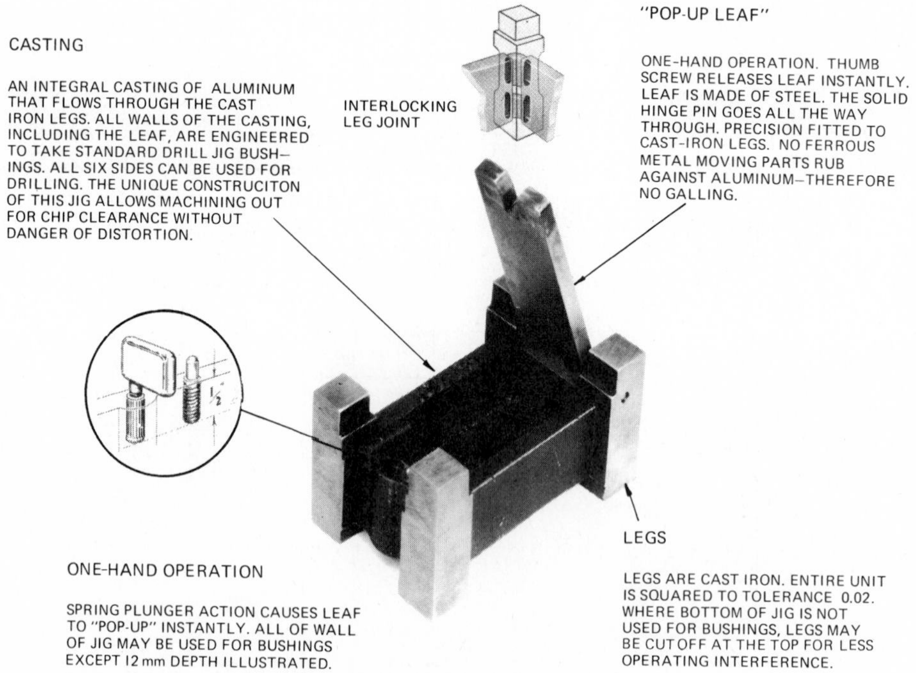
CASTING
AN INTEGRAL CASTING OF ALUMINUM THAT FLOWS THROUGH THE CAST IRON LEGS. ALL WALLS OF THE CASTING, INCLUDING THE LEAF, ARE ENGINEERED TO TAKE STANDARD DRILL JIG BUSHINGS. ALL SIX SIDES CAN BE USED FOR DRILLING. THE UNIQUE CONSTRUCITON OF THIS JIG ALLOWS MACHINING OUT FOR CHIP CLEARANCE WITHOUT DANGER OF DISTORTION.
INTERLOCKING LEG JOINT
"POP-UP LEAF"
ONE-HAND OPERATION. THUMB SCREW RELEASES LEAF INSTANTLY. LEAF IS MADE OF STEEL. THE SOLID HINGE PIN GOES ALL THE WAY THROUGH. PRECISION FITTED TO CAST-IRON LEGS. NO FERROUS METAL MOVING PARTS RUB AGAINST ALUMINUM–THEREFORE NO GALLING.
LEGS
ONE-HAND OPERATION
SPRING PLUNGER ACTION CAUSES LEAF TO "POP-UP" INSTANTLY. ALL OF WALL OF JIG MAY BE USED FOR BUSHINGS EXCEPT 12 mm DEPTH ILLUSTRATED.
LEGS ARE CAST IRON. ENTIRE UNIT IS SQUARED TO TOLERANCE 0.02. WHERE BOTTOM OF JIG IS NOT USED FOR BUSHINGS, LEGS MAY BE CUTOFF AT THE TOP FOR LESS OPERATING INTERFERENCE.

Fig. 26-5-1 Tumble box-type jig body. (Standard Parts Co.)

Assignment

On an A3- or B-size sheet, design a box-type jig for drilling all the holes in one of the parts shown in Figs. 26-5-A to 26-5-D. Select a suitable tumble-box jig from Fig. 26-5-2. Use standard components wherever possible. Scale is 1:1.

REVIEW FOR ASSIGNMENT

Unit 26-1 Jig Design
Unit 26-2 Drill Jig Components
Unit 26-3 Drill Bushings

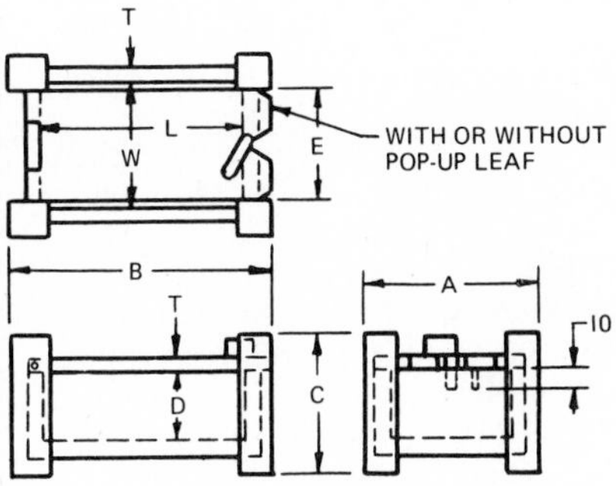
T
L
W
E
WITH OR WITHOUT POP-UP LEAF
B
A
T
D
C
10

DIMENSIONS IN MILLIMETRES

Nominal Capacity	Inside Dim W	Inside Dim L	Inside Dim D	T	E	Outside Dim A	Outside Dim B	Outside Dim C
25 x 50 x 25 25 x 100 x 25	40	60 110	25	8	37	75	95 145	65
50 x 50 x 40 50 x 50 x 60	65	60	40 60	8	62	100	95	80 105
50 x 75 x 40 50 x 75 x 60	65	85	40 60	8	62	100	120	80 105
50 x 100 x 40 50 x 100 x 60	65	110	40 60	8	62	100	145	80 105
75 x 75 x 50 75 x 75 x 75	90	85	50 75	10	87	125	125	95 120
75 x 100 x 50 75 x 100 x 75	90	110	50 75	10	87	125	150	95 120
75 x 125 x 50 75 x 125 x 75	90	135	50 75	10	87	125	175	95 120
75 x 150 x 50 75 x 150 x 75	90	160	50 75	10	87	125	200	95 120
100 x 100 x 50 100 x 100 x 100	115	110	50 100	12	112	160	155	100 150
100 x 125 x 50 100 x 125 x 100	115	135	50 100	12	112	160	180	100 150
100 x 150 x 50 100 x 150 x 100	115	160	50 100	12	112	160	205	100 150
100 x 175 x 50 100 x 175 x 100	115	185	50 100	12	112	160	230	100 150
100 x 200 x 50 100 x 200 x 100	115	210	50 100	12	112	160	255	100 150

DIMENSIONS IN INCHES

Nominal Capacity	Inside Dim W	Inside Dim L	Inside Dim D	T	E	Outside Dim A	Outside Dim B	Outside Dim C
1 x 2 x 1 1 x 4 x 1	1.60	2.50 4.50	1.00	.30	1.50	2.90	3.75 5.75	2.60
2 x 2 x 1.5 2 x 2 x 2.5	2.60	2.50	1.50 2.50	.30	2.50	3.90	3.75	3.10 4.10
2 x 3 x 1.5 2 x 3 x 2.5	2.60	3.50	1.50 2.50	.30	2.50	3.90	4.75	3.10 4.10
2 x 4 x 1.5 2 x 4 x 2.5	2.60	4.50	1.50 2.50	.30	2.50	3.90	5.75	3.10 4.10
3 x 3 x 2 3 x 3 x 3	3.60	3.50	2.00 3.00	.40	3.50	5.00	4.90	3.75 4.75
3 x 4 x 2 3 x 4 x 3	3.60	4.50	2.00 3.00	.40	3.50	5.00	5.90	3.75 4.75
3 x 5 x 2 3 x 5 x 3	3.60	5.50	2.00 3.00	.40	3.50	5.00	6.90	3.75 4.75
3 x 6 x 2 3 x 6 x 3	3.60	6.50	2.00 3.00	.40	3.50	5.00	7.90	3.75 4.75
4 x 4 x 2 4 x 4 x 4	4.60	4.50	2.00 4.00	.50	4.50	6.25	6.10	4.00 6.00
4 x 5 x 2 4 x 5 x 4	4.60	5.50	2.00 4.00	.50	4.50	6.25	7.10	4.00 6.00
4 x 6 x 2 4 x 6 x 4	4.60	6.50	2.00 4.00	.50	4.50	6.25	8.10	4.00 6.00
4 x 7 x 2 4 x 7 x 4	4.60	7.50	2.00 4.00	.50	4.50	6.25	9.10	4.00 6.00
4 x 8 x 2 4 x 8 x 4	4.60	8.50	2.00 4.00	.50	4.50	6.25	10.10	4.00 6.00

Fig. 26-5-2 Nominal sizes for tumble-box jig body. (Standard Parts Co.)

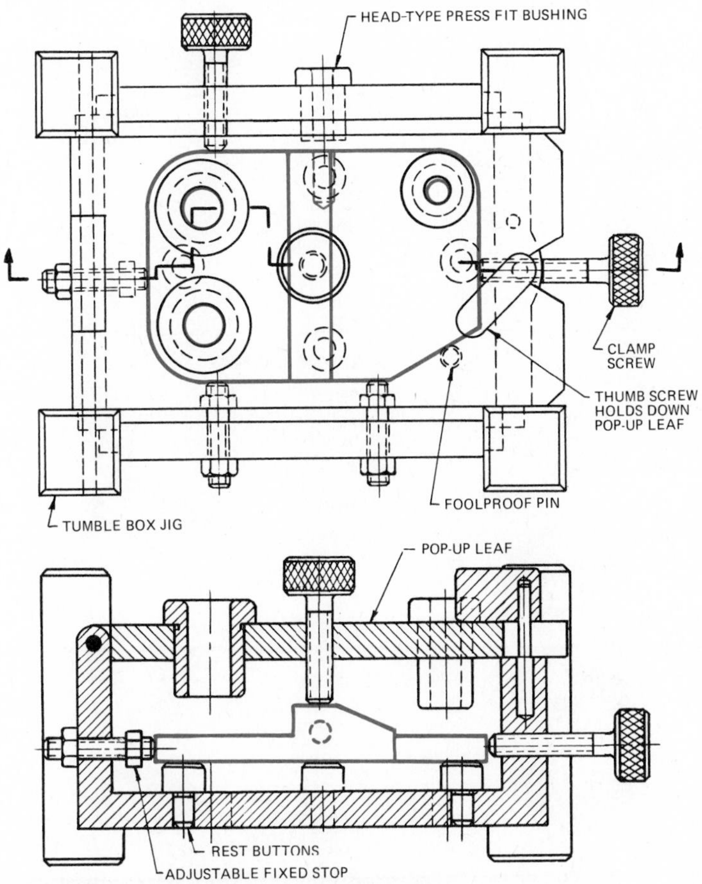
HEAD-TYPE PRESS FIT BUSHING
CLAMP SCREW
THUMB SCREW HOLDS DOWN POP-UP LEAF
FOOLPROOF PIN
TUMBLE BOX JIG
POP-UP LEAF
REST BUTTONS
ADJUSTABLE FIXED STOP

Fig. 26-5-3 Closed or box-type drill jig.

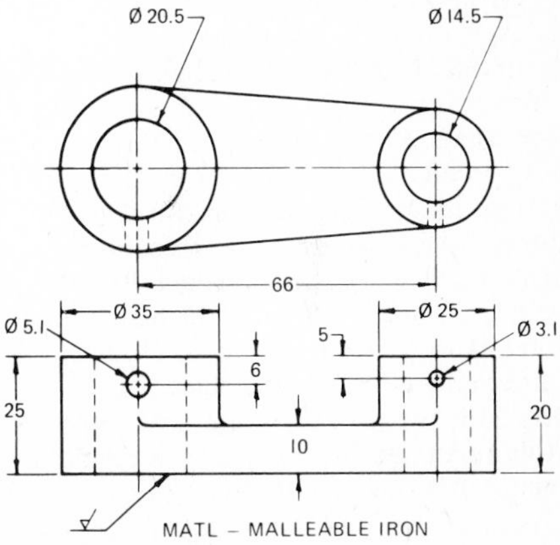
Ø 20.5
Ø 14.5
66
Ø 5.1
Ø 35
Ø 25
Ø 3.1
5
6
25
20
10
MATL – MALLEABLE IRON

Fig. 26-5-A Link.

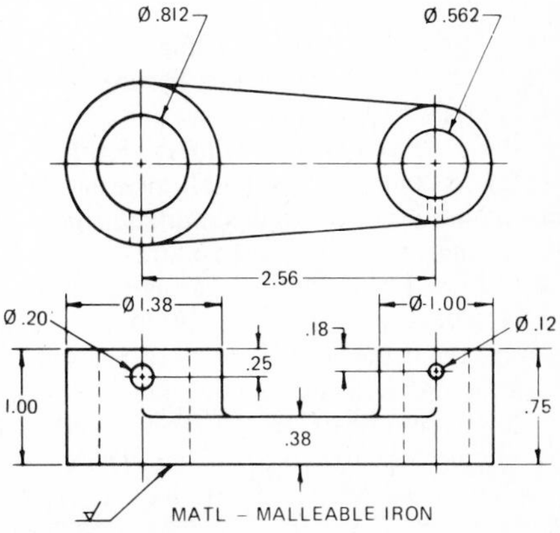
Ø .812
Ø .562
2.56
Ø .20
Ø 1.38
Ø 1.00
Ø .12
.18
.25
1.00
.75
.38
MATL – MALLEABLE IRON

Fig. 26-5-B Link.

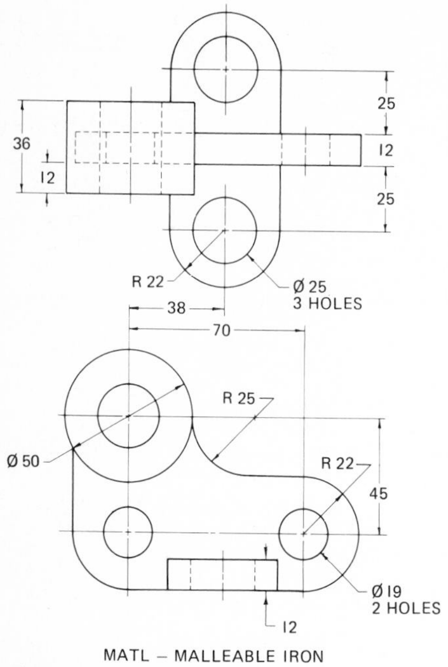

25
36
12
12
25
R 22
Ø 25
3 HOLES
38
70
R 25
Ø 50
R 22
45
Ø 19
2 HOLES
12
MATL – MALLEABLE IRON

Fig. 26-5-C Swivel hanger.

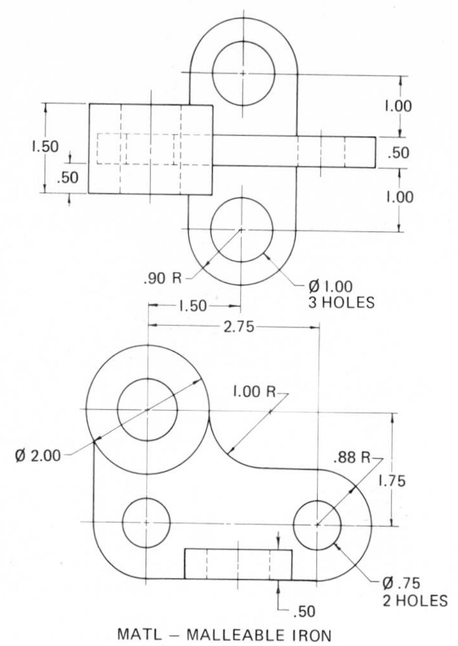

1.00
1.50
.50
.50
1.00
.90 R
Ø 1.00
3 HOLES
1.50
2.75
1.00 R
Ø 2.00
.88 R
1.75
Ø .75
2 HOLES
.50
MATL – MALLEABLE IRON

Fig. 26-5-D Swivel hanger.

Complex Contours. In applications involving complex contours, the foremost advantage of plastic tooling is the ability of the material to conform to a given shape. The methods most commonly used in the fabrication of plastic jigs and fixtures are based on the use of liquid plastic or resin-impregnated glass cloth which is cast or laminated around a master tool. See Fig. 26-6-1. The master tool may be an actual production part or an exact-scale model, usually made of wood, plaster, plastic, or metal. Either method faithfully reproduces the contours of the master tool without complicated and costly machining operations. Tool details, such as drill jig bushings shown in Fig. 26-6-2 and liners, are either embedded at the time of fabrication or potted in place after the tool has cured.

Duplicate Tooling. For industries which must perform simultaneous manufacture of identical parts on several production lines or in separate facilities, the use of plastic techniques permits quantity duplication of a master tool at a low unit cost and with good accuracy.

UNIT 26-6

PLASTIC TOOLING

The expanding use of plastic materials in the fabrication of production tooling has aroused general interest among members of the tooling profession. Continuing improvements in methods and materials have spread the concepts of plastic tooling from its origin in the aircraft and automotive industries to its present-day acceptance by industry in general. As presently foreseen, the role of plastic tooling is that of a supplement to standard metal tooling, rather than replacement. However, the definite advantages of this relatively new medium in certain applications should not be overlooked by the progressive tool designer.

APPLICATIONS FOR PLASTIC TOOLING

The versatility of plastic tooling has proved equally advantageous in very long or very short production runs, in extremely complex or extremely simple production tools, and in highly diversified or highly specialized manufacturing plants. The use of jigs and fixtures fabricated from plastic materials has received wide acceptance in these three general areas: the manufacture of parts having complex contours, the development of prototypes, and the fabrication of duplicate tooling. In addition, plastic counterparts for metal tooling have found specific applications in virtually every industry engaged in manufacturing.

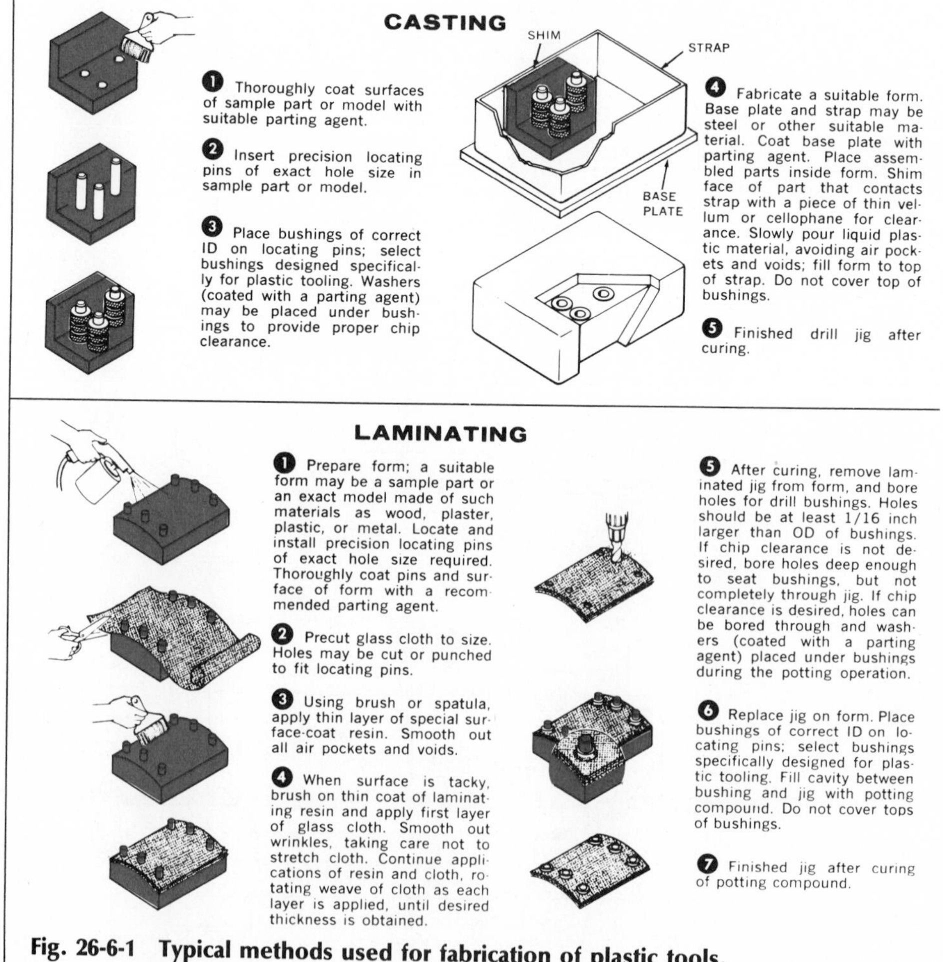

CASTING
1 Thoroughly coat surfaces of sample part or model with suitable parting agent.
2 Insert precision locating pins of exact hole size in sample part or model.
3 Place bushings of correct ID on locating pins; select bushings designed specifically for plastic tooling. Washers (coated with a parting agent) may be placed under bushings to provide proper chip clearance.
SHIM
STRAP
BASE PLATE
4 Fabricate a suitable form. Base plate and strap may be steel or other suitable material. Coat base plate with parting agent. Place assembled parts inside form. Shim face of part that contacts strap with a piece of thin vellum or cellophane for clearance. Slowly pour liquid plastic material, avoiding air pockets and voids; fill form to top of strap. Do not cover top of bushings.
5 Finished drill jig after curing.
LAMINATING
1 Prepare form; a suitable form may be a sample part or an exact model made of such materials as wood, plaster, plastic, or metal. Locate and install precision locating pins of exact hole size required. Thoroughly coat pins and surface of form with a recommended parting agent.
2 Precut glass cloth to size. Holes may be cut or punched to fit locating pins.
3 Using brush or spatula, apply thin layer of special surface-coat resin. Smooth out all air pockets and voids.
4 When surface is tacky, brush on thin coat of laminating resin and apply first layer of glass cloth. Smooth out wrinkles, taking care not to stretch cloth. Continue applications of resin and cloth, rotating weave of cloth as each layer is applied, until desired thickness is obtained.
5 After curing, remove laminated jig from form, and bore holes for drill bushings. Holes should be at least 1/16 inch larger than OD of bushings. If chip clearance is not desired, bore holes deep enough to seat bushings, but not completely through jig. If chip clearance is desired, holes can be bored through and washers (coated with a parting agent) placed under bushings during the potting operation.
6 Replace jig on form. Place bushings of correct ID on locating pins; select bushings specifically designed for plastic tooling. Fill cavity between bushing and jig with potting compound. Do not cover tops of bushings.
7 Finished jig after curing of potting compound.

Fig. 26-6-1 Typical methods used for fabrication of plastic tools. (American Drill Bushing Co.)

EMBEDMENT BUSHINGS HAVE SURFACES THAT CAN BE GRIPPED BY SOFT MATERIALS TO PREVENT SPINOUT.

Fig. 26-6-2 Drill bushing.

Embedment Bushings. Jigs made of laminated glass fiber, potting compounds, and castable and soft metals require special bushings that are designed to prevent spinout. Many such bushing shapes are offered to grip soft tooling.

SOURCE MATERIAL

- American Drill Bushing Co.

Assignment

On an A3- or B-size sheet, prepare a drawing of a drill jig fabricated from plastic materials. The jig is to be designed to drill the holes in one of the parts shown in Figs. 26-6-A to 26-6-D. Scale is 1:1.

REVIEW FOR ASSIGNMENT

Unit 7-5 Assemblies in Section

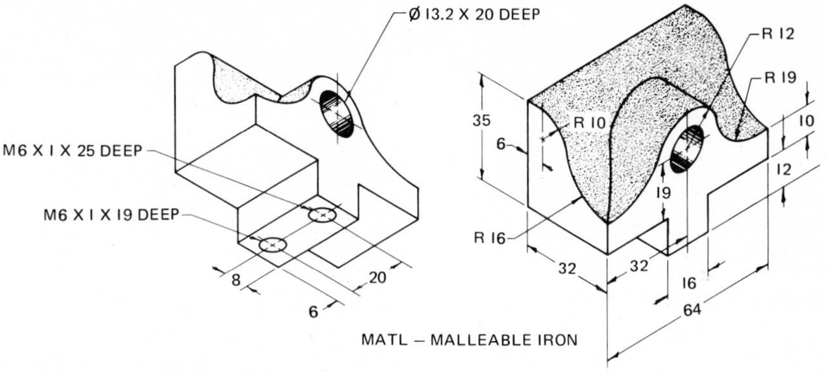

Ø 13.2 X 20 DEEP
M6 X 1 X 25 DEEP
M6 X 1 X 19 DEEP
R 12
R 19
R 10
R 16
35
6
10
12
19
8
6
20
32
32
16
64
MATL – MALLEABLE IRON

Fig. 26-6-C Movable jaw.

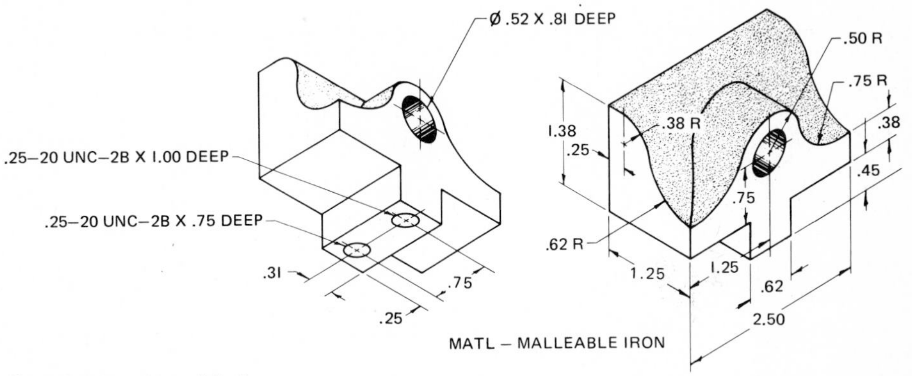

Ø .52 X .81 DEEP
.25–20 UNC–2B X 1.00 DEEP
.25–20 UNC–2B X .75 DEEP
.50 R
.75 R
.38 R
.62 R
1.38
.25
.38
.45
.75
.31
.75
.25
1.25
1.25
.62
2.50
MATL – MALLEABLE IRON

Fig. 26-6-D Movable jaw.

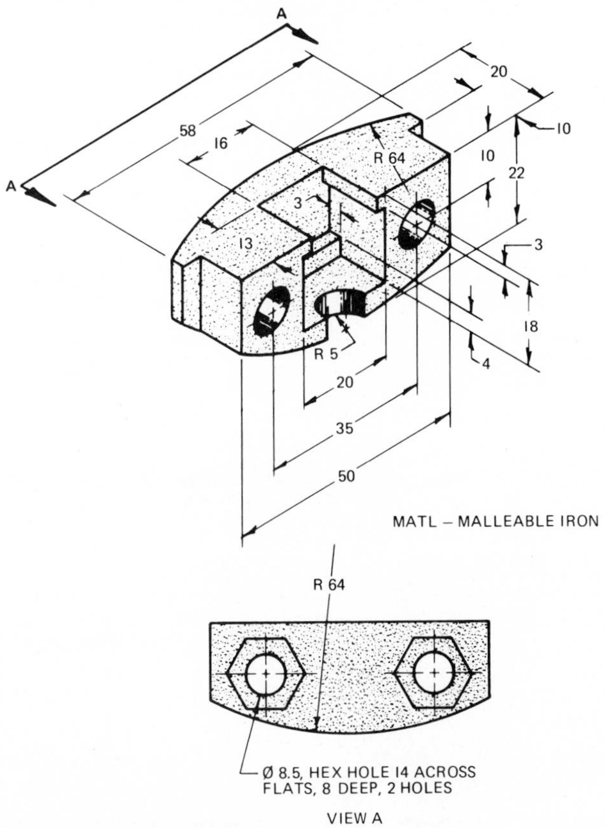

A
A
58
16
20
10
10
22
R 64
3
13
3
18
4
R 5
20
35
50
MATL – MALLEABLE IRON
R 64
Ø 8.5, HEX HOLE 14 ACROSS FLATS, 8 DEEP, 2 HOLES
VIEW A

Fig. 26-6-A Vise guide.

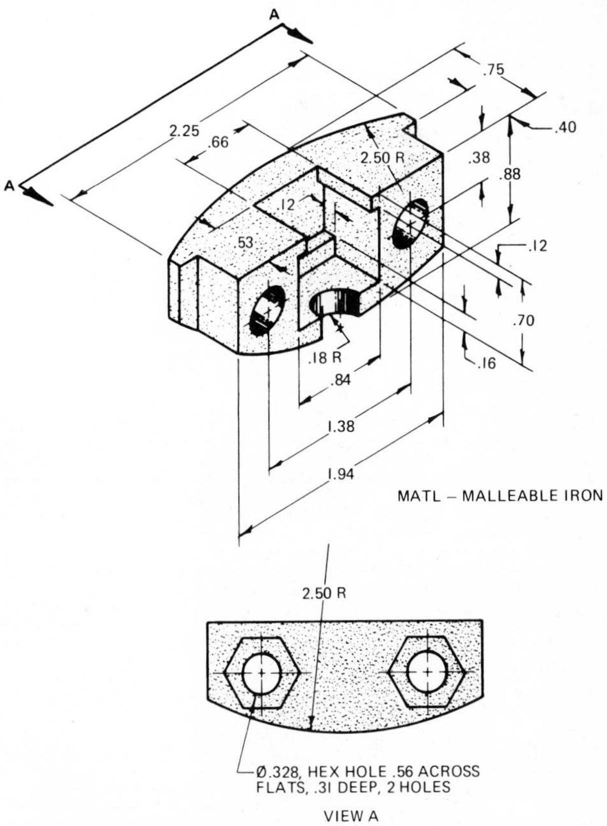

A
A
2.25
.66
.75
.40
.38
.88
2.50 R
.12
.53
.12
.70
.16
.18 R
.84
1.38
1.94
MATL – MALLEABLE IRON
2.50 R
Ø.328, HEX HOLE .56 ACROSS FLATS, .31 DEEP, 2 HOLES
VIEW A

Fig. 26-6-B Vise guide.

UNIT 26-7
FIXTURES

A *fixture* is a device which supports, locates, and holds a workpiece securely in position while machining operations are being performed. It should be noted that the accuracy of the machining depends on the quality of the machine and tools used.

MILLING FIXTURES

The most common type of fixture used is the milling fixture. See Figs. 26-7-1 and 26-7-2. It may be clamped to the milling machine table or held in the milling machine vise. Before a milling fixture is designed, information such as the size and spacing of the T slots, cross-feed, and horizontal traverse of the table must be known. Most drafting offices have this information tabulated in the form of a chart, and the designer may select the most suitable milling machine.

In laying out a fixture, the drawing should be checked to see that no part of the fixture will interfere with the milling arbor or arbor supports. The standard practice of many designers is to show the cutter and arbor on the assembly drawing. A drawing of the cutter and arbor should be done before the clamping is designed, because the necessary clearances may alter the original type of clamping the designer had in mind.

Fixture Components

FIXTURE BASE

Fixture components and the workpiece are usually located on a base, which is securely fastened to the milling machine table with clamping lugs or slots. See Fig. 26-7-3. The size of the lug opening, usually 18 or 20 mm wide, corresponds to the T slot width on the milling machine table. In addition, the base is usually provided with keys or tongues which sit on the table T slots, aligning the fixture so that the workpiece is perpendicular to the cutter arbor axis and parallel to the sides of the cutter. Standard fixture bases in a wide variety of sizes are at the disposal of the designer.

CLAMPS

In milling fixture design, forces resulting from the feed of the table and the rotation of the cutter are encountered. These forces are normally counteracted by the clamp forces. For this reason, fixture clamps must be of heavier design than jig clamps and must be properly located. See Figs. 26-7-4 and 26-7-5. On fixtures where a very large force is encountered, the base rather than the clamps should be designed to take these forces.

SET BLOCKS

Cutter set blocks are mounted on the fixture to properly position the milling cutter in relation to the workpiece. See Figs. 26-7-6 and 26-7-7. The locating surfaces of the set blocks are offset from the finished surfaces on the workpiece that are to be machined. Feeler gages the same thickness as the offset are placed on the located surfaces of the set block, and the position of the milling fixture is adjusted until the cutter touches the feeler gage. The space between the cutter and set block ensures clearance between the cutter and set block during the machining operation. Set blocks are normally fastened to the fixture body with cap screws and dowel pins.

SEQUENCE IN LAYING OUT A FIXTURE

See Figs. 26-7-8 and 26-7-9. The following sequence is recommended in laying out a fixture:

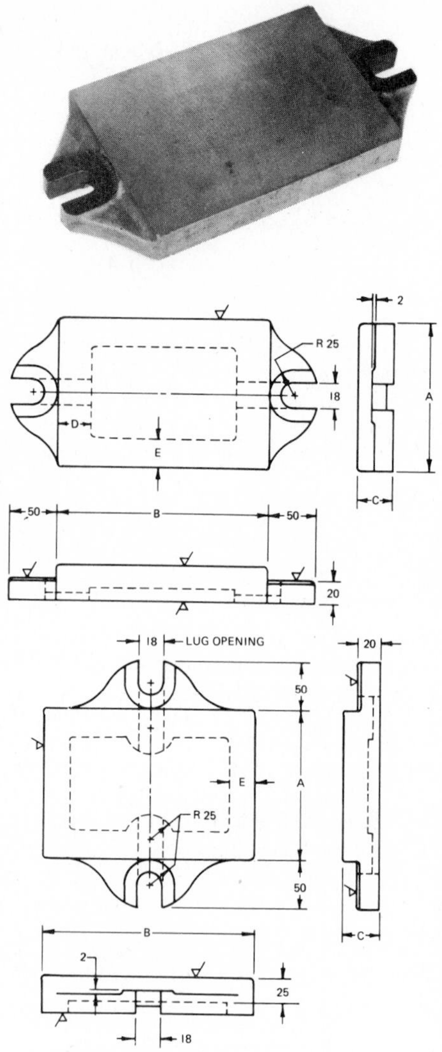
R 25
18
A
C
D
E
50
B
50
20
18
LUG OPENING
20
50
E
A
R 25
50
C
B
2
25
18

A	B	C	D	E
150	225	30	30	30
200	300	35	30	30
250	375	40	35	35

Fig. 26-7-3 Milling fixture base. (Standard Parts Co.)

Fig. 26-7-1 Vertical milling fixture for slotting a hole clamped in a milling vise. (Ex-Cell-O Corp. of Canada Ltd.)

Fig. 26-7-2 Typical milling fixture. (Standard Parts Co.)

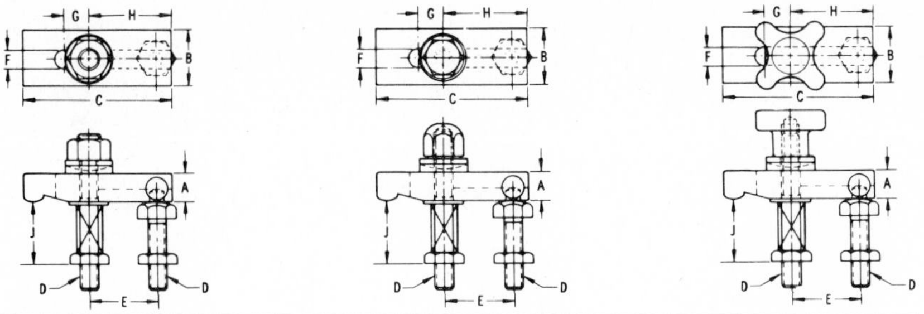

G
H
F
B
C
A
J
D
E

Clamp Assembly	A	B	C	Thread D	E	F	Travel G	H	Capacity J
20005	10	16	50	M6	24	7	12	25	22
20010	12	25	82	M8	32	8	46	38	28
20020	16	32	100	M10	48	35	25	58	42
20030	16	32	125	M12	48	13	38	58	42
20040	22	38	125	M16	50	16	38	64	45
20050	25	45	165	M20	66	20	45	82	50

Fig. 26-7-4 Strap clamp assemblies. (American Drill Bushing Co.)

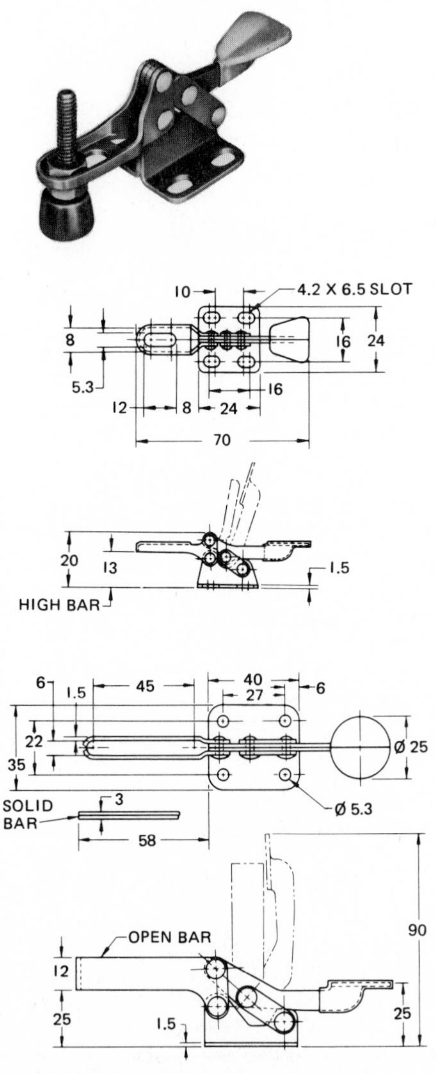

4.2 X 6.5 SLOT
HIGH BAR
SOLID BAR
OPEN BAR
Ø 25
Ø 5.3

Fig. 26-7-5 Toggle clamps.

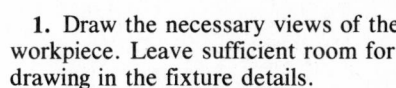

1. Draw the necessary views of the workpiece. Leave sufficient room for drawing in the fixture details.
2. Draw locating devices.
3. Draw the cutter and arbor.
4. Draw the clamping arrangement.
5. Draw the set blocks if required.
6. Draw the fixture base and keying arrangement.

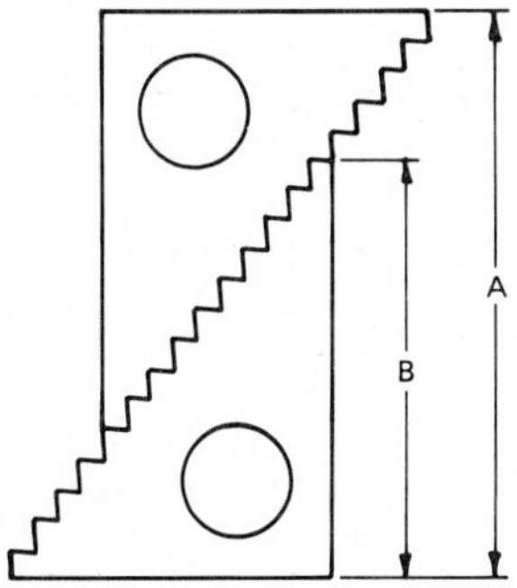

A
B

WIDTH	HEIGHT B	CAPACITY A
25	30	20–40
35	45	30–60
45	90	60–150

STEP BLOCK

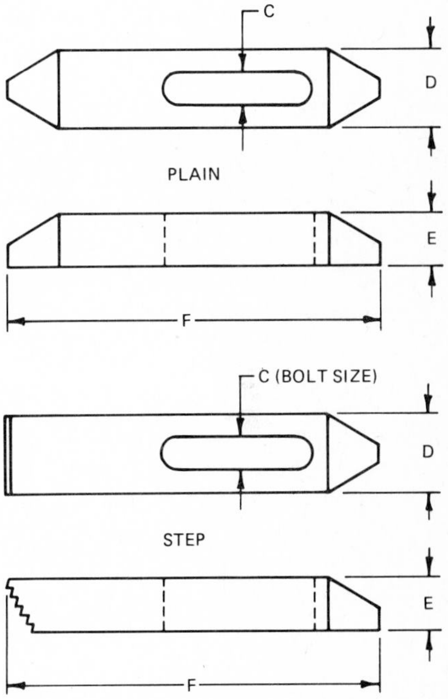

C
D
PLAIN
E
F
C (BOLT SIZE)
D
STEP
E
F

PART	C	D	E	F
A-930	10 TO 12	25	12	60
A-940	12	30	20	150
A-950	18	30	25	150
A-960	20	30	30	200
A-970	25	50	35	250

CLAMP STRAPS

Fig. 26-7-6 Holding components. (American Drill Bushing Co.)

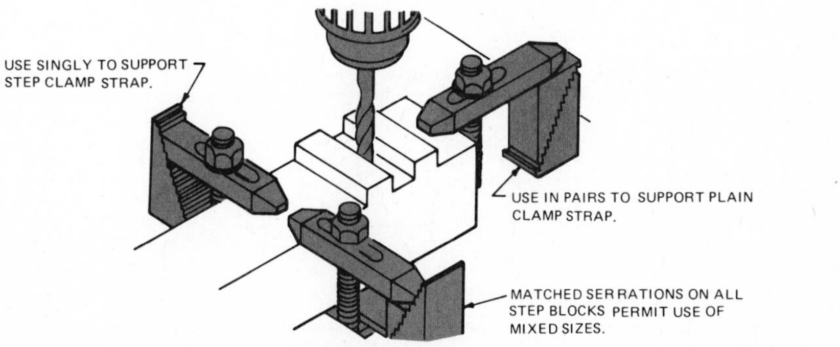

USE SINGLY TO SUPPORT STEP CLAMP STRAP.
USE IN PAIRS TO SUPPORT PLAIN CLAMP STRAP.
MATCHED SERRATIONS ON ALL STEP BLOCKS PERMIT USE OF MIXED SIZES.

Fig. 26-7-7 Application of holding components. (American Drill Bushing Co.)

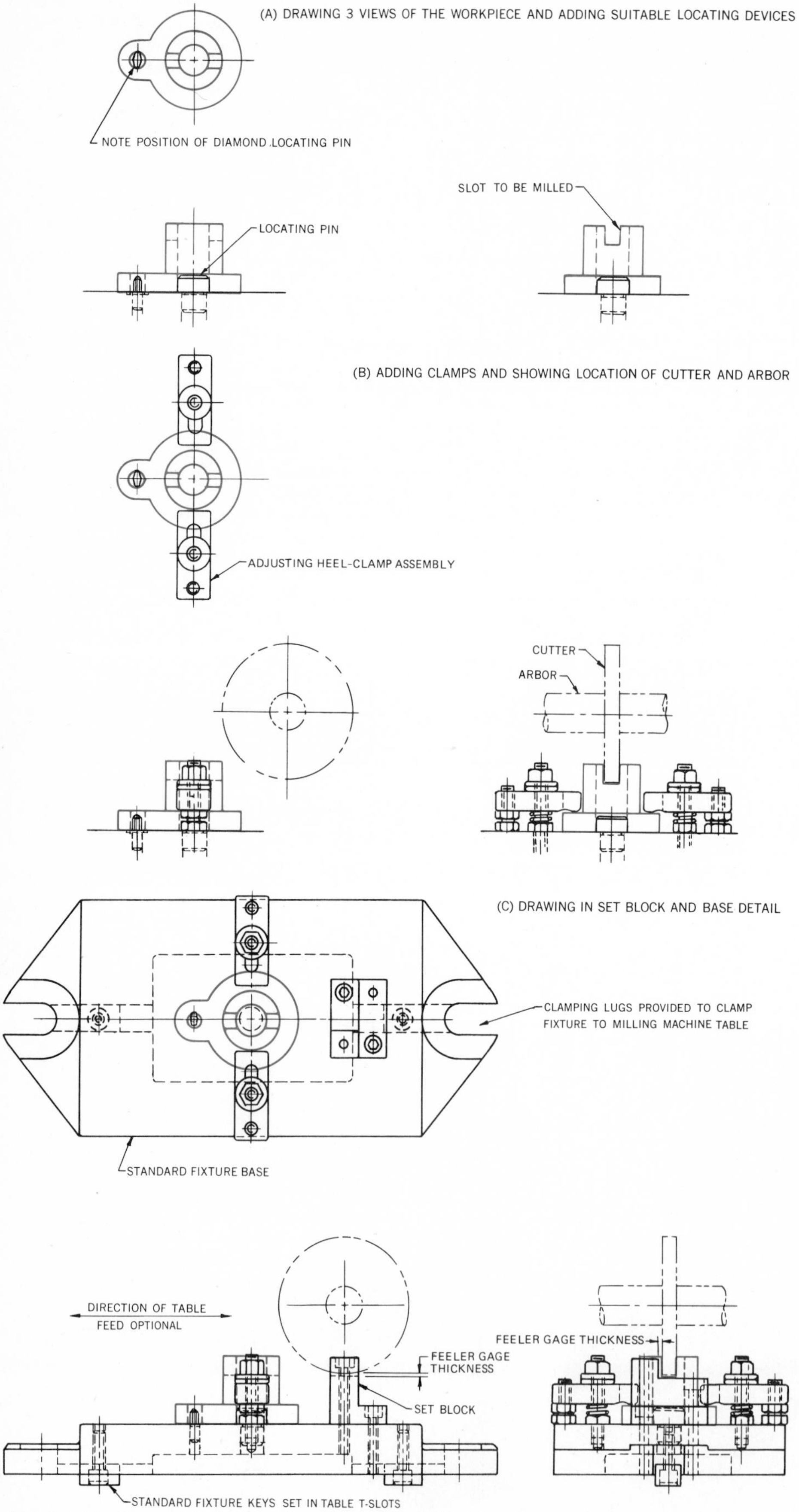
(A) DRAWING 3 VIEWS OF THE WORKPIECE AND ADDING SUITABLE LOCATING DEVICES
NOTE POSITION OF DIAMOND LOCATING PIN
SLOT TO BE MILLED
LOCATING PIN
(B) ADDING CLAMPS AND SHOWING LOCATION OF CUTTER AND ARBOR
ADJUSTING HEEL-CLAMP ASSEMBLY
CUTTER
ARBOR
(C) DRAWING IN SET BLOCK AND BASE DETAIL
CLAMPING LUGS PROVIDED TO CLAMP
FIXTURE TO MILLING MACHINE TABLE
STANDARD FIXTURE BASE
DIRECTION OF TABLE
FEED OPTIONAL
FEELER GAGE THICKNESS
FEELER GAGE
THICKNESS
SET BLOCK
STANDARD FIXTURE KEYS SET IN TABLE T-SLOTS

Fig. 26-7-8 Sequence in drawing a simple milling fixture.

FIXTURE DESIGN CONSIDERATIONS

1. Is the fixture foolproof? Does the design permit only one way of loading?
2. Does the fixture permit rapid loading and unloading?
3. Is ample chip clearance provided?
4. Is the fixture kept as low as possible to avoid chatter and springing of the work?
5. Are the cutting forces taken on the base rather than the clamp?
6. Does any part of the fixture interfere with the milling arbor or supports during the machining operation?
7. Are the clamps located in front of the workpiece?

Assignment

On an A3- or B-size sheet, design a simple milling fixture to mill out the slot in Fig. 26-7-A or 26-7-B or the two outside portions on the top of the part in Fig. 26-7-C or 26-7-D. Use standard components and refer to manufacturers' catalogs wherever possible. Draw the workpiece in red. Scale is 1:2.

REVIEW FOR ASSIGNMENT

Unit 7-5 Assemblies in Section

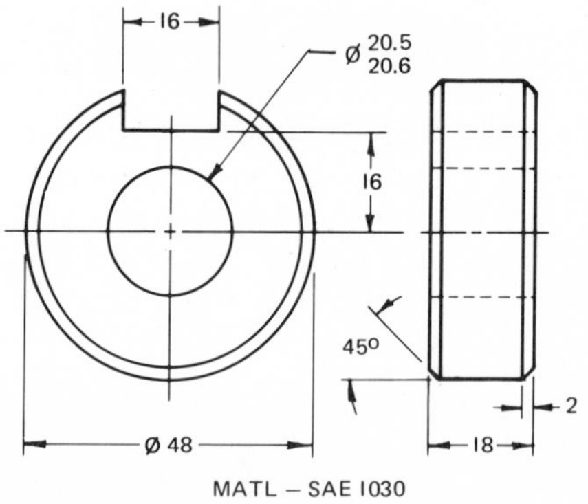
16
Ø 20.5
20.6
16
45°
2
Ø 48
18
MATL – SAE 1030

Fig. 26-7-A Sleeve.

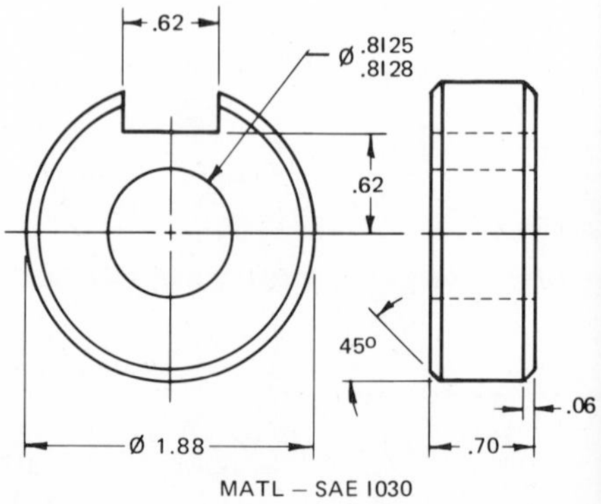
.62
Ø .8125
.8128
.62
45°
.06
Ø 1.88
.70
MATL – SAE 1030

Fig. 26-7-B Sleeve.

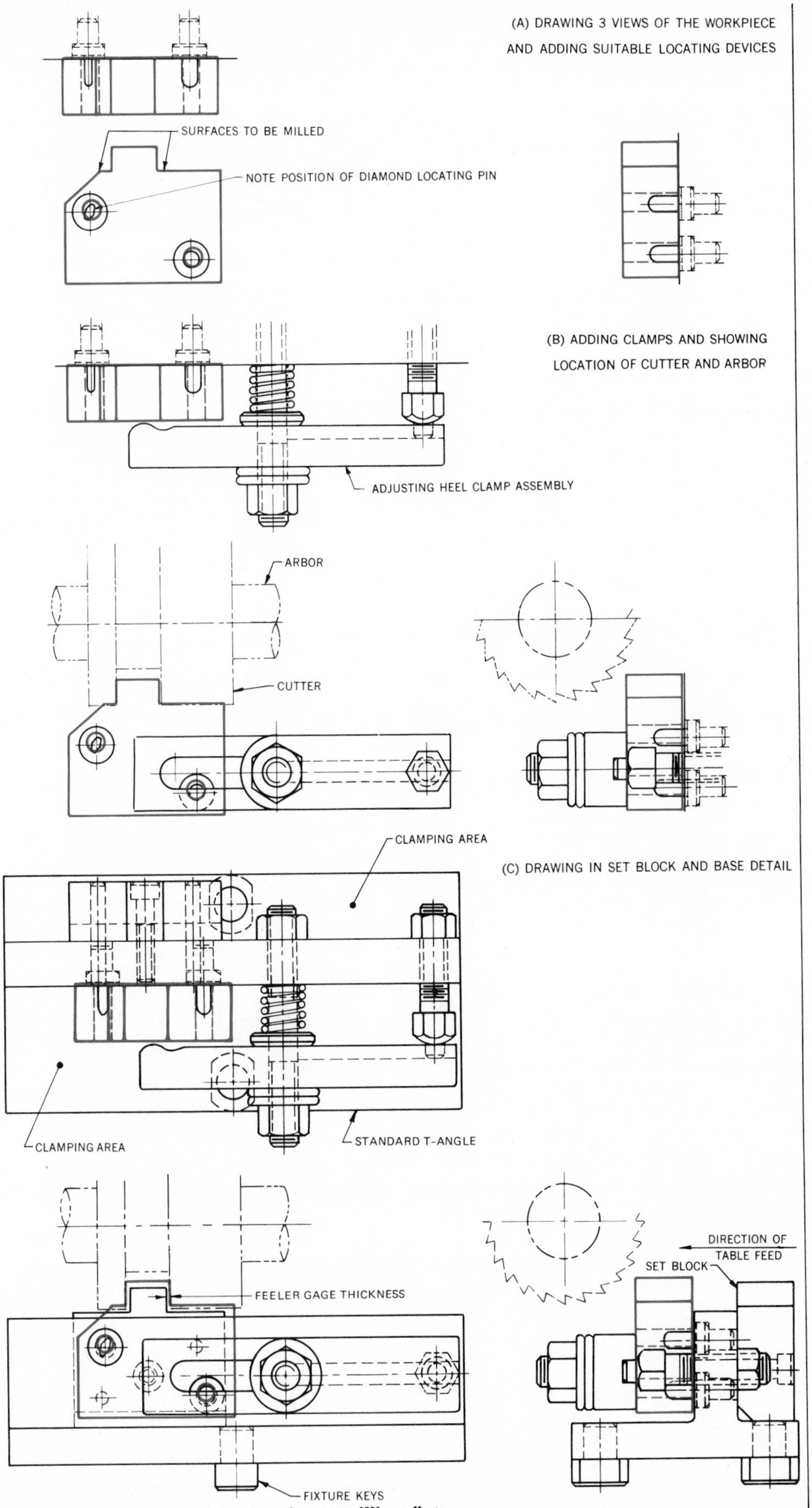
(A) DRAWING 3 VIEWS OF THE WORKPIECE AND ADDING SUITABLE LOCATING DEVICES
SURFACES TO BE MILLED
NOTE POSITION OF DIAMOND LOCATING PIN
(B) ADDING CLAMPS AND SHOWING LOCATION OF CUTTER AND ARBOR
ADJUSTING HEEL CLAMP ASSEMBLY
ARBOR
CUTTER
CLAMPING AREA
(C) DRAWING IN SET BLOCK AND BASE DETAIL
CLAMPING AREA
STANDARD T-ANGLE
FEELER GAGE THICKNESS
DIRECTION OF TABLE FEED
SET BLOCK
FIXTURE KEYS

Fig. 26-7-9 Sequence in drawing a milling fixture.

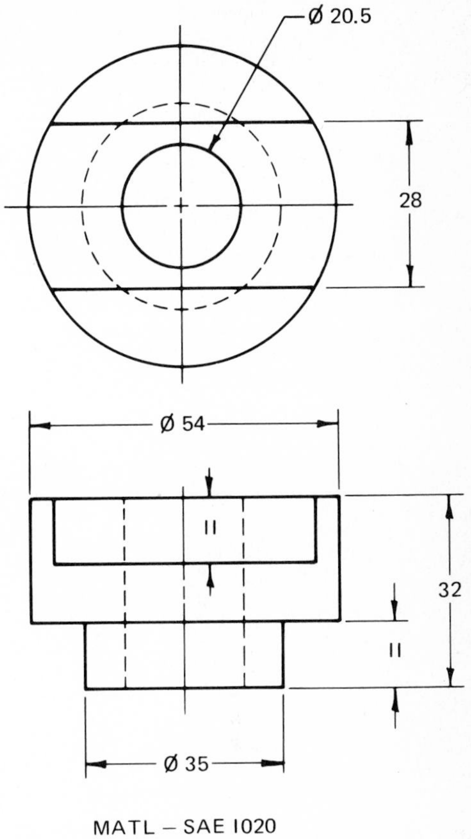
Ø 20.5
28
Ø 54
11
32
11
Ø 35
MATL – SAE 1020

Fig. 26-7-C Drive link.

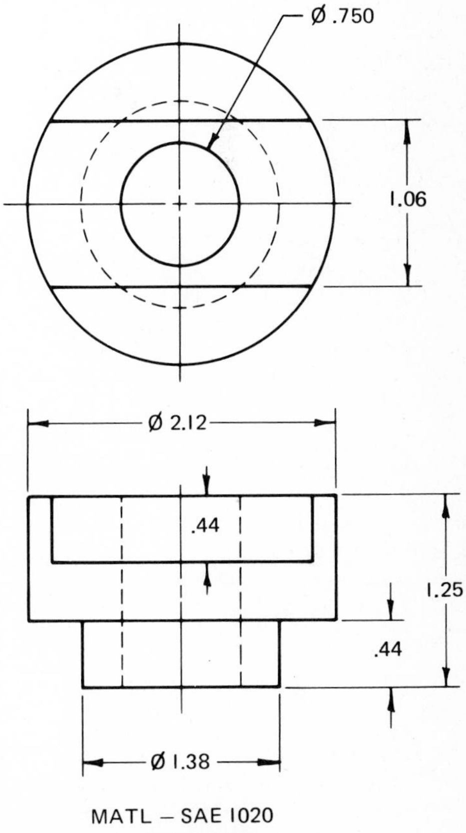
Ø .750
1.06
Ø 2.12
.44
1.25
.44
Ø 1.38
MATL – SAE 1020

Fig. 26-7-D Drive link.

Chapter 27
Die Design

UNIT 27-1
DIE DESIGN

Because of the wide use of sheet metal in the construction of many products, die design and its associated tooling have become indispensable to the engineering industry.

A casual examination of the goods on sale in any hardware store will make it clear that presswork or stamping is the foundation of the mass-production industry. Also, designers will often seek to replace expensive castings and forgings with parts of equal strength constructed from sheet metal for one or more of the following reasons:

1. Faster production, because of the speed at which the presses can operate.
2. Cheaper costs, since many press operations can be carried out by unskilled labor.
3. Lighter construction by skillful arrangement of the sheet-metal components.
4. Accurate parts can be finished completely and good standards and tolerances maintained for an indefinite time, thus saving valuable labor and worker-hours which would otherwise have been expended in the machine shop.

Stamping

Stamping is the art of pressworking sheet metal to change its shape by the use of punches and dies. It may involve punching out a hole or the product itself from a sheet of metal. It may also involve bending or forming, drawing, and coining. See Figs. 27-1-1 and 27-1-2.

Stamping may be divided into the two general classifications: forming and shearing.

Forming. *Forming* includes stampings made by forming sheet metal to the

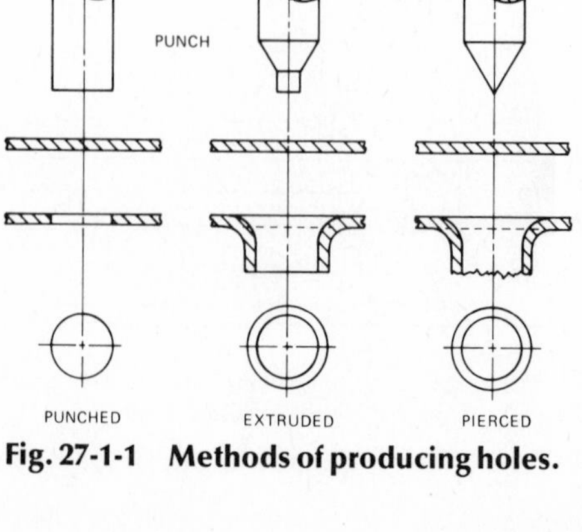
PUNCH
PUNCHED
EXTRUDED
PIERCED

Fig. 27-1-1 Methods of producing holes.

shape desired without cutting or shearing the metal.

Bending and forming are the processes of bending, flanging, folding, twisting, offsetting, or otherwise shaping a blank or a portion of a blank, usually without significantly changing the thickness of the metal.

Drawing is a process in which the metal is stretched or drawn over a form to produce the desired shape from a flat sheet or blank. Drawing usually causes a slight change in metal thickness in the drawn portion.

Coining is the process by which metal is made to ''flow'' by the application of extreme pressure.

Shearing. *Shearing* includes stampings made by shearing the sheet metal either to change the outline or to cut holes in the interior of the part.

Blanking is the process of shearing or cutting out the size and shape of a flat piece necessary to produce the desired finish part.

Punching forms a hole or opening in the part.

Principles of Blanking Dies[1]

Before beginning the study of die design, you must have a clear understanding of the following terms.

PUNCH
DIE SHOE
PRESS BED

Fig. 27-1-2 Punch press used to punch and form metal. (Whitney Metal Tool Co.)

Piece Part. A *piece part* is the product of a die. It may be a complete product in itself (the drafter's erasing shield and a bottle opener are examples of this kind of product), or it may be only one component of a product consisting of many different parts, such as an electrical terminal used in a television set. The die may or may not produce a piece part in a finished state.

Stock Material. This is the general term for any of the various materials from which the piece part is made.

Die. The word *die* has several definitions. This book utilizes two: (1) a complete production tool, the purpose of which is to produce piece parts consistently to required specifications, and (2) the female part of a complete die.

Punch. A *punch* is a male member of a complete die which mates, or acts in conjunction with, the female die to produce a desired effect upon the material being worked. A die can be a simple tool composed of a punch, a die block, and a stripper; or it can be an extremely complex mechanism which performs many and varied operations.

PARTS PRODUCED BY BLANKING

The piece part in Fig. 27-1-3 is produced by blanking. It is shown in relation to the stock material from which it was blanked. The terms used to describe the various aspects of the work are indicated. The *stock strip* is stock material of the required thickness (T) which has been cut to strips of suitable width for the particular job. The stock strip is fed through the die from right to left, as indicated by the feed direction arrow. The *advance* is the distance the stock must be fed (advanced) to allow a clean blanking operation at each press stroke. The *scrap bridge* is the scrap remaining between the openings in the stock strip after blanking. The advance distance is equal to the width of the piece part plus the scrap bridge. The back scrap is so

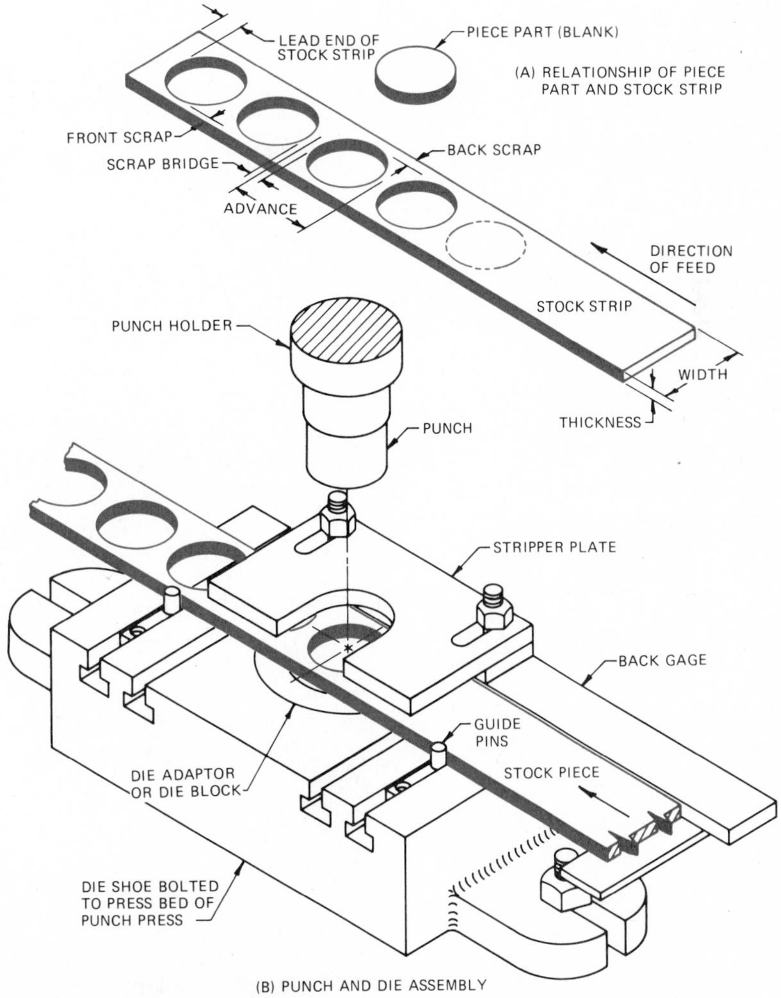
LEAD END OF STOCK STRIP
PIECE PART (BLANK)
(A) RELATIONSHIP OF PIECE PART AND STOCK STRIP
FRONT SCRAP
SCRAP BRIDGE
BACK SCRAP
ADVANCE
DIRECTION OF FEED
STOCK STRIP
WIDTH
THICKNESS
PUNCH HOLDER
PUNCH
STRIPPER PLATE
BACK GAGE
GUIDE PINS
STOCK PIECE
DIE ADAPTOR OR DIE BLOCK
DIE SHOE BOLTED TO PRESS BED OF PUNCH PRESS
(B) PUNCH AND DIE ASSEMBLY

Fig. 27-1-3 Simple blanking die.

called because it is located toward the back of the press (away from the operator). The front scrap is on the side toward the front of the press (toward the operator). The *lead* end of the stock strip is the end that is fed into the die first. The opposite end of the strip is called the *tail*.

The piece part shown in Fig. 27-1-4 is called a *blank* because further work is required to complete it. The slug is the material that is cut out of the blank or piece part by a punching operation.

Whether a die is a blanking die or a punching die depends solely upon its intended use. It is called a *blanking die* if it is meant to produce blanks (B), as shown in Fig. 27-1-3, or a desired contour and size by cutting them out of the required type of material, called the *stock strip*. The blanks are the desired product (piece parts) made by the die. In most cases, the material remaining after the blanks have been cut out is considered scrap.

If the purpose of the die shown in Fig. 27-1-5 is to make openings of a desired contour and size in the required material (A), it is called a *punching die*. In this case, *A* represents the piece part and *B* is called the *slug*. The slug is usually considered scrap.

The basic elements of both dies are identical. The name of the die is derived from its intended use. The physical effects of the tool upon the stock material are the same whether it is punching out slugs in order to produce a desired opening in a piece part or punching out blanks which are the desired product of the die.

DESCRIPTION OF BLANKING DIE

Figures 27-1-3 and 27-1-5 show the basic elements of a standard parts blanking die. These elements are the die block, in which the proper female die opening has been made, an adaptor (may or may not be required depending on the size of hole being made), the punch, and the stripper. In addition, guide and stopping devices are required to control the stock. The die block and stripper are secured to the die shoe, which in turn is fastened to the press bed of the punch press. The punch is locked into the punch holder, which in turn is secured to the ram of the press which moves up and down. Standard punch and die parts are shown in Fig. 27-1-6.

Press Cycle during Blanking. See Fig. 27-1-5*a*. The at-rest position of the press ram is at the top of the stroke. The ram is then at its greatest distance from the bed of the press. When the press is tripped, the clutch allows engagement of the crankshaft with the press flywheel. The crankshaft then rotates through 360°, or one full turn. During the first half of the cycle, the ram is driven toward the press bed. During the last half of the cycle, the ram moves away from the press bed. The distance traveled by the ram during the half cycle (top of stroke to bottom of stroke) is called the *press stroke*.

Action of Blanking or Punching a Die. Refer to Fig. 27-1-5*b*. The stock material (A) is fed or loaded in the proper position on the top surface of the die block. When the press is tripped, the ram drives the punch through the stock strip (A) into the die opening, thereby producing an opening in the stock material by cutting out the blank or slug (B). This blank or slug is withdrawn and is pushed through the die by the blanks or slugs produced subsequently.

Stripping. After the blanking has taken place, the punch is returned to the open or at-rest position by the press ram as it completes its cycle. The stock material clings to the punch and will remain

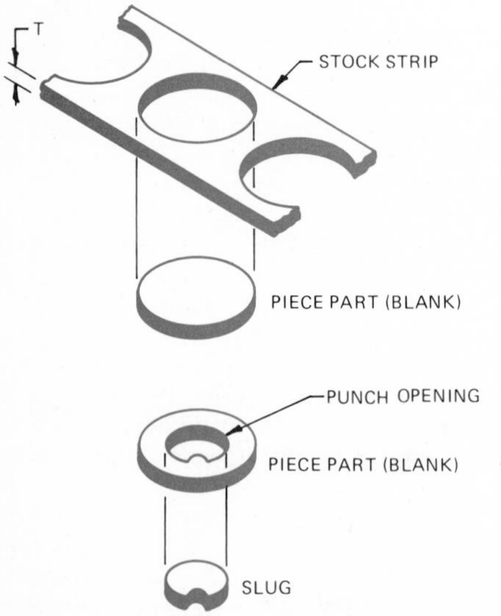
T
STOCK STRIP
PIECE PART (BLANK)
PUNCH OPENING
PIECE PART (BLANK)
SLUG

Fig. 27-1-4 Relationship of stock material, piece part, and slug.

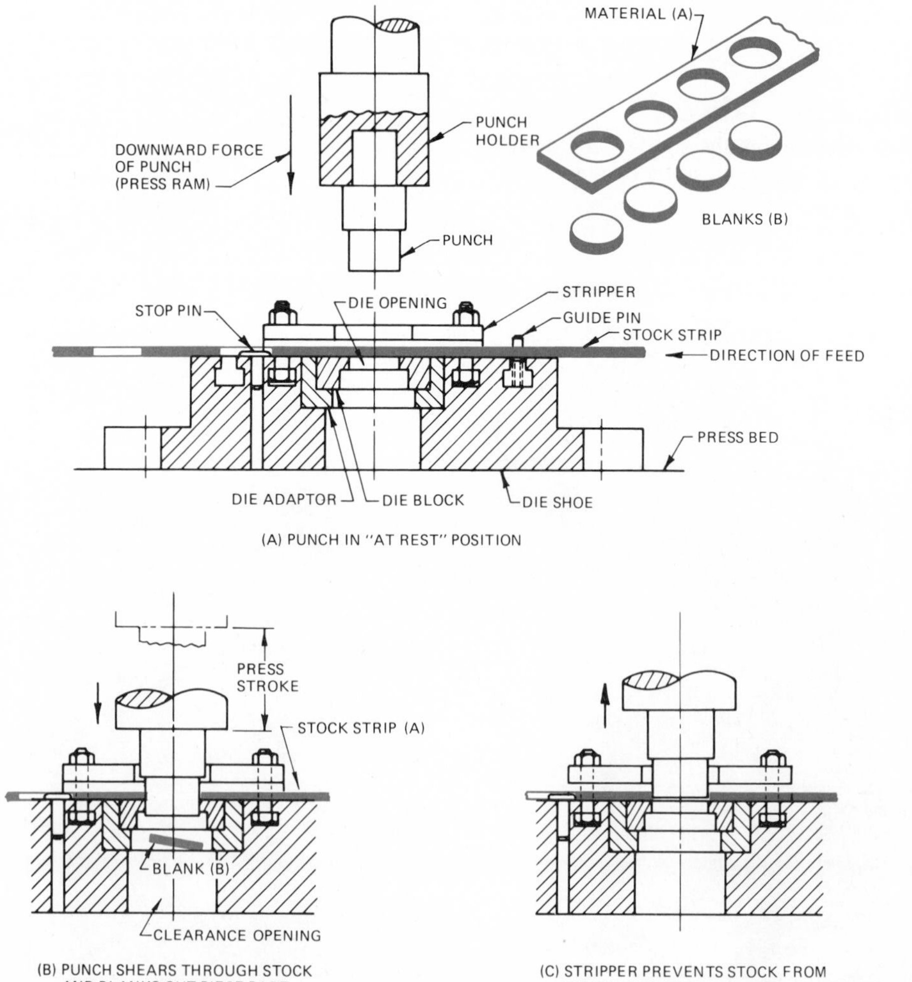
MATERIAL (A)
PUNCH HOLDER
DOWNWARD FORCE OF PUNCH (PRESS RAM)
BLANKS (B)
PUNCH
STOP PIN
DIE OPENING
STRIPPER
GUIDE PIN
STOCK STRIP
DIRECTION OF FEED
PRESS BED
DIE ADAPTOR
DIE BLOCK
DIE SHOE
(A) PUNCH IN "AT REST" POSITION
PRESS STROKE
STOCK STRIP (A)
BLANK (B)
CLEARANCE OPENING
(B) PUNCH SHEARS THROUGH STOCK AND BLANKS OUT PIECE PART
(C) STRIPPER PREVENTS STOCK FROM RISING WITH PUNCH ON RETURN STROKE

Fig. 27-1-5 Simple blanking operation using standard components shown in Fig. 27-1-6.

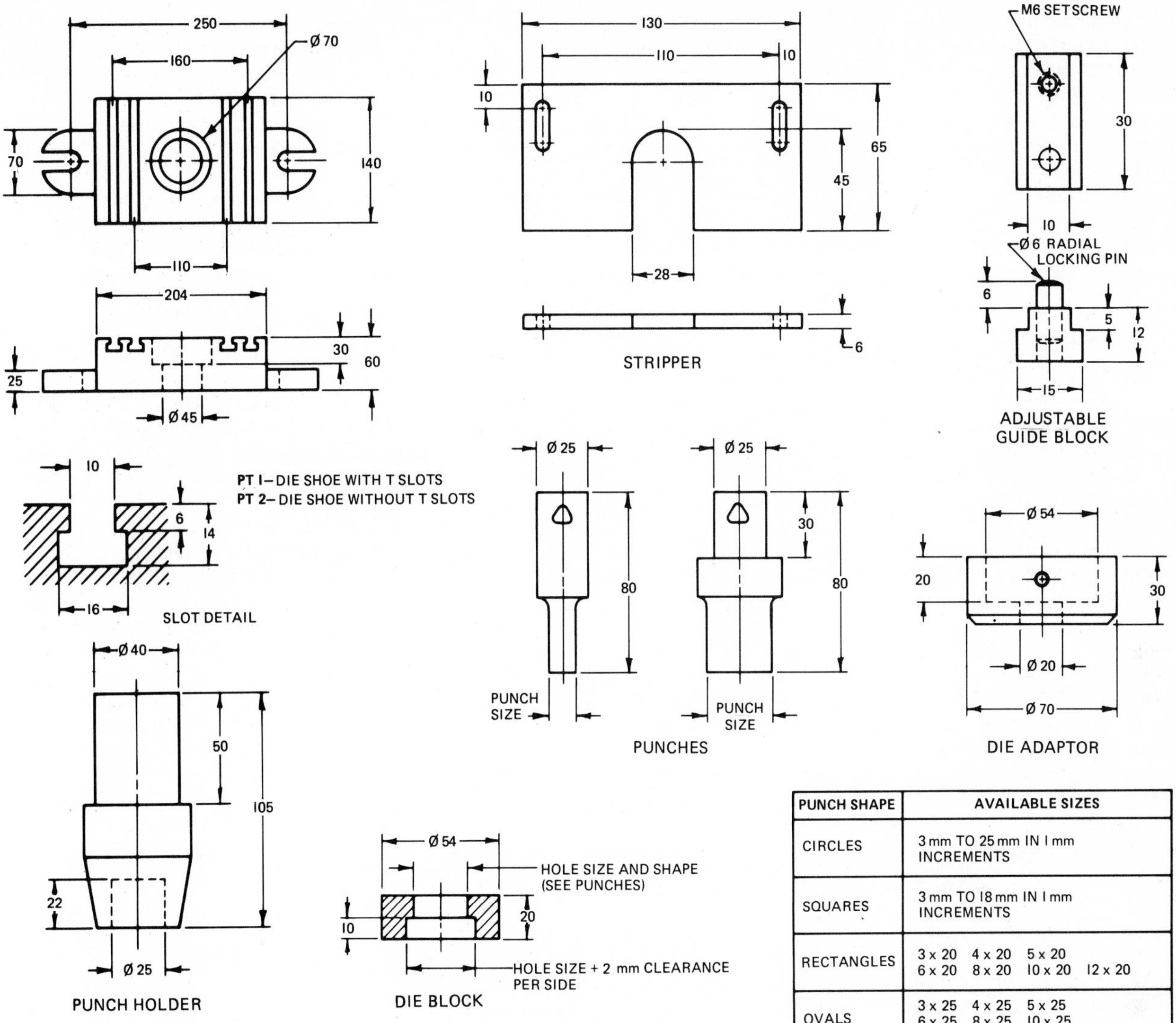
250
160
Ø 70
70
140
110
204
30
60
25
Ø 45
PT 1–DIE SHOE WITH T SLOTS
PT 2–DIE SHOE WITHOUT T SLOTS
10
6
14
16
SLOT DETAIL
Ø 40
50
105
22
Ø 25
PUNCH HOLDER
130
110
10
10
65
45
28
6
STRIPPER
M6 SETSCREW
30
10
Ø 6 RADIAL LOCKING PIN
6
5
12
15
ADJUSTABLE GUIDE BLOCK
Ø 25
Ø 25
80
30
80
PUNCH SIZE
PUNCH SIZE
PUNCHES
Ø 54
20
30
Ø 20
Ø 70
DIE ADAPTOR
Ø 54
HOLE SIZE AND SHAPE (SEE PUNCHES)
10
20
HOLE SIZE + 2 mm CLEARANCE PER SIDE
DIE BLOCK

PUNCH SHAPE	AVAILABLE SIZES
CIRCLES	3 mm TO 25 mm IN 1 mm INCREMENTS
SQUARES	3 mm TO 18 mm IN 1 mm INCREMENTS
RECTANGLES	3 x 20 4 x 20 5 x 20 6 x 20 8 x 20 10 x 20 12 x 20
OVALS	3 x 25 4 x 25 5 x 25 6 x 25 8 x 25 10 x 25 12 x 25

Fig. 27-1-6 Standard punch and die parts. (Whitney Metal Tool Co.)

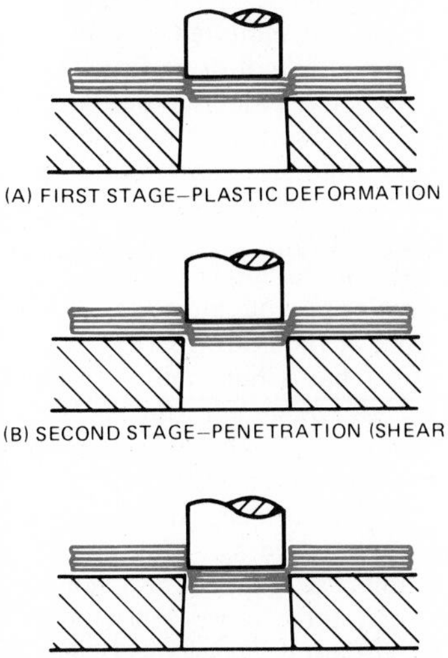
(A) FIRST STAGE–PLASTIC DEFORMATION
(B) SECOND STAGE–PENETRATION (SHEAR)
(C) THIRD STAGE–FRACTURE

Fig. 27-1-7 Critical stages of shearing action on metal.

on the punch unless something is done to prevent it. This is the function of the stripper: it keeps the stock material from traveling with the punch on the return stroke (Fig. 27-1-5*c*). Because the stock material is held back by the stripper, the punch is withdrawn from the material on the return stroke.

REACTION OF THE STOCK MATERIAL: SHEARING ACTION

The result of the forces imposed on the stock material by the working of the blanking die is a shearing action. This shearing action may be considered in three stages, which are important to the die maker because of their direct relationship to the dimensional qualities and appearance of piece parts. They are also related to the effective working and life of the die. These stages showing the reaction of the stock material are illustrated progressively in Fig. 27-1-7.

Critical Stages of Shearing Action of Metal. *First Stage: Plastic Deformation.* The stock material has been placed on the die, the press has been tripped, and the punch is being driven toward the die. The punch contacts the stock material and exerts pressure upon it. When the elastic limit of the stock material is exceeded, plastic deformation takes place.

Second Stage: Penetration. As the driving force of the ram continues, the punch is forced to penetrate the stock material, and the blank or slug is displaced into the die, which opens a corresponding amount. This is the true shearing portion of the cutting cycle, from which the term *shearing action* is derived.

Third Stage: Fracture. Further continuation of the punching pressure then causes fractures to start at the cutting edges of the punch and the die. These are the points of greatest stress concentration. Under the proper cutting conditions, the fractures extend toward one another and meet. When this occurs, the fracture is complete and the blank or slug is separated from the original stock material. The punch then enters the die opening, pushing the blank or slug slightly below the die cutting edge.

These three stages of shear action are responsible for the characteristic appearance of piece parts produced by blanking.

CUTTING CLEARANCE

Proper cutting clearance is necessary to the life of the die and the quality of the piece part. Excessive cutting clearance results in objectionable piece-part characteristics; insufficient cutting clearance causes undue stress and wear on the cutting members of the tool because of the greater punching effort required.

Determining Cutting Clearance. The physical properties and the thickness of the stock material are the factors that determine the amount of cutting clearance. The thickness is easily measured, but the physical properties in relation to cutting clearance are not. See Fig. 27-1-8.

Cutting clearance should be expressed in terms of percentage of stock material thickness per side.

The percentage varies with the properties of the material. A suggested list of percentages for various materials is given in Fig. 27-1-9. Mica, fiber, materials, plastics, etc., generally require less cutting clearance than any of the metals.

There is another result of the necessity for cutting clearance that must be studied and thoroughly understood by the tool and die maker. This is the effect of cutting clearance on the actual dimensions of the piece part, as shown in Fig. 27-1-10.

Relationship of Piece-Part Sizes to Punch and Die Sizes

When blanked piece parts are measured, the measurement is made at the cut

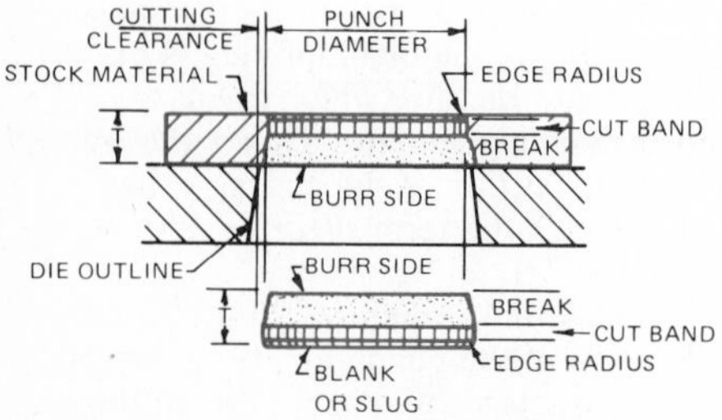

CUTTING CLEARANCE
PUNCH DIAMETER
STOCK MATERIAL
EDGE RADIUS
T
CUT BAND
BREAK
BURR SIDE
DIE OUTLINE
BURR SIDE
BREAK
T
CUT BAND
EDGE RADIUS
BLANK OR SLUG

Fig. 27-1-8 Optimum cutting clearances.

CUTTING CLEARANCE PER SIDE (Percentage of Stock Thickness)		
Material	**Irregular Contours**	**Round**
Aluminum		
Soft, less than 1.5 mm thick	3%	2%
Soft, more than 1.5 mm thick	5%	3%
Hard	5%–8%	4%–6%
Brass & Copper		
Soft	3%	2%
½ Hard	4%	3%
Hard	5%–6%	4%
Steel		
Low Carbon Soft	3%	2%
½ Hard	4%	2%
Hard	5%	3%
Silicon Steel	4%–5%	3%
Stainless Steel	5%–8%	4%–6%

Fig. 27-1-9 Cutting clearances.

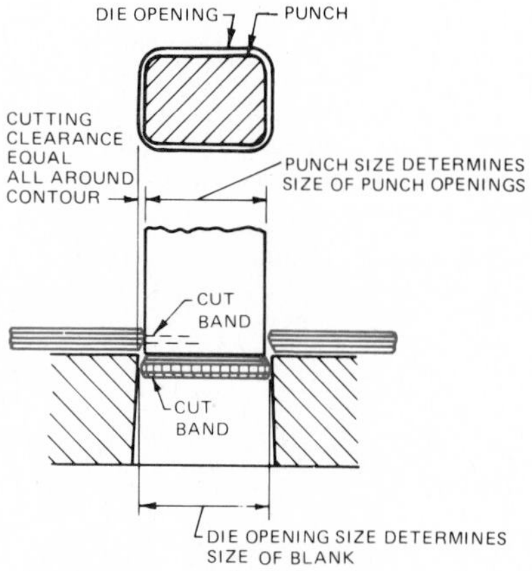

DIE OPENING
PUNCH
CUTTING CLEARANCE EQUAL ALL AROUND CONTOUR
PUNCH SIZE DETERMINES SIZE OF PUNCH OPENINGS
CUT BAND
CUT BAND
DIE OPENING SIZE DETERMINES SIZE OF BLANK

Fig. 27-1-10 Piece part sizes in relation to punch and die size.

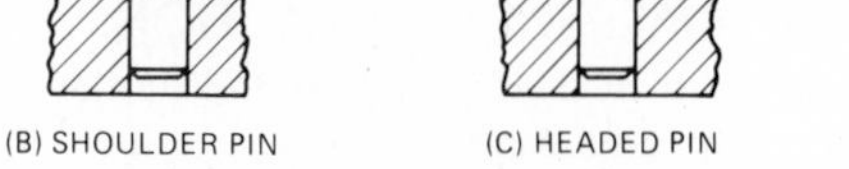

CLEARANCE HOLE IN STRIPPER
PUNCH
STRIPPER PLATE
DIE BLOCK
(A) PLAIN PIN
(B) SHOULDER PIN
(C) HEADED PIN

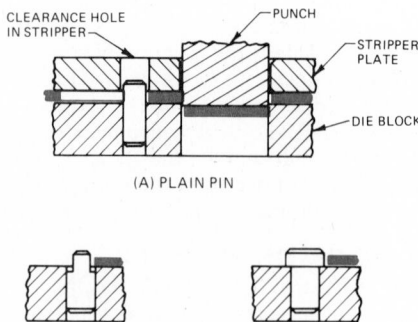

Fig. 27-1-11 Solid pin stops.

band. The blank or slug measures larger than the opening in the stock material from which it was punched. The reasons for this are obvious upon examination of Fig. 27-1-10. The actual cutting of the blank or slug is done by the cutting edge of the die opening. Therefore, the die opening determines the size of a blank or slug. The actual cutting of the opening in the stock material is done by the punch. Therefore, the size of a punched opening is determined by the punch. It is imperative that the designer have a clear understanding of this fact. It enables him or her to know which cutting members must be made to piece-part size and which ones should have the cutting clearance applied to them.

A careful dimensional check of a blank or slug will often reveal that its overall dimensions are slightly larger than the die opening that produced it. This is because the cutting action caused it to be compressed in the die opening. After it passed through the die, the pressure was released; therefore, it expanded a slight amount. Conversely, a punched opening will often be slightly smaller than the punch which produced it. When the punch was withdrawn from the stock material during the stripping stage, the opening closed slightly. Most materials react this way. Satisfactory compensation for this condition is usually achieved by making the punch 0.01 to 0.02 mm larger overall than the desired opening to be punched. If the blank is the desired product, the die opening is then made 0.01 to 0.02 mm smaller overall than the desired blank.

Die Stops

In the majority of applications, stops are installed on dies for the purpose of arresting the feeding movement of the stock strip.

Figure 27-1-11 illustrates three types of solid stop pins. A mounting hole is provided at the desired location in the required die component. The stop pin shown in view A is mounted in a die block. The pin is a light drive fit in the mounting hole. The mounting hole is generally made to suit the stop pin so that a standard pin size can be used. A clearance hole for the pin should be provided in the die shoe.

It is always necessary to provide clearance in any opposing members which cover the site of the stop. In view A, the spring stripper has a hole drilled through it to afford clearance for the stop pin.

The stop pin can be altered as shown in view B to effect a change in the stopping position, if necessary. If a standard dowel pin is employed as a stop pin, do not alter it this way; make a new pin of drill rod. Altering the dowel pin would expose the soft core of the dowel, rendering it unfit for stopping purposes. Changing the stopping position in the other direction may be accomplished by

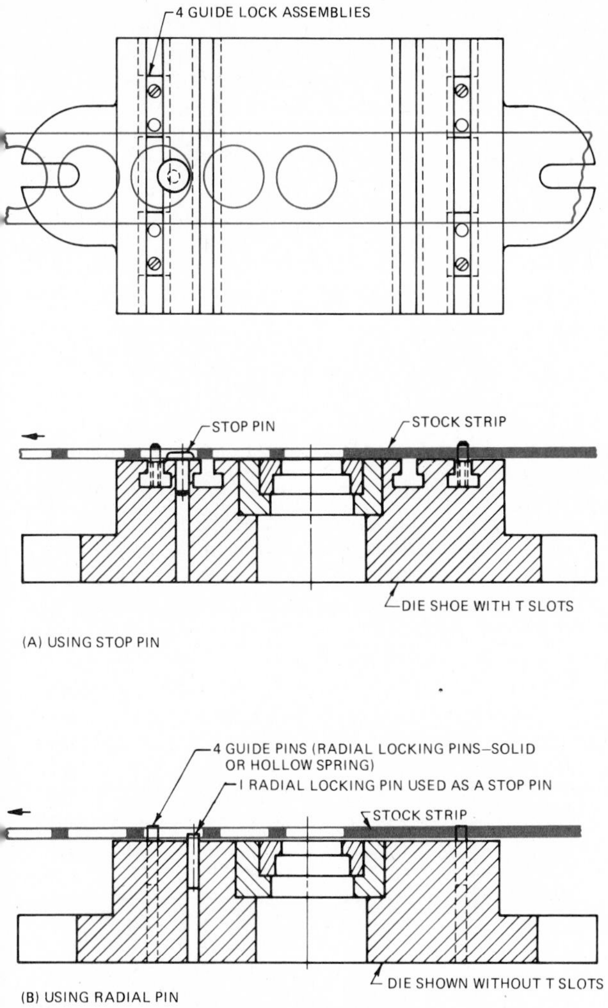
4 GUIDE LOCK ASSEMBLIES
STOP PIN
STOCK STRIP
DIE SHOE WITH T SLOTS
(A) USING STOP PIN
4 GUIDE PINS (RADIAL LOCKING PINS–SOLID OR HOLLOW SPRING)
1 RADIAL LOCKING PIN USED AS A STOP PIN
STOCK STRIP
DIE SHOWN WITHOUT T SLOTS
(B) USING RADIAL PIN

Fig. 27-1-12 Stopping the stock strip.

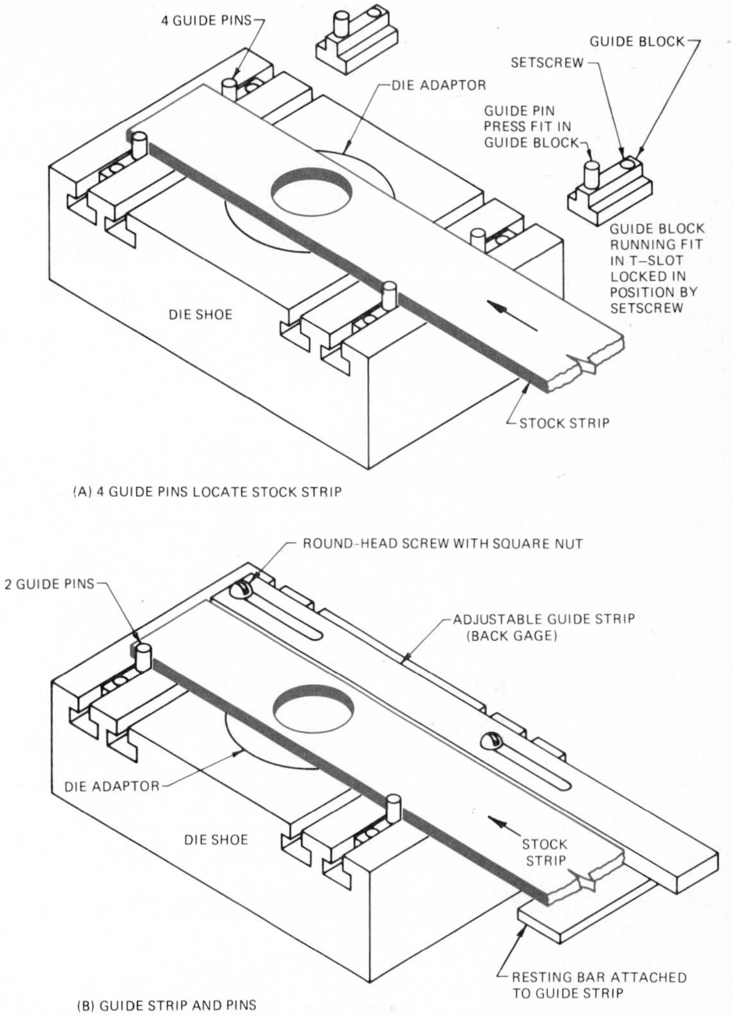
4 GUIDE PINS
DIE ADAPTOR
SETSCREW
GUIDE BLOCK
GUIDE PIN
PRESS FIT IN
GUIDE BLOCK
GUIDE BLOCK
RUNNING FIT
IN T–SLOT
LOCKED IN
POSITION BY
SETSCREW
DIE SHOE
STOCK STRIP
(A) 4 GUIDE PINS LOCATE STOCK STRIP
ROUND-HEAD SCREW WITH SQUARE NUT
2 GUIDE PINS
ADJUSTABLE GUIDE STRIP
(BACK GAGE)
DIE ADAPTOR
DIE SHOE
STOCK
STRIP
RESTING BAR ATTACHED
TO GUIDE STRIP
(B) GUIDE STRIP AND PINS

Fig. 27-1-13 Guiding the stock strip.

substituting a headed stop pin, as shown in view C.

The stock strip shown in Fig. 27-1-12 is located by a simple stop pin pressed into the die shown. In operation, the edge of the hole in the stock strip butts against the stop pin, holding the stock strip at this position until the piece part is punched out. The stock strip is then raised and moved along so that the next hole drops over the stop pin. Two simple forms of stopping the stock strip are shown. The operator should be able to see the stop action. As such, it may be necessary to cut away part of the stripper plate so that the stop pin may be seen. See Fig. 27-1-13.

GUIDE STRIP AND PINS

The stock strip must be guided into the die. This is accomplished by guide pins and guide strips. The two shown in Fig. 27-1-13 are standard parts, adjustable, and, as such, can be used on more than

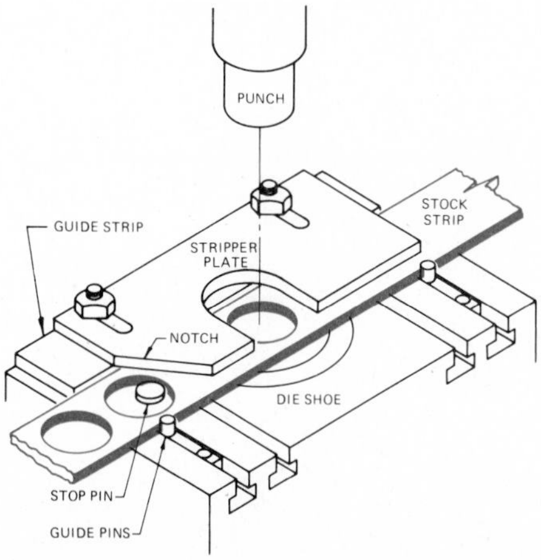
PUNCH
STOCK
STRIP
GUIDE STRIP
STRIPPER
PLATE
NOTCH
DIE SHOE
STOP PIN
GUIDE PINS

Fig. 27-1-14 Cutting notch in stripper plate so operator may see stop action.

one stock strip size. The same screws which hold the stripper plate can be used for the guide strip.

If four guide pins are used, such as shown in Fig. 27-1-13*a*, then a spacer or shim will be required under the stripper plate in order that the stock strip may be raised sufficiently to clear the stop or radial locking pin. Other forms of stops, such as automatic or spring stops, are also used but will not be covered at this time. See Fig. 27-1-14.

Nest Gages

Nest gages are commonly used on dies which perform secondary operations, such as piercing, forming, etc. A typical fixed-pin nest gage is shown in Fig. 27-1-15. Nest gaging is also required for dies in which unit stock is run, since this too requires the feeding of individual pieces into the die.

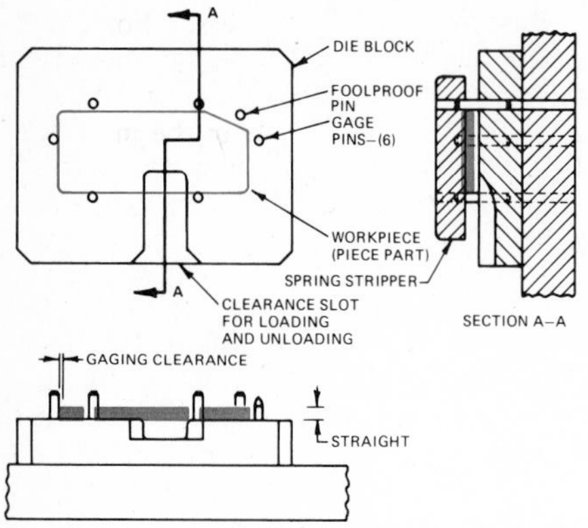
A
DIE BLOCK
FOOLPROOF PIN
GAGE PINS–(6)
WORKPIECE (PIECE PART)
SPRING STRIPPER
A
CLEARANCE SLOT FOR LOADING AND UNLOADING
SECTION A–A
GAGING CLEARANCE
STRAIGHT

Fig. 27-1-15 Typical fixed-pin nest gage.

The purpose of a nest gage is to locate and position the workpiece properly in the die. The workpiece may be unit stock, or it may be an unfinished piece part.

There are four important conditions which must be satisfied in order to provide suitable nest gaging: accuracy, loading ease, unloading ease, and foolproofing.

The number of contact spots (gaging points) required for any given nest gage varies according to size and contour of the workpiece. For round work and triangular work, a minimum of three contact spots will be required. For other contours, the minimum is four.

REFERENCE

1. From *Basic Diemaking* by the National Tool, Die, and Precision Machining Association. Copyright, 1963. McGraw-Hill Book Company. Used by permission.

Assignments

1. On an A3- or B-size sheet, lay out a standard die assembly for each part shown on Fig. 27-1-A or 27-1-B. Use standard parts only as shown in Fig. 27-1-6. Select suitable stops and guides. The punch need not be shown. For No. 1 piece part, the hole is to be punched. For No. 2 piece part, both holes are to be punched, one hole at a time. Show the stock strip and piece part in a second color. Scale is 1:1.

2. On an A3- or B-size sheet, lay out a standard punch and die assembly for one of the piece parts shown in Fig. 27-1-C or 27-1-D. The piece part has already been punched out of the stock strip, and the part must be positioned for the hole to be punched. A nest gage similar to the type shown in Fig. 27-1-15 is recommended. Show the piece part in a second color. Scale is 1:1.

REVIEW FOR ASSIGNMENT

Unit 6-4 Assembly Drawings
Appendix Bolt and Nut Sizes

UNIT 27-2

NESTING OF BLANKS

In die design, one of the first considerations is the layout of the piece parts on the stock strip. Since over 50 percent of the cost of a stamping is for material, the designer must take care to utilize the raw stock to full advantage. The blanks must be positioned so as to utilize the maximum area of the stock strip. See Fig. 27-2-1.

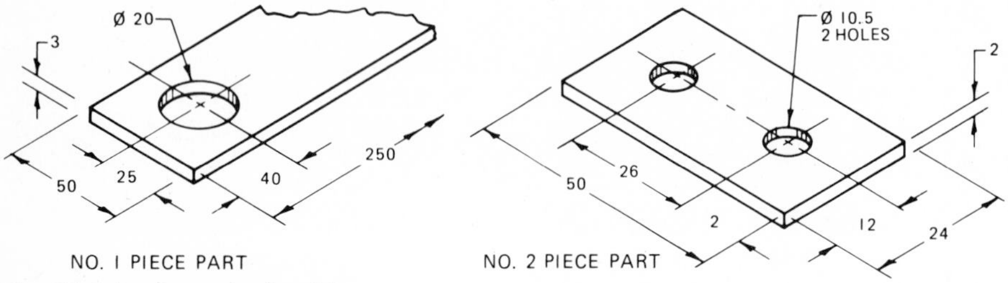
Ø 20
3
50
25
40
250
NO. I PIECE PART
Ø 10.5
2 HOLES
2
50
26
2
12
24
NO. 2 PIECE PART

Fig. 27-1-A Stops and guides.

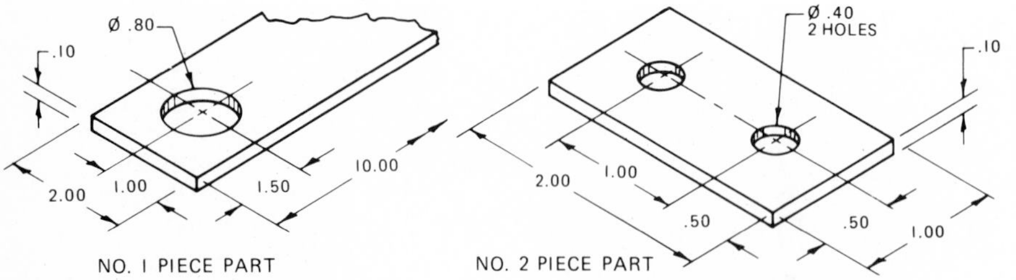
Ø .80
.10
2.00
1.00
1.50
10.00
NO. I PIECE PART
Ø .40
2 HOLES
.10
2.00
1.00
.50
.50
1.00
NO. 2 PIECE PART

Fig. 27-1-B Stops and guides.

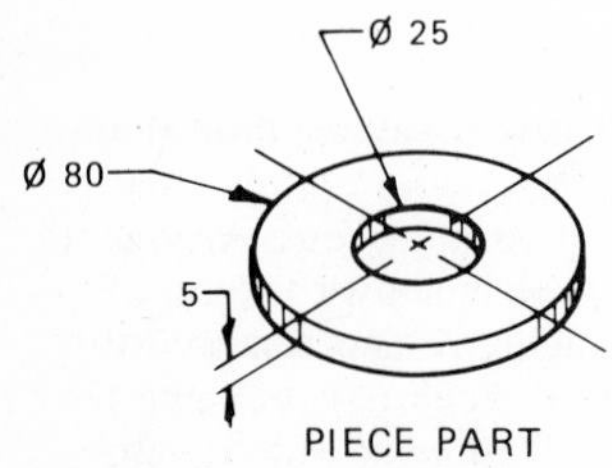
Ø 25
Ø 80
5
PIECE PART

Fig. 27-1-C Nest gage.

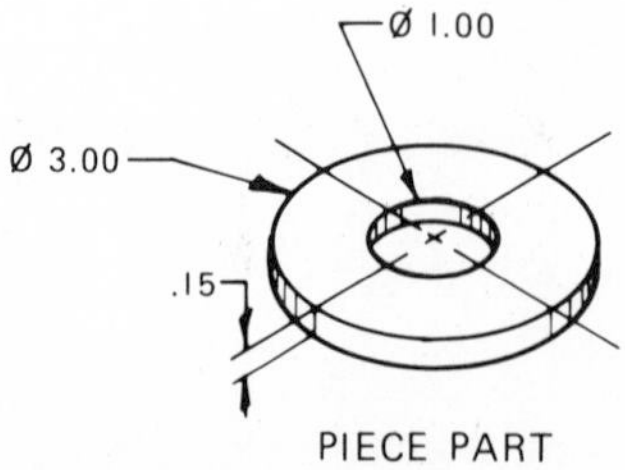
Ø 1.00
Ø 3.00
.15
PIECE PART

Fig. 27-1-D Nest gage.

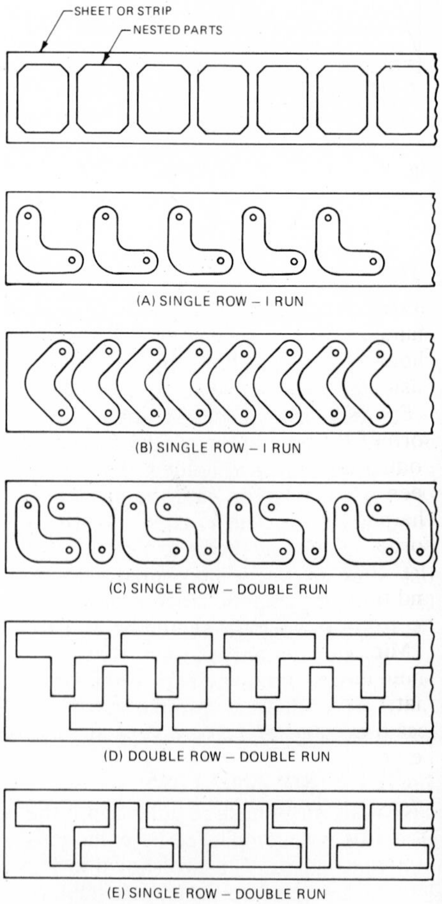
SHEET OR STRIP
NESTED PARTS
(A) SINGLE ROW – I RUN
(B) SINGLE ROW – I RUN
(C) SINGLE ROW – DOUBLE RUN
(D) DOUBLE ROW – DOUBLE RUN
(E) SINGLE ROW – DOUBLE RUN

Fig. 27-2-1 Nesting of blanks.

To accomplish this, a layout of the stock strip is made showing the exact location of each blank. Only a portion of the strip need be drawn if the remaining part of the strip is repetitive.

To reduce the amount of scrap, it may be desirable to alternate or reverse the position of every other blank. However, with this layout method the strip must be fed through the die twice, which involves additional handling costs to produce the same number of parts. As the strip goes through the die the first time, the blanks nesting in one direction are stamped. Then the stock strip is reversed and run through the die the second time to remove the remaining blanks for it. An additional 10 to 15 percent higher labor cost is involved in double-run layouts. If this is more than the cost of the scrap which would be saved, then a single-run strip should be designed. Figure 27-2-2 shows how, by modifying the original design slightly, a stock saving could be achieved.

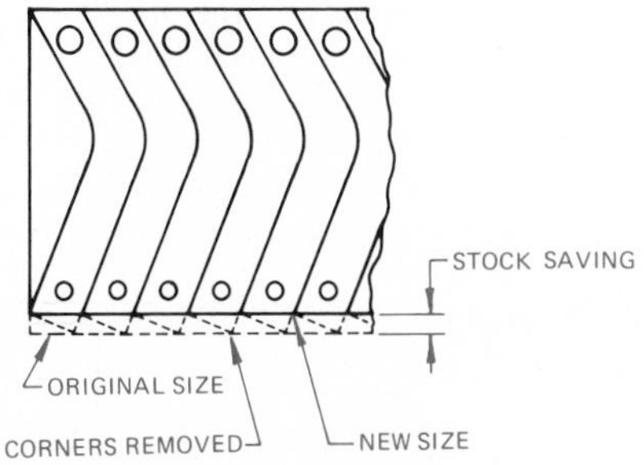
STOCK SAVING
ORIGINAL SIZE
CORNERS REMOVED
NEW SIZE

Fig. 27-2-2 Stock saving.

Design Considerations

In designing parts which are to be stamped, certain design considerations should be adhered to. Some of the more basic design rules follow.

Specifying Cutoffs. There are several correct cutoffs that can be used for economical stampings once the material has been sheared to the correct width. If sharp corners are permissible, the square corner is most economical. Corners along the edge of the strip should be sharp, and those not adjacent to the edge should be round. See Figs. 27-2-3 and 27-2-4.

Minimum Sections. Tabs or slots should never be less than 1.5 to 2 times material thickness in width and never less than 1.0 mm. Its length should not be greater than 5 times its width. See Fig. 27-2-5.

Grain Direction. If the metal grain runs in a direction contrary to the strength requirement of the part, the strength of a stamping may be reduced considerably. Therefore, in designing highly stressed parts, the direction of the grain of the metal should be considered and, when necessary, specified on the drawing. See Fig. 27-2-6.

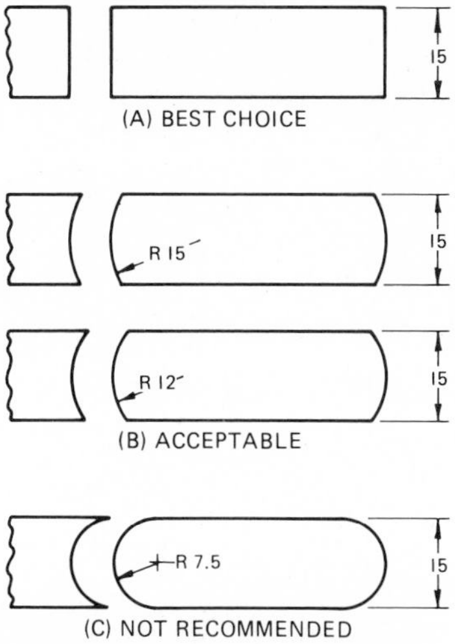
15
(A) BEST CHOICE
R 15
15
R 12
15
(B) ACCEPTABLE
R 7.5
15
(C) NOT RECOMMENDED

Fig. 27-2-3 Specifying cutoffs.

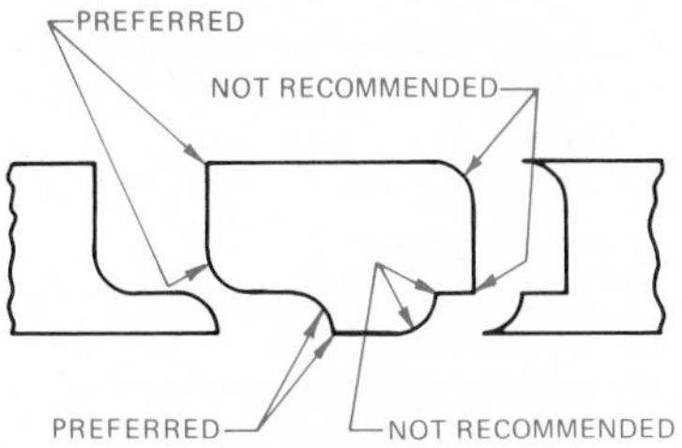
PREFERRED
NOT RECOMMENDED
PREFERRED
NOT RECOMMENDED

Fig. 27-2-4 Corner design.

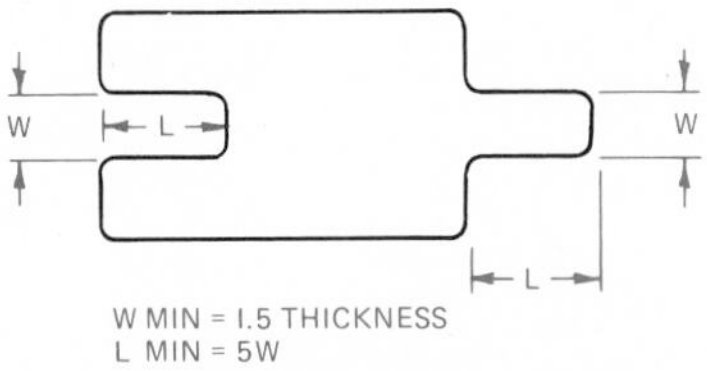
W
L
W
L
W MIN = 1.5 THICKNESS
L MIN = 5W

Fig. 27-2-5 Minimum practical sections.

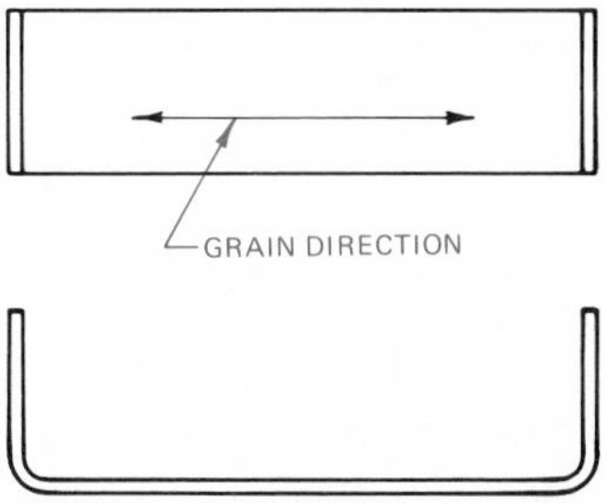
GRAIN DIRECTION

Fig. 27-2-6 Specifying grain direction.

Hole Sizes and Tolerances. For general economy, a hole diameter should not be less than the thickness of the material. If the hole is less than the material thickness or less than 1 mm, the hole should be drilled and the burr removed at added cost.

Unless specified, tolerances shown on hole diameters are considered to apply to the punch side only. See Fig. 27-2-7.

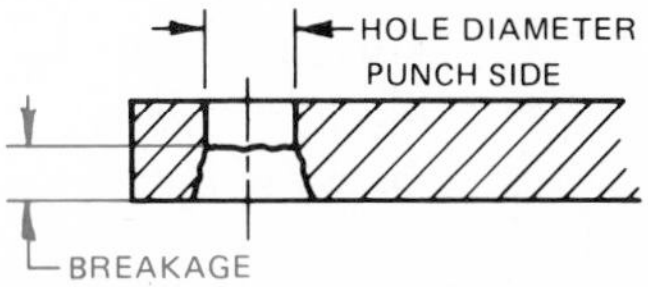
HOLE DIAMETER
PUNCH SIDE
BREAKAGE

Fig. 27-2-7 Punched holes.

Holes near Blank Edge. A hole can be punched without causing a bulge if the web is a minimum of the stock thickness. See Fig. 27-2-8. A bulge will result whenever the web is less than the stock thickness.

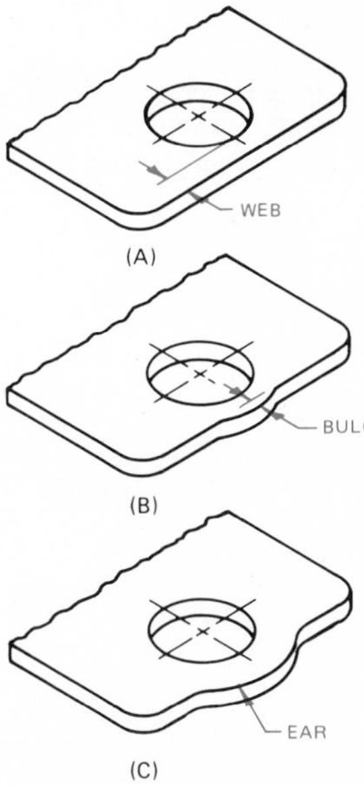
WEB
(A)
BULGE
(B)
EAR
(C)

Fig. 27-2-8 Specifying hole opening near blank edge.

Distance between Holes. The distance between holes or between a hole and the edge of a part should be large enough to prevent tearing of metal and excessive die wear. Recommended minimum distances are shown in Fig. 27-2-9.

Notches with Vertices. Notches in highly stressed parts should be specified

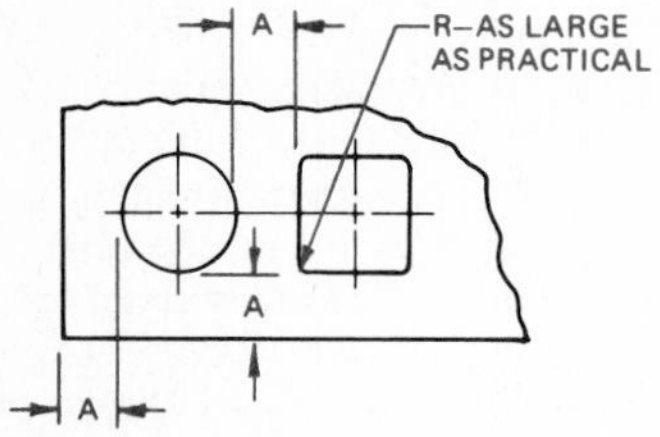
A
R–AS LARGE AS PRACTICAL
A
A

METAL THICKNESS	"A" MINIMUM
UP TO 1.6 INCL	3
OVER 1.6	TWO TIMES METAL THICKNESS

Fig. 27-2-9 Distance between holes.

with a radius at the vertex because a sharp vee might provide the starting point of a tear. The radius should be a minimum of 2 times the metal thickness. A sharp vertex is allowed on lower stress parts when it will aid in lowering design costs. See Fig. 27-2-10.

SOURCE MATERIALS

General Motors Drafting Standards

Assignments

1. On an A3- or B-size sheet, design stock strips for the parts shown on Fig. 27-2-A or 27-2-B. Dimension the width of the stock strip. Scale is 1:1.

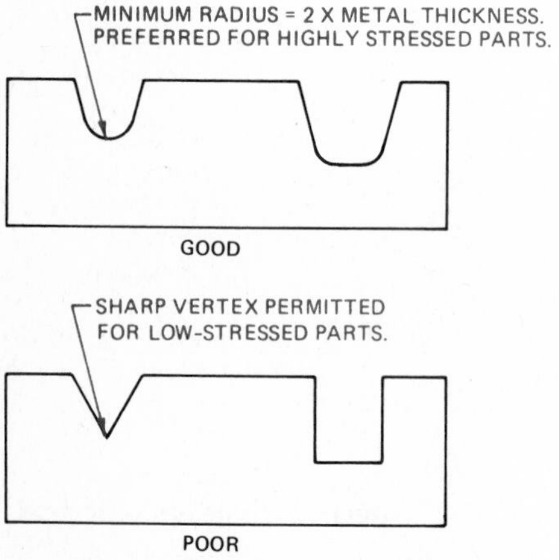
MINIMUM RADIUS = 2 X METAL THICKNESS.
PREFERRED FOR HIGHLY STRESSED PARTS.
GOOD
SHARP VERTEX PERMITTED
FOR LOW-STRESSED PARTS.
POOR

Fig. 27-2-10 Notches with vertices.

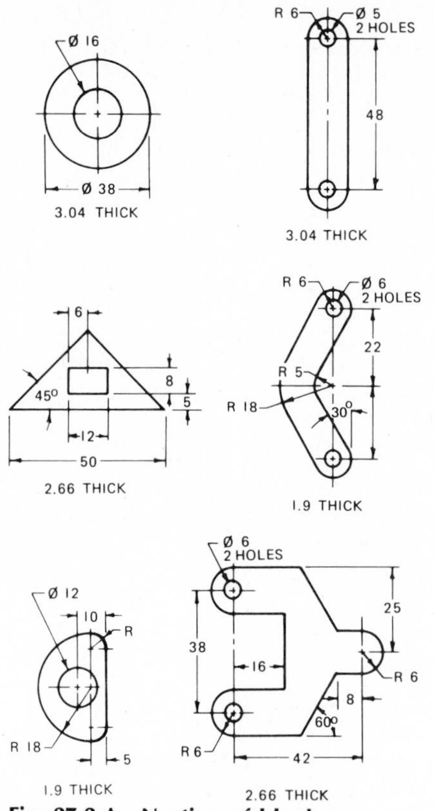
Ø 16
Ø 38
3.04 THICK
R 6
Ø 5
2 HOLES
48
3.04 THICK
6
45°
8
5
12
50
2.66 THICK
R 6
Ø 6
2 HOLES
22
R 5
R 18
30°
1.9 THICK
Ø 12
10
R
R 18
5
1.9 THICK
Ø 6
2 HOLES
25
38
16
R 6
8
60°
R 6
42
2.66 THICK

Fig. 27-2-A Nesting of blanks.

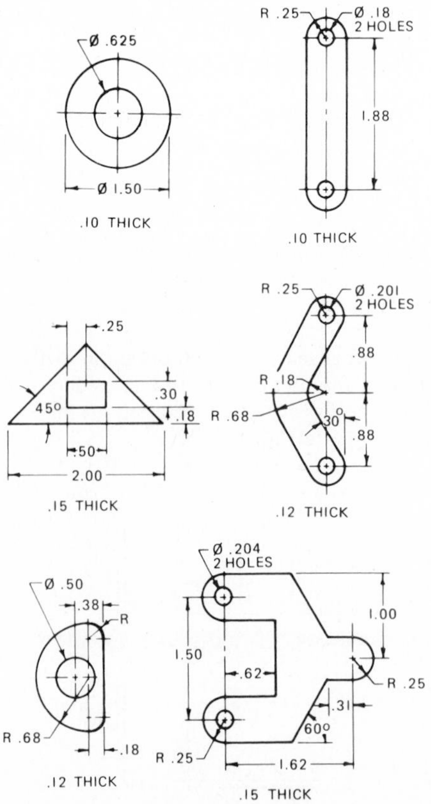
Ø .625
Ø 1.50
.10 THICK
R .25
Ø .18
2 HOLES
1.88
.10 THICK
.25
45°
.30
.18
.50
2.00
.15 THICK
R .25
Ø .201
2 HOLES
.88
R .18
R .68
30°
.88
.12 THICK
Ø .50
.38
R
R .68
.18
.12 THICK
Ø .204
2 HOLES
1.00
1.50
.62
R .25
.31
60°
R .25
1.62
.15 THICK

Fig. 27-2-B Nesting of blanks.

UNIT 27-3

DIE SETS AND COMPONENTS[1]

The punches and dies discussed in Unit 27-1 are relatively simple and inexpensive and have limited use. Many parts produced by punches and dies are more complicated than those already discussed and, as such, require custom-designed die parts.

Die Sets

A *die set* (Fig. 27-3-1) can be defined as a subpress unit consisting of a lower shoe and an upper shoe, together with the guideposts and bushings by which the shoes are aligned.

The following terms are either directly pertinent or closely related to die sets.

Die Shoe. The die-set base is called the *die shoe* (or *die holder*). This remains true even though punches are sometimes mounted on this lower shoe. The great majority of standardized die sets have the guideposts mounted in the die shoe.

Punch Holder. The die-set top member is called the *punch holder* (or *punch shoe*). This remains true even though die blocks are sometimes mounted to this top member. The great majority of standardized die sets have the guide bushings mounted in the punch holder.

Shank. Most punch holders in the smaller sizes are made with a shank which fits the clamping hole in the lower end of the punch press ram. The shank is used to center the die set in the press and to secure the punch holder to the ram.

Guideposts. *Guideposts* (also called *leader pins* or *guide pins*) are cylindrical pins which provide a means of alignment for the die set.

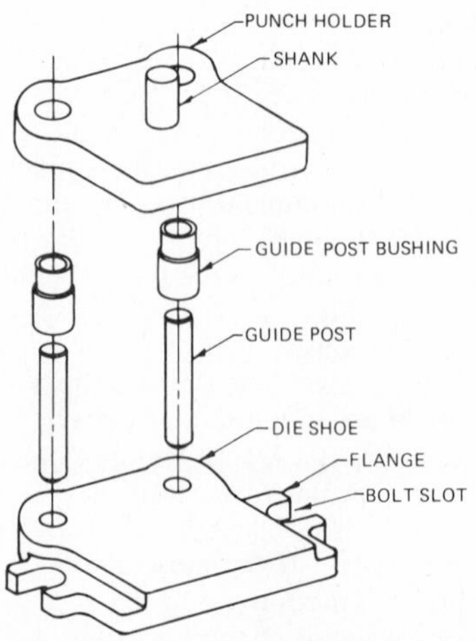
PUNCH HOLDER
SHANK
GUIDE POST BUSHING
GUIDE POST
DIE SHOE
FLANGE
BOLT SLOT

Fig. 27-3-1 Components of typical die set.

Guidepost Bushings. These are installed in the opposing shoe and engage the guideposts with a close sliding fit. The guideposts and bushings, acting together, align the die set.

Flange. This is a ledge which is flush with the bottom surface of a die shoe or the top surface of a punch holder. The ledge extends beyond the die area to provide a means for clamping the shoe member to the bolster plate or press ram, whichever is appropriate.

Die Area. This is the area available on the top surface of the die shoe and the lower surface of the punch holder for the mounting of punch and die components. The die-set guideposts and bushings are normally located outside this area.

Shut Height of Die. This is the distance from the bottom of the die shoe to the top of the punch holder when the die is in its closed working position.

Punches

As previously defined, a punch is a male member of a complete die. It supplements or complements the female die in order to produce a desired effect upon the material being worked. It is both convenient and practical to classify punches in three categories, each of which is composed of two major groups that include the numerous punch types and variations.

PUNCHES MOUNTED IN PUNCH PLATES

A punch plate is so called because its function is to retain and/or position a punch or punches. Punches assembled into punch plates are, for many applications, the most practical method of punch mounting.

One method of assembling a punch and punch plate combination is shown in Fig. 27-3-2. Considered alone (without the punch plate), this is essentially a plain punch except that it does not contain dowels. It does, however, contain mounting screws, which are used to fasten the punch within the confines of the opening provided in the punch plate. With this type of construction, the punch plate is used not to hold the punch during assembly, but only to provide an effective means for accurately positioning and supporting the punch.

PERFORATOR-TYPE PUNCHES

A perforator-type punch can be described as a cutting punch 25 mm or less in diameter, if it is round. If it is not round, then its contour may be circumscribed by a circle whose diameter is 25 mm or less.

For the sake of convenience, these punches are commonly called *perforators* whether or not their function is strictly one of perforating. As a rule, perforators are mounted in punch plates. Perforators whose working contours are other than round are often made with round shanks, and these must be secured against rotation. They are normally keyed to the punch plate in order to ensure proper orientation. Common types of key arrangements are shown in Fig. 27-3-3.

Step-Head Perforators. For general application, the step-head perforator shown in Fig. 27-3-4 is probably the most widely used perforator type.

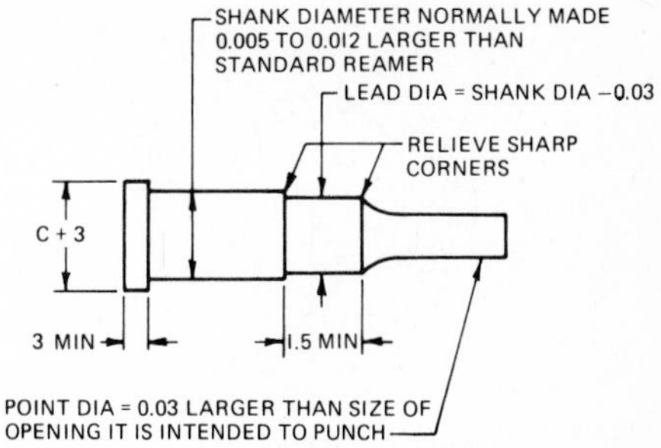
SHANK DIAMETER NORMALLY MADE 0.005 TO 0.012 LARGER THAN STANDARD REAMER
LEAD DIA = SHANK DIA − 0.03
RELIEVE SHARP CORNERS
C + 3
3 MIN
1.5 MIN
POINT DIA = 0.03 LARGER THAN SIZE OF OPENING IT IS INTENDED TO PUNCH

Fig. 27-3-4 Step-head perforator.

Mounting Screws and Dowels

Any construction method that securely holds a die component in assembly and simultaneously ensures accurate positioning of the component throughout the life of the die may be said to be satisfactory. If, in addition, the method provides ease of necessary maintenance and is not needlessly expensive to make originally, it can then be considered good diemaking practice.

It is common procedure to use a combination of screws and dowels to provide for proper mounting (assembling) of the various components that make up a die. The function of the dowels is to provide and maintain accurate positioning of the component. The function of the screws is to retain the component securely in the doweled position.

Die Blocks

The term *die block* refers to the complete unit component which mates with the punch or punches in the manner required to produce the desired effect upon the stock material. It may consist of a single piece, or it may be an assembled unit composed of a number of parts.

Among the simplest of die block constructions is the one-piece ring shown in Fig. 27-3-5. Dowel pins are used to locate and position the die ring. The screw and dowel locations shown are conventionally typical for smaller rings of this kind. Bolt circle diameter C should be such that the screw and dowel positions will not be too close to the wall of the die shoe clearance opening. For rings with minimal cross section W, this usually requires that dimension F be somewhat larger than E. For rings with wider cross sections W, the bolt circle diameter may be made to divide W equally, that is, $E = F$.

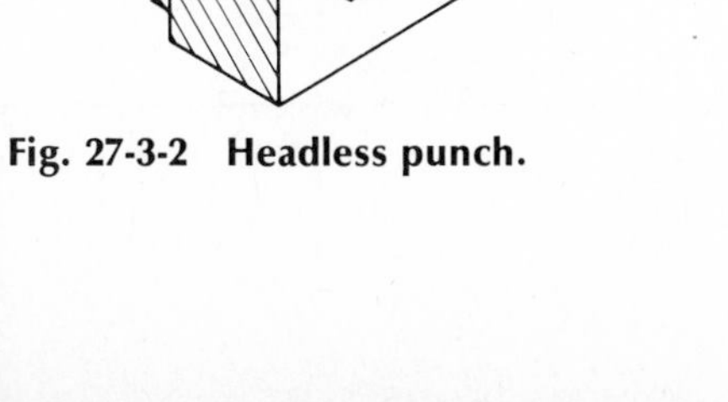
PUNCH PLATE
PUNCH

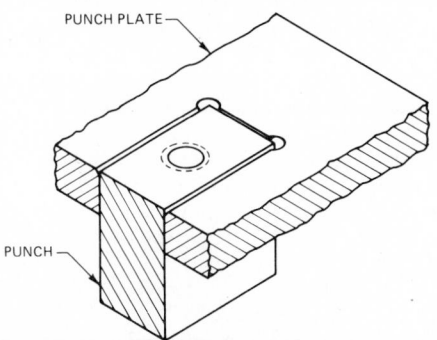

Fig. 27-3-2 Headless punch.

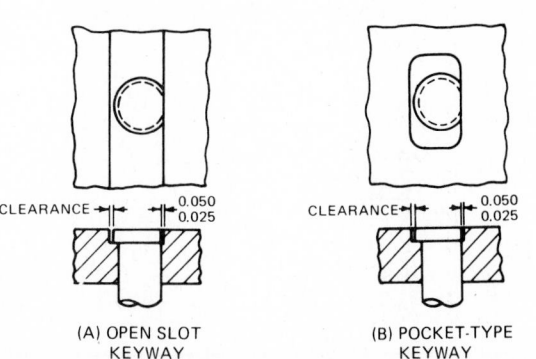

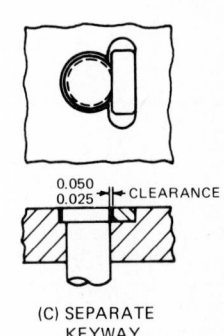

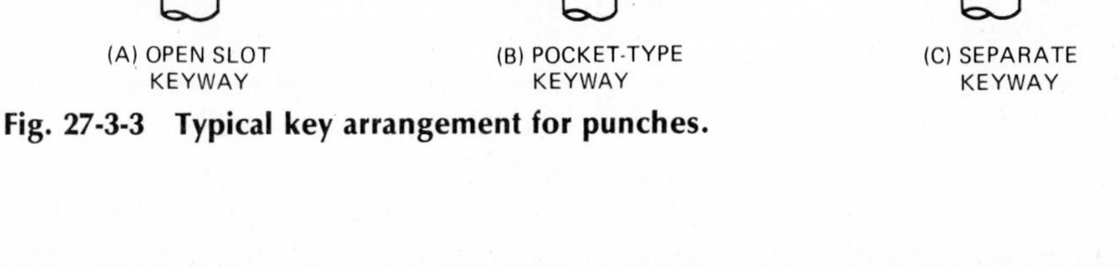
CLEARANCE
0.050
0.025
(A) OPEN SLOT KEYWAY
CLEARANCE
0.050
0.025
(B) POCKET-TYPE KEYWAY
0.050
0.025
CLEARANCE
(C) SEPARATE KEYWAY

Fig. 27-3-3 Typical key arrangement for punches.

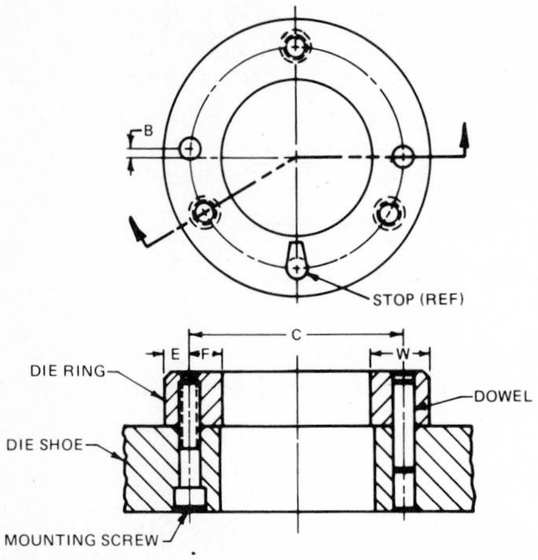
B
STOP (REF)
C
E
F
W
DIE RING
DOWEL
DIE SHOE
MOUNTING SCREW
Fig. 27-3-5 One-piece ring-type die ring—dowel mounted.

Note that one dowel pin *B* is offset. This is a foolproofing procedure; if the dowel were not offset, it would be possible to mount this ring upside down. Foolproofing can, of course, be accomplished in other ways. For example, instead of offsetting, one dowel could be smaller, or one of the screw locations could be offset.

Another common construction for round work is shown in Fig. 27-3-6. Here, the die ring is set in the die shoe.

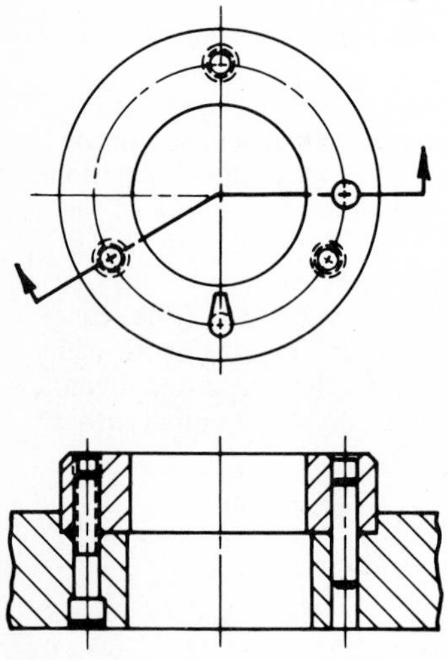
Fig. 27-3-6 One-piece die ring—set in.

Strippers

Stripping is the act of removing the work from the punch or punches. A *stripper* is a device for stripping.

There are two basic stripping actions, depending upon whether the punch is motivated or stationary. With a motivated punch, the stripper acts to arrest the stock material, permitting the moving punch to withdraw from it. This action is depicted in Figs. 27-3-7 and 27-3-8. Figure 27-3-7 illustrates the arresting action of a fixed stripper.

View A. The die is in a closed position, with the punch entered, at the bottom of the press stroke.

View B. The punch is ascending on the return stroke. The stock material (work) clings to the punch, moving upward with it until it contacts the intervening stripper.

View C. The punch, as it continues upward, is withdrawn from the stock material, permitting the stock material to drop toward the die surface.

View D. The working cycle is completed. The punch has returned to its up position. The workpiece has dropped back to the die face and is ready to be removed.

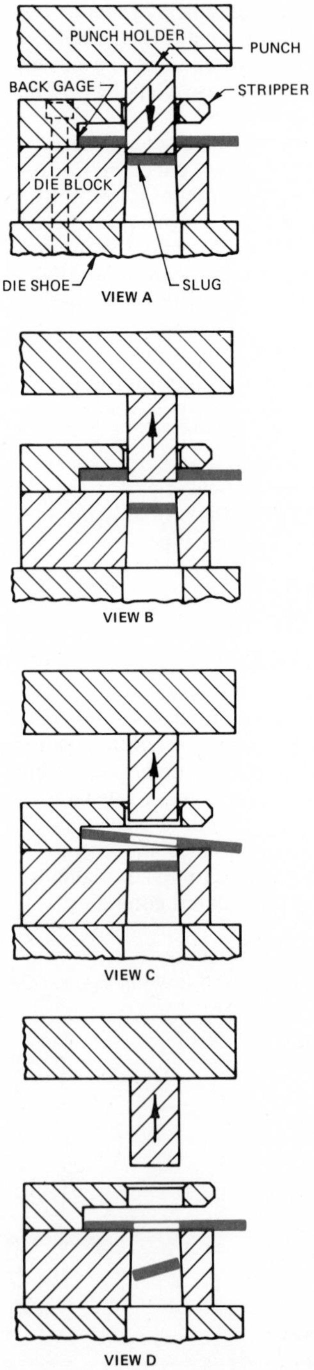
PUNCH HOLDER
PUNCH
BACK GAGE
STRIPPER
DIE BLOCK
DIE SHOE
SLUG
VIEW A
VIEW B
VIEW C
VIEW D
Fig. 27-3-7 Stripping action—fixed stripper, punch moving with ram.

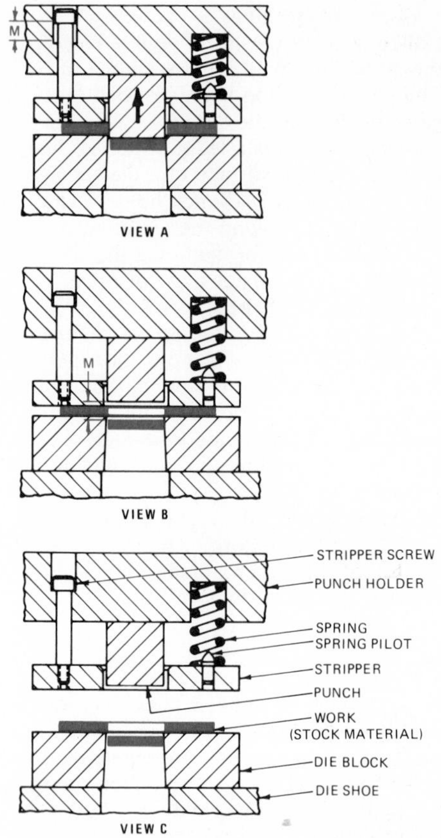
M
VIEW A
M
VIEW B
STRIPPER SCREW
PUNCH HOLDER
SPRING
SPRING PILOT
STRIPPER
PUNCH
WORK
(STOCK MATERIAL)
DIE BLOCK
DIE SHOE
VIEW C
Fig. 27-3-8 Stripping action—pressure-pad stripper, punch moving with ram.

Figure 27-3-8 depicts the arresting action of a pressure-pad-type stripper. The stripper shown is spring-actuated. Gages, spring pins, etc., have been omitted for purposes of clarification.

View A. The die is in a closed position, punch entered, at the bottom of the press stroke. The potential stripper travel for the stripper action is indicated at *M*.

View B. The punch assembly is ascending on the return stroke. The stripper is still in the position shown in view A. The springs have maintained this stripper position, forcing the ascending punch to withdraw from the stock material, as shown. Stripper travel *M* relative to the punch is noted.

View C. The working cycle is completed at the top of the press stroke. The punch holder has carried the punch and stripper assembly with it to the up position. The workpiece has been left behind, lying on the die surface to be retrieved or ejected as required.

Angular Clearance

Angular clearance is a draft or taper applied to the sidewalls of a die opening (Fig. 27-3-9) in order to relieve the internal pressure of the blank or slug as it passes through the opening.

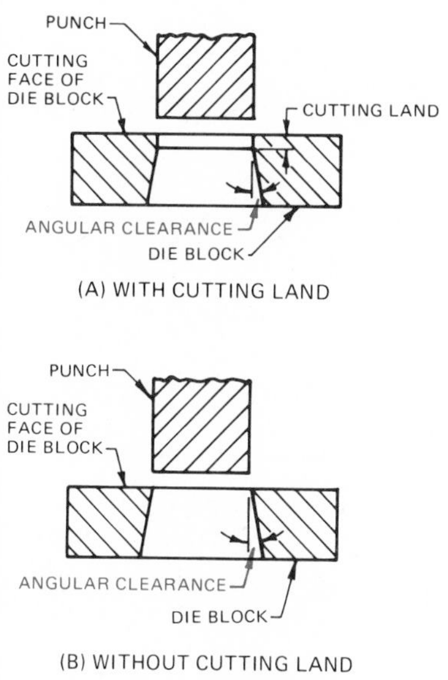
PUNCH
CUTTING FACE OF DIE BLOCK
CUTTING LAND
ANGULAR CLEARANCE
DIE BLOCK
(A) WITH CUTTING LAND
PUNCH
CUTTING FACE OF DIE BLOCK
ANGULAR CLEARANCE
DIE BLOCK
(B) WITHOUT CUTTING LAND

Fig. 27-3-9 Angular clearance.

Specifying Angular Clearance. Angular clearance should be expressed in terms of the amount of clearance per side, not as an overall or included-angle figure. The optimum angular clearance for any given die opening is a variable. It depends upon the type of material to be run, the production requirements, and the method of die construction. The strength of the cutting members also influences the amount of angular clearance to be used.

Generally, soft materials require greater angular clearance than hard ones. Aluminum especially tends to jam or pack in die openings. Heavy gages of material require more angular clearance than thinner material of the same kind.

Maximum and Minimum Angular Clearance. With few exceptions, an angle of 2° per side can be considered a maximum desirable clearance angle. A practical minimum clearance angle is 0.13° per side (1:250 taper).

Methods of Providing Angular Clearance. Figure 27-3-9 illustrates the application of angular clearance to die openings. In view A, the opening is made with a cutting land contiguous to the cutting edge and extending into the die opening. The height of the cutting land should be equal to the thickness of the stock material, but should not be less than 2 mm. A height of 4 mm is considered good average practice for cutting material less than 4 mm thick.

View B shows the angular clearance beginning at the cutting edge. This method is used when it is necessary or desirable to relieve the compression stored within the blanked part as soon as possible. Angular clearance, beginning at the cutting edge, is the preferred method for cutting materials that are abrasive.

Blank and Slug Openings in Die Shoes

After a blank or slug has passed through its die opening, it falls through the clearance opening in the die shoe. The following requirements must be met: (1) the blanks or slugs must fall freely, without interference; (2) the contour of the clearance opening in the die shoe should be made as simple as possible; (3) the opening should not weaken the die shoe any more than necessary; and (4) the contour of the die shoe opening must be such that it provides adequate support for the die block.

Various typical die-shoe clearance openings are shown in Fig. 27-3-10.

Straight and Tapered Die Shoe Openings. At times it is difficult to decide whether die-shoe clearance openings should be made straight or tapered (Fig. 27-3-10*c* and *d*).

The tapered opening is unquestionably the safest. The straight opening, however, is often easier to make, especially when the die shoe is large and/or thick.

Fortunately, the larger blanks cause the least trouble. This usually allows the clearance openings for larger blanks to be made with straight sidewalls. Small slugs or blanks (less than 25 mm across) should have tapered clearance holes in the die shoe. This is true for both round and irregular shapes. The die shoe openings should also be tapered if a die is to be run in an inclined press.

Amount of Taper. The amount of taper on the sidewalls of the drop-through opening in a die shoe is not critical. An angle 0.5° to 2° will satisfy most cases. The angle must begin at the top of the die shoe (next to the die block) and extend clear through the die shoe.

Amount of Offset. The top edge of the die shoe opening is offset from the bottom edge of the die opening. The amount of offset is not critical; it may vary from a step of 0.5 mm to a step of 3 mm or more at times. Straight openings are usually offset more than tapered openings. Illustrations of offsets can be seen in Fig. 27-3-10*b*, *c*, and *d*.

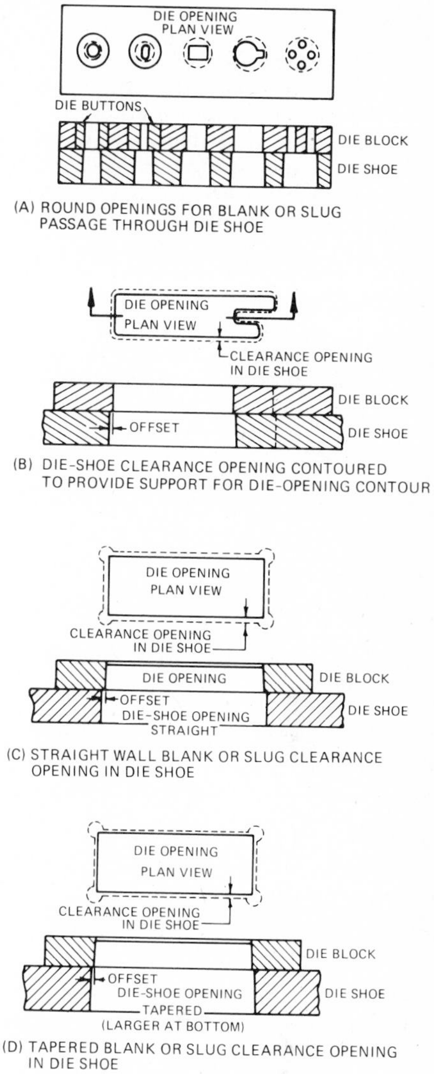
DIE OPENING PLAN VIEW
DIE BUTTONS
DIE BLOCK
DIE SHOE
(A) ROUND OPENINGS FOR BLANK OR SLUG PASSAGE THROUGH DIE SHOE
DIE OPENING
PLAN VIEW
CLEARANCE OPENING IN DIE SHOE
OFFSET
DIE BLOCK
DIE SHOE
(B) DIE-SHOE CLEARANCE OPENING CONTOURED TO PROVIDE SUPPORT FOR DIE-OPENING CONTOUR
DIE OPENING
PLAN VIEW
CLEARANCE OPENING IN DIE SHOE
DIE OPENING
OFFSET
DIE-SHOE OPENING STRAIGHT
DIE BLOCK
DIE SHOE
(C) STRAIGHT WALL BLANK OR SLUG CLEARANCE OPENING IN DIE SHOE
DIE OPENING
PLAN VIEW
CLEARANCE OPENING IN DIE SHOE
OFFSET
DIE-SHOE OPENING TAPERED (LARGER AT BOTTOM)
DIE BLOCK
DIE SHOE
(D) TAPERED BLANK OR SLUG CLEARANCE OPENING IN DIE SHOE

Fig. 27-3-10 Die-shoe openings.

Extra Clearance for Sharp Corners. Since sharp corners on blanks or slugs can dig into the sidewalls of the die-shoe clearance opening, it is a good idea to provide extra clearance for sharp corners. Refer to the plain views in Fig. 27-3-10*c* and *d* and note the extra clearance shown at the corners. In these instances, the extra clearance is provided by drilling holes at the corners far enough out that the sharp corners on the blanks cannot touch the walls of the die shoe opening. These holes also admit the saw blade and make it easier to saw out the opening.

Arrangement of Views

In laying out the design of a punch and die, an approved arrangement of views is recommended over the theoretical arrangement of views, as illustrated in Fig. 27-3-11. The stock strip is normally drawn in red or shown by phantom or broken lines on the assembly drawing. Standard die sets are shown in Fig. 27-3-12.

Fig. 27-3-11 Arrangement of views for punch and die drawings.

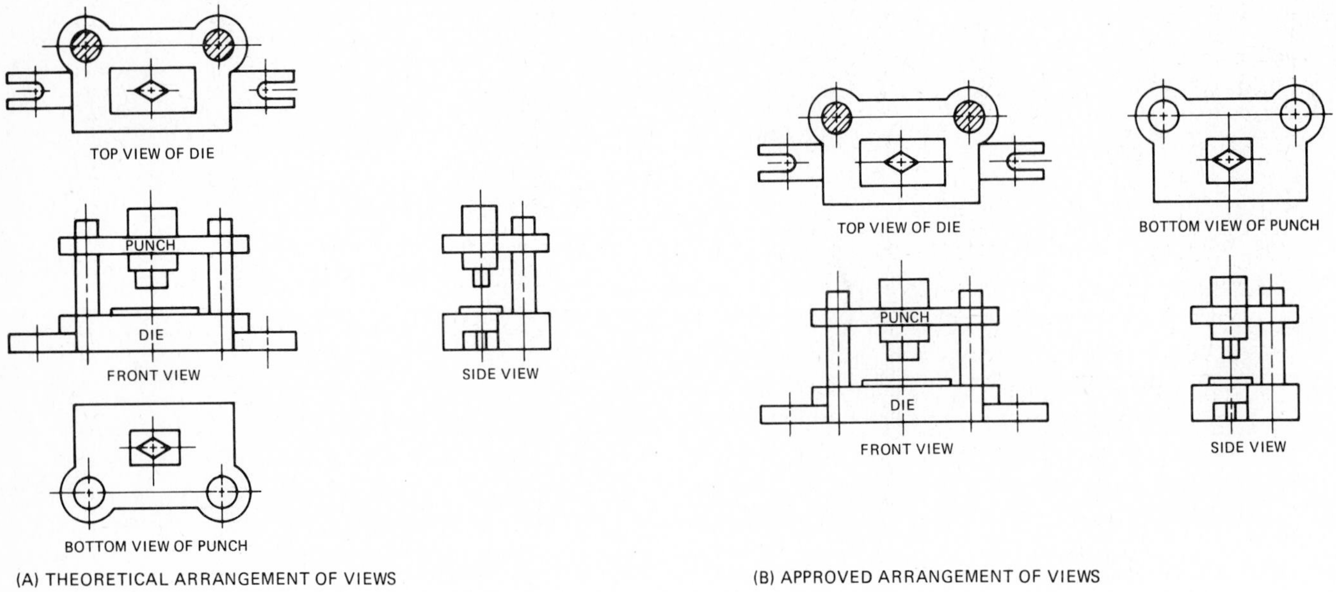
TOP VIEW OF DIE
PUNCH
DIE
FRONT VIEW
SIDE VIEW
BOTTOM VIEW OF PUNCH
(A) THEORETICAL ARRANGEMENT OF VIEWS
TOP VIEW OF DIE
BOTTOM VIEW OF PUNCH
PUNCH
DIE
FRONT VIEW
SIDE VIEW
(B) APPROVED ARRANGEMENT OF VIEWS

Blanking and Punching Dies

BLANKING DIES

Figure 27-3-13 is an assembly drawing of the die which produces the blank for the piece part illustrated. This die does the blanking operation only. The holes in the piece part are punched in a subsequent punching die.

Each component of the die is identified by detail number and name. The number required of each component or detail is also shown. Socket-head cap screws are indicated by the abbreviation SHCS.

Operation of Die. The stock material is furnished in strip form 70 mm wide. It is fed across the die face from right to left until the lead end contacts the stop, at detail 6. The press is then tripped, causing the first piece to be blanked out. After the blanking, the scrap bridge is lifted over the stop. As soon as the bridge passes over the stop, the strip is dropped back to the die face. The head of the stop is now within the previously blanked-out opening. The stock strip is then advanced until the edge of the blanked-out opening contacts the stop. This process is repeated at every press stroke until each strip is used up.

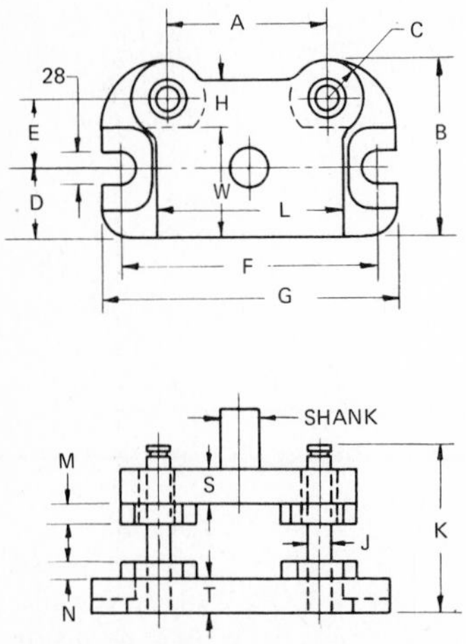
A
C
28
H
E
B
D
W
L
F
G
SHANK
M
S
J
K
T
N

Fig. 27-3-12 **Standard die sets.** (E. A. Baumbach Mfg. Co.)

Width Inches W	Length Inches L	Thickness Die T	Thickness Punch S	Catalog Number	A	B	C	D	E	F	G	H	J	K	M	N
75	100	30	25	B1-341	75	125	30	50	47	165	190	40	20		16	12
		40	25	B1-342												
75	150	40	30	B1-361	125	135	30	50	50	215	250	43	22		20	16
		40	45	B1-362												
		50	30	B1-363												
		50	45	B1-364												
75	200	40	30	B1-381	175	135	30	50	50	260	300	43	22		20	16
		40	45	B1-382												
		50	30	B1-383												
		50	45	B1-384												
75	250	40	30	B1-3101	225	140	35	50	53	315	355	46	25	SPECIFY	20	16
		40	45	B1-3102												
		50	30	B1-3103												
		50	45	B1-3104												
100	100	40	30	B1-442	75	160	30	65	65	165	190	43	22		20	16
		45		B1-443												
100	125	40	30	B1-452	100	160	30	65	65	190	215	43	22		20	16
		45		B1-453												
100	150	40	30	B1-461	125	160	30	65	65	215	250	43	22		20	16
		40	45	B1-462												
		40	55	B1-463												
		45	30	B1-464												
		45	45	B1-465												
		45	55	B1-466												
		70	30	B1-467												
		70	45	B1-468												
		70	55	B1-469												

Fixed Stop. If a plain pin (no head) type of fixed stop was used, it would have to be installed at a safe distance from the cut edge in order to avoid weakening the die block. As a result, the scrap bridge would be wider than necessary, which would mean, of course, that stock material was being wasted. In order to minimize the amount of scrap while still maintaining adequate die block strength, the die shown has a headed fixed-pin stop. The diameter of the head is made rather large in proportion to the shank diameter, which enables the mounting hole to be located a safe distance from the edge of the die opening. The thickness or height of the stop head should be held to a practical minimum in order to make the feeding of the stock strip as easy as possible. The stop is shown in detail in Fig. 27-3-14.

Stripping. A gap-type fixed stripper is used. The gap is made relatively high for ease of feeding. The height of the gap and the open front provide visibility and accessibility for the press operator.

Since the scrap bridge must be lifted over the fixed stop each time the stock strip is advanced, the stripper gap must be high enough to allow free passage of the stock strip during feeding. In this die, the head of the stop is 4 mm high. This combination allows a space of 8 mm between the stop and the stripping surface. The stock material thickness (*T*) is, in this case, 1.21 mm. The feeding clearance between stop and stripper is slightly more than 6*T*, which is ample for ease of feeding.

Because the die is a blanking die, the die opening sizes are derived directly from the piece-part dimensions. The punch dimensions are determined by deducting the cutting clearance on each and every side from the die opening dimensions. Keep in mind that, with blanking dies, the cutting clearance exists all around the cutting periphery.

BLANKING DIE DETAILS

The die details are shown in Fig. 27-3-14. For the sake of clarity, only the overall dimensions and the dimensions pertinent to producing the piece part are shown. Abbreviations used are OHTS for oil-hardening tool steel, CRS for cold-rolled steel, and RC 60 to 62 to indicate the hardness of the hardened two-steel components as measured on the C scale of a Rockwell tester.

Die Block. The die block, part 2, is for a blanking die. Therefore, the piece parts produced will be sized by the die opening. The piece part specifies a width of 40 +0, −0.1 mm. The tolerance is negative (minus). This gives the width

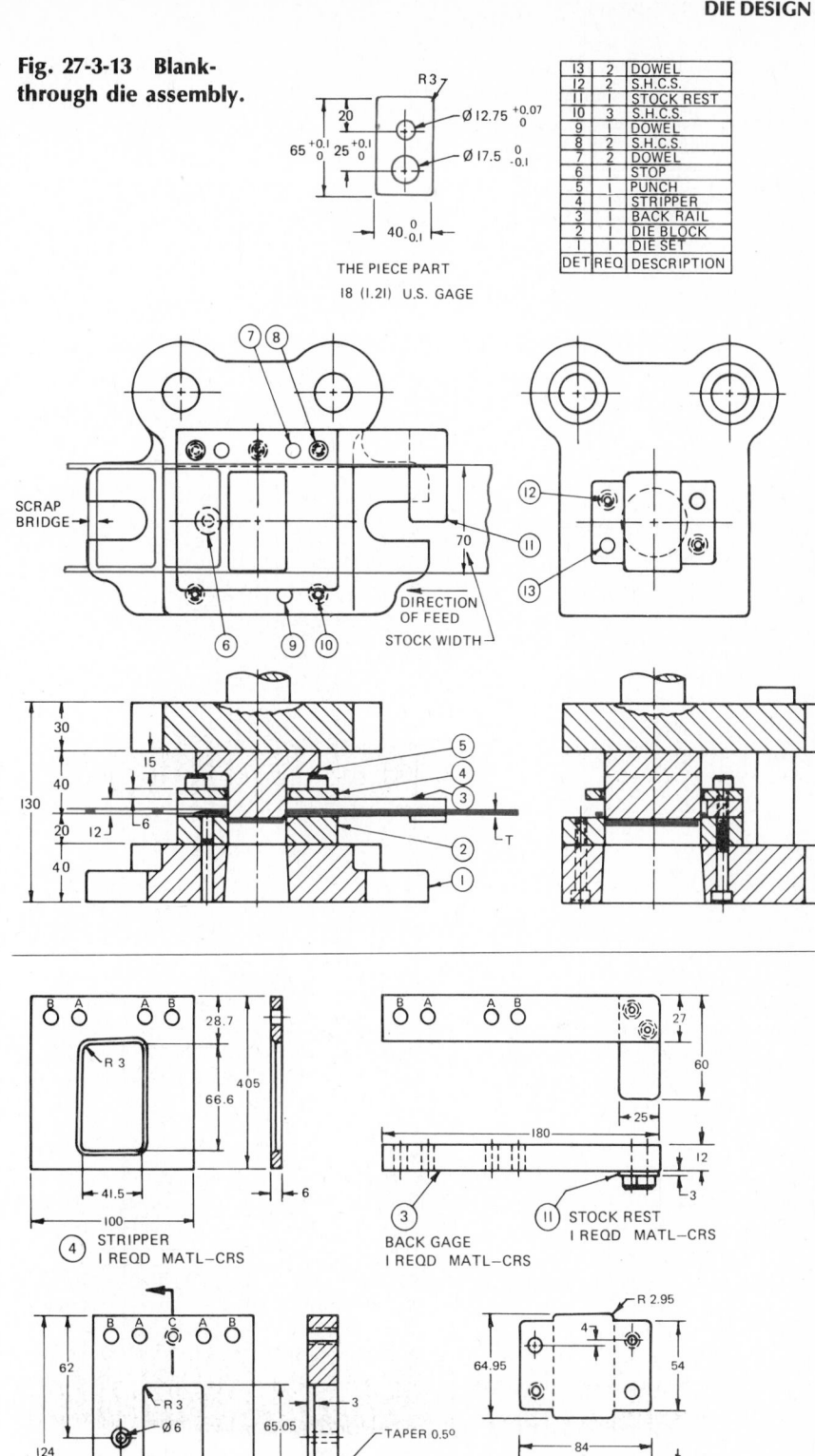

Fig. 27-3-13 Blank-through die assembly.

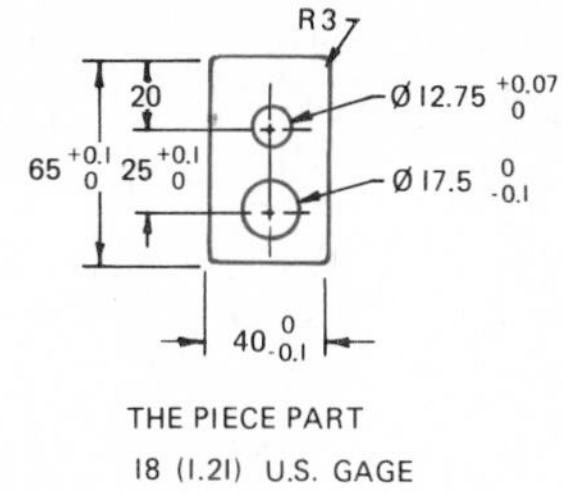
R3
20
65 +0.1 0
25 +0.1 0
Ø 12.75 +0.07 0
Ø 17.5 0 -0.1
40 0 -0.1
THE PIECE PART
18 (1.21) U.S. GAGE

DET	REQ	DESCRIPTION
13	2	DOWEL
12	2	S.H.C.S.
11	1	STOCK REST
10	3	S.H.C.S.
9	1	DOWEL
8	2	S.H.C.S.
7	2	DOWEL
6	1	STOP
5	1	PUNCH
4	1	STRIPPER
3	1	BACK RAIL
2	1	DIE BLOCK
1	1	DIE SET

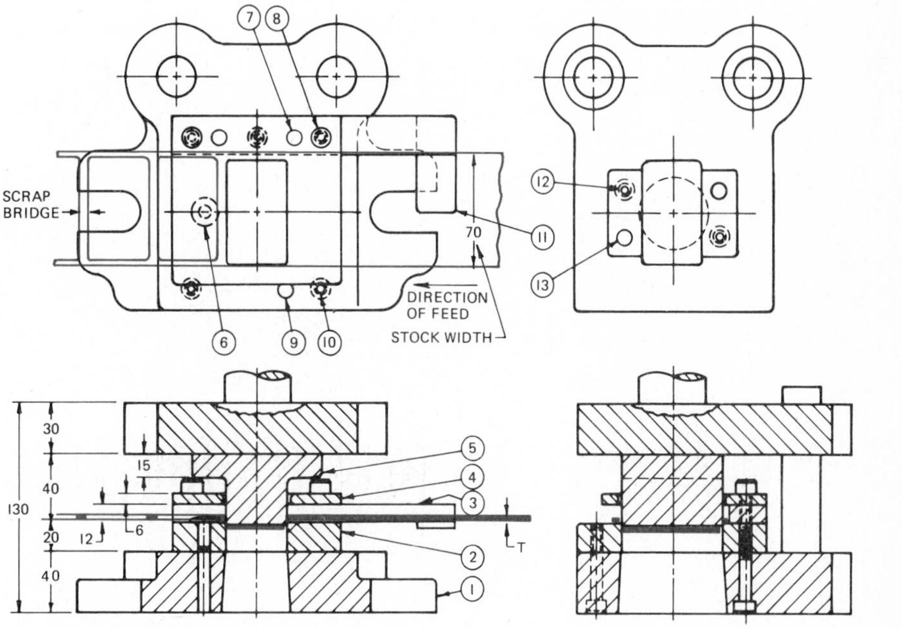
SCRAP BRIDGE
70
DIRECTION OF FEED
STOCK WIDTH
30
15
40
130
20
12
6
40
T

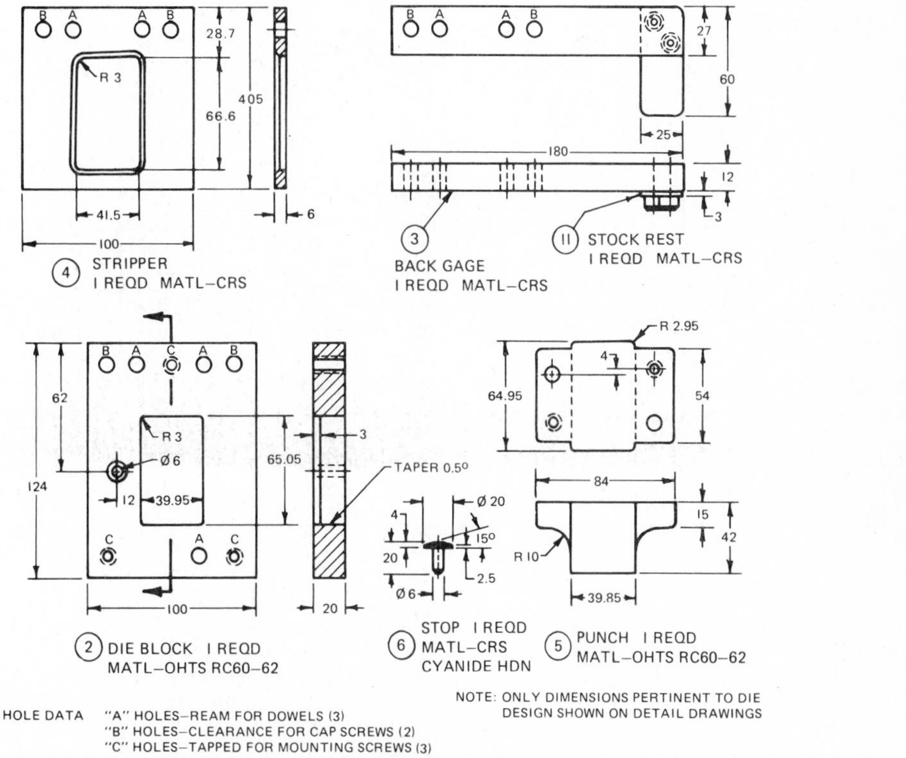
28.7
R 3
405
66.6
41.5
100
6
4 STRIPPER
1 REQD MATL–CRS
27
60
25
180
12
3
3 BACK GAGE
1 REQD MATL–CRS
11 STOCK REST
1 REQD MATL–CRS
62
R 3
Ø 6
65.05
124
12
39.95
100
20
3
TAPER 0.5°
2 DIE BLOCK 1 REQD
MATL–OHTS RC60–62
Ø 20
15°
20
2.5
Ø 6
STOP 1 REQD
6 MATL–CRS
CYANIDE HDN
R 2.95
64.95
54
84
15
42
R 10
39.85
5 PUNCH 1 REQD
MATL–OHTS RC60–62
NOTE: ONLY DIMENSIONS PERTINENT TO DIE DESIGN SHOWN ON DETAIL DRAWINGS
HOLE DATA "A" HOLES–REAM FOR DOWELS (3)
"B" HOLES–CLEARANCE FOR CAP SCREWS (2)
"C" HOLES–TAPPED FOR MOUNTING SCREWS (3)

Fig. 27-3-14 Details of blank-through die.

dimension a range of 39.9 to 40 mm. The mean dimension is 39.95 mm. With this in mind, a figure of 39.95 mm was chosen for the die opening width.

The length of the piece part is 65 +0.1, −0 mm. This tolerance is positive (plus). The range is 65.0 to 65.1 mm, and the mean dimension is, obviously, 65.05. The die opening figure selected for this dimension is 65.05 mm.

The figures chosen give a die opening size of 39.95 × 65.05 mm. These figures are the optimum die opening sizes for the given conditions. A die opening made to these figures will yield a blank (or piece part) that will measure very nearly the exact mean of the dimensions specified on the piece-part drawing.

Tolerances are specified on the piece-part drawing. The purpose of the tolerances is to indicate a range of acceptable sizes: all piece parts produced within the range are acceptable. This does not mean that the die opening can be made to the same tolerance range as the piece part.

A good choice for minimum acceptable die opening size is 39.94 × 65.01 mm. The maximum acceptable die opening should be 39.98 × 65.08 mm.

Punch. The punch sizes shown on part 5 are 39.85 × 64.95 mm, with a corner radius of 2.95. These cutting figures are the result of deducting the cutting clearance from the optimum die opening dimensions. It is standard die design procedure to show punch and die dimensions in this manner.

Back Gage. The back gage, part 3, is 28 mm wide. This dimension is determined as follows. The stock material width is specified as 70 mm. One-half of this stock width is 35 mm. The distance from the center line of the die opening to the back of the die block is 62 mm.

Subtracting 35 from 62 leaves a difference of 27 mm from the edge of the stock strip to the back of the die block. Using this 27 mm dimension for the width of the back gage aids the mounting of the gage of the die block.

Stripper. The stripper opening, part 4, measures 41.5 × 66.5 mm. These dimensions give a clearance space of approximately 0.8 mm on each side between the punch and the stripper.

PUNCH DIE

Figure 27-3-15 is an assembly drawing of a die to punch two holes in the piece part illustrated. This die performs the punching of the holes only; the blank was produced by the blank-through die shown in Fig. 27-3-13. For convenience, the piece part has been shown on each die assembly drawing.

The punch die is set up in the same manner as the blank-through die assembly (Fig. 27-3-13). The components are identified, and the number of each required is stated. Socket-head cap screws are abbreviated SHCS. Socket-head stripper screws are indicated by the abbreviation SHSS.

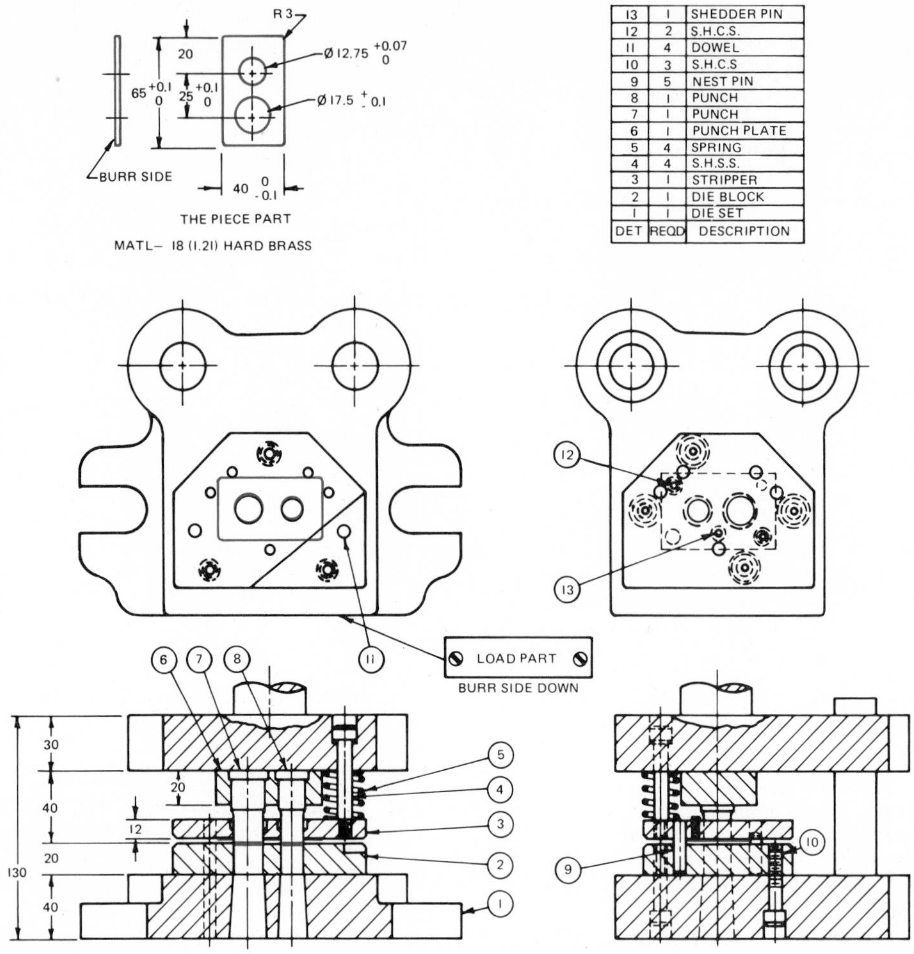
R 3
20
Ø 12.75 +0.07 0
65 +0.1 0
25 +0.1 0
Ø 17.5 + 0.1
BURR SIDE
40 0 -0.1
THE PIECE PART
MATL— 18 (1.21) HARD BRASS
LOAD PART
BURR SIDE DOWN
30
40
20
40
130
12
20

DET	REQD	DESCRIPTION
13	1	SHEDDER PIN
12	2	S.H.C.S.
11	4	DOWEL
10	3	S.H.C.S
9	5	NEST PIN
8	1	PUNCH
7	1	PUNCH
6	1	PUNCH PLATE
5	4	SPRING
4	4	S.H.S.S.
3	1	STRIPPER
2	1	DIE BLOCK
1	1	DIE SET

Fig. 27-3-15 Punch die assembly.

Spring Stripper. This die has a moving stripper actuated by springs. The designer elected to use this type of stripper for reasons of accessibility for the press operator. Using a spring stripper for this die also ensures a flatter piece part after piercing, because the stripper pressure holds the part against the face of the die block while the cutting action is taking place.

Instruction Plate. Note that the drawing calls for an instruction plate. The plate, which reads LOAD PART BURR SIDE DOWN, must be clearly lettered and located at the front of the die where it can be easily seen by the operator. The reason for the plate is that the piece-part drawing specifies a burr side, and because the blank is symmetrical, there is no way to foolproof the loading of the blank in the die. As a result, the press operator must be instructed to feed the part into the die in such a manner that the burr side of the blank and the burr side of the pierced holes will agree, as specified.

Locating and Feeding. Proper location of the pierced holes in the piece part is achieved by a nest area created by the nest pins, part 9. When the blank is placed in the nest area, it is confined between the nest pins, which are correctly located with respect to the die openings and punches.

Cutting Clearance. When the cutting clearance for the blank-through die was determined, the figure was found to be 0.06 mm per side. However, for the punching die, owing to the fact that the punchings are round, a cutting clearance figure of 0.04 mm per side was chosen.

PUNCH DIE DETAILS

Figure 27-3-16 is the detail sheet for the punch die. Major overall sizes and dimensions associated directly with the piece-part requirements are shown.

Punches. Refer to Fig. 27-3-16, parts 7 and 8. One of the openings in the piece part is dimensioned 12.75 +0.07, −0 mm diameter. For this opening, a punch diameter of 12.8 mm was decided upon. A

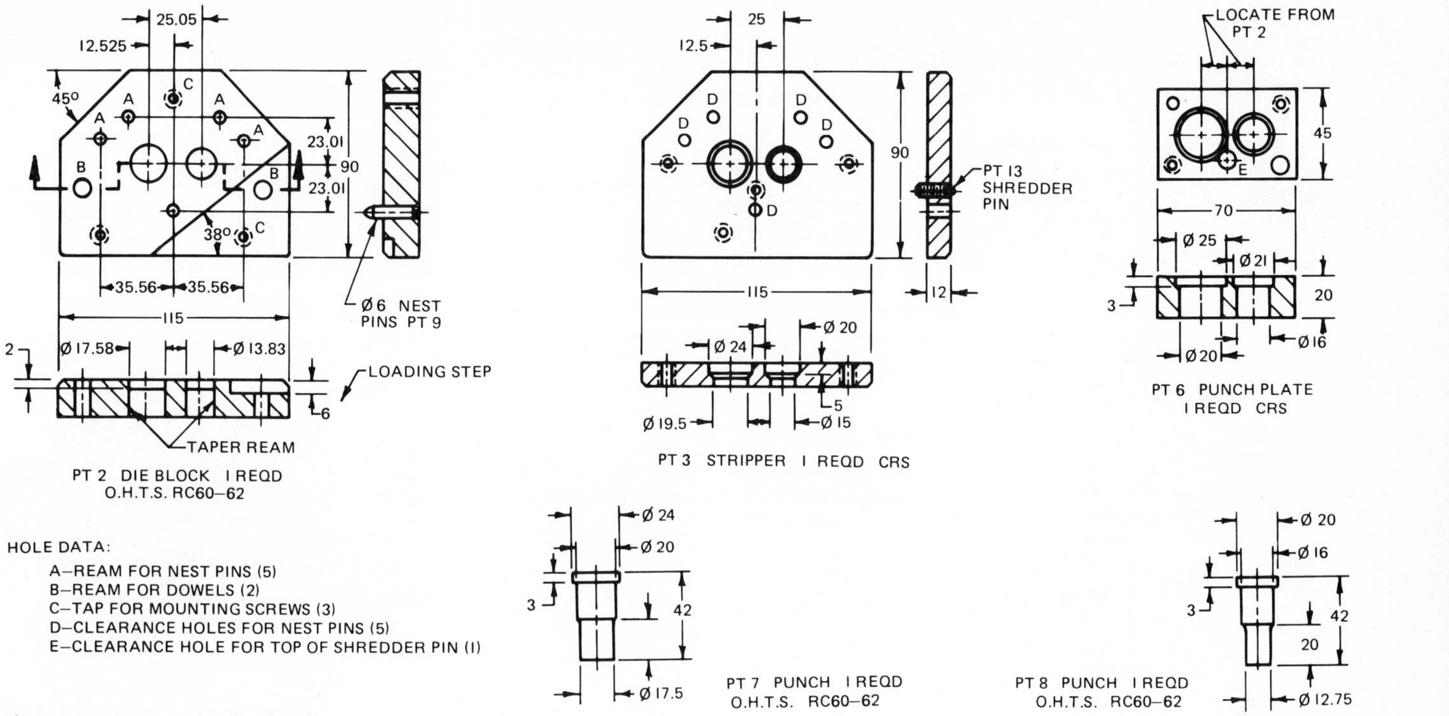

25.05
12.525
45°
23.01
90
23.01
38°
35.56
35.56
115
Ø6 NEST PINS PT 9
2
Ø17.58
Ø13.83
LOADING STEP
6
TAPER REAM
PT 2 DIE BLOCK 1 REQD
O.H.T.S. RC60–62
HOLE DATA:
A–REAM FOR NEST PINS (5)
B–REAM FOR DOWELS (2)
C–TAP FOR MOUNTING SCREWS (3)
D–CLEARANCE HOLES FOR NEST PINS (5)
E–CLEARANCE HOLE FOR TOP OF SHREDDER PIN (1)
25
12.5
90
PT 13 SHREDDER PIN
115
12
Ø24
Ø20
Ø19.5
5
Ø15
PT 3 STRIPPER 1 REQD CRS
LOCATE FROM PT 2
45
70
Ø25
Ø21
20
3
Ø20
Ø16
PT 6 PUNCH PLATE
1 REQD CRS
Ø24
Ø20
3
42
Ø17.5
PT 7 PUNCH 1 REQD
O.H.T.S. RC60–62
PT 8 PUNCH 1 REQD
O.H.T.S. RC60–62
Ø20
Ø16
3
42
20
Ø12.75

Fig. 27-3-16 Punch die details.

punch which has a point or working diameter of 12.8 mm will produce an opening in the piece part of 12.79 to 12.80 mm, which is optimum for this particular opening.

The other opening is specified as 17.5 ± 0.1 diameter. Since the mean dimension for this figure is 17.5 mm, a punch diameter of 17.5 mm is used.

It is an accepted practice to make the shank diameters of round punches in 2 mm increments up to 12 mm diameter and in 4 mm increments for diameters larger than 12 mm. Using diameters of these size increments means that the punch-plate openings can be sized by reamers that are common toolroom equipment. Following this practice led to a choice of 16 and 20 mm diameters for the respective punch shanks.

The punch shanks are fitted to the punch plate. The proper fit is a light tap fit, not a press fit.

Die Block. Since this die block, part 2, is for a punching die, the size of the die openings is derived from the punches. The punch diameters are 17.50 and 12.75 mm. A cutting clearance of 9.04 mm per side is used. The die opening diameters are, therefore, 17.58 and 13.83 mm.

Center-to-center distance between the die openings is taken directly from the piece-part specifications, which give 25 +0.1, −0 mm. Thus the dimensional range is 25.0 to 25.1 mm. For this range the exact mean dimension is 25.05 mm.

After the optimum die opening size and location are decided, the next consideration is to ensure proper location of the pierced openings in relation to the contour of the piece part. This is the function of the nest pins, part 9. In order to serve their purpose, the nest pins must be accurately located. The nest must fit the blank closely enough to ensure the required accuracy of location. Some clearance, however, should exist between the nest pins and the blank to permit ease of loading and unloading. The amount of clearance to allow, of course, depends upon the accuracy requirements of the piece part. In this particular case, a clearance of 0.02 mm between the edges of the blank and the nest pins is satisfactory. Thus the center-to-center distances between the nest pins are 46.02 (40 + 6 + 0.02) and 71.12 (65.1 + 6 + 0.02) mm.

Punch Plate. Refer to Fig. 27-3-16, part 6. The 20- and 14-mm diameters are chosen because a light tap fit is used.

The center-to-center distance is picked up or transferred from the hardened die block. This procedure is followed to eliminate the possibility of minor alignment discrepancies which might result from slight dimensional changes caused by the die-block heat treatment.

The counterbores which receive the heads of the punches are made larger than the punch heads. Clearing for the heads in this way eliminates the possibility of misalignment resulting from contact between the head of the punch and the sidewall of the counterbore.

Foolproofing is accomplished by using dowels of differing diameters.

A shedder or spring pin, part 13, is provided in the stripper. It is a standard purchased item, the length of which cannot be altered.

Stripper. The diameters of the stripper openings (Fig. 27-3-16, part 3) are 15 and 19.5 mm, respectively. These figures give a clearance of approximately 1 mm per side between the punches and the stripper. So much clearance is permissible because of the stock material thickness and is desirable from the die maker's viewpoint because it eliminates the need for precise location and sizing of the stripper openings.

Spring strippers such as this one must have shedder pins to prevent the possibility of a piece part adhering to the stripper. In this stripper, the shedder pin, part 13, is a standard purchased item. It contains a pin which is spring-loaded. A suitable tapped hole is provided in the stripper to receive this unit, which is shown in assembly with the stripper.

REFERENCE

1. From *Basic Diemaking* by the National Tool, Die, and Precision Machining Association. Copyright, 1963. McGraw-Hill Book Company. Used by permission.

Assignments

1. On an A3- or B-size sheet, design suitable punches for the blanking and the punching for the two parts shown in Fig. 27-3-A or 27-3-B. The die set to be used is B1-361, shown in Fig. 27-3-12. Draw only the front and bottom views of the punch. Scale is 1:1.

2. On an A3- or B-size sheet, design a suitable die and stripper plate for one of the two parts shown in Fig. 27-3-C or 27-3-D. The die set to be used is B1-361, shown in Fig. 27-3-15. The die is to be used in conjunction with Fig. 27-3-A or 27-3-B. Draw the front, top, and side views of the die and stripper plate. Scale is 1:1.

3. On an A3- or B-size sheet, draw the details for the blank-through die in assignment 2. Show all the details, but only those dimensions pertinent to the blanking design need be shown on the detail drawings. Scale is to suit.

REVIEW FOR ASSIGNMENTS

Unit 8-3 Cap Screws
Unit 9-2 Pin Fasteners
Unit 9-4 Springs

UNIT 27-4

PROGRESSIVE DIES[1]

A progressive die is a series of dies located on one die shoe, each performing an operation on the strip as it moves along the die from one die station to another. For example, a die of this type may punch at the first station, bend at the second station, and blank at the third station. The completed part is removed from the strip at the last station on the die. An example of a progressive die is shown in Fig. 27-4-1.

Pilots

It is the function of pilots to position workpieces or stock strips accurately for die working. When the work is brought into the required position by the pilots, it is said to be *registered*.

The piloting action, when a manually fed stock strip is registered, is depicted in Fig. 27-4-2. The stop is located to permit overfeeding the strip a slight distance beyond the required registry position. For dies equipped with pilots, all stops except the first (primary) stop are positioned to permit overfeeding. Actual feeding distance is equal to the advance distance plus the overfeed.

It is essential that two basic definitions be associated with the fundamental principles of stops.

Stop Position. This is the location of the actual stopping point of surface against which the stock strip is halted.

Registry Position. This is the exact location in which the stock strip must be established for the work to be dimensionally correct. In a sense, stops can be

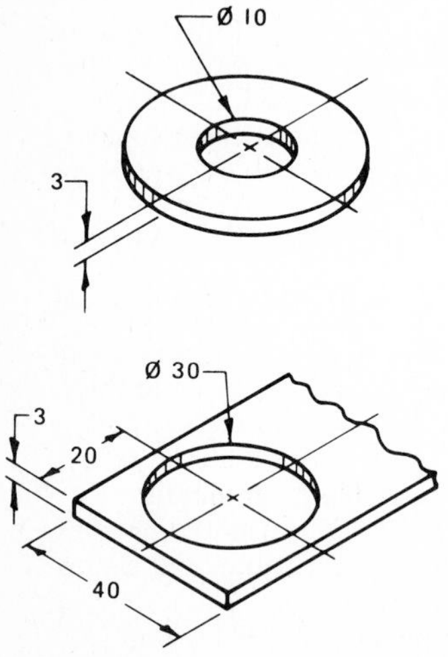
Ø 10
3
Ø 30
3
20
40

Fig. 27-3-A **Punch and blank designs.**

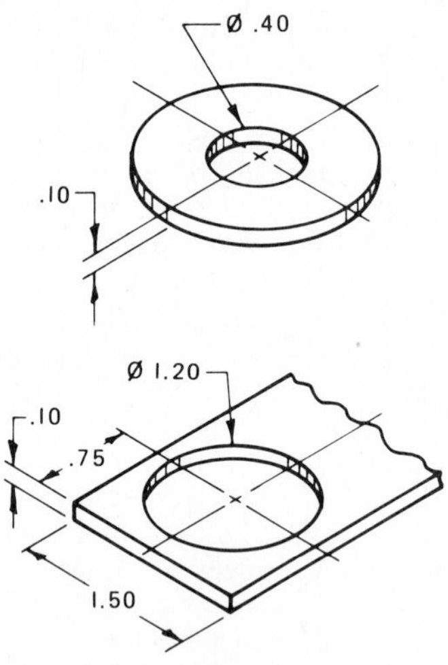
Ø .40
.10
Ø 1.20
.10
.75
1.50

Fig. 27-3-B **Punch and blank designs.**

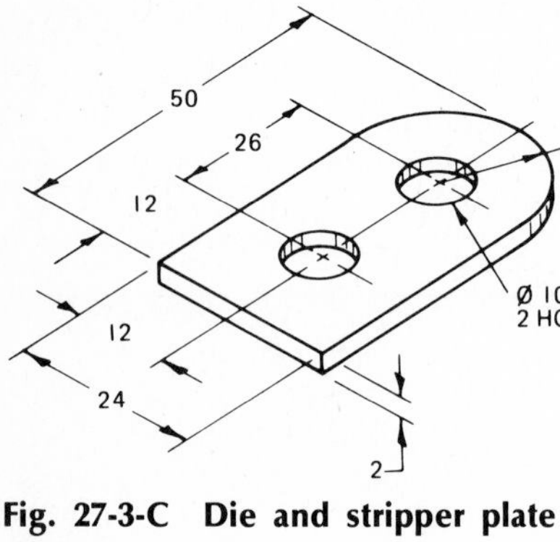
50
26
R
12
Ø 10.5
2 HOLES
12
24
2

Fig. 27-3-C **Die and stripper plate design.**

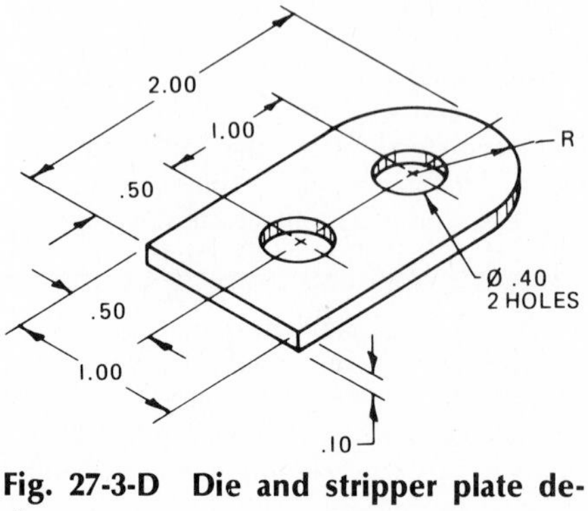
2.00
1.00
R
.50
Ø .40
2 HOLES
.50
1.00
.10

Fig. 27-3-D **Die and stripper plate design.**

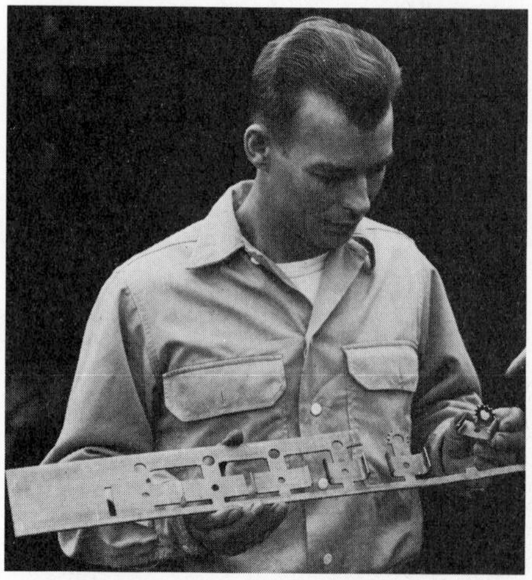

(A) EXAMINING THE STRIP PRODUCED BY THE PROGRESSIVE DIE SHOWN IN (B)

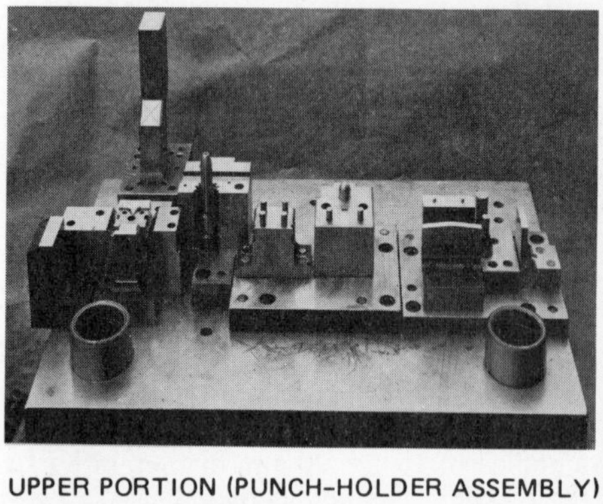

UPPER PORTION (PUNCH-HOLDER ASSEMBLY)

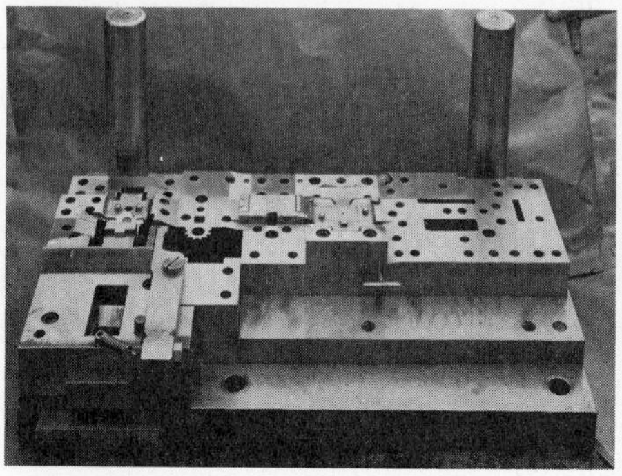

LOWER PORTION (DIE-SHOE ASSEMBLY)
(B) PROGRESSIVE DIE ASSEMBLY

Fig. 27-4-1 **A progressive die.**

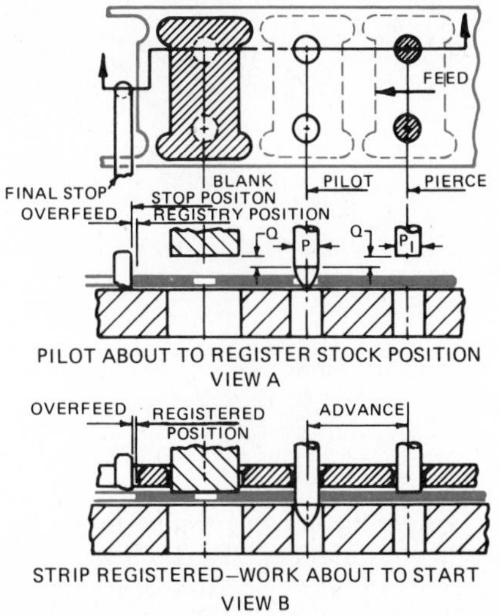
FEED
FINAL STOP
OVERFEED
BLANK
STOP POSITON
REGISTRY POSITION
PILOT
PIERCE
P
P1
Q
Q1
PILOT ABOUT TO REGISTER STOCK POSITION
VIEW A
OVERFEED
REGISTERED POSITION
ADVANCE
STRIP REGISTERED–WORK ABOUT TO START
VIEW B

Fig. 27-4-2 Pilot registry—hand-fed strip.

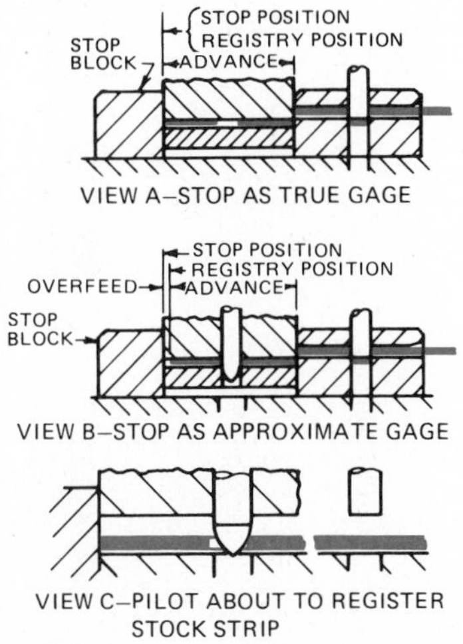
STOP POSITION
REGISTRY POSITION
STOP BLOCK
ADVANCE
VIEW A–STOP AS TRUE GAGE
STOP POSITION
REGISTRY POSITION
OVERFEED
ADVANCE
STOP BLOCK
VIEW B–STOP AS APPROXIMATE GAGE
VIEW C–PILOT ABOUT TO REGISTER STOCK STRIP

Fig. 27-4-3 Fundamental principles of stops.

considered gages as well as stops. Some stops perform a true gaging function. Others act as approximation gages, permitting different components (usually pilots) to perform the final, accurate registry of the stock strip.

The relationship between stop position and registry position depends upon the function of the stop. If a stop acts as a true gage, then the stop position and the registry position are one and the same. If a stop functions as an approximation gage, then the stop position does not coincide with the registry position. It can be said generally that if a stock strip is piloted, it is necessary for the stop to act only as an approximation gage, allowing the strip to be overfed. If a stock strip is not piloted, the stop then functions as a true gage.

Figure 27-4-3 is a comparison drawing illustrating this difference in the function of stops. In each case the stop depicted is a final stop, solid type. However, the same principles apply to all stop categories.

The die shown in view A does not have pilots. The stop position must coincide with the registry position, because the stop is the sole means of positioning the stock strip lengthwise. In operation, the leading end of the strip is advanced

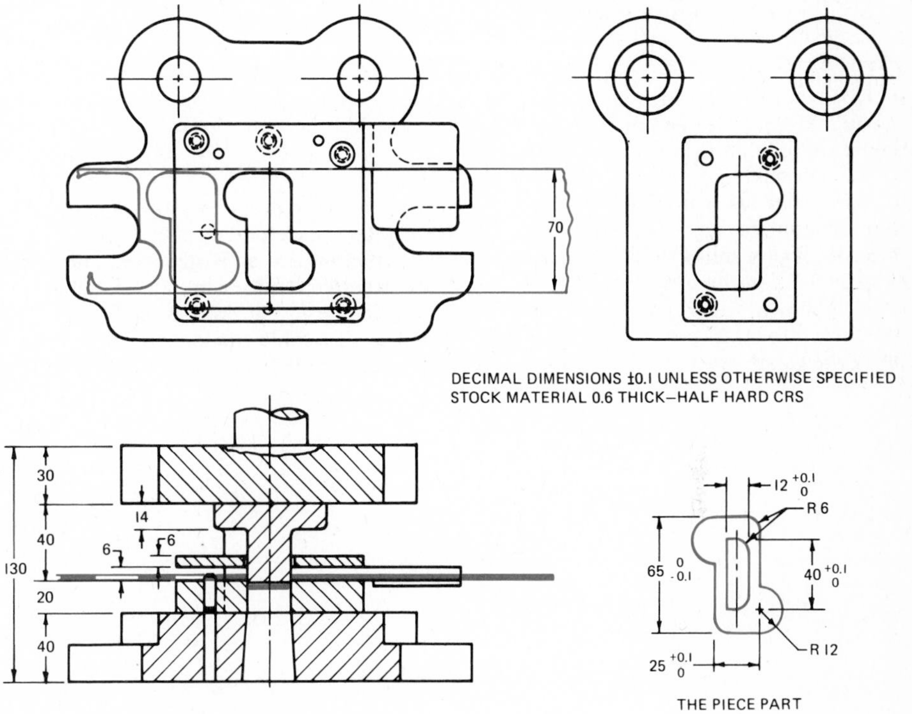
70
DECIMAL DIMENSIONS ±0.1 UNLESS OTHERWISE SPECIFIED
STOCK MATERIAL 0.6 THICK–HALF HARD CRS
30
40
130
20
40
14
6
6
12 +0.1 0
R 6
65 0 -0.1
40 +0.1 0
R 12
25 +0.1 0
THE PIECE PART

Fig. 27-4-4 Blank-through die.

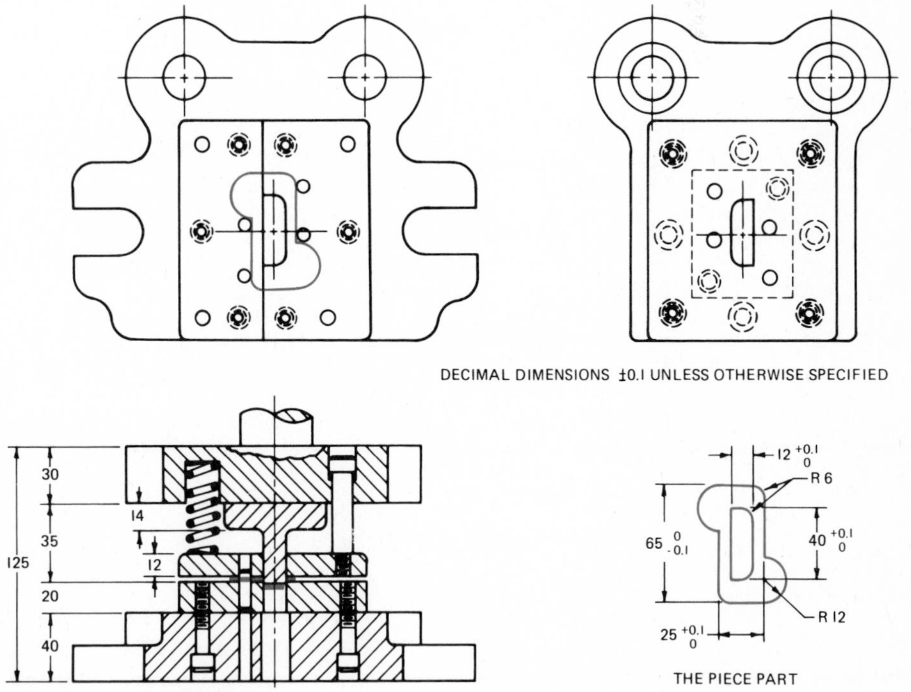
DECIMAL DIMENSIONS ±0.1 UNLESS OTHERWISE SPECIFIED
30
35
125
20
40
14
12
12 +0.1 0
R 6
65 0 -0.1
40 +0.1 0
R 12
25 +0.1 0
THE PIECE PART

Fig. 27-4-5 Punch die.

to the stop and remains against the stop during the press cycle. This stop acts as true gage, registering the stock strip.

REFERENCE

1. *Basic Diemaking* by the National Tool, Die, and Precision Machining Association. Copyright 1963. McGraw-Hill Book Company. Used by permission.

Assignments

1. On an A3- or B-size sheet (assignment 27-4-A), redesign the punch and blank-through dies (Figs. 27-4-4 and 27-4-5) to make a progressive die. The existing punches with minor modification, if required, are to be used on the new die, and a setup similar to that in Fig. 27-4-2 is to be used. The die set B1-462 shown in Fig. 27-3-12 has been purchased for this job. Only the top and front views are required. Scale is 1:1.

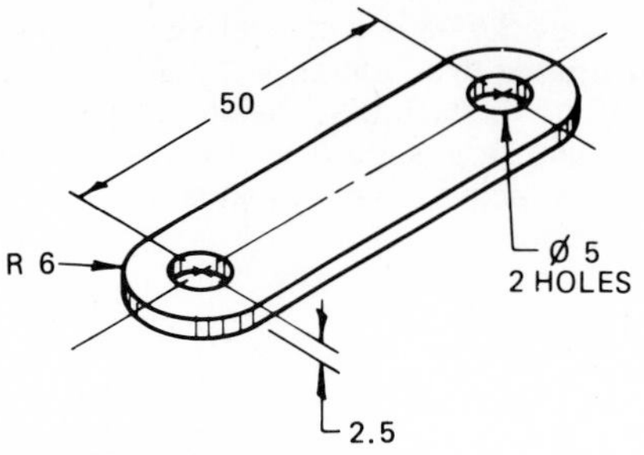
50
R 6
Ø 5
2 HOLES
2.5

Fig. 27-4-B Link.

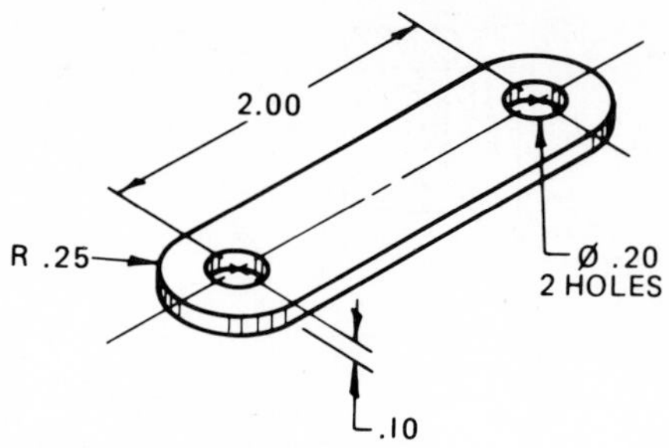
2.00
R .25
Ø .20
2 HOLES
.10

Fig. 27-4-C Link.

Use a second color to show the stock strip and piece part.

2. On an A3- or B-size sheet, design a progressive die to blank and punch the link shown in Fig. 27-4-B or 27-4-C. Select a suitable stock strip width. Use die set B1-3101, shown in Fig. 27-3-12, and set up the drawing similar to the layout shown in Fig. 27-3-11. Scale is 1:2.

REVIEW FOR ASSIGNMENTS

Unit 27-3 Die Sets and Components
Unit 9-4 Springs

Chapter 28 Structural Drafting

UNIT 28-1
STRUCTURAL DRAFTING

The training of the structural steel drafter is of vital importance to the engineering profession, the construction industry, and every structural steel fabricator.

The Building Process

The steps through which a building proceeds from conceptual planning to finished products are, generally speaking, as follows.

1. An owner with the appropriate financing establishes the requirements for a building to fulfill some particular function.

2. A design team (usually an architect and an engineer) studies the owner's needs in reference to set standards and conventions. The following factors may influence the preliminary design: available materials, construction costs, building codes, zoning, health requirements, local bylaws, land condition, fire protection, finance, and setbacks.

3. With these parameters and the owner's requirements, the consultant or design team prepares sketches of the finished building, floor plans, and cost estimates, which are submitted to the owner for approval.

4. The design team then takes over the job. The structural design group designs the building frame, taking into consideration the other factors which influence the type and location of structural members.

5. When the structural arrangements have been finalized, layout drawings are made. These give distances from center line to center line, size and location of structural components, and other specifics of the design. When the layout drawings have been completed, checked, and approved, then they are sent to steel fabricators for tendering. *Tendering* involves quoting a price (usually a price for detailing, supply, fabrication, and erection of the steel members).

6. When the contract has been received by the steel fabricator, he or she makes a list of material required so that the basic shapes can be ordered from the steel producer. The fabricator also begins to detail (draw the individual building members). These are referred to as *shop drawings*.

7. As the shop drawings are completed, they are sent to the shop in order that parts may be fabricated. It is usually during this period that the fabricator will make the erection drawings in conjunction with the structural design group.

8. As the steel is fabricated, it is either stored in the yard or sent to the construction site if it is required immediately.

9. At the site, the steel is erected using the erection drawings. Changes during construction or revisions to steelwork are noted and sent to the fabricator or back to the structural design group. During the steel erection phase, other trades can be working.

10. When the building is completed, all inspections have been made, and the building is certified for occupancy, the owner can take possession.

Fig. 28-1-1 The construction of the Toronto-Dominion Bank Tower, Toronto. (Tisdall, Clark and Lesley Ltd.)

Structural Steel—Plain Material

It is important to remember that the steel produced at the rolling mills and shipped to the fabricating shop comes in a wide variety of shapes (approximately 600) and forms. At this stage it is called *plain material*.

The greatest bulk of this material is shown in Fig. 28-1-2. It can be classified and designated as follows.

1. S shapes (formerly called standard beams or I beams) are rolled in many sizes (75 to 500 mm).

2. C shapes (formerly called standard channels) are available in sizes ranging from 80 to 450 mm.

3. W shapes (formerly called wide-flange shapes) and welded wide-flange (WWF) beams and columns. W shapes are available in sizes ranging from 150 to 900 mm. WWF shapes, sometimes referred to as H shapes, range in size from 350 to 1200 mm.

4. M shapes (formerly called joists and light beams) are similar in contour to the W shapes. They are available in sizes ranging from 150 to 400 mm.

5. Structural tees are produced by splitting S or W shapes, usually through the center of their webs, thus forming two T-shaped pieces from each beam.

6. L shapes, or angles, consisting of two legs set at right angles, are available in sizes ranging from 75 to 200 mm.

7. Hollow structural sections (HSS) consist of round, square, and rectangular sections.

8. Plates, and round and rectangular bars.

When steel shapes such as these are designated on drawings, it is desirable that a standard method of abbreviating be followed that will identify the size and shape of the steel part. See Figs. 28-1-3 and 28-1-4. However, the method for calling for these standard shapes has changed over the last few years. When called upon to revise or modify existing drawings, the drafter must do so in the same convention used previously on that drawing. Therefore, it is important that the drafter not only have the most up-to-date knowledge, but also be familiar with previous standards still in use on old drawings. See Fig. 28-1-5.

Besides having to know the type of shapes available and their drawing designation, one must also be familiar with framing construction terms and where these shapes are used. See Fig. 28-1-6.

The abbreviations shown are intended only for use on design drawings. When lists of materials are being prepared for ordering from the mills, the requirements of the respective mills from which the material is to be ordered should be observed.

All S, C, and MC shapes have a 16.67 percent slope on the inside faces of the flanges. This is equivalent to 9°28′ or a

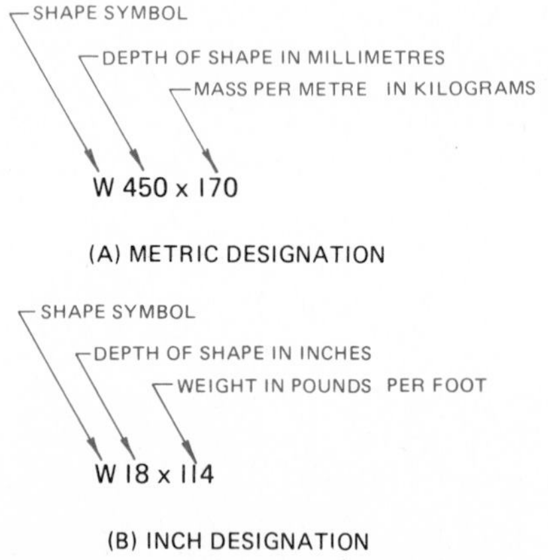
SHAPE SYMBOL
DEPTH OF SHAPE IN MILLIMETRES
MASS PER METRE IN KILOGRAMS
W 450 x 170
(A) METRIC DESIGNATION
SHAPE SYMBOL
DEPTH OF SHAPE IN INCHES
WEIGHT IN POUNDS PER FOOT
W 18 x 114
(B) INCH DESIGNATION

Fig. 28-1-3 Structural steel callouts.

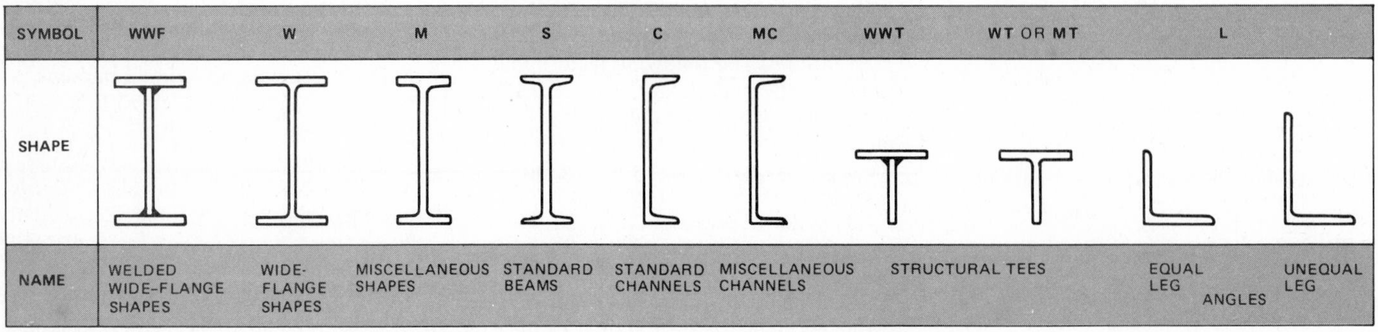
SYMBOL
WWF
W
M
S
C
MC
WWT
WT OR MT
L
SHAPE
NAME
WELDED WIDE-FLANGE SHAPES
WIDE-FLANGE SHAPES
MISCELLANEOUS SHAPES
STANDARD BEAMS
STANDARD CHANNELS
MISCELLANEOUS CHANNELS
STRUCTURAL TEES
EQUAL LEG
UNEQUAL LEG
ANGLES

Fig. 28-1-2 Common structural steel shapes.

Shape	Metric Size Examples See Note 1	Inch Size Examples See Note 2	
		New Designation	Old Designation
Welded Wide Flange Shapes (WWF Shapes)			
—Beam	WWF1000 x 244	WWF48 x 320	48WWF320
—Columns	WWF350 x 315		
Wide Flange Shapes (W Shapes)	W600 x 114	W24 x 76	24WF76
	W160 x 18	W14 x 26	14B26
Miscellaneous Shapes (M Shapes)	M200 x 56	M8 x 18.5	8M18.5
	M160 x 30	M10 x 9	10JR9.0
Standard Beams (S Shapes)	S380 x 64	S24 x 100	24I100
Standard Channels (C Shapes)	C250 x 23	C12 x 20.7	12C20.7
Structural Tees			
—cut from WWF Shapes	WWT280 x 210	WWT24 x 160	ST24WWF160
—cut from W Shapes	WT130 x 16	WT12 x 38	ST12WF38
—cut from M Shapes	MT100 x 14	MT4 x 9.25	ST4M9.25
Bearing Piles (HP Shapes)	HP350 x 109	HP14 x 73	14BP73
Angles (L Shapes) (leg dimensions x thickness)	L75 x 75 x 6	L6 x 6 x .75	L6 x 6 x ¾
	L150 x 100 x 13	L6 x 4 x .62	L6 x 4 x ⅝
Plates (width x thickness)	500 x 12	20 x .50	20 x ½
Square Bar (side)	⌗ 25	⌗ 1.00	Bar 1 ⌗
Round Bar (diameter)	ϕ 30	ϕ 1.25	Bar 1¼ ϕ
Flat Bar (width x thickness)	60 x 6	250 x .25	Bar 2½ x ¼
Round Pipe (type of pipe x OD x wall thickness)	XS 102 OD x 8	12.75 OD x .375	12¾ x ⅜
Square and Rectangular Hollow Structural Sections (outside dimensions x wall thickness)	HSS102 x 102 x 8	HSS4 x 4 x .375	4 x 4RT x ⅜
		HSS8 x 4 x .375	8 x 4RT x ⅜
Steel Pipe Piles (OD x wall thickness)	320 OD x 6		

Note 1—Values shown are nominal depth (millimetres) x mass per metre length (kilograms).
Note 2—Values shown are nominal depth (inches) x weight per foot length (pounds).
Note 3—Metric size examples shown are not necessarily the equivalents of the inch size examples shown.

Fig. 28-1-4 Abbreviations for shapes, plates, bars, and tubes.

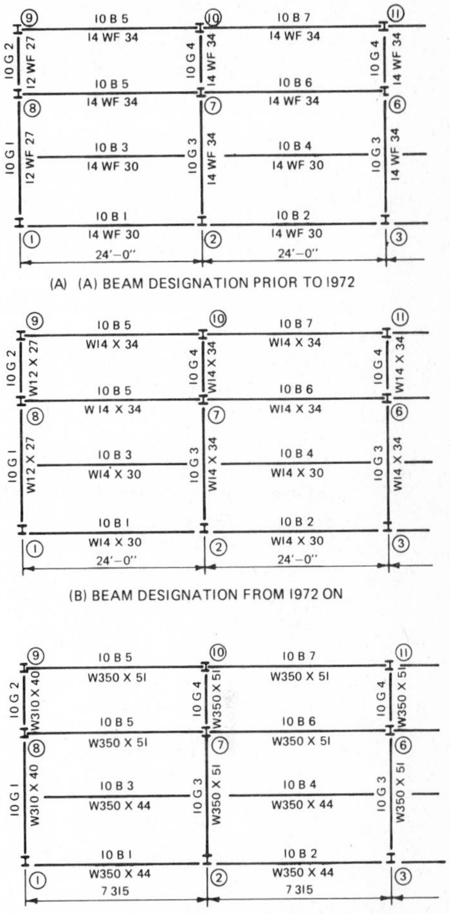

(A) (A) BEAM DESIGNATION PRIOR TO 1972

(B) BEAM DESIGNATION FROM 1972 ON

(C) METRIC BEAM DESIGNATION AND DIMENSIONING

Fig. 28-1-5 Building floor framing plan. (partial view)

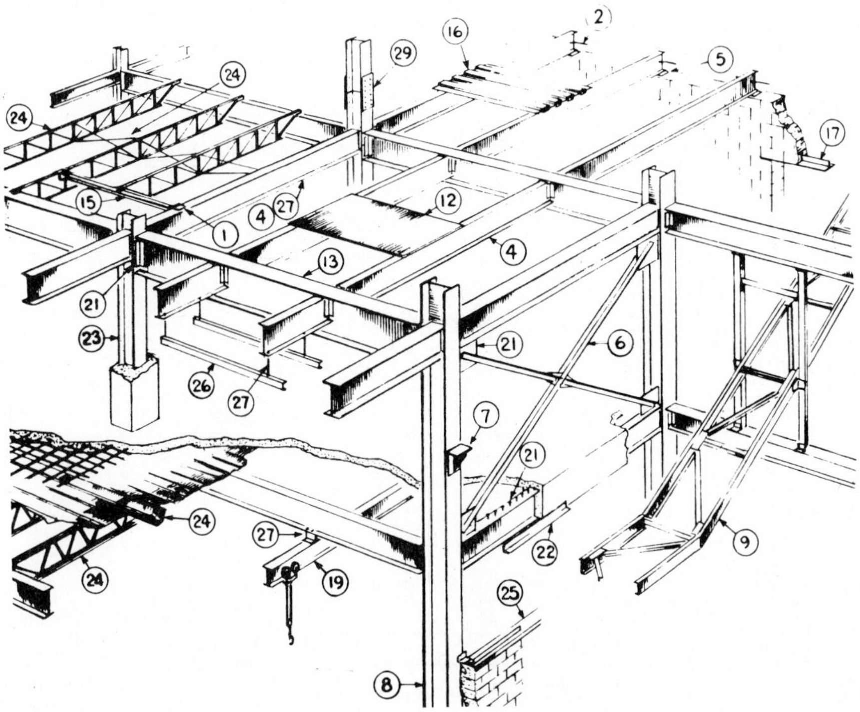

1. Anchors or hangers for open-web steel joists
2. Anchors for structural steel
3. Bases of steel and iron for steel or iron columns
4. Beams, purlins, girts
5. Bearing plates for structural steel
6. Bracing for steel members or frames
7. Brackets attached to the steel frame
8. Columns, concrete-filled pipe, and struts
9. Conveyor structural steel frame work
10. Crane rail beams and stops if size and connections are shown
11. Door frames constituting part of and connected to the steel frame
12. Floor and roof plates (raised pattern or plain (connected to steel frame)
13. Girders
14. Grillage beams of steel
15. Headers or trimmers for support of open-web steel joists where such headers or trimmers frame into structural steel members
16. Light-gage cold-formed steel used to support floor and roofs

Fig. 28-1-6 Structural steel terms.

bevel of 1:6. W-shaped beams and columns are rolled with parallel face flanges or with a 5 percent slope (2°51′) on the inside of the flange, depending on whether they are rolled in Canada or the United States. See Fig. 28-1-7.

In structural steel shape tables, dimensions such as *K* and mean thickness of sloping flanges are given. Since the mean thickness of the sloping flange is given, these dimensions may also be used for all flange shapes. This practice is followed in the interest of standardization.

If it is necessary to have the exact dimensions of a particular shape, they must be obtained from the individual mill's structural shape catalog. These catalogs also give the pertinent radial dimensions in question. See Fig. 28-1-8.

It is customary, on details made to a scale of 1:10 or smaller, for the curve indicating the toes of angles and of flanges, the interior fillets between legs of angles, and the interior fillets between web, or stem, and flanges to be omitted in the drawing. It is usual to exaggerate on detail drawings the thickness of leg, stem, web, or flange.

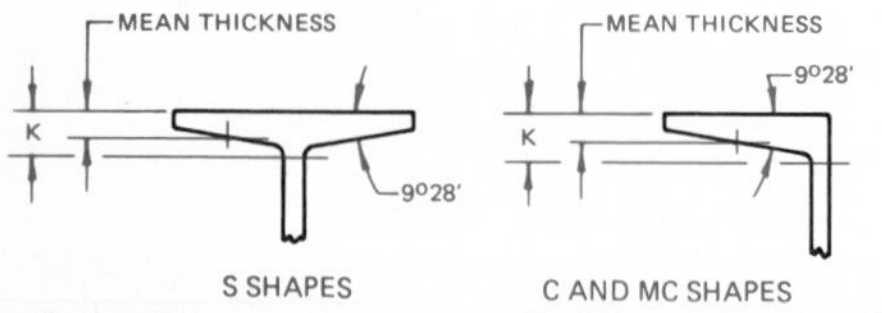
MEAN THICKNESS
MEAN THICKNESS
9°28'
9°28'
K
K
S SHAPES
C AND MC SHAPES

Fig. 28-1-7 Slopes and dimensions of flanges.

STEEL GRADES

There are hundreds of grades of steel produced in mills today. However, only a few of those are suitable for structural applications. The most common structural grade used in the United States is ASTM A-36. In Canada it is G40.21M grade 300W. All structural members discussed in this chapter will be assumed to be fabricated from ASTM A-36, while the bolts are made from A307 or A325 depending on the strength required.

MILL TOLERANCES

There are certain permissible deviations brought about in the manufacture of structural steel that the drafter should understand. These permissible deviations from the published dimensions and contours (as listed in the AISC and CISC manuals, in mill catalogs, and from the lengths specified by the purchaser) are referred to as *mill tolerances*. See Figs. 28-1-9 to 28-1-11.

The factors which contribute to the necessity for a mill tolerance are as follows:

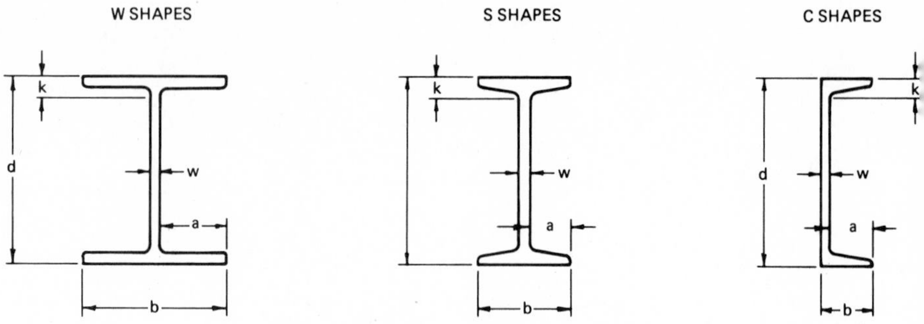
W SHAPES
S SHAPES
C SHAPES
k
d
w
a
b

Metric Designation	Depth d	Flange Width b	Flange Thickness t	Web Thickness W	Distance a	Distance k	Inch Designation
W600 x 140	617	230	22	13.1	109	42	W24 x 94
W600 x 114	607	228	17.3	11.2	109	37	W24 x 76
W450 x 170	469	301	25.2	15.1	143	43	W18 x 114
W450 x 156	465	300	23.1	14.1	143	42	W18 x 105
W450 x 143	461	298	21.1	13	143	38	W18 x 96
W450 x 89	464	192	17.7	10.6	91	30	W18 x 60
W400 x 116	415	218	22.2	13.4	102	42	W16 x 78
W400 x 60	406	178	12.8	7.8	85	29	W16 x 40
W350 x 110	360	256	19.9	11.4	122	38	W14 x 74
W350 x 71	351	204	15.1	8.6	98	32	W14 x 48
W310 x 86	310	254	16.3	9.1	123	35	W12 x 58
W310 x 74	310	205	16.3	9.4	98	35	W12 x 50
W310 x 54	311	167	13.7	7.8	79	27	W12 x 36
W310 x 40	304	165	10.2	6	79	24	W12 x 27
W250 x 72	254	254	14.2	8.6	123	29	W10 x 49
W250 x 49	248	202	11	7.4	97	25	W10 x 33
W250 x 37	256	146	10.9	6.4	70	25	W10 x 25
W250 x 22	254	102	6.8	5.8	48	21	W10 x 15
W200 x 52	206	204	12.5	8	98	25	W8 x 35
W200 x 42	205	166	11.8	7.2	79	24	W8 x 28
W200 x 30	207	134	9.6	6.3	64	22	W8 x 20
W200 x 22	206	102	8	6.2	48	21	W8 x 15
S600 x 149	610	184	22.1	19	83	45	S24 x 100
S600 x 134	610	181	22.1	15.9	83	45	S24 x 90
S500 x 142	508	183	23.7	20.3	83	48	S20 x 95
S500 x 117	508	162	20.1	16.3	73	41	S20 x 75
S450 x 104	457	159	17.6	18.1	70	38	S18 x 70
S380 x 75	381	143	15.8	14	64	35	S15 x 50
S300 x 75	305	139	16.8	17.5	60	37	S12 x 50
S300 x 52	305	129	13.8	10.9	60	30	S12 x 35
S250 x 52	254	126	12.5	15.1	54	29	S10 x 35
S200 x 34	203	106	10.8	11.2	48	26	S8 x 23
C381 x 75	381	95	16.5	18.2	76	36	C15 x 50
C381 x 60	381	90	16.5	13.2	76	36	C15 x 40
C305 x 45	305	81	12.7	13	67	29	C12 x 30
C305 x 31	305	75	12.7	7.2	67	29	C12 x 20.7
C254 x 45	254	77	11.1	17.1	60	26	C10 x 30
C254 x 30	254	70	11.1	9.6	60	26	C10 x 20

Note—Soft converted dimensions from currently used sections adapted from CISC Manual.

Fig. 28-1-8 Properties of common structural steel shapes.

ROLLING TOLERANCES

	Nominal Size	Depth A Over	Depth A Under	Width of Flange B Over	Width of Flange B Under	Out of Square T or T_1 Max	Web Off Center E Max	Maximum Depth of Any Cross Section C Over Nominal Size
Millimetre Sizes	300 and under	4	3	6	5	5	5	6
	Over 300	4	3	6	5	6	5	6

CUTTING TOLERANCES

	W Shapes Nominal Depth	To 9000 Incl Over	To 9000 Incl Under	Over 9000 To 12 000 Incl Over	Over 9000 To 12 000 Incl Under	Over 12 000 To 15 000 Incl Over	Over 12 000 To 15 000 Incl Under	Over 15 000 To 20 000 Incl Over	Over 15 000 To 20 000 Incl Under	Over 20 000 Over	Over 20 000 Under
		Variation from Specified Length for Lengths Given									
Millimetre Sizes	Beams 610 and under	10	10	13	10	16	10	21	10	24	10
	Beams over 610 All Columns	13	13	16	13	19	13	24	13	27	13

Fig. 28-1-9 Rolling and cutting tolerances for W shapes. (CISC)

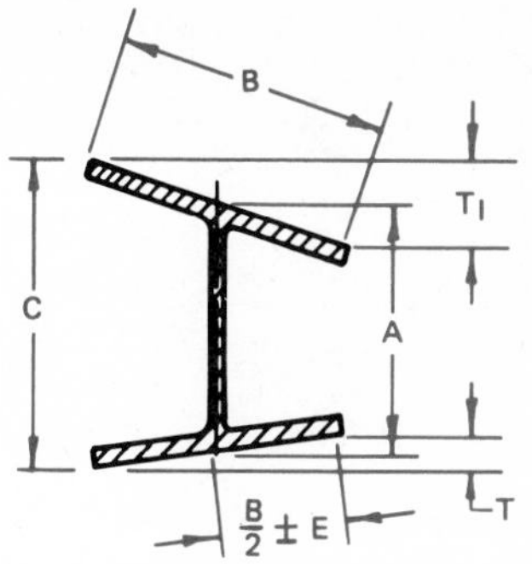
B
C
A
T_1
$\frac{B}{2} \pm E$
T

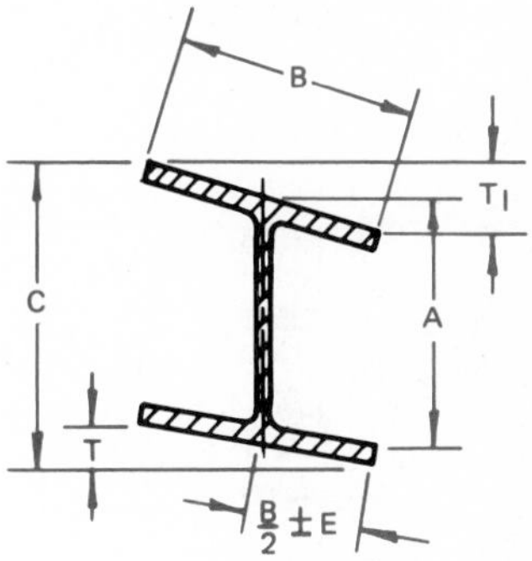
B
C
A
T_1
T
$\frac{B}{2} \pm E$

ROLLING TOLERANCES

Shape	Nominal Specified Size	Depth A Over	Depth A Under	Flange Width B Over	Flange Width B Under	Out of Square $\frac{T}{B}$ or $\frac{T_1}{B}$	Out of Square $\frac{C-D}{B}$
S	75 to 180 incl.	2	2	4	3	0.03	0.03
	181 to 360 incl.	4	2	4	4	0.03	0.03
	361 to 610 incl.	5	3	5	5	0.03	0.03
C	75 to 180 incl.	2	2	4	3	0.03	0.03
	181 to 360 incl.	3	2	3	4	0.03	0.03
	361 to 610 incl.	5	3	3	5	0.03	0.03

CUTTING TOLERANCES

Shape	Up to 9000 Over	Up to 9000 Under	9001 to 12 000 Over	9001 to 12 000 Under	12 001 to 15 000 Over	12 001 to 15 000 Under	15 001 to 20 000 Over	15 001 to 20 000 Under	Over 20 000 Over	Over 20 000 Under
	Variations from Specified Lengths									
S and C Shapes	13	6	19	6	25	6	29	6	32	6

Dimensions in millimetres

Fig. 28-1-10 Rolling and cutting tolerances for S and C shapes. (CISC)

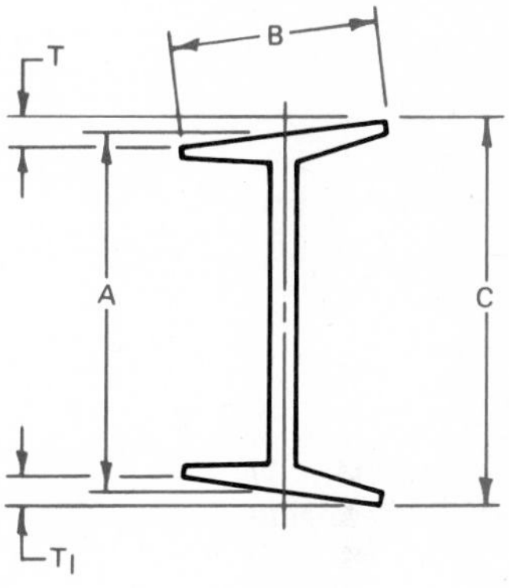
T
B
A
C
T_1

S SHAPES

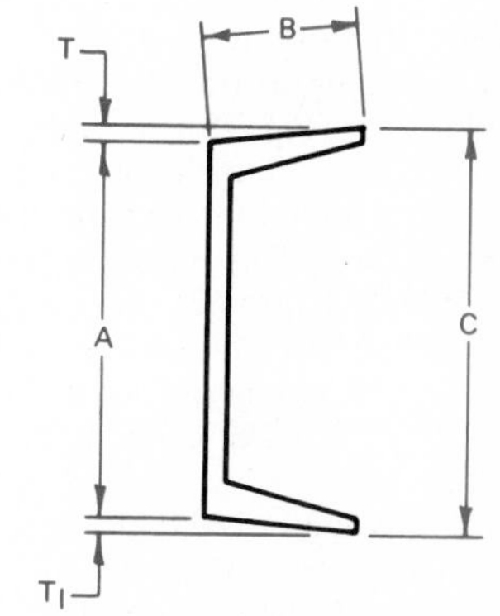
T
B
A
C
T_1

C SHAPES

1. The high speed of the rolling operations required to prevent the metal from cooling before the process has been completed
2. The varying skill of operators in squeezing together the rolls for successive passes of the metal, particularly the final pass (see Fig. 28-1-12).
3. The springing and wearing of the rolls, and other mechanical factors
4. The warping of the steel in the process of cooling

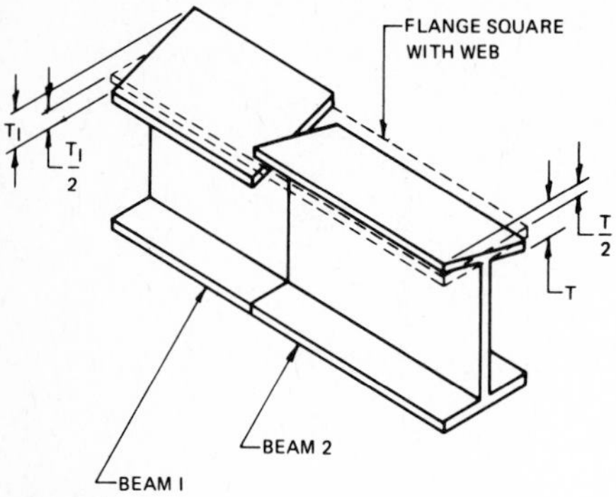
FLANGE SQUARE
WITH WEB
T_1
$\frac{T_1}{2}$
$\frac{T}{2}$
T
BEAM 2
BEAM I

Fig. 28-1-11 Error between mating shapes.

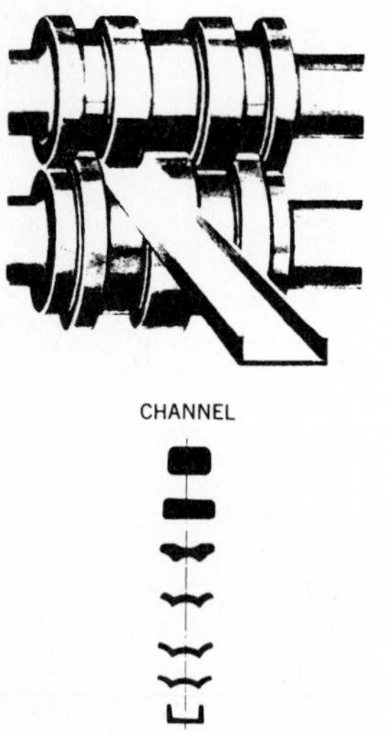
CHANNEL

SMALL STRUCTURAL SHAPES MAY BE FORMED BY A WIDE VARIETY OF PASSING PROCEDURES.

Fig. 28-1-12 The making of a C shape.

5. The subsequent shrinkage in the length of a shape which was cut while the metal was still hot.

Under rolling tolerances, note that the maximum overall depth (*C*) can be 6 mm over the nominal depth. For example, a W600 × 140 beam (Fig. 28-1-8) is shown as having a 617 mm depth. However, its finished actual depth at point *C* after rolling could be 623 mm. The depth at center line *A* could be either 621 mm (617 + 4) or 614 mm (617 − 3). The width of flange *B* could be 236 mm (230 + 6) or 225 mm (230 − 5) in place of the 230 mm width.

Suppose the W600 × 140 is ordered cut to length from the mill as a 15 500-mm length piece. It might be received by the fabricator either as 15 521 mm (15 500 + 21) or 15 490 mm (15 500 − 10) length piece. The fabricators have standards for ordering the plain material that take into consideration these cutting tolerances.

Although this variation of length would not be tolerated in the shop, it is essential that the detailer be aware of its possible occurrence, so that when he or she specifies the required stock, the material obtained will fulfill the purpose for which it was ordered. Another example showing mill tolerances is shown in Fig. 28-1-13.

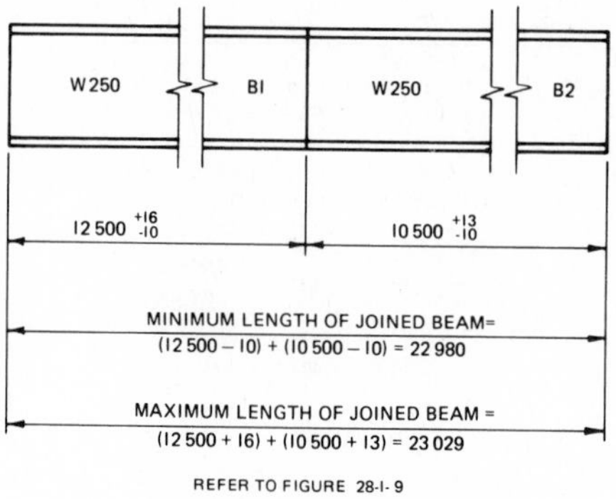
W250
BI
W250
B2
12 500 $^{+16}_{-10}$
10 500 $^{+13}_{-10}$
MINIMUM LENGTH OF JOINED BEAM=
(12 500 – 10) + (10 500 – 10) = 22 980
MAXIMUM LENGTH OF JOINED BEAM =
(12 500 + 16) + (10 500 + 13) = 23 029
REFER TO FIGURE 28-1-9

Fig. 28-1-13 Calculating minimum and maximum length of joined beam.

Detailers can usually disregard mill tolerances when detailing light- and medium-mass trusses, standard beams, standard channels, struts, and most plate girders. However, consideration must be given to the tolerances for all wide-flange beams and other heavy parts.

Handbooks prepared by the American Institute of Steel Construction (AISC) and the Canadian Institute of Steel Construction (CISC) list all the available structural shapes, their properties, and dimensions. However, for the convenience of the student, all the dimensions which are required for examples and problems are reproduced in this text.

Structural Drawing Practices

Figure 28-1-14 is a table indicating the scale and the type of structural drafting in which the scale is most frequently used.

DIMENSIONING

The general practice, in dimensioning structural drawings, is to use the aligned method for dimensions and to place the dimensions above the dimension lines. Otherwise, the same general guidelines as used in mechanical drafting will apply. All dimensions will be in millimetres.

Dimensions should be arranged in a manner most convenient to all who must use the drawing. They should not crowd the sketch and should cross the fewest possible number of other lines. The longest and overall dimensions should be farthest away from the views to which they apply. Dimensioning and descriptions of components (billing), in general, should be placed outside the picture. Dimensions should not be given to the center lines of beams, to the backs of angles, or, except as explained later, to the backs of channels. They should be given to the top or bottom of beams and channels (whichever level is to be held), but never to both top and bottom, because of a possible overrun or underrun, in the depth, resulting from rolling.

As shown in Fig. 28-1-15, when four or more equal spaces between bolts are required, it is convenient to the user of shop drawings for the information to be given as 4 @ 50 = 200 instead of repeating 50 four times. This reduces the possibility of error, both in reading the drawing and in layout of the work in the shop. Do not include in such an equation the distance locating the group itself from some reference point, even though

Scales Millimetre and Metre	Inch and foot	Principal Drawings Used
1:1	Full	Layout
1:5	3 = 1′-0	Layout
1:10	1½ = 1′-0	Layout or detail
1:20	¾ = 1′-0	Detail
	⅜ = 1′-0	Erection or design
	3/16 = 1′-0	Erection or design
	3/32 = 1′-0	Erection or design
	1 = 1′-0	Detail
	½ = 1′-0	Detail or erection
1:50	¼ = 1′-0	Erection or design
1:100	⅛ = 1′-0	Erection or design

Fig. 28-1-14 Scales for structural drawings.

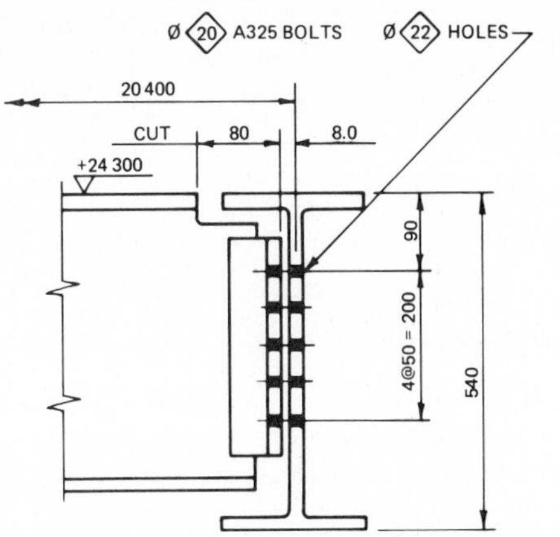

Ø 20 A325 BOLTS
Ø 22 HOLES
20 400
CUT
80
8.0
+24 300
90
4@50 = 200
540

Fig. 28-1-15 Dimensioning structural drawings.

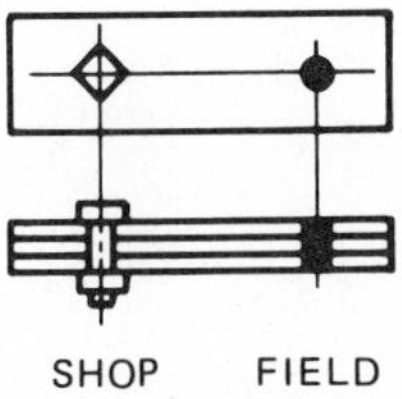

SHOP
FIELD

Fig. 28-1-16 Bolt symbols.

the distance may happen to be the same as the increment of spacing.

Elevation detail dimensions, known as *levels*, are generally furnished by a note on the drawing. When it is desirable to show the level or vertical distance above some established reference point (usually ground level), the value is given in millimetres and placed above the level symbol, as shown in Fig. 28-1-15. A plus or minus precedes the value, indicating that the level indicated is higher or lower than the reference point.

Another dimensioning practice is to enclose bolt and hole sizes in a diamond-shaped frame, as shown in Fig. 28-1-16. This helps to differentiate the circular sizes from the linear dimensions.

The neatness, and hence the legibility, of shop drawings is enhanced by lining up notes and dimensions which have the same common purpose. Thus if the 80 mm cut instructions (Fig. 28-1-15) were required at both ends of the beam, they would be shown at the same elevation, even though the dimension lines were not drawn from end to end of the sketch. Attention paid to these features, resulting in an orderly and systematic presentation of the necessary information, does much to enhance the finished appearance of a shop drawing.

Other dimensioning practices will be explained as the need arises throughout the chapter.

SOURCE MATERIAL

- Canadian Institute of Steel Construction

Assignment

Calculate the limits, tolerances, and sizes of the beams shown in either Fig. 28-1-A or Fig. 28-1-B. Refer to Fig. 28-1-8 for metric sizes of structural steel shapes. For inch-size structural steel shapes, refer to steel handbooks.

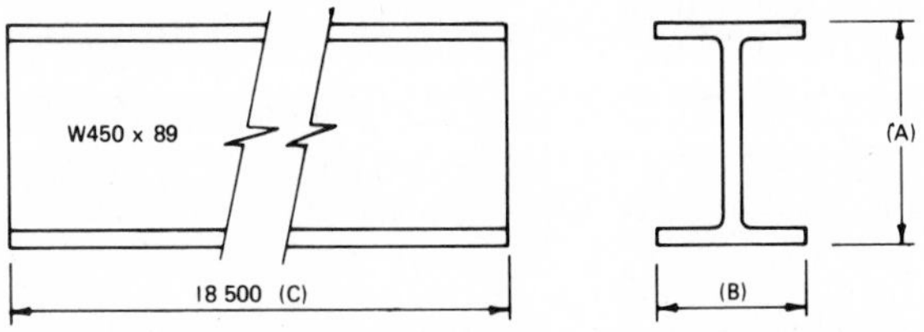

W450 x 89
18 500 (C)
(A)
(B)

W SHAPES

DIM	NOMINAL	TOLERANCE	LIMITS	
			MINIMUM	MAXIMUM
A		+ −		
B		+ −		
C		+ −		

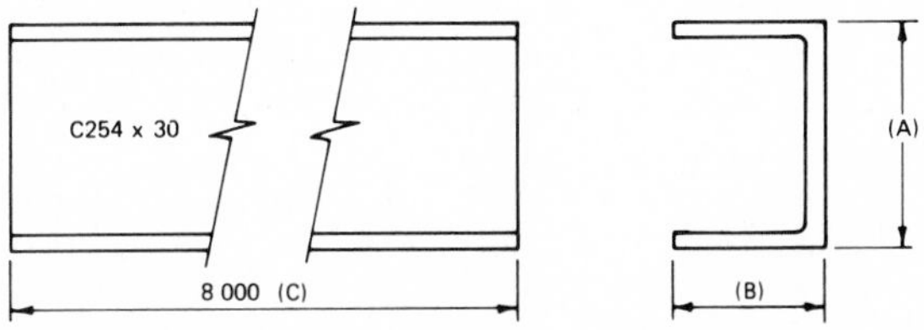

C254 x 30
8 000 (C)
(A)
(B)

C SHAPES

DIM	NOMINAL	TOLERANCE	MINIMUM	MAXIMUM
A		+ −		
B		+ −		
C		+ −		

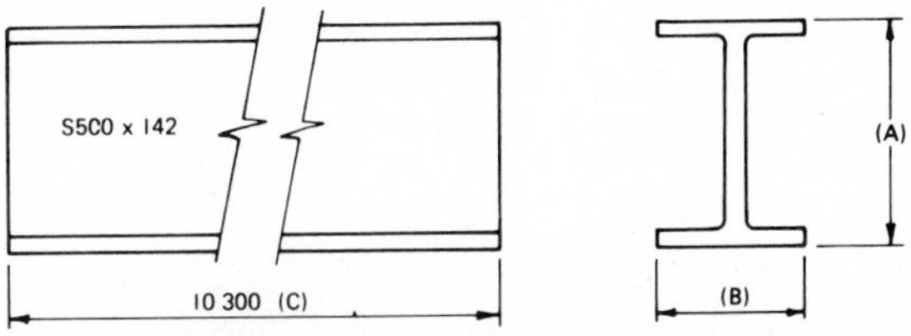

S5C0 x 142
10 300 (C)
(A)
(B)

S SHAPES

DIM	NOMINAL	TOLERANCE	MINIMUM	MAXIMUM
A		+ −		
B		+ −		
C		+ −		

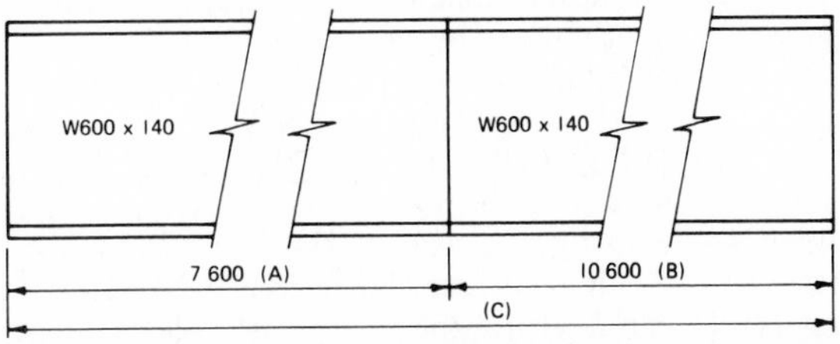

W600 x 140
W600 x 140
7 600 (A)
10 600 (B)
(C)

JOINED W SHAPES

DIM	TOLERANCE	MINIMUM	MAXIMUM
A	+ −		
B	+ −		
C			

Fig. 28-1-A Beam sizes.

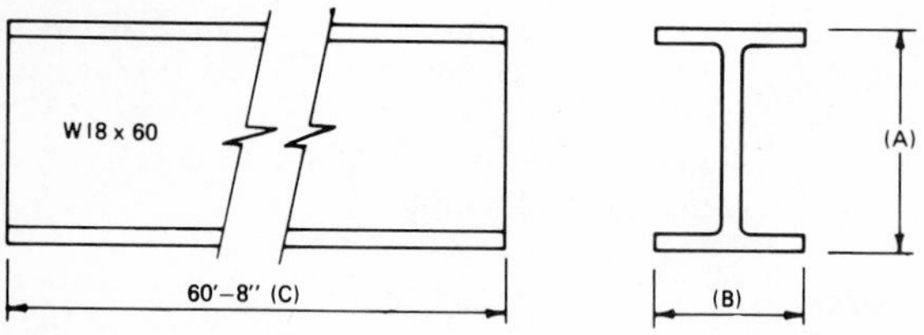
W18 x 60
60'–8" (C)
(A)
(B)

DIM	NOMINAL	TOLERANCE	MINIMUM	MAXIMUM
A		+ –		
B		+ –		
C		+ –		

W SHAPES

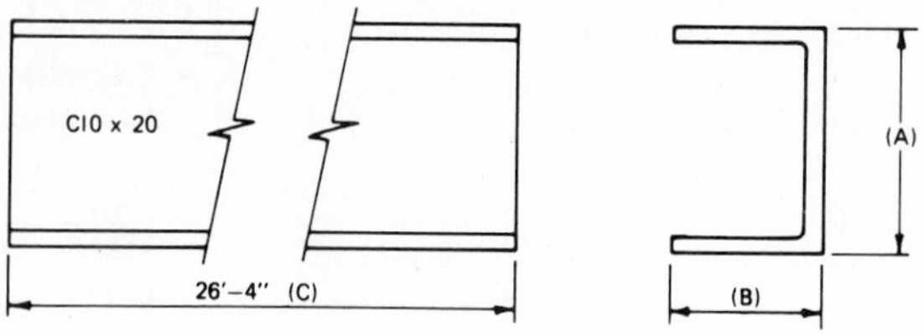
C10 x 20
26'–4" (C)
(A)
(B)

DIM	NOMINAL	TOLERANCE	MINIMUM	MAXIMUM
A		+ –		
B		+ –		
C		+ –		

C SHAPES

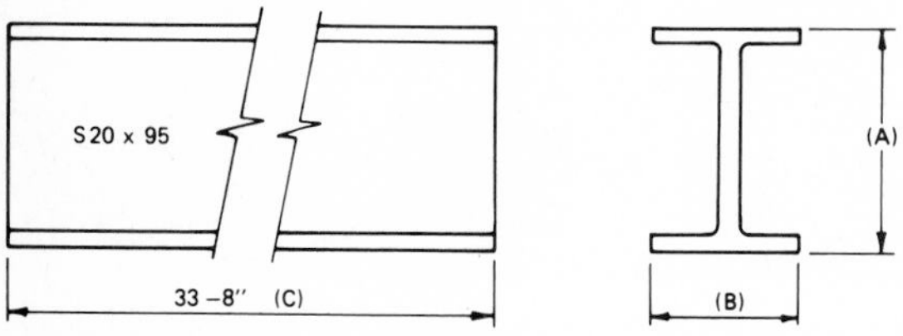
S20 x 95
33 –8" (C)
(A)
(B)

DIM	NOMINAL	TOLERANCE	MINIMUM	MAXIMUM
A		+ –		
B		+ –		
C		+ –		

S SHAPES

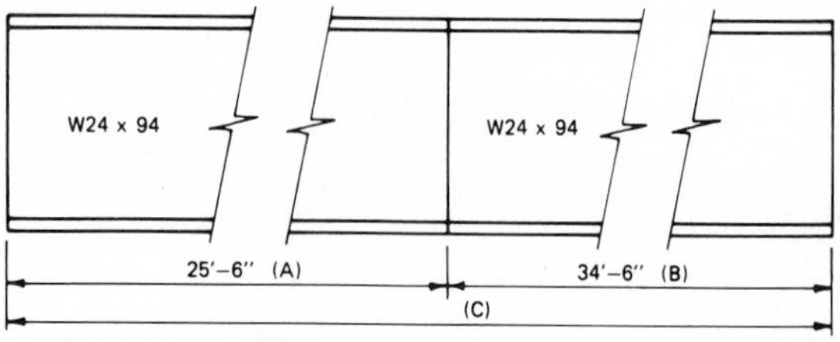
W24 x 94
W24 x 94
25'–6" (A)
34'–6" (B)
(C)

DIM	TOLERANCE	MINIMUM	MAXIMUM
A	+ –		
B	+ –		
C			

JOINED W SHAPES

Fig. 28-1-B Beam sizes.

UNIT 28-2

BEAMS

As a rule, each beam in a system of floor or roof framing makes a convenient erection unit. Hence, the required shop fabrication for each beam is shown on a shop drawing, which provides complete information for that beam. Such a drawing seldom pictures any part of the adjacent members to which this beam will later be joined in the field. However, in the preparation of the beam detail drawing, all the features that have a bearing on the later installation of the beam into its proper location in the frame, as indicated on the design drawing, must be investigated.

The location of the open holes to be provided in the beam for its field connection must match the location of similar holes in the supporting members. Proper clearances must be provided so that the beam can be swung into position after its supporting members have been erected. Any possible interference must be eliminated by cutting away the excess material.

The various fabricating shop drafting rooms do not always agree among themselves on a standard way of making shop drawings. In this text, details will be presented in a manner that all shops could use.

The two principal kinds of beam connections generally used are the framed and the seated types. In the framed type, the beam is connected by means of fittings (generally a pair of short angles) attached to its web. With seated connections, the end of the beam rests on a ledge, or seat, which receives the load from the beam just as if the end of the beam rested upon a wall. See Fig. 28-2-1.

It should be noted that the depth of the beam, dimensions in relation to the depth of the beam, end connections, cuts, and spacing of holes are drawn to scale. Copes, blocks, and cuts are shown in Fig. 28-2-2. It is the practice of the structural detailer to draw the depth dimensions to scale so that the relation of detail is correct and so that the fabricator can interpret the relation of holes to bolts or holes more readily. It is a rule in the fabricating industry that the fabricator is not to scale a drawing; the detailer must have a dimension or note to describe any operation that is required in fabricating the member.

The length of beam and dimensions in relation to the length can be drawn to scale but are usually foreshortened. The reason that the length is usually foreshortened is that the scaled length would, in most cases, take more space on a drawing than is economical and would have no practical value to the fab-

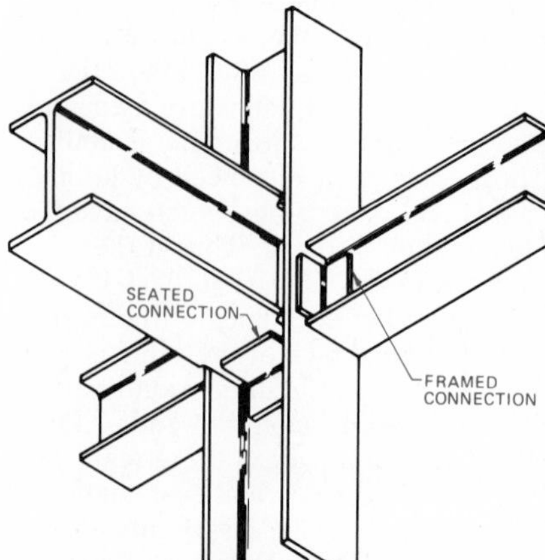
SEATED CONNECTION
FRAMED CONNECTION

Fig. 28-2-1 Beam to column connections.

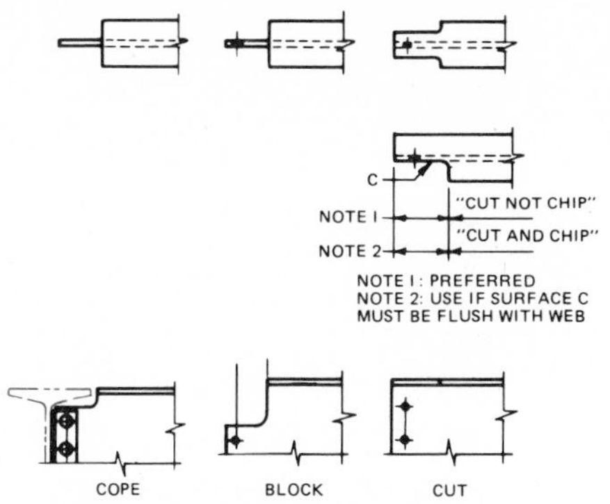
C
NOTE 1
NOTE 2
"CUT NOT CHIP"
"CUT AND CHIP"
NOTE 1: PREFERRED
NOTE 2: USE IF SURFACE C
MUST BE FLUSH WITH WEB
COPE
BLOCK
CUT

Fig. 28-2-2 Copes, blocks, and cuts.

ricator. However, foreshortening the length so much that the holes in the web or some of the detail will appear crowded or ambiguous should be avoided.

The structural detailer does not always draw to the exact scale, but exaggerates the drawings to clarify the object. An example is the two lines that would show the thickness of the top or bottom flange of a beam. If an attempt were made to draw to the exact thickness, in most cases the two lines would run together because of the pencil thickness and would print as one heavy line. To avoid this running together of two parallel lines, the lines are separated more than the actual thickness, so that they clearly define a change in the shape of the object.

Assembly Clearances

In order for members to assemble readily, clearances are required between beams and columns or beams and beams. It may also be necessary to cut or shape the ends of beams for mating parts to fit properly. The recommended clearances are shown in Fig. 28-2-3.

Simple Square-Framed Beams

The information—such as member length, size, and type; number of bolts; or type of fastener—required by the structural detailer is obtained from the design drawing. These drawings usually indicate the type of construction, end loads or loads at support if not normal, type and size of bolts, member shape and size, and any other data that would be required by the detailer.

Figure 28-2-4 represents part of a design drawing for a steel-framed floor system as viewed from above. With its notes, it contains, with the exception of

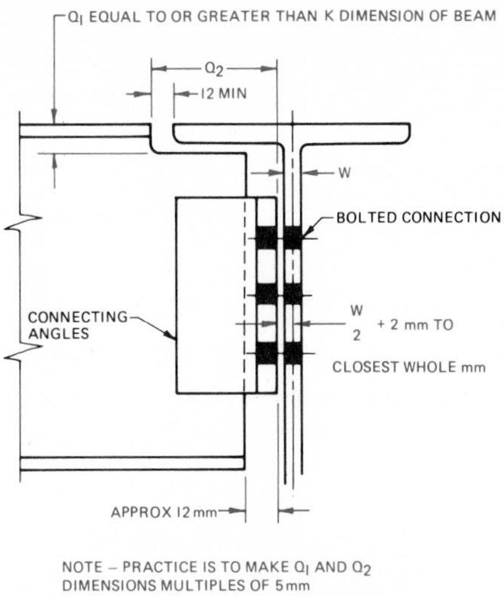
Q1 EQUAL TO OR GREATER THAN K DIMENSION OF BEAM
Q2
12 MIN
W
BOLTED CONNECTION
CONNECTING ANGLES
W/2 + 2 mm TO
CLOSEST WHOLE mm
APPROX 12 mm
NOTE – PRACTICE IS TO MAKE Q1 AND Q2
DIMENSIONS MULTIPLES OF 5 mm

Fig. 28-2-3 Assembly clearances.

connection-angle detail, all the necessary information required by the shop detailer to detail the W450 × 89 beam. Unless otherwise shown by dimensions or notes, members indicated on the design drawing are presumed to be parallel or at right angles to one another, with their webs in a vertical plane, and to be in a level position from end to end. Elevation detail dimensions of beams are generally furnished by a note on the drawing. In Fig. 28-2-4*a* the vertical distance, or elevation, is placed above the level symbol and is shown as +30 020, for the W550 × 109 beam and +30 100 for the W600 × 114 beam. It might have been given by a note reading ALL STEEL FLUSH, TOP AT ELEV. +30 020, as shown by Fig. 28-2-4*b*. Note that the top elevation of W600 × 114 is designated by (+80), meaning that the top of the beam is 80 mm above the reference elevation of + 30 020, or (+ 30 100) as presented in Fig. 28-2-4*a*.

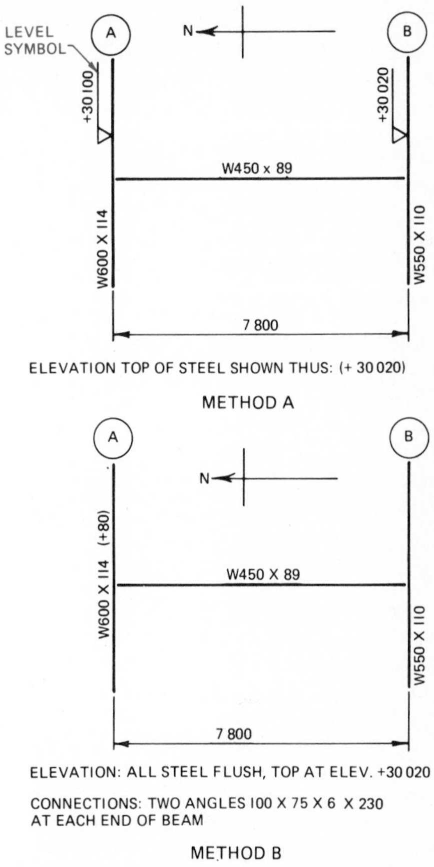
LEVEL SYMBOL
A
B
N
+30 100
+30 020
W450 x 89
W600 X 114
W550 X 110
7 800
ELEVATION TOP OF STEEL SHOWN THUS: (+ 30 020)
METHOD A
W600 X 114 (+80)
W450 X 89
W550 X 110
7 800
ELEVATION: ALL STEEL FLUSH, TOP AT ELEV. +30 020
CONNECTIONS: TWO ANGLES 100 X 75 X 6 X 230
AT EACH END OF BEAM
METHOD B

Fig. 28-2-4 Partial design drawing.

Before starting the drawing, the detailer should first establish what the beam detail is going to look like. This is achieved by making sketches of the connections at both ends. The detailer first investigates the connection at one end, for this example the north end of the W450 × 89 of Fig. 28-2-4. A sketch shown in Fig. 28-2-5*a* is then made of the W450 framing into the W600. This section represents what would be seen if a

Fig. 28-2-5 Detail of W450 x 89 beam.

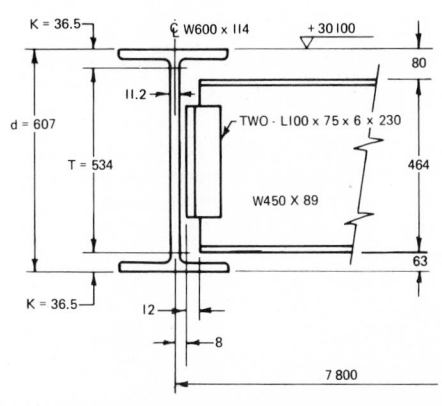

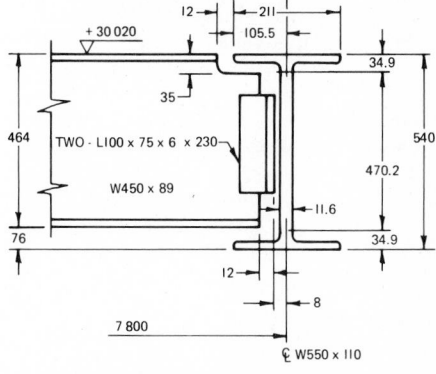

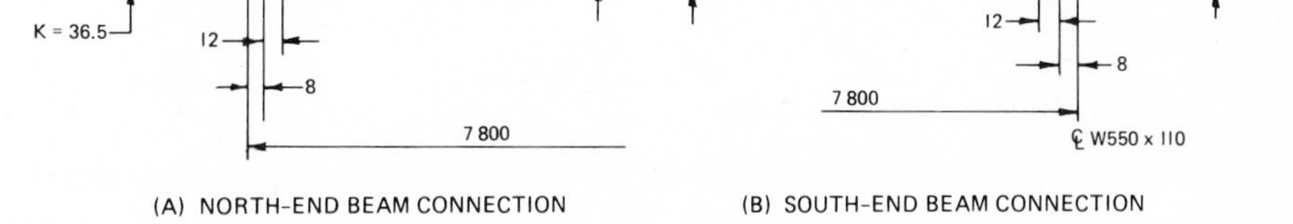
K = 36.5
℄ W600 x 114
+30 100
80
11.2
d = 607
T = 534
TWO - L100 x 75 x 6 x 230
464
W450 X 89
63
K = 36.5
12
8
7 800
(A) NORTH-END BEAM CONNECTION
12
211
105.5
+30 020
34.9
35
464
TWO - L100 x 75 x 6 x 230
540
470.2
W450 x 89
11.6
76
34.9
12
8
7 800
℄ W550 x 110
(B) SOUTH-END BEAM CONNECTION

viewer looked at the connection from the west side of the W450. A sketch is then made of the south end connection of W450. From the sketches the necessary detail requirements can be obtained. Of importance are the number of bolts or size of fillet welds required, and the size of the connecting angles. In this unit, the connecting-angle sizes are given. In Unit 28-3 the calculations for angle size and connections are covered in detail.

Note that the south end flange of the W450 × 89 is flush with that of the supporting W550 × 109 flange. Figure 28-2-5*b* is produced similarly to Fig. 28-2-5*a* for the north end, except we find that the flange of the W450 will interfere with that of the W550. Thus it becomes necessary to notch out or cope the W450.

Some shops would not dimension such a cut, but would give it a standard mark. Others would simply note it on the drawing. Cope to W550 × 109 and let the shop work out its proper shape and size.

Note that the intersection of the horizontal and vertical is not a sharp corner (reentrant-cut), but is burned to a small radius to provide a fillet at this point. However, since the shop has been trained to provide these fillets, they are not generally shown on the detail drawing. The exception to this is in detailing bridge members; then it is essential that all dimensions of the cope be shown. This is to ensure that no fatigue stress raisers will occur.

Since the two beams are flush on top, the minimum depth of the cut Q is made equal to or greater than the K distance for the W550 × 109 beam.

K Distance for the W550 Beam = 34.9. Following the guidelines shown in Fig. 28-2-3, the Q_1 dimension for this beam is 35 mm. The length of the cut, Q_2, as measured from the backs of the connecting angles, should allow for a minimum 12 mm clearance between the toe of the supporting beam and the flange of the supported beam. To determine the length of dimension Q_2, add 12 to half the flange width of the W550 beam. From this value subtract half the web thickness of the W550 beam and 2 mm. The 2 mm dimension is the clearance allowed between the web face and the outer face of the connecting angles.

Therefore $Q_2 = 12 + 105.5 - 5.8 - 2 = 109.7$. As previously used for Q_1, the dimension Q_2 should be raised to the nearest length evenly divisible by 5. Thus the 109.7 dimension (Q_2) is raised to 110.

With the exception of the connecting-angle data, Fig. 28-2-6 is the completed shop drawing of the W450 × 89 beam. Note the following points.

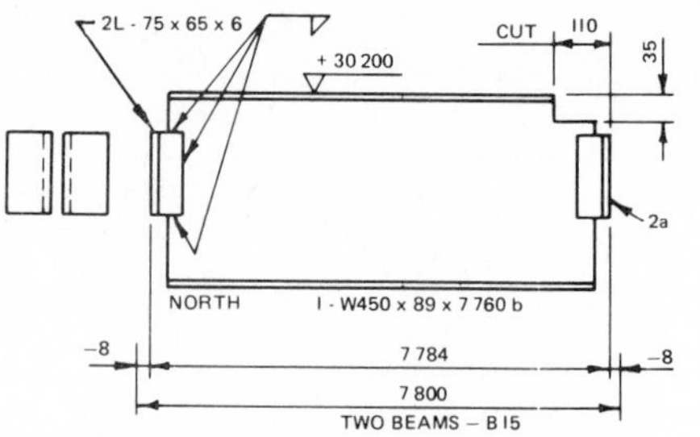

2L - 75 x 65 x 6
CUT
110
+ 30 200
35
2a
NORTH
I - W450 x 89 x 7 760 b
-8
7 784
-8
7 800
TWO BEAMS – B15

NOTE: DETAIL FOR THE CONNECTING ANGLES NOT SHOWN ON THIS DRAWING

Fig. 28-2-6 Detail drawing of W450 x 89 beam.

1. The minus dimensions (−8), shown outside and opposite the dimension line for the back-to-back distance of the end connection angles (7784), are the distances from the center lines of the supporting beams to the back of each connecting angle. For a beam framing to other shapes, the minus dimension (setback distance) is equal to half the web thickness of the supporting member plus 2 mm rounded off to the nearest whole millimetre.

2. The center-to-center distance of 7800 between the two supporting beams is shown for reference purposes.

3. The actual or ordered length of the W450 × 89 should be such that its ends are about 12 mm short of the backs of the connecting angles. This is to allow for inaccurate cutting to specified length at the mill or in the shop and therefore eliminate any extra expense caused by recutting or trimming during fabrication.

4. No top or bottom views are necessary because no holes are required in either flange. In general, the shop should not be required to look at views which convey no necessary instructions.

5. The end connection angles are shown, but not detailed. Information on detailing these angles is given in Unit 28-3. Note the end view of the angles is shown but the W450 beam is not drawn.

6. The complete beam is given a shipping or erection mark, B15, to identify it in the office, shop, and field. There are many systems currently in use for establishing the shipping mark. One of the most common methods is to use a capital letter followed by a sheet number. Each separate shipping piece, detailed on one sheet, has the same number preceded by a different letter. In this example the detail drawing of the W450 × 89 is the second sketch on sheet 15; the third would be C15, the fourth D15; and so on.

7. The connection angles are given assembly or template marks, usually lowercase letters. This is done for two reasons:

(a) It saves the detailing of these angles again, when they are used on the same piece (as on the south end of the beam in this example) or on other beams on the same sheet.

(b) The angles will be punched on a different machine from the one used for the beam. The assembly mark is a guarantee that the correct angle will be assembled on the correct beam. On the given detail, the material required to fabricate one complete shipping piece only is listed, or billed. When duplication of the shipping pieces is required, the shop multiplies the billing for one complete piece by the total number of assemblies required.

SOURCE MATERIAL

- Canadian Institute of Steel Construction

Assignments

1. On an A3- or B-size sheet, make detail drawings of the two connections shown on either Fig. 28-2-A or Fig. 28-2-B. For Fig. 28-2-B refer to inch-size structural steel manuals. The connection angles are welded to the beam web, and the outstanding angles are bolted to the connecting beam. The bolts and holes need not be shown on these drawings. Scale is 1:10 or 1:8.

2. On an A3- or B-size sheet, make sketches of the connections of both ends of the center beam shown on either Fig. 28-2-C or Fig. 28-2-D. After the beam connection sketches have been completed and approved by your instructor, prepare a working drawing of the beam from the sketches and information shown on the drawing. The bolt holes on the outstanding legs of the connection angles need not be shown. Use a conventional break to shorten the length of the beam.

Figure	Connection Sketches	Detail Drawing of Beam
28-2-C	1:8	1:4
28-2-D	1:10	1:5

REVIEW FOR ASSIGNMENT

Unit 7-14 Conventional Breaks
Unit 28-1 Properties of Structural Steel Shapes

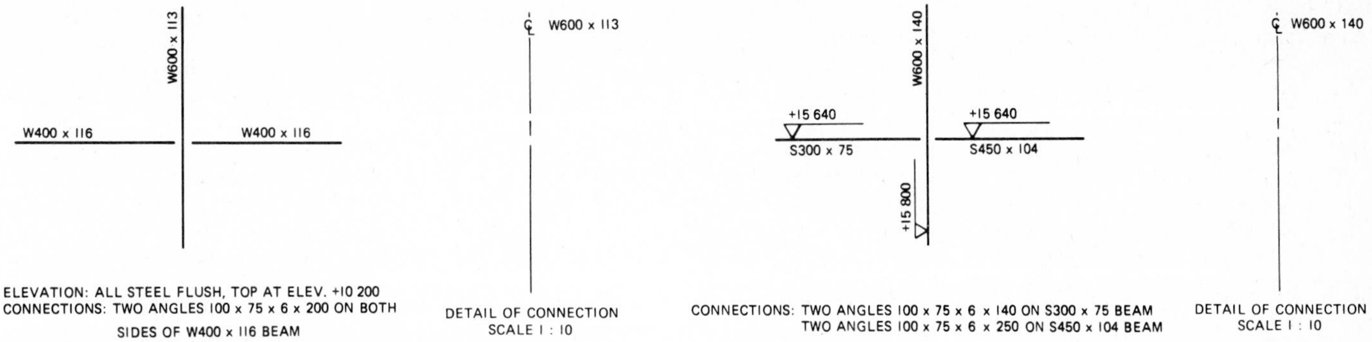
W600 x 113
W400 x 116
W400 x 116
ELEVATION: ALL STEEL FLUSH, TOP AT ELEV. +10 200
CONNECTIONS: TWO ANGLES 100 x 75 x 6 x 200 ON BOTH SIDES OF W400 x 116 BEAM
W600 x 113
DETAIL OF CONNECTION
SCALE 1 : 10
W600 x 140
+15 640
S300 x 75
+15 640
S450 x 104
+15 800
CONNECTIONS: TWO ANGLES 100 x 75 x 6 x 140 ON S300 x 75 BEAM
TWO ANGLES 100 x 75 x 6 x 250 ON S450 x 104 BEAM
W600 x 140
DETAIL OF CONNECTION
SCALE 1 : 10

Fig. 28-2-A Beam and connection details.

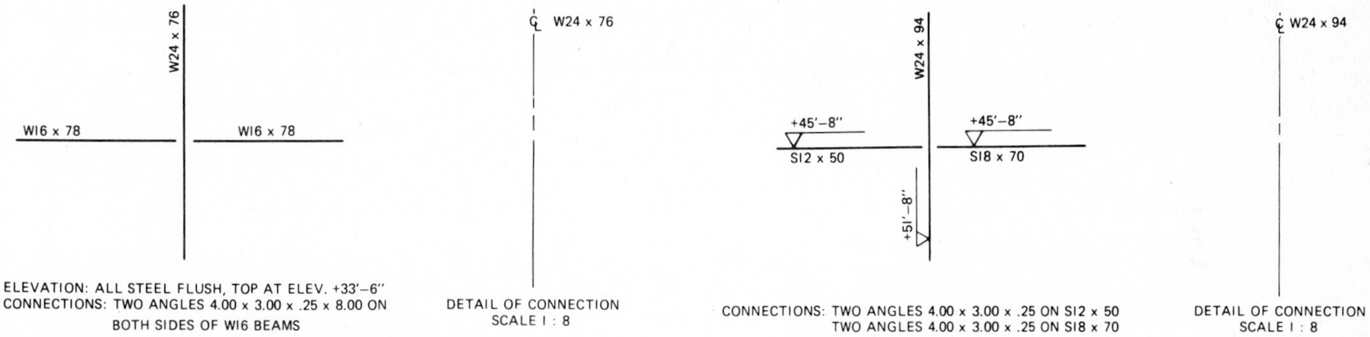
W24 x 76
W16 x 78
W16 x 78
ELEVATION: ALL STEEL FLUSH, TOP AT ELEV. +33'–6"
CONNECTIONS: TWO ANGLES 4.00 x 3.00 x .25 x 8.00 ON BOTH SIDES OF W16 BEAMS
W24 x 76
DETAIL OF CONNECTION
SCALE 1 : 8
W24 x 94
+45'–8"
S12 x 50
+45'–8"
S18 x 70
+51'–8"
CONNECTIONS: TWO ANGLES 4.00 x 3.00 x .25 ON S12 x 50
TWO ANGLES 4.00 x 3.00 x .25 ON S18 x 70
W24 x 94
DETAIL OF CONNECTION
SCALE 1 : 8

Fig. 28-2-B Beam and connection details.

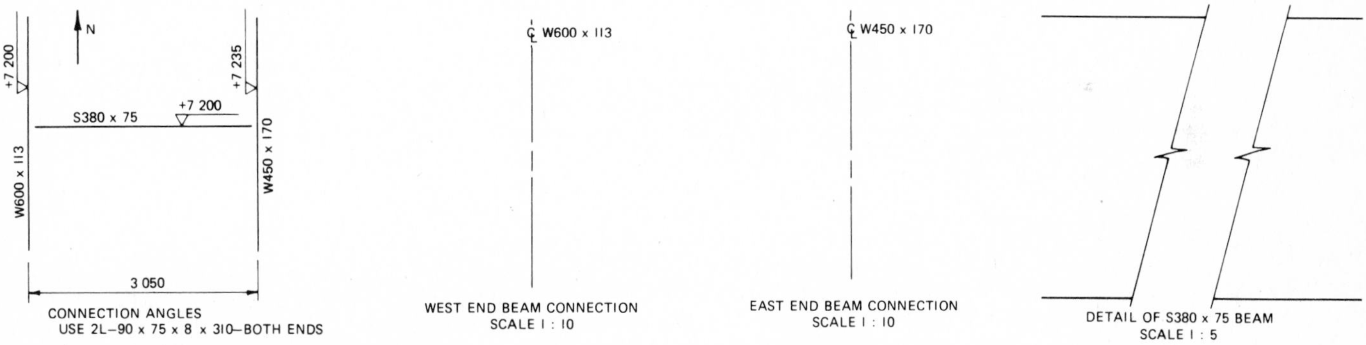
N
+7 200
+7 235
+7 200
S380 x 75
W600 x 113
W450 x 170
3 050
CONNECTION ANGLES
USE 2L–90 x 75 x 8 x 310–BOTH ENDS
W600 x 113
WEST END BEAM CONNECTION
SCALE 1 : 10
W450 x 170
EAST END BEAM CONNECTION
SCALE 1 : 10
DETAIL OF S380 x 75 BEAM
SCALE 1 : 5

Fig. 28-2-C Beam and connection details.

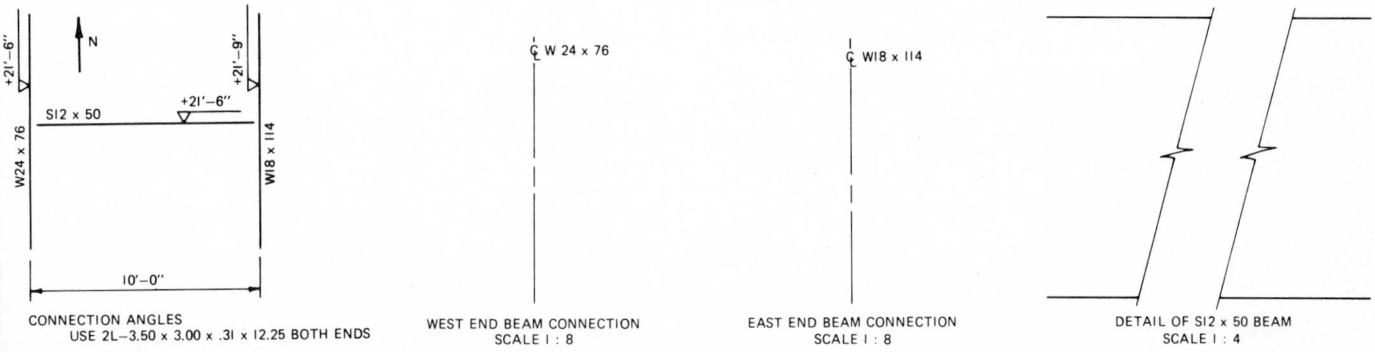
N
+21'–6"
+21'–9"
+21'–6"
S12 x 50
W24 x 76
W18 x 114
10'–0"
CONNECTION ANGLES
USE 2L–3.50 x 3.00 x .31 x 12.25 BOTH ENDS
W 24 x 76
WEST END BEAM CONNECTION
SCALE 1 : 8
W18 x 114
EAST END BEAM CONNECTION
SCALE 1 : 8
DETAIL OF S12 x 50 BEAM
SCALE 1 : 4

Fig. 28-2-D Beam and connection details.

UNIT 28-3

STANDARD CONNECTIONS

Standard framed-beam connections are used for framing structural steel. Since riveting is almost nonexistent in most fabricating plants today, rivets will not be considered in this context. Standard connection angles are shop-welded or bolted to the beam web and field-bolted to their supporting member. Only high-strength bolted connections of the friction type will be considered in this chapter.

When detailing individual members, the shop detailer should bear in mind that each individual member must be joined to other members. The place or location of one member's attachment shape or plate along with the means of fastening is called the *connection plate*, or the *connection* for short.

Bolted Connections

Bolts are placed on standard lines or gages. The distance between bolt holes is referred to as the *bolt pitch*, or *pitch*. The gage and the pitch for a multiple bolt connection detail must be sufficiently large to allow the wrench clearance when a bolt adjacent to a previously installed bolt or adjacent to another part of the shape being joined is tightened. Figure 28-3-1 shows the recommended gages to be used for structural shapes.

Of prime importance is consistency of detail; for example, gages on an individual member should not vary throughout the length of the member. If a connection plate, made from a sheared piece of steel, as shown in Fig. 28-3-2, is to have three holes across its width of 240 mm for 16 mm bolts, then the gage could be 80 mm with edge distances of 30 mm. If the fasteners to be used are 20 mm-diameter bolts, then the edge distance of 30 mm would have to be adjusted. The minimum distance from the center of a bolt hole to any edge should not be less than that given by Fig. 28-3-3. For the 20 mm bolt, the minimum edge distance to the sheared edge is 34 mm. Since the plate is 240 mm wide and a minimum edge distance of 34 mm is required, the gage required would be $(240 - 2 \times 34) \div 2$, or 86 mm. However, 86 mm is considered to be an awkward dimension, and hence it is rounded off to a gage of 85 mm, which produces an edge distance of $(240 - 2 \times 85) \div 2$, or 35 mm which is greater than the minimum edge distance of 34. If the connection plate had been an angle, the same reasoning as used above would still pertain to the connection detail. Another consideration is wrench clearance which is illustrated in Fig. 28-3-4.

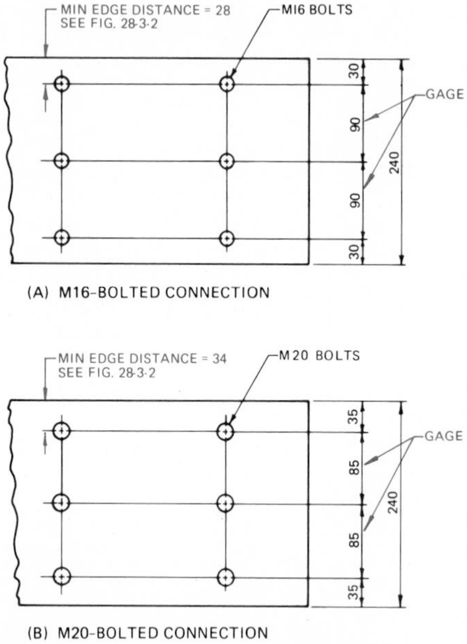
MIN EDGE DISTANCE = 28
SEE FIG. 28-3-2
M16 BOLTS
GAGE
30
90
90
30
240
(A) M16-BOLTED CONNECTION
MIN EDGE DISTANCE = 34
SEE FIG. 28-3-2
M20 BOLTS
GAGE
35
85
85
35
240
(B) M20-BOLTED CONNECTION

Fig. 28-3-2 Establishing gage sizes from edge distance.

BOLT DIAMETER	AT SHEARED EDGE	AT ROLLED OR GAS CUT EDGE
14	26	20
16	28	22
20	34	26
22	38	28
24	42	30
27	46	34
30	52	38
36	64	46

Fig. 28-3-3 Minimum edge distance for bolt holes.

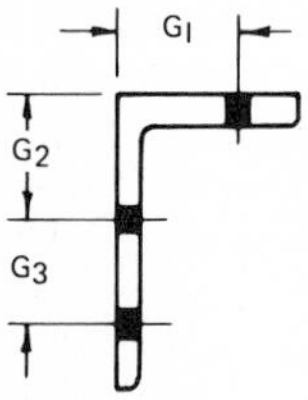
G1
G2
G3

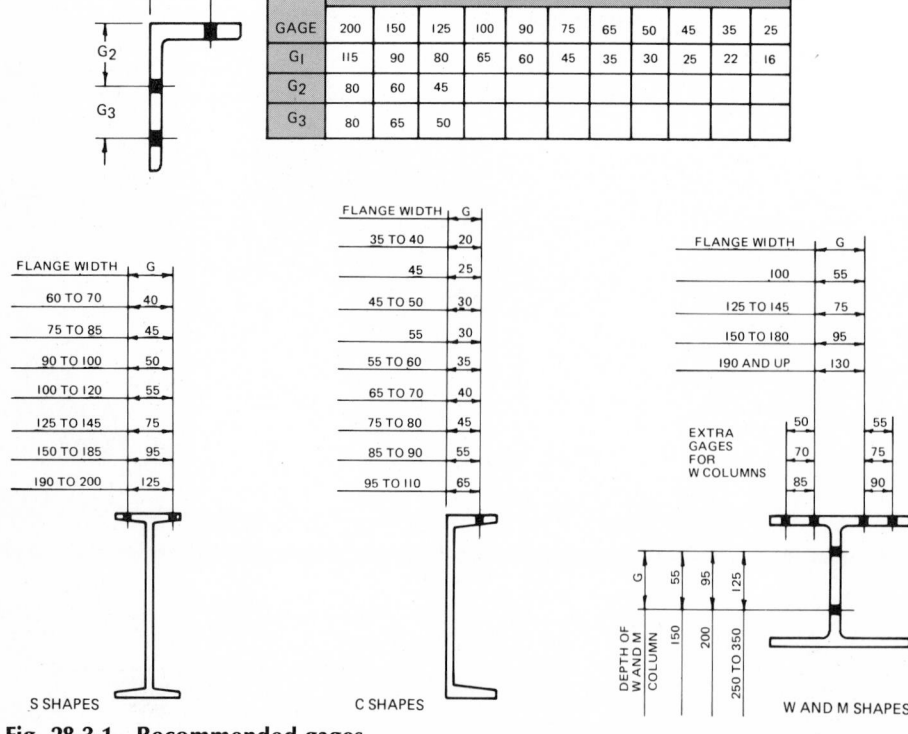

GAGE	LEG SIZE										
	200	150	125	100	90	75	65	50	45	35	25
G1	115	90	80	65	60	45	35	30	25	22	16
G2	80	60	45								
G3	80	65	50								

FLANGE WIDTH	G
60 TO 70	40
75 TO 85	45
90 TO 100	50
100 TO 120	55
125 TO 145	75
150 TO 185	95
190 TO 200	125

S SHAPES

FLANGE WIDTH	G
35 TO 40	20
45	25
45 TO 50	30
55	30
55 TO 60	35
65 TO 70	40
75 TO 80	45
85 TO 90	55
95 TO 110	65

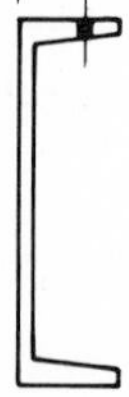

C SHAPES

FLANGE WIDTH	G
100	55
125 TO 145	75
150 TO 180	95
190 AND UP	130

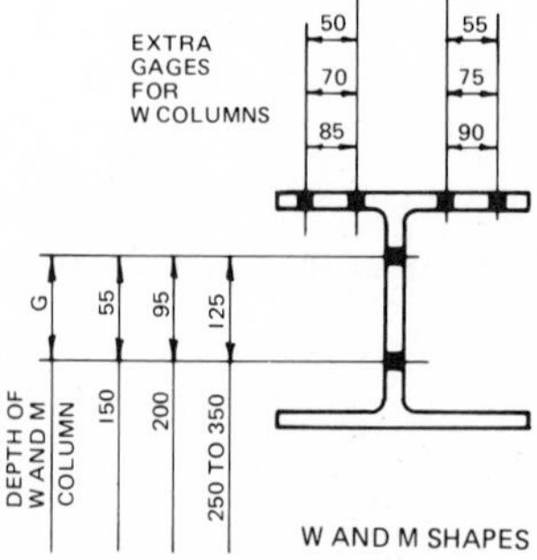
EXTRA GAGES FOR W COLUMNS
50
70
85
55
75
90
G
55
95
125
DEPTH OF W AND M COLUMN
150
200
250 TO 350

W AND M SHAPES

Fig. 28-3-1 Recommended gages.

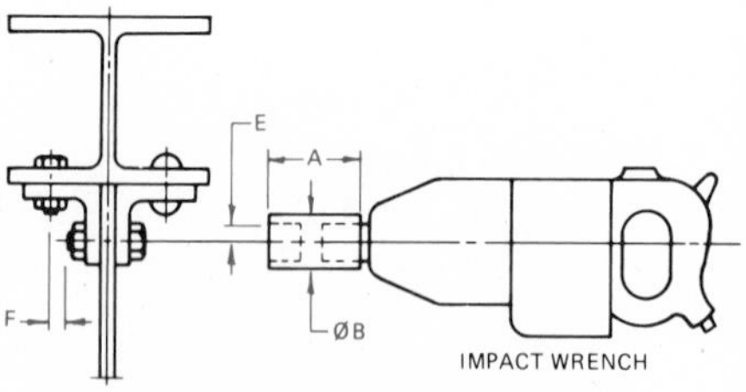
E
A
F
ØB
IMPACT WRENCH

BOLT SIZE	SOCKETS A	SOCKETS B	MIN. CLEAR. E	MIN. CLEAR. F
16	70	55	28	32
20	80	58	29	34
24	90	65	32	36
30	110	75	38	42
36	130	85	43	48

	BOLT SIZE	C	D
LIGHT WRENCHES	16 TO 24	337 TO 356	54
HEAVY WRENCHES	24 TO 36	375 TO 438	64

Fig. 28-3-4 Minimum erection clearances.

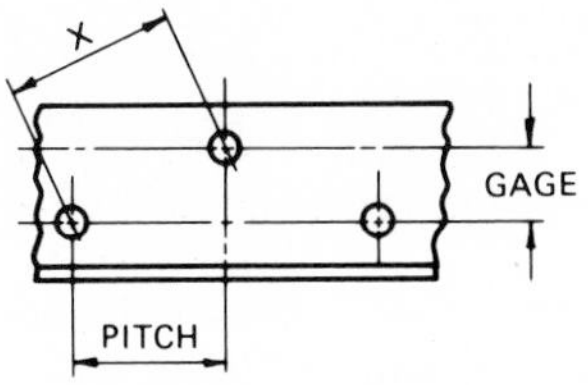
X
GAGE
PITCH

Fig. 28-3-5 **Staggered fasteners.**

Pitch mm	Gage 25	30	35	40	45	50	55	60	65	70	75	80	85	90	95	100	105	110
5	25	30	35	40	45	50	55	60	65	70	75	80	85	90	95	100	105	110
10	27	32	36	41	46	51	56	61	66	71	76	81	86	91	96	100	105	110
15	29	34	38	43	47	52	57	62	67	72	76	81	86	91	96	101	106	111
20	32	36	40	45	49	54	59	63	68	73	78	82	87	92	97	102	107	112
30	35	39	43	47	51	56	60	65	70	74	79	84	89	93	98	103	108	113
30	39	42	46	50	54	58	63	67	72	76	81	85	90	95	100	104	109	114
35	43	46	49	53	57	61	65	69	74	78	83	87	92	97	101	106	111	115
40	47	50	53	57	60	64	68	72	76	81	85	89	94	98	103	108	112	117
45	51	54	57	60	64	67	71	75	79	83	87	92	96	101	105	110	114	119
50	56	58	61	64	67	71	74	78	82	86	90	94	99	103	107	112	116	121
55	60	63	65	68	71	74	78	81	85	89	93	97	101	105	110	114	119	123
60	65	67	69	72	75	78	81	85	88	92	96	100	104	108	112	117	121	125
65	70	72	74	76	79	82	85	88	92	96	99	103	107	111	115	119	123	128
70	74	76	78	81	83	86	89	92	96	99	103	106	110	114	118	122	126	130
75	79	81	83	85	87	90	93	96	99	103	106	110	113	117	121	125	129	133
80	84	85	87	89	92	94	97	100	103	106	110	113	117	120	124	128	132	136
85	89	90	92	94	96	99	101	104	107	110	113	117	120	124	127	131	135	139
90	93	95	97	98	101	103	105	108	111	114	117	120	124	127	131	135	138	142

Occasionally, a gage is too small for bolt holes to be placed adjacent to one another at right angles. When this happens, staggered centers are used as illustrated in Fig. 28-3-5.

Example 1. Given a flat bar 110 mm in width, which is to have a double line of ϕ 16 mm bolts and a gage of 50, calculate the pitch of the bolts. See Fig. 28-3-6.

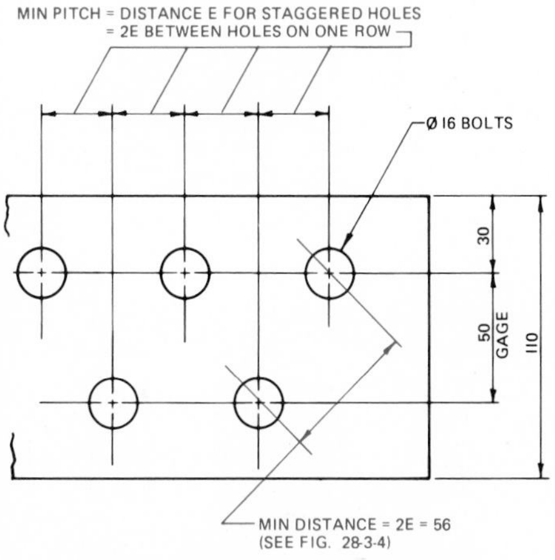
MIN PITCH = DISTANCE E FOR STAGGERED HOLES
= 2E BETWEEN HOLES ON ONE ROW
Ø 16 BOLTS
30
50
GAGE
110
MIN DISTANCE = 2E = 56
(SEE FIG. 28-3-4)

Fig. 28-3-6 Gage and pitch layout.

Solution: From Fig. 28-3-4 the clearance E required for a 16 mm bolt is 28 mm. The minimum recommended distance between holes is $2E$, or 56 mm, which is greater than the gage of 50 mm. Therefore, staggered holes will be required. To determine the minimum pitch, refer to Fig. 28-3-5. Read down the 50 gage column until the dimension 56 is reached. To the extreme left of the 56 value is the pitch required. For this problem, the pitch is 30.

Figure 28-3-7 shows a partial design drawing similar to Fig. 28-2-4 except that it indicates information concerning the connection. It represents part of a design drawing for a steel-framed floor system as viewed from above. With its notes, it contains all the necessary information required by the shop detailer to detail the W450 × 89 beam. With the exception of the connection-angle detail, all information pertaining to this beam was covered in Unit 28-2. Therefore this unit will deal only with the connection-angle detail.

The problem is to select a connection from the W450 beam which will be able to carry the reaction values. Often the reaction values are calculated and given on the design drawing, as shown in Fig. 28-3-8. If the reaction values are not given, the connection is designed to support half the total uniform load capacity. In this case, since the reaction values are not given and there is no indication on the design drawing to indicate other than a uniformly distributed load when the building is complete, we consult Fig. 28-3-9, which states the maximum allowable load for a W450 × 89 beam having a span of 7800 mm lies between the 576 and 540 kilonewtons (kN) values (approximately 555 kN). Connection design load is one-half of this value or 278 kN.

Referring to Fig. 28-3-10 under the column Bolt Capacity in Kilonewtons Friction Connections, we find that four bolts are required to support this load. The length of the connecting angles shown for four bolts is 310 mm, for which the minimum and maximum depth recommendations for beams are 380 and 600 mm. Since these limits bracket the actual depth of the W450 beam, it is acceptable.

Referring to the column Web Framing Leg with Welds, we find the maximum weld capacity of a four-bolt per vertical

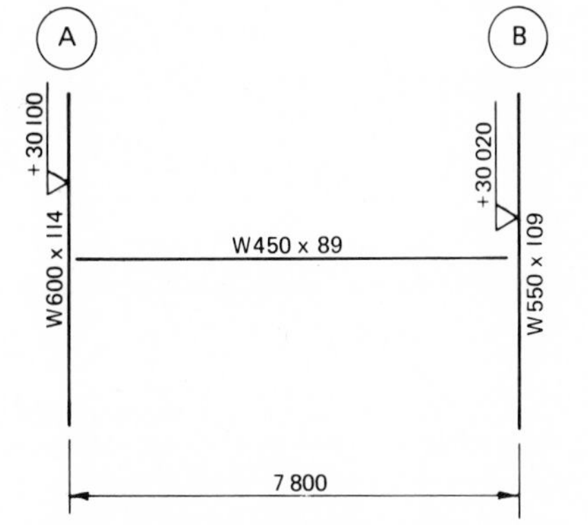
A
B
+30 100
+30 020
W600 x 114
W450 x 89
W550 x 109
7 800

ELEVATION TOP OF STEEL SHOWN THUS (+ 30 020)

NOTES:
ALL HOLES Ø 22
ALL CONNECTIONS TO DEVELOP FULL LENGTH UNLESS OTHERWISE SPECIFIED
BOLTS: M20–A325
CONNECTING ANGLES WELDED TO BEAM, BOLTED TO SUPPORT

Fig. 28-3-7 Partial design drawing.

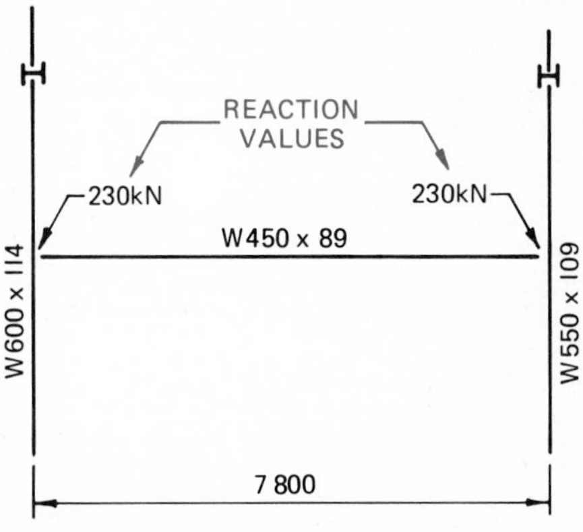
REACTION
VALUES
230kN
230kN
W450 x 89
W600 x 114
W550 x 109
7 800

Fig. 28-3-8 Indicating beam reactions on drawings.

BEAM LOAD TABLE IN KILONEWTONS (kN)								
Span mm	W450 Beam Mass per Metre				W310 Beam Mass per Metre			
	89	82	74	67	60	54	46	40
5000	864	782	708	631	395	361	308	266
5500	785	711	644	573	359	328	280	242
6000	720	652	590	526	329	301	257	222
6500	665	601	545	485	304	278	237	205
7000	617	559	506	451	282	258	220	190
7500	576	521	472	420	263	241	205	177
8000	540	489	443	394	247	226	193	166
8500	508	460	417	371	232	212	181	157

BEAM LOAD TABLE IN KIPS (1000 POUNDS)								
Span Feet	W18 Beam Pounds per Foot				W12 Beam Pounds per Foot			
	60	55	50	45	40	36	31	27
16	177	160	145	130	81	74	63	55
18	161	146	132	118	74	67	57	50
20	148	134	121	108	67	62	53	46
22	136	123	112	100	62	57	49	42
24	126	115	104	92	58	53	45	39
26	118	106	97	86	54	50	42	36
28	111	100	91	81	51	46	40	34
30	104	94	85	76	48	44	37	32

Fig. 28-3-9 Beam load tables.

EITHER LEG WITH BOLTS				WEB FRAMING LEG WITH WELDS				
Angle Width and Gage	Bolts Per Vertical Line	Bolt Capacity in Kilonewtons Friction Connections	Bearing Connections	Weld Capacity in Kilonewtons Fillet Size, D (mm)				Angle Width
				4	6	8	10	mm
	2	169		324	484	646	808	
	3	240		452	678	904	1130	
W = 90	4	320		551	826	1100	1380	
g = 130	5	400		649	974	1300	1620	W = 75
g_1 = 60	6	480		748	1120	1500	1870	
	7	560		846	1270	1690	2120	
	8	640		945	1420	1890	2360	
	2	169		330	484	646	808	
	3	240		428	641	855	1070	
W = 75	4	320		526	789	1050	1320	
G = 100	5	400		625	937	1250	1560	W = 65
G_1 = 45	6	480		723	1080	1450	1810	
	7	560		822	1230	1640	2050	
	8	640		920	1380	1840	2300	
Minimum Required Web Thickness and Angles Where Bolted								
FY = 300			8.6	6.9	10.3	13.8	17.2	

Note 1. Connection angles are assumed to be material with minimum yield strength of 300 MPa.
2. For connections with outstanding legs bolted, the minimum required thickness of the supporting material is one-half the thickness listed above, if beams are attached to one side of the supporting material.

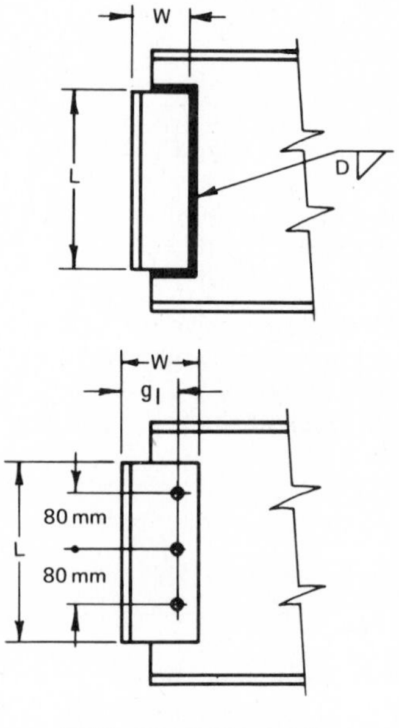
W
L
D
W
g_1
80 mm
80 mm
L
WEB FRAMING LEG

OUTSTANDING LEG WITH WELDS					LENGTH OF CONNECTING ANGLES			
Angle Width	Weld Capacity in Kilonewtons Fillet Size D (mm)				Bolts Per Vertical Line	Suggested Beam Depth Limits (mm)		Angle Length L
mm	4	6	8	10		Min	Max	(mm)
	98	141	186	232	2	200–	300	150
	228	329	438	546	3	300–	450	230
	384	563	755	952	4	380–	600	310
W = 90	490	743	1000	1263	5	450–	800	390
	589	891	1200	1510	6	550–	900	470
	687	1040	1400	1760	7	600–	1100	550
	786	1190	1590	2000	8	700–	1200	630
	125	180	238	297	2	200–	300	150
	276	405	544	681	3	300–	450	230
	386	581	789	1004	4	380–	600	310
W = 75	490	743	1000	1263	5	450–	800	390
	589	891	1200	1510	6	550–	900	470
	687	1040	1400	1760	7	600–	1100	550
	786	1190	1590	2000	8	700–	1200	630
Minimum Required Web or Flange Thickness①								
FY = 300	3.4	5.2	6.9	8.6				

To be used for educational purposes only.

① Thickness listed is for supporting material with beams attached to one side only. If beams are attached to both sides of the supporting material, use double the minimum thickness listed.

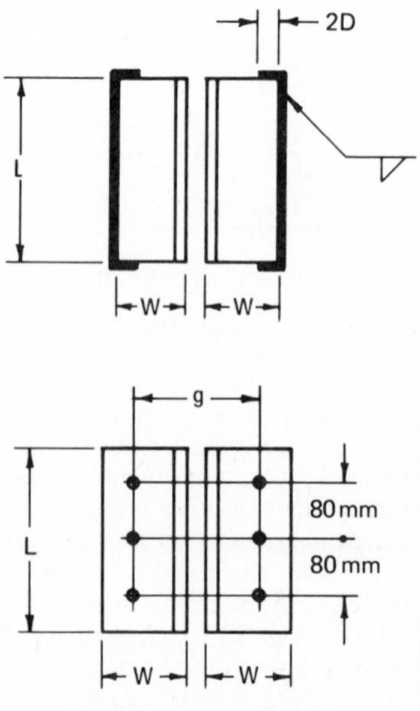
2D
L
W
W
g
80 mm
80 mm
L
W
W
OUTSTANDING LEGS

Fig. 28-3-10 Double angle beam connections for M20-A325 bolts and E480xx fillet welds. (CISC) line angle with a 4 mm weld is 551 kN for a 75 mm angle width and 526 kN for a 65 mm angle width. Both are greater than the 270 kN allowable load and thus are acceptable. Therefore the smaller angle width of 65 is selected for the welded framing leg. Practice dictates that the angle thickness should be 2 mm greater than the weld size; hence the minimum required angle thickness is 4 + 2 = 6 mm thick.

Another check for the minimum thickness of the angle and the minimum permissible web thickness for the beam must be made. In order to determine the minimum angle thickness, refer to the section Minimum Required Web Thickness and Angles Where Bolted. Since connection angles are assumed to be material which has a yield strength of 300 MPa (see note 1 located below the table), the minimum web and angle thickness specified is 8.6 mm. Note 2 states that if the connection is used for outstanding angle legs, then the value as specified can be halved. Therefore, the minimum thickness of angle that can be used is 4.3 mm. This is less than the 6 mm required for welding; and as such, the 6 mm angle thickness is selected. The web thickness of the W450 × 89 beam is 10.16, which is acceptable.

The next step is to select the gage (distance between bolt centers on the two outstanding legs). The recommended gage and angle widths shown are 130 and 90, or 100 and 75, respectively. Both the 90 and 75 mm wide angles are acceptable as far as load capacities are concerned. The clearances shown in Fig. 28-3-11 would be a factor in deciding to use either the 75 or 90 mm wide outstanding length. In this problem, the 75 mm leg was chosen because the use of universal joints is now a widely accepted practice.

The G_1 distance of 45 shown in Fig. 28-3-1 should be used instead of the actual 44.7 dimension shown in Fig. 28-3-11 since a hole clearance of 2 mm (22 mm diameter holes) is used for the M20 bolts.

Some fabricators on occasion choose to use high-strength bolts in the shop to fasten the connection angles to the web of the beam. The combination of the shop and field high-strength bolts in one connection presents other problems, such as additional clearance requirements for entering and tightening the high-strength bolts. These requirements mean larger connection angles, larger gages, and larger spread. An alternate solution to the increased sizes would be to

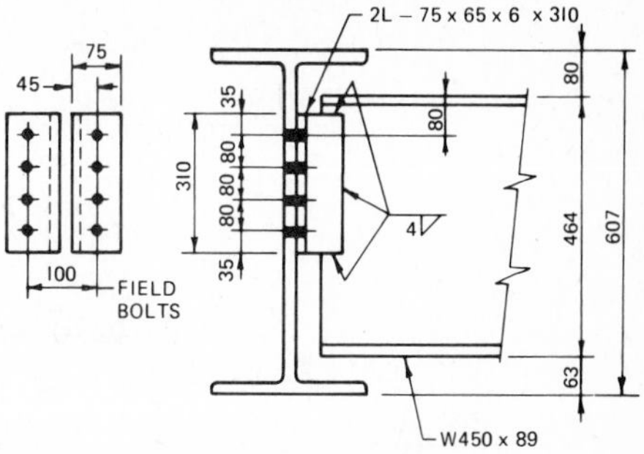
2L – 75 x 65 x 6 x 310
75
45
100
FIELD
BOLTS
310
35
80
80
80
35
80
4
464
607
63
W450 x 89

Fig. 28-3-12 Detail of north end of W450 x 89 beam.

stagger the bolt centers of the outstanding leg. In order not to overcomplicate the subject at this point, all connections discussed in this unit will be angles shop-welded to web and field-bolted to the supporting members.

The location of the connection angles on the horizontal beam must now be set. Practice is to set the distance from the uppermost bolt hole to the top of the beam equal to the pitch distance. From the detailer's sketch, shown in Fig. 28-3-12, the complete detail drawing of the beam (Fig. 28-3-13) is made. Note the following points.

1. The end connection angles are completely detailed. Following their own individual shop standards, many shops could do with less information.
2. The web leg is 65 mm.
3. The outstanding leg of each connection angle is 75 mm, and the recommended gage for this leg size is 45 mm.
4. The open holes, for connection to the web of the supporting beam, are spaced 100 mm apart, center to center. This distance is called the *spread*. It is 2 × G_1 + web thickness of W450 × 89 beam = 2 × 45 + 10.6 = 100.6. Use 100.
5. The pitch, or distance, bolt to bolt, along any gage line is 80 mm.
6. The end distance is equal to half of the remainder left after subtracting the sum of all bolt spaces from the length of the angles. In the case of the four-row connection, it equals (310 – 80 – 80 – 80) ÷ 2 = 35.
7. Instead of noting them on each individual detail, as in Fig. 28-3-13, it is usual practice to call for the sizes of bolts and holes once on each sheet in a general note. Such a note covers all the bolts and holes on the sheet, with exceptions noted on the individual details where they occur.

SOURCE MATERIAL

- Canadian Institute of Steel Construction

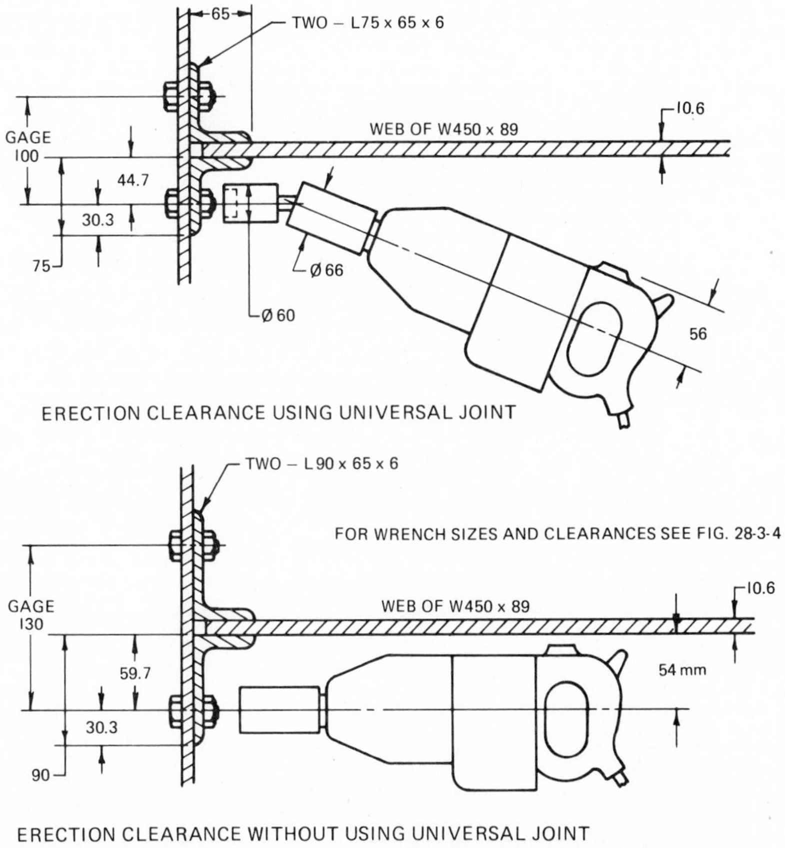
65
TWO – L75 x 65 x 6
GAGE
100
WEB OF W450 x 89
10.6
44.7
30.3
75
Ø 66
Ø 60
56
ERECTION CLEARANCE USING UNIVERSAL JOINT
TWO – L90 x 65 x 6
FOR WRENCH SIZES AND CLEARANCES SEE FIG. 28-3-4
GAGE
130
WEB OF W450 x 89
10.6
59.7
54 mm
30.3
90
ERECTION CLEARANCE WITHOUT USING UNIVERSAL JOINT

Fig. 28-3-11 How erection clearances control gage and connecting angle sizes.

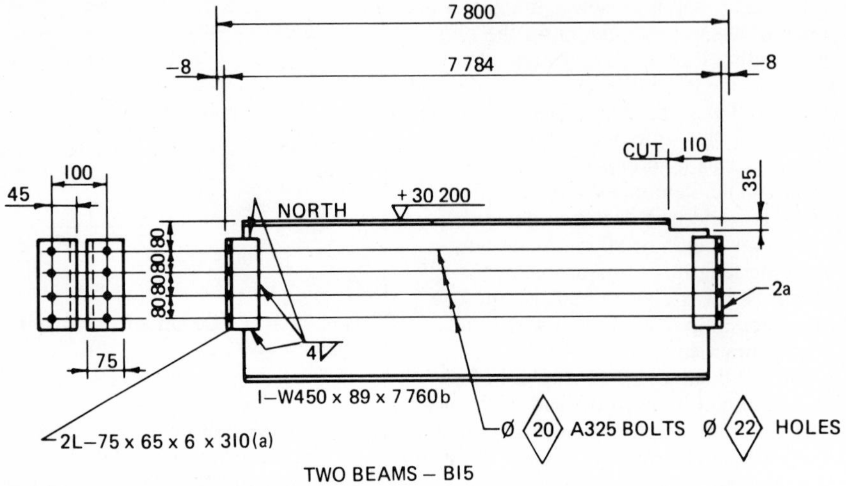

7 800
7 784
–8
–8
CUT
110
35
100
45
+ 30 200
NORTH
80
80
80
80
2a
75
4
I–W450 x 89 x 7 760b
2L–75 x 65 x 6 x 310(a)
Ø 20 A325 BOLTS Ø 22 HOLES
TWO BEAMS – B15

Fig. 28-3-13 Complete beam detail of W450 x 89 from partial design drawing, Fig. 28-3-7.

Assignments

1. On an A3- or B-size sheet, sketch the following beam connections from the design sketch shown in Fig. 28-3-A or Fig. 28-3-B.

(A) North end connection of beam E3.

(B) South end connection of beam K3.

(C) Southwest end connection of beam N3.

Scale is 1:10 or 1:8.

2. On an A3- or B-size sheet, calculate the missing dimensions from the charts and drawings shown in Fig. 28-3-C.

REVIEW FOR ASSIGNMENT

Unit 28-1 Structural Drawing Practices
Unit 28-2 Assembly Clearances

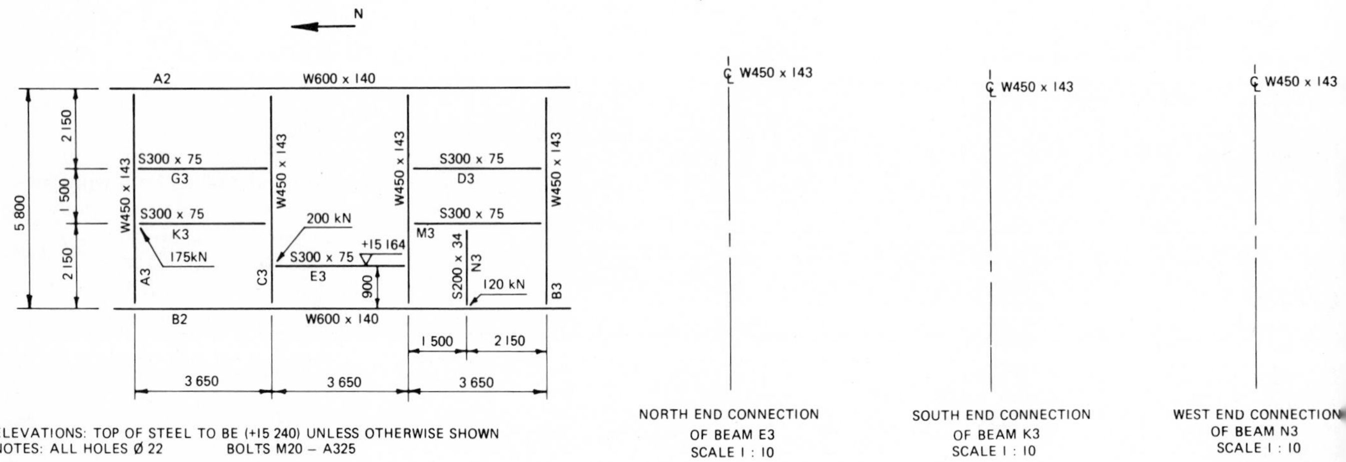

N
A2
W600 x 140
5 800
2 150
1 500
2 150
W450 x 143
S300 x 75
G3
S300 x 75
K3
175kN
A3
W450 x 143
C3
200 kN
+15 164
S300 x 75
E3
900
W450 x 143
S300 x 75
D3
S300 x 75
M3
S200 x 34
N3
120 kN
W450 x 143
B3
B2
W600 x 140
1 500
2 150
3 650
3 650
3 650
ELEVATIONS: TOP OF STEEL TO BE (+15 240) UNLESS OTHERWISE SHOWN
NOTES: ALL HOLES Ø 22 BOLTS M20 – A325
W450 x 143
NORTH END CONNECTION
OF BEAM E3
SCALE 1 : 10
W450 x 143
SOUTH END CONNECTION
OF BEAM K3
SCALE 1 : 10
W450 x 143
WEST END CONNECTION
OF BEAM N3
SCALE 1 : 10

Fig. 28-3-A Beam connections.

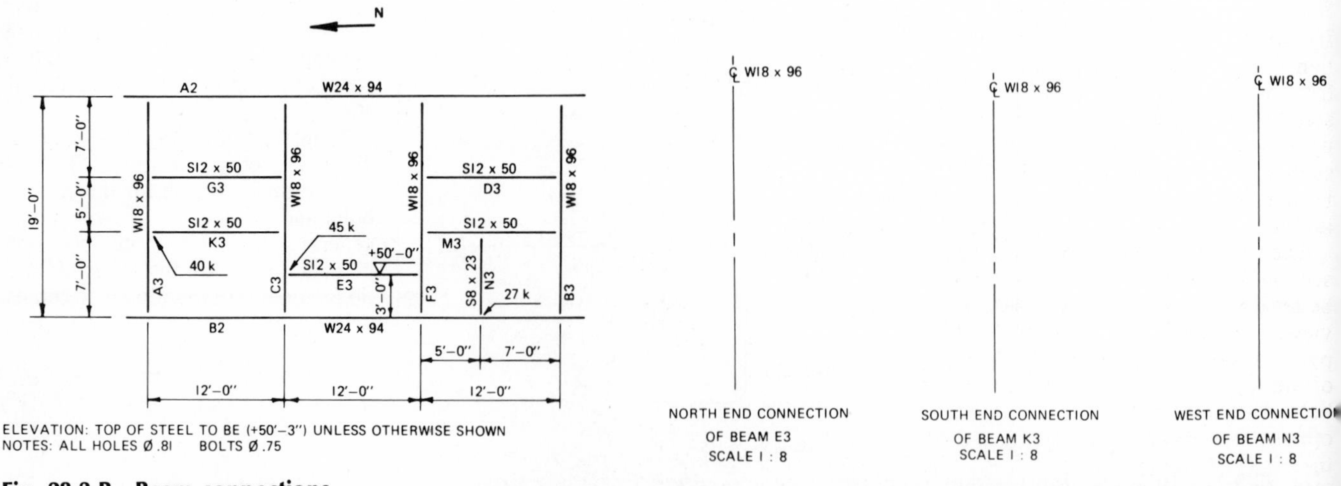

N
A2
W24 x 94
19'–0"
7'–0"
5'–0"
7'–0"
W18 x 96
S12 x 50
G3
S12 x 50
K3
40 k
A3
W18 x 96
C3
45 k
+50'–0"
S12 x 50
E3
3'–0"
W18 x 96
S12 x 50
D3
S12 x 50
M3
F3
S8 x 23
N3
27 k
W18 x 96
B3
B2
W24 x 94
5'–0"
7'–0"
12'–0"
12'–0"
12'–0"
ELEVATION: TOP OF STEEL TO BE (+50'–3") UNLESS OTHERWISE SHOWN
NOTES: ALL HOLES Ø .81 BOLTS Ø .75
W18 x 96
NORTH END CONNECTION
OF BEAM E3
SCALE 1 : 8
W18 x 96
SOUTH END CONNECTION
OF BEAM K3
SCALE 1 : 8
W18 x 96
WEST END CONNECTION
OF BEAM N3
SCALE 1 : 8

Fig. 28-3-B Beam connections.

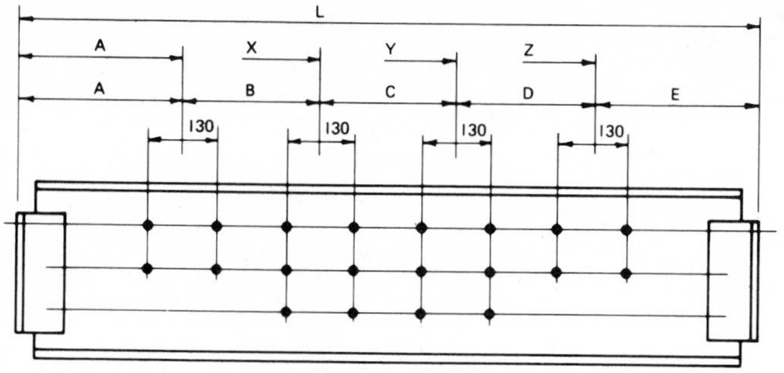

L
A
X
Y
Z
A
B
C
D
E
130
130
130
130

USING THE BEAM SHOWN ABOVE AND DIMENSIONS A–E CALCULATE DIMENSIONS L TO Z AND COMPLETE THE CHART SHOWN BELOW.

DIM	PROBLEM 1	2	3	4	5	6
A	1 907	1 199	1 410	992	2 740	1 661
B	1 029	1 197	1 578	1 330	3 016	1 354
C	911	1 213	1 399	1 659	2 678	1 014
D	2 357	1 206	1 294	1 994	2 956	1 343
E	1 330	1 205	1 708	2 331	2 629	1 688
L						
X						
Y						
Z						

Fig. 28-3-C Calculation of dimensions.

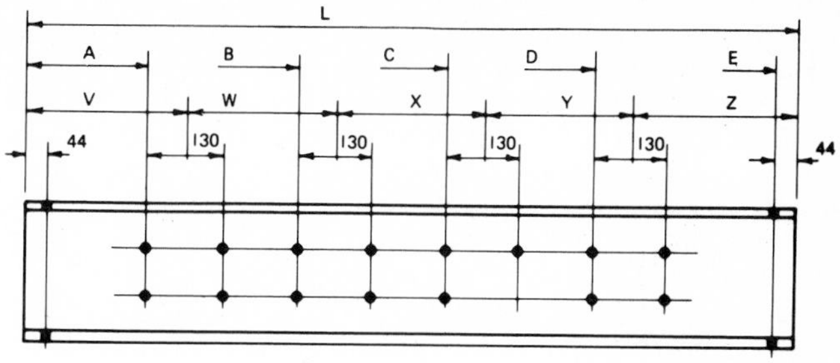

L
A
B
C
D
E
V
W
X
Y
Z
44
130
130
130
130
130
44

USING THE BEAM SHOWN ABOVE AND DIMENSIONS A–E CALCULATE DIMENSIONS L TO Z AND COMPLETE THE CHART SHOWN BELOW.

DIM	PROBLEM 7	8	9
A	1 216	1 548	2 604
B	2 096	3 099	4 672
C	2 986	4 871	6 707
D	3 889	6 256	9 443
E	4 770	7 738	10 862
L			
V			
W			
X			
Y			
Z			

UNIT 28-4

SECTIONING

In many instances the cross-hatching of sections may be omitted on shop drawings because its use is not needed to make the drawing any clearer. In structural detailing, the practice is to omit all lines on a drawing that serve no significant purpose.

Bottom Views

In shop detail drawings, the bottom flange or face of a shape is never viewed from below; that is, the drafter does not stand below the object and look squarely up at it. Rather, she or he cuts a section such as *A-A* in Fig. 28-4-1 and views the bottom flange by looking squarely down on the top side of it.

The reason for substituting a bottom section for a bottom view is to obtain a better correlation between it and the top view. For example, it will be more apparent whether a connection on one side of the top flange and a connection on the bottom flange are on the same side or opposite sides of the member. Note that the cross-hatching is omitted, as previously mentioned.

Elimination of Top and Bottom Views

In the preceding examples, we found that a top and bottom view were not necessary because no holes were required in either flange. Let us now look at an example where there are holes in the bottom and top flanges and see if top and bottom views are required.

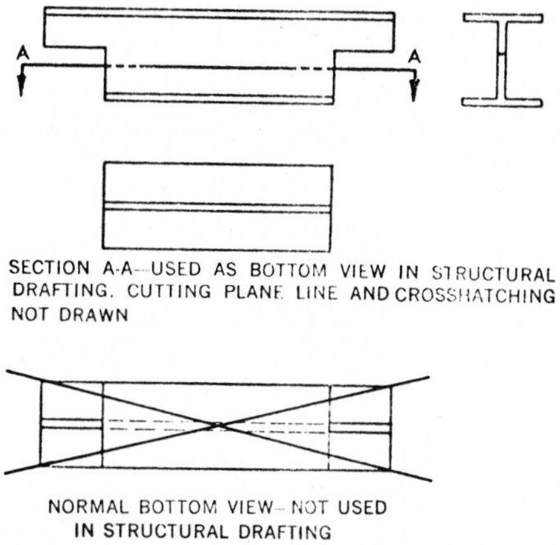

A
A
SECTION A-A—USED AS BOTTOM VIEW IN STRUCTURAL DRAFTING. CUTTING PLANE LINE AND CROSSHATCHING NOT DRAWN
NORMAL BOTTOM VIEW—NOT USED IN STRUCTURAL DRAFTING

Fig. 28-4-1 Bottom view in structural drafting.

In Fig. 28-4-2, the detailing shows a top view and a picture of the bottom flange, taken as a section looking down.

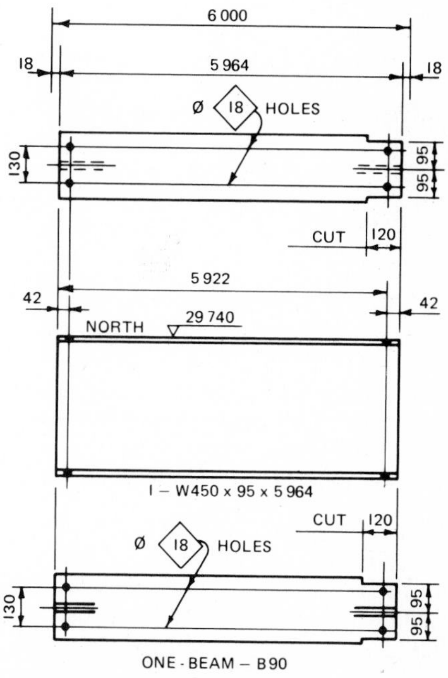

6 000
18
5 964
18
Ø 18 HOLES
130
95
95
CUT
120
5 922
42
42
NORTH
29 740
I – W450 x 95 x 5 964
CUT
120
Ø 18 HOLES
130
95
95
ONE - BEAM – B90

Fig. 28-4-2 Detail of beam.

Note that the dash line in the top view and the solid lines in the bottom view that depict the web are not drawn continuously across the length of the member. Neither is the cut section of the web blackened or cross-hatched. Yet, the drawing is complete as far as being readable and understandable to the fabricator. Remember, use as few lines as possible to describe the object and the shop fabrication.

Detailing in Fig. 28-4-3 eliminates these views. In order to eliminate views of the top and bottom flanges, instructions (including necessary dimensions) for cutting these flanges at the right-hand end have been covered in the note on the web view concerning cutting. The transverse distance between gage lines on the flanges is covered by the note *GA* = 130. In both cases, symmetry about the center line of the beam web is understood. These notes must be explicit in showing what fabrication, if any, is required on each flange. Both methods of presentation are common practice.

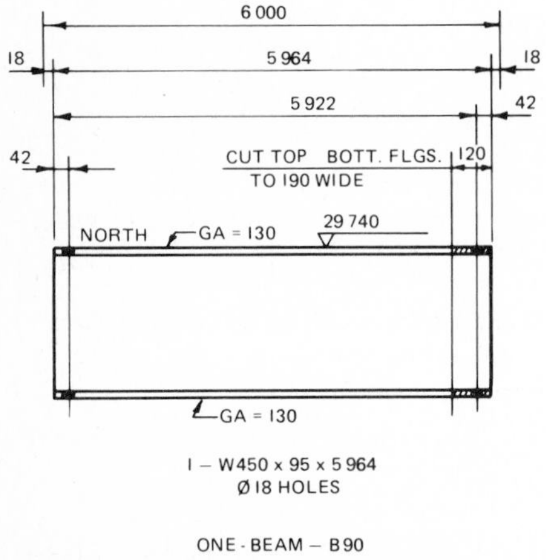
6 000
18
5 964
18
5 922
42
42
CUT TOP BOTT. FLGS. TO 190 WIDE
120
NORTH
GA = 130
29 740
GA = 130
1 – W450 x 95 x 5 964
Ø 18 HOLES
ONE - BEAM – B90

Fig. 28-4-3 Elimination of top and bottom views.

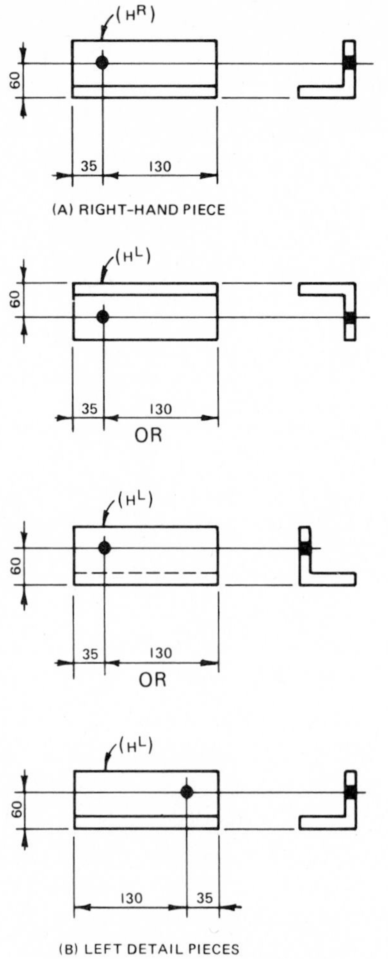
(HR)
60
35
130
(A) RIGHT-HAND PIECE
(HL)
60
35
130
OR
(HL)
60
35
130
OR
(HL)
60
130
35
(B) LEFT DETAIL PIECES

Fig. 28-4-4 Right and left detail pieces.

Right- and Left-Hand Details

Very frequently detail material, such as connection angles and other fittings, is used under conditions where one piece must be the exact opposite of another. In such cases, both the RH (right-hand) and the LH (left-hand) pieces are fabricated from the same sketch; that is, from the detail of the RH piece.

The piece which is made like the drawing is identified by use of the letter R, added to its assembly mark, thus: H^R for right. The one which is made opposite-hand has the letter L added to its assembly mark, thus: H^L for left. No assembly mark should be marked R unless there is an exact opposite, or LH detail piece, needed on the sheet, because all detail pieces are assumed to be RH as shown on the drawing unless otherwise noted. Likewise, no assembly mark should be marked L unless there is also a corresponding right.

If a drawing is placed in front of a mirror, the required RH detail would appear as represented by the drawing, and the required LH detail would appear as reflected in the mirror.

An understanding of rights and lefts, if not innate, may be gotten from Fig. 28-4-4. Note that the two views of H^L to the right of the drawing, even in their rotated position, still picture a fitting which is opposite-hand to H^R.

Pieces which in their assembled position may appear to be rights and lefts often really are alike. Thus the fittings, shown on the left side of Fig. 28-3-11, can be turned upside down and used on the right side of the beam web.

When rights and lefts of whole shipping pieces are encountered, it is the practice in some fabricating shops to note the RH piece AS SHOWN and the LH piece OPPOSITE HAND in the required list. If two shipping pieces are involved, one the exact opposite of the other, the required listing under the single sketch might read (Fig. 28-4-5):

ONE BEAM $A150^R$—AS SHOWN
ONE BEAM $A150^L$—OPPOSITE HAND

If two shipping pieces are involved, one practically but not exactly the oppo-

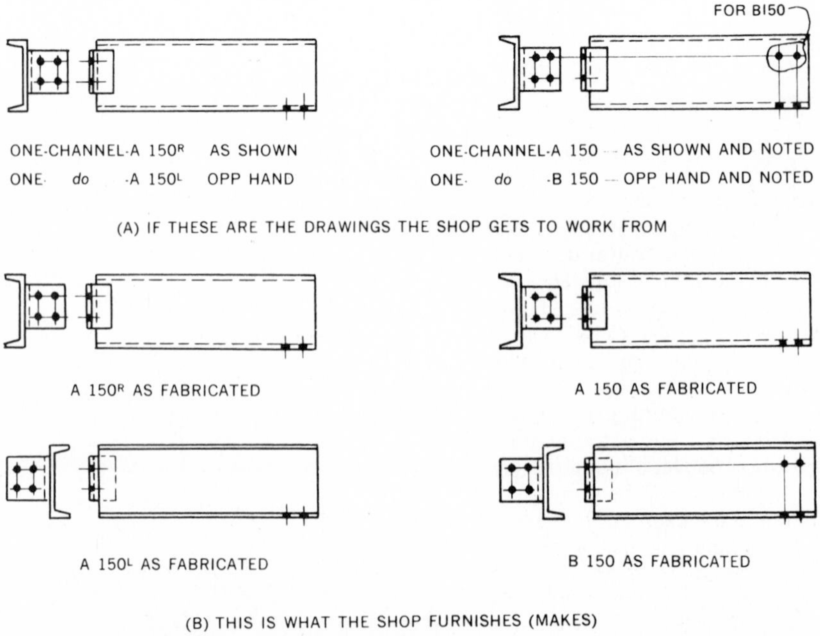
FOR B150
ONE-CHANNEL-A 150R AS SHOWN
ONE- do -A 150L OPP HAND
ONE-CHANNEL-A 150 — AS SHOWN AND NOTED
ONE- do -B 150 — OPP HAND AND NOTED
(A) IF THESE ARE THE DRAWINGS THE SHOP GETS TO WORK FROM
A 150R AS FABRICATED
A 150 AS FABRICATED
A 150L AS FABRICATED
B 150 AS FABRICATED
(B) THIS IS WHAT THE SHOP FURNISHES (MAKES)

Fig. 28-4-5 Right and left shipping pieces.

site of the other, and they are detailed on the same sketch, the required listing under the single sketch would read:

ONE BEAM A150—AS SHOWN AND NOTED
ONE BEAM B150—OPPOSITE HAND AND NOTED

In the case of exact RH and LH shipping pieces, the shipping mark may be the same except for the R and L notation; in the case of combined but different RH and LH shipping pieces, the shipping marks are always different. In both cases only the RH or AS SHOWN shipping piece is detailed. It is the shop which does the reversing, according to the notation OPPOSITE HAND, in the required list.

Before an attempt is made to detail pieces involving combinations of rights and lefts, Fig. 28-4-5 should be studied. If differences between pieces are minor, it is common practice to combine the details of two or more different pieces in a single sketch, by noting the differences, for example, in the case of the two web holes required in B150 but not required in A150.

SOURCE MATERIAL

- Canadian Institute of Steel Construction

Assignments

1. On an A3- or B-size sheet, prepare one drawing for the two beams shown in Fig. 28-4-A (W400 × 60) or 28-4-B (W16 × 40). Eliminate the top and bottom views. Scale is 1:10 or 1:8.

2. On an A3- or B-size sheet, make a complete working drawing of beams B8^R, B8^L, and C6^R, shown on Fig. 28-4-A or 28-4-B. Scale is 1:10 or 1:8.

REVIEW FOR ASSIGNMENTS

Unit 28-1 Properties of Structural Steel Shapes
Unit 28-2 Beams and Assembly Clearances
Unit 28-3 Standard Connections

UNIT 28-5

SEATED BEAM CONNECTIONS

Seated beam connections are used to connect beams to column webs or flanges. There are two types: unstiffened seat connections and stiffened seat connections. Only the unstiffened (angles) will be covered in this unit.

The following procedure is suggested for choosing a seated beam connection. Assume that a W310 × 40 beam is to be placed between two columns, as shown in Fig. 28-5-1. As in the example for the framed beam, Unit 28-3, the beam reactions must first be established. If they are not shown on the drawing, they must be calculated. To do this, the length of the beam must be known. This is found by subtracting half of the nominal depth of each of the two supporting columns from the center-to-center distance. The

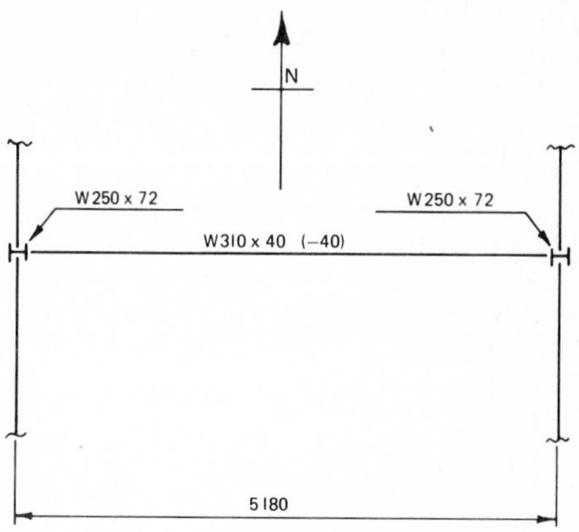
N
W250 x 72
W250 x 72
W310 x 40 (−40)
5180

Fig. 28-5-1 Partial design drawing.

nominal depth of the columns can be found from Fig. 28-5-1. In this example, the length of the span is 5180 − 254 = 4926 mm. From the beam load tables in Fig. 28-3-9 for a W310 × 40 beam having a span of 5000 mm, the total allowable uniform load is 266 kN. The reaction at the end of the beam at each support is half the total load, or 133 kN. The length of the seated beam is now determined. From Fig. 28-1-8, the flange width b for the W310 × 40 beam is 165 mm. Note that Fig. 28-5-2 gives tables for a supporting angle length L of 150 and 200 mm. Since the flange width of 165 mm is greater than the L (length) of 150 mm, the angle length of L = 200 mm for the seat angle will have to be used. The seat angle thickness must now be calculated. The web thickness of the W310 × 40 beam is 6 mm. Refer to Fig. 28-5-2,

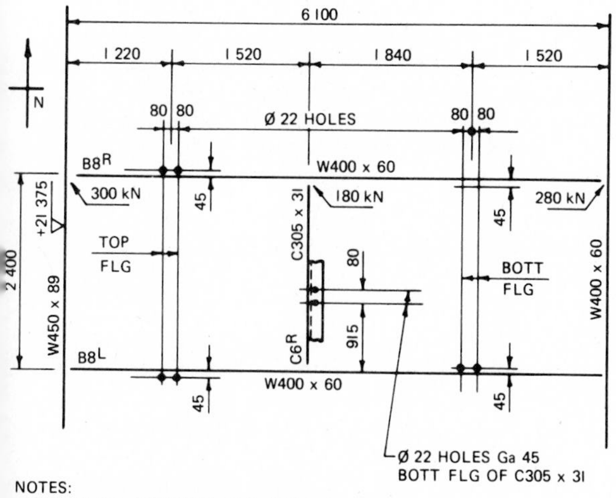
6 100
1 220
1 520
1 840
1 520
N
80 80
Ø 22 HOLES
80 80
B8R
W400 x 60
300 kN
180 kN
280 kN
45
C305 x 31
45
+21 375
TOP
FLG
80
BOTT
FLG
W400 x 60
2 400
W450 x 89
C6R
915
B8L
W400 x 60
45
45
Ø 22 HOLES Ga 45
BOTT FLG OF C305 x 31

NOTES:
- TOP OF ALL MEMBERS AT ELEVATION +21 400 EXCEPT WHERE NOTED
- SHOP CONNECTIONS WELDED
- FIELD CONNECTIONS M20–A325 FRICTION TYPE BOLTS
- USE DOUBLE-ANGLE BEAM CONNECTIONS

Fig. 28-4-A One-view beam drawing.

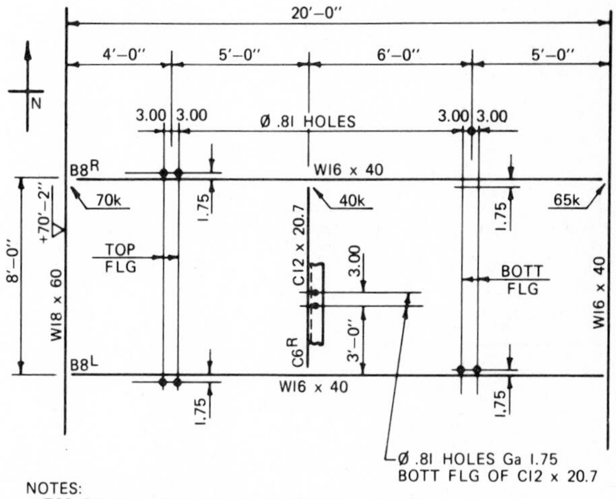
20′–0″
4′–0″
5′–0″
6′–0″
5′–0″
N
3.00 3.00
Ø .81 HOLES
3.00 3.00
B8R
WI6 x 40
70k
40k
65k
1.75
C12 x 20.7
1.75
+70′–2″
TOP
FLG
3.00
BOTT
FLG
8′–0″
WI8 x 60
WI6 x 40
C6R
3′–0″
B8L
WI6 x 40
1.75
1.75
Ø .81 HOLES Ga 1.75
BOTT FLG OF C12 x 20.7

NOTES:
- TOP OF ALL MEMBERS AT ELEVATION +70′–3″ EXCEPT WHERE NOTED
- SHOP CONNECTIONS WELDED
- FIELD CONNECTIONS .75–A325 FRICTION TYPE BOLTS
- USE DOUBLE-ANGLE BEAM CONNECTIONS

Fig. 28-4-B One-view beam drawing.

under the heading Outstanding Leg Capacity = kN, L = 200 mm, and Beam Web Thickness = 6 mm. Read across until a leg capacity of 133 kN or greater is encountered. This occurs at the kilonewton value of 149 where the angle thickness is 16 mm.

It is preferable for most fabricators to shop-weld the seat angle to the column, since the seat will provide support for the beam during erection. Under the heading Vertical Leg Weld Capacity, the angle thickness as previously determined was found to be 16 mm; therefore the maximum permissible weld size would be 16 − 2, or 14 mm. From the table a 14 mm fillet weld will resist a force of 133 kN when a 100 × 100 angle size is used. To the right of this column, the angle thickness range of 6 to 16 mm is specified. The required angle thickness range as determined previously was 16 mm; therefore the dimensions for the seat angle are 100 × 100 × 16 × 200. The next step in the process is to make a sketch of the detail, as shown in Fig. 28-5-3.

The beam will be fastened to the seat angles using M16 = A325 bolts. In order to determine their gage, reference should be made to Fig. 28-3-1. The recommended gage for a W310 which has a flange width of 165 mm is 35 mm. Also from Fig. 28-3-3, the minimum distance for an M16 bolt to a rolled edge is 22 mm. Therefore, the end distance (center of bolt to end of beam) is 100 − 22 − 12 = 66. Use 65 mm.

A top, or cap, angle is used to provide lateral support at the top of the beam. Since it is not required to resist any calculated moment at the end of the beam, this angle can be relatively small. For the top angle 100 × 100 × 6 × 100 is recommended. In this example, no limitations have been specified by the design drawing as to the top angles, and therefore the angle can be placed as shown. However, if the top clearance had been critical, the angle could have been placed in the optional position on either side of the web, whichever provided the most convenient position for field erection purposes.

The top angle is welded to the column and bolted to the beam with two M16 bolts having a gage of 55 mm, as recommended in Fig. 28-3-1. The length of the beam required is equal to the center-to-center distance of the columns minus half of each of the column depths (or the column depth if both columns are the same) minus the 12 mm nominal clearance at each end. For this example, the length of the beam is 5180 − 254 − 2(12) or 4902 mm. The detail drawing of the W310 beam is shown in Fig. 28-5-4.

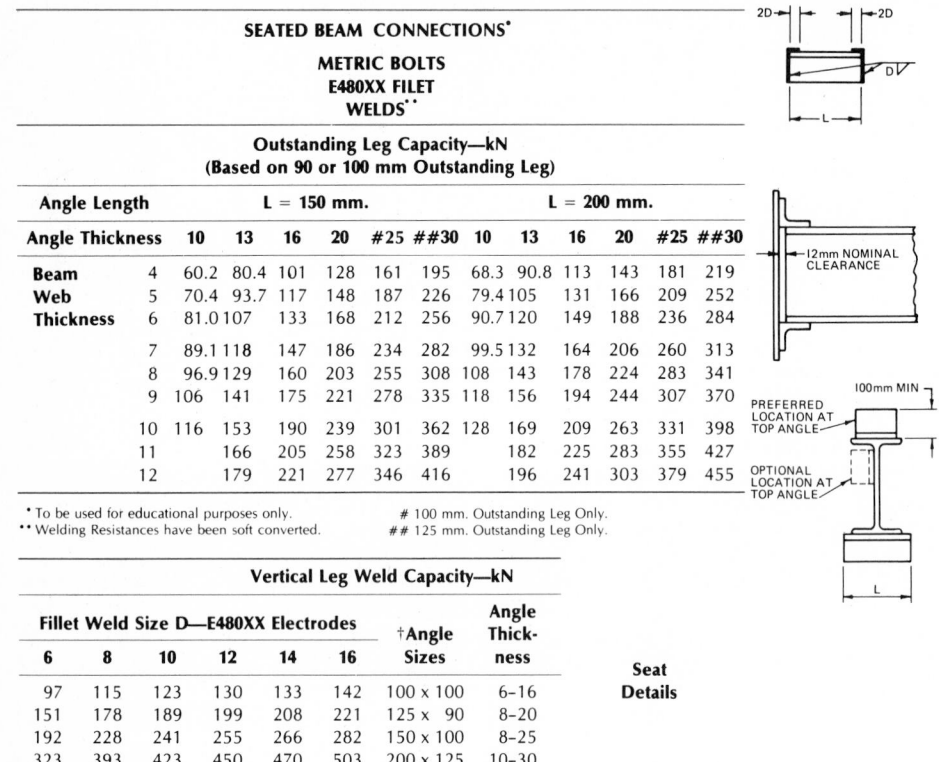

SEATED BEAM CONNECTIONS*

METRIC BOLTS
E480XX FILET WELDS**

Outstanding Leg Capacity—kN
(Based on 90 or 100 mm Outstanding Leg)

Angle Length		L = 150 mm.						L = 200 mm.					
Angle Thickness		10	13	16	20	#25	##30	10	13	16	20	#25	##30
Beam Web Thickness	4	60.2	80.4	101	128	161	195	68.3	90.8	113	143	181	219
	5	70.4	93.7	117	148	187	226	79.4	105	131	166	209	252
	6	81.0	107	133	168	212	256	90.7	120	149	188	236	284
	7	89.1	118	147	186	234	282	99.5	132	164	206	260	313
	8	96.9	129	160	203	255	308	108	143	178	224	283	341
	9	106	141	175	221	278	335	118	156	194	244	307	370
	10	116	153	190	239	301	362	128	169	209	263	331	398
	11		166	205	258	323	389		182	225	283	355	427
	12		179	221	277	346	416		196	241	303	379	455

* To be used for educational purposes only.
** Welding Resistances have been soft converted.
# 100 mm. Outstanding Leg Only.
125 mm. Outstanding Leg Only.

Vertical Leg Weld Capacity—kN

Fillet Weld Size D—E480XX Electrodes						†Angle	Angle Thick-
6	8	10	12	14	16	Sizes	ness
97	115	123	130	133	142	100 x 100	6–16
151	178	189	199	208	221	125 x 90	8–20
192	228	241	255	266	282	150 x 100	8–25
323	393	423	450	470	503	200 x 125	10–30

†Long vertical leg.

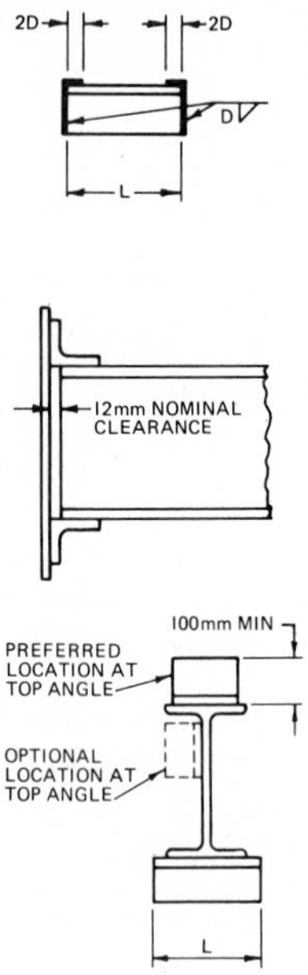

2D
2D
D
L
12mm NOMINAL CLEARANCE
100mm MIN
PREFERRED LOCATION AT TOP ANGLE
OPTIONAL LOCATION AT TOP ANGLE
L

Seat Details

Fig. 28-5-2 Seated beam connections.

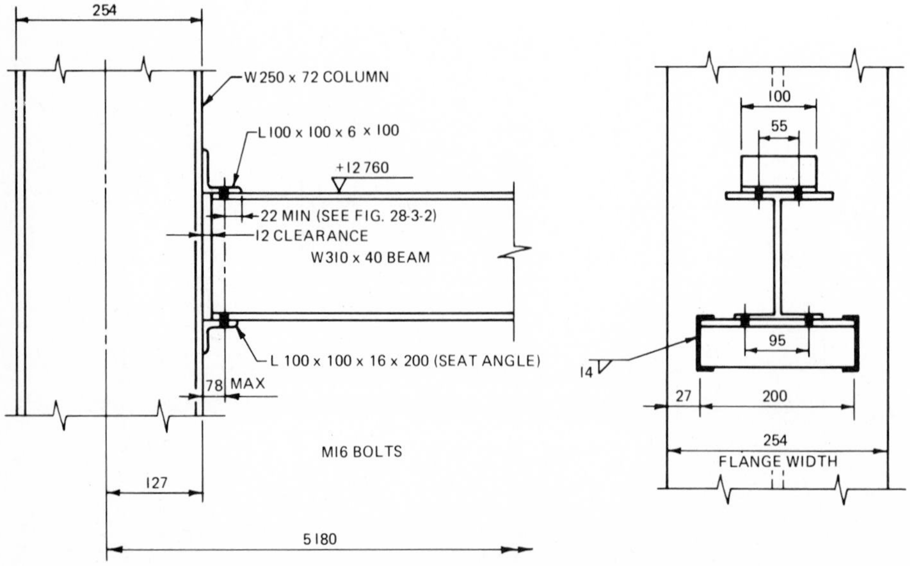

254
W 250 x 72 COLUMN
L 100 x 100 x 6 x 100
+12 760
22 MIN (SEE FIG. 28-3-2)
12 CLEARANCE
W310 x 40 BEAM
L 100 x 100 x 16 x 200 (SEAT ANGLE)
78 MAX
M16 BOLTS
127
5 180
14
100
55
95
27
200
254
FLANGE WIDTH

Fig. 28-5-3 Sketch of seated beam connection for partial design drawing—Fig. 28-5-1.

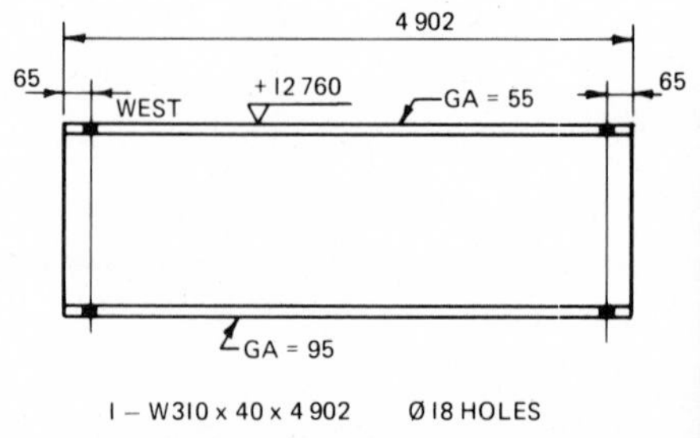

4 902
65
65
WEST
+12 760
GA = 55
GA = 95
1 – W310 x 40 x 4 902
Ø 18 HOLES

Fig. 28-5-4 Detail drawing of W310 beam shown in Fig. 28-5-1.

SOURCE MATERIAL

- Canadian Institute of Steel Construction

Assignments

1. On an A3- or B-size sheet, make detail sketches of the seated beam connections at both ends of beam A shown in Fig. 28-5-A or 28-5-B. Scale is 1:10 or 1:8.

2. On an A3- or B-size sheet, make a one-view detail drawing of beam A in assignment 1. Scale is to suit.

REVIEW FOR ASSIGNMENT

Unit 28-1 Properties of Structural Steel Shapes
Unit 28-2 Beams and Assembly Clearances
Unit 28-4 Elimination of Top and Bottom Views

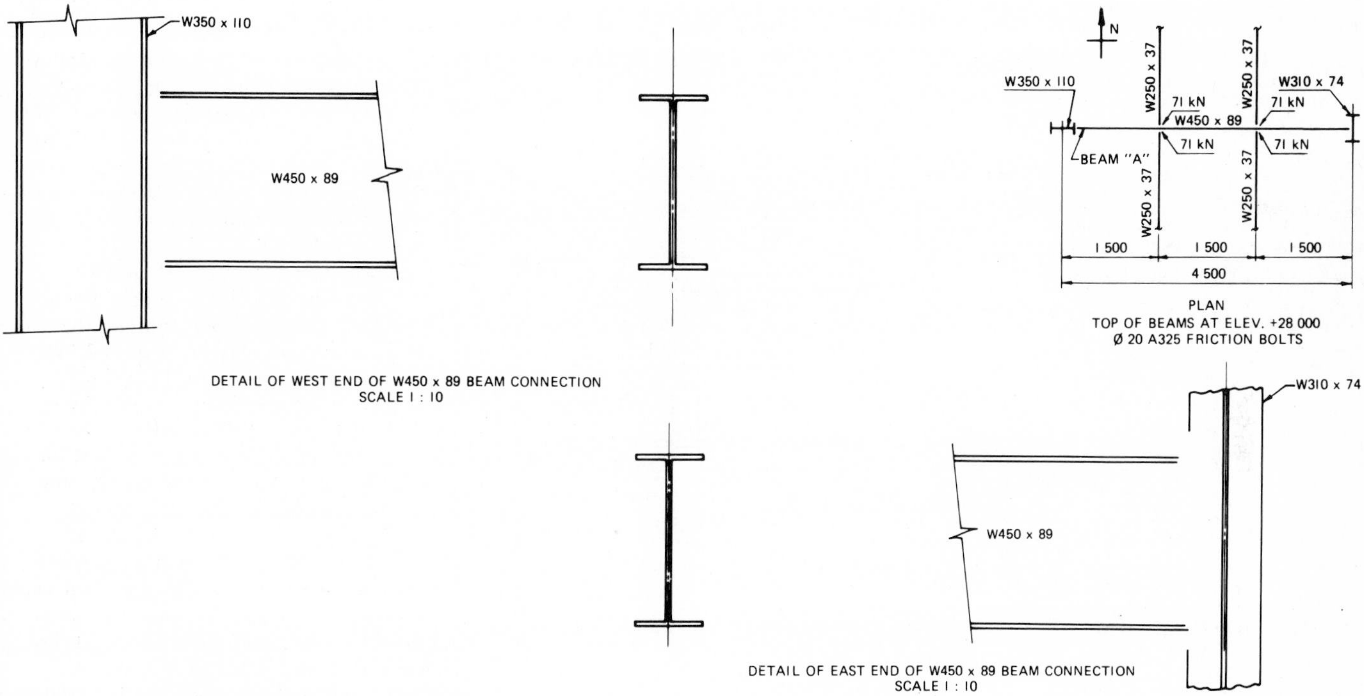
W350 x 110
W450 x 89
DETAIL OF WEST END OF W450 x 89 BEAM CONNECTION
SCALE 1 : 10
N
W350 x 110
W250 x 37
W250 x 37
W310 x 74
71 kN
71 kN
W450 x 89
71 kN
71 kN
BEAM "A"
W250 x 37
W250 x 37
1 500
1 500
1 500
4 500
PLAN
TOP OF BEAMS AT ELEV. +28 000
Ø 20 A325 FRICTION BOLTS
W310 x 74
W450 x 89
DETAIL OF EAST END OF W450 x 89 BEAM CONNECTION
SCALE 1 : 10

Fig. 28-5-A Sketches of seated connection.

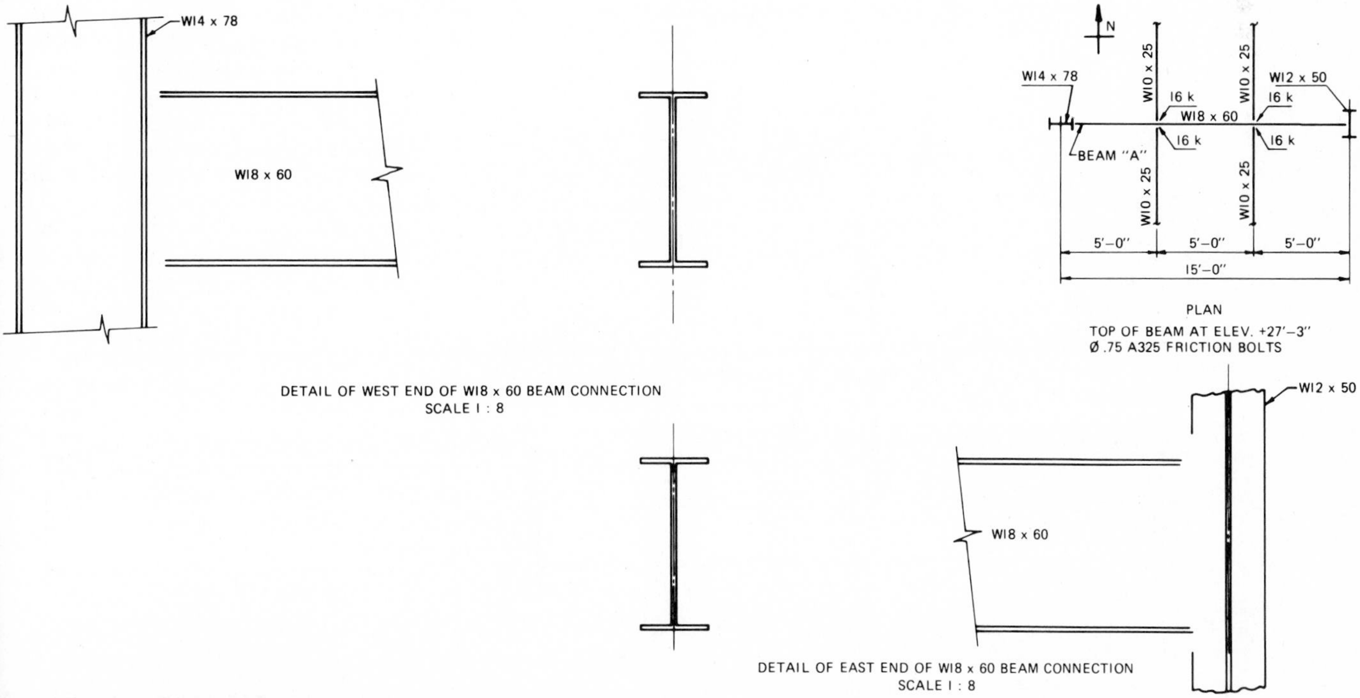
W14 x 78
W18 x 60
DETAIL OF WEST END OF W18 x 60 BEAM CONNECTION
SCALE 1 : 8
N
W14 x 78
W10 x 25
W10 x 25
W12 x 50
16 k
16 k
W18 x 60
16 k
16 k
BEAM "A"
W10 x 25
W10 x 25
5'–0"
5'–0"
5'–0"
15'–0"
PLAN
TOP OF BEAM AT ELEV. +27'–3"
Ø .75 A325 FRICTION BOLTS
W12 x 50
W18 x 60
DETAIL OF EAST END OF W18 x 60 BEAM CONNECTION
SCALE 1 : 8

Fig. 28-5-B Sketches of seated connection.

UNIT 28-6
DIMENSIONING

In addition to the dimensioning techniques discussed in Units 28-1 and 28-3 are the following points for consideration.

Note that longitudinal dimensions along the beams shown in Fig. 28-6-1 are given to the center line of the groups of open holes required for the field connections. This practice can serve two useful purposes. First, it simplifies the dimensioning work for the drafter and later for the checker, since the distances to center lines of the beams are the dimensions given on the design drawing and the erection plan. Second, it is a convenience to the person laying out the beam details who first marks the locations of the center line for a group of holes on the beam and then centers a template at this point, by which he or she can center-punch the location of all the holes required in the group.

In the detail drawing (Fig. 28-6-1), the fabricator preferred to dimension to the center line of the channel webs rather than the backs of the channels. The direction in which the flanges of these channels are to be placed has been indicated on the drawing by the channel symbol, which has been drawn with the web parallel to the line showing the members. When they are installed, the flanges of these channels must point to the same direction as the flanges of the symbol.

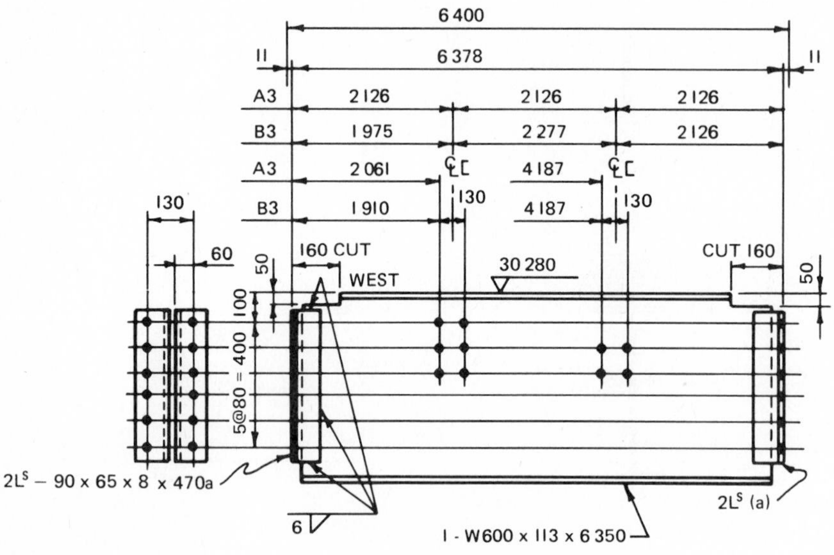
6 400
6 378
11
11
A3
B3
A3
B3
2 126
2 126
2 126
1 975
2 277
2 126
2 061
4 187
1 910
4 187
130
130
130
60
50
160 CUT
WEST
30 280
CUT 160
50
100
5@80 = 400
2L^S – 90 x 65 x 8 x 470a
2L^S (a)
6
I - W600 x 113 x 6 350
ONE - BEAM – A3
ONE - do B3

Fig. 28-6-1 Dimensioning to center line of channel webs.

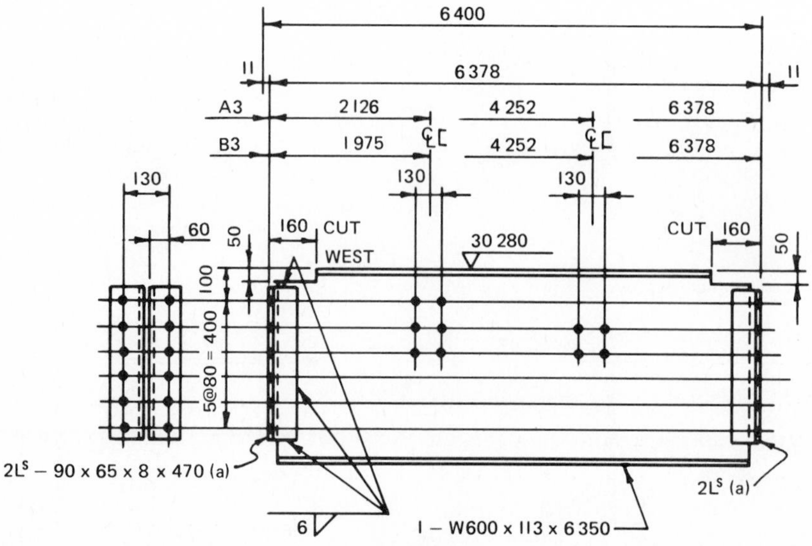
6 400
6 378
11
11
A3
B3
2 126
4 252
6 378
1 975
4 252
6 378
130
130
130
60
50
160
CUT
WEST
30 280
CUT
160
50
100
5@80 = 400
2L^S – 90 x 65 x 8 x 470 (a)
2L^S (a)
6
I – W600 x 113 x 6 350
ONE - BEAM – A3
ONE - do – B3

Fig. 28-6-2 Dimensioning from the left end of beam.

In some shops, however, in addition to locating groups of holes, as noted above, a method of using extension, stub, or running dimensions is employed. This consists of specifying the overall dimension from the left end of the beam to the center line of each group of holes, as shown in Fig. 28-6-2. Note that this practice was also followed in Fig. 28-6-1, to the first line of holes, but here it was done for reference only. In either case, it lessens the shop layout person's work by eliminating the need for calculations and therefore reduces the possibility of an error being made, especially in a shop that uses automated punching equipment.

If the fabricator had preferred to dimension to the backs of the channels, the dimensions would be as shown in Fig. 28-6-3. Dimensions and notes not shown are the same as in Fig. 28-6-1, except that no note is required to identify the dimension reference lines locating the groups of open holes in the beam web. The dimensions 60 mm and 70 mm to each side of the reference lines provide the clue that these open holes are to receive the connection for a channel. It would be understood that the back of the channel will be located toward the smaller of these two dimensions.

Another time-saving device commonly used in the drafting room when pieces are alike, except for some end or intermediate detail, is partial view detailing.

Instead of completely redetailing each piece or trying to combine too many dissimilar details on one sketch by the use of notes indicating to which piece each detail applies, completely detail one piece first. Then, for the second piece, only the difference between it and the first piece needs to be detailed. This partial view is supplemented by a note stating that the parts not shown are THE SAME or possibly OPPOSITE HAND TO the first piece detailed. Figure 28-6-4 represents this practice.

BILLS OF MATERIAL

From the bills of material or Material Bills the workers in the yard, where the structural shapes are stocked, cut the material to length, cut the number of pieces shown on the bills, and send the material into the fabricating shop. From the bills of material the shipping department tallies the number of pieces to be shipped. Therefore, it is extremely important that the drafter include in the bill of material all the material that is shown on the drawing.

The sample shop bill of material (Fig. 28-6-5) shows a typical form that is used in billing the material on shop drawings. The first items are the billing for beam A3 and beam B3 that are shown detailed in Fig. 28-6-1. Note that beam A3 is different from B3 only in the longitudinal spacing of holes in the web, and that the material for the two beams is identical. When the material is the same, the billing of members on a shop drawing is grouped to avoid repetition.

CALCULATION OF MASSES

After the billing operation, it may be necessary to figure the mass of the material on the bill of material. The mass of the material is very important in that the basis of payment for the fabricated steel may be a price per kilogram. For that reason, the mass must be accurate to the nearest kilogram. Also, the shop uses the calculated mass of a member to avoid overloading the cranes or other transporting machinery. The shipping department uses the calculated masses for making up loads and as a basis of payment for shipping. The erection department is interested in masses of members to plan erection procedure and equipment. The masses of structural steel bolts and common structural steel shapes are found in the Appendix.

SOURCE MATERIAL

- Canadian Institute of Steel Construction.

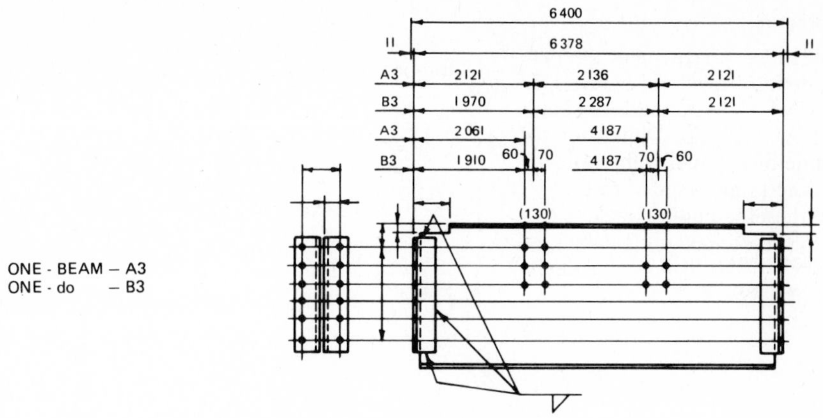
6400
6378
A3
B3
A3
B3
2121
2136
2121
1970
2287
2121
2061
4187
1910
60
70
4187
70
60
(130)
(130)
ONE - BEAM — A3
ONE - do — B3

Fig. 28-6-3 Dimensioning to the backs of channels.

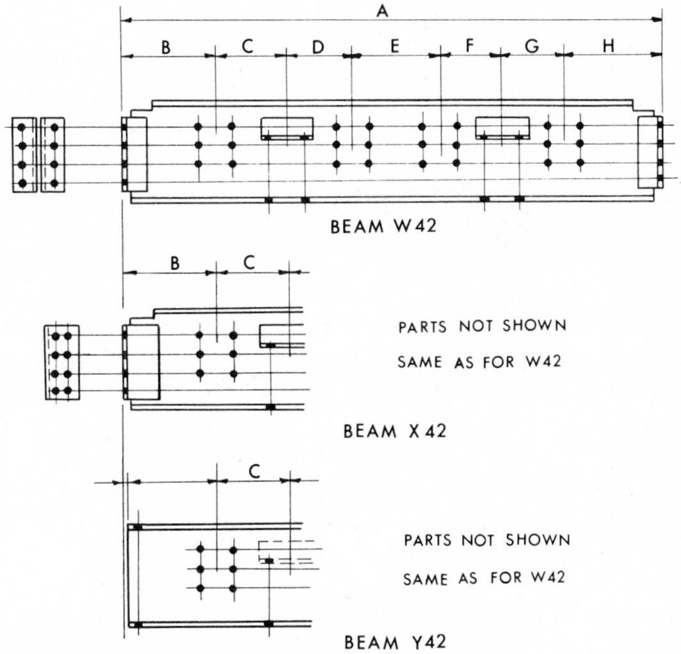
A
B
C
D
E
F
G
H
BEAM W42
PARTS NOT SHOWN
SAME AS FOR W42
BEAM X42
PARTS NOT SHOWN
SAME AS FOR W42
BEAM Y42

Fig. 28-6-4 Partial view detailing.

DATE TO P.A. 26 — CONT. NO.
FG'D BY — DATE — BILL NO. M 3a
CHK'D BY — DATE
CUSTOMER — INSP. AT MILLS — LUMP SUM KILOGRAM PRICE COST PLUS — MADE BY — DATE
DESCRIPTION — SPECN'S — KIND OF MAT'L — CHK'D BY — DATE

BILL OF STRUCTURAL MATERIAL									SHIPPING LIST				
ITEM NO.	NO. PIECES	ASSEMBLY MARK	DESCR.	SIZE	LENGTH	P BILL REFERENCE	EST. MASS	P. A.	NO. PCS.	DESCRIPTION	MARK	ACTUAL MASS	SHIPPING RECORD
1	2		W	600x113	6 350		1435		1	BEAM	A3		733
2	8	a	L	90x65x8	470		31		1	BEAM	B3		733
3							1466						
4													
5													

Fig. 28-6-5 Sample bill of material.

Assignment

1. On an A3- or B-size sheet, prepare a complete detail drawing of beams D3, E3, G3, K3, M3, N3, C3, and F3 shown on Fig. 28-6-A or 28-6-B. Dimension to the center line of channel webs. The connection angles are welded to the beams, and the outstanding angles are bolted to the connecting beams. Make sketches of the beam connections. Scale is 1:10 or 1:8.

2. On an A3- or B-size sheet, prepare a bill of materials and a shipping list for the beams in assignment 1.

REVIEW FOR ASSIGNMENT

Unit 28-1 Properties of Structural Steel Shapes
Unit 28-2 Beams and Assembly Clearances
Unit 28-3 Standard Connections
Unit 28-4 Elimination of Views
Unit 28-5 Seated Beam Connections

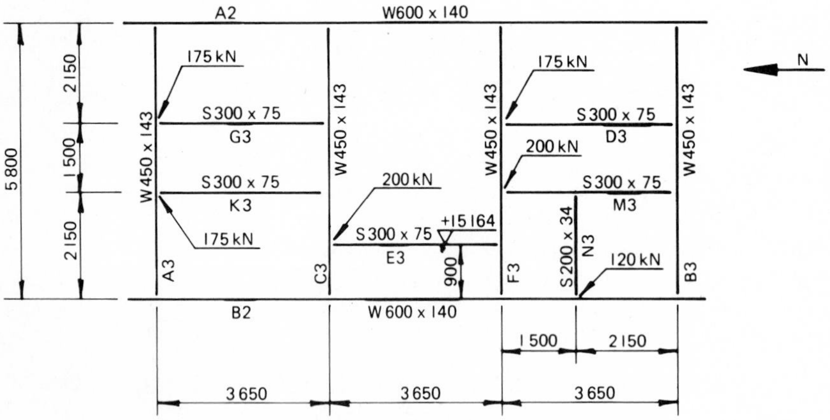
A2
W600 x 140
175 kN
S300 x 75
G3
S300 x 75
K3
175 kN
200 kN
S300 x 75
E3
+15 164
900
S300 x 75
D3
200 kN
S300 x 75
M3
S200 x 34
N3
120 kN
W450 x 143
A3
C3
F3
B3
B2
W 600 x 140
2 150
1 500
5 800
1 500
2150
3 650
3 650
3 650
N

ELEVATIONS: TOP OF STEEL TO BE (+15 240) UNLESS OTHERWISE SHOWN
NOTES: ALL HOLES Ø 22 BOLTS M20 – A325

MAKE A FULLY DIMENSIONED DETAIL DRAWING OF BEAM C3 SHOWN ON THE FLOOR PLAN ABOVE.

Fig. 28-6-A Detail drawings.

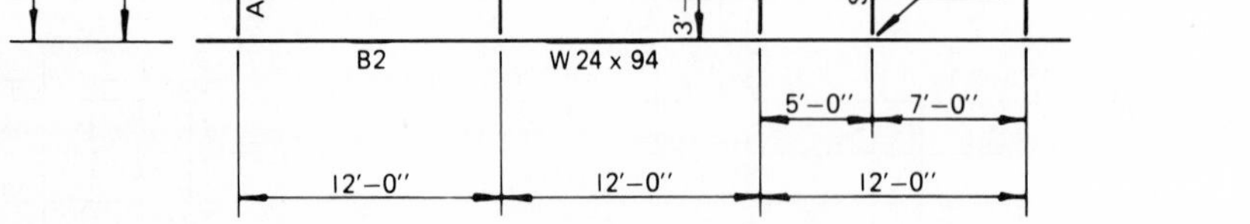
A2
W 24 x 94
40 K
S 12 x 50
G3
S 12 x 50
K3
40 K
45 K
S 12 x 50
E3
+50'–0"
3'–0"
40 K
S 12 x 50
D3
45 K
S 12 x 50
M3
S8 x 23
N3
27 K
W18 x 96
A3
C3
F3
B3
B2
W 24 x 94
7'–0"
5'–0"
19'–0"
5'–0"
7'–0"
12'–0"
12'–0"
12'–0"
N

ELEVATION: TOP OF STEEL TO BE (+50'–3") UNLESS OTHERWISE SHOWN
NOTES: ALL HOLES Ø .81 BOLTS Ø .75–A325

MAKE A FULLY DIMENSIONED DETAIL DRAWING OF BEAM C3 SHOWN ON THE FLOOR PLAN ABOVE.

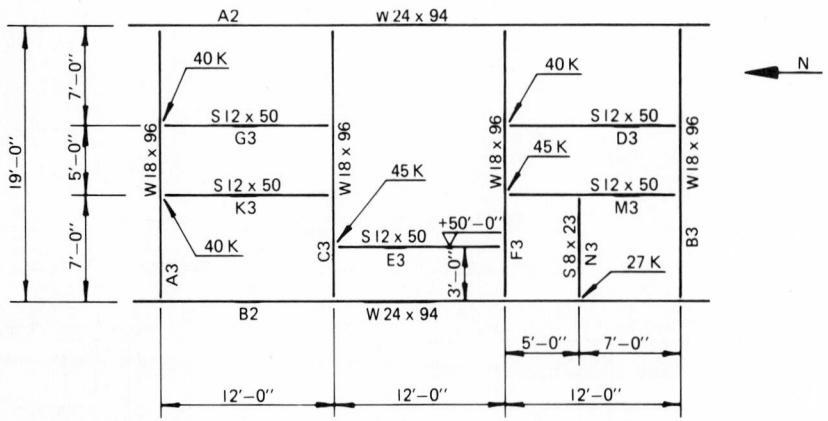

Fig. 28-6-B Detail drawings.

Part 6
Advanced Drafting Design

Chapter 29
Applied Mechanics

UNIT 29-1
FORCES

A *force* is that which changes, or tends to change, the state of rest or uniform motion of a body.

The graphical method of solving mechanical problems involving forces is often used because it is quick and accurate. The force is shown graphically. To describe completely the force, the following particulars must be given:

1. Its magnitude
2. Its point of application
3. Its direction
4. Its sense, i.e., whether it is pushing or pulling

A line is drawn to a given length to represent the magnitude of the force. The direction of this line is parallel to the direction of the force. The sense of the force is indicated by an arrow on the line indicating whether it is acting toward or away from the point of application. The graphical representation of the force is called a *vector*.

Thus a pull of 6N acting at a point A at 45° to the horizon would be represented by the vector AB, as shown in Fig. 29-1-1. By using the scale 10 mm = 1 N, the length of the vector would be 60 mm.

A body is said to be in equilibrium if the forces acting at a point balance one another. If two equal and opposite forces act at a point in a straight line, the body is in equilibrium. Examples are tie bars, which are bars under pull or tension, and struts or columns, which are bars under push or compression (Fig. 29-1-2).

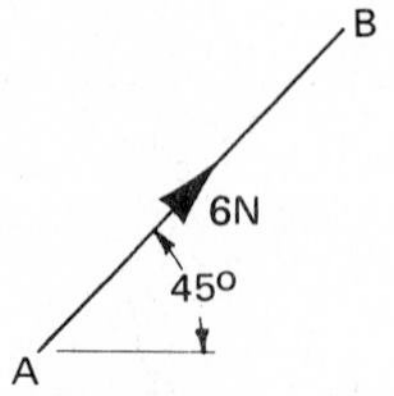

B
6N
45°
A

Fig. 29-1-1 A vector.

Two Forces Acting at a Point

Two or more forces acting at a point may be replaced by one force that will produce the same effect. This force is called the *resultant* of the forces. If two opposite forces of 8 and 5N act at a point O in a straight line, as in Fig. 29-1-3, a resultant force of 3N acting in the same direction as the 8-N force could replace the two original forces.

If two opposite forces F_1 and F_2 act at point O at angles of 120° to each other, as in Fig. 29-1-4, the resultant force R may be found by drawing the two forces to scale and completing the parallelogram. The diagonal OC would be the resultant, and the magnitude of the force

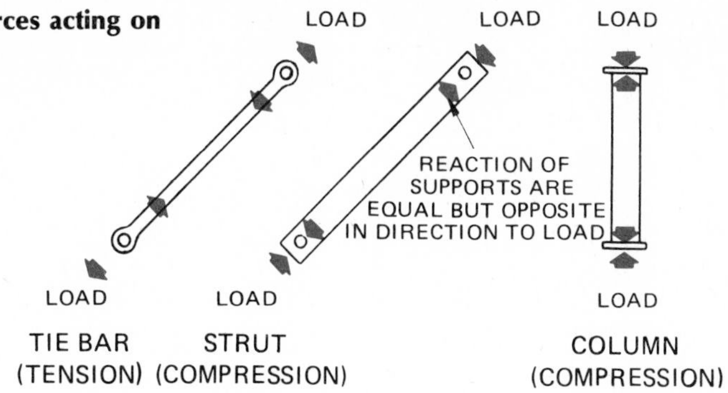
LOAD
LOAD
LOAD
REACTION OF SUPPORTS ARE EQUAL BUT OPPOSITE IN DIRECTION TO LOAD
LOAD
LOAD
LOAD
TIE BAR (TENSION)
STRUT (COMPRESSION)
COLUMN (COMPRESSION)

Fig. 29-1-2 Types of forces acting on supports.

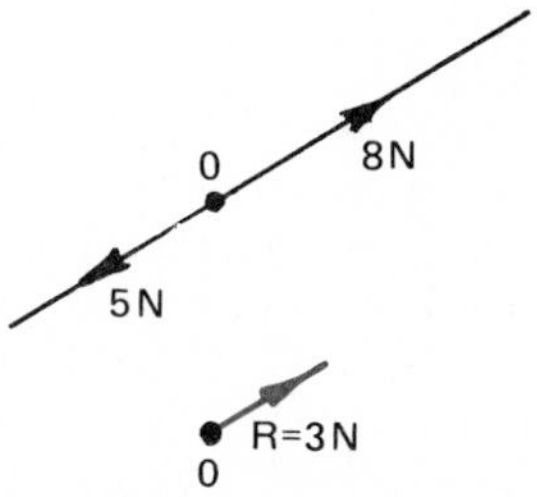
0
8N
5N
R=3N
0

Fig. 29-1-3 Resultant when two forces act in a straight line.

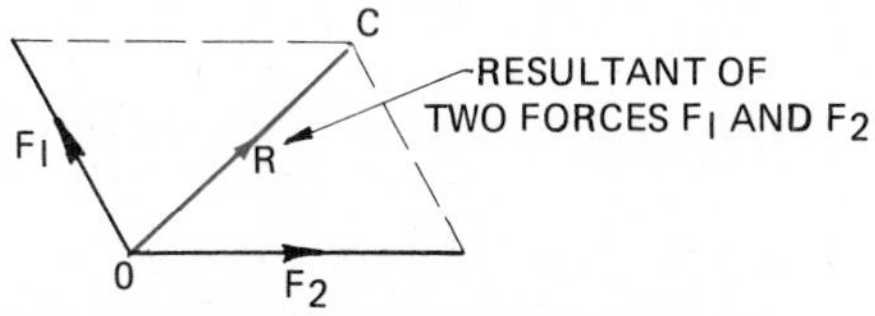
C
RESULTANT OF TWO FORCES F1 AND F2
F1
R
0
F2

Fig. 29-1-4 Parallelogram of forces.

could be measured. This method is called *the parallelogram of forces*. Accuracy of direction and distance is important in laying out forces.

Another way of finding the resultant is the *triangle of force method*. The known force vectors are laid end to end with the forces traveling in the same direction.

The resultant R is found by joining the beginning of the first vector to the end of the last vector, as shown in Fig. 29-1-5*a*, and the direction of the resultant force is in the combined direction of the other two forces.

If a force equal to the resultant of forces F_1 and F_2 but acting in the opposite direction was to act at O, as shown in Fig. 29-1-5*b*, the object would be in equilibrium, since the forces acting at point O tend to balance one another.

This force balancing the other forces is known as the *equilibrant*.

The equilibrant is found in a similar manner to the resultant, by using the triangle of force method. Note that the arrows representing the direction of the forces are pointing the same way around the triangle.

More than Two Forces Acting at a Point

Resultants or equilibrants may be found for any number of forces acting at a point and in one plane. Let A, B, C, and D represent forces acting at a point O, shown in Fig. 29-1-6*a*.

Using the parallelogram of forces method shown in Fig. 29-1-6*b*, we find the resultant R_1 for forces A and B and resultant R_2 for forces C and D. Using resultants R_1 and R_2 in Fig. 29-1-6*c* instead of the forces A, B, C, and D, we find the resultant R of the four forces. The equilibrant or force required to keep the forces A, B, C, and D in equilibrium would be equal to R but would act in the opposite direction.

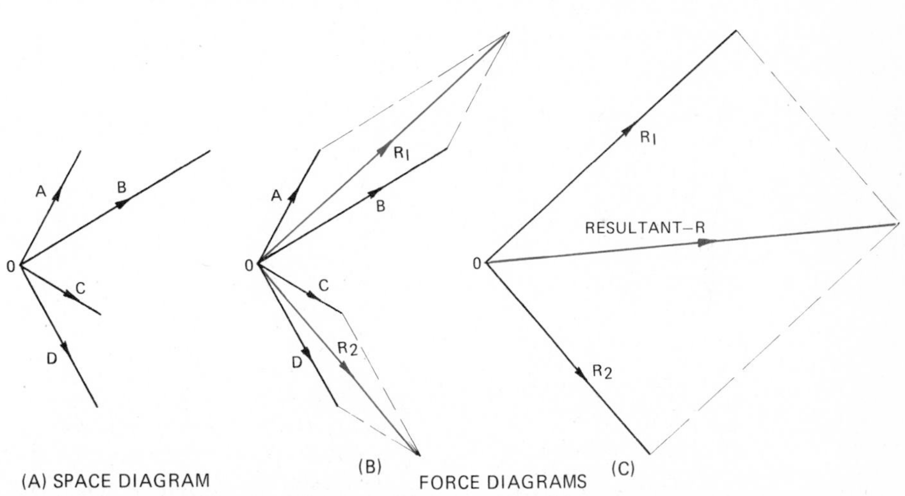
A
B
0
C
D
A
R1
B
0
C
R2
D
R1
RESULTANT–R
0
R2
(A) SPACE DIAGRAM
(B)
FORCE DIAGRAMS
(C)

Fig. 29-1-6 Parallelogram of forces method of finding resultant of *R* for more than two forces acting at a point.

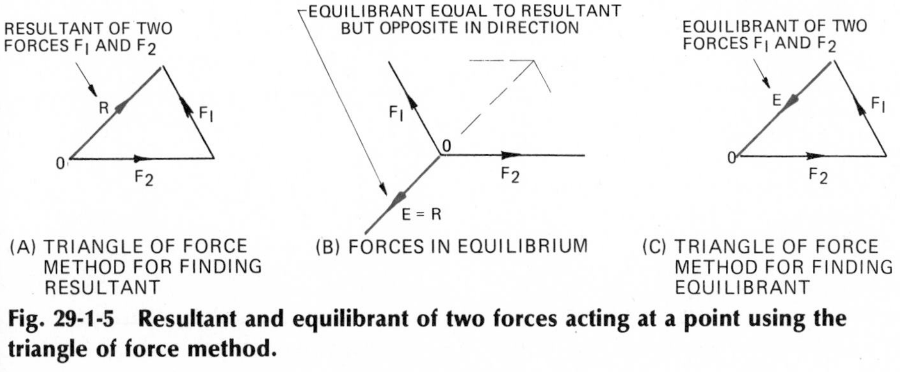
RESULTANT OF TWO FORCES F1 AND F2
R
F1
0
F2
EQUILIBRANT EQUAL TO RESULTANT BUT OPPOSITE IN DIRECTION
F1
0
F2
E = R
EQUILIBRANT OF TWO FORCES F1 AND F2
E
F1
0
F2
(A) TRIANGLE OF FORCE METHOD FOR FINDING RESULTANT
(B) FORCES IN EQUILIBRIUM
(C) TRIANGLE OF FORCE METHOD FOR FINDING EQUILIBRANT

Fig. 29-1-5 Resultant and equilibrant of two forces acting at a point using the triangle of force method.

Polygon of Forces

Using the polygon of forces method shown in Fig. 29-1-7, which is the extension of the triangle method, join the forces A, B, C, and D end to end to form a polygon. Be careful to keep the arrows pointing the same way around. The line joining the beginning of the first force and the end of the last force is the equilibrant.

The following examples illustrate how vector diagrams are applied to practical problems.

Example 1. A crane lifts an 8-metric ton

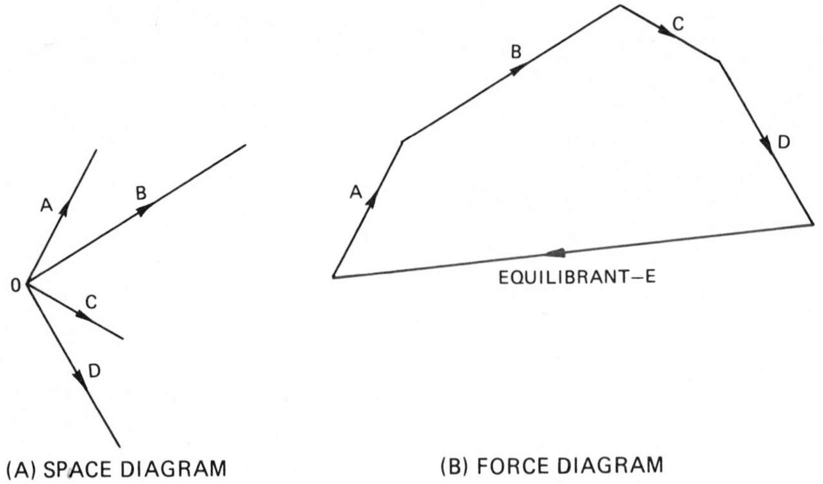
A
B
C
D
0
A
B
C
D
EQUILIBRANT–E
(A) SPACE DIAGRAM
(B) FORCE DIAGRAM

Fig. 29-1-7 Polygon of forces method for finding the equilibrant for more than two forces acting at a point.

(t) steel boiler by means of a chain sling. The sling makes an angle of 30 and 45° with the boiler. Find the forces acting on the sling.

Solution. The force on the crane's chain supporting the sling is equal to the force created by the mass of the 8 t boiler. (A 1 t mass creates a force of 9806.65N.) The angles on the sling indicate the direction of the sling forces F_1 and F_2, as shown in the space diagram in Fig. 29-1-8. The force diagram is then drawn to a suitable scale and the length of lines F_1 and F_2 are measured to find the magnitude of the forces acting on the sling.

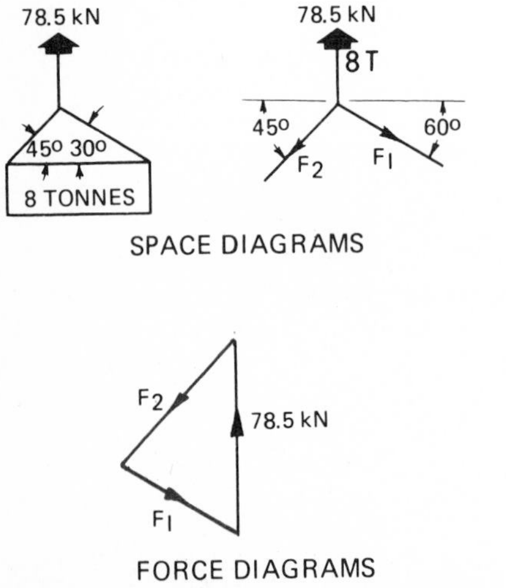
78.5 kN
45° 30°
8 TONNES
78.5 kN
8 T
45°
60°
F2
F1
SPACE DIAGRAMS
F2
78.5 kN
F1
FORCE DIAGRAMS

Fig. 29-1-8 Solution to Example 1 by the force diagram method.

Example 2. In Fig. 29-1-9, a simple wall crane has a load W applied at the end of the jib. Neglecting the mass of the crane parts, calculate the forces acting on the tie and jib.

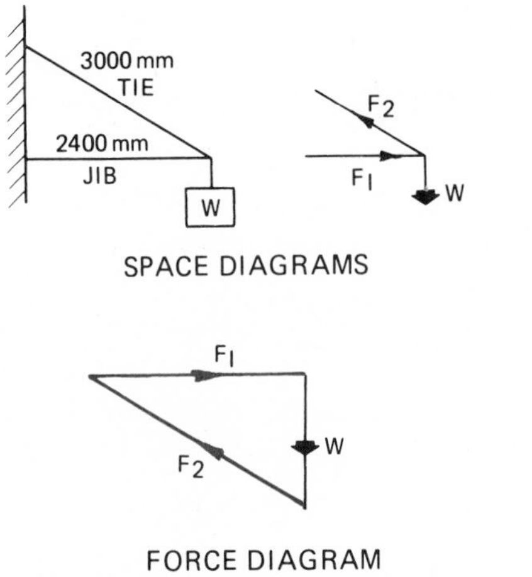
3000 mm
TIE
2400 mm
JIB
W
F2
F1
W
SPACE DIAGRAMS
F1
W
F2
FORCE DIAGRAM

Fig. 29-1-9 Solution to Example 2 by the force diagram method.

Solution. A space diagram is drawn first to find the direction of the force acting on the tie. With the direction of the three forces and the magnitude of one force W known, a force diagram is then drawn to a suitable scale. The length of lines F_1 and F_2 can now be measured to find the magnitude of the forces acting on the jib and tie.

Example 3. In Fig. 29-1-10, a machine having a mass of 12 t is lifted by a jib crane. The length of the jib is 5500 mm and an 8500 mm tie is fastened to a point 4300 mm behind the base of the jib. The chain that lifts the machine passes over a pulley at the top of the jib and connects to a winch located 1800 mm behind the base of the jib. Find the forces acting on the jib and tie.

Solution. The 12 t mass exerts a force of 12 × 9806.65 N, or 117.7 kN, on the chain. A space diagram is drawn first to find the direction of the forces acting on the tie, jib, and chain connected to the winch. With the direction of the four forces and the magnitude of two forces known, a force diagram is then drawn to a suitable scale. The length of lines F_1 and F_2 can now be measured to find the magnitude of the forces acting on the jib and tie.

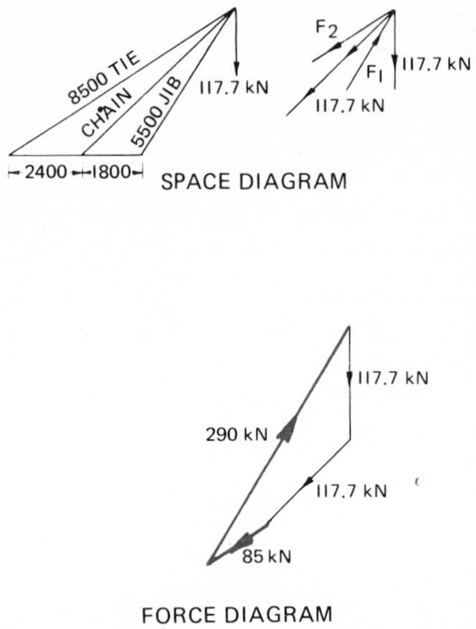
8500 TIE
CHAIN
5500 JIB
117.7 kN
2400
1800
F2
F1
117.7 kN
117.7 kN
SPACE DIAGRAM
117.7 kN
290 kN
117.7 kN
85 kN
FORCE DIAGRAM

Fig. 29-1-10 Solution to Example 3 by the force diagram method.

Example 4. In Fig. 29-1-11, a simple roof truss has a load W applied at the top. Find the forces acting on the rafters, tie bar, and walls.

Solution. A space diagram is drawn first to find the direction of the force acting on the rafters. A force diagram taken

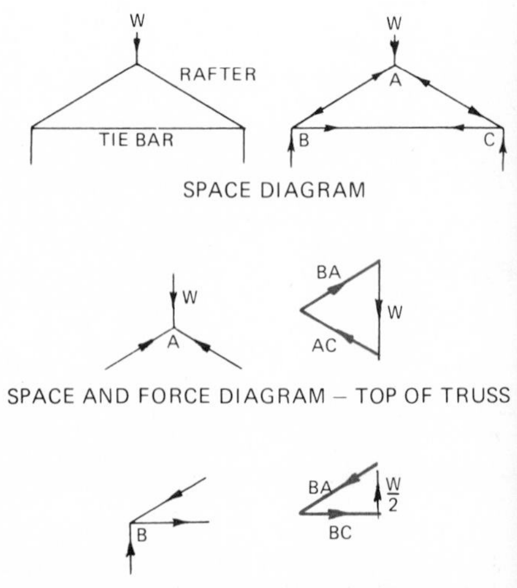
W
RAFTER
TIE BAR
W
A
B
C
SPACE DIAGRAM
W
A
BA
W
AC
SPACE AND FORCE DIAGRAM – TOP OF TRUSS
B
BA
W/2
BC
SPACE AND FORCE DIAGRAM AT TOP OF WALL

Fig. 29-1-11 Solution to Example 4 by the force diagram method.

at point *A* is drawn as the rafters *AB* and *AC* support the load *W*. The forces acting on rafters *AB* and *AC* may be found by making a force diagram taken at point *A*, drawing *W* to scale, and measuring the values of *AB* and *AC*.

Since the force acting on the wall is equal to W/2 (the roof design and load being symmetrical) and the forces acting at points *B* and *C* being equal, only one force diagram need be drawn to find the force acting on the tie.

Assignments

1. On an A3- or B-size sheet, lay out the six problems shown in Fig. 29-1-A, and find their resultant. Scale is to suit. The values shown may be either newtons or pounds.

2. On an A3- or B-size sheet, lay out the six problems shown in Fig. 29-1-B, and find their equilibrants. Scale is to suit. The values shown may be either newtons or pounds.

3. On an A3- or B-size sheet, lay out the four problems shown in Fig. 29-1-C or Fig. 29-1-D, and find the forces acting on each member. Scale is to suit.

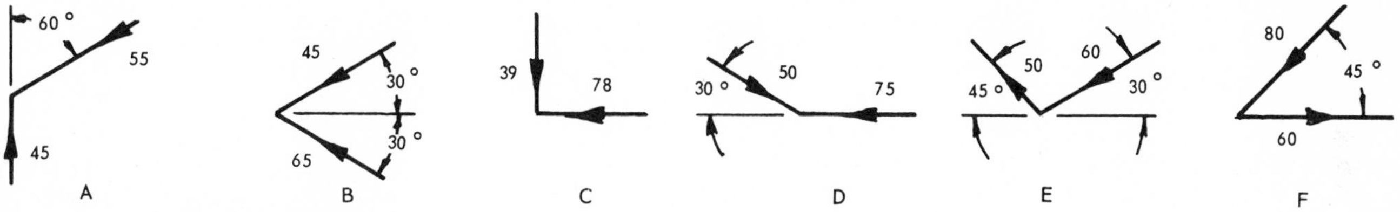
60 °
55
45
A
45
30 °
30 °
65
B
39
78
C
30 °
50
75
D
45 °
50
60
30 °
E
80
45 °
60
F

Fig. 29-1-A **Finding resultants.**

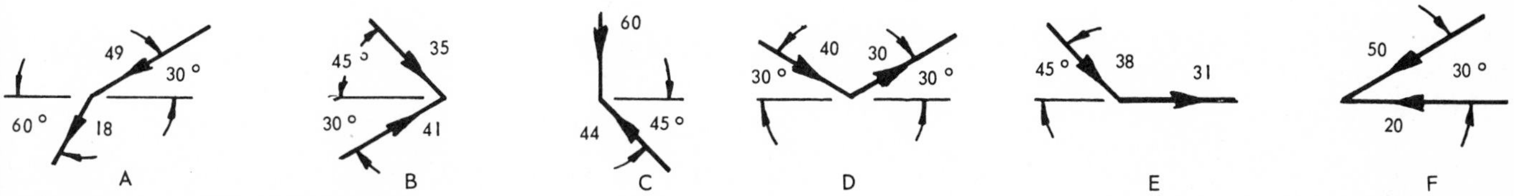
49
30 °
60 °
18
A
45 °
35
30 °
41
B
60
44
45 °
C
40
30
30 °
30 °
D
45 °
38
31
E
50
30 °
20
F

Fig. 29-1-B **Finding equilibrants.**

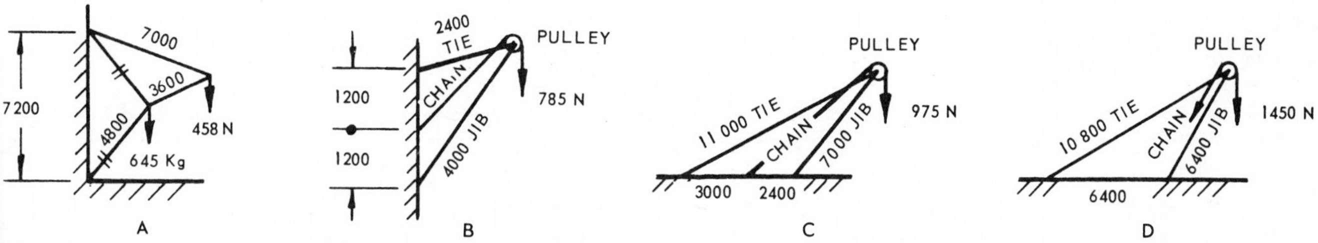
7000
3600
7200
4800
458 N
645 Kg
A
2400
TIE
PULLEY
CHAIN
1200
785 N
4000 JIB
1200
B
PULLEY
11 000 TIE
CHAIN
7000 JIB
975 N
3000
2400
C
PULLEY
10 800 TIE
CHAIN
6400 JIB
1450 N
6400
D

Fig. 29-1-C **Finding forces acting on the tie and jib.**

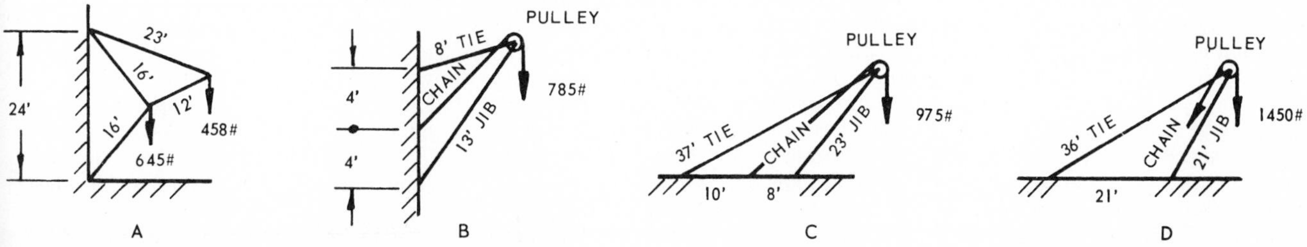
23'
16'
24'
12'
16'
458#
645#
A
PULLEY
8' TIE
CHAIN
4'
785#
13' JIB
4'
B
PULLEY
37' TIE
CHAIN
23' JIB
975#
10'
8'
C
PULLEY
36' TIE
CHAIN
21' JIB
1450#
21'
D

Fig. 29-1-D **Finding forces acting on the tie and jib.**

UNIT 29-2
BEAMS

Bow's Notation

In the previous illustrations, the forces have been identified as F_1, F_2, R, etc. Another system of identifying forces, called *Bow's notation,* is helpful in solving force problems. In the space diagram (Fig. 29-2-1), a boldface capital letter, **A**, **B**, **C**, etc., is placed in the space between two forces and the force is referred to by the two boldface capital letters in the adjoining spaces. The force **AB** in the space diagram is represented by the vector *ab* in the force diagram, the letters *a* and *b* being placed at the beginning and end, respectively, of the vector. The letters in the space diagram are usually given in alphabetical order and in a clockwise direction.

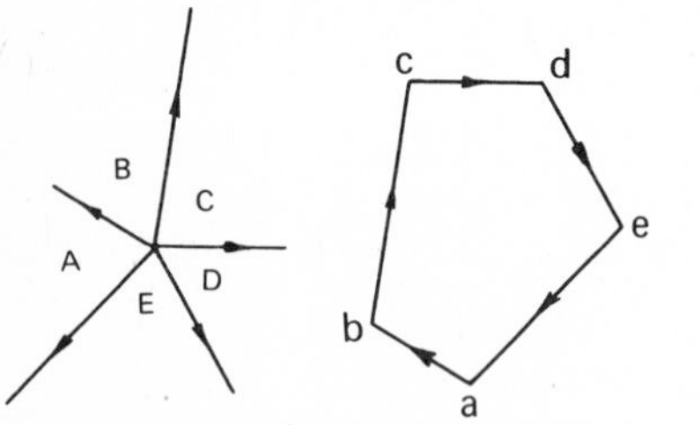

A
B
C
D
E
c
d
e
a
b

(A) SPACE DIAGRAM (B) FORCE DIAGRAM
Fig. 29-2-1 Bow's notation.

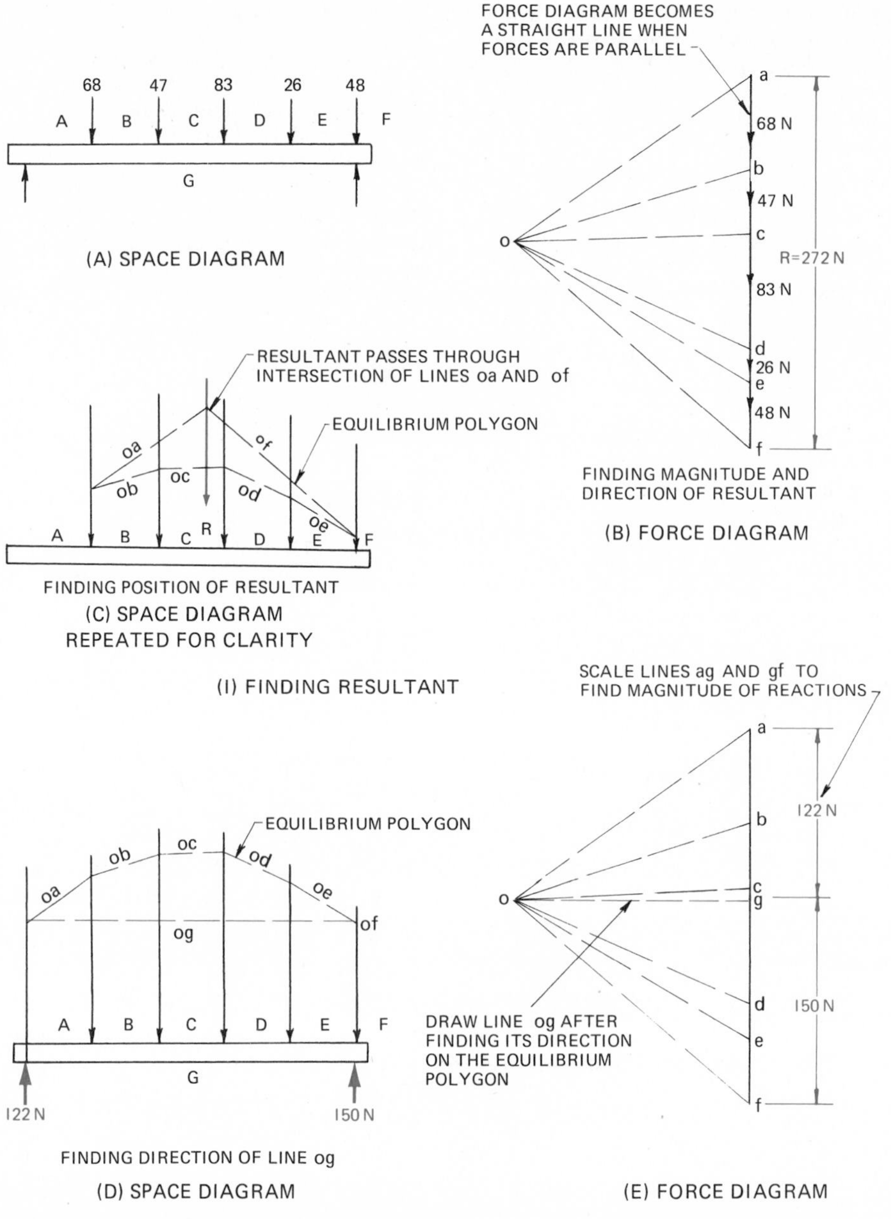

68
47
83
26
48
A
B
C
D
E
F
G
(A) SPACE DIAGRAM
FORCE DIAGRAM BECOMES A STRAIGHT LINE WHEN FORCES ARE PARALLEL
a
68 N
b
47 N
c
o
R=272 N
83 N
d
26 N
e
48 N
f
FINDING MAGNITUDE AND DIRECTION OF RESULTANT
(B) FORCE DIAGRAM
RESULTANT PASSES THROUGH INTERSECTION OF LINES oa AND of
EQUILIBRIUM POLYGON
oa
ob
oc
od
oe
of
R
FINDING POSITION OF RESULTANT
(C) SPACE DIAGRAM REPEATED FOR CLARITY
(I) FINDING RESULTANT
SCALE LINES ag AND gf TO FIND MAGNITUDE OF REACTIONS
122 N
g
150 N
EQUILIBRIUM POLYGON
og
122 N
150 N
DRAW LINE og AFTER FINDING ITS DIRECTION ON THE EQUILIBRIUM POLYGON
FINDING DIRECTION OF LINE og
(D) SPACE DIAGRAM
(E) FORCE DIAGRAM
(II) FINDING REACTIONS

Fig. 29-2-2 Graphic method of finding resultant and reactions of vertical forces acting on a beam.

Equilibrium Polygon

Equilibrium or *funicular polygons* are used in the graphical solutions for finding the magnitude, direction, and point of application of resultants, equilibrants, and reactions. They are also used to check whether or not a number of forces are in equilibrium.

GRAPHIC METHOD OF FINDING RESULTANT AND REACTIONS OF VERTICAL FORCES ACTING ON A BEAM

Example 1. A number of parallel forces are acting on a beam as shown in Fig. 29-2-2*a*. It is required to find, graphically, the magnitude and point of application of the resultant and the magnitude of the reactions.

Solutions. First draw a space and force diagram using Bow's notation. It will be noted that the force diagram is a straight line and not a polygon when all the forces are parallel. The magnitude of the resultant is found by measuring the distance from *a* to *f* on the force diagram. The direction of the resultant is also established but its position with respect to the six forces is still unknown. To find its position, locate a point *o* anywhere on the force diagram and join *o* to each letter with a broken line.

Draw a line anywhere in space *B* of the space diagram (Fig. 29-2-2*c*) parallel to line *ob* in the force diagram, until it intersects forces *AB* and *BC*. Next draw a line parallel to *oc* in space *C* but starting where line *ob* intersects force *BC*. This is continued until the equilibrium polygon is completed. The resultant *R* passes through the intersection of lines *oa* and *of*, thus determining its position. The polygon constructed in the space diagram is called an equilibrium or funicular polygon.

The magnitude of the reaction forces is found by making an equilibrium polygon (see Fig. 29-2-2*d*), which includes reaction forces *AG* and *FG*. These forces were not needed to find the resultant. To construct the equilibrium polygon, draw a line *oa* anywhere in space *A* and parallel to line *oa* in the force diagram, as shown in Fig. 29-2-2*e*. Repeat for lines *ob*, *oc*, *od*, and *oe*, having each of these lines touch each other as shown.

Line *of* will not have any length on the

equilibrium polygon since forces *EF* and *FG* act in the same line. Draw line *og* in the space diagram by joining the start of line *oa* to *of*. Now draw line *og* on the force diagram parallel to line *og* in the space diagram. The magnitude of the reactions (two vertical upward forces supporting the beam) *GA* of 122 kg and *FG* of 150 kg may be found by measuring lines *ag* and *gf* on the force diagram.

GRAPHIC METHOD OF FINDING RESULTANT OF NONPARALLEL FORCES ACTING ON A BEAM

Example 2. A number of nonparallel forces are acting on a beam as shown in Fig. 29-2-3*a*. It is required to find, graphically, the magnitude and the point of application of the resultant.

Solution. To find the resultant, first draw a space and force diagram and label the forces, using Bow's notation. The magnitude and the direction of the resultant are shown by line *ae* on the force diagram (Fig. 29-2-3*b*). The location of the resultant with respect to the four forces must now be found. Locate a point *o* anywhere on the force diagram and join *o* to each letter with a line. Draw a line *ob* anywhere in space *B* of the space diagram (Fig. 29-2-3*c*), parallel to line *ob* in the force diagram. Next draw a line parallel to *oc* in space *C* but starting where line *ob* intersects force *BC*. This is continued until the equilibrium polygon is completed. The resultant, *R* is then drawn through the intersection of lines *oa* and *oe,* thus determining its position.

Assignments

1. On an A3- or B-size sheet, lay out the two beams shown in Fig. 29-2-A or Fig. 29-2-B, and using the graphic method, find the resultant and reactions of the vertical forces. Scale is to suit.

2. On an A3- or B-size sheet, lay out the two beams shown in Fig. 29-2-C or Fig. 29-2-D, and using the graphic method, find the resultant and reactions of the vertical forces. Scale is to suit.

3. On an A3- or B-size sheet, lay out the two beams shown in Fig. 29-2-E or Fig. 29-2-F, and using the graphic method, find the resultant of the nonparallel forces. Scale is to suit.

4. On an A3- or B-size sheet, lay out the two beams shown in Fig. 29-2-G or Fig. 29-2-H, and using the graphic method, find the resultant of the nonparallel forces. Scale is to suit.

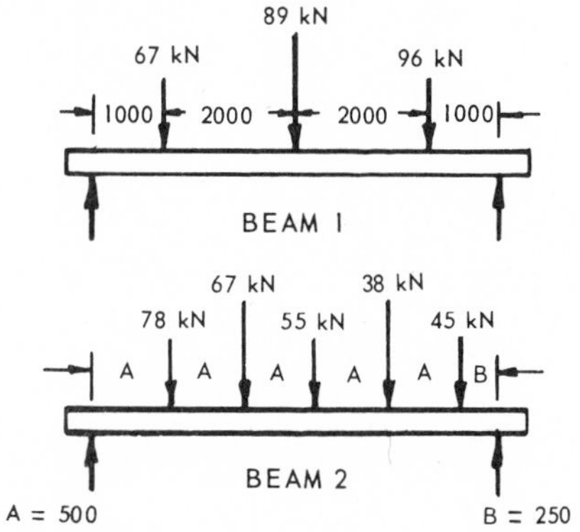
89 kN
67 kN
96 kN
1000
2000
2000
1000
BEAM 1
67 kN
38 kN
78 kN
55 kN
45 kN
A A A A A B
BEAM 2
A = 500
B = 250

Fig. 29-2-A Vertical forces acting on a beam.

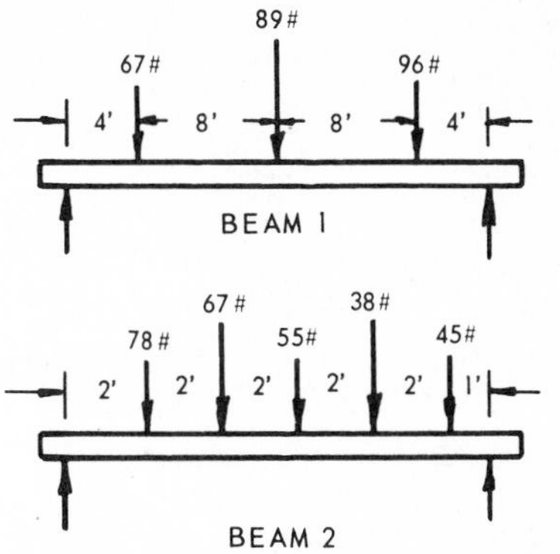
89#
67#
96#
4'
8'
8'
4'
BEAM 1
67#
38#
78#
55#
45#
2'
2'
2'
2'
2'
1'
BEAM 2

Fig. 29-2-B Vertical forces acting on a beam.

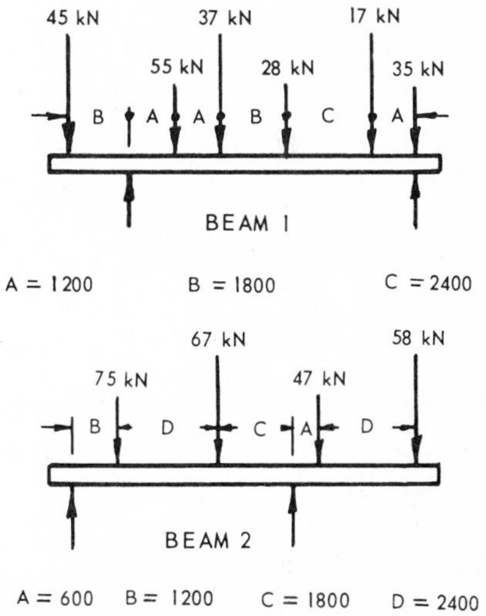
45 kN
37 kN
17 kN
55 kN
28 kN
35 kN
B A A B C A
BEAM 1
A = 1200
B = 1800
C = 2400
67 kN
58 kN
75 kN
47 kN
B D C A D
BEAM 2
A = 600
B = 1200
C = 1800
D = 2400

Fig. 29-2-C Vertical forces acting on a beam.

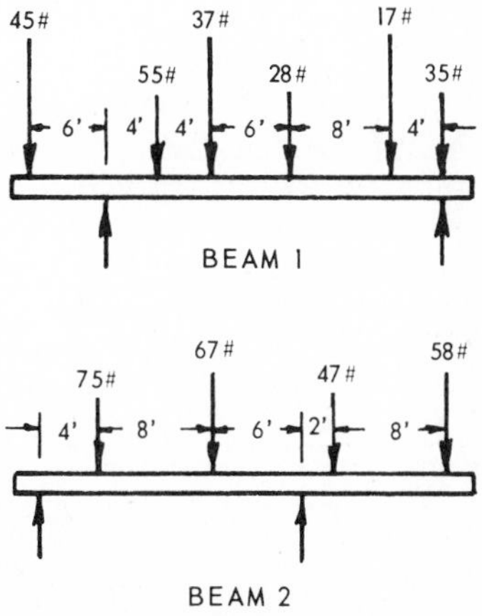
45#
37#
17#
55#
28#
35#
6'
4'
4'
6'
8'
4'
BEAM 1
67#
58#
75#
47#
4'
8'
6'
2'
8'
BEAM 2

Fig. 29-2-D Vertical forces acting on a beam.

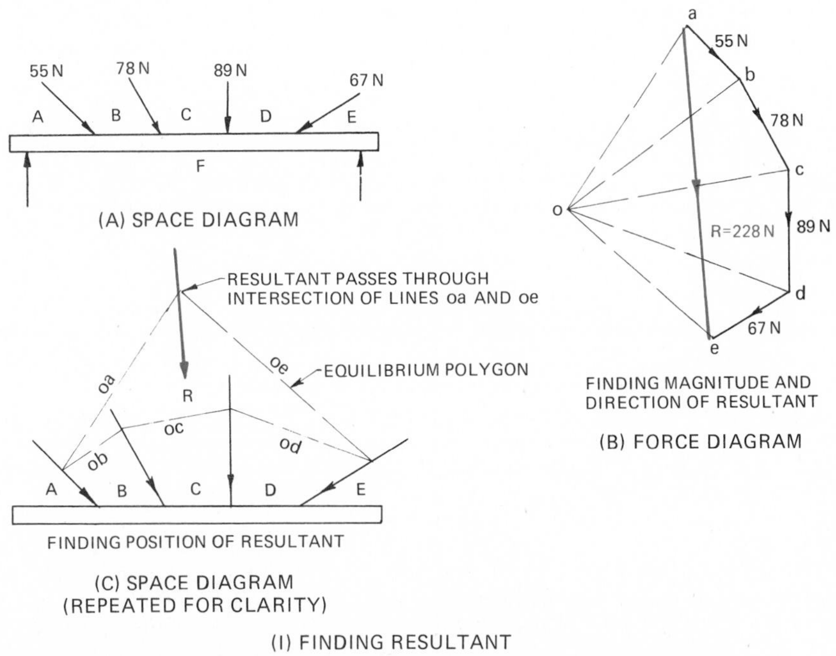
55 N
78 N
89 N
67 N
A B C D E
F
(A) SPACE DIAGRAM
RESULTANT PASSES THROUGH INTERSECTION OF LINES oa AND oe
EQUILIBRIUM POLYGON
oa
oe
R
oc
od
ob
A B C D E
FINDING POSITION OF RESULTANT
(C) SPACE DIAGRAM (REPEATED FOR CLARITY)
a
55 N
b
78 N
c
o
R=228 N
89 N
d
67 N
e
FINDING MAGNITUDE AND DIRECTION OF RESULTANT
(B) FORCE DIAGRAM
(I) FINDING RESULTANT

Fig. 29-2-3 Graphic method of finding resultant of nonparallel forces acting on a beam.

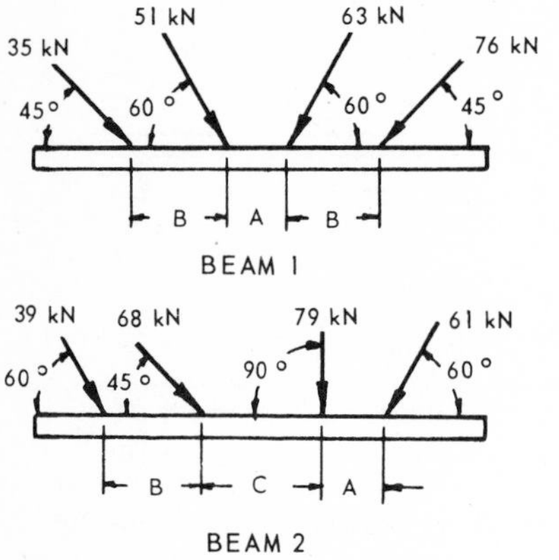
35 kN
51 kN
63 kN
76 kN
45°
60°
60°
45°
B
A
B
BEAM 1
39 kN
68 kN
79 kN
61 kN
60°
45°
90°
60°
B
C
A
BEAM 2

A = 1000 B = 1500 C = 2000

Fig. 29-2-E Nonparallel forces acting on a beam.

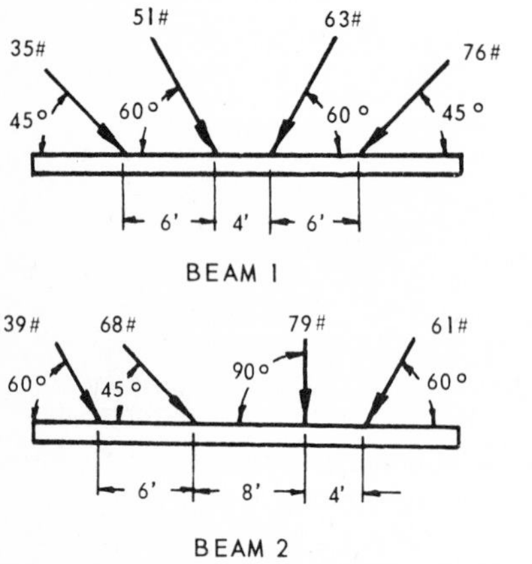
35#
51#
63#
76#
45°
60°
60°
45°
6'
4'
6'
BEAM 1
39#
68#
79#
61#
60°
45°
90°
60°
6'
8'
4'
BEAM 2

Fig. 29-2-F Nonparallel forces acting on a beam.

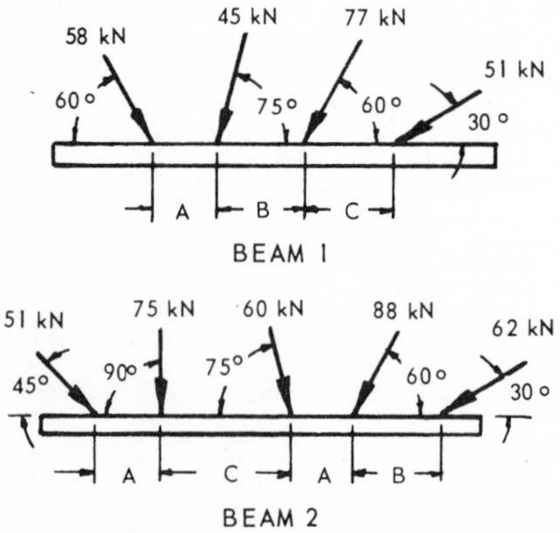
58 kN
45 kN
77 kN
51 kN
60°
75°
60°
30°
A
B
C
BEAM 1
51 kN
75 kN
60 kN
88 kN
62 kN
45°
90°
75°
60°
30°
A
C
A
B
BEAM 2

A = 1000 B = 1500 C = 2000

Fig. 29-2-G Nonparallel forces acting on a beam.

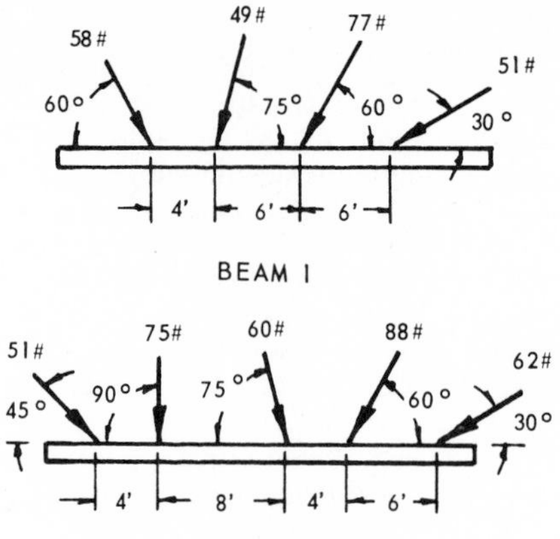
58#
49#
77#
51#
60°
75°
60°
30°
4'
6'
6'
BEAM 1
51#
75#
60#
88#
62#
45°
90°
75°
60°
30°
4'
8'
4'
6'
BEAM 2

Fig. 29-2-H Nonparallel forces acting on a beam.

UNIT 29-3

TRUSS REACTIONS WHEN LOADS ARE PARALLEL

Bridge and Roof Trusses

The graphical solution offers a quick and convenient method of checking or determining truss calculations. Some of the more common types of roof and bridge trusses are shown in Fig. 29-3-1. The loads that the truss and abutments support are combinations of the mass of the truss and the mass of the materials placed on the roof or truss, for example, snow, wind, and live loads such as cars and trains. The upward forces of the abutment or walls are called the *reactions*. Rollers are often used at one support of the truss to allow for expansion and contraction. Truss terminology is shown in Fig. 29-3-2.

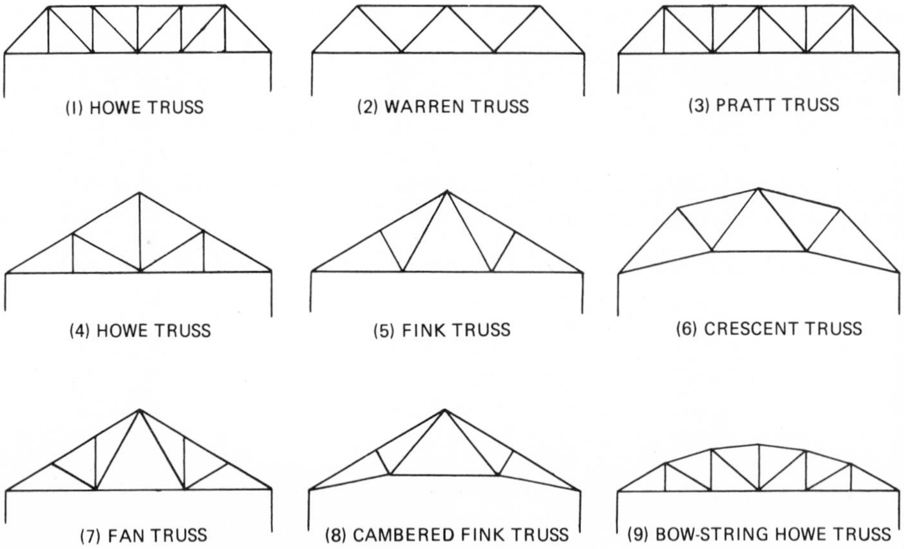
(1) HOWE TRUSS
(2) WARREN TRUSS
(3) PRATT TRUSS
(4) HOWE TRUSS
(5) FINK TRUSS
(6) CRESCENT TRUSS
(7) FAN TRUSS
(8) CAMBERED FINK TRUSS
(9) BOW-STRING HOWE TRUSS

Fig. 29-3-1 Common types of trusses.

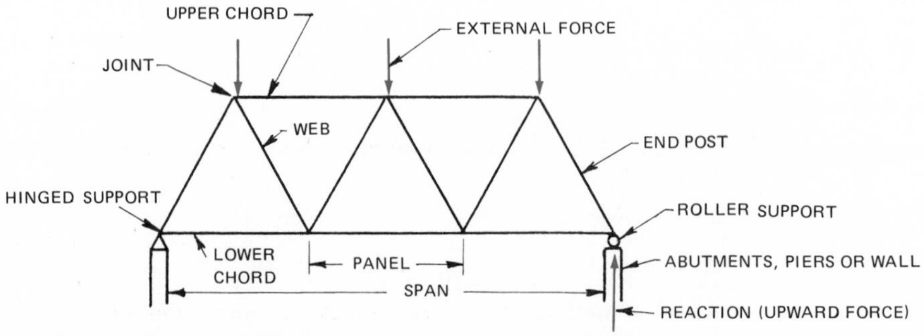
UPPER CHORD
EXTERNAL FORCE
JOINT
WEB
END POST
HINGED SUPPORT
ROLLER SUPPORT
LOWER CHORD
PANEL
ABUTMENTS, PIERS OR WALL
SPAN
REACTION (UPWARD FORCE)

Fig. 29-3-2 Truss terminology.

Graphic Method of Finding Truss Reactions When Loads Are Vertical

Consider the truss in Fig. 29-3-3*a*. The forces acting on the truss are vertical downward forces, while the reaction forces are vertical upward forces, since they must be equal and opposite to the truss forces. The lines of action and directions of all forces are known, and only the magnitude of the reactions must be computed. Draw a space diagram and label the forces using Bow's notation. Next draw the force diagram. It will be noted that the force diagram becomes a straight line when all the forces are parallel. The magnitudes of the reactions *FG* and *GA* are not yet known, but their combined magnitude is equal to 1100 kN, the length of line *af* on the force diagram. Locate a point *o* anywhere on the force diagram and join *o* to points *a* and *f* with a broken line.

Draw a line *ob* parallel to line *ob* in the force diagram, anywhere in space *B* of the space diagram (Fig. 29-3-3*d*) until it intersects forces *AB* and *BC*. Draw a line parallel to *oc* in space *C*, but start where the line *ob* intersects force *BC*.

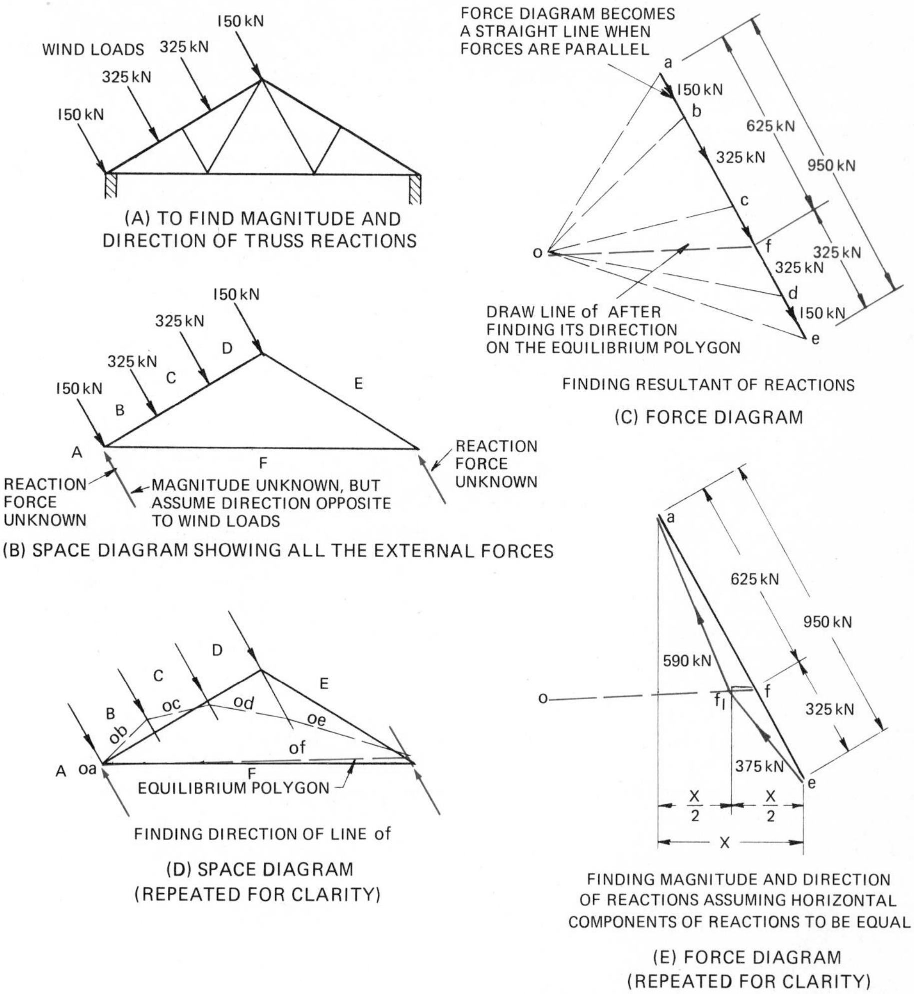
WIND LOADS
150 kN
325 kN
(A) TO FIND MAGNITUDE AND DIRECTION OF TRUSS REACTIONS
REACTION FORCE UNKNOWN
MAGNITUDE UNKNOWN, BUT ASSUME DIRECTION OPPOSITE TO WIND LOADS
(B) SPACE DIAGRAM SHOWING ALL THE EXTERNAL FORCES
EQUILIBRIUM POLYGON
FINDING DIRECTION OF LINE of
(D) SPACE DIAGRAM (REPEATED FOR CLARITY)
FORCE DIAGRAM BECOMES A STRAIGHT LINE WHEN FORCES ARE PARALLEL
625 kN
950 kN
DRAW LINE of AFTER FINDING ITS DIRECTION ON THE EQUILIBRIUM POLYGON
FINDING RESULTANT OF REACTIONS
(C) FORCE DIAGRAM
590 kN
375 kN
FINDING MAGNITUDE AND DIRECTION OF REACTIONS ASSUMING HORIZONTAL COMPONENTS OF REACTIONS TO BE EQUAL
(E) FORCE DIAGRAM (REPEATED FOR CLARITY)

Fig. 29-3-4 Graphic method of finding truss reactions when loads are not vertical.

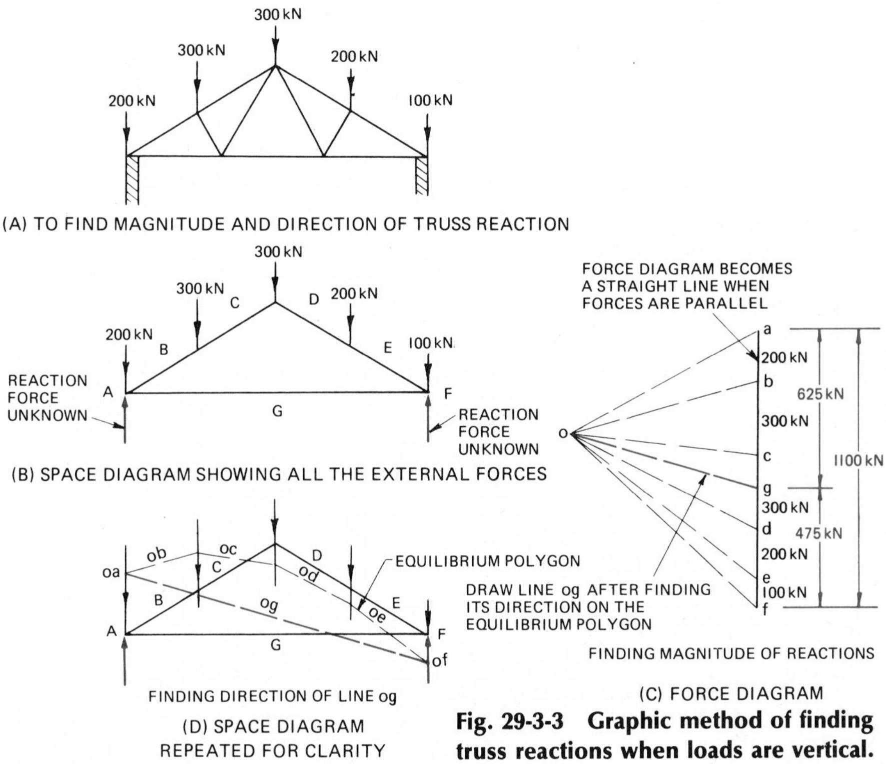
300 kN
200 kN
100 kN
(A) TO FIND MAGNITUDE AND DIRECTION OF TRUSS REACTION
REACTION FORCE UNKNOWN
(B) SPACE DIAGRAM SHOWING ALL THE EXTERNAL FORCES
EQUILIBRIUM POLYGON
DRAW LINE og AFTER FINDING ITS DIRECTION ON THE EQUILIBRIUM POLYGON
FINDING DIRECTION OF LINE og
(D) SPACE DIAGRAM REPEATED FOR CLARITY
FORCE DIAGRAM BECOMES A STRAIGHT LINE WHEN FORCES ARE PARALLEL
625 kN
1100 kN
475 kN
FINDING MAGNITUDE OF REACTIONS
(C) FORCE DIAGRAM

Fig. 29-3-3 Graphic method of finding truss reactions when loads are vertical.

Repeat for lines *od* and *oe*. Line *oa* will not have any length on the equilibrium polygon since forces *AB* and *GA* act in the same line. The same is true for line *of* since forces *EF* and *FG* act in the same line. Close the polygon by joining *oa* to *of* with line *og*. Draw line *og* on the force diagram parallel to the line *og* on the space diagram. The values of the reactions *GA* of 625 kN and *FG* of 475 kN may be found by measuring lines *ga* and *fg* on the force diagram.

Graphic Method of Finding Truss Reactions When Loads Are Not Vertical

When the direction of the resultant of the roof and wind load is not vertical, the reactions are not normally parallel.

The solution for finding the magnitude and direction of the reactions when the truss has fixed supports is based on the assumption that the horizontal components of the reactions are equal, since the truss is assumed to be rigid and rigidly held to the supports.

Consider the truss shown in Fig. 29-3-4*a*. The forces acting on the truss are inclined downward forces, while the reaction forces are inclined upward forces. The lines of action and direction of all the downward forces are known and only the magnitude and direction of the reactions need to be computed. Draw a space diagram and label the forces using Bow's notation. Next draw the force diagram as shown in Fig. 29-3-4*c*. The resultant of the downward forces, line *ae* on the force diagram, is equal but opposite to the resultant of the reaction forces.

The magnitude and direction of the reactions *EF* and *FA* are not yet known, but the magnitude of their resultant is equal to 950 kN. Locate a point *o* anywhere on the force diagram and join *o* to points *a* and *e* with a line. Draw a line *ob* parallel to line *ob* on the force diagram anywhere in space *B* of the space diagram (Fig. 29-3-4*d*) until it intersects forces *AB* and *BC*.

Draw a line parallel to *oc* in space *C* starting where line *ob* intersects force *BC*. Repeat for lines *od* and *oe*. Line *oa* will not have any length on the equilibrium polygon since forces *AB* and *FA* act in the same line. Close the equilibrium polygon by joining *oa* and *oe* with line *of*. Draw line *of* on the force diagram parallel to line *of* on the space diagram. The value of 625 kN and 325 kN may be found by measuring lines *fa* and *ef* on the force diagram. These values are the individual resultants of reactions *EF* and *FA*. Since the horizontal components of the reactions are assumed to be equal, the magnitude and direction of reactions *FA* and *EF* can be found by completing the force diagram as shown in Fig. 29-3-4*e*.

Distance *X* represents the horizontal component of the combined reaction forces. Since the truss is rigidly held to the supports, we may assume that each support will take half the horizontal force. Therefore, X/2 represents the horizontal component of each reaction. From point *f* on line *ae*, extend a horizontal line until it intersects the vertical line bisecting distance *X* at point f_1. The values of the reactions *FA* of 590 kN and *EF* of 375 kN may be found by measuring lines f_1a and ef_1 on the force diagram. The directions of the reaction forces will be parallel to lines f_1a and ef_1.

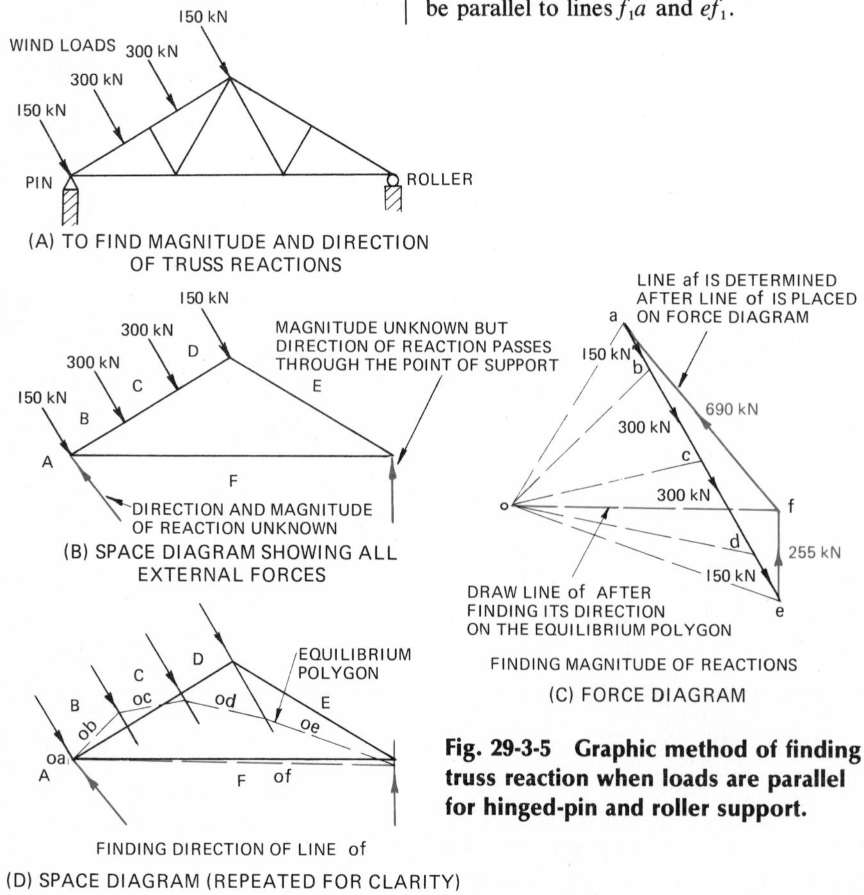
150 kN
WIND LOADS 300 kN
300 kN
150 kN
PIN
ROLLER
(A) TO FIND MAGNITUDE AND DIRECTION OF TRUSS REACTIONS
150 kN
300 kN
300 kN
150 kN
B
C
D
A
MAGNITUDE UNKNOWN BUT DIRECTION OF REACTION PASSES THROUGH THE POINT OF SUPPORT
E
F
DIRECTION AND MAGNITUDE OF REACTION UNKNOWN
(B) SPACE DIAGRAM SHOWING ALL EXTERNAL FORCES
EQUILIBRIUM POLYGON
B
C
D
E
ob
oc
od
oe
oa
A
F
of
FINDING DIRECTION OF LINE of
(D) SPACE DIAGRAM (REPEATED FOR CLARITY)
LINE af IS DETERMINED AFTER LINE of IS PLACED ON FORCE DIAGRAM
a
150 kN
b
690 kN
300 kN
c
300 kN
f
o
d
255 kN
150 kN
e
DRAW LINE of AFTER FINDING ITS DIRECTION ON THE EQUILIBRIUM POLYGON
FINDING MAGNITUDE OF REACTIONS
(C) FORCE DIAGRAM

Fig. 29-3-5 Graphic method of finding truss reaction when loads are parallel for hinged-pin and roller support.

Reaction of Hinged-Pin and Roller Supports

Normally in bridge and roof truss design, one end of the truss is supported by a hinged pin and the other end rests on a roller. This provides for changes in the length of the truss because of temperature changes. The reaction of the roller support is taken through the roller and is usually perpendicular to the path of the roller. The reaction of the hinged-pin support is equal to the direction and size of force required to keep the structure in equilibrium. If the forces acting on the bridge or roof are vertical, then we can assume that the reactions of the hinge and rollers are also vertical. If the resultant of the forces acting on the bridge or roof is inclined due to wind loads and the reaction at the roller support is vertical, then the reaction at the hinged-pin support must be inclined.

Graphic Method of Finding Truss Reactions When Loads Are Parallel for Hinged-Pin and Roller Support

Consider the truss shown in Fig. 29-3-5*a*. The forces acting on the truss are inclined downward forces caused by wind loads, a vertical upward force through the center of the roller, and an inclined upward force at the hinged-pin support. It is required to find the magnitude of the reactions and the direction of the reaction at the hinged-pin support.

Draw a space diagram (Fig. 29-3-5*b*), and label all the forces using Bow's notation. Draw wind forces *AB, BC, CD,* and *DE* on the force diagram. Since the direction of reaction force *EF* is known, its direction can be drawn on the force diagram. Its length or magnitude is not known. Reaction force *FA* cannot be drawn since both its magnitude and direction are not known, but its line of action passes through the point of support. Thus the equilibrium polygon is started at the pin support. Locate a point *o* anywhere on the force diagram and join *o* to to points *a* and *e*. Draw a line *ob* parallel to line *ob* on the force diagram in space *B* of the space diagram (Fig. 29-3-5*d*) starting at the pin support until it intersects force *BC*. Draw a line parallel to *oc* in space *C* but starting where line *ob* intersects force *BC*. Repeat for lines *od* and *oe*. Line *oa* will not have any length on the equilibrium polygon since forces *AB* and *FA* act at the same point. Close the equilibrium polygon by joining *oa* to *oe* with line *of*. Draw line *of* on the force diagram parallel to line *of* on the space diagram until it intersects reaction force *ef* at point *f*. Close the force polygon with line *fa*. The direction of the hinged reaction force *FA* is parallel to line *fa*. The values of the reactions *EF* of 255 kN and *FA* of 690 kN may be found by measuring lines *ef* and *fa* on the force diagram.

Alternative Graphic Method of Finding Truss Reactions When Loads Are Parallel For Hinged-Pin and Roller Support

An alternative method can be employed for finding the truss reactions when the reaction forces at the pin and roller and the load force are not parallel. Consider the forces acting in Fig. 29-3-6*a*. The four wind forces of 150, 300, 300, and 150 kN can be replaced by a resultant force of 900 kN acting midway on the truss since the loads are symmetrical. The direction of the reaction at the hinge support is not known, but the direction of the reaction at the roller support will be vertical. Since the lines of action of any three nonparallel forces in equilibrium intersect at a common point, the direction of the reaction at the hinged support can be found. The lines of action of the resultant wind loads and the vertical reaction force are extended until they intersect at point *O* (Fig. 29-3-6*b*). Since the lines of action of all three forces must pass through point *O*, a line joining point *O* to the point of intersection at the hinged support determines the direction of the left reaction.

Knowing the direction of the three forces and the magnitude of the resultant, the magnitudes of the truss reactions can readily be found by making a force diagram and measuring lines *ca* and *bc*.

Assignments

1. On an A3- or B-size sheet, lay out the two trusses shown in Fig. 29-3-A or Fig. 29-3-B, and by the graphic method find the magnitude of the reactions. Scale is to suit.

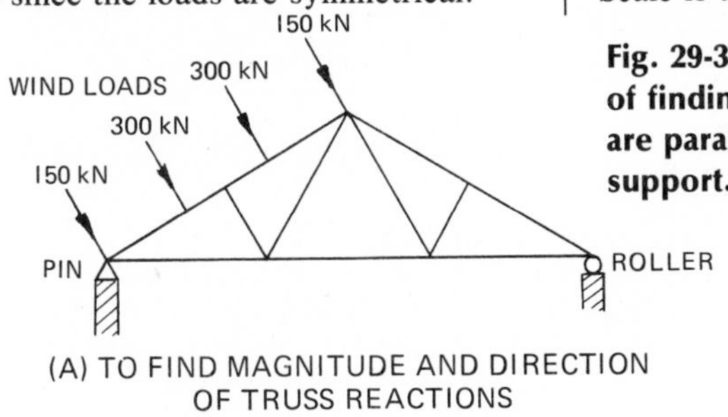
150 kN
WIND LOADS
300 kN
300 kN
150 kN
PIN
ROLLER
(A) TO FIND MAGNITUDE AND DIRECTION OF TRUSS REACTIONS

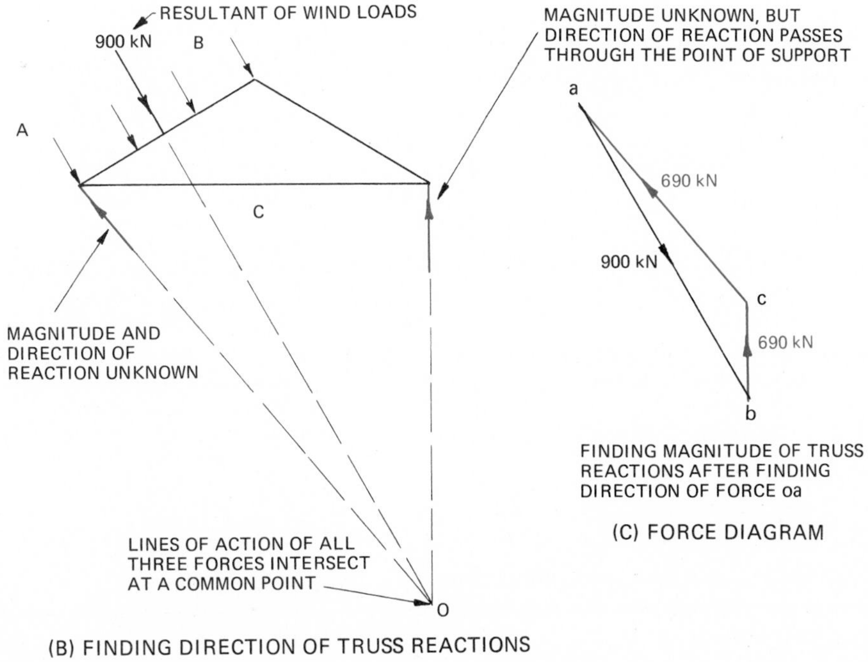
RESULTANT OF WIND LOADS
900 kN
B
A
C
MAGNITUDE AND DIRECTION OF REACTION UNKNOWN
LINES OF ACTION OF ALL THREE FORCES INTERSECT AT A COMMON POINT
O
MAGNITUDE UNKNOWN, BUT DIRECTION OF REACTION PASSES THROUGH THE POINT OF SUPPORT
(B) FINDING DIRECTION OF TRUSS REACTIONS
a
690 kN
900 kN
c
690 kN
b
FINDING MAGNITUDE OF TRUSS REACTIONS AFTER FINDING DIRECTION OF FORCE oa
(C) FORCE DIAGRAM

Fig. 29-3-6 Alternative graphic method of finding truss reactions when loads are parallel for hinged-pin and roller support.

2. On an A3- or B-size sheet, lay out the two trusses shown in Fig. 29-3-C or Fig. 29-3-D, and by the graphic method find the magnitude and direction of the reactions. Scale is to suit.

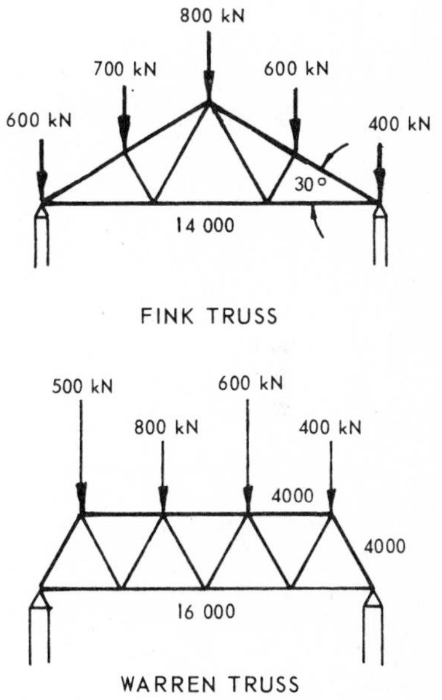
800 kN
700 kN
600 kN
600 kN
400 kN
30°
14 000
FINK TRUSS
500 kN
600 kN
800 kN
400 kN
4000
4000
16 000
WARREN TRUSS

Fig. 29-3-A Vertical loads on truss.

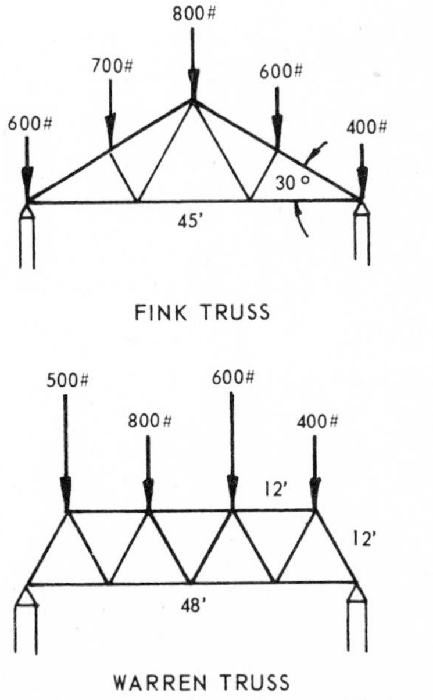
800#
700#
600#
600#
400#
30°
45'
FINK TRUSS
500#
600#
800#
400#
12'
12'
48'
WARREN TRUSS

Fig. 29-3-B Vertical loads on truss.

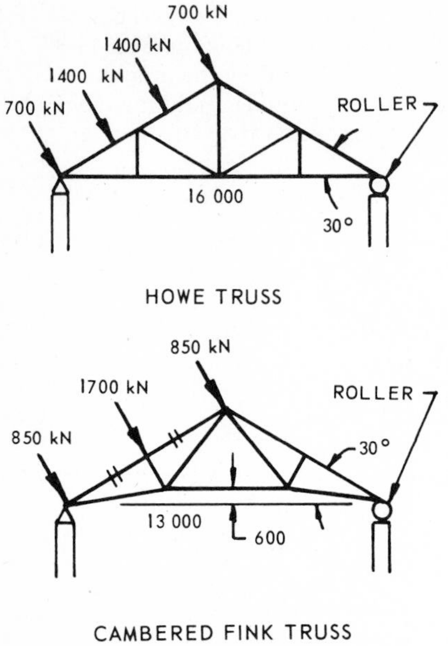
700 kN
1400 kN
1400 kN
700 kN
ROLLER
16 000
30°
HOWE TRUSS
850 kN
1700 kN
850 kN
ROLLER
30°
13 000
600
CAMBERED FINK TRUSS

Fig. 29-3-C Parallel but not vertical loads on truss-hinged and roller supports.

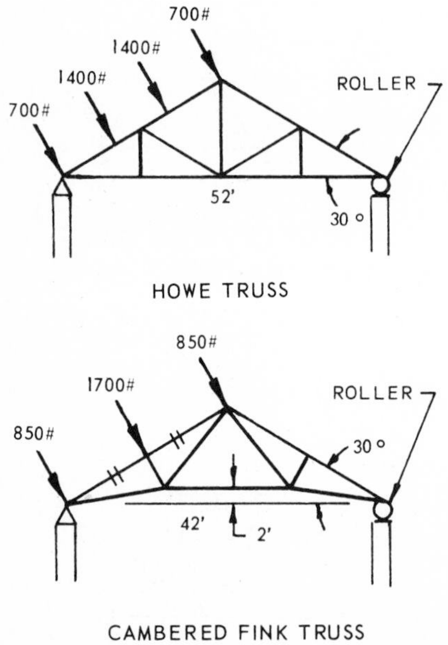
700#
1400#
1400#
700#
ROLLER
52'
30°
HOWE TRUSS
850#
1700#
850#
ROLLER
30°
42'
2'
CAMBERED FINK TRUSS

Fig. 29-3-D Parallel but not vertical loads on truss-hinged and roller supports.

UNIT 29-4

TRUSS REACTIONS WHEN LOADS ARE NOT PARALLEL

Graphic Method of Finding Truss Reactions When Wind and Truss Loads Are Not Parallel

Consider the truss shown in Fig. 29-4-1*a*. The forces acting on the truss are a vertical downward force, an inclined downward force, and the reaction forces, which are inclined upward forces and normally do not act in the same direction. Since the truss is rigid and rigidly held to the supports, we may assume that the horizontal components of the reactions are equal.

The lines of action and direction of the downward forces are known, and only the magnitude and direction of the reactions need be computed. Draw a space diagram and label the forces using Bow's notation. Next draw a force diagram (Fig. 29-4-1*b*). The combined resultant of the two reaction forces is equal to *ac*. The individual magnitudes and directions of the reaction forces *CD* and *DA* are not yet known, but the magnitude of their combined resultant is equal to 970 kN.

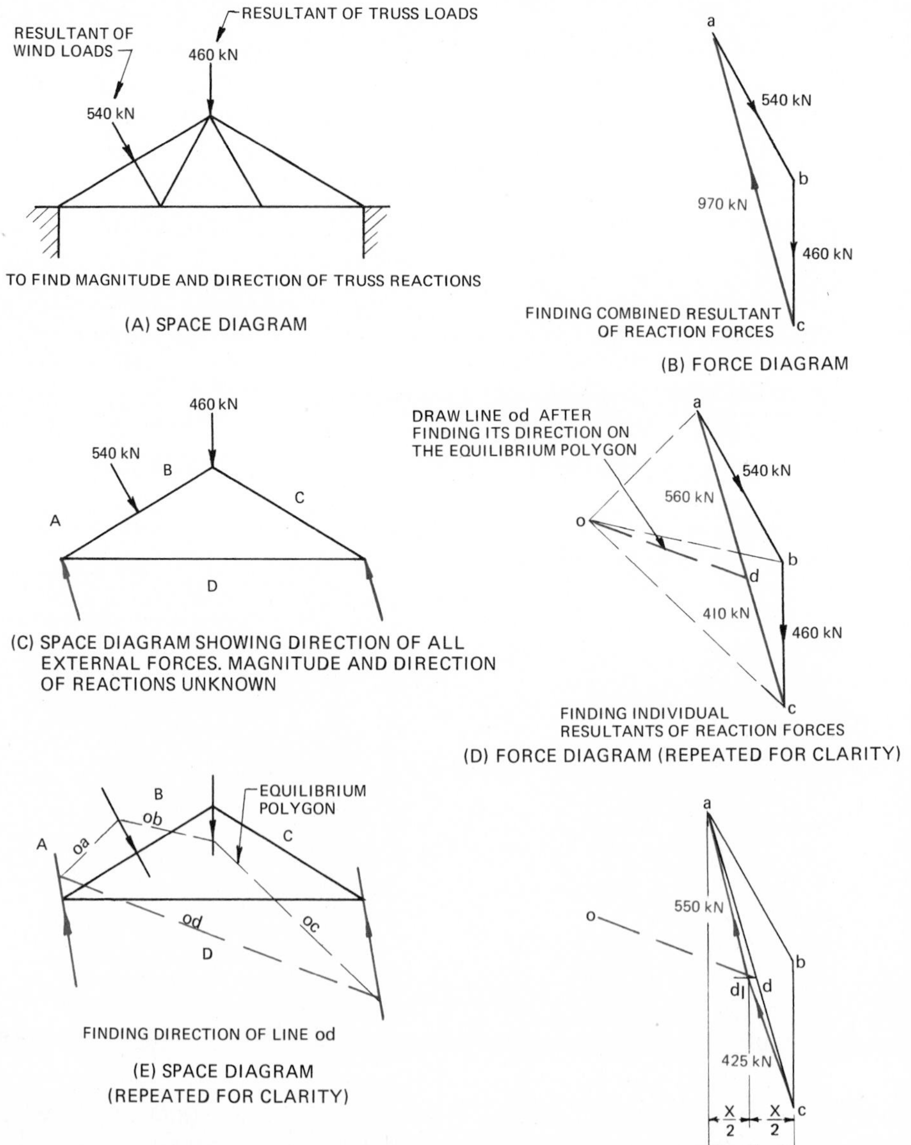
RESULTANT OF TRUSS LOADS
RESULTANT OF WIND LOADS
460 kN
540 kN
TO FIND MAGNITUDE AND DIRECTION OF TRUSS REACTIONS
(A) SPACE DIAGRAM
a
540 kN
b
970 kN
460 kN
c
FINDING COMBINED RESULTANT OF REACTION FORCES
(B) FORCE DIAGRAM
460 kN
540 kN
A
B
C
D
(C) SPACE DIAGRAM SHOWING DIRECTION OF ALL EXTERNAL FORCES. MAGNITUDE AND DIRECTION OF REACTIONS UNKNOWN
DRAW LINE od AFTER FINDING ITS DIRECTION ON THE EQUILIBRIUM POLYGON
a
540 kN
560 kN
o
b
d
410 kN
460 kN
c
FINDING INDIVIDUAL RESULTANTS OF REACTION FORCES
(D) FORCE DIAGRAM (REPEATED FOR CLARITY)
B
ob
EQUILIBRIUM POLYGON
A
oa
C
od
oc
D
FINDING DIRECTION OF LINE od
(E) SPACE DIAGRAM (REPEATED FOR CLARITY)
a
550 kN
o
b
d1
d
425 kN
X/2
X/2
c
X
FINDING MAGNITUDE AND DIRECTION OF REACTIONS
(F) FORCE DIAGRAM (REPEATED FOR CLARITY)

Fig. 29-4-1 Graphic method of finding truss reactions when wind and mass loads are not parallel.

Locate a point *o* anywhere on the force diagram and join *o* to points *a, b,* and *c* with a line, as shown in Fig. 29-4-1*d*. Draw a line *oa* parallel to line *oa* on the force diagram anywhere in space *A* of the space diagram (Fig. 29-4-1*c*) until it intersects forces *DA* and *AB*. Draw a line parallel to *ob* in space *B* but start where line *oa* intersects force *AB*. Repeat for line *oc*. Close the equilibrium polygon with line *od*.

Draw line *od* on the force diagram (Fig. 29-4-1*d*) parallel to line *od* on the equilibrium polygon. The value of 560 kN and 410 kN may be found by measuring lines *ad* and *dc* on the force diagram.

These values are the individual resultants of reactions *CD* and *DA*. Since the horizontal components of the reactions are assumed to be equal, the magnitude and direction of reactions *CD* and *DA* can be found by drawing the force diagram as shown in Fig. 29-4-1*f*. Distance *X* represents the horizontal component of the combined reaction forces, therefore X/2 represents the horizontal component of each reaction. From point *d* on line *ac*, extend a horizontal line until it intersects the vertical line bisecting distance *X* at point d_1. The values of the reactions *CD* of 425 kN and *DA* of 550 kN may be found by measuring lines cd_1 and d_1a on the force diagram. The directions of the reaction forces will be parallel to lines cd_1 and d_1a.

Graphic Method of Finding Truss Reactions, Roller at One End, When Wind and Truss Loads Are Not Parallel

Consider the truss shown in Fig. 29-4-2*a*. The forces acting on the truss are a vertical downward force, an inclined downward force, a vertical upward force through the center of the roller, and an inclined upward force at the hinged pin support. Draw a space diagram and label all the forces using Bow's notation. Next, partially draw the force diagram showing the known forces *AB* and *BC*. Force *CD* is a vertical upward force, but its magnitude is not known. Locate a point *o* anywhere on the force diagram and join *o* to points *a, b,* and *c*. Draw a line *oa* parallel to line *oa* on the force diagram anywhere in space *A* of the space diagram (Fig. 29-4-2*d*), until it intersects forces *DA* and *AB*. Draw a line parallel to *ob* in space *B* but start where line *oa* intersects force *AB*. Repeat for line *oc*. Close the equilibrium polygon by joining *oa* and *oc* with line *od*. Draw line *od* on the force diagram parallel to line *od* on the space diagram until it intersects line *bc* at point *d*. Join *a* to *d* with a line which represents the direction of force *DA*. The values of the reactions *CD* of 360 kg and *DA* of 510 kg may be found by measuring lines *cd* and *da* on the force diagram.

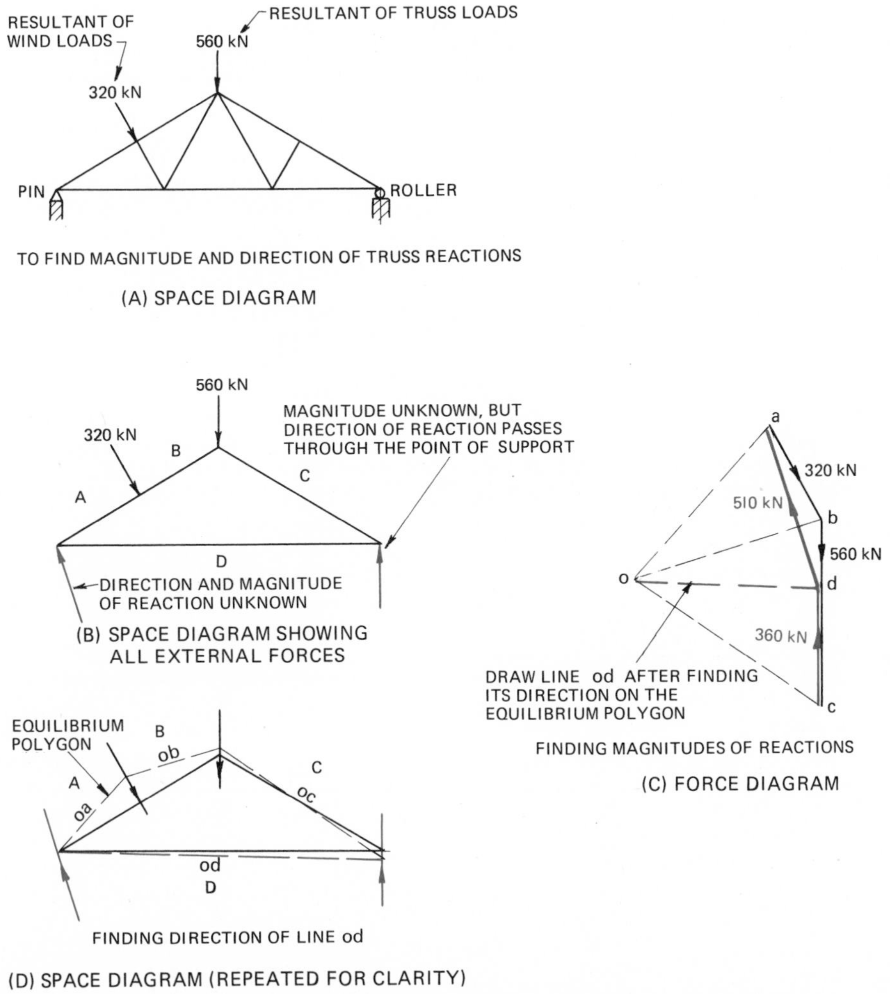
RESULTANT OF WIND LOADS
320 kN
RESULTANT OF TRUSS LOADS
560 kN
PIN
ROLLER
TO FIND MAGNITUDE AND DIRECTION OF TRUSS REACTIONS
(A) SPACE DIAGRAM
560 kN
320 kN
A
B
C
D
MAGNITUDE UNKNOWN, BUT DIRECTION OF REACTION PASSES THROUGH THE POINT OF SUPPORT
DIRECTION AND MAGNITUDE OF REACTION UNKNOWN
(B) SPACE DIAGRAM SHOWING ALL EXTERNAL FORCES
a
320 kN
510 kN
b
560 kN
o
d
360 kN
c
DRAW LINE od AFTER FINDING ITS DIRECTION ON THE EQUILIBRIUM POLYGON
FINDING MAGNITUDES OF REACTIONS
(C) FORCE DIAGRAM
EQUILIBRIUM POLYGON
B
ob
A
oa
C
oc
od
D
FINDING DIRECTION OF LINE od
(D) SPACE DIAGRAM (REPEATED FOR CLARITY)

Fig. 29-4-2 Graphic method of finding truss reactions, roller at one end, when wind and mass loads are not parallel.

Assignments

1. On an A3- or B-size sheet, lay out the two trusses shown in Fig. 29-4-A or Fig. 29-4-B, and by the graphic method find the magnitude and direction of the reactions. Scale is to suit.

2. On an A3- or B-size sheet, lay out the two trusses shown in Fig. 29-4-C or Fig. 29-4-D, and by the graph method find the magnitude and direction of the reactions. Scale is to suit.

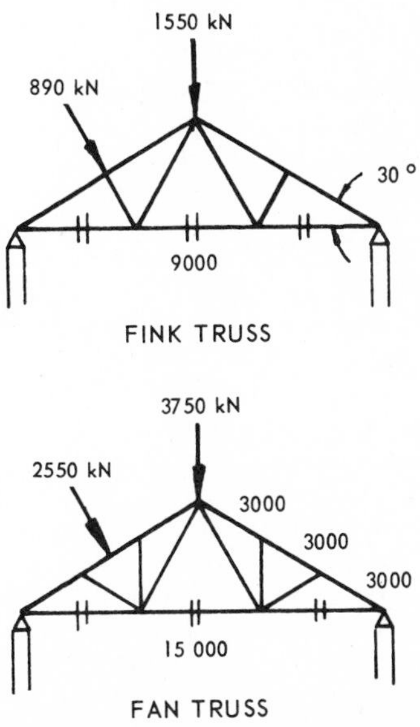
1550 kN
890 kN
30 °
9000
FINK TRUSS
3750 kN
2550 kN
3000
3000
3000
15 000
FAN TRUSS

Fig. 29-4-A Nonparallel loads acting on truss.

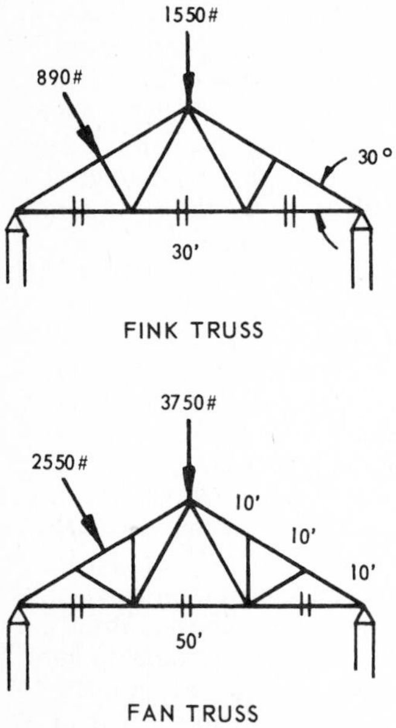
1550#
890#
30°
30'
FINK TRUSS
3750#
2550#
10'
10'
10'
50'
FAN TRUSS

Fig. 29-4-B Nonparallel loads acting on truss.

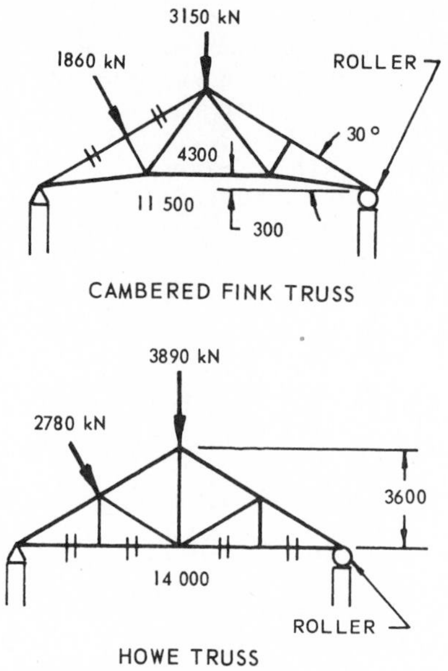
3150 kN
1860 kN
ROLLER
30°
4300
11 500
300
CAMBERED FINK TRUSS
3890 kN
2780 kN
3600
14 000
ROLLER
HOWE TRUSS

Fig. 29-4-C Nonparallel loads acting on truss—roller at one end.

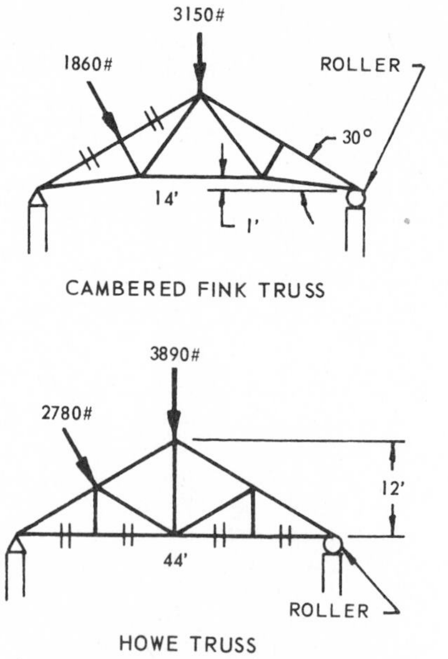
3150#
1860#
ROLLER
30°
14'
1'
CAMBERED FINK TRUSS
3890#
2780#
12'
44'
ROLLER
HOWE TRUSS

Fig. 29-4-D Nonparallel loads acting on truss—roller at one end.

UNIT 29-5

INTERNAL FORCES OR STRESSES IN A TRUSS

Graphic Method of Finding Internal Forces In a Roof Truss

The previous examples relating to roof trusses have been confined to finding the external forces acting on the roof and walls. Once the external forces have been calculated, the forces acting in the truss members can be determined graphically by two methods. Consider the truss shown in Fig. 29-5-1*a*. Since the forces are symmetrical, the reaction forces will be equal; namely 4000 kN each. Using Bow's notation and stating all the forces, draw a space diagram to scale, as in Fig. 29-5-1*b*. Consider joint *ABHG,* the left support. Two of the four forces are known. Draw the force diagram (Fig. 29-5-1*g*) to a convenient scale, starting with vectors *ga* and *ab*. The next force in order is *BH*. Draw a line parallel to *BH* through point *b*. Point *h*, which is one end of vector *bh*, has not yet been established. Draw a line parallel to the last force *HG* through point *g*. Since point *h* is on this line as well as on line *bh, h* must be their point of intersection. The arrows must travel in the same direction around the polygon, thus indicating the direction of the forces acting at the joint. We find that truss member *BH* is under compression and truss member *HG* is under tension. Next, consider joint *BCJH* in Fig. 29-5-1*d*. There are two known forces, *BC* of 2000 kN and *HB* which was found to be 6000 kN and under compression. The directions of forces *CJ* and *JH* are known, but their magnitude is not. The member *CJ* is under compression, but we do not know whether member *JH* is a tie or a strut. Draw the force diagram (Fig. 29-5-1*h*), to a convenient scale by first drawing the vectors *hb* and *bc*. Draw a line parallel to *CJ* through point *c*. Point *J*, which is one end of the vector, has not yet been established. Draw a line parallel to the last force *JH* through point *h*. Since point *j* is on this line as well as on line *cj, j* must be their point of intersection. The arrows must travel in the same direction around the polygon, thus indicating the direction of the force acting at the joint. We find that both truss members *CJ* and *JH* are under compression. Because of this, they are struts. Next consider joint *CDLKJ* (Fig. 29-5-1*e*). The magnitude of three of the five forces is known and the magnitude of the two unknown forces is equal since the loads are symmetrical about the center of the truss. Draw the force diagram (Fig. 29-5-1*j*) to a convenient scale by first drawing the vectors *jc, cd* and *dl*. The next force in order is *LK*. Draw a line parallel to *LK* through point *L*. Point *k*, which is one end of the vector, has not yet been established. Draw a line parallel to the last force *KL* through point *j*. Since point *k* is also on this line as well as on line *jk*, it must be at their point of intersection. The arrows must travel around the polygon in the same direction, thus indicating the direction of the forces acting at the joint. We find that truss members *LK* and *KJ* are under tension and are tie bars. Only one member remains to be calculated. It is not known whether the truss member *GK* is under compression or tension, nor is its magnitude known. From the space diagram shown in Fig. 29-5-1*f*, draw the force diagram in Fig. 29-5-1*k*. *GK* is found to have a magnitude of 3500 kN and is under tension. The second method for finding the forces in the truss members is much faster. A single diagram, called a *loading diagram*, that combines all the separate force polygons is used. The first step in constructing a loading diagram is to draw the force diagram of the external forces, as in Fig. 29-5-1*l*. Now consider a joint where only one or

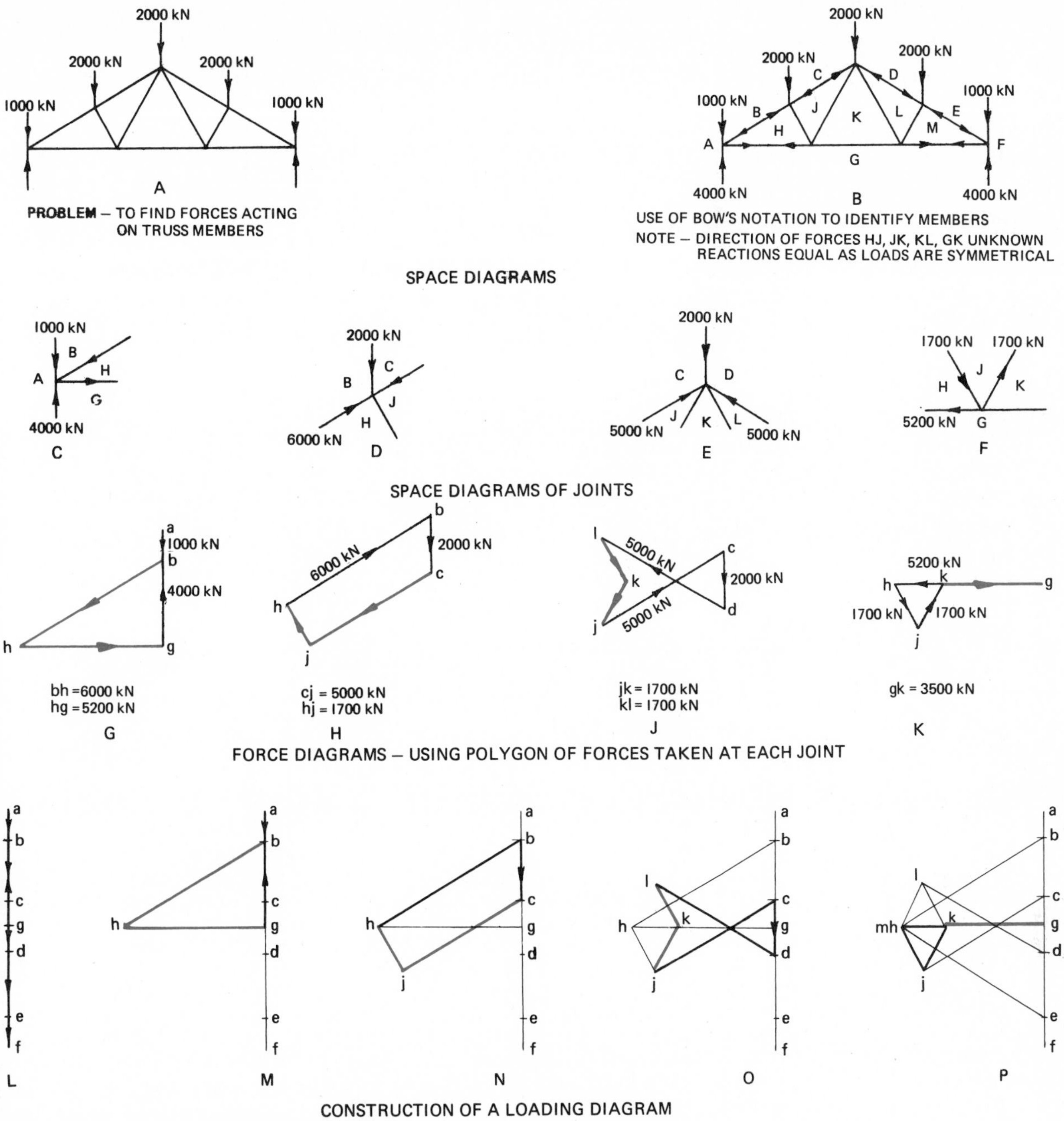
2000 kN
2000 kN
2000 kN
1000 kN
1000 kN
A
PROBLEM – TO FIND FORCES ACTING ON TRUSS MEMBERS
1000 kN
4000 kN
4000 kN
B
USE OF BOW'S NOTATION TO IDENTIFY MEMBERS
NOTE – DIRECTION OF FORCES HJ, JK, KL, GK UNKNOWN
REACTIONS EQUAL AS LOADS ARE SYMMETRICAL
SPACE DIAGRAMS
C
6000 kN
D
5000 kN
5000 kN
E
1700 kN
1700 kN
5200 kN
F
SPACE DIAGRAMS OF JOINTS
bh =6000 kN
hg =5200 kN
G
cj = 5000 kN
hj = 1700 kN
H
jk = 1700 kN
kl = 1700 kN
J
gk = 3500 kN
K
FORCE DIAGRAMS – USING POLYGON OF FORCES TAKEN AT EACH JOINT
L
M
N
O
P
CONSTRUCTION OF A LOADING DIAGRAM

Fig. 29-5-1 Graphic method of finding internal forces in a roof truss.

two forces are unknown, such as joint *ABHG*. The force diagram *ab, bh, hg, ga* can be drawn on the stress diagram in a similar manner to that previously explained. Next, a force diagram similar to Fig. 29-5-1*h* for joint *BCJH* is constructed on the loading diagram, adding vectors *jh* and *cj*. The same procedure is used for joints *CDLKJ* and *GHJK* until the loading diagram is complete. Only the directions of the external forces are shown on the loading diagram, since the direction of the other forces would alternate at the different joints.

Assignments

1. On an A3- or B-size sheet, lay out the fink truss shown in Fig. 29-5-A or Fig. 29-5-B, and by the graphic method find the magnitude of the reactions and the internal forces acting on each member. Scale is to suit.

2. On an A3- or B-size sheet, lay out the fan truss shown in Fig. 29-5-C or Fig. 29-5-D, and by the graphic method find the magnitude of the reactions and the internal forces acting on each member. Scale is to suit.

3. On an A3- or B-size sheet, lay out the crescent truss shown in Fig. 29-5-E or Fig. 29-5-F, and by the graphic method find the magnitude and direction of the reactions and the internal forces acting on each member. Scale is to suit.

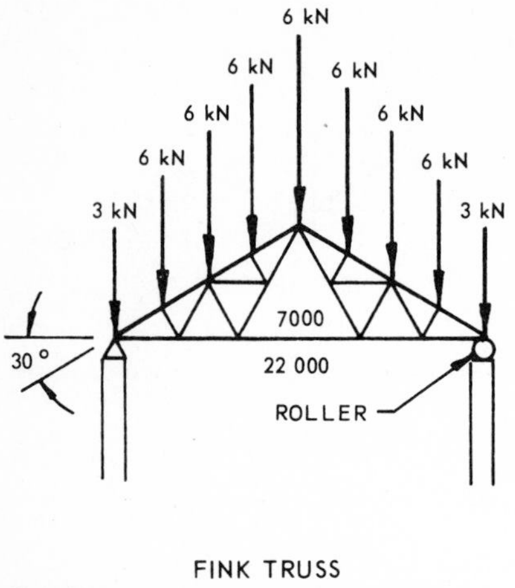
6 kN
6 kN
6 kN
6 kN
6 kN
6 kN
6 kN
3 kN
3 kN
7000
30°
22 000
ROLLER
FINK TRUSS

Fig. 29-5-A Internal roof forces—fink truss.

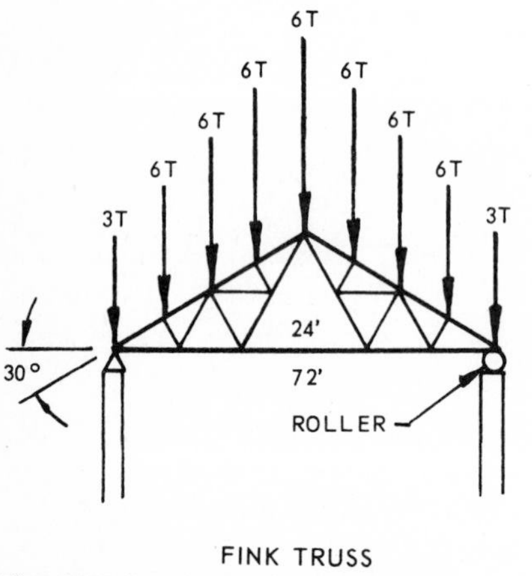
6T
6T
6T
6T
6T
6T
6T
3T
3T
24'
30°
72'
ROLLER
FINK TRUSS

Fig. 29-5-B Internal roof forces—fink truss.

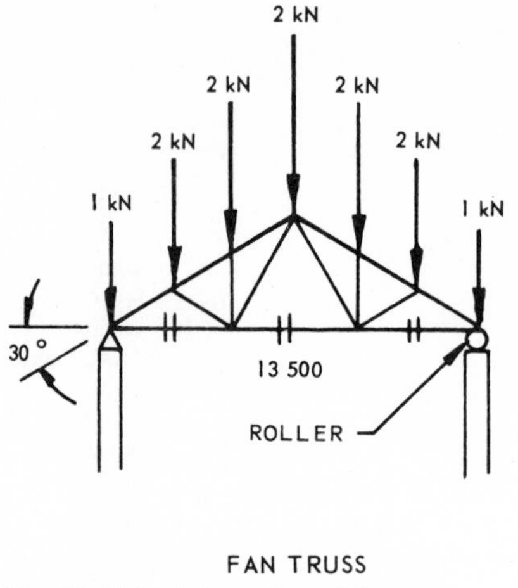
2 kN
2 kN
2 kN
2 kN
2 kN
1 kN
1 kN
30°
13 500
ROLLER
FAN TRUSS

Fig. 29-5-C Internal roof forces—fan truss.

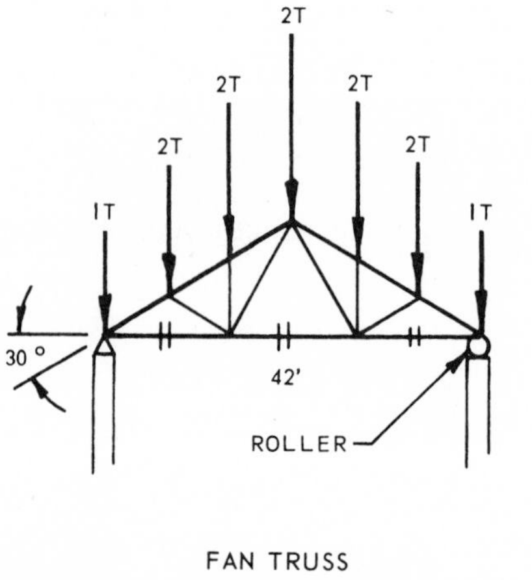
2T
2T
2T
2T
2T
1T
1T
30°
42'
ROLLER
FAN TRUSS

Fig. 29-5-D Internal roof forces—fan truss.

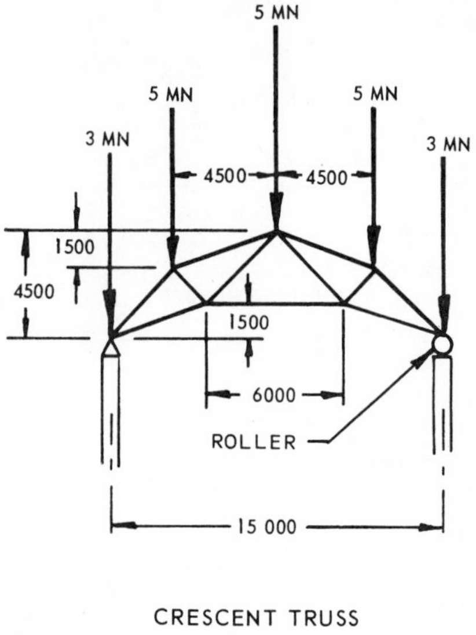
5 MN
5 MN
5 MN
3 MN
3 MN
4500
4500
1500
4500
1500
6000
ROLLER
15 000
CRESCENT TRUSS

Fig. 29-5-E Internal roof forces—crescent truss.

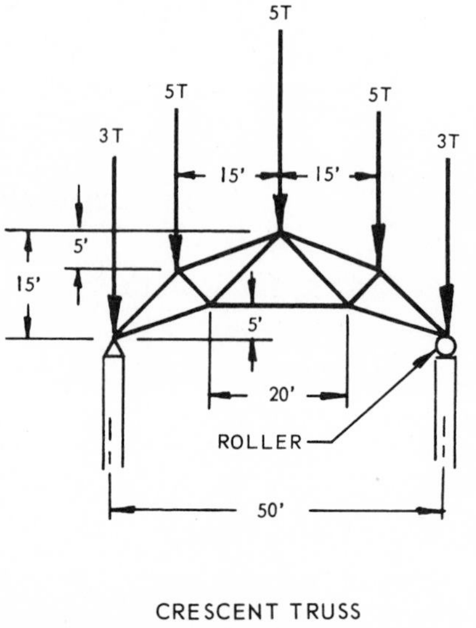
5T
5T
5T
3T
3T
15'
15'
5'
15'
5'
20'
ROLLER
50'
CRESCENT TRUSS

Fig. 29-5-F Internal roof forces—crescent truss.

Chapter 30 Strength of Materials

UNIT 30-1

STRESSES AND STRAIN

RELATIONSHIP BETWEEN MASS AND FORCE

Mass is the quantity of matter in a body. The mass of an object remains constant, regardless of its location on earth. Mass is measured in grams, kilograms, and metric tons, or ounces, pounds, and tons (inch-pound).

Some examples of the use of these units of measurement are in

- Defining quantities of material, packaged in bulk, such as bags of mortar, metric tons of sand
- Defining physical characteristics of material like 95 kg [210 lb] asphalt shingles and 510 g [18 oz], 680 g [24 oz], or 907 g [32 oz] glass
- Defining load capacities for building elements, elevator cranes, hoists, bridges, roads, supports, bearing surfaces
- Specifying application of materials such as 10 kg roofing asphalt per mopping per 100 m² (metric) or 20 lb roofing asphalt per mopping per 100 sq ft (inch-pound)
- Establishing costs for materials, unit prices, and rates on a gram, kilogram, or metric ton basis (metric) or ounce, pound, or ton (inch-pound).

Force is the external agent which changes or tends to change the condition of rest of a body. Force is measured in newtons (N) for light forces, kilonewtons (kN) for intermediate forces, and meganewtons (MN) for high forces (metric). Force units in the inch-pound system are ounce-force, pound-force, ton-force, or kip-force.

Forces related to the design and construction processes are numerous: bearing capacity, applied weight (mass under the influence of gravity) of live, dead, and mobile loads, connection loads, etc. Force may be concentrated on a tiny spot or applied over an immense area.

To convert kilograms to a force value, multiply the mass value (in kilograms) by 9.806 65 to obtain the force in newtons.

Stresses

When a force acts on a piece of material, internal resistance or forces are set up in the material to resist these external forces. These resisting forces are called *stresses* and are measured in pascals (metric) or pounds per square inch or square foot (inch-pound). A *pascal* (Pa) is a pressure or stress produced when a force of one newton (N) is applied to an area of one square metre (m²).

$$Pa = \frac{N}{m^2}$$

The pascal is a very small unit of measure. It is used for very low-stress applications. In most instances the kilopascal (kPa) and megapascal (MPa) are used.

In solving stress problems the following formulas can be used:

$$\text{Stress} = \frac{\text{force}}{\text{area}} \quad \text{or} \quad \text{area} = \frac{\text{force}}{\text{stress}}$$

$$\text{Force} = \text{stress} \times \text{area}$$

There are three kinds of stresses: *tension*, *compression*, and *shear*. Tension or tensile stress is caused by an external force that tends to pull apart or stretch the material. Tie bars supporting heating units or fans from ceiling members are examples of parts subject to tensile stress. See Fig. 30-1-1.

Compression or compressive stress is caused by external forces that tend to crush or push the material together. Basement posts and walls are parts subject to compressive stress.

Shear stress is caused by external forces that tend to cause the particles within the material to slide past one another. Rivets holding metal plates together are subject to shear stress.

These stresses often appear in combination. In a simple beam supporting a load, all three stresses occur. There is a tensile stress along the bottom of the beam, a compressive stress along the top of the beam, and a shear stress at each side at the abutments.

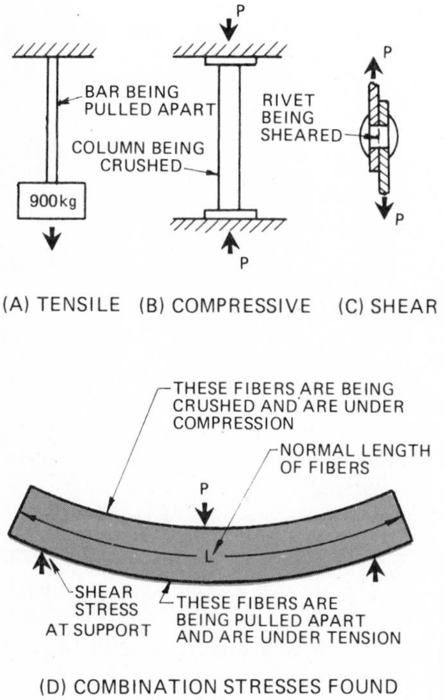
P
BAR BEING PULLED APART
RIVET BEING SHEARED
COLUMN BEING CRUSHED
900kg
P
P
P
(A) TENSILE
(B) COMPRESSIVE
(C) SHEAR
THESE FIBERS ARE BEING CRUSHED AND ARE UNDER COMPRESSION
NORMAL LENGTH OF FIBERS
P
L
SHEAR STRESS AT SUPPORT
THESE FIBERS ARE BEING PULLED APART AND ARE UNDER TENSION
(D) COMBINATION STRESSES FOUND IN A WOODEN BEAM

Fig. 30-1-1 Stresses.

The *ultimate strength* of a material is the highest unit of stress that the material can withstand without breaking.

Loads

The external forces acting on a body, called *loads* and measured in newtons, kilonewtons, and meganewtons (metric) or pounds, tons, and kips (inch-pound), are classified according to the manner in which they are applied.

A *static* load is one that is applied gradually to a part and that remains practically constant once the maximum load is reached. The mass of a building acting on its foundation is an example of a static load. Static loads are also referred to as *dead* loads.

An *impact* or *shock load* is one that is applied suddenly on an object for a short time. When a nail is struck by a hammer or a train passes over a portion of track, the loads resulting from these actions are known as impact loads.

Repeated loads are loads that are alternately applied and removed many times. An example of a part that is subjected to this type of load is a connecting rod in an automobile engine.

Only static and impact loads will be dealt with in this text.

Types of Stresses

Since there are three types of stresses, a material will have three different ultimate strengths. When machine parts are designed, it is not feasible to work precisely to the ultimate strength of the material since the addition of any shock or unforeseen load would cause breaking.

Therefore, when parts are designed, another stress known as the *safe working stress* (allowable unit stress) is used. This stress is obtained by dividing the ultimate strength of the material by a number called the *factor of safety*. Hence

$$\text{Safe working stress (S)} = \frac{\text{ultimate strength (SU)}}{\text{factor of safety (FS)}}$$

Other common terms and formulas are shown in Fig. 30-1-2. The number used for the factor of safety varies according to the material, the proposed location of the part, and the type of force that it must withstand. For example, a wooden part that is subjected to a shock force would have a greater factor of safety than a steel part that is subjected to a dead load. Factors of safety are not used as frequently as they were in the past since many of today's codes for structures and machines list the allowable unit or working stresses to be used. However, in certain applications, such as aircraft design, the ultimate strength and factors of safety are often used. Allowable unit stresses for steel will be covered in greater detail later in the chapter.

TERM	SYMBOL	FORMULA
Force or Load	F	$F = A \times S$
Area	A	$A = \frac{F}{S}$
Stress	S	$S = \frac{F}{A}$
Ultimate Strength	Su	$Su = S \times FS$
Factor of Safety	FS	$FS = \frac{Su}{S}$
Deformation (Unit Strain)	Du	$Du = \frac{Dt}{L} = \frac{S}{E}$
Deformation (Total Strain)	Dt	$Dt = Du \times L$
Coefficients of Linear Expansion	N	
Modulus of Elasticity	E	$E = \frac{S}{Du}$
Length of Part being Deformed	L	$L = \frac{Dt}{Du}$

Fig. 30-1-2 Common terms, symbols, and formulas.

Figure 30-1-3 shows the average values of the ultimate strengths of various materials.

Example 1. A 2 tonne (t) [2000 kg] mass is suspended from a 30 × 50 steel bar. What is the unit stress?

Solution. The force exerted by the 2 t mass is 2000 × 9.807, or 19 614 N. The unit stress will be the force divided by the area:

$$\text{Stress (MPa)} = \frac{\text{force (N)}}{\text{area (m}^2) \times 10^6}$$

$$S = \frac{F}{A}$$

$$= \frac{19\,614}{0.03 \times 0.05 \times 10^6}$$

$$= 13.076 \text{ MPa}$$

Example 2. What tensile force would cause a ϕ50 mm A572M-350 steel rod to fail?

Solution. The cross-sectional area of the rod is equal to $\pi R^2 = 3.1416 \times 25 \times 25 = 1963.5$ mm², or 0.001 964 m². From the table shown in Fig. 30-1-3, A572M-350 steel has an ultimate strength of 450 MPa. Therefore, the tensile force that would cause the rod to fail is

Material		Ultimate Strength			Allowable Unit Stress			Modulus of Elasticity Tension
		Tension	Com-pression	Shear	Tension	Com-pression	Shear	
Aluminum	MPa	103	83	83				78 500
	10^3psi	**15**	**12**	**12**				**11,000**
Brass	MPa	145	207	248				62 000
	10^3psi	**21**	**30**	**36**				**9,000**
Copper	MPa	235	220	248				103 500
	10^3psi	**34**	**32**	**36**				**15,000**
Cast Iron —Gray	MPa	145	620	165		103	28	96 500
	10^3psi	**21**	**90**	**24**		**15**	**4**	**14,000**
Cast Iron —Malleable	MPa	214	317	275	35	52	45	172 000
	10^3psi	**31**	**46**	**40**	**5.2**	**7.5**	**6.6**	**25,000**
Cast Iron —Wrought	MPa	330	330	275	83	83	69	193 000
	10^3psi	**48**	**48**	**40**	**12**	**12**	**10**	**28,000**
Steel— A572M-350	MPa	450	450	345	210	210	140	200 000
G40.2M-350W (See Fig. 30-1-8)	10^3psi	**65**	**65**	**50**	**30**	**30**	**20**	**29,000**

Bold figures denote inch-pound values.

Fig. 30-1-3 Physical properties of common materials.

Area × ultimate strength = 0.001 964 × 450 × 10^6 = 884 kN

Example 3. What is the maximum permissible tensile load an A572M-350 structural steel column having a cross-sectional area of 300 mm², can carry if the factor of safety is given as 5?

Solution. From the table shown in Fig. 30-1-3, A572M-350 structural steel has an ultimate tensile strength of 450 MPa.

Working stress =

$$\frac{\text{ultimate strength}}{\text{factor of safety}} = \frac{450}{5} = 90 \text{ MPa}$$

Therefore, maximum allowable load = $S \times A$ = working stress × area = (90×10^6) Pa × (900×10^{-6}) m² = 81 kN.

Example 4. A 250 × 250 × 2400 Western hemlock construction-grade post supports a mass of 35 Mg. Does this meet the minimum requirements as recommended by the Institute of Timber Construction (ITC)?

Solution. From the table shown in Fig. 30-1-4 under the headings Carrying Load Independently, Compression, and Parallel to Grain, we find the allowable unit stress for Western hemlock, construction grade, is 8 MPa. Therefore

Maximum allowable load =

$$A \times S = \left(\frac{250 \times 250}{10^6}\right) \text{m}^2 \times (8 \times 10^6) \text{ Pa}$$
$$= 500 \text{ kN}$$

A 35 Mg mass exerts a force of 35 000 kg × 9.8 = 343 kN. Therefore, the load is acceptable.

Next check for buckling, using the formula $1/d \leq 10$, where L = length in millimetres and d = least dimension of compression member in millimetres. $1/d$ = 2400/250 = 9.6. Therefore, the post carrying this load would meet ITC requirements.

Example 5. What force is required to punch a 40 mm diameter hole in a no. 12 GS gage sheet?

Solution. The sheared area will be equal to the circumference of the circle multiplied by the thickness of the sheet. Circumference of a 40 mm diameter = 125.7 mm. Thickness of a no. 12 GS plate (see Appendix) = 2.75 mm. Sheared area = 125.7 × 2.75 = 345.7 mm². Ultimate shear strength of steel (see Fig. 30-1-3) is 345 MPa. Therefore the force required to punch hole = $A \times S$ = (345.7×10^{-6}) m² × 345 MPa = 119 kN.

Group	Grade of Lumber		Carrying Load Independently Working Stress				
			Bending		Compression		
			Stress at Extreme Fiber	Longi-tudinal Shear	Parallel to Grain 1/d ≤10	Perpen-dicular to Grain	Tension Parallel to Grain
Douglas Fir	Construction	MPa	10	0.8	8	2.9	10
		psi	**1500**	**120**	**1200**	**415**	**1500**
	Standard	MPa	8	0.7	7	2.7	8
		psi	**1200**	**95**	**1000**	**390**	**1200**
Western Hemlock	Construction	MPa	10	0.7	8	2.5	10
		psi	**1500**	**100**	**1100**	**365**	**1500**
	Standard	MPa	8	0.6	7	2.5	8
		psi	**1200**	**80**	**1000**	**365**	**1200**
Spruce (All)	Structural	MPa	7	0.6	6	2.1	9
		psi	**1050**	**90**	**750**	**300**	**1050**
	Construction	MPa	6	0.5	4	2.1	6
		psi	**840**	**70**	**600**	**300**	**840**
Red Cedar Pine	Structural	MPa	6	0.6	5	1.8	8
		psi	**900**	**80**	**750**	**260**	**900**
	Construction	MPa	5	0.5	4	1.8	5
		psi	**720**	**65**	**600**	**260**	**720**

Note: Values shown are for teaching purposes only. Consult your local building codes for exact values.
Bold figures denote inch-pound values.

Fig. 30-1-4 Allowable unit stresses for sawn timber members.

Deformation

When an object is subjected to a load or force, the shape of the material is changed slightly. This change in length is called *strain*, or *deformation*. See Fig. 30-1-5. The length of an object is shortened by a compressive force or lengthened by a tensile force. The change in size is called *total elongation* and is normally measured in millimetres (metric) or inches (inch-pound) while the change in length per millimetre or inch is called *unit elongation* and is normally measured in millimetres per millimetre (metric) or inches per inch (inch-pound). Normally the deformation is so small that it cannot be detected by the naked eye. The deformation or sag that occurs on a beam when a load is applied is called *deflection*.

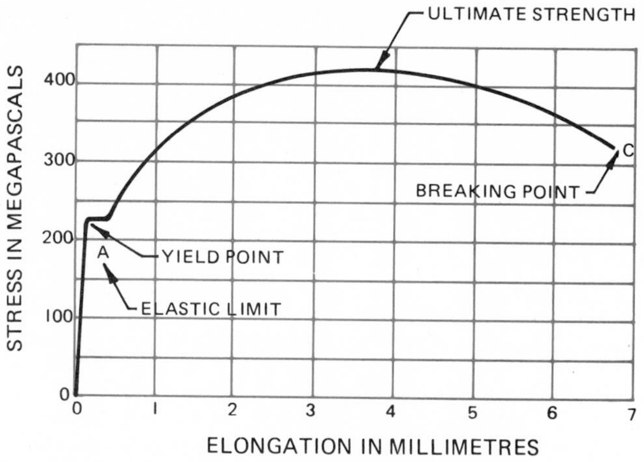
ULTIMATE STRENGTH
BREAKING POINT
C
A
YIELD POINT
ELASTIC LIMIT
STRESS IN MEGAPASCALS
400
300
200
100
0
0
1
2
3
4
5
6
7
ELONGATION IN MILLIMETRES

Fig. 30-1-6 Stress-strain diagram for A572M-290 carbon steel.

STRESS-STRAIN DIAGRAM

The relationship between stress and strain for any material is best shown by a diagram; see Fig. 30-1-6. A piece of carbon steel 25 mm^2 having an area of 0.000 625 m^2 was subjected to tensile loads at 20 kN intervals, and the results were recorded. Up to point *A* on the graph, the elongation of the bar was proportional to the stress. Point *A*, which was recorded at 200 MPa, was the elastic limit for that steel. After point *A*, the elongation increased at a faster rate. At a stress slightly higher than the elastic limit, deformation occurred without an increase in stress. This is known as the *yield point* of the material. As the tension increased, the bar elongated until point *B* was reached. This was the largest load applied, which was recorded at 267 kN. Beyond this point the bar continued to stretch or elongate with less tension. Point *B* was the *ultimate strength* of the material, which was recorded at 410 MPa. The breaking point of the bar was point *C*, which was recorded at 330 MPa.

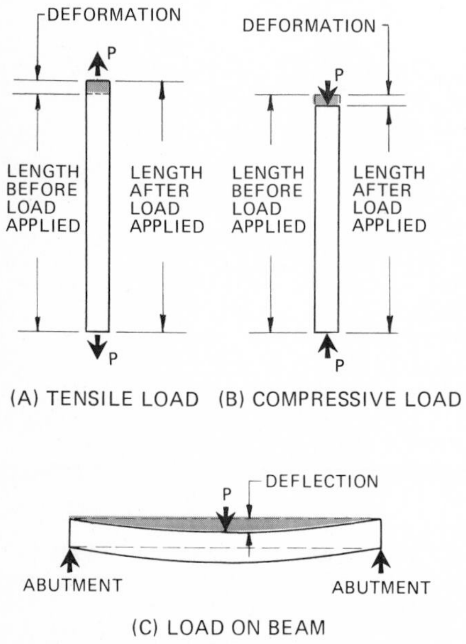
DEFORMATION
DEFORMATION
P
P
LENGTH BEFORE LOAD APPLIED
LENGTH AFTER LOAD APPLIED
LENGTH BEFORE LOAD APPLIED
LENGTH AFTER LOAD APPLIED
P
P
(A) TENSILE LOAD
(B) COMPRESSIVE LOAD
DEFLECTION
P
ABUTMENT
ABUTMENT
(C) LOAD ON BEAM

Fig. 30-1-5 Deformation due to loads.

The strength of any material may be plotted and calculated in a similar manner, although not all materials act in the same way. An example of this would be a cast-iron part. Since the ultimate strength and the breaking point would be the same, the part would break at the maximum load.

If a force acting on an object is not great, the material will return to its original shape when the force is removed. This tendency to return to the original shape after being deformed is called *elasticity* and varies greatly in different materials. For example, lead is said to have little or no elasticity, while spring steel has a great amount. If, however, the material does not return to its original shape after it has been subjected to a force, it is said to be stressed beyond its elastic limit. Up to this elastic limit the deformation is proportional to the load; that is, the unit stress is proportional to the unit strain at any point in a material up to its elastic limit. This is known as *Hooke's law*. Beyond the elastic limit, the deformation ceases to be proportional to the load. The elastic limit of a material is difficult to determine accurately.

The *modulus of elasticity* of a material is defined as the ratio of unit stress to unit deformation (the stress in megapascals divided by the deformation in 1 mm) and is denoted by the letter *E*. It may be used for finding the elongation per millimetre (metric) caused by any given load.

In the inch-pound system, the modulus of elasticity is the stress per square inch divided by the deformation in one inch.

Example 6. A steel bar 3000 mm long elongates 1.91 mm under a tensile force. Calculate the unit deformation.

Solution.

Unit elongation (Du)

$$= \frac{\text{total strain or deformation } (Dt)}{\text{length of part } (L)}$$

$$= \frac{1.91}{3000}$$

$$= 0.000\,636\,6 \text{ mm/mm}$$

Example 7. Find the unit deformation on a piece of steel produced by a stress of 300 MPa.

Solution. From Fig. 30-1-3 we find the modulus of elasticity for steel is 200,000. Therefore,

$$Du = \frac{\mathbf{S}}{\mathbf{E}} = \frac{300}{200\,000} = 0.001\,5 \text{ mm/mm}$$

Example 8. A 5 × 25 mm steel bar 4500 mm in length supports a tensile load of 20 kN. Find the total deformation.

Solution.

$$\mathbf{S} = \frac{\mathbf{P}}{A} = \frac{20\,000}{0.000\,125} \times 10^{-6} = 160 \text{ MPa}$$

The modulus of elasticity for steel (see Fig. 30-1-3) = 200,000. $Du = \mathbf{S}/\mathbf{E}$ = 160/200,000 = 0.0008 mm/mm. Therefore,

$$Dt = Du \times L = 0.0008 \times 4500 = 3.6 \text{ mm}$$

Temperature Stresses. When the temperature of a piece of metal is changed, the length of the metal will be either shortened or lengthened, depending on whether the temperature of the metal is lowered or raised. If, however, the part is rigidly held and is restrained from changing its length, stresses known as *temperature stresses* will result. The main factors concerning temperature stress are (1) amount of heat involved, (2) material undergoing temperature change (aluminum, iron, etc.), and (3) length of part. In order to avoid these stresses, trusses or girders of long spans frequently have one end placed on a roller or a sliding plate.

The linear change per millimetre of length of a part for a degree of change in temperature is called the *coefficient of linear expansion or contraction*. The coefficients of common materials are shown in Fig. 30-1-7. Thus, the total deformation resulting from temperature change can be found as follows. Let total strain or deformation be Dt, the coefficient of linear expansion Ce, temperature change (°C) T, and length of part (in.) L. Then $Dt = Ce \times T \times L$. In the inch-pound system, degrees Fahrenheit is used.

Example 9. A medium steel bar 2500 mm long is raised from 20 to 75°C. How much does it expand?

Solution. The coefficient of linear expansion for medium steel (see Fig. 30-1-7) is 0.000 012 1. Therefore,

$$\text{Total deformation } (Dt) = Ce \times T \times L = 0.000\,012\,1 \times 55 \times 2500 = 1.66 \text{ mm}$$

Example 10. If the steel bar in Example 9 is restrained and is 50 mm², what compressive stress is placed on the bar and what load is placed on the restraining members?

Solution.

$$\text{Total strain} = 1.66 \text{ mm}$$

$$\text{Unit strain} = \frac{1.66}{2500} = 0.000\,664 \text{ mm/mm}$$

$$Du \text{ (unit strain)} = \frac{S}{E}$$

where S = stress

E = modulus of elasticity (see Fig. 30-1-3). Therefore

$$S = Du \times E = 0.000\,664 \times 200\,000 = 132.8 \text{ MPa. Therefore}$$

$$\text{Load} = \text{stress} \times \text{area} = 132.8 \times 10^6 \times 132.8 \times 10^6 \times (0.05 \times 0.05) = 332 \text{ kN}$$

Material	Ce Coefficient of Linear Expansion Millimetres Per Millimetre per 1°C Inches per Inch per 1°F
Aluminum	0.000 023 0 .0000 128
Brass	0.000 018 7 .0000 104
Bronze	0.000 018 2 .0000 101
Copper	0.000 016 7 .0000 093
Iron—Cast	0.000 011 2 .0000 062
Iron—Wrought	0.000 012 2 .0000 068
Steel—Hard	0.000 013 3 .0000 074
Steel—Medium	0.000 012 1 .0000 067
Steel—Soft	0.000 011 0 .0000 061

Bottom figures denote inch-pound values

Fig. 30-1-7 Coefficients of expansion.

Unit Stresses for Steel

The increasing use of high-strength steels no longer permits the continuation of a standard design specification based on the exclusive use of one grade of steel. These high-strength steels afford as much as a 50 percent increase in strength as compared to common structural carbon steel.

To simplify matters, permissible unit stresses for the various grades of steel are given in terms of a percentage of specified minimum yield point. These unit stresses are not to exceed 61 percent of the yield point. For steel having a yield point of 230 MPa, the permissible unit stress would be 140 MPa, which provides for a factor of safety of 1.64. Figure 30-1-8 lists the various grades of steels and their allowable unit stresses. In keeping with the inclusion of steels of several strength grades, a number of corresponding specifications for cast-steel forgings and other materials such as rivets, welding electrodes, and high-strength bolts have been introduced.

Assignments

1. On an A3- or B-size sheet, show the calculations and answers for the problems shown in Assignment 30-1-A or 30-1-B.

ASSIGNMENT 30-1-A

1. A load of 200 kN is suspended from a steel rod. What is the area of the bar if the allowable unit stress is 80 MPa? What is the minimum stock size that can be used (*a*) of a round rod, (*b*) of a square bar?

	METRIC							INCH-POUND						
				Allowable Unit Stress—MPa							Allowable Unit Stress—KSI			
Description	Steel Standard	Minimum Tensile Strength—MPa	Yield Point—MPa	Tension	Compression	Bending	Shear	Steel Standard	Minimum Tensile Strength—KSI	Yield Point—KSI	Tension	Compression	Bending	Shear
General Construction USA	A36	400	250	150	150	165	100	A36	58	36	22	22	24	14.5
	A572M -410	520	410	245	245	270	165	A572 -60	75	60	36	36	40	24
	-380	480	380	230	230	250	150	-55	70	55	33	33	36	22
	-350	450	350	210	210	230	140	-50	65	50	30	30	33	20
	-310	410	310	185	185	205	125	-45	60	45	27	27	30	18
	-290	410	290	175	175	190	115	-42	60	42	25	25	28	16.8
General Construction Canada	G40.21M -410W	520	410	245	245	270	165	G40.21W -70W	75	60	36	36	40	24
	-380W	480	380	230	230	250	150	-60W	70	55	33	33	36	22
	-350W	450	350	210	210	230	140	-50W	65	50	30	30	33	20
	See -300W	450	300	185	185	200	120	-44W	65	44	26	26	29	17.5
	Note 1 -290W	410	290	175	175	190	115	-38W	60	42	25	25	30	16.8
	-230W	380	230	140	140	150	90	-33W	55	33	20	20	20	13.2

Note 1. Metric designations and values were not available at time of printing and as such are soft converted.
2. Values shown are for steel having a maximum thickness of 50 mm (2.00 in.).

Fig. 30-1-8 Allowable working stress for steel.

2. What load may safely be placed on a 200 × 200 mm wooden post if the allowable unit stress is 6 MPa?

3. What is the unit stress on a 100 × 100 × 12 angle iron if the angle is subjected to a 270 kN tensile load? Rounds and fillets need not be considered for calculation purposes.

4. What must be the diameter of an A36 steel rod that safely lifts a load of 70 kN? The factor of safety is 5.

5. What is the maximum tensile load that can be carried by an A36 steel strut having a cross-sectional area of 1200 mm²? The factor of safety is 3. Refer to Fig. 30-1-8.

6. A load of 180 kN is hung from an A36 100 × 10 steel bar. What is the factor of safety?

7. A machine of 12 t rests on four structural-steel legs. If the factor of safety is 5, what is the cross-sectional area of each leg if the load is equally proportioned on each leg? Use A572M-350 steel.

8. What is the maximum load that a 6-in. standard wrought-iron pipe can support using the working stress shown in Fig. 30-1-3?

ASSIGNMENT 30-1-B

1. A load of 40 000 lb is suspended from a steel rod. What is the area of the bar if the allowable unit stress is 12 000 pounds per square inch (psi)? What is the minimum stock size that can be used (*a*) of a round rod, (*b*) of a square bar?

2. What may safely be placed on an 8 × 8 wooden post if the allowable unit stress is 880 psi?

3. What is the unit stress on a 4 × 4 × .50 angle iron if the angle is subjected to a 60 000 lb load? Rounds and fillets need not be considered for calculation purposes.

4. What must be the diameter of an A36 steel rod that safely lifts a load of 15 000 lb? The factor of safety is 5.

5. What is the maximum tensile load that can be carried by an A36 steel strut having a cross-sectional area of 1.75 sq. in.? The factor of safety is 3. Refer to Fig. 30-1-8.

6. A load of 40 000 lb is hung from an A36 steel bar 4 × 0.375 in. What is the factor of safety?

7. A machine of 12 t rests on four structural-steel legs. If the factor of safety is 5, what is the cross-sectional area of each leg if the load is equally proportioned on each leg? Use A572M-350 steel.

8. What is the maximum load a 6-in. standard wrought-iron pipe can support using the working stress shown in Fig. 30-1-3?

2. On an A3- or B-size sheet, show the calculations and answers for the problems shown in Assignment 30-1-C or 30-1-D.

ASSIGNMENT 30-1-C

1. A steel bar 5000 mm long elongates 3.15 under a tensile force. Calculate the unit deformation.

2. What deformation will occur in a 25 mm diameter steel tie rod, 2400 mm long, when a tensile load of 70 kN is applied?

3. What load will elongate a 10 × 25 × 3000 mm long steel tie rod 3 mm?

4. What unit deformation is produced by a tensile load of 150 kN in a 25-mm² steel bar?

5. A 25 × 25 × 3 angle iron 3600 mm long supports a tensile load of 18 kN. Find the total deformation. Ignore the rounds and fillets for calculation purposes.

6. Calculate the allowable load that may be placed on a short 150 × 150 mm post. The material is Eastern spruce, construction grade. Refer to Fig. 30-1-4.

7. A flat, mild-steel bar of rectangular section 75 × 20 mm supports a mass of 12 t. Find the tensile stress and the elongation if the bar length is 6000 mm.

8. Calculate the load required to punch a 50 mm diameter hole in a piece of mild-steel plate 6 mm thick, assuming that the ultimate shear strength of the steel is 415 MPa.

ASSIGNMENT 30-1-D

1. A steel bar 16 ft long elongates .124 in. under a tensile force. Calculate the unit deformation.

2. What deformation will occur in a 1.00 in. diameter steel tie rod, 8 ft long, when a tensile load of 8 tons is applied?

3. What load will elongate a .375 × 1 in. × 10 ft long steel tie rod .125 in.?

4. What unit deformation is produced by a compressive stress of 34,000 lb in a 1.80-sq. in. steel bar?

5. A 1 × 1 × .125 in. angle iron 12 ft long supports a tensile load of 4000 lb. Find the total deformation. Ignore the rounds and fillets for calculation purposes.

6. Calculate the allowable load that may be placed on a short 6 × 6 in. post. The material is Eastern spruce, building grade.

7. A flat, mild-steel bar of rectangular section 3 × .75 in. carries an axial pull of 12 tons. Find the tensile stress and the elongation in inches if the unloaded length is 20 ft.

8. Calculate the load required to punch a 2 in. diameter hole in a piece of mild-steel plate .25 in. thick, assuming that the ultimate shear strength of the steel is 60 000 psi.

UNIT 30-2

BOLTED AND RIVETED JOINTS

It is assumed that the reader has a full understanding of the many advantages of bolted and riveted construction and possesses a knowledge of this type of working drawing and terminology.

The factors of safety for fasteners used in tension are preferably based upon ultimate strength rather than yield point since ultimate strength is of much greater significance for fasteners. The permissible working stresses as shown in Fig. 30-2-1 represent working loads which are approximately one-third to one-half of the value of the ultimate loads observed in tests.

For greater convenience in the proportioning of the bolted connections, permissible stresses for bolts are now given in terms applicable to their normal body area, i.e., the area of the unthreaded shank.

The tension stress permitted for A307 bolts and threaded parts of A36 and G40-21M-230W steel is equivalent to 140 MPa (metric) or 20 000 psi (foot pounds per square inch) applied at the root area of the threads. A similar basis is reflected in the provisions for tension on threaded parts made from other steels.

In recognition of the protection against defects in the threading, the Research Council on Riveted and Bolted Structural Joints has recommended a relatively higher working stress in tension for high-strength bolts.

Permissible stresses for rivets are given in terms applicable to the nominal cross-sectional area of the rivet before driving. See Fig. 30-2-1.

The most common methods of bolting or riveting plates together are by lapping or butting the plates, as shown in Fig. 30-2-2. There are many areas where a failure may occur in this type of connection (Fig. 30-2-3). In the lap joint the rivet may shear between the two plates. Since the rivet would shear in only one plate, it is said to be in *single shear*. The area that would shear would be the cross-sectional area of the rivet.

There is a possibility that the plate may fail by tearing away at its weakest point, the section through the rivet or bolt hole. This would be a tension failure. The area that would fail would be the area of the plate at the center of the holes less the area of the hole.

A third type of failure would be for the rivet or bolt to rip through or crush the plate directly beneath it. This is called a *bearing failure*, and the area that would fail in the plate would be equal to the diameter of the fastener times the plate

TENSION

	140 MPa Rivet Size in Millimetres					20 KSI Rivet Size In Inches				
Rivet Dia. mm (in.)	12	16	20		25	.50	.625	.75	.875	1.00
Area mm² (sq. in.)	113	201	314	380	491	.196	.307	.442	.601	.785
Load kN (kips)	15.82	28.14	43.96	53.2	68.74	3.93	6.14	8.84	12.03	15.71

SHEAR
Check Below to Ensure that the Allowable Load is Not Governed by Bearing

	100 MPa Rivet Size in Millimetres					15 KSI Rivet Size in Inches				
Rivet Dia. mm (in.)	12	16	20	22	25	.50	.625	.75	.875	1.00
Single Shear kN (kips)	11.3	20.1	31.4	38	49.1	2.94	4.60	6.63	9.02	11.78
Double Shear kN (kips)	22.6	40.2	62.8	76	98.2	5.89	9.20	13.25	18.04	23.56

BEARING
Single and Multiple Shear
Check to Ensure that the Allowable Load is Not Governed by Shear

Thickness of Material		310 MPa Rivet Size in Millimetres					45 KSI Rivet Size in Inches				
mm	**in.**	**12**	**16**	**20**	**22**	**25**	**.50**	**.625**	**.75**	**.875**	**1.00**
5	.188	18.6	24.8	31	34.1	38.8	4.22	5.27	6.33	7.38	
6	.250	22.3	29.8	37.2	40.9	46.5	5.62	7.03	8.44	9.84	11.25
8	.312	29.8	39.7	49.6	54.6	62	7.03	8.79	10.55	12.31	14.06
10	.375		49.6	62	68.2	77.5		10.55	12.66	14.77	16.88
12	.500				81.4	93				19.69	22.50
1	1.000	3.72	4.96	6.2	6.82	7.75	22.50	28.13	33.75	39.38	45.00

For material thickness other than those shown, the bearing value is the value for 1 mm (metric) or 1.00 in. (inch-pound) multiplied by the actual thickness. Use values shown in bottom line.

Fig. 30-2-1 Allowable load in kilonewtons (metric) and KSI (inch-pounds) for structural steel.

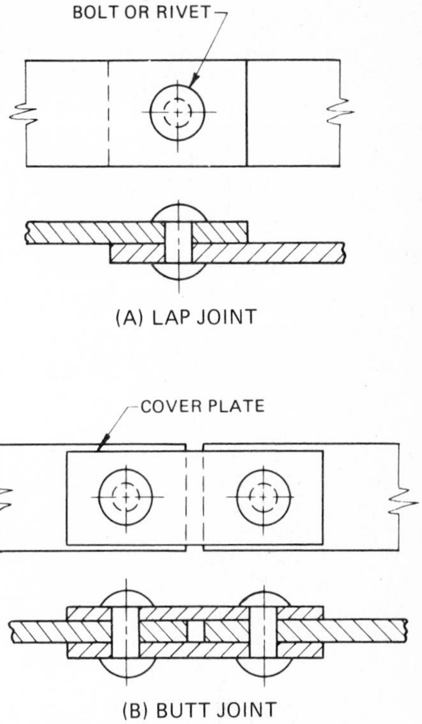
BOLT OR RIVET
(A) LAP JOINT
COVER PLATE
(B) BUTT JOINT

Fig. 30-2-2 Plate connections.

In calculating for bearing failure, a greater allowable working stress is permissible for rivets and high-strength bolts over ordinary bolts.

thickness. If more than one bolt or rivet is used, the load would be divided equally on the fasteners. Since only two pieces of metal are joined, they are said to be in *single bearing*.

In the butt joint shown in Fig. 30-2-3*b*, the rivet would have to be sliced in two sections if the joint were to fail by shear. The rivet is said to be in *double shear*, and twice the area of the rivet is used in the shear calculations.

If the joint fails by tension, that is, pulls or tears away, it will do so at its weakest point—the section through the hole. Since the two outside plates are pulling in one direction and the center plate in the other, the smaller of the two areas must be used in calculating the tensile strength of the joint.

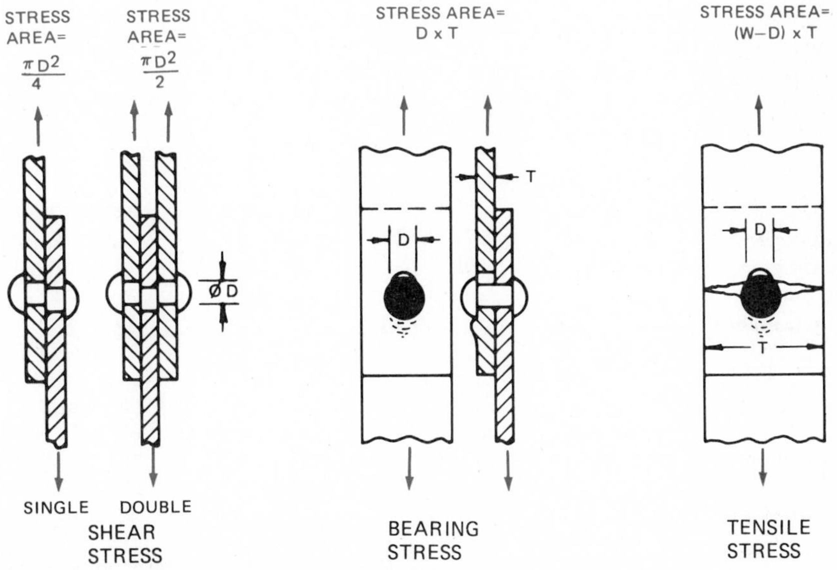
STRESS AREA= $\frac{\pi D^2}{4}$
STRESS AREA= $\frac{\pi D^2}{2}$
STRESS AREA= D x T
STRESS AREA= (W–D) x T
ØD
T
D
SINGLE
DOUBLE
SHEAR STRESS
BEARING STRESS
TENSILE STRESS

Fig. 30-2-3 Stress areas in lap and butt joints.

RIVET HOLES

In calculating the stresses in riveted and bolted joints, a distinction must be made between structural joints and joints in boilers, pipes, and tanks. In structural work, the steel members are generally punched and drilled 1.5 mm [.06 in.] larger than the rivet in the shop and then taken to the site for assembly. In calculating the tensile stress in the joint, the size of the hole is taken as 3 mm [.12 in.] greater than the nominal diameter of the rivet. This is to allow for any unseen damage that may occur around the hole when it is punched and assembled. Areas for shear and bearing are based on the nominal rivet diameter.

In the construction of boilers, etc., where leakage may be a problem, it is essential that the rivet holes line up. These holes are often reamed at assembly. As the finished rivet fills the hole (which is larger than the rivet) completely, the diameter of the hole is used for computing all the stresses.

SPACING OF RIVETS OR BOLTS[1,2]

The minimum distance between the centers of fastener holes is 3 times the diameter of the fastener, but when possible, the distance shall be not less than shown in Fig. 30-2-4.

The maximum pitch of rivets or bolts in line with the stress of compression members composed of plates and shapes does not exceed 16 times the thickness of the thinnest outside plate or shape or 20 times the thickness of the thinnest enclosed plate or shape, with a maximum of 300 mm. When two or more gage center lines are used with rivets and bolts staggered, the maximum pitch of rivets or bolts in the line of stress in each gage line shall not exceed 24 times the thickness of the thinnest plate or shape, with a maximum of 450 mm.

The distance between lines of rivets or bolts measured at right angles to the line of stress shall not exceed 32 times the thickness of the thinnest plate or shape. The minimum distance from the center of any punched hole to any edge shall be that given in Fig. 30-2-4.

Example 1. Lap joint. Two steel plates, 15 × 20 mm are lapped and joined by a 20 mm rivet. What is the allowable tensile load that could be applied to the joint? The holes for rivets are to be punched. Plate material is A36 steel.

Solution. There are three areas that must be checked:

1. The bars failing under a tensile load at the holes
2. The rivet shearing
3. The bearing on the bars directly below the rivet.

As the holes are punched, the diameter of the hole will be taken as 3 mm larger than the rivet diameter for calculation purposes.

1. *Bars failing under a tensile load.* Area of plate at center line of hole = 12 × (50 − 23) = 324 mm². Allowable unit stress = 150 MPa (see Fig. 30-1-8). Therefore

$$\text{Allowable load} = \mathbf{S} \times A = 150 \times 0.000\ 324 = 48.6 \text{ kN}$$

2. *Rivet shearing* (see Fig. 30-2-1). Single shear for 20 mm rivet = 31.4 kN

3. *Bearing on plate below rivet.* 20 mm rivet is bearing on 12 mm thick steel. Allowable load = 74.4 kN (see note at bottom of bearing table, Fig. 30-2-1). The weakest area would be the shear on the rivet. Therefore, allowable tensile load that joint could support = 314 kN.

Example 2. Single-riveted butt joint (Fig. 30-2-5). A boiler has a single riveted butt joint. The boiler plate is 12 mm, and the two cover plates are 8 mm thick. The rivets are 20 mm and are spaced 75 mm apart. Calculate the main stresses that could safely be applied to this joint. Plate material is A36 steel.

Solution. Since the pitch of the rivets is 75 mm, it is assumed that the width of the section taken for calculation purposes is 75 mm. As in Example 1, there are three areas to be checked:

1. The section failing under a tensile load
2. The rivet shearing
3. The bearing on the steel plate below the rivet.

As previously mentioned, for boiler plate construction, the finished rivet is assumed to be the same size as the drilled hole, namely, 21.5 mm.

1. *Plate failing under a tensile load.* Since the area of the two outside plates is greater than the area of the middle plate, the middle plate will fail first. Diameter of rivet hole = 21.5 mm. Area of middle plate = (75 − 21.5) × 12 = 642 mm². Allowable unit stress = 150 MPa (see Fig. 30-1-8). Therefore

$$\text{Allowable load} = \mathbf{S} \times A = 150 \times 10^6 \times 0.000\ 642 = 96.3 \text{ kN}$$

Fastener Diameter		In Sheared Edge		In Rolled Edge of Plates		In Rolled Edge of Structural Shapes		Minimum Spacing of Rivets or Bolts	
mm	inches	mm	inches	mm	inches	mm	inches	mm	inches
12	.50	25	1.00	23	.90	20	.75	50	2.00
16	.625	28	1.10	25	1.00	23	.90	58	2.25
20	.75	32	1.25	28	1.10	25	1.00	65	2.50
22	.875	38	1.50	32	1.25	28	1.10	75	3.00
25	1.00	45	1.75	38	1.50	32	1.25	90	3.50
30	1.125	50	2.00	45	1.75	38	1.50	100	4.00
32	1.25	60	2.25	50	2.00	45	1.75	115	4.50

Fig. 30-2-4 Minimum edge distances and spacings for rivets and bolts.

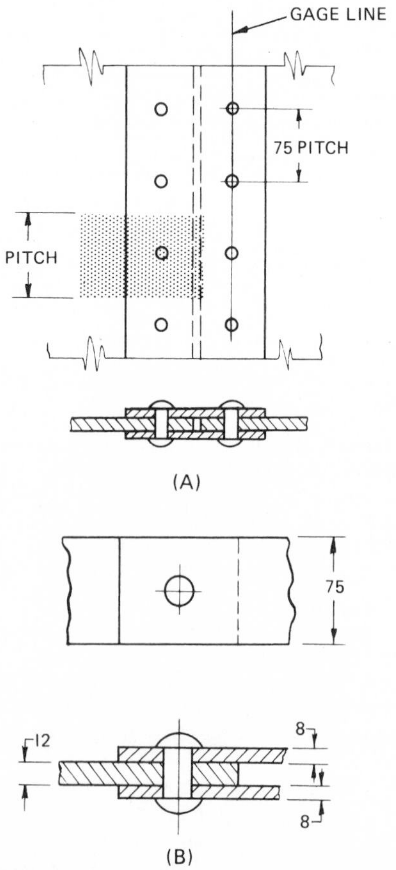
GAGE LINE
75 PITCH
PITCH
(A)
75
12
8
8
(B)

Fig. 30-2-5 Single-riveted butt joint on boiler.

2. *Rivet shear.* Since the rivet would have to shear in two places, it is considered to be in double shear.

Shear area =

$$\frac{\pi \times 21.5^2}{4} \times 2 = 726.1 \text{ mm}^2$$

Allowable shear stress = 100 MPa
Therefore

$$\text{Allowable load} = \mathbf{S} \times A$$
$$= 100 \times 10^6 \times 0.000\,726\,1 = 72.61 \text{ kN}$$

3. *Bearing on plates.* Middle plate (double shear) area = 12 × 21.5 = 258 mm². Outside plates (single shear) has an area equal to

$$\text{Area} = 2 \times 8 \times 21.5 = 344 \text{ mm}^2$$

The weaker area would be the middle plate failing under bearing. Allowable unit stress = 310 MPa. Allowable load = $\mathbf{S} \times A$ = 310 × 10⁶ × 0.000 258 = 79.98 kN. Therefore the weakest area of the three areas checked would be the rivet shearing. Allowable load on joint = 72.61 kN.

Example 3. Roof truss (Fig. 30-2-6). A roof truss has loads of 330 and 285 kN acting on the upper and lower chord members. Calculate the number of 20 mm rivets required to safely carry these loads.

Solution. Since the 10 mm gusset is enclosed by two 8 mm thick angles, the rivets are in double shear. In calculating the bearing stress, it will be noted that the two outer angles having a combined thickness of 16 mm (two 8 mm thick angles) are stronger than the 10 mm thick gusset. Since the problem is one of determining the number of rivets required to carry the load, it can be assumed that the size of the steel is satisfactory for the applied loads. Refer to Fig. 30-2-1:

1. Number of 20 mm rivets in double shear required for:

Upper chord = 330 ÷ 62.8 = 6 rivets
Lower chord = 285 ÷ 62.8 = 5 rivets

2. Number of 20 mm rivets bearing on 10 mm plate required for:

Upper chord = 330 ÷ 62 = 6 rivets
Lower chord = 285 ÷ 62 = 5 rivets

Therefore the minimum allowable rivets required for the upper and lower chords are 6 and 5 rivets, respectively.

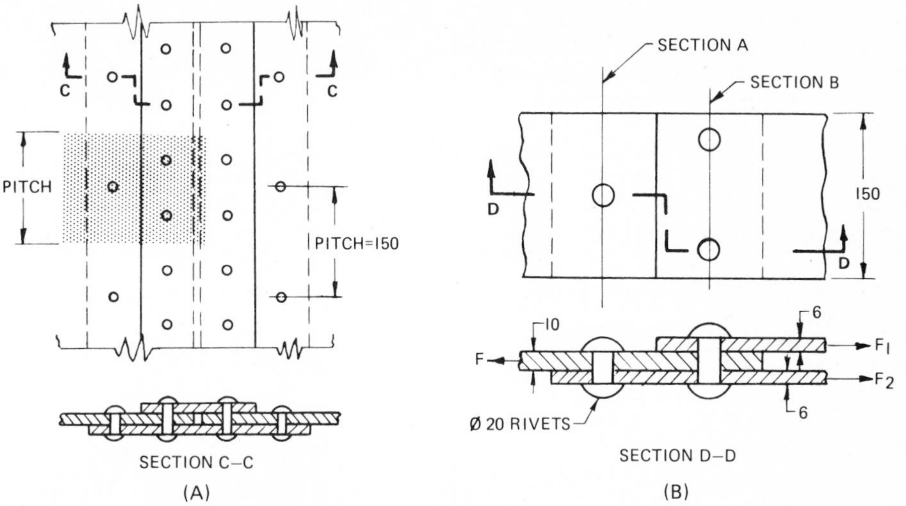

C
PITCH
PITCH=150
SECTION C–C
(A)
SECTION A
SECTION B
D
150
10
6
F
F1
F2
Ø 20 RIVETS
SECTION D–D
(B)

Fig. 30-2-7 Double-riveted butt joint on boiler.

Example 4. Double-riveted butt joint (Fig. 30-2-7). A boiler has a double-riveted butt joint. The boiler plate is 10 mm thick, and the two cover plates are 6 mm. The rivets are 20 mm. A section of the riveted joint is shown. Calculate the main stresses in the joint when the boiler plate is subject to a tensile stress of 40 MPa.

Solution. The length of the repeated section is 150 mm. Since both sides of the joint are the same, only one side of the joint (shown in Fig. 30-2-7*b*) is used in computing the stresses. There are two rivets in double shear and one rivet in single shear. The rivets are 20 mm in diameter and are placed in 21.5 mm drilled holes. As mentioned earlier in boiler work, finished rivets are assumed to be the same size as the drilled holes; thus for calculation purposes the rivets will be 21.5 mm in diameter. The total force exerted on the repeated section is

$$\mathbf{F} = \mathbf{S} \times A$$
$$= 40\,000 \times 150 \times 10 \text{ mm}^2$$
$$= 60 \text{ kN}$$

There are two rivets in double shear and one rivet in single shear, comprising shear areas. It will be assumed that each shear area will carry one-fifth of the load. Shear force on rivets = 60 ÷ 5 = 12 kN. Therefore the unit shear stress on rivets is

$$\mathbf{S} = \frac{\mathbf{F}}{A} = \frac{12}{(\pi \times 21.5^2)/4} = \frac{12\,000}{0.000\,363}$$
$$= 33.06 \text{ MPa}$$

The upper cover plate transmits two-fifths of the load. Therefore $\mathbf{F}_1$ = (0.4)60 kN = 24 kN.

The lower cover plate transmits three-fifths of the load. Therefore $\mathbf{F}_2$ = (0.6)60 kN = 36 kN. Stress on boiler plate taken at section *A*:

$$\mathbf{S} = \frac{\mathbf{F}}{A} = \frac{60 \text{ kN}}{(150 - 21.5) \times 10} = \frac{60\,000}{0.000\,285}$$
$$= 46.69 \text{ MPa}$$

Since one-fifth of the total load has been transmitted to the lower cover plate at section *A*, the load on the boiler plate at section *B* is (0.8)60 = 48 kN. Stress on 10 mm boiler plate taken at section *B* is

$$\mathbf{S} = \frac{\mathbf{F}}{A} = \frac{48 \text{ kN}}{(150 - 43) \times 10} = \frac{48\,000}{0.001\,07}$$
$$= 44.86 \text{ MPa}$$

Since the lower cover plate transmits three-fifths of the total load, the largest stress on the two cover plates will occur on the lower cover plate at section *B*. Stress on bottom cover plate at section *B* is

$$\mathbf{S} = \frac{\mathbf{F}_2}{A} = \frac{36 \text{ kN}}{(150 - 43) \times 6 \times 10^6} = \frac{36\,000}{0.000\,642}$$
$$= 56.07 \text{ MPa}$$

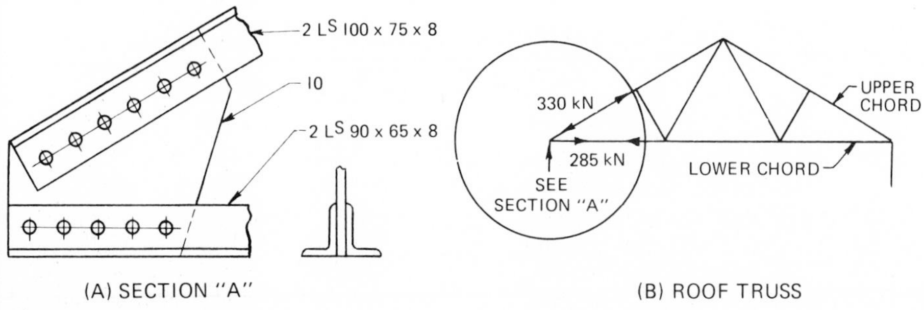

2 LS 100 x 75 x 8
10
2 LS 90 x 65 x 8
(A) SECTION "A"
330 kN
285 kN
SEE SECTION "A"
UPPER CHORD
LOWER CHORD
(B) ROOF TRUSS

Fig. 30-2-6 Roof truss.

In calculating the bearing stresses, the bearing area for the rivet in single shear is the rivet diameter times the thickness of the thinner plate connected (the cover plate). The bearing area for the rivet in double shear is the rivet diameter times the thickness of the boiler plate. The rivets in double shear are subjected to twice the bearing load of those in single shear. Bearing stress at rivet in single shear (section *A*) is

$$S = \frac{F}{A} = \frac{12 \text{ kN}}{21.5 \times 6 \times 10^6} = \frac{12\,000}{0.000\,129}$$
$$= 93.02 \text{ MPa}$$

Bearing stress at rivet in double shear (section *B*) is

$$S = \frac{F_1}{A} = \frac{24 \text{ kN}}{21.5 \times 10 \times 10^6} = \frac{24\,000}{0.000\,215}$$
$$= 111.63 \text{ MPa}$$

Stresses in Thin-Wall Cylinders. An important application of riveted and welded joints is in the construction of boilers and tanks. The pressure of gases or liquids upon the walls of a tank acts outwardly in all directions and uniformly. Therefore, the cylinder shell on a thin-wall vessel is designed with the assumption that the stress is uniform throughout the wall thickness.

The tensile stress in the ends of the cylinder, caused by the pressure inside, is called *longitudinal stress*, or *tension*. The tensile stress acting in the circumferential direction is called *hoop stress*, or *tension*.

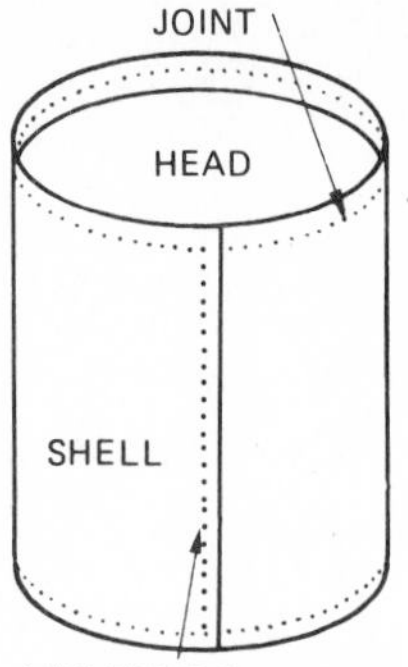

CIRCUMFERENTIAL JOINT
HEAD
SHELL
LONGITUDINAL JOINT
(A) TERMINOLOGY

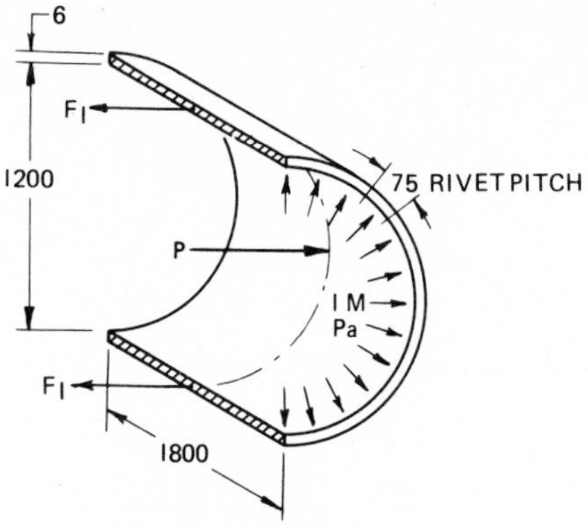

6
1200
F1
P
75 RIVET PITCH
1 MPa
F1
1800
(B) HALF SECTION

Fig. 30-2-8 Thin-wall cylinder.

Example 5. A tank of 1200 mm diameter is made of 6 mm steel plate. The internal pressure is 1 MPa. Calculate the size of rivets required if the pitch on the longitudinal and circumferential joints is 75 mm.

Solution

1. Calculate rivet size for longitudinal seam. Figure 30-2-8 shows a half section of the tank. The internal pressure of 1 MPa acts on the shell surface at every point. The total force acting on the half of the tank shown would be equal to the area of the tank taken at its center times the pressure, or $(1200 \times 1800/10^6)\text{m}^2 \times 1$ MPa $= 2160$ kN. The combined equal pressures of F_1 and F_1 acting on the tank wall are equal in magnitude to **P** but act in opposite directions.

Only the pitch distance of 75 mm, the repeated section, need be used in calculating the size of the rivet along the joint. Therefore

F_1 for repeated section =

$$\frac{1200 \times 75}{10^6} \text{ m}^2 \times (1 \times 10^6) \text{ Pa} \times \frac{1}{2}$$
$$= 45 \text{ kN}$$

or load acting on each rivet. As previously mentioned, in boiler construction the diameter of the rivet hole, which is 1.5 mm larger than the diameter of the rivet, is used in computing all the stresses. Refer to Fig. 30-2-1. The allowable stress in single shear is 100 MPa. The chart shows values of 38 and 49.1 MPa for rivet sizes of 22 and 25 mm, respectively. Since the finished size of the 22 mm diameter rivet will be 23.5 mm in diameter, the allowable load will be computed on the final size. Therefore the allowable load for a 23.5 mm diameter rivet will be:

$$\text{Area} \times \text{stress} =$$
$$0.000\,434 \text{ m}^2 \times 100 \times 10^6 = 43.4 \text{ kN}$$

Since this is less than the load acting on the rivet, the next size larger rivet must be used, and therefore the 25 mm rivet is required. The allowable load in bearing for a 25 mm rivet on 6 mm steel plate is 46.5 MPa. Therefore the size of rivet required along the longitudinal seam is 25 mm.

2. Calculate rivet size for circumferential joint. Number of pitches or repeated sections on circumference equals

$$\frac{\pi \times \text{diameter}}{\text{Pitch}} = \frac{\pi\ (1200)}{075} = 50.2$$

Use 51 rivets. Pressure exerted on head of tank = pressure × area = 1 MPa × $(\pi \times 600^2)$ sq mm = 1131 kN. Therefore

$$\text{Load per rivet} = \frac{1131}{51} = 22.18 \text{ kN}$$

Note: When the pitches on the longitudinal and the circumferential joints are equal, then the load per pitch on the circumferential joint is one-half of the load per pitch on the longitudinal joint.

Refer to Fig. 30-2-1. The allowable load in single shear for a 16 mm rivet (use 17.5 mm for calculations) is 24 kN, and the allowable load in bearing for a 16 mm rivet (use 17.5 mm for calculations) on 6 mm steel is 32.55 kN. Therefore the size of rivet required along circumferential joint is 16 mm.

Bolts, Screws, and Studs

As explained in Chap. 8, metric threaded fasteners are defined by property class numbers which designate their strength. The first number of a two-digit symbol or the first two numbers of a three-digit symbol approximates 1 percent of the minimum tensile stress in megapascals.

The last numeral approximates one-tenth of the ratio expressed as a percentage between minimum yield stress and minimum tensile stress. See Fig. 30-2-9.

For example, a property class 5.8 fastener has a tensile strength of 520 MPa and a yield strength of 420 MPa. The first number, 5, represents approximately 1 percent of the tensile strength of 520 MPa. The yield strength of 420 MPa is approximately four-fifths or 80 percent of the tensile strength of 520 MPa. One-tenth of 80 percent is 8, the last number in the property class. Load capacities and strength of threaded fasteners are shown in Figs. 30-2-9 and 30-2-10.

Example 6. What load can be supported by an M24 × 3 stud, property class 8.8, if a factor of safety of 4 is added to the requirements?

Solution Refer to Fig. 30-2-10. Under the column of 8.8, an M24 × 3 thread has a tensile strength (the maximum force permitted) of 293 kN. Adding a factor of safety to this value, we get that the permissible load is 293 ÷ 4 = 73.25 kN.

Example 7. A platform is supported by four M16 rods which are suspended from the ceiling. The end of the rods are threaded, and plate washers and nuts are attached. If a load of 220 kN is evenly

Property Class	Nominal Diameter	Tensile Strength MPa	Min. Yield Strength MPa
4.6	M5 thru M36	400	240
4.8	M1.6 thru M16	420	340
5.8	M5 thru M24	520	420
8.8	M16 thru M36	830	660
9.8	M1.6 thru M16	900	720
10.9	M5 thru M36	1040	940
12.9	M1.6 thru M36	1220	1100

Fig. 30-2-9 Strength of bolts, screws, and studs.

Thread	Tensile Stress Area mm²	Thread Shear Area mm²	Tensile Strength (kN) for Property Class 4.6	4.8	5.8	8.8	9.8	10.9	12.9
M10 × 1.5	58	15.6	23.2	24.4	30.2		52.2	60.3	70.8
M12 × 1.75	84	19	33.7	35.4	43.8		75.9	87.7	103
M14 × 2	115	22.4	46	48.3	59.8		104	120	140
M16 × 2	157	26.1	62.8	65.9	81.6	130	141	163	192
M20 × 2.5	245	33.3	98		127	203		255	299
M24 × 3	353	40.5	141		184	293		367	431
M30 × 3.5	561	51.6	224			466		583	684
M36 × 4	817	63.1	327			678		850	997

Fig. 30-2-10 Load capacities of threaded fasteners.

distributed on the platform so that each rod is supporting one-fourth of the load, what property class fasteners are required? Use a factor of safety of 2.

Solution The load applied on each rod is 220 kN ÷ 4, or 55 kN. The design load is the applied load times the factor of safety = 55 kN × 2 = 110 kN. Refer to Fig. 30-2-10. An M16 thread, property class 8.8, would be selected because its tensile strength is 130 kN.

REFERENCES AND SOURCE MATERIAL

1. American Institute of Steel Construction.
2. Canadian Institute of Steel Construction.

Assignments

1. On an A3- or B-size sheet, show the calculations for the single-riveted butt joint and the riveted roof truss shown in Fig. 30-2-A or 30-2-B. For the single-riveted butt joint, calculate the main stresses that could safely be applied to the joint. Plate material is A36 steel. For the riveted roof truss, the roof truss has loads as shown acting on the upper and lower chord members. Calculate the number of rivets required to carry these loads.

2. On an A3- or B-size sheet, show the calculations for the double-riveted butt joint and the riveted lap joint shown in Fig. 30-2-C or 30-2-D. For the double-riveted butt joint, calculate the

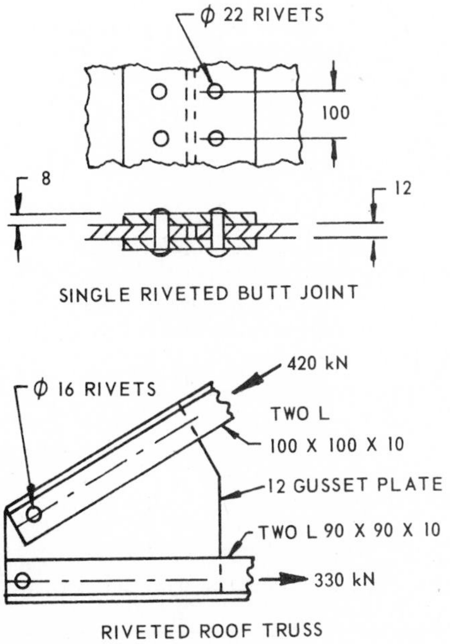

ϕ 22 RIVETS
100
8
12
SINGLE RIVETED BUTT JOINT
ϕ 16 RIVETS
420 kN
TWO L
100 X 100 X 10
12 GUSSET PLATE
TWO L 90 X 90 X 10
330 kN
RIVETED ROOF TRUSS

Fig. 30-2-A Riveted joints.

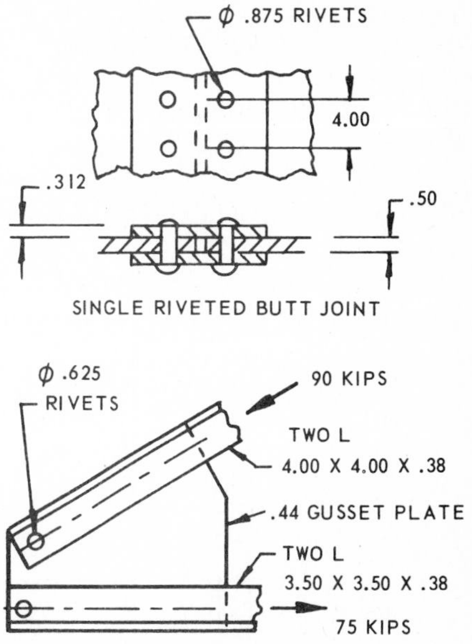

ϕ .875 RIVETS
4.00
.312
.50
SINGLE RIVETED BUTT JOINT
ϕ .625
RIVETS
90 KIPS
TWO L
4.00 X 4.00 X .38
.44 GUSSET PLATE
TWO L
3.50 X 3.50 X .38
75 KIPS
RIVETED ROOF TRUSS

Fig. 30-2-B Riveted joints.

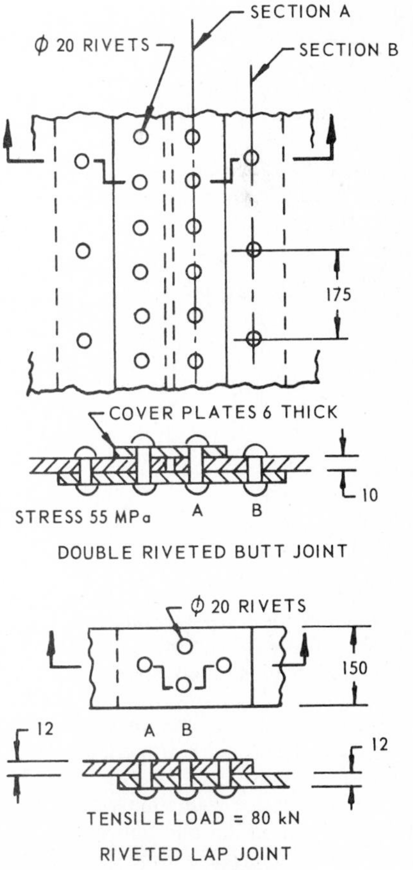

SECTION A
ϕ 20 RIVETS
SECTION B
175
COVER PLATES 6 THICK
10
STRESS 55 MPa
A
B
DOUBLE RIVETED BUTT JOINT
ϕ 20 RIVETS
150
12
A B
12
TENSILE LOAD = 80 kN
RIVETED LAP JOINT

Fig. 30-2-C Riveted joints.

main stresses in the boiler joint when the boiler plate is subjected to the tension stress shown. In the case of the riveted lap joint, a structural joint is fastened by four rivets. Calculate the main stresses when the tensile load shown is applied.

REVIEW FOR ASSIGNMENTS

Unit 30-1 Stresses

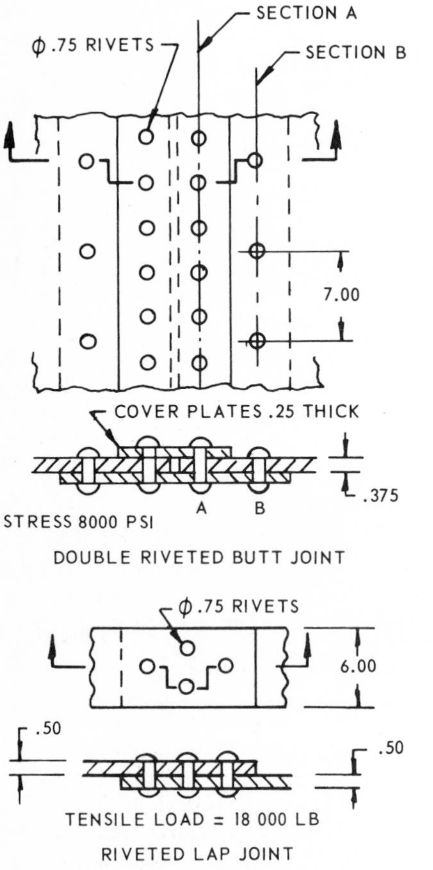
SECTION A
ϕ.75 RIVETS
SECTION B
7.00
COVER PLATES .25 THICK
.375
A
B
STRESS 8000 PSI
DOUBLE RIVETED BUTT JOINT
ϕ.75 RIVETS
6.00
.50
.50
TENSILE LOAD = 18 000 LB
RIVETED LAP JOINT

Fig. 30-2-D Riveted joints.

UNIT 30-3

WELDED JOINTS[1,2]

In addition to riveting, welding is also employed in the joining of structural steel. Fabricated steel construction has also replaced many parts formerly made by castings because of the lower cost and the greater strength at a considerable reduction in mass.

The two types of welds most frequently used are fillet and butt welds. Thus only these types will be covered in this unit.

Fillet Welds

The fillet weld is used to join two parts that either overlap or join at an angle, normally perpendicular, to each other. In the calculations of strength of fillet welds, the *effective area* is considered as the *effective length* of weld times the *effective throat thickness*, as illustrated in Fig. 30-3-1.

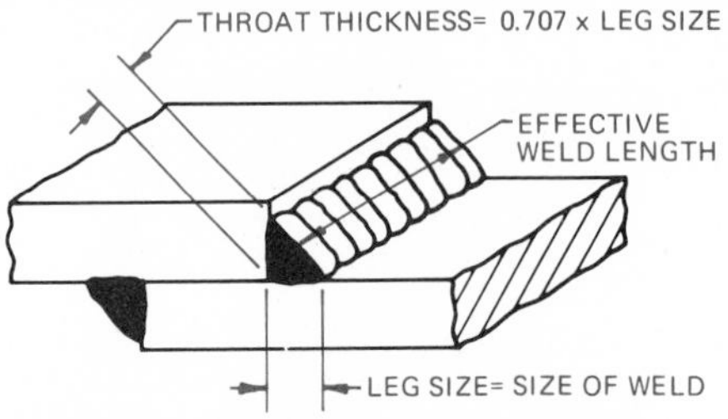
THROAT THICKNESS= 0.707 x LEG SIZE
EFFECTIVE WELD LENGTH
LEG SIZE= SIZE OF WELD

Fig. 30-3-1 Fillet weld nomenclature.

For example, a 10 mm fillet weld 150 mm long (effective length) has an effective area of 10 × 0.7 × 150, or 1050 mm^2.

For purposes of calculating the strength of welds in this unit, the shear stresses shown in Fig. 30-3-3 will be used. The strength of the previous weld would be stress times effective area.

With E410xx electrodes the weld strength is

$$185 \text{ MPa} \times 0.001\ 05 \text{ m}^2 = 194.3 \text{ kN}$$

Using E480xx electrodes, the weld strength is

$$216 \text{ MPa} \times 0.001\ 05 \text{ m}^2 = 226.8 \text{ kN}$$

Another method of calculating the strength of welds is to multiply the leg size of the weld by effective length. The shear resistance factors (SRF) shown in Fig. 30-3-4 are based on the shear stresses shown in Fig. 30-3-3.

The stress of the previous weld using E480xx electrodes would be 1.5 × 150 kN = 225 kN.

The strength value for specified weld sizes is the more convenient method to use.

The following recommendations should be adhered to when welded joints are designed:

Electrodes			
Metric		Inch	Pound
E410XX	E480XX	E60XX	E70XX
185 MPa	216 MPa	13,500	15,700

Fig. 30-3-3 Shear stress for electrodes.

Fillet Weld Size mm	Allowable Load per mm Length in kN Base Metal: A36[1] G40.21M-230W E410XX	Electrodes: A572M[2] G40.21-290W and over E480XX Electrode
4	0.5	0.6
6	0.8	0.9
8	1	1.2
10	1.3	1.5
12	1.6	1.8
16	2.1	2.4
20	2.6	3

1. Based on shear resistance factor of 0.131 kN per mm of weld for 1 mm of weld length (185 MPa shear stress)
2. Based on shear resistance factor of 0.153 kN per mm of weld for 1 mm of weld length (216 MPa shear stress)

Fillet Weld Size Inches	Allowable Load Per Inch kN Length in Kips Base Metal: A36[3] G40.21-33W E60XX Electrode	Metal Electrodes: E70XX[4] Electrode
.18	1.8	2.1
.25	2.4	2.8
.31	3.0	3.5
.38	3.6	4.2
.44	4.2	4.9
.50	4.8	5.6
.62	6.0	7.0
.75	7.2	8.4

3. Based on 600 lb per .062 of weld thickness (13,500 PSI shear stress)
4. Based on 700 lb per .062 of weld thickness (15,800 PSI shear stress)

Fig. 30-3-4 Strength of fillet welds.

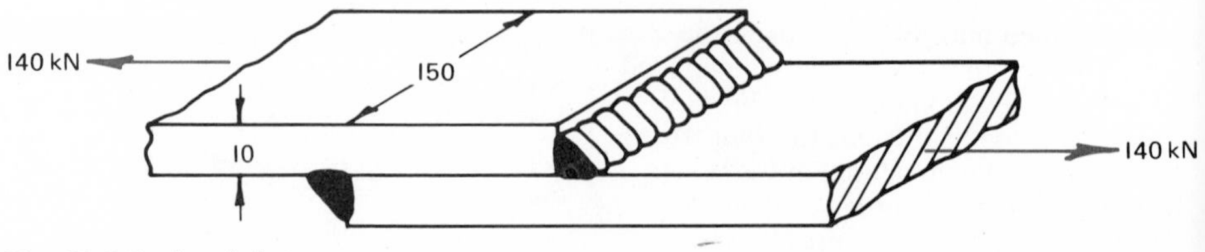
140 kN
150
10
140 kN

Fig. 30-3-2 Lap joint.

- Even-number-size welds, such as shown in Fig. 30-3-4, should be used whenever possible.
- For length of welds use lengths evenly divisible by 5, such as 40, 50, 60, etc.
- Fillet welds should be at least 2 mm less than the thickness of the part being welded.
- Welds should be located on both sides of T-joints evenly spaced around the line of action of the applied load.

Example 1. Two 10 × 150 steel bars are welded with E480 electrodes as shown in Fig. 30-3-2. What size weld is required if a tensile load of 225 kN is applied?

Solution.

$$\text{Weld size} = \frac{\text{load}}{\text{length} \times \text{SRF}}$$

$$= \frac{140}{2 \times 150 \times 0.153} = 3 \text{ mm}$$

Use a 4 mm fillet weld.

Example 2. A 20 × 100 mm A572M-50 bar is welded to a column. What are the size and length of the fillet welds required if a tensile load of 220 kN is applied?

Solution. Since two fillet welds will be used, one on each side of the bar, each fillet weld will be designed to resist a force of 220 kN ÷ 2 = 110 kN.

With a weld running the entire length of 100 mm, a weld having the strength of 110 ÷ 100 or 1.1 kN per millimetre of length is required.

Refer to Fig. 30-3-4: an 8 mm fillet weld is selected. With a 10 mm weld, a weld length of 110 ÷ 1.5 = 73.3 (use 75 mm) would be required.

Example 3. A 250 × 10 A572M-380 steel plate is connected by a pair of fillet welds to the bottom flange of a beam, as shown in Fig. 30-3-5. The plate is subjected to a tensile load of 450 kN. What are the minimum size and length of weld recommended for this connection, if the maximum stress on the plate is 220 MPa?

Solution. Before the weld size is chosen, the minimum plate area at the weld should be established. The minimum plate area, for calculating purposes, at the welded area (see Fig. 30-3-6) is width times plate thickness.

Load = stress × area. Therefore

Minimum width of plate

$$= \frac{\text{load}}{\text{stress} \times \text{width of plate}}$$

$$= \frac{450\ 000}{220 \times 10^6 \times 0.000\ 01}$$

$$= 204.5 \text{ mm}$$

Therefore, the plate width of 250 mm is acceptable. If the minimum weld length of 230 mm per side is used, the load per mm of weld is 450 ÷ (2 × 210) = 1.07 kN.

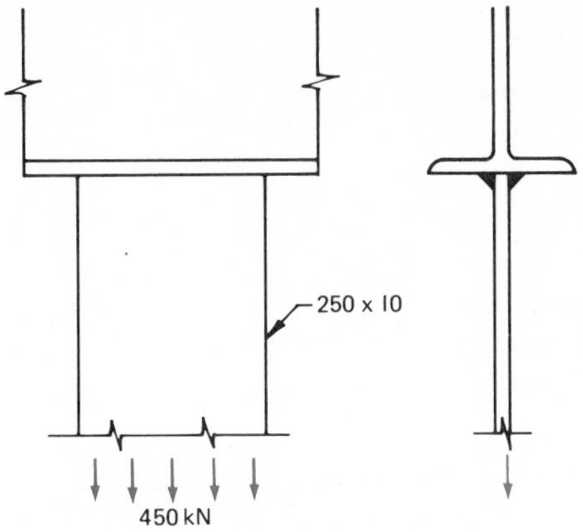
250 x 10
450 kN

Fig. 30-3-5 Bar fillet welded both sides.

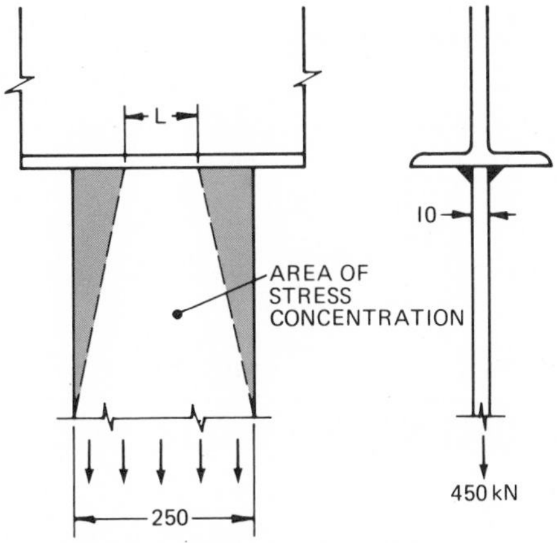
L
10
AREA OF STRESS CONCENTRATION
450 kN
250

Fig. 30-3-6 Plate area at weld.

Refer to Fig. 30-3-4. An 8 mm fillet weld is required. If the fillet weld were to run the entire length of the plate, then the load per mm of weld = 450 ÷ (2 × 250) = 0.9 kN. This would permit a 6 mm weld to be used, which is more economical.

INTERMITTENT FILLET WELDS

Intermittent fillet welds may be used to transfer calculated stress across a joint when the strength required is less than that developed by a continuous fillet weld of the smallest permitted size, and to joint components of built-up members. The effective length of any segment of intermittent fillet welding should be not less than 4 times the weld size, with a minimum of 40 mm.

Example 4. Calculate the size of the intermittent weld shown in Fig. 30-3-7 to safely carry a tensile load of 1.6 MN. Use E480xx electrodes.

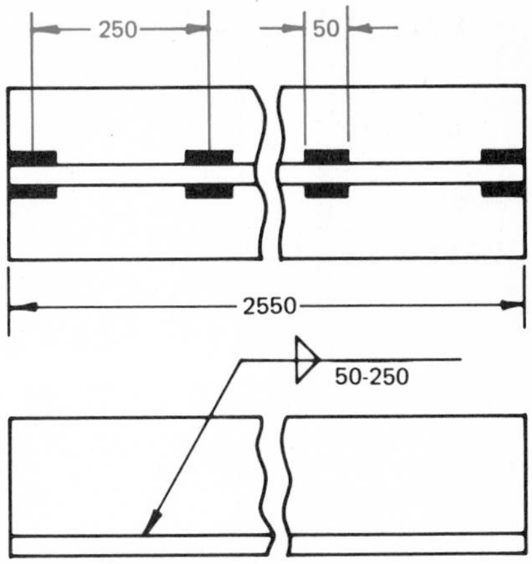
250
50
2550
50-250

Fig. 30-3-7 Intermittent weld.

Solution.

Total length of welds = $11 \times 50 \times 2 = 1100$ mm

Load per millimetre length of weld =

$$\frac{1600}{1100} = 1.45 \text{ kN}$$

Refer to Fig. 30-3-4. Size of fillet weld required = 10 mm.

FILLET WELDS FOR ANGLE IRON

When tension or compression members are connected by two side fillet welds as shown in the previous examples, the weld should be placed in the same line of action as the force being transmitted by the weld. For members having symmetrical cross sections, the length of weld on each side of the member should be equal. For members having unsymmetrical cross sections, as shown in Fig. 30-3-8 where an angle iron is welded to a steel plate, the lengths of welds are so proportioned that the line of action of the force transmitted by the weld will be along the axes of the two members. This is accomplished by assuming that the line of action on the angle member is on the *centroidal* (center of gravity) *axis* and by making the lengths of welds such that $L_1 \times A = L_2 \times B$.

Example 5. A 125 × 90 × 10 A572M-310 angle welded to a steel plate transmits a load of 300 kN. Calculate the length of welds on each side of the angle so that the load acts along the centroidal axis of the angle. See Fig. 30-3-8.

Solution. The maximum fillet weld for 10 mm thick material is 8 mm. The allowable load per millimetre length of 8 mm weld is 1.2 kN. (See Fig. 30-3-4.) Therefore

Minimum permissible length of weld =

$$\frac{300}{1.2} = 250 \text{ mm}$$

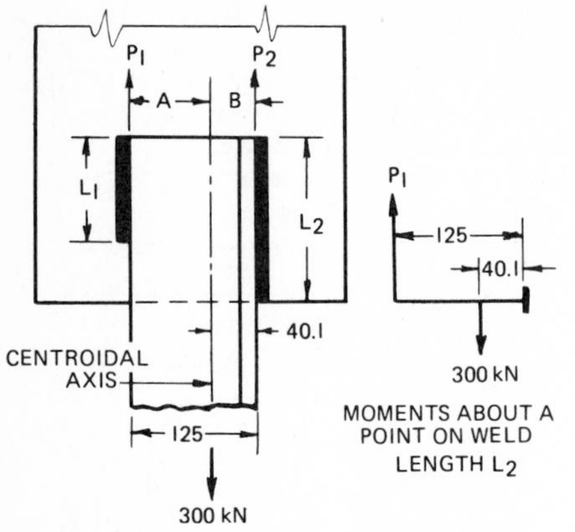
P1
P2
A
B
L1
L2
P1
125
40.1
40.1
CENTROIDAL AXIS
300 kN
MOMENTS ABOUT A POINT ON WELD LENGTH L2
125
300 kN

Fig. 30-3-8 Fillet weld on both sides of angle iron.

Referring to the Appendix, we find the back of the short leg of the angle is 40.1 mm from the centroidal axis.

Consider the 300 kN load being transferred to the plate by loads P_1 and P_2 where $P_1 = 1.2 \times L_1$, $P_2 = 1.2 \times L_2$ and P_1 and P_2 both equal 300 kN. Taking moments about a point on length L_2 (refer to Fig. 30-4-3 for calculation of moments), we have

$$P_1 \times 125 = 300 \times 40.1$$
$$(1.2\ L_1)125 = 300(40.1)$$

Therefore,

$$L_1 = 80 \text{ mm}$$
$$L_1 + L_2 = 250 \text{ mm}$$
$$L_2 = 250 - 80 = 170 \text{ mm}$$

Therefore weld lengths of 80 (L_1) and 170 (L_2) are selected.

Example 6. Use the same members and load as in the previous example, except the fillet weld is welded on three sides, as shown in Fig. 30-3-9.

Solution The design calls for 250 mm of weld to be used; 125 mm of the weld lies along the back of the angle so that the remaining 125 mm of weld is equal to the combined length of welds L_1 and L_2.

The 250 mm of weld should be located so that the line of action of the force transmitted by the weld is along the centroidal axis of the angle.

For calculation purposes, assume there are three welds, P_1, P_2, and P_3, whose combined loads equal 300 kN. P_1 lies along distance L_1 and is equal to 1.2 L_1 kN. P_2 lies along distance L_2 and is equal to 1.2 L_2 kN. P_3 lies midway along the 125 mm width, or 62.5 mm from P_1, and is equal to $125 \times 1.2 = 150$ kN. At a point on line L_2

Clockwise moments

$$= P_1 \times 125 + P_3 \times 62.5$$
$$= L_1 \times 1.2 \times 125 + 150 \times 62.5$$
$$= 150\ L_1 + 9375 \text{ N•m}$$

Counterclockwise moments

$$= 40.1 \times 300 = 12\ 030 \text{ N•m}$$

Clockwise moments = counterclockwise moments

$$150\ L_1 + 9375 \text{ N•m} = 12\ 030 \text{ N•m}$$
$$150\ L_1 = 2655 \text{ N•m}$$
$$L_1 = 17.7 \text{ mm (use 20 mm)}$$
$$L_1 + L_2 = 125 \text{ mm}$$
$$L_2 = 125 - 20 = 105 \text{ mm}$$

Therefore weld lengths of 20 mm (L_1) and 105 mm (L_2) are used on the sides of the angle.

Butt Welds

The butt weld is used to join two pieces of metal that lie on the same plane. In the calculations of strength of butt joints, the effective area of butt welds shall be considered as the effective length of weld times the effective throat thickness. The effective throat thickness depends on the metal thickness, the gap between the adjoining parts, the type of butt weld, and whether the weld is on one or both sides. See Fig. 30-3-10.

Example 7. An open-square butt weld, welded one side, is used to join two A36 steel plates 6 × 250. Compute the safe tensile load that can be applied to the joint.

Solution Refer to Fig. 30-3-10. The effective throat thickness for an open-

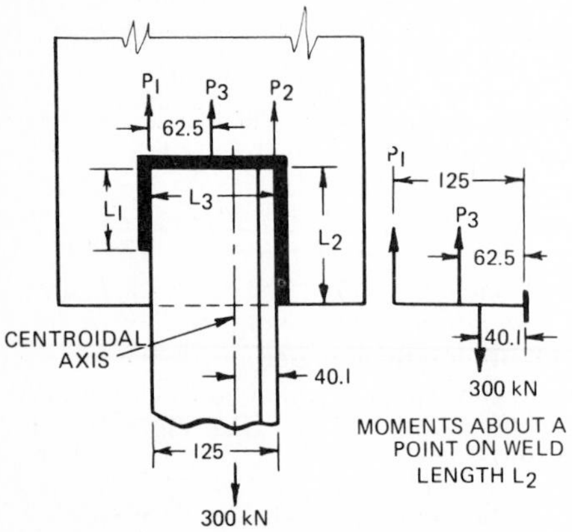
P1
P3
P2
62.5
L1
L3
L2
P1
125
P3
62.5
40.1
CENTROIDAL AXIS
40.1
300 kN
MOMENTS ABOUT A POINT ON WELD LENGTH L2
125
300 kN

Fig. 30-3-9 Fillet weld on sides and end of angle iron.

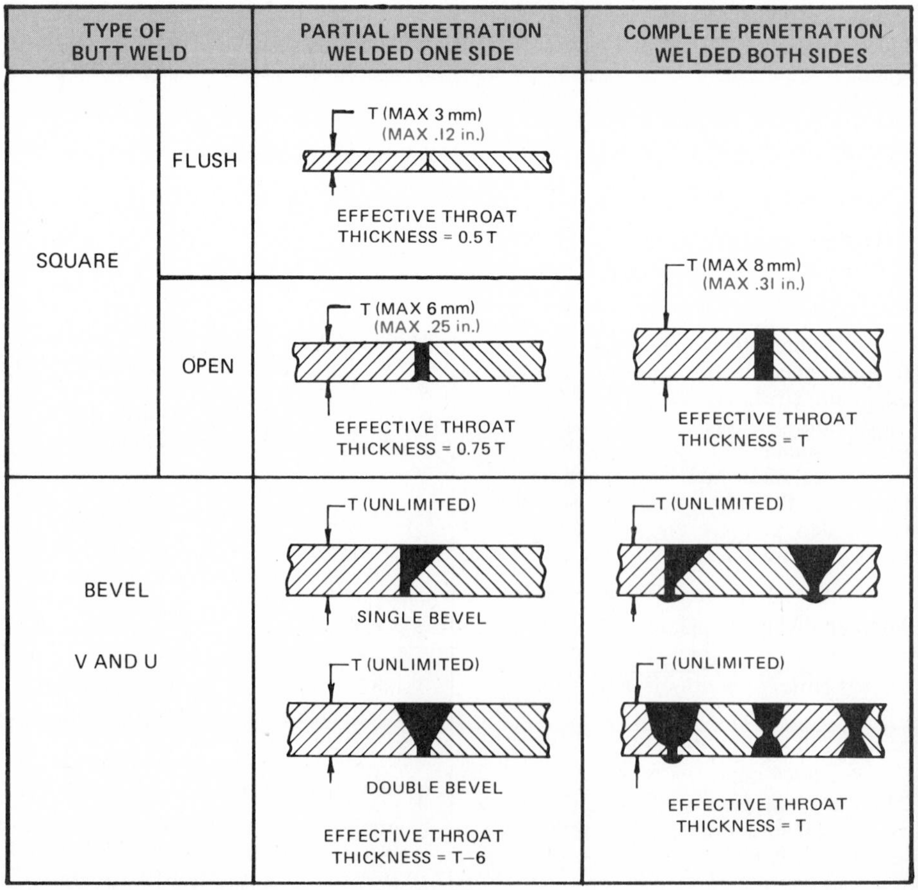
TYPE OF BUTT WELD
PARTIAL PENETRATION WELDED ONE SIDE
COMPLETE PENETRATION WELDED BOTH SIDES
SQUARE
FLUSH
T (MAX 3 mm) (MAX .12 in.)
EFFECTIVE THROAT THICKNESS = 0.5 T
OPEN
T (MAX 6 mm) (MAX .25 in.)
EFFECTIVE THROAT THICKNESS = 0.75 T
T (MAX 8 mm) (MAX .31 in.)
EFFECTIVE THROAT THICKNESS = T
BEVEL
V AND U
T (UNLIMITED)
SINGLE BEVEL
T (UNLIMITED)
DOUBLE BEVEL
EFFECTIVE THROAT THICKNESS = T–6
T (UNLIMITED)
T (UNLIMITED)
EFFECTIVE THROAT THICKNESS = T

Fig. 30-3-10 Strength of butt welds.

square butt weld $= 0.75T = 0.75 \times 6 =$ 4.5 mm. Next (refer to Fig. 30-1-8), the allowable unit tensile stress for A36 steel is 150 MPa. Therefore

$$\text{Safe load} = \text{area} \times \text{unit stress} = (4.5 \times 10^{-6} \times 250)\ \text{m}^2 \times (150 \times 10^6)\ \text{Pa} = 168.75\ \text{kPa}$$

Example 8. A boiler 900 mm in diameter, made of A36 steel, has to withstand a steam pressure of 2.5 MPa. A single-V butt joint, welded one side, is to be used. What is the thickness of boiler plate required?

Solution Figure 30-3-11 shows a half section of the boiler. The total force P acting on the cylinder is the resultant pressure of the internal pressure of 2.5 MPa acting in all directions on the cylinder wall. It will be assumed that for thin-walled cylinders the resultant force P will equal the diameter of the cylinder in metres times the length of the cylinder in metres times the pressure acting within the cylinder.

The total force P is resisted by two equal forces F_1 and F_2. Taking a section of the tank 100 mm in length and calculating the forces P, F_1, and F_2, we have

$$F = S \times A = (2.5 \times 10^6)\ \text{Pa} \times (0.1 \times 0.9)\ \text{m}^2 = 225\ \text{kN}$$

$$F_1 = 225 \div 2 = 112.5\ \text{kN}$$

The single-V butt weld will have to withstand a force of 112.5 kN for every 100 mm of weld. Allowable unit stress for A36 steel is 150 MPa. Weld stress equals plate stress. Therefore

Effective throat thickness of weld

$$= F_1 \div (S \times \text{length of section})$$
$$= 112\ 500 \div (150 \times 10^6 \times 100)$$
$$= 8\ \text{mm}$$

Refer to Fig. 30-3-10; it will be noted that the effective throat thickness of a single-V butt weld, welded one side, is equal to $T - 6$ mm, where T is equal to the thickness of plate. Therefore, minimum plate thickness must be $6 + 8 = 14$ mm to safely carry the load. With a welded-both-sides joint, the plate thickness could be reduced to 8 mm, which would be a considerable saving.

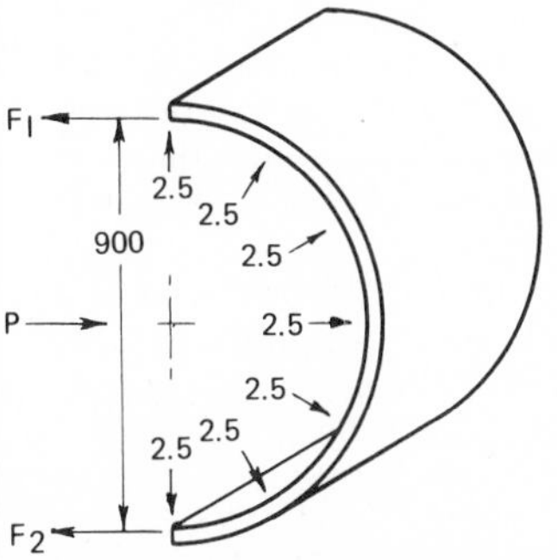

F1
2.5
2.5
900
2.5
P
2.5
2.5
2.5
2.5
F2

Fig. 30-3-11 Welded boiler section.

REFERENCES AND SOURCE MATERIAL

1. American Institute of Steel Construction.
2. Canadian Institute of Steel Construction.

Assignments

1. On an A3- or B-size sheet, show the calculations and answers for the problems shown in Assignment 30-3-A or 30-3-B.

ASSIGNMENT 30-3-A

1. Calculate the safe load in tension on an open-square butt weld, welded one side, joining two 6 × 150 A36 steel plates.
2. Do the same as in problem 1 except use an open square, welded both sides.
3. A 12 × 250 A36 steel bar, welded both sides, is connected to a steel plate. Calculate the length of the two 10 mm side fillet welds required if the plate is subjected to a tensile load of 130 kN.
4. Two A572M-380 structural-steel plates 40 mm thick are connected by a double-U butt weld. Calculate the allowable load the weld will support in a 100-mm-wide section.
5. A ϕ50 A572M-290 steel shaft is welded to a steel plate by a 6 mm fillet weld. Calculate the safe tensile load that could be supported by the weld.
6. A 100× 100× 13 A36 steel angle supporting a load of 400 kN is welded to a steel plate. Calculate the length of side welds required using the largest permissible fillet welds. The line of action on the angle is taken through the centroidal axis.
7. Do the same as in Problem 6 except that the weld is on the sides and end of the angle.

ASSIGNMENT 30-3-B

1. Calculate the safe load in tension on an open-square butt weld, welded one side, joining two .25 × 6 in. A36 steel plates.
2. Do the same as in Problem 1 except use an open square, welded both sides.
3. A .50 × 10 in. A242 steel plate, welded both sides, is connected to a steel plate. Calculate the length of the two .375 in. side fillet welds required if the plate is subject to a tensile load of 15 000 lb.
4. Two A572-55 structural-steel plates 1.50 in. thick are connected by a double-U butt weld. Calculate the allowable load per inch the weld will support.
5. A ϕ2 in. A572-42 steel shaft is welded to a steel plate by a .25 in. fillet weld. Calculate the safe tensile load that could be supported by the weld.
6. A 4× 4× .625 in. A36 steel angle supporting a load of 90 kips is welded to a steel plate. Calculate the length of side welds required if .44 in. fillet welds are used and the line of action on the angle is taken through the centroidal axis.
7. Do the same as in Problem 6 except the weld is on the sides and end of the angle.

2. On an A3- or B-size sheet, show the calculations and answers for the problems shown in Assignment 30-3-C or 30-3-D.

ASSIGNMENT 30-3-C

1. A boiler 1000 mm in diameter, made of A36 steel 12 mm thick, is welded by a single-V butt weld. What is the maximum allowable pressure in the boiler?
2. Do the same as in Problem 8 except use a single-V butt, welded both sides.
3. A boiler 1000 mm in diameter is made of A36 steel and has to withstand a steam pressure of 2.2 MPa. A single-bevel butt joint, welded one side, is used. What is the required thickness of boiler plate?
4. An intermittent fillet weld is used on both sides of a welded Tee A572M-380 steel section 5000 mm long. The weld length is 50 mm on a 200 mm pitch. If a tensile load of 1.3 MN is uniformly distributed along the welded section, what size fillet weld is required?
5. What is the maximum tensile load that can be applied to an A36 steel welded section 2400 mm long? A 12 mm intermittent fillet weld 75 mm long on a 300 mm pitch is used on each side of the plate.

ASSIGNMENT 30-3-D

1. A boiler 42. in. in diameter, made of A36 steel, .50 in. thick welded by a single-V butt weld. What is the allowable pounds per square inch of the boiler?
2. Do the same as in Problem 8 except use a single-V butt, welded both sides.
3. A boiler 42 in. in diameter made of A36 steel, has to withstand a steam pressure of 325 psi. A single-bevel butt joint, welded one side, is used. What is the required thickness of boiler plate?
4. An intermittent fillet weld is used on both sides of a welded Tee A572-55

steel section 16 ft long. The weld length is 3 in. on a pitch of 12 in. If a tensile load of 600 kips is uniformly distributed along the welded section, what size fillet weld is required?

5. What is the maximum tensile load that can be applied to an A36 steel welded section 8 ft long? A .38 in. intermittent fillet weld 3 in. long on a 12 in. pitch is used on each side of the plate.

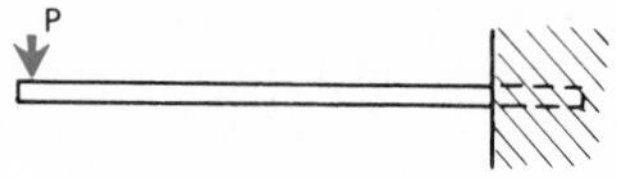
P

(A) CANTILEVER BEAMS

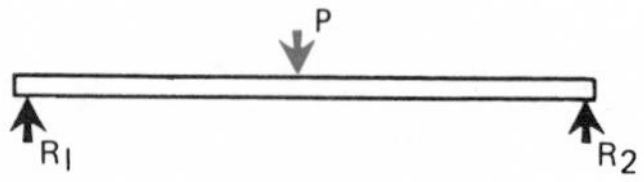
P
R_1
R_2

(B) SIMPLE BEAM—
SUPPORTED AT BOTH ENDS

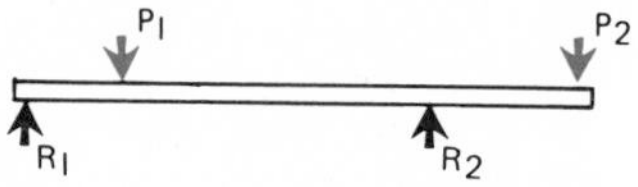
P_1
P_2
R_1
R_2

(C) OVERHANGING BEAM

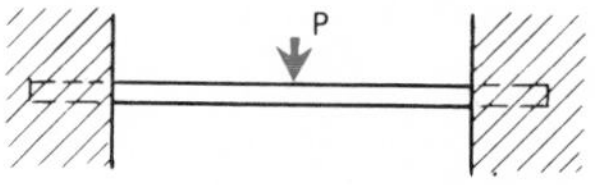
P

(D) BEAM WITH FIXED ENDS

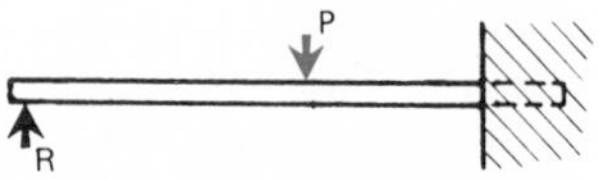
P
R

(E) BEAM FIXED AT ONE END.
SUPPORTED AT OTHER END

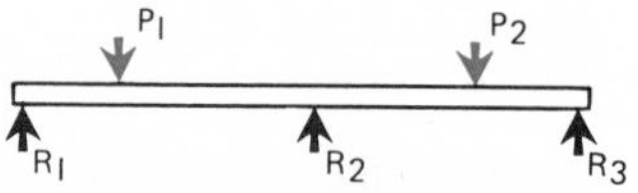
P_1
P_2
R_1
R_2
R_3

(F) CONTINUOUS BEAM

Fig. 30-4-1 Common types of beams.

UNIT 30-4

BEAMS

A *beam* is a structural member or machine part which supports transverse, i.e., perpendicular, loads and reactions. Most beams are placed in a horizontal position with vertical forces acting on them. Examples are floor and ceiling joists, lintels, and floor beams. This unit covers the design of simple beams only where buckling and twisting are not factors and where the beams are of uniform size and shape for the entire length. Forces acting on the beams are assumed to be in the same plane.

Types of Beams. Beams are classified according to the manner in which they are supported. Some of the more common types of beams are shown in Fig. 30-4-1:

1. Cantilever beam: a beam that has one fixed end

2. Simple beam: a beam that is supported at each end

3. Overhanging beam: a beam that has one or both ends projecting beyond its supports

4. Beam with both ends fixed

5. Beam fixed at one end and supported at the other end

6. Continuous beam: a beam supported at more than two points

Kinds of Loads

Two types of loads commonly occur on beams: *concentrated* and *uniformly distributed* loads. A concentrated load extends over a short length of the beam and for calculation purposes is considered as acting at one point. It is usually represented by a line with an arrow indicating its direction of force and the letter *P* as shown in Fig. 30-4-2*a*. Concentrated loads are generally expressed in kilonewtons or meganewtons (metric) or pounds or kips (inch-pound). One kip equals 1000 pounds. A uniformly distributed load is one in which the load is distributed uniformly over a given length or over the entire length of the beam. The mass of the beam is an example of a uniformly distributed load. This type of load is generally expressed in newtons per metre or kilonewtons per metre (metric) or kips per foot (inch-pound). The load is represented in the figure by a rectangular block resting on the beam, as shown in Fig. 30-4-2*b*.

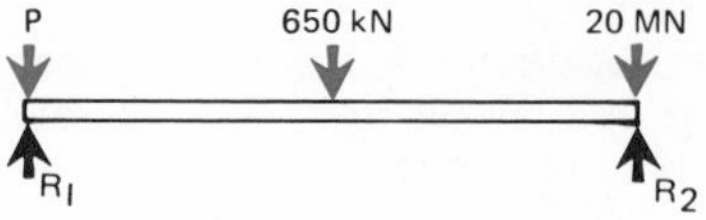
P
650 kN
20 MN
R_1
R_2

(A) THREE METHODS OF INDICATING
CONCENTRATED LOADS

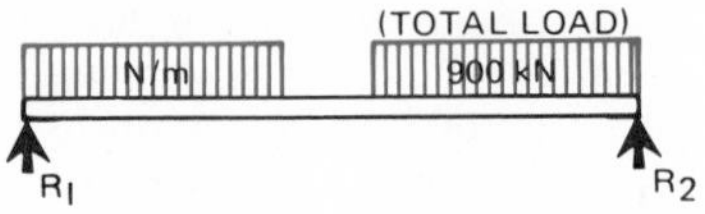
N/m
(TOTAL LOAD)
900 kN
R_1
R_2

(B) UNIFORMLY DISTRIBUTED LOADS

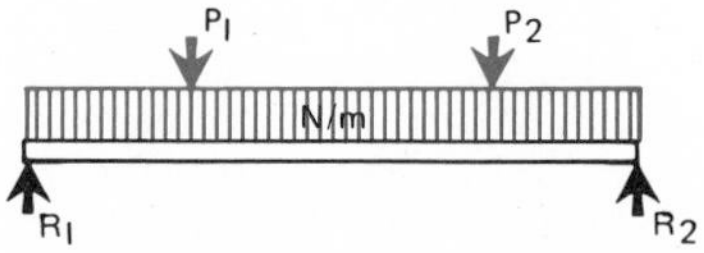
P_1
P_2
N/m
R_1
R_2

(C) COMBINATION OF CONCENTRATED
AND UNIFORMLY DISTRIBUTED LOADS

Fig. 30-4-2 Representation of loads on beam drawings.

The upward forces, or supports, that hold the beam in a state of equilibrium are called the *reactions* and are designated by the letters R_1 (left side) and R_2 (right side). The sum of the reactions $R_1 + R_2$, known as the *forces acting upward,* are equal and opposite to the downward forces or loads.

Moments

When a force acts upon an object at a distance from the object, as through a beam, the force is called a *moment*. See Fig. 30-4-3. A moment is the tendency of a force to cause rotation about a given point or axis. The magnitude of a moment is equal to the magnitude of the force times the perpendicular distance to the point. Since the force is measured in newtons and the distance in metres, moments are measured in newton-metres (N•m). In the inch-pound system the moments are measured in foot-pounds (ft•lb) or inch-pounds (in.•lb).

If a number of forces acting on a point are in equilibrium, the sum of the moments of all the forces about that point is equal to zero.

Therefore the sum of the moments of all forces that tend to produce clockwise moments about a given point is equal and opposite to the sum of all the forces that tend to produce counterclockwise moments at that given point. This *law of equilibrium* is very helpful in solving beam reaction.

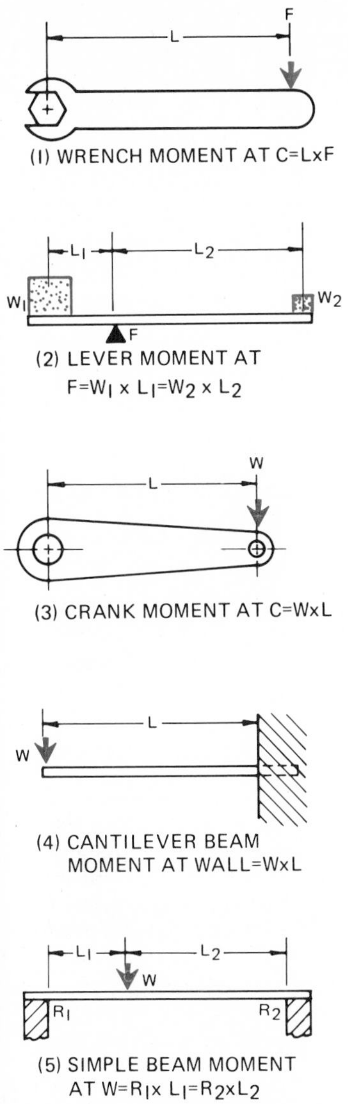
F
L
(1) WRENCH MOMENT AT C=LxF
L1
L2
W1
W2
F
(2) LEVER MOMENT AT
F=W1 x L1=W2 x L2
W
L
(3) CRANK MOMENT AT C=WxL
L
W
(4) CANTILEVER BEAM
MOMENT AT WALL=WxL
L1
L2
W
R1
R2
(5) SIMPLE BEAM MOMENT
AT W=R1x L1=R2xL2

Fig. 30-4-3 Application of moments.

Example 1. A force of 90 N is applied at the end of a wrench 300 mm from the center of the bolt which is being held by the wrench. Calculate the moment.

Solution. The moment may be found by multiplying the force times the distance: $90 \times 0.3 = 2.7$ N•m.

Example 2. A cantilever beam supports a concentrated load of 2 kN located 3600 mm from the support. Neglecting the mass of the beam, calculate the moment at the wall.

Solution. The moment taken at the wall or support may be found by multiplying the force times the distance: $2000 \times 3.6 = 7200$ N•m.

Example 3. A beam 4500 mm long has a concentrated load of 6 kN acting 1500

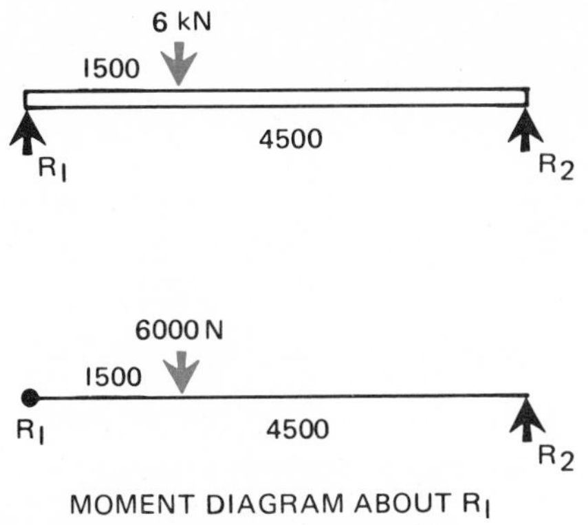
6 kN
1500
4500
R1
R2
6000 N
1500
R1
4500
R2
MOMENT DIAGRAM ABOUT R1

Fig. 30-4-4 Simple beam with concentrated load.

mm away from the left reaction. Neglecting the mass of the beam, calculate the reaction forces (Fig. 30-4-4).

Solution. Taking moments about reactions R_1, we have

Clockwise moments using newtons and metres = $1.5 \times 6000 = 9000$ N•m

Counterclockwise moments = $4.5 \times R_2$

$$\frac{\text{Clockwise}}{\text{moments}} = \frac{\text{Counterclockwise}}{\text{moments}}$$

$$9000 \text{ N•m} = 4.5 \times R_2$$

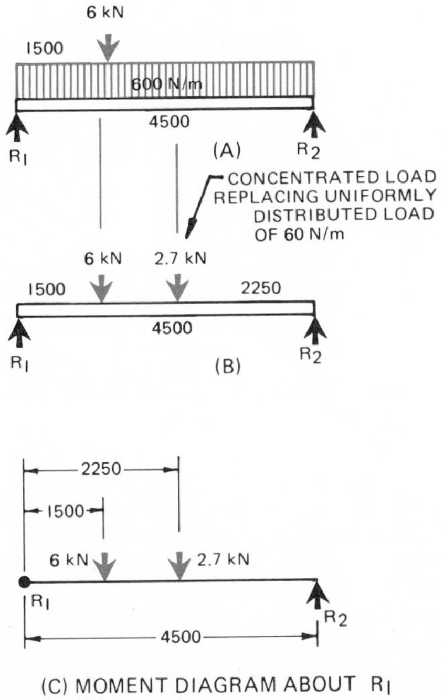
6 kN
1500
600 N/m
4500
R1
(A)
R2
CONCENTRATED LOAD
REPLACING UNIFORMLY
DISTRIBUTED LOAD
OF 60 N/m
6 kN
2.7 kN
1500
2250
4500
R1
(B)
R2
2250
1500
6 kN
2.7 kN
R1
R2
4500
(C) MOMENT DIAGRAM ABOUT R1

Fig. 30-4-5 Simple beam with uniformly distributed and concentrated loads.

Therefore

$$R_2 = \frac{9000}{4.5} = 2 \text{ kN}$$

$$R_1 + R_2 = 6 \text{ kN}$$

Therefore

$$R_1 = 6 - 2 = 4 \text{ kN}$$

Example 4. Use the same data given in Example 3, but include the force of gravity acting on the beam, which is 600 N/m. (See Fig. 30-4-5.)

Solution. Since the force of gravity acting on the beam is uniformly distributed, a concentrated load of 2.7 kN located midway on the beam, as shown in Fig. 30-4-5*b*, would have the same effect on the reactions R_1 and R_2.

Therefore by substituting the 2.7 kN concentrated load 2250 mm from the reactions for the uniformly distributed load, the reaction forces can now be found.

Taking moments about reaction R_1 (Fig. 30-4-5*c*), we have

Clockwise moments using newtons and metres

$$= 6000 \times 1.5 + 2700 \times 2.25$$

$$= 9000 + 6075$$

$$= 15\,075 \text{ N•m}$$

$$\frac{\text{Clockwise}}{\text{moments}} = \frac{\text{Counterclockwise}}{\text{moments}}$$

$$15\,075 \text{ N•m} = 4.5 \times R_2$$

Therefore

$$R_2 = 15\,075 \div 4.5 = 3350 \text{ N}$$

$$R_1 + R_2 = 8.7 \text{ kN}$$

Therefore

$$R_1 = 8.7 - 3.35 = 5.35 \text{ kN}$$

Assignments

1. On an A3- or B-size sheet, calculate the moments and loads for the problems shown in Fig. 30-4-A or 30-4-B.

2. On an A3- or B-size sheet, calculate the moments and reactions for the problems shown in Fig. 30-4-C or 30-4-D.

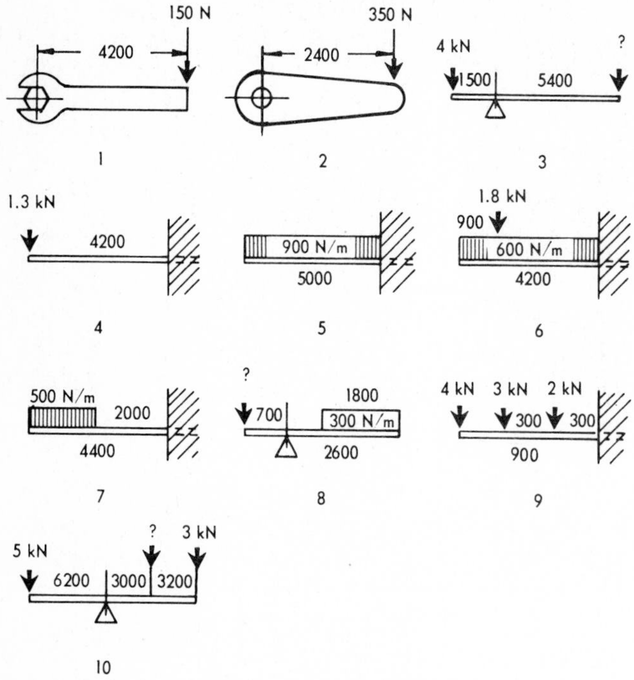
150 N
4200
1
350 N
2400
2
4 kN
?
1500
5400
3
1.3 kN
4200
4
900 N/m
5000
5
1.8 kN
900
600 N/m
4200
6
500 N/m
2000
4400
7
?
700
1800
300 N/m
2600
8
4 kN 3 kN 2 kN
300 300
900
9
5 kN
? 3 kN
6200 3000 3200
10

Fig. 30-4-A Calculating moments and loads.

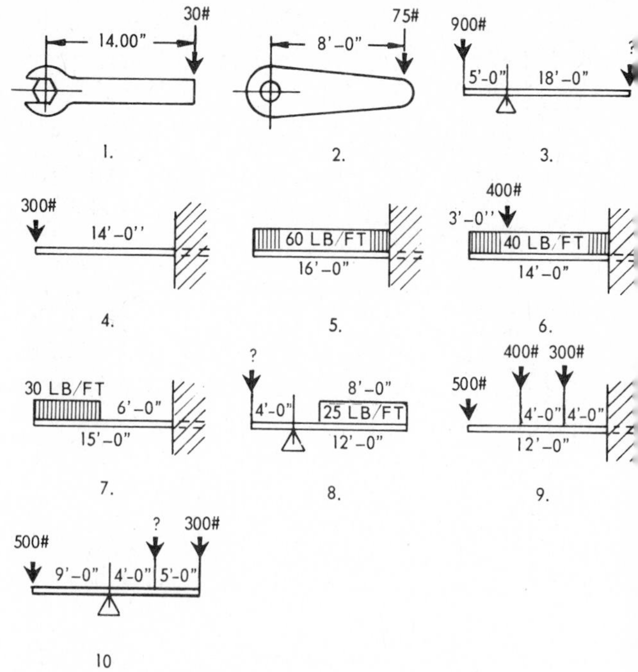
30#
14.00"
1.
75#
8'-0"
2.
900#
5'-0" 18'-0"
3.
300#
14'-0''
4.
60 LB/FT
16'-0"
5.
400#
3'-0''
40 LB/FT
14'-0"
6.
30 LB/FT
6'-0"
15'-0"
7.
?
4'-0"
8'-0"
25 LB/FT
12'-0"
8.
400# 300#
500#
4'-0" 4'-0"
12'-0"
9.
500#
? 300#
9'-0" 4'-0" 5'-0"
10

Fig. 30-4-B Calculating moments and loads.

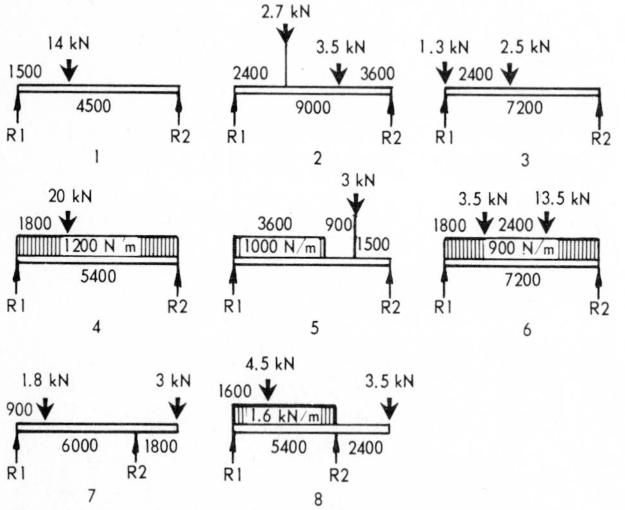
14 kN
1500
4500
R1 R2
1
2.7 kN
3.5 kN
2400 3600
9000
R1 R2
2
1.3 kN 2.5 kN
2400
7200
R1 R2
3
20 kN
1800
1200 N/m
5400
R1 R2
4
3 kN
3600 900
1000 N/m
1500
R1 R2
5
3.5 kN 13.5 kN
1800 2400
900 N/m
7200
R1 R2
6
1.8 kN 3 kN
900
6000 1800
R1 R2
7
4.5 kN
1600
3.5 kN
1.6 kN/m
5400 2400
R1 R2
8

Fig. 30-4-C Calculating moments and reactions.

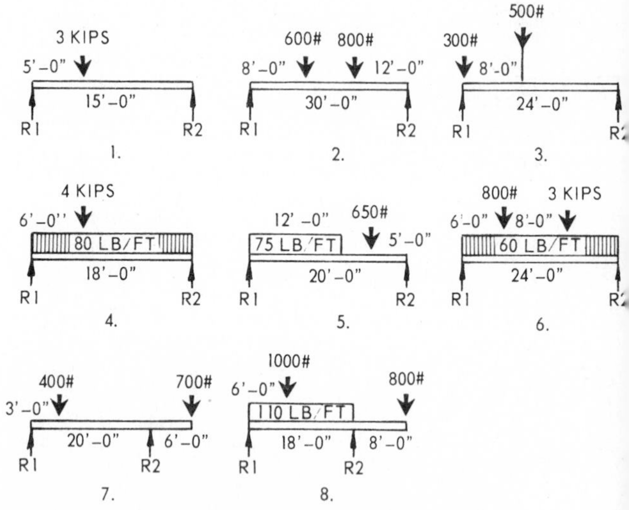
3 KIPS
5'-0"
15'-0"
R1 R2
1.
600# 800#
8'-0" 12'-0"
30'-0"
R1 R2
2.
500#
300#
8'-0"
24'-0"
R1
3.
4 KIPS
6'-0''
80 LB/FT
18'-0"
R1 R2
4.
12' -0" 650#
75 LB/FT
5'-0"
20'-0"
R1 R2
5.
800# 3 KIPS
6'-0" 8'-0"
60 LB/FT
24'-0"
R1
6.
400# 700#
3'-0"
20'-0" 6'-0"
R1 R2
7.
1000#
6'-0"
800#
110 LB/FT
18'-0" 8'-0"
R1 R2
8.

Fig. 30-4-D Calculating moments and reactions.

UNIT 30-5

SHEAR DIAGRAMS

When a beam supports a load, there is a tendency for the beam to fail by shear. In design work, it is essential to know what shear force a beam must resist at any section. The *vertical shear force* at a section of a beam is the algebraic sum of all the external forces acting on either side of the section. This can be further simplified by stating that the vertical shear at any section is equal to the product of the reaction minus the loads. The *section* is the name given to the cross section of the beam where the calculations are made. For simplification, only the forces acting to the left of the section will be calculated in this text.

Shear is designated as either *positive shear* or *negative shear*. When the sum of the vertical forces to the left of the section is upward, the shear is positive. When the sum of the vertical forces to the left of the section is downward, the shear is negative. See Fig. 30-5-1.

This information is represented in a shear force diagram which is normally drawn below the loading diagram of the beam. A horizontal zero base line is

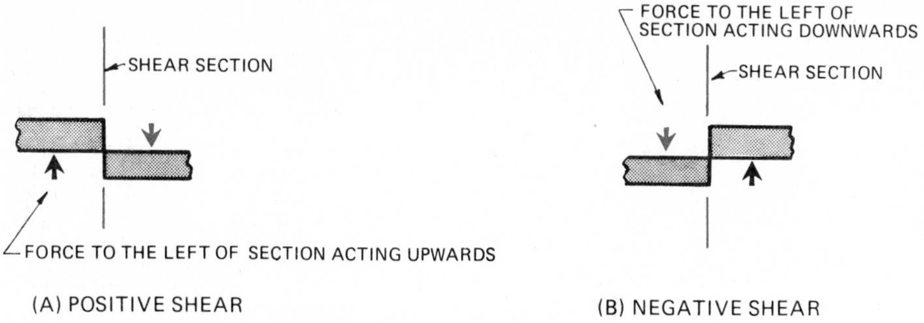
SHEAR SECTION
FORCE TO THE LEFT OF SECTION ACTING UPWARDS
(A) POSITIVE SHEAR
FORCE TO THE LEFT OF SECTION ACTING DOWNWARDS
SHEAR SECTION
(B) NEGATIVE SHEAR

Fig. 30-5-1 Designation of positive and negative shear.

drawn to the same horizontal scale as the loading diagram, and the positive shear is shown above this line while the negative shear is drawn below it. The magnitude of the shear at each section is shown by vertical lines drawn to a convenient scale.

In order to identify the section at which the shear is taken, a symbol (the letter V followed by a number) is used. The letter V refers to the magnitude of the vertical shear, and the number refers to the horizontal distance from the left end of the beam. Thus V_4 refers to the shear force taken at a section 4 m away from the left reaction (R_1) of a simple beam, or 4 m away from the free end of a cantilever beam. The mass of the beam will not be considered unless specified in the examples or problems.

Cantilever Beams

Cantilever beams should be drawn with the support shown at the RH side.

Example 1. Figure 30-5-2*a* represents a cantilever beam with a concentrated load at the free end. Construct the shear diagram.

Solution. Taking sections at various points along the beam and calculating V, the vertical shear to the left of the section, we have

$$V_0 = 0 - 300 = -300 \text{ N}$$
$$V_1 = 0 - 300 = -300 \text{ N}$$
$$V_2 = 0 - 300 = -300 \text{ N}$$
$$V_3 = 0 - 300 = -300 \text{ N}$$
$$V_4 = 0 - 300 = -300 \text{ N}$$

Since there is no reaction to the left of the section, the shear values are all negative and are drawn below the base line.

Example 2. Figure 30-5-2*b* illustrates a cantilever beam with a uniformly distributed load. Construct the shear diagram.

Solution. Taking sections at various points along the beam, starting at the free end, we have

$$V_0 = 0 - 0 = 0 \text{ N}$$
$$V_1 = 0 - 800 = -800 \text{ N}$$
$$V_2 = 0 - 800 \times 2 = -1600 \text{ N}$$
$$V_3 = 0 - 800 \times 3 = -2400 \text{ N}$$
$$V_4 = 0 - 800 \times 4 = -3200 \text{ N}$$

Example 3. Figure 30-5-3*a* illustrates a cantilever beam with a uniformly distributed load at the end of the beam. Construct the shear diagram.

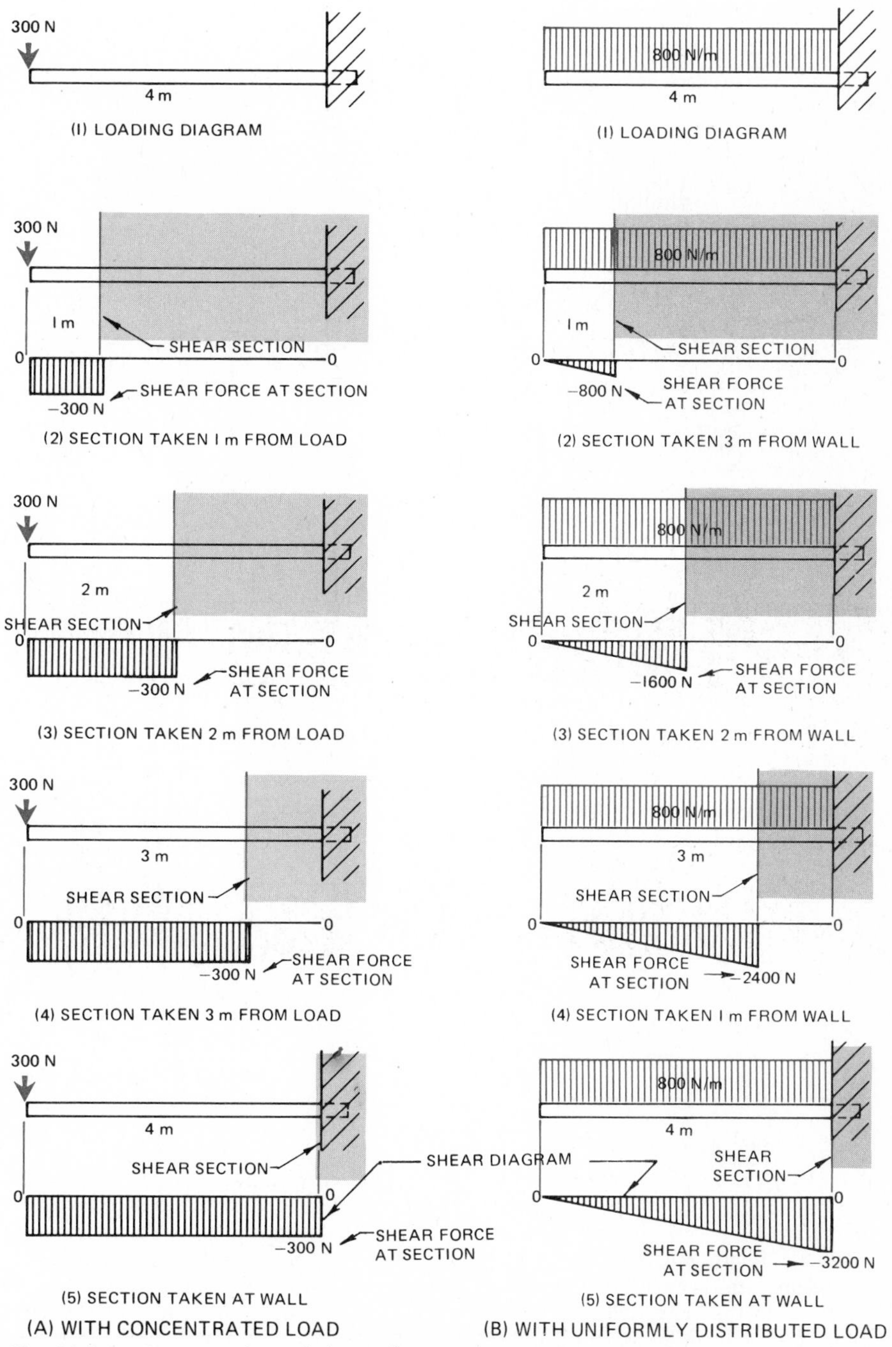
300 N
4 m
(1) LOADING DIAGRAM
300 N
1 m
SHEAR SECTION
SHEAR FORCE AT SECTION
-300 N
(2) SECTION TAKEN 1 m FROM LOAD
300 N
2 m
SHEAR SECTION
SHEAR FORCE AT SECTION
-300 N
(3) SECTION TAKEN 2 m FROM LOAD
300 N
3 m
SHEAR SECTION
SHEAR FORCE AT SECTION
-300 N
(4) SECTION TAKEN 3 m FROM LOAD
300 N
4 m
SHEAR SECTION
SHEAR FORCE AT SECTION
-300 N
(5) SECTION TAKEN AT WALL
(A) WITH CONCENTRATED LOAD
800 N/m
4 m
(1) LOADING DIAGRAM
800 N/m
1 m
SHEAR SECTION
SHEAR FORCE AT SECTION
-800 N
(2) SECTION TAKEN 3 m FROM WALL
800 N/m
2 m
SHEAR SECTION
SHEAR FORCE AT SECTION
-1600 N
(3) SECTION TAKEN 2 m FROM WALL
800 N/m
3 m
SHEAR SECTION
SHEAR FORCE AT SECTION
-2400 N
(4) SECTION TAKEN 1 m FROM WALL
800 N/m
4 m
SHEAR DIAGRAM
SHEAR SECTION
SHEAR FORCE AT SECTION
-3200 N
(5) SECTION TAKEN AT WALL
(B) WITH UNIFORMLY DISTRIBUTED LOAD

Fig. 30-5-2 Construction of shear diagram for cantilever beams.

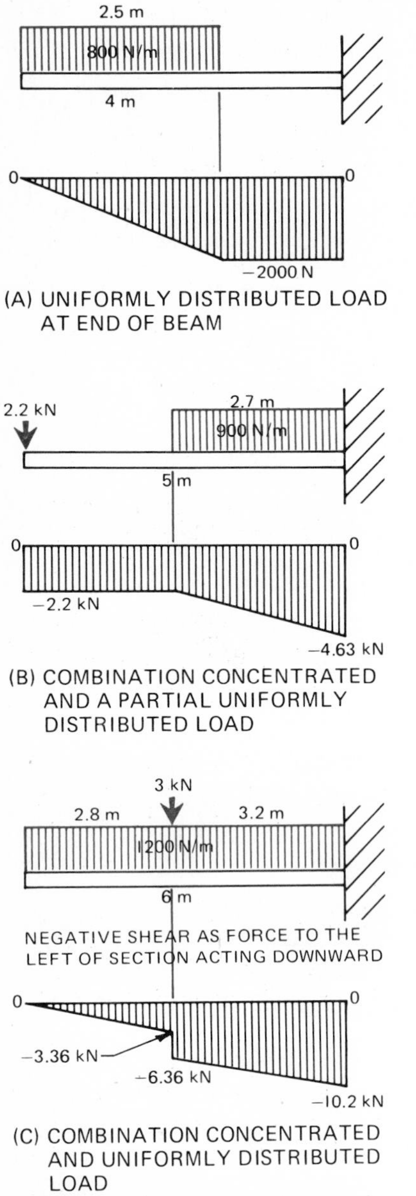
2.5 m
800 N/m
4 m
0
0
−2000 N
(A) UNIFORMLY DISTRIBUTED LOAD AT END OF BEAM
2.2 kN
2.7 m
900 N/m
5 m
0
0
−2.2 kN
−4.63 kN
(B) COMBINATION CONCENTRATED AND A PARTIAL UNIFORMLY DISTRIBUTED LOAD
3 kN
2.8 m
3.2 m
1200 N/m
6 m
NEGATIVE SHEAR AS FORCE TO THE LEFT OF SECTION ACTING DOWNWARD
0
0
−3.36 kN
−6.36 kN
−10.2 kN
(C) COMBINATION CONCENTRATED AND UNIFORMLY DISTRIBUTED LOAD

Fig. 30-5-3 Shear diagrams for cantilever beams.

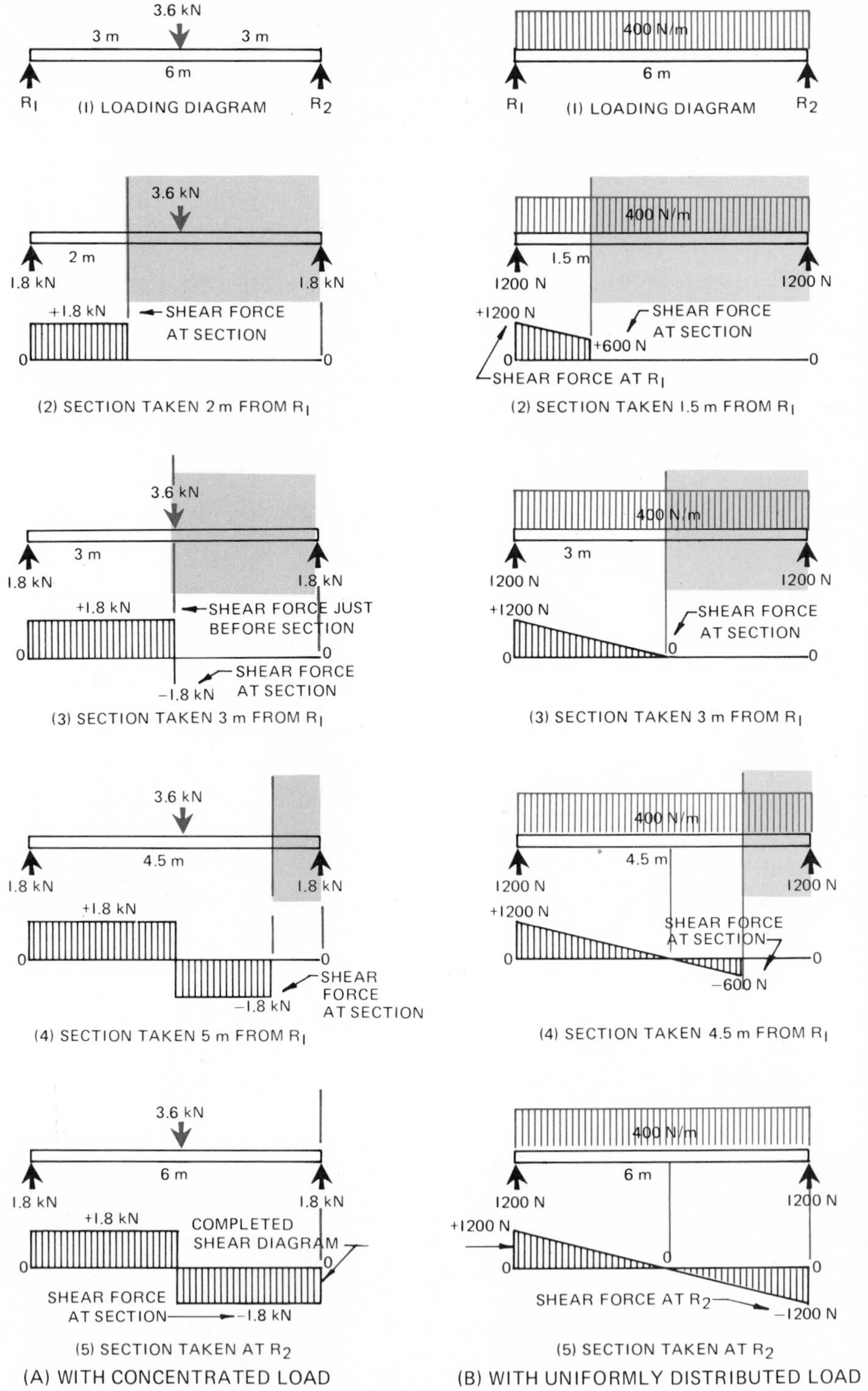
3.6 kN
3 m
3 m
6 m
R1
(1) LOADING DIAGRAM
R2
3.6 kN
2 m
1.8 kN
1.8 kN
+1.8 kN
SHEAR FORCE AT SECTION
(2) SECTION TAKEN 2 m FROM R1
3.6 kN
3 m
1.8 kN
1.8 kN
+1.8 kN
SHEAR FORCE JUST BEFORE SECTION
SHEAR FORCE AT SECTION
−1.8 kN
(3) SECTION TAKEN 3 m FROM R1
3.6 kN
4.5 m
1.8 kN
1.8 kN
+1.8 kN
SHEAR FORCE AT SECTION
−1.8 kN
(4) SECTION TAKEN 5 m FROM R1
3.6 kN
6 m
1.8 kN
1.8 kN
+1.8 kN
COMPLETED SHEAR DIAGRAM
SHEAR FORCE AT SECTION
−1.8 kN
(5) SECTION TAKEN AT R2
(A) WITH CONCENTRATED LOAD
400 N/m
6 m
R1
(1) LOADING DIAGRAM
R2
400 N/m
1.5 m
1200 N
1200 N
+1200 N
SHEAR FORCE AT SECTION
+600 N
SHEAR FORCE AT R1
(2) SECTION TAKEN 1.5 m FROM R1
400 N/m
3 m
1200 N
1200 N
+1200 N
SHEAR FORCE AT SECTION
(3) SECTION TAKEN 3 m FROM R1
400 N/m
4.5 m
1200 N
1200 N
+1200 N
SHEAR FORCE AT SECTION
−600 N
(4) SECTION TAKEN 4.5 m FROM R1
400 N/m
6 m
1200 N
1200 N
+1200 N
SHEAR FORCE AT R2
−1200 N
(5) SECTION TAKEN AT R2
(B) WITH UNIFORMLY DISTRIBUTED LOAD

Fig. 30-5-4 Construction of shear diagram for simple beams.

Solution. Taking sections at various points along the beam, starting at the free end, we have

$$V_0 = 0 - 0 = 0 \text{ N}$$
$$V_1 = 0 - 800 = -800 \text{ N}$$
$$V_2 = 0 - 800 \times 2 = -1600 \text{ N}$$
$$V_{2.5} = 0 - 800 \times 2.5 = -2000 \text{ N}$$
$$V_3 = 0 - 800 \times 2.5 = -2000 \text{ N}$$
$$V_4 = 0 - 800 \times 2.5 = -2000 \text{ N}$$

Example 4. Figure 30-5-3*b* shows a cantilever beam with a concentrated load at the free end of the beam and a uniformly distributed load at the fixed end. Construct the shear diagram.

Solution. Taking sections at various points along the beam, starting at the free end, we have

$$V_0 = 0 - 2.2 = -2.2 \text{ kN}$$
$$V_{2.3} = 0 - 2.2 = -2.2 \text{ kN}$$
$$V_3 = 0 - (2.2 + 0.7 \times 0.9) = -2.83 \text{ kN}$$
$$V_4 = 0 - (2.2 + 1.7 \times 0.9) = -3.73 \text{ kN}$$
$$V_5 = 0 - (2.2 + 2.7 \times 0.9) = -4.63 \text{ kN}$$

Example 5. Figure 30-5-3*c* shows a cantilever beam with a uniformly distributed load and a concentrated load acting in the middle section of the beam. Construct the shear diagram.

Solution. Taking sections at various points along the beam, starting at the free end, we have

$V_0 = 0 - 0 = 0$ kN

$V_1 = 0 - 1.2 \times 1 = -1.2$ kN

$V_{2.7} = 0 - 1.2 \times 2.7 = -3.24$ kN

$V_{2.8} = 0 - (1.2 \times 2.8 + 3) = -6.36$ kN

$V_6 = 0 - (1.2 \times 6 + 3) = -10.2$ kN

Simple Beams

In constructing the shear diagram for a simple beam, the magnitude of the reactions must be calculated first. The shear diagram is constructed in the same manner as for a cantilever beam. For calculation purposes V_0 will be considered as the section where the beam leaves the reaction R_1, and the shear section taken at R_2 will be considered as the section where the beam leaves the reaction R_2.

Example 6. Figure 30-5-4*a* illustrates a simple beam with a concentrated load of 3.6 kN at the center of the span. Construct the shear diagram.

Solution. From the figure it is apparent that the reactions are 1.8 kN each, because of the symmetrical loading. As previously mentioned, only the forces acting to the left of the section will be used for calculating the shear diagram. At the left end of the beam, the only force acting on the beam is the upward force of the reaction R_1. Therefore the shear force at $V_0 = 1.8 - 0 = +1.8$ kN. Taking sections along the beam, we find that

$V_1 = 1.8 - 0 = +1.8$ kN

$V_2 = 1.8 - 0 = +1.8$ kN

$V_{2.5} = 1.8 - 0 = +1.8$ kN

Up to and including the section just to the left of the center of the beam, no new forces are encountered; therefore from V_0 to $V_{2.99}$ the shear force is +1.8 kN. At the center of the beam the downward force of 3.6 kN occurs; thus

$V_3 = 1.8 - 3.6 = -1.8$ kN

Since no new loads are encountered for the remainder of the beam until R_2 is reached,

$V_4 = 1.8 - 3.6 = -1.8$ kN

$V_5 = 1.8 - 3.6 = -1.8$ kN

Just before R_2 is reached, the shear force from V_3 to V_6 is −1.8 kN.

Note that the shear diagram passes through zero, from +1.8 kN to −1.8 kN at the 3.6-kN load.

Example 7. Figure 30-5-4*b* illustrates a simple beam with a uniformly distributed load. Construct the shear diagram.

Solution. Taking sections at intervals along the beam, we have

$V_0 = 1200 - 0 = +1200$ N

$V_{1.5} = 1200 - 1.5 \times 400 = +600$ N

$V_3 = 1200 - 3 \times 400 = +0$ N

$V_{4.5} = 1200 - 4.5 \times 400 = -600$ N

$V_6 = 1200 - 6 \times 400 = -1200$ N

Example 8. Figure 30-5-5 illustrates a simple beam with a partial, uniformly distributed load starting at reaction R_1. Construct the shear diagram.

Solution. The value of reactions R_1 and R_2 must first be found. For calculation purposes, a concentrated load of 200 × 2, or 400 N, acting at the center of the uniformly distributed load will be used in place of the uniformly distributed load. Taking moments about R_1, we have

Clockwise moments = 200 × 2 × 1 = 400 N•m

Counterclockwise moments = $5 \times R_2$

$R_2 = 400 \div 5 = 80$ N

$R_1 = 400 - 80 = 320$ N

Taking sections at intervals along the beam starting at reaction R_1, we have

$V_0 = 320 - 0 = +320$ N

$V_1 = 320 - 1 \times 200 = +120$ N

$V_2 = 320 - 2 \times 200 = -80$ N

$V_3 = 320 - 2 \times 200 = -80$ N

$V_5 = 320 - 2 \times 200 = -80$ N

From the shear diagram it can be seen that the shear passes from positive shear to negative shear between V_1 and V_2. How to locate the position of zero shear will be discussed in Unit 30-6.

Example 9. Figure 30-5-6 illustrates a simple beam with two concentrated loads. Construct the shear diagram.

Solution. The reactions must first be calculated. Taking moments about reaction R_1, we have

Clockwise moments = 1.4 × 3500 + 3.4 × 2700 = 4900 + 9180 = 14 080 N•m

Counterclockwise moments = $6 \times R_2$

$$\text{Clockwise moments} = \text{Counterclockwise moments}$$

$6 \times R_2 = 14\ 080$ N•m

$$R_2 = \frac{14\ 080}{6} = 2347 \text{ N}$$

$R_1 + R_2 = 3500 + 2700 = 6200$ N

Therefore

$R_1 = 6200 - 2347 = 3853$ N

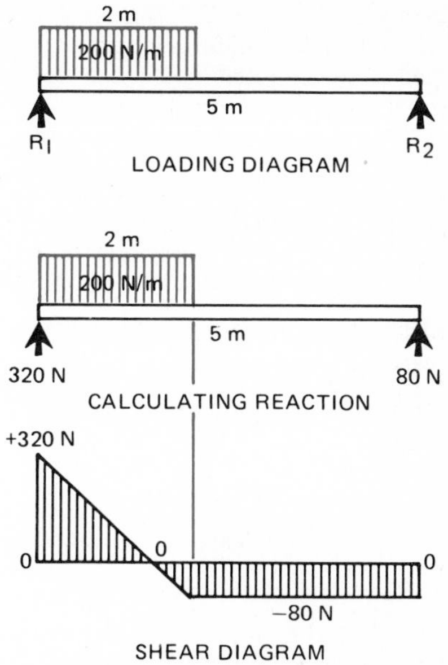
2 m
200 N/m
5 m
R1
R2
LOADING DIAGRAM
2 m
200 N/m
5 m
320 N
80 N
CALCULATING REACTION
+320 N
0
0
0
−80 N
SHEAR DIAGRAM

Fig. 30-5-5 Simple beam with partial uniformly distributed load.

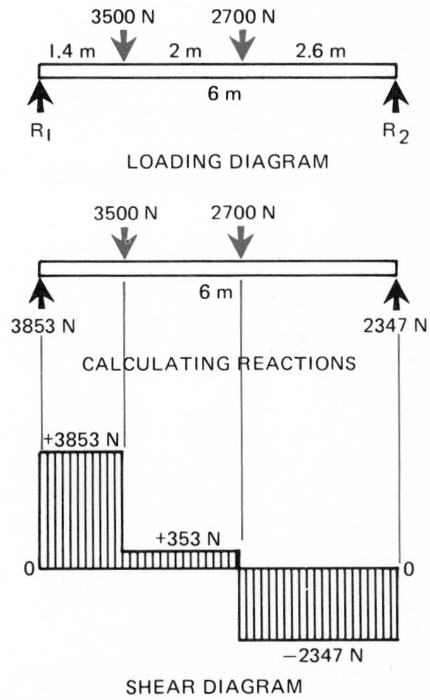
3500 N
2700 N
1.4 m
2 m
2.6 m
6 m
R1
R2
LOADING DIAGRAM
3500 N
2700 N
6 m
3853 N
2347 N
CALCULATING REACTIONS
+3853 N
+353 N
0
0
−2347 N
SHEAR DIAGRAM

Fig. 30-5-6 Simple beam with two concentrated loads.

Taking sections at intervals along the beam, we have

$V_0 = 3853 - 0 = +3853$ N

$V_{1.4} = 3853 - 3500 = +353$ N

$V_{3.4} = 3853 - 3500 - 2700 = -2347$ N

$V_6 = 3853 - 3500 - 2700 = -2347$ N

Example 10. Figure 30-5-7 illustrates a simple beam with a uniformly distributed load and a concentrated load acting on it. Construct the shear diagram.

Solution The reactions must be calculated first. For calculation purposes, a concentrated load of 500 × 6, or 3000 N, acting at the center of the beam will be used in place of the uniformly distributed load. Taking moments about R_1, we have

$$\text{Clockwise moments} = 2 \times 5400 + 3 \times 3000 = 10\,800 + 9000 = 19\,800 \text{ N•m}$$

$$\frac{\text{Clockwise}}{\text{moments}} = \frac{\text{Counterclockwise}}{\text{moments}}$$

$$6 \times R_2 = 19\,800 \text{ N•m}$$

$$R_2 = 19\,800 \div 6 = 3300 \text{ N}$$

$$R_1 + R_2 = 5400 + 3000 = 8400 \text{ N}$$

Therefore

$$R_1 = 8400 - R_2 = 8400 - 3300 = 5100 \text{ N}$$

Taking sections at intervals along the beam gives us

$$V_0 = 5100 - 0 = +5100 \text{ N}$$

$$V_1 = 5100 - 1 \times 500 = +4600 \text{ N}$$

$$V_{1.8} = 5100 - 1.8 \times 500 = +4200 \text{ N}$$

$$V_2 = 5100 - 2 \times 500 - 5400 = -1300 \text{ N}$$

$$V_4 = 5100 - 4 \times 500 - 5400 = -2300 \text{ N}$$

$$V_6 = 5100 - 6 \times 500 - 5400 = -3300 \text{ N}$$

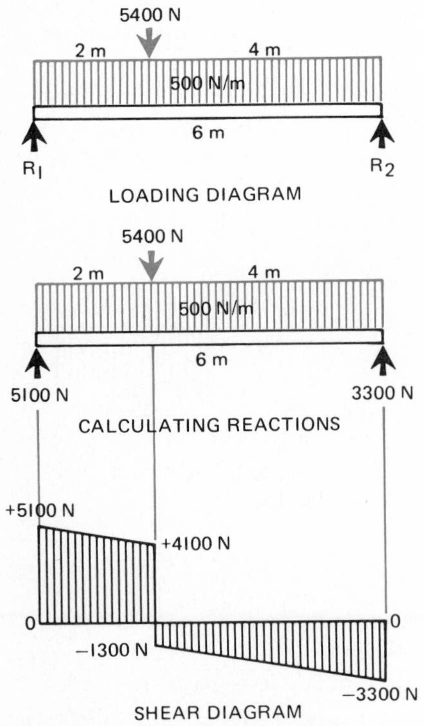
5400 N
2 m
4 m
500 N/m
6 m
R1
R2
LOADING DIAGRAM
5400 N
2 m
4 m
500 N/m
6 m
5100 N
3300 N
CALCULATING REACTIONS
+5100 N
+4100 N
0
0
-1300 N
-3300 N
SHEAR DIAGRAM

Fig. 30-5-7 Simple beam with uniformly distributed load and concentrated load.

Example 11. Figure 30-5-8 illustrates an overhanging beam. Construct the shear diagram.

Solution First the reactions must be calculated. Taking moments about R_1, we have

$$\text{Clockwise moments} = 1.4 \times 24 + 3.6 \times 35 + 7 \times 30$$

$$= 33.6 + 126 + 210$$

$$= 369.6 \text{ kN•m}$$

$$\text{Counterclockwise moments} = 5 \times R_2$$

$$\frac{\text{Clockwise}}{\text{moments}} = \frac{\text{Counterclockwise}}{\text{moments}}$$

$$5 \times R_2 = 369.6 \text{ kN•m}$$

$$R_2 = 369.6 \div 5 = 73.92 \text{ kN}$$

$$R_1 + R_2 = 24 + 35 + 30 = 89 \text{ kN}$$

Therefore

$$R_1 = 89 - 73.92 = 15.08 \text{ kN}$$

Taking sections at intervals along the beam gives us

$$V_0 = 15.08 - 0 = +15.08 \text{ kN}$$

$$V_{1.4} = 15.08 - 24 = -8.92 \text{ kN}$$

$$V_3 = 15.08 - 24 = -8.92 \text{ kN}$$

$$V_{3.6} = 15.08 - 24 - 35 = -43.92 \text{ kN}$$

$$V_5 = 15.08 - 24 - 35 + 73.92 = +30 \text{ kN}$$

$$V_7 = 15.08 - 24 - 35 + 73.92 = +30 \text{ kN}$$

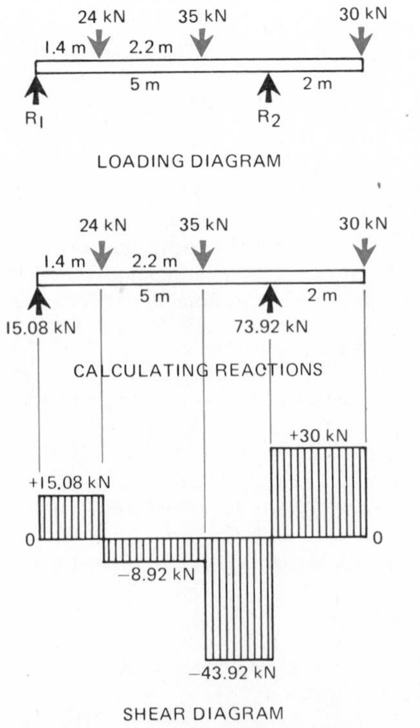
24 kN
35 kN
30 kN
1.4 m
2.2 m
5 m
2 m
R1
R2
LOADING DIAGRAM
24 kN
35 kN
30 kN
1.4 m
2.2 m
5 m
2 m
15.08 kN
73.92 kN
CALCULATING REACTIONS
+30 kN
+15.08 kN
0
0
-8.92 kN
-43.92 kN
SHEAR DIAGRAM

Fig. 30-5-8 Overhanging beam.

CONCLUSION

From the examples given, the following conclusions can be drawn for shear diagrams:

- Where there are concentrated loads, the shear lines are straight horizontal lines changing in value at the loads.
- Where there are uniformly distributed loads, the shear lines are straight inclined lines, the slope of the line being proportional to the load.
- At each concentrated load, including reactions, the shear line rises or drops vertically by an amount equal to the load at that section.

Assignments

1. On an A3- or B-size sheet, draw the four beams shown in Fig. 30-5-A or 30-5-B, showing the loading diagram, the shear diagram, and the calculations. Scale is to suit.

2. On an A3- or B-size sheet, draw the four beams shown in Fig. 30-5-C or 30-5-D, showing the loading diagram, calculated reaction diagram, and the shear diagram plus the calculations. Scale is to suit.

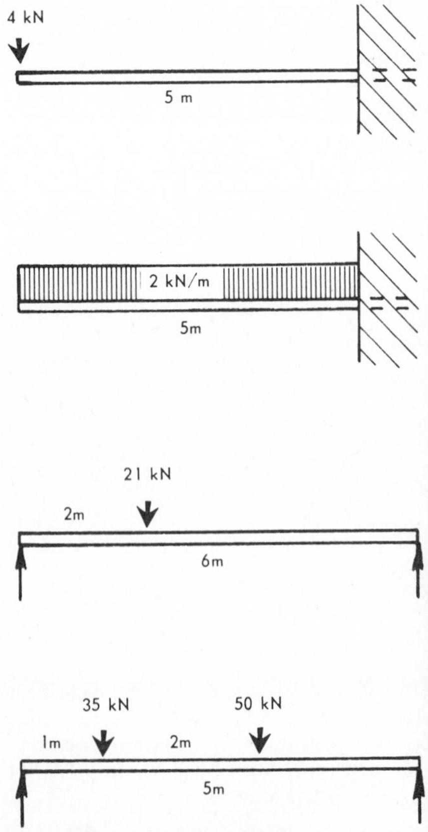
4 kN
5 m
2 kN/m
5m
21 kN
2m
6m
35 kN
50 kN
1m
2m
5m

Fig. 30-5-A Shear problems.

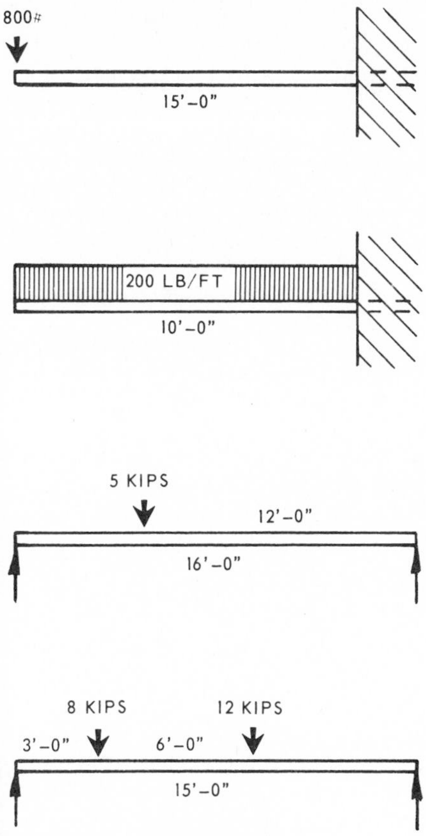
800#
15'-0"
200 LB/FT
10'-0"
5 KIPS
12'-0"
16'-0"
8 KIPS
12 KIPS
3'-0"
6'-0"
15'-0"

Fig. 30-5-B Shear problems.

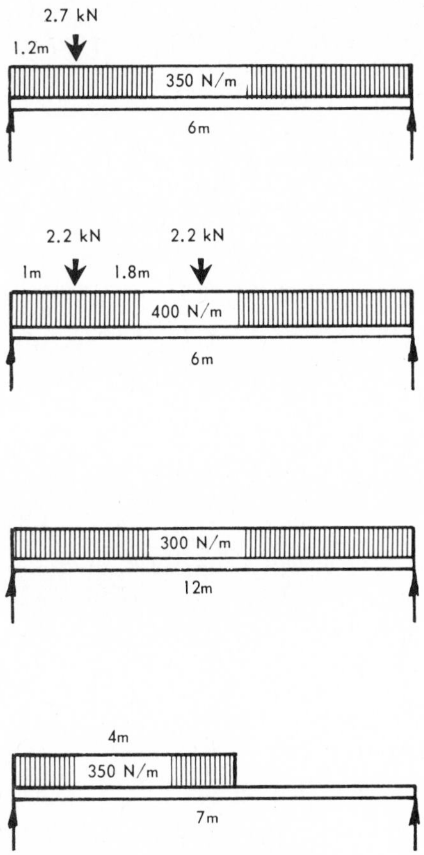
2.7 kN
1.2m
350 N/m
6m
2.2 kN
2.2 kN
1m
1.8m
400 N/m
6m
300 N/m
12m
4m
350 N/m
7m

Fig. 30-5-C Shear problems.

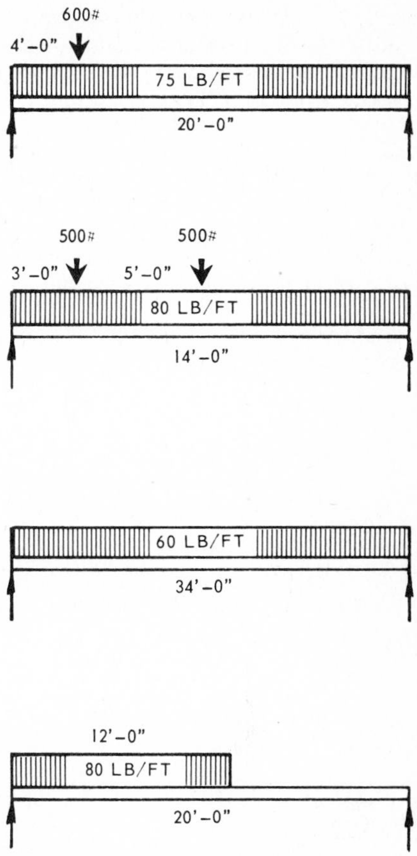
600#
4'-0"
75 LB/FT
20'-0"
500#
500#
3'-0"
5'-0"
80 LB/FT
14'-0"
60 LB/FT
34'-0"
12'-0"
80 LB/FT
20'-0"

Fig. 30-5-D Shear problems.

UNIT 30-6
BENDING MOMENT DIAGRAMS

As previously mentioned, when a load acts on a beam, the force tends to shear the beam. In addition to producing this shearing action, the load tends to *deflect* or *bend* the beam. To determine this deflection, which varies along the beam, the bending stresses must be calculated. Just as the shear diagram shows the shear at any section along the beam, a bending moment diagram is similarly constructed to show the bending moment at any point along the beam and also to indicate where the maximum bending occurs.

The *bending moment* at any section along the beam is equal to the sum of all the moments of the forces acting to the right or left of the beam. In drawing bending moment diagrams, the following points should be noted:

1. Forces are taken to the left of the section.

2. Upward moments are considered positive and are shown above the base line on the bending moment diagram.

3. Downward moments are considered negative and are shown below the base line.

4. The bending moment diagram is drawn directly below the shear diagram and to the same scale.

5. Shear is equal to reaction minus loads.

6. Bending moments are equal to reaction moments minus load moments.

7. In calculating the bending moments at any given section along the beam, the capital letter *M* is used to designate the bending moments. It is followed by a subscript which indicates the distance from the LH end of the beam. Thus, M_4 indicates the bending moments 4 m from the LH end of the beam.

Example 1. Figure 30-6-1*a* shows a cantilever beam with a concentrated load applied to the free end. Figure 30-5-2*a* shows the shear diagram development for this beam. Construct the bending moment diagram.

Solution. Taking moments at intervals along the beam starting at the left-hand end, we find the moments will be negative as the force is acting downward. Thus,

$$M_0 = -300 \times 0 = 0$$

$$M_1 = -300 \times 1 = -300 \text{ N·m}$$

$$M_2 = -300 \times 2 = -600 \text{ N·m}$$

$$M_3 = -300 \times 3 = -900 \text{ N·m}$$

$$M_4 = -300 \times 4 = -1200 \text{ N·m}$$

Example 2. Figure 30-6-1*b* shows a cantilever beam with a uniformly distributed load. Figure 30-5-2*b* shows the shear diagram development for this beam. Construct the bending moment diagram.

Solution. The bending moment at M_0 is zero. The 1 m section to the right of R_1 creates a force of 300 N. The force of any uniform load can be considered as acting at its center of gravity. Thus the 300 N load can be considered as acting 0.5 m away from R_1. Therefore,

$$M_1 = -800 \times 1 \times 0.5 = -400 \text{ N·m}$$

$$M_2 = -800 \times 2 \times 1 = -1600 \text{ N·m}$$

$$M_3 = -800 \times 3 \times 1.5 = -3600 \text{ N·m}$$

$$M_4 = -800 \times 4 \times 2 = -6400 \text{ N·m}$$

Example 3. Figure 30-6-2 shows a cantilever beam with a uniformly distributed load at the end of the beam. The shear diagram development was explained in Unit 30-5, Example 3, Fig. 30-5-3*a*.

Solution. Taking moments at intervals along the beam starting from the LH end, we have

$M_0 = 0$

$M_1 = -800 \times 1 \times 0.5 = -400$ N•m

$M_2 = -800 \times 2 \times 1 = -1600$ N•m

$M_{2.5} = -800 \times 2.5 \times 1.25$
$= -2500$ N•m

$M_3 = -800 \times 2.5 \times (3 - 1.25)$
$= -3500$ N•m

Note that at M_3 the distance from the section to the center of gravity of the load is 3 − 1.25 = 1.75 m, since the center of gravity is 1.25 m from the LH end.

$M_4 = -800 \times 2.5 \times (4 - 1.25) =$
-5500 N•m

Note that from $M_{2.5}$ to M_4 the line on the bending moment diagram is straight.

Example 4. Figure 30-6-3 shows a cantilever beam with a combination concentrated and a partial, uniformly distributed load. The shear diagram development was explained in Unit 30-5, Example 4, Fig. 30-5-3*b*. Construct the bending moment diagram.

Solution. The bending moments from M_0 to $M_{2.3}$ are calculated in the same manners as in Example 1.

$M_0 = -2.2 \times 0 = 0$

$M_1 = -2.2 \times 1 = -2.2$ kN•m

$M_2 = -2.2 \times 2 = -4.4$ kN•m

$M_{2.3} = -2.2 \times 2.3 = -5.06$ kN•m

$M_3 = -2.2 \times 3 + 0.9 \times 0.7 \times 0.35 =$
-6.82 kN•m

$M_4 = -2.2 \times 4 + 0.9 \times 1.7 \times 0.85 =$
-10.25 kN•m

$M_5 = -2.2 \times 5 + 0.9 \times 2.7 \times 1.35 =$
-14.28 kN•m

Example 5. Figure 30-6-4 shows a cantilever beam with a concentrated and uniformly distributed load. The shear diagram development was explained in Unit 30-5, Example 5, Fig. 30-5-3*c*. Construct the bending moment diagram.

Solution. The bending moments from M_0 to $M_{2.8}$ are calculated in the same manner as in Example 2.

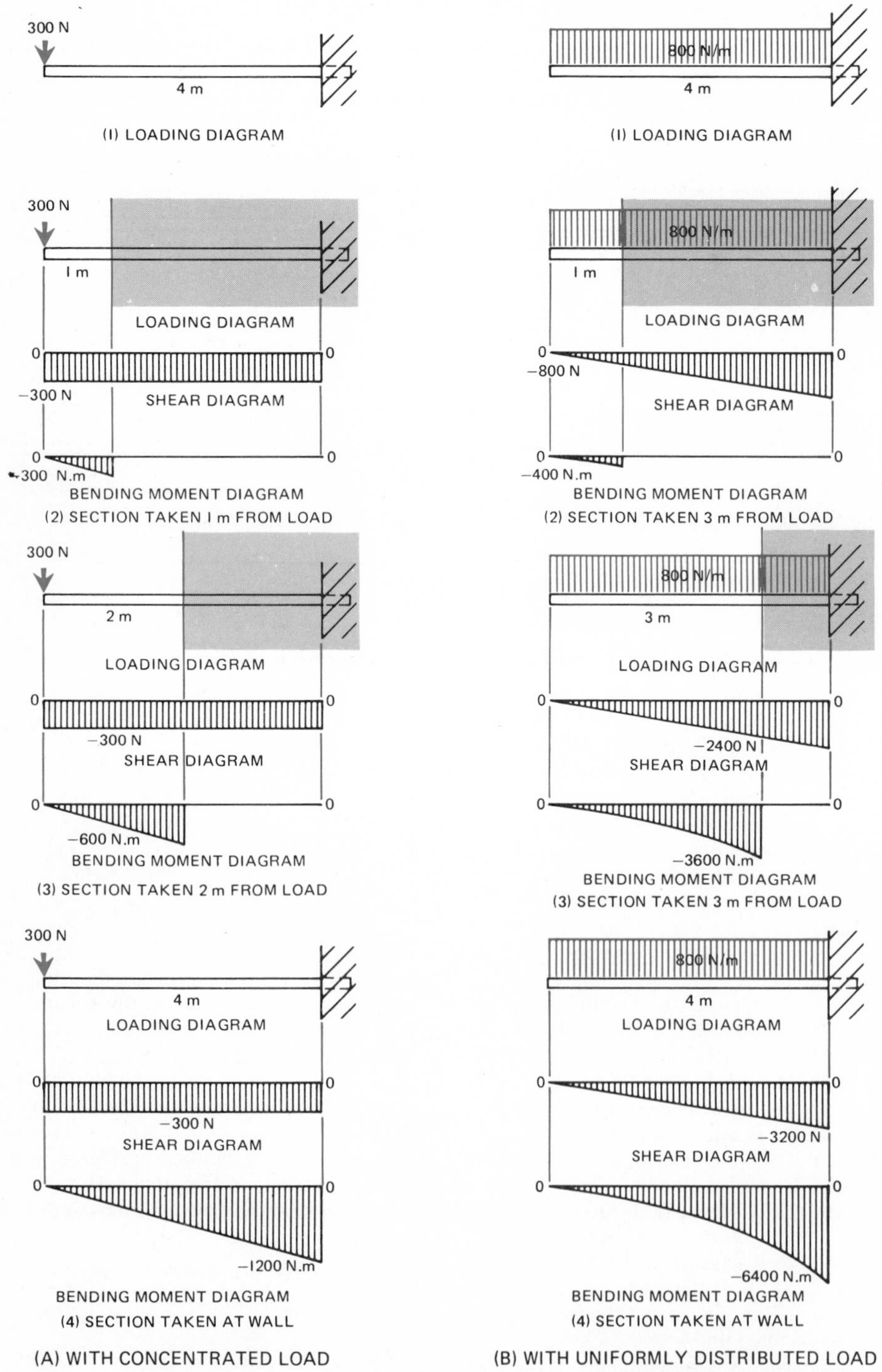
300 N
4 m
(1) LOADING DIAGRAM
800 N/m
LOADING DIAGRAM
SHEAR DIAGRAM
BENDING MOMENT DIAGRAM
-300 N
-300 N.m
-800 N
-400 N.m
(2) SECTION TAKEN I m FROM LOAD
(2) SECTION TAKEN 3 m FROM LOAD
2 m
3 m
-600 N.m
-2400 N
-3600 N.m
(3) SECTION TAKEN 2 m FROM LOAD
(3) SECTION TAKEN 3 m FROM LOAD
-1200 N.m
-3200 N
-6400 N.m
(4) SECTION TAKEN AT WALL
(A) WITH CONCENTRATED LOAD
(B) WITH UNIFORMLY DISTRIBUTED LOAD

Fig. 30-6-1 Construction of bending moment diagram for cantilever beams.

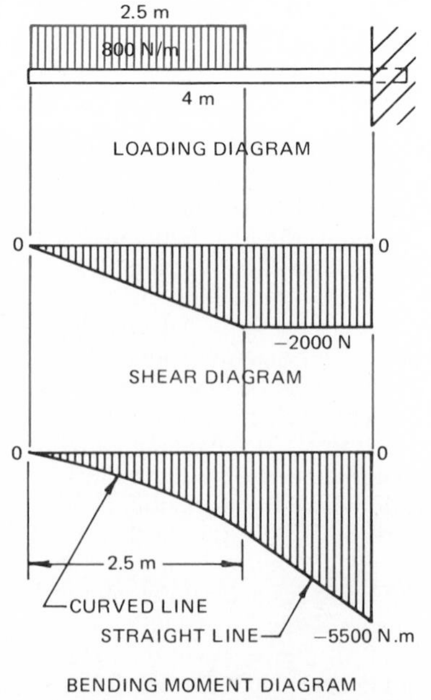
2.5 m
800 N/m
4 m
LOADING DIAGRAM
-2000 N
SHEAR DIAGRAM
2.5 m
CURVED LINE
STRAIGHT LINE
-5500 N.m
BENDING MOMENT DIAGRAM

Fig. 30-6-2 Cantilever beam with uniformly distributed load at end of beam.

$M_0 \ = 0$

$M_1 \ = -1.2 \times 1 \times 0.5 = -0.6$ kN•m

$M_2 \ = -1.2 \times 2 \times 1 = -2.4$ kN•m

$M_{2.8} = -1.2 \times 2.8 \times 1.4 + 3 \times 0 =$
-4.7 kN•m

$M_3 \ = -1.2 \times 3 \times 1.5 + 3 \times 0.2 =$
-6 kN•m

$M_4 \quad -1.2 \times 4 \times 2 + 3 \times 1.2 =$
-13.2 kN•m

$M_5 \ = -1.2 \times 5 \times 2.5 + 3 \times 2.2 =$
-21.6 kN•m

$M_6 \ = -1.2 \times 6 \times 3 + 3 \times 3.2 =$
-31.2 kN•m

Example 6. Figure 30-6-5 shows a simple beam with a partial uniformly distributed load. The development of the shear diagram was explained in Unit 30-5, Example 8, Fig. 30-5-5. Construct the bending moment diagram.

Solution. Taking moments at intervals along the beam, starting at reaction R_1, gives us

$M_0 = +320 \times 0 = 0$

$M_1 = +320 \times 1 - 200 \times 1 \times 0.5 =$
$+220$ N•m

$M_{1.5} = +320 \times 1.5 - 200 \times 1.5 \times 0.75 =$
$+255$ N•m

$M_2 = +320 \times 2 - 200 \times 2 \times 1 =$
$+240$ N•m

From the bending moment calculations it can be seen that the maximum bending moment occurs somewhere between M_1 and M_2 where the zero shear takes place.

The distance between R_1 and zero shear can be found as follows. Let the distance from R_1 to the point at zero shear be X. Thus

$$V_x = +320 - 200 \times X = 0$$
$$X = 320 \div 200 = 1.6 \text{ mm}$$

Maximum bending moment occurs at $M_{1.6}$:

$M_{1.6} = +320 \times 1.6 - 200 \times 1.6 \times 0.8 =$
$+256$ N•m

For calculating the moments for the remaining sections, the uniformly distributed load will be considered as a 400 N concentrated load acting 1 m from R_1.

$M_3 = +320 \times 3 - 400 \times 2 = +160$ N•m

$M_4 = +320 \times 4 - 400 \times 3 = +80$ N•m

$M_5 = +320 \times 5 - 400 \times 4 = 0$

Example 7. Figure 30-6-6*a* shows a simple beam with a concentrated load acting in the center of the beam. The development of the shear diagram was explained in Unit 30-5, Example 6, Fig. 30-5-4*a*. Construct the bending moment diagram.

Solution. Taking moments at intervals along the beam, starting at reaction R_1 which is acting upward, we have

$M_0 = +1.8 \times 0 = 0$

$M_1 = +1.8 \times 1 = +1.8$ kN•m

$M_2 = +1.8 \times 2 = +3.6$ kN•m

$M_3 = +1.8 \times 3 - 3.6 \times 0 = +5.4$ kN•m

$M_4 = +1.8 \times 4 - 3.6 \times 1 = +3.6$ kN•m

$M_5 = +1.8 \times 5 - 3.6 \times 2 = +1.8$ kN•m

$M_6 = +1.8 \times 6 - 3.6 \times 3 = 0$

Note: The maximum bending moment occurs at the point where shear passes through zero.

Example 8. Fig. 30-6-6*b* shows a simple beam with a uniformly distributed load. The development of the shear diagram was explained in Unit 30-5, Example 7, Fig. 30-5-4*b*. Construct the bending moment diagram.

Solution. The bending moment at M_0 is zero. The 1 m section to the right of R_1 creates a force of 400 N. The force of any uniform load can be considered as acting at its center of gravity. Thus, the 400 N load can be considered as acting 0.5 m away from R_1. Therefore,

$M_1 = +1200 \times 1 - 400 \times 1 \times 0.5 =$
$+1000$ N•m

$M_2 = +1200 \times 2 - 400 \times 2 \times 1 =$
$+1600$ N•m

$M_3 = +1200 \times 3 - 400 \times 3 \times 1.5 =$
$+1800$ N•m

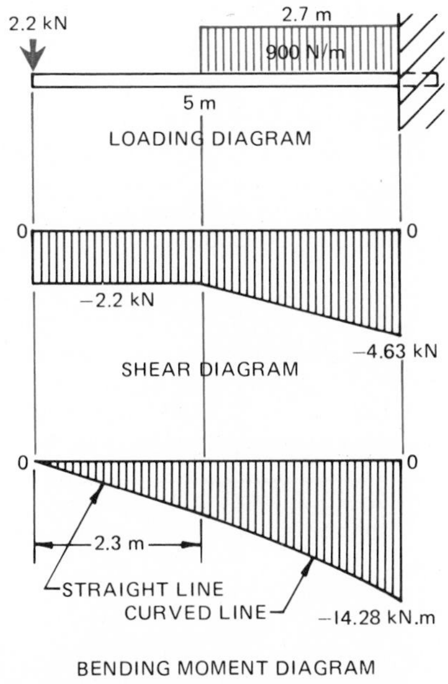
2.2 kN
2.7 m
900 N/m
5 m
LOADING DIAGRAM
0
0
−2.2 kN
−4.63 kN
SHEAR DIAGRAM
0
0
2.3 m
STRAIGHT LINE
CURVED LINE
−14.28 kN.m
BENDING MOMENT DIAGRAM

Fig. 30-6-3 Cantilever beam with concentrated load and partial uniformly distributed loads.

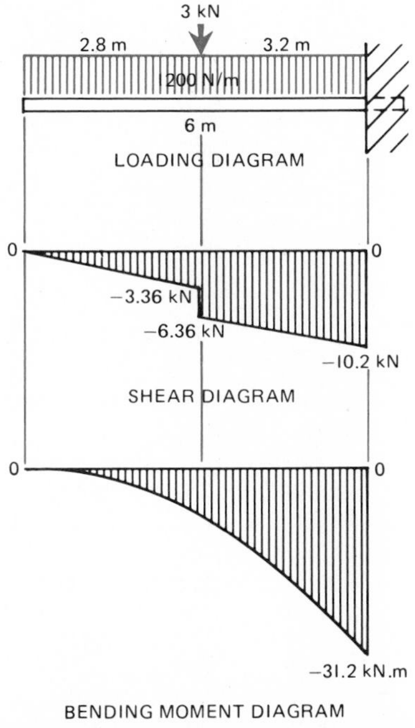
3 kN
2.8 m
3.2 m
1200 N/m
6 m
LOADING DIAGRAM
0
0
−3.36 kN
−6.36 kN
−10.2 kN
SHEAR DIAGRAM
0
0
−31.2 kN.m
BENDING MOMENT DIAGRAM

Fig. 30-6-4 Cantilever beam with concentrated and uniformly distributed loads.

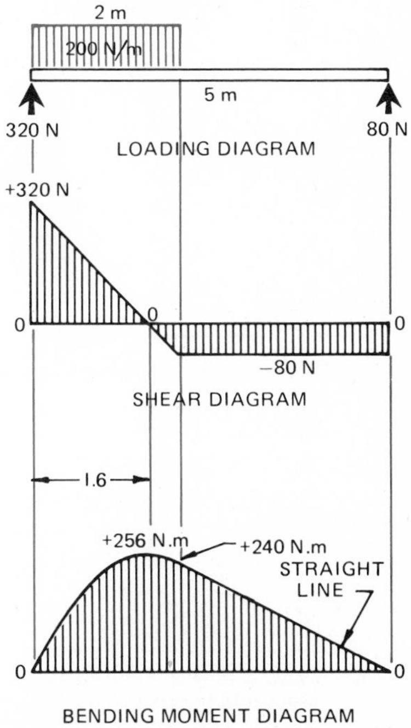
2 m
200 N/m
5 m
320 N
80 N
LOADING DIAGRAM
+320 N
0
0
0
−80 N
SHEAR DIAGRAM
1.6
+256 N.m
+240 N.m
STRAIGHT LINE
0
0
BENDING MOMENT DIAGRAM

Fig. 30-6-5 Simple beam with partial uniformly distributed load.

M_4 = +1200 × 4 − 400 × 4 × 2 = +1600 N•m

M_5 = +1200 × 5 − 400 × 5 × 2.5 = +1000 N•m

M_6 = +1200 × 6 − 400 × 6 × 3 = 0

Note that the maximum bending moment of +1800 N•m occurs at zero shear.

Example 9. Figure 30-6-7 shows a simple beam with two concentrated loads. The development of the shear diagram was explained in Unit 30-5, Example 9, Fig. 30-5-6. Construct the bending moment diagram.

Solution. Taking moments at intervals along the beam, starting at reaction R_1, gives us

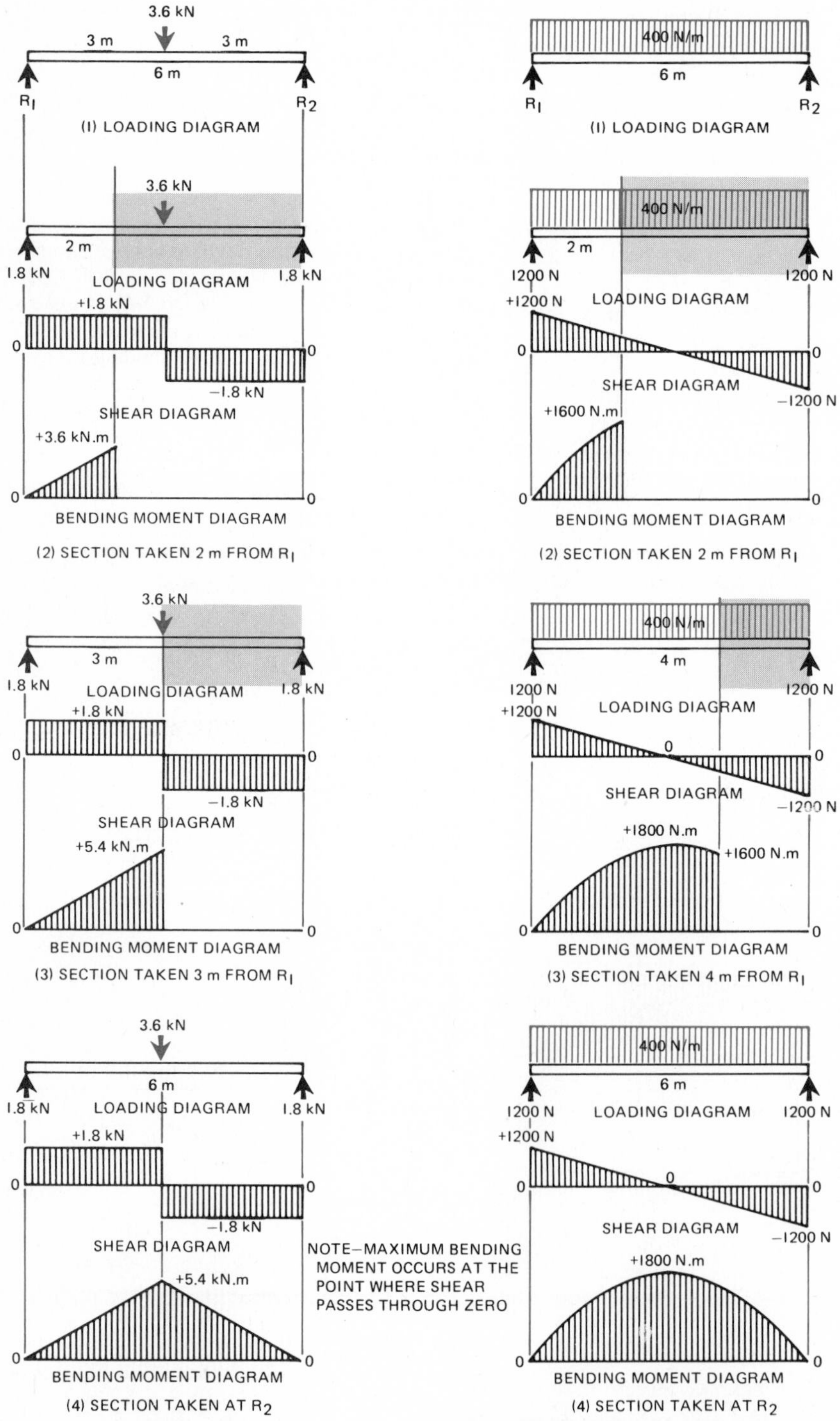

3.6 kN
3 m
3 m
6 m
R1
R2
(1) LOADING DIAGRAM
2 m
1.8 kN
LOADING DIAGRAM
+1.8 kN
−1.8 kN
SHEAR DIAGRAM
+3.6 kN.m
BENDING MOMENT DIAGRAM
(2) SECTION TAKEN 2 m FROM R1
+5.4 kN.m
(3) SECTION TAKEN 3 m FROM R1
(4) SECTION TAKEN AT R2
(A) WITH CONCENTRATED LOAD
400 N/m
1200 N
+1200 N
−1200 N
+1600 N.m
4 m
+1800 N.m
NOTE—MAXIMUM BENDING MOMENT OCCURS AT THE POINT WHERE SHEAR PASSES THROUGH ZERO
(3) SECTION TAKEN 4 m FROM R1
(B) WITH UNIFORMLY DISTRIBUTED LOAD

Fig. 30-6-6 Construction of bending moment diagram for simple beam.

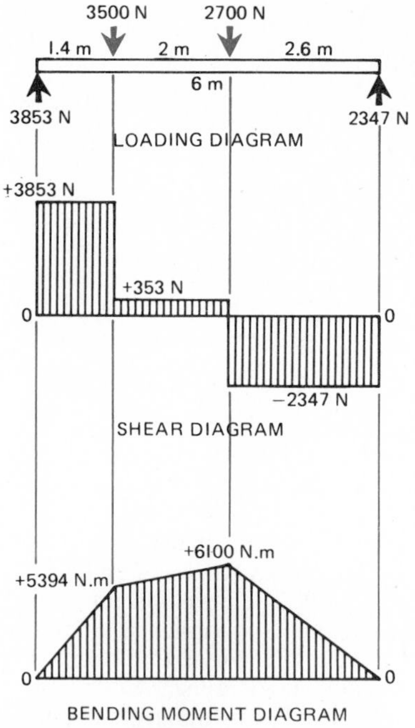

3500 N
2700 N
1.4 m
2 m
2.6 m
6 m
3853 N
2347 N
LOADING DIAGRAM
+3853 N
+353 N
−2347 N
SHEAR DIAGRAM
+6100 N.m
+5394 N.m
BENDING MOMENT DIAGRAM

Fig. 30-6-7 Simple beam with two concentrated loads.

M_0 = +3853 × 0 = 0

M_1 = +3853 × 1 = +3853 N•m

$M_{1.4}$ = +3853 × 1.4 = +5394 N•m

$M_{1.5}$ = +3853 × 1.5 − 3500 × 0.1 = +5430 N•m

M_2 = +3853 × 2 − 3500 × 0.6 = +5606 N•m

$M_{3.4}$ = +3853 × 3.4 − 3500 × 2 − 2700 × 0 = +6100 N•m

M_4 = +3853 × 4 − 3500 × 2.6 − 2700 × 0.6 = +4692 N•m

M_5 = 3853 × 5 − 3500 × 3.6 − 2700 × 1.6 = +2345 N•m

M_6 = 3853 × 6 − 3500 × 4.6 − 2700 × 2.6 = 0

Example 10. Figure 30-6-8 shows a simple beam with a uniformly distributed load and a concentrated load. The development of the shear diagram was explained in Unit 30-5, Example 10, Fig. 30-5-7. Construct the bending moment diagram.

Solution. Taking moments at intervals along the beam, starting at reaction R_1, we have

M_0 = +5100 × 0 = 0

M_1 = +5100 × 1 − 500 × 1 × 0.5 = +4850 N•m

M_2 = +5100 × 2 − 500 × 2 × 1 − 5400 × 0 = +9200 N•m

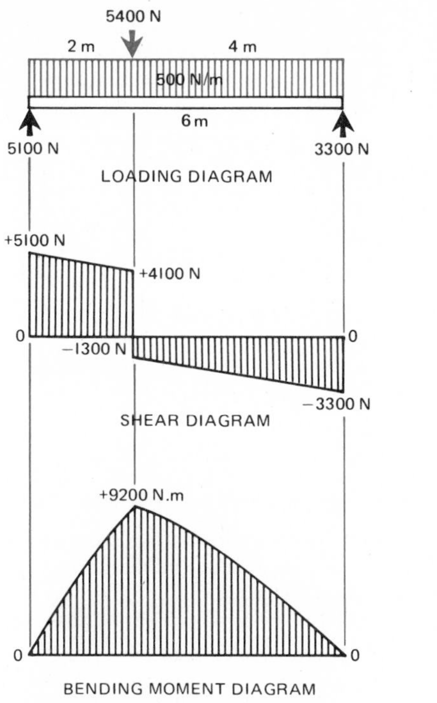
5400 N
2 m
4 m
500 N/m
6 m
5100 N
3300 N
LOADING DIAGRAM
+5100 N
+4100 N
0
-1300 N
0
-3300 N
SHEAR DIAGRAM
+9200 N.m
0
0
BENDING MOMENT DIAGRAM

Fig. 30-6-8 Simple beam with uniformly distributed and concentrated loads.

$$M_3 = +5100 \times 3 - 500 \times 3 \times 1.5 - 5400 \times 1 = +7650 \text{ N•m}$$

$$M_4 = +5100 \times 4 - 500 \times 4 \times 2 - 5400 \times 2 = +6000 \text{ N•m}$$

$$M_5 = +5100 \times 5 - 500 \times 5 \times 2.5 - 5400 \times 3 = +3050 \text{ N•m}$$

$$M_6 = +5100 \times 6 - 500 \times 6 \times 3 - 5400 \times 4 = 0$$

Example 11. Figure 30-6-9 shows an overhanging beam with three concentrated loads. The development of the shear diagram was explained in Unit 30-5, Example 11, Fig. 30-5-8. Construct the bending moment diagram.

Solution. Taking moments at intervals along the beam, starting at reaction R_1, we have

$$M_0 = +15.08 \times 0 = 0$$

$$M_1 = +15.08 \times 1 = +15.08 \text{ kN•m}$$

$$M_{1.4} = +15.08 \times 1.4 - 24 \times 0 = +21.1 \text{ kN•m}$$

$$M_2 = +15.08 \times 2 - 24 \times 0.6 = +16.12 \text{ kN•m}$$

$$M_3 = +15.08 \times 3 - 24 \times 1.6 = +6.84 \text{ kN•m}$$

$$M_{3.6} = +15.08 \times 3.6 - 24 \times 2.2 - 35 \times 0 = +1.39 \text{ kN•m}$$

$$M_4 = +15.08 \times 4 - 24 \times 2.6 - 35 \times 0.4 = -16.08 \text{ kN•m}$$

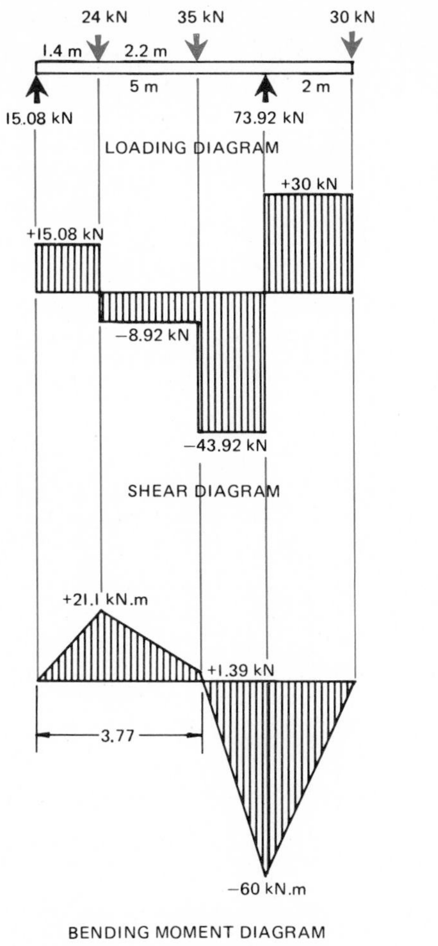
24 kN
35 kN
30 kN
1.4 m
2.2 m
5 m
2 m
15.08 kN
73.92 kN
LOADING DIAGRAM
+30 kN
+15.08 kN
-8.92 kN
-43.92 kN
SHEAR DIAGRAM
+21.1 kN.m
+1.39 kN
3.77
-60 kN.m
BENDING MOMENT DIAGRAM

Fig. 30-6-9 Overhanging beam.

$$M_5 = +15.08 \times 5 - 24 \times 3.6 - 35 \times 1.4 + 73.92 \times 0 = -60 \text{ kN•m}$$

$$M_6 = +15.08 \times 6 - 24 \times 4.6 - 35 \times 2.4 + 73.92 \times 1 = -30 \text{ kN•m}$$

$$M_7 = +15.08 \times 7 - 24 \times 5.6 - 35 \times 3.4 + 73.92 \times 2 = 0$$

From the bending moment diagram it can be seen that zero bending moment occurs to the right of the 35-kN load. Its exact location can be found as follows.

Let the distance between R_1 and the point where zero takes place be X. Then

$$M_x = 0$$

$$M_x = +15.08 \times X - 24(X - 1.4) = 0$$

$$= +15.08X - 24X + 33.6$$

$$8.92X = 33.6$$

$$X = 3.77 \text{ mm}$$

Example 12. Figure 30-6-10 shows a simple beam with a partial uniformly distributed load and a concentrated load. Find the position and magnitude of the maximum bending moment.

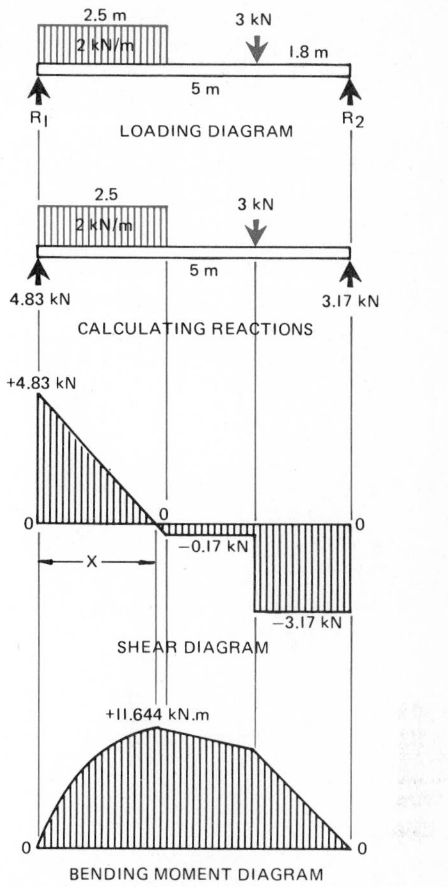
2.5 m
2 kN/m
3 kN
1.8 m
5 m
R1
R2
LOADING DIAGRAM
2.5
2 kN/m
3 kN
5 m
4.83 kN
3.17 kN
CALCULATING REACTIONS
+4.83 kN
0
0
0
-0.17 kN
X
-3.17 kN
SHEAR DIAGRAM
+11.644 kN.m
0
0
BENDING MOMENT DIAGRAM

Fig. 30-6-10 Simple beam with a partial uniformly distributed load and a concentrated load.

Solution. Reactions R_1 and R_2 must be calculated first. Taking moments about R_1, we have

$$\text{Clockwise moments} = 1.25 \times 2 \times 2.5 + 3.2 \times 3 = 15.85 \text{ kN•m}$$

$$\text{Counterclockwise moments} = 5 \times R_2$$

$$R_2 = 15.85 \div 5 = 3.17 \text{ kN}$$

$$R_1 = 2.5 \times 2 + 3 - 3.17 = 4.83 \text{ kN}$$

Next construct the shear diagram, taking sections at intervals along the beam, starting at reaction R_1:

$$V_0 = 4.83 - 0 = +4.83 \text{ kN}$$

$$V_1 = 4.83 - 1 \times 2 = +2.83 \text{ kN}$$

$$V_2 = 4.83 - 2 \times 2 = +0.83 \text{ kN}$$

$$V_{2.5} = 4.83 - 2.5 \times 2 = -0.17 \text{ kN}$$

$$V_{3.2} = 4.83 - 2.5 \times 2 - 3 = -3.17 \text{ kN}$$

$$V_5 = 4.83 - 2.5 \times 2 - 3 = -3.17 \text{ kN}$$

From the shear diagram, it is noted that the section having zero shear lies somewhere between R_1 and the end of the 2.5 m uniformly distributed load. Its exact position, X distance from R_1, may be found by taking the shear at Vx,

BEAM AND LOADING L IN METRES	MAXIMUM BENDING MOMENTS	MAXIMUM SHEAR	MAXIMUM DEFLECTION
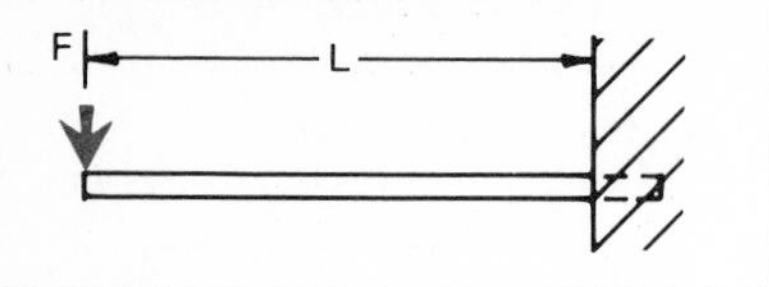 F L	FL	F	$\frac{FL^3}{3EI}$
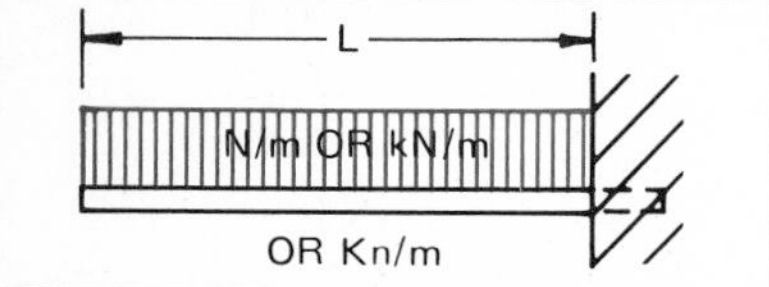 L N/m OR kN/m OR Kn/m	$\frac{NL^2}{2}$	NL	$\frac{NL^4}{8EI}$
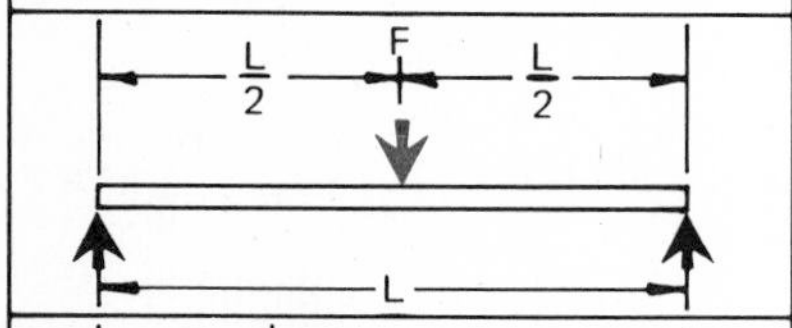 L/2 F L/2 L	$\frac{FL}{4}$	$\frac{F}{2}$	$\frac{FL^3}{48EI}$
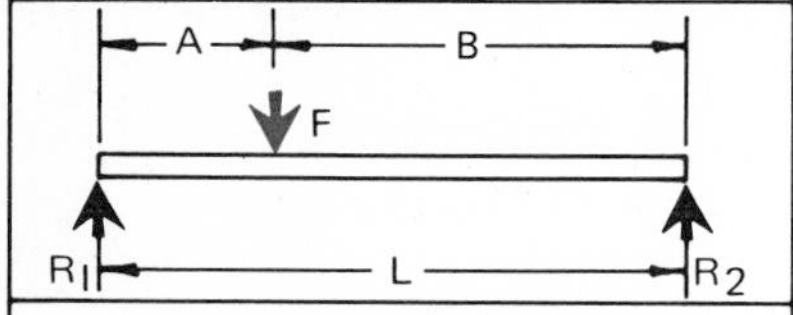 A B F R1 L R2	$\frac{FAB}{L}$	WHEN B IS GREATER THAN A $\frac{FB}{L}$	$\frac{FA^2B^2}{3EIL}$
 N/m OR kN/m L	$\frac{NL^2}{8}$	$\frac{NL}{2}$	$\frac{5NL^3}{384EI}$

E = MODULUS OF ELASTICITY I = MOMENT OF INERTIA

Fig. 30-6-11 Maximum bending moments, shear, and deflection for commonly occurring loads on beams.

which is zero. Thus $Vx = 4.83 - X \times 2 = 0$.
Therefore, $X = 4.83 \div 2 = 2.415$.

The maximum bending moment will occur where the shear passes through zero or 2.415 m from R_1. Thus the maximum bending moment is

$$M_2\bullet 415 = 4.83 \times 2.415 - 2 \times 2.415 \times 1.2075 = 5.8 \text{ kN}\bullet\text{m}$$

Example 13. A simple beam 8 m long carries a 3600 N concentrated load 2 m from the left abutment. Calculate the maximum bending moment and shear.

Solution The maximum bending moment for a simple beam with a concentrated load at any point is

$$\frac{FAB}{L} = \frac{3600 \times 2 \times 6}{8} = 5400 \text{ N}\bullet\text{m}$$

Maximum shear =

$$\frac{FB}{L} = \frac{3600 \times 6}{8} = 2700 \text{ N}$$

CONCLUSION

From the examples given, the following conclusions can be drawn from bending moment diagrams:

1. Where there are no loads on a part of a beam, the bending moment line is a straight, sloping line.

2. Where there is a uniformly distributed load, the bending moment line is a curve.

3. The maximum bending moment occurs at a section on the beam at which the shear passes through zero.

Standard beam formulas are shown in Fig. 30-6-11.

Assignments

1. On an A3- or B-size sheet, draw the bending moment diagrams for the four cantilevered beams shown in Fig. 30-6-A or 30-6-B. Below each diagram show the calculations. Scale is to suit.

2. On an A3- or B-size sheet, draw the bending moment diagrams for the four concentrated loaded beams shown in Fig. 30-6-C or 30-6-D. Below each diagram show the bending moment calculations.

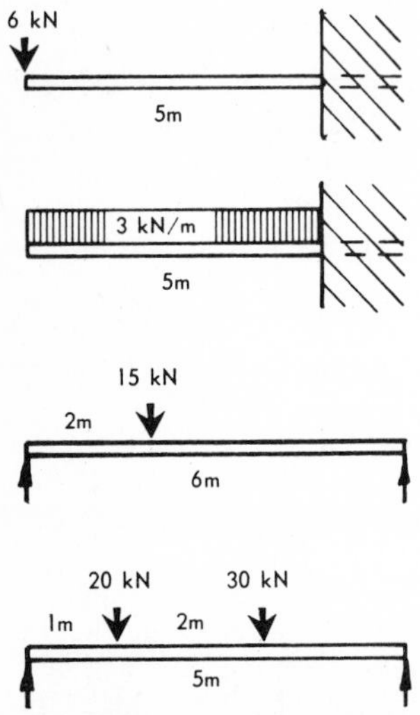
6 kN
5m
3 kN/m
5m
15 kN
2m
6m
20 kN
30 kN
1m
2m
5m

Fig. 30-6-A Bending moment problems.

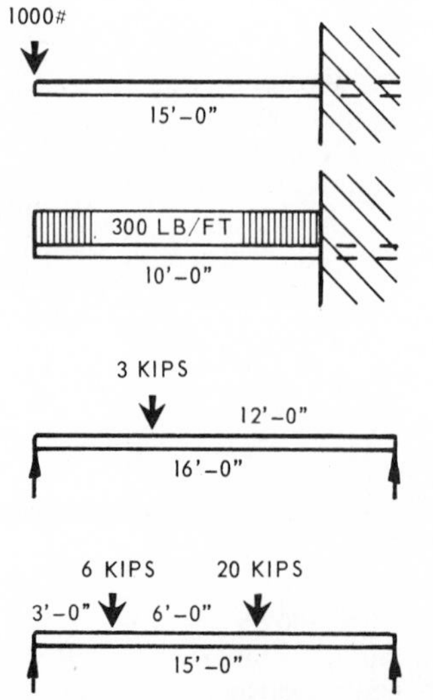
1000#
15'-0"
300 LB/FT
10'-0"
3 KIPS
12'-0"
16'-0"
6 KIPS
20 KIPS
3'-0"
6'-0"
15'-0"

Fig. 30-6-B Bending moment problems.

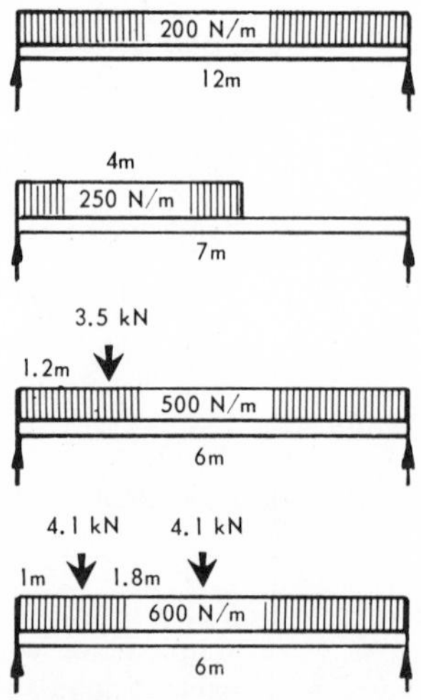
200 N/m
12m
4m
250 N/m
7m
3.5 kN
1.2m
500 N/m
6m
4.1 kN
4.1 kN
1m
1.8m
600 N/m
6m

Fig. 30-6-C Bending moment problems.

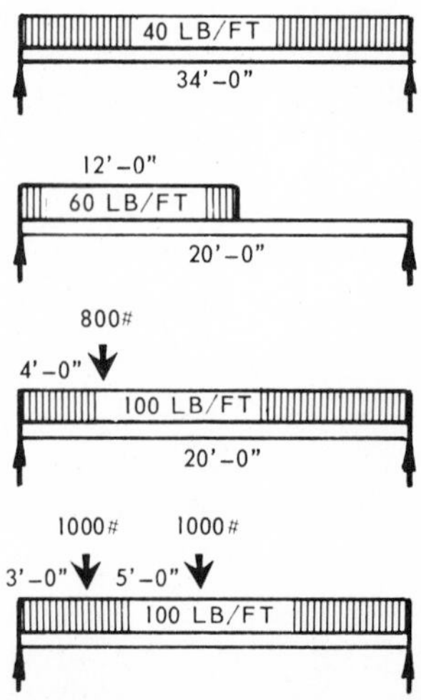
40 LB/FT
34'-0"
12'-0"
60 LB/FT
20'-0"
800#
4'-0"
100 LB/FT
20'-0"
1000#
1000#
3'-0"
5'-0"
100 LB/FT

Fig. 30-6-D Bending moment problems.

UNIT 30-7

BEAM DESIGN

It has been found from experience that beams normally fail at the section where the bending moment is maximum, rather than by shearing at the supports. Therefore, in beam design, it is customary first to select a suitable beam size to withstand the bending forces and then to check it for shear and deflection. The ability of a beam to resist bending depends on such factors as the material used, the shape of its cross section, and the way the cross section is turned with respect to the load. To illustrate this last point, one may bend a flat steel rule across its thin axis; but if the steel rule is set on its edge, then it is virtually impossible to bend the rule in the direction of its width. This resistance to bending can be measured in terms of a quantity called the *section modulus* of the section concerned. The theory and the mathematics behind the development of the section modulus of beams and shapes will not be covered in this text.

Thus the ability of any beam to resist bending is directly related to its section modulus, which is expressed in cubic millimetres and is denoted as Z in calculations. The bending stress S, the bending moment M_0, and the section modulus are related by the formula $M_0 = Z \times S \div 10^6$ in which the quantities are in newton-millimetres, cubic millimetres, and pascals, respectively. The stress in pascals is divided by 10^6 to obtain the stress per square millimetre. The section modulus for certain regular sections can be found from the formulas given in Fig. 30-7-1. The values of Z for structural-steel shapes and many common circular and rectangular sizes are tabulated in most engineers' handbooks.

The letter S is frequently used in textbooks to designate section modulus. However, to avoid confusion with the letter S for stress, the letter Z will be used to designate section modulus throughout this chapter.

Structural shapes may be placed in two general positions, as shown in Fig. 30-7-2. Since the resistance to bending will depend on the position of the beam with regard to its neutral axis, two section modulus values are generally shown in engineering tables. One value is used when the beam is in the upright position, as shown in Fig. 30-7-2*c*(1) where the XX axis is the neutral axis; the other is used when the beam is in the flat position, as shown in Fig. 30-7-2*c*(2), where the YY axis is the neutral axis. The *neutral axis* is defined as the axis which passes through the centroid of the cross-sectional area.

The majority of engineering handbooks show only one illustration of the structural shape with both the XX and YY axes shown as illustrated in Fig. 30-7-2*c*(3).

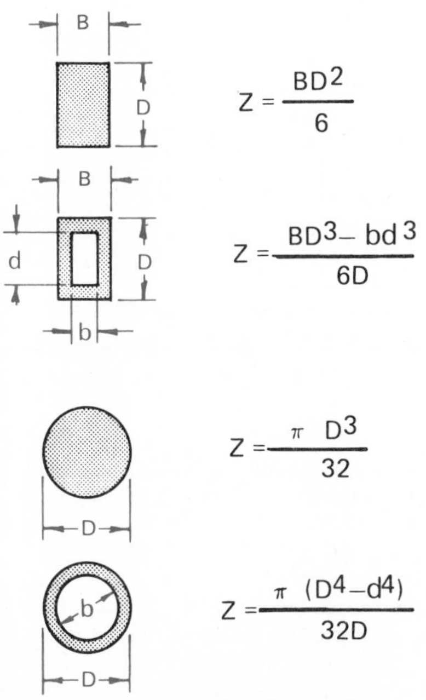
B
D
$Z = \frac{BD^2}{6}$
B
d
D
b
$Z = \frac{BD^3 - bd^3}{6D}$
D
$Z = \frac{\pi D^3}{32}$
b
D
$Z = \frac{\pi (D^4 - d^4)}{32D}$

Fig. 30-7-1 Formulas for section moduli for common shapes.

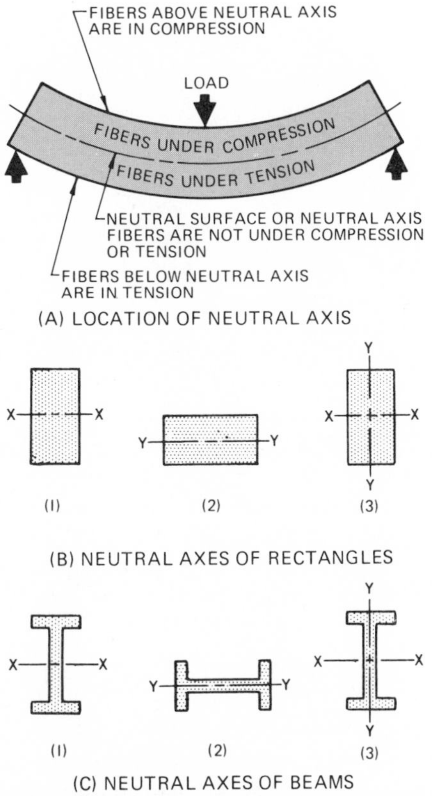
FIBERS ABOVE NEUTRAL AXIS ARE IN COMPRESSION
LOAD
FIBERS UNDER COMPRESSION
FIBERS UNDER TENSION
NEUTRAL SURFACE OR NEUTRAL AXIS FIBERS ARE NOT UNDER COMPRESSION OR TENSION
FIBERS BELOW NEUTRAL AXIS ARE IN TENSION
(A) LOCATION OF NEUTRAL AXIS
X
Y
(1)
(2)
(3)
(B) NEUTRAL AXES OF RECTANGLES
(1)
(2)
(3)
(C) NEUTRAL AXES OF BEAMS

Fig. 30-7-2 Neutral axis.

Shearing Stresses in Beams

In designing beams for vertical shear, it is customary to consider only the full height of the webs of S, C, WT, and WWT beams to carry the full load; the flanges are not considered.

Example 1. A cantilever beam 3 m long supports a 20 kN load at the end of the beam. What size A36 beam is required?

Solution. First select a beam to withstand the bending forces. Refer to Fig. 30-6-11; maximum bending moments = $FL = 20 \times 3 = 60$ kN•m, or 60×10^6 N•mm. The allowable bend stress for A36 steel (see Fig. 30-1-8) is 165 MPa.

Section modulus required = Z =

$$M \div \frac{S}{10^6} = (60 \times 10^6) \div \frac{165 \times 10^6}{10^6}$$

$$= 363\ 600 \text{ mm}^3$$

Refer to Fig. 30-7-3. A W200 × 42 has a section modulus of 398 000 mm³, which is acceptable. If the depth of the beam is not an important design factor, then the S250 × 38 beam, which is lighter and has a section modulus of 406 000, would be the most economical.

Next the beam must be checked for vertical shear. Maximum shear force = 20 kN. Web area of a W200 × 42 beam (refer to the structural-steel handbook) = 8 × 204 = 1632 mm². Average vertical shear stress = 20 kN ÷ 0.001 632 = 12.25 MPa. Permissible vertical shear stress for steel (see Fig. 30-1-8) is 100 MPa. Therefore the W200 × 42 beam is acceptable.

Example 2. A simple beam 6 m long supports a uniformly distributed load of 6 kN•m. Neglecting the mass of the beam, select the lightest A572M-310 beam to safely carry this load.

Solution. First select a beam to withstand the bending forces. Refer to Fig. 30-6-10. Maximum bending moments = $NL^2 \div 8 = 6000 \times 10^6 \div 8 = 27\ 000$ N•m, or 27×10^6 N•mm. The allowable bending stress for A572M-310 steel (see Fig. 30-1-8) is 205 MPa. Section modulus required = $Z = M \div (S \div 10^6) = 27 \times 10^6 \times 10^6 \div (205 \times 10^6) = 131\ 700$ mm³.

Referring to Fig. 30-7-3, we find that a W150 × 24 beam has a section modulus of 165 000, which is acceptable.

Next the beam must be checked for vertical shear.

Maximum shear force =

$$\frac{NL}{2} = \frac{6 \text{ kN} \times 6}{2} = 18 \text{ kN}$$

Web area of a W150 × 24 beam (refer to the structural-steel handbook) = 159 × 6 = 954 mm². Average vertical shear stress = 18 000 ÷ 0.000 954 = 18.9 MPa.

METRIC			INCH-POUND		
Section Modulus 10^3 mm^3	Shape	Moment of Inertia = I 10^6 mm^4**	Section Modulus in^3	Shape	Moment of Inertia = I in^4**
2573	W450×127	599	157	W18×85	1440
2474	W400×1311	508	151	W16×88	1220
2474	W550×109	666	151	W21×73	1600
2327	W450×115	537	142	W18×77	1290
2294	W550×101	616	140	W21×68	1480
2147	W350×125	386	131	W14×84	928
2048	W300×137	328	125	W12×92	789
1983	W350×116	354	121	W14×78	851
1934	W450×95	437	118	W18×64	1050
1901	W400×106	392	116	W16×71	941
1901	W300×106	301	116	W12×85	723
1835	W350×110	332	112	W14×74	797
1753	W300×118	276	107	W12×79	663
1721	W400×95	348	105	W16×64	836
1598	W300×107	248	97.5	W12×72	597
1545	W400×86	311	94.3	W16×58	748
1511	W350×91	267	92.2	W14×61	641
1442	W300×97	222	88	W12×65	533
1326	W400×75	273	80.9	W16×50	657
1206	W250×98	159	73.6	W10×66	382
1157	W300×79	177	70.6	W12×53	426
1062	W300×75	164	64.8	W12×50	395
1062	S300×75	202	64.8	S15×50	486
1027	W350×64	179	62.7	W14×43	429
991	W250×80	127	60.5	W10×54	306
977	S375×64	186	59.6	S15×42.9	447
896	W350×56	161	54.7	W14×38	386
850	W300×60	129	51.9	W12×40	310
832	S300×75	127	50.8	S12×50	305
806	W250×31	104	49.2	W10×45	249
710	W200×71	76.6	43.3	W8×48	184
685	W350×44	121	41.8	W14×30	290
595	S300×47.4	90.7	36.3	S12×31.8	218
575	W250×49	71.2	35.1	W10×33	171
506	W250×43	65.8	30.9	W10×29	158
451	W200×46	45.8	27.5	W8×31	110
406	S250×38	51.6	24.8	S10×25.4	124
398	W200×42	40.7	24.3	W8×28	97.8
354	W250×31	44.5	21.6	W10×21	107
341	W200×36	34.3	20.8	W8×24	82.5
265	S200×34	27	16.2	S8×23	64.9
233	W200×26	23.6	14.2	W8×17	56.6
220	W150×30	17.3	13.4	W6×20	41.5
197	M300×17.6	29.9	12	M12×11.8	71.9
165	W150×24	13.2	10.1	W6×16	31.7
118	W150×18	9.03	7.23	W6×12	21.7
99	W100×19.4	4.7	5.43	W4×13	11.3
83	W150×12.7	6.16	5.08	W6×8.5	14.8

*Soft Converted **Taken at X-X axis

Fig. 30-7-3 Section modulus and moment of inertia for shapes used as beams.

The permissible vertical shear stress for A572M-310 steel (see Fig. 30-1-8) is 125 MPa. Therefore the W150 × 24 beam is acceptable.

Example 3. A simple beam 5 m long supports a 130 kN concentrated load 1.2 m from the left abutment. What size A36 beam is required?

Solution. First select a beam to withstand the bending forces. Refer to Fig. 30-6-11. Maximum bending moments = $FAB \div L = 130\,000 \times 1.2 \times 3.8 \div 5 = 118\,560$ N•m, or 118 560 000 N•mm. Allowable bending stress for A36 steel (see Fig. 30-1-8) is 165 MPa.

Section modulus required = Z

$$= M \div \frac{S}{10^6}$$

$$= 118\,560\,000 \times \frac{10^6}{165 \times 10^6}$$

$$= 718\,500 \text{ mm}^3$$

Referring to Fig. 30-7-3, we find that a W250 × 67 beam has a section modulus of 806 000.

Next the beam must be checked for vertical shear.

Maximum shear force =

$$\frac{130\,000 \times 3.8}{5} = 98\,800 \text{ N}$$

Web area of a W250 × 67 beam (refer to the structural-steel handbook) = 257 × 9.5 = 2441 mm^2.

Average vertical shear stress =

$$\frac{98\,800}{0.002\,441} = 40.4 \text{ MPa}$$

The permissible vertical shear stress for A36 steel (see Fig. 30-1-8) is 100 MPa. Therefore the W250 × 67 beam is acceptable.

Example 4. A floor 5 m wide has a uniformly distributed load of 3000 N•m. The floor joists are 38 mm wide and are spaced 400 mm center to center. If the allowable bending stress is not to exceed 9600 kPa, what depth of floor joists must be used?

Solution. On each floor joist, the uniformly distributed load is 3000 × (400 ÷ 1000) = 1200 N•m. For a simple beam with a uniformly distributed load, the maximum bending moment = $FL^2 \div 8$ = (1200 × 5 × 5) ÷ 8 = 3750 N•m, or 375 × 10^4 N•mm.

Allowable bending stress = 10 MPa. Therefore

Section modulus required

$$= M \div \frac{S}{10^6}$$

$$= 375 \times 10^4 \times \frac{10^6}{10 \times 10^6}$$

$$= 375\,000 \text{ mm}^3$$

The section modulus for the joist = $bd^2 \div 6$ where b = width = 38 mm and d = depth, which is unknown. Therefore

$$d = \sqrt{\frac{375\,000 \times 6}{38}} = 243 \text{ mm}$$

Since standard joist sizes are 38 × 184, 38 × 235, and 38 × 286, the joist size of 38 × 286 is selected.

Deflection of Beams

The vertical distance a horizontally placed beam moves when it bends under an applied load is called *deflection*. Since deflection may cause cracking in plastered ceilings or buckling of floors, the limitations placed on the deflection of the beam may be the governing factor in its selection. In building construction the maximum deflection of beams is limited to 1/360 of the span, or, for cantilever beams, 1/180 of the span. After the beam is selected to withstand the bending and shearing stresses, it must then be checked for deflection. The theory and the mathematics behind the development of beam deflection will not be covered in this text. The formulas using the *double-integration method* for finding the deflection of simple beams are shown in Fig. 30-6-11.

Example 5. A W200 × 42 cantilever beam 2.5 m long has a concentrated load of 20 kN at its free end. Is the deflection excessive?

Solution. Refer to Fig. 30-6-11. The maximum deflection for a cantilever beam having a concentrated load at its free end is $FL^3 \div (3\ EI)$, where $F = 20$ kN, $L = 2500$ mm, E = 200 000 Pa (see Fig. 30-1-3), and $I = 40.7 \times 10^6$ mm^4 (Fig. 30-7-3). Therefore

$$\text{Maximum deflection} = \frac{20\ 000 \times 2500^4}{3 \times 200\ 000 \times 40.7 \times 10^6} = 12.8 \text{ mm}$$

Allowable deflection = 1/180 of the span (for cantilever beams) = 2500 ÷ 180 = 13.9 mm. Therefore, since the maximum deflection is less than the allowable, the W200 × 42 beam is acceptable.

Example 6. A W300 × 60 simple beam has a concentrated load of 27 kN acting at the center of the beam. The beam span is 6 m. Check for deflection.

Solution. Refer to Fig. 30-6-11. The maximum deflection for a simple beam with a concentrated load is $FL^3 \div (48\ EI)$ where $F = 27$ kN, $L = 6000$ mm, $E =$ 200 000 Pa (Fig. 30-1-3), and $I = 129 \times 10^6$ mm^4 (Fig. 30-7-3). Therefore

$$\text{Maximum deflection} = \frac{27\ 000 \times 6000^3}{48 \times 200\ 000 \times 128 \times 10^6} = 4.7 \text{ mm}$$

Allowable deflection = span ÷ 360 = 6000 ÷ 360 = 16.7 mm. Since the maximum deflection is less than the allowable, the W300 × 60 beam is acceptable.

Assignments

1. On an A3- or B-size sheet, calculate the section modulus and select the proper steel section for the beam problems shown in Assignment 30-7-A or 30-7-B.

ASSIGNMENT 30-7-A

1. Select the lightest beam from Fig. 30-7-3 that can be used to safely carry an 18 kN load suspended from the end of a 3 m cantilever beam. Use A572M 310 steel.

2. A simple beam 10 m long supports a uniformly distributed load of 5 kN•m. Neglecting the mass of the beam, select the lightest A36 steel beam that will safely carry this load.

3. A simple beam 8 m long supports a 200 kN concentrated load 3 m from the left abutment. Neglecting the mass of the beam, select the lightest A36 beam that will safely carry this load.

4. A cantilever beam 5 m long carries a uniformly distributed load of 8 kN•m. Neglecting the mass of the beam, select the lightest A572 M-350 beam that will safely carry this load.

ASSIGNMENT 30-7-B

1. Select the lightest beam from Fig. 30-7-3 that can be used to safely carry a 4000 lb load suspended from the end of a 9 ft cantilever beam. Use A572-45 steel.

2. A simple beam 30 ft long supports a uniformly distributed load of 300 lb/ft. Neglecting the mass of the beam, select the lightest A36 steel beam that will safely carry this load.

3. A simple beam 24 ft long supports a 44 000 lb concentrated load 10 ft from the left abutment. Neglecting the mass of the beam, select the lightest A36 steel beam that will safely carry this load.

4. A cantilever beam 16 ft long carries a uniformly distributed load of 500 lb/ft. Neglecting the mass of the beam, select the lightest A572-50 steel beam that will safely carry this load.

2. On an A3- or B-size sheet, calculate the section modulus and select the proper steel section for the beam problems shown in Assignment 30-7-C or 30-7-D.

ASSIGNMENT 30-7-C

1. What is the maximum uniformly distributed load that can be safely placed on a simple W300 × 60 beam 6 m long made of A572 M-310 steel?

2. A simple beam 6 m long has a uniformly distributed load of 5 kN•m and a 10 kN concentrated load 1.2 m from the left abutment. Select the lightest A36 steel beam from Fig. 30-1-8 that will safely carry this load.

3. How many 38 × 286 wood beams must be joined to form a simple beam that will carry safely a concentrated load of 25 kN acting at the center of the beam? The beam is to be 3.5 m long, and the allowable bending stress must not exceed 8000 kPa.

4. A simple W250 × 98 beam 6 m long is made of A36 steel. What are the largest concentrated load the beam can support at the middle of the span and the largest uniformly distributed load the beam can support?

ASSIGNMENT 30-7-D

1. What is the maximum uniformly distributed load that can be safely placed on a simple W12 × 40 beam 20 ft long made of A572-45 steel?

2. A simple beam 20 ft long has a uniformly distributed load of 300 lb/ft and a 2400 lb concentrated load 4 ft from the left abutment. Select the lightest A36 steel beam that will safely carry this load.

3. How must 2 × 12 in. (1⅝ × 11⅝ in. actual size) wood beams be joined to form a simple beam that will carry safely a concentrated load of 6000 lb acting at the center of the beam? The beam is to be 12 ft long, and the allowable bending stress must not exceed 1200 psi.

4. A simple W10 × 66 beam 20 ft long is made of A36 steel. What are the largest concentrated load the beam can support at the middle of the span and the largest uniformly distributed load the beam can support?

3. On an A3- or B-size sheet, calculate the deflection for the beam problems shown in Assignment 30-7-E or 30-7-F.

ASSIGNMENT 30-7-E

1. A W250 × 43 beam has a 6-m span. What is the largest concentrated load, located at the center, that the beam can carry if the maximum deflection is the span ÷ 360?

2. Do the same as in Problem 1, except substitute a uniformly distributed load for a concentrated load.

3. An S300 × 47.4 cantilever beam 4 m long has a concentrated load of 35 kN at its free end. Is the deflection excessive?

4. A 250 mm wide × 400 mm deep simple wooden beam has a span of 5 m and a uniformly distributed load of 14 kN•m. Check the beam for deflection using values of 9400 for E and 1226 for I.

ASSIGNMENT 30-7-F

1. A W10 × 29 beam has a 20-ft span. What is the largest concentrated load, located at the center, that the beam can carry if the maximum deflection is the span ÷ 360?

2. Do the same as in Problem 1, except substitute a uniformly distributed load for a concentrated load.

3. An S12 × 31.8 cantilever beam 12 ft long has a concentrated load of 8000 lb at its free end. Is the deflection excessive?

4. A 10 in. wide × 16 in. deep simple wooden beam has a span of 16 ft and a uniformly distributed load of 1000 lb/ft. Check the beam for deflection, using values of 1,320,000 for E and 2948 for I.

UNIT 30-8

TORSION AND POWER TRANSMISSION OF SHAFTS

Torsion of Shafts

Since many machines and mechanical devices are largely composed of rotating parts, their natural method of construction is to mount the components on circular shafts. The selected size of these shafts is dependent on several factors, one being the resistance of the shaft to bending loads and another being the resistance of the shaft to twisting. This twisting action in the shaft is known as *torsion* and is denoted by the term *torque* which is measured in newton-metres. The abbreviation **T** is used in the following formulas to denote torque.

In the SI system, the derived unit for torque is the newton-metre. Since the force and distance vectors are perpendicular, the value of newton-metres is not multiplied to obtain **J** for joule.

Consider the shaft *B* shown in Fig. 30-8-1. If a force *F* is applied at the end of the lever *A* at a distance *R* from the center of the shaft, then the shaft would be subject to a twisting moment or torque of $F \times R$ N•m. For example, if the force *F* were 50 N and the distance *R* were 0.125 m, then the torque on the shaft would be $50 \times 0.125 = 6.25$ N•m. Note that the length of the shaft is not considered at this point.

Example 1. What torque is necessary to shear a key 90 mm long × 25 mm wide on a shaft whose diameter is 100 mm? The ultimate shear stress for the key steel is 330 MPa.

Solution. Area of key in shear = 90 × 25 = 0.002 25 m². Force required to shear key = area × ultimate shear stress = 0.002 25 × 330 × 10⁶ = 742 500 N. Torque on shaft produced by this force at the circumference = 742 500 × 0.050 = 37 125 N•m.

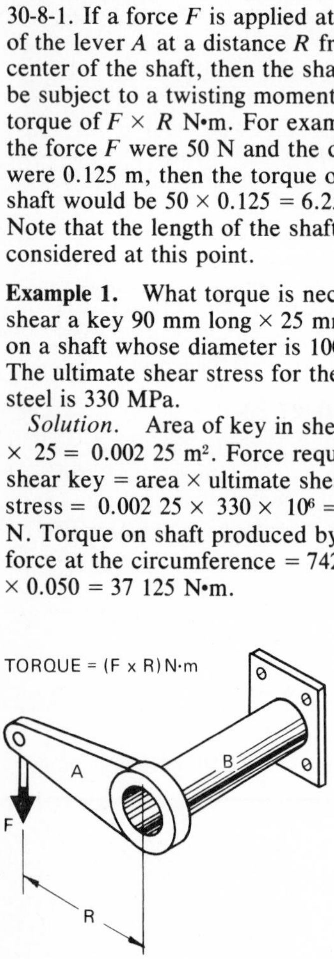
TORQUE = (F x R) N·m
A
B
F
R

Fig. 30-8-1 Torsion shafts.

Power Transmission by Shafts

Power can be supplied in a multitude of forms. Its common applications are in the forms of energy, namely, *mechanical power, heat power,* and *electrical power.* Only applications dealing with mechanical power will be discussed in this unit. The watt is the SI unit for all forms of power. A watt is energy per unit, measured in newton-metres per second (N•m/s). This derived unit has been given the special name *watt*. The preferred units are *watt* (W) for small values, *kilowatt* (kW) for intermediate values, and *megawatt* (MW) for large values.

Since power is the rate at which work is done (the work done per second), one watt of power is equal to one newton-metre of work done in one second.

In the inch-pound system, power is expressed in horsepower (hp). One horsepower is the amount of power required to raise a pound 33,000 feet in one minute. Thus, 1 hp = 33,000 ft•lb/min of work.

The watt (SI) and the horsepower (inch-pound) are the basic units used in engineering calculation associated with power. Consider a disk of radius R_M having a force *F* applied at the rim. The work done by the force *F* in one revolution is the force times the distance traveled by the force. The work done per revolution = $F \times 2\pi R$ N•m. If the shaft rotates at *N* r/min, then the work done would be $F \times 2\pi R \times N \div 60$ *W*. Thus power in watts can be calculated by using the formula *W* = 0.104 72 × load (*N*) × radius of shaft (*R*) × revolutions per minute.

Example 2. How many watts could be safely transmitted at 65 r/min by the key in Example 1 if a factor of safety of 4 was to be used?

Solution. Safe working stress = 330 ÷ 4 = 82.5 MPa. Load = area × stress = 0.000 25 m² × 82.5 × 10⁶ = 185 625 N. Torque due to this load = 185 625 × 0.05 = 9281.25 N•m. Therefore *W* = 0.104 72 × 185 625 × 0.05 × 65 = 63 175.6 *W*, or 63.2 kW.

Example 3. Calculate the torque in a shaft rotating at 100 r/min transmitting 20 kW.

Solution.

$$\text{Torque} = \frac{W}{0.104\,72 \times \text{r/min}}$$

$$= \frac{20\,000}{0.104\,72 \times 100} = 1910 \text{ N•m}$$

Example 4. How many 20 mm bolts on a 250 mm pitch diameter would be required to transmit 75 kW at 65 r/min if the allowable shear stress for the bolts is 70 MPa?

Solution.

$$\text{Torque} = \frac{W}{0.104\,72 \times \text{r/min}}$$

$$= \frac{75\,000}{0.104\,72 \times 65} = 11\,018 \text{ N•m}$$

Allowable shear load per bolt = area × stress = $\pi r^2 \times 70\,000$

$$= \frac{3.1416 \times 10 \times 10 \times 70\,000\,000}{10^6}$$

= 21 991 N

Torque per bolt = load × distance = 21 991 × 0.125 = 2749 N•m

Therefore number of bolts required is

$$\frac{\text{Total torque}}{\text{Torque per bolt}} = \frac{11\,018}{2749} = 4.008 \text{ bolts}$$

Thus four bolts are required.

SHEARING STRESSES

The complete relationship among torque, shear stress, and diameter of shaft is given in the following formula, the mathematics of which is not covered in this text:

$$\frac{T}{J} = \frac{S}{R}$$

where *T* = torque, in newton-metres
J = polar moment of inertia (see Fig. 30-8-2)
S = shear stress, in pascals
R = radius of shaft, in metres

Example 5. Calculate the torque that can be safely carried by a ϕ24 shaft if the shear stress is not to exceed 55 MPa. How many kilowatts can be transmitted by this shaft turning at 1000 r/min?

TYPE OF SHAFT	POLAR MOMENT OF INERTIA (J)
SOLID 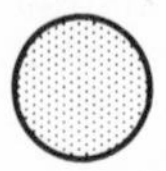	$\frac{\pi D^4}{32}$ OR $\frac{\pi R^4}{2}$
HOLLOW 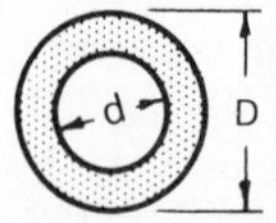 d D	$\frac{\pi (D^4 - d^4)}{32}$ OR $\frac{\pi (R^4 - r^4)}{2}$

Fig. 30-8-2 Polar moment of inertia.

Solution.

$$\text{Torque} = \frac{J \times S}{R} = \frac{\pi R^4/2 \times 55 \times 10^6}{R}$$

$$= \frac{3.1416 \times 0.012^4 \times 55 \times 10^6}{0.012 \times 2}$$

$$= 145.7 \text{ N·m}$$

$$\text{Kilowatts} = \frac{0.104\,72 \times 145.7 \times 1000}{1000}$$

$$= 15.26 \text{ kW}$$

Example 6. A shaft is required to transmit 40 kW from an electric motor. What must its diameter be if it is rotating at 700 r/min and the material is steel with an allowable shear stress of 60 MPa?

Solution.

$$W = 0.104\,72 \times T \times \text{r/min}$$

$$T = \frac{40\,000}{0.104\,72 \times 700} = 546 \text{ N·m}$$

$$\frac{T}{J} = \frac{S}{R}$$

$$\frac{546}{\pi R^4/2} = \frac{60 \times 10^6}{R}$$

$$\pi R^3 \times 60 \times 10^6 = 546 \times 2$$

$$R^3 = 5.8 \times 10^{-6}$$

$$R = \sqrt[3]{5.8 \times 10^{-6}}$$

$$= \sqrt[3]{0.000\,005\,8}$$

$$= 0.018 \text{ m}$$

Therefore, diameter = 0.018 × 2 = 0.036 m, or 36 mm.

Assignments

1. On an A3- or B-size sheet, calculate the answers to the problems shown in Assignment 30-8-A or 30-8-B.

ASSIGNMENT 30-8-A

1. What torque in newton-metres would shear a 20 × 60 mm long square key on a ϕ75 shaft if the ultimate shear stress for the key is 400 MPa?

2. A 150 PD gear is keyed to a 25 mm shaft by a 6 mm square key 25 mm long. If the ultimate shear stress for the key is 330 MPa and the load is assumed to act on the pitch diameter of the gear, what load acting on the gear would shear the key?

3. How many kilowatts could be safely transmitted at 250 r/min by a 12 mm square key 50 mm long on a 50 mm diameter shaft? Allowable shear stress = 90 MPa.

4. A flanged coupling is held by four 16 mm bolts seated on a 200 mm pitch circle. The flanges are keyed to a 60 mm diameter shaft with 10 mm square keys 80 mm long. The allowable shear stress for the bolts is 70 MPa. Find the total torque transmitted by the bolts, the shear stress on the key, and the kilowatts transmitted at 200 r/min.

ASSIGNMENT 30-8-B

1. What torque in pound-inches would shear a .75 × 2.50 in. long square key on a 3 in. diameter shaft if the ultimate shear stress for the key were 60 000 psi?

2. A 6 PD gear is keyed to a 1 in. diameter shaft by a .25 in. square key 1 in. long. If the ultimate shear stress for the key is 48 000 psi and the load is assumed to act on the pitch diameter of the gear, what load acting on the gear would shear the key?

3. What horsepower could be safely transmitted at 250 r/min by a .50 in. square key 2 in. long on a 2 in. diameter shaft? Allowable shear stress = 13 200 psi.

4. A flanged coupling is held by four .625 in. bolts located on an 8 in. pitch circle. The flanges are keyed to a 2.50 in. diameter shaft with .375 in. square keys 3 in. long. The allowable shear stress for the bolts is 10 000 psi. Find the total torque transmitted by the bolts, the shear stress in each key, and the horsepower transmitted at 200 r/min.

2. On an A3- or B-size sheet, calculate the answers to the problems shown in Assignment 30-8-C or 30-8-D.

ASSIGNMENT 30-8-C

1. Determine the maximum allowable kilowatts which can be transmitted by a 150 mm diameter shaft running at 240 r/min when the permissible shear stress is 55 MPa. The shaft has a coupling on it having six bolts on a 300 mm pitch circle. Determine the diameter of the bolts if the maximum shear stress in the bolts must not exceed 100 MPa.

2. A pulley is driven by an 80 mm long × 6 mm wide key. The allowable unit shear stress of the steel key is 80 MPa. How much torque will the key transmit safely if the shaft has a diameter of 75 mm?

3. How many kilowatts can be transmitted by a flanged coupling consisting of six 20 mm diameter bolts arranged on the circumference of a 150 mm bolt circle? The allowable shear stress in the bolt material is 70 MPa, and the coupling rotates at 300 r/min.

4. A flanged coupling is used to connect two shafts, the left and right shafts being 75 mm and 50 mm in diameter respectively. The key, keying the left shaft to its flange is 40 mm long and 6 mm wide. The key, keying the right shaft to its flange, is 40 mm long and 8 mm wide. The key material has an allowable shearing strength of 110 MPa. What is the limiting torque which can be transmitted by the assembly, as determined by the capacity of the keys?

ASSIGNMENT 30-8-D

1. Determine the maximum allowable horsepower which can be transmitted by a 6 in. diameter shaft running at 240 r/min when the permissible shear stress is 8000 psi. The shaft has a coupling on it having six bolts on a 10.50 in. diameter pitch circle. Determine the diameter of the bolts if the maximum shear stress in the bolts must not exceed 14 600 psi.

2. A pulley is driven by a 3 in. long × .28 in. wide key. The allowable unit shearing stress of the steel key is 12 000 psi. How much torque will the key transmit safely if the shaft has a diameter of 3 in.?

3. What horsepower can be transmitted by a flanged coupling consisting of six .75 in. diameter bolts arranged on the circumference of a 6 in. bolt circle? The allowable shearing stress in the bolt material is 10 000 psi, and the coupling rotates at 300 r/min.

4. A flanged coupling is used to connect two shafts, the left and right shafts being 3 and 2 in. in diameter respectively. The key, keying the left shaft to its flange, is 1.50 in. long and .24 in. wide. The key, keying the right shaft to its flange, is 1.50 in. long and .30 in. wide. The key material has an allowable shearing strength of 16 000 psi. What is the limiting torque which can be transmitted by the assembly, as determined by the capacity of the keys?

Chapter 31 Engineering Tolerancing

UNIT 31-1 MODERN ENGINEERING TOLERANCING

An engineering drawing of a manufactured part is intended to convey information from the designer to the manufacturer and inspector. It must contain all information necessary for the part to be correctly manufactured. It must also enable the inspector to make a precise determination of whether the parts are acceptable.

Therefore each drawing must convey three essential types of information:

1. The material to be used
2. The size or dimensions of the part
3. The shape or geometric characteristics

The drawing must also indicate permissible variations for each of these aspects, in the form of tolerance or limits.

Materials are usually covered by separate specifications or supplementary documents, and the drawings need only make reference to these.

Size is specified by linear and angular dimensions. Tolerances may be applied directly to these dimensions or may be specified by means of a general tolerance note.

Shape and geometric characteristics, such as orientation and position, are indicated by views on the drawing, supplemented to some extent by dimensions.

In the past tolerances were often shown for which no precise interpretation existed, such as on dimensions which originated at nonexistent center lines. Specification of datum features was often omitted, resulting in measurements being made from actual surfaces when datums were intended. There was confusion concerning the precise effect of various methods of expressing tolerances and of the number of decimal places used. While tolerancing of geometric characteristics was sometimes specified in the form of simple notes, no precise methods or interpretations were established. Straight or circular lines were drawn, without indicating how straight or round they should be. Square corners were drawn, without indicating how much the 90° angle could vary.

Modern systems of tolerancing, which include geometric and positional tolerancing, use of datum and datum targets, and more precise interpretations of linear and angular tolerances, provide designers and drafters with a means of expressing permissible variations in a very precise manner. Furthermore, the methods and symbols are international in scope and therefore break down language barriers.

It is not necessary to use geometric tolerances for every feature on a part drawing. In most cases it is to be expected that if each feature meets all dimensional tolerances, form variations will be adequately controlled by the accuracy of the manufacturing process and equipment used.

This chapter covers the application of modern tolerancing methods on drawings. For a more in-depth coverage of this subject refer to *Modern Engineering Tolerances* by Hill and Jensen, published by McGraw-Hill Book Co.

National and International Standards. References are made to technical drawing standards published by various national and international standardizing bodies. These bodies are generally referred to by their acronyms, as shown in Fig. 31-1-1.

Most of the symbols in all these standards are identical, but there are some variations. These are chiefly in the methods of indicating datum features and of applying the symbols to drawings. In view of the exchange of drawings among the United States, Canada, Great Britain, and other countries, it would be advantageous for drafters and designers to become acquainted with these different symbols.

For this reason whenever differences between United States and the other countries occur, two methods are shown in some of the illustrations, and each is labeled with the acronym of the appropriate standardizing body, ANSI or ISO (ISO, CSA, and BSI [British Standards Institute] all use identical symbols). However, differences in symbols or methods of application do not in any way affect the principles or interpretation of tolerances, unless specifically noted.

ILLUSTRATIONS

Most of the drawings in this chapter are not complete working drawings. They are intended only to illustrate a principle. Therefore, to avoid distraction from the information being presented, most of the details that are not essential to explain the principle have been omitted.

Definitions of Basic Terms

Definitions of some of the basic terms used in dimensioning and tolerancing of drawings follow. While these terms are not new, their exact meaning warrants special attention in order that there be no ambiguity in the precise interpretation of tolerancing methods described in this chapter.

DIMENSION

A *dimension* is a geometric characteristic, of which the size is specified, such as diameter, length, angle, location, or center distance. The term is also used for convenience to indicate the size or value of a dimension, as specified on a drawing. See Fig. 31-1-2.

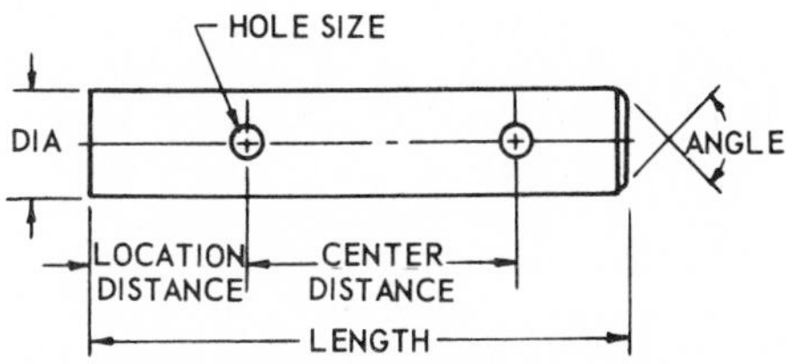

HOLE SIZE
DIA
ANGLE
LOCATION DISTANCE
CENTER DISTANCE
LENGTH

Fig. 31-1-2 Dimensions of a part.

TOLERANCE

The *tolerance* on a dimension is the total permissible variation in its size, which is equal to the difference between the limits of size. The plural term *tolerances* is sometimes used to denote the permissible variations from the specified size when the tolerance is expressed bilaterally.

For example, in Fig. 31-1-3*a* the tolerance on the center distance dimension 40 ± 0.2 is 0.4 mm, but in common parlance the values +0.2 and −0.2 are often referred to as the tolerances.

A *bilateral tolerance* is a tolerance which is expressed as plus and minus values, where neither is zero, to denote permissible variations in both directions from the specified size.

A *unilateral tolerance* is one which applies in only one direction from the specified size, so that the permissible variation in the other direction is zero.

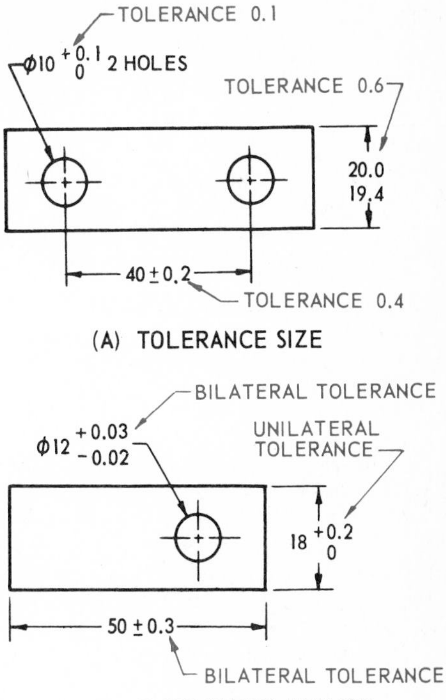

TOLERANCE 0.1
Ø10 +0.1 0 2 HOLES
TOLERANCE 0.6
20.0
19.4
40 ± 0.2
TOLERANCE 0.4
(A) TOLERANCE SIZE
BILATERAL TOLERANCE
Ø12 +0.03 -0.02
UNILATERAL TOLERANCE
18 +0.2 0
50 ± 0.3
BILATERAL TOLERANCE
(B) TYPE OF TOLERANCE

Fig. 31-1-3 Tolerances.

SIZE OF DIMENSIONS

In theory, it is impossible to produce a part to an exact size, because every part, if measured with sufficient accuracy, would be found to be a slightly different size. However, for purposes of discussion and interpretation, a number of distinct sizes for each dimension have to be recognized.

Actual Size. In ordinary practice, *actual size* simply means the measured size of an individual part.

Nominal Size. The *nominal size* is the designation of size used for purposes of general identification.

The nominal size is used in referring to a part in an assembly drawing stocklist, in a specification, or in other such documents. It is very often identical to the basic size but may differ widely; for example, the diameter of a ½-in. steel pipe is 21.34 mm [0.84 in.]. The nominal size is ½ in.

ACRONYM	STANDARDIZING BODY	STANDARD FOR DIMENSIONING AND TOLERANCING
ANSI	American National Standards Institute	ANSI Y14.5
ISO	International Organization for Standardization	ISO R1101
CSA	Canadian Standards Association	CSA B78.2
BSI	British Standards Institution	BS308 Part III
SAA	Standards Association of Australia	AS CZ1 Sect. 8

Fig. 31-1-1 Standardizing bodies.

Specified Size. This is the size specified on the drawing when the size is associated with a tolerance. The specified size is usually identical to the design size or, if no allowance is involved, to the basic size.

Design Size. The *design size* of a dimension is the size in relation to which the tolerance for that dimension is assigned.

Theoretically, it is the size on which the design of the individual feature is based, and therefore it is the size which should be specified on the drawing. For dimensions of mating features it is derived from the basic size by the application of the allowance, but when there is no allowance, it is identical to the basic size.

Basic Size. The *basic size* of a dimension is the theoretical size from which the limits for that dimension are derived by the application of the allowance and the tolerance.

Figure 31-1-4 shows two mating features with the tolerance and allowance zones exaggerated, to illustrate the sizes, tolerances, and allowances. This figure also illustrates the origin of tolerance block diagrams, as shown in Fig. 31-1-5, which are commonly used to show the relationships among part limits, gage or inspection limits, and gage tolerances.

DEVIATIONS

The differences between the basic, or zero, line and the maximum and minimum sizes are called the *upper* and *lower deviations*, respectively.

Thus in Fig. 31-1-6 the upper deviation of the external part is −0.1, and the

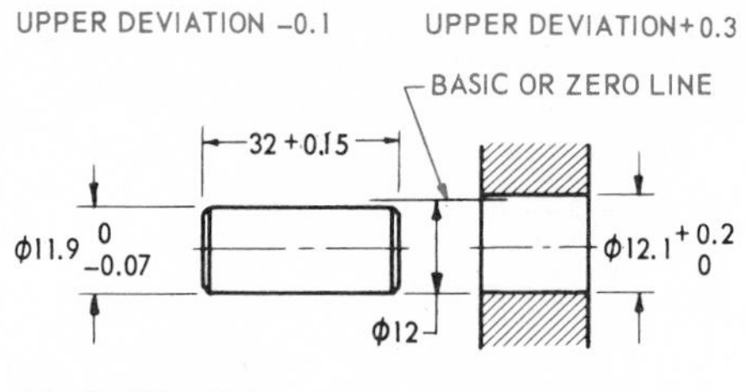

UPPER DEVIATION −0.1
UPPER DEVIATION+0.3
BASIC OR ZERO LINE
32 +0.15
ϕ11.9 0 −0.07
ϕ12.1 +0.2 0
ϕ12
LOWER DEVIATION −0.17
LOWER DEVIATION+0.1

Fig. 31-1-6 Deviations.

lower deviation is −0.17. For the hole diameter, the upper deviation is +0.3, and the lower deviation is +0.1, whereas for the length of the pin the upper and lower deviations are +0.15 and −0.15, respectively.

EXACT DIMENSIONS

True-Position Dimension. A *true-position dimension* is a dimension which specified the mean position of a feature or features.

Datum Dimension. A *datum dimension* is a dimension which establishes the true position of a datum or datum target.

Basic Dimension. A *basic dimension* represents the basic size of a feature.

These terms are called *exact dimensions*. See Fig. 31-1-7. They are shown without direct tolerances, and each is enclosed in a rectangular frame to indicate that the tolerances in the general tolerance note do not apply.

FEATURE

A *feature* is a specific, characteristic por-

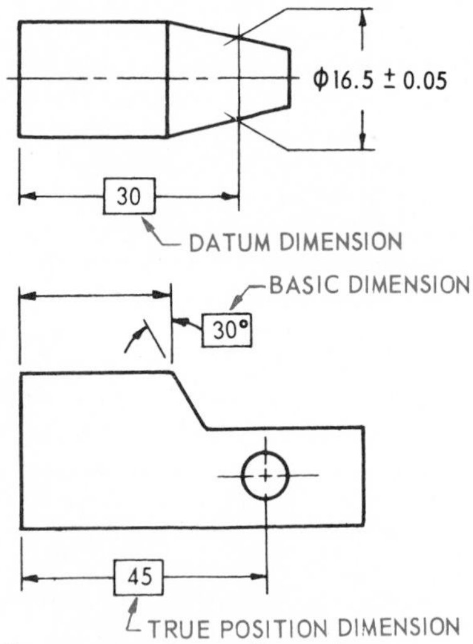

ϕ16.5 ± 0.05
30
DATUM DIMENSION
BASIC DIMENSION
30°
45
TRUE POSITION DIMENSION

Fig. 31-1-7 Exact dimensions.

tion of a part, such as a surface, hole, slot, screw thread, or profile.

While a feature may include one or more surfaces, the term is generally used in geometric tolerancing in a more restricted sense, to indicate a specific point, line, or surface. Some examples are the axis of a hole, the edge of a part, or a single flat or curved surface, to which reference is being made or which forms the basis for a datum.

AXIS

An *axis* is a theoretical straight line about which a part or circular feature revolves or could be considered to revolve. See Fig. 31-1-8.

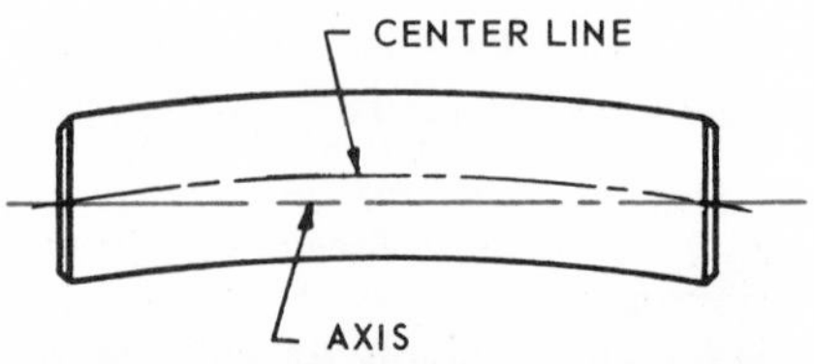

CENTER LINE
AXIS

Fig. 31-1-8 Divergence of axis and center line when part is deformed.

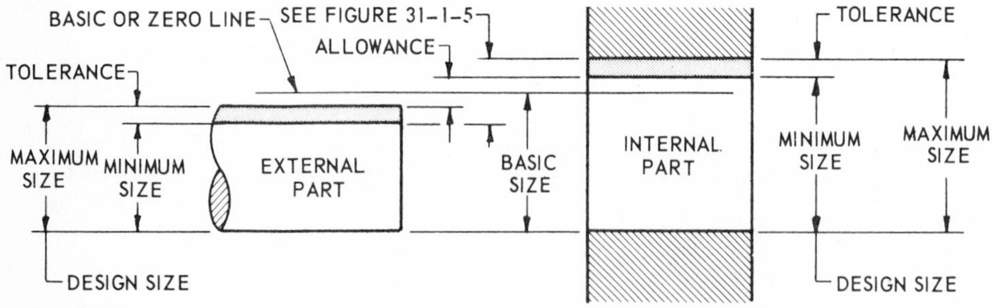

BASIC OR ZERO LINE
SEE FIGURE 31-1-5
ALLOWANCE
TOLERANCE
TOLERANCE
MAXIMUM SIZE
MINIMUM SIZE
EXTERNAL PART
BASIC SIZE
INTERNAL PART
MINIMUM SIZE
MAXIMUM SIZE
DESIGN SIZE
DESIGN SIZE

Fig. 31-1-4 Sizes of mating parts.

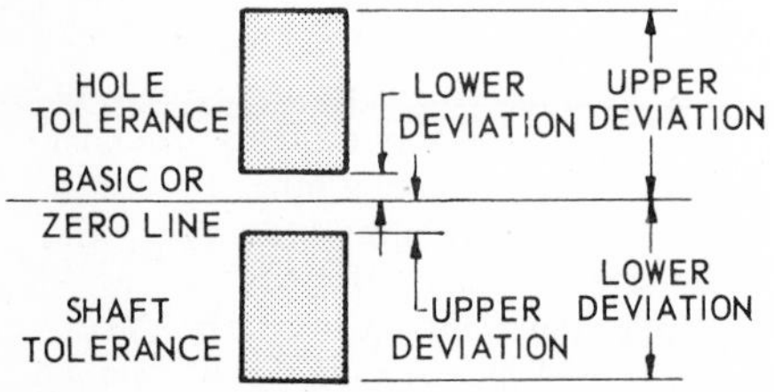

HOLE TOLERANCE
LOWER DEVIATION
UPPER DEVIATION
BASIC OR ZERO LINE
LOWER DEVIATION
SHAFT TOLERANCE
UPPER DEVIATION

Fig. 31-1-5 Tolerance block diagram.

In drafting practice there is often confusion between the use of the terms *axis* and *center line*. The drawing of a part represents the ideal or perfect form of the part, in which the longitudinal center line of circular features, such as holes and shafts, is coincident with the axis.

Interpretation of Drawings and Dimensions

It should not be necessary to specify the geometric shape of a feature, unless some particular precision is required. Lines which appear to be straight imply straightness; those that appear to be round imply circularity; those that appear to be parallel imply parallelism; those that appear to be square imply perpendicularity; center lines imply symmetry; and features that appear to be concentric about a common center line imply concentricity.

Therefore it is not necessary to add angular dimensions of 90° to corners of rectangular parts nor to specify that opposite sides are parallel.

However, if a particular departure from the illustrated form is permissible, or if a certain degree of precision of form is required, it must be specified. If a slight departure from the true geometric form or position is required, it should be exaggerated pictorially in order to indicate clearly where the dimensions apply. Figure 31-1-9 gives some examples. Dimensions which are not to scale should be underlined freehand.

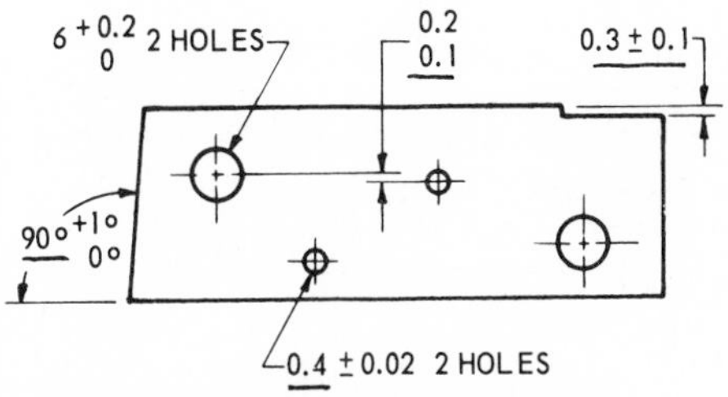

6 +0.2 0 2 HOLES
0.2 0.1
0.3 ± 0.1
90° +1° 0°
0.4 ± 0.02 2 HOLES

Fig. 31-1-9 Exaggeration of small dimensions.

POINT-TO-POINT DIMENSIONS

When datums are not specified, linear dimensions are intended to apply on a point-to-point basis, either between opposing points on the indicated surfaces or directly between the points indicated on the drawing.

The examples shown in Fig. 31-1-10 should help to clarify this principle of point-to-point dimensions.

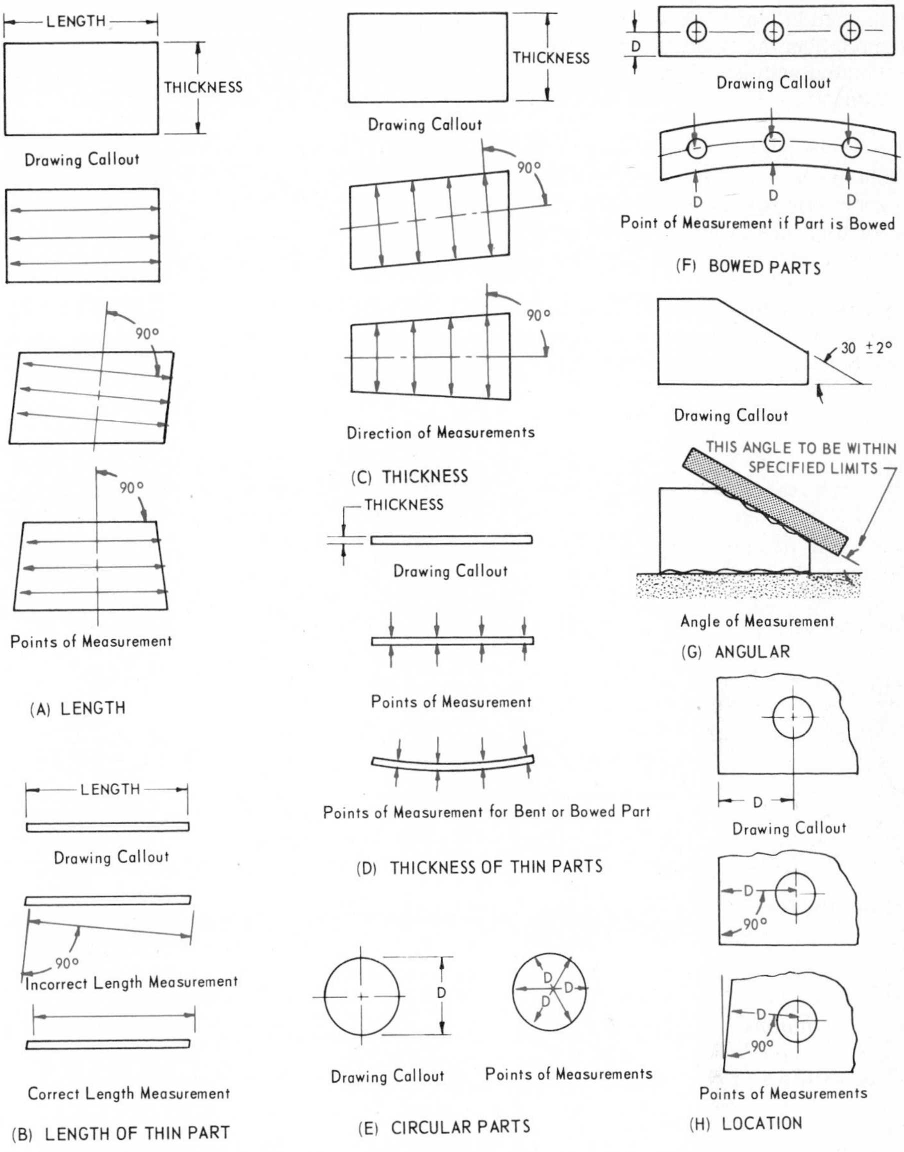

LENGTH
THICKNESS
Drawing Callout
90°
Points of Measurement
(A) LENGTH
LENGTH
Drawing Callout
90°
Incorrect Length Measurement
Correct Length Measurement
(B) LENGTH OF THIN PART
THICKNESS
Drawing Callout
90°
Direction of Measurements
(C) THICKNESS
THICKNESS
Drawing Callout
Points of Measurement
Points of Measurement for Bent or Bowed Part
(D) THICKNESS OF THIN PARTS
D
Drawing Callout
Points of Measurements
(E) CIRCULAR PARTS
D
Drawing Callout
Point of Measurement if Part is Bowed
(F) BOWED PARTS
30 ±2°
Drawing Callout
THIS ANGLE TO BE WITHIN SPECIFIED LIMITS
Angle of Measurement
(G) ANGULAR
D
Drawing Callout
90°
Points of Measurements
(H) LOCATION

Fig. 31-1-10 Point-to-point dimensions when datums not used.

LOCATION DIMENSIONS WITH DATUMS

When location dimensions originate from a feature or surface indicated as a datum, measurement is made from the theoretical datum, not from the actual feature or surface of the part.

There will be many cases where a curved center line, as shown in Fig. 31-1-10*f*, would not meet functional requirements or where the position of the hole in Fig. 31-1-10*h* would be required to be measured parallel to the base. This can easily be specified by referring the dimension to a datum feature, as shown in Fig. 31-1-11. This will be more fully

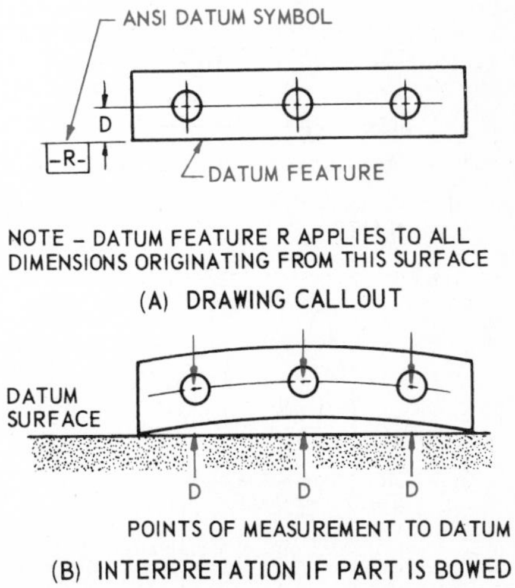

ANSI DATUM SYMBOL
D
-R-
DATUM FEATURE
NOTE – DATUM FEATURE R APPLIES TO ALL DIMENSIONS ORIGINATING FROM THIS SURFACE
(A) DRAWING CALLOUT
DATUM SURFACE
D
POINTS OF MEASUREMENT TO DATUM
(B) INTERPRETATION IF PART IS BOWED

Fig. 31-1-11 Dimension referred to a datum.

explained in Unit 31-7, when the interpretation of coordinate tolerances is compared with geometric and positional tolerances.

ASSUMED DATUMS

There are often cases where the basic rules for measurements on a point-to-point basis cannot be applied, because the originating point, lines, or surfaces are offset in relation to the features located by the dimensions. It is then necessary to assume a suitable datum, which is usually the theoretical extension of one of the lines or surfaces involved.

The following general rules cover three types of dimensioning procedures commonly encountered.

1. If a dimension refers to two parallel edges or planes, the larger edge or surface is assumed to be the datum feature. For example, if the surfaces of the part shown in Fig. 31-1-12*a* were not quite parallel, as shown in the lower view, dimension *D* would be acceptable if the top surface were within limits when measured at *a* and *b*, but need not be within limits if measured at *c*.

2. If only one of the extension lines refers to a straight edge or surface, the extension of that edge or surface is assumed to be the datum. Thus in Fig. 31-1-12*b* measurement of dimension *A* is made to a datum surface as shown at *a* in the bottom view.

3. If both extension lines refer to offset points rather than to edges or surfaces, generally it should be assumed that the datum is a line running through one of these points and parallel to the line or surface to which it is dimensionally related. Thus in Fig. 31-1-12*c* dimension *A* is measured from the center of hole *D* to a line through the center of hole *C* which is parallel to the base line, as at *a*.

PERMISSIBLE FORM VARIATIONS

The actual size of a feature must be within the limits of size, as specified on the drawing, at all points of measurement. This means that each measurement, made at any cross section of the feature, must be not greater than the maximum limit of size nor smaller than the minimum limit of size. See Fig. 31-1-13.

By themselves, toleranced linear dimensions, or limits of size, do not give specific control over many other variations of form, orientation, and, to some extent, position, such as errors of squareness of related features or deviations caused by bending of parts, lobing, eccentricity, and the like. Therefore features may actually cross the boundaries of perfect form at the maximum material size and at the minimum material size.

In order to meet functional requirements, it is often necessary to control

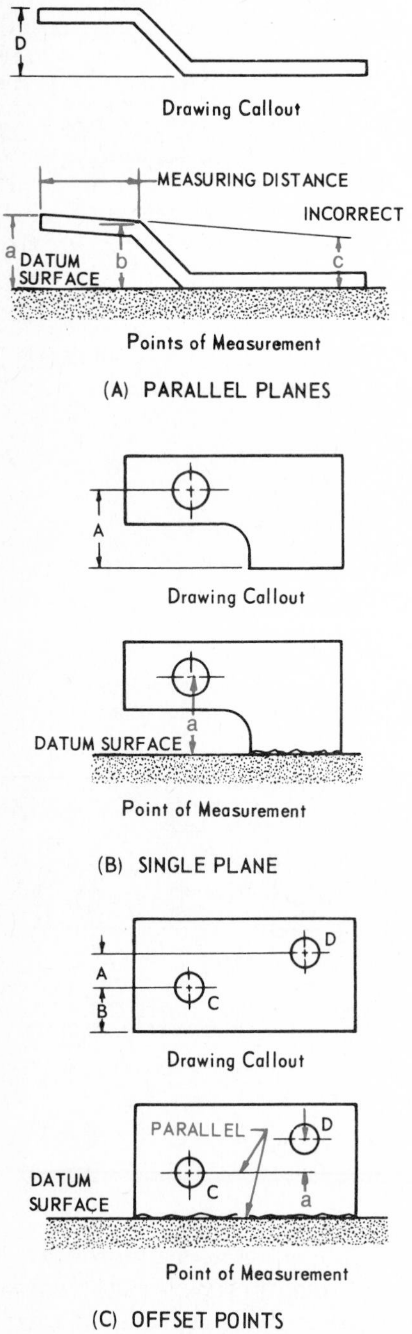
D
Drawing Callout
MEASURING DISTANCE
INCORRECT
a
DATUM
SURFACE
b
c
Points of Measurement
(A) PARALLEL PLANES
A
Drawing Callout
a
DATUM SURFACE
Point of Measurement
(B) SINGLE PLANE
D
A
C
B
Drawing Callout
PARALLEL
D
DATUM
SURFACE
C
a
Point of Measurement
(C) OFFSET POINTS

Fig. 31-1-12 Assumed datums.

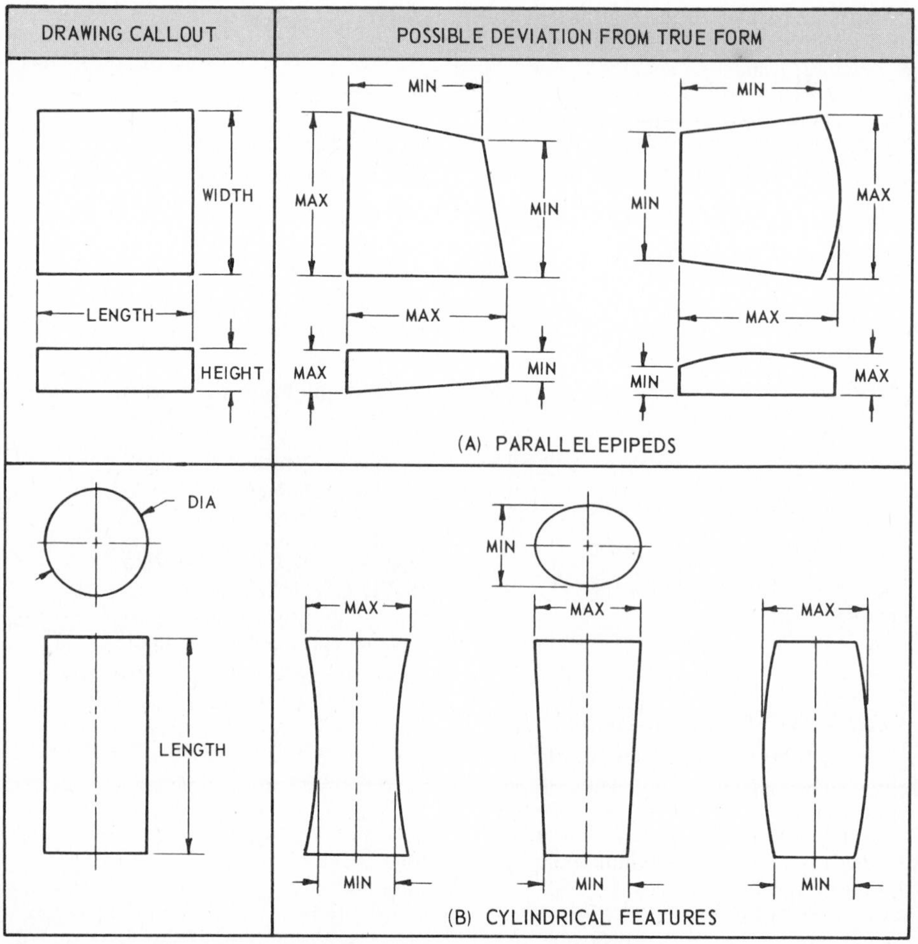
DRAWING CALLOUT
POSSIBLE DEVIATION FROM TRUE FORM
WIDTH
LENGTH
HEIGHT
MIN
MAX
(A) PARALLELEPIPEDS
DIA
LENGTH
(B) CYLINDRICAL FEATURES

Fig. 31-1-13 Deviations permitted by toleranced dimensions.

such deviations. This is done to ensure that parts are not only within their limits of size but also within specified limits of geometric form, orientation, and position. In the case of mating parts, such as holes and shafts, it is usually necessary to ensure that they do not cross the boundary of perfect form at the maximum material size, by reason of being bent or otherwise deformed. This condition is shown in Fig. 31-1-14, where features do not cross the maximum material boundary but are permitted to cross the boundary of perfect form at the minimum material condition.

If only size tolerances or limits of size are specified for an individual feature and no geometric tolerance is given, it might be expected that no element of the feature would actually extend beyond the maximum material boundary of perfect form. Examples are shown in Fig. 31-1-15. Note, however, that this does not prevent features from crossing the least material perfect-form boundary. According to ANSI rules, all parts are expected to have perfect form of individual features at the maximum material condition, when not otherwise specified.

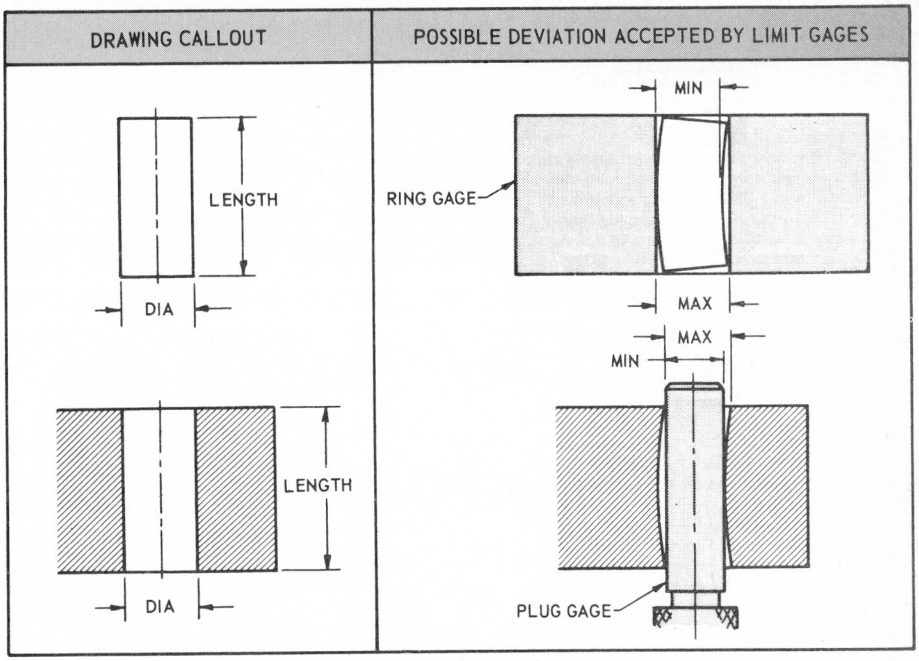
DRAWING CALLOUT
POSSIBLE DEVIATION ACCEPTED BY LIMIT GAGES
MIN
LENGTH
RING GAGE
DIA
MAX
MAX
MIN
LENGTH
DIA
PLUG GAGE

Fig. 31-1-15 Form variations accepted by limit gages.

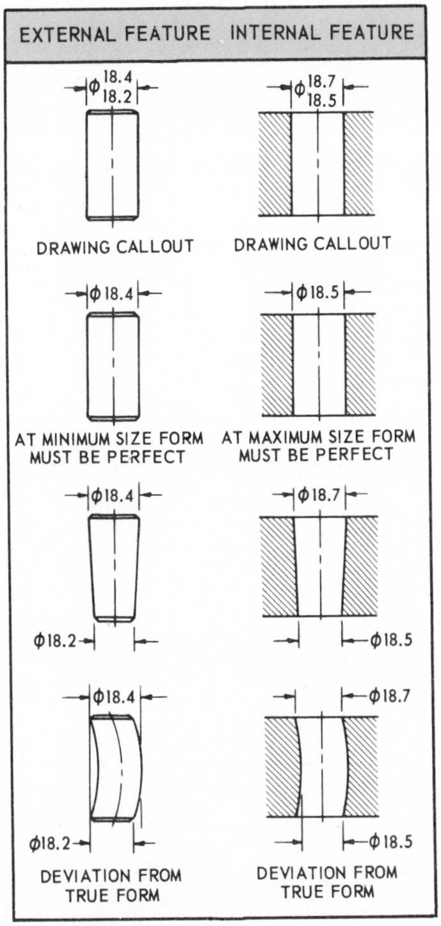
EXTERNAL FEATURE
INTERNAL FEATURE
ϕ 18.4 18.2
ϕ 18.7 18.5
DRAWING CALLOUT
DRAWING CALLOUT
ϕ18.4
ϕ18.5
AT MINIMUM SIZE FORM MUST BE PERFECT
AT MAXIMUM SIZE FORM MUST BE PERFECT
ϕ18.4
ϕ18.7
ϕ18.2
ϕ18.5
ϕ18.4
ϕ18.7
ϕ18.2
ϕ18.5
DEVIATION FROM TRUE FORM
DEVIATION FROM TRUE FORM

Fig. 31-1-14 Examples of deviation of form when perfect form at the maximum material condition is required.

Assignments

1. Parts may deviate from true form and still be acceptable providing the measurements lie within the limits of size. Show by means of a sketch with dimensions two acceptable form variations for each part shown in Fig. 31-1-A.

2. On an A3- or B-size sheet, prepare sketches from the drawings shown in Fig. 31-1-B or 31-1-C and the following information.

(*a*) Using illustration (A) make a tolerance block diagram similar to Fig. 31-1-5. Show the deviations and limits of size.

(*b*) Draw illustration (B) and shade in and dimension the tolerance zone.

(*c*) The exaggeration of sizes is used when it improves the clarity of the drawing. Draw illustration (C) and exaggerate the sizes which would improve the readability of the drawing. Dimension the exaggerated features.

(*d*) With reference to illustration (D), is the part acceptable? State your reason.

(*e*) With reference to the drawing callout shown in illustration (E), what parts would pass inspection?

(*f*) In the drawing callout in illustration (F), what parts in illustration (E) would pass inspection?

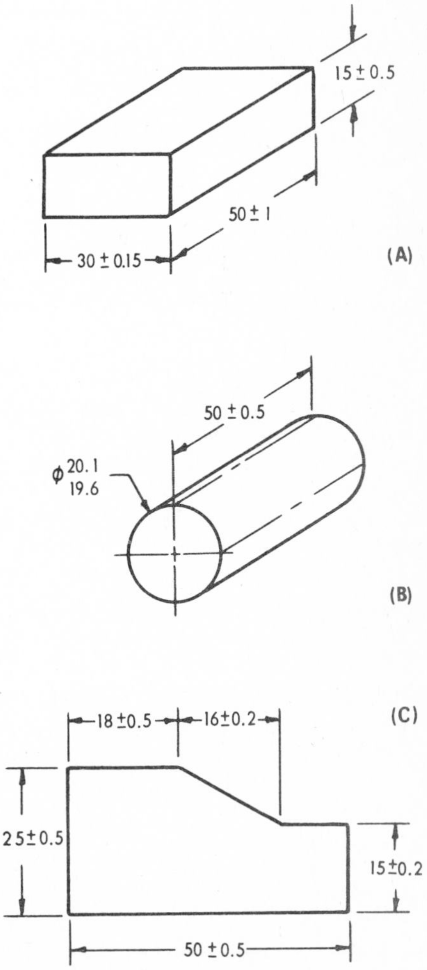
15 ± 0.5
50 ± 1
30 ± 0.15
(A)
50 ± 0.5
ϕ 20.1 19.6
(B)
18 ±0.5
16±0.2
(C)
25±0.5
15±0.2
50 ±0.5

Fig. 31-1-A Assignments.

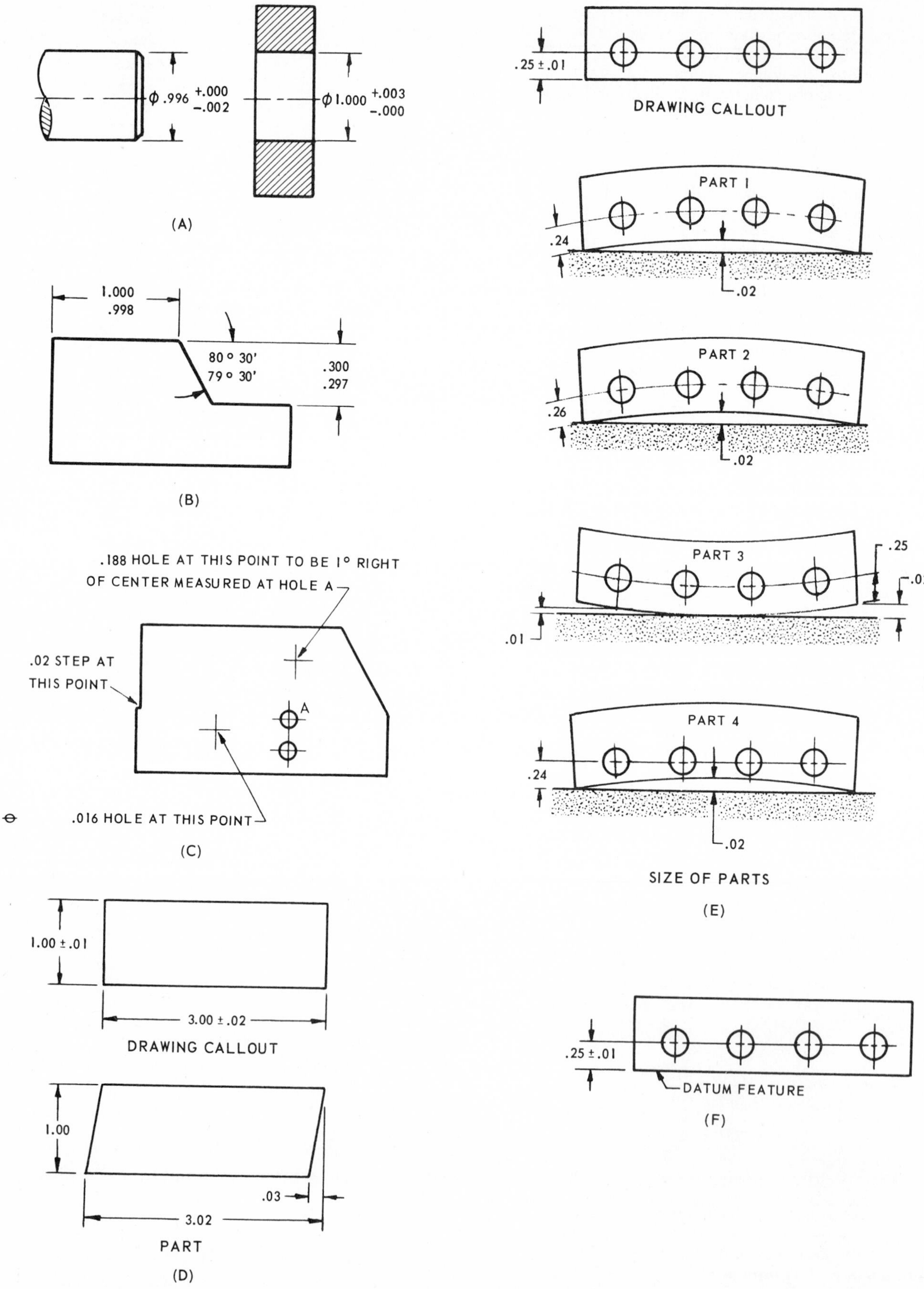

ϕ .996 +.000 −.002
ϕ 1.000 +.003 −.000
(A)
1.000
.998
80° 30'
79° 30'
.300
.297
(B)
.188 HOLE AT THIS POINT TO BE 1° RIGHT OF CENTER MEASURED AT HOLE A
.02 STEP AT THIS POINT
A
.016 HOLE AT THIS POINT
(C)
1.00 ±.01
3.00 ±.02
DRAWING CALLOUT
1.00
.03
3.02
PART
(D)
.25 ±.01
DRAWING CALLOUT
PART 1
.24
.02
PART 2
.26
.02
PART 3
.25
.02
.01
PART 4
.24
.02
SIZE OF PARTS
(E)
.25 ±.01
DATUM FEATURE
(F)

Fig. 31-1-B Assignments.

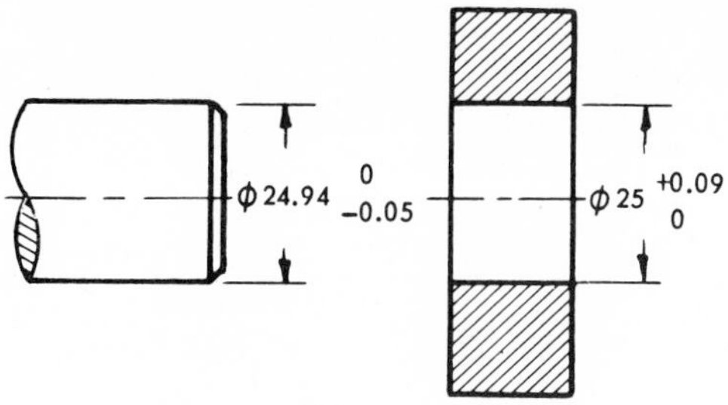
ϕ 24.94 0 −0.05
ϕ 25 +0.09 0

(A)

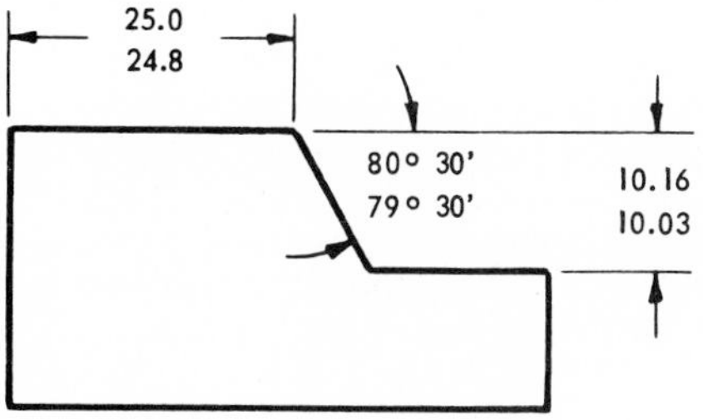
25.0
24.8
80° 30'
79° 30'
10.16
10.03

(B)

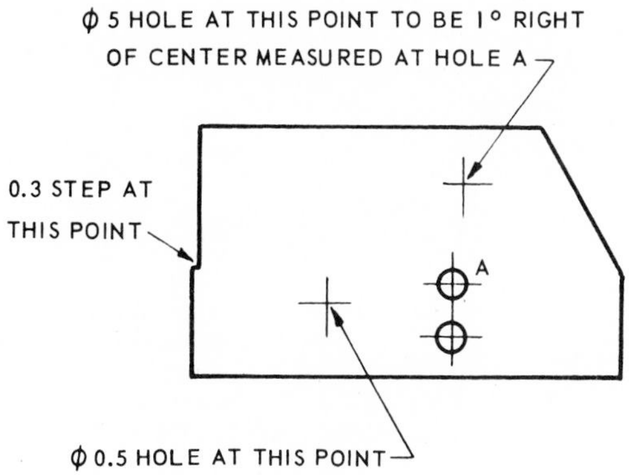
ϕ 5 HOLE AT THIS POINT TO BE 1° RIGHT OF CENTER MEASURED AT HOLE A
0.3 STEP AT THIS POINT
A
ϕ 0.5 HOLE AT THIS POINT

(C)

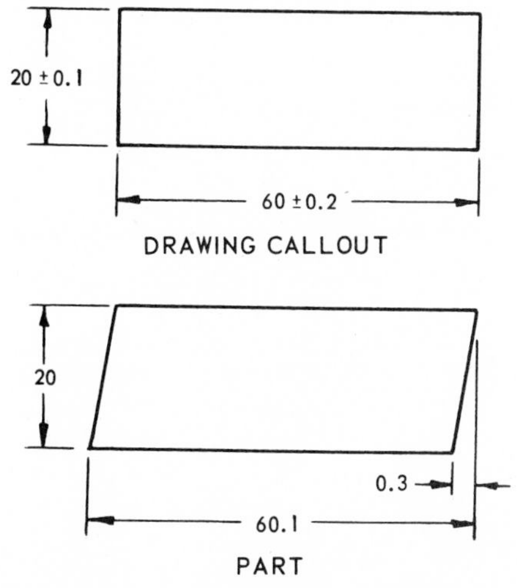
20 ±0.1
60 ±0.2
DRAWING CALLOUT
20
0.3
60.1
PART

(D)

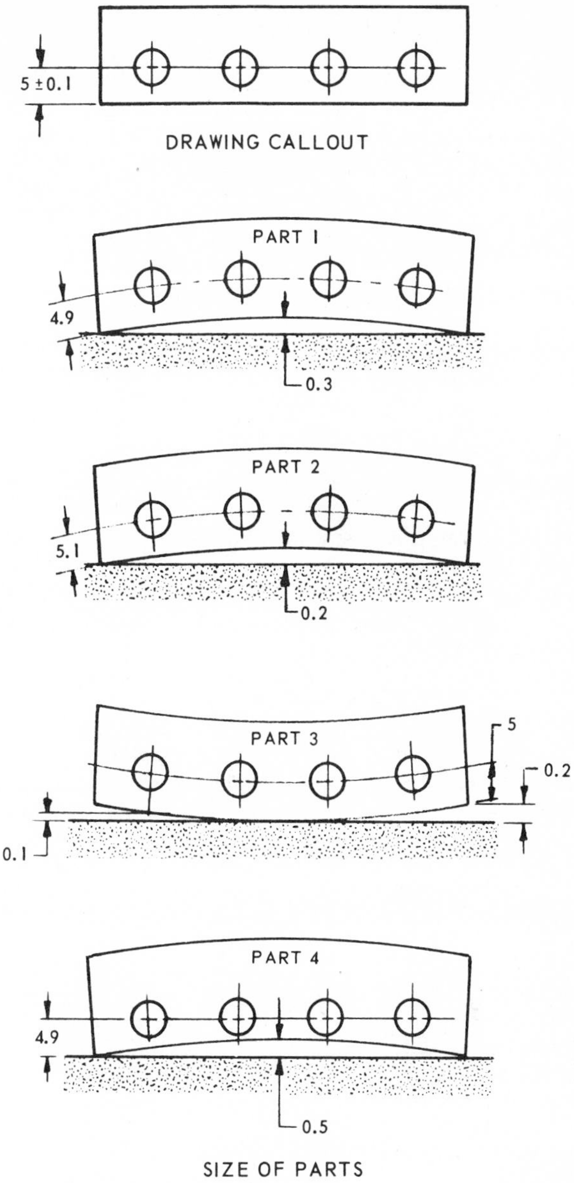
5 ±0.1
DRAWING CALLOUT
PART 1
4.9
0.3
PART 2
5.1
0.2
PART 3
5
0.2
0.1
PART 4
4.9
0.5
SIZE OF PARTS

(E)

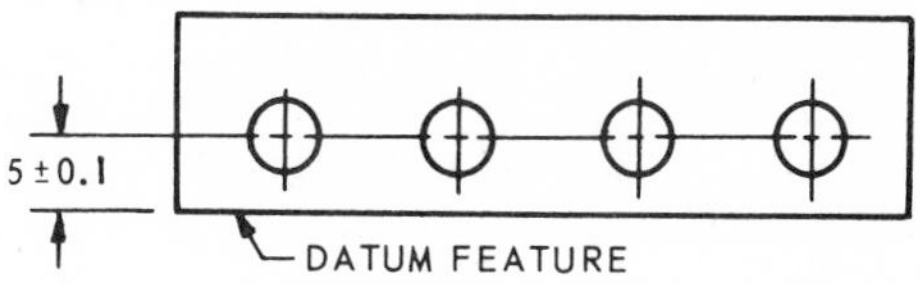
5 ±0.1
DATUM FEATURE

(F)

Fig. 31-1-C Assignments.

UNIT 31-2

GEOMETRIC TOLERANCING

A geometric tolerance is the maximum permissible variation of form, orientation, or location of a feature from that indicated or specified on the drawing. The tolerance value represents the width or diameter of the tolerance zone, within which the point, line, or surface of the feature shall lie.

From this definition it follows that a feature would be permitted to have any variation of form, or take up any position, within the specified geometric tolerance zone.

For example, a line controlled by a straightness tolerance of 0.15 mm must be contained within a tolerance zone 0.15 mm wide. See Fig. 31-2-1.

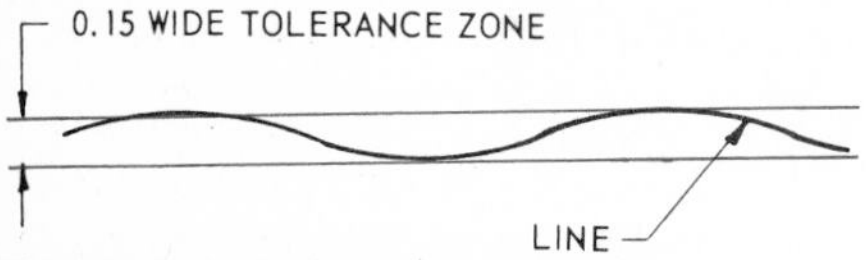
0.15 WIDE TOLERANCE ZONE
LINE

Fig. 31-2-1 Tolerance zone for straightness of a line.

Points, Lines, and Surfaces

The production and measurement of engineering parts deals, in most cases, with surfaces of objects. These surfaces may be flat, cylindrical, conical, or spherical or have some more or less irregular shape or contour. Lines might enter the picture as edges of parts or axes of cylinders, and points might refer to the intersection of two axes.

However, measurement usually has to take place at specific points. A line or surface is evaluated dimensionally by making a series of measurements at various points along its length.

Therefore, geometric tolerances are chiefly concerned with points and lines, while surfaces are considered to be composed of a series of line elements running in two or more directions.

Points have position but no size, and therefore position is the only characteristic that requires control. Lines and surfaces have to be controlled for form, orientation, and location. Therefore geometric tolerances provide for control of the characteristics shown in Fig. 31-2-2.

Feature Control Notes and Symbols

Some geometric tolerances have been used for many years in the form of notes such as "parallel with surface *A* within 0.001" and "Straight within 0.005." While such notes are now obsolete, the reader should be prepared to recognize them on older drawings.

The modern method is to specify geometric tolerances by means of *feature control symbols*. A feature control symbol consists of a rectangular frame, having a suitable height to accommodate the lettering used, and the length required to contain the necessary information. This frame is divided into two or more compartments. See Fig. 31-2-3.

The first compartment contains a symbol representing the geometric characteristic to be controlled. The second compartment contains the required tolerance value. When necessary, other compartments are added to contain datum references, as explained in Unit 31-5.

CONTROLLING	GEOMETRICAL SYMBOL
Form of a line	Straightness, roundness and profile
Form of a surface	Flatness, cylindricity, and profile
Orientation	Angularity, parallelism and perpendicularity
Location	Position, concentricity and symmetry
Runout	This is a special composite characteristic, which will be explained in unit 31–11

Fig. 31-2-2 Geometrical characteristics.

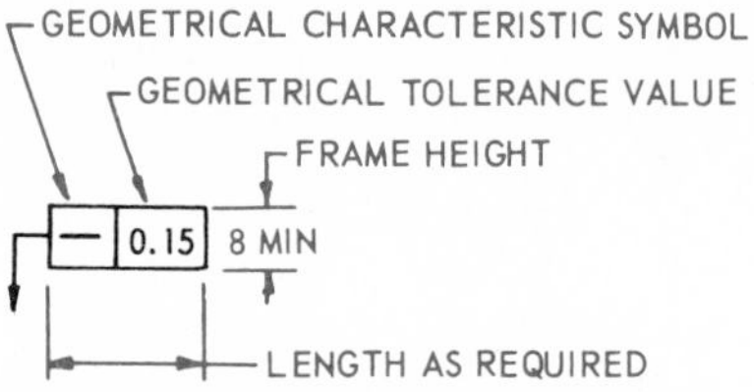
GEOMETRICAL CHARACTERISTIC SYMBOL
GEOMETRICAL TOLERANCE VALUE
FRAME HEIGHT
0.15
8 MIN
LENGTH AS REQUIRED

Fig. 31-2-3 Feature control symbol.

GEOMETRICAL CHARACTERISTIC	SYMBOL	SIZE OF SYMBOL
Profile (of a line)	⌒	length equal to frame height
Straightness	—	length equal to frame height
Roundness	○	diameter equal to 75% of frame height

Fig. 31-2-4 Geometrical symbols for form of a line.

Geometric characteristic symbols used for form of a line are shown in Fig. 31-2-4. Other symbols will be introduced as required, but all are shown for reference purposes in the Appendix.

Application to Drawings

The feature control symbol is normally connected by means of a leader line from either end of the symbol to the feature to be controlled. This leader terminates in an arrowhead, which is positioned as follows:

1. It is on the outline of the feature when the tolerance refers to the line itself or to the surface represented by the line, as shown in Fig. 31-2-5*a*.

2. It is on an extension line from the feature, in line with the size dimension, when the tolerance refers to the axis, center line, or median plane of the feature, as shown in Fig. 31-2-5*b*.

3. The feature control symbol may also be associated directly with the size dimension of the feature being controlled, as shown in Fig. 31-2-5*c*.

ANSI also permits the feature control symbol to be related to the feature by attaching the side or end of the symbol directly to an extension line from the feature or to a dimension line pertaining to the size of the feature. These methods are shown in Fig. 31-2-6 but are not used in any other national standard.

Application to Surfaces

The leader and arrowhead from the feature control symbol should touch the surface of the feature or the extension line

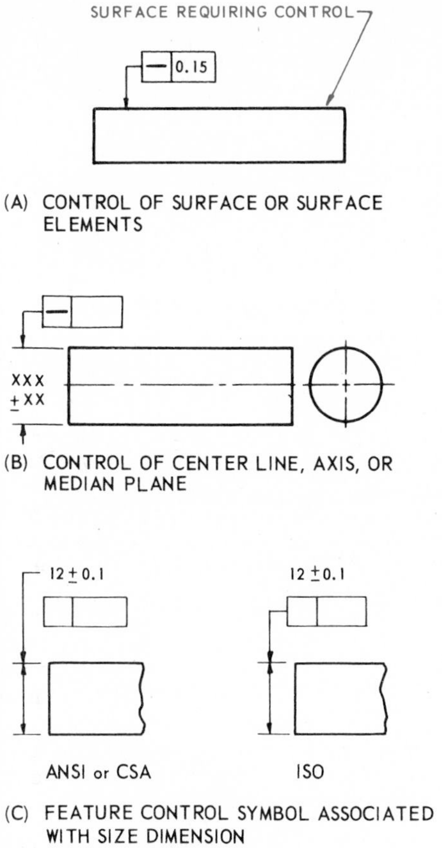
SURFACE REQUIRING CONTROL
0.15
(A) CONTROL OF SURFACE OR SURFACE ELEMENTS
XXX
±XX
(B) CONTROL OF CENTER LINE, AXIS, OR MEDIAN PLANE
12 ± 0.1
12 ± 0.1
ANSI or CSA
ISO
(C) FEATURE CONTROL SYMBOL ASSOCIATED WITH SIZE DIMENSION

Fig. 31-2-5 Application of feature control symbols.

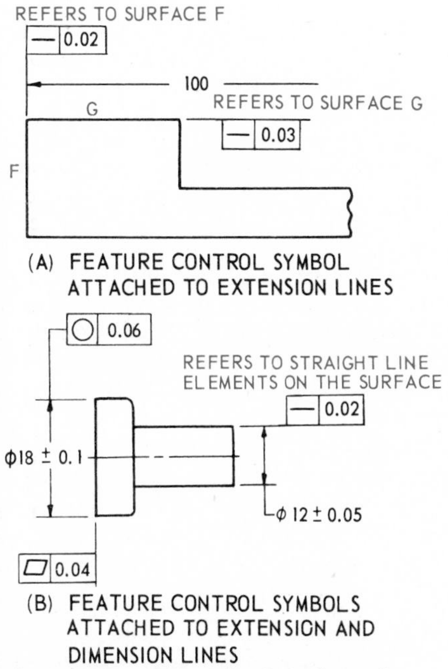
REFERS TO SURFACE F
0.02
100
G
REFERS TO SURFACE G
0.03
F
(A) FEATURE CONTROL SYMBOL ATTACHED TO EXTENSION LINES
0.06
REFERS TO STRAIGHT LINE ELEMENTS ON THE SURFACE
0.02
ϕ18 ± 0.1
ϕ 12 ± 0.05
0.04
(B) FEATURE CONTROL SYMBOLS ATTACHED TO EXTENSION AND DIMENSION LINES

Fig. 31-2-6 Alternative location of feature control symbol for ANSI drawings.

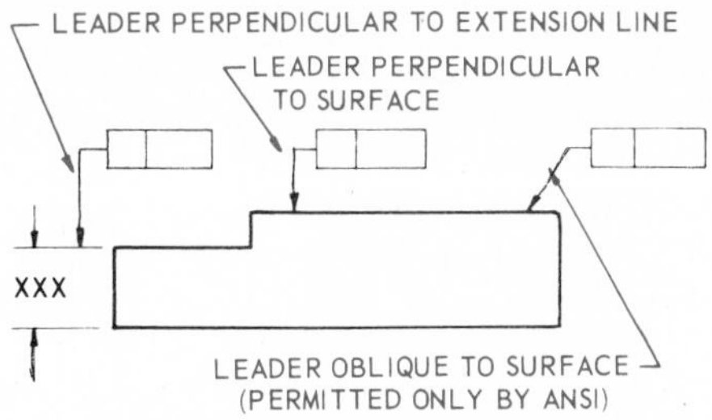
LEADER PERPENDICULAR TO EXTENSION LINE
LEADER PERPENDICULAR TO SURFACE
XXX
LEADER OBLIQUE TO SURFACE (PERMITTED ONLY BY ANSI)

Fig. 31-2-7 Direction of leaders for feature control.

in the direction of the width or diameter of the tolerance zone. This is usually perpendicular to the line or surface. However, ANSI permits the leader to touch the surface outline at an angle other than 90°, as shown in Fig. 31-2-7.

Two or more feature control symbols, which apply to the same feature, are drawn together with a single leader and arrowhead, as shown in Fig. 31-2-8.

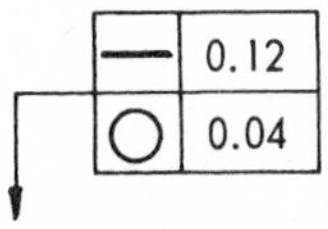
0.12
0.04

Fig. 31-2-8 Two controls of one surface.

Preferably the leader from the feature control symbol should be directed to the line on the drawing representing the surface or feature in its characteristic profile. Thus, in Fig. 31-2-9 the straightness tolerance is directed to the side view, and the roundness tolerance to the end view. This is not always possible, and tolerances connected to an alternative view, such as a roundness tolerance connected to a side view, are acceptable.

When space is limited, or when it is more convenient, the leader and arrowhead may be directed to an extension line from the surface, as shown in Fig. 31-2-10, but it must not be in line with the dimension.

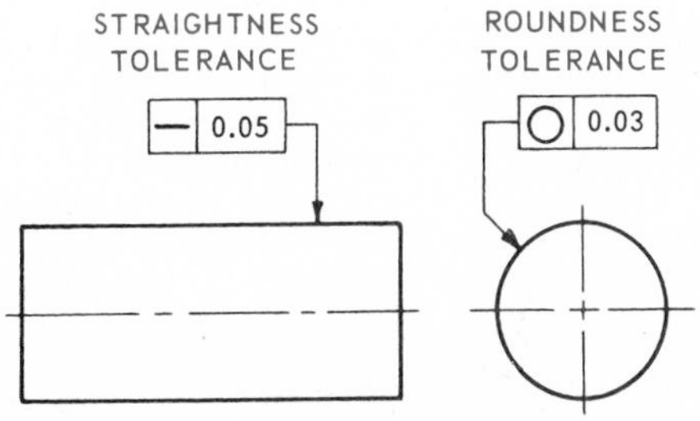
STRAIGHTNESS TOLERANCE
ROUNDNESS TOLERANCE
0.05
0.03

Fig. 31-2-9 Preferred location of feature control symbol.

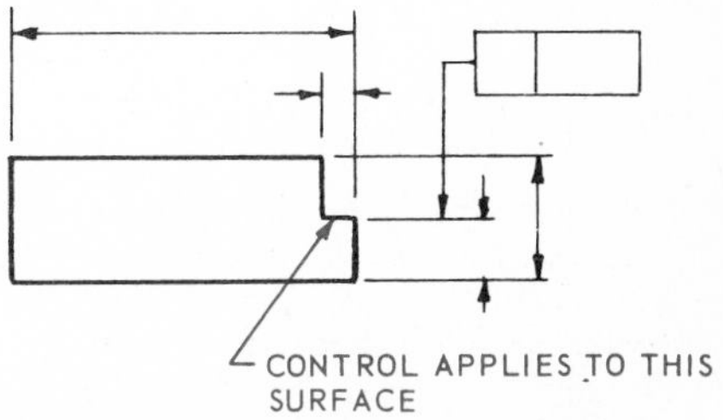
CONTROL APPLIES TO THIS SURFACE

Fig. 31-2-10 Acceptable location of feature control leader when space is limited.

Circular Tolerance Zones

When the resulting tolerance zone is circular or cylindrical, such as when straightness of the center line of a cylindrical feature is specified, a diameter symbol precedes the tolerance value in the feature control symbol. See Fig. 31-2-11.

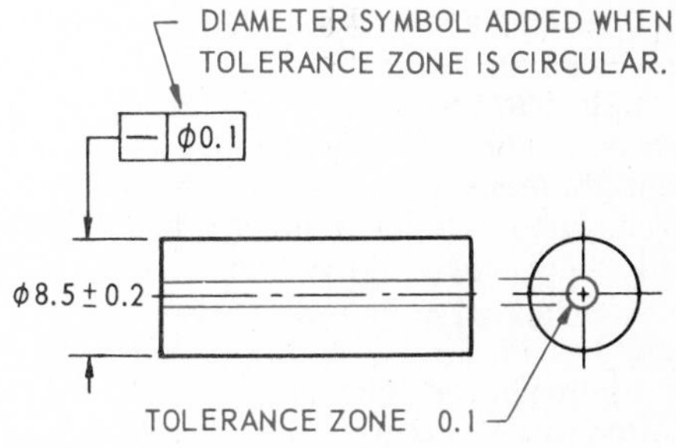
DIAMETER SYMBOL ADDED WHEN TOLERANCE ZONE IS CIRCULAR.
ϕ0.1
ϕ8.5 ± 0.2
TOLERANCE ZONE 0.1

Fig. 31-2-11 Circular tolerance zone.

Straightness

Lines. Straightness is fundamentally a characteristic of a line, such as the edge of a part or a line scribed on a surface. A straightness tolerance is specified on a drawing by means of a feature control symbol, which is directed by a leader to the line requiring control, as shown in Fig. 31-2-12.

It states in symbolic form that the line shall be straight within 0.15 mm. This

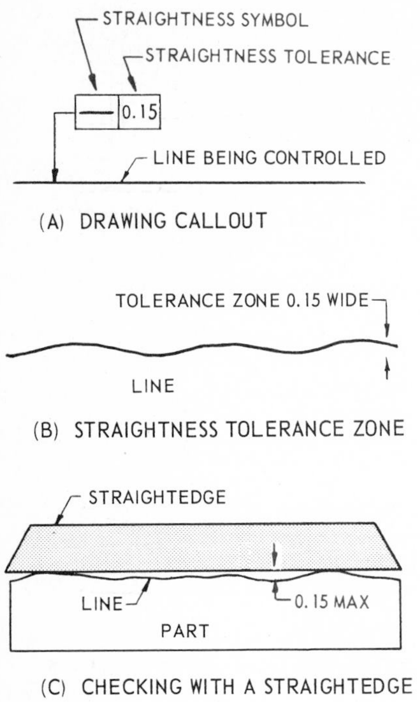
STRAIGHTNESS SYMBOL
STRAIGHTNESS TOLERANCE
0.15
LINE BEING CONTROLLED
(A) DRAWING CALLOUT
TOLERANCE ZONE 0.15 WIDE
LINE
(B) STRAIGHTNESS TOLERANCE ZONE
STRAIGHTEDGE
LINE
0.15 MAX
PART
(C) CHECKING WITH A STRAIGHTEDGE

Fig. 31-2-12 Straightness symbol and application.

means that the line shall be contained within a tolerance zone consisting of the area between two parallel straight lines in the same plane, separated by the specified tolerance.

Theoretically, straightness could be measured by bringing a straightedge into contact with the line and determining that any space between the straightedge and the line does not exceed the specified tolerance.

Cylindrical Surfaces. For cylindrical parts, or curved surfaces which are straight in one direction, the feature control symbol should be directed to the side view, where line elements appear as a straight line, as shown in Figs. 31-2-13 and 31-2-14.

A straightness tolerance thus applied to the surface controls surface elements only. Therefore it would control bending or a wavy condition of the surface or a barrel-shaped part, but it would not necessarily control the straightness of the center line or the conicity of the cylinder.

Straightness of a cylindrical surface is interpreted to mean that each line element of the surface shall be contained within a tolerance zone consisting of the space between two parallel planes, separated by the width of the specified tolerance, when the part is rolled along one of the planes. All circular elements of the surface must be within the specified size tolerance. No error in straightness would

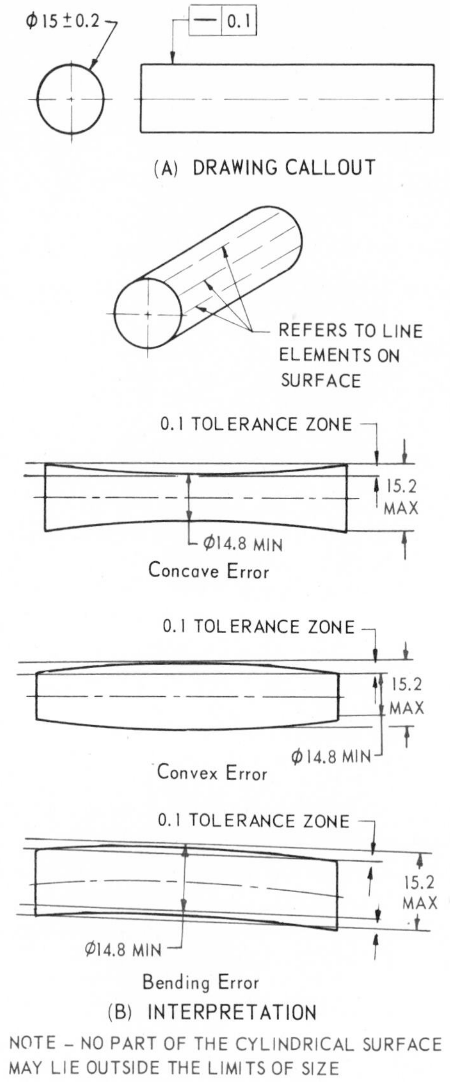
ϕ15 ± 0.2
0.1
(A) DRAWING CALLOUT
REFERS TO LINE ELEMENTS ON SURFACE
0.1 TOLERANCE ZONE
15.2 MAX
ϕ14.8 MIN
Concave Error
0.1 TOLERANCE ZONE
15.2 MAX
ϕ14.8 MIN
Convex Error
0.1 TOLERANCE ZONE
15.2 MAX
ϕ14.8 MIN
Bending Error
(B) INTERPRETATION
NOTE – NO PART OF THE CYLINDRICAL SURFACE MAY LIE OUTSIDE THE LIMITS OF SIZE

Fig. 31-2-13 Straightness errors in line elements of a cylindrical surface.

be permitted if the diameter were at its maximum material size.

Conical Surfaces. A straightness tolerance can be applied to a conical surface in the same manner as for a cylindrical surface, as shown in Fig. 31-2-15, and will ensure that the rate of taper is uniform. The actual rate of taper, or the taper angle, must be separately toleranced.

Flat Surfaces. A straightness tolerance applied to a flat surface indicates straightness control in one direction only and must be directed to the line on the drawing representing the surface to be controlled and the direction in which control is required, as shown in Fig. 31-2-16*a*. It is then interpreted to mean that each line element on the surface in the indicated direction shall lie within a tolerance zone.

Different straightness tolerances may be specified in two or more directions

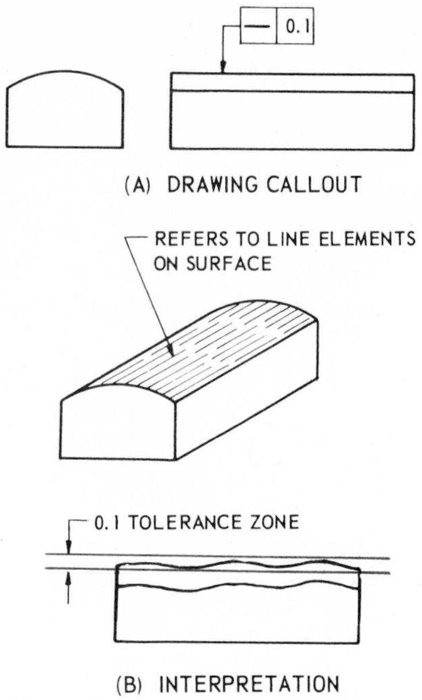
0.1
(A) DRAWING CALLOUT
REFERS TO LINE ELEMENTS ON SURFACE
0.1 TOLERANCE ZONE
(B) INTERPRETATION

Fig. 31-2-14 Straightness of surface line elements.

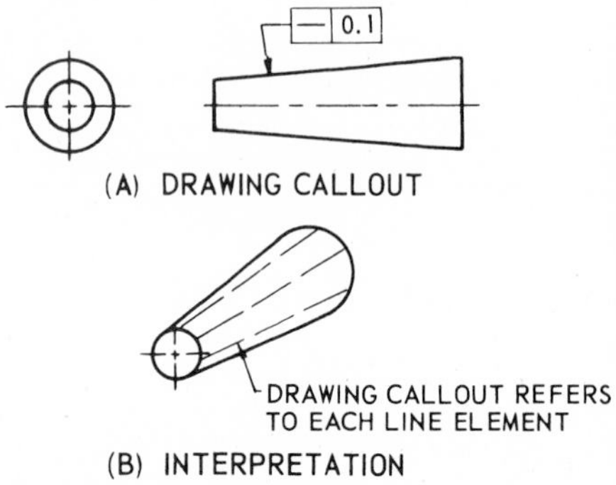
0.1
(A) DRAWING CALLOUT
DRAWING CALLOUT REFERS TO EACH LINE ELEMENT
(B) INTERPRETATION

Fig. 31-2-15 Straightness of a conical surface.

when required, as shown in Fig. 31-2-16*b*. However, if the same straightness tolerance is required in two coordinate directions on the same surface, a flatness tolerance rather than a straightness tolerance is used.

If it is not otherwise necessary to draw all three views, the straightness tolerances may all be shown on a single view by indicating the direction with short lines terminated by arrowheads, as shown in Fig. 31-2-16*c*.

Straightness in a Specified Length. It is often desirable on long parts to specify a straightness tolerance over a specific length, either with or without a maximum overall tolerance. For example, a certain amount of bow may be

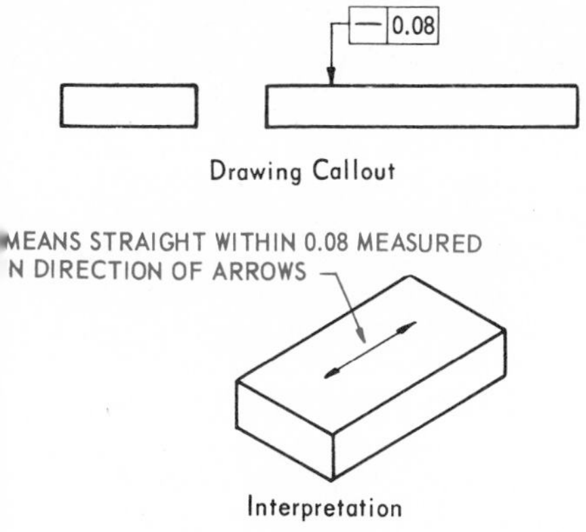

— 0.08
Drawing Callout
MEANS STRAIGHT WITHIN 0.08 MEASURED
N DIRECTION OF ARROWS
Interpretation

(A) STRAIGHTNESS IN ONE DIRECTION

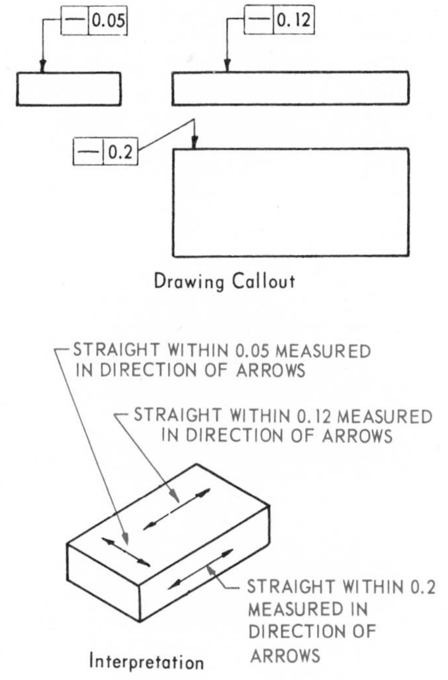

— 0.05
— 0.12
— 0.2
Drawing Callout
STRAIGHT WITHIN 0.05 MEASURED IN DIRECTION OF ARROWS
STRAIGHT WITHIN 0.12 MEASURED IN DIRECTION OF ARROWS
STRAIGHT WITHIN 0.2 MEASURED IN DIRECTION OF ARROWS
Interpretation

(B) STRAIGHTNESS IN SEVERAL DIRECTIONS

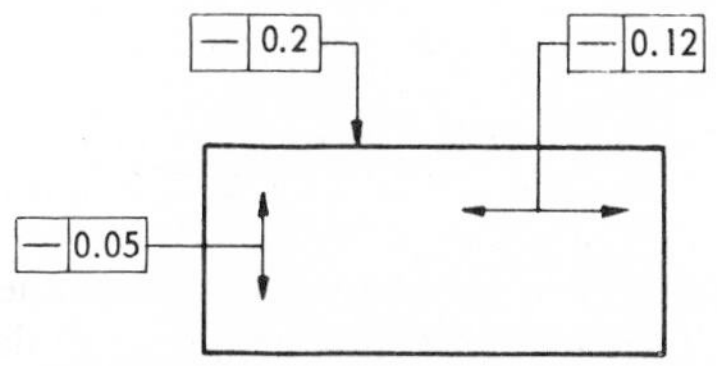

— 0.2
— 0.12
— 0.05

(C) THREE STRAIGHTNESS TOLERANCES ON ONE VIEW

Fig. 31-2-16 Three straightness tolerances on one view.

quite acceptable if spread over the entire length, but quite undesirable if concentrated at some point along the length.

This requirement is specified on the drawing by including the specified length with the tolerance in the feature control symbol and separating them with a diagonal line, as shown in Fig. 31-2-17.

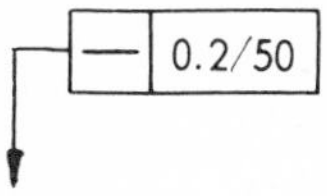

— 0.2/50

Fig. 31-2-17 Tolerance in a specified length.

Note, however, that the expression 0.2/50 does not mean 0.2 mm per 50 mm of length, but 0.2 mm in any 50 mm long portion of the part.

It can be readily shown that for a part which is uniformly bowed in a circular arc, a bow of 0.01 mm in 10 mm would be 0.04 mm in 20 mm, or 0.16 mm in 40 mm, etc., as illustrated in Fig. 31-2-18.

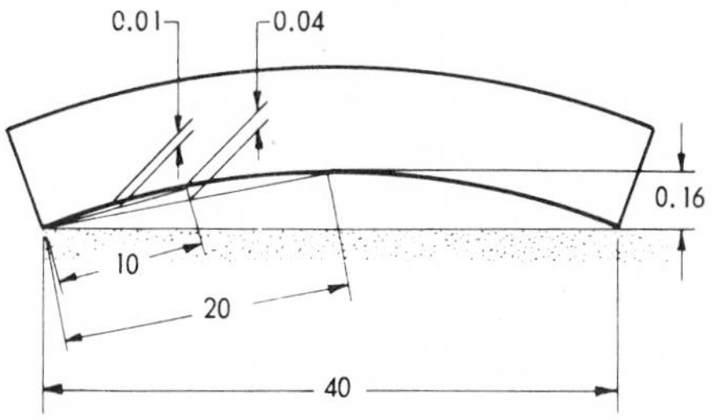

0.01
0.04
0.16
10
20
40

Fig. 31-2-18 Resultant in specifying straightness per unit length with no specified total.

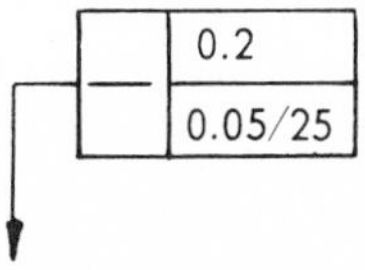

— 0.2
0.05/25

Fig. 31-2-19 Overall tolerance combined with a tolerance in a specified length.

When a maximum overall tolerance is to be combined with a tolerance in a specified length, the tolerances are shown in a double feature control symbol, as in Fig. 31-2-19. This means: STRAIGHT WITHIN 0.2 mm FOR THE FULL LENGTH, BUT NOT TO EXCEED 0.05 mm IN ANY 25 mm LENGTH.

Assignment

On an A3- or B-size sheet, prepare sketches and solutions for the drawings shown in Fig. 31-2-A or 31-2-B and the following information.

1. Apply two feature control symbols, one to denote roundness, the other to control straightness to the surface of the cylindrical feature *R* shown in illustration (A).

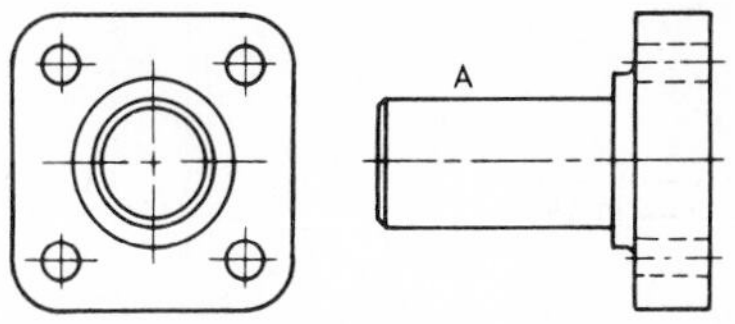

A

CONTROL	TOLERANCE
STRAIGHTNESS	0.1
ROUNDNESS	0.02

(A)

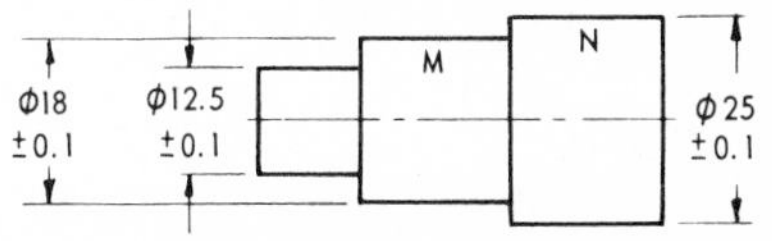

⌀18 ±0.1
⌀12.5 ±0.1
M
N
⌀25 ±0.1

FEATURE	STRAIGHTNESS TOLERANCE
CENTER LINE OF ⌀M	0.15
SURFACE OF ⌀N (SHOW TWO METHODS)	0.2

(B)

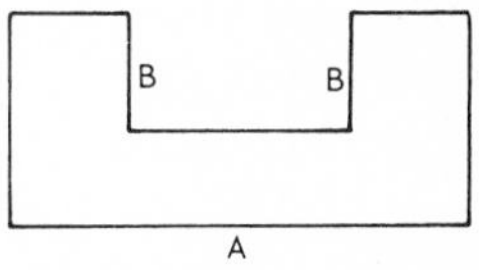

B
B
A

SURFACE A TO BE STRAIGHT WITHIN 0.05
SURFACES B TO BE STRAIGHT WITHIN 0.1
(C)

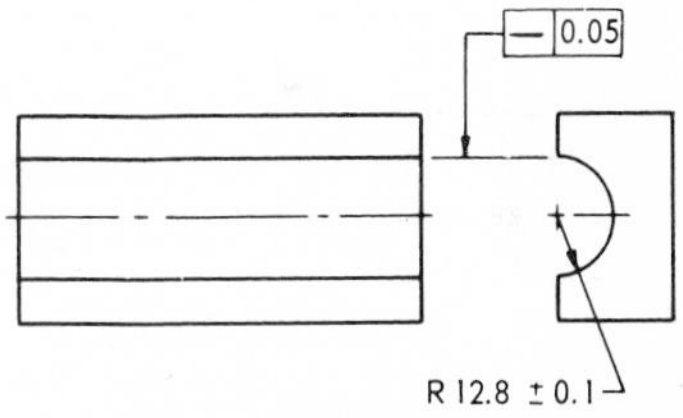

— 0.05
R 12.8 ± 0.1

FEATURE SIZE	PERMISSIBLE STRAIGHTNESS ERROR
12.77	
12.82	
12.9	

(D)

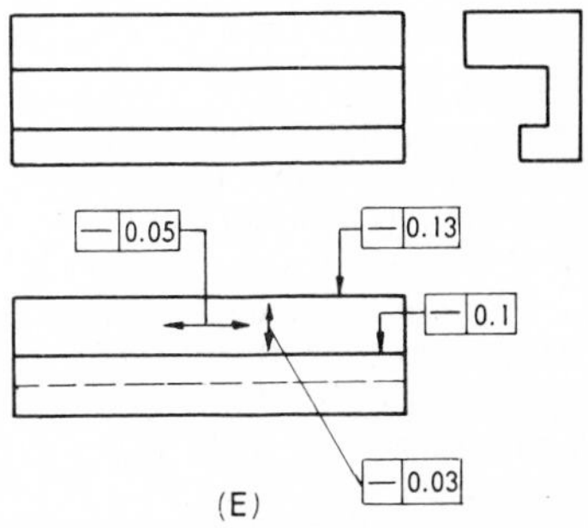

— 0.05
— 0.13
— 0.1
— 0.03

(E)

Fig. 31-2-A Assignments.

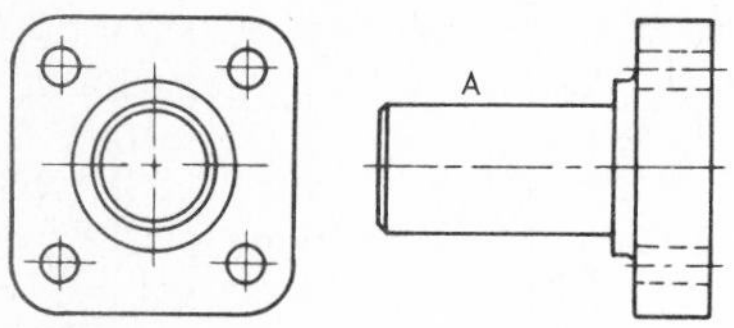
A

CONTROL	TOLERANCE
STRAIGHTNESS	.02
ROUNDNESS	.005

(A)

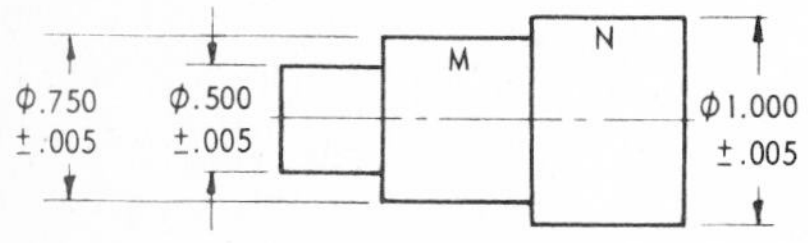
M
N
ϕ.750
±.005
ϕ.500
±.005
ϕ1.000
±.005

FEATURE	STRAIGHTNESS TOLERANCE
CENTER LINE OF ϕM	
SURFACE OF ϕN (SHOW TWO METHODS)	

(B)

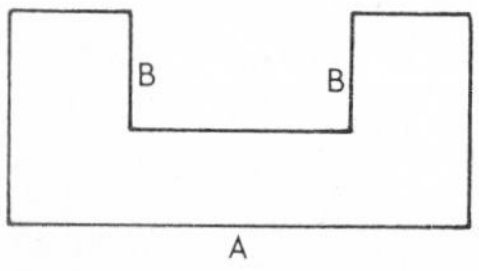
B
B
A

SURFACE A TO BE STRAIGHT WITHIN .002

SURFACES B TO BE STRAIGHT WITHIN .004

(C)

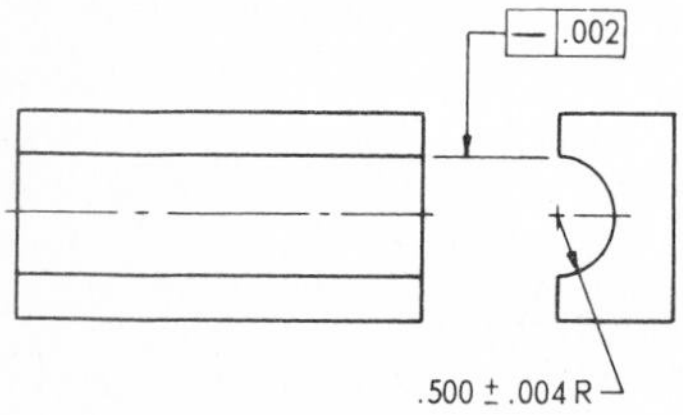
— .002
.500 ± .004 R

FEATURE SIZE	PERMISSIBLE STRAIGHTNESS ERROR
.496	
.500	
.503	

(D)

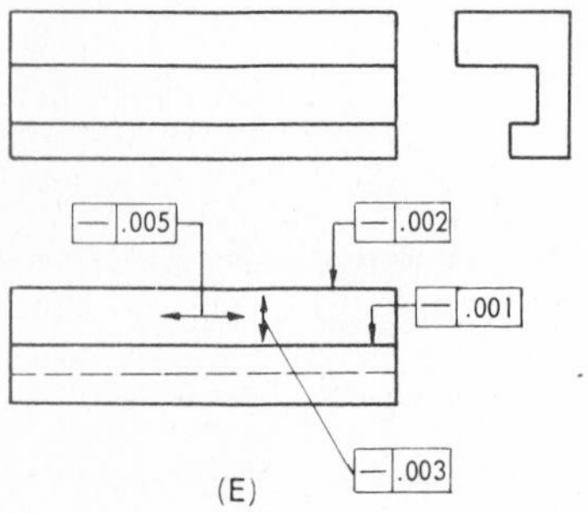
— .005
— .002
— .001
— .003

(E)

Fig. 31-2-B Assignments.

2. It is essential that the axis of features M and N be straight within the tolerances shown in illustration (B). Add the feature control symbols to the drawing which would control the form and tolerance prescribed.

3. ANSI permits the feature control symbol to be connected to extension and dimension lines. Using these methods, apply feature control symbols to illustration (C), which would indicate the form and tolerances prescribed.

4. With straightness specified as shown in illustration (D), what is the maximum permissible deviation from straightness if the radius was as shown in the chart?

5. With reference to illustration (E), eliminate the lower view and show the straightness tolerances on the upper views.

UNIT 31-3
RELATIONSHIP TO FEATURE OF SIZE

Definitions

Maximum Material Condition (MMC). This term refers to that limit of size of a feature which results in the part containing the maximum amount of material. Thus it is the maximum limit of size for an external feature, such as a shaft, or the minimum limit of size for an internal feature, such as a hole. See Fig. 31-3-1.

Virtual Condition (Size). *Virtual condition* refers to the overall envelope of perfect form, within which the feature would just fit. For an external feature such as a shaft, it is the maximum measured size plus the effect of actual form variations, such as straightness, flatness, or roundness. For an internal feature such as a hole, it is the minimum measured size minus the effect of such form variations.

Least Material Condition (LMC). This term refers to that size of a feature which results in the part containing the minimum amount of material. Thus it is the minimum limit of size for an external feature and the maximum limit of size for an internal feature.

Regardless of Feature Size (RFS). This term indicates that a form or positional tolerance applies to any size of a feature which lies within its size tolerance.

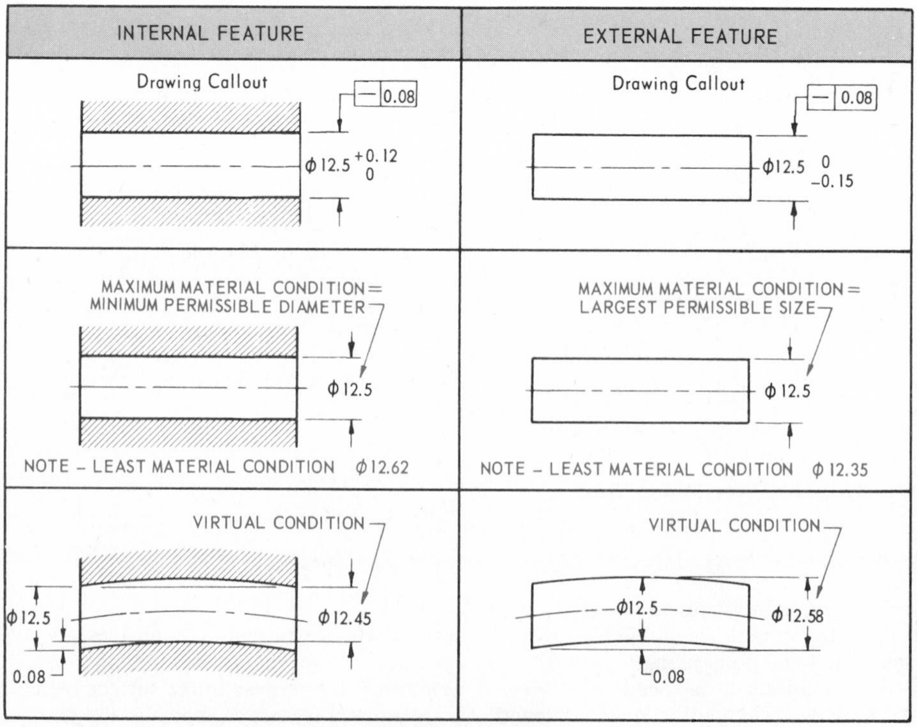
INTERNAL FEATURE
EXTERNAL FEATURE
Drawing Callout
— 0.08
ϕ12.5 +0.12 0
Drawing Callout
— 0.08
ϕ12.5 0 −0.15
MAXIMUM MATERIAL CONDITION = MINIMUM PERMISSIBLE DIAMETER
ϕ12.5
NOTE – LEAST MATERIAL CONDITION ϕ12.62
MAXIMUM MATERIAL CONDITION = LARGEST PERMISSIBLE SIZE
ϕ12.5
NOTE – LEAST MATERIAL CONDITION ϕ12.35
VIRTUAL CONDITION
ϕ12.5
ϕ12.45
0.08
VIRTUAL CONDITION
ϕ12.5
ϕ12.58
0.08

Fig. 31-3-1 Maximum material and virtual condition.

Features of Size

Geometric tolerances so far considered concern only lines, line elements, and single surfaces. These are features having no diameter or thickness, and tolerances applied to them cannot be affected by feature size.

Features of size are features which do have diameter or thickness. These may be cylinders, such as shafts and holes. They may also be slots, tabs, or rectangular or flat parts, where two parallel, flat surfaces are considered to form a single feature.

If freedom of assembly of mating parts is the chief criterion for establishing a geometric tolerance for a feature of size, the least favorable assembly condition exists when the parts are made to the maximum material condition. Further geometric variations can then be permitted, without jeopardizing assembly, as the features approach their least material condition.

Example 1. The effect of a form tolerance is shown in Fig. 31-3-2, wherein a cylindrical pin of 7.62 +0, −0.12 diameter is intended to assemble into a round hole of 7.62 +0.12, −0 diameter. If both parts are at their maximum material condition of 7.62, it is evident that both would have to be perfectly round and straight in order to assemble. However, if the pin was at its least material condition of 7.5, it could be bent up to 0.12 and still assemble in the smallest permissible hole.

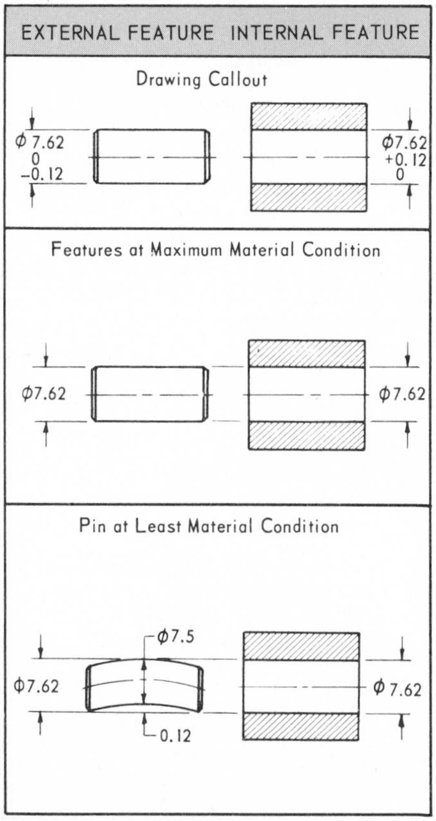
EXTERNAL FEATURE
INTERNAL FEATURE
Drawing Callout
Φ7.62 0 −0.12
Φ7.62 +0.12 0
Features at Maximum Material Condition
Φ7.62
Φ7.62
Pin at Least Material Condition
Φ7.5
Φ7.62
Φ7.62
0.12

Fig. 31-3-2 Effect of form variations.

Example 2. Another example based on the location of features is shown in Fig. 31-3-3. This shows a part with two projecting pins required to assemble into a mating part having two holes at the same center distance.

The worst assembly condition exists when the pins and holes are at their maximum material condition, which is 6.4 mm diameter. Theoretically, these parts would just assemble if their form, orientation (squareness to the surface), and center distances were perfect. However, if the pins and holes were at their least material condition of 6.3 and 6.5, it is evident that one center distance could be increased and the other decreased by 0.1 without jeopardizing the assembly condition.

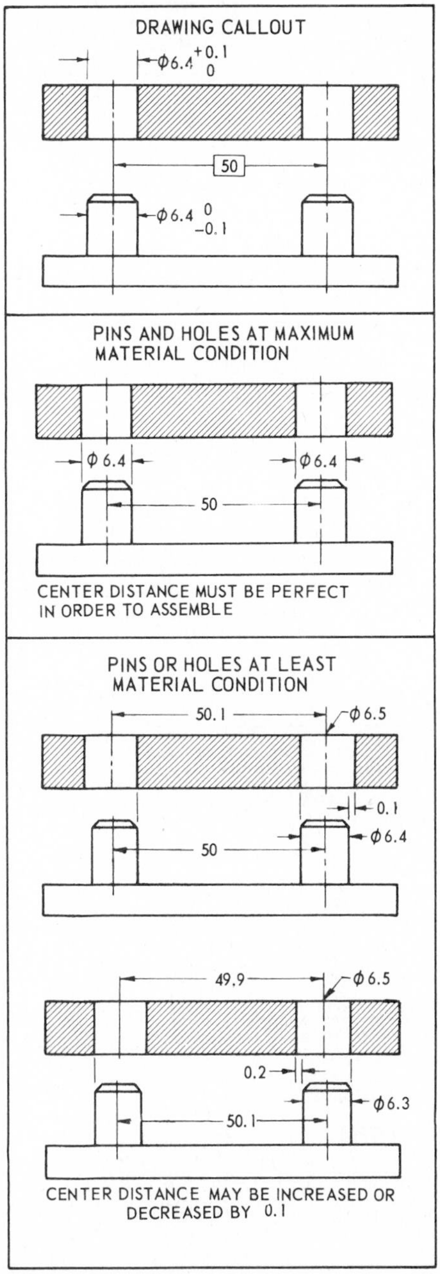
DRAWING CALLOUT
Φ6.4 +0.1 0
50
Φ6.4 0 −0.1
PINS AND HOLES AT MAXIMUM MATERIAL CONDITION
Φ6.4
Φ6.4
50
CENTER DISTANCE MUST BE PERFECT IN ORDER TO ASSEMBLE
PINS OR HOLES AT LEAST MATERIAL CONDITION
50.1
Φ6.5
0.1
Φ6.4
50
49.9
Φ6.5
0.2
Φ6.3
50.1
CENTER DISTANCE MAY BE INCREASED OR DECREASED BY 0.1

Fig. 31-3-3 Effect on location.

Maximum Material Condition

The symbol for maximum material condition is shown in Fig. 31-3-4. The symbol dimensions are based on percentages of the feature control symbol frame height. These are 75 percent of the frame height for the diameter of the circle and 40 percent for the height of the letter M.

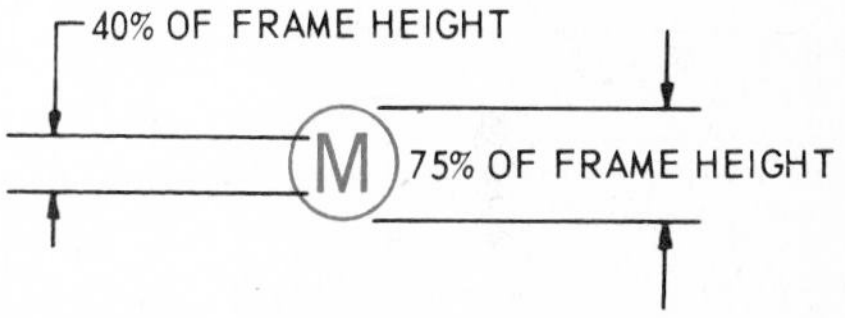
40% OF FRAME HEIGHT
M
75% OF FRAME HEIGHT

Fig. 31-3-4 MMC symbol.

If a geometric tolerance is required to be modified on an MMC basis, it is specified on the drawing by including the symbol Ⓜ immediately after the tolerance value in the feature control symbol, as shown in Fig. 31-3-5.

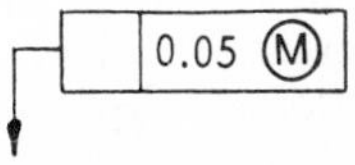
0.05 Ⓜ

Fig. 31-3-5 Application of MMC symbol.

A form tolerance modified in this way can be applied only to a feature of size; it cannot be applied to a single surface. It controls the boundary of the feature, such as a complete cylindrical surface, or two parallel surfaces of a flat feature. This permits the feature surface or surfaces to cross the maximum material boundary by the amount of the form tolerance. If it is required that the virtual condition be kept within the maximum material boundary, the form tolerance must be specified as zero at MMC, as shown in Fig. 31-3-6.

Application with Maximum Value. It is sometimes necessary to ensure that the geometric tolerance does not vary

0 Ⓜ

Fig. 31-3-6 MMC symbol with zero tolerance.

over the full range permitted by the size variations. For such applications a maximum limit may be applied to the geometric tolerance, in addition to the tolerance permitted at the maximum material condition, as shown in Fig. 31-3-7.

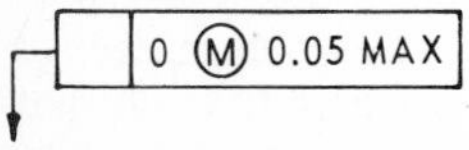
0 Ⓜ 0.05 MAX

EXAMPLE 1

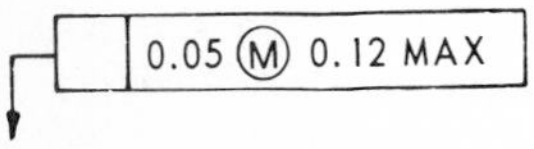
0.05 Ⓜ 0.12 MAX

EXAMPLE 2

Fig. 31-3-7 Tolerance with a maximum value.

Regardless of Feature Size

When MMC is not specified with a geometric tolerance for a feature of size, no relationship is intended to exist between the feature size and the geometric tolerance. In other words, the tolerance applies regardless of feature size.

In this case, the geometric tolerance controls the form, orientation, or location of the center line, axis, or median plane of the feature.

Prior to publication of the 1973 issue of ANSI Y14.5, the ANSI standard stated that positional tolerances, unless otherwise specified, were intended to apply on an MMC basis. Therefore, if positional tolerancing was used on a RFS basis, it became necessary to add a symbol after the tolerance. This symbol is Ⓢ, and it is applied as shown in Fig. 31-3-8. With publication of Y14.5 (1973), the rule was changed, so that all geometric tolerances for feature of size, whether of form, orientation, or position, apply regardless of feature of size, unless

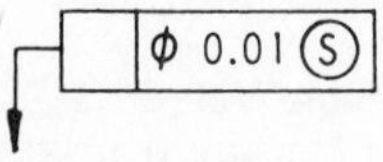
ϕ 0.01 Ⓢ

Fig. 31-3-8 Application of RFS symbol.

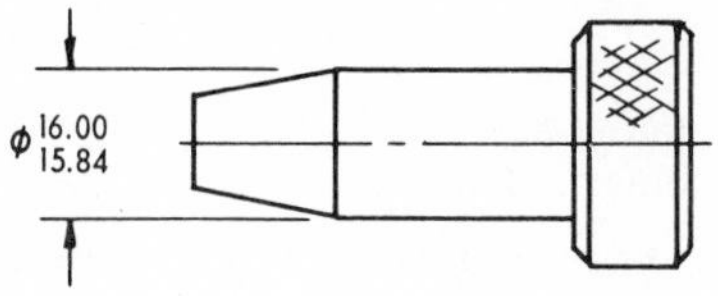
ϕ 16.00
15.84

SIZE	ALLOWABLE TOLERANCE
ϕ15.9	
MMC	
LMC	

(A)

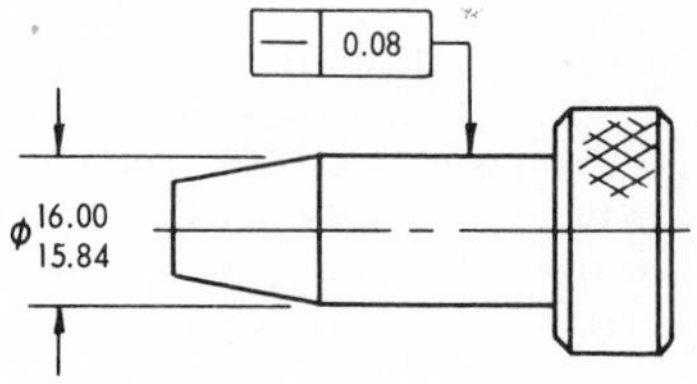
0.08
ϕ 16.00
15.84

SIZE	DIAMETER
VIRTUAL CONDITION	
ϕ16 AND BENT 0.08	
ϕ15.92 AND BENT 0.06	
ϕ15.84 AND PERFECTLY STRAIGHT	

(B)

ADD A STRAIGHTNESS TOLERANCE OF 0.08 TO THE SURFACES OF THE THREE DIMENSIONED DIAMETERS.

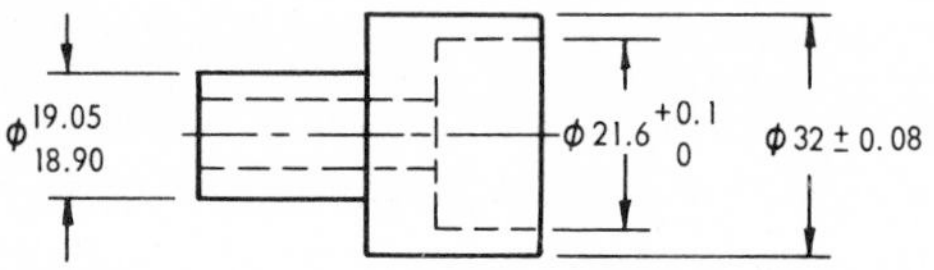
ϕ 19.05
18.90
ϕ 21.6 +0.1
0
ϕ 32 ± 0.08

SIZE	DIAMETER		
	ϕ 19.05 / 18.90	ϕ 21.6	ϕ 32
(A) LMC			
(B) MMC			
(C) VIRTUAL CONDITION			
(D) FEATURE SIZE			

(C)

ADD A STRAIGHTNESS TOLERANCE OF 0.08 TO THE DIMENSIONED FLAT SURFACES.

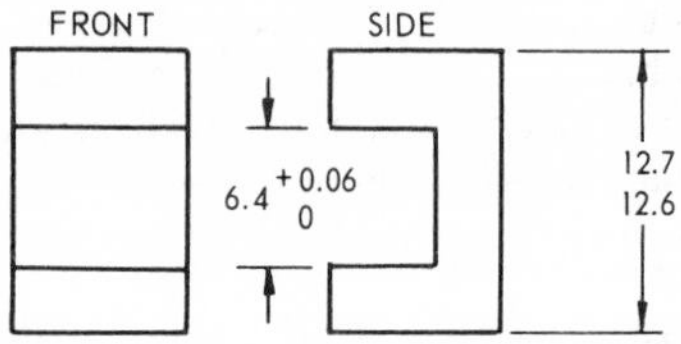
FRONT
SIDE
6.4 +0.06
0
12.7
12.6

SIZE	FEATURE	
	6.4	12.7 / 12.6
(A) LMC		
(B) MMC		
(C) VIRTUAL CONDITION		
(D) FEATURE SIZE		

(D)

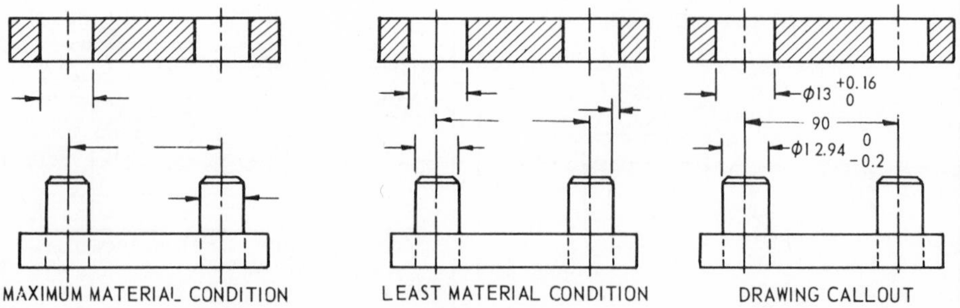
ϕ13 +0.16
0
90
ϕ12.94 0
-0.2
MAXIMUM MATERIAL CONDITION
LEAST MATERIAL CONDITION
DRAWING CALLOUT

(E)

Fig. 31-3-A Assignments.

modified by Ⓜ. Thus the use of the Ⓢ symbol is no longer necessary. This new interpretation agrees with ISO, CSA, and BSI standards. However, the symbol Ⓢ may still be used when desired to avoid ambiguity with older ANSI drawings and when modifications to existing drawings are necessary.

Assignment

On an A3- or B-size sheet, make drawings, complete with dimensions and geometric tolerances, of the problems shown in Fig. 31-3-A or 31-3-B and the following information.

1. If perfect form at the maximum material condition is required, what deviations from straightness would be permitted for the sizes shown in the table for the diameter shown in illustration (A)? Add the feature control symbol.
2. With reference to illustration (B), what is the virtual condition? Calculate the feature size of the individual parts having the sizes and variations shown in the table. Assume that the parts are perfectly round and uniform in diameter.
3. With reference to illustrations (C) and (D), complete the tables and add the straightness tolerance specified to the dimensioned features when Ⓜ is specified and perfect form at MMC is not required. Include at the bottom of the table the feature size when the feature is at the least material size and has the maximum permissible form variation.
4. With reference to illustration (E), calculate equal tolerances for the distances between the holes at MMC and LMC.

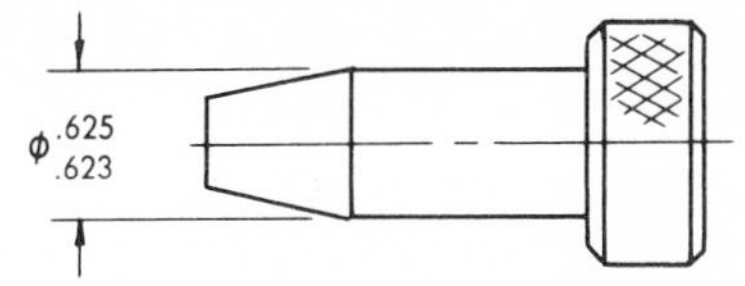

ϕ .625 .623

SIZE	ALLOWABLE TOLERANCE
ϕ.624	
MMC	
LMC	

(A)

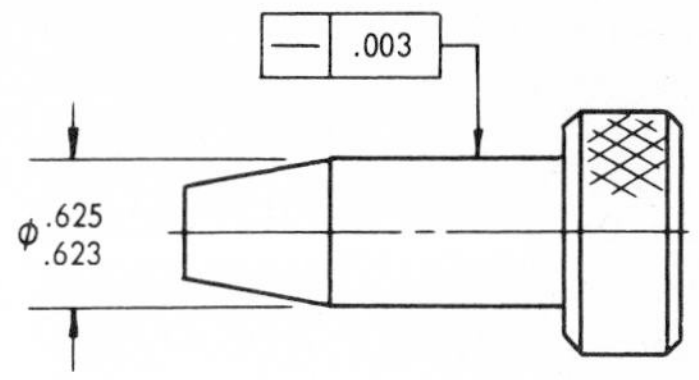

— .003
ϕ .625 .623

SIZE	DIAMETER
VIRTUAL CONDITION	
ϕ.625 AND BENT .002	
ϕ.624 AND BENT .003	
ϕ.623 AND PERFECTLY STRAIGHT	

(B)

ADD A STRAIGHTNESS TOLERANCE OF .003 TO THE SURFACES OF THE THREE DIMENSIONED DIAMETERS.

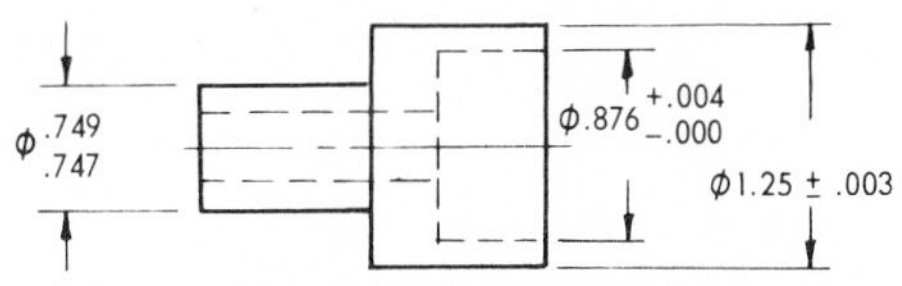

ϕ .749 .747
ϕ.876 +.004 −.000
ϕ1.25 ± .003

SIZE	DIAMETER ϕ .749/.747	ϕ.876	ϕ1.25
(A) LMC			
(B) MMC			
(C) VIRTUAL CONDITION			
(D) FEATURE SIZE			

(C)

ADD A STRAIGHTNESS TOLERANCE OF .003 TO THE DIMENSIONED FLAT SURFACES.

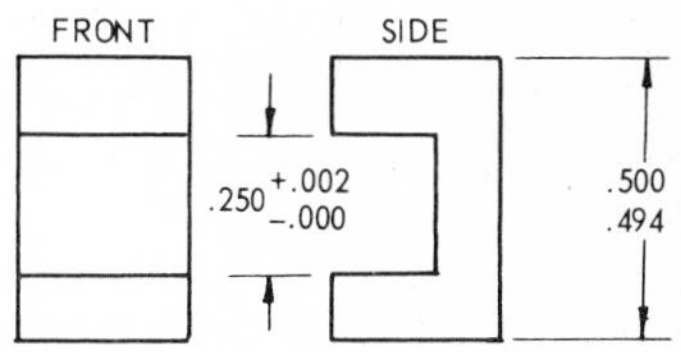

FRONT
SIDE
.250 +.002 −.000
.500 .494

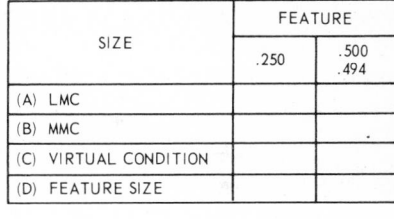

SIZE	FEATURE .250	.500/.494
(A) LMC		
(B) MMC		
(C) VIRTUAL CONDITION		
(D) FEATURE SIZE		

(D)

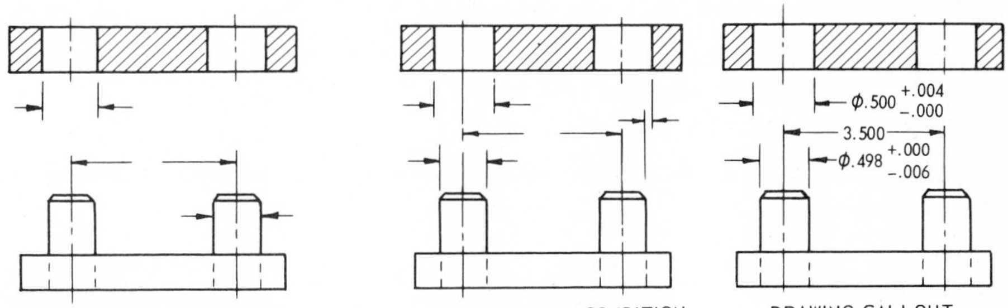

ϕ.500 +.004 −.000
3.500
ϕ.498 +.000 −.006
MAXIMUM MATERIAL CONDITION
LEAST MATERIAL CONDITION
DRAWING CALLOUT

(E)

Fig. 31-3-B Assignments.

UNIT 31-4

STRAIGHTNESS OF FEATURES AND FLATNESS

In applying a geometric tolerance to a feature of size, it is often desirable to control the feature as a whole, rather than merely its surface elements. Such control may be intended to control the center line or median plane, in which case it applies regardless of feature size (RFS). Alternatively it may be intended to control the bounding surfaces of the feature, in which case the tolerance is modified on an MMC basis. Straightness tolerances using both of these methods are described in this unit.

The same straightness symbol is used in the feature control symbol as for straightness of surface elements in Unit 31-3. However, when not modified by MMC, the feature control symbol may be directed to extension lines from the diameter or thickness, as shown in Fig. 31-4-1. Any of the variations shown in Fig. 31-2-5 may be used when it is more convenient.

Positioning the feature control symbol between the views should be avoided unless it is perfectly clear whether it refers to the lengthwise or widthwise straightness (see Fig. 31-4-2).

A straightness tolerance, not modified by MMC, may be applied to parts or features of any size or shape, providing they have a center line or median plane which is intended to be straight in the direction indicated. Examples are parts having a cross section which is circular, hexagonal, square, or rectangular.

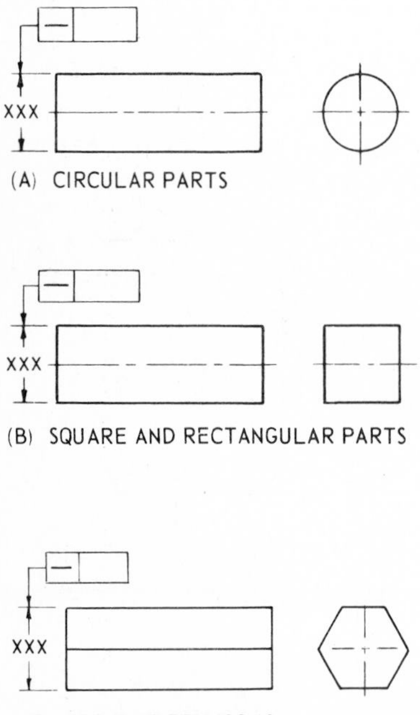
XXX
(A) CIRCULAR PARTS
XXX
(B) SQUARE AND RECTANGULAR PARTS
XXX
(C) REGULAR POLYGONS

Fig. 31-4-1 Straightness of the center line—RFS.

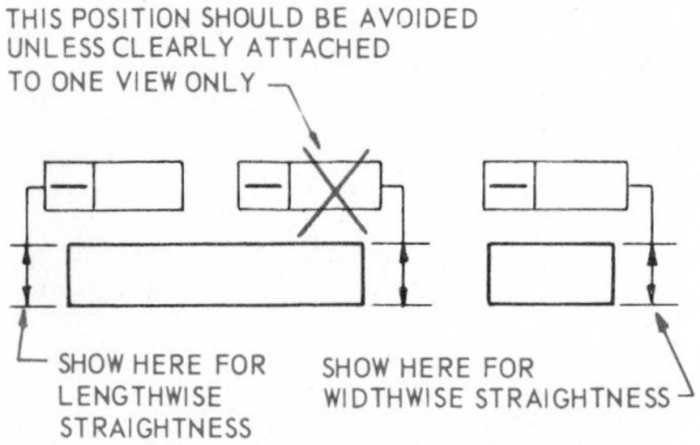
THIS POSITION SHOULD BE AVOIDED UNLESS CLEARLY ATTACHED TO ONE VIEW ONLY
SHOW HERE FOR LENGTHWISE STRAIGHTNESS
SHOW HERE FOR WIDTHWISE STRAIGHTNESS

Fig. 31-4-2 Straightness of center plane—RFS.

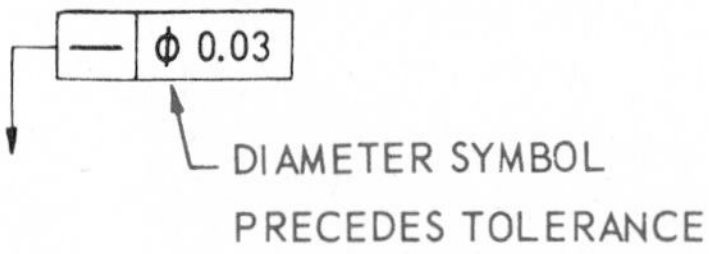
ϕ 0.03
DIAMETER SYMBOL PRECEDES TOLERANCE

Fig. 31-4-3 Diameter symbol added when tolerance zone is circular or cylindrical.

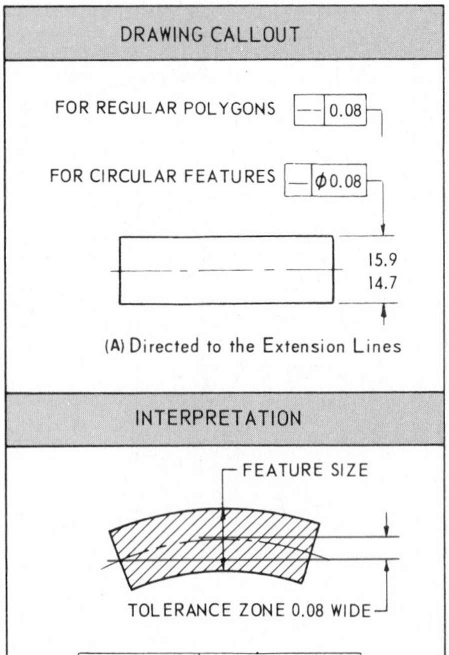
DRAWING CALLOUT
FOR REGULAR POLYGONS
0.08
FOR CIRCULAR FEATURES
ϕ0.08
15.9
14.7
(A) Directed to the Extension Lines
INTERPRETATION
FEATURE SIZE
TOLERANCE ZONE 0.08 WIDE

FEATURE SIZE	WIDTH OF TOLERANCE ZONE
15.9	0.08
15.6	0.08
15.3	0.08
15.0	0.08
14.7	0.08

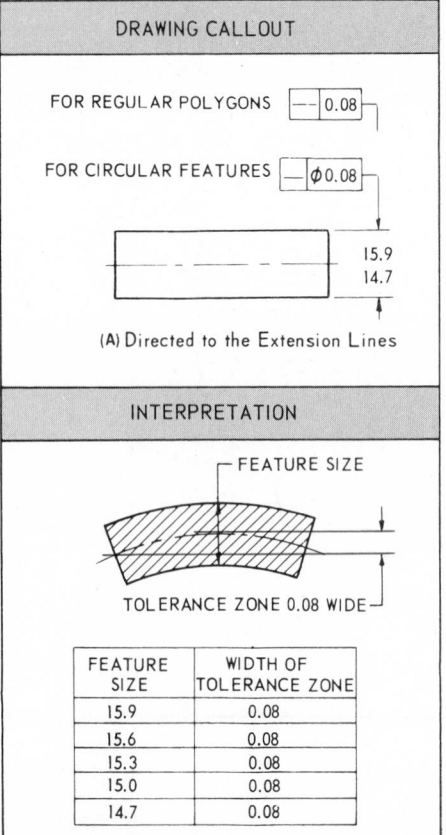

Fig. 31-4-4 Tolerance zones for cylindrical features.

Tolerances directed in this manner apply to straightness of the center line or center plane between all opposing line elements of the surfaces in the longitudinal direction to which the control is directed. The width of the tolerance zone is in the direction of the arrowhead. If the cross section forms a regular polygon, such as a hexagon or square, the tolerance applies to the center lines between each pair of sides, without its being necessary to so state on the drawing. If the cross section is circular, the tolerance zone becomes circular and a diameter symbol then precedes the tolerance, as shown in Figs. 31-4-3 and 31-4-4.

Control in Specific Directions

As already stated, straightness of a center line or median plane applies only to center lines that run in the direction of the line or line elements to which the straightness tolerance is directed. If there could be some ambiguity, a note should be added, such as THIS DIRECTION ONLY, as shown in Fig. 31-4-5*a*. If the part is circular and it is intended that the tolerance apply in all directions, a diameter symbol should precede the tolerance value, as shown in Fig. 31-4-5*b*.

If a tolerance is shown in two directions, it is measured in these two directions, and the tolerance zone is then a parallelepiped, as shown in Fig. 31-4-6.

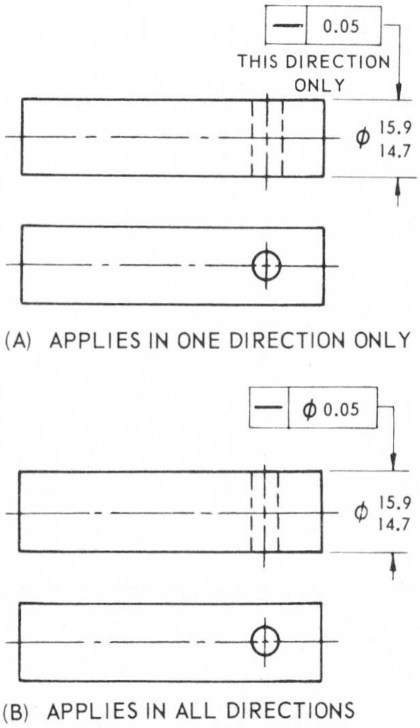
0.05
THIS DIRECTION ONLY
ϕ 15.9
14.7
(A) APPLIES IN ONE DIRECTION ONLY
ϕ 0.05
ϕ 15.9
14.7
(B) APPLIES IN ALL DIRECTIONS

Fig. 31-4-5 Direction of application of straightness.

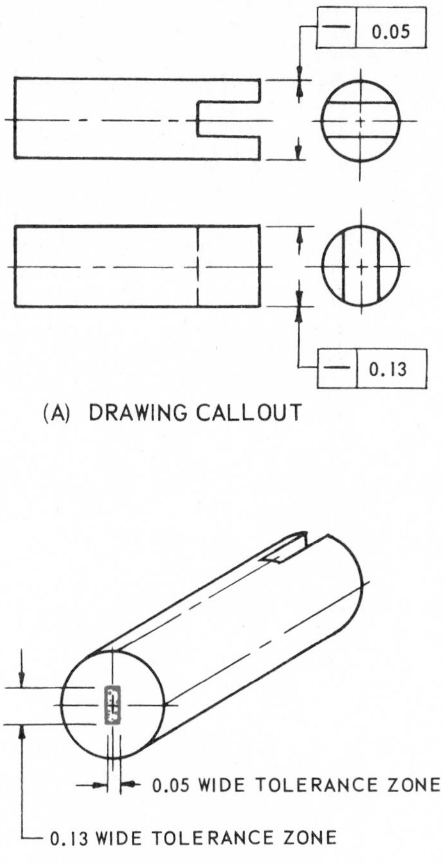

0.05
0.13
(A) DRAWING CALLOUT
0.05 WIDE TOLERANCE ZONE
0.13 WIDE TOLERANCE ZONE
(B) TOLERANCE ZONE

Fig. 31-4-6 Straightness in two directions.

Straightness—MMC

Straightness on an MMC basis can be considered to be an extension of the dimensional tolerance. This is because it actually specifies the virtual condition.

It is quite permissible to specify a geometric tolerance of zero MMC, which means that the virtual condition coincides with the maximum material size. Therefore, if a feature is at its maximum material limit everywhere, no errors of straightness are permitted.

Straightness on an MMC basis can be applied to any part or feature having straight-line elements in a plane which includes the diameter or thickness. This includes practically all the parts already shown on an RFS basis. However, it should not be used for features which do not have a uniform cross section.

To signify that the maximum material principle applies to a straightness tolerance, the symbol Ⓜ is placed immediately after the tolerance value in the feature control symbol. Since there must always be a feature size dimension associated with such a tolerance, it is useful and convenient to establish this relationship by directing the feature control symbol in line with the dimension, or associating it with the dimension by a common leader, as shown in Fig. 31-4-7.

A straightness tolerance modified by Ⓜ means that the feature shall lie within a tolerance zone consisting of the space between two parallel lines, in the same plane as the longitudinal section of the feature being evaluated. For external features, these zone lines are separated by the specified straightness tolerance, plus the maximum material condition of the feature, as shown in Fig. 31-4-8.

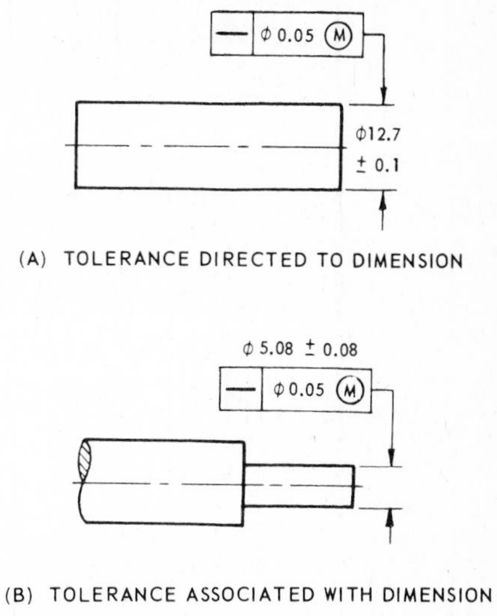

Φ 0.05 Ⓜ
Φ12.7
± 0.1
(A) TOLERANCE DIRECTED TO DIMENSION
Φ 5.08 ± 0.08
Φ 0.05 Ⓜ
(B) TOLERANCE ASSOCIATED WITH DIMENSION

Fig. 31-4-7 Straightness—MMC.

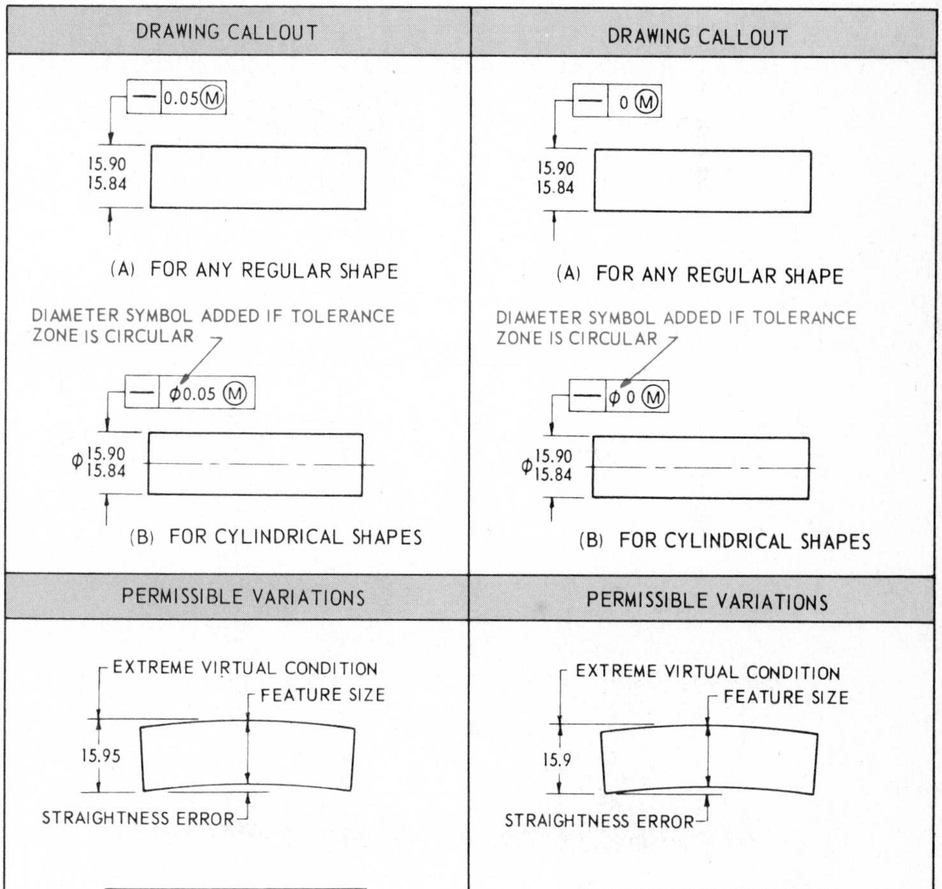

DRAWING CALLOUT
0.05Ⓜ
15.90
15.84
(A) FOR ANY REGULAR SHAPE
DIAMETER SYMBOL ADDED IF TOLERANCE ZONE IS CIRCULAR
Φ0.05 Ⓜ
Φ15.90
15.84
(B) FOR CYLINDRICAL SHAPES
PERMISSIBLE VARIATIONS
EXTREME VIRTUAL CONDITION
FEATURE SIZE
15.95
STRAIGHTNESS ERROR
DRAWING CALLOUT
0 Ⓜ
15.90
15.84
(A) FOR ANY REGULAR SHAPE
DIAMETER SYMBOL ADDED IF TOLERANCE ZONE IS CIRCULAR
Φ 0 Ⓜ
Φ15.90
15.84
(B) FOR CYLINDRICAL SHAPES
PERMISSIBLE VARIATIONS
EXTREME VIRTUAL CONDITION
FEATURE SIZE
15.9
STRAIGHTNESS ERROR

FEATURE SIZE	PERMISSIBLE STRAIGHTNESS ERROR
15.90	0.05
15.89	0.06
15.88	0.07
15.87	0.08
15.86	0.09
15.85	0.10
15.84	0.11

FEATURE SIZE	PERMISSIBLE STRAIGHTNESS ERROR
15.90	0.00
15.89	0.01
15.88	0.02
15.87	0.03
15.86	0.04
15.85	0.05
15.84	0.06

Fig. 31-4-8 Straightness—MMC.

Straightness with a Maximum Value

If it is desired to ensure that the straightness error does not become too great when the part approaches the least material condition, a maximum value may be added, as shown in Fig. 31-4-9.

Flatness

The symbol for flatness is a parallelogram, with angles of 60 and 120°, as shown in Fig. 31-4-10. The length and height are based on a percentage of the height of the feature control symbol frame.

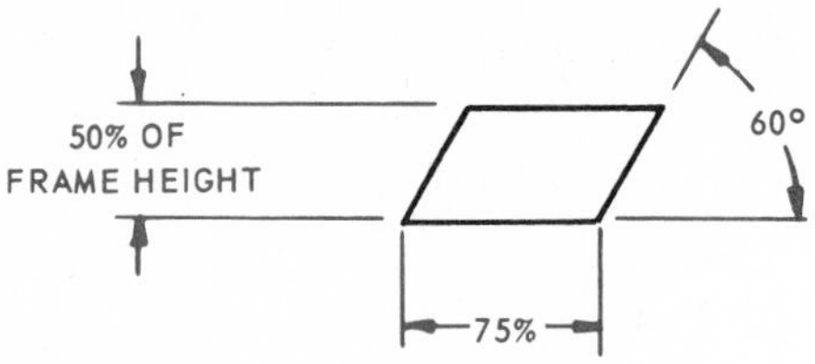

50% OF
FRAME HEIGHT
60°
75%

Fig. 31-4-10 Flatness symbol.

FLATNESS OF A SURFACE

Flatness of a surface is a condition in which all surface elements are in one plane. On such a surface all line elements in two or more directions are straight.

A flatness tolerance is applied to a line representing the surface of a part by means of a feature control symbol, as shown in Fig. 31-4-11.

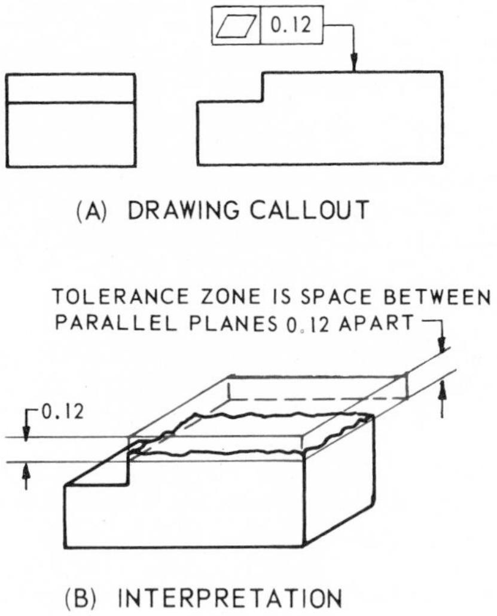

0.12
(A) DRAWING CALLOUT
TOLERANCE ZONE IS SPACE BETWEEN
PARALLEL PLANES 0.12 APART
0.12
(B) INTERPRETATION

Fig. 31-4-11 Flatness of a surface.

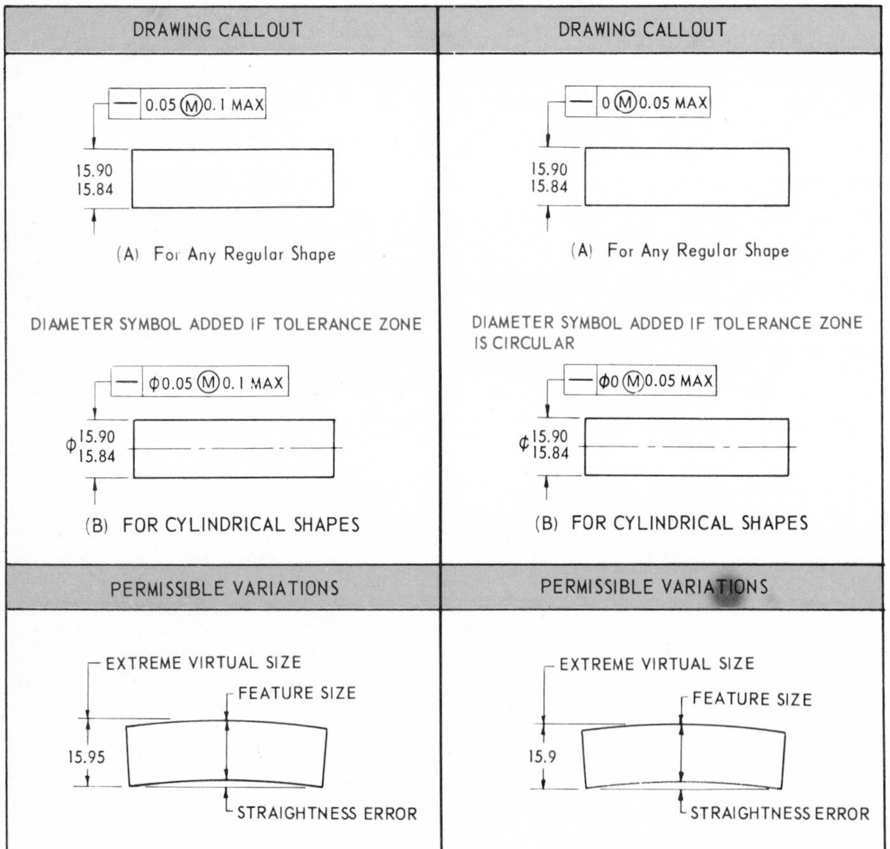

DRAWING CALLOUT
0.05 Ⓜ 0.1 MAX
15.90
15.84
(A) For Any Regular Shape
DIAMETER SYMBOL ADDED IF TOLERANCE ZONE
φ0.05 Ⓜ 0.1 MAX
φ 15.90
15.84
(B) FOR CYLINDRICAL SHAPES
DRAWING CALLOUT
0 Ⓜ 0.05 MAX
15.90
15.84
(A) For Any Regular Shape
DIAMETER SYMBOL ADDED IF TOLERANCE ZONE
IS CIRCULAR
φ0 Ⓜ 0.05 MAX
φ 15.90
15.84
(B) FOR CYLINDRICAL SHAPES
PERMISSIBLE VARIATIONS
EXTREME VIRTUAL SIZE
FEATURE SIZE
15.95
STRAIGHTNESS ERROR
PERMISSIBLE VARIATIONS
EXTREME VIRTUAL SIZE
FEATURE SIZE
15.9
STRAIGHTNESS ERROR

FEATURE SIZE	PERMISSIBLE STRAIGHTNESS ERROR
15.90	0.05
15.89	0.06
15.88	0.07
15.87	0.08
15.86	0.09
15.85	0.10
15.84	0.10

FEATURE SIZE	PERMISSIBLE STRAIGHTNESS ERROR
15.90	0.00
15.89	0.01
15.88	0.02
15.87	0.03
15.86	0.04
15.85	0.05
15.84	0.06

Fig. 31-4-9 Straightness of internal feature—MMC.

A flatness tolerance means that all points on the surface shall be contained within a tolerance zone consisting of the space between two parallel planes which are separated by the specified tolerance. These planes may be oriented in any manner to contain the surface; that is, they are not necessarily parallel to the base.

If the same control is desired on two or more surfaces, a suitable note may be added instead of repeating the symbol, as shown in Fig. 31-4-12.

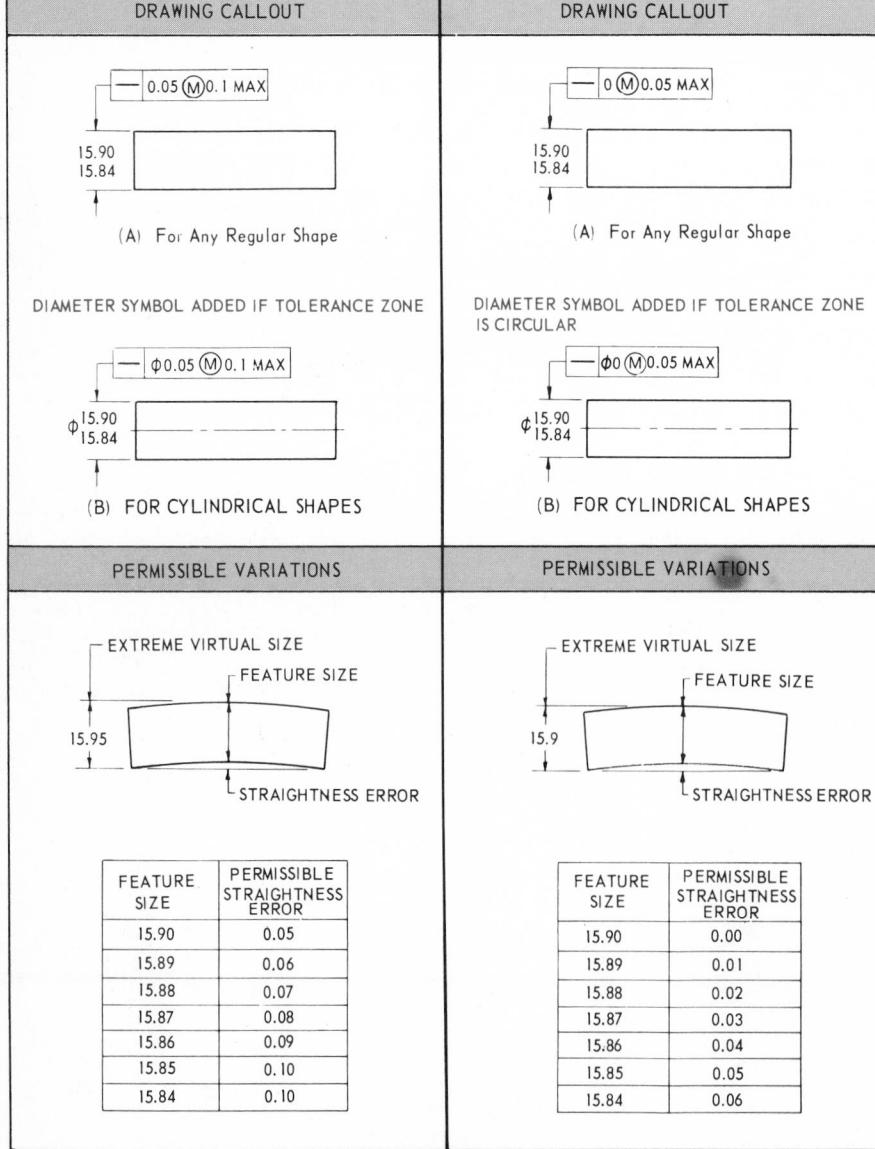

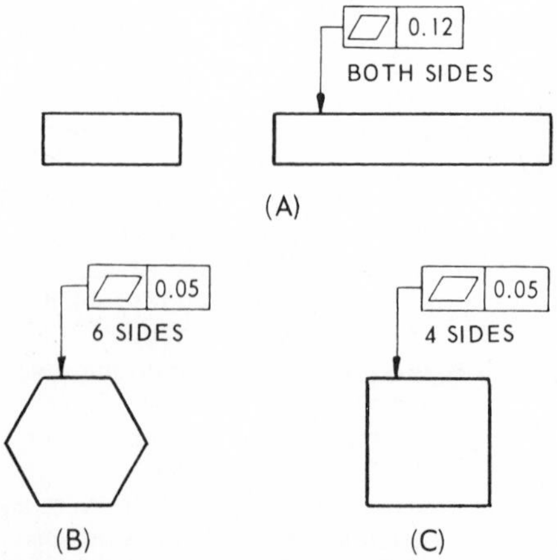

0.12
BOTH SIDES
(A)
0.05
6 SIDES
0.05
4 SIDES
(B)
(C)

Fig. 31-4-12 Controlling flatness on two or more surfaces.

Assignment

On an A3- or B-size sheet, prepare sketches and solutions for the drawings shown in Fig. 31-4-A or 31-4-B and the following information.

1. With reference to illustration (A), are the parts *A* to *E* acceptable?

2. Design a ring and a snap gage to check the pins shown in illustration (B). The ring gage should be of such a size as to check the entire length of pin. The snap gage has two open ends, the openings to measure the minimum and maximum acceptable pin diameters.

3. In illustration (C), part 1 is required to fit into part 2 so that there will not be any interference and the maximum clearance will never exceed 0.1 mm. Show suitable flatness tolerances for both parts and a size with the largest possible tolerance for part 2.

4. Show the tolerance zones, and widths, for the parts shown in illustration (D). Determine the extreme virtual condition for parts 2 and 3.

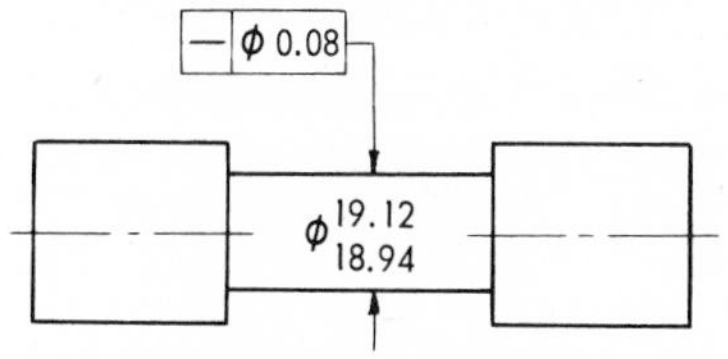

— ϕ 0.08
ϕ 19.12
18.94

PART	FEATURE SIZE	STRAIGHTNESS DEVIATION	ACCEPTABLE
A	19.14	0.02	
B	19.06	0.14	
C	19.00	0.2	
D	18.92	0.08	
E	18.94	0.26	

(A)

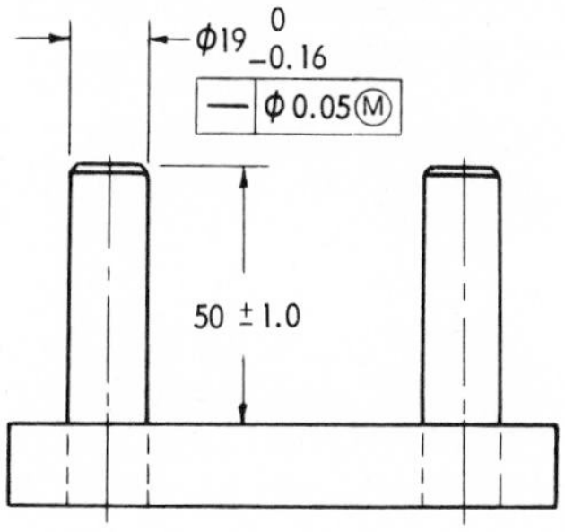

ϕ19 0 −0.16
— ϕ 0.05 Ⓜ
50 ± 1.0

(B)

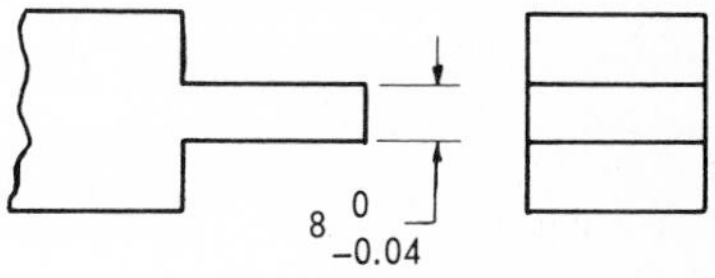

8 0 −0.04

PART 1

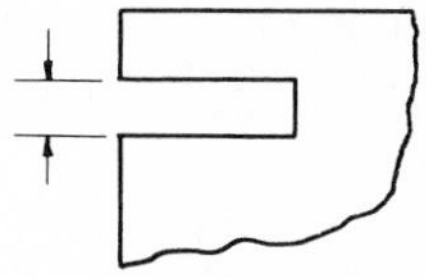

PART 2

(C)

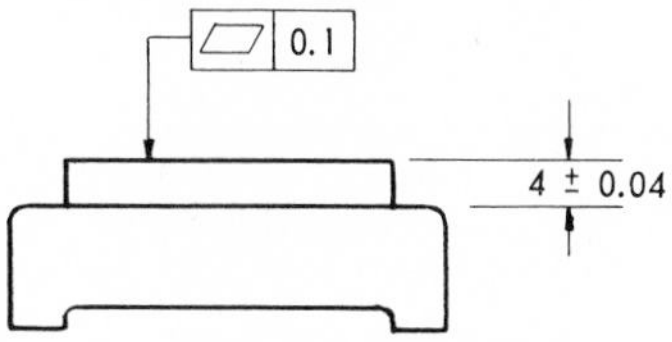

⏥ 0.1
4 ± 0.04

PART 1

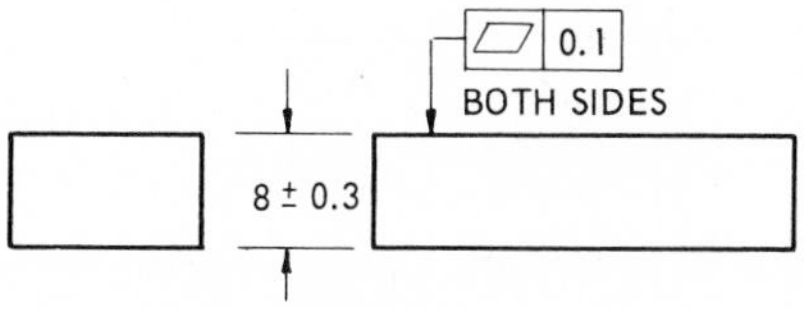

⏥ 0.1
BOTH SIDES
8 ± 0.3

PART 2

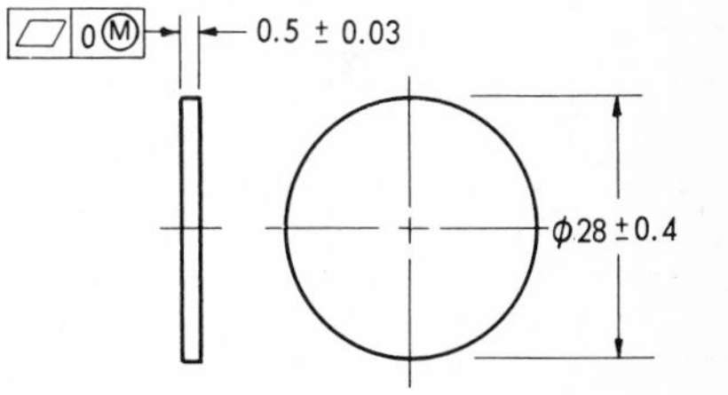

⏥ 0 Ⓜ
0.5 ± 0.03
ϕ28 ± 0.4

PART 3

(D)

Fig. 31-4-A Assignments.

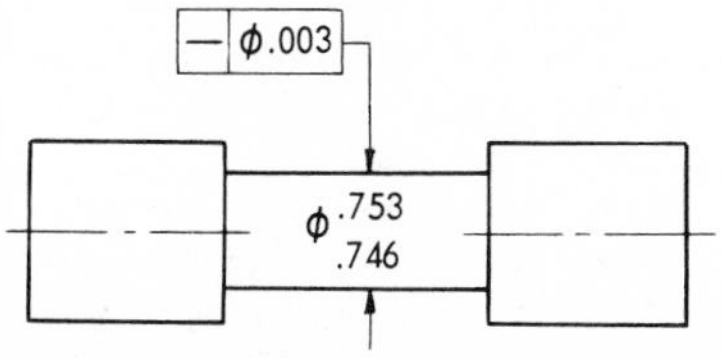

— ϕ.003
ϕ .753
.746

PART	FEATURE SIZE	STRAIGHTNESS DEVIATION	ACCEPTABLE
A	.754	.001	
B	.750	.002	
C	.748	.0035	
D	.745	.003	
E	.746	.004	

(A)

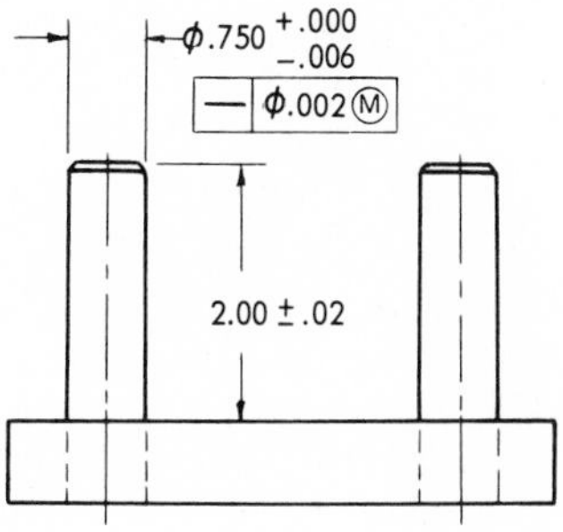

ϕ.750 +.000 −.006
— ϕ.002 Ⓜ
2.00 ± .02

(B)

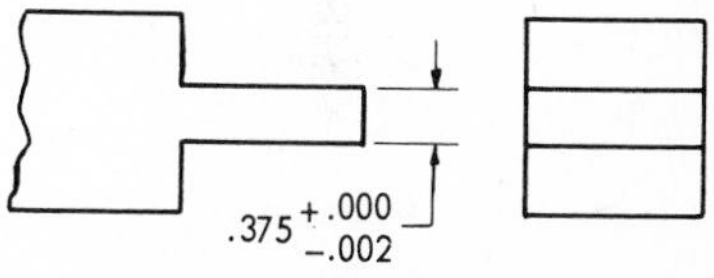

.375 +.000 −.002

PART 1

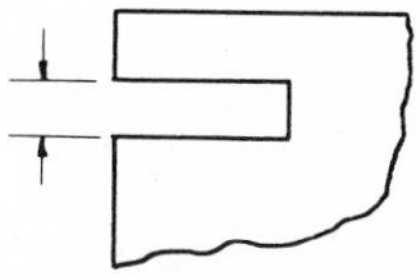

PART 2

(C)

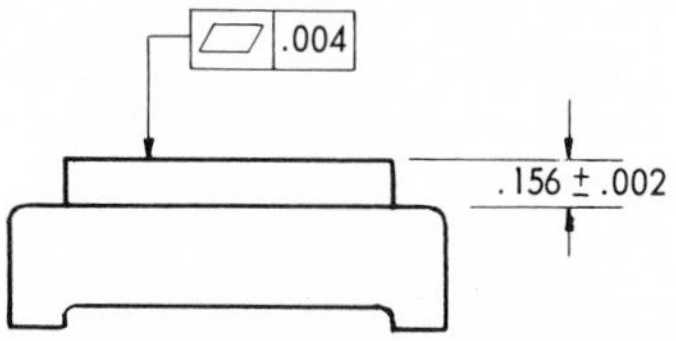

⏥ .004
.156 ± .002

PART 1

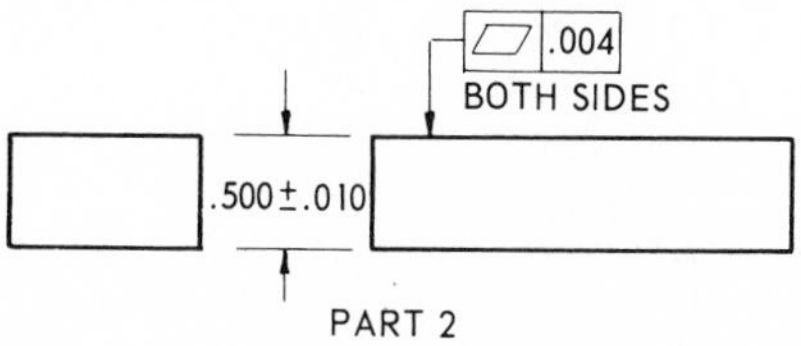

⏥ .004
BOTH SIDES
.500 ± .010

PART 2

(D)

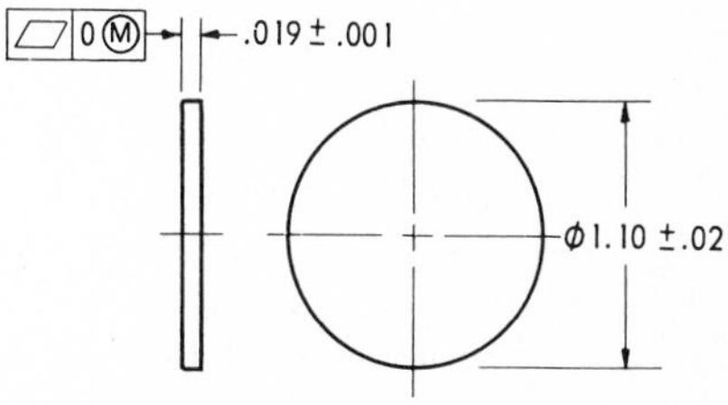

⏥ 0 Ⓜ
.019 ± .001
ϕ1.10 ± .02

PART 3

Fig. 31-4-B Assignments.

UNIT 31-5

DATUMS AND THE THREE-PLANE CONCEPT

Datums

Datum. A *datum* is a point, line, plane, or other geometric surface from which dimensions are measured when so specified or to which geometric tolerances are referenced. A datum has an exact form and represents an exact or fixed location, for purposes of manufacture or measurement.

Datum Feature. A datum feature is a feature of a part, such as an edge, a surface, or a hole, which forms the basis for a datum or is used to establish its location.

Datums for Geometric Tolerancing

As defined, datums are exact geometric points, lines, or surfaces, each based on one or more datum features of the part. Surfaces are usually either flat or cylindrical, but other shapes are used when necessary. The datum features, being physical surfaces of the part, are subject to manufacturing errors and variations. For example, a flat surface of a part, if greatly magnified, will show some irregularity. If brought into contact with a perfect plane, it will touch only at the highest points, as shown in Fig. 31-5-1. The true datums are theoretical but are considered to exist in the form of locating surfaces of machines, fixtures, and gaging equipment, on which the part rests or with which it makes contact during manufacture and measurement.

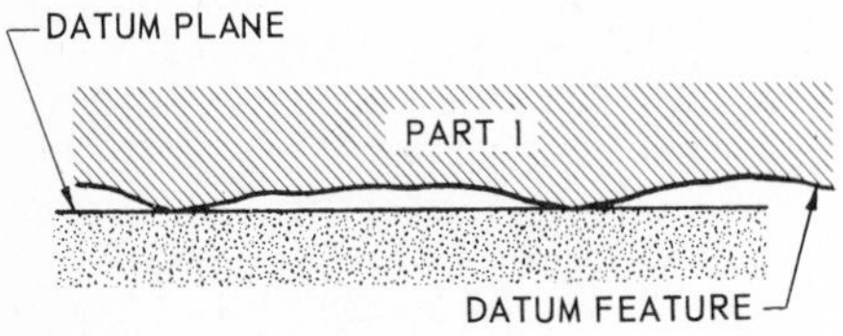
DATUM PLANE
PART 1
DATUM FEATURE

Fig. 31-5-1 Magnified section of a flat surface.

Three-Plane System

Geometric tolerances, such as straightness and flatness, refer to unrelated lines and surfaces and do not require the use of data.

Orientation and locational tolerances refer to related features; that is, they control the relationship of features to one another or to a datum or datum system. Such datum features must be properly identified on the drawing.

Usually only one datum is required for orientation purposes, but positional relationships may require a datum system consisting of two or three datums. These datums are designated as *primary*, *secondary*, and *tertiary*. When these datums are plane surfaces that are mutually perpendicular, they are commonly referred to as a *three-plane datum system*, or a *datum reference frame*.

Primary Datum. If the primary datum feature is a flat surface, it could lay on a suitable plane surface, such as the surface of a gage, which would then become a primary datum, as shown in Fig. 31-5-2. Theoretically, there will be a minimum of three high spots on the flat surface which would come in contact with the surface of the gage.

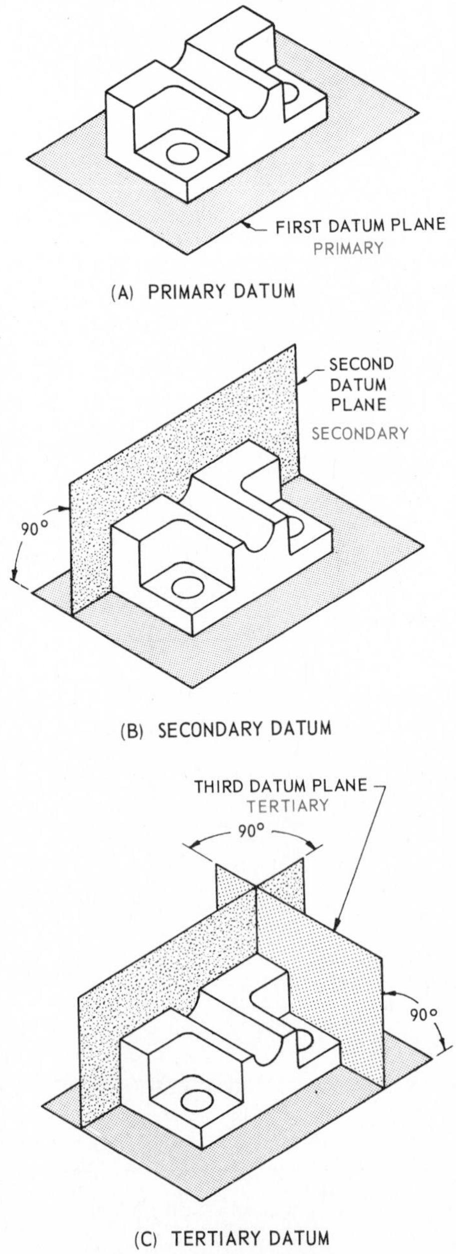
FIRST DATUM PLANE
PRIMARY
(A) PRIMARY DATUM
SECOND DATUM PLANE
SECONDARY
90°
(B) SECONDARY DATUM
THIRD DATUM PLANE
TERTIARY
90°
90°
(C) TERTIARY DATUM

Fig. 31-5-2 The datum planes.

Secondary Datum. If the part, while lying on this primary plane, is brought into contact with a secondary plane, it will theoretically touch at a minimum of two points.

Tertiary Datum. The part can now be slid along, while maintaining contact with the primary and secondary planes, until it contacts a third plane. This plane then becomes the tertiary datum, and the part will theoretically touch it at only one point.

These three planes constitute a datum system from which measurements can be taken. They will appear on the drawing, as shown in Fig. 31-5-3, except that the datum features would be identified in their correct sequence by the methods described later in this unit.

It must be remembered that the majority of parts are not of the simple rectangular shape, and considerably more ingenuity may be required to establish suitable data for more complex shapes.

IDENTIFICATION OF DATUMS

Datum symbols are required to serve two purposes:

1. To indicate the datum surface or feature on the drawing
2. To identify the datum feature so that it can easily be referred to in other requirements.

There are two methods of datum symbolization in general use for such purposes: one is shown and used in ANSI standards; the other, the ISO method, is used in most other countries of the world.

ANSI DATUM SYMBOLIZATION

In the current ANSI system, every datum feature is identified by a capital letter, enclosed in a rectangular box.

A dash is placed before and after the letter, to identify it as applying to a datum feature, as shown in Fig. 31-5-4.

This identifying symbol may be directed to the datum feature in any one of the following ways.

1. By attaching a side, end, or corner of the symbol frame to an extension line from the feature
2. By running a leader with arrowhead from the symbol frame to the feature
3. By adding the symbol to a note, a dimension, or a feature control symbol pertaining to the feature.

These methods are illustrated in Fig. 31-5-5.

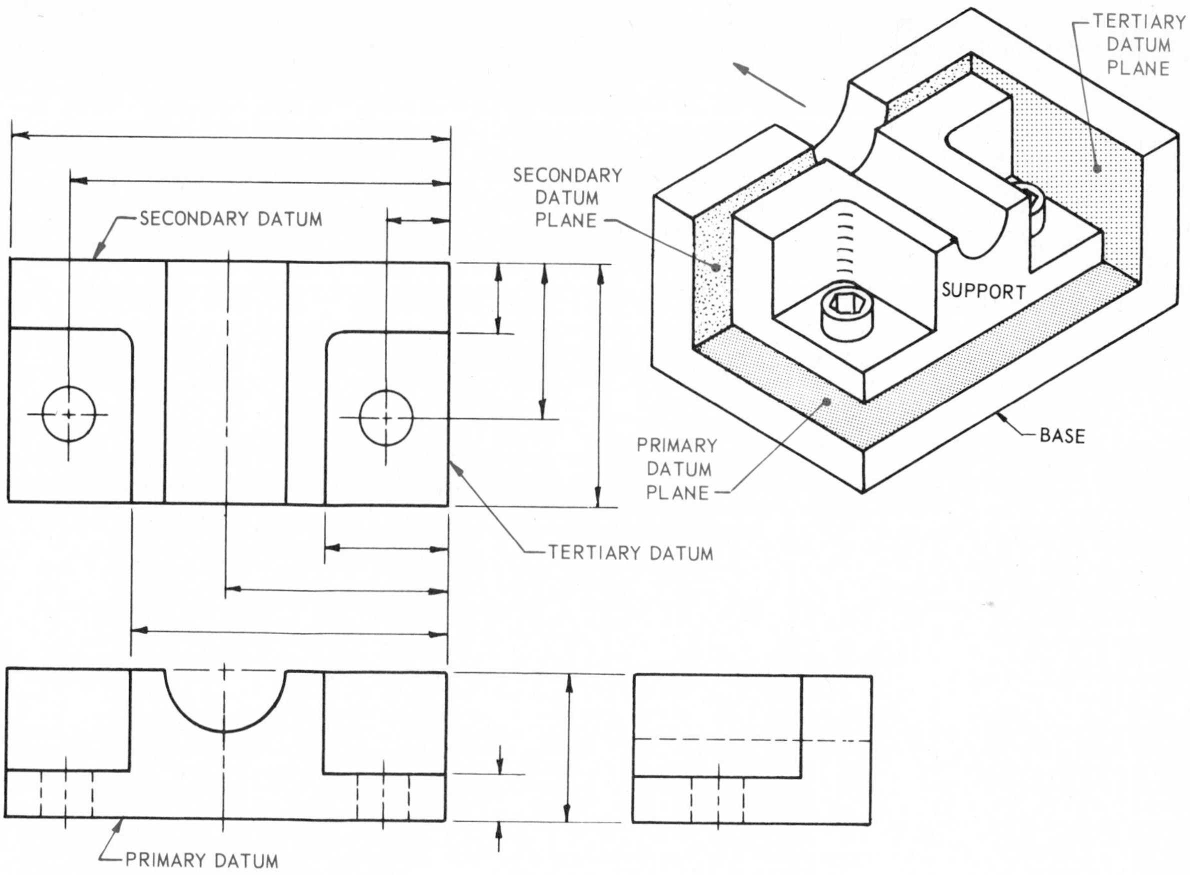

TERTIARY DATUM PLANE
SECONDARY DATUM PLANE
SECONDARY DATUM
SUPPORT
PRIMARY DATUM PLANE
BASE
TERTIARY DATUM
PRIMARY DATUM

Fig. 31-5-3 Three-plane datum system.

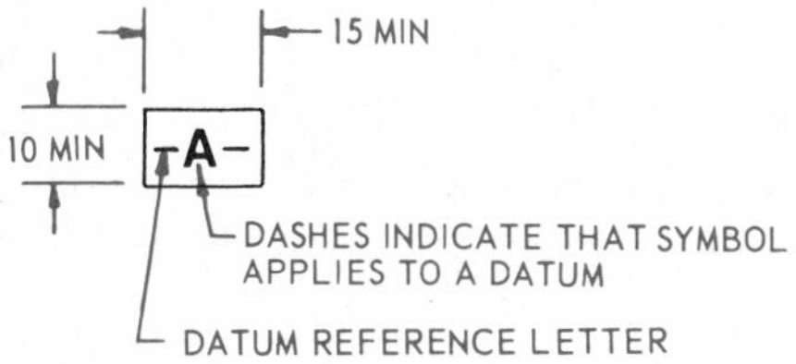

15 MIN
10 MIN
-A-
DASHES INDICATE THAT SYMBOL APPLIES TO A DATUM
DATUM REFERENCE LETTER

Fig. 31-5-4 Datum feature identification—ANSI.

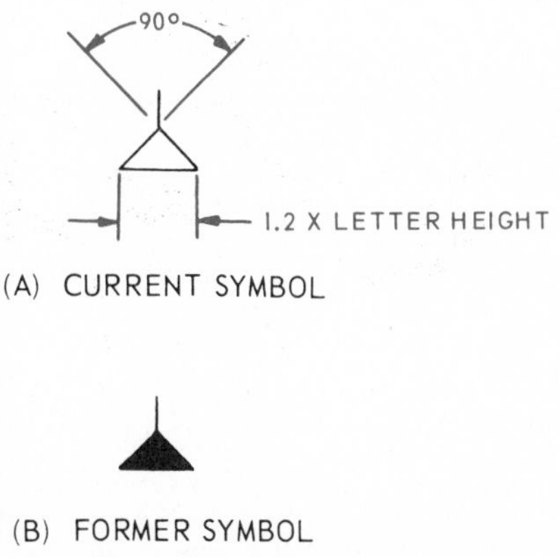

90°
1.2 X LETTER HEIGHT
(A) CURRENT SYMBOL
(B) FORMER SYMBOL

Fig. 31-5-5 Datum feature indicator symbol.

ISO DATUM SYMBOLIZATION

The ISO method is used in Canadian, British, and most other national standards. The ISO datum feature indicator symbol is a right-angled triangle, with a leader projecting from the 90° apex, as shown in Fig. 31-5-5. The base of the triangle should be slightly greater than the height of the lettering used on the drawing. The triangle was formerly filled in, but because of microfilming and cost reduction it was changed to a hollow triangle.

The datum is identified by a capital letter placed in a square frame and connected to the datum indicator symbol, as shown in Fig. 31-5-6.

APPLICATION OF DATUM IDENTIFICATION TO DRAWINGS

The datum feature indicator symbol is placed on or directed to the datum feature in one of the following ways.

1. The datum feature indicator symbol is placed on or directed to the outline of the feature when the datum feature is the surface itself or line elements of the surface. This is the preferred method.

2. It is on an extension line from the datum feature surface, but not in line with the dimension. This method is useful when the datum feature surface is very small or if, for other reasons, it is difficult to place the symbol on the line representing the surface.

3. It is on a leader ending in a dot on the datum feature surface. This method is useful when it is not convenient to apply it to the line representing the surface or when the view showing the surface as a line is not otherwise required. This method is not included in the ANSI standard.

4. It is on the center line when the datum is the datum axis of the feature or of all features on the common center line. This method is also not included in the ANSI standard.

5. It is on an extension line, in line with the feature dimension, when the datum is the datum axis of the cylindrical

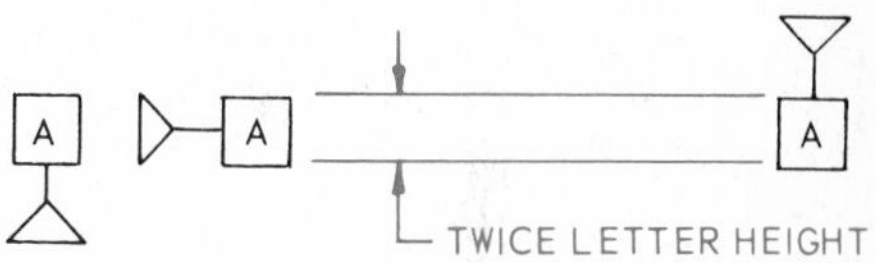

A
A
A
TWICE LETTER HEIGHT

Fig. 31-5-6 Datum identification symbol.

surface represented by the extension line, especially when applied on an MMC basis.

6. It is on two extension lines, using two datum symbols on the same leader, when two surfaces form a single datum, such as two sides of a slot or of a rectangular part.

7. It is on a leader line to a small feature, such as a hole, when the surface is too small to place the symbol on the surface and extension lines are not used.

All these methods are illustrated in Figs. 31-5-7 and 31-5-8. The symbol should not be placed on a hidden-feature line, but on the extension line to it or on the surface line in a sectional view.

ASSOCIATION WITH GEOMETRIC TOLERANCES

The datum letter is placed in the feature control symbol frame by adding an extra compartment for the datum reference, as shown in Fig. 31-5-9.

If two or more datum references are involved, then additional frames are added and the datum references are placed in these frames in the correct order, that is, primary, secondary, and tertiary datums, as shown in Fig. 31-5-10.

MULTIPLE DATUM FEATURES

If a single datum is established by two datum features, such as two ends of a shaft, the features are each identified by separate letters. Both letters are then placed in the same compartment of the feature control symbol, with a dash between them, as shown in Fig. 31-5-11. The datum, in this case, is the common line between the two datum features.

DATUMS BASED ON FEATURES OF SIZE

When a feature of size is specified as a datum feature, the datum has to be established from the full surface of a cylindrical feature or from two opposing surfaces of other features of size. However, the true datum is a datum axis, center line, or median plane of the feature.

There are two methods of establishing such datums: on an MMC basis or without MMC.

Datum Features—Without MMC

When a feature of size is specified as a datum feature and is not modified on an MMC basis, the datum axis or median plane is established from the virtual size of each individual part.

When there are other features on a common center line, the symbol should be directed to extension lines and be as-

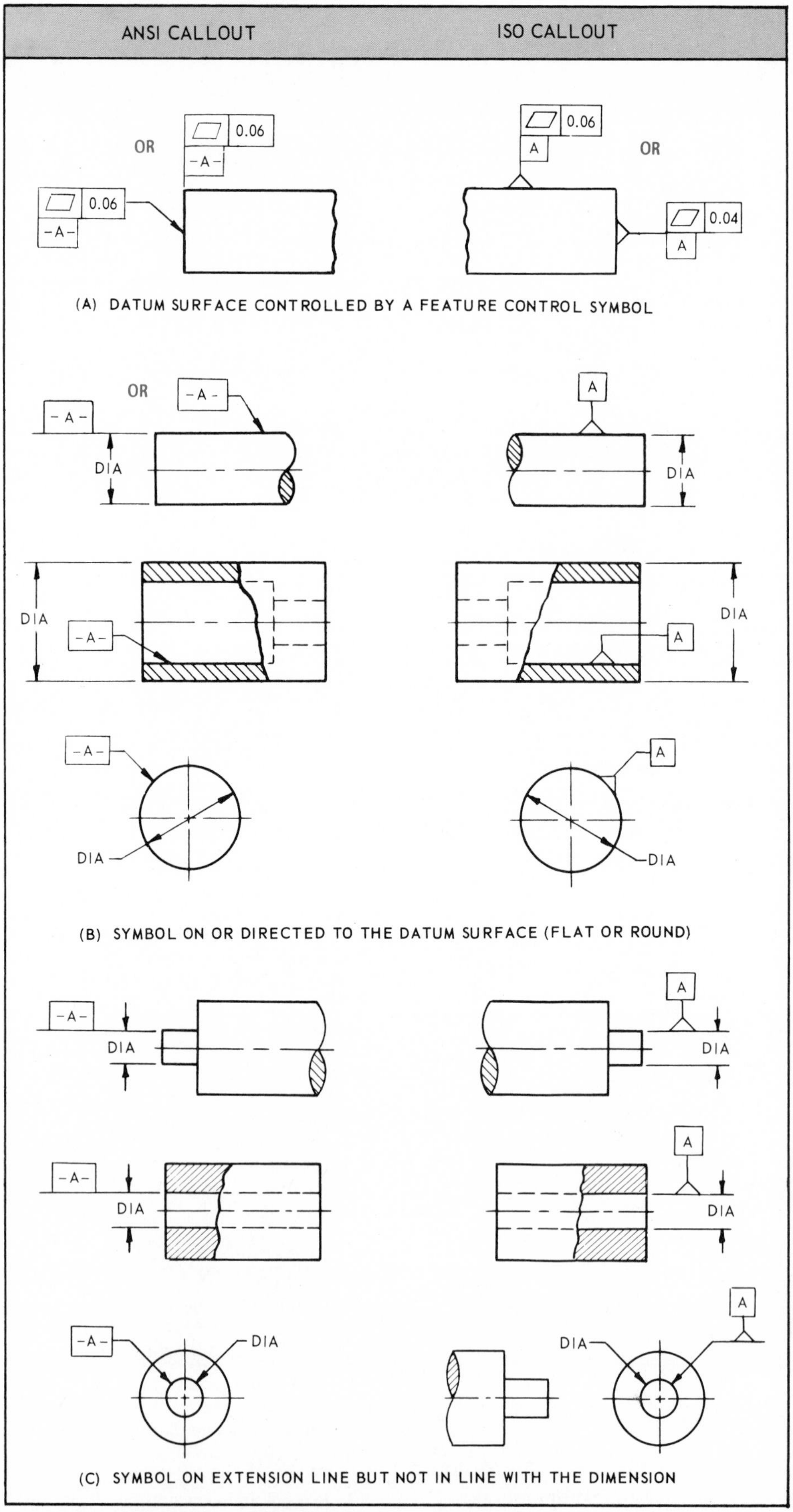
ANSI CALLOUT
ISO CALLOUT
OR
0.06
-A-
0.06
-A-
0.06
A
OR
0.04
A
(A) DATUM SURFACE CONTROLLED BY A FEATURE CONTROL SYMBOL
OR
-A-
-A-
DIA
A
DIA
DIA
-A-
DIA
A
-A-
DIA
A
DIA
(B) SYMBOL ON OR DIRECTED TO THE DATUM SURFACE (FLAT OR ROUND)
-A-
DIA
A
DIA
-A-
DIA
A
DIA
-A-
DIA
A
DIA
(C) SYMBOL ON EXTENSION LINE BUT NOT IN LINE WITH THE DIMENSION

Fig. 31-5-7 Placement of datum symbol for single features.

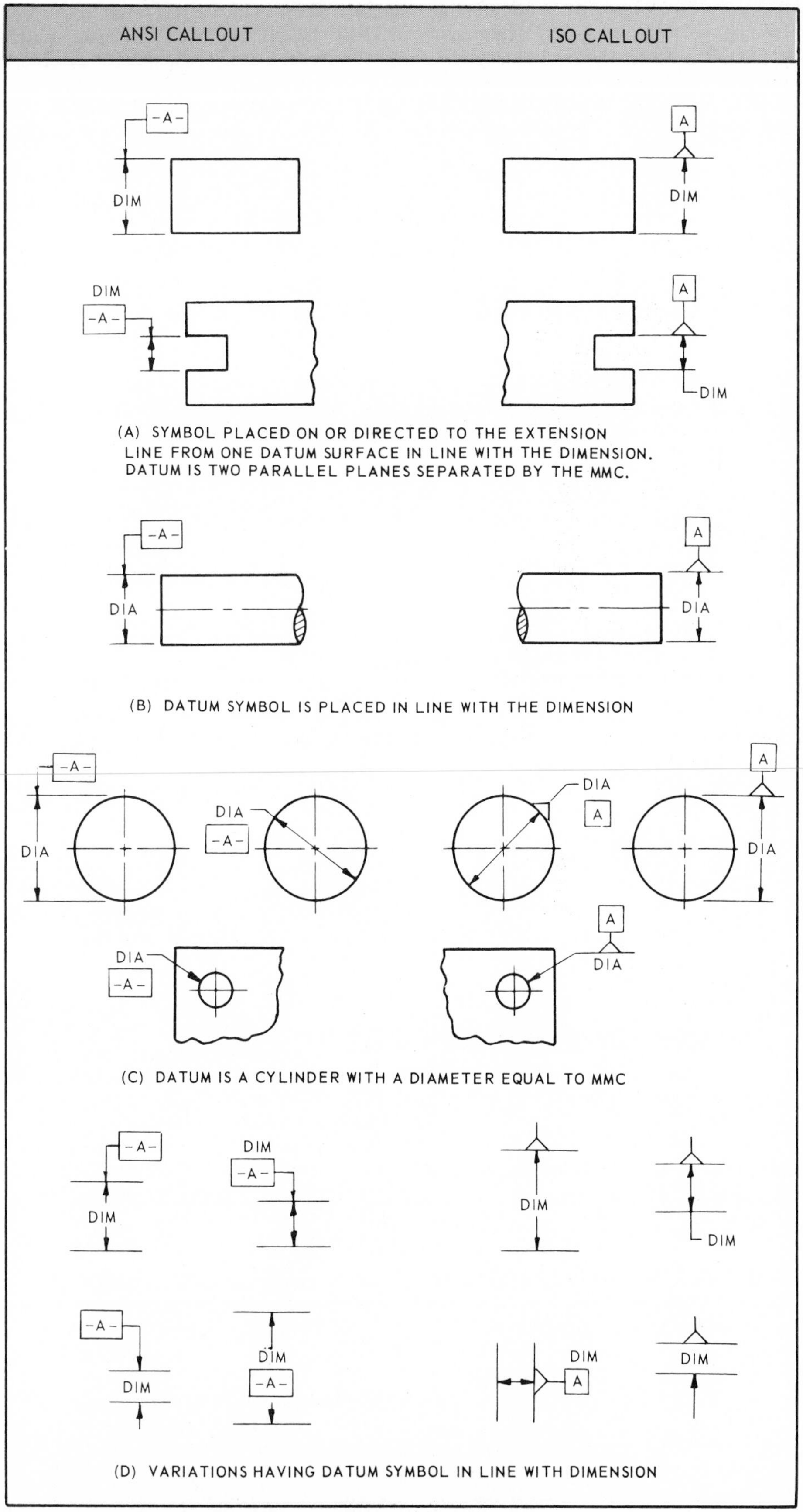
ANSI CALLOUT
ISO CALLOUT
-A-
A
DIM
(A) SYMBOL PLACED ON OR DIRECTED TO THE EXTENSION LINE FROM ONE DATUM SURFACE IN LINE WITH THE DIMENSION. DATUM IS TWO PARALLEL PLANES SEPARATED BY THE MMC.
DIA
(B) DATUM SYMBOL IS PLACED IN LINE WITH THE DIMENSION
(C) DATUM IS A CYLINDER WITH A DIAMETER EQUAL TO MMC
(D) VARIATIONS HAVING DATUM SYMBOL IN LINE WITH DIMENSION

Fig. 31-5-8 Placement of datum symbol for feature of size.

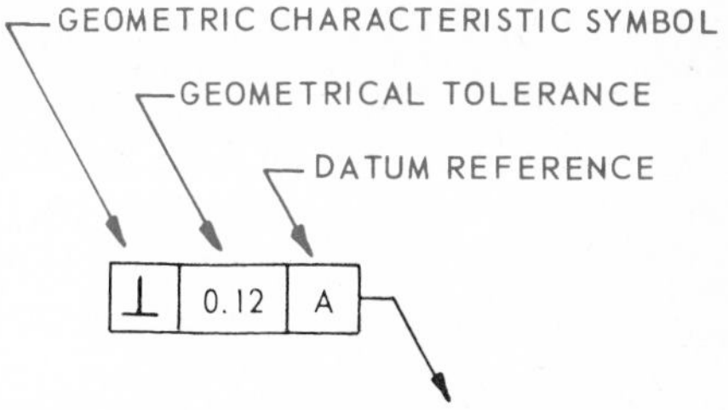
GEOMETRIC CHARACTERISTIC SYMBOL
GEOMETRICAL TOLERANCE
DATUM REFERENCE
⊥ 0.12 A

Fig. 31-5-9 Feature control symbol referenced to a datum.

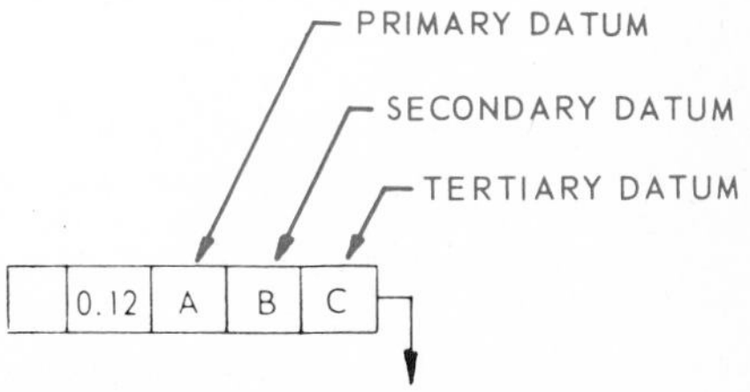
PRIMARY DATUM
SECONDARY DATUM
TERTIARY DATUM
0.12 A B C

Fig. 31-5-10 Multiple datum references.

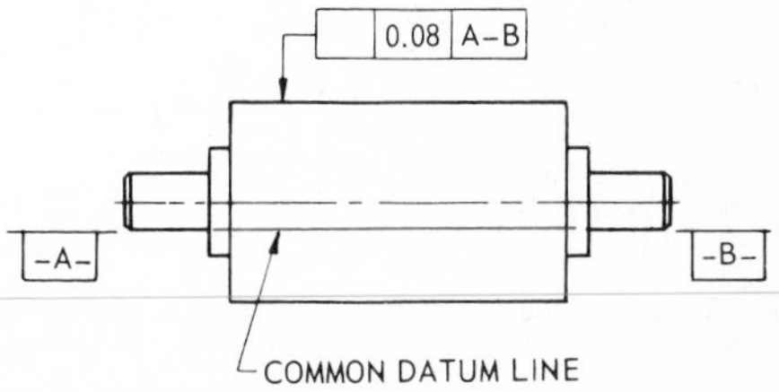
0.08 A–B
-A-
-B-
COMMON DATUM LINE

Fig. 31-5-11 Two datum features for one datum.

sociated with or in line with the dimension, as shown in Fig. 31-5-12.

External Cylindrical Datum Features. If the axis of an external cylindrical feature is designated as a primary datum feature, the datum is the axis of an imaginary perfect cylinder. This is the smallest such cylinder which can be circumscribed around the feature, so that it just contacts the highest points on the surface. Geometric tolerances which are referenced to such a datum refer theoretically to the axis of this datum cylinder, as shown in Fig. 31-5-13.

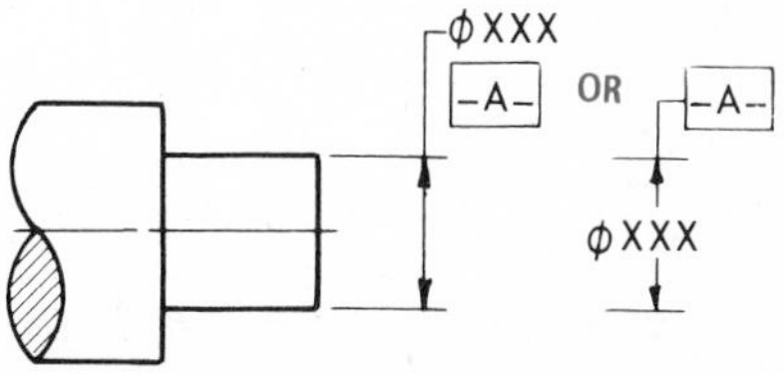
ϕXXX
-A-
OR
-A-
ϕXXX

Fig. 31-5-12 Symbols on extension lines when center line is common to other features.

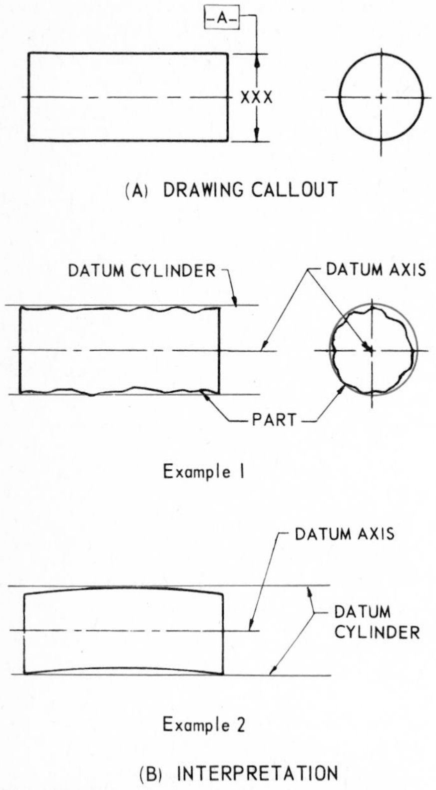

-A-
XXX
(A) DRAWING CALLOUT
DATUM CYLINDER
DATUM AXIS
PART
Example 1
DATUM AXIS
DATUM CYLINDER
Example 2
(B) INTERPRETATION

Fig. 31-5-13 Datum passed on center line of part.

Cylindrical Features as Secondary Datums. It is often desirable to specify an end face or flange of a cylindrical feature as the primary datum, using the cylindrical feature as the secondary datum. In such cases the secondary datum symbol may be directed to the cylindrical surface rather than to the center line.

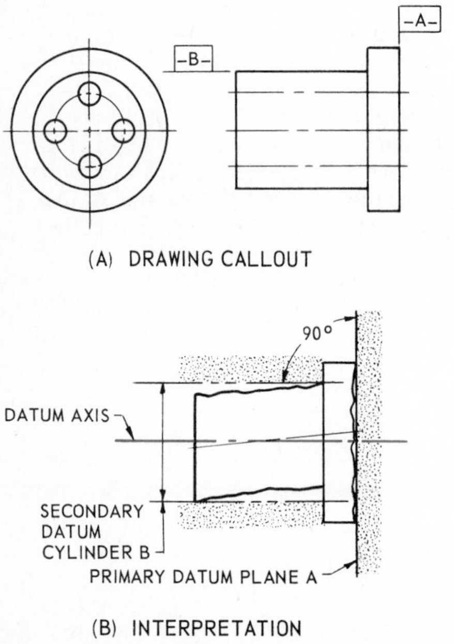

-A-
-B-
(A) DRAWING CALLOUT
90°
DATUM AXIS
SECONDARY DATUM CYLINDER B
PRIMARY DATUM PLANE A
(B) INTERPRETATION

Fig. 31-5-14 Cylindrical feature as secondary datum.

The primary datum is then a perfect plane on which the part would normally rest. The secondary datum is still the axis of an imaginary perfect cylinder, but also one that is perpendicular to the primary datum. This cylinder would theoretically touch the feature at only two points. This is illustrated in Fig. 31-5-14, where the part has been purposely drawn out of square to show the effect of such deviations.

Internal Cylindrical Datum Features. If an internal cylindrical feature, such as a hole, is specified as a primary datum feature, the datum is again the axis of an imaginary perfect cylinder. In this case it is the largest such cylinder which can be inscribed within the feature. It will therefore contact the inner surface of the feature at the highest points on the surface, as shown in Fig. 31-5-15.

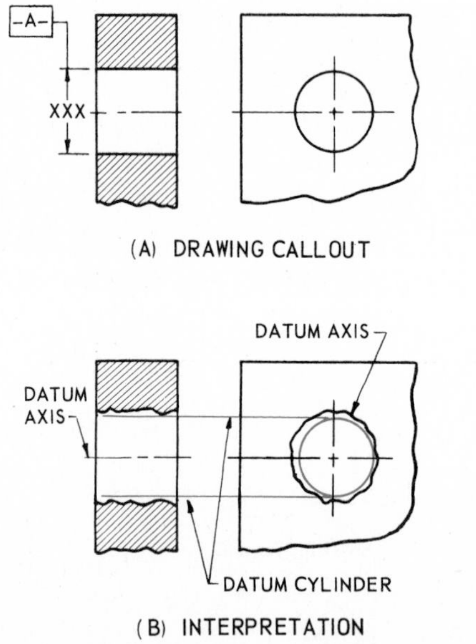

-A-
XXX
(A) DRAWING CALLOUT
DATUM AXIS
DATUM AXIS
DATUM CYLINDER
(B) INTERPRETATION

Fig. 31-5-15 Internal cylindrical feature as datum.

Datum Features—MMC

The measurement of geometric tolerances, which refer to datums that are features of size, is greatly simplified if the datums are specified on an MMC basis. This permits the use of fixed GO gages.

Datum features on an MMC basis always apply at their virtual condition. *Virtual condition* refers to the potential boundary of a feature, as specified on a drawing, derived from the collective effect of the maximum material limit of size and the specified form or orientation tolerance. These are added for external features, such as shafts, and subtracted for internal features, such as holes and slots.

If no form or orientation tolerance is specified, it is assumed, for datum reference purposes, that the form tolerance is zero MMC.

The fact that a datum applies on an MMC basis is indicated in the feature control symbol by the addition of the MMC symbol Ⓜ immediately following the datum reference, as shown in Fig. 31-5-16. When there is more than one datum reference, the MMC symbol must be added for each datum where this modification is required.

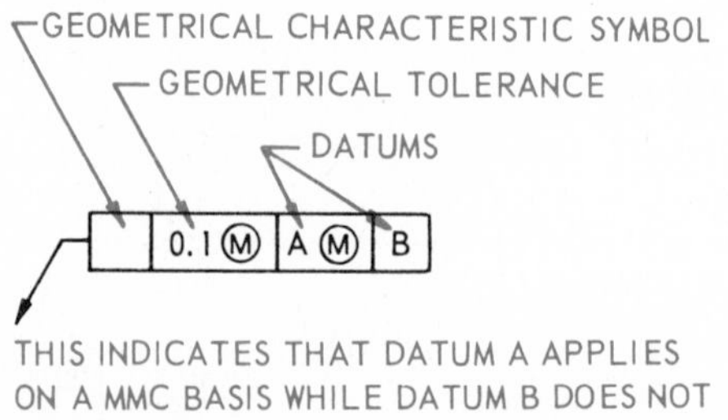

GEOMETRICAL CHARACTERISTIC SYMBOL
GEOMETRICAL TOLERANCE
DATUMS
0.1 Ⓜ | A Ⓜ | B
THIS INDICATES THAT DATUM A APPLIES ON A MMC BASIS WHILE DATUM B DOES NOT

Fig. 31-5-16 Reference to datums—MMC.

The datum identification symbol could be shown in any of the ways used for data that are not on an MMC basis. However, it is convenient to direct the symbol in line or in association with the feature size dimension. It may also be coupled to the form tolerance if one is used, as shown in Fig. 31-5-17. This is permissible because an MMC datum is based on the maximum material size of the feature and the applicable form tolerance.

External Features with Regular Cross Sections—MMC. For external features having a cross section which is circular or which comprises a regular polygon, the datum will be of the same shape as the datum feature. The width or diameter of the datum will be equal to the maximum material limit of size, plus the specified form tolerance.

In Fig. 31-5-18, because no form tolerance is specified, the cylindrical gaging elements are made to the maximum material size of 14.38 mm. This provides an exact location of the part when it is made to the maximum material condition. However, it allows a deviation of 0.08 mm in any direction from true position if the part is everywhere at the

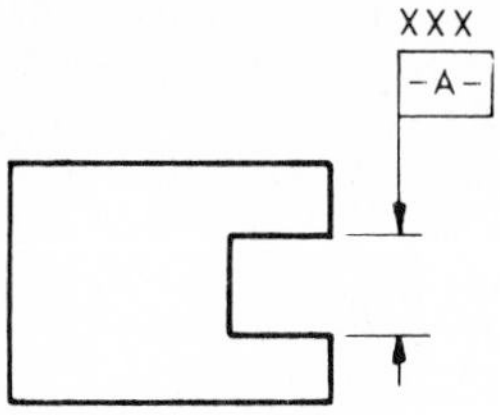
XXX
-A-

(A) APPLIED ON MMC BASIS

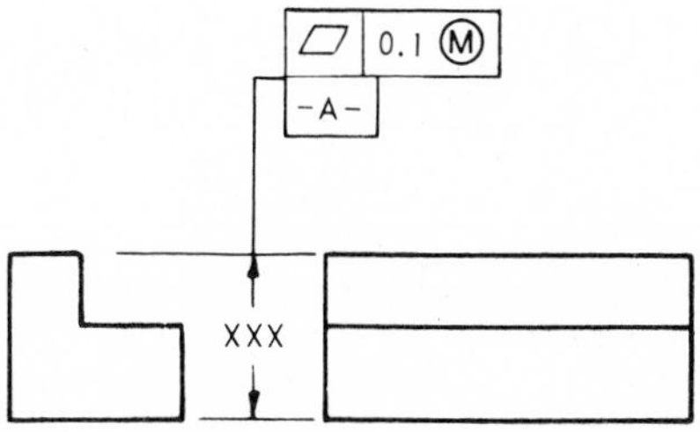
0.1 Ⓜ
-A-
XXX

(B) COUPLED WITH FORM TOLERANCES

Fig. 31-5-17 Specifying datums.

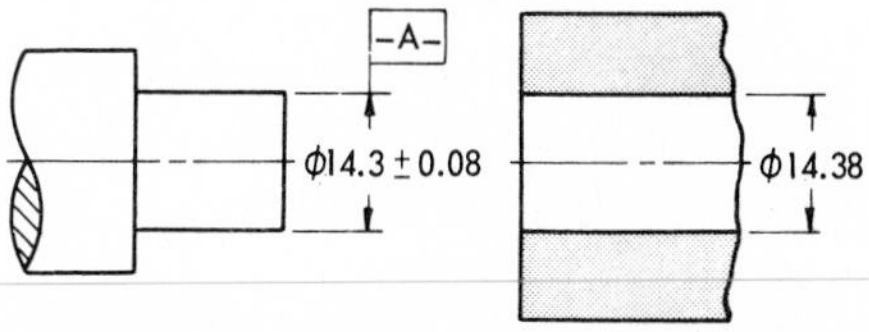
-A-
ϕ14.3 ± 0.08
ϕ14.38

Fig. 31-5-18 Gage element for circular datum.

minimum material size of 14.22 mm and there are no form errors.

Rectangular Features—MMC. If the specified datum feature consists of two flat surfaces and the cross section is not a regular polygon, then the datum consists of two parallel planes. These are separated by a distance equal to the maximum material condition, plus the specified form tolerance.

Figure 31-5-19 shows a flatness tolerance which applies separately to both datum feature surfaces. It should be noted that in such cases the form tolerance is not doubled in calculating the virtual condition. This is because if the part is everywhere at its maximum material condition, as shown in the gaging position, then a convex point on one side can only be offset by a concave point on the other side. If this were not true, the size dimension would be exceeded.

It should be noted that Fig. 31-5-19*b* shows the element of a gage which locates on the datum on an MMC basis. It does not check the flatness requirement.

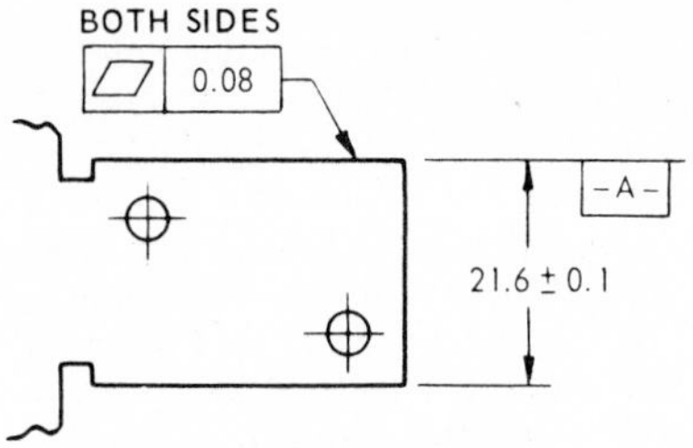
BOTH SIDES
0.08
-A-
21.6 ± 0.1

(A) DRAWING CALLOUT

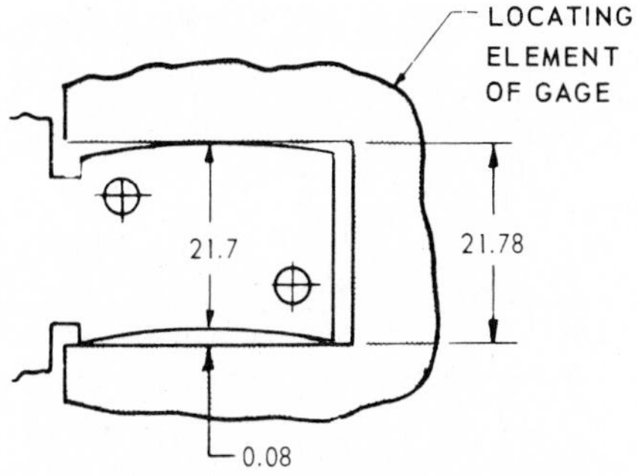
LOCATING
ELEMENT
OF GAGE
21.7
21.78
0.08

(B) LOCATING ELEMENT OF GAGE

Fig. 31-5-19 Gage element for rectangular datum feature.

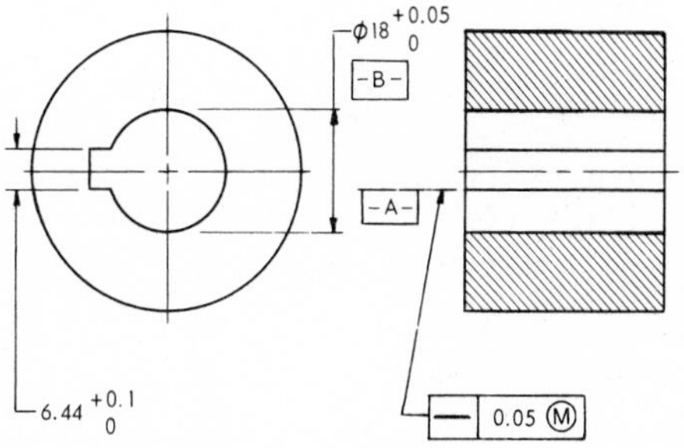
ϕ18 +0.05 0
-B-
-A-
6.44 +0.1 0
0.05 Ⓜ

(A) ANSI CALLOUT

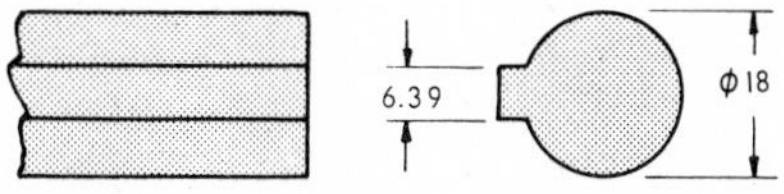
6.39
ϕ18

(B) ISO CALLOUT

Fig. 31-5-20 Gage element for internal datum feature.

Internal Features—MMC. The same rules apply to internal features, except that the maximum material condition is the minimum limit of size, and the specified form tolerance is subtracted from this limit to obtain the virtual condition. The size of the locating element of the gage is identical to the virtual condition, as shown in Fig. 31-5-20.

Assignment

On an A3- or B-size sheet, prepare sketches and solutions for the drawings shown in Fig. 31-5-A or 31-5-B and the following information.

1. Draw the top and front views of illustration (A) and show the following information:

- Surface *A* is datum *A*.
- Surface *B* is datum *B*.
- Surfaces *C* and *D* are datum features *C* and *D*, respectively.
- Add the geometric tolerances from the information shown on the drawing.

2. With reference to illustration (B):

- Pins 1, 2, and 3 are used to establish the secondary and tertiary datums for the part shown.
- Make a two-view drawing of the part shown and identify the primary, secondary, and tertiary datum planes as *A*, *B*, and *C*, respectively.
- Place the flatness tolerance specified for the back of the slot on the drawing.
- With reference to the slot tolerance and assuming that the top surface is smooth and flat, are the three parts shown acceptable?

3. With reference to illustration (C), diameter *M* is to be used as datum *A*, the end face of diameter *N* is to be used as datum *B*, and the two sides of the slot in diameter *N* are to be used as datum *C*. Prepare two drawings, one with ANSI drawing standards, the other with ISO drawing standards which will identify these datums.

4. Prepare a three-view drawing showing the datums and feature control symbols for illustration (D).

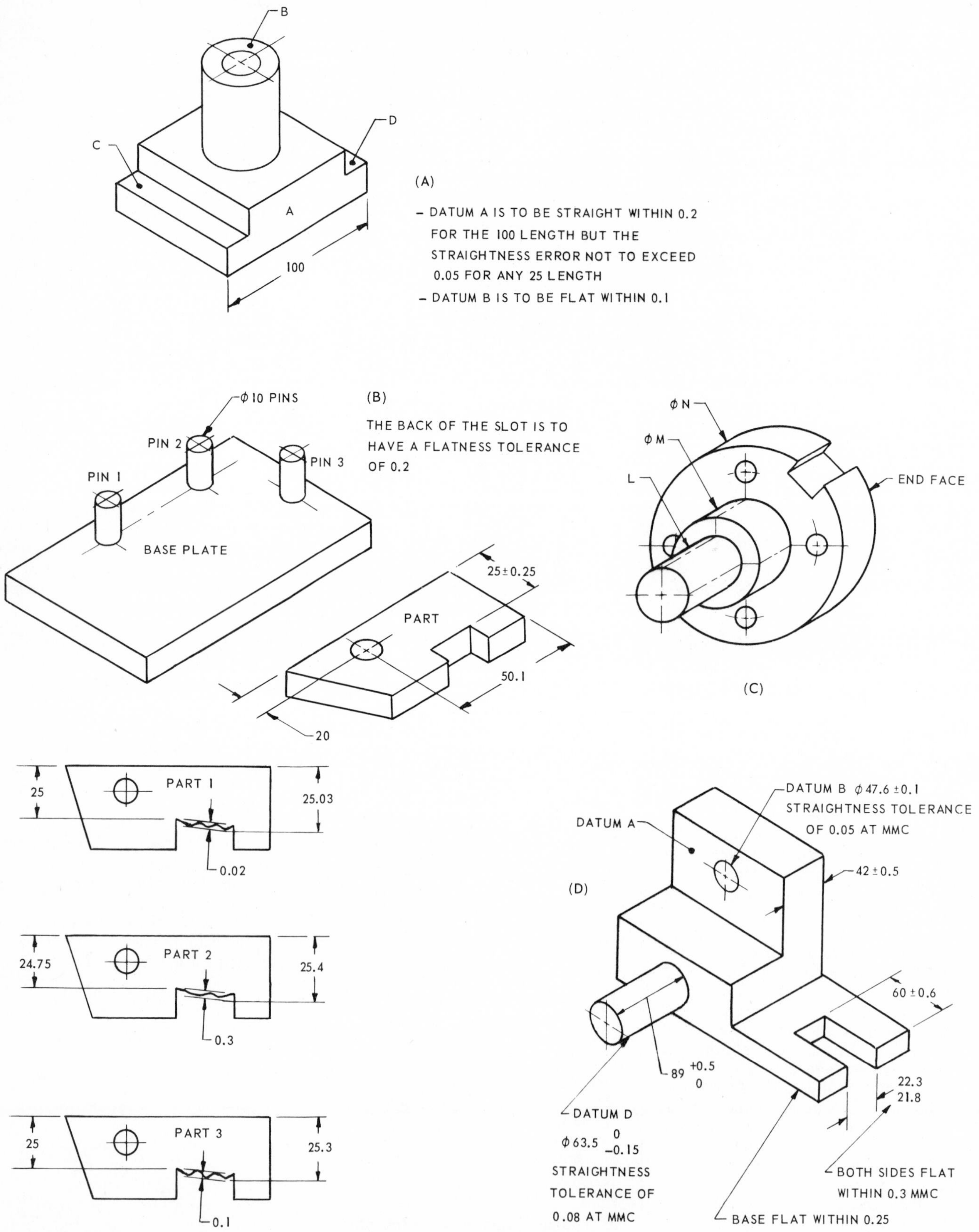
B
C
D
A
100
(A)
- DATUM A IS TO BE STRAIGHT WITHIN 0.2 FOR THE 100 LENGTH BUT THE STRAIGHTNESS ERROR NOT TO EXCEED 0.05 FOR ANY 25 LENGTH
- DATUM B IS TO BE FLAT WITHIN 0.1
ϕ 10 PINS
PIN 1
PIN 2
PIN 3
BASE PLATE
(B)
THE BACK OF THE SLOT IS TO HAVE A FLATNESS TOLERANCE OF 0.2
25±0.25
PART
50.1
20
PART 1
25
25.03
0.02
PART 2
24.75
25.4
0.3
PART 3
25
25.3
0.1
ϕ N
ϕ M
L
END FACE
(C)
DATUM A
DATUM B ϕ 47.6 ±0.1
STRAIGHTNESS TOLERANCE OF 0.05 AT MMC
42 ±0.5
(D)
60 ±0.6
89 +0.5 0
22.3
21.8
DATUM D
ϕ 63.5 0 −0.15
STRAIGHTNESS TOLERANCE OF 0.08 AT MMC
BOTH SIDES FLAT WITHIN 0.3 MMC
BASE FLAT WITHIN 0.25

Fig. 31-5-A Assignments.

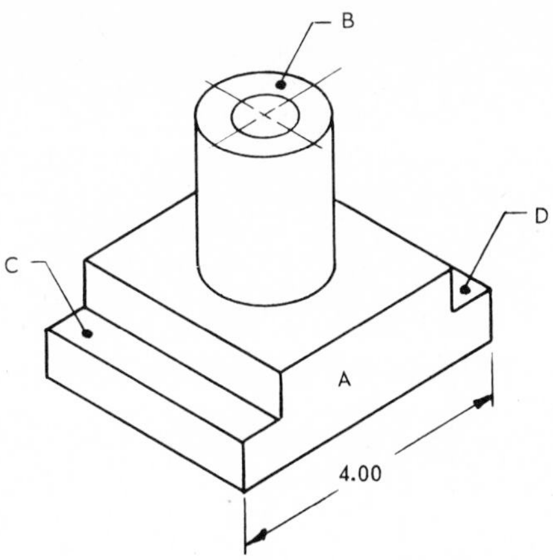
B
C
D
A
4.00

DATUM A IS TO BE STRAIGHT WITHIN .008 FOR THE 4.00 LENGTH BUT THE STRAIGHTNESS ERROR NOT TO EXCEED .002 FOR ANY 1.00 LENGTH.
DATUM B IS TO BE FLAT WITHIN .004 .

(A)

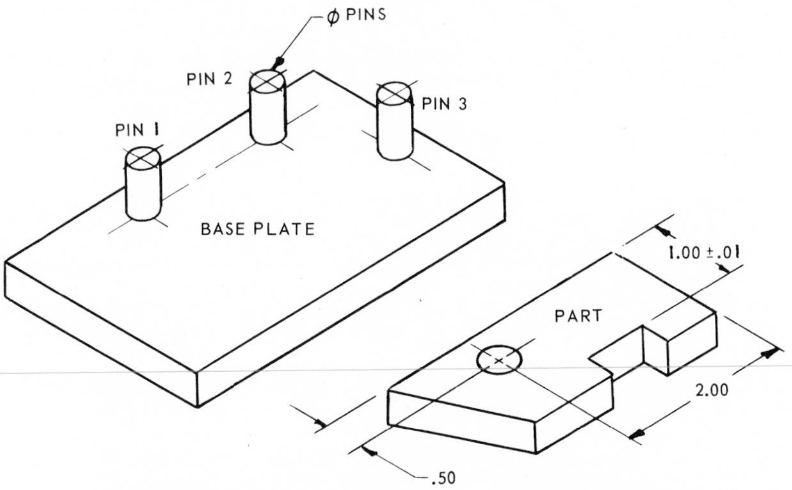
ϕ PINS
PIN 2
PIN 3
PIN 1
BASE PLATE
1.00 ±.01
PART
2.00
.50

THE BACK OF THE SLOT IS TO HAVE A FLATNESS TOLERANCE OF .01

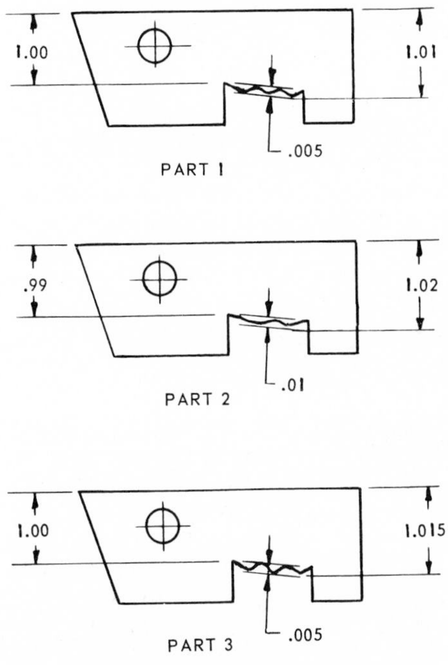
1.00
1.01
.005
PART 1
.99
1.02
.01
PART 2
1.00
1.015
.005
PART 3

(B)

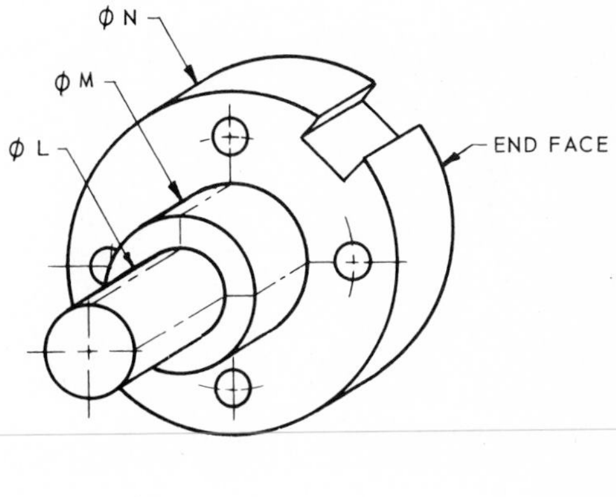
ϕ N
ϕ M
ϕ L
END FACE

(C)

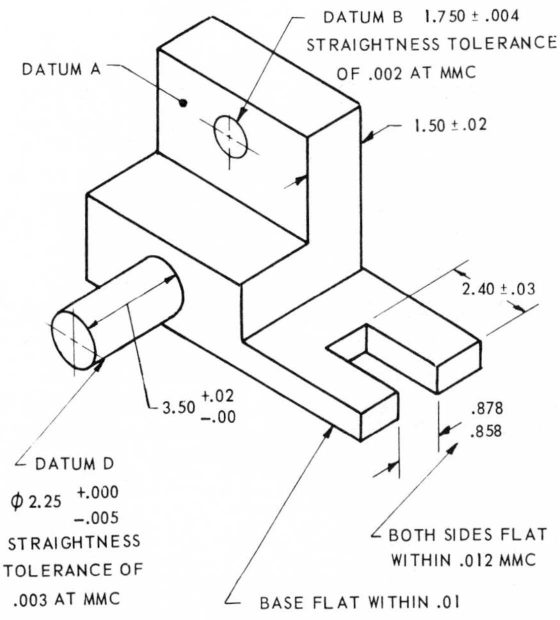
DATUM B 1.750 ± .004
STRAIGHTNESS TOLERANCE
OF .002 AT MMC
DATUM A
1.50 ± .02
2.40 ± .03
3.50 +.02 −.00
.878
.858
DATUM D
ϕ 2.25 +.000 −.005
STRAIGHTNESS
TOLERANCE OF
.003 AT MMC
BOTH SIDES FLAT
WITHIN .012 MMC
BASE FLAT WITHIN .01

(D)

Fig. 31-5-B Assignments.

UNIT 31-6
ORIENTATION

Angular Relationships of Flat Surfaces

Orientation refers to the angular relationship which exists between two or more lines, surfaces, or other features.

The general geometric characteristic for orientation is termed *angularity*. This term may be used to describe angular relationships, of any angle, between straight lines or surfaces with straight-line elements, such as flat or cylindrical surfaces. For two particular types of angularity special terms are used. These are *perpendicularity*, or squareness, for features related by a 90° angle, and *parallelism* for features related by an angle of 0°.

An orientation tolerance indicates a relationship between two or more features. Whenever possible, the feature to which the controlled feature is related should be designated as a datum.

Symbols for Geometric Characteristics

There are three geometric symbols for these characteristics, as shown in Fig. 31-6-1. The proportions are based on the height of the feature control symbol frame.

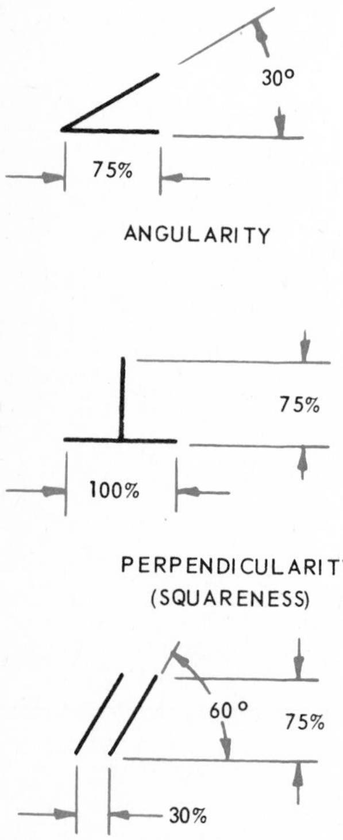
30°
75%
ANGULARITY
75%
100%
PERPENDICULARITY
(SQUARENESS)
60°
75%
30%
PARALLELISM

Fig. 31-6-1 Orientation symbols.

For geometric tolerancing of angularity, the angle between the datum and the controlled feature should be stated as a basic angle. Therefore it should be enclosed in a rectangular frame, as shown in Fig. 31-6-2, to indicate that the general tolerance note does not apply. However, the angle need not be stated for either perpendicularity (90°) or parallelism (0°).

The datum is identified with a capital letter, and the same letter is then used in the feature control symbol, as shown in Fig. 31-6-2.

For the tolerancing of angularity the same characteristic symbols are used for both lines and surfaces.

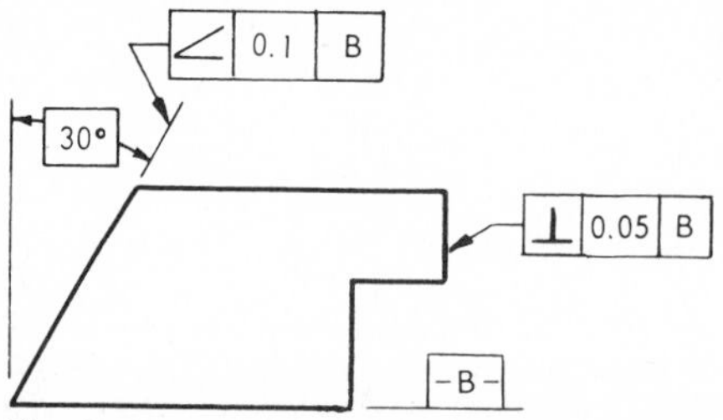
0.1
B
30°
0.05
B
-B-

Fig. 31-6-2 Orientation tolerances with identified datum.

Orientation Tolerancing of Flat Surfaces

Figure 31-6-3 shows three simple parts in which one flat surface is designated as a datum feature and another flat surface is related to it by one of the orientation tolerances.

Each of these tolerances is interpreted to mean that the designated surface shall be contained within a tolerance zone consisting of the space between two parallel planes, separated by the specified tolerance (0.05 mm) and related to the datum by the basic angle specified (30, 90, or 0°).

CONTROL IN TWO DIRECTIONS

The measuring principles for angularity indicate the method of aligning the part prior to making angularity measurements. Proper alignment ensures that line elements of the surface perpendicular to the angular line elements are parallel to the datum.

For example, the part in Fig. 31-6-4 will be aligned so that line elements running horizontally in the left-hand view will be parallel to the datum. However, these line elements will bear a proper relationship with the sides, ends, and top faces only if these surfaces are true and square with the primary datum.

It may be functionally more important to measure the angle in a direction parallel to a side or perpendicular to the front or back face. In this case, one side or face must be chosen as a secondary datum, as shown in Fig. 31-6-4.

Under these circumstances, the part is aligned on the angle plate so that the secondary datum *B* is exactly parallel to the side of the angle plate.

Lines Related to Surfaces

INTERNAL CYLINDRICAL FEATURES

Figure 31-6-5 shows some simple parts in which the axis or center line of a hole is related by an orientation tolerance to a flat surface. The flat surface is designated as the datum feature.

The center line of the hole must be contained within a tolerance zone consisting of the space between two parallel planes. These planes are separated by a specified tolerance of 0.15 mm and are related to the datum by one of the basic angles 45, 90, or 0°. Figure 31-6-6 clearly illustrates the tolerance zone for angularity.

When the tolerance is one of perpendicularity, the tolerance-zone planes can be revolved around the feature axis without the angle being affected. The tolerance zone therefore becomes a cylinder. This cylindrical zone is perpendicular to the datum and has a diameter equal to the specified tolerance, as shown in Fig. 31-6-7.

Control of Direction of Angularity. The angularity tolerance in Fig. 31-6-5 controls angularity of the hole in the direction in which it has its greatest slope, that is, the minimum angle. In some cases, it may be functionally more important to control the angle in a specific direction, such as parallel to a side. In this case, a secondary datum should be added, as shown in Fig. 31-6-8.

Control of Parallelism in Two Directions. The feature control symbol shown in Fig. 31-6-5*c* controls parallelism with a base only. If control with a side is also required, the side should be designated as a secondary datum, as shown in Fig. 31-6-9. The tolerance zone will be a parallelepiped, and two separate measurements will have to be made.

Controls on MMC Basis. Since a hole is a feature of size, any of the tolerances shown in Fig. 31-6-5 can be modified on an MMC basis. This is specified by adding the symbol Ⓜ after the tolerance. Figure 31-6-10 shows an example.

Control of Center Lines. Tolerances intended to control orientation of the center line of a feature are applied to drawings as shown in Fig. 31-6-11.

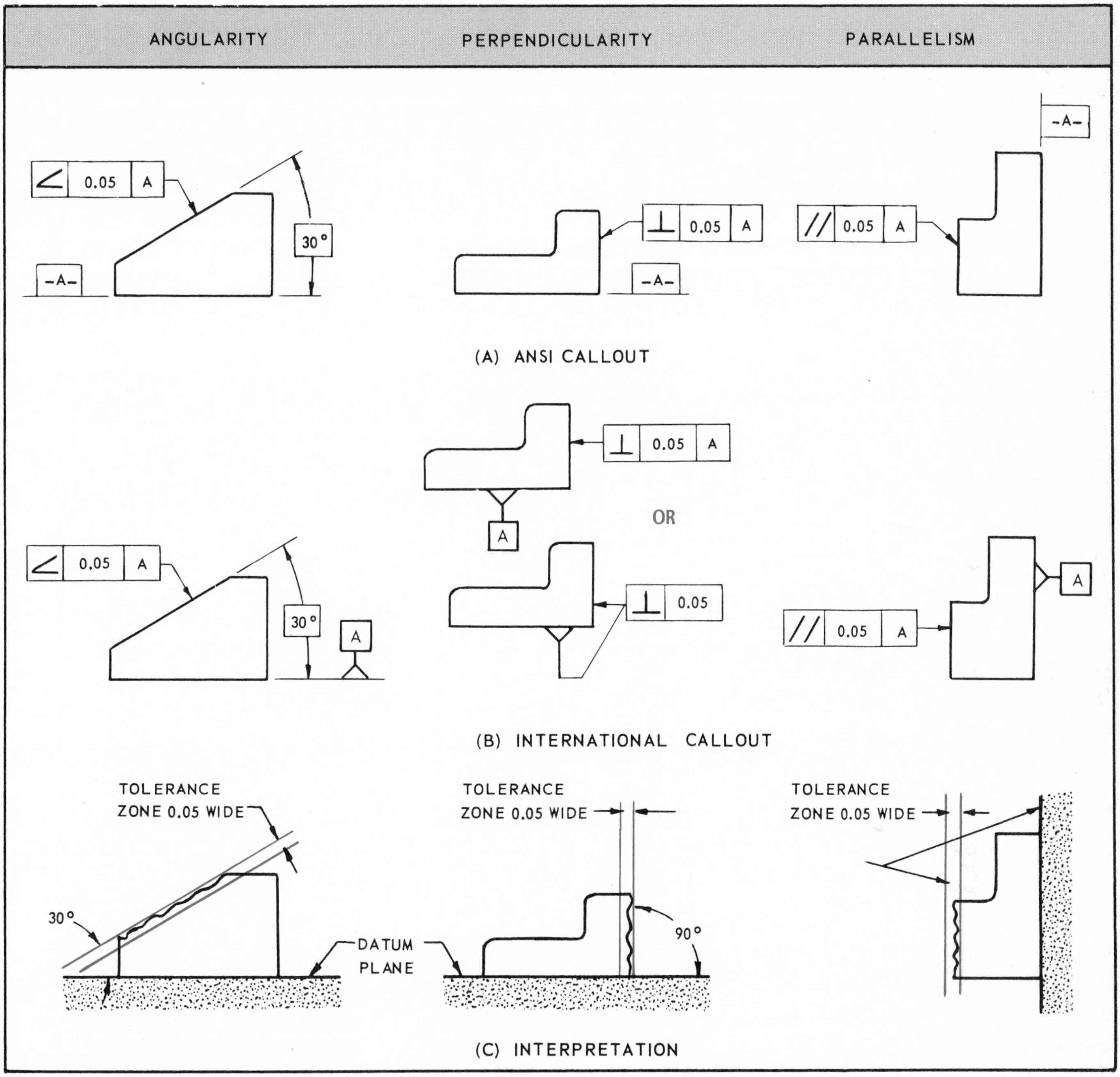
ANGULARITY
PERPENDICULARITY
PARALLELISM
0.05 A
30°
-A-
(A) ANSI CALLOUT
OR
0.05
A
(B) INTERNATIONAL CALLOUT
TOLERANCE ZONE 0.05 WIDE
30°
DATUM PLANE
90°
(C) INTERPRETATION

Fig. 31-6-3 Orientation tolerancing.

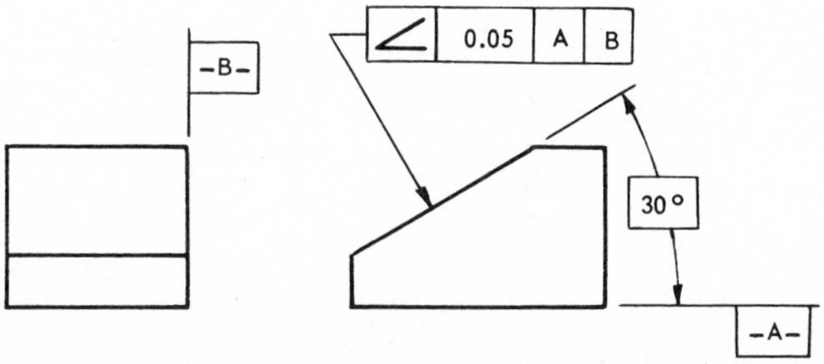
-B-
0.05 A B
30°
-A-

Fig. 31-6-4 Angularity referred to a datum system.

The center line of the cylindrical feature must be contained within a tolerance zone consisting of the space between two parallel planes separated by the specified tolerance. The parallel planes are related to the datum by the basic angle of 75, 90, or 0°.

Since the tolerance planes for perpendicularity can be revolved around the feature axis, the tolerance zone effectively becomes a cylinder. The diameter of this cylinder is equal to the specified tolerance.

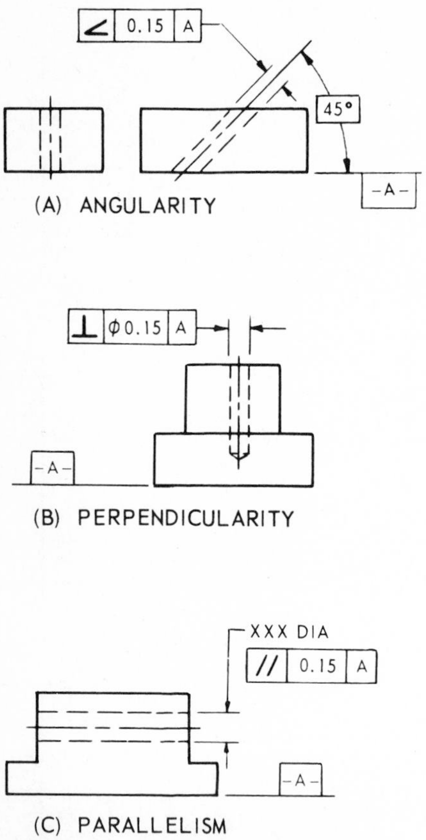
0.15 A
45°
-A-
(A) ANGULARITY
⌀0.15 A
-A-
(B) PERPENDICULARITY
XXX DIA
// 0.15 A
-A-
(C) PARALLELISM

Fig. 31-6-5 Orientation of lines and surfaces.

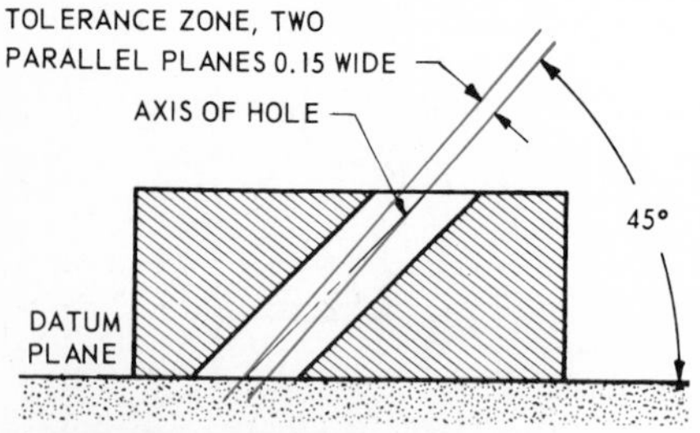
TOLERANCE ZONE, TWO
PARALLEL PLANES 0.15 WIDE
AXIS OF HOLE
45°
DATUM
PLANE

Fig. 31-6-6 Tolerance zone for angularity—Fig. 31-6-5.

TOLERANCES APPLIED TO LINE ELEMENTS OF THE SURFACE

In some cases it may be more important to control line elements of the cylindrical surface, instead of its axis or center line. For this purpose the tolerance is directed to the surface, with a single arrowhead, as shown in Fig. 31-6-12.

Except for perpendicularity, such controls apply to only two line elements of the surface. These are the lines on opposite sides of the feature, which lie in a plane perpendicular to the datum and

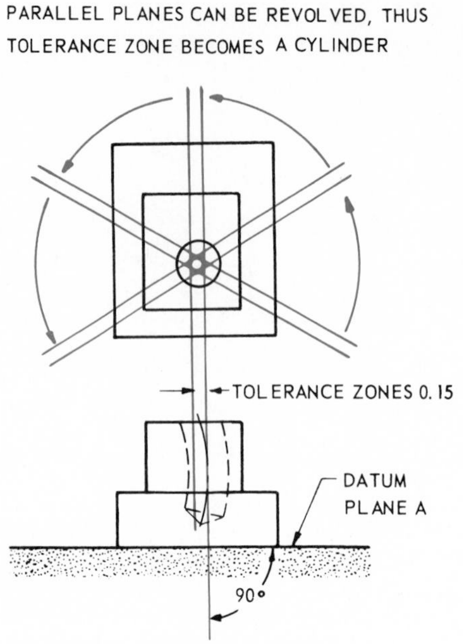
PARALLEL PLANES CAN BE REVOLVED, THUS
TOLERANCE ZONE BECOMES A CYLINDER
TOLERANCE ZONES 0.15
DATUM
PLANE A
90°

Fig. 31-6-7 Tolerance zone for perpendicularity—Fig. 31-6-5.

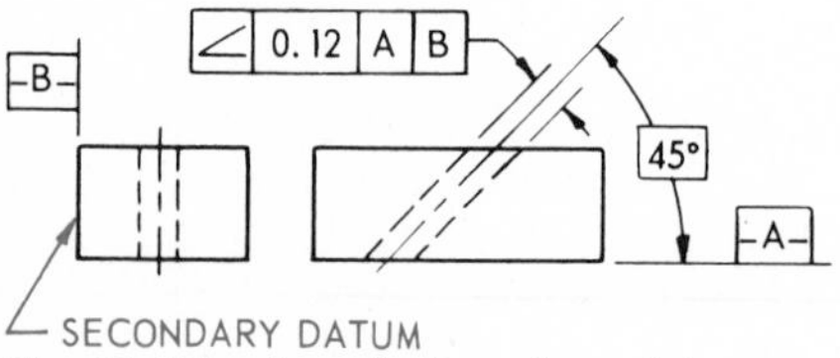
0.12 A B
-B-
45°
-A-
SECONDARY DATUM

Fig. 31-6-8 Angularity referred to two datums.

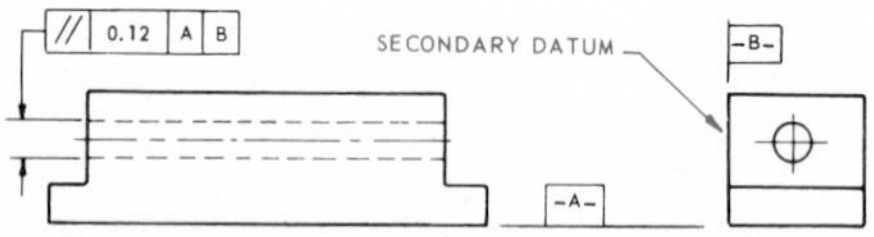
// 0.12 A B
SECONDARY DATUM
-B-
-A-

Fig. 31-6-9 Parallelism controlled in two directions.

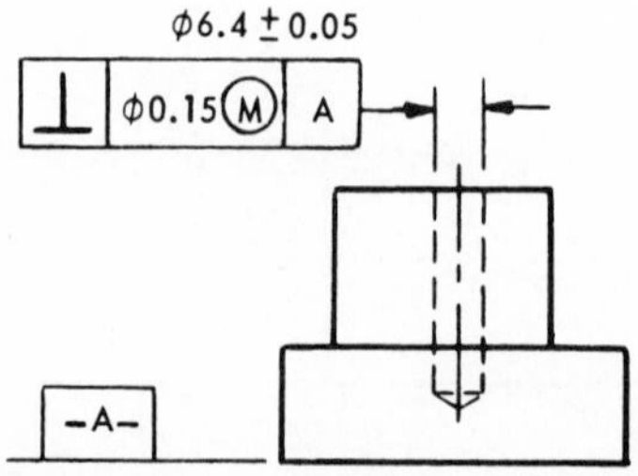
⌀6.4 ± 0.05
⌀0.15Ⓜ A
-A-

Fig. 31-6-10 Feature controlled on MMC basis.

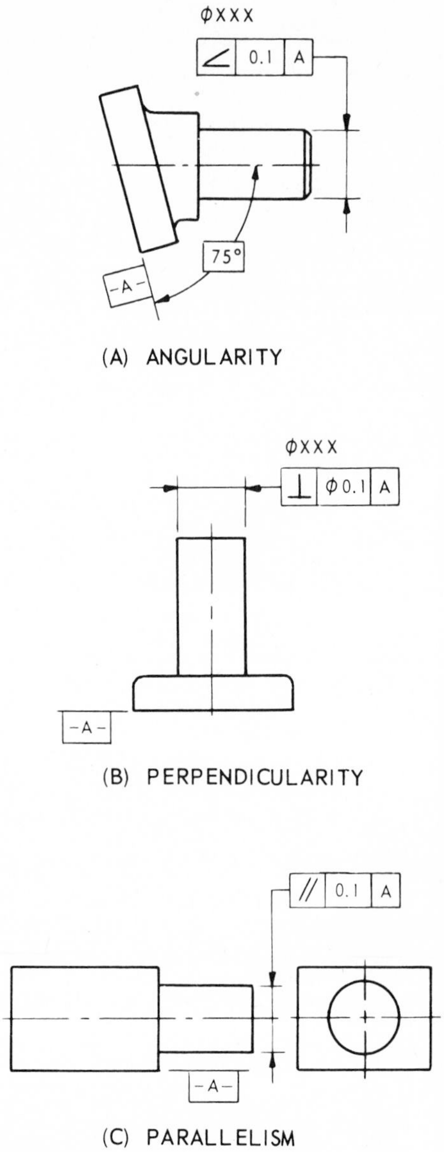
⌀XXX
0.1 A
75°
-A-
(A) ANGULARITY
⌀XXX
⌀0.1 A
-A-
(B) PERPENDICULARITY
// 0.1 A
-A-
(C) PARALLELISM

Fig. 31-6-11 Orientation of external cylindrical features.

which include the axis of the cylindrical feature. There are two tolerance zones, each of which may be considered to be the area between two parallel lines, which lie in this measurement plane. These lines are separated by the

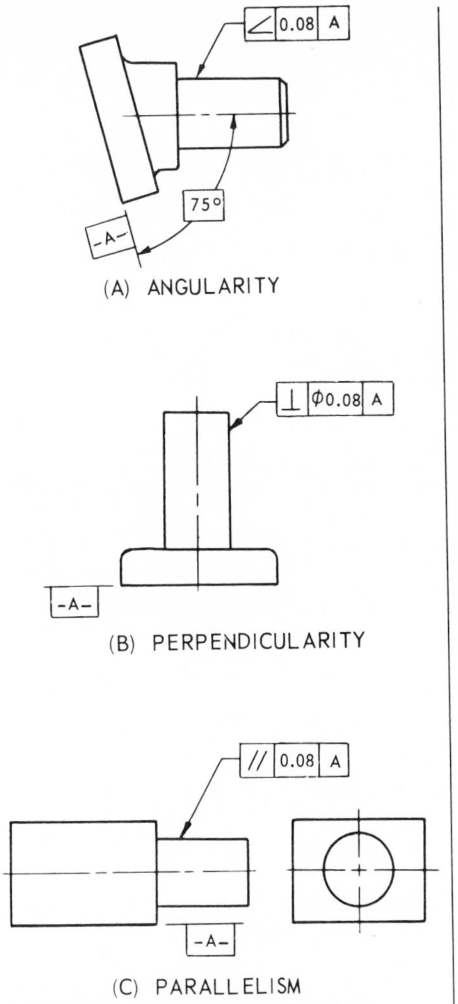
0.08 A
75°
-A-
(A) ANGULARITY
Φ0.08 A
-A-
(B) PERPENDICULARITY
// 0.08 A
-A-
(C) PARALLELISM

Fig. 31-6-12 Orientation of surface elements.

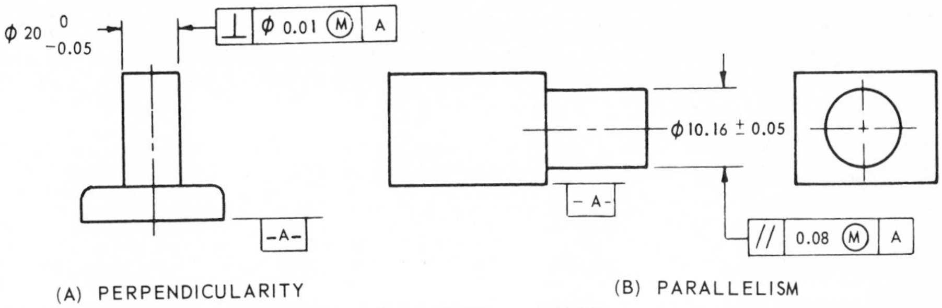
Φ 20 0 −0.05
⊥ Φ 0.01 (M) A
-A-
(A) PERPENDICULARITY
Φ 10.16 ± 0.05
- A -
// 0.08 (M) A
(B) PARALLELISM

Fig. 31-6-14 Perpendicularity and parallelism—MMC.

specified tolerance and are related to the datum by the basic angle. Figure 31-6-13 shows these zones for a parallelism tolerance. Figure 31-6-14 illustrates perpendicularity and parallelism on an MMC basis.

For perpendicularity the measurement plane can be revolved around the axis without altering its relationship to the datum. Therefore, unless otherwise specified, perpendicularity applies to line elements in all positions of the measuring plane.

ORIENTATION USING LINES AS DATUMS

The following examples of orientation tolerancing involve internal or external cylindrical datum features. These features establish datums which are either line elements of the datum feature surface or the axes of datum cylinders. Many of the methods are also applicable to the center lines of noncylindrical features, such as those having square or hexagonal cross sections.

WHEN MMC IS NOT SPECIFIED

The tolerance zone is always the space between two parallel planes separated by the specified tolerance and related to the datum by the basic angle. The controlled feature, whether it be a flat surface, the center line of the feature, or a line element of the surface, must be contained within this tolerance zone.

Example 1. Figure 31-6-15 shows a flat surface related to an external cylindrical datum feature. Since the MMC is not specified, the real datum is the axis of a datum cylinder, which is the smallest perfect cylinder that can be circumscribed around the feature. The tolerance zone is the space between two parallel planes 0.01 mm apart, which are related to the datum axis by the basic angle.

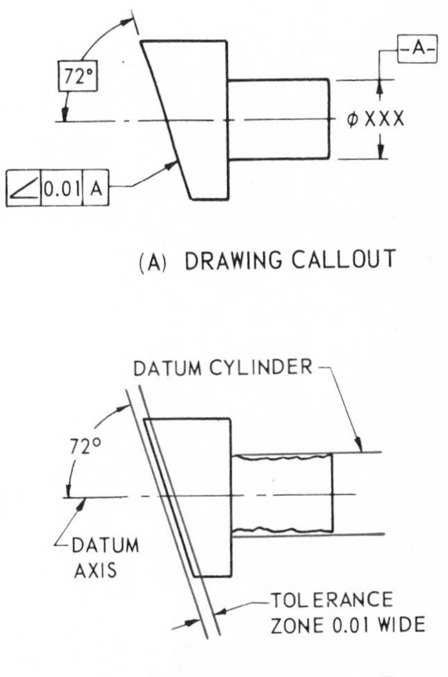
72°
-A-
Φ XXX
0.01 A
(A) DRAWING CALLOUT
DATUM CYLINDER
72°
DATUM AXIS
TOLERANCE ZONE 0.01 WIDE
(B) TOLERANCE ZONE

Fig. 31-6-15 Angularity of flat surface with an external cylinder.

Example 2. Figure 31-6-16 is similar to Fig. 31-6-15 except that the flat surface is related to an internal cylindrical feature. The datum is therefore the axis of the largest perfect cylinder which will fit within the center hole of the part. The tolerance zone is the space between two parallel planes, 0.1 mm apart, related to this datum axis by the specified basic angle.

Example 3. Figure 31-6-17 shows a simple parallelism requirement of a flat surface in relation to a cylindrical hole. The tolerance zone is the space between two parallel planes, 0.1 mm apart and parallel to the datum axis.

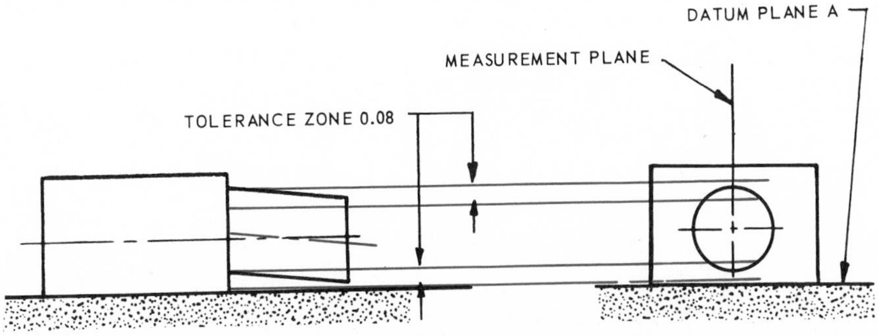
DATUM PLANE A
MEASUREMENT PLANE
TOLERANCE ZONE 0.08

Fig. 31-6-13 Tolerance zones for parallelism in Fig. 31-6-12.

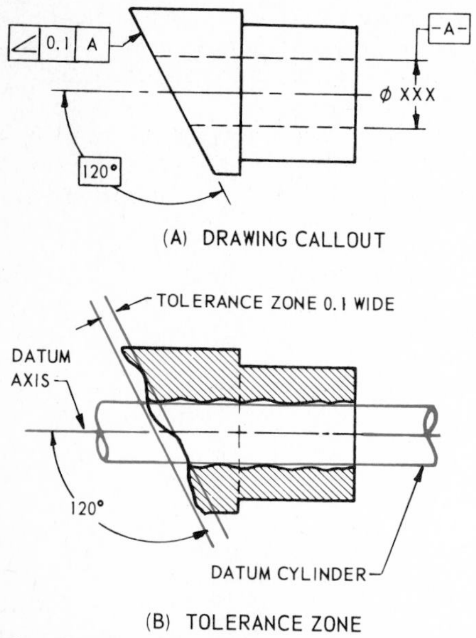
0.1 A
-A-
Ø XXX
120°
(A) DRAWING CALLOUT
TOLERANCE ZONE 0.1 WIDE
DATUM AXIS
120°
DATUM CYLINDER
(B) TOLERANCE ZONE

Fig. 31-6-16 Angularity of flat surface with an internal cylinder.

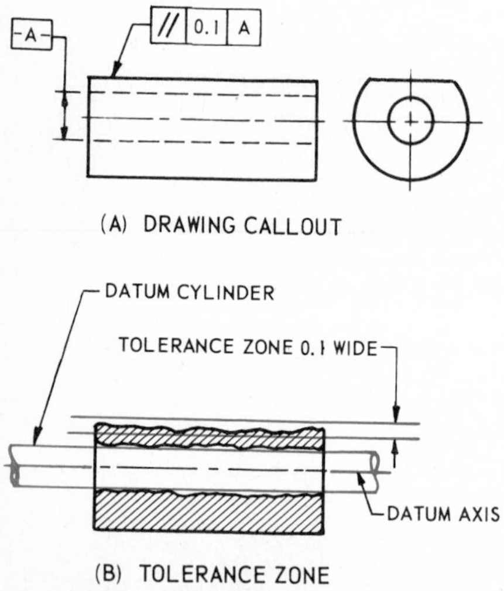
-A-
// 0.1 A
(A) DRAWING CALLOUT
DATUM CYLINDER
TOLERANCE ZONE 0.1 WIDE
DATUM AXIS
(B) TOLERANCE ZONE

Fig. 31-6-17 Parallelism of a flat surface with a cylindrical hole.

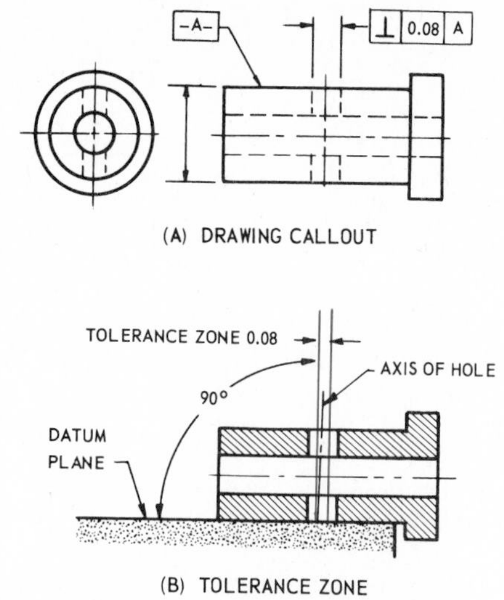
-A-
⊥ 0.08 A
(A) DRAWING CALLOUT
TOLERANCE ZONE 0.08
AXIS OF HOLE
90°
DATUM PLANE
(B) TOLERANCE ZONE

Fig. 31-6-18 Perpendicularity of hole with line elements of surface.

Example 4. Figure 31-6-18 shows a requirement for perpendicularity of the axis of a hole with line elements of a cylindrical surface. There are actually two line elements of the cylindrical surface perpendicular to the hole, represented by the upper and lower solid lines in the illustration.

Assignments

1. On an A3- or B-size sheet, prepare sketches and solutions for the drawings shown in Fig. 31-6-A or 31-6-B and the following information.

(*a*) Today's drafters must be capable of interpreting and preparing drawings for use in other countries as well as in their own locality. From the information shown in illustration (A) prepare a three-view sketch showing the datums and geometric tolerancing for use in the United States and in Canada or Europe.

(*b*) The surfaces shown in illustration (B) are required to be controlled in the following manner. Surfaces *A*, *B*, *C*, and *D* are datums *A*, *B*, *C*, *D*, respectively. Prepare a three-view drawing showing the datums, and feature control symbols from the information supplied.

2. On an A3- or B-size sheet, prepare sketches and solutions for the drawings shown in Fig. 31-6-C or 31-6-D and the following information.

(*a*) In illustration (A) it is required that the hole be parallel with datum *A* within ±0.5°. Given that the tangent of 0.5° is 0.0087, show the drawing callout and describe the tolerance zone.

(*b*) In illustration (B) it is functionally necessary that the shaft portion of the part not depart from perpendicularity

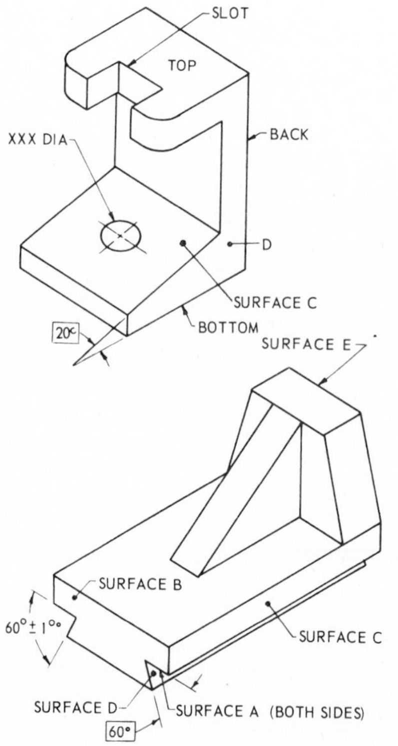
SLOT
TOP
BACK
XXX DIA
D
SURFACE C
BOTTOM
20°
SURFACE E
SURFACE B
60° ± 1°
SURFACE C
SURFACE D
SURFACE A (BOTH SIDES)
60°

BRACKET

- BOTTOM TO BE DATUM A
- BACK TO BE DATUM B
- HOLE TO BE PERPENDICULAR TO BOTTOM WITHIN 0.08
- BACK TO BE PERPENDICULAR TO BOTTOM WITHIN 0.1
- TOP TO BE PARALLEL WITH BOTTOM WITHIN 0.12
- SURFACE C TO HAVE AN ANGULARITY TOLERANCE OF 0.16 WITH THE BOTTOM. SURFACE D TO BE THE SECONDARY DATUM FOR THIS FEATURE.
- THE SIDES OF THE SLOT TO BE PARALLEL WITH EACH OTHER WITHIN 0.05

DOVETAIL SLIDE

- SURFACE D OF THE DOVETAIL MUST HAVE AN ANGULARITY TOLERANCE OF 0.05 mm WITH DATUM A.
- SURFACE C SHOULD BE PERPENDICULAR TO DATUM A WITHIN 0.03 mm.
- SURFACE E MUST BE PERPENDICULAR TO DATUMS A AND D WITHIN 0.02 mm.

Fig. 31-6-A Assignments.

with the holes by more than the tolerance specified. Show the drawing callout for this, and indicate the shape and size of the tolerance zone.

(*c*) Sketch the basic elements of a gage to check the perpendicularity requirement in illustration (C). If the large hole were round and straight and at the size measured, what would be the greatest angularity error of the hole axis?

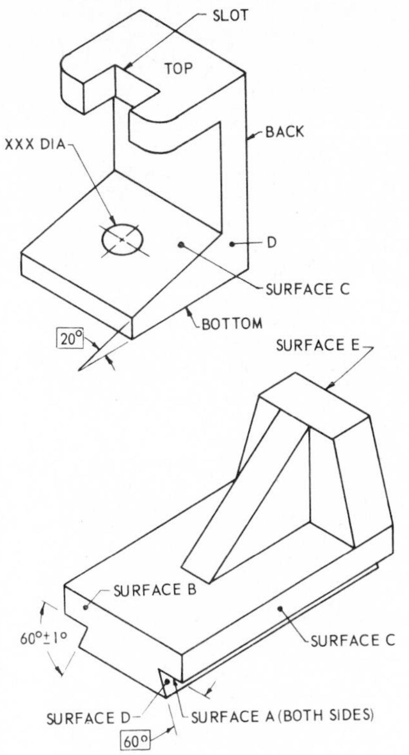
SLOT
TOP
BACK
XXX DIA
D
SURFACE C
BOTTOM
20°
SURFACE E
SURFACE B
60°±1°
SURFACE C
SURFACE D
SURFACE A (BOTH SIDES)
60°

BRACKET
- BOTTOM TO BE DATUM A
- BACK TO BE DATUM B
- HOLE TO BE PERPENDICULAR TO BOTTOM WITHIN .003
- BACK TO BE PERPENDICULAR TO BOTTOM WITHIN .004
- TOP TO BE PARALLEL WITH BOTTOM WITHIN .005
- SURFACE C TO HAVE AN ANGULARITY TOLERANCE OF .005 WITH THE BOTTOM. SURFACE D TO BE THE SECONDARY DATUM FOR THIS FEATURE.
- THE SIDES OF THE SLOT TO BE PARALLEL WITH EACH OTHER WITHIN .002

DOVETAIL SLIDE
- SURFACE D OF THE DOVETAIL MUST HAVE AN ANGULARITY TOLERANCE OF .003 WITH DATUM A.
- SURFACE C SHOULD BE PERPENDICULAR TO DATUM A WITHIN .001 .
- SURFACE E MUST BE PERPENDICULAR TO DATUMS A AND D WITHIN .002 .

Fig. 31-6-B Assignments.

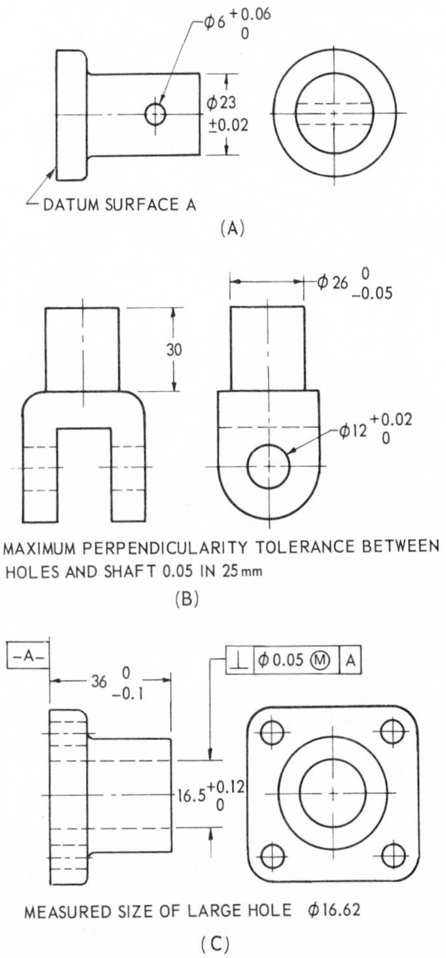
φ6 +0.06 0
φ23 ±0.02
DATUM SURFACE A
(A)
φ26 0 −0.05
30
φ12 +0.02 0
MAXIMUM PERPENDICULARITY TOLERANCE BETWEEN HOLES AND SHAFT 0.05 IN 25 mm
(B)
-A-
36 0 −0.1
⊥ φ0.05 Ⓜ A
16.5 +0.12 0
MEASURED SIZE OF LARGE HOLE φ16.62
(C)

Fig. 31-6-C Assignments.

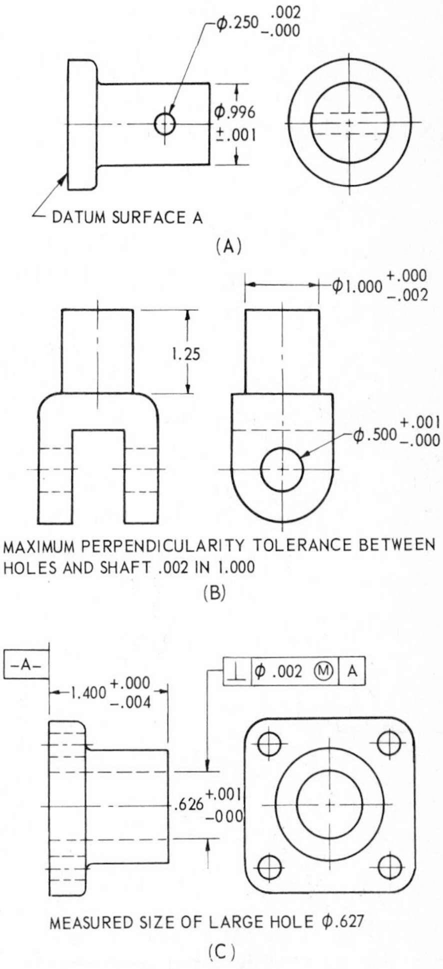
φ.250 .002 −.000
φ.996 ±.001
DATUM SURFACE A
(A)
φ1.000 +.000 −.002
1.25
φ.500 +.001 −.000
MAXIMUM PERPENDICULARITY TOLERANCE BETWEEN HOLES AND SHAFT .002 IN 1.000
(B)
-A-
1.400 +.000 −.004
⊥ φ .002 Ⓜ A
.626 +.001 −000
MEASURED SIZE OF LARGE HOLE φ.627
(C)

Fig. 31-6-D Assignments.

UNIT 31-7

TOLERANCING FOR LOCATION OF FEATURES

The location of features is one of the most frequently used applications of dimensions on technical drawings. Tolerancing may be accomplished either by *coordinate tolerances* applied to the dimensions or by *geometric (positional) tolerancing*.

Positional tolerancing is especially useful when applied on an MMC basis to groups or patterns of holes or other small features in the mass production of parts. This method meets functional requirements in most cases and permits assessment with simple gaging procedures.

Most examples in this unit are devoted to the principles involved in the location of small, round holes, because they represent the most commonly used applications. The same principles apply, however, to the location of other features, such as slots, tabs, bosses, and noncircular holes.

Tolerancing Methods

The location of a single hole is usually indicated by means of rectangular coordinate dimensions, extending from suitable edges or other features of the part to the axis of the hole. Other dimensioning methods, such as polar coordinates, may be used when circumstances warrant.

There are two standard methods of tolerancing the location of holes, as illustrated in Fig. 31-7-1:

1. *Coordinate tolerancing*, which refers to tolerances applied directly to the coordinate dimensions or to applicable tolerances specified in a general tolerance note.

2. (*a*) *Positional tolerancing*, regardless of feature size.

(*b*) *Positional tolerancing*, MMC basis. These positional tolerancing methods are part of the system of geometric tolerancing.

Any of these tolerancing methods can be substituted one for the other, although with differing results. It is necessary, however, to first analyze the widely used method of coordinate tolerancing in order to explain and understand the advantages and disadvantages of the positional tolerancing methods.

Coordinate Tolerancing

Coordinate dimensions and tolerances may be applied to the location of a single hole, as shown in Fig. 31-7-2. They indicate the location of the hole axis and result in a rectangular or wedge-shaped tolerance zone within which the axis of the hole must lie.

If the two coordinate tolerances are equal, the tolerance zone formed will be a square. Unequal tolerances result in a rectangular tolerance zone. Where one of the locating dimensions is a radius, polar dimensioning gives a circular ring section tolerance zone. For simplicity, square tolerance zones are used in the analyses

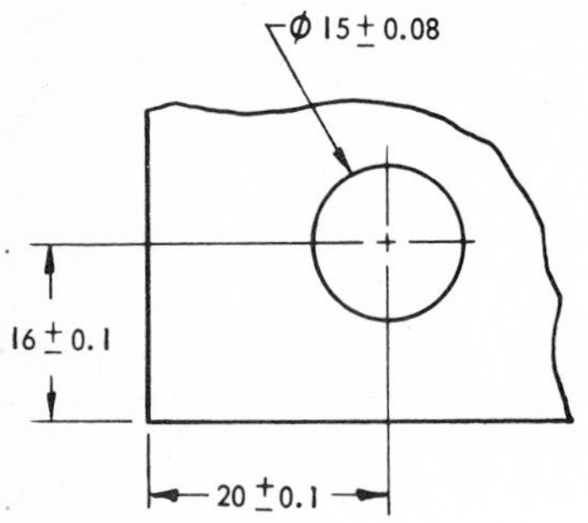

ϕ 15 ± 0.08
16 ± 0.1
20 ± 0.1

(A) COORDINATE TOLERANCING

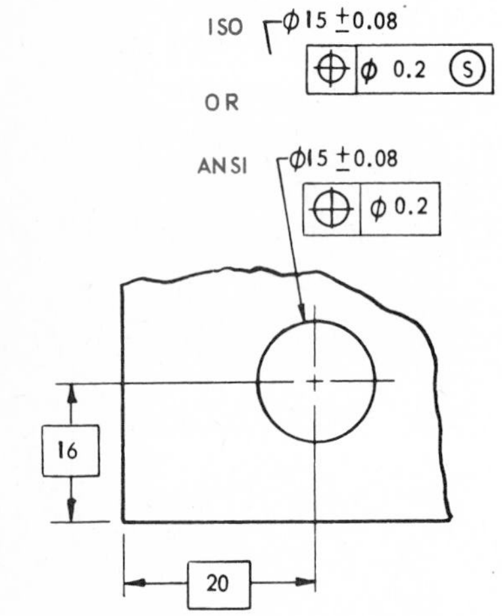

ISO
ϕ 15 ± 0.08
ϕ 0.2 Ⓢ
OR
ANSI
ϕ15 ± 0.08
ϕ 0.2
16
20

(B) POSITIONAL TOLERANCING – RFS

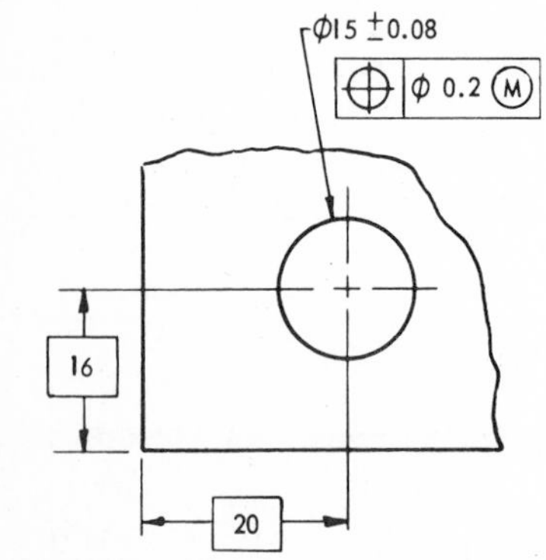

ϕ15 ± 0.08
ϕ 0.2 Ⓜ
16
20

(C) POSITIONAL TOLERANCING – MMC

Fig. 31-7-1 Comparison of tolerancing methods.

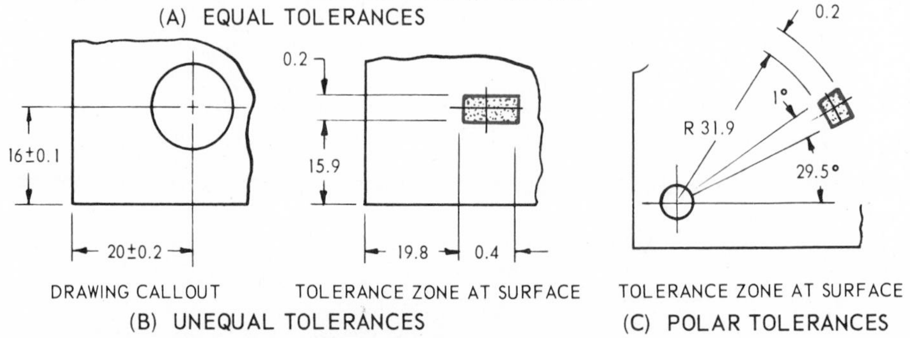

16±0.1
20±0.1
0.2
15.9
19.9
0.2
DRAWING CALLOUT
TOLERANCE ZONE AT SURFACE
(A) EQUAL TOLERANCES
16±0.1
20±0.2
0.2
15.9
19.8
0.4
DRAWING CALLOUT
TOLERANCE ZONE AT SURFACE
(B) UNEQUAL TOLERANCES
R 32±0.1
30°±0.5°
DRAWING CALLOUT
0.2
1°
R 31.9
29.5°
TOLERANCE ZONE AT SURFACE
(C) POLAR TOLERANCES

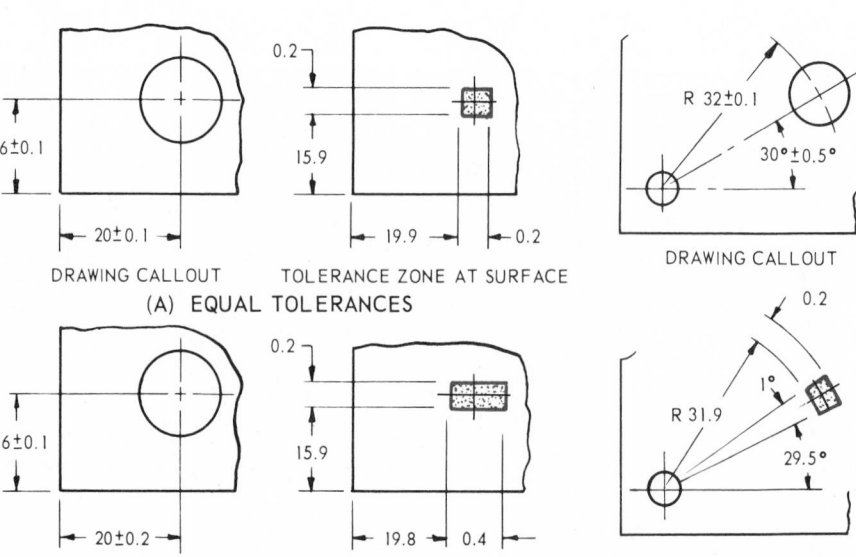

Fig. 31-7-2 Tolerance zones for coordinate tolerances.

of most of the examples in this section.

It should be noted that the tolerance zone extends for the full depth of the hole, that is, the whole length of the axis. This is illustrated in Fig. 31-7-3 and explained in more detail in a later unit. In most of the illustrations, tolerances will be analyzed as they apply at the surface of the part, where the axis is represented by a point.

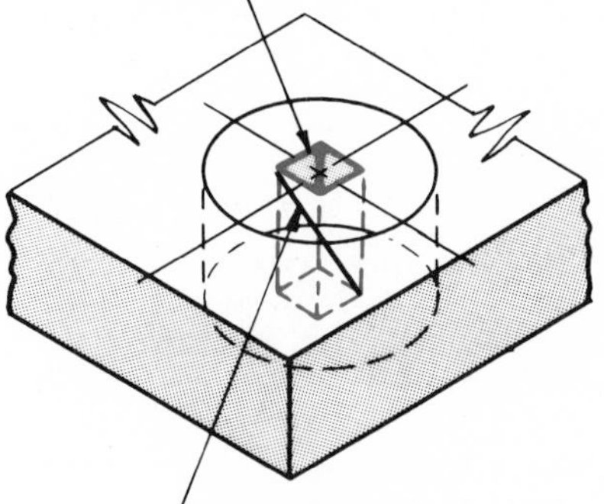

SQUARE TOLERANCE ZONE
EXTREME PERMISSIBLE VARIATION IN POSITION OF AXIS

Fig. 31-7-3 Tolerance zone extending through part.

MAXIMUM PERMISSIBLE ERROR

The actual position of the feature axis may be anywhere within the rectangular tolerance zone. For square tolerance zones, the maximum allowable variation from the desired position occurs in a direction of 45° from the direction of the coordinate dimensions. See Fig. 31-7-4.

For rectangular tolerance zones this maximum tolerance is the square root of the sum of the squares of the individual tolerances, or expressed mathematically

$$\sqrt{X^2 + Y^2}$$

For the examples shown in Fig. 31-7-2, the tolerance zones are shown in Fig.

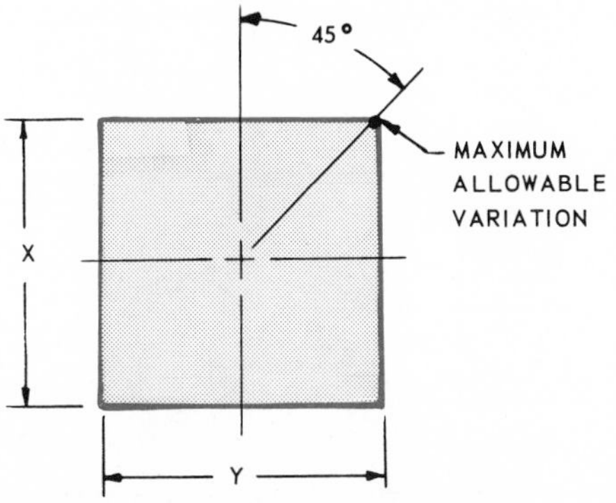

45°
MAXIMUM ALLOWABLE VARIATION
X
Y

Fig. 31-7-4 Square tolerance zone.

31-7-5, and the maximum tolerance values are as shown in the following examples.

Example 1.

$$\sqrt{0.2^2 + 0.2^2} = 0.28$$

Example 2.

$$\sqrt{0.2^2 + 0.4^2} = 0.45$$

Example 3. For polar coordinates the extreme variation is

$$\sqrt{A^2 + T^2}$$

where $A = R \text{ Tan } a$
T = tolerance on radius
R = mean radius
a = angular tolerance

Thus, the extreme variation in the third example is

$$\sqrt{(32 \times 0.017\ 45)^2 + 0.2^2} = 0.59$$

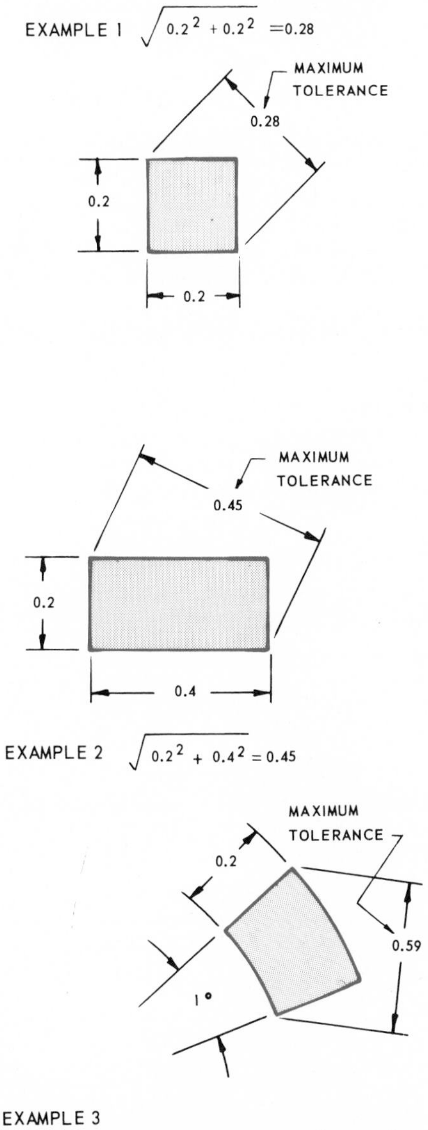

EXAMPLE 1 $\sqrt{0.2^2 + 0.2^2}$ =0.28
MAXIMUM TOLERANCE
0.28
0.2
0.2
MAXIMUM TOLERANCE
0.45
0.2
0.4
EXAMPLE 2 $\sqrt{0.2^2 + 0.4^2}$ = 0.45
MAXIMUM TOLERANCE
0.2
0.59
1°
EXAMPLE 3

Fig. 31-7-5 Tolerance zones for parts shown in Fig. 31-7-2.

Note: Mathematically, A in the above formula should be $2R \tan a/2$, instead of $R \tan a$, and T should be $T \cos a/2$; but the difference in results is quite insignificant for the tolerances normally used.

Some values of tan A for commonly used angular tolerances are as follows.

A	Tan a	A	Tan a	A	Tan a
0° 5′	.00145	0°25′	.00727	0°45′	.01309
0°10′	.00291	0°30′	.00873	0°50′	.01454
0°15′	.00436	0°35′	.01018	0°55′	.01600
0°20′	.00582	0°40′	.01164	1° 0′	.01745

USE OF CHART

A quick and easy method of finding the maximum positional error permitted with coordinate tolerancing, without having to calculate squares and square roots, is by use of a chart like that shown in Fig. 31-7-6.

In the first example shown in Fig. 31-7-2, the tolerance in both directions is 0.2 mm. The extensions of the horizontal and vertical lines of 0.2 in the chart intersect at point A, which lies between the radii of 0.275 and 0.3 mm. When interpolated and rounded to two decimal places, a maximum permissible variation of position is 0.28 mm.

In the second example shown in Fig. 31-7-2, the tolerances are 0.2 mm in one direction and 0.4 mm in the other. The extensions of the vertical and horizontal lines at 0.2 and 0.4 mm respectively in the chart intersect at point B, which lies between the radii of 0.425 and 0.45 mm. When interpolated and rounded to two decimal places, a maximum variation of position is 0.45 mm. Figure 31-7-6 also shows a chart for use with tolerances in inches.

ADVANTAGES OF COORDINATE TOLERANCING

The advantages claimed for direct coordinate tolerancing are as follows.

1. It is simple and easily understood, and therefore it is commonly used.

2. It permits direct measurements to be made with standard instruments and does not require the use of special-purpose functional gages or other calculations.

DISADVANTAGES OF COORDINATE TOLERANCING

There are a number of disadvantages to the direct tolerancing method:

1. It results in a square or rectangular tolerance zone within which the axis must lie. For a square zone this permits

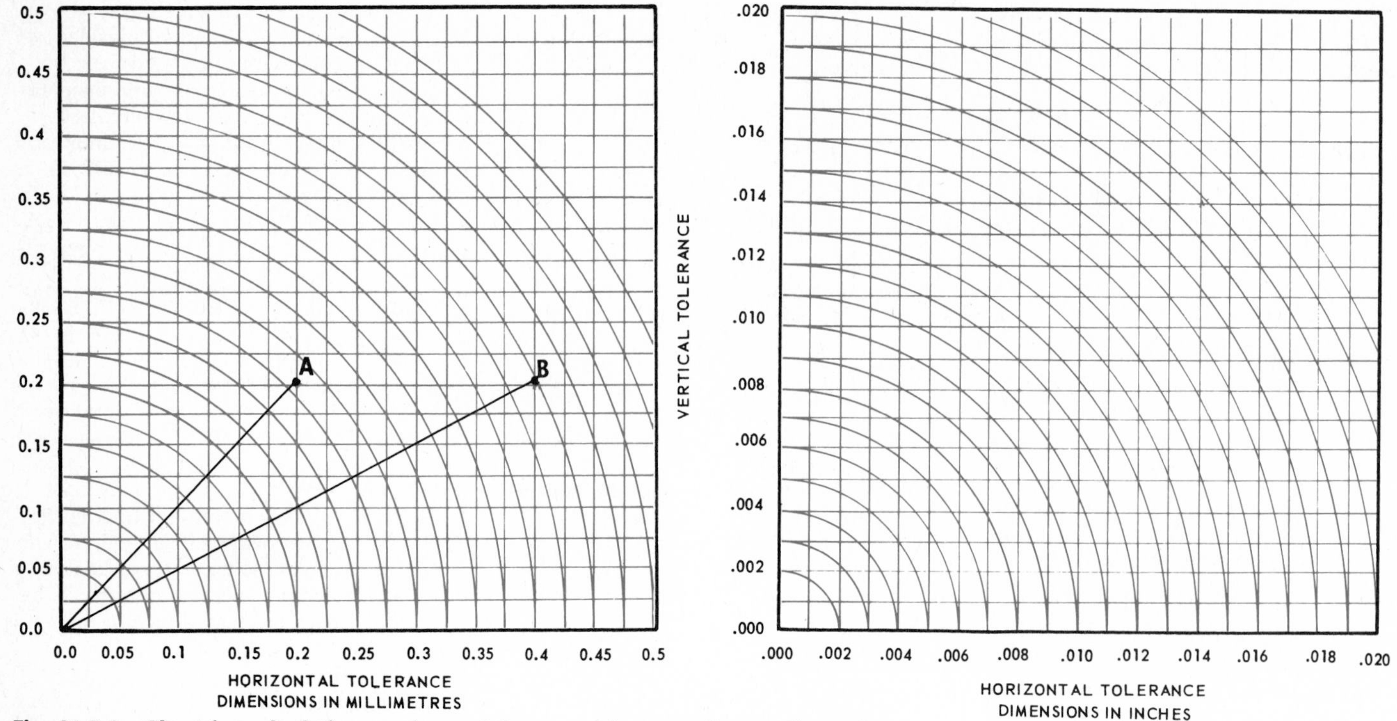
VERTICAL TOLERANCE
HORIZONTAL TOLERANCE
DIMENSIONS IN MILLIMETRES
HORIZONTAL TOLERANCE
DIMENSIONS IN INCHES
A
B

Fig. 31-7-6 Chart for calculating maximum tolerance using coordinate tolerancing.

a variation in a 45° direction of approximately 1.4 times the specified tolerance. This amount of variation may necessitate the specification of tolerances which are only 70 percent of those that are functionally acceptable.

2. It may result in an undesirable accumulation of tolerances when several features are involved, especially when chain dimensioning is used.

3. It is more difficult to assess clearances between mating features and components than when positional tolerancing is used, especially when a group or a pattern of features is involved.

4. It does not correspond to the control exercised by fixed functional GO gages, often desirable in mass production of parts. This becomes particularly important in dealing with a group of holes. With direct coordinate tolerancing, the location of each hole has to be measured separately in two directions, whereas with positional tolerancing on an MMC basis one functional gage checks all holes in one operation.

Positional Tolerancing

Positional tolerancing is part of the system of geometric tolerancing. In this system the location of features is shown either by coordinate dimensions or by polar and angular dimensions, except that the dimensions are shown without direct tolerances. These dimensions represent the basic sizes and are commonly known as *true-position* dimensions. Each such dimension is enclosed in a rectangular frame to indicate that it represents an exact value, to which tolerances shown in the general tolerance note do not apply. See Fig. 31-7-7. The frame size need not be any larger than that necessary to enclose the dimension. Permissible deviations from true position are then indicated by a positional tolerance as described in this unit.

Symbol for Position

The geometric characteristic symbol for position is a circle, whose diameter is approximately 60 percent of the feature control symbol frame height, and two solid center lines whose length is 75 percent of the frame height, as shown in Fig. 31-7-8. This symbol is used in the feature control symbol in the same manner as for other geometric tolerances.

Positional tolerances may be specified on an MMC basis; but when MMC is not specified, they apply to the position of the axis, regardless of the size of the feature. The ANSI standard shows the use of the symbol Ⓢ, meaning regardless of feature size, for all positional tolerances which are not modified by Ⓜ, maximum material condition.

Positional tolerancing on this basis controls the position of the axis of the hole. For this reason the feature control symbol is normally attached to the size of the feature, as shown in Fig. 31-7-9.

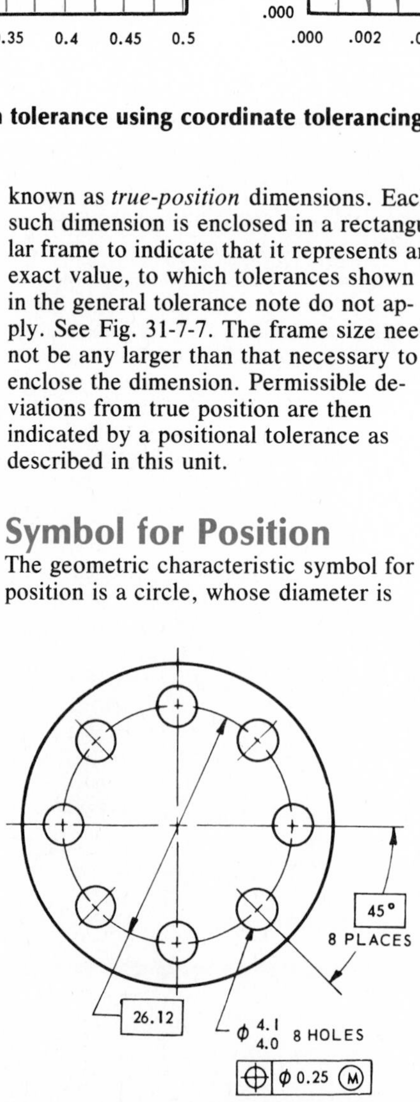
45°
8 PLACES
26.12
ϕ 4.1 4.0 8 HOLES
ϕ 0.25 Ⓜ

Fig. 31-7-7 Identifying true-position dimensions.

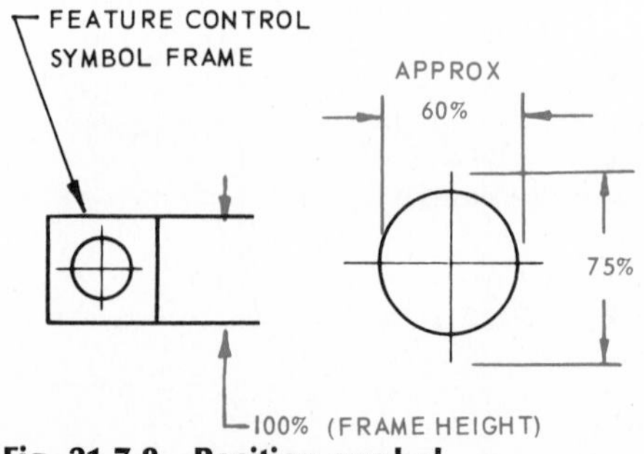
FEATURE CONTROL SYMBOL FRAME
APPROX 60%
75%
100% (FRAME HEIGHT)

Fig. 31-7-8 Position symbol.

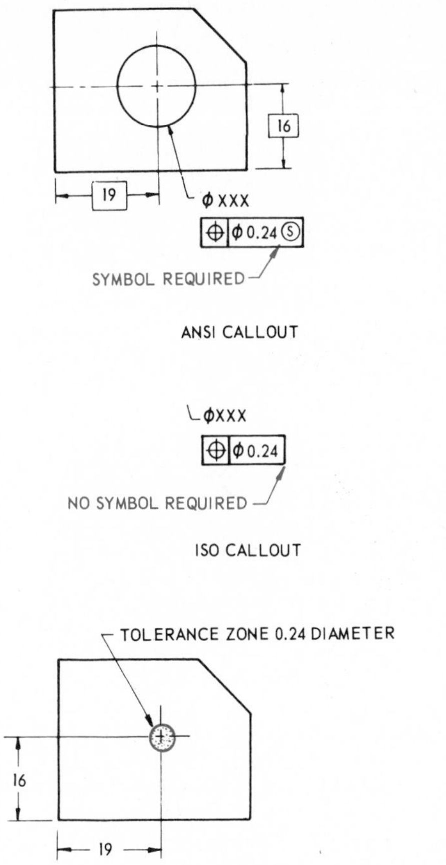
16
19
ϕXXX
ϕ0.24 Ⓢ
SYMBOL REQUIRED
ANSI CALLOUT
ϕXXX
ϕ0.24
NO SYMBOL REQUIRED
ISO CALLOUT
TOLERANCE ZONE 0.24 DIAMETER
16
19
(B) TOLERANCE ZONE

Fig. 31-7-9 Positional tolerancing—regardless of feature size.

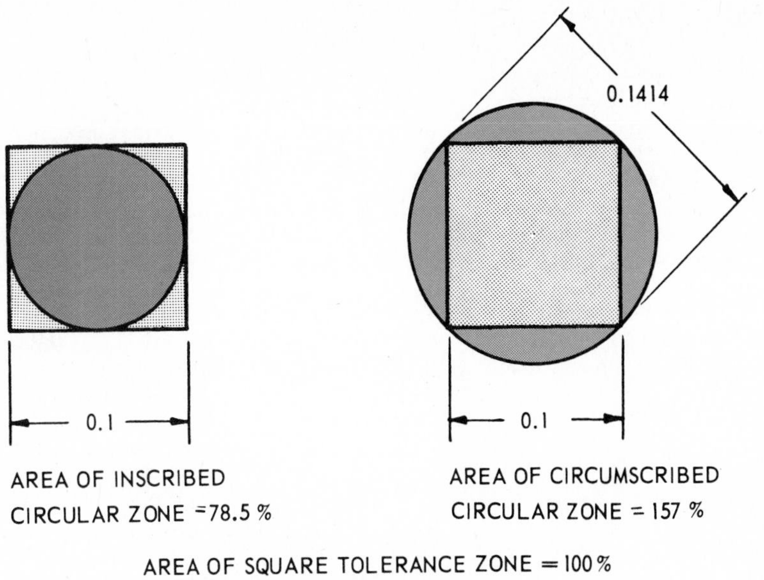
0.1414
0.1
0.1
AREA OF INSCRIBED CIRCULAR ZONE ≈78.5 %
AREA OF CIRCUMSCRIBED CIRCULAR ZONE = 157 %
AREA OF SQUARE TOLERANCE ZONE = 100 %

Fig. 31-7-10 Relationship of tolerance zones.

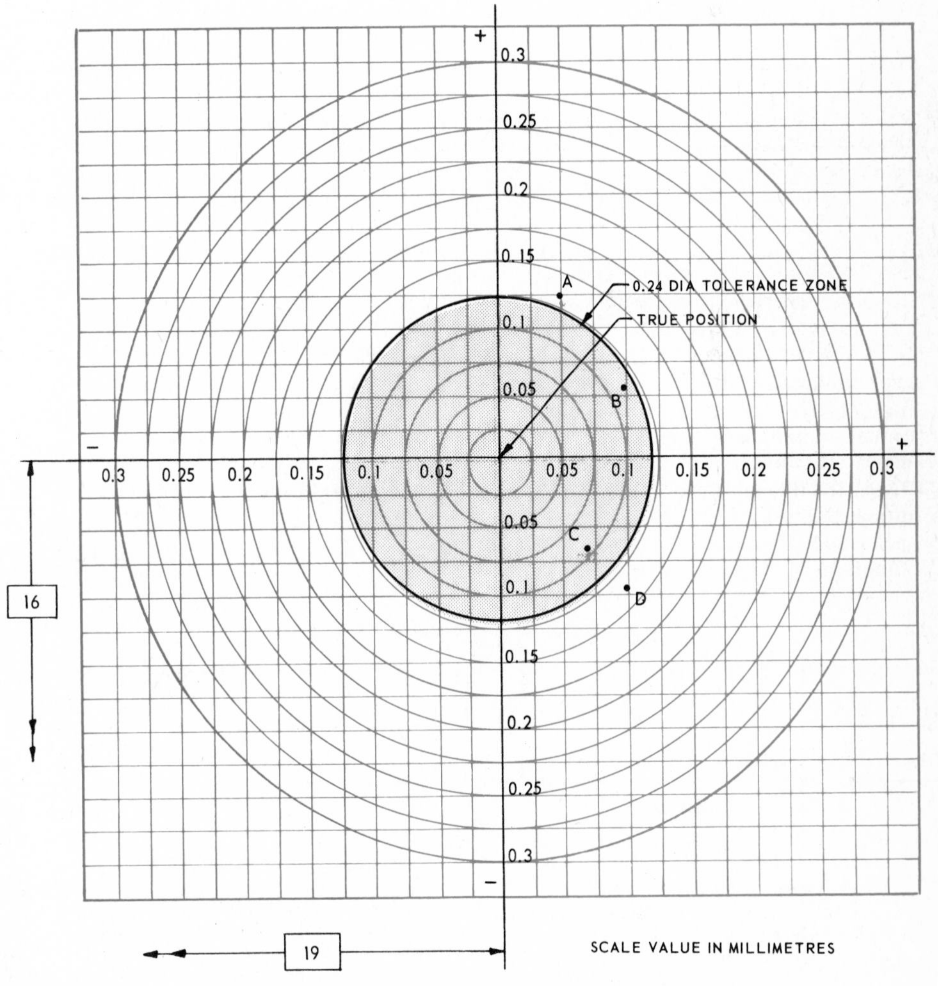
0.24 DIA TOLERANCE ZONE
TRUE POSITION
A
B
C
D
16
19
SCALE VALUE IN MILLIMETRES
NOTE – RED CIRCULAR AREA REPRESENTS ϕ0.24 TOLERANCE ZONE

Fig. 31-7-11 Chart for evaluating positional tolerancing—RFS.

The positional tolerance represents the diameter of a cylindrical tolerance zone, located at true position as determined by the true-position dimensions on the drawing, within which the axis or center line of the hole must lie.

Except for the fact that the tolerance zone is circular instead of square, a positional tolerance on this basis has exactly the same meaning as direct coordinate tolerancing with equal tolerances in both directions.

It has already been shown that with rectangular coordinate tolerancing the maximum permissible error in location is not the value indicated by the horizontal and vertical tolerances, but rather is equivalent to the length of the diagonal between the two tolerances. For square tolerance zones this amount is 1.4 times the specified tolerance values. If the same tolerance as shown for a square coordinate tolerance is specified for a positional tolerance (circular), the total area of the tolerance zone is reduced by

21.5 percent. The specified tolerance can therefore be increased to an amount equal to the diagonal of the coordinate tolerance zone without affecting the clearance between the hole and its mating part.

It is quite practical, however, to replace coordinate tolerances with a positional tolerance having a value equal to the diagonal of the coordinate tolerance zone. This does not affect the clearance between the hole and its mating part, yet it offers 57 percent more tolerance area, as shown in Fig. 31-7-10. Such a change would most likely result in a reduction in the number of parts rejected for positional errors.

A simpler method is to make coordinate measurements and evaluate them on a chart, as shown in Fig. 31-7-11. For example, if measurements of four parts are as shown in the table below, only two are acceptable. These positions are shown on the chart.

Part	Measurements		Acceptability
A	16.12	19.05	Rejected
B	16.05	19.10	Accepted
C	15.93	19.07	Accepted
D	15.90	19.10	Rejected

Positional Tolerancing—Maximum Material Condition

The problems of tolerancing for the position of holes are simplified when positional tolerancing is applied on an MMC basis. This overcomes the disadvantages listed for the coordinate method. Positional tolerancing simplifies measuring procedures because it permits the use of functional GO gages. It also permits an increase in positional variations as the size departs from the maximum material size without jeopardizing free assembly of mating features.

A positional tolerance on an MMC basis is specified on a drawing, on either the front or the side view, as shown in Fig. 31-7-12. The MMC symbol Ⓜ is added in the feature control symbol immediately after the tolerance.

A positional tolerance applied to a hole on an MMC basis means that the boundary of the hole must fall outside a perfect cylinder having a diameter equal to the maximum material condition of the feature minus the positional tolerance. This cylinder is located with its axis at true position. The hole must, of course, meet its diameter limits.

The effect is illustrated in Fig. 31-7-13, where the gage cylinder is shown at true position and the minimum and maximum diameter holes are drawn to show the extreme permissible variations in position in one direction.

Therefore, if a hole is at its maximum material condition (minimum diameter), the position of its axis must lie within a circular tolerance zone having a diameter equal to the specified tolerance. If the hole is at its maximum diameter (least material condition), the diameter of the tolerance zone is increased by the amount of the feature tolerance. The greatest deviation in one direction from true position is therefore

$$\frac{H + P}{2}$$

where H = hole diameter tolerance
P = positional tolerance

An example is illustrated in Fig. 31-7-14. The hole at its least material condition has a 13.8 mm diameter.

$$H = 0.1$$
$$P = 0.12$$

Therefore, the greatest deviation in any one direction is

$$\frac{H + P}{2} = \frac{0.1 + 0.12}{2} = 0.11$$

It must be emphasized that positional tolerancing, even on an MMC basis, is not a cure-all for positional tolerancing problems; each method of tolerancing has its own area of usefulness. In each application a method must be selected which best suits the particular case. The preferred methods for various types of applications are as follows.

Positional Tolerancing on an MMC Basis. This method is preferred when production quantities warrant the provision of functional GO gages, because gaging is then limited to one simple operation, even when a group of holes is involved. This method also facilitates manufacture by permitting larger variations in position when the diameter departs from the maximum material condition. It cannot be used when it is essential that variations in location of the axis be observed regardless of feature size.

Coordinate Tolerancing Method. This method is preferred in many applications where it is not economical to provide functional gages. Coordinate tolerancing may require a reduction in the specified tolerance from the estimated permissible variation in order to compensate for a square tolerance zone. Also, it does not permit an increase in the permissible variation as the size approaches the least material condition.

Positional Tolerancing, RFS. This method is very similar to the coordinate

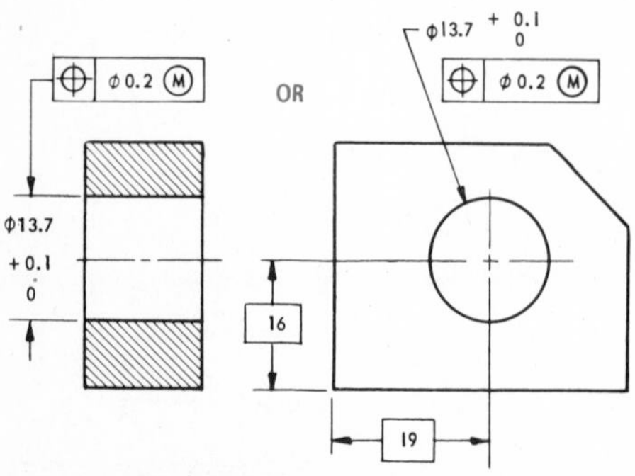

⌖ ⌀0.2 Ⓜ
OR
⌀13.7 +0.1 0
⌀13.7
+0.1
0
16
19

Fig. 31-7-12 Positional tolerancing—MMC.

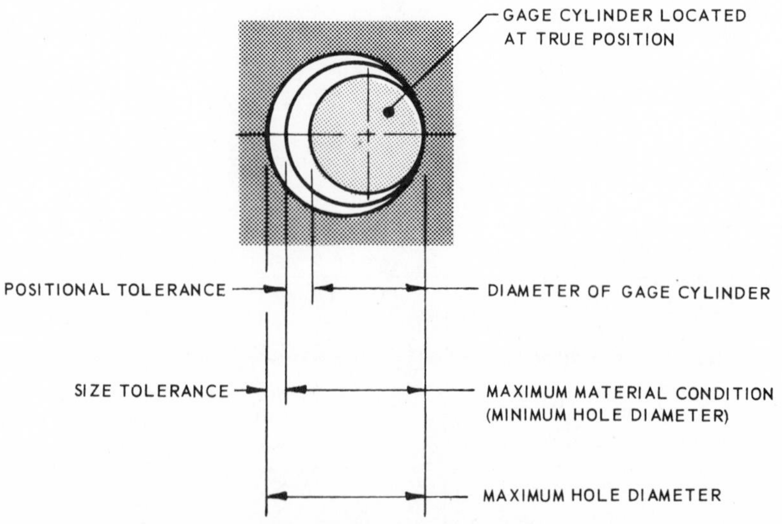

GAGE CYLINDER LOCATED AT TRUE POSITION
POSITIONAL TOLERANCE
DIAMETER OF GAGE CYLINDER
SIZE TOLERANCE
MAXIMUM MATERIAL CONDITION (MINIMUM HOLE DIAMETER)
MAXIMUM HOLE DIAMETER

Fig. 31-7-13 Positional variations for tolerancing—MMC.

tolerancing method except that it results in a round, instead of a square, tolerance zone. The specified tolerance value can therefore be increased by 40 percent without affecting free assembly of mating parts. However, it is probably the least used method since measurement entails the use of mathematics or a chart.

Assignment

On an A3- or B-size sheet, prepare drawings and solutions for the illustrations in Fig. 31-7-A or 31-7-B and the following information.

1. If coordinate tolerances as shown in illustration (A) are given, what are the shape of the tolerance zone and the distance from the specified position to the extreme permissible position?
2. If coordinate tolerances as shown in illustration (B) are given, what are the shape of the tolerance zone and the distance between extreme permissible positions of the hole?
3. In illustration (C) add the largest equal tolerances so that if two such parts are assembled with the edges aligned, the distance between their hole centers could never be more than that shown.
4. If a tolerance shown in illustration (D) is specified for the vertical dimension, what tolerance should be added to the horizontal dimension to meet the same requirement for problem no. 3 mentioned above?
5. In order to assemble correctly, the hole in the part in illustration (E) must not vary from its true position by more than that shown on the drawing when the

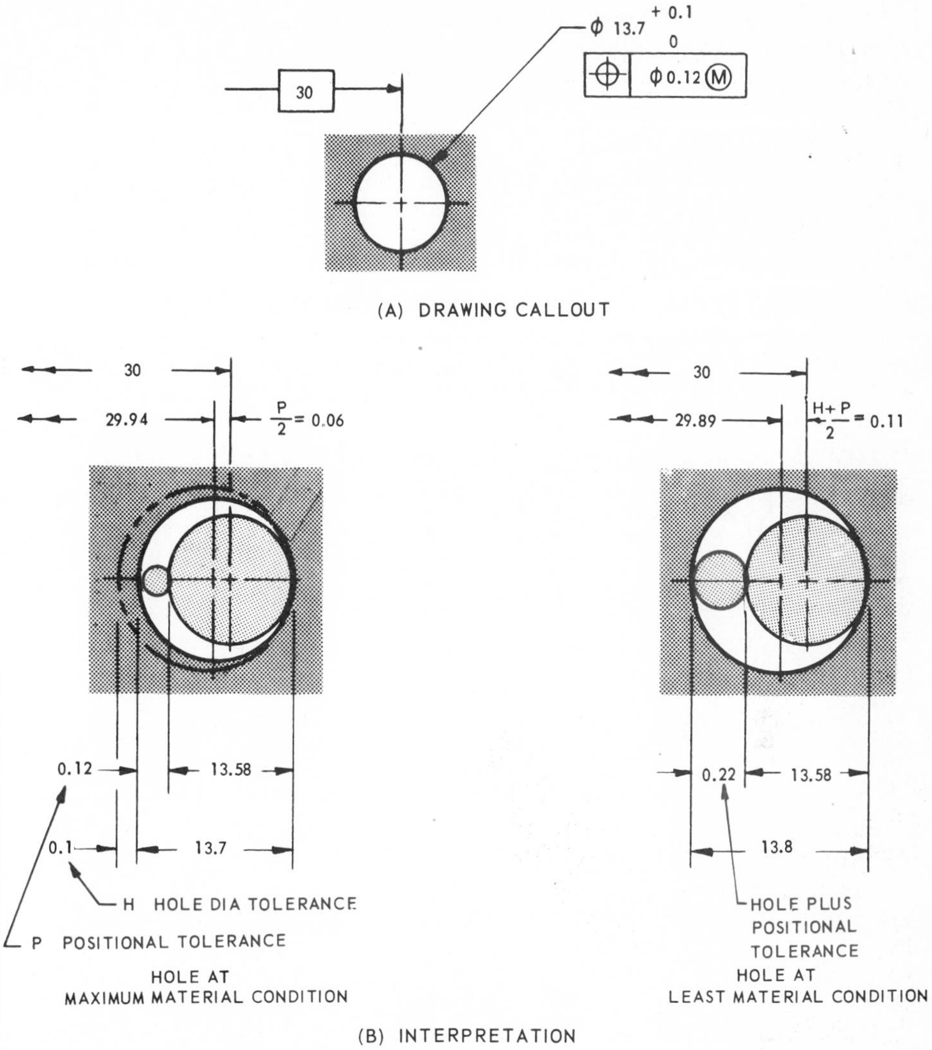

30
ϕ 13.7 +0.1 0
ϕ 0.12 Ⓜ
(A) DRAWING CALLOUT
30
29.94
P/2 = 0.06
0.12
13.58
0.1
13.7
H HOLE DIA TOLERANCE
P POSITIONAL TOLERANCE
HOLE AT
MAXIMUM MATERIAL CONDITION
30
29.89
(H+P)/2 = 0.11
0.22
13.58
13.8
HOLE PLUS
POSITIONAL
TOLERANCE
HOLE AT
LEAST MATERIAL CONDITION
(B) INTERPRETATION

Fig. 31-7-14 Hole with an MMC positional tolerance.

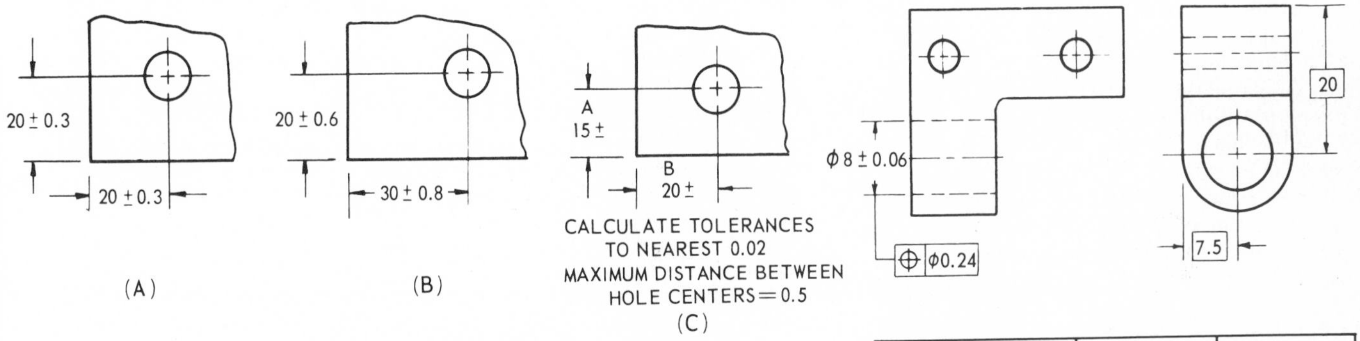

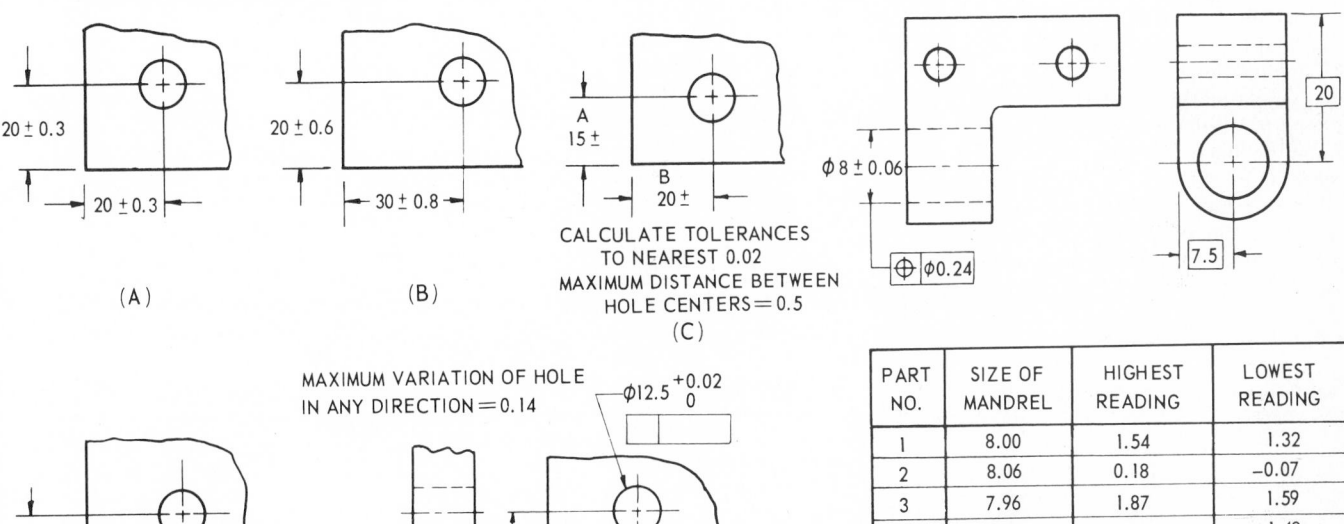

20 ± 0.3
20 ± 0.3
(A)
20 ± 0.6
30 ± 0.8
(B)
A
15 ±
B
20 ±
CALCULATE TOLERANCES
TO NEAREST 0.02
MAXIMUM DISTANCE BETWEEN
HOLE CENTERS = 0.5
(C)
ϕ 8 ± 0.06
ϕ0.24
20
7.5

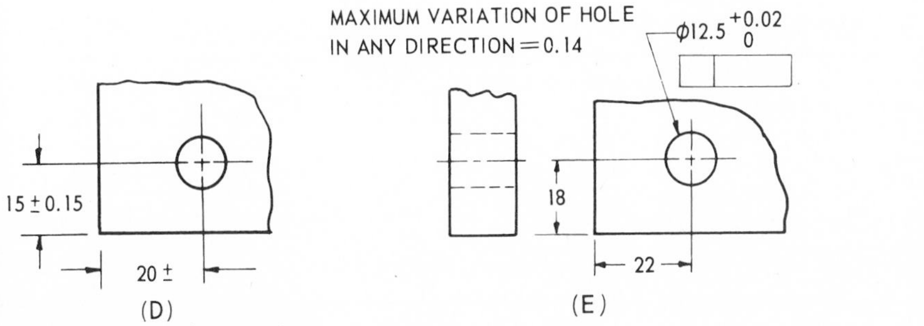

15 ± 0.15
20 ±
(D)
MAXIMUM VARIATION OF HOLE
IN ANY DIRECTION = 0.14
ϕ12.5 +0.02 0
18
22
(E)

PART NO.	SIZE OF MANDREL	HIGHEST READING	LOWEST READING
1	8.00	1.54	1.32
2	8.06	0.18	−0.07
3	7.96	1.87	1.59
4	7.94	1.72	1.48
5	8.00	1.95	1.85
6	8.05	1.24	1.02

(F)

Fig. 31-7-A Assignments.

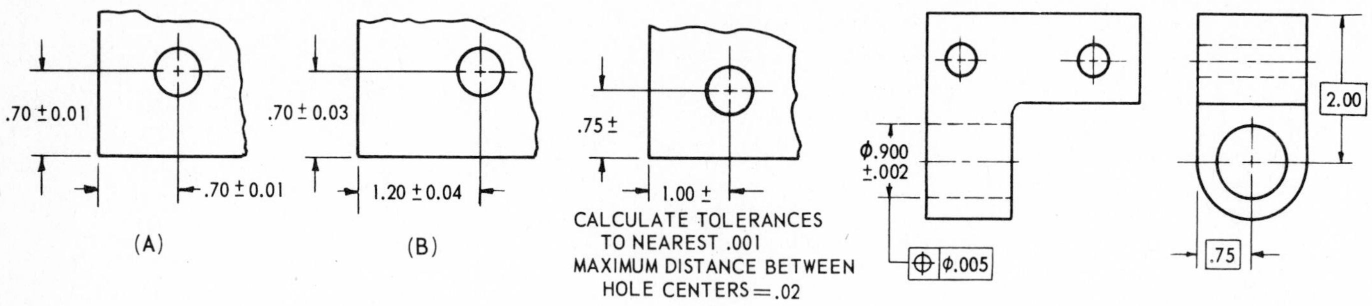

.70 ±0.01
.70 ±0.01
(A)
.70 ± 0.03
1.20 ±0.04
(B)
.75 ±
1.00 ±
CALCULATE TOLERANCES
TO NEAREST .001
MAXIMUM DISTANCE BETWEEN
HOLE CENTERS = .02
(C)
Φ.900
±.002
Φ.005
2.00
.75

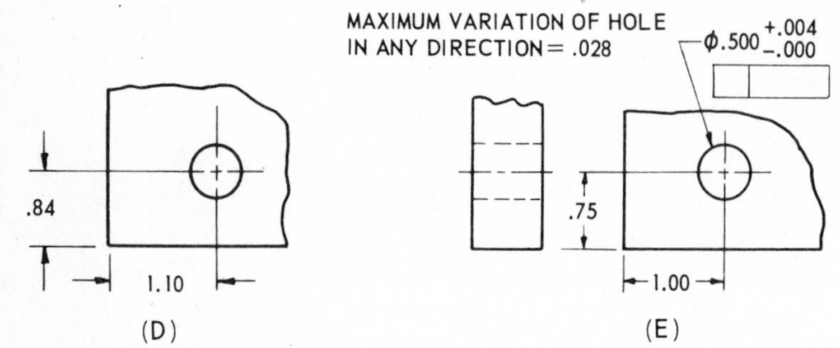

.84
1.10
(D)
MAXIMUM VARIATION OF HOLE
IN ANY DIRECTION = .028
Φ.500 +.004 −.000
.75
1.00
(E)

PART NO.	SIZE OF MANDREL	HIGHEST READING	LOWEST READING
1	.901	1.060	1.052
2	.900	1.007	.998
3	.898	1.074	1.063
4	.899	1.068	1.058
5	.902	1.077	1.073
6	.901	1.049	1.040

(F)

Fig. 31-7-B Assignments.

hole is at its smallest size. Show suitable tolerancing to achieve this:

- By means of coordinate tolerancing
- By positional tolerancing without MMC
- By positional tolerancing on an MMC basis.

What would be the maximum permissible departure from true position if the hole were at its maximum diameter, using positional tolerancing RFS?

What would be the maximum permissible departure from true position with a maximum-diameter hole and a positional tolerance on an MMC basis?

6. The part shown in illustration (F) is set on a revolving table, adjusted so that the part revolves about the true-position center of the large hole. If both indicators give identical readings and the results shown are obtained, which parts are acceptable?

What is the positional error for each part?

UNIT 31-8

DATUMS FOR POSITIONAL TOLERANCING

In the examples given so far, the position of the axis of the hole was established by using dimensions from the actual surfaces of the part. These surfaces were not designated as datums. The true or mean position of the axis was therefore a line parallel to a surface line element on each of the surfaces from which the dimensions were drawn, as illustrated in Fig. 31-8-1.

If these sides are off-square with one another or with other surfaces of the part, the true position of the axis would be similarly off-square, as shown in a somewhat exaggerated format in Fig. 31-8-2.

In some applications this may be the desired requirement, but in most cases it is preferable to have the hole either related to other surfaces or features or related to a full side rather than a line on the surface. It is then necessary to specify the desired datum feature or features in the required order of priority.

The first consideration in such applications is to decide on the primary datum feature. The usual course of action is to specify as the primary datum the surface

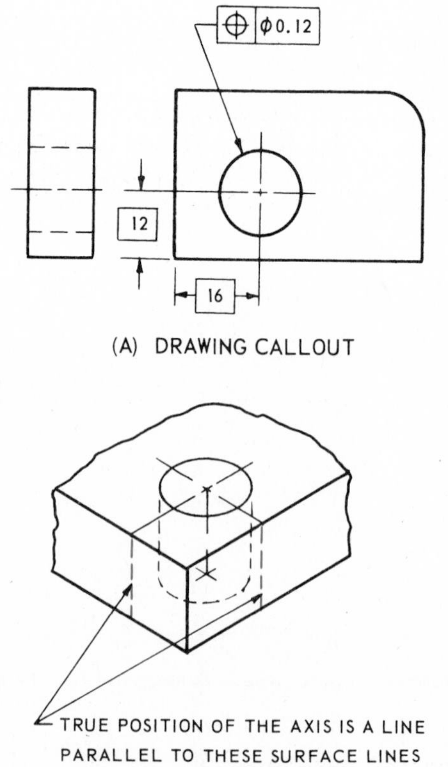

Φ0.12
12
16
(A) DRAWING CALLOUT
TRUE POSITION OF THE AXIS IS A LINE
PARALLEL TO THESE SURFACE LINES
(B) INTERPRETATION OF TRUE POSITION

Fig. 31-8-1 Lines from which measurements are made.

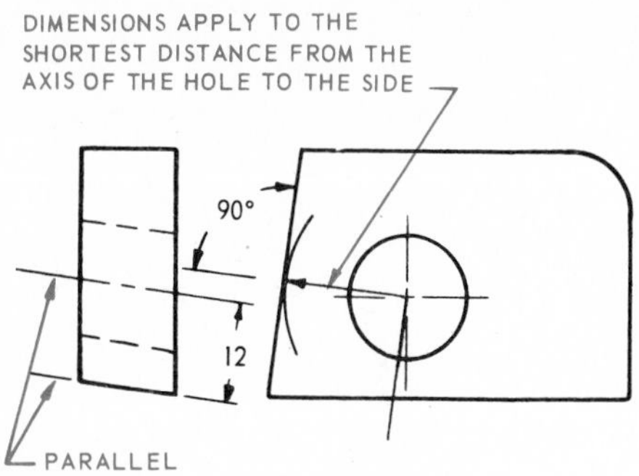

DIMENSIONS APPLY TO THE SHORTEST DISTANCE FROM THE AXIS OF THE HOLE TO THE SIDE
90°
12
PARALLEL

Fig. 31-8-2 Results when sides are off-square.

into which the hole is produced. This will ensure that the true position of the axis is perpendicular to this surface or at the basic angle, if it is other than 90°. Secondary and tertiary datum features are then selected and identified, if required.

Figure 31-8-3 shows a part similar to that shown in Fig. 31-8-1 but with the addition of a primary datum feature and the MMC modifier. Figure 31-8-4 shows the same part with three datum features specified.

Long Holes. It is not always essential to have the true position of a hole perpendicular to the face into which the hole is produced. It may be functionally more important, especially with long holes, to have it parallel to one of the sides. Figure 31-8-5 is a case in point. In this example the sides are designated as primary and secondary datums. A tertiary datum is not required.

Circular Datums. Circular features, such as holes or external cylindrical features, can be used as datums just as readily as flat surfaces. In the simple part shown in Fig. 31-8-6, it is quite evident that the true position of the small hole is established from the axis of the large hole. In cases like these, it may not be necessary to specify one of the holes as a datum, although it would facilitate gaging if the datum were specified on an MMC basis.

In other cases, such as that shown in Fig. 31-8-7, it is essential to specify the

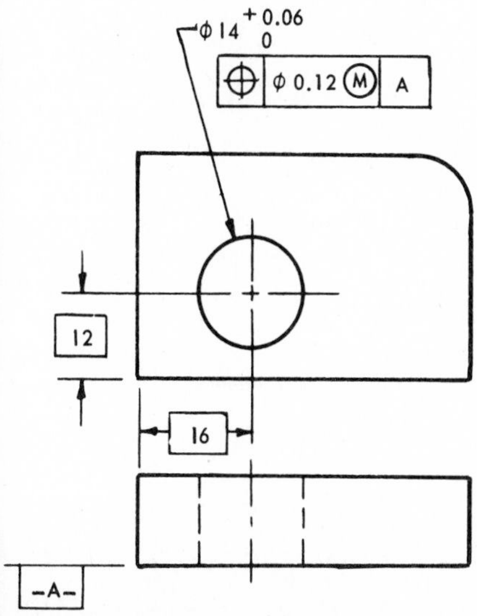

Φ14 +0.06 0
⌖ Φ 0.12 Ⓜ A
12
16
-A-

Fig. 31-8-3 Part with one datum feature specified.

Fig. 31-8-5 Datum system for a long hole.

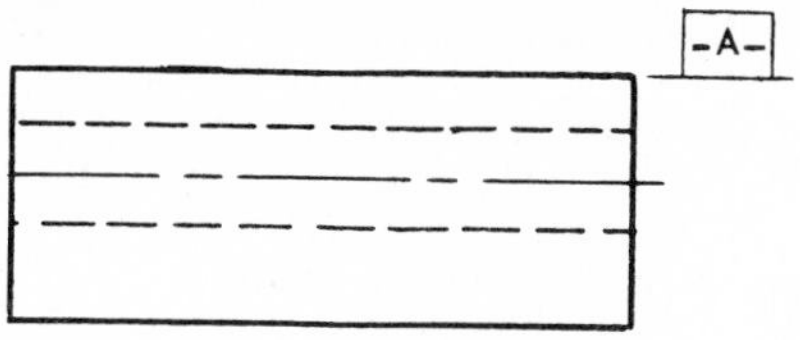

-A-

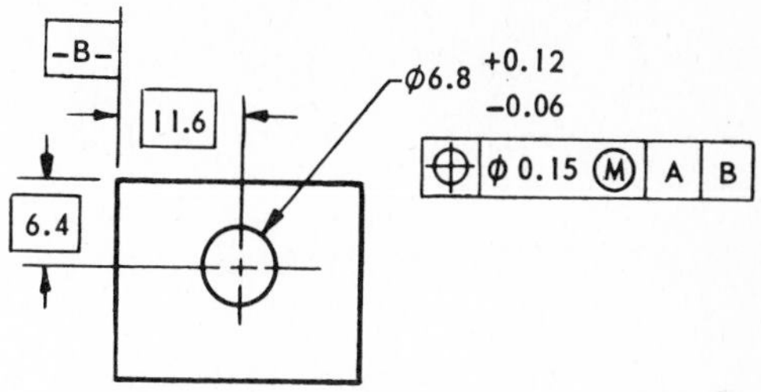

-B-
11.6
6.4
Φ6.8 +0.12 -0.06
⌖ Φ 0.15 Ⓜ A B

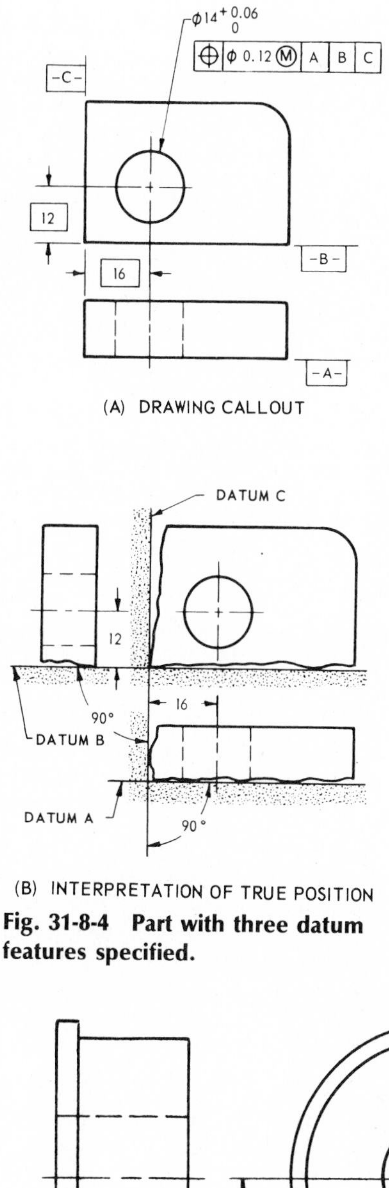

Φ14 +0.06 0
⌖ Φ 0.12 Ⓜ A B C
-C-
12
16
-B-
-A-
(A) DRAWING CALLOUT
DATUM C
12
16
90°
DATUM B
DATUM A
90°
(B) INTERPRETATION OF TRUE POSITION

Fig. 31-8-4 Part with three datum features specified.

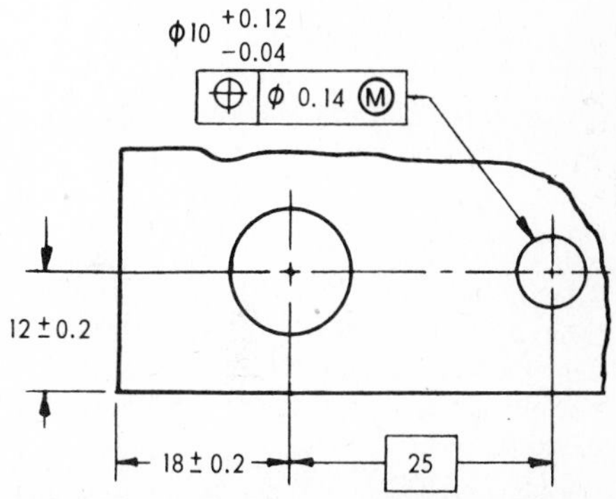

Φ10 +0.12 -0.04
⌖ Φ 0.14 Ⓜ
12 ± 0.2
18 ± 0.2
25

Fig. 31-8-6 Part where specification of the datum feature may not be required.

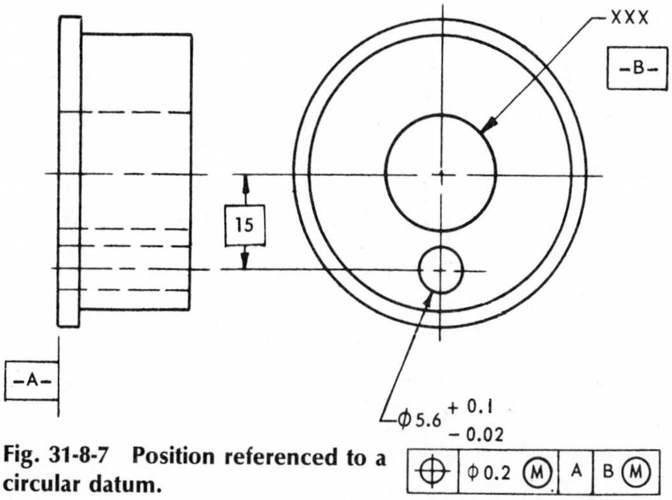

XXX
-B-
15
-A-
Φ5.6 + 0.1 - 0.02
⌖ Φ 0.2 Ⓜ A B Ⓜ

Fig. 31-8-7 Position referenced to a circular datum.

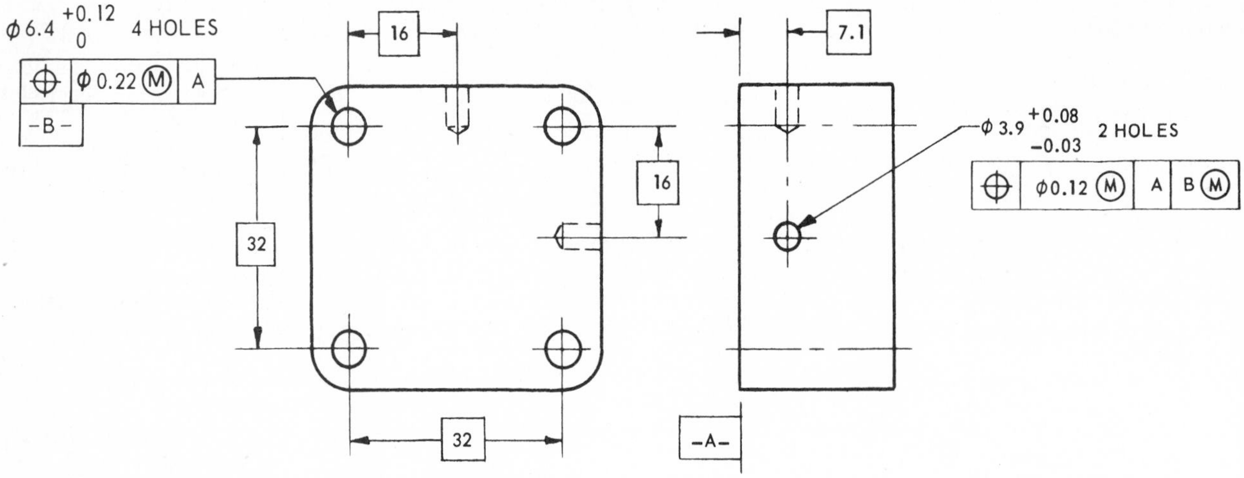
⌀6.4 +0.12 0 4 HOLES
⌀0.22 Ⓜ A
-B-
16
7.1
32
16
32
-A-
⌀3.9 +0.08 −0.03 2 HOLES
⌀0.12 Ⓜ A B Ⓜ

Fig. 31-8-8 Group of holes forming a single datum.

datum; otherwise, the origin of the true-position dimension would be left in doubt. It could be either the axis of the hole or the axis of the outside cylindrical surface.

Multiple-Hole Datums. The axis of holes is sometimes specified as a datum feature with MMC being specified for the datum reference. On an MMC basis, any number of holes or similar features which form a group or pattern may be specified as a single datum. All features forming such a datum must be related with a positional tolerance on an MMC basis. See Fig. 31-8-8.

Datum Targets

The full feature surface was used to establish a datum for the features so far designated as datum features. This may not always be practical for the following reasons.

1. The surface of a feature may be so large that a gage designed to make contact with the full surface may be too expensive or too cumbersome to use.

2. Functional requirements of the part may necessitate the use of only a portion of a surface as a datum feature, for example, the portion which contacts a mating part in assembly.

3. A surface selected as a datum feature may not be sufficiently true, and a flat datum feature may rock when placed on a datum plane, so that accurate and repeatable measurements from the surface would not be possible. This is particularly so for surfaces of castings, forgings, weldments, and some sheet-metal and formed parts. Such parts may be subject to bowing, warping, or other imperfections, on or adjacent to the surface.

A useful technique to overcome such problems is the datum target method. In this method certain points, lines, or small areas on the surfaces are selected as the bases for establishment of datums. For flat surfaces, this usually requires three target points or areas for a primary datum, two for a secondary datum, and one for a tertiary datum. In many cases the use of such target areas will eliminate the need for costly machining which might otherwise be required to produce surfaces suitable for use as datum features.

It is not necessary to use targets for all datums. It is quite logical, for example, to use targets for the primary datum and other surfaces or features for secondary and tertiary datums if required; or to use a flat surface of a part as the primary datum and to indicate fixed points or lines on the edges as secondary and tertiary datums.

Datum targets should be spaced as far apart as possible to provide maximum rigidity for making measurements. At the same time, they must be kept clear of parting lines, risers, rough edges, welds, and other irregularities.

IDENTIFICATION OF TARGETS

Each datum target is shown on a view of the part in its desired location by means of a datum target symbol. These symbols are shown in Fig. 31-8-9.

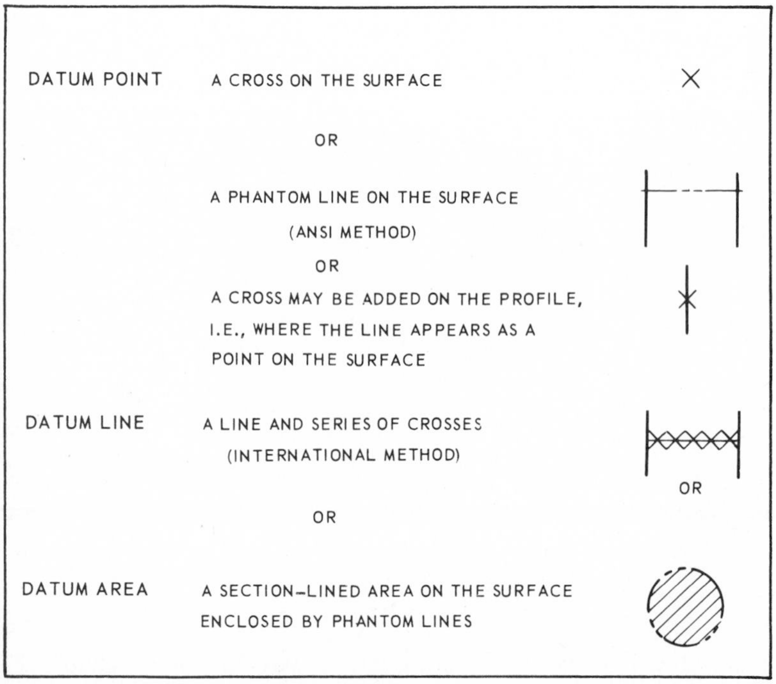
DATUM POINT
A CROSS ON THE SURFACE
OR
A PHANTOM LINE ON THE SURFACE
(ANSI METHOD)
OR
A CROSS MAY BE ADDED ON THE PROFILE,
I.E., WHERE THE LINE APPEARS AS A
POINT ON THE SURFACE
DATUM LINE
A LINE AND SERIES OF CROSSES
(INTERNATIONAL METHOD)
OR
OR
DATUM AREA
A SECTION-LINED AREA ON THE SURFACE
ENCLOSED BY PHANTOM LINES

Fig. 31-8-9 Symbols for datum targets.

Each datum target is then identified by means of a datum target identification symbol. This symbol consists of a circle with a diameter approximately 1.75 times the frame height of the feature control symbols used on the drawing. This circle is divided by a horizontal line, as shown in Fig. 31-8-10. The upper half contains a letter that identifies the datum feature, and the lower half contains a number that identifies that particular target in the datum system. Targets should be numbered consecutively; for example, in a three-plane, six-point datum system, if the datums are A, B, and C, the datum targets would be A_1, A_2, A_3, B_4, B_5, and C_6.

Each datum target identification symbol is connected to its datum target by a leader which ends with a dot or arrowhead, except when the symbol is attached to an extension line.

The datum surface may also be identified in the normal manner.

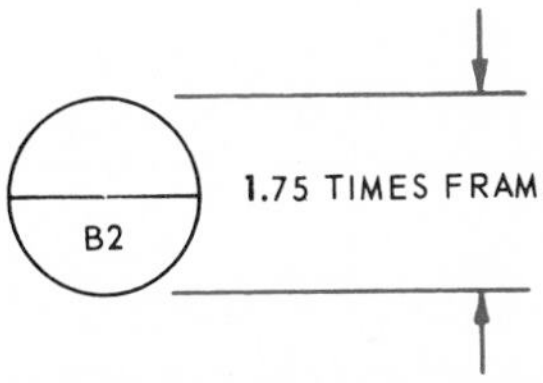
1.75 TIMES FRAME HEIGHT
B2

Fig. 31-8-10 Datum target identification symbol.

DIMENSIONING FOR TARGET LOCATION

The location of datum targets and the size of target areas are shown by means of datum dimensions. Each dimension is shown, without tolerances, enclosed in a rectangular frame, indicating that the general tolerance does not apply. Dimensions locating a set of datum targets should be dimensionally related or have a common origin.

The extension lines for these dimensions may also serve as leader lines for the target identification symbols.

TARGET POINTS

Each target point is shown on the surface, in its desired location, by means of a cross, drawn at approximately 45° to the coordinate dimensions. The cross consists of lines equal in length to the height of the lettering used, as shown in Fig. 31-8-11. Figure 31-8-12 illustrates two such points as they would appear on a drawing.

Target points may be represented on tools, fixtures, and gages by spherically ended pins, as shown in Fig. 31-8-13.

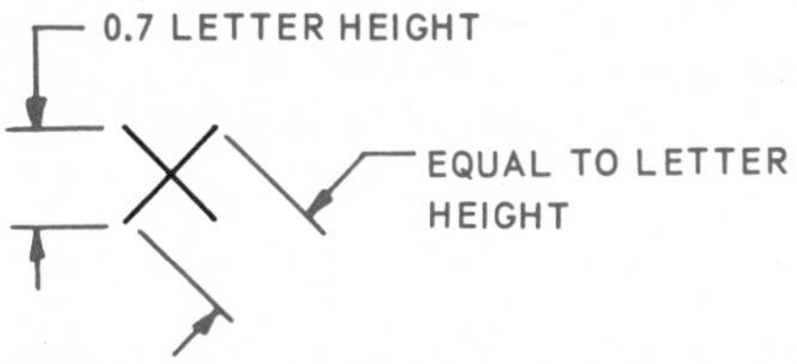
0.7 LETTER HEIGHT
EQUAL TO LETTER HEIGHT

Fig. 31-8-11 Symbol for a datum target point.

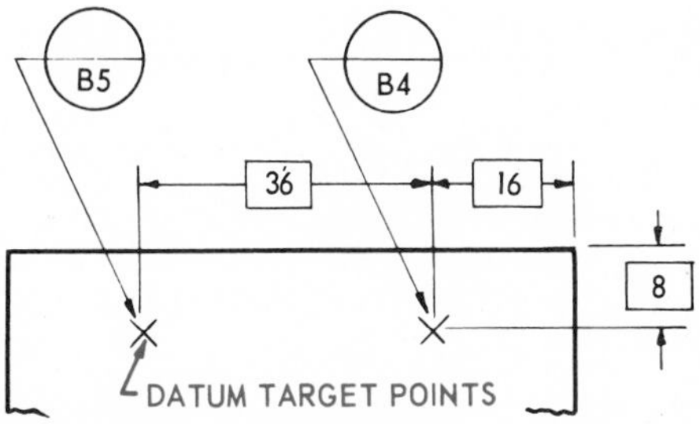
B5
B4
36
16
8
DATUM TARGET POINTS

Fig. 31-8-12 Datum target points on a drawing.

TARGETS NOT IN THE SAME PLANE

In most applications datum target points which form a single datum are all located on the same surface, as shown in Fig. 31-8-12. However, this is not essential. They may be located on different surfaces, to meet functional requirements, as shown, for example, in Fig. 31-8-14. In some cases the datum plane may be located in space, that is, not actually touching the part, as shown in Fig. 31-8-15. In such applications the controlled features must be dimensioned from the specified datum, and the position of the datum from the datum targets must be shown by means of exact datum dimensions. For example, in Fig. 31-8-15 datum B is positioned by means of datum dimensions 9, 12, and 28. The top surface is controlled from this datum by means of a toleranced dimension, and the hole is positioned by means of a true-position dimension [26] and a positional tolerance.

TARGET LINES

When a datum feature consists of a line on a surface, it is shown on the drawing by a series of crosses, as if it consisted of a series of target points, with a line drawn through the crosses or by a phantom line, as shown in Fig. 31-8-16.

A datum line on the edge of a part may also be shown on the plan view, by means of a single cross drawn on the line representing the datum edge, as shown in Fig. 31-8-17.

Figure 31-8-18 is interpreted to mean that the true position of the hole axis is parallel to datum A, rather than perpendicular to the face of the part. For gaging purposes the part is supported in contact with three gage pins, or similar surfaces, as shown in Fig. 31-8-19.

DATUM TARGET AREAS

Datum target areas are indicated by drawing the datum target boundary on the plan view in phantom lines and section-lining the area, as shown in Fig.

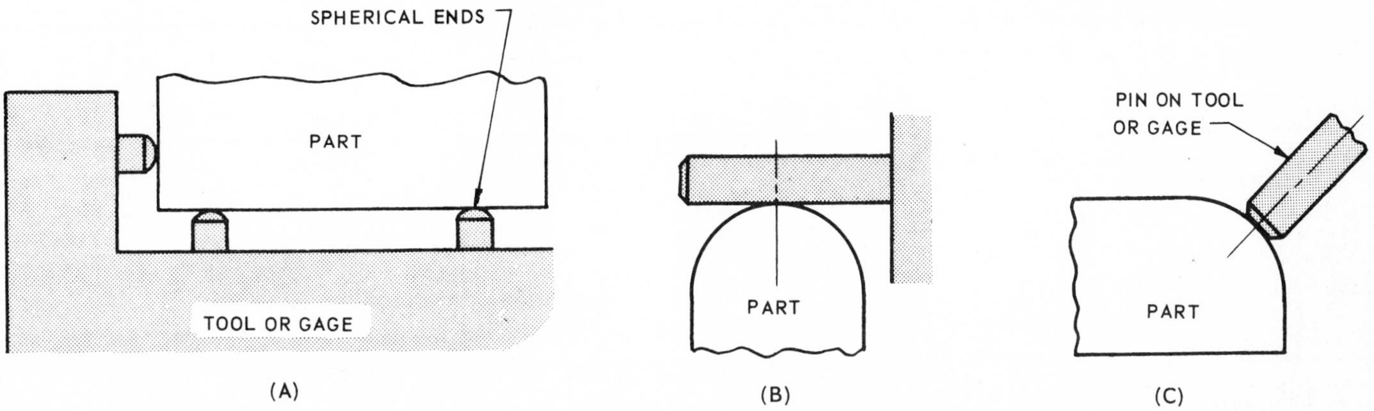
SPHERICAL ENDS
PART
TOOL OR GAGE
(A)
PART
(B)
PIN ON TOOL OR GAGE
PART
(C)

Fig. 31-8-13 Location of part on datum target points.

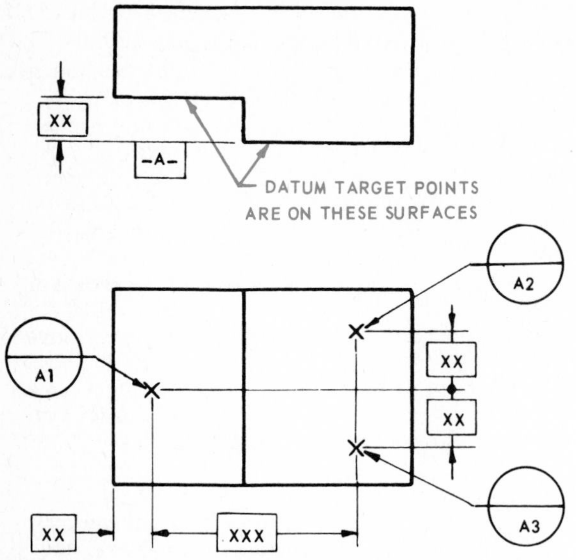

XX
-A-
DATUM TARGET POINTS
ARE ON THESE SURFACES
A1
A2
A3
XX
XX
XX
XXX

Fig. 31-8-14 **Datum target points on different planes.**

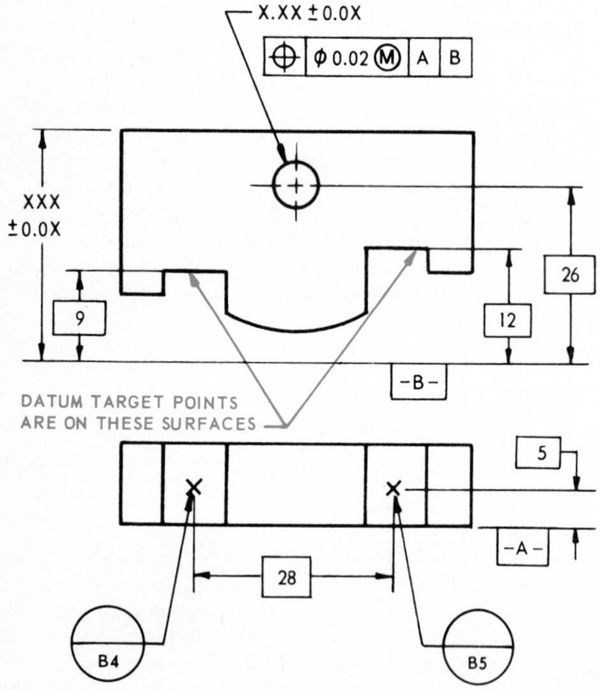

X.XX ± 0.0X
⊕ ϕ 0.02 Ⓜ A B
XXX
±0.0X
9
12
26
-B-
DATUM TARGET POINTS
ARE ON THESE SURFACES
5
-A-
28
B4
B5

Fig. 31-8-15 **Datum outside of part profile.**

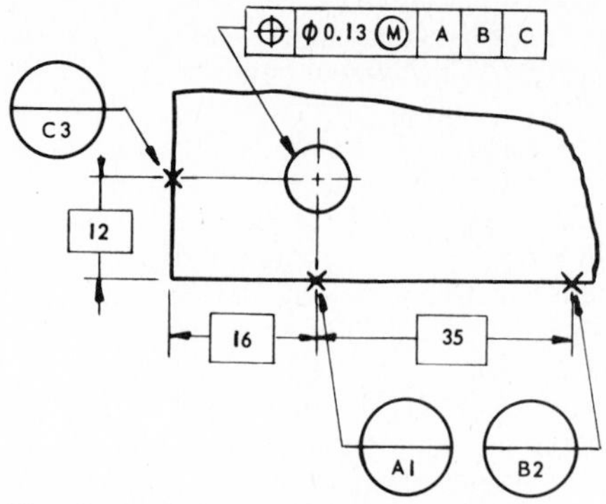

⊕ ϕ0.13 Ⓜ A B C
C3
12
16
35
A1
B2

Fig. 31-8-18 **Part with three datum lines.**

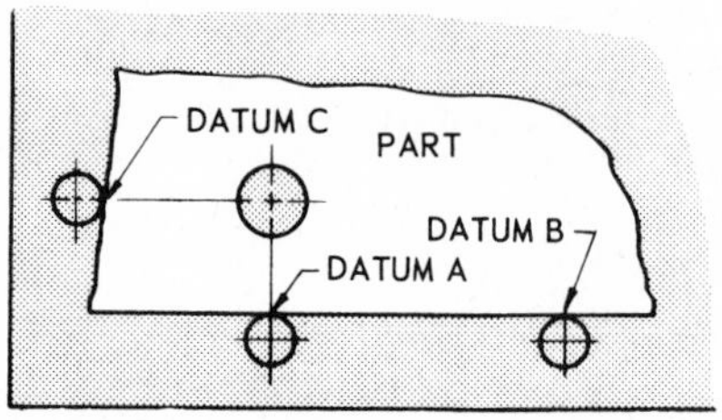

DATUM C
PART
DATUM B
DATUM A

Fig. 31-8-19 **Location of part in Fig. 31-8-18 in a gage.**

31-8-20. The datum target symbol is drawn on an extension line or connected to the boundary of the area by a leader with an arrowhead. The size of the target areas may be placed adjacent to the target symbol or be covered by a suitable note, such as TARGET AREAS ϕ8.

Datum target areas may have any desired shape, a few of which are shown in Fig. 31-8-21. Target areas should be kept as small as possible, consistent with functional requirements, to avoid having large, section-lined areas on the drawing.

Assignments

1. On an A3- or B-size sheet, make a two-view drawing of the part shown in

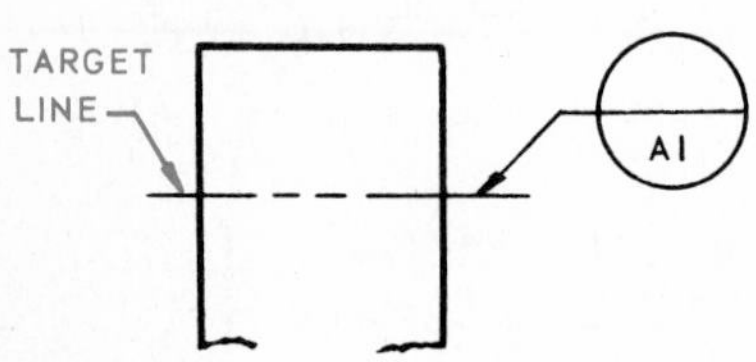

TARGET
LINE
A1

Fig. 31-8-16 **Target line on surface.**

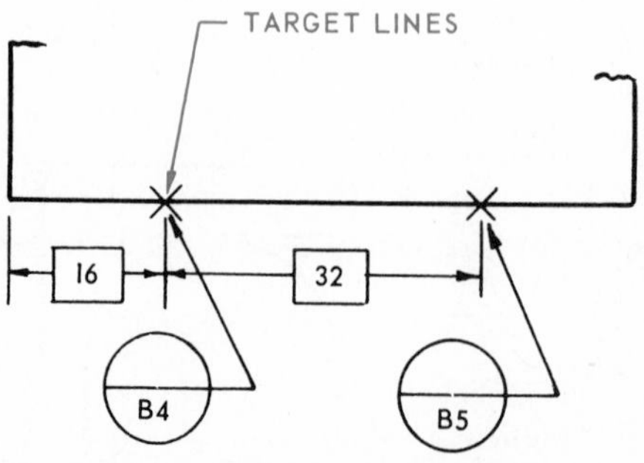

TARGET LINES
16
32
B4
B5

Fig. 31-8-17 **Target lines on edge of part.**

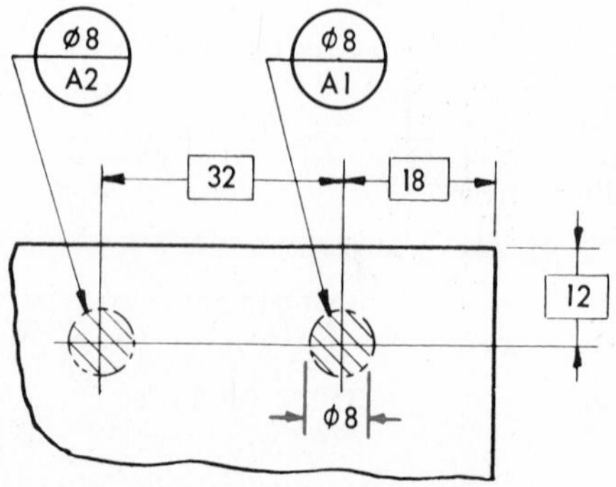

ϕ8
A2
ϕ8
A1
32
18
12
ϕ8

Fig. 31-8-20 **Datum target areas.**

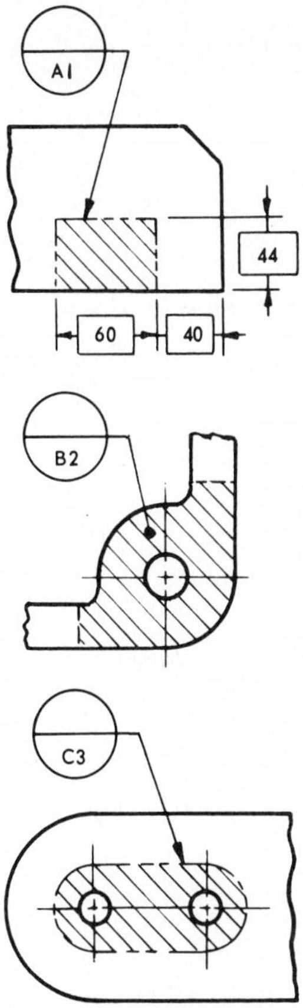
A1
44
60
40
B2
C3

Fig. 31-8-21 Typical target areas.

Fig. 31-8-A or 31-8-B showing the datum features. Only the dimensions related to the datums need be shown. Scale is 1:1. Datum information is as follows:

- *Primary datum* (three points). Points A_1 and A_2 are located on center of surface M, one-fifth the height distance from the top and bottom, respectively. Point A_3 is located on the center of surface N midway between the center of hole and the right end.
- *Secondary datum* is a datum line located on the center of surface D.
- *Tertiary datum* is a datum point located on the center of surface E.

2. On an A3- or B-size sheet, make a three-view drawing of the bearing housing shown in Fig. 31-8-C or 31-8-D showing the datum features. Only the dimensions related to the datums need be shown. Scale is 1:2.

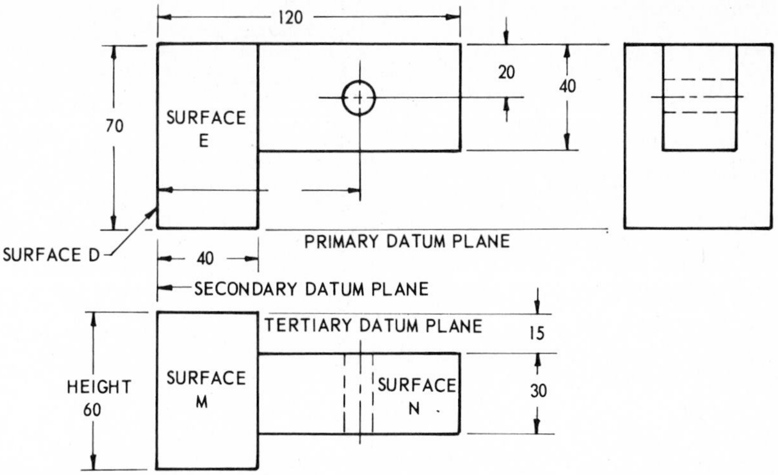
120
70
SURFACE E
20
40
SURFACE D
40
PRIMARY DATUM PLANE
SECONDARY DATUM PLANE
TERTIARY DATUM PLANE
15
HEIGHT 60
SURFACE M
SURFACE N
30

Fig. 31-8-A Assignments.

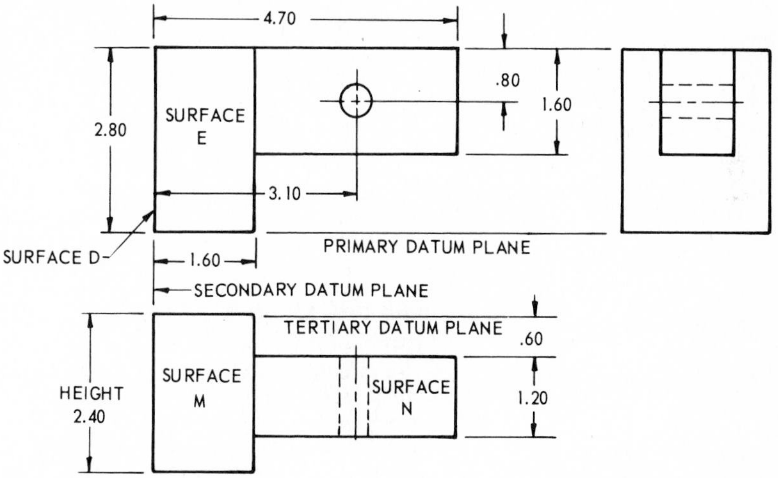
4.70
2.80
SURFACE E
.80
1.60
3.10
SURFACE D
1.60
PRIMARY DATUM PLANE
SECONDARY DATUM PLANE
TERTIARY DATUM PLANE
.60
HEIGHT 2.40
SURFACE M
SURFACE N
1.20

Fig. 31-8-B Assignments.

ROUNDS & FILLETS R4

DATUM AND LOCATION				
DATUM DESCRIPTION		LOCATION FROM		
		PRIMARY DATUM PLANE	SECONDARY DATUM PLANE	TERTIARY DATUM PLANE
DATUM A TARGET AREAS ϕ16	A1		14	30
	A2		14	130
	A3		94	80
DATUM B TARGET LINES	B4			20
	B5			140
DATUM C TARGET POINT	C6	20	44	

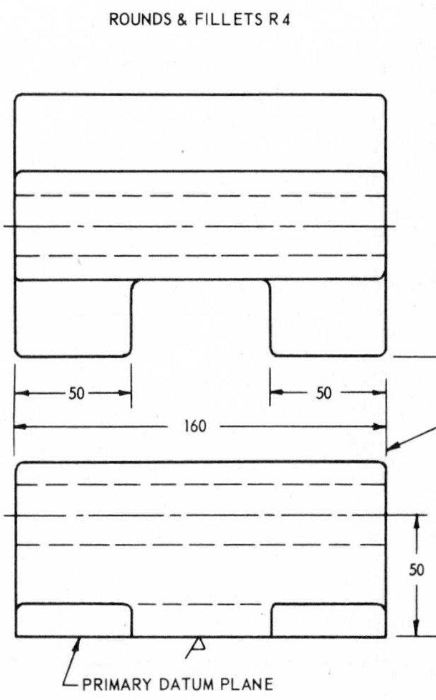
50
50
160
50
14
PRIMARY DATUM PLANE

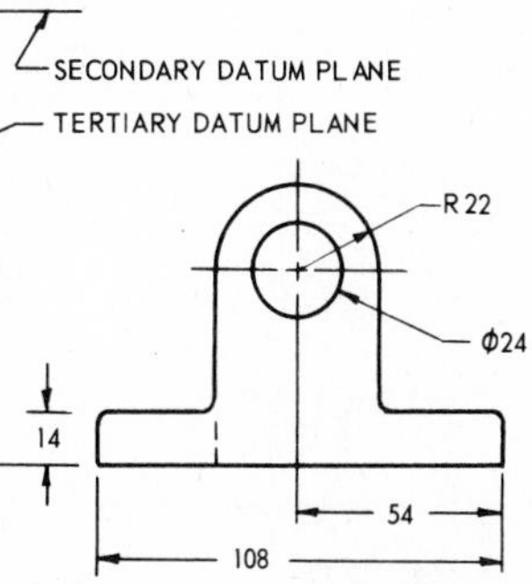
SECONDARY DATUM PLANE
TERTIARY DATUM PLANE
R22
ϕ24
54
108

Fig. 31-8-C Assignments.

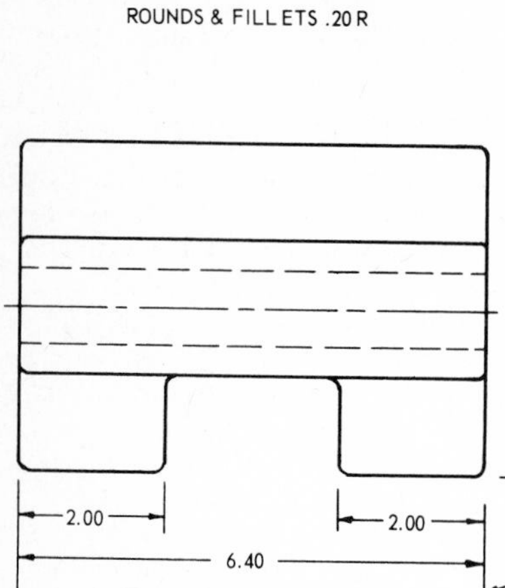
ROUNDS & FILLETS .20 R
2.00
2.00
6.40

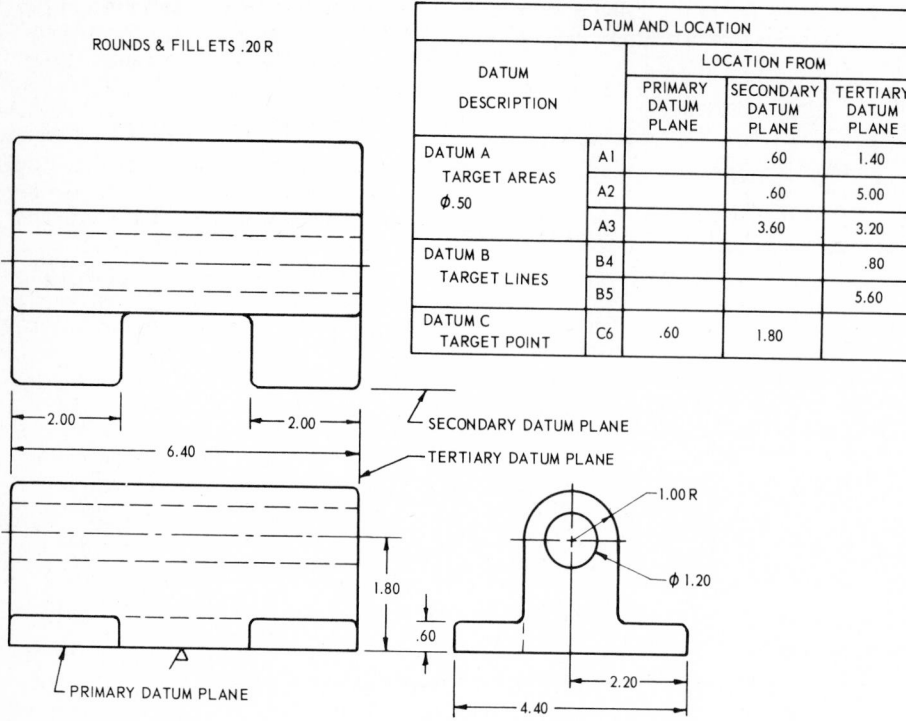

DATUM DESCRIPTION		DATUM AND LOCATION		
		LOCATION FROM		
		PRIMARY DATUM PLANE	SECONDARY DATUM PLANE	TERTIARY DATUM PLANE
DATUM A TARGET AREAS ϕ.50	A1		.60	1.40
	A2		.60	5.00
	A3		3.60	3.20
DATUM B TARGET LINES	B4			.80
	B5			5.60
DATUM C TARGET POINT	C6	.60	1.80	

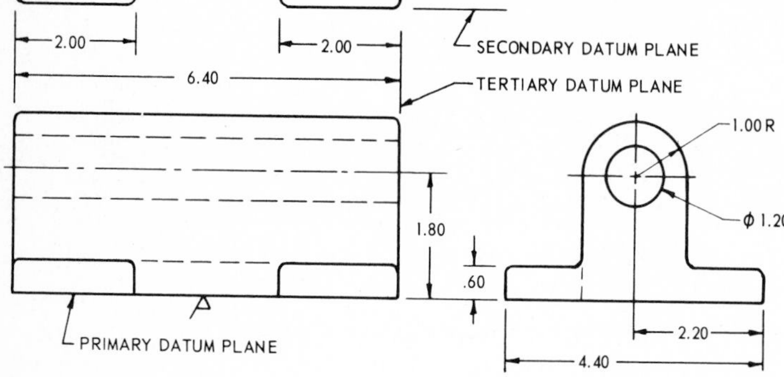
SECONDARY DATUM PLANE
TERTIARY DATUM PLANE
1.00 R
ϕ 1.20
1.80
.60
2.20
4.40
PRIMARY DATUM PLANE

Fig. 31-8-D Assignments.

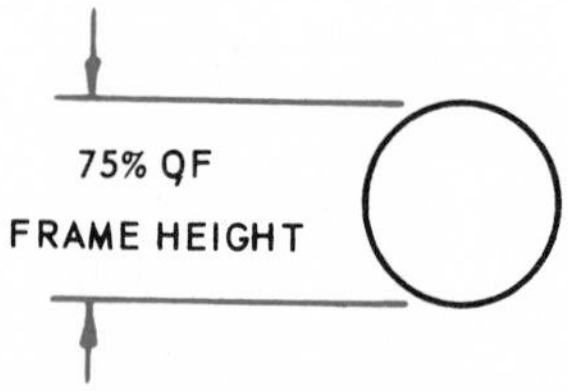
75% OF
FRAME HEIGHT

Fig. 31-9-2 Roundness symbol.

A roundness tolerance specifies the width of an annular tolerance zone, bounded by two concentric circles in the same plane, within which the circular line or the periphery of the feature in that plane shall lie, as shown in Fig. 31-9-3. A roundness tolerance cannot be modified on an MMC basis since it controls surface elements only.

Roundness of Noncylindrical Parts. Noncylindrical parts refer to conical parts and other features which are circular in cross section but which have variable diameters, such as those shown in Fig. 31-9-4. Since many sizes of circles may be involved, it is usually best to direct the roundness tolerance to the longitudinal surfaces as shown.

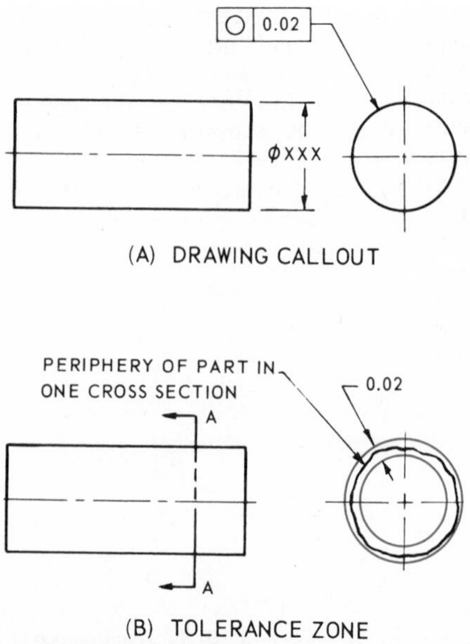
0.02
ϕXXX
(A) DRAWING CALLOUT
PERIPHERY OF PART IN ONE CROSS SECTION
0.02
A
A
(B) TOLERANCE ZONE

Fig. 31-9-3 Roundness tolerance.

UNIT 31-9

ROUNDNESS AND CYLINDRICITY

Roundness

Roundness refers to a condition of a circular line or the surface of a circular feature where all points on the line, or on the periphery of a plane cross section of the feature, are equidistant from a common center point.

Examples of circular features would include disks, spheres, cylinders, and cones. The measurement plane for a sphere is any plane which passes through a section of maximum diameter. For a cylinder, cone, or other nonspherical feature, the measurement plane is any plane perpendicular to the axis or center line.

ERRORS OF ROUNDNESS

Errors of roundness (out-of-roundness) of a circular line or the periphery of a cross section of a circular feature may occur as *ovality*, where differences appear between the major and minor axes; as *lobing*, where the diametral values may be constant or nearly so; or as *random irregularities* from a true circle. All these errors are illustrated in Fig. 31-9-1. The geometric characteristic symbol for roundness is simply a circle, having a diameter 75 percent of the feature control symbol frame height, as shown in Fig. 31-9-2.

ROUNDNESS TOLERANCE—RFS

A roundness tolerance may be specified by using this symbol in the feature control symbol. It is expressed on an RFS basis.

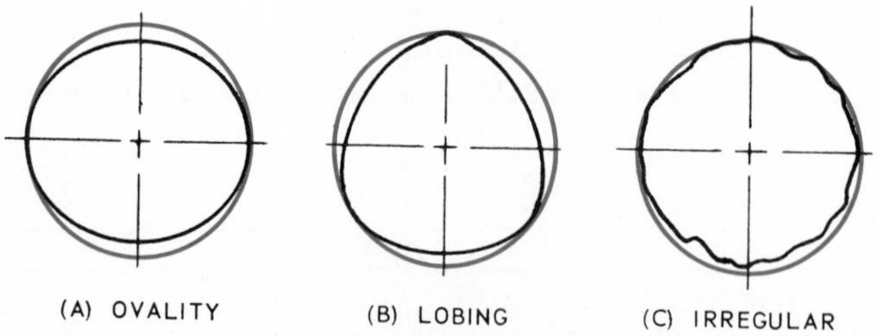
(A) OVALITY
(B) LOBING
(C) IRREGULAR

Fig. 31-9-1 Common types of roundness errors.

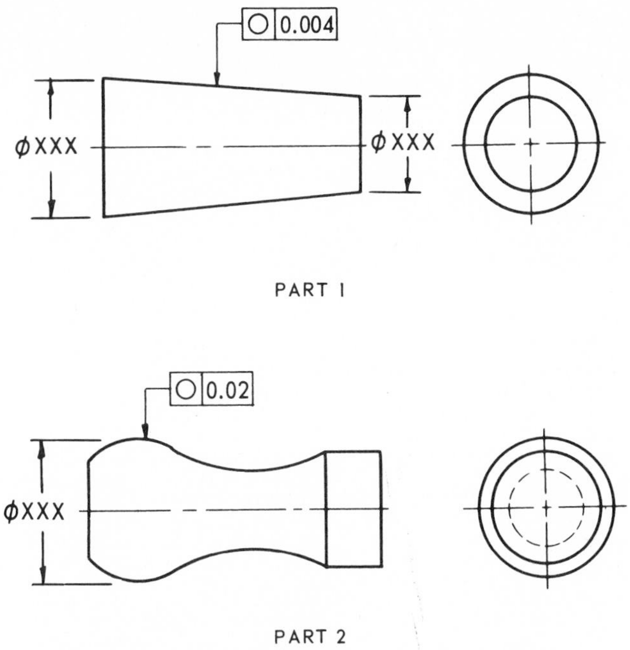
0.004
ϕXXX
ϕXXX
PART 1
0.02
ϕXXX
PART 2

Fig. 31-9-4 Roundness tolerance on noncylindrical parts.

Cylindricity

Cylindricity refers to a condition of a surface which forms a cylinder where the surface elements in cross sections parallel to the axis are straight and parallel and in cross sections perpendicular to the axis are round. Cylindricity thus combines in one term geometric form tolerances for roundness, straightness, and parallelism of the surface elements.

Cylindricity tolerances can be applied only to cylindrical surfaces, such as round holes and shafts. No specific geometric tolerances have been devised for other circular forms, which require the use of several geometric tolerances. A conical surface, for example, must be controlled by a combination of tolerances for roundness, straightness, and angularity.

Errors of cylindricity may be caused by out-of-roundness, like ovality or lobing, by errors of straightness caused by bending or by diametral variation, by errors of parallelism like conicity or taper, and by random irregularities from a true cylindrical form.

The geometric characteristic symbol for cylindricity consists of a circle with two tangent lines at 60°, as shown in Fig. 31-9-5. The circle diameter is 62 percent of the height of the feature control symbol frame.

CYLINDRICITY TOLERANCE—RFS

The cylindricity tolerance symbol is used in a feature control symbol in the usual manner, and it is directed to the cylindrical surface, in either the side or end view, as shown in Fig. 31-9-6.

Since cylindricity is a form tolerance controlling surface elements only, it cannot be modified on an MMC basis.

Assignment

On an A3- or B-size sheet, prepare drawings and solutions for the illustrations shown in Fig. 31-9-A or 31-9-B and the following information.

1. Sketch the tolerance zone for the roundness tolerance in illustration (A). If measurements made at cross sections *AA*, *BB*, and *CC*, indicate that all points on the periphery fall within the annular rings shown, would you conclude that the part met the specified roundness tolerance? If not, which cross section is not acceptable?

2. Add roundness tolerances to the diameters shown in illustration (B). The roundness tolerances are to be one-fifth of the size tolerances for each diameter.

3. Show on each part in illustration (C) a cylindrical tolerance. The size of the cylindrical tolerance is to equal one-quarter the size tolerance for each diameter.

4. Sketch the tolerance zone for the cylindrical tolerance in illustration (D), indicating its size and shape for a part shown.

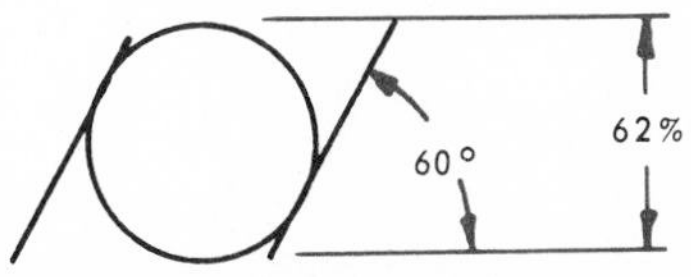
60°
62%

Fig. 31-9-5 Cylindricity symbol.

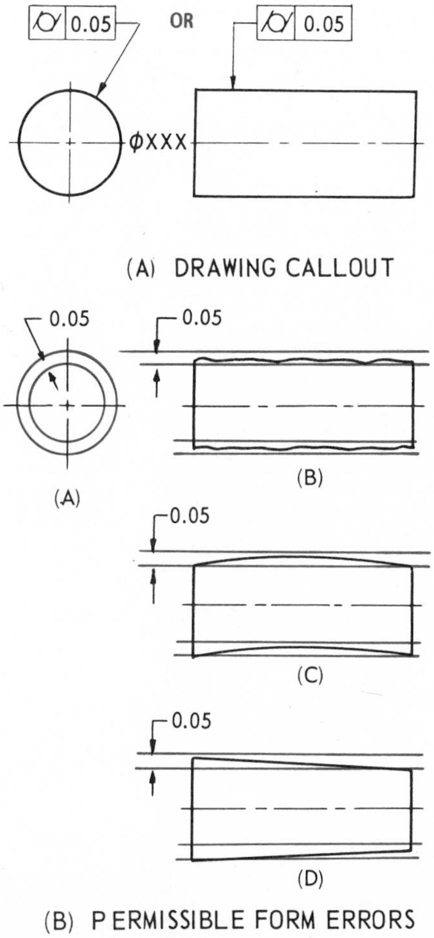
0.05
OR
0.05
ϕXXX
(A) DRAWING CALLOUT
0.05
0.05
(A)
(B)
0.05
(C)
0.05
(D)
(B) PERMISSIBLE FORM ERRORS

Fig. 31-9-6 Cylindricity tolerance directed to either view.

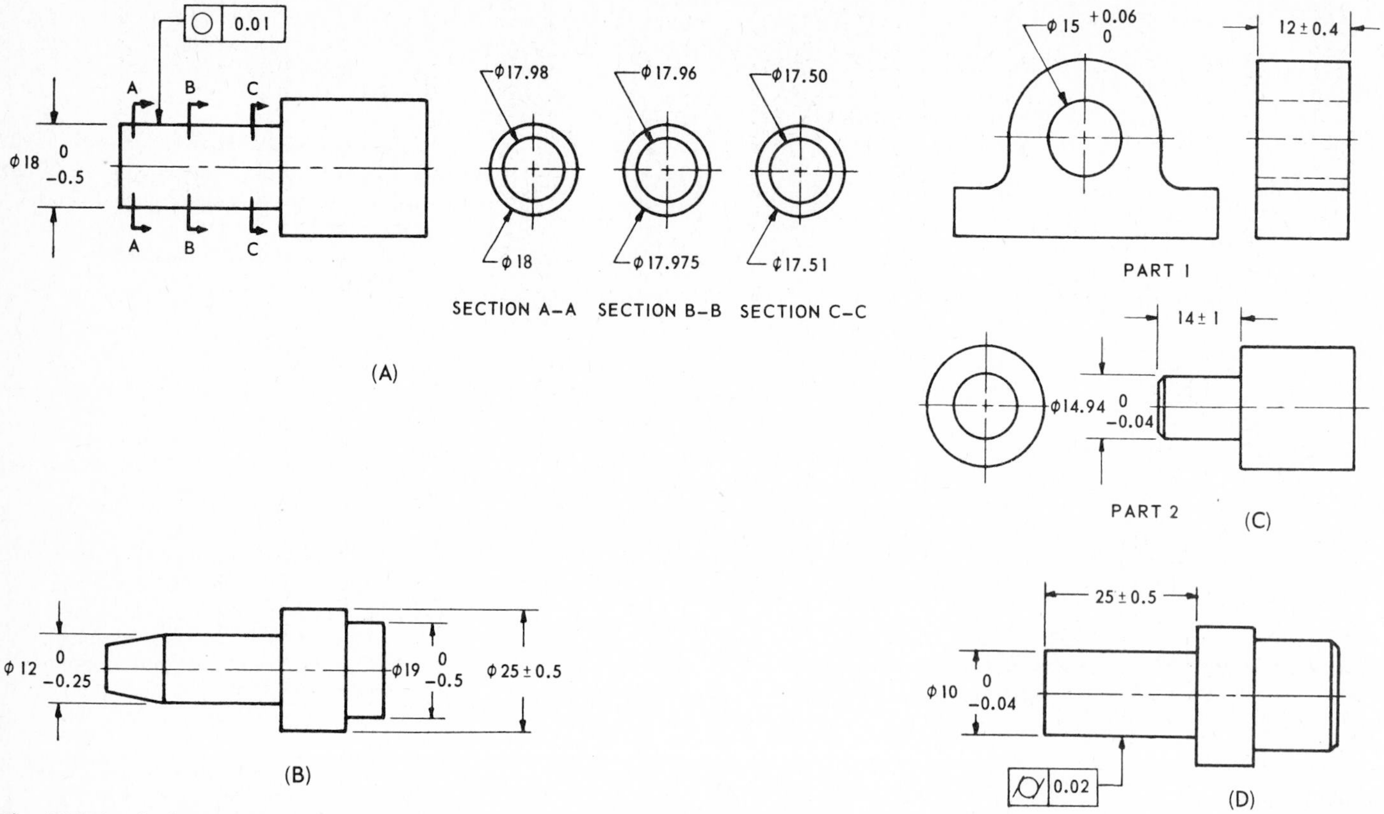
0.01
A B C
A B C
Φ18 0 −0.5
Φ17.98
Φ18
Φ17.96
Φ17.975
Φ17.50
Φ17.51
SECTION A–A
SECTION B–B
SECTION C–C
(A)
Φ15 +0.06 0
12 ± 0.4
PART 1
14 ± 1
Φ14.94 0 −0.04
PART 2
(C)
Φ12 0 −0.25
Φ19 0 −0.5
Φ25 ± 0.5
(B)
25 ± 0.5
Φ10 0 −0.04
0.02
(D)

Fig. 31-9-A Assignments.

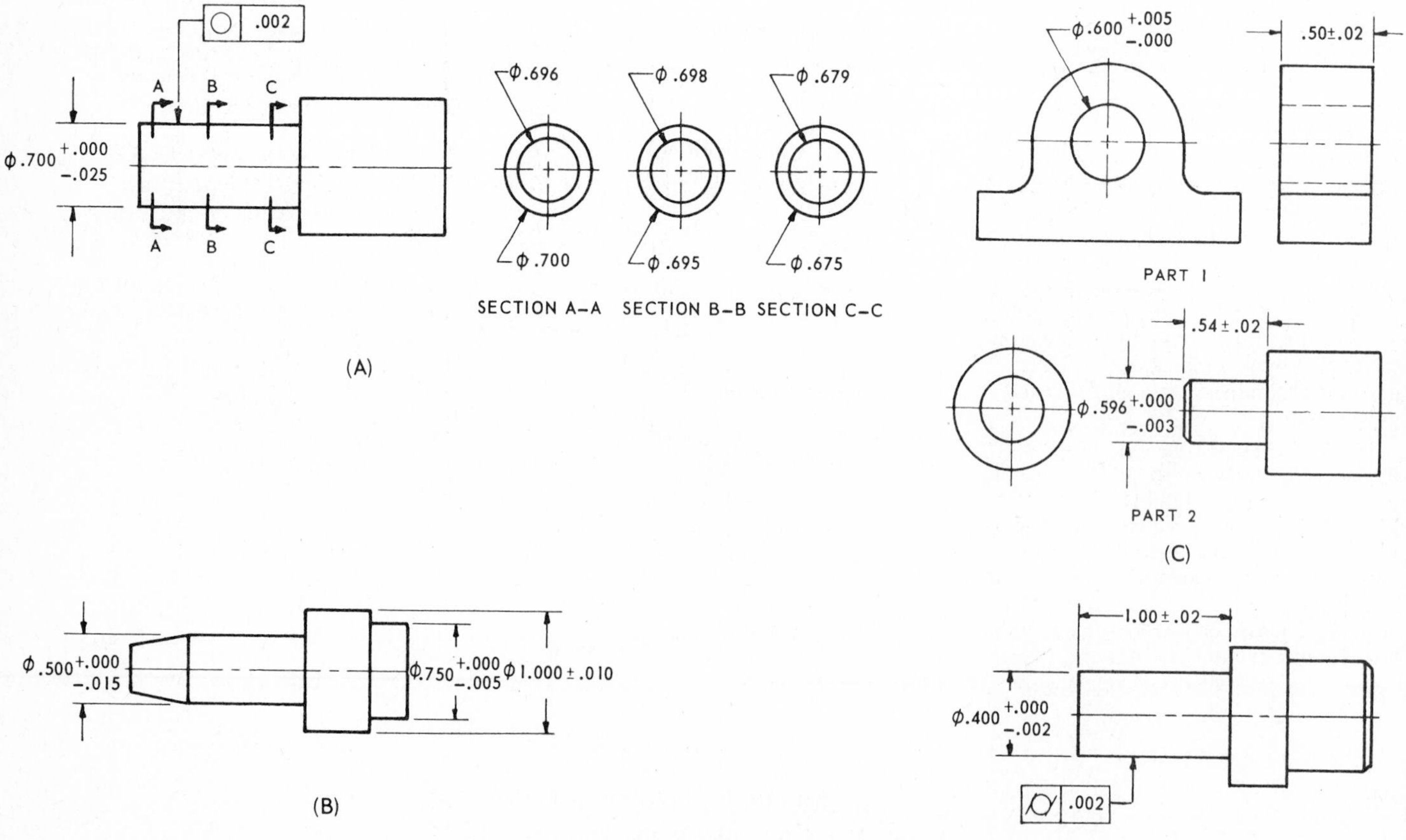
.002
A B C
A B C
Φ.700 +.000 −.025
Φ.696
Φ.700
Φ.698
Φ.695
Φ.679
Φ.675
SECTION A–A
SECTION B–B
SECTION C–C
(A)
Φ.600 +.005 −.000
.50±.02
PART 1
.54 ± .02
Φ.596 +.000 −.003
PART 2
(C)
Φ.500 +.000 −.015
Φ.750 +.000 −.005
Φ1.000 ± .010
(B)
1.00 ± .02
Φ.400 +.000 −.002
.002

Fig. 31-9-B Assignments.

(D)

UNIT 31-10

PROFILE TOLERANCING

Profiles

A *profile* is the outline form or shape of a line or surface. A *line profile* may be the outline of a part or feature as depicted in a view on a drawing. It may represent the edge of a part, or it may refer to line elements of a surface in a single direction, such as the outline of cross sections through the part. In contrast, a *surface profile* outlines the form or shape of a complete surface.

The elements of a line profile may be straight lines, arcs, or other curved lines. The elements of a surface profile may be flat surfaces, spherical surfaces, cylindrical surfaces, or surfaces composed of various line profiles in two or more directions.

A profile tolerance may be applied to a single, independent line or surface, or any part of a complex line or surface. It may be applied to a feature of size, in which case MMC is specified. It may be related to a datum, in which case it also controls orientation and, in some cases, position of the line or surface.

Profile Symbols

There are two geometric characteristic symbols for profiles, one for lines and one for surfaces. Separate symbols are required, because it is often necessary to distinguish between line elements of a surface and the complete surface itself. The symbol for profile of a line consists of a semicircle with a diameter equal to the height of the feature control symbol frame. The symbol for profile of a surface is identical except that the semicircle is closed by a straight line at the bottom, as shown in Fig. 31-10-1. All other geometric tolerances of form and orientation are merely special cases of profile tolerancing.

Profile tolerances are used to control the position of lines and surfaces which are neither flat nor cylindrical.

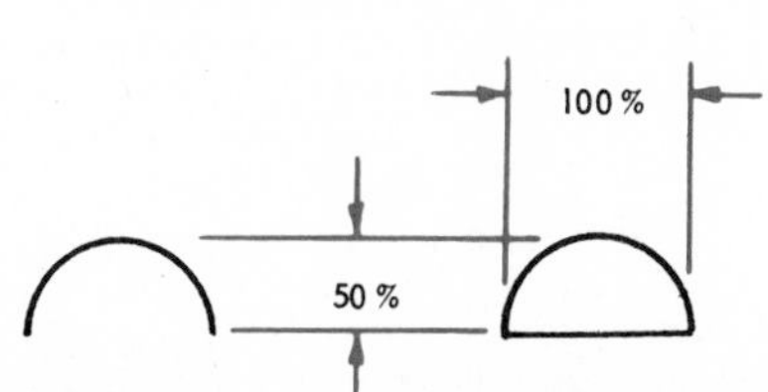
100 %
50 %
PROFILE OF
A LINE
PROFILE OF
A SURFACE

Fig. 31-10-1 Profile symbols.

Profile of a Line

A profile-of-a-line tolerance may be directed to a line of any length or shape. If a datum is not referenced and the line is intended to be straight, the profile tolerance is identical to a straightness tolerance in meaning and interpretation. If the line is intended to be circular, the profile tolerance is identical to a roundness tolerance. The form of lines of any other shape or combinations of straight and curved lines must be controlled by a profile tolerance.

Similarly, a profile-of-a-surface tolerance may be directed to a surface of any size or shape. If a datum is not referenced and the surface is intended to be flat, the profile tolerance is identical to a flatness tolerance. If the surface is intended to be cylindrical, the profile tolerance is identical to a cylindricity tolerance. To control the form of all other surfaces, or combinations of straight, flat, and curved surfaces, a profile tolerance must be used.

A profile tolerance may also be directed to a line or a surface with a datum reference in order to control the orientation of the line or surface as well as its form. In this case, the tolerance zone is oriented to the datum in the manner shown on the drawing. If the feature being controlled is identical to orientation tolerances of angularity, parallelism, or perpendicularity, the orientation of any lines or surfaces not straight or flat must be controlled by a profile tolerance referenced to a suitable datum.

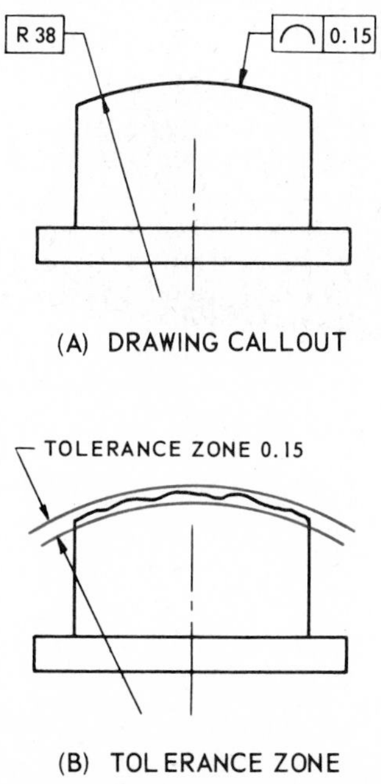
R 38
0.15
(A) DRAWING CALLOUT
TOLERANCE ZONE 0.15
(B) TOLERANCE ZONE

Fig. 31-10-2 Simple profile with profile tolerance.

If a profile tolerance is directed to a line or surface and is referenced to a datum, and if the datum is also associated with the profile by means of a basic or true-position dimension, then the profile tolerance is identical to a positional tolerance. It then controls the position of the profile as well as its orientation and form.

A profile-of-a-line tolerance is specified in the usual manner, by including the symbol and tolerance in a feature control symbol directed to the line to be controlled, as shown in Fig. 31-10-2.

If the line on the drawing to which the tolerance is directed represents a surface, the tolerance applies to all line elements of the surface parallel to the plane of the view on the drawing, unless otherwise specified.

The tolerance indicates a tolerance zone consisting of the area between two parallel lines, separated by the specified tolerance, which are themselves parallel to the basic form of the line in a plane parallel to the view on the drawing.

BILATERAL AND UNILATERAL TOLERANCES

The profile tolerance zone is normally equally disposed about the basic profile in a form known as a *bilateral tolerance zone*. The width of this zone is always measured normal to the profile surface. The tolerance zone may be considered to be bounded by two lines enveloping a series of circles, each having a diameter equal to the specified profile tolerance, with their centers on the theoretical, basic profile, as shown in Fig. 31-10-3.

Occasionally it is desirable to have the tolerance zone wholly on one side of the basic profile instead of equally divided on both sides. Such zones are called *unilateral tolerance zones*. They are specified by showing a thick chain or zone line close to the profile surface. The tolerance is directed to this line, as shown in Fig. 31-10-4.

In many cases, unilateral tolerancing has very little or no significance, especially if the surface is nearly flat or if the radius is separately controlled. Therefore unilateral tolerances should not be used unless there is a definite need.

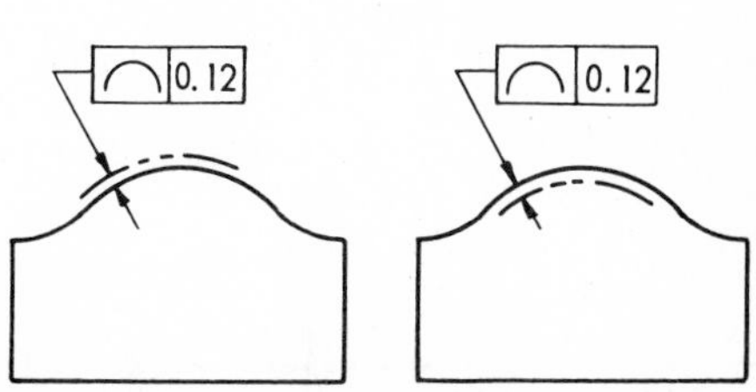
0.12
0.12

Fig. 31-10-3 Bilateral tolerance zones.

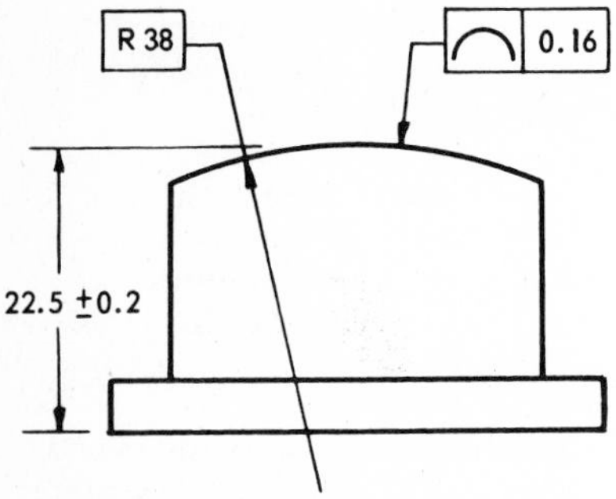
R 38
0.16
22.5 ±0.2

Fig. 31-10-4 Position and form as separate requirements.

METHOD OF DIMENSIONING

The true or mean profile is established by means of basic dimensions, each of which is enclosed in a rectangular frame to indicate that the tolerance in the general tolerance note does not apply.

When the profile tolerance is not intended to control the position of the profile, there must be a clear distinction between dimensions which control the position of the profile and those which control the form or shape of the profile.

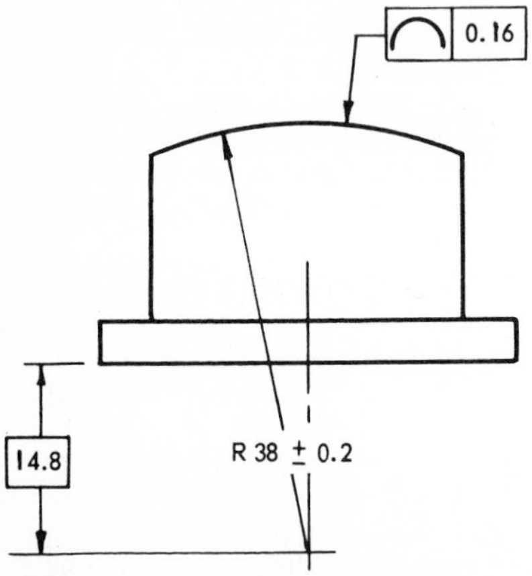
0.16
14.8
R 38 ± 0.2

Fig. 31-10-5 Position and radius separate from form.

The position of the profile is shown by one or more toleranced dimensions, all of which must be individually measured. The basic profile dimensions must relate all points on the profile to one another or to the ends or center line of the profile, so that the complete basic profile may be established without reference to any other surface or features and without reference to, or inclusion of, any toleranced dimensions.

Any convenient method of dimensioning may be used to establish the basic profile. Examples are chain or common-point dimensions, dimensioning to points on a surface or to the intersection of lines, dimensioning located on tangent radii, and angles.

To illustrate, the simple part in Fig. 31-10-4 shows a dimension of 22.5 ± 0.2 controlling the height of the profile. This dimension must be separately measured. The radius of 38 mm is a basic dimension, and it becomes part of the profile. Therefore the profile tolerance zone has radii of 37.92 and 38.08, but is free to flat in any direction within the limits of the positional tolerance zone in order to enclose the curved profile.

If the radius had been shown as a toleranced dimension, without the rectangular frame, as in Fig. 31-10-5, it would become a separate measurement. The profile tolerance would only control the finer irregularities of the surface and would have to be checked against tolerance zones of various radii in exactly the same manner as roundness.

Figure 31-10-6 shows a more complex profile, where the profile is located by a single toleranced dimension. There are, however, five basic dimensions defining the true profile.

In this case, the tolerance on the height indicates a tolerance zone 1 mm wide extending the full length of the profile. This is because the profile is established by basic dimensions. No other dimension exists to affect the orientation or height. The profile tolerance specifies a 0.2 mm wide tolerance zone, which may lie anywhere within the 1 mm tolerance zone.

EXTENT OF CONTROLLED PROFILE

The profile is generally intended to extend to the first abrupt change or sharp corner. For example, in Fig. 31-10-7 it extends from the upper left- to the upper right-hand corners, unless otherwise specified. If the extent of the profile is not clearly identified by sharp corners or by basic profile dimensions, it must be indicated by a note under the feature control symbol, such as FROM A TO B.

If the controlled profile includes a sharp corner, the corner represents a

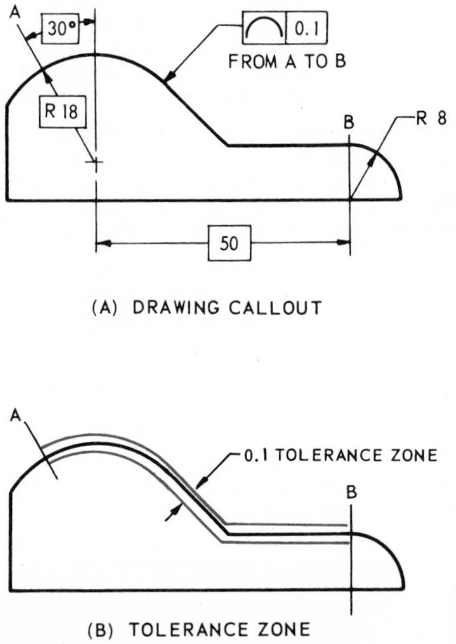
A
30°
0.1
FROM A TO B
R 18
B
R 8
50
(A) DRAWING CALLOUT
A
0.1 TOLERANCE ZONE
B
(B) TOLERANCE ZONE

Fig. 31-10-7 Specifying extent of profile.

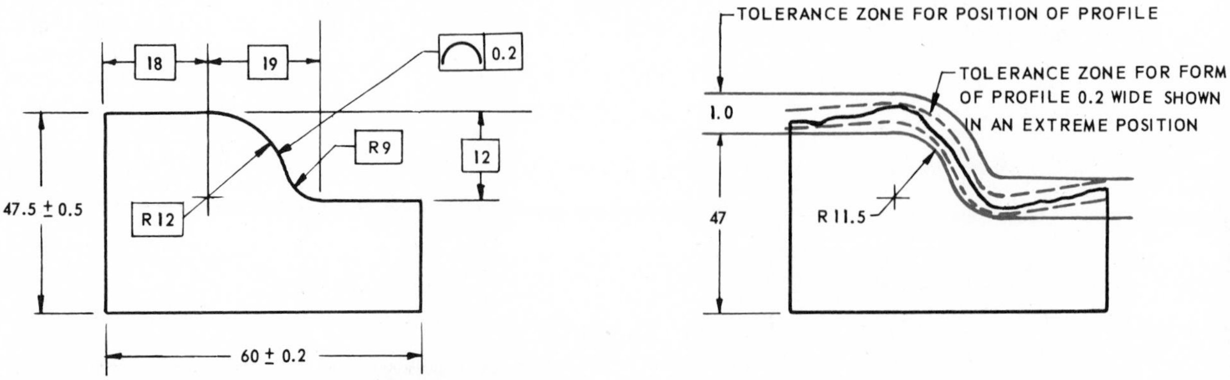
18
19
0.2
R9
12
47.5 ± 0.5
R 12
60 ± 0.2
TOLERANCE ZONE FOR POSITION OF PROFILE
TOLERANCE ZONE FOR FORM OF PROFILE 0.2 WIDE SHOWN IN AN EXTREME POSITION
1.0
47
R 11.5

Fig. 31-10-6 Profile defined by basic dimensions.

discontinuity of the tolerance boundary, and the boundary is considered to extend to the intersection of the boundary lines, as shown in Fig. 31-10-8. Since the part boundary could be anywhere within the profile tolerance zone, such corners could become somewhat rounded. If this is undesirable, the drawing must indicate the design requirements, such as by specifying a maximum radius for each individual corner.

If the profile tolerance is not intended to apply to a radius within the length of the profile, then the radius dimension should be separately toleranced, instead of being shown as a basic dimension.

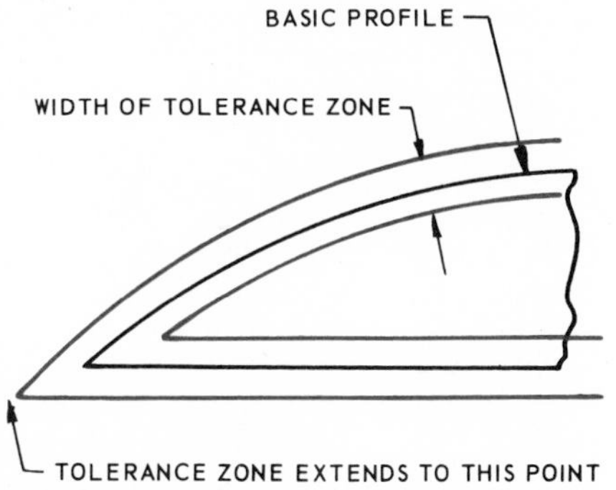
BASIC PROFILE
WIDTH OF TOLERANCE ZONE
TOLERANCE ZONE EXTENDS TO THIS POINT

Fig. 31-10-8 Tolerance zone at a sharp corner.

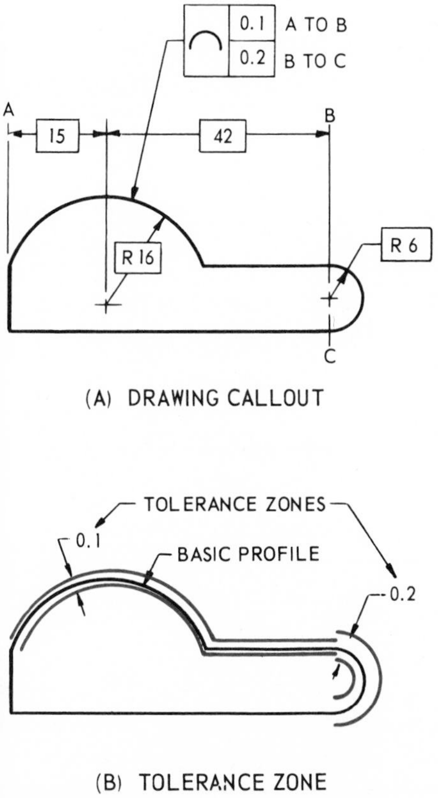
0.1 A TO B
0.2 B TO C
A
B
C
15
42
R 16
R 6
(A) DRAWING CALLOUT
TOLERANCE ZONES
0.1
BASIC PROFILE
0.2
(B) TOLERANCE ZONE

Fig. 31-10-9 Dual tolerance zones.

DUAL TOLERANCES

If different profile tolerances are to be applied to different parts of a profile, the tolerances may be shown in a combined feature control symbol, but the extent of each must be clearly stated, as shown in Fig. 31-10-9. In this instance tolerance zones will then appear as shown.

ALL-AROUND PROFILE TOLERANCES

It is sometimes desirable to show basic dimensions for the complete outline of a part and to add a profile tolerance or tolerances which apply all around the part. If more than one tolerance is used, the extent to which each applies must be clearly indicated. When only two tolerances are used, such as shown in Fig. 31-10-10, there may be some doubt as to which direction from point *A* the tolerances apply. In such instances, a means of indicating the direction the tolerances are taken from point *A* is required. Any of the methods shown may be used to indicate direction of readings from point *A*.

If the same tolerance applies to the complete profile, the words ALL AROUND should be added, as shown in Fig. 31-10-11.

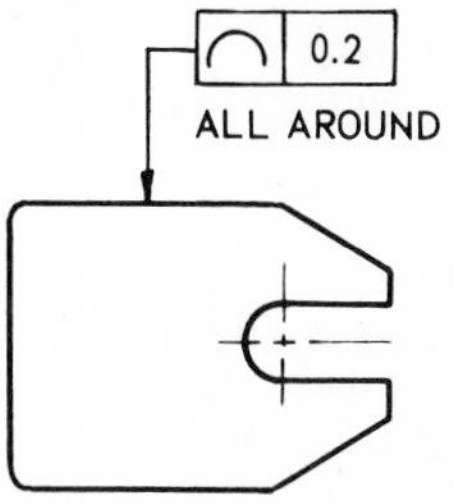
0.2
ALL AROUND

Fig. 31-10-11 All-around tolerance.

PROFILE TOLERANCE ON AN MMC BASIS

A profile tolerance may be specified on an MMC basis for any profile that includes opposing surfaces. This method is especially suitable when the tolerance applies all around the part. However, then the tolerance controls the profile only at the maximum material size. It must be possible to measure the part to ensure that the minimum material limits of size are also met. Specification on an MMC basis is particularly useful for parts of simple or regular cross sections, such as square, rectangular, hexagonal, or round bars, as shown in Fig. 31-10-12.

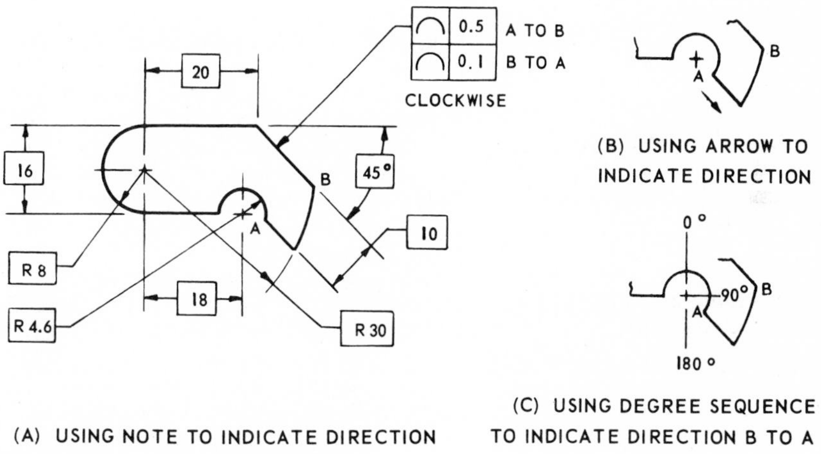
0.5 A TO B
0.1 B TO A
CLOCKWISE
20
16
45°
B
A
10
R 8
18
R 4.6
R 30
(A) USING NOTE TO INDICATE DIRECTION
(B) USING ARROW TO INDICATE DIRECTION
0°
90°
180°
(C) USING DEGREE SEQUENCE TO INDICATE DIRECTION B TO A

Fig. 31-10-10 All-around profile tolerances.

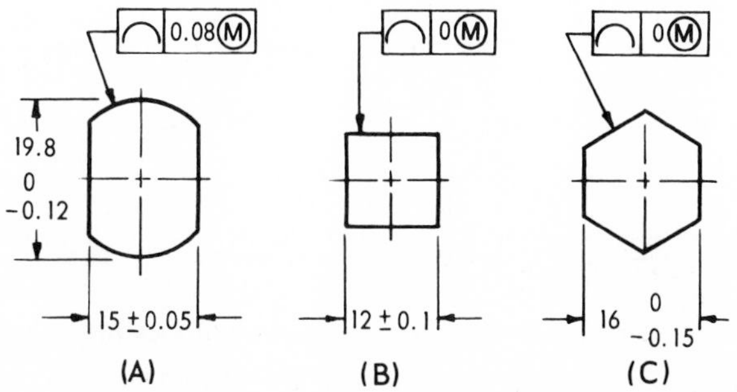
0.08Ⓜ
0Ⓜ
0Ⓜ
19.8
0
-0.12
15 ± 0.05
12 ± 0.1
16
0
-0.15
(A)
(B)
(C)

Fig. 31-10-12 Profiles tolerance—MMC.

Profile of a Surface

As already stated, a profile-of-a-line tolerance, when directed to a line on a drawing which represents a surface, applies to the profile of all cross sections parallel to the view on the drawing, unless otherwise specified. This is illustrated by the cross sections *AA* and *BB* in Fig. 31-10-13.

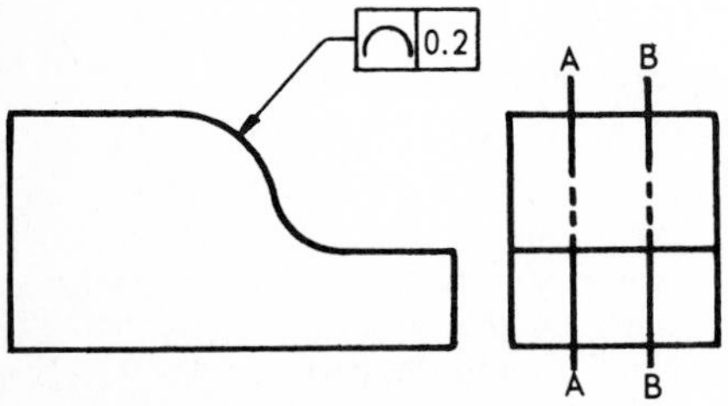
0.2
A
B
A
B

Fig. 31-10-13 Tolerance applies to profile of cross sections.

If the profile in a plane normal to the plane of the drawing requires tolerancing, a separate tolerance may be added to the side view. Frequently this profile will be straight, in which case a straightness tolerance may be substituted, as shown in Fig. 31-10-14.

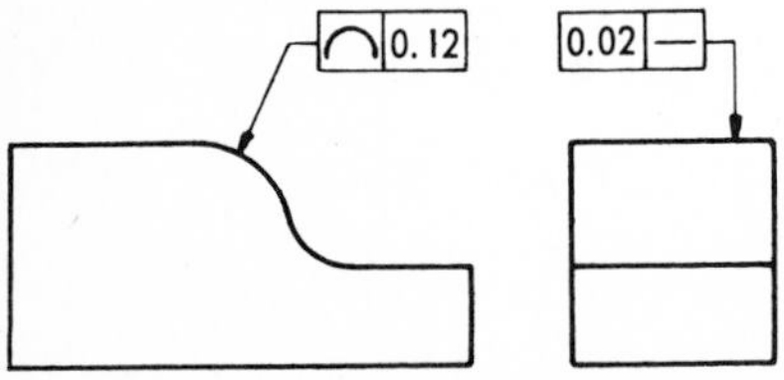
0.12
0.02

Fig. 31-10-14 Profile and straightness tolerance to same surface.

If the same tolerance is intended to apply over the whole surface, instead of to lines or line elements in specific directions, the profile of a surface symbol is used, as shown in Fig. 31-10-15. While the profile tolerance may be directed to the surface in either view, it is usually directed to the view showing the more characteristic profile.

The profile-of-a-surface tolerance indicates a tolerance zone having the same form as the basic surface, with a uniform width equal to the specified tolerance within which the entire surface must lie.

ORIENTATION

In the examples so far, the profile tolerances have been shown without reference to a datum, and it has been permissible to move evaluating charts in any direction in an effort to encompass the line or surface profile. More often than not, requirements call for the profile to be oriented to some other surface or feature of the part. This is specified simply by indicating suitable data. Figure 31-10-16 shows a simple part where the base is designated as a datum.

POSITION OF LINES AND SURFACES

Position of features, such as holes and bosses, is normally controlled by a positional tolerance. Such a tolerance automatically controls the orientation and form of the feature, since all points on the feature surface must lie within the positional tolerance zone.

Positional tolerances may also be applied to surfaces, such as the position of the flat surface in Fig. 31-10-17. The surface must be dimensioned from the datum feature or features by means of basic dimensions, as shown. The tolerance zone is then established at an exact position from the datum, and all points on the surface must lie within this tolerance zone without further adjustment.

Positional tolerancing may also be applied to lines or surfaces of other shapes. It is usually confined, however,

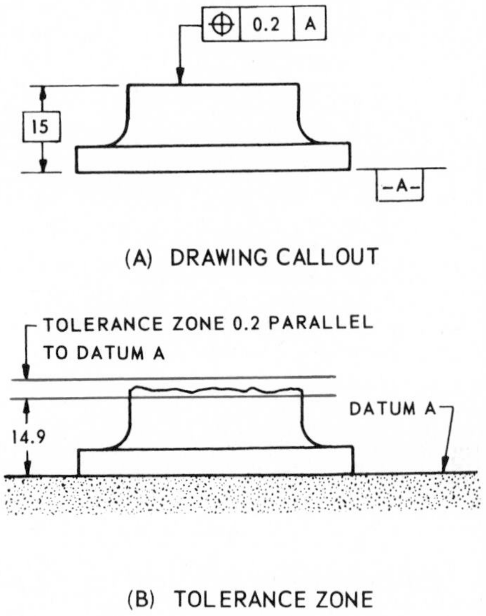
0.2
A
15
-A-
(A) DRAWING CALLOUT
TOLERANCE ZONE 0.2 PARALLEL
TO DATUM A
DATUM A
14.9
(B) TOLERANCE ZONE

Fig. 31-10-17 Position of a surface.

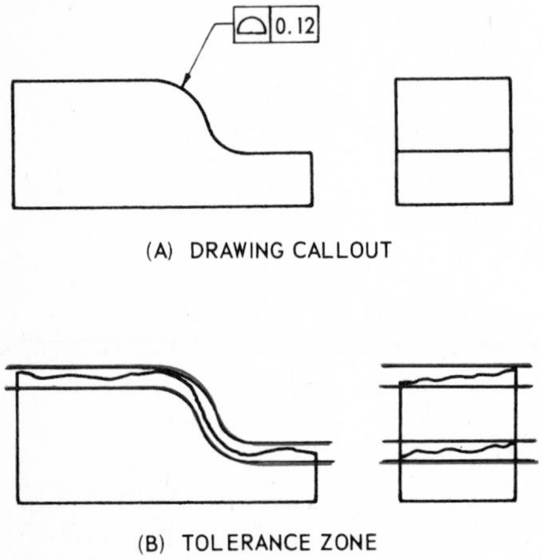
0.12
(A) DRAWING CALLOUT
(B) TOLERANCE ZONE

Fig. 31-10-15 Profile of a surface.

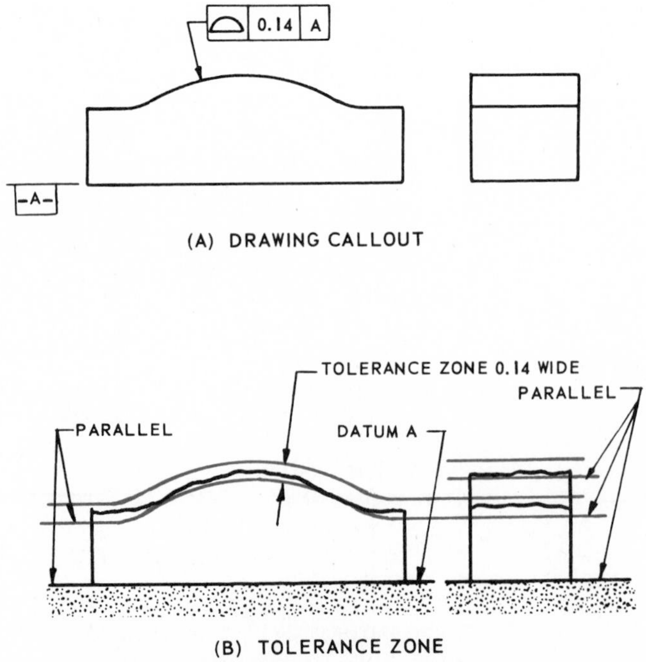
0.14
A
-A-
(A) DRAWING CALLOUT
TOLERANCE ZONE 0.14 WIDE
PARALLEL
PARALLEL
DATUM A
(B) TOLERANCE ZONE

Fig. 31-10-16 Profile referenced to a datum.

to straight lines, circular lines, flat or cylindrical surfaces, or sometimes regular shapes such as hexagons. For example, in Fig. 31-10-18, it is immaterial whether the geometric symbol is position or profile—the interpretation is identical.

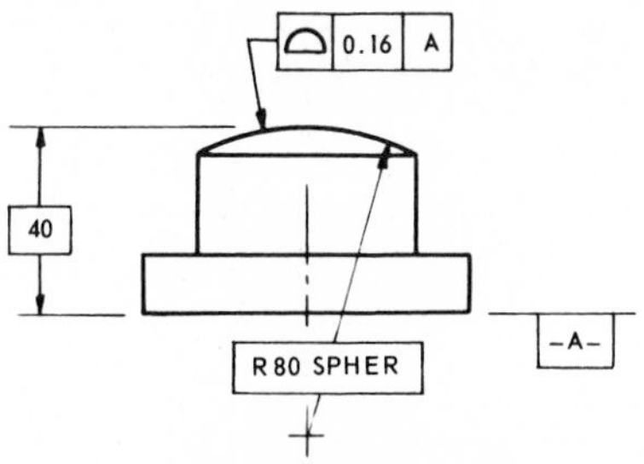
0.16
A
40
R 80 SPHER
-A-

(A) DRAWING CALLOUT

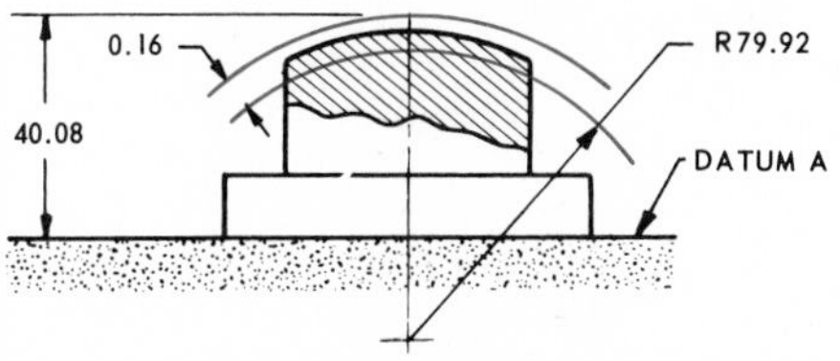
0.16
R79.92
40.08
DATUM A

(B) TOLERANCE ZONE

Fig. 31-10-18 Profile tolerance for position.

The criterion which distinguishes a profile tolerance as applying to position or to orientation is whether the profile is related to the datum by a basic dimension or by a toleranced dimension. This is illustrated by Fig. 31-10-19.

CYLINDRICAL DATA

Profile tolerances controlling position are very useful for parts which can be revolved around a cylindrical datum feature, such as screw-machine cams. The position or size as well as the form can easily be assessed by revolving the part on the datum axis in conjunction with a dividing head, with an indicator gage to make direct measurements on the periphery. Figure 31-10-20 shows an example.

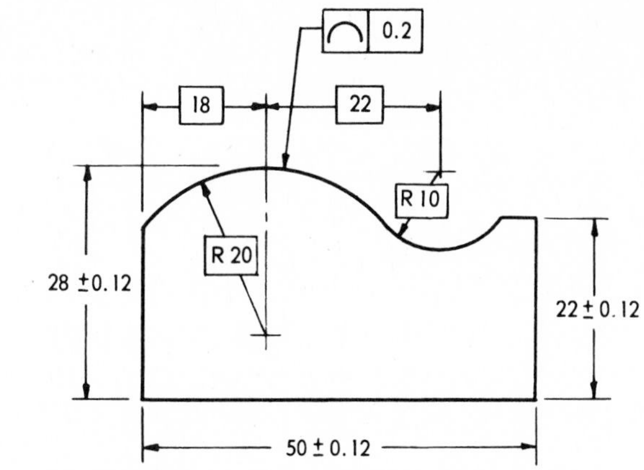
0.2
18
22
R 10
R 20
28 ±0.12
22±0.12
50±0.12

(A) PROFILE TOLERANCE CONTROLS FORM OF PROFILE ONLY

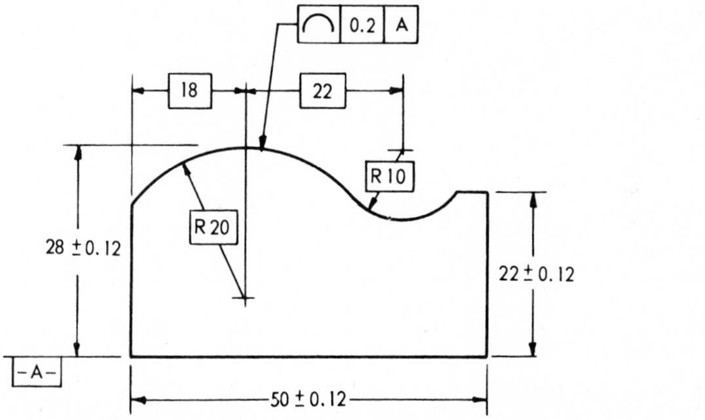
0.2
A
18
22
R 10
R 20
28 ±0.12
22±0.12
-A-
50±0.12

(B) PROFILE TOLERANCE CONTROLS FORM AND ORIENTATION OF PROFILE

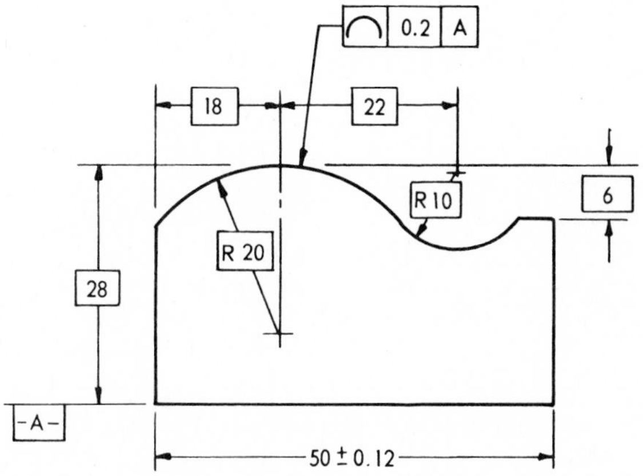
0.2
A
18
22
R 10
6
R 20
28
-A-
50±0.12

(C) PROFILE TOLERANCE CONTROLS FORM, ORIENTATION, AND POSITION OF PROFILE

Fig. 31-10-19 Comparison of profile tolerances.

Assignments

On an A3- or B-size sheet, prepare drawings and solutions for the illustrations shown in Fig. 31-10-A or Fig. 31-10-B and the following information.

1. In illustration (A) it is required to have the form of the indented portion controlled by the line profile tolerance shown. Show the tolerance and sketch the resulting tolerance zone on the drawing.

2. It is required to control the profile in illustration (B) with the tolerance described on the drawing. Add the line profile tolerance to the drawing and sketch the resulting tolerance zone.

3. Draw the tolerance zone, showing its relationship to the datums, for the profile tolerance shown in illustration (C).

4. A cam is dimensioned as shown in illustration (D). If parts are measured with an indicator which was set to zero and the following readings were obtained, which parts shown in the chart would be acceptable? Of the nonacceptable parts, which could be made acceptable by regrinding?

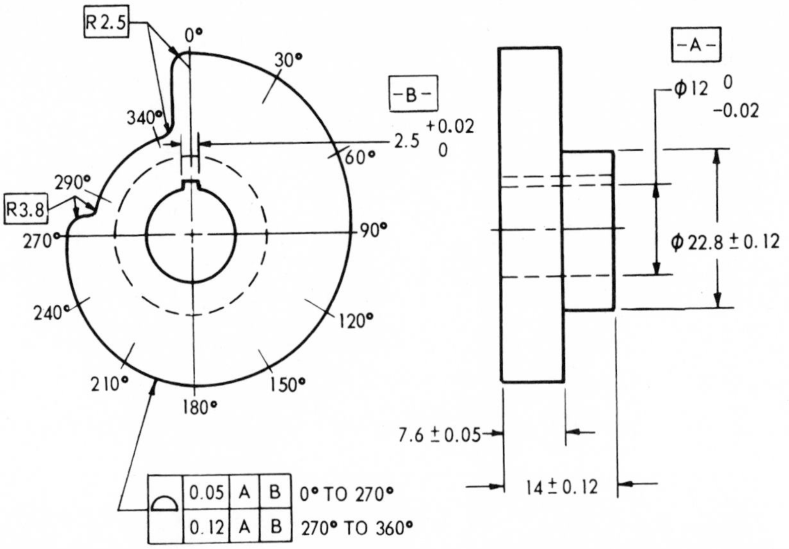
R2.5
0°
30°
340°
-B-
2.5 +0.02 0
60°
290°
R3.8
270°
90°
240°
120°
210°
150°
180°
0.05 A B 0° TO 270°
0.12 A B 270° TO 360°
-A-
ϕ12 0 -0.02
ϕ22.8 ± 0.12
7.6 ± 0.05
14 ± 0.12

BASIC VALUES	
DEGREES	RADIUS
0	25
30	25
60	24
90	23
120	22
150	21
180	20
210	19
240	18
270	17
290	14
340	14

Fig. 31-10-20 Profile tolerances control form and size of cam profile.

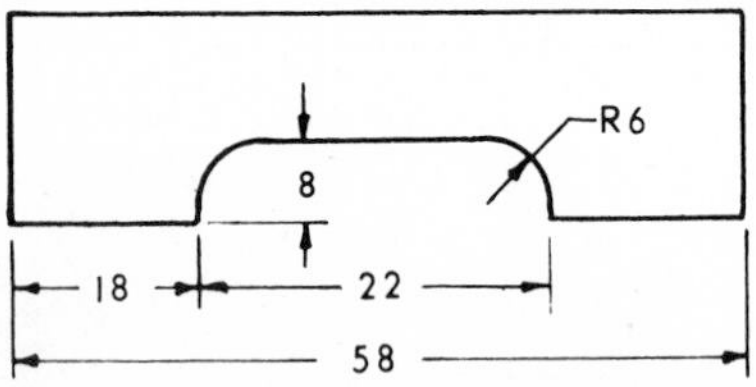
R6
8
18
22
58

PLACE A LINE PROFILE TOLERANCE OF 0.05 ON THE INDENTED PORTION OF THE PART

(A)

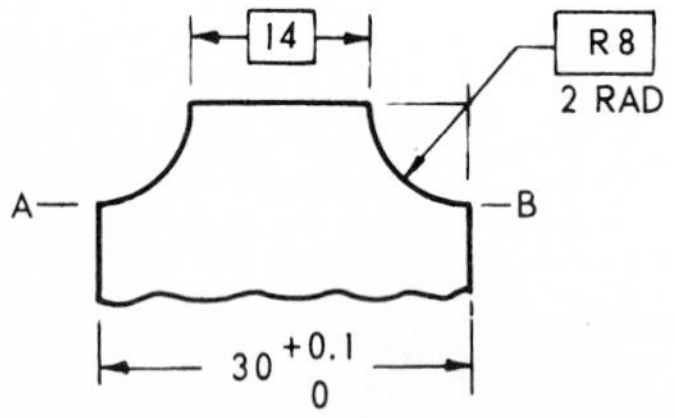
14
R8
2 RAD
A
B
30 +0.1 0

CONTROL THE PROFILE A TO B WITH A LINE PROFILE TOLERANCE OF 0.08 EXCEPT THAT THE 14 STRAIGHT PORTION CAN BE PERMITTED TO VARY VERTICALLY BY ±0.1 .

(B)

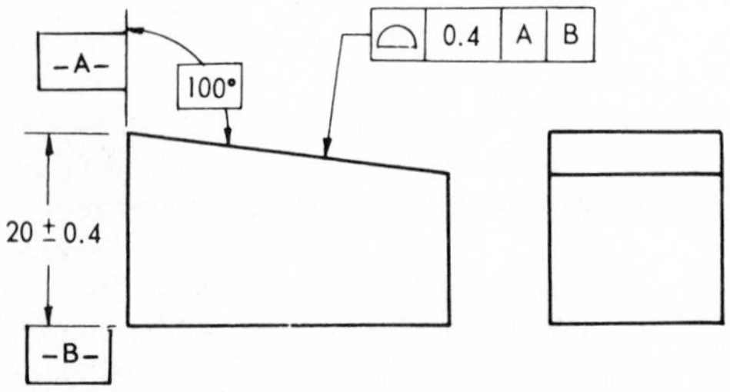
-A-
100°
0.4 A B
20 ± 0.4
-B-

(C)

(D)

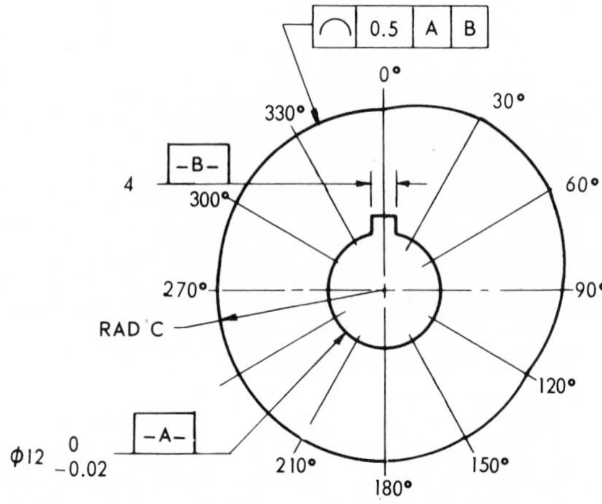
0.5 A B
0°
30°
330°
-B-
4
300°
60°
270°
90°
RAD C
120°
-A-
ϕ12 0 -0.02
210°
150°
180°

ANGLE	C RAD
0	24.5
30	27.5
60	27.5
90	25
120	22
150	20.5
180 TO 270	20
300	20.5
330	22

ANGLE	DISTANCE FROM CENTER			
	PART 1	PART 2	PART 3	PART 4
0	24.6	24.7	24.5	24.4
30	27.8	27.7	27.4	27.4
60	27.6	27.6	27.25	27.5
90	25.1	25	24.6	25.1
120	22.4	21.9	21.7	22.1
150	20.6	20.4	20.2	20.5
180 TO 270	20.2 TO 20.3	19.8 TO 19.75	19.6 TO 20.3	20.1 TO 20
300	20.4	20.3	20.2	20.4
330	21.9	22	21.7	21.9

Fig. 31-10-A Assignments.

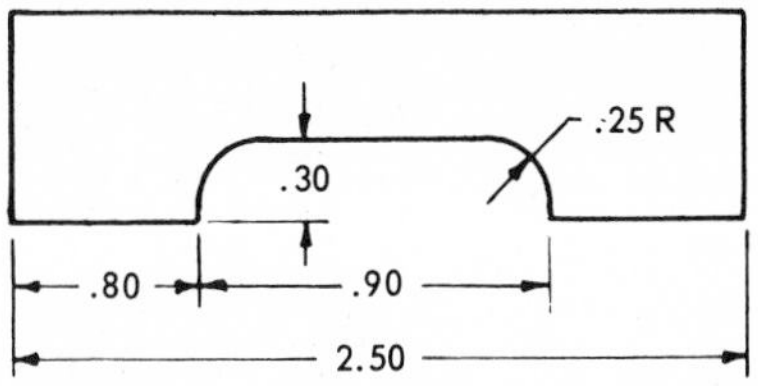
.25 R
.30
.80
.90
2.50

PLACE A LINE PROFILE TOLERANCE OF .002
ON THE INDENTED PORTION OF THE PART

(A)

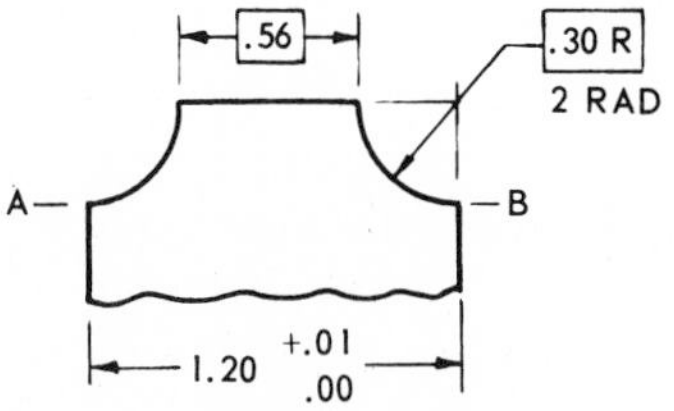
.56
.30 R
2 RAD
A
B
1.20 +.01 .00

CONTROL THE PROFILE A TO B WITH A LINE PROFILE TOLERANCE OF .003 EXCEPT THAT THE .56 STRAIGHT PORTION CAN BE PERMITTED TO VARY VERTICALLY BY ±.004

(B)

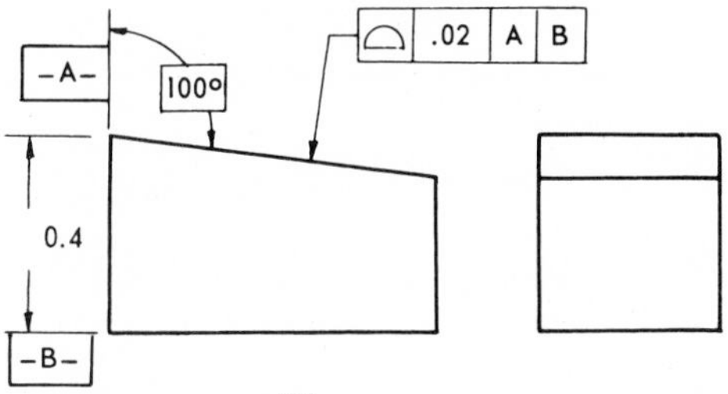
-A-
100°
.02 A B
0.4
-B-

(C)

(D)

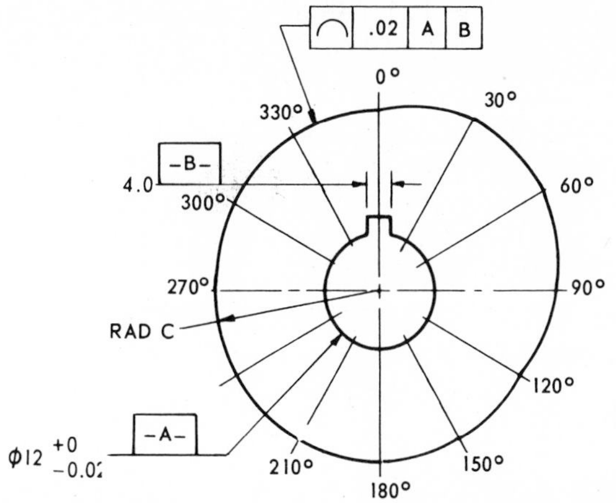
.02 A B
0°
30°
60°
90°
120°
150°
180°
210°
270°
300°
330°
-B-
4.0
RAD C
-A-
ϕ12 +0 -0.0

ANGLE	C RAD
0	.965
30	1.083
60	1.083
90	.984
120	.866
150	.807
180 TO 270	.787
300	.807
330	.866

ANGLE	DISTANCE FROM CENTER			
	PART 1	PART 2	PART 3	PART 4
0	.968	.972	.965	1.079
30	1.094	1.091	1.079	1.079
60	1.087	1.087	1.073	1.083
90	.988	.984	.969	.988
120	.882	.862	.854	.870
150	.811	.803	.795	.807
180 TO 270	.795 TO .799	.780 TO .778	.772 TO .799	.791 TO .787
300	.803	.799	.795	.803
330	.862	.866	.854	.862

Fig. 31-10-B Assignments.

UNIT 31-11
CORRELATIVE TOLERANCES

Correlative geometric tolerancing refers to tolerancing for the control of two or more features intended to be correlated in position or attitude. Examples of such correlated tolerancing include coplanarity, for control of two or more flat surfaces; symmetry, for control of features equally disposed about a center line; concentricity and coaxiality, for control of features having common axes or center lines; and runout, for control of surfaces related to an axis. These are all tolerances of location, for which the positional symbol and positional tolerances could be used. Special symbols, however, have been provided for some of them to clarify and simplify drawing callout requirements.

When position is to be separately controlled, other form or orientation tolerances may be applied to control the correlation of features.

Coplanarity

Coplanarity refers to the relative position of two or more flat surfaces, which are intended to lie in the same geometric plane. No special symbol exists for coplanarity. There are several methods of tolerancing. Either the symbol for position or the symbol for flatness may be used, depending on the type of control required.

Figure 31-11-1 illustrates a suitable method of tolerancing when one or more surfaces are required to be coplanar with a principal surface, which is then specified as a datum feature. Here the tolerance controls orientation and flatness of the toleranced surfaces within the same limits, but not flatness of the datum feature surface. The position of the surfaces in relation to the base or other features of the part must be separately dimensioned and toleranced, as shown by the 25 mm dimension.

Figure 31-11-2 shows a case where coplanar surfaces are also required to be accurately located and parallel to another surface of a part, which is then designated as a datum feature. In this case, the datum and the controlled surfaces must be associated by a basic dimension.

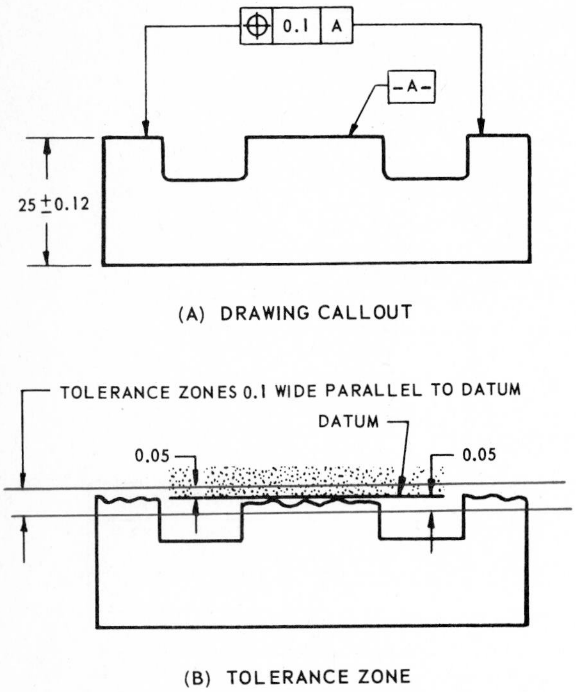
0.1
A
-A-
25 ± 0.12
(A) DRAWING CALLOUT
TOLERANCE ZONES 0.1 WIDE PARALLEL TO DATUM
DATUM
0.05
0.05
(B) TOLERANCE ZONE

Fig. 31-11-1 Surfaces coplanar with datum surface.

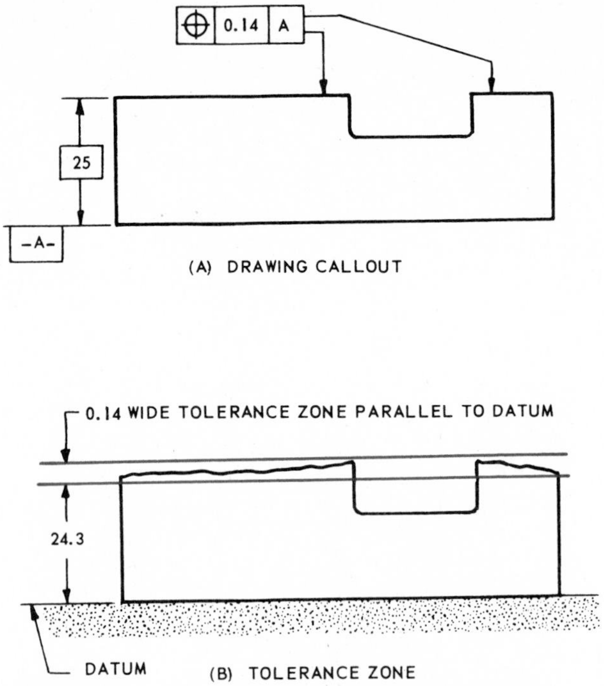
0.14
A
25
-A-
(A) DRAWING CALLOUT
0.14 WIDE TOLERANCE ZONE PARALLEL TO DATUM
24.3
DATUM
(B) TOLERANCE ZONE

Fig. 31-11-2 Coplanar surfaces parallel to a datum.

Although the positional tolerance applies separately to each surface, the tolerance zones are of equal width and are each parallel to and at equal distances from the datum surface. This condition results essentially in one tolerance zone within which the surfaces must lie simultaneously. The positional tolerance controls orientation and form (parallelism and flatness) of both surfaces within the same limits.

It is sometimes necessary to reference surfaces to a datum system instead of to a single datum surface. This occurs when it is necessary to control coplanar surfaces perpendicular to a datum instead of parallel to it. Figure 31-11-3 shows a case where the coplanar surfaces are required to be perpendicular to the axis of a hole.

Simultaneous Requirements. If it is necessary to hold coplanarity of two or more surfaces within close limits but permit a greater variation in the position of the surfaces in relation to other features of the part, two methods of tolerancing may be used. Both give the same results.

The surfaces may be positioned relative to one another, using the symbol for position, as shown in the upper view of Fig. 31-11-4. Alternatively, they may be treated as one flat surface, using a flatness tolerance and symbol as shown in the lower view. The word SIMULTANEOUS must be added beneath the feature control symbol. Otherwise the tolerance would apply separately to each surface and would control their flatness, but would not ensure that they were in the same plane. A profile symbol may also be used instead of the flatness symbol without affecting the interpretation in the least.

A parallelism tolerance is sometimes used for coplanar surfaces, as shown in Fig. 31-11-5. Such a tolerance controls only parallelism and flatness of the surfaces. Their positions may vary anywhere within the tolerance zone for position specified by the toleranced dimension.

Other Correlated Surfaces. Many features other than flat surfaces may be correlated if they have line elements in a single direction in the same straight or circular line. It is often possible to control such features using controls similar to those applied to coplanar surfaces. Either a positional tolerance or a simultaneous form tolerance may be used, depending on the type of control required. Figure 31-11-6 gives a few examples.

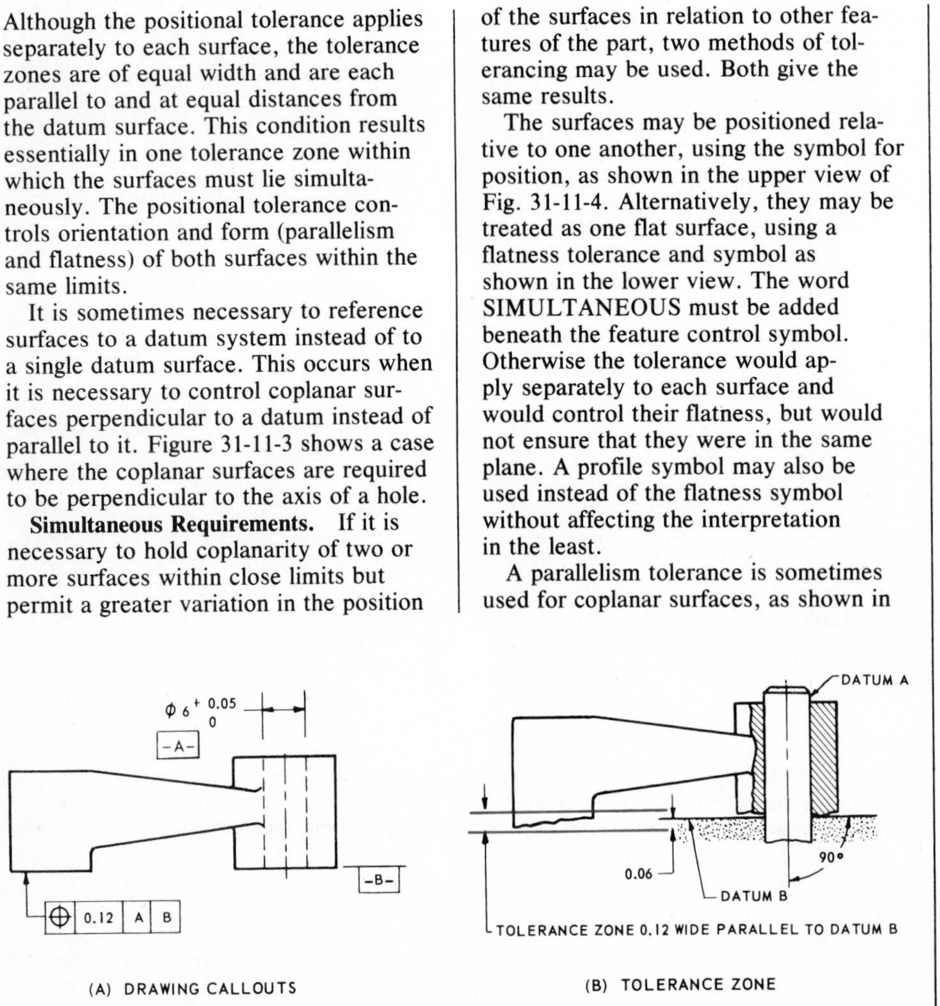
ϕ 6 $^{+0.05}_{0}$
-A-
-B-
0.12
A
B
(A) DRAWING CALLOUTS
DATUM A
90°
0.06
DATUM B
TOLERANCE ZONE 0.12 WIDE PARALLEL TO DATUM B
(B) TOLERANCE ZONE

Fig. 31-11-3 Surface referenced to a datum system.

Symmetry

Symmetry is a condition in which a feature or features are symmetrically disposed about a center line or center plane of another feature. The center line or plane of the second feature is usually specified as a datum.

A symmetry tolerance specifies the width of a tolerance zone. This width is the area between two parallel lines or the space between two parallel planes equally disposed about the datum axis or median plane.

Symmetry is therefore a special case of position. The advantage of using the symmetry symbol rather than the position symbol is that it indicates that the

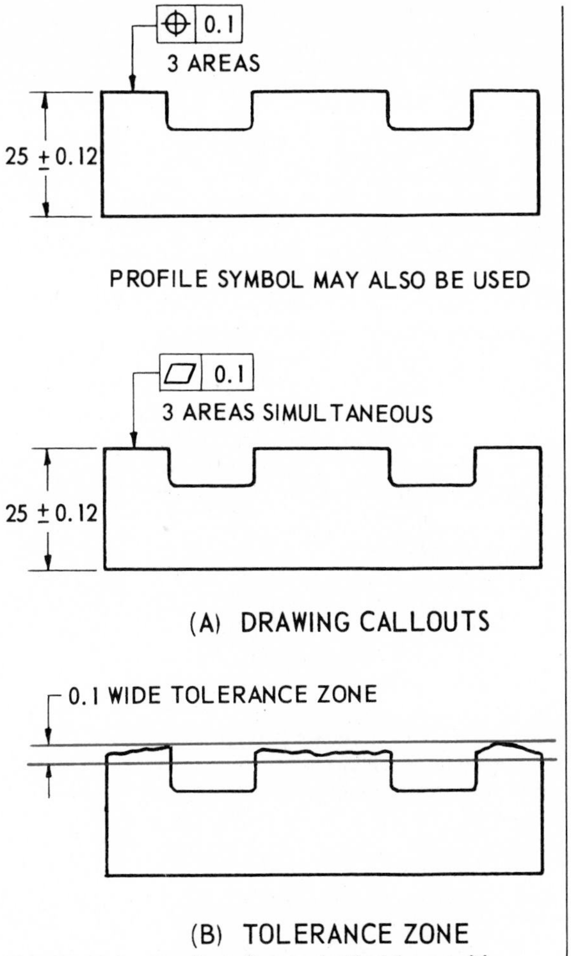
0.1
3 AREAS
25 ± 0.12
PROFILE SYMBOL MAY ALSO BE USED
0.1
3 AREAS SIMULTANEOUS
25 ± 0.12
(A) DRAWING CALLOUTS
0.1 WIDE TOLERANCE ZONE
(B) TOLERANCE ZONE

Fig. 31-11-4 Coplanarity controlled by position or flatness.

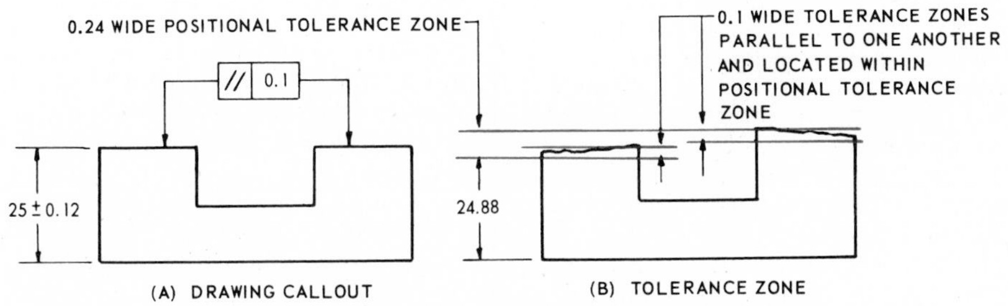
0.24 WIDE POSITIONAL TOLERANCE ZONE
0.1 WIDE TOLERANCE ZONES PARALLEL TO ONE ANOTHER AND LOCATED WITHIN POSITIONAL TOLERANCE ZONE
// 0.1
25 ± 0.12
24.88
(A) DRAWING CALLOUT
(B) TOLERANCE ZONE

Fig. 31-11-5 Parallelism tolerance for coplanar surfaces.

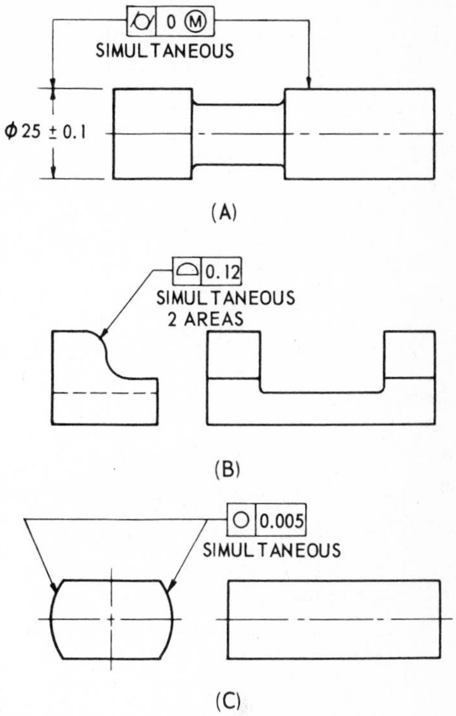
0 Ⓜ
SIMULTANEOUS
ϕ25 ± 0.1
(A)
0.12
SIMULTANEOUS
2 AREAS
(B)
0.005
SIMULTANEOUS
(C)

Fig. 31-11-6 Tolerancing correlated surfaces.

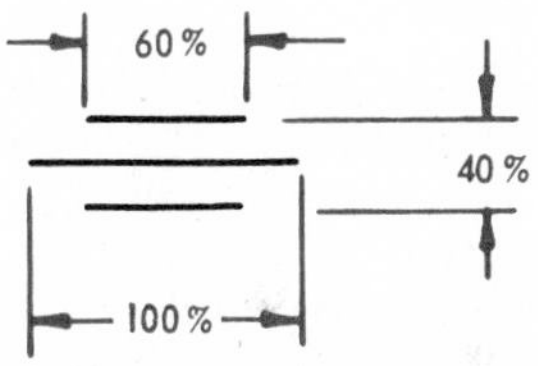
60 %
40 %
100 %

Fig. 31-11-7 Symmetry symbol.

true position is symmetrical and often eliminates the need for basic dimensions to correlate the position of the features. Thus it serves the same purpose for noncylindrical features as concentricity serves for circular features.

The geometric characteristic symbol for symmetry consists of three horizontal lines, as shown in Fig. 31-11-7. The dimensions refer to percentages of the feature control symbol frame height.

SYMMETRY—RFS

Figure 31-11-8 shows a simple example in which a slot is intended to be symmetrical with the overall width of the part, which in turn is specified as the datum. Note that it is not necessary to show a true-position dimension for the slot, or equal dimensions from the sides of the slot to the sides of the part, which might be deemed necessary if a positional tolerance were used.

The tolerance zone is equally disposed about the center line or median plane of the datum feature.

SYMMETRY—MMC

Whenever possible, symmetry should be specified on an MMC basis. Some examples of parts toleranced on this basis, with suitable gaging principles, are shown in Fig. 31-11-9.

Concentricity

Concentricity is a condition in which two or more features, such as circles, spheres, cylinders, cones, or hexagons, share a common center or axis. An example would be a round hole through the center of a cylindrical part.

A concentricity tolerance is a particular case of a positional tolerance. It controls the permissible variation in position, or eccentricity, of the center line of the controlled feature in relation to the

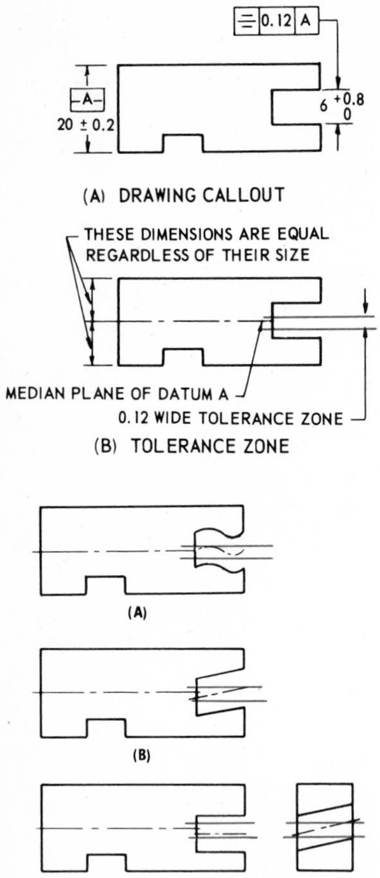
0.12 A
-A-
20 ± 0.2
6 +0.8 0
(A) DRAWING CALLOUT
THESE DIMENSIONS ARE EQUAL REGARDLESS OF THEIR SIZE
MEDIAN PLANE OF DATUM A
0.12 WIDE TOLERANCE ZONE
(B) TOLERANCE ZONE
(A)
(B)
(C) PERMISSIBLE VARIATIONS

Fig. 31-11-8 Symmetry tolerance—RFS.

axis of the datum feature when the feature and datum are intended to be concentric or coaxial.

The tolerance zone for circles is a circle; for other circular features it is a cylinder, concentric with the datum axis, having a diameter equal to the specified tolerance. The center of all cross sections normal to the axis of the controlled feature must lie within this tolerance zone.

The geometric characteristic symbol used for both concentricity and coaxiality consists of two concentric circles, having diameters equal to 75 and 50 percent, respectively, of the feature control symbol frame height, as shown in Fig. 31-11-10.

CONCENTRICITY—RFS

Concentricity tolerance, because of its unique characteristic, is always used on an RFS basis.

Figure 31-11-11 shows two concentric circles. The small circle is designated as a datum feature, and a concentricity tolerance is applied to the outer circle. To meet drawing requirements, the center of all diameters of the larger circle must lie within the tolerance zone.

Figure 31-11-12 shows a common type of part where the outer diameter is required to be concentric with the center bore, which is designated as a datum feature.

Figure 31-11-13 shows an example of a part where two cylindrical portions are intended to be coaxial. This figure also illustrates the extreme errors of eccentricity and parallelism that the concentricity tolerance would permit.

A concentricity tolerance may be referenced to a datum system, instead of to a single datum, to meet certain functional requirements. Figure 31-11-14 gives an example in which the tolerance zone is perpendicular to datum *A* and also concentric with the axis of datum *B* in the plane of datum *A*.

Fig. 31-11-9 Symmetry tolerance—MMC.

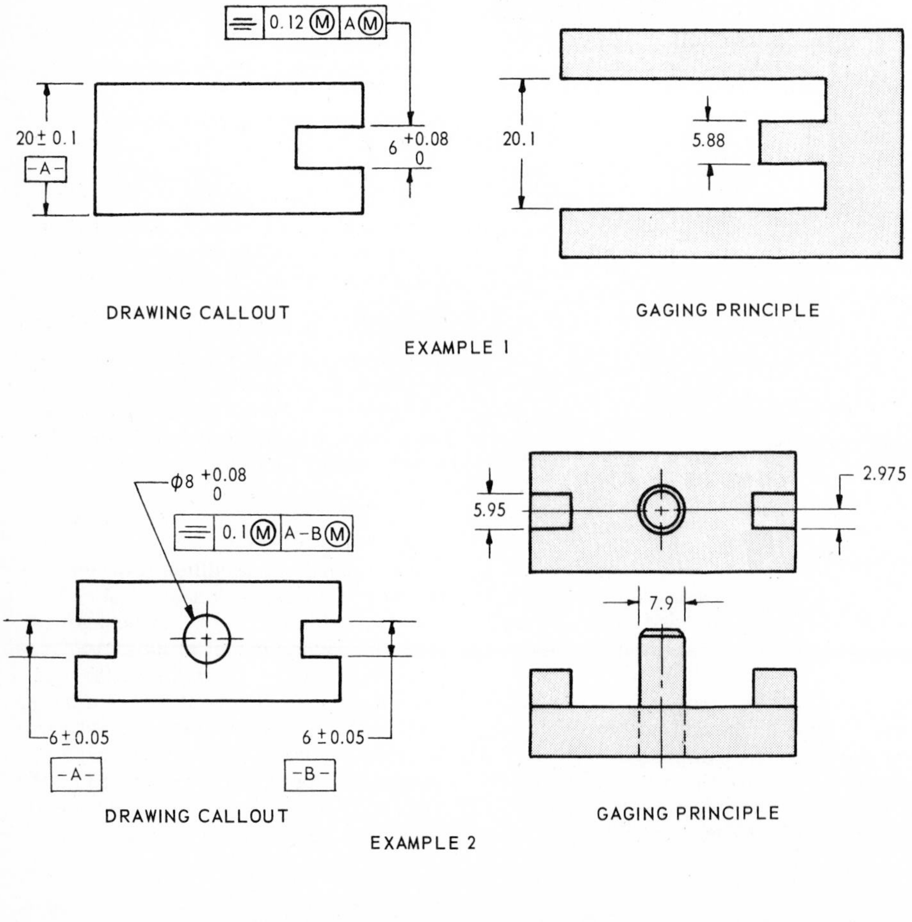
0.12 Ⓜ A Ⓜ
20 ± 0.1
-A-
6 +0.08 0
20.1
5.88
DRAWING CALLOUT
GAGING PRINCIPLE
EXAMPLE 1
ϕ8 +0.08 0
0.1 Ⓜ A-B Ⓜ
5.95
2.975
7.9
6 ± 0.05
6 ± 0.05
-A-
-B-
DRAWING CALLOUT
GAGING PRINCIPLE
EXAMPLE 2

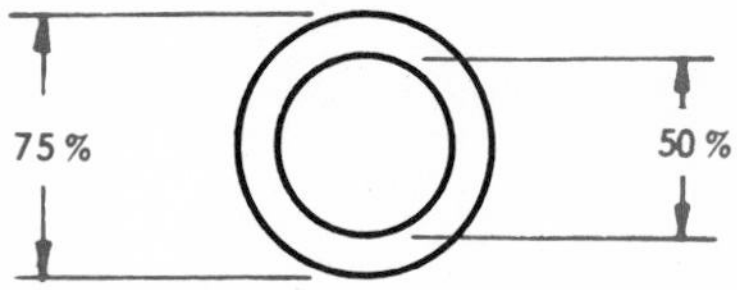
75 %
50 %

Fig. 31-11-10 Concentricity symbol.

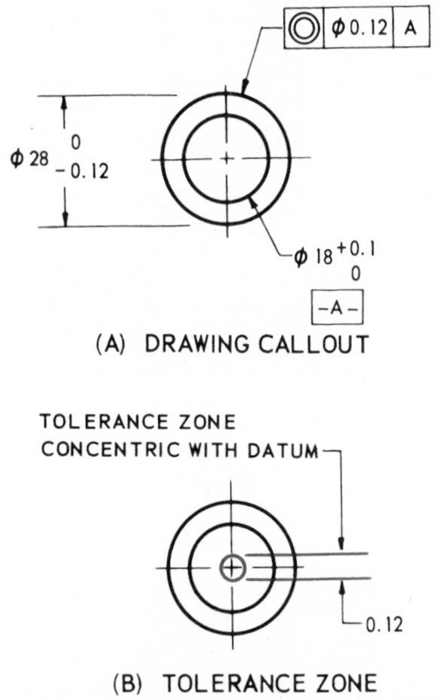
ϕ 0.12 A
ϕ 28 0 -0.12
ϕ 18 +0.1 0
-A-
(A) DRAWING CALLOUT
TOLERANCE ZONE CONCENTRIC WITH DATUM
0.12
(B) TOLERANCE ZONE

Fig. 31-11-11 Concentricity of circles.

Runout

Runout is the deviation in position of a surface of revolution as a part is revolved about a datum axis.

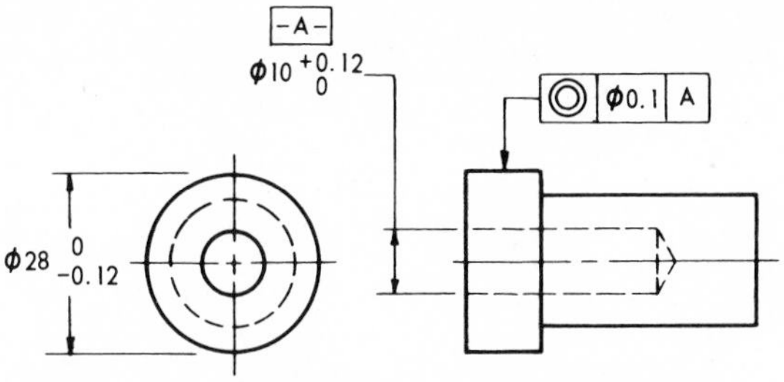
-A-
ϕ10 +0.12 0
ϕ0.1 A
ϕ28 0 -0.12

Fig. 31-11-12 Cylindrical part with cylindricity tolerance.

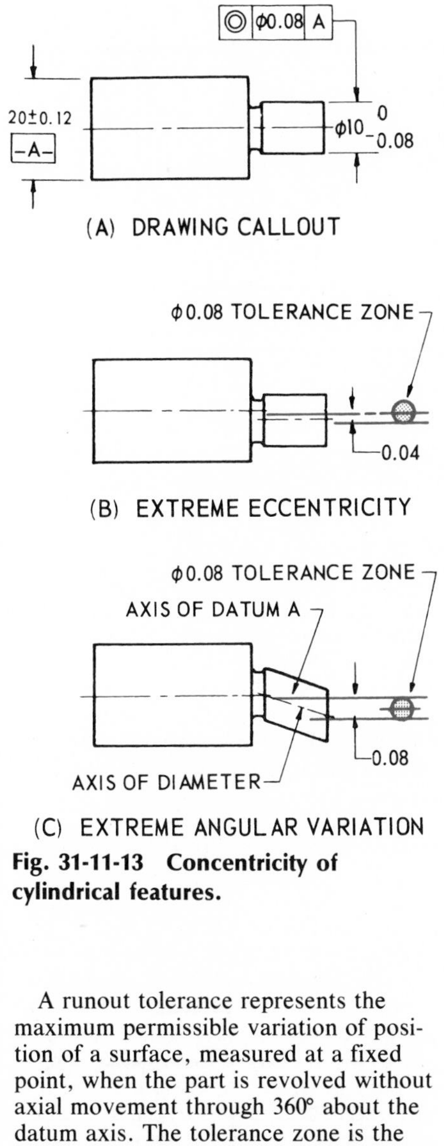
ϕ0.08 A
20±0.12
-A-
ϕ10 0 -0.08
(A) DRAWING CALLOUT
ϕ0.08 TOLERANCE ZONE
0.04
(B) EXTREME ECCENTRICITY
ϕ0.08 TOLERANCE ZONE
AXIS OF DATUM A
0.08
AXIS OF DIAMETER
(C) EXTREME ANGULAR VARIATION

Fig. 31-11-13 Concentricity of cylindrical features.

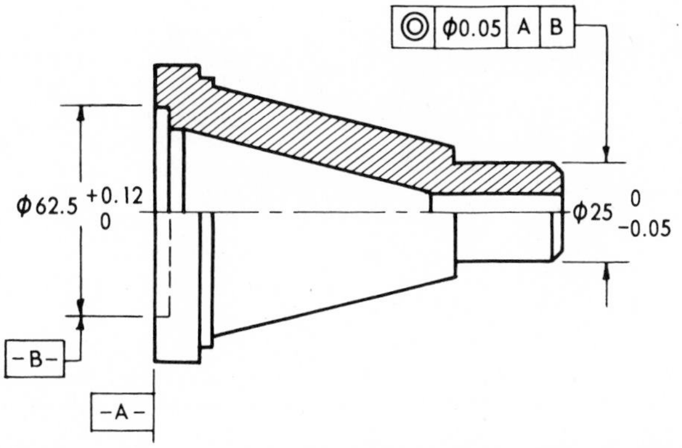
ϕ0.05 A B
ϕ62.5 +0.12 0
ϕ25 0 -0.05
-B-
-A-

Fig. 31-11-14 Concentricity referenced to datum system.

A runout tolerance represents the maximum permissible variation of position of a surface, measured at a fixed point, when the part is revolved without axial movement through 360° about the datum axis. The tolerance zone is the area, or space, of uniform thickness between two shapes coaxial with the datum axis and parallel to the true profile of the controlled surface.

Unless otherwise specified, the runout variation is measured normal to the surface being controlled by contacting the surface with an indicator gage while the part is revolved on its datum axis. The runout error is the full movement of the indicator.

Runout is sometimes described as a locational tolerance but is more often listed as a composite tolerance. This is because runout controls not the absolute position of a feature but rather the variations in position. In so doing, it also controls many other variations of form and orientation in relation to the datum axis.

There are two concepts in runout tolerancing. One is known as *circular runout* and concerns runout of each circular element or cross section. The other is called *total runout* and provides composite control of all surface elements simultaneously. Circular runout is less complex than total runout and is generally easier to measure.

There are two geometric characteristic symbols for runout, one for circular and one for total runout. These are shown in Fig. 31-11-15, where the dimensions are given as percentages of the feature control symbol frame height. These symbols are used with suitable tolerances in a feature control symbol in the usual manner.

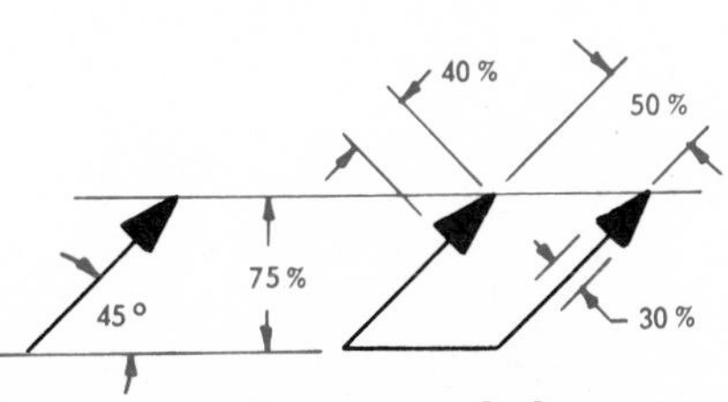
40 %
50 %
75 %
45°
30 %

Fig. 31-11-15 Runout symbols.

CIRCULAR RUNOUT

Circular runout means that the runout tolerance must be met in each circular element and that a sufficient number of measurements must be made at different positions along the surface to ensure that the tolerance is met at all cross sections of the part.

Thus in Fig. 31-11-16 the surface is measured at several positions along the surface, as shown by the three indicator positions. At each position the indicator movement during one revolution of the part must not exceed the specified tolerance, in this case 0.1 mm. For a cylindrical feature such as this, runout error is caused by eccentricity and errors of roundness. It is not affected by taper (conicity) or errors of straightness of the straight-line elements such as barrel shaping.

Circular runout may be applied to surfaces of revolution which are at any desired angle in relation to the datum axis.

Figure 31-11-17 shows a part where the tolerance is applied to a surface at right angles to the axis. In this case, an error will be shown, generally referred to as *wobble*, if the surface is flat but not perpendicular to the axis, as shown at (A).

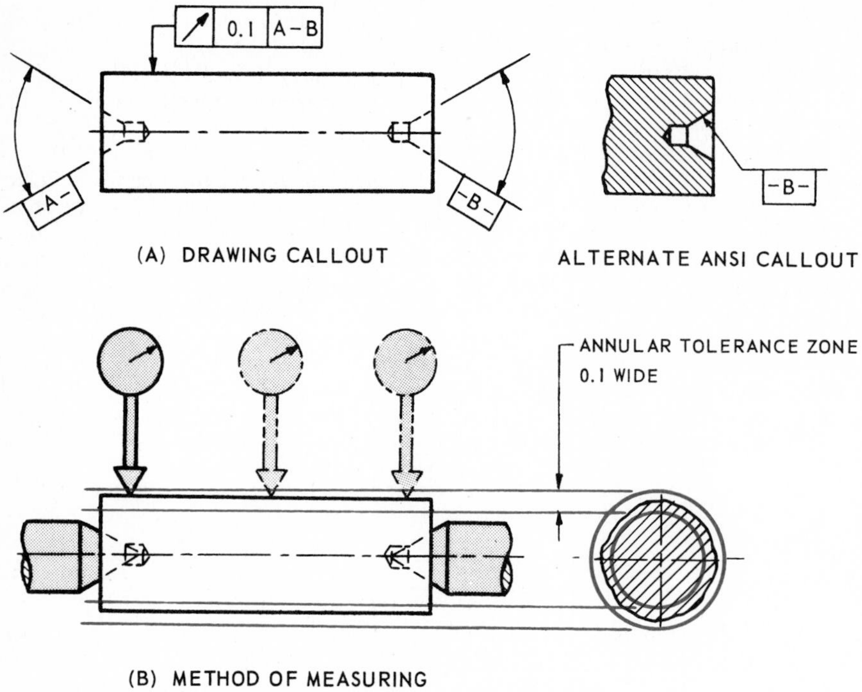
0.1
A−B
-A-
-B-
(A) DRAWING CALLOUT
ALTERNATE ANSI CALLOUT
ANNULAR TOLERANCE ZONE
0.1 WIDE
(B) METHOD OF MEASURING

Fig. 31-11-16 Circular runout of cylindrical features.

No error will be indicated if the surface is convex or concave but otherwise perfect, as shown at (B).

Circular runout can also be applied to curved surfaces, as shown in Fig. 31-11-18. Unless otherwise specified, measurement is always made normal to the surface, as shown by the arrows in the figure.

A runout tolerance directed to a surface applies to the full length of the surface up to an abrupt change in direction. Thus in Fig. 31-11-18 the tolerance applies to the curved portion but not to the cylindrical portion. If a control is intended to apply to more than one portion of a surface, additional leaders and arrowheads may be used, as in Fig. 31-11-19, where the same tolerance applies. If different tolerance values are required, separate tolerances must be specified.

If only part of a surface or several consecutive portions require the same tolerance, the length to which the tolerance applies may be indicated as shown in Fig. 31-11-20.

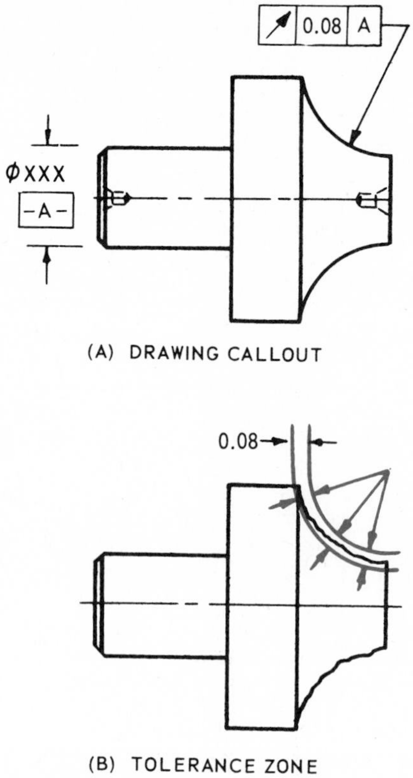
0.08
A
ϕXXX
-A-
(A) DRAWING CALLOUT
0.08
(B) TOLERANCE ZONE

Fig. 31-11-18 Runout of curved surface.

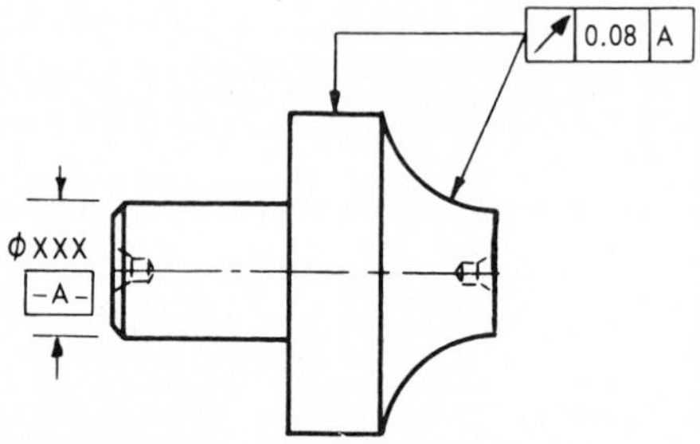
0.08
A
ϕXXX
-A-

Fig. 31-11-19 Tolerance applicable to two surfaces.

Fig. 31-11-17 Runout perpendicular to axis.

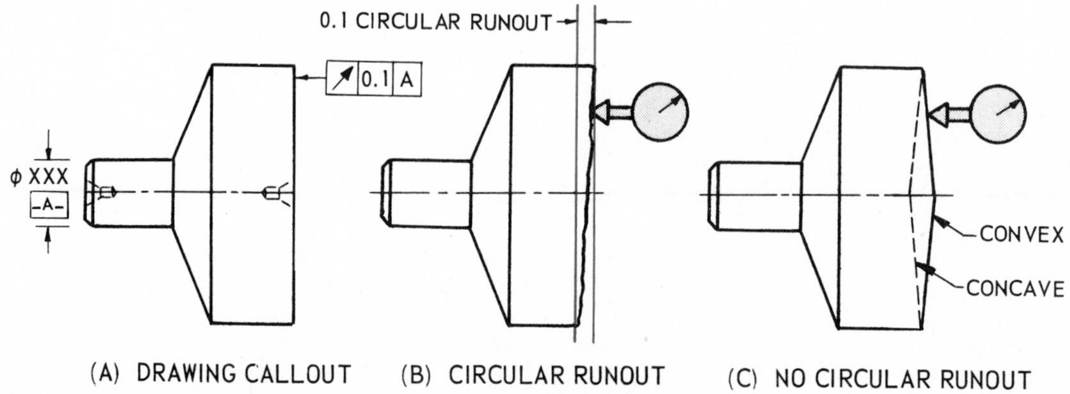
0.1 CIRCULAR RUNOUT
0.1
A
ϕ XXX
-A-
CONVEX
CONCAVE
(A) DRAWING CALLOUT
(B) CIRCULAR RUNOUT
(C) NO CIRCULAR RUNOUT

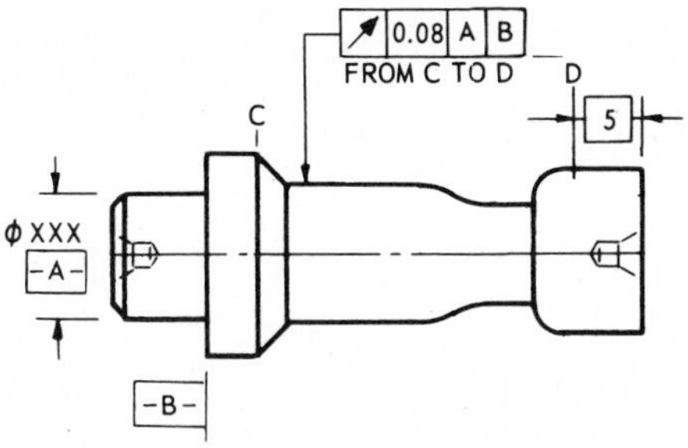
0.08 A B
FROM C TO D
C
D
5
ϕ XXX
-A-
-B-

Fig. 31-11-20 Tolerance applicable to part of a surface.

TOTAL RUNOUT

Total runout concerns the runout of a complete surface, not merely the runout of each circular element. For measurement purposes the checking indicator must traverse the full length or extent of the surface while the part is revolved about its datum axis. Measurements are made over the whole surface without resetting the indicator. Total runout is the difference between the lowest indicator reading in any position and the highest reading in that or in any other position on the same surface. Thus in Fig. 31-11-21 the tolerance zone is the space between two concentric cylinders separated by the specified tolerance and coaxial with the datum axis. All points on the controlled surface must lie within this zone simultaneously.

METHODS OF ESTABLISHING DATUMS

In many examples the datum axis has been established from centers drilled in the two ends of the part, in which case the part is mounted between centers for measurement purposes. This is an ideal method of mounting and revolving the part when such centers have been provided for manufacturing purposes. When centers are not provided, any cylindrical or conical surface may be used to establish the datum axis if it is chosen on the basis of the functional requirements of the part. In some cases, a runout tolerance may also be applied to the datum feature.

Figure 31-11-22 shows a simple, external cylindrical feature specified as the datum feature.

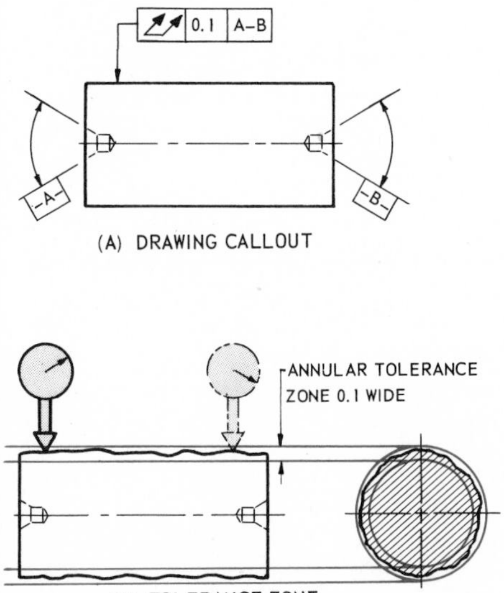
0.1 A–B
-A-
-B-
(A) DRAWING CALLOUT
ANNULAR TOLERANCE
ZONE 0.1 WIDE
(B) TOLERANCE ZONE

Fig. 31-11-21 Tolerance zone for total runout.

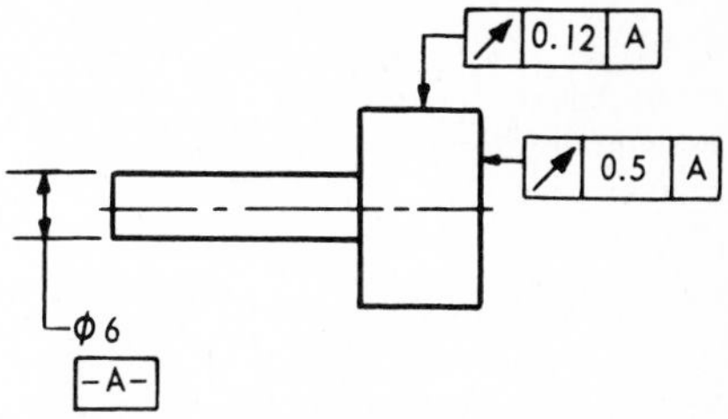
0.12 A
0.5 A
ϕ6
-A-

Fig. 31-11-22 Cylindrical datum feature.

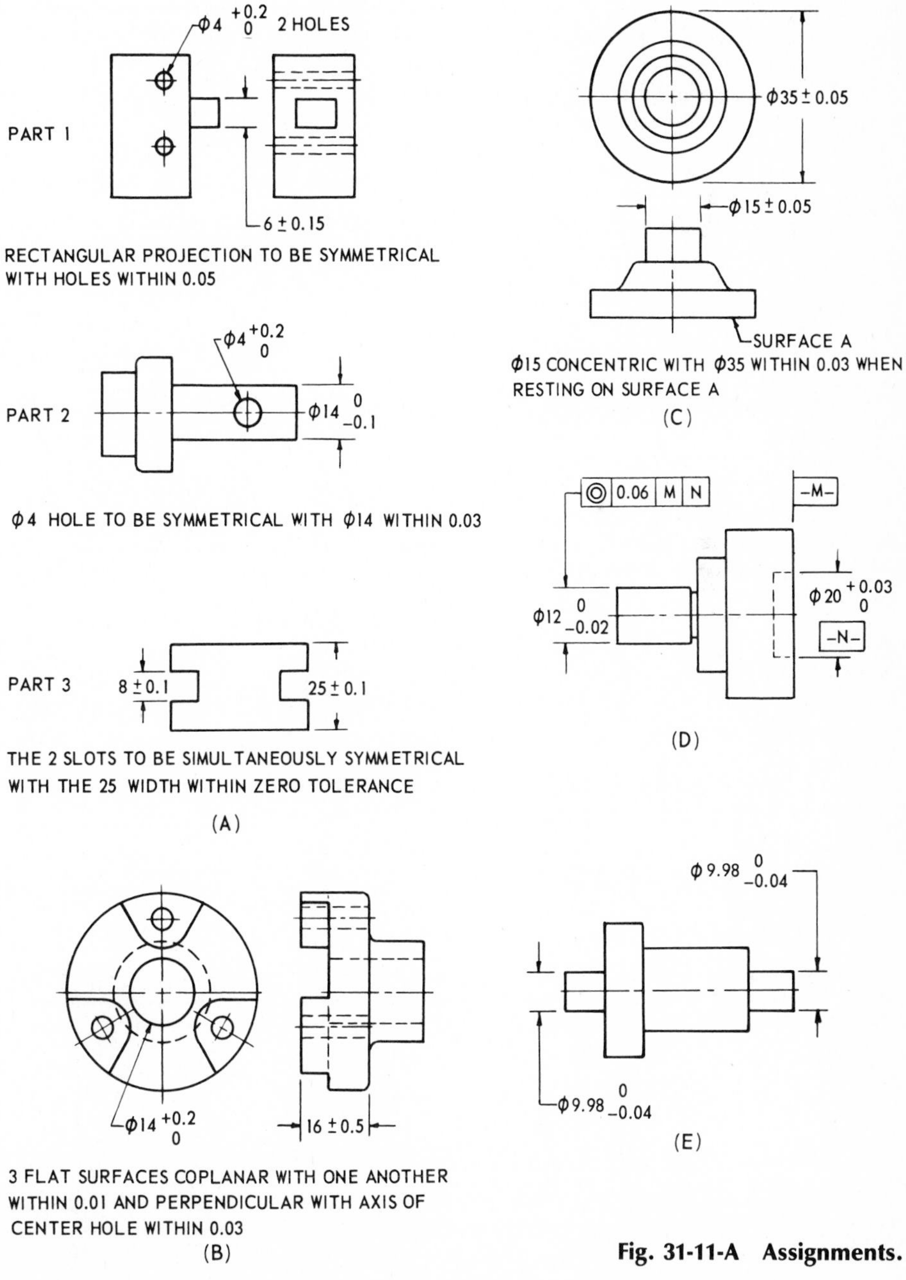
ϕ4 +0.2 0 2 HOLES
PART 1
6 ± 0.15
RECTANGULAR PROJECTION TO BE SYMMETRICAL
WITH HOLES WITHIN 0.05
ϕ4 +0.2 0
PART 2
ϕ14 0 -0.1
ϕ4 HOLE TO BE SYMMETRICAL WITH ϕ14 WITHIN 0.03
PART 3
8 ± 0.1
25 ± 0.1
THE 2 SLOTS TO BE SIMULTANEOUSLY SYMMETRICAL
WITH THE 25 WIDTH WITHIN ZERO TOLERANCE
(A)
ϕ14 +0.2 0
16 ± 0.5
3 FLAT SURFACES COPLANAR WITH ONE ANOTHER
WITHIN 0.01 AND PERPENDICULAR WITH AXIS OF
CENTER HOLE WITHIN 0.03
(B)
ϕ35 ± 0.05
ϕ15 ± 0.05
SURFACE A
ϕ15 CONCENTRIC WITH ϕ35 WITHIN 0.03 WHEN
RESTING ON SURFACE A
(C)
0.06 M N
-M-
ϕ12 0 -0.02
ϕ20 +0.03 0
-N-
(D)
ϕ9.98 0 -0.04
ϕ9.98 0 -0.04
(E)

Fig. 31-11-A Assignments.

Assignment

On an A3- or B-size sheet, prepare drawings and solutions for the illustrations shown in Fig. 31-11-A or 31-11-B and the following information.

1. In illustration (A) show tolerances which will ensure that features are symmetrical at their maximum material conditions as shown on drawings.

2. It is required to have the three flat surfaces in illustration (B) coplanar with one another and perpendicular with the axis of the center hole within the tolerances specified. Add suitable geometric tolerances to the drawing. Show the tolerance zones for the part.

3. It is required to have the top diameter in illustration (C) concentric with the bottom diameter, when resting on surface *A*. Show how this should be specified on the drawing of an RFS basis.

4. Show the size and form of the tolerance zone for the concentricity tolerance shown in illustration (D).

5. The part shown in illustration (E) is intended to function by rotating with the two end diameters supported in bearings. The two larger diameters are required to run true within the tolerance specified. Show how this would be toleranced.

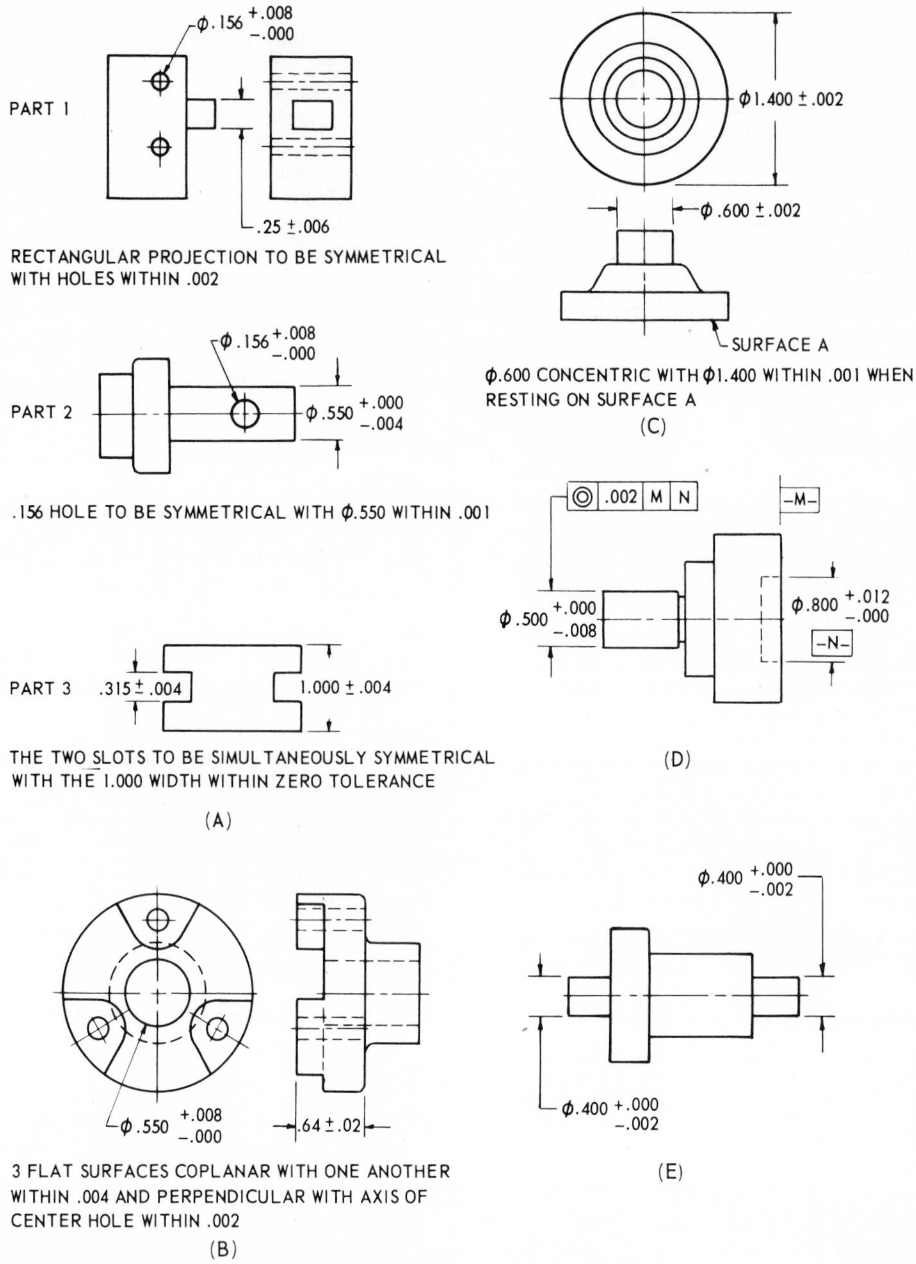

PART 1
ϕ.156 +.008 −.000
.25 ±.006
RECTANGULAR PROJECTION TO BE SYMMETRICAL WITH HOLES WITHIN .002
PART 2
ϕ.156 +.008 −.000
ϕ.550 +.000 −.004
.156 HOLE TO BE SYMMETRICAL WITH ϕ.550 WITHIN .001
PART 3
.315 ± .004
1.000 ± .004
THE TWO SLOTS TO BE SIMULTANEOUSLY SYMMETRICAL WITH THE 1.000 WIDTH WITHIN ZERO TOLERANCE
(A)
ϕ1.400 ±.002
ϕ.600 ±.002
SURFACE A
ϕ.600 CONCENTRIC WITH ϕ1.400 WITHIN .001 WHEN RESTING ON SURFACE A
(C)
.002 M N
-M-
ϕ.500 +.000 −.008
ϕ.800 +.012 −.000
-N-
(D)
ϕ.550 +.008 −.000
.64 ±.02
3 FLAT SURFACES COPLANAR WITH ONE ANOTHER WITHIN .004 AND PERPENDICULAR WITH AXIS OF CENTER HOLE WITHIN .002
(B)
ϕ.400 +.000 −.002
ϕ.400 +.000 −.002
(E)

Fig. 31-11-B Assignments.

Chapter 32
Descriptive Geometry

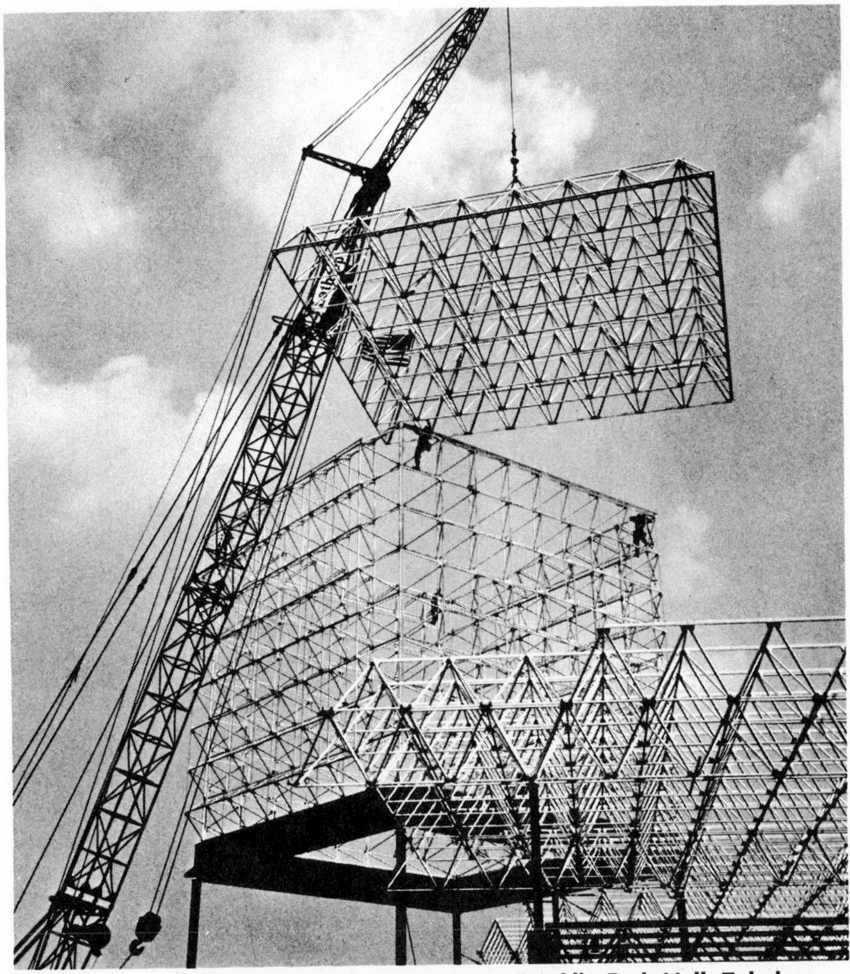

Fig. 32-1-1 Geometric space-frame structure, Franklin Park Mall, Toledo, Ohio. (Unistrut Corp.)

UNIT 32-1

GRAPHIC SOLUTIONS

A major problem in technical drawing and design is the projection for finding the true views of lines and planes. The following is a brief review of the principles of descriptive geometry involved in the solution of such problems. The designer working along with an engineering team can solve problems graphically with geometric elements. Structures that occupy space have three-dimensional forms made up of a combination of geometric elements (Fig. 32-1-1). The graphic solutions of three-dimensional forms require an understanding of the space relations which points, lines, and planes share in forming any given shape. Problems which many times require mathematical solutions can often be solved graphically with the accuracy that will allow manufacturing and construction. *Basic descriptive geometry* is one of the designer's methods of thinking through and solving problems. In the 18th century a French mathematician, Gaspard Monge, developed the principles of graphically solving spatial problems related to military structures. Descriptive geometry was introduced to the U.S. Military Academy at West Point by Claude Crozet in 1816. The Mongean method of presentation has changed, but the basic principles are still taught in engineering schools throughout the world. The visual studies required in descriptive geometry assist in the development of the reasoning powers used in graphically solving problems.

In this chapter a graphic technique is used to analyze all geometric elements. The visual examination of geometric elements will assist in the description of structures of every possible shape. The basic shape of most structures designed by human beings is rectangular. This is a result of the relatively simple planning needed to form this type of structure. Figure 32-1-2 shows the basic geometric elements and some of the common and unusual geometric features of engineering designs.

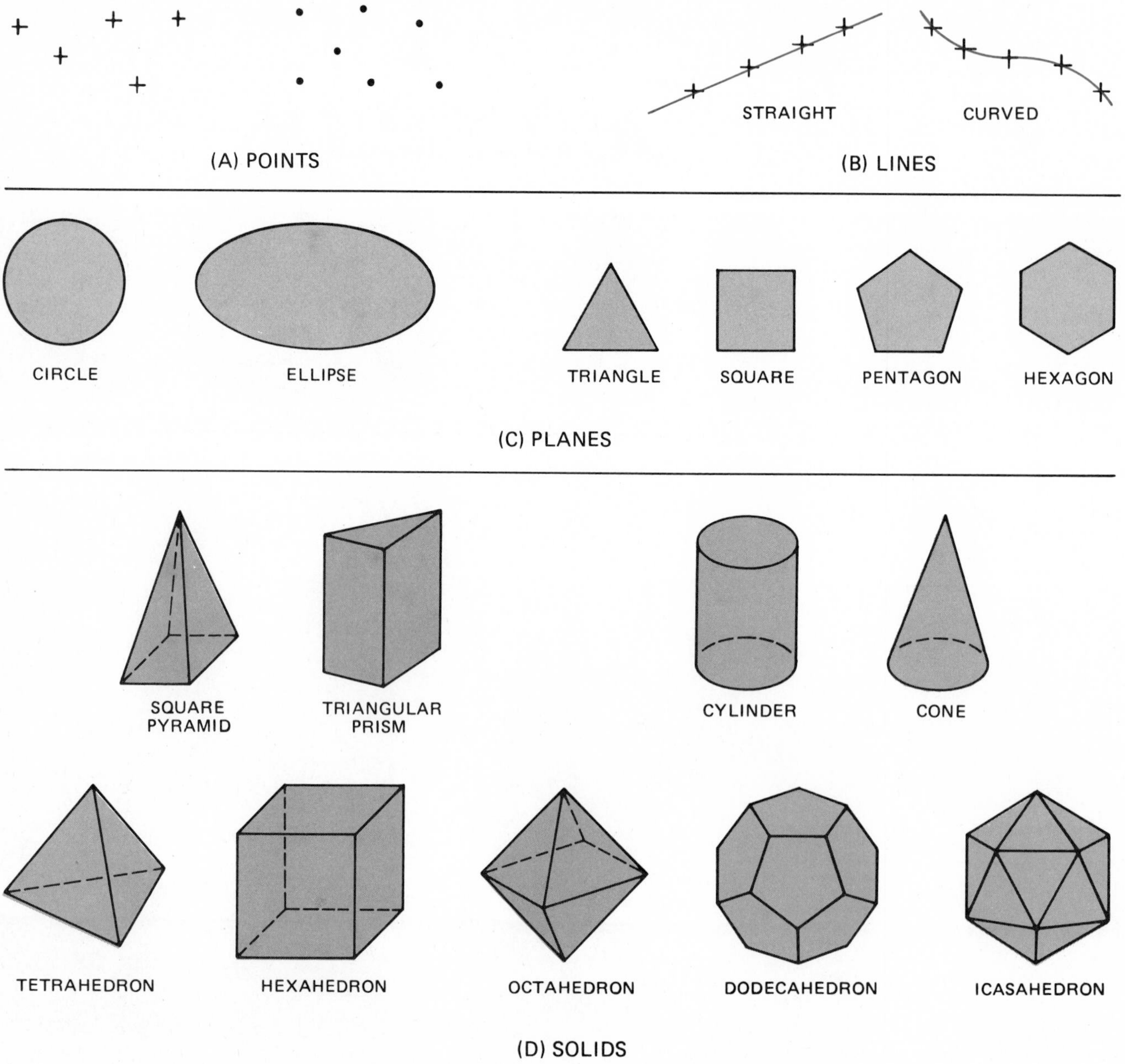
(A) POINTS
STRAIGHT
CURVED
(B) LINES
CIRCLE
ELLIPSE
TRIANGLE
SQUARE
PENTAGON
HEXAGON
(C) PLANES
SQUARE PYRAMID
TRIANGULAR PRISM
CYLINDER
CONE
TETRAHEDRON
HEXAHEDRON
OCTAHEDRON
DODECAHEDRON
ICASAHEDRON
(D) SOLIDS

Fig. 32-1-2 Basic geometric elements and shapes. (a) Points. (b) Lines. (c) Planes. (d) Solids.

Reference Planes

Unfolding of the reference planes forms a two-dimensional surface which a drafter uses to construct and solve problems (see Fig. 32-1-3). The planes are labeled so that *T* represents the top or horizontal reference plane, *F* represents the front or vertical reference plane, and *S* represents the side (end) or profile reference plane. Thus a point 1 on the part, line, or plane would be identified as 1_F on the front reference plane, 1_T on the top reference plane, and 1_S on the side reference plane.

The folding lines shown on the box are referred to as *reference lines* on the drawing. Other reference planes and reference lines are drawn and labeled as required.

Points

A point can be considered physically real and can be located by a small dot or a small cross. It is normally identified by two or more projections. In Fig. 32-1-4*a*, points *A* and *B* are located on all three reference planes. Notice that the unfolding of the three planes forms a two-dimensional surface with the fold lines remaining. The folding lines are labeled as shown to indicate that *F* represents the front view, *T* represents the top view, and *S* represents the profile or right-side view. In Fig. 32-1-4*b* the planes are replaced with reference lines RL_1 and RL_2, which are placed in the same position as the fold lines in Fig. 32-1-4*a*.

Lines in Space

Lines in descriptive geometry are grouped into three classes depending on how they are positioned with relationship to the reference lines.

Normal Lines. A line which is perpendicular to the reference plane will project as a point on that plane. In Fig. 32-1-5*a*, line *AB* is perpendicular to the front reference plane. As such, it is shown as a point ($A_F B_F$) in the front view and as a true-length line in the top and

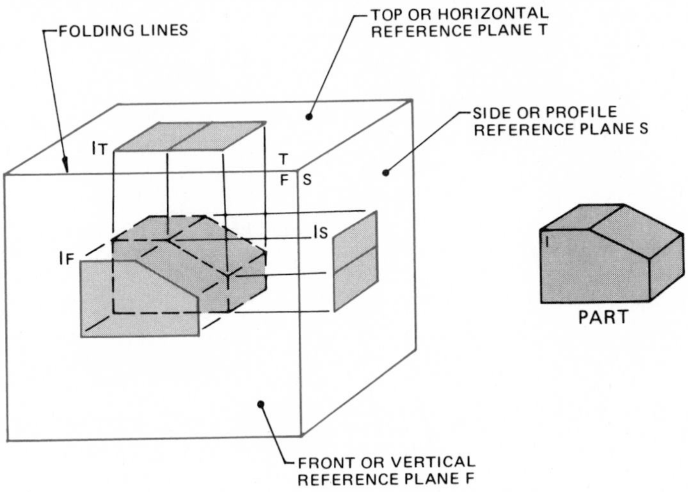
FOLDING LINES
TOP OR HORIZONTAL REFERENCE PLANE T
SIDE OR PROFILE REFERENCE PLANE S
1T
T
F S
1S
1F
1
PART
FRONT OR VERTICAL REFERENCE PLANE F

(A) PICTORIAL VIEW OF REFERENCE PLANES

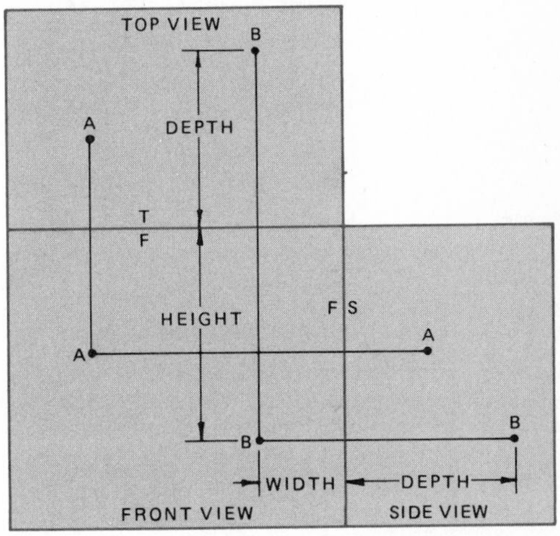
TOP VIEW
B
A
DEPTH
T
F
HEIGHT
F S
A
A
B
B
WIDTH
DEPTH
FRONT VIEW
SIDE VIEW

(A) POINTS A AND B IDENTIFIED ON UNFOLDED REFERENCE PLANE

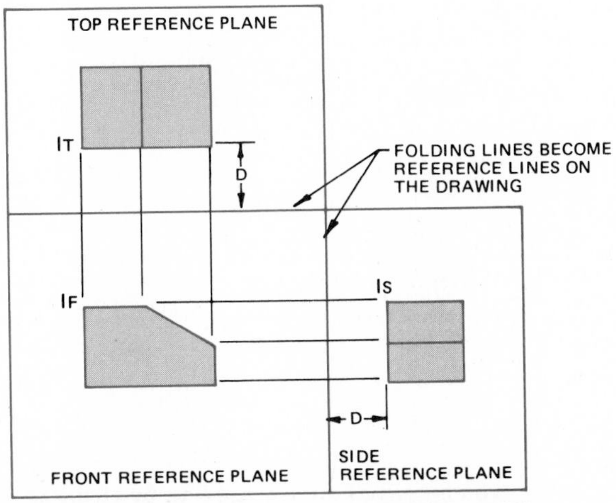
TOP REFERENCE PLANE
1T
D
FOLDING LINES BECOME REFERENCE LINES ON THE DRAWING
1S
1F
D
SIDE REFERENCE PLANE
FRONT REFERENCE PLANE

(B) UNFOLDING OF THE THREE REFERENCE PLANES

Fig. 32-1-3 Reference planes. (a) Pictorial view of reference planes. (b) Unfolding of the three reference planes.

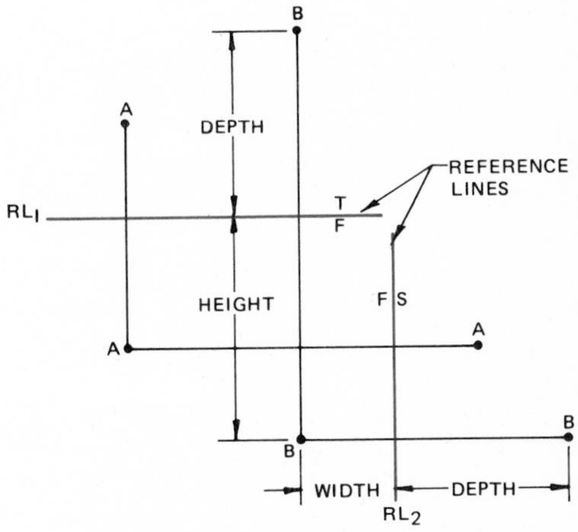
B
A
DEPTH
REFERENCE LINES
RL1
T
F
HEIGHT
F S
A
A
B
B
WIDTH
DEPTH
RL2

(B) POINTS A AND B IDENTIFIED TO REFERENCE LINES

Fig. 32-1-4 Points in space. (a) Points A and B identified on unfolded reference plane. (b) Points A and B identified to reference lines.

side views (lines A_TB_T and A_SB_S, respectively).

Inclined Lines. A line that appears inclined in one plane, as shown in Fig. 32-1-5*b*, and is parallel to one on the other two principal planes will appear foreshortened in the other two views. The inclined line shown in the front view will be the true length of line *AB*.

Oblique Lines. A line that appears inclined in all three views is an oblique line. It is neither parallel nor perpendicular to any of the three planes. The true length of the line is not shown in any of these views. See Fig. 32-1-5*c*.

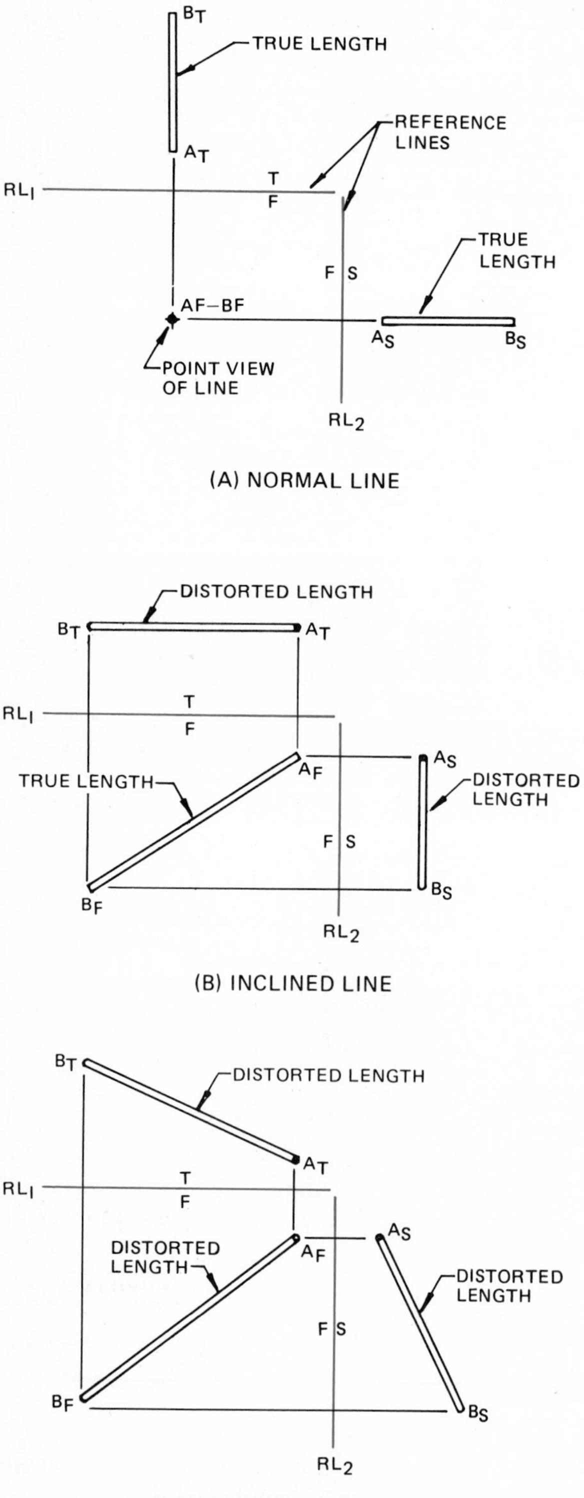
TRUE LENGTH
REFERENCE LINES
RL1
T
F
F S
TRUE LENGTH
AF–BF
POINT VIEW OF LINE
AS
BS
RL2
(A) NORMAL LINE
DISTORTED LENGTH
BT
AT
TRUE LENGTH
AF
AS
DISTORTED LENGTH
BF
BS
(B) INCLINED LINE
DISTORTED LENGTH
DISTORTED LENGTH
(C) OBLIQUE LINES

Fig. 32-1-5 **Lines in space. (a) Normal line. (b) Inclined line. (c) Oblique lines.**

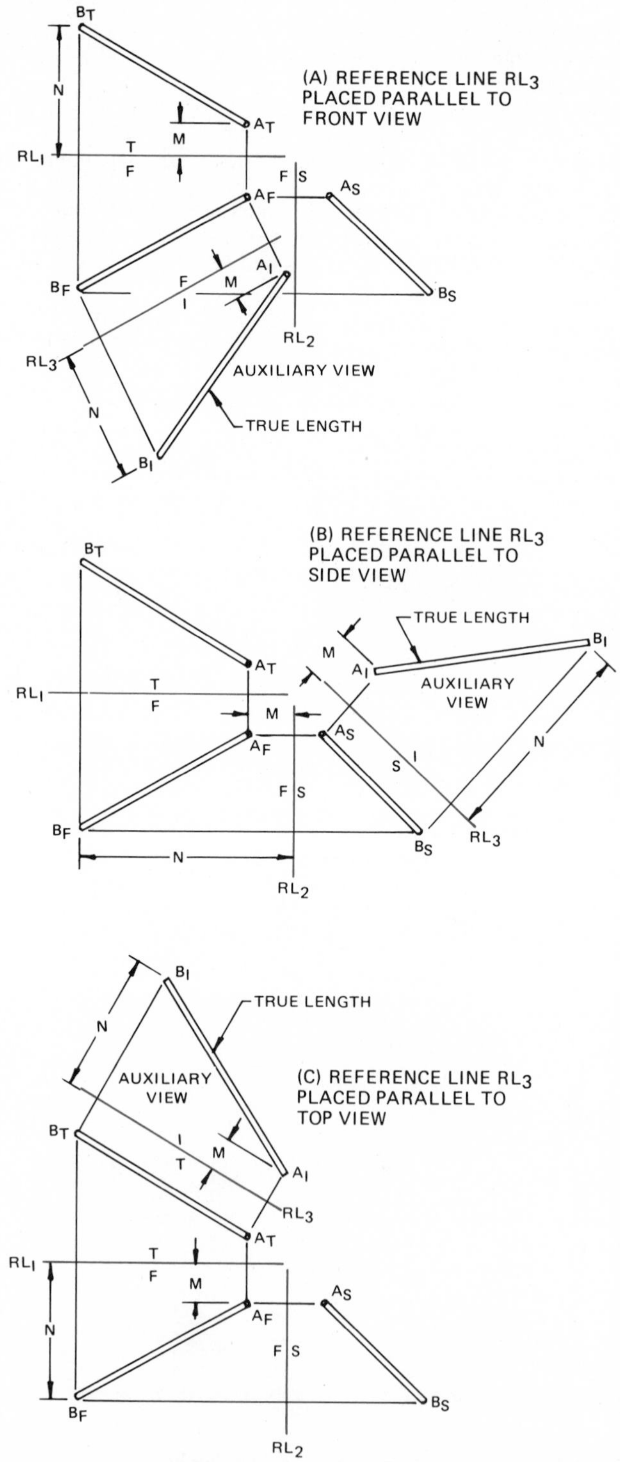
(A) REFERENCE LINE RL3 PLACED PARALLEL TO FRONT VIEW
AUXILIARY VIEW
TRUE LENGTH
(B) REFERENCE LINE RL3 PLACED PARALLEL TO SIDE VIEW
TRUE LENGTH
AUXILIARY VIEW
(C) REFERENCE LINE RL3 PLACED PARALLEL TO TOP VIEW
TRUE LENGTH
AUXILIARY VIEW
RL1
RL2
RL3
N
M

Fig. 32-1-6 True length of an oblique line by auxiliary projection. (a) Reference line RL_3 placed parallel to front view. (b) Reference line RL_3 placed parallel to side view. (c) Reference line RL_3 placed parallel to top view.

True Length of an Oblique Line by Auxiliary Projection

The true length of a normal line and the inclined line have line projections which are parallel to a principal plane of projection. It is therefore concluded that a line parallel to a plane of projection shows true length in that projection. Since an oblique line is not parallel to any of the three principal reference planes, an auxiliary reference line RL_3 can be placed parallel to any one of the oblique lines, as shown in Fig. 32-1-6. Transfer distances designated as M and N shown in the regular views to the auxiliary view locating points A_1 and B_1, respectively. Join points A_1 and B_1 with a line, obtaining the true length of line AB.

Point on a Line

The line $A_F B_F$ in the front view of Fig. 32-1-7*a* contains a point C. To place point C on the line in the other two views, it is necessary to project construction lines perpendicular to the reference lines RL_1 and RL_2, as shown in Fig. 32-1-7. The construction lines are projected to line $A_T B_T$ in the top view and line $A_S B_S$ in the side view, locating point C on the line in these views.

If point C was to be located on the true length of line AB, then another reference line, such as RL_1, is required, and the distances L and M in the front view are then used to locate the true length of line $A_1 B_1$ in the auxiliary view. Position C is then projected perpendicular to line $A_S B_S$ in the side view to locate C on the true-length line.

Point-on-Point View of a Line

When the front view $A_F B_F$ and top view $A_T B_T$ are given, as in Fig. 32-1-8, and the point-on-point view of line AB is required, the procedure is as follows:

- Draw a true-length view of AB by the method described in True Length of an Oblique Line by Auxiliary Projection (Fig. 32-1-6).
- Draw a reference line perpendicular to the true length of line $A_1 B_1$, and label it RL_3.
- The next adjacent, secondary auxiliary view $A_2 B_2$ will be a point-on-point view of line AB.

Assignments

1. Divide an A3- or B-size sheet into four parts. With the use of reference lines, locate the points in the third view for the drawings shown in Fig. 32-1-A. Scale is to suit.

2. Divide an A3- or B-size sheet into four parts. With the use of reference lines, locate the lines in the other views for the drawings shown in Fig. 32-1-B. Scale is to suit.

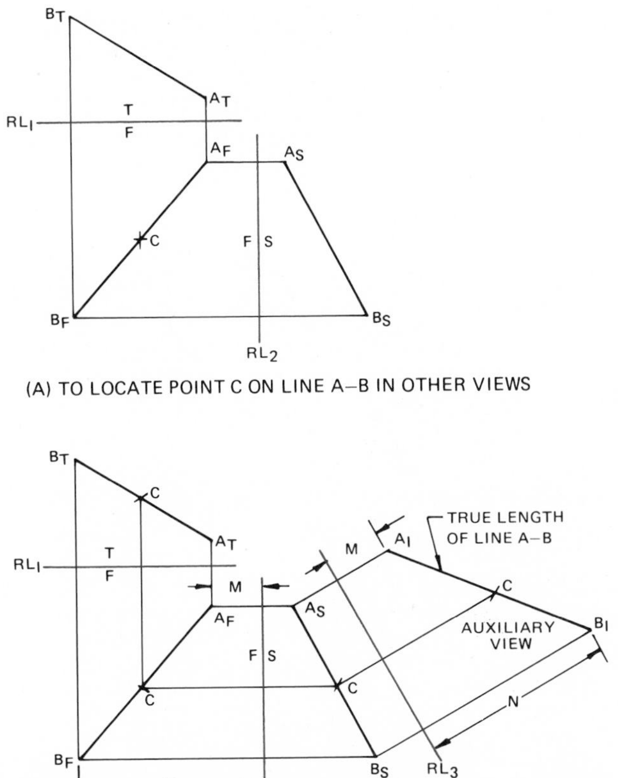
BT
AT
RL1
T
F
AF
AS
C
F S
BF
BS
RL2
(A) TO LOCATE POINT C ON LINE A–B IN OTHER VIEWS
TRUE LENGTH
OF LINE A–B
M
A1
C
AUXILIARY
VIEW
B1
N
RL3

Fig. 32-1-7 Point on a line. (a) To locate point C on line AB in other views.

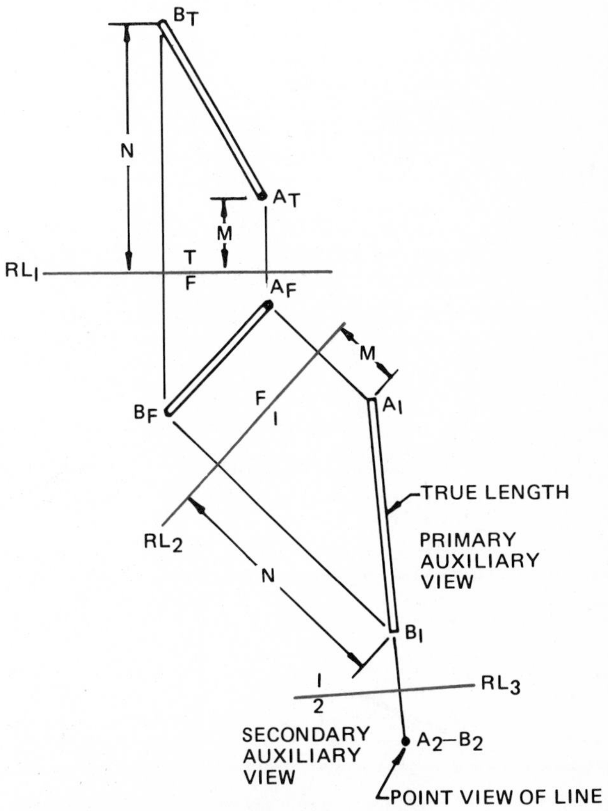
BT
N
AT
M
T
F
RL1
AF
M
F
1
A1
BF
TRUE LENGTH
PRIMARY
AUXILIARY
VIEW
RL2
N
B1
1
2
RL3
SECONDARY
AUXILIARY
VIEW
A2–B2
POINT VIEW OF LINE

Fig. 32-1-8 Point-on-point view of a line.

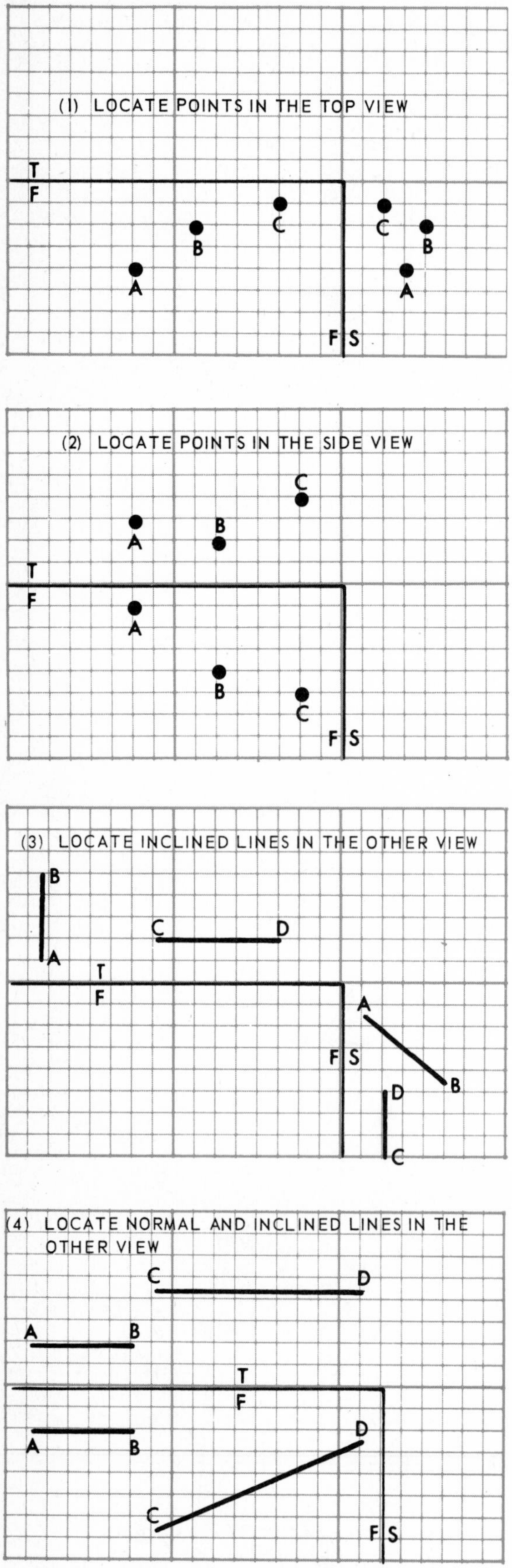
(1) LOCATE POINTS IN THE TOP VIEW
T
F
A B C C B A
F S
(2) LOCATE POINTS IN THE SIDE VIEW
C A B
T
F
A B C
F S
(3) LOCATE INCLINED LINES IN THE OTHER VIEW
B A C D
T
F
A B D C
F S
(4) LOCATE NORMAL AND INCLINED LINES IN THE OTHER VIEW
C D
A B
T
F
A B D C
F S

Fig. 32-1-A Assignments.

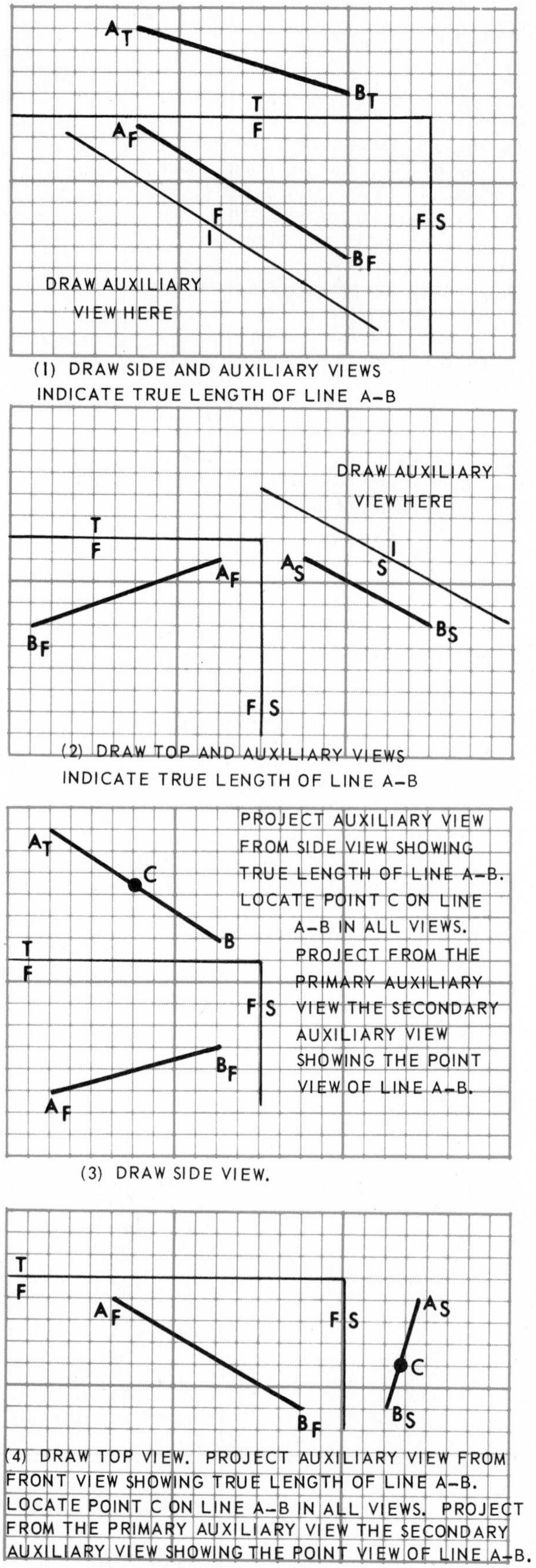
A_T B_T
T
F
A_F B_F
F
1
F S
DRAW AUXILIARY VIEW HERE
(1) DRAW SIDE AND AUXILIARY VIEWS
INDICATE TRUE LENGTH OF LINE A–B
DRAW AUXILIARY VIEW HERE
T
F
A_F B_F
A_S B_S
S
1
F S
(2) DRAW TOP AND AUXILIARY VIEWS
INDICATE TRUE LENGTH OF LINE A–B
A_T C B
T
F
F S
A_F B_F
PROJECT AUXILIARY VIEW FROM SIDE VIEW SHOWING TRUE LENGTH OF LINE A–B. LOCATE POINT C ON LINE A–B IN ALL VIEWS. PROJECT FROM THE PRIMARY AUXILIARY VIEW THE SECONDARY AUXILIARY VIEW SHOWING THE POINT VIEW OF LINE A–B.
(3) DRAW SIDE VIEW.
T
F
A_F B_F
F S
A_S C B_S
(4) DRAW TOP VIEW. PROJECT AUXILIARY VIEW FROM FRONT VIEW SHOWING TRUE LENGTH OF LINE A–B. LOCATE POINT C ON LINE A–B IN ALL VIEWS. PROJECT FROM THE PRIMARY AUXILIARY VIEW THE SECONDARY AUXILIARY VIEW SHOWING THE POINT VIEW OF LINE A–B.

Fig. 32-1-B Assignments.

UNIT 32-2

PLANES IN SPACE

Planes for practical studies are considered to be without thickness and can be extended without limit. A plane may be represented or determined by intersecting lines, two parallel lines, a line and a point, three points, or a triangle.

The three basic planes, referred to as the *normal plane*, *inclined plane*, and *oblique plane*, are identified by their relationship to the three principal reference planes. Figure 32-2-1 illustrates the three basic planes, each plane being triangular in shape.

Normal Plane. This is a plane whose surface, in this case a triangular surface, appears in its true shape in the front view and as a line in the other two views.

Inclined Plane. This results when the shape of the triangular plane appears distorted in two views and as a line in the other view.

Oblique Plane. This is a plane whose shape appears distorted in all three views.

Locating a Line in a Plane

The top and front views shown in Fig. 32-2-2*a* show a triangular plane *ABC* and lines *RS* and *MN*, each located in one of the views. To find their location in the other views, refer to Fig. 32-2-2*b* and the following procedures.

To locate line RS in the front view:

- Line R_TS_T crosses over lines A_TB_T and A_TC_T at points D_T and E_T, respectively.
- Project points D_T and E_T to front view, locating points D_F and E_F.
- Extend a line through points D_F and E_F.
- The length of the line can be found by projecting points R_T and S_T to the front view, locating the end points R_F and S_F.

To locate line MN in the top view:

- Extend line M_FN_F in the front view, locating points H_F and G_F on lines A_FB_F and A_FC_F, respectively.
- Project points H_F and G_F to the top view, locating points H_T and G_T.
- Draw a line through points H_T and G_T.
- Project points M_F and N_F to top view, locating points on line M_TN_T.

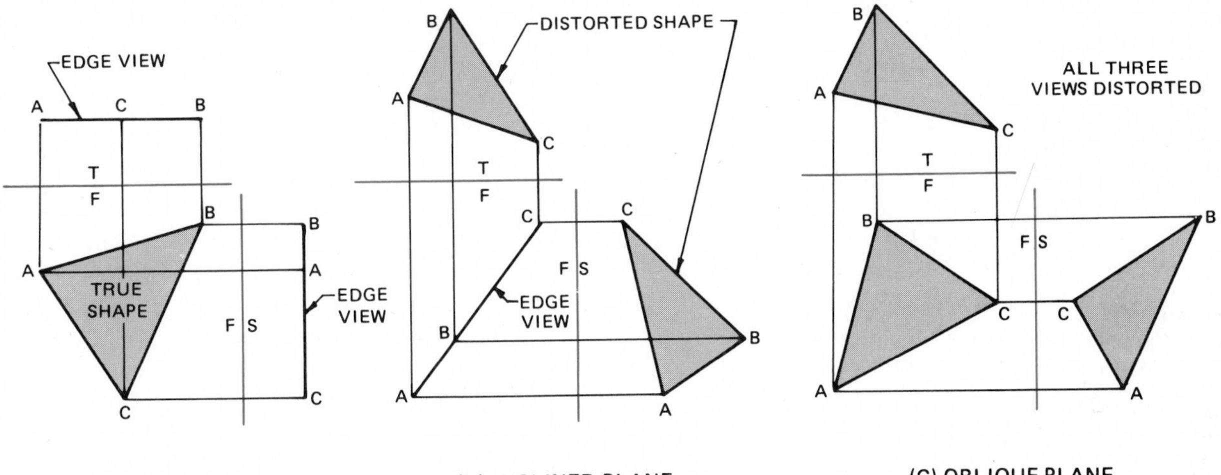

EDGE VIEW
TRUE SHAPE
DISTORTED SHAPE
ALL THREE VIEWS DISTORTED
T
F
F S

(A) NORMAL PLANE (B) INCLINED PLANE (C) OBLIQUE PLANE

Fig. 32-2-1 Planes in space. (a) Normal plane. (b) Inclined plane. (c) Oblique plane.

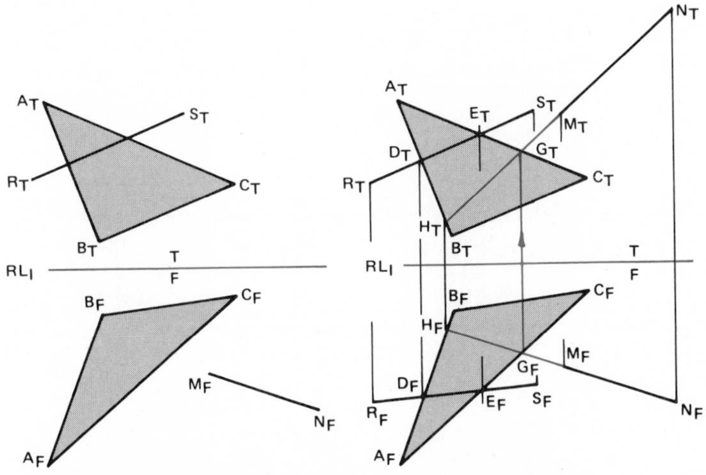

RL1
T
F

(A) PROBLEM—TO LOCATE LINE IN THE OTHER VIEW (B) SOLUTION

Fig. 32-2-2 Locating a line in a plane. (a) Problem—to locate line in the other view. (b) Solution.

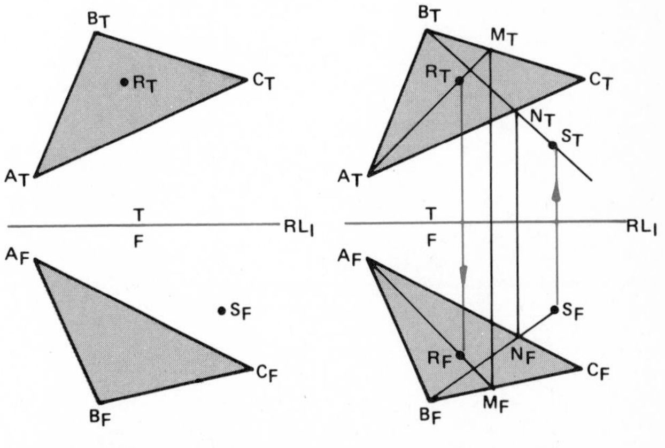

RL1
T
F

(A) PROBLEM—TO LOCATE POINT IN OTHER VIEW (B) SOLUTION

Fig. 32-2-3 Locating a point in a plane. (a) Problem—to locate point in other view. (b) Solution.

Locating a Point on a Plane

The top and front views shown in Fig. 32-2-3*a* show a triangular plane *ABC* and points *R* and *S* each located in one of the views. To find their location in the other views, refer to Fig. 32-2-3*b* and the following procedures.

To locate point R in the front view:

- Draw a line from A_T passing through point R_T to a point M_T on line B_TC_T.
- Project point M_T to front view, locating point M_F.
- Join points A_F and M_F with a line.
- Project point R_T to front view, locating point R_F.

To locate point S in the top view:

- Draw a line between points B_F to S_F, locating point N_F on line A_FC_F.
- Project point N_F to top view, locating point N_T.
- Draw a line through points B_T and N_T.
- Project point S_F to top view, locating point S_T.

Locating the Piercing Point of a Line and a Plane— Cutting-Plane Method

The top and front views shown in Fig. 32-2-4 show a line *UV* passing somewhere through plane *ABC*. The piercing point of the line through the plane is found as follows:

- Locate points D_T and E_T in the top view.
- Project points D_T and E_T to the front view, locating points D_F and E_F.
- The intersection of lines D_FE_F and U_FV_F is the piercing point.
- Project point O_F to top view, locating point O_T.

Locating the Piercing Point of a Line and a Plane— Auxiliary-View Method

The top and front views in Fig. 32-2-5 show a line *UV* passing somewhere through plane *ABC*. The piercing point of the line through the plane is found as follows:

- Draw line A_TD_T in the top view parallel to reference line RL_1.
- Project point D_T to front view, locating point D_F.
- Draw reference line RL_2 perpendicular to a line intersecting points A_F and D_F in the front view.
- Draw an auxiliary view which shows the edge view of the plane. The intersection of the plane and line locates the piercing point O_1.
- Project the piercing point to front and top views.

Assignments

1. *Normal and Inclined Planes in Space*. Divide an A3- or B-size sheet into four parts. With the use of reference lines, draw the front views (true shape) of triangle *ABC* shown in illustrations 1 and 2 of Fig. 32-2-A. Draw the side view in illustration 3 and the top view in illustration 4. Scale is to suit.

2. *Oblique Planes in Space*. Divide an A3- or B-size sheet into four parts. With the use of reference lines, draw the third view as indicated for the four planes shown in Fig. 32-2-B. Scale is to suit.

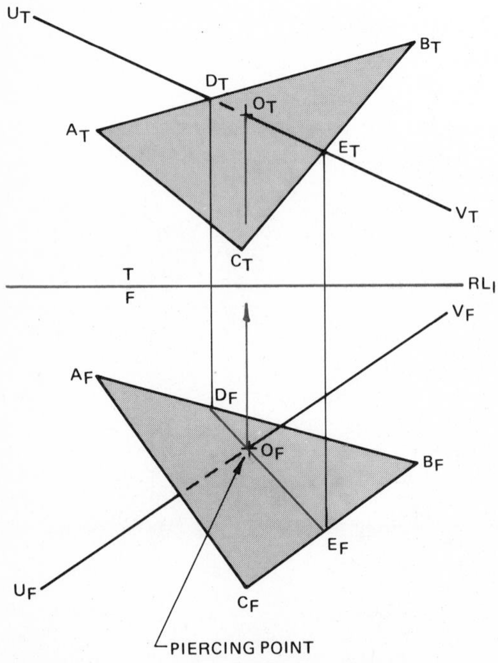
UT
BT
DT
OT
AT
ET
VT
CT
T
F
RL1
VF
AF
DF
OF
BF
EF
UF
CF
PIERCING POINT

Fig. 32-2-4 Locating piercing point of a line and a plane-cutting plane method.

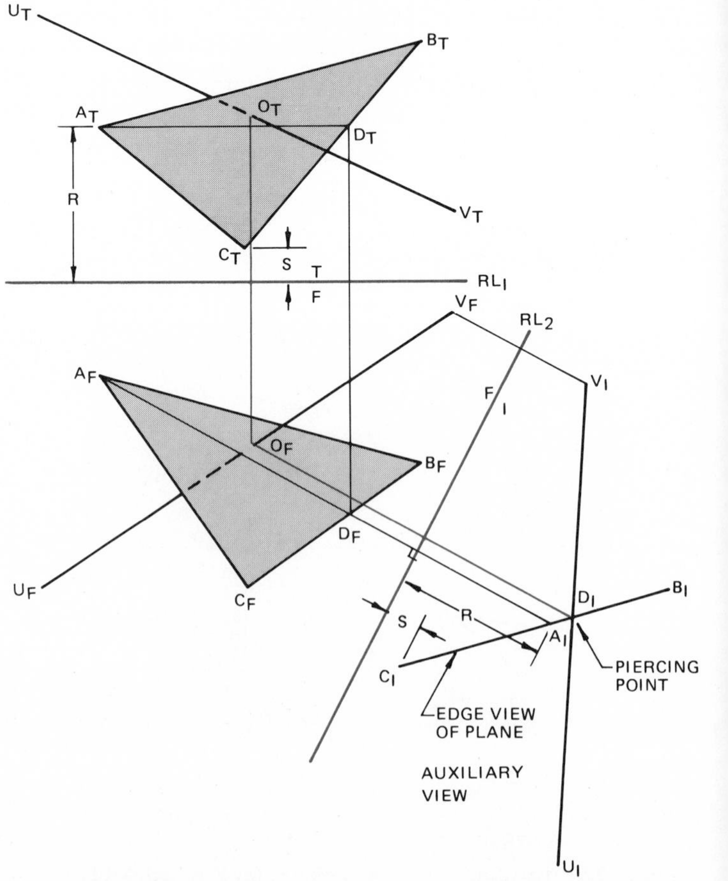
UT
BT
AT
OT
DT
R
VT
CT
S
T
F
RL1
VF
RL2
AF
V1
F
1
OF
BF
DF
UF
CF
D1
B1
S
R
A1
C1
PIERCING POINT
EDGE VIEW OF PLANE
AUXILIARY VIEW
U1

Fig. 32-2-5 Piercing point of a line and a plane-auxiliary view method.

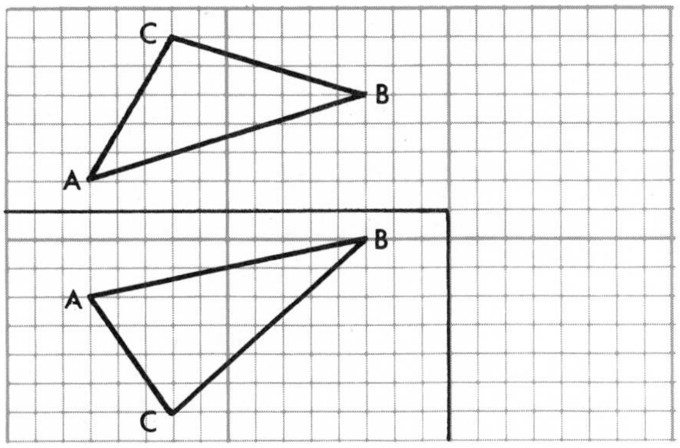
C
B
A
B
A
C

(1) DRAW THE SIDE VIEW OF TRIANGLE ABC

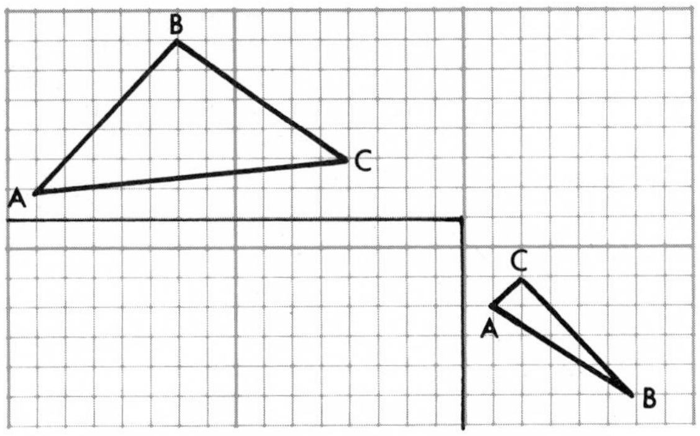
B
C
A
C
A
B

(2) DRAW THE FRONT VIEW OF TRIANGLE ABC

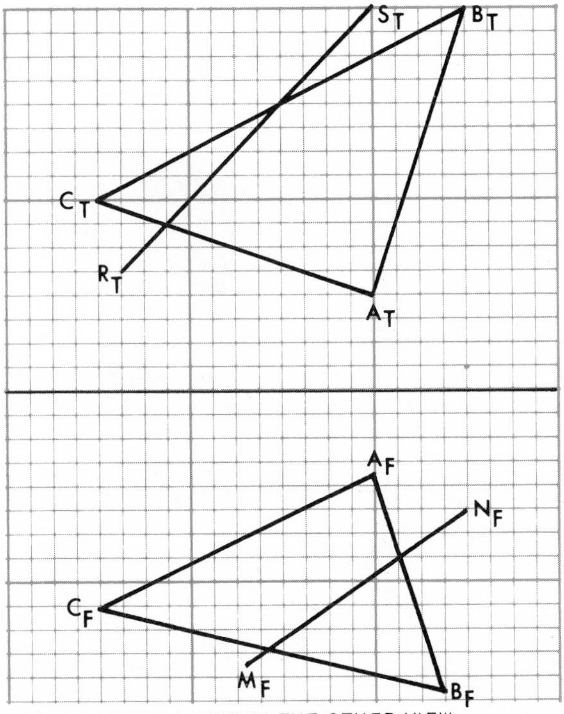
S_T
B_T
C_T
R_T
A_T
A_F
N_F
C_F
M_F
B_F

(3) LOCATING A LINE IN THE OTHER VIEW

Fig. 32-2-A Assignments.

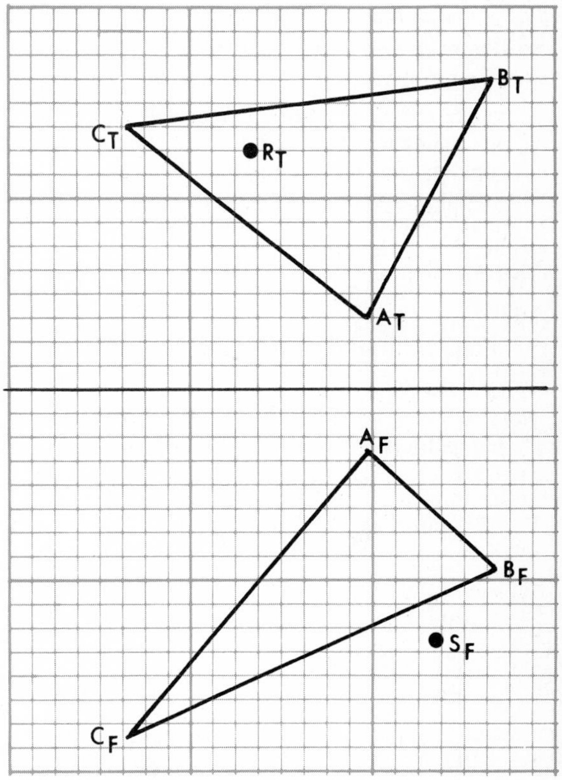
B_T
C_T
R_T
A_T
A_F
B_F
S_F
C_F

(1) LOCATING A POINT IN SPACE

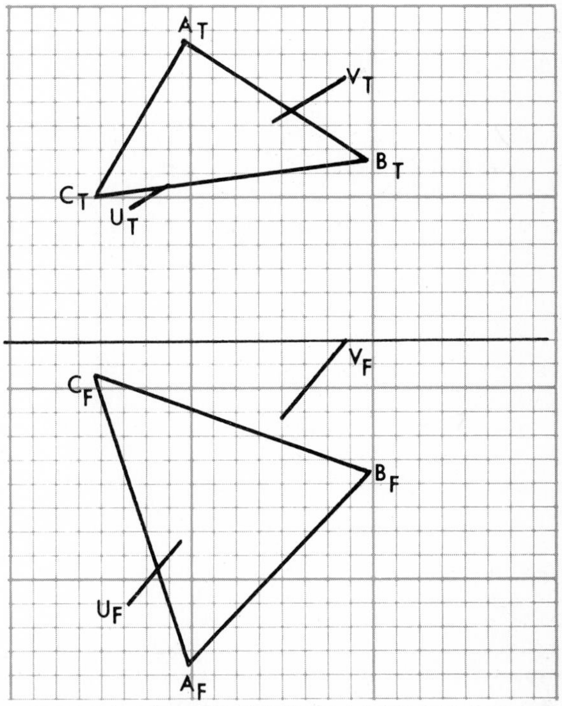
A_T
V_T
B_T
C_T
U_T
V_F
C_F
B_F
U_F
A_F

(2) LOCATING PIERCING POINT OF A LINE AND A PLANE

Fig. 32-2-B Assignments.

UNIT 32-3

ESTABLISHING VISIBILITY OF LINES IN SPACE

Visibility of Skew Lines by Testing

In the example of the two nonintersecting pipes shown in Fig. 32-3-1*a*, it is not apparent which pipe is nearest the viewer at the crossing points in the two views. To establish which of the pipes lies in front of the other, the following procedure is used.

To establish the visible pipe at the crossing shown in the top view (Fig. 32-3-1*b*):

- Label the crossing of lines A_TB_T and C_TD_T as 1 , 2 .
- Project the crossing point to the front view, establishing points 1 and 2 .
- Point 1 is closer to reference line RL_1, which means that line A_TB_T is nearer when the top view is being observed and thus is visible.

To establish the visible pipe at the crossing shown in the front view:

- Label the crossing of lines A_FB_F and C_FD_F as 3 , 4 .
- Project the crossing point to the top view, establishing points 3 and 4 .
- Point 4 is closer to reference line RL_1, which means that line C_FD_F is nearer when the front view is being observed and thus is visible. Figure 32-3-1*c* shows the correct crossings of the pipes.

Visibility of Lines and Surfaces by Testing

In cases where points or lines are approximately the same distance away from the viewer, it may be necessary to

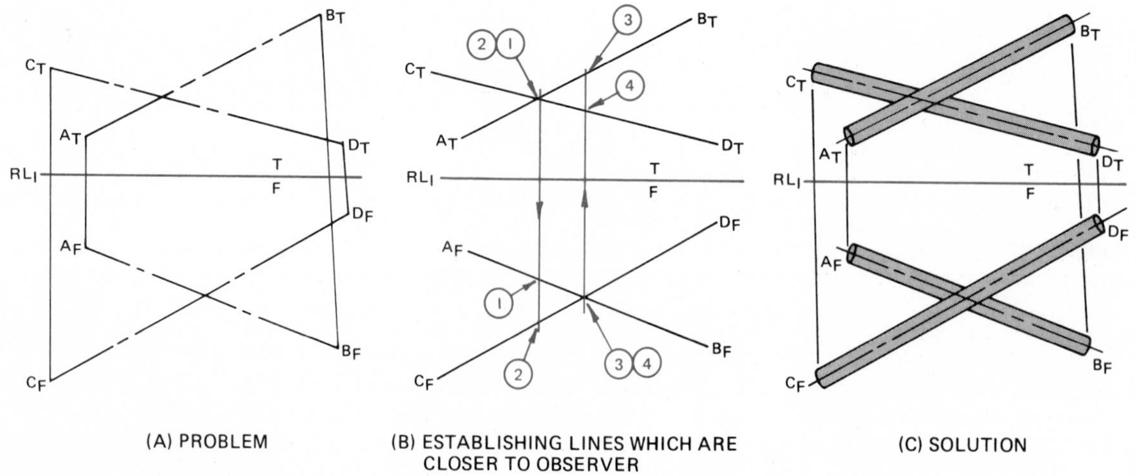
(A) PROBLEM
(B) ESTABLISHING LINES WHICH ARE CLOSER TO OBSERVER
(C) SOLUTION

Fig. 32-3-1 Visibility of skew lines by testing. (a) Problem. (b) Establishing lines which are closer to observer. (c) Solution.

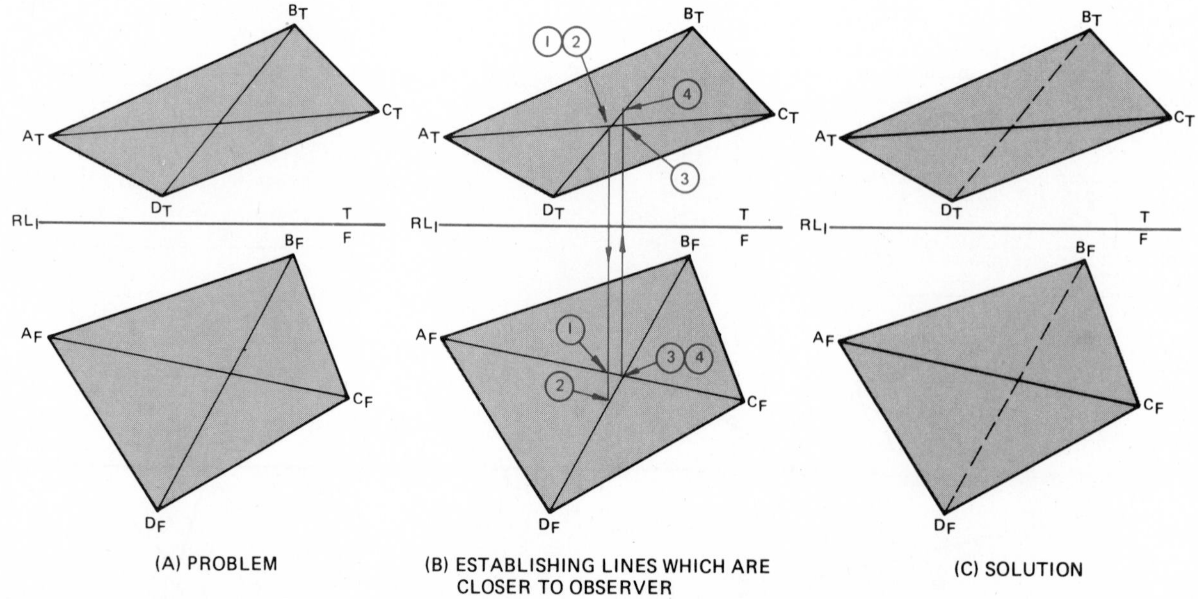
(A) PROBLEM
(B) ESTABLISHING LINES WHICH ARE CLOSER TO OBSERVER
(C) SOLUTION

Fig. 32-3-2 Establishing visibility of lines and surfaces by testing. (a) Problem. (b) Establishing lines which are closer to observer. (c) Solution.

graphically check the visibility of lines and points, as in Fig. 32-3-2.

To check visibility of lines A_TC_T and B_TD_T in the top view:

- Label intersection of lines A_TC_T and B_TD_T as 1 , 2 .
- Project the point of intersection to the front view, establishing point 1 on line A_FC_F and point 2 on line B_FD_F.
- Point 1 is closer to reference line RL_1, which means that line A_TC_T is nearer when the top view is being observed and thus is visible.
- Point 2 is farther away from reference line RL_1, which means that line B_TD_T would not be seen when one is viewing from the top.

To check visibility of lines A_FC_F and B_FD_F in the front view:

- Label the intersection of lines A_FC_F and B_FD_F as 3 , 4 .
- Project the point of intersection to the top view, establishing point 3 on line A_TC_T and point 4 on line B_TD_T.
- Point 3 is closer to reference line RL_1, which means that line A_FC_F is nearer when the front view is being observed and thus is visible.
- Point 4 is farther away from reference line RL_1, which means that line B_FD_F would not be seen when one is viewing from the front. Figure 32-3-2*c* shows the completed top and front views of the part.

Visibility of Lines and Surfaces by Observation

In order to fully understand the shape of an object, it is necessary to know which lines and surfaces are visible in each of the views. Determining their visibility can, in most cases, be done by inspection. With reference to Fig. 32-3-3*a*, the outline of the part is obviously visible. However, the visibility of lines and surfaces within the outline must be determined. This is accomplished by determining the position of O_F in the front view. Since position O is the closest point to the reference line, RL_1 in the front view, it must be the point which is closest to the observer when viewing the top view. Thus it can be seen, and the lines converging to point O_T are visible.

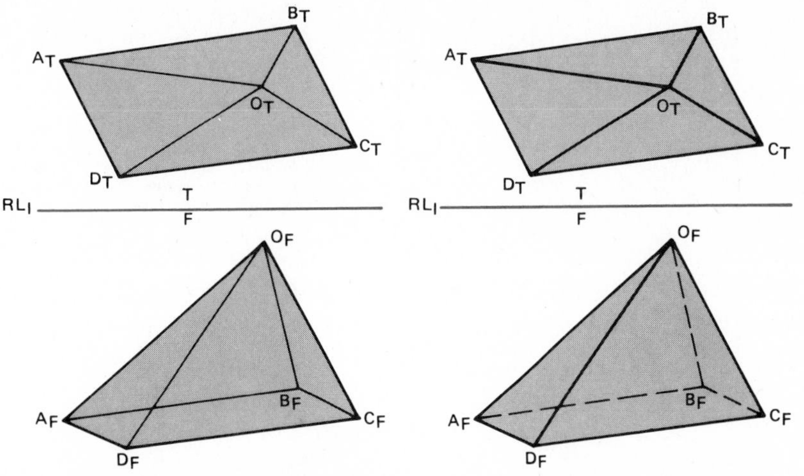

AT
BT
OT
CT
DT
RL1
T
F
OF
AF
BF
CF
DF

(A) PROBLEM—TO DETERMINE VISIBILITY OF LINES

(B) SOLUTION

Fig. 32-3-3 Visibility of lines and surfaces by observation. (a) Problem—to determine visibility of lines. (b) Solution.

With reference to determining the visibility of the lines in the front view, see the top view. Plane $O_TC_TD_T$ is closest to reference line RL_1. Therefore it must be the closest surface when the observer is looking at the front view, and it must be visible. Since point B_T in the top view is farthest away from reference line RL_1, it is the point which is farthest away when the observer is looking at the front view. Since it lies behind surface $O_FC_FD_F$, it cannot be seen.

From this example it may be stated that lines or points closest to the observer will be visible, and lines and points farthest away from the viewer but lying within the outline of the view will be hidden.

Assignments

1. *Establishing Visibility of Skew Lines by Observation.* Divide an A3- or B-size sheet into four parts, as shown in Fig. 32-3-A. By observation, sketch in the circular pipes in a manner similar to that shown in Fig. 32-3-1, showing the direction in which the pipes are sloping and which pipe is closer to the observer in the two views. Scale is to suit.

2. *Establishing Visibility of Lines and Surfaces by Testing.* Divide an A3- or B-size sheet into four parts, as shown in Fig. 32-3-B. By testing in the manner shown in Fig. 32-3-2, complete the two-view drawings, showing the visible and hidden lines. Scale is to suit.

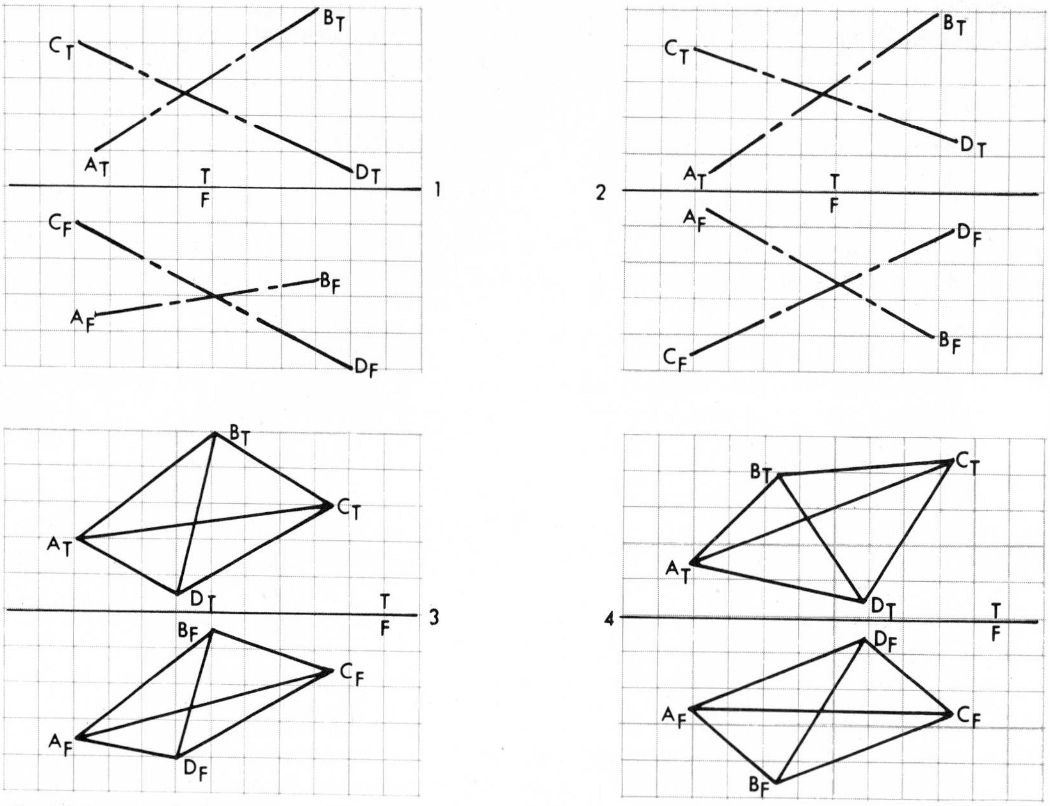

B_T C_T A_T D_T T F 1 C_F B_F A_F D_F
2 B_T C_T D_T A_T T F A_F D_F B_F C_F
B_T C_T A_T D_T T F 3 B_F C_F A_F D_F
B_T C_T A_T D_T 4 T F D_F A_F C_F B_F

Fig. 32-3-A Assignment.

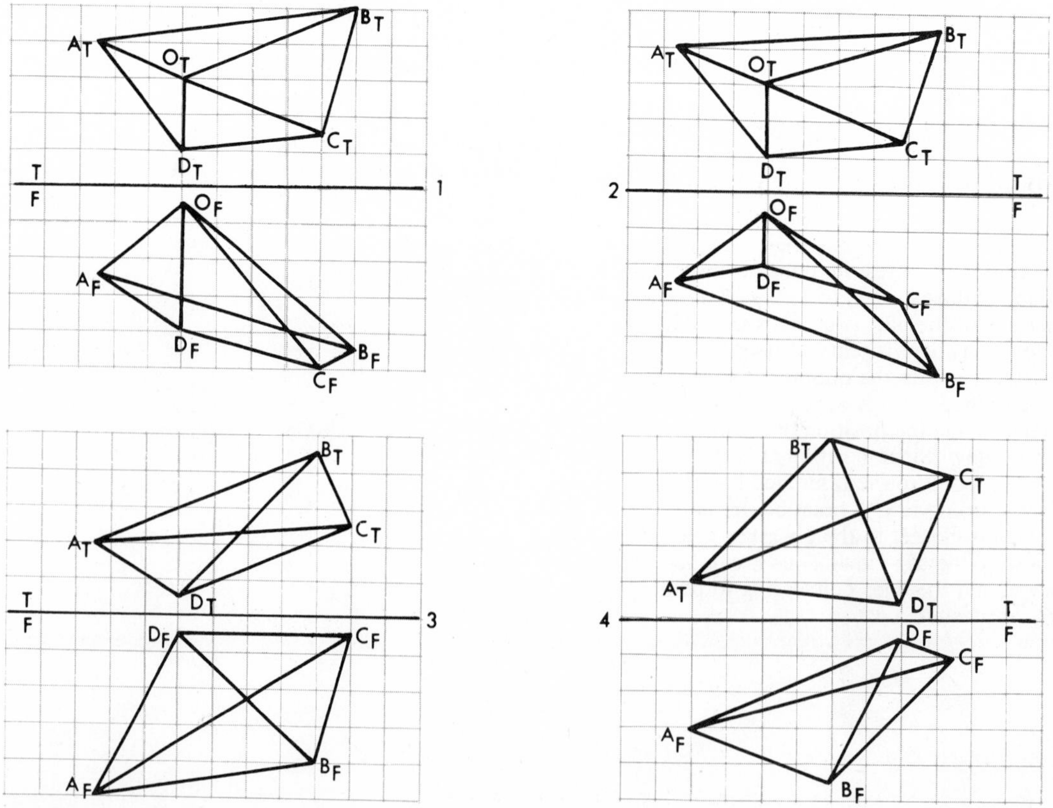

A_T O_T B_T C_T D_T T F 1 O_F A_F D_F B_F C_F
A_T O_T B_T C_T D_T 2 T F O_F A_F D_F C_F B_F
B_T C_T A_T D_T T F 3 D_F C_F B_F A_F
B_T C_T A_T D_T 4 T F D_F C_F A_F B_F

Fig. 32-3-B Assignment.

UNIT 32-4

DISTANCES BETWEEN LINES AND POINTS

Distance from a Point to a Line

When the front and side views are given, as in Fig. 32-4-1, and the shortest distance between line *AB* and point *P* is required, the procedure is as follows:

- Draw reference line RL_2 parallel to line A_SB_S in the side view.
- Transfer distances designated as *R*, *S*, and *U* in the front view to the primary auxiliary view. The resulting line A_1B_1 in the auxiliary view is the true length of line *AB*.
- Next draw reference line RL_3 perpendicular to line A_1B_1.
- Transfer distances designated as *L*, *M*, and *N* in the side view to the secondary auxiliary view, establishing points P_2 and A_2B_2, the latter being the point view of line *AB*.
- The shortest distance between point *P* and line *AB* is shown in the secondary auxiliary view.

Figure 32-4-2 illustrates the application of the point-on-point view of a line to determine the clearance between a hydraulic cylinder and a clip on the wheel housing.

Shortest Distance between Two Skew Lines

When the front and top views are given, as in Fig. 32-4-3, and the shortest distance between the two lines *AB* and *CD* is required, the procedure is as follows:

- Draw reference line RL_2 parallel to line A_FB_F in the front view.
- Transfer the distances designated as *R*, *S*, *U*, and *V* in the top view to the primary auxiliary view. The resulting line A_1B_1 in the auxiliary view is the true length of line *AB*.
- Next draw reference line RL_3 perpendicular to line A_1B_1.
- Transfer the distances designated as *L*,

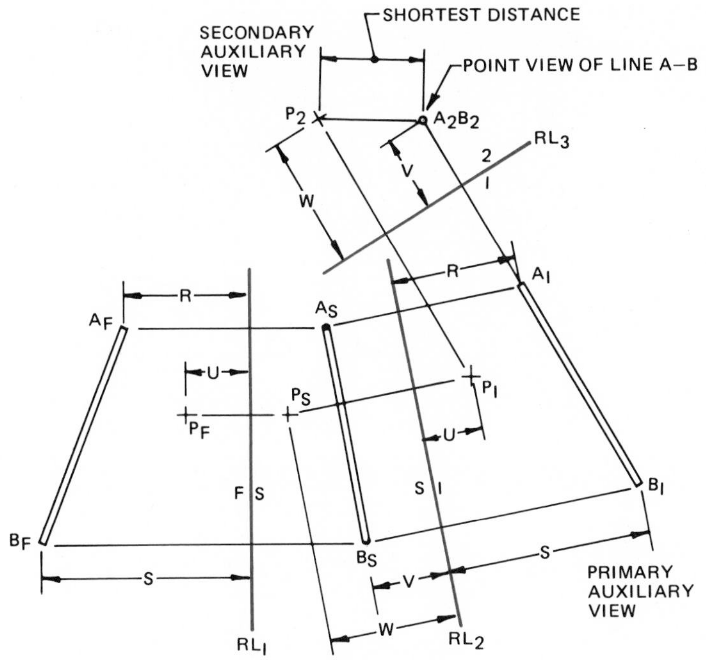

SHORTEST DISTANCE
SECONDARY AUXILIARY VIEW
POINT VIEW OF LINE A–B
P2
A2B2
RL3
PRIMARY AUXILIARY VIEW
RL1
RL2

Fig. 32-4-1 Distance from a point to a line.

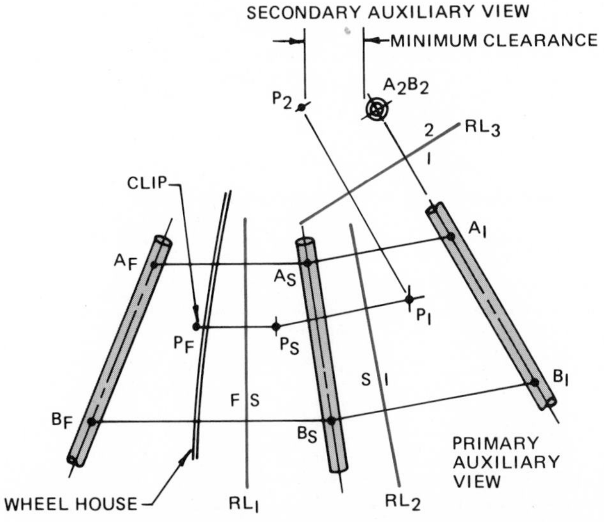

SECONDARY AUXILIARY VIEW
MINIMUM CLEARANCE
CLIP
WHEEL HOUSE
PRIMARY AUXILIARY VIEW
RL1
RL2
RL3

Fig. 32-4-2 Design application of distance from a point to a line.

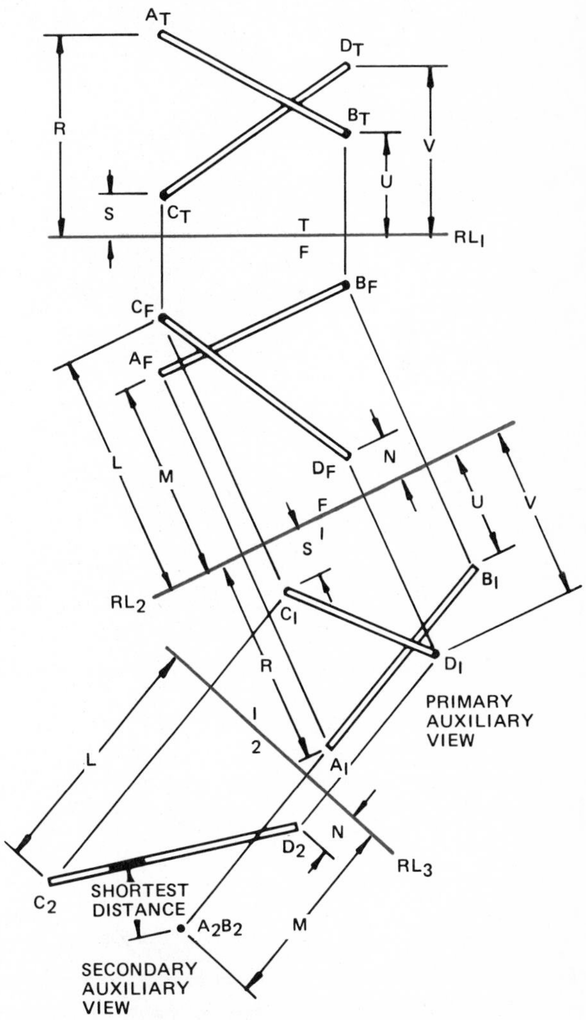

PRIMARY AUXILIARY VIEW
SHORTEST DISTANCE
SECONDARY AUXILIARY VIEW
RL1
RL2
RL3

Fig. 32-4-3 Shortest distance between skew lines.

M, and N in the front view to the secondary auxiliary view. The resulting line C_2D_2 is the true length of line CD. Point A_2B_2 is a point view of line AB.

- The shortest distance between the two lines AB and CD is shown in the secondary auxiliary view.

Assignments

1. *Finding Distance from a Point to a Line.* Divide an A3- or B-size sheet into two parts. With the use of reference lines and auxiliary views, find the distance from a point to a line in the two problems shown in Fig. 32-4-A. Scale is to suit.

2. *Finding the Shortest Distance between Skew Lines.* Divide an A3- or B-size sheet into two parts. With the use of reference lines and auxiliary views, find the shortest distance between the skew lines for the two problems shown in Fig. 32-4-B. Scale is to suit.

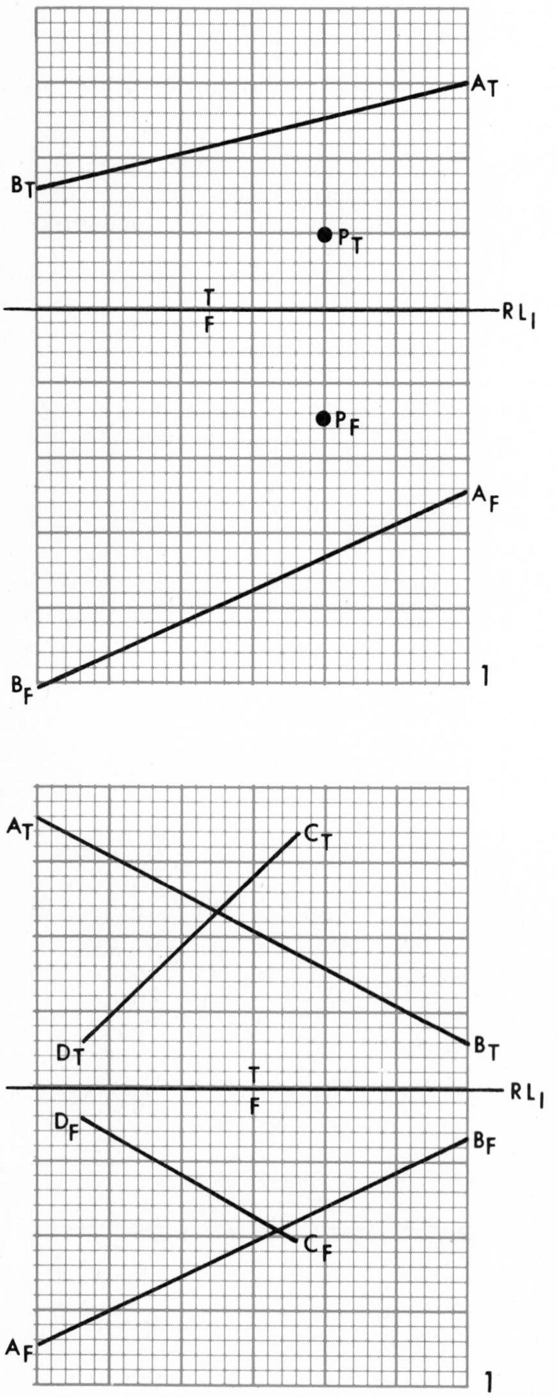

AT
BT
PT
T
F
RL1
PF
AF
BF
1
AT
CT
DT
BT
T
F
RL1
DF
BF
CF
AF
1

Fig. 32-4-A Assignment.

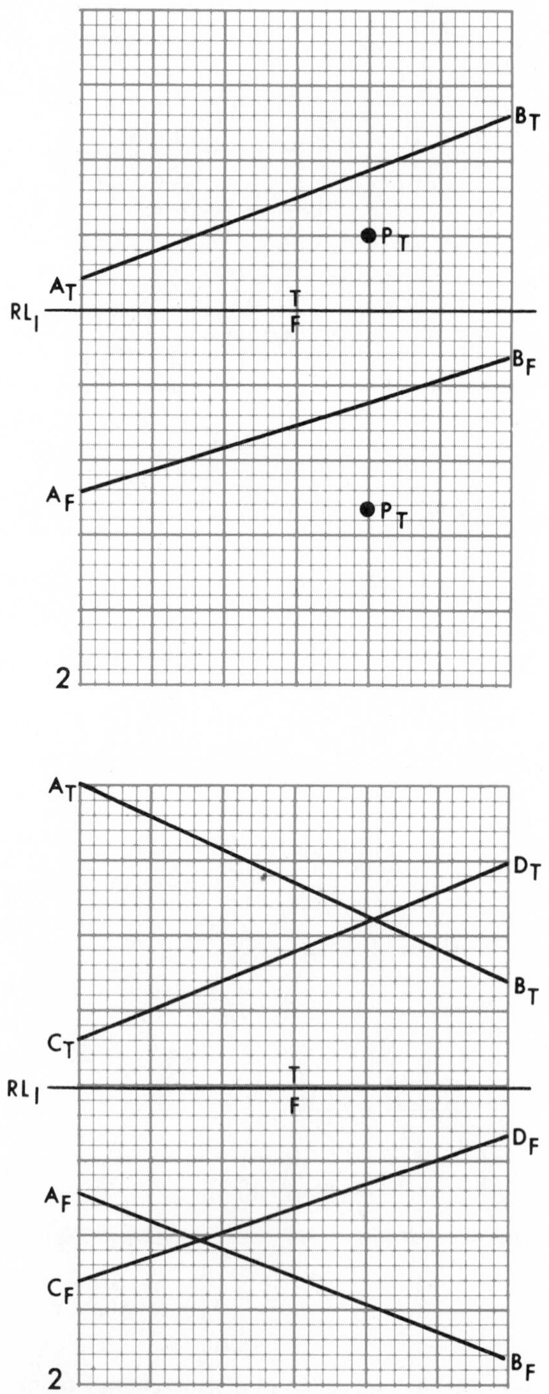

BT
PT
AT
RL1
T
F
BF
AF
PT
2
AT
DT
BT
CT
RL1
T
F
DF
AF
CF
BF
2

Fig. 32-4-B Assignment.

UNIT 32-5

EDGE AND TRUE VIEW OF PLANES

The primary planes of projection are horizontal, vertical (or frontal), and profile. A plane that is not parallel to a primary plane is not visible in its true dimensions. To show a plane in true view, it must be revolved until it is parallel to a projection plane. Figure 32-5-1 shows an oblique plane *ABC* in the top and front views. The object is to find the true view of this plane. When the top and front views are examined carefully, no line is parallel to the reference line in either view. However, a line C_TD_T can be drawn on the plane parallel to the reference line RL_1 and projected to the front view for its true length D_FC_F. Next,

- Draw reference line RL_2 perpendicular to line D_FC_F in the front view.
- Transfer the distances designated *R*, *S*, and *U* in the top view to the primary auxiliary view. The resulting line A_1B_1 is the edge view of the plane.
- Draw reference line RL_3 parallel to line A_1B_1.
- Transfer the distances *L*, *M*, and *N* in the front view to construct the true shape of plane *ABC* in the secondary auxiliary view.

Design Application. Figure 32-5-2 shows the application of the procedure followed for Fig. 32-5-1. Points *A*, *B*, *C*, and *D* correspond in both drawings, but line *AC* is omitted in Fig. 32-5-2 since it serves no practical purpose in the design.

Planes in Combination

Figure 32-5-3 demonstrates a solution where a combination of planes is involved. Note that $A_TB_TC_T$ and $A_FB_FC_F$ form one plane while $B_TC_TD_T$ and $B_FC_FD_F$ form another. Also line *BC* is common to both planes. The objective in the problem is to find the true bends at the angles *ABC* and *BCD*. The procedure is as follows:

- Construct the primary auxiliary view which shows the true length of line *BC*.
- Construct a secondary auxiliary view which shows *BC* as a point-on-point view. The result is the edge view of both planes *ABC* and *BCD* in the secondary auxiliary view.
- Since any view adjacent to a point-on-point view of a line must show the line in its true length, *BC* will be in true length in secondary auxiliary views 2 and 3. Therefore, projecting perpendicularly from the edge views in the secondary auxiliary view 1 to the secondary auxiliary views 2 and 3

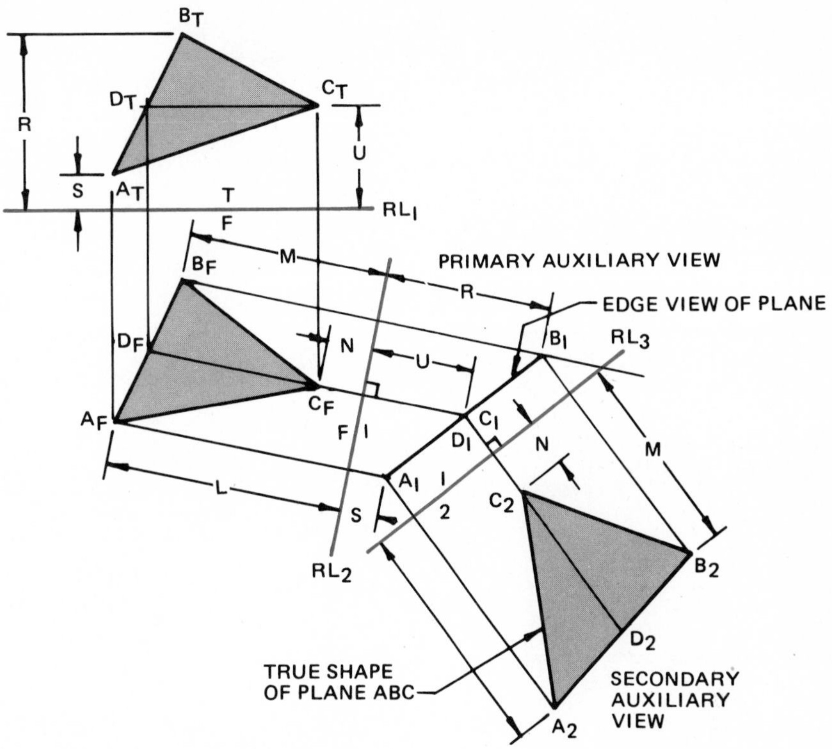
BT
CT
DT
AT
R
S
U
T
F
RL1
BF
DF
AF
CF
M
N
L
R
U
PRIMARY AUXILIARY VIEW
EDGE VIEW OF PLANE
B1
RL3
C1
D1
A1
F 1
1 2
C2
B2
D2
A2
M
N
S
RL2
TRUE SHAPE OF PLANE ABC
SECONDARY AUXILIARY VIEW

Fig. 32-5-1 True view of a plane.

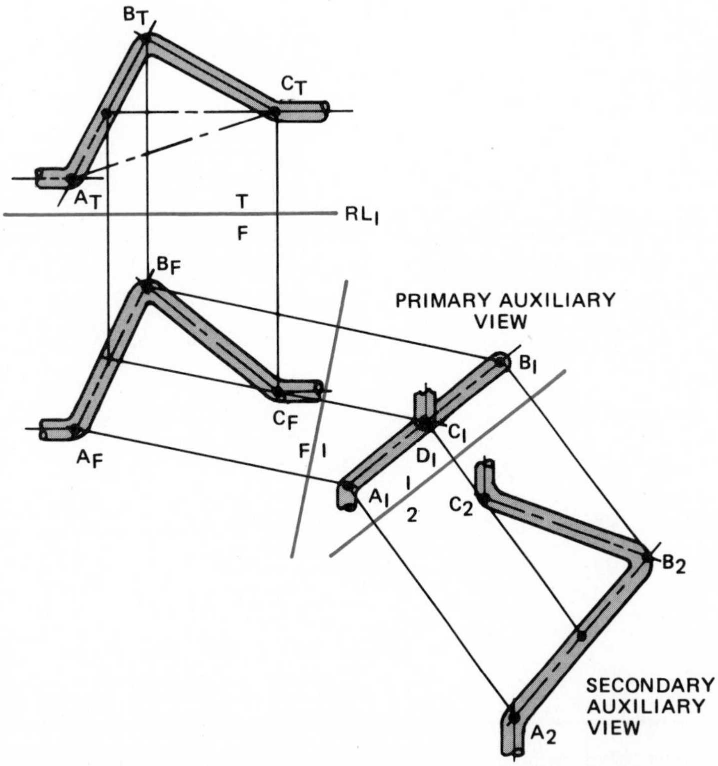
BT
CT
AT
T
F
RL1
BF
CF
AF
PRIMARY AUXILIARY VIEW
B1
C1
D1
A1
F 1
1 2
C2
B2
A2
SECONDARY AUXILIARY VIEW

Fig. 32-5-2 Design application of true view of a plane (Fig. 32-5-1).

gives not only *BC* in true length in auxiliary views 2 and 3, but also the true angles *ABC* and *BCD*.

Assignments

1. *True View of a Plane.* Divide an A3- or B-size sheet into two parts. With the use of reference lines and auxiliary views, find the true view of the plane for the two problems shown in Fig. 32-5-A. Scale is to suit.

2. *Finding True Angles of Intersecting Planes.* Divide an A3- or B-size sheet into two parts. With the use of reference lines and auxiliary views, find the true angles of the intersecting planes for the two problems shown in Fig. 32-5-B. Scale to suit.

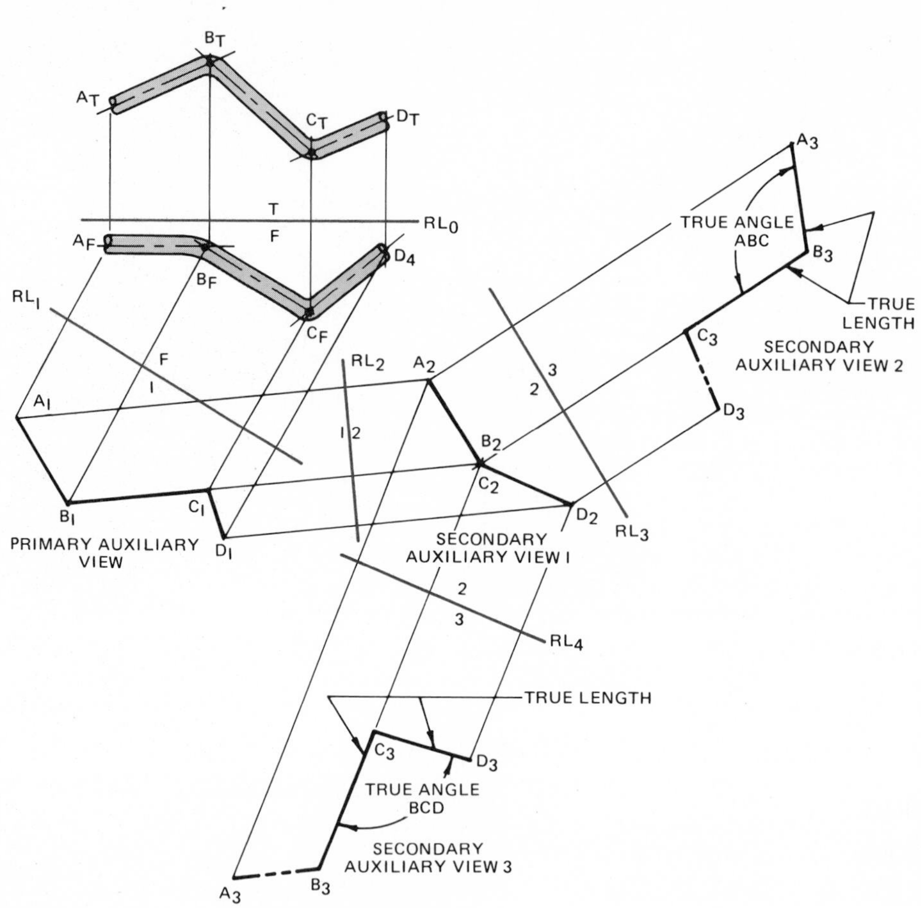

B_T
A_T
C_T
D_T
T
F
RL_0
A_F
B_F
C_F
D_4
RL_1
F
1
A_1
B_1
C_1
D_1
PRIMARY AUXILIARY VIEW
RL_2
1 2
A_2
B_2
C_2
D_2
SECONDARY AUXILIARY VIEW 1
RL_3
2
3
A_3
TRUE ANGLE ABC
B_3
TRUE LENGTH
C_3
D_3
SECONDARY AUXILIARY VIEW 2
RL_4
2
3
TRUE LENGTH
C_3
D_3
TRUE ANGLE BCD
SECONDARY AUXILIARY VIEW 3
A_3
B_3

Fig. 32-5-3 Use of planes in combination.

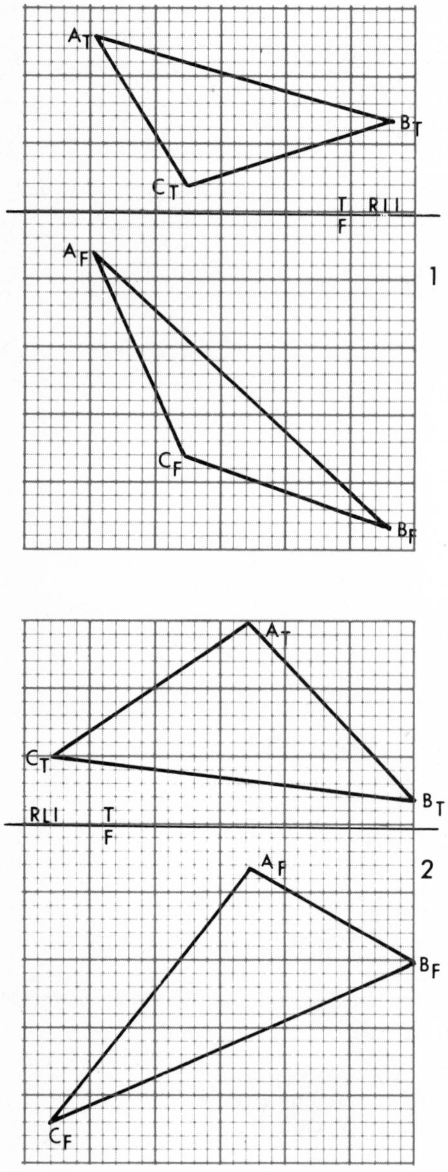

A_T
B_T
C_T
T RL1
F
A_F
1
C_F
B_F
A_T
C_T
B_T
RL1 T
F
A_F
2
B_F
C_F

Fig. 32-5-A Assignment.

Fig. 32-5-B Assignment.

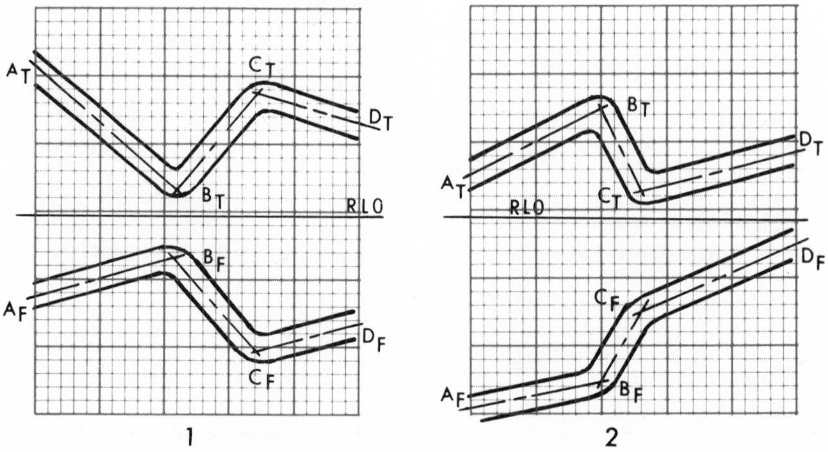

A_T
C_T
D_T
B_T
RL0
B_F
A_F
D_F
C_F
1
B_T
D_T
A_T
RL0
C_T
D_F
C_F
A_F
B_F
2

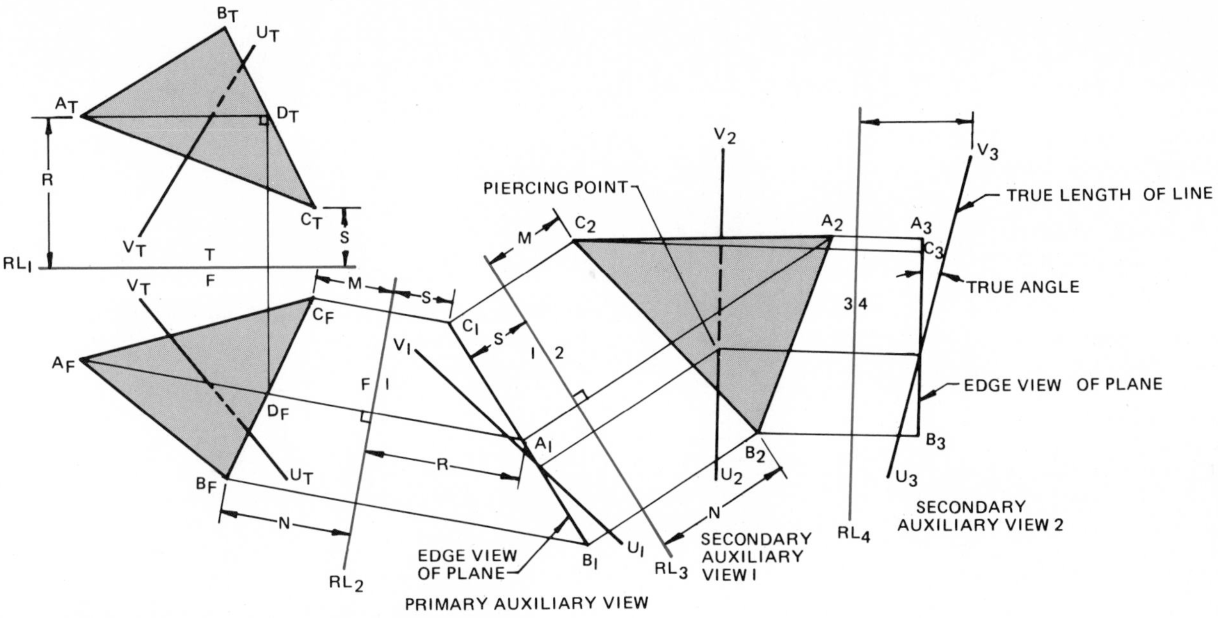

PIERCING POINT
TRUE LENGTH OF LINE
TRUE ANGLE
EDGE VIEW OF PLANE
SECONDARY AUXILIARY VIEW 2
SECONDARY AUXILIARY VIEW I
EDGE VIEW OF PLANE
PRIMARY AUXILIARY VIEW
RL1
RL2
RL3
RL4

Fig. 32-6-1 The angle a line makes with a plane.

UNIT 32-6

ANGLES BETWEEN LINES AND PLANES

The Angle a Line Makes with a Plane

The top and front views in Fig. 32-6-1 show a line *UV* passing somewhere through plane *ABC*. The true angle between the line and the plane will be shown in the view which shows the edge view of the plane and the true length of the line. This view is found as follows:

- Draw line A_TD_T in the top view parallel to reference line RL_1.
- Project point D_T to front view, locating point D_F.
- Draw reference line RL_2 perpendicular to a line intersecting points A_F and D_F in the front view.
- Draw the primary auxiliary view which shows the edge view of the plane but distorted length of line *UV*.
- Draw reference line RL_3 parallel to edge view of the plane shown in the auxiliary view.
- Draw the secondary auxiliary view 1 which shows the true view of the plane and location of the piercing point.
- Draw reference line RL_4 parallel to line V_2U_2.
- Draw the secondary auxiliary view 2 which shows the true length of line *UV* and the true angle between line and edge view of plane.

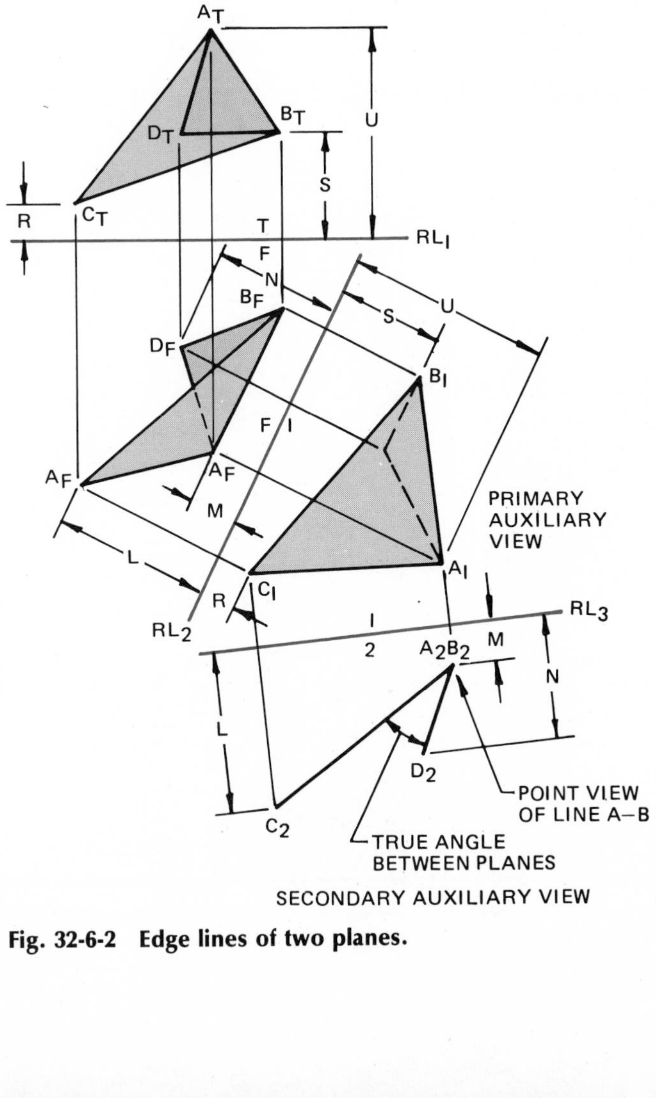

RL1
RL2
RL3
PRIMARY AUXILIARY VIEW
POINT VIEW OF LINE A–B
TRUE ANGLE BETWEEN PLANES
SECONDARY AUXILIARY VIEW

Fig. 32-6-2 Edge lines of two planes.

Edge Lines of Two Planes

Figure 32-6-2 shows a line of intersection *AB* made by two planes, triangles *ABC* and *ABD*. When the top and front views are given, the point-on-point view of line *AB* and the true angle between the planes are found as follows:

- Draw reference line RL_2 parallel to line A_FB_F in the front view.
- Transfer the distances designated R, S, and U in the top view to the primary auxiliary view. The resulting line A_1B_1 in the auxiliary view is the true length of line *AB*.
- Next draw reference line RL_2 perpendicular to line A_1B_1.
- Transfer the distances designated as L, M, and N in the front view to the secondary auxiliary view.
- Point A_2B_2 is a point-on-point view of line *AB*. The true angle between the two planes is seen in the secondary auxiliary view.

Assignments

1. *The Angle a Line Makes with a Plane.* On an A3- or B-size sheet, find the angle the line makes with a plane for the layout shown in Fig. 32-6-A. Scale is to suit.

2. *The Angle a Line Makes with a Plane.* On an A3- or B-size sheet, find the angle the line makes with a plane for the layout shown in Fig. 32-6-B. Scale is to suit.

3. *Edge Line of Two Planes.* Divide an A3- or B-size sheet into two parts. With the use of reference lines and auxiliary views, find the edge line of the two planes shown in the two problems in Fig. 32-6-C. Scale is to suit.

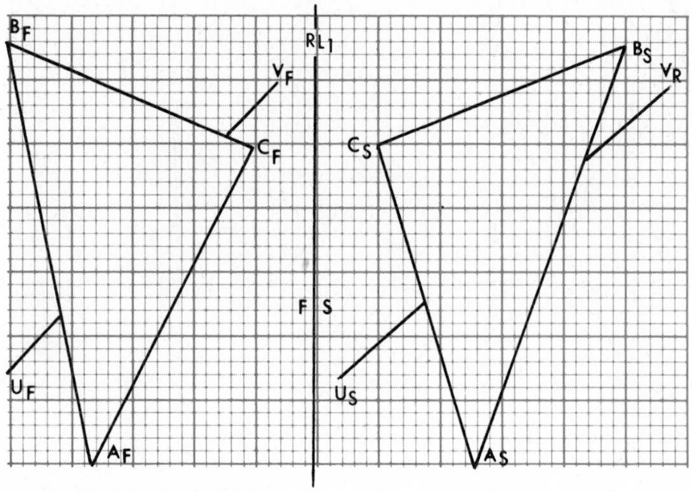
BF
RL1
VF
CF
CS
BS
VR
F S
UF
US
AF
AS

Fig. 32-6-A Assignment.

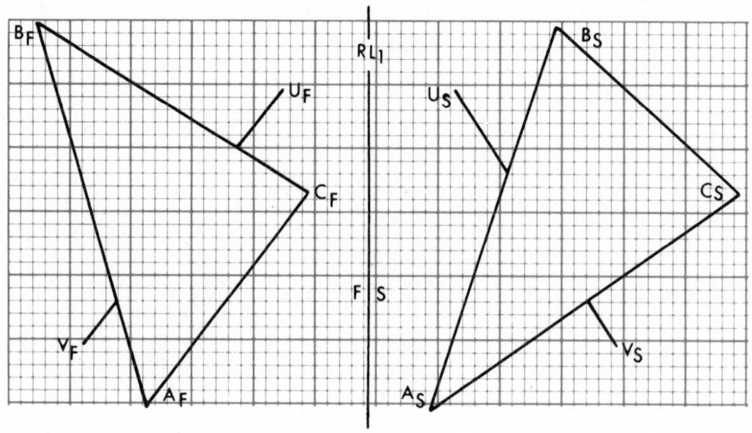
BF
RL1
UF
CF
VF
AF
BS
US
CS
F S
VS
AS

Fig. 32-6-B Assignment.

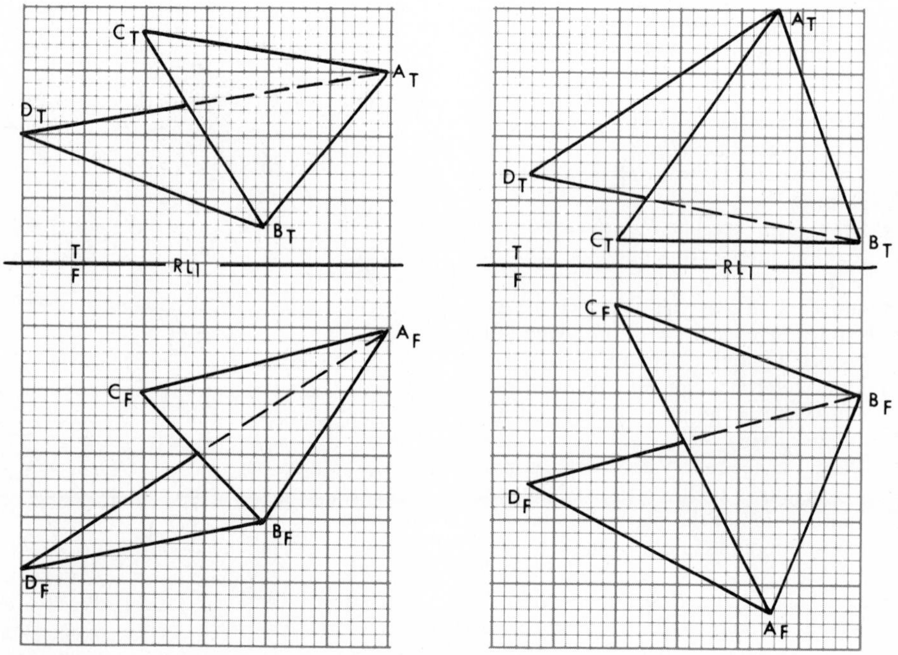
CT
AT
DT
BT
T
F
RL1
AF
CF
BF
DF

Fig. 32-6-C Assignment.

Chapter 33
Design

UNIT 33-1
THE DESIGN PROCESS

The purpose of any design department is to create a product which not only will function efficiently but also will be a financial success. Although most designs are more complicated than the examples given in this chapter, the main steps in designing a product follow a similar pattern.

In the design process, considerations should be given to each of the steps shown in Fig. 33-1-1. The design can be treated as a process in which the input is the problem and the output is the solution.

A detailer may work from a complete set of instructions such as a complete assembly drawing or may have a free hand in the design of the part. If a detailer extracts the information from a complete assembly drawing, then many of the decisions have already been made by the designer or engineer. If the designer or engineer has made the decisions, there are several factors the detailer must consider before starting the detail drawing. Normally, the final design is a compromise of many factors.

The Engineering Approach to Successful Design

Each field of engineering has its techniques and rules—and its standards for the use of the construction materials peculiar to that field.

The steps from idea to production are based on logical and well-known design principles. These principles apply equally to the manufacture of gears, optical systems, industrial components, consumer articles, or rockets—and regardless of the material of construction. These steps are not necessarily in order, but all are essential for a successful application.

DEFINING THE END-USE REQUIREMENTS

As an initial step, the product designer must anticipate the conditions of use and the performance requirements of the product. Consideration must be given to such things as environment, load, speed of production, life expectancy, optimum size, maintenance, shape, color, strength, and stiffness.

These end-use requirements can be ascertained through market analyses, surveys, examinations of similar products,

DESIGN PROBLEM
ANALYZE NEEDS
SET OBJECTIVES
CREATE ALTERNATIVES
CHECK FOR FEASIBILITY
SELECT THE SOLUTION
DRAW THE DESIGN

Fig. 33-1-1 Steps in the design process.

testing, general experience, and frequently material supplies.

A clear definition of product requirements leads directly to choice of the material of construction.

SELECTING THE MATERIAL

Material is a very important factor to consider in designing a part. Perhaps plastic is the better choice of material over wood or metal. Would one choice of metal be better than another? What elements will come into contact with water? If the part is to be immersed in an oil solution, will the choice of plastic be minimized or ruled out? Is strength a factor? If so, what materials will meet the stresses required? What material is in stock or easily obtainable? Is the material *the* correct choice if a plating or coating is required?

There are thousands of engineering materials available, yet no single one will exhibit all desired properties in their proper relationships. Therefore, a compromise among properties, cost, and manufacturing process determines the material of construction. Even within one series, materials differ because of varying formulations. Just as steel compositions vary—tool steel and stainless steel, for example—so do the plastics.

The designer needs a firm set of properties and engineering data upon which to base the first design. The data can come from handbooks or, more likely, from the published literature provided by the manufacturer of the material.

DRAFTING THE PRELIMINARY DESIGN

The designer blends the end-use requirements and the properties of the selected material into a preliminary design.

Engineering techniques and formulas are used to achieve the three requirements of design success:

- Economic feasibility
- Functional feasibility
- Attractive appearance

The production method to be used will often set limitations on design. The designer should be aware of the strengths and weaknesses of the method selected. The material supplier and the processor, with their experience in hundreds of applications, can assist here.

PROTOTYPING THE DESIGN

The prototype is the opportunity for the designer to see the product as a three-dimensional object. This, too, is the first opportunity for checking the engineering design. The quality of the prototype is quite important. The method used in producing the prototype may not be the same as that planned for the final production line, but the design must be identical to that expected on the production line—otherwise tests may be misleading and analysis false.

If the search for the right material has been narrowed only to two or three, prototyping will help spotlight one.

TESTING THE DESIGN

Every design should be given an actual or simulated service test while in the prototype stage to ensure that the obvious is not overlooked and that the not so obvious is taken into account. The end-use requirements dictate the design testing program. An engine part might be given temperature, vibration, and hydrocarbon-resistance tests; a luggage fixture might be subjected to abrasion tests; and a toaster knob might be checked for electrical and heat insulation. Other tests, such as field testing or consumer reactions, are part of the necessary procedure for completely evaluating any design.

TAKING A SECOND LOOK

The second look at the design provides an answer to the basic question, Is the product doing the right job at the right price?

At this point, most products can be improved by redesigning for better production economies or for important functional or aesthetic changes. Weak sections can be strengthened, colors changed, and new features added. Substantial changes in design will require retesting. Now is the time to set up production. The first step is to write the specification.

WRITING A MEANINGFUL SPECIFICATION

The purpose of the specification is to eliminate any variations in the product that will not satisfy the functional, aesthetic, or economic requirements. The specification is a complete set of written requirements which the part must meet. The specification for the part should include such things as the material of construction by brand and generic name, method of fabrication, dimensions, color, surface finish, packaging, printing, and every other detail of production to which there could be more than one answer.

SETTING UP PRODUCTION

Is one or one thousand parts required? When a large quantity is required, then the number of methods of producing the parts increases. Perhaps a casting, a forging, or a stamping may be the most sound choice. If only a few parts are required, then prefabricating or machining may be the better choice.

Production. Is the part to be manufactured within the plant, or can it be sent out to be produced? In many cases company policy may be to produce the part within the plant. If this is the case, then the production choices are limited to the methods available within the plant.

After the specification is written but before the production line can start, tooling must be designed, built, and integrated with the processing equipment. (In some cases, dies and molds can be started while testing is in progress.)

Production efficiency and economy can be realized through proper design of tools. The processor is an important source of aid in this area.

TIME FACTOR

In some instances, such as when a breakdown of a machine is holding up production within the plant, the best method of producing a part may take second place if it involves too much time.

WORKFORCE

This factor ties in with time. A machine breaks down at 1400 hours. It is essential that the machine be back in operation by 800 hours the following day. What personnel are available to produce the part, with overtime, or is there a night shift?

CONTROLLING QUALITY

Good inspection practice requires a checklist to maintain a consistently good product. The inspection checklist, for the most part, will conform to the end-use requirements set forth in the specification.

Here, too, it is beneficial to consult with the supplier or the molder, who knows the processing and finishing characteristics of the material.

Part Specifications

All material applications start out as ideas in someone's mind. From this point the idea must be developed into a production item. The transition is accomplished in a series of logical steps. Ensuring the quality of the final production item starts with the writing of a specification.

The specification is a complete set of written requirements; the purpose is to ensure that the finished part will perform as intended. The scope of the specification depends on the performance required of the part. In general, as a specification becomes more complex, the cost of the part increases.

Let's take a look at a typical, although hypothetical, part (shown in Fig. 33-1-2). Assume that it has been developed through the steps of the engineering approach and is ready for a meaningful specification.

Any good specification should contain three basic portions: (1) the raw material, (2) the design of the part, and (3) the performance of the part in use.

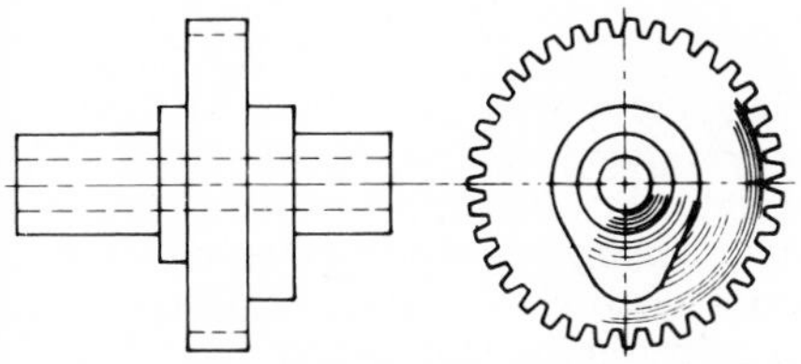

Fig. 33-1-2 Combined gear and cam.

RAW MATERIAL

The raw material for any part is selected for its physical properties, with due regard for economic and engineering requirements. The section of a specification dealing with raw material should be divided into two major parts: identification and quality.

DESIGN OF THE PART

The second major portion of the specification involves the design of the part. The design specification includes tolerances and surface finish. If the part is going to be cast, then parting lines, flash, gate location, and warpage must be considered.

Dimensional Tolerance. The dimensional tolerances should be as close as required for functioning. As tolerances become tighter than this, the cost of both tooling and fabrication rises very rapidly (see Fig. 33-1-3). Such tight dimensional tolerances require very close control of the processing variables and necessitate extra inspection, which contributes to a high unit cost.

Other important considerations are:

1. Critical dimensions should be identified with specific tolerances; let overall tolerances control less important dimensions.

2. If parts are to be machined, allow generous tolerances in these areas.

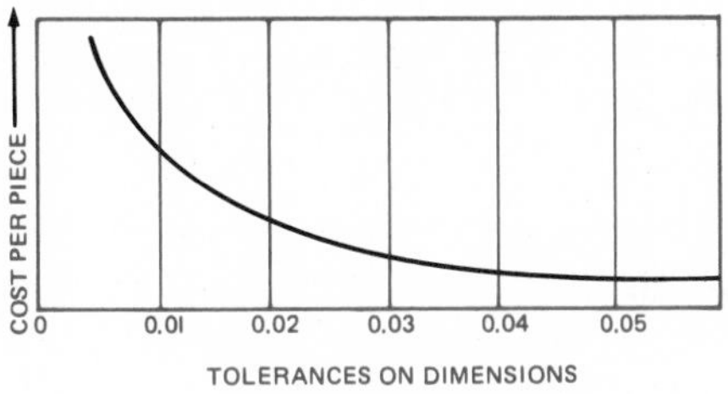
COST PER PIECE
0
0.01
0.02
0.03
0.04
0.05
TOLERANCES ON DIMENSIONS

Fig. 33-1-3 Cost of part increases rapidly as tolerances get smaller.

Surface Finish. The type and degree of surface finish required should be clearly indicated. If a highly polished finish is necessary, it should be specified. On other surfaces, finish need not be covered except in general terms. Surfaces which must be clear of imperfections such as tool marks, sinks, blisters and flow lines should be clearly indicated.

Parting Line. The location of a parting line of a mold may be influenced by flash, part appearance, and structural design. In such cases, parting-line location should be specified on the drawings. Experienced processors can assist in altering part design to simplify tooling or molding.

The parting line in the gear-cam part could be located in several places. One end of the gear simplifies the tooling requirement.

Flash. Where flash is undesirable, the drawing should so indicate. The length of allowable flash may be given in a measurable dimension.

Gating. The gate should be located where it will cause the least difficulty. The specification on gating should define areas to avoid, such as cam or bearing surfaces. To permit maximum concentricity in the hypothetical part, the gate should be located as shown in Fig. 33-1-4.

Warpage. The allowable warpage of parts after molding should be specified, even though all dimensions may be met.

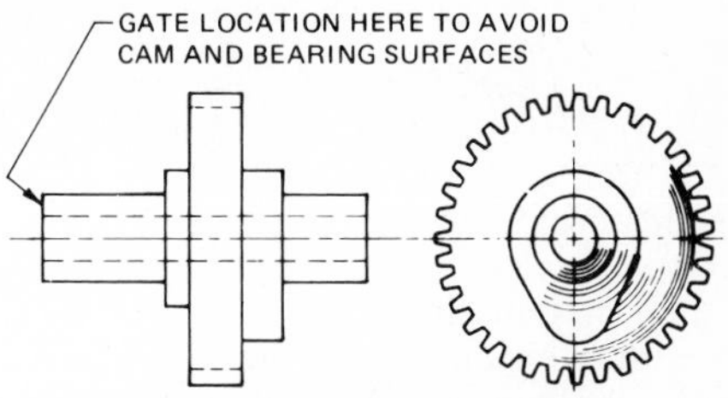
GATE LOCATION HERE TO AVOID
CAM AND BEARING SURFACES

Fig. 33-1-4 Recommended gate location for maximum concentricity.

When minimum warpage is desired, the cooling of parts in a jig or fixture may be necessary. It should be remembered that post-molding operations (annealing, moisture conditioning, etc.) may cause warpage.

Minimum warpage depends on a proper balance of several factors. These include uniform part thickness, location of knockout pins, and optimum molding conditions.

Performance Specifications

The third major portion of a suggested specification concerns the quality of the part. Earlier work on design, material selection, and end-use testing has given assurance of part performance. Performance specifications are concerned with two factors: the first is visual inspection, and the second is simulated service tests.

Visual Tests. This kind of test is concerned with color, weld lines, etc. The test is established by considering the effect of each of these on the final performance of the part. Not all the factors may be necessary for every part.

Simulated Service Test. The second test for a part should simulate the ultimate use for the part. Care must be taken to use a meaningful test. Excessive speeds, loads, or impacts, well beyond ultimate requirement, can frequently cause rejection of good parts.

A simulated service test on a part like the cam-gear assembly of Fig. 33-1-2 might consist of impact and/or torque loading.

Do's and Don'ts for Designers

Designers want reliably functioning parts that are dependably procurable at the lowest installed cost. They can best meet their needs by consulting with vendors and understanding custom-metal-part manufacturing and pricing. Here is a checklist to use in the design of a part or assembly.

DON'TS

1. Don't specify tolerances tighter than essential to mechanism functioning.
2. Don't specify every dimension as mandatory; mark noncritical ones as reference only.
3. Don't specify material that is too good (too expensive) for the service.
4. Don't specify material that is available only on special purchase unless there is no alternative. If in doubt, ask your vendor.

DO'S

1. Do leave adequate space for assembly, i.e., bolt clearance, finger grip, etc.
2. Do consider manufacturing economics.
3. Do consider utilizing stock items when you need only a small quantity of parts. Your savings in design time, procurement costs, and delivery time may be appreciable.
4. Do realize that for small quantities or one part, the cost of raw material is not important; material availability and minimum-quantity purchase restrictions are important.
5. Do realize that for large-quantity purchases, precise specification of raw material can be extremely important.
6. Do realize that the total cost of a custom part is not the purchase cost but the installed cost.
7. Do consider in your product reliability, the relation between part cost, part reliability, the cost of replacement of a broken part.

REFERENCES AND SOURCE MATERIAL

1. E. I. duPont de Nemours & Co. (Inc.)
2. The Wallace Barnes Company Ltd.

Assignments

1. Your drafting supervisor has assigned to you the responsibility of designing an attractive single toggle switch plate for use in kitchens and bathrooms. The toggle plate and clearance hole requirements are shown in Fig. 33-1-A. The production run will exceed 25,000 and four different color plates are required. On an A3- or B-size sheet, lay out the design of the plate and include on the drawing the production and specifications data which you would submit with your design.
2. Design a suitable handle for a metal storage cabinet from the following data and the information shown in Fig. 33-1-B. The latch to be released with a quarter turn (anticlockwise direction) of the square latch shaft. The shaft is 10 mm² and protrudes beyond the cover

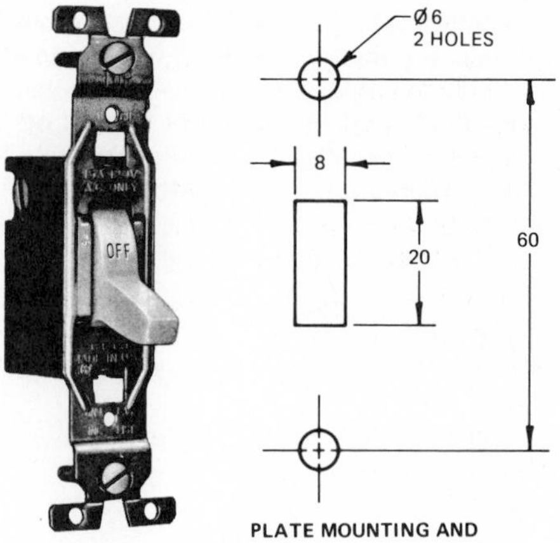

Ø6
2 HOLES
8
20
60
OFF
PLATE MOUNTING AND
TOGGLE SWITCH CLEARANCE

Fig. 33-1-A Assignment.

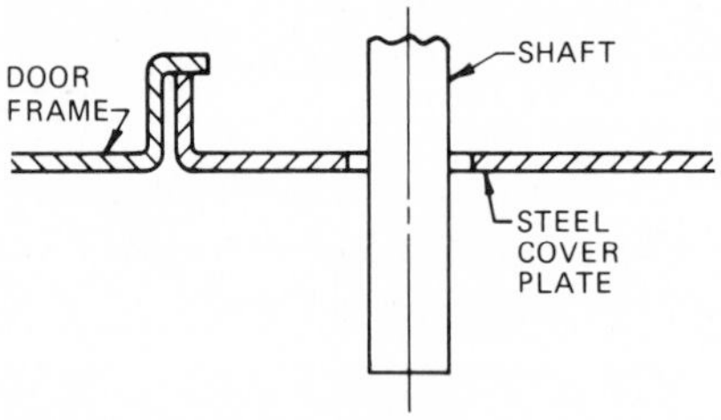

DOOR
FRAME
SHAFT
STEEL
COVER
PLATE

Fig. 33-1-B Assignment.

by 25 mm. The lock and shaft are securely fastened to the inside face of the cover and as such do not require any support on the handle side of the cover plate. A 16-mm diameter hole is punched in the 2-mm thick cover plate. Quantity 5000. On an A3- or B-size sheet, lay out the design of the handle and include on the drawing the production and specification data which you would submit with the design.

3. *Fig. 33-1-C.* Your drafting supervisor has asked you to design a table-mounted holder for a cassette microphone. The material is to be of such quality as to give an attractive finish as well as supporting the mass of the microphone. A θ20 mm tapered hole (large end) with a 3 mm wide slot on top is required for attaching the microphone to the holder. On an A3- or B-size sheet, lay out the holder and show the microphone in phantom lines.
4. Design a personalized key holder for a recreation vehicle that is used occasionally. When not in use it must hang on a wall or key rack. The material is to be lightweight and colorful.
5. Due to increased drafting room costs, your supervisor has asked you to design a holder to use up the short-ends of wooden pencils that have been too small to hold in the hand. The material

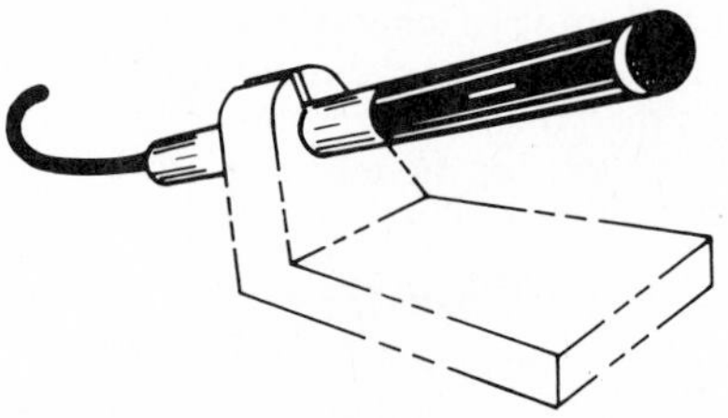

Fig. 33-1-C Assignment.

you choose should have good characteristics for grip and color. A quantity of 1000 is required.

6. Design a small tic-tac-toe board that could be given away as a souvenir or award. Your school crest or name must appear on the board or be an intricate part of the design. The material selected should be one that could be produced in one of your shops on campus.

UNIT 33-2

ASSEMBLY CONSIDERATIONS

An *assembly* is a combination of two or more parts which are joined by any of a number of different methods. A *subassembly* is made to facilitate the production of a larger assembly.

This unit describes various methods of attaching component parts to produce an assembly and some of the problems concerning assembly cost, tools, and practicability.

Although various examples of assembly methods and attachments are presented, they are not to be considered the only methods nor are they to have any preference over other means. Product design, volume, cost, and facilities are the determining factors influencing the need for an assembly.

The quality of the finished product as an assembly depends on effective attachment or fastening methods, regardless of the quality of the individual parts.

All assemblies, regardless of size, shape, or design, should be given the following considerations, since, in many cases, an analysis will dictate changes in design which will effect a cost savings.

Cost of Assembly

The following points should be carefully noted in determining the least expensive method of assembly.

PRODUCT VOLUME

Careful design will reduce costs of assembly at any volume level. However, many times the greatest savings are realized when the volume is high enough to justify the capital expenditure necessary for time-saving equipment that could not be justified at lower volumes. In many cases, it can be readily demonstrated that the procurement of highly specialized machinery may be justified by the increased efficiency made possible by such equipment.

PRODUCT DESIGN

Cost of any part or assembly is the responsibility of the design engineer and engineering management. To effectively control costs, it is essential that the design, fabrication, and assembly costs be continuously foremost in the minds of all responsible personnel. It is generally true that the simpler the design, the lower the cost of producing the finished product. In considering new designs or methods, all factors involved must be taken into account in calculating the savings or increased cost, including equipment obsolescence.

There is no general solution for any assembly problem. For example, an assembly may be made in one plant following a set sequence, while in another plant it may be made quite differently—the difference being due to established plant practices, equipment and facilities, tooling fixture design, volume, and labor costs.

Methods of assembly also have an important bearing on cost. Figure 33-2-1 shows a typical cost analysis covering seven possible methods for assembling a simple bracket to its carrying member, as illustrated in Fig. 33-2-2.

Sometimes methods of attachment are designed into the product without careful consideration of the economics involved. Obviously there is considerable difference in the effectiveness of the various attaching means. However, with cost comparisons at hand similar to those shown in Fig. 33-2-1, it becomes only a question of simple economics to choose the least expensive method that will do the job satisfactorily.

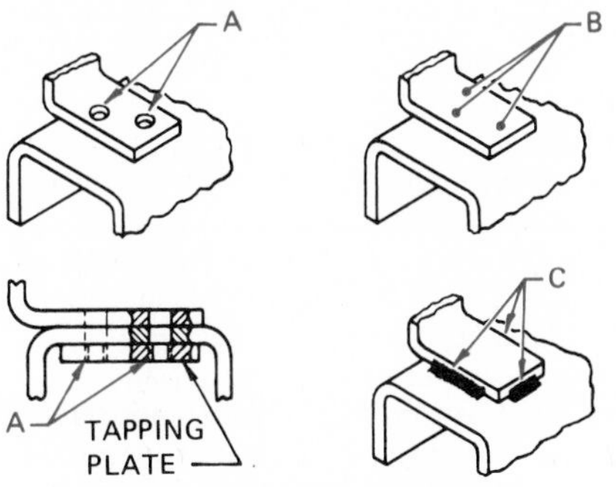
A
B
C
A
TAPPING
PLATE
Fig. 33-2-2 Bracket assembly.

Operation or Material	Cost Factor	Qty & Pos	COST COMPARISONS						
			Spot Welds	Proj Welds	Rivets	Arc Welds	Bolts & Nuts	Bolts & Tapping Plate	Blind Rivets
Method			1	2	3	4	5	6	7
Spotwelding	100	3B	300					200	
Projection Welding	100	3B		300					
Forming Weld Proj	89	3B		267					
Punching Hole	89	4A			356		356	356	356
Rivet	70	2A			140				
Driving Rivets	96	2A			192				192
Arc Welding (25 mm)	250	3C				750			
Bolt	115	2A					230	230	
Nut	106	2A					212		
Lockwasher	18	2A					36	36	
Assembling Bolts	136	2A					272	272	
Tapping Plate (Matl)	321	1A						321	
Drilling Hole	89	2A						178	
Tapping Hole	89	2A						178	
Blind Rivet	742	2A							1484
TOTAL COST			300	567	688	750	1106	1771	2032

The above table is for illustrative purposes only and its application should be adjusted to costs prevailing at the time of its use. Cost comparisons are based on spotwelding as Unit 100. The table is not intended to indicate that the least costly method is the best; function and strength of assembly must also be considered.

Fig. 33-2-1 Assembly methods cost analysis chart with reference to Figure 33-2-2.

EASE OF ASSEMBLY

Cost of assembly labor and equipment, as well as space requirements for assembly operations and equipment, is directly dependent on the ease and speed with which the assembly can be made.

Intricate assemblies require careful, slow hand fitting or expensive jigs and fixtures, or both. If the assembly is made from simple components which can be rapidly assembled, the cost will be lower.

Hole tolerances should be as liberal as is commensurate with the functional requirements of the assembly in order to facilitate the assembly operations. In many instances, a redesign may be justified to eliminate tight fits and unnecessarily close clearances which slow up the operation and make the assembly difficult. Not only should each subassembly be reviewed from this standpoint, but accessibility of all parts used to attach the subassembly to the main assembly should also be investigated to determine whether the tools normally used by the production department can reach the points of attachment.

QUALITY

Thought must be given to the finished appearance, the functional limitations, and the sales appeal of the completed product. Lack of attention to refinements in the assembly will otherwise completely offset the closest attention to the details.

SERVICE

One of the most frequently heard criticisms of assembled products is the difficulty and cost of removing and replacing some minor part of the unit. Often the labor cost of replacing a bearing, gasket, or minor assembly exceeds the cost of the replaced parts. The cost of replacing parts can be minimized if consideration is given during the design period to providing for rapid and easy disassembly of functioning parts.

It must also be remembered that automobiles, trucks, industrial engines, airplanes, locomotives, etc., have to function in all types of weather, and parts may be subject to moisture and rust. Likewise, products made to handle corrosive vapors and liquids must be given special consideration, and the designer should use fastenings which will be least affected by such exposure.

Attachments

Attaching means used in assemblies are broadly divided into three categories: permanent, semipermanent, and quick detachable. Each has an important function in the assembly of component parts.

PERMANENT ATTACHMENTS

Permanent attachments include welding, brazing, soldering, riveting, peening, staking, crimping, spinning, stapling, stitching, pressing, and shrinking. Welding is the most popular because of its satisfactory attachment and because it can be accomplished by many different processes.

Welding. *Welding* is the process of joining metallic parts by fusing them at their junction with heat and with or without pressure. For a more complete discussion, see the chapter on welding. In considering resistance welded attachments, select electrode shapes from a clearance standpoint (see Fig. 33-2-3).

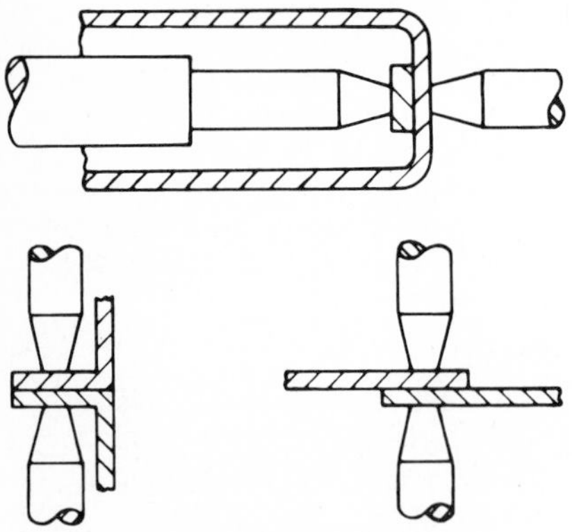

Fig. 33-2-3 Resistance welding.

Brazing. *Brazing* is the process of joining metallic parts by heating them at the junction points to a suitable temperature using a nonferrous filler metal that has a melting point below that of the base metals.

Soft Soldering. *Soft soldering* is the process of joining metal parts by melting into their heated joints an alloy of nonferrous metal. The silver brazing alloys, which are often called *hard solders*, have a much higher melting point and fall within the field of brazing.

Soft soldering may be used to fabricate low-stressed assemblies, to provide liquid- and gas-tight joints, or to ensure electrical contact. The strength and rigidity of a joint can be greatly increased by mechanically securing the joint before soldering, such as by staking, crimping, or folding. How soldered joints are indicated on drawings and a more complete discussion of assemblies of thin-gage metals can be found in the chapter on developments.

Rivets. *Rivets* are a permanent type of fastener for attaching parts of an assembly. The most common types are solid, blind, split, and tubular.

Solid rivets are used in assemblies that are not intended to be taken apart. Riveting may be done hot or cold. The cold method has replaced hot riveting in most assemblies because of the speed and efficiency of new power tools and the freedom from scale.

The use of these power tools facilitates the cold riveting process, which has the advantage of a more complete filling of holes as a result of the absence of scale and contraction of the rivets upon cooling.

Blind rivets are designed for use where it is impossible to have access to both ends of the rivet, such as riveting a bracket to a box section. In other applications, they may also be used in place of solid rivets; however, they are more costly than solid rivets. The cost of installation time plus the unit price of both methods should be considered before a process is chosen. Blind rivets are available in various designs to suit both light- and heavy-duty applications.

Tubular rivets do not make as strong joints as solid rivets, but they can be easily installed by either a spinning or a squeezing process. Although spinning is considered more desirable from a strength standpoint, the squeezing operation is used more generally because of the simplicity of equipment.

Split rivets are somewhat limited in their applications. They are usually installed with the same type of squeezer equipment used for tubular rivets. An application of tubular and spit rivets is shown in Fig. 33-2-4.

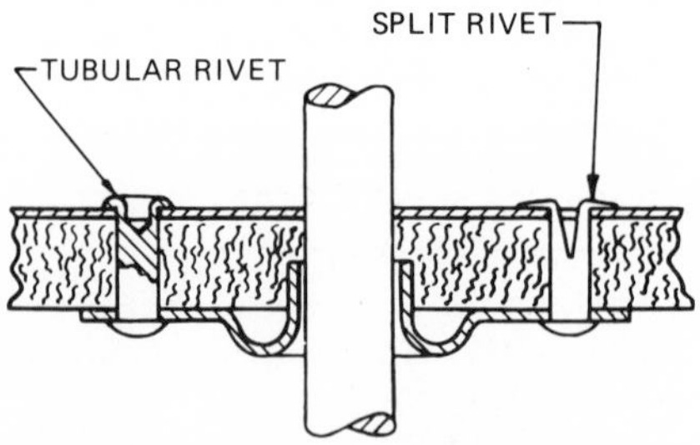
SPLIT RIVET
TUBULAR RIVET

Fig. 33-2-4 Tubular and split rivet applications.

Since many manufacturers supply the various types of riveting equipment, it is impractical to give detailed specifications for any particular tool. The designer should contact either the tool engineer or the suppliers to determine what equipment is available or can be used.

Impact riveting, known also as *peening*, is used to secure a shoulder pin or rivet in an assembly of two or more parts. Impact riveting can be used to advantage where stock thickness or hardness of parts varies; the operator can control the force and number of blows required to produce a secure assembly. In impact riveting, a round shaft is often swaged into contact with the sides of a hexagonal hole to solidly lock the pin and eliminate any possibility of rotation of the pin in the part (see Fig. 33-2-5).

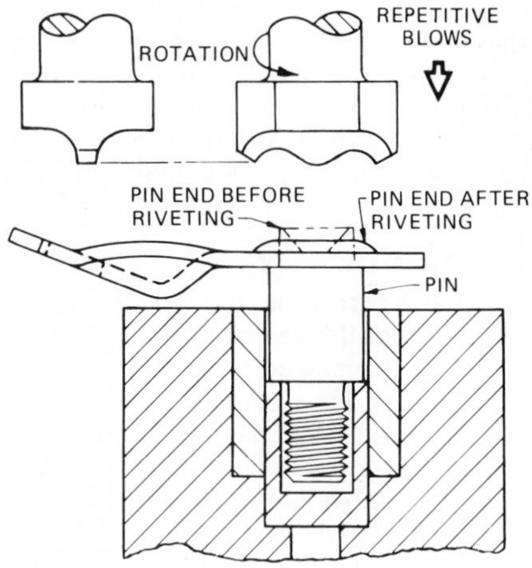
ROTATION
REPETITIVE BLOWS
PIN END BEFORE RIVETING
PIN END AFTER RIVETING
PIN

Fig. 33-2-5 Impact riveting.

Spin riveting results in a better head-bearing surface than impact or squeeze riveting, and has less tendency to cause shaft distortion than squeeze riveting. However, it is usually slower in operation, and the tool cost is normally much higher. Spin riveting can be used to advantage where one of the assembled parts must be free to move.

Squeeze riveting can be used to advantage in fastening two or more parts where the holes for the rivet may be slightly mismatched, as well as in true matched holes. Squeeze riveting of pins into parts tends to distort the pin below the riveting point and, by so doing, fill the hole in the plates even though they may be slightly mismatched. This method of riveting can be done with either hot or cold rivets (see Fig. 33-2-6).

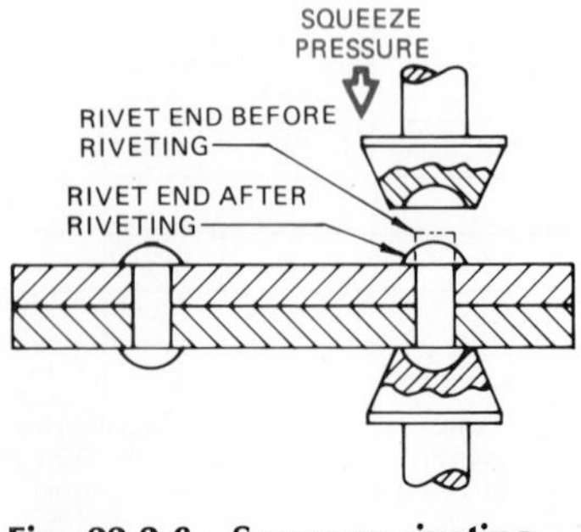
SQUEEZE PRESSURE
RIVET END BEFORE RIVETING
RIVET END AFTER RIVETING

Fig. 33-2-6 Squeeze riveting.

Crimping. This method is used to secure two or more pieces of metal in assembly by folding over the metal of one part to squeeze or clinch the other part or parts.

In crimping, the part must be designed to allow enough extruded metal on the crimped part to fold over in complete contact with its assembly mates but without excess metal which may be forced out from under the crimping punch. Successful crimping requires a die or tool designed for the specific crimping operation, as illustrated in Fig. 33-2-7. Crimping is less expensive than riveting or welding and can be used when the metal of one part is ductile enough to allow folding over without cracking.

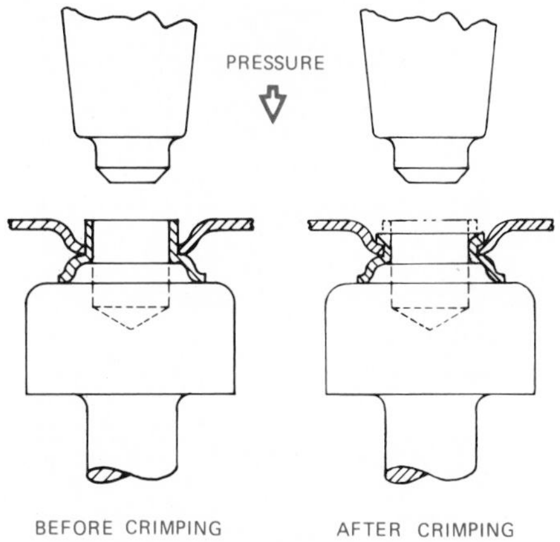
PRESSURE
BEFORE CRIMPING
AFTER CRIMPING

Fig. 33-2-7 Crimping.

Stitching. This method is used to secure metal to metal, fabric to metal, rubber to metal, etc., as shown in Fig. 33-2-8. Stitching is economical because of the speed of assembly and the low cost of assembly material. In stitching, a continuous wire is fed into a machine which forms the staples and then drives them through the material and clinches over the ends to secure the assembly. No prepunched holes are required.

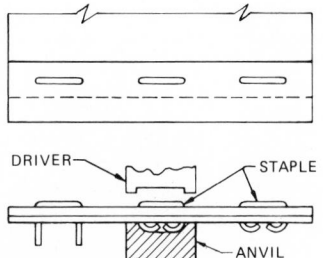

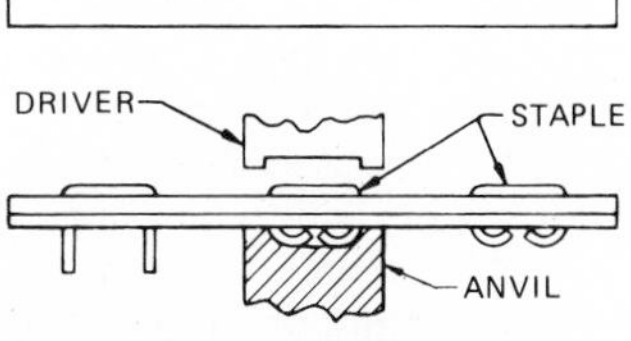
DRIVER
STAPLE
ANVIL

Fig. 33-2-8 Stitching.

Press Fit. The term *press fit* applies to the assembling of a part, such as a shaft, into a hole which is slightly smaller in diameter than the shaft (see Fig. 33-2-9). The degree of interference depends on the size of the hole, the mass of material around the hole, and the kind and quality of material.

Press-fit components are assembled by means of a hand or power press or a series of blows. This method of assembly is less costly than securing with setscrews, threads, or welding.

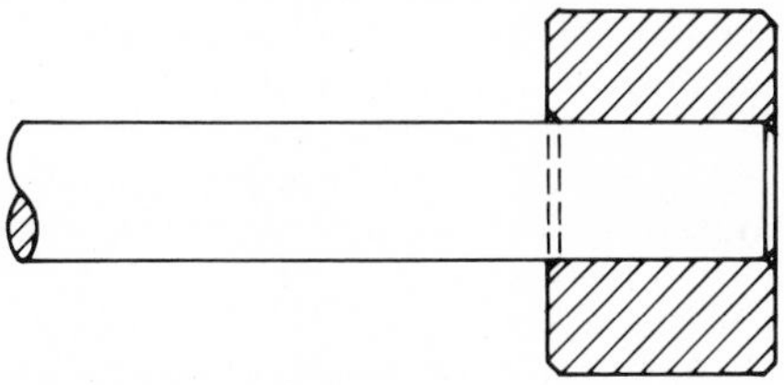

Fig. 33-2-9 Press fit.

Shrink Fit. This method is a modification of a press fit, adapted particularly to large diameters. Diameters that would provide sufficient interference to hold the two parts together permanently could not be pressed together cold. In these cases, the ring is preheated, then slipped over the shaft or wheel and allowed to cool in place. As the ring cools, it tends to shrink to its normal diameter, thereby producing a pressure on the shaft sufficient to hold it.

Cementing. Cementing or bonding with a suitable adhesive agent is another method used in production to make permanent or semipermanent assemblies. Some good examples of the use of adhesive agents for making permanent assemblies are brake linings to brake shoes and door weather strips to doors.

SEMIPERMANENT ATTACHMENTS

Semipermanent attachments include pins and threaded fastenings, such as bolts, screws, studs, nuts, washers, nails, and pins. Many factors must be taken into consideration when a fastener selection is made, such as strength, appearance, permanence, corrosion resistance, materials to be joined, cost, assembling, and disassembling.

Bolts. The proper diameter for a bolt is usually determined by design requirement and controlled by the engineer or designer. The factors which govern this decision are the strength requirement of the assembled unit and the material and heat treatment of the bolt. The type of head is also determined by design requirements such as unit pressure exerted by the bolt head, space limitations, and driving torque.

Hexagonal bolts are in most general use. They have a washer face, or the underside of the head is chamfered. They may be used in a threaded hole or with a nut. A typical application of a hexagonal bolt is shown in Fig. 33-2-10. Hexagonal bolts are available with conical points for piloting purposes and with slotted heads for special applications.

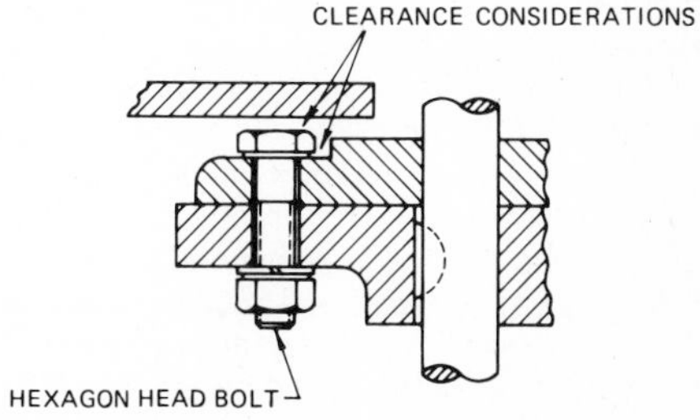
CLEARANCE CONSIDERATIONS
HEXAGON HEAD BOLT

Fig. 33-2-10 Hexagon bolt application.

Flanged hex-head bolts are usually specified where a bolt is to be used against a material that has a relatively low compressive strength, such as aluminum. The flanged head is also advantageous where an oversized hole or slotted hole is necessary.

Round-head bolts are made with variously shaped necks under the head for such specific purposes as embedding in wood or metal to prevent rotation or as a means of retention in thin metal, as shown in Fig. 33-2-11.

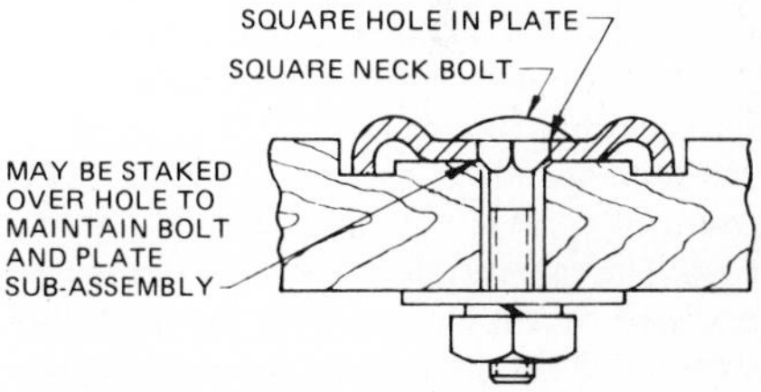
SQUARE HOLE IN PLATE
SQUARE NECK BOLT
MAY BE STAKED
OVER HOLE TO
MAINTAIN BOLT
AND PLATE
SUB-ASSEMBLY

Fig. 33-2-11 Round-head square-neck bolt assembly.

Square-head bolts are better adapted to heavy machinery, conveyors, and fixtures.

In selecting thread pitches, the bolt material strength and internal thread material strength must be considered, since a coarse pitch produces a stronger internal thread and a fine pitch produces a stronger external thread. The following two factors should also be considered:

1. A fine-thread bolt has a larger tensile-stress area and therefore a stronger core.

2. There is less chance of cross-threading coarse threads and of damaging them with rough handling.

In general, coarse threads should be used in materials which have relatively low shear strength, such as castings and soft metals, and for applications requiring rapid assembly or disassembly. Fine threads should be used where fine adjustment is necessary and where thin walls may be encountered. Also, fine threads may be required to obtain adequate tool life when high-strength forgings are tapped.

Studs. These are sometimes called *stud bolts*. Studs have threads on both ends, to be screwed permanently into a fixed part at one end and receive a nut on the exposed end. They are made of different materials depending on their use. Normally, they are made with coarse threads on the stud end and fine threads on the nut end, as shown in Fig. 33-2-12.

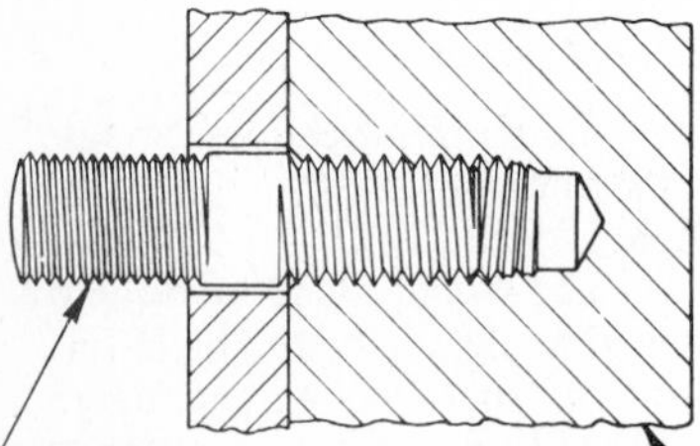
NUT END
CASTING
Fig. 33-2-12 Stud application.

Machine Screws. Machine screws generally differ from bolts in range of diameters, head shapes, and driver provisions. Their use is restricted to light assemblies such as instrument-panel mountings, moldings, and wire and pipe clips. The size of screw to be used is determined by the tightness required of the parts to be fastened. The smallest screws that will withstand sufficient torque to prevent loosening from vibration are generally used. Machine screws can be assembled into a nut or a threaded hole in a functional part.

The *flat head* is used where a flush surface is required. The *oval head* is generally used for reasons of appearance. Other head types are used for functional reasons; for example, *pan* and *truss heads* are used to cover large clearance holes and elongated holes.

The *hexagonal heads* are preferable from a driving standpoint; however, they are not suitable for appearance in many locations. For appearance, the *cross-recess head* is popular. In addition to its good appearance, it has the advantage of better driving characteristics than the slotted heads, which are susceptible to driver slippage and resultant marred surfaces.

For applications where mating parts are subject to misalignment, it may be desirable to specify special "dog points" on machine screws to provide a piloting action and avoid cross-threading. The use of header points will also facilitate assembly in many instances.

Setscrews. Setscrews are used extensively in tools, jigs and fixtures, control knobs, hand wheels, cam levers, and collars. In order to avoid accidents to operators, setscrews with the end protruding above the hole should never be used on power-rotated or oscillating parts. A typical setscrew installation is shown in Fig. 33-2-13.

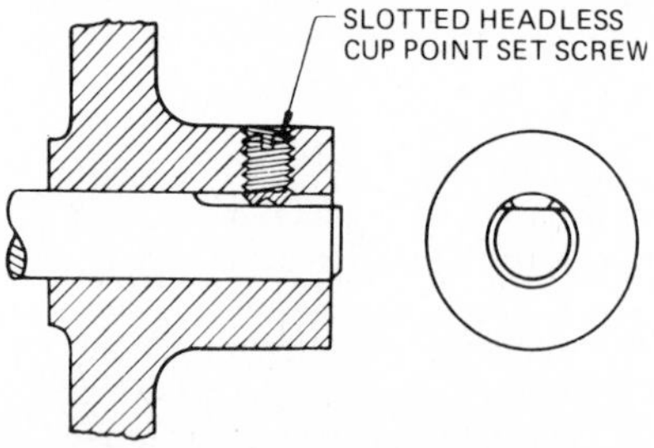
SLOTTED HEADLESS
CUP POINT SET SCREW
Fig. 33-2-13 Setscrew application.

Screw-and-Washer Assemblies. A preassembled screw and washer comprise a unit assembly, as shown in Fig. 33-2-14. The washer is free to rotate relative to the screw and is held in place under the head of the screw by the threads, which are rolled after the washer is assembled. Screw-and-washer assemblies result in a labor savings, since only one part need be handled. In addition, they ensure that a washer will be included in the assembly. Procurement and stock control are also simplified. These are factors which bear consideration in specifying screws and washers for attachments and should be weighed against the added unit cost for screw-and-washer assemblies.

Wood Screws. Wood screws are generally cold-headed parts with cut threads and are used for joining metal or plastic to wood or wood to wood. They are available in flat-, oval-, and round-head styles with slotted and cross-recess driver provisions; and they are made from various metals with standard finishes. Since the type of wood used and the direction of grain affect the assembly, little can be said regarding the proper size and quantity of screws required for a given assembly. Experience or practical tests should govern the selection of size and quantity of screws to be used.

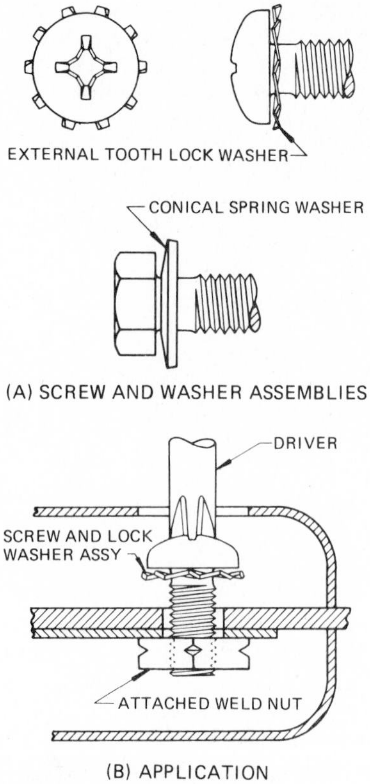
EXTERNAL TOOTH LOCK WASHER
CONICAL SPRING WASHER
(A) SCREW AND WASHER ASSEMBLIES
DRIVER
SCREW AND LOCK
WASHER ASSY
ATTACHED WELD NUT
(B) APPLICATION
Fig. 33-2-14 Screw-and-washer assemblies.

Drive Screws for Wood. These screws are modified wood screws, having a longer thread lead which permits them to be driven into the wood with a mechanical or hand hammer. Driver impressions are provided in the heads for removal of the screw with conventional tools. These screws are desirable for high-speed assembly of metal parts to wood members where a large quantity of screws are to be driven.

Screw Nails. This type of fastener is generally used for joining sheet metal to wood. The spiral threads cut into the burr formed in the sheet metal by the pilot and force their way into the wood to provide the fastening. They are driven in a manner similar to ordinary nails. A pierced hole should be provided in heavy sheet metal to facilitate assembly.

Drive Screws for Metal. Hardened metallic drive screws provide a permanent fastening for heavy sheet metal, castings, plastics, etc., and may be used in place of tapping screws or machine screws. Drive screws are hammered or otherwise forced into holes of suitable size. The unthreaded pilot guides the drive screw in straight, and the hardened spiral thread, which extends to the head, forms the required mating thread in the hole.

The thickness of metal into which the screw is to driven must at least be approximately the same as the outside diameter of the drive screw to ensure adequate thread engagement. An advantage in using these screws in place of machine screws is the elimination of tapped holes; however, a pilot hole is necessary (see Fig. 33-2-15).

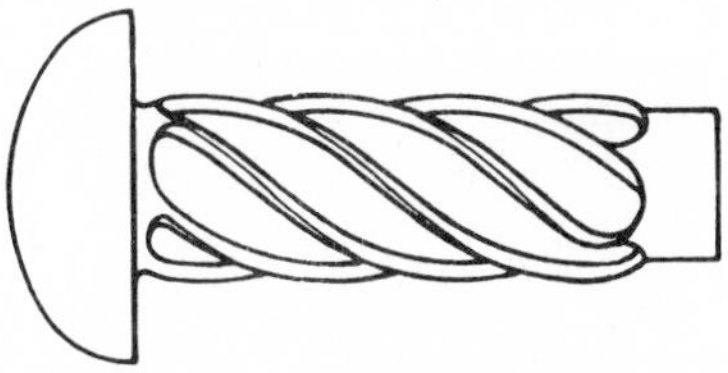

Fig. 33-2-15 Metallic drive screw.

Tapping Screws. These screws were developed primarily to eliminate tapping operations or nuts in certain assemblies of sheet-metal parts, plastics, and soft castings. There are four basic categories of tapping screws:

1. Thread-forming tapping screws which displace material around the screw thread

2. Thread-cutting tapping screws which cut threads into the mating material

3. Thread-swaging tapping screws which form the mating material into a thread

4. Drilling and thread-forming tapping screws which have a point designed to drill a hole in the mating material before the threads engage

Nuts. Many types of nuts are available for specific requirements. It is desirable to minimize the use of special designs in favor of the more commonly used nuts.

Slotted nuts with cotter pins or wire can be used to help retain the nut on the bolt.

Jam nuts are used where height is restricted or as a means of locking the working nut, if assembled as shown in Fig. 33-2-16.

A *locknut* is a nut having a special means for gripping an externally threaded member so that relative back-off rotation between the nut and the companion member is impeded.

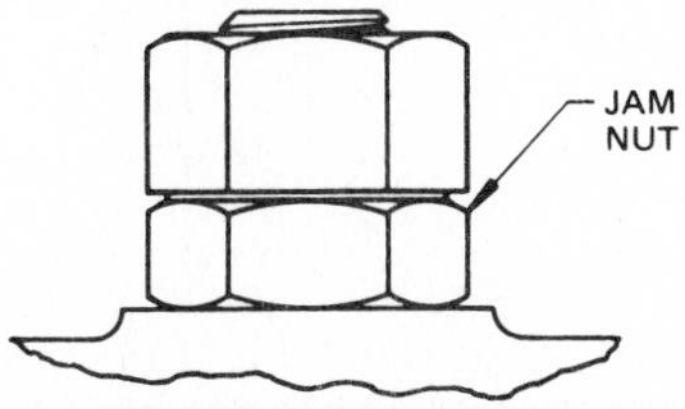
JAM
NUT

Fig. 33-2-16 Jam nut application.

Prevailing-torque-type locknuts employ a self-contained locking feature such as deformed or undersize threads, variable lead angle, plastic or fiber washers, or plug inserts. This type of nut resists screwing on, as well as unscrewing, and does not depend on bolt load for locking (see Fig. 33-2-17).

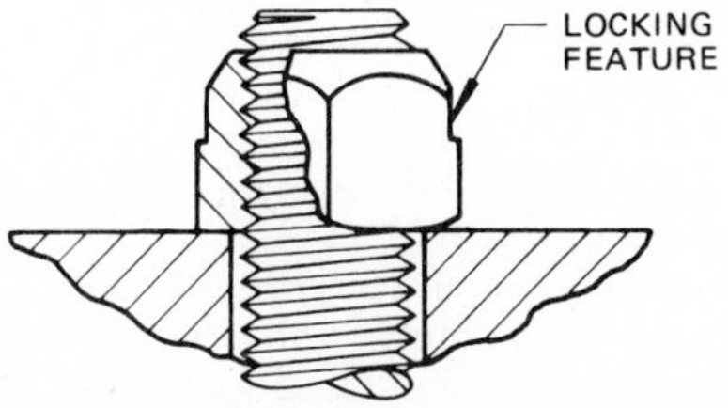
LOCKING
FEATURE

Fig. 33-2-17 Prevailing torque locknut.

Free-running locknuts develop their locking action after the nut has been seated by reactive spring force against the threads or by friction against the bearing surface.

The necessity and cost of locknuts as compared with other types of locking devices should be considered in their selection.

Spring nuts are made of thin spring metal and have arched prongs or formed embossments to fit a single lead of a mating screw thread (see Fig. 33-2-18). Spring nuts are used extensively for sheet-metal construction where relatively high torques and strength are not required.

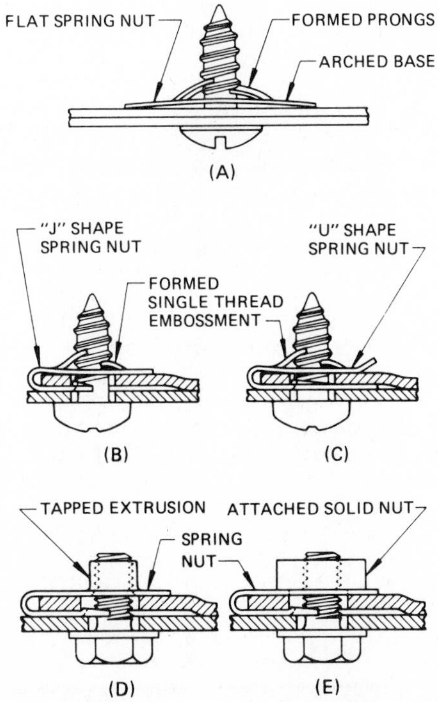
FLAT SPRING NUT
FORMED PRONGS
ARCHED BASE
(A)
"J" SHAPE
SPRING NUT
"U" SHAPE
SPRING NUT
FORMED
SINGLE THREAD
EMBOSSMENT
(B)
(C)
TAPPED EXTRUSION
ATTACHED SOLID NUT
SPRING
NUT
(D)
(E)

Fig. 33-2-18 Spring-nut assemblies.

Another type of spring nut is available which can be pushed on over rivets, tubing, nails, or other unthreaded parts and provides a positive bite that grips securely even on very smooth surfaces. Figure 33-2-19 shows a typical application.

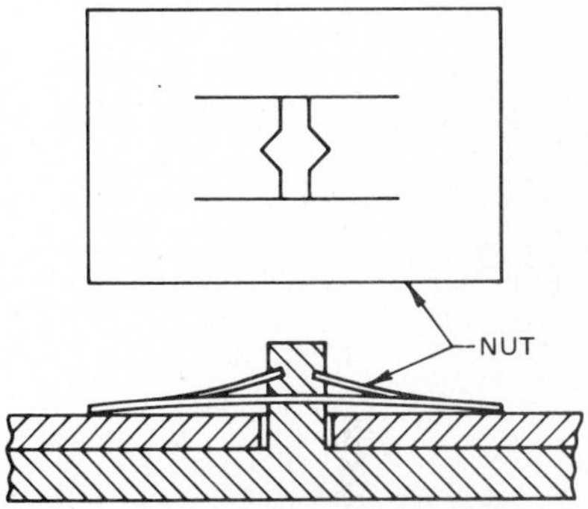
NUT

Fig. 33-2-19 Push-on spring nut.

Stamped nuts are usually fabricated from thin spring steel and have arched prongs formed to fit a single lead of a mating screw thread. They have the same functional usage as a spring nut, with the additional advantage of provisions for turning the nut (see Fig. 33-2-20).

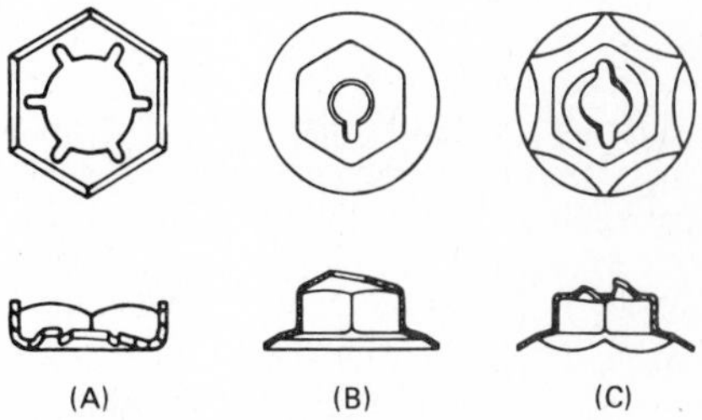
(A)
(B)
(C)

Fig. 33-2-20 Stamped nuts.

Crown nuts are generally used where it is desirable to cover the end of the externally threaded part for purposes of appearance or protection from sharp edges.

Wing nuts, as the name implies, are provided with two wings to facilitate hand tightening and loosening. They are used where high torque is not required and where the nuts are to be disassembled and reassembled frequently.

Barrel and *sleeve nuts* are usually made to resemble a screw head at the outer or exposed end. They are used in assemblies where any other type of nut would present a less favorable appearance.

Clinch nuts were developed for sheet-metal assemblies where the nut is inaccessible for wrenching. They are provided on one side with a shoulder and smaller pilot, which is inserted into a preformed hole in the sheet metal, and are permanently attached by spinning or staking the portion of pilot extending through the hole (see Fig. 33-2-21).

Weld nuts are similar to the clinch nuts in function. However, they are supplied with weld projections and are either spot- or projection-welded to the sheet metal instead of being peened.

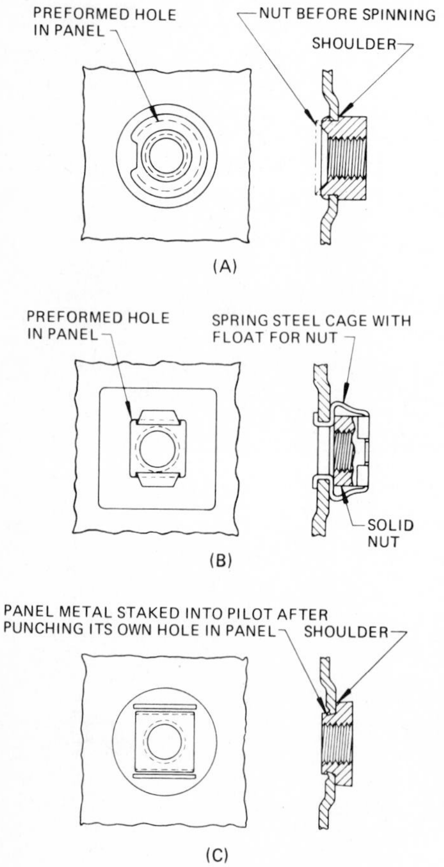
PREFORMED HOLE IN PANEL
NUT BEFORE SPINNING
SHOULDER
(A)
PREFORMED HOLE IN PANEL
SPRING STEEL CAGE WITH FLOAT FOR NUT
SOLID NUT
(B)
PANEL METAL STAKED INTO PILOT AFTER PUNCHING ITS OWN HOLE IN PANEL
SHOULDER
(C)

Fig. 33-2-21 Assembled clinch nuts.

Washers. The four basic types of washers are flat (plain) washers, conical-spring washers, helical-spring lockwashers, and tooth lockwashers. All four types are available in standard sizes to suit standard bolts and screws.

Flat washers are used under the head of a screw or bolt, or under a nut, for four principal purposes:

1. To spread the load over a greater area
2. To reduce frictional variations during assembly
3. To provide bearing surface over large clearance holes or slots
4. To prevent marring of parts during assembly.

Conical-spring washers are made of steel, hardened and tempered. The relatively high supporting load and spring return make this washer effective where bolt tension may be lost because of such factors as thermal expansion or compression set of gaskets.

Helical-spring lockwashers are usually used as a hardened thrust washer or as a spacer.

Three commonly used types of *tooth lockwashers* are shown in Fig. 33-2-22. The hardened teeth of these washers are twisted offset to bite both the bolt head or nut and the respective work surface to resist loosening of the assembly. They are also used to ensure good electrical contact.

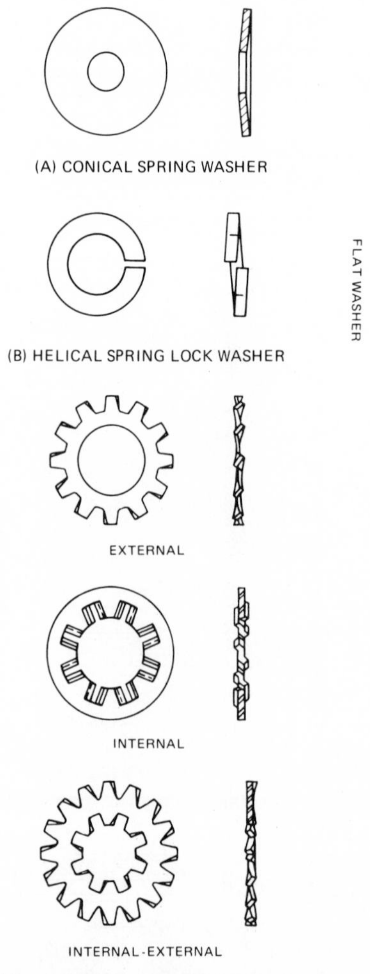
(A) CONICAL SPRING WASHER
FLAT WASHER
(B) HELICAL SPRING LOCK WASHER
EXTERNAL
INTERNAL
INTERNAL-EXTERNAL

Fig. 33-2-22 Washer types.

Pins. Cotter pins, spring pins, groove pins, taper pins, and clevis pins are used to retain parts of an assembly in relative position.

Cotter pins are used for retaining slotted nuts, moveable links or rods, etc., as shown in Figs. 33-2-23 and 33-2-25.

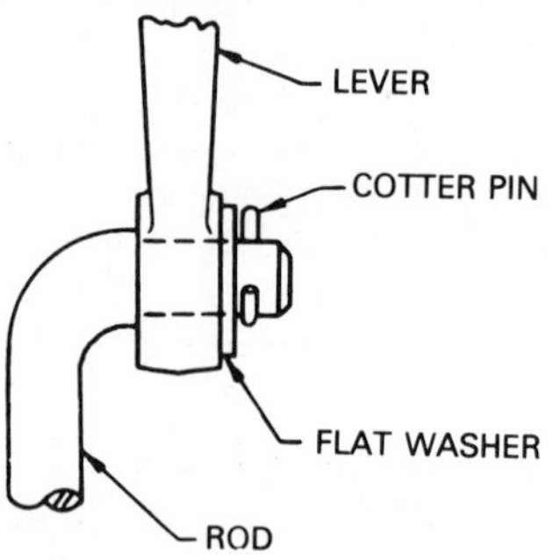
LEVER
COTTER PIN
FLAT WASHER
ROD

Fig. 33-2-23 Cotter-pin application.

Groove pins are straight pins having longitudinal grooves rolled or pressed into the body which provide a reactive expansion effect when the pin is driven into a drilled hole (see Fig. 33-2-24). *Spring pins* provide a spring effect which serves to retain the pin when it is driven into a drilled hole of diameter slightly

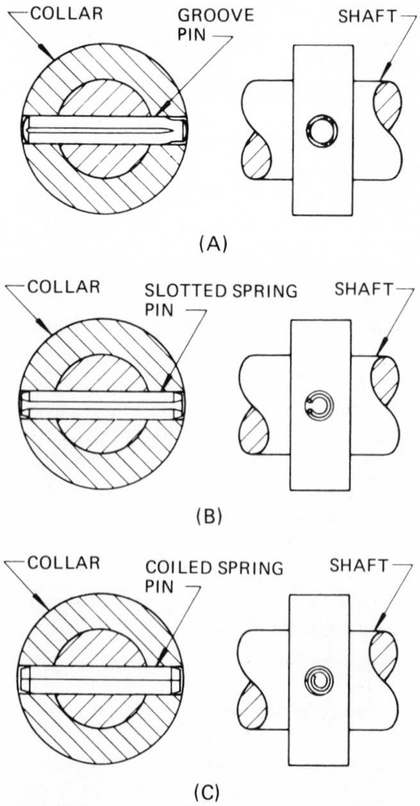
COLLAR
GROOVE PIN
SHAFT
(A)
COLLAR
SLOTTED SPRING PIN
SHAFT
(B)
COLLAR
COILED SPRING PIN
SHAFT
(C)

Fig. 33-2-24 Groove- and spring-pin application.

smaller than that of the pin. These types of pins eliminate reaming or peening and can be disassembled a number of times without serious loss of holding power. They are used for semipermanent fastening of levers, collars, gears, cams, etc., to shafts. They may also be used as guides or locating pins.

Taper pins serve the same functional purpose as groove pins. However, they require taper-reamed holes at assembly and are retained only by taper lock, which can totally disengage when minor displacement occurs. Assemblies using taper pins are more costly than grooved pins.

The primary purpose of *clevis pins* is to attach clevises to rod ends and levers and to serve as bearings. They are held in place by cotter pins, as shown in Fig. 33-2-25.

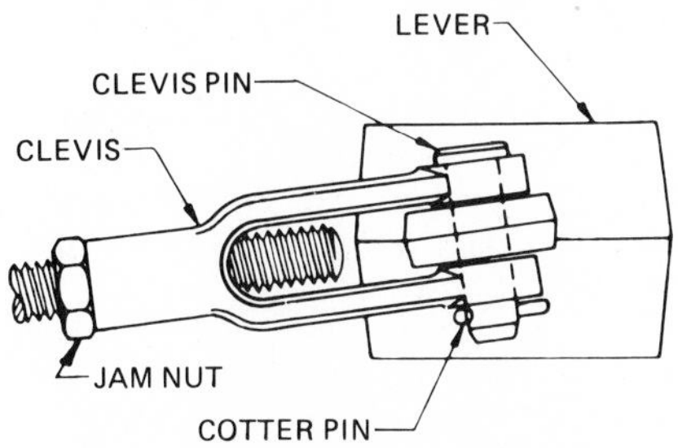
LEVER
CLEVIS PIN
CLEVIS
JAM NUT
COTTER PIN

Fig. 33-2-25 Clevis-pin application.

Quick Detachable Attachments. Quick detachable attachments include clips or certain kinds of snap fasteners. These attaching means are convenient and timesaving by providing for quick assembly and disassembly.

Design Checklist

The following design checklist will serve as a helpful guide in reviewing a design.

1. Keep the number of separate pieces in a subassembly as low as practicable by combining single components into one assembly if functional requirements permit.

2. Check particularly left- and right-hand parts to determine whether they can be made identical and so avoid carrying an extra part.

3. Check advisability of using lockwashers assembled to bolts as purchased assemblies.

4. Check to determine if similar parts between models can be standardized.

5. If subassemblies require alignment, the parts should be designed to secure such alignment without the use of special jigs by providing tabs, shoulders, notches, or contour locators.

6. Provide adequate clearances between parts for assembly tools.

7. Provide clearances between parts to allow for tolerance stackup variations.

8. Specify the simplest, most effective, and cheapest type of attachment practicable and commensurate with the functional requirements.

9. Avoid blind-riveting operations wherever possible.

10. Where rivets are used, provide sufficient clearances between parts and from flanges to permit use of a standard riveting gun.

11. Avoid use of slotted nuts and cotter pins wherever possible.

12. Standardize as far as practicable on bolt and thread sizes. Hold number of bolt lengths to a minimum and recheck frequently to reduce number of lengths in use. Avoid smaller sizes since they are difficult to handle and losses on this class of material are high.

13. Try to avoid riveting or welding operations in shop areas where this type of equipment is not normally used.

Assignments

1. *Fig. 33-2-A*. Design a combined pencil holder and masking tape dispenser which can be mounted on the edge of a drafting table. It should hold four pencils and one roll of masking tape. The tape has a 75 mm diameter center hole and is 25 mm wide. Prepare an assembly drawing and include a bill of material.

The material selected should be able to be used in mass production methods.

2. A 20 mm diameter steel rotary shaft must be supported at two metre intervals along a ceiling. Split journal bearings, 40 mm long, are recommended by the engineering department. A maximum of eight brackets are required. On an A3- or B-size sheet, design a suitable bracket complete with dimensions.

3. Same as assignment no. 2 except the quantity is 2000.

4. Two vertical 8 mm diameter electrical conductors are to be supported on the inside wall of an oil-filled metal tank. The conductors, made of copper, have no insulation on them. The voltage they carry is such that there must be a minimum distance of 50 mm between conductors or any other metal when supported by nonconductive material, such as plastic, wood, etc. Although there are no external forces acting on these conductors, they should be supported every 600 mm. On an A3- or B-size sheet, prepare an assembly drawing showing the conductors, support, and tank wall. Include a bill of material. On a second sheet prepare the details of the parts required. Scale to suit.

5. Design a container to store twelve 8-track tape cassettes. This container should have a lid or cover and look attractive when installed on the bottom edge of an automobile instrument panel.

6. Design a napkin holder to sit on the table. The napkins can be stored either flat 170 mm square or folded in half 85 x 170 mm. Their position for dispensing is your choice as long as one napkin can be taken at a time. The holder must hold a minimum of 12 napkins and come in a variety of colors or a selection of wood grains to complement the modern kitchen.

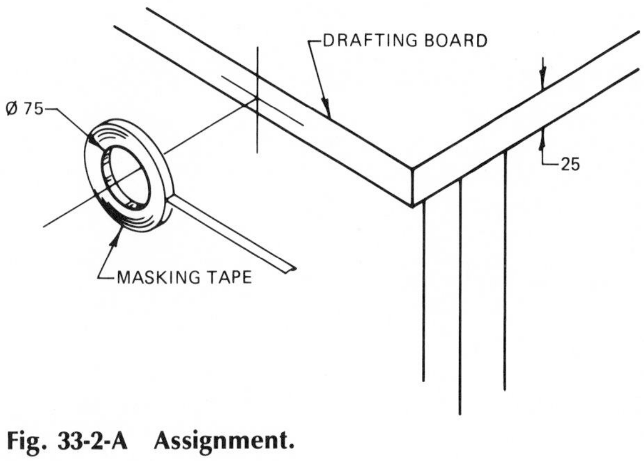
DRAFTING BOARD
Ø 75
25
MASKING TAPE

Fig. 33-2-A Assignment.

Appendix

ANSI AND CSA PUBLICATIONS

ANSI—American National Standards Institute

ANSI	Y14.1	Size and Format
*ANSI	Y14.2-1957	Line Conventions, Sectioning and Lettering
*ANSI	Y14.3-1957	Projections
*ANSI	Y14.4-1957	Pictorial Drawing
*ANSI	Y14.5-1966	Dimensioning and Tolerancing for Engineering Drawings
*ANSI	Y14.6-1957	Screw Threads
ANSI	Y14.7	Gears, Splines, and Serrations
ANSI	Y14.7.1	Gear Drawing Standards—Part 1, for Spur, Helical, Double Helical and Rack
ANSI	Y14.9	Forgings
ANSI	Y14.10	Metal Stampings
ANSI	Y14.11	Plastics
ANSI	Y14.14	Mechanical Assemblies
ANSI	Y14.15	Electrical and Electronics Diagrams
ANSI	Y14.15A	Interconnection Diagrams
ANSI	Y14.17	Fluid Power Diagrams
ANSI	Y32.2	Graphic Symbols for Electrical and Electronics Diagrams
ANSI	Y32.9	Graphic Electrical Wiring Symbols for Architectural and Electrical Layout Drawings
ANSI	B1.1	Unified Screw Threads
ANSI	B18.2.1	Square and Hex Bolts and Screws
ANSI	B18.2.2	Square and Hex Nuts
ANSI	B18.3	Socket Cap, Shoulder, and Setscrews
ANSI	B18.6.2	Slotted-Head Cap Screws, Square-Head Setscrews, Slotted-Headless Setscrews
ANSI	B18.6.3	Machine Screws and Machine Screw Nuts
ANSI	B17.2	Woodruff Key and Keyslot Dimensions
ANSI	B17.1	Keys and Keyseats
ANSI	B18.21.1	Lock Washers
ANSI	B27.2	Plain Washers
ANSI	B46.1	Surface Texture

CSA—Canadian Standards

CSA	B1.1	Unified and American Screw Threads
CSA	B19.1	Plain Washers
CSA	B33.1	Square and Hexagon Bolts and Nuts, Studs and Wrench Openings
CSA	B35.1	Machine Screws, Stove Bolts and Associated Nuts
CSA	B78.1	Drawing Standard—General Principles
CSA	B78.2	Drawing Standard—Dimensioning and Tolerancing
CSA	B95	Surface Texture
CSA	B97.1	Limits and Fits for Engineering and Manufacturing
CSA	Z85	Abbreviations for Scientific and Engineering Terms
33-GP-7		Architectural Drawing Practices (National Research Council, Ottawa, Canada)

Note: The above standards may be purchased from:
AMERICAN NATIONAL STANDARDS INSTITUTE, INC.
1430 BROADWAY,
NEW YORK, N.Y. 10018

CANADIAN STANDARDS ASSOCIATION
178 REXDALE BOULEVARD
REXDALE, ONTARIO, CANADA, M9W 1R3

THE AMERICAN SOCIETY OF MECHANICAL ENGINEERS
UNITED ENGINEERING CENTER
345 EAST 47TH STREET,
NEW YORK, N.Y. 10017

Table 1 ANSI and CSA publications.

Quantity	Metric Unit	Symbol	Metric to Inch-Pound Unit	Inch-Pound to Metric Unit
Length	millimetre centimetre metre kilometre	mm cm m km	1 mm = 0.0394 in. 1 cm = 0.394 in. 1 m = 39.37 in. = 3.28 ft 1 km = 0.62 mile	1 in. = 25.4 mm 1 ft. = 30.5 cm 1 yd. = 0.914 m = 914 mm 1 mile = 1.61 km
Area	square millimetre square centimetre square metre	mm^2 cm^2 m^2	1 mm^2 = 0.001 55 sq. in. 1 cm^2 = 0.155 sq. in. 1 m^2 = 10.8 sq. ft. = 1.2 sq. yd.	1 sq. in. = 6 452 mm^2 1 sq. ft. = 0.093 m^2 1 sq. yd. = 0.836 m^2
Mass	milligram gram kilogram tonne	mg g kg t	1 g = 0.035 oz. 1 kg = 2.205 lb. 1 tonne = 1.102 tons	1 oz. = 28.3 g 1 lb. = 0.454 kg 1 ton = 907.2 kg = 0.907 tonnes
Volume	cubic centimetre cubic metre millilitre	cm^3 m^3 m	1 mm^3 = 0.000 061 cu. in. 1 cm^3 = 0.061 cu. in. 1 m^3 = 35.3 cu ft. = 1.308 cu. yd. 1 $m\ell$ = 0.035 fl. oz.	1 fl. oz. = 28.4 cm^3 1 cu. in. = 16.387 cm^3 1 cu. ft. = 0.028 m^3 1 cu. yd. = 0.756 m^3
Capacity	litre	L	U.S. Measure 1 pt. = 0.473 L 1 qt. = 0.946 L 1 gal = 3.785 L Imperial Measure 1 pt. = 0.568 L 1 qt. = 1.137 L 1 gal = 4.546 L	U.S. Measure 1 L = 2.113 pt. = 1.057 qt. = 0.264 gal. Imperial Measure 1 L = 1.76 pt. = 0.88 qt. = 0.22 gal.
Temperature	Celsius degree	°C	°C = $\frac{5}{9}$(°F-32)	°F = $\frac{9}{5}$ × °C +32
Force	newton kilonewton	N kN	1 N = 0.225 lb (f) 1 kN = 0.225 kip (f) = 0.112 ton (f)	1 lb (f) = 4.45N = 0.004 448 kN
Energy/Work	joule kilojoule megajoule	J kJ MJ	1 J = 0.737 ft · lb 1 J = 0.948 Btu 1 MJ = 0.278 kWh	1 ft · lb = 1.355 J 1 Btu = 1.055 J 1 kWh = 3.6 MJ
Power	kilowatt	kW	1 kW = 1.34 hp 1 W = 0.0226 ft · lb/min.	1 hp (550 ft · lb/s) = 0.746 kW 1 ft · lb/min = 44.2537 W
Pressure	kilopascal	kPa	1 kPa = 0.145 psi = 20.885 psf = 0.01 ton-force per sq. ft.	1 psi = 6.895 kPa 1 lb-force/sq. ft. = 47.88 Pa 1 ton-force/sq. ft. = 95.76 kPa
	*kilogram per square centimetre	kg/cm^2	1 kg/cm^2 = 13.780 psi	
Torque	newton metre *kilogram metre *kilogram per centimetre	N · m kg/m kg/cm	1 N · m = 0.74 lb · ft 1 kg/m = 7.24 lb · ft 1 kg/cm = 0.86 lb · in	1 lb · ft = 1.36 N · m 1 lb · ft = 0.14 kg/m 1 lb · in = 1.2 kg/cm
Speed/Velocity	metres per second kilometres per hour	m/s km/h	1 m/s = 3.28 ft/s 1 km/h = 0.62 mph	1 ft/s = 0.305 m/s 1 mph = 1.61 km/h

*Not SI units, but included here because they are employed on some of the gages and indicators currently in use in industry.

Table 2 Metric conversion tables.

ONE HUNDREDTH OF AN INCH INCREMENTS TO ONE INCH

Inch	.00	.01	.02	.03	.04	.05	.06	.07	.08	.09
.00	0.00	0.25	0.51	0.76	1.02	1.27	1.52	1.78	2.03	2.29
.10	2.54	2.79	3.05	3.30	3.56	3.81	4.06	4.32	4.57	4.83
.20	5.08	5.33	5.59	5.84	6.10	6.35	6.60	6.86	7.11	7.37
.30	7.62	7.87	8.13	8.38	8.64	8.89	9.14	9.40	9.65	9.91
.40	10.16	10.41	10.67	10.92	11.18	11.43	11.68	11.94	12.19	12.45
.50	12.70	12.95	13.21	13.46	13.72	13.97	14.22	14.48	14.73	14.99
.60	15.24	15.49	15.75	16.00	16.26	16.51	16.76	17.02	17.27	17.53
.70	17.78	18.03	18.29	18.54	18.80	19.05	19.30	19.56	19.81	20.07
.80	20.32	20.57	20.83	21.08	21.34	21.59	21.84	22.10	22.35	22.61
.90	22.86	23.11	23.37	23.62	23.88	24.13	24.38	24.64	24.89	25.15

ONE TENTH OF AN INCH INCREMENTS TO TWENTY INCHES

Inches	0	.10	.20	.30	.40	.50	.60	.70	.80	.90
0	0.0	2.5	5.1	7.6	10.2	12.7	15.2	17.8	20.3	22.9
1	25.4	27.9	30.5	33.0	35.6	38.1	40.6	43.2	45.7	48.3
2	50.8	53.3	55.9	58.4	61.0	63.5	66.0	68.6	71.1	73.7
3	76.2	78.7	81.3	83.8	86.4	88.9	91.4	94.0	96.5	99.1
4	101.6	104.1	106.7	109.2	111.8	114.3	116.8	119.4	121.9	124.5
5	127.0	129.5	132.1	134.6	137.2	139.7	142.2	144.8	147.3	149.9
6	152.4	154.9	157.5	160.0	162.6	165.1	167.6	170.2	172.7	175.3
7	177.8	180.3	182.9	185.4	188.0	190.5	193.0	195.6	198.1	200.7
8	203.2	205.7	208.3	210.8	213.4	215.9	218.4	221.0	223.5	226.1
9	228.6	231.1	233.7	236.2	238.8	241.3	243.8	246.4	248.9	251.5
10	254.0	256.5	259.1	261.6	264.2	266.7	269.2	271.8	274.3	276.9
11	279.4	281.9	284.5	287.0	289.6	292.1	294.6	297.2	299.7	302.3
12	304.8	307.3	309.9	312.4	315.0	317.5	320.0	322.6	325.1	327.7
13	330.2	332.7	335.3	337.8	340.4	342.9	345.4	348.0	350.5	353.1
14	355.6	358.1	360.7	363.2	365.8	368.3	370.8	373.4	375.9	378.5
15	381.0	383.5	386.1	388.6	391.2	393.7	396.2	398.8	401.3	403.9
16	406.4	408.9	411.5	414.0	416.6	419.1	421.6	424.2	426.7	429.3
17	431.8	434.3	436.9	439.4	442.0	444.5	447.0	449.6	452.1	454.7
18	457.2	459.7	462.3	464.8	467.4	469.9	472.4	475.0	477.5	480.1
19	482.6	485.1	487.7	490.2	492.8	495.3	497.8	500.4	502.9	505.5
20	508.0	510.5	513.1	515.6	518.2	520.7	523.2	525.8	528.3	530.9

Table 3 Conversion of decimals of an inch to millimetres.

IN.	0	1/16	1/8	3/16	1/4	5/16	3/8	7/16	1/2	9/16	5/8	11/16	3/4	13/16	7/8	15/16
0	.0	1.6	3.2	4.8	6.4	7.9	9.5	11.1	12.7	14.3	15.9	17.5	19.1	20.6	22.2	23.8
1	25.4	27.0	28.6	30.2	31.8	33.3	34.9	36.5	38.1	39.7	41.3	42.9	44.5	46.0	47.6	49.2
2	50.8	52.4	54.0	55.6	57.2	58.7	60.3	61.9	63.5	65.1	66.7	68.3	69.9	71.4	73.0	74.6
3	76.2	77.8	79.4	81.0	82.6	84.1	85.7	87.3	88.9	90.5	92.1	93.7	95.3	96.8	98.4	100.0
4	101.6	103.2	104.8	106.4	108.0	109.5	111.1	112.7	114.3	115.9	117.5	119.1	120.7	122.2	123.8	125.4
5	127.0	128.6	130.2	131.8	133.4	134.9	136.5	138.1	139.7	141.3	142.9	144.5	146.1	147.6	149.2	150.8
6	152.4	154.0	155.6	157.2	158.8	160.3	161.9	163.5	165.1	166.7	168.3	169.9	171.5	173.0	174.6	176.2
7	177.8	179.4	181.0	182.6	184.2	185.7	187.3	188.9	190.5	192.1	193.7	195.3	196.9	198.4	200.0	201.6
8	203.2	204.8	206.4	208.0	209.6	211.1	212.7	214.3	215.9	217.5	219.1	220.7	222.3	223.8	225.4	227.0
9	228.6	230.2	231.8	233.4	235.0	236.5	238.1	239.7	241.3	242.9	244.5	246.1	247.7	249.2	250.8	252.4
10	254.0	255.6	257.2	258.8	260.4	261.9	263.5	265.1	266.7	268.3	269.9	271.5	273.1	274.6	276.2	277.8
11	279.4	281.0	282.6	284.2	285.8	287.3	288.9	290.5	292.1	293.7	295.3	296.9	298.5	300.0	301.6	303.2
12	304.8	306.4	308.0	309.6	311.2	312.7	314.3	315.9	317.5	319.1	320.7	322.3	323.9	325.4	327.0	328.6
13	330.2	331.8	333.4	335.0	336.6	338.1	339.7	341.3	342.9	344.5	346.1	347.7	349.3	350.8	352.4	354.0
14	355.6	357.2	358.8	360.4	362.0	363.5	365.1	366.7	368.3	369.9	371.5	373.1	374.7	376.2	377.8	379.4

Table 4 Conversion of fractions of an inch to millimetres.

NUMBER	SQUARE	SQUARE ROOT	CIRCUMFERENCE OF CIRCLE	AREA OF CIRCLE	NUMBER	SQUARE	SQUARE ROOT	CIRCUMFERENCE OF CIRCLE	AREA OF CIRCLE	NUMBER	SQUARE	SQUARE ROOT	CIRCUMFERENCE OF CIRCLE	AREA OF CIRCLE
1	1	1	3.14	0.78	36	1296	6.0000	113.10	1017.88	71	5041	8.4261	223.05	3959.19
2	4	1.41	6.28	3.14	37	1369	6.0828	116.24	1075.21	72	5184	8.4853	226.19	4071.50
3	9	1.73	9.43	7.07	38	1444	6.1644	119.38	1134.11	73	5329	8.5440	229.34	4185.39
4	16	2.00	12.57	12.57	39	1521	6.2450	122.52	1194.59	74	5476	8.6023	232.48	4300.84
5	25	2.34	15.71	19.64	40	1600	6.3246	125.66	1256.64	75	5625	8.6603	235.62	4417.88
6	36	2.4495	18.85	28.27	41	1681	6.4031	128.81	1320.25	76	5776	8.7178	238.76	4536.47
7	49	2.6458	21.99	38.48	42	1764	6.4807	131.95	1385.44	77	5929	8.7750	241.90	4656.64
8	64	2.8284	25.13	50.27	43	1849	6.5574	135.09	1452.20	78	6084	8.8318	245.04	4778.37
9	81	3.0000	28.27	63.62	44	1936	6.6332	138.23	1520.53	79	6241	8.8882	248.19	4901.68
10	100	3.1623	31.46	78.54	45	2025	6.7082	141.37	1590.43	80	6400	8.9443	251.33	5026.56
11	121	3.3166	34.56	95.03	46	2116	6.7823	144.51	1661.90	81	6561	9.0000	254.47	5183.01
12	144	3.4641	37.70	113.09	47	2209	6.8557	147.65	1734.94	82	6724	9.0554	257.61	5281.03
13	169	3.6056	40.84	132.73	48	2304	6.9282	150.80	1809.56	83	6889	9.1104	260.75	5410.62
14	196	3.7417	43.98	153.94	49	2401	7.0000	153.94	1885.74	84	7056	9.1652	263.89	5541.78
15	225	3.8730	47.12	176.72	50	2500	7.0711	157.08	1963.50	85	7225	9.2200	267.04	5674.52
16	256	4.0000	50.27	201.06	51	2601	7.1414	160.22	2042.82	86	7396	9.2736	270.18	5808.82
17	289	4.1231	53.41	226.98	52	2704	7.2111	163.36	2123.72	87	7569	9.3274	273.32	5944.69
18	324	4.2426	56.55	254.47	53	2809	7.2801	166.50	2206.18	88	7744	9.3808	276.46	6082.14
19	361	4.3589	59.69	283.53	54	2916	7.3485	169.65	2290.22	89	7921	9.4340	279.60	6221.15
20	400	4.4721	62.83	314.16	55	3025	7.4162	172.79	2375.83	90	8100	9.4868	282.74	6361.74
21	441	4.5826	65.97	346.36	56	3136	7.4833	175.93	2463.01	91	8281	9.5393	285.89	6503.90
22	484	4.6904	69.12	380.13	57	3249	7.5498	179.07	2551.76	92	8464	9.5917	289.03	6647.63
23	529	4.7958	72.26	415.48	58	3364	7.6158	182.21	2642.08	93	8649	9.6437	292.17	6792.92
24	576	4.8990	75.39	452.39	59	3481	7.6811	185.35	2733.97	94	8836	9.6954	295.31	6939.79
25	625	5.0000	78.54	490.87	60	3600	7.7460	188.50	2827.43	95	9025	9.7468	298.45	7088.24
26	676	5.0990	81.68	530.93	61	3721	7.8102	191.64	3922.47	96	9216	9.7979	301.59	7238.25
27	729	5.1962	84.82	572.56	62	3844	7.8740	194.78	3019.07	97	9409	9.8489	304.74	7389.83
28	784	5.2915	87.97	615.75	63	3969	7.9373	197.92	3117.25	98	9604	9.8995	307.88	7542.98
29	841	5.3852	91.11	660.52	64	4096	8.0000	201.06	3216.99	99	9801	9.9509	311.02	7697.71
30	900	5.4772	94.25	706.86	65	4225	8.0623	204.20	3318.31	100	10 000	10.000	314.16	7854.00
31	961	5.5678	97.39	754.77	66	4356	8.1240	207.35	3421.19	101	10 201	10.0499	317.30	8011.87
32	1024	5.6569	100.53	804.25	67	4489	8.1854	210.49	3525.65	102	10 404	10.0995	320.44	8171.30
33	1089	5.7446	103.67	855.30	68	4624	8.2462	213.63	3631.68	103	10 609	10.1489	323.58	8332.31
34	1156	5.8310	106.81	907.92	69	4761	8.3066	216.77	3739.28	104	10 816	10.1980	326.73	8494.89
35	1225	5.9161	109.96	962.113	70	4900	8.3666	219.91	3848.50	105	11 025	10.2470	329.87	8659.04

Table 5 Function of numbers.

ANGLE	SINE	COSINE	TAN	COTAN	ANGLE
0°	**.0000**	**1.0000**	**.0000**	θ	**90°**
1°	0.0175	0.9998	0.0175	57.290	89°
2°	0.0349	0.9994	0.0349	28.636	88°
3°	0.0523	0.9986	0.0524	19.081	87°
4°	0.0698	0.9976	0.0699	14.301	86°
5°	0.0872	0.9962	0.0875	11.430	85°
6°	0.1045	0.9945	0.1051	9.5144	84°
7°	0.1219	0.9925	0.1228	8.1443	83°
8°	0.1392	0.9903	0.1405	7.1154	82°
9°	0.1564	0.9877	0.1584	6.3138	81°
10°	0.1736	0.9848	0.1763	5.6713	80°
11°	0.1908	0.9816	0.1944	5.1446	79°
12°	0.2079	0.9781	0.2126	4.7046	78°
13°	0.2250	0.9744	0.2309	4.3315	77°
14°	0.2419	0.9703	0.2493	4.0108	76°
15°	0.2588	0.9659	0.2679	3.7321	75°
16°	0.2756	0.9613	0.2867	3.4874	74°
17°	0.2924	0.9563	0.3057	3.2709	73°
18°	0.3090	0.9511	0.3249	3.0777	72°
19°	0.3256	0.9455	0.3443	2.9042	71°
20°	0.3420	0.9397	0.3640	2.7475	70°
21°	0.3584	0.9336	0.3839	2.6051	69°
22°	0.3746	0.9272	0.4040	2.4751	68°
23°	0.3907	0.9205	0.4245	2.3559	67°
24°	0.4067	0.9135	0.4452	2.2460	66°
25°	0.4226	0.9063	0.4663	2.1445	65°
26°	0.4384	0.8988	0.4877	2.0503	64°
27°	0.4540	0.8910	0.5095	1.9626	63°
28°	0.4695	0.8829	0.5317	1.8807	62°
29°	0.4848	0.8746	0.5543	1.8040	61°
30°	0.5000	0.8660	0.5774	1.7321	60°
31°	0.5150	0.8572	0.6009	1.6643	59°
32°	0.5299	0.8480	0.6249	1.6003	58°
33°	0.5446	0.8387	0.6494	1.5399	57°
34°	0.5592	0.8290	0.6745	1.4826	56°
35°	0.5736	0.8192	0.7002	1.4281	55°
36°	0.5878	0.8090	0.7265	1.3764	54°
37°	0.6018	0.7986	0.7536	1.3270	53°
38°	0.6157	0.7880	0.7813	1.2799	52°
39°	0.6293	0.7771	0.8098	1.2349	51°
40°	0.6428	0.7660	0.8391	1.1918	50°
41°	0.6561	0.7547	0.8693	1.1504	49°
42°	0.6691	0.7431	0.9004	1.1106	48°
43°	0.6820	0.7314	0.9325	1.0724	47°
44°	0.6947	0.7193	0.9657	1.0355	46°
45°	0.7071	0.7071	0.0000	1.0000	45°
ANGLE	**COSINE**	**SINE**	**COTAN**	**TAN**	**ANGLE**

Table 6 Trigonometric functions.

And	&
Across Flats	A/F
American National Standards Institute	ANSI
Angular	ANG
Approximate	APPROX
Assembly	ASSY
Basic	BASIC
Bill of Material	B/M
Bolt Circle	BC
Brass	BR
Brown and Sharpe Gauge	B & S GA
Bushing	BUSH
Canada Standards Institute	CSI
Casting	CSTG
Cast Iron	CI
Centimetre	cm
Center Line	℄
Center to Center	C to C
Chamfered	CHAM
Circularity	CIR
Cold-Rolled Steel	CRS
Concentric	CONC
Counterbore	CBORE
Countersink	CSK
Cubic Centimetre	cm^3
Cubic Metre	m^3
Datum	DATUM
Degree (Angle)	° or DEG
Diameter	ϕ or DIA
Diametral Pitch	DP
Dimension	DIM
Drawing	DWG
Eccentric	ECC
Figure	FIG
Finish All Over	FAO
Gage	GA
Heat Treat	HT TR
Head	HD
Heavy	HY
Hexagon	HEX
Hydraulic	HYD
Inside Diameter	ID
International Organization for Standardization	ISO
Iron Pipe Size	IPS
Kilogram	kg
Kilometre	km
Large End	LE
Left Hand	LH
Litre	L
Machined	√ or ∀
Machine Steel	MS or MACH ST
Material	MATL
Maximum	MAX
Maximum Material Condition	M or MMC
Metre	m
Metric Thread	M
Micrometre	μm
Millimetre	mm
Minimum	MIN
Minute (Angle)	MIN
Newton	N
Nominal	NOM
Not to Scale	NTS or ___
Number	NO
Outside Diameter	OD
Parallel	PAR
Pascal	Pa
Perpendicular	PERP
Pitch	P
Pitch Circle Diameter	PCD
Pitch Diameter	PD
Plate	PL
Radian	rad
Radius	R
Reference or Reference Dimension	REF
Revolutions per Minute	rev/min
Right Hand	RH
Second (Arc)	(″)
Second (Time)	SEC
Section	SECT
Slotted	SLOT
Socket	SOCK
Spherical	SPHER
Spotface	SFACE
Square	SQ
Square Centimetre	cm^2
Square Metre	m^2
Steel	ST
Straight	STR
Symmetrical	SYM
Thread	THD
Through	THRU
Tolerance	TOL
True Profile	TP
Undercut	UCUT
U.S. Sheet-Metal Gage	USS GA
Watt	W
Wrought iron	WI

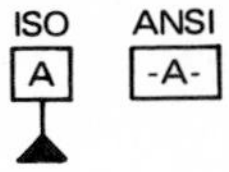
ISO
A
ANSI
-A-

Table 7 Abbreviations and symbols.

METRIC DRILL SIZES Preferred	METRIC DRILL SIZES Available	Reference Decimal Equivalent (Inches)
—	0.40	.0157
—	0.42	.0165
—	0.45	.0177
—	0.48	.0189
0.50	—	.0197
—	0.52	.0205
0.55	—	.0217
—	0.58	.0228
0.60	—	.0236
—	0.62	.0244
0.65	—	.0256
—	0.68	.0268
0.70	—	.0276
—	0.72	.0283
0.75	—	.0295
—	0.78	.0307
0.80	—	.0315
—	0.82	.0323
0.85	—	.0335
—	0.88	.0346
0.90	—	.0354
—	0.92	.0362
0.95	—	.0374
—	0.98	.0386
1.00	—	.0394
—	1.03	.0406
1.05	—	.0413
—	1.08	.0425
1.10	—	.0433
—	1.15	.0453
1.20	—	.0472
1.25	—	.0492
1.3	—	.0512
—	1.35	.0531
1.4	—	.0551
—	1.45	.0571
1.5	—	.0591
—	1.55	.0610
1.6	—	.0630
—	1.65	.0650
1.7	—	.0669
—	1.75	.0689
1.8	—	.0709
—	1.85	.0728
1.9	—	.0748
—	1.95	.0768
2.0	—	.0787
—	2.05	.0807
2.1	—	.0827
—	2.15	.0846
2.2	—	.0866
—	2.3	.0906
2.4	—	.0945
2.5	—	.0984
2.6	—	.1024

METRIC DRILL SIZES Preferred	METRIC DRILL SIZES Available	Reference Decimal Equivalent (Inches)
—	2.7	.1063
2.8	—	.1102
—	2.9	.1142
3.0	—	.1181
—	3.1	.1220
3.2	—	.1260
—	3.3	.1299
3.4	—	.1339
—	3.5	.1378
3.6	—	.1417
—	3.7	.1457
3.8	—	.1496
—	3.9	.1535
4.0	—	.1575
—	4.1	.1614
4.2	—	.1654
—	4.4	.1732
4.5	—	.1772
—	4.6	.1811
4.8	—	.1890
5.0	—	.1969
—	5.2	.2047
5.3	—	.2087
—	5.4	.2126
5.6	—	.2205
—	5.8	.2283
6.0	—	.2362
—	6.2	.2441
6.3	—	.2480
—	6.5	.2559
6.7	—	.2638
—	6.8	.2677
—	6.9	.2717
7.1	—	.2795
—	7.3	.2874
7.5	—	.2953
—	7.8	.3071
8.0	—	.3150
—	8.2	.3228
8.5	—	.3346
—	8.8	.3465
9.0	—	.3543
—	9.2	.3622
9.5	—	.3740
—	9.8	.3858
10	—	.3937
—	10.3	.4055
10.5	—	.4134
—	10.8	.4252
11	—	.4331
—	11.5	.4528
12	—	.4724
12.5	—	.4921
13	—	.5118
—	13.5	.5315

METRIC DRILL SIZES Preferred	METRIC DRILL SIZES Available	Reference Decimal Equivalent (Inches)
14	—	.5512
—	14.5	.5709
15	—	.5906
—	15.5	.6102
16	—	.6299
—	16.5	.6496
17	—	.6693
—	17.5	.6890
18	—	.7087
—	18.5	.7283
19	—	.7480
—	19.5	.7677
20	—	.7874
—	20.5	.8071
21	—	.8268
—	21.5	.8465
22	—	.8661
—	23	.9055
24	—	.9449
25	—	.9843
26	—	1.0236
—	27	1.0630
28	—	1.1024
—	29	1.1417
30	—	1.1811
—	31	1.2205
32	—	1.2598
—	33	1.2992
34	—	1.3386
—	35	1.3780
36	—	1.4173
—	37	1.4567
38	—	1.4361
—	39	1.5354
40	—	1.5748
—	41	1.6142
42	—	1.6535
—	43.5	1.7126
45	—	1.7717
—	46.5	1.8307
48	—	1.8898
50	—	1.9685
—	51.5	2.0276
53	—	2.0866
—	54	2.1260
56	—	2.2047
—	58	2.2835
60	—	2.3622

Table 8 Metric twist drill sizes.

NUMBER OR LETTER SIZE DRILL	SIZE		NUMBER OR LETTER SIZE DRILL	SIZE		NUMBER OR LETTER SIZE DRILL	SIZE		NUMBER OR LETTER SIZE DRILL	SIZE	
	mm	INCHES		mm	INCHES		mm	INCHES		mm	INCHES
80	0.343	.014	50	1.778	.070	20	4.089	.161	K	7.137	.821
79	0.368	.015	49	1.854	.073	19	4.216	.166	L	7.366	.290
78	0.406	.016	48	1.930	.076	18	4.305	.170	M	7.493	.295
77	0.457	.018	47	1.994	.079	17	4.394	.173	N	7.671	.302
76	0.508	.020	46	2.057	.081	16	4.496	.177	0	8.026	.316
75	0.533	.021	45	2.083	.082	15	4.572	.180	P	8.204	.323
74	0.572	.023	44	2.184	.086	14	4.623	.182	Q	8.433	.332
73	0.610	.024	43	2.261	.089	13	4.700	.185	R	8.611	.339
72	0.635	.025	42	2.375	.094	12	4.800	.189	S	8.839	.348
71	0.660	.026	41	2.438	.096	11	4.851	.191	T	9.093	.358
70	0.711	.028	40	2.489	.098	10	4.915	.194	U	9.347	.368
69	0.742	.029	39	2.527	.100	9	4.978	.196	V	9.576	.377
68	0.787	.031	38	2.578	.102	8	5.080	.199	W	9.804	.386
67	0.813	.032	37	2.642	.104	7	5.105	.201	X	10.084	.397
66	0.838	.033	36	2.705	.107	6	5.182	.204	Y	10.262	.404
65	0.889	.035	35	2.794	.110	5	5.220	.206	Z	10.490	.413
64	0.914	.036	34	2.819	.111	4	5.309	.209			
63	0.940	.037	33	2.870	.113	3	5.410	.213			
62	0.965	.038	32	2.946	.116	2	5.613	.221			
61	0.991	.039	31	3.048	.120	1	5.791	.228			
60	1.016	.040	30	3.264	.129	A	5.944	.234			
59	1.041	.041	29	3.354	.136	B	6.045	.238			
58	1.069	.042	28	3.569	.141	C	6.147	.242			
57	1.092	.043	27	3.658	.144	D	6.248	.246			
56	1.181	.047	26	3.734	.147	E	6.350	.250			
55	1.321	.052	25	3.797	.150	F	6.528	.257			
54	1.397	.055	24	3.861	.152	G	6.629	.261			
53	1.511	.060	23	3.912	.154	H	6.756	.266			
52	1.613	.064	22	3.988	.157	I	6.909	.272			
51	1.702	.067	21	4.039	.159	J	7.036	.277			

Table 9 Number and letter-size drills.

NOMINAL SIZE DIA (mm)		SERIES WITH GRADED PITCHES				SERIES WITH CONSTANT PITCHES																	
		COARSE		FINE		4		3		2		1.5		1.25		1		0.75		0.5		0.35	
Preferred		Thread Pitch	Tap Drill Size	Thread Pitch	Tap Drill Size	Thread Pitch	Tap Drill Size	Thread Pitch	Tap Drill Size	Thread Pitch	Tap Drill Size	Thread Pitch	Tap Drill Size	Thread Pitch	Tap Drill Size	Thread Pitch	Tap Drill Size	Thread Pitch	Tap Drill Size	Thread Pitch	Tap Drill Size	Thread Pitch	Tap Drill Size
1.6		0.35	1.25																				
	1.8	0.35	1.45																				
2		0.4	1.6																				
	2.2	0.45	1.75																				
2.5		0.45	2.05																			0.35	2.15
3		0.5	2.5																			0.35	2.65
	3.5	0.6	2.9																			0.35	3.15
4		0.7	3.3																	0.5	3.5		
	4.5	0.75	3.7																	0.5	4.0		
5		0.8	4.2																	0.5	4.5		
6		1	5.0															0.75	5.2				
8		1.25	6.7	1	7.0											1	7.0	0.75	7.2				
10		1.5	8.5	1.25	8.7									1.25	8.7	1	9.0	0.75	9.2				
12		1.75	10.2	1.25	10.8							1.5	10.5	1.25	10.7	1	11						
	14	2	12	1.5	12.5							1.5	12.5	1.25	12.7	1	13						
16		2	14	1.5	14.5							1.5	14.5			1	15						
	18	2.5	15.5	1.5	16.5					2	16	1.5	16.5			1	17						
20		2.5	17.5	1.5	18.5					2	18	1.5	18.5			1	19						
	22	2.5	19.5	1.5	20.5					2	20	1.5	20.5			1	21						
24		3	21	2	22					2	22	1.5	22.5			1	23						
	27	3	24	2	25					2	25	1.5	25.5			1	26						
30		3.5	26.5	2	28					2	28	1.5	28.5			1	29						
	33	3.5	29.5	2	31					2	31	1.5	31.5										
36		4	32	3	33					2	34	1.5	34.5										
	39	4	35	3	36					2	37	1.5	37.5										
42		4.5	37.5	3	39	4	38	3	39	2	40	1.5	40.5										
	45	4.5	39	3	42	4	41	3	42	2	43	1.5	43.5										
48		5	43	3	45	4	44	3	45	2	46	1.5	46.5										

Table 10 IS0 metric screw threads.

THREADS PER INCH AND TAP DRILL SIZES

SIZE INCHES		Graded Pitch Series						Constant Pitch Series					
		Coarse UNC		Fine UNF		Extra Fine UNEF		8 UN		12 UN		16 UN	
Number or Fraction	Deci-mal	Threads per Inch	Tap Drill Dia.	Threads per Inch	Tap Drill Dia.	Threads per Inch	Tap Drill Dia.	Threads per Inch	Tap Drill Dia.	Threads per Inch	Tap Drill Dia.	Threads per Inch	Tap Drill Dia.
0	.060	—	—	80	3/64	—	—	—	—	—	—	—	—
2	.086	56	No. 50	64	No. 49	—	—	—	—	—	—	—	—
4	.112	40	No. 43	48	No. 42	—	—	—	—	—	—	—	—
5	.125	40	No. 38	44	No. 37	—	—	—	—	—	—	—	—
6	.138	32	No. 36	40	No. 33	—	—	—	—	—	—	—	—
8	.164	32	No. 29	36	No. 29	—	—	—	—	—	—	—	—
10	.190	24	No. 25	32	No. 21	—	—	—	—	—	—	—	—
1/4	.250	20	7	28	3	32	.219	—	—	—	—	—	—
5/16	.312	18	F	24	1	32	.281	—	—	—	—	—	—
3/8	.375	16	.312	24	Q	32	.344	—	—	—	—	UNC	—
7/16	.438	14	U	20	.391	28	Y	—	—	—	—	16	V
1/2	.500	13	.422	20	.453	28	.469	—	—	—	—	16	.438
9/16	.562	12	.484	18	.516	24	.516	—	—	UNC	—	16	.500
5/8	.625	11	.531	18	.578	24	.578	—	—	12	.547	16	.562
3/4	.750	10	.656	16	.688	20	.703	—	—	12	.672	UNF	—
7/8	.875	9	.766	14	.812	20	.828	—	—	12	.797	16	.812
1	1.000	8	.875	12	.922	20	.953	UNC	—	UNF	—	16	.938
1 1/8	1.125	7	.984	12	1.047	18	1.078	8	1.000	UNF	—	16	1.062
1 1/4	1.250	7	1.109	12	1.172	18	1.188	8	1.125	UNF	—	16	1.188
1 3/8	1.375	6	1.219	12	1.297	18	1.312	8	1.250	UNF	—	16	1.312
1 1/2	1.500	6	1.344	12	1.422	18	1.438	8	1.375	UNF	—	16	1.438
1 5/8	1.625	—	—	—	—	18	—	8	1.500	12	1.547	16	1.562
1 3/4	1.750	5	1.562	—	—	—	—	8	1.625	12	1.672	16	1.688
1 7/8	1.875	—	—	—	—	—	—	8	1.750	12	1.797	16	1.812
2	2.000	4.5	1.781	—	—	—	—	8	1.875	12	1.922	16	1.938
2 1/4	2.250	4.5	2.031	—	—	—	—	8	2.125	12	2.172	16	2.188
2 1/2	2.500	4	2.250	—	—	—	—	8	2.375	12	2.422	16	2.438
2 3/4	2.750	4	2.500	—	—	—	—	8	2.625	12	2.672	16	2.688
3	3.000	4	2.750	—	—	—	—	8	2.875	12	2.922	16	2.938
3 1/4	3.250	4	3.000	—	—	—	—	8	3.125	12	3.172	16	3.188
3 1/2	3.500	4	3.250	—	—	—	—	8	3.375	12	3.422	16	3.438
3 3/4	3.750	4	3.500	—	—	—	—	8	3.625	12	3.668	16	3.688
4	4.000	4	3.750	—	—	—	—	8	3.875	12	3.922	16	3.938

Note: The tap diameter sizes shown are nominal. The class and length of thread will govern the limits on the tapped hole size.

Table 11 Inch screw threads.

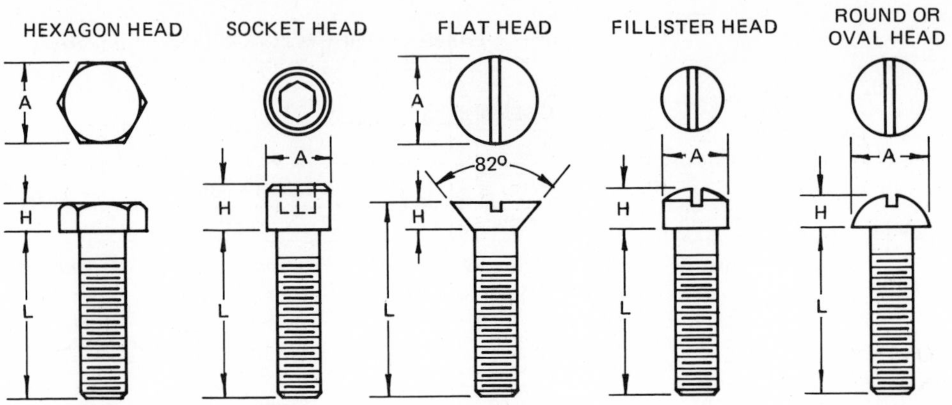
HEXAGON HEAD
SOCKET HEAD
FLAT HEAD
FILLISTER HEAD
ROUND OR OVAL HEAD
A
H
L
82°

NOMINAL SIZE	HEXAGON HEAD		SOCKET HEAD			FLAT HEAD		FILLISTER HEAD		ROUND OR OVAL HEAD	
	A	H	A	H	KEY SIZE	A	H	A	H	A	H
M3	5.5	2	5.5	3	2.5	5.6	1.6	6	2.4	5.6	
4	7	2.8	7	4	3	7.5	2.2	8	3.1	7.5	
5	8.5	3.5	9	5	4	9.2	2.5	10	3.8	9.2	
6	10	4	10	6	5	11	3	12	4.6	11	
8	13	5.5	13	8	6	14.5	4	16	6	14.5	
10	17	7	16	10	8	18	5	20	7.5	18	
12	19	8	18	12	10						
14	22	9	22	14	12						
16	24	10	24	16	14						

Table 12 Common cap screws.

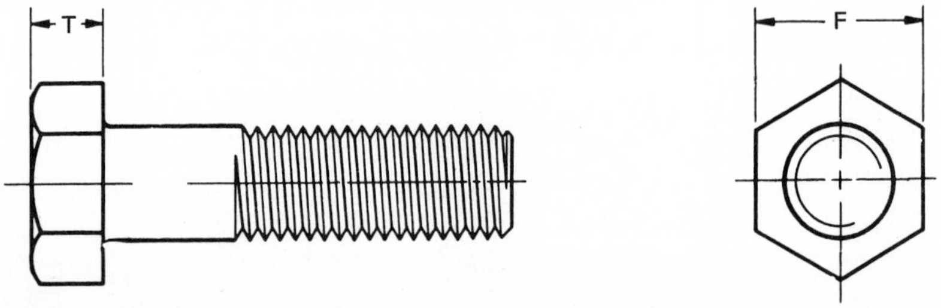
T
F

NOMINAL BOLT SIZE AND THREAD PITCH	WIDTH ACROSS FLATS F	THICKNESS T
M5 x 0.8	8	3.9
M6 x 1	10	4.7
M8 x 1.25	13	5.7
M10 x 1.5	15	6.8
M12 x 1.75	18	8
M14 x 2	21	9.3
M16 x 2	24	10.5
M20 x 2.5	30	13.1
M24 x 3	36	15.6
M30 x 3.5	46	19.5
M36 x 4	55	23.4

Table 13 Hexagon-head bolts and cap screws.

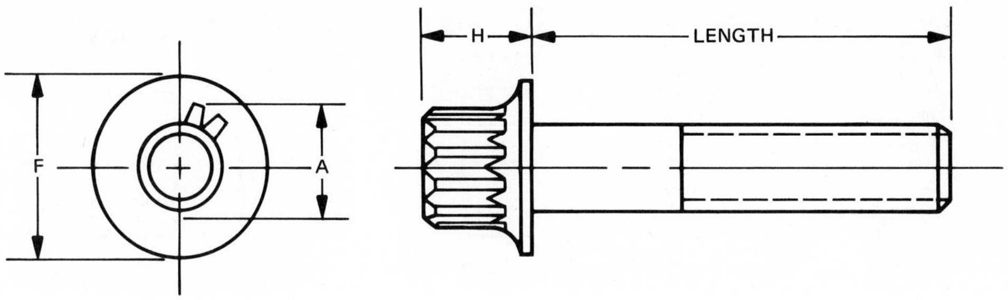
H
LENGTH
F
A

NOMINAL BOLT SIZE AND THREAD PITCH	HEAD SIZES		
	F	A	H
M5 x 0.8	9.4	5.9	5
M6 x 1	11.8	7.4	6.3
M8 x 1.25	15	9.4	8
M10 x 1.5	18.6	11.7	10
M12 x 1.75	22.8	14	12
M14 x 2	26.4	16.3	14
M16 x 2	30.3	18.7	16
M20 x 2.5	37.4	23.4	20

Table 14 Twelve-spline flange screws.

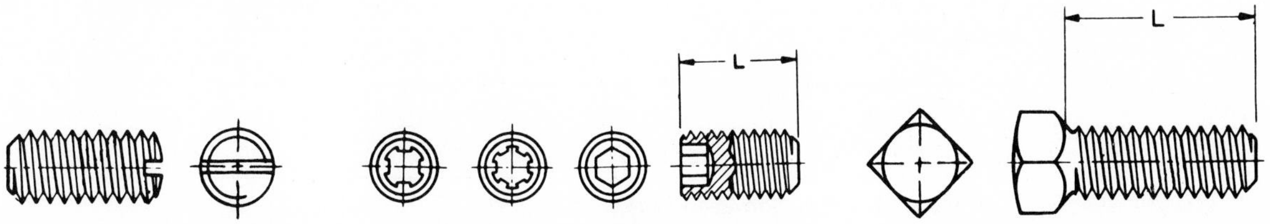
L
L
SLOTTED HEADLESS
SPLINE
HEX SOCKET
SQUARE HEAD

SETSCREW HEADS

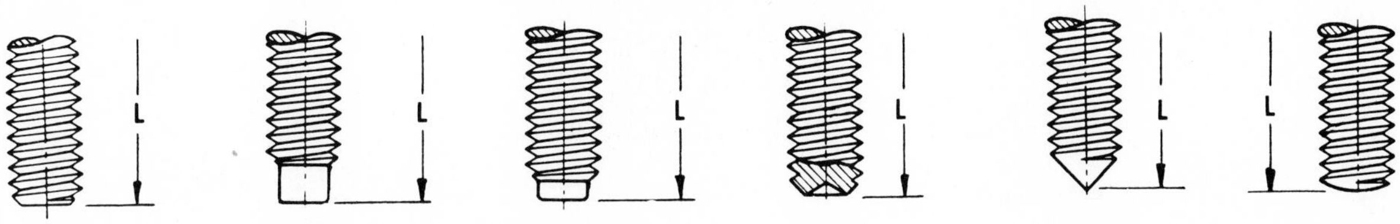
L
L
L
L
L
L
FLAT
DOG
HALF DOG
CUP
CONE
OVAL

SETSCREW POINTS

NOMINAL SIZE	KEY SIZE
M1.4	0.7
2	0.9
3	1.5
4	2
5	2
6	3
8	4
10	5
12	6
16	8

Table 15 Setscrews.

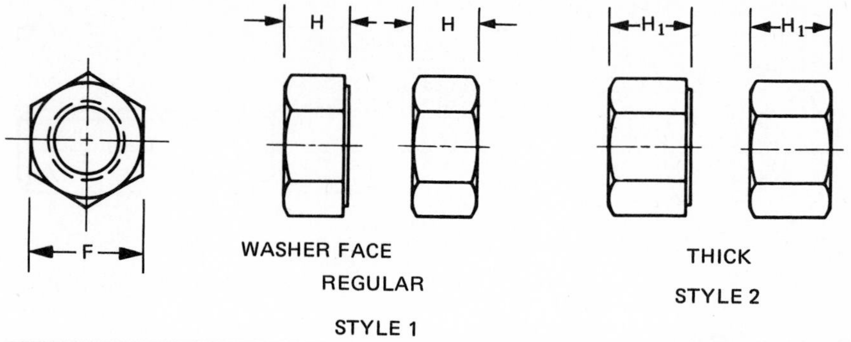
H
H
H1
H1
F
WASHER FACE
REGULAR
STYLE 1
THICK
STYLE 2

NOMINAL NUT SIZE AND THREAD PITCH	DISTANCE ACROSS FLATS F	THICKNESS MAX. Style 1 H	Style 2 H_1
M4 x 0.7	7	—	3.2
M5 x 0.8	8	4.5	5.3
M6 x 1	10	5.6	6.5
M8 x 1.25	13	6.6	7.8
M10 x 1.5	15	9	10.7
M12 x 1.75	18	10.7	12.8
M14 x 2	21	12.5	14.9
M16 x 2	24	14.5	17.4
M20 x 2.5	30	18.4	21.2
M24 x 3	36	22	25.4
M30 x 3.5	46	26.7	31
M36 x 4	55	32	37.6

Table 16 Hexagon-head nuts.

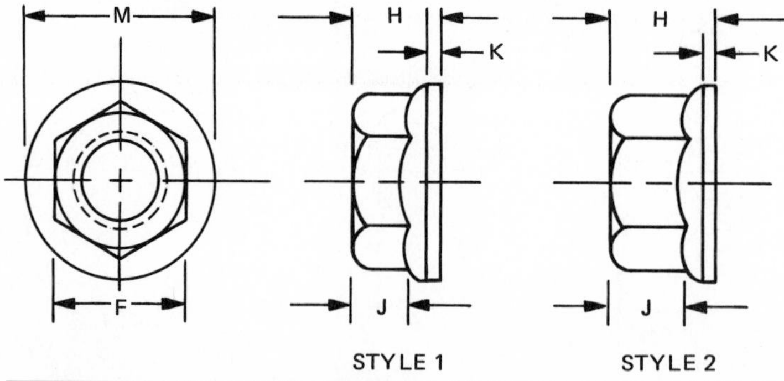
M
F
H
K
J
H
K
J
STYLE 1
STYLE 2

NOMINAL NUT SIZE AND THREAD PITCH	WIDTH ACROSS FLATS F	STYLE 1 H	STYLE 1 J	K	M	STYLE 2 H	STYLE 2 J
M6 x 1	10	5.8	3	1	14.2	6.7	3.7
M8 x 1.25	13	6.8	3.7	1.3	17.6	8	4.5
M10 x 1.5	15	9.6	5.5	1.5	21.5	11.2	6.7
M12 x 1.75	18	11.6	6.7	2	25.6	13.5	8.2
M14 x 2	21	13.4	7.8	2.3	29.6	15.7	9.6
M16 x 2	24	15.9	9.5	2.5	34.2	18.4	11.7
M20 x 2.5	30	19.2	11.1	2.8	42.3	22	12.6
M24 x 3							
M30 x 3.5							
M36 x 4							

Table 17 Hex flange nuts.

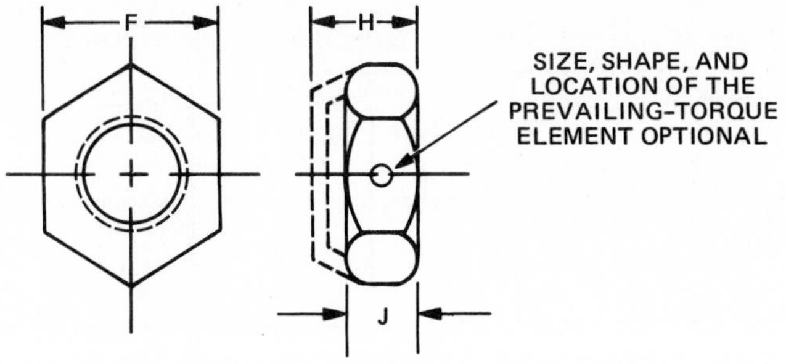
F
H
J
SIZE, SHAPE, AND LOCATION OF THE PREVAILING-TORQUE ELEMENT OPTIONAL

HEX NUTS

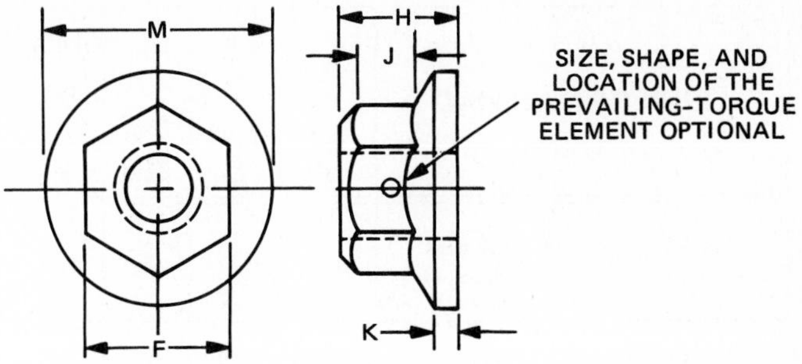
M
H
J
F
K
SIZE, SHAPE, AND LOCATION OF THE PREVAILING-TORQUE ELEMENT OPTIONAL

HEX FLANGE NUTS

NOMINAL NUT SIZE AND THREAD PITCH	WIDTH ACROSS FLATS F	HEX NUTS Style 1 H max.	HEX NUTS Style 1 J max.	HEX NUTS Style 2 H max.	HEX NUTS Style 2 J max.	HEX FLANGE NUTS Style 1 H	HEX FLANGE NUTS Style 1 J	K	M	HEX FLANGE NUTS Style 2 H	HEX FLANGE NUTS Style 2 J
M5 × 0.8	8.0	6.1	2.3	7.6	2.9						
M6 × 1	10	7.6	3	8.8	3.7	7.6	3	1	14.2	8.8	3.7
M8 × 1.25	13	9.1	3.7	10.3	4.5	9.1	3.7	1.3	17.6	10.3	4.5
M10 × 1.5	15	12	5.5	14	6.7	12	5.5	1.5	21.5	14	6.7
M12 × 1.75	18	14.2	6.7	16.8	8.2	14.4	6.7	2	25.6	16.8	8.2
M14 × 2	21	16.5	7.8	18.9	9.6	16.6	7.8	2.3	29.6	18.9	9.6
M16 × 2	24	18.5	9.5	21.4	11.7	18.9	9.5	2.5	34.2	21.4	11.7
M20 × 2.5	30	23.4	11.1	26.5	12.6	23.4	11.1	2.8	42.3	26.5	12.6
M24 × 3	36	28	13.3	31.4	15.1						
M30 × 3.5	46	33.7	16.4	38	18.5						
M36 × 4	55	40	20.1	45.6	22.8						

Table 18 Prevailing-torque insert-type nuts.

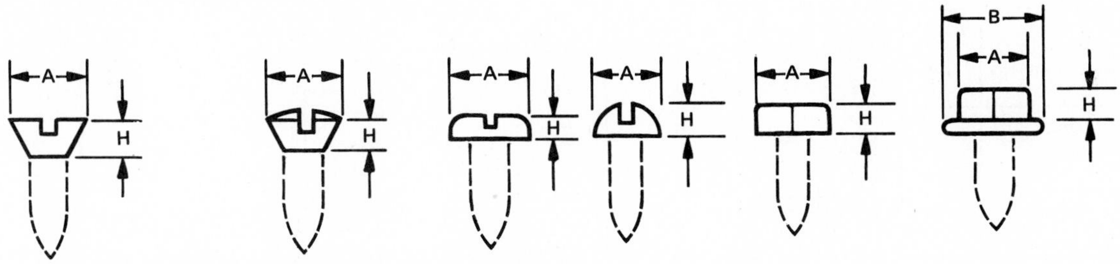
A
H
A
H
A
H
A
H
A
H
B
A
H

NOMINAL SIZE	SLOTTED FLAT COUNTERSUNK HEAD A	SLOTTED FLAT COUNTERSUNK HEAD H	SLOTTED OVAL COUNTERSUNK HEAD A	SLOTTED OVAL COUNTERSUNK HEAD H	PAN HEAD A	PAN HEAD Slot H	PAN HEAD Recess H	HEX HEAD A	HEX HEAD H	HEX WASHER HEAD A	HEX WASHER HEAD B	HEX WASHER HEAD H
2	3.6	1.2	3.6	1.2	3.9	1.4	1.6	3.2	1.3	3.2	4.2	1.3
2.5	4.6	1.5	4.6	1.5	4.9	1.7	2	4	1.4	4	5.3	1.4
3	5.5	1.8	5.5	1.8	5.8	1.9	1.3	5	1.5	5	6.2	1.5
3.5	6.5	2.1	6.5	2.1	6.8	2.3	2.5	5.5	2.4	5.5	7.5	2.4
4	7.5	2.3	7.5	2.3	7.8	2.6	2.8	7	2.8	7	9.2	2.8
5	9.5	2.9	9.5	2.9	9.8	3.1	3.5	8	3	8	10.5	3
6	11.9	3.6	11.9	3.6	12	3.9	4.3	10	4.8	10	13.2	4.8
8	15.2	4.4	15.2	4.4	15.6	5	5.6	13	5.8	13	17.2	5.8
10	1.9	5.4	19	5.4	19.5	6.2	7	15	7.5	15	19.8	7.5
12	22.9	6.4	22.9	6.4	23.4	7.5	8.3	18	9.5	18	23.8	9.5

Table 19 Tapping screws.

KIND OF MATERIAL	THREAD-FORMING								THREAD-CUTTING			SELF DRILLING	
	Type A	Type B	Type AB	HEX HEAD B	SWAGE FORM*	SWAGE FORM* B	Type U	Type 21	Type F*	Type L	Type B-F*	DRIL-KWICK	TAPITS*
SHEET METAL 0.4 to 1.2 thick (Steel, Brass, Aluminum, Monel, etc.)	●	●	●	●	●	●		●				●	●
SHEET STAINLESS STEEL 0.4 to 1.2 thick	●	●	●	●	●	●		●	●			●	●
SHEET METAL 1.2 to 5 thick (Steel, Brass, Aluminum, etc.)		●	●	●	●	●	●	●	●			●	
STRUCTURAL STEEL 5 to 12 thick				●	●		●		●				
CASTINGS (Aluminum, Magnesium, Zinc, Brass, Bronze, etc.)		●	●	●	●	●	●		●				
CASTINGS (Grey Iron, Malleable Iron, Steel, etc.)					●		●		●				
FORGINGS (Steel, Brass, Bronze, etc.)					●		●		●				
PLYWOOD, Resin Impregnated: Compreg, Pregwood, etc. **NATURAL WOODS**	●	●	●	●			●		●		●	●	●
ASBESTOS and other compositions: Ebony, Asbestos, Transite, Fiberglas, Insurok, etc.	●	●	●	●		●						●	●
PHENOL FORMALDEHYDE: Molded: Bakelite, Durez, etc. **Cast:** Catalin, Marblette, etc. **Laminated:** Formica, Textolite, etc.		●	●	●		●	●		●		●		
UREA FORMALDEHYDE: Molded: Plaskon, Beetle, etc. **MELAMINE FORMALDEHYDE:** Melantite, Melamac						●	●				●		
CELLULOSE ACETATES and NITRATES: Tenite, Lumarith, Plastacele Pyralin, Celanese, etc. **ACRYLATE & STYRENE RESINS:** Lucite, Plexiglas, Styron, etc.		●	●	●	●	●	●				●		
NYLON PLASTICS: Nylon, Zytel					●	●	●			●			

Table 20 Selector guide to thread cutting screws.

BOLT SIZE	FLAT WASHERS ID	FLAT WASHERS OD	FLAT WASHERS THICK	LOCKWASHERS ID	LOCKWASHERS OD	LOCKWASHERS THICK	SPRING LOCKWASHERS ID	SPRING LOCKWASHERS OD	SPRING LOCKWASHERS THICK
2	2.2	5.5	0.5	2.1	3.3	0.5			
3	3.2	7	0.5	3.1	5.7	0.8			
4	4.3	9	0.8	4.1	7.1	0.9	4.2	8	0.3 0.4
5	5.3	11	1	5.1	8.7	1.2	5.2	10	0.4 0.5
6	6.4	12	1.5	6.1	11.1	1.6	6.2	12.5	0.5 0.7
7	7.4	14	1.5	7.1	12.1	1.6	7.2	14	0.5 0.8
8	8.4	17	2	8.2	14.2	2	8.2	16	0.6 0.9
10	10.5	21	2.5	10.2	17.2	2.2	10.2	20	0.8 1.1
12	13	24	2.5	12.3	20.2	2.5	12.2	25	0.9 1.5
14	15	28	2.5	14.2	23.2	3	14.2	28	1.0 1.5
16	17	30	3	16.2	26.2	3.5	16.3	31.5	1.2 1.7
18	19	34	3	18.2	28.2	3.5	18.3	35.5	1.2 2.0
20	21	36	3	20.2	32.2	4	20.4	40	1.5 2.25
22	23	39	4	22.5	34.5	4	22.4	45	1.75 2.5
24	25	44	4	24.5	38.5	5			
27	28	50	4	27.5	41.5	5			
30	31	56	4	30.5	46.5	6			

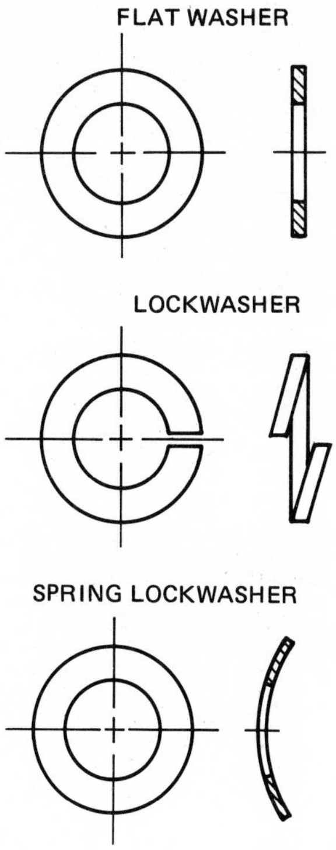
FLAT WASHER
LOCKWASHER
SPRING LOCKWASHER

Table 21 Common washer sizes.

OUTSIDE DIAMETER MAX.	INSIDE DIAMETER MIN.	STOCK THICKNESS t	H APPROX.	LOAD IN FLAT POSITION kg	OUTSIDE DIAMETER MAX.	INSIDE DIAMETER MIN.	STOCK THICKNESS t	H APPROX.	LOAD IN FLAT POSITION kg
4.8	2.4	0.16 0.25	0.33 0.38	4.5 12.6	25.4	12.8	0.89 1.27 1.85	1.7 1.9 2.3	118 272 605
6.4	3.2	0.22 0.34	0.44 0.51	7.7 21.8	28.6	14.4	0.97 1.42	1.9 2.1	130 334
7.9	4	0.27 0.42	0.55 0.64	12.3 34.	31.8	16	1.02 1.58	2.1 2.3	150 395
9.5	4.8	0.38 0.51	0.69 0.76	25 50	34.9	17.6	1.12 1.70	2.2 2.6	170 465
12.7	6.5	0.46 0.64 0.97	0.86 0.97 1.20	32 72 181	38.1	19.2	1.14 1.83	2.4 2.7	182 535
15.9	8.1	0.56 0.81	1.07 1.22	48 118	44.5	22.4	1.45 2.16	2.9 3.3	295 740
19.1	9.7	0.71 1.02 1.42	1.3 1.5 1.8	80 188 383	50.8	25.4	1.65 2.46	3.3 3.7	390 970
22.2	11.2	0.79 1.14	1.5 1.7	93 225	63.5	31.8	2.03 3.05	4.1 4.6	572 1452

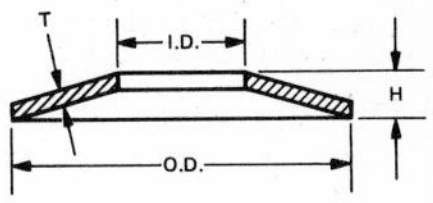
T
I.D.
H
O.D.

Table 22 Belleville washers. (Wallace Barnes Co. Ltd.)

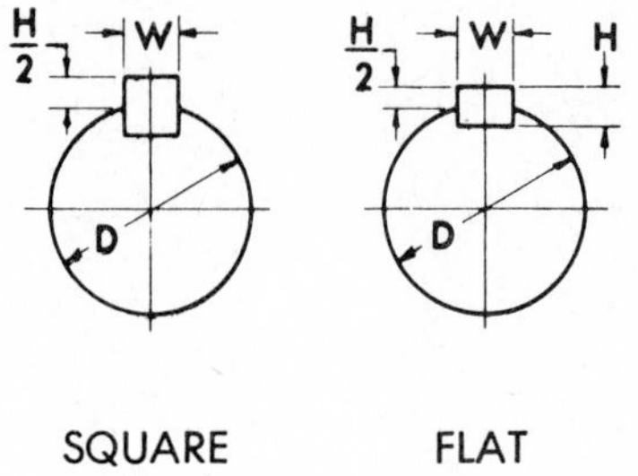

H/2
W
D
H
SQUARE
FLAT

DIAMETER OF SHAFT (mm)		SQUARE KEY		FLAT KEY	
		NOMINAL SIZE		NOMINAL SIZE	
OVER	UP TO	W	H	W	H
6	8	2	2		
8	10	3	3		
10	12	4	4		
12	17	5	5		
17	22	6	6		
22	30	7	7	8	7
30	38	8	8	10	8
38	44	9	9	12	8
44	50	10	10	14	9
50	58	12	12	16	10

Table 23 Square and flat stock keys.

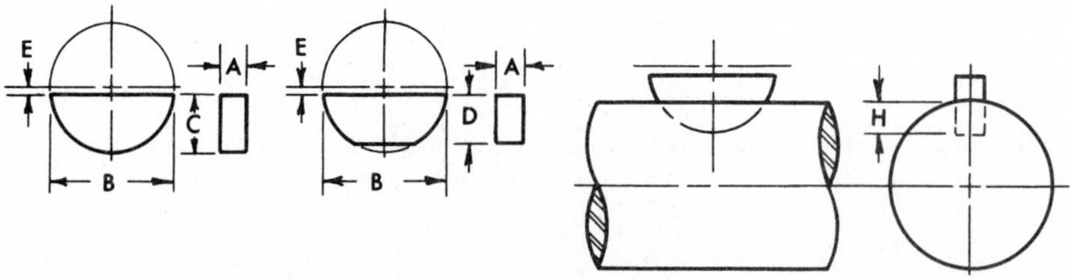

E
A
C
B
D
H

KEY	NOMINAL (A × B)		KEY			KEY SEAT
NO.	MILLIMETRES	INCHES	E	C	D	H
204	1.6 × 6.4	0.062 × 0.250	0.5	2.8	2.8	4.3
304	2.4 × 12.7	0.094 × 0.500	1.3	5.1	4.8	3.8
305	2.4 × 15.9	0.094 × 0.625	1.5	6.4	6.1	5.1
404	3.2 × 12.7	0.125 × 0.500	1.3	5.1	4.8	3.6
405	3.2 × 15.9	0.125 × 0.625	1.5	6.4	6.1	4.6
406	3.2 × 19.1	0.125 × 0.750	1.5	7.9	7.6	6.4
505	4.0 × 15.9	0.156 × 0.625	1.5	6.4	6.1	4.3
506	4.0 × 19.1	0.156 × 0.750	1.5	7.9	7.6	5.8
507	4.0 × 22.2	0.156 × 0.875	1.5	9.7	9.1	7.4
606	4.8 × 19.1	0.188 × 0.750	1.5	7.9	7.6	5.3
607	4.8 × 22.2	0.188 × 0.875	1.5	9.7	9.1	7.1
608	4.8 × 25.4	0.188 × 1.000	1.5	11.2	10.9	8.6
609	4.8 × 28.6	0.188 × 1.250	2.0	12.2	11.9	9.9
807	6.4 × 22.2	0.250 × 0.875	1.5	9.7	9.1	6.4
808	6.4 × 25.4	0.250 × 1.000	1.5	11.2	10.9	7.9

NOTE: METRIC KEY SIZES WERE NOT AVAILABLE AT THE TIME OF PUBLICATION. SIZES SHOWN ARE INCH-DESIGNED KEY-SIZES SOFT CONVERTED TO MILLIMETRES. CONVERSION WAS NECESSARY TO ALLOW THE STUDENT TO COMPARE KEYS WITH SLOT SIZES GIVEN IN MILLIMETRES.

Table 24 Woodruff keys.

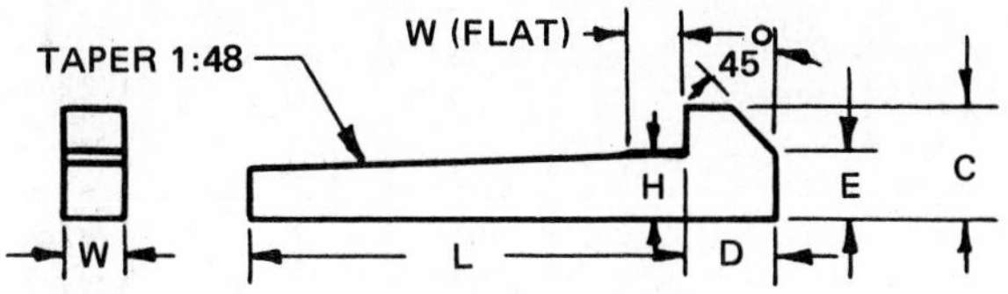
TAPER 1:48
W (FLAT)
45°
H
E
C
W
L
D
L (MIN) = 4W
L (MAX) = 16W

SHAFT DIAMETER	SQUARE TYPE					FLAT TYPE				
	W	H	C	D	E	W	H	C	D	E
12–14	3.2	3.2	6.4	5.4	4	3.2	2.4	5	3.2	3.2
16–22	4.8	4.8	10	7	5.4	4.8	3.2	6.4	5	4
24–32	6.4	6.4	11	8.6	8.6	6.4	5	8	6.4	5
34–35	8	8	14	10	10	8	6.4	10	8	6.4
36–44	10	10	18	12	12	10	6.4	11	10	8
46–58	13	13	22	15	16	13	10	16	13	11
60–70	16	16	27	19	20	16	11	20	16	13
72–82	20	20	32	22	22	20	13	22	20	16

Note: Metric standards governing key sizes were not available at the time of publication. The sizes given in the above chart are "soft conversion" from current standards and are not representative of the precise metric key sizes which may be available in the future. Metric sizes are given only to allow the student to complete the drawing assignment.

Table 25 Square and flat gib-head keys.

L (MIN) = 2W

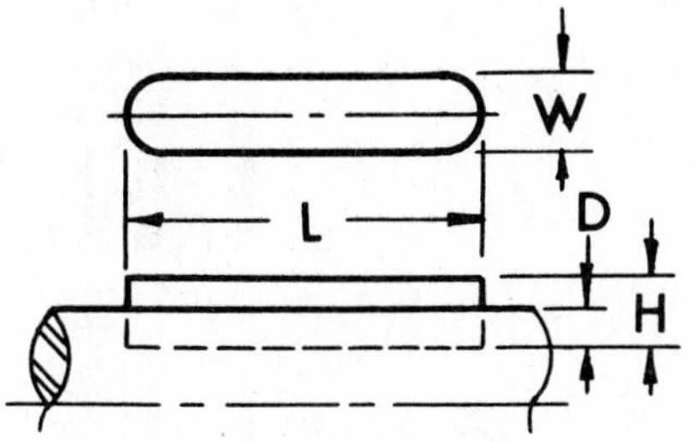
W
L
D
H

KEY NO.	L	W	H	D	KEY NO.	L	W	H	D	KEY NO.	L	W	H	D
1	12	1.6	2.4	1.6	17	28	7	8	5	27	50	6	10	6.5
2	12	2.4	3.6	2.4	18	28	6.5	10	6	28	50	8	12	8
3	12	3.2	5	3.2	C	28	10	12	8	29	50	10	14	10
4	16	2.4	3.6	2.4	19	32	5	7	5	54	55	6	10	6.5
5	16	3.2	5	3.2	20	32	7	8	5	55	55	8	12	8
6	16	4	6	4	21	32	6.5	10	6	56	55	10	14	10
7	20	3.2	5	3.2	D	32	10	12	8	57	55	11	16	12
8	20	4	6	4	E	32	12	14	10	58	65	8	12	8
9	20	5	7	5	22	35	6.5	10	6	59	65	10	14	10
10	22	4	6	4	23	35	10	12	8	60	65	11	16	12
11	22	5	7	5	F	35	12	14	10	61	65	12	20	13
12	22	6	8.4	7	24	38	6.5	10	6	30	75	10	14	10
A	22	6.5	10	6.5	25	38	10	12	8	31	75	11	16	12
13	25	5	7	5	G	38	12	14	10	32	75	12	20	13
14	25	6	8.4	6	51	45	6.5	10	6	33	75	14	22	14
15	25	6.5	10	6.5	52	45	10	12	8	34	75	16	24	16
B	25	8	12	8	53	45	12	14	10					
16	28	5	7	5	26	50	5	7	5					

Note: Metric standards governing key sizes were not available at the time of publication. The sizes given in the above chart are "soft conversion" from current standards and are not representative of the precise metric key sizes which may be available in the future. Metric sizes are given only to allow the student to complete the drawing assignments.

Table 26 Pratt and Whitney keys.

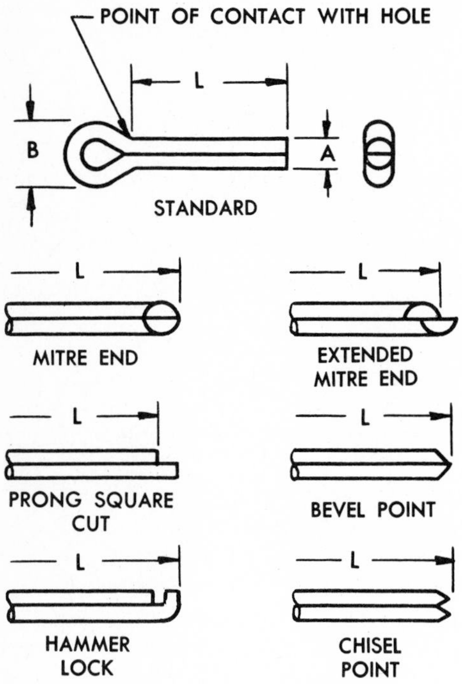
POINT OF CONTACT WITH HOLE
L
B
A
STANDARD
L
MITRE END
L
EXTENDED
MITRE END
L
PRONG SQUARE
CUT
L
BEVEL POINT
L
HAMMER
LOCK
L
CHISEL
POINT

NOMINAL BOLT OR THREAD-SIZE RANGE	NOMINAL COTTER-PIN SIZE	COTTER-PIN HOLE	MIN. END CLEARANCE*
– 2.5	0.6	0.8	1.5
2.5– 3.5	0.8	1.0	2.0
3.5– 4.5	1.0	1.2	2.0
4.5– 5.5	1.2	1.4	2.5
5.5– 7.0	1.6	1.8	2.5
7.0– 9.0	2.0	2.2	3.0
9.0–11	2.5	2.8	3.5
11–14	3.2	3.6	5
14–20	4	4.5	6
20–27	5	5.6	7
27–39	6.3	6.7	10
39–56	8.0	8.5	15
56–80	10	10.5	20

*End of bolt to center of hole

Table 27 Cotter pins.

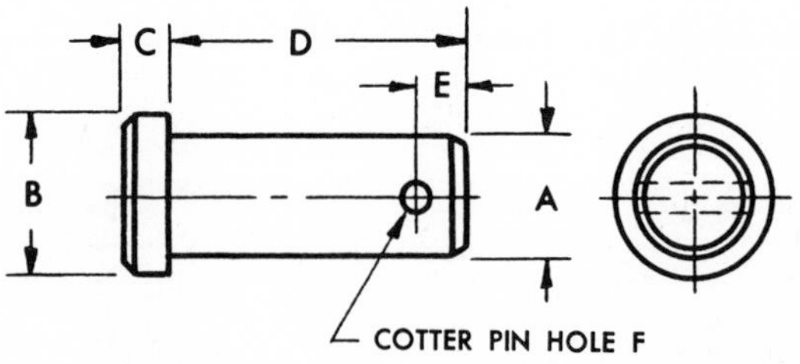
C
D
E
B
A
COTTER PIN HOLE F

PIN DIA. A	B	C	MIN. D	E	DRILL SIZE F
4	6	1		2.2	1
6	10	2		3.2	1.6
8	14	3		3.5	2
10	18	4		4.5	3.2
12	20	4		5.5	3.2
16	25	4.5		6	4
20	30	5		8	5
24	36	6		9	6.3

Table 28 Clevis pins.

NUMBER	7/0	6/0	5/0	4/0	3/0	2/0	0	1	2	3	4	5	6	7	8	9	10
SIZE (LARGE END)	1.6	2	2.4	2.8	3.2	3.6	4	4.4	4.9	5.6	6.4	7.4	8	10.4	12.5	15	18
LENGTH																	
10	x	x															
12	x	x	x	x	x	x	x										
16	x	x	x	x	x	x	x										
20		x	x	x	x	x	x	x	x	x							
22					x	x	x	x	x	x							
25			x	x	x	x	x	x	x	x	x	x					
30						x	x	x	x	x	x	x	x				
40							x	x	x	x	x	x	x				
45								x	x	x	x	x	x				
50								x	x	x	x	x	x	x	x		
55									x	x	x	x	x	x	x		
65									x	x	x	x	x	x	x		
70										x	x	x	x	x	x	x	
75										x	x	x	x	x	x	x	

Table 29 Taper pins.

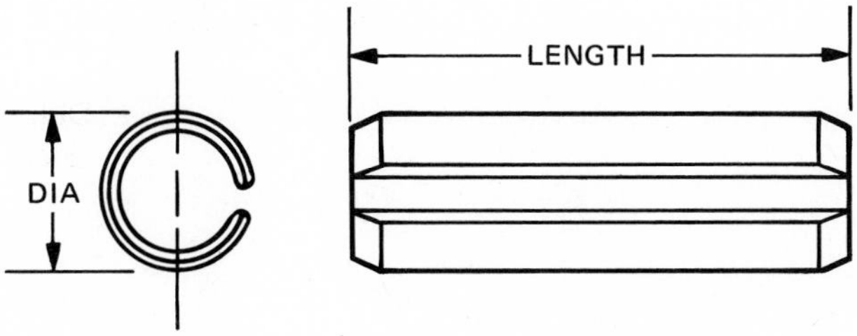
LENGTH
DIA

	PIN DIAMETER												
Length	1.5	2	2.5	3	4	5	6	7	8	9	10	11	12
5	x	x											
10	x	x	x	x									
15	x	x	x	x	x	x							
20	x	x	x	x	x	x	x	x					
25	x	x	x	x	x	x	x	x	x	x			
30		x	x	x	x	x	x	x	x	x	x	x	x
35		x	x	x	x	x	x	x	x	x	x	x	x
40		x	x	x	x	x	x	x	x	x	x	x	x
45				x	x	x	x	x	x	x	x	x	x
50				x	x	x	x	x	x	x	x	x	x
55					x	x	x	x	x	x	x	x	x
60					x	x	x	x	x	x	x	x	x
70							x	x	x	x	x	x	x
75							x	x	x	x	x	x	x
80								x	x	x	x	x	x

Table 30 Spring pins.

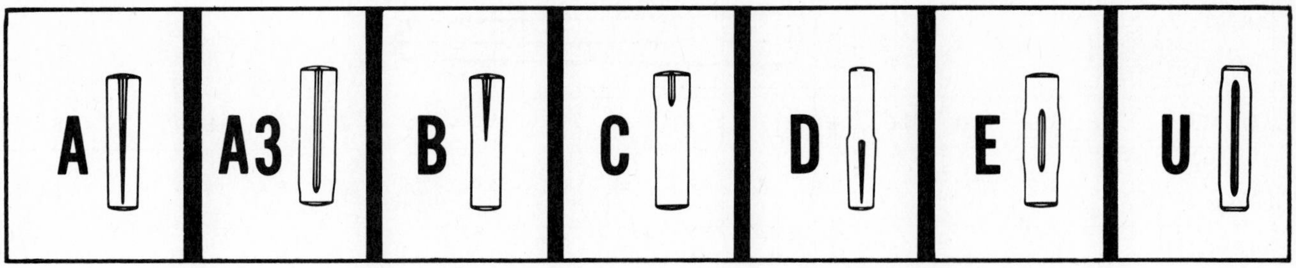
A
A3
B
C
D
E
U

	PIN DIAMETER												
Length	**1.5**	**2**	**2.5**	**3**	**4**	**5**	**6**	**7**	**8**	**9**	**10**	**11**	**12**
5	X	X	X	X	X								
10	X	X	X	X	X	X	X						
15	X	X	X	X	X	X	X	X	X				
20	X	X	X	X	X	X	X	X	X	X	X		
25	X	X	X	X	X	X	X	X	X	X	X	X	X
30		X	X	X	X	X	X	X	X	X	X	X	X
35				X	X	X	X	X	X	X	X	X	X
40					X	X	X	X	X	X	X	X	X
45						X	X	X	X	X	X	X	X
50						X	X	X	X	X	X	X	X
55							X	X	X	X	X	X	X
60							X	X	X	X	X	X	X
65							X	X	X	X	X	X	X
70								X	X	X	X	X	X
75								X	X	X	X	X	X
80								X	X	X	X	X	X

Note: Metric size pins were not available at the time of publication. Sizes were soft converted to allow students to complete drawing assignments.

Table 31 Groove pins.

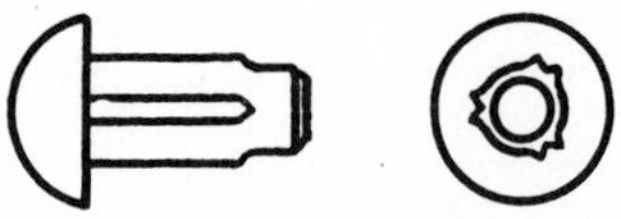

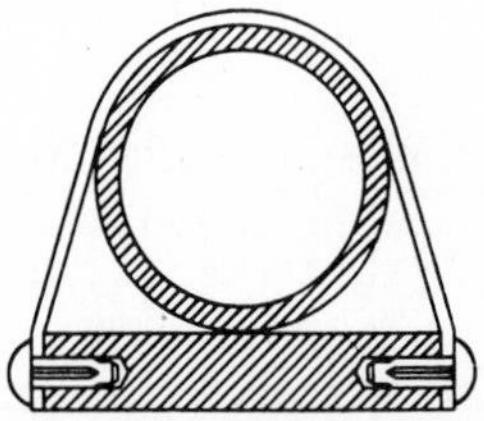

WIDELY USED FOR FASTENING BRACKETS

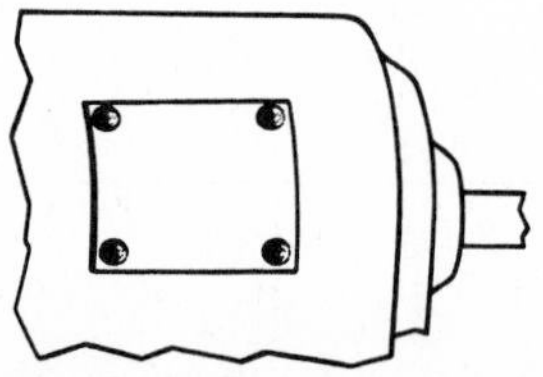

ATTACHING NAMEPLATES, INSTRUCTION PANELS

STUD NUMBER	SHANK DIA.	RECOMMENDED DRILL SIZE	MAXIMUM HEAD DIA.	MAXIMUM HEAD HEIGHT	STANDARD LENGTHS					
					4	6	8	10	12	14
0	1.7	1.7	3.3	1.3	•	•	•			
2	2.2	2.2	4.1	1.8	•	•	•			
4	2.6	2.6	5.4	2.2		•	•	•		
6	3.0	3.0	6.6	2.6			•	•	•	
7	3.4	3.4	7.8	3.0				•	•	•
8	3.8	3.8	7.8	3.0					•	•
10	4.1	4.1	9.1	3.5					•	•
12	5.0	5.0	10.4	3.9						•
14	5.6	5.6	11.6	4.3						•
16	6.3	6.3	12	4.4						•

Note: Metric size studs were not available at the time of publication. Sizes were soft converted to allow students to complete drawing assignments.

Table 32 Grooved studs. (Drive-Lok)

Use these columns first to locate your correct

GRIP LENGTH

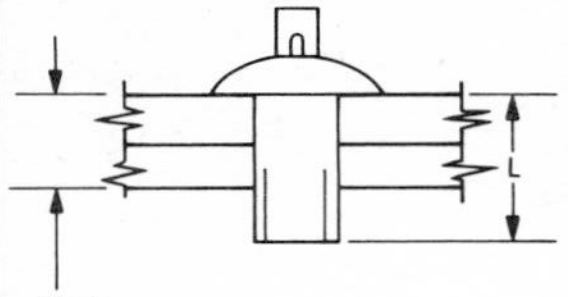
L

Grip =

Total thickness of all sheets fastened together

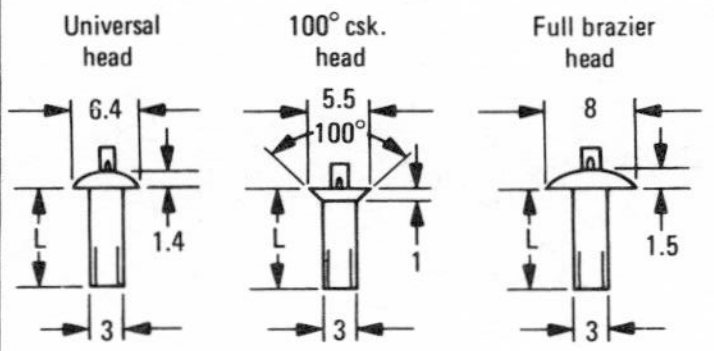
Universal head
100° csk. head
Full brazier head
6.4
1.4
3
5.5
100°
1
3
8
1.5
3
L

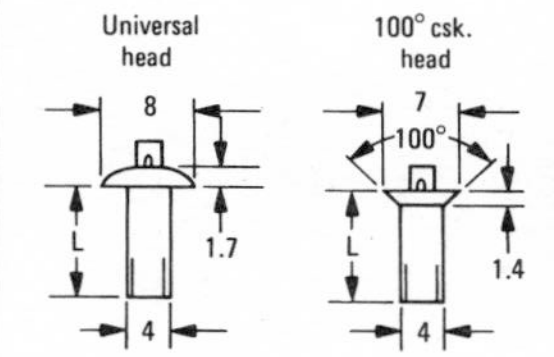
Universal head
100° csk. head
8
1.7
4
7
100°
1.4
4
L

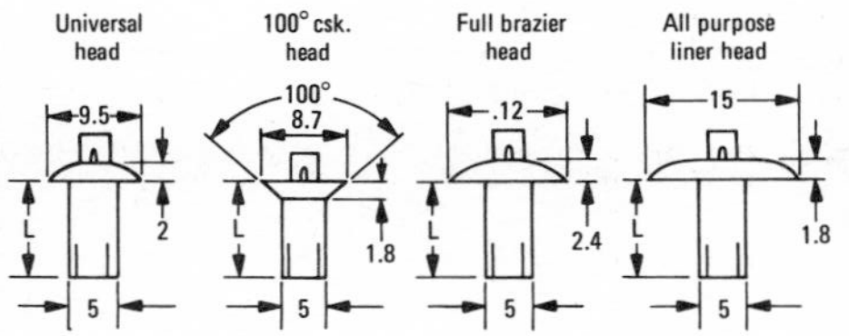
Universal head
100° csk. head
Full brazier head
All purpose liner head
9.5
2
5
100°
8.7
1.8
5
.12
2.4
5
15
1.8
5
L

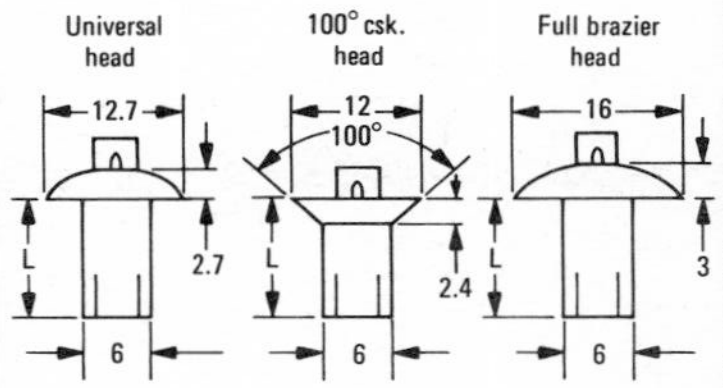
Universal head
100° csk. head
Full brazier head
12.7
2.7
6
12
100°
2.4
6
16
3
6
L

Minimum Grip	Nominal Grip	Maximum Grip	Length Under Head L	3 mm Dia. Part Numbers			Length Under Head L	4 mm Dia. Part Numbers		Length Under Head L	5 mm Dia. Part Numbers				Length Under Head L	6 mm Dia. Part Numbers		
				Universal Head	100° Csk. Head	Full Brazier Head		Universal Head	100° Csk. Head		Universal Head	100° Csk. Head	Full Brazier Head	All Purpose Liner Head		Universal Head	100° Csk. Head	Full Brazier Head
1	2	3	5	✓	✓	✓	5	✓		5	✓		✓	✓	5	✓	✓	✓
2	3	4	6	✓	✓	✓	6	✓	✓	6	✓		✓	✓	6	✓	✓	✓
4	5	6	8	✓	✓	✓	8	✓	✓	8	✓	✓	✓	✓	8	✓	✓	✓
5	6	7	10	✓	✓	✓	10	✓	✓	10	✓	✓	✓	✓	10	✓	✓	✓
7	8	9	12	✓	✓	✓	12	✓	✓	12	✓	✓	✓	✓	12	✓	✓	✓
9	10	11	14	✓	✓	✓	14	✓	✓	14	✓	✓	✓	✓	14	✓	✓	✓
11	12	13	16	✓			16	✓	✓	16	✓	✓	✓	✓	16	✓	✓	✓
13	14	15	18				18	✓	✓	18	✓	✓	✓	✓	18	✓	✓	✓
15	16	17	20				20	✓	✓	20	✓	✓	✓	✓	20	✓	✓	✓
19	20	21	24				24			24	✓		✓	✓	24	✓	✓	✓
23	24	25	28				28			28	✓		✓		28	✓	✓	✓
27	28	29	32				32			32	✓		✓		32	✓	✓	✓

Note: Metric drive rivets were not available at the time of publication. Sizes were soft converted to allow students to complete drawing assignments.

HIT THE PIN

Drive pin flush with rivet head

Expanding prongs clinch sheets tightly, eliminating gaps.

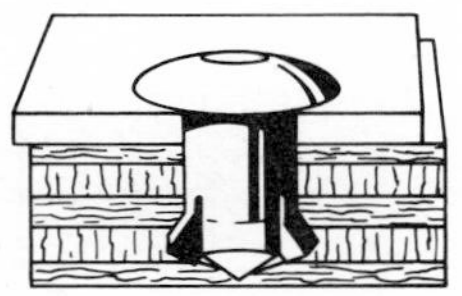

Metal and wood pulled tightly together. Nothing protrudes through wood.

IN METAL

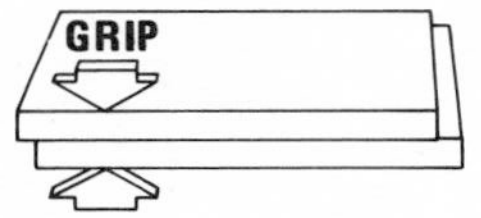
GRIP

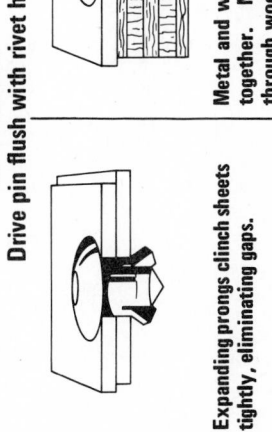

Grip Length = Total thickness of sheets to be fastened.

IN WOOD

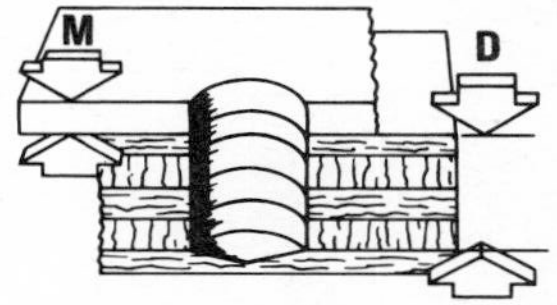
M
D

Use "L" Dimension (length under head) instead of grip length. L = M (thickness of metal) + D (hole depth in wood).

Table 33 Aluminum drive rivets.

Southco Lion Fasteners

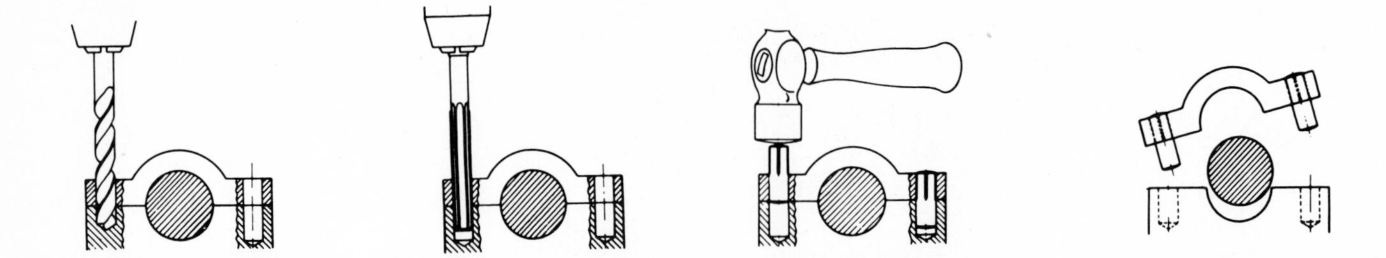

DRILL HOLE SLIGHTLY UNDERSIZE

REAM FULL SIZE

DRIVE OR PRESS *LOK DOWELS* INTO PLACE

LOK DOWELS LOCK SECURELY AND PARTS SEPARATE EASILY

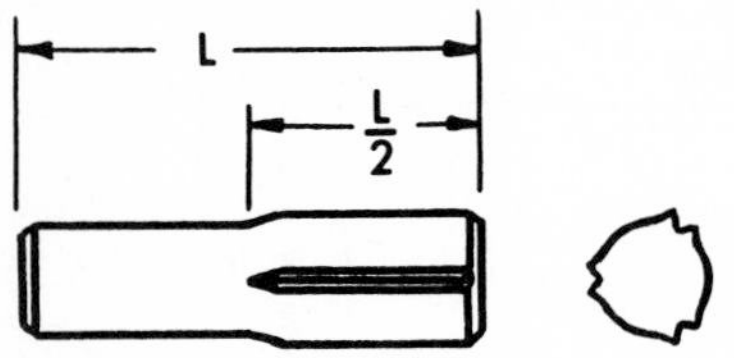
L
L/2

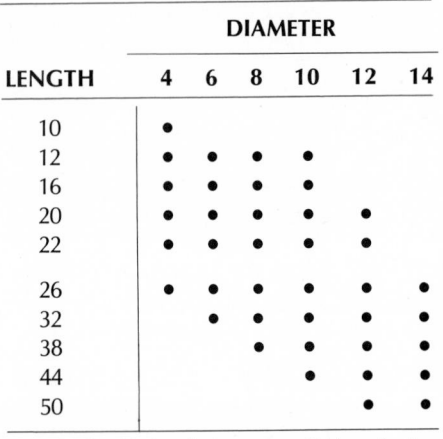

LENGTH	DIAMETER 4	6	8	10	12	14
10	•					
12	•	•	•	•		
16	•	•	•	•		
20	•	•	•	•	•	
22	•	•	•	•	•	
26	•	•	•	•	•	•
32		•	•	•	•	•
38			•	•	•	•
44				•	•	•
50					•	•

Note: Metric size dowels were not available at the time of publication. Sizes were soft converted to allow students to complete drawing assignments.

Table 34 Lok dowels. (Drive-Lok)

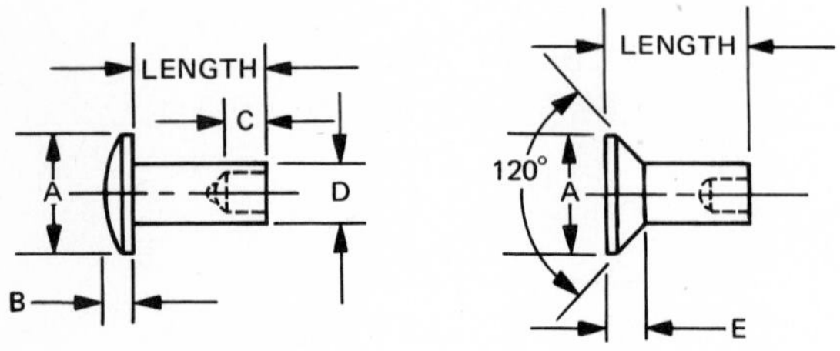
LENGTH
C
A
D
B
LENGTH
120°
A
E

(A) SEMITUBULAR

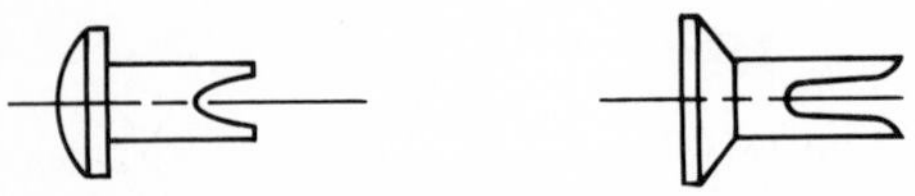

(B) SPLIT

D	A	B	C	E	MIN. LENGTH
1.5	2.8	0.4	1.2		1.6
2.2	3.7	0.6	1.6	0.8	2
2.5	4.7	0.7	2.0		2
3.1	5.5	0.9	2.4	1.0	2.4
3.6	5.9	1	3.2	1.2	3.2
4.7	7.9	1.5	3.9	1.6	4
5.4	11.1	1.7	4.8	1.8	4.8
6.3	12.7	2	5.6	2.2	5.6
7.7	14.3	2.4	6.2		6.4

Table 35 Semitubular and split rivets.

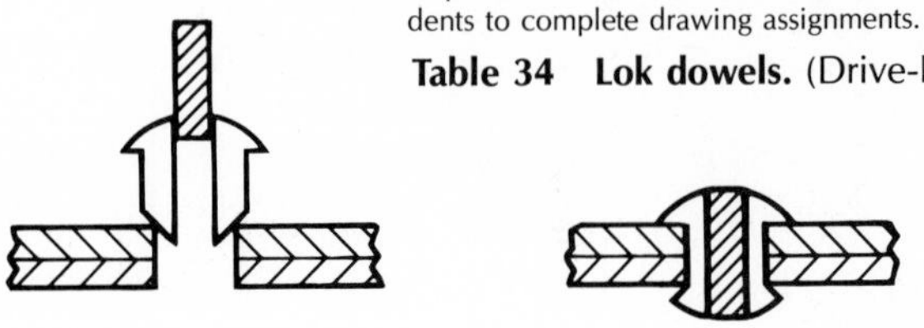

HOLE DIA.	PANEL RANGE	HEAD Dia.	HEAD Height	MAXIMUM LOAD KILOGRAMS Tension	MAXIMUM LOAD KILOGRAMS Shear
3.18	0.8– 3.6 0.8– 3.2	4.8 5.5	1.2 1.5	4	7
4.01	5.9– 9.4	5.5	1.3	7	11
4.75	1.6– 4.0 4.0– 7.1	9.5 11.1	3.2 1.9	11	18
5.54	1.6– 3.2 2.4– 8.0	9.5	2.4 2.0	16 11	25 18
6.35	2.3– 5.6 3.2– 9.5	16 19	3.2 1.3	22 16	34 34
7.14	3.4– 8.1	12.3	1.9	34	44
9.53	6.4–12.7	11.1	2.6	36	72
12.7	8.1– 9.4	19	2.5	45	90

Table 36 Plastic rivets.

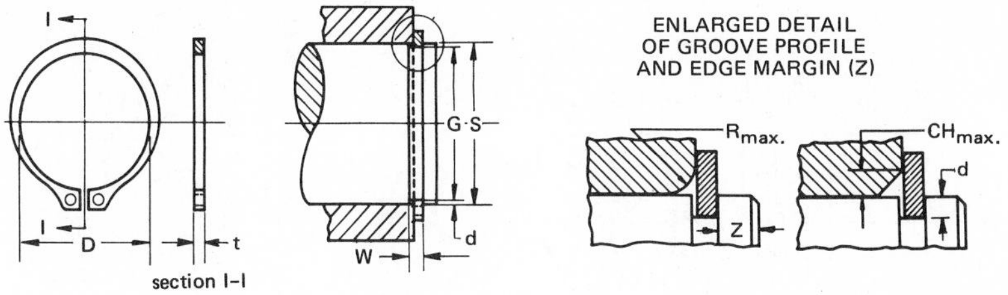
ENLARGED DETAIL
OF GROOVE PROFILE
AND EDGE MARGIN (Z)
Rmax.
CHmax.
d
Z
G
S
d
W
D
t
section I-I

SHAFT DIA.	EXTERNAL SERIES	RETAINING RING DIMENSIONS		GROOVE DIMENSIONS Diameter		GROOVE DIMENSIONS Width		MAXIMUM ALLOWABLE CORNER RADII AND CHAMFER OF RETAINED PARTS		EDGE MARGIN	NOMINAL GROOVE DEPTH (REF)
S	Size—No.	D	t	G	Tol.	W	Tol.	R Max.	Ch. Max.	z	d
4	M5100-4	3.6	0.25	3.80	−0.08	0.32	+0.05	0.35	0.25	0.3	0.10
6	M5100-6	5.5	0.4	5.70	−0.08	0.5	+0.1	0.35	0.25	0.5	0.15
8	M5100-8	7.2	0.6	7.50	−0.1	0.7	+0.15	0.5	0.35	0.8	0.25
10	M5100-10	9.0	0.6	9.40	−0.1	0.7	+0.15	0.7	0.4	0.9	0.30
12	M5100-12	10.9	0.6	11.35	−0.12	0.7	+0.15	0.8	0.45	1.0	0.33
14	M5100-14	12.9	0.9	13.25	−0.12	1.0	+0.15	0.9	0.5	1.2	0.38
16	M5100-16	14.7	0.9	15.10	−0.15	1.0	+0.15	1.1	0.6	1.4	0.45
18	M5100-18	16.7	1.1	17.00	−0.15	1.2	+0.15	1.2	0.7	1.5	0.50
20	M5100-20	18.4	1.1	18.85	−0.15	1.2	+0.15	1.2	0.7	1.7	0.58
22	M5100-22	20.3	1.1	20.70	−0.15	1.2	+0.15	1.3	0.8	1.9	0.65
24	M5100-24	22.2	1.1	22.60	−0.15	1.2	+0.15	1.4	0.8	2.1	0.70
25	M5100-25	23.1	1.1	23.50	−0.15	1.2	+0.15	1.4	0.8	2.3	0.75
30	M5100-30	27.9	1.3	28.35	−0.2	1.4	+0.15	1.6	1.0	2.5	0.83
35	M5100-35	32.3	1.3	32.9	−0.2	1.4	+0.15	1.8	1.1	3.1	1.05
40	M5100-40	36.8	1.6	37.7	−0.3	1.75	+0.2	2.1	1.2	3.4	1.15
45	M5100-45	41.6	1.6	42.4	−0.3	1.75	+0.2	2.3	1.4	3.9	1.3
50	M5100-50	46.2	1.6	47.2	−0.3	1.75	+0.2	2.4	1.4	4.2	1.4

Table 37 Retaining rings—external. (© 1965, 1958 Waldes Koh-i-noor, Inc. Reprinted with permission.)

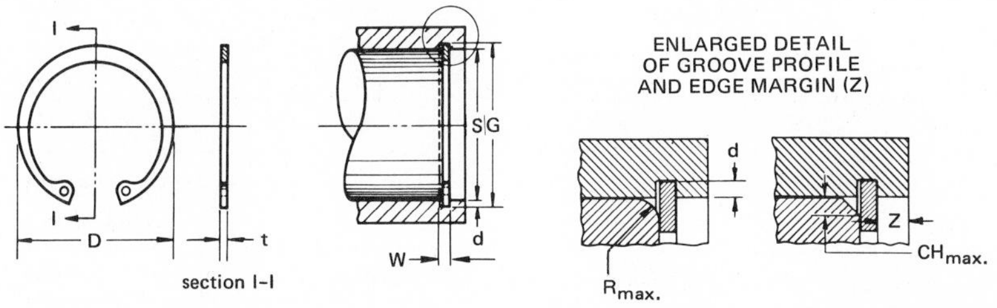
ENLARGED DETAIL
OF GROOVE PROFILE
AND EDGE MARGIN (Z)
d
Z
CHmax.
Rmax.
S
G
d
W
D
t
section I-I

HOUSING DIA.	INTERNAL SERIES	RETAINING RING DIMENSIONS		GROOVE DIMENSIONS Diameter		GROOVE DIMENSIONS Width		MAXIMUM ALLOWABLE CORNER RADII AND CHAMFER OF RETAINED PARTS		EDGE MARGIN	NOMINAL GROOVE DEPTH
S	Size—No.	D	t	G	Tol.	W	Tol.	R Max.	Ch. Max.	z	d
8	MN5000-8	8.80	0.4	8.40	+0.6	0.5	+0.1	0.4	0.3	0.6	0.2
10	MN5000-10	11.10	0.6	10.50	+0.1	0.7	+0.15	0.5	0.35	0.8	0.25
12	MN5000-12	13.30	0.6	12.65	+0.1	0.7	+0.15	0.6	0.4	1.0	0.33
14	MN5000-14	15.45	0.9	14.80	+0.1	1.0	+0.15	0.7	0.5	1.2	0.40
16	MN5000-16	17.70	0.9	16.90	+0.1	1.0	+0.15	0.7	0.5	1.4	0.45
18	MN5000-18	20.05	0.9	19.05	+0.1	1.0	+0.15	0.75	0.6	1.6	0.53
20	MN5000-20	22.25	0.9	21.15	+0.15	1.0	+0.15	0.9	0.7	1.7	0.57
22	MN5000-22	24.40	1.1	23.30	+0.15	1.2	+0.15	0.9	0.7	1.9	0.65
24	MN5000-24	26.55	1.1	25.4	+0.15	1.2	+0.15	1.0	0.8	2.1	0.70
25	MN5000-25	27.75	1.1	26.6	+0.15	1.2	+0.15	1.0	0.8	2.4	0.80
30	MN5000-30	33.40	1.3	31.9	+0.2	1.4	+0.15	1.2	1.0	2.9	0.95
35	MN5000-35	38.75	1.3	37.2	+0.2	1.4	+0.15	1.2	1.0	3.3	1.10
40	MN5000-40	44.25	1.6	42.4	+0.2	1.75	+0.2	1.7	1.3	3.6	1.20
45	MN5000-45	49.95	1.6	47.6	+0.2	1.75	+0.2	1.7	1.3	3.9	1.30
50	MN5000-50	55.35	1.6	53.1	+0.2	1.75	+0.2	1.7	1.3	4.6	1.55

Table 38 Retaining rings—internal. (© 1965, 1958 Waldes Koh-i-noor, Inc. Reprinted with permission.)

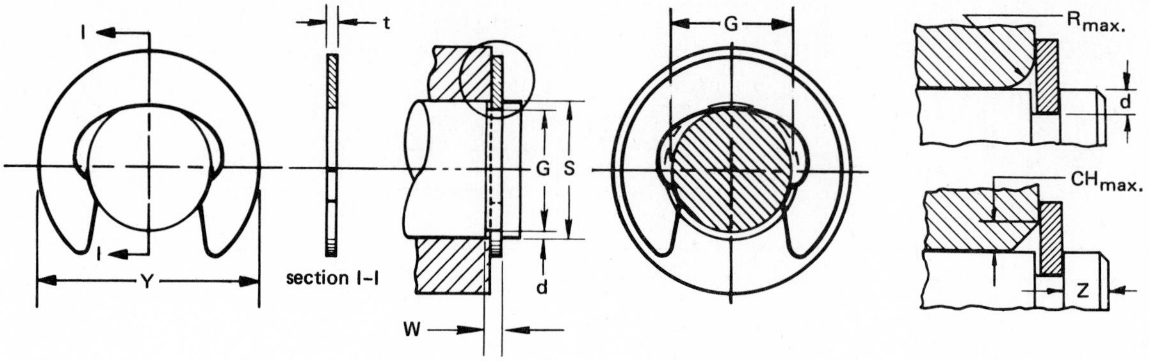

I
t
I
section I-I
Y
G
S
d
W
G
$R_{max.}$
d
$CH_{max.}$
Z

SHAFT DIA.	EXTERNAL SERIES 11-410	RETAINING RING DIMENSIONS		GROOVE DIMENSIONS: Diameter		GROOVE DIMENSIONS: Width		MAXIMUM ALLOWABLE CORNER RADII AND CHAMFER OF RETAINED PARTS		EDGE MARGIN	NOMINAL GROOVE DEPTH (REF)
S	Size—No.	Y	t	G	Tol.	W	Tol.	R Max.	Ch. Max.	z	d
8	11-410-080	10	0.6	7	−0.1	0.7	+0.15	0.6	0.45	1.5	0.5
10	11-410-100	12.2	0.6	9	−0.1	0.7	+0.15	0.6	0.45	1.5	0.5
12	11-410-120	14.4	0.6	10.9	−0.1	0.7	+0.15	0.6	0.45	1.7	0.5
14	11-410-140	16.3	1	12.7	−0.1	1.1	+0.15	1	0.8	2	0.65
16	11-410-160	18.5	1	14.5	−0.1	1.1	+0.15	1	0.8	2.3	0.75
18	11-410-180	20.4	1.2	16.3	−0.1	1.3	+0.15	1.2	0.9	2.6	0.85
20	11-410-200	22.6	1.2	18.1	−0.2	1.3	+0.15	1.2	0.9	2.9	0.95
22	11-410-220	25	1.2	19.9	−0.2	1.3	+0.15	1.2	0.9	3.2	1.05
24	11-410-240	27.1	1.2	21.7	−0.2	1.3	+0.15	1.2	0.9	3.5	1.15
25	11-410-250	28.3	1.2	22.6	−0.2	1.3	+0.15	1.2	0.9	3.6	1.2
30	11-410-300	33.7	1.5	27	−0.2	1.3	+0.2	1.5	1.15	4.5	1.5
35	11-410-350	39.4	1.5	31.5	−0.25	1.6	+0.2	1.5	1.15	5.3	1.75
40	11-410-400	45	1.5	36	−0.25	1.6	+0.2	1.5	1.15	6	2
45	11-410-450	50.6	1.5	40.5	−0.25	1.6	+0.2	1.5	1.15	6.8	2.25
50	11-410-500	56.4	2	45	−0.25	2.2	+0.2	2	1.5	7.5	2.5

Table 39 Retaining rings—radial assembly. (© 1965, 1958 Waldes Koh-i-noor, Inc. Reprinted with permission.)

MORSE TAPERS			BROWN AND SHARPE TAPERS					
No. of Taper	TAPER: mm per 100 mm	TAPER: inches per Foot	No. of Taper	TAPER: mm per 100 mm	TAPER: inches per Foot	No. of Taper	TAPER: mm per 100 mm	TAPER: inches per Foot
0	5.21	.625	1	4.18	.502	9	4.18	.501
1	4.99	.599	2	4.18	.502	10	4.3	.516
2	4.99	.599	3	4.18	.502	11	4.18	.501
3	5.02	.602	4	4.18	.502	12	4.17	.500
4	5.19	.623	5	4.18	.502	13	4.17	.500
5	5.26	.631	6	4.19	.503	14	4.17	.500
6	5.22	.626	7	4.18	.502	15	4.17	.500
7	5.20	.624	8	4.18	.501	16	4.17	.500

Table 40 Machine tapers.

PRODUCT	OUTSTANDING FEATURES	APPLICATION METHOD	COLOR
1357	A high performance adhesive with long bonding range, excellent initial strength. Meets specification requirements of MMM-A-121 (supersedes MIL-A-1154 C), MIL-A-5092 B, Type II, and MIL-A-21366. Bonds rubber, cloth, wood, foamed glass, paper honeycomb, decorative plastic laminates. Also used with metal-to-metal for bonds of moderate strength.	Spray or Brush	Gray/ Green or Olive
2210	Fast drying, exhibits aggressive tack that allows coated surfaces to knit easily under hand roller pressure. Excellent water and oil resistance. Meets specification requirements of MMM-A-121 (supersedes MIL-A-1154 C), MIL-A-21366, and MMM-A-00130a. Bonds a wide range of materials including rubber, leather, cloth, aluminum, wood, hardboard. Used extensively for bonding decorative plastic laminates.	Brush, Roller, or Trowel	Yellow
2215	Fast drying and has a rapid rate of strength build-up. Its aggressive tack permits adhesive coated surfaces to bond easily with moderate pressure. Bonds decorative plastic laminates to metal, wood, and particle board. Also used for general bonding of rubber, leather, cloth, aluminum, wood, hardboard, etc.	Spray	Light Yellow
2218	Offers rapid strength build-up, high-ultimate strength. Has a high softening point and excellent resistance to plastic flow. Adhesion to steel is especially good. Meets specification requirements of MMM-A-121 (supersedes MIL-A-1154 C). Bonds high density decorative plastic laminates to metal or wood. Widely used to fabricate honeycomb and sandwich-type building panels with various face sheets, including porcelain enamel steel.	Spray	Green
2226	Water dispersed, has high immediate bond strength, long bonding range. Changes color from blue to green while drying. Used to bond foamed plastics, plastic laminate, wood, rubber, plywood, wallboard, wood veneer, plaster, and canvas to themselves and to each other.	Spray or Brush	Wet: Lt. Blue Dry: Green
4420	High performance, fast-drying adhesive designed for application by pressure curtain coating and mechanical roll coating. Bonds decorative plastic laminates to plywood or particle board and is suitable for conveyor line production of laminated panels of various types such as aluminum to wood or hardboard.	Roll Coating	Yellow
4488	Lower viscosity version of Cement 4420 for use specifically with flow-over or Weir-type curtain coaters. Bonds decorative plastic laminates to plywood or particle board and is suitable for conveyor line production of laminated panels of various types, such as aluminum to wood or hardboard.	Curtain and Roll Coating	Yellow
4518	Designed for spray application with automatic or production line equipment. Dries very fast, requires pressure from a niproll (rotary press) or platen press to assure proper bonding. Bonds decorative plastic laminates to plywood or particle board on both flat work and postforming. Also used for conveyor line production of laminated panels of various types such as aluminum to wood or hardboard. Meets requirements of MIL-A-5092 B, Type II.	Spray	Green
4729	Similar to Cement 2218 except that it is formulated with a nonflammable solvent. Requires force drying to prevent blushing.	Spray	Red
5034	Water dispersed, has high immediate bond strength and long bonding range. Bonds foamed plastics, plastic laminate, wood, rubber, plywood, wallboard, wood veneer, plaster, and canvas to themselves and to each other.	Spray or Brush	Neutral

Table 41 Physical properties and application data of adhesives. (3M Company.)

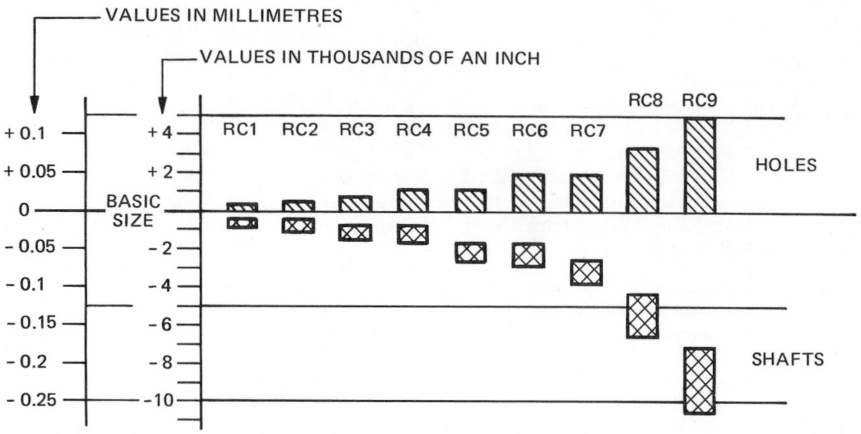
VALUES IN MILLIMETRES
VALUES IN THOUSANDS OF AN INCH
+ 0.1
+ 0.05
0
- 0.05
- 0.1
- 0.15
- 0.2
- 0.25
+ 4
+ 2
BASIC SIZE
- 2
- 4
- 6
- 8
-10
RC1 RC2 RC3 RC4 RC5 RC6 RC7 RC8 RC9
HOLES
SHAFTS

VALUES IN MILLIMETRES

Nominal Size Range Millimetres		Class RC1 Precision Sliding			Class RC2 Sliding Fit			Class RC3 Precision Running			Class RC4 Close Running		
Over	To	Hole Tol. H5 −0	Minimum Clearance	Shaft Tol. g4 +0	Hole Tol. H6 −0	Minimum Clearance	Shaft Tol. g5 +0	Hole Tol. H7 −0	Minimum Clearance	Shaft Tol. f6 +0	Hole Tol. H8 −0	Minimum Clearance	Shaft Tol. f7 +0
0	3	+0.004	0.003	−0.003	+0.006	0.003	−0.004	+0.010	0.008	−0.006	+0.015	0.008	−0.010
3	6	+0.005	0.004	−0.004	+0.008	0.004	−0.005	+0.013	0.010	−0.008	+0.018	0.010	−0.013
6	10	+0.006	0.005	−0.004	+0.010	0.005	−0.006	+0.015	0.013	−0.010	+0.023	0.013	−0.015
10	18	+0.008	0.006	−0.005	+0.010	0.006	−0.008	+0.018	0.015	−0.010	+0.025	0.015	−0.018
18	30	+0.010	0.008	−0.006	+0.013	0.008	−0.010	+0.020	0.020	−0.013	+0.03	0.02	−0.02
30	50	+0.010	0.010	−0.008	+0.015	0.010	−0.010	+0.03	0.03	−0.015	+0.04	0.03	−0.03
50	80	+0.013	0.010	−0.008	+0.018	0.010	−0.013	+0.03	0.03	−0.02	+0.05	0.03	−0.03
80	120	+0.015	0.013	−0.010	+0.023	0.013	−0.015	+0.04	0.04	−0.02	+0.06	0.04	−0.04
120	180	+0.018	0.015	−0.013	+0.025	0.015	−0.018	+0.04	0.04	−0.03	+0.06	0.04	−0.04
180	250	+0.020	0.015	−0.015	+0.030	0.015	−0.020	+0.05	0.05	−0.03	+0.07	0.05	−0.05
250	315	+0.023	0.020	−0.015	+0.030	0.020	−0.023	+0.05	0.06	−0.03	+0.08	0.06	−0.05
315	400	+0.025	0.025	−0.018	+0.036	0.025	−0.025	+0.06	0.08	−0.04	+0.09	0.08	−0.06

VALUES IN THOUSANDTHS OF AN INCH

Nominal Size Range Inches		Class RC1 Precision Sliding			Class RC2 Sliding Fit			Class RC3 Precision Running			Class RC4 Close Running		
Over	To	Hole Tol. GR5 −0	Minimum Clearance	Shaft Tol. GR4 +0	Hole Tol. GR6 −0	Minimum Clearance	Shaft Tol. GR5 +0	Hole Tol. GR7 −0	Minimum Clearance	Shaft Tol. GR6 +0	Hole Tol. GR8 −0	Minimum Clearance	Shaft Tol. GR7 +0
0	.12	+0.15	0.1	−0.12	+0.25	0.1	−0.15	+0.4	0.3	−0.25	+0.6	0.3	−0.4
.12	.24	+0.2	0.15	−0.15	+0.3	0.15	−0.2	+0.5	0.4	−0.3	+0.7	0.4	−0.5
.24	.40	+0.25	0.2	−0.15	+0.4	0.2	−0.25	+0.6	0.5	−0.4	+0.9	0.5	−0.6
.40	.71	+0.3	0.25	−0.2	+0.4	0.25	−0.3	+0.7	0.6	−0.4	+1.0	0.6	−0.7
.71	1.19	+0.4	0.3	−0.25	+0.5	0.3	−0.4	+0.8	0.8	−0.5	+1.2	0.8	−0.8
1.19	1.97	+0.4	0.4	−0.3	+0.6	0.4	−0.4	+1.0	1.0	−0.6	+1.6	1.0	−1.0
1.97	3.15	+0.5	0.4	−0.3	+0.7	0.4	−0.5	+1.2	1.2	−0.7	+1.8	1.2	−1.2
3.15	4.73	+0.6	0.5	−0.4	+0.9	0.5	−0.6	+1.4	1.4	−0.9	+2.2	1.4	−1.4
4.73	7.09	+0.7	0.6	−0.5	+1.0	0.6	−0.7	+1.6	1.6	−1.0	+2.5	1.6	−1.6
7.09	9.85	+0.8	0.6	−0.6	+1.2	0.6	−0.8	+1.8	2.0	−1.2	+2.8	2.0	−1.8
9.85	12.41	+0.9	0.8	−0.6	+1.2	0.8	−0.9	+2.0	2.5	−1.2	+3.0	2.5	−2.0
12.41	15.75	+1.0	1.0	−0.7	+1.4	1.0	−1.0	+2.2	3.0	−1.4	+3.5	3.0	−2.2

Note—Metric sizes were not available at time of publication. Valves have been soft converted for problem solving.

Table 42 Running and sliding fits.

VALUES IN MILLIMETRES

Class RC5 Medium Running			Class RC6 Medium Running			Class RC7 Free Running			Class RC8 Loose Running			Class RC9 Loose Running		
Hole Tol. H8	Minimum Clearance	Shaft Tol. e7	Hole Tol. H9	Minimum Clearance	Shaft Tol. e8	Hole Tol. H9	Minimum Clearance	Shaft Tol. d8	Hole Tol. H10	Minimum Clearance	Shaft Tol. e9	Hole Tol. GR11	Minimum Clearance	Shaft Tol. GR10
−0		+0	−0		+0	−0		+0	−0		+0	−0		+0
+0.015	0.015	−0.010	+0.025	0.015	−0.015	+0.025	0.025	−0.015	+0.041	0.064	−0.025	+0.06	0.10	−0.04
+0.018	0.020	−0.013	+0.030	0.015	−0.018	+0.030	0.030	−0.018	+0.046	0.071	−0.030	+0.08	0.11	−0.05
+0.023	0.025	−0.015	+0.036	0.025	−0.023	+0.036	0.040	−0.023	+0.056	0.076	−0.036	+0.09	0.13	−0.06
+0.025	0.03	−0.018	+0.04	0.03	−0.025	+0.04	0.05	−0.025	+0.07	0.09	−0.04	+0.10	0.15	−0.07
+0.03	0.04	−0.02	+0.05	0.04	−0.03	+0.05	0.06	−0.03	+0.09	0.11	−0.05	+0.13	0.18	−0.09
+0.04	0.05	−0.03	+0.06	0.05	−0.04	+0.06	0.08	−0.04	+0.10	0.13	−0.06	+0.15	0.20	−0.10
+0.05	0.06	−0.03	+0.08	0.06	−0.05	+0.08	0.10	−0.05	+0.11	0.15	−0.08	+0.18	0.23	−0.12
+0.06	0.08	−0.04	+0.09	0.08	−0.06	+0.09	0.13	−0.06	+0.13	0.18	−0.09	+0.23	0.25	−0.13
+0.06	0.09	−0.04	+0.10	0.09	−0.06	+0.10	0.15	−0.06	+0.15	0.20	−0.10	+0.25	0.30	−0.15
+0.07	0.11	−0.05	+0.11	0.10	−0.07	+0.11	0.18	−0.07	+0.18	0.25	−0.11	+0.30	0.38	−0.18
+0.08	0.13	−0.05	+0.13	0.13	−0.08	+0.13	0.20	−0.08	+0.20	0.30	−0.13	+0.30	0.46	−0.20
+0.09	0.15	−0.06	+0.15	0.15	−0.09	+0.15	0.25	−0.09	+0.23	0.36	−0.15	+0.36	0.56	−0.23

VALUES IN THOUSANDTHS OF AN INCH

Class RC5 Medium Running			Class RC6 Medium Running			Class RC7 Free Running			Class RC8 Loose Running			Class RC9 Loose Running		
Hole Tol. GR8	Minimum Clearance	Shaft Tol. GR7	Hole Tol. GR9	Minimum Clearance	Shaft Tol. GR8	Hole Tol. GR9	Minimum Clearance	Shaft Tol. GR8	Hole Tol. GR10	Minimum Clearance	Shaft Tol. GR9	Hole Tol. GR11	Minimum Clearance	Shaft Tol. GR10
−0		+0	−0		+0	−0		+0	−0		+0	−0		+0
+0.6	0.6	−0.4	+1.0	0.6	−0.6	+1.0	1.0	−0.6	+1.6	2.5	−1.0	+2.5	4.0	−1.6
+0.7	0.8	−0.5	+1.2	0.8	−0.7	+1.2	1.2	−0.7	+1.8	2.8	−1.2	+3.0	4.5	−1.8
+0.9	1.0	−0.6	+1.4	1.0	−0.9	+1.4	1.6	−0.9	+2.2	3.0	−1.4	+3.5	5.0	−2.2
+1.0	1.2	−0.7	+1.6	1.2	−1.0	+1.6	2.0	−1.0	+2.8	3.5	−1.6	+4.0	6.0	−2.8
+1.2	1.6	−0.8	+2.0	1.6	−1.2	+2.0	2.5	−1.2	+3.5	4.5	−2.0	+5.0	7.0	−3.5
+1.6	2.0	−1.0	+2.5	2.0	−1.6	+2.5	3.0	−1.6	+4.0	5.0	−2.5	+6.0	8.0	−4.0
+1.8	2.5	−1.2	+3.0	2.5	−1.8	+3.0	4.0	−1.8	+4.5	6.0	−3.0	+7.0	9.0	−4.5
+2.2	3.0	−1.4	+3.5	3.0	−2.2	+3.5	5.0	−2.2	+5.0	7.0	−3.5	+9.0	10.0	−5.0
+2.5	3.5	−1.6	+4.0	3.5	−2.5	+4.0	6.0	−2.5	+6.0	8.0	−4.0	+10.0	12.0	−6.0
+2.8	4.5	−1.8	+4.5	4.0	−2.8	+4.5	7.0	−2.8	+7.0	10.0	−4.5	+12.0	15.0	−7.0
+3.0	5.0	−2.0	+5.0	5.0	−3.0	+5.0	8.0	−3.0	+8.0	12.0	−5.0	+12.0	18.0	−8.0
+3.5	6.0	−2.2	+6.0	6.0	−3.5	+6.0	10.0	−3.5	+9.0	14.0	−6.0	+14.0	22.0	−9.0

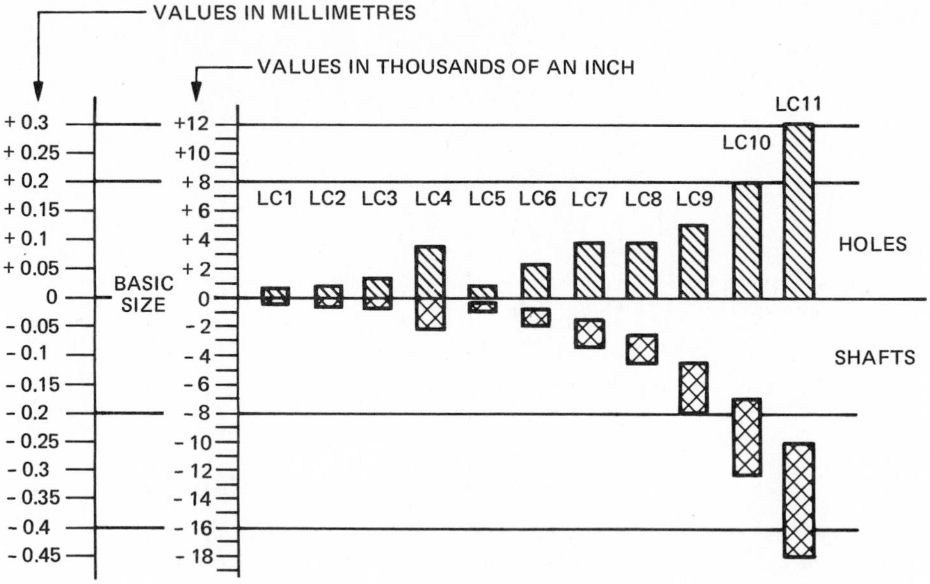
VALUES IN MILLIMETRES
VALUES IN THOUSANDS OF AN INCH
+ 0.3
+ 0.25
+ 0.2
+ 0.15
+ 0.1
+ 0.05
0
- 0.05
- 0.1
- 0.15
- 0.2
- 0.25
- 0.3
- 0.35
- 0.4
- 0.45
BASIC SIZE
+12
+10
+ 8
+ 6
+ 4
+ 2
0
- 2
- 4
- 6
- 8
- 10
- 12
- 14
- 16
- 18
LC1 LC2 LC3 LC4 LC5 LC6 LC7 LC8 LC9
LC10
LC11
HOLES
SHAFTS

VALUES IN MILLIMETRES

Nominal Size Range Millimetres		Class LC1			Class LC2			Class LC3			Class LC4			Class LC5		
		Hole Tol. H6	Minimum Clearance	Shaft Tol. h5	Hole Tol. H7	Minimum Clearance	Shaft Tol. h6	Hole Tol. H8	Minimum Clearance	Shaft Tol. h7	Hole Tol. H10	Minimum Clearance	Shaft Tol. h9	Hole Tol. H7	Minimum Clearance	Shaft Tol. g6
Over	To	−0		+0	−0		+0	−0		+0	−0		+0	−0		+0
0	3	+0.006	0	−0.004	+0.010	0	−0.006	+0.015	0	−0.010	+0.041	0	−0.025	+0.010	0.002	−0.006
3	6	+0.008	0	−0.005	+0.013	0	−0.008	+0.018	0	−0.013	+0.046	0	−0.030	+0.013	0.004	−0.008
6	10	+0.010	0	−0.006	+0.015	0	−0.010	+0.023	0	−0.015	+0.056	0	−0.036	+0.015	0.005	−0.010
10	18	+0.010	0	−0.008	+0.018	0	−0.010	+0.025	0	−0.018	+0.07	0	−0.04	+0.018	0.006	−0.010
18	30	+0.013	0	−0.010	+0.020	0	−0.013	+0.030	0	−0.020	+0.09	0	−0.05	+0.020	0.008	−0.013
30	50	+0.015	0	−0.010	+0.025	0	−0.015	+0.041	0	−0.025	+0.10	0	−0.06	+1.025	0.010	−0.015
50	80	+0.018	0	−0.013	+0.030	0	−0.018	+0.046	0	−0.030	+0.11	0	−0.08	+0.030	0.010	−0.018
80	120	+0.023	0	−0.015	+0.036	0	−0.023	+0.056	0	−0.036	+0.13	0	−0.09	+0.036	0.013	−0.023
120	180	+0.025	0	−0.018	+0.041	0	−0.025	+0.064	0	−0.041	+0.15	0	−0.10	+0.041	0.015	−0.025
180	250	+0.030	0	−0.020	+0.046	0	−0.030	+0.071	0	−0.046	+0.18	0	−0.11	+0.046	0.015	−0.030
250	315	+0.030	0	−0.023	+0.051	0	−0.030	+0.076	0	−0.051	+0.20	0	−0.13	+0.051	0.018	−0.030
315	400	+0.036	0	−0.025	+0.056	0	−0.036	+0.089	0	−0.056	+0.23	0	−0.15	+0.056	0.018	−0.036

VALUES IN THOUSANDTHS OF AN INCH

Nominal Size Range Inches		Class LC1			Class LC2			Class LC3			Class LC4			Class LC5		
		Hole Tol. GR6	Minimum Clearance	Shaft Tol. GR5	Hole Tol. GR7	Minimum Clearance	Shaft Tol. GR6	Hole Tol. GR8	Minimum Clearance	Shaft Tol. GR7	Hole Tol. GR10	Minimum Clearance	Shaft Tol. GR9	Hole Tol. GR7	Minimum Clearance	Shaft Tol. GR6
Over	To	−0		+0	−0		+0	−0		+0	−0		+0	−0		+0
0	.12	+0.25	0	−0.15	+0.4	0	−0.25	+0.6	0	−0.4	+1.6	0	−1.0	+0.4	0.1	−0.25
.12	.24	+0.3	0	−0.2	+0.5	0	−0.3	+0.7	0	−0.5	+1.8	0	−1.2	+0.5	0.15	−0.3
.24	.40	+0.4	0	−0.25	+0.6	0	−0.4	+0.9	0	−0.6	+2.2	0	−1.4	+0.6	0.2	−0.4
.40	.71	+0.4	0	−0.3	+0.7	0	−0.4	+1.0	0	−0.7	+2.8	0	−1.6	+0.7	0.25	−0.4
.71	1.19	+0.5	0	−0.4	+0.8	0	−0.5	+1.2	0	−0.8	+3.5	0	−2.0	+0.8	0.3	−0.5
1.19	1.97	+0.6	0	−0.4	+1.0	0	−0.6	+1.6	0	−1.0	+4.0	0	−2.5	+1.0	0.4	−0.6
1.97	3.15	+0.7	0	−0.5	+1.2	0	−0.7	+1.8	0	−1.2	+4.5	0	−3.0	+1.2	0.4	−0.7
3.15	4.73	+0.9	0	−0.6	+1.4	0	−0.9	+2.2	0	−1.4	+5.0	0	−3.5	+1.4	0.5	−0.9
4.73	7.09	+1.0	0	−0.7	+1.6	0	−1.0	+2.5	0	−1.6	+6.0	0	−4.0	+1.6	0.6	−1.0
7.09	9.85	+1.2	0	−0.8	+1.8	0	−1.2	+2.8	0	−1.8	+7.0	0	−4.5	+1.8	0.6	−1.2
9.85	12.41	+1.2	0	−0.9	+2.0	0	−1.2	+3.0	0	−2.0	+8.0	0	−5.0	+2.0	0.7	−1.2
12.41	15.75	+1.4	0	−1.0	+2.2	0	−1.4	+3.5	0	−2.2	+9.0	0	−6.0	+2.2	0.7	−1.4

Note: Metric sizes were not available at time of publication. Values have been soft converted for problem solving.

Table 43 Locational clearance fits.

VALUES IN MILLIMETRES

Class LC6			Class LC7			Class LC8			Class LC9			Class LC10			Class LC11		
Hole Tol. H9 −0	Minimum Clearance	Shaft Tol. f8 +0	Hole Tol. H10 −0	Minimum Clearance	Shaft Tol. e9 +0	Hole Tol. H10 −0	Minimum Clearance	Shaft Tol. d9 +0	Hole Tol. H11 −0	Minimum Clearance	Shaft Tol. c10 +0	Hole Tol. GR12 −0	Minimum Clearance	Shaft Tol. GR11 +0	Hole Tol. GR13 −0	Minimum Clearance	Shaft Tol. GR12 +0
+0.025	0.008	−0.015	+0.041	0.015	−0.025	+0.041	0.025	−0.025	+0.064	0.064	−0.041	+0.10	0.10	−0.06	+0.15	0.13	−0.10
+0.030	0.010	−0.018	+0.046	0.020	−0.030	+0.046	0.030	−0.030	+0.076	0.071	−0.046	+0.13	0.11	−0.08	+0.18	0.15	−0.13
+0.036	0.013	−0.023	+0.056	0.025	−0.036	+0.056	0.041	−0.036	+0.089	0.076	−0.056	+0.15	0.13	−0.09	+0.23	0.18	−0.15
+0.041	0.015	−0.025	+0.07	0.03	−0.04	+0.07	0.05	−0.04	+0.10	0.09	−0.07	+0.18	0.15	−0.10	+0.25	0.20	−0.18
+0.05	0.02	−0.03	+0.09	0.04	−0.05	+0.09	0.06	−0.05	+0.13	0.11	−0.09	+0.20	0.18	−0.13	+0.31	0.25	−0.20
+0.06	0.03	−0.04	+0.10	0.05	−0.06	+0.10	0.09	−0.06	+0.15	0.13	−0.10	+0.25	0.20	−0.15	+0.41	0.31	−0.25
+0.08	0.03	−0.05	+0.11	0.06	−0.08	+0.11	0.10	−0.08	+0.18	0.15	−0.11	+0.31	0.25	−0.18	+0.46	0.36	−0.31
+0.09	0.04	−0.06	+0.13	0.08	−0.09	+0.13	0.13	−0.09	+0.23	0.18	−0.13	+0.36	0.28	−0.23	+0.56	0.41	−0.36
+0.10	0.04	−0.06	+0.15	0.09	−0.10	+0.15	0.15	−0.10	+0.25	0.20	−0.15	+0.41	0.31	−0.25	+0.64	0.46	−0.41
+0.11	0.05	−0.07	+0.18	0.10	−0.11	+0.18	0.18	−0.11	+0.31	0.25	−0.18	+0.46	0.41	−0.31	+0.71	0.56	−0.46
+0.13	0.06	−0.08	+0.20	0.11	−0.13	+0.20	0.18	−0.13	+0.31	0.31	−0.20	+0.51	0.51	−0.31	+0.76	0.71	−0.51
+0.15	0.06	−0.09	+0.23	0.13	−0.15	+0.23	0.20	−0.15	+0.36	0.36	−0.23	+0.56	0.56	−0.36	+0.89	0.76	−0.56

VALUES IN THOUSANDTHS OF AN INCH

Class LC6			Class LC7			Class LC8			Class LC9			Class LC10			Class LC11		
Hole Tol. GR9 −0	Minimum Clearance	Shaft Tol. GR8 +0	Hole Tol. GR10 −0	Minimum Clearance	Shaft Tol. GR9 +0	Hole Tol. GR10 −0	Minimum Clearance	Shaft Tol. GR9 +0	Hole Tol. GR11 −0	Minimum Clearance	Shaft Tol. GR10 +0	Hole Tol. GR12 −0	Minimum Clearance	Shaft Tol. GR11 +0	Hole Tol. GR13 −0	Minimum Clearance	Shaft Tol. GR12 +0
+1.0	0.3	−0.6	+1.6	0.6	−1.0	+1.6	1.0	−1.0	+2.5	2.5	−1.6	+4.0	4.0	−2.5	+6.0	5.0	−4.0
+1.2	0.4	−0.7	+1.8	0.8	−1.2	+1.8	1.2	−1.2	+3.0	2.8	−1.8	+5.0	4.5	−3.0	+7.0	6.0	−5.0
+1.4	0.5	−0.9	+2.2	1.0	−1.4	+2.2	1.6	−1.4	+3.5	3.0	−2.2	+6.0	5.0	−3.5	+9.0	7.0	−6.0
+1.6	0.6	−1.0	+2.8	1.2	−1.6	+2.8	2.0	−1.6	−4.0	3.5	−2.8	+7.0	6.0	−4.0	+10.0	8.0	−7.0
+2.0	0.8	−1.2	+3.5	1.6	−2.0	+3.5	2.5	−2.0	+5.0	4.5	−3.5	+8.0	7.0	−5.0	+12.0	10.0	−8.0
+2.5	1.0	−1.6	+4.0	2.0	−2.5	+4.0	3.6	−2.5	+6.0	5.0	−4.0	+10.0	8.0	−6.0	+16.0	12.0	−10.0
+3.0	1.2	−1.8	+4.5	2.5	−3.0	+4.5	4.0	−3.0	+7.0	6.0	−4.5	+12.0	10.0	−7.0	+18.0	14.0	−12.0
+3.5	1.4	−2.2	+5.0	3.0	−3.5	+5.0	5.0	−3.5	+9.0	7.0	−5.0	+14.0	11.0	−9.0	+22.0	16.0	−14.0
+4.0	1.6	−2.5	+6.0	3.5	−4.0	+6.0	6.0	−4.0	+10.0	8.0	−6.0	+16.0	12.0	−10.0	+25.0	18.0	−16.0
+4.5	2.0	−2.8	+7.0	4.0	−4.5	+7.0	7.0	−4.5	+12.0	10.0	−7.0	+18.0	16.0	−12.0	+28.0	22.0	−18.0
+5.0	2.2	−3.0	+8.0	4.5	−5.0	+8.0	7.0	−5.0	+12.0	12.0	−8.0	+20.0	20.0	−12.0	+30.0	28.0	−20.0
+6.0	2.5	−3.5	+9.0	5.0	−6.0	+9.0	8.0	−6.0	+14.0	14.0	−9.0	+22.0	22.0	−14.0	+35.0	30.0	−22.0

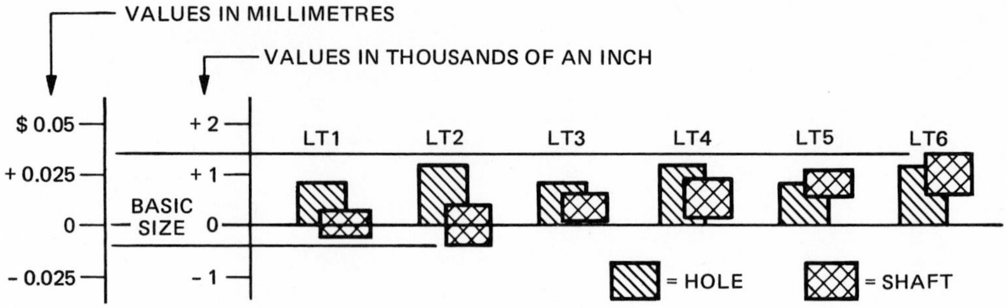
VALUES IN MILLIMETRES
VALUES IN THOUSANDS OF AN INCH
$ 0.05
+ 0.025
0
- 0.025
+ 2
+ 1
BASIC SIZE 0
- 1
LT1
LT2
LT3
LT4
LT5
LT6
= HOLE
= SHAFT

VALUES IN MILLIMETRES

Nominal Size Range Millimetres		Class LT1			Class LT2		
		Hole Tol. H7	Maximum Interference	Shaft Tol. Js6	Hole Tol. H8	Maximum Interference	Shaft Tol. Js7
Over	To	−0		+0	−0		+0
0	3	+0.010	0.002	−0.006	+0.015	0.005	−0.010
3	6	+0.013	0.004	−0.008	+0.018	0.006	−0.013
6	10	+0.015	0.005	−0.010	+0.023	0.008	−0.015
10	18	+0.018	0.005	−0.010	+0.025	0.008	−0.018
18	30	+0.020	0.006	−0.013	+0.030	0.010	−0.020
30	50	+0.025	0.008	−0.015	+0.041	0.013	−0.025
50	80	+0.030	0.008	−0.018	+0.046	0.015	−0.030
80	120	+0.036	0.010	−0.023	+0.056	0.018	−0.036
120	180	+0.041	0.013	−0.025	+0.064	0.020	−0.041
180	250	+0.046	0.015	−0.030	+0.071	0.023	−0.046
250	315	+0.051	0.015	−0.030	+0.076	0.025	−0.051
315	400	+0.056	0.018	−0.036	+0.089	0.025	−0.056

VALUES IN THOUSANDTHS OF AN INCH

Nominal Size Range Inches		Class LT1			Class LT2		
		Hole Tol. GR7	Maximum Interference	Shaft Tol. GR6	Hole Tol. GR8	Maximum Interference	Shaft Tol. GR7
Over	To	−0		+0	−0		+0
0	.12	+0.4	0.1	−0.25	+0.6	0.2	−0.4
.12	.24	+0.5	0.15	−0.3	+0.7	0.25	−0.5
.24	.40	+0.6	0.2	−0.4	+0.9	0.3	−0.6
.40	.71	+0.7	0.2	−0.4	+1.0	0.3	−0.7
.71	1.19	+0.8	0.25	−0.5	+1.2	0.4	−0.8
1.19	1.97	+1.0	0.3	−0.6	+1.6	0.5	−1.0
1.97	3.15	+1.2	0.3	−0.7	+1.8	0.6	−1.2
3.15	4.73	+1.4	0.4	−0.9	+2.2	0.7	−1.4
4.73	7.09	+1.6	0.5	−1.0	+2.5	0.8	−1.6
7.09	9.85	+1.8	0.6	−1.2	+2.8	0.9	−1.8
9.85	12.41	+2.0	0.6	−1.2	+3.0	1.0	−2.0
12.41	15.75	+2.2	0.7	−1.4	+3.5	1.0	−2.2

Note: Metric sizes were not available at time of publication. Values have been soft converted for problem solving.

Table 44 Transition Fits

VALUES IN MILLIMETRES

Class LT3			Class LT4			Class LT5			Class LT6		
Hole Tol. H7	Maximum Interference	Shaft Tol. K6	Hole Tol. H8	Maximum Interference	Shaft Tol. K7	Hole Tol. H7	Maximum Interference	Shaft Tol. n6	Hole Tol. H8	Maximum Interference	Shaft Tol. n7
−0		+0	−0		+0	−0		+0	−0		+0
+0.010	0.006	−0.006	+0.015	0.010	−0.010	+0.010	0.013	−0.006	+0.015	0.016	−0.010
+0.013	0.010	−0.008	+0.018	0.015	−0.013	+0.013	0.015	−0.008	+0.018	0.020	−0.013
+0.015	0.013	−0.010	+0.023	0.018	−0.015	+0.015	0.020	−0.010	+0.023	0.025	−0.015
+0.018	0.013	−0.010	+0.025	0.020	−0.018	+0.018	0.023	−0.010	+0.025	0.030	−0.018
+0.020	0.015	−0.013	+0.030	0.023	−0.020	+0.020	0.028	−0.013	+0.030	0.036	−0.020
+0.025	0.018	−0.015	+0.041	0.028	−0.025	+0.025	0.033	−0.015	+0.041	0.044	−0.025
+0.030	0.020	−0.018	+0.046	0.033	−0.030	+0.030	0.038	−0.018	+0.046	0.051	−0.030
+0.036	0.025	−0.023	+0.056	0.038	−0.036	+0.036	0.048	−0.023	+0.056	0.062	−0.036
+0.041	0.028	−0.025	+0.064	0.044	−0.041	+0.041	0.056	−0.025	+0.064	0.071	−0.041
+0.046	0.036	−0.030	+0.071	0.051	−0.046	+0.046	0.066	−0.030	+0.071	0.081	−0.046
+0.051	0.036	−0.030	+0.076	0.056	−0.051	+0.051	0.066	−0.030	+0.076	0.086	−0.051
+0.056	0.041	−0.036	+0.089	0.062	−0.056	+0.056	0.076	−0.036	+0.089	0.096	−0.056

VALUES IN THOUSANDTHS OF AN INCH

Class LT3			Class LT4			Class LT5			Class LT6		
Hole Tol. GR7	Maximum Interference	Shaft Tol. GR6	Hole Tol. GR8	Maximum Interference	Shaft Tol. GR7	Hole Tol. GR7	Maximum Interference	Shaft Tol. GR6	Hole Tol. GR8	Maximum Interference	Shaft Tol. GR7
−0		+0	−0		+0	−0		+0	−0		+0
+0.4	0.25	−0.25	+0.6	0.4	−0.4	+0.4	0.5	−0.25	+0.6	0.65	−0.4
+0.5	0.4	−0.3	+0.7	0.6	−0.5	+0.5	0.6	−0.3	+0.7	0.8	−0.5
+0.6	0.5	−0.4	+0.9	0.7	−0.6	+0.6	0.8	−0.4	+0.9	1.0	−0.6
+0.7	0.5	−0.4	+1.0	0.8	−0.7	+0.7	0.9	−0.4	+1.0	1.2	−0.7
+0.8	0.6	−0.5	+1.2	0.9	−0.8	+0.8	1.1	−0.5	+1.2	1.4	−0.8
+1.0	0.7	−0.6	+1.6	1.1	−1.0	+1.0	1.3	−0.6	+1.6	1.7	−1.0
+1.2	0.8	−0.7	+1.8	1.3	−1.2	+1.2	1.5	−0.7	+1.8	2.0	−1.2
+1.4	1.0	−0.9	+2.2	1.5	−1.4	+1.4	1.9	−0.9	+2.2	2.4	−1.4
+1.6	1.1	−1.0	+2.5	1.7	−1.6	+1.6	2.2	−1.0	+2.5	2.8	−1.6
+1.8	1.4	−1.2	+2.8	2.0	−1.8	+1.8	2.6	−1.2	+2.8	3.2	−1.8
+2.0	1.4	−1.2	+3.0	2.2	−2.0	+2.0	2.6	−1.2	+3.0	3.4	−2.0
+2.2	1.6	−1.4	+3.5	2.4	−2.2	+2.2	3.0	−1.4	+3.5	3.8	−2.2

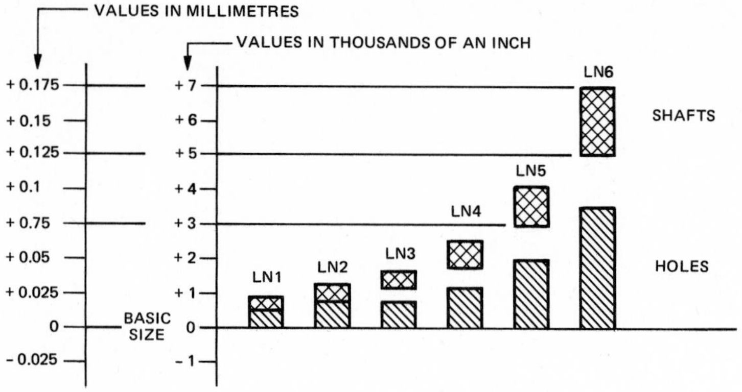
VALUES IN MILLIMETRES
VALUES IN THOUSANDS OF AN INCH
+ 0.175
+ 0.15
+ 0.125
+ 0.1
+ 0.75
+ 0.05
+ 0.025
0
- 0.025
BASIC SIZE
+ 7
+ 6
+ 5
+ 4
+ 3
+ 2
+ 1
0
- 1
LN1
LN2
LN3
LN4
LN5
LN6
SHAFTS
HOLES

VALUES IN MILLIMETRES

Nominal Size Range Millimetres		Class LN1 Light Press Fit			Class LN2 Medium Press Fit		
		Hole Tol. GR6	Maximum Interference	Shaft Tol. GR5	Hole Tol. H7	Maximum Interference	Shaft Tol. P6
Over	To	−0		+0	−0		+0
0	3	+0.006	0	−0.004	+0.010	0.016	−0.006
3	6	+0.008	0	−0.005	+0.013	0.020	−0.008
6	10	+0.010	0	−0.006	+0.015	0.025	−0.010
10	18	+0.010	0	−0.008	+0.018	0.028	−0.010
18	30	+0.013	0	−0.010	+0.020	0.033	−0.013
30	50	+0.015	0	−0.010	+0.025	0.041	−0.015
50	80	+0.018	0	−0.013	+0.030	0.054	−0.018
80	120	+0.023	0	−0.015	+0.036	0.064	−0.023
120	180	+0.025	0	−0.018	+0.041	0.071	−0.025
180	250	+0.030	0	−0.020	+0.046	0.081	−0.030
250	315	+0.030	0	−0.023	+0.051	0.086	−0.030
315	400	+0.036	0	−0.025	+0.056	0.099	−0.036

VALUES IN THOUSANDTHS OF AN INCH

Nominal Size Range Inches		Class LN1 Light Press Fit			Class LN2 Medium Press Fit		
		Hole Tol. GR6	Maximum Interference	Shaft Tol. GR5	Hole Tol. GR7	Maximum Interference	Shaft Tol. GR6
Over	To	−0		+0	−0		+0
0	.12	+0.25	0.4	−0.15	+0.4	0.65	−0.25
.12	.24	+0.3	0.5	−0.2	+0.5	0.8	−0.3
.24	.40	+0.4	0.65	−0.25	+0.6	1.0	−0.4
.40	.71	+0.4	0.7	−0.3	+0.7	1.1	−0.4
.71	1.19	+0.5	0.9	−0.4	+0.8	1.3	−0.5
1.19	1.97	+0.6	1.0	−0.4	+1.0	1.6	−0.6
1.97	3.15	+0.7	1.3	−0.5	+1.2	2.1	−0.7
3.15	4.73	+0.9	1.6	−0.6	+1.4	2.5	−0.9
4.73	7.09	+1.0	1.9	−0.7	+1.6	2.8	−1.0
7.09	9.85	+1.2	2.2	−0.8	+1.8	3.2	−1.2
9.85	12.41	+1.2	2.3	−0.9	+2.0	3.4	−1.2
12.41	15.75	+1.4	2.6	−1.0	+2.2	3.9	−1.4

Note: Metric sizes were not available at time of publication. Values have been soft converted for problem solving.

Table 45 Locational interference fits

VALUES IN MILLIMETRES

Class LN3 Heavy Press Fit			Class LN4			Class LN5			Class LN6		
Hole Tol. H7	Maximum Interference	Shaft Tol. r6	Hole Tol. GR8	Maximum Interference	Shaft Tol. GR7	Hole Tol. GR9	Maximum Interference	Shaft Tol. GR8	Hole Tol. GR10	Maximum Interference	Shaft Tol. GR9
−0		+0	−0		+0	−0		+0	−0		+0
+0.010	0.019	−0.006	+0.015	0.030	−0.010	+0.025	0.046	−0.015	+0.041	0.076	−0.025
+0.013	0.023	−0.008	+0.018	0.038	−0.013	+0.030	0.059	−0.018	+0.046	0.091	−0.030
+0.015	0.030	−0.010	+0.023	0.046	−0.015	+0.036	0.071	−0.023	+0.056	0.112	−0.036
+0.018	0.036	−0.010	+0.025	0.056	−0.018	+0.041	0.086	−0.025	+0.071	0.142	−0.041
+0.020	0.044	−0.013	+0.030	0.066	−0.020	+0.051	0.107	−0.030	+0.089	0.178	−0.051
+0.025	0.051	−0.015	+0.041	0.086	−0.025	+0.064	0.135	−0.041	+0.102	0.216	−0.064
+0.030	0.059	−0.018	+0.046	0.102	−0.030	+0.076	0.160	−0.046	+0.114	0.254	−0.076
+0.036	0.074	−0.023	+0.056	0.122	−0.036	+0.102	0.196	−0.056	+0.127	0.292	−0.102
+0.041	0.089	−0.025	+0.064	0.142	−0.041	+0.114	0.221	−0.064	+0.152	0.343	−0.114
+0.046	0.107	−0.030	+0.071	0.168	−0.046	+0.127	0.262	−0.071	+0.178	0.419	−0.127
+0.051	0.119	−0.030	+0.076	0.191	−0.051	+0.152	0.305	−0.076	+0.203	0.483	−0.152
+0.056	0.150	−0.036	+0.089	0.221	−0.056	+0.152	0.368	−0.089	+0.229	0.584	−0.152

VALUES IN THOUSANDTHS OF AN INCH

Class LN3 Heavy Press Fit			Class LN4			Class LN5			Class LN6		
Hole Tol. GR7	Maximum Interference	Shaft Tol. GR6	Hole Tol. GR8	Maximum Interference	Shaft Tol. GR7	Hole Tol. GR9	Maximum Interference	Shaft Tol. GR8	Hole Tol. GR10	Maximum Interference	Shaft Tol. GR9
−0		+0	−0		+0	−0		+0	−0		+0
+0.4	0.75	−0.25	+0.6	1.2	−0.4	+1.0	1.8	−0.6	+1.6	3.0	−1.0
+0.5	0.9	−0.3	+0.7	1.5	−0.5	+1.2	2.3	−0.7	+1.8	3.6	−1.2
+0.6	1.2	−0.4	+0.9	1.8	−0.6	+1.4	2.8	−0.9	+2.2	4.4	−1.4
+0.7	1.4	−0.4	+1.0	2.2	−0.7	+1.6	3.4	−1.0	+2.8	5.6	−1.6
+0.8	1.7	−0.5	+1.2	2.6	−0.8	+2.0	4.2	−1.2	+3.5	7.0	−2.0
+1.0	2.0	−0.6	+1.6	3.4	−1.0	+2.5	5.3	−1.6	+4.0	8.5	−2.5
+1.2	2.3	−0.7	+1.8	4.0	−1.2	+3.0	6.3	−1.8	+4.5	10.0	−3.0
+1.4	2.9	−0.9	+2.2	4.8	−1.4	+4.0	7.7	−2.2	+5.0	11.5	−3.5
+1.6	3.5	−1.0	+2.5	5.6	−1.6	+4.5	8.7	−2.5	+6.0	13.5	−4.0
+1.8	4.2	−1.2	+2.8	6.6	−1.8	+5.0	10.3	−2.8	+7.0	16.5	−4.5
+2.0	4.7	−1.2	+3.0	7.5	−2.0	+6.0	12.0	−3.0	+8.0	19	−5.0
+2.2	5.9	−1.4	+3.5	8.7	−2.2	+6.0	14.5	−3.5	+9.0	23	−6.0

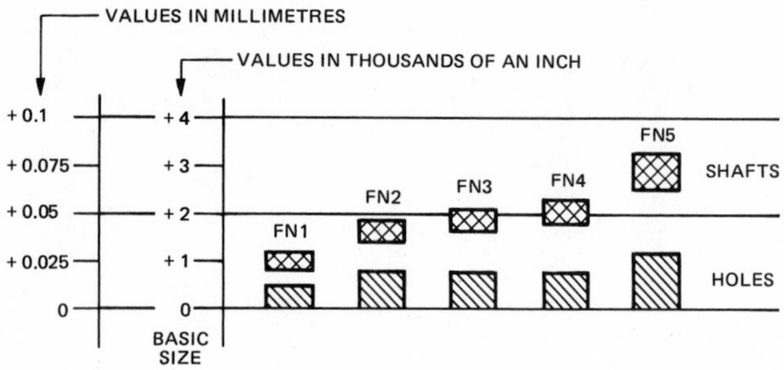
VALUES IN MILLIMETRES
VALUES IN THOUSANDS OF AN INCH
+0.1
+0.075
+0.05
+0.025
0
+4
+3
+2
+1
0
BASIC SIZE
FN1
FN2
FN3
FN4
FN5
SHAFTS
HOLES

VALUES IN MILLIMETRES

Nominal Size Range Millimetres		Class FN1 Light Drive Fit			Class FN2 Medium Drive Fit		
		Hole Tol. GR6	Maximum Interference	Shaft Tol. GR5	Hole Tol. H7	Maximum Interference	Shaft Tol. s6
Over	To	−0		+0	−0		+0
0	3	+0.006	0.013	−0.004	+0.010	0.021	−0.006
3	6	+0.007	0.015	−0.005	+0.013	0.025	−0.007
6	10	+0.010	0.019	−0.006	+0.015	0.036	−0.010
10	14	+0.010	0.020	−0.008	+0.018	0.041	−0.010
14	18	+0.010	0.023	−0.008	+0.018	0.041	−0.010
18	24	+0.013	0.028	−0.010	+0.020	0.048	−0.013
24	30	+0.013	0.030	−0.010	+0.020	0.048	−0.013
30	40	+0.015	0.033	−0.010	+0.025	0.061	−0.015
40	50	+0.015	0.036	−0.010	+0.025	0.061	−0.015
50	65	+0.018	0.046	−0.013	+0.030	0.069	−0.018
65	80	+0.018	0.048	−0.013	+0.030	0.074	−0.018
80	100	+0.023	0.061	−0.015	+0.036	0.094	−0.023

VALUES IN THOUSANDTHS OF AN INCH

Nominal Size Range Inches		Class FN1 Light Drive Fit			Class FN2 Medium Drive Fit		
		Hole Tol. GR6	Maximum Interference	Shaft Tol. GR5	Hole Tol. GR7	Maximum Interference	Shaft Tol. GR6
Over	To	−0		+0	−0		+0
0	.12	+0.25	0.5	−0.15	+0.4	0.85	−0.25
.12	.24	+0.3	0.6	−0.2	+0.5	1.0	−0.3
.24	.40	+0.4	0.75	−0.25	+0.6	1.4	−0.4
.40	.56	+0.4	0.8	−0.3	+0.7	1.6	−0.4
.56	.71	+0.4	0.9	−0.3	+0.7	1.6	−0.4
.71	.95	+0.5	1.1	−0.4	+0.8	1.9	−0.5
.95	1.19	+0.5	1.2	−0.4	+0.8	1.9	−0.5
1.19	1.58	+0.6	1.3	−0.4	+1.0	2.4	−0.6
1.58	1.97	+0.6	1.4	−0.4	+1.0	2.4	−0.6
1.97	2.56	+0.7	1.8	−0.5	+1.2	2.7	−0.7
2.56	3.15	+0.7	1.9	−0.5	+1.2	2.9	−0.7
3.15	3.94	+0.9	2.4	−0.6	+1.4	3.7	−0.9

Note: Metric sizes were not available at time of publication. Values have been soft converted for problem solving.

Table 46 Force and shrink fits.

VALUES IN MILLIMETRES

Class FN3 Heavy Drive Fit			Class FN4 Shrink Fit			Class FN5 Heavy Shrink Fit		
Hole Tol. H7	Maximum Interference	Shaft Tol. t6	Hole Tol. H7	Maximum Interference	Shaft Tol. u6	Hole Tol. H8	Maximum Interference	Shaft Tol. x7
−0		+0	−0		+0	−0		+0
			+0.010	0.024	−0.006	+0.015	0.033	−0.010
			+0.013	0.030	−0.007	+0.018	0.043	−0.013
			+0.015	0.041	−0.010	+0.023	0.051	−0.015
			+0.018	0.046	−0.010	+0.025	0.058	−0.018
			+0.018	0.046	−0.010	+0.025	0.064	−0.018
			+0.020	0.053	−0.013	+0.030	0.076	−0.020
+0.020	0.053	−0.013	+0.020	0.058	−0.013	+0.030	0.084	−0.020
+0.025	0.066	−0.015	+0.025	0.079	−0.015	+0.041	0.102	−0.025
+0.025	0.071	−0.015	+0.025	0.086	−0.015	+0.041	0.127	−0.025
+0.030	0.082	−0.018	+0.030	0.107	−0.018	+0.046	0.157	−0.030
+0.030	0.094	−0.018	+0.030	0.119	−0.018	+0.046	0.183	−0.030
+0.036	0.112	−0.023	+0.036	0.150	−0.023	+0.056	0.213	−0.036

VALUES IN THOUSANDTHS OF AN INCH

Class FN3 Heavy Drive Fit			Class FN4 Shrink Fit			Class FN5 Heavy Shrink Fit		
Hole Tol. GR7	Maximum Interference	Shaft Tol. GR6	Hole Tol. GR7	Maximum Interference	Shaft Tol. GR6	Hole Tol. GR8	Maximum Interference	Shaft Tol. GR7
−0		+0	−0		+0	−0		+0
			+0.4	0.95	−0.25	+0.6	1.3	−0.4
			+0.5	1.2	−0.3	+0.7	1.7	−0.5
			+0.6	1.6	−0.4	+0.9	2.0	−0.6
			+0.7	1.8	−0.4	+1.0	2.3	−0.7
			+0.7	1.8	−0.4	+1.0	2.5	−0.7
			+0.8	2.1	−0.5	+1.2	3.0	−0.8
+0.8	2.1	−0.5	+0.8	2.3	−0.5	+1.2	3.3	−0.8
+1.0	2.6	−0.6	+1.0	3.1	−0.6	+1.6	4.0	−1.0
+1.0	2.8	−0.6	+1.0	3.4	−0.6	+1.6	5.0	−1.0
+1.2	3.2	−0.7	+1.2	4.2	−0.7	+1.8	6.2	−1.2
+1.2	3.7	−0.7	+1.2	4.7	−0.7	+1.8	7.2	−1.2
+1.4	4.4	−0.9	+1.4	5.9	−0.9	+2.2	8.4	−1.4

NORTH AMERICAN GAGES								EUROPEAN GAGES					
Ferrous metals, such as galvanized steel, tin plate				Nonferrous metals, such as copper, brass, aluminum		Steel and iron wire and bare copper piano wire		Galvanized steel, tin plate, copper, strip steel and steel, copper and aluminum tubes				Nonferrous	
U.S. Standard (USS)		U.S. Standard (Revised) Formerly Manufactures Standard		American Standard or Brown and Sharpe (B & S)		American Steel and Wire (Steel W.G.)		Birmingham (BWG)		New Birmingham (BG)		Imperial Wire Gage Imperial Standard (SWG)	
Gage	mm	Gage	mm	Gage	mm	Gage	mm	Gage	mm	Gage	mm	Gage	mm
		3	6.01	3	5.83								
4	5.95	4	5.70	4	5.19	4	0.33	4	6.05	4	6.35	4	5.89
5	5.56	5	5.31	5	4.62	5	0.36	5	5.59	5	5.65	5	5.39
6	5.16	6	4.94	6	4.12	6	0.41	6	5.16	6	5.03	6	4.88
7	4.76	7	4.55	7	3.67	7	0.46	7	4.57	7	4.48	7	4.47
8	4.37	8	4.18	8	3.26	8	0.51	8	4.19	8	3.99	8	4.06
9	3.97	9	3.80	9	2.91	9	0.56	9	3.76	9	3.55	9	3.66
10	3.57	10	3.42	10	2.59	10	0.61	10	3.40	10	3.18	10	3.25
11	3.18	11	3.04	11	2.30	11	0.66	11	3.05	11	2.83	11	2.95
12	2.78	12	2.66	12	2.05	12	0.74	12	2.77	12	2.52	12	2.64
13	2.38	13	2.78	13	1.83	13	0.79	13	2.41	13	2.24	13	2.34
14	1.98	14	1.90	14	1.63	14	0.84	14	2.11	14	1.99	14	2.03
15	1.79	15	1.71	15	1.45	15	0.89	15	1.83	15	1.78	15	1.83
16	1.59	16	1.52	16	1.29	16	0.94	16	1.65	16	1.59	16	1.63
17	1.43	17	1.37	17	1.15	17	0.99	17	1.47	17	1.41	17	1.42
18	1.27	18	1.21	18	1.02	18	1.04	18	1.25	18	2.58	18	1.22
19	1.11	19	1.06	19	0.91	19	1.09	19	1.07	19	1.19	19	1.02
20	0.95	20	0.91	20	0.81	20	1.14	20	0.89	20	1.00	20	0.91
21	0.87	21	0.84	21	0.72	21	1.19	21	0.81	21	0.89	21	0.81
22	0.79	22	0.76	22	0.65	22	1.24	22	0.71	22	0.79	22	0.71
23	0.71	23	0.68	23	0.57	23	1.30	23	0.64	23	0.71	23	0.61
24	0.64	24	0.61	24	0.51	24	1.40	24	0.56	24	0.63	24	0.56
25	0.56	25	0.53	25	0.46	25	1.50	25	0.51	25	0.56	25	0.51
26	0.48	26	0.46	26	0.40	26	1.60	26	0.46	26	0.50	26	0.46
27	0.44	27	0.42	27	0.36	27	1.70	27	0.41	27	0.44	27	0.42
28	0.40	28	0.38	28	0.32	28	1.80	28	0.36	28	0.40	28	0.38
29	0.36	29	0.34	29	0.29	29	1.90	29	0.33	29	0.35	29	0.35
30	0.32	30	0.31	30	0.25	30	2.03	30	0.31	30	0.31	30	0.32
31	0.28	31	0.27	31	0.23	31	2.16	31	0.25	31	0.28		
32	0.26	32	0.25	32	0.20	32	2.29	32	0.23			32	0.27
33	0.24	33	0.23	33	0.18	33	2.41	33	0.20	33	0.22	33	0.25
34	0.22	34	0.21	34	0.16	34	2.54	34	0.18	34	0.20	34	0.23
						35	2.69	35	0.13	35	0.18	35	0.21
36	0.18	36	0.17	36	0.13	36	2.84	36	0.10	36	0.16		
						37	3.00					37	0.17
38	0.16	38	0.15	38	0.10	38	3.15			38	0.12	38	0.15
						39	3.30						
						40	3.51			40	0.10	40	0.12
						41	3.71					42	0.10

Note: Metric standards governing gage sizes were not available at the time of publication. The sizes given in the above chart are "soft conversion" from current inch standards and are not meant to be representative of the precise metric gage sizes which may be available in the future. Conversions are given only to allow the student to compare gage sizes readily with the metric drill sizes.

Table 47 Wire and sheet-metal gages and thicknesses.

Millimetres Thickness	FLAT SHEET				
	Cold Rolled Sheet	Hot Rolled Sheet	Galvanized Steel Sheet	Aluminum Coated Sheet	Mass in kg/m²
12					94.2
11					86.3
10		•			78.5
9		•			70.6
8		•			62.8
7.5		•			58.9
7		•			54.9
6.5		•			51.0
6		•			47.1
5.5		•			43.2
5		•			39.2
4.8		•			37.7
4.5		•			35.3
4.2		•			33.0
4		•	•		31.4
3.8		•	•		29.8
3.6		•	•		28.6
3.5		•	•		27.5
3.4		•	•		26.7
3.2	•	•	•		25.1
3	•	•	•		23.5
2.8	•	•	•		22.0
2.6	•	•	•		20.4
2.5	•	•	•		19.6
2.4	•	•	•		18.8
2.2	•	•	•		17.3
2.1	•	•	•		16.5
2	•	•	•		15.7
1.9	•	•	•		14.9
1.8	•	•	•		14.1
1.7	•	•	•		13.3
1.6	•	•	•		12.6
1.5	•		•	•	11.8
1.4	•		•	•	11.0
1.3	•		•	•	10.2
1.2	•		•	•	9.4
1.1	•		•	•	8.6
1.05	•		•	•	8.2
1.0	•		•	•	7.8
0.95	•		•	•	7.5
0.9	•		•	•	7.1
0.85	•		•	•	6.7
0.8	•		•	•	6.3
0.75	•		•	•	5.9
0.7	•		•	•	5.5
0.65	•		•	•	5.1
0.6	•		•	•	4.7
0.55	•		•	•	4.3
0.5	•		•	•	3.9
0.45	•		•		3.5
0.4	•		•		3.1
0.35					2.7

Size mm	Area mm²		Mass kg/m²	
	Round	Square	Round	Square
1.6	2.01	2.56	0.016	0.020
1.8	2.54	3.24	0.020	0.025
2	3.14	4	0.025	0.031
2.2	3.8	4.84	0.030	0.038
2.5	4.91	6.25	0.039	0.049
2.8	6.16	7.84	0.048	0.061
3	7.07	9	0.055	0.071
3.5	9.62	12.25	0.075	0.096
4	12.57	16	0.099	0.125
4.5	15.9	20.25	0.125	0.159
5	19.64	25	0.154	0.196
5.5	23.76	30.25	0.186	0.237
6	28.27	36	0.222	0.282
6.5	33.18	42.25	0.260	0.331
7	38.48	49	0.302	0.384
7.5	44.18	56.25	0.346	0.441
8	50.27	64	0.394	0.502
9	63.62	81	0.499	0.635
10	78.54	100	0.616	0.784
11	95.03	121	0.745	0.949
12	113.1	144	0.887	1.129
13	132.73	169	1.041	1.325
14	153.94	196	1.207	1.537
15	176.72	225	1.386	1.764
16	201.06	256	1.577	2.007
17	226.98	289	1.780	2.266
18	254.47	324	1.996	2.541
19	283.53	361	2.223	2.831
20	314.16	400	2.464	3.137
21	346.36	441	2.716	3.458
22	380.13	484	2.981	3.795
23	415.48	529	3.258	4.148
24	452.39	576	3.548	4.517
25	490.88	625	3.850	4.901

Table 48 Mass and areas of flat, square, and round steel

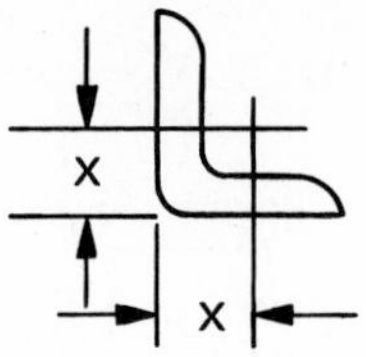

X
X

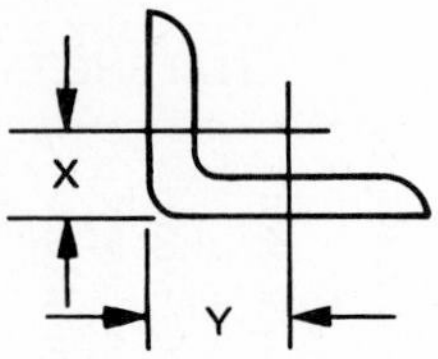

X
Y

EQUAL ANGLES			UNEQUAL ANGLES			
Section Size	Mass Per Metre	Centroidal Axis X	Section Size	Mass Per Metre	Centroidal Axis X	Centroidal Axis Y
mm	kg	mm	mm	kg	mm	mm
200 x 200 x 20	59	57.4	200 x 125 x 20	47.9	31.5	69
16	48.2	55.9	16	38.8	30	67.5
13	39.5	54.8	13	31.8	28.9	66.4
150 x 150 x 20	44	44.8	150 x 100 x 20	36.1	27.4	52.4
16	35.7	43.4	16	29.4	25.9	50.9
13	29.3	42.3	13	24.2	24.9	49.9
10	22.8	41.2	10	18.8	23.8	48.8
125 x 125 x 16	29.4	37.1	125 x 90 x 16	25	24.7	42.2
13	24.2	36	13	20.6	23.7	41.2
10	18.8	34.9	10	16.1	22.6	40.1
8	15.2	34.2	8	13	21.8	39.3
100 x 100 x 13	19.1	29.8	100 x 75 x 13	16.5	20.9	33.4
10	14.9	28.7	10	13	19.8	32.3
8	12.1	28	8	10.5	19	31.5
6	9.14	27.2	6	7.96	18.3	30.8
90 x 90 x 13	17	27.2	90 x 75 x 16	18.7	22.8	30.3
10	13.3	26.2	13	15.5	21.8	29.3
8	10.8	25.5	10	12.2	20.7	28.2
6	8.2	24.7	8	9.86	20	27.5
75 x 75 x 13	14	23.5	6	7.49	19.3	26.8
10	11	22.4	90 x 65 x 16	17.5	19.5	32
8	8.92	21.7	13	14.5	18.4	30.9
6	6.78	21	10	11.4	17.3	29.8
65 x 65 x 10	9.42	19.9	8	9.23	16.6	29.1
8	7.66	19.2	6	7.12	15.9	28.4
6	5.84	18.5	75 x 50 x 13	11.4	14.8	27.3
50 x 50 x 10	7.07	16.1	10	9.03	13.7	26.2
8	5.78	15.4	8	7.35	13	25.5
6	4.43	14.7	6	5.6	12.2	24.7
4	3.01	14	65 x 45 x 10	7.85	12.9	22.9
45 x 45 x 8	5.15	14.2	8	6.41	12.2	22.2
6	3.96	13.4	6	4.9	11.4	21.4
4	2.7	12.7	4	3.33	10.7	20.7
35 x 35 x 8	3.89	11.6	50 x 35 x 8	4.84	10.1	17.6
6	3.01	10.9	6	3.72	9.42	16.9
4	2.07	10.2	4	2.54	8.7	16.2
25 x 25 x 6	2.07	8.4	45 x 25 x 6	3.01	6.71	16.7
4	1.44	7.7	4	2.07	5.98	16

Table 49 Angle iron sizes

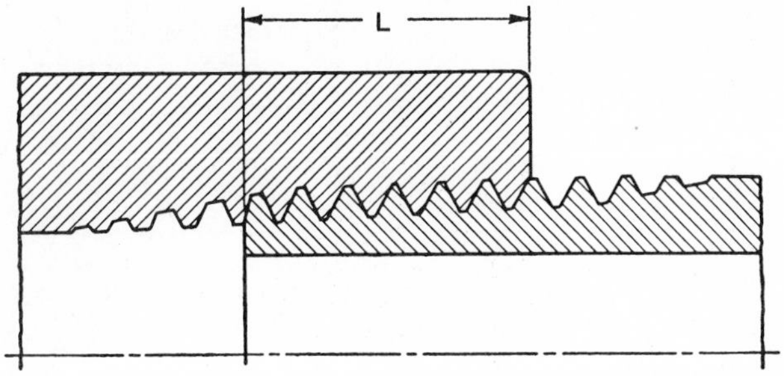
L

Nominal Pipe Size Inches	Outside Diameter mm	Wall Thickness (mm)			Approx. Distance Pipe Enters Fitting L	Mass (kg/m)		
		Standard	Extra Strong	Double Extra Strong		Standard	Extra Strong	Double Extra Strong
⅛ (.125)	10.3	1.7	2.4	—	5	0.36	0.46	—
¼ (.250)	13.7	2.9	3.1	—	7	0.63	0.80	—
⅜ (.375)	17.1	2.4	3.3	—	8	0.85	1.10	—
½ (.500)	21.3	2.8	3.8	7.8	10	1.26	1.62	2.54
¾ (1.00)	26.7	2.9	4	8.1	11	1.68	2.19	3.63
1.00	33.4	3.5	4.6	9.4	13	2.50	3.23	5.45
1.25	42.1	3.6	5	10	14	3.38	4.46	7.75
1.50	48.3	3.8	5.2	10.4	14	4.05	5.40	9.54
2.00	60.3	4	5.7	11.4	15	5.43	7.47	13.44
2.50	73	5.3	7.2	14.4	22	8.91	11.40	20.39
3.00	88.9	5.6	7.8	15.6	24	11.28	15.25	27.65
3.50	101.6	5.9	8.3	16.5	25	13.54	18.62	34.00
4	114.3	6.1	8.7	17.5	27	16.06	22.30	40.98
5	141.3	6.7	9.7	19.5	29	21.74	30.92	57.37
6	168.3	7.3	11.2	22.5	32	28.23	42.52	79.11
8	219	8.4	13	22.7	38	42.49	64.57	107.77

Table 50 American Standard wrought iron pipe

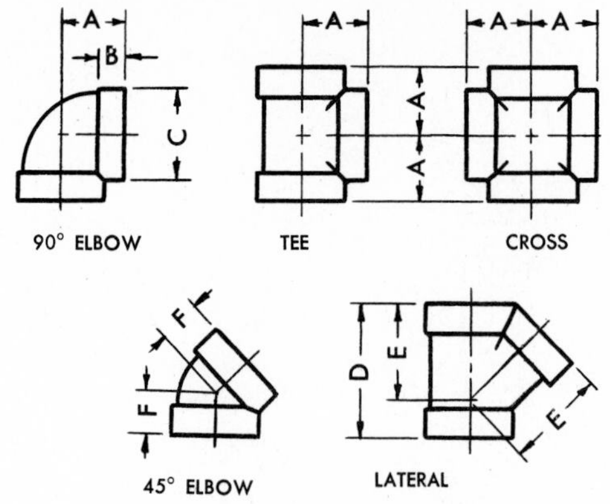
A
B
C
90° ELBOW
A
A
A
TEE
A
A
CROSS
F
F
45° ELBOW
D
E
E
LATERAL

DIMENSIONS IN MILLIMETRES						
Nominal Pipe Size in Inches	A	Min B	Min C	D	E	F
.25	21	10	24	—	—	19
.375	24	11	28	—	—	20
.50	28	13	34	64	47	22
.75	33	14	41	76	57	25
1.00	38	16	50	89	70	28
1.25	44	18	61	108	83	33
1.50	49	19	68	124	97	36
2.00	57	21	83	146	108	43
2.50	69	24	98	171	132	50
3.00	78	25	117	200	155	55
3.50	87	27	132	225	174	61
4.00	96	28	147	248	194	66
5	114	30	179	295	235	77
6	130	33	210	341	273	88
8	167	37	270	430	346	109
10	205	43	333	613	425	131

Table 51 **American standard (125 lb) cast-iron screwed-pipe fittings.**

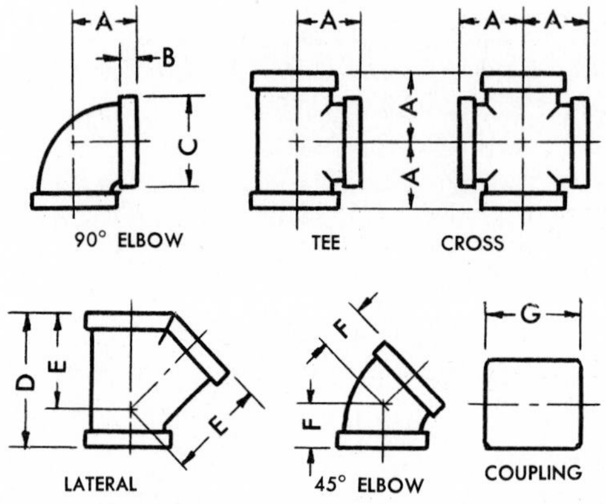
A
B
C
90° ELBOW
A
A
A
TEE
A
A
CROSS
D
E
E
LATERAL
F
F
45° ELBOW
G
COUPLING

DIMENSIONS IN MILLIMETRES							
Nominal Pipe Size in Inches	A	B	C	D	E	F	G
.125	18	5.0	18	—	—	—	24
.250	21	5	21	—	—	19	27
.375	24	6	26	49	36	20	29
.500	28	6	30	59	43	22	34
.750	33	7	37	70	52	25	39
1.00	38	8	45	83	62	28	42
1.25	44	9	55	100	74	33	49
1.50	49	9	62	111	83	36	55
2.00	57	11	75	131	100	43	64
2.50	69	12	91	159	120	50	73
3.00	78	14	109	184	141	55	81
3.50	87	15	123	—	—	61	87
4.00	96	17	137	228	177	66	94
5.00	114	20	167	—	—	77	—
6.00	130	23	197	—	—	88	—

Table 52 **American Standard (150 lb) Malleable-iron screwed-pipe fittings.**

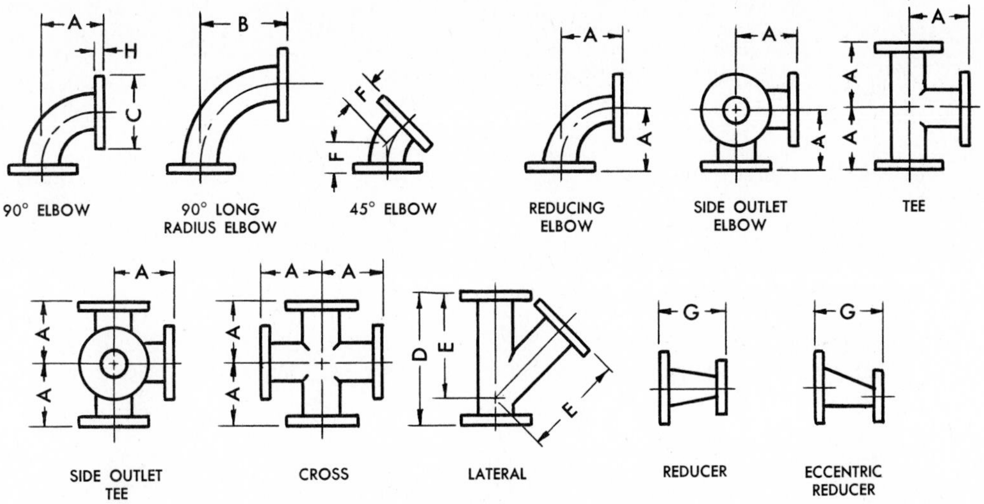
90° ELBOW
90° LONG RADIUS ELBOW
45° ELBOW
REDUCING ELBOW
SIDE OUTLET ELBOW
TEE
SIDE OUTLET TEE
CROSS
LATERAL
REDUCER
ECCENTRIC REDUCER

DIMENSIONS IN MILLIMETRES

Nominal Pipe Size in Inches	A	B	C	D	E	F	G	H
1.00	89	127	108	190	146	45	—	11
1.25	95	140	118	203	159	51	—	13
1.50	102	152	127	229	178	57	—	14
2.00	114	165	152	267	203	64	127	16
2.50	127	178	178	305	241	76	140	18
3.00	140	197	190	330	254	76	152	19
3.50	153	216	216	368	292	89	165	21
4.00	165	229	229	381	305	102	178	24
5.00	190	260	254	432	343	114	203	24
6.00	203	292	280	457	368	127	229	25

Table 53 American Standard flanged fittings.

DIMENSIONS IN MILLIMETRES

Nominal Pipe Size in Inches	Center to End of Short Radius Elbow A	Center to End of Long Radius Elbow B	Center to End of 45° Long Radius Elbow C	Center to End of Tees D	Center to End of Cross E	Length of Reducer Largest End F
1.00	25	38	22	38	38	51
1.25	32	48	25	48	48	51
1.50	38	57	28	57	57	64
2.00	51	76	35	64	64	76
2.50	64	95	44	76	76	89
3.00	76	114	51	86	86	89
3.50	89	133	57	95	95	102
4.00	102	152	64	105	105	102
5.00	127	190	79	124	124	127
6.00	152	229	95	143	143	140

Table 54 American Standard steel butt-welding fittings.

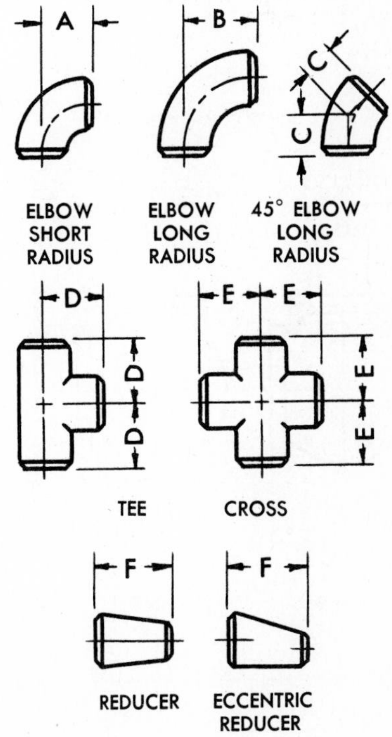
ELBOW SHORT RADIUS
ELBOW LONG RADIUS
45° ELBOW LONG RADIUS
TEE
CROSS
REDUCER
ECCENTRIC REDUCER

CHECK VALVES

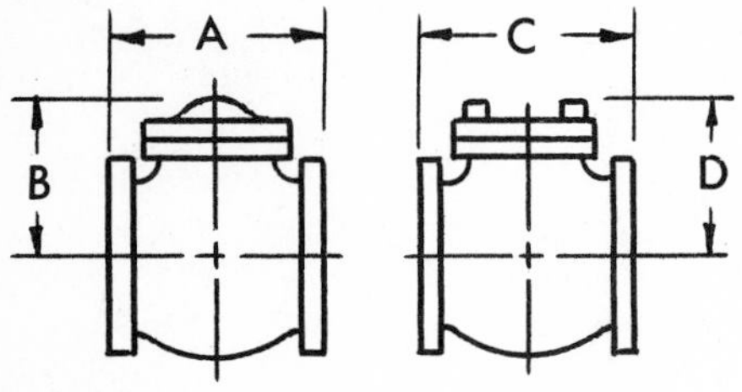
A
C
B
D

LIFT SWING

Nominal Pipe Size in Inches	Lift Check			Swing Check		
	A			C		
	Screwed	Flanged	B	Screwed	Flanged	D
2.00	165	—	89	165	203	108
2.50	178	—	108	178	216	122
3.00	203	241	127	203	241	129
3.50	—	—	—	229	267	148
4.00	254	292	159	254	292	157
5.00	—	330	178	286	330	183
6.00	—	356	210	318	356	190
8.00	—	—	—	—	495	259
10.00	—	—	—	—	622	308

DIMENSIONS IN MILLIMETRES

GATE VALVES

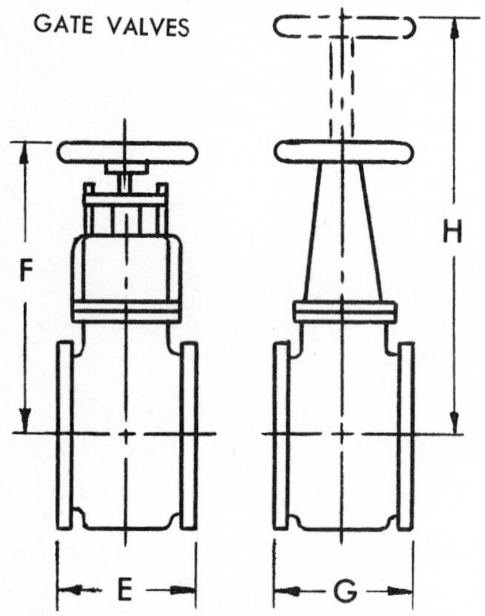
F
H
E
G

Nominal Pipe Size in Inches	Nonrising Spindle			Rising Spindle		
	E			G		
	Screwed	Flanged	F	Screwed	Flanged	H
1.50	121	—	267	121	—	333
2.00	121	178	267	121	178	333
2.50	140	190	284	140	190	368
3.00	152	203	321	152	203	422
3.50	168	216	338	168	216	468
4.00	181	229	387	181	229	535
5.00	206	254	454	206	254	635
6.00	229	267	513	229	267	743
8.00	254	292	610	254	292	946
10.00	—	330	716	—	330	1121

DIMENSIONS IN MILLIMETRES

GLOBE AND ANGLE VALVES

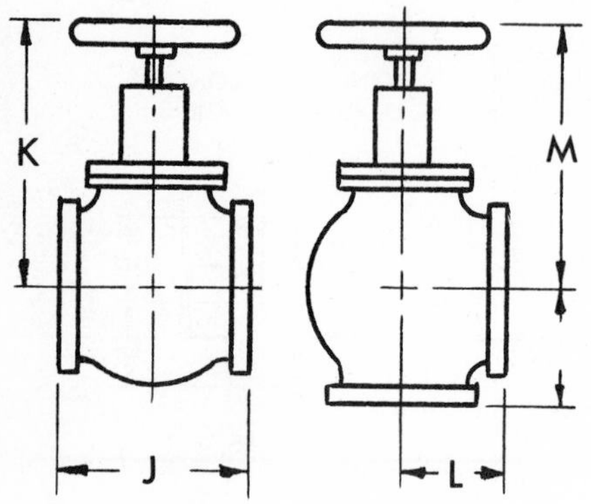
K
M
J
L

GLOBE ANGLE

Nominal Pipe Size in Inches	Globe			Angle		
	J			L		
	Screwed	Flanged	K	Screwed	Flanged	M
2.00	121	178	240	89	98	264
2.50	140	190	281	98	114	281
3.00	152	203	314	119	118	346
3.50	168	216	335	127	136	348
4.00	181	229	387	152	149	378
5.00	206	254	438	160	165	449
6.00	229	267	478	203	203	484
8.00	254	292	562	—	235	578
10.00	—	330	629	—	270	633

Dimensions taken from manufacturer's catalogs for drawing purposes.

Table 55 Common values.

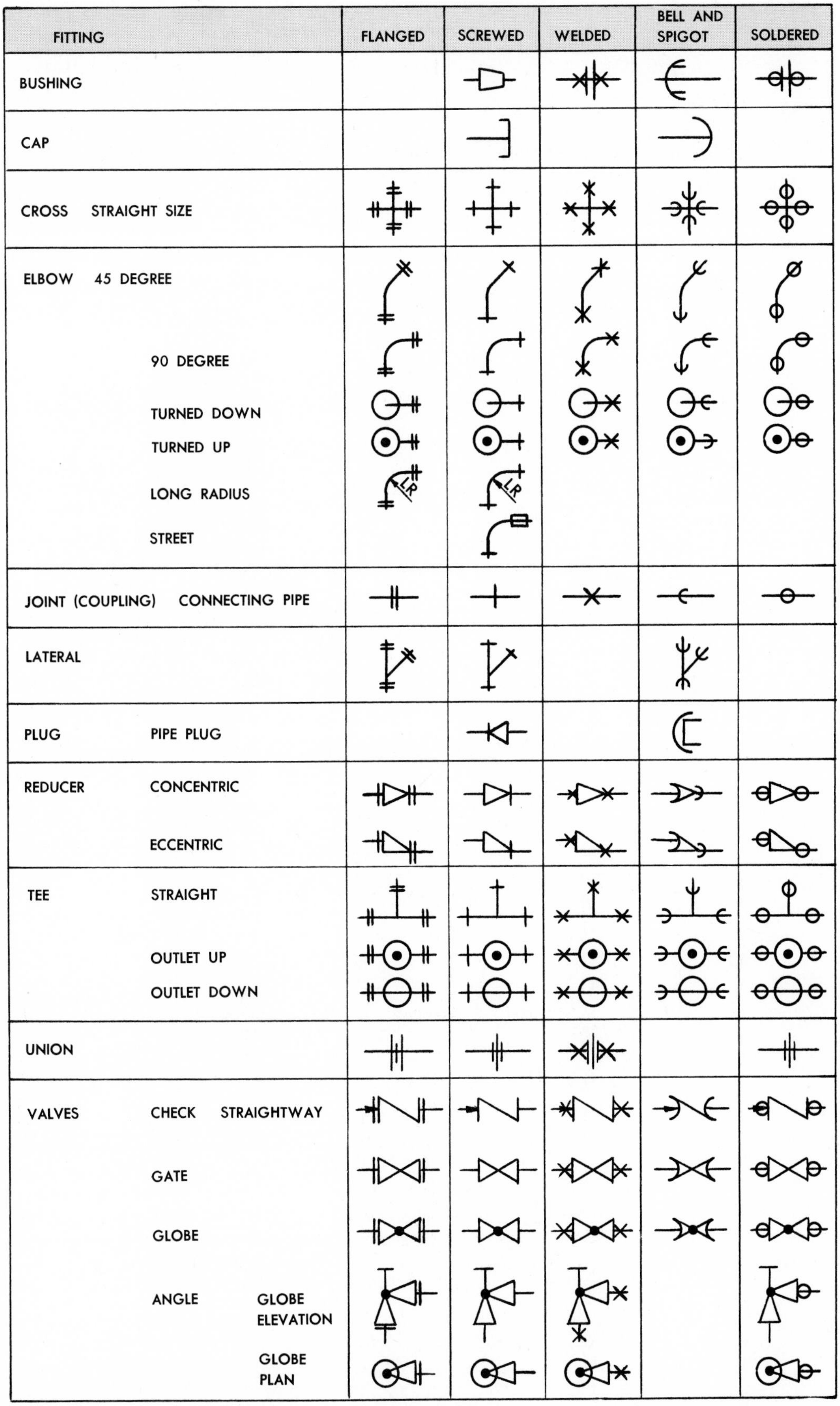
FITTING
FLANGED
SCREWED
WELDED
BELL AND SPIGOT
SOLDERED
BUSHING
CAP
CROSS STRAIGHT SIZE
ELBOW 45 DEGREE
90 DEGREE
TURNED DOWN
TURNED UP
LONG RADIUS
STREET
JOINT (COUPLING) CONNECTING PIPE
LATERAL
PLUG PIPE PLUG
REDUCER CONCENTRIC
ECCENTRIC
TEE STRAIGHT
OUTLET UP
OUTLET DOWN
UNION
VALVES CHECK STRAIGHTWAY
GATE
GLOBE
ANGLE GLOBE ELEVATION
GLOBE PLAN

Table 56 Graphic symbols formerly used for pipe fittings.

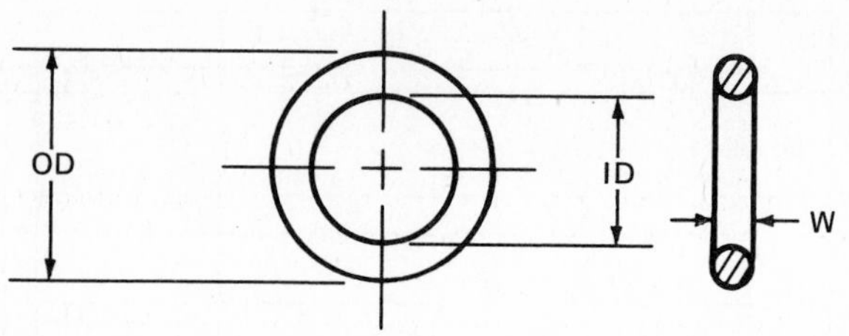
OD
ID
W

	W = 1.5	W = 2	W = 3	W = 4	W = 5	W = 6
ID	OD	OD	OD	OD	OD	OD
3	6					
4	7					
5	8					
6	9					
7	10					
8	11	12	14			
9	12	13	15			
10	13	14	16	18	20	
12	15	16	18	20	22	24
14	17	18	20	22	24	26
15	18	19	21	23	25	27
16	19	20	22	24	26	28
18	21	22	24	26	28	30
20	23	24	26	28	30	32
22	25	26	28	30	32	34
24	27	28	30	32	34	36
25	28	29	31	33	35	37
26	29	30	32	34	36	38
28	31	32	34	36	38	40
30	33	34	36	38	40	42
32	35	36	38	40	42	44
34	37	38	40	42	44	46
35	38	39	41	43	45	47
36	39	40	42	44	46	48
38	41	42	44	46	48	50
40	43	44	46	48	50	52
42	45	46	48	50	52	54
44	47	48	50	52	54	56
45	48	49	51	53	55	57
46	49	50	52	54	56	58
48	51	52	54	56	58	60
50	53	54	56	58	60	62
52	55	56	58	60	62	64
54	57	58	60	62	64	66
55	58	59	61	63	65	67
56	59	60	62	64	66	68
58	61	62	64	66	68	70
60	63	64	66	68	70	72
65	68	69	71	73	75	77
70	73	74	76	78	80	82
75	78	79	81	83	85	87
80	83	84	86	88	90	92

Table 57 O rings

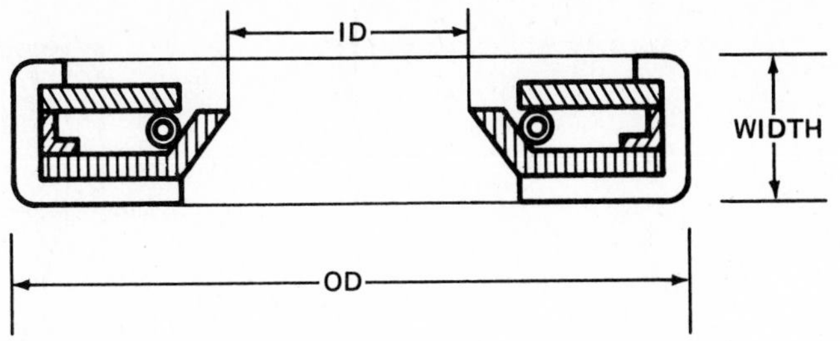
ID
WIDTH
OD

Inside Dia		Outside Dia		Width	
mm	in.	mm	in.	mm	in.
10	.375	19	.753	6	.25
10	.375	21	.840	8	.31
11	.438	26	1.003	8	.31
11	.438	28	1.128	8	.31
12	.500	26	1.003	8	.31
12	.500	28	1.128	8	.31
12	.500	32	1.254	10	.38
14	.562	26	1.003	8	.31
		28	1.128	8	.31
14	.562	32	1.250	10	.38
16	.625	28	1.128	8	.31
16	.625	32	1.250	10	.38
16	.625	35	1.379	10	.38
18	.688	28	1.128	8	.31
18	.688	32	1.254	10	.38
18	.688	35	1.379	10	.38
20	.750	32	1.254	10	.38
20	.750	35	1.379	10	
20	.750	38	1.503	10	.38
20	.750	44	1.756	12	.44
21	.8125	32	1.254	10	.38
21	.8125	35	1.379	10	.38
21	.8125	38	1.503	10	.38
21	.8125	44	1.756	12	.44
22	.875	35	1.379	10	.38
22	.875	38	1.503	10	.38
22	.875	42	1.628	12	.44
22	.875	44	1.756	12	.44
24	.938	38	1.503	10	.38
24	.938	42	1.628	10	.38
24	.938	44	1.756	12	.44
25	1.000	38	1.503	10	.38
25	1.000	44	1.756	12	.44
25	1.000	48	1.878	12	.44
25	1.000	50	2.004	12	.44
25	1.000				
26	1.062	38	1.503	10	.38
26		42	1.628	12	.44
26		44	1.756	12	.44
28	1.125	42	1.628	12	.44
28		44	1.756	12	.44
28		50	1.987	12	.50
30	1.188	46	1.832	12	.44
30		50	1.987	12	.50
30		58	2.254	12	.50
32	1.25	44	1.756	12	.44
32		48	1.878	12	.44
32		52	2.066	12	.50
34	1.312	52	2.060	12	.44
34		58	2.254	12	.50
34		60	2.378	12	.50
35	1.375	52	2.066	12	.44
35		58	2.254	12	.50
35		62	2.441	12	.50
36	1.438	58	2.254	12	.50
36		64	2.506	12	.50
36		66	2.627	12	.50
38	1.500	58	2.254	10	.38
38		62	2.410	12	.50
38		70	2.720	12	.50
40	1.562	62	2.441	12	.50
40		68	2.690	12	.50
40		74	2.879	12	.50
42	1.625	62	2.441	10	.38
42		74	2.879	10	.38
42		66	2.627	12	.50
42		74	2.879	12	.50
42		78	3.066	12	.50
44	1.75	58	2.254	12	.50
44		62	2.441	12	.50
44		64	2.506	12	.50

Table 58 Oil seals.

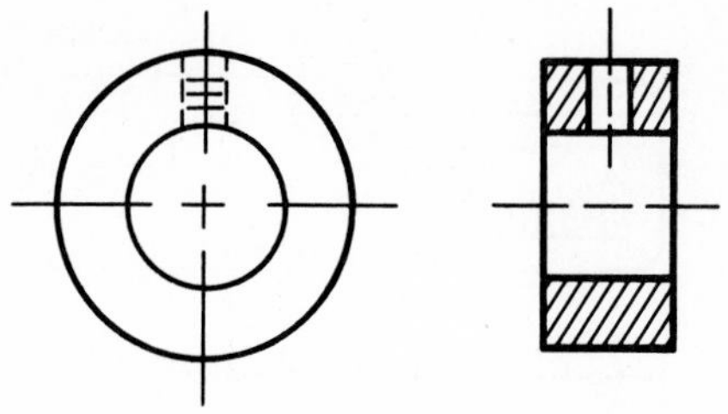

Shaft Dia		Outside Dia		Width		Size of Set Screw	
mm	in · lb	mm	in · lb	mm	in · lb	mm	in · lb
10	.375	20	.75	10	.40	M6	.250
12	.500	25	1.00	10	.44	M6	.250
16	.625	28	1.10	12	.50	M8	.3125
20	.750	30	1.20	14	.56	M8	.3125
22	.875	40	1.50	14	.56	M8	.3125
24	1.000	40	1.60	16	.60	M8	.3125

Table 59 Setscrew collars.

	Bolt Size									
Bolt Length	M16 × 2		M20 × 2.5		M24 × 3		M30 × 3.5		M36 × 4	
	Min	Max	Min	Max	Min	Max	Min	Max	Min	Max
45	12	29								
50	17	34								
55	22	39	15	35						
60	27	44	20	40						
65	32	49	25	45	20	41				
70	37	54	30	50	25	46				
75	42	59	35	55	30	51	22	45		
80	47	64	40	60	35	56	27	50		
85	52	69	45	65	40	61	32	55	25	49
90	57	74	50	70	45	66	37	60	30	54
95	62	79	55	75	50	71	42	65	35	59
100	67	84	60	80	55	76	47	70	40	64
110	72	94	65	90	60	86	52	80	45	74
120	82	104	75	100	70	96	62	90	55	84
130	92	114	85	110	80	106	72	100	65	94
140	102	124	95	120	90	116	82	110	75	104
150	112	134	105	130	100	126	92	120	85	114
160	122	144	115	140	110	136	102	130	95	124
170	132	154	125	150	120	146	112	140	105	134

Table 60 Minimum and maximum grips for metric bolts. (CISC)

	KILOGRAMS PER 100 UNITS				
Bolt Length	M16	M20	M24	M30	M36
45	23.5	36.8	53	83	121
50	24.5	38.2	55	86	125
55	25.4	39.6	57	89	130
60	26.3	41.0	59	92	135
65	27.1	42.4	61	95	139
70	28.0	43.8	63	98	144
75	28.9	45.2	65	101	148
80	29.8	46.6	67	104	153
85	30.7	48.0	69	108	157
90	31.6	49.3	71	111	162
95	32.5	50.7	73	114	166
100	33.4	52.1	75	117	171
110	35.2	54.9	79	123	180
120	36.9	57.7	83	129	189
130	38.7	60.5	87	136	198
140	40.5	63.2	91	142	207
150	42.3	66.0	95	148	217
160	44.1	68.8	99	154	226
170	45.8	71.6	103	161	235
Plain Round Washer	3.0	4.7	6.7	10.5	15.3
Hex Nut	8.6	13.4	19.3	30	44

Table 61 Approximate mass of bolts, washers, and hexagon nuts.

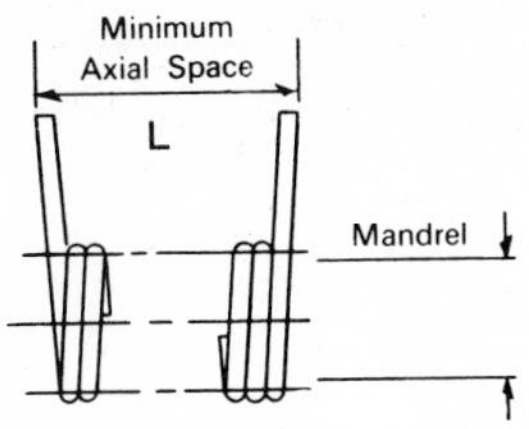
Minimum
Axial Space
L
Mandrel

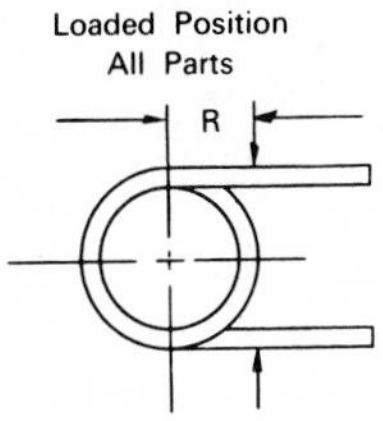
Loaded Position
All Parts
R

Figures show springs wound left-hand

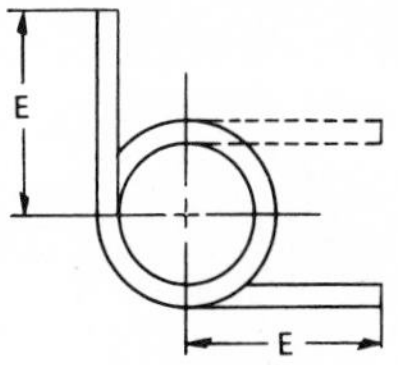
E
E

Fig. 1
90° Deflection

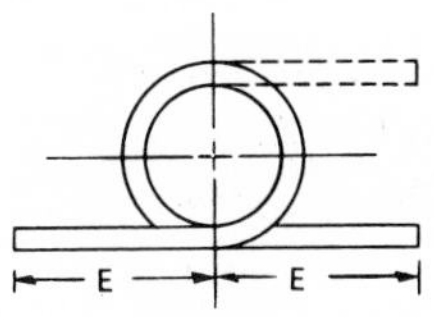
E
E

Fig. 2
180° Deflection

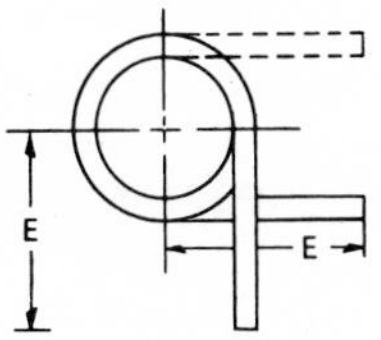
E
E

Fig. 3
270° Deflection

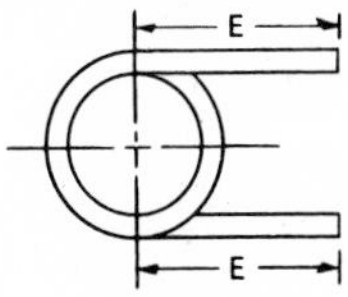
E
E

Fig. 4
360° Deflection

Wire Dia	Outside Dia	Pos. of Ends — Fig.	Def, Deg.	Torque — Kg = mm	Radius R	Suggested Mandrel Size	E	Min. Axial Space — L
0.35	3.2	1	90	0.81	6.4	1.6	13	1.7
	3.4	2	180		6.4	2.0	13	2.7
	3.2	3	270		6.4	1.6	13	4.0
	4.9	2	180		9.5	2.8	19	2.0
	5.1	3	270		9.5	3.2	19	2.6
	5.2	4	360		9.5	3.2	19	3.2
0.43	4.1	1	90	1.35	6.4	2.4	13	2.1
	4.4	2	180		6.4	2.4	13	3.2
	4.1	3	270		6.4	2.4	13	4.9
	6.3	2	180		9.5	4.0	19	2.4
	6.6	3	270		9.5	4.0	19	3.1
	6.0	4	360		9.5	3.6	19	4.3
0.51	4.9	1	90	2.15	9.5	2.8	19	2.4
	4.6	2	180		9.5	2.8	19	4.3
	4.5	3	270		9.5	2.4	19	6.2
	6.2	2	180		12.7	3.6	25	3.3
	6.8	3	270		12.7	4.4	25	4.2
	6.5	4	360		12.7	4.0	25	6.4
0.59	5.2	1	90	3.54	9.5	2.8	19	2.8
	4.9	2	180		9.5	2.8	19	5.0
	4.8	3	270		9.5	2.8	19	7.2
	6.6	2	180		12.7	4.0	25	3.8
	6.4	3	270		12.7	4.0	25	5.4
	6.9	4	360		12.7	4.4	25	6.5
0.71	6.8	1	90	5.93	12.7	4.0	25	3.4
	6.3	2	180		12.7	3.6	25	6.0
	6.2	3	270		12.7	3.6	25	8.8
	8.7	2	180		12.7	5.2	25	4.6
	8.4	3	270		12.7	5.2	25	6.6
	9.0	4	360		12.7	5.6	25	7.9

Table 62 Torsion springs. (Wallace Barnes Co. Ltd.)

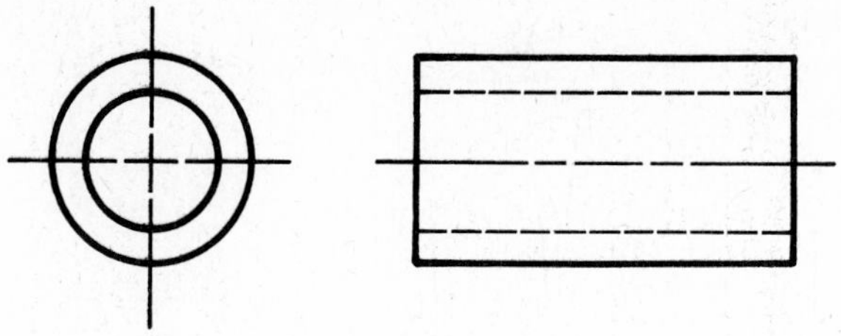

Inside Dia		Outside Dia		Length—Millimetres and Inches								
mm	in.	mm	in.	15 .50	20 .75	25 1.00	30 1.25	35 1.50	40 1.75	50 2.00	60 2.50	75 3.00
10	.375	16	.625	•	•	•	•					
		20	.750	•	•	•	•					
12	.500	16	.625	•	•	•	•	•				
		20	.750	•	•	•	•	•	•	•		
16	.625	22	.875	•	•	•	•	•	•	•		
		25	1.000	•	•	•	•	•	•	•		
20	.750	25	1.000	•	•	•	•	•	•	•		
		28	1.125	•	•	•	•	•	•	•		
22	.875	28	1.125		•	•	•	•	•	•	•	
		32	1.250		•	•	•	•	•	•	•	
25	1.000	32	1.250			•	•	•	•	•	•	•
		35	1.375			•	•	•	•	•	•	•

Table 63 Standard plain (journal) bearings.

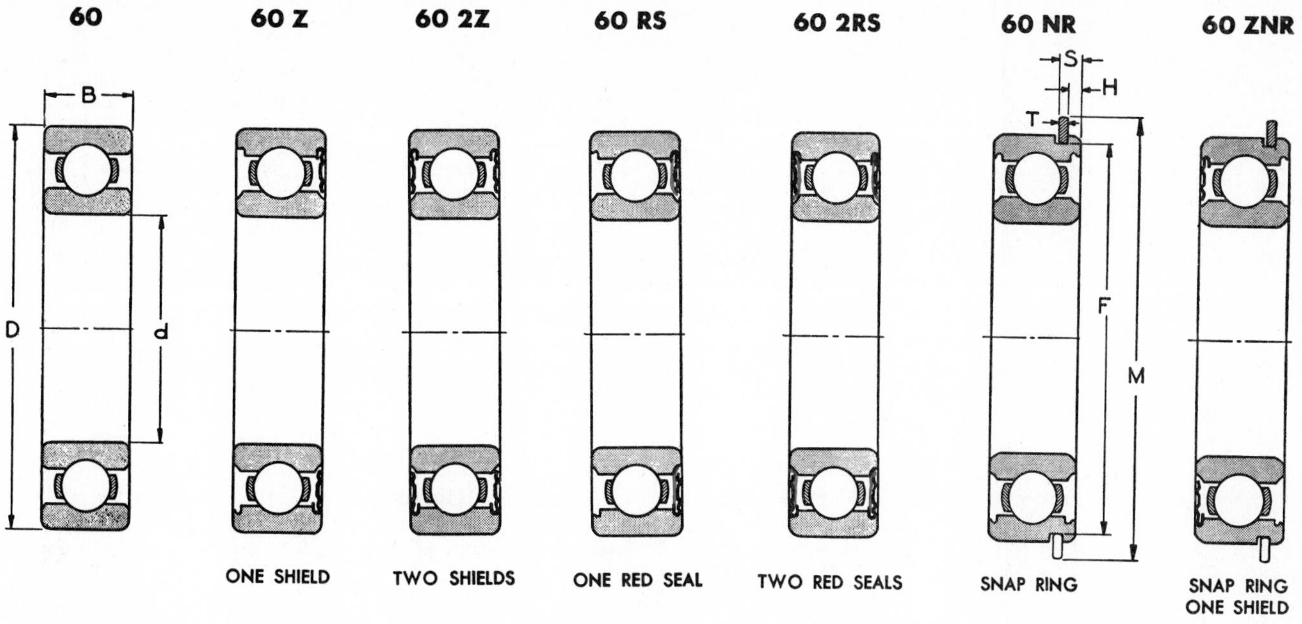
60
60 Z
60 2Z
60 RS
60 2RS
60 NR
60 ZNR
B
D
d
S
H
T
F
M
ONE SHIELD
TWO SHIELDS
ONE RED SEAL
TWO RED SEALS
SNAP RING
SNAP RING ONE SHIELD

Bearing Number							Nominal Bearing Dimensions										
							d		D		B		F	M	S	H	T
							mm	in.	mm	in.	mm	in.	Millimetres				
6000	6000 Z	6000 2Z	6000 RS	6000 2RS			10	.3937	26	1.0236	8	.3150					
6001	6001 Z	6001 2Z	6001 RS	6001 2RS			12	.4724	28	1.1024	8	.3150					
6002	6002 Z	6002 2Z	6002 RS	6002 2RS	6002 NR	6002 ZNR	15	.5906	32	1.2598	9	.3543	30	36.5	3	2	1
6003	6003 Z	6003 2Z	6003 RS	6003 2RS			17	.6693	35	1.3780	10	.3937					
6004	6004 Z	6004 2Z	6004 RS	6004 2RS			20	.7874	42	1.6535	12	.4724					
6005	6005 Z	6005 2Z	6005 RS	6005 2RS			25	.9843	47	1.8504	12	.4724					
6006	6006 Z	6006 2Z	6006 RS	6006 2RS			30	1.1811	55	2.1654	13	.5118					
6007	6007 Z	6007 2Z	6007 RS	6007 2RS			35	1.3780	62	2.4409	14	.5512					
6008	6008 Z	6008 2Z	6008 RS	6008 2RS			40	1.5748	68	2.6772	15	.5906					
6009							45	1.7717	75	2.9528	16	.6299					
6010	6010 Z	6010 2Z	6010 RS	6010 2RS	6010 NR	6010 ZNR	50	1.9685	80	3.1496	16	.6299	76.8	86	4	2.5	1.6
6011	6011 Z	6011 2Z	6011 RS	6011 2RS	6011 NR	6011 ZNR	55	2.1654	90	3.5433	18	.7087	86.8	96.5	5	3	2.4
6012							60	2.3622	95	3.7402	18	.7087					
6013	6013 Z	6013 2Z			6013 NR	6013 ZNR	65	2.5591	100	3.9370	18	.7087	96.8	106	5	3	2.4
6014							70	2.7559	110	4.3307	20	.7874					
6015	6015 Z	6015 2Z			6015 NR	6015 ZNR	75	2.9528	115	4.5276	20	.7874	112	121	5	3	2.4

Note: When shield is required on same side as snap ring, order as ZNBR.

Table 64 Radial ball bearings. (Canadian SKF Co. Ltd.)

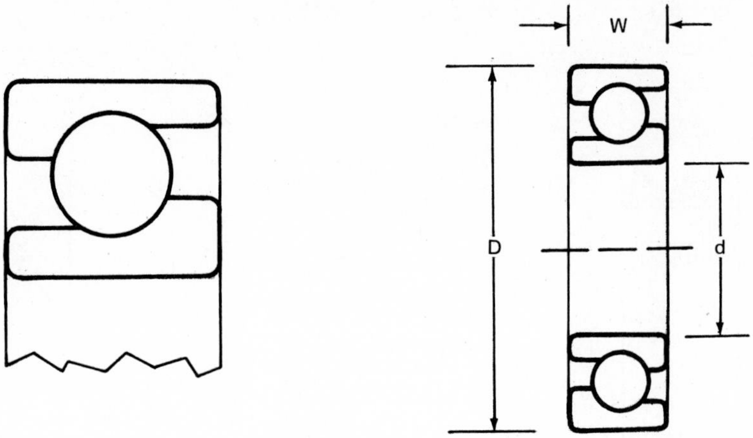
W
D
d

Series	Bearing Number	Bore d (mm)	Bore d (in.)	Outside Diameter D (mm)	Outside Diameter D (in.)	Width W (mm)	Width W (in.)	Basic Load Rating (kN)	Basic Load Rating (Kips)
Light	7205B	25	0.984	52	2.047	15	0.591	11.4	2.5
	7206B	30	1.181	62	2.441	16	0.630	15.6	3.4
	7207B	35	1.378	72	2.835	17	0.670	20.8	4.6
	7208B	40	1.575	80	3.150	18	0.709	24.5	5.5
	7209B	45	1.772	85	3.347	19	0.748	27.5	6.2
	7210B	50	1.969	90	3.543	20	0.787	28.5	6.4
	7211B	55	2.165	100	3.937	21	0.827	36	8.2
	7212B	60	2.362	110	4.331	22	0.866	43	9.7
Medium	7304B	20	0.787	52	2.047	15	0.591	13.4	3.1
	7305B	25	0.984	62	2.441	17	0.670	19	4.3
	7306B	30	1.181	72	2.835	19	0.748	24	5.4
	7307B	35	1.378	80	3.150	21	0.827	28	6.3
	7308B	40	1.575	90	3.543	23	0.906	34.5	7.8
	7309B	45	1.772	100	3.937	25	0.984	45	10
	7310B	50	1.969	110	4.331	27	1.063	52	11.6
	7311B	55	2.165	120	4.724	29	1.142	61	13.4
	7312B	60	2.362	130	5.118	31	1.221	69.5	15.6
Heavy	7405B	25	0.984	80	3.150	21	0.827	30.5	6.8
	7406B	30	1.181	90	3.543	23	0.906	36.5	8.3
	7407B	35	1.378	100	3.937	25	0.984	46.5	10.4
	7408B	40	1.575	110	4.331	27	1.063	54	12
	7409B	45	1.772	120	4.724	29	1.142	65.5	14.6
	7410B	50	1.969	130	5.118	31	1.221	73.5	16.6
	7411B	55	2.165	140	5.512	33	1.299	85	19
	7412B	60	2.362	150	5.906	35	1.378	91.5	20.4

Table 65 Angular contact bearings. (Canadian SKC Co. Ltd.)

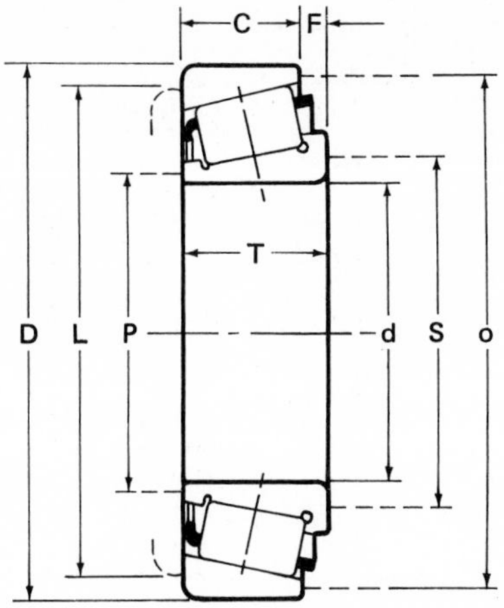
C
F
T
D
L
P
d
S
o

Bearing Number	Nominal External Sizes: Bore mm	Bore in.	Outside Dia mm	Outside Dia in.	Width mm	Width in.	Reference Ratings: Kilonewtons Radial F	Kilonewtons Thrust C	Pounds Radial F	Pounds Thrust C	Nominal Sizes (mm) F	C	S	Preferred Shoulder D/A (mm) P	L	O
4059	15	.590	35	1.375	11	.430	2	9	325	290	2	9	20	18	28	32
03062	16	.625	41	1.625	14	.562	3	11	625	381	3	11	21	20	33	37
5062	16	.625	46	1.850	14	.566	3	11	715	500	3	11	24	21	40	43
1749	17	.688	40	1.570	14	.545	3	11	660	371	3	11	23	20	37	36
9195	17.6	.695	49	1.938	20	.906	6	14	1190	620	6	14	27	24	41	44
9070	17.6	.695	49	1.938	23	.906	6	17	1190	620	6	17	27	24	39	44
1774	19	.748	57	2.240	20	.753	4	16	1300	785	4	16	27	24	48	51
1910	20	.750	45	1.781	16	.610	4	12	885	525	4	12	25	59	38	41
9067	20	.750	50	1.938	21	.835	4	17	1190	620	4	17	25	23	39	44
9078	20	.750	50	1.938	23	.906	6	17	1190	620	6	17	25	23	39	44
7075	20	.750	51	2.000	15	.591	3	12	795	630	3	12	25	24	44	47
7087	22	.857	50	1.969	14	.531	4	10	795	630	4	10	28	27	44	47
1380	22	.875	54	2.125	20	.762	5	15	1220	700	5	15	28	28	44	48
1280	22	.875	57	2.250	22	.875	5	17	1580	1080	5	17	28	28	48	52
4643	25	1.000	48	1.980	14	.560	4	10	850	625	4	10	33	29	44	47
1780	25	1.000	57	2.240	20	.762	4	16	1300	785	4	16	30	30	48	52

Table 66 Tapered roller bearings. (Canadian SKF Co. Ltd.)

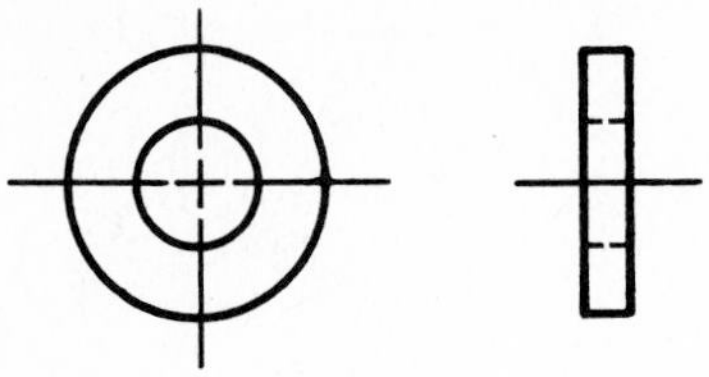

Inside Dia		Thickness		Outside Dia					
mm	in.	mm	in.	mm	in.	mm	in.	mm	in.
12	.500	3 5	.12 .18	20	.80	25	1.10	30	1.25
16	.625	3 5	.12 .18	30	1.25	40	1.50	46	1.75
20	.750	3 5	.12 .18	34	1.30	40	1.60	50	2.00
22	.875	3 5	.12 .18	34	1.30	40	1.60	50	2.00
25	1.000	3 5 6	.12 .18 .25	40	1.60	50	2.00	60	2.25

Table 67 Thrust plain bearings.

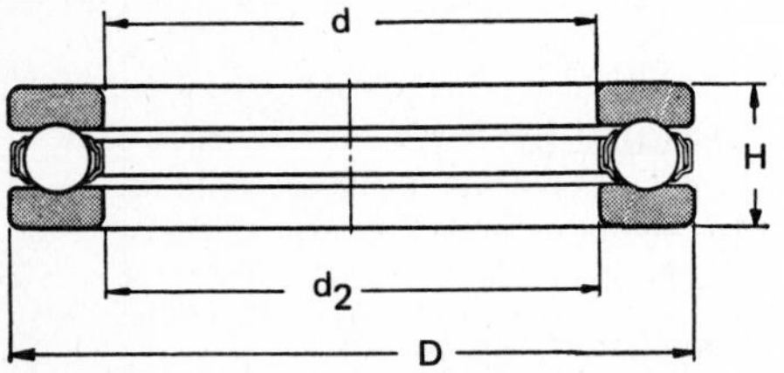
d
H
d_2
D

BEARING NUMBER	Nominal Bearing Dimensions: d		d_2		D		H		Max. Fillet Radius*	Shoulder Diameter in.	
	mm	in.	mm	in.	mm	in.	mm	in.	in.	Shaft Min.	Housing Max.
51100	10	.3937	10.2	.402	24	.9449	9	.354	.012	.750	.563
51101	12	.4724	12.2	.480	26	1.0236	9	.354	.012	.844	.656
51102	15	.5906	15.2	.598	28	1.1024	9	.354	.012	.938	.750
51103	17	.6693	17.2	.677	30	1.1811	9	.354	.012	1.031	.844
51104	20	.7874	20.2	.795	35	1.3780	10	.394	.012	1.188	.969
51105	25	.9843	25.2	.992	42	1.6535	11	.433	.024	1.438	1.219
51106	30	1.1811	30.2	1.189	47	1.8504	11	.433	.024	1.563	1.469
51107X	35	1.3780	35.2	1.386	52	2.0472	12	.472	.024	1.813	1.656
51108	40	1.5748	40.2	1.583	60	2.3622	13	.512	.024	2.063	1.875
51109	45	1.7717	45.2	1.780	65	2.5591	14	.551	.024	2.281	2.063
51110	50	1.9685	50.2	1.976	70	2.7559	14	.551	.024	2.500	2.250
51111	55	2.1654	55.2	2.173	78	3.0709	16	.630	.024	2.750	2.500
51112	60	2.3622	60.2	2.370	85	3.3465	17	.669	.039	3.000	2.719
51113	65	2.5591	65.2	2.567	90	3.5433	18	.709	.039	3.188	2.875
51114	70	2.7559	70.2	2.764	95	3.7402	18	.709	.039	3.375	3.063
51115	75	2.9528	75.2	2.961	100	3.9370	19	.748	.039	3.625	3.250
51116	80	3.1496	80.2	3.157	105	4.1339	19	.748	.039	3.813	3.438
51117	85	3.3465	85.2	3.354	110	4.3307	19	.748	.039	4.000	3.625

* The maximum fillet on the shaft or in the housing, which will be cleared by the bearing corner.

Table 68 Thrust roller bearings. (Canadian SKF Co. Ltd.)

Style	No. of Teeth	Metric Sizes (mm) 3.18 Module 30 Face: Pitch Dia	Hole	Hub Dia	Hub Proj.	Inch Sizes 8 Pitch 1.25 Face: Pitch Dia	Hole	Hub Dia	Hub Proj.
Steel Plain	12	38.2	20	28	20	1.500	.750	1.12	.75
	14	44.5	20	35	20	1.750	.750	1.38	.75
	15	47.7	22	40	20	1.875	.875	1.50	.75
	16	50.9	22	40	20	2.000	.875	1.62	.75
	18	57.2	22	48	20	2.250	.875	1.88	.75
	20	63.6	22	54	20	2.500	.875	2.12	.75
	22	70.0	22	60	20	2.750	.875	2.38	.75
Cast Iron Spoke Web	24	76.3	22	54	25	3.000	.875	2.12	1.00
	28	89.0	22	56	25	3.500	.875	2.25	1.00
	30	95.4	22	56	25	3.750	.875	2.25	1.00
	32	101.8	25	56	25	4.000	1.000	2.25	1.00
	36	114.5	25	64	25	4.500	1.000	2.50	1.00
	40	127.2	25	64	25	5.000	1.000	2.50	1.00
	42	133.6	25	64	25	5.250	1.000	2.50	1.00
	44	139.9	25	64	25	5.500	1.000	2.50	1.00
	48	152.6	25	64	25	6.000	1.000	2.50	1.00
	54	171.7	25	64	25	6.750	1.000	2.50	1.00
	56	178.1	25	64	25	7.000	1.000	2.50	1.00
	60	190.8	25	64	25	7.500	1.000	2.50	1.00
	64	203.5	25	64	25	8.000	1.000	2.50	1.00
	72	229.0	25	64	25	9.000	1.000	2.50	1.00
	80	254.4	28	76	28	10.000	1.125	3.00	1.12
	84	267.1	28	76	28	10.500	1.125	3.00	1.12
	88	279.8	28	76	28	11.000	1.125	3.00	1.12
	96	305.3	28	76	28	12.000	1.125	3.00	1.12

Note: Metric size gears were not available at time of publication. Values were soft converted for problem solving only.

Table 69 3.18 Module (eight-pitch) spur-gear data. (Boston Gear Works)

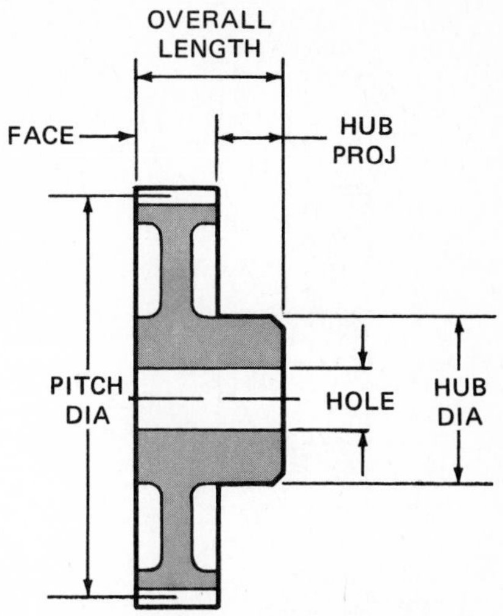
OVERALL LENGTH
FACE
HUB PROJ
PITCH DIA
HOLE
HUB DIA

3.18 Module (8 Pitch)

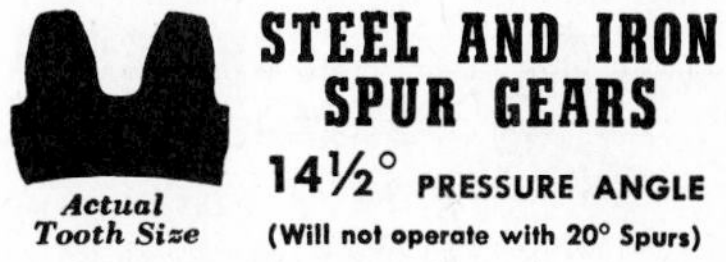
STEEL AND IRON
SPUR GEARS
14½° PRESSURE ANGLE
(Will not operate with 20° Spurs)
Actual Tooth Size

SPUR GEARS
STEEL AND IRON
20° PRESSURE ANGLE
(Will not operate with 14½° Spurs)
Actual Tooth Size

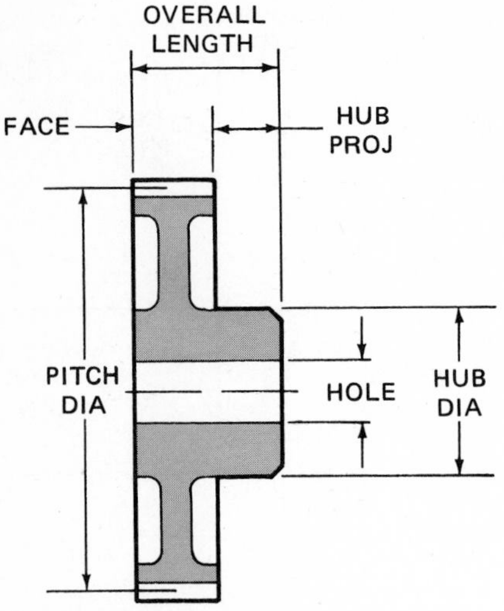
OVERALL
LENGTH
FACE
HUB
PROJ
PITCH
DIA
HOLE
HUB
DIA

4.23 Module (6 Pitch)

SPUR GEARS
STEEL AND IRON
14½° PRESSURE ANGLE
(Will not operate with 20° Spurs)
Actual
Tooth Size
SPUR GEARS
STEEL AND IRON
20° PRESSURE ANGLE
(Will not operate with 14½° Spurs)
Actual
Tooth Size

		Metric Sizes (mm) 4.23 Module 40 Face				Inch Sizes 6 Pitch 1.50 Face			
				Hub				Hub	
Style	No. of Teeth	Pitch Dia	Hole	Dia	Proj.	Pitch Dia	Hole	Dia	Proj.
Steel Plain	12	50.8	25	38	22	2.000	1.00	1.50	.88
	14	59.2	25	46	22	2.333	1.00	1.81	.88
	15	63.5	25	50	22	2.500	1.00	2.00	.88
	16	67.7	25	55	22	2.667	1.00	2.16	.88
	18	76.1	25	64	22	3.000	1.00	2.50	.88
	20	84.6	25	72	22	3.333	1.00	2.84	.88
	21	88.8	25	76	22	3.500	1.00	3.00	.88
Cast Iron Spoke Web	24	101.5	28	64	25	4.000	1.12	2.50	1.00
	27	114.2	28	64	25	4.500	1.12	2.50	1.00
	30	126.9	28	64	25	5.000	1.12	2.50	1.00
	32	135.4	28	64	25	5.333	1.12	2.50	1.00
	33	139.6	28	64	25	5.500	1.12	2.50	1.00
	36	152.3	28	64	25	6.000	1.12	2.50	1.00
	40	169.2	28	64	25	6.667	1.12	2.50	1.00
	42	177.7	28	64	25	7.000	1.12	2.50	1.00
	48	203.0	28	64	25	8.000	1.12	2.50	1.00
	54	228.4	28	64	25	9.000	1.12	2.50	1.00
	60	253.8	30	76	30	10.000	1.25	3.00	1.25
	64	270.7	30	76	30	10.667	1.25	3.00	1.25
	66	279.2	30	76	30	11.000	1.25	3.00	1.25
	72	304.6	30	76	30	12.000	1.25	3.00	1.25
	84	355.3	30	82	30	14.000	1.25	3.25	1.25

Note: Metric size gears were not available at time of publication. Values were soft converted for problem solving only.

Table 70 4.23 module (six-pitch) spur-gear data. (Boston Gear Works)

Style	No. of Teeth	Metric Sizes (mm) 5.08 Module 50 Face: Pitch Dia	Hole	Hub Dia	Hub Proj.	Inch Sizes 5 Pitch 2.00 Face: Pitch Dia	Hole	Hub Dia	Hub Proj.
	12	61.0	26	45	22	2.400	1.06	1.78	.88
	14	71.1	26	55	22	2.800	1.06	2.18	.88
Steel Plain	15	76.2	26	60	22	3.000	1.06	2.38	.88
	16	81.3	26	65	22	3.200	1.06	2.59	.88
	18	91.4	26	75	22	3.600	1.06	3.00	.88
	20	101.6	26	85	22	4.000	1.06	3.38	.88
	24	121.9	26	75	30	4.800	1.06	3.00	1.25
	25	127.0	26	75	30	5.000	1.06	3.00	1.25
	30	152.4	26	75	30	6.000	1.06	3.00	1.25
	35	177.8	30	75	30	7.000	1.18	3.00	1.25
	40	203.2	30	75	30	8.000	1.18	3.00	1.25
Cast Iron Web	45	228.6	30	75	30	9.000	1.18	3.00	1.25
	50	254.0	30	90	30	10.000	1.18	3.50	1.25
	55	279.4	30	90	30	11.000	1.18	3.50	1.25
	60	304.8	30	90	30	12.000	1.18	3.50	1.25
Cast Iron Spoke	70	355.6	30	90	30	14.000	1.18	3.50	1.25
	80	406.4	30	90	30	16.000	1.18	3.50	1.25
	90	457.2	30	90	30	18.000	1.18	3.50	1.25
	100	508.0	32	95	38	20.000	1.31	3.75	1.50
	110	558.8	32	95	38	22.000	1.31	3.75	1.50
	120	609.6	32	100	38	24.000	1.31	4.00	1.50

Note: Metric size gears were not available at the time of publication. Values were soft converted for problem solving only.

Table 71 5.08 module (five-pitch) spur-gear data. (Boston Gear Works)

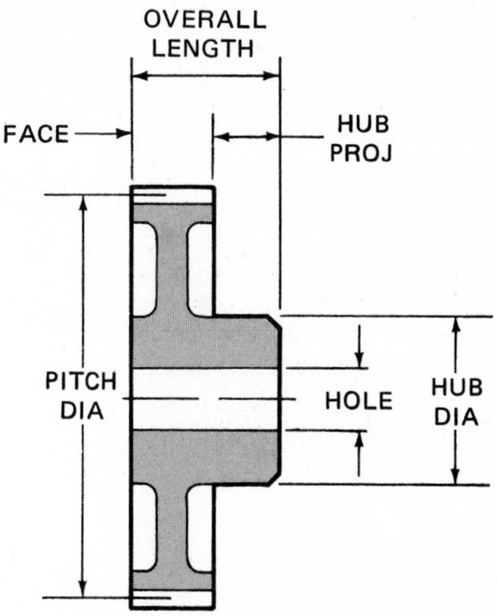

OVERALL LENGTH
FACE
HUB PROJ
PITCH DIA
HOLE
HUB DIA

5.08 Module (5 Pitch)

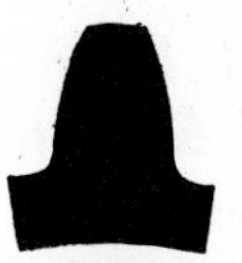

Actual Tooth Size

SPUR GEARS STEEL AND IRON

14½° PRESSURE ANGLE

(Will not operate with 20° Spurs)

SPUR GEARS STEEL AND IRON

20° PRESSURE ANGLE

(Will not operate with 14½° Spurs)

Actual Tooth Size

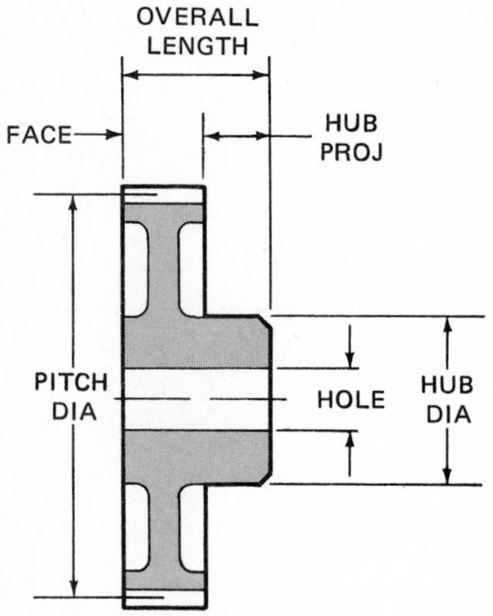
OVERALL LENGTH
FACE
HUB PROJ
PITCH DIA
HOLE
HUB DIA

6.35 Module (4 Pitch)

SPUR GEARS
STEEL AND IRON
14½° PRESSURE ANGLE
(Will not operate with 20° Spurs)
Actual Tooth Size

SPUR GEARS
STEEL AND IRON
20° PRESSURE ANGLE
(Will not operate with 14½° Spurs)
Actual Tooth Size

		Metric Sizes (mm) 6.35 Module		70 Face		Inch Sizes 4 Pitch		2.75 Face	
				Hub				Hub	
Style	No. of Teeth	Pitch Dia	Hole	Dia	Proj.	Pitch Dia	Hole	Dia	Proj.
Steel Plain	12	76.2	28	58	22	3.000	1.12	2.25	.88
	14	88.9	28	70	22	3.500	1.12	2.75	.88
	15	95.3	28	76	22	3.750	1.12	3.00	.88
	16	101.6	28	82	22	4.000	1.12	3.25	.88
	18	114.3	28	96	22	4.500	1.12	3.75	.88
	20	127.0	28	108	22	5.000	1.12	4.25	.88
	22	139.7	28	120	22	5.500	1.12	4.75	.88
	24	152.4	28	90	38	6.000	1.12	3.50	1.50
	28	177.8	30	90	38	7.000	1.25	3.50	1.50
	30	190.5	30	90	38	7.500	1.25	3.50	1.50
	32	203.2	30	90	38	8.000	1.25	3.50	1.50
Cast Iron Web	36	228.6	30	90	38	9.000	1.25	3.50	1.50
	40	254.0	30	100	38	10.000	1.25	4.00	1.50
	42	266.7	30	100	38	10.500	1.25	4.00	1.50
	44	279.4	30	100	38	11.000	1.25	4.00	1.50
Cast Iron Spoke	48	304.8	30	100	38	12.000	1.25	4.00	1.50
	54	342.9	30	100	38	13.500	1.25	4.00	1.50
	56	355.6	30	100	38	14.000	1.25	4.00	1.50
	60	381.0	30	100	38	15.000	1.25	4.00	1.50
	64	406.4	30	100	38	16.000	1.25	4.00	1.50
	72	457.2	30	100	38	18.000	1.25	4.00	1.50
	80	508.0	35	115	38	20.000	1.38	4.50	1.50

Note: Metric size gears were not available at the time of publication. Values were soft converted for problem solving only.

Table 72 6.35 module (four-pitch) spur-gear data. (Boston Gear Works)

20° PRESSURE ANGLE—STRAIGHT TOOTH

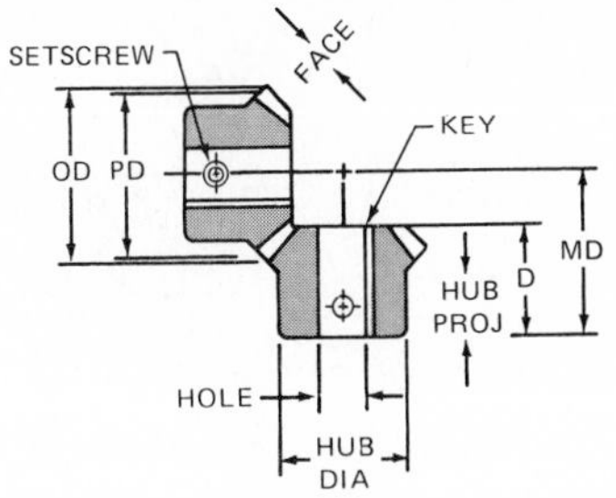
SETSCREW
FACE
KEY
OD
PD
MD
D
HUB
PROJ
HOLE
HUB
DIA

Module Pitch	Metric Sizes: Teeth	Pitch Dia	Face	OD Approx.	D	MD	Hub Dia	Hub Prov. Approx.	Hole	Keyway	Inch Sizes: Pitch Dia	Face	OD Approx.	D	MD	Hub Dia	Hub Prov. Approx.	Hole	Keyway
									12	3 × 1.5								.500	.12 × .06
2.54	20	50	11.2	54	34	50	42	20	16	5 × 2.5	**2.000**	**.44**	**2.15**	**1.36**	**2.00**	**1.62**	**.81**	**.625**	**.18 × .09**
									20	5 × 2.5								**.750**	**.18 × .09**
									20	5 × 2.5								**.750**	**.18 × .09**
10	25	62	14	68	41	62	50	24	22	5 × 2.5	**2.500**	**.55**	**2.65**	**1.62**	**2.44**	**2.00**	**.94**	**.875**	**.18 × .09**
									25	6 × 3								**1.000**	**.25 × .12**
					40	65	45	20	20	5 × 2.5				**1.58**	**2.56**	**1.75**	**.81**	**.750**	**.18 × .09**
3.18	24	76	16	80	45	70	58	27	25	6 × 3	**3.000**	**.64**	**3.18**	**1.76**	**2.75**	**2.25**	**1.06**	**1.000**	**.25 × .12**
					45	70	64	28	30	6 × 3				**1.76**	**2.75**	**2.50**	**1.12**	**1.250**	**.25 × .12**
									25									**1.000**	
8	28	88	16	93	53	82	64	32	30	6 × 3	**3.500**	**.75**	**3.68**	**2.09**	**3.25**	**2.50**	**1.25**	**1.188**	**.25 × .12**
									32									**1.250**	
4.23	24	100	22	108	58	92	76	33	32	6 × 3	**4.000**	**.86**	**4.24**	**2.31**	**3.62**	**3.00**	**1.31**	**1.25**	**.25 × .12**
									38	10 × 5								**1.50**	**.38 × .18**
6	27	114	24	120	66	104	82	38	38	10 × 5	**4.500**	**.96**	**4.74**	**2.62**	**4.12**	**3.25**	**1.50**	**1.50**	**.38 × .18**
									35	8 × 4								**1.38**	**.31 × .16**
5.08	25	128	28	134	76	117	88	44	38	10 × 5	**5.000**	**1.10**	**5.29**	**3.00**	**4.62**	**3.50**	**1.75**	**1.50**	**.38 × .18**
5									44	10 × 5								**1.75**	**.38 × .18**

NOTE: Metric size gears were not available at time of publication. Values were soft converted for problem solving only.

Table 73 Miter gears. (Boston Gear Works)

20° PRESSURE ANGLE

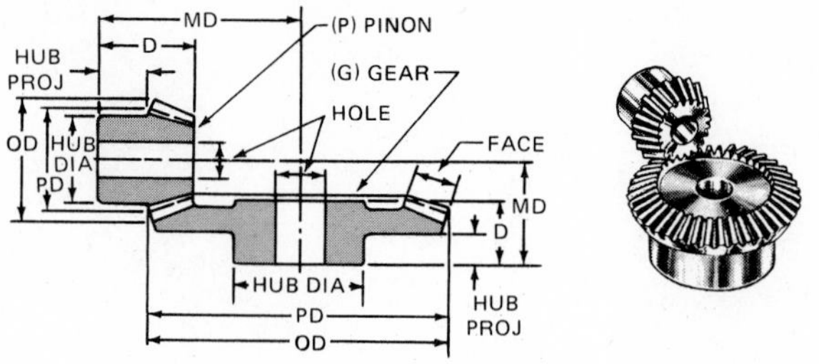
MD
D
(P) PINON
HUB PROJ
(G) GEAR
HOLE
OD
HUB DIA
PD
FACE
MD
D
HUB DIA
PD
OD
HUB PROJ

Module Pitch	Teeth	Metric Sizes: Pitch Dia	Face	Hole	D	MD	Hub Dia	Hub Proj. Approx.	OD Approx.	Inch Sizes: Pitch Dia	Face	Hole	D	MD	Hub Dia	Hub Proj. Approx.	OD Approx.
	50	127	18	20	33.0	66.6	50	25	128.5	5.000	.70	.750	1.30	2.62	2.00	1.00	5.06
	25	63.5			39.4	85.8	50	20	69.8	2.500			1.55	3.38	2.00	.75	2.75
2.54	60	152.4	20	25	47.2	69.8	76	35	153.4	6.000	.78	1.00	1.86	2.75	3.00	1.38	6.04
10	20	50.8		22	54.9	111.2	45	33	57.6	2.000		.875	2.16	4.38	1.75	1.30	2.27
	60	152.4	18.3	22	41.1	57.2	64	28	153.2	6.000	.72	.875	1.62	2.25	2.50	1.12	6.03
	15	38.1		16	40.6	98.6	36	22	45.2	1.500		.625	1.60	3.88	1.44	.84	1.78
	40	127.2	20.8	25	46.7	67.6	76	32	129	5.000	.82	1.000	1.84	2.88	3.00	1.25	5.08
	20	63.6			57.9	101.6	54	35	71.4	2.500			2.28	4.00	2.12	1.40	2.81
3.18	48	152.6	21.4	22	41.2	60.4	70	25	153.7	6.000	.84	.875	1.62	2.38	2.75	1.00	6.05
8	16	50.9		20	52.8	108	45	48	59.7	2.000		.750	2.08	4.25	1.75	1.88	2.35
	64	203.5	21.4	25	47.7	69.8	70	32	204.2	8.000	.84	1.000	1.88	2.75	2.75	1.25	8.04
	16	50.9		22	53.1	133.4	48	30	60	2.000		.875	2.09	5.25	1.88	1.22	2.36
		127	20.8	25	46.7	67.6	76	32	129	5.000	.82	1.000	1.84	2.88	3.00	1.25	5.08
		63.5			57.9	101.6	54	35	71.4	2.500			2.28	4.00	2.12	1.40	2.81
		152.4	21.4	22	41.1	60.4	70	25	153.7	6.000	.84	.875	1.62	2.38	2.75	1.00	6.05
		50.8		20	52.8	108	45	30	59.7	2.000		.750	2.08	4.25	1.75	1.18	2.35
4.23		203.2	21.4	25	47.7	69.8	70	32	204.2	8.000	.84	1.000	1.88	2.75	2.75	1.25	8.04
6		50.8		22	53.1	133.4	48	30	60	2.000		.875	2.09	5.25	1.88	1.22	2.36
		152.4	27	28	57.2	88.9	82	38	155	6.000	1.06	1.125	2.25	3.50	3.25	1.50	6.10
		76.2			70.1	120.6	64	40	86.6	3.000			2.76	4.75	2.50	1.60	3.41
		190.5	27	28	53.8	76.2	82	32	192.3	7.500	1.06	1.125	2.12	3.00	3.25	1.25	7.57
		63.5		22	65.0	133.4	54	36	75	2.500		.875	2.56	5.25	2.12	1.44	2.95

NOTE: Metric size gears were not available at time of publication. Values were soft converted for problem solving only.

Table 74 Bevel gears. (Boston Gear Works)

TYPE P
HEADLESS PRESS FIT BUSHINGS

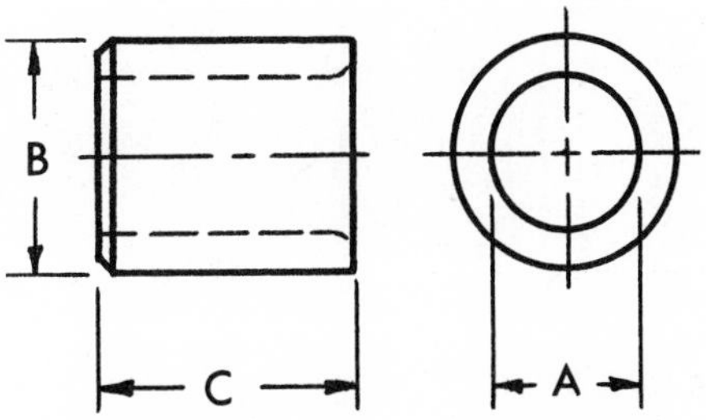
B
C
A

TYPE H
HEAD PRESS FIT BUSHINGS

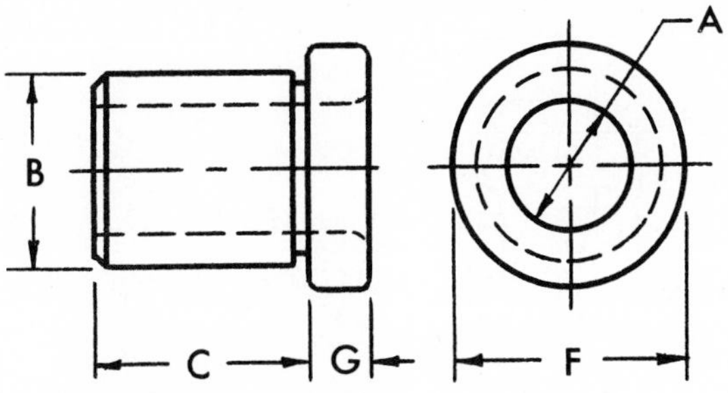
A
B
C
G
F

	Inside Dia (A)		Outside Dia (B)		Bushing Length C											Technical Data	
	From	To	Max.	Min.	6	8	10	12	16	20	25	30	35	40	45	F	G
Millimetre Sizes	3	5	7.978	7.971	•	•	•	•	•	•	•	•	•	•		10	3
	5	6.5	10.358	10.351	•	•	•	•	•	•	•	•	•	•		12	4
	5	8	12.972	12.736	•	•	•	•	•	•	•	•	•	•		15	5
	8	11	15.918	15.911	•	•	•	•	•	•	•	•	•	•		20	6
	8	13.5	19.096	19.088	•	•	•	•	•	•	•	•	•	•	•	24	6
	12	16	22.271	22.263		•	•	•	•	•	•	•	•	•	•	28	6
	12	20	25.446	25.438			•	•	•	•	•	•	•	•	•	32	8
	16	26	34.981	34.971			•	•	•	•	•	•	•	•	•	40	10
	25	35	44.508	44.498			•	•	•	•	•	•	•	•	•	50	10
					.25	.31	.38	.50	.75	1.00	1.38	1.75	2.12	2.50	3.00	F	G
Inch Sizes	.125	.194	.3141	.3138	•	•	•	•	•	•	•	•				.42	.12
	.188	.257	.4078	.4075	•	•	•	•	•	•	•	•	•			.50	.16
	.188	.316	.5017	.5014	•	•	•	•	•	•	•	•	•			.60	.22
	.312	.438	.6267	.6264	•	•	•	•	•	•	•	•	•	•		.80	.22
	.312	.531	.7518	.7515	•	•	•	•	•	•	•	•	•	•	•	.92	.22
	.500	.656	.8768	.8765		•	•	•	•	•	•	•	•	•	•	1.10	.25
	.500	.766	1.0018	1.0015				•	•	•	•	•	•	•	•	1.24	.32
	.625	1.031	1.3772	1.3768				•	•	•	•	•	•	•	•	1.60	.38
	1.000	1.390	1.7523	1.7519					•	•	•	•	•	•	•	1.98	.38

Note: Metric size bushings were not available at time of publication. Values shown were soft converted for problem solving only.

Table 75 Press-fit drill jig bushings. (American Drill Bushing Co.).

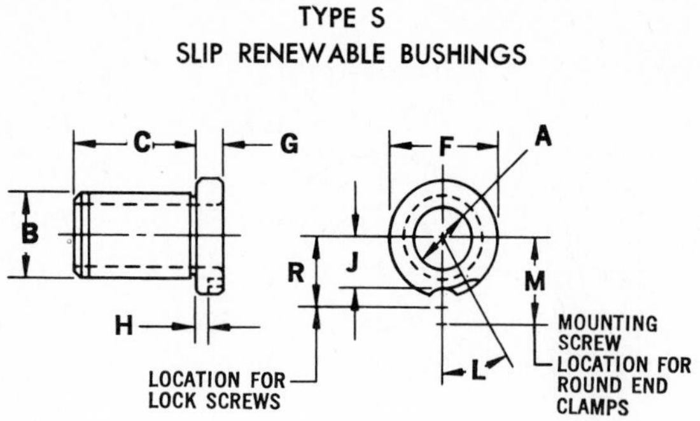
TYPE S
SLIP RENEWABLE BUSHINGS
C
G
F
A
B
R
J
M
H
L
MOUNTING SCREW LOCATION FOR ROUND END CLAMPS
LOCATION FOR LOCK SCREWS

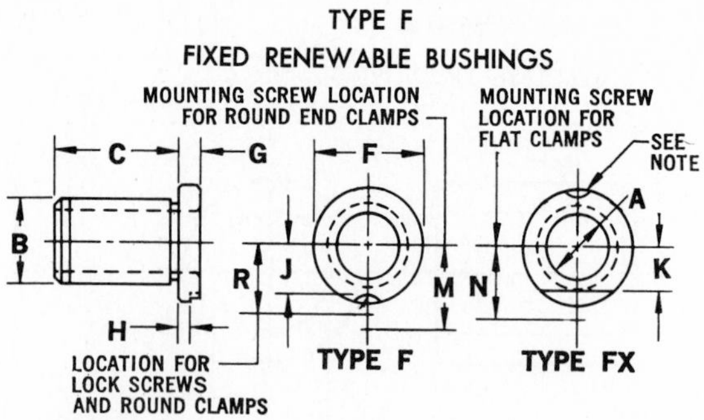
TYPE F
FIXED RENEWABLE BUSHINGS
MOUNTING SCREW LOCATION FOR ROUND END CLAMPS
MOUNTING SCREW LOCATION FOR FLAT CLAMPS
SEE NOTE
C
G
F
A
B
R
J
M
N
K
H
LOCATION FOR LOCK SCREWS AND ROUND CLAMPS
TYPE F
TYPE FX

NOTE: Type "FX" bushings (at the manufacturer's option) may be supplied with lock screw slot opposite the flat clamp milling.

	Inside Dia (A)		Outside Dia (B)		Bushing Length C									Technical Data								
	From	To	From	To	10	12	16	20	25	30	40	50	60	F	G	H	J	K	L	M	N	R
Millimetre Sizes	1.6	2.3	7.938	7.932	•	•	•	•	•	•				14	10	3	4.3	4.3	65°	16	17	13
	2.4	4.9	7.938	7.932	•	•	•	•	•	•				14	10	3	4.3	4.3	65°	16	17	13
	3.6	4.8	12.700	12.695	•	•	•	•	•	•	•	•		20	11	3	7.6	7.6	65°	20	20	16
	4.8	8.7	12.700	12.695	•	•	•	•	•	•	•	•		20	11	3	7.6	7.6	65°	20	20	16
	7.1	14.2	19.050	19.045		•	•	•	•	•	•	•	•	26	11	3	10.7	10.7	50°	22	20	20
	11.9	19.8	25.400	25.350		•	•	•	•	•	•	•	•	36	11	4.5	15.2	15.2	35°	28	29	24
	18.2	26.9	34.925	34.917				•	•	•	•	•	•	46	11	4.5	19.8	17.5	30°	33	34	28
	24.6	35.7	44.450	44.442				•	•	•	•	•	•	58	16	4.5	25.4	22.4	30°	42	38	35
					.38	.50	.62	.75	1.00	1.38	1.75	2.12	2.50	F	G	H	J	K	L	M	N	R
Inch Sizes	.064	.089	.3125	.3123	•	•	•	•	•	•				.55	.38	.12	.17	.17	65°	.62	.68	.50
	.094	.194	.3125	.3123	•	•	•	•	•	•				.55	.38	.12	.17	.17	65°	.62	.68	.50
	.141	.188	.5000	.4998	•	•	•	•	•	•	•	•		.80	.44	.12	.30	.26	65°	.75	.78	.62
	.189	.344	.5000	.4998	•	•	•	•	•	•	•	•		.80	.44	.12	.30	.26	65°	.75	.78	.62
	.281	.562	.7500	.7498		•	•	•	•	•	•	•	•	1.05	.44	.12	.42	.39	50°	.88	.80	.75
	.469	.781	1.0000	.9998		•	•	•	•	•	•	•	•	1.42	.44	.18	.60	.50	35°	1.11	1.14	.92
	.719	1.062	1.3750	1.3747				•	•	•	•	•	•	1.80	.44	.18	.78	.69	30°	1.30	1.33	1.10
	.969	1.406	1.7500	1.7497				•	•	•	•	•	•	2.30	.62	.18	1.00	.88	30°	1.64	1.52	1.39

Note: Metric size bushings were not available at time of publication. Values shown were soft converted for problem solving only.

Table 76 Renewable drill jig bushings. (American Drill Bushing Co.)

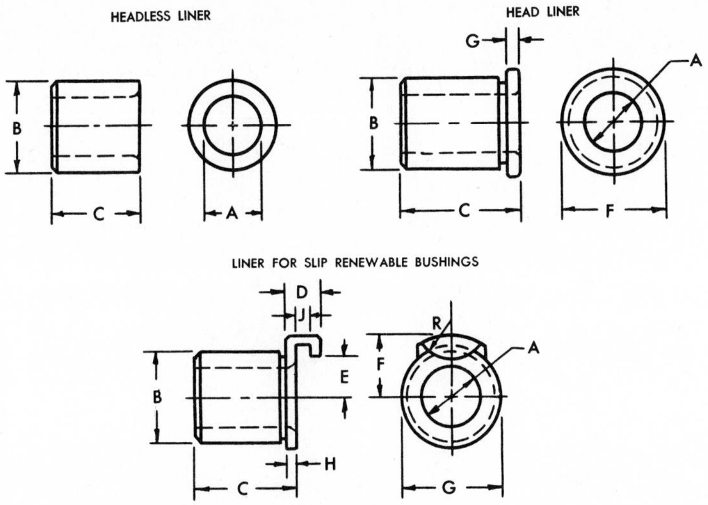
HEADLESS LINER
HEAD LINER
LINER FOR SLIP RENEWABLE BUSHINGS
A
B
C
D
E
F
G
H
J
R

	Nominal Inside Dia	Outside Dia (B)		Liner Length C										Technical Data	
	(A)	Max.	Min.	8	10	12	16	20	25	35	45	55	65	F	G
Millimetre Sizes	5	7.978	7.971	•	•	•	•	•						—	—
	8	12.743	12.736	•	•	•	•	•	•	•				16	2.3
	12	19.096	19.088	•	•	•	•	•	•	•	•	•		22	2.3
	20	25.446	25.438			•	•	•	•	•	•	•	•	28	3.0
	25	34.981	34.971			•	•	•	•	•	•	•	•	38	3.0
	35	44.508	44.498					•	•	•	•	•	•	48	4.5
				.31	.38	.50	.62	.75	1.00	1.38	1.75	2.12	2.50	F	G
Inch Sizes	.188	.3141	.3138	•	•	•	•	•						—	—
	.312	.5017	.5014	•	•	•	•	•	•	•				.62	.09
	.500	.7518	.7515	•	•	•	•	•	•	•	•	•		.88	.09
	.750	1.0018	1.0015			•	•	•	•	•	•	•	•	1.12	.12
	1.000	1.3772	1.3768			•	•	•	•	•	•	•	•	1.50	.12
	1.375	1.7523	1.7519					•	•	•	•	•	•	1.88	.18

	Nominal Inside Dia	Outside Dia (B)		Liner Length C						Technical Data						
	(A)	Max.	Min.	8	10	12	20	25	35	D	E	F	G	H	J	R
Millimetre Sizes	5	7.978	7.971	•	•	•	•			5.5	3	6	9	1.5	2.5	4
	8	12.743	12.736		•	•	•			8.1	1.5	10	14	2.3	3.5	8
	12	19.096	19.088			•	•	•		8.1	8	12	20	2.3	3.5	8
	20	25.446	25.438			•	•	•		9.5	11	16	27	3	3.5	8
	25	34.981	34.971			•	•	•	•	11	15	22	36	3	5	8
				.31	.38	.50	.75	1.00	1.38	D	E	F	G	H	J	R
Inch Sizes	.188	.3141	.3138	•	•	•	•			.23	.11	.25	.36	.06	.10	.16
	.312	.5017	.5014		•	•	•			.31	.18	.38	.56	.09	.14	.31
	.500	.7518	.7515			•	•	•		.31	.31	.50	.81	.09	.14	.31
	.750	1.0018	1.0015			•	•	•		.38	.44	.66	1.06	.12	.14	.31
	1.000	1.3772	1.3768			•	•	•	•	.38	.61	.88	1.44	.12	.20	.31

Note: Metric size drill bushing liners were not available at time of publication. Values shown were soft converted for problem solving only.

Table 77 Drill bushing liners. (American Drill Bushing Co.)

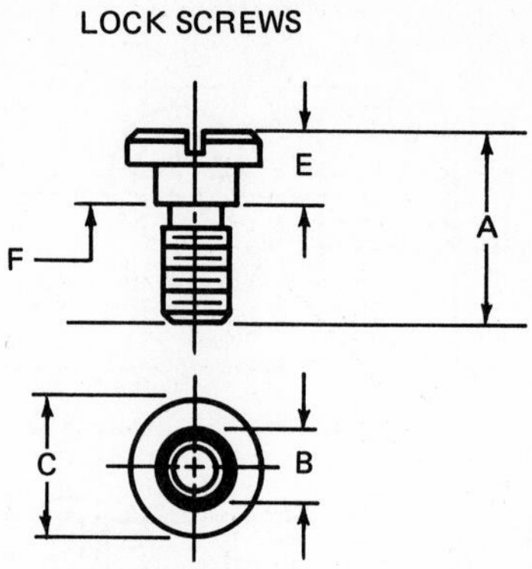
LOCK SCREWS
E
A
F
C
B

Part No.	A	B	C	E	F	Thread
LS-1	16	10	16	6	2.5	M8
LS-2	22	10	16	10	3.5	M8
LS-3	25	11	20	10	5.0	M10
LS-4	27	11	20	11	5.8	M10

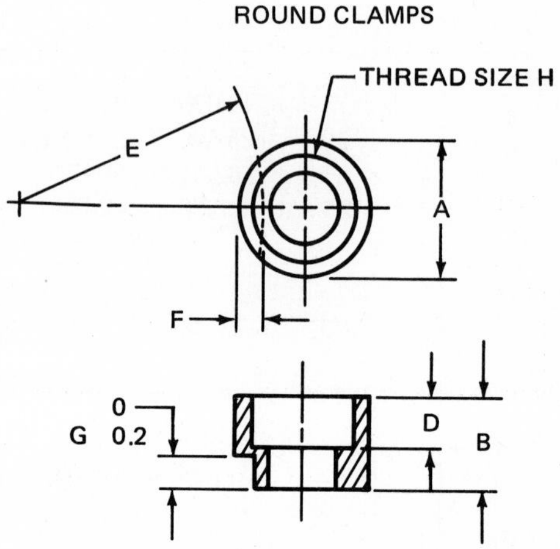
ROUND CLAMPS
THREAD SIZE H
E
A
F
0
G 0.2
D
B

Part No.	A	B	D	E	F	G	H	Replaces Lock Screw
CL-1	16	8	4	14	2.7	3	M8	LS-1
CL-2	16	12	5.5	23	3.2	4.5	M8	LS-2
CL-3	20	12	7	35	4.0	4.5	M10	LS-3
CL-4	20	13	6	64	10.5	4.5	M10	LS-4

Table 78 Locking devices for renewable bushings. (American Drill Bushing Co.)

ROUND END CLAMPS

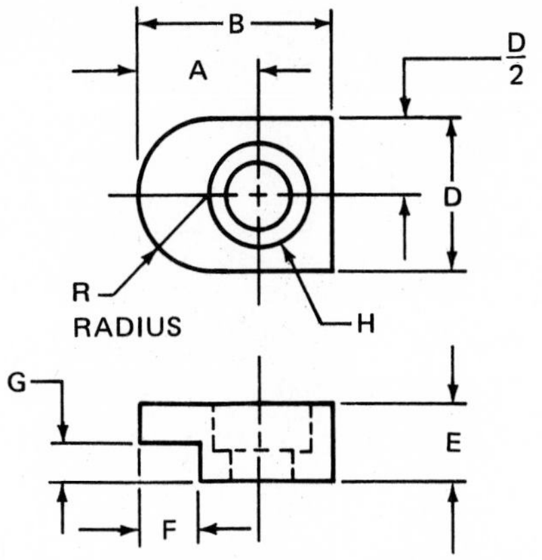
B
A
D
2
D
R
RADIUS
H
G
E
F

FOR FLUSH MOUNTING OF LINERS

Part No.	A	B	D	E	F	G	H
RE-1	11	20	16	8	5	3	M6
RE-2	12	20	16	10	6	4.5	M6
RE-3	16	25	20	10	7	4.5	M8

FOR FLUSH MOUNTING OF LINERS

Part No.	A	B	D	E	F	G	H
FC-1	12	22	16	8	5	3	M6
FC-2	16	25	25	10	6	4.5	M8
FC-3	16	28	25	10	7	4.5	M8

FOR PROJECTED MOUNTING OF SHOULDER LINERS

Part No.	A	B	D	E	F	G	H
FC-10	12	22	16	10	5	5	M6
FC-11	12	22	16	10	5	6	M6

FLAT CLAMPS

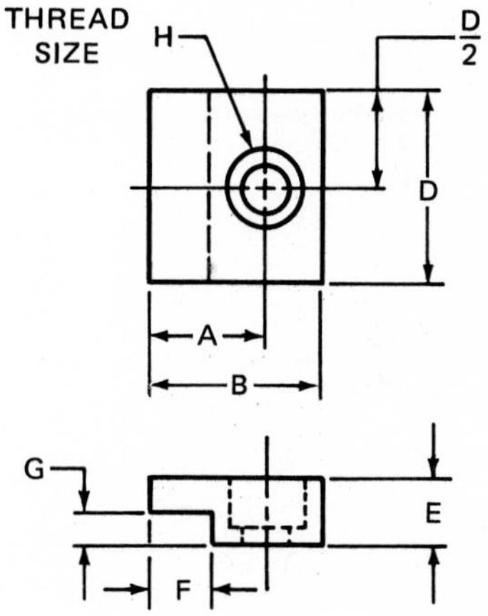
THREAD SIZE
H
D
2
D
A
B
G
E
F

Note: Metric sizes were not available at time of publication. Values shown were soft converted for problem solving only.

Table 78 Continued.

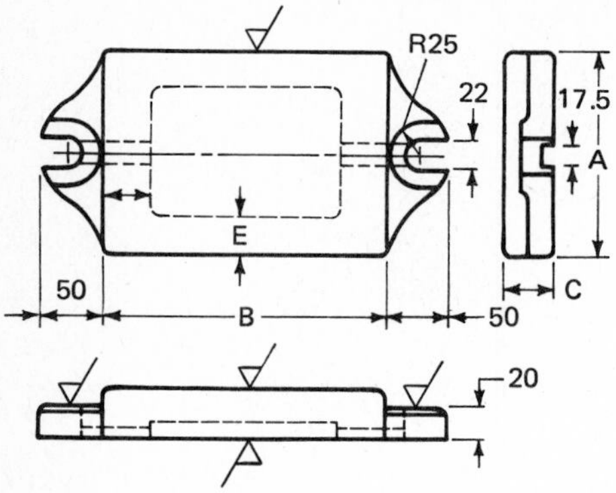
R25
22
17.5
A
E
50
B
50
C
20

A	B	C	D	E
150	225	30	45	35
200	300	35	50	40
250	375	40	50	45
300	450	45	55	50

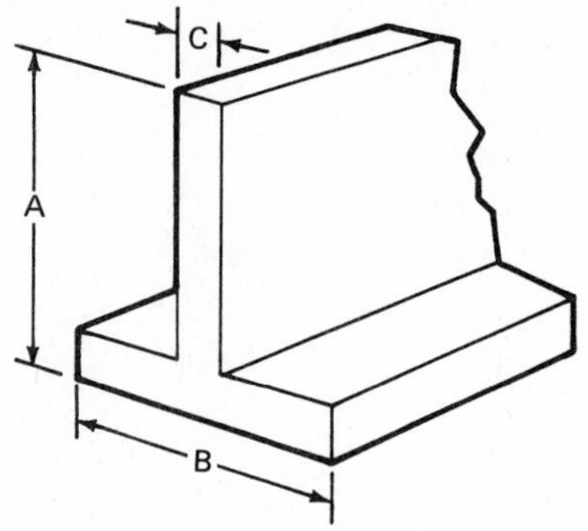
C
A
B

Cat. No.	A	B	C
ET-75	75	75	15
ET-100	100	100	20
ET-125	125	125	20
ET-150	150	150	25
ET-200	200	200	30

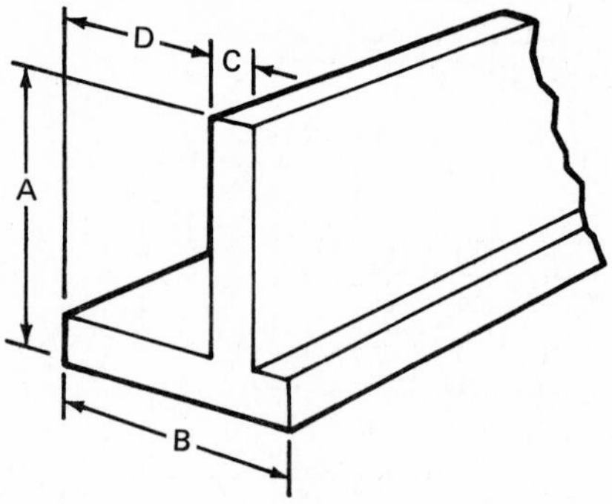
D
C
A
B

Cat. No.	AB	C	D
UT-75	75	15	45
UT-100	100	20	65
UT-125	125	20	85
UT-150	150	25	90
UT-200	200	30	120

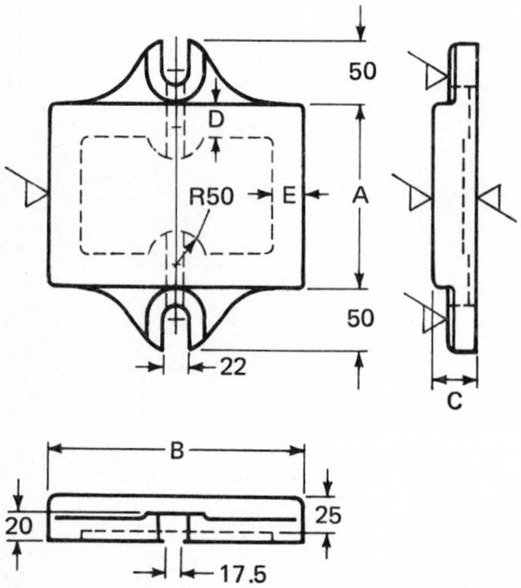
50
D
R50
E
A
50
22
C
B
20
25
17.5

A	B	C	D	E
150	225	30	30	30
200	300	35	30	30
250	375	40	35	35

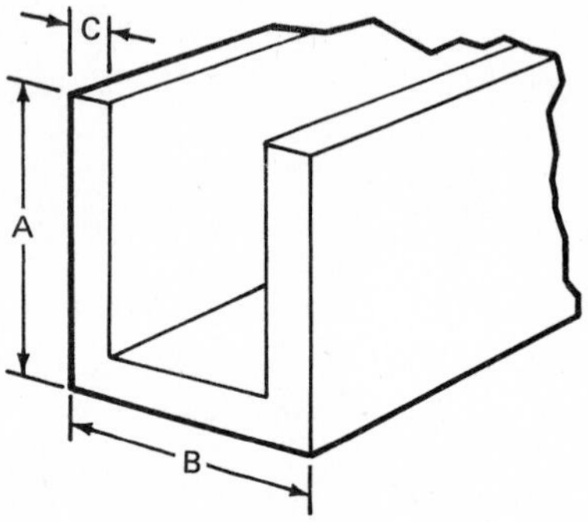
C
A
B

Cat. No.	A	B	C
U-75	75	75	15
U-100	100	100	20
U-125	125	125	20
U-150	150	150	25
U-200	200	200	30

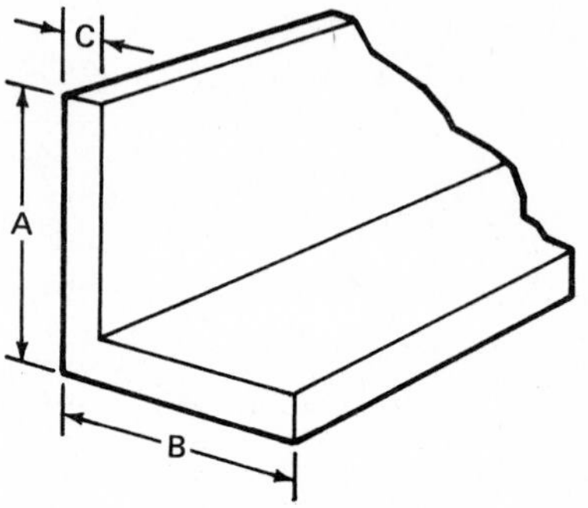
C
A
B

Cat. No.	A	B	C
L-75	75	75	15
L-100	100	100	20
L-125	125	125	20
L-150	150	150	25
L-200	200	200	30

Table 79 Fixture bases and microsections.

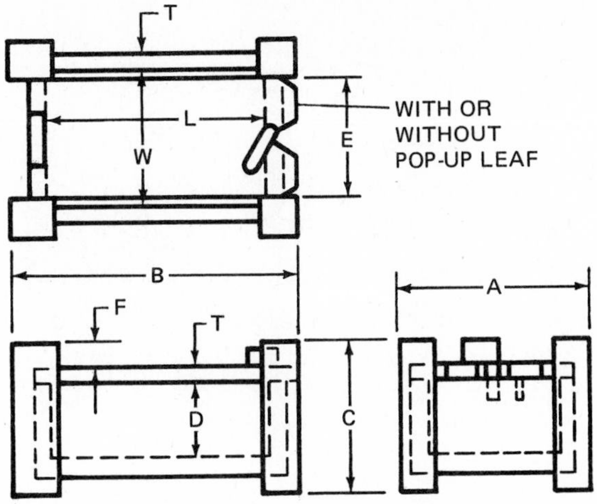
T
L
W
E
WITH OR
WITHOUT
POP-UP LEAF
B
A
F
T
D
C

Nominal Capacity	Inside Dimensions			Outside Dimensions					
	W	L	D	A	B	C	T	E	F
40			40			80			
50 × 50 × 50	65	60	50	100	95	90	8	60	15
60			60			100			
40			40			80			
50 × 75 × 50	65	85	50	100	120	90	8	60	15
60			60			100			
40			40			80			
50 × 100 × 50	65	110	50	100	145	90	8	60	15
60			60			100			
40			40			85			
75 × 75 × 50	90	85	50	130	125	95	10	85	15
75			75			120			
50			50			95			
75 × 100 × 60	90	110	60	130	150	105	10	85	15
75			75			120			
50			50			95			
75 × 125 × 60	90	135	60	130	175	105	10	85	15
75			75			120			
50			50			95			
75 × 150 × 60	90	160	60	130	200	105	10	85	15
75			75			120			
50			50			100			
100 × 100 × 75	115	110	75	160	155	125	12	110	15
100			100			150			
50			50			100			
100 × 125 × 75	115	135	75	160	180	125	12	110	15
100			100			150			
50			50			100			
100 × 150 × 75	115	160	75	160	205	125	12	110	15
100			100			150			
50			50			100			
100 × 200 × 75	115	210	75	160	255	125	12	110	15
100			100			150			

Note: Metric sizes were not available at time of publication. Values were soft converted for problem solving only.

Table 80 Tumble box jig. (Standard Parts Co.)

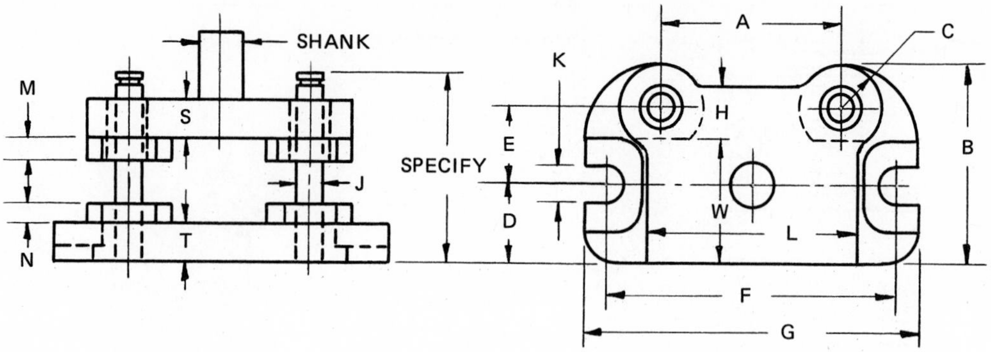
SHANK
M
S
J
T
N
SPECIFY
E
D
K
A
C
H
B
W
L
F
G

	Width	Length	Thickness													
			Die	Punch												
	W	L	T	S	A	B	C	D	E	F	G	H	J	K	M	N
Millimetre Sizes	75	100	30	25	75	125	25	50	50	165	190	40	20	28	16	12
	75	150	40	30	125	135	35	50	50	215	250	45	22	28	20	16
	75	200	40	30	175	135	35	50	50	265	300	45	22	28	20	16
	75	250	40	30	225	140	40	50	50	320	350	45	25	28	20	16
	100	100	40	30	75	160	30	65	65	165	190	45	22	28	20	16
	100	125	40	30	100	160	30	65	65	190	215	45	22	28	20	16
	100	150	40	30	125	160	30	65	65	215	250	45	22	28	20	16
	100	200	40	30	175	160	30	65	65	265	300	45	22	28	20	16
Inch Sizes	3.00	4.00	1.25	1.00	3.00	5.00	1.12	2.00	1.88	6.50	7.50	1.50	.75	1.06	.62	.50
	3.00	6.00	1.50	1.25	5.00	5.25	1.25	2.00	2.00	8.50	10.00	1.68	.88	1.06	.75	.62
	3.00	8.00	1.50	1.25	7.00	5.25	1.25	2.00	2.00	10.50	12.00	1.68	.88	1.06	.75	.62
	3.00	10.00	1.50	1.25	9.00	5.44	1.38	2.00	2.06	12.50	14.00	1.81	1.00	1.06	.75	.62
	4.00	4.00	1.50	1.25	3.00	6.25	1.25	2.50	2.50	6.50	7.50	1.68	.88	1.06	.75	.62
	4.00	5.00	1.50	1.25	4.00	6.25	1.25	2.50	2.50	7.50	8.50	1.68	.88	1.06	.75	.62
	4.00	6.00	1.50	1.25	5.00	6.25	1.25	2.50	2.50	8.50	10.00	1.68	.88	1.06	.75	.62
	4.00	8.00	1.50	1.25	7.00	6.25	1.25	2.50	2.50	10.50	12.00	1.68	.88	1.06	.75	.62

Note: Metric size die sets were not available at time of publication. Values shown were soft converted for problem solving only.

Table 81 Die set. (E.A. Baumbach Mfg. Co.)

GEOMETRICAL CHARACTERISTIC				SYMBOL
Individual Features	Form of a Line	Straightness		⏤
		Roundness		○
		Profile of a Line		⌒
	Form of a Surface	Flatness		▱
		Cylindricity		⌭
		Profile of a Surface		⌓
Related Features	Orientation	Angularity		∠
		Parallelism		//
		Perpendicularity		⟂
	Location	Position		⌖
		Concentricity		◎
		Symmetry		⌯
	Runout	Circular		↗
		Total	ANSI	↗ TOTAL
			ISO	⌰
Symbols for Features of Size		Basic Dimension		[120]
		Maximum Material Condition		Ⓜ
		Regardless of Feature Size	ANSI	Ⓢ
			ISO	NONE
Datums		Datum Identification	ANSI	[A]
			ISO	[A] with triangle
		Datum Target		A4 (in circle)

Table 82 Geometrical tolerancing symbols.

Index